P9-CDJ-294

Chemistry 2e

SENIOR CONTRIBUTING AUTHORS
PAUL FLOWERS, UNIVERSITY OF NORTH CAROLINA AT PEMBROKE
KLAUS THEOPOLD, UNIVERSITY OF DELAWARE
RICHARD LANGLEY, STEPHEN F. AUSTIN STATE UNIVERSITY
WILLIAM R. ROBINSON, PHD

ISBN: 978-1-947172-62-3

OpenStax
Rice University
6100 Main Street MS-375
Houston, Texas 77005

To learn more about OpenStax, visit https://openstax.org.
Individual print copies and bulk orders can be purchased through our website.

HARDCOVER BOOK ISBN-13	**978-1-947172-62-3**
B&W PAPERBACK BOOK ISBN-13	**978-1-59399-578-2**
DIGITAL VERSION ISBN-13	**978-1-947172-61-6**

2 3 4 5 6 7 8 9 10 RS 22 19

Printed by

XanEdu

17177 Laurel Park Dr., Suite 233
Livonia, MI 48152
800-562-2147
www.xanedu.com

OpenStax

OpenStax provides free, peer-reviewed, openly licensed textbooks for introductory college and Advanced Placement® courses and low-cost, personalized courseware that helps students learn. A nonprofit ed tech initiative based at Rice University, we're committed to helping students access the tools they need to complete their courses and meet their educational goals.

Rice University

OpenStax, OpenStax CNX, and OpenStax Tutor are initiatives of Rice University. As a leading research university with a distinctive commitment to undergraduate education, Rice University aspires to path-breaking research, unsurpassed teaching, and contributions to the betterment of our world. It seeks to fulfill this mission by cultivating a diverse community of learning and discovery that produces leaders across the spectrum of human endeavor.

Philanthropic Support

OpenStax is grateful for the generous philanthropic partners who advance our mission to improve educational access and learning for everyone. To see the impact of our supporter community and our most updated list of partners, please visit openstax.org/impact.

Arnold Ventures
Chan Zuckerberg Initiative
Chegg, Inc.
Arthur and Carlyse Ciocca Charitable Foundation
Digital Promise
Ann and John Doerr
Bill & Melinda Gates Foundation
Girard Foundation
Google Inc.
The William and Flora Hewlett Foundation
The Hewlett-Packard Company
Intel Inc.
Rusty and John Jaggers
The Calvin K. Kazanjian Economics Foundation
Charles Koch Foundation
Leon Lowenstein Foundation, Inc.
The Maxfield Foundation
Burt and Deedee McMurtry
Michelson 20MM Foundation
National Science Foundation
The Open Society Foundations
Jumee Yhu and David E. Park III
Brian D. Patterson USA-International Foundation
The Bill and Stephanie Sick Fund
Steven L. Smith & Diana T. Go
Stand Together
Robin and Sandy Stuart Foundation
The Stuart Family Foundation
Tammy and Guillermo Treviño
Valhalla Charitable Foundation
White Star Education Foundation
Schmidt Futures
William Marsh Rice University

Contents

Preface 1

CHAPTER 1
Essential Ideas 9
Introduction 9
1.1 Chemistry in Context 9
1.2 Phases and Classification of Matter 14
1.3 Physical and Chemical Properties 23
1.4 Measurements 27
1.5 Measurement Uncertainty, Accuracy, and Precision 33
1.6 Mathematical Treatment of Measurement Results 41
Key Terms 48
Key Equations 49
Summary 49
Exercises 50

CHAPTER 2
Atoms, Molecules, and Ions 61
Introduction 61
2.1 Early Ideas in Atomic Theory 62
2.2 Evolution of Atomic Theory 66
2.3 Atomic Structure and Symbolism 71
2.4 Chemical Formulas 79
2.5 The Periodic Table 84
2.6 Ionic and Molecular Compounds 89
2.7 Chemical Nomenclature 96
Key Terms 104
Key Equations 105
Summary 105
Exercises 107

CHAPTER 3
Composition of Substances and Solutions 117
Introduction 117
3.1 Formula Mass and the Mole Concept 118
3.2 Determining Empirical and Molecular Formulas 129
3.3 Molarity 136
3.4 Other Units for Solution Concentrations 144
Key Terms 150
Key Equations 150
Summary 150
Exercises 151

CHAPTER 4
Stoichiometry of Chemical Reactions

Stoichiometry of Chemical Reactions 159
Introduction 159
4.1 Writing and Balancing Chemical Equations 160
4.2 Classifying Chemical Reactions 166
4.3 Reaction Stoichiometry 180
4.4 Reaction Yields 185
4.5 Quantitative Chemical Analysis 190
Key Terms 198
Key Equations 199
Summary 199
Exercises 200

CHAPTER 5
Thermochemistry

Thermochemistry 211
Introduction 211
5.1 Energy Basics 212
5.2 Calorimetry 221
5.3 Enthalpy 233
Key Terms 247
Key Equations 248
Summary 248
Exercises 248

CHAPTER 6
Electronic Structure and Periodic Properties of Elements

Electronic Structure and Periodic Properties of Elements 257
Introduction 257
6.1 Electromagnetic Energy 258
6.2 The Bohr Model 270
6.3 Development of Quantum Theory 274
6.4 Electronic Structure of Atoms (Electron Configurations) 287
6.5 Periodic Variations in Element Properties 295
Key Terms 304
Key Equations 305
Summary 305
Exercises 307

CHAPTER 7
Chemical Bonding and Molecular Geometry

Chemical Bonding and Molecular Geometry 313
Introduction 313
7.1 Ionic Bonding 313
7.2 Covalent Bonding 317
7.3 Lewis Symbols and Structures 322
7.4 Formal Charges and Resonance 332
7.5 Strengths of Ionic and Covalent Bonds 336
7.6 Molecular Structure and Polarity 343
Key Terms 358
Key Equations 359
Summary 359

Exercises 360

CHAPTER 8
Advanced Theories of Covalent Bonding 375
Introduction 375
8.1 Valence Bond Theory 376
8.2 Hybrid Atomic Orbitals 379
8.3 Multiple Bonds 390
8.4 Molecular Orbital Theory 393
Key Terms 408
Key Equations 408
Summary 409
Exercises 409

CHAPTER 9
Gases 415
Introduction 415
9.1 Gas Pressure 416
9.2 Relating Pressure, Volume, Amount, and Temperature: The Ideal Gas Law 425
9.3 Stoichiometry of Gaseous Substances, Mixtures, and Reactions 437
9.4 Effusion and Diffusion of Gases 449
9.5 The Kinetic-Molecular Theory 454
9.6 Non-Ideal Gas Behavior 458
Key Terms 462
Key Equations 462
Summary 463
Exercises 464

CHAPTER 10
Liquids and Solids 475
Introduction 475
10.1 Intermolecular Forces 476
10.2 Properties of Liquids 487
10.3 Phase Transitions 493
10.4 Phase Diagrams 503
10.5 The Solid State of Matter 510
10.6 Lattice Structures in Crystalline Solids 516
Key Terms 534
Key Equations 535
Summary 535
Exercises 536

CHAPTER 11
Solutions and Colloids 547
Introduction 547
11.1 The Dissolution Process 548
11.2 Electrolytes 552
11.3 Solubility 555
11.4 Colligative Properties 564

11.5 Colloids 583
Key Terms 592
Key Equations 593
Summary 593
Exercises 594

CHAPTER 12
Kinetics 599

Introduction 599
12.1 Chemical Reaction Rates 600
12.2 Factors Affecting Reaction Rates 605
12.3 Rate Laws 607
12.4 Integrated Rate Laws 614
12.5 Collision Theory 625
12.6 Reaction Mechanisms 630
12.7 Catalysis 635
Key Terms 642
Key Equations 642
Summary 643
Exercises 644

CHAPTER 13
Fundamental Equilibrium Concepts 657

Introduction 657
13.1 Chemical Equilibria 657
13.2 Equilibrium Constants 660
13.3 Shifting Equilibria: Le Châtelier's Principle 669
13.4 Equilibrium Calculations 675
Key Terms 683
Key Equations 683
Summary 683
Exercises 684

CHAPTER 14
Acid-Base Equilibria 693

Introduction 693
14.1 Brønsted-Lowry Acids and Bases 693
14.2 pH and pOH 697
14.3 Relative Strengths of Acids and Bases 702
14.4 Hydrolysis of Salts 716
14.5 Polyprotic Acids 721
14.6 Buffers 724
14.7 Acid-Base Titrations 730
Key Terms 738
Key Equations 738
Summary 739
Exercises 740

CHAPTER 15
Equilibria of Other Reaction Classes 749

Introduction 749
15.1 Precipitation and Dissolution 749
15.2 Lewis Acids and Bases 763
15.3 Coupled Equilibria 767
Key Terms 772
Key Equations 772
Summary 772
Exercises 773

CHAPTER 16
Thermodynamics 783

Introduction 783
16.1 Spontaneity 783
16.2 Entropy 787
16.3 The Second and Third Laws of Thermodynamics 793
16.4 Free Energy 797
Key Terms 809
Key Equations 809
Summary 809
Exercises 810

CHAPTER 17
Electrochemistry 817

Introduction 817
17.1 Review of Redox Chemistry 818
17.2 Galvanic Cells 821
17.3 Electrode and Cell Potentials 824
17.4 Potential, Free Energy, and Equilibrium 830
17.5 Batteries and Fuel Cells 834
17.6 Corrosion 840
17.7 Electrolysis 843
Key Terms 849
Key Equations 850
Summary 850
Exercises 851

CHAPTER 18
Representative Metals, Metalloids, and Nonmetals 857

Introduction 857
18.1 Periodicity 858
18.2 Occurrence and Preparation of the Representative Metals 867
18.3 Structure and General Properties of the Metalloids 870
18.4 Structure and General Properties of the Nonmetals 877
18.5 Occurrence, Preparation, and Compounds of Hydrogen 885
18.6 Occurrence, Preparation, and Properties of Carbonates 891
18.7 Occurrence, Preparation, and Properties of Nitrogen 893
18.8 Occurrence, Preparation, and Properties of Phosphorus 897

18.9 Occurrence, Preparation, and Compounds of Oxygen 899
18.10 Occurrence, Preparation, and Properties of Sulfur 913
18.11 Occurrence, Preparation, and Properties of Halogens 915
18.12 Occurrence, Preparation, and Properties of the Noble Gases 920
Key Terms 922
Summary 923
Exercises 924

CHAPTER 19
Transition Metals and Coordination Chemistry 935
Introduction 935
19.1 Occurrence, Preparation, and Properties of Transition Metals and Their Compounds 935
19.2 Coordination Chemistry of Transition Metals 948
19.3 Spectroscopic and Magnetic Properties of Coordination Compounds 962
Key Terms 971
Summary 972
Exercises 973

CHAPTER 20
Organic Chemistry 977
Introduction 977
20.1 Hydrocarbons 978
20.2 Alcohols and Ethers 995
20.3 Aldehydes, Ketones, Carboxylic Acids, and Esters 999
20.4 Amines and Amides 1005
Key Terms 1014
Summary 1014
Exercises 1015

CHAPTER 21
Nuclear Chemistry 1021
Introduction 1021
21.1 Nuclear Structure and Stability 1022
21.2 Nuclear Equations 1028
21.3 Radioactive Decay 1031
21.4 Transmutation and Nuclear Energy 1042
21.5 Uses of Radioisotopes 1055
21.6 Biological Effects of Radiation 1059
Key Terms 1067
Key Equations 1068
Summary 1069
Exercises 1070

Appendix A The Periodic Table 1077

Appendix B Essential Mathematics 1079

Appendix C Units and Conversion Factors 1087

Appendix D Fundamental Physical Constants 1091

Appendix E Water Properties 1093

Appendix F Composition of Commercial Acids and Bases 1099

Appendix G Standard Thermodynamic Properties for Selected Substances 1101

Appendix H Ionization Constants of Weak Acids 1119

Appendix I Ionization Constants of Weak Bases 1123

Appendix J Solubility Products 1125

Appendix K Formation Constants for Complex Ions 1131

Appendix L Standard Electrode (Half-Cell) Potentials 1133

Appendix M Half-Lives for Several Radioactive Isotopes 1139

Answer Key 1141

Index 1201

PREFACE

Welcome to *Chemistry 2e*, an OpenStax resource. This textbook was written to increase student access to high-quality learning materials, maintaining highest standards of academic rigor at little to no cost.

About OpenStax

OpenStax is a nonprofit based at Rice University, and it's our mission to improve student access to education. Our first openly licensed college textbook was published in 2012, and our library has since scaled to over 25 books for college and AP® courses used by hundreds of thousands of students. OpenStax Tutor, our low-cost personalized learning tool, is being used in college courses throughout the country. Through our partnerships with philanthropic foundations and our alliance with other educational resource organizations, OpenStax is breaking down the most common barriers to learning and empowering students and instructors to succeed.

About OpenStax resources

Customization

Chemistry 2e is licensed under a Creative Commons Attribution 4.0 International (CC BY) license, which means that you can distribute, remix, and build upon the content, as long as you provide attribution to OpenStax and its content contributors.

Because our books are openly licensed, you are free to use the entire book or pick and choose the sections that are most relevant to the needs of your course. Feel free to remix the content by assigning your students certain chapters and sections in your syllabus, in the order that you prefer. You can even provide a direct link in your syllabus to the sections in the web view of your book.

Instructors also have the option of creating a customized version of their OpenStax book. The custom version can be made available to students in low-cost print or digital form through their campus bookstore. Visit the Instructor Resources section of your book page on OpenStax.org for more information.

Errata

All OpenStax textbooks undergo a rigorous review process. However, like any professional-grade textbook, errors sometimes occur. Since our books are web based, we can make updates periodically when deemed pedagogically necessary. If you have a correction to suggest, submit it through the link on your book page on OpenStax.org. Subject matter experts review all errata suggestions. OpenStax is committed to remaining transparent about all updates, so you will also find a list of past errata changes on your book page on OpenStax.org.

Format

You can access this textbook for free in web view or PDF through OpenStax.org, and for a low cost in print.

About *Chemistry 2e*

Chemistry 2e is designed to meet the scope and sequence requirements of the two-semester general chemistry course. The textbook provides an important opportunity for students to learn the core concepts of chemistry and understand how those concepts apply to their lives and the world around them. The book also includes a number of innovative features, including interactive exercises and real-world applications, designed to enhance student learning. The second edition has been revised to incorporate clearer, more current, and more dynamic explanations, while maintaining the same organization as the first edition. Substantial improvements have been made in the figures, illustrations, and example exercises that support the text narrative.

Coverage and scope

Our *Chemistry 2e* textbook adheres to the scope and sequence of most general chemistry courses nationwide. We strive to make chemistry, as a discipline, interesting and accessible to students. With this objective in mind, the content of this textbook has been developed and arranged to provide a logical progression from fundamental to more advanced concepts of chemical science. Topics are introduced within the context of familiar experiences whenever possible, treated with an appropriate rigor to satisfy the intellect of the learner, and reinforced in subsequent discussions of related content. The organization and pedagogical features were developed and vetted with feedback from chemistry educators dedicated to the project.

Changes to the second edition

OpenStax only undertakes second editions when significant modifications to the text are necessary. In the case of *Chemistry 2e*, user feedback indicated that we needed to focus on a few key areas, which we have done in the following ways:

Content revisions for clarity and accuracy. The revision plan varied by chapter based on need. About five chapters were extensively rewritten and another twelve chapters were substantially revised to improve the readability and clarity of the narrative.

Example and end-of-chapter exercises. The example and end-of-chapter exercises in several chapters were subjected to a rigorous accuracy check and revised to correct any errors, and additional exercises were added to several chapters to more fully support chapter content.

Art and illustrations. Under the guidance of the authors and expert scientific illustrators, especially those well-versed in creating accessible art, the OpenStax team made changes to much of the art in the first edition of *Chemistry*. The revisions included correcting errors, redesigning illustrations to improve understanding, and recoloring for overall consistency.

Accessibility improvements. As with all OpenStax books, the first edition of *Chemistry* was created with a focus on accessibility. We have emphasized and improved that approach in the second edition. To accommodate users of specific assistive technologies, all alternative text was reviewed and revised for comprehensiveness and clarity. Many illustrations were revised to improve the color contrast, which is important for some visually impaired students. Overall, the OpenStax platform has been continually upgraded to improve accessibility.

Pedagogical foundation and features

Throughout *Chemistry 2e*, you will find features that draw the students into scientific inquiry by taking selected topics a step further. Students and educators alike will appreciate discussions in these feature boxes.

- **Chemistry in Everyday Life** ties chemistry concepts to everyday issues and real-world applications of science that students encounter in their lives. Topics include cell phones, solar thermal energy power plants, plastics recycling, and measuring blood pressure.
- **How Sciences Interconnect** feature boxes discuss chemistry in context of its interconnectedness with other scientific disciplines. Topics include neurotransmitters, greenhouse gases and climate change, and proteins and enzymes.
- **Portrait of a Chemist** presents a short bio and an introduction to the work of prominent figures from history and present day so that students can see the "faces" of contributors in this field as well as science in action.

Comprehensive art program

Our art program is designed to enhance students' understanding of concepts through clear, effective illustrations, diagrams, and photographs.

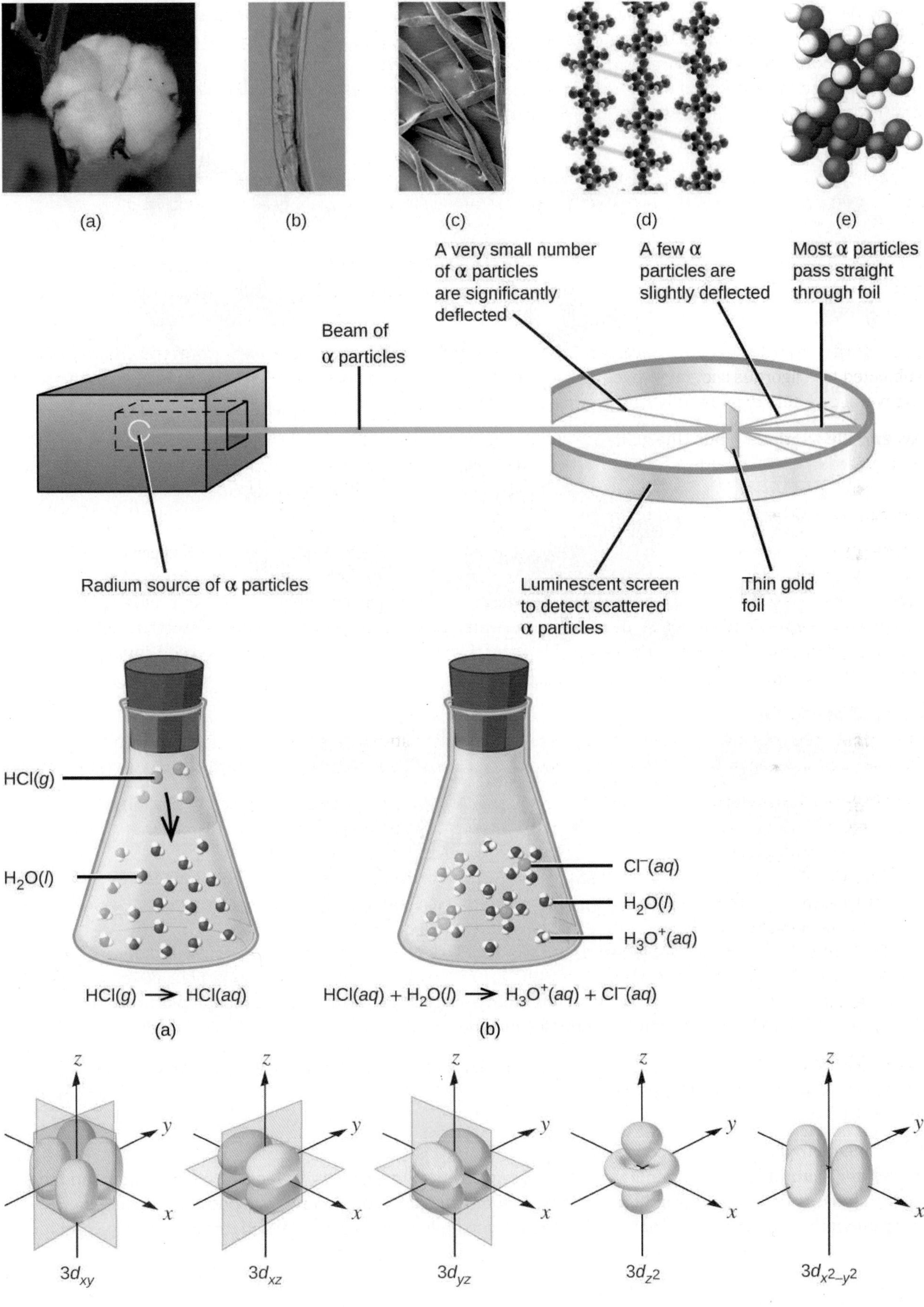
(a)
(b)
(c)
(d)
(e)
A very small number of α particles are significantly deflected
A few α particles are slightly deflected
Most α particles pass straight through foil
Beam of α particles
Radium source of α particles
Luminescent screen to detect scattered α particles
Thin gold foil
HCl(g)
H2O(l)
Cl−(aq)
H2O(l)
H3O+(aq)
HCl(g) → HCl(aq)
HCl(aq) + H2O(l) → H3O+(aq) + Cl−(aq)
(a)
(b)
z
y
x
3dxy
3dxz
3dyz
3dz2
3dx2−y2

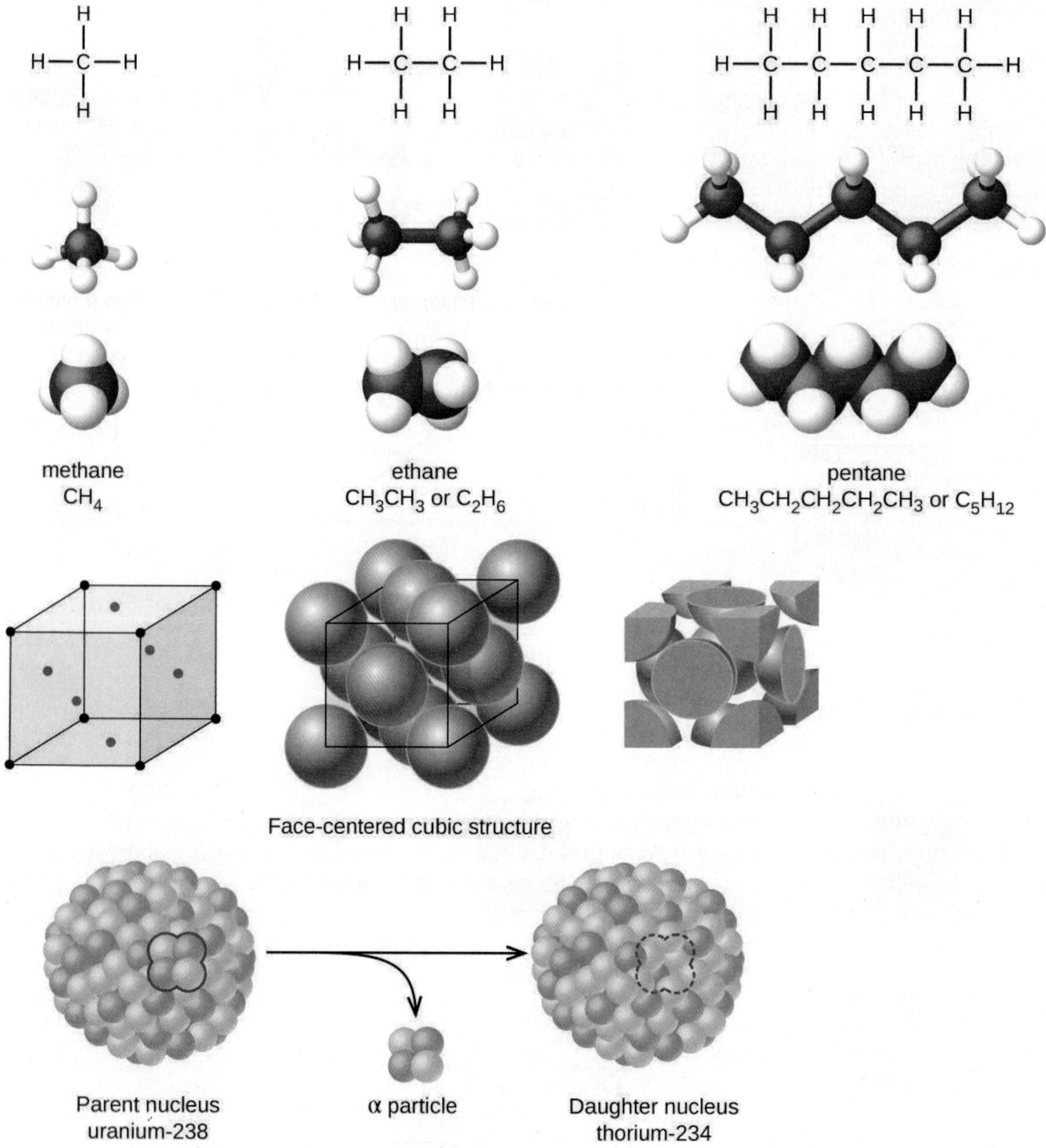

Interactives that engage

Chemistry 2e incorporates links to relevant interactive exercises and animations that help bring topics to life through our **Link to Learning** feature. Examples include:

- PhET simulations
- IUPAC data and interactives
- TED Talks

Assessments that reinforce key concepts

In-chapter **Examples** walk students through problems by posing a question, stepping out a solution, and then asking students to practice the skill with a "Check Your Learning" component. The book also includes assessments at the end of each chapter so students can apply what they've learned through practice problems.

Additional resources

Student and instructor resources

We've compiled additional resources for both students and instructors, including Getting Started Guides, an instructor solutions manual, and PowerPoint slides. Instructor resources require a verified instructor account, which you can apply for when you log in or create your account on OpenStax.org. Take advantage of these resources to supplement your OpenStax book.

Community Hubs

OpenStax partners with the Institute for the Study of Knowledge Management in Education (ISKME) to offer Community Hubs on OER Commons — a platform for instructors to share community-created resources that support OpenStax books, free of charge. Through our Community Hubs, instructors can upload their own materials or download resources to use in their own courses, including additional ancillaries, teaching material, multimedia, and relevant course content. We encourage instructors to join the hubs for the subjects most relevant to your teaching and research as an opportunity both to enrich your courses and to engage with other faculty.

To reach the Community Hubs, visit www.oercommons.org/hubs/OpenStax (https://www.oercommons.org/hubs/OpenStax).

Technology partners

As allies in making high-quality learning materials accessible, our technology partners offer optional low-cost tools that are integrated with OpenStax books. To access the technology options for your text, visit your book page on OpenStax.org.

About the authors

Senior contributing authors

Paul Flowers, University of North Carolina at Pembroke

Dr. Paul Flowers earned a BS in Chemistry from St. Andrews Presbyterian College in 1983 and a PhD in Analytical Chemistry from the University of Tennessee in 1988. After a one-year postdoctoral appointment at Los Alamos National Laboratory, he joined the University of North Carolina at Pembroke in the fall of 1989. Dr. Flowers teaches courses in general and analytical chemistry, and conducts experimental research involving the development of new devices and methods for microscale chemical analysis.

Klaus Theopold, University of Delaware

Dr. Klaus Theopold (born in Berlin, Germany) received his Vordiplom from the Universität Hamburg in 1977. He then decided to pursue his graduate studies in the United States, where he received his PhD in inorganic chemistry from UC Berkeley in 1982. After a year of postdoctoral research at MIT, he joined the faculty at Cornell University. In 1990, he moved to the University of Delaware, where he is a Professor in the Department of Chemistry and Biochemistry and serves as an Associate Director of the University's Center for Catalytic Science and Technology. Dr. Theopold regularly teaches graduate courses in inorganic and organometallic chemistry as well as general chemistry.

Richard Langley, Stephen F. Austin State University

Dr. Richard Langley earned BS degrees in Chemistry and Mineralogy from Miami University of Ohio in the early 1970s and went on to receive his PhD in Chemistry from the University of Nebraska in 1977. After a postdoctoral fellowship at the Arizona State University Center for Solid State Studies, Dr. Langley taught in the University of Wisconsin system and participated in research at Argonne National Laboratory. Moving to Stephen F. Austin State University in 1982, Dr. Langley today serves as Professor of Chemistry. His areas of specialization are solid state chemistry, synthetic inorganic chemistry, fluorine chemistry, and chemical education.

William R. Robinson, PhD

Contributing authors

Mark Blaser, Shasta College
Simon Bott, University of Houston

Donald Carpenetti, Craven Community College
Andrew Eklund, Alfred University
Emad El-Giar, University of Louisiana at Monroe
Don Frantz, Wilfrid Laurier University
Paul Hooker, Westminster College
Jennifer Look, Mercer University
George Kaminski, Worcester Polytechnic Institute
Carol Martinez, Central New Mexico Community College
Troy Milliken, Jackson State University
Vicki Moravec, Trine University
Jason Powell, Ferrum College
Thomas Sorensen, University of Wisconsin–Milwaukee
Allison Soult, University of Kentucky

Reviewers

Casey Akin, College Station Independent School District
Lara AL-Hariri, University of Massachusetts–Amherst
Sahar Atwa, University of Louisiana at Monroe
Todd Austell, University of North Carolina–Chapel Hill
Bobby Bailey, University of Maryland–University College
Robert Baker, Trinity College
Jeffrey Bartz, Kalamazoo College
Greg Baxley, Cuesta College
Ashley Beasley Green, National Institute of Standards and Technology
Patricia Bianconi, University of Massachusetts
Lisa Blank, Lyme Central School District
Daniel Branan, Colorado Community College System
Dorian Canelas, Duke University
Emmanuel Chang, York College
Carolyn Collins, College of Southern Nevada
Colleen Craig, University of Washington
Yasmine Daniels, Montgomery College–Germantown
Patricia Dockham, Grand Rapids Community College
Erick Fuoco, Richard J. Daley College
Andrea Geyer, University of Saint Francis
Daniel Goebbert, University of Alabama
John Goodwin, Coastal Carolina University
Stephanie Gould, Austin College
Patrick Holt, Bellarmine University
George A. Kaminski, Worcester Polytechnic Institute
Kevin Kolack, Queensborough Community College
Amy Kovach, Roberts Wesleyan College
Judit Kovacs Beagle, University of Dayton
Krzysztof Kuczera, University of Kansas
Marcus Lay, University of Georgia
Pamela Lord, University of Saint Francis
Oleg Maksimov, Excelsior College
John Matson, Virginia Tech
Katrina Miranda, University of Arizona
Douglas Mulford, Emory University
Mark Ott, Jackson College
Adrienne Oxley, Columbia College
Richard Pennington, Georgia Gwinnett College
Rodney Powell, Coastal Carolina Community College

Jeanita Pritchett, Montgomery College–Rockville
Aheda Saber, University of Illinois at Chicago
Raymond Sadeghi, University of Texas at San Antonio
Nirmala Shankar, Rutgers University
Jonathan Smith, Temple University
Bryan Spiegelberg, Rider University
Ron Sternfels, Roane State Community College
Cynthia Strong, Cornell College
Kris Varazo, Francis Marion University
Victor Vilchiz, Virginia State University
Alex Waterson, Vanderbilt University
JuchaoYan, Eastern New Mexico University
Mustafa Yatin, Salem State University
Kazushige Yokoyama, State University of New York at Geneseo
Curtis Zaleski, Shippensburg University
Wei Zhang, University of Colorado–Boulder

CHAPTER 1
Essential Ideas

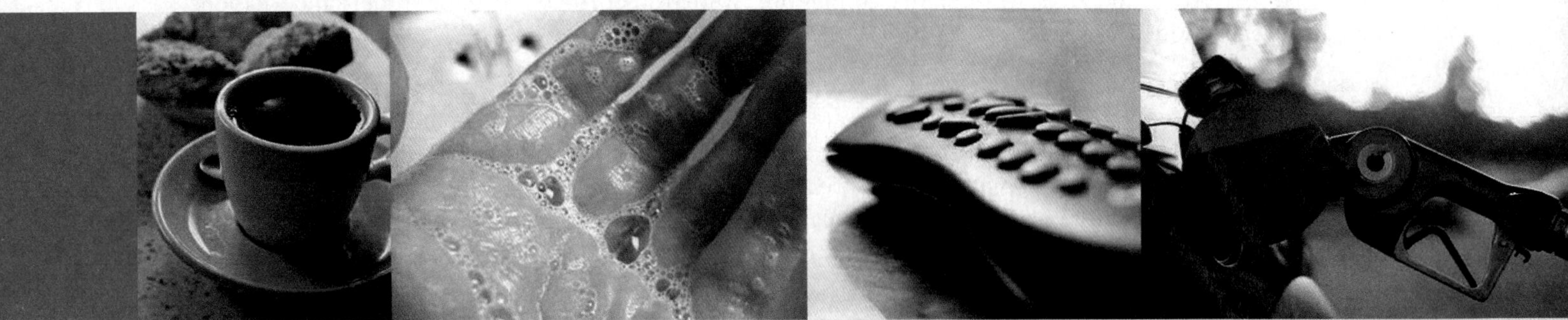

Figure 1.1 Chemical substances and processes are essential for our existence, providing sustenance, keeping us clean and healthy, fabricating electronic devices, enabling transportation, and much more. (credit "left": modification of work by "vxla"/Flickr; credit "left middle": modification of work by "the Italian voice"/Flickr; credit "right middle": modification of work by Jason Trim; credit "right": modification of work by "gosheshe"/Flickr)

CHAPTER OUTLINE

1.1 Chemistry in Context
1.2 Phases and Classification of Matter
1.3 Physical and Chemical Properties
1.4 Measurements
1.5 Measurement Uncertainty, Accuracy, and Precision
1.6 Mathematical Treatment of Measurement Results

INTRODUCTION Your alarm goes off and, after hitting "snooze" once or twice, you pry yourself out of bed. You make a cup of coffee to help you get going, and then you shower, get dressed, eat breakfast, and check your phone for messages. On your way to school, you stop to fill your car's gas tank, almost making you late for the first day of chemistry class. As you find a seat in the classroom, you read the question projected on the screen: "Welcome to class! Why should we study chemistry?"

Do you have an answer? You may be studying chemistry because it fulfills an academic requirement, but if you consider your daily activities, you might find chemistry interesting for other reasons. Most everything you do and encounter during your day involves chemistry. Making coffee, cooking eggs, and toasting bread involve chemistry. The products you use—like soap and shampoo, the fabrics you wear, the electronics that keep you connected to your world, the gasoline that propels your car—all of these and more involve chemical substances and processes. Whether you are aware or not, chemistry is part of your everyday world. In this course, you will learn many of the essential principles underlying the chemistry of modern-day life.

1.1 Chemistry in Context

LEARNING OBJECTIVES

By the end of this section, you will be able to:

- Outline the historical development of chemistry
- Provide examples of the importance of chemistry in everyday life
- Describe the scientific method
- Differentiate among hypotheses, theories, and laws
- Provide examples illustrating macroscopic, microscopic, and symbolic domains

Throughout human history, people have tried to convert matter into more useful forms. Our Stone Age ancestors chipped pieces of flint into useful tools and carved wood into statues and toys. These endeavors involved changing the shape of a substance without changing the substance itself. But as our knowledge increased, humans began to change the composition of the substances as well—clay was converted into pottery, hides were cured to make garments, copper ores were transformed into copper tools and weapons, and grain was made into bread.

Humans began to practice chemistry when they learned to control fire and use it to cook, make pottery, and smelt metals. Subsequently, they began to separate and use specific components of matter. A variety of drugs such as aloe, myrrh, and opium were isolated from plants. Dyes, such as indigo and Tyrian purple, were extracted from plant and animal matter. Metals were combined to form alloys—for example, copper and tin were mixed together to make bronze—and more elaborate smelting techniques produced iron. Alkalis were extracted from ashes, and soaps were prepared by combining these alkalis with fats. Alcohol was produced by fermentation and purified by distillation.

Attempts to understand the behavior of matter extend back for more than 2500 years. As early as the sixth century BC, Greek philosophers discussed a system in which water was the basis of all things. You may have heard of the Greek postulate that matter consists of four elements: earth, air, fire, and water. Subsequently, an amalgamation of chemical technologies and philosophical speculations was spread from Egypt, China, and the eastern Mediterranean by alchemists, who endeavored to transform "base metals" such as lead into "noble metals" like gold, and to create elixirs to cure disease and extend life (Figure 1.2).

FIGURE 1.2 (a) This portrayal shows an alchemist's workshop circa 1580. Although alchemy made some useful contributions to how to manipulate matter, it was not scientific by modern standards. (b) While the equipment used by Alma Levant Hayden in this 1952 picture might not seem as sleek as you might find in a lab today, her approach was highly methodical and carefully recorded. A department head at the FDA, Hayden is most famous for exposing an aggressively marketed anti-cancer drug as nothing more than an unhelpful solution of common substances. (credit a: Chemical Heritage Foundation; b: NIH History Office)

From alchemy came the historical progressions that led to modern chemistry: the isolation of drugs from natural sources, such as plants and animals. But while many of the substances extracted or processed from those natural sources were critical in the treatment of diseases, many were scarce. For example, progesterone, which is critical to women's health, became available as a medicine in 1935, but its animal sources produced extremely small quantities, limiting its availability and increasing its expense. Likewise, in the 1940s, cortisone came into use to treat arthritis and other disorders and injuries, but it took a 36-step process to synthesize. Chemist Percy Lavon Julian turned to a more plentiful source: soybeans. Previously, Julian had developed a lab to isolate soy protein, which was used in firefighting among other applications. He focused on using the soy sterols—substances mostly used in plant membranes—and was able to quickly produce progesterone and later testosterone and other hormones. He later developed a process to do the same for cortisone, and laid the groundwork for modern drug design. Since soybeans and similar plant sources were extremely plentiful, the drugs soon became widely available, saving many lives.

Chemistry: The Central Science

Chemistry is sometimes referred to as "the central science" due to its interconnectedness with a vast array of other STEM disciplines (STEM stands for areas of study in the science, technology, engineering, and math fields). Chemistry and the language of chemists play vital roles in biology, medicine, materials science, forensics, environmental science, and many other fields (Figure 1.3). The basic principles of physics are essential for understanding many aspects of chemistry, and there is extensive overlap between many subdisciplines within the two fields, such as chemical physics and nuclear chemistry. Mathematics, computer science, and information theory provide important tools that help us calculate, interpret, describe, and generally make sense of the chemical world. Biology and chemistry converge in biochemistry, which is crucial to understanding the many complex factors and processes that keep living organisms (such as us) alive. Chemical engineering, materials science, and nanotechnology combine chemical principles and empirical findings to produce useful substances, ranging from gasoline to fabrics to electronics. Agriculture, food science, veterinary science, and brewing and wine making help provide sustenance in the form of food and drink to the world's population. Medicine, pharmacology, biotechnology, and botany identify and produce substances that help keep us healthy. Environmental science, geology, oceanography, and atmospheric science incorporate many chemical ideas to help us better understand and protect our physical world. Chemical ideas are used to help understand the universe in astronomy and cosmology.

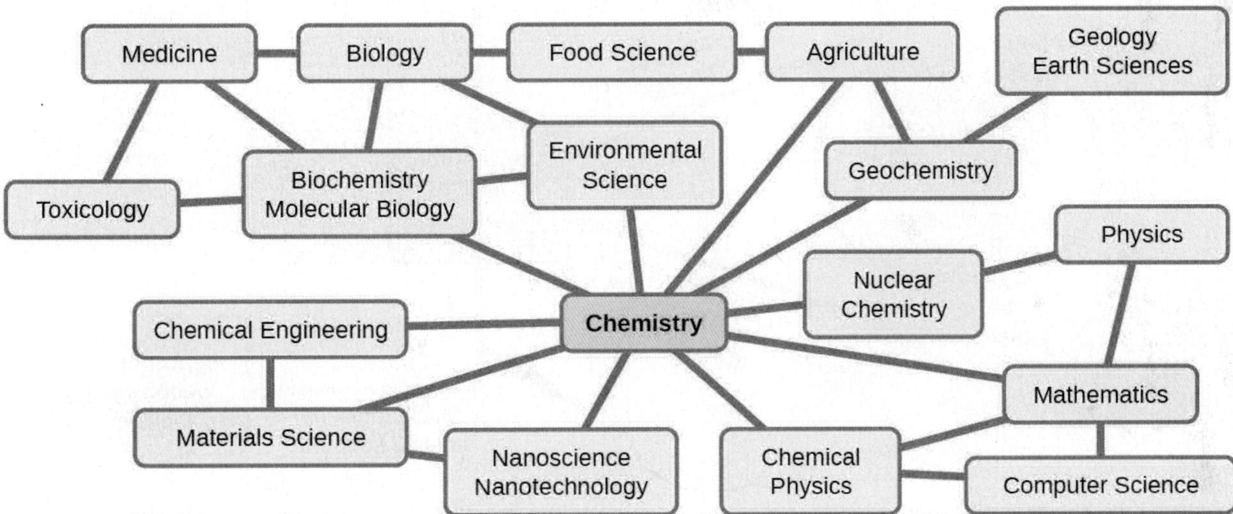

FIGURE 1.3 Knowledge of chemistry is central to understanding a wide range of scientific disciplines. This diagram shows just some of the interrelationships between chemistry and other fields.

What are some changes in matter that are essential to daily life? Digesting and assimilating food, synthesizing polymers that are used to make clothing, containers, cookware, and credit cards, and refining crude oil into gasoline and other products are just a few examples. As you proceed through this course, you will discover many different examples of changes in the composition and structure of matter, how to classify these changes and how they occurred, their causes, the changes in energy that accompany them, and the principles and laws involved. As you learn about these things, you will be learning **chemistry**, the study of the composition, properties, and interactions of matter. The practice of chemistry is not limited to chemistry books or laboratories: It happens whenever someone is involved in changes in matter or in conditions that may lead to such changes.

The Scientific Method

Chemistry is a science based on observation and experimentation. Doing chemistry involves attempting to answer questions and explain observations in terms of the laws and theories of chemistry, using procedures that are accepted by the scientific community. There is no single route to answering a question or explaining an observation, but there is an aspect common to every approach: Each uses knowledge based on experiments that can be reproduced to verify the results. Some routes involve a **hypothesis**, a tentative explanation of observations that acts as a guide for gathering and checking information. A hypothesis is tested by experimentation, calculation, and/or comparison with the experiments of others and then refined as needed.

Some hypotheses are attempts to explain the behavior that is summarized in laws. The **laws** of science summarize a vast number of experimental observations, and describe or predict some facet of the natural world. If such a hypothesis turns out to be capable of explaining a large body of experimental data, it can reach the status of a theory. Scientific **theories** are well-substantiated, comprehensive, testable explanations of particular aspects of nature. Theories are accepted because they provide satisfactory explanations, but they can be modified if new data become available. The path of discovery that leads from question and observation to law or hypothesis to theory, combined with experimental verification of the hypothesis and any necessary modification of the theory, is called the **scientific method** (Figure 1.4).

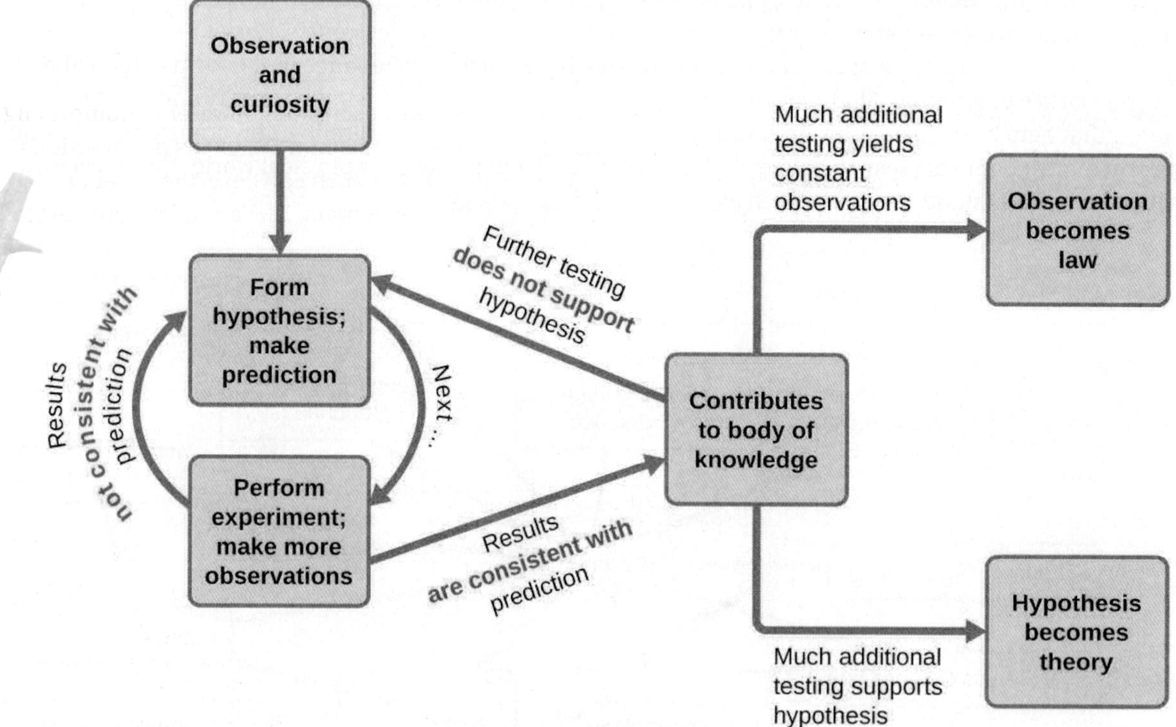

FIGURE 1.4 The scientific method follows a process similar to the one shown in this diagram. All the key components are shown, in roughly the right order. Scientific progress is seldom neat and clean: It requires open inquiry and the reworking of questions and ideas in response to findings.

The Domains of Chemistry

Chemists study and describe the behavior of matter and energy in three different domains: macroscopic, microscopic, and symbolic. These domains provide different ways of considering and describing chemical behavior.

Macro is a Greek word that means "large." The **macroscopic domain** is familiar to us: It is the realm of everyday things that are large enough to be sensed directly by human sight or touch. In daily life, this includes the food you eat and the breeze you feel on your face. The macroscopic domain includes everyday and laboratory chemistry, where we observe and measure physical and chemical properties such as density, solubility, and flammability.

Micro comes from Greek and means "small." The **microscopic domain** of chemistry is often visited in the imagination. Some aspects of the microscopic domain are visible through standard optical microscopes, for example, many biological cells. More sophisticated instruments are capable of imaging even smaller entities such as molecules and atoms (see Figure 1.5 (**b**)).

However, most of the subjects in the microscopic domain of chemistry are too small to be seen even with the most advanced microscopes and may only be pictured in the mind. Other components of the microscopic domain include ions and electrons, protons and neutrons, and chemical bonds, each of which is far too small to see.

The **symbolic domain** contains the specialized language used to represent components of the macroscopic and microscopic domains. Chemical symbols (such as those used in the periodic table), chemical formulas, and chemical equations are part of the symbolic domain, as are graphs, drawings, and calculations. These symbols play an important role in chemistry because they help interpret the behavior of the macroscopic domain in terms of the components of the microscopic domain. One of the challenges for students learning chemistry is recognizing that the same symbols can represent different things in the macroscopic and microscopic domains, and one of the features that makes chemistry fascinating is the use of a domain that must be imagined to explain behavior in a domain that can be observed.

A helpful way to understand the three domains is via the essential and ubiquitous substance of water. That water is a liquid at moderate temperatures, will freeze to form a solid at lower temperatures, and boil to form a gas at higher temperatures (Figure 1.5) are macroscopic observations. But some properties of water fall into the microscopic domain—what cannot be observed with the naked eye. The description of water as comprising two hydrogen atoms and one oxygen atom, and the explanation of freezing and boiling in terms of attractions between these molecules, is within the microscopic arena. The formula H_2O, which can describe water at either the macroscopic or microscopic levels, is an example of the symbolic domain. The abbreviations (*g*) for gas, (*s*) for solid, and (*l*) for liquid are also symbolic.

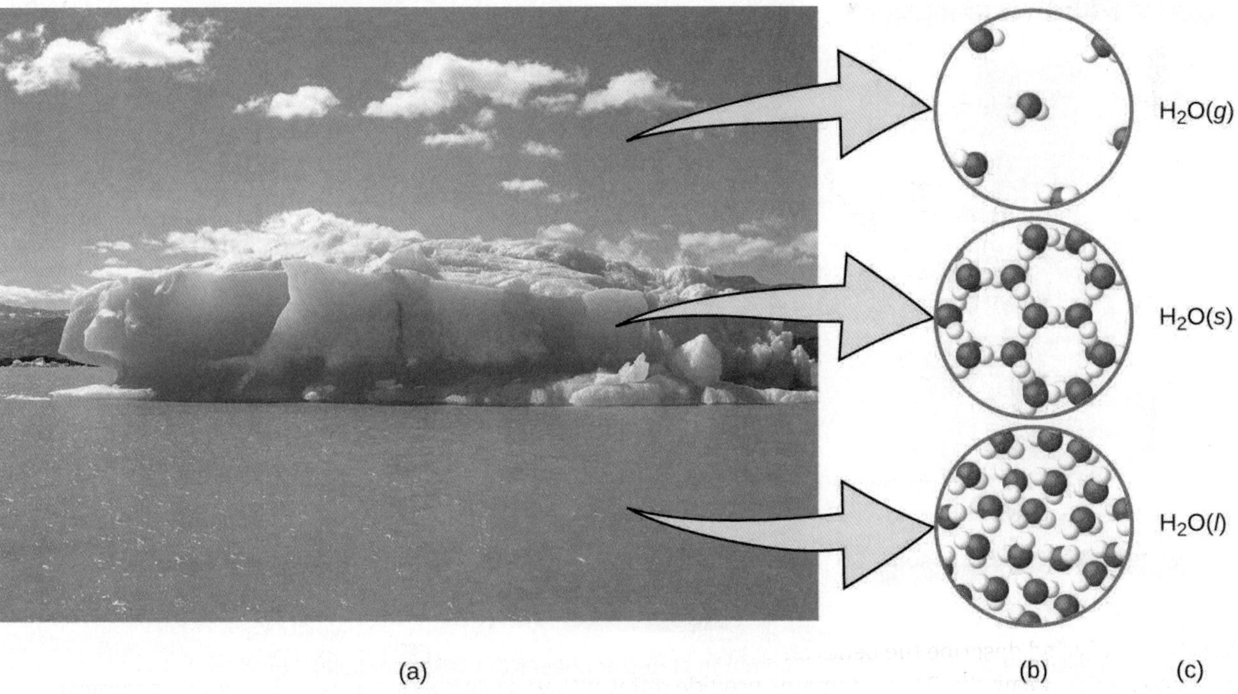

FIGURE 1.5 (a) Moisture in the air, icebergs, and the ocean represent water in the macroscopic domain. (b) At the molecular level (microscopic domain), gas molecules are far apart and disorganized, solid water molecules are close together and organized, and liquid molecules are close together and disorganized. (c) The formula H_2O symbolizes water, and (*g*), (*s*), and (*l*) symbolize its phases. Note that clouds are actually comprised of either very small liquid water droplets or solid water crystals; gaseous water in our atmosphere is not visible to the naked eye, although it may be sensed as humidity. (credit a: modification of work by "Gorkaazk"/Wikimedia Commons)

1.2 Phases and Classification of Matter

LEARNING OBJECTIVES

By the end of this section, you will be able to:

- Describe the basic properties of each physical state of matter: solid, liquid, and gas
- Distinguish between mass and weight
- Apply the law of conservation of matter
- Classify matter as an element, compound, homogeneous mixture, or heterogeneous mixture with regard to its physical state and composition
- Define and give examples of atoms and molecules

Matter is defined as anything that occupies space and has mass, and it is all around us. Solids and liquids are more obviously matter: We can see that they take up space, and their weight tells us that they have mass. Gases are also matter; if gases did not take up space, a balloon would not inflate (increase its volume) when filled with gas.

Solids, liquids, and gases are the three states of matter commonly found on earth (Figure 1.6). A **solid** is rigid and possesses a definite shape. A **liquid** flows and takes the shape of its container, except that it forms a flat or slightly curved upper surface when acted upon by gravity. (In zero gravity, liquids assume a spherical shape.) Both liquid and solid samples have volumes that are very nearly independent of pressure. A **gas** takes both the shape and volume of its container.

FIGURE 1.6 The three most common states or phases of matter are solid, liquid, and gas.

A fourth state of matter, plasma, occurs naturally in the interiors of stars. A **plasma** is a gaseous state of matter that contains appreciable numbers of electrically charged particles (Figure 1.7). The presence of these charged particles imparts unique properties to plasmas that justify their classification as a state of matter distinct from gases. In addition to stars, plasmas are found in some other high-temperature environments (both natural and man-made), such as lightning strikes, certain television screens, and specialized analytical instruments used to detect trace amounts of metals.

FIGURE 1.7 A plasma torch can be used to cut metal. (credit: “Hypertherm”/Wikimedia Commons)

LINK TO LEARNING

In a tiny cell in a plasma television, the plasma emits ultraviolet light, which in turn causes the display at that location to appear a specific color. The composite of these tiny dots of color makes up the image that you see. Watch this video (http://openstax.org/l/16plasma) to learn more about plasma and the places you encounter it.

Some samples of matter appear to have properties of solids, liquids, and/or gases at the same time. This can occur when the sample is composed of many small pieces. For example, we can pour sand as if it were a liquid because it is composed of many small grains of solid sand. Matter can also have properties of more than one state when it is a mixture, such as with clouds. Clouds appear to behave somewhat like gases, but they are actually mixtures of air (gas) and tiny particles of water (liquid or solid).

The **mass** of an object is a measure of the amount of matter in it. One way to measure an object’s mass is to measure the force it takes to accelerate the object. It takes much more force to accelerate a car than a bicycle because the car has much more mass. A more common way to determine the mass of an object is to use a balance to compare its mass with a standard mass.

Although weight is related to mass, it is not the same thing. **Weight** refers to the force that gravity exerts on an object. This force is directly proportional to the mass of the object. The weight of an object changes as the force of gravity changes, but its mass does not. An astronaut’s mass does not change just because she goes to the moon. But her weight on the moon is only one-sixth her earth-bound weight because the moon’s gravity is only one-sixth that of the earth’s. She may feel “weightless” during her trip when she experiences negligible external forces (gravitational or any other), although she is, of course, never “massless.”

The **law of conservation of matter** summarizes many scientific observations about matter: It states that *there is no detectable change in the total quantity of matter present when matter converts from one type to another (a chemical change) or changes among solid, liquid, or gaseous states (a physical change).* Brewing beer and the operation of batteries provide examples of the conservation of matter (Figure 1.8). During the brewing of beer, the ingredients (water, yeast, grains, malt, hops, and sugar) are converted into beer (water, alcohol, carbonation, and flavoring substances) with no actual loss of substance. This is most clearly seen during the bottling process, when glucose turns into ethanol and carbon dioxide, and the total mass of the substances does not change. This can also be seen in a lead-acid car battery: The original substances (lead, lead oxide, and sulfuric acid), which are capable of producing electricity, are changed into other substances (lead sulfate and water) that do not produce electricity, with no change in the actual amount of matter.

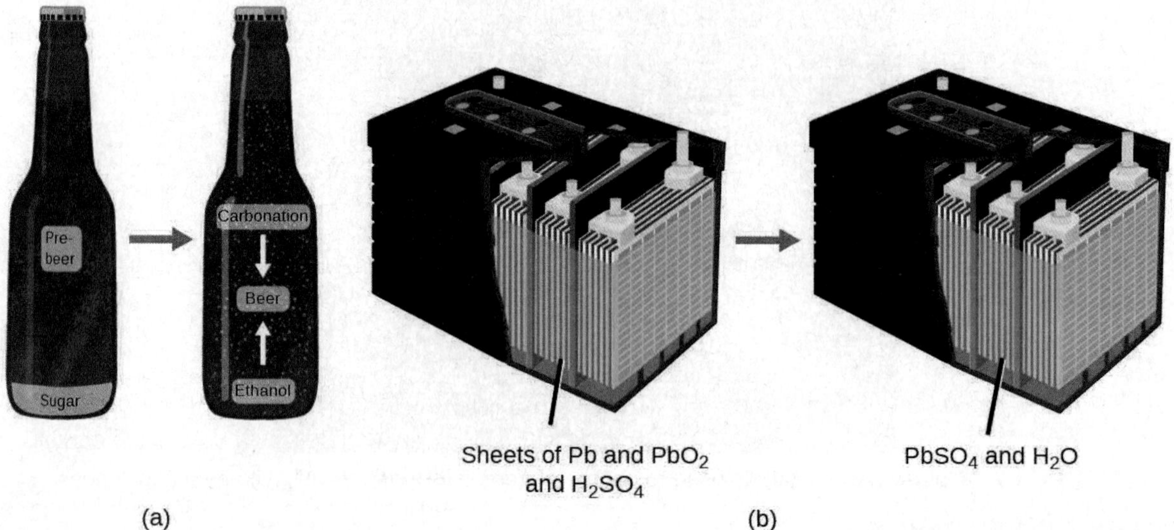

FIGURE 1.8 (a) The mass of beer precursor materials is the same as the mass of beer produced: Sugar has become alcohol and carbon dioxide. (b) The mass of the lead, lead oxide, and sulfuric acid consumed by the production of electricity is exactly equal to the mass of lead sulfate and water that is formed.

Although this conservation law holds true for all conversions of matter, convincing examples are few and far between because, outside of the controlled conditions in a laboratory, we seldom collect all of the material that is produced during a particular conversion. For example, when you eat, digest, and assimilate food, all of the matter in the original food is preserved. But because some of the matter is incorporated into your body, and much is excreted as various types of waste, it is challenging to verify by measurement.

Classifying Matter

Matter can be classified into several categories. Two broad categories are mixtures and pure substances. A **pure substance** has a constant composition. All specimens of a pure substance have exactly the same makeup and properties. Any sample of sucrose (table sugar) consists of 42.1% carbon, 6.5% hydrogen, and 51.4% oxygen by mass. Any sample of sucrose also has the same physical properties, such as melting point, color, and sweetness, regardless of the source from which it is isolated.

Pure substances may be divided into two classes: elements and compounds. Pure substances that cannot be broken down into simpler substances by chemical changes are called **elements**. Iron, silver, gold, aluminum, sulfur, oxygen, and copper are familiar examples of the more than 100 known elements, of which about 90 occur naturally on the earth, and two dozen or so have been created in laboratories.

Pure substances that are comprised of two or more elements are called **compounds**. Compounds may be broken down by chemical changes to yield either elements or other compounds, or both. Mercury(II) oxide, an orange, crystalline solid, can be broken down by heat into the elements mercury and oxygen (Figure 1.9). When heated in the absence of air, the compound sucrose is broken down into the element carbon and the compound water. (The initial stage of this process, when the sugar is turning brown, is known as caramelization—this is what imparts the characteristic sweet and nutty flavor to caramel apples, caramelized onions, and caramel). Silver(I) chloride is a white solid that can be broken down into its elements, silver and chlorine, by absorption of light. This property is the basis for the use of this compound in photographic films and photochromic eyeglasses (those with lenses that darken when exposed to light).

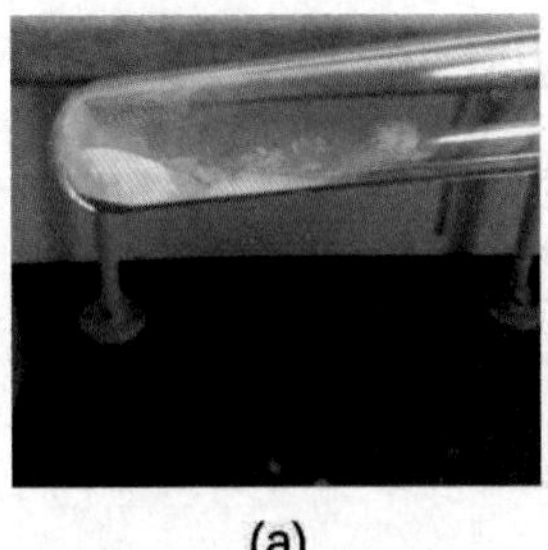
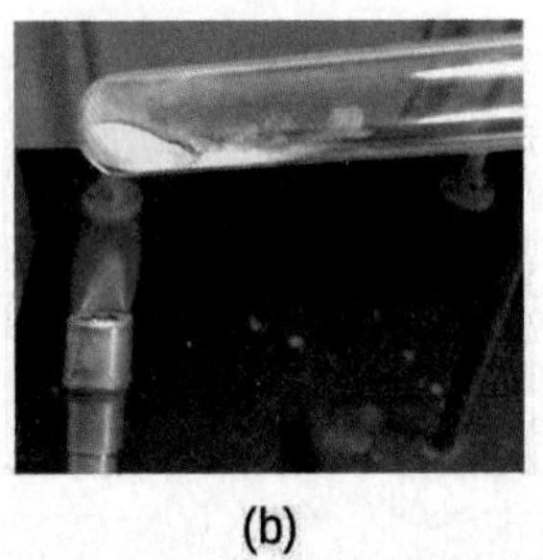
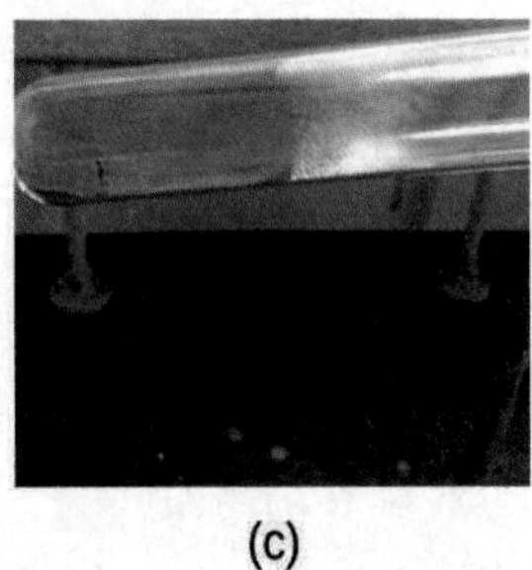

(a) (b) (c)

FIGURE 1.9 (a) The compound mercury(II) oxide, (b) when heated, (c) decomposes into silvery droplets of liquid mercury and invisible oxygen gas. (credit: modification of work by Paul Flowers)

LINK TO LEARNING

Many compounds break down when heated. This site (http://openstax.org/l/16mercury) shows the breakdown of mercury oxide, HgO. You can also view an example of the photochemical decomposition of silver chloride (http://openstax.org/l/16silvchloride) (AgCl), the basis of early photography.

The properties of combined elements are different from those in the free, or uncombined, state. For example, white crystalline sugar (sucrose) is a compound resulting from the chemical combination of the element carbon, which is a black solid in one of its uncombined forms, and the two elements hydrogen and oxygen, which are colorless gases when uncombined. Free sodium, an element that is a soft, shiny, metallic solid, and free chlorine, an element that is a yellow-green gas, combine to form sodium chloride (table salt), a compound that is a white, crystalline solid.

A **mixture** is composed of two or more types of matter that can be present in varying amounts and can be separated by physical changes, such as evaporation (you will learn more about this later). A mixture with a composition that varies from point to point is called a **heterogeneous mixture**. Italian dressing is an example of a heterogeneous mixture (Figure 1.10). Its composition can vary because it may be prepared from varying amounts of oil, vinegar, and herbs. It is not the same from point to point throughout the mixture—one drop may be mostly vinegar, whereas a different drop may be mostly oil or herbs because the oil and vinegar separate and the herbs settle. Other examples of heterogeneous mixtures are chocolate chip cookies (we can see the separate bits of chocolate, nuts, and cookie dough) and granite (we can see the quartz, mica, feldspar, and more).

A **homogeneous mixture**, also called a **solution**, exhibits a uniform composition and appears visually the same throughout. An example of a solution is a sports drink, consisting of water, sugar, coloring, flavoring, and electrolytes mixed together uniformly (Figure 1.10). Each drop of a sports drink tastes the same because each drop contains the same amounts of water, sugar, and other components. Note that the composition of a sports drink can vary—it could be made with somewhat more or less sugar, flavoring, or other components, and still be a sports drink. Other examples of homogeneous mixtures include air, maple syrup, gasoline, and a solution of salt in water.

FIGURE 1.10 (a) Oil and vinegar salad dressing is a heterogeneous mixture because its composition is not uniform throughout. (b) A commercial sports drink is a homogeneous mixture because its composition is uniform throughout. (credit a "left": modification of work by John Mayer; credit a "right": modification of work by Umberto Salvagnin; credit b "left: modification of work by Jeff Bedford)

Although there are just over 100 elements, tens of millions of chemical compounds result from different combinations of these elements. Each compound has a specific composition and possesses definite chemical and physical properties that distinguish it from all other compounds. And, of course, there are innumerable ways to combine elements and compounds to form different mixtures. A summary of how to distinguish between the various major classifications of matter is shown in (Figure 1.11).

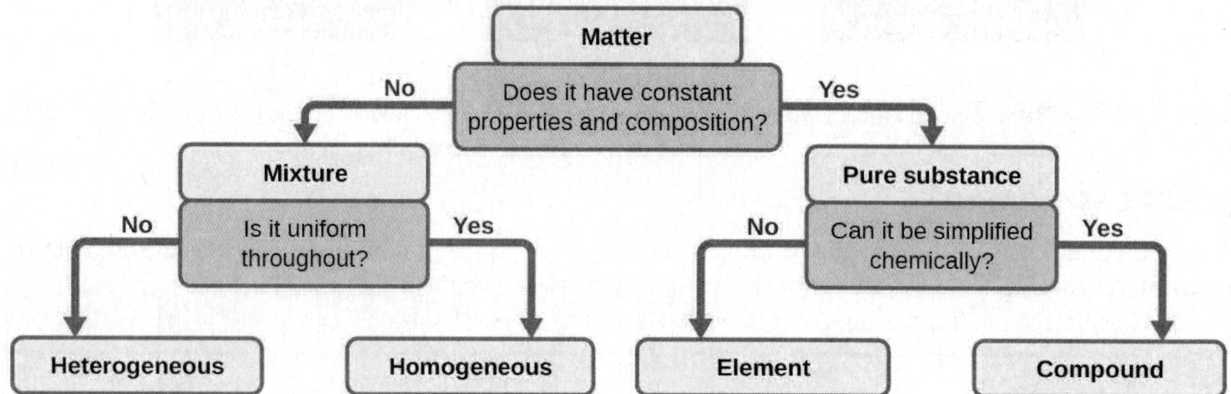

FIGURE 1.11 Depending on its properties, a given substance can be classified as a homogeneous mixture, a heterogeneous mixture, a compound, or an element.

Eleven elements make up about 99% of the earth's crust and atmosphere (Table 1.1). Oxygen constitutes nearly one-half and silicon about one-quarter of the total quantity of these elements. A majority of elements on earth are found in chemical combinations with other elements; about one-quarter of the elements are also found in the free state.

Elemental Composition of Earth

Element	Symbol	Percent Mass	Element	Symbol	Percent Mass
oxygen	O	49.20	chlorine	Cl	0.19
silicon	Si	25.67	phosphorus	P	0.11
aluminum	Al	7.50	manganese	Mn	0.09
iron	Fe	4.71	carbon	C	0.08
calcium	Ca	3.39	sulfur	S	0.06
sodium	Na	2.63	barium	Ba	0.04
potassium	K	2.40	nitrogen	N	0.03
magnesium	Mg	1.93	fluorine	F	0.03
hydrogen	H	0.87	strontium	Sr	0.02
titanium	Ti	0.58	all others	-	0.47

TABLE 1.1

Atoms and Molecules

An **atom** is the smallest particle of an element that has the properties of that element and can enter into a chemical combination. Consider the element gold, for example. Imagine cutting a gold nugget in half, then cutting one of the halves in half, and repeating this process until a piece of gold remained that was so small that it could not be cut in half (regardless of how tiny your knife may be). This minimally sized piece of gold is an atom (from the Greek *atomos*, meaning "indivisible") (Figure 1.12). This atom would no longer be gold if it were divided any further.

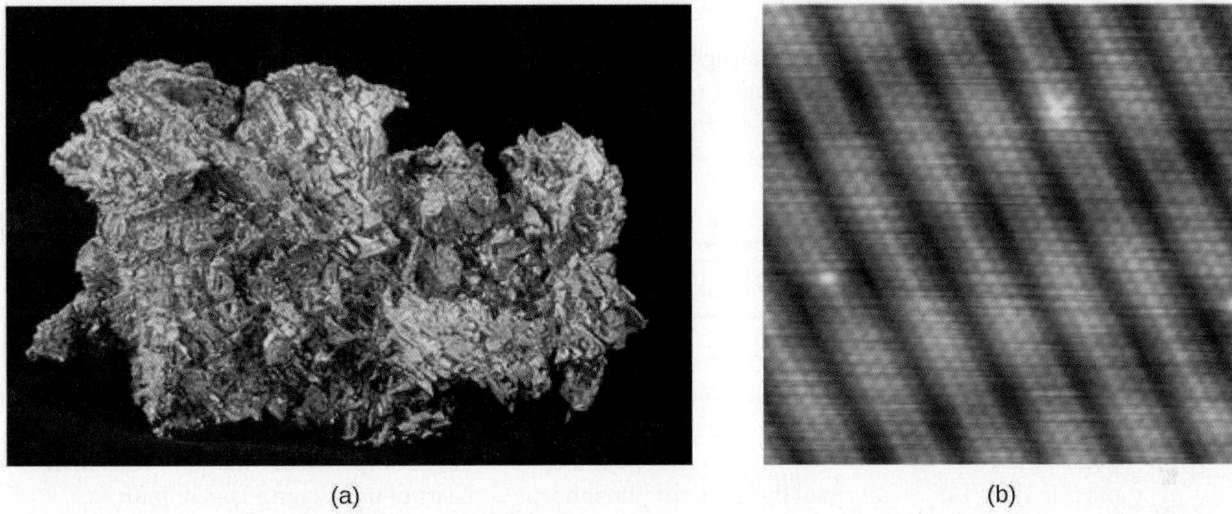

FIGURE 1.12 (a) This photograph shows a gold nugget. (b) A scanning-tunneling microscope (STM) can generate views of the surfaces of solids, such as this image of a gold crystal. Each sphere represents one gold atom. (credit a: modification of work by United States Geological Survey; credit b: modification of work by "Erwinrossen"/Wikimedia Commons)

The first suggestion that matter is composed of atoms is attributed to the Greek philosophers Leucippus and Democritus, who developed their ideas in the 5th century BCE. However, it was not until the early nineteenth century that John Dalton (1766–1844), a British schoolteacher with a keen interest in science, supported this hypothesis with quantitative measurements. Since that time, repeated experiments have confirmed many aspects of this hypothesis, and it has become one of the central theories of chemistry. Other aspects of Dalton's atomic theory are still used but with minor revisions (details of Dalton's theory are provided in the chapter on atoms and molecules).

An atom is so small that its size is difficult to imagine. One of the smallest things we can see with our unaided eye is a single thread of a spider web: These strands are about 1/10,000 of a centimeter (0.0001 cm) in diameter. Although the cross-section of one strand is almost impossible to see without a microscope, it is huge on an atomic scale. A single carbon atom in the web has a diameter of about 0.000000015 centimeter, and it would take about 7000 carbon atoms to span the diameter of the strand. To put this in perspective, if a carbon atom were the size of a dime, the cross-section of one strand would be larger than a football field, which would require about 150 million carbon atom "dimes" to cover it. (Figure 1.13) shows increasingly close microscopic and atomic-level views of ordinary cotton.

FIGURE 1.13 These images provide an increasingly closer view: (a) a cotton boll, (b) a single cotton fiber viewed under an optical microscope (magnified 40 times), (c) an image of a cotton fiber obtained with an electron microscope (much higher magnification than with the optical microscope); and (d and e) atomic-level models of the fiber (spheres of different colors represent atoms of different elements). (credit c: modification of work by "Featheredtar"/Wikimedia Commons)

An atom is so light that its mass is also difficult to imagine. A billion lead atoms (1,000,000,000 atoms) weigh about 3×10^{-13} grams, a mass that is far too light to be weighed on even the world's most sensitive balances. It would require over 300,000,000,000,000 lead atoms (300 trillion, or 3×10^{14}) to be weighed, and they would weigh only 0.0000001 gram.

It is rare to find collections of individual atoms. Only a few elements, such as the gases helium, neon, and argon, consist of a collection of individual atoms that move about independently of one another. Other elements, such as the gases hydrogen, nitrogen, oxygen, and chlorine, are composed of units that consist of pairs of atoms (Figure 1.14). One form of the element phosphorus consists of units composed of four phosphorus atoms. The element sulfur exists in various forms, one of which consists of units composed of eight sulfur atoms. These units are called molecules. A **molecule** consists of two or more atoms joined by strong forces called chemical bonds. The atoms in a molecule move around as a unit, much like the cans of soda in a six-pack or a bunch of keys joined together on a single key ring. A molecule may consist of two or more identical atoms, as in the molecules found in the elements hydrogen, oxygen, and sulfur, or it may consist of two or more different atoms, as in the molecules found in water. Each water molecule is a unit that contains two hydrogen atoms and one oxygen atom. Each glucose molecule is a unit that contains 6 carbon atoms, 12 hydrogen atoms, and 6 oxygen atoms. Like atoms, molecules are incredibly small and light. If an ordinary glass of water were enlarged to the size of the earth, the water molecules inside it would be about the size of golf balls.

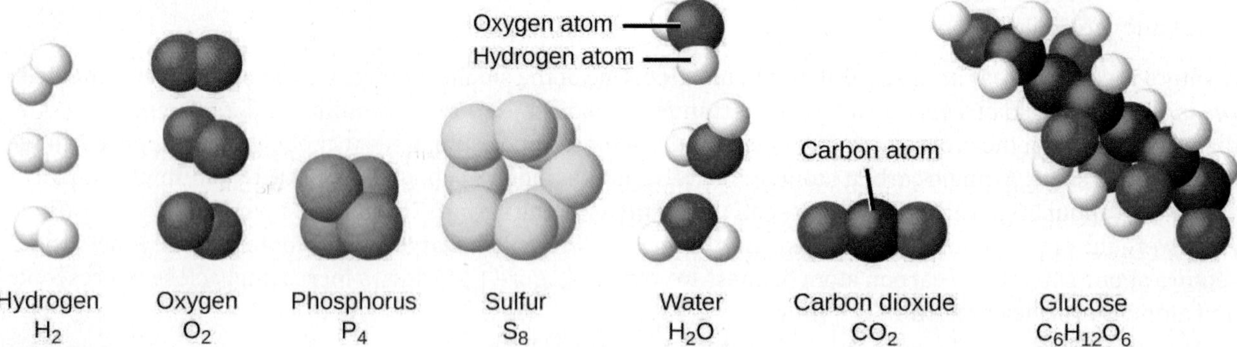

FIGURE 1.14 The elements hydrogen, oxygen, phosphorus, and sulfur form molecules consisting of two or more atoms of the same element. The compounds water, carbon dioxide, and glucose consist of combinations of atoms of different elements.

Chemistry in Everyday Life

Decomposition of Water / Production of Hydrogen

Water consists of the elements hydrogen and oxygen combined in a 2 to 1 ratio. Water can be broken down into hydrogen and oxygen gases by the addition of energy. One way to do this is with a battery or power supply, as shown in (Figure 1.15).

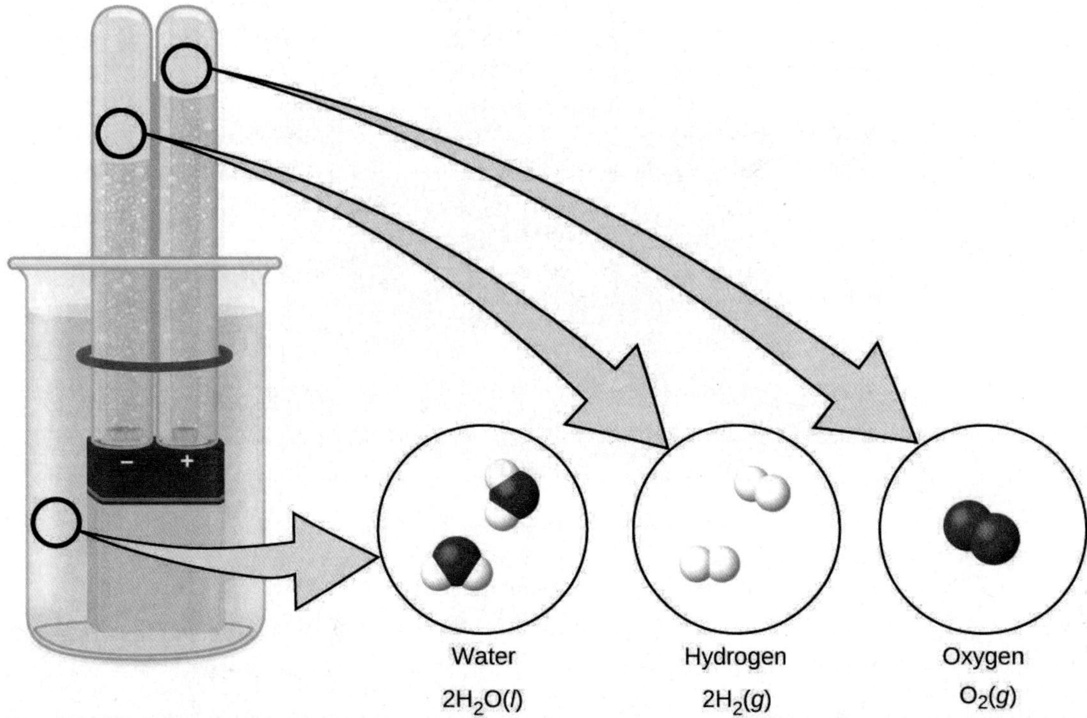

FIGURE 1.15 The decomposition of water is shown at the macroscopic, microscopic, and symbolic levels. The battery provides an electric current (microscopic) that decomposes water. At the macroscopic level, the liquid separates into the gases hydrogen (on the left) and oxygen (on the right). Symbolically, this change is presented by showing how liquid H_2O separates into H_2 and O_2 gases.

The breakdown of water involves a rearrangement of the atoms in water molecules into different molecules, each composed of two hydrogen atoms and two oxygen atoms, respectively. Two water molecules form one oxygen molecule and two hydrogen molecules. The representation for what occurs, $2H_2O(l) \longrightarrow 2H_2(g) + O_2(g)$, will be explored in more depth in later chapters.

The two gases produced have distinctly different properties. Oxygen is not flammable but is required for combustion of a fuel, and hydrogen is highly flammable and a potent energy source. How might this knowledge be applied in our world? One application involves research into more fuel-efficient transportation. Fuel-cell vehicles (FCV) run on hydrogen instead of gasoline (Figure 1.16). They are more efficient than vehicles with internal combustion engines, are nonpolluting, and reduce greenhouse gas emissions, making us less dependent on fossil fuels. FCVs are not yet economically viable, however, and current hydrogen production depends on natural gas. If we can develop a process to economically decompose water, or produce hydrogen in another environmentally sound way, FCVs may be the way of the future.

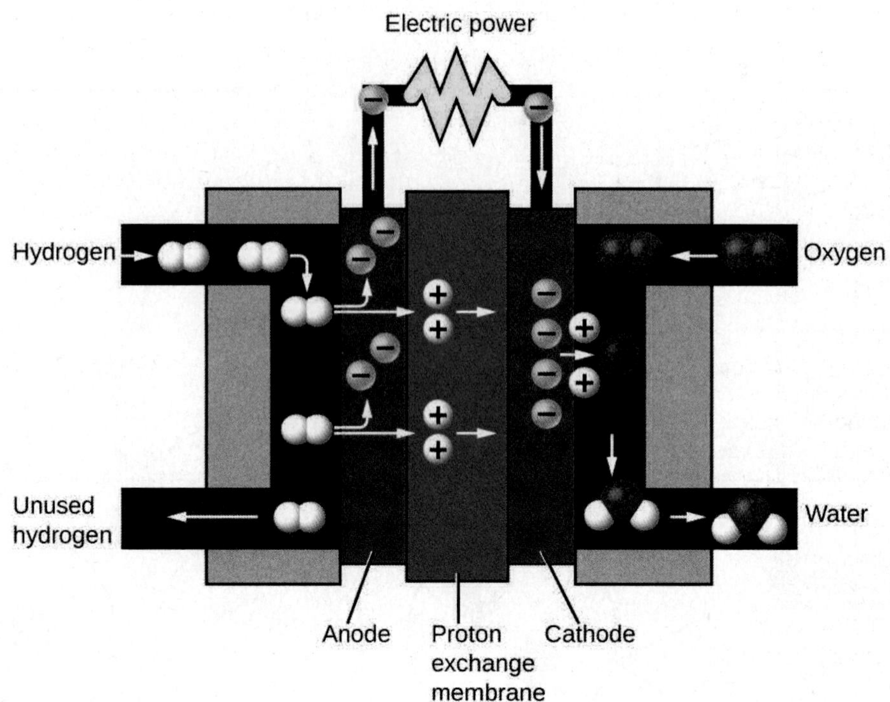

FIGURE 1.16 A fuel cell generates electrical energy from hydrogen and oxygen via an electrochemical process and produces only water as the waste product.

Chemistry in Everyday Life

Chemistry of Cell Phones

Imagine how different your life would be without cell phones (Figure 1.17) and other smart devices. Cell phones are made from numerous chemical substances, which are extracted, refined, purified, and assembled using an extensive and in-depth understanding of chemical principles. About 30% of the elements that are found in nature are found within a typical smart phone. The case/body/frame consists of a combination of sturdy, durable polymers composed primarily of carbon, hydrogen, oxygen, and nitrogen [acrylonitrile butadiene styrene (ABS) and polycarbonate thermoplastics], and light, strong, structural metals, such as aluminum, magnesium, and iron. The display screen is made from a specially toughened glass (silica glass strengthened by the addition of aluminum, sodium, and potassium) and coated with a material to make it conductive (such as indium tin oxide). The circuit board uses a semiconductor material (usually silicon); commonly used metals like copper, tin, silver, and gold; and more unfamiliar elements such as yttrium, praseodymium, and gadolinium. The battery relies upon lithium ions and a variety of other materials, including iron, cobalt, copper, polyethylene oxide, and polyacrylonitrile.

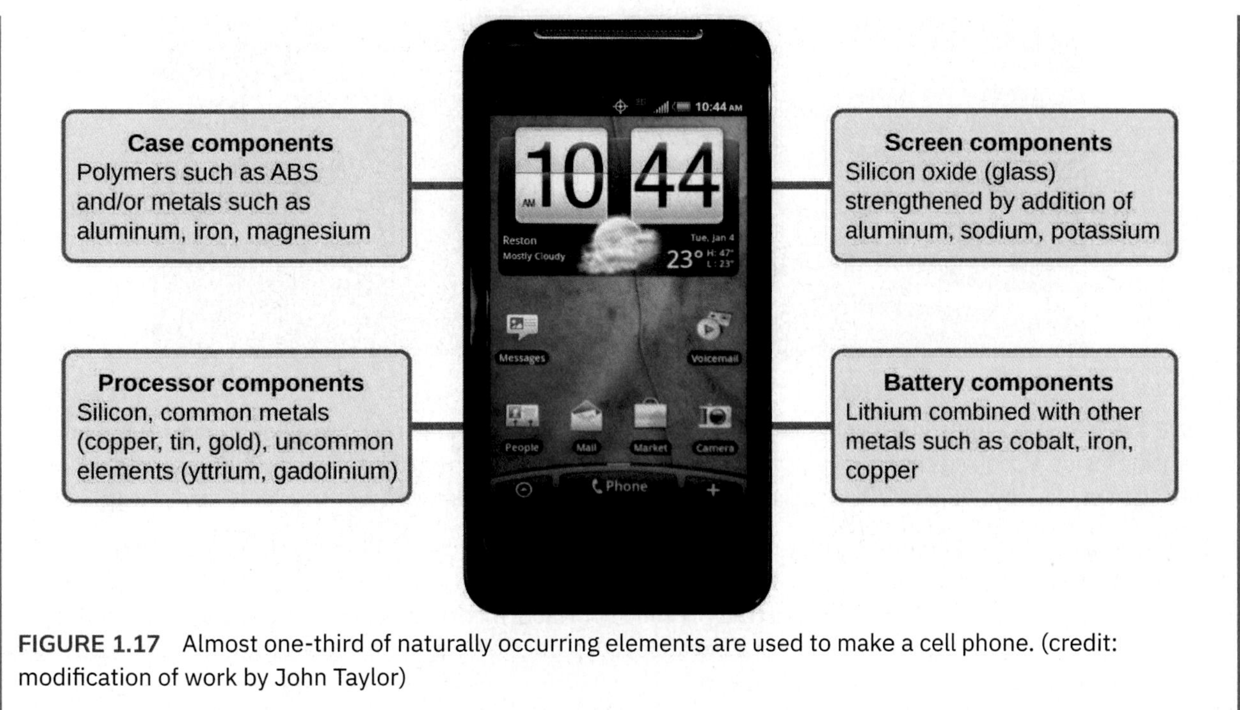

FIGURE 1.17 Almost one-third of naturally occurring elements are used to make a cell phone. (credit: modification of work by John Taylor)

1.3 Physical and Chemical Properties

LEARNING OBJECTIVES

By the end of this section, you will be able to:

- Identify properties of and changes in matter as physical or chemical
- Identify properties of matter as extensive or intensive

The characteristics that distinguish one substance from another are called properties. A **physical property** is a characteristic of matter that is not associated with a change in its chemical composition. Familiar examples of physical properties include density, color, hardness, melting and boiling points, and electrical conductivity. Some physical properties, such as density and color, may be observed without changing the physical state of the matter. Other physical properties, such as the melting temperature of iron or the freezing temperature of water, can only be observed as matter undergoes a physical change. A **physical change** is a change in the state or properties of matter without any accompanying change in the chemical identities of the substances contained in the matter. Physical changes are observed when wax melts, when sugar dissolves in coffee, and when steam condenses into liquid water (Figure 1.18). Other examples of physical changes include magnetizing and demagnetizing metals (as is done with common antitheft security tags) and grinding solids into powders (which can sometimes yield noticeable changes in color). In each of these examples, there is a change in the physical state, form, or properties of the substance, but no change in its chemical composition.

(a) (b)

FIGURE 1.18 (a) Wax undergoes a physical change when solid wax is heated and forms liquid wax. (b) Steam condensing inside a cooking pot is a physical change, as water vapor is changed into liquid water. (credit a: modification of work by “95jb14”/Wikimedia Commons; credit b: modification of work by “mjneuby”/Flickr)

The change of one type of matter into another type (or the inability to change) is a **chemical property**. Examples of chemical properties include flammability, toxicity, acidity, and many other types of reactivity. Iron, for example, combines with oxygen in the presence of water to form rust; chromium does not oxidize (Figure 1.19). Nitroglycerin is very dangerous because it explodes easily; neon poses almost no hazard because it is very unreactive.

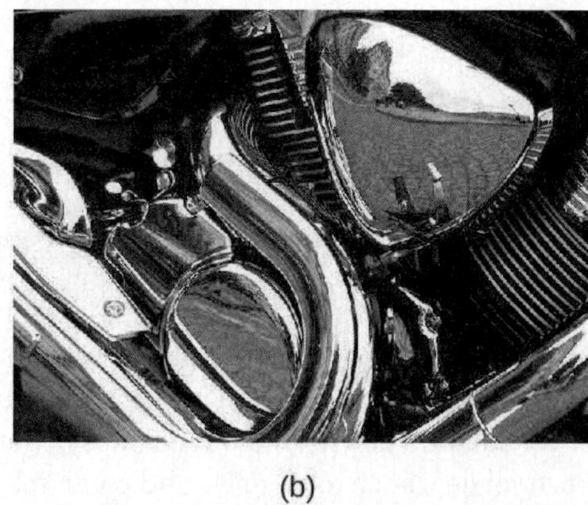

(a) (b)

FIGURE 1.19 (a) One of the chemical properties of iron is that it rusts; (b) one of the chemical properties of chromium is that it does not. (credit a: modification of work by Tony Hisgett; credit b: modification of work by “Atoma”/Wikimedia Commons)

A **chemical change** always produces one or more types of matter that differ from the matter present before the change. The formation of rust is a chemical change because rust is a different kind of matter than the iron, oxygen, and water present before the rust formed. The explosion of nitroglycerin is a chemical change because the gases produced are very different kinds of matter from the original substance. Other examples of chemical changes include reactions that are performed in a lab (such as copper reacting with nitric acid), all forms of combustion (burning), and food being cooked, digested, or rotting (Figure 1.20).

FIGURE 1.20 (a) Copper and nitric acid undergo a chemical change to form copper nitrate and brown, gaseous nitrogen dioxide. (b) During the combustion of a match, cellulose in the match and oxygen from the air undergo a chemical change to form carbon dioxide and water vapor. (c) Cooking red meat causes a number of chemical changes, including the oxidation of iron in myoglobin that results in the familiar red-to-brown color change. (d) A banana turning brown is a chemical change as new, darker (and less tasty) substances form. (credit b: modification of work by Jeff Turner; credit c: modification of work by Gloria Cabada-Leman; credit d: modification of work by Roberto Verzo)

Properties of matter fall into one of two categories. If the property depends on the amount of matter present, it is an **extensive property**. The mass and volume of a substance are examples of extensive properties; for instance, a gallon of milk has a larger mass than a cup of milk. The value of an extensive property is directly proportional to the amount of matter in question. If the property of a sample of matter does not depend on the amount of matter present, it is an **intensive property**. Temperature is an example of an intensive property. If the gallon and cup of milk are each at 20 °C (room temperature), when they are combined, the temperature remains at 20 °C. As another example, consider the distinct but related properties of heat and temperature. A drop of hot cooking oil spattered on your arm causes brief, minor discomfort, whereas a pot of hot oil yields severe burns. Both the drop and the pot of oil are at the same temperature (an intensive property), but the pot clearly contains much more heat (extensive property).

Chemistry in Everyday Life

Hazard Diamond

You may have seen the symbol shown in Figure 1.21 on containers of chemicals in a laboratory or workplace. Sometimes called a “fire diamond” or “hazard diamond,” this chemical hazard diamond provides valuable information that briefly summarizes the various dangers of which to be aware when working with a particular substance.

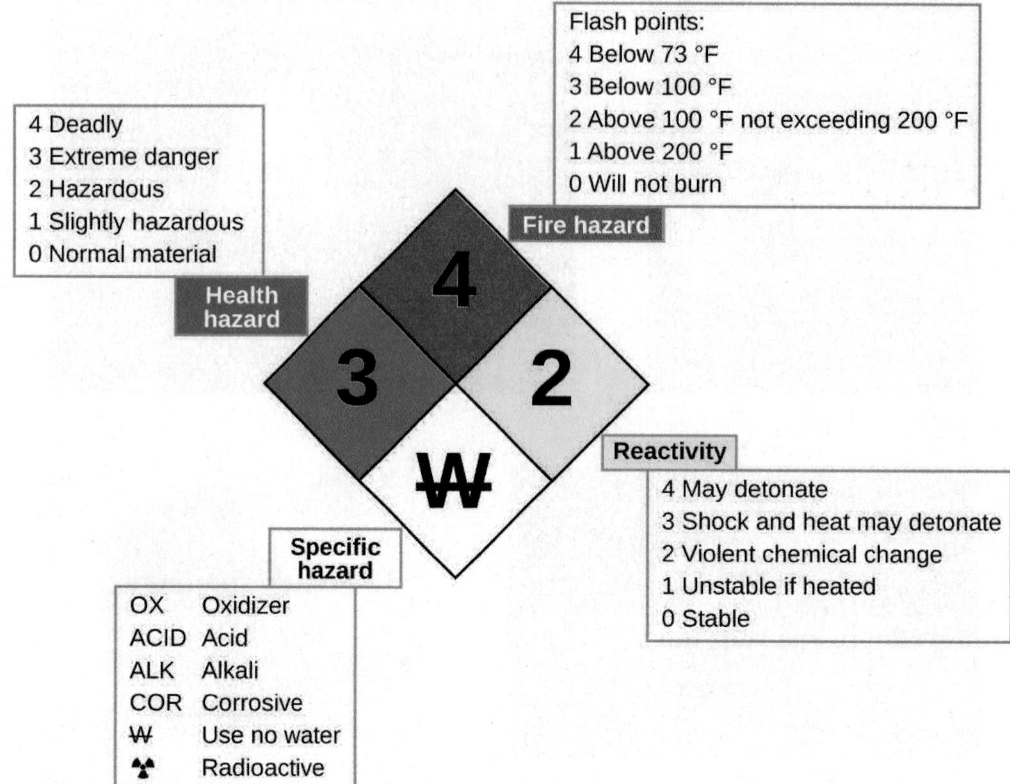

FIGURE 1.21 The National Fire Protection Agency (NFPA) hazard diamond summarizes the major hazards of a chemical substance.

The National Fire Protection Agency (NFPA) 704 Hazard Identification System was developed by NFPA to provide safety information about certain substances. The system details flammability, reactivity, health, and other hazards. Within the overall diamond symbol, the top (red) diamond specifies the level of fire hazard (temperature range for flash point). The blue (left) diamond indicates the level of health hazard. The yellow (right) diamond describes reactivity hazards, such as how readily the substance will undergo detonation or a violent chemical change. The white (bottom) diamond points out special hazards, such as if it is an oxidizer (which allows the substance to burn in the absence of air/oxygen), undergoes an unusual or dangerous reaction with water, is corrosive, acidic, alkaline, a biological hazard, radioactive, and so on. Each hazard is rated on a scale from 0 to 4, with 0 being no hazard and 4 being extremely hazardous.

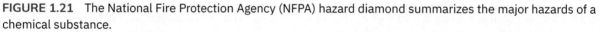

While many elements differ dramatically in their chemical and physical properties, some elements have similar properties. For example, many elements conduct heat and electricity well, whereas others are poor conductors. These properties can be used to sort the elements into three classes: metals (elements that conduct well), nonmetals (elements that conduct poorly), and metalloids (elements that have intermediate conductivities).

The periodic table is a table of elements that places elements with similar properties close together (Figure 1.22). You will learn more about the periodic table as you continue your study of chemistry.

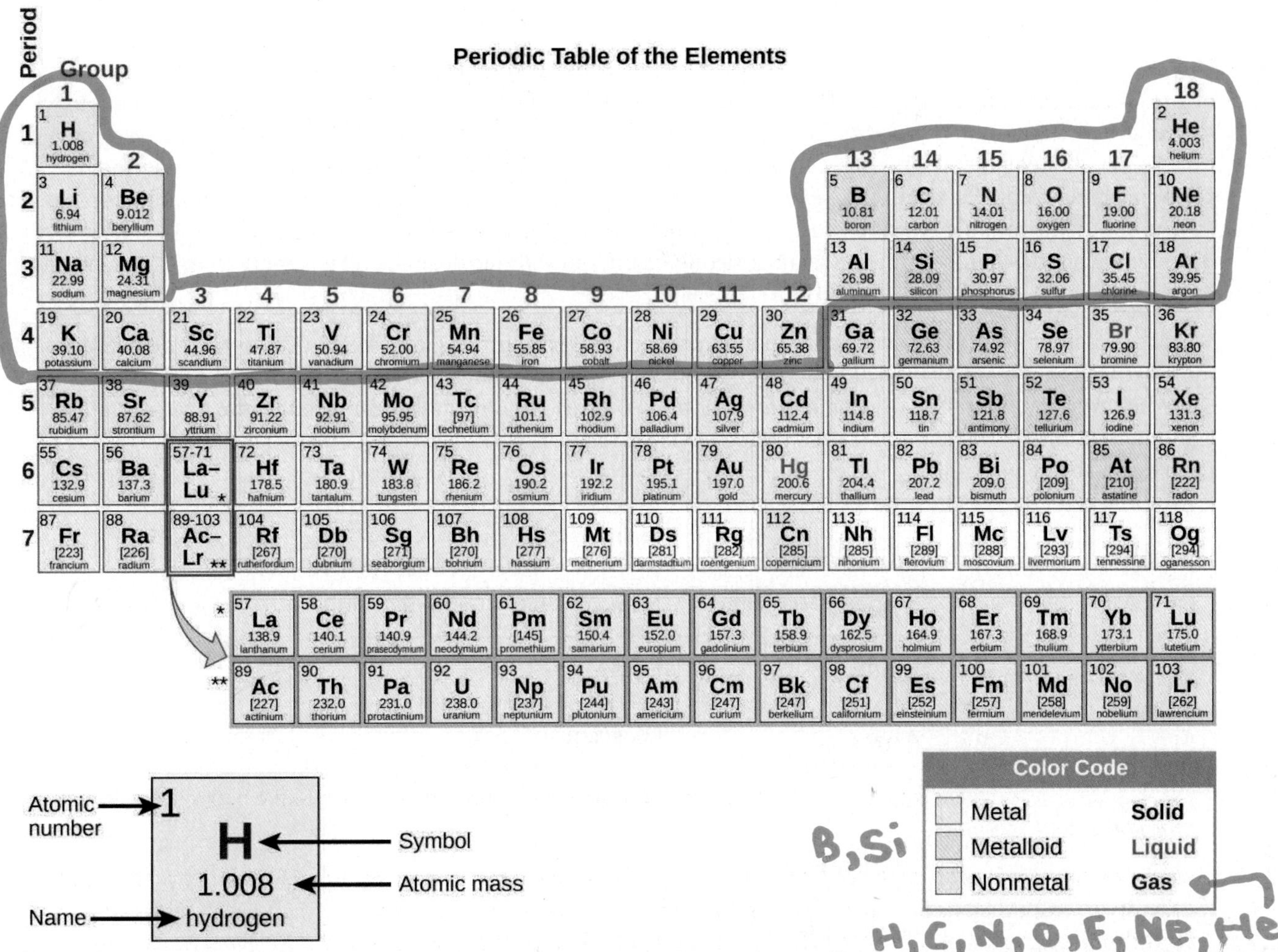

FIGURE 1.22 The periodic table shows how elements may be grouped according to certain similar properties. Note the background color denotes whether an element is a metal, metalloid, or nonmetal, whereas the element symbol color indicates whether it is a solid, liquid, or gas.

1.4 Measurements

LEARNING OBJECTIVES

By the end of this section, you will be able to:

- Explain the process of measurement
- Identify the three basic parts of a quantity
- Describe the properties and units of length, mass, volume, density, temperature, and time
- Perform basic unit calculations and conversions in the metric and other unit systems

Measurements provide much of the information that informs the hypotheses, theories, and laws describing the behavior of matter and energy in both the macroscopic and microscopic domains of chemistry. Every measurement provides three kinds of information: the size or magnitude of the measurement (a number); a standard of comparison for the measurement (a unit); and an indication of the uncertainty of the measurement. While the number and unit are explicitly represented when a quantity is written, the uncertainty is an aspect of the measurement result that is more implicitly represented and will be discussed later.

The number in the measurement can be represented in different ways, including decimal form and scientific notation. (Scientific notation is also known as exponential notation; a review of this topic can be found in Appendix B.) For example, the maximum takeoff weight of a Boeing 777-200ER airliner is 298,000 kilograms, which can also be written as 2.98×10^5 kg. The mass of the average mosquito is about 0.0000025 kilograms, which can be written as 2.5×10^{-6} kg.

Units, such as liters, pounds, and centimeters, are standards of comparison for measurements. A 2-liter bottle of a soft drink contains a volume of beverage that is twice that of the accepted volume of 1 liter. The meat used to prepare a 0.25-pound hamburger weighs one-fourth as much as the accepted weight of 1 pound. Without units, a number can be meaningless, confusing, or possibly life threatening. Suppose a doctor prescribes phenobarbital to control a patient's seizures and states a dosage of "100" without specifying units. Not only will this be confusing to the medical professional giving the dose, but the consequences can be dire: 100 mg given three times per day can be effective as an anticonvulsant, but a single dose of 100 g is more than 10 times the lethal amount.

The measurement units for seven fundamental properties ("base units") are listed in Table 1.2. The standards for these units are fixed by international agreement, and they are called the **International System of Units** or **SI Units** (from the French, *Le Système International d'Unités*). SI units have been used by the United States National Institute of Standards and Technology (NIST) since 1964. Units for other properties may be derived from these seven base units.

Base Units of the SI System

Property Measured	Name of Unit	Symbol of Unit
length	meter	m
mass	kilogram	kg
time	second	s
temperature	kelvin	K
electric current	ampere	A
amount of substance	mole	mol
luminous intensity	candela	cd

TABLE 1.2

Everyday measurement units are often defined as fractions or multiples of other units. Milk is commonly packaged in containers of 1 gallon (4 quarts), 1 quart (0.25 gallon), and one pint (0.5 quart). This same approach is used with SI units, but these fractions or multiples are always powers of 10. Fractional or multiple SI units are named using a prefix and the name of the base unit. For example, a length of 1000 meters is also called a kilometer because the prefix *kilo* means "one thousand," which in scientific notation is 10^3 (1 kilometer = 1000 m = 10^3 m). The prefixes used and the powers to which 10 are raised are listed in Table 1.3.

Common Unit Prefixes

Prefix	Symbol	Factor	Example
femto	f	10^{-15}	1 femtosecond (fs) = 1×10^{-15} s (0.000000000000001 s)
pico	p	10^{-12}	1 picometer (pm) = 1×10^{-12} m (0.000000000001 m)

TABLE 1.3

Prefix	Symbol	Factor	Example
nano	n	10^{-9}	4 nanograms (ng) = 4×10^{-9} g (0.000000004 g)
micro	µ	10^{-6}	1 microliter (µL) = 1×10^{-6} L (0.000001 L)
milli	m	10^{-3}	2 millimoles (mmol) = 2×10^{-3} mol (0.002 mol)
centi	c	10^{-2}	7 centimeters (cm) = 7×10^{-2} m (0.07 m)
deci	d	10^{-1}	1 deciliter (dL) = 1×10^{-1} L (0.1 L)
kilo	k	10^{3}	1 kilometer (km) = 1×10^{3} m (1000 m)
mega	M	10^{6}	3 megahertz (MHz) = 3×10^{6} Hz (3,000,000 Hz)
giga	G	10^{9}	8 gigayears (Gyr) = 8×10^{9} yr (8,000,000,000 yr)
tera	T	10^{12}	5 terawatts (TW) = 5×10^{12} W (5,000,000,000,000 W)

TABLE 1.3

LINK TO LEARNING

Need a refresher or more practice with scientific notation? Visit this site (http://openstax.org/l/16notation) to go over the basics of scientific notation.

SI Base Units

The initial units of the metric system, which eventually evolved into the SI system, were established in France during the French Revolution. The original standards for the meter and the kilogram were adopted there in 1799 and eventually by other countries. This section introduces four of the SI base units commonly used in chemistry. Other SI units will be introduced in subsequent chapters.

Length

The standard unit of **length** in both the SI and original metric systems is the **meter (m)**. A meter was originally specified as 1/10,000,000 of the distance from the North Pole to the equator. It is now defined as the distance light in a vacuum travels in 1/299,792,458 of a second. A meter is about 3 inches longer than a yard (Figure 1.23); one meter is about 39.37 inches or 1.094 yards. Longer distances are often reported in kilometers (1 km = 1000 m = 10^3 m), whereas shorter distances can be reported in centimeters (1 cm = 0.01 m = 10^{-2} m) or millimeters (1 mm = 0.001 m = 10^{-3} m).

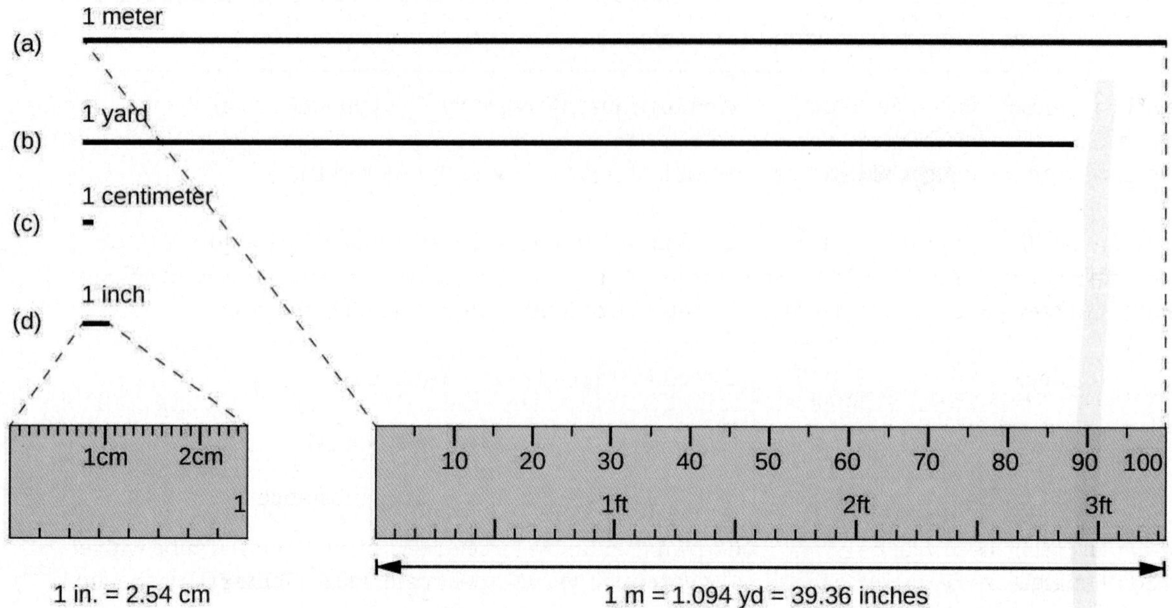

FIGURE 1.23 The relative lengths of 1 m, 1 yd, 1 cm, and 1 in. are shown (not actual size), as well as comparisons of 2.54 cm and 1 in., and of 1 m and 1.094 yd.

Mass

The standard unit of mass in the SI system is the **kilogram (kg)**. The kilogram was previously defined by the International Union of Pure and Applied Chemistry (IUPAC) as the mass of a specific reference object. This object was originally one liter of pure water, and more recently it was a metal cylinder made from a platinum-iridium alloy with a height and diameter of 39 mm (Figure 1.24). In May 2019, this definition was changed to one that is based instead on precisely measured values of several fundamental physical constants.[1]. One kilogram is about 2.2 pounds. The gram (g) is exactly equal to 1/1000 of the mass of the kilogram (10^{-3} kg).

FIGURE 1.24 This replica prototype kilogram as previously defined is housed at the National Institute of Standards and Technology (NIST) in Maryland. (credit: National Institutes of Standards and Technology)

Temperature

Temperature is an intensive property. The SI unit of temperature is the **kelvin (K)**. The IUPAC convention is to use kelvin (all lowercase) for the word, K (uppercase) for the unit symbol, and neither the word "degree" nor the degree symbol (°). The degree **Celsius (°C)** is also allowed in the SI system, with both the word "degree" and the degree symbol used for Celsius measurements. Celsius degrees are the same magnitude as those of kelvin, but the two scales place their zeros in different places. Water freezes at 273.15 K (0 °C) and boils at 373.15 K (100 °C) by definition, and normal human body temperature is approximately 310 K (37 °C). The conversion

1 For details see https://www.nist.gov/pml/weights-and-measures/si-units-mass

between these two units and the Fahrenheit scale will be discussed later in this chapter.

Time

The SI base unit of time is the **second (s)**. Small and large time intervals can be expressed with the appropriate prefixes; for example, 3 microseconds = 0.000003 s = 3×10^{-6} and 5 megaseconds = 5,000,000 s = 5×10^{6} s. Alternatively, hours, days, and years can be used.

Derived SI Units

We can derive many units from the seven SI base units. For example, we can use the base unit of length to define a unit of volume, and the base units of mass and length to define a unit of density.

Volume

Volume is the measure of the amount of space occupied by an object. The standard SI unit of volume is defined by the base unit of length (Figure 1.25). The standard volume is a **cubic meter (m^3)**, a cube with an edge length of exactly one meter. To dispense a cubic meter of water, we could build a cubic box with edge lengths of exactly one meter. This box would hold a cubic meter of water or any other substance.

A more commonly used unit of volume is derived from the decimeter (0.1 m, or 10 cm). A cube with edge lengths of exactly one decimeter contains a volume of one cubic decimeter (dm^3). A **liter (L)** is the more common name for the cubic decimeter. One liter is about 1.06 quarts.

A **cubic centimeter (cm^3)** is the volume of a cube with an edge length of exactly one centimeter. The abbreviation **cc** (for **c**ubic **c**entimeter) is often used by health professionals. A cubic centimeter is equivalent to a **milliliter (mL)** and is 1/1000 of a liter.

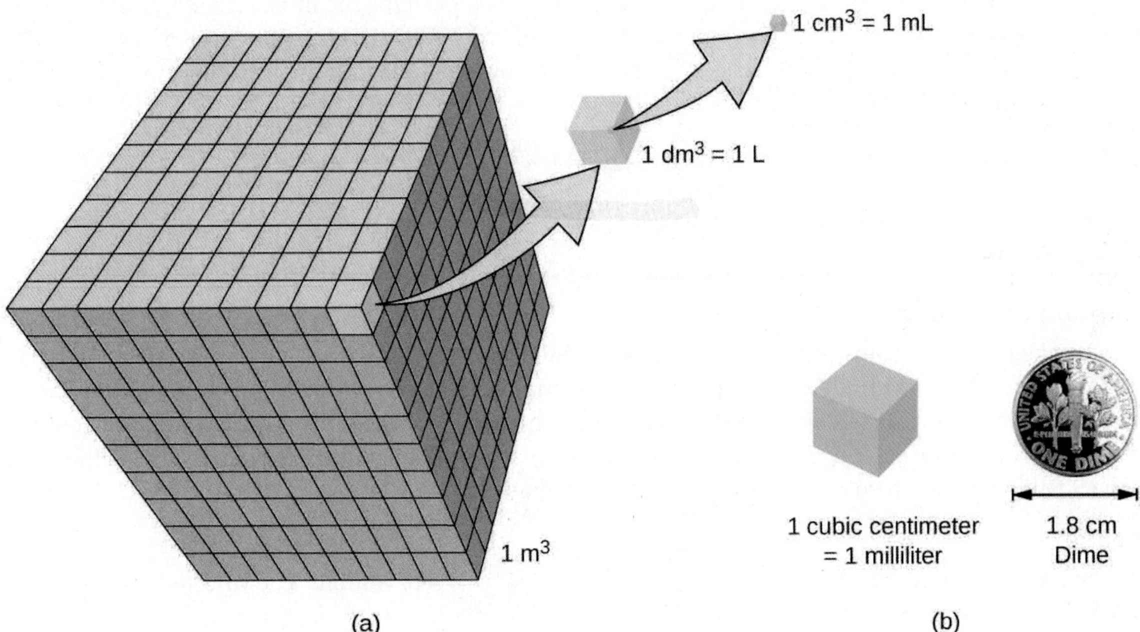

FIGURE 1.25 (a) The relative volumes are shown for cubes of 1 m^3, 1 dm^3 (1 L), and 1 cm^3 (1 mL) (not to scale). (b) The diameter of a dime is compared relative to the edge length of a 1-cm^3 (1-mL) cube.

Density

We use the mass and volume of a substance to determine its density. Thus, the units of density are defined by the base units of mass and length.

The **density** of a substance is the ratio of the mass of a sample of the substance to its volume. The SI unit for density is the kilogram per cubic meter (kg/m^3). For many situations, however, this is an inconvenient unit, and we often use grams per cubic centimeter (g/cm^3) for the densities of solids and liquids, and grams per liter (g/L) for gases. Although there are exceptions, most liquids and solids have densities that range from about 0.7 g/cm^3 (the density of gasoline) to 19 g/cm^3 (the density of gold). The density of air is about 1.2 g/L. Table 1.4

shows the densities of some common substances.

Densities of Common Substances

Solids	Liquids	Gases (at 25 °C and 1 atm)
ice (at 0 °C) 0.92 g/cm^3	water 1.0 g/cm^3	dry air 1.20 g/L
oak (wood) 0.60–0.90 g/cm^3	ethanol 0.79 g/cm^3	oxygen 1.31 g/L
iron 7.9 g/cm^3	acetone 0.79 g/cm^3	nitrogen 1.14 g/L
copper 9.0 g/cm^3	glycerin 1.26 g/cm^3	carbon dioxide 1.80 g/L
lead 11.3 g/cm^3	olive oil 0.92 g/cm^3	helium 0.16 g/L
silver 10.5 g/cm^3	gasoline 0.70–0.77 g/cm^3	neon 0.83 g/L
gold 19.3 g/cm^3	mercury 13.6 g/cm^3	radon 9.1 g/L

TABLE 1.4

While there are many ways to determine the density of an object, perhaps the most straightforward method involves separately finding the mass and volume of the object, and then dividing the mass of the sample by its volume. In the following example, the mass is found directly by weighing, but the volume is found indirectly through length measurements.

$$\text{density} = \frac{\text{mass}}{\text{volume}}$$

 **EXAMPLE 1.1**

Calculation of Density

Gold—in bricks, bars, and coins—has been a form of currency for centuries. In order to swindle people into paying for a brick of gold without actually investing in a brick of gold, people have considered filling the centers of hollow gold bricks with lead to fool buyers into thinking that the entire brick is gold. It does not work: Lead is a dense substance, but its density is not as great as that of gold, 19.3 g/cm^3. What is the density of lead if a cube of lead has an edge length of 2.00 cm and a mass of 90.7 g?

Solution

The density of a substance can be calculated by dividing its mass by its volume. The volume of a cube is calculated by cubing the edge length.

$$\text{volume of lead cube} = 2.00 \text{ cm} \times 2.00 \text{ cm} \times 2.00 \text{ cm} = 8.00 \text{ cm}^3$$

$$\text{density} = \frac{\text{mass}}{\text{volume}} = \frac{90.7 \text{ g}}{8.00 \text{ cm}^3} = 11.3 \text{ g/cm}^3$$

(We will discuss the reason for rounding to the first decimal place in the next section.)

Check Your Learning

(a) To three decimal places, what is the volume of a cube (cm^3) with an edge length of 0.843 cm?

(b) If the cube in part (a) is copper and has a mass of 5.34 g, what is the density of copper to two decimal places?

Answer:
(a) 0.599 cm^3; (b) 8.91 g/cm^3

LINK TO LEARNING

To learn more about the relationship between mass, volume, and density, use this interactive simulator (http://openstax.org/l/16phetmasvolden) to explore the density of different materials.

EXAMPLE 1.2

Using Displacement of Water to Determine Density

This exercise uses a simulation (http://openstax.org/l/16phetmasvolden) to illustrate an alternative approach to the determination of density that involves measuring the object's volume via displacement of water. Use the simulator to determine the densities iron and wood.

Solution

Click the "turn fluid into water" button in the simulator to adjust the density of liquid in the beaker to 1.00 g/mL. Remove the red block from the beaker and note the volume of water is 25.5 mL. Select the iron sample by clicking "iron" in the table of materials at the bottom of the screen, place the iron block on the balance pan, and observe its mass is 31.48 g. Transfer the iron block to the beaker and notice that it sinks, displacing a volume of water equal to its own volume and causing the water level to rise to 29.5 mL. The volume of the iron block is therefore:

$$v_{\text{iron}} = 29.5 \text{ mL} - 25.5 \text{ mL} = 4.0 \text{ mL}$$

The density of the iron is then calculated to be:

$$\text{density} = \frac{\text{mass}}{\text{volume}} = \frac{31.48 \text{ g}}{4.0 \text{ mL}} = 7.9 \text{ g/mL}$$

Remove the iron block from the beaker, change the block material to wood, and then repeat the mass and volume measurements. Unlike iron, the wood block does not sink in the water but instead floats on the water's surface. To measure its volume, drag it beneath the water's surface so that it is fully submerged.

$$\text{density} = \frac{\text{mass}}{\text{volume}} = \frac{1.95 \text{ g}}{3.0 \text{ mL}} = 0.65 \text{ g/mL}$$

Note: The sink versus float behavior illustrated in this example demonstrates the property of "buoyancy" (see end of chapter Exercise 1.42 and Exercise 1.43).

Check Your Learning

Following the water displacement approach, use the simulator to measure the density of the foam sample.

Answer:
0.230 g/mL

1.5 Measurement Uncertainty, Accuracy, and Precision

LEARNING OBJECTIVES

By the end of this section, you will be able to:

- Define accuracy and precision
- Distinguish exact and uncertain numbers
- Correctly represent uncertainty in quantities using significant figures
- Apply proper rounding rules to computed quantities

Counting is the only type of measurement that is free from uncertainty, provided the number of objects being counted does not change while the counting process is underway. The result of such a counting measurement is an example of an **exact number**. By counting the eggs in a carton, one can determine *exactly* how many eggs the carton contains. The numbers of defined quantities are also exact. By definition, 1 foot is exactly 12 inches, 1 inch is exactly 2.54 centimeters, and 1 gram is exactly 0.001 kilogram. Quantities derived from measurements other than counting, however, are uncertain to varying extents due to practical limitations of the measurement process used.

Significant Figures in Measurement

The numbers of measured quantities, unlike defined or directly counted quantities, are not exact. To measure the volume of liquid in a graduated cylinder, you should make a reading at the bottom of the meniscus, the lowest point on the curved surface of the liquid.

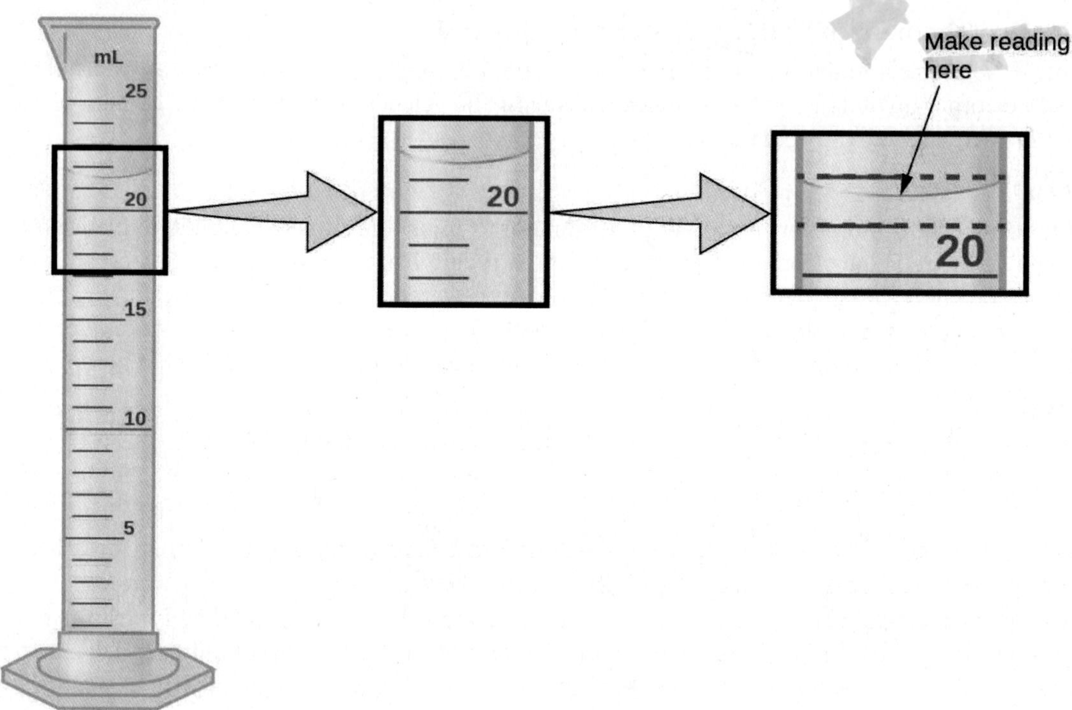

FIGURE 1.26 To measure the volume of liquid in this graduated cylinder, you must mentally subdivide the distance between the 21 and 22 mL marks into tenths of a milliliter, and then make a reading (estimate) at the bottom of the meniscus.

Refer to the illustration in Figure 1.26. The bottom of the meniscus in this case clearly lies between the 21 and 22 markings, meaning the liquid volume is *certainly* greater than 21 mL but less than 22 mL. The meniscus appears to be a bit closer to the 22-mL mark than to the 21-mL mark, and so a reasonable estimate of the liquid's volume would be 21.6 mL. In the number 21.6, then, the digits 2 and 1 are certain, but the 6 is an estimate. Some people might estimate the meniscus position to be equally distant from each of the markings and estimate the tenth-place digit as 5, while others may think it to be even closer to the 22-mL mark and estimate this digit to be 7. Note that it would be pointless to attempt to estimate a digit for the hundredths place, given that the tenths-place digit is uncertain. In general, numerical scales such as the one on this graduated cylinder will permit measurements to one-tenth of the smallest scale division. The scale in this case has 1-mL divisions, and so volumes may be measured to the nearest 0.1 mL.

This concept holds true for all measurements, even if you do not actively make an estimate. If you place a quarter on a standard electronic balance, you may obtain a reading of 6.72 g. The digits 6 and 7 are certain, and the 2 indicates that the mass of the quarter is likely between 6.71 and 6.73 grams. The quarter weighs *about* 6.72 grams, with a nominal uncertainty in the measurement of ± 0.01 gram. If the coin is weighed on a more sensitive balance, the mass might be 6.723 g. This means its mass lies between 6.722 and 6.724 grams, an uncertainty of 0.001 gram. Every measurement has some **uncertainty**, which depends on the device used (and the user's ability). All of the digits in a measurement, including the uncertain last digit, are called **significant figures** or **significant digits**. Note that zero may be a measured value; for example, if you stand on a scale that shows weight to the nearest pound and it shows "120," then the 1 (hundreds), 2 (tens) and 0 (ones) are all significant (measured) values.

A measurement result is properly reported when its significant digits accurately represent the certainty of the measurement process. But what if you were analyzing a reported value and trying to determine what is significant and what is not? Well, for starters, all nonzero digits are significant, and it is only zeros that require some thought. We will use the terms "leading," "trailing," and "captive" for the zeros and will consider how to deal with them.

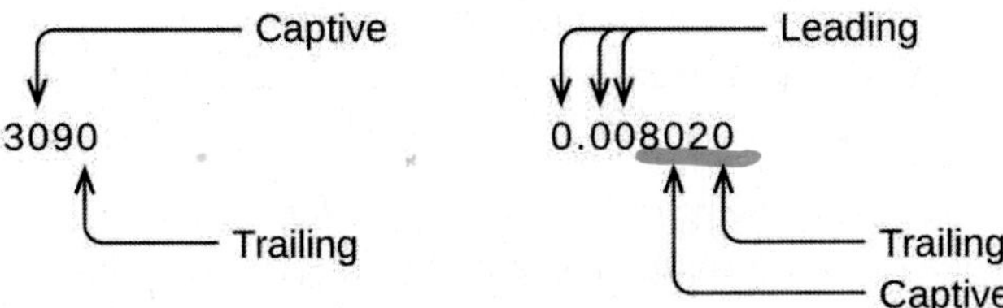

Starting with the first nonzero digit on the left, count this digit and all remaining digits to the right. This is the number of significant figures in the measurement unless the last digit is a trailing zero lying to the left of the decimal point.

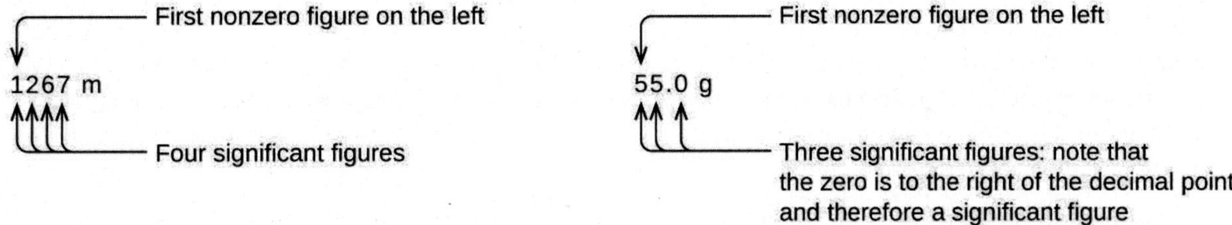

Captive zeros result from measurement and are therefore always significant. Leading zeros, however, are never significant—they merely tell us where the decimal point is located.

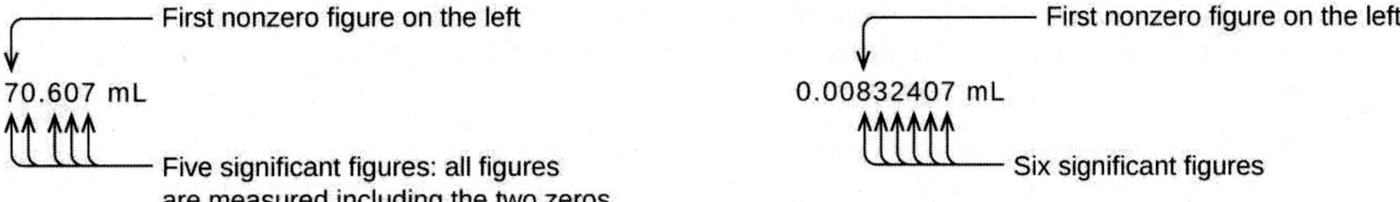

The leading zeros in this example are not significant. We could use exponential notation (as described in Appendix B) and express the number as 8.32407×10^{-3}; then the number 8.32407 contains all of the significant figures, and 10^{-3} locates the decimal point.

The number of significant figures is uncertain in a number that ends with a zero to the left of the decimal point location. The zeros in the measurement 1,300 grams could be significant or they could simply indicate where the decimal point is located. The ambiguity can be resolved with the use of exponential notation: 1.3×10^3 (two significant figures), 1.30×10^3 (three significant figures, if the tens place was measured), or 1.300×10^3 (four significant figures, if the ones place was also measured). In cases where only the decimal-formatted number is available, it is prudent to assume that all trailing zeros are not significant.

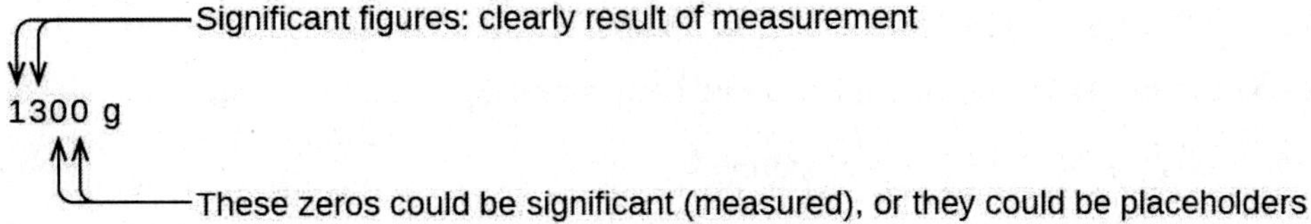

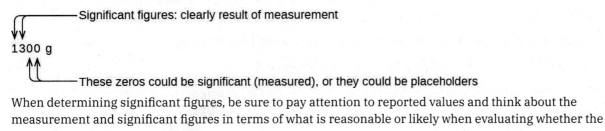

When determining significant figures, be sure to pay attention to reported values and think about the measurement and significant figures in terms of what is reasonable or likely when evaluating whether the value makes sense. For example, the official January 2014 census reported the resident population of the US as 317,297,725. Do you think the US population was correctly determined to the reported nine significant figures, that is, to the exact number of people? People are constantly being born, dying, or moving into or out of the country, and assumptions are made to account for the large number of people who are not actually counted. Because of these uncertainties, it might be more reasonable to expect that we know the population to within perhaps a million or so, in which case the population should be reported as 3.17×10^8 people.

Significant Figures in Calculations

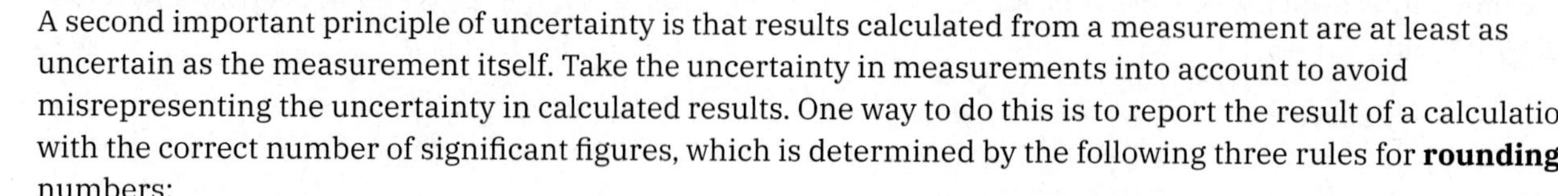

A second important principle of uncertainty is that results calculated from a measurement are at least as uncertain as the measurement itself. Take the uncertainty in measurements into account to avoid misrepresenting the uncertainty in calculated results. One way to do this is to report the result of a calculation with the correct number of significant figures, which is determined by the following three rules for **rounding** numbers:

1. When adding or subtracting numbers, round the result to the same number of decimal places as the number with the least number of decimal places (the least certain value in terms of addition and subtraction).
2. When multiplying or dividing numbers, round the result to the same number of digits as the number with the least number of significant figures (the least certain value in terms of multiplication and division).
3. If the digit to be dropped (the one immediately to the right of the digit to be retained) is less than 5, "round down" and leave the retained digit unchanged; if it is more than 5, "round up" and increase the retained digit by 1. If the dropped digit is 5, and it's either the last digit in the number or it's followed only by zeros, round up or down, whichever yields an even value for the retained digit. If any nonzero digits follow the dropped 5, round up. (The last part of this rule may strike you as a bit odd, but it's based on reliable statistics and is aimed at avoiding any bias when dropping the digit "5," since it is equally close to both possible values of the retained digit.)

The following examples illustrate the application of this rule in rounding a few different numbers to three significant figures:

- 0.028675 rounds "up" to 0.0287 (the dropped digit, 7, is greater than 5)
- 18.3384 rounds "down" to 18.3 (the dropped digit, 3, is less than 5)
- 6.8752 rounds "up" to 6.88 (the dropped digit is 5, and a nonzero digit follows it)
- 92.85 rounds "down" to 92.8 (the dropped digit is 5, and the retained digit is even)

Let's work through these rules with a few examples.

EXAMPLE 1.3

Rounding Numbers

Round the following to the indicated number of significant figures:

(a) 31.57 (to two significant figures)

(b) 8.1649 (to three significant figures)

(c) 0.051065 (to four significant figures)

(d) 0.90275 (to four significant figures)

Solution

(a) 31.57 rounds "up" to 32 (the dropped digit is 5, and the retained digit is even)

(b) 8.1649 rounds "down" to 8.16 (the dropped digit, 4, is less than 5)

(c) 0.051065 rounds "down" to 0.05106 (the dropped digit is 5, and the retained digit is even)

(d) 0.90275 rounds "up" to 0.9028 (the dropped digit is 5, and the retained digit is even)

Check Your Learning

Round the following to the indicated number of significant figures:

(a) 0.424 (to two significant figures)

(b) 0.0038661 (to three significant figures)

(c) 421.25 (to four significant figures)

(d) 28,683.5 (to five significant figures)

Answer:

(a) 0.42; (b) 0.00387; (c) 421.2; (d) 28,684

 **EXAMPLE 1.4**

Addition and Subtraction with Significant Figures

Rule: When adding or subtracting numbers, round the result to the same number of decimal places as the number with the fewest decimal places (i.e., the least certain value in terms of addition and subtraction).

(a) Add 1.0023 g and 4.383 g.

(b) Subtract 421.23 g from 486 g.

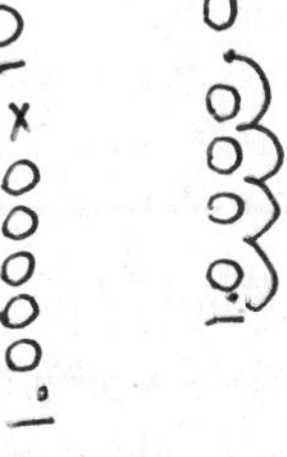

Solution

(a)

$$\begin{array}{r} 1.0023 \text{ g} \\ +\ 4.383 \text{ g} \\ \hline 5.3853 \text{ g} \end{array}$$

Answer is 5.385 g (round to the thousandths place; three decimal places)

(b)

$$\begin{array}{r} 486 \text{ g} \\ -421.23 \text{ g} \\ \hline 64.77 \text{ g} \end{array}$$

Answer is 65 g (round to the ones place; no decimal places)

1.0023 ← Ten thousandths place
+ 4.383 ← Thousandths place: least precise
5.3853
Round to thousandths
(a)

486 g
− 421.23 g
64.77 g → Answer is 65 g
Round to ones
(b)

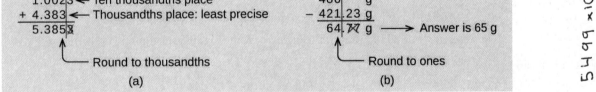

Check Your Learning

(a) Add 2.334 mL and 0.31 mL.

(b) Subtract 55.8752 m from 56.533 m.

Answer:

(a) 2.64 mL; (b) 0.658 m

EXAMPLE 1.5

Multiplication and Division with Significant Figures

Rule: When multiplying or dividing numbers, round the result to the same number of digits as the number with the fewest significant figures (the least certain value in terms of multiplication and division).

(a) Multiply 0.6238 cm by 6.6 cm.

(b) Divide 421.23 g by 486 mL.

Solution

(a)
$$\begin{array}{l} 0.6238 \text{ cm} \times 6.6 \text{ cm} = 4.11708 \text{ cm}^2 \longrightarrow \text{result is } 4.1 \text{ cm}^2 \text{ (round to two significant figures)} \\ \text{four significant figures} \times \text{two significant figures} \longrightarrow \text{two significant figures answer} \end{array}$$

(b)
$$\begin{array}{l} \frac{421.23 \text{ g}}{486 \text{ mL}} = 0.866728... \text{ g/mL} \longrightarrow \text{result is } 0.867 \text{ g/mL (round to three significant figures)} \\ \frac{\text{five significant figures}}{\text{three significant figures}} \longrightarrow \text{three significant figures answer} \end{array}$$

Check Your Learning

(a) Multiply 2.334 cm and 0.320 cm.

(b) Divide 55.8752 m by 56.53 s.

Answer:

(a) 0.747 cm^2 (b) 0.9884 m/s

In the midst of all these technicalities, it is important to keep in mind the reason for these rules about significant figures and rounding—to correctly represent the certainty of the values reported and to ensure that a calculated result is not represented as being more certain than the least certain value used in the calculation.

EXAMPLE 1.6

Calculation with Significant Figures

One common bathtub is 13.44 dm long, 5.920 dm wide, and 2.54 dm deep. Assume that the tub is rectangular and calculate its approximate volume in liters.

Solution

$$\begin{aligned} V &= l \times w \times d \\ &= 13.44 \text{ dm} \times 5.920 \text{ dm} \times 2.54 \text{ dm} \\ &= 202.09459... \text{ dm}^3 \text{ (value from calculator)} \\ &= 202 \text{ dm}^3\text{, or } 202 \text{ L (answer rounded to three significant figures)} \end{aligned}$$

Check Your Learning

What is the density of a liquid with a mass of 31.1415 g and a volume of 30.13 cm^3?

Answer:

1.034 g/mL

EXAMPLE 1.7

Experimental Determination of Density Using Water Displacement

A piece of rebar is weighed and then submerged in a graduated cylinder partially filled with water, with results as shown.

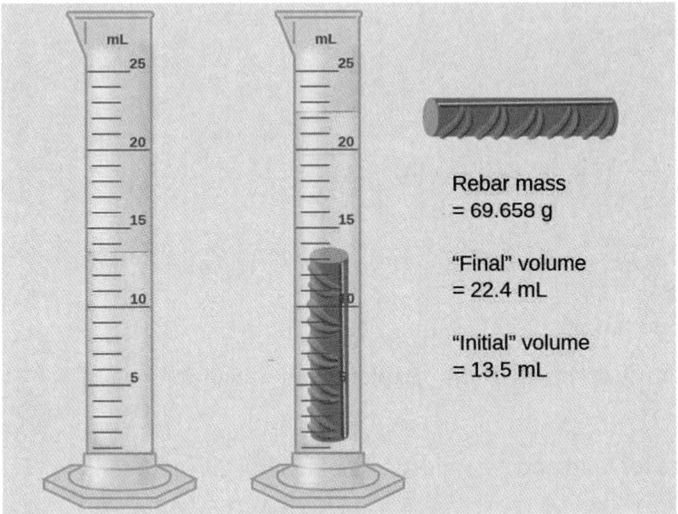

(a) Use these values to determine the density of this piece of rebar.

(b) Rebar is mostly iron. Does your result in (a) support this statement? How?

Solution

The volume of the piece of rebar is equal to the volume of the water displaced:

$$\text{volume} = 22.4\text{ mL} - 13.5\text{ mL} = 8.9\text{ mL} = 8.9\text{ cm}^3$$

(rounded to the nearest 0.1 mL, per the rule for addition and subtraction)

The density is the mass-to-volume ratio:

$$\text{density} = \frac{\text{mass}}{\text{volume}} = \frac{69.658\text{ g}}{8.9\text{ cm}^3} = 7.8\text{ g/cm}^3$$

(rounded to two significant figures, per the rule for multiplication and division)

From Table 1.4, the density of iron is 7.9 g/cm^3, very close to that of rebar, which lends some support to the fact that rebar is mostly iron.

Check Your Learning

An irregularly shaped piece of a shiny yellowish material is weighed and then submerged in a graduated cylinder, with results as shown.

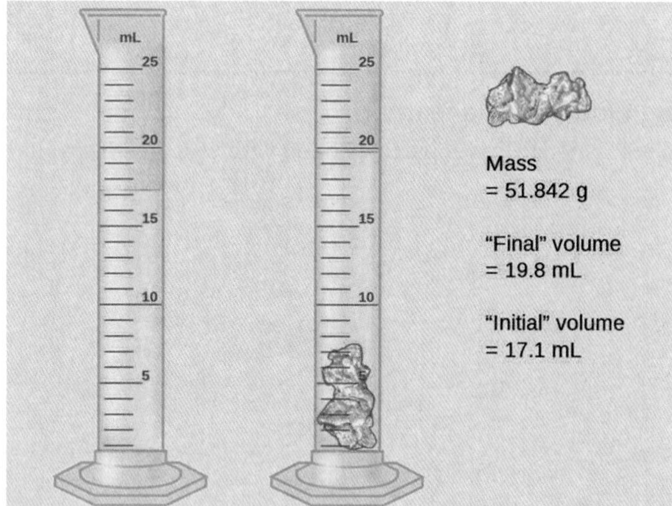

(a) Use these values to determine the density of this material.

(b) Do you have any reasonable guesses as to the identity of this material? Explain your reasoning.

Answer:
(a) 19 g/cm^3; (b) It is likely gold; the right appearance for gold and very close to the density given for gold in Table 1.4.

Accuracy and Precision

Scientists typically make repeated measurements of a quantity to ensure the quality of their findings and to evaluate both the **precision** and the **accuracy** of their results. Measurements are said to be precise if they yield very similar results when repeated in the same manner. A measurement is considered accurate if it yields a result that is very close to the true or accepted value. Precise values agree with each other; accurate values agree with a true value. These characterizations can be extended to other contexts, such as the results of an archery competition (Figure 1.27).

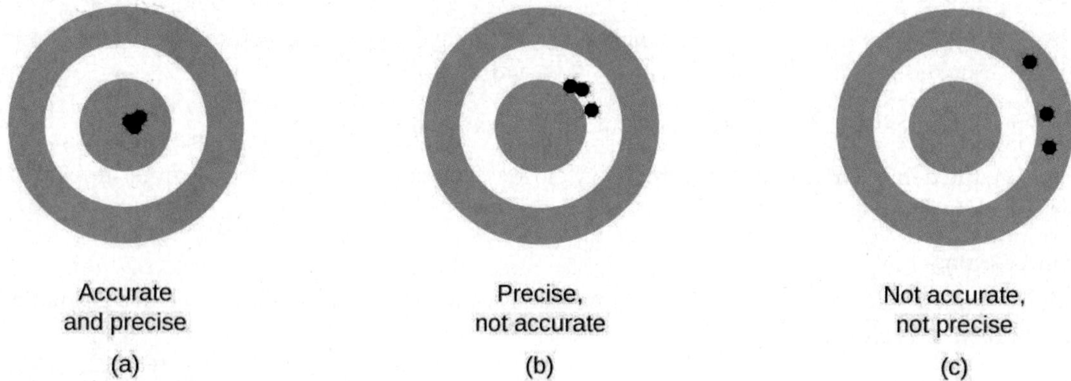

FIGURE 1.27 (a) These arrows are close to both the bull's eye and one another, so they are both accurate and precise. (b) These arrows are close to one another but not on target, so they are precise but not accurate. (c) These arrows are neither on target nor close to one another, so they are neither accurate nor precise.

Suppose a quality control chemist at a pharmaceutical company is tasked with checking the accuracy and precision of three different machines that are meant to dispense 10 ounces (296 mL) of cough syrup into storage bottles. She proceeds to use each machine to fill five bottles and then carefully determines the actual volume dispensed, obtaining the results tabulated in Table 1.5.

Volume (mL) of Cough Medicine Delivered by 10-oz (296 mL) Dispensers

Dispenser #1	Dispenser #2	Dispenser #3
283.3	298.3	296.1
284.1	294.2	295.9
283.9	296.0	296.1
284.0	297.8	296.0
284.1	293.9	296.1

TABLE 1.5

Considering these results, she will report that dispenser #1 is precise (values all close to one another, within a few tenths of a milliliter) but not accurate (none of the values are close to the target value of 296 mL, each being more than 10 mL too low). Results for dispenser #2 represent improved accuracy (each volume is less than 3 mL away from 296 mL) but worse precision (volumes vary by more than 4 mL). Finally, she can report that dispenser #3 is working well, dispensing cough syrup both accurately (all volumes within 0.1 mL of the target volume) and precisely (volumes differing from each other by no more than 0.2 mL).

1.6 Mathematical Treatment of Measurement Results

LEARNING OBJECTIVES

By the end of this section, you will be able to:

- Explain the dimensional analysis (factor label) approach to mathematical calculations involving quantities
- Use dimensional analysis to carry out unit conversions for a given property and computations involving two or more properties

It is often the case that a quantity of interest may not be easy (or even possible) to measure directly but instead must be calculated from other directly measured properties and appropriate mathematical relationships. For example, consider measuring the average speed of an athlete running sprints. This is typically accomplished by measuring the *time* required for the athlete to run from the starting line to the finish line, and the *distance* between these two lines, and then computing *speed* from the equation that relates these three properties:

$$\text{speed} = \frac{\text{distance}}{\text{time}}$$

An Olympic-quality sprinter can run 100 m in approximately 10 s, corresponding to an average speed of

$$\frac{100\text{ m}}{10\text{ s}} = 10\text{ m/s}$$

Note that this simple arithmetic involves dividing the numbers of each measured quantity to yield the number of the computed quantity (100/10 = 10) *and likewise* dividing the units of each measured quantity to yield the unit of the computed quantity (m/s = m/s). Now, consider using this same relation to predict the time required for a person running at this speed to travel a distance of 25 m. The same relation among the three properties is used, but in this case, the two quantities provided are a speed (10 m/s) and a distance (25 m). To yield the sought property, time, the equation must be rearranged appropriately:

$$\text{time} = \frac{\text{distance}}{\text{speed}}$$

The time can then be computed as:

$$\frac{25\text{ m}}{10\text{ m/s}} = 2.5\text{ s}$$

Again, arithmetic on the numbers (25/10 = 2.5) was accompanied by the same arithmetic on the units (m/(m/s) = s) to yield the number and unit of the result, 2.5 s. Note that, just as for numbers, when a unit is divided by an identical unit (in this case, m/m), the result is "1"—or, as commonly phrased, the units "cancel."

These calculations are examples of a versatile mathematical approach known as **dimensional analysis** (or the **factor-label method**). Dimensional analysis is based on this premise: *the units of quantities must be subjected to the same mathematical operations as their associated numbers.* This method can be applied to computations ranging from simple unit conversions to more complex, multi-step calculations involving several different quantities.

Conversion Factors and Dimensional Analysis

A ratio of two equivalent quantities expressed with different measurement units can be used as a **unit conversion factor**. For example, the lengths of 2.54 cm and 1 in. are equivalent (by definition), and so a unit conversion factor may be derived from the ratio,

$$\frac{2.54\text{ cm}}{1\text{ in.}}\ (2.54\text{ cm} = 1\text{ in.})\text{ or }2.54\ \frac{\text{cm}}{\text{in.}}$$

Several other commonly used conversion factors are given in Table 1.6.

Common Conversion Factors

Length	Volume	Mass
1 m = 1.0936 yd	1 L = 1.0567 qt	1 kg = 2.2046 lb
1 in. = 2.54 cm (exact)	1 qt = 0.94635 L	1 lb = 453.59 g
1 km = 0.62137 mi	1 ft^3 = 28.317 L	1 (avoirdupois) oz = 28.349 g
1 mi = 1609.3 m	1 tbsp = 14.787 mL	1 (troy) oz = 31.103 g

TABLE 1.6

When a quantity (such as distance in inches) is multiplied by an appropriate unit conversion factor, the quantity is converted to an equivalent value with different units (such as distance in centimeters). For example, a basketball player's vertical jump of 34 inches can be converted to centimeters by:

$$34\ \cancel{\text{in.}} \times \frac{2.54\text{ cm}}{1\ \cancel{\text{in.}}} = 86\text{ cm}$$

Since this simple arithmetic involves *quantities*, the premise of dimensional analysis requires that we multiply both *numbers and units*. The numbers of these two quantities are multiplied to yield the number of the product quantity, 86, whereas the units are multiplied to yield $\frac{\text{in.} \times \text{cm}}{\text{in.}}$. Just as for numbers, a ratio of identical units is also numerically equal to one, $\frac{\text{in.}}{\text{in.}} = 1$, and the unit product thus simplifies to *cm*. (When identical units divide to yield a factor of 1, they are said to "cancel.") Dimensional analysis may be used to confirm the proper application of unit conversion factors as demonstrated in the following example.

EXAMPLE 1.8

Using a Unit Conversion Factor

The mass of a competition frisbee is 125 g. Convert its mass to ounces using the unit conversion factor derived from the relationship 1 oz = 28.349 g (Table 1.6).

Solution

Given the conversion factor, the mass in ounces may be derived using an equation similar to the one used for converting length from inches to centimeters.

$$x \text{ oz} = 125 \text{ g} \times \text{unit conversion factor}$$

The unit conversion factor may be represented as:

$$\frac{1 \text{ oz}}{28.349 \text{ g}} \text{ and } \frac{28.349 \text{ g}}{1 \text{ oz}}$$

The correct unit conversion factor is the ratio that cancels the units of grams and leaves ounces.

$$\begin{aligned} x \text{ oz} &= 125 \cancel{\text{g}} \times \frac{1 \text{ oz}}{28.349 \cancel{\text{g}}} \\ &= \left(\frac{125}{28.349}\right) \text{oz} \\ &= 4.41 \text{ oz (three significant figures)} \end{aligned}$$

Check Your Learning

Convert a volume of 9.345 qt to liters.

Answer:

8.844 L

Beyond simple unit conversions, the factor-label method can be used to solve more complex problems involving computations. Regardless of the details, the basic approach is the same—all the *factors* involved in the calculation must be appropriately oriented to ensure that their *labels* (units) will appropriately cancel and/or combine to yield the desired unit in the result. As your study of chemistry continues, you will encounter many opportunities to apply this approach.

EXAMPLE 1.9

Computing Quantities from Measurement Results and Known Mathematical Relations

What is the density of common antifreeze in units of g/mL? A 4.00-qt sample of the antifreeze weighs 9.26 lb.

Solution

Since density $= \frac{\text{mass}}{\text{volume}}$, we need to divide the mass in grams by the volume in milliliters. In general: the number of units of B = the number of units of A × unit conversion factor. The necessary conversion factors are given in Table 1.6: 1 lb = 453.59 g; 1 L = 1.0567 qt; 1 L = 1,000 mL. Mass may be converted from pounds to grams as follows:

$$9.26 \cancel{\text{lb}} \times \frac{453.59 \text{ g}}{1 \cancel{\text{lb}}} = 4.20 \times 10^3 \text{ g}$$

Volume may be converted from quarts to milliliters via two steps:

Step 1. *Convert quarts to liters.*

$$4.00 \cancel{\text{qt}} \times \frac{1 \text{ L}}{1.0567 \cancel{\text{qt}}} = 3.78 \text{ L}$$

Step 2. *Convert liters to milliliters.*

$$3.78 \cancel{\text{L}} \times \frac{1000 \text{ mL}}{1 \cancel{\text{L}}} = 3.78 \times 10^3 \text{ mL}$$

Then,

$$\text{density} = \frac{4.20 \times 10^3 \text{ g}}{3.78 \times 10^3 \text{ mL}} = 1.11 \text{ g/mL}$$

Alternatively, the calculation could be set up in a way that uses three unit conversion factors sequentially as follows:

$$\frac{9.26 \cancel{\text{lb}}}{4.00 \cancel{\text{qt}}} \times \frac{453.59 \text{ g}}{1 \cancel{\text{lb}}} \times \frac{1.0567 \cancel{\text{qt}}}{1 \cancel{\text{L}}} \times \frac{1 \cancel{\text{L}}}{1000 \text{ mL}} = 1.11 \text{ g/mL}$$

Check Your Learning

What is the volume in liters of 1.000 oz, given that 1 L = 1.0567 qt and 1 qt = 32 oz (exactly)?

Answer:

2.956×10^{-2} L

 **EXAMPLE 1.10**

Computing Quantities from Measurement Results and Known Mathematical Relations

While being driven from Philadelphia to Atlanta, a distance of about 1250 km, a 2014 Lamborghini Aventador Roadster uses 213 L gasoline.

(a) What (average) fuel economy, in miles per gallon, did the Roadster get during this trip?

(b) If gasoline costs $3.80 per gallon, what was the fuel cost for this trip?

Solution

(a) First convert distance from kilometers to miles:

$$1250 \cancel{\text{km}} \times \frac{0.62137 \text{ mi}}{1 \cancel{\text{km}}} = 777 \text{ mi}$$

and then convert volume from liters to gallons:

$$213 \cancel{\text{L}} \times \frac{1.0567 \cancel{\text{qt}}}{1 \cancel{\text{L}}} \times \frac{1 \text{ gal}}{4 \cancel{\text{qt}}} = 56.3 \text{ gal}$$

Finally,

$$\text{(average) mileage} = \frac{777 \text{ mi}}{56.3 \text{ gal}} = 13.8 \text{ miles/gallon} = 13.8 \text{ mpg}$$

Alternatively, the calculation could be set up in a way that uses all the conversion factors sequentially, as follows:

$$\frac{1250 \cancel{\text{km}}}{213 \cancel{\text{L}}} \times \frac{0.62137 \text{ mi}}{1 \cancel{\text{km}}} \times \frac{1 \cancel{\text{L}}}{1.0567 \cancel{\text{qt}}} \times \frac{4 \cancel{\text{qt}}}{1 \text{ gal}} = 13.8 \text{ mpg}$$

(b) Using the previously calculated volume in gallons, we find:

$$56.3 \cancel{\text{gal}} \times \frac{\$3.80}{1 \cancel{\text{gal}}} = \$214$$

Check Your Learning

A Toyota Prius Hybrid uses 59.7 L gasoline to drive from San Francisco to Seattle, a distance of 1300 km (two significant digits).

(a) What (average) fuel economy, in miles per gallon, did the Prius get during this trip?

(b) If gasoline costs $3.90 per gallon, what was the fuel cost for this trip?

Answer:

(a) 51 mpg; (b) $62

Conversion of Temperature Units

We use the word **temperature** to refer to the hotness or coldness of a substance. One way we measure a change in temperature is to use the fact that most substances expand when their temperature increases and contract when their temperature decreases. The liquid in a common glass thermometer changes its volume as the temperature changes, and the position of the trapped liquid's surface along a printed scale may be used as a measure of temperature.

Temperature scales are defined relative to selected reference temperatures: Two of the most commonly used are the freezing and boiling temperatures of water at a specified atmospheric pressure. On the Celsius scale, 0 °C is defined as the freezing temperature of water and 100 °C as the boiling temperature of water. The space between the two temperatures is divided into 100 equal intervals, which we call degrees. On the **Fahrenheit** scale, the freezing point of water is defined as 32 °F and the boiling temperature as 212 °F. The space between these two points on a Fahrenheit thermometer is divided into 180 equal parts (degrees).

Defining the Celsius and Fahrenheit temperature scales as described in the previous paragraph results in a slightly more complex relationship between temperature values on these two scales than for different units of measure for other properties. Most measurement units for a given property are directly proportional to one another (y = mx). Using familiar length units as one example:

$$\text{length in feet} = \left(\frac{1\ \text{ft}}{12\ \text{in.}}\right) \times \text{length in inches}$$

where y = length in feet, x = length in inches, and the proportionality constant, m, is the conversion factor. The Celsius and Fahrenheit temperature scales, however, do not share a common zero point, and so the relationship between these two scales is a linear one rather than a proportional one (y = mx + b). Consequently, converting a temperature from one of these scales into the other requires more than simple multiplication by a conversion factor, m, it also must take into account differences in the scales' zero points (b).

The linear equation relating Celsius and Fahrenheit temperatures is easily derived from the two temperatures used to define each scale. Representing the Celsius temperature as x and the Fahrenheit temperature as y, the slope, m, is computed to be:

$$m = \frac{\Delta y}{\Delta x} = \frac{212\ °\text{F} - 32\ °\text{F}}{100\ °\text{C} - 0\ °\text{C}} = \frac{180\ °\text{F}}{100\ °\text{C}} = \frac{9\ °\text{F}}{5\ °\text{C}}$$

The y-intercept of the equation, b, is then calculated using either of the equivalent temperature pairs, (100 °C, 212 °F) or (0 °C, 32 °F), as:

$$b = y - mx = 32\ °\text{F} - \frac{9\ °\text{F}}{5\ °\text{C}} \times 0\ °\text{C} = 32\ °\text{F}$$

The equation relating the temperature (T) scales is then:

$$T_{°\text{F}} = \left(\frac{9\ °\text{F}}{5\ °\text{C}} \times T_{°\text{C}}\right) + 32\ °\text{F}$$

An abbreviated form of this equation that omits the measurement units is:

$$T_{°\text{F}} = \left(\frac{9}{5} \times T_{°\text{C}}\right) + 32$$

Rearrangement of this equation yields the form useful for converting from Fahrenheit to Celsius:

$$T_{°\text{C}} = \frac{5}{9}(T_{°\text{F}} - 32)$$

As mentioned earlier in this chapter, the SI unit of temperature is the kelvin (K). Unlike the Celsius and Fahrenheit scales, the kelvin scale is an absolute temperature scale in which 0 (zero) K corresponds to the lowest temperature that can theoretically be achieved. Since the kelvin temperature scale is absolute, a degree symbol is not included in the unit abbreviation, K. The early 19th-century discovery of the relationship between a gas's volume and temperature suggested that the volume of a gas would be zero at −273.15 °C. In 1848, British physicist William Thompson, who later adopted the title of Lord Kelvin, proposed an absolute temperature scale based on this concept (further treatment of this topic is provided in this text's chapter on gases).

The freezing temperature of water on this scale is 273.15 K and its boiling temperature is 373.15 K. Notice the numerical difference in these two reference temperatures is 100, the same as for the Celsius scale, and so the linear relation between these two temperature scales will exhibit a slope of $1\ \frac{\text{K}}{°\text{C}}$. Following the same approach, the equations for converting between the kelvin and Celsius temperature scales are derived to be:

$$T_{\text{K}} = T_{°\text{C}} + 273.15$$

$$T_{°\text{C}} = T_{\text{K}} - 273.15$$

The 273.15 in these equations has been determined experimentally, so it is not exact. Figure 1.28 shows the relationship among the three temperature scales.

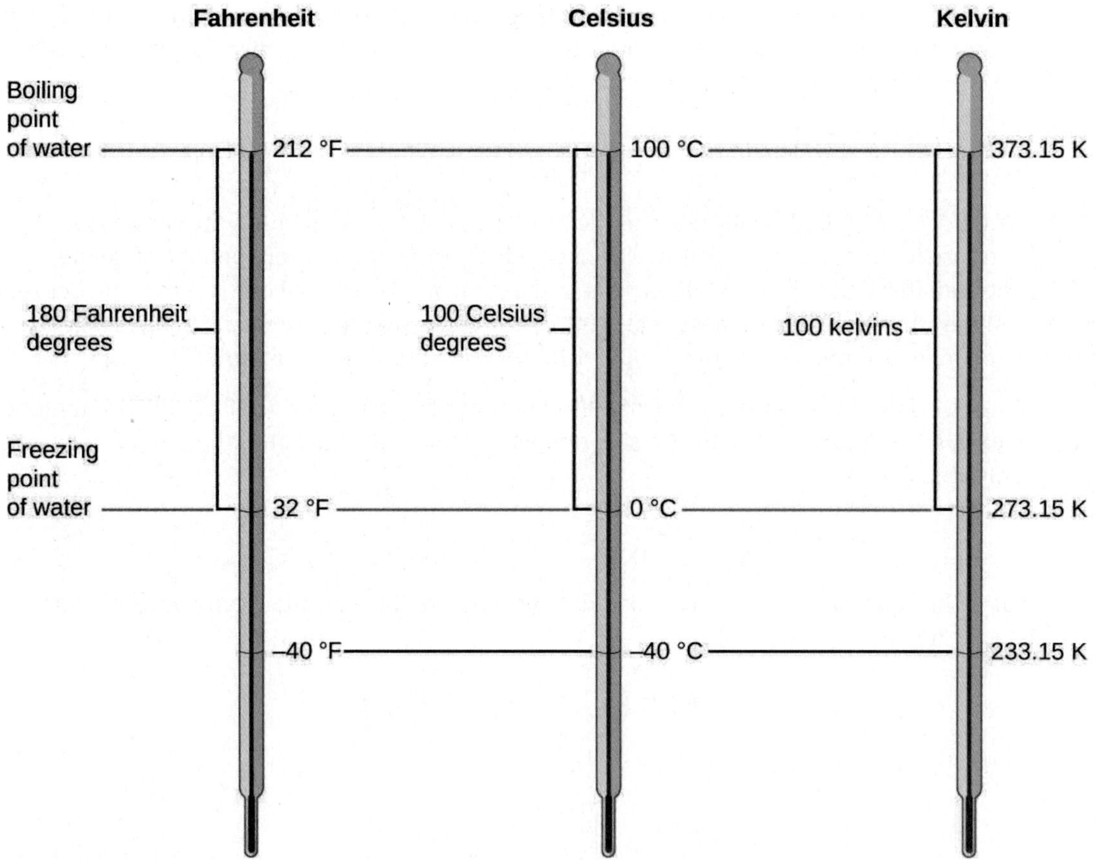

FIGURE 1.28 The Fahrenheit, Celsius, and kelvin temperature scales are compared.

Although the kelvin (absolute) temperature scale is the official SI temperature scale, Celsius is commonly used in many scientific contexts and is the scale of choice for nonscience contexts in almost all areas of the world. Very few countries (the U.S. and its territories, the Bahamas, Belize, Cayman Islands, and Palau) still use Fahrenheit for weather, medicine, and cooking.

EXAMPLE 1.11

Conversion from Celsius

Normal body temperature has been commonly accepted as 37.0 °C (although it varies depending on time of day and method of measurement, as well as among individuals). What is this temperature on the kelvin scale and on the Fahrenheit scale?

Solution

$$\text{K} = °\text{C} + 273.15 = 37.0 + 273.2 = 310.2\ \text{K}$$

$$°\text{F} = \frac{9}{5}°\text{C} + 32.0 = \left(\frac{9}{5} \times 37.0\right) + 32.0 = 66.6 + 32.0 = 98.6\ °\text{F}$$

Check Your Learning

Convert 80.92 °C to K and °F.

Answer:

354.07 K, 177.7 °F

EXAMPLE 1.12

Conversion from Fahrenheit

Baking a ready-made pizza calls for an oven temperature of 450 °F. If you are in Europe, and your oven thermometer uses the Celsius scale, what is the setting? What is the kelvin temperature?

Solution

$$°\text{C} = \frac{5}{9}(°\text{F} - 32) = \frac{5}{9}(450 - 32) = \frac{5}{9} \times 418 = 232\ °\text{C} \longrightarrow \text{set oven to } 230\ °\text{C} \qquad \text{(two significant figures)}$$

$$\text{K} = °\text{C} + 273.15 = 230 + 273 = 503\ \text{K} \longrightarrow 5.0 \times 10^2\ \text{K} \qquad \text{(two significant figures)}$$

Check Your Learning

Convert 50 °F to °C and K.

Answer:

10 °C, 280 K

Key Terms

accuracy how closely a measurement aligns with a correct value

atom smallest particle of an element that can enter into a chemical combination

Celsius (°C) unit of temperature; water freezes at 0 °C and boils at 100 °C on this scale

chemical change change producing a different kind of matter from the original kind of matter

chemical property behavior that is related to the change of one kind of matter into another kind of matter

chemistry study of the composition, properties, and interactions of matter

compound pure substance that can be decomposed into two or more elements

cubic centimeter (cm^3 or cc) volume of a cube with an edge length of exactly 1 cm

cubic meter (m^3) SI unit of volume

density ratio of mass to volume for a substance or object

dimensional analysis (also, factor-label method) versatile mathematical approach that can be applied to computations ranging from simple unit conversions to more complex, multi-step calculations involving several different quantities

element substance that is composed of a single type of atom; a substance that cannot be decomposed by a chemical change

exact number number derived by counting or by definition

extensive property property of a substance that depends on the amount of the substance

Fahrenheit unit of temperature; water freezes at 32 °F and boils at 212 °F on this scale

gas state in which matter has neither definite volume nor shape

heterogeneous mixture combination of substances with a composition that varies from point to point

homogeneous mixture (also, solution) combination of substances with a composition that is uniform throughout

hypothesis tentative explanation of observations that acts as a guide for gathering and checking information

intensive property property of a substance that is independent of the amount of the substance

kelvin (K) SI unit of temperature; 273.15 K = 0 °C

kilogram (kg) standard SI unit of mass; 1 kg = approximately 2.2 pounds

law statement that summarizes a vast number of experimental observations, and describes or predicts some aspect of the natural world

law of conservation of matter when matter converts from one type to another or changes form, there is no detectable change in the total amount of matter present

length measure of one dimension of an object

liquid state of matter that has a definite volume but indefinite shape

liter (L) (also, cubic decimeter) unit of volume; 1 L = 1,000 cm^3

macroscopic domain realm of everyday things that are large enough to sense directly by human sight and touch

mass fundamental property indicating amount of matter

matter anything that occupies space and has mass

meter (m) standard metric and SI unit of length; 1 m = approximately 1.094 yards

microscopic domain realm of things that are much too small to be sensed directly

milliliter (mL) 1/1,000 of a liter; equal to 1 cm^3

mixture matter that can be separated into its components by physical means

molecule bonded collection of two or more atoms of the same or different elements

physical change change in the state or properties of matter that does not involve a change in its chemical composition

physical property characteristic of matter that is not associated with any change in its chemical composition

plasma gaseous state of matter containing a large number of electrically charged atoms and/or molecules

precision how closely a measurement matches the same measurement when repeated

pure substance homogeneous substance that has a constant composition

rounding procedure used to ensure that calculated results properly reflect the uncertainty in the measurements used in the calculation

scientific method path of discovery that leads from question and observation to law or hypothesis to theory, combined with experimental verification of the hypothesis and any necessary modification of the theory

second (s) SI unit of time

SI units (International System of Units) standards fixed by international agreement in the International System of Units (*Le Système International d'Unités*)

significant figures (also, significant digits) all of

the measured digits in a determination, including the uncertain last digit

solid state of matter that is rigid, has a definite shape, and has a fairly constant volume

symbolic domain specialized language used to represent components of the macroscopic and microscopic domains, such as chemical symbols, chemical formulas, chemical equations, graphs, drawings, and calculations

temperature intensive property representing the hotness or coldness of matter

theory well-substantiated, comprehensive, testable explanation of a particular aspect of nature

uncertainty estimate of amount by which measurement differs from true value

unit standard of comparison for measurements

unit conversion factor ratio of equivalent quantities expressed with different units; used to convert from one unit to a different unit

volume amount of space occupied by an object

weight force that gravity exerts on an object

Key Equations

$$\text{density} = \frac{\text{mass}}{\text{volume}}$$

$$T_{°C} = \frac{5}{9} \times \left(T_{°F} - 32\right)$$

$$T_{°F} = \left(\frac{9}{5} \times T_{°C}\right) + 32$$

$$T_K = °C + 273.15$$

$$T_{°C} = K - 273.15$$

Summary

1.1 Chemistry in Context

Chemistry deals with the composition, structure, and properties of matter, and the ways by which various forms of matter may be interconverted. Thus, it occupies a central place in the study and practice of science and technology. Chemists use the scientific method to perform experiments, pose hypotheses, and formulate laws and develop theories, so that they can better understand the behavior of the natural world. To do so, they operate in the macroscopic, microscopic, and symbolic domains. Chemists measure, analyze, purify, and synthesize a wide variety of substances that are important to our lives.

1.2 Phases and Classification of Matter

Matter is anything that occupies space and has mass. The basic building block of matter is the atom, the smallest unit of an element that can enter into combinations with atoms of the same or other elements. In many substances, atoms are combined into molecules. On earth, matter commonly exists in three states: solids, of fixed shape and volume; liquids, of variable shape but fixed volume; and gases, of variable shape and volume. Under high-temperature conditions, matter also can exist as a plasma. Most matter is a mixture: It is composed of two or more types of matter that can be present in varying amounts and can be separated by physical means. Heterogeneous mixtures vary in composition from point to point; homogeneous mixtures have the same composition from point to point. Pure substances consist of only one type of matter. A pure substance can be an element, which consists of only one type of atom and cannot be broken down by a chemical change, or a compound, which consists of two or more types of atoms.

1.3 Physical and Chemical Properties

All substances have distinct physical and chemical properties, and may undergo physical or chemical changes. Physical properties, such as hardness and boiling point, and physical changes, such as melting or freezing, do not involve a change in the composition of matter. Chemical properties, such flammability and acidity, and chemical changes, such as rusting, involve production of matter that differs from that present beforehand.

Measurable properties fall into one of two categories. Extensive properties depend on the amount of matter present, for example, the mass of gold. Intensive properties do not depend on the amount of matter present, for example, the density of gold. Heat is an example of an extensive property, and temperature is an example of an intensive property.

1.4 Measurements

Measurements provide quantitative information that is critical in studying and practicing chemistry. Each measurement has an amount, a unit for comparison, and an uncertainty. Measurements can be represented in either decimal or scientific notation. Scientists primarily use SI (International System) units such as meters, seconds, and kilograms, as well as derived units, such as liters (for volume) and g/cm^3 (for density). In many cases, it is convenient to use prefixes that yield fractional and multiple units, such as microseconds (10^{-6} seconds) and megahertz (10^6 hertz), respectively.

1.5 Measurement Uncertainty, Accuracy, and Precision

Quantities can be defined or measured. Measured quantities have an associated uncertainty that is represented by the number of significant figures in the quantity's number. The uncertainty of a calculated quantity depends on the uncertainties in the quantities used in the calculation and is reflected in how the value is rounded. Quantities are characterized with regard to accuracy (closeness to a true or accepted value) and precision (variation among replicate measurement results).

1.6 Mathematical Treatment of Measurement Results

Measurements are made using a variety of units. It is often useful or necessary to convert a measured quantity from one unit into another. These conversions are accomplished using unit conversion factors, which are derived by simple applications of a mathematical approach called the factor-label method or dimensional analysis. This strategy is also employed to calculate sought quantities using measured quantities and appropriate mathematical relations.

Exercises

1.1 Chemistry in Context

1. Explain how you could experimentally determine whether the outside temperature is higher or lower than 0 °C (32 °F) without using a thermometer.
2. Identify each of the following statements as being most similar to a hypothesis, a law, or a theory. Explain your reasoning.
(a) Falling barometric pressure precedes the onset of bad weather.
(b) All life on earth has evolved from a common, primitive organism through the process of natural selection.
(c) My truck's gas mileage has dropped significantly, probably because it's due for a tune-up.
3. Identify each of the following statements as being most similar to a hypothesis, a law, or a theory. Explain your reasoning.
(a) The pressure of a sample of gas is directly proportional to the temperature of the gas.
(b) Matter consists of tiny particles that can combine in specific ratios to form substances with specific properties.
(c) At a higher temperature, solids (such as salt or sugar) will dissolve better in water.
4. Identify each of the underlined items as a part of either the macroscopic domain, the microscopic domain, or the symbolic domain of chemistry. For any in the symbolic domain, indicate whether they are symbols for a macroscopic or a microscopic feature.
(a) The mass of a lead pipe is 14 lb.
(b) The mass of a certain chlorine atom is 35 amu.
(c) A bottle with a label that reads Al contains aluminum metal.
(d) Al is the symbol for an aluminum atom.
5. Identify each of the underlined items as a part of either the macroscopic domain, the microscopic domain, or the symbolic domain of chemistry. For those in the symbolic domain, indicate whether they are symbols for a macroscopic or a microscopic feature.
(a) A certain molecule contains one H atom and one Cl atom.
(b) Copper wire has a density of about 8 g/cm^3.
(c) The bottle contains 15 grams of Ni powder.
(d) A sulfur molecule is composed of eight sulfur atoms.
6. According to one theory, the pressure of a gas increases as its volume decreases because the molecules in the gas have to move a shorter distance to hit the walls of the container. Does this theory follow a macroscopic or microscopic description of chemical behavior? Explain your answer.

7. The amount of heat required to melt 2 lbs of ice is twice the amount of heat required to melt 1 lb of ice. Is this observation a macroscopic or microscopic description of chemical behavior? Explain your answer.

1.2 Phases and Classification of Matter

8. Why is an object's mass, rather than its weight, used to indicate the amount of matter it contains?
9. What properties distinguish solids from liquids? Liquids from gases? Solids from gases?
10. How does a heterogeneous mixture differ from a homogeneous mixture? How are they similar?
11. How does a homogeneous mixture differ from a pure substance? How are they similar?
12. How does an element differ from a compound? How are they similar?
13. How do molecules of elements and molecules of compounds differ? In what ways are they similar?
14. How does an atom differ from a molecule? In what ways are they similar?
15. Many of the items you purchase are mixtures of pure compounds. Select three of these commercial products and prepare a list of the ingredients that are pure compounds.
16. Classify each of the following as an element, a compound, or a mixture:
    (a) copper
    (b) water
    (c) nitrogen
    (d) sulfur
    (e) air
    (f) sucrose
    (g) a substance composed of molecules each of which contains two iodine atoms
    (h) gasoline
17. Classify each of the following as an element, a compound, or a mixture:
    (a) iron
    (b) oxygen
    (c) mercury oxide
    (d) pancake syrup
    (e) carbon dioxide
    (f) a substance composed of molecules each of which contains one hydrogen atom and one chlorine atom
    (g) baking soda
    (h) baking powder
18. A sulfur atom and a sulfur molecule are not identical. What is the difference?
19. How are the molecules in oxygen gas, the molecules in hydrogen gas, and water molecules similar? How do they differ?
20. Why are astronauts in space said to be "weightless," but not "massless"?
21. Prepare a list of the principal chemicals consumed and produced during the operation of an automobile.
22. Matter is everywhere around us. Make a list by name of fifteen different kinds of matter that you encounter every day. Your list should include (and label at least one example of each) the following: a solid, a liquid, a gas, an element, a compound, a homogenous mixture, a heterogeneous mixture, and a pure substance.
23. When elemental iron corrodes it combines with oxygen in the air to ultimately form red brown iron(III) oxide called rust. (a) If a shiny iron nail with an initial mass of 23.2 g is weighed after being coated in a layer of rust, would you expect the mass to have increased, decreased, or remained the same? Explain. (b) If the mass of the iron nail increases to 24.1 g, what mass of oxygen combined with the iron?

24. As stated in the text, convincing examples that demonstrate the law of conservation of matter outside of the laboratory are few and far between. Indicate whether the mass would increase, decrease, or stay the same for the following scenarios where chemical reactions take place:
(a) Exactly one pound of bread dough is placed in a baking tin. The dough is cooked in an oven at 350 °F releasing a wonderful aroma of freshly baked bread during the cooking process. Is the mass of the baked loaf less than, greater than, or the same as the one pound of original dough? Explain.
(b) When magnesium burns in air a white flaky ash of magnesium oxide is produced. Is the mass of magnesium oxide less than, greater than, or the same as the original piece of magnesium? Explain.
(c) Antoine Lavoisier, the French scientist credited with first stating the law of conservation of matter, heated a mixture of tin and air in a sealed flask to produce tin oxide. Did the mass of the sealed flask and contents decrease, increase, or remain the same after the heating?

25. Yeast converts glucose to ethanol and carbon dioxide during anaerobic fermentation as depicted in the simple chemical equation here:

$$\text{glucose} \longrightarrow \text{ethanol} + \text{carbon dioxide}$$

(a) If 200.0 g of glucose is fully converted, what will be the total mass of ethanol and carbon dioxide produced?
(b) If the fermentation is carried out in an open container, would you expect the mass of the container and contents after fermentation to be less than, greater than, or the same as the mass of the container and contents before fermentation? Explain.
(c) If 97.7 g of carbon dioxide is produced, what mass of ethanol is produced?

1.3 Physical and Chemical Properties

26. Classify the six underlined properties in the following paragraph as chemical or physical:
Fluorine is a pale yellow gas that reacts with most substances. The free element melts at −220 °C and boils at −188 °C. Finely divided metals burn in fluorine with a bright flame. Nineteen grams of fluorine will react with 1.0 gram of hydrogen.

27. Classify each of the following changes as physical or chemical:
(a) condensation of steam
(b) burning of gasoline
(c) souring of milk
(d) dissolving of sugar in water
(e) melting of gold

28. Classify each of the following changes as physical or chemical:
(a) coal burning
(b) ice melting
(c) mixing chocolate syrup with milk
(d) explosion of a firecracker
(e) magnetizing of a screwdriver

29. The volume of a sample of oxygen gas changed from 10 mL to 11 mL as the temperature changed. Is this a chemical or physical change?

30. A 2.0-liter volume of hydrogen gas combined with 1.0 liter of oxygen gas to produce 2.0 liters of water vapor. Does oxygen undergo a chemical or physical change?

31. Explain the difference between extensive properties and intensive properties.

32. Identify the following properties as either extensive or intensive.
(a) volume
(b) temperature
(c) humidity
(d) heat
(e) boiling point

33. The density (d) of a substance is an intensive property that is defined as the ratio of its mass (m) to its volume (V).

$$\text{density} = \frac{\text{mass}}{\text{volume}} \qquad d = \frac{m}{V}$$

Considering that mass and volume are both extensive properties, explain why their ratio, density, is intensive.

1.4 Measurements

34. Is one liter about an ounce, a pint, a quart, or a gallon?
35. Is a meter about an inch, a foot, a yard, or a mile?
36. Indicate the SI base units or derived units that are appropriate for the following measurements:
(a) the length of a marathon race (26 miles 385 yards)
(b) the mass of an automobile
(c) the volume of a swimming pool
(d) the speed of an airplane
(e) the density of gold
(f) the area of a football field
(g) the maximum temperature at the South Pole on April 1, 1913
37. Indicate the SI base units or derived units that are appropriate for the following measurements:
(a) the mass of the moon
(b) the distance from Dallas to Oklahoma City
(c) the speed of sound
(d) the density of air
(e) the temperature at which alcohol boils
(f) the area of the state of Delaware
(g) the volume of a flu shot or a measles vaccination
38. Give the name and symbol of the prefixes used with SI units to indicate multiplication by the following exact quantities.
(a) 10^3
(b) 10^{-2}
(c) 0.1
(d) 10^{-3}
(e) 1,000,000
(f) 0.000001
39. Give the name of the prefix and the quantity indicated by the following symbols that are used with SI base units.
(a) c
(b) d
(c) G
(d) k
(e) m
(f) n
(g) p
(h) T
40. A large piece of jewelry has a mass of 132.6 g. A graduated cylinder initially contains 48.6 mL water. When the jewelry is submerged in the graduated cylinder, the total volume increases to 61.2 mL.
(a) Determine the density of this piece of jewelry.
(b) Assuming that the jewelry is made from only one substance, what substance is it likely to be? Explain.

41. Visit this density simulation (http://openstax.org/l/16phetmasvolden) and click the "turn fluid into water" button to adjust the density of liquid in the beaker to 1.00 g/mL.
(a) Use the water displacement approach to measure the mass and volume of the unknown material (select the green block with question marks).
(b) Use the measured mass and volume data from step (a) to calculate the density of the unknown material.
(c) Link out to the link provided.
(d) Assuming this material is a copper-containing gemstone, identify its three most likely identities by comparing the measured density to the values tabulated at this gemstone density guide (https://www.ajsgem.com/articles/gemstone-density-definitive-guide.html).
(e) How are mass and density related for blocks of the same volume?

42. Visit this density simulation (http://openstax.org/l/16phetmasvolden) and click the "reset" button to ensure all simulator parameters are at their default values.
(a) Use the water displacement approach to measure the mass and volume of the red block.
(b) Use the measured mass and volume data from step (a) to calculate the density of the red block.
(c) Use the vertical green slide control to adjust the fluid density to values well above, then well below, and finally nearly equal to the density of the red block, reporting your observations.

43. Visit this density simulation (http://openstax.org/l/16phetmasvolden) and click the "turn fluid into water" button to adjust the density of liquid in the beaker to 1.00 g/mL. Change the block material to foam, and then wait patiently until the foam block stops bobbing up and down in the water.
(a) The foam block should be floating on the surface of the water (that is, only partially submerged). What is the volume of water displaced?
(b) Use the water volume from part (a) and the density of water (1.00 g/mL) to calculate the mass of water displaced.
(c) Remove and weigh the foam block. How does the block's mass compare to the mass of displaced water from part (b)?

1.5 Measurement Uncertainty, Accuracy, and Precision

44. Express each of the following numbers in scientific notation with correct significant figures:
(a) 711.0
(b) 0.239
(c) 90743
(d) 134.2
(e) 0.05499
(f) 10000.0
(g) 0.000000738592

45. Express each of the following numbers in exponential notation with correct significant figures:
(a) 704
(b) 0.03344
(c) 547.9
(d) 22086
(e) 1000.00
(f) 0.0000000651
(g) 0.007157

46. Indicate whether each of the following can be determined exactly or must be measured with some degree of uncertainty:
(a) the number of eggs in a basket
(b) the mass of a dozen eggs
(c) the number of gallons of gasoline necessary to fill an automobile gas tank
(d) the number of cm in 2 m
(e) the mass of a textbook
(f) the time required to drive from San Francisco to Kansas City at an average speed of 53 mi/h

47. Indicate whether each of the following can be determined exactly or must be measured with some degree of uncertainty:
(a) the number of seconds in an hour
(b) the number of pages in this book
(c) the number of grams in your weight
(d) the number of grams in 3 kilograms
(e) the volume of water you drink in one day
(f) the distance from San Francisco to Kansas City

48. How many significant figures are contained in each of the following measurements?
(a) 38.7 g
(b) 2×10^{18} m
(c) 3,486,002 kg
(d) 9.74150×10^{-4} J
(e) 0.0613 cm^3
(f) 17.0 kg
(g) 0.01400 g/mL

49. How many significant figures are contained in each of the following measurements?
(a) 53 cm
(b) 2.05×10^{8} m
(c) 86,002 J
(d) 9.740×10^{4} m/s
(e) 10.0613 m^3
(f) 0.17 g/mL
(g) 0.88400 s

50. The following quantities were reported on the labels of commercial products. Determine the number of significant figures in each.
(a) 0.0055 g active ingredients
(b) 12 tablets
(c) 3% hydrogen peroxide
(d) 5.5 ounces
(e) 473 mL
(f) 1.75% bismuth
(g) 0.001% phosphoric acid
(h) 99.80% inert ingredients

51. Round off each of the following numbers to two significant figures:
(a) 0.436
(b) 9.000
(c) 27.2
(d) 135
(e) 1.497×10^{-3}
(f) 0.445

52. Round off each of the following numbers to two significant figures:
(a) 517
(b) 86.3
(c) 6.382×10^{3}
(d) 5.0008
(e) 22.497
(f) 0.885

53. Perform the following calculations and report each answer with the correct number of significant figures.
(a) 628×342
(b) $(5.63 \times 10^2) \times (7.4 \times 10^3)$
(c) $\frac{28.0}{13.483}$
(d) 8119×0.000023
(e) $14.98 + 27,340 + 84.7593$
(f) $42.7 + 0.259$

54. Perform the following calculations and report each answer with the correct number of significant figures.
(a) 62.8×34
(b) $0.147 + 0.0066 + 0.012$
(c) $38 \times 95 \times 1.792$
(d) $15 - 0.15 - 0.6155$
(e) $8.78 \times \left(\frac{0.0500}{0.478}\right)$
(f) $140 + 7.68 + 0.014$
(g) $28.7 - 0.0483$
(h) $\frac{(88.5-87.57)}{45.13}$

55. Consider the results of the archery contest shown in this figure.
(a) Which archer is most precise?
(b) Which archer is most accurate?
(c) Who is both least precise and least accurate?

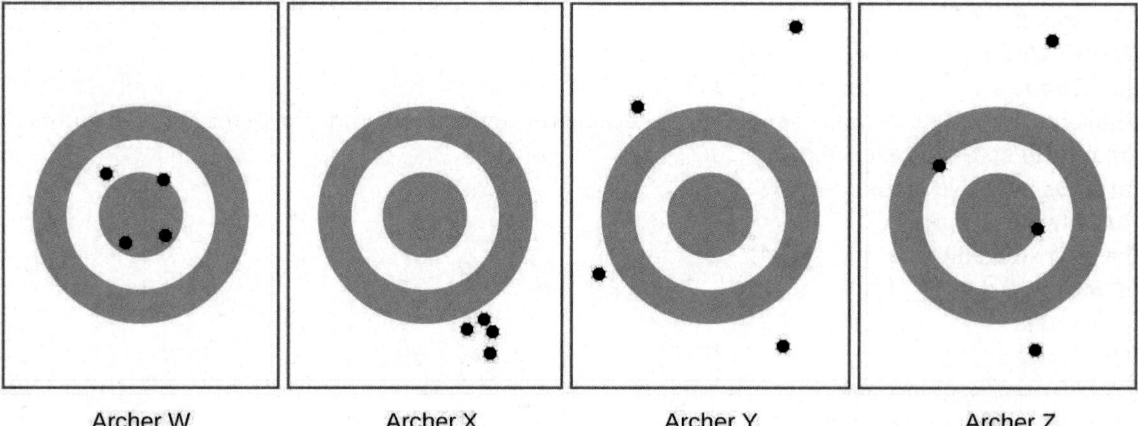

56. Classify the following sets of measurements as accurate, precise, both, or neither.
(a) Checking for consistency in the weight of chocolate chip cookies: 17.27 g, 13.05 g, 19.46 g, 16.92 g
(b) Testing the volume of a batch of 25-mL pipettes: 27.02 mL, 26.99 mL, 26.97 mL, 27.01 mL
(c) Determining the purity of gold: 99.9999%, 99.9998%, 99.9998%, 99.9999%

1.6 Mathematical Treatment of Measurement Results

57. Write conversion factors (as ratios) for the number of:
(a) yards in 1 meter
(b) liters in 1 liquid quart
(c) pounds in 1 kilogram

58. Write conversion factors (as ratios) for the number of:
(a) kilometers in 1 mile
(b) liters in 1 cubic foot
(c) grams in 1 ounce

59. The label on a soft drink bottle gives the volume in two units: 2.0 L and 67.6 fl oz. Use this information to derive a conversion factor between the English and metric units. How many significant figures can you justify in your conversion factor?

60. The label on a box of cereal gives the mass of cereal in two units: 978 grams and 34.5 oz. Use this information to find a conversion factor between the English and metric units. How many significant figures can you justify in your conversion factor?
61. Soccer is played with a round ball having a circumference between 27 and 28 in. and a weight between 14 and 16 oz. What are these specifications in units of centimeters and grams?
62. A woman's basketball has a circumference between 28.5 and 29.0 inches and a maximum weight of 20 ounces (two significant figures). What are these specifications in units of centimeters and grams?
63. How many milliliters of a soft drink are contained in a 12.0-oz can?
64. A barrel of oil is exactly 42 gal. How many liters of oil are in a barrel?
65. The diameter of a red blood cell is about 3×10^{-4} in. What is its diameter in centimeters?
66. The distance between the centers of the two oxygen atoms in an oxygen molecule is 1.21×10^{-8} cm. What is this distance in inches?
67. Is a 197-lb weight lifter light enough to compete in a class limited to those weighing 90 kg or less?
68. A very good 197-lb weight lifter lifted 192 kg in a move called the clean and jerk. What was the mass of the weight lifted in pounds?
69. Many medical laboratory tests are run using 5.0 μL blood serum. What is this volume in milliliters?
70. If an aspirin tablet contains 325 mg aspirin, how many grams of aspirin does it contain?
71. Use scientific (exponential) notation to express the following quantities in terms of the SI base units in Table 1.2:
    (a) 0.13 g
    (b) 232 Gg
    (c) 5.23 pm
    (d) 86.3 mg
    (e) 37.6 cm
    (f) 54 μm
    (g) 1 Ts
    (h) 27 ps
    (i) 0.15 mK
72. Complete the following conversions between SI units.
    (a) 612 g = ________ mg
    (b) 8.160 m = ________ cm
    (c) 3779 μg = ________ g
    (d) 781 mL = ________ L
    (e) 4.18 kg = ________ g
    (f) 27.8 m = ________ km
    (g) 0.13 mL = ________ L
    (h) 1738 km = ________ m
    (i) 1.9 Gg = ________ g
73. Gasoline is sold by the liter in many countries. How many liters are required to fill a 12.0-gal gas tank?
74. Milk is sold by the liter in many countries. What is the volume of exactly 1/2 gal of milk in liters?
75. A long ton is defined as exactly 2240 lb. What is this mass in kilograms?
76. Make the conversion indicated in each of the following:
    (a) the men's world record long jump, 29 ft 4¼ in., to meters
    (b) the greatest depth of the ocean, about 6.5 mi, to kilometers
    (c) the area of the state of Oregon, 96,981 mi^2, to square kilometers
    (d) the volume of 1 gill (exactly 4 oz) to milliliters
    (e) the estimated volume of the oceans, 330,000,000 mi^3, to cubic kilometers.
    (f) the mass of a 3525-lb car to kilograms
    (g) the mass of a 2.3-oz egg to grams

77. Make the conversion indicated in each of the following:
(a) the length of a soccer field, 120 m (three significant figures), to feet
(b) the height of Mt. Kilimanjaro, at 19,565 ft, the highest mountain in Africa, to kilometers
(c) the area of an 8.5- × 11-inch sheet of paper in cm^2
(d) the displacement volume of an automobile engine, 161 $in.^3$, to liters
(e) the estimated mass of the atmosphere, 5.6×10^{15} tons, to kilograms
(f) the mass of a bushel of rye, 32.0 lb, to kilograms
(g) the mass of a 5.00-grain aspirin tablet to milligrams (1 grain = 0.00229 oz)

78. Many chemistry conferences have held a 50-Trillion Angstrom Run (two significant figures). How long is this run in kilometers and in miles? (1 Å = 1×10^{-10} m)

79. A chemist's 50-Trillion Angstrom Run (see Exercise 1.78) would be an archeologist's 10,900 cubit run. How long is one cubit in meters and in feet? (1 Å = 1×10^{-8} cm)

80. The gas tank of a certain luxury automobile holds 22.3 gallons according to the owner's manual. If the density of gasoline is 0.8206 g/mL, determine the mass in kilograms and pounds of the fuel in a full tank.

81. As an instructor is preparing for an experiment, he requires 225 g phosphoric acid. The only container readily available is a 150-mL Erlenmeyer flask. Is it large enough to contain the acid, whose density is 1.83 g/mL?

82. To prepare for a laboratory period, a student lab assistant needs 125 g of a compound. A bottle containing 1/4 lb is available. Did the student have enough of the compound?

83. A chemistry student is 159 cm tall and weighs 45.8 kg. What is her height in inches and weight in pounds?

84. In a recent Grand Prix, the winner completed the race with an average speed of 229.8 km/h. What was his speed in miles per hour, meters per second, and feet per second?

85. Solve these problems about lumber dimensions.
(a) To describe to a European how houses are constructed in the US, the dimensions of "two-by-four" lumber must be converted into metric units. The thickness × width × length dimensions are 1.50 in. × 3.50 in. × 8.00 ft in the US. What are the dimensions in cm × cm × m?
(b) This lumber can be used as vertical studs, which are typically placed 16.0 in. apart. What is that distance in centimeters?

86. The mercury content of a stream was believed to be above the minimum considered safe—1 part per billion (ppb) by weight. An analysis indicated that the concentration was 0.68 parts per billion. What quantity of mercury in grams was present in 15.0 L of the water, the density of which is 0.998 g/ml? $\left(1 \text{ ppb Hg} = \frac{1 \text{ ng Hg}}{1 \text{ g water}}\right)$

87. Calculate the density of aluminum if 27.6 cm^3 has a mass of 74.6 g.

88. Osmium is one of the densest elements known. What is its density if 2.72 g has a volume of 0.121 cm^3?

89. Calculate these masses.
(a) What is the mass of 6.00 cm^3 of mercury, density = 13.5939 g/cm^3?
(b) What is the mass of 25.0 mL octane, density = 0.702 g/cm^3?

90. Calculate these masses.
(a) What is the mass of 4.00 cm^3 of sodium, density = 0.97 g/cm^3 ?
(b) What is the mass of 125 mL gaseous chlorine, density = 3.16 g/L?

91. Calculate these volumes.
(a) What is the volume of 25 g iodine, density = 4.93 g/cm^3?
(b) What is the volume of 3.28 g gaseous hydrogen, density = 0.089 g/L?

92. Calculate these volumes.
(a) What is the volume of 11.3 g graphite, density = 2.25 g/cm^3?
(b) What is the volume of 39.657 g bromine, density = 2.928 g/cm^3?

93. Convert the boiling temperature of gold, 2966 °C, into degrees Fahrenheit and kelvin.

94. Convert the temperature of scalding water, 54 °C, into degrees Fahrenheit and kelvin.

95. Convert the temperature of the coldest area in a freezer, −10 °F, to degrees Celsius and kelvin.

96. Convert the temperature of dry ice, −77 °C, into degrees Fahrenheit and kelvin.

97. Convert the boiling temperature of liquid ammonia, −28.1 °F, into degrees Celsius and kelvin.

98. The label on a pressurized can of spray disinfectant warns against heating the can above 130 °F. What are the corresponding temperatures on the Celsius and kelvin temperature scales?

99. The weather in Europe was unusually warm during the summer of 1995. The TV news reported temperatures as high as 45 °C. What was the temperature on the Fahrenheit scale?

CHAPTER 2

Atoms, Molecules, and Ions

Figure 2.1 Analysis of molecules in an exhaled breath can provide valuable information, leading to early diagnosis of diseases or detection of environmental exposure to harmful substances. (credit: modification of work by Paul Flowers)

CHAPTER OUTLINE

2.1 Early Ideas in Atomic Theory
2.2 Evolution of Atomic Theory
2.3 Atomic Structure and Symbolism
2.4 Chemical Formulas
2.5 The Periodic Table
2.6 Ionic and Molecular Compounds
2.7 Chemical Nomenclature

INTRODUCTION Lung diseases and lung cancers are among the world's most devastating illnesses partly due to delayed detection and diagnosis. Most noninvasive screening procedures aren't reliable, and patients often resist more accurate methods due to discomfort with the procedures or with the potential danger that the procedures cause. But what if you could be accurately diagnosed through a simple breath test?

Early detection of biomarkers, substances that indicate an organism's disease or physiological state, could allow diagnosis and treatment before a condition becomes serious or irreversible. Recent studies have shown that your exhaled breath can contain molecules that may be biomarkers for recent exposure to environmental contaminants or for pathological conditions ranging from asthma to lung cancer. Scientists are working to develop biomarker "fingerprints" that could be used to diagnose a specific disease based on the amounts and identities of certain molecules in a patient's exhaled breath. In Sangeeta Bhatia's lab at MIT, a team used substances that react specifically inside diseased lung tissue; the products of the reactions will be present as biomarkers that can be identified through mass spectrometry (an analytical method discussed later in the chapter). A potential application would allow patients with early symptoms to inhale or ingest a "sensor" substance, and, minutes later, to breathe into a detector for diagnosis. Similar research by scientists such as Laura López-Sánchez has provided similar processes for lung cancer. An essential concept underlying this goal is that of a molecule's identity, which is determined by the numbers and types of atoms it contains, and how they are bonded together. This chapter will describe some of the fundamental chemical principles related to the composition of matter, including those central to the concept of molecular identity.

2.1 Early Ideas in Atomic Theory

LEARNING OBJECTIVES

By the end of this section, you will be able to:

- State the postulates of Dalton's atomic theory
- Use postulates of Dalton's atomic theory to explain the laws of definite and multiple proportions

The earliest recorded discussion of the basic structure of matter comes from ancient Greek philosophers, the scientists of their day. In the fifth century BC, Leucippus and Democritus argued that all matter was composed of small, finite particles that they called *atomos*, a term derived from the Greek word for "indivisible." They thought of atoms as moving particles that differed in shape and size, and which could join together. Later, Aristotle and others came to the conclusion that matter consisted of various combinations of the four "elements"—fire, earth, air, and water—and could be infinitely divided. Interestingly, these philosophers thought about atoms and "elements" as philosophical concepts, but apparently never considered performing experiments to test their ideas.

The Aristotelian view of the composition of matter held sway for over two thousand years, until English schoolteacher John Dalton helped to revolutionize chemistry with his hypothesis that the behavior of matter could be explained using an atomic theory. First published in 1807, many of Dalton's hypotheses about the microscopic features of matter are still valid in modern atomic theory. Here are the postulates of **Dalton's atomic theory**.

1. Matter is composed of exceedingly small particles called atoms. An atom is the smallest unit of an element that can participate in a chemical change.
2. An element consists of only one type of atom, which has a mass that is characteristic of the element and is the same for all atoms of that element (Figure 2.2). A macroscopic sample of an element contains an incredibly large number of atoms, all of which have identical chemical properties.

FIGURE 2.2 A pre-1982 copper penny (left) contains approximately 3×10^{22} copper atoms (several dozen are represented as brown spheres at the right), each of which has the same chemical properties. (credit: modification of work by "slgckgc"/Flickr)

3. Atoms of one element differ in properties from atoms of all other elements.
4. A compound consists of atoms of two or more elements combined in a small, whole-number ratio. In a given compound, the numbers of atoms of each of its elements are always present in the same ratio (Figure 2.3).

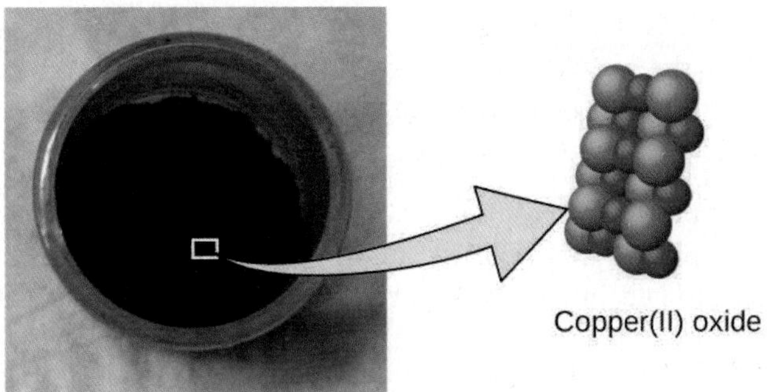

FIGURE 2.3 Copper(II) oxide, a powdery, black compound, results from the combination of two types of atoms—copper (brown spheres) and oxygen (red spheres)—in a 1:1 ratio. (credit: modification of work by "Chemicalinterest"/Wikimedia Commons)

5. Atoms are neither created nor destroyed during a chemical change, but are instead rearranged to yield substances that are different from those present before the change (Figure 2.4).

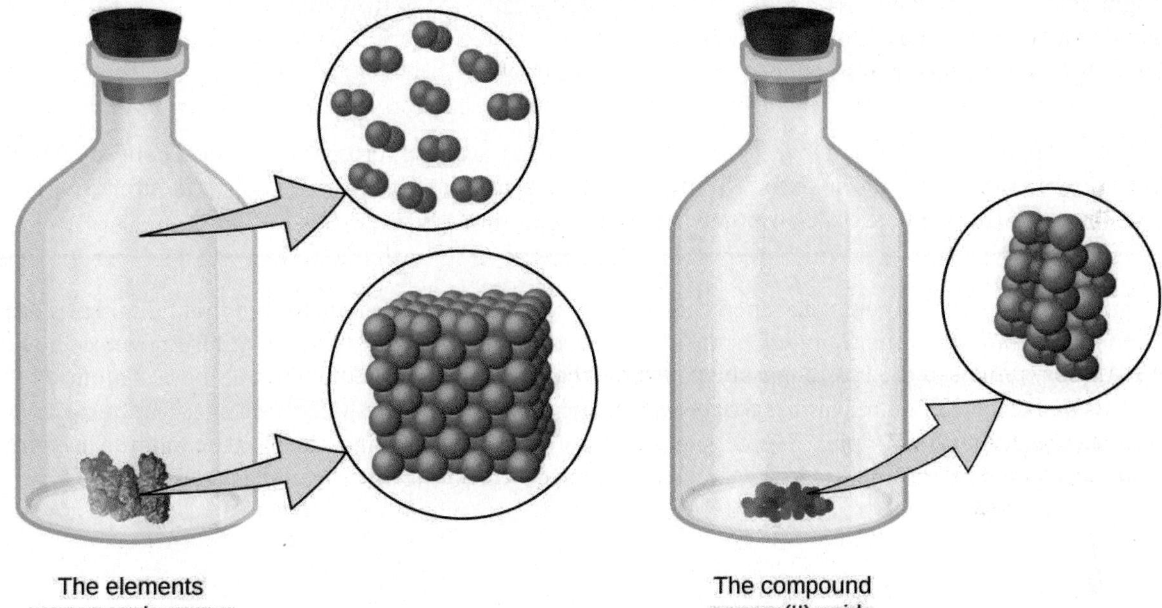

FIGURE 2.4 When the elements copper (a shiny, red-brown solid, shown here as brown spheres) and oxygen (a clear and colorless gas, shown here as red spheres) react, their atoms rearrange to form a compound containing copper and oxygen (a powdery, black solid). (credit copper: modification of work by http://images-of-elements.com/copper.php)

Dalton's atomic theory provides a microscopic explanation of the many macroscopic properties of matter that you've learned about. For example, if an element such as copper consists of only one kind of atom, then it cannot be broken down into simpler substances, that is, into substances composed of fewer types of atoms. And if atoms are neither created nor destroyed during a chemical change, then the total mass of matter present when matter changes from one type to another will remain constant (the law of conservation of matter).

Testing Dalton's Atomic Theory

In the following drawing, the green spheres represent atoms of a certain element. The purple spheres represent atoms of another element. If the spheres touch, they are part of a single unit of a compound. Does the following chemical change represented by these symbols violate any of the ideas of Dalton's atomic theory? If so, which one?

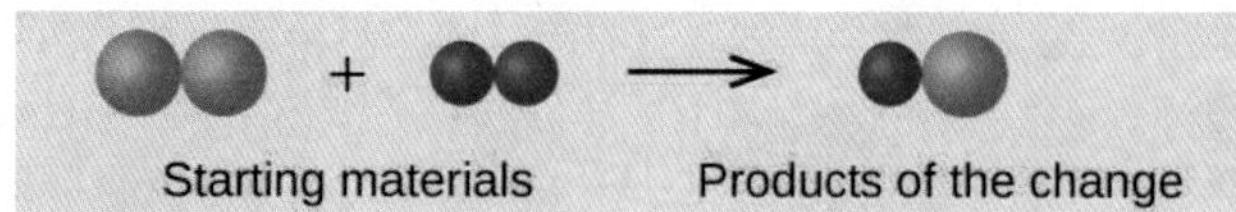

Solution

The starting materials consist of two green spheres and two purple spheres. The products consist of only one green sphere and one purple sphere. This violates Dalton's postulate that atoms are neither created nor destroyed during a chemical change, but are merely redistributed. (In this case, atoms appear to have been destroyed.)

Check Your Learning

In the following drawing, the green spheres represent atoms of a certain element. The purple spheres represent atoms of another element. If the spheres touch, they are part of a single unit of a compound. Does the following chemical change represented by these symbols violate any of the ideas of Dalton's atomic theory? If so, which one?

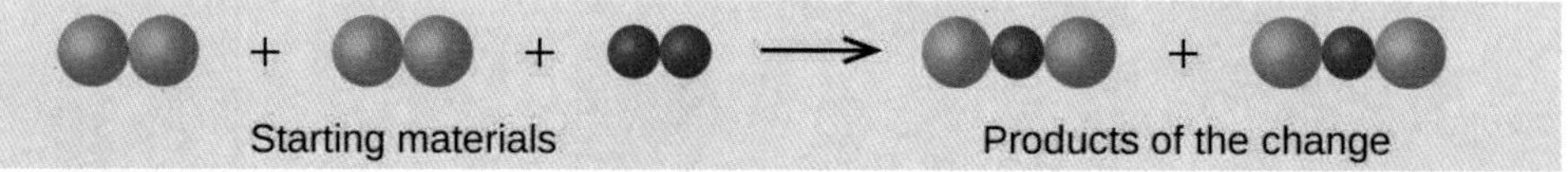

Answer:

The starting materials consist of four green spheres and two purple spheres. The products consist of four green spheres and two purple spheres. This does not violate any of Dalton's postulates: Atoms are neither created nor destroyed, but are redistributed in small, whole-number ratios.

Dalton knew of the experiments of French chemist Joseph Proust, who demonstrated that *all samples of a pure compound contain the same elements in the same proportion by mass*. This statement is known as the **law of definite proportions** or the **law of constant composition**. The suggestion that the numbers of atoms of the elements in a given compound always exist in the same ratio is consistent with these observations. For example, when different samples of isooctane (a component of gasoline and one of the standards used in the octane rating system) are analyzed, they are found to have a carbon-to-hydrogen mass ratio of 5.33:1, as shown in Table 2.1.

Constant Composition of Isooctane

Sample	Carbon	Hydrogen	Mass Ratio
A	14.82 g	2.78 g	$\frac{14.82 \text{ g carbon}}{2.78 \text{ g hydrogen}} = \frac{5.33 \text{ g carbon}}{1.00 \text{ g hydrogen}}$
B	22.33 g	4.19 g	$\frac{22.33 \text{ g carbon}}{4.19 \text{ g hydrogen}} = \frac{5.33 \text{ g carbon}}{1.00 \text{ g hydrogen}}$
C	19.40 g	3.64 g	$\frac{19.40 \text{ g carbon}}{3.63 \text{ g hydrogen}} = \frac{5.33 \text{ g carbon}}{1.00 \text{ g hydrogen}}$

TABLE 2.1

It is worth noting that although all samples of a particular compound have the same mass ratio, the converse is not true in general. That is, samples that have the same mass ratio are not necessarily the same substance. For example, there are many compounds other than isooctane that also have a carbon-to-hydrogen mass ratio of 5.33:1.00.

Dalton also used data from Proust, as well as results from his own experiments, to formulate another interesting law. The **law of multiple proportions** states that *when two elements react to form more than one compound, a fixed mass of one element will react with masses of the other element in a ratio of small, whole* *numbers*. For example, copper and chlorine can form a green, crystalline solid with a mass ratio of 0.558 g chlorine to 1 g copper, as well as a brown crystalline solid with a mass ratio of 1.116 g chlorine to 1 g copper. These ratios by themselves may not seem particularly interesting or informative; however, if we take a ratio of these ratios, we obtain a useful and possibly surprising result: a small, whole-number ratio.

$$\frac{\frac{1.116 \text{ g Cl}}{1 \text{ g Cu}}}{\frac{0.558 \text{ g Cl}}{1 \text{ g Cu}}} = \frac{2}{1}$$

This 2-to-1 ratio means that the brown compound has twice the amount of chlorine per amount of copper as the green compound.

This can be explained by atomic theory if the copper-to-chlorine ratio in the brown compound is 1 copper atom to 2 chlorine atoms, and the ratio in the green compound is 1 copper atom to 1 chlorine atom. The ratio of chlorine atoms (and thus the ratio of their masses) is therefore 2 to 1 (Figure 2.5).

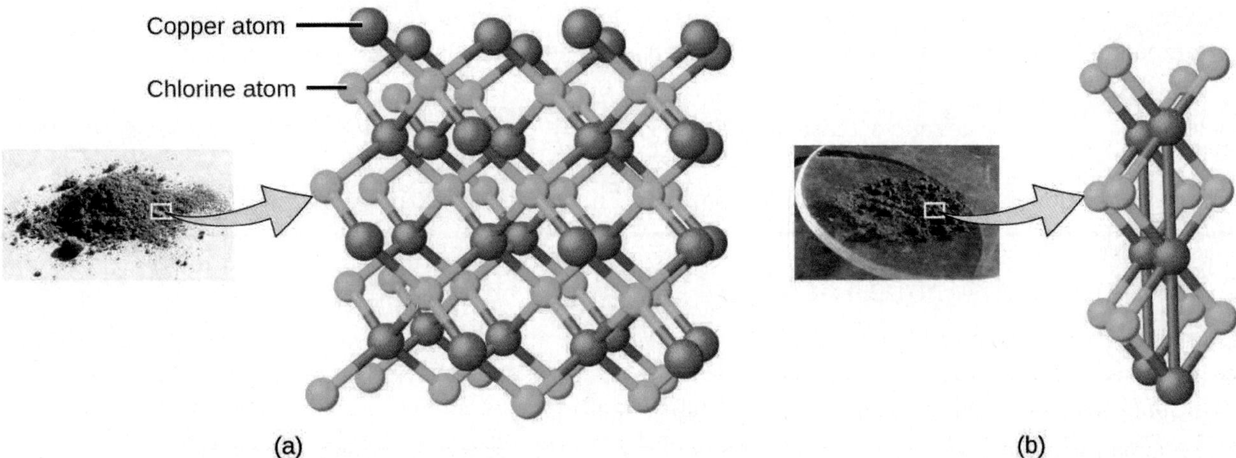

FIGURE 2.5 Compared to the copper chlorine compound in (a), where copper is represented by brown spheres and chlorine by green spheres, the copper chlorine compound in (b) has twice as many chlorine atoms per copper atom. (credit a: modification of work by "Benjah-bmm27"/Wikimedia Commons; credit b: modification of work by "Walkerma"/Wikimedia Commons)

EXAMPLE 2.2

Laws of Definite and Multiple Proportions

A sample of compound A (a clear, colorless gas) is analyzed and found to contain 4.27 g carbon and 5.69 g oxygen. A sample of compound B (also a clear, colorless gas) is analyzed and found to contain 5.19 g carbon and 13.84 g oxygen. Are these data an example of the law of definite proportions, the law of multiple proportions, or neither? What do these data tell you about substances A and B?

Solution

In compound A, the mass ratio of oxygen to carbon is:

$$\frac{1.33 \text{ g O}}{1 \text{ g C}}$$

In compound B, the mass ratio of oxygen to carbon is:

$$\frac{2.67 \text{ g O}}{1 \text{ g C}}$$

The ratio of these ratios is:

$$\frac{\frac{1.33 \text{ g O}}{1 \text{ g C}}}{\frac{2.67 \text{ g O}}{1 \text{ g C}}} = \frac{1}{2}$$

This supports the law of multiple proportions. This means that A and B are different compounds, with A having one-half as much oxygen per amount of carbon (or twice as much carbon per amount of oxygen) as B. A possible pair of compounds that would fit this relationship would be A = CO and B = CO_2.

Check Your Learning

A sample of compound X (a clear, colorless, combustible liquid with a noticeable odor) is analyzed and found to contain 14.13 g carbon and 2.96 g hydrogen. A sample of compound Y (a clear, colorless, combustible liquid with a noticeable odor that is slightly different from X's odor) is analyzed and found to contain 19.91 g carbon and 3.34 g hydrogen. Are these data an example of the law of definite proportions, the law of multiple proportions, or neither? What do these data tell you about substances X and Y?

Answer:

In compound X, the mass ratio of carbon to hydrogen is $\frac{14.13 \text{ g C}}{2.96 \text{ g H}}$. In compound Y, the mass ratio of carbon to hydrogen is $\frac{19.91 \text{ g C}}{3.34 \text{ g H}}$. The ratio of these ratios is $\frac{\frac{14.13 \text{ g C}}{2.96 \text{ g H}}}{\frac{19.91 \text{ g C}}{3.34 \text{ g H}}} = \frac{4.77 \text{ g C/g H}}{5.96 \text{ g C/g H}} = 0.800 = \frac{4}{5}$. This small, whole-number ratio supports the law of multiple proportions. This means that X and Y are different compounds.

2.2 Evolution of Atomic Theory

LEARNING OBJECTIVES

By the end of this section, you will be able to:

- Outline milestones in the development of modern atomic theory
- Summarize and interpret the results of the experiments of Thomson, Millikan, and Rutherford
- Describe the three subatomic particles that compose atoms
- Define isotopes and give examples for several elements

If matter is composed of atoms, what are atoms composed of? Are they the smallest particles, or is there something smaller? In the late 1800s, a number of scientists interested in questions like these investigated the electrical discharges that could be produced in low-pressure gases, with the most significant discovery made by English physicist J. J. Thomson using a cathode ray tube. This apparatus consisted of a sealed glass tube from which almost all the air had been removed; the tube contained two metal electrodes. When high voltage was applied across the electrodes, a visible beam called a cathode ray appeared between them. This beam was deflected toward the positive charge and away from the negative charge, and was produced in the same way with identical properties when different metals were used for the electrodes. In similar experiments, the ray was simultaneously deflected by an applied magnetic field, and measurements of the extent of deflection and the magnetic field strength allowed Thomson to calculate the charge-to-mass ratio of the cathode ray particles. The results of these measurements indicated that these particles were much lighter than atoms (Figure 2.6).

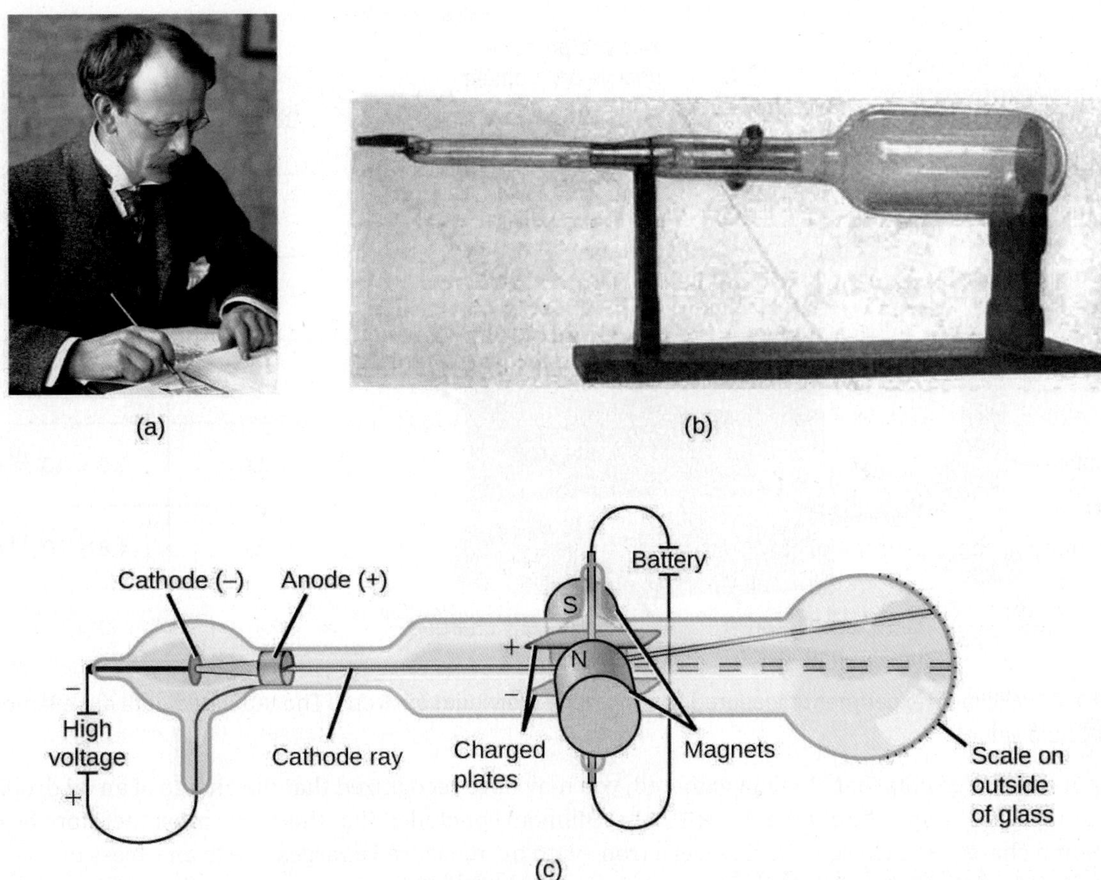

FIGURE 2.6 (a) J. J. Thomson produced a visible beam in a cathode ray tube. (b) This is an early cathode ray tube, invented in 1897 by Ferdinand Braun. (c) In the cathode ray, the beam (shown in yellow) comes from the cathode and is accelerated past the anode toward a fluorescent scale at the end of the tube. Simultaneous deflections by applied electric and magnetic fields permitted Thomson to calculate the mass-to-charge ratio of the particles composing the cathode ray. (credit a: modification of work by Nobel Foundation; credit b: modification of work by Eugen Nesper; credit c: modification of work by "Kurzon"/Wikimedia Commons)

Based on his observations, here is what Thomson proposed and why: The particles are attracted by positive (+) charges and repelled by negative (–) charges, so they must be negatively charged (like charges repel and unlike charges attract); they are less massive than atoms and indistinguishable, regardless of the source material, so they must be fundamental, subatomic constituents of all atoms. Although controversial at the time, Thomson's idea was gradually accepted, and his cathode ray particle is what we now call an **electron**, a negatively charged, subatomic particle with a mass more than one thousand-times less that of an atom. The term "electron" was coined in 1891 by Irish physicist George Stoney, from "*electr*ic i*on*."

LINK TO LEARNING

Click here (http://openstax.org/l/16JJThomson) to hear Thomson describe his discovery in his own voice.

In 1909, more information about the electron was uncovered by American physicist Robert A. Millikan via his "oil drop" experiments. Millikan created microscopic oil droplets, which could be electrically charged by friction as they formed or by using X-rays. These droplets initially fell due to gravity, but their downward progress could be slowed or even reversed by an electric field lower in the apparatus. By adjusting the electric field strength and making careful measurements and appropriate calculations, Millikan was able to determine the charge on individual drops (Figure 2.7).

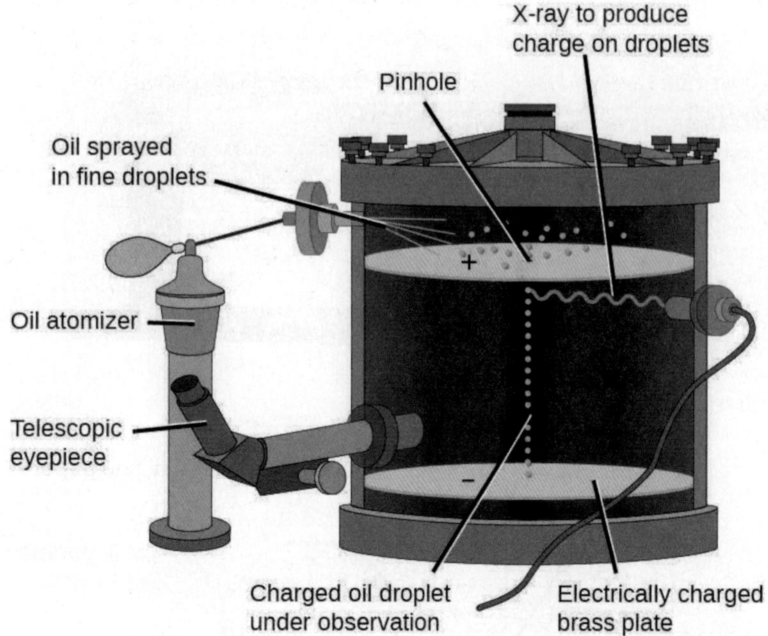

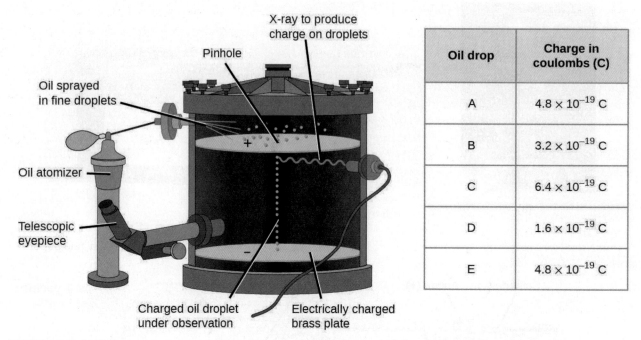

Oil drop	Charge in coulombs (C)
A	4.8×10^{-19} C
B	3.2×10^{-19} C
C	6.4×10^{-19} C
D	1.6×10^{-19} C
E	4.8×10^{-19} C

FIGURE 2.7 Millikan's experiment measured the charge of individual oil drops. The tabulated data are examples of a few possible values.

Looking at the charge data that Millikan gathered, you may have recognized that the charge of an oil droplet is always a multiple of a specific charge, 1.6×10^{-19} C. Millikan concluded that this value must therefore be a fundamental charge—the charge of a single electron—with his measured charges due to an excess of one electron (1 times 1.6×10^{-19} C), two electrons (2 times 1.6×10^{-19} C), three electrons (3 times 1.6×10^{-19} C), and so on, on a given oil droplet. Since the charge of an electron was now known due to Millikan's research, and the charge-to-mass ratio was already known due to Thomson's research (1.759×10^{11} C/kg), it only required a simple calculation to determine the mass of the electron as well.

$$\text{Mass of electron} = 1.602 \times 10^{-19}\ \text{C} \times \frac{1\ \text{kg}}{1.759 \times 10^{11}\ \text{C}} = 9.107 \times 10^{-31}\ \text{kg}$$

Scientists had now established that the atom was not indivisible as Dalton had believed, and due to the work of Thomson, Millikan, and others, the charge and mass of the negative, subatomic particles—the electrons—were known. However, the positively charged part of an atom was not yet well understood. In 1904, Thomson proposed the "plum pudding" model of atoms, which described a positively charged mass with an equal amount of negative charge in the form of electrons embedded in it, since all atoms are electrically neutral. A competing model had been proposed in 1903 by Hantaro Nagaoka, who postulated a Saturn-like atom, consisting of a positively charged sphere surrounded by a halo of electrons (Figure 2.8).

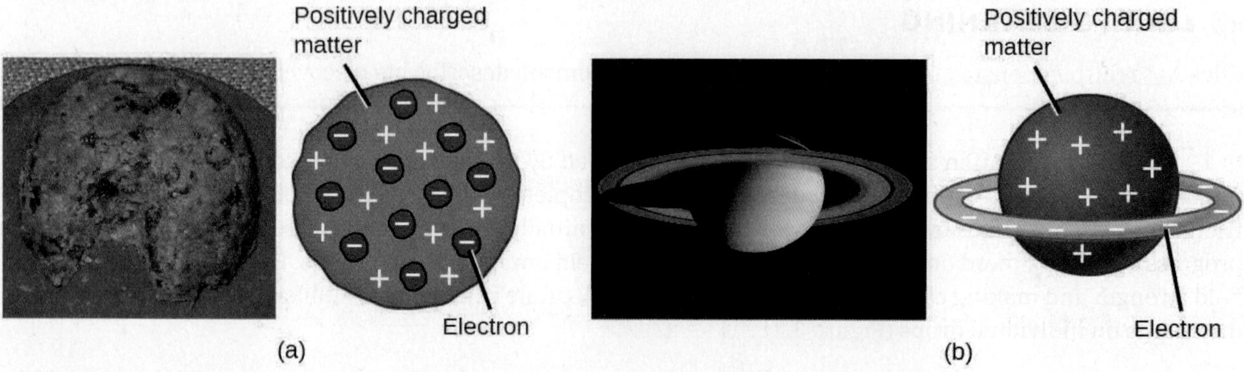

FIGURE 2.8 (a) Thomson suggested that atoms resembled plum pudding, an English dessert consisting of moist cake with embedded raisins ("plums"). (b) Nagaoka proposed that atoms resembled the planet Saturn, with a ring of

electrons surrounding a positive "planet." (credit a: modification of work by "Man vyi"/Wikimedia Commons; credit b: modification of work by "NASA"/Wikimedia Commons)

The next major development in understanding the atom came from Ernest Rutherford, a physicist from New Zealand who largely spent his scientific career in Canada and England. He performed a series of experiments using a beam of high-speed, positively charged **alpha particles (α particles)** that were produced by the radioactive decay of radium; α particles consist of two protons and two neutrons (you will learn more about radioactive decay in the chapter on nuclear chemistry). Rutherford and his colleagues Hans Geiger (later famous for the Geiger counter) and Ernest Marsden aimed a beam of α particles, the source of which was embedded in a lead block to absorb most of the radiation, at a very thin piece of gold foil and examined the resultant scattering of the α particles using a luminescent screen that glowed briefly where hit by an α particle.

What did they discover? Most particles passed right through the foil without being deflected at all. However, some were diverted slightly, and a very small number were deflected almost straight back toward the source (Figure 2.9). Rutherford described finding these results: "It was quite the most incredible event that has ever happened to me in my life. It was almost as incredible as if you fired a 15-inch shell at a piece of tissue paper and it came back and hit you."[1]

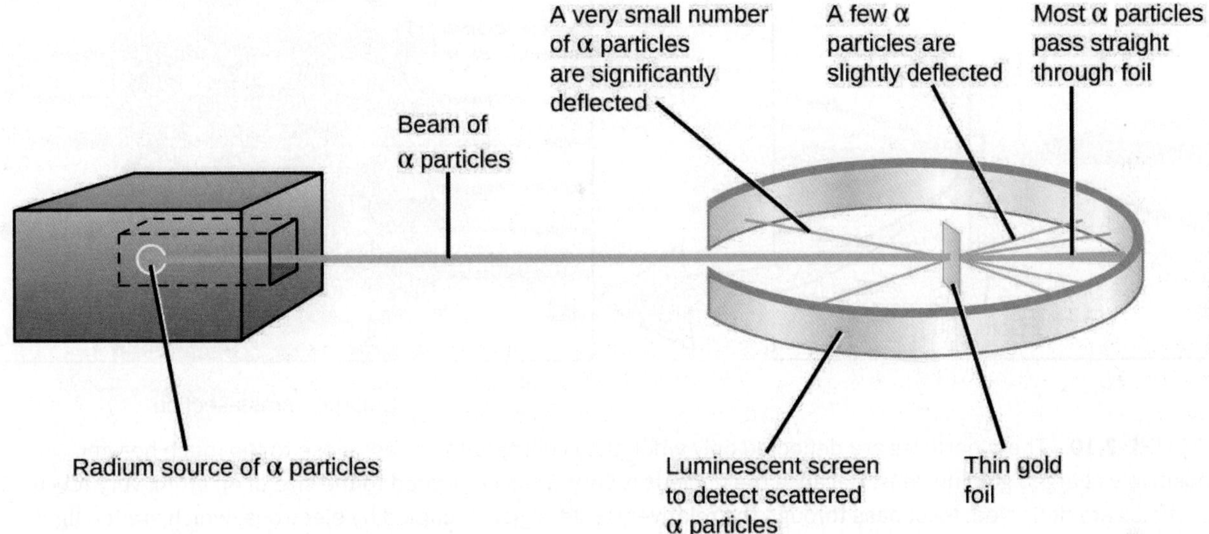

FIGURE 2.9 Geiger and Rutherford fired α particles at a piece of gold foil and detected where those particles went, as shown in this schematic diagram of their experiment. Most of the particles passed straight through the foil, but a few were deflected slightly and a very small number were significantly deflected.

Here is what Rutherford deduced: Because most of the fast-moving α particles passed through the gold atoms undeflected, they must have traveled through essentially empty space inside the atom. Alpha particles are positively charged, so deflections arose when they encountered another positive charge (like charges repel each other). Since like charges repel one another, the few positively charged α particles that changed paths abruptly must have hit, or closely approached, another body that also had a highly concentrated, positive charge. Since the deflections occurred a small fraction of the time, this charge only occupied a small amount of the space in the gold foil. Analyzing a series of such experiments in detail, Rutherford drew two conclusions:

1. The volume occupied by an atom must consist of a large amount of empty space.
2. A small, relatively heavy, positively charged body, the **nucleus**, must be at the center of each atom.

LINK TO LEARNING

View this simulation (http://openstax.org/l/16Rutherford) of the Rutherford gold foil experiment. Adjust the slit width to produce a narrower or broader beam of α particles to see how that affects the scattering pattern.

1 Ernest Rutherford, "The Development of the Theory of Atomic Structure," ed. J. A. Ratcliffe, in *Background to Modern Science*, eds. Joseph Needham and Walter Pagel, (Cambridge, UK: Cambridge University Press, 1938), 61–74. Accessed September 22, 2014, https://ia600508.us.archive.org/3/items/backgroundtomode032734mbp/backgroundtomode032734mbp.pdf.

This analysis led Rutherford to propose a model in which an atom consists of a very small, positively charged nucleus, in which most of the mass of the atom is concentrated, surrounded by the negatively charged electrons, so that the atom is electrically neutral (Figure 2.10). After many more experiments, Rutherford also discovered that the nuclei of other elements contain the hydrogen nucleus as a "building block," and he named this more fundamental particle the **proton**, the positively charged, subatomic particle found in the nucleus. With one addition, which you will learn next, this nuclear model of the atom, proposed over a century ago, is still used today.

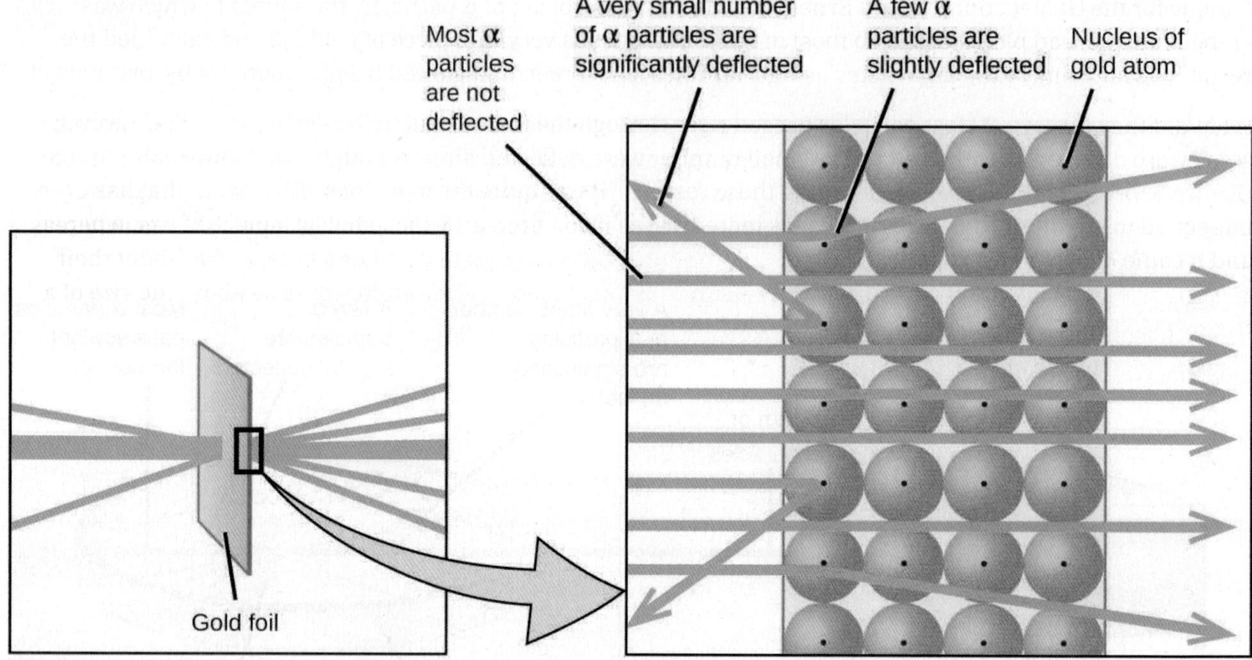

FIGURE 2.10 The α particles are deflected only when they collide with or pass close to the much heavier, positively charged gold nucleus. Because the nucleus is very small compared to the size of an atom, very few α particles are deflected. Most pass through the relatively large region occupied by electrons, which are too light to deflect the rapidly moving particles.

LINK TO LEARNING

The Rutherford Scattering simulation (http://openstax.org/l/16PhetScatter) allows you to investigate the differences between a "plum pudding" atom and a Rutherford atom by firing α particles at each type of atom.

Another important finding was the discovery of isotopes. During the early 1900s, scientists identified several substances that appeared to be new elements, isolating them from radioactive ores. For example, a "new element" produced by the radioactive decay of thorium was initially given the name mesothorium. However, a more detailed analysis showed that mesothorium was chemically identical to radium (another decay product), despite having a different atomic mass. This result, along with similar findings for other elements, led the English chemist Frederick Soddy to realize that an element could have types of atoms with different masses that were chemically indistinguishable. These different types are called **isotopes**—atoms of the same element that differ in mass. Soddy was awarded the Nobel Prize in Chemistry in 1921 for this discovery.

One puzzle remained: The nucleus was known to contain almost all of the mass of an atom, with the number of protons only providing half, or less, of that mass. Different proposals were made to explain what constituted the remaining mass, including the existence of neutral particles in the nucleus. As you might expect, detecting uncharged particles is very challenging, and it was not until 1932 that James Chadwick found evidence of **neutrons**, uncharged, subatomic particles with a mass approximately the same as that of protons. The existence of the neutron also explained isotopes: They differ in mass because they have different numbers of

neutrons, but they are chemically identical because they have the same number of protons. This will be explained in more detail later in this chapter.

2.3 Atomic Structure and Symbolism

LEARNING OBJECTIVES

By the end of this section, you will be able to:

- Write and interpret symbols that depict the atomic number, mass number, and charge of an atom or ion
- Define the atomic mass unit and average atomic mass
- Calculate average atomic mass and isotopic abundance

The development of modern atomic theory revealed much about the inner structure of atoms. It was learned that an atom contains a very small nucleus composed of positively charged protons and uncharged neutrons, surrounded by a much larger volume of space containing negatively charged electrons. The nucleus contains the majority of an atom's mass because protons and neutrons are much heavier than electrons, whereas electrons occupy almost all of an atom's volume. The diameter of an atom is on the order of 10^{-10} m, whereas the diameter of the nucleus is roughly 10^{-15} m—about 100,000 times smaller. For a perspective about their relative sizes, consider this: If the nucleus were the size of a blueberry, the atom would be about the size of a football stadium (Figure 2.11).

FIGURE 2.11 If an atom could be expanded to the size of a football stadium, the nucleus would be the size of a single blueberry. (credit middle: modification of work by "babyknight"/Wikimedia Commons; credit right: modification of work by Paxson Woelber)

Atoms—and the protons, neutrons, and electrons that compose them—are extremely small. For example, a carbon atom weighs less than 2×10^{-23} g, and an electron has a charge of less than 2×10^{-19} C (coulomb). When describing the properties of tiny objects such as atoms, we use appropriately small units of measure, such as the **atomic mass unit (amu)** and the **fundamental unit of charge (e)**. The amu was originally defined based on hydrogen, the lightest element, then later in terms of oxygen. Since 1961, it has been defined with regard to the most abundant isotope of carbon, atoms of which are assigned masses of exactly 12 amu. (This isotope is known as "carbon-12" as will be discussed later in this module.) Thus, one amu is exactly $\frac{1}{12}$ of the mass of one carbon-12 atom: 1 amu = 1.6605×10^{-24} g. (The **Dalton (Da)** and the **unified atomic mass unit (u)** are alternative units that are equivalent to the amu.) The fundamental unit of charge (also called the elementary charge) equals the magnitude of the charge of an electron (e) with e = 1.602×10^{-19} C.

A proton has a mass of 1.0073 amu and a charge of 1+. A neutron is a slightly heavier particle with a mass 1.0087 amu and a charge of zero; as its name suggests, it is neutral. The electron has a charge of 1− and is a much lighter particle with a mass of about 0.00055 amu (it would take about 1800 electrons to equal the mass of one proton). The properties of these fundamental particles are summarized in Table 2.2. (An observant student might notice that the sum of an atom's subatomic particles does not equal the atom's actual mass: The total mass of six protons, six neutrons, and six electrons is 12.0993 amu, slightly larger than 12.00 amu. This "missing" mass is known as the mass defect, and you will learn about it in the chapter on nuclear chemistry.)

Properties of Subatomic Particles

Name	Location	Charge (C)	Unit Charge	Mass (amu)	Mass (g)
electron	outside nucleus	-1.602×10^{-19}	1−	0.00055	0.00091×10^{-24}
proton	nucleus	1.602×10^{-19}	1+	1.00727	1.67262×10^{-24}
neutron	nucleus	0	0	1.00866	1.67493×10^{-24}

TABLE 2.2

The number of protons in the nucleus of an atom is its **atomic number (Z)**. This is the defining trait of an element: Its value determines the identity of the atom. For example, any atom that contains six protons is the element carbon and has the atomic number 6, regardless of how many neutrons or electrons it may have. A neutral atom must contain the same number of positive and negative charges, so the number of protons equals the number of electrons. Therefore, the atomic number also indicates the number of electrons in an atom. The total number of protons and neutrons in an atom is called its **mass number (A)**. The number of neutrons is therefore the difference between the mass number and the atomic number: A – Z = number of neutrons.

$$\begin{aligned} \text{atomic number (Z)} &= \text{number of protons} \\ \text{mass number (A)} &= \text{number of protons} + \text{number of neutrons} \\ \text{A} - \text{Z} &= \text{number of neutrons} \end{aligned}$$

Atoms are electrically neutral if they contain the same number of positively charged protons and negatively charged electrons. When the numbers of these subatomic particles are *not* equal, the atom is electrically charged and is called an **ion**. The charge of an atom is defined as follows:

Atomic charge = number of protons − number of electrons

As will be discussed in more detail later in this chapter, atoms (and molecules) typically acquire charge by gaining or losing electrons. An atom that gains one or more electrons will exhibit a negative charge and is called an **anion**. Positively charged atoms called **cations** are formed when an atom loses one or more electrons. For example, a neutral sodium atom (Z = 11) has 11 electrons. If this atom loses one electron, it will become a cation with a 1+ charge (11 − 10 = 1+). A neutral oxygen atom (Z = 8) has eight electrons, and if it gains two electrons it will become an anion with a 2− charge (8 − 10 = 2−).

EXAMPLE 2.3

Composition of an Atom

Iodine is an essential trace element in our diet; it is needed to produce thyroid hormone. Insufficient iodine in the diet can lead to the development of a goiter, an enlargement of the thyroid gland (Figure 2.12).

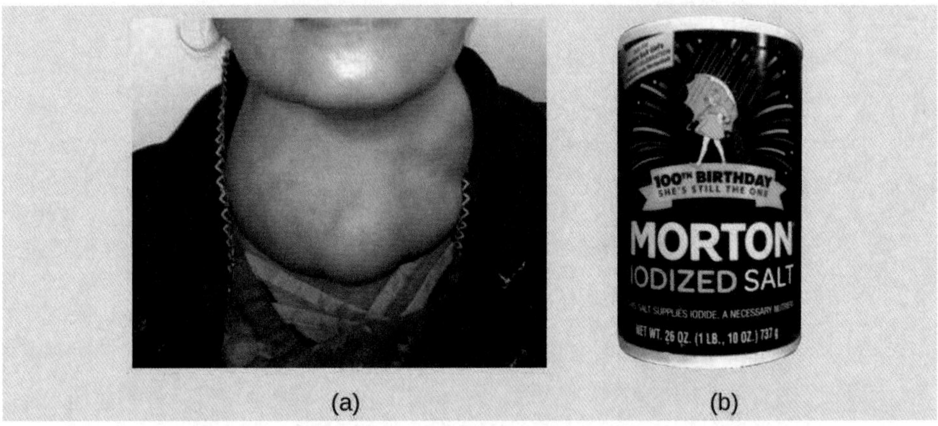

FIGURE 2.12 (a) Insufficient iodine in the diet can cause an enlargement of the thyroid gland called a goiter. (b) The addition of small amounts of iodine to salt, which prevents the formation of goiters, has helped eliminate this concern in the US where salt consumption is high. (credit a: modification of work by "Almazi"/Wikimedia Commons; credit b: modification of work by Mike Mozart)

The addition of small amounts of iodine to table salt (iodized salt) has essentially eliminated this health concern in the United States, but as much as 40% of the world's population is still at risk of iodine deficiency. The iodine atoms are added as anions, and each has a 1− charge and a mass number of 127. Determine the numbers of protons, neutrons, and electrons in one of these iodine anions.

Solution

The atomic number of iodine (53) tells us that a neutral iodine atom contains 53 protons in its nucleus and 53 electrons outside its nucleus. Because the sum of the numbers of protons and neutrons equals the mass number, 127, the number of neutrons is 74 (127 − 53 = 74). Since the iodine is added as a 1− anion, the number of electrons is 54 [53 – (1–) = 54].

Check Your Learning

An ion of platinum has a mass number of 195 and contains 74 electrons. How many protons and neutrons does it contain, and what is its charge?

Answer:

78 protons; 117 neutrons; charge is 4+

Chemical Symbols

A **chemical symbol** is an abbreviation that we use to indicate an element or an atom of an element. For example, the symbol for mercury is Hg (Figure 2.13). We use the same symbol to indicate one atom of mercury (microscopic domain) or to label a container of many atoms of the element mercury (macroscopic domain).

FIGURE 2.13 The symbol Hg represents the element mercury regardless of the amount; it could represent one atom of mercury or a large amount of mercury.

The symbols for several common elements and their atoms are listed in Table 2.3. Some symbols are derived from the common name of the element; others are abbreviations of the name in another language. Most symbols have one or two letters, but three-letter symbols have been used to describe some elements that have atomic numbers greater than 112. To avoid confusion with other notations, only the first letter of a symbol is capitalized. For example, Co is the symbol for the element cobalt, but CO is the notation for the compound carbon monoxide, which contains atoms of the elements carbon (C) and oxygen (O). All known elements and their symbols are in the periodic table in Figure 2.26 (also found in Appendix A).

Some Common Elements and Their Symbols

Element	Symbol	Element	Symbol
aluminum	Al	iron	Fe (from *ferrum*)
bromine	Br	lead	Pb (from *plumbum*)
calcium	Ca	magnesium	Mg
carbon	C	mercury	Hg (from *hydrargyrum*)
chlorine	Cl	nitrogen	N
chromium	Cr	oxygen	O
cobalt	Co	potassium	K (from *kalium*)
copper	Cu (from *cuprum*)	silicon	Si
fluorine	F	silver	Ag (from *argentum*)
gold	Au (from *aurum*)	sodium	Na (from *natrium*)
helium	He	sulfur	S

TABLE 2.3

Element	Symbol	Element	Symbol
hydrogen	H	tin	Sn (from *stannum*)
iodine	I	zinc	Zn

TABLE 2.3

Traditionally, the discoverer (or discoverers) of a new element names the element. However, until the name is recognized by the International Union of Pure and Applied Chemistry (IUPAC), the recommended name of the new element is based on the Latin word(s) for its atomic number. For example, element 106 was called unnilhexium (Unh), element 107 was called unnilseptium (Uns), and element 108 was called unniloctium (Uno) for several years. These elements are now named after scientists (or occasionally locations); for example, element 106 is now known as *seaborgium* (Sg) in honor of Glenn Seaborg, a Nobel Prize winner who was active in the discovery of several heavy elements. Element 109 was named in honor of Lise Meitner, who discovered nuclear fission, a phenomenon that would have world-changing impacts; Meitner also contributed to the discovery of some major isotopes, discussed immediately below.

LINK TO LEARNING

Visit this site (http://openstax.org/l/16IUPAC) to learn more about IUPAC, the International Union of Pure and Applied Chemistry, and explore its periodic table.

Isotopes

The symbol for a specific isotope of any element is written by placing the mass number as a superscript to the left of the element symbol (Figure 2.14). The atomic number is sometimes written as a subscript preceding the symbol, but since this number defines the element's identity, as does its symbol, it is often omitted. For example, magnesium exists as a mixture of three isotopes, each with an atomic number of 12 and with mass numbers of 24, 25, and 26, respectively. These isotopes can be identified as ^{24}Mg, ^{25}Mg, and ^{26}Mg. These isotope symbols are read as "element, mass number" and can be symbolized consistent with this reading. For instance, ^{24}Mg is read as "magnesium 24," and can be written as "magnesium-24" or "Mg-24." ^{25}Mg is read as "magnesium 25," and can be written as "magnesium-25" or "Mg-25." All magnesium atoms have 12 protons in their nucleus. They differ only because a ^{24}Mg atom has 12 neutrons in its nucleus, a ^{25}Mg atom has 13 neutrons, and a ^{26}Mg has 14 neutrons.

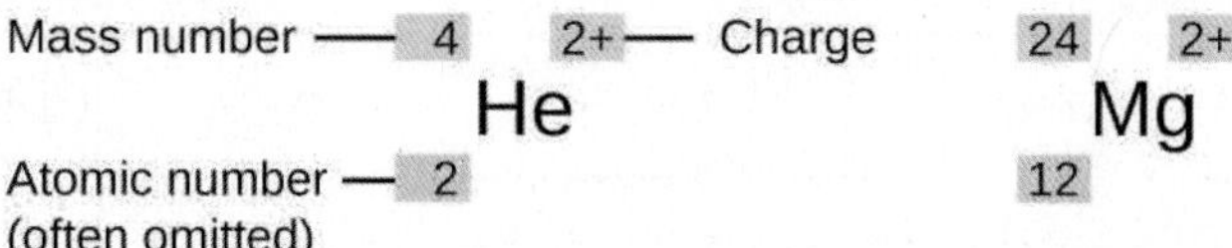

FIGURE 2.14 The symbol for an atom indicates the element via its usual two-letter symbol, the mass number as a left superscript, the atomic number as a left subscript (sometimes omitted), and the charge as a right superscript.

Information about the naturally occurring isotopes of elements with atomic numbers 1 through 10 is given in Table 2.4. Note that in addition to standard names and symbols, the isotopes of hydrogen are often referred to using common names and accompanying symbols. Hydrogen-2, symbolized ^{2}H, is also called deuterium and sometimes symbolized D. Hydrogen-3, symbolized ^{3}H, is also called tritium and sometimes symbolized T.

Nuclear Compositions of Atoms of the Very Light Elements

Element	Symbol	Atomic Number	Number of Protons	Number of Neutrons	Mass (amu)	% Natural Abundance
hydrogen	$^{1}_{1}H$ (protium)	1	1	0	1.0078	99.989
	$^{2}_{1}H$ (deuterium)	1	1	1	2.0141	0.0115
	$^{3}_{1}H$ (tritium)	1	1	2	3.01605	— (trace)
helium	$^{3}_{2}He$	2	2	1	3.01603	0.00013
	$^{4}_{2}He$	2	2	2	4.0026	100
lithium	$^{6}_{3}Li$	3	3	3	6.0151	7.59
	$^{7}_{3}Li$	3	3	4	7.0160	92.41
beryllium	$^{9}_{4}Be$	4	4	5	9.0122	100
boron	$^{10}_{5}B$	5	5	5	10.0129	19.9
	$^{11}_{5}B$	5	5	6	11.0093	80.1
carbon	$^{12}_{6}C$	6	6	6	12.0000	98.89
	$^{13}_{6}C$	6	6	7	13.0034	1.11
	$^{14}_{6}C$	6	6	8	14.0032	— (trace)
nitrogen	$^{14}_{7}N$	7	7	7	14.0031	99.63
	$^{15}_{7}N$	7	7	8	15.0001	0.37
oxygen	$^{16}_{8}O$	8	8	8	15.9949	99.757
	$^{17}_{8}O$	8	8	9	16.9991	0.038
	$^{18}_{8}O$	8	8	10	17.9992	0.205
fluorine	$^{19}_{9}F$	9	9	10	18.9984	100
neon	$^{20}_{10}Ne$	10	10	10	19.9924	90.48

TABLE 2.4

Element	Symbol	Atomic Number	Number of Protons	Number of Neutrons	Mass (amu)	% Natural Abundance
	$^{21}_{10}\text{Ne}$	10	10	11	20.9938	0.27
	$^{22}_{10}\text{Ne}$	10	10	12	21.9914	9.25

TABLE 2.4

LINK TO LEARNING

Use this Build an Atom simulator (http://openstax.org/l/16PhetAtomBld) to build atoms of the first 10 elements, see which isotopes exist, check nuclear stability, and gain experience with isotope symbols.

Atomic Mass

Because each proton and each neutron contribute approximately one amu to the mass of an atom, and each electron contributes far less, the **atomic mass** of a single atom is approximately equal to its mass number (a whole number). However, the average masses of atoms of most elements are not whole numbers because most elements exist naturally as mixtures of two or more isotopes.

The mass of an element shown in a periodic table or listed in a table of atomic masses is a weighted, average mass of all the isotopes present in a naturally occurring sample of that element. This is equal to the sum of each individual isotope's mass multiplied by its fractional abundance.

$$\text{average mass} = \sum_i (\text{fractional abundance} \times \text{isotopic mass})_i$$

For example, the element boron is composed of two isotopes: About 19.9% of all boron atoms are ^{10}B with a mass of 10.0129 amu, and the remaining 80.1% are ^{11}B with a mass of 11.0093 amu. The average atomic mass for boron is calculated to be:

$$\begin{aligned}\text{boron average mass} &= (0.199 \times 10.0129\text{ amu}) + (0.801 \times 11.0093\text{ amu}) \\ &= 1.99\text{ amu} + 8.82\text{ amu} \\ &= 10.81\text{ amu}\end{aligned}$$

It is important to understand that no single boron atom weighs exactly 10.8 amu; 10.8 amu is the average mass of all boron atoms, and individual boron atoms weigh either approximately 10 amu or 11 amu.

EXAMPLE 2.4

Calculation of Average Atomic Mass

A meteorite found in central Indiana contains traces of the noble gas neon picked up from the solar wind during the meteorite's trip through the solar system. Analysis of a sample of the gas showed that it consisted of 91.84% ^{20}Ne (mass 19.9924 amu), 0.47% ^{21}Ne (mass 20.9940 amu), and 7.69% ^{22}Ne (mass 21.9914 amu). What is the average mass of the neon in the solar wind?

Solution

$$\begin{aligned}\text{average mass} &= (0.9184 \times 19.9924\text{ amu}) + (0.0047 \times 20.9940\text{ amu}) + (0.0769 \times 21.9914\text{ amu}) \\ &= (18.36 + 0.099 + 1.69)\text{ amu} \\ &= 20.15\text{ amu}\end{aligned}$$

The average mass of a neon atom in the solar wind is 20.15 amu. (The average mass of a terrestrial neon atom is 20.1796 amu. This result demonstrates that we may find slight differences in the natural abundance of isotopes, depending on their origin.)

Check Your Learning

A sample of magnesium is found to contain 78.70% of ^{24}Mg atoms (mass 23.98 amu), 10.13% of ^{25}Mg atoms (mass 24.99 amu), and 11.17% of ^{26}Mg atoms (mass 25.98 amu). Calculate the average mass of a Mg atom.

Answer:

24.31 amu

We can also do variations of this type of calculation, as shown in the next example.

EXAMPLE 2.5

Calculation of Percent Abundance

Naturally occurring chlorine consists of ^{35}Cl (mass 34.96885 amu) and ^{37}Cl (mass 36.96590 amu), with an average mass of 35.453 amu. What is the percent composition of Cl in terms of these two isotopes?

Solution

The average mass of chlorine is the fraction that is ^{35}Cl times the mass of ^{35}Cl plus the fraction that is ^{37}Cl times the mass of ^{37}Cl.

$$\text{average mass} = (\text{fraction of } ^{35}\text{Cl} \times \text{mass of } ^{35}\text{Cl}) + (\text{fraction of } ^{37}\text{Cl} \times \text{mass of } ^{37}\text{Cl})$$

If we let x represent the fraction that is ^{35}Cl, then the fraction that is ^{37}Cl is represented by 1.00 – x.

(The fraction that is ^{35}Cl + the fraction that is ^{37}Cl must add up to 1, so the fraction of ^{37}Cl must equal 1.00 – the fraction of ^{35}Cl.)

Substituting this into the average mass equation, we have:

$$\begin{aligned} 35.453\text{ amu} &= (x \times 34.96885\text{ amu}) + [(1.00 - x) \times 36.96590\text{ amu}] \\ 35.453 &= 34.96885x + 36.96590 - 36.96590x \\ 1.99705x &= 1.513 \\ x &= \frac{1.513}{1.99705} = 0.7576 \end{aligned}$$

So solving yields: x = 0.7576, which means that 1.00 – 0.7576 = 0.2424. Therefore, chlorine consists of 75.76% ^{35}Cl and 24.24% ^{37}Cl.

Check Your Learning

Naturally occurring copper consists of ^{63}Cu (mass 62.9296 amu) and ^{65}Cu (mass 64.9278 amu), with an average mass of 63.546 amu. What is the percent composition of Cu in terms of these two isotopes?

Answer:

69.15% Cu-63 and 30.85% Cu-65

LINK TO LEARNING

Visit this site (http://openstax.org/l/16PhetAtomMass) to make mixtures of the main isotopes of the first 18 elements, gain experience with average atomic mass, and check naturally occurring isotope ratios using the Isotopes and Atomic Mass simulation.

As you will learn, isotopes are important in nature and especially in human understanding of science and medicine. Let's consider just one natural, stable isotope: Oxygen-18, which is noted in the table above and is referred to as one of the environmental isotopes. It is important in paleoclimatology, for example, because scientists can use the ratio between Oxygen-18 and Oxygen-16 in an ice core to determine the temperature of precipitation over time. Oxygen-18 was also critical to the discovery of metabolic pathways and the mechanisms of enzymes. Mildred Cohn pioneered the usage of these isotopes to act as tracers, so that researchers could follow their path through reactions and gain a better understanding of what is happening. One of her first discoveries provided insight into the phosphorylation of glucose that takes place in mitochondria. And the methods of using isotopes for this research contributed to entire fields of study.

The occurrence and natural abundances of isotopes can be experimentally determined using an instrument called a mass spectrometer. Mass spectrometry (MS) is widely used in chemistry, forensics, medicine, environmental science, and many other fields to analyze and help identify the substances in a sample of material. In a typical mass spectrometer (Figure 2.15), the sample is vaporized and exposed to a high-energy electron beam that causes the sample's atoms (or molecules) to become electrically charged, typically by losing one or more electrons. These cations then pass through a (variable) electric or magnetic field that deflects each cation's path to an extent that depends on both its mass and charge (similar to how the path of a large steel ball rolling past a magnet is deflected to a lesser extent that that of a small steel ball). The ions are detected, and a plot of the relative number of ions generated versus their mass-to-charge ratios (a *mass spectrum*) is made. The height of each vertical feature or peak in a mass spectrum is proportional to the fraction of cations with the specified mass-to-charge ratio. Since its initial use during the development of modern atomic theory, MS has evolved to become a powerful tool for chemical analysis in a wide range of applications.

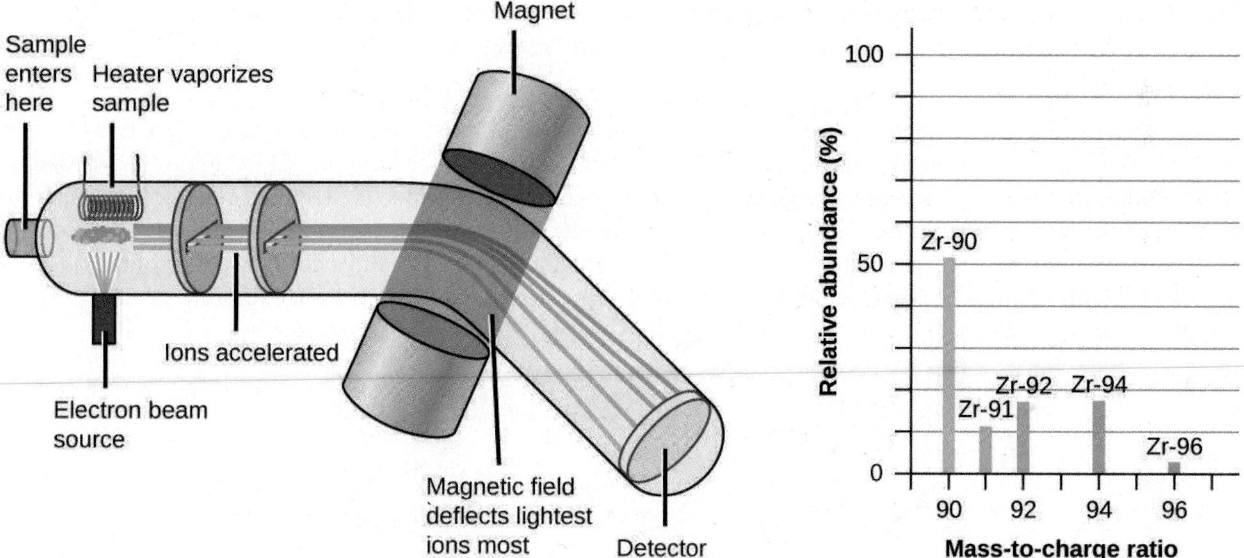

FIGURE 2.15 Analysis of zirconium in a mass spectrometer produces a mass spectrum with peaks showing the different isotopes of Zr.

LINK TO LEARNING

See an animation (http://openstax.org/l/16MassSpec) that explains mass spectrometry. Watch this video (http://openstax.org/l/16RSChemistry) from the Royal Society for Chemistry for a brief description of the rudiments of mass spectrometry.

2.4 Chemical Formulas

LEARNING OBJECTIVES

By the end of this section, you will be able to:

- Symbolize the composition of molecules using molecular formulas and empirical formulas
- Represent the bonding arrangement of atoms within molecules using structural formulas

A **molecular formula** is a representation of a molecule that uses chemical symbols to indicate the types of atoms followed by subscripts to show the number of atoms of each type in the molecule. (A subscript is used only when more than one atom of a given type is present.) Molecular formulas are also used as abbreviations for the names of compounds.

The **structural formula** for a compound gives the same information as its molecular formula (the types and numbers of atoms in the molecule) but also shows how the atoms are connected in the molecule. The structural formula for methane contains symbols for one C atom and four H atoms, indicating the number of atoms in the molecule (Figure 2.16). The lines represent bonds that hold the atoms together. (A chemical bond is an attraction between atoms or ions that holds them together in a molecule or a crystal.) We will discuss chemical bonds and see how to predict the arrangement of atoms in a molecule later. For now, simply know that the lines are an indication of how the atoms are connected in a molecule. A ball-and-stick model shows the geometric arrangement of the atoms with atomic sizes not to scale, and a space-filling model shows the relative sizes of the atoms.

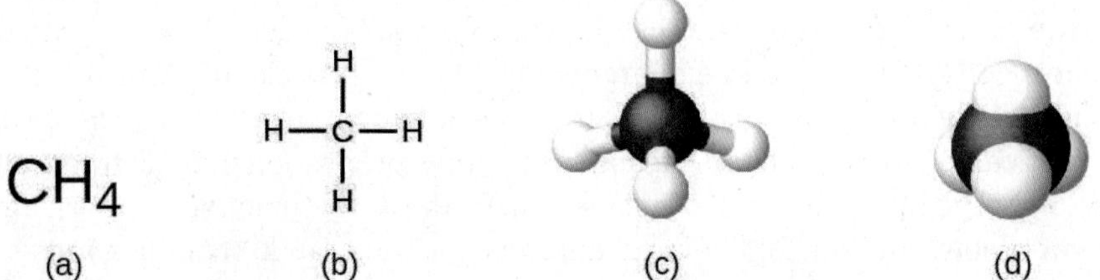

FIGURE 2.16 A methane molecule can be represented as (a) a molecular formula, (b) a structural formula, (c) a ball-and-stick model, and (d) a space-filling model. Carbon and hydrogen atoms are represented by black and white spheres, respectively.

Although many elements consist of discrete, individual atoms, some exist as molecules made up of two or more atoms of the element chemically bonded together. For example, most samples of the elements hydrogen, oxygen, and nitrogen are composed of molecules that contain two atoms each (called diatomic molecules) and thus have the molecular formulas H_2, O_2, and N_2, respectively. Other elements commonly found as diatomic molecules are fluorine (F_2), chlorine (Cl_2), bromine (Br_2), and iodine (I_2). The most common form of the element sulfur is composed of molecules that consist of eight atoms of sulfur; its molecular formula is S_8 (Figure 2.17).

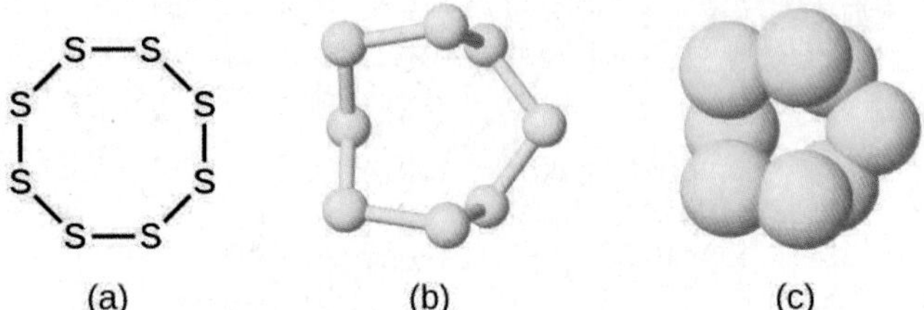

FIGURE 2.17 A molecule of sulfur is composed of eight sulfur atoms and is therefore written as S_8. It can be represented as (a) a structural formula, (b) a ball-and-stick model, and (c) a space-filling model. Sulfur atoms are represented by yellow spheres.

It is important to note that a subscript following a symbol and a number in front of a symbol do not represent the same thing; for example, H_2 and 2H represent distinctly different species. H_2 is a molecular formula; it represents a diatomic molecule of hydrogen, consisting of two atoms of the element that are chemically bonded together. The expression 2H, on the other hand, indicates two separate hydrogen atoms that are not combined as a unit. The expression $2H_2$ represents two molecules of diatomic hydrogen (Figure 2.18).

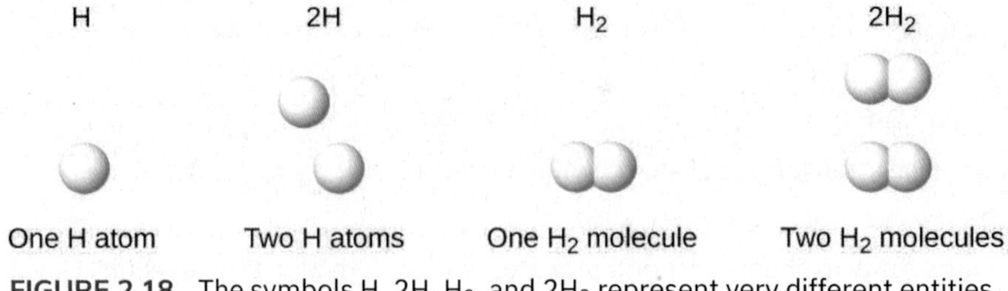

FIGURE 2.18 The symbols H, 2H, H_2, and $2H_2$ represent very different entities.

Compounds are formed when two or more elements chemically combine, resulting in the formation of bonds. For example, hydrogen and oxygen can react to form water, and sodium and chlorine can react to form table salt. We sometimes describe the composition of these compounds with an **empirical formula**, which indicates the types of atoms present and *the simplest whole-number ratio of the number of atoms (or ions) in the compound.* For example, titanium dioxide (used as pigment in white paint and in the thick, white, blocking type of sunscreen) has an empirical formula of TiO_2. This identifies the elements titanium (Ti) and oxygen (O) as the constituents of titanium dioxide, and indicates the presence of twice as many atoms of the element oxygen as atoms of the element titanium (Figure 2.19).

FIGURE 2.19 (a) The white compound titanium dioxide provides effective protection from the sun. (b) A crystal of titanium dioxide, TiO_2, contains titanium and oxygen in a ratio of 1 to 2. The titanium atoms are gray and the oxygen atoms are red. (credit a: modification of work by "osseous"/Flickr)

As discussed previously, we can describe a compound with a molecular formula, in which the subscripts indicate the *actual numbers of atoms* of each element in a molecule of the compound. In many cases, the molecular formula of a substance is derived from experimental determination of both its empirical formula and its molecular mass (the sum of atomic masses for all atoms composing the molecule). For example, it can be determined experimentally that benzene contains two elements, carbon (C) and hydrogen (H), and that for every carbon atom in benzene, there is one hydrogen atom. Thus, the empirical formula is CH. An experimental determination of the molecular mass reveals that a molecule of benzene contains six carbon atoms and six hydrogen atoms, so the molecular formula for benzene is C_6H_6 (Figure 2.20).

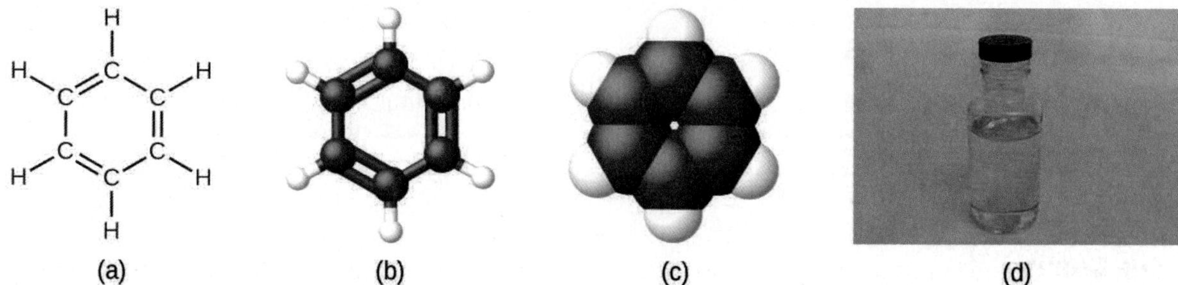

FIGURE 2.20 Benzene, C_6H_6, is produced during oil refining and has many industrial uses. A benzene molecule can be represented as (a) a structural formula, (b) a ball-and-stick model, and (c) a space-filling model. (d) Benzene is a clear liquid. (credit d: modification of work by Sahar Atwa)

If we know a compound's formula, we can easily determine the empirical formula. (This is somewhat of an academic exercise; the reverse chronology is generally followed in actual practice.) For example, the molecular formula for acetic acid, the component that gives vinegar its sharp taste, is $C_2H_4O_2$. This formula indicates that a molecule of acetic acid (Figure 2.21) contains two carbon atoms, four hydrogen atoms, and two oxygen atoms. The ratio of atoms is 2:4:2. Dividing by the lowest common denominator (2) gives the simplest, whole-number ratio of atoms, 1:2:1, so the empirical formula is CH_2O. Note that a molecular formula is always a whole-number multiple of an empirical formula.

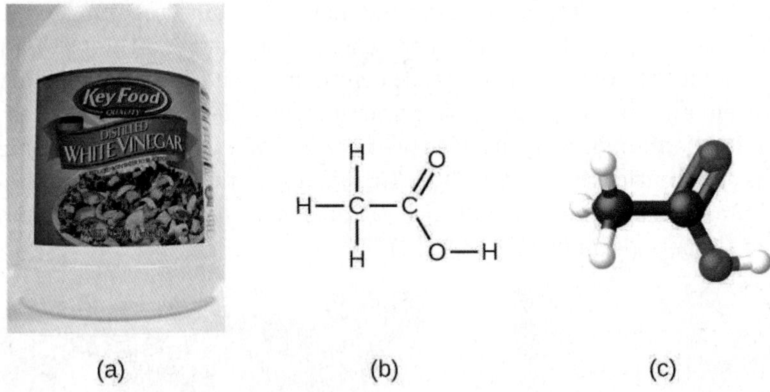

FIGURE 2.21 (a) Vinegar contains acetic acid, $C_2H_4O_2$, which has an empirical formula of CH_2O. It can be represented as (b) a structural formula and (c) as a ball-and-stick model. (credit a: modification of work by "HomeSpot HQ"/Flickr)

EXAMPLE 2.6

Empirical and Molecular Formulas

Molecules of glucose (blood sugar) contain 6 carbon atoms, 12 hydrogen atoms, and 6 oxygen atoms. What are the molecular and empirical formulas of glucose?

Solution

The molecular formula is $C_6H_{12}O_6$ because one molecule actually contains 6 C, 12 H, and 6 O atoms. The simplest whole-number ratio of C to H to O atoms in glucose is 1:2:1, so the empirical formula is CH_2O.

Check Your Learning

A molecule of metaldehyde (a pesticide used for snails and slugs) contains 8 carbon atoms, 16 hydrogen atoms, and 4 oxygen atoms. What are the molecular and empirical formulas of metaldehyde?

Answer:

Molecular formula, $C_8H_{16}O_4$; empirical formula, C_2H_4O

LINK TO LEARNING

You can explore molecule building (http://openstax.org/l/16molbuilding) using an online simulation.

Portrait of a Chemist

Lee Cronin

What is it that chemists do? According to Lee Cronin (Figure 2.22), chemists make very complicated molecules by "chopping up" small molecules and "reverse engineering" them. He wonders if we could "make a really cool universal chemistry set" by what he calls "app-ing" chemistry. Could we "app" chemistry?

In a 2012 TED talk, Lee describes one fascinating possibility: combining a collection of chemical "inks" with a 3D printer capable of fabricating a reaction apparatus (tiny test tubes, beakers, and the like) to fashion a "universal toolkit of chemistry." This toolkit could be used to create custom-tailored drugs to fight a new superbug or to "print" medicine personally configured to your genetic makeup, environment, and health situation. Says Cronin, "What Apple did for music, I'd like to do for the discovery and distribution of prescription drugs."[2] View his full talk (http://openstax.org/l/16LeeCronin) at the TED website.

2 Lee Cronin, "Print Your Own Medicine," Talk presented at TED Global 2012, Edinburgh, Scotland, June 2012.

FIGURE 2.22 Chemist Lee Cronin has been named one of the UK's 10 most inspirational scientists. The youngest chair at the University of Glasgow, Lee runs a large research group, collaborates with many scientists worldwide, has published over 250 papers in top scientific journals, and has given more than 150 invited talks. His research focuses on complex chemical systems and their potential to transform technology, but also branches into nanoscience, solar fuels, synthetic biology, and even artificial life and evolution. (credit: image courtesy of Lee Cronin)

It is important to be aware that it may be possible for the same atoms to be arranged in different ways: Compounds with the same molecular formula may have different atom-to-atom bonding and therefore different structures. For example, could there be another compound with the same formula as acetic acid, $C_2H_4O_2$? And if so, what would be the structure of its molecules?

If you predict that another compound with the formula $C_2H_4O_2$ could exist, then you demonstrated good chemical insight and are correct. Two C atoms, four H atoms, and two O atoms can also be arranged to form a methyl formate, which is used in manufacturing, as an insecticide, and for quick-drying finishes. Methyl formate molecules have one of the oxygen atoms between the two carbon atoms, differing from the arrangement in acetic acid molecules. Acetic acid and methyl formate are examples of **isomers**—compounds with the same chemical formula but different molecular structures (Figure 2.23). Note that this small difference in the arrangement of the atoms has a major effect on their respective chemical properties. You would certainly not want to use a solution of methyl formate as a substitute for a solution of acetic acid (vinegar) when you make salad dressing.

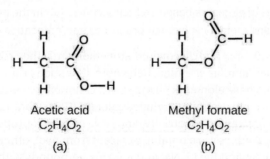

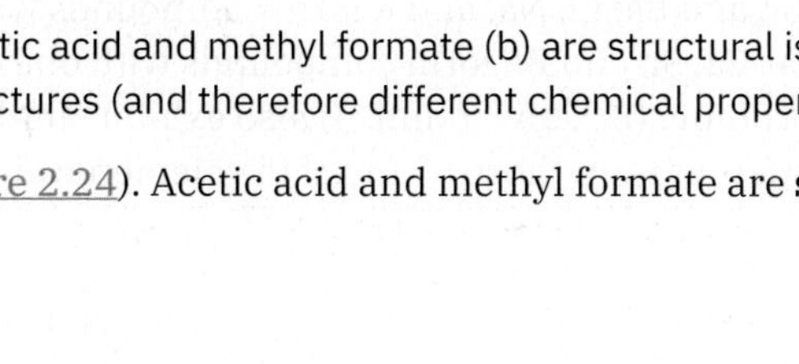

Acetic acid
$C_2H_4O_2$
(a)

Methyl formate
$C_2H_4O_2$
(b)

FIGURE 2.23 Molecules of (a) acetic acid and methyl formate (b) are structural isomers; they have the same formula ($C_2H_4O_2$) but different structures (and therefore different chemical properties).

Many types of isomers exist (Figure 2.24). Acetic acid and methyl formate are **structural isomers**, compounds in which the molecules differ in how the atoms are connected to each other. There are also various types of **spatial isomers**, in which the relative orientations of the atoms in space can be different. For example, the compound carvone (found in caraway seeds, spearmint, and mandarin orange peels) consists of two isomers that are mirror images of each other. *S*-(+)-carvone smells like caraway, and *R*-(−)-carvone smells like spearmint.

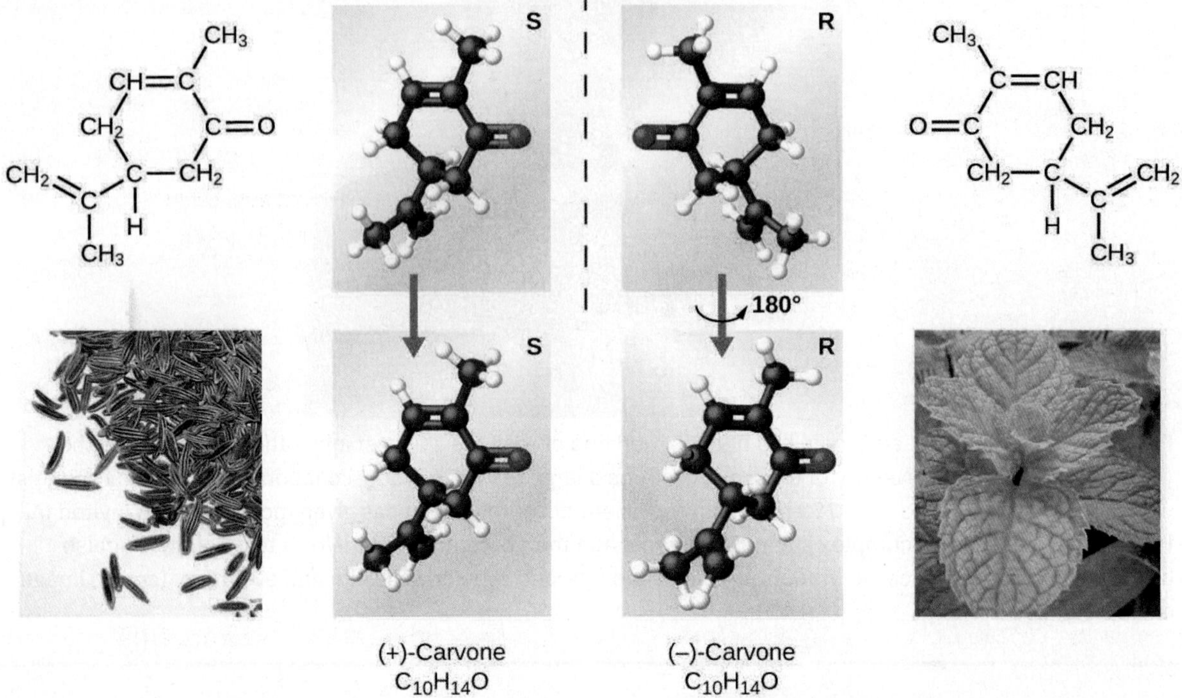

FIGURE 2.24 Molecules of carvone are spatial isomers; they only differ in the relative orientations of the atoms in space. (credit bottom left: modification of work by "Miansari66"/Wikimedia Commons; credit bottom right: modification of work by Forest & Kim Starr)

LINK TO LEARNING

Select this link (http://openstax.org/l/16isomers) to view an explanation of isomers, spatial isomers, and why they have different smells (select the video titled "Mirror Molecule: Carvone").

2.5 The Periodic Table

LEARNING OBJECTIVES

By the end of this section, you will be able to:

- State the periodic law and explain the organization of elements in the periodic table
- Predict the general properties of elements based on their location within the periodic table
- Identify metals, nonmetals, and metalloids by their properties and/or location on the periodic table

As early chemists worked to purify ores and discovered more elements, they realized that various elements could be grouped together by their similar chemical behaviors. One such grouping includes lithium (Li), sodium (Na), and potassium (K): These elements all are shiny, conduct heat and electricity well, and have similar chemical properties. A second grouping includes calcium (Ca), strontium (Sr), and barium (Ba), which also are shiny, good conductors of heat and electricity, and have chemical properties in common. However, the specific properties of these two groupings are notably different from each other. For example: Li, Na, and K are much more reactive than are Ca, Sr, and Ba; Li, Na, and K form compounds with oxygen in a ratio of two of their atoms to one oxygen atom, whereas Ca, Sr, and Ba form compounds with one of their atoms to one oxygen atom. Fluorine (F), chlorine (Cl), bromine (Br), and iodine (I) also exhibit similar properties to each other, but these properties are drastically different from those of any of the elements above.

Dimitri Mendeleev in Russia (1869) and Lothar Meyer in Germany (1870) independently recognized that there was a periodic relationship among the properties of the elements known at that time. Both published tables with the elements arranged according to increasing atomic mass. But Mendeleev went one step further than Meyer: He used his table to predict the existence of elements that would have the properties similar to aluminum and silicon, but were yet unknown. The discoveries of gallium (1875) and germanium (1886) provided great support for Mendeleev's work. Although Mendeleev and Meyer had a long dispute over priority, Mendeleev's contributions to the development of the periodic table are now more widely recognized (Figure 2.25).

(a)

Reihen	Gruppo I. — R^2O	Gruppo II. — RO	Gruppo III. — R^2O^3	Gruppo IV. RH^4 RO^2	Gruppo V. RH^3 R^2O^5	Gruppo VI. RH^2 RO^3	Gruppo VII. RH R^2O^7	Gruppo VIII. — RO^4
1	H=1							
2	Li=7	Be=9,4	B=11	C=12	N=14	O=16	F=19	
3	Na=23	Mg=24	Al=27,3	Si=28	P=31	S=32	Cl=35,5	
4	K=39	Ca=40	—=44	Ti=48	V=51	Cr=52	Mn=55	Fe=56, Co=59, Ni=59, Cu=63.
5	(Cu=63)	Zn=65	—=68	—=72	As=75	Se=78	Br=80	
6	Rb=85	Sr=87	?Yt=88	Zr=90	Nb=94	Mo=96	—=100	Ru=104, Rh=104, Pd=106, Ag=108.
7	(Ag=108)	Cd=112	In=113	Sn=118	Sb=122	Te=125	J=127	
8	Cs=133	Ba=137	?Di=138	?Ce=140	—	—	—	— — — —
9	(—)	—	—	—	—	—	—	
10	—	—	?Er=178	?La=180	Ta=182	W=184	—	Os=195, Ir=197, Pt=198, Au=199.
11	(Au=199)	Hg=200	Tl=204	Pb=207	Bi=208	—	—	
12	—	—	—	Th=231	—	U=240	—	— — — —

(b)

FIGURE 2.25 (a) Dimitri Mendeleev is widely credited with creating (b) the first periodic table of the elements. (credit a: modification of work by Serge Lachinov; credit b: modification of work by "Den fjättrade ankan"/Wikimedia Commons)

By the twentieth century, it became apparent that the periodic relationship involved atomic numbers rather than atomic masses. The modern statement of this relationship, the **periodic law**, is as follows: *the properties of the elements are periodic functions of their atomic numbers.* A modern **periodic table** arranges the elements in increasing order of their atomic numbers and groups atoms with similar properties in the same vertical column (Figure 2.26). Each box represents an element and contains its atomic number, symbol, average atomic mass, and (sometimes) name. The elements are arranged in seven horizontal rows, called **periods** or **series**, and 18 vertical columns, called **groups**. Groups are labeled at the top of each column. In the United States, the labels traditionally were numerals with capital letters. However, IUPAC recommends that the numbers 1 through 18 be used, and these labels are more common. For the table to fit on a single page, parts of two of the rows, a total of 14 columns, are usually written below the main body of the table.

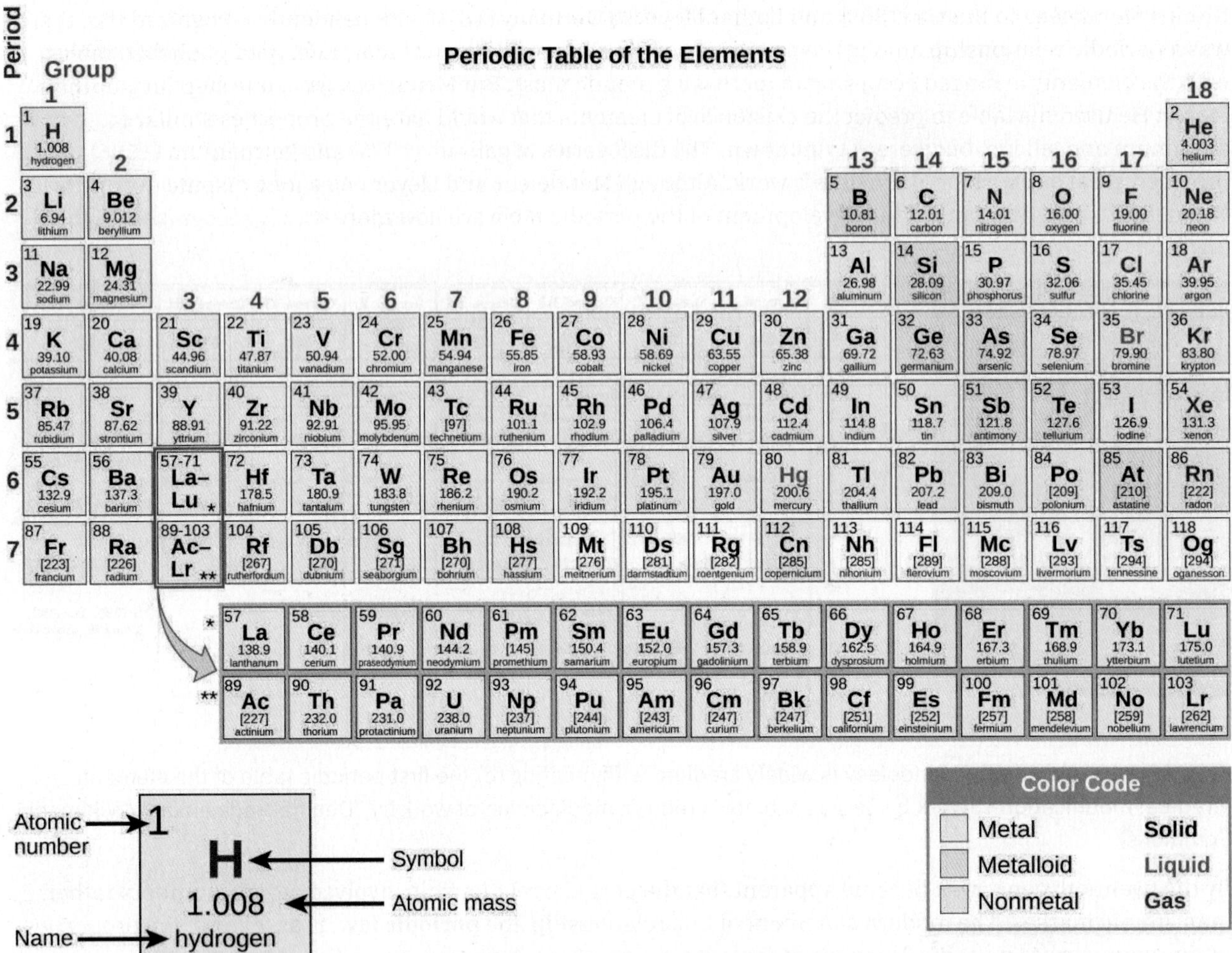

FIGURE 2.26 Elements in the periodic table are organized according to their properties.

Even after the periodic nature of elements and the table itself were widely accepted, gaps remained. Mendeleev had predicted, and others including Henry Moseley had later confirmed, that there should be elements below Manganese in Group 7. German chemists Ida Tacke and Walter Noddack set out to find the elements, a quest being pursued by scientists around the world. Their method was unique in that they did not only consider the properties of manganese, but also the elements horizontally adjacent to the missing elements 43 and 75 on the table. Thus, by investigating ores containing minerals of ruthenium (Ru), tungsten (W), osmium (Os), and so on, they were able to identify naturally occurring elements that helped complete the table. Rhenium, one of their discoveries, was one of the last natural elements to be discovered and is the last stable element to be discovered. (Francium, the last natural element to be discovered, was identified by Marguerite Perey in 1939.)

Many elements differ dramatically in their chemical and physical properties, but some elements are similar in their behaviors. For example, many elements appear shiny, are malleable (able to be deformed without breaking) and ductile (can be drawn into wires), and conduct heat and electricity well. Other elements are not shiny, malleable, or ductile, and are poor conductors of heat and electricity. We can sort the elements into large classes with common properties: **metals** (elements that are shiny, malleable, good conductors of heat and electricity—shaded yellow); **nonmetals** (elements that appear dull, poor conductors of heat and electricity—shaded green); and **metalloids** (elements that conduct heat and electricity moderately well, and possess some properties of metals and some properties of nonmetals—shaded purple).

The elements can also be classified into the **main-group elements** (or **representative elements**) in the columns labeled 1, 2, and 13–18; the **transition metals** in the columns labeled 3–12[3]; and **inner transition metals** in the two rows at the bottom of the table (the top-row elements are called **lanthanides** and the bottom-row elements are **actinides**; Figure 2.27). The elements can be subdivided further by more specific properties, such as the composition of the compounds they form. For example, the elements in group 1 (the first column) form compounds that consist of one atom of the element and one atom of hydrogen. These elements (except hydrogen) are known as **alkali metals**, and they all have similar chemical properties. The elements in group 2 (the second column) form compounds consisting of one atom of the element and two atoms of hydrogen: These are called **alkaline earth metals**, with similar properties among members of that group. Other groups with specific names are the **pnictogens** (group 15), **chalcogens** (group 16), **halogens** (group 17), and the **noble gases** (group 18, also known as **inert gases**). The groups can also be referred to by the first element of the group: For example, the chalcogens can be called the oxygen group or oxygen family. Hydrogen is a unique, nonmetallic element with properties similar to both group 1 and group 17 elements. For that reason, hydrogen may be shown at the top of both groups, or by itself.

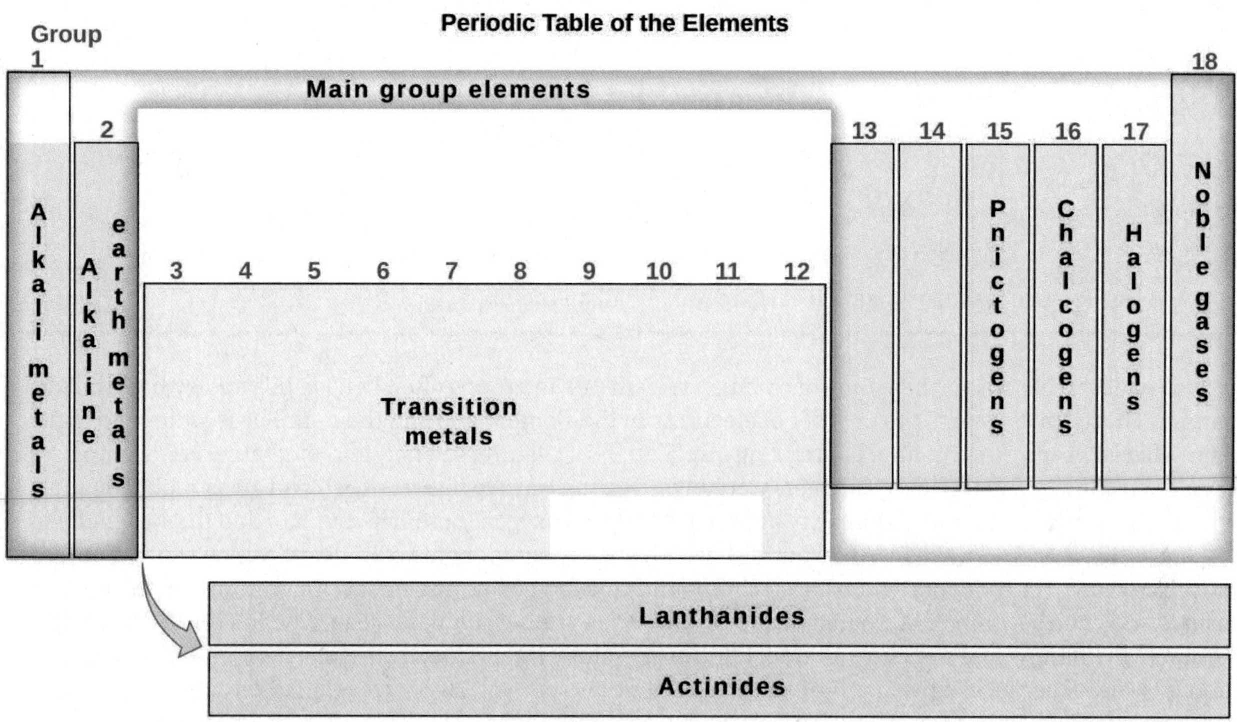

FIGURE 2.27 The periodic table organizes elements with similar properties into groups.

LINK TO LEARNING

Click on this link (https://openstax.org/l/16Periodic) for an interactive periodic table, which you can use to explore the properties of the elements (includes podcasts and videos of each element). You may also want to try this one (https://openstax.org/l/16Periodic2) that shows photos of all the elements.

EXAMPLE 2.7

Naming Groups of Elements

Atoms of each of the following elements are essential for life. Give the group name for the following elements:

(a) chlorine

3 Per the IUPAC definition, group 12 elements are not transition metals, though they are often referred to as such. Additional details on this group's elements are provided in a chapter on transition metals and coordination chemistry.

(b) calcium

(c) sodium

(d) sulfur

Solution

The family names are as follows:

(a) halogen

(b) alkaline earth metal

(c) alkali metal

(d) chalcogen

Check Your Learning

Give the group name for each of the following elements:

(a) krypton

(b) selenium

(c) barium

(d) lithium

Answer:

(a) noble gas; (b) chalcogen; (c) alkaline earth metal; (d) alkali metal

---

As you will learn in your further study of chemistry, elements in groups often behave in a somewhat similar manner. This is partly due to the number of electrons in their outer shell and their similar readiness to bond. These shared properties can have far-ranging implications in nature, science, and medicine. For example, when Gertrude Elion and George Hitchens were investigating ways to interrupt cell and virus replication to fight diseases, they utilized the similarity between sulfur and oxygen (both in Group 16) and their capacity to bond in similar ways. Elion focused on purines, which are key components of DNA and which contain oxygen. She found that by introducing sulfur-based compounds (called purine analogues) that mimic the structure of purines, molecules within DNA would bond to the analogues rather than the "regular" DNA purine. With the normal DNA bonding and structure altered, Elion successfully interrupted cell replication. At its core, the strategy worked because of the similarity between sulfur and oxygen. Her discovery led directly to important treatments for leukemia. Overall, Elion's work with George Hitchens not only led to more treatments, but also changed the entire methodology of drug development. By using specific elements and compounds to target specific aspects of tumor cells, viruses, and bacteria, they laid the groundwork for many of today's most common and important medicines, used to help millions of people each year. They were awarded the Nobel Prize in 1988.

In studying the periodic table, you might have noticed something about the atomic masses of some of the elements. Element 43 (technetium), element 61 (promethium), and most of the elements with atomic number 84 (polonium) and higher have their atomic mass given in square brackets. This is done for elements that consist entirely of unstable, radioactive isotopes (you will learn more about radioactivity in the nuclear chemistry chapter). An average atomic weight cannot be determined for these elements because their radioisotopes may vary significantly in relative abundance, depending on the source, or may not even exist in nature. The number in square brackets is the atomic mass number (an approximate atomic mass) of the most stable isotope of that element.

2.6 Ionic and Molecular Compounds

LEARNING OBJECTIVES

By the end of this section, you will be able to:

- Define ionic and molecular (covalent) compounds
- Predict the type of compound formed from elements based on their location within the periodic table
- Determine formulas for simple ionic compounds

In ordinary chemical reactions, the nucleus of each atom (and thus the identity of the element) remains unchanged. Electrons, however, can be added to atoms by transfer from other atoms, lost by transfer to other atoms, or shared with other atoms. The transfer and sharing of electrons among atoms govern the chemistry of the elements. During the formation of some compounds, atoms gain or lose electrons, and form electrically charged particles called ions (Figure 2.28).

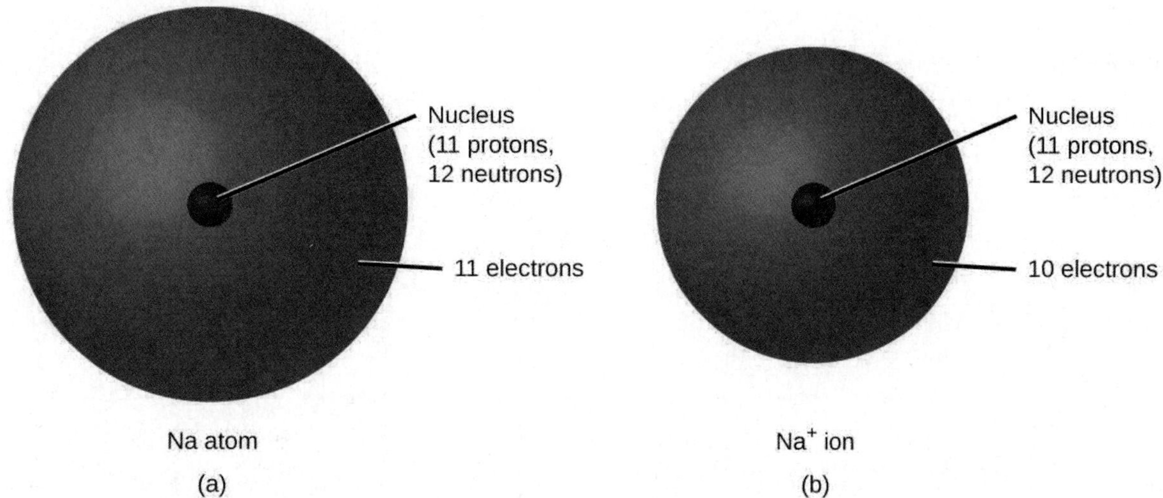

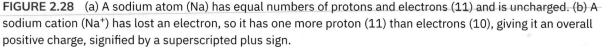

FIGURE 2.28 (a) A sodium atom (Na) has equal numbers of protons and electrons (11) and is uncharged. (b) A sodium cation (Na^+) has lost an electron, so it has one more proton (11) than electrons (10), giving it an overall positive charge, signified by a superscripted plus sign.

You can use the periodic table to predict whether an atom will form an anion or a cation, and you can often predict the charge of the resulting ion. Atoms of many main-group metals lose enough electrons to leave them with the same number of electrons as an atom of the preceding noble gas. To illustrate, an atom of an alkali metal (group 1) loses one electron and forms a cation with a 1+ charge; an alkaline earth metal (group 2) loses two electrons and forms a cation with a 2+ charge, and so on. For example, a neutral calcium atom, with 20 protons and 20 electrons, readily loses two electrons. This results in a cation with 20 protons, 18 electrons, and a 2+ charge. It has the same number of electrons as atoms of the preceding noble gas, argon, and is symbolized Ca^{2+}. The name of a metal ion is the same as the name of the metal atom from which it forms, so Ca^{2+} is called a calcium ion.

When atoms of nonmetal elements form ions, they generally gain enough electrons to give them the same number of electrons as an atom of the next noble gas in the periodic table. Atoms of group 17 gain one electron and form anions with a 1− charge; atoms of group 16 gain two electrons and form ions with a 2− charge, and so on. For example, the neutral bromine atom, with 35 protons and 35 electrons, can gain one electron to provide it with 36 electrons. This results in an anion with 35 protons, 36 electrons, and a 1− charge. It has the same number of electrons as atoms of the next noble gas, krypton, and is symbolized Br^-. (A discussion of the theory supporting the favored status of noble gas electron numbers reflected in these predictive rules for ion formation is provided in a later chapter of this text.)

Note the usefulness of the periodic table in predicting likely ion formation and charge (Figure 2.29). Moving from the far left to the right on the periodic table, main-group elements tend to form cations with a charge equal to the group number. That is, group 1 elements form 1+ ions; group 2 elements form 2+ ions, and so on. Moving from the far right to the left on the periodic table, elements often form anions with a negative charge equal to the number of groups moved left from the noble gases. For example, group 17 elements (one group left of the noble gases) form 1– ions; group 16 elements (two groups left) form 2– ions, and so on. This trend can be used as a guide in many cases, but its predictive value decreases when moving toward the center of the periodic table. In fact, transition metals and some other metals often exhibit variable charges that are not predictable by their location in the table. For example, copper can form ions with a 1+ or 2+ charge, and iron can form ions with a 2+ or 3+ charge.

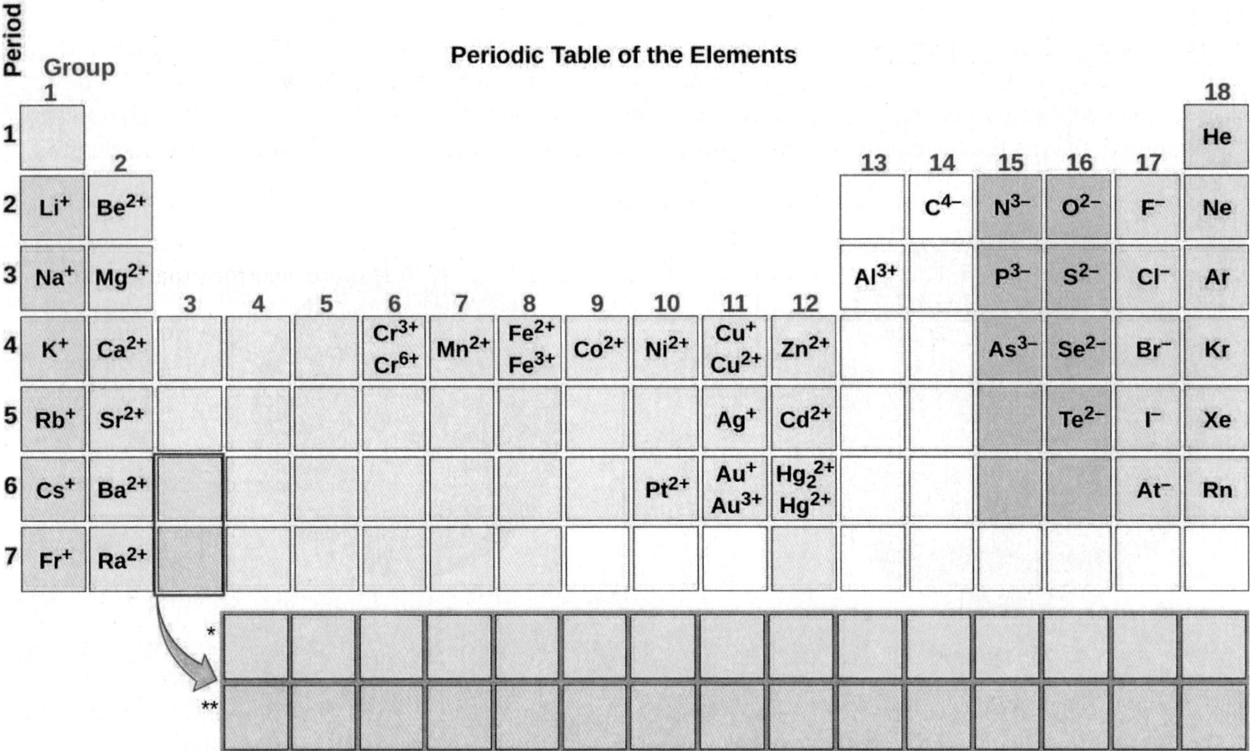

FIGURE 2.29 Some elements exhibit a regular pattern of ionic charge when they form ions.

EXAMPLE 2.8

Composition of Ions

An ion found in some compounds used as antiperspirants contains 13 protons and 10 electrons. What is its symbol?

Solution

Because the number of protons remains unchanged when an atom forms an ion, the atomic number of the element must be 13. Knowing this lets us use the periodic table to identify the element as Al (aluminum). The Al atom has lost three electrons and thus has three more positive charges (13) than it has electrons (10). This is the aluminum cation, Al^{3+}.

Check Your Learning

Give the symbol and name for the ion with 34 protons and 36 electrons.

Answer:

Se^{2-}, the selenide ion

EXAMPLE 2.9

Formation of Ions

Magnesium and nitrogen react to form an ionic compound. Predict which forms an anion, which forms a cation, and the charges of each ion. Write the symbol for each ion and name them.

Solution

Magnesium's position in the periodic table (group 2) tells us that it is a metal. Metals form positive ions (cations). A magnesium atom must lose two electrons to have the same number electrons as an atom of the previous noble gas, neon. Thus, a magnesium atom will form a cation with two fewer electrons than protons and a charge of 2+. The symbol for the ion is Mg^{2+}, and it is called a magnesium ion.

Nitrogen's position in the periodic table (group 15) reveals that it is a nonmetal. Nonmetals form negative ions (anions). A nitrogen atom must gain three electrons to have the same number of electrons as an atom of the following noble gas, neon. Thus, a nitrogen atom will form an anion with three more electrons than protons and a charge of 3−. The symbol for the ion is N^{3-}, and it is called a nitride ion.

Check Your Learning

Aluminum and carbon react to form an ionic compound. Predict which forms an anion, which forms a cation, and the charges of each ion. Write the symbol for each ion and name them.

Answer:

Al will form a cation with a charge of 3+: Al^{3+}, an aluminum ion. Carbon will form an anion with a charge of 4−: C^{4-}, a carbide ion.

The ions that we have discussed so far are called **monatomic ions**, that is, they are ions formed from only one atom. We also find many **polyatomic ions**. These ions, which act as discrete units, are electrically charged molecules (a group of bonded atoms with an overall charge). Some of the more important polyatomic ions are listed in Table 2.5. **Oxyanions** are polyatomic ions that contain one or more oxygen atoms. At this point in your study of chemistry, you should memorize the names, formulas, and charges of the most common polyatomic ions. Because you will use them repeatedly, they will soon become familiar.

Common Polyatomic Ions

Name	Formula	Related Acid	Formula
ammonium	$NH_4{}^+$		
hydronium	H_3O^+		
peroxide	$O_2{}^{2-}$		
hydroxide	OH^-		
acetate	CH_3COO^-	acetic acid	CH_3COOH
cyanide	CN^-	hydrocyanic acid	HCN
azide	$N_3{}^-$	hydrazoic acid	HN_3
carbonate	$CO_3{}^{2-}$	carbonic acid	H_2CO_3

TABLE 2.5

Name	Formula	Related Acid	Formula
bicarbonate	HCO_3^-		
nitrate	NO_3^-	nitric acid	HNO_3
nitrite	NO_2^-	nitrous acid	HNO_2
sulfate	SO_4^{2-}	sulfuric acid	H_2SO_4
hydrogen sulfate	HSO_4^-		
sulfite	SO_3^{2-}	sulfurous acid	H_2SO_3
hydrogen sulfite	HSO_3^-		
phosphate	PO_4^{3-}	phosphoric acid	H_3PO_4
hydrogen phosphate	HPO_4^{2-}		
dihydrogen phosphate	$H_2PO_4^-$		
perchlorate	ClO_4^-	perchloric acid	$HClO_4$
chlorate	ClO_3^-	chloric acid	$HClO_3$
chlorite	ClO_2^-	chlorous acid	$HClO_2$
hypochlorite	ClO^-	hypochlorous acid	$HClO$
chromate	CrO_4^{2-}	chromic acid	H_2CrO_4
dichromate	$Cr_2O_7^{2-}$	dichromic acid	$H_2Cr_2O_7$
permanganate	MnO_4^-	permanganic acid	$HMnO_4$

TABLE 2.5

Note that there is a system for naming some polyatomic ions; *-ate* and *-ite* are suffixes designating polyatomic ions containing more or fewer oxygen atoms. *Per-* (short for "hyper") and *hypo-* (meaning "under") are prefixes meaning more oxygen atoms than *-ate* and fewer oxygen atoms than *-ite*, respectively. For example, perchlorate is ClO_4^-, chlorate is ClO_3^-, chlorite is ClO_2^- and hypochlorite is ClO^-. Unfortunately, the number of oxygen atoms corresponding to a given suffix or prefix is not consistent; for example, nitrate is NO_3^- while sulfate is SO_4^{2-}. This will be covered in more detail in the next module on nomenclature.

The nature of the attractive forces that hold atoms or ions together within a compound is the basis for classifying chemical bonding. When electrons are transferred and ions form, **ionic bonds** result. Ionic bonds are electrostatic forces of attraction, that is, the attractive forces experienced between objects of opposite electrical charge (in this case, cations and anions). When electrons are "shared" and molecules form, **covalent bonds** result. Covalent bonds are the attractive forces between the positively charged nuclei of the bonded atoms and one or more pairs of electrons that are located between the atoms. Compounds are classified as ionic or molecular (covalent) on the basis of the bonds present in them.

Ionic Compounds

When an element composed of atoms that readily lose electrons (a metal) reacts with an element composed of atoms that readily gain electrons (a nonmetal), a transfer of electrons usually occurs, producing ions. The compound formed by this transfer is stabilized by the electrostatic attractions (ionic bonds) between the ions of opposite charge present in the compound. For example, when each sodium atom in a sample of sodium metal (group 1) gives up one electron to form a sodium cation, Na^+, and each chlorine atom in a sample of chlorine gas (group 17) accepts one electron to form a chloride anion, Cl^-, the resulting compound, NaCl, is composed of sodium ions and chloride ions in the ratio of one Na^+ ion for each Cl^- ion. Similarly, each calcium atom (group 2) can give up two electrons and transfer one to each of two chlorine atoms to form $CaCl_2$, which is composed of Ca^{2+} and Cl^- ions in the ratio of one Ca^{2+} ion to two Cl^- ions.

A compound that contains ions and is held together by ionic bonds is called an **ionic compound**. The periodic table can help us recognize many of the compounds that are ionic: When a metal is combined with one or more nonmetals, the compound is usually ionic. This guideline works well for predicting ionic compound formation for most of the compounds typically encountered in an introductory chemistry course. However, it is not always true (for example, aluminum chloride, $AlCl_3$, is not ionic).

You can often recognize ionic compounds because of their properties. Ionic compounds are solids that typically melt at high temperatures and boil at even higher temperatures. For example, sodium chloride melts at 801 °C and boils at 1413 °C. (As a comparison, the molecular compound water melts at 0 °C and boils at 100 °C.) In solid form, an ionic compound is not electrically conductive because its ions are unable to flow ("electricity" is the flow of charged particles). When molten, however, it can conduct electricity because its ions are able to move freely through the liquid (Figure 2.30).

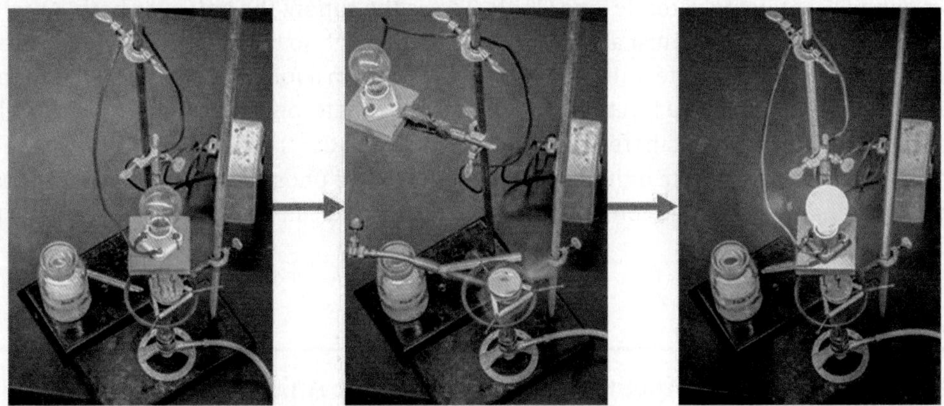

FIGURE 2.30 Sodium chloride melts at 801 °C and conducts electricity when molten. (credit: modification of work by Mark Blaser and Matt Evans)

LINK TO LEARNING

Watch this video (http://openstax.org/l/16moltensalt) to see a mixture of salts melt and conduct electricity.

In every ionic compound, the total number of positive charges of the cations equals the total number of negative charges of the anions. Thus, ionic compounds are electrically neutral overall, even though they contain positive and negative ions. We can use this observation to help us write the formula of an ionic compound. The formula of an ionic compound must have a ratio of ions such that the numbers of positive and negative charges are equal.

EXAMPLE 2.10

Predicting the Formula of an Ionic Compound

The gemstone sapphire (Figure 2.31) is mostly a compound of aluminum and oxygen that contains aluminum cations, Al^{3+}, and oxygen anions, O^{2-}. What is the formula of this compound?

FIGURE 2.31 Although pure aluminum oxide is colorless, trace amounts of iron and titanium give blue sapphire its characteristic color. (credit: modification of work by Stanislav Doronenko)

Solution

Because the ionic compound must be electrically neutral, it must have the same number of positive and negative charges. Two aluminum ions, each with a charge of 3+, would give us six positive charges, and three oxide ions, each with a charge of 2−, would give us six negative charges. The formula would be Al_2O_3.

Check Your Learning

Predict the formula of the ionic compound formed between the sodium cation, Na^+, and the sulfide anion, S^{2-}.

Answer:

Na_2S

Many ionic compounds contain polyatomic ions (Table 2.5) as the cation, the anion, or both. As with simple ionic compounds, these compounds must also be electrically neutral, so their formulas can be predicted by treating the polyatomic ions as discrete units. We use parentheses in a formula to indicate a group of atoms that behave as a unit. For example, the formula for calcium phosphate, one of the minerals in our bones, is $Ca_3(PO_4)_2$. This formula indicates that there are three calcium ions (Ca^{2+}) for every two phosphate ($PO_4{}^{3-}$) groups. The $PO_4{}^{3-}$ groups are discrete units, each consisting of one phosphorus atom and four oxygen atoms, and having an overall charge of 3−. The compound is electrically neutral, and its formula shows a total count of three Ca, two P, and eight O atoms.

EXAMPLE 2.11

Predicting the Formula of a Compound with a Polyatomic Anion

Baking powder contains calcium dihydrogen phosphate, an ionic compound composed of the ions Ca^{2+} and $H_2PO_4{}^-$. What is the formula of this compound?

Solution

The positive and negative charges must balance, and this ionic compound must be electrically neutral. Thus, we must have two negative charges to balance the 2+ charge of the calcium ion. This requires a ratio of one Ca^{2+} ion to two $H_2PO_4{}^-$ ions. We designate this by enclosing the formula for the dihydrogen phosphate ion in parentheses and adding a subscript 2. The formula is $Ca(H_2PO_4)_2$.

Check Your Learning

Predict the formula of the ionic compound formed between the lithium ion and the peroxide ion, $O_2{}^{2-}$ (Hint: Use the periodic table to predict the sign and the charge on the lithium ion.)

Answer:

Li_2O_2

Because an ionic compound is not made up of single, discrete molecules, it may not be properly symbolized using a *molecular* formula. Instead, ionic compounds must be symbolized by a formula indicating the *relative numbers* of its constituent ions. For compounds containing only monatomic ions (such as NaCl) and for many compounds containing polyatomic ions (such as $CaSO_4$), these formulas are just the empirical formulas introduced earlier in this chapter. However, the formulas for some ionic compounds containing polyatomic ions are not empirical formulas. For example, the ionic compound sodium oxalate is comprised of Na^+ and $C_2O_4{}^{2-}$ ions combined in a 2:1 ratio, and its formula is written as $Na_2C_2O_4$. The subscripts in this formula are not the smallest-possible whole numbers, as each can be divided by 2 to yield the empirical formula, $NaCO_2$. This is not the accepted formula for sodium oxalate, however, as it does not accurately represent the compound's polyatomic anion, $C_2O_4{}^{2-}$.

Molecular Compounds

Many compounds do not contain ions but instead consist solely of discrete, neutral molecules. These **molecular compounds** (covalent compounds) result when atoms share, rather than transfer (gain or lose), electrons. Covalent bonding is an important and extensive concept in chemistry, and it will be treated in considerable detail in a later chapter of this text. We can often identify molecular compounds on the basis of their physical properties. Under normal conditions, molecular compounds often exist as gases, low-boiling liquids, and low-melting solids, although many important exceptions exist.

Whereas ionic compounds are usually formed when a metal and a nonmetal combine, covalent compounds are usually formed by a combination of nonmetals. Thus, the periodic table can help us recognize many of the compounds that are covalent. While we can use the positions of a compound's elements in the periodic table to predict whether it is ionic or covalent at this point in our study of chemistry, you should be aware that this is a very simplistic approach that does not account for a number of interesting exceptions. Shades of gray exist between ionic and molecular compounds, and you'll learn more about those later.

EXAMPLE 2.12

Predicting the Type of Bonding in Compounds

Predict whether the following compounds are ionic or molecular:

(a) KI, the compound used as a source of iodine in table salt

(b) H_2O_2, the bleach and disinfectant hydrogen peroxide

(c) $CHCl_3$, the anesthetic chloroform

(d) Li_2CO_3, a source of lithium in antidepressants

Solution

(a) Potassium (group 1) is a metal, and iodine (group 17) is a nonmetal; KI is predicted to be ionic.

(b) Hydrogen (group 1) is a nonmetal, and oxygen (group 16) is a nonmetal; H_2O_2 is predicted to be molecular.

(c) Carbon (group 14) is a nonmetal, hydrogen (group 1) is a nonmetal, and chlorine (group 17) is a nonmetal; $CHCl_3$ is predicted to be molecular.

(d) Lithium (group 1) is a metal, and carbonate is a polyatomic ion; Li_2CO_3 is predicted to be ionic.

Check Your Learning

Using the periodic table, predict whether the following compounds are ionic or covalent:

(a) SO_2

(b) CaF_2

(c) N_2H_4

(d) $Al_2(SO_4)_3$

Answer:
(a) molecular; (b) ionic; (c) molecular; (d) ionic

2.7 Chemical Nomenclature

LEARNING OBJECTIVES

By the end of this section, you will be able to:

- Derive names for common types of inorganic compounds using a systematic approach

Nomenclature, a collection of rules for naming things, is important in science and in many other situations. This module describes an approach that is used to name simple ionic and molecular compounds, such as NaCl, $CaCO_3$, and N_2O_4. The simplest of these are **binary compounds**, those containing only two elements, but we will also consider how to name ionic compounds containing polyatomic ions, and one specific, very important class of compounds known as acids (subsequent chapters in this text will focus on these compounds in great detail). We will limit our attention here to inorganic compounds, compounds that are composed principally of elements other than carbon, and will follow the nomenclature guidelines proposed by IUPAC. The rules for organic compounds, in which carbon is the principle element, will be treated in a later chapter on organic chemistry.

Ionic Compounds

To name an inorganic compound, we need to consider the answers to several questions. First, is the compound ionic or molecular? If the compound is ionic, does the metal form ions of only one type (fixed charge) or more than one type (variable charge)? Are the ions monatomic or polyatomic? If the compound is molecular, does it contain hydrogen? If so, does it also contain oxygen? From the answers we derive, we place the compound in an appropriate category and then name it accordingly.

Compounds Containing Only Monatomic Ions

The name of a binary compound containing monatomic ions consists of the name of the cation (the name of the metal) followed by the name of the anion (the name of the nonmetallic element with its ending replaced by the suffix –*ide*). Some examples are given in Table 2.6.

Names of Some Ionic Compounds

NaCl, sodium chloride	Na_2O, sodium oxide
KBr, potassium bromide	CdS, cadmium sulfide
CaI_2, calcium iodide	Mg_3N_2, magnesium nitride
CsF, cesium fluoride	Ca_3P_2, calcium phosphide
LiCl, lithium chloride	Al_4C_3, aluminum carbide

TABLE 2.6

Compounds Containing Polyatomic Ions

Compounds containing polyatomic ions are named similarly to those containing only monatomic ions, i.e. by naming first the cation and then the anion. Examples are shown in Table 2.7.

Names of Some Polyatomic Ionic Compounds

$KC_2H_3O_2$, potassium acetate	NH_4Cl, ammonium chloride
$NaHCO_3$, sodium bicarbonate	$CaSO_4$, calcium sulfate
$Al_2(CO_3)_3$, aluminum carbonate	$Mg_3(PO_4)_2$, magnesium phosphate

TABLE 2.7

Chemistry in Everyday Life

Ionic Compounds in Your Cabinets

Every day you encounter and use a large number of ionic compounds. Some of these compounds, where they are found, and what they are used for are listed in Table 2.8. Look at the label or ingredients list on the various products that you use during the next few days, and see if you run into any of those in this table, or find other ionic compounds that you could now name or write as a formula.

Everyday Ionic Compounds

Ionic Compound	Use
NaCl, sodium chloride	ordinary table salt
KI, potassium iodide	added to “iodized” salt for thyroid health
NaF, sodium fluoride	ingredient in toothpaste
$NaHCO_3$, sodium bicarbonate	baking soda; used in cooking (and as antacid)
Na_2CO_3, sodium carbonate	washing soda; used in cleaning agents
NaOCl, sodium hypochlorite	active ingredient in household bleach
$CaCO_3$ calcium carbonate	ingredient in antacids
$Mg(OH)_2$, magnesium hydroxide	ingredient in antacids
$Al(OH)_3$, aluminum hydroxide	ingredient in antacids
NaOH, sodium hydroxide	lye; used as drain cleaner
K_3PO_4, potassium phosphate	food additive (many purposes)
$MgSO_4$, magnesium sulfate	added to purified water
Na_2HPO_4, sodium hydrogen phosphate	anti-caking agent; used in powdered products

TABLE 2.8

Ionic Compound	Use
Na_2SO_3, sodium sulfite	preservative

TABLE 2.8

Compounds Containing a Metal Ion with a Variable Charge

Most of the transition metals and some main group metals can form two or more cations with different charges. Compounds of these metals with nonmetals are named with the same method as compounds in the first category, except the charge of the metal ion is specified by a Roman numeral in parentheses after the name of the metal. The charge of the metal ion is determined from the formula of the compound and the charge of the anion. For example, consider binary ionic compounds of iron and chlorine. Iron typically exhibits a charge of either 2+ or 3+ (see Figure 2.29), and the two corresponding compound formulas are $FeCl_2$ and $FeCl_3$. The simplest name, "iron chloride," will, in this case, be ambiguous, as it does not distinguish between these two compounds. In cases like this, the charge of the metal ion is included as a Roman numeral in parentheses immediately following the metal name. These two compounds are then unambiguously named iron(II) chloride and iron(III) chloride, respectively. Other examples are provided in Table 2.9.

Some Ionic Compounds with Variably Charged Metal Ions

Compound	Name
$FeCl_2$	iron(II) chloride
$FeCl_3$	iron(III) chloride
Hg_2O	mercury(I) oxide
HgO	mercury(II) oxide
SnF_2	tin(II) fluoride
SnF_4	tin(IV) fluoride

TABLE 2.9

Out-of-date nomenclature used the suffixes *–ic* and *–ous* to designate metals with higher and lower charges, respectively: Iron(III) chloride, $FeCl_3$, was previously called ferric chloride, and iron(II) chloride, $FeCl_2$, was known as ferrous chloride. Though this naming convention has been largely abandoned by the scientific community, it remains in use by some segments of industry. For example, you may see the words *stannous fluoride* on a tube of toothpaste. This represents the formula SnF_2, which is more properly named tin(II) fluoride. The other fluoride of tin is SnF_4, which was previously called stannic fluoride but is now named tin(IV) fluoride.

Ionic Hydrates

Ionic compounds that contain water molecules as integral components of their crystals are called **hydrates**. The name for an ionic hydrate is derived by adding a term to the name for the anhydrous (meaning "not hydrated") compound that indicates the number of water molecules associated with each formula unit of the compound. The added word begins with a Greek prefix denoting the number of water molecules (see Table 2.10) and ends with "hydrate." For example, the anhydrous compound copper(II) sulfate also exists as a hydrate containing five water molecules and named copper(II) sulfate pentahydrate. Washing soda is the common name for a hydrate of sodium carbonate containing 10 water molecules; the systematic name is sodium carbonate decahydrate.

Formulas for ionic hydrates are written by appending a vertically centered dot, a coefficient representing the number of water molecules, and the formula for water. The two examples mentioned in the previous paragraph are represented by the formulas

$$\text{copper(II) sulfate pentahydrate } CuSO_4{\bullet}5H_2O$$
$$\text{sodium carbonate decahydrate } Na_2CO_3{\bullet}10H_2O$$

Nomenclature Prefixes

Number	Prefix	Number	Prefix
1 (sometimes omitted)	mono-	6	hexa-
2	di-	7	hepta-
3	tri-	8	octa-
4	tetra-	9	nona-
5	penta-	10	deca-

TABLE 2.10

EXAMPLE 2.13

Naming Ionic Compounds

Name the following ionic compounds

(a) Fe_2S_3

(b) CuSe

(c) GaN

(d) $MgSO_4{\cdot}7H_2O$

(e) $Ti_2(SO_4)_3$

Solution

The anions in these compounds have a fixed negative charge (S^{2-}, Se^{2-}, N^{3-}, and $SO_4{}^{2-}$), and the compounds must be neutral. Because the total number of positive charges in each compound must equal the total number of negative charges, the positive ions must be Fe^{3+}, Cu^{2+}, Ga^{3+}, Mg^{2+}, and Ti^{3+}. These charges are used in the names of the metal ions:

(a) iron(III) sulfide

(b) copper(II) selenide

(c) gallium(III) nitride

(d) magnesium sulfate heptahydrate

(e) titanium(III) sulfate

Check Your Learning

Write the formulas of the following ionic compounds:

(a) chromium(III) phosphide

(b) mercury(II) sulfide

(c) manganese(II) phosphate

(d) copper(I) oxide

(e) iron(III) chloride dihydrate

Answer:

(a) CrP; (b) HgS; (c) $Mn_3(PO_4)_2$; (d) Cu_2O; (e) $FeCl_3{\cdot}2H_2O$

Chemistry in Everyday Life

Erin Brockovich and Chromium Contamination

In the early 1990s, legal file clerk Erin Brockovich (Figure 2.32) discovered a high rate of serious illnesses in the small town of Hinckley, California. Her investigation eventually linked the illnesses to groundwater contaminated by Cr(VI) used by Pacific Gas & Electric (PG&E) to fight corrosion in a nearby natural gas pipeline. As dramatized in the film *Erin Brockovich* (for which Julia Roberts won an Oscar), Erin and lawyer Edward Masry sued PG&E for contaminating the water near Hinckley in 1993. The settlement they won in 1996—$333 million—was the largest amount ever awarded for a direct-action lawsuit in the US at that time.

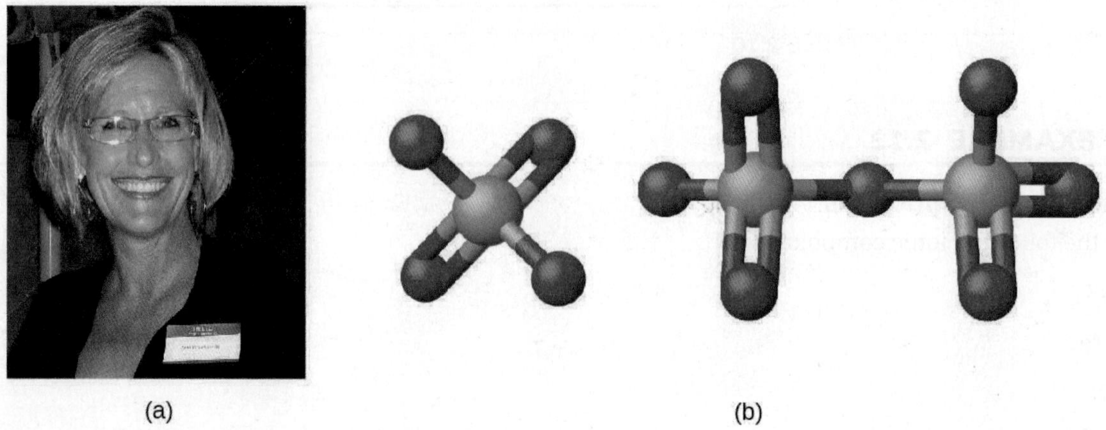

FIGURE 2.32 (a) Erin Brockovich found that Cr(VI), used by PG&E, had contaminated the Hinckley, California, water supply. (b) The Cr(VI) ion is often present in water as the polyatomic ions chromate, $CrO_4{}^{2-}$ (left), and dichromate, $Cr_2O_7{}^{2-}$ (right).

Chromium compounds are widely used in industry, such as for chrome plating, in dye-making, as preservatives, and to prevent corrosion in cooling tower water, as occurred near Hinckley. In the environment, chromium exists primarily in either the Cr(III) or Cr(VI) forms. Cr(III), an ingredient of many vitamin and nutritional supplements, forms compounds that are not very soluble in water, and it has low toxicity. But Cr(VI) is much more toxic and forms compounds that are reasonably soluble in water. Exposure to small amounts of Cr(VI) can lead to damage of the respiratory, gastrointestinal, and immune systems, as well as the kidneys, liver, blood, and skin.

Despite cleanup efforts, Cr(VI) groundwater contamination remains a problem in Hinckley and other locations across the globe. A 2010 study by the Environmental Working Group found that of 35 US cities tested, 31 had higher levels of Cr(VI) in their tap water than the public health goal of 0.02 parts per billion set by the California Environmental Protection Agency.

Molecular (Covalent) Compounds

The bonding characteristics of inorganic molecular compounds are different from ionic compounds, and they are named using a different system as well. The charges of cations and anions dictate their ratios in ionic compounds, so specifying the names of the ions provides sufficient information to determine chemical formulas. However, because covalent bonding allows for significant variation in the combination ratios of the atoms in a molecule, the names for molecular compounds must explicitly identify these ratios.

Compounds Composed of Two Elements

When two nonmetallic elements form a molecular compound, several combination ratios are often possible. For example, carbon and oxygen can form the compounds CO and CO_2. Since these are different substances with different properties, they cannot both have the same name (they cannot both be called carbon oxide). To deal with this situation, we use a naming method that is somewhat similar to that used for ionic compounds, but with added prefixes to specify the numbers of atoms of each element. The name of the more metallic element (the one farther to the left and/or bottom of the periodic table) is first, followed by the name of the more nonmetallic element (the one farther to the right and/or top) with its ending changed to the suffix *–ide*. The numbers of atoms of each element are designated by the Greek prefixes shown in Table 2.10.

When only one atom of the first element is present, the prefix *mono-* is usually deleted from that part. Thus, CO is named carbon monoxide, and CO_2 is called carbon dioxide. When two vowels are adjacent, the *a* in the Greek prefix is usually dropped. Some other examples are shown in Table 2.11.

Names of Some Molecular Compounds Composed of Two Elements

Compound	Name	Compound	Name
SO_2	sulfur dioxide	BCl_3	boron trichloride
SO_3	sulfur trioxide	SF_6	sulfur hexafluoride
NO_2	nitrogen dioxide	PF_5	phosphorus pentafluoride
N_2O_4	dinitrogen tetroxide	P_4O_{10}	tetraphosphorus decaoxide
N_2O_5	dinitrogen pentoxide	IF_7	iodine heptafluoride

TABLE 2.11

There are a few common names that you will encounter as you continue your study of chemistry. For example, although NO is often called nitric oxide, its proper name is nitrogen monoxide. Similarly, N_2O is known as nitrous oxide even though our rules would specify the name dinitrogen monoxide. (And H_2O is usually called water, not dihydrogen monoxide.) You should commit to memory the common names of compounds as you encounter them.

EXAMPLE 2.14

Naming Covalent Compounds

Name the following covalent compounds:

(a) SF_6

(b) N_2O_3

(c) Cl_2O_7

(d) P_4O_6

Solution

Because these compounds consist solely of nonmetals, we use prefixes to designate the number of atoms of each element:

(a) sulfur hexafluoride

(b) dinitrogen trioxide

(c) dichlorine heptoxide

(d) tetraphosphorus hexoxide

Check Your Learning

Write the formulas for the following compounds:

(a) phosphorus pentachloride

(b) dinitrogen monoxide

(c) iodine heptafluoride

(d) carbon tetrachloride

Answer:

(a) PCl_5; (b) N_2O; (c) IF_7; (d) CCl_4

LINK TO LEARNING

The following website (http://openstax.org/l/16chemcompname) provides practice with naming chemical compounds and writing chemical formulas. You can choose binary, polyatomic, and variable charge ionic compounds, as well as molecular compounds.

Binary Acids

Some compounds containing hydrogen are members of an important class of substances known as acids. The chemistry of these compounds is explored in more detail in later chapters of this text, but for now, it will suffice to note that many acids release hydrogen ions, H^+, when dissolved in water. To denote this distinct chemical property, a mixture of water with an acid is given a name derived from the compound's name. If the compound is a **binary acid** (comprised of hydrogen and one other nonmetallic element):

1. The word "hydrogen" is changed to the prefix *hydro-*
2. The other nonmetallic element name is modified by adding the suffix *-ic*
3. The word "acid" is added as a second word

For example, when the gas HCl (hydrogen chloride) is dissolved in water, the solution is called *hydrochloric acid*. Several other examples of this nomenclature are shown in Table 2.12.

Names of Some Simple Acids

Name of Gas	Name of Acid
HF(*g*), hydrogen fluoride	HF(*aq*), hydrofluoric acid
HCl(*g*), hydrogen chloride	HCl(*aq*), hydrochloric acid

TABLE 2.12

Name of Gas	Name of Acid
$HBr(g)$, hydrogen bromide	$HBr(aq)$, hydrobromic acid
$HI(g)$, hydrogen iodide	$HI(aq)$, hydroiodic acid
$H_2S(g)$, hydrogen sulfide	$H_2S(aq)$, hydrosulfuric acid

TABLE 2.12

Oxyacids

Many compounds containing three or more elements (such as organic compounds or coordination compounds) are subject to specialized nomenclature rules that you will learn later. However, we will briefly discuss the important compounds known as **oxyacids**, compounds that contain hydrogen, oxygen, and at least one other element, and are bonded in such a way as to impart acidic properties to the compound (you will learn the details of this in a later chapter). Typical oxyacids consist of hydrogen combined with a polyatomic, oxygen-containing ion. To name oxyacids:

1. Omit "hydrogen"
2. Start with the root name of the anion
3. Replace *–ate* with *–ic*, or *–ite* with *–ous*
4. Add "acid"

For example, consider H_2CO_3 (which you might be tempted to call "hydrogen carbonate"). To name this correctly, "hydrogen" is omitted; the *–ate* of carbonate is replace with *–ic*; and acid is added—so its name is carbonic acid. Other examples are given in Table 2.13. There are some exceptions to the general naming method (e.g., H_2SO_4 is called sulfuric acid, not sulfic acid, and H_2SO_3 is sulfurous, not sulfous, acid).

Names of Common Oxyacids

Formula	Anion Name	Acid Name
$HC_2H_3O_2$	acetate	acetic acid
HNO_3	nitrate	nitric acid
HNO_2	nitrite	nitrous acid
$HClO_4$	perchlorate	perchloric acid
H_2CO_3	carbonate	carbonic acid
H_2SO_4	sulfate	sulfuric acid
H_2SO_3	sulfite	sulfurous acid
H_3PO_4	phosphate	phosphoric acid

TABLE 2.13

Key Terms

actinide inner transition metal in the bottom of the bottom two rows of the periodic table

alkali metal element in group 1

alkaline earth metal element in group 2

alpha particle (α particle) positively charged particle consisting of two protons and two neutrons

anion negatively charged atom or molecule (contains more electrons than protons)

atomic mass average mass of atoms of an element, expressed in amu

atomic mass unit (amu) (also, unified atomic mass unit, u, or Dalton, Da) unit of mass equal to $\frac{1}{12}$ of the mass of a ^{12}C atom

atomic number (Z) number of protons in the nucleus of an atom

binary acid compound that contains hydrogen and one other element, bonded in a way that imparts acidic properties to the compound (ability to release H^+ ions when dissolved in water)

binary compound compound containing two different elements.

cation positively charged atom or molecule (contains fewer electrons than protons)

chalcogen element in group 16

chemical symbol one-, two-, or three-letter abbreviation used to represent an element or its atoms

covalent bond attractive force between the nuclei of a molecule's atoms and pairs of electrons between the atoms

covalent compound (also, molecular compound) composed of molecules formed by atoms of two or more different elements

Dalton (Da) alternative unit equivalent to the atomic mass unit

Dalton's atomic theory set of postulates that established the fundamental properties of atoms

electron negatively charged, subatomic particle of relatively low mass located outside the nucleus

empirical formula formula showing the composition of a compound given as the simplest whole-number ratio of atoms

fundamental unit of charge (also called the elementary charge) equals the magnitude of the charge of an electron (e) with $e = 1.602 \times 10^{-19}$ C

group vertical column of the periodic table

halogen element in group 17

hydrate compound containing one or more water molecules bound within its crystals

inert gas (also, noble gas) element in group 18

inner transition metal (also, lanthanide or actinide) element in the bottom two rows; if in the first row, also called lanthanide, or if in the second row, also called actinide

ion electrically charged atom or molecule (contains unequal numbers of protons and electrons)

ionic bond electrostatic forces of attraction between the oppositely charged ions of an ionic compound

ionic compound compound composed of cations and anions combined in ratios, yielding an electrically neutral substance

isomers compounds with the same chemical formula but different structures

isotopes atoms that contain the same number of protons but different numbers of neutrons

lanthanide inner transition metal in the top of the bottom two rows of the periodic table

law of constant composition (also, law of definite proportions) all samples of a pure compound contain the same elements in the same proportions by mass

law of definite proportions (also, law of constant composition) all samples of a pure compound contain the same elements in the same proportions by mass

law of multiple proportions when two elements react to form more than one compound, a fixed mass of one element will react with masses of the other element in a ratio of small whole numbers

main-group element (also, representative element) element in groups 1, 2, and 13–18

mass number (A) sum of the numbers of neutrons and protons in the nucleus of an atom

metal element that is shiny, malleable, good conductor of heat and electricity

metalloid element that conducts heat and electricity moderately well, and possesses some properties of metals and some properties of nonmetals

molecular compound (also, covalent compound) composed of molecules formed by atoms of two or more different elements

molecular formula formula indicating the composition of a molecule of a compound and giving the actual number of atoms of each element in a molecule of the compound.

monatomic ion ion composed of a single atom

neutron uncharged, subatomic particle located in the nucleus

noble gas (also, inert gas) element in group 18

nomenclature system of rules for naming objects

of interest

nonmetal element that appears dull, poor conductor of heat and electricity

nucleus massive, positively charged center of an atom made up of protons and neutrons

oxyacid compound that contains hydrogen, oxygen, and one other element, bonded in a way that imparts acidic properties to the compound (ability to release H^+ ions when dissolved in water)

oxyanion polyatomic anion composed of a central atom bonded to oxygen atoms

period (also, series) horizontal row of the periodic table

periodic law properties of the elements are periodic function of their atomic numbers.

periodic table table of the elements that places elements with similar chemical properties close together

pnictogen element in group 15

polyatomic ion ion composed of more than one atom

proton positively charged, subatomic particle located in the nucleus

representative element (also, main-group element) element in columns 1, 2, and 12–18

series (also, period) horizontal row of the period table

spatial isomers compounds in which the relative orientations of the atoms in space differ

structural formula shows the atoms in a molecule and how they are connected

structural isomer one of two substances that have the same molecular formula but different physical and chemical properties because their atoms are bonded differently

transition metal element in groups 3–12 (more strictly defined, 3–11; see chapter on transition metals and coordination chemistry)

unified atomic mass unit (u) alternative unit equivalent to the atomic mass unit

Key Equations

$$\text{average mass} = \sum_i (\text{fractional abundance} \times \text{isotopic mass})_i$$

Summary

2.1 Early Ideas in Atomic Theory

The ancient Greeks proposed that matter consists of extremely small particles called atoms. Dalton postulated that each element has a characteristic type of atom that differs in properties from atoms of all other elements, and that atoms of different elements can combine in fixed, small, whole-number ratios to form compounds. Samples of a particular compound all have the same elemental proportions by mass. When two elements form different compounds, a given mass of one element will combine with masses of the other element in a small, whole-number ratio. During any chemical change, atoms are neither created nor destroyed.

2.2 Evolution of Atomic Theory

Although no one has actually seen the inside of an atom, experiments have demonstrated much about atomic structure. Thomson's cathode ray tube showed that atoms contain small, negatively charged particles called electrons. Millikan discovered that there is a fundamental electric charge—the charge of an electron. Rutherford's gold foil experiment showed that atoms have a small, dense, positively charged nucleus; the positively charged particles within the nucleus are called protons. Chadwick discovered that the nucleus also contains neutral particles called neutrons. Soddy demonstrated that atoms of the same element can differ in mass; these are called isotopes.

2.3 Atomic Structure and Symbolism

An atom consists of a small, positively charged nucleus surrounded by electrons. The nucleus contains protons and neutrons; its diameter is about 100,000 times smaller than that of the atom. The mass of one atom is usually expressed in atomic mass units (amu), which is referred to as the atomic mass. An amu is defined as exactly $\frac{1}{12}$ of the mass of a carbon-12 atom and is equal to 1.6605×10^{-24} g.

Protons are relatively heavy particles with a charge of 1+ and a mass of 1.0073 amu. Neutrons are relatively heavy particles with no charge and a mass of 1.0087 amu. Electrons are light particles with a charge of 1− and a mass of 0.00055 amu. The number of protons in the nucleus is called the atomic number (Z) and is the property that defines an atom's elemental identity. The sum of the numbers of protons and neutrons in the nucleus is called the mass number and, expressed in amu, is approximately equal to the mass of the atom. An atom is neutral when it contains equal numbers of electrons and protons.

Isotopes of an element are atoms with the same atomic number but different mass numbers; isotopes of an element, therefore, differ from each other only in the number of neutrons within the nucleus. When a naturally occurring element is composed of several isotopes, the atomic mass of the element represents the average of the masses of the isotopes involved. A chemical symbol identifies the atoms in a substance using symbols, which are one-, two-, or three-letter abbreviations for the atoms.

2.4 Chemical Formulas

A molecular formula uses chemical symbols and subscripts to indicate the exact numbers of different atoms in a molecule or compound. An empirical formula gives the simplest, whole-number ratio of atoms in a compound. A structural formula indicates the bonding arrangement of the atoms in the molecule. Ball-and-stick and space-filling models show the geometric arrangement of atoms in a molecule. Isomers are compounds with the same molecular formula but different arrangements of atoms.

2.5 The Periodic Table

The discovery of the periodic recurrence of similar properties among the elements led to the formulation of the periodic table, in which the elements are arranged in order of increasing atomic number in rows known as periods and columns known as groups. Elements in the same group of the periodic table have similar chemical properties. Elements can be classified as metals, metalloids, and nonmetals, or as a main-group elements, transition metals, and inner transition metals. Groups are numbered 1–18 from left to right. The elements in group 1 are known as the alkali metals; those in group 2 are the alkaline earth metals; those in 15 are the pnictogens; those in 16 are the chalcogens; those in 17 are the halogens; and those in 18 are the noble gases.

2.6 Ionic and Molecular Compounds

Metals (particularly those in groups 1 and 2) tend to lose the number of electrons that would leave them with the same number of electrons as in the preceding noble gas in the periodic table. By this means, a positively charged ion is formed. Similarly, nonmetals (especially those in groups 16 and 17, and, to a lesser extent, those in Group 15) can gain the number of electrons needed to provide atoms with the same number of electrons as in the next noble gas in the periodic table. Thus, nonmetals tend to form negative ions. Positively charged ions are called cations, and negatively charged ions are called anions. Ions can be either monatomic (containing only one atom) or polyatomic (containing more than one atom).

Compounds that contain ions are called ionic compounds. Ionic compounds generally form from metals and nonmetals. Compounds that do not contain ions, but instead consist of atoms bonded tightly together in molecules (uncharged groups of atoms that behave as a single unit), are called covalent compounds. Covalent compounds usually form from two nonmetals.

2.7 Chemical Nomenclature

Chemists use nomenclature rules to clearly name compounds. Ionic and molecular compounds are named using somewhat-different methods. Binary ionic compounds typically consist of a metal and a nonmetal. The name of the metal is written first, followed by the name of the nonmetal with its ending changed to *–ide*. For example, K_2O is called potassium oxide. If the metal can form ions with different charges, a Roman numeral in parentheses follows the name of the metal to specify its charge. Thus, $FeCl_2$ is iron(II) chloride and $FeCl_3$ is iron(III) chloride. Some compounds contain polyatomic ions; the names of common polyatomic ions should be memorized. Molecular compounds can form compounds with different ratios of their elements, so prefixes are used to specify the numbers of atoms of each element in a molecule of the compound. Examples include SF_6, sulfur hexafluoride, and N_2O_4, dinitrogen tetroxide. Acids are an important class of compounds containing hydrogen and having special nomenclature rules. Binary acids are named using the prefix *hydro-*, changing the *–ide* suffix to *–ic*, and adding "acid;" HCl is hydrochloric acid. Oxyacids are named by changing the ending of the anion (*–ate* to *–ic* and *–ite* to *–ous*), and adding "acid;" H_2CO_3 is carbonic acid.

Exercises

2.1 Early Ideas in Atomic Theory

1. In the following drawing, the green spheres represent atoms of a certain element. The purple spheres represent atoms of another element. If the spheres of different elements touch, they are part of a single unit of a compound. The following chemical change represented by these spheres may violate one of the ideas of Dalton's atomic theory. Which one?

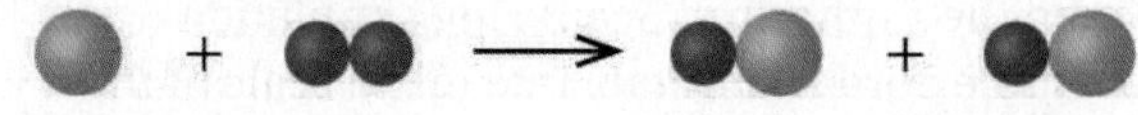

2. Which postulate of Dalton's theory is consistent with the following observation concerning the weights of reactants and products? When 100 grams of solid calcium carbonate is heated, 44 grams of carbon dioxide and 56 grams of calcium oxide are produced.
3. Identify the postulate of Dalton's theory that is violated by the following observations: 59.95% of one sample of titanium dioxide is titanium; 60.10% of a different sample of titanium dioxide is titanium.
4. Samples of compound X, Y, and Z are analyzed, with results shown here.

Compound	Description	Mass of Carbon	Mass of Hydrogen
X	clear, colorless, liquid with strong odor	1.776 g	0.148 g
Y	clear, colorless, liquid with strong odor	1.974 g	0.329 g
Z	clear, colorless, liquid with strong odor	7.812 g	0.651 g

Do these data provide example(s) of the law of definite proportions, the law of multiple proportions, neither, or both? What do these data tell you about compounds X, Y, and Z?

2.2 Evolution of Atomic Theory

5. The existence of isotopes violates one of the original ideas of Dalton's atomic theory. Which one?
6. How are electrons and protons similar? How are they different?
7. How are protons and neutrons similar? How are they different?
8. Predict and test the behavior of α particles fired at a "plum pudding" model atom.
(a) Predict the paths taken by α particles that are fired at atoms with a Thomson's plum pudding model structure. Explain why you expect the α particles to take these paths.
(b) If α particles of higher energy than those in (a) are fired at plum pudding atoms, predict how their paths will differ from the lower-energy α particle paths. Explain your reasoning.
(c) Now test your predictions from (a) and (b). Open the Rutherford Scattering simulation (http://openstax.org/l/16PhetScatter) and select the "Plum Pudding Atom" tab. Set "Alpha Particles Energy" to "min," and select "show traces." Click on the gun to start firing α particles. Does this match your prediction from (a)? If not, explain why the actual path would be that shown in the simulation. Hit the pause button, or "Reset All." Set "Alpha Particles Energy" to "max," and start firing α particles. Does this match your prediction from (b)? If not, explain the effect of increased energy on the actual paths as shown in the simulation.

9. Predict and test the behavior of α particles fired at a Rutherford atom model.
   (a) Predict the paths taken by α particles that are fired at atoms with a Rutherford atom model structure. Explain why you expect the α particles to take these paths.
   (b) If α particles of higher energy than those in (a) are fired at Rutherford atoms, predict how their paths will differ from the lower-energy α particle paths. Explain your reasoning.
   (c) Predict how the paths taken by the α particles will differ if they are fired at Rutherford atoms of elements other than gold. What factor do you expect to cause this difference in paths, and why?
   (d) Now test your predictions from (a), (b), and (c). Open the Rutherford Scattering simulation (http://openstax.org/l/16PhetScatter) and select the "Rutherford Atom" tab. Due to the scale of the simulation, it is best to start with a small nucleus, so select "20" for both protons and neutrons, "min" for energy, show traces, and then start firing α particles. Does this match your prediction from (a)? If not, explain why the actual path would be that shown in the simulation. Pause or reset, set energy to "max," and start firing α particles. Does this match your prediction from (b)? If not, explain the effect of increased energy on the actual path as shown in the simulation. Pause or reset, select "40" for both protons and neutrons, "min" for energy, show traces, and fire away. Does this match your prediction from (c)? If not, explain why the actual path would be that shown in the simulation. Repeat this with larger numbers of protons and neutrons. What generalization can you make regarding the type of atom and effect on the path of α particles? Be clear and specific.

2.3 Atomic Structure and Symbolism

10. In what way are isotopes of a given element always different? In what way(s) are they always the same?
11. Write the symbol for each of the following ions:
    (a) the ion with a 1+ charge, atomic number 55, and mass number 133
    (b) the ion with 54 electrons, 53 protons, and 74 neutrons
    (c) the ion with atomic number 15, mass number 31, and a 3– charge
    (d) the ion with 24 electrons, 30 neutrons, and a 3+ charge
12. Write the symbol for each of the following ions:
    (a) the ion with a 3+ charge, 28 electrons, and a mass number of 71
    (b) the ion with 36 electrons, 35 protons, and 45 neutrons
    (c) the ion with 86 electrons, 142 neutrons, and a 4+ charge
    (d) the ion with a 2+ charge, atomic number 38, and mass number 87
13. Open the Build an Atom simulation (http://openstax.org/l/16PhetAtomBld) and click on the Atom icon.
    (a) Pick any one of the first 10 elements that you would like to build and state its symbol.
    (b) Drag protons, neutrons, and electrons onto the atom template to make an atom of your element. State the numbers of protons, neutrons, and electrons in your atom, as well as the net charge and mass number.
    (c) Click on "Net Charge" and "Mass Number," check your answers to (b), and correct, if needed.
    (d) Predict whether your atom will be stable or unstable. State your reasoning.
    (e) Check the "Stable/Unstable" box. Was your answer to (d) correct? If not, first predict what you can do to make a stable atom of your element, and then do it and see if it works. Explain your reasoning.
14. Open the Build an Atom simulation (http://openstax.org/l/16PhetAtomBld).
    (a) Drag protons, neutrons, and electrons onto the atom template to make a neutral atom of Oxygen-16 and give the isotope symbol for this atom.
    (b) Now add two more electrons to make an ion and give the symbol for the ion you have created.
15. Open the Build an Atom simulation (http://openstax.org/l/16PhetAtomBld).
    (a) Drag protons, neutrons, and electrons onto the atom template to make a neutral atom of Lithium-6 and give the isotope symbol for this atom.
    (b) Now remove one electron to make an ion and give the symbol for the ion you have created.

16. Determine the number of protons, neutrons, and electrons in the following isotopes that are used in medical diagnoses:
(a) atomic number 9, mass number 18, charge of 1–
(b) atomic number 43, mass number 99, charge of 7+
(c) atomic number 53, atomic mass number 131, charge of 1–
(d) atomic number 81, atomic mass number 201, charge of 1+
(e) Name the elements in parts (a), (b), (c), and (d).
17. The following are properties of isotopes of two elements that are essential in our diet. Determine the number of protons, neutrons and electrons in each and name them.
(a) atomic number 26, mass number 58, charge of 2+
(b) atomic number 53, mass number 127, charge of 1–
18. Give the number of protons, electrons, and neutrons in neutral atoms of each of the following isotopes:
(a) ${}^{10}_{5}B$
(b) ${}^{199}_{80}Hg$
(c) ${}^{63}_{29}Cu$
(d) ${}^{13}_{6}C$
(e) ${}^{77}_{34}Se$
19. Give the number of protons, electrons, and neutrons in neutral atoms of each of the following isotopes:
(a) ${}^{7}_{3}Li$
(b) ${}^{125}_{52}Te$
(c) ${}^{109}_{47}Ag$
(d) ${}^{15}_{7}N$
(e) ${}^{31}_{15}P$
20. Click on the site (http://openstax.org/l/16PhetAtomMass) and select the “Mix Isotopes” tab, hide the “Percent Composition” and “Average Atomic Mass” boxes, and then select the element boron.
(a) Write the symbols of the isotopes of boron that are shown as naturally occurring in significant amounts.
(b) Predict the relative amounts (percentages) of these boron isotopes found in nature. Explain the reasoning behind your choice.
(c) Add isotopes to the black box to make a mixture that matches your prediction in (b). You may drag isotopes from their bins or click on “More” and then move the sliders to the appropriate amounts.
(d) Reveal the “Percent Composition” and “Average Atomic Mass” boxes. How well does your mixture match with your prediction? If necessary, adjust the isotope amounts to match your prediction.
(e) Select “Nature's” mix of isotopes and compare it to your prediction. How well does your prediction compare with the naturally occurring mixture? Explain. If necessary, adjust your amounts to make them match “Nature's” amounts as closely as possible.
21. Repeat Exercise 2.20 using an element that has three naturally occurring isotopes.
22. An element has the following natural abundances and isotopic masses: 90.92% abundance with 19.99 amu, 0.26% abundance with 20.99 amu, and 8.82% abundance with 21.99 amu. Calculate the average atomic mass of this element.
23. Average atomic masses listed by IUPAC are based on a study of experimental results. Bromine has two isotopes, ${}^{79}Br$ and ${}^{81}Br$, whose masses (78.9183 and 80.9163 amu, respectively) and abundances (50.69% and 49.31%, respectively) were determined in earlier experiments. Calculate the average atomic mass of bromine based on these experiments.
24. Variations in average atomic mass may be observed for elements obtained from different sources. Lithium provides an example of this. The isotopic composition of lithium from naturally occurring minerals is 7.5% ${}^{6}Li$ and 92.5% ${}^{7}Li$, which have masses of 6.01512 amu and 7.01600 amu, respectively. A commercial source of lithium, recycled from a military source, was 3.75% ${}^{6}Li$ (and the rest ${}^{7}Li$). Calculate the average atomic mass values for each of these two sources.

25. The average atomic masses of some elements may vary, depending upon the sources of their ores. Naturally occurring boron consists of two isotopes with accurately known masses (^{10}B, 10.0129 amu and ^{11}B, 11.00931 amu). The actual atomic mass of boron can vary from 10.807 to 10.819, depending on whether the mineral source is from Turkey or the United States. Calculate the percent abundances leading to the two values of the average atomic masses of boron from these two countries.
26. The ^{18}O:^{16}O abundance ratio in some meteorites is greater than that used to calculate the average atomic mass of oxygen on earth. Is the average mass of an oxygen atom in these meteorites greater than, less than, or equal to that of a terrestrial oxygen atom?

2.4 Chemical Formulas

27. Explain why the symbol for an atom of the element oxygen and the formula for a molecule of oxygen differ.
28. Explain why the symbol for the element sulfur and the formula for a molecule of sulfur differ.
29. Write the molecular and empirical formulas of the following compounds:

(a)

O=C=O

(b)

H—C≡C—H

(c)

H H
\ /
C=C
/ \
H H

(d)

O
|
O—S—O—H
|
O—H

30. Write the molecular and empirical formulas of the following compounds:

(a)

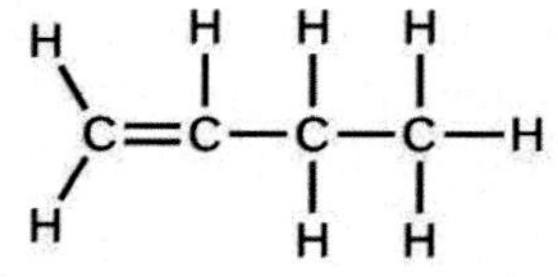

```
H       H  H  H
 \      |  |  |
  C == C - C - C - H
 /         |  |
H          H  H
```

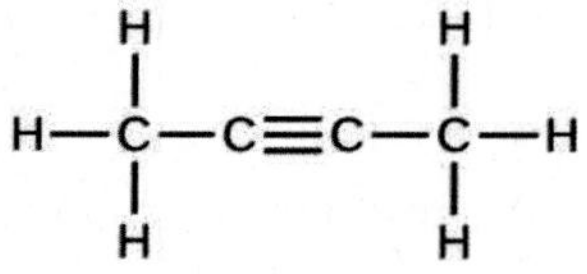

(b)

```
    H              H
    |              |
H - C - C ≡ C  -   C - H
    |              |
    H              H
```

(c)

```
      Cl   Cl
      |    |
Cl - Si - Si - Cl
      |    |
      H    H
```

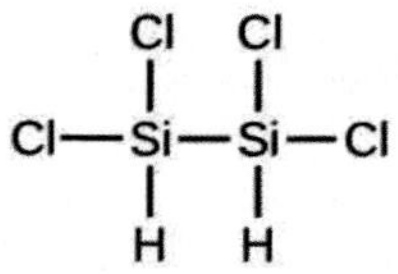

(d)

```
      O - H
      |
O  -  P - O - H
      |
      O - H
```

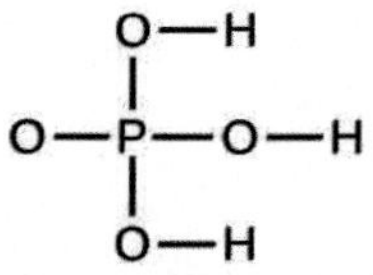

31. Determine the empirical formulas for the following compounds:

(a) caffeine, $C_8H_{10}N_4O_2$
(b) sucrose, $C_{12}H_{22}O_{11}$
(c) hydrogen peroxide, H_2O_2
(d) glucose, $C_6H_{12}O_6$
(e) ascorbic acid (vitamin C), $C_6H_8O_6$

32. Determine the empirical formulas for the following compounds:

(a) acetic acid, $C_2H_4O_2$
(b) citric acid, $C_6H_8O_7$
(c) hydrazine, N_2H_4
(d) nicotine, $C_{10}H_{14}N_2$
(e) butane, C_4H_{10}

33. Write the empirical formulas for the following compounds:

(a)

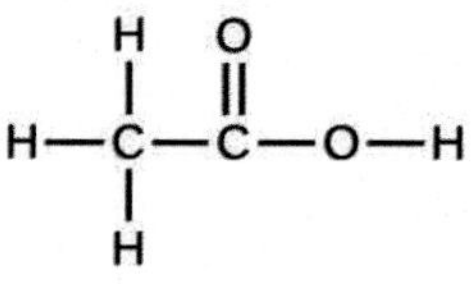

```
    H   O
    |   ||
H - C - C - O - H
    |
    H
```

(b)

```
    H   O       H   H
    |   ||      |   |
H - C - C - O - C - C - H
    |           |   |
    H           H   H
```

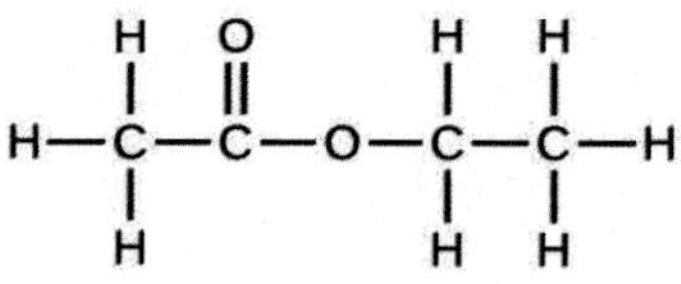

34. Open the Build a Molecule simulation (http://openstax.org/l/16molbuilding) and select the “Larger Molecules” tab. Select an appropriate atom’s “Kit” to build a molecule with two carbon and six hydrogen atoms. Drag atoms into the space above the “Kit” to make a molecule. A name will appear when you have made an actual molecule that exists (even if it is not the one you want). You can use the scissors tool to separate atoms if you would like to change the connections. Click on “3D” to see the molecule, and look at both the space-filling and ball-and-stick possibilities.
(a) Draw the structural formula of this molecule and state its name.
(b) Can you arrange these atoms in any way to make a different compound?

35. Use the Build a Molecule simulation (http://openstax.org/l/16molbuilding) to repeat Exercise 2.34, but build a molecule with two carbons, six hydrogens, and one oxygen.
(a) Draw the structural formula of this molecule and state its name.
(b) Can you arrange these atoms to make a different molecule? If so, draw its structural formula and state its name.
(c) How are the molecules drawn in (a) and (b) the same? How do they differ? What are they called (the type of relationship between these molecules, not their names).?

36. Use the Build a Molecule simulation (http://openstax.org/l/16molbuilding) to repeat Exercise 2.34, but build a molecule with three carbons, seven hydrogens, and one chlorine.
(a) Draw the structural formula of this molecule and state its name.
(b) Can you arrange these atoms to make a different molecule? If so, draw its structural formula and state its name.
(c) How are the molecules drawn in (a) and (b) the same? How do they differ? What are they called (the type of relationship between these molecules, not their names)?

2.5 The Periodic Table

37. Using the periodic table, classify each of the following elements as a metal or a nonmetal, and then further classify each as a main-group (representative) element, transition metal, or inner transition metal:
(a) uranium
(b) bromine
(c) strontium
(d) neon
(e) gold
(f) americium
(g) rhodium
(h) sulfur
(i) carbon
(j) potassium

38. Using the periodic table, classify each of the following elements as a metal or a nonmetal, and then further classify each as a main-group (representative) element, transition metal, or inner transition metal:
(a) cobalt
(b) europium
(c) iodine
(d) indium
(e) lithium
(f) oxygen
(g) cadmium
(h) terbium
(i) rhenium

39. Using the periodic table, identify the lightest member of each of the following groups:
(a) noble gases
(b) alkaline earth metals
(c) alkali metals
(d) chalcogens

40. Using the periodic table, identify the heaviest member of each of the following groups:
(a) alkali metals
(b) chalcogens
(c) noble gases
(d) alkaline earth metals

41. Use the periodic table to give the name and symbol for each of the following elements:
(a) the noble gas in the same period as germanium
(b) the alkaline earth metal in the same period as selenium
(c) the halogen in the same period as lithium
(d) the chalcogen in the same period as cadmium

42. Use the periodic table to give the name and symbol for each of the following elements:
(a) the halogen in the same period as the alkali metal with 11 protons
(b) the alkaline earth metal in the same period with the neutral noble gas with 18 electrons
(c) the noble gas in the same row as an isotope with 30 neutrons and 25 protons
(d) the noble gas in the same period as gold

43. Write a symbol for each of the following neutral isotopes. Include the atomic number and mass number for each.
(a) the alkali metal with 11 protons and a mass number of 23
(b) the noble gas element with 75 neutrons in its nucleus and 54 electrons in the neutral atom
(c) the isotope with 33 protons and 40 neutrons in its nucleus
(d) the alkaline earth metal with 88 electrons and 138 neutrons

44. Write a symbol for each of the following neutral isotopes. Include the atomic number and mass number for each.
(a) the chalcogen with a mass number of 125
(b) the halogen whose longest-lived isotope is radioactive
(c) the noble gas, used in lighting, with 10 electrons and 10 neutrons
(d) the lightest alkali metal with three neutrons

2.6 Ionic and Molecular Compounds

45. Using the periodic table, predict whether the following chlorides are ionic or covalent: KCl, NCl_3, ICl, $MgCl_2$, PCl_5, and CCl_4.

46. Using the periodic table, predict whether the following chlorides are ionic or covalent: $SiCl_4$, PCl_3, $CaCl_2$, CsCl, $CuCl_2$, and $CrCl_3$.

47. For each of the following compounds, state whether it is ionic or covalent. If it is ionic, write the symbols for the ions involved:
(a) NF_3
(b) BaO
(c) $(NH_4)_2CO_3$
(d) $Sr(H_2PO_4)_2$
(e) IBr
(f) Na_2O

48. For each of the following compounds, state whether it is ionic or covalent, and if it is ionic, write the symbols for the ions involved:
(a) $KClO_4$
(b) $Mg(C_2H_3O_2)_2$
(c) H_2S
(d) Ag_2S
(e) N_2Cl_4
(f) $Co(NO_3)_2$

49. For each of the following pairs of ions, write the formula of the compound they will form:
(a) Ca^{2+}, S^{2-}
(b) $NH_4{}^+$, $SO_4{}^{2-}$
(c) Al^{3+}, Br^-
(d) Na^+, $HPO_4{}^{2-}$
(e) Mg^{2+}, $PO_4{}^{3-}$

50. For each of the following pairs of ions, write the formula of the compound they will form:
(a) K^+, O^{2-}
(b) $NH_4{}^+$, $PO_4{}^{3-}$
(c) Al^{3+}, O^{2-}
(d) Na^+, $CO_3{}^{2-}$
(e) Ba^{2+}, $PO_4{}^{3-}$

2.7 Chemical Nomenclature

51. Name the following compounds:
(a) CsCl
(b) BaO
(c) K_2S
(d) $BeCl_2$
(e) HBr
(f) AlF_3

52. Name the following compounds:
(a) NaF
(b) Rb_2O
(c) BCl_3
(d) H_2Se
(e) P_4O_6
(f) ICl_3

53. Write the formulas of the following compounds:
(a) rubidium bromide
(b) magnesium selenide
(c) sodium oxide
(d) calcium chloride
(e) hydrogen fluoride
(f) gallium phosphide
(g) aluminum bromide
(h) ammonium sulfate

54. Write the formulas of the following compounds:
(a) lithium carbonate
(b) sodium perchlorate
(c) barium hydroxide
(d) ammonium carbonate
(e) sulfuric acid
(f) calcium acetate
(g) magnesium phosphate
(h) sodium sulfite

55. Write the formulas of the following compounds:
(a) chlorine dioxide
(b) dinitrogen tetraoxide
(c) potassium phosphide
(d) silver(I) sulfide
(e) aluminum fluoride trihydrate
(f) silicon dioxide

56. Write the formulas of the following compounds:
(a) barium chloride
(b) magnesium nitride
(c) sulfur dioxide
(d) nitrogen trichloride
(e) dinitrogen trioxide
(f) tin(IV) chloride

57. Each of the following compounds contains a metal that can exhibit more than one ionic charge. Name these compounds:
(a) Cr_2O_3
(b) $FeCl_2$
(c) CrO_3
(d) $TiCl_4$
(e) $CoCl_2{\cdot}6H_2O$
(f) MoS_2

58. Each of the following compounds contains a metal that can exhibit more than one ionic charge. Name these compounds:
(a) $NiCO_3$
(b) MoO_3
(c) $Co(NO_3)_2$
(d) V_2O_5
(e) MnO_2
(f) Fe_2O_3

59. The following ionic compounds are found in common household products. Write the formulas for each compound:
(a) potassium phosphate
(b) copper(II) sulfate
(c) calcium chloride
(d) titanium(IV) oxide
(e) ammonium nitrate
(f) sodium bisulfate (the common name for sodium hydrogen sulfate)

60. The following ionic compounds are found in common household products. Name each of the compounds:
(a) $Ca(H_2PO_4)_2$
(b) $FeSO_4$
(c) $CaCO_3$
(d) MgO
(e) $NaNO_2$
(f) KI

61. What are the IUPAC names of the following compounds?
(a) manganese dioxide
(b) mercurous chloride (Hg_2Cl_2)
(c) ferric nitrate [$Fe(NO_3)_3$]
(d) titanium tetrachloride
(e) cupric bromide ($CuBr_2$)

CHAPTER 3
Composition of Substances and Solutions

Figure 3.1 The water in a swimming pool is a complex mixture of substances whose relative amounts must be carefully maintained to ensure the health and comfort of people using the pool. (credit: modification of work by Vic Brincat)

CHAPTER OUTLINE

3.1 Formula Mass and the Mole Concept
3.2 Determining Empirical and Molecular Formulas
3.3 Molarity
3.4 Other Units for Solution Concentrations

INTRODUCTION Swimming pools have long been a popular means of recreation, exercise, and physical therapy. Since it is impractical to refill large pools with fresh water on a frequent basis, pool water is regularly treated with chemicals to prevent the growth of harmful bacteria and algae. Proper pool maintenance requires regular additions of various chemical compounds in carefully measured amounts. For example, the relative amount of calcium ion, Ca^{2+}, in the water should be maintained within certain limits to prevent eye irritation and avoid damage to the pool bed and plumbing. To maintain proper calcium levels, calcium cations are added to the water in the form of an ionic compound that also contains anions; thus, it is necessary to know both the relative amount of Ca^{2+} in the compound and the volume of water in the pool in order to achieve the proper calcium level. Quantitative aspects of the composition of substances (such as the calcium-containing compound) and mixtures (such as the pool water) are the subject of this chapter.

3.1 Formula Mass and the Mole Concept

LEARNING OBJECTIVES

By the end of this section, you will be able to:

- Calculate formula masses for covalent and ionic compounds
- Define the amount unit mole and the related quantity Avogadro's number Explain the relation between mass, moles, and numbers of atoms or molecules, and perform calculations deriving these quantities from one another

Many argue that modern chemical science began when scientists started exploring the quantitative as well as the qualitative aspects of chemistry. For example, Dalton's atomic theory was an attempt to explain the results of measurements that allowed him to calculate the relative masses of elements combined in various compounds. Understanding the relationship between the masses of atoms and the chemical formulas of compounds allows us to quantitatively describe the composition of substances.

Formula Mass

An earlier chapter of this text described the development of the atomic mass unit, the concept of average atomic masses, and the use of chemical formulas to represent the elemental makeup of substances. These ideas can be extended to calculate the **formula mass** of a substance by summing the average atomic masses of all the atoms represented in the substance's formula.

Formula Mass for Covalent Substances

For covalent substances, the formula represents the numbers and types of atoms composing a single molecule of the substance; therefore, the formula mass may be correctly referred to as a molecular mass. Consider chloroform ($CHCl_3$), a covalent compound once used as a surgical anesthetic and now primarily used in the production of tetrafluoroethylene, the building block for the "anti-stick" polymer, Teflon. The molecular formula of chloroform indicates that a single molecule contains one carbon atom, one hydrogen atom, and three chlorine atoms. The average molecular mass of a chloroform molecule is therefore equal to the sum of the average atomic masses of these atoms. Figure 3.2 outlines the calculations used to derive the molecular mass of chloroform, which is 119.37 amu.

Element	Quantity		Average atomic mass (amu)		Subtotal (amu)
C	1	×	12.01	=	12.01
H	1	×	1.008	=	1.008
Cl	3	×	35.45	=	106.35
			Molecular mass		119.37

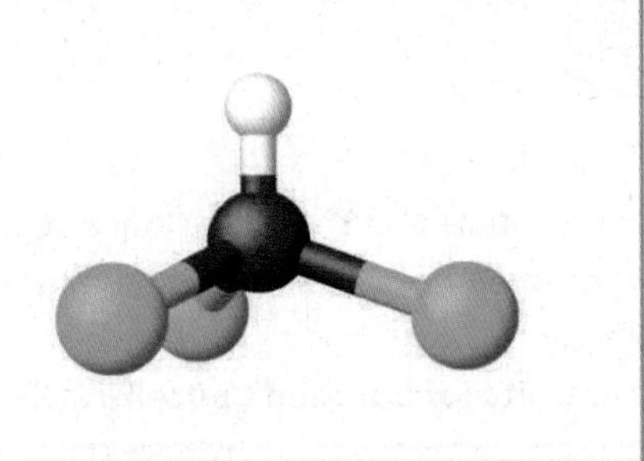

FIGURE 3.2 The average mass of a chloroform molecule, $CHCl_3$, is 119.37 amu, which is the sum of the average atomic masses of each of its constituent atoms. The model shows the molecular structure of chloroform.

Likewise, the molecular mass of an aspirin molecule, $C_9H_8O_4$, is the sum of the atomic masses of nine carbon atoms, eight hydrogen atoms, and four oxygen atoms, which amounts to 180.15 amu (Figure 3.3).

Element	Quantity		Average atomic mass (amu)		Subtotal (amu)
C	9	×	12.01	=	108.09
H	8	×	1.008	=	8.064
O	4	×	16.00	=	64.00
			Molecular mass		180.15

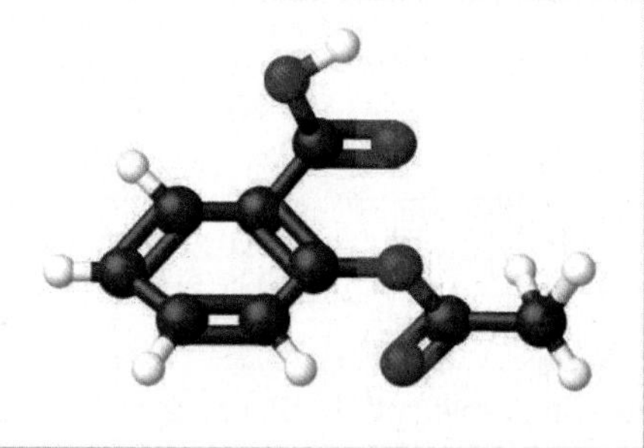

FIGURE 3.3 The average mass of an aspirin molecule is 180.15 amu. The model shows the molecular structure of aspirin, $C_9H_8O_4$.

EXAMPLE 3.1

Computing Molecular Mass for a Covalent Compound

Ibuprofen, $C_{13}H_{18}O_2$, is a covalent compound and the active ingredient in several popular nonprescription pain medications, such as Advil and Motrin. What is the molecular mass (amu) for this compound?

Solution

Molecules of this compound are composed of 13 carbon atoms, 18 hydrogen atoms, and 2 oxygen atoms. Following the approach described above, the average molecular mass for this compound is therefore:

Element	Quantity		Average atomic mass (amu)		Subtotal (amu)
C	13	×	12.01	=	156.13
H	18	×	1.008	=	18.144
O	2	×	16.00	=	32.00
			Molecular mass		206.27

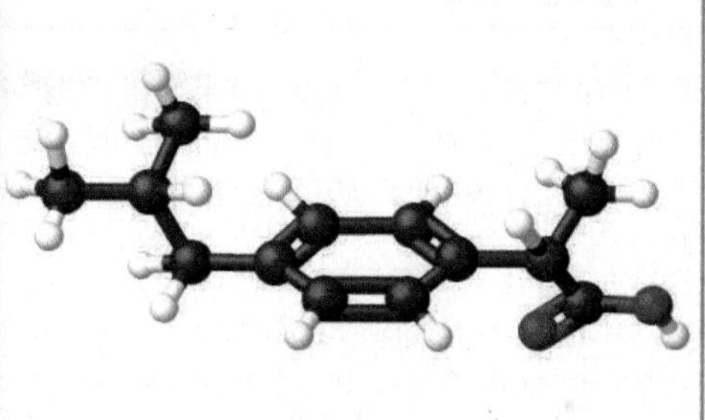

Check Your Learning

Acetaminophen, $C_8H_9NO_2$, is a covalent compound and the active ingredient in several popular nonprescription pain medications, such as Tylenol. What is the molecular mass (amu) for this compound?

Answer:

151.16 amu

Formula Mass for Ionic Compounds

Ionic compounds are composed of discrete cations and anions combined in ratios to yield electrically neutral bulk matter. The formula mass for an ionic compound is calculated in the same way as the formula mass for covalent compounds: by summing the average atomic masses of all the atoms in the compound's formula. Keep in mind, however, that the formula for an ionic compound does not represent the composition of a discrete molecule, so it may not correctly be referred to as the "molecular mass."

As an example, consider sodium chloride, NaCl, the chemical name for common table salt. Sodium chloride is an ionic compound composed of sodium cations, Na^+, and chloride anions, Cl^-, combined in a 1:1 ratio. The formula mass for this compound is computed as 58.44 amu (see Figure 3.4).

Element	Quantity		Average atomic mass (amu)		Subtotal
Na	1	×	22.99	=	22.99
Cl	1	×	35.45	=	35.45
			Formula mass		58.44

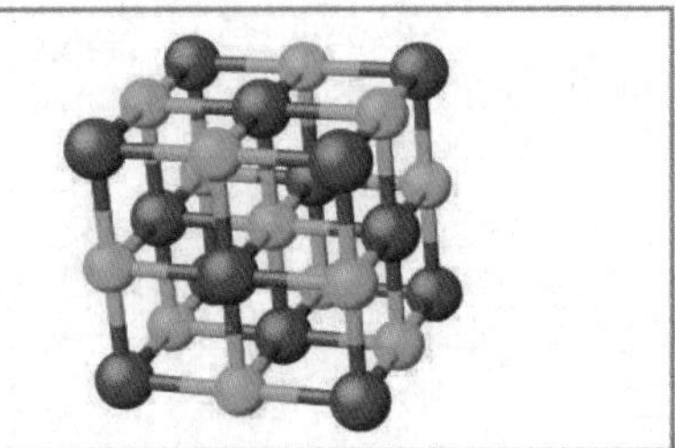

FIGURE 3.4 Table salt, NaCl, contains an array of sodium and chloride ions combined in a 1:1 ratio. Its formula mass is 58.44 amu.

Note that the average masses of neutral sodium and chlorine atoms were used in this computation, rather than the masses for sodium cations and chlorine anions. This approach is perfectly acceptable when computing the formula mass of an ionic compound. Even though a sodium cation has a slightly smaller mass than a sodium atom (since it is missing an electron), this difference will be offset by the fact that a chloride anion is slightly more massive than a chloride atom (due to the extra electron). Moreover, the mass of an electron is negligibly small with respect to the mass of a typical atom. Even when calculating the mass of an isolated ion, the missing or additional electrons can generally be ignored, since their contribution to the overall mass is negligible, reflected only in the nonsignificant digits that will be lost when the computed mass is properly rounded. The few exceptions to this guideline are very light ions derived from elements with precisely known atomic masses.

EXAMPLE 3.2

Computing Formula Mass for an Ionic Compound

Aluminum sulfate, $Al_2(SO_4)_3$, is an ionic compound that is used in the manufacture of paper and in various water purification processes. What is the formula mass (amu) of this compound?

Solution

The formula for this compound indicates it contains Al^{3+} and $SO_4{}^{2-}$ ions combined in a 2:3 ratio. For purposes of computing a formula mass, it is helpful to rewrite the formula in the simpler format, $Al_2S_3O_{12}$. Following the approach outlined above, the formula mass for this compound is calculated as follows:

Element	Quantity		Average atomic mass (amu)		Subtotal (amu)
Al	2	×	26.98	=	53.96
S	3	×	32.06	=	96.18
O	12	×	16.00	=	192.00
			Molecular mass		342.14

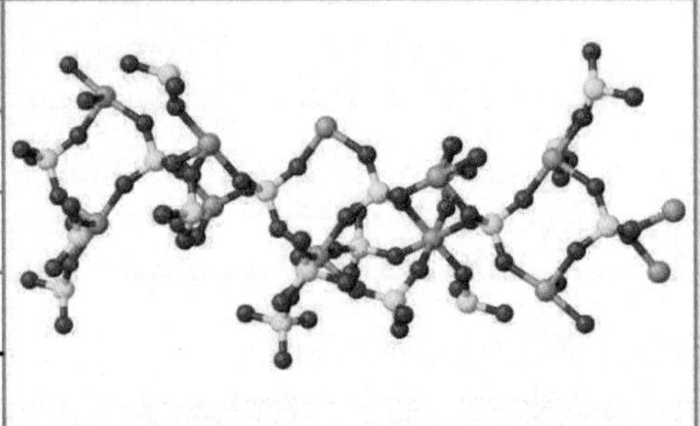

Check Your Learning

Calcium phosphate, $Ca_3(PO_4)_2$, is an ionic compound and a common anti-caking agent added to food products. What is the formula mass (amu) of calcium phosphate?

Answer:

310.18 amu

The Mole

The identity of a substance is defined not only by the types of atoms or ions it contains, but by the quantity of each type of atom or ion. For example, water, H_2O, and hydrogen peroxide, H_2O_2, are alike in that their respective molecules are composed of hydrogen and oxygen atoms. However, because a hydrogen peroxide molecule contains two oxygen atoms, as opposed to the water molecule, which has only one, the two substances exhibit very different properties. Today, sophisticated instruments allow the direct measurement of these defining microscopic traits; however, the same traits were originally derived from the measurement of macroscopic properties (the masses and volumes of bulk quantities of matter) using relatively simple tools (balances and volumetric glassware). This experimental approach required the introduction of a new unit for amount of substances, the *mole*, which remains indispensable in modern chemical science.

The *mole* is an amount unit similar to familiar units like pair, dozen, gross, etc. It provides a specific measure of *the number* of atoms or molecules in a sample of matter. One Latin connotation for the word "mole" is "large mass" or "bulk," which is consistent with its use as the name for this unit. The mole provides a link between an easily measured macroscopic property, bulk mass, and an extremely important fundamental property, number of atoms, molecules, and so forth. A **mole** of substance is that amount in which there are 6.02214076 $\times 10^{23}$ discrete entities (atoms or molecules). This large number is a fundamental constant known as **Avogadro's number (N_A)** or the Avogadro constant in honor of Italian scientist Amedeo Avogadro. This constant is properly reported with an explicit unit of "per mole," a conveniently rounded version being 6.022×10^{23}/mol.

Consistent with its definition as an amount unit, 1 mole of any element contains the same number of atoms as 1 mole of any other element. The masses of 1 mole of different elements, however, are different, since the masses of the individual atoms are drastically different. The **molar mass** of an element (or compound) is the mass in grams of 1 mole of that substance, a property expressed in units of grams per mole (g/mol) (see Figure 3.5).

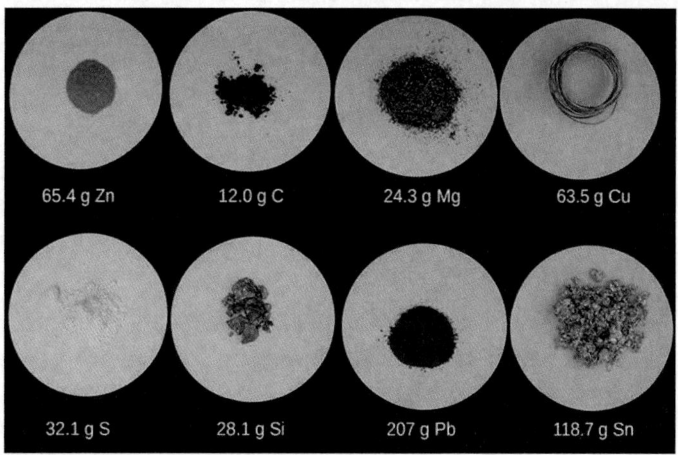

FIGURE 3.5 Each sample contains 6.022×10^{23} atoms —1.00 mol of atoms. From left to right (top row): 65.4 g zinc, 12.0 g carbon, 24.3 g magnesium, and 63.5 g copper. From left to right (bottom row): 32.1 g sulfur, 28.1 g silicon, 207 g lead, and 118.7 g tin. (credit: modification of work by Mark Ott)

The molar mass of any substance is numerically equivalent to its atomic or formula weight in amu. Per the amu definition, a single ^{12}C atom weighs 12 amu (its atomic mass is 12 amu). A mole of ^{12}C weighs 12 g (its molar mass is 12 g/mol). This relationship holds for all elements, since their atomic masses are measured relative to that of the amu-reference substance, ^{12}C. Extending this principle, the molar mass of a compound in grams is likewise numerically equivalent to its formula mass in amu (Figure 3.6).

FIGURE 3.6 Each sample contains 6.02×10^{23} molecules or formula units—1.00 mol of the compound or element. Clock-wise from the upper left: 130.2 g of $C_8H_{17}OH$ (1-octanol, formula mass 130.2 amu), 454.4 g of HgI_2 (mercury(II) iodide, formula mass 454.4 amu), 32.0 g of CH_3OH (methanol, formula mass 32.0 amu) and 256.5 g of S_8 (sulfur, formula mass 256.5 amu). (credit: Sahar Atwa)

Element	Average Atomic Mass (amu)	Molar Mass (g/mol)	Atoms/Mole
C	12.01	12.01	6.022×10^{23}
H	1.008	1.008	6.022×10^{23}
O	16.00	16.00	6.022×10^{23}
Na	22.99	22.99	6.022×10^{23}
Cl	35.45	35.45	6.022×10^{23}

While atomic mass and molar mass are numerically equivalent, keep in mind that they are vastly different in terms of scale, as represented by the vast difference in the magnitudes of their respective units (amu versus g). To appreciate the enormity of the mole, consider a small drop of water weighing about 0.03 g (see Figure 3.7). Although this represents just a tiny fraction of 1 mole of water (~18 g), it contains more water molecules than can be clearly imagined. If the molecules were distributed equally among the roughly seven billion people on earth, each person would receive more than 100 billion molecules.

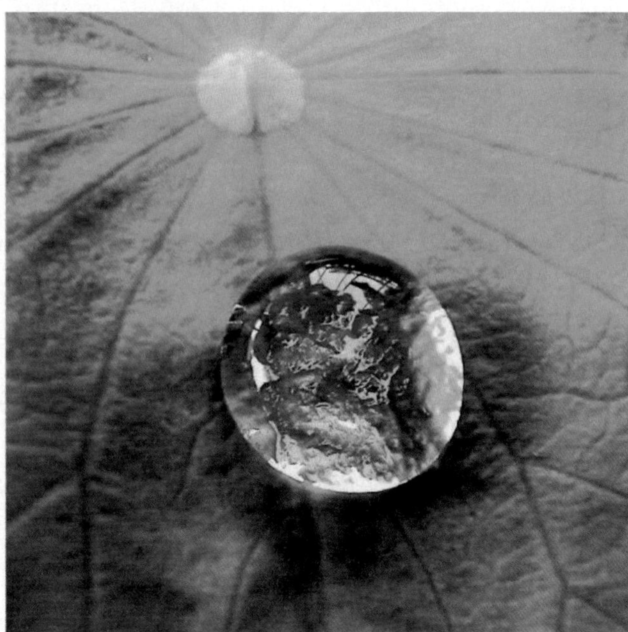

FIGURE 3.7 The number of molecules in a single droplet of water is roughly 100 billion times greater than the number of people on earth. (credit: "tanakawho"/Wikimedia commons)

LINK TO LEARNING

The mole is used in chemistry to represent 6.022×10^{23} of something, but it can be difficult to conceptualize such a large number. Watch this video (http://openstax.org/l/16molevideo) and then complete the "Think" questions that follow. Explore more about the mole by reviewing the information under "Dig Deeper."

The relationships between formula mass, the mole, and Avogadro's number can be applied to compute various quantities that describe the composition of substances and compounds, as demonstrated in the next several example problems.

EXAMPLE 3.3

Deriving Moles from Grams for an Element

According to nutritional guidelines from the US Department of Agriculture, the estimated average requirement for dietary potassium is 4.7 g. What is the estimated average requirement of potassium in moles?

Solution

The mass of K is provided, and the corresponding amount of K in moles is requested. Referring to the periodic table, the atomic mass of K is 39.10 amu, and so its molar mass is 39.10 g/mol. The given mass of K (4.7 g) is a bit more than one-tenth the molar mass (39.10 g), so a reasonable "ballpark" estimate of the number of moles would be slightly greater than 0.1 mol.

The molar amount of a substance may be calculated by dividing its mass (g) by its molar mass (g/mol):

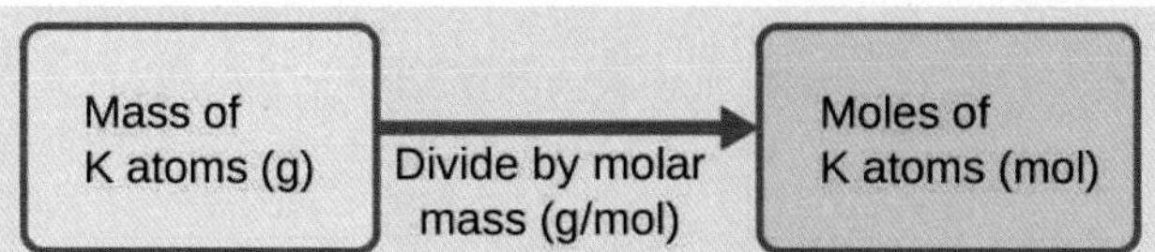

The factor-label method supports this mathematical approach since the unit "g" cancels and the answer has units of "mol:"

$$4.7\ \cancel{\text{g K}}\left(\frac{\text{mol K}}{39.10\ \cancel{\text{g K}}}\right) = 0.12\ \text{mol K}$$

The calculated magnitude (0.12 mol K) is consistent with our ballpark expectation, since it is a bit greater than 0.1 mol.

Check Your Learning

Beryllium is a light metal used to fabricate transparent X-ray windows for medical imaging instruments. How many moles of Be are in a thin-foil window weighing 3.24 g?

Answer:

0.360 mol

EXAMPLE 3.4

Deriving Grams from Moles for an Element

A liter of air contains 9.2×10^{-4} mol argon. What is the mass of Ar in a liter of air?

Solution

The molar amount of Ar is provided and must be used to derive the corresponding mass in grams. Since the amount of Ar is less than 1 mole, the mass will be less than the mass of 1 mole of Ar, approximately 40 g. The molar amount in question is approximately one-one thousandth (~10^{-3}) of a mole, and so the corresponding mass should be roughly one-one thousandth of the molar mass (~0.04 g):

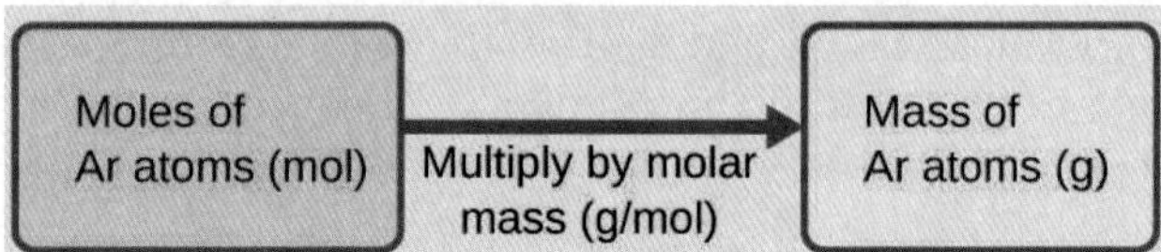

In this case, logic dictates (and the factor-label method supports) multiplying the provided amount (mol) by the molar mass (g/mol):

$$9.2 \times 10^{-4}\ \cancel{\text{mol Ar}}\left(\frac{39.95\ \text{g Ar}}{\cancel{\text{mol Ar}}}\right) = 0.037\ \text{g Ar}$$

The result is in agreement with our expectations, around 0.04 g Ar.

Check Your Learning

What is the mass of 2.561 mol of gold?

Answer:

504.4 g

EXAMPLE 3.5

Deriving Number of Atoms from Mass for an Element

Copper is commonly used to fabricate electrical wire (Figure 3.8). How many copper atoms are in 5.00 g of copper wire?

FIGURE 3.8 Copper wire is composed of many, many atoms of Cu. (credit: Emilian Robert Vicol)

Solution

The number of Cu atoms in the wire may be conveniently derived from its mass by a two-step computation: first calculating the molar amount of Cu, and then using Avogadro's number (N_A) to convert this molar amount to number of Cu atoms:

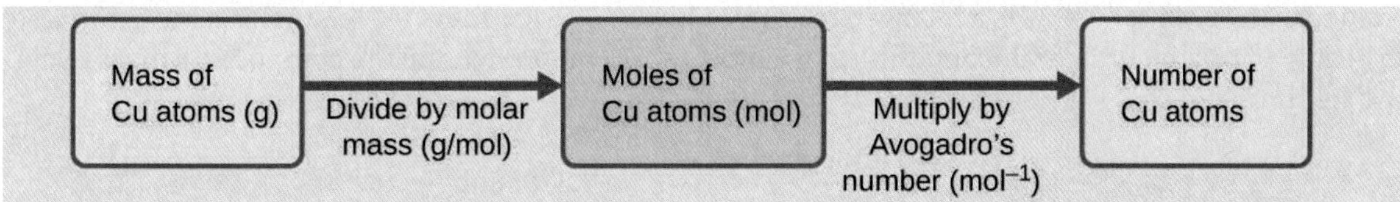

Considering that the provided sample mass (5.00 g) is a little less than one-tenth the mass of 1 mole of Cu (~64 g), a reasonable estimate for the number of atoms in the sample would be on the order of one-tenth N_A, or approximately 10^{22} Cu atoms. Carrying out the two-step computation yields:

$$5.00 \cancel{\text{g Cu}} \left(\frac{\cancel{\text{mol Cu}}}{63.55 \cancel{\text{g Cu}}}\right)\left(\frac{6.022 \times 10^{23} \text{ Cu atoms}}{\cancel{\text{mol Cu}}}\right) = 4.74 \times 10^{22} \text{Cu atoms}$$

The factor-label method yields the desired cancellation of units, and the computed result is on the order of 10^{22} as expected.

Check Your Learning

A prospector panning for gold in a river collects 15.00 g of pure gold. How many Au atoms are in this quantity of gold?

Answer:

4.586×10^{22} Au atoms

EXAMPLE 3.6

Deriving Moles from Grams for a Compound

Our bodies synthesize protein from amino acids. One of these amino acids is glycine, which has the molecular formula $C_2H_5O_2N$. How many moles of glycine molecules are contained in 28.35 g of glycine?

Solution

Derive the number of moles of a compound from its mass following the same procedure used for an element in Example 3.3:

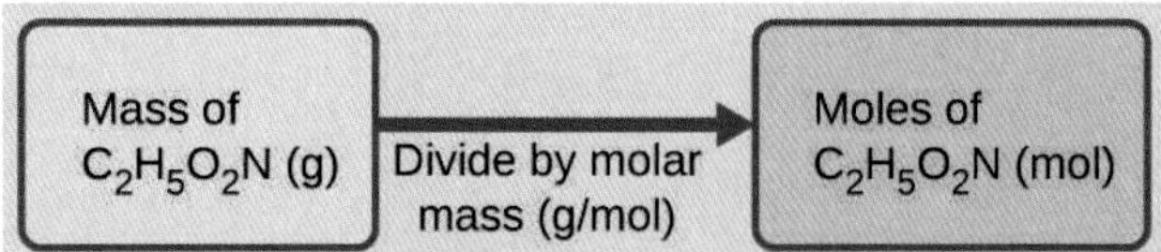

The molar mass of glycine is required for this calculation, and it is computed in the same fashion as its molecular mass. One mole of glycine, $C_2H_5O_2N$, contains 2 moles of carbon, 5 moles of hydrogen, 2 moles of oxygen, and 1 mole of nitrogen:

Element	Quantity (mol element/ mol compound)		Molar mass (g/mol element)		Subtotal (g/mol compound)
C	2	×	12.01	=	24.02
H	5	×	1.008	=	5.040
O	2	×	16.00	=	32.00
N	1	×	14.007	=	14.007
			Molecular mass (g/mol compound)		75.07

The provided mass of glycine (~28 g) is a bit more than one-third the molar mass (~75 g/mol), so the computed result is expected to be a bit greater than one-third of a mole (~0.33 mol). Dividing the compound's mass by its molar mass yields:

$$28.35 \cancel{\text{g glycine}} \left(\frac{\text{mol glycine}}{75.07 \cancel{\text{g glycine}}} \right) = 0.378 \text{ mol glycine}$$

This result is consistent with the rough estimate.

Check Your Learning

How many moles of sucrose, $C_{12}H_{22}O_{11}$, are in a 25-g sample of sucrose?

Answer:

0.073 mol

EXAMPLE 3.7

Deriving Grams from Moles for a Compound

Vitamin C is a covalent compound with the molecular formula $C_6H_8O_6$. The recommended daily dietary allowance of vitamin C for children aged 4–8 years is 1.42×10^{-4} mol. What is the mass of this allowance in grams?

Solution

As for elements, the mass of a compound can be derived from its molar amount as shown:

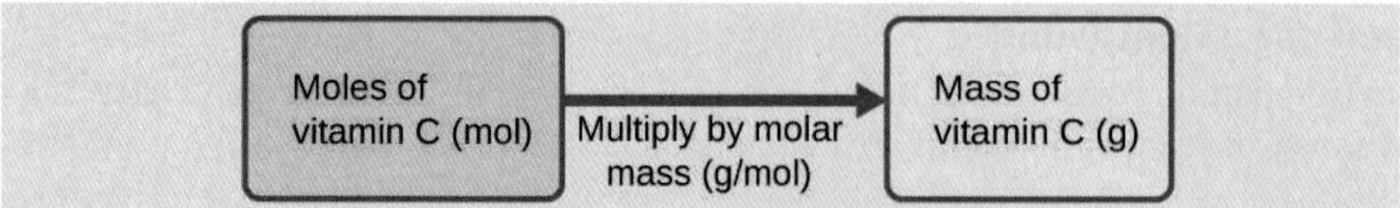

The molar mass for this compound is computed to be 176.124 g/mol. The given number of moles is a very small fraction of a mole (~10^{-4} or one-ten thousandth); therefore, the corresponding mass is expected to be about one-ten thousandth of the molar mass (~0.02 g). Performing the calculation yields:

$$1.42 \times 10^{-4}\ \cancel{\text{mol vitamin C}} \left(\frac{176.124\ \text{g vitamin C}}{\cancel{\text{mol vitamin C}}} \right) = 0.0250\ \text{g vitamin C}$$

This is consistent with the anticipated result.

Check Your Learning

What is the mass of 0.443 mol of hydrazine, N_2H_4?

Answer:

14.2 g

EXAMPLE 3.8

Deriving the Number of Atoms and Molecules from the Mass of a Compound

A packet of an artificial sweetener contains 40.0 mg of saccharin ($C_7H_5NO_3S$), which has the structural formula:

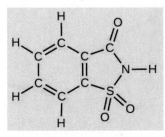

Given that saccharin has a molar mass of 183.18 g/mol, how many saccharin molecules are in a 40.0-mg (0.0400-g) sample of saccharin? How many carbon atoms are in the same sample?

Solution

The number of molecules in a given mass of compound is computed by first deriving the number of moles, as demonstrated in Example 3.6, and then multiplying by Avogadro's number:

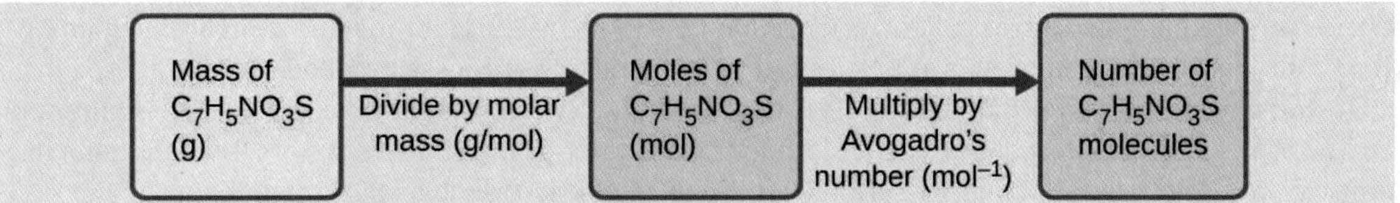

Using the provided mass and molar mass for saccharin yields:

$$0.0400\ \cancel{\text{g } C_7H_5NO_3S} \left(\frac{\cancel{\text{mol } C_7H_5NO_3S}}{183.18\ \cancel{\text{g } C_7H_5NO_3S}} \right) \left(\frac{6.022 \times 10^{23}\ C_7H_5NO_3S\ \text{molecules}}{1\ \cancel{\text{mol } C_7H_5NO_3S}} \right)$$
$$= 1.31 \times 10^{20}\ C_7H_5NO_3S\ \text{molecules}$$

The compound's formula shows that each molecule contains seven carbon atoms, and so the number of C atoms in the provided sample is:

$$1.31 \times 10^{20}\ C_7H_5NO_3S\ \text{molecules} \left(\frac{7\ \text{C atoms}}{1\ C_7H_5NO_3S\ \text{molecule}} \right) = 9.17 \times 10^{20}\ \text{C atoms}$$

Check Your Learning

How many C_4H_{10} molecules are contained in 9.213 g of this compound? How many hydrogen atoms?

Answer:

9.545×10^{22} molecules C_4H_{10}; 9.545×10^{23} atoms H

HOW SCIENCES INTERCONNECT

Counting Neurotransmitter Molecules in the Brain

The brain is the control center of the central nervous system (Figure 3.9). It sends and receives signals to and from muscles and other internal organs to monitor and control their functions; it processes stimuli detected by sensory organs to guide interactions with the external world; and it houses the complex physiological processes that give rise to our intellect and emotions. The broad field of neuroscience spans all aspects of the structure and function of the central nervous system, including research on the anatomy and physiology of the brain. Great progress has been made in brain research over the past few decades, and the BRAIN Initiative, a federal initiative announced in 2013, aims to accelerate and capitalize on these advances through the concerted efforts of various industrial, academic, and government agencies (more details available at www.whitehouse.gov/share/brain-initiative).

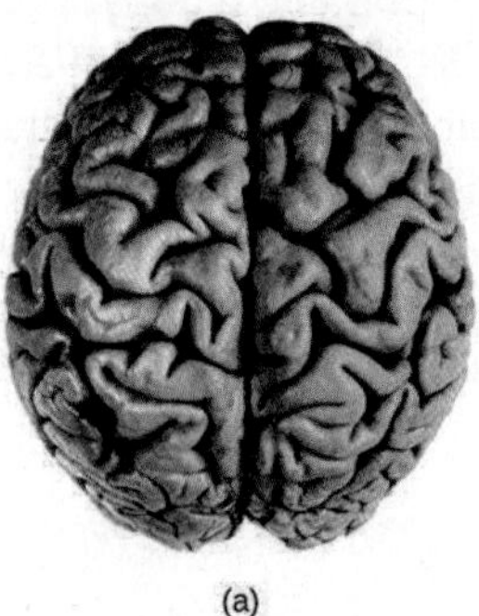

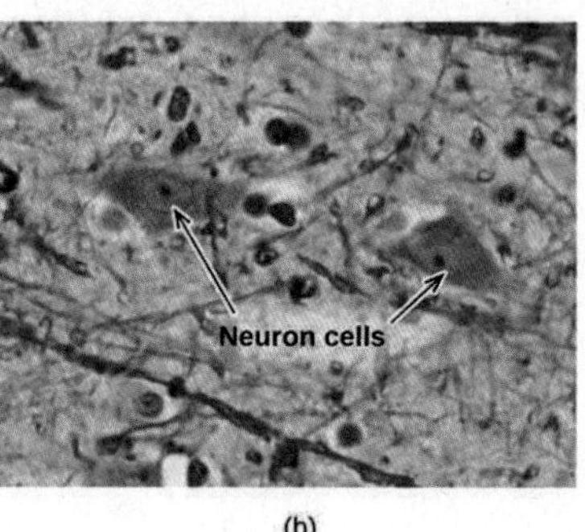

FIGURE 3.9 (a) A typical human brain weighs about 1.5 kg and occupies a volume of roughly 1.1 L. (b) Information is transmitted in brain tissue and throughout the central nervous system by specialized cells called neurons (micrograph shows cells at 1600× magnification).

Specialized cells called neurons transmit information between different parts of the central nervous system by way of electrical and chemical signals. Chemical signaling occurs at the interface between different neurons when one of the cells releases molecules (called neurotransmitters) that diffuse across the small gap between the cells (called the synapse) and bind to the surface of the other cell. These neurotransmitter molecules are stored in small intracellular structures called vesicles that fuse to the cell membrane and then break open to release their contents when the neuron is appropriately stimulated. This process is called exocytosis (see Figure 3.10). One neurotransmitter that has been very extensively studied is dopamine, $C_8H_{11}NO_2$. Dopamine is involved in various neurological processes that impact a wide variety of human behaviors. Dysfunctions in the dopamine systems of the brain underlie serious neurological diseases such as Parkinson's and schizophrenia.

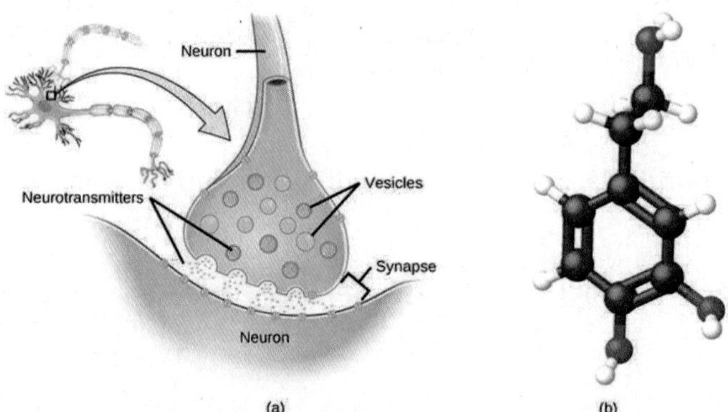

FIGURE 3.10 (a) Chemical signals are transmitted from neurons to other cells by the release of neurotransmitter molecules into the small gaps (synapses) between the cells. (b) Dopamine, $C_8H_{11}NO_2$, is a neurotransmitter involved in a number of neurological processes.

One important aspect of the complex processes related to dopamine signaling is the number of neurotransmitter molecules released during exocytosis. Since this number is a central factor in determining neurological response (and subsequent human thought and action), it is important to know how this number changes with certain controlled stimulations, such as the administration of drugs. It is also important to understand the mechanism responsible for any changes in the number of neurotransmitter molecules released—for example, some dysfunction in exocytosis, a change in the number of vesicles in the neuron, or a change in the number of neurotransmitter molecules in each vesicle.

Significant progress has been made recently in directly measuring the number of dopamine molecules stored in individual vesicles and the amount actually released when the vesicle undergoes exocytosis. Using miniaturized probes that can selectively detect dopamine molecules in very small amounts, scientists have determined that the vesicles of a certain type of mouse brain neuron contain an average of 30,000 dopamine molecules per vesicle (about 5×10^{-20} mol or 50 zmol). Analysis of these neurons from mice subjected to various drug therapies shows significant changes in the average number of dopamine molecules contained in individual vesicles, increasing or decreasing by up to three-fold, depending on the specific drug used. These studies also indicate that not all of the dopamine in a given vesicle is released during exocytosis, suggesting that it may be possible to regulate the fraction released using pharmaceutical therapies.[1]

3.2 Determining Empirical and Molecular Formulas

LEARNING OBJECTIVES

By the end of this section, you will be able to:

- Compute the percent composition of a compound
- Determine the empirical formula of a compound
- Determine the molecular formula of a compound

The previous section discussed the relationship between the bulk mass of a substance and the number of atoms or molecules it contains (moles). Given the chemical formula of the substance, one may determine the amount of the substance (moles) from its mass, and vice versa. But what if the chemical formula of a substance is unknown? In this section, these same principles will be applied to derive the chemical formulas of unknown substances from experimental mass measurements.

Percent Composition

The elemental makeup of a compound defines its chemical identity, and chemical formulas are the most succinct way of representing this elemental makeup. When a compound's formula is unknown, measuring the mass of each of its constituent elements is often the first step in the process of determining the formula experimentally. The results of these measurements permit the calculation of the compound's **percent composition**, defined as the percentage by mass of each element in the compound. For example, consider a gaseous compound composed solely of carbon and hydrogen. The percent composition of this compound could be represented as follows:

$$\%\text{H} = \frac{\text{mass H}}{\text{mass compound}} \times 100\%$$

$$\%\text{C} = \frac{\text{mass C}}{\text{mass compound}} \times 100\%$$

If analysis of a 10.0-g sample of this gas showed it to contain 2.5 g H and 7.5 g C, the percent composition would be calculated to be 25% H and 75% C:

$$\%\text{H} = \frac{2.5 \text{ g H}}{10.0 \text{ g compound}} \times 100\% = 25\%$$

1 Omiatek, Donna M., Amanda J. Bressler, Ann-Sofie Cans, Anne M. Andrews, Michael L. Heien, and Andrew G. Ewing. "The Real Catecholamine Content of Secretory Vesicles in the CNS Revealed by Electrochemical Cytometry." *Scientific Report* 3 (2013): 1447, accessed January 14, 2015, doi:10.1038/srep01447.

$$\% \text{C} = \frac{7.5 \text{ g C}}{10.0 \text{ g compound}} \times 100\% = 75\%$$

EXAMPLE 3.9

Calculation of Percent Composition

Analysis of a 12.04-g sample of a liquid compound composed of carbon, hydrogen, and nitrogen showed it to contain 7.34 g C, 1.85 g H, and 2.85 g N. What is the percent composition of this compound?

Solution

To calculate percent composition, divide the experimentally derived mass of each element by the overall mass of the compound, and then convert to a percentage:

$$\% \text{C} = \frac{7.34 \text{ g C}}{12.04 \text{ g compound}} \times 100\% = 61.0\%$$
$$\% \text{H} = \frac{1.85 \text{ g H}}{12.04 \text{ g compound}} \times 100\% = 15.4\%$$
$$\% \text{N} = \frac{2.85 \text{ g N}}{12.04 \text{ g compound}} \times 100\% = 23.7\%$$

The analysis results indicate that the compound is 61.0% C, 15.4% H, and 23.7% N by mass.

Check Your Learning

A 24.81-g sample of a gaseous compound containing only carbon, oxygen, and chlorine is determined to contain 3.01 g C, 4.00 g O, and 17.81 g Cl. What is this compound's percent composition?

Answer:

12.1% C, 16.1% O, 71.79% Cl

Determining Percent Composition from Molecular or Empirical Formulas

Percent composition is also useful for evaluating the relative abundance of a given element in different compounds of known formulas. As one example, consider the common nitrogen-containing fertilizers ammonia (NH_3), ammonium nitrate (NH_4NO_3), and urea (CH_4N_2O). The element nitrogen is the active ingredient for agricultural purposes, so the mass percentage of nitrogen in the compound is a practical and economic concern for consumers choosing among these fertilizers. For these sorts of applications, the percent composition of a compound is easily derived from its formula mass and the atomic masses of its constituent elements. A molecule of NH_3 contains one N atom weighing 14.01 amu and three H atoms weighing a total of (3 × 1.008 amu) = 3.024 amu. The formula mass of ammonia is therefore (14.01 amu + 3.024 amu) = 17.03 amu, and its percent composition is:

$$\% \text{N} = \frac{14.01 \text{ amu N}}{17.03 \text{ amu NH}_3} \times 100\% = 82.27\%$$
$$\% \text{H} = \frac{3.024 \text{ amu H}}{17.03 \text{ amu NH}_3} \times 100\% = 17.76\%$$

This same approach may be taken considering a pair of molecules, a dozen molecules, or a mole of molecules, etc. The latter amount is most convenient and would simply involve the use of molar masses instead of atomic and formula masses, as demonstrated Example 3.10. As long as the molecular or empirical formula of the compound in question is known, the percent composition may be derived from the atomic or molar masses of the compound's elements.

EXAMPLE 3.10

Determining Percent Composition from a Molecular Formula

Aspirin is a compound with the molecular formula $C_9H_8O_4$. What is its percent composition?

Solution

To calculate the percent composition, the masses of C, H, and O in a known mass of $C_9H_8O_4$ are needed. It is convenient to consider 1 mol of $C_9H_8O_4$ and use its molar mass (180.159 g/mole, determined from the chemical formula) to calculate the percentages of each of its elements:

$$\%\,C = \frac{9\text{ mol C} \times \text{molar mass C}}{\text{molar mass }C_9H_8O_4} \times 100 = \frac{9 \times 12.01\text{ g/mol}}{180.159\text{ g/mol}} \times 100 = \frac{108.09\text{ g/mol}}{180.159\text{ g/mol}} \times 100$$
$$\%\,C = 60.00\%\,C$$

$$\%\,H = \frac{8\text{ mol H} \times \text{molar mass H}}{\text{molar mass }C_9H_8O_4} \times 100 = \frac{8 \times 1.008\text{ g/mol}}{180.159\text{ g/mol}} \times 100 = \frac{8.064\text{ g/mol}}{180.159\text{ g/mol}} \times 100$$
$$\%\,H = 4.476\%\,H$$

$$\%\,O = \frac{4\text{ mol O} \times \text{molar mass O}}{\text{molar mass }C_9H_8O_4} \times 100 = \frac{4 \times 16.00\text{ g/mol}}{180.159\text{ g/mol}} \times 100 = \frac{64.00\text{ g/mol}}{180.159\text{ g/mol}} \times 100$$
$$\%\,O = 35.52\%$$

Note that these percentages sum to equal 100.00% when appropriately rounded.

Check Your Learning

To three significant digits, what is the mass percentage of iron in the compound Fe_2O_3?

Answer:

69.9% Fe

Determination of Empirical Formulas

As previously mentioned, the most common approach to determining a compound's chemical formula is to first measure the masses of its constituent elements. However, keep in mind that chemical formulas represent the relative *numbers*, not masses, of atoms in the substance. Therefore, any experimentally derived data involving mass must be used to derive the corresponding numbers of atoms in the compound. This is accomplished using molar masses to convert the mass of each element to a number of moles. These molar amounts are used to compute whole-number ratios that can be used to derive the empirical formula of the substance. Consider a sample of compound determined to contain 1.71 g C and 0.287 g H. The corresponding numbers of atoms (in moles) are:

$$1.71\text{ g C} \times \frac{1\text{ mol C}}{12.01\text{ g C}} = 0.142\text{ mol C}$$
$$0.287\text{ g H} \times \frac{1\text{ mol H}}{1.008\text{ g H}} = 0.284\text{ mol H}$$

Thus, this compound may be represented by the formula $C_{0.142}H_{0.284}$. Per convention, formulas contain whole-number subscripts, which can be achieved by dividing each subscript by the smaller subscript:

$$C_{\frac{0.142}{0.142}}H_{\frac{0.284}{0.142}}\text{ or }CH_2$$

(Recall that subscripts of "1" are not written but rather assumed if no other number is present.)

The empirical formula for this compound is thus CH_2. This may or not be the compound's *molecular formula* as well; however, additional information is needed to make that determination (as discussed later in this section).

Consider as another example a sample of compound determined to contain 5.31 g Cl and 8.40 g O. Following the same approach yields a tentative empirical formula of:

$$Cl_{0.150}O_{0.525} = Cl_{\frac{0.150}{0.150}}O_{\frac{0.525}{0.150}} = ClO_{3.5}$$

In this case, dividing by the smallest subscript still leaves us with a decimal subscript in the empirical formula. To convert this into a whole number, multiply each of the subscripts by two, retaining the same atom ratio and yielding Cl_2O_7 as the final empirical formula.

In summary, empirical formulas are derived from experimentally measured element masses by:

1. Deriving the number of moles of each element from its mass
2. Dividing each element's molar amount by the smallest molar amount to yield subscripts for a tentative empirical formula
3. Multiplying all coefficients by an integer, if necessary, to ensure that the smallest whole-number ratio of subscripts is obtained

Figure 3.11 outlines this procedure in flow chart fashion for a substance containing elements A and X.

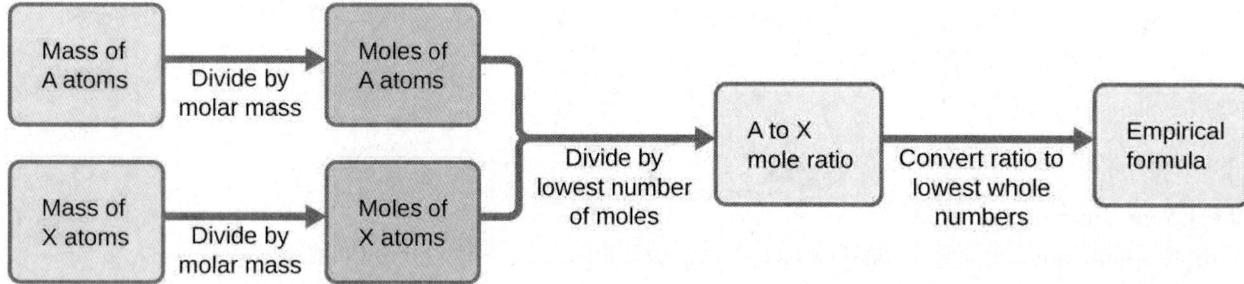

FIGURE 3.11 The empirical formula of a compound can be derived from the masses of all elements in the sample.

EXAMPLE 3.11

Determining a Compound's Empirical Formula from the Masses of Its Elements

A sample of the black mineral hematite (Figure 3.12), an oxide of iron found in many iron ores, contains 34.97 g of iron and 15.03 g of oxygen. What is the empirical formula of hematite?

FIGURE 3.12 Hematite is an iron oxide that is used in jewelry. (credit: Mauro Cateb)

Solution

This problem provides the mass in grams of each element. Begin by finding the moles of each:

$$34.97 \text{ g Fe} \left(\frac{\text{mol Fe}}{55.85 \text{ g}}\right) = 0.6261 \text{ mol Fe}$$
$$15.03 \text{ g O} \left(\frac{\text{mol O}}{16.00 \text{ g}}\right) = 0.9394 \text{ mol O}$$

Next, derive the iron-to-oxygen molar ratio by dividing by the lesser number of moles:

$$\frac{0.6261}{0.6261} = 1.000 \text{ mol Fe}$$
$$\frac{0.9394}{0.6261} = 1.500 \text{ mol O}$$

The ratio is 1.000 mol of iron to 1.500 mol of oxygen ($Fe_1O_{1.5}$). Finally, multiply the ratio by two to get the smallest possible whole number subscripts while still maintaining the correct iron-to-oxygen ratio:

$$2\,(Fe_1O_{1.5}) = Fe_2O_3$$

The empirical formula is Fe_2O_3.

Check Your Learning

What is the empirical formula of a compound if a sample contains 0.130 g of nitrogen and 0.370 g of oxygen?

Answer:

N_2O_5

LINK TO LEARNING

For additional worked examples illustrating the derivation of empirical formulas, watch the brief video (http://openstax.org/l/16empforms) clip.

Deriving Empirical Formulas from Percent Composition

Finally, with regard to deriving empirical formulas, consider instances in which a compound's percent composition is available rather than the absolute masses of the compound's constituent elements. In such cases, the percent composition can be used to calculate the masses of elements present in any convenient mass of compound; these masses can then be used to derive the empirical formula in the usual fashion.

EXAMPLE 3.12

Determining an Empirical Formula from Percent Composition

The bacterial fermentation of grain to produce ethanol forms a gas with a percent composition of 27.29% C and 72.71% O (Figure 3.13). What is the empirical formula for this gas?

FIGURE 3.13 An oxide of carbon is removed from these fermentation tanks through the large copper pipes at the top. (credit: "Dual Freq"/Wikimedia Commons)

Solution

Since the scale for percentages is 100, it is most convenient to calculate the mass of elements present in a sample weighing 100 g. The calculation is "most convenient" because, per the definition for percent composition, the mass of a given element in grams is numerically equivalent to the element's mass percentage. This numerical equivalence results from the definition of the "percentage" unit, whose name is derived from the Latin phrase *per centum* meaning "by the hundred." Considering this definition, the mass percentages provided may be more conveniently expressed as fractions:

$$\begin{aligned} 27.29\%\ \text{C} &= \frac{27.29\ \text{g C}}{100\ \text{g compound}} \\ 72.71\%\ \text{O} &= \frac{72.71\ \text{g O}}{100\ \text{g compound}} \end{aligned}$$

The molar amounts of carbon and oxygen in a 100-g sample are calculated by dividing each element's mass by its molar mass:

$$\begin{aligned} 27.29\ \text{g C}\left(\frac{\text{mol C}}{12.01\ \text{g}}\right) &= 2.272\ \text{mol C} \\ 72.71\ \text{g O}\left(\frac{\text{mol O}}{16.00\ \text{g}}\right) &= 4.544\ \text{mol O} \end{aligned}$$

Coefficients for the tentative empirical formula are derived by dividing each molar amount by the lesser of the two:

$$\begin{aligned} \frac{2.272\ \text{mol C}}{2.272} &= 1 \\ \frac{4.544\ \text{mol O}}{2.272} &= 2 \end{aligned}$$

Since the resulting ratio is one carbon to two oxygen atoms, the empirical formula is CO_2.

Check Your Learning

What is the empirical formula of a compound containing 40.0% C, 6.71% H, and 53.28% O?

Answer:

CH_2O

Derivation of Molecular Formulas

Recall that empirical formulas are symbols representing the *relative* numbers of a compound's elements. Determining the *absolute* numbers of atoms that compose a single molecule of a covalent compound requires knowledge of both its empirical formula and its molecular mass or molar mass. These quantities may be determined experimentally by various measurement techniques. Molecular mass, for example, is often derived from the mass spectrum of the compound (see discussion of this technique in the previous chapter on atoms and molecules). Molar mass can be measured by a number of experimental methods, many of which will be introduced in later chapters of this text.

Molecular formulas are derived by comparing the compound's molecular or molar mass to its **empirical formula mass**. As the name suggests, an empirical formula mass is the sum of the average atomic masses of all the atoms represented in an empirical formula. If the molecular (or molar) mass of the substance is known, it may be divided by the empirical formula mass to yield the number of empirical formula units per molecule (*n*):

$$\frac{\text{molecular or molar mass}\left(\text{amu or }\frac{\text{g}}{\text{mol}}\right)}{\text{empirical formula mass}\left(\text{amu or }\frac{\text{g}}{\text{mol}}\right)} = n \text{ formula units/molecule}$$

The molecular formula is then obtained by multiplying each subscript in the empirical formula by *n*, as shown by the generic empirical formula A_xB_y:

$$(A_xB_y)_n = A_{nx}B_{ny}$$

For example, consider a covalent compound whose empirical formula is determined to be CH_2O. The empirical formula mass for this compound is approximately 30 amu (the sum of 12 amu for one C atom, 2 amu for two H atoms, and 16 amu for one O atom). If the compound's molecular mass is determined to be 180 amu, this indicates that molecules of this compound contain six times the number of atoms represented in the empirical formula:

$$\frac{180 \text{ amu/molecule}}{30 \frac{\text{amu}}{\text{formula unit}}} = 6 \text{ formula units/molecule}$$

Molecules of this compound are then represented by molecular formulas whose subscripts are six times greater than those in the empirical formula:

$$(CH_2O)_6 = C_6H_{12}O_6$$

Note that this same approach may be used when the molar mass (g/mol) instead of the molecular mass (amu) is used. In this case, *one mole* of empirical formula units and molecules is considered, as opposed to single units and molecules.

 **EXAMPLE 3.13**

Determination of the Molecular Formula for Nicotine

Nicotine, an alkaloid in the nightshade family of plants that is mainly responsible for the addictive nature of cigarettes, contains 74.02% C, 8.710% H, and 17.27% N. If 40.57 g of nicotine contains 0.2500 mol nicotine, what is the molecular formula?

Solution

Determining the molecular formula from the provided data will require comparison of the compound's empirical formula mass to its molar mass. As the first step, use the percent composition to derive the compound's empirical formula. Assuming a convenient, a 100-g sample of nicotine yields the following molar amounts of its elements:

$$\begin{aligned}(74.02 \text{ g C})\left(\frac{1 \text{ mol C}}{12.01 \text{ g C}}\right) &= 6.163 \text{ mol C}\\(8.710 \text{ g H})\left(\frac{1 \text{ mol H}}{1.01 \text{ g H}}\right) &= 8.624 \text{ mol H}\\(17.27 \text{ g N})\left(\frac{1 \text{ mol N}}{14.01 \text{ g N}}\right) &= 1.233 \text{ mol N}\end{aligned}$$

Next, calculate the molar ratios of these elements relative to the least abundant element, N.

$$6.163 \text{ mol C} / 1.233 \text{ mol N} = 5$$
$$8.264 \text{ mol H} / 1.233 \text{ mol N} = 7$$
$$1.233 \text{ mol N} / 1.233 \text{ mol N} = 1$$

$$\begin{aligned}\frac{1.233}{1.233} &= 1.000 \text{ mol N}\\\frac{6.163}{1.233} &= 4.998 \text{ mol C}\\\frac{8.624}{1.233} &= 6.994 \text{ mol H}\end{aligned}$$

The C-to-N and H-to-N molar ratios are adequately close to whole numbers, and so the empirical formula is C_5H_7N. The empirical formula mass for this compound is therefore 81.13 amu/formula unit, or 81.13 g/mol formula unit.

Calculate the molar mass for nicotine from the given mass and molar amount of compound:

$$\frac{40.57 \text{ g nicotine}}{0.2500 \text{ mol nicotine}} = \frac{162.3 \text{ g}}{\text{mol}}$$

Comparing the molar mass and empirical formula mass indicates that each nicotine molecule contains two formula units:

$$\frac{162.3 \text{ g/mol}}{81.13 \frac{\text{g}}{\text{formula unit}}} = 2 \text{ formula units/molecule}$$

Finally, derive the molecular formula for nicotine from the empirical formula by multiplying each subscript by two:

$$(C_5H_7N)_2 = C_{10}H_{14}N_2$$

Check Your Learning

What is the molecular formula of a compound with a percent composition of 49.47% C, 5.201% H, 28.84% N, and 16.48% O, and a molecular mass of 194.2 amu?

Answer:

$C_8H_{10}N_4O_2$

3.3 Molarity

LEARNING OBJECTIVES

By the end of this section, you will be able to:

- Describe the fundamental properties of solutions
- Calculate solution concentrations using molarity
- Perform dilution calculations using the dilution equation

Preceding sections of this chapter focused on the composition of substances: samples of matter that contain only one type of element or compound. However, mixtures—samples of matter containing two or more substances physically combined—are more commonly encountered in nature than are pure substances. Similar to a pure substance, the relative composition of a mixture plays an important role in determining its properties. The relative amount of oxygen in a planet's atmosphere determines its ability to sustain aerobic life. The relative amounts of iron, carbon, nickel, and other elements in steel (a mixture known as an "alloy") determine its physical strength and resistance to corrosion. The relative amount of the active ingredient in a medicine determines its effectiveness in achieving the desired pharmacological effect. The relative amount of sugar in a beverage determines its sweetness (see Figure 3.14). This section will describe one of the most common ways in which the relative compositions of mixtures may be quantified.

FIGURE 3.14 Sugar is one of many components in the complex mixture known as coffee. The amount of sugar in a given amount of coffee is an important determinant of the beverage's sweetness. (credit: Jane Whitney)

Solutions

Solutions have previously been defined as homogeneous mixtures, meaning that the composition of the mixture (and therefore its properties) is uniform throughout its entire volume. Solutions occur frequently in nature and have also been implemented in many forms of manmade technology. A more thorough treatment of solution properties is provided in the chapter on solutions and colloids, but provided here is an introduction to some of the basic properties of solutions.

The relative amount of a given solution component is known as its **concentration**. Often, though not always, a solution contains one component with a concentration that is significantly greater than that of all other components. This component is called the **solvent** and may be viewed as the medium in which the other components are dispersed, or **dissolved**. Solutions in which water is the solvent are, of course, very common on our planet. A solution in which water is the solvent is called an **aqueous solution**.

A **solute** is a component of a solution that is typically present at a much lower concentration than the solvent. Solute concentrations are often described with qualitative terms such as **dilute** (of relatively low concentration) and **concentrated** (of relatively high concentration).

Concentrations may be quantitatively assessed using a wide variety of measurement units, each convenient for particular applications. **Molarity (*M*)** is a useful concentration unit for many applications in chemistry. Molarity is defined as the number of moles of solute in exactly 1 liter (1 L) of the solution:

$$M = \frac{\text{mol solute}}{\text{L solution}}$$

EXAMPLE 3.14

Calculating Molar Concentrations

A 355-mL soft drink sample contains 0.133 mol of sucrose (table sugar). What is the molar concentration of sucrose in the beverage?

Solution

Since the molar amount of solute and the volume of solution are both given, the molarity can be calculated using the definition of molarity. Per this definition, the solution volume must be converted from mL to L:

$$M = \frac{\text{mol solute}}{\text{L solution}} = \frac{0.133\ \text{mol}}{355\ \text{mL} \times \frac{1\ \text{L}}{1000\ \text{mL}}} = 0.375\ M$$

Check Your Learning

A teaspoon of table sugar contains about 0.01 mol sucrose. What is the molarity of sucrose if a teaspoon of sugar has been dissolved in a cup of tea with a volume of 200 mL?

Answer:

0.05 *M*

 **EXAMPLE 3.15**

Deriving Moles and Volumes from Molar Concentrations

How much sugar (mol) is contained in a modest sip (~10 mL) of the soft drink from Example 3.14?

Solution

Rearrange the definition of molarity to isolate the quantity sought, moles of sugar, then substitute the value for molarity derived in Example 3.14, 0.375 *M*:

$$M = \frac{\text{mol solute}}{\text{L solution}}$$

$$\text{mol solute} = M \times \text{L solution}$$

$$\text{mol solute} = 0.375\ \frac{\text{mol sugar}}{\text{L}} \times \left(10\ \text{mL} \times \frac{1\ \text{L}}{1000\ \text{mL}}\right) = 0.004\ \text{mol sugar}$$

Check Your Learning

What volume (mL) of the sweetened tea described in Example 3.14 contains the same amount of sugar (mol) as 10 mL of the soft drink in this example?

Answer:

80 mL

 **EXAMPLE 3.16**

Calculating Molar Concentrations from the Mass of Solute

Distilled white vinegar (Figure 3.15) is a solution of acetic acid, CH_3CO_2H, in water. A 0.500-L vinegar solution contains 25.2 g of acetic acid. What is the concentration of the acetic acid solution in units of molarity?

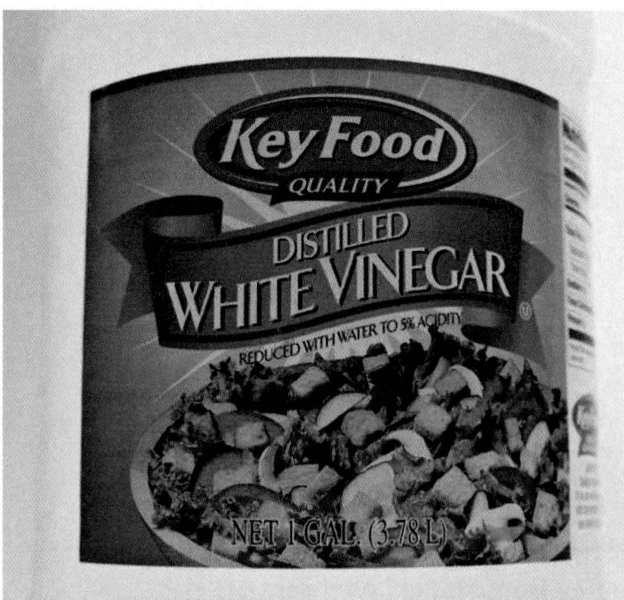

FIGURE 3.15 Distilled white vinegar is a solution of acetic acid in water.

Solution

As in previous examples, the definition of molarity is the primary equation used to calculate the quantity sought. Since the mass of solute is provided instead of its molar amount, use the solute's molar mass to obtain the amount of solute in moles:

$$M = \frac{\text{mol solute}}{\text{L solution}} = \frac{25.2\text{ g CH}_3\text{CO}_2\text{H} \times \frac{1\text{ mol CH}_3\text{CO}_2\text{H}}{60.052\text{ g CH}_3\text{CO}_2\text{H}}}{0.500\text{ L solution}} = 0.839\ M$$

$$M = \frac{\text{mol solute}}{\text{L solution}} = 0.839\ M$$
$$M = \frac{0.839\text{ mol solute}}{1.00\text{ L solution}}$$

Check Your Learning

Calculate the molarity of 6.52 g of $CoCl_2$ (128.9 g/mol) dissolved in an aqueous solution with a total volume of 75.0 mL.

Answer:

0.674 *M*

EXAMPLE 3.17

Determining the Mass of Solute in a Given Volume of Solution

How many grams of NaCl are contained in 0.250 L of a 5.30-*M* solution?

Solution

The volume and molarity of the solution are specified, so the amount (mol) of solute is easily computed as demonstrated in Example 3.15:

$$M = \frac{\text{mol solute}}{\text{L solution}}$$
$$\text{mol solute} = M \times \text{L solution}$$

$$\text{mol solute} = 5.30\ \frac{\text{mol NaCl}}{\text{L}} \times 0.250\ \text{L} = 1.325\ \text{mol NaCl}$$

Finally, this molar amount is used to derive the mass of NaCl:

$$1.325\ \text{mol NaCl} \times \frac{58.44\ \text{g NaCl}}{\text{mol NaCl}} = 77.4\ \text{g NaCl}$$

Check Your Learning

How many grams of $CaCl_2$ (110.98 g/mol) are contained in 250.0 mL of a 0.200-*M* solution of calcium chloride?

Answer:

5.55 g $CaCl_2$

When performing calculations stepwise, as in Example 3.17, it is important to refrain from rounding any intermediate calculation results, which can lead to rounding errors in the final result. In Example 3.17, the molar amount of NaCl computed in the first step, 1.325 mol, would be properly rounded to 1.32 mol if it were to be reported; however, although the last digit (5) is not significant, it must be retained as a guard digit in the intermediate calculation. If the guard digit had not been retained, the final calculation for the mass of NaCl would have been 77.1 g, a difference of 0.3 g.

In addition to retaining a guard digit for intermediate calculations, rounding errors may also be avoided by performing computations in a single step (see Example 3.18). This eliminates intermediate steps so that only the final result is rounded.

EXAMPLE 3.18

Determining the Volume of Solution Containing a Given Mass of Solute

In Example 3.16, the concentration of acetic acid in white vinegar was determined to be 0.839 *M*. What volume of vinegar contains 75.6 g of acetic acid?

Solution

First, use the molar mass to calculate moles of acetic acid from the given mass:

$$\text{g solute} \times \frac{\text{mol solute}}{\text{g solute}} = \text{mol solute}$$

Then, use the molarity of the solution to calculate the volume of solution containing this molar amount of solute:

$$\text{mol solute} \times \frac{\text{L solution}}{\text{mol solute}} = \text{L solution}$$

Combining these two steps into one yields:

$$\text{g solute} \times \frac{\text{mol solute}}{\text{g solute}} \times \frac{\text{L solution}}{\text{mol solute}} = \text{L solution}$$

$$75.6\ \text{g}\ CH_3CO_2H \left(\frac{\text{mol}\ CH_3CO_2H}{60.05\ \text{g}}\right)\left(\frac{\text{L solution}}{0.839\ \text{mol}\ CH_3CO_2H}\right) = 1.50\ \text{L solution}$$

Check Your Learning

What volume of a 1.50-*M* KBr solution contains 66.0 g KBr?

Answer:

0.370 L

Dilution of Solutions

Dilution is the process whereby the concentration of a solution is lessened by the addition of solvent. For example, a glass of iced tea becomes increasingly diluted as the ice melts. The water from the melting ice increases the volume of the solvent (water) and the overall volume of the solution (iced tea), thereby reducing the relative concentrations of the solutes that give the beverage its taste (Figure 3.16).

FIGURE 3.16 Both solutions contain the same mass of copper nitrate. The solution on the right is more dilute because the copper nitrate is dissolved in more solvent. (credit: Mark Ott)

Dilution is also a common means of preparing solutions of a desired concentration. By adding solvent to a measured portion of a more concentrated *stock solution*, a solution of lesser concentration may be prepared. For example, commercial pesticides are typically sold as solutions in which the active ingredients are far more concentrated than is appropriate for their application. Before they can be used on crops, the pesticides must be diluted. This is also a very common practice for the preparation of a number of common laboratory reagents.

A simple mathematical relationship can be used to relate the volumes and concentrations of a solution before and after the dilution process. According to the definition of molarity, the number of moles of solute in a solution (n) is equal to the product of the solution's molarity (M) and its volume in liters (L):

$$n = ML$$

Expressions like these may be written for a solution before and after it is diluted:

$$n_1 = M_1 L_1$$
$$n_2 = M_2 L_2$$

where the subscripts "1" and "2" refer to the solution before and after the dilution, respectively. Since the dilution process *does not change the amount of solute in the solution,* $n_1 = n_2$. Thus, these two equations may be set equal to one another:

$$M_1 L_1 = M_2 L_2$$

This relation is commonly referred to as the dilution equation. Although this equation uses molarity as the unit of concentration and liters as the unit of volume, other units of concentration and volume may be used as long as the units properly cancel per the factor-label method. Reflecting this versatility, the dilution equation is often written in the more general form:

$$C_1 V_1 = C_2 V_2$$

where C and V are concentration and volume, respectively.

LINK TO LEARNING

Use the simulation (http://openstax.org/l/16Phetsolvents) to explore the relations between solute amount, solution volume, and concentration and to confirm the dilution equation.

EXAMPLE 3.19

Determining the Concentration of a Diluted Solution

If 0.850 L of a 5.00-*M* solution of copper nitrate, $Cu(NO_3)_2$, is diluted to a volume of 1.80 L by the addition of water, what is the molarity of the diluted solution?

Solution

The stock concentration, C_1, and volume, V_1, are provided as well as the volume of the diluted solution, V_2. Rearrange the dilution equation to isolate the unknown property, the concentration of the diluted solution, C_2:

$$C_1V_1 = C_2V_2$$

$$C_2 = \frac{C_1V_1}{V_2}$$

Since the stock solution is being diluted by more than two-fold (volume is increased from 0.85 L to 1.80 L), the diluted solution's concentration is expected to be less than one-half 5 *M*. This ballpark estimate will be compared to the calculated result to check for any gross errors in computation (for example, such as an improper substitution of the given quantities). Substituting the given values for the terms on the right side of this equation yields:

$$C_2 = \frac{0.850\ \text{L} \times 5.00\ \frac{\text{mol}}{\text{L}}}{1.80\ \text{L}} = 2.36\ M$$

This result compares well to our ballpark estimate (it's a bit less than one-half the stock concentration, 5 *M*).

Check Your Learning

What is the concentration of the solution that results from diluting 25.0 mL of a 2.04-*M* solution of CH_3OH to 500.0 mL?

Answer:

0.102 *M* CH_3OH

EXAMPLE 3.20

Volume of a Diluted Solution

What volume of 0.12 *M* HBr can be prepared from 11 mL (0.011 L) of 0.45 *M* HBr?

Solution

Provided are the volume and concentration of a stock solution, V_1 and C_1, and the concentration of the resultant diluted solution, C_2. Find the volume of the diluted solution, V_2 by rearranging the dilution equation to isolate V_2:

$$C_1V_1 = C_2V_2$$

$$V_2 = \frac{C_1V_1}{C_2}$$

Since the diluted concentration (0.12 *M*) is slightly more than one-fourth the original concentration (0.45 *M*), the volume of the diluted solution is expected to be roughly four times the original volume, or around 44 mL. Substituting the given values and solving for the unknown volume yields:

$$V_2 = \frac{(0.45\ M)(0.011\ \text{L})}{(0.12\ M)}$$
$$V_2 = 0.041\ \text{L}$$

The volume of the 0.12-*M* solution is 0.041 L (41 mL). The result is reasonable and compares well with the rough estimate.

Check Your Learning

A laboratory experiment calls for 0.125 *M* HNO_3. What volume of 0.125 *M* HNO_3 can be prepared from 0.250 L of 1.88 *M* HNO_3?

Answer:

3.76 L

EXAMPLE 3.21

Volume of a Concentrated Solution Needed for Dilution

What volume of 1.59 *M* KOH is required to prepare 5.00 L of 0.100 *M* KOH?

Solution

Given are the concentration of a stock solution, C_1, and the volume and concentration of the resultant diluted solution, V_2 and C_2. Find the volume of the stock solution, V_1 by rearranging the dilution equation to isolate V_1:

$$C_1V_1 = C_2V_2$$

$$V_1 = \frac{C_2V_2}{C_1}$$

Since the concentration of the diluted solution 0.100 *M* is roughly one-sixteenth that of the stock solution (1.59 *M*), the volume of the stock solution is expected to be about one-sixteenth that of the diluted solution, or around 0.3 liters. Substituting the given values and solving for the unknown volume yields:

$$V_1 = \frac{(0.100\ M)(5.00\ \text{L})}{1.59\ M}$$
$$V_1 = 0.314\ \text{L}$$

Thus, 0.314 L of the 1.59-*M* solution is needed to prepare the desired solution. This result is consistent with the rough estimate.

Check Your Learning

What volume of a 0.575-*M* solution of glucose, $C_6H_{12}O_6$, can be prepared from 50.00 mL of a 3.00-*M* glucose solution?

Answer:

0.261 L

3.4 Other Units for Solution Concentrations

LEARNING OBJECTIVES

By the end of this section, you will be able to:

- Define the concentration units of mass percentage, volume percentage, mass-volume percentage, parts-per-million (ppm), and parts-per-billion (ppb)
- Perform computations relating a solution's concentration and its components' volumes and/or masses using these units

The previous section introduced molarity, a very useful measurement unit for evaluating the concentration of solutions. However, molarity is only one measure of concentration. This section will describe some other units of concentration that are commonly used in various applications, either for convenience or by convention.

Mass Percentage

Earlier in this chapter, percent composition was introduced as a measure of the relative amount of a given element in a compound. Percentages are also commonly used to express the composition of mixtures, including solutions. The **mass percentage** of a solution component is defined as the ratio of the component's mass to the solution's mass, expressed as a percentage:

$$\text{mass percentage} = \frac{\text{mass of component}}{\text{mass of solution}} \times 100\%$$

Mass percentage is also referred to by similar names such as *percent mass, percent weight, weight/weight percent,* and other variations on this theme. The most common symbol for mass percentage is simply the percent sign, %, although more detailed symbols are often used including %mass, %weight, and (w/w)%. Use of these more detailed symbols can prevent confusion of mass percentages with other types of percentages, such as volume percentages (to be discussed later in this section).

Mass percentages are popular concentration units for consumer products. The label of a typical liquid bleach bottle (Figure 3.17) cites the concentration of its active ingredient, sodium hypochlorite (NaOCl), as being 7.4%. A 100.0-g sample of bleach would therefore contain 7.4 g of NaOCl.

FIGURE 3.17 Liquid bleach is an aqueous solution of sodium hypochlorite (NaOCl). This brand has a concentration of 7.4% NaOCl by mass.

EXAMPLE 3.22

Calculation of Percent by Mass

A 5.0-g sample of spinal fluid contains 3.75 mg (0.00375 g) of glucose. What is the percent by mass of glucose in spinal fluid?

Solution

The spinal fluid sample contains roughly 4 mg of glucose in 5000 mg of fluid, so the mass fraction of glucose should be a bit less than one part in 1000, or about 0.1%. Substituting the given masses into the equation defining mass percentage yields:

$$\%\text{ glucose} = \frac{3.75\text{ mg glucose} \times \frac{1\text{ g}}{1000\text{ mg}}}{5.0\text{ g spinal fluid}} = 0.075\%$$

The computed mass percentage agrees with our rough estimate (it's a bit less than 0.1%).

Note that while any mass unit may be used to compute a mass percentage (mg, g, kg, oz, and so on), the same unit must be used for both the solute and the solution so that the mass units cancel, yielding a dimensionless ratio. In this case, the solute mass unit in the numerator was converted from mg to g to match the units in the denominator. Alternatively, the spinal fluid mass unit in the denominator could have been converted from g to mg instead. As long as identical mass units are used for both solute and solution, the computed mass percentage will be correct.

Check Your Learning

A bottle of a tile cleanser contains 135 g of HCl and 775 g of water. What is the percent by mass of HCl in this cleanser?

Answer:

14.8%

EXAMPLE 3.23

Calculations using Mass Percentage

"Concentrated" hydrochloric acid is an aqueous solution of 37.2% HCl that is commonly used as a laboratory reagent. The density of this solution is 1.19 g/mL. What mass of HCl is contained in 0.500 L of this solution?

Solution

The HCl concentration is near 40%, so a 100-g portion of this solution would contain about 40 g of HCl. Since the solution density isn't greatly different from that of water (1 g/mL), a reasonable estimate of the HCl mass in 500 g (0.5 L) of the solution is about five times greater than that in a 100 g portion, or 5 × 40 = 200 g. In order to derive the mass of solute in a solution from its mass percentage, the mass of the solution must be known. Using the solution density given, convert the solution's volume to mass, and then use the given mass percentage to calculate the solute mass. This mathematical approach is outlined in this flowchart:

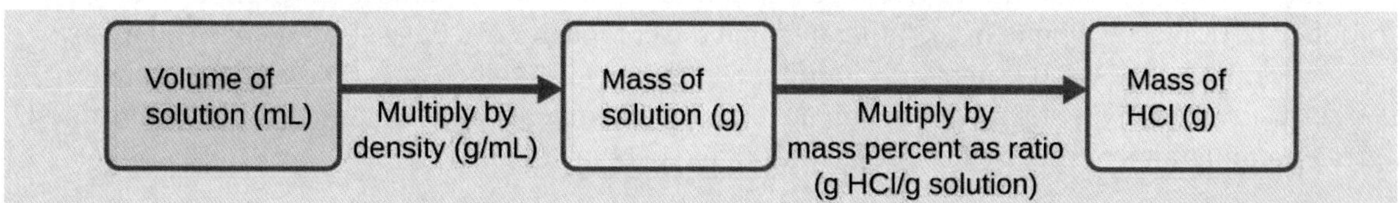

For proper unit cancellation, the 0.500-L volume is converted into 500 mL, and the mass percentage is expressed as a ratio, 37.2 g HCl/g solution:

$$500\ \text{mL solution}\left(\frac{1.19\ \text{g solution}}{\text{mL solution}}\right)\left(\frac{37.2\ \text{g HCl}}{100\ \text{g solution}}\right) = 221\ \text{g HCl}$$

This mass of HCl is consistent with our rough estimate of approximately 200 g.

Check Your Learning

What volume of concentrated HCl solution contains 125 g of HCl?

Answer:

282 mL

Volume Percentage

Liquid volumes over a wide range of magnitudes are conveniently measured using common and relatively inexpensive laboratory equipment. The concentration of a solution formed by dissolving a liquid solute in a liquid solvent is therefore often expressed as a **volume percentage**, %vol or (v/v)%:

$$\text{volume percentage} = \frac{\text{volume solute}}{\text{volume solution}} \times 100\%$$

EXAMPLE 3.24

Calculations using Volume Percentage

Rubbing alcohol (isopropanol) is usually sold as a 70%vol aqueous solution. If the density of isopropyl alcohol is 0.785 g/mL, how many grams of isopropyl alcohol are present in a 355 mL bottle of rubbing alcohol?

Solution

Per the definition of volume percentage, the isopropanol volume is 70% of the total solution volume. Multiplying the isopropanol volume by its density yields the requested mass:

$$(355\ \text{mL solution})\left(\frac{70\ \text{mL isopropyl alcohol}}{100\ \text{mL solution}}\right)\left(\frac{0.785\ \text{g isopropyl alcohol}}{1\ \text{mL isopropyl alcohol}}\right) = 195\ \text{g isopropyl alchol}$$

Check Your Learning

Wine is approximately 12% ethanol (CH_3CH_2OH) by volume. Ethanol has a molar mass of 46.06 g/mol and a density 0.789 g/mL. How many moles of ethanol are present in a 750-mL bottle of wine?

Answer:

1.5 mol ethanol

Mass-Volume Percentage

“Mixed” percentage units, derived from the mass of solute and the volume of solution, are popular for certain biochemical and medical applications. A **mass-volume percent** is a ratio of a solute’s mass to the solution’s volume expressed as a percentage. The specific units used for solute mass and solution volume may vary, depending on the solution. For example, physiological saline solution, used to prepare intravenous fluids, has a concentration of 0.9% mass/volume (m/v), indicating that the composition is 0.9 g of solute per 100 mL of solution. The concentration of glucose in blood (commonly referred to as “blood sugar”) is also typically expressed in terms of a mass-volume ratio. Though not expressed explicitly as a percentage, its concentration is usually given in milligrams of glucose per deciliter (100 mL) of blood (Figure 3.18).

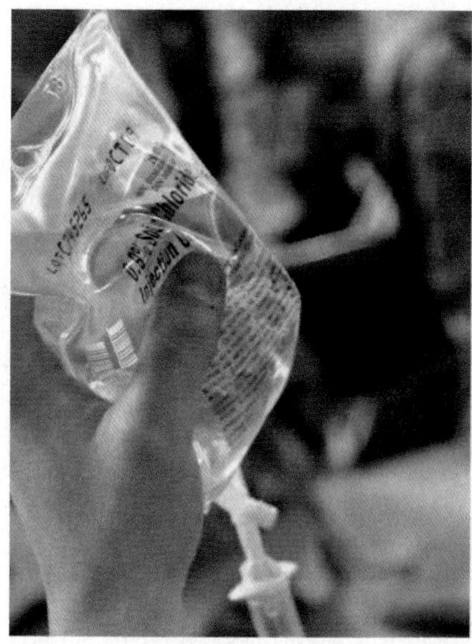

(a)

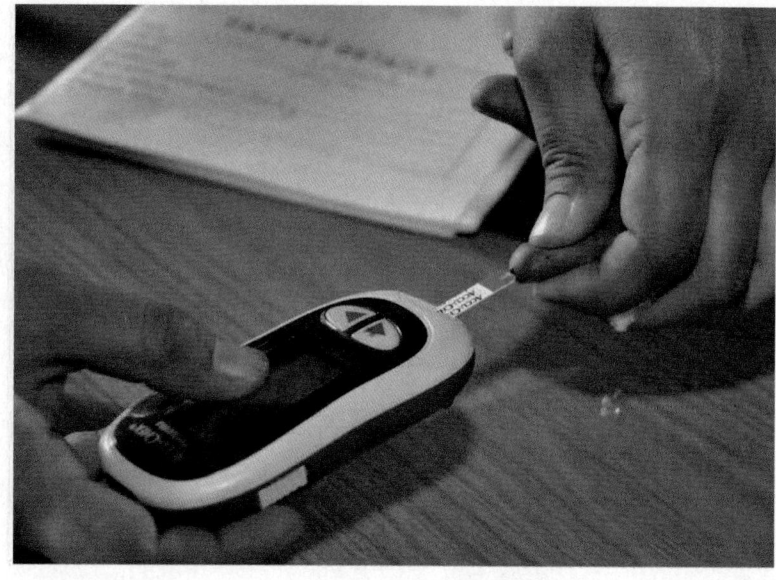

(b)

FIGURE 3.18 "Mixed" mass-volume units are commonly encountered in medical settings. (a) The NaCl concentration of physiological saline is 0.9% (m/v). (b) This device measures glucose levels in a sample of blood. The normal range for glucose concentration in blood (fasting) is around 70–100 mg/dL. (credit a: modification of work by "The National Guard"/Flickr; credit b: modification of work by Biswarup Ganguly)

Parts per Million and Parts per Billion

Very low solute concentrations are often expressed using appropriately small units such as **parts per million (ppm)** or **parts per billion (ppb)**. Like percentage ("part per hundred") units, ppm and ppb may be defined in terms of masses, volumes, or mixed mass-volume units. There are also ppm and ppb units defined with respect to numbers of atoms and molecules.

The mass-based definitions of ppm and ppb are given here:

$$\text{ppm} = \frac{\text{mass solute}}{\text{mass solution}} \times 10^6 \text{ ppm}$$
$$\text{ppb} = \frac{\text{mass solute}}{\text{mass solution}} \times 10^9 \text{ ppb}$$

Both ppm and ppb are convenient units for reporting the concentrations of pollutants and other trace contaminants in water. Concentrations of these contaminants are typically very low in treated and natural waters, and their levels cannot exceed relatively low concentration thresholds without causing adverse effects on health and wildlife. For example, the EPA has identified the maximum safe level of fluoride ion in tap water to be 4 ppm. Inline water filters are designed to reduce the concentration of fluoride and several other trace-level contaminants in tap water (Figure 3.19).

(a) (b)

FIGURE 3.19 (a) In some areas, trace-level concentrations of contaminants can render unfiltered tap water unsafe for drinking and cooking. (b) Inline water filters reduce the concentration of solutes in tap water. (credit a: modification of work by Jenn Durfey; credit b: modification of work by "vastateparkstaff"/Wikimedia commons)

EXAMPLE 3.25

Calculation of Parts per Million and Parts per Billion Concentrations

According to the EPA, when the concentration of lead in tap water reaches 15 ppb, certain remedial actions must be taken. What is this concentration in ppm? At this concentration, what mass of lead (µg) would be contained in a typical glass of water (300 mL)?

Solution

The definitions of the ppm and ppb units may be used to convert the given concentration from ppb to ppm. Comparing these two unit definitions shows that ppm is 1000 times greater than ppb (1 ppm = 10^3 ppb). Thus:

$$15 \text{ ppb} \times \frac{1 \text{ ppm}}{10^3 \text{ ppb}} = 0.015 \text{ ppm}$$

The definition of the ppb unit may be used to calculate the requested mass if the mass of the solution is provided. Since the volume of solution (300 mL) is given, its density must be used to derive the corresponding mass. Assume the density of tap water to be roughly the same as that of pure water (~1.00 g/mL), since the concentrations of any dissolved substances should not be very large. Rearranging the equation defining the ppb unit and substituting the given quantities yields:

$$\text{ppb} = \frac{\text{mass solute}}{\text{mass solution}} \times 10^9 \text{ ppb}$$

$$\text{mass solute} = \frac{\text{ppb} \times \text{mass solution}}{10^9 \text{ ppb}}$$

$$\text{mass solute} = \frac{15 \text{ ppb} \times 300 \text{ mL} \times \frac{1.00 \text{ g}}{\text{mL}}}{10^9 \text{ ppb}} = 4.5 \times 10^{-6} \text{ g}$$

Finally, convert this mass to the requested unit of micrograms:

$$4.5 \times 10^{-6}\ \text{g} \times \frac{1\ \mu\text{g}}{10^{-6}\ \text{g}} = 4.5\ \mu\text{g}$$

Check Your Learning

A 50.0-g sample of industrial wastewater was determined to contain 0.48 mg of mercury. Express the mercury concentration of the wastewater in ppm and ppb units.

Answer:

9.6 ppm, 9600 ppb

Key Terms

aqueous solution solution for which water is the solvent

Avogadro's number (N_A) experimentally determined value of the number of entities comprising 1 mole of substance, equal to 6.022×10^{23} mol^{-1}

concentrated qualitative term for a solution containing solute at a relatively high concentration

concentration quantitative measure of the relative amounts of solute and solvent present in a solution

dilute qualitative term for a solution containing solute at a relatively low concentration

dilution process of adding solvent to a solution in order to lower the concentration of solutes

dissolved describes the process by which solute components are dispersed in a solvent

empirical formula mass sum of average atomic masses for all atoms represented in an empirical formula

formula mass sum of the average masses for all atoms represented in a chemical formula; for covalent compounds, this is also the molecular mass

mass percentage ratio of solute-to-solution mass expressed as a percentage

mass-volume percent ratio of solute mass to solution volume, expressed as a percentage

molar mass mass in grams of 1 mole of a substance

molarity (M) unit of concentration, defined as the number of moles of solute dissolved in 1 liter of solution

mole amount of substance containing the same number of atoms, molecules, ions, or other entities as the number of atoms in exactly 12 grams of ^{12}C

parts per billion (ppb) ratio of solute-to-solution mass multiplied by 10^9

parts per million (ppm) ratio of solute-to-solution mass multiplied by 10^6

percent composition percentage by mass of the various elements in a compound

solute solution component present in a concentration less than that of the solvent

solvent solution component present in a concentration that is higher relative to other components

volume percentage ratio of solute-to-solution volume expressed as a percentage

Key Equations

$$\%\,X = \frac{\text{mass X}}{\text{mass compound}} \times 100\%$$

$$\%\,X = \frac{\text{mass X}}{\text{mass compound}} \times 100\%$$

$$\frac{\text{molecular or molar mass}\left(\text{amu or } \frac{\text{g}}{\text{mol}}\right)}{\text{empirical formula mass}\left(\text{amu or } \frac{\text{g}}{\text{mol}}\right)} = n \text{ formula units/molecule}$$

$$(A_xB_y)_n = A_{nx}B_{ny}$$

$$M = \frac{\text{mol solute}}{\text{L solution}}$$

$$C_1V_1 = C_2V_2$$

$$\text{Percent by mass} = \frac{\text{mass of solute}}{\text{mass of solution}} \times 100$$

$$\text{ppm} = \frac{\text{mass solute}}{\text{mass solution}} \times 10^6 \text{ ppm}$$

$$\text{ppb} = \frac{\text{mass solute}}{\text{mass solution}} \times 10^9 \text{ ppb}$$

Summary

3.1 Formula Mass and the Mole Concept

The formula mass of a substance is the sum of the average atomic masses of each atom represented in the chemical formula and is expressed in atomic mass units. The formula mass of a covalent compound is also called the molecular mass. A convenient amount unit for expressing very large numbers of atoms or molecules is the mole. Experimental measurements have determined the number of entities composing 1 mole of substance to be 6.022×10^{23}, a quantity called Avogadro's number. The mass in grams of 1 mole of substance is its molar mass. Due to the use of the same reference substance in defining the atomic mass unit and the mole, the formula mass (amu) and molar mass (g/mol) for any substance are numerically equivalent (for example, one H_2O molecule weighs approximately18 amu and 1 mole of H_2O molecules weighs approximately 18 g).

3.2 Determining Empirical and Molecular Formulas

The chemical identity of a substance is defined by the types and relative numbers of atoms composing its fundamental entities (molecules in the case of covalent compounds, ions in the case of ionic compounds). A compound's percent composition provides the mass percentage of each element in the compound, and it is often experimentally determined and used to derive the compound's empirical formula. The empirical formula mass of a covalent compound may be compared to the compound's molecular or molar mass to derive a molecular formula.

3.3 Molarity

Solutions are homogeneous mixtures. Many solutions contain one component, called the solvent, in which other components, called solutes, are dissolved. An aqueous solution is one for which the solvent is water. The concentration of a solution is a measure of the relative amount of solute in a given amount of solution. Concentrations may be measured using various units, with one very useful unit being molarity, defined as the number of moles of solute per liter of solution. The solute concentration of a solution may be decreased by adding solvent, a process referred to as dilution. The dilution equation is a simple relation between concentrations and volumes of a solution before and after dilution.

3.4 Other Units for Solution Concentrations

In addition to molarity, a number of other solution concentration units are used in various applications. Percentage concentrations based on the solution components' masses, volumes, or both are useful for expressing relatively high concentrations, whereas lower concentrations are conveniently expressed using ppm or ppb units. These units are popular in environmental, medical, and other fields where mole-based units such as molarity are not as commonly used.

Exercises

3.1 Formula Mass and the Mole Concept

1. What is the total mass (amu) of carbon in each of the following molecules?
   (a) CH_4
   (b) $CHCl_3$
   (c) $C_{12}H_{10}O_6$
   (d) $CH_3CH_2CH_2CH_2CH_3$
2. What is the total mass of hydrogen in each of the molecules?
   (a) CH_4
   (b) $CHCl_3$
   (c) $C_{12}H_{10}O_6$
   (d) $CH_3CH_2CH_2CH_2CH_3$
3. Calculate the molecular or formula mass of each of the following:
   (a) P_4
   (b) H_2O
   (c) $Ca(NO_3)_2$
   (d) CH_3CO_2H (acetic acid)
   (e) $C_{12}H_{22}O_{11}$ (sucrose, cane sugar)

4. Determine the molecular mass of the following compounds:

(a)

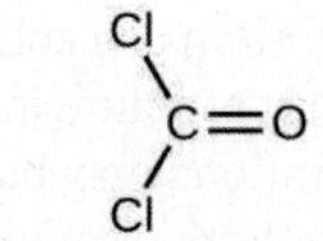

```
  Cl
    \
     C=O
    /
  Cl
```

(b)

```
H—C≡C—H
```

(c)

```
Br      Br
  \    /
   C=C
  /    \
H        H
```

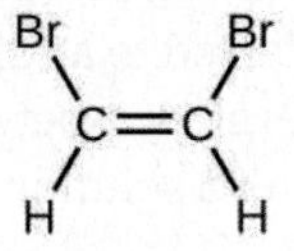

(d)

```
      O
      ||
O=S—O—H
      |
      O—H
```

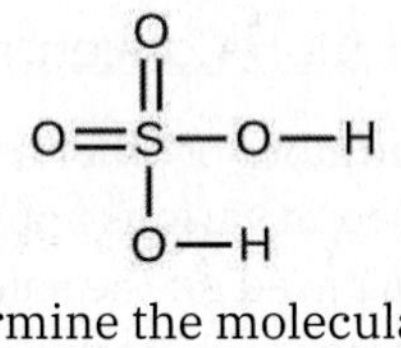

5. Determine the molecular mass of the following compounds:

(a)

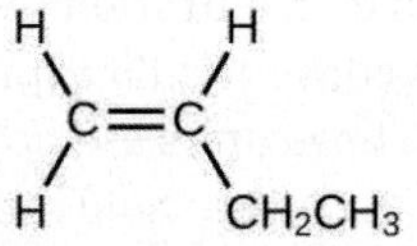

```
H        H
  \    /
   C=C
  /    \
H        CH2CH3
```

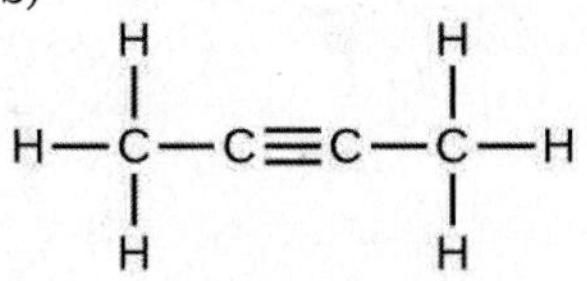

(b)

```
    H            H
    |            |
H—C—C≡C—C—H
    |            |
    H            H
```

(c)

```
     Cl   Cl
     |    |
Cl—Si—Si—Cl
     |    |
     H    H
```

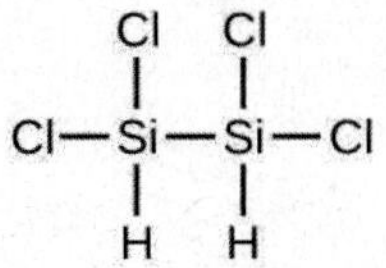

(d)

```
     O—H
     |
O=P—O—H
     |
     O—H
```

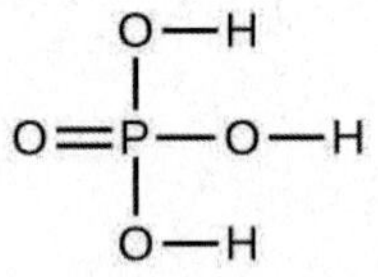

6. Which molecule has a molecular mass of 28.05 amu?

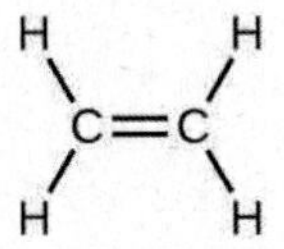

(a)

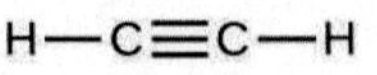

```
H—C≡C—H
```

(b)

```
H      H
  \  /
   C=C
  /  \
H      H
```

(c)

```
    H  H
    |  |
H—C—C—H
    |  |
    H  H
```

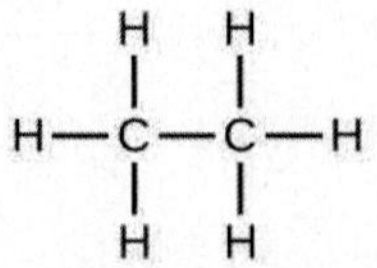

7. Write a sentence that describes how to determine the number of moles of a compound in a known mass of the compound using its molecular formula.

8. Compare 1 mole of H_2, 1 mole of O_2, and 1 mole of F_2.
   (a) Which has the largest number of molecules? Explain why.
   (b) Which has the greatest mass? Explain why.
9. Which contains the greatest mass of oxygen: 0.75 mol of ethanol (C_2H_5OH), 0.60 mol of formic acid (HCO_2H), or 1.0 mol of water (H_2O)? Explain why.
10. Which contains the greatest number of moles of oxygen atoms: 1 mol of ethanol (C_2H_5OH), 1 mol of formic acid (HCO_2H), or 1 mol of water (H_2O)? Explain why.
11. How are the molecular mass and the molar mass of a compound similar and how are they different?
12. Calculate the molar mass of each of the following compounds:
   (a) hydrogen fluoride, HF
   (b) ammonia, NH_3
   (c) nitric acid, HNO_3
   (d) silver sulfate, Ag_2SO_4
   (e) boric acid, $B(OH)_3$
13. Calculate the molar mass of each of the following:
   (a) S_8
   (b) C_5H_{12}
   (c) $Sc_2(SO_4)_3$
   (d) CH_3COCH_3 (acetone)
   (e) $C_6H_{12}O_6$ (glucose)
14. Calculate the empirical or molecular formula mass and the molar mass of each of the following minerals:
   (a) limestone, $CaCO_3$
   (b) halite, NaCl
   (c) beryl, $Be_3Al_2Si_6O_{18}$
   (d) malachite, $Cu_2(OH)_2CO_3$
   (e) turquoise, $CuAl_6(PO_4)_4(OH)_8(H_2O)_4$
15. Calculate the molar mass of each of the following:
   (a) the anesthetic halothane, $C_2HBrClF_3$
   (b) the herbicide paraquat, $C_{12}H_{14}N_2Cl_2$
   (c) caffeine, $C_8H_{10}N_4O_2$
   (d) urea, $CO(NH_2)_2$
   (e) a typical soap, $C_{17}H_{35}CO_2Na$
16. Determine the number of moles of compound and the number of moles of each type of atom in each of the following:
   (a) 25.0 g of propylene, C_3H_6
   (b) 3.06×10^{-3} g of the amino acid glycine, $C_2H_5NO_2$
   (c) 25 lb of the herbicide Treflan, $C_{13}H_{16}N_2O_4F$ (1 lb = 454 g)
   (d) 0.125 kg of the insecticide Paris Green, $Cu_4(AsO_3)_2(CH_3CO_2)_2$
   (e) 325 mg of aspirin, $C_6H_4(CO_2H)(CO_2CH_3)$
17. Determine the mass of each of the following:
   (a) 0.0146 mol KOH
   (b) 10.2 mol ethane, C_2H_6
   (c) 1.6×10^{-3} mol $Na_2\ SO_4$
   (d) 6.854×10^3 mol glucose, $C_6\ H_{12}\ O_6$
   (e) 2.86 mol $Co(NH_3)_6Cl_3$
18. Determine the number of moles of the compound and determine the number of moles of each type of atom in each of the following:
   (a) 2.12 g of potassium bromide, KBr
   (b) 0.1488 g of phosphoric acid, H_3PO_4
   (c) 23 kg of calcium carbonate, $CaCO_3$
   (d) 78.452 g of aluminum sulfate, $Al_2(SO_4)_3$
   (e) 0.1250 mg of caffeine, $C_8H_{10}N_4O_2$

19. Determine the mass of each of the following:
(a) 2.345 mol LiCl
(b) 0.0872 mol acetylene, C_2H_2
(c) 3.3×10^{-2} mol Na_2CO_3
(d) 1.23×10^3 mol fructose, $C_6H_{12}O_6$
(e) 0.5758 mol $FeSO_4(H_2O)_7$
20. The approximate minimum daily dietary requirement of the amino acid leucine, $C_6H_{13}NO_2$, is 1.1 g. What is this requirement in moles?
21. Determine the mass in grams of each of the following:
(a) 0.600 mol of oxygen atoms
(b) 0.600 mol of oxygen molecules, O_2
(c) 0.600 mol of ozone molecules, O_3
22. A 55-kg woman has 7.5×10^{-3} mol of hemoglobin (molar mass = 64,456 g/mol) in her blood. How many hemoglobin molecules is this? What is this quantity in grams?
23. Determine the number of atoms and the mass of zirconium, silicon, and oxygen found in 0.3384 mol of zircon, $ZrSiO_4$, a semiprecious stone.
24. Determine which of the following contains the greatest mass of hydrogen: 1 mol of CH_4, 0.6 mol of C_6H_6, or 0.4 mol of C_3H_8.
25. Determine which of the following contains the greatest mass of aluminum: 122 g of $AlPO_4$, 266 g of Al_2Cl_6, or 225 g of Al_2S_3.
26. Diamond is one form of elemental carbon. An engagement ring contains a diamond weighing 1.25 carats (1 carat = 200 mg). How many atoms are present in the diamond?
27. The Cullinan diamond was the largest natural diamond ever found (January 25, 1905). It weighed 3104 carats (1 carat = 200 mg). How many carbon atoms were present in the stone?
28. One 55-gram serving of a particular cereal supplies 270 mg of sodium, 11% of the recommended daily allowance. How many moles and atoms of sodium are in the recommended daily allowance?
29. A certain nut crunch cereal contains 11.0 grams of sugar (sucrose, $C_{12}H_{22}O_{11}$) per serving size of 60.0 grams. How many servings of this cereal must be eaten to consume 0.0278 moles of sugar?
30. A tube of toothpaste contains 0.76 g of sodium monofluorophosphate (Na_2PO_3F) in 100 mL.
(a) What mass of fluorine atoms in mg was present?
(b) How many fluorine atoms were present?
31. Which of the following represents the least number of molecules?
(a) 20.0 g of H_2O (18.02 g/mol)
(b) 77.0 g of CH_4 (16.06 g/mol)
(c) 68.0 g of C_3H_6 (42.08 g/mol)
(d) 100.0 g of N_2O (44.02 g/mol)
(e) 84.0 g of HF (20.01 g/mol)

3.2 Determining Empirical and Molecular Formulas

32. What information is needed to determine the molecular formula of a compound from the empirical formula?
33. Calculate the following to four significant figures:
(a) the percent composition of ammonia, NH_3
(b) the percent composition of photographic fixer solution ("hypo"), $Na_2S_2O_3$
(c) the percent of calcium ion in $Ca_3(PO_4)_2$
34. Determine the following to four significant figures:
(a) the percent composition of hydrazoic acid, HN_3
(b) the percent composition of TNT, $C_6H_2(CH_3)(NO_2)_3$
(c) the percent of SO_4^{2-} in $Al_2(SO_4)_3$
35. Determine the percent ammonia, NH_3, in $Co(NH_3)_6Cl_3$, to three significant figures.
36. Determine the percent water in $CuSO_4 \cdot 5H_2O$ to three significant figures.

37. Determine the empirical formulas for compounds with the following percent compositions:
(a) 15.8% carbon and 84.2% sulfur
(b) 40.0% carbon, 6.7% hydrogen, and 53.3% oxygen

38. Determine the empirical formulas for compounds with the following percent compositions:
(a) 43.6% phosphorus and 56.4% oxygen
(b) 28.7% K, 1.5% H, 22.8% P, and 47.0% O

39. A compound of carbon and hydrogen contains 92.3% C and has a molar mass of 78.1 g/mol. What is its molecular formula?

40. Dichloroethane, a compound that is often used for dry cleaning, contains carbon, hydrogen, and chlorine. It has a molar mass of 99 g/mol. Analysis of a sample shows that it contains 24.3% carbon and 4.1% hydrogen. What is its molecular formula?

41. Determine the empirical and molecular formula for chrysotile asbestos. Chrysotile has the following percent composition: 28.03% Mg, 21.60% Si, 1.16% H, and 49.21% O. The molar mass for chrysotile is 520.8 g/mol.

42. Polymers are large molecules composed of simple units repeated many times. Thus, they often have relatively simple empirical formulas. Calculate the empirical formulas of the following polymers:
(a) Lucite (Plexiglas); 59.9% C, 8.06% H, 32.0% O
(b) Saran; 24.8% C, 2.0% H, 73.1% Cl
(c) polyethylene; 86% C, 14% H
(d) polystyrene; 92.3% C, 7.7% H
(e) Orlon; 67.9% C, 5.70% H, 26.4% N

43. A major textile dye manufacturer developed a new yellow dye. The dye has a percent composition of 75.95% C, 17.72% N, and 6.33% H by mass with a molar mass of about 240 g/mol. Determine the molecular formula of the dye.

3.3 Molarity

44. Explain what changes and what stays the same when 1.00 L of a solution of NaCl is diluted to 1.80 L.

45. What information is needed to calculate the molarity of a sulfuric acid solution?

46. A 200-mL sample and a 400-mL sample of a solution of salt have the same molarity. In what ways are the two samples identical? In what ways are these two samples different?

47. Determine the molarity for each of the following solutions:
(a) 0.444 mol of $CoCl_2$ in 0.654 L of solution
(b) 98.0 g of phosphoric acid, H_3PO_4, in 1.00 L of solution
(c) 0.2074 g of calcium hydroxide, $Ca(OH)_2$, in 40.00 mL of solution
(d) 10.5 kg of $Na_2SO_4{\cdot}10H_2O$ in 18.60 L of solution
(e) 7.0×10^{-3} mol of I_2 in 100.0 mL of solution
(f) 1.8×10^4 mg of HCl in 0.075 L of solution

48. Determine the molarity of each of the following solutions:
(a) 1.457 mol KCl in 1.500 L of solution
(b) 0.515 g of H_2SO_4 in 1.00 L of solution
(c) 20.54 g of $Al(NO_3)_3$ in 1575 mL of solution
(d) 2.76 kg of $CuSO_4{\cdot}5H_2O$ in 1.45 L of solution
(e) 0.005653 mol of Br_2 in 10.00 mL of solution
(f) 0.000889 g of glycine, $C_2H_5NO_2$, in 1.05 mL of solution

49. Consider this question: What is the mass of the solute in 0.500 L of 0.30 *M* glucose, $C_6H_{12}O_6$, used for intravenous injection?
(a) Outline the steps necessary to answer the question.
(b) Answer the question.

50. Consider this question: What is the mass of solute in 200.0 L of a 1.556-*M* solution of KBr?
(a) Outline the steps necessary to answer the question.
(b) Answer the question.

51. Calculate the number of moles and the mass of the solute in each of the following solutions:
(a) 2.00 L of 18.5 *M* H_2SO_4, concentrated sulfuric acid
(b) 100.0 mL of 3.8×10^{-6} *M* NaCN, the minimum lethal concentration of sodium cyanide in blood serum
(c) 5.50 L of 13.3 *M* H_2CO, the formaldehyde used to "fix" tissue samples
(d) 325 mL of 1.8×10^{-6} *M* $FeSO_4$, the minimum concentration of iron sulfate detectable by taste in drinking water
52. Calculate the number of moles and the mass of the solute in each of the following solutions:
(a) 325 mL of 8.23×10^{-5} *M* KI, a source of iodine in the diet
(b) 75.0 mL of 2.2×10^{-5} *M* H_2SO_4, a sample of acid rain
(c) 0.2500 L of 0.1135 *M* K_2CrO_4, an analytical reagent used in iron assays
(d) 10.5 L of 3.716 *M* $(NH_4)_2SO_4$, a liquid fertilizer
53. Consider this question: What is the molarity of $KMnO_4$ in a solution of 0.0908 g of $KMnO_4$ in 0.500 L of solution?
(a) Outline the steps necessary to answer the question.
(b) Answer the question.
54. Consider this question: What is the molarity of HCl if 35.23 mL of a solution of HCl contain 0.3366 g of HCl?
(a) Outline the steps necessary to answer the question.
(b) Answer the question.
55. Calculate the molarity of each of the following solutions:
(a) 0.195 g of cholesterol, $C_{27}H_{46}O$, in 0.100 L of serum, the average concentration of cholesterol in human serum
(b) 4.25 g of NH_3 in 0.500 L of solution, the concentration of NH_3 in household ammonia
(c) 1.49 kg of isopropyl alcohol, C_3H_7OH, in 2.50 L of solution, the concentration of isopropyl alcohol in rubbing alcohol
(d) 0.029 g of I_2 in 0.100 L of solution, the solubility of I_2 in water at 20 °C
56. Calculate the molarity of each of the following solutions:
(a) 293 g HCl in 666 mL of solution, a concentrated HCl solution
(b) 2.026 g $FeCl_3$ in 0.1250 L of a solution used as an unknown in general chemistry laboratories
(c) 0.001 mg Cd^{2+} in 0.100 L, the maximum permissible concentration of cadmium in drinking water
(d) 0.0079 g $C_7H_5SNO_3$ in one ounce (29.6 mL), the concentration of saccharin in a diet soft drink.
57. There is about 1.0 g of calcium, as Ca^{2+}, in 1.0 L of milk. What is the molarity of Ca^{2+} in milk?
58. What volume of a 1.00-*M* $Fe(NO_3)_3$ solution can be diluted to prepare 1.00 L of a solution with a concentration of 0.250 *M*?
59. If 0.1718 L of a 0.3556-*M* C_3H_7OH solution is diluted to a concentration of 0.1222 *M*, what is the volume of the resulting solution?
60. If 4.12 L of a 0.850 *M*-H_3PO_4 solution is be diluted to a volume of 10.00 L, what is the concentration of the resulting solution?
61. What volume of a 0.33-*M* $C_{12}H_{22}O_{11}$ solution can be diluted to prepare 25 mL of a solution with a concentration of 0.025 *M*?
62. What is the concentration of the NaCl solution that results when 0.150 L of a 0.556-*M* solution is allowed to evaporate until the volume is reduced to 0.105 L?
63. What is the molarity of the diluted solution when each of the following solutions is diluted to the given final volume?
(a) 1.00 L of a 0.250-*M* solution of $Fe(NO_3)_3$ is diluted to a final volume of 2.00 L
(b) 0.5000 L of a 0.1222-*M* solution of C_3H_7OH is diluted to a final volume of 1.250 L
(c) 2.35 L of a 0.350-*M* solution of H_3PO_4 is diluted to a final volume of 4.00 L
(d) 22.50 mL of a 0.025-*M* solution of $C_{12}H_{22}O_{11}$ is diluted to 100.0 mL
64. What is the final concentration of the solution produced when 225.5 mL of a 0.09988-*M* solution of Na_2CO_3 is allowed to evaporate until the solution volume is reduced to 45.00 mL?
65. A 2.00-L bottle of a solution of concentrated HCl was purchased for the general chemistry laboratory. The solution contained 868.8 g of HCl. What is the molarity of the solution?

66. An experiment in a general chemistry laboratory calls for a 2.00-*M* solution of HCl. How many mL of 11.9 *M* HCl would be required to make 250 mL of 2.00 *M* HCl?
67. What volume of a 0.20-*M* K_2SO_4 solution contains 57 g of K_2SO_4?
68. The US Environmental Protection Agency (EPA) places limits on the quantities of toxic substances that may be discharged into the sewer system. Limits have been established for a variety of substances, including hexavalent chromium, which is limited to 0.50 mg/L. If an industry is discharging hexavalent chromium as potassium dichromate ($K_2Cr_2O_7$), what is the maximum permissible molarity of that substance?

3.4 Other Units for Solution Concentrations

69. Consider this question: What mass of a concentrated solution of nitric acid (68.0% HNO_3 by mass) is needed to prepare 400.0 g of a 10.0% solution of HNO_3 by mass?
    (a) Outline the steps necessary to answer the question.
    (b) Answer the question.
70. What mass of a 4.00% NaOH solution by mass contains 15.0 g of NaOH?
71. What mass of solid NaOH (97.0% NaOH by mass) is required to prepare 1.00 L of a 10.0% solution of NaOH by mass? The density of the 10.0% solution is 1.109 g/mL.
72. What mass of HCl is contained in 45.0 mL of an aqueous HCl solution that has a density of 1.19 g cm^{-3} and contains 37.21% HCl by mass?
73. The hardness of water (hardness count) is usually expressed in parts per million (by mass) of $CaCO_3$, which is equivalent to milligrams of $CaCO_3$ per liter of water. What is the molar concentration of Ca^{2+} ions in a water sample with a hardness count of 175 mg $CaCO_3$/L?
74. The level of mercury in a stream was suspected to be above the minimum considered safe (1 part per billion by weight). An analysis indicated that the concentration was 0.68 parts per billion. Assume a density of 1.0 g/mL and calculate the molarity of mercury in the stream.
75. In Canada and the United Kingdom, devices that measure blood glucose levels provide a reading in millimoles per liter. If a measurement of 5.3 m*M* is observed, what is the concentration of glucose ($C_6H_{12}O_6$) in mg/dL?
76. A throat spray is 1.40% by mass phenol, C_6H_5OH, in water. If the solution has a density of 0.9956 g/mL, calculate the molarity of the solution.
77. Copper(I) iodide (CuI) is often added to table salt as a dietary source of iodine. How many moles of CuI are contained in 1.00 lb (454 g) of table salt containing 0.0100% CuI by mass?
78. A cough syrup contains 5.0% ethyl alcohol, C_2H_5OH, by mass. If the density of the solution is 0.9928 g/mL, determine the molarity of the alcohol in the cough syrup.
79. D5W is a solution used as an intravenous fluid. It is a 5.0% by mass solution of dextrose ($C_6H_{12}O_6$) in water. If the density of D5W is 1.029 g/mL, calculate the molarity of dextrose in the solution.
80. Find the molarity of a 40.0% by mass aqueous solution of sulfuric acid, H_2SO_4, for which the density is 1.3057 g/mL.

CHAPTER 4
Stoichiometry of Chemical Reactions

Figure 4.1 Many modern rocket fuels are solid mixtures of substances combined in carefully measured amounts and ignited to yield a thrust-generating chemical reaction. (credit: modification of work by NASA)

CHAPTER OUTLINE

4.1 Writing and Balancing Chemical Equations
4.2 Classifying Chemical Reactions
4.3 Reaction Stoichiometry
4.4 Reaction Yields
4.5 Quantitative Chemical Analysis

INTRODUCTION Solid-fuel rockets are a central feature in the world's space exploration programs, including the new Space Launch System being developed by the National Aeronautics and Space Administration (NASA) to replace the retired Space Shuttle fleet (Figure 4.1). The engines of these rockets rely on carefully prepared solid mixtures of chemicals combined in precisely measured amounts. Igniting the mixture initiates a vigorous chemical reaction that rapidly generates large amounts of gaseous products. These gases are ejected from the rocket engine through its nozzle, providing the thrust needed to propel heavy payloads into space. Both the nature of this chemical reaction and the relationships between the amounts of the substances being consumed and produced by the reaction are critically important considerations that determine the success of the technology. This chapter will describe how to symbolize chemical reactions using chemical equations, how to classify some common chemical reactions by identifying patterns of reactivity, and how to determine the quantitative relations between the amounts of substances involved in chemical reactions—that is, the reaction *stoichiometry*.

4.1 Writing and Balancing Chemical Equations

LEARNING OBJECTIVES

By the end of this section, you will be able to:

- Derive chemical equations from narrative descriptions of chemical reactions.
- Write and balance chemical equations in molecular, total ionic, and net ionic formats.

An earlier chapter of this text introduced the use of element symbols to represent individual atoms. When atoms gain or lose electrons to yield ions, or combine with other atoms to form molecules, their symbols are modified or combined to generate chemical formulas that appropriately represent these species. Extending this symbolism to represent both the identities and the relative quantities of substances undergoing a chemical (or physical) change involves writing and balancing a **chemical equation**. Consider as an example the reaction between one methane molecule (CH_4) and two diatomic oxygen molecules (O_2) to produce one carbon dioxide molecule (CO_2) and two water molecules (H_2O). The chemical equation representing this process is provided in the upper half of Figure 4.2, with space-filling molecular models shown in the lower half of the figure.

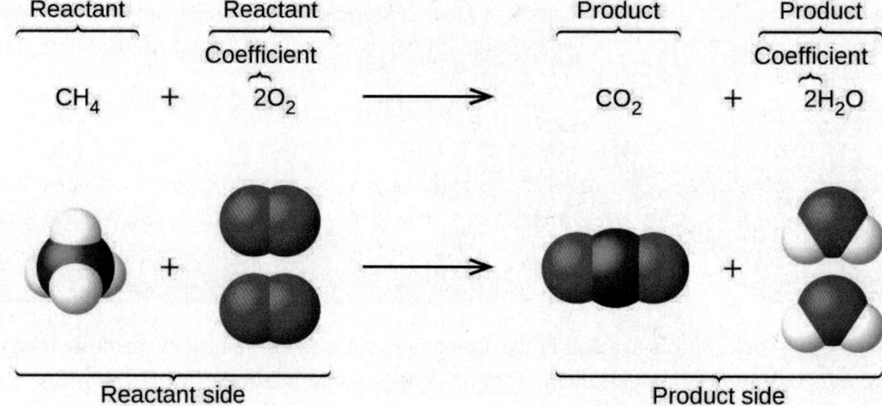

FIGURE 4.2 The reaction between methane and oxygen to yield carbon dioxide and water (shown at bottom) may be represented by a chemical equation using formulas (top).

This example illustrates the fundamental aspects of any chemical equation:

1. The substances undergoing reaction are called **reactants**, and their formulas are placed on the left side of the equation.
2. The substances generated by the reaction are called **products**, and their formulas are placed on the right side of the equation.
3. Plus signs (+) separate individual reactant and product formulas, and an arrow ($\longrightarrow$) separates the reactant and product (left and right) sides of the equation.
4. The relative numbers of reactant and product species are represented by **coefficients** (numbers placed immediately to the left of each formula). A coefficient of 1 is typically omitted.

It is common practice to use the smallest possible whole-number coefficients in a chemical equation, as is done in this example. Realize, however, that these coefficients represent the *relative* numbers of reactants and products, and, therefore, they may be correctly interpreted as ratios. Methane and oxygen react to yield carbon dioxide and water in a 1:2:1:2 ratio. This ratio is satisfied if the numbers of these molecules are, respectively, 1-2-1-2, or 2-4-2-4, or 3-6-3-6, and so on (Figure 4.3). Likewise, these coefficients may be interpreted with regard to any amount (number) unit, and so this equation may be correctly read in many ways, including:

- *One* methane molecule and *two* oxygen molecules react to yield *one* carbon dioxide molecule and *two* water molecules.
- *One dozen* methane molecules and *two dozen* oxygen molecules react to yield *one dozen* carbon dioxide molecules and *two dozen* water molecules.
- *One mole* of methane molecules and *2 moles* of oxygen molecules react to yield *1 mole* of carbon dioxide

molecules and 2 *moles* of water molecules.

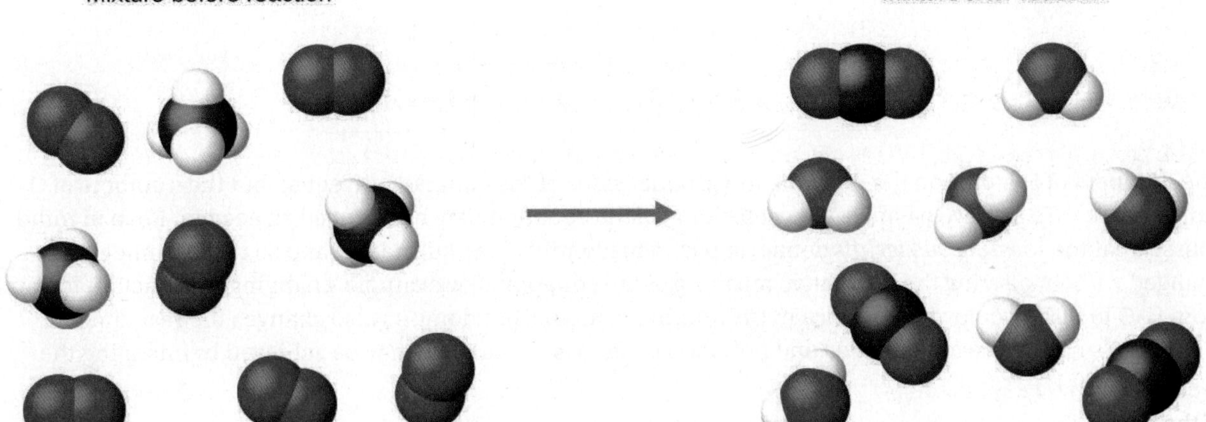

FIGURE 4.3 Regardless of the absolute numbers of molecules involved, the ratios between numbers of molecules of each species that react (the reactants) and molecules of each species that form (the products) are the same and are given by the chemical reaction equation.

Balancing Equations

The chemical equation described in section 4.1 is **balanced**, meaning that equal numbers of atoms for each element involved in the reaction are represented on the reactant and product sides. This is a requirement the equation must satisfy to be consistent with the law of conservation of matter. It may be confirmed by simply summing the numbers of atoms on either side of the arrow and comparing these sums to ensure they are equal. Note that the number of atoms for a given element is calculated by multiplying the coefficient of any formula containing that element by the element's subscript in the formula. If an element appears in more than one formula on a given side of the equation, the number of atoms represented in each must be computed and then added together. For example, both product species in the example reaction, CO_2 and H_2O, contain the element oxygen, and so the number of oxygen atoms on the product side of the equation is

$$\left(1\ \text{CO}_2\ \text{molecule} \times \frac{2\ \text{O atoms}}{\text{CO}_2\ \text{molecule}}\right) + \left(2\ \text{H}_2\text{O molecules} \times \frac{1\ \text{O atom}}{\text{H}_2\text{O molecule}}\right) = 4\ \text{O atoms}$$

The equation for the reaction between methane and oxygen to yield carbon dioxide and water is confirmed to be balanced per this approach, as shown here:

$$\text{CH}_4 + 2\text{O}_2 \longrightarrow \text{CO}_2 + 2\text{H}_2\text{O}$$

Element	Reactants	Products	Balanced?
C	1 × 1 = 1	1 × 1 = 1	1 = 1, yes
H	4 × 1 = 4	2 × 2 = 4	4 = 4, yes
O	2 × 2 = 4	(1 × 2) + (2 × 1) = 4	4 = 4, yes

A balanced chemical equation often may be derived from a qualitative description of some chemical reaction by a fairly simple approach known as balancing by inspection. Consider as an example the decomposition of water to yield molecular hydrogen and oxygen. This process is represented qualitatively by an *unbalanced* chemical equation:

$$\text{H}_2\text{O} \longrightarrow \text{H}_2 + \text{O}_2 \qquad \text{(unbalanced)}$$

Comparing the number of H and O atoms on either side of this equation confirms its imbalance:

Element	Reactants	Products	Balanced?
H	$1 \times 2 = 2$	$1 \times 2 = 2$	$2 = 2$, yes
O	$1 \times 1 = 1$	$1 \times 2 = 2$	$1 \neq 2$, no

The numbers of H atoms on the reactant and product sides of the equation are equal, but the numbers of O atoms are not. To achieve balance, the *coefficients* of the equation may be changed as needed. Keep in mind, of course, that the *formula subscripts* define, in part, the identity of the substance, and so these cannot be changed without altering the qualitative meaning of the equation. For example, changing the reactant formula from H_2O to H_2O_2 would yield balance in the number of atoms, but doing so also changes the reactant's identity (it's now hydrogen peroxide and not water). The O atom balance may be achieved by changing the coefficient for H_2O to 2.

$$\mathbf{2}H_2O \longrightarrow H_2 + O_2 \qquad \text{(unbalanced)}$$

Element	Reactants	Products	Balanced?
H	$\mathbf{2} \times 2 = 4$	$1 \times 2 = 2$	$4 \neq 2$, no
O	$2 \times 1 = 2$	$1 \times 2 = 2$	$2 = 2$, yes

The H atom balance was upset by this change, but it is easily reestablished by changing the coefficient for the H_2 product to 2.

$$2H_2O \longrightarrow \mathbf{2}H_2 + O_2 \qquad \text{(balanced)}$$

Element	Reactants	Products	Balanced?
H	$2 \times 2 = 4$	$\mathbf{2} \times 2 = 4$	$4 = 4$, yes
O	$2 \times 1 = 2$	$1 \times 2 = 2$	$2 = 2$, yes

These coefficients yield equal numbers of both H and O atoms on the reactant and product sides, and the balanced equation is, therefore:

$$2H_2O \longrightarrow 2H_2 + O_2$$

 **EXAMPLE 4.1**

Balancing Chemical Equations

Write a balanced equation for the reaction of molecular nitrogen (N_2) and oxygen (O_2) to form dinitrogen pentoxide.

Solution

First, write the unbalanced equation.

$$N_2 + O_2 \longrightarrow N_2O_5 \qquad \text{(unbalanced)}$$

Next, count the number of each type of atom present in the unbalanced equation.

Element	Reactants	Products	Balanced?
N	$1 \times 2 = 2$	$1 \times 2 = 2$	$2 = 2$, yes
O	$1 \times 2 = 2$	$1 \times 5 = 5$	$2 \neq 5$, no

Though nitrogen is balanced, changes in coefficients are needed to balance the number of oxygen atoms. To balance the number of oxygen atoms, a reasonable first attempt would be to change the coefficients for the O_2 and N_2O_5 to integers that will yield 10 O atoms (the least common multiple for the O atom subscripts in these two formulas).

$$N_2 + \mathbf{5}O_2 \longrightarrow \mathbf{2}N_2O_5 \qquad \text{(unbalanced)}$$

Element	Reactants	Products	Balanced?
N	$1 \times 2 = 2$	$\mathbf{2} \times 2 = 4$	$2 \neq 4$, no
O	$\mathbf{5} \times 2 = 10$	$\mathbf{2} \times 5 = 10$	$10 = 10$, yes

The N atom balance has been upset by this change; it is restored by changing the coefficient for the reactant N_2 to 2.

$$2N_2 + 5O_2 \longrightarrow 2N_2O_5$$

Element	Reactants	Products	Balanced?
N	$\mathbf{2} \times 2 = 4$	$2 \times 2 = 4$	$4 = 4$, yes
O	$5 \times 2 = 10$	$2 \times 5 = 10$	$10 = 10$, yes

The numbers of N and O atoms on either side of the equation are now equal, and so the equation is balanced.

Check Your Learning

Write a balanced equation for the decomposition of ammonium nitrate to form molecular nitrogen, molecular oxygen, and water. (Hint: Balance oxygen last, since it is present in more than one molecule on the right side of the equation.)

Answer:

$2NH_4NO_3 \longrightarrow 2N_2 + O_2 + 4H_2O$

It is sometimes convenient to use fractions instead of integers as intermediate coefficients in the process of balancing a chemical equation. When balance is achieved, all the equation's coefficients may then be multiplied by a whole number to convert the fractional coefficients to integers without upsetting the atom balance. For example, consider the reaction of ethane (C_2H_6) with oxygen to yield H_2O and CO_2, represented by the unbalanced equation:

$$C_2H_6 + O_2 \longrightarrow H_2O + CO_2 \qquad \text{(unbalanced)}$$

Following the usual inspection approach, one might first balance C and H atoms by changing the coefficients for the two product species, as shown:

$$C_2H_6 + O_2 \longrightarrow 3H_2O + 2CO_2 \qquad \text{(unbalanced)}$$

This results in seven O atoms on the product side of the equation, an odd number—no integer coefficient can be used with the O_2 reactant to yield an odd number, so a fractional coefficient, $\frac{7}{2}$, is used instead to yield a provisional balanced equation:

$$C_2H_6 + \frac{7}{2}O_2 \longrightarrow 3H_2O + 2CO_2$$

A conventional balanced equation with integer-only coefficients is derived by multiplying each coefficient by 2:

$$2C_2H_6 + 7O_2 \longrightarrow 6H_2O + 4CO_2$$

Finally with regard to balanced equations, recall that convention dictates use of the *smallest whole-number coefficients*. Although the equation for the reaction between molecular nitrogen and molecular hydrogen to produce ammonia is, indeed, balanced,

$$3N_2 + 9H_2 \longrightarrow 6NH_3$$

the coefficients are not the smallest possible integers representing the relative numbers of reactant and product molecules. Dividing each coefficient by the greatest common factor, 3, gives the preferred equation:

$$N_2 + 3H_2 \longrightarrow 2NH_3$$

LINK TO LEARNING

Use this interactive tutorial (http://openstax.org/l/16BalanceEq) for additional practice balancing equations.

Additional Information in Chemical Equations

The physical states of reactants and products in chemical equations very often are indicated with a parenthetical abbreviation following the formulas. Common abbreviations include *s* for solids, *l* for liquids, *g* for gases, and *aq* for substances dissolved in water (*aqueous solutions*, as introduced in the preceding chapter). These notations are illustrated in the example equation here:

$$2Na(s) + 2H_2O(l) \longrightarrow 2NaOH(aq) + H_2(g)$$

This equation represents the reaction that takes place when sodium metal is placed in water. The solid sodium reacts with liquid water to produce molecular hydrogen gas and the ionic compound sodium hydroxide (a solid in pure form, but readily dissolved in water).

Special conditions necessary for a reaction are sometimes designated by writing a word or symbol above or below the equation's arrow. For example, a reaction carried out by heating may be indicated by the uppercase Greek letter delta (Δ) over the arrow.

$$CaCO_3(s) \xrightarrow{\Delta} CaO(s) + CO_2(g)$$

Other examples of these special conditions will be encountered in more depth in later chapters.

Equations for Ionic Reactions

Given the abundance of water on earth, it stands to reason that a great many chemical reactions take place in aqueous media. When ions are involved in these reactions, the chemical equations may be written with various levels of detail appropriate to their intended use. To illustrate this, consider a reaction between ionic compounds taking place in an aqueous solution. When aqueous solutions of $CaCl_2$ and $AgNO_3$ are mixed, a reaction takes place producing aqueous $Ca(NO_3)_2$ and solid AgCl:

$$CaCl_2(aq) + 2AgNO_3(aq) \longrightarrow Ca(NO_3)_2(aq) + 2AgCl(s)$$

This balanced equation, derived in the usual fashion, is called a **molecular equation** because it doesn't explicitly represent the ionic species that are present in solution. When ionic compounds dissolve in water, they may *dissociate* into their constituent ions, which are subsequently dispersed homogenously throughout the resulting solution (a thorough discussion of this important process is provided in the chapter on solutions). Ionic compounds dissolved in water are, therefore, more realistically represented as dissociated ions, in this case:

$$\begin{aligned} CaCl_2(aq) &\longrightarrow Ca^{2+}(aq) + 2Cl^-(aq) \\ 2AgNO_3(aq) &\longrightarrow 2Ag^+(aq) + 2NO_3{}^-(aq) \\ Ca(NO_3)_2(aq) &\longrightarrow Ca^{2+}(aq) + 2NO_3{}^-(aq) \end{aligned}$$

Unlike these three ionic compounds, AgCl does not dissolve in water to a significant extent, as signified by its physical state notation, *s*.

Explicitly representing all dissolved ions results in a **complete ionic equation**. In this particular case, the formulas for the dissolved ionic compounds are replaced by formulas for their dissociated ions:

$$Ca^{2+}(aq) + 2Cl^-(aq) + 2Ag^+(aq) + 2NO_3{}^-(aq) \longrightarrow Ca^{2+}(aq) + 2NO_3{}^-(aq) + 2AgCl(s)$$

Examining this equation shows that two chemical species are present in identical form on both sides of the arrow, $Ca^{2+}(aq)$ and $NO_3{}^-(aq)$. These **spectator ions**—ions whose presence is required to maintain charge neutrality—are neither chemically nor physically changed by the process, and so they may be eliminated from the equation to yield a more succinct representation called a **net ionic equation**:

$$\begin{aligned} \cancel{Ca^{2+}(aq)} + 2Cl^-(aq) + 2Ag^+(aq) + \cancel{2NO_3{}^-(aq)} &\longrightarrow \cancel{Ca^{2+}(aq)} + \cancel{2NO_3{}^-(aq)} + 2AgCl(s) \\ 2Cl^-(aq) + 2Ag^+(aq) &\longrightarrow 2AgCl(s) \end{aligned}$$

Following the convention of using the smallest possible integers as coefficients, this equation is then written:

$$Cl^-(aq) + Ag^+(aq) \longrightarrow AgCl(s)$$

This net ionic equation indicates that solid silver chloride may be produced from dissolved chloride and silver(I) ions, regardless of the source of these ions. These molecular and complete ionic equations provide additional information, namely, the ionic compounds used as sources of Cl^- and Ag^+.

EXAMPLE 4.2

Ionic and Molecular Equations

When carbon dioxide is dissolved in an aqueous solution of sodium hydroxide, the mixture reacts to yield aqueous sodium carbonate and liquid water. Write balanced molecular, complete ionic, and net ionic equations for this process.

Solution

Begin by identifying formulas for the reactants and products and arranging them properly in chemical equation form:

$$CO_2(aq) + NaOH(aq) \longrightarrow Na_2CO_3(aq) + H_2O(l) \qquad \text{(unbalanced)}$$

Balance is achieved easily in this case by changing the coefficient for NaOH to 2, resulting in the molecular equation for this reaction:

$$CO_2(aq) + 2NaOH(aq) \longrightarrow Na_2CO_3(aq) + H_2O(l)$$

The two dissolved ionic compounds, NaOH and Na_2CO_3, can be represented as dissociated ions to yield the complete ionic equation:

$$CO_2(aq) + 2Na^+(aq) + 2OH^-(aq) \longrightarrow 2Na^+(aq) + CO_3{}^{2-}(aq) + H_2O(l)$$

Finally, identify the spectator ion(s), in this case $Na^+(aq)$, and remove it from each side of the equation to generate the net ionic equation:

$$CO_2(aq) + \cancel{2Na^+(aq)} + 2OH^-(aq) \longrightarrow \cancel{2Na^+(aq)} + {CO_3}^{2-}(aq) + H_2O(l)$$
$$CO_2(aq) + 2OH^-(aq) \longrightarrow {CO_3}^{2-}(aq) + H_2O(l)$$

Check Your Learning

Diatomic chlorine and sodium hydroxide (lye) are commodity chemicals produced in large quantities, along with diatomic hydrogen, via the electrolysis of brine, according to the following unbalanced equation:

$$NaCl(aq) + H_2O(l) \xrightarrow{\text{electricity}} NaOH(aq) + H_2(g) + Cl_2(g)$$

Write balanced molecular, complete ionic, and net ionic equations for this process.

Answer:

$2NaCl(aq) + 2H_2O(l) \longrightarrow 2NaOH(aq) + H_2(g) + Cl_2(g)$ (molecular)
$2Na^+(aq) + 2Cl^-(aq) + 2H_2O(l) \longrightarrow 2Na^+(aq) + 2OH^-(aq) + H_2(g) + Cl_2(g)$ (complete ionic)
$2Cl^-(aq) + 2H_2O(l) \longrightarrow 2OH^-(aq) + H_2(g) + Cl_2(g)$ (net ionic)

4.2 Classifying Chemical Reactions

LEARNING OBJECTIVES

By the end of this section, you will be able to:

- Define three common types of chemical reactions (precipitation, acid-base, and oxidation-reduction)
- Classify chemical reactions as one of these three types given appropriate descriptions or chemical equations
- Identify common acids and bases
- Predict the solubility of common inorganic compounds by using solubility rules
- Compute the oxidation states for elements in compounds

Humans interact with one another in various and complex ways, and we classify these interactions according to common patterns of behavior. When two humans exchange information, we say they are communicating. When they exchange blows with their fists or feet, we say they are fighting. Faced with a wide range of varied interactions between chemical substances, scientists have likewise found it convenient (or even necessary) to classify chemical interactions by identifying common patterns of reactivity. This module will provide an introduction to three of the most prevalent types of chemical reactions: precipitation, acid-base, and oxidation-reduction.

Precipitation Reactions and Solubility Rules

A **precipitation reaction** is one in which dissolved substances react to form one (or more) solid products. Many reactions of this type involve the exchange of ions between ionic compounds in aqueous solution and are sometimes referred to as *double displacement, double replacement,* or *metathesis* reactions. These reactions are common in nature and are responsible for the formation of coral reefs in ocean waters and kidney stones in animals. They are used widely in industry for production of a number of commodity and specialty chemicals. Precipitation reactions also play a central role in many chemical analysis techniques, including spot tests used to identify metal ions and *gravimetric methods* for determining the composition of matter (see the last module of this chapter).

The extent to which a substance may be dissolved in water, or any solvent, is quantitatively expressed as its **solubility**, defined as the maximum concentration of a substance that can be achieved under specified conditions. Substances with relatively large solubilities are said to be **soluble**. A substance will **precipitate** when solution conditions are such that its concentration exceeds its solubility. Substances with relatively low solubilities are said to be **insoluble**, and these are the substances that readily precipitate from solution. More information on these important concepts is provided in a later chapter on solutions. For purposes of predicting the identities of solids formed by precipitation reactions, one may simply refer to patterns of solubility that have been observed for many ionic compounds (Table 4.1).

	contain these ions	exceptions
Soluble Ionic Compounds	NH_4^+ group I cations: Li^+ Na^+ K^+ Rb^+ Cs^+	none
	Cl^- Br^- I^-	compounds with Ag^+, Hg_2^{2+}, and Pb^{2+}
	F^-	compounds with group 2 metal cations, Pb^{2+}, and Fe^{3+}
	$C_2H_3O_2^-$ HCO_3^- NO_3^- ClO_3^-	none
	SO_4^{2-}	compounds with Ag^+, Ba^{2+}, Ca^{2+}, Hg_2^{2+}, Pb^{2+} and Sr^{2+}
	contain these ions	exceptions
Insoluble Ionic Compounds	CO_3^{2-} CrO_4^{2-} PO_4^{3-} S^{2-}	compounds with group 1 cations and NH_4^+
	OH^-	compounds with group 1 cations and Ba^{2+}

TABLE 4.1

A vivid example of precipitation is observed when solutions of potassium iodide and lead nitrate are mixed, resulting in the formation of solid lead iodide:

$$2KI(aq) + Pb(NO_3)_2(aq) \longrightarrow PbI_2(s) + 2KNO_3(aq)$$

This observation is consistent with the solubility guidelines: The only insoluble compound among all those involved is lead iodide, one of the exceptions to the general solubility of iodide salts.

The net ionic equation representing this reaction is:

$$Pb^{2+}(aq) + 2I^-(aq) \longrightarrow PbI_2(s)$$

Lead iodide is a bright yellow solid that was formerly used as an artist's pigment known as iodine yellow (Figure 4.4). The properties of pure PbI_2 crystals make them useful for fabrication of X-ray and gamma ray detectors.

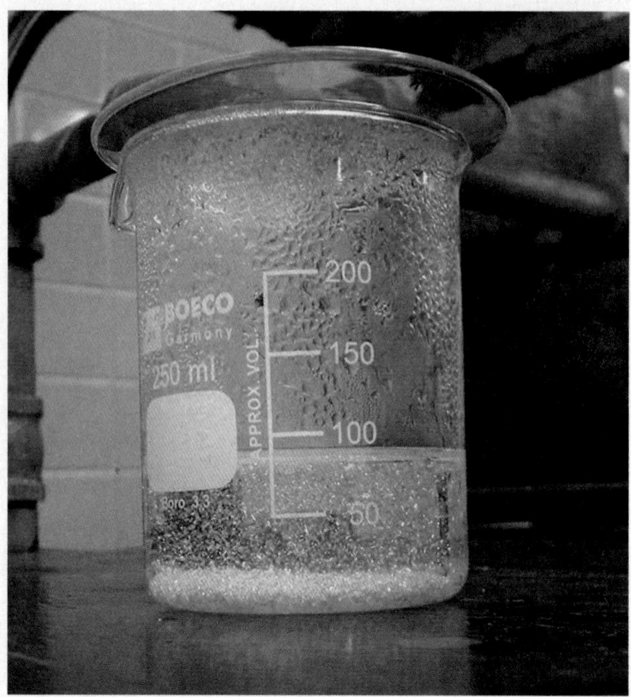

FIGURE 4.4 A precipitate of PbI_2 forms when solutions containing Pb^{2+} and I^- are mixed. (credit: Der Kreole/ Wikimedia Commons)

The solubility guidelines in Table 4.1 may be used to predict whether a precipitation reaction will occur when solutions of soluble ionic compounds are mixed together. One merely needs to identify all the ions present in the solution and then consider if possible cation/anion pairing could result in an insoluble compound. For example, mixing solutions of silver nitrate and sodium chloride will yield a solution containing Ag^+, $NO_3{}^-$, Na^+, and Cl^- ions. Aside from the two ionic compounds originally present in the solutions, $AgNO_3$ and NaCl, two additional ionic compounds may be derived from this collection of ions: $NaNO_3$ and AgCl. The solubility guidelines indicate all nitrate salts are soluble but that AgCl is one of insoluble. A precipitation reaction, therefore, is predicted to occur, as described by the following equations:

$$NaCl(aq) + AgNO_3(aq) \longrightarrow AgCl(s) + NaNO_3(aq) \qquad \text{(molecular)}$$

$$Ag^+(aq) + Cl^-(aq) \longrightarrow AgCl(s) \qquad \text{(net ionic)}$$

EXAMPLE 4.3

Predicting Precipitation Reactions

Predict the result of mixing reasonably concentrated solutions of the following ionic compounds. If precipitation is expected, write a balanced net ionic equation for the reaction.

(a) potassium sulfate and barium nitrate

(b) lithium chloride and silver acetate

(c) lead nitrate and ammonium carbonate

Solution

(a) The two possible products for this combination are KNO_3 and $BaSO_4$. The solubility guidelines indicate

$BaSO_4$ is insoluble, and so a precipitation reaction is expected. The net ionic equation for this reaction, derived in the manner detailed in the previous module, is

$$Ba^{2+}(aq) + SO_4^{2-}(aq) \longrightarrow BaSO_4(s)$$

(b) The two possible products for this combination are $LiC_2H_3O_2$ and AgCl. The solubility guidelines indicate AgCl is insoluble, and so a precipitation reaction is expected. The net ionic equation for this reaction, derived in the manner detailed in the previous module, is

$$Ag^+(aq) + Cl^-(aq) \longrightarrow AgCl(s)$$

(c) The two possible products for this combination are $PbCO_3$ and NH_4NO_3. The solubility guidelines indicate $PbCO_3$ is insoluble, and so a precipitation reaction is expected. The net ionic equation for this reaction, derived in the manner detailed in the previous module, is

$$Pb^{2+}(aq) + CO_3^{2-}(aq) \longrightarrow PbCO_3(s)$$

Check Your Learning

Which solution could be used to precipitate the barium ion, Ba^{2+}, in a water sample: sodium chloride, sodium hydroxide, or sodium sulfate? What is the formula for the expected precipitate?

Answer:

sodium sulfate, $BaSO_4$

Acid-Base Reactions

An **acid-base reaction** is one in which a hydrogen ion, H^+, is transferred from one chemical species to another. Such reactions are of central importance to numerous natural and technological processes, ranging from the chemical transformations that take place within cells and the lakes and oceans, to the industrial-scale production of fertilizers, pharmaceuticals, and other substances essential to society. The subject of acid-base chemistry, therefore, is worthy of thorough discussion, and a full chapter is devoted to this topic later in the text.

For purposes of this brief introduction, we will consider only the more common types of acid-base reactions that take place in aqueous solutions. In this context, an **acid** is a substance that will dissolve in water to yield hydronium ions, H_3O^+. As an example, consider the equation shown here:

$$HCl(aq) + H_2O(aq) \longrightarrow Cl^-(aq) + H_3O^+(aq)$$

The process represented by this equation confirms that hydrogen chloride is an acid. When dissolved in water, H_3O^+ ions are produced by a chemical reaction in which H^+ ions are transferred from HCl molecules to H_2O molecules (Figure 4.5).

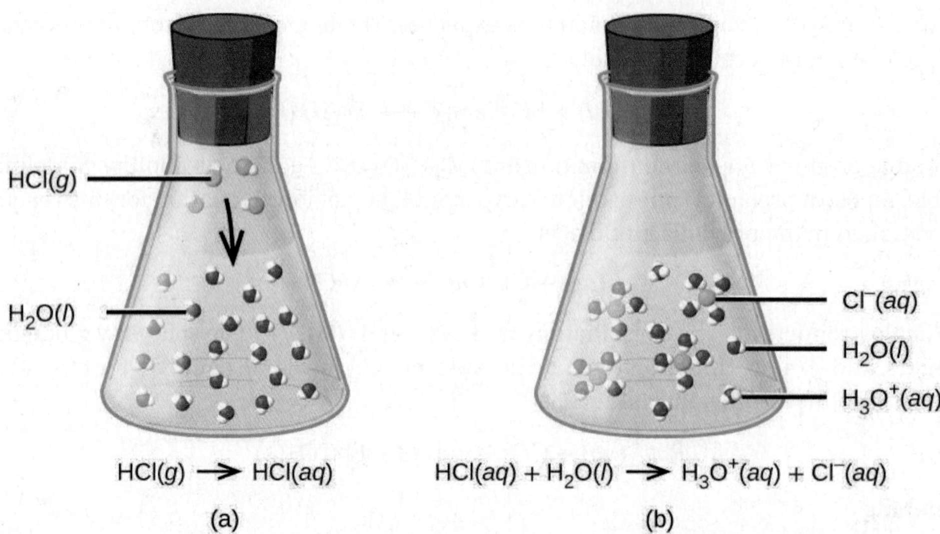

FIGURE 4.5 When hydrogen chloride gas dissolves in water, (a) it reacts as an acid, transferring protons to water molecules to yield (b) hydronium ions (and solvated chloride ions).

The nature of HCl is such that its reaction with water as just described is essentially 100% efficient: Virtually every HCl molecule that dissolves in water will undergo this reaction. Acids that completely react in this fashion are called **strong acids**, and HCl is one among just a handful of common acid compounds that are classified as strong (Table 4.2). A far greater number of compounds behave as **weak acids** and only partially react with water, leaving a large majority of dissolved molecules in their original form and generating a relatively small amount of hydronium ions. Weak acids are commonly encountered in nature, being the substances partly responsible for the tangy taste of citrus fruits, the stinging sensation of insect bites, and the unpleasant smells associated with body odor. A familiar example of a weak acid is acetic acid, the main ingredient in food vinegars:

$$CH_3CO_2H(aq) + H_2O(l) \rightleftharpoons CH_3CO_2^-(aq) + H_3O^+(aq)$$

When dissolved in water under typical conditions, only about 1% of acetic acid molecules are present in the ionized form, $CH_3CO_2^-$ (Figure 4.6). (The use of a double-arrow in the equation above denotes the partial reaction aspect of this process, a concept addressed fully in the chapters on chemical equilibrium.)

FIGURE 4.6 (a) Fruits such as oranges, lemons, and grapefruit contain the weak acid citric acid. (b) Vinegars contain the weak acid acetic acid. (credit a: modification of work by Scott Bauer; credit b: modification of work by Brücke-Osteuropa/Wikimedia Commons)

Common Strong Acids

Compound Formula	Name in Aqueous Solution
HBr	hydrobromic acid
HCl	hydrochloric acid
HI	hydroiodic acid
HNO_3	nitric acid
$HClO_4$	perchloric acid
H_2SO_4	sulfuric acid

TABLE 4.2

A **base** is a substance that will dissolve in water to yield hydroxide ions, OH^-. The most common bases are ionic compounds composed of alkali or alkaline earth metal cations (groups 1 and 2) combined with the hydroxide ion—for example, NaOH and $Ca(OH)_2$. Unlike the acid compounds discussed previously, these compounds do not react chemically with water; instead they dissolve and dissociate, releasing hydroxide ions directly into the solution. For example, KOH and $Ba(OH)_2$ dissolve in water and dissociate completely to produce cations (K^+ and Ba^{2+}, respectively) and hydroxide ions, OH^-. These bases, along with other hydroxides that completely dissociate in water, are considered **strong bases**.

Consider as an example the dissolution of lye (sodium hydroxide) in water:

$$\text{NaOH}(s) \longrightarrow \text{Na}^+(aq) + \text{OH}^-(aq)$$

This equation confirms that sodium hydroxide is a base. When dissolved in water, NaOH dissociates to yield Na^+ and OH^- ions. This is also true for any other ionic compound containing hydroxide ions. Since the dissociation process is essentially complete when ionic compounds dissolve in water under typical conditions, NaOH and other ionic hydroxides are all classified as strong bases.

Unlike ionic hydroxides, some compounds produce hydroxide ions when dissolved by chemically reacting with water molecules. In all cases, these compounds react only partially and so are classified as **weak bases**. These types of compounds are also abundant in nature and important commodities in various technologies. For example, global production of the weak base ammonia is typically well over 100 metric tons annually, being widely used as an agricultural fertilizer, a raw material for chemical synthesis of other compounds, and an active ingredient in household cleaners (Figure 4.7). When dissolved in water, ammonia reacts partially to yield hydroxide ions, as shown here:

$$\text{NH}_3(aq) + \text{H}_2\text{O}(l) \rightleftharpoons \text{NH}_4{}^+(aq) + \text{OH}^-(aq)$$

This is, by definition, an acid-base reaction, in this case involving the transfer of H^+ ions from water molecules to ammonia molecules. Under typical conditions, only about 1% of the dissolved ammonia is present as $NH_4{}^+$ ions.

(a)

(b)

FIGURE 4.7 Ammonia is a weak base used in a variety of applications. (a) Pure ammonia is commonly applied as an agricultural fertilizer. (b) Dilute solutions of ammonia are effective household cleansers. (credit a: modification of work by National Resources Conservation Service; credit b: modification of work by pat00139)

A **neutralization reaction** is a specific type of acid-base reaction in which the reactants are an acid and a base (but not water), and the products are often a **salt** and water

$$\text{acid} + \text{base} \longrightarrow \text{salt} + \text{water}$$

To illustrate a neutralization reaction, consider what happens when a typical antacid such as milk of magnesia (an aqueous suspension of solid $Mg(OH)_2$) is ingested to ease symptoms associated with excess stomach acid (HCl):

$$\text{Mg(OH)}_2(s) + 2\text{HCl}(aq) \longrightarrow \text{MgCl}_2(aq) + 2\text{H}_2\text{O}(l).$$

Note that in addition to water, this reaction produces a salt, magnesium chloride.

EXAMPLE 4.4

Writing Equations for Acid-Base Reactions

Write balanced chemical equations for the acid-base reactions described here:

(a) the weak acid hydrogen hypochlorite reacts with water

(b) a solution of barium hydroxide is neutralized with a solution of nitric acid

Solution

(a) The two reactants are provided, HOCl and H_2O. Since the substance is reported to be an acid, its reaction with water will involve the transfer of H^+ from HOCl to H_2O to generate hydronium ions, H_3O^+ and hypochlorite ions, OCl^-.

$$\text{HOCl}(aq) + \text{H}_2\text{O}(l) \rightleftharpoons \text{OCl}^-(aq) + \text{H}_3\text{O}^+(aq)$$

A double-arrow is appropriate in this equation because it indicates the HOCl is a weak acid that has not reacted completely.

(b) The two reactants are provided, $Ba(OH)_2$ and HNO_3. Since this is a neutralization reaction, the two products will be water and a salt composed of the cation of the ionic hydroxide (Ba^{2+}) and the anion generated when the acid transfers its hydrogen ion (NO_3^-).

$$\text{Ba(OH)}_2(aq) + 2\text{HNO}_3(aq) \longrightarrow \text{Ba(NO}_3)_2(aq) + 2\text{H}_2\text{O}(l)$$

Check Your Learning

Write the net ionic equation representing the neutralization of any strong acid with an ionic hydroxide. (Hint: Consider the ions produced when a strong acid is dissolved in water.)

Answer:

$$\text{H}_3\text{O}^+(aq) + \text{OH}^-(aq) \longrightarrow 2\text{H}_2\text{O}(l)$$

Chemistry in Everyday Life

Stomach Antacids

Our stomachs contain a solution of roughly 0.03 *M* HCl, which helps us digest the food we eat. The burning sensation associated with heartburn is a result of the acid of the stomach leaking through the muscular valve at the top of the stomach into the lower reaches of the esophagus. The lining of the esophagus is not protected from the corrosive effects of stomach acid the way the lining of the stomach is, and the results can be very painful. When we have heartburn, it feels better if we reduce the excess acid in the esophagus by taking an antacid. As you may have guessed, antacids are bases. One of the most common antacids is calcium carbonate, $CaCO_3$. The reaction,

$$CaCO_3(s) + 2HCl(aq) \rightleftharpoons CaCl_2(aq) + H_2O(l) + CO_2(g)$$

not only neutralizes stomach acid, it also produces $CO_2(g)$, which may result in a satisfying belch.

Milk of Magnesia is a suspension of the sparingly soluble base magnesium hydroxide, $Mg(OH)_2$. It works according to the reaction:

$$Mg(OH)_2(s) \rightleftharpoons Mg^{2+}(aq) + 2OH^-(aq)$$

The hydroxide ions generated in this equilibrium then go on to react with the hydronium ions from the stomach acid, so that:

$$H_3O^+ + OH^- \rightleftharpoons 2H_2O(l)$$

This reaction does not produce carbon dioxide, but magnesium-containing antacids can have a laxative effect. Several antacids have aluminum hydroxide, $Al(OH)_3$, as an active ingredient. The aluminum hydroxide tends to cause constipation, and some antacids use aluminum hydroxide in concert with magnesium hydroxide to balance the side effects of the two substances.

Chemistry in Everyday Life

Culinary Aspects of Chemistry

Examples of acid-base chemistry are abundant in the culinary world. One example is the use of baking soda, or sodium bicarbonate in baking. $NaHCO_3$ is a base. When it reacts with an acid such as lemon juice, buttermilk, or sour cream in a batter, bubbles of carbon dioxide gas are formed from decomposition of the resulting carbonic acid, and the batter "rises." Baking powder is a combination of sodium bicarbonate, and one or more acid salts that react when the two chemicals come in contact with water in the batter.

Many people like to put lemon juice or vinegar, both of which are acids, on cooked fish (Figure 4.8). It turns out that fish have volatile amines (bases) in their systems, which are neutralized by the acids to yield involatile ammonium salts. This reduces the odor of the fish, and also adds a "sour" taste that we seem to enjoy.

$$CH_3COOH + NH_2CH_2CH_2CH_2CH_2NH_2 \longrightarrow CH_3COO^- + NH_3^+CH_2CH_2CH_2CH_2NH_2$$

Acetic acid + Putrescine ⟶ Acetate ion + Putrescinium ion

FIGURE 4.8 A neutralization reaction takes place between citric acid in lemons or acetic acid in vinegar, and the bases in the flesh of fish.

Pickling is a method used to preserve vegetables using a naturally produced acidic environment. The vegetable, such as a cucumber, is placed in a sealed jar submerged in a brine solution. The brine solution favors the growth of beneficial bacteria and suppresses the growth of harmful bacteria. The beneficial bacteria feed on starches in the cucumber and produce lactic acid as a waste product in a process called fermentation. The lactic acid eventually increases the acidity of the brine to a level that kills any harmful bacteria, which require a basic environment. Without the harmful bacteria consuming the cucumbers they are able to last much longer than if they were unprotected. A byproduct of the pickling process changes the flavor of the vegetables with the acid making them taste sour.

LINK TO LEARNING

Explore the microscopic view (http://openstax.org/l/16AcidsBases) of strong and weak acids and bases.

Oxidation-Reduction Reactions

Earth's atmosphere contains about 20% molecular oxygen, O_2, a chemically reactive gas that plays an essential role in the metabolism of aerobic organisms and in many environmental processes that shape the world. The term **oxidation** was originally used to describe chemical reactions involving O_2, but its meaning has evolved to refer to a broad and important reaction class known as *oxidation-reduction (redox) reactions*. A few examples of such reactions will be used to develop a clear picture of this classification.

Some redox reactions involve the transfer of electrons between reactant species to yield ionic products, such as the reaction between sodium and chlorine to yield sodium chloride:

$$2Na(s) + Cl_2(g) \longrightarrow 2NaCl(s)$$

It is helpful to view the process with regard to each individual reactant, that is, to represent the fate of each reactant in the form of an equation called a **half-reaction**:

$$2Na(s) \longrightarrow 2Na^+(s) + 2e^-$$
$$Cl_2(g) + 2e^- \longrightarrow 2Cl^-(s)$$

These equations show that Na atoms *lose electrons* while Cl atoms (in the Cl_2 molecule) *gain electrons*, the "*s*" subscripts for the resulting ions signifying they are present in the form of a solid ionic compound. For redox reactions of this sort, the loss and gain of electrons define the complementary processes that occur:

$$\textbf{oxidation} = \text{loss of electrons}$$
$$\textbf{reduction} = \text{gain of electrons}$$

In this reaction, then, sodium is *oxidized* and chlorine undergoes **reduction**. Viewed from a more active perspective, sodium functions as a **reducing agent (reductant)**, since it provides electrons to (or reduces) chlorine. Likewise, chlorine functions as an **oxidizing agent (oxidant)**, as it effectively removes electrons from (oxidizes) sodium.

$$\textbf{reducing agent} = \text{species that is oxidized}$$
$$\textbf{oxidizing agent} = \text{species that is reduced}$$

Some redox processes, however, do not involve the transfer of electrons. Consider, for example, a reaction similar to the one yielding NaCl:

$$H_2(g) + Cl_2(g) \longrightarrow 2HCl(g)$$

The product of this reaction is a covalent compound, so transfer of electrons in the explicit sense is not involved. To clarify the similarity of this reaction to the previous one and permit an unambiguous definition of redox reactions, a property called *oxidation number* has been defined. The **oxidation number** (or **oxidation state**) of an element in a compound is the charge its atoms would possess *if the compound were ionic*. The following guidelines are used to assign oxidation numbers to each element in a molecule or ion.

1. The oxidation number of an atom in an elemental substance is zero.
2. The oxidation number of a monatomic ion is equal to the ion's charge.
3. Oxidation numbers for common nonmetals are usually assigned as follows:
   - Hydrogen: +1 when combined with nonmetals, −1 when combined with metals
   - Oxygen: −2 in most compounds, sometimes −1 (so-called peroxides, $O_2^{\,2-}$), very rarely $-\frac{1}{2}$ (so-called superoxides, $O_2^{\,-}$), positive values when combined with F (values vary)
   - Halogens: −1 for F always, −1 for other halogens except when combined with oxygen or other halogens (positive oxidation numbers in these cases, varying values)
4. The sum of oxidation numbers for all atoms in a molecule or polyatomic ion equals the charge on the molecule or ion.

Note: The proper convention for reporting charge is to write the number first, followed by the sign (e.g., 2+), while oxidation number is written with the reversed sequence, sign followed by number (e.g., +2). This convention aims to emphasize the distinction between these two related properties.

 **EXAMPLE 4.5**

Assigning Oxidation Numbers

Follow the guidelines in this section of the text to assign oxidation numbers to all the elements in the following species:

(a) H_2S

(b) $SO_3^{\,2-}$

(c) Na_2SO_4

Solution

(a) According to guideline 3, the oxidation number for H is +1.

Using this oxidation number and the compound's formula, guideline 4 may then be used to calculate the oxidation number for sulfur:

$$\text{charge on } H_2S = 0 = (2 \times +1) + (1 \times x)$$
$$x = 0 - (2 \times +1) = -2$$

(b) Guideline 3 suggests the oxidation number for oxygen is −2.

Using this oxidation number and the ion's formula, guideline 4 may then be used to calculate the oxidation number for sulfur:

$$\text{charge on } SO_3{}^{2-} = -2 = (3 \times -2) + (1 \times x)$$
$$x = -2 - (3 \times -2) = +4$$

(c) For ionic compounds, it's convenient to assign oxidation numbers for the cation and anion separately.

According to guideline 2, the oxidation number for sodium is +1.

Assuming the usual oxidation number for oxygen (−2 per guideline 3), the oxidation number for sulfur is calculated as directed by guideline 4:

$$\text{charge on } SO_4{}^{2-} = -2 = (4 \times -2) + (1 \times x)$$
$$x = -2 - (4 \times -2) = +6$$

Check Your Learning

Assign oxidation states to the elements whose atoms are underlined in each of the following compounds or ions:

(a) $K\underline{N}O_3$

(b) $\underline{Al}H_3$

(c) $\underline{N}H_4{}^+$

(d) $H_2\underline{P}O_4{}^-$

Answer:

(a) N, +5; (b) Al, +3; (c) N, −3; (d) P, +5

Using the oxidation number concept, an all-inclusive definition of redox reaction has been established. **Oxidation-reduction (redox) reactions** are those in which one or more elements involved undergo a change in oxidation number. (While the vast majority of redox reactions involve changes in oxidation number for two or more elements, a few interesting exceptions to this rule do exist Example 4.6.) Definitions for the complementary processes of this reaction class are correspondingly revised as shown here:

$$\begin{aligned} \textbf{oxidation} &= \text{increase in oxidation number} \\ \textbf{reduction} &= \text{decrease in oxidation number} \end{aligned}$$

Returning to the reactions used to introduce this topic, they may now both be identified as redox processes. In the reaction between sodium and chlorine to yield sodium chloride, sodium is oxidized (its oxidation number increases from 0 in Na to +1 in NaCl) and chlorine is reduced (its oxidation number decreases from 0 in Cl_2 to −1 in NaCl). In the reaction between molecular hydrogen and chlorine, hydrogen is oxidized (its oxidation number increases from 0 in H_2 to +1 in HCl) and chlorine is reduced (its oxidation number decreases from 0 in

Cl_2 to −1 in HCl).

Several subclasses of redox reactions are recognized, including **combustion reactions** in which the reductant (also called a *fuel*) and oxidant (often, but not necessarily, molecular oxygen) react vigorously and produce significant amounts of heat, and often light, in the form of a flame. Solid rocket-fuel reactions such as the one depicted in Figure 4.1 are combustion processes. A typical propellant reaction in which solid aluminum is oxidized by ammonium perchlorate is represented by this equation:

$$10\text{Al}(s) + 6\text{NH}_4\text{ClO}_4(s) \longrightarrow 4\text{Al}_2\text{O}_3(s) + 2\text{AlCl}_3(s) + 12\text{H}_2\text{O}(g) + 3\text{N}_2(g)$$

LINK TO LEARNING

Watch a brief video (http://openstax.org/l/16hybridrocket) showing the test firing of a small-scale, prototype, hybrid rocket engine planned for use in the new Space Launch System being developed by NASA. The first engines firing at
3 s (green flame) use a liquid fuel/oxidant mixture, and the second, more powerful engines firing at 4 s (yellow flame) use a solid mixture.

Single-displacement (replacement) reactions are redox reactions in which an ion in solution is displaced (or replaced) via the oxidation of a metallic element. One common example of this type of reaction is the acid oxidation of certain metals:

$$\text{Zn}(s) + 2\text{HCl}(aq) \longrightarrow \text{ZnCl}_2(aq) + \text{H}_2(g)$$

Metallic elements may also be oxidized by solutions of other metal salts; for example:

$$\text{Cu}(s) + 2\text{AgNO}_3(aq) \longrightarrow \text{Cu}(\text{NO}_3)_2(aq) + 2\text{Ag}(s)$$

This reaction may be observed by placing copper wire in a solution containing a dissolved silver salt. Silver ions in solution are reduced to elemental silver at the surface of the copper wire, and the resulting Cu^{2+} ions dissolve in the solution to yield a characteristic blue color (Figure 4.9).

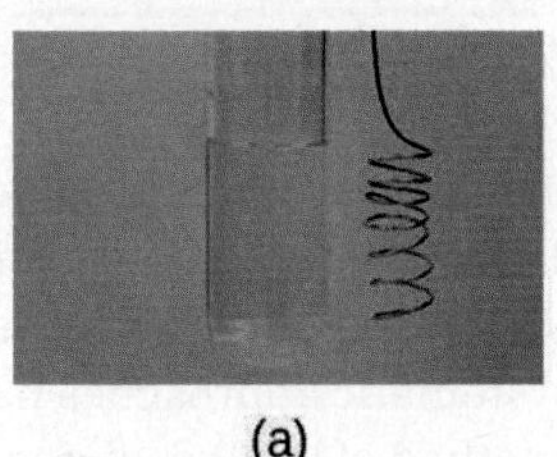

(a)

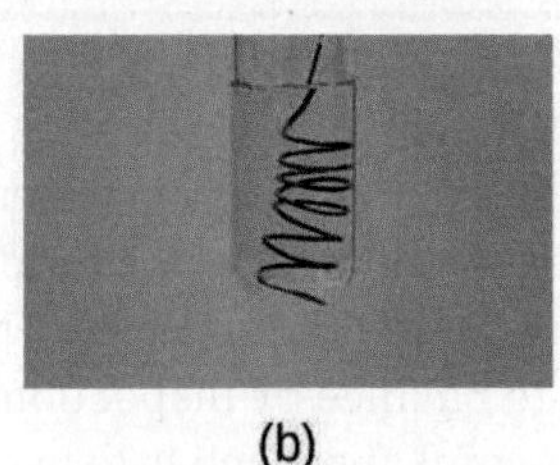

(b)

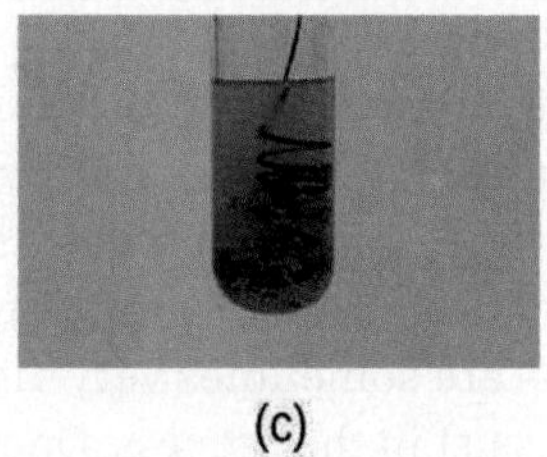

(c)

FIGURE 4.9 (a) A copper wire is shown next to a solution containing silver(I) ions. (b) Displacement of dissolved silver ions by copper ions results in (c) accumulation of gray-colored silver metal on the wire and development of a blue color in the solution, due to dissolved copper ions. (credit: modification of work by Mark Ott)

EXAMPLE 4.6

Describing Redox Reactions

Identify which equations represent redox reactions, providing a name for the reaction if appropriate. For those reactions identified as redox, name the oxidant and reductant.

(a) $\text{ZnCO}_3(s) \longrightarrow \text{ZnO}(s) + \text{CO}_2(g)$

(b) $2\text{Ga}(l) + 3\text{Br}_2(l) \longrightarrow 2\text{GaBr}_3(s)$

(c) $2\text{H}_2\text{O}_2(aq) \longrightarrow 2\text{H}_2\text{O}(l) + \text{O}_2(g)$

(d) $\text{BaCl}_2(aq) + \text{K}_2\text{SO}_4(aq) \longrightarrow \text{BaSO}_4(s) + 2\text{KCl}(aq)$

(e) $\text{C}_2\text{H}_4(g) + 3\text{O}_2(g) \longrightarrow 2\text{CO}_2(g) + 2\text{H}_2\text{O}(l)$

Solution

Redox reactions are identified per definition if one or more elements undergo a change in oxidation number.

(a) This is not a redox reaction, since oxidation numbers remain unchanged for all elements.

(b) This is a redox reaction. Gallium is oxidized, its oxidation number increasing from 0 in Ga(*l*) to +3 in $GaBr_3(s)$. The reducing agent is Ga(*l*). Bromine is reduced, its oxidation number decreasing from 0 in $Br_2(l)$ to −1 in $GaBr_3(s)$. The oxidizing agent is $Br_2(l)$.

(c) This is a redox reaction. It is a particularly interesting process, as it involves the same element, oxygen, undergoing both oxidation and reduction (a so-called *disproportionation reaction)*. Oxygen is oxidized, its oxidation number increasing from −1 in $H_2O_2(aq)$ to 0 in $O_2(g)$. Oxygen is also reduced, its oxidation number decreasing from −1 in $H_2O_2(aq)$ to −2 in $H_2O(l)$. For disproportionation reactions, the same substance functions as an oxidant and a reductant.

(d) This is not a redox reaction, since oxidation numbers remain unchanged for all elements.

(e) This is a redox reaction (combustion). Carbon is oxidized, its oxidation number increasing from −2 in $C_2H_4(g)$ to +4 in $CO_2(g)$. The reducing agent (fuel) is $C_2H_4(g)$. Oxygen is reduced, its oxidation number decreasing from 0 in $O_2(g)$ to −2 in $H_2O(l)$. The oxidizing agent is $O_2(g)$.

Check Your Learning

This equation describes the production of tin(II) chloride:

$$\text{Sn}(s) + 2\text{HCl}(g) \longrightarrow \text{SnCl}_2(s) + \text{H}_2(g)$$

Is this a redox reaction? If so, provide a more specific name for the reaction if appropriate, and identify the oxidant and reductant.

Answer:

Yes, a single-replacement reaction. Sn(*s*)is the reductant, HCl(*g*) is the oxidant.

Balancing Redox Reactions via the Half-Reaction Method

Redox reactions that take place in aqueous media often involve water, hydronium ions, and hydroxide ions as reactants or products. Although these species are not oxidized or reduced, they do participate in chemical change in other ways (e.g., by providing the elements required to form oxyanions). Equations representing these reactions are sometimes very difficult to balance by inspection, so systematic approaches have been developed to assist in the process. One very useful approach is to use the method of half-reactions, which involves the following steps:

1. Write the two half-reactions representing the redox process.
2. Balance all elements except oxygen and hydrogen.
3. Balance oxygen atoms by adding H_2O molecules.
4. Balance hydrogen atoms by adding H^+ ions.
5. Balance charge by adding electrons.
6. If necessary, multiply each half-reaction's coefficients by the smallest possible integers to yield equal numbers of electrons in each.
7. Add the balanced half-reactions together and simplify by removing species that appear on both sides of the equation.
8. For reactions occurring in basic media (excess hydroxide ions), carry out these additional steps:
   a. Add OH^- ions to both sides of the equation in numbers equal to the number of H^+ ions.
   b. On the side of the equation containing both H^+ and OH^- ions, combine these ions to yield water molecules.
   c. Simplify the equation by removing any redundant water molecules.

9. Finally, check to see that both the number of atoms and the total charges[1] are balanced.

Balancing Redox Reactions in Acidic Solution

Write a balanced equation for the reaction between dichromate ion and iron(II) to yield iron(III) and chromium(III) in acidic solution.

$$Cr_2O_7{}^{2-} + Fe^{2+} \longrightarrow Cr^{3+} + Fe^{3+}$$

Solution

Step 1.
Write the two half-reactions.

Each half-reaction will contain one reactant and one product with one element in common.

$$Fe^{2+} \longrightarrow Fe^{3+}$$
$$Cr_2O_7{}^{2-} \longrightarrow Cr^{3+}$$

Step 2.
Balance all elements except oxygen and hydrogen. The iron half-reaction is already balanced, but the chromium half-reaction shows two Cr atoms on the left and one Cr atom on the right. Changing the coefficient on the right side of the equation to 2 achieves balance with regard to Cr atoms.

$$Fe^{2+} \longrightarrow Fe^{3+}$$
$$Cr_2O_7{}^{2-} \longrightarrow 2Cr^{3+}$$

Step 3.
Balance oxygen atoms by adding H_2O *molecules.* The iron half-reaction does not contain O atoms. The chromium half-reaction shows seven O atoms on the left and none on the right, so seven water molecules are added to the right side.

$$Fe^{2+} \longrightarrow Fe^{3+}$$

$$Cr_2O_7{}^{2-} \longrightarrow 2Cr^{3+} + 7H_2O$$

Step 4.
Balance hydrogen atoms by adding H^+ *ions.* The iron half-reaction does not contain H atoms. The chromium half-reaction shows 14 H atoms on the right and none on the left, so 14 hydrogen ions are added to the left side.

$$Fe^{2+} \longrightarrow Fe^{3+}$$
$$Cr_2O_7{}^{2-} + 14H^+ \longrightarrow 2Cr^{3+} + 7H_2O$$

Step 5.
Balance charge by adding electrons. The iron half-reaction shows a total charge of 2+ on the left side (1 Fe^{2+} ion) and 3+ on the right side (1 Fe^{3+} ion). Adding one electron to the right side brings that side's total charge to (3+) + (1−) = 2+, and charge balance is achieved.

The chromium half-reaction shows a total charge of (1 × 2−) + (14 × 1+) = 12+ on the left side (1 $Cr_2O_7{}^{2-}$ ion and 14 H^+ ions). The total charge on the right side is (2 × 3+) = 6 + (2 Cr^{3+} ions). Adding six electrons to

1 The requirement of "charge balance" is just a specific type of "mass balance" in which the species in question are electrons. An equation must represent equal numbers of electrons on the reactant and product sides, and so both atoms and charges must be balanced.

the left side will bring that side's total charge to (12+ + 6−) = 6+, and charge balance is achieved.

$$Fe^{2+} \longrightarrow Fe^{3+} + e^-$$
$$Cr_2O_7{}^{2-} + 14H^+ + 6e^- \longrightarrow 2Cr^{3+} + 7H_2O$$

Step 6.
Multiply the two half-reactions so the number of electrons in one reaction equals the number of electrons in the other reaction. To be consistent with mass conservation, and the idea that redox reactions involve the transfer (not creation or destruction) of electrons, the iron half-reaction's coefficient must be multiplied by 6.

$$6Fe^{2+} \longrightarrow 6Fe^{3+} + 6e^-$$
$$Cr_2O_7{}^{2-} + 6e^- + 14H^+ \longrightarrow 2Cr^{3+} + 7H_2O$$

Step 7.
Add the balanced half-reactions and cancel species that appear on both sides of the equation.

$$6Fe^{2+} + Cr_2O_7{}^{2-} + 6e^- + 14H^+ \longrightarrow 6Fe^{3+} + 6e^- + 2Cr^{3+} + 7H_2O$$

Only the six electrons are redundant species. Removing them from each side of the equation yields the simplified, balanced equation here:

$$6Fe^{2+} + Cr_2O_7{}^{2-} + 14H^+ \longrightarrow 6Fe^{3+} + 2Cr^{3+} + 7H_2O$$

A final check of atom and charge balance confirms the equation is balanced.

	Reactants	Products
Fe	6	6
Cr	2	2
O	7	7
H	14	14
charge	24+	24+

Check Your Learning

In basic solution, molecular chlorine, Cl_2, reacts with hydroxide ions, OH^-, to yield chloride ions, Cl^-. and chlorate ions, $ClO_3{}^-$. HINT: This is a *disproportionation reaction* in which the element chlorine is both oxidized and reduced. Write a balanced equation for this reaction.

Answer:

$3Cl_2(aq) + 6OH^-(aq) \longrightarrow 5Cl^-(aq) + ClO_3{}^-(aq) + 3H_2O(l)$

4.3 Reaction Stoichiometry

LEARNING OBJECTIVES

By the end of this section, you will be able to:

- Explain the concept of stoichiometry as it pertains to chemical reactions
- Use balanced chemical equations to derive stoichiometric factors relating amounts of reactants and products
- Perform stoichiometric calculations involving mass, moles, and solution molarity

A balanced chemical equation provides a great deal of information in a very succinct format. Chemical formulas provide the identities of the reactants and products involved in the chemical change, allowing classification of the reaction. Coefficients provide the relative numbers of these chemical species, allowing a quantitative assessment of the relationships between the amounts of substances consumed and produced by the reaction. These quantitative relationships are known as the reaction's **stoichiometry**, a term derived from the Greek words *stoicheion* (meaning "element") and *metron* (meaning "measure"). In this module, the use of balanced chemical equations for various stoichiometric applications is explored.

The general approach to using stoichiometric relationships is similar in concept to the way people go about many common activities. Food preparation, for example, offers an appropriate comparison. A recipe for making eight pancakes calls for 1 cup pancake mix, $\frac{3}{4}$ cup milk, and one egg. The "equation" representing the preparation of pancakes per this recipe is

$$1\text{ cup mix} + \frac{3}{4}\text{ cup milk} + 1\text{ egg} \longrightarrow 8\text{ pancakes}$$

If two dozen pancakes are needed for a big family breakfast, the ingredient amounts must be increased proportionally according to the amounts given in the recipe. For example, the number of eggs required to make 24 pancakes is

$$24\ \cancel{\text{pancakes}} \times \frac{1\text{ egg}}{8\ \cancel{\text{pancakes}}} = 3\text{ eggs}$$

Balanced chemical equations are used in much the same fashion to determine the amount of one reactant required to react with a given amount of another reactant, or to yield a given amount of product, and so forth. The coefficients in the balanced equation are used to derive **stoichiometric factors** that permit computation of the desired quantity. To illustrate this idea, consider the production of ammonia by reaction of hydrogen and nitrogen:

$$N_2(g) + 3H_2(g) \longrightarrow 2NH_3(g)$$

This equation shows ammonia molecules are produced from hydrogen molecules in a 2:3 ratio, and stoichiometric factors may be derived using any amount (number) unit:

$$\frac{2\text{ NH}_3\text{ molecules}}{3\text{ H}_2\text{ molecules}} \text{ or } \frac{2\text{ doz NH}_3\text{ molecules}}{3\text{ doz H}_2\text{ molecules}} \text{ or } \frac{2\text{ mol NH}_3\text{ molecules}}{3\text{ mol H}_2\text{ molecules}}$$

These stoichiometric factors can be used to compute the number of ammonia molecules produced from a given number of hydrogen molecules, or the number of hydrogen molecules required to produce a given number of ammonia molecules. Similar factors may be derived for any pair of substances in any chemical equation.

EXAMPLE 4.8

Moles of Reactant Required in a Reaction

How many moles of I_2 are required to react with 0.429 mol of Al according to the following equation (see Figure 4.10)?

$$2Al + 3I_2 \longrightarrow 2AlI_3$$

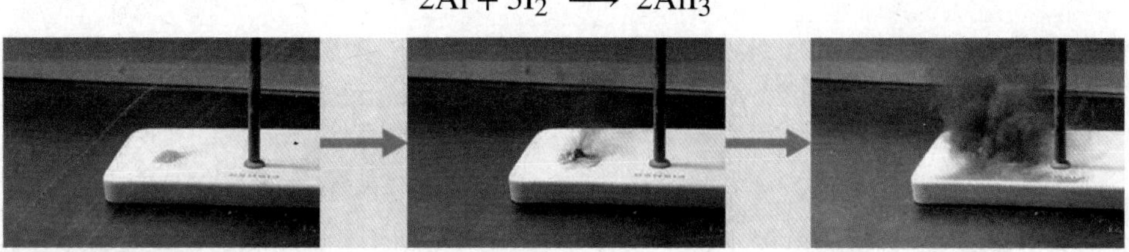

FIGURE 4.10 Aluminum and iodine react to produce aluminum iodide. The heat of the reaction vaporizes some of the solid iodine as a purple vapor. (credit: modification of work by Mark Ott)

Solution

Referring to the balanced chemical equation, the stoichiometric factor relating the two substances of interest is $\frac{3 \text{ mol } I_2}{2 \text{ mol Al}}$. The molar amount of iodine is derived by multiplying the provided molar amount of aluminum by this factor:

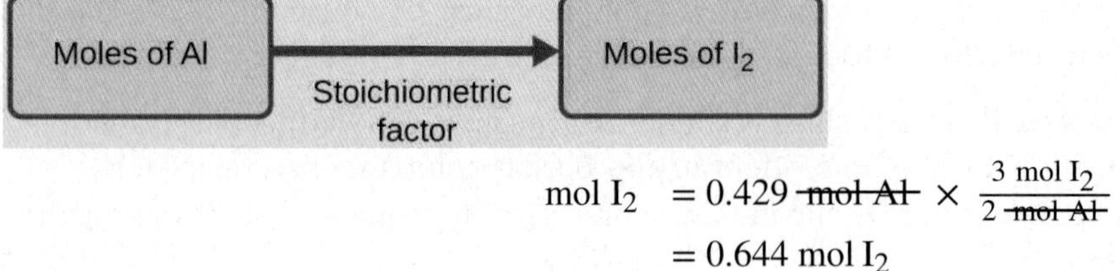

$$\begin{aligned} \text{mol } I_2 &= 0.429 \text{ mol Al} \times \frac{3 \text{ mol } I_2}{2 \text{ mol Al}} \\ &= 0.644 \text{ mol } I_2 \end{aligned}$$

Check Your Learning

How many moles of $Ca(OH)_2$ are required to react with 1.36 mol of H_3PO_4 to produce $Ca_3(PO_4)_2$ according to the equation $3Ca(OH)_2 + 2H_3PO_4 \longrightarrow Ca_3(PO_4)_2 + 6H_2O$?

Answer:

2.04 mol

EXAMPLE 4.9

Number of Product Molecules Generated by a Reaction

How many carbon dioxide molecules are produced when 0.75 mol of propane is combusted according to this equation?

$$C_3H_8 + 5O_2 \longrightarrow 3CO_2 + 4H_2O$$

Solution

The approach here is the same as for Example 4.8, though the absolute number of molecules is requested, not the number of moles of molecules. This will simply require use of the moles-to-numbers conversion factor, Avogadro's number.

The balanced equation shows that carbon dioxide is produced from propane in a 3:1 ratio:

$$\frac{3 \text{ mol } CO_2}{1 \text{ mol } C_3H_8}$$

Using this stoichiometric factor, the provided molar amount of propane, and Avogadro's number,

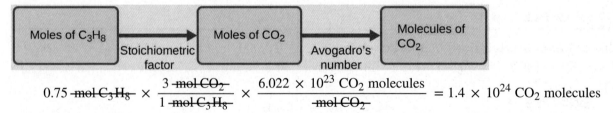

Moles of C_3H_8 → (Stoichiometric factor) → Moles of CO_2 → (Avogadro's number) → Molecules of CO_2

$$0.75 \text{ mol } C_3H_8 \times \frac{3 \text{ mol } CO_2}{1 \text{ mol } C_3H_8} \times \frac{6.022 \times 10^{23} \text{ } CO_2 \text{ molecules}}{\text{mol } CO_2} = 1.4 \times 10^{24} \text{ } CO_2 \text{ molecules}$$

Check Your Learning

How many NH_3 molecules are produced by the reaction of 4.0 mol of $Ca(OH)_2$ according to the following equation:

$$(NH_4)_2SO_4 + Ca(OH)_2 \longrightarrow 2NH_3 + CaSO_4 + 2H_2O$$

Answer:

4.8×10^{24} NH_3 molecules

These examples illustrate the ease with which the amounts of substances involved in a chemical reaction of known stoichiometry may be related. Directly measuring numbers of atoms and molecules is, however, not an easy task, and the practical application of stoichiometry requires that we use the more readily measured property of mass.

EXAMPLE 4.10

Relating Masses of Reactants and Products

What mass of sodium hydroxide, NaOH, would be required to produce 16 g of the antacid milk of magnesia [magnesium hydroxide, $Mg(OH)_2$] by the following reaction?

$$MgCl_2(aq) + 2NaOH(aq) \longrightarrow Mg(OH)_2(s) + 2NaCl(aq)$$

Solution

The approach used previously in Example 4.8 and Example 4.9 is likewise used here; that is, we must derive an appropriate stoichiometric factor from the balanced chemical equation and use it to relate the amounts of the two substances of interest. In this case, however, masses (not molar amounts) are provided and requested, so additional steps of the sort learned in the previous chapter are required. The calculations required are outlined in this flowchart:

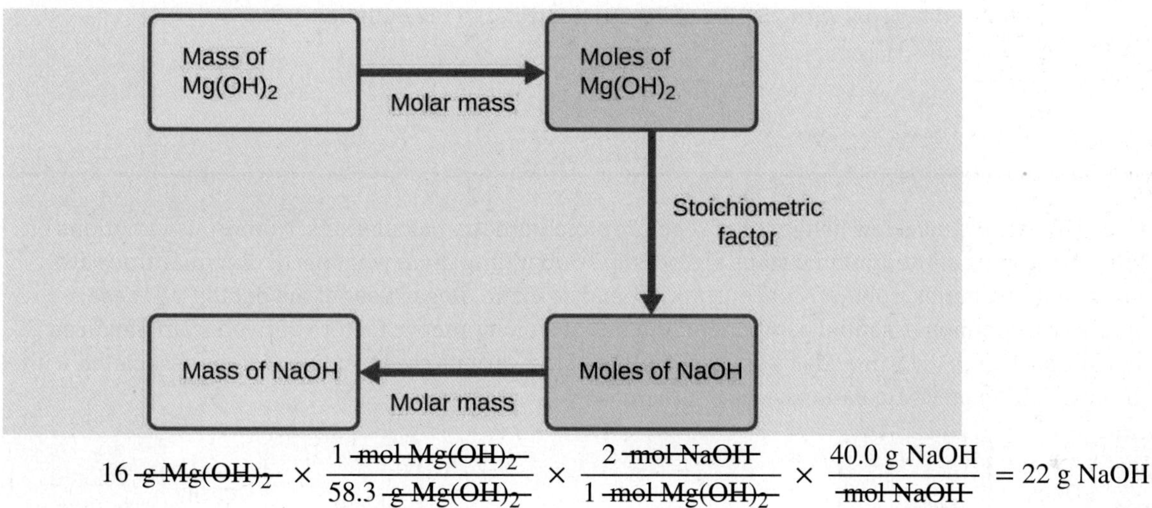

$$16 \cancel{\text{g Mg(OH)}_2} \times \frac{1 \cancel{\text{mol Mg(OH)}_2}}{58.3 \cancel{\text{g Mg(OH)}_2}} \times \frac{2 \cancel{\text{mol NaOH}}}{1 \cancel{\text{mol Mg(OH)}_2}} \times \frac{40.0 \text{ g NaOH}}{\cancel{\text{mol NaOH}}} = 22 \text{ g NaOH}$$

Check Your Learning

What mass of gallium oxide, Ga_2O_3, can be prepared from 29.0 g of gallium metal? The equation for the reaction is $4Ga + 3O_2 \longrightarrow 2Ga_2O_3$.

Answer:

39.0 g

EXAMPLE 4.11

Relating Masses of Reactants

What mass of oxygen gas, O_2, from the air is consumed in the combustion of 702 g of octane, C_8H_{18}, one of the principal components of gasoline?

$$2C_8H_{18} + 25O_2 \longrightarrow 16CO_2 + 18H_2O$$

Solution

The approach required here is the same as for the Example 4.10, differing only in that the provided and requested masses are both for reactant species.

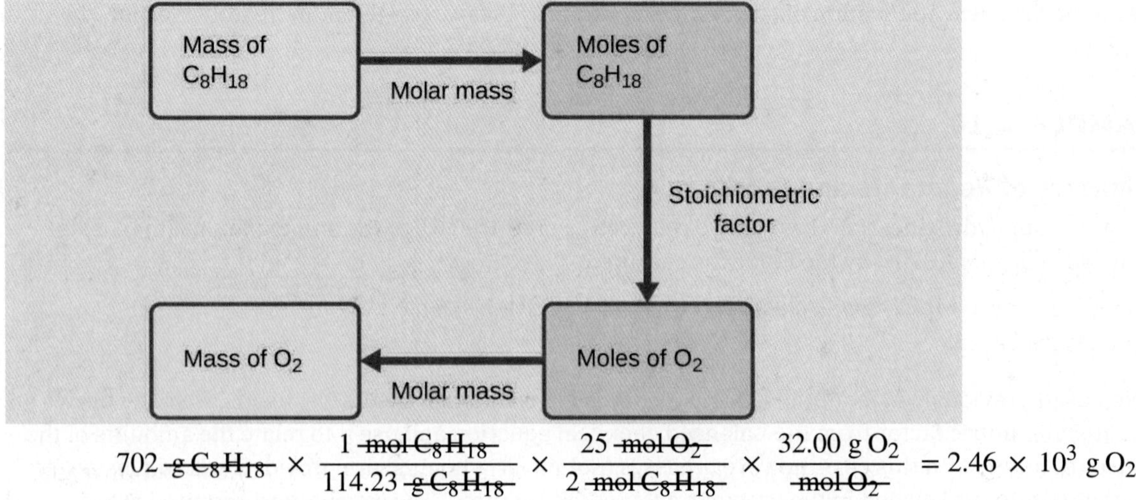

$$702 \ \cancel{g\ C_8H_{18}} \times \frac{1 \ \cancel{mol\ C_8H_{18}}}{114.23 \ \cancel{g\ C_8H_{18}}} \times \frac{25 \ \cancel{mol\ O_2}}{2 \ \cancel{mol\ C_8H_{18}}} \times \frac{32.00 \ g\ O_2}{\cancel{mol\ O_2}} = 2.46 \times 10^3 \ g\ O_2$$

Check Your Learning

What mass of CO is required to react with 25.13 g of Fe_2O_3 according to the equation $Fe_2O_3 + 3CO \longrightarrow 2Fe + 3CO_2$?

Answer:

13.22 g

These examples illustrate just a few instances of reaction stoichiometry calculations. Numerous variations on the beginning and ending computational steps are possible depending upon what particular quantities are provided and sought (volumes, solution concentrations, and so forth). Regardless of the details, all these calculations share a common essential component: the use of stoichiometric factors derived from balanced chemical equations. Figure 4.11 provides a general outline of the various computational steps associated with many reaction stoichiometry calculations.

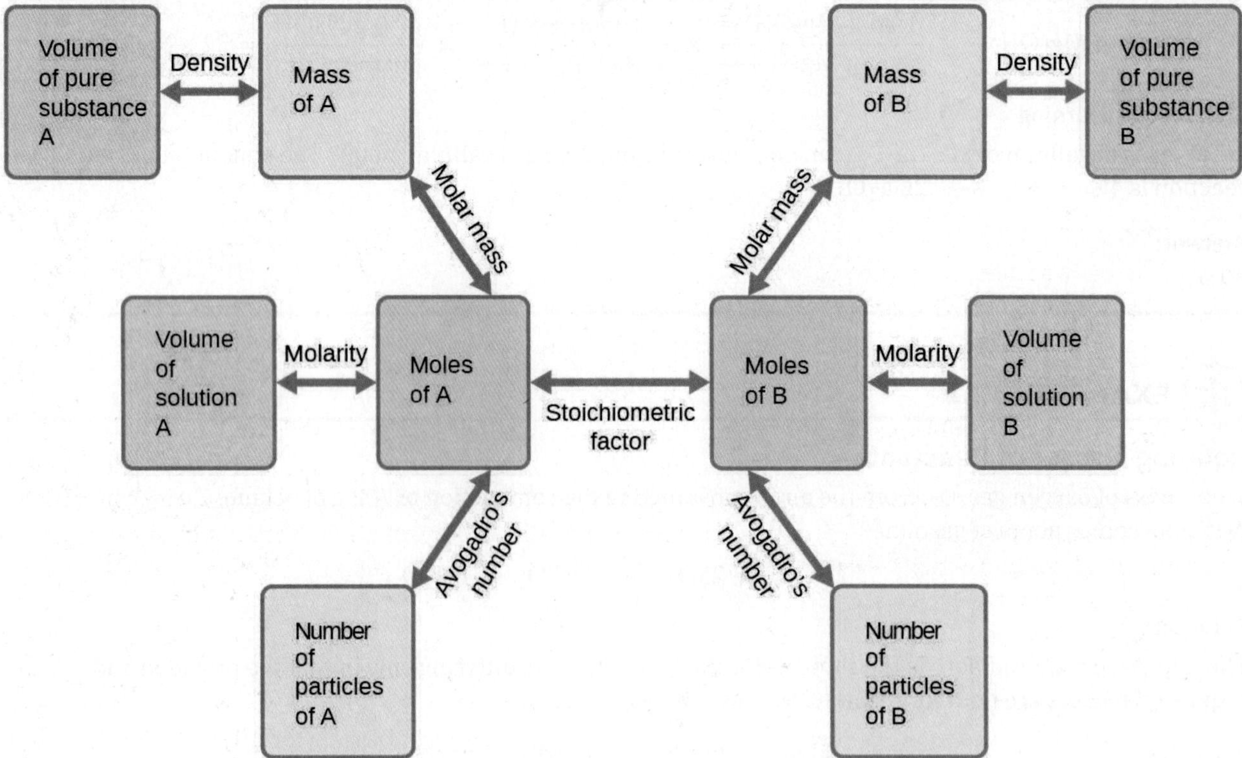

FIGURE 4.11 The flowchart depicts the various computational steps involved in most reaction stoichiometry

calculations.

Chemistry in Everyday Life

Airbags

Airbags (Figure 4.12) are a safety feature provided in most automobiles since the 1990s. The effective operation of an airbag requires that it be rapidly inflated with an appropriate amount (volume) of gas when the vehicle is involved in a collision. This requirement is satisfied in many automotive airbag systems through use of explosive chemical reactions, one common choice being the decomposition of sodium azide, NaN_3. When sensors in the vehicle detect a collision, an electrical current is passed through a carefully measured amount of NaN_3 to initiate its decomposition:

$$2NaN_3(s) \longrightarrow 3N_2(g) + 2Na(s)$$

This reaction is very rapid, generating gaseous nitrogen that can deploy and fully inflate a typical airbag in a fraction of a second (~0.03–0.1 s). Among many engineering considerations, the amount of sodium azide used must be appropriate for generating enough nitrogen gas to fully inflate the air bag and ensure its proper function. For example, a small mass (~100 g) of NaN_3 will generate approximately 50 L of N_2.

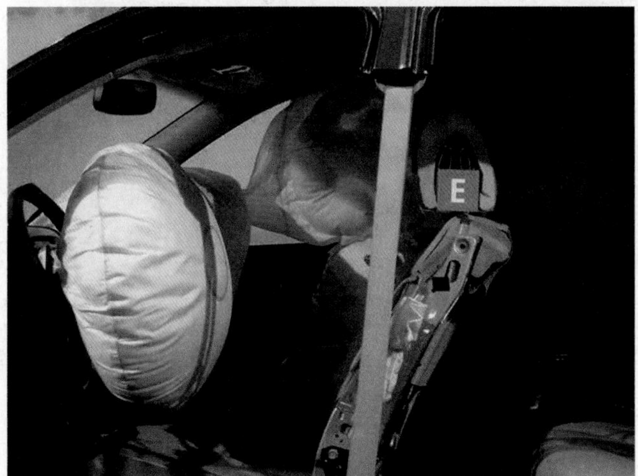

FIGURE 4.12 Airbags deploy upon impact to minimize serious injuries to passengers. (credit: Jon Seidman)

4.4 Reaction Yields

LEARNING OBJECTIVES

By the end of this section, you will be able to:

- Explain the concepts of theoretical yield and limiting reactants/reagents.
- Derive the theoretical yield for a reaction under specified conditions.
- Calculate the percent yield for a reaction.

The relative amounts of reactants and products represented in a balanced chemical equation are often referred to as *stoichiometric amounts*. All the exercises of the preceding module involved stoichiometric amounts of reactants. For example, when calculating the amount of product generated from a given amount of reactant, it was assumed that any other reactants required were available in stoichiometric amounts (or greater). In this module, more realistic situations are considered, in which reactants are not present in stoichiometric amounts.

Limiting Reactant

Consider another food analogy, making grilled cheese sandwiches (Figure 4.13):

$$\text{1 slice of cheese} + \text{2 slices of bread} \longrightarrow \text{1 sandwich}$$

Stoichiometric amounts of sandwich ingredients for this recipe are bread and cheese slices in a 2:1 ratio.

Provided with 28 slices of bread and 11 slices of cheese, one may prepare 11 sandwiches per the provided recipe, using all the provided cheese and having six slices of bread left over. In this scenario, the number of sandwiches prepared has been *limited* by the number of cheese slices, and the bread slices have been provided in *excess*.

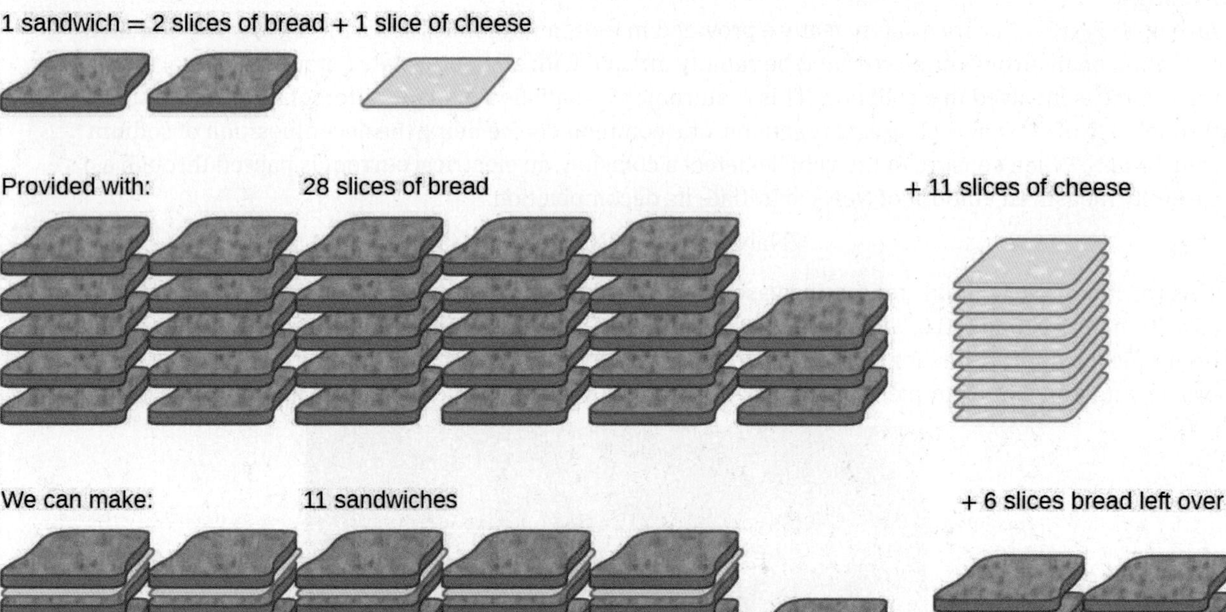

FIGURE 4.13 Sandwich making can illustrate the concepts of limiting and excess reactants.

Consider this concept now with regard to a chemical process, the reaction of hydrogen with chlorine to yield hydrogen chloride:

$$H_2(g) + Cl_2(g) \longrightarrow 2HCl(g)$$

The balanced equation shows the hydrogen and chlorine react in a 1:1 stoichiometric ratio. If these reactants are provided in any other amounts, one of the reactants will nearly always be entirely consumed, thus limiting the amount of product that may be generated. This substance is the **limiting reactant**, and the other substance is the **excess reactant**. Identifying the limiting and excess reactants for a given situation requires computing the molar amounts of each reactant provided and comparing them to the stoichiometric amounts represented in the balanced chemical equation. For example, imagine combining 3 moles of H_2 and 2 moles of Cl_2. This represents a 3:2 (or 1.5:1) ratio of hydrogen to chlorine present for reaction, which is greater than the stoichiometric ratio of 1:1. Hydrogen, therefore, is present in excess, and chlorine is the limiting reactant. Reaction of all the provided chlorine (2 mol) will consume 2 mol of the 3 mol of hydrogen provided, leaving 1 mol of hydrogen unreacted.

An alternative approach to identifying the limiting reactant involves comparing the amount of product expected for the complete reaction of each reactant. Each reactant amount is used to separately calculate the amount of product that would be formed per the reaction's stoichiometry. The reactant yielding the lesser amount of product is the limiting reactant. For the example in the previous paragraph, complete reaction of the hydrogen would yield

$$\text{mol HCl produced} = 3 \text{ mol } H_2 \times \frac{2 \text{ mol HCl}}{1 \text{ mol } H_2} = 6 \text{ mol HCl}$$

Complete reaction of the provided chlorine would produce

$$\text{mol HCl produced} = 2 \text{ mol } Cl_2 \times \frac{2 \text{ mol HCl}}{1 \text{ mol } Cl_2} = 4 \text{ mol HCl}$$

The chlorine will be completely consumed once 4 moles of HCl have been produced. Since enough hydrogen was provided to yield 6 moles of HCl, there will be unreacted hydrogen remaining once this reaction is complete. Chlorine, therefore, is the limiting reactant and hydrogen is the excess reactant (Figure 4.14).

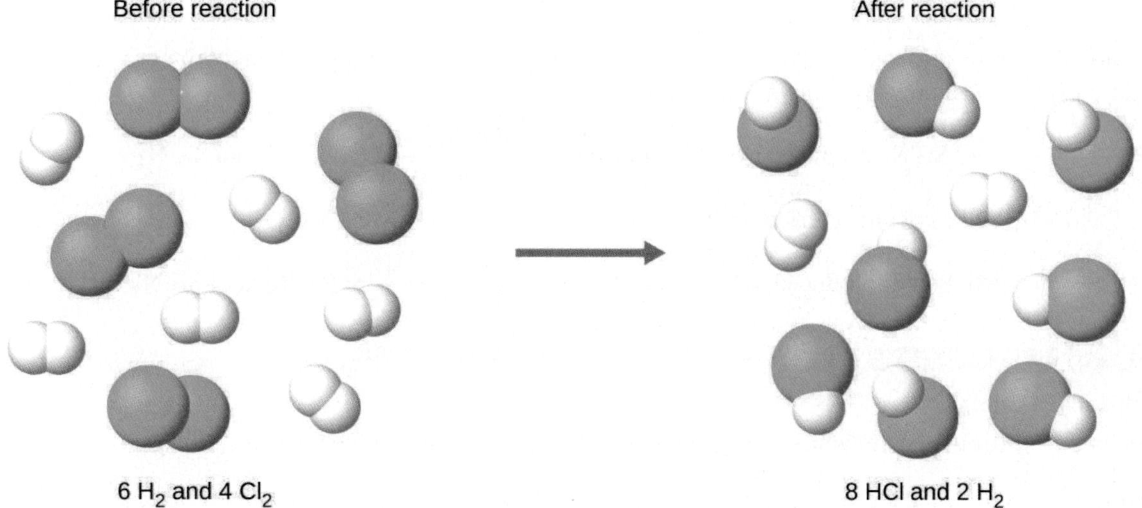

FIGURE 4.14 When H_2 and Cl_2 are combined in nonstoichiometric amounts, one of these reactants will limit the amount of HCl that can be produced. This illustration shows a reaction in which hydrogen is present in excess and chlorine is the limiting reactant.

LINK TO LEARNING

View this interactive simulation (http://openstax.org/l/16reactantprod) illustrating the concepts of limiting and excess reactants.

EXAMPLE 4.12

Identifying the Limiting Reactant

Silicon nitride is a very hard, high-temperature-resistant ceramic used as a component of turbine blades in jet engines. It is prepared according to the following equation:

$$3\text{Si}\,(s) + 2\text{N}_2(g) \longrightarrow \text{Si}_3\text{N}_4(s)$$

Which is the limiting reactant when 2.00 g of Si and 1.50 g of N_2 react?

Solution

Compute the provided molar amounts of reactants, and then compare these amounts to the balanced equation to identify the limiting reactant.

$$\text{mol Si} = 2.00\,\cancel{\text{g Si}} \times \frac{1\text{ mol Si}}{28.09\,\cancel{\text{g Si}}} = 0.0712\text{ mol Si}$$

$$\text{mol N}_2 = 1.50\,\cancel{\text{g N}_2} \times \frac{1\text{ mol N}_2}{28.02\,\cancel{\text{g N}_2}} = 0.0535\text{ mol N}_2$$

The provided Si:N_2 molar ratio is:

$$\frac{0.0712\text{ mol Si}}{0.0535\text{ mol N}_2} = \frac{1.33\text{ mol Si}}{1\text{ mol N}_2}$$

The stoichiometric Si:N_2 ratio is:

$$\frac{3 \text{ mol Si}}{2 \text{ mol N}_2} = \frac{1.5 \text{ mol Si}}{1 \text{ mol N}_2}$$

Comparing these ratios shows that Si is provided in a less-than-stoichiometric amount, and so is the limiting reactant.

Alternatively, compute the amount of product expected for complete reaction of each of the provided reactants. The 0.0712 moles of silicon would yield

$$\text{mol Si}_3\text{N}_4 \text{ produced} = 0.0712 \text{ mol Si} \times \frac{1 \text{ mol Si}_3\text{N}_4}{3 \text{ mol Si}} = 0.0237 \text{ mol Si}_3\text{N}_4$$

while the 0.0535 moles of nitrogen would produce

$$\text{mol Si}_3\text{N}_4 \text{ produced} = 0.0535 \text{ mol N}_2 \times \frac{1 \text{ mol Si}_3\text{N}_4}{2 \text{ mol N}_2} = 0.0268 \text{ mol Si}_3\text{N}_4$$

Since silicon yields the lesser amount of product, it is the limiting reactant.

Check Your Learning

Which is the limiting reactant when 5.00 g of H_2 and 10.0 g of O_2 react and form water?

Answer:

O_2

Percent Yield

The amount of product that *may be* produced by a reaction under specified conditions, as calculated per the stoichiometry of an appropriate balanced chemical equation, is called the **theoretical yield** of the reaction. In practice, the amount of product obtained is called the **actual yield**, and it is often less than the theoretical yield for a number of reasons. Some reactions are inherently inefficient, being accompanied by *side reactions* that generate other products. Others are, by nature, incomplete (consider the partial reactions of weak acids and bases discussed earlier in this chapter). Some products are difficult to collect without some loss, and so less than perfect recovery will reduce the actual yield. The extent to which a reaction's theoretical yield is achieved is commonly expressed as its **percent yield**:

$$\text{percent yield} = \frac{\text{actual yield}}{\text{theoretical yield}} \times 100\%$$

Actual and theoretical yields may be expressed as masses or molar amounts (or any other appropriate property; e.g., volume, if the product is a gas). As long as both yields are expressed using the same units, these units will cancel when percent yield is calculated.

 **EXAMPLE 4.13**

Calculation of Percent Yield

Upon reaction of 1.274 g of copper sulfate with excess zinc metal, 0.392 g copper metal was obtained according to the equation:

$$\text{CuSO}_4(aq) + \text{Zn}(s) \longrightarrow \text{Cu}(s) + \text{ZnSO}_4(aq)$$

What is the percent yield?

Solution

The provided information identifies copper sulfate as the limiting reactant, and so the theoretical yield is found by the approach illustrated in the previous module, as shown here:

$$1.274 \cancel{\text{g CuSO}_4} \times \frac{1 \cancel{\text{mol CuSO}_4}}{159.62 \cancel{\text{g CuSO}_4}} \times \frac{1 \cancel{\text{mol Cu}}}{1 \cancel{\text{mol CuSO}_4}} \times \frac{63.55 \text{ g Cu}}{1 \cancel{\text{mol Cu}}} = 0.5072 \text{ g Cu}$$

Using this theoretical yield and the provided value for actual yield, the percent yield is calculated to be

$$\text{percent yield} = \left(\frac{\text{actual yield}}{\text{theoretical yield}}\right) \times 100$$

$$\begin{aligned}\text{percent yield} &= \left(\frac{0.392\text{ g Cu}}{0.5072\text{ g Cu}}\right) \times 100 \\ &= 77.3\%\end{aligned}$$

Check Your Learning

What is the percent yield of a reaction that produces 12.5 g of the gas Freon CF_2Cl_2 from 32.9 g of CCl_4 and excess HF?

$$CCl_4 + 2HF \longrightarrow CF_2Cl_2 + 2HCl$$

Answer:

48.3%

HOW SCIENCES INTERCONNECT

Green Chemistry and Atom Economy

The purposeful design of chemical products and processes that minimize the use of environmentally hazardous substances and the generation of waste is known as *green chemistry*. Green chemistry is a philosophical approach that is being applied to many areas of science and technology, and its practice is summarized by guidelines known as the "Twelve Principles of Green Chemistry" (see details at this website (http://openstax.org/l/16greenchem)). One of the 12 principles is aimed specifically at maximizing the efficiency of processes for synthesizing chemical products. The *atom economy* of a process is a measure of this efficiency, defined as the percentage by mass of the final product of a synthesis relative to the masses of *all* the reactants used:

$$\text{atom economy} = \frac{\text{mass of product}}{\text{mass of reactants}} \times 100\%$$

Though the definition of atom economy at first glance appears very similar to that for percent yield, be aware that this property represents a difference in the *theoretical* efficiencies of *different* chemical processes. The percent yield of a given chemical process, on the other hand, evaluates the efficiency of a process by comparing the yield of product actually obtained to the maximum yield predicted by stoichiometry.

The synthesis of the common nonprescription pain medication, ibuprofen, nicely illustrates the success of a green chemistry approach (Figure 4.15). First marketed in the early 1960s, ibuprofen was produced using a six-step synthesis that required 514 g of reactants to generate each mole (206 g) of ibuprofen, an atom economy of 40%. In the 1990s, an alternative process was developed by the BHC Company (now BASF Corporation) that requires only three steps and has an atom economy of ~80%, nearly twice that of the original process. The BHC process generates significantly less chemical waste; uses less-hazardous and recyclable materials; and provides significant cost-savings to the manufacturer (and, subsequently, the consumer). In recognition of the positive environmental impact of the BHC process, the company received the Environmental Protection Agency's Greener Synthetic Pathways Award in 1997.

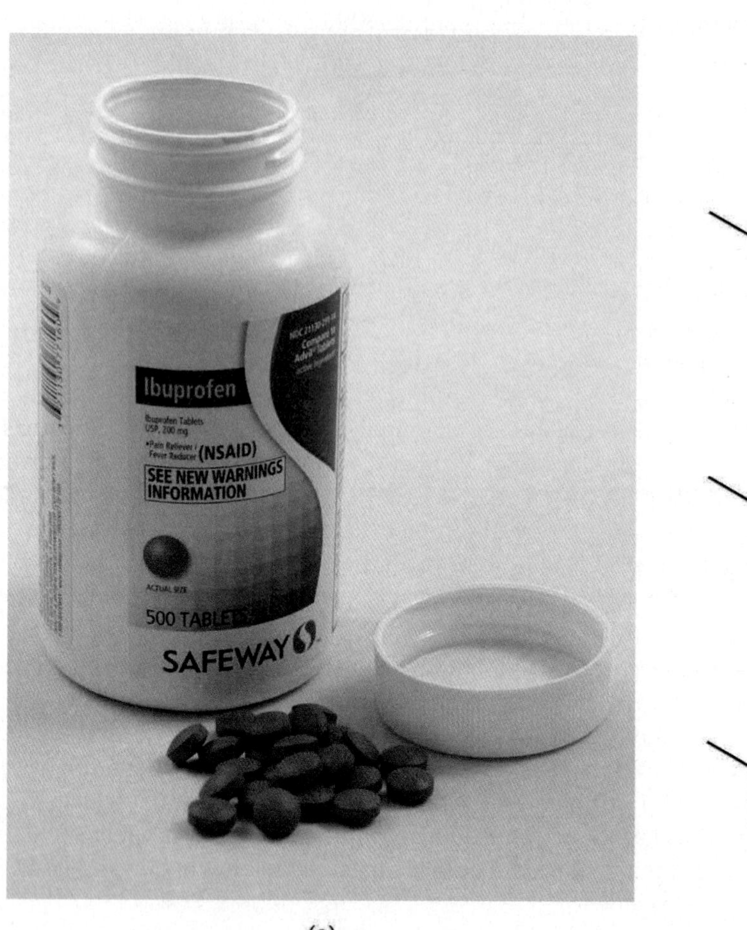

(a) (b)

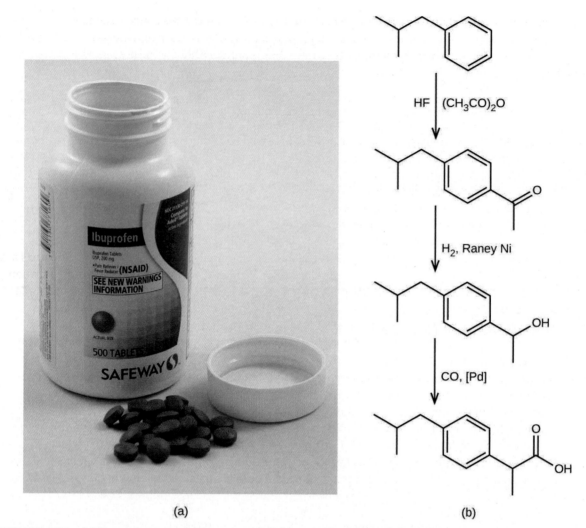

FIGURE 4.15 (a) Ibuprofen is a popular nonprescription pain medication commonly sold as 200 mg tablets. (b) The BHC process for synthesizing ibuprofen requires only three steps and exhibits an impressive atom economy. (credit a: modification of work by Derrick Coetzee)

4.5 Quantitative Chemical Analysis

LEARNING OBJECTIVES

By the end of this section, you will be able to:

- Describe the fundamental aspects of titrations and gravimetric analysis.
- Perform stoichiometric calculations using typical titration and gravimetric data.

In the 18th century, the strength (actually the concentration) of vinegar samples was determined by noting the amount of potassium carbonate, K_2CO_3, which had to be added, a little at a time, before bubbling ceased. The greater the weight of potassium carbonate added to reach the point where the bubbling ended, the more concentrated the vinegar.

We now know that the effervescence that occurred during this process was due to reaction with acetic acid, CH_3CO_2H, the compound primarily responsible for the odor and taste of vinegar. Acetic acid reacts with potassium carbonate according to the following equation:

$$2CH_3CO_2H(aq) + K_2CO_3(s) \longrightarrow 2CH_3CO_2K(aq) + CO_2(g) + H_2O(l)$$

The bubbling was due to the production of CO_2.

The test of vinegar with potassium carbonate is one type of **quantitative analysis**—the determination of the amount or concentration of a substance in a sample. In the analysis of vinegar, the concentration of the solute (acetic acid) was determined from the amount of reactant that combined with the solute present in a known volume of the solution. In other types of chemical analyses, the amount of a substance present in a sample is determined by measuring the amount of product that results.

Titration

The described approach to measuring vinegar strength was an early version of the analytical technique known as **titration analysis**. A typical titration analysis involves the use of a **buret** (Figure 4.16) to make incremental additions of a solution containing a known concentration of some substance (the **titrant**) to a sample solution containing the substance whose concentration is to be measured (the **analyte**). The titrant and analyte undergo a chemical reaction of known stoichiometry, and so measuring the volume of titrant solution required for complete reaction with the analyte (the **equivalence point** of the titration) allows calculation of the analyte concentration. The equivalence point of a titration may be detected visually if a distinct change in the appearance of the sample solution accompanies the completion of the reaction. The halt of bubble formation in the classic vinegar analysis is one such example, though, more commonly, special dyes called **indicators** are added to the sample solutions to impart a change in color at or very near the equivalence point of the titration. Equivalence points may also be detected by measuring some solution property that changes in a predictable way during the course of the titration. Regardless of the approach taken to detect a titration's equivalence point, the volume of titrant actually measured is called the **end point**. Properly designed titration methods typically ensure that the difference between the equivalence and end points is negligible. Though any type of chemical reaction may serve as the basis for a titration analysis, the three described in this chapter (precipitation, acid-base, and redox) are most common. Additional details regarding titration analysis are provided in the chapter on acid-base equilibria.

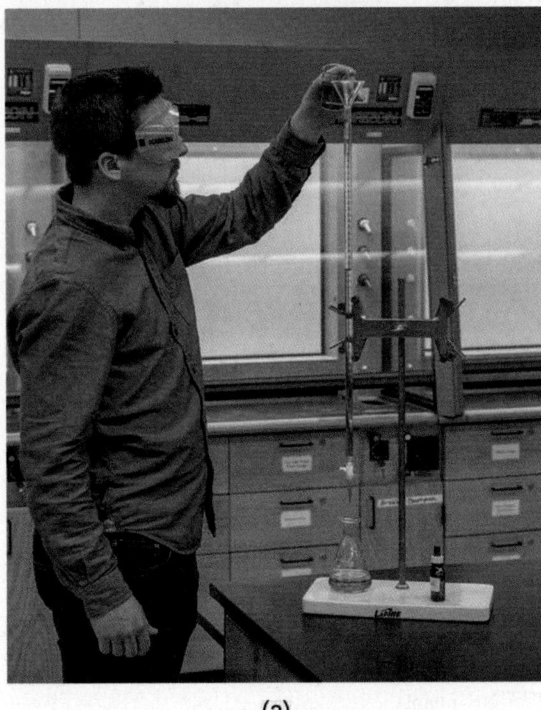

(a)

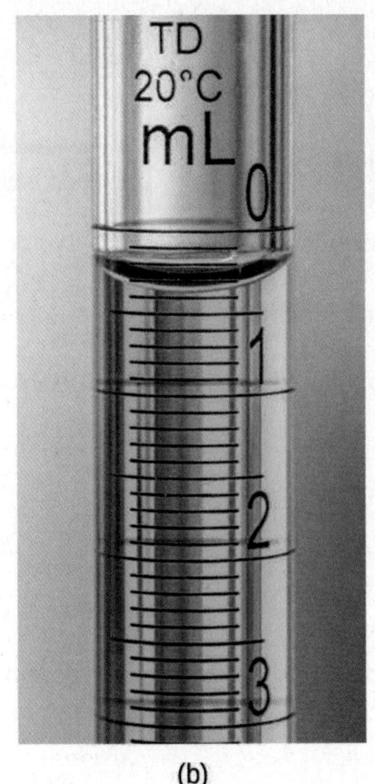

(b)

FIGURE 4.16 (a) A student fills a buret in preparation for a titration analysis. (b) A typical buret permits volume measurements to the nearest 0.01 mL. (credit a: modification of work by Mark Blaser and Matt Evans; credit b: modification of work by Mark Blaser and Matt Evans)

EXAMPLE 4.14

Titration Analysis

The end point in a titration of a 50.00-mL sample of aqueous HCl was reached by addition of 35.23 mL of 0.250 M NaOH titrant. The titration reaction is:

$$\text{HCl}\,(aq) + \text{NaOH}\,(aq) \longrightarrow \text{NaCl}\,(aq) + \text{H}_2\text{O}\,(l)$$

What is the molarity of the HCl?

Solution

As for all reaction stoichiometry calculations, the key issue is the relation between the molar amounts of the chemical species of interest as depicted in the balanced chemical equation. The approach outlined in previous modules of this chapter is followed, with additional considerations required, since the amounts of reactants provided and requested are expressed as solution concentrations.

For this exercise, the calculation will follow the following outlined steps:

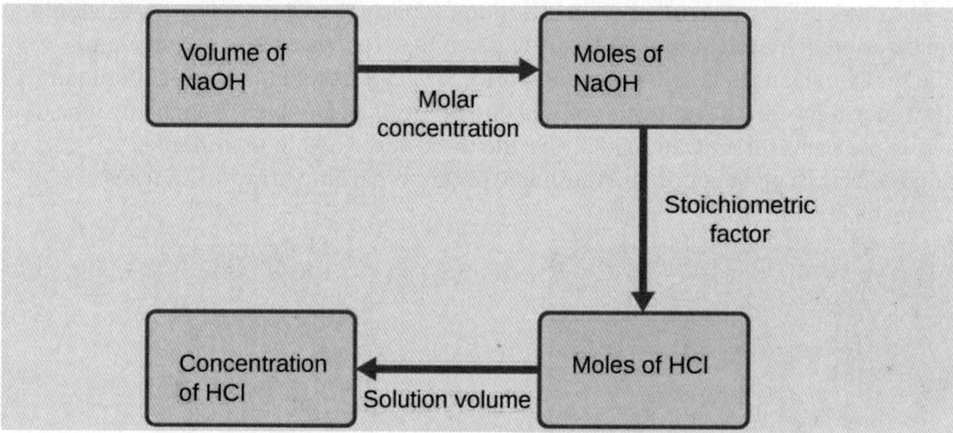

The molar amount of HCl is calculated to be:

$$35.23\ \cancel{\text{mL NaOH}} \times \frac{1\ \cancel{\text{L}}}{1000\ \cancel{\text{mL}}} \times \frac{0.250\ \cancel{\text{mol NaOH}}}{1\ \cancel{\text{L}}} \times \frac{1\ \text{mol HCl}}{1\ \cancel{\text{mol NaOH}}} = 8.81 \times 10^{-3}\ \text{mol HCl}$$

Using the provided volume of HCl solution and the definition of molarity, the HCl concentration is:

$$M = \frac{\text{mol HCl}}{\text{L solution}}$$

$$M = \frac{8.81 \times 10^{-3}\ \text{mol HCl}}{50.00\ \text{mL} \times \frac{1\ \text{L}}{1000\ \text{mL}}}$$

$$M = 0.176\ M$$

Note: For these types of titration calculations, it is convenient to recognize that solution molarity is also equal to the number of *milli*moles of solute per *milli*liter of solution:

$$M = \frac{\text{mol solute}}{\text{L solution}} \times \frac{\frac{10^3\ \text{mmol}}{\text{mol}}}{\frac{10^3\ \text{mL}}{\text{L}}} = \frac{\text{mmol solute}}{\text{mL solution}}$$

Using this version of the molarity unit will shorten the calculation by eliminating two conversion factors:

$$\frac{35.23\ \text{mL NaOH} \times \frac{0.250\ \text{mmol NaOH}}{\text{mL NaOH}} \times \frac{1\ \text{mmol HCl}}{1\ \text{mmol NaOH}}}{50.00\ \text{mL solution}} = 0.176\ M\ \text{HCl}$$

Check Your Learning

A 20.00-mL sample of aqueous oxalic acid, $H_2C_2O_4$, was titrated with a 0.09113-*M* solution of potassium permanganate, $KMnO_4$ (see net ionic equation below).

$$2MnO_4^-(aq) + 5H_2C_2O_4(aq) + 6H^+(aq) \longrightarrow 10CO_2(g) + 2Mn^{2+}(aq) + 8H_2O(l)$$

A volume of 23.24 mL was required to reach the end point. What is the oxalic acid molarity?

Answer:

0.2648 *M*

Gravimetric Analysis

A **gravimetric analysis** is one in which a sample is subjected to some treatment that causes a change in the physical state of the analyte that permits its separation from the other components of the sample. Mass measurements of the sample, the isolated analyte, or some other component of the analysis system, used along with the known stoichiometry of the compounds involved, permit calculation of the analyte concentration. Gravimetric methods were the first techniques used for quantitative chemical analysis, and they remain important tools in the modern chemistry laboratory.

The required change of state in a gravimetric analysis may be achieved by various physical and chemical processes. For example, the moisture (water) content of a sample is routinely determined by measuring the mass of a sample before and after it is subjected to a controlled heating process that evaporates the water. Also common are gravimetric techniques in which the analyte is subjected to a precipitation reaction of the sort described earlier in this chapter. The precipitate is typically isolated from the reaction mixture by filtration, carefully dried, and then weighed (Figure 4.17). The mass of the precipitate may then be used, along with relevant stoichiometric relationships, to calculate analyte concentration.

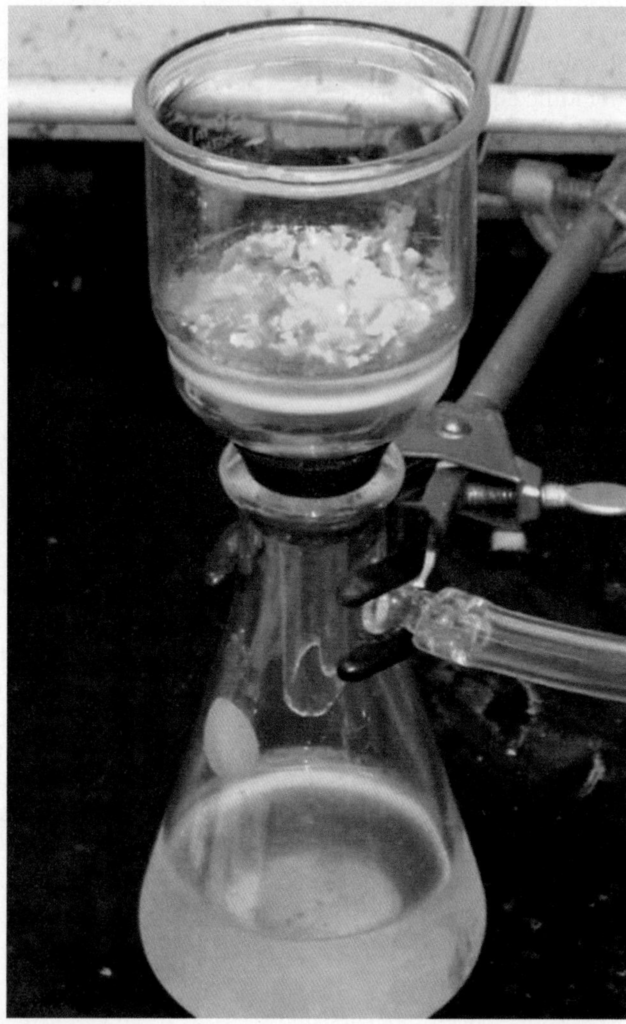

FIGURE 4.17 Precipitate may be removed from a reaction mixture by filtration.

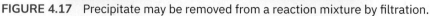

EXAMPLE 4.15

Gravimetric Analysis

A 0.4550-g solid mixture containing $MgSO_4$ is dissolved in water and treated with an excess of $Ba(NO_3)_2$, resulting in the precipitation of 0.6168 g of $BaSO_4$.

$$MgSO_4(aq) + Ba(NO_3)_2(aq) \longrightarrow BaSO_4(s) + Mg(NO_3)_2(aq)$$

What is the concentration (mass percent) of $MgSO_4$ in the mixture?

Solution

The plan for this calculation is similar to others used in stoichiometric calculations, the central step being the connection between the moles of $BaSO_4$ and $MgSO_4$ through their stoichiometric factor. Once the mass of $MgSO_4$ is computed, it may be used along with the mass of the sample mixture to calculate the requested percentage concentration.

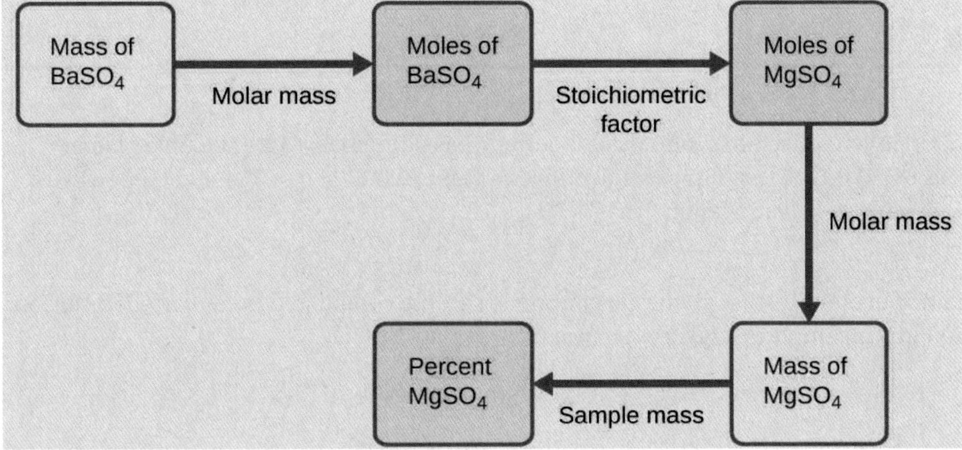

The mass of $MgSO_4$ that would yield the provided precipitate mass is

$$0.6168 \cancel{\text{g BaSO}_4} \times \frac{1 \cancel{\text{mol BaSO}_4}}{233.43 \cancel{\text{g BaSO}_4}} \times \frac{1 \cancel{\text{mol MgSO}_4}}{1 \cancel{\text{mol BaSO}_4}} \times \frac{120.37 \text{ g MgSO}_4}{1 \cancel{\text{mol MgSO}_4}} = 0.3181 \text{ g MgSO}_4$$

The concentration of $MgSO_4$ in the sample mixture is then calculated to be

$$\text{percent MgSO}_4 = \frac{\text{mass MgSO}_4}{\text{mass sample}} \times 100\%$$

$$\frac{0.3181 \text{ g}}{0.4550 \text{ g}} \times 100\% = 69.91\%$$

Check Your Learning

What is the percent of chloride ion in a sample if 1.1324 g of the sample produces 1.0881 g of AgCl when treated with excess Ag^+?

$$\text{Ag}^+(aq) + \text{Cl}^-(aq) \longrightarrow \text{AgCl}(s)$$

Answer:

23.76%

The elemental composition of hydrocarbons and related compounds may be determined via a gravimetric method known as **combustion analysis**. In a combustion analysis, a weighed sample of the compound is heated to a high temperature under a stream of oxygen gas, resulting in its complete combustion to yield gaseous products of known identities. The complete combustion of hydrocarbons, for example, will yield carbon dioxide and water as the only products. The gaseous combustion products are swept through separate, preweighed collection devices containing compounds that selectively absorb each product (Figure 4.18). The mass increase of each device corresponds to the mass of the absorbed product and may be used in an appropriate stoichiometric calculation to derive the mass of the relevant element.

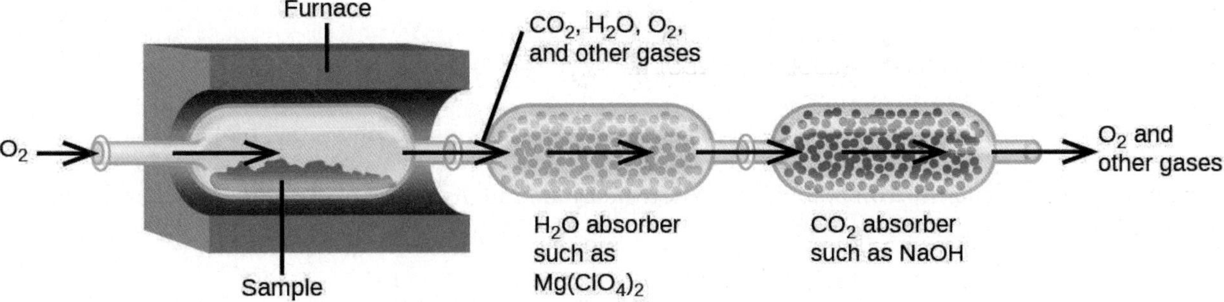

FIGURE 4.18 This schematic diagram illustrates the basic components of a combustion analysis device for determining the carbon and hydrogen content of a sample.

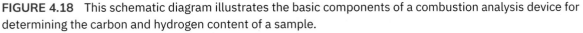

Combustion Analysis

Polyethylene is a hydrocarbon polymer used to produce food-storage bags and many other flexible plastic items. A combustion analysis of a 0.00126-g sample of polyethylene yields 0.00394 g of CO_2 and 0.00161 g of H_2O. What is the empirical formula of polyethylene?

Solution

The primary assumption in this exercise is that all the carbon in the sample combusted is converted to carbon dioxide, and all the hydrogen in the sample is converted to water:

$$C_xH_y(s) + \text{excess } O_2(g) \longrightarrow xCO_2(g) + \frac{y}{2}H_2O(g)$$

Note that a balanced equation is not necessary for the task at hand. To derive the empirical formula of the compound, only the subscripts x and y are needed.

First, calculate the molar amounts of carbon and hydrogen in the sample, using the provided masses of the carbon dioxide and water, respectively. With these molar amounts, the empirical formula for the compound may be written as described in the previous chapter of this text. An outline of this approach is given in the following flow chart:

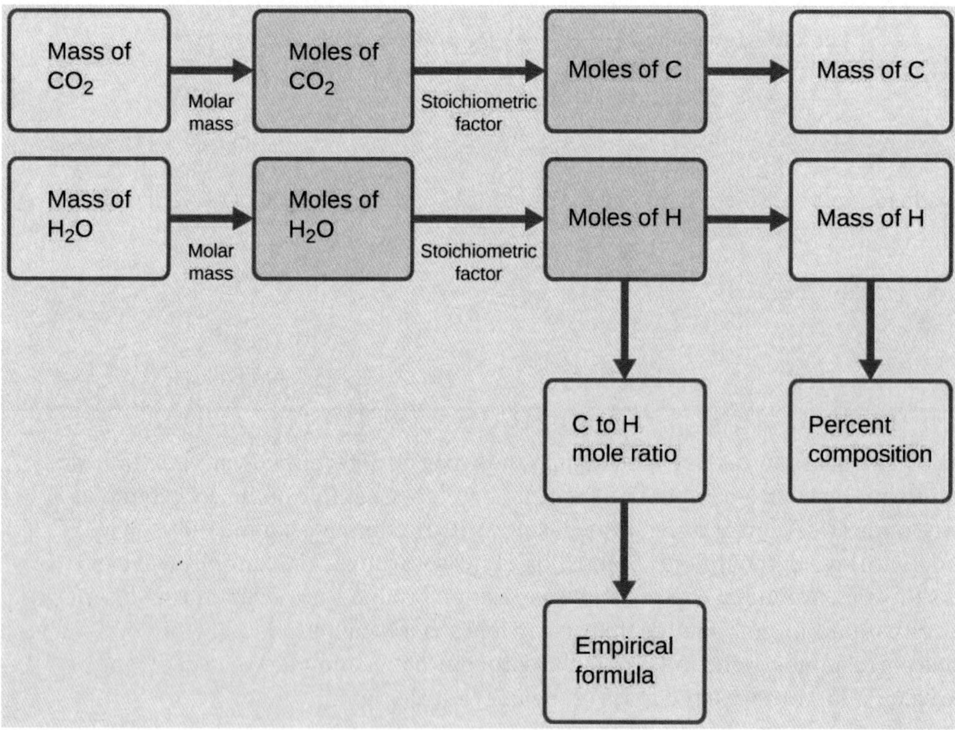

$$\text{mol C} = 0.00394 \text{ g } CO_2 \times \frac{1 \text{ mol } CO_2}{44.01 \text{ g}} \times \frac{1 \text{ mol C}}{1 \text{ mol } CO_2} = 8.95 \times 10^{-5} \text{ mol C}$$

$$\text{mol H} = 0.00161 \text{ g } H_2O \times \frac{1 \text{ mol } H_2O}{18.02 \text{ g}} \times \frac{2 \text{ mol H}}{1 \text{ mol } H_2O} = 1.79 \times 10^{-4} \text{ mol H}$$

The empirical formula for the compound is then derived by identifying the smallest whole-number multiples for these molar amounts. The H-to-C molar ratio is

$$\frac{\text{mol H}}{\text{mol C}} = \frac{1.79 \times 10^{-4} \text{ mol H}}{8.95 \times 10^{-5} \text{ mol C}} = \frac{2 \text{ mol H}}{1 \text{ mol C}}$$

and the empirical formula for polyethylene is CH_2.

Check Your Learning

A 0.00215-g sample of polystyrene, a polymer composed of carbon and hydrogen, produced 0.00726 g of CO_2 and 0.00148 g of H_2O in a combustion analysis. What is the empirical formula for polystyrene?

Answer:

CH

Key Terms

acid substance that produces H_3O^+ when dissolved in water

acid-base reaction reaction involving the transfer of a hydrogen ion between reactant species

actual yield amount of product formed in a reaction

analyte chemical species of interest

balanced equation chemical equation with equal numbers of atoms for each element in the reactant and product

base substance that produces OH^- when dissolved in water

buret device used for the precise delivery of variable liquid volumes, such as in a titration analysis

chemical equation symbolic representation of a chemical reaction

coefficient number placed in front of symbols or formulas in a chemical equation to indicate their relative amount

combustion analysis gravimetric technique used to determine the elemental composition of a compound via the collection and weighing of its gaseous combustion products

combustion reaction vigorous redox reaction producing significant amounts of energy in the form of heat and, sometimes, light

complete ionic equation chemical equation in which all dissolved ionic reactants and products, including spectator ions, are explicitly represented by formulas for their dissociated ions

end point measured volume of titrant solution that yields the change in sample solution appearance or other property expected for stoichiometric equivalence (see *equivalence point*)

equivalence point volume of titrant solution required to react completely with the analyte in a titration analysis; provides a stoichiometric amount of titrant for the sample's analyte according to the titration reaction

excess reactant reactant present in an amount greater than required by the reaction stoichiometry

gravimetric analysis quantitative chemical analysis method involving the separation of an analyte from a sample by a physical or chemical process and subsequent mass measurements of the analyte, reaction product, and/or sample

half-reaction an equation that shows whether each reactant loses or gains electrons in a reaction.

indicator substance added to the sample in a titration analysis to permit visual detection of the end point

insoluble of relatively low solubility; dissolving only to a slight extent

limiting reactant reactant present in an amount lower than required by the reaction stoichiometry, thus limiting the amount of product generated

molecular equation chemical equation in which all reactants and products are represented as neutral substances

net ionic equation chemical equation in which only those dissolved ionic reactants and products that undergo a chemical or physical change are represented (excludes spectator ions)

neutralization reaction reaction between an acid and a base to produce salt and water

oxidation process in which an element's oxidation number is increased by loss of electrons

oxidation number (also, oxidation state) the charge each atom of an element would have in a compound if the compound were ionic

oxidation-reduction reaction (also, redox reaction) reaction involving a change in oxidation number for one or more reactant elements

oxidizing agent (also, oxidant) substance that brings about the oxidation of another substance, and in the process becomes reduced

percent yield measure of the efficiency of a reaction, expressed as a percentage of the theoretical yield

precipitate insoluble product that forms from reaction of soluble reactants

precipitation reaction reaction that produces one or more insoluble products; when reactants are ionic compounds, sometimes called double-displacement or metathesis

product substance formed by a chemical or physical change; shown on the right side of the arrow in a chemical equation

quantitative analysis the determination of the amount or concentration of a substance in a sample

reactant substance undergoing a chemical or physical change; shown on the left side of the arrow in a chemical equation

reducing agent (also, reductant) substance that brings about the reduction of another substance, and in the process becomes oxidized

reduction process in which an element's oxidation number is decreased by gain of electrons

salt ionic compound that can be formed by the

reaction of an acid with a base that contains a cation and an anion other than hydroxide or oxide

single-displacement reaction (also, replacement) redox reaction involving the oxidation of an elemental substance by an ionic species

solubility the extent to which a substance may be dissolved in water, or any solvent

soluble of relatively high solubility; dissolving to a relatively large extent

spectator ion ion that does not undergo a chemical or physical change during a reaction, but its presence is required to maintain charge neutrality

stoichiometric factor ratio of coefficients in a balanced chemical equation, used in computations relating amounts of reactants and products

stoichiometry relationships between the amounts of reactants and products of a chemical reaction

strong acid acid that reacts completely when dissolved in water to yield hydronium ions

strong base base that reacts completely when dissolved in water to yield hydroxide ions

theoretical yield amount of product that may be produced from a given amount of reactant(s) according to the reaction stoichiometry

titrant solution containing a known concentration of substance that will react with the analyte in a titration analysis

titration analysis quantitative chemical analysis method that involves measuring the volume of a reactant solution required to completely react with the analyte in a sample

weak acid acid that reacts only to a slight extent when dissolved in water to yield hydronium ions

weak base base that reacts only to a slight extent when dissolved in water to yield hydroxide ions

Key Equations

$$\text{percent yield} = \left(\frac{\text{actual yield}}{\text{theoretical yield}}\right) \times 100$$

Summary

4.1 Writing and Balancing Chemical Equations

Chemical equations are symbolic representations of chemical and physical changes. Formulas for the substances undergoing the change (reactants) and substances generated by the change (products) are separated by an arrow and preceded by integer coefficients indicating their relative numbers. Balanced equations are those whose coefficients result in equal numbers of atoms for each element in the reactants and products. Chemical reactions in aqueous solution that involve ionic reactants or products may be represented more realistically by complete ionic equations and, more succinctly, by net ionic equations.

4.2 Classifying Chemical Reactions

Chemical reactions are classified according to similar patterns of behavior. A large number of important reactions are included in three categories: precipitation, acid-base, and oxidation-reduction (redox). Precipitation reactions involve the formation of one or more insoluble products. Acid-base reactions involve the transfer of hydrogen ions between reactants. Redox reactions involve a change in oxidation number for one or more reactant elements. Writing balanced equations for some redox reactions that occur in aqueous solutions is simplified by using a systematic approach called the half-reaction method.

4.3 Reaction Stoichiometry

A balanced chemical equation may be used to describe a reaction's stoichiometry (the relationships between amounts of reactants and products). Coefficients from the equation are used to derive stoichiometric factors that subsequently may be used for computations relating reactant and product masses, molar amounts, and other quantitative properties.

4.4 Reaction Yields

When reactions are carried out using less-than-stoichiometric quantities of reactants, the amount of product generated will be determined by the limiting reactant. The amount of product generated by a chemical reaction is its actual yield. This yield is often less than the amount of product predicted by the stoichiometry of the balanced chemical equation representing the reaction (its theoretical yield). The extent to which a reaction generates the theoretical amount of product is expressed as its percent yield.

4.5 Quantitative Chemical Analysis

The stoichiometry of chemical reactions may serve as the basis for quantitative chemical analysis methods. Titrations involve measuring the volume of a titrant solution required to completely react with a sample solution. This volume is then used to calculate the concentration of analyte in the sample using the stoichiometry of the titration reaction. Gravimetric analysis involves separating the analyte from the sample by a physical or chemical process, determining its mass, and then calculating its concentration in the sample based on the stoichiometry of the relevant process. Combustion analysis is a gravimetric method used to determine the elemental composition of a compound by collecting and weighing the gaseous products of its combustion.

Exercises

4.1 Writing and Balancing Chemical Equations

1. What does it mean to say an equation is balanced? Why is it important for an equation to be balanced?
2. Consider molecular, complete ionic, and net ionic equations.
(a) What is the difference between these types of equations?
(b) In what circumstance would the complete and net ionic equations for a reaction be identical?
3. Balance the following equations:
(a) $PCl_5(s) + H_2O(l) \longrightarrow POCl_3(l) + HCl(aq)$
(b) $Cu(s) + HNO_3(aq) \longrightarrow Cu(NO_3)_2(aq) + H_2O(l) + NO(g)$
(c) $H_2(g) + I_2(s) \longrightarrow HI(s)$
(d) $Fe(s) + O_2(g) \longrightarrow Fe_2O_3(s)$
(e) $Na(s) + H_2O(l) \longrightarrow NaOH(aq) + H_2(g)$
(f) $(NH_4)_2Cr_2O_7(s) \longrightarrow Cr_2O_3(s) + N_2(g) + H_2O(g)$
(g) $P_4(s) + Cl_2(g) \longrightarrow PCl_3(l)$
(h) $PtCl_4(s) \longrightarrow Pt(s) + Cl_2(g)$
4. Balance the following equations:
(a) $Ag(s) + H_2S(g) + O_2(g) \longrightarrow Ag_2S(s) + H_2O(l)$
(b) $P_4(s) + O_2(g) \longrightarrow P_4O_{10}(s)$
(c) $Pb(s) + H_2O(l) + O_2(g) \longrightarrow Pb(OH)_2(s)$
(d) $Fe(s) + H_2O(l) \longrightarrow Fe_3O_4(s) + H_2(g)$
(e) $Sc_2O_3(s) + SO_3(l) \longrightarrow Sc_2(SO_4)_3(s)$
(f) $Ca_3(PO_4)_2(aq) + H_3PO_4(aq) \longrightarrow Ca(H_2PO_4)_2(aq)$
(g) $Al(s) + H_2SO_4(aq) \longrightarrow Al_2(SO_4)_3(s) + H_2(g)$
(h) $TiCl_4(s) + H_2O(g) \longrightarrow TiO_2(s) + HCl(g)$
5. Write a balanced molecular equation describing each of the following chemical reactions.
(a) Solid calcium carbonate is heated and decomposes to solid calcium oxide and carbon dioxide gas.
(b) Gaseous butane, C_4H_{10}, reacts with diatomic oxygen gas to yield gaseous carbon dioxide and water vapor.
(c) Aqueous solutions of magnesium chloride and sodium hydroxide react to produce solid magnesium hydroxide and aqueous sodium chloride.
(d) Water vapor reacts with sodium metal to produce solid sodium hydroxide and hydrogen gas.
6. Write a balanced equation describing each of the following chemical reactions.
(a) Solid potassium chlorate, $KClO_3$, decomposes to form solid potassium chloride and diatomic oxygen gas.
(b) Solid aluminum metal reacts with solid diatomic iodine to form solid Al_2I_6.
(c) When solid sodium chloride is added to aqueous sulfuric acid, hydrogen chloride gas and aqueous sodium sulfate are produced.
(d) Aqueous solutions of phosphoric acid and potassium hydroxide react to produce aqueous potassium dihydrogen phosphate and liquid water.

7. Colorful fireworks often involve the decomposition of barium nitrate and potassium chlorate and the reaction of the metals magnesium, aluminum, and iron with oxygen.
(a) Write the formulas of barium nitrate and potassium chlorate.
(b) The decomposition of solid potassium chlorate leads to the formation of solid potassium chloride and diatomic oxygen gas. Write an equation for the reaction.
(c) The decomposition of solid barium nitrate leads to the formation of solid barium oxide, diatomic nitrogen gas, and diatomic oxygen gas. Write an equation for the reaction.
(d) Write separate equations for the reactions of the solid metals magnesium, aluminum, and iron with diatomic oxygen gas to yield the corresponding metal oxides. (Assume the iron oxide contains Fe^{3+} ions.)
8. Fill in the blank with a single chemical formula for a covalent compound that will balance the equation:

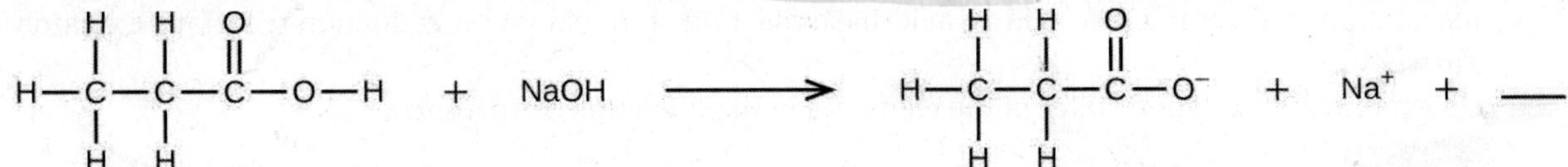

9. Aqueous hydrogen fluoride (hydrofluoric acid) is used to etch glass and to analyze minerals for their silicon content. Hydrogen fluoride will also react with sand (silicon dioxide).
(a) Write an equation for the reaction of solid silicon dioxide with hydrofluoric acid to yield gaseous silicon tetrafluoride and liquid water.
(b) The mineral fluorite (calcium fluoride) occurs extensively in Illinois. Solid calcium fluoride can also be prepared by the reaction of aqueous solutions of calcium chloride and sodium fluoride, yielding aqueous sodium chloride as the other product. Write complete and net ionic equations for this reaction.
10. A novel process for obtaining magnesium from sea water involves several reactions. Write a balanced chemical equation for each step of the process.
(a) The first step is the decomposition of solid calcium carbonate from seashells to form solid calcium oxide and gaseous carbon dioxide.
(b) The second step is the formation of solid calcium hydroxide as the only product from the reaction of the solid calcium oxide with liquid water.
(c) Solid calcium hydroxide is then added to the seawater, reacting with dissolved magnesium chloride to yield solid magnesium hydroxide and aqueous calcium chloride.
(d) The solid magnesium hydroxide is added to a hydrochloric acid solution, producing dissolved magnesium chloride and liquid water.
(e) Finally, the magnesium chloride is melted and electrolyzed to yield liquid magnesium metal and diatomic chlorine gas.
11. From the balanced molecular equations, write the complete ionic and net ionic equations for the following:
(a) $K_2C_2O_4(aq) + Ba(OH)_2(aq) \longrightarrow 2KOH(aq) + BaC_2O_4(s)$
(b) $Pb(NO_3)_2(aq) + H_2SO_4(aq) \longrightarrow PbSO_4(s) + 2HNO_3(aq)$
(c) $CaCO_3(s) + H_2SO_4(aq) \longrightarrow CaSO_4(s) + CO_2(g) + H_2O(l)$

4.2 Classifying Chemical Reactions

12. Use the following equations to answer the next four questions:
i. $H_2O(s) \longrightarrow H_2O(l)$
ii. $Na^+(aq) + Cl^-(aq) + Ag^+(aq) + NO_3{}^-(aq) \longrightarrow AgCl(s) + Na^+(aq) + NO_3{}^-(aq)$
iii. $CH_3OH(g) + O_2(g) \longrightarrow CO_2(g) + H_2O(g)$
iv. $2H_2O(l) \longrightarrow 2H_2(g) + O_2(g)$
v. $H^+(aq) + OH^-(aq) \longrightarrow H_2O(l)$
(a) Which equation describes a physical change?
(b) Which equation identifies the reactants and products of a combustion reaction?
(c) Which equation is not balanced?
(d) Which is a net ionic equation?

13. Indicate what type, or types, of reaction each of the following represents:
(a) $Ca(s) + Br_2(l) \longrightarrow CaBr_2(s)$
(b) $Ca(OH)_2(aq) + 2HBr(aq) \longrightarrow CaBr_2(aq) + 2H_2O(l)$
(c) $C_6H_{12}(l) + 9O_2(g) \longrightarrow 6CO_2(g) + 6H_2O(g)$

14. Indicate what type, or types, of reaction each of the following represents:
(a) $H_2O(g) + C(s) \longrightarrow CO(g) + H_2(g)$
(b) $2KClO_3(s) \longrightarrow 2KCl(s) + 3O_2(g)$
(c) $Al(OH)_3(aq) + 3HCl(aq) \longrightarrow AlCl_3(aq) + 3H_2O(l)$
(d) $Pb(NO_3)_2(aq) + H_2SO_4(aq) \longrightarrow PbSO_4(s) + 2HNO_3(aq)$

15. Silver can be separated from gold because silver dissolves in nitric acid while gold does not. Is the dissolution of silver in nitric acid an acid-base reaction or an oxidation-reduction reaction? Explain your answer.

16. Determine the oxidation states of the elements in the following compounds:
(a) NaI
(b) $GdCl_3$
(c) $LiNO_3$
(d) H_2Se
(e) Mg_2Si
(f) RbO_2, rubidium superoxide
(g) HF

17. Determine the oxidation states of the elements in the compounds listed. None of the oxygen-containing compounds are peroxides or superoxides.
(a) H_3PO_4
(b) $Al(OH)_3$
(c) SeO_2
(d) KNO_2
(e) In_2S_3
(f) P_4O_6

18. Determine the oxidation states of the elements in the compounds listed. None of the oxygen-containing compounds are peroxides or superoxides.
(a) H_2SO_4
(b) $Ca(OH)_2$
(c) BrOH
(d) $ClNO_2$
(e) $TiCl_4$
(f) NaH

19. Classify the following as acid-base reactions or oxidation-reduction reactions:
(a) $Na_2S(aq) + 2HCl(aq) \longrightarrow 2NaCl(aq) + H_2S(g)$
(b) $2Na(s) + 2HCl(aq) \longrightarrow 2NaCl(aq) + H_2(g)$
(c) $Mg(s) + Cl_2(g) \longrightarrow MgCl_2(s)$
(d) $MgO(s) + 2HCl(aq) \longrightarrow MgCl_2(aq) + H_2O(l)$
(e) $K_3P(s) + 2O_2(g) \longrightarrow K_3PO_4(s)$
(f) $3KOH(aq) + H_3PO_4(aq) \longrightarrow K_3PO_4(aq) + 3H_2O(l)$

20. Identify the atoms that are oxidized and reduced, the change in oxidation state for each, and the oxidizing and reducing agents in each of the following equations:
(a) $Mg(s) + NiCl_2(aq) \longrightarrow MgCl_2(aq) + Ni(s)$
(b) $PCl_3(l) + Cl_2(g) \longrightarrow PCl_5(s)$
(c) $C_2H_4(g) + 3O_2(g) \longrightarrow 2CO_2(g) + 2H_2O(g)$
(d) $Zn(s) + H_2SO_4(aq) \longrightarrow ZnSO_4(aq) + H_2(g)$
(e) $2K_2S_2O_3(s) + I_2(s) \longrightarrow K_2S_4O_6(s) + 2KI(s)$
(f) $3Cu(s) + 8HNO_3(aq) \longrightarrow 3Cu(NO_3)_2(aq) + 2NO(g) + 4H_2O(l)$

21. Complete and balance the following acid-base equations:
(a) HCl gas reacts with solid $Ca(OH)_2(s)$.
(b) A solution of $Sr(OH)_2$ is added to a solution of HNO_3.
22. Complete and balance the following acid-base equations:
(a) A solution of $HClO_4$ is added to a solution of LiOH.
(b) Aqueous H_2SO_4 reacts with NaOH.
(c) $Ba(OH)_2$ reacts with HF gas.
23. Complete and balance the following oxidation-reduction reactions, which give the highest possible oxidation state for the oxidized atoms.
(a) $Al(s) + F_2(g) \longrightarrow$
(b) $Al(s) + CuBr_2(aq) \longrightarrow$ (single displacement)
(c) $P_4(s) + O_2(g) \longrightarrow$
(d) $Ca(s) + H_2O(l) \longrightarrow$ (products are a strong base and a diatomic gas)
24. Complete and balance the following oxidation-reduction reactions, which give the highest possible oxidation state for the oxidized atoms.
(a) $K(s) + H_2O(l) \longrightarrow$
(b) $Ba(s) + HBr(aq) \longrightarrow$
(c) $Sn(s) + I_2(s) \longrightarrow$
25. Complete and balance the equations for the following acid-base neutralization reactions. If water is used as a solvent, write the reactants and products as aqueous ions. In some cases, there may be more than one correct answer, depending on the amounts of reactants used.
(a) $Mg(OH)_2(s) + HClO_4(aq) \longrightarrow$
(b) $SO_3(g) + H_2O(l) \longrightarrow$ (assume an excess of water and that the product dissolves)
(c) $SrO(s) + H_2SO_4(l) \longrightarrow$
26. When heated to 700–800 °C, diamonds, which are pure carbon, are oxidized by atmospheric oxygen. (They burn!) Write the balanced equation for this reaction.
27. The military has experimented with lasers that produce very intense light when fluorine combines explosively with hydrogen. What is the balanced equation for this reaction?
28. Write the molecular, total ionic, and net ionic equations for the following reactions:
(a) $Ca(OH)_2(aq) + HC_2H_3O_2(aq) \longrightarrow$
(b) $H_3PO_4(aq) + CaCl_2(aq) \longrightarrow$
29. Great Lakes Chemical Company produces bromine, Br_2, from bromide salts such as NaBr, in Arkansas brine by treating the brine with chlorine gas. Write a balanced equation for the reaction of NaBr with Cl_2.
30. In a common experiment in the general chemistry laboratory, magnesium metal is heated in air to produce MgO. MgO is a white solid, but in these experiments it often looks gray, due to small amounts of Mg_3N_2, a compound formed as some of the magnesium reacts with nitrogen. Write a balanced equation for each reaction.
31. Lithium hydroxide may be used to absorb carbon dioxide in enclosed environments, such as manned spacecraft and submarines. Write an equation for the reaction that involves 2 mol of LiOH per 1 mol of CO_2. (Hint: Water is one of the products.)
32. Calcium propionate is sometimes added to bread to retard spoilage. This compound can be prepared by the reaction of calcium carbonate, $CaCO_3$, with propionic acid, $C_2H_5CO_2H$, which has properties similar to those of acetic acid. Write the balanced equation for the formation of calcium propionate.
33. Complete and balance the equations of the following reactions, each of which could be used to remove hydrogen sulfide from natural gas:
(a) $Ca(OH)_2(s) + H_2S(g) \longrightarrow$
(b) $Na_2CO_3(aq) + H_2S(g) \longrightarrow$
34. Copper(II) sulfide is oxidized by molecular oxygen to produce gaseous sulfur trioxide and solid copper(II) oxide. The gaseous product then reacts with liquid water to produce liquid dihydrogen sulfate as the only product. Write the two equations which represent these reactions.

35. Write balanced chemical equations for the reactions used to prepare each of the following compounds from the given starting material(s). In some cases, additional reactants may be required.
(a) solid ammonium nitrate from gaseous molecular nitrogen via a two-step process (first reduce the nitrogen to ammonia, then neutralize the ammonia with an appropriate acid)
(b) gaseous hydrogen bromide from liquid molecular bromine via a one-step redox reaction
(c) gaseous H_2S from solid Zn and S via a two-step process (first a redox reaction between the starting materials, then reaction of the product with a strong acid)

36. Calcium cyclamate $Ca(C_6H_{11}NHSO_3)_2$ is an artificial sweetener used in many countries around the world but is banned in the United States. It can be purified industrially by converting it to the barium salt through reaction of the acid $C_6H_{11}NHSO_3H$ with barium carbonate, treatment with sulfuric acid (barium sulfate is very insoluble), and then neutralization with calcium hydroxide. Write the balanced equations for these reactions.

37. Complete and balance each of the following half-reactions (steps 2–5 in half-reaction method):
(a) $Sn^{4+}(aq) \longrightarrow Sn^{2+}(aq)$
(b) $\left[Ag(NH_3)_2\right]^+(aq) \longrightarrow Ag(s) + NH_3(aq)$
(c) $Hg_2Cl_2(s) \longrightarrow Hg(l) + Cl^-(aq)$
(d) $H_2O(l) \longrightarrow O_2(g)$ (in acidic solution)
(e) $IO_3{}^-(aq) \longrightarrow I_2(s)$ (in basic solution)
(f) $SO_3{}^{2-}(aq) \longrightarrow SO_4{}^{2-}(aq)$ (in acidic solution)
(g) $MnO_4{}^-(aq) \longrightarrow Mn^{2+}(aq)$ (in acidic solution)
(h) $Cl^-(aq) \longrightarrow ClO_3{}^-(aq)$ (in basic solution)

38. Complete and balance each of the following half-reactions (steps 2–5 in half-reaction method):
(a) $Cr^{2+}(aq) \longrightarrow Cr^{3+}(aq)$
(b) $Hg(l) + Br^-(aq) \longrightarrow HgBr_4{}^{2-}(aq)$
(c) $ZnS(s) \longrightarrow Zn(s) + S^{2-}(aq)$
(d) $H_2(g) \longrightarrow H_2O(l$ (in basic solution)
(e) $H_2(g) \longrightarrow H_3O^+(aq)$ (in acidic solution)
(f) $NO_3{}^-(aq) \longrightarrow HNO_2(aq)$ (in acidic solution)
(g) $MnO_2(s) \longrightarrow MnO_4{}^-(aq)$ (in basic solution)
(h) $Cl^-(aq) \longrightarrow ClO_3{}^-(aq)$ (in acidic solution)

39. Balance each of the following equations according to the half-reaction method:
(a) $Sn^{2+}(aq) + Cu^{2+}(aq) \longrightarrow Sn^{4+}(aq) + Cu^+(aq)$
(b) $H_2S(g) + Hg_2{}^{2+}(aq) \longrightarrow Hg(l) + S(s)$ (in acid)
(c) $CN^-(aq) + ClO_2(aq) \longrightarrow CNO^-(aq) + Cl^-(aq)$ (in acid)
(d) $Fe^{2+}(aq) + Ce^{4+}(aq) \longrightarrow Fe^{3+}(aq) + Ce^{3+}(aq)$
(e) $HBrO(aq) \longrightarrow Br^-(aq) + O_2(g)$ (in acid)

40. Balance each of the following equations according to the half-reaction method:
(a) $Zn(s) + NO_3{}^-(aq) \longrightarrow Zn^{2+}(aq) + N_2(g)$ (in acid)
(b) $Zn(s) + NO_3{}^-(aq) \longrightarrow Zn^{2+}(aq) + NH_3(aq)$ (in base)
(c) $CuS(s) + NO_3{}^-(aq) \longrightarrow Cu^{2+}(aq) + S(s) + NO(g)$ (in acid)
(d) $NH_3(aq) + O_2(g) \longrightarrow NO_2(g)$ (gas phase)
(e) $H_2O_2(aq) + MnO_4{}^-(aq) \longrightarrow Mn^{2+}(aq) + O_2(g)$ (in acid)
(f) $NO_2(g) \longrightarrow NO_3{}^-(aq) + NO_2{}^-(aq)$ (in base)
(g) $Fe^{3+}(aq) + I^-(aq) \longrightarrow Fe^{2+}(aq) + I_2(aq)$

41. Balance each of the following equations according to the half-reaction method:
(a) $MnO_4{}^-(aq) + NO_2{}^-(aq) \longrightarrow MnO_2(s) + NO_3{}^-(aq)$ (in base)
(b) $MnO_4{}^{2-}(aq) \longrightarrow MnO_4{}^-(aq) + MnO_2(s)$ (in base)
(c) $Br_2(l) + SO_2(g) \longrightarrow Br^-(aq) + SO_4{}^{2-}(aq)$ (in acid)

4.3 Reaction Stoichiometry

42. Write the balanced equation, then outline the steps necessary to determine the information requested in each of the following:
(a) The number of moles and the mass of chlorine, Cl_2, required to react with 10.0 g of sodium metal, Na, to produce sodium chloride, NaCl.
(b) The number of moles and the mass of oxygen formed by the decomposition of 1.252 g of mercury(II) oxide.
(c) The number of moles and the mass of sodium nitrate, $NaNO_3$, required to produce 128 g of oxygen. ($NaNO_2$ is the other product.)
(d) The number of moles and the mass of carbon dioxide formed by the combustion of 20.0 kg of carbon in an excess of oxygen.
(e) The number of moles and the mass of copper(II) carbonate needed to produce 1.500 kg of copper(II) oxide. (CO_2 is the other product.)
(f)

The number of moles and the mass of $\begin{array}{c}\mathrm{H}\quad\mathrm{H}\\ \mathrm{Br{-}\overset{|}{\underset{|}{C}}{-}\overset{|}{\underset{|}{C}}{-}Br}\\ \mathrm{H}\quad\mathrm{H}\end{array}$ formed by the reaction of 12.85 g of $\begin{array}{c}\mathrm{H}\quad\quad\mathrm{H}\\ \mathrm{C{=}C}\\ \mathrm{H}\quad\quad\mathrm{H}\end{array}$ with an excess of Br_2.

43. Determine the number of moles and the mass requested for each reaction in Exercise 4.42.

44. Write the balanced equation, then outline the steps necessary to determine the information requested in each of the following:
(a) The number of moles and the mass of Mg required to react with 5.00 g of HCl and produce $MgCl_2$ and H_2.
(b) The number of moles and the mass of oxygen formed by the decomposition of 1.252 g of silver(I) oxide.
(c) The number of moles and the mass of magnesium carbonate, $MgCO_3$, required to produce 283 g of carbon dioxide. (MgO is the other product.)
(d) The number of moles and the mass of water formed by the combustion of 20.0 kg of acetylene, C_2H_2, in an excess of oxygen.
(e) The number of moles and the mass of barium peroxide, BaO_2, needed to produce 2.500 kg of barium oxide, BaO (O_2 is the other product.)
(f)

The number of moles and the mass of $\begin{array}{c}\mathrm{H}\quad\quad\mathrm{H}\\ \mathrm{C{=}C}\\ \mathrm{H}\quad\quad\mathrm{H}\end{array}$ required to react with H_2O to produce 9.55 g of $\begin{array}{c}\mathrm{H}\quad\mathrm{H}\\ \mathrm{H{-}\overset{|}{\underset{|}{C}}{-}\overset{|}{\underset{|}{C}}{-}O{-}H}\\ \mathrm{H}\quad\mathrm{H}\end{array}$.

45. Determine the number of moles and the mass requested for each reaction in Exercise 4.44.

46. H_2 is produced by the reaction of 118.5 mL of a 0.8775-M solution of H_3PO_4 according to the following equation: $2Cr + 2H_3PO_4 \longrightarrow 3H_2 + 2CrPO_4$.
(a) Outline the steps necessary to determine the number of moles and mass of H_2.
(b) Perform the calculations outlined.

47. Gallium chloride is formed by the reaction of 2.6 L of a 1.44 M solution of HCl according to the following equation: $2Ga + 6HCl \longrightarrow 2GaCl_3 + 3H_2$.
(a) Outline the steps necessary to determine the number of moles and mass of gallium chloride.
(b) Perform the calculations outlined.

48. I_2 is produced by the reaction of 0.4235 mol of $CuCl_2$ according to the following equation: $2CuCl_2 + 4KI \longrightarrow 2CuI + 4KCl + I_2$.
(a) How many molecules of I_2 are produced?
(b) What mass of I_2 is produced?

49. Silver is often extracted from ores such as $K[Ag(CN)_2]$ and then recovered by the reaction
$2K\left[Ag(CN)_2\right](aq) + Zn(s) \longrightarrow 2Ag(s) + Zn(CN)_2(aq) + 2KCN(aq)$
(a) How many molecules of $Zn(CN)_2$ are produced by the reaction of 35.27 g of $K[Ag(CN)_2]$?
(b) What mass of $Zn(CN)_2$ is produced?

50. What mass of silver oxide, Ag_2O, is required to produce 25.0 g of silver sulfadiazine, $AgC_{10}H_9N_4SO_2$, from the reaction of silver oxide and sulfadiazine?
$2C_{10}H_{10}N_4SO_2 + Ag_2O \longrightarrow 2AgC_{10}H_9N_4SO_2 + H_2O$

51. Carborundum is silicon carbide, SiC, a very hard material used as an abrasive on sandpaper and in other applications. It is prepared by the reaction of pure sand, SiO_2, with carbon at high temperature. Carbon monoxide, CO, is the other product of this reaction. Write the balanced equation for the reaction, and calculate how much SiO_2 is required to produce 3.00 kg of SiC.

52. Automotive air bags inflate when a sample of sodium azide, NaN_3, is very rapidly decomposed.
$2NaN_3(s) \longrightarrow 2Na(s) + 3N_2(g)$
What mass of sodium azide is required to produce 2.6 ft^3 (73.6 L) of nitrogen gas with a density of 1.25 g/L?

53. Urea, $CO(NH_2)_2$, is manufactured on a large scale for use in producing urea-formaldehyde plastics and as a fertilizer. What is the maximum mass of urea that can be manufactured from the CO_2 produced by combustion of 1.00×10^3 kg of carbon followed by the reaction?
$CO_2(g) + 2NH_3(g) \longrightarrow CO(NH_2)_2(s) + H_2O(l)$

54. In an accident, a solution containing 2.5 kg of nitric acid was spilled. Two kilograms of Na_2CO_3 was quickly spread on the area and CO_2 was released by the reaction. Was sufficient Na_2CO_3 used to neutralize all of the acid?

55. A compact car gets 37.5 miles per gallon on the highway. If gasoline contains 84.2% carbon by mass and has a density of 0.8205 g/mL, determine the mass of carbon dioxide produced during a 500-mile trip (3.785 liters per gallon).

56. What volume of 0.750 M hydrochloric acid solution can be prepared from the HCl produced by the reaction of 25.0 g of NaCl with excess sulfuric acid?
$NaCl(s) + H_2SO_4(l) \longrightarrow HCl(g) + NaHSO_4(s)$

57. What volume of a 0.2089 *M* KI solution contains enough KI to react exactly with the $Cu(NO_3)_2$ in 43.88 mL of a 0.3842 *M* solution of $Cu(NO_3)_2$?
$2Cu(NO_3)_2 + 4KI \longrightarrow 2CuI + I_2 + 4KNO_3$

58. A mordant is a substance that combines with a dye to produce a stable fixed color in a dyed fabric. Calcium acetate is used as a mordant. It is prepared by the reaction of acetic acid with calcium hydroxide.
$2CH_3CO_2H + Ca(OH)_2 \longrightarrow Ca(CH_3CO_2)_2 + 2H_2O$
What mass of $Ca(OH)_2$ is required to react with the acetic acid in 25.0 mL of a solution having a density of 1.065 g/mL and containing 58.0% acetic acid by mass?

59. The toxic pigment called white lead, $Pb_3(OH)_2(CO_3)_2$, has been replaced in white paints by rutile, TiO_2. How much rutile (g) can be prepared from 379 g of an ore that contains 88.3% ilmenite ($FeTiO_3$) by mass?
$2FeTiO_3 + 4HCl + Cl_2 \longrightarrow 2FeCl_3 + 2TiO_2 + 2H_2O$

4.4 Reaction Yields

60. The following quantities are placed in a container: 1.5×10^{24} atoms of hydrogen, 1.0 mol of sulfur, and 88.0 g of diatomic oxygen.
(a) What is the total mass in grams for the collection of all three elements?
(b) What is the total number of moles of atoms for the three elements?
(c) If the mixture of the three elements formed a compound with molecules that contain two hydrogen atoms, one sulfur atom, and four oxygen atoms, which substance is consumed first?
(d) How many atoms of each remaining element would remain unreacted in the change described in (c)?

61. What is the limiting reactant in a reaction that produces sodium chloride from 8 g of sodium and 8 g of diatomic chlorine?

62. Which of the postulates of Dalton's atomic theory explains why we can calculate a theoretical yield for a chemical reaction?

63. A student isolated 25 g of a compound following a procedure that would theoretically yield 81 g. What was his percent yield?

64. A sample of 0.53 g of carbon dioxide was obtained by heating 1.31 g of calcium carbonate. What is the percent yield for this reaction?
$CaCO_3(s) \longrightarrow CaO(s) + CO_2(s)$

65. Freon-12, CCl_2F_2, is prepared from CCl_4 by reaction with HF. The other product of this reaction is HCl. Outline the steps needed to determine the percent yield of a reaction that produces 12.5 g of CCl_2F_2 from 32.9 g of CCl_4. Freon-12 has been banned and is no longer used as a refrigerant because it catalyzes the decomposition of ozone and has a very long lifetime in the atmosphere. Determine the percent yield.

66. Citric acid, $C_6H_8O_7$, a component of jams, jellies, and fruity soft drinks, is prepared industrially via fermentation of sucrose by the mold *Aspergillus niger*. The equation representing this reaction is

$$C_{12}H_{22}O_{11} + H_2O + 3O_2 \longrightarrow 2C_6H_8O_7 + 4H_2O$$

What mass of citric acid is produced from exactly 1 metric ton (1.000×10^3 kg) of sucrose if the yield is 92.30%?

67. Toluene, $C_6H_5CH_3$, is oxidized by air under carefully controlled conditions to benzoic acid, $C_6H_5CO_2H$, which is used to prepare the food preservative sodium benzoate, $C_6H_5CO_2Na$. What is the percent yield of a reaction that converts 1.000 kg of toluene to 1.21 kg of benzoic acid?

$$2C_6H_5CH_3 + 3O_2 \longrightarrow 2C_6H_5CO_2H + 2H_2O$$

68. In a laboratory experiment, the reaction of 3.0 mol of H_2 with 2.0 mol of I_2 produced 1.0 mol of HI. Determine the theoretical yield in grams and the percent yield for this reaction.

69. Outline the steps needed to solve the following problem, then do the calculations. Ether, $(C_2H_5)_2O$, which was originally used as an anesthetic but has been replaced by safer and more effective medications, is prepared by the reaction of ethanol with sulfuric acid.

$$2C_2H_5OH + H_2SO_4 \longrightarrow (C_2H_5)_2O + H_2SO_4{\cdot}H_2O$$

What is the percent yield of ether if 1.17 L (d = 0.7134 g/mL) is isolated from the reaction of 1.500 L of C_2H_5OH
(d = 0.7894 g/mL)?

70. Outline the steps needed to determine the limiting reactant when 30.0 g of propane, C_3H_8, is burned with 75.0 g of oxygen.
Determine the limiting reactant.

71. Outline the steps needed to determine the limiting reactant when 0.50 mol of Cr and 0.75 mol of H_3PO_4 react according to the following chemical equation.

$$2Cr + 2H_3PO_4 \longrightarrow 2CrPO_4 + 3H_2$$

Determine the limiting reactant.

72. What is the limiting reactant when 1.50 g of lithium and 1.50 g of nitrogen combine to form lithium nitride, a component of advanced batteries, according to the following unbalanced equation?

$$Li + N_2 \longrightarrow Li_3N$$

73. Uranium can be isolated from its ores by dissolving it as $UO_2(NO_3)_2$, then separating it as solid $UO_2(C_2O_4){\cdot}3H_2O$. Addition of 0.4031 g of sodium oxalate, $Na_2C_2O_4$, to a solution containing 1.481 g of uranyl nitrate, $UO_2(NO_3)_2$, yields 1.073 g of solid $UO_2(C_2O_4){\cdot}3H_2O$.

$$Na_2C_2O_4 + UO_2(NO_3)_2 + 3H_2O \longrightarrow UO_2(C_2O_4){\cdot}3H_2O + 2NaNO_3$$

Determine the limiting reactant and the percent yield of this reaction.

74. How many molecules of $C_2H_4Cl_2$ can be prepared from 15 C_2H_4 molecules and 8 Cl_2 molecules?

75. How many molecules of the sweetener saccharin can be prepared from 30 C atoms, 25 H atoms, 12 O atoms, 8 S atoms, and 14 N atoms?

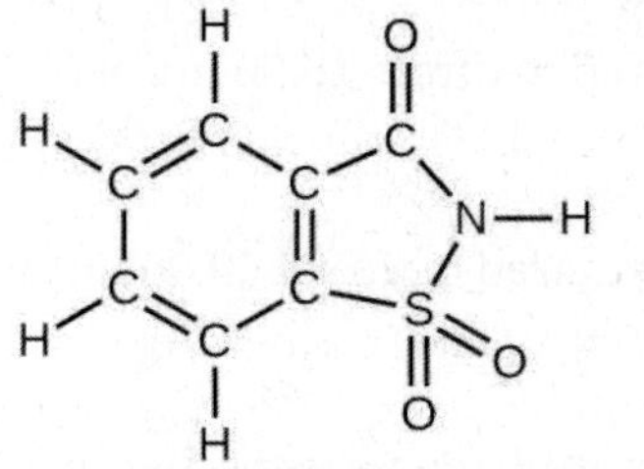

76. The phosphorus pentoxide used to produce phosphoric acid for cola soft drinks is prepared by burning phosphorus in oxygen.

(a) What is the limiting reactant when 0.200 mol of P_4 and 0.200 mol of O_2 react according to

$$P_4 + 5O_2 \longrightarrow P_4O_{10}$$

(b) Calculate the percent yield if 10.0 g of P_4O_{10} is isolated from the reaction.

77. Would you agree to buy 1 trillion (1,000,000,000,000) gold atoms for $5? Explain why or why not. Find the current price of gold at http://money.cnn.com/data/commodities/ (1 troy ounce = 31.1 g)

4.5 Quantitative Chemical Analysis

78. What volume of 0.0105-M HBr solution is required to titrate 125 mL of a 0.0100-*M* $Ca(OH)_2$ solution?
$Ca(OH)_2(aq) + 2HBr(aq) \longrightarrow CaBr_2(aq) + 2H_2O(l)$
79. Titration of a 20.0-mL sample of acid rain required 1.7 mL of 0.0811 *M* NaOH to reach the end point. If we assume that the acidity of the rain is due to the presence of sulfuric acid, what was the concentration of sulfuric acid in this sample of rain?
80. What is the concentration of NaCl in a solution if titration of 15.00 mL of the solution with 0.2503 *M* $AgNO_3$ requires 20.22 mL of the $AgNO_3$ solution to reach the end point?
$AgNO_3(aq) + NaCl(aq) \longrightarrow AgCl(s) + NaNO_3(aq)$
81. In a common medical laboratory determination of the concentration of free chloride ion in blood serum, a serum sample is titrated with a $Hg(NO_3)_2$ solution.
$2Cl^-(aq) + Hg(NO_3)_2(aq) \longrightarrow 2NO_3^-(aq) + HgCl_2(s)$
What is the Cl^- concentration in a 0.25-mL sample of normal serum that requires 1.46 mL of 8.25×10^{-4} *M* $Hg(NO_3)_2(aq)$ to reach the end point?
82. Potatoes can be peeled commercially by soaking them in a 3-M to 6-M solution of sodium hydroxide, then removing the loosened skins by spraying them with water. Does a sodium hydroxide solution have a suitable concentration if titration of 12.00 mL of the solution requires 30.6 mL of 1.65 M HCI to reach the end point?
83. A sample of gallium bromide, $GaBr_3$, weighing 0.165 g was dissolved in water and treated with silver nitrate, $AgNO_3$, resulting in the precipitation of 0.299 g AgBr. Use these data to compute the %Ga (by mass) $GaBr_3$.
84. The principal component of mothballs is naphthalene, a compound with a molecular mass of about 130 amu, containing only carbon and hydrogen. A 3.000-mg sample of naphthalene burns to give 10.3 mg of CO_2. Determine its empirical and molecular formulas.
85. A 0.025-g sample of a compound composed of boron and hydrogen, with a molecular mass of ~28 amu, burns spontaneously when exposed to air, producing 0.063 g of B_2O_3. What are the empirical and molecular formulas of the compound?
86. Sodium bicarbonate (baking soda), $NaHCO_3$, can be purified by dissolving it in hot water (60 °C), filtering to remove insoluble impurities, cooling to 0 °C to precipitate solid $NaHCO_3$, and then filtering to remove the solid, leaving soluble impurities in solution. Any $NaHCO_3$ that remains in solution is not recovered. The solubility of $NaHCO_3$ in hot water of 60 °C is 164 g/L. Its solubility in cold water of 0 °C is 69 g/L. What is the percent yield of $NaHCO_3$ when it is purified by this method?
87. What volume of 0.600 *M* HCl is required to react completely with 2.50 g of sodium hydrogen carbonate?
$NaHCO_3(aq) + HCl(aq) \longrightarrow NaCl(aq) + CO_2(g) + H_2O(l)$
88. What volume of 0.08892 *M* HNO_3 is required to react completely with 0.2352 g of potassium hydrogen phosphate?
$2HNO_3(aq) + K_2HPO_4(aq) \longrightarrow H_3PO_4(aq) + 2KNO_3(aq)$
89. What volume of a 0.3300-*M* solution of sodium hydroxide would be required to titrate 15.00 mL of 0.1500 *M* oxalic acid?
$C_2O_4H_2(aq) + 2NaOH(aq) \longrightarrow Na_2C_2O_4(aq) + 2H_2O(l)$
90. What volume of a 0.00945-*M* solution of potassium hydroxide would be required to titrate 50.00 mL of a sample of acid rain with a H_2SO_4 concentration of 1.23×10^{-4} *M*.
$H_2SO_4(aq) + 2KOH(aq) \longrightarrow K_2SO_4(aq) + 2H_2O(l)$
91. A sample of solid calcium hydroxide, $Ca(OH)_2$, is allowed to stand in water until a saturated solution is formed. A titration of 75.00 mL of this solution with 5.00×10^{-2} *M* HCl requires 36.6 mL of the acid to reach the end point.
$Ca(OH)_2(aq) + 2HCl(aq) \longrightarrow CaCl_2(aq) + 2H_2O(l)$
What is the molarity?

92. What mass of $Ca(OH)_2$ will react with 25.0 g of butanoic to form the preservative calcium butanoate according to the equation?

$$CH_3CH_2CH_2COOH + Ca(OH)_2 \longrightarrow 2Ca(CH_3CH_2CH_2COO)_2 + 2H_2O$$

93. How many milliliters of a 0.1500-M solution of KOH will be required to titrate 40.00 mL of a 0.0656-M solution of H_3PO_4?

$H_3PO_4(aq) + 2KOH(aq) \longrightarrow K_2HPO_4(aq) + 2H_2O(l)$

94. Potassium hydrogen phthalate, $KHC_8H_4O_4$, or KHP, is used in many laboratories, including general chemistry laboratories, to standardize solutions of base. KHP is one of only a few stable solid acids that can be dried by warming and weighed. A 0.3420-g sample of $KHC_8H_4O_4$ reacts with 35.73 mL of a NaOH solution in a titration. What is the molar concentration of the NaOH?

$KHC_8H_4O_4(aq) + NaOH(aq) \longrightarrow KNaC_8H_4O_4(aq) + H_2O(aq)$

95. The reaction of WCl_6 with Al at ~400 °C gives black crystals of a compound containing only tungsten and chlorine. A sample of this compound, when reduced with hydrogen, gives 0.2232 g of tungsten metal and hydrogen chloride, which is absorbed in water. Titration of the hydrochloric acid thus produced requires 46.2 mL of 0.1051 M NaOH to reach the end point. What is the empirical formula of the black tungsten chloride?

CHAPTER 5
Thermochemistry

Figure 5.1 Sliding a match head along a rough surface initiates a combustion reaction that produces energy in the form of heat and light. (credit: modification of work by Laszlo Ilyes)

CHAPTER OUTLINE

5.1 Energy Basics
5.2 Calorimetry
5.3 Enthalpy

INTRODUCTION Chemical reactions, such as those that occur when you light a match, involve changes in energy as well as matter. Societies at all levels of development could not function without the energy released by chemical reactions. In 2012, about 85% of US energy consumption came from the combustion of petroleum products, coal, wood, and garbage. We use this energy to produce electricity (38%); to transport food, raw materials, manufactured goods, and people (27%); for industrial production (21%); and to heat and power our homes and businesses (10%).[1] While these combustion reactions help us meet our essential energy needs, they are also recognized by the majority of the scientific community as a major contributor to global climate change.

Useful forms of energy are also available from a variety of chemical reactions other than combustion. For example, the energy produced by the batteries in a cell phone, car, or flashlight results from chemical reactions. This chapter introduces many of the basic ideas necessary to explore the relationships between chemical changes and energy, with a focus on thermal energy.

1 US Energy Information Administration, *Primary Energy Consumption by Source and Sector, 2012*, http://www.eia.gov/totalenergy/data/monthly/pdf/flow/css_2012_energy.pdf. Data derived from US Energy Information Administration, *Monthly Energy Review* (January 2014).

5.1 Energy Basics

LEARNING OBJECTIVES

By the end of this section, you will be able to:

- Define energy, distinguish types of energy, and describe the nature of energy changes that accompany chemical and physical changes
- Distinguish the related properties of heat, thermal energy, and temperature
- Define and distinguish specific heat and heat capacity, and describe the physical implications of both
- Perform calculations involving heat, specific heat, and temperature change

Chemical changes and their accompanying changes in energy are important parts of our everyday world (Figure 5.2). The macronutrients in food (proteins, fats, and carbohydrates) undergo metabolic reactions that provide the energy to keep our bodies functioning. We burn a variety of fuels (gasoline, natural gas, coal) to produce energy for transportation, heating, and the generation of electricity. Industrial chemical reactions use enormous amounts of energy to produce raw materials (such as iron and aluminum). Energy is then used to manufacture those raw materials into useful products, such as cars, skyscrapers, and bridges.

(a)

(b)

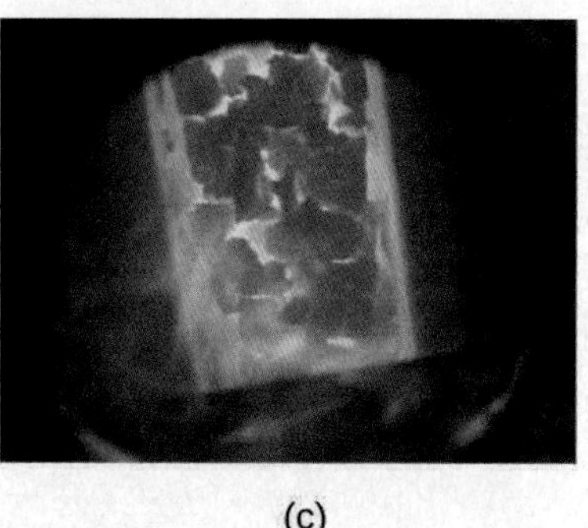
(c)

FIGURE 5.2 The energy involved in chemical changes is important to our daily lives: (a) A cheeseburger for lunch provides the energy you need to get through the rest of the day; (b) the combustion of gasoline provides the energy that moves your car (and you) between home, work, and school; and (c) coke, a processed form of coal, provides the energy needed to convert iron ore into iron, which is essential for making many of the products we use daily. (credit a: modification of work by "Pink Sherbet Photography"/Flickr; credit b: modification of work by Jeffery Turner)

Over 90% of the energy we use comes originally from the sun. Every day, the sun provides the earth with almost 10,000 times the amount of energy necessary to meet all of the world's energy needs for that day. Our challenge is to find ways to convert and store incoming solar energy so that it can be used in reactions or chemical processes that are both convenient and nonpolluting. Plants and many bacteria capture solar energy through photosynthesis. We release the energy stored in plants when we burn wood or plant products such as ethanol. We also use this energy to fuel our bodies by eating food that comes directly from plants or from animals that got their energy by eating plants. Burning coal and petroleum also releases stored solar energy: These fuels are fossilized plant and animal matter.

This chapter will introduce the basic ideas of an important area of science concerned with the amount of heat absorbed or released during chemical and physical changes—an area called **thermochemistry**. The concepts introduced in this chapter are widely used in almost all scientific and technical fields. Food scientists use them to determine the energy content of foods. Biologists study the energetics of living organisms, such as the metabolic combustion of sugar into carbon dioxide and water. The oil, gas, and transportation industries, renewable energy providers, and many others endeavor to find better methods to produce energy for our commercial and personal needs. Engineers strive to improve energy efficiency, find better ways to heat and cool our homes, refrigerate our food and drinks, and meet the energy and cooling needs of computers and electronics, among other applications. Understanding thermochemical principles is essential for chemists, physicists, biologists, geologists, every type of engineer, and just about anyone who studies or does any kind of science.

Energy

Energy can be defined as the capacity to supply heat or do work. One type of **work (*w*)** is the process of causing matter to move against an opposing force. For example, we do work when we inflate a bicycle tire—we move matter (the air in the pump) against the opposing force of the air already in the tire.

Like matter, energy comes in different types. One scheme classifies energy into two types: **potential energy**, the energy an object has because of its relative position, composition, or condition, and **kinetic energy**, the energy that an object possesses because of its motion. Water at the top of a waterfall or dam has potential energy because of its position; when it flows downward through generators, it has kinetic energy that can be used to do work and produce electricity in a hydroelectric plant (Figure 5.3). A battery has potential energy because the chemicals within it can produce electricity that can do work.

(a)

(b)

FIGURE 5.3 (a) Water at a higher elevation, for example, at the top of Victoria Falls, has a higher potential energy than water at a lower elevation. As the water falls, some of its potential energy is converted into kinetic energy. (b) If the water flows through generators at the bottom of a dam, such as the Hoover Dam shown here, its kinetic energy is converted into electrical energy. (credit a: modification of work by Steve Jurvetson; credit b: modification of work by "curimedia"/Wikimedia commons)

Energy can be converted from one form into another, but all of the energy present before a change occurs always exists in some form after the change is completed. This observation is expressed in the law of conservation of energy: during a chemical or physical change, energy can be neither created nor destroyed, although it can be changed in form. (This is also one version of the first law of thermodynamics, as you will learn later.)

When one substance is converted into another, there is always an associated conversion of one form of energy into another. Heat is usually released or absorbed, but sometimes the conversion involves light, electrical energy, or some other form of energy. For example, chemical energy (a type of potential energy) is stored in the molecules that compose gasoline. When gasoline is combusted within the cylinders of a car's engine, the rapidly expanding gaseous products of this chemical reaction generate mechanical energy (a type of kinetic energy) when they move the cylinders' pistons.

According to the law of conservation of matter (seen in an earlier chapter), there is no detectable change in the total amount of matter during a chemical change. When chemical reactions occur, the energy changes are relatively modest and the mass changes are too small to measure, so the laws of conservation of matter and energy hold well. However, in nuclear reactions, the energy changes are much larger (by factors of a million or so), the mass changes are measurable, and matter-energy conversions are significant. This will be examined in more detail in a later chapter on nuclear chemistry.

Thermal Energy, Temperature, and Heat

Thermal energy is kinetic energy associated with the random motion of atoms and molecules. **Temperature** is a quantitative measure of "hot" or "cold." When the atoms and molecules in an object are moving or vibrating quickly, they have a higher average kinetic energy (KE), and we say that the object is "hot." When the atoms and molecules are moving slowly, they have lower average KE, and we say that the object is "cold" (Figure 5.4). Assuming that no chemical reaction or phase change (such as melting or vaporizing) occurs, increasing the amount of thermal energy in a sample of matter will cause its temperature to increase. And, assuming that no chemical reaction or phase change (such as condensation or freezing) occurs, decreasing the amount of thermal energy in a sample of matter will cause its temperature to decrease.

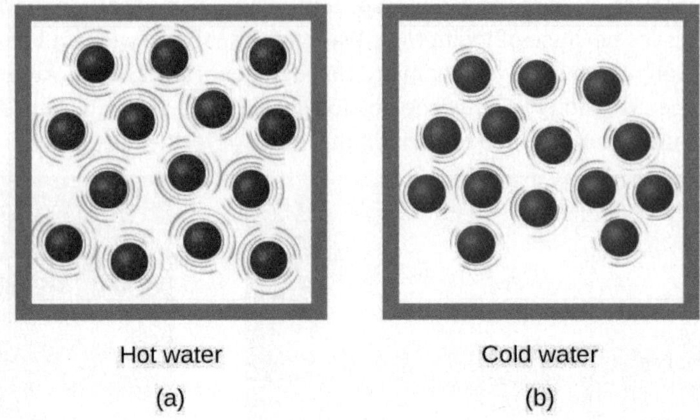

FIGURE 5.4 (a) The molecules in a sample of hot water move more rapidly than (b) those in a sample of cold water.

LINK TO LEARNING

Click on this interactive simulation (http://openstax.org/l/16PHETtempFX) to view the effects of temperature on molecular motion.

Most substances expand as their temperature increases and contract as their temperature decreases. This property can be used to measure temperature changes, as shown in Figure 5.5. The operation of many thermometers depends on the expansion and contraction of substances in response to temperature changes.

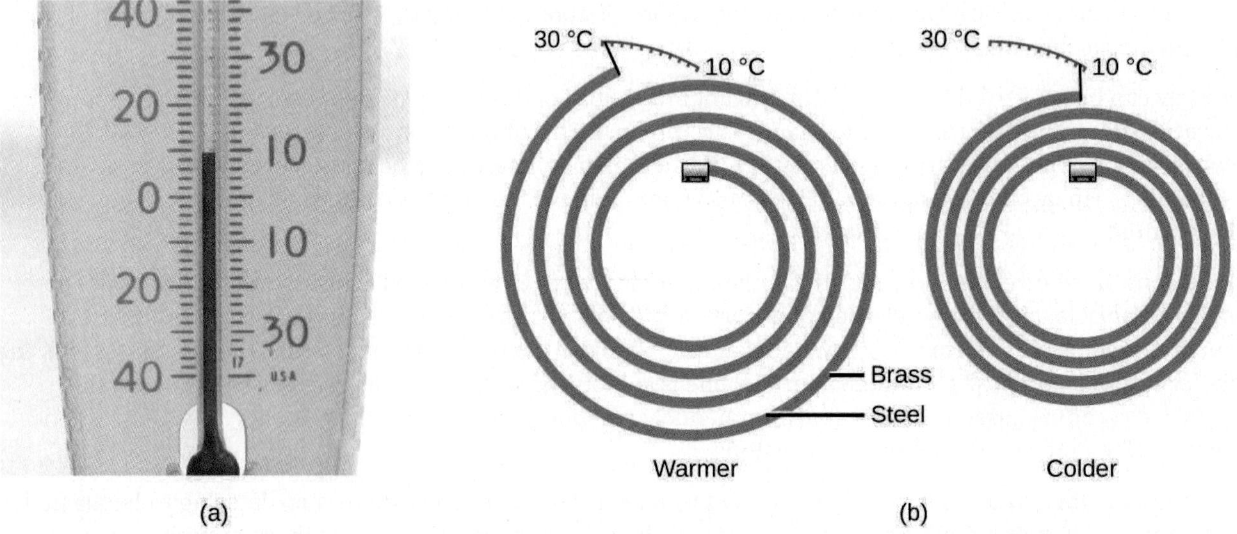

FIGURE 5.5 (a) In an alcohol or mercury thermometer, the liquid (dyed red for visibility) expands when heated and contracts when cooled, much more so than the glass tube that contains the liquid. (b) In a bimetallic thermometer, two different metals (such as brass and steel) form a two-layered strip. When heated or cooled, one of the metals (brass) expands or contracts more than the other metal (steel), causing the strip to coil or uncoil. Both types of thermometers have a calibrated scale that indicates the temperature. (credit a: modification of work by "dwstucke"/Flickr)

LINK TO LEARNING

The following demonstration (http://openstax.org/l/16Bimetallic) allows one to view the effects of heating and cooling a coiled bimetallic strip.

Heat (*q*) is the transfer of thermal energy between two bodies at different temperatures. Heat flow (a redundant term, but one commonly used) increases the thermal energy of one body and decreases the thermal energy of the other. Suppose we initially have a high temperature (and high thermal energy) substance (H) and a low temperature (and low thermal energy) substance (L). The atoms and molecules in H have a higher average KE than those in L. If we place substance H in contact with substance L, the thermal energy will flow spontaneously from substance H to substance L. The temperature of substance H will decrease, as will the average KE of its molecules; the temperature of substance L will increase, along with the average KE of its molecules. Heat flow will continue until the two substances are at the same temperature (Figure 5.6).

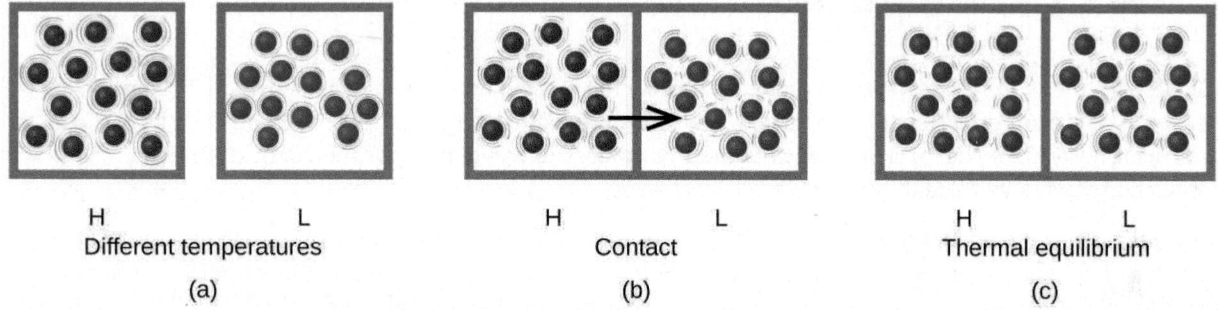

FIGURE 5.6 (a) Substances H and L are initially at different temperatures, and their atoms have different average kinetic energies. (b) When they contact each other, collisions between the molecules result in the transfer of kinetic (thermal) energy from the hotter to the cooler matter. (c) The two objects reach "thermal equilibrium" when both substances are at the same temperature and their molecules have the same average kinetic energy.

LINK TO LEARNING

Click on the PhET simulation (http://openstax.org/l/16PHETenergy) to explore energy forms and changes. Visit the Energy Systems tab to create combinations of energy sources, transformation methods, and outputs. Click on Energy Symbols to visualize the transfer of energy.

Matter undergoing chemical reactions and physical changes can release or absorb heat. A change that releases heat is called an **exothermic process**. For example, the combustion reaction that occurs when using an oxyacetylene torch is an exothermic process—this process also releases energy in the form of light as evidenced by the torch's flame (Figure 5.7). A reaction or change that absorbs heat is an **endothermic process**. A cold pack used to treat muscle strains provides an example of an endothermic process. When the substances in the cold pack (water and a salt like ammonium nitrate) are brought together, the resulting process absorbs heat, leading to the sensation of cold.

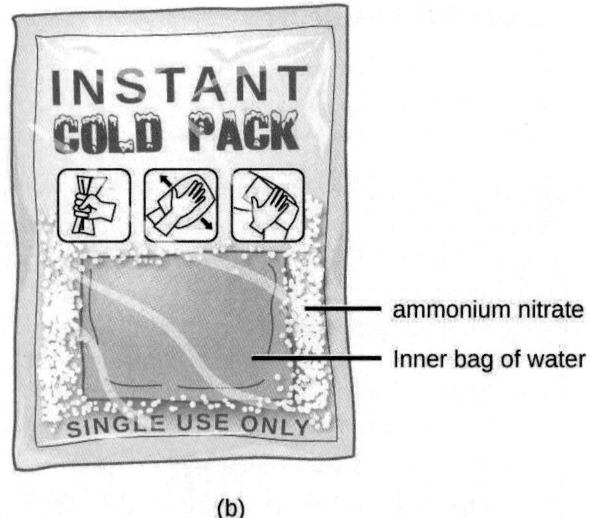

(a) (b)

FIGURE 5.7 (a) An oxyacetylene torch produces heat by the combustion of acetylene in oxygen. The energy released by this exothermic reaction heats and then melts the metal being cut. The sparks are tiny bits of the molten metal flying away. (b) A cold pack uses an endothermic process to create the sensation of cold. (credit a: modification of work by "Skatebiker"/Wikimedia commons)

Historically, energy was measured in units of **calories (cal)**. A calorie is the amount of energy required to raise one gram of water by 1 degree C (1 kelvin). However, this quantity depends on the atmospheric pressure and the starting temperature of the water. The ease of measurement of energy changes in calories has meant that the calorie is still frequently used. The Calorie (with a capital C), or large calorie, commonly used in quantifying food energy content, is a kilocalorie. The SI unit of heat, work, and energy is the joule. A **joule (J)** is defined as the amount of energy used when a force of 1 newton moves an object 1 meter. It is named in honor of the English physicist James Prescott Joule. One joule is equivalent to 1 kg m^2/s^2, which is also called 1 newton–meter. A kilojoule (kJ) is 1000 joules. To standardize its definition, 1 calorie has been set to equal 4.184 joules.

We now introduce two concepts useful in describing heat flow and temperature change. The **heat capacity (*C*)** of a body of matter is the quantity of heat (q) it absorbs or releases when it experiences a temperature change (ΔT) of 1 degree Celsius (or equivalently, 1 kelvin):

$$C = \frac{q}{\Delta T}$$

Heat capacity is determined by both the type and amount of substance that absorbs or releases heat. It is therefore an extensive property—its value is proportional to the amount of the substance.

For example, consider the heat capacities of two cast iron frying pans. The heat capacity of the large pan is five times greater than that of the small pan because, although both are made of the same material, the mass of the large pan is five times greater than the mass of the small pan. More mass means more atoms are present in the larger pan, so it takes more energy to make all of those atoms vibrate faster. The heat capacity of the small cast iron frying pan is found by observing that it takes 18,150 J of energy to raise the temperature of the pan by 50.0 °C:

$$C_{\text{small pan}} = \frac{18{,}140\ \text{J}}{50.0\ °\text{C}} = 363\ \text{J/°C}$$

The larger cast iron frying pan, while made of the same substance, requires 90,700 J of energy to raise its temperature by 50.0 °C. The larger pan has a (proportionally) larger heat capacity because the larger amount of material requires a (proportionally) larger amount of energy to yield the same temperature change:

$$C_{\text{large pan}} = \frac{90{,}700\ \text{J}}{50.0\ °\text{C}} = 1814\ \text{J/°C}$$

The **specific heat capacity (*c*)** of a substance, commonly called its "specific heat," is the quantity of heat required to raise the temperature of 1 gram of a substance by 1 degree Celsius (or 1 kelvin):

$$c = \frac{q}{\text{m}\Delta\text{T}}$$

Specific heat capacity depends only on the kind of substance absorbing or releasing heat. It is an intensive property—the type, but not the amount, of the substance is all that matters. For example, the small cast iron frying pan has a mass of 808 g. The specific heat of iron (the material used to make the pan) is therefore:

$$c_{\text{iron}} = \frac{18{,}140\text{ J}}{(808\text{ g})(50.0\text{ °C})} = 0.449\text{ J/g °C}$$

The large frying pan has a mass of 4040 g. Using the data for this pan, we can also calculate the specific heat of iron:

$$c_{\text{iron}} = \frac{90{,}700\text{ J}}{(4040\text{ g})(50.0\text{ °C})} = 0.449\text{ J/g °C}$$

Although the large pan is more massive than the small pan, since both are made of the same material, they both yield the same value for specific heat (for the material of construction, iron). Note that specific heat is measured in units of energy per temperature per mass and is an intensive property, being derived from a ratio of two extensive properties (heat and mass). The molar heat capacity, also an intensive property, is the heat capacity per mole of a particular substance and has units of J/mol °C (Figure 5.8).

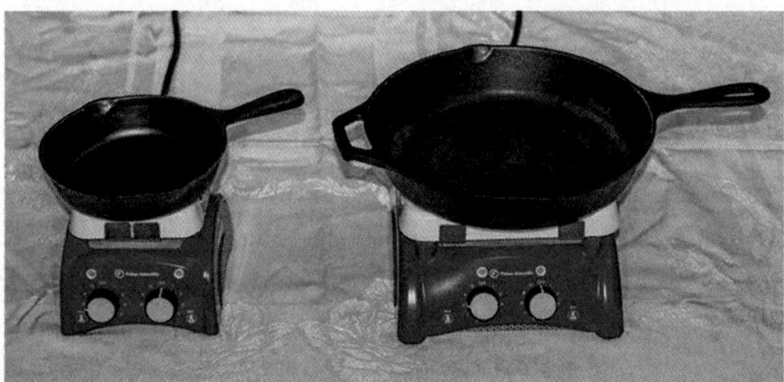

FIGURE 5.8 Because of its larger mass, a large frying pan has a larger heat capacity than a small frying pan. Because they are made of the same material, both frying pans have the same specific heat. (credit: Mark Blaser)

Water has a relatively high specific heat (about 4.2 J/g °C for the liquid and 2.09 J/g °C for the solid); most metals have much lower specific heats (usually less than 1 J/g °C). The specific heat of a substance varies somewhat with temperature. However, this variation is usually small enough that we will treat specific heat as constant over the range of temperatures that will be considered in this chapter. Specific heats of some common substances are listed in Table 5.1.

Specific Heats of Common Substances at 25 °C and 1 bar

Substance	Symbol (*state*)	Specific Heat (J/g °C)
helium	He(*g*)	5.193
water	$H_2O(l)$	4.184
ethanol	$C_2H_6O(l)$	2.376
ice	$H_2O(s)$	2.093 (at −10 °C)
water vapor	$H_2O(g)$	1.864

TABLE 5.1

Substance	Symbol (*state*)	Specific Heat (J/g °C)
nitrogen	$N_2(g)$	1.040
air		1.007
oxygen	$O_2(g)$	0.918
aluminum	$Al(s)$	0.897
carbon dioxide	$CO_2(g)$	0.853
argon	$Ar(g)$	0.522
iron	$Fe(s)$	0.449
copper	$Cu(s)$	0.385
lead	$Pb(s)$	0.130
gold	$Au(s)$	0.129
silicon	$Si(s)$	0.712

TABLE 5.1

If we know the mass of a substance and its specific heat, we can determine the amount of heat, q, entering or leaving the substance by measuring the temperature change before and after the heat is gained or lost:

$$q = \text{(specific heat)} \times \text{(mass of substance)} \times \text{(temperature change)}$$
$$q = c \times m \times \Delta T = c \times m \times (T_{\text{final}} - T_{\text{initial}})$$

In this equation, c is the specific heat of the substance, m is its mass, and ΔT (which is read "delta T") is the temperature change, $T_{\text{final}} - T_{\text{initial}}$. If a substance gains thermal energy, its temperature increases, its final temperature is higher than its initial temperature, $T_{\text{final}} - T_{\text{initial}}$ has a positive value, and the value of q is positive. If a substance loses thermal energy, its temperature decreases, the final temperature is lower than the initial temperature, $T_{\text{final}} - T_{\text{initial}}$ has a negative value, and the value of q is negative.

Measuring Heat

A flask containing 8.0×10^2 g of water is heated, and the temperature of the water increases from 21 °C to 85 °C. How much heat did the water absorb?

Solution

To answer this question, consider these factors:

- the specific heat of the substance being heated (in this case, water)
- the amount of substance being heated (in this case, 8.0×10^2 g)
- the magnitude of the temperature change (in this case, from 21 °C to 85 °C).

The specific heat of water is 4.184 J/g °C, so to heat 1 g of water by 1 °C requires 4.184 J. We note that since 4.184 J is required to heat 1 g of water by 1 °C, we will need *800 times as much* to heat 8.0×10^2 g of water by 1

°C. Finally, we observe that since 4.184 J are required to heat 1 g of water by 1 °C, we will need *64 times as much* to heat it by 64 °C (that is, from 21 °C to 85 °C).

This can be summarized using the equation:

$$q = c \times m \times \Delta T = c \times m \times (T_{\text{final}} - T_{\text{initial}})$$

$$\begin{aligned} &= (4.184\ \text{J}/\cancel{\text{g}}\ °\text{C}) \times (8.0 \text{ x } 10^2\ \cancel{\text{g}}) \times (85 - 21)\ °\text{C} \\ &= (4.184\ \text{J}/\cancel{\text{g}}\ °\cancel{\text{C}}) \times (8.0 \text{ x } 10^2\ \cancel{\text{g}}) \times (64)\ °\cancel{\text{C}} \\ &= 210{,}000\ \text{J}\ (= 2.1 \times 10^2\ \text{kJ}) \end{aligned}$$

Because the temperature increased, the water absorbed heat and q is positive.

Check Your Learning

How much heat, in joules, must be added to a 502 g iron skillet to increase its temperature from 25 °C to 250 °C? The specific heat of iron is 0.449 J/g °C.

Answer:

5.07×10^4 J

Note that the relationship between heat, specific heat, mass, and temperature change can be used to determine any of these quantities (not just heat) if the other three are known or can be deduced.

EXAMPLE 5.2

Determining Other Quantities

A piece of unknown metal weighs 348 g. When the metal piece absorbs 6.64 kJ of heat, its temperature increases from 22.4 °C to 43.6 °C. Determine the specific heat of this metal (which might provide a clue to its identity).

Solution

Since mass, heat, and temperature change are known for this metal, we can determine its specific heat using the relationship:

$$q = c \times m \times \Delta T = c \times m \times (T_{\text{final}} - T_{\text{initial}})$$

Substituting the known values:

$$6640\ \text{J} = c \times (348\ \text{g}) \times (43.6 - 22.4)\ °\text{C}$$

Solving:

$$c = \frac{6640\ \text{J}}{(348\ \text{g}) \times (21.2\ °\text{C})} = 0.900\ \text{J/g}\ °\text{C}$$

Comparing this value with the values in Table 5.1, this value matches the specific heat of aluminum, which suggests that the unknown metal may be aluminum.

Check Your Learning

A piece of unknown metal weighs 217 g. When the metal piece absorbs 1.43 kJ of heat, its temperature increases from 24.5 °C to 39.1 °C. Determine the specific heat of this metal, and predict its identity.

Answer:

c = 0.451 J/g °C; the metal is likely to be iron

Chemistry in Everyday Life

Solar Thermal Energy Power Plants

The sunlight that reaches the earth contains thousands of times more energy than we presently capture. Solar thermal systems provide one possible solution to the problem of converting energy from the sun into energy we can use. Large-scale solar thermal plants have different design specifics, but all concentrate sunlight to heat some substance; the heat "stored" in that substance is then converted into electricity.

The Solana Generating Station in Arizona's Sonora Desert produces 280 megawatts of electrical power. It uses parabolic mirrors that focus sunlight on pipes filled with a heat transfer fluid (HTF) (Figure 5.9). The HTF then does two things: It turns water into steam, which spins turbines, which in turn produces electricity, and it melts and heats a mixture of salts, which functions as a thermal energy storage system. After the sun goes down, the molten salt mixture can then release enough of its stored heat to produce steam to run the turbines for 6 hours. Molten salts are used because they possess a number of beneficial properties, including high heat capacities and thermal conductivities.

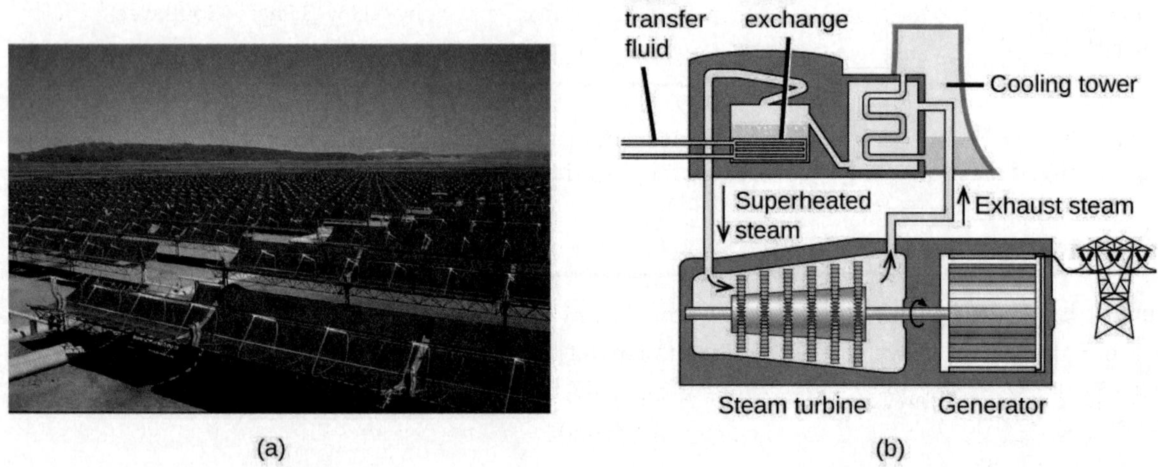

FIGURE 5.9 This solar thermal plant uses parabolic trough mirrors to concentrate sunlight. (credit a: modification of work by Bureau of Land Management)

The 377-megawatt Ivanpah Solar Generating System, located in the Mojave Desert in California, is the largest solar thermal power plant in the world (Figure 5.10). Its 170,000 mirrors focus huge amounts of sunlight on three water-filled towers, producing steam at over 538 °C that drives electricity-producing turbines. It produces enough energy to power 140,000 homes. Water is used as the working fluid because of its large heat capacity and heat of vaporization.

(a)

(b)

FIGURE 5.10 (a) The Ivanpah solar thermal plant uses 170,000 mirrors to concentrate sunlight on water-filled towers. (b) It covers 4000 acres of public land near the Mojave Desert and the California-Nevada border. (credit a: modification of work by Craig Dietrich; credit b: modification of work by "USFWS Pacific Southwest Region"/Flickr)

5.2 Calorimetry

LEARNING OBJECTIVES

By the end of this section, you will be able to:

- Explain the technique of calorimetry
- Calculate and interpret heat and related properties using typical calorimetry data

One technique we can use to measure the amount of heat involved in a chemical or physical process is known as **calorimetry**. Calorimetry is used to measure amounts of heat transferred to or from a substance. To do so, the heat is exchanged with a calibrated object (calorimeter). The temperature change measured by the calorimeter is used to derive the amount of heat transferred by the process under study. The measurement of heat transfer using this approach requires the definition of a **system** (the substance or substances undergoing the chemical or physical change) and its **surroundings** (all other matter, including components of the measurement apparatus, that serve to either provide heat to the system or absorb heat from the system).

A **calorimeter** is a device used to measure the amount of heat involved in a chemical or physical process. For example, when an exothermic reaction occurs in solution in a calorimeter, the heat produced by the reaction is absorbed by the solution, which increases its temperature. When an endothermic reaction occurs, the heat required is absorbed from the thermal energy of the solution, which decreases its temperature (Figure 5.11). The temperature change, along with the specific heat and mass of the solution, can then be used to calculate the amount of heat involved in either case.

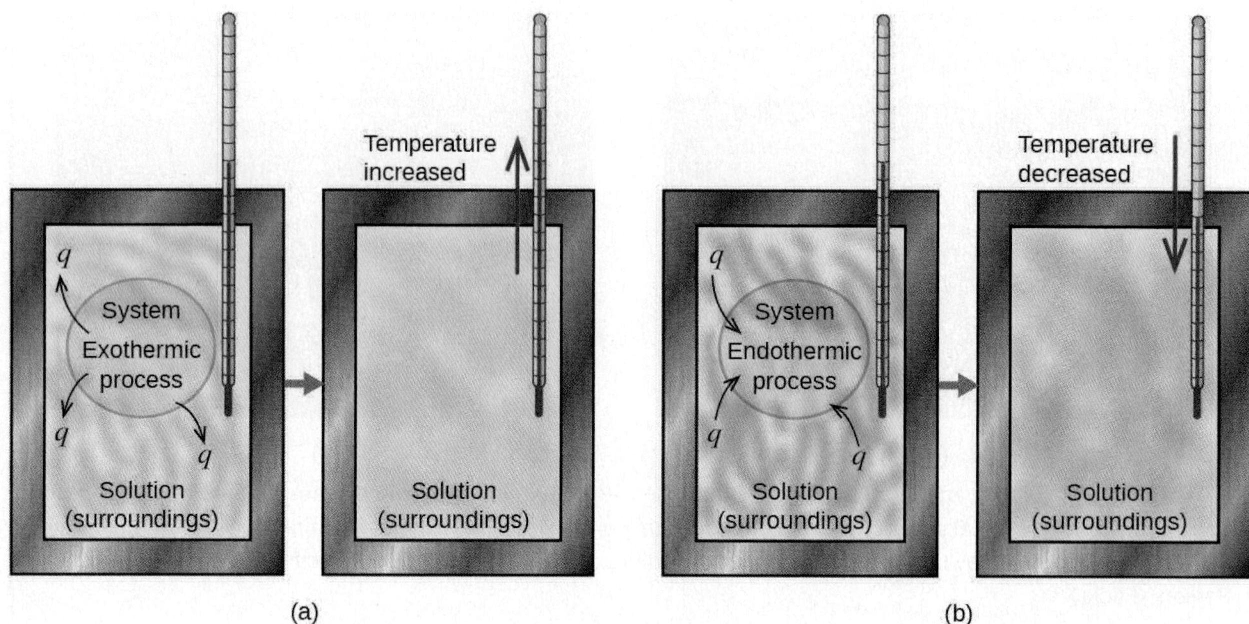

FIGURE 5.11 In a calorimetric determination, either (a) an exothermic process occurs and heat, q, is negative, indicating that thermal energy is transferred from the system to its surroundings, or (b) an endothermic process occurs and heat, q, is positive, indicating that thermal energy is transferred from the surroundings to the system.

Calorimetry measurements are important in understanding the heat transferred in reactions involving everything from microscopic proteins to massive machines. During her time at the National Bureau of Standards, research chemist Reatha Clark King performed calorimetric experiments to understand the precise heats of various flourine compounds. Her work was important to NASA in their quest for better rocket fuels.

Scientists use well-insulated calorimeters that all but prevent the transfer of heat between the calorimeter and its environment, which effectively limits the "surroundings" to the nonsystem components with the calorimeter (and the calorimeter itself). This enables the accurate determination of the heat involved in chemical processes, the energy content of foods, and so on. General chemistry students often use simple calorimeters constructed from polystyrene cups (Figure 5.12). These easy-to-use "coffee cup" calorimeters allow more heat exchange with the outside environment, and therefore produce less accurate energy values.

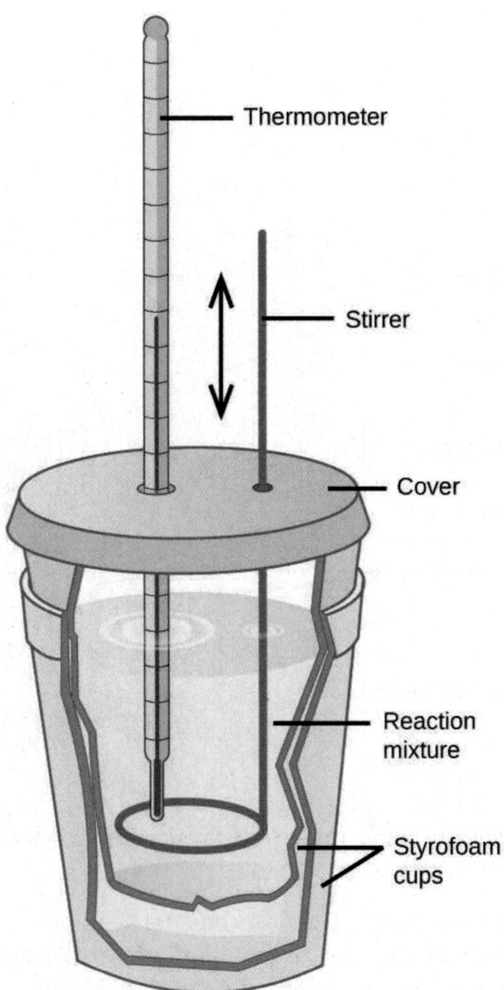

FIGURE 5.12 A simple calorimeter can be constructed from two polystyrene cups. A thermometer and stirrer extend through the cover into the reaction mixture.

Commercial solution calorimeters are also available. Relatively inexpensive calorimeters often consist of two thin-walled cups that are nested in a way that minimizes thermal contact during use, along with an insulated cover, handheld stirrer, and simple thermometer. More expensive calorimeters used for industry and research typically have a well-insulated, fully enclosed reaction vessel, motorized stirring mechanism, and a more accurate temperature sensor (Figure 5.13).

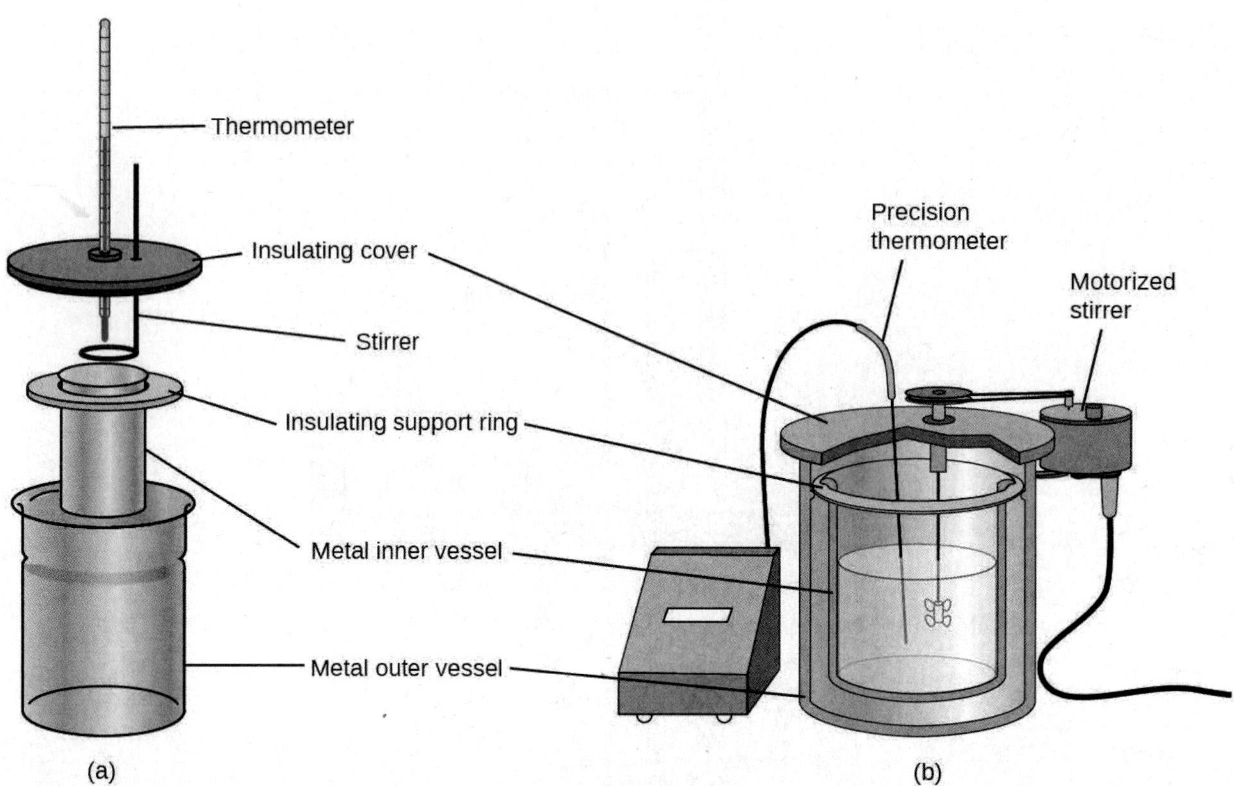

FIGURE 5.13 Commercial solution calorimeters range from (a) simple, inexpensive models for student use to (b) expensive, more accurate models for industry and research.

Before discussing the calorimetry of chemical reactions, consider a simpler example that illustrates the core idea behind calorimetry. Suppose we initially have a high-temperature substance, such as a hot piece of metal (M), and a low-temperature substance, such as cool water (W). If we place the metal in the water, heat will flow from M to W. The temperature of M will decrease, and the temperature of W will increase, until the two substances have the same temperature—that is, when they reach thermal equilibrium (Figure 5.14). If this occurs in a calorimeter, ideally all of this heat transfer occurs between the two substances, with no heat gained or lost by either its external environment. Under these ideal circumstances, the net heat change is zero:

$$q_{\text{substance M}} + q_{\text{substance W}} = 0$$

This relationship can be rearranged to show that the heat gained by substance M is equal to the heat lost by substance W:

$$q_{\text{substance M}} = -q_{\text{substance W}}$$

The magnitude of the heat (change) is therefore the same for both substances, and the negative sign merely shows that $q_{\text{substance M}}$ and $q_{\text{substance W}}$ are opposite in direction of heat flow (gain or loss) but does not indicate the arithmetic sign of either q value (that is determined by whether the matter in question gains or loses heat, per definition). In the specific situation described, $q_{\text{substance M}}$ is a negative value and $q_{\text{substance W}}$ is positive, since heat is transferred from M to W.

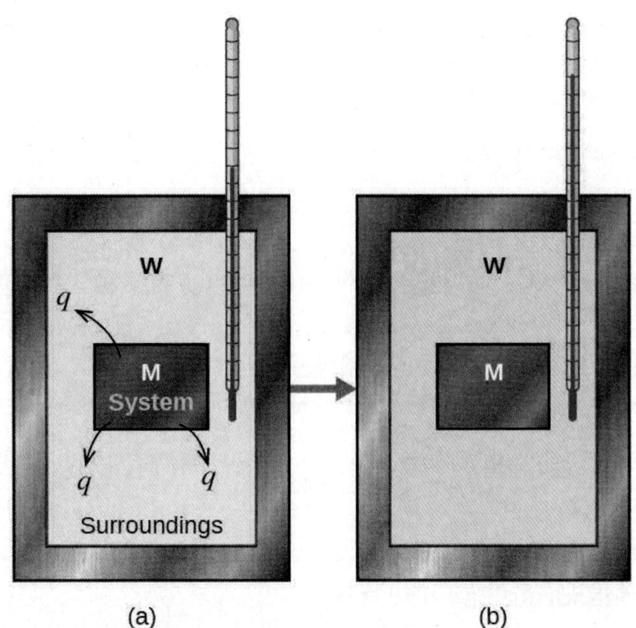

FIGURE 5.14 In a simple calorimetry process, (a) heat, *q*, is transferred from the hot metal, M, to the cool water, W, until (b) both are at the same temperature.

EXAMPLE 5.3

Heat Transfer between Substances at Different Temperatures

A 360.0-g piece of rebar (a steel rod used for reinforcing concrete) is dropped into 425 mL of water at 24.0 °C. The final temperature of the water was measured as 42.7 °C. Calculate the initial temperature of the piece of rebar. Assume the specific heat of steel is approximately the same as that for iron (Table 5.1), and that all heat transfer occurs between the rebar and the water (there is no heat exchange with the surroundings).

Solution

The temperature of the water increases from 24.0 °C to 42.7 °C, so the water absorbs heat. That heat came from the piece of rebar, which initially was at a higher temperature. Assuming that all heat transfer was between the rebar and the water, with no heat "lost" to the outside environment, then *heat given off by rebar = −heat taken in by water*, or:

$$q_{\text{rebar}} = -q_{\text{water}}$$

Since we know how heat is related to other measurable quantities, we have:

$$(c \times m \times \Delta T)_{\text{rebar}} = -(c \times m \times \Delta T)_{\text{water}}$$

Letting f = final and i = initial, in expanded form, this becomes:

$$c_{\text{rebar}} \times m_{\text{rebar}} \times (T_{\text{f,rebar}} - T_{\text{i,rebar}}) = -c_{\text{water}} \times m_{\text{water}} \times (T_{\text{f,water}} - T_{\text{i,water}})$$

The density of water is 1.0 g/mL, so 425 mL of water = 425 g. Noting that the final temperature of both the rebar and water is 42.7 °C, substituting known values yields:

$$\left(0.449 \text{ J/g °C}\right)\left(360.0 \text{ g}\right)\left(42.7 \text{ °C} - T_{\text{i,rebar}}\right) = -\left(4.184 \text{ J/g °C}\right)\left(425\text{g}\right)\left(42.7 \text{ °C} - 24.0 \text{ °C}\right)$$

$$T_{i,rebar} = \frac{(4.184 \text{ J/g °C})(425 \text{ g})(42.7 \text{ °C} - 24.0 \text{ °C})}{(0.449 \text{ J/g °C})(360.0 \text{ g})} + 42.7 \text{ °C}$$

Solving this gives $T_{\text{i,rebar}}$= 248 °C, so the initial temperature of the rebar was 248 °C.

Check Your Learning

A 248-g piece of copper is dropped into 390 mL of water at 22.6 °C. The final temperature of the water was measured as 39.9 °C. Calculate the initial temperature of the piece of copper. Assume that all heat transfer occurs between the copper and the water.

Answer:

The initial temperature of the copper was 335.6 °C.

Check Your Learning

A 248-g piece of copper initially at 314 °C is dropped into 390 mL of water initially at 22.6 °C. Assuming that all heat transfer occurs between the copper and the water, calculate the final temperature.

Answer:

The final temperature (reached by both copper and water) is 38.7 °C.

This method can also be used to determine other quantities, such as the specific heat of an unknown metal.

EXAMPLE 5.4

Identifying a Metal by Measuring Specific Heat

A 59.7 g piece of metal that had been submerged in boiling water was quickly transferred into 60.0 mL of water initially at 22.0 °C. The final temperature is 28.5 °C. Use these data to determine the specific heat of the metal. Use this result to identify the metal.

Solution

Assuming perfect heat transfer, *heat given off by metal = −heat taken in by water*, or:

$$q_{\text{metal}} = -q_{\text{water}}$$

In expanded form, this is:

$$c_{\text{metal}} \times m_{\text{metal}} \times (T_{\text{f,metal}} - T_{\text{i, metal}}) = -c_{\text{water}} \times m_{\text{water}} \times (T_{\text{f,water}} - T_{\text{i,water}})$$

Noting that since the metal was submerged in boiling water, its initial temperature was 100.0 °C; and that for water, 60.0 mL = 60.0 g; we have:

$$(c_{\text{metal}})(59.7\text{ g})(28.5\text{ °C} - 100.0\text{ °C}) = -(4.18\text{ J/g °C})(60.0\text{ g})(28.5\text{ °C} - 22.0\text{ °C})$$

Solving this:

$$c_{\text{metal}} = \frac{-(4.184\text{ J/g °C})(60.0\text{ g})(6.5\text{ °C})}{(59.7\text{ g})(-71.5\text{ °C})} = 0.38\text{ J/g °C}$$

Comparing this with values in Table 5.1, our experimental specific heat is closest to the value for copper (0.39 J/g °C), so we identify the metal as copper.

Check Your Learning

A 92.9-g piece of a silver/gray metal is heated to 178.0 °C, and then quickly transferred into 75.0 mL of water initially at 24.0 °C. After 5 minutes, both the metal and the water have reached the same temperature: 29.7 °C. Determine the specific heat and the identity of the metal. (Note: You should find that the specific heat is close to that of two different metals. Explain how you can confidently determine the identity of the metal).

Answer:

c_{metal}= 0.13 J/g °C

This specific heat is close to that of either gold or lead. It would be difficult to determine which metal this was based solely on the numerical values. However, the observation that the metal is silver/gray in addition to the value for the specific heat indicates that the metal is lead.

When we use calorimetry to determine the heat involved in a chemical reaction, the same principles we have been discussing apply. The amount of heat absorbed by the calorimeter is often small enough that we can neglect it (though not for highly accurate measurements, as discussed later), and the calorimeter minimizes energy exchange with the outside environment. Because energy is neither created nor destroyed during a chemical reaction, the heat produced or consumed in the reaction (the "system"), q_{reaction}, plus the heat absorbed or lost by the solution (the "surroundings"), q_{solution}, must add up to zero:

$$q_{\text{reaction}} + q_{\text{solution}} = 0$$

This means that the amount of heat produced or consumed in the reaction equals the amount of heat absorbed or lost by the solution:

$$q_{\text{reaction}} = -q_{\text{solution}}$$

This concept lies at the heart of all calorimetry problems and calculations.

Heat Produced by an Exothermic Reaction

When 50.0 mL of 1.00 M HCl(*aq*) and 50.0 mL of 1.00 M NaOH(*aq*), both at 22.0 °C, are added to a coffee cup calorimeter, the temperature of the mixture reaches a maximum of 28.9 °C. What is the approximate amount of heat produced by this reaction?

$$\text{HCl}(aq) + \text{NaOH}(aq) \longrightarrow \text{NaCl}(aq) + \text{H}_2\text{O}(l)$$

Solution

To visualize what is going on, imagine that you could combine the two solutions so quickly that no reaction took place while they mixed; then after mixing, the reaction took place. At the instant of mixing, you have 100.0 mL of a mixture of HCl and NaOH at 22.0 °C. The HCl and NaOH then react until the solution temperature reaches 28.9 °C.

The heat given off by the reaction is equal to that taken in by the solution. Therefore:

$$q_{\text{reaction}} = -q_{\text{solution}}$$

(It is important to remember that this relationship only holds if the calorimeter does not absorb any heat from the reaction, and there is no heat exchange between the calorimeter and the outside environment.)

Next, we know that the heat absorbed by the solution depends on its specific heat, mass, and temperature change:

$$q_{\text{solution}} = (c \times m \times \Delta T)_{\text{solution}}$$

To proceed with this calculation, we need to make a few more reasonable assumptions or approximations. Since the solution is aqueous, we can proceed as if it were water in terms of its specific heat and mass values. The density of water is approximately 1.0 g/mL, so 100.0 mL has a mass of about 1.0×10^2 g (two significant figures). The specific heat of water is approximately 4.184 J/g °C, so we use that for the specific heat of the solution. Substituting these values gives:

$$q_{\text{solution}} = (4.184\ \text{J/g °C})(1.0 \times 10^2\ \text{g})(28.9\ \text{°C} - 22.0\ \text{°C}) = 2.9 \times 10^3\ \text{J}$$

Finally, since we are trying to find the heat of the reaction, we have:

$$q_{\text{reaction}} = -q_{\text{solution}} = -2.9 \times 10^3\ \text{J}$$

The negative sign indicates that the reaction is exothermic. It produces 2.9 kJ of heat.

Check Your Learning

When 100 mL of 0.200 M NaCl(*aq*) and 100 mL of 0.200 M $AgNO_3$(*aq*), both at 21.9 °C, are mixed in a coffee cup calorimeter, the temperature increases to 23.5 °C as solid AgCl forms. How much heat is produced by this precipitation reaction? What assumptions did you make to determine your value?

Answer:
1.34 × 1.3 kJ; assume no heat is absorbed by the calorimeter, no heat is exchanged between the calorimeter and its surroundings, and that the specific heat and mass of the solution are the same as those for water

Chemistry in Everyday Life

Thermochemistry of Hand Warmers

When working or playing outdoors on a cold day, you might use a hand warmer to warm your hands (Figure 5.15). A common reusable hand warmer contains a supersaturated solution of $NaC_2H_3O_2$ (sodium acetate) and a metal disc. Bending the disk creates nucleation sites around which the metastable $NaC_2H_3O_2$ quickly crystallizes (a later chapter on solutions will investigate saturation and supersaturation in more detail).

The process $NaC_2H_3O_2(aq) \longrightarrow NaC_2H_3O_2(s)$ is exothermic, and the heat produced by this process is absorbed by your hands, thereby warming them (at least for a while). If the hand warmer is reheated, the $NaC_2H_3O_2$ redissolves and can be reused.

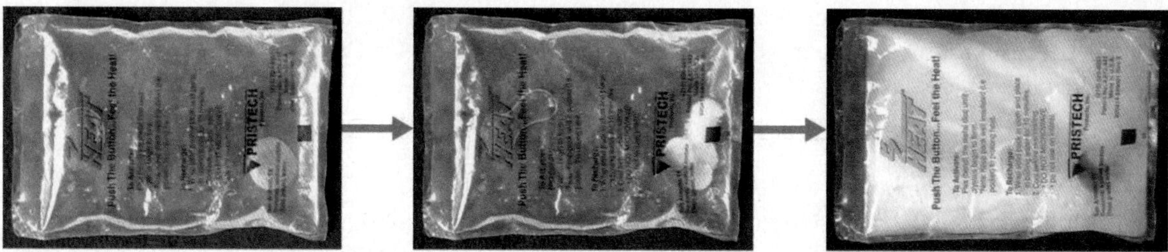

FIGURE 5.15 Chemical hand warmers produce heat that warms your hand on a cold day. In this one, you can see the metal disc that initiates the exothermic precipitation reaction. (credit: modification of work by Science Buddies TV/YouTube)

Another common hand warmer produces heat when it is ripped open, exposing iron and water in the hand warmer to oxygen in the air. One simplified version of this exothermic reaction is $2Fe(s) + \frac{3}{2}O_2(g) \longrightarrow Fe_2O_3(s)$. Salt in the hand warmer catalyzes the reaction, so it produces heat more rapidly; cellulose, vermiculite, and activated carbon help distribute the heat evenly. Other types of hand warmers use lighter fluid (a platinum catalyst helps lighter fluid oxidize exothermically), charcoal (charcoal oxidizes in a special case), or electrical units that produce heat by passing an electrical current from a battery through resistive wires.

LINK TO LEARNING

This link (http://openstax.org/l/16Handwarmer) shows the precipitation reaction that occurs when the disk in a chemical hand warmer is flexed.

EXAMPLE 5.6

Heat Flow in an Instant Ice Pack

When solid ammonium nitrate dissolves in water, the solution becomes cold. This is the basis for an "instant ice pack" (Figure 5.16). When 3.21 g of solid NH_4NO_3 dissolves in 50.0 g of water at 24.9 °C in a calorimeter, the temperature decreases to 20.3 °C.

Calculate the value of q for this reaction and explain the meaning of its arithmetic sign. State any assumptions that you made.

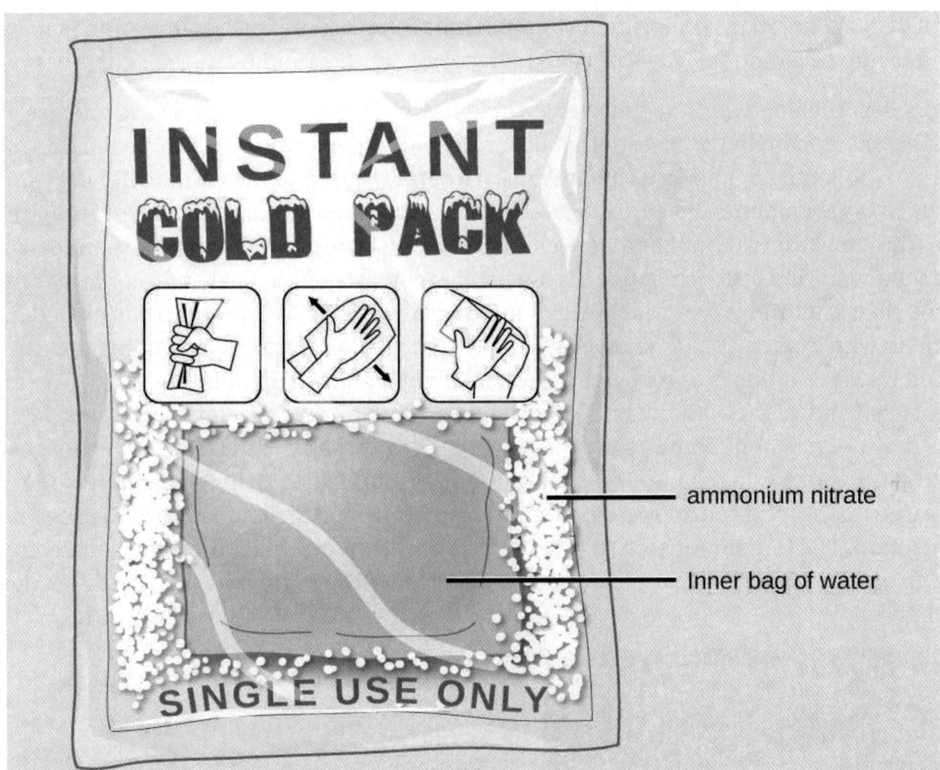

FIGURE 5.16 An instant cold pack consists of a bag containing solid ammonium nitrate and a second bag of water. When the bag of water is broken, the pack becomes cold because the dissolution of ammonium nitrate is an endothermic process that removes thermal energy from the water. The cold pack then removes thermal energy from your body.

Solution

We assume that the calorimeter prevents heat transfer between the solution and its external environment (including the calorimeter itself), in which case:

$$q_{\text{rxn}} = -q_{\text{soln}}$$

with "rxn" and "soln" used as shorthand for "reaction" and "solution," respectively.

Assuming also that the specific heat of the solution is the same as that for water, we have:

$$\begin{aligned} q_{\text{rxn}} &= -q_{\text{soln}} = -(c \times m \times \Delta T)_{\text{soln}} \\ &= -[(4.184\ \text{J/g °C}) \times (53.2\ \text{g}) \times (20.3\ \text{°C} - 24.9\ \text{°C})] \\ &= -[(4.184\ \text{J/g °C}) \times (53.2\ \text{g}) \times (-4.6\ \text{°C})] \\ &+1.0 \times 10^3\ \text{J} = +1.0\ \text{kJ} \end{aligned}$$

The positive sign for q indicates that the dissolution is an endothermic process.

Check Your Learning

When a 3.00-g sample of KCl was added to 3.00×10^2 g of water in a coffee cup calorimeter, the temperature decreased by 1.05 °C. How much heat is involved in the dissolution of the KCl? What assumptions did you make?

Answer:

1.33 kJ; assume that the calorimeter prevents heat transfer between the solution and its external environment (including the calorimeter itself) and that the specific heat of the solution is the same as that for water

If the amount of heat absorbed by a calorimeter is too large to neglect or if we require more accurate results, then we must take into account the heat absorbed both by the solution and by the calorimeter.

The calorimeters described are designed to operate at constant (atmospheric) pressure and are convenient to measure heat flow accompanying processes that occur in solution. A different type of calorimeter that operates at constant volume, colloquially known as a **bomb calorimeter**, is used to measure the energy produced by reactions that yield large amounts of heat and gaseous products, such as combustion reactions. (The term “bomb” comes from the observation that these reactions can be vigorous enough to resemble explosions that would damage other calorimeters.) This type of calorimeter consists of a robust steel container (the “bomb”) that contains the reactants and is itself submerged in water (Figure 5.17). The sample is placed in the bomb, which is then filled with oxygen at high pressure. A small electrical spark is used to ignite the sample. The energy produced by the reaction is absorbed by the steel bomb and the surrounding water. The temperature increase is measured and, along with the known heat capacity of the calorimeter, is used to calculate the energy produced by the reaction. Bomb calorimeters require calibration to determine the heat capacity of the calorimeter and ensure accurate results. The calibration is accomplished using a reaction with a known q, such as a measured quantity of benzoic acid ignited by a spark from a nickel fuse wire that is weighed before and after the reaction. The temperature change produced by the known reaction is used to determine the heat capacity of the calorimeter. The calibration is generally performed each time before the calorimeter is used to gather research data.

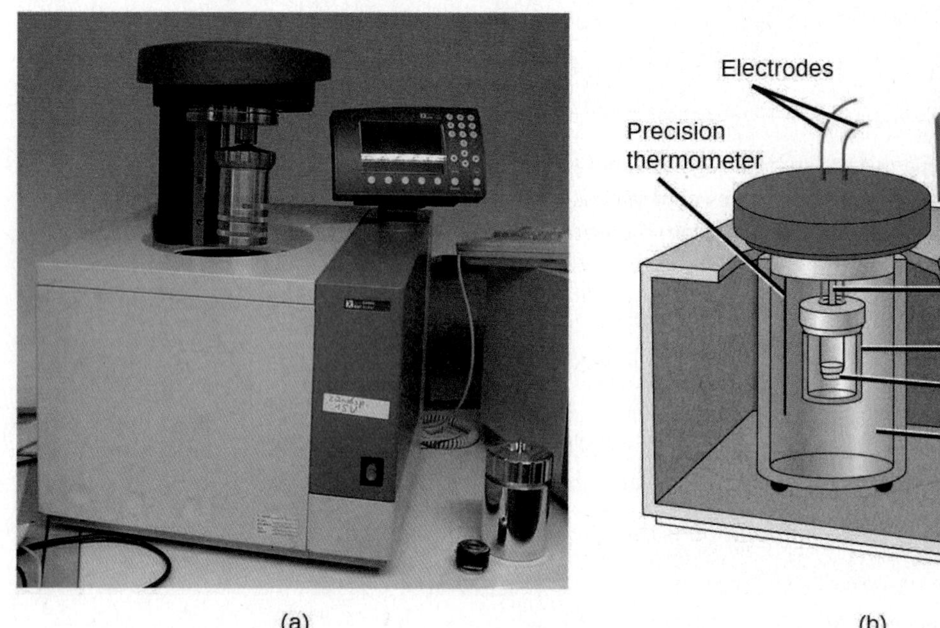

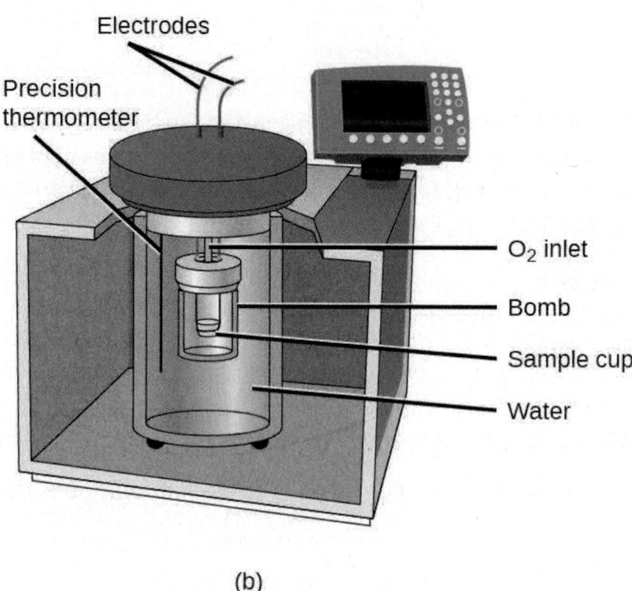

(a) (b)

FIGURE 5.17 (a) A bomb calorimeter is used to measure heat produced by reactions involving gaseous reactants or products, such as combustion. (b) The reactants are contained in the gas-tight “bomb,” which is submerged in water and surrounded by insulating materials. (credit a: modification of work by “Harbor1”/Wikimedia commons)

LINK TO LEARNING

Click on this link (http://openstax.org/l/16BombCal) to view how a bomb calorimeter is prepared for action.

This site (http://openstax.org/l/16Calorcalcs) shows calorimetric calculations using sample data.

EXAMPLE 5.7

Bomb Calorimetry

When 3.12 g of glucose, $C_6H_{12}O_6$, is burned in a bomb calorimeter, the temperature of the calorimeter increases from 23.8 °C to 35.6 °C. The calorimeter contains 775 g of water, and the bomb itself has a heat capacity of 893 J/°C. How much heat was produced by the combustion of the glucose sample?

Solution

The combustion produces heat that is primarily absorbed by the water and the bomb. (The amounts of heat absorbed by the reaction products and the unreacted excess oxygen are relatively small and dealing with them is beyond the scope of this text. We will neglect them in our calculations.)

The heat produced by the reaction is absorbed by the water and the bomb:

$$\begin{aligned} q_{\text{rxn}} &= -(q_{\text{water}} + q_{\text{bomb}}) \\ &= -[(4.184\ \text{J/g °C}) \times (775\ \text{g}) \times (35.6\ \text{°C} - 23.8\ \text{°C}) + 893\ \text{J/°C} \times (35.6\ \text{°C} - 23.8\ \text{°C})] \\ &= -(38{,}300\ \text{J} + 10{,}500\ \text{J}) \\ &= -48{,}800\ \text{J} = -48.8\ \text{kJ} \end{aligned}$$

This reaction released 48.7 kJ of heat when 3.12 g of glucose was burned.

Check Your Learning

When 0.963 g of benzene, C_6H_6, is burned in a bomb calorimeter, the temperature of the calorimeter increases by 8.39 °C. The bomb has a heat capacity of 784 J/°C and is submerged in 925 mL of water. How much heat was produced by the combustion of the benzene sample?

Answer:

q_{rx} = −39.0 kJ (the reaction produced 39.0 kJ of heat)

Since the first one was constructed in 1899, 35 calorimeters have been built to measure the heat produced by a living person.[2] These whole-body calorimeters of various designs are large enough to hold an individual human being. More recently, whole-room calorimeters allow for relatively normal activities to be performed, and these calorimeters generate data that more closely reflect the real world. These calorimeters are used to measure the metabolism of individuals under different environmental conditions, different dietary regimes, and with different health conditions, such as diabetes.

For example Carla Prado's team at University of Alberta undertook whole-body calorimetry to understand the energy expenditures of women who had recently given birth. Studies like this help develop better recommendations and regimens for nutrition, exercise, and general wellbeing during this period of significant physiological change. In humans, metabolism is typically measured in Calories per day. A **nutritional calorie (Calorie)** is the energy unit used to quantify the amount of energy derived from the metabolism of foods; one Calorie is equal to 1000 calories (1 kcal), the amount of energy needed to heat 1 kg of water by 1 °C.

Chemistry in Everyday Life

Measuring Nutritional Calories

In your day-to-day life, you may be more familiar with energy being given in Calories, or nutritional calories, which are used to quantify the amount of energy in foods. One calorie (cal) = exactly 4.184 joules, and one Calorie (note the capitalization) = 1000 cal, or 1 kcal. (This is approximately the amount of energy needed to heat 1 kg of water by 1 °C.)

The macronutrients in food are proteins, carbohydrates, and fats or oils. Proteins provide about 4 Calories per gram, carbohydrates also provide about 4 Calories per gram, and fats and oils provide about 9 Calories/g. Nutritional labels on food packages show the caloric content of one serving of the food, as well as the breakdown into Calories from each of the three macronutrients (Figure 5.18).

2 Francis D. Reardon et al. "The Snellen human calorimeter revisited, re-engineered and upgraded: Design and performance characteristics." *Medical and Biological Engineering and Computing* 8 (2006)721–28, http://link.springer.com/article/10.1007/s11517-006-0086-5.

FIGURE 5.18 (a) Macaroni and cheese contain energy in the form of the macronutrients in the food. (b) The food's nutritional information is shown on the package label. In the US, the energy content is given in Calories (per serving); the rest of the world usually uses kilojoules. (credit a: modification of work by "Rex Roof"/Flickr)

For the example shown in (b), the total energy per 228-g portion is calculated by:

$$(5\text{ g protein} \times 4\text{ Calories/g}) + (31\text{ g carb} \times 4\text{ Calories/g}) + (12\text{ g fat} \times 9\text{ Calories/g}) = 252\text{ Calories}$$

So, you can use food labels to count your Calories. But where do the values come from? And how accurate are they? The caloric content of foods can be determined by using bomb calorimetry; that is, by burning the food and measuring the energy it contains. A sample of food is weighed, mixed in a blender, freeze-dried, ground into powder, and formed into a pellet. The pellet is burned inside a bomb calorimeter, and the measured temperature change is converted into energy per gram of food.

Today, the caloric content on food labels is derived using a method called the Atwater system that uses the average caloric content of the different chemical constituents of food, protein, carbohydrate, and fats. The average amounts are those given in the equation and are derived from the various results given by bomb calorimetry of whole foods. The carbohydrate amount is discounted a certain amount for the fiber content, which is indigestible carbohydrate. To determine the energy content of a food, the quantities of carbohydrate, protein, and fat are each multiplied by the average Calories per gram for each and the products summed to obtain the total energy.

LINK TO LEARNING

Click on this link (http://openstax.org/l/16USDA) to access the US Department of Agriculture (USDA) National Nutrient Database, containing nutritional information on over 8000 foods.

5.3 Enthalpy

LEARNING OBJECTIVES

By the end of this section, you will be able to:

- State the first law of thermodynamics
- Define enthalpy and explain its classification as a state function
- Write and balance thermochemical equations
- Calculate enthalpy changes for various chemical reactions
- Explain Hess's law and use it to compute reaction enthalpies

Thermochemistry is a branch of **chemical thermodynamics**, the science that deals with the relationships between heat, work, and other forms of energy in the context of chemical and physical processes. As we concentrate on thermochemistry in this chapter, we need to consider some widely used concepts of thermodynamics.

Substances act as reservoirs of energy, meaning that energy can be added to them or removed from them. Energy is stored in a substance when the kinetic energy of its atoms or molecules is raised. The greater kinetic energy may be in the form of increased translations (travel or straight-line motions), vibrations, or rotations of the atoms or molecules. When thermal energy is lost, the intensities of these motions decrease and the kinetic energy falls. The total of all possible kinds of energy present in a substance is called the **internal energy (*U*)**, sometimes symbolized as *E*.

As a system undergoes a change, its internal energy can change, and energy can be transferred from the system to the surroundings, or from the surroundings to the system. Energy is transferred into a system when it absorbs heat (q) from the surroundings or when the surroundings do work (w) on the system. For example, energy is transferred into room-temperature metal wire if it is immersed in hot water (the wire absorbs heat from the water), or if you rapidly bend the wire back and forth (the wire becomes warmer because of the work done on it). Both processes increase the internal energy of the wire, which is reflected in an increase in the wire's temperature. Conversely, energy is transferred out of a system when heat is lost from the system, or when the system does work on the surroundings.

The relationship between internal energy, heat, and work can be represented by the equation:

$$\Delta U = q + w$$

as shown in Figure 5.19. This is one version of the **first law of thermodynamics**, and it shows that the internal energy of a system changes through heat flow into or out of the system (positive q is heat flow in; negative q is heat flow out) or work done on or by the system. The work, w, is positive if it is done on the system and negative if it is done by the system.

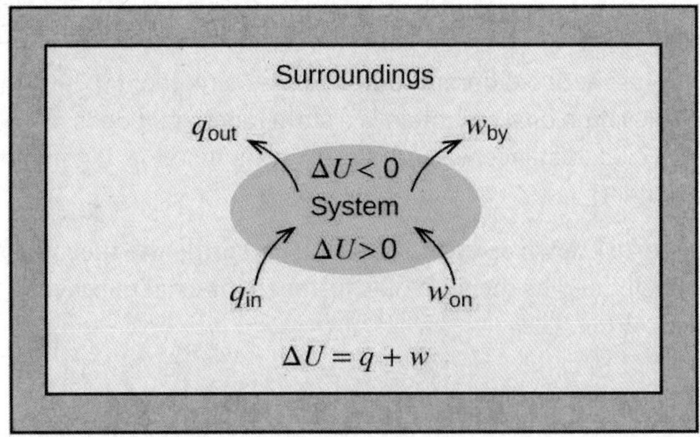

FIGURE 5.19 The internal energy, U, of a system can be changed by heat flow and work. If heat flows into the system, q_{in}, or work is done on the system, w_{on}, its internal energy increases, $\Delta U > 0$. If heat flows out of the system, q_{out}, or work is done by the system, w_{by}, its internal energy decreases, $\Delta U < 0$.

A type of work called **expansion work** (or pressure-volume work) occurs when a system pushes back the surroundings against a restraining pressure, or when the surroundings compress the system. An example of this occurs during the operation of an internal combustion engine. The reaction of gasoline and oxygen is exothermic. Some of this energy is given off as heat, and some does work pushing the piston in the cylinder. The substances involved in the reaction are the system, and the engine and the rest of the universe are the surroundings. The system loses energy by both heating and doing work on the surroundings, and its internal energy decreases. (The engine is able to keep the car moving because this process is repeated many times per second while the engine is running.) We will consider how to determine the amount of work involved in a chemical or physical change in the chapter on thermodynamics.

LINK TO LEARNING

This view of an internal combustion engine (http://openstax.org/l/16combustion) illustrates the conversion of energy produced by the exothermic combustion reaction of a fuel such as gasoline into energy of motion.

As discussed, the relationship between internal energy, heat, and work can be represented as $\Delta U = q + w$. Internal energy is an example of a **state function** (or state variable), whereas heat and work are not state functions. The value of a state function depends only on the state that a system is in, and not on how that state is reached. If a quantity is not a state function, then its value *does* depend on how the state is reached. An example of a state function is altitude or elevation. If you stand on the summit of Mt. Kilimanjaro, you are at an altitude of 5895 m, and it does not matter whether you hiked there or parachuted there. The distance you traveled to the top of Kilimanjaro, however, is not a state function. You could climb to the summit by a direct route or by a more roundabout, circuitous path (Figure 5.20). The distances traveled would differ (distance is not a state function) but the elevation reached would be the same (altitude is a state function).

FIGURE 5.20 Paths X and Y represent two different routes to the summit of Mt. Kilimanjaro. Both have the same change in elevation (altitude or elevation on a mountain is a state function; it does not depend on path), but they have very different distances traveled (distance walked is not a state function; it depends on the path). (credit: modification of work by Paul Shaffner)

Chemists ordinarily use a property known as **enthalpy (*H*)** to describe the thermodynamics of chemical and physical processes. Enthalpy is defined as the sum of a system's internal energy (U) and the mathematical product of its pressure (P) and volume (V):

$$H = U + PV$$

Enthalpy is also a state function. Enthalpy values for specific substances cannot be measured directly; only enthalpy *changes* for chemical or physical processes can be determined. For processes that take place at constant pressure (a common condition for many chemical and physical changes), the **enthalpy change (Δ*H*)** is:

$$\Delta H = \Delta U + P\Delta V$$

The mathematical product $P\Delta V$ represents work (w), namely, expansion or pressure-volume work as noted. By their definitions, the arithmetic signs of ΔV and w will always be opposite:

$$P\Delta V = -w$$

Substituting this equation and the definition of internal energy into the enthalpy-change equation yields:

$$\begin{aligned}\Delta H &= \Delta U + P\Delta V \\ &= q_\text{p} + w - w \\ &= q_\text{p}\end{aligned}$$

where q_p is the heat of reaction under conditions of constant pressure.

And so, if a chemical or physical process is carried out at constant pressure with the only work done caused by expansion or contraction, then the heat flow (q_p) and enthalpy change (ΔH) for the process are equal.

The heat given off when you operate a Bunsen burner is equal to the enthalpy change of the methane combustion reaction that takes place, since it occurs at the essentially constant pressure of the atmosphere. On the other hand, the heat produced by a reaction measured in a bomb calorimeter (Figure 5.17) is not equal to ΔH because the closed, constant-volume metal container prevents the pressure from remaining constant (it may increase or decrease if the reaction yields increased or decreased amounts of gaseous species). Chemists usually perform experiments under normal atmospheric conditions, at constant external pressure with $q = \Delta H$, which makes enthalpy the most convenient choice for determining heat changes for chemical reactions.

The following conventions apply when using ΔH:

- A negative value of an enthalpy change, $\Delta H < 0$, indicates an exothermic reaction; a positive value, $\Delta H > 0$, indicates an endothermic reaction. If the direction of a chemical equation is reversed, the arithmetic sign of its ΔH is changed (a process that is endothermic in one direction is exothermic in the opposite direction).
- Chemists use a thermochemical equation to represent the changes in both matter and energy. In a thermochemical equation, the enthalpy change of a reaction is shown as a ΔH value following the equation for the reaction. This ΔH value indicates the amount of heat associated with the reaction involving the number of moles of reactants and products *as shown in the chemical equation*. For example, consider this equation:

  $$\text{H}_2(g) + \frac{1}{2}\text{O}_2(g) \longrightarrow \text{H}_2\text{O}(l) \qquad \Delta H = -286 \text{ kJ}$$

  This equation indicates that when 1 mole of hydrogen gas and $\frac{1}{2}$ mole of oxygen gas at some temperature and pressure change to 1 mole of liquid water at the same temperature and pressure, 286 kJ of heat are released to the surroundings. If the coefficients of the chemical equation are multiplied by some factor, the enthalpy change must be multiplied by that same factor (ΔH is an extensive property):

  $$\begin{array}{ll}\text{(two-fold increase in amounts)} & \\ 2\text{H}_2(g) + \text{O}_2(g) \longrightarrow 2\text{H}_2\text{O}(l) & \Delta H = 2 \times (-286 \text{ kJ}) = -572 \text{ kJ} \\ \text{(two-fold decrease in amounts)} & \\ \frac{1}{2}\text{H}_2(g) + \frac{1}{4}\text{O}_2(g) \longrightarrow \frac{1}{2}\text{H}_2\text{O}(l) & \Delta H = \frac{1}{2} \times (-286 \text{ kJ}) = -143 \text{ kJ}\end{array}$$

- The enthalpy change of a reaction depends on the physical states of the reactants and products, so these must be shown. For example, when 1 mole of hydrogen gas and $\frac{1}{2}$ mole of oxygen gas change to 1 mole of liquid water at the same temperature and pressure, 286 kJ of heat are released. If gaseous water forms, only 242 kJ of heat are released.

  $$\text{H}_2(g) + \frac{1}{2}\text{O}_2(g) \longrightarrow \text{H}_2\text{O}(g) \qquad \Delta H = -242 \text{ kJ}$$

EXAMPLE 5.8

Writing Thermochemical Equations

When 0.0500 mol of HCl(*aq*) reacts with 0.0500 mol of NaOH(*aq*) to form 0.0500 mol of NaCl(*aq*), 2.9 kJ of heat are produced. Write a balanced thermochemical equation for the reaction of one mole of HCl.

$$\text{HCl}(aq) + \text{NaOH}(aq) \longrightarrow \text{NaCl}(aq) + \text{H}_2\text{O}(l)$$

Solution

For the reaction of 0.0500 mol acid (HCl), $q = -2.9$ kJ. The reactants are provided in stoichiometric amounts (same molar ratio as in the balanced equation), and so the amount of acid may be used to calculate a molar enthalpy change. Since ΔH is an extensive property, it is proportional to the amount of acid neutralized:

$$\Delta H = 1 \cancel{\text{mol HCl}} \times \frac{-2.9 \text{ kJ}}{0.0500 \cancel{\text{mol HCl}}} = -58 \text{ kJ}$$

The thermochemical equation is then

$$\text{HCl}(aq) + \text{NaOH}(aq) \longrightarrow \text{NaCl}(aq) + \text{H}_2\text{O}(l) \qquad \Delta H = -58 \text{ kJ}$$

Check Your Learning

When 1.34 g Zn(*s*) reacts with 60.0 mL of 0.750 M HCl(*aq*), 3.14 kJ of heat are produced. Determine the enthalpy change per mole of zinc reacting for the reaction:

$$\text{Zn}(s) + 2\text{HCl}(aq) \longrightarrow \text{ZnCl}_2(aq) + \text{H}_2(g)$$

Answer:

$\Delta H = -153$ kJ

Be sure to take both stoichiometry and limiting reactants into account when determining the ΔH for a chemical reaction.

EXAMPLE 5.9

Writing Thermochemical Equations

A gummy bear contains 2.67 g sucrose, $C_{12}H_{22}O_{11}$. When it reacts with 7.19 g potassium chlorate, $KClO_3$, 43.7 kJ of heat are produced. Write a thermochemical equation for the reaction of one mole of sucrose:

$$\text{C}_{12}\text{H}_{22}\text{O}_{11}(aq) + 8\text{KClO}_3(aq) \longrightarrow 12\text{CO}_2(g) + 11\text{H}_2\text{O}(l) + 8\text{KCl}(aq).$$

Solution

Unlike the previous example exercise, this one does not involve the reaction of stoichiometric amounts of reactants, and so the *limiting reactant* must be identified (it limits the yield of the reaction and the amount of thermal energy produced or consumed).

The provided amounts of the two reactants are

$$(2.67 \text{ g})\,(1 \text{ mol}/342.3 \text{ g}) = 0.00780 \text{ mol C}_{12}\text{ H}_{22}\text{ O}_{11}$$
$$(7.19 \text{ g})\,(1 \text{ mol}/122.5 \text{ g}) = 0.0587 \text{ mol KClO}_3$$

The provided molar ratio of perchlorate-to-sucrose is then

$$0.0587 \text{ mol KClO}_3/0.00780 \text{ mol C}_{12}\text{H}_{22}\text{O}_{11} = 7.52$$

The balanced equation indicates 8 mol $KClO_3$ are required for reaction with 1 mol $C_{12}H_{22}O_{11}$. Since the provided amount of $KClO_3$ is less than the stoichiometric amount, it is the limiting reactant and may be used to compute the enthalpy change:

$$\Delta H = -43.7 \quad kJ/0.0587 \text{ mol } KClO_3 = 744 \text{ kJ/mol } KClO_3$$

Because the equation, as written, represents the reaction of 8 mol $KClO_3$, the enthalpy change is

$$(744 \text{ kJ/mol } KClO_3)(8 \text{ mol } KClO_3) = 5960 \text{ kJ}$$

The enthalpy change for this reaction is −5960 kJ, and the thermochemical equation is:

$$C_{12}H_{22}O_{11} + 8KClO_3 \longrightarrow 12CO_2 + 11H_2O + 8KCl \qquad \Delta H = -5960 \text{ kJ}$$

Check Your Learning

When 1.42 g of iron reacts with 1.80 g of chlorine, 3.22 g of $FeCl_2(s)$ and 8.60 kJ of heat is produced. What is the enthalpy change for the reaction when 1 mole of $FeCl_2(s)$ is produced?

Answer:

$\Delta H = -338$ kJ

Enthalpy changes are typically tabulated for reactions in which both the reactants and products are at the same conditions. A **standard state** is a commonly accepted set of conditions used as a reference point for the determination of properties under other different conditions. For chemists, the IUPAC standard state refers to materials under a pressure of 1 bar and solutions at 1 M, and does not specify a temperature. Many thermochemical tables list values with a standard state of 1 atm. Because the ΔH of a reaction changes very little with such small changes in pressure (1 bar = 0.987 atm), ΔH values (except for the most precisely measured values) are essentially the same under both sets of standard conditions. We will include a superscripted "o" in the enthalpy change symbol to designate standard state. Since the usual (but not technically standard) temperature is 298.15 K, this temperature will be assumed unless some other temperature is specified. Thus, the symbol (ΔH°) is used to indicate an enthalpy change for a process occurring under these conditions. (The symbol ΔH is used to indicate an enthalpy change for a reaction occurring under nonstandard conditions.)

The enthalpy changes for many types of chemical and physical processes are available in the reference literature, including those for combustion reactions, phase transitions, and formation reactions. As we discuss these quantities, it is important to pay attention to the *extensive* nature of enthalpy and enthalpy changes. Since the enthalpy change for a given reaction is proportional to the amounts of substances involved, it may be reported on that basis (i.e., as the ΔH for specific amounts of reactants). However, we often find it more useful to divide one extensive property (ΔH) by another (amount of substance), and report a per-amount *intensive* value of ΔH, often "normalized" to a per-mole basis. (Note that this is similar to determining the intensive property specific heat from the extensive property heat capacity, as seen previously.)

Standard Enthalpy of Combustion

Standard enthalpy of combustion (ΔH_C°) is the enthalpy change when 1 mole of a substance burns (combines vigorously with oxygen) under standard state conditions; it is sometimes called "heat of combustion." For example, the enthalpy of combustion of ethanol, −1366.8 kJ/mol, is the amount of heat produced when one mole of ethanol undergoes complete combustion at 25 °C and 1 atmosphere pressure, yielding products also at 25 °C and 1 atm.

$$C_2H_5OH(l) + 3O_2(g) \longrightarrow 2CO_2 + 3H_2O(l) \qquad \Delta H^\circ = -1366.8 \text{ kJ}$$

Enthalpies of combustion for many substances have been measured; a few of these are listed in Table 5.2. Many readily available substances with large enthalpies of combustion are used as fuels, including hydrogen, carbon (as coal or charcoal), and **hydrocarbons** (compounds containing only hydrogen and carbon), such as methane, propane, and the major components of gasoline.

Standard Molar Enthalpies of Combustion

Substance	Combustion Reaction	Enthalpy of Combustion, ΔH_c° $\left(\frac{\text{kJ}}{\text{mol}}\text{ at 25 °C}\right)$
carbon	$C(s) + O_2(g) \longrightarrow CO_2(g)$	−393.5
hydrogen	$H_2(g) + \frac{1}{2}O_2(g) \longrightarrow H_2O(l)$	−285.8
magnesium	$Mg(s) + \frac{1}{2}O_2(g) \longrightarrow MgO(s)$	−601.6
sulfur	$S(s) + O_2(g) \longrightarrow SO_2(g)$	−296.8
carbon monoxide	$CO(g) + \frac{1}{2}O_2(g) \longrightarrow CO_2(g)$	−283.0
methane	$CH_4(g) + 2O_2(g) \longrightarrow CO_2(g) + 2H_2O(l)$	−890.8
acetylene	$C_2H_2(g) + \frac{5}{2}O_2(g) \longrightarrow 2CO_2(g) + H_2O(l)$	−1301.1
ethanol	$C_2H_5OH(l) + 3O_2(g) \longrightarrow 2CO_2(g) + 3H_2O(l)$	−1366.8
methanol	$CH_3OH(l) + \frac{3}{2}O_2(g) \longrightarrow CO_2(g) + 2H_2O(l)$	−726.1
isooctane	$C_8H_{18}(l) + \frac{25}{2}O_2(g) \longrightarrow 8CO_2(g) + 9H_2O(l)$	−5461

TABLE 5.2

EXAMPLE 5.10

Using Enthalpy of Combustion

As Figure 5.21 suggests, the combustion of gasoline is a highly exothermic process. Let us determine the approximate amount of heat produced by burning 1.00 L of gasoline, assuming the enthalpy of combustion of gasoline is the same as that of isooctane, a common component of gasoline. The density of isooctane is 0.692 g/mL.

FIGURE 5.21 The combustion of gasoline is very exothermic. (credit: modification of work by “AlexEagle”/Flickr)

Solution

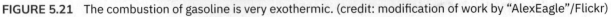

Starting with a known amount (1.00 L of isooctane), we can perform conversions between units until we arrive at the desired amount of heat or energy. The enthalpy of combustion of isooctane provides one of the necessary conversions. Table 5.2 gives this value as −5460 kJ per 1 mole of isooctane (C_8H_{18}).

Using these data,

$$1.00\ \cancel{\text{L}\ C_8H_{18}} \times \frac{1000\ \cancel{\text{mL}\ C_8H_{18}}}{1\ \cancel{\text{L}\ C_8H_{18}}} \times \frac{0.692\ \cancel{\text{g}\ C_8H_{18}}}{1\ \cancel{\text{mL}\ C_8H_{18}}} \times \frac{1\ \cancel{\text{mol}\ C_8H_{18}}}{114\ \cancel{\text{g}\ C_8H_{18}}} \times \frac{-5460\ \text{kJ}}{1\ \cancel{\text{mol}\ C_8H_{18}}} = -3.31 \times 10^4\ \text{kJ}$$

The combustion of 1.00 L of isooctane produces 33,100 kJ of heat. (This amount of energy is enough to melt 99.2 kg, or about 218 lbs, of ice.)

Note: If you do this calculation one step at a time, you would find:

$$1.00\ \text{L}\ C_8H_{18} \longrightarrow 1.00 \times 10^3\ \text{mL}\ C_8H_{18}$$
$$1.00 \times 10^3\ \text{mL}\ C_8H_{18} \longrightarrow 692\ \text{g}\ C_8H_{18}$$
$$692\ \text{g}\ C_8H_{18} \longrightarrow 6.07\ \text{mol}\ C_8H_{18}$$
$$6.07\ \text{mol}\ C_8H_{18} \longrightarrow -3.31 \times 10^4\ \text{kJ}$$

Check Your Learning

How much heat is produced by the combustion of 125 g of acetylene?

Answer:

6.25×10^3 kJ

Chemistry in Everyday Life

Emerging Algae-Based Energy Technologies (Biofuels)

As reserves of fossil fuels diminish and become more costly to extract, the search is ongoing for replacement fuel sources for the future. Among the most promising biofuels are those derived from algae (Figure 5.22). The species of algae used are nontoxic, biodegradable, and among the world's fastest growing organisms. About 50% of algal weight is oil, which can be readily converted into fuel such as biodiesel. Algae can yield 26,000 gallons of biofuel per hectare—much more energy per acre than other crops. Some strains of algae can flourish in brackish water that is not usable for growing other crops. Algae can produce biodiesel, biogasoline, ethanol, butanol, methane, and even jet fuel.

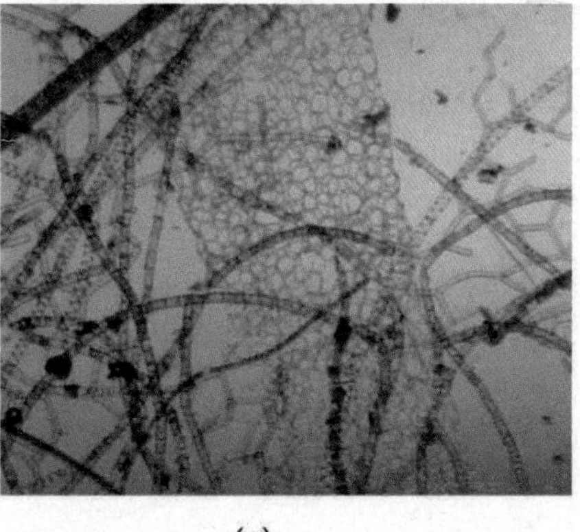

(a)

(b)

(c)

FIGURE 5.22 (a) Tiny algal organisms can be (b) grown in large quantities and eventually (c) turned into a useful fuel such as biodiesel. (credit a: modification of work by Micah Sittig; credit b: modification of work by Robert Kerton; credit c: modification of work by John F. Williams)

According to the US Department of Energy, only 39,000 square kilometers (about 0.4% of the land mass of the US or less than $\frac{1}{7}$ of the area used to grow corn) can produce enough algal fuel to replace all the petroleum-based fuel used in the US. The cost of algal fuels is becoming more competitive—for instance, the US Air Force is producing jet fuel from algae at a total cost of under $5 per gallon.[3] The process used to produce algal fuel is as follows: grow the algae (which use sunlight as their energy source and CO_2 as a raw material); harvest the algae; extract the fuel compounds (or precursor compounds); process as necessary (e.g., perform a transesterification reaction to make biodiesel); purify; and distribute (Figure 5.23).

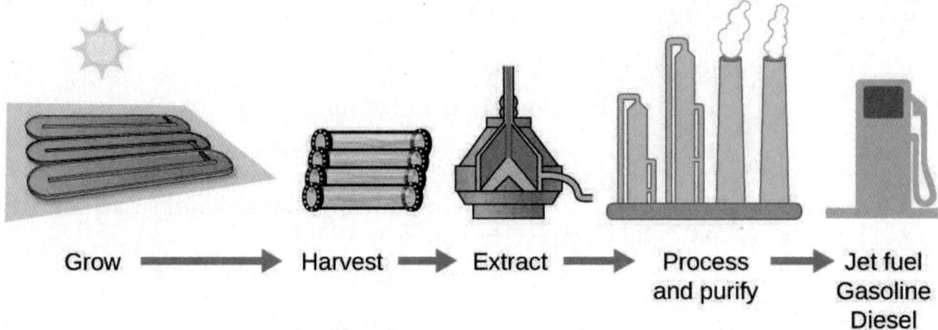

FIGURE 5.23 Algae convert sunlight and carbon dioxide into oil that is harvested, extracted, purified, and transformed into a variety of renewable fuels.

LINK TO LEARNING

Click here (http://openstax.org/l/16biofuel) to learn more about the process of creating algae biofuel.

Standard Enthalpy of Formation

A **standard enthalpy of formation** ΔH_f° is an enthalpy change for a reaction in which exactly 1 mole of a pure substance is formed from free elements in their most stable states under standard state conditions. These values are especially useful for computing or predicting enthalpy changes for chemical reactions that are impractical or dangerous to carry out, or for processes for which it is difficult to make measurements. If we have values for the appropriate standard enthalpies of formation, we can determine the enthalpy change for any reaction, which we will practice in the next section on Hess's law.

The standard enthalpy of formation of $CO_2(g)$ is −393.5 kJ/mol. This is the enthalpy change for the exothermic reaction:

$$\text{C}(s) + \text{O}_2(g) \longrightarrow \text{CO}_2(g) \qquad \Delta H_\text{f}^\circ = \Delta H^\circ = -393.5 \text{ kJ}$$

starting with the reactants at a pressure of 1 atm and 25 °C (with the carbon present as graphite, the most stable form of carbon under these conditions) and ending with one mole of CO_2, also at 1 atm and 25 °C. For nitrogen dioxide, $NO_2(g)$, ΔH_f° is 33.2 kJ/mol. This is the enthalpy change for the reaction:

$$\frac{1}{2}\text{N}_2(g) + \text{O}_2(g) \longrightarrow \text{NO}_2(g) \qquad \Delta H_\text{f}^\circ = \Delta H^\circ = +33.2 \text{ kJ}$$

A reaction equation with $\frac{1}{2}$ mole of N_2 and 1 mole of O_2 is correct in this case because the standard enthalpy of formation always refers to 1 mole of product, $NO_2(g)$.

You will find a table of standard enthalpies of formation of many common substances in Appendix G. These values indicate that formation reactions range from highly exothermic (such as −2984 kJ/mol for the formation of P_4O_{10}) to strongly endothermic (such as +226.7 kJ/mol for the formation of acetylene, C_2H_2). By definition, the standard enthalpy of formation of an element in its most stable form is equal to zero under

3 For more on algal fuel, see http://www.theguardian.com/environment/2010/feb/13/algae-solve-pentagon-fuel-problem.

standard conditions, which is 1 atm for gases and 1 M for solutions.

EXAMPLE 5.11

Evaluating an Enthalpy of Formation

Ozone, $O_3(g)$, forms from oxygen, $O_2(g)$, by an endothermic process. Ultraviolet radiation is the source of the energy that drives this reaction in the upper atmosphere. Assuming that both the reactants and products of the reaction are in their standard states, determine the standard enthalpy of formation, ΔH_f° of ozone from the following information:

$$3O_2(g) \longrightarrow 2O_3(g) \qquad \Delta H^\circ = +286 \text{ kJ}$$

Solution

ΔH_f° is the enthalpy change for the formation of one mole of a substance in its standard state from the elements in their standard states. Thus, ΔH_f° for $O_3(g)$ is the enthalpy change for the reaction:

$$\frac{3}{2}O_2(g) \longrightarrow O_3(g)$$

For the formation of 2 mol of $O_3(g)$, $\Delta H^\circ = +286$ kJ. This ratio, $\left(\frac{286 \text{ kJ}}{2 \text{ mol } O_3}\right)$, can be used as a conversion factor to find the heat produced when 1 mole of $O_3(g)$ is formed, which is the enthalpy of formation for $O_3(g)$:

$$\Delta H^\circ \text{ for 1 mole of } O_3(g) = 1 \cancel{\text{mol } O_3} \times \frac{286 \text{ kJ}}{2 \cancel{\text{mol } O_3}} = 143 \text{ kJ}$$

Therefore, $\Delta H_f^\circ [O_3(g)] = +143$ kJ/mol.

Check Your Learning

Hydrogen gas, H_2, reacts explosively with gaseous chlorine, Cl_2, to form hydrogen chloride, $HCl(g)$. What is the enthalpy change for the reaction of 1 mole of $H_2(g)$ with 1 mole of $Cl_2(g)$ if both the reactants and products are at standard state conditions? The standard enthalpy of formation of $HCl(g)$ is −92.3 kJ/mol.

Answer:

For the reaction $H_2(g) + Cl_2(g) \longrightarrow 2HCl(g) \qquad \Delta H^\circ = -184.6 \text{ kJ}$

EXAMPLE 5.12

Writing Reaction Equations for ΔH_f°

Write the heat of formation reaction equations for:

(a) $C_2H_5OH(l)$

(b) $Ca_3(PO_4)_2(s)$

Solution

Remembering that ΔH_f° reaction equations are for forming 1 mole of the compound from its constituent elements under standard conditions, we have:

(a) $2C(s, \text{ graphite}) + 3H_2(g) + \frac{1}{2}O_2(g) \longrightarrow C_2H_5OH(l)$

(b) $3Ca(s) + \frac{1}{2}P_4(s) + 4O_2(g) \longrightarrow Ca_3(PO_4)_2(s)$

Note: The standard state of carbon is graphite, and phosphorus exists as P_4.

Check Your Learning

Write the heat of formation reaction equations for:

(a) $C_2H_5OC_2H_5(l)$

(b) $Na_2CO_3(s)$

Answer:

(a) $4C(s,\ \text{graphite}) + 5H_2(g) + \frac{1}{2}O_2(g) \longrightarrow C_2H_5OC_2H_5(l)$; (b) $2Na(s) + C(s,\ \text{graphite}) + \frac{3}{2}O_2(g) \longrightarrow Na_2CO_3(s)$

Hess's Law

There are two ways to determine the amount of heat involved in a chemical change: measure it experimentally, or calculate it from other experimentally determined enthalpy changes. Some reactions are difficult, if not impossible, to investigate and make accurate measurements for experimentally. And even when a reaction is not hard to perform or measure, it is convenient to be able to determine the heat involved in a reaction without having to perform an experiment.

This type of calculation usually involves the use of **Hess's law**, which states: *If a process can be written as the sum of several stepwise processes, the enthalpy change of the total process equals the sum of the enthalpy changes of the various steps.* Hess's law is valid because enthalpy is a state function: Enthalpy changes depend only on where a chemical process starts and ends, but not on the path it takes from start to finish. For example, we can think of the reaction of carbon with oxygen to form carbon dioxide as occurring either directly or by a two-step process. The direct process is written:

$$C(s) + O_2(g) \longrightarrow CO_2(g) \qquad \Delta H° = -394 \text{ kJ}$$

In the two-step process, first carbon monoxide is formed:

$$C(s) + \frac{1}{2}O_2(g) \longrightarrow CO(g) \qquad \Delta H° = -111 \text{ kJ}$$

Then, carbon monoxide reacts further to form carbon dioxide:

$$CO(g) + \frac{1}{2}O_2(g) \longrightarrow CO_2(g) \qquad \Delta H° = -283 \text{ kJ}$$

The equation describing the overall reaction is the sum of these two chemical changes:

$$\begin{array}{l} \text{Step 1: } C(s) + \frac{1}{2}O_2(g) \longrightarrow CO(g) \\ \underline{\text{Step 2: } CO(g) + \frac{1}{2}O_2(g) \longrightarrow CO_2(g)} \\ \text{Sum: } C(s) + \frac{1}{2}O_2(g) + CO(g) + \frac{1}{2}O_2(g) \longrightarrow CO(g) + CO_2(g) \end{array}$$

Because the CO produced in Step 1 is consumed in Step 2, the net change is:

$$C(s) + O_2(g) \longrightarrow CO_2(g)$$

According to Hess's law, the enthalpy change of the reaction will equal the sum of the enthalpy changes of the steps.

$$\begin{array}{ll} C(s) + \frac{1}{2}O_2(g) \longrightarrow CO(g) & \Delta H° = -111 \text{ kJ} \\ \underline{CO(g) + \frac{1}{2}O_2(g) \longrightarrow CO_2(g)} & \underline{\Delta H° = -283 \text{ kJ}} \\ C(s) + O_2(g) \longrightarrow CO_2(g) & \Delta H° = -394 \text{ kJ} \end{array}$$

The result is shown in Figure 5.24. We see that ΔH of the overall reaction is the same whether it occurs in one step or two. This finding (overall ΔH for the reaction = sum of ΔH values for reaction "steps" in the overall reaction) is true in general for chemical and physical processes.

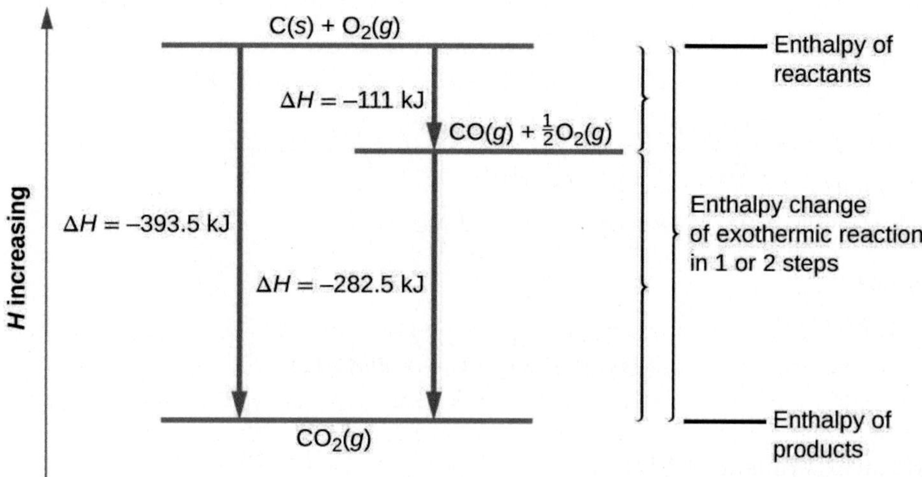

FIGURE 5.24 The formation of $CO_2(g)$ from its elements can be thought of as occurring in two steps, which sum to the overall reaction, as described by Hess's law. The horizontal blue lines represent enthalpies. For an exothermic process, the products are at lower enthalpy than are the reactants.

Before we further practice using Hess's law, let us recall two important features of ΔH.

1. ΔH is directly proportional to the quantities of reactants or products. For example, the enthalpy change for the reaction forming 1 mole of $NO_2(g)$ is +33.2 kJ:

$$\frac{1}{2}\text{N}_2(g) + \text{O}_2(g) \longrightarrow \text{NO}_2(g) \qquad \Delta H = +33.2 \text{ kJ}$$

When 2 moles of NO_2 (twice as much) are formed, the ΔH will be twice as large:

$$\text{N}_2(g) + 2\text{O}_2(g) \longrightarrow 2\text{NO}_2(g) \qquad \Delta H = +66.4 \text{ kJ}$$

In general, if we multiply or divide an equation by a number, then the enthalpy change should also be multiplied or divided by the same number.

2. ΔH for a reaction in one direction is equal in magnitude and opposite in sign to ΔH for the reaction in the reverse direction. For example, given that:

$$\text{H}_2(g) + \text{Cl}_2(g) \longrightarrow 2\text{HCl}(g) \qquad \Delta H = -184.6 \text{ kJ}$$

Then, for the "reverse" reaction, the enthalpy change is also "reversed":

$$2\text{HCl}(g) \longrightarrow \text{H}_2(g) + \text{Cl}_2(g) \qquad \Delta H = +184.6 \text{ kJ}$$

EXAMPLE 5.13

Stepwise Calculation of ΔH_f° Using Hess's Law

Determine the enthalpy of formation, ΔH_f°, of $FeCl_3(s)$ from the enthalpy changes of the following two-step process that occurs under standard state conditions:

$$\text{Fe}(s) + \text{Cl}_2(g) \longrightarrow \text{FeCl}_2(s) \qquad \Delta H^\circ = -341.8 \text{ kJ}$$

$$\text{FeCl}_2(s) + \frac{1}{2}\text{Cl}_2(g) \longrightarrow \text{FeCl}_3(s) \qquad \Delta H^\circ = -57.7 \text{ kJ}$$

Solution

We are trying to find the standard enthalpy of formation of $FeCl_3(s)$, which is equal to ΔH° for the reaction:

$$\text{Fe}(s) + \frac{3}{2}\text{Cl}_2(g) \longrightarrow \text{FeCl}_3(s) \qquad \Delta H_f^\circ = ?$$

Looking at the reactions, we see that the reaction for which we want to find $\Delta H°$ is the sum of the two reactions with known ΔH values, so we must sum their ΔHs:

$$\begin{array}{lr} Fe(s) + Cl_2(g) \longrightarrow FeCl_2(s) & \Delta H° = -341.8\ kJ \\ FeCl_2(s) + \frac{1}{2}Cl_2(g) \longrightarrow FeCl_3(s) & \Delta H° = -57.7\ kJ \\ \hline Fe(s) + \frac{3}{2}Cl_2(g) \longrightarrow FeCl_3(s) & \Delta H° = -399.5\ kJ \end{array}$$

The enthalpy of formation, $\Delta H_f°$, of $FeCl_3(s)$ is −399.5 kJ/mol.

Check Your Learning

Calculate ΔH for the process:

$$N_2(g) + 2O_2(g) \longrightarrow 2NO_2(g)$$

from the following information:

$$N_2(g) + O_2(g) \longrightarrow 2NO(g) \qquad \Delta H = 180.5\ kJ$$

$$NO(g) + \frac{1}{2}O_2(g) \longrightarrow NO_2(g) \qquad \Delta H = -57.06\ kJ$$

Answer:

66.4 kJ

Here is a less straightforward example that illustrates the thought process involved in solving many Hess's law problems. It shows how we can find many standard enthalpies of formation (and other values of ΔH) if they are difficult to determine experimentally.

EXAMPLE 5.14

A More Challenging Problem Using Hess's Law

Chlorine monofluoride can react with fluorine to form chlorine trifluoride:

$(i)\ ClF(g) + F_2(g) \longrightarrow ClF_3(g) \qquad \Delta H° = ?$

Use the reactions here to determine the $\Delta H°$ for reaction *(i)*:

$(ii)\ 2OF_2(g) \longrightarrow O_2(g) + 2F_2(g) \qquad \Delta H°_{(ii)} = -49.4\ kJ$

$(iii)\ 2ClF(g) + O_2(g) \longrightarrow Cl_2O(g) + OF_2(g) \qquad \Delta H°_{(iii)} = +214.0\ kJ$

$(iv)\ ClF_3(g) + O_2(g) \longrightarrow \frac{1}{2}Cl_2O(g) + \frac{3}{2}OF_2(g) \qquad \Delta H°_{(iv)} = +236.2\ kJ$

Solution

Our goal is to manipulate and combine reactions *(ii)*, *(iii)*, and *(iv)* such that they add up to reaction *(i)*. Going from left to right in *(i)*, we first see that $ClF(g)$ is needed as a reactant. This can be obtained by multiplying reaction *(iii)* by $\frac{1}{2}$, which means that the $\Delta H°$ change is also multiplied by $\frac{1}{2}$:

$$ClF(g) + \frac{1}{2}O_2(g) \longrightarrow \frac{1}{2}Cl_2O(g) + \frac{1}{2}OF_2(g) \qquad \Delta H° = \frac{1}{2}(214.0) = +107.0\ kJ$$

Next, we see that F_2 is also needed as a reactant. To get this, reverse and halve reaction *(ii)*, which means that the $\Delta H°$ changes sign and is halved:

$$\frac{1}{2}O_2(g) + F_2(g) \longrightarrow OF_2(g) \qquad \Delta H° = +24.7\ kJ$$

To get ClF_3 as a product, reverse *(iv)*, changing the sign of $\Delta H°$:

$$\frac{1}{2}Cl_2O(g) + \frac{3}{2}OF_2(g) \longrightarrow ClF_3(g) + O_2(g) \qquad \Delta H° = -236.2 \text{ kJ}$$

Now check to make sure that these reactions add up to the reaction we want:

$$ClF(g) + \frac{1}{2}O_2(g) \longrightarrow \frac{1}{2}Cl_2O(g) + \frac{1}{2}OF_2(g) \qquad \Delta H° = +107.0 \text{ kJ}$$

$$\frac{1}{2}O_2(g) + F_2(g) \longrightarrow OF_2(g) \qquad \Delta H° = +24.7 \text{ kJ}$$

$$\frac{1}{2}Cl_2O(g) + \frac{3}{2}OF_2(g) \longrightarrow ClF_3(g) + O_2(g) \qquad \Delta H° = -236.2 \text{ kJ}$$

$$ClF(g) + F_2 \longrightarrow ClF_3(g) \qquad \Delta H° = -104.5 \text{ kJ}$$

Reactants $\frac{1}{2}O_2$ and $\frac{1}{2}O_2$ cancel out product O_2; product $\frac{1}{2}Cl_2O$ cancels reactant $\frac{1}{2}Cl_2O$; and reactant $\frac{3}{2}OF_2$ is cancelled by products $\frac{1}{2}OF_2$ and OF_2. This leaves only reactants $ClF(g)$ and $F_2(g)$ and product $ClF_3(g)$, which are what we want. Since summing these three modified reactions yields the reaction of interest, summing the three modified $\Delta H°$ values will give the desired $\Delta H°$:

$$\Delta H° = (+107.0 \text{ kJ}) + (24.7 \text{ kJ}) + (-236.2 \text{ kJ}) = -104.5 \text{ kJ}$$

Check Your Learning

Aluminum chloride can be formed from its elements:

$$(i)\ 2Al(s) + 3Cl_2(g) \longrightarrow 2AlCl_3(s) \qquad \Delta H° = ?$$

Use the reactions here to determine the $\Delta H°$ for reaction *(i)*:

$$(ii)\ HCl(g) \longrightarrow HCl(aq) \qquad \Delta H°_{(ii)} = -74.8 \text{ kJ}$$

$$(iii)\ H_2(g) + Cl_2(g) \longrightarrow 2HCl(g) \qquad \Delta H°_{(iii)} = -185 \text{ kJ}$$

$$(iv)\ AlCl_3(aq) \longrightarrow AlCl_3(s) \qquad \Delta H°_{(iv)} = +323 \text{ kJ/mol}$$

$$(v)\ 2Al(s) + 6HCl(aq) \longrightarrow 2AlCl_3(aq) + 3H_2(g) \qquad \Delta H°_{(v)} = -1049 \text{ kJ}$$

Answer:

−1407 kJ

We also can use Hess's law to determine the enthalpy change of any reaction if the corresponding enthalpies of formation of the reactants and products are available. The stepwise reactions we consider are: (i) decompositions of the reactants into their component elements (for which the enthalpy changes are proportional to the negative of the enthalpies of formation of the reactants), followed by (ii) re-combinations of the elements to give the products (with the enthalpy changes proportional to the enthalpies of formation of the products). The standard enthalpy change of the overall reaction is therefore equal to: (ii) the sum of the standard enthalpies of formation of all the products plus (i) the sum of the negatives of the standard enthalpies of formation of the reactants. This is usually rearranged slightly to be written as follows, with $\sum$ representing "the sum of" and n standing for the stoichiometric coefficients:

$$\Delta H°_{\text{reaction}} = \sum n \times \Delta H°_{\text{f}}(\text{products}) - \sum n \times \Delta H°_{\text{f}}(\text{reactants})$$

The following example shows in detail why this equation is valid, and how to use it to calculate the enthalpy change for a reaction of interest.

 **EXAMPLE 5.15**

Using Hess's Law

What is the standard enthalpy change for the reaction:

$$3NO_2(g) + H_2O(l) \longrightarrow 2HNO_3(aq) + NO(g) \qquad \Delta H° = ?$$

Solution: Using the Equation

Use the special form of Hess's law given previously, and values from Appendix G:

$$\Delta H^\circ_{\text{reaction}} = \sum n \times \Delta H^\circ_{\text{f}}(\text{products}) - \sum n \times \Delta H^\circ_{\text{f}}(\text{reactants})$$

$$\begin{aligned}
&= \left[2\,\cancel{\text{mol HNO}_3(aq)} \times \frac{-207.4\text{ kJ}}{\cancel{\text{mol HNO}_3(aq)}} + 1\,\cancel{\text{mol NO}(g)} \times \frac{+90.2\text{ kJ}}{\cancel{\text{mol NO}(g)}}\right] \\
&- \left[3\,\cancel{\text{mol NO}_2(g)} \times \frac{+33.2\text{ kJ}}{\cancel{\text{mol NO}_2(g)}} + 1\,\cancel{\text{mol H}_2\text{O}(l)} \times \frac{-285.8\text{ kJ}}{\cancel{\text{mol H}_2\text{O}(l)}}\right] \\
&= [2 \times (-206.64) + 90.25] - [3 \times 33.2 + -(-285.83)] \\
&= -323.03 + 186.23 \\
&= -136.80\text{ kJ}
\end{aligned}$$

Solution: Supporting Why the General Equation Is Valid

Alternatively, we can write this reaction as the sum of the decompositions of $3NO_2(g)$ and $1H_2O(l)$ into their constituent elements, and the formation of $2HNO_3(aq)$ and $1NO(g)$ from their constituent elements. Writing out these reactions, and noting their relationships to the ΔH°_f values for these compounds (from Appendix G), we have:

$$3\text{NO}_2(g) \longrightarrow 3/2\text{N}_2(g) + 3\text{O}_2(g) \qquad \Delta H^\circ_1 = -99.6\text{ kJ}$$

$$\text{H}_2\text{O}(l) \longrightarrow \text{H}_2(g) + \frac{1}{2}\text{O}_2(g) \qquad \Delta H^\circ_2 = +285.8\text{ kJ }[-1 \times \Delta H^\circ_f(\text{H}_2\text{O})]$$

$$\text{H}_2(g) + \text{N}_2(g) + 3\text{O}_2(g) \longrightarrow 2\text{HNO}_3(aq) \qquad \Delta H^\circ_3 = -414.8\text{ kJ }[2 \times \Delta H^\circ_f(\text{HNO}_3)]$$

$$\frac{1}{2}\text{N}_2(g) + \frac{1}{2}\text{O}_2(g) \longrightarrow \text{NO}(g) \qquad \Delta H^\circ_4 = +90.2\text{ kJ }[1 \times (\text{NO})]$$

Summing these reaction equations gives the reaction we are interested in:

$$3\text{NO}_2(g) + \text{H}_2\text{O}(l) \longrightarrow 2\text{HNO}_3(aq) + \text{NO}(g)$$

Summing their enthalpy changes gives the value we want to determine:

$$\begin{aligned}
\Delta H^\circ_{\text{rxn}} &= \Delta H^\circ_1 + \Delta H^\circ_2 + \Delta H^\circ_3 + \Delta H^\circ_4 = (-99.6\text{ kJ}) + (+285.8\text{ kJ}) + (-414.8\text{ kJ}) + (+90.2\text{ kJ}) \\
&= -138.4\text{ kJ}
\end{aligned}$$

So the standard enthalpy change for this reaction is $\Delta H° = -138.4$ kJ.

Note that this result was obtained by (1) multiplying the ΔH°_f of each product by its stoichiometric coefficient and summing those values, (2) multiplying the ΔH°_f of each reactant by its stoichiometric coefficient and summing those values, and then (3) subtracting the result found in (2) from the result found in (1). This is also the procedure in using the general equation, as shown.

Check Your Learning

Calculate the heat of combustion of 1 mole of ethanol, $C_2H_5OH(l)$, when $H_2O(l)$ and $CO_2(g)$ are formed. Use the following enthalpies of formation: $C_2H_5OH(l)$, −278 kJ/mol; $H_2O(l)$, −286 kJ/mol; and $CO_2(g)$, −394 kJ/mol.

Answer:

−1368 kJ/mol

Key Terms

bomb calorimeter device designed to measure the energy change for processes occurring under conditions of constant volume; commonly used for reactions involving solid and gaseous reactants or products

calorie (cal) unit of heat or other energy; the amount of energy required to raise 1 gram of water by 1 degree Celsius; 1 cal is defined as 4.184 J

calorimeter device used to measure the amount of heat absorbed or released in a chemical or physical process

calorimetry process of measuring the amount of heat involved in a chemical or physical process

chemical thermodynamics area of science that deals with the relationships between heat, work, and all forms of energy associated with chemical and physical processes

endothermic process chemical reaction or physical change that absorbs heat

energy capacity to supply heat or do work

enthalpy (*H*) sum of a system's internal energy and the mathematical product of its pressure and volume

enthalpy change (Δ*H*) heat released or absorbed by a system under constant pressure during a chemical or physical process

exothermic process chemical reaction or physical change that releases heat

expansion work (pressure-volume work) work done as a system expands or contracts against external pressure

first law of thermodynamics internal energy of a system changes due to heat flow in or out of the system or work done on or by the system

heat (*q*) transfer of thermal energy between two bodies

heat capacity (*C*) extensive property of a body of matter that represents the quantity of heat required to increase its temperature by 1 degree Celsius (or 1 kelvin)

Hess's law if a process can be represented as the sum of several steps, the enthalpy change of the process equals the sum of the enthalpy changes of the steps

hydrocarbon compound composed only of hydrogen and carbon; the major component of fossil fuels

internal energy (*U*) total of all possible kinds of energy present in a substance or substances

joule (J) SI unit of energy; 1 joule is the kinetic energy of an object with a mass of 2 kilograms moving with a velocity of 1 meter per second, 1 J = 1 kg m^2/s and 4.184 J = 1 cal

kinetic energy energy of a moving body, in joules, equal to $\frac{1}{2}mv^2$ (where m = mass and v = velocity)

nutritional calorie (Calorie) unit used for quantifying energy provided by digestion of foods, defined as 1000 cal or 1 kcal

potential energy energy of a particle or system of particles derived from relative position, composition, or condition

specific heat capacity (*c*) intensive property of a substance that represents the quantity of heat required to raise the temperature of 1 gram of the substance by 1 degree Celsius (or 1 kelvin)

standard enthalpy of combustion (ΔH_c°) heat released when one mole of a compound undergoes complete combustion under standard conditions

standard enthalpy of formation (ΔH_f°) enthalpy change of a chemical reaction in which 1 mole of a pure substance is formed from its elements in their most stable states under standard state conditions

standard state set of physical conditions as accepted as common reference conditions for reporting thermodynamic properties; 1 bar of pressure, and solutions at 1 molar concentrations, usually at a temperature of 298.15 K

state function property depending only on the state of a system, and not the path taken to reach that state

surroundings all matter other than the system being studied

system portion of matter undergoing a chemical or physical change being studied

temperature intensive property of matter that is a quantitative measure of "hotness" and "coldness"

thermal energy kinetic energy associated with the random motion of atoms and molecules

thermochemistry study of measuring the amount of heat absorbed or released during a chemical reaction or a physical change

work (*w*) energy transfer due to changes in external, macroscopic variables such as pressure and volume; or causing matter to move against an opposing force

Key Equations

$$q = c \times m \times \Delta T = c \times m \times (T_{\text{final}} - T_{\text{initial}})$$

$$\Delta U = q + w$$

$$\Delta H^\circ_{\text{reaction}} = \sum n \times \Delta H^\circ_{\text{f}}(\text{products}) - \sum n \times \Delta H^\circ_{\text{f}}(\text{reactants})$$

Summary

5.1 Energy Basics

Energy is the capacity to supply heat or do work (applying a force to move matter). Kinetic energy (KE) is the energy of motion; potential energy is energy due to relative position, composition, or condition. When energy is converted from one form into another, energy is neither created nor destroyed (law of conservation of energy or first law of thermodynamics).

The thermal energy of matter is due to the kinetic energies of its constituent atoms or molecules. Temperature is an intensive property of matter reflecting hotness or coldness that increases as the average kinetic energy increases. Heat is the transfer of thermal energy between objects at different temperatures. Chemical and physical processes can absorb heat (endothermic) or release heat (exothermic). The SI unit of energy, heat, and work is the joule (J).

Specific heat and heat capacity are measures of the energy needed to change the temperature of a substance or object. The amount of heat absorbed or released by a substance depends directly on the type of substance, its mass, and the temperature change it undergoes.

5.2 Calorimetry

Calorimetry is used to measure the amount of thermal energy transferred in a chemical or physical process. This requires careful measurement of the temperature change that occurs during the process and the masses of the system and surroundings. These measured quantities are then used to compute the amount of heat produced or consumed in the process using known mathematical relations.

Calorimeters are designed to minimize energy exchange between their contents and the external environment. They range from simple coffee cup calorimeters used by introductory chemistry students to sophisticated bomb calorimeters used to determine the energy content of food.

5.3 Enthalpy

If a chemical change is carried out at constant pressure and the only work done is caused by expansion or contraction, q for the change is called the enthalpy change with the symbol ΔH, or ΔH° for reactions occurring under standard state conditions at 298 K. The value of ΔH for a reaction in one direction is equal in magnitude, but opposite in sign, to ΔH for the reaction in the opposite direction, and ΔH is directly proportional to the quantity of reactants and products. The standard enthalpy of formation, $\Delta H^\circ_{\text{f}}$, is the enthalpy change accompanying the formation of 1 mole of a substance from the elements in their most stable states at 1 bar and 298.15 K. If the enthalpies of formation are available for the reactants and products of a reaction, the enthalpy change can be calculated using Hess's law: If a process can be written as the sum of several stepwise processes, the enthalpy change of the total process equals the sum of the enthalpy changes of the various steps.

Exercises

5.1 Energy Basics

1. A burning match and a bonfire may have the same temperature, yet you would not sit around a burning match on a fall evening to stay warm. Why not?
2. Prepare a table identifying several energy transitions that take place during the typical operation of an automobile.
3. Explain the difference between heat capacity and specific heat of a substance.
4. Calculate the heat capacity, in joules and in calories per degree, of the following:
   (a) 28.4 g of water
   (b) 1.00 oz of lead

5. Calculate the heat capacity, in joules and in calories per degree, of the following:
(a) 45.8 g of nitrogen gas
(b) 1.00 pound of aluminum metal
6. How much heat, in joules and in calories, must be added to a 75.0–g iron block with a specific heat of 0.449 J/g °C to increase its temperature from 25 °C to its melting temperature of 1535 °C?
7. How much heat, in joules and in calories, is required to heat a 28.4-g (1-oz) ice cube from −23.0 °C to −1.0 °C?
8. How much would the temperature of 275 g of water increase if 36.5 kJ of heat were added?
9. If 14.5 kJ of heat were added to 485 g of liquid water, how much would its temperature increase?
10. A piece of unknown substance weighs 44.7 g and requires 2110 J to increase its temperature from 23.2 °C to 89.6 °C.
(a) What is the specific heat of the substance?
(b) If it is one of the substances found in Table 5.1, what is its likely identity?
11. A piece of unknown solid substance weighs 437.2 g, and requires 8460 J to increase its temperature from 19.3 °C to 68.9 °C.
(a) What is the specific heat of the substance?
(b) If it is one of the substances found in Table 5.1, what is its likely identity?
12. An aluminum kettle weighs 1.05 kg.
(a) What is the heat capacity of the kettle?
(b) How much heat is required to increase the temperature of this kettle from 23.0 °C to 99.0 °C?
(c) How much heat is required to heat this kettle from 23.0 °C to 99.0 °C if it contains 1.25 L of water (density of 0.997 g/mL and a specific heat of 4.184 J/g °C)?
13. Most people find waterbeds uncomfortable unless the water temperature is maintained at about 85 °F. Unless it is heated, a waterbed that contains 892 L of water cools from 85 °F to 72 °F in 24 hours. Estimate the amount of electrical energy required over 24 hours, in kWh, to keep the bed from cooling. Note that 1 kilowatt-hour (kWh) = 3.6×10^6 J, and assume that the density of water is 1.0 g/mL (independent of temperature). What other assumptions did you make? How did they affect your calculated result (i.e., were they likely to yield "positive" or "negative" errors)?

5.2 Calorimetry

14. A 500-mL bottle of water at room temperature and a 2-L bottle of water at the same temperature were placed in a refrigerator. After 30 minutes, the 500-mL bottle of water had cooled to the temperature of the refrigerator. An hour later, the 2-L of water had cooled to the same temperature. When asked which sample of water lost the most heat, one student replied that both bottles lost the same amount of heat because they started at the same temperature and finished at the same temperature. A second student thought that the 2-L bottle of water lost more heat because there was more water. A third student believed that the 500-mL bottle of water lost more heat because it cooled more quickly. A fourth student thought that it was not possible to tell because we do not know the initial temperature and the final temperature of the water. Indicate which of these answers is correct and describe the error in each of the other answers.
15. Would the amount of heat measured for the reaction in Example 5.5 be greater, lesser, or remain the same if we used a calorimeter that was a poorer insulator than a coffee cup calorimeter? Explain your answer.
16. Would the amount of heat absorbed by the dissolution in Example 5.6 appear greater, lesser, or remain the same if the experimenter used a calorimeter that was a poorer insulator than a coffee cup calorimeter? Explain your answer.
17. Would the amount of heat absorbed by the dissolution in Example 5.6 appear greater, lesser, or remain the same if the heat capacity of the calorimeter were taken into account? Explain your answer.
18. How many milliliters of water at 23 °C with a density of 1.00 g/mL must be mixed with 180 mL (about 6 oz) of coffee at 95 °C so that the resulting combination will have a temperature of 60 °C? Assume that coffee and water have the same density and the same specific heat.
19. How much will the temperature of a cup (180 g) of coffee at 95 °C be reduced when a 45 g silver spoon (specific heat 0.24 J/g °C) at 25 °C is placed in the coffee and the two are allowed to reach the same temperature? Assume that the coffee has the same density and specific heat as water.

20. A 45-g aluminum spoon (specific heat 0.88 J/g °C) at 24 °C is placed in 180 mL (180 g) of coffee at 85 °C and the temperature of the two become equal.
(a) What is the final temperature when the two become equal? Assume that coffee has the same specific heat as water.
(b) The first time a student solved this problem she got an answer of 88 °C. Explain why this is clearly an incorrect answer.
21. The temperature of the cooling water as it leaves the hot engine of an automobile is 240 °F. After it passes through the radiator it has a temperature of 175 °F. Calculate the amount of heat transferred from the engine to the surroundings by one gallon of water with a specific heat of 4.184 J/g °C.
22. A 70.0-g piece of metal at 80.0 °C is placed in 100 g of water at 22.0 °C contained in a calorimeter like that shown in Figure 5.12. The metal and water come to the same temperature at 24.6 °C. How much heat did the metal give up to the water? What is the specific heat of the metal?
23. If a reaction produces 1.506 kJ of heat, which is trapped in 30.0 g of water initially at 26.5 °C in a calorimeter like that in Figure 5.12, what is the resulting temperature of the water?
24. A 0.500-g sample of KCl is added to 50.0 g of water in a calorimeter (Figure 5.12). If the temperature decreases by 1.05 °C, what is the approximate amount of heat involved in the dissolution of the KCl, assuming the specific heat of the resulting solution is 4.18 J/g °C? Is the reaction exothermic or endothermic?
25. Dissolving 3.0 g of $CaCl_2(s)$ in 150.0 g of water in a calorimeter (Figure 5.12) at 22.4 °C causes the temperature to rise to 25.8 °C. What is the approximate amount of heat involved in the dissolution, assuming the specific heat of the resulting solution is 4.18 J/g °C? Is the reaction exothermic or endothermic?
26. When 50.0 g of 0.200 M $NaCl(aq)$ at 24.1 °C is added to 100.0 g of 0.100 M $AgNO_3(aq)$ at 24.1 °C in a calorimeter, the temperature increases to 25.2 °C as $AgCl(s)$ forms. Assuming the specific heat of the solution and products is 4.20 J/g °C, calculate the approximate amount of heat in joules produced.
27. The addition of 3.15 g of $Ba(OH)_2{\cdot}8H_2O$ to a solution of 1.52 g of NH_4SCN in 100 g of water in a calorimeter caused the temperature to fall by 3.1 °C. Assuming the specific heat of the solution and products is 4.20 J/g °C, calculate the approximate amount of heat absorbed by the reaction, which can be represented by the following equation:
$Ba(OH)_2{\cdot}8H_2O(s) + 2NH_4SCN(aq) \longrightarrow Ba(SCN)_2(aq) + 2NH_3(aq) + 10H_2O(l)$
28. The reaction of 50 mL of acid and 50 mL of base described in Example 5.5 increased the temperature of the solution by 6.9 °C. How much would the temperature have increased if 100 mL of acid and 100 mL of base had been used in the same calorimeter starting at the same temperature of 22.0 °C? Explain your answer.
29. If the 3.21 g of NH_4NO_3 in Example 5.6 were dissolved in 100.0 g of water under the same conditions, how much would the temperature change? Explain your answer.
30. When 1.0 g of fructose, $C_6H_{12}O_6(s)$, a sugar commonly found in fruits, is burned in oxygen in a bomb calorimeter, the temperature of the calorimeter increases by 1.58 °C. If the heat capacity of the calorimeter and its contents is 9.90 kJ/°C, what is q for this combustion?
31. When a 0.740-g sample of trinitrotoluene (TNT), $C_7H_5N_2O_6$, is burned in a bomb calorimeter, the temperature increases from 23.4 °C to 26.9 °C. The heat capacity of the calorimeter is 534 J/°C, and it contains 675 mL of water. How much heat was produced by the combustion of the TNT sample?
32. One method of generating electricity is by burning coal to heat water, which produces steam that drives an electric generator. To determine the rate at which coal is to be fed into the burner in this type of plant, the heat of combustion per ton of coal must be determined using a bomb calorimeter. When 1.00 g of coal is burned in a bomb calorimeter (Figure 5.17), the temperature increases by 1.48 °C. If the heat capacity of the calorimeter is 21.6 kJ/°C, determine the heat produced by combustion of a ton of coal (2.000×10^3 pounds).
33. The amount of fat recommended for someone with a daily diet of 2000 Calories is 65 g. What percent of the calories in this diet would be supplied by this amount of fat if the average number of Calories for fat is 9.1 Calories/g?
34. A teaspoon of the carbohydrate sucrose (common sugar) contains 16 Calories (16 kcal). What is the mass of one teaspoon of sucrose if the average number of Calories for carbohydrates is 4.1 Calories/g?

35. What is the maximum mass of carbohydrate in a 6-oz serving of diet soda that contains less than 1 Calorie per can if the average number of Calories for carbohydrates is 4.1 Calories/g?
36. A pint of premium ice cream can contain 1100 Calories. What mass of fat, in grams and pounds, must be produced in the body to store an extra 1.1×10^3 Calories if the average number of Calories for fat is 9.1 Calories/g?
37. A serving of a breakfast cereal contains 3 g of protein, 18 g of carbohydrates, and 6 g of fat. What is the Calorie content of a serving of this cereal if the average number of Calories for fat is 9.1 Calories/g, for carbohydrates is 4.1 Calories/g, and for protein is 4.1 Calories/g?
38. Which is the least expensive source of energy in kilojoules per dollar: a box of breakfast cereal that weighs 32 ounces and costs $4.23, or a liter of isooctane (density, 0.6919 g/mL) that costs $0.45? Compare the nutritional value of the cereal with the heat produced by combustion of the isooctane under standard conditions. A 1.0-ounce serving of the cereal provides 130 Calories.

5.3 Enthalpy

39. Explain how the heat measured in Example 5.5 differs from the enthalpy change for the exothermic reaction described by the following equation:
$HCl(aq) + NaOH(aq) \longrightarrow NaCl(aq) + H_2O(l)$
40. Using the data in the check your learning section of Example 5.5, calculate ΔH in kJ/mol of $AgNO_3(aq)$ for the reaction: $NaCl(aq) + AgNO_3(aq) \longrightarrow AgCl(s) + NaNO_3(aq)$
41. Calculate the enthalpy of solution (ΔH for the dissolution) per mole of NH_4NO_3 under the conditions described in Example 5.6.
42. Calculate ΔH for the reaction described by the equation. (*Hint*: Use the value for the approximate amount of heat absorbed by the reaction that you calculated in a previous exercise.)
$Ba(OH)_2{\cdot}8H_2O(s) + 2NH_4SCN(aq) \longrightarrow Ba(SCN)_2(aq) + 2NH_3(aq) + 10H_2O(l)$
43. Calculate the enthalpy of solution (ΔH for the dissolution) per mole of $CaCl_2$ (refer to Exercise 5.25).
44. Although the gas used in an oxyacetylene torch (Figure 5.7) is essentially pure acetylene, the heat produced by combustion of one mole of acetylene in such a torch is likely not equal to the enthalpy of combustion of acetylene listed in Table 5.2. Considering the conditions for which the tabulated data are reported, suggest an explanation.
45. How much heat is produced by burning 4.00 moles of acetylene under standard state conditions?
46. How much heat is produced by combustion of 125 g of methanol under standard state conditions?
47. How many moles of isooctane must be burned to produce 100 kJ of heat under standard state conditions?
48. What mass of carbon monoxide must be burned to produce 175 kJ of heat under standard state conditions?
49. When 2.50 g of methane burns in oxygen, 125 kJ of heat is produced. What is the enthalpy of combustion per mole of methane under these conditions?
50. How much heat is produced when 100 mL of 0.250 M HCl (density, 1.00 g/mL) and 200 mL of 0.150 M NaOH (density, 1.00 g/mL) are mixed?
$HCl(aq) + NaOH(aq) \longrightarrow NaCl(aq) + H_2O(l) \qquad \Delta H^\circ = -58\ kJ$
If both solutions are at the same temperature and the specific heat of the products is 4.19 J/g °C, how much will the temperature increase? What assumption did you make in your calculation?
51. A sample of 0.562 g of carbon is burned in oxygen in a bomb calorimeter, producing carbon dioxide. Assume both the reactants and products are under standard state conditions, and that the heat released is directly proportional to the enthalpy of combustion of graphite. The temperature of the calorimeter increases from 26.74 °C to 27.93 °C. What is the heat capacity of the calorimeter and its contents?
52. Before the introduction of chlorofluorocarbons, sulfur dioxide (enthalpy of vaporization, 6.00 kcal/mol) was used in household refrigerators. What mass of SO_2 must be evaporated to remove as much heat as evaporation of 1.00 kg of CCl_2F_2 (enthalpy of vaporization is 17.4 kJ/mol)?
The vaporization reactions for SO_2 and CCl_2F_2 are $SO_2(l) \longrightarrow SO_2(g)$ and $CCl_2F(l) \longrightarrow CCl_2F_2(g)$, respectively.
53. Homes may be heated by pumping hot water through radiators. What mass of water will provide the same amount of heat when cooled from 95.0 to 35.0 °C, as the heat provided when 100 g of steam is cooled from 110 °C to 100 °C.

54. Which of the enthalpies of combustion in Table 5.2 the table are also standard enthalpies of formation?
55. Does the standard enthalpy of formation of $H_2O(g)$ differ from $\Delta H°$ for the reaction
$2H_2(g) + O_2(g) \longrightarrow 2H_2O(g)$?
56. Joseph Priestly prepared oxygen in 1774 by heating red mercury(II) oxide with sunlight focused through a lens. How much heat is required to decompose exactly 1 mole of red $HgO(s)$ to $Hg(l)$ and $O_2(g)$ under standard conditions?
57. How many kilojoules of heat will be released when exactly 1 mole of manganese, Mn, is burned to form $Mn_3O_4(s)$ at standard state conditions?
58. How many kilojoules of heat will be released when exactly 1 mole of iron, Fe, is burned to form $Fe_2O_3(s)$ at standard state conditions?
59. The following sequence of reactions occurs in the commercial production of aqueous nitric acid:
$4NH_3(g) + 5O_2(g) \longrightarrow 4NO(g) + 6H_2O(l) \quad \Delta H = -907 \text{ kJ}$
$2NO(g) + O_2(g) \longrightarrow 2NO_2(g) \quad \Delta H = -113 \text{ kJ}$
$3NO_2 + H_2O(l) \longrightarrow 2HNO_3(aq) + NO(g) \quad \Delta H = -139 \text{ kJ}$
Determine the total energy change for the production of one mole of aqueous nitric acid by this process.
60. Both graphite and diamond burn.
$C(s, \text{diamond}) + O_2(g) \longrightarrow CO_2(g)$
For the conversion of graphite to diamond:
$C(s, \text{graphite}) \longrightarrow C(s, \text{diamond}) \quad \Delta H° = 1.90 \text{ kJ}$
Which produces more heat, the combustion of graphite or the combustion of diamond?
61. From the molar heats of formation in Appendix G, determine how much heat is required to evaporate one mole of water: $H_2O(l) \longrightarrow H_2O(g)$
62. Which produces more heat?
$Os(s) \longrightarrow 2O_2(g) \longrightarrow OsO_4(s)$
or
$Os(s) \longrightarrow 2O_2(g) \longrightarrow OsO_4(g)$
for the phase change $OsO_4(s) \longrightarrow OsO_4(g) \quad \Delta H = 56.4 \text{ kJ}$
63. Calculate $\Delta H°$ for the process
$Sb(s) + \frac{5}{2}Cl_2(g) \longrightarrow SbCl_5(s)$
from the following information:

$Sb(s) + \frac{3}{2}Cl_2(g) \longrightarrow SbCl_3(s) \quad \Delta H° = -314 \text{ kJ}$
$SbCl_3(s) + Cl_2(g) \longrightarrow SbCl_5(s) \quad \Delta H° = -80 \text{ kJ}$
64. Calculate $\Delta H°$ for the process $Zn(s) + S(s) + 2O_2(g) \longrightarrow ZnSO_4(s)$
from the following information:
$Zn(s) + S(s) \longrightarrow ZnS(s) \quad \Delta H° = -206.0 \text{ kJ}$
$ZnS(s) + 2O_2(g) \longrightarrow ZnSO_4(s) \quad \Delta H° = -776.8 \text{ kJ}$
65. Calculate ΔH for the process $Hg_2Cl_2(s) \longrightarrow 2Hg(l) + Cl_2(g)$
from the following information:
$Hg(l) + Cl_2(g) \longrightarrow HgCl_2(s) \quad \Delta H = -224 \text{ kJ}$
$Hg(l) + HgCl_2(s) \longrightarrow Hg_2Cl_2(s) \quad \Delta H = -41.2 \text{ kJ}$
66. Calculate $\Delta H°$ for the process $Co_3O_4(s) \longrightarrow 3Co(s) + 2O_2(g)$
from the following information:

$Co(s) + \frac{1}{2}O_2(g) \longrightarrow CoO(s) \quad \Delta H° = -237.9 \text{kJ}$
$3CoO(s) + \frac{1}{2}O_2(g) \longrightarrow Co_3O_4(s) \quad \Delta H° = -177.5 \text{kJ}$
67. Calculate the standard molar enthalpy of formation of $NO(g)$ from the following data:
$N_2(g) + 2O_2 \longrightarrow 2NO_2(g) \quad \Delta H° = 66.4 \text{ kJ}$
$2NO(g) + O_2 \longrightarrow 2NO_2(g) \quad \Delta H° = -114.1 \text{ kJ}$

68. Using the data in Appendix G, calculate the standard enthalpy change for each of the following reactions:
(a) $N_2(g) + O_2(g) \longrightarrow 2NO(g)$
(b) $Si(s) + 2Cl_2(g) \longrightarrow SiCl_4(g)$
(c) $Fe_2O_3(s) + 3H_2(g) \longrightarrow 2Fe(s) + 3H_2O(l)$
(d) $2LiOH(s) + CO_2(g) \longrightarrow Li_2CO_3(s) + H_2O(g)$

69. Using the data in Appendix G, calculate the standard enthalpy change for each of the following reactions:
(a) $Si(s) + 2F_2(g) \longrightarrow SiF_4(g)$
(b) $2C(s) + 2H_2(g) + O_2(g) \longrightarrow CH_3CO_2H(l)$
(c) $CH_4(g) + N_2(g) \longrightarrow HCN(g) + NH_3(g)$;
(d) $CS_2(g) + 3Cl_2(g) \longrightarrow CCl_4(g) + S_2Cl_2(g)$

70. The following reactions can be used to prepare samples of metals. Determine the enthalpy change under standard state conditions for each.
(a) $2Ag_2O(s) \longrightarrow 4Ag(s) + O_2(g)$
(b) $SnO(s) + CO(g) \longrightarrow Sn(s) + CO_2(g)$
(c) $Cr_2O_3(s) + 3H_2(g) \longrightarrow 2Cr(s) + 3H_2O(l)$
(d) $2Al(s) + Fe_2O_3(s) \longrightarrow Al_2O_3(s) + 2Fe(s)$

71. The decomposition of hydrogen peroxide, H_2O_2, has been used to provide thrust in the control jets of various space vehicles. Using the data in Appendix G, determine how much heat is produced by the decomposition of exactly 1 mole of H_2O_2 under standard conditions.
$2H_2O_2(l) \longrightarrow 2H_2O(g) + O_2(g)$

72. Calculate the enthalpy of combustion of propane, $C_3H_8(g)$, for the formation of $H_2O(g)$ and $CO_2(g)$. The enthalpy of formation of propane is −104 kJ/mol.

73. Calculate the enthalpy of combustion of butane, $C_4H_{10}(g)$ for the formation of $H_2O(g)$ and $CO_2(g)$. The enthalpy of formation of butane is −126 kJ/mol.

74. Both propane and butane are used as gaseous fuels. Which compound produces more heat per gram when burned?

75. The white pigment TiO_2 is prepared by the reaction of titanium tetrachloride, $TiCl_4$, with water vapor in the gas phase: $TiCl_4(g) + 2H_2O(g) \longrightarrow TiO_2(s) + 4HCl(g)$.
How much heat is evolved in the production of exactly 1 mole of $TiO_2(s)$ under standard state conditions?

76. Water gas, a mixture of H_2 and CO, is an important industrial fuel produced by the reaction of steam with red hot coke, essentially pure carbon: $C(s) + H_2O(g) \longrightarrow CO(g) + H_2(g)$.
(a) Assuming that coke has the same enthalpy of formation as graphite, calculate $\Delta H°$ for this reaction.
(b) Methanol, a liquid fuel that could possibly replace gasoline, can be prepared from water gas and additional hydrogen at high temperature and pressure in the presence of a suitable catalyst:
$2H_2(g) + CO(g) \longrightarrow CH_3OH(g)$.
Under the conditions of the reaction, methanol forms as a gas. Calculate $\Delta H°$ for this reaction and for the condensation of gaseous methanol to liquid methanol.
(c) Calculate the heat of combustion of 1 mole of liquid methanol to $H_2O(g)$ and $CO_2(g)$.

77. In the early days of automobiles, illumination at night was provided by burning acetylene, C_2H_2. Though no longer used as auto headlamps, acetylene is still used as a source of light by some cave explorers. The acetylene is (was) prepared in the lamp by the reaction of water with calcium carbide, CaC_2:
$CaC_2(s) + 2H_2O(l) \longrightarrow Ca(OH)_2(s) + C_2H_2(g)$.
Calculate the standard enthalpy of the reaction. The ΔH_f° of CaC_2 is −15.14 kcal/mol.

78. From the data in Table 5.2, determine which of the following fuels produces the greatest amount of heat per gram when burned under standard conditions: $CO(g)$, $CH_4(g)$, or $C_2H_2(g)$.

79. The enthalpy of combustion of hard coal averages −35 kJ/g, that of gasoline, 1.28×10^5 kJ/gal. How many kilograms of hard coal provide the same amount of heat as is available from 1.0 gallon of gasoline? Assume that the density of gasoline is 0.692 g/mL (the same as the density of isooctane).

80. Ethanol, C_2H_5OH, is used as a fuel for motor vehicles, particularly in Brazil.
(a) Write the balanced equation for the combustion of ethanol to $CO_2(g)$ and $H_2O(g)$, and, using the data in Appendix G, calculate the enthalpy of combustion of 1 mole of ethanol.
(b) The density of ethanol is 0.7893 g/mL. Calculate the enthalpy of combustion of exactly 1 L of ethanol.
(c) Assuming that an automobile's mileage is directly proportional to the heat of combustion of the fuel, calculate how much farther an automobile could be expected to travel on 1 L of gasoline than on 1 L of ethanol. Assume that gasoline has the heat of combustion and the density of n–octane, C_8H_{18} ($\Delta H_f^\circ = -208.4$ kJ/mol; density = 0.7025 g/mL).

81. Among the substances that react with oxygen and that have been considered as potential rocket fuels are diborane [B_2H_6, produces $B_2O_3(s)$ and $H_2O(g)$], methane [CH_4, produces $CO_2(g)$ and $H_2O(g)$], and hydrazine [N_2H_4, produces $N_2(g)$ and $H_2O(g)$]. On the basis of the heat released by 1.00 g of each substance in its reaction with oxygen, which of these compounds offers the best possibility as a rocket fuel? The ΔH_f° of $B_2H_6(g)$, $CH_4(g)$, and $N_2H_4(l)$ may be found in Appendix G.

82. How much heat is produced when 1.25 g of chromium metal reacts with oxygen gas under standard conditions?

83. Ethylene, C_2H_4, a byproduct from the fractional distillation of petroleum, is fourth among the 50 chemical compounds produced commercially in the largest quantities. About 80% of synthetic ethanol is manufactured from ethylene by its reaction with water in the presence of a suitable catalyst.

$$C_2H_4(g) + H_2O(g) \longrightarrow C_2H_5OH(l)$$

Using the data in the table in Appendix G, calculate ΔH° for the reaction.

84. The oxidation of the sugar glucose, $C_6H_{12}O_6$, is described by the following equation:

$$C_6H_{12}O_6(s) + 6O_2(g) \longrightarrow 6CO_2(g) + 6H_2O(l) \qquad \Delta H = -2816\text{ kJ}$$

The metabolism of glucose gives the same products, although the glucose reacts with oxygen in a series of steps in the body.
(a) How much heat in kilojoules can be produced by the metabolism of 1.0 g of glucose?
(b) How many Calories can be produced by the metabolism of 1.0 g of glucose?

85. Propane, C_3H_8, is a hydrocarbon that is commonly used as a fuel.
(a) Write a balanced equation for the complete combustion of propane gas.
(b) Calculate the volume of air at 25 °C and 1.00 atmosphere that is needed to completely combust 25.0 grams of propane. Assume that air is 21.0 percent O_2 by volume. (Hint: We will see how to do this calculation in a later chapter on gases—for now use the information that 1.00 L of air at 25 °C and 1.00 atm contains 0.275 g of O_2.)
(c) The heat of combustion of propane is −2,219.2 kJ/mol. Calculate the heat of formation, ΔH_f° of propane given that ΔH_f° of $H_2O(l) = -285.8$ kJ/mol and ΔH_f° of $CO_2(g) = -393.5$ kJ/mol.
(d) Assuming that all of the heat released in burning 25.0 grams of propane is transferred to 4.00 kilograms of water, calculate the increase in temperature of the water.

86. During a recent winter month in Sheboygan, Wisconsin, it was necessary to obtain 3500 kWh of heat provided by a natural gas furnace with 89% efficiency to keep a small house warm (the efficiency of a gas furnace is the percent of the heat produced by combustion that is transferred into the house).
(a) Assume that natural gas is pure methane and determine the volume of natural gas in cubic feet that was required to heat the house. The average temperature of the natural gas was 56 °F; at this temperature and a pressure of 1 atm, natural gas has a density of 0.681 g/L.
(b) How many gallons of LPG (liquefied petroleum gas) would be required to replace the natural gas used? Assume the LPG is liquid propane [C_3H_8: density, 0.5318 g/mL; enthalpy of combustion, 2219 kJ/mol for the formation of $CO_2(g)$ and $H_2O(l)$] and the furnace used to burn the LPG has the same efficiency as the gas furnace.
(c) What mass of carbon dioxide is produced by combustion of the methane used to heat the house?
(d) What mass of water is produced by combustion of the methane used to heat the house?
(e) What volume of air is required to provide the oxygen for the combustion of the methane used to heat the house? Air contains 23% oxygen by mass. The average density of air during the month was 1.22 g/L.
(f) How many kilowatt–hours (1 kWh = 3.6×10^6 J) of electricity would be required to provide the heat necessary to heat the house? Note electricity is 100% efficient in producing heat inside a house.
(g) Although electricity is 100% efficient in producing heat inside a house, production and distribution of electricity is not 100% efficient. The efficiency of production and distribution of electricity produced in a coal-fired power plant is about 40%. A certain type of coal provides 2.26 kWh per pound upon combustion. What mass of this coal in kilograms will be required to produce the electrical energy necessary to heat the house if the efficiency of generation and distribution is 40%?

CHAPTER 6
Electronic Structure and Periodic Properties of Elements

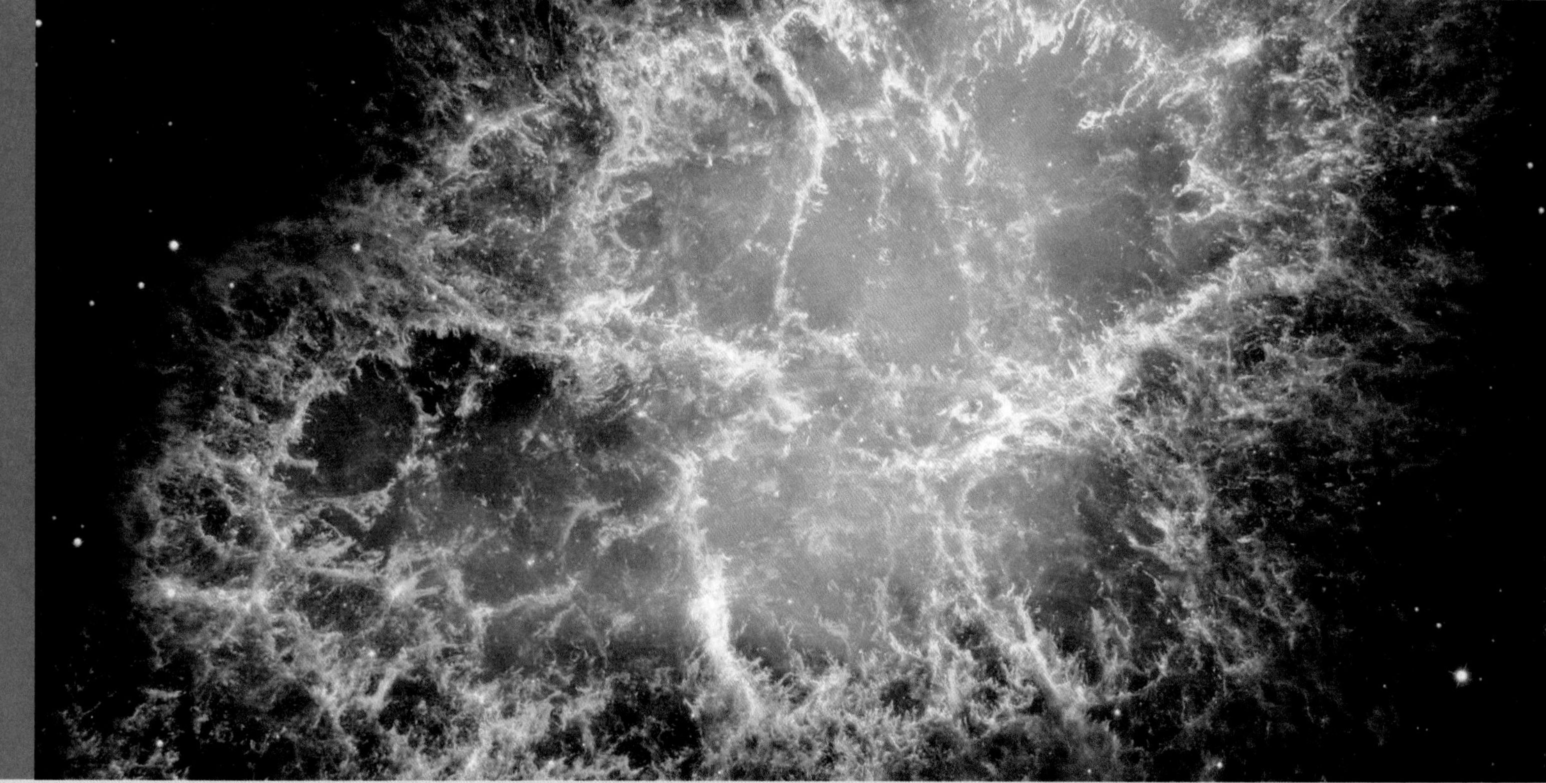

Figure 6.1 The Crab Nebula consists of remnants of a supernova (the explosion of a star). NASA's Hubble Space Telescope produced this composite image. Measurements of the emitted light wavelengths enabled astronomers to identify the elements in the nebula, determining that it contains specific ions including S^+ (green filaments) and O^{2+} (red filaments). (credit: modification of work by NASA and ESA)

CHAPTER OUTLINE

6.1 Electromagnetic Energy
6.2 The Bohr Model
6.3 Development of Quantum Theory
6.4 Electronic Structure of Atoms (Electron Configurations)
6.5 Periodic Variations in Element Properties

INTRODUCTION In 1054, Chinese astronomers recorded the appearance of a "guest star" in the sky, visible even during the day, which then disappeared slowly over the next two years. The sudden appearance was due to a supernova explosion, which was much brighter than the original star. Even though this supernova was observed almost a millennium ago, the remaining Crab Nebula (Figure 6.1) continues to release energy today. It emits not only visible light but also infrared light, X-rays, and other forms of electromagnetic radiation. The nebula emits both continuous spectra (the blue-white glow) and atomic emission spectra (the colored filaments). In this chapter, we will discuss light and other forms of electromagnetic radiation and how they are related to the electronic structure of atoms. We will also see how this radiation can be used to identify elements, even from thousands of light years away.

6.1 Electromagnetic Energy

LEARNING OBJECTIVES

By the end of this section, you will be able to:

- Explain the basic behavior of waves, including travelling waves and standing waves
- Describe the wave nature of light
- Use appropriate equations to calculate related light-wave properties such as frequency, wavelength, and energy
- Distinguish between line and continuous emission spectra
- Describe the particle nature of light

The nature of light has been a subject of inquiry since antiquity. In the seventeenth century, Isaac Newton performed experiments with lenses and prisms and was able to demonstrate that white light consists of the individual colors of the rainbow combined together. Newton explained his optics findings in terms of a "corpuscular" view of light, in which light was composed of streams of extremely tiny particles travelling at high speeds according to Newton's laws of motion. Others in the seventeenth century, such as Christiaan Huygens, had shown that optical phenomena such as reflection and refraction could be equally well explained in terms of light as waves travelling at high speed through a medium called "luminiferous aether" that was thought to permeate all space. Early in the nineteenth century, Thomas Young demonstrated that light passing through narrow, closely spaced slits produced interference patterns that could not be explained in terms of Newtonian particles but could be easily explained in terms of waves. Later in the nineteenth century, after James Clerk Maxwell developed his theory of **electromagnetic radiation** and showed that light was the visible part of a vast spectrum of electromagnetic waves, the particle view of light became thoroughly discredited. By the end of the nineteenth century, scientists viewed the physical universe as roughly comprising two separate domains: matter composed of particles moving according to Newton's laws of motion, and electromagnetic radiation consisting of waves governed by Maxwell's equations. Today, these domains are referred to as classical mechanics and classical electrodynamics (or classical electromagnetism). Although there were a few physical phenomena that could not be explained within this framework, scientists at that time were so confident of the overall soundness of this framework that they viewed these aberrations as puzzling paradoxes that would ultimately be resolved somehow within this framework. As we shall see, these paradoxes led to a contemporary framework that intimately connects particles and waves at a fundamental level called wave-particle duality, which has superseded the classical view.

Visible light and other forms of electromagnetic radiation play important roles in chemistry, since they can be used to infer the energies of electrons within atoms and molecules. Much of modern technology is based on electromagnetic radiation. For example, radio waves from a mobile phone, X-rays used by dentists, the energy used to cook food in your microwave, the radiant heat from red-hot objects, and the light from your television screen are forms of electromagnetic radiation that all exhibit wavelike behavior.

Waves

A **wave** is an oscillation or periodic movement that can transport energy from one point in space to another. Common examples of waves are all around us. Shaking the end of a rope transfers energy from your hand to the other end of the rope, dropping a pebble into a pond causes waves to ripple outward along the water's surface, and the expansion of air that accompanies a lightning strike generates sound waves (thunder) that can travel outward for several miles. In each of these cases, kinetic energy is transferred through matter (the rope, water, or air) while the matter remains essentially in place. An insightful example of a wave occurs in sports stadiums when fans in a narrow region of seats rise simultaneously and stand with their arms raised up for a few seconds before sitting down again while the fans in neighboring sections likewise stand up and sit down in sequence. While this wave can quickly encircle a large stadium in a few seconds, none of the fans actually travel with the wave-they all stay in or above their seats.

Waves need not be restricted to travel through matter. As Maxwell showed, electromagnetic waves consist of an electric field oscillating in step with a perpendicular magnetic field, both of which are perpendicular to the direction of travel. These waves can travel through a vacuum at a constant speed of 2.998×10^8 m/s, the speed of light (denoted by *c*).

All waves, including forms of electromagnetic radiation, are characterized by, a **wavelength** (denoted by λ, the lowercase Greek letter lambda), a **frequency** (denoted by ν, the lowercase Greek letter nu), and an **amplitude**. As can be seen in Figure 6.2, the wavelength is the distance between two consecutive peaks or troughs in a wave (measured in meters in the SI system). Electromagnetic waves have wavelengths that fall within an enormous range-wavelengths of kilometers (10^3 m) to picometers (10^{-12} m) have been observed. The frequency is the number of wave cycles that pass a specified point in space in a specified amount of time (in the SI system, this is measured in seconds). A cycle corresponds to one complete wavelength. The unit for frequency, expressed as cycles per second [s^{-1}], is the **hertz (Hz)**. Common multiples of this unit are megahertz, (1 MHz = 1×10^6 Hz) and gigahertz (1 GHz = 1×10^9 Hz). The amplitude corresponds to the magnitude of the wave's displacement and so, in Figure 6.2, this corresponds to one-half the height between the peaks and troughs. The amplitude is related to the intensity of the wave, which for light is the brightness, and for sound is the loudness.

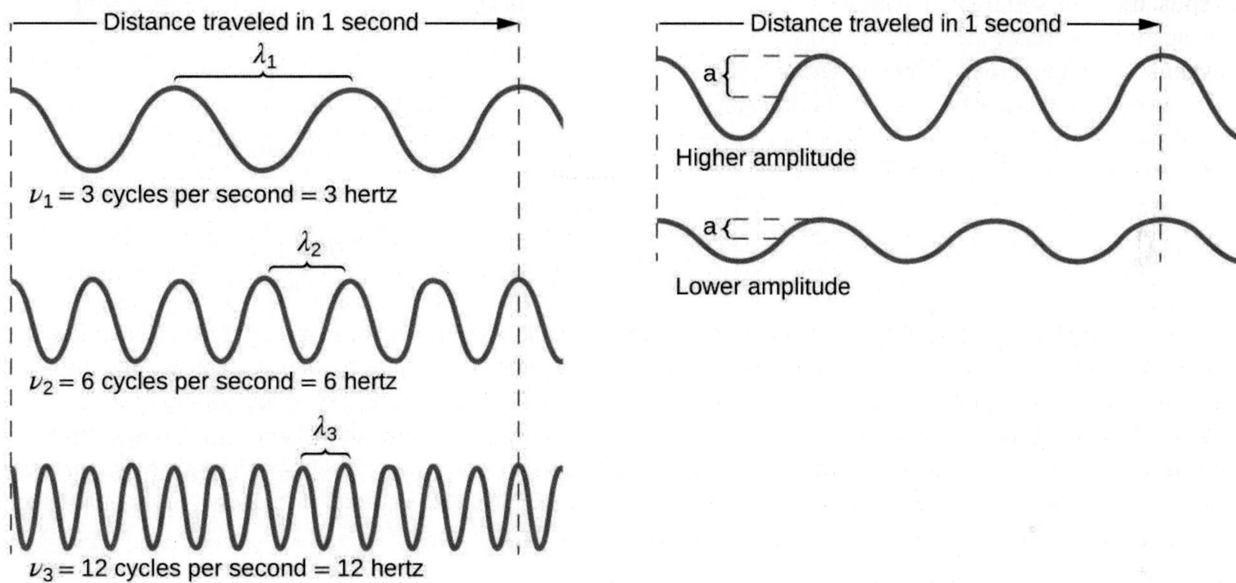

FIGURE 6.2 One-dimensional sinusoidal waves show the relationship among wavelength, frequency, and speed. The wave with the shortest wavelength has the highest frequency. Amplitude is one-half the height of the wave from peak to trough.

The product of a wave's wavelength (λ) and its frequency (ν), $\lambda\nu$, is the speed of the wave. Thus, for electromagnetic radiation in a vacuum, speed is equal to the fundamental constant, *c*:

$$c = 2.998 \times 10^8 \text{ms}^{-1} = \lambda\nu$$

Wavelength and frequency are inversely proportional: As the wavelength increases, the frequency decreases. The inverse proportionality is illustrated in Figure 6.3. This figure also shows the **electromagnetic spectrum**, the range of all types of electromagnetic radiation. Each of the various colors of visible light has specific frequencies and wavelengths associated with them, and you can see that visible light makes up only a small portion of the electromagnetic spectrum. Because the technologies developed to work in various parts of the electromagnetic spectrum are different, for reasons of convenience and historical legacies, different units are typically used for different parts of the spectrum. For example, radio waves are usually specified as frequencies (typically in units of MHz), while the visible region is usually specified in wavelengths (typically in units of nm or angstroms).

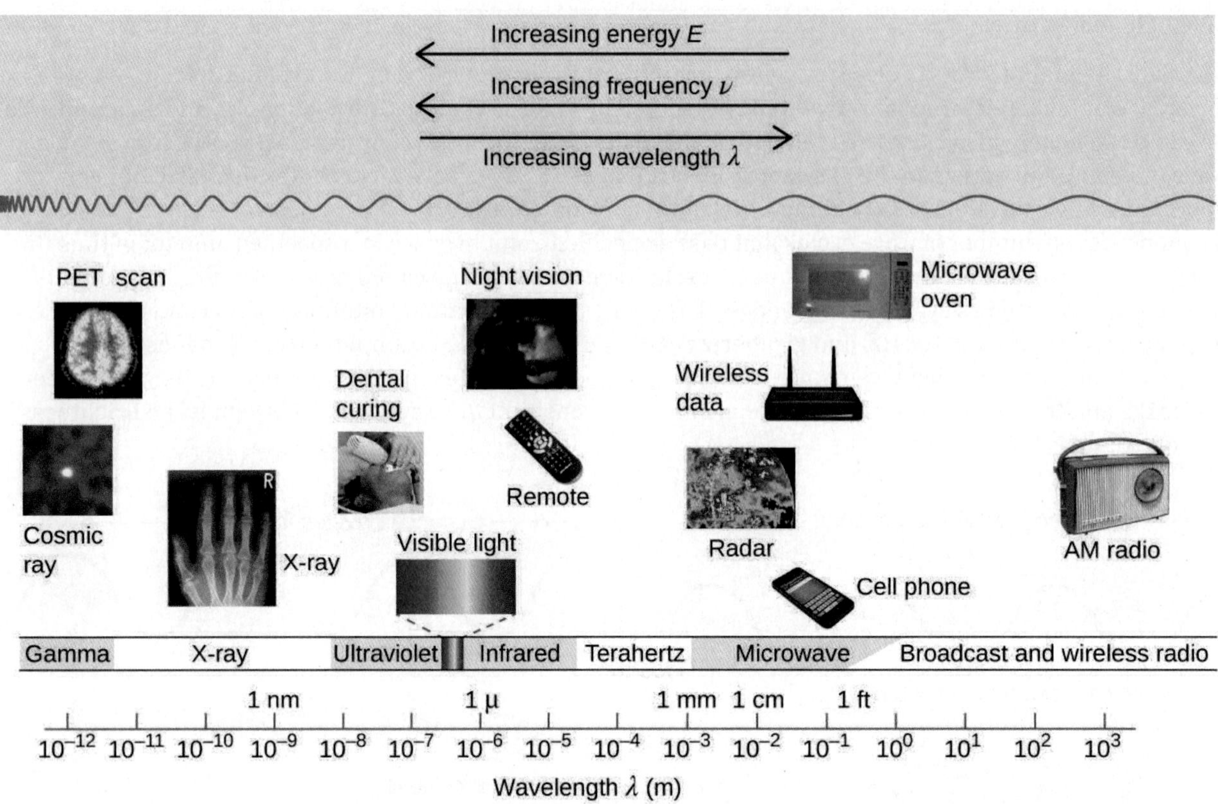

FIGURE 6.3 Portions of the electromagnetic spectrum are shown in order of decreasing frequency and increasing wavelength. (credit "Cosmic ray": modification of work by NASA; credit "PET scan": modification of work by the National Institute of Health; credit "X-ray": modification of work by Dr. Jochen Lengerke; credit "Dental curing": modification of work by the Department of the Navy; credit "Night vision": modification of work by the Department of the Army; credit "Remote": modification of work by Emilian Robert Vicol; credit "Cell phone": modification of work by Brett Jordan; credit "Microwave oven": modification of work by Billy Mabray; credit "Ultrasound": modification of work by Jane Whitney; credit "AM radio": modification of work by Dave Clausen)

EXAMPLE 6.1

Determining the Frequency and Wavelength of Radiation

A sodium streetlight gives off yellow light that has a wavelength of 589 nm (1 nm = 1×10^{-9} m). What is the frequency of this light?

Solution

We can rearrange the equation $c = \lambda\nu$ to solve for the frequency:

$$\nu = \frac{c}{\lambda}$$

Since c is expressed in meters per second, we must also convert 589 nm to meters.

$$\nu = \left(\frac{2.998 \times 10^{8} \cancel{\text{m}}\,\text{s}^{-1}}{589 \cancel{\text{nm}}}\right)\left(\frac{1 \times 10^{9} \cancel{\text{nm}}}{1 \cancel{\text{m}}}\right) = 5.09 \times 10^{14}\,\text{s}^{-1}$$

Check Your Learning

One of the frequencies used to transmit and receive cellular telephone signals in the United States is 850 MHz. What is the wavelength in meters of these radio waves?

Answer:

0.353 m = 35.3 cm

Chemistry in Everyday Life

Wireless Communication

FIGURE 6.4 Radio and cell towers are typically used to transmit long-wavelength electromagnetic radiation. Increasingly, cell towers are designed to blend in with the landscape, as with the Tucson, Arizona, cell tower (right) disguised as a palm tree. (credit left: modification of work by Sir Mildred Pierce; credit middle: modification of work by M.O. Stevens)

Many valuable technologies operate in the radio (3 kHz-300 GHz) frequency region of the electromagnetic spectrum. At the low frequency (low energy, long wavelength) end of this region are AM (amplitude modulation) radio signals (540-2830 kHz) that can travel long distances. FM (frequency modulation) radio signals are used at higher frequencies (87.5-108.0 MHz). In AM radio, the information is transmitted by varying the amplitude of the wave (Figure 6.5). In FM radio, by contrast, the amplitude is constant and the instantaneous frequency varies.

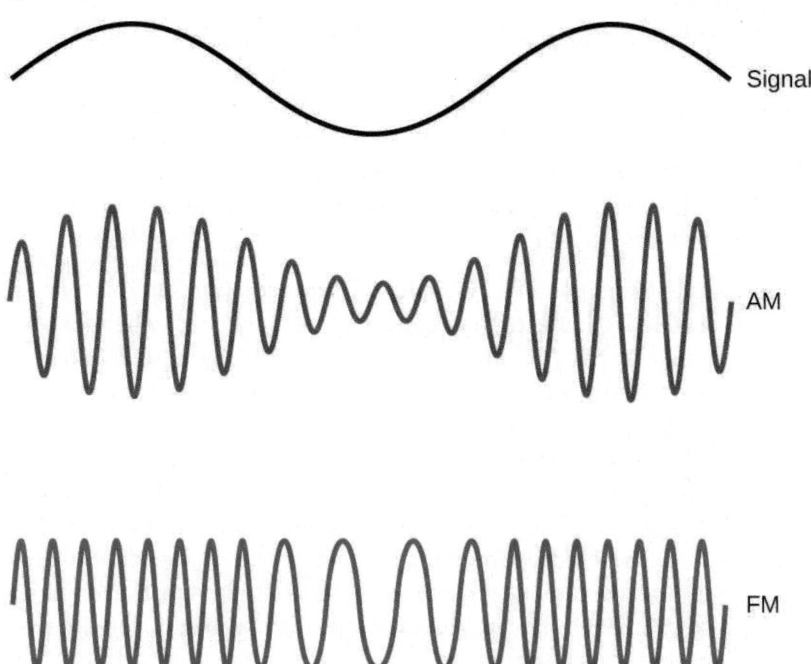

FIGURE 6.5 This schematic depicts how amplitude modulation (AM) and frequency modulation (FM) can be used to transmit a radio wave.

Other technologies also operate in the radio-wave portion of the electromagnetic spectrum. For example, 4G cellular telephone signals are approximately 880 MHz, while Global Positioning System (GPS) signals operate at 1.228 and 1.575 GHz, local area wireless technology (Wi-Fi) networks operate at 2.4 to 5 GHz, and highway toll sensors operate at 5.8 GHz. The frequencies associated with these applications are convenient because such waves tend not to be absorbed much by common building materials.

One particularly characteristic phenomenon of waves results when two or more waves come into contact: They interfere with each other. Figure 6.6 shows the **interference patterns** that arise when light passes through narrow slits closely spaced about a wavelength apart. The fringe patterns produced depend on the wavelength, with the fringes being more closely spaced for shorter wavelength light passing through a given set of slits. When the light passes through the two slits, each slit effectively acts as a new source, resulting in two closely spaced waves coming into contact at the detector (the camera in this case). The dark regions in Figure 6.6 correspond to regions where the peaks for the wave from one slit happen to coincide with the troughs for the wave from the other slit (destructive interference), while the brightest regions correspond to the regions where the peaks for the two waves (or their two troughs) happen to coincide (constructive interference). Likewise, when two stones are tossed close together into a pond, interference patterns are visible in the interactions between the waves produced by the stones. Such interference patterns cannot be explained by particles moving according to the laws of classical mechanics.

FIGURE 6.6 Interference fringe patterns are shown for light passing through two closely spaced, narrow slits. The spacing of the fringes depends on the wavelength, with the fringes being more closely spaced for the shorter-wavelength blue light. (credit: PASCO)

Portrait of a Chemist

Dorothy Crowfoot Hodgkin

X-rays exhibit wavelengths of approximately 0.01–10 nm. Since these wavelengths are comparable to the spaces between atoms in a crystalline solid, X-rays are scattered when they pass through crystals. The scattered rays undergo constructive and destructive interference that creates a specific diffraction pattern that may be measured and used to precisely determine the positions of atoms within the crystal. This phenomenon of X-ray diffraction is the basis for very powerful techniques enabling the determination of molecular structure. One of the pioneers who applied this powerful technology to important biochemical substances was Dorothy Crowfoot Hodgkin.

Born in Cairo, Egypt, in 1910 to British parents, Dorothy's fascination with chemistry was fostered early in her life. At age 11 she was enrolled in a prestigious English grammar school where she was one of just two girls allowed to study chemistry. On her 16th birthday, her mother, Molly, gifted her a book on X-ray crystallography, which had a profound impact on the trajectory of her career. She studied chemistry at Oxford University, graduating with first-class honors in 1932 and directly entering Cambridge University to pursue a doctoral degree. At Cambridge, Dorothy recognized the promise of X-ray crystallography for protein structure determinations, conducting research that earned her a PhD in 1937. Over the course of a very productive career, Dr. Hodgkin was credited with determining structures for several important biomolecules, including cholesterol iodide, penicillin, and vitamin B12. In recognition of her achievements in the use of X-ray techniques to elucidate the structures of biochemical substances, she was awarded the 1964 Nobel Prize in Chemistry. In 1969, she led a team of scientists who deduced the structure of insulin, facilitating the mass production of this hormone and greatly advancing the treatment of diabetic patients worldwide. Dr. Hodgkin continued working with the international scientific community, earning numerous distinctions and awards prior to her death in 1993.

Not all waves are travelling waves. **Standing waves** (also known as **stationary waves**) remain constrained within some region of space. As we shall see, standing waves play an important role in our understanding of the electronic structure of atoms and molecules. The simplest example of a standing wave is a one-dimensional wave associated with a vibrating string that is held fixed at its two end points. Figure 6.7 shows the four lowest-energy standing waves (the fundamental wave and the lowest three harmonics) for a vibrating string at a particular amplitude. Although the string's motion lies mostly within a plane, the wave itself is considered to be one dimensional, since it lies along the length of the string. The motion of string segments in a direction perpendicular to the string length generates the waves and so the amplitude of the waves is visible as the maximum displacement of the curves seen in Figure 6.7. The key observation from the figure is *that only those waves having an integer number, n, of half-wavelengths between the end points can form.* A system with fixed end points such as this restricts the number and type of the possible waveforms. This is an example of **quantization**, in which only discrete values from a more general set of continuous values of some property are observed. Another important observation is that the harmonic waves (those waves displaying more than one-half wavelength) all have one or more points between the two end points that are not in motion. These special points are **nodes**. The energies of the standing waves with a given amplitude in a vibrating string increase with the number of half-wavelengths n. Since the number of nodes is $n - 1$, the energy can also be said to depend on the number of nodes, generally increasing as the number of nodes increases.

FIGURE 6.7 A vibrating string shows some one-dimensional standing waves. Since the two end points of the string are held fixed, only waves having an integer number of half-wavelengths can form. The points on the string between the end points that are not moving are called the nodes.

An example of two-dimensional standing waves is shown in Figure 6.8, which shows the vibrational patterns on a flat surface. Although the vibrational amplitudes cannot be seen like they could in the vibrating string, the nodes have been made visible by sprinkling the drum surface with a powder that collects on the areas of the surface that have minimal displacement. For one-dimensional standing waves, the nodes were points on the line, but for two-dimensional standing waves, the nodes are lines on the surface (for three-dimensional standing waves, the nodes are two-dimensional surfaces within the three-dimensional volume).

FIGURE 6.8 Two-dimensional standing waves can be visualized on a vibrating surface. The surface has been sprinkled with a powder that collects near the nodal lines. There are two types of nodes visible: radial nodes

(circles) and angular nodes (radii).

LINK TO LEARNING

You can watch the formation of various radial nodes here (http://openstax.org/l/16radnodes) as singer Imogen Heap projects her voice across a kettle drum.

Blackbody Radiation and the Ultraviolet Catastrophe

The last few decades of the nineteenth century witnessed intense research activity in commercializing newly discovered electric lighting. This required obtaining a better understanding of the distributions of light emitted from various sources being considered. Artificial lighting is usually designed to mimic natural sunlight within the limitations of the underlying technology. Such lighting consists of a range of broadly distributed frequencies that form a **continuous spectrum**. Figure 6.9 shows the wavelength distribution for sunlight. The most intense radiation is in the visible region, with the intensity dropping off rapidly for shorter wavelength ultraviolet (UV) light, and more slowly for longer wavelength infrared (IR) light.

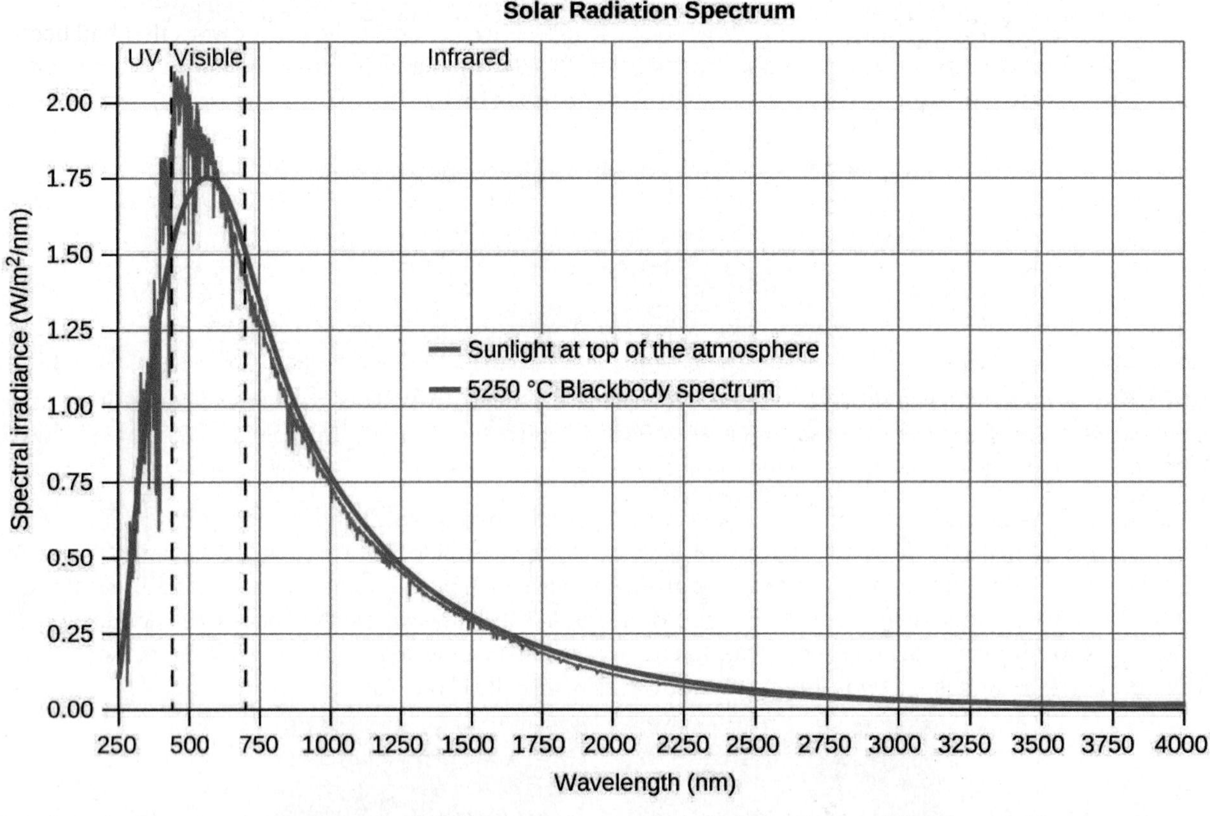

FIGURE 6.9 The spectral distribution (light intensity vs. wavelength) of sunlight reaches the Earth's atmosphere as UV light, visible light, and IR light. The unabsorbed sunlight at the top of the atmosphere has a distribution that approximately matches the theoretical distribution of a blackbody at 5250 °C, represented by the blue curve. (credit: modification of work by American Society for Testing and Materials (ASTM) Terrestrial Reference Spectra for Photovoltaic Performance Evaluation)

In Figure 6.9, the solar distribution is compared to a representative distribution, called a blackbody spectrum, that corresponds to a temperature of 5250 °C. The blackbody spectrum matches the solar spectrum quite well. A **blackbody** is a convenient, ideal emitter that approximates the behavior of many materials when heated. It is "ideal" in the same sense that an ideal gas is a convenient, simple representation of real gases that works well, provided that the pressure is not too high nor the temperature too low. A good approximation of a blackbody that can be used to observe blackbody radiation is a metal oven that can be heated to very high temperatures. The oven has a small hole allowing for the light being emitted within the oven to be observed with a spectrometer so that the wavelengths and their intensities can be measured. Figure 6.10 shows the resulting curves for some representative temperatures. Each distribution depends only on a single parameter: the temperature. The maxima in the blackbody curves, λ_{max}, shift to shorter wavelengths as the temperature increases, reflecting the observation that metals being heated to high temperatures begin to glow a darker red that becomes brighter as the temperature increases, eventually becoming white hot at very high temperatures as the intensities of all of the visible wavelengths become appreciable. This common observation was at the heart of the first paradox that showed the fundamental limitations of classical physics that we will examine.

Physicists derived mathematical expressions for the blackbody curves using well-accepted concepts from the theories of classical mechanics and classical electromagnetism. The theoretical expressions as functions of temperature fit the observed experimental blackbody curves well at longer wavelengths, but showed significant discrepancies at shorter wavelengths. Not only did the theoretical curves not show a peak, they absurdly showed the intensity becoming infinitely large as the wavelength became smaller, which would imply that everyday objects at room temperature should be emitting large amounts of UV light. This became known as the "ultraviolet catastrophe" because no one could find any problems with the theoretical treatment that could lead to such unrealistic short-wavelength behavior. Finally, around 1900, Max Planck derived a theoretical expression for blackbody radiation that fit the experimental observations exactly (within experimental error). Planck developed his theoretical treatment by extending the earlier work that had been based on the premise that the atoms composing the oven vibrated at increasing frequencies (or decreasing wavelengths) as the temperature increased, with these vibrations being the source of the emitted electromagnetic radiation. But where the earlier treatments had allowed the vibrating atoms to have any energy values obtained from a continuous set of energies (perfectly reasonable, according to classical physics), Planck found that by restricting the vibrational energies to discrete values for each frequency, he could derive an expression for blackbody radiation that correctly had the intensity dropping rapidly for the short wavelengths in the UV region.

$$E = nh\nu,\ n = 1,\ 2,\ 3,\ \ldots$$

The quantity h is a constant now known as Planck's constant, in his honor. Although Planck was pleased he had resolved the blackbody radiation paradox, he was disturbed that to do so, he needed to assume the vibrating atoms required quantized energies, which he was unable to explain. The value of Planck's constant is very small, 6.626×10^{-34} joule seconds (J s), which helps explain why energy quantization had not been observed previously in macroscopic phenomena.

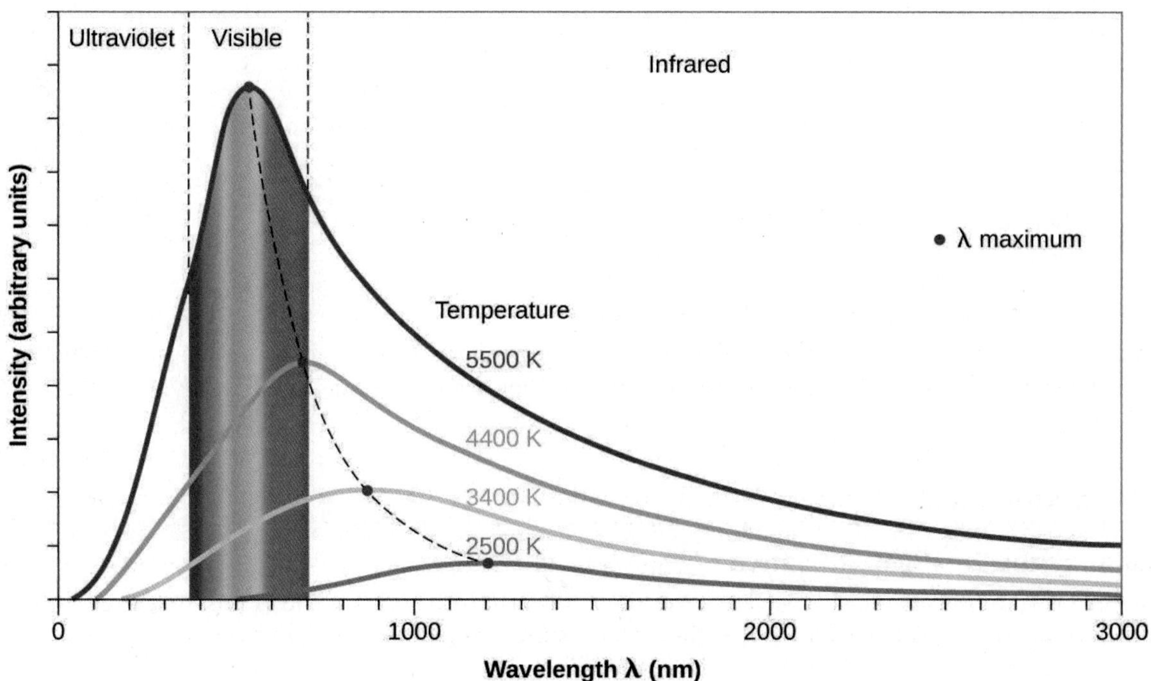

FIGURE 6.10 Blackbody spectral distribution curves are shown for some representative temperatures.

The Photoelectric Effect

The next paradox in the classical theory to be resolved concerned the photoelectric effect (Figure 6.11). It had been observed that electrons could be ejected from the clean surface of a metal when light having a frequency greater than some threshold frequency was shone on it. Surprisingly, the kinetic energy of the ejected electrons did not depend on the brightness of the light, but increased with increasing frequency of the light. Since the electrons in the metal had a certain amount of binding energy keeping them there, the incident light needed to have more energy to free the electrons. According to classical wave theory, a wave's energy depends on its intensity (which depends on its amplitude), not its frequency. One part of these observations was that the number of electrons ejected within in a given time period was seen to increase as the brightness increased. In 1905, Albert Einstein was able to resolve the paradox by incorporating Planck's quantization findings into the discredited particle view of light (Einstein actually won his Nobel prize for this work, and not for his theories of relativity for which he is most famous).

Einstein argued that the quantized energies that Planck had postulated in his treatment of blackbody radiation could be applied to the light in the photoelectric effect so that the light striking the metal surface should not be viewed as a wave, but instead as a stream of particles (later called **photons**) whose energy depended on their frequency, according to Planck's formula, $E = h\nu$ (or, in terms of wavelength using $c = \nu\lambda$, $E = \frac{hc}{\lambda}$). Electrons were ejected when hit by photons having sufficient energy (a frequency greater than the threshold). The greater the frequency, the greater the kinetic energy imparted to the escaping electrons by the collisions. Einstein also argued that the light intensity did not depend on the amplitude of the incoming wave, but instead corresponded to the number of photons striking the surface within a given time period. This explains why the number of ejected electrons increased with increasing brightness, since the greater the number of incoming photons, the greater the likelihood that they would collide with some of the electrons.

With Einstein's findings, the nature of light took on a new air of mystery. Although many light phenomena could be explained either in terms of waves or particles, certain phenomena, such as the interference patterns obtained when light passed through a double slit, were completely contrary to a particle view of light, while other phenomena, such as the photoelectric effect, were completely contrary to a wave view of light. Somehow, at a deep fundamental level still not fully understood, light is both wavelike and particle-like. This is known as **wave-particle duality**.

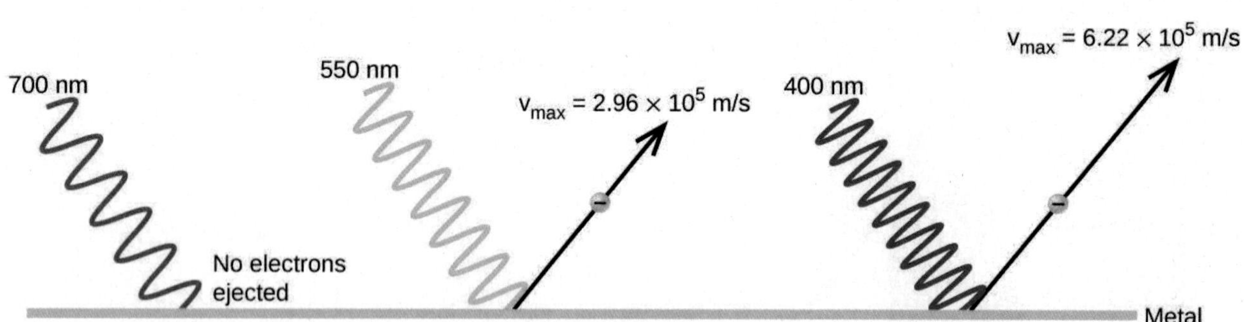

FIGURE 6.11 Photons with low frequencies do not have enough energy to cause electrons to be ejected via the photoelectric effect. For any frequency of light above the threshold frequency, the kinetic energy of an ejected electron will increase linearly with the energy of the incoming photon.

EXAMPLE 6.2

Calculating the Energy of Radiation

When we see light from a neon sign, we are observing radiation from excited neon atoms. If this radiation has a wavelength of 640 nm, what is the energy of the photon being emitted?

Solution

We use the part of Planck's equation that includes the wavelength, λ, and convert units of nanometers to meters so that the units of λ and c are the same.

$$E = \frac{hc}{\lambda}$$

$$E = \frac{(6.626 \times 10^{-34} \text{J}\cancel{\text{s}})(2.998 \times 10^{8} \text{m}\cancel{\text{s}}^{-1})}{(640 \cancel{\text{nm}})\left(\frac{1 \text{ m}}{10^{9} \cancel{\text{nm}}}\right)}$$

$$E = 3.10 \times 10^{-19} \text{J}$$

Check Your Learning

The microwaves in an oven are of a specific frequency that will heat the water molecules contained in food. (This is why most plastics and glass do not become hot in a microwave oven-they do not contain water molecules.) This frequency is about 3×10^9 Hz. What is the energy of one photon in these microwaves?

Answer:

2×10^{-24} J

LINK TO LEARNING

Use this simulation program (http://openstax.org/l/16photelec) to experiment with the photoelectric effect to see how intensity, frequency, type of metal, and other factors influence the ejected photons.

EXAMPLE 6.3

Photoelectric Effect

Identify which of the following statements are false and, where necessary, change the italicized word or phrase to make them true, consistent with Einstein's explanation of the photoelectric effect.

(a) Increasing the brightness of incoming light *increases* the kinetic energy of the ejected electrons.

(b) Increasing the *wavelength* of incoming light increases the kinetic energy of the ejected electrons.

(c) Increasing the brightness of incoming light *increases* the number of ejected electrons.

(d) Increasing the *frequency* of incoming light can increase the number of ejected electrons.

Solution

(a) False. Increasing the brightness of incoming light *has no effect* on the kinetic energy of the ejected electrons. Only energy, not the number or amplitude, of the photons influences the kinetic energy of the electrons.

(b) False. Increasing the *frequency* of incoming light increases the kinetic energy of the ejected electrons. Frequency is proportional to energy and inversely proportional to wavelength. Frequencies above the threshold value transfer the excess energy into the kinetic energy of the electrons.

(c) True. Because the number of collisions with photons increases with brighter light, the number of ejected electrons increases.

(d) True with regard to the threshold energy binding the electrons to the metal. Below this threshold, electrons are not emitted and above it they are. Once over the threshold value, further increasing the frequency does not increase the number of ejected electrons

Check Your Learning

Calculate the threshold energy in kJ/mol of electrons in aluminum, given that the lowest frequency photon for which the photoelectric effect is observed is 9.87×10^{14} Hz.

Answer:

394 kJ/mol

Line Spectra

Another paradox within the classical electromagnetic theory that scientists in the late nineteenth century struggled with concerned the light emitted from atoms and molecules. When solids, liquids, or condensed gases are heated sufficiently, they radiate some of the excess energy as light. Photons produced in this manner have a range of energies, and thereby produce a continuous spectrum in which an unbroken series of wavelengths is present. Most of the light generated from stars (including our sun) is produced in this fashion. You can see all the visible wavelengths of light present in sunlight by using a prism to separate them. As can be seen in Figure 6.9, sunlight also contains UV light (shorter wavelengths) and IR light (longer wavelengths) that can be detected using instruments but that are invisible to the human eye. Incandescent (glowing) solids such as tungsten filaments in incandescent lights also give off light that contains all wavelengths of visible light. These continuous spectra can often be approximated by blackbody radiation curves at some appropriate temperature, such as those shown in Figure 6.10.

In contrast to continuous spectra, light can also occur as discrete or **line spectra** having very narrow line widths interspersed throughout the spectral regions such as those shown in Figure 6.13. Exciting a gas at low partial pressure using an electrical current, or heating it, will produce line spectra. Fluorescent light bulbs and neon signs operate in this way (Figure 6.12). Each element displays its own characteristic set of lines, as do molecules, although their spectra are generally much more complicated.

FIGURE 6.12 Neon signs operate by exciting a gas at low partial pressure using an electrical current. This sign shows the elaborate artistic effects that can be achieved. (credit: Dave Shaver)

Each emission line consists of a single wavelength of light, which implies that the light emitted by a gas consists of a set of discrete energies. For example, when an electric discharge passes through a tube containing hydrogen gas at low pressure, the H_2 molecules are broken apart into separate H atoms and we see a blue-pink color. Passing the light through a prism produces a line spectrum, indicating that this light is composed of photons of four visible wavelengths, as shown in Figure 6.13.

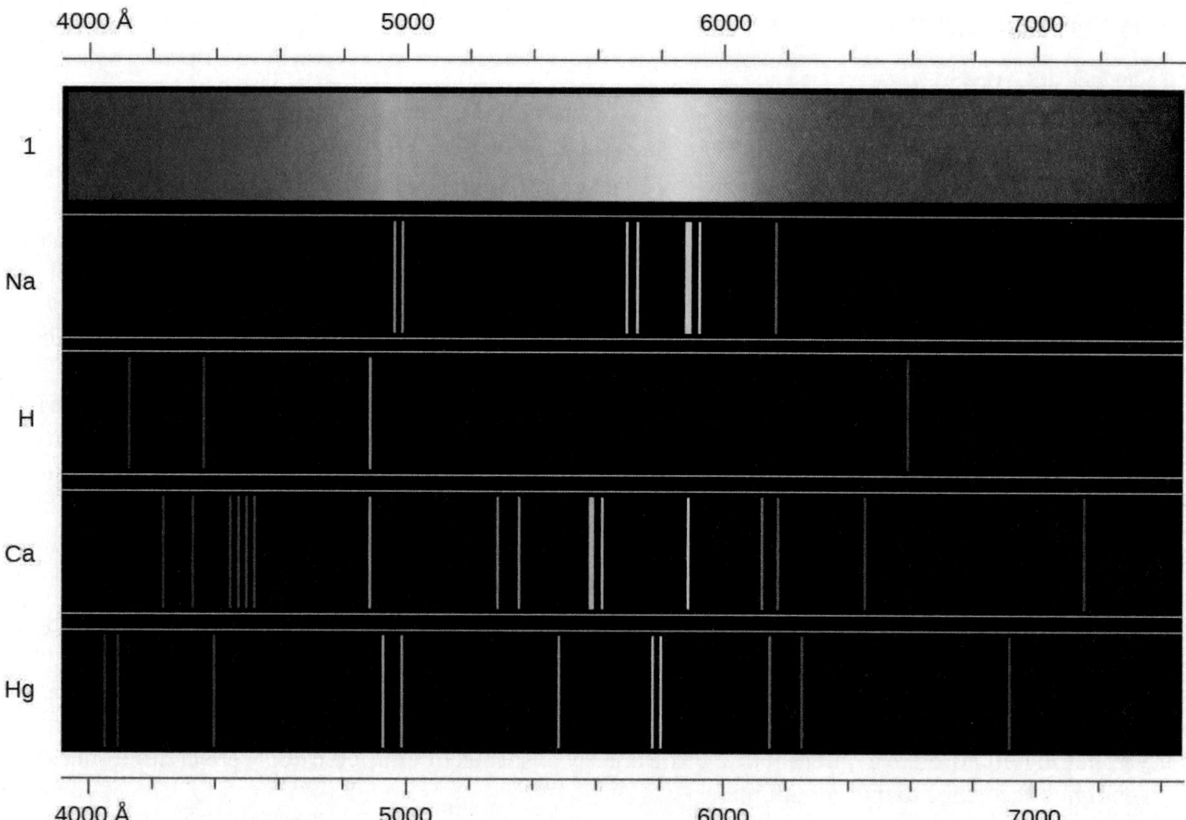

FIGURE 6.13 Compare the two types of emission spectra: continuous spectrum of white light (top) and the line spectra of the light from excited sodium, hydrogen, calcium, and mercury atoms.

The origin of discrete spectra in atoms and molecules was extremely puzzling to scientists in the late nineteenth century, since according to classical electromagnetic theory, only continuous spectra should be observed. Even more puzzling, in 1885, Johann Balmer was able to derive an empirical equation that related the four visible wavelengths of light emitted by hydrogen atoms to whole integers. That equation is the following one, in which k is a constant:

$$\frac{1}{\lambda} = k\left(\frac{1}{4} - \frac{1}{n^2}\right), \; n = 3, 4, 5, 6$$

Other discrete lines for the hydrogen atom were found in the UV and IR regions. Johannes Rydberg generalized Balmer's work and developed an empirical formula that predicted all of hydrogen's emission lines, not just those restricted to the visible range, where, n_1 and n_2 are integers, $n_1 < n_2$, and R_∞ is the Rydberg constant (1.097×10^7 m^{-1}).

$$\frac{1}{\lambda} = R_\infty\left(\frac{1}{n_1^2} - \frac{1}{n_2^2}\right)$$

Even in the late nineteenth century, spectroscopy was a very precise science, and so the wavelengths of hydrogen were measured to very high accuracy, which implied that the Rydberg constant could be determined very precisely as well. That such a simple formula as the Rydberg formula could account for such precise measurements seemed astounding at the time, but it was the eventual explanation for emission spectra by Neils Bohr in 1913 that ultimately convinced scientists to abandon classical physics and spurred the development of modern quantum mechanics.

6.2 The Bohr Model

LEARNING OBJECTIVES

By the end of this section, you will be able to:

- Describe the Bohr model of the hydrogen atom
- Use the Rydberg equation to calculate energies of light emitted or absorbed by hydrogen atoms

Following the work of Ernest Rutherford and his colleagues in the early twentieth century, the picture of atoms consisting of tiny dense nuclei surrounded by lighter and even tinier electrons continually moving about the nucleus was well established. This picture was called the planetary model, since it pictured the atom as a miniature "solar system" with the electrons orbiting the nucleus like planets orbiting the sun. The simplest atom is hydrogen, consisting of a single proton as the nucleus about which a single electron moves. The electrostatic force attracting the electron to the proton depends only on the distance between the two particles. This classical mechanics description of the atom is incomplete, however, since an electron moving in an elliptical orbit would be accelerating (by changing direction) and, according to classical electromagnetism, it should continuously emit electromagnetic radiation. This loss in orbital energy should result in the electron's orbit getting continually smaller until it spirals into the nucleus, implying that atoms are inherently unstable.

In 1913, Niels Bohr attempted to resolve the atomic paradox by ignoring classical electromagnetism's prediction that the orbiting electron in hydrogen would continuously emit light. Instead, he incorporated into the classical mechanics description of the atom Planck's ideas of quantization and Einstein's finding that light consists of photons whose energy is proportional to their frequency. Bohr assumed that the electron orbiting the nucleus would not normally emit any radiation (the stationary state hypothesis), but it would emit or absorb a photon if it moved to a different orbit. The energy absorbed or emitted would reflect differences in the orbital energies according to this equation:

$$|\Delta E| = |E_{\text{f}} - E_{\text{i}}| = h\nu = \frac{hc}{\lambda}$$

In this equation, h is Planck's constant and E_i and E_f are the initial and final orbital energies, respectively. The absolute value of the energy difference is used, since frequencies and wavelengths are always positive. Instead of allowing for continuous values of energy, Bohr assumed the energies of these electron orbitals were quantized:

$$E_n = -\frac{k}{n^2}, \; n = 1, 2, 3, \ldots$$

In this expression, k is a constant comprising fundamental constants such as the electron mass and charge and Planck's constant. Inserting the expression for the orbit energies into the equation for ΔE gives

$$\Delta E = k\left(\frac{1}{n_1^2} - \frac{1}{n_2^2}\right) = \frac{hc}{\lambda}$$

or

$$\frac{1}{\lambda} = \frac{k}{hc}\left(\frac{1}{n_1^2} - \frac{1}{n_2^2}\right)$$

which is identical to the Rydberg equation in which $R_\infty = \frac{k}{hc}$. When Bohr calculated his theoretical value for the Rydberg constant, R_∞, and compared it with the experimentally accepted value, he got excellent agreement. Since the Rydberg constant was one of the most precisely measured constants at that time, this level of agreement was astonishing and meant that **Bohr's model** was taken seriously, despite the many assumptions that Bohr needed to derive it.

The lowest few energy levels are shown in Figure 6.14. One of the fundamental laws of physics is that matter is most stable with the lowest possible energy. Thus, the electron in a hydrogen atom usually moves in the $n = 1$ orbit, the orbit in which it has the lowest energy. When the electron is in this lowest energy orbit, the atom is said to be in its **ground electronic state** (or simply ground state). If the atom receives energy from an outside source, it is possible for the electron to move to an orbit with a higher n value and the atom is now in an **excited electronic state** (or simply an excited state) with a higher energy. When an electron transitions from an excited state (higher energy orbit) to a less excited state, or ground state, the difference in energy is emitted as a photon. Similarly, if a photon is absorbed by an atom, the energy of the photon moves an electron from a lower energy orbit up to a more excited one. We can relate the energy of electrons in atoms to what we learned previously about energy. The law of conservation of energy says that we can neither create nor destroy energy. Thus, if a certain amount of external energy is required to excite an electron from one energy level to another, that same amount of energy will be liberated when the electron returns to its initial state (Figure 6.15).

Since Bohr's model involved only a single electron, it could also be applied to the single electron ions He^{+}, Li^{2+}, Be^{3+}, and so forth, which differ from hydrogen only in their nuclear charges, and so one-electron atoms and ions are collectively referred to as hydrogen-like atoms. The energy expression for hydrogen-like atoms is a generalization of the hydrogen atom energy, in which Z is the nuclear charge (+1 for hydrogen, +2 for He, +3 for Li, and so on) and k has a value of 2.179×10^{-18} J.

$$E_n = -\frac{kZ^2}{n^2}$$

The sizes of the circular orbits for hydrogen-like atoms are given in terms of their radii by the following expression, in which a_0 is a constant called the Bohr radius, with a value of 5.292×10^{-11} m:

$$r = \frac{n^2}{Z}a_0$$

The equation also shows us that as the electron's energy increases (as n increases), the electron is found at greater distances from the nucleus. This is implied by the inverse dependence of electrostatic attraction on distance, since, as the electron moves away from the nucleus, the electrostatic attraction between it and the nucleus decreases and it is held less tightly in the atom. Note that as n gets larger and the orbits get larger, their energies get closer to zero, and so the limits $n \longrightarrow \infty$ and $r \longrightarrow \infty$ imply that $E = 0$ corresponds to the ionization limit where the electron is completely removed from the nucleus. Thus, for hydrogen in the ground state $n = 1$, the ionization energy would be:

$$\Delta E = E_{n \longrightarrow \infty} - E_1 = 0 + k = k$$

With three extremely puzzling paradoxes now solved (blackbody radiation, the photoelectric effect, and the hydrogen atom), and all involving Planck's constant in a fundamental manner, it became clear to most physicists at that time that the classical theories that worked so well in the macroscopic world were fundamentally flawed and could not be extended down into the microscopic domain of atoms and molecules. Unfortunately, despite Bohr's remarkable achievement in deriving a theoretical expression for the Rydberg constant, he was unable to extend his theory to the next simplest atom, He, which only has two electrons. Bohr's model was severely flawed, since it was still based on the classical mechanics notion of precise orbits, a concept that was later found to be untenable in the microscopic domain, when a proper model of quantum mechanics was developed to supersede classical mechanics.

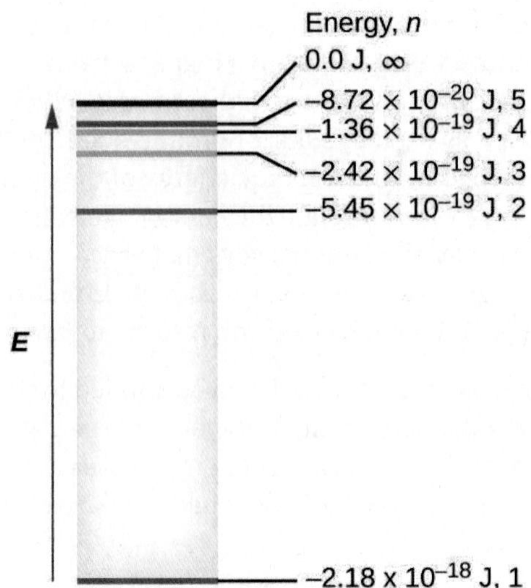

FIGURE 6.14 Quantum numbers and energy levels in a hydrogen atom. The more negative the calculated value, the lower the energy.

EXAMPLE 6.4

Calculating the Energy of an Electron in a Bohr Orbit

Early researchers were very excited when they were able to predict the energy of an electron at a particular distance from the nucleus in a hydrogen atom. If a spark promotes the electron in a hydrogen atom into an orbit with $n = 3$, what is the calculated energy, in joules, of the electron?

Solution

The energy of the electron is given by this equation:

$$E = \frac{-kZ^2}{n^2}$$

The atomic number, Z, of hydrogen is 1; $k = 2.179 \times 10^{-18}$ J; and the electron is characterized by an n value of 3. Thus,

$$E = \frac{-(2.179 \times 10^{-18}\text{ J}) \times (1)^2}{(3)^2} = -2.421 \times 10^{-19}\text{ J}$$

Check Your Learning

The electron in Figure 6.15 is promoted even further to an orbit with $n = 6$. What is its new energy?

Answer:

-6.053×10^{-20} J

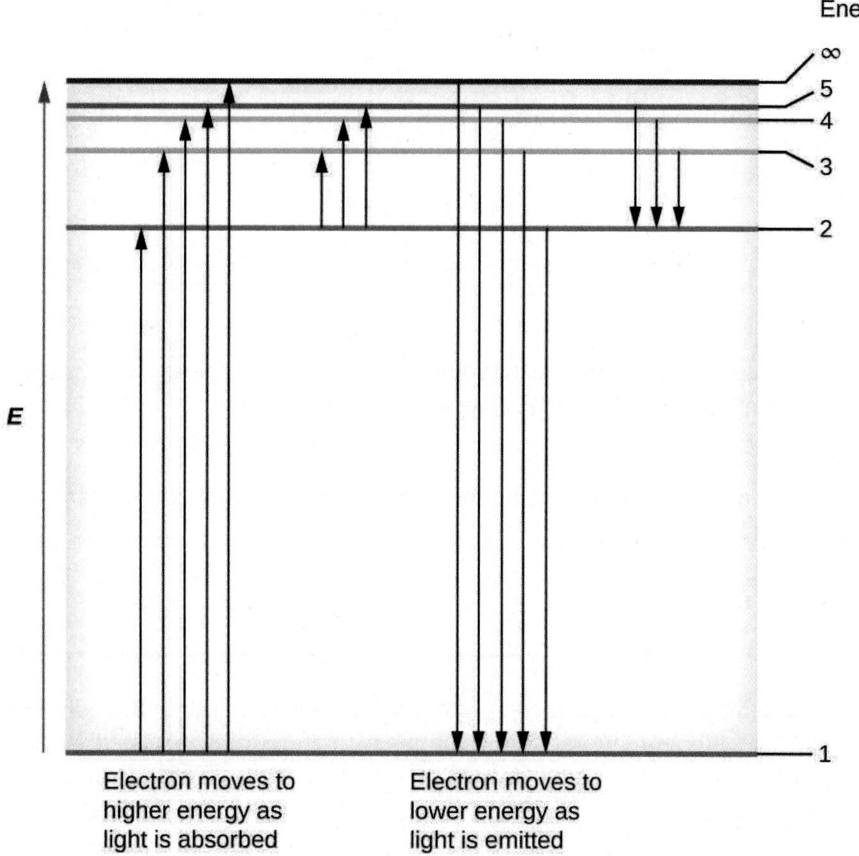

FIGURE 6.15 The horizontal lines show the relative energy of orbits in the Bohr model of the hydrogen atom, and the vertical arrows depict the energy of photons absorbed (left) or emitted (right) as electrons move between these orbits.

EXAMPLE 6.5

Calculating the Energy and Wavelength of Electron Transitions in a One–electron (Bohr) System

What is the energy (in joules) and the wavelength (in meters) of the line in the spectrum of hydrogen that represents the movement of an electron from Bohr orbit with $n = 4$ to the orbit with $n = 6$? In what part of the electromagnetic spectrum do we find this radiation?

Solution

In this case, the electron starts out with $n = 4$, so $n_1 = 4$. It comes to rest in the $n = 6$ orbit, so $n_2 = 6$. The difference in energy between the two states is given by this expression:

$$\Delta E = E_1 - E_2 = 2.179 \times 10^{-18} \left(\frac{1}{n_1^2} - \frac{1}{n_2^2} \right)$$

$$\Delta E = 2.179 \times 10^{-18} \left(\frac{1}{4^2} - \frac{1}{6^2} \right) \text{ J}$$

$$\Delta E = 2.179 \times 10^{-18} \left(\frac{1}{16} - \frac{1}{36} \right) \text{ J}$$

$$\Delta E = 7.566 \times 10^{-20} \text{ J}$$

This energy difference is positive, indicating a photon enters the system (is absorbed) to excite the electron from the $n = 4$ orbit up to the $n = 6$ orbit. The wavelength of a photon with this energy is found by the expression $E = \frac{hc}{\lambda}$. Rearrangement gives:

$$\lambda = \frac{hc}{E}$$

$$= \left(6.626 \times 10^{-34}\ \cancel{\text{J}}\ \cancel{\text{s}}\right) \times \frac{2.998 \times 10^{8}\ \text{m}\ \cancel{\text{s}}^{-1}}{7.566 \times 10^{-20}\ \cancel{\text{J}}}$$

$$= 2.626 \times 10^{-6}\ \text{m}$$

From the illustration of the electromagnetic spectrum in Electromagnetic Energy, we can see that this wavelength is found in the infrared portion of the electromagnetic spectrum.

Check Your Learning

What is the energy in joules and the wavelength in meters of the photon produced when an electron falls from the $n = 5$ to the $n = 3$ level in a He^+ ion ($Z = 2$ for He^+)?

Answer:

6.198×10^{-19} J; 3.205×10^{-7} m

Bohr's model of the hydrogen atom provides insight into the behavior of matter at the microscopic level, but it does not account for electron–electron interactions in atoms with more than one electron. It does introduce several important features of all models used to describe the distribution of electrons in an atom. These features include the following:

- The energies of electrons (energy levels) in an atom are quantized, described by **quantum numbers**: integer numbers having only specific allowed value and used to characterize the arrangement of electrons in an atom.
- An electron's energy increases with increasing distance from the nucleus.
- The discrete energies (lines) in the spectra of the elements result from quantized electronic energies.

Of these features, the most important is the postulate of quantized energy levels for an electron in an atom. As a consequence, the model laid the foundation for the quantum mechanical model of the atom. Bohr won a Nobel Prize in Physics for his contributions to our understanding of the structure of atoms and how that is related to line spectra emissions.

6.3 Development of Quantum Theory

LEARNING OBJECTIVES

By the end of this section, you will be able to:

- Extend the concept of wave–particle duality that was observed in electromagnetic radiation to matter as well
- Understand the general idea of the quantum mechanical description of electrons in an atom, and that it uses the notion of three-dimensional wave functions, or orbitals, that define the distribution of probability to find an electron in a particular part of space
- List and describe traits of the four quantum numbers that form the basis for completely specifying the state of an electron in an atom

Bohr's model explained the experimental data for the hydrogen atom and was widely accepted, but it also raised many questions. Why did electrons orbit at only fixed distances defined by a single quantum number $n = 1, 2, 3$, and so on, but never in between? Why did the model work so well describing hydrogen and one-electron ions, but could not correctly predict the emission spectrum for helium or any larger atoms? To answer these questions, scientists needed to completely revise the way they thought about matter.

Behavior in the Microscopic World

We know how matter behaves in the macroscopic world—objects that are large enough to be seen by the naked eye follow the rules of classical physics. A billiard ball moving on a table will behave like a particle: It will continue in a straight line unless it collides with another ball or the table cushion, or is acted on by some other force (such as friction). The ball has a well-defined position and velocity (or a well-defined momentum, $p = mv$, defined by mass m and velocity v) at any given moment. In other words, the ball is moving in a classical trajectory. This is the typical behavior of a classical object.

When waves interact with each other, they show interference patterns that are not displayed by macroscopic particles such as the billiard ball. For example, interacting waves on the surface of water can produce interference patterns similar to those shown on Figure 6.16. This is a case of wave behavior on the macroscopic scale, and it is clear that particles and waves are very different phenomena in the macroscopic realm.

FIGURE 6.16 An interference pattern on the water surface is formed by interacting waves. The waves are caused by reflection of water from the rocks. (credit: modification of work by Sukanto Debnath)

As technological improvements allowed scientists to probe the microscopic world in greater detail, it became increasingly clear by the 1920s that very small pieces of matter follow a different set of rules from those we observe for large objects. The unquestionable separation of waves and particles was no longer the case for the microscopic world.

One of the first people to pay attention to the special behavior of the microscopic world was Louis de Broglie. He asked the question: If electromagnetic radiation can have particle-like character, can electrons and other submicroscopic particles exhibit wavelike character? In his 1925 doctoral dissertation, de Broglie extended the wave–particle duality of light that Einstein used to resolve the photoelectric-effect paradox to material particles. He predicted that a particle with mass m and velocity v (that is, with linear momentum p) should also exhibit the behavior of a wave with a wavelength value λ, given by this expression in which h is the familiar Planck's constant:

$$\lambda = \frac{h}{mv} = \frac{h}{p}$$

This is called the *de Broglie wavelength*. Unlike the other values of λ discussed in this chapter, the de Broglie wavelength is a characteristic of particles and other bodies, not electromagnetic radiation (note that this equation involves velocity [v, m/s], not frequency [ν, Hz]. Although these two symbols appear nearly identical, they mean very different things). Where Bohr had postulated the electron as being a particle orbiting the nucleus in quantized orbits, de Broglie argued that Bohr's assumption of quantization can be explained if the electron is considered not as a particle, but rather as a circular standing wave such that only an integer number of wavelengths could fit exactly within the orbit (Figure 6.17).

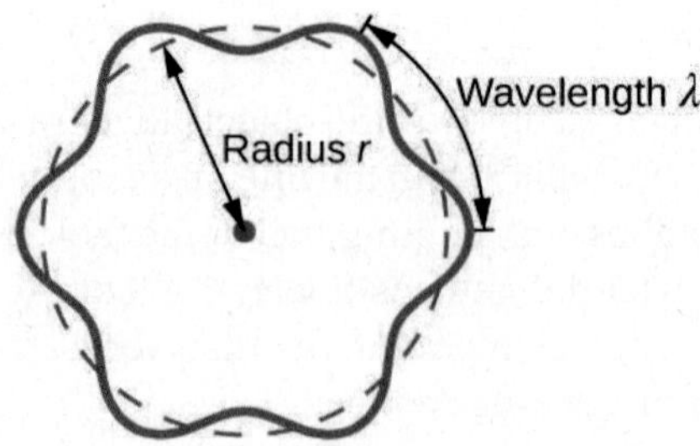

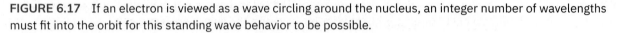

FIGURE 6.17 If an electron is viewed as a wave circling around the nucleus, an integer number of wavelengths must fit into the orbit for this standing wave behavior to be possible.

For a circular orbit of radius r, the circumference is $2\pi r$, and so de Broglie's condition is:

$$2\pi r = n\lambda, \; n = 1, 2, 3, \ldots$$

Shortly after de Broglie proposed the wave nature of matter, two scientists at Bell Laboratories, C. J. Davisson and L. H. Germer, demonstrated experimentally that electrons can exhibit wavelike behavior by showing an interference pattern for electrons travelling through a regular atomic pattern in a crystal. The regularly spaced atomic layers served as slits, as used in other interference experiments. Since the spacing between the layers serving as slits needs to be similar in size to the wavelength of the tested wave for an interference pattern to form, Davisson and Germer used a crystalline nickel target for their "slits," since the spacing of the atoms within the lattice was approximately the same as the de Broglie wavelengths of the electrons that they used. Figure 6.18 shows an interference pattern. It is strikingly similar to the interference patterns for light shown in Electromagnetic Energy for light passing through two closely spaced, narrow slits. The wave–particle duality of matter can be seen in Figure 6.18 by observing what happens if electron collisions are recorded over a long period of time. Initially, when only a few electrons have been recorded, they show clear particle-like behavior, having arrived in small localized packets that appear to be random. As more and more electrons arrived and were recorded, a clear interference pattern that is the hallmark of wavelike behavior emerged. Thus, it appears that while electrons are small localized particles, their motion does not follow the equations of motion implied by classical mechanics, but instead it is governed by some type of a wave equation. Thus the wave–particle duality first observed with photons is actually a fundamental behavior intrinsic to all quantum particles.

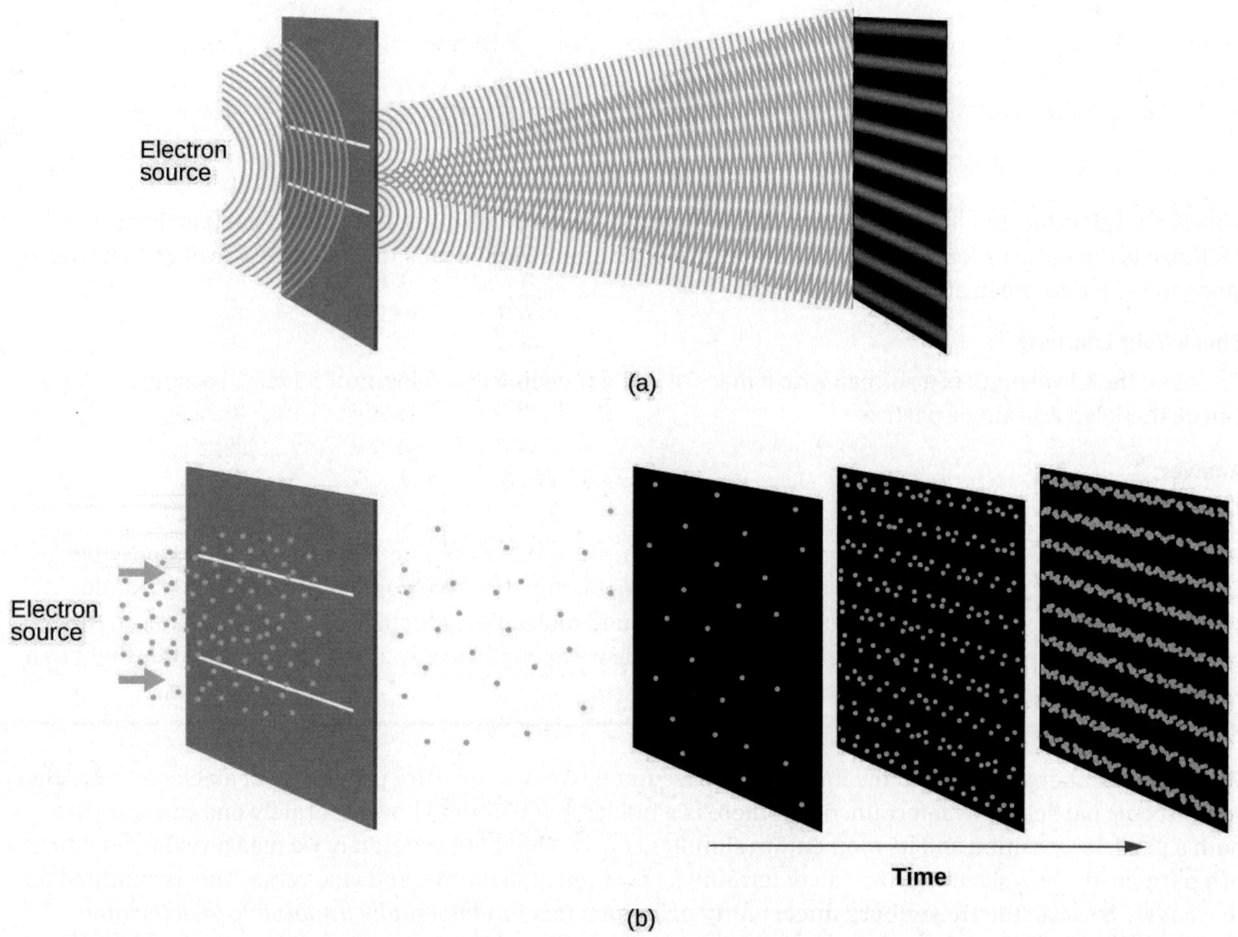

FIGURE 6.18 (a) The interference pattern for electrons passing through very closely spaced slits demonstrates that quantum particles such as electrons can exhibit wavelike behavior. (b) The experimental results illustrated here demonstrate the wave–particle duality in electrons.

LINK TO LEARNING

View the Dr. Quantum – Double Slit Experiment cartoon (http://openstax.org/l/16duality) for an easy-to-understand description of wave–particle duality and the associated experiments.

EXAMPLE 6.6

Calculating the Wavelength of a Particle

If an electron travels at a velocity of 1.000×10^7 m s^{-1} and has a mass of 9.109×10^{-28} g, what is its wavelength?

Solution

We can use de Broglie's equation to solve this problem, but we first must do a unit conversion of Planck's constant. You learned earlier that 1 J = 1 kg m^2/s^2. Thus, we can write $h = 6.626 \times 10^{-34}$ J s as 6.626×10^{-34} kg m^2/s.

$$\lambda = \frac{h}{mv}$$

$$= \frac{6.626 \times 10^{-34} \text{ kg m}^2/\text{s}}{(9.109 \times 10^{-31} \text{ kg})(1.000 \times 10^7 \text{ m/s})}$$
$$= 7.274 \times 10^{-11} \text{m}$$

This is a small value, but it is significantly larger than the size of an electron in the classical (particle) view. This size is the same order of magnitude as the size of an atom. This means that electron wavelike behavior is going to be noticeable in an atom.

Check Your Learning

Calculate the wavelength of a softball with a mass of 100 g traveling at a velocity of 35 m s^{-1}, assuming that it can be modeled as a single particle.

Answer:

1.9×10^{-34} m.

We never think of a thrown softball having a wavelength, since this wavelength is so small it is impossible for our senses or any known instrument to detect (strictly speaking, the wavelength of a real baseball would correspond to the wavelengths of its constituent atoms and molecules, which, while much larger than this value, would still be microscopically tiny). The de Broglie wavelength is only appreciable for matter that has a very small mass and/or a very high velocity.

Werner Heisenberg considered the limits of how accurately we can measure properties of an electron or other microscopic particles. He determined that there is a fundamental limit to how accurately one can measure both a particle's position and its momentum simultaneously. The more accurately we measure the momentum of a particle, the less accurately we can determine its position at that time, and vice versa. This is summed up in what we now call the **Heisenberg uncertainty principle**: *It is fundamentally impossible to determine simultaneously and exactly both the momentum and the position of a particle.* For a particle of mass m moving with velocity v_x in the x direction (or equivalently with momentum p_x), the product of the uncertainty in the position, Δx, and the uncertainty in the momentum, Δp_x, must be greater than or equal to $\frac{\hbar}{2}$ (where $\hbar = \frac{h}{2\pi}$, the value of Planck's constant divided by 2π).

$$\Delta x \times \Delta p_x = (\Delta x)(m\Delta v) \geq \frac{\hbar}{2}$$

This equation allows us to calculate the limit to how precisely we can know both the simultaneous position of an object and its momentum. For example, if we improve our measurement of an electron's position so that the uncertainty in the position (Δx) has a value of, say, 1 pm (10^{-12} m, about 1% of the diameter of a hydrogen atom), then our determination of its momentum must have an uncertainty with a value of at least

$$\left[\Delta p = m\Delta v = \frac{\hbar}{(2\Delta x)}\right] = \frac{(1.055 \times 10^{-34} \text{ kg m}^2/\text{s})}{(2 \times 1 \times 10^{-12} \text{ m})} = 5 \times 10^{-23} \text{ kg m/s}.$$

The value of $\hbar$ is not large, so the uncertainty in the position or momentum of a macroscopic object like a baseball is too insignificant to observe. However, the mass of a microscopic object such as an electron is small enough that the uncertainty can be large and significant.

It should be noted that Heisenberg's uncertainty principle is not just limited to uncertainties in position and momentum, but it also links other dynamical variables. For example, when an atom absorbs a photon and makes a transition from one energy state to another, the uncertainty in the energy and the uncertainty in the time required for the transition are similarly related, as $\Delta E \, \Delta t \geq \frac{\hbar}{2}$.

Heisenberg's principle imposes ultimate limits on what is knowable in science. The uncertainty principle can be shown to be a consequence of wave–particle duality, which lies at the heart of what distinguishes modern quantum theory from classical mechanics.

LINK TO LEARNING

Read this article (http://openstax.org/l/16uncertainty) that describes a recent macroscopic demonstration of the uncertainty principle applied to microscopic objects.

The Quantum–Mechanical Model of an Atom

Shortly after de Broglie published his ideas that the electron in a hydrogen atom could be better thought of as being a circular standing wave instead of a particle moving in quantized circular orbits, Erwin Schrödinger extended de Broglie's work by deriving what is today known as the Schrödinger equation. When Schrödinger applied his equation to hydrogen-like atoms, he was able to reproduce Bohr's expression for the energy and, thus, the Rydberg formula governing hydrogen spectra. Schrödinger described electrons as three-dimensional stationary waves, or **wavefunctions**, represented by the Greek letter psi, ψ. A few years later, Max Born proposed an interpretation of the wavefunction ψ that is still accepted today: Electrons are still particles, and so the waves represented by ψ are not physical waves but, instead, are complex probability amplitudes. The square of the magnitude of a wavefunction $|\psi|^2$ describes the probability of the quantum particle being present near a certain location in space. This means that wavefunctions can be used to determine the distribution of the electron's density with respect to the nucleus in an atom. In the most general form, the Schrödinger equation can be written as:

$$\hat{H}\psi = E\psi$$

$\hat{H}$ is the Hamiltonian operator, a set of mathematical operations representing the total energy of the quantum particle (such as an electron in an atom), ψ is the wavefunction of this particle that can be used to find the special distribution of the probability of finding the particle, and E is the actual value of the total energy of the particle.

Schrödinger's work, as well as that of Heisenberg and many other scientists following in their footsteps, is generally referred to as **quantum mechanics**.

LINK TO LEARNING

You may also have heard of Schrödinger because of his famous thought experiment. This story (http://openstax.org/l/16superpos) explains the concepts of superposition and entanglement as related to a cat in a box with poison.

Understanding Quantum Theory of Electrons in Atoms

The goal of this section is to understand the electron orbitals (location of electrons in atoms), their different energies, and other properties. The use of quantum theory provides the best understanding to these topics. This knowledge is a precursor to chemical bonding.

As was described previously, electrons in atoms can exist only on discrete energy levels but not between them. It is said that the energy of an electron in an atom is quantized, that is, it can be equal only to certain specific values and can jump from one energy level to another but not transition smoothly or stay between these levels.

The energy levels are labeled with an n value, where $n = 1, 2, 3, \ldots$. Generally speaking, the energy of an electron in an atom is greater for greater values of n. This number, n, is referred to as the principal quantum number. The **principal quantum number** defines the location of the energy level. It is essentially the same concept as the n in the Bohr atom description. Another name for the principal quantum number is the shell number. The **shells** of an atom can be thought of concentric circles radiating out from the nucleus. The electrons that belong to a specific shell are most likely to be found within the corresponding circular area. The further we proceed from the nucleus, the higher the shell number, and so the higher the energy level (Figure 6.19). The positively charged protons in the nucleus stabilize the electronic orbitals by electrostatic attraction between the positive charges of the protons and the negative charges of the electrons. So the further away the electron is from the nucleus, the greater the energy it has.

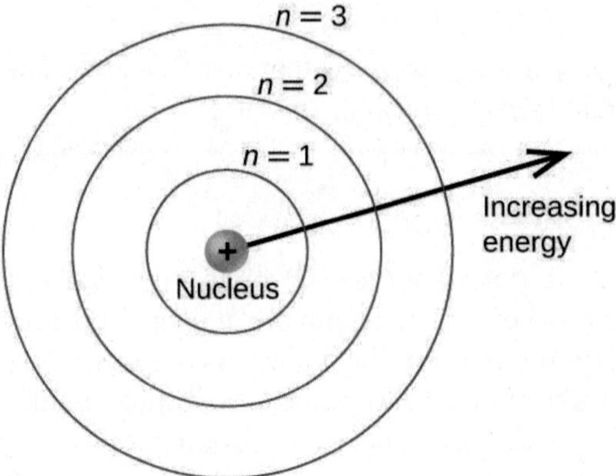

FIGURE 6.19 Different shells are numbered by principal quantum numbers.

This quantum mechanical model for where electrons reside in an atom can be used to look at electronic transitions, the events when an electron moves from one energy level to another. If the transition is to a higher energy level, energy is absorbed, and the energy change has a positive value. To obtain the amount of energy necessary for the transition to a higher energy level, a photon is absorbed by the atom. A transition to a lower energy level involves a release of energy, and the energy change is negative. This process is accompanied by emission of a photon by the atom. The following equation summarizes these relationships and is based on the hydrogen atom:

$$\Delta E = E_{\text{final}} - E_{\text{initial}}$$

$$= -2.18 \times 10^{-18} \left(\frac{1}{n_{\text{f}}^2} - \frac{1}{n_{\text{i}}^2} \right) \text{J}$$

The values n_{f} and n_{i} are the final and initial energy states of the electron. Example 6.5 in the previous section of the chapter demonstrates calculations of such energy changes.

The principal quantum number is one of three quantum numbers used to characterize an orbital. An **atomic orbital** is a general region in an atom within which an electron is most probable to reside. The quantum mechanical model specifies the probability of finding an electron in the three-dimensional space around the nucleus and is based on solutions of the Schrödinger equation. In addition, the principal quantum number defines the energy of an electron in a hydrogen or hydrogen-like atom or an ion (an atom or an ion with only one electron) and the general region in which discrete energy levels of electrons in a multi-electron atoms and ions are located.

Another quantum number is *l*, the **secondary (angular momentum) quantum number**. It is an integer that may take the values, $l = 0, 1, 2, \ldots, n - 1$. This means that an orbital with $n = 1$ can have only one value of *l*, $l = 0$, whereas $n = 2$ permits $l = 0$ and $l = 1$, and so on. Whereas the principal quantum number, *n*, defines the general size and energy of the orbital, the secondary quantum number *l* specifies the shape of the orbital. Orbitals with the same value of *l* define a **subshell**.

Orbitals with $l = 0$ are called ***s* orbitals** and they make up the *s* subshells. The value $l = 1$ corresponds to the *p* orbitals. For a given *n*, ***p* orbitals** constitute a *p* subshell (e.g., 3*p* if $n = 3$). The orbitals with $l = 2$ are called the ***d* orbitals**, followed by the *f-, g-, and h*-orbitals for $l = 3$, 4, and 5.

There are certain distances from the nucleus at which the probability density of finding an electron located at a particular orbital is zero. In other words, the value of the wavefunction ψ is zero at this distance for this orbital. Such a value of radius *r* is called a radial node. The number of radial nodes in an orbital is $n - l - 1$.

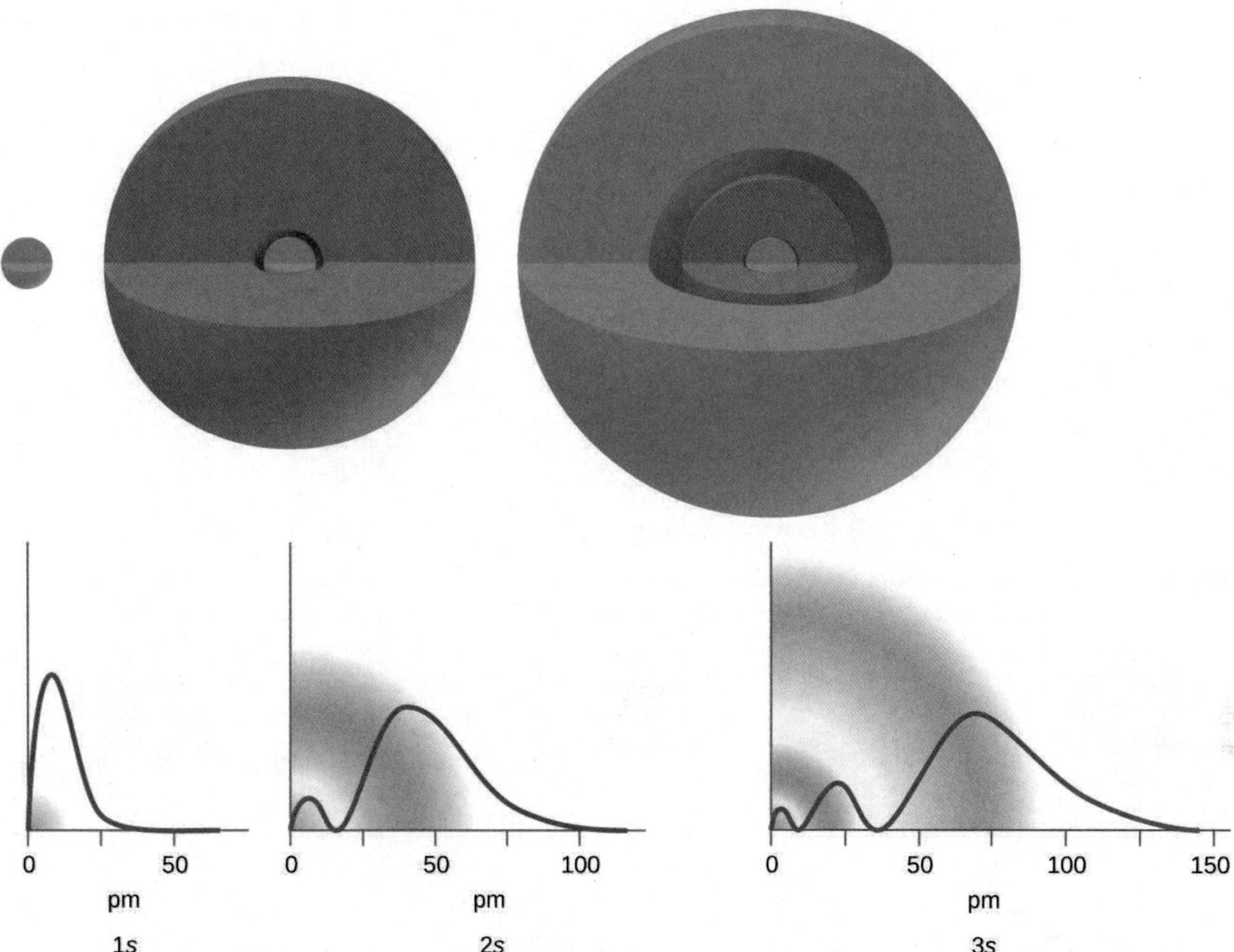

FIGURE 6.20 The graphs show the probability (*y* axis) of finding an electron for the 1*s*, 2*s*, 3*s* orbitals as a function of distance from the nucleus.

Consider the examples in Figure 6.20. The orbitals depicted are of the *s* type, thus $l = 0$ for all of them. It can be seen from the graphs of the probability densities that there are $1 - 0 - 1 = 0$ places where the density is zero (nodes) for 1*s* ($n = 1$), $2 - 0 - 1 = 1$ node for 2*s*, and $3 - 0 - 1 = 2$ nodes for the 3*s* orbitals.

The *s* subshell electron density distribution is spherical and the *p* subshell has a dumbbell shape. The *d* and ***f* orbitals** are more complex. These shapes represent the three-dimensional regions within which the electron is likely to be found.

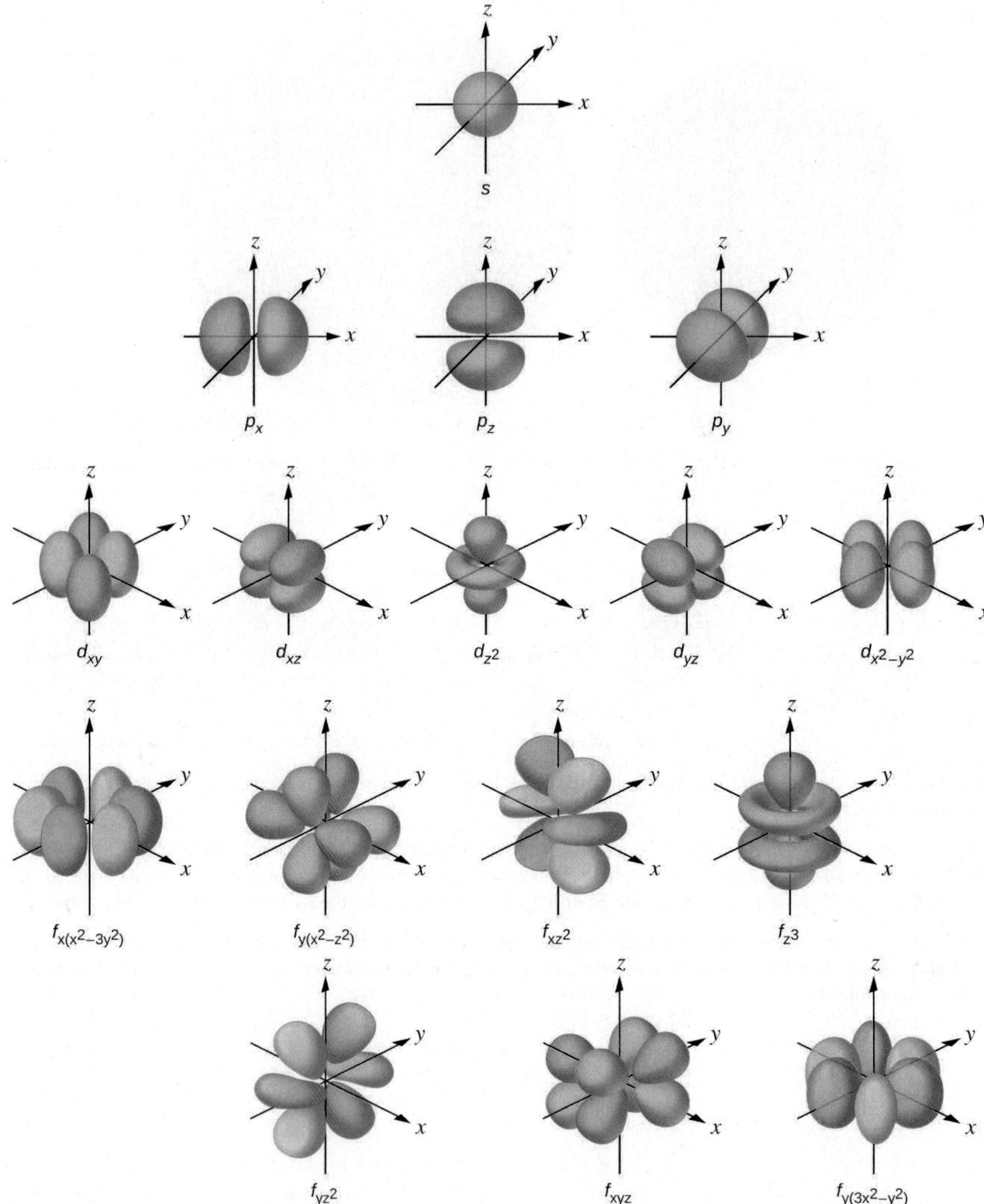

FIGURE 6.21 Shapes of *s*, *p*, *d*, and *f* orbitals.

The **magnetic quantum number**, m_l, specifies the relative spatial orientation of a particular orbital. Generally speaking, m_l can be equal to $-l, -(l-1), \ldots, 0, \ldots, (l-1), l$. The total number of possible orbitals with the same value of l (that is, in the same subshell) is $2l + 1$. Thus, there is one *s*-orbital in an *s* subshell ($l = 0$), there are three *p*-orbitals in a *p* subshell ($l = 1$), five *d*-orbitals in a *d* subshell ($l = 2$), seven *f*-orbitals in an *f* subshell ($l = 3$), and so forth. The principal quantum number defines the general value of the electronic energy. The angular momentum quantum number determines the shape of the orbital. And the magnetic quantum number specifies orientation of the orbital in space, as can be seen in Figure 6.21.

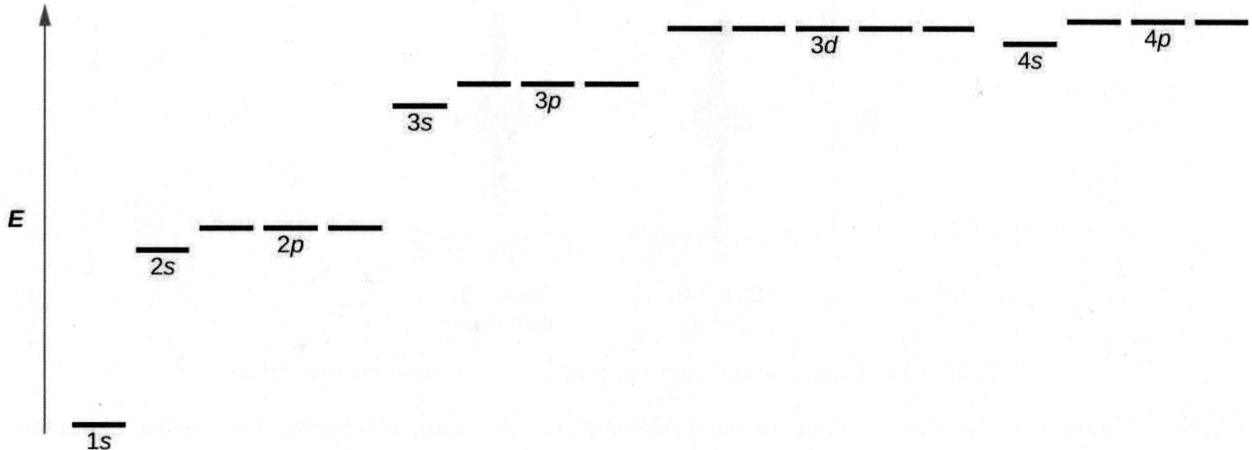

FIGURE 6.22 The chart shows the energies of electron orbitals in a multi-electron atom.

Figure 6.22 illustrates the energy levels for various orbitals. The number before the orbital name (such as 2*s*, 3*p*, and so forth) stands for the principal quantum number, *n*. The letter in the orbital name defines the subshell with a specific angular momentum quantum number $l = 0$ for *s* orbitals, 1 for *p* orbitals, 2 for *d* orbitals. Finally, there are more than one possible orbitals for $l \geq 1$, each corresponding to a specific value of m_l. In the case of a hydrogen atom or a one-electron ion (such as He^+, Li^{2+}, and so on), energies of all the orbitals with the same *n* are the same. This is called a degeneracy, and the energy levels for the same principal quantum number, *n*, are called **degenerate orbitals**. However, in atoms with more than one electron, this degeneracy is eliminated by the electron–electron interactions, and orbitals that belong to different subshells have different energies, as shown on Figure 6.22. Orbitals within the same subshell are still degenerate and have the same energy.

While the three quantum numbers discussed in the previous paragraphs work well for describing electron orbitals, some experiments showed that they were not sufficient to explain all observed results. It was demonstrated in the 1920s that when hydrogen-line spectra are examined at extremely high resolution, some lines are actually not single peaks but, rather, pairs of closely spaced lines. This is the so-called fine structure of the spectrum, and it implies that there are additional small differences in energies of electrons even when they are located in the same orbital. These observations led Samuel Goudsmit and George Uhlenbeck to propose that electrons have a fourth quantum number. They called this the **spin quantum number**, or $\boldsymbol{m_s}$.

The other three quantum numbers, *n*, *l*, and m_l, are properties of specific atomic orbitals that also define in what part of the space an electron is most likely to be located. Orbitals are a result of solving the Schrödinger equation for electrons in atoms. The electron spin is a different kind of property. It is a completely quantum phenomenon with no analogues in the classical realm. In addition, it cannot be derived from solving the Schrödinger equation and is not related to the normal spatial coordinates (such as the Cartesian *x*, *y*, and *z*). Electron spin describes an intrinsic electron "rotation" or "spinning." Each electron acts as a tiny magnet or a tiny rotating object with an angular momentum, or as a loop with an electric current, even though this rotation or current cannot be observed in terms of spatial coordinates.

The magnitude of the overall electron spin can only have one value, and an electron can only "spin" in one of two quantized states. One is termed the α state, with the *z* component of the spin being in the positive direction of the *z* axis. This corresponds to the spin quantum number $m_s = \frac{1}{2}$. The other is called the β state, with the *z* component of the spin being negative and $m_s = -\frac{1}{2}$. Any electron, regardless of the atomic orbital it is located in, can only have one of those two values of the spin quantum number. The energies of electrons having $m_s = -\frac{1}{2}$ and $m_s = \frac{1}{2}$ are different if an external magnetic field is applied.

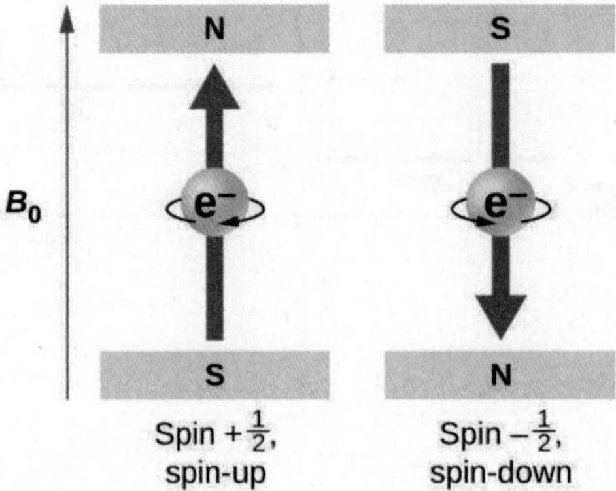

FIGURE 6.23 Electrons with spin values $\pm\frac{1}{2}$ in an external magnetic field.

Figure 6.23 illustrates this phenomenon. An electron acts like a tiny magnet. Its moment is directed up (in the positive direction of the *z* axis) for the $\frac{1}{2}$ spin quantum number and down (in the negative *z* direction) for the spin quantum number of $-\frac{1}{2}$. A magnet has a lower energy if its magnetic moment is aligned with the external magnetic field (the left electron on Figure 6.23) and a higher energy for the magnetic moment being opposite to the applied field. This is why an electron with $m_s = \frac{1}{2}$ has a slightly lower energy in an external field in the positive *z* direction, and an electron with $m_s = -\frac{1}{2}$ has a slightly higher energy in the same field. This is true even for an electron occupying the same orbital in an atom. A spectral line corresponding to a transition for electrons from the same orbital but with different spin quantum numbers has two possible values of energy; thus, the line in the spectrum will show a fine structure splitting.

The Pauli Exclusion Principle

An electron in an atom is completely described by four quantum numbers: *n*, *l*, m_l, and m_s. The first three quantum numbers define the orbital and the fourth quantum number describes the intrinsic electron property called spin. An Austrian physicist Wolfgang Pauli formulated a general principle that gives the last piece of information that we need to understand the general behavior of electrons in atoms. The **Pauli exclusion principle** can be formulated as follows: No two electrons in the same atom can have exactly the same set of all the four quantum numbers. What this means is that two electrons can share the same orbital (the same set of the quantum numbers *n*, *l*, and m_l) only if their spin quantum numbers m_s have different values. Since the spin quantum number can only have two values $\left(\pm\frac{1}{2}\right)$, no more than two electrons can occupy the same orbital (and if two electrons are located in the same orbital, they must have opposite spins). Therefore, any atomic orbital can be populated by only zero, one, or two electrons.

The properties and meaning of the quantum numbers of electrons in atoms are briefly summarized in Table 6.1.

Quantum Numbers, Their Properties, and Significance

Name	Symbol	Allowed values	Physical meaning
principal quantum number	n	1, 2, 3, 4, ...	shell, the general region for the value of energy for an electron on the orbital
angular momentum or azimuthal quantum number	l	$0 \le l \le n - 1$	subshell, the shape of the orbital
magnetic quantum number	m_l	$-l \le m_l \le l$	orientation of the orbital

Quantum Numbers, Their Properties, and Significance			
Name	Symbol	Allowed values	Physical meaning
spin quantum number	m_s	$\frac{1}{2}, -\frac{1}{2}$	direction of the intrinsic quantum "spinning" of the electron

TABLE 6.1

EXAMPLE 6.7

Working with Shells and Subshells

Indicate the number of subshells, the number of orbitals in each subshell, and the values of l and m_l for the orbitals in the $n = 4$ shell of an atom.

Solution

For $n = 4$, l can have values of 0, 1, 2, and 3. Thus, *s*, *p*, *d*, and *f* subshells are found in the $n = 4$ shell of an atom. For $l = 0$ (the *s* subshell), m_l can only be 0. Thus, there is only one 4*s* orbital. For $l = 1$ (*p*-type orbitals), m can have values of –1, 0, +1, so we find three 4*p* orbitals. For $l = 2$ (*d*-type orbitals), m_l can have values of –2, –1, 0, +1, +2, so we have five 4*d* orbitals. When $l = 3$ (*f*-type orbitals), m_l can have values of –3, –2, –1, 0, +1, +2, +3, and we can have seven 4*f* orbitals. Thus, we find a total of 16 orbitals in the $n = 4$ shell of an atom.

Check Your Learning

Identify the subshell in which electrons with the following quantum numbers are found: (a) $n = 3, l = 1$; (b) $n = 5, l = 3$; (c) $n = 2, l = 0$.

Answer:

(a) 3*p* (b) 5*f* (c) 2s

EXAMPLE 6.8

Maximum Number of Electrons

Calculate the maximum number of electrons that can occupy a shell with (a) $n = 2$, (b) $n = 5$, and (c) n as a variable. Note you are only looking at the orbitals with the specified n value, not those at lower energies.

Solution

(a) When $n = 2$, there are four orbitals (a single 2*s* orbital, and three orbitals labeled 2*p*). These four orbitals can contain eight electrons.

(b) When $n = 5$, there are five subshells of orbitals that we need to sum:

$$\begin{array}{l} 1 \text{ orbital labeled } 5s \\ 3 \text{ orbitals labeled } 5p \\ 5 \text{ orbitals labeled } 5d \\ 7 \text{ orbitals labeled } 5f \\ \underline{+9 \text{ orbitals labeled } 5g} \\ 25 \text{ orbitals total} \end{array}$$

Again, each orbital holds two electrons, so 50 electrons can fit in this shell.

(c) The number of orbitals in any shell n will equal n^2. There can be up to two electrons in each orbital, so the maximum number of electrons will be $2 \times n^2$.

Check Your Learning

If a shell contains a maximum of 32 electrons, what is the principal quantum number, n?

Answer:

$n = 4$

EXAMPLE 6.9

Working with Quantum Numbers

Complete the following table for atomic orbitals:

Orbital	n	l	m_l degeneracy	Radial nodes (no.)
4f				
	4	1		
	7		7	3
5d				

Solution

The table can be completed using the following rules:

- The orbital designation is nl, where l = 0, 1, 2, 3, 4, 5, ... is mapped to the letter sequence s, p, d, f, g, h, ...,
- The m_l degeneracy is the number of orbitals within an l subshell, and so is $2l + 1$ (there is one s orbital, three p orbitals, five d orbitals, seven f orbitals, and so forth).
- The number of radial nodes is equal to $n - l - 1$.

Orbital	n	l	m_l degeneracy	Radial nodes (no.)
4f	4	3	7	0
4p	4	1	3	2
7f	7	3	7	3
5d	5	2	5	2

Check Your Learning

How many orbitals have $l = 2$ and $n = 3$?

Answer:

The five degenerate 3d orbitals

6.4 Electronic Structure of Atoms (Electron Configurations)

LEARNING OBJECTIVES

By the end of this section, you will be able to:

- Derive the predicted ground-state electron configurations of atoms
- Identify and explain exceptions to predicted electron configurations for atoms and ions
- Relate electron configurations to element classifications in the periodic table

Having introduced the basics of atomic structure and quantum mechanics, we can use our understanding of quantum numbers to determine how atomic orbitals relate to one another. This allows us to determine which orbitals are occupied by electrons in each atom. The specific arrangement of electrons in orbitals of an atom determines many of the chemical properties of that atom.

Orbital Energies and Atomic Structure

The energy of atomic orbitals increases as the principal quantum number, *n*, increases. In any atom with two or more electrons, the repulsion between the electrons makes energies of subshells with different values of *l* differ so that the energy of the orbitals increases within a shell in the order $s < p < d < f$. Figure 6.24 depicts how these two trends in increasing energy relate. The 1*s* orbital at the bottom of the diagram is the orbital with electrons of lowest energy. The energy increases as we move up to the 2*s* and then 2*p*, 3*s*, and 3*p* orbitals, showing that the increasing *n* value has more influence on energy than the increasing *l* value for small atoms. However, this pattern does not hold for larger atoms. The 3*d* orbital is higher in energy than the 4*s* orbital. Such overlaps continue to occur frequently as we move up the chart.

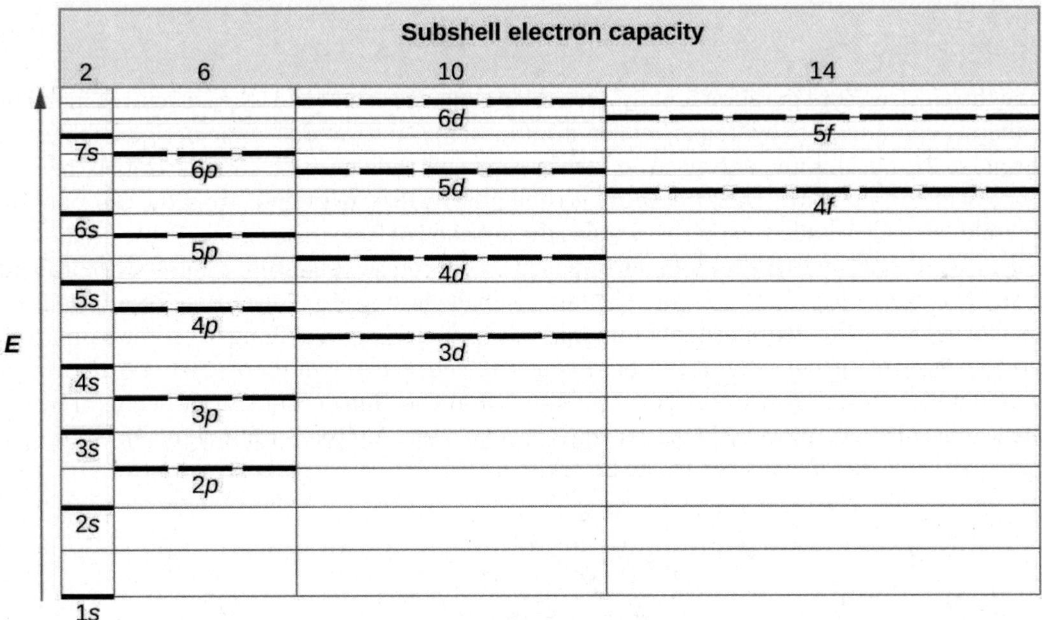

FIGURE 6.24 Generalized energy-level diagram for atomic orbitals in an atom with two or more electrons (not to scale).

Electrons in successive atoms on the periodic table tend to fill low-energy orbitals first. Thus, many students find it confusing that, for example, the 5*p* orbitals fill immediately after the 4*d*, and immediately before the 6*s*. The filling order is based on observed experimental results, and has been confirmed by theoretical calculations. As the principal quantum number, *n*, increases, the size of the orbital increases and the electrons spend more time farther from the nucleus. Thus, the attraction to the nucleus is weaker and the energy associated with the orbital is higher (less stabilized). But this is not the only effect we have to take into account. Within each shell, as the value of *l* increases, the electrons are less penetrating (meaning there is less electron density found close to the nucleus), in the order $s > p > d > f$. Electrons that are closer to the nucleus slightly repel electrons that are farther out, offsetting the more dominant electron–nucleus attractions slightly (recall that all electrons have −1 charges, but nuclei have +*Z* charges). This phenomenon is called shielding and will be discussed in more detail in the next section. Electrons in orbitals that experience more shielding are less stabilized and thus higher in energy. For small orbitals (1*s* through 3*p*), the increase in energy due to *n* is more significant than the increase due to *l*; however, for larger orbitals the two trends are comparable and cannot be simply predicted. We will discuss methods for remembering the observed order.

The arrangement of electrons in the orbitals of an atom is called the **electron configuration** of the atom. We describe an electron configuration with a symbol that contains three pieces of information (Figure 6.25):

1. The number of the principal quantum shell, *n*,
2. The letter that designates the orbital type (the subshell, *l*), and
3. A superscript number that designates the number of electrons in that particular subshell.

For example, the notation $2p^4$ (read "two–p–four") indicates four electrons in a *p* subshell ($l = 1$) with a principal quantum number (*n*) of 2. The notation $3d^8$ (read "three–d–eight") indicates eight electrons in the *d* subshell (i.e., $l = 2$) of the principal shell for which $n = 3$.

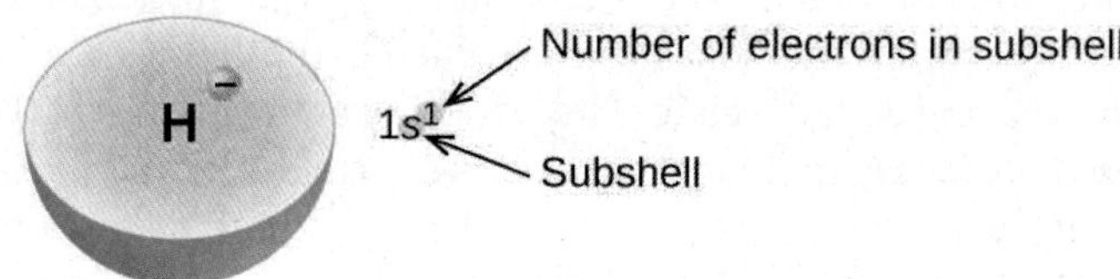

FIGURE 6.25 The diagram of an electron configuration specifies the subshell (*n* and *l* value, with letter symbol) and superscript number of electrons.

The Aufbau Principle

To determine the electron configuration for any particular atom, we can "build" the structures in the order of atomic numbers. Beginning with hydrogen, and continuing across the periods of the periodic table, we add one proton at a time to the nucleus and one electron to the proper subshell until we have described the electron configurations of all the elements. This procedure is called the **Aufbau principle**, from the German word *Aufbau* ("to build up"). Each added electron occupies the subshell of lowest energy available (in the order shown in Figure 6.24), subject to the limitations imposed by the allowed quantum numbers according to the Pauli exclusion principle. Electrons enter higher-energy subshells only after lower-energy subshells have been filled to capacity. Figure 6.26 illustrates the traditional way to remember the filling order for atomic orbitals. Since the arrangement of the periodic table is based on the electron configurations, Figure 6.27 provides an alternative method for determining the electron configuration. The filling order simply begins at hydrogen and includes each subshell as you proceed in increasing *Z* order. For example, after filling the 3*p* block up to Ar, we see the orbital will be 4s (K, Ca), followed by the 3*d* orbitals.

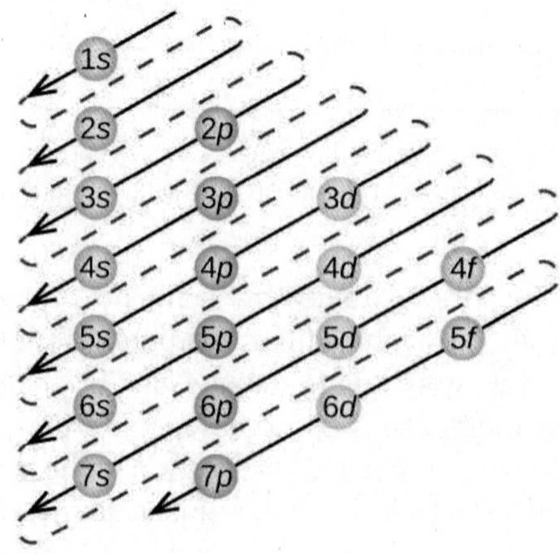

FIGURE 6.26 This diagram depicts the energy order for atomic orbitals and is useful for deriving ground-state electron configurations.

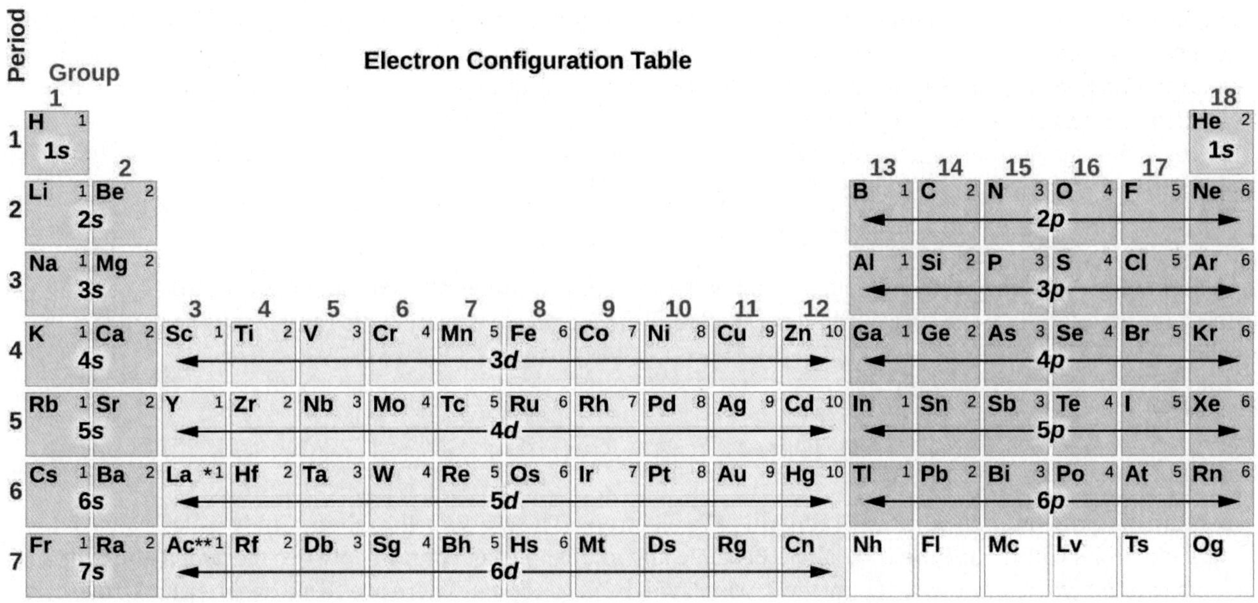

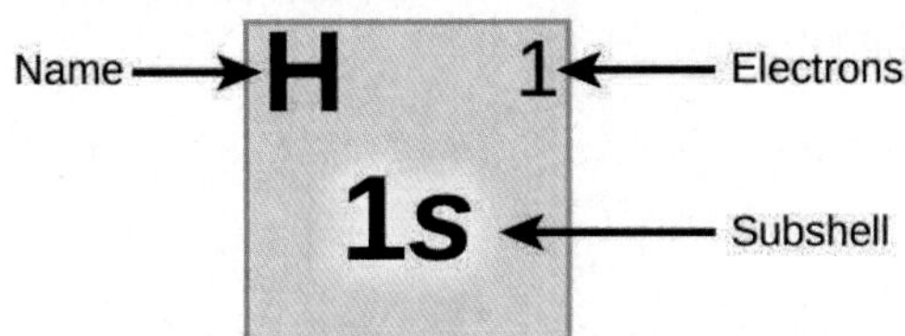

FIGURE 6.27 This partial periodic table shows electron configurations for the valence subshells of atoms. By "building up" from hydrogen, this table can be used to determine the electron configuration for atoms of most elements in the periodic table. (Electron configurations of the lanthanides and actinides are not accurately predicted by this simple approach. See Figure 6.29

We will now construct the ground-state electron configuration and orbital diagram for a selection of atoms in the first and second periods of the periodic table. **Orbital diagrams** are pictorial representations of the electron configuration, showing the individual orbitals and the pairing arrangement of electrons. We start with a single hydrogen atom (atomic number 1), which consists of one proton and one electron. Referring to Figure 6.26 or Figure 6.27, we would expect to find the electron in the 1*s* orbital. By convention, the $m_s = +\frac{1}{2}$ value is usually filled first. The electron configuration and the orbital diagram are:

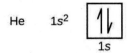

H $1s^1$ [↿]
1s

Following hydrogen is the noble gas helium, which has an atomic number of 2. The helium atom contains two protons and two electrons. The first electron has the same four quantum numbers as the hydrogen atom electron ($n = 1$, $l = 0$, $m_l = 0$, $m_s = +\frac{1}{2}$). The second electron also goes into the 1*s* orbital and fills that orbital. The second electron has the same *n*, *l*, and m_l quantum numbers, but must have the opposite spin quantum number, $m_s = -\frac{1}{2}$. This is in accord with the Pauli exclusion principle: No two electrons in the same atom can have the same set of four quantum numbers. For orbital diagrams, this means two arrows go in each box (representing two electrons in each orbital) and the arrows must point in opposite directions (representing paired spins). The electron configuration and orbital diagram of helium are:

He $1s^2$ [↿⇂]
1s

The $n = 1$ shell is completely filled in a helium atom.

The next atom is the alkali metal lithium with an atomic number of 3. The first two electrons in lithium fill the 1*s* orbital and have the same sets of four quantum numbers as the two electrons in helium. The remaining electron must occupy the orbital of next lowest energy, the 2*s* orbital (Figure 6.26 or Figure 6.27). Thus, the electron configuration and orbital diagram of lithium are:

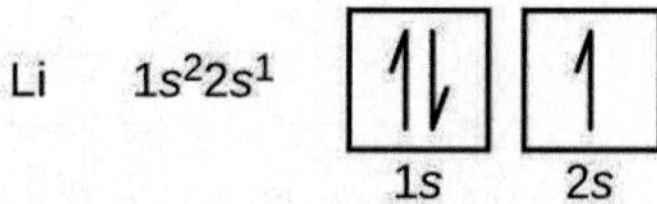

An atom of the alkaline earth metal beryllium, with an atomic number of 4, contains four protons in the nucleus and four electrons surrounding the nucleus. The fourth electron fills the remaining space in the 2*s* orbital.

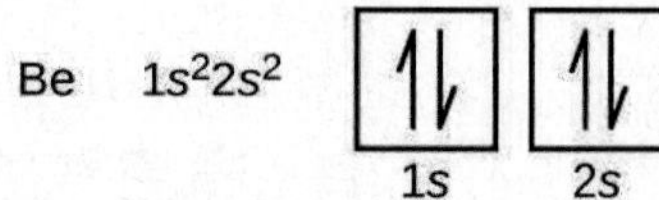

An atom of boron (atomic number 5) contains five electrons. The $n = 1$ shell is filled with two electrons and three electrons will occupy the $n = 2$ shell. Because any *s* subshell can contain only two electrons, the fifth electron must occupy the next energy level, which will be a 2*p* orbital. There are three degenerate 2*p* orbitals ($m_l = -1, 0, +1$) and the electron can occupy any one of these *p* orbitals. When drawing orbital diagrams, we include empty boxes to depict any empty orbitals in the same subshell that we are filling.

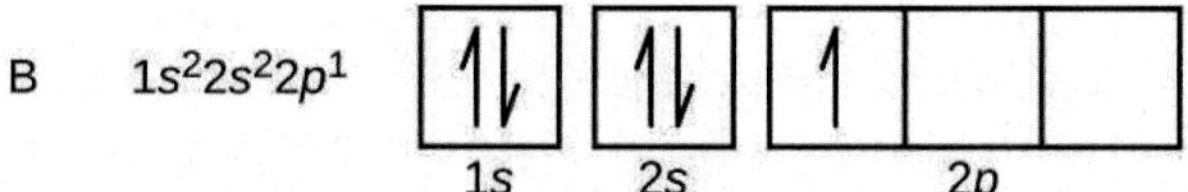

Carbon (atomic number 6) has six electrons. Four of them fill the 1*s* and 2*s* orbitals. The remaining two electrons occupy the 2*p* subshell. We now have a choice of filling one of the 2*p* orbitals and pairing the electrons or of leaving the electrons unpaired in two different, but degenerate, *p* orbitals. The orbitals are filled as described by **Hund's rule**: the lowest-energy configuration for an atom with electrons within a set of degenerate orbitals is that having the maximum number of unpaired electrons. Thus, the two electrons in the carbon 2*p* orbitals have identical n, l, and m_s quantum numbers and differ in their m_l quantum number (in accord with the Pauli exclusion principle). The electron configuration and orbital diagram for carbon are:

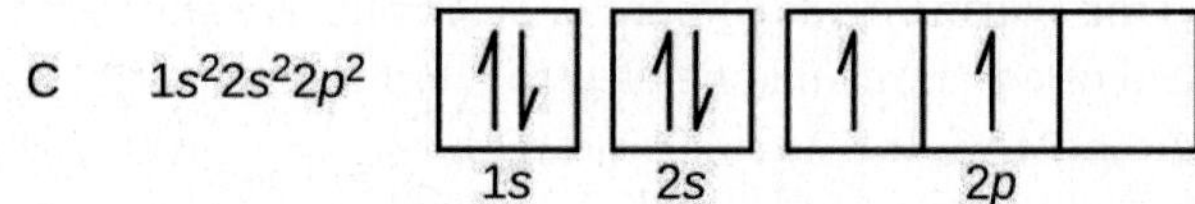

Nitrogen (atomic number 7) fills the 1*s* and 2*s* subshells and has one electron in each of the three 2*p* orbitals, in accordance with Hund's rule. These three electrons have unpaired spins. Oxygen (atomic number 8) has a pair of electrons in any one of the 2*p* orbitals (the electrons have opposite spins) and a single electron in each of the other two. Fluorine (atomic number 9) has only one 2*p* orbital containing an unpaired electron. All of the electrons in the noble gas neon (atomic number 10) are paired, and all of the orbitals in the $n = 1$ and the $n = 2$ shells are filled. The electron configurations and orbital diagrams of these four elements are:

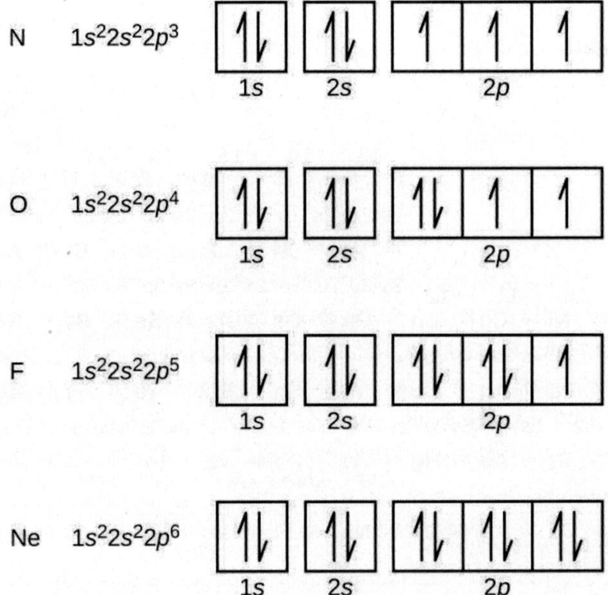

The alkali metal sodium (atomic number 11) has one more electron than the neon atom. This electron must go into the lowest-energy subshell available, the 3*s* orbital, giving a $1s^2 2s^2 2p^6 3s^1$ configuration. The electrons occupying the outermost shell orbital(s) (highest value of *n*) are called **valence electrons**, and those occupying the inner shell orbitals are called **core electrons** (Figure 6.28). Since the core electron shells correspond to noble gas electron configurations, we can abbreviate electron configurations by writing the noble gas that matches the core electron configuration, along with the valence electrons in a condensed format. For our sodium example, the symbol [Ne] represents core electrons, ($1s^2 2s^2 2p^6$) and our abbreviated or condensed configuration is [Ne]$3s^1$.

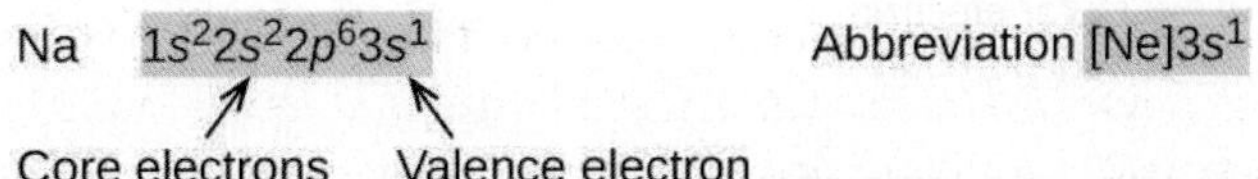

FIGURE 6.28 A core-abbreviated electron configuration (right) replaces the core electrons with the noble gas symbol whose configuration matches the core electron configuration of the other element.

Similarly, the abbreviated configuration of lithium can be represented as [He]$2s^1$, where [He] represents the configuration of the helium atom, which is identical to that of the filled inner shell of lithium. Writing the configurations in this way emphasizes the similarity of the configurations of lithium and sodium. Both atoms, which are in the alkali metal family, have only one electron in a valence *s* subshell outside a filled set of inner shells.

$$\text{Li: [He]}\,2s^1$$
$$\text{Na: [Ne]}\,3s^1$$

The alkaline earth metal magnesium (atomic number 12), with its 12 electrons in a [Ne]$3s^2$ configuration, is analogous to its family member beryllium, [He]$2s^2$. Both atoms have a filled *s* subshell outside their filled inner shells. Aluminum (atomic number 13), with 13 electrons and the electron configuration [Ne]$3s^2 3p^1$, is analogous to its family member boron, [He]$2s^2 2p^1$.

The electron configurations of silicon (14 electrons), phosphorus (15 electrons), sulfur (16 electrons), chlorine (17 electrons), and argon (18 electrons) are analogous in the electron configurations of their outer shells to their corresponding family members carbon, nitrogen, oxygen, fluorine, and neon, respectively, except that the principal quantum number of the outer shell of the heavier elements has increased by one to $n = 3$. Figure 6.29 shows the lowest energy, or ground-state, electron configuration for these elements as well as that for atoms of each of the known elements.

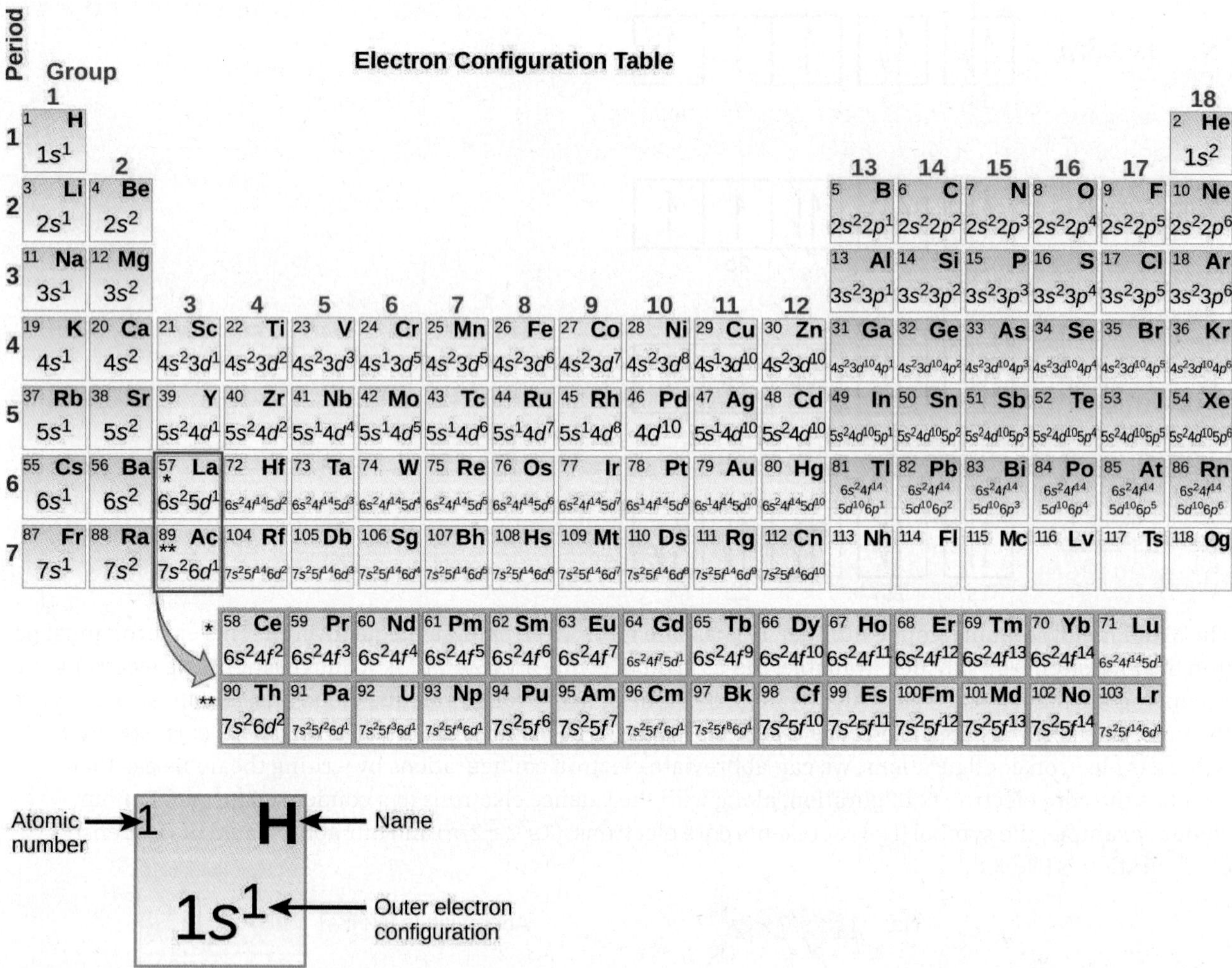

FIGURE 6.29 This version of the periodic table shows the outer-shell electron configuration of each element. Note that down each group, the configuration is often similar.

When we come to the next element in the periodic table, the alkali metal potassium (atomic number 19), we might expect that we would begin to add electrons to the 3d subshell. However, all available chemical and physical evidence indicates that potassium is like lithium and sodium, and that the next electron is not added to the 3d level but is, instead, added to the 4s level (Figure 6.29). As discussed previously, the 3d orbital with no radial nodes is higher in energy because it is less penetrating and more shielded from the nucleus than the 4s, which has three radial nodes. Thus, potassium has an electron configuration of [Ar]4s^1. Hence, potassium corresponds to Li and Na in its valence shell configuration. The next electron is added to complete the 4s subshell and calcium has an electron configuration of [Ar]4s^2. This gives calcium an outer-shell electron configuration corresponding to that of beryllium and magnesium.

Beginning with the transition metal scandium (atomic number 21), additional electrons are added successively to the 3d subshell. This subshell is filled to its capacity with 10 electrons (remember that for $l = 2$ [d orbitals], there are $2l + 1 = 5$ values of m_l, meaning that there are five d orbitals that have a combined capacity of 10 electrons). The 4p subshell fills next. Note that for three series of elements, scandium (Sc) through copper (Cu), yttrium (Y) through silver (Ag), and lutetium (Lu) through gold (Au), a total of 10 d electrons are successively added to the (n – 1) shell next to the n shell to bring that (n – 1) shell from 8 to 18 electrons. For two series, lanthanum (La) through lutetium (Lu) and actinium (Ac) through lawrencium (Lr), 14 f electrons ($l = 3$, $2l + 1 = 7$ m_l values; thus, seven orbitals with a combined capacity of 14 electrons) are successively added to the (n – 2) shell to bring that shell from 18 electrons to a total of 32 electrons.

EXAMPLE 6.10

Quantum Numbers and Electron Configurations

What is the electron configuration and orbital diagram for a phosphorus atom? What are the four quantum numbers for the last electron added?

Solution

The atomic number of phosphorus is 15. Thus, a phosphorus atom contains 15 electrons. The order of filling of the energy levels is 1s, 2s, 2p, 3s, 3p, 4s, . . . The 15 electrons of the phosphorus atom will fill up to the 3p orbital, which will contain three electrons:

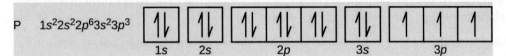

P	$1s^2 2s^2 2p^6 3s^2 3p^3$	↑↓	↑↓	↑↓ ↑↓ ↑↓	↑↓	↑ ↑ ↑
		1s	2s	2p	3s	3p

The last electron added is a 3p electron. Therefore, n = 3 and, for a p-type orbital, l = 1. The m_l value could be –1, 0, or +1. The three p orbitals are degenerate, so any of these m_l values is correct. For unpaired electrons, convention assigns the value of $+\frac{1}{2}$ for the spin quantum number; thus, $m_s = +\frac{1}{2}$.

Check Your Learning

Identify the atoms from the electron configurations given:

(a) $[Ar]4s^2 3d^5$

(b) $[Kr]5s^2 4d^{10} 5p^6$

Answer:

(a) Mn (b) Xe

The periodic table can be a powerful tool in predicting the electron configuration of an element. However, we do find exceptions to the order of filling of orbitals that are shown in Figure 6.26 or Figure 6.27. For instance, the electron configurations (shown in Figure 6.29) of the transition metals chromium (Cr; atomic number 24) and copper (Cu; atomic number 29), among others, are not those we would expect. In general, such exceptions involve subshells with very similar energy, and small effects can lead to changes in the order of filling.

In the case of Cr and Cu, we find that half-filled and completely filled subshells apparently represent conditions of preferred stability. This stability is such that an electron shifts from the 4s into the 3d orbital to gain the extra stability of a half-filled 3d subshell (in Cr) or a filled 3d subshell (in Cu). Other exceptions also occur. For example, niobium (Nb, atomic number 41) is predicted to have the electron configuration $[Kr]5s^2 4d^3$. Experimentally, we observe that its ground-state electron configuration is actually $[Kr]5s^1 4d^4$. We can rationalize this observation by saying that the electron–electron repulsions experienced by pairing the electrons in the 5s orbital are larger than the gap in energy between the 5s and 4d orbitals. There is no simple method to predict the exceptions for atoms where the magnitude of the repulsions between electrons is greater than the small differences in energy between subshells.

Electron Configurations and the Periodic Table

As described earlier, the periodic table arranges atoms based on increasing atomic number so that elements with the same chemical properties recur periodically. When their electron configurations are added to the table (Figure 6.29), we also see a periodic recurrence of similar electron configurations in the outer shells of these elements. Because they are in the outer shells of an atom, valence electrons play the most important role in chemical reactions. The outer electrons have the highest energy of the electrons in an atom and are more easily lost or shared than the core electrons. Valence electrons are also the determining factor in some physical properties of the elements.

Elements in any one group (or column) have the same number of valence electrons; the alkali metals lithium and sodium each have only one valence electron, the alkaline earth metals beryllium and magnesium each have two, and the halogens fluorine and chlorine each have seven valence electrons. The similarity in chemical properties among elements of the same group occurs because they have the same number of valence electrons. It is the loss, gain, or sharing of valence electrons that defines how elements react.

It is important to remember that the periodic table was developed on the basis of the chemical behavior of the elements, well before any idea of their atomic structure was available. Now we can understand why the periodic table has the arrangement it has—the arrangement puts elements whose atoms have the same number of valence electrons in the same group. This arrangement is emphasized in Figure 6.29, which shows in periodic-table form the electron configuration of the last subshell to be filled by the Aufbau principle. The colored sections of Figure 6.29 show the three categories of elements classified by the orbitals being filled: main group, transition, and inner transition elements. These classifications determine which orbitals are counted in the **valence shell**, or highest energy level orbitals of an atom.

1. **Main group elements** (sometimes called **representative elements**) are those in which the last electron added enters an *s* or a *p* orbital in the outermost shell, shown in blue and red in Figure 6.29. This category includes all the nonmetallic elements, as well as many metals and the metalloids. The valence electrons for main group elements are those with the highest *n* level. For example, gallium (Ga, atomic number 31) has the electron configuration $[Ar]\underline{4s^2}3d^{10}\underline{4p^1}$, which contains three valence electrons (underlined). The completely filled *d* orbitals count as core, not valence, electrons.
2. **Transition elements or transition metals**. These are metallic elements in which the last electron added enters a *d* orbital. The valence electrons (those added after the last noble gas configuration) in these elements include the *ns* and $(n-1)$ *d* electrons. The official IUPAC definition of transition elements specifies those with partially filled *d* orbitals. Thus, the elements with completely filled orbitals (Zn, Cd, Hg, as well as Cu, Ag, and Au in Figure 6.29) are not technically transition elements. However, the term is frequently used to refer to the entire *d* block (colored yellow in Figure 6.29), and we will adopt this usage in this textbook.
3. **Inner transition elements** are metallic elements in which the last electron added occupies an *f* orbital. They are shown in green in Figure 6.29. The valence shells of the inner transition elements consist of the $(n-2)f$, the $(n-1)d$, and the *ns* subshells. There are two inner transition series:

a. The lanthanide series: lanthanum (La) through lutetium (Lu)
b. The actinide series: actinium (Ac) through lawrencium (Lr)

Lanthanum and actinium, because of their similarities to the other members of the series, are included and used to name the series, even though they are transition metals with no *f* electrons.

Electron Configurations of Ions

Ions are formed when atoms gain or lose electrons. A cation (positively charged ion) forms when one or more electrons are removed from a parent atom. For main group elements, the electrons that were added last are the first electrons removed. For transition metals and inner transition metals, however, electrons in the *s* orbital are easier to remove than the *d* or *f* electrons, and so the highest *ns* electrons are lost, and then the $(n-1)d$ or $(n-2)f$ electrons are removed. An anion (negatively charged ion) forms when one or more electrons are added to a parent atom. The added electrons fill in the order predicted by the Aufbau principle.

 **EXAMPLE 6.11**

Predicting Electron Configurations of Ions

What is the electron configuration of:

(a) Na^+

(b) P^{3-}

(c) Al^{2+}

(d) Fe^{2+}

(e) Sm^{3+}

Solution

First, write out the electron configuration for each parent atom. We have chosen to show the full, unabbreviated configurations to provide more practice for students who want it, but listing the core-abbreviated electron configurations is also acceptable.

Next, determine whether an electron is gained or lost. Remember electrons are negatively charged, so ions with a positive charge have *lost* an electron. For main group elements, the last orbital gains or loses the electron. For transition metals, the last *s* orbital loses an electron before the *d* orbitals.

(a) Na: $1s^22s^22p^63s^1$. Sodium cation loses one electron, so Na^+: $1s^22s^22p^63s^1$ = Na^+: $1s^22s^22p^6$.

(b) P: $1s^22s^22p^63s^23p^3$. Phosphorus trianion gains three electrons, so P^{3-}: $1s^22s^22p^63s^23p^6$.

(c) Al: $1s^22s^22p^63s^23p^1$. Aluminum dication loses two electrons Al^{2+}: $1s^22s^22p^63s^23p^1$ =

Al^{2+}: $1s^22s^22p^63s^1$.

(d) Fe: $1s^22s^22p^63s^23p^64s^23d^6$. Iron(II) loses two electrons and, since it is a transition metal, they are removed from the 4*s* orbital Fe^{2+}: $1s^22s^22p^63s^23p^64s^23d^6 = 1s^22s^22p^63s^23p^63d^6$.

(e). Sm: $1s^22s^22p^63s^23p^64s^23d^{10}4p^65s^24d^{10}5p^66s^24f^6$. Samarium trication loses three electrons. The first two will be lost from the 6*s* orbital, and the final one is removed from the 4*f* orbital. Sm^{3+}: $1s^22s^22p^63s^23p^64s^23d^{10}4p^65s^24d^{10}5p^66s^24f^6 = 1s^22s^22p^63s^23p^64s^23d^{10}4p^65s^24d^{10}5p^64f^5$.

Check Your Learning

Which ion with a +2 charge has the electron configuration $1s^22s^22p^63s^23p^64s^23d^{10}4p^64d^5$? Which ion with a +3 charge has this configuration?

Answer:

Tc^{2+}, Ru^{3+}

6.5 Periodic Variations in Element Properties

LEARNING OBJECTIVES

By the end of this section, you will be able to:

- Describe and explain the observed trends in atomic size, ionization energy, and electron affinity of the elements

The elements in groups (vertical columns) of the periodic table exhibit similar chemical behavior. This similarity occurs because the members of a group have the same number and distribution of electrons in their valence shells. However, there are also other patterns in chemical properties on the periodic table. For example, as we move down a group, the metallic character of the atoms increases. Oxygen, at the top of group 16 (6A), is a colorless gas; in the middle of the group, selenium is a semiconducting solid; and, toward the bottom, polonium is a silver-grey solid that conducts electricity.

As we go across a period from left to right, we add a proton to the nucleus and an electron to the valence shell with each successive element. As we go down the elements in a group, the number of electrons in the valence shell remains constant, but the principal quantum number increases by one each time. An understanding of the electronic structure of the elements allows us to examine some of the properties that govern their chemical behavior. These properties vary periodically as the electronic structure of the elements changes. They are (1) size (radius) of atoms and ions, (2) ionization energies, and (3) electron affinities.

LINK TO LEARNING

Explore visualizations (http://openstax.org/l/16pertrends) of the periodic trends discussed in this section (and many more trends). With just a few clicks, you can create three-dimensional versions of the periodic table showing atomic size or graphs of ionization energies from all measured elements.

Variation in Covalent Radius

The quantum mechanical picture makes it difficult to establish a definite size of an atom. However, there are several practical ways to define the radius of atoms and, thus, to determine their relative sizes that give roughly similar values. We will use the **covalent radius** (Figure 6.30), which is defined as one-half the distance between the nuclei of two identical atoms when they are joined by a covalent bond (this measurement is possible because atoms within molecules still retain much of their atomic identity). We know that as we scan down a group, the principal quantum number, *n*, increases by one for each element. Thus, the electrons are being added to a region of space that is increasingly distant from the nucleus. Consequently, the size of the atom (and its covalent radius) must increase as we increase the distance of the outermost electrons from the nucleus. This trend is illustrated for the covalent radii of the halogens in Table 6.2 and Figure 6.30. The trends for the entire periodic table can be seen in Figure 6.30.

Covalent Radii of the Halogen Group Elements

Atom	Covalent radius (pm)	Nuclear charge
F	64	+9
Cl	99	+17
Br	114	+35
I	133	+53
At	148	+85

TABLE 6.2

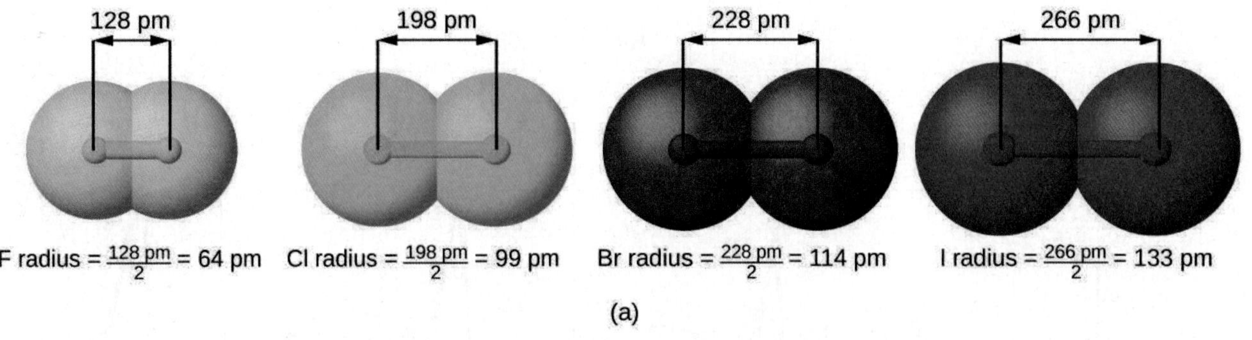

FIGURE 6.30 (a) The radius of an atom is defined as one-half the distance between the nuclei in a molecule consisting of two identical atoms joined by a covalent bond. The atomic radius for the halogens increases down the group as *n* increases. (b) Covalent radii of the elements are shown to scale. The general trend is that radii increase down a group and decrease across a period.

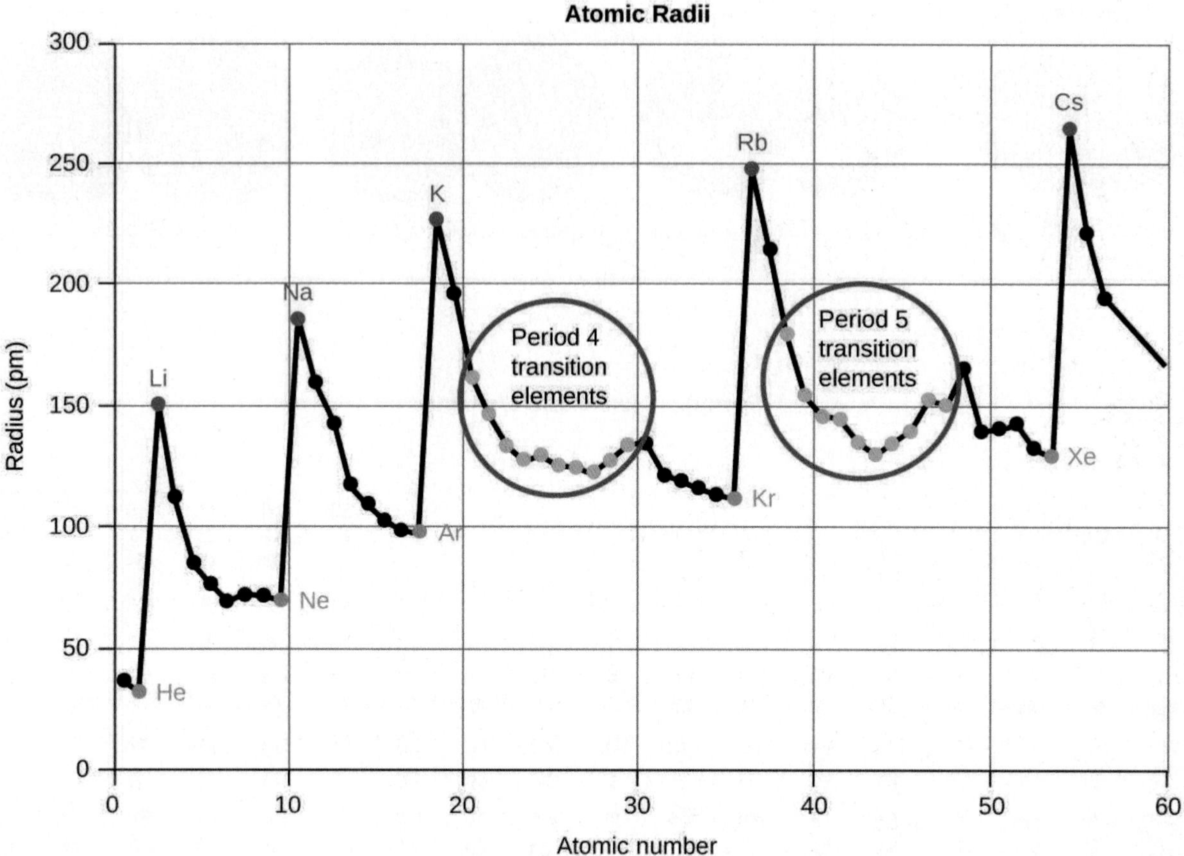

FIGURE 6.31 Within each period, the trend in atomic radius decreases as *Z* increases; for example, from K to Kr. Within each group (e.g., the alkali metals shown in purple), the trend is that atomic radius increases as *Z* increases.

As shown in Figure 6.31, as we move across a period from left to right, we generally find that each element has a smaller covalent radius than the element preceding it. This might seem counterintuitive because it implies that atoms with more electrons have a smaller atomic radius. This can be explained with the concept of **effective nuclear charge, Z_{eff}**. This is the pull exerted on a specific electron by the nucleus, taking into account any electron–electron repulsions. For hydrogen, there is only one electron and so the nuclear charge (Z) and the effective nuclear charge (Z_{eff}) are equal. For all other atoms, the inner electrons partially *shield* the outer electrons from the pull of the nucleus, and thus:

$$Z_{eff} = Z - shielding$$

Shielding is determined by the probability of another electron being between the electron of interest and the nucleus, as well as by the electron–electron repulsions the electron of interest encounters. Core electrons are adept at shielding, while electrons in the same valence shell do not block the nuclear attraction experienced by each other as efficiently. Thus, each time we move from one element to the next across a period, *Z* increases by one, but the shielding increases only slightly. Thus, Z_{eff} increases as we move from left to right across a period. The stronger pull (higher effective nuclear charge) experienced by electrons on the right side of the periodic table draws them closer to the nucleus, making the covalent radii smaller.

Thus, as we would expect, the outermost or valence electrons are easiest to remove because they have the highest energies, are shielded more, and are farthest from the nucleus. As a general rule, when the representative elements form cations, they do so by the loss of the *ns* or *np* electrons that were added last in the Aufbau process. The transition elements, on the other hand, lose the *ns* electrons before they begin to lose the $(n - 1)d$ electrons, even though the *ns* electrons are added first, according to the Aufbau principle.

EXAMPLE 6.12

Sorting Atomic Radii

Predict the order of increasing covalent radius for Ge, Fl, Br, Kr.

Solution

Radius increases as we move down a group, so Ge < Fl (Note: Fl is the symbol for flerovium, element 114, NOT fluorine). Radius decreases as we move across a period, so Kr < Br < Ge. Putting the trends together, we obtain Kr < Br < Ge < Fl.

Check Your Learning

Give an example of an atom whose size is smaller than fluorine.

Answer:

Ne or He

Variation in Ionic Radii

Ionic radius is the measure used to describe the size of an ion. A cation always has fewer electrons and the same number of protons as the parent atom; it is smaller than the atom from which it is derived (Figure 6.32). For example, the covalent radius of an aluminum atom ($1s^22s^22p^63s^23p^1$) is 118 pm, whereas the ionic radius of an Al^{3+} ($1s^22s^22p^6$) is 68 pm. As electrons are removed from the outer valence shell, the remaining core electrons occupying smaller shells experience a greater effective nuclear charge Z_{eff} (as discussed) and are drawn even closer to the nucleus.

Al 118
Al^{3+} 68
S 104
S^{2-} 170

FIGURE 6.32 The radius for a cation is smaller than the parent atom (Al), due to the lost electrons; the radius for an anion is larger than the parent (S), due to the gained electrons.

Cations with larger charges are smaller than cations with smaller charges (e.g., V^{2+} has an ionic radius of 79 pm, while that of V^{3+} is 64 pm). Proceeding down the groups of the periodic table, we find that cations of successive elements with the same charge generally have larger radii, corresponding to an increase in the principal quantum number, *n*.

An anion (negative ion) is formed by the addition of one or more electrons to the valence shell of an atom. This results in a greater repulsion among the electrons and a decrease in Z_{eff} per electron. Both effects (the increased number of electrons and the decreased Z_{eff}) cause the radius of an anion to be larger than that of the parent atom (Figure 6.32). For example, a sulfur atom ($[Ne]3s^23p^4$) has a covalent radius of 104 pm, whereas the ionic radius of the sulfide anion ($[Ne]3s^23p^6$) is 170 pm. For consecutive elements proceeding down any group, anions have larger principal quantum numbers and, thus, larger radii.

Atoms and ions that have the same electron configuration are said to be **isoelectronic**. Examples of isoelectronic species are N^{3-}, O^{2-}, F^-, Ne, Na^+, Mg^{2+}, and Al^{3+} ($1s^22s^22p^6$). Another isoelectronic series is P^{3-}, S^{2-}, Cl^-, Ar, K^+, Ca^{2+}, and Sc^{3+} ($[Ne]3s^23p^6$). For atoms or ions that are isoelectronic, the number of protons determines the size. The greater the nuclear charge, the smaller the radius in a series of isoelectronic ions and atoms.

Variation in Ionization Energies

The amount of energy required to remove the most loosely bound electron from a gaseous atom in its ground state is called its first **ionization energy** (IE_1). The first ionization energy for an element, X, is the energy required to form a cation with +1 charge:

$$X(g) \longrightarrow X^+(g) + e^- \qquad IE_1$$

The energy required to remove the second most loosely bound electron is called the second ionization energy (IE_2).

$$X^+(g) \longrightarrow X^{2+}(g) + e^- \qquad IE_2$$

The energy required to remove the third electron is the third ionization energy, and so on. Energy is always required to remove electrons from atoms or ions, so ionization processes are endothermic and IE values are always positive. For larger atoms, the most loosely bound electron is located farther from the nucleus and so is easier to remove. Thus, as size (atomic radius) increases, the ionization energy should decrease. Relating this logic to what we have just learned about radii, we would expect first ionization energies to decrease down a group and to increase across a period.

Figure 6.33 graphs the relationship between the first ionization energy and the atomic number of several elements. The values of first ionization energy for the elements are given in Figure 6.34. Within a period, the IE_1 generally increases with increasing *Z*. Down a group, the IE_1 value generally decreases with increasing *Z*. There are some systematic deviations from this trend, however. Note that the ionization energy of boron (atomic number 5) is less than that of beryllium (atomic number 4) even though the nuclear charge of boron is greater by one proton. This can be explained because the energy of the subshells increases as *l* increases, due to penetration and shielding (as discussed previously in this chapter). Within any one shell, the *s* electrons are lower in energy than the *p* electrons. This means that an *s* electron is harder to remove from an atom than a *p* electron in the same shell. The electron removed during the ionization of beryllium ([He]$2s^2$) is an *s* electron, whereas the electron removed during the ionization of boron ([He]$2s^22p^1$) is a *p* electron; this results in a lower first ionization energy for boron, even though its nuclear charge is greater by one proton. Thus, we see a small deviation from the predicted trend occurring each time a new subshell begins.

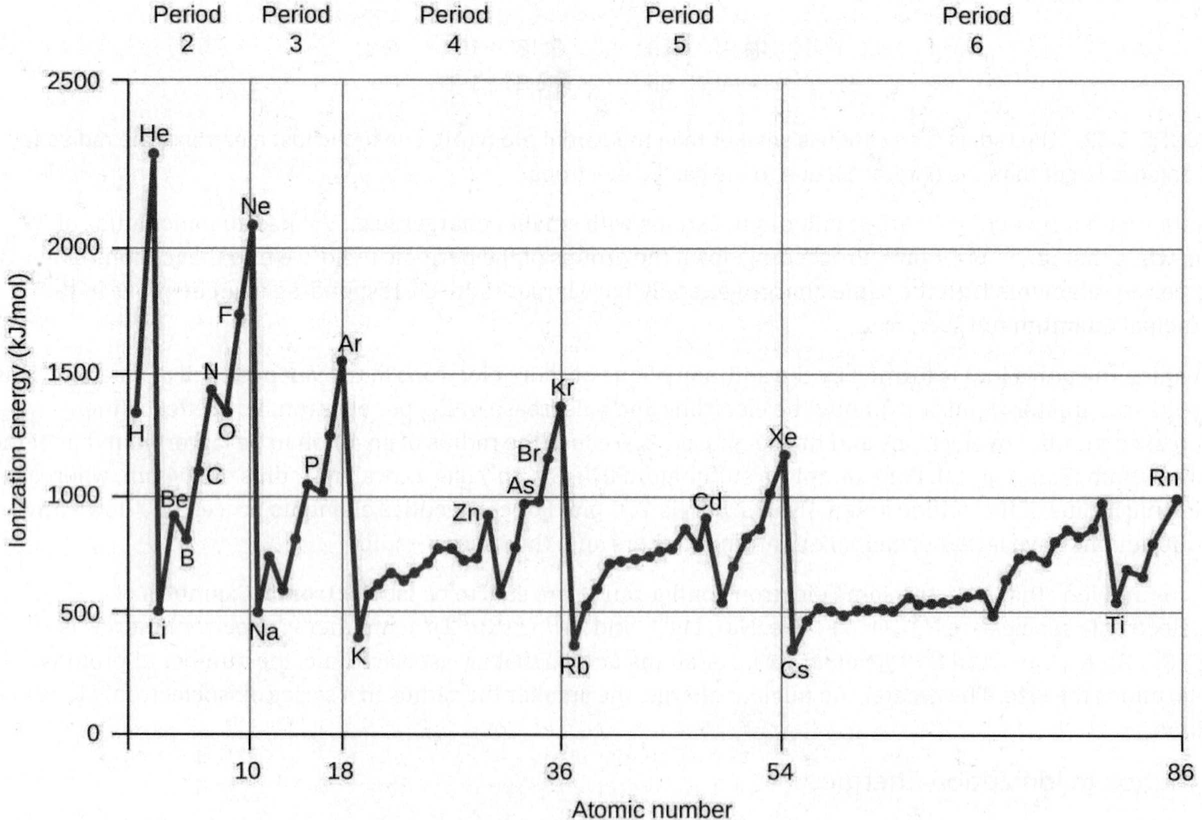

FIGURE 6.33 The first ionization energy of the elements in the first five periods are plotted against their atomic number.

First Ionization Energies of Some Elements (kJ/mol)

Period \ Group	1	2	3	4	5	6	7	8	9	10	11	12	13	14	15	16	17	18
1	H 1310																	He 2370
2	Li 520	Be 900											B 800	C 1090	N 1400	O 1310	F 1680	Ne 2080
3	Na 490	Mg 730											Al 580	Si 780	P 1060	S 1000	Cl 1250	Ar 1520
4	K 420	Ca 590	Sc 630	Ti 660	V 650	Cr 660	Mn 710	Fe 760	Co 760	Ni 730	Cu 740	Zn 910	Ga 580	Ge 780	As 960	Se 950	Br 1140	Kr 1350
5	Rb 400	Sr 550	Y 620	Zr 660	Nb 670	Mo 680	Tc 700	Ru 710	Rh 720	Pd 800	Ag 730	Cd 870	In 560	Sn 700	Sb 830	Te 870	I 1010	Xe 1170
6	Cs 380	Ba 500	La 540	Hf 700	Ta 760	W 770	Re 760	Os 840	Ir 890	Pt 870	Au 890	Hg 1000	Tl 590	Pb 710	Bi 800	Po 810	At ...	Rn 1030
7	Fr ...	Ra 510																

FIGURE 6.34 This version of the periodic table shows the first ionization energy (IE_1), in kJ/mol, of selected elements.

Another deviation occurs as orbitals become more than one-half filled. The first ionization energy for oxygen is slightly less than that for nitrogen, despite the trend in increasing IE_1 values across a period. Looking at the orbital diagram of oxygen, we can see that removing one electron will eliminate the electron–electron repulsion caused by pairing the electrons in the 2*p* orbital and will result in a half-filled orbital (which is energetically favorable). Analogous changes occur in succeeding periods (note the dip for sulfur after phosphorus in Figure 6.34).

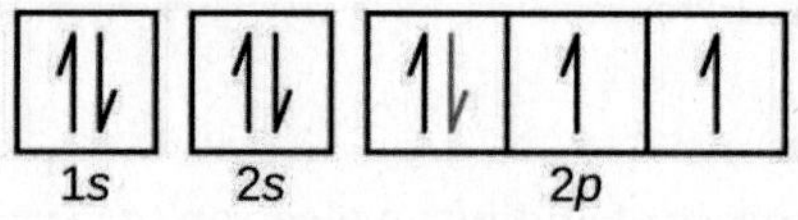

O $1s^2 2s^2 2p^4$ [↑↓] 1*s* [↑↓] 2*s* [↑↓][↑][↑] 2*p*

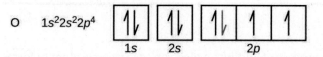

Removing an electron from a cation is more difficult than removing an electron from a neutral atom because of the greater electrostatic attraction to the cation. Likewise, removing an electron from a cation with a higher positive charge is more difficult than removing an electron from an ion with a lower charge. Thus, successive ionization energies for one element always increase. As seen in Table 6.3, there is a large increase in the ionization energies for each element. This jump corresponds to removal of the core electrons, which are harder to remove than the valence electrons. For example, Sc and Ga both have three valence electrons, so the rapid increase in ionization energy occurs after the third ionization.

Successive Ionization Energies for Selected Elements (kJ/mol)

Element	IE_1	IE_2	IE_3	IE_4	IE_5	IE_6	IE_7
K	418.8	3051.8	4419.6	5876.9	7975.5	9590.6	11343
Ca	589.8	1145.4	4912.4	6490.6	8153.0	10495.7	12272.9
Sc	633.1	1235.0	2388.7	7090.6	8842.9	10679.0	13315.0
Ga	578.8	1979.4	2964.6	6180	8298.7	10873.9	13594.8

TABLE 6.3

Element	IE_1	IE_2	IE_3	IE_4	IE_5	IE_6	IE_7
Ge	762.2	1537.5	3302.1	4410.6	9021.4	Not available	Not available
As	944.5	1793.6	2735.5	4836.8	6042.9	12311.5	Not available

TABLE 6.3

EXAMPLE 6.13

Ranking Ionization Energies

Predict the order of increasing energy for the following processes: IE_1 for Al, IE_1 for Tl, IE_2 for Na, IE_3 for Al.

Solution

Removing the $6p^1$ electron from Tl is easier than removing the $3p^1$ electron from Al because the higher *n* orbital is farther from the nucleus, so $IE_1(Tl) < IE_1(Al)$. Ionizing the third electron from Al $(Al^{2+} \longrightarrow Al^{3+} + e^-)$ requires more energy because the cation Al^{2+} exerts a stronger pull on the electron than the neutral Al atom, so $IE_1(Al) < IE_3(Al)$. The second ionization energy for sodium removes a core electron, which is a much higher energy process than removing valence electrons. Putting this all together, we obtain: $IE_1(Tl) < IE_1(Al) < IE_3(Al) < IE_2(Na)$.

Check Your Learning

Which has the lowest value for IE_1: O, Po, Pb, or Ba?

Answer:

Ba

Variation in Electron Affinities

The **electron affinity** (EA) is the energy change for the process of adding an electron to a gaseous atom to form an anion (negative ion).

$$X(g) + e^- \longrightarrow X^-(g) \qquad EA_1$$

This process can be either endothermic or exothermic, depending on the element. The EA of some of the elements is given in Figure 6.35. You can see that many of these elements have negative values of EA, which means that energy is released when the gaseous atom accepts an electron. However, for some elements, energy is required for the atom to become negatively charged and the value of their EA is positive. Just as with ionization energy, subsequent EA values are associated with forming ions with more charge. The second EA is the energy associated with adding an electron to an anion to form a –2 ion, and so on.

As we might predict, it becomes easier to add an electron across a series of atoms as the effective nuclear charge of the atoms increases. We find, as we go from left to right across a period, EAs tend to become more negative. The exceptions found among the elements of group 2 (2A), group 15 (5A), and group 18 (8A) can be understood based on the electronic structure of these groups. The noble gases, group 18 (8A), have a completely filled shell and the incoming electron must be added to a higher *n* level, which is more difficult to do. Group 2 (2A) has a filled *ns* subshell, and so the next electron added goes into the higher energy *np*, so, again, the observed EA value is not as the trend would predict. Finally, group 15 (5A) has a half-filled *np* subshell and the next electron must be paired with an existing *np* electron. In all of these cases, the initial relative stability of the electron configuration disrupts the trend in EA.

We also might expect the atom at the top of each group to have the most negative EA; their first ionization potentials suggest that these atoms have the largest effective nuclear charges. However, as we move down a group, we see that the *second* element in the group most often has the most negative EA. This can be attributed to the small size of the $n = 2$ shell and the resulting large electron–electron repulsions. For example, chlorine, with an EA value of −348 kJ/mol, has the highest value of any element in the periodic table. The EA of fluorine is −322 kJ/mol. When we add an electron to a fluorine atom to form a fluoride anion (F^-), we add an electron to the $n = 2$ shell. The electron is attracted to the nucleus, but there is also significant repulsion from the other electrons already present in this small valence shell. The chlorine atom has the same electron configuration in the valence shell, but because the entering electron is going into the $n = 3$ shell, it occupies a considerably larger region of space and the electron–electron repulsions are reduced. The entering electron does not experience as much repulsion and the chlorine atom accepts an additional electron more readily, resulting in a more negative EA.

Electron Affinity Values for Selected Elements (kJ/mol)

Period \ Group	1	2	3	4	5	6	7	8	9	10	11	12	13	14	15	16	17	18
1	H –72																	He +20*
2	Li –60	Be +240*											B –23	C –123	N 0	O –141	F –322	Ne –30
3	Na –53	Mg +230*											Al –44	Si –120	P –74	S –200	Cl –348	Ar +35*
4	K –48	Ca +150*	Sc	Ti	V	Cr	Mn	Fe	Co	Ni	Cu	Zn	Ga –40*	Ge –115	As –7	Se –195	Br –324	Kr +40*
5	Rb –46	Sr +160*	Y	Zr	Nb	Mo	Tc	Ru	Rh	Pd	Ag	Cd	In –40*	Sn –121	Sb –101	Te –190	I –295	Xe +40*
6	Cs –45	Ba +50*	La	Hf	Ta	W	Re	Os	Ir	Pt	Au	Hg	Tl –50	Pb –101	Bi –101	Po –170	At –270*	Rn +40*
7	Fr	Ra																

* Calculated value

FIGURE 6.35 This version of the periodic table displays the electron affinity values (in kJ/mol) for selected elements.

The properties discussed in this section (size of atoms and ions, effective nuclear charge, ionization energies, and electron affinities) are central to understanding chemical reactivity. For example, because fluorine has an energetically favorable EA and a large energy barrier to ionization (IE), it is much easier to form fluorine anions than cations. Metallic properties including conductivity and malleability (the ability to be formed into sheets) depend on having electrons that can be removed easily. Thus, metallic character increases as we move down a group and decreases across a period in the same trend observed for atomic size because it is easier to remove an electron that is farther away from the nucleus.

Key Terms

amplitude extent of the displacement caused by a wave

atomic orbital mathematical function that describes the behavior of an electron in an atom (also called the wavefunction)

Aufbau principle procedure in which the electron configuration of the elements is determined by "building" them in order of atomic numbers, adding one proton to the nucleus and one electron to the proper subshell at a time

blackbody idealized perfect absorber of all incident electromagnetic radiation; such bodies emit electromagnetic radiation in characteristic continuous spectra called blackbody radiation

Bohr's model of the hydrogen atom structural model in which an electron moves around the nucleus only in circular orbits, each with a specific allowed radius

continuous spectrum electromagnetic radiation given off in an unbroken series of wavelengths (e.g., white light from the sun)

core electron electron in an atom that occupies the orbitals of the inner shells

covalent radius one-half the distance between the nuclei of two identical atoms when they are joined by a covalent bond

***d* orbital** region of space with high electron density that is either four lobed or contains a dumbbell and torus shape; describes orbitals with $l = 2$.

degenerate orbitals orbitals that have the same energy

effective nuclear charge charge that leads to the Coulomb force exerted by the nucleus on an electron, calculated as the nuclear charge minus shielding

electromagnetic radiation energy transmitted by waves that have an electric-field component and a magnetic-field component

electromagnetic spectrum range of energies that electromagnetic radiation can comprise, including radio, microwaves, infrared, visible, ultraviolet, X-rays, and gamma rays

electron affinity energy change associated with addition of an electron to a gaseous atom or ion

electron configuration listing that identifies the electron occupancy of an atom's shells and subshells

electron density a measure of the probability of locating an electron in a particular region of space, it is equal to the squared absolute value of the wave function ψ

excited state state having an energy greater than the ground-state energy

***f* orbital** multilobed region of space with high electron density, describes orbitals with $l = 3$

frequency (ν) number of wave cycles (peaks or troughs) that pass a specified point in space per unit time

ground state state in which the electrons in an atom, ion, or molecule have the lowest energy possible

Heisenberg uncertainty principle rule stating that it is impossible to exactly determine both certain conjugate dynamical properties such as the momentum and the position of a particle at the same time. The uncertainty principle is a consequence of quantum particles exhibiting wave–particle duality

hertz (Hz) the unit of frequency, which is the number of cycles per second, s^{-1}

Hund's rule every orbital in a subshell is singly occupied with one electron before any one orbital is doubly occupied, and all electrons in singly occupied orbitals have the same spin

intensity property of wave-propagated energy related to the amplitude of the wave, such as brightness of light or loudness of sound

interference pattern pattern typically consisting of alternating bright and dark fringes; it results from constructive and destructive interference of waves

ionization energy energy required to remove an electron from a gaseous atom or ion

isoelectronic group of ions or atoms that have identical electron configurations

line spectrum electromagnetic radiation emitted at discrete wavelengths by a specific atom (or atoms) in an excited state

magnetic quantum number (m_l) quantum number signifying the orientation of an atomic orbital around the nucleus

node any point of a standing wave with zero amplitude

orbital diagram pictorial representation of the electron configuration showing each orbital as a box and each electron as an arrow

***p* orbital** dumbbell-shaped region of space with high electron density, describes orbitals with $l = 1$

Pauli exclusion principle specifies that no two electrons in an atom can have the same value for all four quantum numbers

photon smallest possible packet of electromagnetic radiation, a particle of light

principal quantum number (n) quantum number

specifying the shell an electron occupies in an atom

quantization limitation of some property to specific discrete values, not continuous

quantum mechanics field of study that includes quantization of energy, wave-particle duality, and the Heisenberg uncertainty principle to describe matter

quantum number number having only specific allowed values and used to characterize the arrangement of electrons in an atom

***s* orbital** spherical region of space with high electron density, describes orbitals with $l = 0$

secondary (angular momentum) quantum number (*l*) quantum number distinguishing the different shapes of orbitals; it is also a measure of the orbital angular momentum

shell atomic orbitals with the same principal quantum number, n

spin quantum number (m_s) number specifying the electron spin direction, either $+\frac{1}{2}$ or $-\frac{1}{2}$

standing wave (also, stationary wave) localized wave phenomenon characterized by discrete wavelengths determined by the boundary conditions used to generate the waves; standing waves are inherently quantized

subshell atomic orbitals with the same values of n and l

valence electrons electrons in the high energy outer shell(s) of an atom

valence shell high energy outer shell(s) of an atom

wave oscillation of a property over time or space; can transport energy from one point to another

wave-particle duality observation that elementary particles can exhibit both wave-like and particle-like properties

wavefunction (ψ) mathematical description of an atomic orbital that describes the shape of the orbital; it can be used to calculate the probability of finding the electron at any given location in the orbital, as well as dynamical variables such as the energy and the angular momentum

wavelength (λ) distance between two consecutive peaks or troughs in a wave

Key Equations

$$c = \lambda\nu$$

$$E = h\nu = \frac{hc}{\lambda}, \text{ where } h = 6.626 \times 10^{-34} \text{ J s}$$

$$\frac{1}{\lambda} = R_\infty\left(\frac{1}{n_1^2} - \frac{1}{n_2^2}\right)$$

$$E_n = -\frac{kZ^2}{n^2}, \; n = 1, 2, 3, \ldots$$

$$\Delta E = kZ^2\left(\frac{1}{n_1^2} - \frac{1}{n_2^2}\right)$$

$$r = \frac{n^2}{Z} a_0$$

Summary

6.1 Electromagnetic Energy

Light and other forms of electromagnetic radiation move through a vacuum with a constant speed, c, of 2.998×10^8 m s^{-1}. This radiation shows wavelike behavior, which can be characterized by a frequency, ν, and a wavelength, λ, such that $c = \lambda\nu$. Light is an example of a travelling wave. Other important wave phenomena include standing waves, periodic oscillations, and vibrations. Standing waves exhibit quantization, since their wavelengths are limited to discrete integer multiples of some characteristic lengths. Electromagnetic radiation that passes through two closely spaced narrow slits having dimensions roughly similar to the wavelength will show an interference pattern that is a result of constructive and destructive interference of the waves. Electromagnetic radiation also demonstrates properties of particles called photons. The energy of a photon is related to the frequency (or alternatively, the wavelength) of the radiation as $E = h\nu$ (or $E = \frac{hc}{\lambda}$), where h is Planck's constant. That light demonstrates both wavelike and particle-like behavior is known as wave-particle duality. All forms of electromagnetic radiation share these properties, although various forms including X-rays, visible light, microwaves, and radio waves interact differently with matter and have very different practical applications. Electromagnetic radiation can be generated by exciting matter to higher energies, such as by heating it. The emitted light can be either continuous (incandescent sources like the sun) or discrete (from specific types of excited atoms). Continuous spectra often have distributions that can be approximated as blackbody radiation at some appropriate temperature. The line spectrum of hydrogen can be obtained by passing the light from an electrified tube of hydrogen gas through a prism. This line spectrum was simple enough that an empirical formula called the Rydberg formula could be derived from the spectrum. Three historically important paradoxes from the late 19th and early 20th centuries that could not be explained within the existing framework of classical mechanics and classical electromagnetism were the blackbody problem, the photoelectric effect, and the discrete spectra of atoms. The resolution of these paradoxes ultimately led to quantum theories that superseded the classical theories.

6.2 The Bohr Model

Bohr incorporated Planck's and Einstein's quantization ideas into a model of the hydrogen atom that resolved the paradox of atom stability and discrete spectra. The Bohr model of the hydrogen atom explains the connection between the quantization of photons and the quantized emission from atoms. Bohr described the hydrogen atom in terms of an electron moving in a circular orbit about a nucleus. He postulated that the electron was restricted to certain orbits characterized by discrete energies. Transitions between these allowed orbits result in the absorption or emission of photons. When an electron moves from a higher-energy orbit to a more stable one, energy is emitted in the form of a photon. To move an electron from a stable orbit to a more excited one, a photon of energy must be absorbed. Using the Bohr model, we can calculate the energy of an electron and the radius of its orbit in any one-electron system.

6.3 Development of Quantum Theory

Macroscopic objects act as particles. Microscopic objects (such as electrons) have properties of both a particle and a wave. Their exact trajectories cannot be determined. The quantum mechanical model of atoms describes the three-dimensional position of the electron in a *probabilistic* manner according to a mathematical function called a wavefunction, often denoted as ψ. Atomic wavefunctions are also called orbitals. The squared magnitude of the wavefunction describes the distribution of the probability of finding the electron in a particular region in space. Therefore, atomic orbitals describe the areas in an atom where electrons are most likely to be found.

An atomic orbital is characterized by three quantum numbers. The principal quantum number, n, can be any positive integer. The general region for value of energy of the orbital and the average distance of an electron from the nucleus are related to n. Orbitals having the same value of n are said to be in the same shell. The secondary (angular momentum) quantum number, l, can have any integer value from 0 to $n - 1$. This quantum number describes the shape or type of the orbital. Orbitals with the same principal quantum number and the same l value belong to the same subshell. The magnetic quantum number, m_l, with $2l + 1$ values ranging from $-l$ to $+l$, describes the orientation of the orbital in space. In addition, each electron has a spin quantum number, m_s, that can be equal to $\pm \frac{1}{2}$. No two electrons in the same atom can have the same set of values for all the four quantum numbers.

6.4 Electronic Structure of Atoms (Electron Configurations)

The relative energy of the subshells determine the order in which atomic orbitals are filled ($1s$, $2s$, $2p$, $3s$, $3p$, $4s$, $3d$, $4p$, and so on). Electron configurations and orbital diagrams can be determined by applying the Pauli exclusion principle (no two electrons can have the same set of four quantum numbers) and Hund's rule (whenever possible, electrons retain unpaired spins in degenerate orbitals).

Electrons in the outermost orbitals, called valence electrons, are responsible for most of the chemical behavior of elements. In the periodic table, elements with analogous valence electron configurations usually occur within the same group. There are some exceptions to the predicted filling order, particularly when half-filled or completely filled orbitals can be formed. The periodic table can be divided into three categories based on the orbital in which the last electron to be added is placed: main group elements (s and p orbitals), transition elements (d orbitals), and inner transition elements (f orbitals).

6.5 Periodic Variations in Element Properties

Electron configurations allow us to understand many periodic trends. Covalent radius increases as we move down a group because the n level (orbital size) increases. Covalent radius mostly decreases as we move left to right across a period because the effective nuclear charge experienced by the electrons increases, and the electrons are pulled in tighter to the nucleus. Anionic radii are larger than the parent atom, while cationic radii are smaller, because the number of valence electrons has changed while the nuclear charge has remained constant. Ionization energy (the energy associated with forming a cation) decreases down a group and mostly increases across a period because it is easier to remove an electron from a larger, higher energy orbital. Electron affinity (the energy associated with forming an anion) is more favorable (exothermic) when electrons are placed into lower energy orbitals, closer to the nucleus. Therefore, electron affinity becomes increasingly negative as we move left to right across the periodic table and decreases as we move down a group. For both IE and electron affinity data, there are exceptions to the trends when dealing with completely filled or half-filled subshells.

Exercises

6.1 Electromagnetic Energy

1. The light produced by a red neon sign is due to the emission of light by excited neon atoms. Qualitatively describe the spectrum produced by passing light from a neon lamp through a prism.
2. An FM radio station found at 103.1 on the FM dial broadcasts at a frequency of 1.031×10^8 s^{-1} (103.1 MHz). What is the wavelength of these radio waves in meters?
3. FM-95, an FM radio station, broadcasts at a frequency of 9.51×10^7 s^{-1} (95.1 MHz). What is the wavelength of these radio waves in meters?
4. A bright violet line occurs at 435.8 nm in the emission spectrum of mercury vapor. What amount of energy, in joules, must be released by an electron in a mercury atom to produce a photon of this light?
5. Light with a wavelength of 614.5 nm looks orange. What is the energy, in joules, per photon of this orange light? What is the energy in eV (1 eV = 1.602×10^{-19} J)?
6. Heated lithium atoms emit photons of light with an energy of 2.961×10^{-19} J. Calculate the frequency and wavelength of one of these photons. What is the total energy in 1 mole of these photons? What is the color of the emitted light?
7. A photon of light produced by a surgical laser has an energy of 3.027×10^{-19} J. Calculate the frequency and wavelength of the photon. What is the total energy in 1 mole of photons? What is the color of the emitted light?
8. When rubidium ions are heated to a high temperature, two lines are observed in its line spectrum at wavelengths (a) 7.9×10^{-7} m and (b) 4.2×10^{-7} m. What are the frequencies of the two lines? What color do we see when we heat a rubidium compound?
9. The emission spectrum of cesium contains two lines whose frequencies are (a) 3.45×10^{14} Hz and (b) 6.53 $\times 10^{14}$ Hz. What are the wavelengths and energies per photon of the two lines? What color are the lines?
10. Photons of infrared radiation are responsible for much of the warmth we feel when holding our hands before a fire. These photons will also warm other objects. How many infrared photons with a wavelength of 1.5×10^{-6} m must be absorbed by the water to warm a cup of water (175 g) from 25.0 °C to 40 °C?
11. One of the radiographic devices used in a dentist's office emits an X-ray of wavelength 2.090×10^{-11} m. What is the energy, in joules, and frequency of this X-ray?
12. The eyes of certain reptiles pass a single visual signal to the brain when the visual receptors are struck by photons of a wavelength of 850 nm. If a total energy of 3.15×10^{-14} J is required to trip the signal, what is the minimum number of photons that must strike the receptor?
13. RGB color television and computer displays use cathode ray tubes that produce colors by mixing red, green, and blue light. If we look at the screen with a magnifying glass, we can see individual dots turn on and off as the colors change. Using a spectrum of visible light, determine the approximate wavelength of each of these colors. What is the frequency and energy of a photon of each of these colors?

14. Answer the following questions about a Blu-ray laser:
(a) The laser on a Blu-ray player has a wavelength of 405 nm. In what region of the electromagnetic spectrum is this radiation? What is its frequency?
(b) A Blu-ray laser has a power of 5 milliwatts (1 watt = 1 J s^{-1}). How many photons of light are produced by the laser in 1 hour?
(c) The ideal resolution of a player using a laser (such as a Blu-ray player), which determines how close together data can be stored on a compact disk, is determined using the following formula: Resolution = $0.60(\lambda/NA)$, where λ is the wavelength of the laser and NA is the numerical aperture. Numerical aperture is a measure of the size of the spot of light on the disk; the larger the NA, the smaller the spot. In a typical Blu-ray system, NA = 0.95. If the 405-nm laser is used in a Blu-ray player, what is the closest that information can be stored on a Blu-ray disk?
(d) The data density of a Blu-ray disk using a 405-nm laser is 1.5×10^7 bits mm^{-2}. Disks have an outside diameter of 120 mm and a hole of 15-mm diameter. How many data bits can be contained on the disk? If a Blu-ray disk can hold 9,400,000 pages of text, how many data bits are needed for a typed page? (Hint: Determine the area of the disk that is available to hold data. The area inside a circle is given by $A = \pi r^2$, where the radius r is one-half of the diameter.)
15. What is the threshold frequency for sodium metal if a photon with frequency 6.66×10^{14} s^{-1} ejects an electron with 7.74×10^{-20} J kinetic energy? Will the photoelectric effect be observed if sodium is exposed to orange light?

6.2 The Bohr Model

16. Why is the electron in a Bohr hydrogen atom bound less tightly when it has a quantum number of 3 than when it has a quantum number of 1?
17. What does it mean to say that the energy of the electrons in an atom is quantized?
18. Using the Bohr model, determine the energy, in joules, necessary to ionize a ground-state hydrogen atom. Show your calculations.
19. The electron volt (eV) is a convenient unit of energy for expressing atomic-scale energies. It is the amount of energy that an electron gains when subjected to a potential of 1 volt; 1 eV = 1.602×10^{-19} J. Using the Bohr model, determine the energy, in electron volts, of the photon produced when an electron in a hydrogen atom moves from the orbit with $n = 5$ to the orbit with $n = 2$. Show your calculations.
20. Using the Bohr model, determine the lowest possible energy, in joules, for the electron in the Li^{2+} ion.
21. Using the Bohr model, determine the lowest possible energy for the electron in the He^+ ion.
22. Using the Bohr model, determine the energy of an electron with $n = 6$ in a hydrogen atom.
23. Using the Bohr model, determine the energy of an electron with $n = 8$ in a hydrogen atom.
24. How far from the nucleus in angstroms (1 angstrom = 1×10^{-10} m) is the electron in a hydrogen atom if it has an energy of -8.72×10^{-20} J?
25. What is the radius, in angstroms, of the orbital of an electron with $n = 8$ in a hydrogen atom?
26. Using the Bohr model, determine the energy in joules of the photon produced when an electron in a He^+ ion moves from the orbit with $n = 5$ to the orbit with $n = 2$.
27. Using the Bohr model, determine the energy in joules of the photon produced when an electron in a Li^{2+} ion moves from the orbit with $n = 2$ to the orbit with $n = 1$.
28. Consider a large number of hydrogen atoms with electrons randomly distributed in the $n = 1, 2, 3$, and 4 orbits.
(a) How many different wavelengths of light are emitted by these atoms as the electrons fall into lower-energy orbits?
(b) Calculate the lowest and highest energies of light produced by the transitions described in part (a).
(c) Calculate the frequencies and wavelengths of the light produced by the transitions described in part (b).
29. How are the Bohr model and the Rutherford model of the atom similar? How are they different?

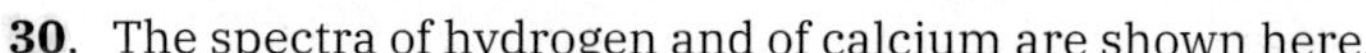

30. The spectra of hydrogen and of calcium are shown here.

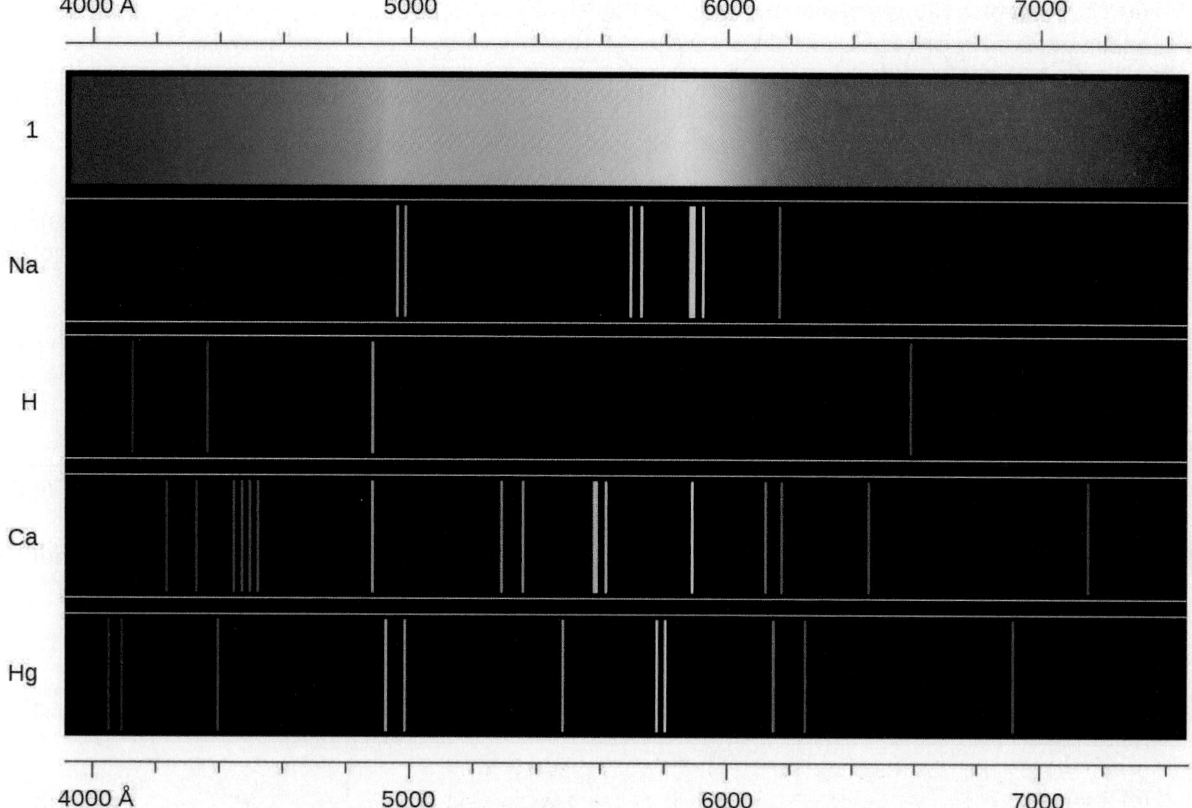

What causes the lines in these spectra? Why are the colors of the lines different? Suggest a reason for the observation that the spectrum of calcium is more complicated than the spectrum of hydrogen.

6.3 Development of Quantum Theory

31. How are the Bohr model and the quantum mechanical model of the hydrogen atom similar? How are they different?

32. What are the allowed values for each of the four quantum numbers: n, l, m_l, and m_s?

33. Describe the properties of an electron associated with each of the following four quantum numbers: n, l, m_l, and m_s.

34. Answer the following questions:
(a) Without using quantum numbers, describe the differences between the shells, subshells, and orbitals of an atom.
(b) How do the quantum numbers of the shells, subshells, and orbitals of an atom differ?

35. Identify the subshell in which electrons with the following quantum numbers are found:
(a) $n = 2, l = 1$
(b) $n = 4, l = 2$
(c) $n = 6, l = 0$

36. Which of the subshells described in the previous question contain degenerate orbitals? How many degenerate orbitals are in each?

37. Identify the subshell in which electrons with the following quantum numbers are found:
(a) $n = 3, l = 2$
(b) $n = 1, l = 0$
(c) $n = 4, l = 3$

38. Which of the subshells described in the previous question contain degenerate orbitals? How many degenerate orbitals are in each?

39. Sketch the boundary surface of a $d_{x^2-y^2}$ and a p_y orbital. Be sure to show and label the axes.

40. Sketch the p_x and d_{xz} orbitals. Be sure to show and label the coordinates.

41. Consider the orbitals shown here in outline.

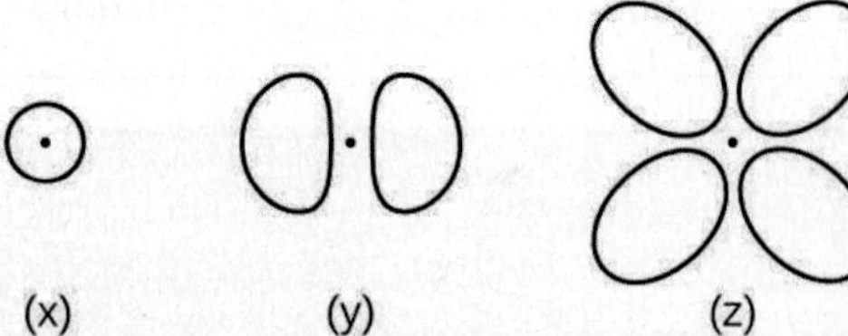

(a) What is the maximum number of electrons contained in an orbital of type (x)? Of type (y)? Of type (z)?
(b) How many orbitals of type (x) are found in a shell with $n = 2$? How many of type (y)? How many of type (z)?
(c) Write a set of quantum numbers for an electron in an orbital of type (x) in a shell with $n = 4$. Of an orbital of type (y) in a shell with $n = 2$. Of an orbital of type (z) in a shell with $n = 3$.
(d) What is the smallest possible n value for an orbital of type (x)? Of type (y)? Of type (z)?
(e) What are the possible l and m_l values for an orbital of type (x)? Of type (y)? Of type (z)?

42. State the Heisenberg uncertainty principle. Describe briefly what the principle implies.
43. How many electrons could be held in the second shell of an atom if the spin quantum number m_s could have three values instead of just two? (Hint: Consider the Pauli exclusion principle.)
44. Which of the following equations describe particle-like behavior? Which describe wavelike behavior? Do any involve both types of behavior? Describe the reasons for your choices.
(a) $c = \lambda\nu$
(b) $E = \frac{mv^2}{2}$
(c) $r = \frac{n^2 a_0}{Z}$
(d) $E = h\nu$
(e) $\lambda = \frac{h}{mv}$
45. Write a set of quantum numbers for each of the electrons with an n of 4 in a Se atom.

6.4 Electronic Structure of Atoms (Electron Configurations)

46. Read the labels of several commercial products and identify monatomic ions of at least four transition elements contained in the products. Write the complete electron configurations of these cations.
47. Read the labels of several commercial products and identify monatomic ions of at least six main group elements contained in the products. Write the complete electron configurations of these cations and anions.
48. Using complete subshell notation (not abbreviations, $1s^2 2s^2 2p^6$, and so forth), predict the electron configuration of each of the following atoms:
(a) C
(b) P
(c) V
(d) Sb
(e) Sm
49. Using complete subshell notation ($1s^2 2s^2 2p^6$, and so forth), predict the electron configuration of each of the following atoms:
(a) N
(b) Si
(c) Fe
(d) Te
(e) Tb
50. Is $1s^2 2s^2 2p^6$ the symbol for a macroscopic property or a microscopic property of an element? Explain your answer.
51. What additional information do we need to answer the question “Which ion has the electron configuration $1s^2 2s^2 2p^6 3s^2 3p^6$”?

52. Draw the orbital diagram for the valence shell of each of the following atoms:
    (a) C
    (b) P
    (c) V
    (d) Sb
    (e) Ru
53. Use an orbital diagram to describe the electron configuration of the valence shell of each of the following atoms:
    (a) N
    (b) Si
    (c) Fe
    (d) Te
    (e) Mo
54. Using complete subshell notation ($1s^2 2s^2 2p^6$, and so forth), predict the electron configurations of the following ions.
    (a) N^{3-}
    (b) Ca^{2+}
    (c) S^-
    (d) Cs^{2+}
    (e) Cr^{2+}
    (f) Gd^{3+}
55. Which atom has the electron configuration $1s^2 2s^2 2p^6 3s^2 3p^6 4s^2 3d^{10} 4p^6 5s^2 4d^2$?
56. Which atom has the electron configuration $1s^2 2s^2 2p^6 3s^2 3p^6 3d^7 4s^2$?
57. Which ion with a +1 charge has the electron configuration $1s^2 2s^2 2p^6 3s^2 3p^6 3d^{10} 4s^2 4p^6$? Which ion with a −2 charge has this configuration?
58. Which of the following atoms contains only three valence electrons: Li, B, N, F, Ne?
59. Which of the following has two unpaired electrons?
    (a) Mg
    (b) Si
    (c) S
    (d) Both Mg and S
    (e) Both Si and S.
60. Which atom would be expected to have a half-filled $6p$ subshell?
61. Which atom would be expected to have a half-filled $4s$ subshell?
62. In one area of Australia, the cattle did not thrive despite the presence of suitable forage. An investigation showed the cause to be the absence of sufficient cobalt in the soil. Cobalt forms cations in two oxidation states, Co^{2+} and Co^{3+}. Write the electron structure of the two cations.
63. Thallium was used as a poison in the Agatha Christie mystery story "The Pale Horse." Thallium has two possible cationic forms, +1 and +3. The +1 compounds are the more stable. Write the electron structure of the +1 cation of thallium.
64. Write the electron configurations for the following atoms or ions:
    (a) B^{3+}
    (b) O^-
    (c) Cl^{3+}
    (d) Ca^{2+}
    (e) Ti
65. Cobalt–60 and iodine–131 are radioactive isotopes commonly used in nuclear medicine. How many protons, neutrons, and electrons are in atoms of these isotopes? Write the complete electron configuration for each isotope.
66. Write a set of quantum numbers for each of the electrons with an n of 3 in a Sc atom.

6.5 Periodic Variations in Element Properties

67. Based on their positions in the periodic table, predict which has the smallest atomic radius: Mg, Sr, Si, Cl, I.

68. Based on their positions in the periodic table, predict which has the largest atomic radius: Li, Rb, N, F, I.

69. Based on their positions in the periodic table, predict which has the largest first ionization energy: Mg, Ba, B, O, Te.

70. Based on their positions in the periodic table, predict which has the smallest first ionization energy: Li, Cs, N, F, I.

71. Based on their positions in the periodic table, rank the following atoms in order of increasing first ionization energy: F, Li, N, Rb

72. Based on their positions in the periodic table, rank the following atoms in order of increasing first ionization energy: Mg, O, S, Si

73. Atoms of which group in the periodic table have a valence shell electron configuration of ns^2np^3?

74. Atoms of which group in the periodic table have a valence shell electron configuration of ns^2?

75. Based on their positions in the periodic table, list the following atoms in order of increasing radius: Mg, Ca, Rb, Cs.

76. Based on their positions in the periodic table, list the following atoms in order of increasing radius: Sr, Ca, Si, Cl.

77. Based on their positions in the periodic table, list the following ions in order of increasing radius: K^+, Ca^{2+}, Al^{3+}, Si^{4+}.

78. List the following ions in order of increasing radius: Li^+, Mg^{2+}, Br^-, Te^{2-}.

79. Which atom and/or ion is (are) isoelectronic with Br^+: Se^{2+}, Se, As^-, Kr, Ga^{3+}, Cl^-?

80. Which of the following atoms and ions is (are) isoelectronic with S^{2+}: Si^{4+}, Cl^{3+}, Ar, As^{3+}, Si, Al^{3+}?

81. Compare both the numbers of protons and electrons present in each to rank the following ions in order of increasing radius: As^{3-}, Br^-, K^+, Mg^{2+}.

82. Of the five elements Al, Cl, I, Na, Rb, which has the most exothermic reaction? (E represents an atom.) What name is given to the energy for the reaction? Hint: Note the process depicted does *not* correspond to electron affinity.)

$$E^+(g) + e^- \longrightarrow E(g)$$

83. Of the five elements Sn, Si, Sb, O, Te, which has the most endothermic reaction? (E represents an atom.) What name is given to the energy for the reaction?

$$E(g) \longrightarrow E^+(g) + e^-$$

84. The ionic radii of the ions S^{2-}, Cl^-, and K^+ are 184, 181, 138 pm respectively. Explain why these ions have different sizes even though they contain the same number of electrons.

85. Which main group atom would be expected to have the lowest second ionization energy?

86. Explain why Al is a member of group 13 rather than group 3?

CHAPTER 7
Chemical Bonding and Molecular Geometry

Figure 7.1 Nicknamed "buckyballs," buckminsterfullerene molecules (C_{60}) contain only carbon atoms (left) arranged to form a geometric framework of hexagons and pentagons, similar to the pattern on a soccer ball (center). This molecular structure is named after architect R. Buckminster Fuller, whose innovative designs combined simple geometric shapes to create large, strong structures such as this weather radar dome near Tucson, Arizona (right). (credit middle: modification of work by "Petey21"/Wikimedia Commons; credit right: modification of work by Bill Morrow)

CHAPTER OUTLINE

7.1 Ionic Bonding
7.2 Covalent Bonding
7.3 Lewis Symbols and Structures
7.4 Formal Charges and Resonance
7.5 Strengths of Ionic and Covalent Bonds
7.6 Molecular Structure and Polarity

INTRODUCTION It has long been known that pure carbon occurs in different forms (allotropes) including graphite and diamonds. But it was not until 1985 that a new form of carbon was recognized: buckminsterfullerene. This molecule was named after the architect and inventor R. Buckminster Fuller (1895–1983), whose signature architectural design was the geodesic dome, characterized by a lattice shell structure supporting a spherical surface. Experimental evidence revealed the formula, C_{60}, and then scientists determined how 60 carbon atoms could form one symmetric, stable molecule. They were guided by bonding theory—the topic of this chapter—which explains how individual atoms connect to form more complex structures.

7.1 Ionic Bonding

LEARNING OBJECTIVES

By the end of this section, you will be able to:

- Explain the formation of cations, anions, and ionic compounds
- Predict the charge of common metallic and nonmetallic elements, and write their electron configurations

As you have learned, ions are atoms or molecules bearing an electrical charge. A cation (a positive ion) forms when a neutral atom loses one or more electrons from its valence shell, and an anion (a negative ion) forms when a neutral atom gains one or more electrons in its valence shell.

Compounds composed of ions are called ionic compounds (or salts), and their constituent ions are held together by **ionic bonds**: electrostatic forces of attraction between oppositely charged cations and anions. The properties of ionic compounds shed some light on the nature of ionic bonds. Ionic solids exhibit a crystalline structure and tend to be rigid and brittle; they also tend to have high melting and boiling points, which suggests that ionic bonds are very strong. Ionic solids are also poor conductors of electricity for the same reason—the strength of ionic bonds prevents ions from moving freely in the solid state. Most ionic solids, however, dissolve readily in water. Once dissolved or melted, ionic compounds are excellent conductors of electricity and heat because the ions can move about freely.

Neutral atoms and their associated ions have very different physical and chemical properties. Sodium *atoms* form sodium metal, a soft, silvery-white metal that burns vigorously in air and reacts explosively with water. Chlorine *atoms* form chlorine gas, Cl_2, a yellow-green gas that is extremely corrosive to most metals and very poisonous to animals and plants. The vigorous reaction between the elements sodium and chlorine forms the white, crystalline compound sodium chloride, common table salt, which contains sodium *cations* and chloride *anions* (Figure 7.2). The compound composed of these ions exhibits properties entirely different from the properties of the elements sodium and chlorine. Chlorine is poisonous, but sodium chloride is essential to life; sodium atoms react vigorously with water, but sodium chloride simply dissolves in water.

FIGURE 7.2 (a) Sodium is a soft metal that must be stored in mineral oil to prevent reaction with air or water. (b) Chlorine is a pale yellow-green gas. (c) When combined, they form white crystals of sodium chloride (table salt). (credit a: modification of work by "Jurii"/Wikimedia Commons)

The Formation of Ionic Compounds

Binary ionic compounds are composed of just two elements: a metal (which forms the cations) and a nonmetal (which forms the anions). For example, NaCl is a binary ionic compound. We can think about the formation of such compounds in terms of the periodic properties of the elements. Many metallic elements have relatively low ionization potentials and lose electrons easily. These elements lie to the left in a period or near the bottom of a group on the periodic table. Nonmetal atoms have relatively high electron affinities and thus readily gain electrons lost by metal atoms, thereby filling their valence shells. Nonmetallic elements are found in the upper-right corner of the periodic table.

As all substances must be electrically neutral, the total number of positive charges on the cations of an ionic compound must equal the total number of negative charges on its anions. The formula of an ionic compound represents the simplest ratio of the numbers of ions necessary to give identical numbers of positive and negative charges. For example, the formula for aluminum oxide, Al_2O_3, indicates that this ionic compound contains two aluminum cations, Al^{3+}, for every three oxide anions, O^{2-} [thus, $(2 \times +3) + (3 \times -2) = 0$].

It is important to note, however, that the formula for an ionic compound does *not* represent the physical arrangement of its ions. It is incorrect to refer to a sodium chloride (NaCl) "molecule" because there is not a single ionic bond, per se, between any specific pair of sodium and chloride ions. The attractive forces between ions are isotropic—the same in all directions—meaning that any particular ion is equally attracted to all of the nearby ions of opposite charge. This results in the ions arranging themselves into a tightly bound, three-dimensional lattice structure. Sodium chloride, for example, consists of a regular arrangement of equal numbers of Na^+ cations and Cl^- anions (Figure 7.3).

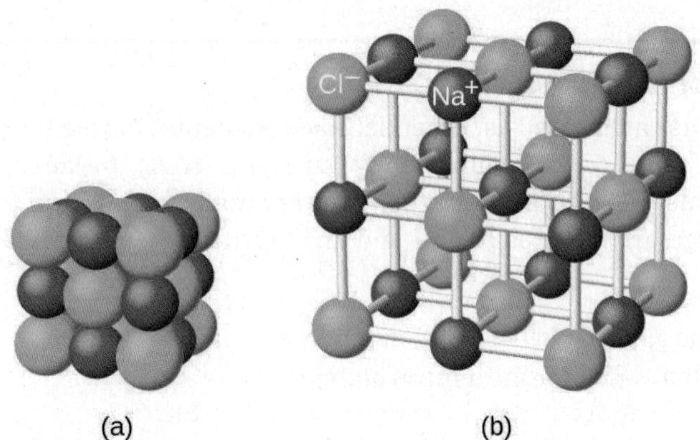

FIGURE 7.3 The atoms in sodium chloride (common table salt) are arranged to (a) maximize opposite charges interacting. The smaller spheres represent sodium ions, the larger ones represent chloride ions. In the expanded view (b), the geometry can be seen more clearly. Note that each ion is "bonded" to all of the surrounding ions—six in this case.

The strong electrostatic attraction between Na^+ and Cl^- ions holds them tightly together in solid NaCl. It requires 769 kJ of energy to dissociate one mole of solid NaCl into separate gaseous Na^+ and Cl^- ions:

$$\text{NaCl}(s) \longrightarrow \text{Na}^+(g) + \text{Cl}^-(g) \qquad \Delta H = 769 \text{ kJ}$$

Electronic Structures of Cations

When forming a cation, an atom of a main group element tends to lose all of its valence electrons, thus assuming the electronic structure of the noble gas that precedes it in the periodic table. For groups 1 (the alkali metals) and 2 (the alkaline earth metals), the group numbers are equal to the numbers of valence shell electrons and, consequently, to the charges of the cations formed from atoms of these elements when all valence shell electrons are removed. For example, calcium is a group 2 element whose neutral atoms have 20 electrons and a ground state electron configuration of $1s^2 2s^2 2p^6 3s^2 3p^6 4s^2$. When a Ca atom loses both of its valence electrons, the result is a cation with 18 electrons, a 2+ charge, and an electron configuration of $1s^2 2s^2 2p^6 3s^2 3p^6$. The Ca^{2+} ion is therefore isoelectronic with the noble gas Ar.

For groups 13–17, the group numbers exceed the number of valence electrons by 10 (accounting for the possibility of full *d* subshells in atoms of elements in the fourth and greater periods). Thus, the charge of a cation formed by the loss of all valence electrons is equal to the group number minus 10. For example, aluminum (in group 13) forms 3+ ions (Al^{3+}).

Exceptions to the expected behavior involve elements toward the bottom of the groups. In addition to the expected ions Tl^{3+}, Sn^{4+}, Pb^{4+}, and Bi^{5+}, a partial loss of these atoms' valence shell electrons can also lead to the formation of Tl^+, Sn^{2+}, Pb^{2+}, and Bi^{3+} ions. The formation of these 1+, 2+, and 3+ cations is ascribed to the **inert pair effect**, which reflects the relatively low energy of the valence *s*-electron pair for atoms of the heavy elements of groups 13, 14, and 15. Mercury (group 12) also exhibits an unexpected behavior: it forms a diatomic ion, ${Hg_2}^{2+}$ (an ion formed from two mercury atoms, with an Hg-Hg bond), in addition to the expected monatomic ion Hg^{2+} (formed from only one mercury atom).

Transition and inner transition metal elements behave differently than main group elements. Most transition metal cations have 2+ or 3+ charges that result from the loss of their outermost *s* electron(s) first, sometimes followed by the loss of one or two *d* electrons from the next-to-outermost shell. For example, iron ($1s^2 2s^2 2p^6 3s^2 3p^6 3d^6 4s^2$) forms the ion Fe^{2+} ($1s^2 2s^2 2p^6 3s^2 3p^6 3d^6$) by the loss of the 4*s* electrons and the ion Fe^{3+} ($1s^2 2s^2 2p^6 3s^2 3p^6 3d^5$) by the loss of the 4*s* electrons and one of the 3*d* electrons. Although the *d* orbitals of the transition elements are—according to the Aufbau principle—the last to fill when building up electron configurations, the outermost *s* electrons are the first to be lost when these atoms ionize. When the inner transition metals form ions, they usually have a 3+ charge, resulting from the loss of their outermost *s* electrons and a *d* or *f* electron.

EXAMPLE 7.1

Determining the Electronic Structures of Cations

There are at least 14 elements categorized as “essential trace elements” for the human body. They are called “essential” because they are required for healthy bodily functions, “trace” because they are required only in small amounts, and “elements” in spite of the fact that they are really ions. Two of these essential trace elements, chromium and zinc, are required as Cr^{3+} and Zn^{2+}. Write the electron configurations of these cations.

Solution

First, write the electron configuration for the neutral atoms:

Zn: $[Ar]3d^{10}4s^2$

Cr: $[Ar]3d^54s^1$

Next, remove electrons from the highest energy orbital. For the transition metals, electrons are removed from the *s* orbital first and then from the *d* orbital. For the *p*-block elements, electrons are removed from the *p* orbitals and then from the *s* orbital. Zinc is a member of group 12, so it should have a charge of 2+, and thus loses only the two electrons in its *s* orbital. Chromium is a transition element and should lose its *s* electrons and then its *d* electrons when forming a cation. Thus, we find the following electron configurations of the ions:

Zn^{2+}: $[Ar]3d^{10}$

Cr^{3+}: $[Ar]3d^3$

Check Your Learning

Potassium and magnesium are required in our diet. Write the electron configurations of the ions expected from these elements.

Answer:

K^+: [Ar], Mg^{2+}: [Ne]

Electronic Structures of Anions

Most monatomic anions form when a neutral nonmetal atom gains enough electrons to completely fill its outer *s* and *p* orbitals, thereby reaching the electron configuration of the next noble gas. Thus, it is simple to determine the charge on such a negative ion: The charge is equal to the number of electrons that must be gained to fill the *s* and *p* orbitals of the parent atom. Oxygen, for example, has the electron configuration $1s^22s^22p^4$, whereas the oxygen anion has the electron configuration of the noble gas neon (Ne), $1s^22s^22p^6$. The two additional electrons required to fill the valence orbitals give the oxide ion the charge of 2– (O^{2-}).

EXAMPLE 7.2

Determining the Electronic Structure of Anions

Selenium and iodine are two essential trace elements that form anions. Write the electron configurations of the anions.

Solution

Se^{2-}: $[Ar]3d^{10}4s^24p^6$

I^-: $[Kr]4d^{10}5s^25p^6$

Check Your Learning

Write the electron configurations of a phosphorus atom and its negative ion. Give the charge on the anion.

Answer:

P: $[Ne]3s^23p^3$; P^{3-}: $[Ne]3s^23p^6$

7.2 Covalent Bonding

LEARNING OBJECTIVES

By the end of this section, you will be able to:

- Describe the formation of covalent bonds
- Define electronegativity and assess the polarity of covalent bonds

Ionic bonding results from the electrostatic attraction of oppositely charged ions that are typically produced by the transfer of electrons between metallic and nonmetallic atoms. A different type of bonding results from the mutual attraction of atoms for a "shared" pair of electrons. Such bonds are called **covalent bonds**. Covalent bonds are formed between two atoms when both have similar tendencies to attract electrons to themselves (i.e., when both atoms have identical or fairly similar ionization energies and electron affinities). For example, two hydrogen atoms bond covalently to form an H_2 molecule; each hydrogen atom in the H_2 molecule has two electrons stabilizing it, giving each atom the same number of valence electrons as the noble gas He.

Compounds that contain covalent bonds exhibit different physical properties than ionic compounds. Because the attraction between molecules, which are electrically neutral, is weaker than that between electrically charged ions, covalent compounds generally have much lower melting and boiling points than ionic compounds. In fact, many covalent compounds are liquids or gases at room temperature, and, in their solid states, they are typically much softer than ionic solids. Furthermore, whereas ionic compounds are good conductors of electricity when dissolved in water, most covalent compounds are insoluble in water; since they are electrically neutral, they are poor conductors of electricity in any state.

Formation of Covalent Bonds

Nonmetal atoms frequently form covalent bonds with other nonmetal atoms. For example, the hydrogen molecule, H_2, contains a covalent bond between its two hydrogen atoms. Figure 7.4 illustrates why this bond is formed. Starting on the far right, we have two separate hydrogen atoms with a particular potential energy, indicated by the red line. Along the x-axis is the distance between the two atoms. As the two atoms approach each other (moving left along the x-axis), their valence orbitals (1s) begin to overlap. The single electrons on each hydrogen atom then interact with both atomic nuclei, occupying the space around both atoms. The strong attraction of each shared electron to both nuclei stabilizes the system, and the potential energy decreases as the bond distance decreases. If the atoms continue to approach each other, the positive charges in the two nuclei begin to repel each other, and the potential energy increases. The **bond length** is determined by the distance at which the lowest potential energy is achieved.

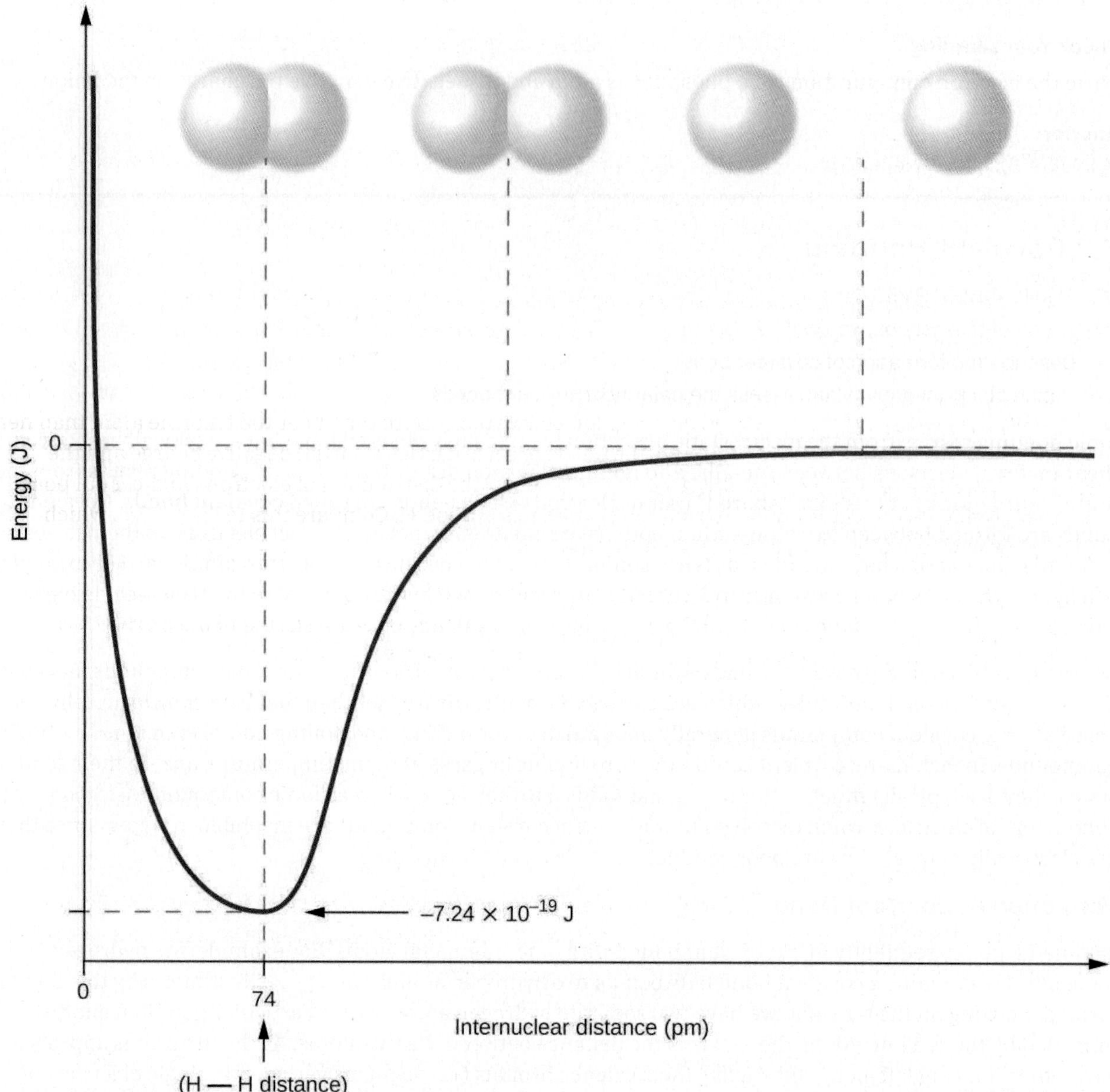

FIGURE 7.4 The potential energy of two separate hydrogen atoms (right) decreases as they approach each other, and the single electrons on each atom are shared to form a covalent bond. The bond length is the internuclear distance at which the lowest potential energy is achieved.

It is essential to remember that energy must be added to break chemical bonds (an endothermic process), whereas forming chemical bonds releases energy (an exothermic process). In the case of H_2, the covalent bond is very strong; a large amount of energy, 436 kJ, must be added to break the bonds in one mole of hydrogen molecules and cause the atoms to separate:

$$H_2(g) \longrightarrow 2H(g) \qquad \Delta H = 436 \text{ kJ}$$

Conversely, the same amount of energy is released when one mole of H_2 molecules forms from two moles of H atoms:

$$2H(g) \longrightarrow H_2(g) \qquad \Delta H = -436 \text{ kJ}$$

Pure vs. Polar Covalent Bonds

If the atoms that form a covalent bond are identical, as in H_2, Cl_2, and other diatomic molecules, then the electrons in the bond must be shared equally. We refer to this as a **pure covalent bond**. Electrons shared in pure covalent bonds have an equal probability of being near each nucleus.

In the case of Cl_2, each atom starts off with seven valence electrons, and each Cl shares one electron with the other, forming one covalent bond:

$$\text{Cl} + \text{Cl} \longrightarrow \text{Cl}_2$$

The total number of electrons around each individual atom consists of six nonbonding electrons and two shared (i.e., bonding) electrons for eight total electrons, matching the number of valence electrons in the noble gas argon. Since the bonding atoms are identical, Cl_2 also features a pure covalent bond.

When the atoms linked by a covalent bond are different, the bonding electrons are shared, but no longer equally. Instead, the bonding electrons are more attracted to one atom than the other, giving rise to a shift of electron density toward that atom. This unequal distribution of electrons is known as a **polar covalent bond**, characterized by a partial positive charge on one atom and a partial negative charge on the other. The atom that attracts the electrons more strongly acquires the partial negative charge and vice versa. For example, the electrons in the H–Cl bond of a hydrogen chloride molecule spend more time near the chlorine atom than near the hydrogen atom. Thus, in an HCl molecule, the chlorine atom carries a partial negative charge and the hydrogen atom has a partial positive charge. Figure 7.5 shows the distribution of electrons in the H–Cl bond. Note that the shaded area around Cl is much larger than it is around H. Compare this to Figure 7.4, which shows the even distribution of electrons in the H_2 nonpolar bond.

We sometimes designate the positive and negative atoms in a polar covalent bond using a lowercase Greek letter "delta," δ, with a plus sign or minus sign to indicate whether the atom has a partial positive charge (δ+) or a partial negative charge (δ–). This symbolism is shown for the H–Cl molecule in Figure 7.5.

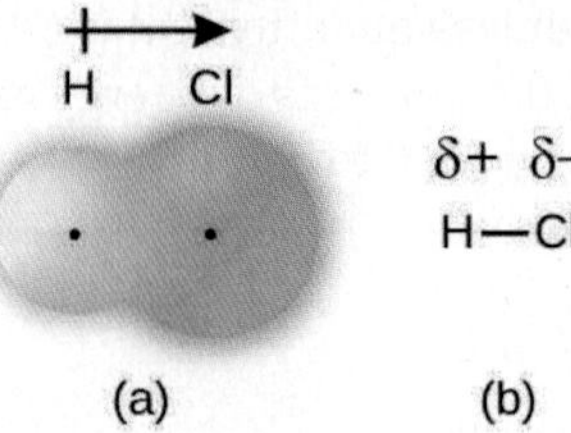

FIGURE 7.5 (a) The distribution of electron density in the HCl molecule is uneven. The electron density is greater around the chlorine nucleus. The small, black dots indicate the location of the hydrogen and chlorine nuclei in the molecule. (b) Symbols δ+ and δ– indicate the polarity of the H–Cl bond.

Electronegativity

Whether a bond is nonpolar or polar covalent is determined by a property of the bonding atoms called **electronegativity**. Electronegativity is a measure of the tendency of an atom to attract electrons (or electron density) towards itself. It determines how the shared electrons are distributed between the two atoms in a bond. The more strongly an atom attracts the electrons in its bonds, the larger its electronegativity. Electrons in a polar covalent bond are shifted toward the more electronegative atom; thus, the more electronegative atom is the one with the partial negative charge. The greater the difference in electronegativity, the more polarized the electron distribution and the larger the partial charges of the atoms.

Figure 7.6 shows the electronegativity values of the elements as proposed by one of the most famous chemists of the twentieth century: Linus Pauling (Figure 7.7). In general, electronegativity increases from left to right across a period in the periodic table and decreases down a group. Thus, the nonmetals, which lie in the upper right, tend to have the highest electronegativities, with fluorine the most electronegative element of all (EN = 4.0). Metals tend to be less electronegative elements, and the group 1 metals have the lowest electronegativities. Note that noble gases are excluded from this figure because these atoms usually do not share electrons with others atoms since they have a full valence shell. (While noble gas compounds such as XeO_2 do exist, they can only be formed under extreme conditions, and thus they do not fit neatly into the general model of electronegativity.)

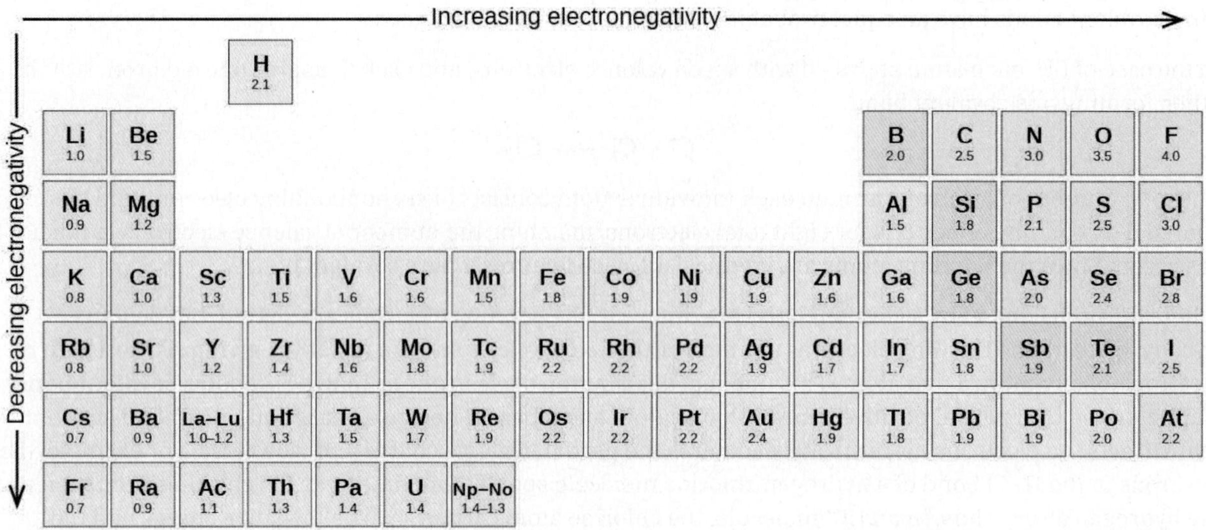

FIGURE 7.6 The electronegativity values derived by Pauling follow predictable periodic trends, with the higher electronegativities toward the upper right of the periodic table.

Electronegativity versus Electron Affinity

We must be careful not to confuse electronegativity and electron affinity. The electron affinity of an element is a measurable physical quantity, namely, the energy released or absorbed when an isolated gas-phase atom acquires an electron, measured in kJ/mol. Electronegativity, on the other hand, describes how tightly an atom attracts electrons in a bond. It is a dimensionless quantity that is calculated, not measured. Pauling derived the first electronegativity values by comparing the amounts of energy required to break different types of bonds. He chose an arbitrary relative scale ranging from 0 to 4.

Portrait of a Chemist

Linus Pauling

Linus Pauling, shown in Figure 7.7, is the only person to have received two unshared (individual) Nobel Prizes: one for chemistry in 1954 for his work on the nature of chemical bonds and one for peace in 1962 for his opposition to weapons of mass destruction. He developed many of the theories and concepts that are foundational to our current understanding of chemistry, including electronegativity and resonance structures.

FIGURE 7.7 Linus Pauling (1901–1994) made many important contributions to the field of chemistry. He was also a prominent activist, publicizing issues related to health and nuclear weapons.

Pauling also contributed to many other fields besides chemistry. His research on sickle cell anemia revealed the cause of the disease—the presence of a genetically inherited abnormal protein in the blood—and paved the way for the field of molecular genetics. His work was also pivotal in curbing the testing of nuclear weapons; he proved that radioactive fallout from nuclear testing posed a public health risk.

Electronegativity and Bond Type

The absolute value of the difference in electronegativity (ΔEN) of two bonded atoms provides a rough measure of the polarity to be expected in the bond and, thus, the bond type. When the difference is very small or zero, the bond is covalent and nonpolar. When it is large, the bond is polar covalent or ionic. The absolute values of the electronegativity differences between the atoms in the bonds H–H, H–Cl, and Na–Cl are 0 (nonpolar), 0.9 (polar covalent), and 2.1 (ionic), respectively. The degree to which electrons are shared between atoms varies from completely equal (pure covalent bonding) to not at all (ionic bonding). Figure 7.8 shows the relationship between electronegativity difference and bond type.

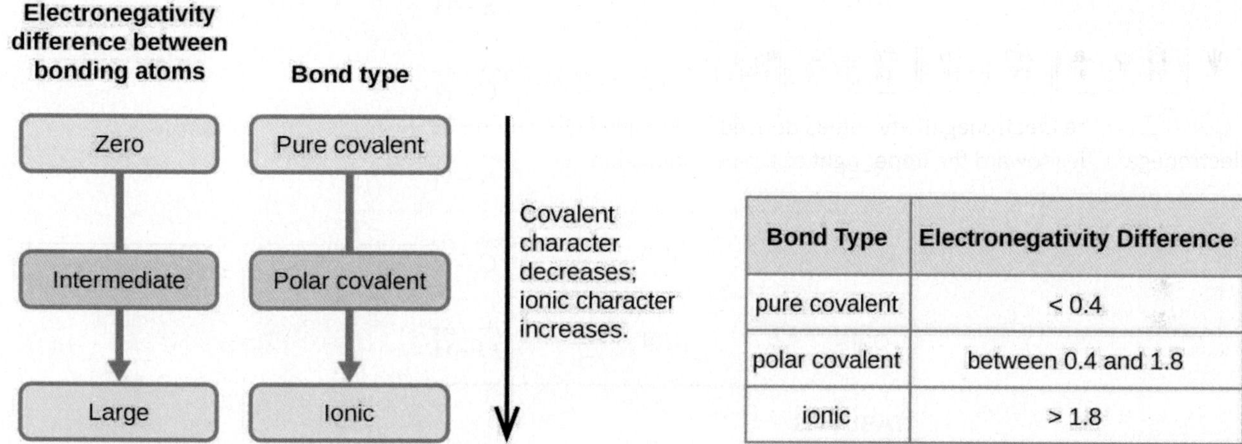

Bond Type	Electronegativity Difference
pure covalent	< 0.4
polar covalent	between 0.4 and 1.8
ionic	> 1.8

FIGURE 7.8 As the electronegativity difference increases between two atoms, the bond becomes more ionic.

A rough approximation of the electronegativity differences associated with covalent, polar covalent, and ionic bonds is shown in Figure 7.8. This table is just a general guide, however, with many exceptions. For example, the H and F atoms in HF have an electronegativity difference of 1.9, and the N and H atoms in NH_3 a difference of 0.9, yet both of these compounds form bonds that are considered polar covalent. Likewise, the Na and Cl atoms in NaCl have an electronegativity difference of 2.1, and the Mn and I atoms in MnI_2 have a difference of 1.0, yet both of these substances form ionic compounds.

The best guide to the covalent or ionic character of a bond is to consider the types of atoms involved and their relative positions in the periodic table. Bonds between two nonmetals are generally covalent; bonding between a metal and a nonmetal is often ionic.

Some compounds contain both covalent and ionic bonds. The atoms in polyatomic ions, such as OH^-, $NO_3{}^-$, and $NH_4{}^+$, are held together by polar covalent bonds. However, these polyatomic ions form ionic compounds by combining with ions of opposite charge. For example, potassium nitrate, KNO_3, contains the K^+ cation and the polyatomic $NO_3{}^-$ anion. Thus, bonding in potassium nitrate is ionic, resulting from the electrostatic attraction between the ions K^+ and $NO_3{}^-$, as well as covalent between the nitrogen and oxygen atoms in $NO_3{}^-$.

EXAMPLE 7.3

Electronegativity and Bond Polarity

Bond polarities play an important role in determining the structure of proteins. Using the electronegativity values in Figure 7.6, arrange the following covalent bonds—all commonly found in amino acids—in order of increasing polarity. Then designate the positive and negative atoms using the symbols δ+ and δ–:

C–H, C–N, C–O, N–H, O–H, S–H

Solution

The polarity of these bonds increases as the absolute value of the electronegativity difference increases. The atom with the δ– designation is the more electronegative of the two. Table 7.1 shows these bonds in order of increasing polarity.

Bond Polarity and Electronegativity Difference

Bond	ΔEN	Polarity
C–H	0.4	$\overset{\delta-}{C}-\overset{\delta+}{H}$
S–H	0.4	$\overset{\delta-}{S}-\overset{\delta+}{H}$
C–N	0.5	$\overset{\delta+}{C}-\overset{\delta-}{N}$
N–H	0.9	$\overset{\delta-}{N}-\overset{\delta+}{H}$
C–O	1.0	$\overset{\delta+}{C}-\overset{\delta-}{O}$
O–H	1.4	$\overset{\delta-}{O}-\overset{\delta+}{H}$

TABLE 7.1

Check Your Learning

Silicones are polymeric compounds containing, among others, the following types of covalent bonds: Si–O, Si–C, C–H, and C–C. Using the electronegativity values in Figure 7.6, arrange the bonds in order of increasing polarity and designate the positive and negative atoms using the symbols δ+ and δ–.

Answer:

Bond	Electronegativity Difference	Polarity
C–C	0.0	nonpolar
C–H	0.4	$\overset{\delta-}{C}-\overset{\delta+}{H}$
Si–C	0.7	$\overset{\delta+}{Si}-\overset{\delta-}{C}$
Si–O	1.7	$\overset{\delta+}{Si}-\overset{\delta-}{O}$

7.3 Lewis Symbols and Structures

LEARNING OBJECTIVES

By the end of this section, you will be able to:

- Write Lewis symbols for neutral atoms and ions
- Draw Lewis structures depicting the bonding in simple molecules

Thus far in this chapter, we have discussed the various types of bonds that form between atoms and/or ions. In all cases, these bonds involve the sharing or transfer of valence shell electrons between atoms. In this section, we will explore the typical method for depicting valence shell electrons and chemical bonds, namely Lewis symbols and Lewis structures.

Lewis Symbols

We use Lewis symbols to describe valence electron configurations of atoms and monatomic ions. A **Lewis symbol** consists of an elemental symbol surrounded by one dot for each of its valence electrons:

•Ca•

Figure 7.9 shows the Lewis symbols for the elements of the third period of the periodic table.

Atoms	Electronic Configuration	Lewis Symbol
sodium	[Ne]$3s^1$	Na•
magnesium	[Ne]$3s^2$	•Mg•
aluminum	[Ne]$3s^23p^1$	•Al•
silicon	[Ne]$3s^23p^2$	•Si•
phosphorus	[Ne]$3s^23p^3$	•P•
sulfur	[Ne]$3s^23p^4$	:S•
chlorine	[Ne]$3s^23p^5$	:Cl•
argon	[Ne]$3s^23p^6$	:Ar:

FIGURE 7.9 Lewis symbols illustrating the number of valence electrons for each element in the third period of the periodic table.

Lewis symbols can also be used to illustrate the formation of cations from atoms, as shown here for sodium and calcium:

Na• ⟶ Na^+ + e^- (sodium atom ⟶ sodium cation)

•Ca• ⟶ Ca^{2+} + $2e^-$ (calcium atom ⟶ calcium cation)

Likewise, they can be used to show the formation of anions from atoms, as shown here for chlorine and sulfur:

:Cl• + e^- ⟶ :Cl:$^-$ (chlorine atom ⟶ chlorine anion)

:S• + $2e^-$ ⟶ :S:$^{2-}$ (sulfur atom ⟶ sulfide anion)

Figure 7.10 demonstrates the use of Lewis symbols to show the transfer of electrons during the formation of ionic compounds.

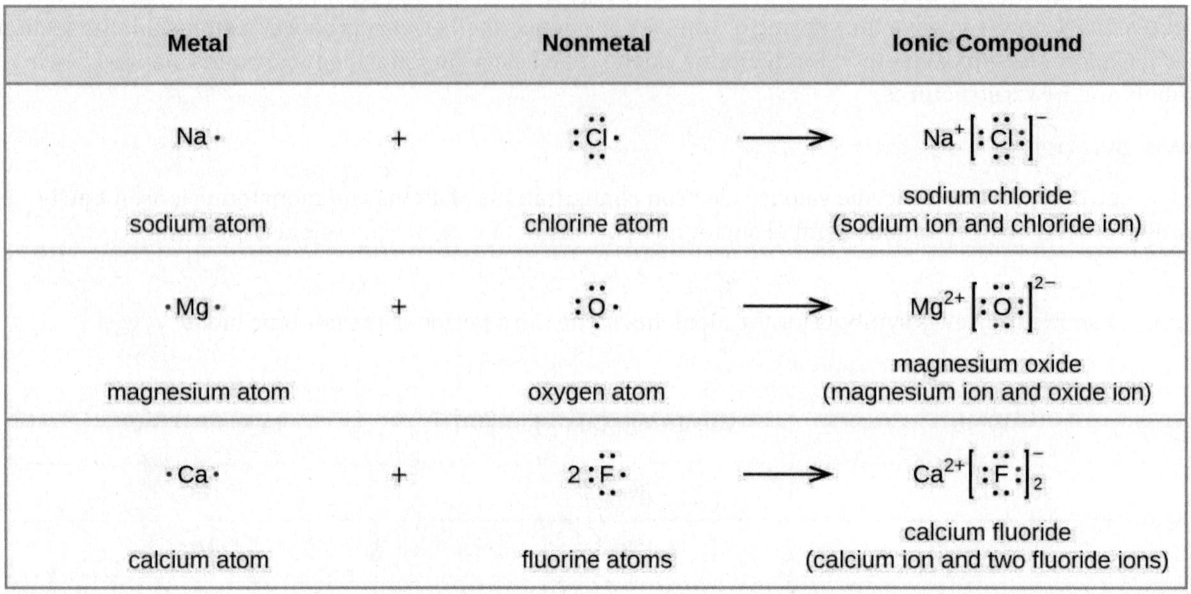

Metal		Nonmetal		Ionic Compound
Na· sodium atom	+	:Cl· chlorine atom	⟶	$Na^+[:Cl:]^-$ sodium chloride (sodium ion and chloride ion)
·Mg· magnesium atom	+	:O· oxygen atom	⟶	$Mg^{2+}[:O:]^{2-}$ magnesium oxide (magnesium ion and oxide ion)
·Ca· calcium atom	+	2 :F· fluorine atoms	⟶	$Ca^{2+}[:F:]^-_2$ calcium fluoride (calcium ion and two fluoride ions)

FIGURE 7.10 Cations are formed when atoms lose electrons, represented by fewer Lewis dots, whereas anions are formed by atoms gaining electrons. The total number of electrons does not change.

Lewis Structures

We also use Lewis symbols to indicate the formation of covalent bonds, which are shown in **Lewis structures**, drawings that describe the bonding in molecules and polyatomic ions. For example, when two chlorine atoms form a chlorine molecule, they share one pair of electrons:

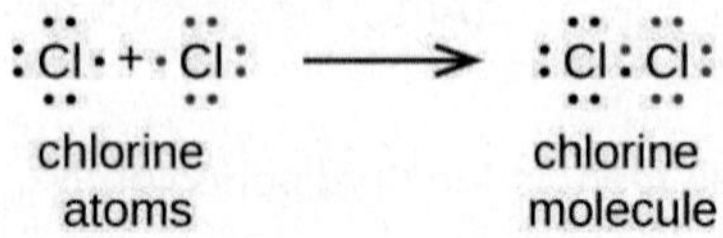

:Cl· + ·Cl: ⟶ :Cl:Cl:

chlorine atoms — chlorine molecule

The Lewis structure indicates that each Cl atom has three pairs of electrons that are not used in bonding (called **lone pairs**) and one shared pair of electrons (written between the atoms). A dash (or line) is sometimes used to indicate a shared pair of electrons:

H—H   :Cl—Cl:

A single shared pair of electrons is called a **single bond**. Each Cl atom interacts with eight valence electrons: the six in the lone pairs and the two in the single bond.

The Octet Rule

The other halogen molecules (F_2, Br_2, I_2, and At_2) form bonds like those in the chlorine molecule: one single bond between atoms and three lone pairs of electrons per atom. This allows each halogen atom to have a noble gas electron configuration. The tendency of main group atoms to form enough bonds to obtain eight valence electrons is known as the **octet rule**.

The number of bonds that an atom can form can often be predicted from the number of electrons needed to reach an octet (eight valence electrons); this is especially true of the nonmetals of the second period of the periodic table (C, N, O, and F). For example, each atom of a group 14 element has four electrons in its outermost shell and therefore requires four more electrons to reach an octet. These four electrons can be gained by forming four covalent bonds, as illustrated here for carbon in CCl_4 (carbon tetrachloride) and silicon in SiH_4 (silane). Because hydrogen only needs two electrons to fill its valence shell, it is an exception to the octet rule. The transition elements and inner transition elements also do not follow the octet rule:

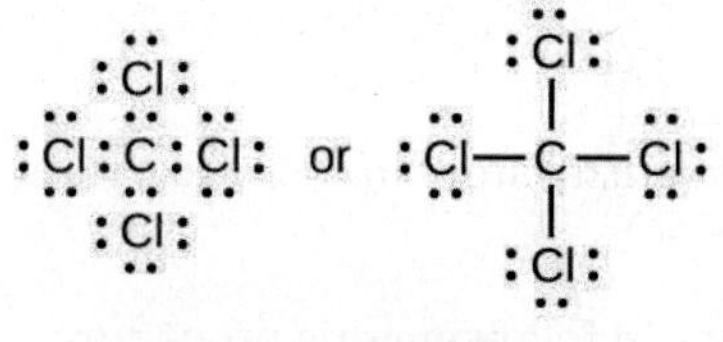

carbon tetrachloride 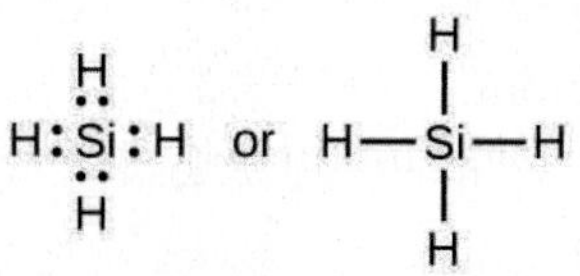   silane

Group 15 elements such as nitrogen have five valence electrons in the atomic Lewis symbol: one lone pair and three unpaired electrons. To obtain an octet, these atoms form three covalent bonds, as in NH_3 (ammonia). Oxygen and other atoms in group 16 obtain an octet by forming two covalent bonds:

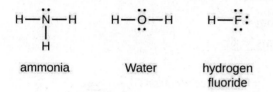

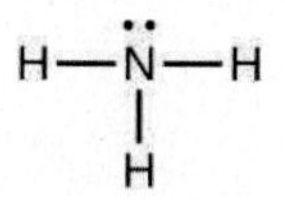

ammonia 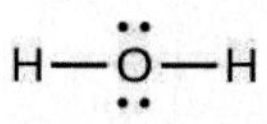   Water 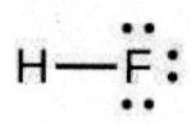   hydrogen fluoride

Double and Triple Bonds

As previously mentioned, when a pair of atoms shares one pair of electrons, we call this a single bond. However, a pair of atoms may need to share more than one pair of electrons in order to achieve the requisite octet. A **double bond** forms when two pairs of electrons are shared between a pair of atoms, as between the carbon and oxygen atoms in CH_2O (formaldehyde) and between the two carbon atoms in C_2H_4 (ethylene):

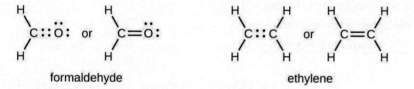

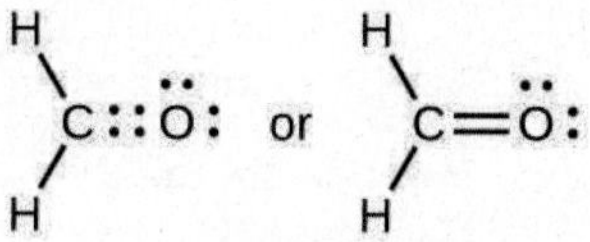

formaldehyde 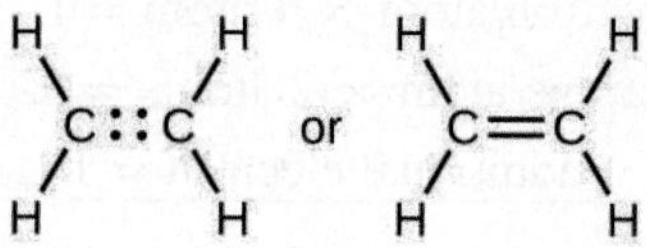   ethylene

A **triple bond** forms when three electron pairs are shared by a pair of atoms, as in carbon monoxide (CO) and the cyanide ion (CN^-):

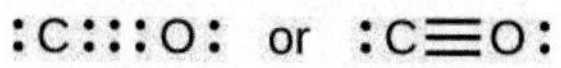

:C:::O: or :C≡O: :C:::N:$^-$ or :C≡N:$^-$

carbon monoxide 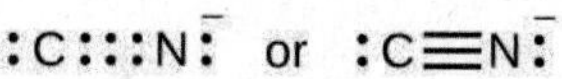   cyanide ion

Writing Lewis Structures with the Octet Rule

For very simple molecules and molecular ions, we can write the Lewis structures by merely pairing up the unpaired electrons on the constituent atoms. See these examples:

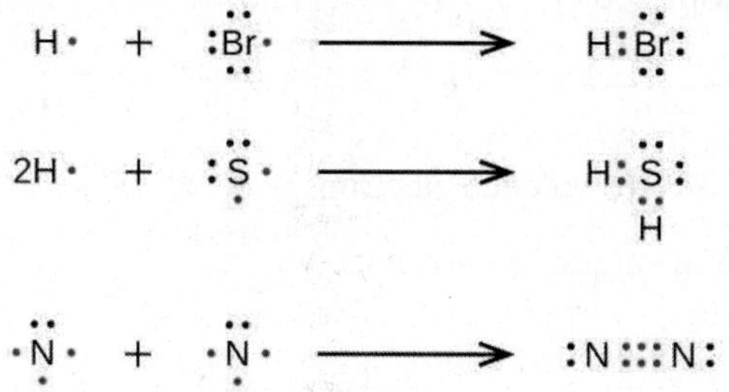

H· + :Br· ⟶ H:Br:

2H· + :S· ⟶ H:S:H

·N· + ·N· ⟶ :N:::N:

For more complicated molecules and molecular ions, it is helpful to follow the step-by-step procedure outlined here:

1. Determine the total number of valence (outer shell) electrons. For cations, subtract one electron for each positive charge. For anions, add one electron for each negative charge.
2. Draw a skeleton structure of the molecule or ion, arranging the atoms around a central atom. (Generally, the least electronegative element should be placed in the center.) Connect each atom to the central atom with a single bond (one electron pair).
3. Distribute the remaining electrons as lone pairs on the terminal atoms (except hydrogen), completing an

octet around each atom.

4. Place all remaining electrons on the central atom.
5. Rearrange the electrons of the outer atoms to make multiple bonds with the central atom in order to obtain octets wherever possible.

Let us determine the Lewis structures of SiH_4, CHO_2^-, NO^+, and OF_2 as examples in following this procedure:

1. Determine the total number of valence (outer shell) electrons in the molecule or ion.
   - For a molecule, we add the number of valence electrons on each atom in the molecule:

$$\begin{aligned} &SiH_4 \\ &\text{Si: 4 valence electrons/atom} \times 1 \text{ atom} = 4 \\ +&\underline{\text{H: 1 valence electron/atom} \times 4 \text{ atoms} = 4} \\ &= 8 \text{ valence electrons} \end{aligned}$$

   - For a *negative ion,* such as CHO_2^-, we add the number of valence electrons on the atoms to the number of negative charges on the ion (one electron is gained for each single negative charge):

$$\begin{aligned} &CHO_2^- \\ &\text{C: 4 valence electrons/atom} \times 1 \text{ atom} = 4 \\ &\text{H: 1 valence electron/atom} \times 1 \text{ atom} = 1 \\ &\text{O: 6 valence electrons/atom} \times 2 \text{ atoms} = 12 \\ +&\underline{\qquad\qquad 1 \text{ additional electron} = 1} \\ &= 18 \text{ valence electrons} \end{aligned}$$

   - For a *positive ion,* such as NO^+, we add the number of valence electrons on the atoms in the ion and then subtract the number of positive charges on the ion (one electron is lost for each single positive charge) from the total number of valence electrons:

$$\begin{aligned} &NO^+ \\ &\text{N: 5 valence electrons/atom} \times 1 \text{ atom} = 5 \\ &\text{O: 6 valence electron/atom} \times 1 \text{ atom} = 6 \\ +&\underline{-1 \text{ electron (positive charge)} = -1} \\ &= 10 \text{ valence electrons} \end{aligned}$$

   - Since OF_2 is a neutral molecule, we simply add the number of valence electrons:

$$\begin{aligned} &OF_2 \\ &\text{O: 6 valence electrons/atom} \times 1 \text{ atom} = 6 \\ +&\underline{\text{F: 7 valence electrons/atom} \times 2 \text{ atoms} = 14} \\ &= 20 \text{ valence electrons} \end{aligned}$$

2. Draw a skeleton structure of the molecule or ion, arranging the atoms around a central atom and connecting each atom to the central atom with a single (one electron pair) bond. (Note that we denote ions with brackets around the structure, indicating the charge outside the brackets:)

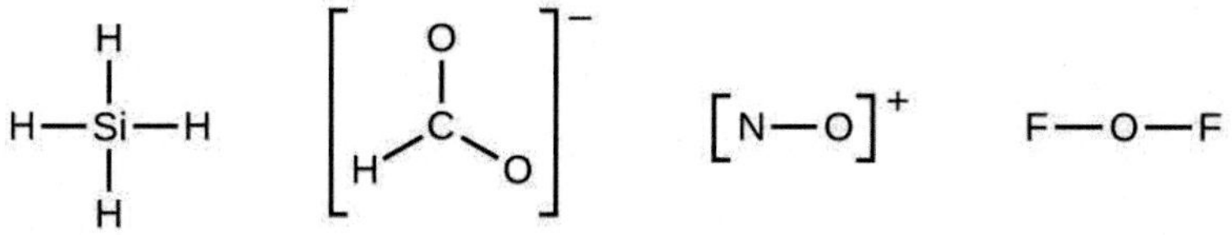

When several arrangements of atoms are possible, as for CHO_2^-, we must use experimental evidence to choose the correct one. In general, the less electronegative elements are more likely to be central atoms. In CHO_2^-, the less electronegative carbon atom occupies the central position with the oxygen and hydrogen atoms surrounding it. Other examples include P in $POCl_3$, S in SO_2, and Cl in ClO_4^-. An exception is that hydrogen is almost never a central atom. As the most electronegative element, fluorine also cannot be a central atom.

3. Distribute the remaining electrons as lone pairs on the terminal atoms (except hydrogen) to complete their valence shells with an octet of electrons.
   - There are no remaining electrons on SiH_4, so it is unchanged:

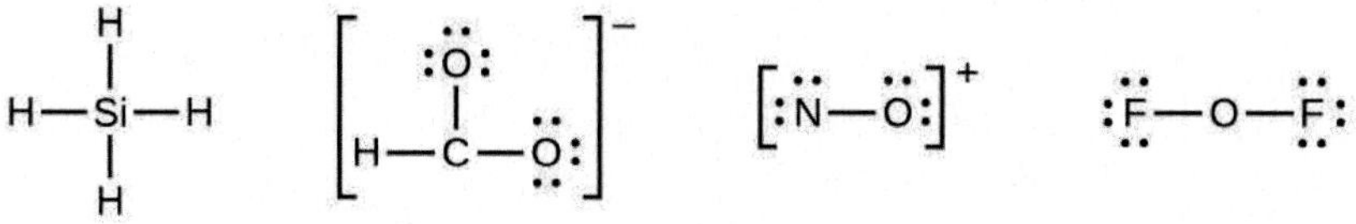

4. Place all remaining electrons on the central atom.
   - For SiH_4, CHO_2^-, and NO^+, there are no remaining electrons; we already placed all of the electrons determined in Step 1.
   - For OF_2, we had 16 electrons remaining in Step 3, and we placed 12, leaving 4 to be placed on the central atom:

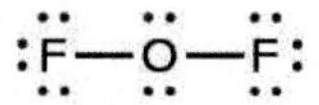

5. Rearrange the electrons of the outer atoms to make multiple bonds with the central atom in order to obtain octets wherever possible.
   - SiH_4: Si already has an octet, so nothing needs to be done.
   - CHO_2^- : We have distributed the valence electrons as lone pairs on the oxygen atoms, but the carbon atom lacks an octet:

     gives

   - NO^+: For this ion, we added eight valence electrons, but neither atom has an octet. We cannot add any more electrons since we have already used the total that we found in Step 1, so we must move electrons to form a multiple bond:

     gives

     This still does not produce an octet, so we must move another pair, forming a triple bond:

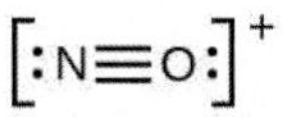

   - In OF_2, each atom has an octet as drawn, so nothing changes.

EXAMPLE 7.4

Writing Lewis Structures

NASA's Cassini-Huygens mission detected a large cloud of toxic hydrogen cyanide (HCN) on Titan, one of Saturn's moons. Titan also contains ethane (H_3CCH_3), acetylene (HCCH), and ammonia (NH_3). What are the Lewis structures of these molecules?

Solution

Step 1. *Calculate the number of valence electrons.*
HCN: (1 × 1) + (4 × 1) + (5 × 1) = 10
H_3CCH_3: (1 × 3) + (2 × 4) + (1 × 3) = 14
HCCH: (1 × 1) + (2 × 4) + (1 × 1) = 10
NH_3: (5 × 1) + (3 × 1) = 8
Step 2. *Draw a skeleton and connect the atoms with single bonds.* Remember that H is never a central atom:

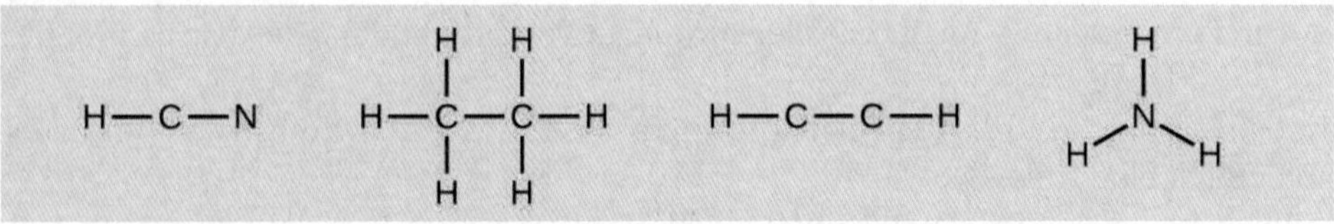

Step 3. *Where needed, distribute electrons to the terminal atoms:*

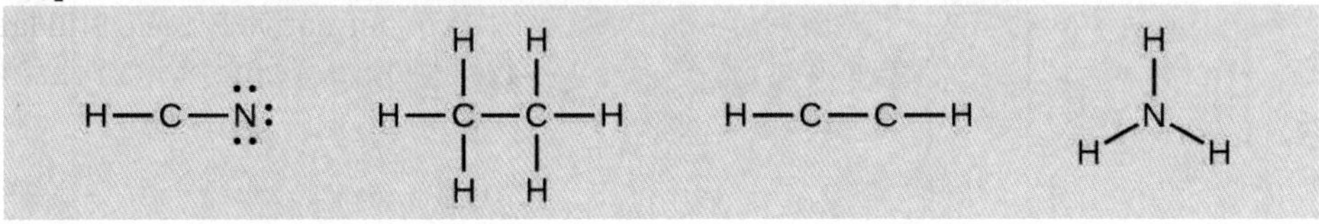

HCN: six electrons placed on N
H_3CCH_3: no electrons remain
HCCH: no terminal atoms capable of accepting electrons
NH_3: no terminal atoms capable of accepting electrons
Step 4. *Where needed, place remaining electrons on the central atom:*

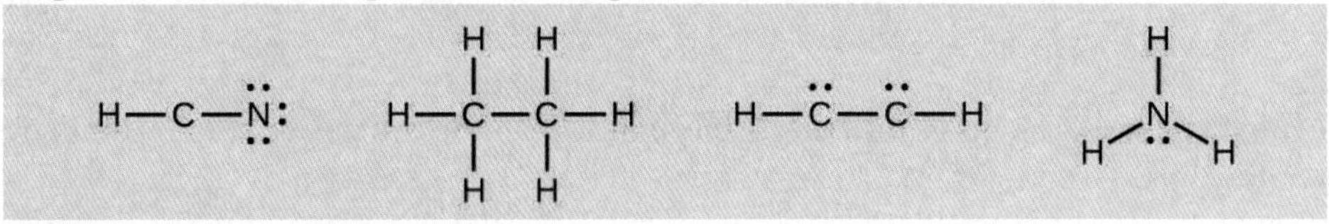

HCN: no electrons remain
H_3CCH_3: no electrons remain
HCCH: four electrons placed on carbon
NH_3: two electrons placed on nitrogen
Step 5. *Where needed, rearrange electrons to form multiple bonds in order to obtain an octet on each atom:*
HCN: form two more C–N bonds
H_3CCH_3: all atoms have the correct number of electrons
HCCH: form a triple bond between the two carbon atoms
NH_3: all atoms have the correct number of electrons

gives

gives

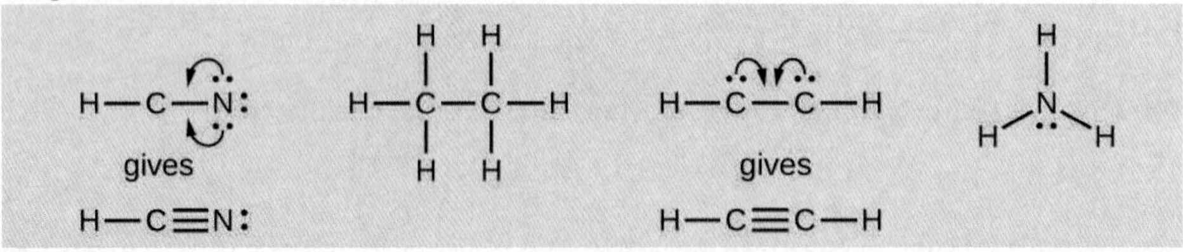

Check Your Learning

Both carbon monoxide, CO, and carbon dioxide, CO_2, are products of the combustion of fossil fuels. Both of these gases also cause problems: CO is toxic and CO_2 has been implicated in global climate change. What are the Lewis structures of these two molecules?

Answer:

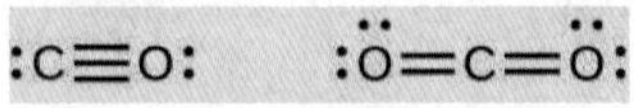

HOW SCIENCES INTERCONNECT

Fullerene Chemistry

Carbon, in various forms and compounds, has been known since prehistoric times, . Soot has been used as a pigment (often called carbon black) for thousands of years. Charcoal, high in carbon content, has likewise been critical to human development. Carbon is the key additive to iron in the steelmaking process, and diamonds have a unique place in both culture and industry. With all this usage came significant study, particularly with the emergence of organic chemistry. And even with all the known forms and functions of the element, scientists began to uncover the potential for even more varied and extensive carbon structures.

As early as the 1960s, chemists began to observe complex carbon structures, but they had little evidence to support their concepts, or their work did not make it into the mainstream. Eiji Osawa predicted a spherical form based on observations of a similar structure, but his work was not widely known outside Japan. In a similar manner, the most comprehensive advance was likely computational chemist Elena Galpern's, who in 1973 predicted a highly stable, 60-carbon molecule; her work was also isolated to her native Russia. Still later, Harold Kroto, working with Canadian radio astronomers, sought to uncover the nature of long carbon chains that had been discovered in interstellar space.

Kroto sought to use a machine developed by Richard Smalley's team at Rice University to learn more about these structures. Together with Robert Curl, who had introduced them, and three graduate students—James Heath, Sean O'Brien, and Yuan Liu—they performed an intensive series of experiments that led to a major discovery.

In 1996, the Nobel Prize in Chemistry was awarded to Richard Smalley (Figure 7.11), Robert Curl, and Harold Kroto for their work in discovering a new form of carbon, the C_{60} buckminsterfullerene molecule (Figure 7.1). An entire class of compounds, including spheres and tubes of various shapes, were discovered based on C_{60}. This type of molecule, called a fullerene, shows promise in a variety of applications. Because of their size and shape, fullerenes can encapsulate other molecules, so they have shown potential in various applications from hydrogen storage to targeted drug delivery systems. They also possess unique electronic and optical properties that have been put to good use in solar powered devices and chemical sensors.

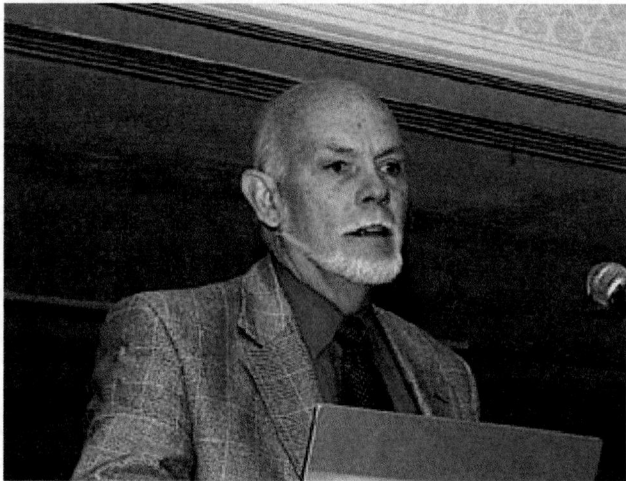

FIGURE 7.11 Richard Smalley (1943–2005), a professor of physics, chemistry, and astronomy at Rice University, was one of the leading advocates for fullerene chemistry. Upon his death in 2005, the US Senate honored him as the “Father of Nanotechnology.” (credit: United States Department of Energy)

Exceptions to the Octet Rule

Many covalent molecules have central atoms that do not have eight electrons in their Lewis structures. These molecules fall into three categories:

- Odd-electron molecules have an odd number of valence electrons, and therefore have an unpaired electron.
- Electron-deficient molecules have a central atom that has fewer electrons than needed for a noble gas configuration.
- Hypervalent molecules have a central atom that has more electrons than needed for a noble gas configuration.

Odd-electron Molecules

We call molecules that contain an odd number of electrons **free radicals**. Nitric oxide, NO, is an example of an odd-electron molecule; it is produced in internal combustion engines when oxygen and nitrogen react at high temperatures.

To draw the Lewis structure for an odd-electron molecule like NO, we follow the same five steps we would for other molecules, but with a few minor changes:

1. *Determine the total number of valence (outer shell) electrons.* The sum of the valence electrons is 5 (from N) + 6 (from O) = 11. The odd number immediately tells us that we have a free radical, so we know that not every atom can have eight electrons in its valence shell.
2. *Draw a skeleton structure of the molecule.* We can easily draw a skeleton with an N–O single bond: N–O
3. *Distribute the remaining electrons as lone pairs on the terminal atoms.* In this case, there is no central atom, so we distribute the electrons around both atoms. We give eight electrons to the more electronegative atom in these situations; thus oxygen has the filled valence shell:
   :N—O:
4. *Place all remaining electrons on the central atom.* Since there are no remaining electrons, this step does not apply.
5. *Rearrange the electrons to make multiple bonds with the central atom in order to obtain octets wherever possible.* We know that an odd-electron molecule cannot have an octet for every atom, but we want to get each atom as close to an octet as possible. In this case, nitrogen has only five electrons around it. To move closer to an octet for nitrogen, we take one of the lone pairs from oxygen and use it to form a NO double bond. (We cannot take another lone pair of electrons on oxygen and form a triple bond because nitrogen would then have nine electrons:)
   :N=O:

Electron-deficient Molecules

We will also encounter a few molecules that contain central atoms that do not have a filled valence shell. Generally, these are molecules with central atoms from groups 2 and 13, outer atoms that are hydrogen, or other atoms that do not form multiple bonds. For example, in the Lewis structures of beryllium dihydride, BeH_2, and boron trifluoride, BF_3, the beryllium and boron atoms each have only four and six electrons, respectively. It is possible to draw a structure with a double bond between a boron atom and a fluorine atom in BF_3, satisfying the octet rule, but experimental evidence indicates the bond lengths are closer to that expected for B–F single bonds. This suggests the best Lewis structure has three B–F single bonds and an electron deficient boron. The reactivity of the compound is also consistent with an electron deficient boron. However, the B–F bonds are slightly shorter than what is actually expected for B–F single bonds, indicating that some double bond character is found in the actual molecule.

H—Be—H   :F: | B with :F: and :F:

An atom like the boron atom in BF_3, which does not have eight electrons, is very reactive. It readily combines with a molecule containing an atom with a lone pair of electrons. For example, NH_3 reacts with BF_3 because the lone pair on nitrogen can be shared with the boron atom:

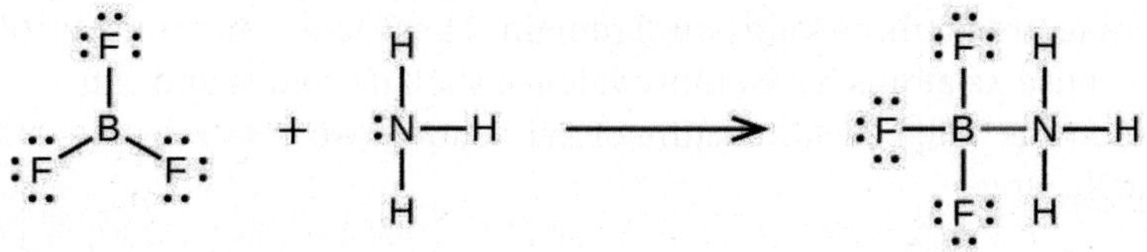

Hypervalent Molecules

Elements in the second period of the periodic table ($n = 2$) can accommodate only eight electrons in their valence shell orbitals because they have only four valence orbitals (one $2s$ and three $2p$ orbitals). Elements in the third and higher periods ($n \geq 3$) have more than four valence orbitals and can share more than four pairs of electrons with other atoms because they have empty d orbitals in the same shell. Molecules formed from these elements are sometimes called **hypervalent molecules**. Figure 7.12 shows the Lewis structures for two hypervalent molecules, PCl_5 and SF_6.

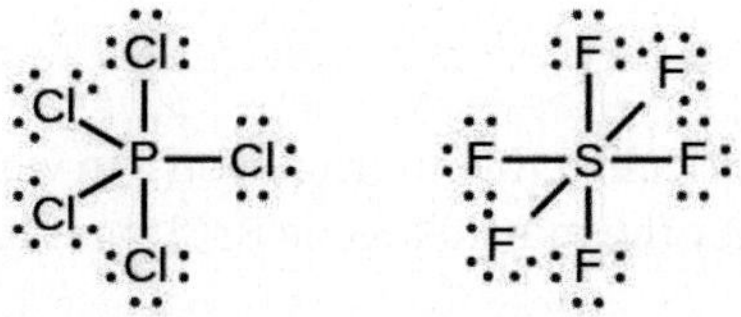

FIGURE 7.12 In PCl_5, the central atom phosphorus shares five pairs of electrons. In SF_6, sulfur shares six pairs of electrons.

In some hypervalent molecules, such as IF_5 and XeF_4, some of the electrons in the outer shell of the central atom are lone pairs:

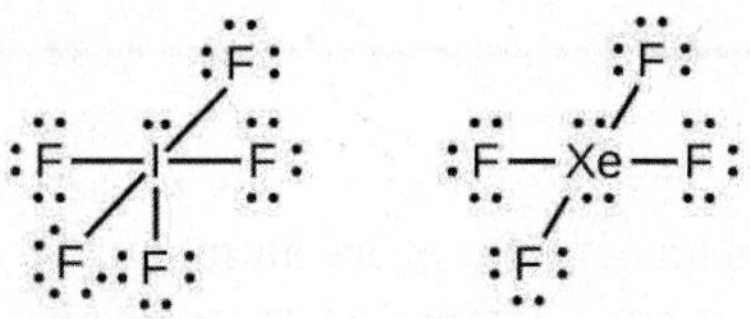

When we write the Lewis structures for these molecules, we find that we have electrons left over after filling the valence shells of the outer atoms with eight electrons. These additional electrons must be assigned to the central atom.

EXAMPLE 7.5

Writing Lewis Structures: Octet Rule Violations

Xenon is a noble gas, but it forms a number of stable compounds. We examined XeF_4 earlier. What are the Lewis structures of XeF_2 and XeF_6?

Solution

We can draw the Lewis structure of any covalent molecule by following the six steps discussed earlier. In this case, we can condense the last few steps, since not all of them apply.

Step 1. *Calculate the number of valence electrons:*
XeF_2: $8 + (2 \times 7) = 22$
XeF_6: $8 + (6 \times 7) = 50$

Step 2. *Draw a skeleton joining the atoms by single bonds.* Xenon will be the central atom because fluorine cannot be a central atom:

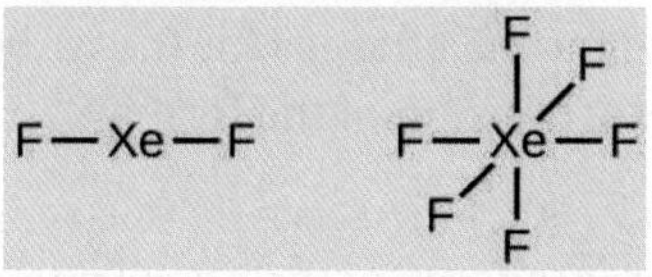

Step 3. *Distribute the remaining electrons.*
XeF_2: We place three lone pairs of electrons around each F atom, accounting for 12 electrons and giving

each F atom 8 electrons. Thus, six electrons (three lone pairs) remain. These lone pairs must be placed on the Xe atom. This is acceptable because Xe atoms have empty valence shell *d* orbitals and can accommodate more than eight electrons. The Lewis structure of XeF_2 shows two bonding pairs and three lone pairs of electrons around the Xe atom:

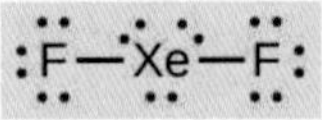

XeF_6: We place three lone pairs of electrons around each F atom, accounting for 36 electrons. Two electrons remain, and this lone pair is placed on the Xe atom:

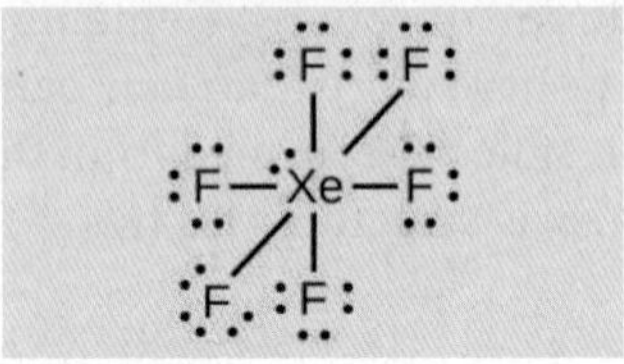

Check Your Learning

The halogens form a class of compounds called the interhalogens, in which halogen atoms covalently bond to each other. Write the Lewis structures for the interhalogens $BrCl_3$ and $ICl_4{}^-$.

Answer:

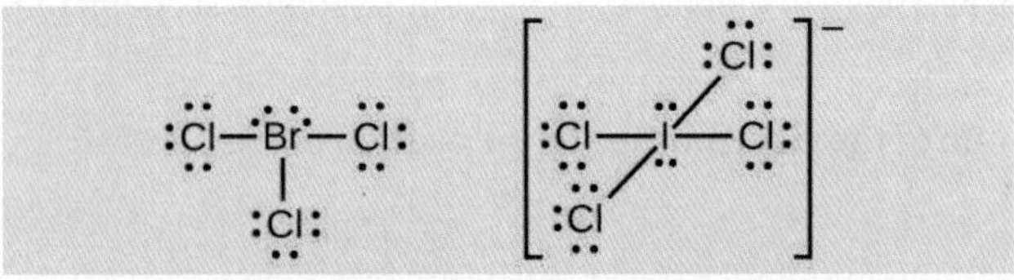

7.4 Formal Charges and Resonance

LEARNING OBJECTIVES

By the end of this section, you will be able to:

- Compute formal charges for atoms in any Lewis structure
- Use formal charges to identify the most reasonable Lewis structure for a given molecule
- Explain the concept of resonance and draw Lewis structures representing resonance forms for a given molecule

In the previous section, we discussed how to write Lewis structures for molecules and polyatomic ions. As we have seen, however, in some cases, there is seemingly more than one valid structure for a molecule. We can use the concept of formal charges to help us predict the most appropriate Lewis structure when more than one is reasonable.

Calculating Formal Charge

The **formal charge** of an atom in a molecule is the *hypothetical* charge the atom would have if we could redistribute the electrons in the bonds evenly between the atoms. Another way of saying this is that formal charge results when we take the number of valence electrons of a neutral atom, subtract the nonbonding electrons, and then subtract the number of bonds connected to that atom in the Lewis structure.

Thus, we calculate formal charge as follows:

$$\text{formal charge} = \text{\# valence shell electrons (free atom)} - \text{\# lone pair electrons} - \frac{1}{2}\text{\# bonding electrons}$$

We can double-check formal charge calculations by determining the sum of the formal charges for the whole structure. The sum of the formal charges of all atoms in a molecule must be zero; the sum of the formal charges in an ion should equal the charge of the ion.

We must remember that the formal charge calculated for an atom is not the *actual* charge of the atom in the molecule. Formal charge is only a useful bookkeeping procedure; it does not indicate the presence of actual charges.

EXAMPLE 7.6

Calculating Formal Charge from Lewis Structures

Assign formal charges to each atom in the interhalogen ion $ICl_4{}^{-}$.

Solution

Step 1. *We divide the bonding electron pairs equally for all I–Cl bonds:*

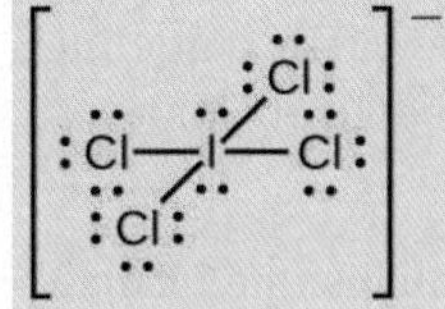

Step 2. *We assign lone pairs of electrons to their atoms.* Each Cl atom now has seven electrons assigned to it, and the I atom has eight.

Step 3. *Subtract this number from the number of valence electrons for the neutral atom:*

I: 7 – 8 = –1

Cl: 7 – 7 = 0

The sum of the formal charges of all the atoms equals –1, which is identical to the charge of the ion (–1).

Check Your Learning

Calculate the formal charge for each atom in the carbon monoxide molecule:

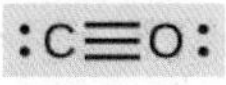

:C≡O:

Answer:

C −1, O +1

EXAMPLE 7.7

Calculating Formal Charge from Lewis Structures

Assign formal charges to each atom in the interhalogen molecule $BrCl_3$.

Solution

Step 1. *Assign one of the electrons in each Br–Cl bond to the Br atom and one to the Cl atom in that bond:*

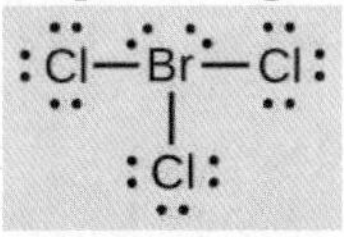

Step 2. *Assign the lone pairs to their atom.* Now each Cl atom has seven electrons and the Br atom has seven electrons.

Step 3. *Subtract this number from the number of valence electrons for the neutral atom.* This gives the formal charge:

Br: 7 – 7 = 0

Cl: 7 – 7 = 0

All atoms in $BrCl_3$ have a formal charge of zero, and the sum of the formal charges totals zero, as it must in a neutral molecule.

Check Your Learning

Determine the formal charge for each atom in NCl_3.

Answer:

N: 0; all three Cl atoms: 0

Cl—N—Cl
|
Cl

Using Formal Charge to Predict Molecular Structure

The arrangement of atoms in a molecule or ion is called its **molecular structure**. In many cases, following the steps for writing Lewis structures may lead to more than one possible molecular structure—different multiple bond and lone-pair electron placements or different arrangements of atoms, for instance. A few guidelines involving formal charge can be helpful in deciding which of the possible structures is most likely for a particular molecule or ion:

1. A molecular structure in which all formal charges are zero is preferable to one in which some formal charges are not zero.
2. If the Lewis structure must have nonzero formal charges, the arrangement with the smallest nonzero formal charges is preferable.
3. Lewis structures are preferable when adjacent formal charges are zero or of the opposite sign.
4. When we must choose among several Lewis structures with similar distributions of formal charges, the structure with the negative formal charges on the more electronegative atoms is preferable.

To see how these guidelines apply, let us consider some possible structures for carbon dioxide, CO_2. We know from our previous discussion that the less electronegative atom typically occupies the central position, but formal charges allow us to understand *why* this occurs. We can draw three possibilities for the structure: carbon in the center and double bonds, carbon in the center with a single and triple bond, and oxygen in the center with double bonds:

O=C=O	O≡C—O	O=O=C	Structure
0 0 0	+1 0 −1	0 +2 −2	Formal charge

Comparing the three formal charges, we can definitively identify the structure on the left as preferable because it has only formal charges of zero (Guideline 1).

As another example, the thiocyanate ion, an ion formed from a carbon atom, a nitrogen atom, and a sulfur atom, could have three different molecular structures: NCS^-, CNS^-, or CSN^-. The formal charges present in each of these molecular structures can help us pick the most likely arrangement of atoms. Possible Lewis structures and the formal charges for each of the three possible structures for the thiocyanate ion are shown here:

Structure	[N=C=S]⁻	[C=N=S]⁻	[C=S=N]⁻
Formal charge	−1 0 0	−2 +1 0	−2 +2 −1

Note that the sum of the formal charges in each case is equal to the charge of the ion (–1). However, the first arrangement of atoms is preferred because it has the lowest number of atoms with nonzero formal charges (Guideline 2). Also, it places the least electronegative atom in the center, and the negative charge on the more electronegative element (Guideline 4).

EXAMPLE 7.8

Using Formal Charge to Determine Molecular Structure

Nitrous oxide, N_2O, commonly known as laughing gas, is used as an anesthetic in minor surgeries, such as the routine extraction of wisdom teeth. Which is the more likely structure for nitrous oxide?

Solution

Determining formal charge yields the following:

:N≡N=Ö:	:N̈=O=N̈:
0 +1 −1	−1 +2 −1

The structure with a terminal oxygen atom best satisfies the criteria for the most stable distribution of formal charge:

:N≡N—Ö:

The number of atoms with formal charges are minimized (Guideline 2), there is no formal charge with a magnitude greater than one (Guideline 2), the negative formal charge is on the more electronegative element (Guideline 4), and the less electronegative atom is in the center position.

Check Your Learning

Which is the most likely molecular structure for the nitrite $\left(NO_2{}^-\right)$ ion?

[:N̈=Ö—Ö:]⁻ or [:Ö=N̈—Ö:]⁻

Answer:

ONO⁻

Resonance

Notice that the more likely structure for the nitrite anion in Example 7.8 may actually be drawn in two different ways, distinguished by the locations of the N-O and N=O bonds:

[:Ö—N̈=Ö]⁻ [Ö=N̈—Ö:]⁻

If nitrite ions do indeed contain a single and a double bond, we would expect for the two bond lengths to be different. A double bond between two atoms is shorter (and stronger) than a single bond between the same two atoms. Experiments show, however, that both N–O bonds in $NO_2{}^-$ have the same strength and length, and are identical in all other properties.

It is not possible to write a single Lewis structure for $NO_2{}^-$ in which nitrogen has an octet and both bonds are equivalent. Instead, we use the concept of **resonance**: if two or more Lewis structures with the same arrangement of atoms can be written for a molecule or ion, the actual distribution of electrons is an *average* of that shown by the various Lewis structures. The actual distribution of electrons in each of the nitrogen-oxygen bonds in $NO_2{}^-$ is the average of a double bond and a single bond. We call the individual Lewis structures **resonance forms**. The actual electronic structure of the molecule (the average of the resonance forms) is called a **resonance hybrid** of the individual resonance forms. A double-headed arrow between Lewis structures indicates that they are resonance forms.

[:Ö—N̈=Ö]⁻ ⟷ [Ö=N̈—Ö:]⁻

We should remember that a molecule described as a resonance hybrid *never* possesses an electronic structure described by either resonance form. It does not fluctuate between resonance forms; rather, the actual electronic structure is *always* the average of that shown by all resonance forms. George Wheland, one of the pioneers of resonance theory, used a historical analogy to describe the relationship between resonance forms and resonance hybrids. A medieval traveler, having never before seen a rhinoceros, described it as a hybrid of a dragon and a unicorn because it had many properties in common with both. Just as a rhinoceros is neither a dragon sometimes nor a unicorn at other times, a resonance hybrid is neither of its resonance forms at any given time. Like a rhinoceros, it is a real entity that experimental evidence has shown to exist. It has some characteristics in common with its resonance forms, but the resonance forms themselves are convenient, imaginary images (like the unicorn and the dragon).

The carbonate anion, $CO_3{}^{2-}$, provides a second example of resonance:

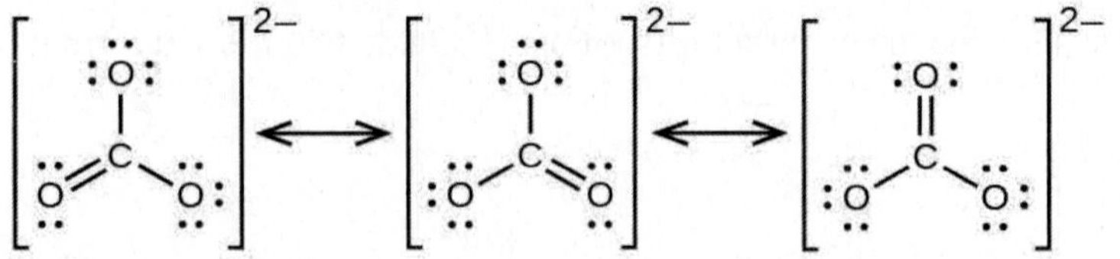

One oxygen atom must have a double bond to carbon to complete the octet on the central atom. All oxygen atoms, however, are equivalent, and the double bond could form from any one of the three atoms. This gives rise to three resonance forms of the carbonate ion. Because we can write three identical resonance structures, we know that the actual arrangement of electrons in the carbonate ion is the average of the three structures. Again, experiments show that all three C–O bonds are exactly the same.

LINK TO LEARNING

Use this online quiz (http://openstax.org/l/16LewisMake) to practice your skills in drawing resonance structures and estimating formal charges.

7.5 Strengths of Ionic and Covalent Bonds

LEARNING OBJECTIVES

By the end of this section, you will be able to:

- Describe the energetics of covalent and ionic bond formation and breakage
- Use the Born-Haber cycle to compute lattice energies for ionic compounds
- Use average covalent bond energies to estimate enthalpies of reaction

A bond's strength describes how strongly each atom is joined to another atom, and therefore how much energy is required to break the bond between the two atoms. In this section, you will learn about the bond strength of covalent bonds, and then compare that to the strength of ionic bonds, which is related to the lattice energy of a compound.

Bond Strength: Covalent Bonds

Stable molecules exist because covalent bonds hold the atoms together. We measure the strength of a covalent bond by the energy required to break it, that is, the energy necessary to separate the bonded atoms. Separating any pair of bonded atoms requires energy (see Figure 7.4). The stronger a bond, the greater the energy required to break it.

The energy required to break a specific covalent bond in one mole of gaseous molecules is called the bond energy or the bond dissociation energy. The bond energy for a diatomic molecule, D_{X-Y}, is defined as the standard enthalpy change for the endothermic reaction:

$$XY(g) \longrightarrow X(g) + Y(g) \qquad D_{X-Y} = \Delta H^\circ$$

For example, the bond energy of the pure covalent H–H bond, D_{H-H}, is 436 kJ per mole of H–H bonds broken:

$$H_2(g) \longrightarrow 2H(g) \qquad D_{H-H} = \Delta H^\circ = 436 \text{ kJ}$$

Molecules with three or more atoms have two or more bonds. The sum of all bond energies in such a molecule is equal to the standard enthalpy change for the endothermic reaction that breaks all the bonds in the molecule. For example, the sum of the four C–H bond energies in CH_4, 1660 kJ, is equal to the standard enthalpy change of the reaction:

$$\text{H}{-}\underset{\text{H}}{\overset{\text{H}}{\text{C}}}{-}\text{H}(g) \longrightarrow \text{C}(g) + 4\text{H}(g) \qquad \Delta H^\circ = 1660 \text{ kJ}$$

The average C–H bond energy, D_{C-H}, is 1660/4 = 415 kJ/mol because there are four moles of C–H bonds broken per mole of the reaction. Although the four C–H bonds are equivalent in the original molecule, they do not each require the same energy to break; once the first bond is broken (which requires 439 kJ/mol), the remaining bonds are easier to break. The 415 kJ/mol value is the average, not the exact value required to break any one bond.

The strength of a bond between two atoms increases as the number of electron pairs in the bond increases. Generally, as the bond strength increases, the bond length decreases. Thus, we find that triple bonds are stronger and shorter than double bonds between the same two atoms; likewise, double bonds are stronger and shorter than single bonds between the same two atoms. Average bond energies for some common bonds appear in Table 7.2, and a comparison of bond lengths and bond strengths for some common bonds appears in Table 7.3. When one atom bonds to various atoms in a group, the bond strength typically decreases as we move down the group. For example, C–F is 439 kJ/mol, C–Cl is 330 kJ/mol, and C–Br is 275 kJ/mol.

Bond Energies (kJ/mol)

Bond	Bond Energy	Bond	Bond Energy	Bond	Bond Energy
H–H	436	C–S	260	F–Cl	255
H–C	415	C–Cl	330	F–Br	235
H–N	390	C–Br	275	Si–Si	230
H–O	464	C–I	240	Si–P	215
H–F	569	N–N	160	Si–S	225
H–Si	395	$\text{N} = \text{N}$	418	Si–Cl	359
H–P	320	$\text{N} \equiv \text{N}$	946	Si–Br	290
H–S	340	N–O	200	Si–I	215
H–Cl	432	N–F	270	P–P	215
H–Br	370	N–P	210	P–S	230
H–I	295	N–Cl	200	P–Cl	330
C–C	345	N–Br	245	P–Br	270
$\text{C} = \text{C}$	611	O–O	140	P–I	215
$\text{C} \equiv \text{C}$	837	$\text{O} = \text{O}$	498	S–S	215
C–N	290	O–F	160	S–Cl	250

Bond Energies (kJ/mol)					
Bond	**Bond Energy**	**Bond**	**Bond Energy**	**Bond**	**Bond Energy**
C = N	615	O–Si	370	S–Br	215
C ≡ N	891	O–P	350	Cl–Cl	243
C–O	350	O–Cl	205	Cl–Br	220
C = O	741	O–I	200	Cl–I	210
C ≡ O	1080	F–F	160	Br–Br	190
C–F	439	F–Si	540	Br–I	180
C–Si	360	F–P	489	I–I	150
C–P	265	F–S	285		

TABLE 7.2

Average Bond Lengths and Bond Energies for Some Common Bonds

Bond	Bond Length (Å)	Bond Energy (kJ/mol)
C–C	1.54	345
C = C	1.34	611
C ≡ C	1.20	837
C–N	1.43	290
C = N	1.38	615
C ≡ N	1.16	891
C–O	1.43	350
C = O	1.23	741
C ≡ O	1.13	1080

TABLE 7.3

We can use bond energies to calculate approximate enthalpy changes for reactions where enthalpies of formation are not available. Calculations of this type will also tell us whether a reaction is exothermic or endothermic. An exothermic reaction (ΔH negative, heat produced) results when the bonds in the products are stronger than the bonds in the reactants. An endothermic reaction (ΔH positive, heat absorbed) results when the bonds in the products are weaker than those in the reactants.

The enthalpy change, ΔH, for a chemical reaction is approximately equal to the sum of the energy required to break all bonds in the reactants (energy "in", positive sign) plus the energy released when all bonds are formed in the products (energy "out," negative sign). This can be expressed mathematically in the following way:

$$\Delta H = \Sigma \text{D}_{\text{bonds broken}} - \Sigma \text{D}_{\text{bonds formed}}$$

In this expression, the symbol Σ means "the sum of" and D represents the bond energy in kilojoules per mole, which is always a positive number. The bond energy is obtained from a table (like Table 7.3) and will depend on whether the particular bond is a single, double, or triple bond. Thus, in calculating enthalpies in this manner, it is important that we consider the bonding in all reactants and products. Because D values are typically averages for one type of bond in many different molecules, this calculation provides a rough estimate, not an exact value, for the enthalpy of reaction.

Consider the following reaction:

$$\text{H}_2(g) + \text{Cl}_2(g) \longrightarrow 2\text{HCl}(g)$$

or

$$\text{H–H}(g) + \text{Cl–Cl}(g) \longrightarrow 2\text{H–Cl}(g)$$

To form two moles of HCl, one mole of H–H bonds and one mole of Cl–Cl bonds must be broken. The energy required to break these bonds is the sum of the bond energy of the H–H bond (436 kJ/mol) and the Cl–Cl bond (243 kJ/mol). During the reaction, two moles of H–Cl bonds are formed (bond energy = 432 kJ/mol), releasing 2 × 432 kJ; or 864 kJ. Because the bonds in the products are stronger than those in the reactants, the reaction releases more energy than it consumes:

$$\begin{aligned} \Delta H &= \Sigma \text{D}_{\text{bonds broken}} - \Sigma \text{D}_{\text{bonds formed}} \\ \Delta H &= \left[\text{D}_{\text{H–H}} + \text{D}_{\text{Cl–Cl}}\right] - 2\text{D}_{\text{H–Cl}} \\ &= [436 + 243] - 2(432) = -185 \text{ kJ} \end{aligned}$$

This excess energy is released as heat, so the reaction is exothermic. Appendix G gives a value for the standard molar enthalpy of formation of HCl(g), ΔH_f°, of –92.307 kJ/mol. Twice that value is –184.6 kJ, which agrees well with the answer obtained earlier for the formation of two moles of HCl.

EXAMPLE 7.9

Using Bond Energies to Calculate Approximate Enthalpy Changes

Methanol, CH_3OH, may be an excellent alternative fuel. The high-temperature reaction of steam and carbon produces a mixture of the gases carbon monoxide, CO, and hydrogen, H_2, from which methanol can be produced. Using the bond energies in Table 7.3, calculate the approximate enthalpy change, ΔH, for the reaction here:

$$\text{CO}(g) + 2\text{H}_2(g) \longrightarrow \text{CH}_3\text{OH}(g)$$

Solution

First, we need to write the Lewis structures of the reactants and the products:

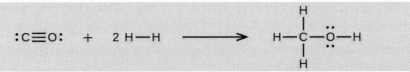

:C≡O: + 2 H—H ⟶ H—C(H)(H)—Ö—H

From this, we see that ΔH for this reaction involves the energy required to break a C–O triple bond and two H–H single bonds, as well as the energy produced by the formation of three C–H single bonds, a C–O single bond, and an O–H single bond. We can express this as follows:

$$\Delta H = \Sigma D_{\text{bonds broken}} - \Sigma D_{\text{bonds formed}}$$
$$\Delta H = \left[D_{C\equiv O} + 2(D_{H-H})\right] - \left[3\left(D_{C-H}\right) + D_{C-O} + D_{O-H}\right]$$

Using the bond energy values in Table 7.3, we obtain:

$$\begin{aligned}\Delta H &= [1080 + 2(436)] - [3(415) + 350 + 464] \\ &= -107 \text{ kJ}\end{aligned}$$

We can compare this value to the value calculated based on ΔH_f° data from Appendix G:

$$\begin{aligned}\Delta H &= \left[\Delta H_f^\circ CH_3OH(g)\right] - \left[\Delta H_f^\circ CO(g) + 2 \times \Delta H_f^\circ H_2\right] \\ &= [-201.0] - [-110.52 + 2 \times 0] \\ &= -90.5 \text{ kJ}\end{aligned}$$

Note that there is a fairly significant gap between the values calculated using the two different methods. This occurs because D values are the *average* of different bond strengths; therefore, they often give only rough agreement with other data.

Check Your Learning

Ethyl alcohol, CH_3CH_2OH, was one of the first organic chemicals deliberately synthesized by humans. It has many uses in industry, and it is the alcohol contained in alcoholic beverages. It can be obtained by the fermentation of sugar or synthesized by the hydration of ethylene in the following reaction:

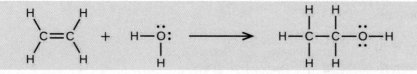

Using the bond energies in Table 7.3, calculate an approximate enthalpy change, ΔH, for this reaction.

Answer:

–35 kJ

Ionic Bond Strength and Lattice Energy

An ionic compound is stable because of the electrostatic attraction between its positive and negative ions. The lattice energy of a compound is a measure of the strength of this attraction. The **lattice energy ($\Delta H_{\text{lattice}}$)** of an ionic compound is defined as the energy required to separate one mole of the solid into its component gaseous ions. For the ionic solid MX, the lattice energy is the enthalpy change of the process:

$$MX(s) \longrightarrow M^{n+}(g) + X^{n-}(g) \qquad \Delta H_{\text{lattice}}$$

Note that we are using the convention where the ionic solid is separated into ions, so our lattice energies will be *endothermic* (positive values). Some texts use the equivalent but opposite convention, defining lattice energy as the energy released when separate ions combine to form a lattice and giving negative (exothermic) values. Thus, if you are looking up lattice energies in another reference, be certain to check which definition is being used. In both cases, a larger magnitude for lattice energy indicates a more stable ionic compound. For sodium chloride, $\Delta H_{\text{lattice}}$ = 769 kJ. Thus, it requires 769 kJ to separate one mole of solid NaCl into gaseous Na^+ and Cl^- ions. When one mole each of gaseous Na^+ and Cl^- ions form solid NaCl, 769 kJ of heat is released.

The lattice energy $\Delta H_{\text{lattice}}$ of an ionic crystal can be expressed by the following equation (derived from Coulomb's law, governing the forces between electric charges):

$$\Delta H_{\text{lattice}} = \frac{C\left(Z^+\right)\left(Z^-\right)}{R_o}$$

in which C is a constant that depends on the type of crystal structure; Z^+ and Z^- are the charges on the ions; and R_o is the interionic distance (the sum of the radii of the positive and negative ions). Thus, the lattice energy of an ionic crystal increases rapidly as the charges of the ions increase and the sizes of the ions decrease. When all other parameters are kept constant, doubling the charge of both the cation and anion quadruples the lattice energy. For example, the lattice energy of LiF (Z^+ and $Z^- = 1$) is 1023 kJ/mol, whereas that of MgO (Z^+ and $Z^- = 2$) is 3900 kJ/mol (R_o is nearly the same—about 200 pm for both compounds).

Different interatomic distances produce different lattice energies. For example, we can compare the lattice energy of MgF_2 (2957 kJ/mol) to that of MgI_2 (2327 kJ/mol) to observe the effect on lattice energy of the smaller ionic size of F^- as compared to I^-.

Lattice Energy Comparisons

The precious gem ruby is aluminum oxide, Al_2O_3, containing traces of Cr^{3+}. The compound Al_2Se_3 is used in the fabrication of some semiconductor devices. Which has the larger lattice energy, Al_2O_3 or Al_2Se_3?

Solution

In these two ionic compounds, the charges Z^+ and Z^- are the same, so the difference in lattice energy will depend upon R_o. The O^{2-} ion is smaller than the Se^{2-} ion. Thus, Al_2O_3 would have a shorter interionic distance than Al_2Se_3, and Al_2O_3 would have the larger lattice energy.

Check Your Learning

Zinc oxide, ZnO, is a very effective sunscreen. How would the lattice energy of ZnO compare to that of NaCl?

Answer:

ZnO would have the larger lattice energy because the Z values of both the cation and the anion in ZnO are greater, and the interionic distance of ZnO is smaller than that of NaCl.

The Born-Haber Cycle

It is not possible to measure lattice energies directly. However, the lattice energy can be calculated using the equation given in the previous section or by using a thermochemical cycle. The **Born-Haber cycle** is an application of Hess's law that breaks down the formation of an ionic solid into a series of individual steps:

- ΔH_f°, the standard enthalpy of formation of the compound
- *IE*, the ionization energy of the metal
- *EA*, the electron affinity of the nonmetal
- ΔH_s°, the enthalpy of sublimation of the metal
- *D*, the bond dissociation energy of the nonmetal
- $\Delta H_{lattice}$, the lattice energy of the compound

Figure 7.13 diagrams the Born-Haber cycle for the formation of solid cesium fluoride.

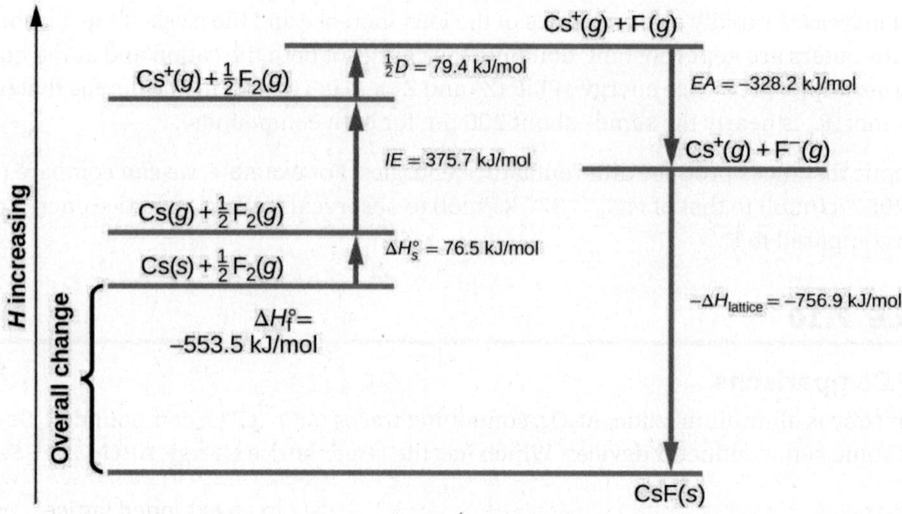

FIGURE 7.13 The Born-Haber cycle shows the relative energies of each step involved in the formation of an ionic solid from the necessary elements in their reference states.

We begin with the elements in their most common states, Cs(*s*) and F_2(*g*). The ΔH_s° represents the conversion of solid cesium into a gas, and then the ionization energy converts the gaseous cesium atoms into cations. In the next step, we account for the energy required to break the F–F bond to produce fluorine atoms. Converting one mole of fluorine atoms into fluoride ions is an exothermic process, so this step gives off energy (the electron affinity) and is shown as decreasing along the *y*-axis. We now have one mole of Cs cations and one mole of F anions. These ions combine to produce solid cesium fluoride. The enthalpy change in this step is the negative of the lattice energy, so it is also an exothermic quantity. The total energy involved in this conversion is equal to the experimentally determined enthalpy of formation, ΔH_f°, of the compound from its elements. In this case, the overall change is exothermic.

Hess's law can also be used to show the relationship between the enthalpies of the individual steps and the enthalpy of formation. Table 7.4 shows this for fluoride, CsF.

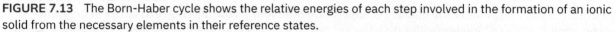

Enthalpy of sublimation of Cs(*s*)	$\text{Cs}(s) \longrightarrow \text{Cs}(g)$	$\Delta H = \Delta H_s^\circ = 76.5\text{kJ/mol}$
One-half of the bond energy of F_2	$\frac{1}{2}\text{F}_2(g) \longrightarrow \text{F}(g)$	$\Delta H = \frac{1}{2}D = 79.4\text{kJ/mol}$
Ionization energy of Cs(*g*)	$\text{Cs}(g) \longrightarrow \text{Cs}^+(g) + \text{e}^-$	$\Delta H = IE = 375.7\text{kJ/mol}$
Electron affinity of F	$\text{F}(g) + \text{e}^- \longrightarrow \text{F}^-(g)$	$\Delta H = EA = -328.2\text{kJ/mol}$
Negative of the lattice energy of CsF(*s*)	$\text{Cs}^+(g) + \text{F}^-(g) \longrightarrow \text{CsF}(s)$	$\Delta H = -\Delta H_{\text{lattice}} = ?$
Enthalpy of formation of CsF(*s*), add steps 1–5	$\Delta H = \Delta H_f^\circ = \Delta H_s^\circ + \frac{1}{2}D + IE + (EA) + (-\Delta H_{\text{lattice}})$ $\text{Cs}(s) + \frac{1}{2}\text{F}_2(g) \longrightarrow \text{CsF}(s)$	$\Delta H = -553.5\text{kJ/mol}$

TABLE 7.4

Thus, the lattice energy can be calculated from other values. For cesium fluoride, using this data, the lattice energy is:

$$\Delta H_{\text{lattice}} = 76.5 + 79.4 + 375.7 + (-328.2) - (-553.5) = 756.9\ \text{kJ/mol}$$

The Born-Haber cycle may also be used to calculate any one of the other quantities in the equation for lattice energy, provided that the remainder is known. For example, if the relevant enthalpy of sublimation ΔH_s°, ionization energy (IE), bond dissociation enthalpy (D), lattice energy $\Delta H_{\text{lattice}}$, and standard enthalpy of formation ΔH_f° are known, the Born-Haber cycle can be used to determine the electron affinity of an atom.

Lattice energies calculated for ionic compounds are typically much higher than bond dissociation energies measured for covalent bonds. Whereas lattice energies typically fall in the range of 600–4000 kJ/mol (some even higher), covalent bond dissociation energies are typically between 150–400 kJ/mol for single bonds. Keep in mind, however, that these are not directly comparable values. For ionic compounds, lattice energies are associated with many interactions, as cations and anions pack together in an extended lattice. For covalent bonds, the bond dissociation energy is associated with the interaction of just two atoms.

7.6 Molecular Structure and Polarity

LEARNING OBJECTIVES

By the end of this section, you will be able to:

- Predict the structures of small molecules using valence shell electron pair repulsion (VSEPR) theory
- Explain the concepts of polar covalent bonds and molecular polarity
- Assess the polarity of a molecule based on its bonding and structure

Thus far, we have used two-dimensional Lewis structures to represent molecules. However, molecular structure is actually three-dimensional, and it is important to be able to describe molecular bonds in terms of their distances, angles, and relative arrangements in space (Figure 7.14). A **bond angle** is the angle between any two bonds that include a common atom, usually measured in degrees. A **bond distance** (or bond length) is the distance between the nuclei of two bonded atoms along the straight line joining the nuclei. Bond distances are measured in Ångstroms (1 Å = 10^{-10} m) or picometers (1 pm = 10^{-12} m, 100 pm = 1 Å).

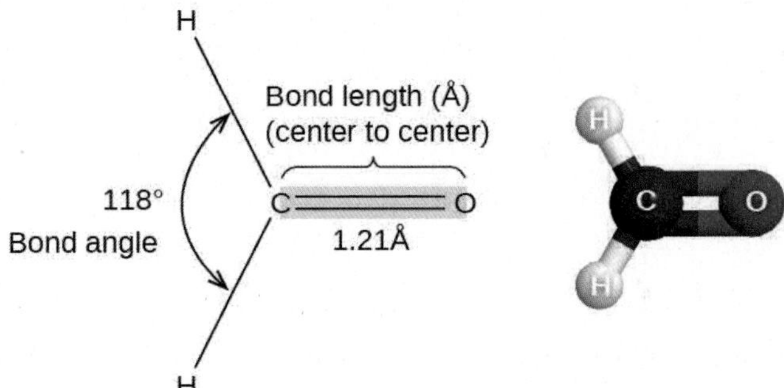

FIGURE 7.14 Bond distances (lengths) and angles are shown for the formaldehyde molecule, H_2CO.

VSEPR Theory

Valence shell electron-pair repulsion theory (VSEPR theory) enables us to predict the molecular structure, including approximate bond angles around a central atom, of a molecule from an examination of the number of bonds and lone electron pairs in its Lewis structure. The VSEPR model assumes that electron pairs in the valence shell of a central atom will adopt an arrangement that minimizes repulsions between these electron pairs by maximizing the distance between them. The electrons in the valence shell of a central atom form either bonding pairs of electrons, located primarily between bonded atoms, or lone pairs. The electrostatic repulsion of these electrons is reduced when the various regions of high electron density assume positions as far from each other as possible.

VSEPR theory predicts the arrangement of electron pairs around each central atom and, usually, the correct arrangement of atoms in a molecule. We should understand, however, that the theory only considers electron-pair repulsions. Other interactions, such as nuclear-nuclear repulsions and nuclear-electron attractions, are also involved in the final arrangement that atoms adopt in a particular molecular structure.

As a simple example of VSEPR theory, let us predict the structure of a gaseous BeF_2 molecule. The Lewis structure of BeF_2 (Figure 7.15) shows only two electron pairs around the central beryllium atom. With two bonds and no lone pairs of electrons on the central atom, the bonds are as far apart as possible, and the electrostatic repulsion between these regions of high electron density is reduced to a minimum when they are on opposite sides of the central atom. The bond angle is 180° (Figure 7.15).

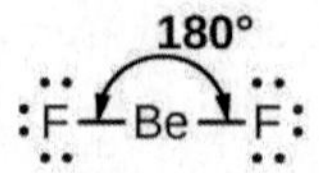

180°

F–Be–F

FIGURE 7.15 The BeF_2 molecule adopts a linear structure in which the two bonds are as far apart as possible, on opposite sides of the Be atom.

Figure 7.16 illustrates this and other electron-pair geometries that minimize the repulsions among regions of high electron density (bonds and/or lone pairs). Two regions of electron density around a central atom in a molecule form a **linear** geometry; three regions form a **trigonal planar** geometry; four regions form a **tetrahedral** geometry; five regions form a **trigonal bipyramidal** geometry; and six regions form an **octahedral** geometry.

Number of regions	Two regions of high electron density (bonds and/or unshared pairs)	Three regions of high electron density (bonds and/or unshared pairs)	Four regions of high electron density (bonds and/or unshared pairs)	Five regions of high electron density (bonds and/or unshared pairs)	Six regions of high electron density (bonds and/or unshared pairs)
Spatial arrangement	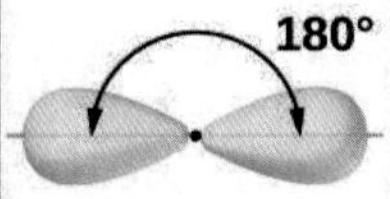 180°	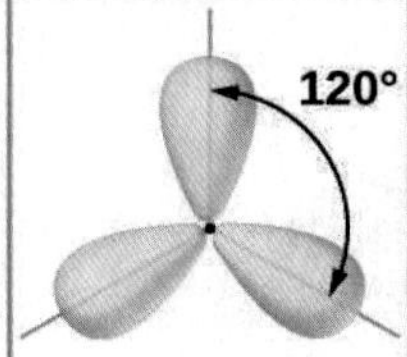 120°	 109.5°	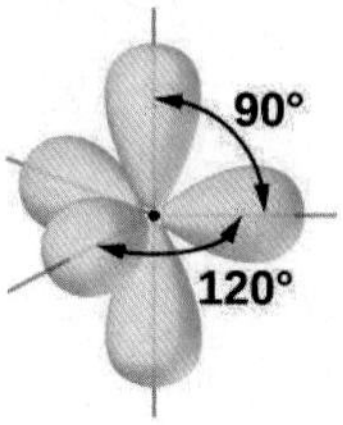 90°, 120°	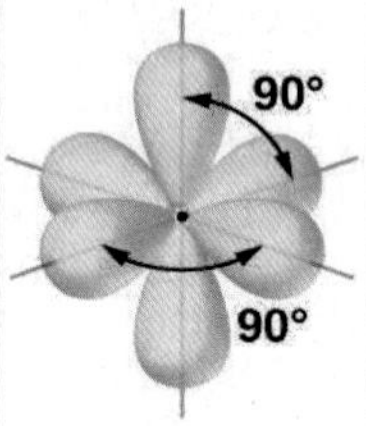 90°, 90°
Line-dash-wedge notation	H—Be—H	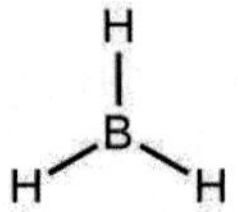 BH_3	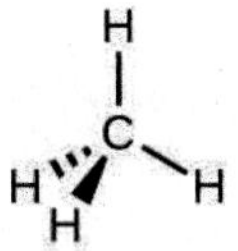 CH_4	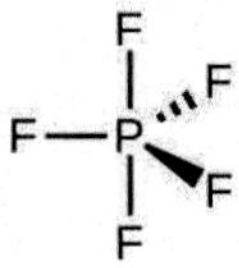 PF_5	 SF_6
Electron region geometry	Linear; 180° angle	Trigonal planar; all angles 120°	Tetrahedral; all angles 109.5°	Trigonal bipyramidal; angles of 90° or 120° An attached atom may be equatorial (in the plane of the triangle) or axial (above or below the plane of the triangle).	Octahedral; all angles 90° or 180°

FIGURE 7.16 The basic electron-pair geometries predicted by VSEPR theory maximize the space around any region of electron density (bonds or lone pairs).

Electron-pair Geometry versus Molecular Structure

It is important to note that electron-pair geometry around a central atom is *not* the same thing as its molecular structure. The electron-pair geometries shown in Figure 7.16 describe all regions where electrons are located, bonds as well as lone pairs. Molecular structure describes the location of the *atoms*, not the electrons.

We differentiate between these two situations by naming the geometry that includes *all* electron pairs the **electron-pair geometry**. The structure that includes only the placement of the atoms in the molecule is called the **molecular structure**. The electron-pair geometries will be the same as the molecular structures when there are no lone electron pairs around the central atom, but they will be different when there are lone pairs present on the central atom.

For example, the methane molecule, CH_4, which is the major component of natural gas, has four bonding pairs of electrons around the central carbon atom; the electron-pair geometry is tetrahedral, as is the molecular structure (Figure 7.17). On the other hand, the ammonia molecule, NH_3, also has four electron pairs associated with the nitrogen atom, and thus has a tetrahedral electron-pair geometry. One of these regions, however, is a lone pair, which is not included in the molecular structure, and this lone pair influences the shape of the molecule (Figure 7.18).

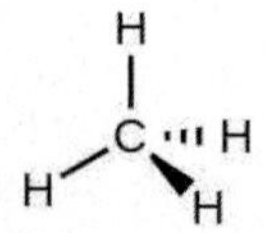

FIGURE 7.17 The molecular structure of the methane molecule, CH_4, is shown with a tetrahedral arrangement of the hydrogen atoms. VSEPR structures like this one are often drawn using the wedge and dash notation, in which solid lines represent bonds in the plane of the page, solid wedges represent bonds coming up out of the plane, and dashed lines represent bonds going down into the plane.

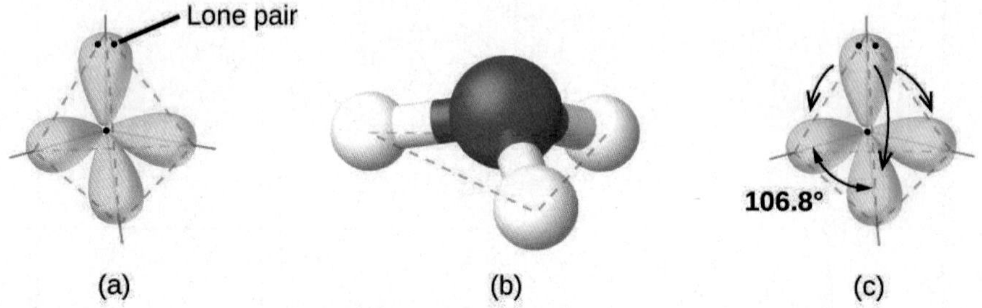

FIGURE 7.18 (a) The electron-pair geometry for the ammonia molecule is tetrahedral with one lone pair and three single bonds. (b) The trigonal pyramidal molecular structure is determined from the electron-pair geometry. (c) The actual bond angles deviate slightly from the idealized angles because the lone pair takes up a larger region of space than do the single bonds, causing the HNH angle to be slightly smaller than 109.5°.

As seen in Figure 7.18, small distortions from the ideal angles in Figure 7.16 can result from differences in repulsion between various regions of electron density. VSEPR theory predicts these distortions by establishing an order of repulsions and an order of the amount of space occupied by different kinds of electron pairs. The order of electron-pair repulsions from greatest to least repulsion is:

$$\text{lone pair-lone pair} > \text{lone pair-bonding pair} > \text{bonding pair-bonding pair}$$

This order of repulsions determines the amount of space occupied by different regions of electrons. A lone pair of electrons occupies a larger region of space than the electrons in a triple bond; in turn, electrons in a triple bond occupy more space than those in a double bond, and so on. The order of sizes from largest to smallest is:

$$\text{lone pair} > \text{triple bond} > \text{double bond} > \text{single bond}$$

Consider formaldehyde, H_2CO, which is used as a preservative for biological and anatomical specimens (Figure 7.14). This molecule has regions of high electron density that consist of two single bonds and one double bond. The basic geometry is trigonal planar with 120° bond angles, but we see that the double bond causes slightly larger angles (121°), and the angle between the single bonds is slightly smaller (118°).

In the ammonia molecule, the three hydrogen atoms attached to the central nitrogen are not arranged in a flat, trigonal planar molecular structure, but rather in a three-dimensional trigonal pyramid (Figure 7.18) with the nitrogen atom at the apex and the three hydrogen atoms forming the base. The ideal bond angles in a trigonal pyramid are based on the tetrahedral electron pair geometry. Again, there are slight deviations from the ideal because lone pairs occupy larger regions of space than do bonding electrons. The H–N–H bond angles in NH_3 are slightly smaller than the 109.5° angle in a regular tetrahedron (Figure 7.16) because the lone pair-bonding pair repulsion is greater than the bonding pair-bonding pair repulsion (Figure 7.18). Figure 7.19 illustrates the ideal molecular structures, which are predicted based on the electron-pair geometries for various combinations of lone pairs and bonding pairs.

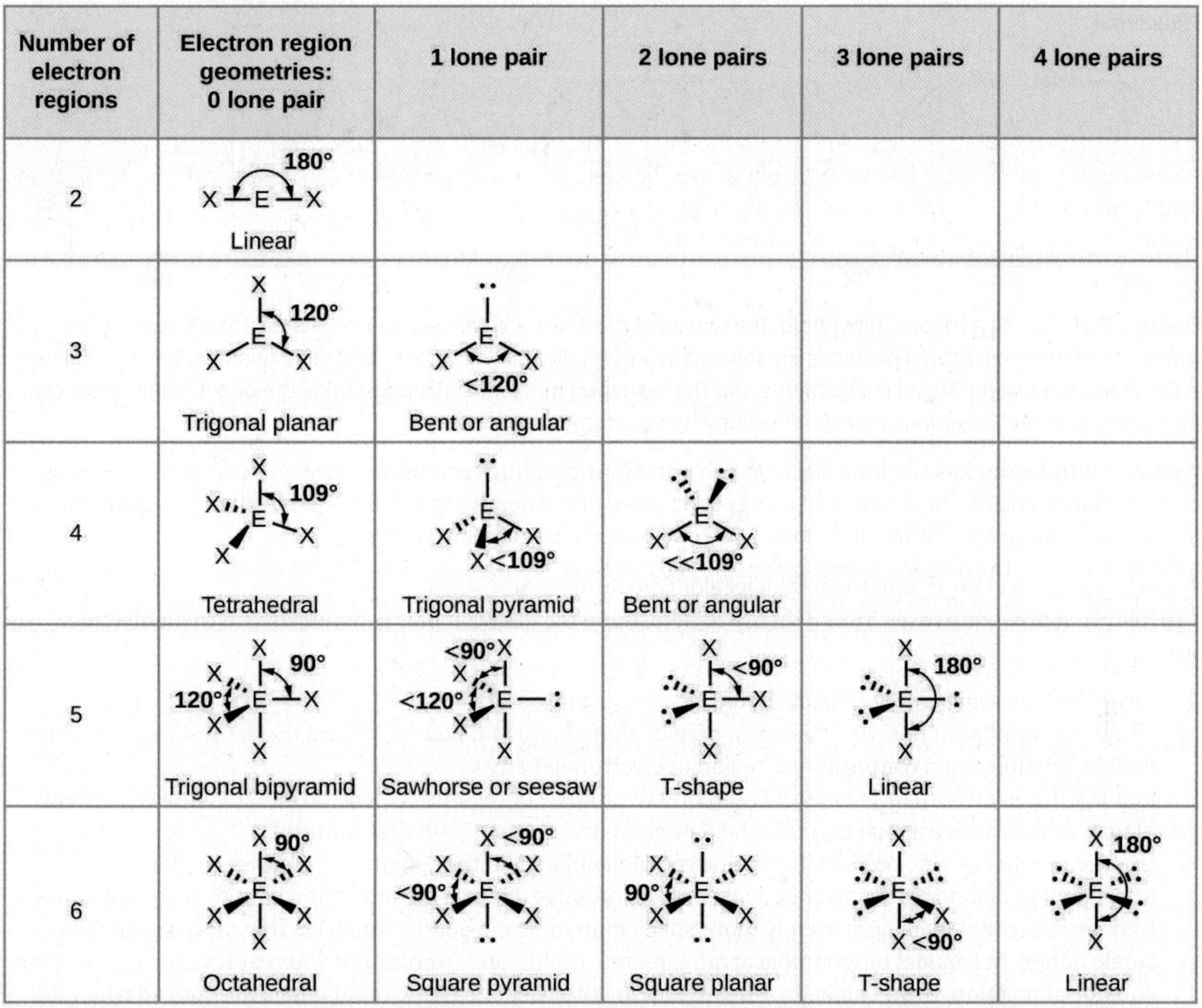

Number of electron regions	Electron region geometries: 0 lone pair	1 lone pair	2 lone pairs	3 lone pairs	4 lone pairs
2	180° Linear				
3	120° Trigonal planar	<120° Bent or angular			
4	109° Tetrahedral	<109° Trigonal pyramid	<<109° Bent or angular		
5	90° 120° Trigonal bipyramid	<90° <120° Sawhorse or seesaw	<90° T-shape	180° Linear	
6	90° Octahedral	<90° <90° Square pyramid	90° Square planar	<90° T-shape	180° Linear

FIGURE 7.19 The molecular structures are identical to the electron-pair geometries when there are no lone pairs present (first column). For a particular number of electron pairs (row), the molecular structures for one or more lone pairs are determined based on modifications of the corresponding electron-pair geometry.

According to VSEPR theory, the terminal atom locations (Xs in Figure 7.19) are equivalent within the linear, trigonal planar, and tetrahedral electron-pair geometries (the first three rows of the table). It does not matter which X is replaced with a lone pair because the molecules can be rotated to convert positions. For trigonal bipyramidal electron-pair geometries, however, there are two distinct X positions, as shown in Figure 7.20: an **axial position** (if we hold a model of a trigonal bipyramid by the two axial positions, we have an axis around which we can rotate the model) and an **equatorial position** (three positions form an equator around the middle of the molecule). As shown in Figure 7.19, the axial position is surrounded by bond angles of 90°, whereas the equatorial position has more space available because of the 120° bond angles. In a trigonal bipyramidal electron-pair geometry, lone pairs always occupy equatorial positions because these more spacious positions can more easily accommodate the larger lone pairs.

Theoretically, we can come up with three possible arrangements for the three bonds and two lone pairs for the ClF_3 molecule (Figure 7.20). The stable structure is the one that puts the lone pairs in equatorial locations, giving a T-shaped molecular structure.

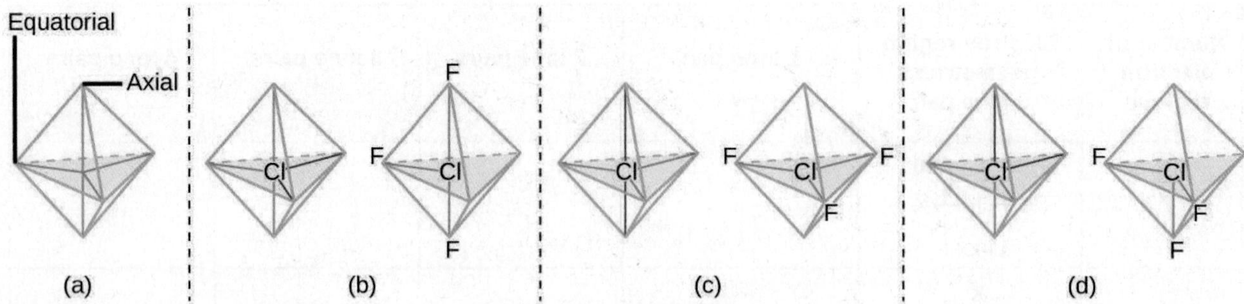

FIGURE 7.20 (a) In a trigonal bipyramid, the two axial positions are located directly across from one another, whereas the three equatorial positions are located in a triangular arrangement. (b–d) The two lone pairs (red lines) in ClF_3 have several possible arrangements, but the T-shaped molecular structure (b) is the one actually observed, consistent with the larger lone pairs both occupying equatorial positions.

When a central atom has two lone electron pairs and four bonding regions, we have an octahedral electron-pair geometry. The two lone pairs are on opposite sides of the octahedron (180° apart), giving a square planar molecular structure that minimizes lone pair-lone pair repulsions (Figure 7.19).

Predicting Electron Pair Geometry and Molecular Structure

The following procedure uses VSEPR theory to determine the electron pair geometries and the molecular structures:

1. Write the Lewis structure of the molecule or polyatomic ion.
2. Count the number of regions of electron density (lone pairs and bonds) around the central atom. A single, double, or triple bond counts as one region of electron density.
3. Identify the electron-pair geometry based on the number of regions of electron density: linear, trigonal planar, tetrahedral, trigonal bipyramidal, or octahedral (Figure 7.19, first column).
4. Use the number of lone pairs to determine the molecular structure (Figure 7.19). If more than one arrangement of lone pairs and chemical bonds is possible, choose the one that will minimize repulsions, remembering that lone pairs occupy more space than multiple bonds, which occupy more space than single bonds. In trigonal bipyramidal arrangements, repulsion is minimized when every lone pair is in an equatorial position. In an octahedral arrangement with two lone pairs, repulsion is minimized when the lone pairs are on opposite sides of the central atom.

The following examples illustrate the use of VSEPR theory to predict the molecular structure of molecules or ions that have no lone pairs of electrons. In this case, the molecular structure is identical to the electron pair geometry.

EXAMPLE 7.11

Predicting Electron-pair Geometry and Molecular Structure: CO_2 and BCl_3

Predict the electron-pair geometry and molecular structure for each of the following:

(a) carbon dioxide, CO_2, a molecule produced by the combustion of fossil fuels

(b) boron trichloride, BCl_3, an important industrial chemical

Solution

(a) We write the Lewis structure of CO_2 as:

This shows us two regions of high electron density around the carbon atom—each double bond counts as one region, and there are no lone pairs on the carbon atom. Using VSEPR theory, we predict that the two regions of electron density arrange themselves on opposite sides of the central atom with a bond angle of 180°. The electron-pair geometry and molecular structure are identical, and CO_2 molecules are linear.

(b) We write the Lewis structure of BCl_3 as:

Thus we see that BCl_3 contains three bonds, and there are no lone pairs of electrons on boron. The arrangement of three regions of high electron density gives a trigonal planar electron-pair geometry. The B–Cl bonds lie in a plane with 120° angles between them. BCl_3 also has a trigonal planar molecular structure (Figure 7.21).

FIGURE 7.21

The electron-pair geometry and molecular structure of BCl_3 are both trigonal planar. Note that the VSEPR geometry indicates the correct bond angles (120°), unlike the Lewis structure shown above.

Check Your Learning

Carbonate, $CO_3{}^{2-}$, is a common polyatomic ion found in various materials from eggshells to antacids. What are the electron-pair geometry and molecular structure of this polyatomic ion?

Answer:

The electron-pair geometry is trigonal planar and the molecular structure is trigonal planar. Due to resonance, all three C–O bonds are identical. Whether they are single, double, or an average of the two, each bond counts as one region of electron density.

EXAMPLE 7.12

Predicting Electron-pair Geometry and Molecular Structure: Ammonium

Two of the top 50 chemicals produced in the United States, ammonium nitrate and ammonium sulfate, both used as fertilizers, contain the ammonium ion. Predict the electron-pair geometry and molecular structure of the $NH_4{}^+$ cation.

Solution

We write the Lewis structure of $NH_4{}^+$ as:

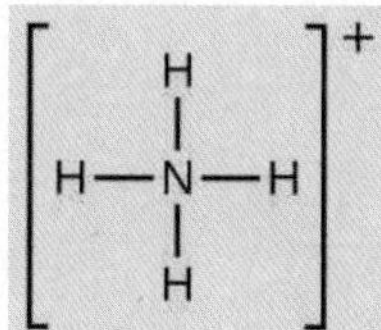

We can see that $NH_4{}^+$ contains four bonds from the nitrogen atom to hydrogen atoms and no lone pairs. We expect the four regions of high electron density to arrange themselves so that they point to the corners of a tetrahedron with the central nitrogen atom in the middle (Figure 7.19). Therefore, the electron pair geometry of $NH_4{}^+$ is tetrahedral, and the molecular structure is also tetrahedral (Figure 7.22).

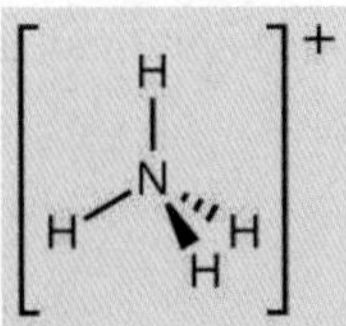

FIGURE 7.22 The ammonium ion displays a tetrahedral electron-pair geometry as well as a tetrahedral molecular structure.

Check Your Learning

Identify a molecule with trigonal bipyramidal molecular structure.

Answer:

Any molecule with five electron pairs around the central atoms including no lone pairs will be trigonal bipyramidal. PF_5 is a common example.

The next several examples illustrate the effect of lone pairs of electrons on molecular structure.

EXAMPLE 7.13

Predicting Electron-pair Geometry and Molecular Structure: Lone Pairs on the Central Atom

Predict the electron-pair geometry and molecular structure of a water molecule.

Solution

The Lewis structure of H_2O indicates that there are four regions of high electron density around the oxygen atom: two lone pairs and two chemical bonds:

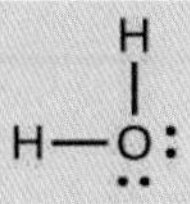

We predict that these four regions are arranged in a tetrahedral fashion (Figure 7.23), as indicated in Figure 7.19. Thus, the electron-pair geometry is tetrahedral and the molecular structure is bent with an angle slightly less than 109.5°. In fact, the bond angle is 104.5°.

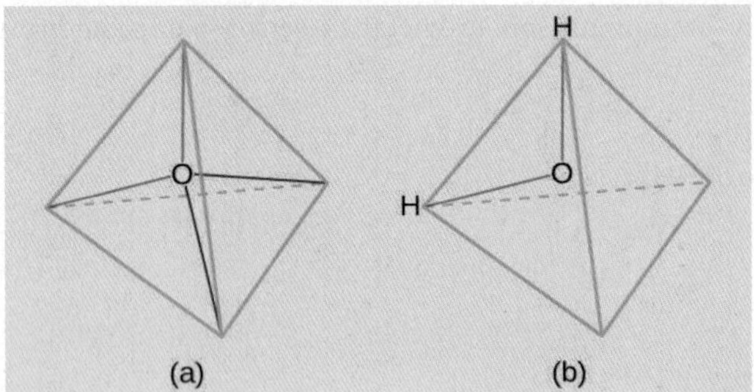

FIGURE 7.23 (a) H_2O has four regions of electron density around the central atom, so it has a tetrahedral electron-pair geometry. (b) Two of the electron regions are lone pairs, so the molecular structure is bent.

Check Your Learning

The hydronium ion, H_3O^+, forms when acids are dissolved in water. Predict the electron-pair geometry and molecular structure of this cation.

Answer:

electron pair geometry: tetrahedral; molecular structure: trigonal pyramidal

---

EXAMPLE 7.14

Predicting Electron-pair Geometry and Molecular Structure: SF_4

Sulfur tetrafluoride, SF_4, is extremely valuable for the preparation of fluorine-containing compounds used as herbicides (i.e., SF_4 is used as a fluorinating agent). Predict the electron-pair geometry and molecular structure of a SF_4 molecule.

Solution

The Lewis structure of SF_4 indicates five regions of electron density around the sulfur atom: one lone pair and four bonding pairs:

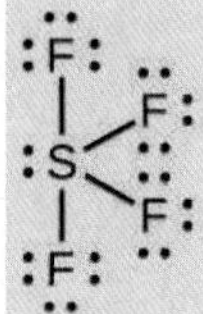

We expect these five regions to adopt a trigonal bipyramidal electron-pair geometry. To minimize lone pair repulsions, the lone pair occupies one of the equatorial positions. The molecular structure (Figure 7.24) is that of a seesaw (Figure 7.19).

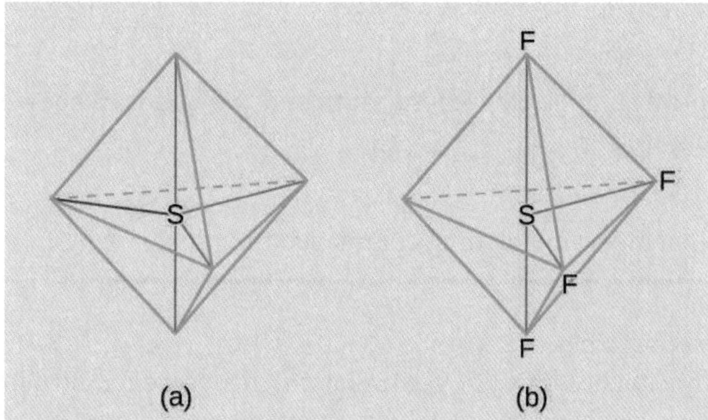

FIGURE 7.24 (a) SF4 has a trigonal bipyramidal arrangement of the five regions of electron density. (b) One of the regions is a lone pair, which results in a seesaw-shaped molecular structure.

Check Your Learning

Predict the electron pair geometry and molecular structure for molecules of XeF_2.

Answer:

The electron-pair geometry is trigonal bipyramidal. The molecular structure is linear.

EXAMPLE 7.15

Predicting Electron-pair Geometry and Molecular Structure: XeF_4

Of all the noble gases, xenon is the most reactive, frequently reacting with elements such as oxygen and fluorine. Predict the electron-pair geometry and molecular structure of the XeF_4 molecule.

Solution

The Lewis structure of XeF_4 indicates six regions of high electron density around the xenon atom: two lone pairs and four bonds:

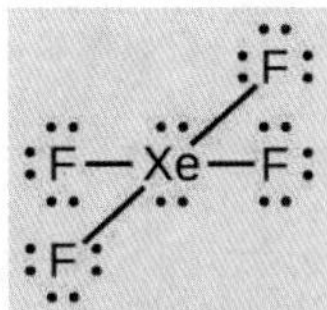

These six regions adopt an octahedral arrangement (Figure 7.19), which is the electron-pair geometry. To minimize repulsions, the lone pairs should be on opposite sides of the central atom (Figure 7.25). The five atoms are all in the same plane and have a square planar molecular structure.

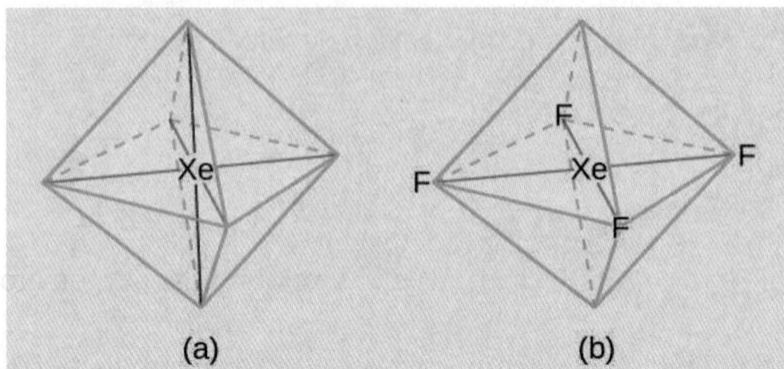

FIGURE 7.25 (a) XeF_4 adopts an octahedral arrangement with two lone pairs (red lines) and four bonds in the electron-pair geometry. (b) The molecular structure is square planar with the lone pairs directly across from one another.

Check Your Learning

In a certain molecule, the central atom has three lone pairs and two bonds. What will the electron pair geometry and molecular structure be?

Answer:

electron pair geometry: trigonal bipyramidal; molecular structure: linear

---

Molecular Structure for Multicenter Molecules

When a molecule or polyatomic ion has only one central atom, the molecular structure completely describes the shape of the molecule. Larger molecules do not have a single central atom, but are connected by a chain of interior atoms that each possess a "local" geometry. The way these local structures are oriented with respect to each other also influences the molecular shape, but such considerations are largely beyond the scope of this introductory discussion. For our purposes, we will only focus on determining the local structures.

 **EXAMPLE 7.16**

Predicting Structure in Multicenter Molecules

The Lewis structure for the simplest amino acid, glycine, $H_2NCH_2CO_2H$, is shown here. Predict the local geometry for the nitrogen atom, the two carbon atoms, and the oxygen atom with a hydrogen atom attached:

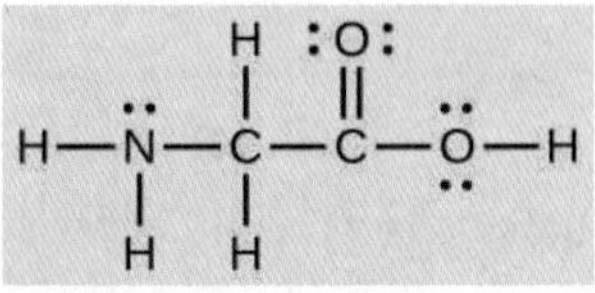

Solution

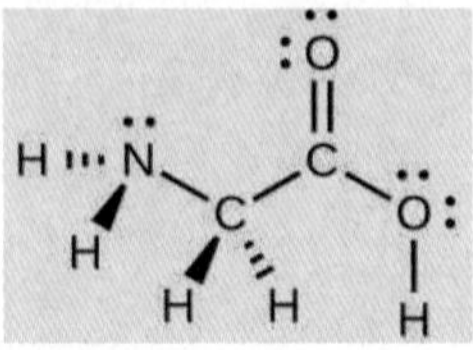

Consider each central atom independently. The electron-pair geometries:

- nitrogen––four regions of electron density; tetrahedral
- carbon ($\underline{C}H_2$)––four regions of electron density; tetrahedral
- carbon ($\underline{C}O_2$)—three regions of electron density; trigonal planar
- oxygen ($\underline{O}H$)—four regions of electron density; tetrahedral

The local structures:

- nitrogen––three bonds, one lone pair; trigonal pyramidal
- carbon ($\underline{C}H_2$)—four bonds, no lone pairs; tetrahedral
- carbon ($\underline{C}O_2$)—three bonds (double bond counts as one bond), no lone pairs; trigonal planar
- oxygen ($\underline{O}H$)—two bonds, two lone pairs; bent (109°)

Check Your Learning

Another amino acid is alanine, which has the Lewis structure shown here. Predict the electron-pair geometry and local structure of the nitrogen atom, the three carbon atoms, and the oxygen atom with hydrogen attached:

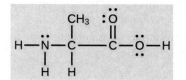

Answer:

electron-pair geometries: nitrogen––tetrahedral; carbon ($\underline{C}H$)—tetrahedral; carbon ($\underline{C}H_3$)—tetrahedral; carbon ($\underline{C}O_2$)—trigonal planar; oxygen ($\underline{O}H$)—tetrahedral; local structures: nitrogen—trigonal pyramidal; carbon ($\underline{C}H$)—tetrahedral; carbon ($\underline{C}H_3$)—tetrahedral; carbon ($\underline{C}O_2$)—trigonal planar; oxygen ($\underline{O}H$)—bent (109°)

LINK TO LEARNING

The molecular shape simulator (http://openstax.org/l/16MolecShape) lets you build various molecules and practice naming their electron-pair geometries and molecular structures.

EXAMPLE 7.17

Molecular Simulation

Using molecular shape simulator (http://openstax.org/l/16MolecShape) allows us to control whether bond angles and/or lone pairs are displayed by checking or unchecking the boxes under "Options" on the right. We can also use the "Name" checkboxes at bottom-left to display or hide the electron pair geometry (called "electron geometry" in the simulator) and/or molecular structure (called "molecular shape" in the simulator).

Build the molecule HCN in the simulator based on the following Lewis structure:

$$H–C\equiv N$$

Click on each bond type or lone pair at right to add that group to the central atom. Once you have the complete molecule, rotate it to examine the predicted molecular structure. What molecular structure is this?

Solution

The molecular structure is linear.

Check Your Learning

Build a more complex molecule in the simulator. Identify the electron-group geometry, molecular structure, and bond angles. Then try to find a chemical formula that would match the structure you have drawn.

Answer:

Answers will vary. For example, an atom with four single bonds, a double bond, and a lone pair has an octahedral electron-group geometry and a square pyramidal molecular structure. $XeOF_4$ is a molecule that adopts this structure.

Molecular Polarity and Dipole Moment

As discussed previously, polar covalent bonds connect two atoms with differing electronegativities, leaving one atom with a partial positive charge (δ+) and the other atom with a partial negative charge (δ–), as the electrons are pulled toward the more electronegative atom. This separation of charge gives rise to a **bond dipole moment**. The magnitude of a bond dipole moment is represented by the Greek letter mu (μ) and is given by the formula shown here, where Q is the magnitude of the partial charges (determined by the electronegativity difference) and r is the distance between the charges:

$$\mu = \text{Qr}$$

This bond moment can be represented as a **vector**, a quantity having both direction and magnitude (Figure 7.26). Dipole vectors are shown as arrows pointing along the bond from the less electronegative atom toward the more electronegative atom. A small plus sign is drawn on the less electronegative end to indicate the partially positive end of the bond. The length of the arrow is proportional to the magnitude of the electronegativity difference between the two atoms.

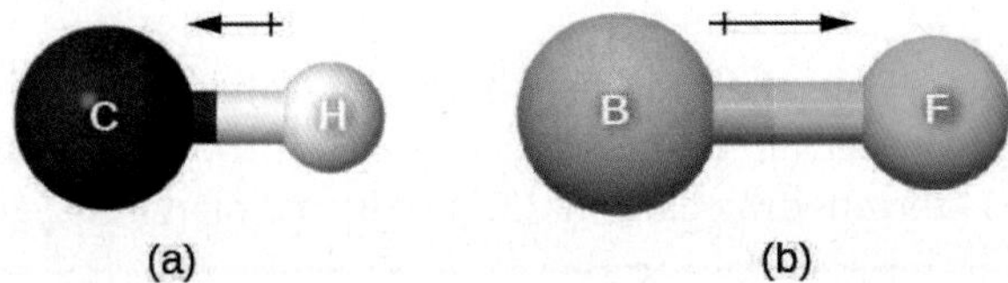

FIGURE 7.26 (a) There is a small difference in electronegativity between C and H, represented as a short vector. (b) The electronegativity difference between B and F is much larger, so the vector representing the bond moment is much longer.

A whole molecule may also have a separation of charge, depending on its molecular structure and the polarity of each of its bonds. If such a charge separation exists, the molecule is said to be a **polar molecule** (or dipole); otherwise the molecule is said to be nonpolar. The **dipole moment** measures the extent of net charge separation in the molecule as a whole. We determine the dipole moment by adding the bond moments in three-dimensional space, taking into account the molecular structure.

For diatomic molecules, there is only one bond, so its bond dipole moment determines the molecular polarity. Homonuclear diatomic molecules such as Br_2 and N_2 have no difference in electronegativity, so their dipole moment is zero. For heteronuclear molecules such as CO, there is a small dipole moment. For HF, there is a larger dipole moment because there is a larger difference in electronegativity.

When a molecule contains more than one bond, the geometry must be taken into account. If the bonds in a molecule are arranged such that their bond moments cancel (vector sum equals zero), then the molecule is nonpolar. This is the situation in CO_2 (Figure 7.27). Each of the bonds is polar, but the molecule as a whole is nonpolar. From the Lewis structure, and using VSEPR theory, we determine that the CO_2 molecule is linear with polar C=O bonds on opposite sides of the carbon atom. The bond moments cancel because they are pointed in opposite directions. In the case of the water molecule (Figure 7.27), the Lewis structure again shows that there are two bonds to a central atom, and the electronegativity difference again shows that each of these bonds has a nonzero bond moment. In this case, however, the molecular structure is bent because of the lone pairs on O, and the two bond moments do not cancel. Therefore, water does have a net dipole moment and is a polar molecule (dipole).

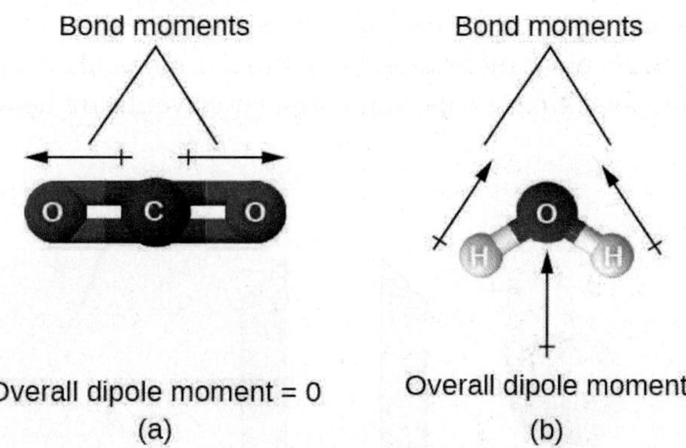

FIGURE 7.27 The overall dipole moment of a molecule depends on the individual bond dipole moments and how they are arranged. (a) Each CO bond has a bond dipole moment, but they point in opposite directions so that the net CO_2 molecule is nonpolar. (b) In contrast, water is polar because the OH bond moments do not cancel out.

The OCS molecule has a structure similar to CO_2, but a sulfur atom has replaced one of the oxygen atoms. To determine if this molecule is polar, we draw the molecular structure. VSEPR theory predicts a linear molecule:

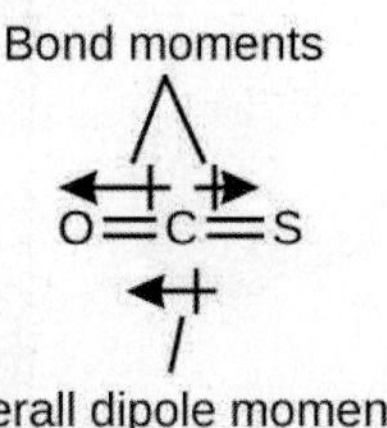

The C-O bond is considerably polar. Although C and S have very similar electronegativity values, S is slightly more electronegative than C, and so the C-S bond is just slightly polar. Because oxygen is more electronegative than sulfur, the oxygen end of the molecule is the negative end.

Chloromethane, CH_3Cl, is a tetrahedral molecule with three slightly polar C-H bonds and a more polar C-Cl bond. The relative electronegativities of the bonded atoms is H < C < Cl, and so the bond moments all point toward the Cl end of the molecule and sum to yield a considerable dipole moment (the molecules are relatively polar).

For molecules of high symmetry such as BF_3 (trigonal planar), CH_4 (tetrahedral), PF_5 (trigonal bipyramidal), and SF_6 (octahedral), all the bonds are of identical polarity (same bond moment) and they are oriented in geometries that yield nonpolar molecules (dipole moment is zero). Molecules of less geometric symmetry, however, may be polar even when all bond moments are identical. For these molecules, the directions of the equal bond moments are such that they sum to give a nonzero dipole moment and a polar molecule. Examples of such molecules include hydrogen sulfide, H_2S (nonlinear), and ammonia, NH_3 (trigonal pyramidal).

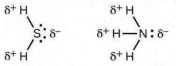

δ^+ H　　　　δ^+ H
S: δ^-　　δ^+ H—N: δ^-
δ^+ H　　　　δ^+ H

To summarize, to be polar, a molecule must:

1. Contain at least one polar covalent bond.
2. Have a molecular structure such that the sum of the vectors of each bond dipole moment does not cancel.

Properties of Polar Molecules

Polar molecules tend to align when placed in an electric field with the positive end of the molecule oriented toward the negative plate and the negative end toward the positive plate (Figure 7.28). We can use an electrically charged object to attract polar molecules, but nonpolar molecules are not attracted. Also, polar solvents are better at dissolving polar substances, and nonpolar solvents are better at dissolving nonpolar substances.

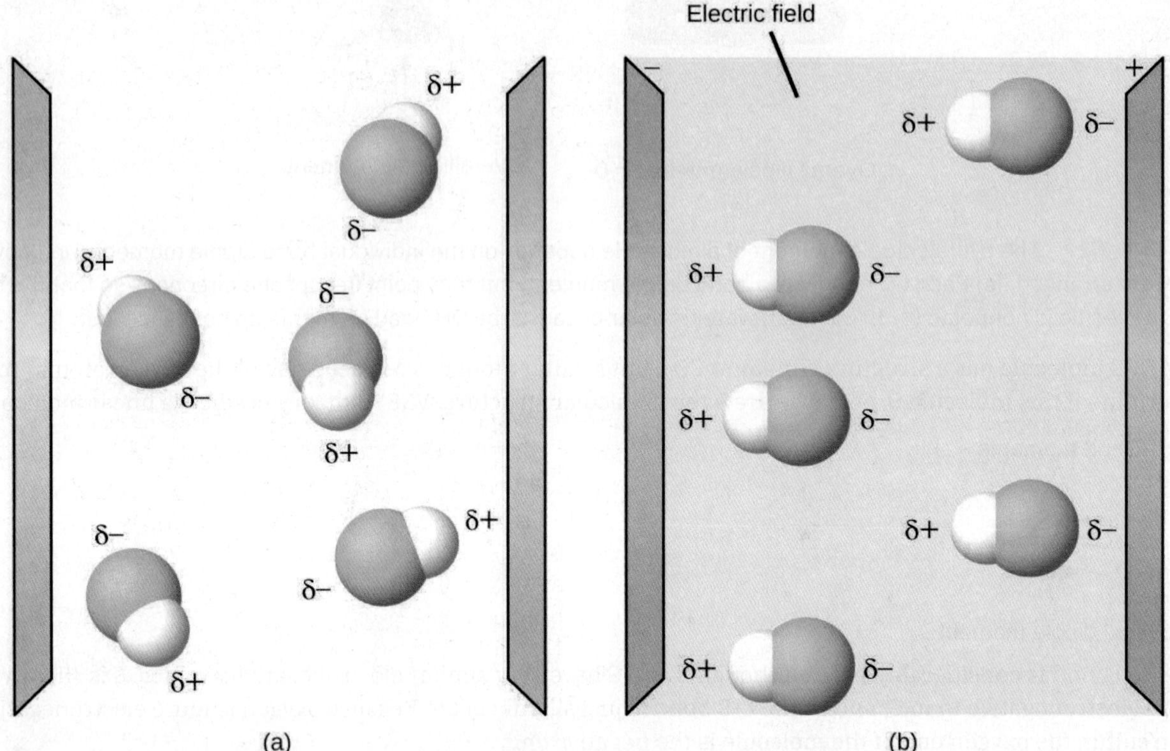

FIGURE 7.28 (a) Molecules are always randomly distributed in the liquid state in the absence of an electric field. (b) When an electric field is applied, polar molecules like HF will align to the dipoles with the field direction.

LINK TO LEARNING

The molecule polarity simulation (http://openstax.org/l/16MolecPolarity) provides many ways to explore dipole moments of bonds and molecules.

EXAMPLE 7.18

Polarity Simulations

Open the molecule polarity simulation (http://openstax.org/l/16MolecPolarity) and select the "Three Atoms" tab at the top. This should display a molecule ABC with three electronegativity adjustors. You can display or hide the bond moments, molecular dipoles, and partial charges at the right. Turning on the Electric Field will show whether the molecule moves when exposed to a field, similar to Figure 7.28.

Use the electronegativity controls to determine how the molecular dipole will look for the starting bent molecule if:

(a) A and C are very electronegative and B is in the middle of the range.

(b) A is very electronegative, and B and C are not.

Solution

(a) Molecular dipole moment points immediately between A and C.

(b) Molecular dipole moment points along the A–B bond, toward A.

Check Your Learning

Determine the partial charges that will give the largest possible bond dipoles.

Answer:

The largest bond moments will occur with the largest partial charges. The two solutions above represent how unevenly the electrons are shared in the bond. The bond moments will be maximized when the electronegativity difference is greatest. The controls for A and C should be set to one extreme, and B should be set to the opposite extreme. Although the magnitude of the bond moment will not change based on whether B is the most electronegative or the least, the direction of the bond moment will.

Key Terms

axial position location in a trigonal bipyramidal geometry in which there is another atom at a 180° angle and the equatorial positions are at a 90° angle

bond angle angle between any two covalent bonds that share a common atom

bond dipole moment separation of charge in a bond that depends on the difference in electronegativity and the bond distance represented by partial charges or a vector

bond distance (also, bond length) distance between the nuclei of two bonded atoms

bond energy (also, bond dissociation energy) energy required to break a covalent bond in a gaseous substance

bond length distance between the nuclei of two bonded atoms at which the lowest potential energy is achieved

Born-Haber cycle thermochemical cycle relating the various energetic steps involved in the formation of an ionic solid from the relevant elements

covalent bond bond formed when electrons are shared between atoms

dipole moment property of a molecule that describes the separation of charge determined by the sum of the individual bond moments based on the molecular structure

double bond covalent bond in which two pairs of electrons are shared between two atoms

electron-pair geometry arrangement around a central atom of all regions of electron density (bonds, lone pairs, or unpaired electrons)

electronegativity tendency of an atom to attract electrons in a bond to itself

equatorial position one of the three positions in a trigonal bipyramidal geometry with 120° angles between them; the axial positions are located at a 90° angle

formal charge charge that would result on an atom by taking the number of valence electrons on the neutral atom and subtracting the nonbonding electrons and the number of bonds (one-half of the bonding electrons)

free radical molecule that contains an odd number of electrons

hypervalent molecule molecule containing at least one main group element that has more than eight electrons in its valence shell

inert pair effect tendency of heavy atoms to form ions in which their valence *s* electrons are not lost

ionic bond strong electrostatic force of attraction between cations and anions in an ionic compound

lattice energy ($\Delta H_{\text{lattice}}$) energy required to separate one mole of an ionic solid into its component gaseous ions

Lewis structure diagram showing lone pairs and bonding pairs of electrons in a molecule or an ion

Lewis symbol symbol for an element or monatomic ion that uses a dot to represent each valence electron in the element or ion

linear shape in which two outside groups are placed on opposite sides of a central atom

lone pair two (a pair of) valence electrons that are not used to form a covalent bond

molecular structure arrangement of atoms in a molecule or ion

molecular structure structure that includes only the placement of the atoms in the molecule

octahedral shape in which six outside groups are placed around a central atom such that a three-dimensional shape is generated with four groups forming a square and the other two forming the apex of two pyramids, one above and one below the square plane

octet rule guideline that states main group atoms will form structures in which eight valence electrons interact with each nucleus, counting bonding electrons as interacting with both atoms connected by the bond

polar covalent bond covalent bond between atoms of different electronegativities; a covalent bond with a positive end and a negative end

polar molecule (also, dipole) molecule with an overall dipole moment

pure covalent bond (also, nonpolar covalent bond) covalent bond between atoms of identical electronegativities

resonance situation in which one Lewis structure is insufficient to describe the bonding in a molecule and the average of multiple structures is observed

resonance forms two or more Lewis structures that have the same arrangement of atoms but different arrangements of electrons

resonance hybrid average of the resonance forms shown by the individual Lewis structures

single bond bond in which a single pair of electrons is shared between two atoms

tetrahedral shape in which four outside groups are placed around a central atom such that a three-dimensional shape is generated with four corners and 109.5° angles between each pair and the central atom

trigonal bipyramidal shape in which five outside groups are placed around a central atom such that three form a flat triangle with 120° angles between each pair and the central atom, and the other two form the apex of two pyramids, one above and one below the triangular plane

trigonal planar shape in which three outside groups are placed in a flat triangle around a central atom with 120° angles between each pair and the central atom

triple bond bond in which three pairs of electrons are shared between two atoms

valence shell electron-pair repulsion theory (VSEPR) theory used to predict the bond angles in a molecule based on positioning regions of high electron density as far apart as possible to minimize electrostatic repulsion

vector quantity having magnitude and direction

Key Equations

$$\text{formal charge} = \#\text{ valence shell electrons (free atom)} - \#\text{ lone pair electrons} - \frac{1}{2}\#\text{ bonding electrons}$$

Bond energy for a diatomic molecule: $XY(g) \longrightarrow X(g) + Y(g) \qquad D_{X–Y} = \Delta H°$

Enthalpy change: $\Delta H = \Sigma D_{\text{bonds broken}} - \Sigma D_{\text{bonds formed}}$

Lattice energy for a solid MX: $MX(s) \longrightarrow M^{n+}(g) + X^{n-}(g) \qquad \Delta H_{\text{lattice}}$

Lattice energy for an ionic crystal: $\Delta H_{\text{lattice}} = \frac{C(Z^+)(Z^-)}{R_o}$

Summary

7.1 Ionic Bonding

Atoms gain or lose electrons to form ions with particularly stable electron configurations. The charges of cations formed by the representative metals may be determined readily because, with few exceptions, the electronic structures of these ions have either a noble gas configuration or a completely filled electron shell. The charges of anions formed by the nonmetals may also be readily determined because these ions form when nonmetal atoms gain enough electrons to fill their valence shells.

7.2 Covalent Bonding

Covalent bonds form when electrons are shared between atoms and are attracted by the nuclei of both atoms. In pure covalent bonds, the electrons are shared equally. In polar covalent bonds, the electrons are shared unequally, as one atom exerts a stronger force of attraction on the electrons than the other. The ability of an atom to attract a pair of electrons in a chemical bond is called its electronegativity. The difference in electronegativity between two atoms determines how polar a bond will be. In a diatomic molecule with two identical atoms, there is no difference in electronegativity, so the bond is nonpolar or pure covalent. When the electronegativity difference is very large, as is the case between metals and nonmetals, the bonding is characterized as ionic.

7.3 Lewis Symbols and Structures

Valence electronic structures can be visualized by drawing Lewis symbols (for atoms and monatomic ions) and Lewis structures (for molecules and polyatomic ions). Lone pairs, unpaired electrons, and single, double, or triple bonds are used to indicate where the valence electrons are located around each atom in a Lewis structure. Most structures—especially those containing second row elements—obey the octet rule, in which every atom (except H) is surrounded by eight electrons. Exceptions to the octet rule occur for odd-electron molecules (free radicals), electron-deficient molecules, and hypervalent molecules.

7.4 Formal Charges and Resonance

In a Lewis structure, formal charges can be assigned to each atom by treating each bond as if one-half of the electrons are assigned to each atom. These hypothetical formal charges are a guide to determining the most appropriate Lewis structure. A structure in which the formal charges are as close to zero as possible is preferred. Resonance occurs in cases where two or more Lewis structures with identical arrangements of atoms but different distributions of electrons can be written. The actual distribution of electrons (the resonance hybrid) is an average of the distribution indicated by the individual Lewis structures (the resonance forms).

7.5 Strengths of Ionic and Covalent Bonds

The strength of a covalent bond is measured by its bond dissociation energy, that is, the amount of energy required to break that particular bond in a mole of molecules. Multiple bonds are stronger than single bonds between the same atoms. The enthalpy of a reaction can be estimated based on the energy input required to break bonds and the energy released when new bonds are formed. For ionic bonds, the lattice energy is the energy required to separate one mole of a compound into its gas phase ions. Lattice energy increases for ions with higher charges and shorter distances between ions. Lattice energies are often calculated using the Born-Haber cycle, a thermochemical cycle including all of the energetic steps involved in converting elements into an ionic compound.

7.6 Molecular Structure and Polarity

VSEPR theory predicts the three-dimensional arrangement of atoms in a molecule. It states that valence electrons will assume an electron-pair geometry that minimizes repulsions between areas of high electron density (bonds and/or lone pairs). Molecular structure, which refers only to the placement of atoms in a molecule and not the electrons, is equivalent to electron-pair geometry only when there are no lone electron pairs around the central atom. A dipole moment measures a separation of charge. For one bond, the bond dipole moment is determined by the difference in electronegativity between the two atoms. For a molecule, the overall dipole moment is determined by both the individual bond moments and how these dipoles are arranged in the molecular structure. Polar molecules (those with an appreciable dipole moment) interact with electric fields, whereas nonpolar molecules do not.

Exercises

7.1 Ionic Bonding

1. Does a cation gain protons to form a positive charge or does it lose electrons?
2. Iron(III) sulfate [$Fe_2(SO_4)_3$] is composed of Fe^{3+} and $SO_4{}^{2-}$ ions. Explain why a sample of iron(III) sulfate is uncharged.
3. Which of the following atoms would be expected to form negative ions in binary ionic compounds and which would be expected to form positive ions: P, I, Mg, Cl, In, Cs, O, Pb, Co?
4. Which of the following atoms would be expected to form negative ions in binary ionic compounds and which would be expected to form positive ions: Br, Ca, Na, N, F, Al, Sn, S, Cd?
5. Predict the charge on the monatomic ions formed from the following atoms in binary ionic compounds:
   (a) P
   (b) Mg
   (c) Al
   (d) O
   (e) Cl
   (f) Cs
6. Predict the charge on the monatomic ions formed from the following atoms in binary ionic compounds:
   (a) I
   (b) Sr
   (c) K
   (d) N
   (e) S
   (f) In

7. Write the electron configuration for each of the following ions:
(a) As^{3-}
(b) I^{-}
(c) Be^{2+}
(d) Cd^{2+}
(e) O^{2-}
(f) Ga^{3+}
(g) Li^{+}
(h) N^{3-}
(i) Sn^{2+}
(j) Co^{2+}
(k) Fe^{2+}
(l) As^{3+}
8. Write the electron configuration for the monatomic ions formed from the following elements (which form the greatest concentration of monatomic ions in seawater):
(a) Cl
(b) Na
(c) Mg
(d) Ca
(e) K
(f) Br
(g) Sr
(h) F
9. Write out the full electron configuration for each of the following atoms and for the monatomic ion found in binary ionic compounds containing the element:
(a) Al
(b) Br
(c) Sr
(d) Li
(e) As
(f) S
10. From the labels of several commercial products, prepare a list of six ionic compounds in the products. For each compound, write the formula. (You may need to look up some formulas in a suitable reference.)

7.2 Covalent Bonding

11. Why is it incorrect to speak of a molecule of solid NaCl?
12. What information can you use to predict whether a bond between two atoms is covalent or ionic?
13. Predict which of the following compounds are ionic and which are covalent, based on the location of their constituent atoms in the periodic table:
(a) Cl_2CO
(b) MnO
(c) NCl_3
(d) $CoBr_2$
(e) K_2S
(f) CO
(g) CaF_2
(h) HI
(i) CaO
(j) IBr
(k) CO_2
14. Explain the difference between a nonpolar covalent bond, a polar covalent bond, and an ionic bond.

15. From its position in the periodic table, determine which atom in each pair is more electronegative:
(a) Br or Cl
(b) N or O
(c) S or O
(d) P or S
(e) Si or N
(f) Ba or P
(g) N or K

16. From its position in the periodic table, determine which atom in each pair is more electronegative:
(a) N or P
(b) N or Ge
(c) S or F
(d) Cl or S
(e) H or C
(f) Se or P
(g) C or Si

17. From their positions in the periodic table, arrange the atoms in each of the following series in order of increasing electronegativity:
(a) C, F, H, N, O
(b) Br, Cl, F, H, I
(c) F, H, O, P, S
(d) Al, H, Na, O, P
(e) Ba, H, N, O, As

18. From their positions in the periodic table, arrange the atoms in each of the following series in order of increasing electronegativity:
(a) As, H, N, P, Sb
(b) Cl, H, P, S, Si
(c) Br, Cl, Ge, H, Sr
(d) Ca, H, K, N, Si
(e) Cl, Cs, Ge, H, Sr

19. Which atoms can bond to sulfur so as to produce a positive partial charge on the sulfur atom?

20. Which is the most polar bond?
(a) C–C
(b) C–H
(c) N–H
(d) O–H
(e) Se–H

21. Identify the more polar bond in each of the following pairs of bonds:
(a) HF or HCl
(b) NO or CO
(c) SH or OH
(d) PCl or SCl
(e) CH or NH
(f) SO or PO
(g) CN or NN

22. Which of the following molecules or ions contain polar bonds?
(a) O_3
(b) S_8
(c) $O_2{}^{2-}$
(d) $NO_3{}^{-}$
(e) CO_2
(f) H_2S
(g) $BH_4{}^{-}$

7.3 Lewis Symbols and Structures

23. Write the Lewis symbols for each of the following ions:
(a) As^{3-}
(b) I^-
(c) Be^{2+}
(d) O^{2-}
(e) Ga^{3+}
(f) Li^+
(g) N^{3-}

24. Many monatomic ions are found in seawater, including the ions formed from the following list of elements. Write the Lewis symbols for the monatomic ions formed from the following elements:
(a) Cl
(b) Na
(c) Mg
(d) Ca
(e) K
(f) Br
(g) Sr
(h) F

25. Write the Lewis symbols of the ions in each of the following ionic compounds and the Lewis symbols of the atom from which they are formed:
(a) MgS
(b) Al_2O_3
(c) $GaCl_3$
(d) K_2O
(e) Li_3N
(f) KF

26. In the Lewis structures listed here, M and X represent various elements in the third period of the periodic table. Write the formula of each compound using the chemical symbols of each element:
(a)

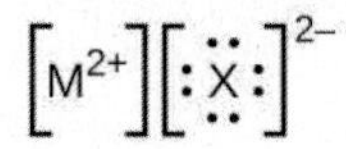

$$[M^{2+}][:\ddot{\underset{..}{X}}:]^{2-}$$

(b)

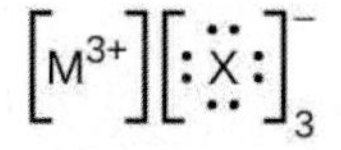

$$[M^{3+}][:\ddot{\underset{..}{X}}:]^{-}_{3}$$

(c)

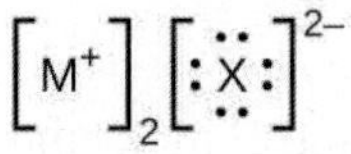

$$[M^{+}]_2[:\ddot{\underset{..}{X}}:]^{2-}$$

(d)

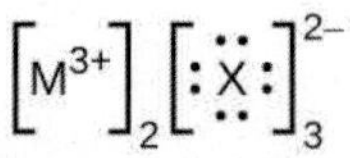

$$[M^{3+}]_2[:\ddot{\underset{..}{X}}:]^{2-}_{3}$$

27. Write the Lewis structure for the diatomic molecule P_2, an unstable form of phosphorus found in high-temperature phosphorus vapor.

28. Write Lewis structures for the following:
    (a) H_2
    (b) HBr
    (c) PCl_3
    (d) SF_2
    (e) H_2CCH_2
    (f) HNNH
    (g) H_2CNH
    (h) NO^-
    (i) N_2
    (j) CO
    (k) CN^-
29. Write Lewis structures for the following:
    (a) O_2
    (b) H_2CO
    (c) AsF_3
    (d) ClNO
    (e) $SiCl_4$
    (f) H_3O^+
    (g) $NH_4{}^+$
    (h) $BF_4{}^-$
    (i) HCCH
    (j) ClCN
    (k) $C_2{}^{2+}$
30. Write Lewis structures for the following:
    (a) ClF_3
    (b) PCl_5
    (c) BF_3
    (d) $PF_6{}^-$
31. Write Lewis structures for the following:
    (a) SeF_6
    (b) XeF_4
    (c) $SeCl_3{}^+$
    (d) Cl_2BBCl_2 (contains a B–B bond)
32. Write Lewis structures for:
    (a) $PO_4{}^{3-}$
    (b) $ICl_4{}^-$
    (c) $SO_3{}^{2-}$
    (d) HONO
33. Correct the following statement: "The bonds in solid $PbCl_2$ are ionic; the bond in a HCl molecule is covalent. Thus, all of the valence electrons in $PbCl_2$ are located on the Cl^- ions, and all of the valence electrons in a HCl molecule are shared between the H and Cl atoms."
34. Write Lewis structures for the following molecules or ions:
    (a) SbH_3
    (b) XeF_2
    (c) Se_8 (a cyclic molecule with a ring of eight Se atoms)
35. Methanol, H_3COH, is used as the fuel in some race cars. Ethanol, C_2H_5OH, is used extensively as motor fuel in Brazil. Both methanol and ethanol produce CO_2 and H_2O when they burn. Write the chemical equations for these combustion reactions using Lewis structures instead of chemical formulas.
36. Many planets in our solar system contain organic chemicals including methane (CH_4) and traces of ethylene (C_2H_4), ethane (C_2H_6), propyne (H_3CCCH), and diacetylene (HCCCCH). Write the Lewis structures for each of these molecules.

37. Carbon tetrachloride was formerly used in fire extinguishers for electrical fires. It is no longer used for this purpose because of the formation of the toxic gas phosgene, Cl_2CO. Write the Lewis structures for carbon tetrachloride and phosgene.

38. Identify the atoms that correspond to each of the following electron configurations. Then, write the Lewis symbol for the common ion formed from each atom:

(a) $1s^2 2s^2 2p^5$

(b) $1s^2 2s^2 2p^6 3s^2$

(c) $1s^2 2s^2 2p^6 3s^2 3p^6 4s^2 3d^{10}$

(d) $1s^2 2s^2 2p^6 3s^2 3p^6 4s^2 3d^{10} 4p^4$

(e) $1s^2 2s^2 2p^6 3s^2 3p^6 4s^2 3d^{10} 4p^1$

39. The arrangement of atoms in several biologically important molecules is given here. Complete the Lewis structures of these molecules by adding multiple bonds and lone pairs. Do not add any more atoms.

(a) the amino acid serine:

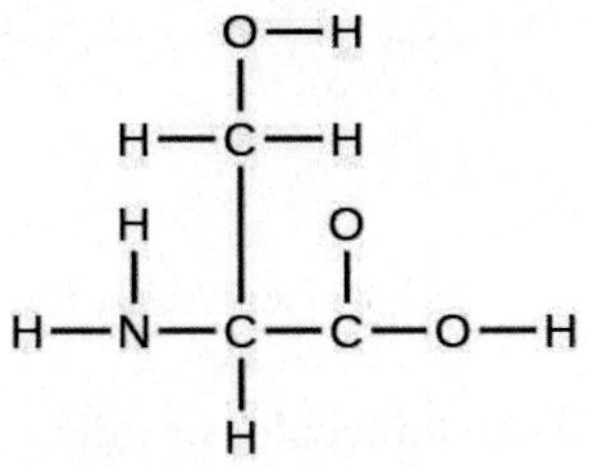

```
          O—H
          |
       H—C—H
     H    |    O
     |    |    |
H—N—C—C—O—H
          |
          H
```

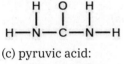

(b) urea:

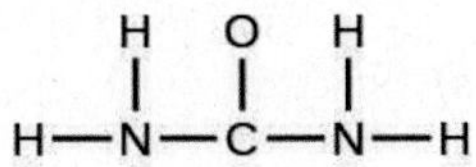

```
   H   O   H
   |   |   |
H—N—C—N—H
```

(c) pyruvic acid:

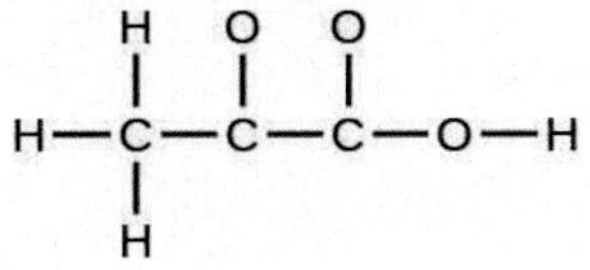

```
   H   O   O
   |   |   |
H—C—C—C—O—H
   |
   H
```

(d) uracil:

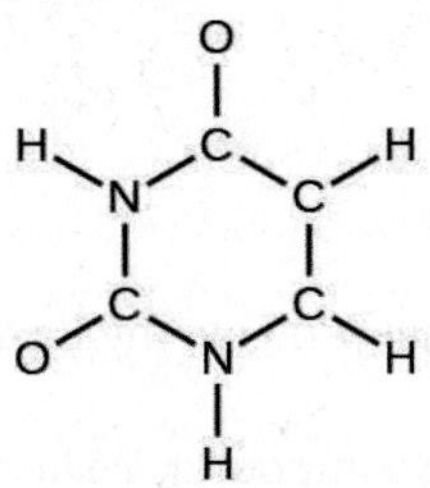

```
         O
         |
  H      C      H
    \  /   \  /
     N       C
     |       |
     C       C
   /   \   /   \
  O      N      H
         |
         H
```

(e) carbonic acid:

```
        O
        |
H—O—C—O—H
```

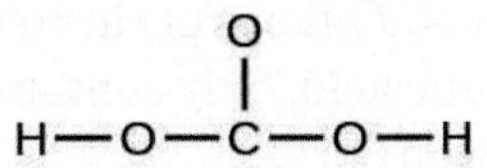

40. A compound with a molar mass of about 28 g/mol contains 85.7% carbon and 14.3% hydrogen by mass. Write the Lewis structure for a molecule of the compound.

41. A compound with a molar mass of about 42 g/mol contains 85.7% carbon and 14.3% hydrogen by mass. Write the Lewis structure for a molecule of the compound.

42. Two arrangements of atoms are possible for a compound with a molar mass of about 45 g/mol that contains 52.2% C, 13.1% H, and 34.7% O by mass. Write the Lewis structures for the two molecules.

43. How are single, double, and triple bonds similar? How do they differ?

7.4 Formal Charges and Resonance

44. Write resonance forms that describe the distribution of electrons in each of these molecules or ions.
(a) selenium dioxide, OSeO
(b) nitrate ion, NO_3^-
(c) nitric acid, HNO_3 (N is bonded to an OH group and two O atoms)
(d) benzene, C_6H_6: 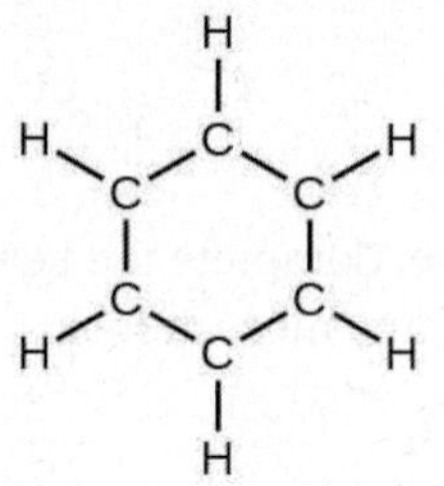
(e) the formate ion:

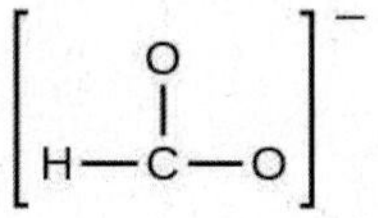

45. Write resonance forms that describe the distribution of electrons in each of these molecules or ions.
(a) sulfur dioxide, SO_2
(b) carbonate ion, CO_3^{2-}
(c) hydrogen carbonate ion, HCO_3^- (C is bonded to an OH group and two O atoms)
(d) pyridine: 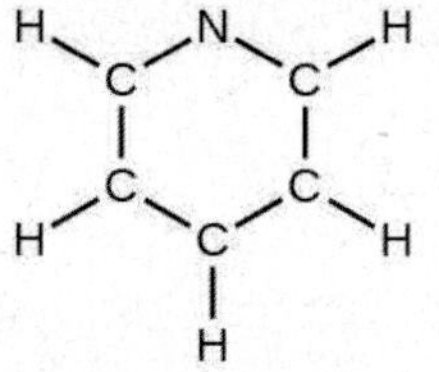
(e) the allyl ion:

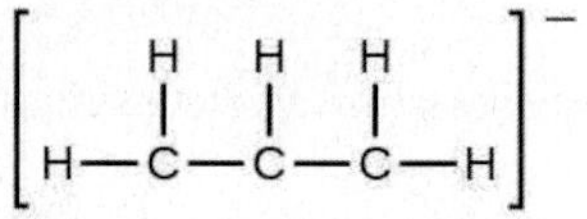

46. Write the resonance forms of ozone, O_3, the component of the upper atmosphere that protects the Earth from ultraviolet radiation.

47. Sodium nitrite, which has been used to preserve bacon and other meats, is an ionic compound. Write the resonance forms of the nitrite ion, NO_2^-.

48. In terms of the bonds present, explain why acetic acid, CH_3CO_2H, contains two distinct types of carbon-oxygen bonds, whereas the acetate ion, formed by loss of a hydrogen ion from acetic acid, only contains one type of carbon-oxygen bond. The skeleton structures of these species are shown:

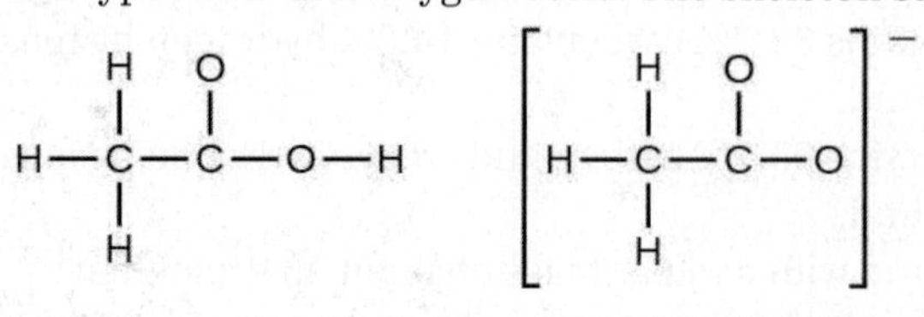

49. Write the Lewis structures for the following, and include resonance structures where appropriate. Indicate which has the strongest carbon-oxygen bond.
(a) CO_2
(b) CO

50. Toothpastes containing sodium hydrogen carbonate (sodium bicarbonate) and hydrogen peroxide are widely used. Write Lewis structures for the hydrogen carbonate ion and hydrogen peroxide molecule, with resonance forms where appropriate.

51. Determine the formal charge of each element in the following:
(a) HCl
(b) CF_4
(c) PCl_3
(d) PF_5

52. Determine the formal charge of each element in the following:
(a) H_3O^+
(b) $SO_4{}^{2-}$
(c) NH_3
(d) $O_2{}^{2-}$
(e) H_2O_2

53. Calculate the formal charge of chlorine in the molecules Cl_2, $BeCl_2$, and ClF_5.

54. Calculate the formal charge of each element in the following compounds and ions:
(a) F_2CO
(b) NO^-
(c) $BF_4{}^-$
(d) $SnCl_3{}^-$
(e) H_2CCH_2
(f) ClF_3
(g) SeF_6
(h) $PO_4{}^{3-}$

55. Draw all possible resonance structures for each of these compounds. Determine the formal charge on each atom in each of the resonance structures:
(a) O_3
(b) SO_2
(c) $NO_2{}^-$
(d) $NO_3{}^-$

56. Based on formal charge considerations, which of the following would likely be the correct arrangement of atoms in nitrosyl chloride: ClNO or ClON?

57. Based on formal charge considerations, which of the following would likely be the correct arrangement of atoms in hypochlorous acid: HOCl or OClH?

58. Based on formal charge considerations, which of the following would likely be the correct arrangement of atoms in sulfur dioxide: OSO or SOO?

59. Draw the structure of hydroxylamine, H_3NO, and assign formal charges; look up the structure. Is the actual structure consistent with the formal charges?

60. Iodine forms a series of fluorides (listed here). Write Lewis structures for each of the four compounds and determine the formal charge of the iodine atom in each molecule:
(a) IF
(b) IF_3
(c) IF_5
(d) IF_7

61. Write the Lewis structure and chemical formula of the compound with a molar mass of about 70 g/mol that contains 19.7% nitrogen and 80.3% fluorine by mass, and determine the formal charge of the atoms in this compound.

62. Which of the following structures would we expect for nitrous acid? Determine the formal charges:

H
|
:Ö—N=Ö or H—Ö—N̈=Ö

63. Sulfuric acid is the industrial chemical produced in greatest quantity worldwide. About 90 billion pounds are produced each year in the United States alone. Write the Lewis structure for sulfuric acid, H_2SO_4, which has two oxygen atoms and two OH groups bonded to the sulfur.

7.5 Strengths of Ionic and Covalent Bonds

64. Which bond in each of the following pairs of bonds is the strongest?
(a) C–C or $C = C$
(b) C–N or $C \equiv N$
(c)$C \equiv O$ or $C = O$
(d) H–F or H–Cl
(e) C–H or O–H
(f) C–N or C–O

65. Using the bond energies in Table 7.2, determine the approximate enthalpy change for each of the following reactions:
(a) $H_2(g) + Br_2(g) \longrightarrow 2HBr(g)$
(b) $CH_4(g) + I_2(g) \longrightarrow CH_3I(g) + HI(g)$
(c) $C_2H_4(g) + 3O_2(g) \longrightarrow 2CO_2(g) + 2H_2O(g)$

66. Using the bond energies in Table 7.2, determine the approximate enthalpy change for each of the following reactions:
(a) $Cl_2(g) + 3F_2(g) \longrightarrow 2ClF_3(g)$
(b) $H_2C = CH_2(g) + H_2(g) \longrightarrow H_3CCH_3(g)$
(c) $2C_2H_6(g) + 7O_2(g) \longrightarrow 4CO_2(g) + 6H_2O(g)$

67. When a molecule can form two different structures, the structure with the stronger bonds is usually the more stable form. Use bond energies to predict the correct structure of the hydroxylamine molecule:

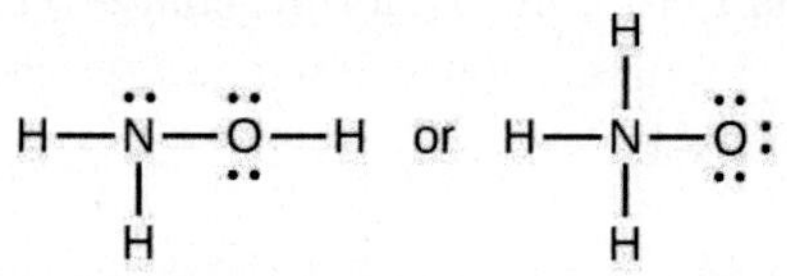

H–N(H)–O–H or H–N(H)(H)–O

68. How does the bond energy of HCl(*g*) differ from the standard enthalpy of formation of HCl(*g*)?

69. Using the standard enthalpy of formation data in Appendix G, show how the standard enthalpy of formation of HCl(*g*) can be used to determine the bond energy.

70. Using the standard enthalpy of formation data in Appendix G, calculate the bond energy of the carbon-sulfur double bond in CS_2.

71. Using the standard enthalpy of formation data in Appendix G, determine which bond is stronger: the S–F bond in $SF_4(g)$ or in $SF_6(g)$?

72. Using the standard enthalpy of formation data in Appendix G, determine which bond is stronger: the P–Cl bond in $PCl_3(g)$ or in $PCl_5(g)$?

73. Complete the following Lewis structure by adding bonds (not atoms), and then indicate the longest bond:

```
   H  H        H  H

H  C  C  C  C  C  C  H

   H  H
```

74. Use the bond energy to calculate an approximate value of ΔH for the following reaction. Which is the more stable form of FNO_2?

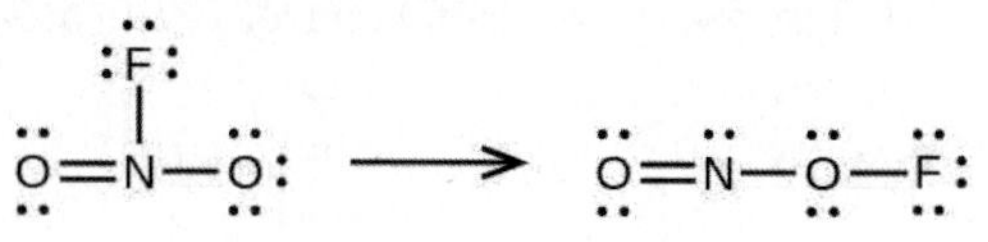

O=N(F)–O → O=N–O–F

75. Use principles of atomic structure to answer each of the following:[1]
(a) The radius of the Ca atom is 197 pm; the radius of the Ca^{2+} ion is 99 pm. Account for the difference.
(b) The lattice energy of $CaO(s)$ is –3460 kJ/mol; the lattice energy of K_2O is –2240 kJ/mol. Account for the difference.
(c) Given these ionization values, explain the difference between Ca and K with regard to their first and second ionization energies.

Element	First Ionization Energy (kJ/mol)	Second Ionization Energy (kJ/mol)
K	419	3050
Ca	590	1140

(d) The first ionization energy of Mg is 738 kJ/mol and that of Al is 578 kJ/mol. Account for this difference.

76. The lattice energy of LiF is 1023 kJ/mol, and the Li–F distance is 200.8 pm. NaF crystallizes in the same structure as LiF but with a Na–F distance of 231 pm. Which of the following values most closely approximates the lattice energy of NaF: 510, 890, 1023, 1175, or 4090 kJ/mol? Explain your choice.

77. For which of the following substances is the least energy required to convert one mole of the solid into separate ions?
(a) MgO
(b) SrO
(c) KF
(d) CsF
(e) MgF_2

78. The reaction of a metal, M, with a halogen, X_2, proceeds by an exothermic reaction as indicated by this equation: $\mathrm{M}(s) + \mathrm{X}_2(g) \longrightarrow \mathrm{MX}_2(s)$. For each of the following, indicate which option will make the reaction more exothermic. Explain your answers.
(a) a large radius vs. a small radius for M^{+2}
(b) a high ionization energy vs. a low ionization energy for M
(c) an increasing bond energy for the halogen
(d) a decreasing electron affinity for the halogen
(e) an increasing size of the anion formed by the halogen

79. The lattice energy of LiF is 1023 kJ/mol, and the Li–F distance is 201 pm. MgO crystallizes in the same structure as LiF but with a Mg–O distance of 205 pm. Which of the following values most closely approximates the lattice energy of MgO: 256 kJ/mol, 512 kJ/mol, 1023 kJ/mol, 2046 kJ/mol, or 4008 kJ/mol? Explain your choice.

80. Which compound in each of the following pairs has the larger lattice energy? Note: Mg^{2+} and Li^+ have similar radii; O^{2-} and F^- have similar radii. Explain your choices.
(a) MgO or MgSe
(b) LiF or MgO
(c) Li_2O or LiCl
(d) Li_2Se or MgO

81. Which compound in each of the following pairs has the larger lattice energy? Note: Ba^{2+} and K^+ have similar radii; S^{2-} and Cl^- have similar radii. Explain your choices.
(a) K_2O or Na_2O
(b) K_2S or BaS
(c) KCl or BaS
(d) BaS or $BaCl_2$

1 This question is taken from the Chemistry Advanced Placement Examination and is used with the permission of the Educational Testing Service.

82. Which of the following compounds requires the most energy to convert one mole of the solid into separate ions?
(a) MgO
(b) SrO
(c) KF
(d) CsF
(e) MgF_2

83. Which of the following compounds requires the most energy to convert one mole of the solid into separate ions?
(a) K_2S
(b) K_2O
(c) CaS
(d) Cs_2S
(e) CaO

84. The lattice energy of KF is 794 kJ/mol, and the interionic distance is 269 pm. The Na–F distance in NaF, which has the same structure as KF, is 231 pm. Which of the following values is the closest approximation of the lattice energy of NaF: 682 kJ/mol, 794 kJ/mol, 924 kJ/mol, 1588 kJ/mol, or 3175 kJ/mol? Explain your answer.

7.6 Molecular Structure and Polarity

85. Explain why the HOH molecule is bent, whereas the HBeH molecule is linear.

86. What feature of a Lewis structure can be used to tell if a molecule's (or ion's) electron-pair geometry and molecular structure will be identical?

87. Explain the difference between electron-pair geometry and molecular structure.

88. Why is the H–N–H angle in NH_3 smaller than the H–C–H bond angle in CH_4? Why is the H–N–H angle in NH_4^+ identical to the H–C–H bond angle in CH_4?

89. Explain how a molecule that contains polar bonds can be nonpolar.

90. As a general rule, MX_n molecules (where M represents a central atom and X represents terminal atoms; n = 2 – 5) are polar if there is one or more lone pairs of electrons on M. NH_3 (M = N, X = H, n = 3) is an example. There are two molecular structures with lone pairs that are exceptions to this rule. What are they?

91. Predict the electron pair geometry and the molecular structure of each of the following molecules or ions:
(a) SF_6
(b) PCl_5
(c) BeH_2
(d) CH_3^+

92. Identify the electron pair geometry and the molecular structure of each of the following molecules or ions:
(a) IF_6^+
(b) CF_4
(c) BF_3
(d) SiF_5^-
(e) $BeCl_2$

93. What are the electron-pair geometry and the molecular structure of each of the following molecules or ions?
(a) ClF_5
(b) ClO_2^-
(c) $TeCl_4^{2-}$
(d) PCl_3
(e) SeF_4
(f) PH_2^-

94. Predict the electron pair geometry and the molecular structure of each of the following ions:
(a) H_3O^+
(b) PCl_4^-
(c) $SnCl_3^+$
(d) $BrCl_4^-$
(e) ICl_3
(f) XeF_4
(g) SF_2

95. Identify the electron pair geometry and the molecular structure of each of the following molecules:
(a) ClNO (N is the central atom)
(b) CS_2
(c) Cl_2CO (C is the central atom)
(d) Cl_2SO (S is the central atom)
(e) SO_2F_2 (S is the central atom)
(f) XeO_2F_2 (Xe is the central atom)
(g) $ClOF_2^+$ (Cl is the central atom)

96. Predict the electron pair geometry and the molecular structure of each of the following:
(a) IOF_5 (I is the central atom)
(b) $POCl_3$ (P is the central atom)
(c) Cl_2SeO (Se is the central atom)
(d) $ClSO^+$ (S is the central atom)
(e) F_2SO (S is the central atom)
(f) NO_2^-
(g) SiO_4^{4-}

97. Which of the following molecules and ions contain polar bonds? Which of these molecules and ions have dipole moments?
(a) ClF_5
(b) ClO_2^-
(c) $TeCl_4^{2-}$
(d) PCl_3
(e) SeF_4
(f) PH_2^-
(g) XeF_2

98. Which of these molecules and ions contain polar bonds? Which of these molecules and ions have dipole moments?
(a) H_3O^+
(b) PCl_4^-
(c) $SnCl_3^-$
(d) $BrCl_4^-$
(e) ICl_3
(f) XeF_4
(g) SF_2

99. Which of the following molecules have dipole moments?
(a) CS_2
(b) SeS_2
(c) CCl_2F_2
(d) PCl_3 (P is the central atom)
(e) ClNO (N is the central atom)

100. Identify the molecules with a dipole moment:
(a) SF_4
(b) CF_4
(c) Cl_2CCBr_2
(d) CH_3Cl
(e) H_2CO

101. The molecule XF_3 has a dipole moment. Is X boron or phosphorus?

102. The molecule XCl_2 has a dipole moment. Is X beryllium or sulfur?

103. Is the Cl_2BBCl_2 molecule polar or nonpolar?

104. There are three possible structures for PCl_2F_3 with phosphorus as the central atom. Draw them and discuss how measurements of dipole moments could help distinguish among them.

105. Describe the molecular structure around the indicated atom or atoms:
(a) the sulfur atom in sulfuric acid, H_2SO_4 [$(HO)_2SO_2$]
(b) the chlorine atom in chloric acid, $HClO_3$ [$HOClO_2$]
(c) the oxygen atom in hydrogen peroxide, HOOH
(d) the nitrogen atom in nitric acid, HNO_3 [$HONO_2$]
(e) the oxygen atom in the OH group in nitric acid, HNO_3 [$HONO_2$]
(f) the central oxygen atom in the ozone molecule, O_3
(g) each of the carbon atoms in propyne, CH_3CCH
(h) the carbon atom in Freon, CCl_2F_2
(i) each of the carbon atoms in allene, H_2CCCH_2

106. Draw the Lewis structures and predict the shape of each compound or ion:
(a) CO_2
(b) $NO_2{}^-$
(c) SO_3
(d) $SO_3{}^{2-}$

107. A molecule with the formula AB_2, in which A and B represent different atoms, could have one of three different shapes. Sketch and name the three different shapes that this molecule might have. Give an example of a molecule or ion for each shape.

108. A molecule with the formula AB_3, in which A and B represent different atoms, could have one of three different shapes. Sketch and name the three different shapes that this molecule might have. Give an example of a molecule or ion that has each shape.

109. Draw the Lewis electron dot structures for these molecules, including resonance structures where appropriate:
(a) $CS_3{}^{2-}$
(b) CS_2
(c) CS
(d) predict the molecular shapes for $CS_3{}^{2-}$ and CS_2 and explain how you arrived at your predictions

110. What is the molecular structure of the stable form of FNO_2? (N is the central atom.)

111. A compound with a molar mass of about 42 g/mol contains 85.7% carbon and 14.3% hydrogen. What is its molecular structure?

112. Use the simulation (http://openstax.org/l/16MolecPolarity) to perform the following exercises for a two-atom molecule:
(a) Adjust the electronegativity value so the bond dipole is pointing toward B. Then determine what the electronegativity values must be to switch the dipole so that it points toward A.
(b) With a partial positive charge on A, turn on the electric field and describe what happens.
(c) With a small partial negative charge on A, turn on the electric field and describe what happens.
(d) Reset all, and then with a large partial negative charge on A, turn on the electric field and describe what happens.

113. Use the simulation (http://openstax.org/l/16MolecPolarity) to perform the following exercises for a real molecule. You may need to rotate the molecules in three dimensions to see certain dipoles.
(a) Sketch the bond dipoles and molecular dipole (if any) for O_3. Explain your observations.
(b) Look at the bond dipoles for NH_3. Use these dipoles to predict whether N or H is more electronegative.
(c) Predict whether there should be a molecular dipole for NH_3 and, if so, in which direction it will point. Check the molecular dipole box to test your hypothesis.

114. Use the Molecule Shape simulator (http://openstax.org/l/16MolecShape) to build a molecule. Starting with the central atom, click on the double bond to add one double bond. Then add one single bond and one lone pair. Rotate the molecule to observe the complete geometry. Name the electron group geometry and molecular structure and predict the bond angle. Then click the check boxes at the bottom and right of the simulator to check your answers.

115. Use the Molecule Shape simulator (http://openstax.org/l/16MolecShape) to explore real molecules. On the Real Molecules tab, select H_2O. Switch between the "real" and "model" modes. Explain the difference observed.

116. Use the Molecule Shape simulator (http://openstax.org/l/16MolecShape) to explore real molecules. On the Real Molecules tab, select "model" mode and S_2O. What is the model bond angle? Explain whether the "real" bond angle should be larger or smaller than the ideal model angle.

CHAPTER 8
Advanced Theories of Covalent Bonding

Figure 8.1 Oxygen molecules orient randomly most of the time, as shown in the top magnified view. However, when we pour liquid oxygen through a magnet, the molecules line up with the magnetic field, and the attraction allows them to stay suspended between the poles of the magnet where the magnetic field is strongest. Other diatomic molecules (like N_2) flow past the magnet. The detailed explanation of bonding described in this chapter allows us to understand this phenomenon. (credit: modification of work by Jefferson Lab)

CHAPTER OUTLINE

8.1 Valence Bond Theory
8.2 Hybrid Atomic Orbitals
8.3 Multiple Bonds
8.4 Molecular Orbital Theory

INTRODUCTION We have examined the basic ideas of bonding, showing that atoms share electrons to form molecules with stable Lewis structures and that we can predict the shapes of those molecules by valence shell electron pair repulsion (VSEPR) theory. These ideas provide an important starting point for understanding chemical bonding. But these models sometimes fall short in their abilities to predict the behavior of real substances. How can we reconcile the geometries of *s, p,* and *d* atomic orbitals with molecular shapes that show angles like 120° and 109.5°? Furthermore, we know that electrons and magnetic behavior are related through electromagnetic fields. Both N_2 and O_2 have fairly similar Lewis structures that contain lone pairs of electrons.

:N≡N: :Ö=Ö:

Yet oxygen demonstrates very different magnetic behavior than nitrogen. We can pour liquid nitrogen through a magnetic field with no visible interactions, while liquid oxygen (shown in Figure 8.1) is attracted to the magnet and floats in the magnetic field. We need to understand the additional concepts of valence bond theory, orbital hybridization, and molecular orbital theory to understand these observations.

8.1 Valence Bond Theory

LEARNING OBJECTIVES

By the end of this section, you will be able to:

- Describe the formation of covalent bonds in terms of atomic orbital overlap
- Define and give examples of σ and π bonds

As we know, a scientific theory is a strongly supported explanation for observed natural laws or large bodies of experimental data. For a theory to be accepted, it must explain experimental data and be able to predict behavior. For example, VSEPR theory has gained widespread acceptance because it predicts three-dimensional molecular shapes that are consistent with experimental data collected for thousands of different molecules. However, VSEPR theory does not provide an explanation of chemical bonding.

There are successful theories that describe the electronic structure of atoms. We can use quantum mechanics to predict the specific regions around an atom where electrons are likely to be located: A spherical shape for an *s* orbital, a dumbbell shape for a *p* orbital, and so forth. However, these predictions only describe the orbitals around free atoms. When atoms bond to form molecules, atomic orbitals are not sufficient to describe the regions where electrons will be located in the molecule. A more complete understanding of electron distributions requires a model that can account for the electronic structure of molecules. One popular theory holds that a covalent bond forms when a pair of electrons is shared by two atoms and is simultaneously attracted by the nuclei of both atoms. In the following sections, we will discuss how such bonds are described by valence bond theory and hybridization.

Valence bond theory describes a covalent bond as the overlap of half-filled atomic orbitals (each containing a single electron) that yield a pair of electrons shared between the two bonded atoms. We say that orbitals on two different atoms **overlap** when a portion of one orbital and a portion of a second orbital occupy the same region of space. According to valence bond theory, a covalent bond results when two conditions are met: (1) an orbital on one atom overlaps an orbital on a second atom and (2) the single electrons in each orbital combine to form an electron pair. The mutual attraction between this negatively charged electron pair and the two atoms' positively charged nuclei serves to physically link the two atoms through a force we define as a covalent bond. The strength of a covalent bond depends on the extent of overlap of the orbitals involved. Orbitals that overlap extensively form bonds that are stronger than those that have less overlap.

The energy of the system depends on how much the orbitals overlap. Figure 8.2 illustrates how the sum of the energies of two hydrogen atoms (the colored curve) changes as they approach each other. When the atoms are far apart there is no overlap, and by convention we set the sum of the energies at zero. As the atoms move together, their orbitals begin to overlap. Each electron begins to feel the attraction of the nucleus in the other atom. In addition, the electrons begin to repel each other, as do the nuclei. While the atoms are still widely separated, the attractions are slightly stronger than the repulsions, and the energy of the system decreases. (A bond begins to form.) As the atoms move closer together, the overlap increases, so the attraction of the nuclei for the electrons continues to increase (as do the repulsions among electrons and between the nuclei). At some specific distance between the atoms, which varies depending on the atoms involved, the energy reaches its lowest (most stable) value. This optimum distance between the two bonded nuclei is the bond distance between the two atoms. The bond is stable because at this point, the attractive and repulsive forces combine to create the lowest possible energy configuration. If the distance between the nuclei were to decrease further, the repulsions between nuclei and the repulsions as electrons are confined in closer proximity to each other would become stronger than the attractive forces. The energy of the system would then rise (making the system destabilized), as shown at the far left of Figure 8.2.

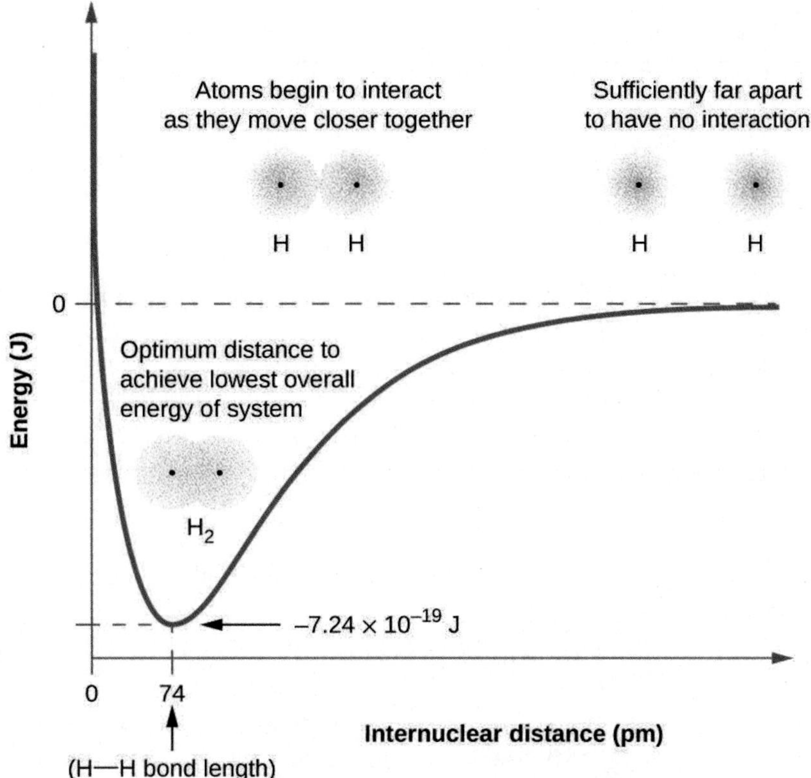

FIGURE 8.2 (a) The interaction of two hydrogen atoms changes as a function of distance. (b) The energy of the system changes as the atoms interact. The lowest (most stable) energy occurs at a distance of 74 pm, which is the bond length observed for the H_2 molecule.

The bond energy is the difference between the energy minimum (which occurs at the bond distance) and the energy of the two separated atoms. This is the quantity of energy released when the bond is formed. Conversely, the same amount of energy is required to break the bond. For the H_2 molecule shown in Figure 8.2, at the bond distance of 74 pm the system is 7.24×10^{-19} J lower in energy than the two separated hydrogen atoms. This may seem like a small number. However, we know from our earlier description of thermochemistry that bond energies are often discussed on a per-mole basis. For example, it requires 7.24×10^{-19} J to break one H–H bond, but it takes 4.36×10^5 J to break 1 mole of H–H bonds. A comparison of some bond lengths and energies is shown in Table 8.1. We can find many of these bonds in a variety of molecules, and this table provides average values. For example, breaking the first C–H bond in CH_4 requires 439.3 kJ/mol, while breaking the first C–H bond in $H–CH_2C_6H_5$ (a common paint thinner) requires 375.5 kJ/mol.

Representative Bond Energies and Lengths

Bond	Length (pm)	Energy (kJ/mol)	Bond	Length (pm)	Energy (kJ/mol)
H–H	74	436	C–O	140.1	358
H–C	106.8	413	$C=O$	119.7	745
H–N	101.5	391	$C\equiv O$	113.7	1072
H–O	97.5	467	H–Cl	127.5	431

TABLE 8.1

Bond	Length (pm)	Energy (kJ/mol)	Bond	Length (pm)	Energy (kJ/mol)
C–C	150.6	347	H–Br	141.4	366
$C=C$	133.5	614	H–I	160.9	298
$C\equiv C$	120.8	839	O–O	148	146
C–N	142.1	305	$O=O$	120.8	498
$C=N$	130.0	615	F–F	141.2	159
$C\equiv N$	116.1	891	Cl–Cl	198.8	243

TABLE 8.1

In addition to the distance between two orbitals, the orientation of orbitals also affects their overlap (other than for two *s* orbitals, which are spherically symmetric). Greater overlap is possible when orbitals are oriented such that they overlap on a direct line between the two nuclei. Figure 8.3 illustrates this for two *p* orbitals from different atoms; the overlap is greater when the orbitals overlap end to end rather than at an angle.

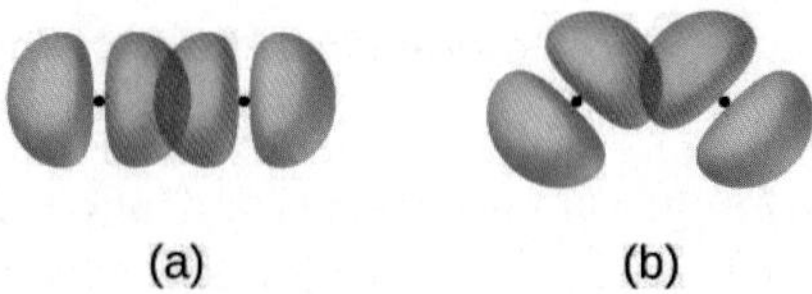

FIGURE 8.3 (a) The overlap of two *p* orbitals is greatest when the orbitals are directed end to end. (b) Any other arrangement results in less overlap. The dots indicate the locations of the nuclei.

The overlap of two *s* orbitals (as in H_2), the overlap of an *s* orbital and a *p* orbital (as in HCl), and the end-to-end overlap of two *p* orbitals (as in Cl_2) all produce **sigma bonds (σ bonds)**, as illustrated in Figure 8.4. A σ bond is a covalent bond in which the electron density is concentrated in the region along the internuclear axis; that is, a line between the nuclei would pass through the center of the overlap region. Single bonds in Lewis structures are described as σ bonds in valence bond theory.

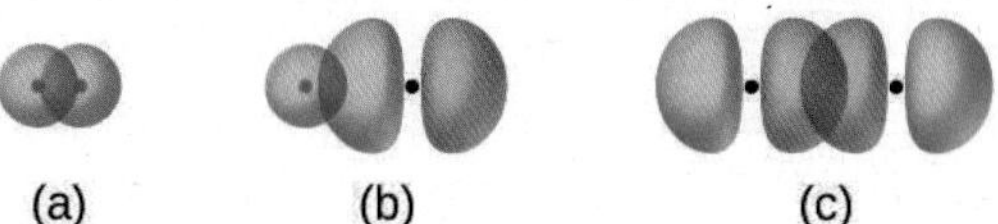

FIGURE 8.4 Sigma (σ) bonds form from the overlap of the following: (a) two *s* orbitals, (b) an *s* orbital and a *p* orbital, and (c) two *p* orbitals. The dots indicate the locations of the nuclei.

A **pi bond (π bond)** is a type of covalent bond that results from the side-by-side overlap of two *p* orbitals, as illustrated in Figure 8.5. In a π bond, the regions of orbital overlap lie on opposite sides of the internuclear axis. Along the axis itself, there is a **node**, that is, a plane with no probability of finding an electron.

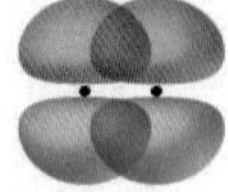

FIGURE 8.5 Pi (π) bonds form from the side-by-side overlap of two *p* orbitals. The dots indicate the location of the nuclei.

While all single bonds are σ bonds, multiple bonds consist of both σ and π bonds. As the Lewis structures below suggest, O_2 contains a double bond, and N_2 contains a triple bond. The double bond consists of one σ bond and one π bond, and the triple bond consists of one σ bond and two π bonds. Between any two atoms, the first bond formed will always be a σ bond, but there can only be one σ bond in any one location. In any multiple bond, there will be one σ bond, and the remaining one or two bonds will be π bonds. These bonds are described in more detail later in this chapter.

H—Cl:	:O=O:	:N≡N:
One σ bond No π bonds	One σ bond One π bond	One σ bond Two π bonds

As seen in Table 8.1, an average carbon-carbon single bond is 347 kJ/mol, while in a carbon-carbon double bond, the π bond increases the bond strength by 267 kJ/mol. Adding an additional π bond causes a further increase of 225 kJ/mol. We can see a similar pattern when we compare other σ and π bonds. Thus, each individual π bond is generally weaker than a corresponding σ bond between the same two atoms. In a σ bond, there is a greater degree of orbital overlap than in a π bond.

EXAMPLE 8.1

Counting σ and π Bonds

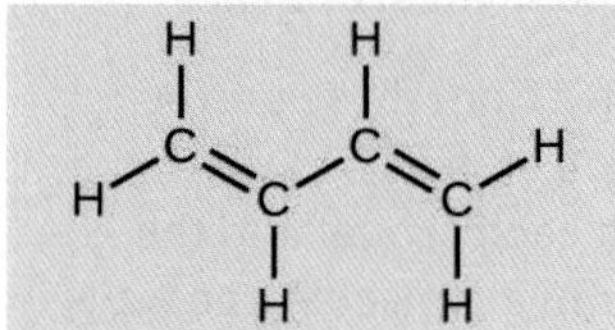

Butadiene, C_4H_6, is used to make synthetic rubber. Identify the number of σ and π bonds contained in this molecule.

Solution

There are six σ C–H bonds and one σ C–C bond, for a total of seven from the single bonds. There are two double bonds that each have a π bond in addition to the σ bond. This gives a total nine σ and two π bonds overall.

Check Your Learning

Identify each illustration as depicting a σ or π bond:

(a) side-by-side overlap of a 4*p* and a 2*p* orbital

(b) end-to-end overlap of a 4*p* and 4*p* orbital

(c) end-to-end overlap of a 4*p* and a 2*p* orbital

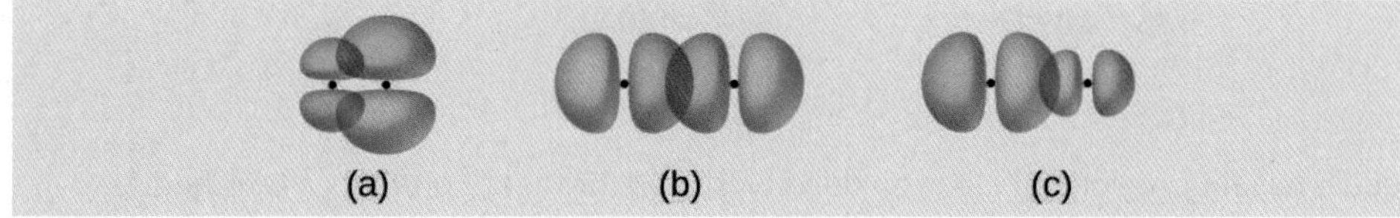

Answer:

(a) is a π bond with a node along the axis connecting the nuclei while (b) and (c) are σ bonds that overlap along the axis.

8.2 Hybrid Atomic Orbitals

LEARNING OBJECTIVES

By the end of this section, you will be able to:

- Explain the concept of atomic orbital hybridization
- Determine the hybrid orbitals associated with various molecular geometries

Thinking in terms of overlapping atomic orbitals is one way for us to explain how chemical bonds form in diatomic molecules. However, to understand how molecules with more than two atoms form stable bonds, we require a more detailed model. As an example, let us consider the water molecule, in which we have one oxygen atom bonding to two hydrogen atoms. Oxygen has the electron configuration $1s^2 2s^2 2p^4$, with two unpaired electrons (one in each of the two $2p$ orbitals). Valence bond theory would predict that the two O–H bonds form from the overlap of these two $2p$ orbitals with the $1s$ orbitals of the hydrogen atoms. If this were the case, the bond angle would be 90°, as shown in Figure 8.6, because p orbitals are perpendicular to each other. Experimental evidence shows that the bond angle is 104.5°, not 90°. The prediction of the valence bond theory model does not match the real-world observations of a water molecule; a different model is needed.

FIGURE 8.6 The hypothetical overlap of two of the $2p$ orbitals on an oxygen atom (red) with the $1s$ orbitals of two hydrogen atoms (blue) would produce a bond angle of 90°. This is not consistent with experimental evidence.[1]

Quantum-mechanical calculations suggest why the observed bond angles in H_2O differ from those predicted by the overlap of the $1s$ orbital of the hydrogen atoms with the $2p$ orbitals of the oxygen atom. The mathematical expression known as the wave function, ψ, contains information about each orbital and the wavelike properties of electrons in an isolated atom. When atoms are bound together in a molecule, the wave functions combine to produce new mathematical descriptions that have different shapes. This process of combining the wave functions for atomic orbitals is called **hybridization** and is mathematically accomplished by the *linear combination of atomic orbitals*, LCAO, (a technique that we will encounter again later). The new orbitals that result are called **hybrid orbitals**. The valence orbitals in an *isolated* oxygen atom are a $2s$ orbital and three $2p$ orbitals. The valence orbitals in an oxygen atom in a water molecule differ; they consist of four equivalent hybrid orbitals that point approximately toward the corners of a tetrahedron (Figure 8.7). Consequently, the overlap of the O and H orbitals should result in a tetrahedral bond angle (109.5°). The observed angle of 104.5° is experimental evidence for which quantum-mechanical calculations give a useful explanation: Valence bond theory must include a hybridization component to give accurate predictions.

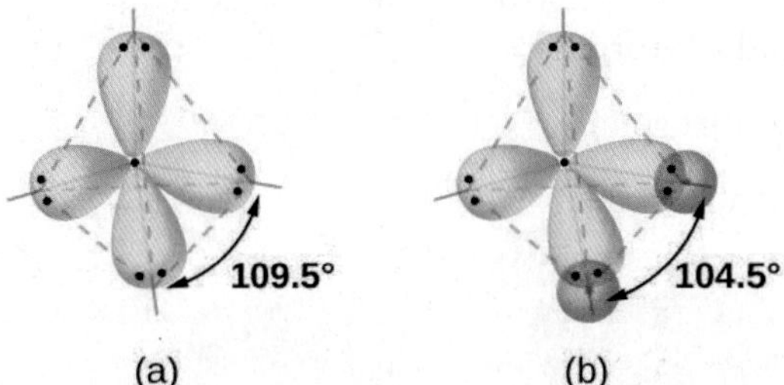

FIGURE 8.7 (a) A water molecule has four regions of electron density, so VSEPR theory predicts a tetrahedral arrangement of hybrid orbitals. (b) Two of the hybrid orbitals on oxygen contain lone pairs, and the other two overlap with the $1s$ orbitals of hydrogen atoms to form the O–H bonds in H_2O. This description is more consistent with the experimental structure.

The following ideas are important in understanding hybridization:

1. Hybrid orbitals do not exist in isolated atoms. They are formed only in covalently bonded atoms.
2. Hybrid orbitals have shapes and orientations that are very different from those of the atomic orbitals in isolated atoms.
3. A set of hybrid orbitals is generated by combining atomic orbitals. The number of hybrid orbitals in a set is equal to the number of atomic orbitals that were combined to produce the set.

1 Note that orbitals may sometimes be drawn in an elongated "balloon" shape rather than in a more realistic "plump" shape in order to make the geometry easier to visualize.

4. All orbitals in a set of hybrid orbitals are equivalent in shape and energy.
5. The type of hybrid orbitals formed in a bonded atom depends on its electron-pair geometry as predicted by the VSEPR theory.
6. Hybrid orbitals overlap to form σ bonds. Unhybridized orbitals overlap to form π bonds.

In the following sections, we shall discuss the common types of hybrid orbitals.

sp Hybridization

The beryllium atom in a gaseous $BeCl_2$ molecule is an example of a central atom with no lone pairs of electrons in a linear arrangement of three atoms. There are two regions of valence electron density in the $BeCl_2$ molecule that correspond to the two covalent Be–Cl bonds. To accommodate these two electron domains, two of the Be atom's four valence orbitals will mix to yield two hybrid orbitals. This hybridization process involves mixing of the valence *s* orbital with one of the valence *p* orbitals to yield two equivalent ***sp* hybrid orbitals** that are oriented in a linear geometry (Figure 8.8). In this figure, the set of *sp* orbitals appears similar in shape to the original *p* orbital, but there is an important difference. The number of atomic orbitals combined always equals the number of hybrid orbitals formed. The *p* orbital is one orbital that can hold up to two electrons. The *sp* set is two equivalent orbitals that point 180° from each other. The two electrons that were originally in the *s* orbital are now distributed to the two *sp* orbitals, which are half filled. In gaseous $BeCl_2$, these half-filled hybrid orbitals will overlap with orbitals from the chlorine atoms to form two identical σ bonds.

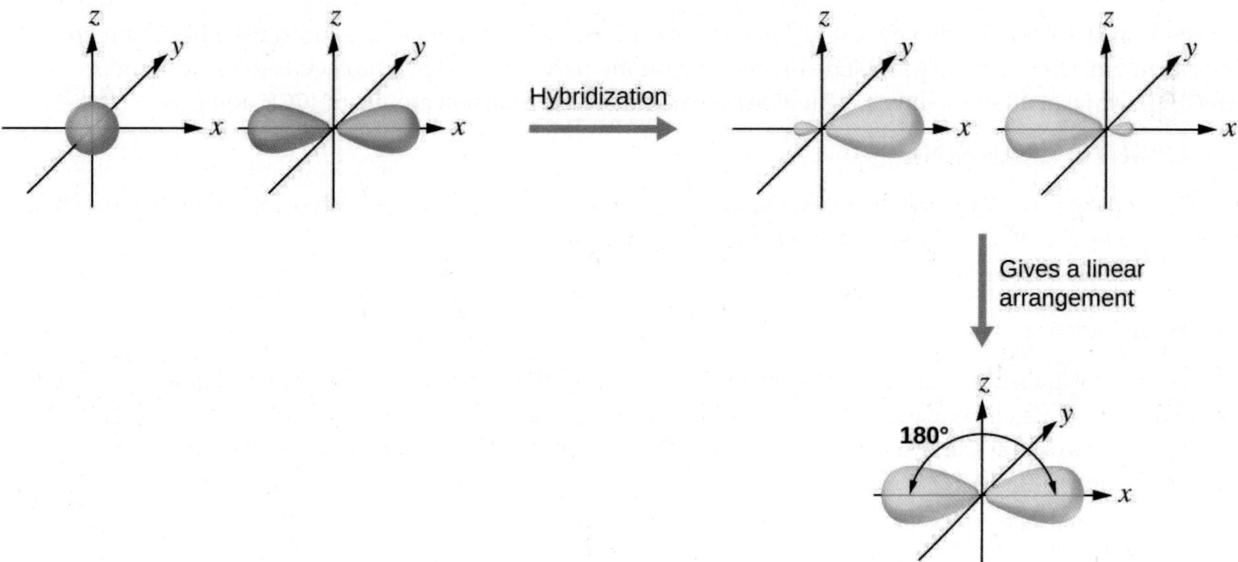

FIGURE 8.8 Hybridization of an *s* orbital (blue) and a *p* orbital (red) of the same atom produces two *sp* hybrid orbitals (yellow). Each hybrid orbital is oriented primarily in just one direction. Note that each *sp* orbital contains one lobe that is significantly larger than the other. The set of two *sp* orbitals are oriented at 180°, which is consistent with the geometry for two domains.

We illustrate the electronic differences in an isolated Be atom and in the bonded Be atom in the orbital energy-level diagram in Figure 8.9. These diagrams represent each orbital by a horizontal line (indicating its energy) and each electron by an arrow. Energy increases toward the top of the diagram. We use one upward arrow to indicate one electron in an orbital and two arrows (up and down) to indicate two electrons of opposite spin.

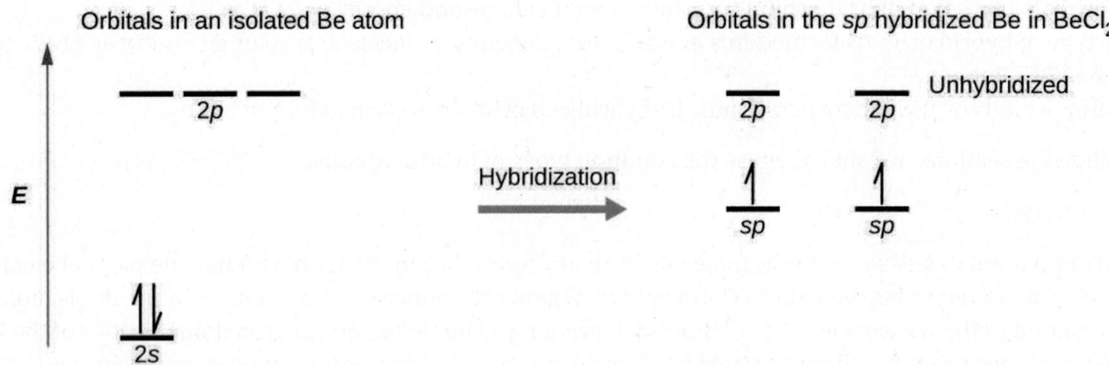

FIGURE 8.9 This orbital energy-level diagram shows the *sp* hybridized orbitals on Be in the linear $BeCl_2$ molecule. Each of the two *sp* hybrid orbitals holds one electron and is thus half filled and available for bonding via overlap with a Cl 3*p* orbital.

When atomic orbitals hybridize, the valence electrons occupy the newly created orbitals. The Be atom had two valence electrons, so each of the *sp* orbitals gets one of these electrons. Each of these electrons pairs up with the unpaired electron on a chlorine atom when a hybrid orbital and a chlorine orbital overlap during the formation of the Be–Cl bonds.

Any central atom surrounded by just two regions of valence electron density in a molecule will exhibit *sp* hybridization. Other examples include the mercury atom in the linear $HgCl_2$ molecule, the zinc atom in $Zn(CH_3)_2$, which contains a linear C–Zn–C arrangement, and the carbon atoms in HCCH and CO_2.

LINK TO LEARNING

Check out the University of Wisconsin-Oshkosh website (http://openstax.org/l/16hybridorbital) to learn about visualizing hybrid orbitals in three dimensions.

sp^2 Hybridization

The valence orbitals of a central atom surrounded by three regions of electron density consist of a set of three ***sp*^{2} hybrid orbitals** and one unhybridized *p* orbital. This arrangement results from *sp*^{2} hybridization, the mixing of one *s* orbital and two *p* orbitals to produce three identical hybrid orbitals oriented in a trigonal planar geometry (Figure 8.10).

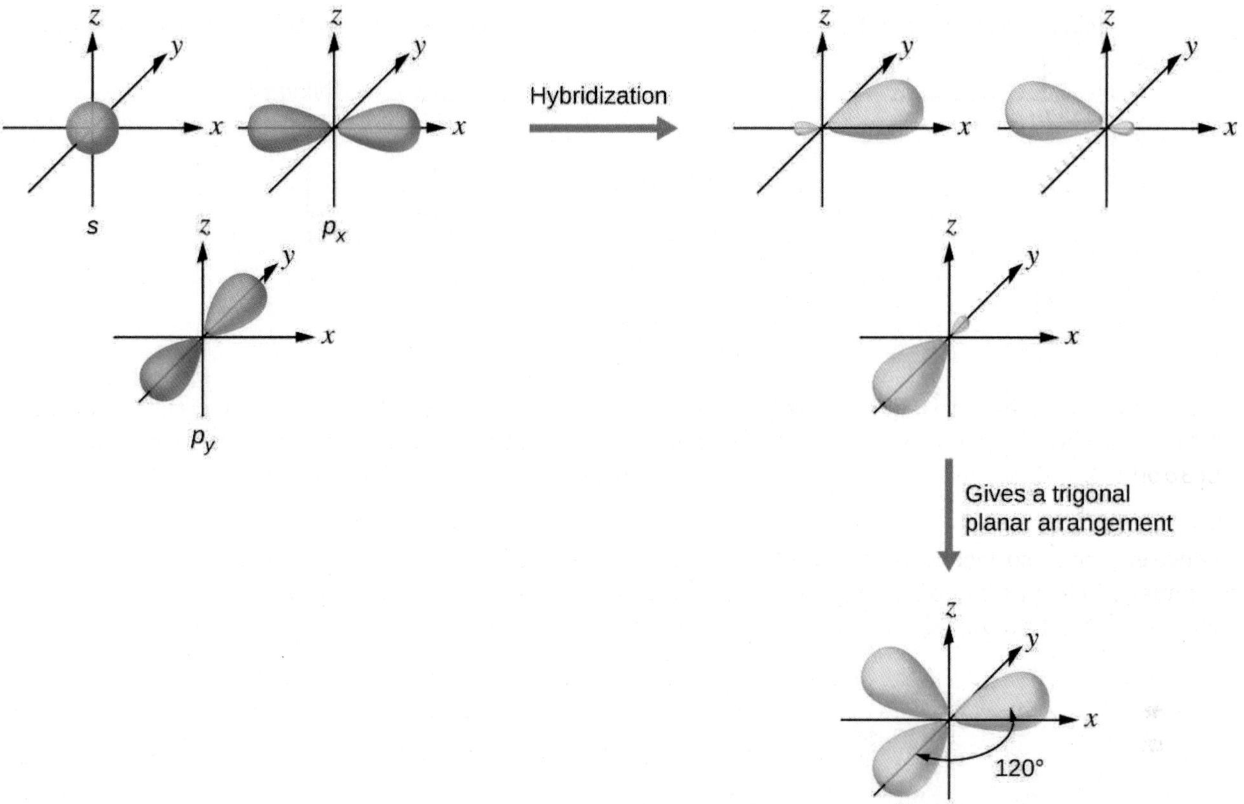

FIGURE 8.10 The hybridization of an *s* orbital (blue) and two *p* orbitals (red) produces three equivalent sp^2 hybridized orbitals (yellow) oriented at 120° with respect to each other. The remaining unhybridized *p* orbital is not shown here, but is located along the z axis.

Although quantum mechanics yields the "plump" orbital lobes as depicted in Figure 8.10, sometimes for clarity these orbitals are drawn thinner and without the minor lobes, as in Figure 8.11, to avoid obscuring other features of a given illustration. We will use these "thinner" representations whenever the true view is too crowded to easily visualize.

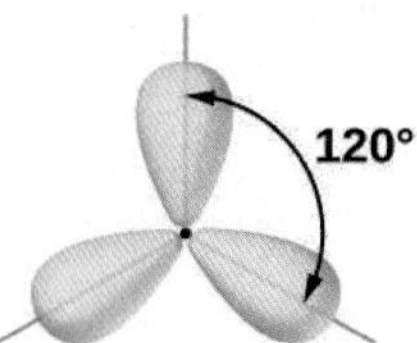

FIGURE 8.11 This alternate way of drawing the trigonal planar sp^2 hybrid orbitals is sometimes used in more crowded figures.

The observed structure of the borane molecule, BH_3, suggests sp^2 hybridization for boron in this compound. The molecule is trigonal planar, and the boron atom is involved in three bonds to hydrogen atoms (Figure 8.12). We can illustrate the comparison of orbitals and electron distribution in an isolated boron atom and in the bonded atom in BH_3 as shown in the orbital energy level diagram in Figure 8.13. We redistribute the three valence electrons of the boron atom in the three sp^2 hybrid orbitals, and each boron electron pairs with a hydrogen electron when B–H bonds form.

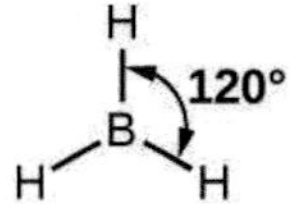

FIGURE 8.12 BH_3 is an electron-deficient molecule with a trigonal planar structure.

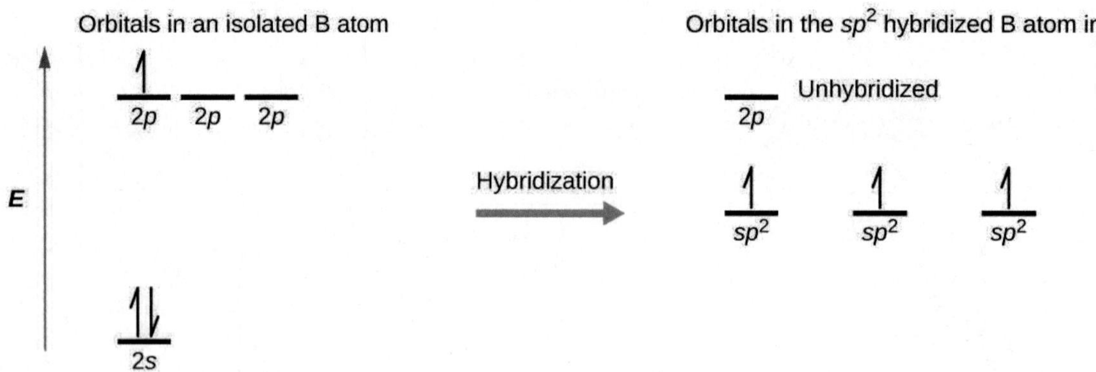

FIGURE 8.13 In an isolated B atom, there are one 2*s* and three 2*p* valence orbitals. When boron is in a molecule with three regions of electron density, three of the orbitals hybridize and create a set of three sp^2 orbitals and one unhybridized 2*p* orbital. The three half-filled hybrid orbitals each overlap with an orbital from a hydrogen atom to form three σ bonds in BH_3.

Any central atom surrounded by three regions of electron density will exhibit sp^2 hybridization. This includes molecules with a lone pair on the central atom, such as ClNO (Figure 8.14), or molecules with two single bonds and a double bond connected to the central atom, as in formaldehyde, CH_2O, and ethene, H_2CCH_2.

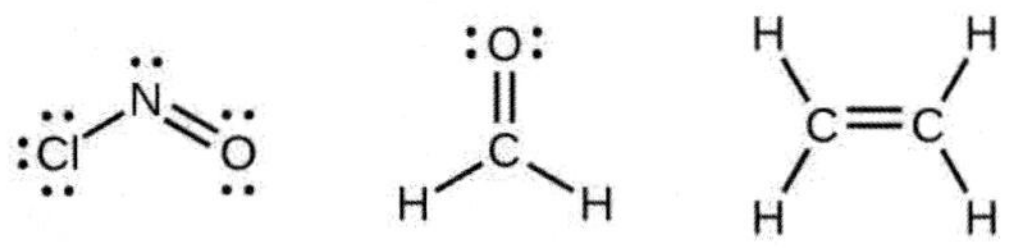

FIGURE 8.14 The central atom(s) in each of the structures shown contain three regions of electron density and are sp^2 hybridized. As we know from the discussion of VSEPR theory, a region of electron density contains all of the electrons that point in one direction. A lone pair, an unpaired electron, a single bond, or a multiple bond would each count as one region of electron density.

sp^3 Hybridization

The valence orbitals of an atom surrounded by a tetrahedral arrangement of bonding pairs and lone pairs consist of a set of four ***sp*^3^ hybrid orbitals**. The hybrids result from the mixing of one *s* orbital and all three *p* orbitals that produces four identical sp^3 hybrid orbitals (Figure 8.15). Each of these hybrid orbitals points toward a different corner of a tetrahedron.

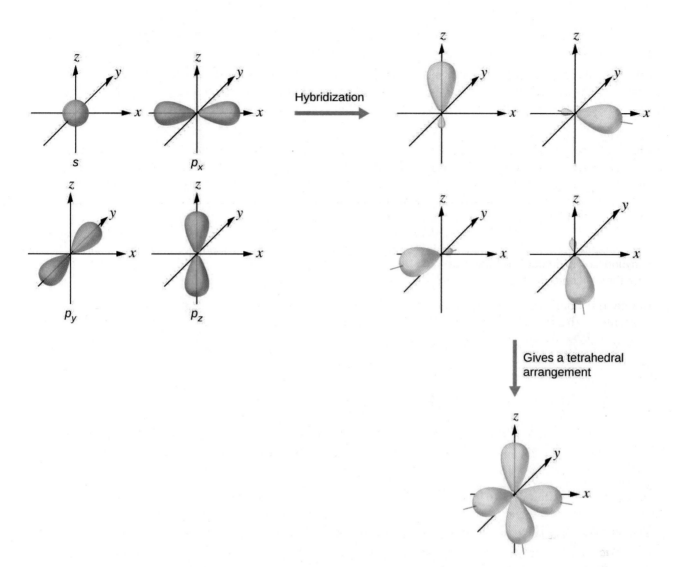

FIGURE 8.15 The hybridization of an *s* orbital (blue) and three *p* orbitals (red) produces four equivalent sp^3 hybridized orbitals (yellow) oriented at 109.5° with respect to each other.

A molecule of methane, CH_4, consists of a carbon atom surrounded by four hydrogen atoms at the corners of a tetrahedron. The carbon atom in methane exhibits sp^3 hybridization. We illustrate the orbitals and electron distribution in an isolated carbon atom and in the bonded atom in CH_4 in Figure 8.16. The four valence electrons of the carbon atom are distributed equally in the hybrid orbitals, and each carbon electron pairs with a hydrogen electron when the C–H bonds form.

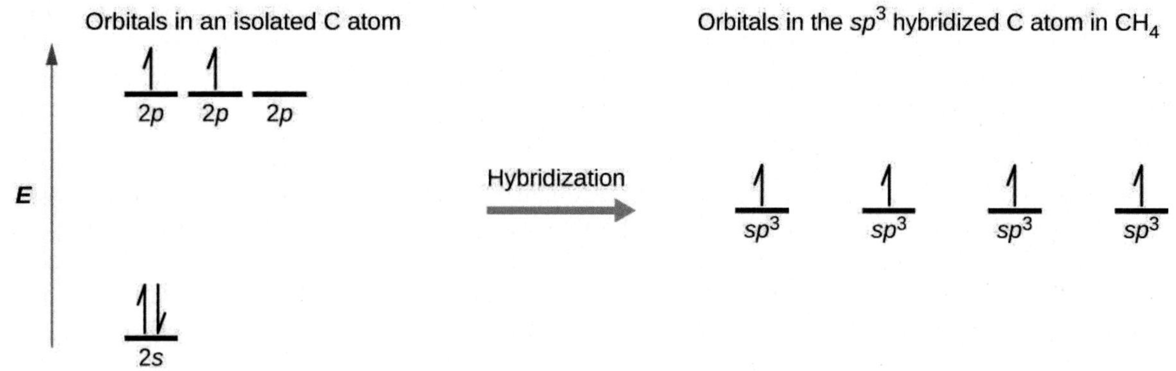

FIGURE 8.16 The four valence atomic orbitals from an isolated carbon atom all hybridize when the carbon bonds

in a molecule like CH_4 with four regions of electron density. This creates four equivalent sp^3 hybridized orbitals. Overlap of each of the hybrid orbitals with a hydrogen orbital creates a C–H σ bond.

In a methane molecule, the 1*s* orbital of each of the four hydrogen atoms overlaps with one of the four sp^3 orbitals of the carbon atom to form a sigma (σ) bond. This results in the formation of four strong, equivalent covalent bonds between the carbon atom and each of the hydrogen atoms to produce the methane molecule, CH_4.

The structure of ethane, C_2H_6, is similar to that of methane in that each carbon in ethane has four neighboring atoms arranged at the corners of a tetrahedron—three hydrogen atoms and one carbon atom (Figure 8.17). However, in ethane an sp^3 orbital of one carbon atom overlaps end to end with an sp^3 orbital of a second carbon atom to form a σ bond between the two carbon atoms. Each of the remaining sp^3 hybrid orbitals overlaps with an *s* orbital of a hydrogen atom to form carbon–hydrogen σ bonds. The structure and overall outline of the bonding orbitals of ethane are shown in Figure 8.17. The orientation of the two CH_3 groups is not fixed relative to each other. Experimental evidence shows that rotation around σ bonds occurs easily.

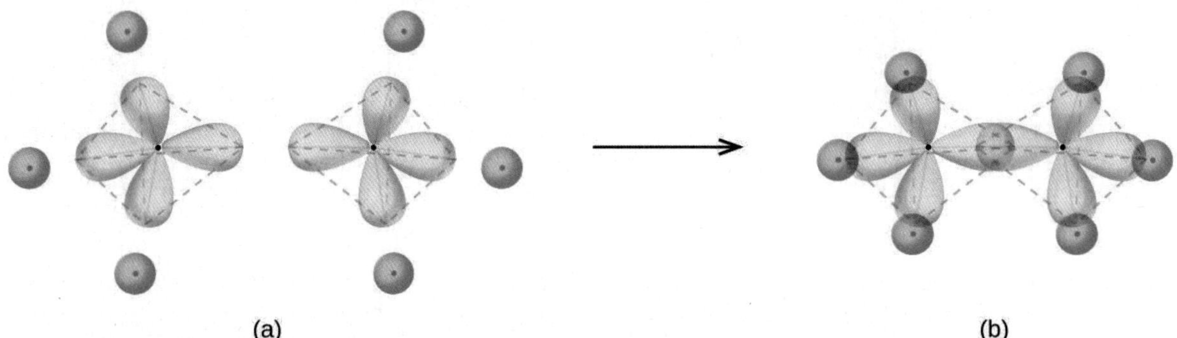

FIGURE 8.17 (a) In the ethane molecule, C_2H_6, each carbon has four sp^3 orbitals. (b) These four orbitals overlap to form seven σ bonds.

An sp^3 hybrid orbital can also hold a lone pair of electrons. For example, the nitrogen atom in ammonia is surrounded by three bonding pairs and a lone pair of electrons directed to the four corners of a tetrahedron. The nitrogen atom is sp^3 hybridized with one hybrid orbital occupied by the lone pair.

The molecular structure of water is consistent with a tetrahedral arrangement of two lone pairs and two bonding pairs of electrons. Thus we say that the oxygen atom is sp^3 hybridized, with two of the hybrid orbitals occupied by lone pairs and two by bonding pairs. Since lone pairs occupy more space than bonding pairs, structures that contain lone pairs have bond angles slightly distorted from the ideal. Perfect tetrahedra have angles of 109.5°, but the observed angles in ammonia (107.3°) and water (104.5°) are slightly smaller. Other examples of sp^3 hybridization include CCl_4, PCl_3, and NCl_3.

sp^3d and sp^3d^2 Hybridization

To describe the five bonding orbitals in a trigonal bipyramidal arrangement, we must use five of the valence shell atomic orbitals (the *s* orbital, the three *p* orbitals, and one of the *d* orbitals), which gives five **sp^3d hybrid orbitals**. With an octahedral arrangement of six hybrid orbitals, we must use six valence shell atomic orbitals (the *s* orbital, the three *p* orbitals, and two of the *d* orbitals in its valence shell), which gives six **sp^3d^2 hybrid orbitals**. These hybridizations are only possible for atoms that have *d* orbitals in their valence subshells (that is, not those in the first or second period).

In a molecule of phosphorus pentachloride, PCl_5, there are five P–Cl bonds (thus five pairs of valence electrons around the phosphorus atom) directed toward the corners of a trigonal bipyramid. We use the 3*s* orbital, the three 3*p* orbitals, and one of the 3*d* orbitals to form the set of five sp^3d hybrid orbitals (Figure 8.19) that are involved in the P–Cl bonds. Other atoms that exhibit sp^3d hybridization include the sulfur atom in SF_4 and the chlorine atoms in ClF_3 and in $ClF_4{}^+$. (The electrons on fluorine atoms are omitted for clarity.)

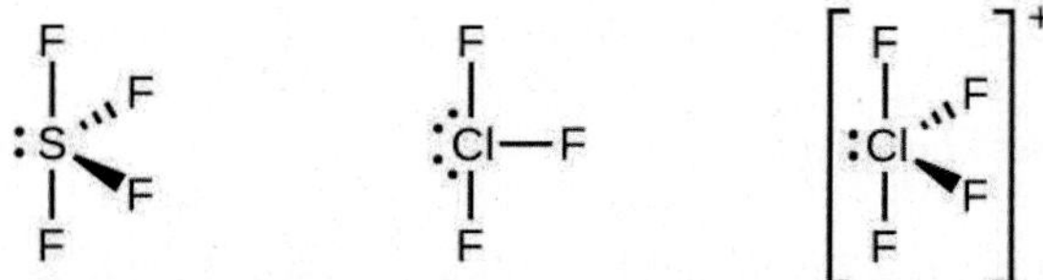

FIGURE 8.18 The three compounds pictured exhibit sp^3d hybridization in the central atom and a trigonal bipyramid form. SF_4 and $ClF_4{}^+$ have one lone pair of electrons on the central atom, and ClF_3 has two lone pairs giving it the T-shape shown.

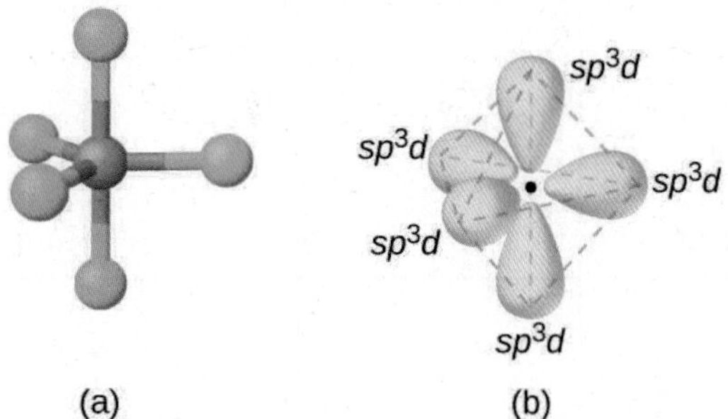

FIGURE 8.19 (a) The five regions of electron density around phosphorus in PCl_5 require five hybrid sp^3d orbitals. (b) These orbitals combine to form a trigonal bipyramidal structure with each large lobe of the hybrid orbital pointing at a vertex. As before, there are also small lobes pointing in the opposite direction for each orbital (not shown for clarity).

The sulfur atom in sulfur hexafluoride, SF_6, exhibits sp^3d^2 hybridization. A molecule of sulfur hexafluoride has six bonding pairs of electrons connecting six fluorine atoms to a single sulfur atom. There are no lone pairs of electrons on the central atom. To bond six fluorine atoms, the 3*s* orbital, the three 3*p* orbitals, and two of the 3*d* orbitals form six equivalent sp^3d^2 hybrid orbitals, each directed toward a different corner of an octahedron. Other atoms that exhibit sp^3d^2 hybridization include the phosphorus atom in $PCl_6{}^-$, the iodine atom in the interhalogens $IF_6{}^+$, IF_5, $ICl_4{}^-$, $IF_4{}^-$ and the xenon atom in XeF_4.

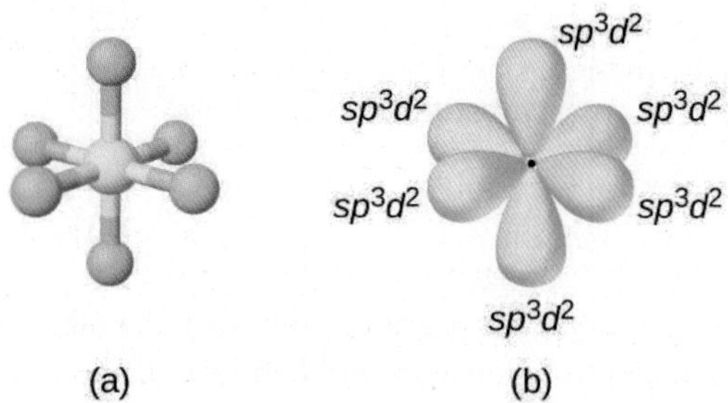

FIGURE 8.20 (a) Sulfur hexafluoride, SF_6, has an octahedral structure that requires sp^3d^2 hybridization. (b) The six sp^3d^2 orbitals form an octahedral structure around sulfur. Again, the minor lobe of each orbital is not shown for clarity.

Assignment of Hybrid Orbitals to Central Atoms

The hybridization of an atom is determined based on the number of regions of electron density that surround it. The geometrical arrangements characteristic of the various sets of hybrid orbitals are shown in Figure 8.21. These arrangements are identical to those of the electron-pair geometries predicted by VSEPR theory. VSEPR theory predicts the shapes of molecules, and hybrid orbital theory provides an explanation for how those shapes are formed. To find the hybridization of a central atom, we can use the following guidelines:

1. Determine the Lewis structure of the molecule.
2. Determine the number of regions of electron density around an atom using VSEPR theory, in which single bonds, multiple bonds, radicals, and lone pairs each count as one region.
3. Assign the set of hybridized orbitals from Figure 8.21 that corresponds to this geometry.

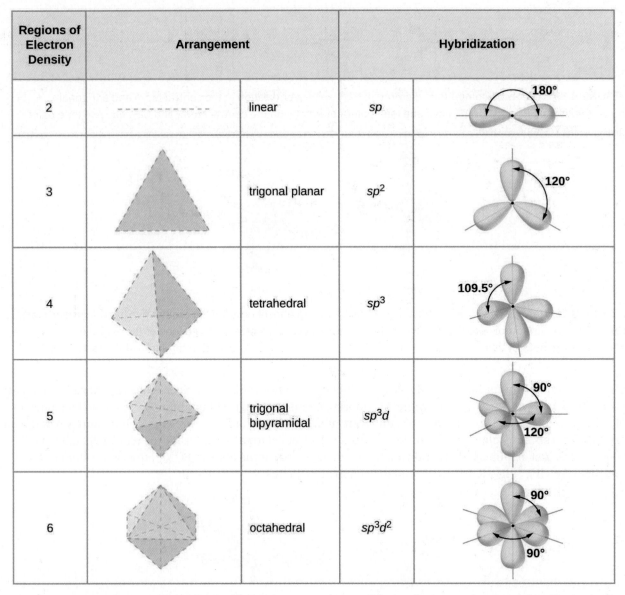

Regions of Electron Density	Arrangement		Hybridization	
2		linear	*sp*	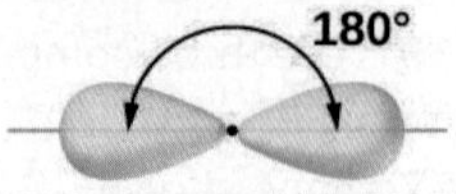 180°
3		trigonal planar	sp^2	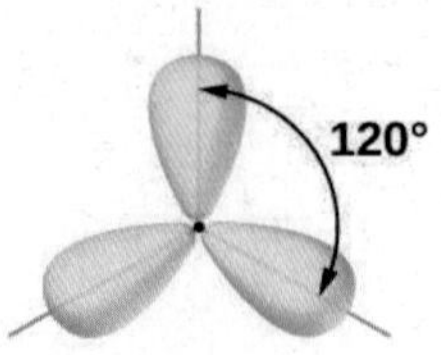 120°
4	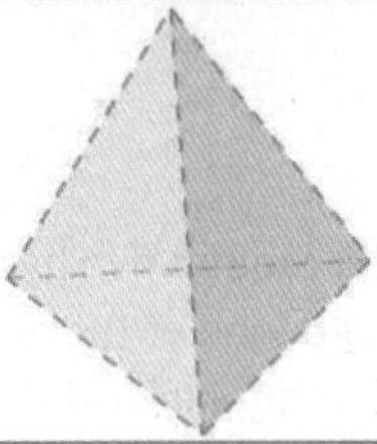	tetrahedral	sp^3	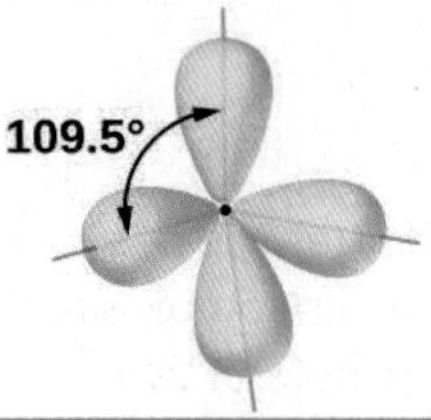 109.5°
5	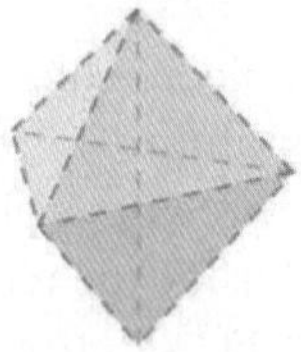	trigonal bipyramidal	sp^3d	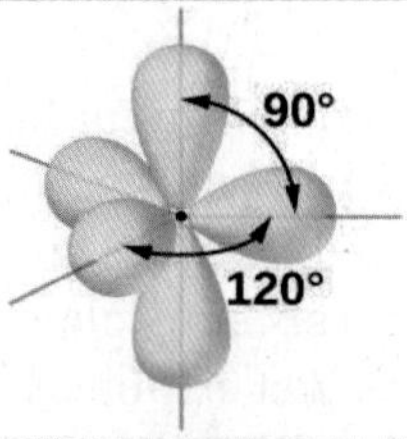 90° 120°
6	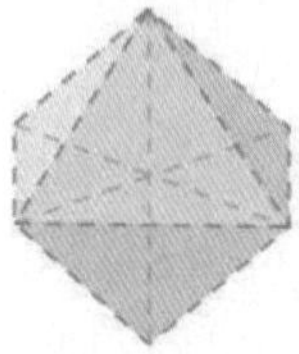	octahedral	sp^3d^2	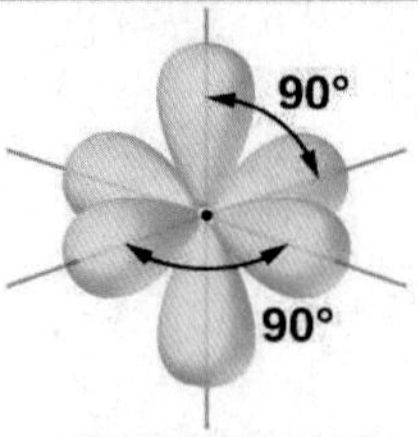 90° 90°

FIGURE 8.21 The shapes of hybridized orbital sets are consistent with the electron-pair geometries. For example, an atom surrounded by three regions of electron density is sp^2 hybridized, and the three sp^2 orbitals are arranged in a trigonal planar fashion.

It is important to remember that hybridization was devised to rationalize experimentally observed molecular geometries. The model works well for molecules containing small central atoms, in which the valence electron pairs are close together in space. However, for larger central atoms, the valence-shell electron pairs are farther from the nucleus, and there are fewer repulsions. Their compounds exhibit structures that are often not consistent with VSEPR theory, and hybridized orbitals are not necessary to explain the observed data. For example, we have discussed the H–O–H bond angle in H_2O, 104.5°, which is more consistent with sp^3 hybrid orbitals (109.5°) on the central atom than with $2p$ orbitals (90°). Sulfur is in the same group as oxygen, and H_2S has a similar Lewis structure. However, it has a much smaller bond angle (92.1°), which indicates much less hybridization on sulfur than oxygen. Continuing down the group, tellurium is even larger than sulfur, and for H_2Te, the observed bond angle (90°) is consistent with overlap of the $5p$ orbitals, without invoking hybridization. We invoke hybridization where it is necessary to explain the observed structures.

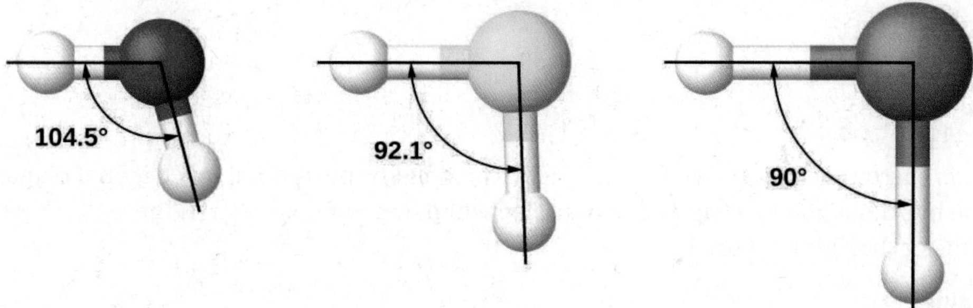

EXAMPLE 8.2

Assigning Hybridization

Ammonium sulfate is important as a fertilizer. What is the hybridization of the sulfur atom in the sulfate ion, SO_4^{2-}?

Solution

The Lewis structure of sulfate shows there are four regions of electron density. The hybridization is *sp*^{3}.

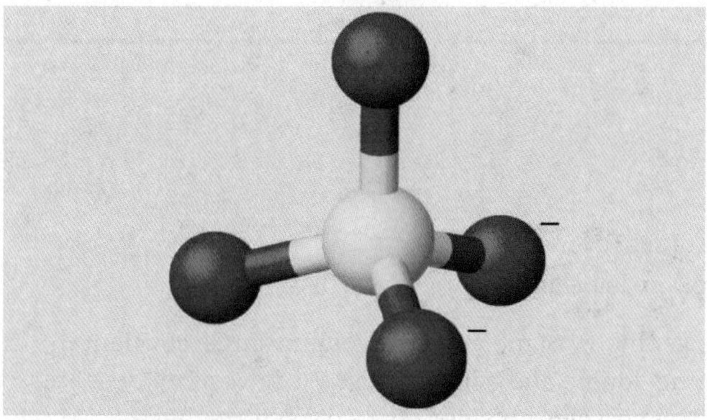

Check Your Learning

What is the hybridization of the selenium atom in SeF_4?

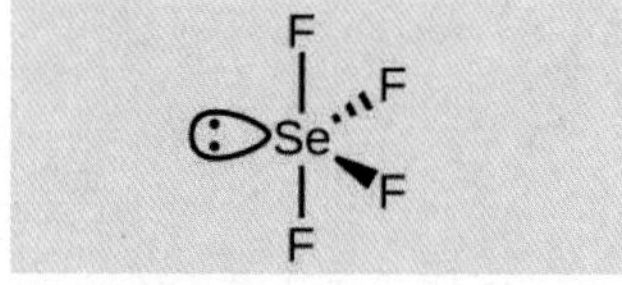

Answer:

The selenium atom is *sp*^{3}*d* hybridized.

EXAMPLE 8.3

Assigning Hybridization

Urea, $NH_2C(O)NH_2$, is sometimes used as a source of nitrogen in fertilizers. What is the hybridization of the carbon atom in urea?

Solution

The Lewis structure of urea is

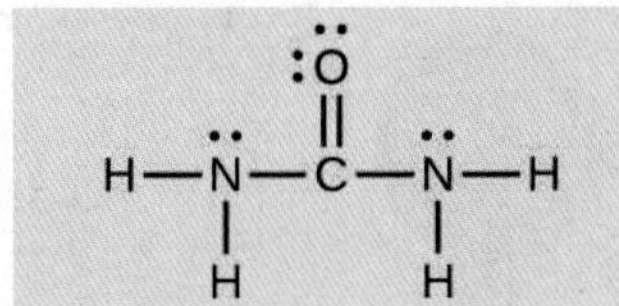

The carbon atom is surrounded by three regions of electron density, positioned in a trigonal planar arrangement. The hybridization in a trigonal planar electron pair geometry is *sp*^{2} (Figure 8.21), which is the hybridization of the carbon atom in urea.

Check Your Learning

Acetic acid, $H_3CC(O)OH$, is the molecule that gives vinegar its odor and sour taste. What is the hybridization of the two carbon atoms in acetic acid?

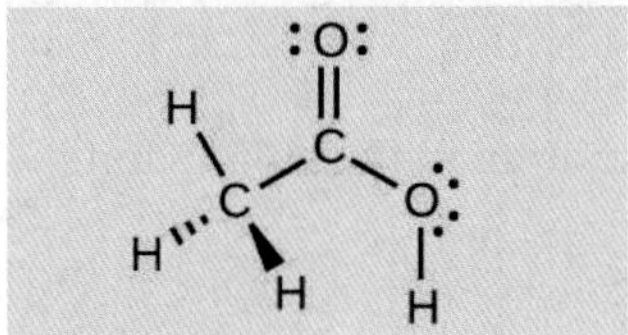

Answer:

$H_3\underline{C}$, *sp*^{3}; $\underline{C}(O)OH$, *sp*^{2}

8.3 Multiple Bonds

LEARNING OBJECTIVES

By the end of this section, you will be able to:

- Describe multiple covalent bonding in terms of atomic orbital overlap
- Relate the concept of resonance to π-bonding and electron delocalization

The hybrid orbital model appears to account well for the geometry of molecules involving single covalent bonds. Is it also capable of describing molecules containing double and triple bonds? We have already discussed that multiple bonds consist of σ and π bonds. Next we can consider how we visualize these components and how they relate to hybrid orbitals. The Lewis structure of ethene, C_2H_4, shows us that each carbon atom is surrounded by one other carbon atom and two hydrogen atoms.

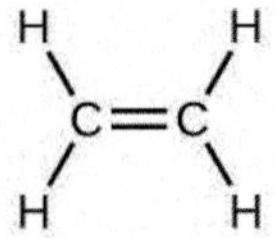

H H
C=C
H H

The three bonding regions form a trigonal planar electron-pair geometry. Thus we expect the σ bonds from each carbon atom are formed using a set of *sp*^{2} hybrid orbitals that result from hybridization of two of the 2*p* orbitals and the 2*s* orbital (Figure 8.22). These orbitals form the C–H single bonds and the σ bond in the $C = C$ double bond (Figure 8.23). The π bond in the $C = C$ double bond results from the overlap of the third (remaining) 2*p* orbital on each carbon atom that is not involved in hybridization. This unhybridized *p* orbital (lobes shown in red and blue in Figure 8.23) is perpendicular to the plane of the *sp*^{2} hybrid orbitals. Thus the unhybridized 2*p* orbitals overlap in a side-by-side fashion, above and below the internuclear axis (Figure 8.23) and form a π bond.

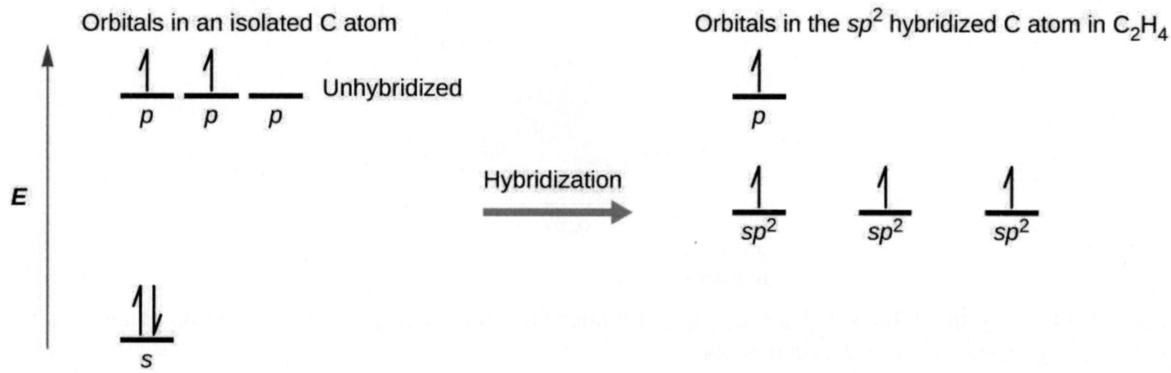

FIGURE 8.22 In ethene, each carbon atom is sp^2 hybridized, and the sp^2 orbitals and the *p* orbital are singly occupied. The hybrid orbitals overlap to form σ bonds, while the *p* orbitals on each carbon atom overlap to form a π bond.

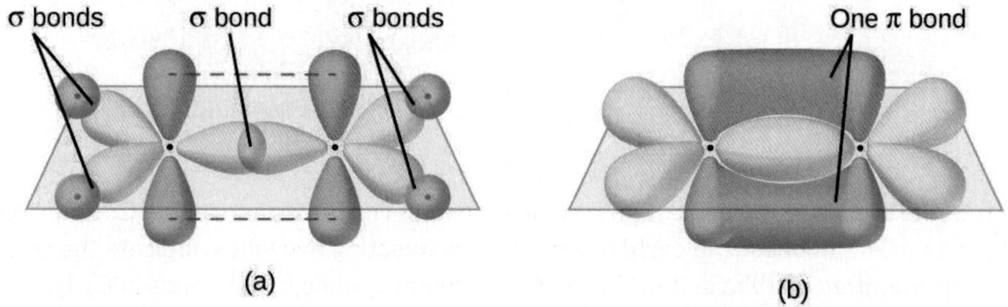

FIGURE 8.23 In the ethene molecule, C_2H_4, there are (a) five σ bonds. One C–C σ bond results from overlap of sp^2 hybrid orbitals on the carbon atom with one sp^2 hybrid orbital on the other carbon atom. Four C–H bonds result from the overlap between the C atoms' sp^2 orbitals with *s* orbitals on the hydrogen atoms. (b) The π bond is formed by the side-by-side overlap of the two unhybridized *p* orbitals in the two carbon atoms. The two lobes of the π bond are above and below the plane of the σ system.

In an ethene molecule, the four hydrogen atoms and the two carbon atoms are all in the same plane. If the two planes of sp^2 hybrid orbitals tilted relative to each other, the *p* orbitals would not be oriented to overlap efficiently to create the π bond. The planar configuration for the ethene molecule occurs because it is the most stable bonding arrangement. This is a significant difference between σ and π bonds; rotation around single (σ) bonds occurs easily because the end-to-end orbital overlap does not depend on the relative orientation of the orbitals on each atom in the bond. In other words, rotation around the internuclear axis does not change the extent to which the σ bonding orbitals overlap because the bonding electron density is symmetric about the axis. Rotation about the internuclear axis is much more difficult for multiple bonds; however, this would drastically alter the off-axis overlap of the π bonding orbitals, essentially breaking the π bond.

In molecules with *sp* hybrid orbitals, two unhybridized *p* orbitals remain on the atom (Figure 8.24). We find this situation in acetylene, H–C≡C–H, which is a linear molecule. The *sp* hybrid orbitals of the two carbon atoms overlap end to end to form a σ bond between the carbon atoms (Figure 8.25). The remaining *sp* orbitals form σ bonds with hydrogen atoms. The two unhybridized *p* orbitals per carbon are positioned such that they overlap side by side and, hence, form two π bonds. The two carbon atoms of acetylene are thus bound together by one σ bond and two π bonds, giving a triple bond.

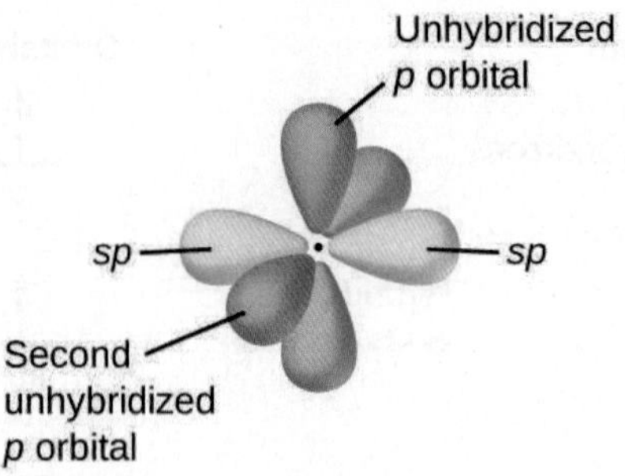

FIGURE 8.24 Diagram of the two linear *sp* hybrid orbitals of a carbon atom, which lie in a straight line, and the two unhybridized *p* orbitals at perpendicular angles.

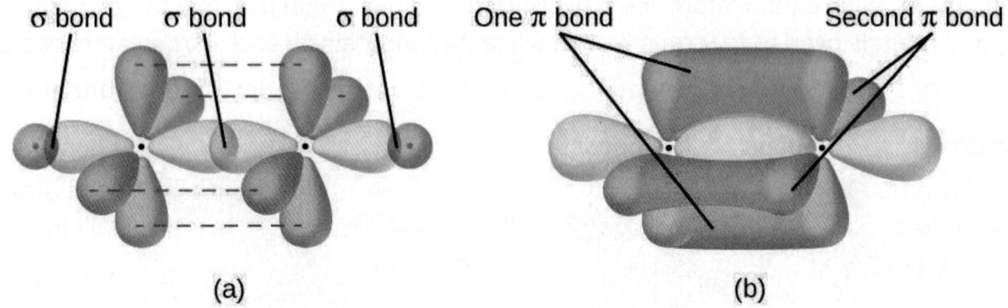

FIGURE 8.25 (a) In the acetylene molecule, C_2H_2, there are two C–H σ bonds and a $C \equiv C$ triple bond involving one C–C σ bond and two C–C π bonds. The dashed lines, each connecting two lobes, indicate the side-by-side overlap of the four unhybridized *p* orbitals. (b) This shows the overall outline of the bonds in C_2H_2. The two lobes of each of the π bonds are positioned across from each other around the line of the C–C σ bond.

Hybridization involves only σ bonds, lone pairs of electrons, and single unpaired electrons (radicals). Structures that account for these features describe the correct hybridization of the atoms. However, many structures also include resonance forms. Remember that resonance forms occur when various arrangements of π bonds are possible. Since the arrangement of π bonds involves only the unhybridized orbitals, resonance does not influence the assignment of hybridization.

For example, molecule benzene has two resonance forms (Figure 8.26). We can use either of these forms to determine that each of the carbon atoms is bonded to three other atoms with no lone pairs, so the correct hybridization is sp^2. The electrons in the unhybridized *p* orbitals form π bonds. Neither resonance structure completely describes the electrons in the π bonds. They are not located in one position or the other, but in reality are delocalized throughout the ring. Valence bond theory does not easily address delocalization. Bonding in molecules with resonance forms is better described by molecular orbital theory. (See the next module.)

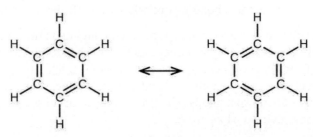

FIGURE 8.26 Each carbon atom in benzene, C_6H_6, is sp^2 hybridized, independently of which resonance form is considered. The electrons in the π bonds are not located in one set of *p* orbitals or the other, but rather delocalized throughout the molecule.

EXAMPLE 8.4

Assignment of Hybridization Involving Resonance

Some acid rain results from the reaction of sulfur dioxide with atmospheric water vapor, followed by the formation of sulfuric acid. Sulfur dioxide, SO_2, is a major component of volcanic gases as well as a product of the combustion of sulfur-containing coal. What is the hybridization of the S atom in SO_2?

Solution

The resonance structures of SO_2 are

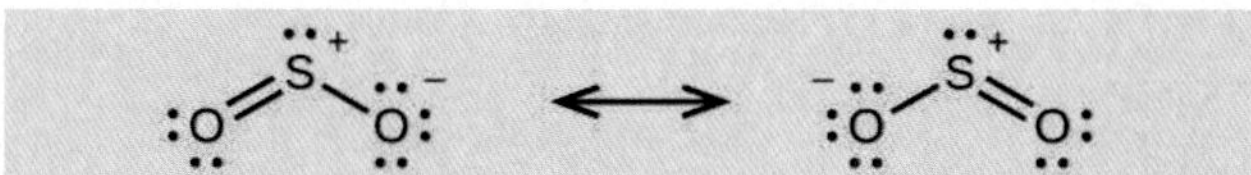

The sulfur atom is surrounded by two bonds and one lone pair of electrons in either resonance structure. Therefore, the electron-pair geometry is trigonal planar, and the hybridization of the sulfur atom is sp^2.

Check Your Learning

Another acid in acid rain is nitric acid, HNO_3, which is produced by the reaction of nitrogen dioxide, NO_2, with atmospheric water vapor. What is the hybridization of the nitrogen atom in NO_2? (Note: the lone electron on nitrogen occupies a hybridized orbital just as a lone pair would.)

Answer:

sp^2

8.4 Molecular Orbital Theory

LEARNING OBJECTIVES

By the end of this section, you will be able to:

- Outline the basic quantum-mechanical approach to deriving molecular orbitals from atomic orbitals
- Describe traits of bonding and antibonding molecular orbitals
- Calculate bond orders based on molecular electron configurations
- Write molecular electron configurations for first- and second-row diatomic molecules
- Relate these electron configurations to the molecules' stabilities and magnetic properties

For almost every covalent molecule that exists, we can now draw the Lewis structure, predict the electron-pair geometry, predict the molecular geometry, and come close to predicting bond angles. However, one of the most important molecules we know, the oxygen molecule O_2, presents a problem with respect to its Lewis structure. We would write the following Lewis structure for O_2:

$$:\ddot{O}=\ddot{O}:$$

This electronic structure adheres to all the rules governing Lewis theory. There is an O=O double bond, and each oxygen atom has eight electrons around it. However, this picture is at odds with the magnetic behavior of oxygen. By itself, O_2 is not magnetic, but it is attracted to magnetic fields. Thus, when we pour liquid oxygen past a strong magnet, it collects between the poles of the magnet and defies gravity, as in Figure 8.1. Such attraction to a magnetic field is called **paramagnetism**, and it arises in molecules that have unpaired electrons. And yet, the Lewis structure of O_2 indicates that all electrons are paired. How do we account for this discrepancy?

Magnetic susceptibility measures the force experienced by a substance in a magnetic field. When we compare the weight of a sample to the weight measured in a magnetic field (Figure 8.27), paramagnetic samples that are attracted to the magnet will appear heavier because of the force exerted by the magnetic field. We can calculate the number of unpaired electrons based on the increase in weight.

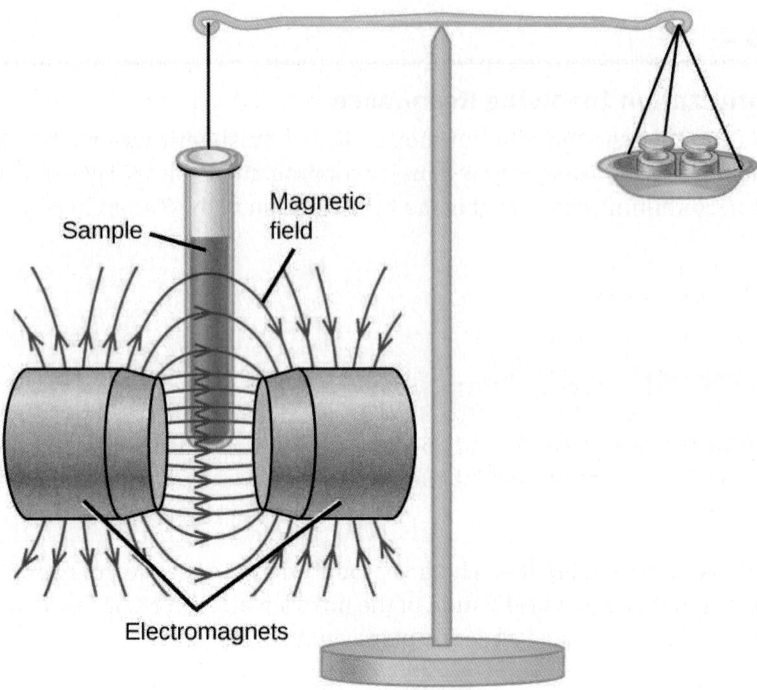

FIGURE 8.27 A Gouy balance compares the mass of a sample in the presence of a magnetic field with the mass with the electromagnet turned off to determine the number of unpaired electrons in a sample.

Experiments show that each O_2 molecule has two unpaired electrons. The Lewis-structure model does not predict the presence of these two unpaired electrons. Unlike oxygen, the apparent weight of most molecules decreases slightly in the presence of an inhomogeneous magnetic field. Materials in which all of the electrons are paired are **diamagnetic** and weakly repel a magnetic field. Paramagnetic and diamagnetic materials do not act as permanent magnets. Only in the presence of an applied magnetic field do they demonstrate attraction or repulsion.

LINK TO LEARNING

Water, like most molecules, contains all paired electrons. Living things contain a large percentage of water, so they demonstrate diamagnetic behavior. If you place a frog near a sufficiently large magnet, it will levitate. You can see videos (http://openstax.org/l/16diamagnetic) of diamagnetic floating frogs, strawberries, and more.

Molecular orbital theory (MO theory) provides an explanation of chemical bonding that accounts for the paramagnetism of the oxygen molecule. It also explains the bonding in a number of other molecules, such as violations of the octet rule and more molecules with more complicated bonding (beyond the scope of this text) that are difficult to describe with Lewis structures. Additionally, it provides a model for describing the energies of electrons in a molecule and the probable location of these electrons. Unlike valence bond theory, which uses hybrid orbitals that are assigned to one specific atom, MO theory uses the combination of atomic orbitals to yield molecular orbitals that are *delocalized* over the entire molecule rather than being localized on its constituent atoms. MO theory also helps us understand why some substances are electrical conductors, others are semiconductors, and still others are insulators. Table 8.2 summarizes the main points of the two complementary bonding theories. Both theories provide different, useful ways of describing molecular structure.

Comparison of Bonding Theories

Valence Bond Theory	Molecular Orbital Theory
considers bonds as localized between one pair of atoms	considers electrons delocalized throughout the entire molecule
creates bonds from overlap of atomic orbitals (*s, p, d*...) and hybrid orbitals (*sp*, sp^2, sp^3...)	combines atomic orbitals to form molecular orbitals (σ, σ*, π, π*)
forms σ or π bonds	creates bonding and antibonding interactions based on which orbitals are filled
predicts molecular shape based on the number of regions of electron density	predicts the arrangement of electrons in molecules
needs multiple structures to describe resonance	

TABLE 8.2

Molecular orbital theory describes the distribution of electrons in molecules in much the same way that the distribution of electrons in atoms is described using atomic orbitals. Using quantum mechanics, the behavior of an electron in a molecule is still described by a wave function, Ψ, analogous to the behavior in an atom. Just like electrons around isolated atoms, electrons around atoms in molecules are limited to discrete (quantized) energies. The region of space in which a valence electron in a molecule is likely to be found is called a **molecular orbital (Ψ^2)**. Like an atomic orbital, a molecular orbital is full when it contains two electrons with opposite spin.

We will consider the molecular orbitals in molecules composed of two identical atoms (H_2 or Cl_2, for example). Such molecules are called **homonuclear diatomic molecules**. In these diatomic molecules, several types of molecular orbitals occur.

The mathematical process of combining atomic orbitals to generate molecular orbitals is called the **linear combination of atomic orbitals (LCAO)**. The wave function describes the wavelike properties of an electron. Molecular orbitals are combinations of atomic orbital wave functions. Combining waves can lead to constructive interference, in which peaks line up with peaks, or destructive interference, in which peaks line up with troughs (Figure 8.28). In orbitals, the waves are three dimensional, and they combine with in-phase waves producing regions with a higher probability of electron density and out-of-phase waves producing nodes, or regions of no electron density.

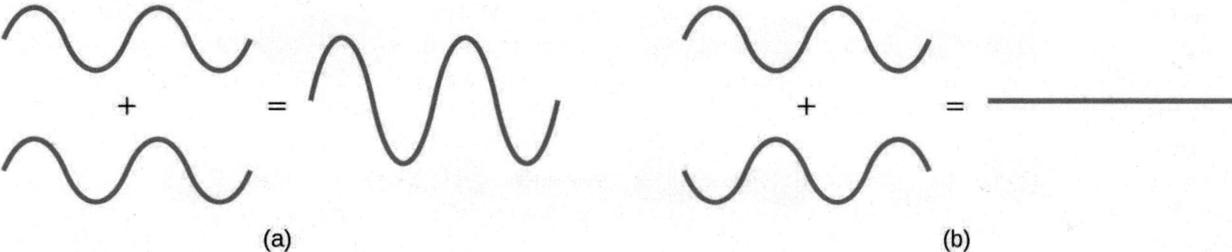

FIGURE 8.28 (a) When in-phase waves combine, constructive interference produces a wave with greater amplitude. (b) When out-of-phase waves combine, destructive interference produces a wave with less (or no) amplitude.

There are two types of molecular orbitals that can form from the overlap of two atomic *s* orbitals on adjacent atoms. The two types are illustrated in Figure 8.29. The in-phase combination produces a lower energy $\boldsymbol{\sigma_s}$

molecular orbital (read as "sigma-s") in which most of the electron density is directly between the nuclei. The out-of-phase addition (which can also be thought of as subtracting the wave functions) produces a higher energy σ_s^* **molecular orbital** (read as "sigma-s-star") molecular orbital in which there is a node between the nuclei. The asterisk signifies that the orbital is an antibonding orbital. Electrons in a σ_s orbital are attracted by both nuclei at the same time and are more stable (of lower energy) than they would be in the isolated atoms. Adding electrons to these orbitals creates a force that holds the two nuclei together, so we call these orbitals **bonding orbitals**. Electrons in the σ_s^* orbitals are located well away from the region between the two nuclei. The attractive force between the nuclei and these electrons pulls the two nuclei apart. Hence, these orbitals are called **antibonding orbitals**. Electrons fill the lower-energy bonding orbital before the higher-energy antibonding orbital, just as they fill lower-energy atomic orbitals before they fill higher-energy atomic orbitals.

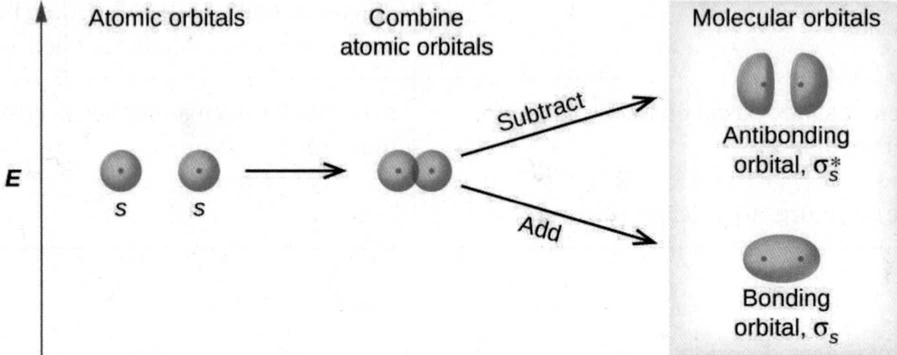

FIGURE 8.29 Sigma (σ) and sigma-star (σ*) molecular orbitals are formed by the combination of two *s* atomic orbitals. The dots (·) indicate the locations of nuclei.

LINK TO LEARNING

You can watch animations (http://openstax.org/l/16molecorbital) visualizing the calculated atomic orbitals combining to form various molecular orbitals at the Orbitron website.

In *p* orbitals, the wave function gives rise to two lobes with opposite phases, analogous to how a two-dimensional wave has both parts above and below the average. We indicate the phases by shading the orbital lobes different colors. When orbital lobes of the same phase overlap, constructive wave interference increases the electron density. When regions of opposite phase overlap, the destructive wave interference decreases electron density and creates nodes. When *p* orbitals overlap end to end, they create σ and σ* orbitals (Figure 8.30). If two atoms are located along the *x*-axis in a Cartesian coordinate system, the two p_x orbitals overlap end to end and form σ_{px} (bonding) and σ_{px}^* (antibonding) (read as "sigma-p-x" and "sigma-p-x star," respectively). Just as with *s*-orbital overlap, the asterisk indicates the orbital with a node between the nuclei, which is a higher-energy, antibonding orbital.

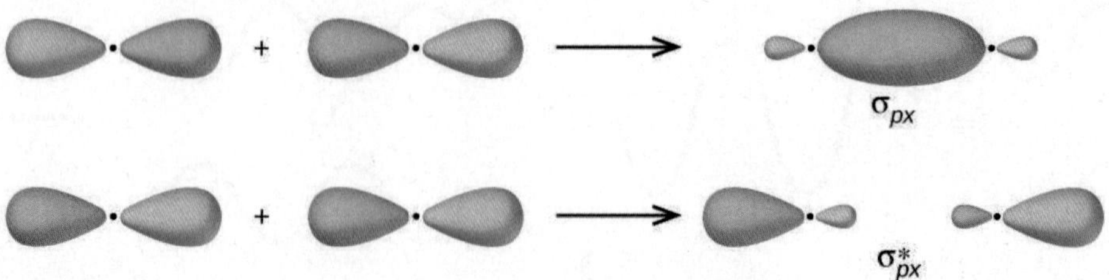

FIGURE 8.30 Combining wave functions of two *p* atomic orbitals along the internuclear axis creates two molecular orbitals, σ_p and σ_p^*.

The side-by-side overlap of two *p* orbitals gives rise to a **pi (π) bonding molecular orbital** and a **π* antibonding molecular orbital**, as shown in Figure 8.31. In valence bond theory, we describe π bonds as containing a nodal plane containing the internuclear axis and perpendicular to the lobes of the *p* orbitals, with electron density on either side of the node. In molecular orbital theory, we describe the π orbital by this same shape, and a π bond exists when this orbital contains electrons. Electrons in this orbital interact with both nuclei and help hold the two atoms together, making it a bonding orbital. For the out-of-phase combination, there are two nodal planes created, one along the internuclear axis and a perpendicular one between the nuclei.

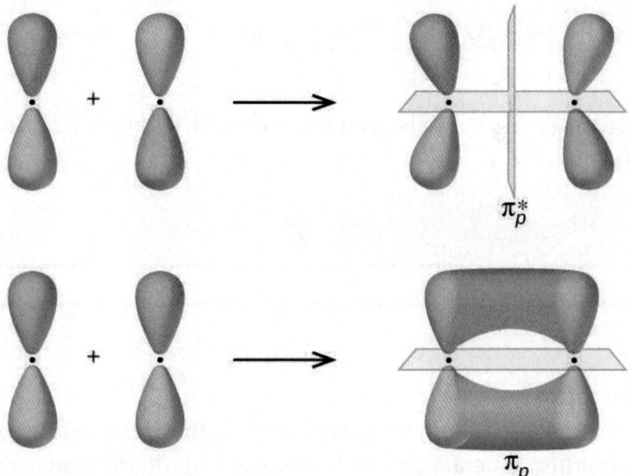

FIGURE 8.31 Side-by-side overlap of each two *p* orbitals results in the formation of two π molecular orbitals. Combining the out-of-phase orbitals results in an antibonding molecular orbital with two nodes. One contains the internuclear axis, and one is perpendicular to the axis. Combining the in-phase orbitals results in a bonding orbital. There is a node (blue) containing the internuclear axis with the two lobes of the orbital located above and below this node.

In the molecular orbitals of diatomic molecules, each atom also has two sets of *p* orbitals oriented side by side (p_y and p_z), so these four atomic orbitals combine pairwise to create two π orbitals and two π* orbitals. The π_{py} and π_{py}^* orbitals are oriented at right angles to the π_{pz} and π_{pz}^* orbitals. Except for their orientation, the π_{py} and π_{pz} orbitals are identical and have the same energy; they are **degenerate orbitals**. The π_{py}^* and π_{pz}^* antibonding orbitals are also degenerate and identical except for their orientation. A total of six molecular orbitals results from the combination of the six atomic *p* orbitals in two atoms: σ_{px} and σ_{px}^*, π_{py} and π_{py}^*, π_{pz} and π_{pz}^*.

EXAMPLE 8.5

Molecular Orbitals

Predict what type (if any) of molecular orbital would result from adding the wave functions so each pair of orbitals shown overlap. The orbitals are all similar in energy.

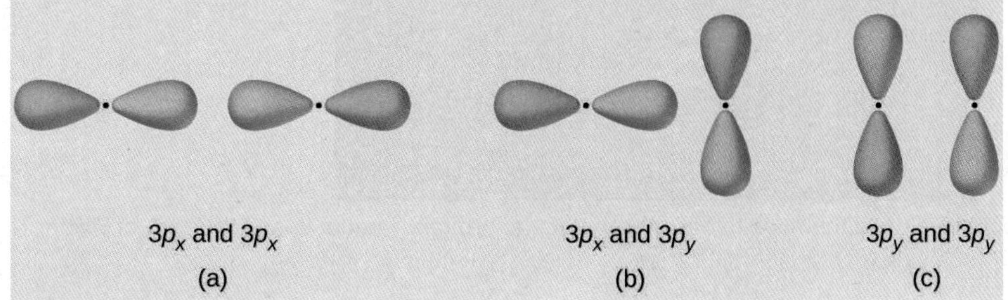

Solution

(a) is an in-phase combination, resulting in a σ_{3p} orbital

(b) will not result in a new orbital because the in-phase component (bottom) and out-of-phase component (top) cancel out. Only orbitals with the correct alignment can combine.

(c) is an out-of-phase combination, resulting in a π^*_{3p} orbital.

Check Your Learning

Label the molecular orbital shown as σ or π, bonding or antibonding and indicate where the node occurs.

Answer:

The orbital is located along the internuclear axis, so it is a σ orbital. There is a node bisecting the internuclear axis, so it is an antibonding orbital.

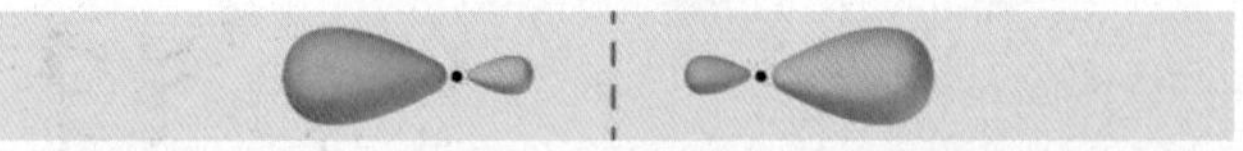

Portrait of a Chemist

Walter Kohn: Nobel Laureate

Walter Kohn (Figure 8.32) is a theoretical physicist who studies the electronic structure of solids. His work combines the principles of quantum mechanics with advanced mathematical techniques. This technique, called density functional theory, makes it possible to compute properties of molecular orbitals, including their shape and energies. Kohn and mathematician John Pople were awarded the Nobel Prize in Chemistry in 1998 for their contributions to our understanding of electronic structure. Kohn also made significant contributions to the physics of semiconductors.

FIGURE 8.32 Walter Kohn developed methods to describe molecular orbitals. (credit: image courtesy of Walter Kohn)

Kohn's biography has been remarkable outside the realm of physical chemistry as well. He was born in Austria, and during World War II he was part of the Kindertransport program that rescued 10,000 children from the Nazi regime. His summer jobs included discovering gold deposits in Canada and helping Polaroid explain how its instant film worked. Dr. Kohn passed away in 2016 at the age of 93.

HOW SCIENCES INTERCONNECT

Computational Chemistry in Drug Design

While the descriptions of bonding described in this chapter involve many theoretical concepts, they also have many practical, real-world applications. For example, drug design is an important field that uses our understanding of chemical bonding to develop pharmaceuticals. This interdisciplinary area of study uses biology (understanding diseases and how they operate) to identify specific targets, such as a binding site that is involved in a disease pathway. By modeling the structures of the binding site and potential drugs, computational chemists can predict which structures can fit together and how effectively they will bind (see Figure 8.33). Thousands of potential candidates can be narrowed down to a few of the most promising candidates. These candidate molecules are then carefully tested to determine side effects, how effectively they can be transported through the body, and other factors. Dozens of important new pharmaceuticals have been discovered with the aid of computational chemistry, and new research projects are underway.

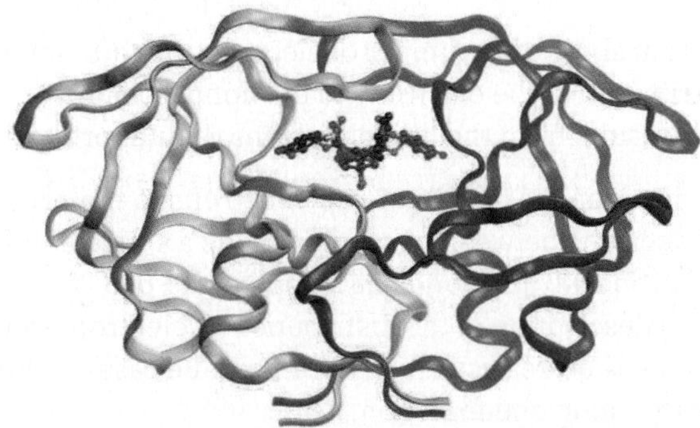

FIGURE 8.33 The molecule shown, HIV-1 protease, is an important target for pharmaceutical research. By designing molecules that bind to this protein, scientists are able to drastically inhibit the progress of the disease.

Molecular Orbital Energy Diagrams

The relative energy levels of atomic and molecular orbitals are typically shown in a **molecular orbital diagram** (Figure 8.34). For a diatomic molecule, the atomic orbitals of one atom are shown on the left, and those of the other atom are shown on the right. Each horizontal line represents one orbital that can hold two electrons. The molecular orbitals formed by the combination of the atomic orbitals are shown in the center. Dashed lines show which of the atomic orbitals combine to form the molecular orbitals. For each pair of atomic orbitals that combine, one lower-energy (bonding) molecular orbital and one higher-energy (antibonding) orbital result. Thus we can see that combining the six $2p$ atomic orbitals results in three bonding orbitals (one σ and two π) and three antibonding orbitals (one σ^* and two π^*).

We predict the distribution of electrons in these molecular orbitals by filling the orbitals in the same way that we fill atomic orbitals, by the Aufbau principle. Lower-energy orbitals fill first, electrons spread out among degenerate orbitals before pairing, and each orbital can hold a maximum of two electrons with opposite spins (Figure 8.34). Just as we write electron configurations for atoms, we can write the molecular electronic configuration by listing the orbitals with superscripts indicating the number of electrons present. For clarity, we place parentheses around molecular orbitals with the same energy. In this case, each orbital is at a different energy, so parentheses separate each orbital. Thus we would expect a diatomic molecule or ion containing seven electrons (such as $Be_2{}^+$) would have the molecular electron configuration $(\sigma_{1s})^2(\sigma_{1s}^*)^2(\sigma_{2s})^2(\sigma_{2s}^*)^1$. It is common to omit the core electrons from molecular orbital diagrams and configurations and include only the valence electrons.

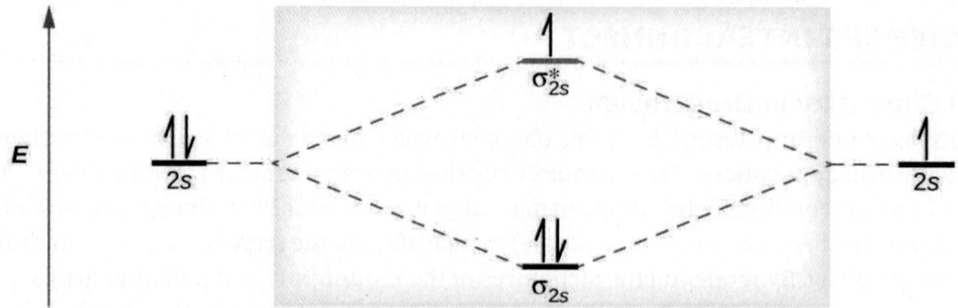

FIGURE 8.34 This is the molecular orbital diagram for the homonuclear diatomic $Be_2{}^+$, showing the molecular orbitals of the valence shell only. The molecular orbitals are filled in the same manner as atomic orbitals, using the Aufbau principle and Hund's rule.

Bond Order

The filled molecular orbital diagram shows the number of electrons in both bonding and antibonding molecular orbitals. The net contribution of the electrons to the bond strength of a molecule is identified by determining the **bond order** that results from the filling of the molecular orbitals by electrons.

When using Lewis structures to describe the distribution of electrons in molecules, we define bond order as the number of bonding pairs of electrons between two atoms. Thus a single bond has a bond order of 1, a double bond has a bond order of 2, and a triple bond has a bond order of 3. We define bond order differently when we use the molecular orbital description of the distribution of electrons, but the resulting bond order is usually the same. The MO technique is more accurate and can handle cases when the Lewis structure method fails, but both methods describe the same phenomenon.

In the molecular orbital model, an electron contributes to a bonding interaction if it occupies a bonding orbital and it contributes to an antibonding interaction if it occupies an antibonding orbital. The bond order is calculated by subtracting the destabilizing (antibonding) electrons from the stabilizing (bonding) electrons. Since a bond consists of two electrons, we divide by two to get the bond order. We can determine bond order with the following equation:

$$\text{bond order} = \frac{(\text{number of bonding electrons}) - (\text{number of antibonding electrons})}{2}$$

The order of a covalent bond is a guide to its strength; a bond between two given atoms becomes stronger as the bond order increases (Table 8.1). If the distribution of electrons in the molecular orbitals between two atoms is such that the resulting bond would have a bond order of zero, a stable bond does not form. We next look at some specific examples of MO diagrams and bond orders.

Bonding in Diatomic Molecules

A dihydrogen molecule (H_2) forms from two hydrogen atoms. When the atomic orbitals of the two atoms combine, the electrons occupy the molecular orbital of lowest energy, the σ_{1s} bonding orbital. A dihydrogen molecule, H_2, readily forms because the energy of a H_2 molecule is lower than that of two H atoms. The σ_{1s} orbital that contains both electrons is lower in energy than either of the two 1*s* atomic orbitals.

A molecular orbital can hold two electrons, so both electrons in the H_2 molecule are in the σ_{1s} bonding orbital; the electron configuration is $(\sigma_{1s})^2$. We represent this configuration by a molecular orbital energy diagram (Figure 8.35) in which a single upward arrow indicates one electron in an orbital, and two (upward and downward) arrows indicate two electrons of opposite spin.

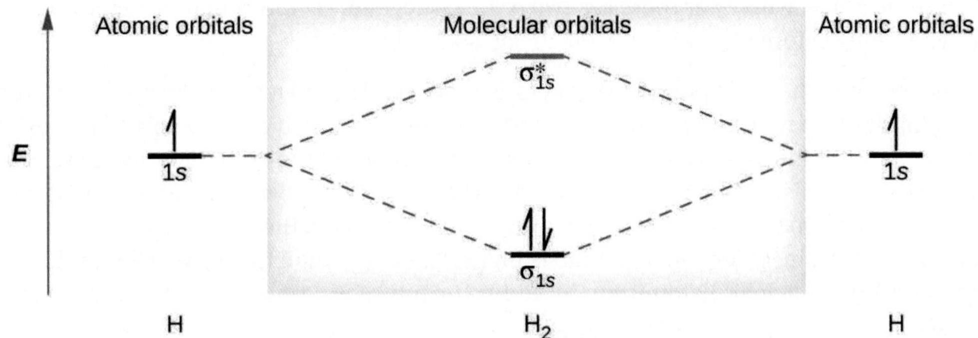

FIGURE 8.35 The molecular orbital energy diagram predicts that H_2 will be a stable molecule with lower energy than the separated atoms.

A dihydrogen molecule contains two bonding electrons and no antibonding electrons so we have

$$\text{bond order in } H_2 = \frac{(2-0)}{2} = 1$$

Because the bond order for the H–H bond is equal to 1, the bond is a single bond.

A helium atom has two electrons, both of which are in its 1*s* orbital. Two helium atoms do not combine to form a dihelium molecule, He_2, with four electrons, because the stabilizing effect of the two electrons in the lower-energy bonding orbital would be offset by the destabilizing effect of the two electrons in the higher-energy antibonding molecular orbital. We would write the hypothetical electron configuration of He_2 as $(\sigma_{1s})^2(\sigma^*_{1s})^2$ as in Figure 8.36. The net energy change would be zero, so there is no driving force for helium atoms to form the diatomic molecule. In fact, helium exists as discrete atoms rather than as diatomic molecules. The bond order in a hypothetical dihelium molecule would be zero.

$$\text{bond order in } He_2 = \frac{(2-2)}{2} = 0$$

A bond order of zero indicates that no bond is formed between two atoms.

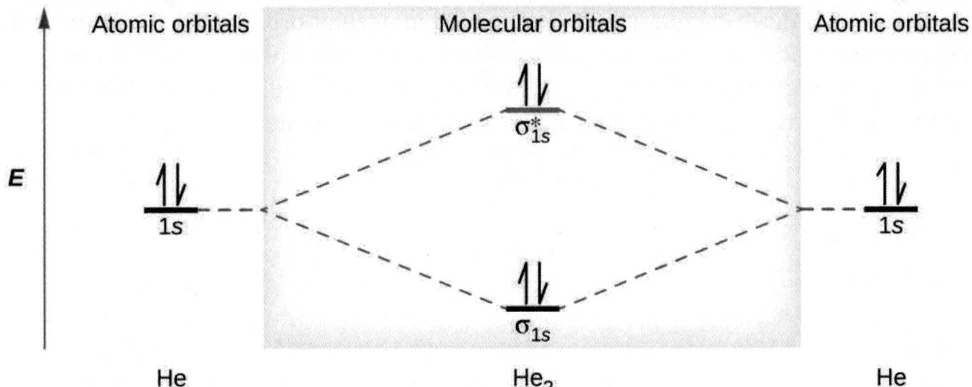

FIGURE 8.36 The molecular orbital energy diagram predicts that He_2 will not be a stable molecule, since it has equal numbers of bonding and antibonding electrons.

The Diatomic Molecules of the Second Period

Eight possible homonuclear diatomic molecules might be formed by the atoms of the second period of the periodic table: Li_2, Be_2, B_2, C_2, N_2, O_2, F_2, and Ne_2. However, we can predict that the Be_2 molecule and the Ne_2 molecule would not be stable. We can see this by a consideration of the molecular electron configurations (Table 8.3).

We predict valence molecular orbital electron configurations just as we predict electron configurations of atoms. Valence electrons are assigned to valence molecular orbitals with the lowest possible energies.

Consistent with Hund's rule, whenever there are two or more degenerate molecular orbitals, electrons fill each orbital of that type singly before any pairing of electrons takes place.

As we saw in valence bond theory, σ bonds are generally more stable than π bonds formed from degenerate atomic orbitals. Similarly, in molecular orbital theory, σ orbitals are usually more stable than π orbitals. However, this is not always the case. The MOs for the valence orbitals of the second period are shown in Figure 8.37. Looking at Ne_2 molecular orbitals, we see that the order is consistent with the generic diagram shown in the previous section. However, for atoms with three or fewer electrons in the *p* orbitals (Li through N) we observe a different pattern, in which the σ_p orbital is higher in energy than the π_p set. Obtain the molecular orbital diagram for a homonuclear diatomic ion by adding or subtracting electrons from the diagram for the neutral molecule.

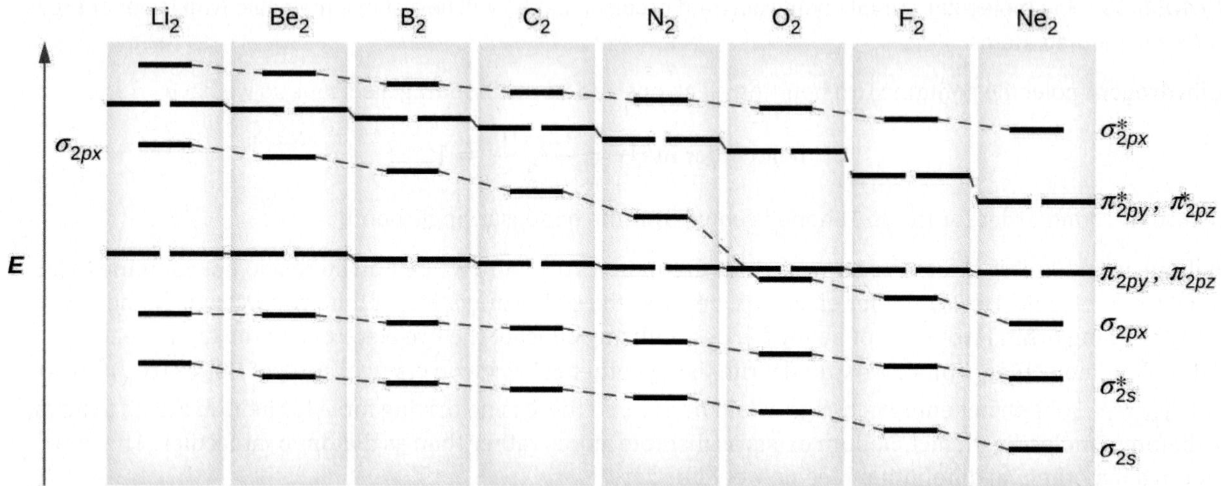

FIGURE 8.37 This shows the MO diagrams for each homonuclear diatomic molecule in the second period. The orbital energies decrease across the period as the effective nuclear charge increases and atomic radius decreases. Between N_2 and O_2, the order of the orbitals changes.

This switch in orbital ordering occurs because of a phenomenon called **s-p mixing**. s-p mixing does not create new orbitals; it merely influences the energies of the existing molecular orbitals. The σ_s wavefunction mathematically combines with the σ_p wavefunction, with the result that the σ_s orbital becomes more stable, and the σ_p orbital becomes less stable (Figure 8.38). Similarly, the antibonding orbitals also undergo s-p mixing, with the σ_{s*} becoming more stable and the σ_{p*} becoming less stable.

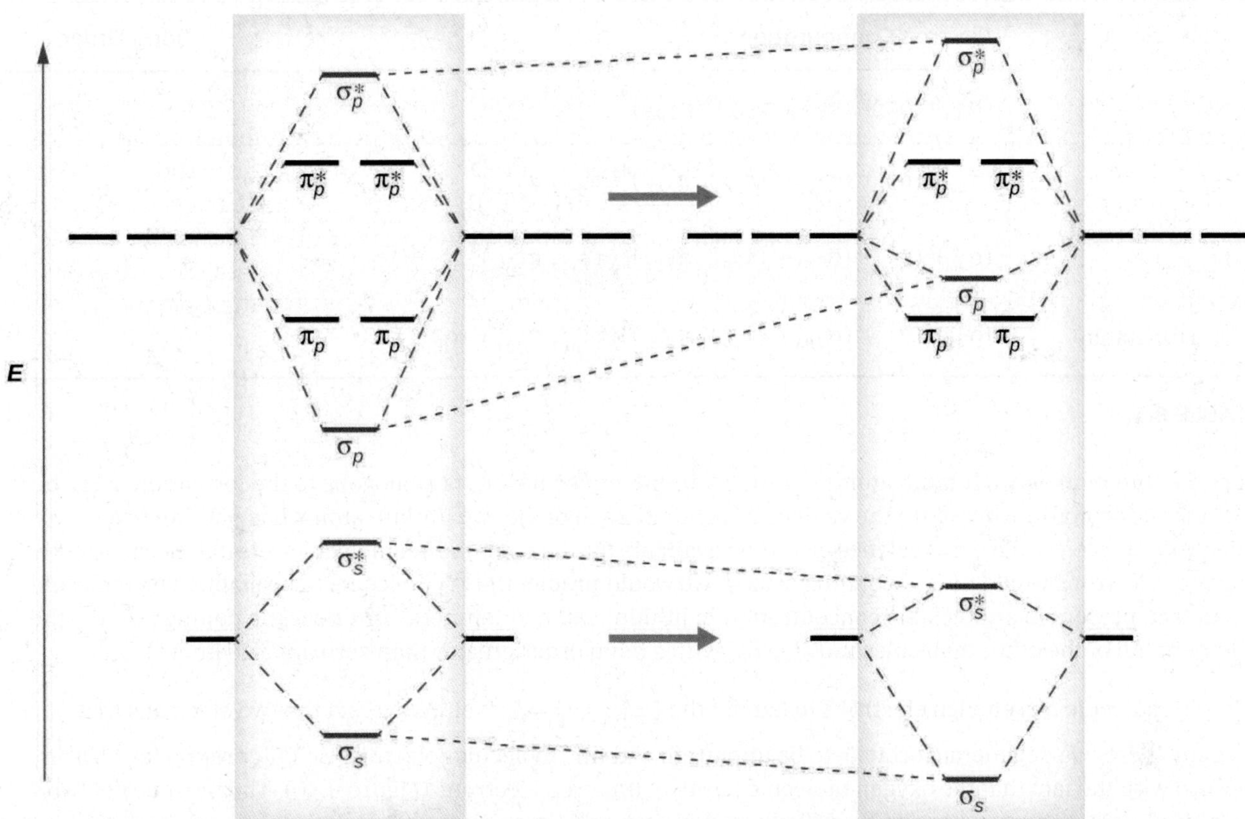

FIGURE 8.38 Without mixing, the MO pattern occurs as expected, with the σ_p orbital lower in energy than the π_p orbitals. When s-p mixing occurs, the orbitals shift as shown, with the σ_p orbital higher in energy than the π_p orbitals.

s-p mixing occurs when the *s* and *p* orbitals have similar energies. The energy difference between 2*s* and 2*p* orbitals in O, F, and Ne is greater than that in Li, Be, B, C, and N. Because of this, O_2, F_2, and Ne_2 exhibit negligible s-p mixing (not sufficient to change the energy ordering), and their MO diagrams follow the normal pattern, as shown in Figure 8.37. All of the other period 2 diatomic molecules do have s-p mixing, which leads to the pattern where the σ_p orbital is raised above the π_p set.

Using the MO diagrams shown in Figure 8.37, we can add in the electrons and determine the molecular electron configuration and bond order for each of the diatomic molecules. As shown in Table 8.3, Be_2 and Ne_2 molecules would have a bond order of 0, and these molecules do not exist.

Electron Configuration and Bond Order for Molecular Orbitals in Homonuclear Diatomic Molecules of Period Two Elements

Molecule	Electron Configuration	Bond Order
Li_2	$(\sigma_{2s})^2$	1
Be_2 (unstable)	$(\sigma_{2s})^2(\sigma_{2s}^*)^2$	0
B_2	$(\sigma_{2s})^2(\sigma_{2s}^*)^2(\pi_{2py}, \pi_{2pz})^2$	1
C_2	$(\sigma_{2s})^2(\sigma_{2s}^*)^2(\pi_{2py}, \pi_{2pz})^4$	2

TABLE 8.3

Molecule	Electron Configuration	Bond Order
N_2	$(\sigma_{2s})^2(\sigma_{2s}^*)^2(\pi_{2py}, \pi_{2pz})^4(\sigma_{2px})^2$	3
O_2	$(\sigma_{2s})^2(\sigma_{2s}^*)^2(\sigma_{2px})^2(\pi_{2py}, \pi_{2pz})^4(\pi_{2py}^*, \pi_{2pz}^*)^2$	2
F_2	$(\sigma_{2s})^2(\sigma_{2s}^*)^2(\sigma_{2px})^2(\pi_{2py}, \pi_{2pz})^4(\pi_{2py}^*, \pi_{2pz}^*)^4$	1
Ne_2 (unstable)	$(\sigma_{2s})^2(\sigma_{2s}^*)^2(\sigma_{2px})^2(\pi_{2py}, \pi_{2pz})^4(\pi_{2py}^*, \pi_{2pz}^*)^4(\sigma_{2px}^*)^2$	0

TABLE 8.3

The combination of two lithium atoms to form a lithium molecule, Li_2, is analogous to the formation of H_2, but the atomic orbitals involved are the valence 2*s* orbitals. Each of the two lithium atoms has one valence electron. Hence, we have two valence electrons available for the σ_{2s} bonding molecular orbital. Because both valence electrons would be in a bonding orbital, we would predict the Li_2 molecule to be stable. The molecule is, in fact, present in appreciable concentration in lithium vapor at temperatures near the boiling point of the element. All of the other molecules in Table 8.3 with a bond order greater than zero are also known.

The O_2 molecule has enough electrons to half fill the $\left(\pi_{2py}^*, \pi_{2pz}^*\right)$ level. We expect the two electrons that occupy these two degenerate orbitals to be unpaired, and this molecular electronic configuration for O_2 is in accord with the fact that the oxygen molecule has two unpaired electrons (Figure 8.40). The presence of two unpaired electrons has proved to be difficult to explain using Lewis structures, but the molecular orbital theory explains it quite well. In fact, the unpaired electrons of the oxygen molecule provide a strong piece of support for the molecular orbital theory.

HOW SCIENCES INTERCONNECT

Band Theory

When two identical atomic orbitals on different atoms combine, two molecular orbitals result (see Figure 8.29). The bonding orbital is lower in energy than the original atomic orbitals because the atomic orbitals are in-phase in the molecular orbital. The antibonding orbital is higher in energy than the original atomic orbitals because the atomic orbitals are out-of-phase.

In a solid, similar things happen, but on a much larger scale. Remember that even in a small sample there are a huge number of atoms (typically > 10^{23} atoms), and therefore a huge number of atomic orbitals that may be combined into molecular orbitals. When *N* valence atomic orbitals, all of the same energy and each containing one (1) electron, are combined, *N*/2 (filled) bonding orbitals and *N*/2 (empty) antibonding orbitals will result. Each bonding orbital will show an energy lowering as the atomic orbitals are *mostly* in-phase, but each of the bonding orbitals will be a little different and have slightly different energies. The antibonding orbitals will show an increase in energy as the atomic orbitals are *mostly* out-of-phase, but each of the antibonding orbitals will also be a little different and have slightly different energies. The allowed energy levels for all the bonding orbitals are so close together that they form a band, called the valence band. Likewise, all the antibonding orbitals are very close together and form a band, called the conduction band. Figure 8.39 shows the bands for three important classes of materials: insulators, semiconductors, and conductors.

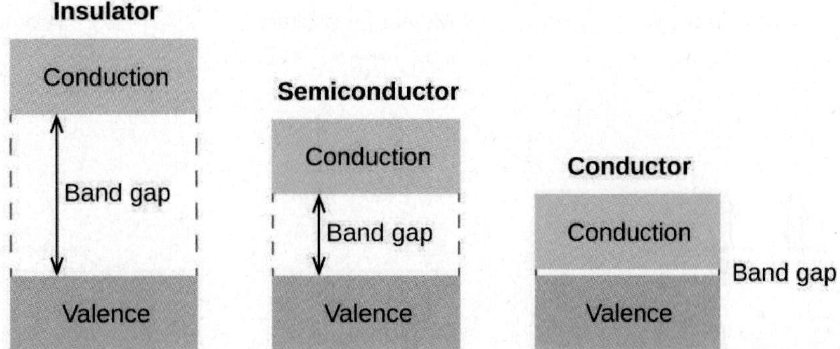

FIGURE 8.39 Molecular orbitals in solids are so closely spaced that they are described as bands. The valence band is lower in energy and the conduction band is higher in energy. The type of solid is determined by the size of the "band gap" between the valence and conduction bands. Only a very small amount of energy is required to move electrons from the valence band to the conduction band in a conductor, and so they conduct electricity well. In an insulator, the band gap is large, so that very few electrons move, and they are poor conductors of electricity. Semiconductors are in between: they conduct electricity better than insulators, but not as well as conductors.

In order to conduct electricity, electrons must move from the filled valence band to the empty conduction band where they can move throughout the solid. The size of the band gap, or the energy difference between the top of the valence band and the bottom of the conduction band, determines how easy it is to move electrons between the bands. Only a small amount of energy is required in a conductor because the band gap is very small. This small energy difference is "easy" to overcome, so they are good conductors of electricity. In an insulator, the band gap is so "large" that very few electrons move into the conduction band; as a result, insulators are poor conductors of electricity. Semiconductors conduct electricity when "moderate" amounts of energy are provided to move electrons out of the valence band and into the conduction band. Semiconductors, such as silicon, are found in many electronics.

Semiconductors are used in devices such as computers, smartphones, and solar cells. Solar cells produce electricity when light provides the energy to move electrons out of the valence band. The electricity that is generated may then be used to power a light or tool, or it can be stored for later use by charging a battery. As of December 2014, up to 46% of the energy in sunlight could be converted into electricity using solar cells.

EXAMPLE 8.6

Molecular Orbital Diagrams, Bond Order, and Number of Unpaired Electrons

Draw the molecular orbital diagram for the oxygen molecule, O_2. From this diagram, calculate the bond order for O_2. How does this diagram account for the paramagnetism of O_2?

Solution

We draw a molecular orbital energy diagram similar to that shown in Figure 8.37. Each oxygen atom contributes six electrons, so the diagram appears as shown in Figure 8.40.

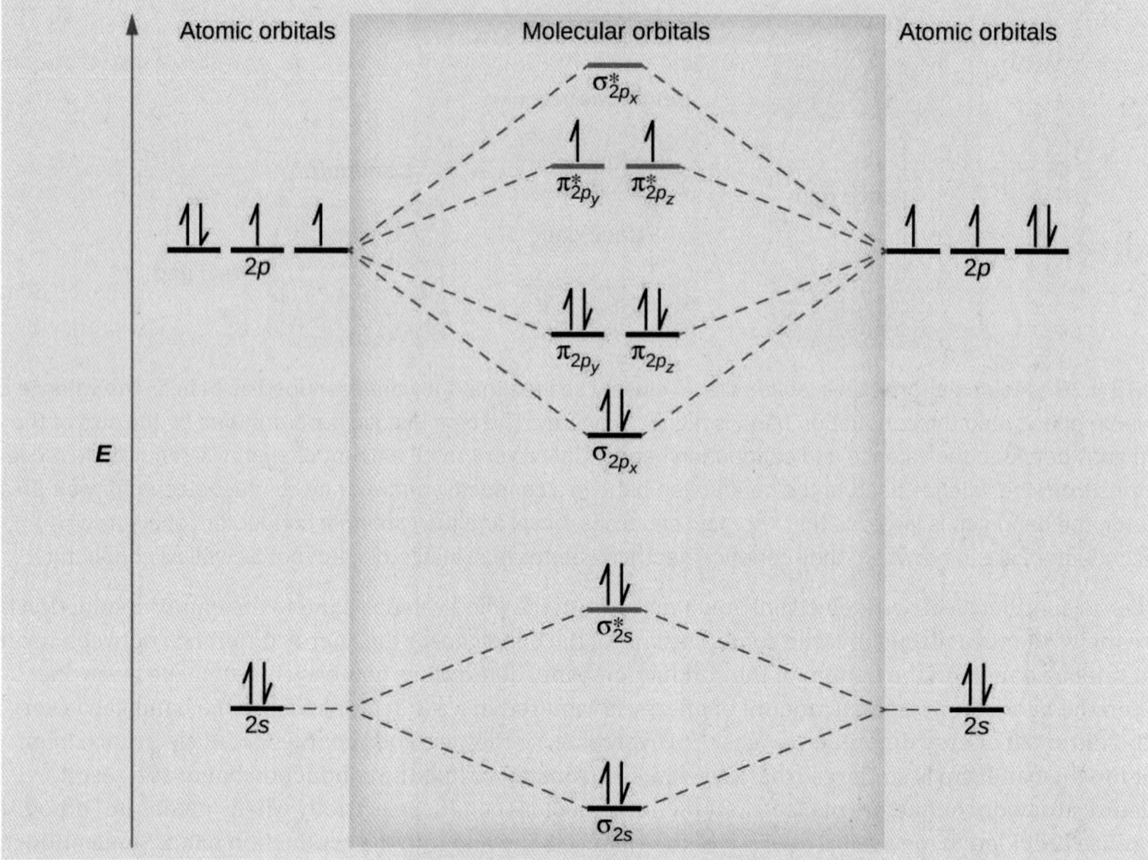

FIGURE 8.40 The molecular orbital energy diagram for O_2 predicts two unpaired electrons.

We calculate the bond order as

$$O_2 = \frac{(8-4)}{2} = 2$$

Oxygen's paramagnetism is explained by the presence of two unpaired electrons in the $(\pi_{2py}, \pi_{2pz})^*$ molecular orbitals.

Check Your Learning

The main component of air is N_2. From the molecular orbital diagram of N_2, predict its bond order and whether it is diamagnetic or paramagnetic.

Answer:

N_2 has a bond order of 3 and is diamagnetic.

EXAMPLE 8.7

Ion Predictions with MO Diagrams

Give the molecular orbital configuration for the valence electrons in $C_2{}^{2-}$. Will this ion be stable?

Solution

Looking at the appropriate MO diagram, we see that the π orbitals are lower in energy than the σ_p orbital. The valence electron configuration for C_2 is $(\sigma_{2s})^2(\sigma^*_{2s})^2(\pi_{2py}, \pi_{2pz})^4$. Adding two more electrons to generate the $C_2{}^{2-}$ anion will give a valence electron configuration of $(\sigma_{2s})^2(\sigma^*_{2s})^2(\pi_{2py}, \pi_{2pz})^4(\sigma_{2px})^2$. Since this has six more bonding electrons than antibonding, the bond order will be 3, and the ion should be stable.

Check Your Learning

How many unpaired electrons would be present on a $Be_2{}^{2-}$ ion? Would it be paramagnetic or diamagnetic?

Answer:

two, paramagnetic

LINK TO LEARNING

Creating molecular orbital diagrams for molecules with more than two atoms relies on the same basic ideas as the diatomic examples presented here. However, with more atoms, computers are required to calculate how the atomic orbitals combine. See three-dimensional drawings (http://openstax.org/l/16orbitaldiag) of the molecular orbitals for C_6H_6.

Key Terms

antibonding orbital molecular orbital located outside of the region between two nuclei; electrons in an antibonding orbital destabilize the molecule

bond order number of pairs of electrons between two atoms; it can be found by the number of bonds in a Lewis structure or by the difference between the number of bonding and antibonding electrons divided by two

bonding orbital molecular orbital located between two nuclei; electrons in a bonding orbital stabilize a molecule

degenerate orbitals orbitals that have the same energy

diamagnetism phenomenon in which a material is not magnetic itself but is repelled by a magnetic field; it occurs when there are only paired electrons present

homonuclear diatomic molecule molecule consisting of two identical atoms

hybrid orbital orbital created by combining atomic orbitals on a central atom

hybridization model that describes the changes in the atomic orbitals of an atom when it forms a covalent compound

linear combination of atomic orbitals technique for combining atomic orbitals to create molecular orbitals

molecular orbital region of space in which an electron has a high probability of being found in a molecule

molecular orbital diagram visual representation of the relative energy levels of molecular orbitals

molecular orbital theory model that describes the behavior of electrons delocalized throughout a molecule in terms of the combination of atomic wave functions

node plane separating different lobes of orbitals, where the probability of finding an electron is zero

overlap coexistence of orbitals from two different atoms sharing the same region of space, leading to the formation of a covalent bond

paramagnetism phenomenon in which a material is not magnetic itself but is attracted to a magnetic field; it occurs when there are unpaired electrons present

pi bond (π bond) covalent bond formed by side-by-side overlap of atomic orbitals; the electron density is found on opposite sides of the internuclear axis

s-p mixing change that causes σ_p orbitals to be less stable than π_p orbitals due to the mixing of *s* and *p*-based molecular orbitals of similar energies.

sigma bond (σ bond) covalent bond formed by overlap of atomic orbitals along the internuclear axis

***sp* hybrid orbital** one of a set of two orbitals with a linear arrangement that results from combining one *s* and one *p* orbital

***sp*^{2} hybrid orbital** one of a set of three orbitals with a trigonal planar arrangement that results from combining one *s* and two *p* orbitals

***sp*^{3} hybrid orbital** one of a set of four orbitals with a tetrahedral arrangement that results from combining one *s* and three *p* orbitals

***sp*^{3}*d* hybrid orbital** one of a set of five orbitals with a trigonal bipyramidal arrangement that results from combining one *s*, three *p*, and one *d* orbital

***sp*^{3}*d*^{2} hybrid orbital** one of a set of six orbitals with an octahedral arrangement that results from combining one *s*, three *p*, and two *d* orbitals

valence bond theory description of bonding that involves atomic orbitals overlapping to form σ or π bonds, within which pairs of electrons are shared

π bonding orbital molecular orbital formed by side-by-side overlap of atomic orbitals, in which the electron density is found on opposite sides of the internuclear axis

π* bonding orbital antibonding molecular orbital formed by out of phase side-by-side overlap of atomic orbitals, in which the electron density is found on both sides of the internuclear axis, and there is a node between the nuclei

σ bonding orbital molecular orbital in which the electron density is found along the axis of the bond

σ* bonding orbital antibonding molecular orbital formed by out-of-phase overlap of atomic orbital along the axis of the bond, generating a node between the nuclei

Key Equations

$$\text{bond order} = \frac{(\text{number of bonding electron}) - (\text{number of antibonding electrons})}{2}$$

Summary

8.1 Valence Bond Theory

Valence bond theory describes bonding as a consequence of the overlap of two separate atomic orbitals on different atoms that creates a region with one pair of electrons shared between the two atoms. When the orbitals overlap along an axis containing the nuclei, they form a σ bond. When they overlap in a fashion that creates a node along this axis, they form a π bond.

8.2 Hybrid Atomic Orbitals

We can use hybrid orbitals, which are mathematical combinations of some or all of the valence atomic orbitals, to describe the electron density around covalently bonded atoms. These hybrid orbitals either form sigma (σ) bonds directed toward other atoms of the molecule or contain lone pairs of electrons. We can determine the type of hybridization around a central atom from the geometry of the regions of electron density about it. Two such regions imply *sp* hybridization; three, sp^2 hybridization; four, sp^3 hybridization; five, sp^3d hybridization; and six, sp^3d^2 hybridization. Pi (π) bonds are formed from unhybridized atomic orbitals (*p* or *d* orbitals).

8.3 Multiple Bonds

Multiple bonds consist of a σ bond located along the axis between two atoms and one or two π bonds. The σ bonds are usually formed by the overlap of hybridized atomic orbitals, while the π bonds are formed by the side-by-side overlap of unhybridized orbitals. Resonance occurs when there are multiple unhybridized orbitals with the appropriate alignment to overlap, so the placement of π bonds can vary.

8.4 Molecular Orbital Theory

Molecular orbital (MO) theory describes the behavior of electrons in a molecule in terms of combinations of the atomic wave functions. The resulting molecular orbitals may extend over all the atoms in the molecule. Bonding molecular orbitals are formed by in-phase combinations of atomic wave functions, and electrons in these orbitals stabilize a molecule. Antibonding molecular orbitals result from out-of-phase combinations of atomic wave functions and electrons in these orbitals make a molecule less stable. Molecular orbitals located along an internuclear axis are called σ MOs. They can be formed from *s* orbitals or from *p* orbitals oriented in an end-to-end fashion. Molecular orbitals formed from *p* orbitals oriented in a side-by-side fashion have electron density on opposite sides of the internuclear axis and are called π orbitals.

We can describe the electronic structure of diatomic molecules by applying molecular orbital theory to the valence electrons of the atoms. Electrons fill molecular orbitals following the same rules that apply to filling atomic orbitals; Hund's rule and the Aufbau principle tell us that lower-energy orbitals will fill first, electrons will spread out before they pair up, and each orbital can hold a maximum of two electrons with opposite spins. Materials with unpaired electrons are paramagnetic and attracted to a magnetic field, while those with all-paired electrons are diamagnetic and repelled by a magnetic field. Correctly predicting the magnetic properties of molecules is in advantage of molecular orbital theory over Lewis structures and valence bond theory.

Exercises

8.1 Valence Bond Theory

1. Explain how σ and π bonds are similar and how they are different.
2. Draw a curve that describes the energy of a system with H and Cl atoms at varying distances. Then, find the minimum energy of this curve two ways.
   (a) Use the bond energy found in Table 8.1 to calculate the energy for one single HCl bond (Hint: How many bonds are in a mole?)
   (b) Use the enthalpy of reaction and the bond energies for H_2 and Cl_2 to solve for the energy of one mole of HCl bonds.
   $$H_2(g) + Cl_2(g) \rightleftharpoons 2HCl(g) \qquad \Delta H^\circ_{rxn} = -184.7 \text{ kJ/mol}$$
3. Explain why bonds occur at specific average bond distances instead of the atoms approaching each other infinitely close.

4. Use valence bond theory to explain the bonding in F_2, HF, and ClBr. Sketch the overlap of the atomic orbitals involved in the bonds.
5. Use valence bond theory to explain the bonding in O_2. Sketch the overlap of the atomic orbitals involved in the bonds in O_2.
6. How many σ and π bonds are present in the molecule HCN?
7. A friend tells you N_2 has three π bonds due to overlap of the three *p*-orbitals on each N atom. Do you agree?
8. Draw the Lewis structures for CO_2 and CO, and predict the number of σ and π bonds for each molecule.
   (a) CO_2
   (b) CO

8.2 Hybrid Atomic Orbitals

9. Why is the concept of hybridization required in valence bond theory?
10. Give the shape that describes each hybrid orbital set:
    (a) sp^2
    (b) sp^3d
    (c) sp
    (d) sp^3d^2
11. Explain why a carbon atom cannot form five bonds using sp^3d hybrid orbitals.
12. What is the hybridization of the central atom in each of the following?
    (a) BeH_2
    (b) SF_6
    (c) $PO_4{}^{3-}$
    (d) PCl_5
13. A molecule with the formula AB_3 could have one of four different shapes. Give the shape and the hybridization of the central A atom for each.
14. Methionine, $CH_3SCH_2CH_2CH(NH_2)CO_2H$, is an amino acid found in proteins. The Lewis structure of this compound is shown below. What is the hybridization type of each carbon, oxygen, the nitrogen, and the sulfur?

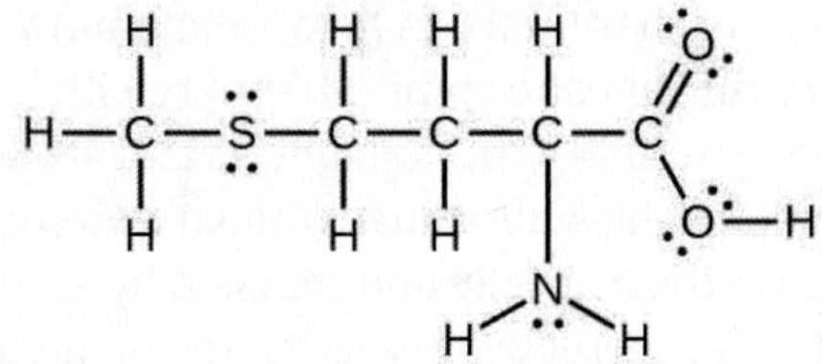

```
    H       H   H   H   O
    |   ..  |   |   |  //
H — C — S — C — C — C — C
    |   ..  |   |   |    \
    H       H   H   |     O — H
                    N
                  H   H
```

15. Sulfuric acid is manufactured by a series of reactions represented by the following equations:
    $S_8(s) + 8O_2(g) \longrightarrow 8SO_2(g)$
    $2SO_2(g) + O_2(g) \longrightarrow 2SO_3(g)$
    $SO_3(g) + H_2O(l) \longrightarrow H_2SO_4(l)$
    Draw a Lewis structure, predict the molecular geometry by VSEPR, and determine the hybridization of sulfur for the following:
    (a) circular S_8 molecule
    (b) SO_2 molecule
    (c) SO_3 molecule
    (d) H_2SO_4 molecule (the hydrogen atoms are bonded to oxygen atoms)
16. Two important industrial chemicals, ethene, C_2H_4, and propene, C_3H_6, are produced by the steam (or thermal) cracking process:

$$2C_3H_8(g) \longrightarrow C_2H_4(g) + C_3H_6(g) + CH_4(g) + H_2(g)$$

    For each of the four carbon compounds, do the following:
    (a) Draw a Lewis structure.
    (b) Predict the geometry about the carbon atom.
    (c) Determine the hybridization of each type of carbon atom.

17. Analysis of a compound indicates that it contains 77.55% Xe and 22.45% F by mass.
(a) What is the empirical formula for this compound? *(Assume this is also the molecular formula in responding to the remaining parts of this exercise).*
(b) Write a Lewis structure for the compound.
(c) Predict the shape of the molecules of the compound.
(d) What hybridization is consistent with the shape you predicted?

18. Consider nitrous acid, HNO_2 (HONO).
(a) Write a Lewis structure.
(b) What are the electron pair and molecular geometries of the internal oxygen and nitrogen atoms in the HNO_2 molecule?
(c) What is the hybridization on the internal oxygen and nitrogen atoms in HNO_2?

19. Strike-anywhere matches contain a layer of $KClO_3$ and a layer of P_4S_3. The heat produced by the friction of striking the match causes these two compounds to react vigorously, which sets fire to the wooden stem of the match. $KClO_3$ contains the $ClO_3{}^-$ ion. P_4S_3 is an unusual molecule with the skeletal structure.

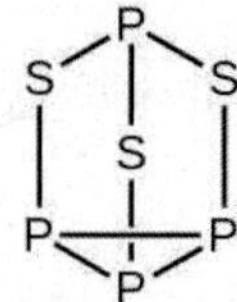

(a) Write Lewis structures for P_4S_3 and the $ClO_3{}^-$ ion.
(b) Describe the geometry about the P atoms, the S atom, and the Cl atom in these species.
(c) Assign a hybridization to the P atoms, the S atom, and the Cl atom in these species.
(d) Determine the oxidation states and formal charge of the atoms in P_4S_3 and the $ClO_3{}^-$ ion.

20. Identify the hybridization of each carbon atom in the following molecule. (The arrangement of atoms is given; you need to determine how many bonds connect each pair of atoms.)

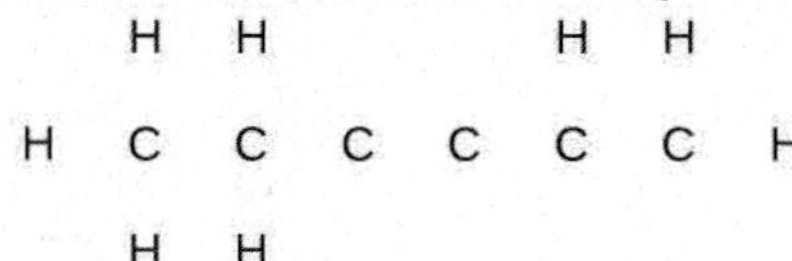

21. Write Lewis structures for NF_3 and PF_5. On the basis of hybrid orbitals, explain the fact that NF_3, PF_3, and PF_5 are stable molecules, but NF_5 does not exist.

22. In addition to NF_3, two other fluoro derivatives of nitrogen are known: N_2F_4 and N_2F_2. What shapes do you predict for these two molecules? What is the hybridization for the nitrogen in each molecule?

8.3 Multiple Bonds

23. The bond energy of a C–C single bond averages 347 kJ mol^{-1}; that of a $C \equiv C$ triple bond averages 839 kJ mol^{-1}. Explain why the triple bond is not three times as strong as a single bond.

24. For the carbonate ion, $CO_3{}^{2-}$, draw all of the resonance structures. Identify which orbitals overlap to create each bond.

25. A useful solvent that will dissolve salts as well as organic compounds is the compound acetonitrile, H_3CCN. It is present in paint strippers.
(a) Write the Lewis structure for acetonitrile, and indicate the direction of the dipole moment in the molecule.
(b) Identify the hybrid orbitals used by the carbon atoms in the molecule to form σ bonds.
(c) Describe the atomic orbitals that form the π bonds in the molecule. Note that it is not necessary to hybridize the nitrogen atom.

26. For the molecule allene, $H_2C = C = CH_2$, give the hybridization of each carbon atom. Will the hydrogen atoms be in the same plane or perpendicular planes?

27. Identify the hybridization of the central atom in each of the following molecules and ions that contain multiple bonds:
(a) ClNO (N is the central atom)
(b) CS_2
(c) Cl_2CO (C is the central atom)
(d) Cl_2SO (S is the central atom)
(e) SO_2F_2 (S is the central atom)
(f) XeO_2F_2 (Xe is the central atom)
(g) $ClOF_2^+$ (Cl is the central atom)

28. Describe the molecular geometry and hybridization of the N, P, or S atoms in each of the following compounds.
(a) H_3PO_4, phosphoric acid, used in cola soft drinks
(b) NH_4NO_3, ammonium nitrate, a fertilizer and explosive
(c) S_2Cl_2, disulfur dichloride, used in vulcanizing rubber
(d) $K_4[O_3POPO_3]$, potassium pyrophosphate, an ingredient in some toothpastes

29. For each of the following molecules, indicate the hybridization requested and whether or not the electrons will be delocalized:
(a) ozone (O_3) central O hybridization
(b) carbon dioxide (CO_2) central C hybridization
(c) nitrogen dioxide (NO_2) central N hybridization
(d) phosphate ion (PO_4^{3-}) central P hybridization

30. For each of the following structures, determine the hybridization requested and whether the electrons will be delocalized:
(a) Hybridization of each carbon

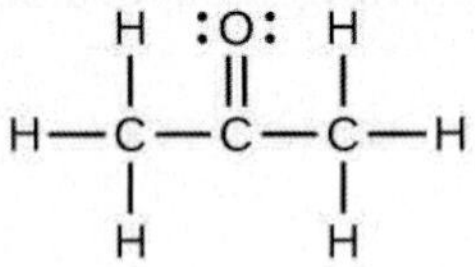

(b) Hybridization of sulfur

(c) All atoms

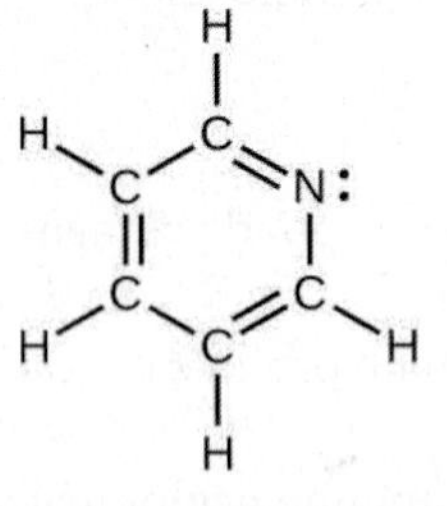

31. Draw the orbital diagram for carbon in CO_2 showing how many carbon atom electrons are in each orbital.

8.4 Molecular Orbital Theory

32. Sketch the distribution of electron density in the bonding and antibonding molecular orbitals formed from two *s* orbitals and from two *p* orbitals.

33. How are the following similar, and how do they differ?
(a) σ molecular orbitals and π molecular orbitals
(b) ψ for an atomic orbital and ψ for a molecular orbital
(c) bonding orbitals and antibonding orbitals

34. If molecular orbitals are created by combining five atomic orbitals from atom A and five atomic orbitals from atom B combine, how many molecular orbitals will result?

35. Can a molecule with an odd number of electrons ever be diamagnetic? Explain why or why not.

36. Can a molecule with an even number of electrons ever be paramagnetic? Explain why or why not.

37. Why are bonding molecular orbitals lower in energy than the parent atomic orbitals?

38. Calculate the bond order for an ion with this configuration:

$$(\sigma_{2s})^2(\sigma_{2s}^*)^2(\sigma_{2px})^2(\pi_{2py}, \pi_{2pz})^4(\pi_{2py}^*, \pi_{2pz}^*)^3$$

39. Explain why an electron in the bonding molecular orbital in the H_2 molecule has a lower energy than an electron in the 1*s* atomic orbital of either of the separated hydrogen atoms.

40. Predict the valence electron molecular orbital configurations for the following, and state whether they will be stable or unstable ions.

(a) ${Na_2}^{2+}$
(b) ${Mg_2}^{2+}$
(c) ${Al_2}^{2+}$
(d) ${Si_2}^{2+}$
(e) ${P_2}^{2+}$
(f) ${S_2}^{2+}$
(g) ${F_2}^{2+}$
(h) ${Ar_2}^{2+}$

41. Determine the bond order of each member of the following groups, and determine which member of each group is predicted by the molecular orbital model to have the strongest bond.

(a) H_2, ${H_2}^+$, ${H_2}^-$
(b) O_2, ${O_2}^{2+}$, ${O_2}^{2-}$
(c) Li_2, ${Be_2}^+$, Be_2
(d) F_2, ${F_2}^+$, ${F_2}^-$
(e) N_2, ${N_2}^+$, ${N_2}^-$

42. For the first ionization energy for an N_2 molecule, what molecular orbital is the electron removed from?

43. Compare the atomic and molecular orbital diagrams to identify the member of each of the following pairs that has the highest first ionization energy (the most tightly bound electron) in the gas phase:

(a) H and H_2
(b) N and N_2
(c) O and O_2
(d) C and C_2
(e) B and B_2

44. Which of the period 2 homonuclear diatomic molecules are predicted to be paramagnetic?

45. A friend tells you that the 2*s* orbital for fluorine starts off at a much lower energy than the 2*s* orbital for lithium, so the resulting σ_{2s} molecular orbital in F_2 is more stable than in Li_2. Do you agree?

46. True or false: Boron contains $2s^22p^1$ valence electrons, so only one *p* orbital is needed to form molecular orbitals.

47. What charge would be needed on F_2 to generate an ion with a bond order of 2?

48. Predict whether the MO diagram for S_2 would show s-p mixing or not.

49. Explain why ${N_2}^{2+}$ is diamagnetic, while ${O_2}^{4+}$, which has the same number of valence electrons, is paramagnetic.

50. Using the MO diagrams, predict the bond order for the stronger bond in each pair:

(a) B_2 or ${B_2}^+$
(b) F_2 or ${F_2}^+$
(c) O_2 or ${O_2}^{2+}$
(d) ${C_2}^+$ or ${C_2}^-$

CHAPTER 9
Gases

Figure 9.1 The hot air inside these balloons is less dense than the surrounding cool air. This results in a buoyant force that causes the balloons to rise when their guy lines are untied. (credit: modification of work by Anthony Quintano)

CHAPTER OUTLINE

9.1 Gas Pressure
9.2 Relating Pressure, Volume, Amount, and Temperature: The Ideal Gas Law
9.3 Stoichiometry of Gaseous Substances, Mixtures, and Reactions
9.4 Effusion and Diffusion of Gases
9.5 The Kinetic-Molecular Theory
9.6 Non-Ideal Gas Behavior

INTRODUCTION We are surrounded by an ocean of gas—the atmosphere—and many of the properties of gases are familiar to us from our daily activities. Heated gases expand, which can make a hot air balloon rise (Figure 9.1) or cause a blowout in a bicycle tire left in the sun on a hot day.

Gases have played an important part in the development of chemistry. In the seventeenth and eighteenth centuries, many scientists investigated gas behavior, providing the first mathematical descriptions of the behavior of matter.

In this chapter, we will examine the relationships between gas temperature, pressure, amount, and volume. We will study a simple theoretical model and use it to analyze the experimental behavior of gases. The results of these analyses will show us the limitations of the theory and how to improve on it.

9.1 Gas Pressure

LEARNING OBJECTIVES

By the end of this section, you will be able to:

- Define the property of pressure
- Define and convert among the units of pressure measurements
- Describe the operation of common tools for measuring gas pressure
- Calculate pressure from manometer data

The earth's atmosphere exerts a pressure, as does any other gas. Although we do not normally notice atmospheric pressure, we are sensitive to pressure changes—for example, when your ears "pop" during take-off and landing while flying, or when you dive underwater. Gas pressure is caused by the force exerted by gas molecules colliding with the surfaces of objects (Figure 9.2). Although the force of each collision is very small, any surface of appreciable area experiences a large number of collisions in a short time, which can result in a high pressure. In fact, normal air pressure is strong enough to crush a metal container when not balanced by equal pressure from inside the container.

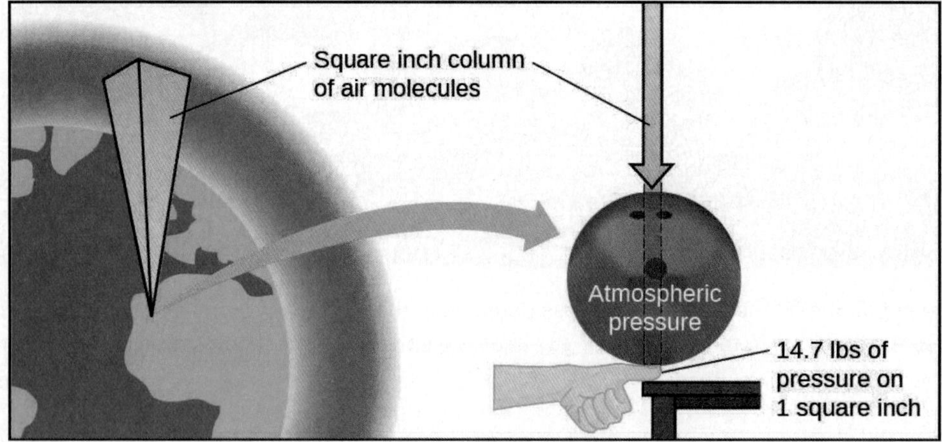

FIGURE 9.2 The atmosphere above us exerts a large pressure on objects at the surface of the earth, roughly equal to the weight of a bowling ball pressing on an area the size of a human thumbnail.

LINK TO LEARNING

A dramatic illustration (http://openstax.org/l/16atmospressur1) of atmospheric pressure is provided in this brief video, which shows a railway tanker car imploding when its internal pressure is decreased.

A smaller scale demonstration (http://openstax.org/l/16atmospressur2) of this phenomenon is briefly explained.

Atmospheric pressure is caused by the weight of the column of air molecules in the atmosphere above an object, such as the tanker car. At sea level, this pressure is roughly the same as that exerted by a full-grown African elephant standing on a doormat, or a typical bowling ball resting on your thumbnail. These may seem like huge amounts, and they are, but life on earth has evolved under such atmospheric pressure. If you actually perch a bowling ball on your thumbnail, the pressure experienced is *twice* the usual pressure, and the sensation is unpleasant.

In general, **pressure** is defined as the force exerted on a given area: $P = \frac{F}{A}$. Note that pressure is directly proportional to force and inversely proportional to area. Thus, pressure can be increased either by increasing the amount of force or by decreasing the area over which it is applied; pressure can be decreased by decreasing the force or increasing the area.

Let's apply this concept to determine which exerts a greater pressure in Figure 9.3—the elephant or the figure skater? A large African elephant can weigh 7 tons, supported on four feet, each with a diameter of about 1.5 ft (footprint area of 250 in^2), so the pressure exerted by each foot is about 14 lb/in^2:

$$\text{pressure per elephant foot} = 14{,}000\ \frac{\text{lb}}{\text{elephant}} \times \frac{1\ \text{elephant}}{4\ \text{feet}} \times \frac{1\ \text{foot}}{250\ \text{in}^2} = 14\ \text{lb/in}^2$$

The figure skater weighs about 120 lbs, supported on two skate blades, each with an area of about 2 in^2, so the pressure exerted by each blade is about 30 lb/in^2:

$$\text{pressure per skate blade} = 120\ \frac{\text{lb}}{\text{skater}} \times \frac{1\ \text{skater}}{2\ \text{blades}} \times \frac{1\ \text{blade}}{2\ \text{in}^2} = 30\ \text{lb/in}^2$$

Even though the elephant is more than one hundred-times heavier than the skater, it exerts less than one-half of the pressure. On the other hand, if the skater removes their skates and stands with bare feet (or regular footwear) on the ice, the larger area over which their weight is applied greatly reduces the pressure exerted:

$$\text{pressure per human foot} = 120\ \frac{\text{lb}}{\text{skater}} \times \frac{1\ \text{skater}}{2\ \text{feet}} \times \frac{1\ \text{foot}}{30\ \text{in}^2} = 2\ \text{lb/in}^2$$

(a)

(b)

FIGURE 9.3 Although (a) an elephant's weight is large, creating a very large force on the ground, (b) the figure skater exerts a much higher pressure on the ice due to the small surface area of the skates. (credit a: modification of work by Guido da Rozze; credit b: modification of work by Ryosuke Yagi)

The SI unit of pressure is the **pascal (Pa)**, with 1 Pa = 1 N/m^2, where N is the newton, a unit of force defined as 1 kg m/s^2. One pascal is a small pressure; in many cases, it is more convenient to use units of kilopascal (1 kPa = 1000 Pa) or **bar** (1 bar = 100,000 Pa). In the United States, pressure is often measured in pounds of force on an area of one square inch—**pounds per square inch (psi)**—for example, in car tires. Pressure can also be measured using the unit **atmosphere (atm)**, which originally represented the average sea level air pressure at the approximate latitude of Paris (45°). Table 9.1 provides some information on these and a few other common units for pressure measurements

Pressure Units

Unit Name and Abbreviation	Definition or Relation to Other Unit
pascal (Pa)	1 Pa = 1 N/m^2 recommended IUPAC unit
kilopascal (kPa)	1 kPa = 1000 Pa
pounds per square inch (psi)	air pressure at sea level is ~14.7 psi
atmosphere (atm)	1 atm = 101,325 Pa = 760 torr air pressure at sea level is ~1 atm

TABLE 9.1

Unit Name and Abbreviation	Definition or Relation to Other Unit
bar (bar, or b)	1 bar = 100,000 Pa (exactly) commonly used in meteorology
millibar (mbar, or mb)	1000 mbar = 1 bar
inches of mercury (in. Hg)	1 in. Hg = 3386 Pa used by aviation industry, also some weather reports
torr	1 torr = $\frac{1}{760}$ atm named after Evangelista Torricelli, inventor of the barometer
millimeters of mercury (mm Hg)	1 mm Hg ~1 torr

TABLE 9.1

EXAMPLE 9.1

Conversion of Pressure Units

The United States National Weather Service reports pressure in both inches of Hg and millibars. Convert a pressure of 29.2 in. Hg into:

(a) torr

(b) atm

(c) kPa

(d) mbar

Solution

This is a unit conversion problem. The relationships between the various pressure units are given in Table 9.1.

(a) $29.2 \cancel{\text{in Hg}} \times \frac{25.4 \cancel{\text{mm}}}{1 \cancel{\text{in}}} \times \frac{1 \text{ torr}}{1 \cancel{\text{mm Hg}}} = 742 \text{ torr}$

(b) $742 \cancel{\text{torr}} \times \frac{1 \text{ atm}}{760 \cancel{\text{torr}}} = 0.976 \text{ atm}$

(c) $742 \cancel{\text{torr}} \times \frac{101.325 \text{ kPa}}{760 \cancel{\text{torr}}} = 98.9 \text{ kPa}$

(d) $98.9 \cancel{\text{kPa}} \times \frac{1000 \cancel{\text{Pa}}}{1 \cancel{\text{kPa}}} \times \frac{1 \cancel{\text{bar}}}{100{,}000 \cancel{\text{Pa}}} \times \frac{1000 \text{ mbar}}{1 \cancel{\text{bar}}} = 989 \text{ mbar}$

Check Your Learning

A typical barometric pressure in Kansas City is 740 torr. What is this pressure in atmospheres, in millimeters of mercury, in kilopascals, and in bar?

Answer:

0.974 atm; 740 mm Hg; 98.7 kPa; 0.987 bar

We can measure atmospheric pressure, the force exerted by the atmosphere on the earth's surface, with a **barometer** (Figure 9.4). A barometer is a glass tube that is closed at one end, filled with a nonvolatile liquid such as mercury, and then inverted and immersed in a container of that liquid. The atmosphere exerts pressure on the liquid outside the tube, the column of liquid exerts pressure inside the tube, and the pressure at the liquid surface is the same inside and outside the tube. The height of the liquid in the tube is therefore proportional to the pressure exerted by the atmosphere.

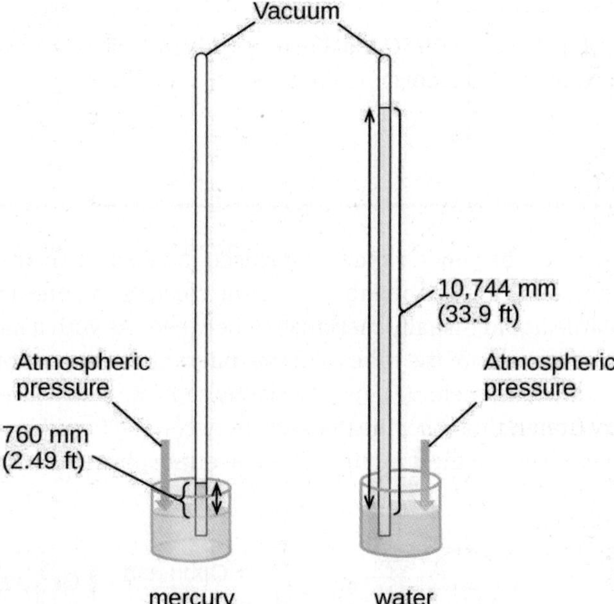

FIGURE 9.4 In a barometer, the height, *h*, of the column of liquid is used as a measurement of the air pressure. Using very dense liquid mercury (left) permits the construction of reasonably sized barometers, whereas using water (right) would require a barometer more than 30 feet tall.

If the liquid is water, normal atmospheric pressure will support a column of water over 10 meters high, which is rather inconvenient for making (and reading) a barometer. Because mercury (Hg) is about 13.6-times denser than water, a mercury barometer only needs to be $\frac{1}{13.6}$ as tall as a water barometer—a more suitable size. Standard atmospheric pressure of 1 atm at sea level (101,325 Pa) corresponds to a column of mercury that is about 760 mm (29.92 in.) high. The **torr** was originally intended to be a unit equal to one millimeter of mercury, but it no longer corresponds exactly. The pressure exerted by a fluid due to gravity is known as **hydrostatic pressure**, *p*:

$$p = h\rho g$$

where h is the height of the fluid, ρ is the density of the fluid, and g is acceleration due to gravity.

EXAMPLE 9.2

Calculation of Barometric Pressure

Show the calculation supporting the claim that atmospheric pressure near sea level corresponds to the pressure exerted by a column of mercury that is about 760 mm high. The density of mercury = 13.6 g/cm^3.

Solution

The hydrostatic pressure is given by $p = h\rho g$, with h = 760 mm, ρ = 13.6 g/cm^3, and g = 9.81 m/s^2. Plugging these values into the equation and doing the necessary unit conversions will give us the value we seek. (Note: We are expecting to find a pressure of ~101,325 Pa.)

$$101{,}325\ N/\text{m}^2 = 101{,}325\ \frac{\text{kg·m/s}^2}{\text{m}^2} = 101{,}325\ \frac{\text{kg}}{\text{m·s}^2}$$

$$p = \left(760\text{ mm} \times \frac{1\text{ m}}{1000\text{ mm}}\right) \times \left(\frac{13.6\text{ g}}{1\text{ cm}^3} \times \frac{1\text{ kg}}{1000\text{ g}} \times \frac{(100\text{ cm})^3}{(1\text{ m})^3}\right) \times \left(\frac{9.81\text{ m}}{1\text{ s}^2}\right)$$

$$= (0.760\text{ m})\left(13{,}600\text{ kg/m}^3\right)\left(9.81\text{ m/s}^2\right) = 1.01 \times 10^5\text{ kg/ms}^2 = 1.01 \times 10^5\ N/\text{m}^2$$

$$= 1.01 \times 10^5 \text{ Pa}$$

Check Your Learning

Calculate the height of a column of water at 25 °C that corresponds to normal atmospheric pressure. The density of water at this temperature is 1.0 g/cm^3.

Answer:

10.3 m

A **manometer** is a device similar to a barometer that can be used to measure the pressure of a gas trapped in a container. A closed-end manometer is a U-shaped tube with one closed arm, one arm that connects to the gas to be measured, and a nonvolatile liquid (usually mercury) in between. As with a barometer, the distance between the liquid levels in the two arms of the tube (h in the diagram) is proportional to the pressure of the gas in the container. An open-end manometer (Figure 9.5) is the same as a closed-end manometer, but one of its arms is open to the atmosphere. In this case, the distance between the liquid levels corresponds to the difference in pressure between the gas in the container and the atmosphere.

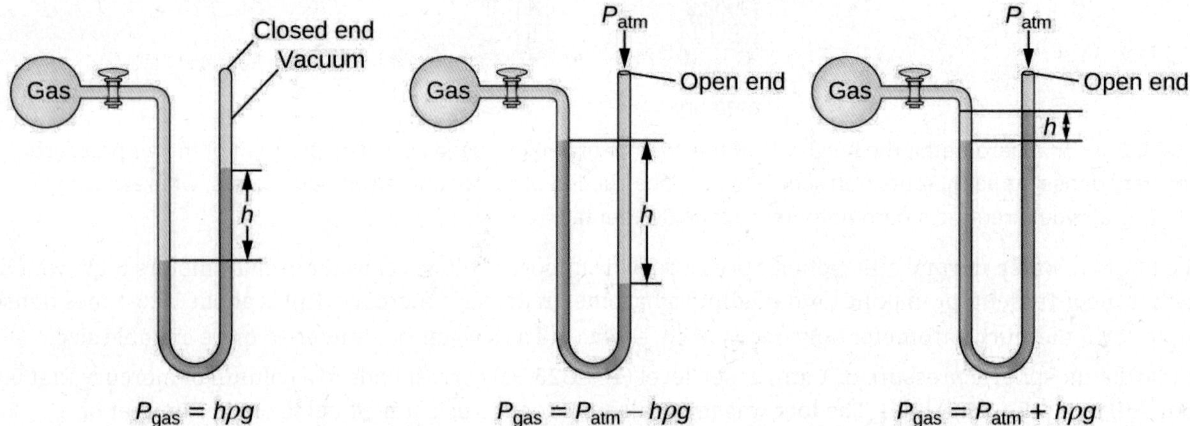

FIGURE 9.5 A manometer can be used to measure the pressure of a gas. The (difference in) height between the liquid levels (h) is a measure of the pressure. Mercury is usually used because of its large density.

EXAMPLE 9.3

Calculation of Pressure Using a Closed-End Manometer

The pressure of a sample of gas is measured with a closed-end manometer, as shown to the right. The liquid in the manometer is mercury. Determine the pressure of the gas in:

(a) torr

(b) Pa

(c) bar

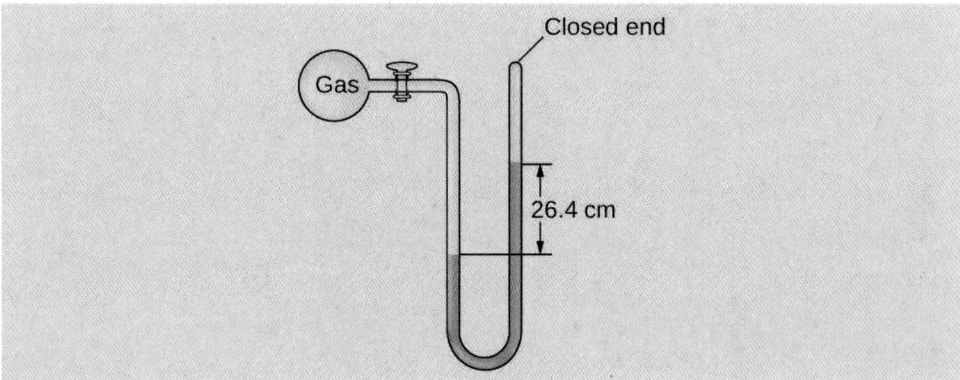

Solution

The pressure of the gas is equal to a column of mercury of height 26.4 cm. (The pressure at the bottom horizontal line is equal on both sides of the tube. The pressure on the left is due to the gas and the pressure on the right is due to 26.4 cm Hg, or mercury.) We could use the equation $p = h\rho g$ as in Example 9.2, but it is simpler to just convert between units using Table 9.1.

$$\text{(a) } 26.4\ \cancel{\text{cm Hg}} \times \frac{10\ \cancel{\text{mm Hg}}}{1\ \cancel{\text{cm Hg}}} \times \frac{1\ \text{torr}}{1\ \cancel{\text{mm Hg}}} = 264\ \text{torr}$$

$$\text{(b) } 264\ \cancel{\text{torr}} \times \frac{1\ \cancel{\text{atm}}}{760\ \cancel{\text{torr}}} \times \frac{101{,}325\ \text{Pa}}{1\ \cancel{\text{atm}}} = 35{,}200\ \text{Pa}$$

$$\text{(c) } 35{,}200\ \cancel{\text{Pa}} \times \frac{1\ \text{bar}}{100{,}000\ \cancel{\text{Pa}}} = 0.352\ \text{bar}$$

Check Your Learning

The pressure of a sample of gas is measured with a closed-end manometer. The liquid in the manometer is mercury. Determine the pressure of the gas in:

(a) torr

(b) Pa

(c) bar

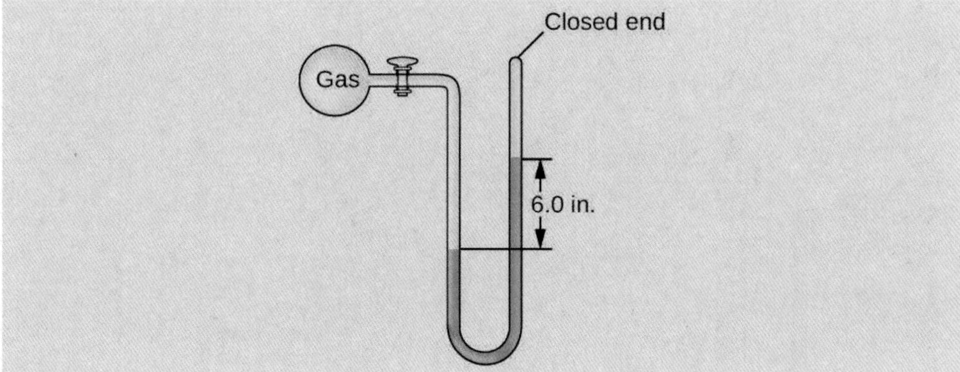

Answer:

(a) ~150 torr; (b) ~20,000 Pa; (c) ~0.20 bar

EXAMPLE 9.4

Calculation of Pressure Using an Open-End Manometer

The pressure of a sample of gas is measured at sea level with an open-end Hg (mercury) manometer, as shown to the right. Determine the pressure of the gas in:

(a) mm Hg

(b) atm

(c) kPa

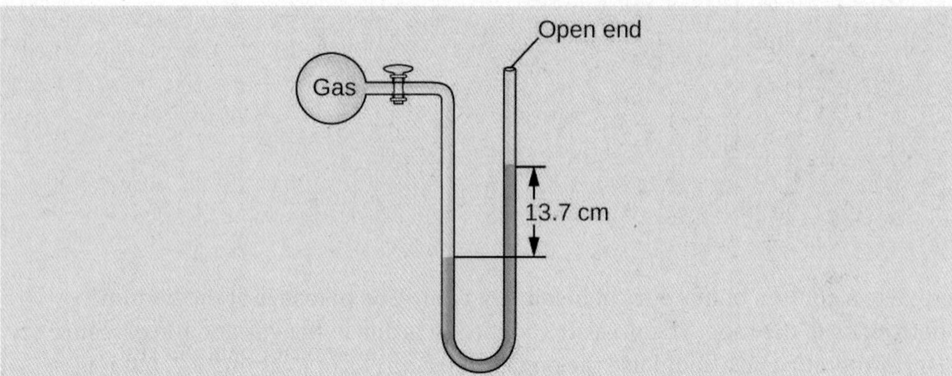

Solution

The pressure of the gas equals the hydrostatic pressure due to a column of mercury of height 13.7 cm plus the pressure of the atmosphere at sea level. (The pressure at the bottom horizontal line is equal on both sides of the tube. The pressure on the left is due to the gas and the pressure on the right is due to 13.7 cm of Hg plus atmospheric pressure.)

(a) In mm Hg, this is: 137 mm Hg + 760 mm Hg = 897 mm Hg

(b) $897 \cancel{\text{mm Hg}} \times \frac{1 \text{ atm}}{760 \cancel{\text{mm Hg}}} = 1.18 \text{ atm}$

(c) $1.18 \cancel{\text{atm}} \times \frac{101.325 \text{ kPa}}{1 \cancel{\text{atm}}} = 1.20 \times 10^2 \text{ kPa}$

Check Your Learning

The pressure of a sample of gas is measured at sea level with an open-end Hg manometer, as shown to the right. Determine the pressure of the gas in:

(a) mm Hg

(b) atm

(c) kPa

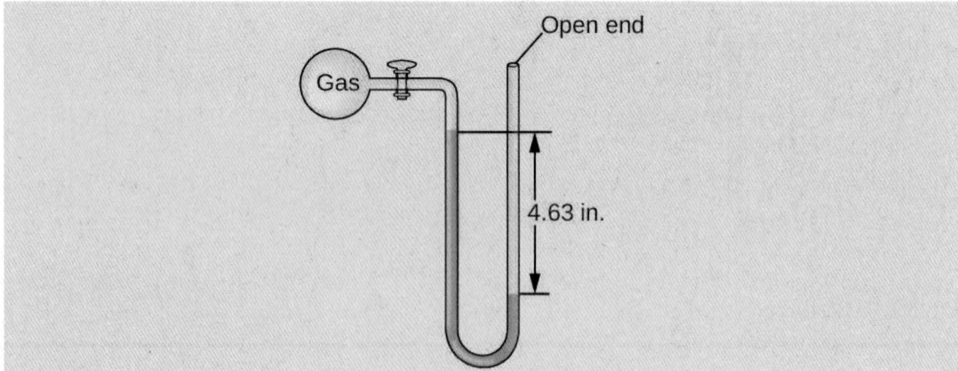

Answer:

(a) 642 mm Hg; (b) 0.845 atm; (c) 85.6 kPa

Chemistry in Everyday Life

Measuring Blood Pressure

Blood pressure is measured using a device called a sphygmomanometer (Greek *sphygmos* = "pulse"). It consists of an inflatable cuff to restrict blood flow, a manometer to measure the pressure, and a method of determining when blood flow begins and when it becomes impeded (Figure 9.6). Since its invention in 1881, it has been an essential medical device. There are many types of sphygmomanometers: manual ones that require a stethoscope and are used by medical professionals; mercury ones, used when the most accuracy is required; less accurate mechanical ones; and digital ones that can be used with little training but that have limitations. When using a sphygmomanometer, the cuff is placed around the upper arm and inflated until blood flow is completely blocked, then slowly released. As the heart beats, blood forced through the arteries causes a rise in pressure. This rise in pressure at which blood flow begins is the *systolic pressure*—the peak pressure in the cardiac cycle. When the cuff's pressure equals the arterial systolic pressure, blood flows past the cuff, creating audible sounds that can be heard using a stethoscope. This is followed by a decrease in pressure as the heart's ventricles prepare for another beat. As cuff pressure continues to decrease, eventually sound is no longer heard; this is the *diastolic pressure*—the lowest pressure (resting phase) in the cardiac cycle. Blood pressure units from a sphygmomanometer are in terms of millimeters of mercury (mm Hg).

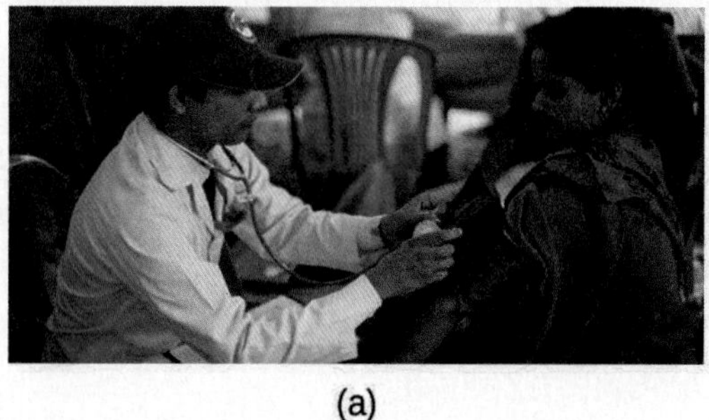

(a)

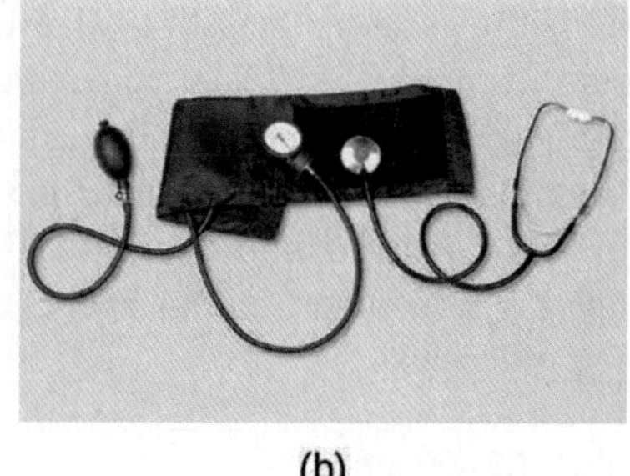

(b)

FIGURE 9.6 (a) A medical technician prepares to measure a patient's blood pressure with a sphygmomanometer. (b) A typical sphygmomanometer uses a valved rubber bulb to inflate the cuff and a diaphragm gauge to measure pressure. (credit a: modification of work by Master Sgt. Jeffrey Allen)

HOW SCIENCES INTERCONNECT

Meteorology, Climatology, and Atmospheric Science

Throughout the ages, people have observed clouds, winds, and precipitation, trying to discern patterns and make predictions: when it is best to plant and harvest; whether it is safe to set out on a sea voyage; and much more. We now face complex weather and atmosphere-related challenges that will have a major impact on our civilization and the ecosystem. Several different scientific disciplines use chemical principles to help us better understand weather, the atmosphere, and climate. These are meteorology, climatology, and atmospheric science. Meteorology is the study of the atmosphere, atmospheric phenomena, and atmospheric effects on earth's weather. Meteorologists seek to understand and predict the weather in the short term, which can save lives and benefit the economy. Weather forecasts (Figure 9.7) are the result of thousands of measurements of air pressure, temperature, and the like, which are compiled, modeled, and analyzed in weather centers worldwide.

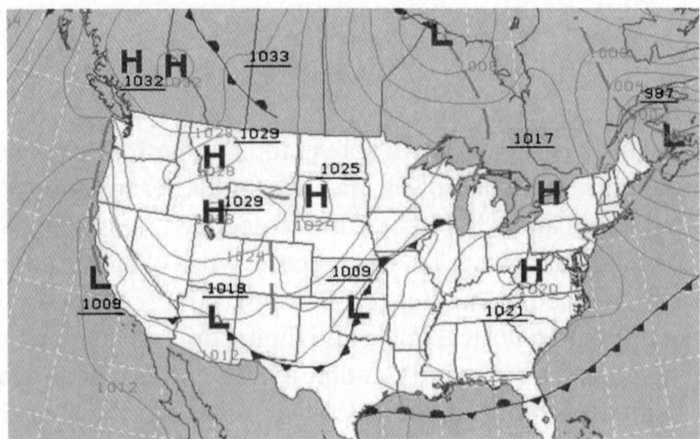

FIGURE 9.7 Meteorologists use weather maps to describe and predict weather. Regions of high (H) and low (L) pressure have large effects on weather conditions. The gray lines represent locations of constant pressure known as isobars. (credit: modification of work by National Oceanic and Atmospheric Administration)

In terms of weather, low-pressure systems occur when the earth's surface atmospheric pressure is lower than the surrounding environment: Moist air rises and condenses, producing clouds. Movement of moisture and air within various weather fronts instigates most weather events.

The atmosphere is the gaseous layer that surrounds a planet. Earth's atmosphere, which is roughly 100–125 km thick, consists of roughly 78.1% nitrogen and 21.0% oxygen, and can be subdivided further into the regions shown in Figure 9.8: the exosphere (furthest from earth, > 700 km above sea level), the thermosphere (80–700 km), the mesosphere (50–80 km), the stratosphere (second lowest level of our atmosphere, 12–50 km above sea level), and the troposphere (up to 12 km above sea level, roughly 80% of the earth's atmosphere by mass and the layer where most weather events originate). As you go higher in the troposphere, air density and temperature both decrease.

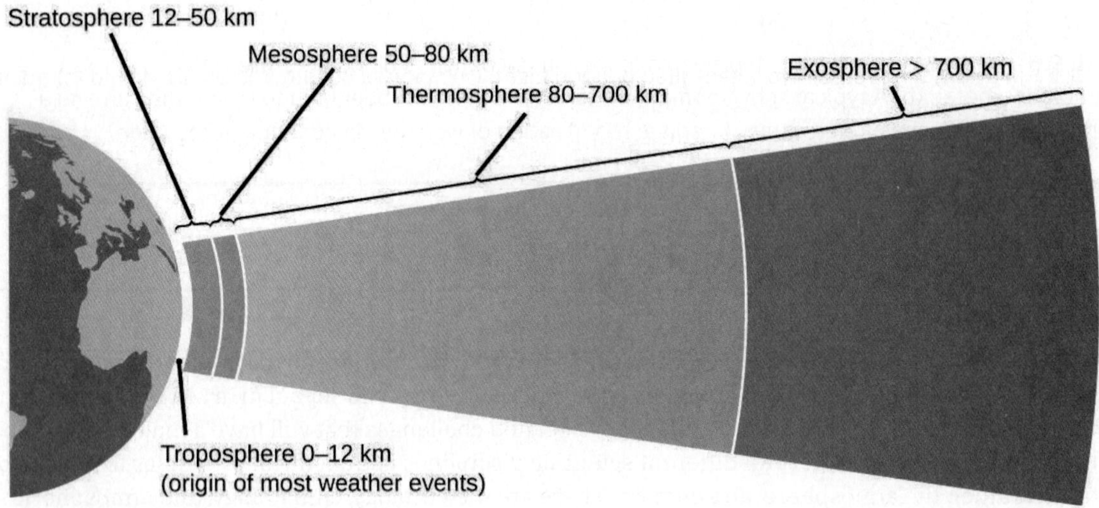

FIGURE 9.8 Earth's atmosphere has five layers: the troposphere, the stratosphere, the mesosphere, the thermosphere, and the exosphere.

Climatology is the study of the climate, averaged weather conditions over long time periods, using atmospheric data. However, climatologists study patterns and effects that occur over decades, centuries, and millennia, rather than shorter time frames of hours, days, and weeks like meteorologists. Atmospheric science is an even broader field, combining meteorology, climatology, and other scientific disciplines that study the atmosphere.

9.2 Relating Pressure, Volume, Amount, and Temperature: The Ideal Gas Law

LEARNING OBJECTIVES

By the end of this section, you will be able to:

- Identify the mathematical relationships between the various properties of gases
- Use the ideal gas law, and related gas laws, to compute the values of various gas properties under specified conditions

During the seventeenth and especially eighteenth centuries, driven both by a desire to understand nature and a quest to make balloons in which they could fly (Figure 9.9), a number of scientists established the relationships between the macroscopic physical properties of gases, that is, pressure, volume, temperature, and amount of gas. Although their measurements were not precise by today's standards, they were able to determine the mathematical relationships between pairs of these variables (e.g., pressure and temperature, pressure and volume) that hold for an *ideal* gas—a hypothetical construct that real gases approximate under certain conditions. Eventually, these individual laws were combined into a single equation—the *ideal gas law*—that relates gas quantities for gases and is quite accurate for low pressures and moderate temperatures. We will consider the key developments in individual relationships (for pedagogical reasons not quite in historical order), then put them together in the ideal gas law.

(a)

(b)

(c)

FIGURE 9.9 In 1783, the first (a) hydrogen-filled balloon flight, (b) manned hot air balloon flight, and (c) manned hydrogen-filled balloon flight occurred. When the hydrogen-filled balloon depicted in (a) landed, the frightened villagers of Gonesse reportedly destroyed it with pitchforks and knives. The launch of the latter was reportedly viewed by 400,000 people in Paris.

Pressure and Temperature: Amontons's Law

Imagine filling a rigid container attached to a pressure gauge with gas and then sealing the container so that no gas may escape. If the container is cooled, the gas inside likewise gets colder and its pressure is observed to decrease. Since the container is rigid and tightly sealed, both the volume and number of moles of gas remain constant. If we heat the sphere, the gas inside gets hotter (Figure 9.10) and the pressure increases.

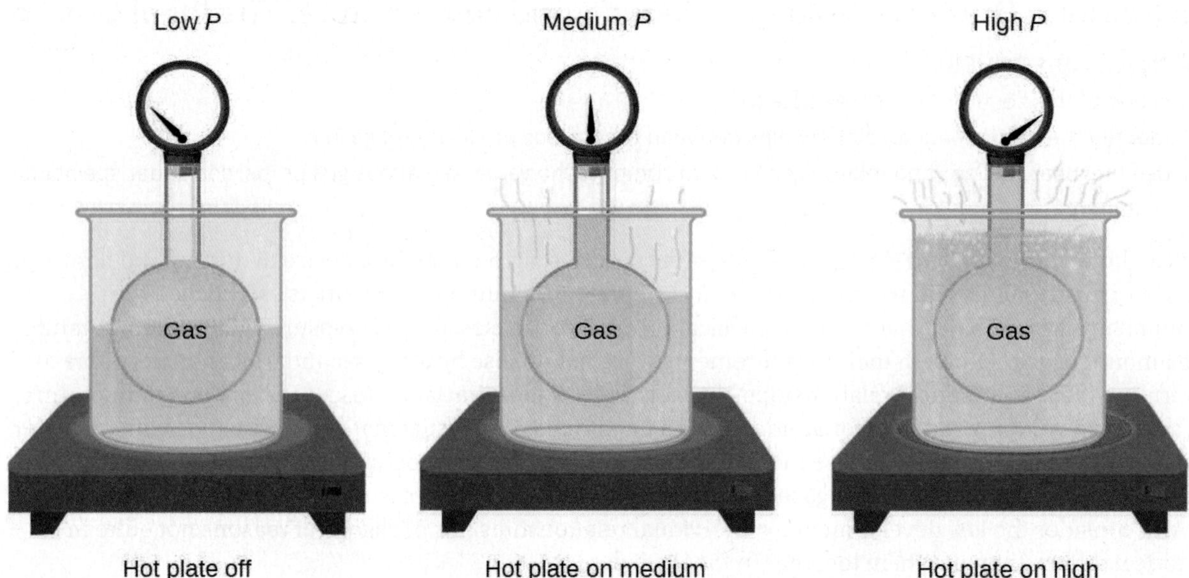

FIGURE 9.10 The effect of temperature on gas pressure: When the hot plate is off, the pressure of the gas in the sphere is relatively low. As the gas is heated, the pressure of the gas in the sphere increases.

This relationship between temperature and pressure is observed for any sample of gas confined to a constant volume. An example of experimental pressure-temperature data is shown for a sample of air under these conditions in Figure 9.11. We find that temperature and pressure are linearly related, and if the temperature is on the kelvin scale, then *P* and *T* are directly proportional (again, when *volume and moles of gas are held constant*); if the temperature on the kelvin scale increases by a certain factor, the gas pressure increases by the same factor.

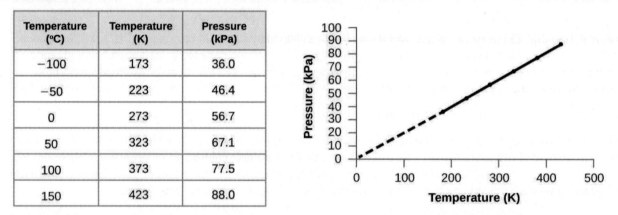

Temperature (°C)	Temperature (K)	Pressure (kPa)
−100	173	36.0
−50	223	46.4
0	273	56.7
50	323	67.1
100	373	77.5
150	423	88.0

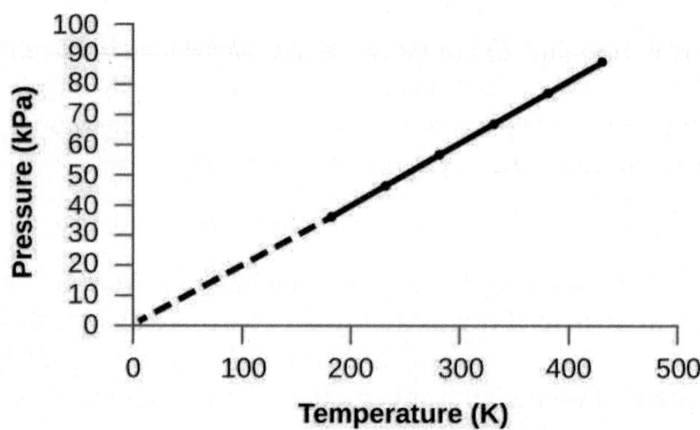

FIGURE 9.11 For a constant volume and amount of air, the pressure and temperature are directly proportional, provided the temperature is in kelvin. (Measurements cannot be made at lower temperatures because of the condensation of the gas.) When this line is extrapolated to lower pressures, it reaches a pressure of 0 at −273 °C, which is 0 on the kelvin scale and the lowest possible temperature, called absolute zero.

Guillaume Amontons was the first to empirically establish the relationship between the pressure and the temperature of a gas (~1700), and Joseph Louis Gay-Lussac determined the relationship more precisely (~1800). Because of this, the *P*-*T* relationship for gases is known as either **Amontons's law** or **Gay-Lussac's law**. Under either name, it states that *the pressure of a given amount of gas is directly proportional to its temperature on the kelvin scale when the volume is held constant.* Mathematically, this can be written:

$$P \propto T \text{ or } P = \text{constant} \times T \text{ or } P = k \times T$$

where ∝ means "is proportional to," and *k* is a proportionality constant that depends on the identity, amount, and volume of the gas.

For a confined, constant volume of gas, the ratio $\frac{P}{T}$ is therefore constant (i.e., $\frac{P}{T} = k$). If the gas is initially in "Condition 1" (with $P = P_1$ and $T = T_1$), and then changes to "Condition 2" (with $P = P_2$ and $T = T_2$), we have that $\frac{P_1}{T_1} = k$ and $\frac{P_2}{T_2} = k$, which reduces to $\frac{P_1}{T_1} = \frac{P_2}{T_2}$. This equation is useful for pressure-temperature calculations for a confined gas at constant volume. Note that temperatures must be on the kelvin scale for any gas law calculations (0 on the kelvin scale and the lowest possible temperature is called **absolute zero**). (Also note that there are at least three ways we can describe how the pressure of a gas changes as its temperature changes: We can use a table of values, a graph, or a mathematical equation.)

EXAMPLE 9.5

Predicting Change in Pressure with Temperature

A can of hair spray is used until it is empty except for the propellant, isobutane gas.

(a) On the can is the warning "Store only at temperatures below 120 °F (48.8 °C). Do not incinerate." Why?

(b) The gas in the can is initially at 24 °C and 360 kPa, and the can has a volume of 350 mL. If the can is left in a car that reaches 50 °C on a hot day, what is the new pressure in the can?

Solution

(a) The can contains an amount of isobutane gas at a constant volume, so if the temperature is increased by heating, the pressure will increase proportionately. High temperature could lead to high pressure, causing the can to burst. (Also, isobutane is combustible, so incineration could cause the can to explode.)

(b) We are looking for a pressure change due to a temperature change at constant volume, so we will use Amontons's/Gay-Lussac's law. Taking P_1 and T_1 as the initial values, T_2 as the temperature where the pressure is unknown and P_2 as the unknown pressure, and converting °C to K, we have:

$$\frac{P_1}{T_1} = \frac{P_2}{T_2} \text{ which means that } \frac{360 \text{ kPa}}{297 \text{ K}} = \frac{P_2}{323 \text{ K}}$$

Rearranging and solving gives: $P_2 = \frac{360 \text{ kPa} \times 323 \cancel{\text{K}}}{297 \cancel{\text{K}}} = 390 \text{ kPa}$

Check Your Learning

A sample of nitrogen, N_2, occupies 45.0 mL at 27 °C and 600 torr. What pressure will it have if cooled to –73 °C while the volume remains constant?

Answer:

400 torr

Volume and Temperature: Charles's Law

If we fill a balloon with air and seal it, the balloon contains a specific amount of air at atmospheric pressure, let's say 1 atm. If we put the balloon in a refrigerator, the gas inside gets cold and the balloon shrinks (although both the amount of gas and its pressure remain constant). If we make the balloon very cold, it will shrink a great deal, and it expands again when it warms up.

LINK TO LEARNING

This video (http://openstax.org/l/16CharlesLaw) shows how cooling and heating a gas causes its volume to decrease or increase, respectively.

These examples of the effect of temperature on the volume of a given amount of a confined gas at constant pressure are true in general: The volume increases as the temperature increases, and decreases as the temperature decreases. Volume-temperature data for a 1-mole sample of methane gas at 1 atm are listed and graphed in Figure 9.12.

Temperature (°C)	Temperature (K)	Volume (L)
–3	270	22
–23	250	21
–53	220	18
–162	111	9

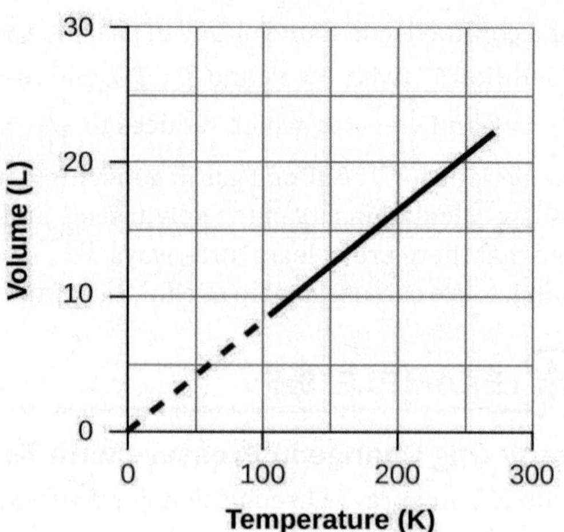

FIGURE 9.12 The volume and temperature are linearly related for 1 mole of methane gas at a constant pressure of 1 atm. If the temperature is in kelvin, volume and temperature are directly proportional. The line stops at 111 K because methane liquefies at this temperature; when extrapolated, it intersects the graph's origin, representing a temperature of absolute zero.

The relationship between the volume and temperature of a given amount of gas at constant pressure is known as Charles's law in recognition of the French scientist and balloon flight pioneer Jacques Alexandre César Charles. **Charles's law** states that *the volume of a given amount of gas is directly proportional to its temperature on the kelvin scale when the pressure is held constant.*

Mathematically, this can be written as:

$$V \propto T \text{ or } V = \text{constant}\cdot T \text{ or } V = k\cdot T \text{ or } V_1/T_1 = V_2/T_2$$

with k being a proportionality constant that depends on the amount and pressure of the gas.

For a confined, constant pressure gas sample, $\frac{V}{T}$ is constant (i.e., the ratio = k), and as seen with the P-T relationship, this leads to another form of Charles's law: $\frac{V_1}{T_1} = \frac{V_2}{T_2}$.

 **EXAMPLE 9.6**

Predicting Change in Volume with Temperature

A sample of carbon dioxide, CO_2, occupies 0.300 L at 10 °C and 750 torr. What volume will the gas have at 30 °C and 750 torr?

Solution

Because we are looking for the volume change caused by a temperature change at constant pressure, this is a job for Charles's law. Taking V_1 and T_1 as the initial values, T_2 as the temperature at which the volume is unknown and V_2 as the unknown volume, and converting °C into K we have:

$$\frac{V_1}{T_1} = \frac{V_2}{T_2} \text{ which means that } \frac{0.300\text{ L}}{283\text{ K}} = \frac{V_2}{303\text{ K}}$$

Rearranging and solving gives: $V_2 = \frac{0.300\text{ L} \times 303 \cancel{\text{K}}}{283 \cancel{\text{K}}} = 0.321\text{ L}$

This answer supports our expectation from Charles's law, namely, that raising the gas temperature (from 283 K to 303 K) at a constant pressure will yield an increase in its volume (from 0.300 L to 0.321 L).

Check Your Learning

A sample of oxygen, O_2, occupies 32.2 mL at 30 °C and 452 torr. What volume will it occupy at –70 °C and the same pressure?

Answer:

21.6 mL

EXAMPLE 9.7

Measuring Temperature with a Volume Change

Temperature is sometimes measured with a gas thermometer by observing the change in the volume of the gas as the temperature changes at constant pressure. The hydrogen in a particular hydrogen gas thermometer has a volume of 150.0 cm^3 when immersed in a mixture of ice and water (0.00 °C). When immersed in boiling liquid ammonia, the volume of the hydrogen, at the same pressure, is 131.7 cm^3. Find the temperature of boiling ammonia on the kelvin and Celsius scales.

Solution

A volume change caused by a temperature change at constant pressure means we should use Charles's law. Taking V_1 and T_1 as the initial values, T_2 as the temperature at which the volume is unknown and V_2 as the unknown volume, and converting °C into K we have:

$$\frac{V_1}{T_1} = \frac{V_2}{T_2} \text{ which means that } \frac{150.0 \text{ cm}^3}{273.15 \text{ K}} = \frac{131.7 \text{ cm}^3}{T_2}$$

Rearrangement gives $T_2 = \frac{131.7 \cancel{\text{cm}^3} \times 273.15 \text{ K}}{150.0 \cancel{\text{cm}^3}} = 239.8 \text{ K}$

Subtracting 273.15 from 239.8 K, we find that the temperature of the boiling ammonia on the Celsius scale is −33.4 °C.

Check Your Learning

What is the volume of a sample of ethane at 467 K and 1.1 atm if it occupies 405 mL at 298 K and 1.1 atm?

Answer:

635 mL

Volume and Pressure: Boyle's Law

If we partially fill an airtight syringe with air, the syringe contains a specific amount of air at constant temperature, say 25 °C. If we slowly push in the plunger while keeping temperature constant, the gas in the syringe is compressed into a smaller volume and its pressure increases; if we pull out the plunger, the volume increases and the pressure decreases. This example of the effect of volume on the pressure of a given amount of a confined gas is true in general. Decreasing the volume of a contained gas will increase its pressure, and increasing its volume will decrease its pressure. In fact, if the volume increases by a certain factor, the pressure decreases by the same factor, and vice versa. Volume-pressure data for an air sample at room temperature are graphed in Figure 9.13.

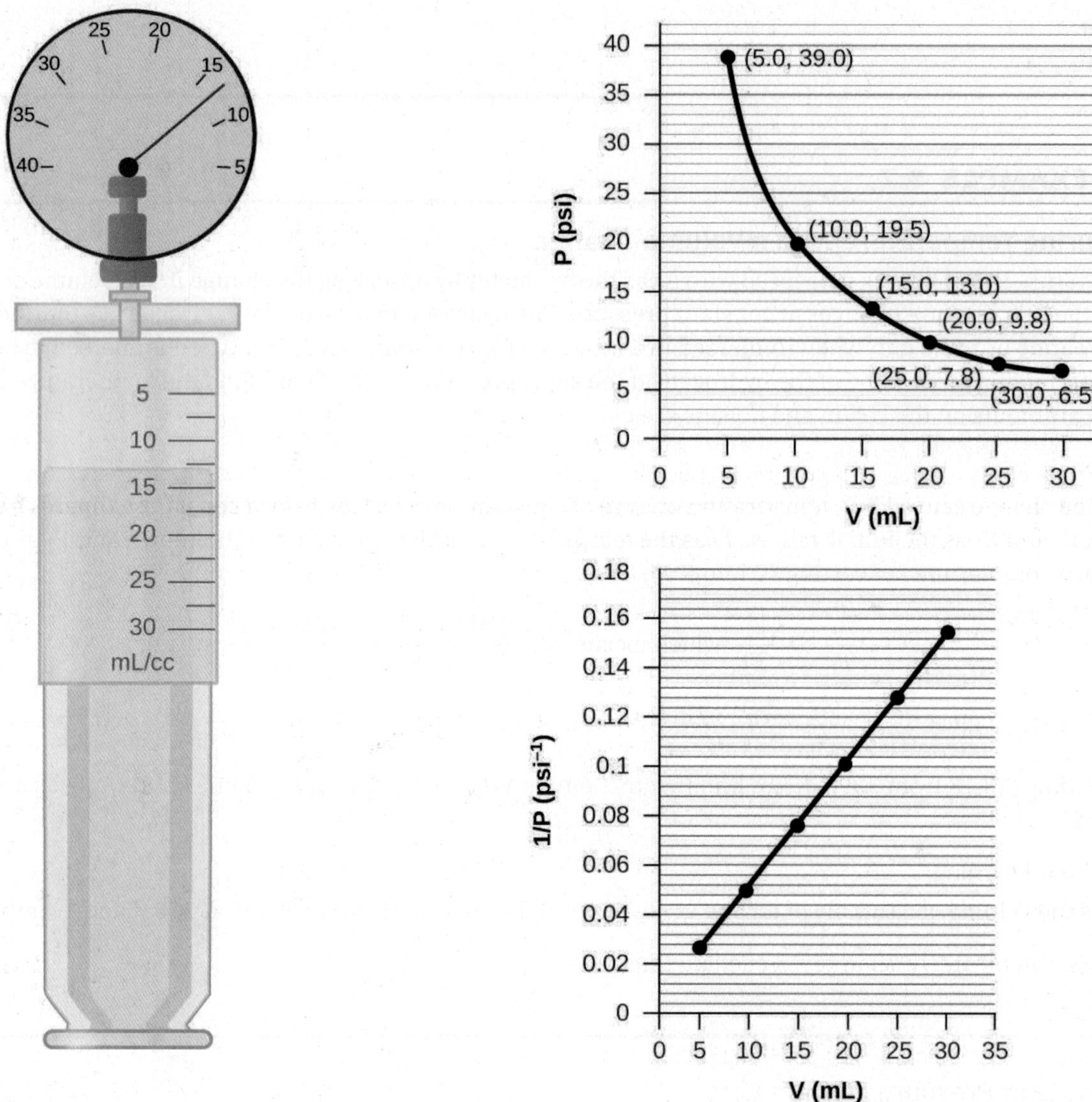

FIGURE 9.13 When a gas occupies a smaller volume, it exerts a higher pressure; when it occupies a larger volume, it exerts a lower pressure (assuming the amount of gas and the temperature do not change). Since *P* and *V* are inversely proportional, a graph of $\frac{1}{P}$ vs. *V* is linear.

Unlike the *P*-*T* and *V*-*T* relationships, pressure and volume are not directly proportional to each other. Instead, *P* and *V* exhibit inverse proportionality: Increasing the pressure results in a decrease of the volume of the gas. Mathematically this can be written:

$$P \propto 1/V \text{ or } P = k \cdot 1/V \text{ or } P \cdot V = k \text{ or } P_1V_1 = P_2V_2$$

with *k* being a constant. Graphically, this relationship is shown by the straight line that results when plotting the inverse of the pressure $\left(\frac{1}{P}\right)$ versus the volume (*V*), or the inverse of volume $\left(\frac{1}{V}\right)$ versus the pressure (*P*). Graphs with curved lines are difficult to read accurately at low or high values of the variables, and they are more difficult to use in fitting theoretical equations and parameters to experimental data. For those reasons, scientists often try to find a way to "linearize" their data. If we plot *P* versus *V*, we obtain a hyperbola (see Figure 9.14).

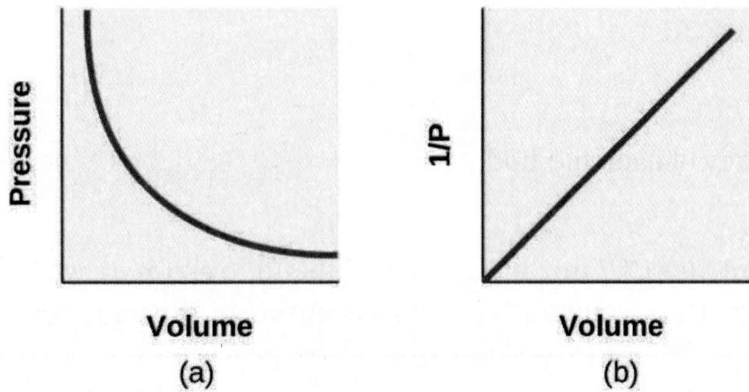

FIGURE 9.14 The relationship between pressure and volume is inversely proportional. (a) The graph of *P* vs. *V* is a hyperbola, whereas (b) the graph of $\left(\frac{1}{P}\right)$ vs. *V* is linear.

The relationship between the volume and pressure of a given amount of gas at constant temperature was first published by the English natural philosopher Robert Boyle over 300 years ago. It is summarized in the statement now known as **Boyle's law**: *The volume of a given amount of gas held at constant temperature is inversely proportional to the pressure under which it is measured.*

EXAMPLE 9.8

Volume of a Gas Sample

The sample of gas in Figure 9.13 has a volume of 15.0 mL at a pressure of 13.0 psi. Determine the pressure of the gas at a volume of 7.5 mL, using:

(a) the *P*-*V* graph in Figure 9.13

(b) the $\frac{1}{P}$ vs. *V* graph in Figure 9.13

(c) the Boyle's law equation

Comment on the likely accuracy of each method.

Solution

(a) Estimating from the *P*-*V* graph gives a value for *P* somewhere around 27 psi.

(b) Estimating from the $\frac{1}{P}$ versus *V* graph give a value of about 26 psi.

(c) From Boyle's law, we know that the product of pressure and volume (*PV*) for a given sample of gas at a constant temperature is always equal to the same value. Therefore we have $P_1V_1 = k$ and $P_2V_2 = k$ which means that $P_1V_1 = P_2V_2$.

Using P_1 and V_1 as the known values 13.0 psi and 15.0 mL, P_2 as the pressure at which the volume is unknown, and V_2 as the unknown volume, we have:

$$P_1V_1 = P_2V_2 \text{ or } 13.0\text{ psi } \times 15.0\text{ mL} = P_2 \times 7.5\text{ mL}$$

Solving:

$$P_2 = \frac{13.0\text{ psi } \times 15.0\ \cancel{\text{mL}}}{7.5\ \cancel{\text{mL}}} = 26\text{ psi}$$

It was more difficult to estimate well from the *P*-*V* graph, so (a) is likely more inaccurate than (b) or (c). The calculation will be as accurate as the equation and measurements allow.

Check Your Learning

The sample of gas in Figure 9.13 has a volume of 30.0 mL at a pressure of 6.5 psi. Determine the volume of the gas at a pressure of 11.0 psi, using:

(a) the *P*-*V* graph in Figure 9.13

(b) the $\frac{1}{P}$ vs. V graph in Figure 9.13

(c) the Boyle's law equation

Comment on the likely accuracy of each method.

Answer:

(a) about 17–18 mL; (b) ~18 mL; (c) 17.7 mL; it was more difficult to estimate well from the P-V graph, so (a) is likely more inaccurate than (b); the calculation will be as accurate as the equation and measurements allow

Chemistry in Everyday Life

Breathing and Boyle's Law

What do you do about 20 times per minute for your whole life, without break, and often without even being aware of it? The answer, of course, is respiration, or breathing. How does it work? It turns out that the gas laws apply here. Your lungs take in gas that your body needs (oxygen) and get rid of waste gas (carbon dioxide). Lungs are made of spongy, stretchy tissue that expands and contracts while you breathe. When you inhale, your diaphragm and intercostal muscles (the muscles between your ribs) contract, expanding your chest cavity and making your lung volume larger. The increase in volume leads to a decrease in pressure (Boyle's law). This causes air to flow into the lungs (from high pressure to low pressure). When you exhale, the process reverses: Your diaphragm and rib muscles relax, your chest cavity contracts, and your lung volume decreases, causing the pressure to increase (Boyle's law again), and air flows out of the lungs (from high pressure to low pressure). You then breathe in and out again, and again, repeating this Boyle's law cycle for the rest of your life (Figure 9.15).

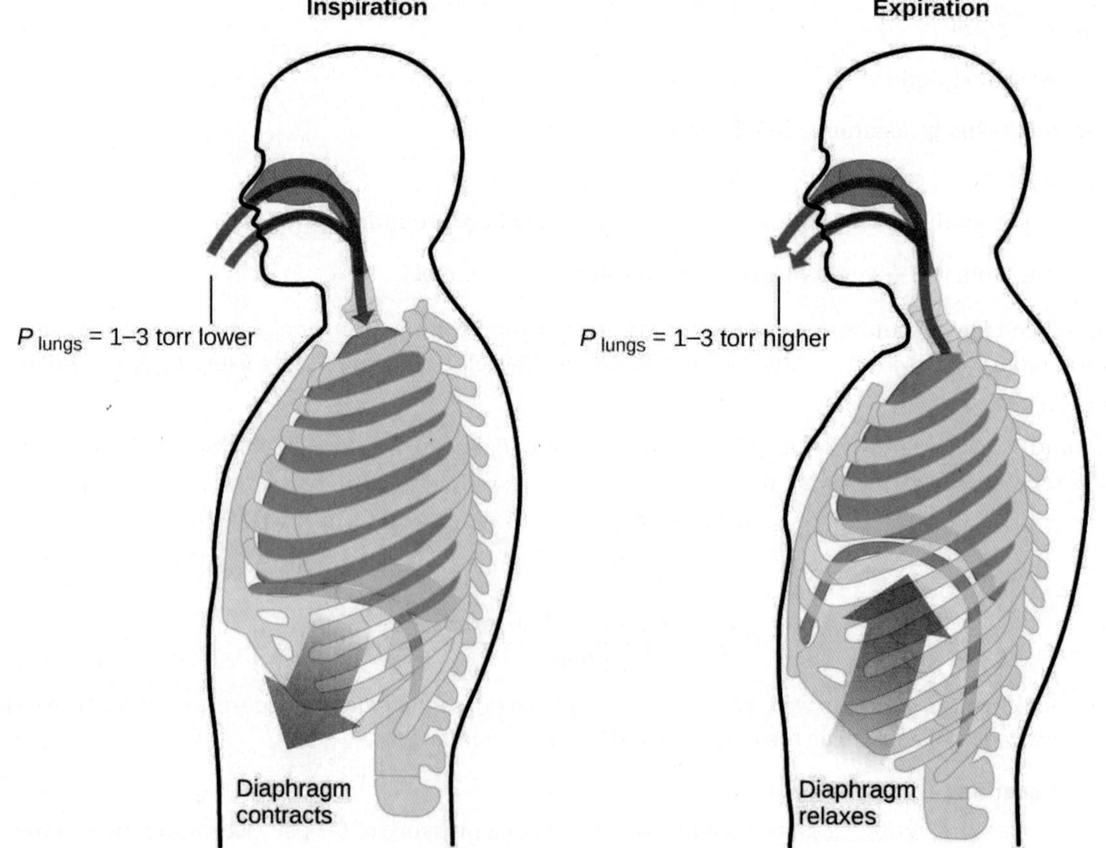

FIGURE 9.15 Breathing occurs because expanding and contracting lung volume creates small pressure differences between your lungs and your surroundings, causing air to be drawn into and forced out of your lungs.

Moles of Gas and Volume: Avogadro's Law

The Italian scientist Amedeo Avogadro advanced a hypothesis in 1811 to account for the behavior of gases, stating that equal volumes of all gases, measured under the same conditions of temperature and pressure, contain the same number of molecules. Over time, this relationship was supported by many experimental observations as expressed by **Avogadro's law**: *For a confined gas, the volume (V) and number of moles (n) are directly proportional if the pressure and temperature both remain constant.*

In equation form, this is written as:

$$V \propto n \quad \text{or} \quad V = k \times n \quad \text{or} \quad \frac{V_1}{n_1} = \frac{V_2}{n_2}$$

Mathematical relationships can also be determined for the other variable pairs, such as P versus n, and n versus T.

LINK TO LEARNING

Visit this interactive PhET simulation (http://openstax.org/l/16IdealGasLaw) to investigate the relationships between pressure, volume, temperature, and amount of gas. Use the simulation to examine the effect of changing one parameter on another while holding the other parameters constant (as described in the preceding sections on the various gas laws).

The Ideal Gas Law

To this point, four separate laws have been discussed that relate pressure, volume, temperature, and the number of moles of the gas:

- Boyle's law: PV = constant at constant T and n
- Amontons's law: $\frac{P}{T}$ = constant at constant V and n
- Charles's law: $\frac{V}{T}$ = constant at constant P and n
- Avogadro's law: $\frac{V}{n}$ = constant at constant P and T

Combining these four laws yields the **ideal gas law**, a relation between the pressure, volume, temperature, and number of moles of a gas:

$$PV = nRT$$

where P is the pressure of a gas, V is its volume, n is the number of moles of the gas, T is its temperature on the kelvin scale, and R is a constant called the **ideal gas constant** or the universal gas constant. The units used to express pressure, volume, and temperature will determine the proper form of the gas constant as required by dimensional analysis, the most commonly encountered values being 0.08206 L atm mol^{-1} K^{-1} and 8.314 kPa L mol^{-1} K^{-1}.

Gases whose properties of P, V, and T are accurately described by the ideal gas law (or the other gas laws) are said to exhibit *ideal behavior* or to approximate the traits of an **ideal gas**. An ideal gas is a hypothetical construct that may be used along with *kinetic molecular theory* to effectively explain the gas laws as will be described in a later module of this chapter. Although all the calculations presented in this module assume ideal behavior, this assumption is only reasonable for gases under conditions of relatively low pressure and high temperature. In the final module of this chapter, a modified gas law will be introduced that accounts for the *non-ideal* behavior observed for many gases at relatively high pressures and low temperatures.

The ideal gas equation contains five terms, the gas constant R and the variable properties P, V, n, and T. Specifying any four of these terms will permit use of the ideal gas law to calculate the fifth term as demonstrated in the following example exercises.

EXAMPLE 9.9

Using the Ideal Gas Law

Methane, CH_4, is being considered for use as an alternative automotive fuel to replace gasoline. One gallon of gasoline could be replaced by 655 g of CH_4. What is the volume of this much methane at 25 °C and 745 torr?

Solution

We must rearrange $PV = nRT$ to solve for V: $V = \frac{nRT}{P}$

If we choose to use $R = 0.08206$ L atm mol^{-1} K^{-1}, then the amount must be in moles, temperature must be in kelvin, and pressure must be in atm.

Converting into the "right" units:

$$n = 655 \cancel{\text{g CH}_4} \times \frac{1\text{ mol}}{16.043 \cancel{\text{g CH}_4}} = 40.8\text{ mol}$$

$$T = 25\text{ °C} + 273 = 298\text{ K}$$

$$P = 745 \cancel{\text{torr}} \times \frac{1\text{ atm}}{760 \cancel{\text{torr}}} = 0.980\text{ atm}$$

$$V = \frac{nRT}{P} = \frac{(40.8 \cancel{\text{mol}})(0.08206\text{ L} \cancel{\text{atm mol}^{-1}\text{ K}^{-1}})(298 \cancel{\text{K}})}{0.980 \cancel{\text{atm}}} = 1.02 \times 10^3\text{L}$$

It would require 1020 L (269 gal) of gaseous methane at about 1 atm of pressure to replace 1 gal of gasoline. It requires a large container to hold enough methane at 1 atm to replace several gallons of gasoline.

Check Your Learning

Calculate the pressure in bar of 2520 moles of hydrogen gas stored at 27 °C in the 180-L storage tank of a modern hydrogen-powered car.

Answer:

350 bar

If the number of moles of an ideal gas are kept constant under two different sets of conditions, a useful mathematical relationship called the combined gas law is obtained: $\frac{P_1V_1}{T_1} = \frac{P_2V_2}{T_2}$ using units of atm, L, and K. Both sets of conditions are equal to the product of $n \times R$ (where n = the number of moles of the gas and R is the ideal gas law constant).

EXAMPLE 9.10

Using the Combined Gas Law

When filled with air, a typical scuba tank with a volume of 13.2 L has a pressure of 153 atm (Figure 9.16). If the water temperature is 27 °C, how many liters of air will such a tank provide to a diver's lungs at a depth of approximately 70 feet in the ocean where the pressure is 3.13 atm?

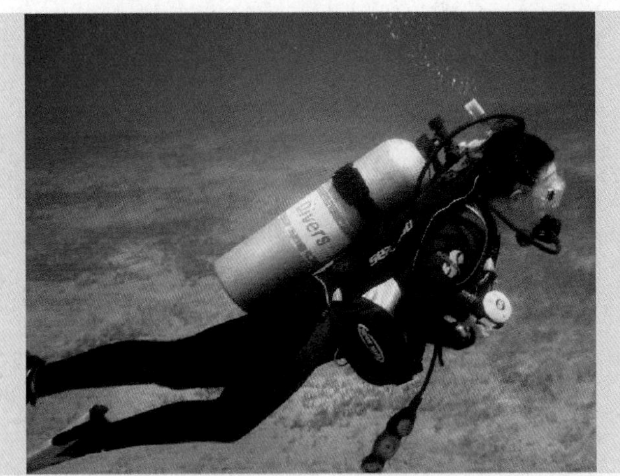

FIGURE 9.16 Scuba divers use compressed air to breathe while underwater. (credit: modification of work by Mark Goodchild)

Letting *1* represent the air in the scuba tank and *2* represent the air in the lungs, and noting that body temperature (the temperature the air will be in the lungs) is 37 °C, we have:

$$\frac{P_1 V_1}{T_1} = \frac{P_2 V_2}{T_2} \longrightarrow \frac{(153 \text{ atm})(13.2 \text{ L})}{(300 \text{ K})} = \frac{(3.13 \text{ atm})(V_2)}{(310 \text{ K})}$$

Solving for V_2:

$$V_2 = \frac{(153 \cancel{\text{atm}})(13.2 \text{ L})(310 \cancel{\text{K}})}{(300 \cancel{\text{K}})(3.13 \cancel{\text{atm}})} = 667 \text{ L}$$

(Note: Be advised that this particular example is one in which the assumption of ideal gas behavior is not very reasonable, since it involves gases at relatively high pressures and low temperatures. Despite this limitation, the calculated volume can be viewed as a good “ballpark” estimate.)

Check Your Learning

A sample of ammonia is found to occupy 0.250 L under laboratory conditions of 27 °C and 0.850 atm. Find the volume of this sample at 0 °C and 1.00 atm.

Answer:

0.193 L

Chemistry in Everyday Life

The Interdependence between Ocean Depth and Pressure in Scuba Diving

Whether scuba diving at the Great Barrier Reef in Australia (shown in Figure 9.17) or in the Caribbean, divers must understand how pressure affects a number of issues related to their comfort and safety.

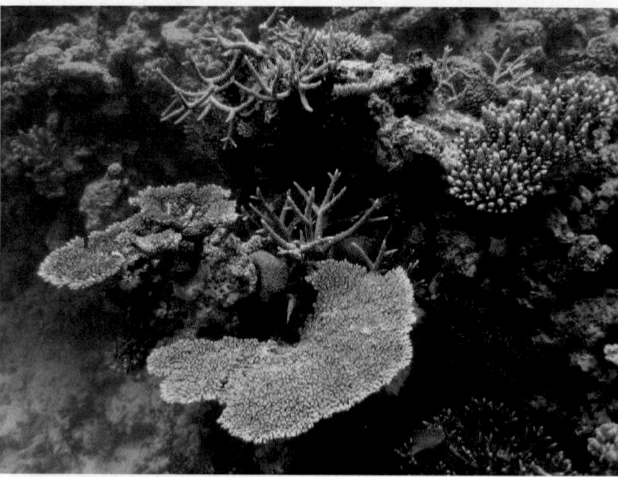

FIGURE 9.17 Scuba divers, whether at the Great Barrier Reef or in the Caribbean, must be aware of buoyancy, pressure equalization, and the amount of time they spend underwater, to avoid the risks associated with pressurized gases in the body. (credit: Kyle Taylor)

Pressure increases with ocean depth, and the pressure changes most rapidly as divers reach the surface. The pressure a diver experiences is the sum of all pressures above the diver (from the water and the air). Most pressure measurements are given in units of atmospheres, expressed as "atmospheres absolute" or ATA in the diving community: Every 33 feet of salt water represents 1 ATA of pressure in addition to 1 ATA of pressure from the atmosphere at sea level. As a diver descends, the increase in pressure causes the body's air pockets in the ears and lungs to compress; on the ascent, the decrease in pressure causes these air pockets to expand, potentially rupturing eardrums or bursting the lungs. Divers must therefore undergo equalization by adding air to body airspaces on the descent by breathing normally and adding air to the mask by breathing out of the nose or adding air to the ears and sinuses by equalization techniques; the corollary is also true on ascent, divers must release air from the body to maintain equalization. Buoyancy, or the ability to control whether a diver sinks or floats, is controlled by the buoyancy compensator (BCD). If a diver is ascending, the air in their BCD expands because of lower pressure according to Boyle's law (decreasing the pressure of gases increases the volume). The expanding air increases the buoyancy of the diver, and they begin to ascend. The diver must vent air from the BCD or risk an uncontrolled ascent that could rupture the lungs. In descending, the increased pressure causes the air in the BCD to compress and the diver sinks much more quickly; the diver must add air to the BCD or risk an uncontrolled descent, facing much higher pressures near the ocean floor. The pressure also impacts how long a diver can stay underwater before ascending. The deeper a diver dives, the more compressed the air that is breathed because of increased pressure: If a diver dives 33 feet, the pressure is 2 ATA and the air would be compressed to one-half of its original volume. The diver uses up available air twice as fast as at the surface.

Standard Conditions of Temperature and Pressure

We have seen that the volume of a given quantity of gas and the number of molecules (moles) in a given volume of gas vary with changes in pressure and temperature. Chemists sometimes make comparisons against a **standard temperature and pressure (STP)** for reporting properties of gases: 273.15 K and 1 atm (101.325 kPa).[1] At STP, one mole of an ideal gas has a volume of about 22.4 L—this is referred to as the **standard molar volume** (Figure 9.18).

1 The IUPAC definition of standard pressure was changed from 1 atm to 1 bar (100 kPa) in 1982, but the prior definition remains in use by many literature resources and will be used in this text.

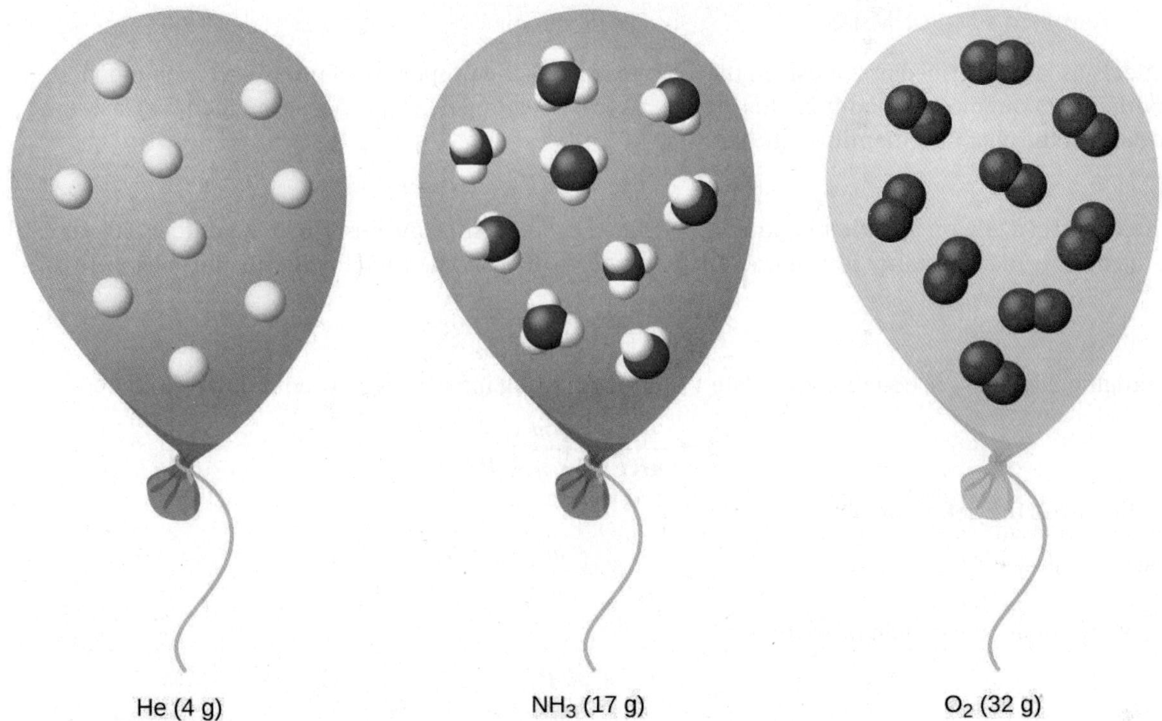

FIGURE 9.18 Regardless of its chemical identity, one mole of gas behaving ideally occupies a volume of ~22.4 L at STP.

9.3 Stoichiometry of Gaseous Substances, Mixtures, and Reactions

LEARNING OBJECTIVES

By the end of this section, you will be able to:

- Use the ideal gas law to compute gas densities and molar masses
- Perform stoichiometric calculations involving gaseous substances
- State Dalton's law of partial pressures and use it in calculations involving gaseous mixtures

The study of the chemical behavior of gases was part of the basis of perhaps the most fundamental chemical revolution in history. French nobleman Antoine Lavoisier, widely regarded as the "father of modern chemistry," changed chemistry from a qualitative to a quantitative science through his work with gases. He discovered the law of conservation of matter, discovered the role of oxygen in combustion reactions, determined the composition of air, explained respiration in terms of chemical reactions, and more. He was a casualty of the French Revolution, guillotined in 1794. Of his death, mathematician and astronomer Joseph-Louis Lagrange said, "It took the mob only a moment to remove his head; a century will not suffice to reproduce it."[2] Much of the knowledge we do have about Lavoisier's contributions is due to his wife, Marie-Anne Paulze Lavoisier, who worked with him in his lab. A trained artist fluent in several languages, she created detailed illustrations of the equipment in his lab, and translated texts from foreign scientists to complement his knowledge. After his execution, she was instrumental in publishing Lavoisier's major treatise, which unified many concepts of chemistry and laid the groundwork for significant further study.

As described in an earlier chapter of this text, we can turn to chemical stoichiometry for answers to many of the questions that ask "How much?" The essential property involved in such use of stoichiometry is the amount of substance, typically measured in moles (n). For gases, molar amount can be derived from convenient experimental measurements of pressure, temperature, and volume. Therefore, these measurements are useful in assessing the stoichiometry of pure gases, gas mixtures, and chemical reactions involving gases. This section will not introduce any new material or ideas, but will provide examples of applications and ways to integrate concepts already discussed.

2 "Quotations by Joseph-Louis Lagrange," last modified February 2006, accessed February 10, 2015, http://www-history.mcs.st-andrews.ac.uk/Quotations/Lagrange.html

Gas Density and Molar Mass

The ideal gas law described previously in this chapter relates the properties of pressure *P*, volume *V*, temperature *T*, and molar amount *n*. This law is universal, relating these properties in identical fashion regardless of the chemical identity of the gas:

$$PV = nRT$$

The density *d* of a gas, on the other hand, is determined by its identity. As described in another chapter of this text, the density of a substance is a characteristic property that may be used to identify the substance.

$$d = \frac{m}{V}$$

Rearranging the ideal gas equation to isolate *V* and substituting into the density equation yields

$$d = \frac{mP}{nRT} = \left(\frac{m}{n}\right)\frac{P}{RT}$$

The ratio m/n is the definition of molar mass, $\mathcal{M}$:

$$\mathcal{M} = \frac{m}{n}$$

The density equation can then be written

$$d = \frac{\mathcal{M}P}{RT}$$

This relation may be used for calculating the densities of gases of known identities at specified values of pressure and temperature as demonstrated in Example 9.11.

 **EXAMPLE 9.11**

Measuring Gas Density

What is the density of molecular nitrogen gas at STP?

Solution

The molar mass of molecular nitrogen, N_2, is 28.01 g/mol. Substituting this value along with standard temperature and pressure into the gas density equation yields

$$d = \frac{\mathcal{M}P}{RT} = \frac{(28.01\ \text{g/mol})(1.00\ \text{atm})}{(0.0821\ \text{L·atm·mol}^{-1}\text{K}^{-1})(273\ \text{K})} = 1.25\ \text{g/L}$$

Check Your Learning

What is the density of molecular hydrogen gas at 17.0 °C and a pressure of 760 torr?

Answer:

d = 0.0847 g/L

When the identity of a gas is unknown, measurements of the mass, pressure, volume, and temperature of a sample can be used to calculate the molar mass of the gas (a useful property for identification purposes). Combining the ideal gas equation

$$PV = nRT$$

and the definition of molar mass

$$\mathcal{M} = \frac{m}{n}$$

yields the following equation:

$$\mathcal{M} = \frac{mRT}{PV}$$

Determining the molar mass of a gas via this approach is demonstrated in Example 9.12.

EXAMPLE 9.12

Determining the Molecular Formula of a Gas from its Molar Mass and Empirical Formula

Cyclopropane, a gas once used with oxygen as a general anesthetic, is composed of 85.7% carbon and 14.3% hydrogen by mass. Find the empirical formula. If 1.56 g of cyclopropane occupies a volume of 1.00 L at 0.984 atm and 50 °C, what is the molecular formula for cyclopropane?

Solution

First determine the empirical formula of the gas. Assume 100 g and convert the percentage of each element into grams. Determine the number of moles of carbon and hydrogen in the 100-g sample of cyclopropane. Divide by the smallest number of moles to relate the number of moles of carbon to the number of moles of hydrogen. In the last step, realize that the smallest whole number ratio is the empirical formula:

$$85.7 \text{ g C} \times \frac{1 \text{ mol C}}{12.01 \text{ g C}} = 7.136 \text{ mol C} \qquad \frac{7.136}{7.136} = 1.00 \text{ mol C}$$

$$14.3 \text{ g H} \times \frac{1 \text{ mol H}}{1.01 \text{ g H}} = 14.158 \text{ mol H} \qquad \frac{14.158}{7.136} = 1.98 \text{ mol H}$$

Empirical formula is CH_2 [empirical mass (EM) of 14.03 g/empirical unit].

Next, use the provided values for mass, pressure, temperature and volume to compute the molar mass of the gas:

$$\mathcal{M} = \frac{mRT}{PV} = \frac{(1.56 \text{ g})(0.0821 \text{ L·atm·mol}^{-1}\text{K}^{-1})(323 \text{ K})}{(0.984 \text{ atm})(1.00 \text{ L})} = 42.0 \text{ g/mol}$$

Comparing the molar mass to the empirical formula mass shows how many empirical formula units make up a molecule:

$$\frac{\mathcal{M}}{EM} = \frac{42.0 \text{ g/mol}}{14.0 \text{ g/mol}} = 3$$

The molecular formula is thus derived from the empirical formula by multiplying each of its subscripts by three:

$$(CH_2)_3 = C_3H_6$$

Check Your Learning

Acetylene, a fuel used welding torches, is composed of 92.3% C and 7.7% H by mass. Find the empirical formula. If 1.10 g of acetylene occupies of volume of 1.00 L at 1.15 atm and 59.5 °C, what is the molecular formula for acetylene?

Answer:

Empirical formula, CH; Molecular formula, C_2H_2

EXAMPLE 9.13

Determining the Molar Mass of a Volatile Liquid

The approximate molar mass of a volatile liquid can be determined by:

1. Heating a sample of the liquid in a flask with a tiny hole at the top, which converts the liquid into gas that

may escape through the hole
2. Removing the flask from heat at the instant when the last bit of liquid becomes gas, at which time the flask will be filled with only gaseous sample at ambient pressure
3. Sealing the flask and permitting the gaseous sample to condense to liquid, and then weighing the flask to determine the sample's mass (see Figure 9.19)

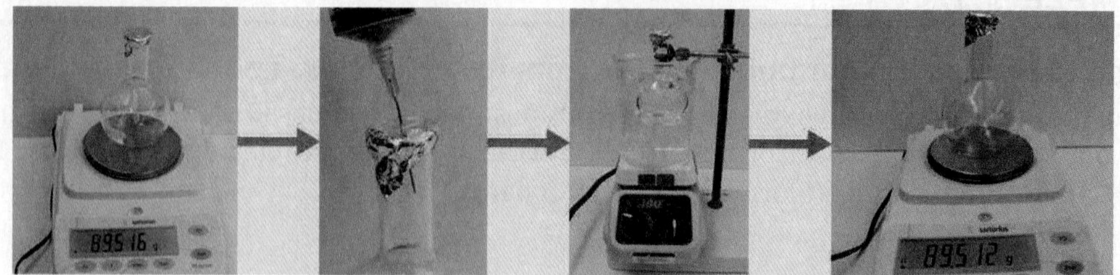

FIGURE 9.19 When the volatile liquid in the flask is heated past its boiling point, it becomes gas and drives air out of the flask. At $t_{l \longrightarrow g}$, the flask is filled with volatile liquid gas at the same pressure as the atmosphere. If the flask is then cooled to room temperature, the gas condenses and the mass of the gas that filled the flask, and is now liquid, can be measured. (credit: modification of work by Mark Ott)

Using this procedure, a sample of chloroform gas weighing 0.494 g is collected in a flask with a volume of 129 cm^3 at 99.6 °C when the atmospheric pressure is 742.1 mm Hg. What is the approximate molar mass of chloroform?

Solution

Since $\mathcal{M} = \frac{m}{n}$ and $n = \frac{PV}{RT}$, substituting and rearranging gives $\mathcal{M} = \frac{mRT}{PV}$,

then

$$\mathcal{M} = \frac{mRT}{PV} = \frac{(0.494\text{ g}) \times 0.08206\text{ L·atm/mol K} \times 372.8\text{ K}}{0.976\text{ atm} \times 0.129\text{ L}} = 120\text{ g/mol}.$$

Check Your Learning

A sample of phosphorus that weighs 3.243×10^{-2} g exerts a pressure of 31.89 kPa in a 56.0-mL bulb at 550 °C. What are the molar mass and molecular formula of phosphorus vapor?

Answer:
124 g/mol P_4

The Pressure of a Mixture of Gases: Dalton's Law

Unless they chemically react with each other, the individual gases in a mixture of gases do not affect each other's pressure. Each individual gas in a mixture exerts the same pressure that it would exert if it were present alone in the container (Figure 9.20). The pressure exerted by each individual gas in a mixture is called its **partial pressure**. This observation is summarized by **Dalton's law of partial pressures**: *The total pressure of a mixture of ideal gases is equal to the sum of the partial pressures of the component gases*:

$$P_{Total} = P_A + P_B + P_C + ... = \Sigma_i P_i$$

In the equation P_{Total} is the total pressure of a mixture of gases, P_A is the partial pressure of gas A; P_B is the partial pressure of gas B; P_C is the partial pressure of gas C; and so on.

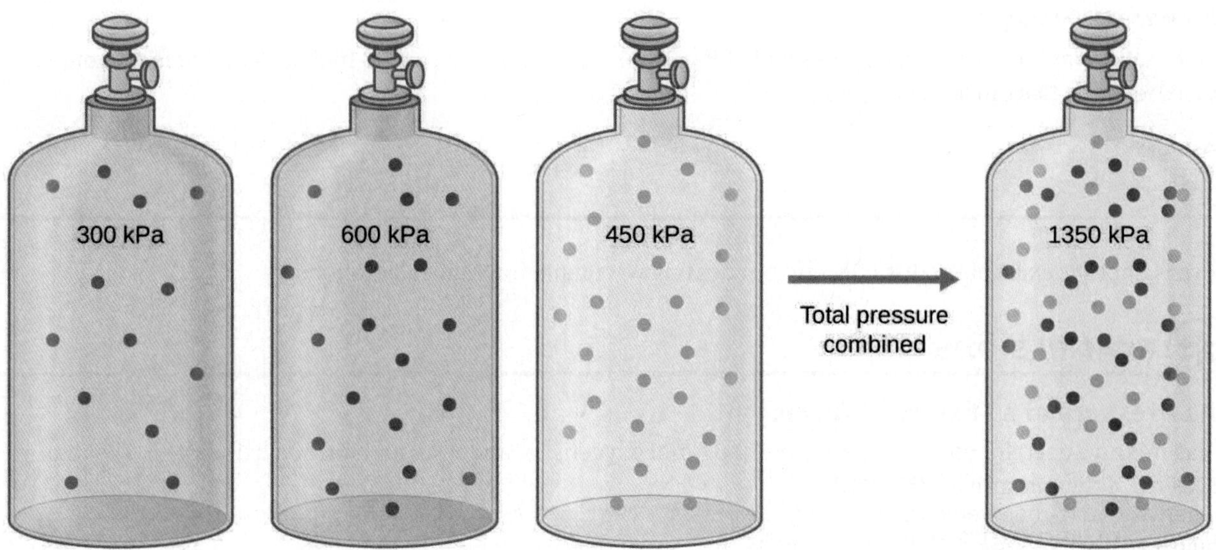

FIGURE 9.20 If equal-volume cylinders containing gasses at pressures of 300 kPa, 450 kPa, and 600 kPa are all combined in the same-size cylinder, the total pressure of the gas mixture is 1350 kPa.

The partial pressure of gas A is related to the total pressure of the gas mixture via its **mole fraction (*X*)**, a unit of concentration defined as the number of moles of a component of a solution divided by the total number of moles of all components:

$$P_A = X_A \times P_{Total} \qquad \text{where} \qquad X_A = \frac{n_A}{n_{Total}}$$

where P_A, X_A, and n_A are the partial pressure, mole fraction, and number of moles of gas A, respectively, and n_{Total} is the number of moles of all components in the mixture.

EXAMPLE 9.14

The Pressure of a Mixture of Gases

A 10.0-L vessel contains 2.50×10^{-3} mol of H_2, 1.00×10^{-3} mol of He, and 3.00×10^{-4} mol of Ne at 35 °C.

(a) What are the partial pressures of each of the gases?

(b) What is the total pressure in atmospheres?

Solution

The gases behave independently, so the partial pressure of each gas can be determined from the ideal gas equation, using $P = \frac{nRT}{V}$:

$$P_{H_2} = \frac{(2.50 \times 10^{-3}\ \cancel{\text{mol}})(0.08206\ \cancel{\text{L}}\ \text{atm}\ \cancel{\text{mol}^{-1}\ \text{K}^{-1}})(308\ \cancel{\text{K}})}{10.0\ \cancel{\text{L}}} = 6.32 \times 10^{-3}\ \text{atm}$$

$$P_{He} = \frac{(1.00 \times 10^{-3}\ \cancel{\text{mol}})(0.08206\ \cancel{\text{L}}\ \text{atm}\ \cancel{\text{mol}^{-1}\ \text{K}^{-1}})(308\ \cancel{\text{K}})}{10.0\ \cancel{\text{L}}} = 2.53 \times 10^{-3}\ \text{atm}$$

$$P_{Ne} = \frac{(3.00 \times 10^{-4}\ \cancel{\text{mol}})(0.08206\ \cancel{\text{L}}\ \text{atm}\ \cancel{\text{mol}^{-1}\ \text{K}^{-1}})(308\ \cancel{\text{K}})}{10.0\ \cancel{\text{L}}} = 7.58 \times 10^{-4}\ \text{atm}$$

The total pressure is given by the sum of the partial pressures:

$$P_T = P_{H_2} + P_{He} + P_{Ne} = (0.00632 + 0.00253 + 0.00076)\ \text{atm} = 9.61 \times 10^{-3}\ \text{atm}$$

Check Your Learning

A 5.73-L flask at 25 °C contains 0.0388 mol of N_2, 0.147 mol of CO, and 0.0803 mol of H_2. What is the total pressure in the flask in atmospheres?

Answer:

1.137 atm

Here is another example of this concept, but dealing with mole fraction calculations.

EXAMPLE 9.15

The Pressure of a Mixture of Gases

A gas mixture used for anesthesia contains 2.83 mol oxygen, O_2, and 8.41 mol nitrous oxide, N_2O. The total pressure of the mixture is 192 kPa.

(a) What are the mole fractions of O_2 and N_2O?

(b) What are the partial pressures of O_2 and N_2O?

Solution

The mole fraction is given by $X_A = \frac{n_A}{n_{Total}}$ and the partial pressure is $P_A = X_A \times P_{Total}$.

For O_2,

$$X_{O_2} = \frac{n_{O_2}}{n_{Total}} = \frac{2.83 \text{ mol}}{(2.83 + 8.41) \text{ mol}} = 0.252$$

and $P_{O_2} = X_{O_2} \times P_{Total} = 0.252 \times 192 \text{ kPa} = 48.4 \text{ kPa}$

For N_2O,

$$X_{N_2O} = \frac{n_{N_2O}}{n_{\text{Total}}} = \frac{8.41 \text{ mol}}{(2.83 + 8.41) \text{ mol}} = 0.748$$

and

$$P_{N_2O} = X_{N_2O} \times P_{\text{Total}} = 0.748 \times 192 \text{ kPa} = 144 \text{ kPa}$$

Check Your Learning

What is the pressure of a mixture of 0.200 g of H_2, 1.00 g of N_2, and 0.820 g of Ar in a container with a volume of 2.00 L at 20 °C?

Answer:

1.87 atm

Collection of Gases over Water

A simple way to collect gases that do not react with water is to capture them in a bottle that has been filled with water and inverted into a dish filled with water. The pressure of the gas inside the bottle can be made equal to the air pressure outside by raising or lowering the bottle. When the water level is the same both inside and outside the bottle (Figure 9.21), the pressure of the gas is equal to the atmospheric pressure, which can be measured with a barometer.

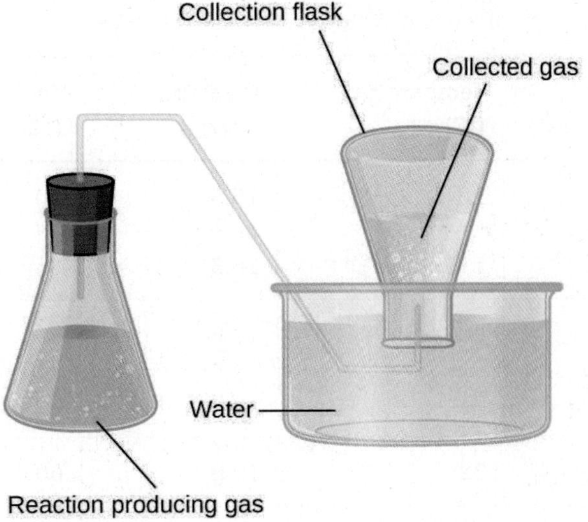

FIGURE 9.21 When a reaction produces a gas that is collected above water, the trapped gas is a mixture of the gas produced by the reaction and water vapor. If the collection flask is appropriately positioned to equalize the water levels both within and outside the flask, the pressure of the trapped gas mixture will equal the atmospheric pressure outside the flask (see the earlier discussion of manometers).

However, there is another factor we must consider when we measure the pressure of the gas by this method. Water evaporates and there is always gaseous water (water vapor) above a sample of liquid water. As a gas is collected over water, it becomes saturated with water vapor and the total pressure of the mixture equals the partial pressure of the gas plus the partial pressure of the water vapor. The pressure of the pure gas is therefore equal to the total pressure minus the pressure of the water vapor—this is referred to as the "dry" gas pressure, that is, the pressure of the gas only, without water vapor. The **vapor pressure of water**, which is the pressure exerted by water vapor in equilibrium with liquid water in a closed container, depends on the temperature (Figure 9.22); more detailed information on the temperature dependence of water vapor can be found in Table 9.2, and vapor pressure will be discussed in more detail in the next chapter on liquids.

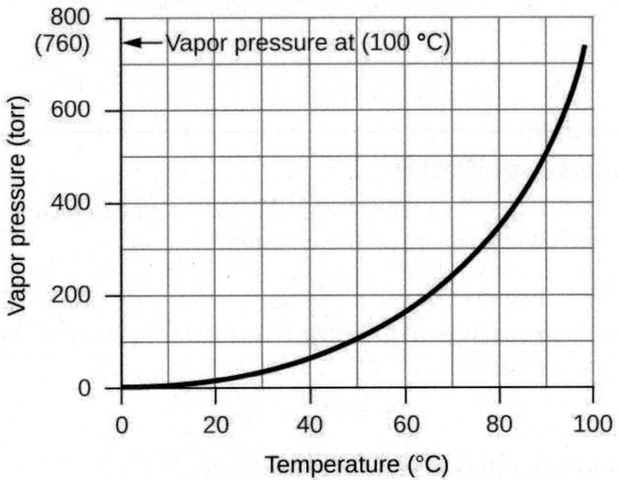

FIGURE 9.22 This graph shows the vapor pressure of water at sea level as a function of temperature.

Vapor Pressure of Ice and Water in Various Temperatures at Sea Level

Temperature (°C)	Pressure (torr)	Temperature (°C)	Pressure (torr)	Temperature (°C)	Pressure (torr)
–10	1.95	18	15.5	30	31.8
–5	3.0	19	16.5	35	42.2
–2	3.9	20	17.5	40	55.3
0	4.6	21	18.7	50	92.5
2	5.3	22	19.8	60	149.4
4	6.1	23	21.1	70	233.7
6	7.0	24	22.4	80	355.1
8	8.0	25	23.8	90	525.8
10	9.2	26	25.2	95	633.9
12	10.5	27	26.7	99	733.2
14	12.0	28	28.3	100.0	760.0
16	13.6	29	30.0	101.0	787.6

TABLE 9.2

EXAMPLE 9.16

Pressure of a Gas Collected Over Water

If 0.200 L of argon is collected over water at a temperature of 26 °C and a pressure of 750 torr in a system like that shown in Figure 9.21, what is the partial pressure of argon?

Solution

According to Dalton's law, the total pressure in the bottle (750 torr) is the sum of the partial pressure of argon and the partial pressure of gaseous water:

$$P_T = P_{Ar} + P_{H_2O}$$

Rearranging this equation to solve for the pressure of argon gives:

$$P_{Ar} = P_T - P_{H_2O}$$

The pressure of water vapor above a sample of liquid water at 26 °C is 25.2 torr (Appendix E), so:

$$P_{Ar} = 750 \text{ torr} - 25.2 \text{ torr} = 725 \text{ torr}$$

Check Your Learning

A sample of oxygen collected over water at a temperature of 29.0 °C and a pressure of 764 torr has a volume of 0.560 L. What volume would the dry oxygen from this sample have under the same conditions of temperature and pressure?

Answer:
0.537 L

Chemical Stoichiometry and Gases

Chemical stoichiometry describes the quantitative relationships between reactants and products in chemical reactions.

We have previously measured quantities of reactants and products using masses for solids and volumes in conjunction with the molarity for solutions; now we can also use gas volumes to indicate quantities. If we know the volume, pressure, and temperature of a gas, we can use the ideal gas equation to calculate how many moles of the gas are present. If we know how many moles of a gas are involved, we can calculate the volume of a gas at any temperature and pressure.

Avogadro's Law Revisited

Sometimes we can take advantage of a simplifying feature of the stoichiometry of gases that solids and solutions do not exhibit: All gases that show ideal behavior contain the same number of molecules in the same volume (at the same temperature and pressure). Thus, the ratios of volumes of gases involved in a chemical reaction are given by the coefficients in the equation for the reaction, provided that the gas volumes are measured at the same temperature and pressure.

We can extend Avogadro's law (that the volume of a gas is directly proportional to the number of moles of the gas) to chemical reactions with gases: Gases combine, or react, in definite and simple proportions by volume, provided that all gas volumes are measured at the same temperature and pressure. For example, since nitrogen and hydrogen gases react to produce ammonia gas according to $N_2(g) + 3H_2(g) \longrightarrow 2NH_3(g)$, a given volume of nitrogen gas reacts with three times that volume of hydrogen gas to produce two times that volume of ammonia gas, if pressure and temperature remain constant.

The explanation for this is illustrated in Figure 9.23. According to Avogadro's law, equal volumes of gaseous N_2, H_2, and NH_3, at the same temperature and pressure, contain the same number of molecules. Because one molecule of N_2 reacts with three molecules of H_2 to produce two molecules of NH_3, the volume of H_2 required is three times the volume of N_2, and the volume of NH_3 produced is two times the volume of N_2.

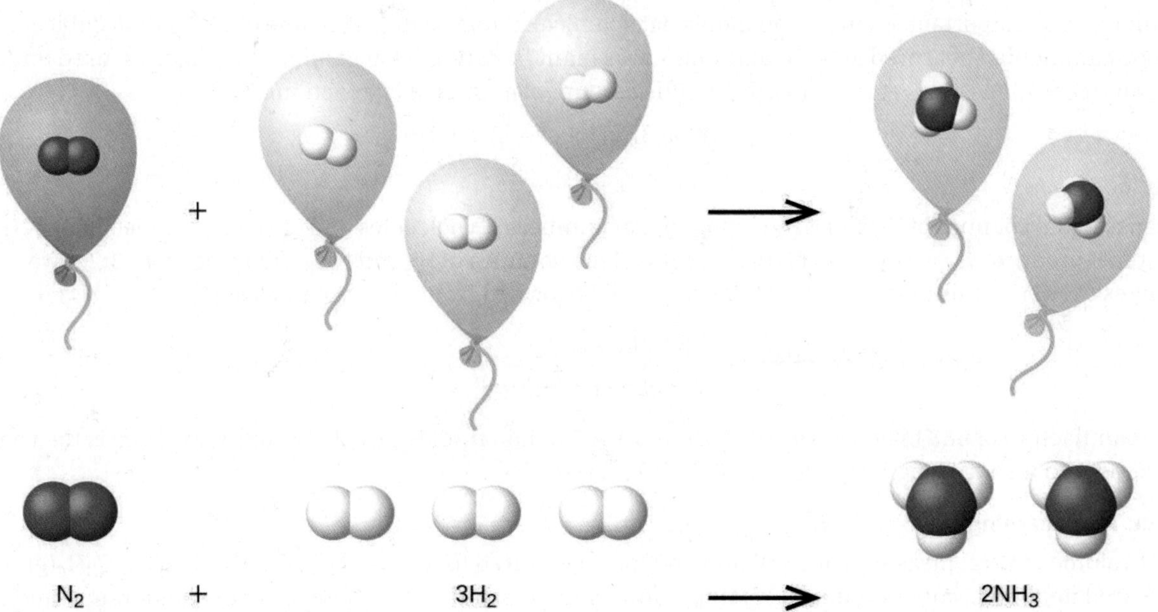

FIGURE 9.23 One volume of N_2 combines with three volumes of H_2 to form two volumes of NH_3.

EXAMPLE 9.17

Reaction of Gases

Propane, $C_3H_8(g)$, is used in gas grills to provide the heat for cooking. What volume of $O_2(g)$ measured at 25 °C and 760 torr is required to react with 2.7 L of propane measured under the same conditions of temperature and pressure? Assume that the propane undergoes complete combustion.

Solution

The ratio of the volumes of C_3H_8 and O_2 will be equal to the ratio of their coefficients in the balanced equation for the reaction:

$$\begin{array}{ccc} C_3H_8(g) + 5O_2(g) & \longrightarrow & 3CO_2(g) + 4H_2O(l) \\ 1 \text{ volume} + 5 \text{ volumes} & & 3 \text{ volumes} + 4 \text{ volumes} \end{array}$$

From the equation, we see that one volume of C_3H_8 will react with five volumes of O_2:

$$2.7 \cancel{\text{L C}_3\text{H}_8} \times \frac{5 \text{ L } O_2}{1 \cancel{\text{L C}_3\text{H}_8}} = 13.5 \text{ L } O_2$$

A volume of 13.5 L of O_2 will be required to react with 2.7 L of C_3H_8.

Check Your Learning

An acetylene tank for an oxyacetylene welding torch provides 9340 L of acetylene gas, C_2H_2, at 0 °C and 1 atm. How many tanks of oxygen, each providing 7.00×10^3 L of O_2 at 0 °C and 1 atm, will be required to burn the acetylene?

$$2C_2H_2 + 5O_2 \longrightarrow 4CO_2 + 2H_2O$$

Answer:

3.34 tanks (2.34×10^4 L)

EXAMPLE 9.18

Volumes of Reacting Gases

Ammonia is an important fertilizer and industrial chemical. Suppose that a volume of 683 billion cubic feet of gaseous ammonia, measured at 25 °C and 1 atm, was manufactured. What volume of $H_2(g)$, measured under the same conditions, was required to prepare this amount of ammonia by reaction with N_2?

$$N_2(g) + 3H_2(g) \longrightarrow 2NH_3(g)$$

Solution

Because equal volumes of H_2 and NH_3 contain equal numbers of molecules and each three molecules of H_2 that react produce two molecules of NH_3, the ratio of the volumes of H_2 and NH_3 will be equal to 3:2. Two volumes of NH_3, in this case in units of billion ft^3, will be formed from three volumes of H_2:

$$683 \cancel{\text{billion ft}^3 \text{ NH}_3} \times \frac{3 \text{ billion ft}^3 \text{ H}_2}{2 \cancel{\text{billion ft}^3 \text{ NH}_3}} = 1.02 \times 10^3 \text{ billion ft}^3 \text{ H}_2$$

The manufacture of 683 billion ft^3 of NH_3 required 1020 billion ft^3 of H_2. (At 25 °C and 1 atm, this is the volume of a cube with an edge length of approximately 1.9 miles.)

Check Your Learning

What volume of $O_2(g)$ measured at 25 °C and 760 torr is required to react with 17.0 L of ethylene, $C_2H_4(g)$, measured under the same conditions of temperature and pressure? The products are CO_2 and water vapor.

Answer:

51.0 L

EXAMPLE 9.19

Volume of Gaseous Product

What volume of hydrogen at 27 °C and 723 torr may be prepared by the reaction of 8.88 g of gallium with an excess of hydrochloric acid?

$$2\text{Ga}\,(s) + 6\text{HCl}\,(aq) \longrightarrow 2\text{GaCl}_3\,(aq) + 3\text{H}_2(g)$$

Solution

Convert the provided mass of the limiting reactant, Ga, to moles of hydrogen produced:

$$8.88\ \cancel{\text{g Ga}} \times \frac{1\ \cancel{\text{mol Ga}}}{69.723\ \cancel{\text{g Ga}}} \times \frac{3\ \text{mol H}_2}{2\ \cancel{\text{mol Ga}}} = 0.191\ \text{mol H}_2$$

Convert the provided temperature and pressure values to appropriate units (K and atm, respectively), and then use the molar amount of hydrogen gas and the ideal gas equation to calculate the volume of gas:

$$V = \left(\frac{nRT}{P}\right) = \frac{0.191\ \cancel{\text{mol}} \times 0.08206\ \text{L}\ \cancel{\text{atm mol}^{-1}\ \text{K}^{-1}} \times 300\ \text{K}}{0.951\ \cancel{\text{atm}}} = 4.94\ \text{L}$$

Check Your Learning

Sulfur dioxide is an intermediate in the preparation of sulfuric acid. What volume of SO_2 at 343 °C and 1.21 atm is produced by burning 1.00 kg of sulfur in excess oxygen?

Answer:

1.30×10^3 L

HOW SCIENCES INTERCONNECT

Greenhouse Gases and Climate Change

The thin skin of our atmosphere keeps the earth from being an ice planet and makes it habitable. In fact, this is due to less than 0.5% of the air molecules. Of the energy from the sun that reaches the earth, almost $\frac{1}{3}$ is reflected back into space, with the rest absorbed by the atmosphere and the surface of the earth. Some of the energy that the earth absorbs is re-emitted as infrared (IR) radiation, a portion of which passes back out through the atmosphere into space. Most if this IR radiation, however, is absorbed by certain atmospheric gases, effectively trapping heat within the atmosphere in a phenomenon known as the *greenhouse effect*. This effect maintains global temperatures within the range needed to sustain life on earth. Without our atmosphere, the earth's average temperature would be lower by more than 30 °C (nearly 60 °F). The major greenhouse gases (GHGs) are water vapor, carbon dioxide, methane, and ozone. Since the Industrial Revolution, human activity has been increasing the concentrations of GHGs, which have changed the energy balance and are significantly altering the earth's climate (Figure 9.24).

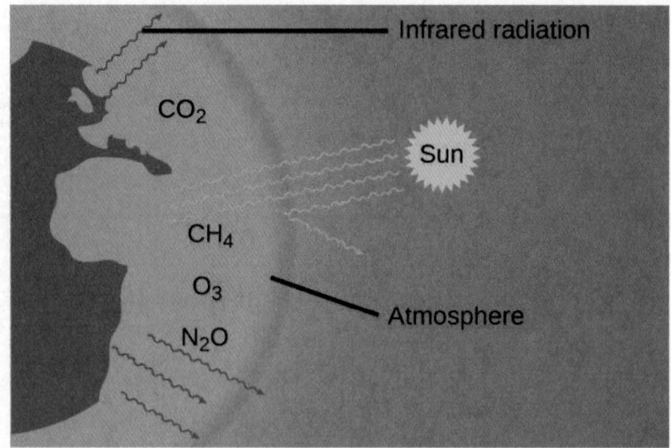

FIGURE 9.24 Greenhouse gases trap enough of the sun's energy to make the planet habitable—this is known as the greenhouse effect. Human activities are increasing greenhouse gas levels, warming the planet and causing more extreme weather events.

There is strong evidence from multiple sources that higher atmospheric levels of CO_2 are caused by human activity, with fossil fuel burning accounting for about $\frac{3}{4}$ of the recent increase in CO_2. Reliable data from ice cores reveals that CO_2 concentration in the atmosphere is at the highest level in the past 800,000 years; other evidence indicates that it may be at its highest level in 20 million years. In recent years, the CO_2 concentration has increased preindustrial levels of ~280 ppm to more than 400 ppm today (Figure 9.25).

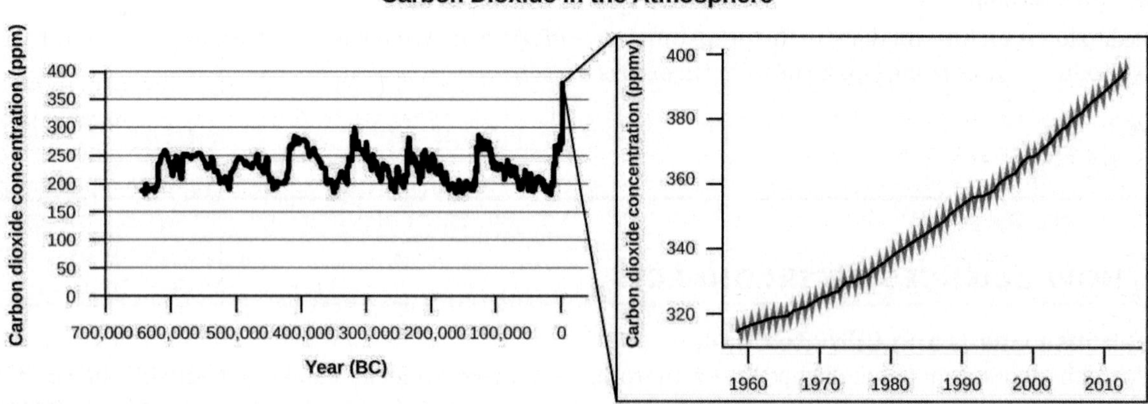

FIGURE 9.25 CO_2 levels over the past 700,000 years were typically from 200–300 ppm, with a steep, unprecedented increase over the past 50 years.

LINK TO LEARNING

Click here (http://openstax.org/l/16GlobalWarming) to see a 2-minute video explaining greenhouse gases and global warming.

Portrait of a Chemist

Susan Solomon

Atmospheric and climate scientist Susan Solomon (Figure 9.26) is the author of one of *The New York Times* books of the year (*The Coldest March*, 2001), one of Time magazine's 100 most influential people in the world (2008), and a working group leader of the Intergovernmental Panel on Climate Change (IPCC), which was the recipient of the 2007 Nobel Peace Prize. She helped determine and explain the cause of the formation of the ozone hole over Antarctica, and has authored many important papers on climate change.

She has been awarded the top scientific honors in the US and France (the National Medal of Science and the Grande Medaille, respectively), and is a member of the National Academy of Sciences, the Royal Society, the French Academy of Sciences, and the European Academy of Sciences. Formerly a professor at the University of Colorado, she is now at MIT, and continues to work at NOAA.

For more information, watch this video (http://openstax.org/l/16SusanSolomon) about Susan Solomon.

FIGURE 9.26 Susan Solomon's research focuses on climate change and has been instrumental in determining the cause of the ozone hole over Antarctica. (credit: National Oceanic and Atmospheric Administration)

9.4 Effusion and Diffusion of Gases

LEARNING OBJECTIVES

By the end of this section, you will be able to:

- Define and explain effusion and diffusion
- State Graham's law and use it to compute relevant gas properties

If you have ever been in a room when a piping hot pizza was delivered, you have been made aware of the fact that gaseous molecules can quickly spread throughout a room, as evidenced by the pleasant aroma that soon reaches your nose. Although gaseous molecules travel at tremendous speeds (hundreds of meters per second), they collide with other gaseous molecules and travel in many different directions before reaching the desired target. At room temperature, a gaseous molecule will experience billions of collisions per second. The **mean free path** is the average distance a molecule travels between collisions. The mean free path increases with decreasing pressure; in general, the mean free path for a gaseous molecule will be hundreds of times the diameter of the molecule

In general, we know that when a sample of gas is introduced to one part of a closed container, its molecules very quickly disperse throughout the container; this process by which molecules disperse in space in response to differences in concentration is called **diffusion** (shown in Figure 9.27). The gaseous atoms or molecules are, of course, unaware of any concentration gradient, they simply move randomly—regions of higher concentration have more particles than regions of lower concentrations, and so a net movement of species from high to low concentration areas takes place. In a closed environment, diffusion will ultimately result in equal concentrations of gas throughout, as depicted in Figure 9.27. The gaseous atoms and molecules continue to move, but since their concentrations are the same in both bulbs, the rates of transfer between the bulbs are equal (no *net* transfer of molecules occurs).

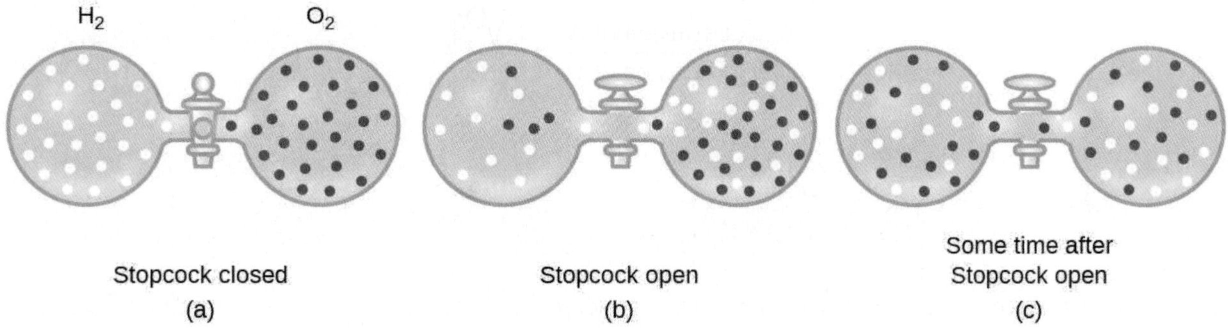

FIGURE 9.27 (a) Two gases, H_2 and O_2, are initially separated. (b) When the stopcock is opened, they mix together.

The lighter gas, H_2, passes through the opening faster than O_2, so just after the stopcock is opened, more H_2 molecules move to the O_2 side than O_2 molecules move to the H_2 side. (c) After a short time, both the slower-moving O_2 molecules and the faster-moving H_2 molecules have distributed themselves evenly on both sides of the vessel.

We are often interested in the **rate of diffusion**, the amount of gas passing through some area per unit time:

$$\text{rate of diffusion} = \frac{\text{amount of gas passing through an area}}{\text{unit of time}}$$

The diffusion rate depends on several factors: the concentration gradient (the increase or decrease in concentration from one point to another); the amount of surface area available for diffusion; and the distance the gas particles must travel. Note also that the time required for diffusion to occur is inversely proportional to the rate of diffusion, as shown in the rate of diffusion equation.

A process involving movement of gaseous species similar to diffusion is **effusion**, the escape of gas molecules through a tiny hole such as a pinhole in a balloon into a vacuum (Figure 9.28). Although diffusion and effusion rates both depend on the molar mass of the gas involved, their rates are not equal; however, the ratios of their rates are the same.

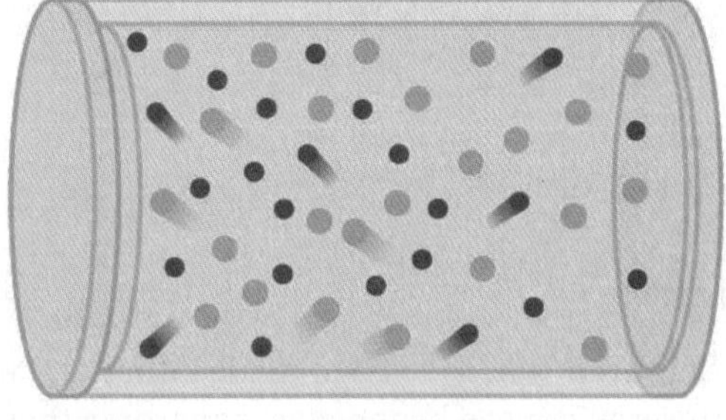

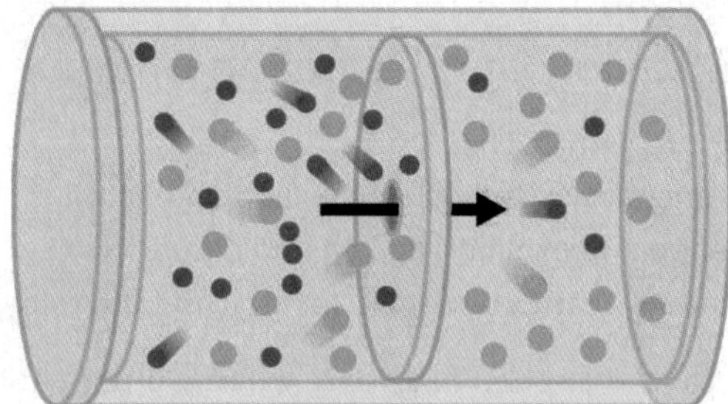

FIGURE 9.28 Diffusion involves the unrestricted dispersal of molecules throughout space due to their random motion. When this process is restricted to passage of molecules through very small openings in a physical barrier, the process is called effusion.

If a mixture of gases is placed in a container with porous walls, the gases effuse through the small openings in the walls. The lighter gases pass through the small openings more rapidly (at a higher rate) than the heavier ones (Figure 9.29). In 1832, Thomas Graham studied the rates of effusion of different gases and formulated **Graham's law of effusion**: *The rate of effusion of a gas is inversely proportional to the square root of the mass of its particles*:

$$\text{rate of effusion} \propto \frac{1}{\sqrt{\mathcal{M}}}$$

This means that if two gases A and B are at the same temperature and pressure, the ratio of their effusion rates is inversely proportional to the ratio of the square roots of the masses of their particles:

$$\frac{\text{rate of effusion of A}}{\text{rate of effusion of B}} = \frac{\sqrt{\mathcal{M}_B}}{\sqrt{\mathcal{M}_A}}$$

FIGURE 9.29 The left photograph shows two balloons inflated with different gases, helium (orange) and argon (blue).The right-side photograph shows the balloons approximately 12 hours after being filled, at which time the helium balloon has become noticeably more deflated than the argon balloon, due to the greater effusion rate of the lighter helium gas. (credit: modification of work by Paul Flowers)

EXAMPLE 9.20

Applying Graham's Law to Rates of Effusion

Calculate the ratio of the rate of effusion of hydrogen to the rate of effusion of oxygen.

Solution

From Graham's law, we have:

$$\frac{\text{rate of effusion of hydrogen}}{\text{rate of effusion of oxygen}} = \frac{\sqrt{32\ \cancel{\text{g mol}^{-1}}}}{\sqrt{2\ \cancel{\text{g mol}^{-1}}}} = \frac{\sqrt{16}}{\sqrt{1}} = \frac{4}{1}$$

Hydrogen effuses four times as rapidly as oxygen.

Check Your Learning

At a particular pressure and temperature, nitrogen gas effuses at the rate of 79 mL/s. Under the same conditions, at what rate will sulfur dioxide effuse?

Answer:

52 mL/s

EXAMPLE 9.21

Effusion Time Calculations

It takes 243 s for 4.46×10^{-5} mol Xe to effuse through a tiny hole. Under the same conditions, how long will it take 4.46×10^{-5} mol Ne to effuse?

Solution

It is important to resist the temptation to use the times directly, and to remember how rate relates to time as well as how it relates to mass. Recall the definition of rate of effusion:

$$\text{rate of effusion} = \frac{\text{amount of gas transferred}}{\text{time}}$$

and combine it with Graham's law:

$$\frac{\text{rate of effusion of gas Xe}}{\text{rate of effusion of gas Ne}} = \frac{\sqrt{\mathcal{M}_{\text{Ne}}}}{\sqrt{\mathcal{M}_{\text{Xe}}}}$$

To get:

$$\frac{\frac{\text{amount of Xe transferred}}{\text{time for Xe}}}{\frac{\text{amount of Ne transferred}}{\text{time for Ne}}} = \frac{\sqrt{\mathcal{M}_{Ne}}}{\sqrt{\mathcal{M}_{Xe}}}$$

Noting that *amount of A* = *amount of B*, and solving for *time for Ne*:

$$\frac{\frac{\cancel{\text{amount of Xe}}}{\text{time for Xe}}}{\frac{\cancel{\text{amount of Ne}}}{\text{time for Ne}}} = \frac{\text{time for Ne}}{\text{time for Xe}} = \frac{\sqrt{\mathcal{M}_{Ne}}}{\sqrt{\mathcal{M}_{Xe}}} = \frac{\sqrt{\mathcal{M}_{Ne}}}{\sqrt{\mathcal{M}_{Xe}}}$$

and substitute values:

$$\frac{\text{time for Ne}}{243\text{ s}} = \sqrt{\frac{20.2\ \cancel{\text{g mol}}}{131.3\ \cancel{\text{g mol}}}} = 0.392$$

Finally, solve for the desired quantity:

$$\text{time for Ne} = 0.392 \times 243\text{ s} = 95.3\text{ s}$$

Note that this answer is reasonable: Since Ne is lighter than Xe, the effusion rate for Ne will be larger than that for Xe, which means the time of effusion for Ne will be smaller than that for Xe.

Check Your Learning

A party balloon filled with helium deflates to $\frac{2}{3}$ of its original volume in 8.0 hours. How long will it take an identical balloon filled with the same number of moles of air ($\mathcal{M}$ = 28.2 g/mol) to deflate to $\frac{1}{2}$ of its original volume?

Answer:

32 h

EXAMPLE 9.22

Determining Molar Mass Using Graham's Law

An unknown gas effuses 1.66 times more rapidly than CO_2. What is the molar mass of the unknown gas? Can you make a reasonable guess as to its identity?

Solution

From Graham's law, we have:

$$\frac{\text{rate of effusion of Unknown}}{\text{rate of effusion of CO}_2} = \frac{\sqrt{\mathcal{M}_{CO_2}}}{\sqrt{\mathcal{M}_{Unknown}}}$$

Plug in known data:

$$\frac{1.66}{1} = \frac{\sqrt{44.0\text{ g/mol}}}{\sqrt{\mathcal{M}_{Unknown}}}$$

Solve:

$$\mathcal{M}_{Unknown} = \frac{44.0\text{ g/mol}}{(1.66)^2} = 16.0\text{ g/mol}$$

The gas could well be CH_4, the only gas with this molar mass.

Check Your Learning

Hydrogen gas effuses through a porous container 8.97-times faster than an unknown gas. Estimate the molar mass of the unknown gas.

Answer:

163 g/mol

HOW SCIENCES INTERCONNECT

Use of Diffusion for Nuclear Energy Applications: Uranium Enrichment

Gaseous diffusion has been used to produce enriched uranium for use in nuclear power plants and weapons. Naturally occurring uranium contains only 0.72% of ^{235}U, the kind of uranium that is "fissile," that is, capable of sustaining a nuclear fission chain reaction. Nuclear reactors require fuel that is 2–5% ^{235}U, and nuclear bombs need even higher concentrations. One way to enrich uranium to the desired levels is to take advantage of Graham's law. In a gaseous diffusion enrichment plant, uranium hexafluoride (UF_6, the only uranium compound that is volatile enough to work) is slowly pumped through large cylindrical vessels called diffusers, which contain porous barriers with microscopic openings. The process is one of diffusion because the other side of the barrier is not evacuated. The $^{235}UF_6$ molecules have a higher average speed and diffuse through the barrier a little faster than the heavier $^{238}UF_6$ molecules. The gas that has passed through the barrier is slightly enriched in $^{235}UF_6$ and the residual gas is slightly depleted. The small difference in molecular weights between $^{235}UF_6$ and $^{238}UF_6$ only about 0.4% enrichment, is achieved in one diffuser (Figure 9.30). But by connecting many diffusers in a sequence of stages (called a cascade), the desired level of enrichment can be attained.

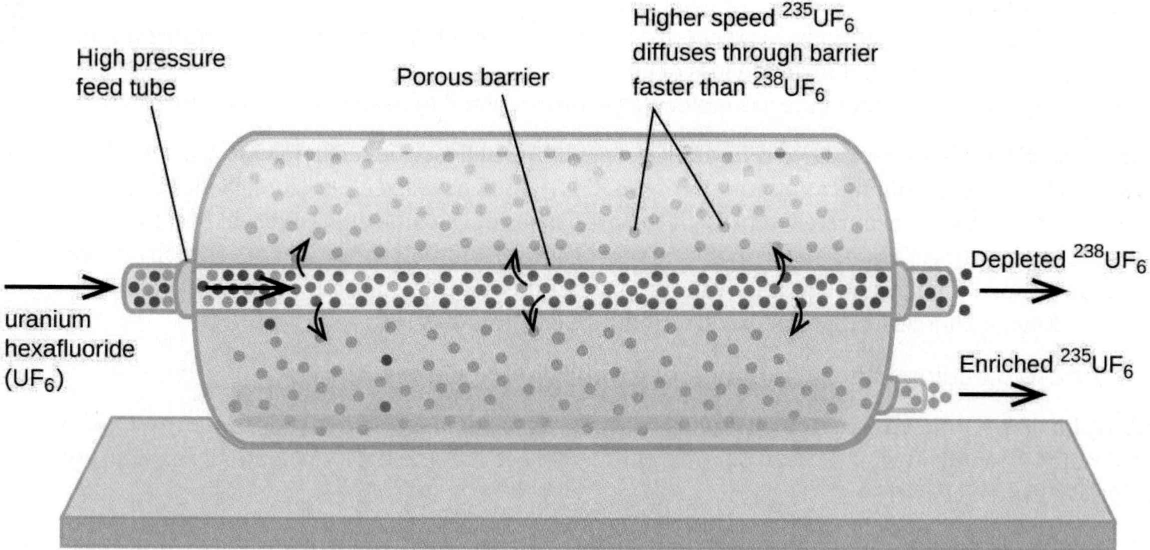

FIGURE 9.30 In a diffuser, gaseous UF_6 is pumped through a porous barrier, which partially separates $^{235}UF_6$ from $^{238}UF_6$ The UF_6 must pass through many large diffuser units to achieve sufficient enrichment in ^{235}U.

The large scale separation of gaseous $^{235}UF_6$ from $^{238}UF_6$ was first done during the World War II, at the atomic energy installation in Oak Ridge, Tennessee, as part of the Manhattan Project (the development of the first atomic bomb). Although the theory is simple, this required surmounting many daunting technical challenges to make it work in practice. The barrier must have tiny, uniform holes (about 10^{-6} cm in diameter) and be porous enough to produce high flow rates. All materials (the barrier, tubing, surface coatings, lubricants, and gaskets) need to be able to contain, but not react with, the highly reactive and corrosive UF_6.

Because gaseous diffusion plants require very large amounts of energy (to compress the gas to the high pressures required and drive it through the diffuser cascade, to remove the heat produced during compression, and so on), it is now being replaced by gas centrifuge technology, which requires far less energy. A current hot political issue is how to deny this technology to Iran, to prevent it from producing enough enriched uranium for them to use to make nuclear weapons.

9.5 The Kinetic-Molecular Theory

LEARNING OBJECTIVES

- State the postulates of the kinetic-molecular theory
- Use this theory's postulates to explain the gas laws

The gas laws that we have seen to this point, as well as the ideal gas equation, are empirical, that is, they have been derived from experimental observations. The mathematical forms of these laws closely describe the macroscopic behavior of most gases at pressures less than about 1 or 2 atm. Although the gas laws describe relationships that have been verified by many experiments, they do not tell us why gases follow these relationships.

The **kinetic molecular theory** (KMT) is a simple microscopic model that effectively explains the gas laws described in previous modules of this chapter. This theory is based on the following five postulates described here. (Note: The term "molecule" will be used to refer to the individual chemical species that compose the gas, although some gases are composed of atomic species, for example, the noble gases.)

1. Gases are composed of molecules that are in continuous motion, travelling in straight lines and changing direction only when they collide with other molecules or with the walls of a container.
2. The molecules composing the gas are negligibly small compared to the distances between them.
3. The pressure exerted by a gas in a container results from collisions between the gas molecules and the container walls.
4. Gas molecules exert no attractive or repulsive forces on each other or the container walls; therefore, their collisions are *elastic* (do not involve a loss of energy).
5. The average kinetic energy of the gas molecules is proportional to the kelvin temperature of the gas.

The test of the KMT and its postulates is its ability to explain and describe the behavior of a gas. The various gas laws can be derived from the assumptions of the KMT, which have led chemists to believe that the assumptions of the theory accurately represent the properties of gas molecules. We will first look at the individual gas laws (Boyle's, Charles's, Amontons's, Avogadro's, and Dalton's laws) conceptually to see how the KMT explains them. Then, we will more carefully consider the relationships between molecular masses, speeds, and kinetic energies with temperature, and explain Graham's law.

The Kinetic-Molecular Theory Explains the Behavior of Gases, Part I

Recalling that gas pressure is exerted by rapidly moving gas molecules and depends directly on the number of molecules hitting a unit area of the wall per unit of time, we see that the KMT conceptually explains the behavior of a gas as follows:

- *Amontons's law.* If the temperature is increased, the average speed and kinetic energy of the gas molecules increase. If the volume is held constant, the increased speed of the gas molecules results in more frequent and more forceful collisions with the walls of the container, therefore increasing the pressure (Figure 9.31).
- *Charles's law.* If the temperature of a gas is increased, a constant pressure may be maintained only if the volume occupied by the gas increases. This will result in greater average distances traveled by the molecules to reach the container walls, as well as increased wall surface area. These conditions will decrease the both the frequency of molecule-wall collisions and the number of collisions per unit area, the combined effects of which balance the effect of increased collision forces due to the greater kinetic energy at the higher temperature.
- *Boyle's law.* If the gas volume volume of a given amount of gas at a given temperature is decreased (that is, if the gas is *compressed*), the molecules will be exposed to a decreased container wall area. Collisions with

the container wall will therefore occur more frequently and the pressure exerted by the gas will increase (Figure 9.31).

- *Avogadro's law.* At constant pressure and temperature, the frequency and force of molecule-wall collisions are constant. Under such conditions, increasing the number of gaseous molecules will require a proportional increase in the container volume in order to yield a decrease in the number of collisions per unit area to compensate for the increased frequency of collisions (Figure 9.31).
- *Dalton's Law.* Because of the large distances between them, the molecules of one gas in a mixture bombard the container walls with the same frequency whether other gases are present or not, and the total pressure of a gas mixture equals the sum of the (partial) pressures of the individual gases.

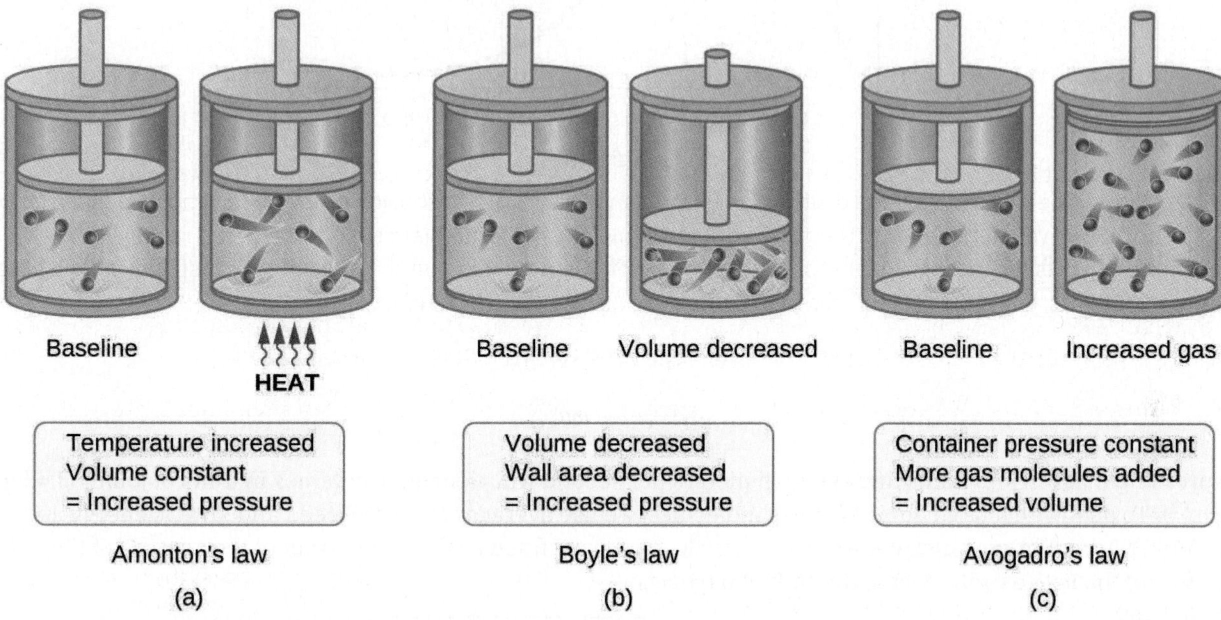

FIGURE 9.31 (a) When gas temperature increases, gas pressure increases due to increased force and frequency of molecular collisions. (b) When volume decreases, gas pressure increases due to increased frequency of molecular collisions. (c) When the amount of gas increases at a constant pressure, volume increases to yield a constant number of collisions per unit wall area per unit time.

molecular speeds and Kinetic Energy

The previous discussion showed that the KMT qualitatively explains the behaviors described by the various gas laws. The postulates of this theory may be applied in a more quantitative fashion to derive these individual laws. To do this, we must first look at speeds and kinetic energies of gas molecules, and the temperature of a gas sample.

In a gas sample, individual molecules have widely varying speeds; however, because of the *vast* number of molecules and collisions involved, the molecular speed distribution and average speed are constant. This molecular speed distribution is known as a Maxwell-Boltzmann distribution, and it depicts the relative numbers of molecules in a bulk sample of gas that possesses a given speed (Figure 9.32).

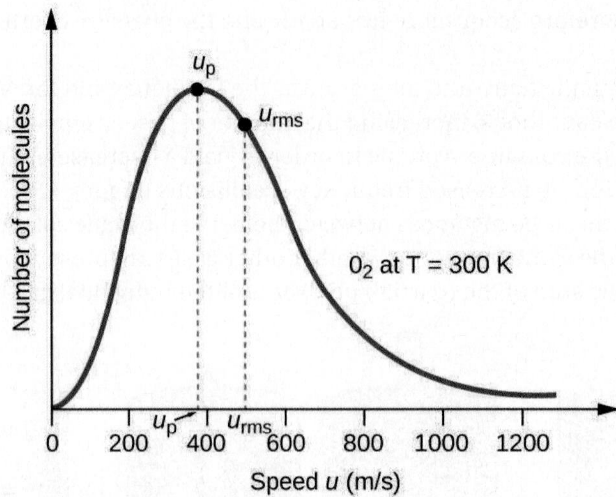

FIGURE 9.32 The molecular speed distribution for oxygen gas at 300 K is shown here. Very few molecules move at either very low or very high speeds. The number of molecules with intermediate speeds increases rapidly up to a maximum, which is the most probable speed, then drops off rapidly. Note that the most probable speed, v_p, is a little less than 400 m/s, while the root mean square speed, u_{rms}, is closer to 500 m/s.

The kinetic energy (KE) of a particle of mass (m) and speed (u) is given by:

$$\text{KE} = \frac{1}{2}mu^2$$

Expressing mass in kilograms and speed in meters per second will yield energy values in units of joules (J = kg m^2 s^{-2}). To deal with a large number of gas molecules, we use averages for both speed and kinetic energy. In the KMT, the **root mean square speed** of a particle, $\boldsymbol{u_{rms}}$, is defined as the square root of the average of the squares of the speeds with n = the number of particles:

$$u_{\text{rms}} = \sqrt{\overline{u^2}} = \sqrt{\frac{u_1^2 + u_2^2 + u_3^2 + u_4^2 + \ldots}{n}}$$

The average kinetic energy for a mole of particles, KE_{avg}, is then equal to:

$$\text{KE}_{\text{avg}} = \frac{1}{2}Mu_{\text{rms}}^2$$

where M is the molar mass expressed in units of kg/mol. The KE_{avg} of a mole of gas molecules is also directly proportional to the temperature of the gas and may be described by the equation:

$$\text{KE}_{\text{avg}} = \frac{3}{2}RT$$

where R is the gas constant and T is the kelvin temperature. When used in this equation, the appropriate form of the gas constant is 8.314 J/mol·K (8.314 kg m^{2}s^{-2}mol^{-1}K^{-1}). These two separate equations for KE_{avg} may be combined and rearranged to yield a relation between molecular speed and temperature:

$$\frac{1}{2}Mu_{\text{rms}}^2 = \frac{3}{2}RT$$

$$u_{\text{rms}} = \sqrt{\frac{3RT}{M}}$$

EXAMPLE 9.23

Calculation of u_{rms}

Calculate the root-mean-square speed for a nitrogen molecule at 30 °C.

Solution

Convert the temperature into Kelvin:

$$30\ °\text{C} + 273 = 303\ \text{K}$$

Determine the molar mass of nitrogen in kilograms:

$$\frac{28.0\ \cancel{\text{g}}}{1\ \text{mol}} \times \frac{1\ \text{kg}}{1000\ \cancel{\text{g}}} = 0.028\ \text{kg/mol}$$

Replace the variables and constants in the root-mean-square speed equation, replacing Joules with the equivalent kg m^2s^{-2}:

$$u_{\text{rms}} = \sqrt{\frac{3RT}{M}}$$

$$u_{\text{rms}} = \sqrt{\frac{3(8.314\ \text{J/mol K})(303\ \text{K})}{(0.028\ \text{kg/mol})}} = \sqrt{2.70 \times 10^5\ \text{m}^2\text{s}^{-2}} = 519\ \text{m/s}$$

Check Your Learning

Calculate the root-mean-square speed for a mole of oxygen molecules at −23 °C.

Answer:

441 m/s

If the temperature of a gas increases, its KE_{avg} increases, more molecules have higher speeds and fewer molecules have lower speeds, and the distribution shifts toward higher speeds overall, that is, to the right. If temperature decreases, KE_{avg} decreases, more molecules have lower speeds and fewer molecules have higher speeds, and the distribution shifts toward lower speeds overall, that is, to the left. This behavior is illustrated for nitrogen gas in Figure 9.33.

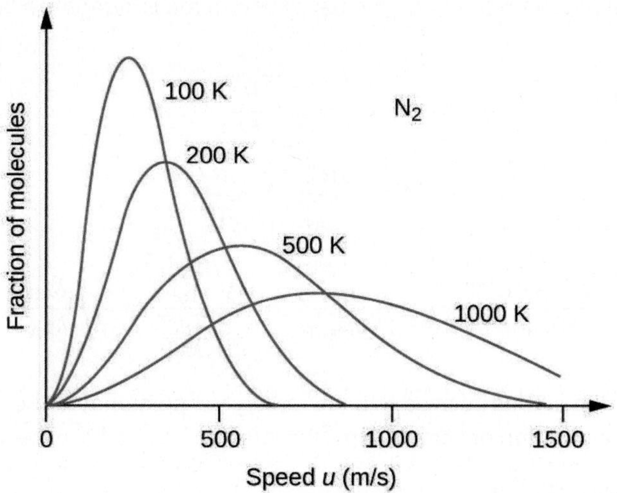

FIGURE 9.33 The molecular speed distribution for nitrogen gas (N_2) shifts to the right and flattens as the temperature increases; it shifts to the left and heightens as the temperature decreases.

At a given temperature, all gases have the same KE_{avg} for their molecules. Gases composed of lighter molecules have more high-speed particles and a higher u_{rms}, with a speed distribution that peaks at relatively higher speeds. Gases consisting of heavier molecules have more low-speed particles, a lower u_{rms}, and a speed distribution that peaks at relatively lower speeds. This trend is demonstrated by the data for a series of noble gases shown in Figure 9.34.

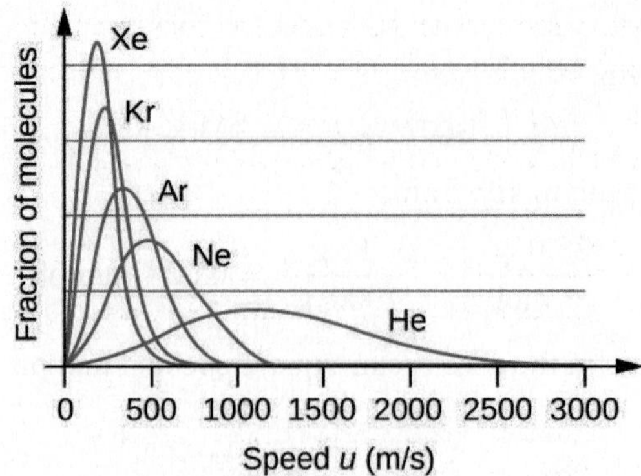

FIGURE 9.34 molecular speed is directly related to molecular mass. At a given temperature, lighter molecules move faster on average than heavier molecules.

LINK TO LEARNING

The gas simulator (http://openstax.org/l/16MolecVelocity) may be used to examine the effect of temperature on molecular speeds. Examine the simulator's "energy histograms" (molecular speed distributions) and "species information" (which gives average speed values) for molecules of different masses at various temperatures.

The Kinetic-Molecular Theory Explains the Behavior of Gases, Part II

According to Graham's law, the molecules of a gas are in rapid motion and the molecules themselves are small. The average distance between the molecules of a gas is large compared to the size of the molecules. As a consequence, gas molecules can move past each other easily and diffuse at relatively fast rates.

The rate of effusion of a gas depends directly on the (average) speed of its molecules:

$$\text{effusion rate} \propto u_{\text{rms}}$$

Using this relation, and the equation relating molecular speed to mass, Graham's law may be easily derived as shown here:

$$u_{\text{rms}} = \sqrt{\frac{3RT}{M}}$$

$$M = \frac{3RT}{u_{\text{rms}}^2} = \frac{3RT}{\bar{u}^2}$$

$$\frac{\text{effusion rate A}}{\text{effusion rate B}} = \frac{u_{\text{rms A}}}{u_{\text{rms B}}} = \frac{\sqrt{\frac{3RT}{M_{\text{A}}}}}{\sqrt{\frac{3RT}{M_{\text{B}}}}} = \sqrt{\frac{M_{\text{B}}}{M_{\text{A}}}}$$

The ratio of the rates of effusion is thus derived to be inversely proportional to the ratio of the square roots of their masses. This is the same relation observed experimentally and expressed as Graham's law.

9.6 Non-Ideal Gas Behavior

LEARNING OBJECTIVES

By the end of this section, you will be able to:

- Describe the physical factors that lead to deviations from ideal gas behavior
- Explain how these factors are represented in the van der Waals equation
- Define compressibility (Z) and describe how its variation with pressure reflects non-ideal behavior
- Quantify non-ideal behavior by comparing computations of gas properties using the ideal gas law and the van der Waals equation

Thus far, the ideal gas law, $PV = nRT$, has been applied to a variety of different types of problems, ranging from reaction stoichiometry and empirical and molecular formula problems to determining the density and molar mass of a gas. As mentioned in the previous modules of this chapter, however, the behavior of a gas is often non-ideal, meaning that the observed relationships between its pressure, volume, and temperature are not accurately described by the gas laws. In this section, the reasons for these deviations from ideal gas behavior are considered.

One way in which the accuracy of $PV = nRT$ can be judged is by comparing the actual volume of 1 mole of gas (its molar volume, V_m) to the molar volume of an ideal gas at the same temperature and pressure. This ratio is called the **compressibility factor (Z)** with:

$$Z = \frac{\text{molar volume of gas at same } T \text{ and } P}{\text{molar volume of ideal gas at same } T \text{ and } P} = \left(\frac{PV_m}{RT}\right)_{\text{measured}}$$

Ideal gas behavior is therefore indicated when this ratio is equal to 1, and any deviation from 1 is an indication of non-ideal behavior. Figure 9.35 shows plots of Z over a large pressure range for several common gases.

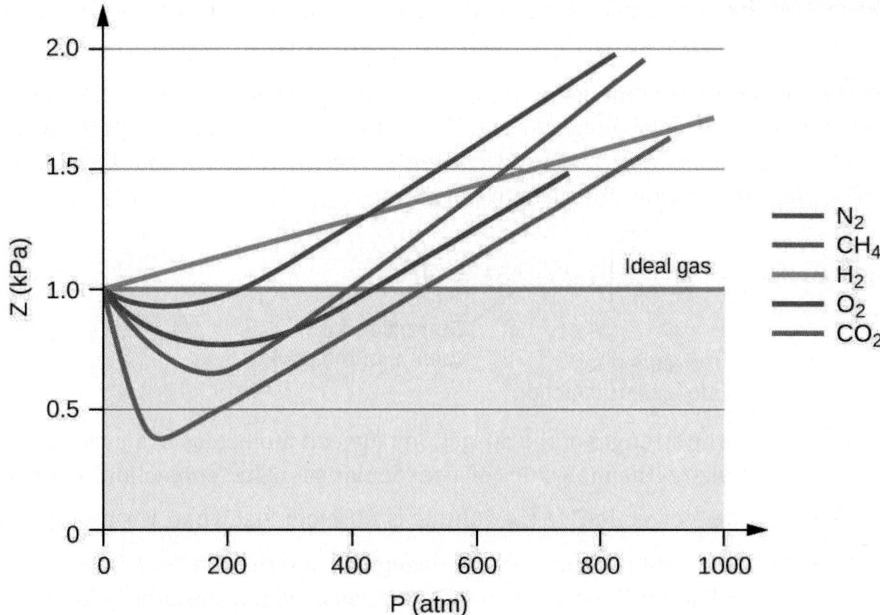

FIGURE 9.35 A graph of the compressibility factor (Z) vs. pressure shows that gases can exhibit significant deviations from the behavior predicted by the ideal gas law.

As is apparent from Figure 9.35, the ideal gas law does not describe gas behavior well at relatively high pressures. To determine why this is, consider the differences between real gas properties and what is expected of a hypothetical ideal gas.

Particles of a hypothetical ideal gas have no significant volume and do not attract or repel each other. In general, real gases approximate this behavior at relatively low pressures and high temperatures. However, at high pressures, the molecules of a gas are crowded closer together, and the amount of empty space between the molecules is reduced. At these higher pressures, the volume of the gas molecules themselves becomes appreciable relative to the total volume occupied by the gas. The gas therefore becomes less compressible at these high pressures, and although its volume continues to decrease with increasing pressure, this decrease is not *proportional* as predicted by Boyle's law.

At relatively low pressures, gas molecules have practically no attraction for one another because they are (on average) so far apart, and they behave almost like particles of an ideal gas. At higher pressures, however, the force of attraction is also no longer insignificant. This force pulls the molecules a little closer together, slightly decreasing the pressure (if the volume is constant) or decreasing the volume (at constant pressure) (Figure 9.36). This change is more pronounced at low temperatures because the molecules have lower KE relative to the attractive forces, and so they are less effective in overcoming these attractions after colliding with one another.

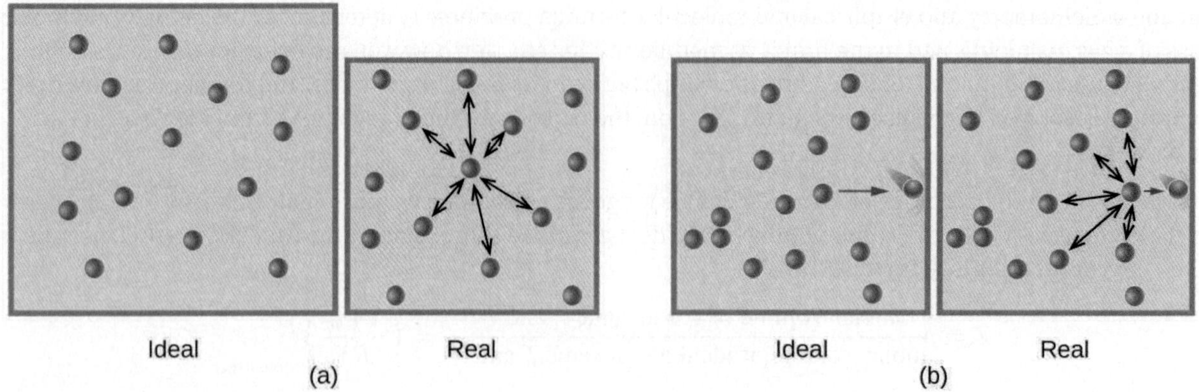

FIGURE 9.36 (a) Attractions between gas molecules serve to decrease the gas volume at constant pressure compared to an ideal gas whose molecules experience no attractive forces. (b) These attractive forces will decrease the force of collisions between the molecules and container walls, therefore reducing the pressure exerted at constant volume compared to an ideal gas.

There are several different equations that better approximate gas behavior than does the ideal gas law. The first, and simplest, of these was developed by the Dutch scientist Johannes van der Waals in 1879. The **van der Waals equation** improves upon the ideal gas law by adding two terms: one to account for the volume of the gas molecules and another for the attractive forces between them.

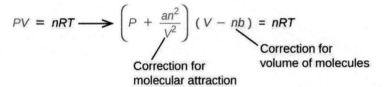

$$PV = nRT \longrightarrow \left(P + \frac{an^2}{V^2}\right)(V - nb) = nRT$$

Correction for molecular attraction

Correction for volume of molecules

The constant a corresponds to the strength of the attraction between molecules of a particular gas, and the constant b corresponds to the size of the molecules of a particular gas. The "correction" to the pressure term in the ideal gas law is $\frac{n^2 a}{V^2}$, and the "correction" to the volume is nb. Note that when V is relatively large and n is relatively small, both of these correction terms become negligible, and the van der Waals equation reduces to the ideal gas law, $PV = nRT$. Such a condition corresponds to a gas in which a relatively low number of molecules is occupying a relatively large volume, that is, a gas at a relatively low pressure. Experimental values for the van der Waals constants of some common gases are given in Table 9.3.

Values of van der Waals Constants for Some Common Gases

Gas	a (L^2 atm/mol^2)	b (L/mol)
N_2	1.39	0.0391
O_2	1.36	0.0318
CO_2	3.59	0.0427
H_2O	5.46	0.0305
He	0.0342	0.0237
CCl_4	20.4	0.1383

TABLE 9.3

At low pressures, the correction for intermolecular attraction, *a*, is more important than the one for molecular volume, *b*. At high pressures and small volumes, the correction for the volume of the molecules becomes important because the molecules themselves are incompressible and constitute an appreciable fraction of the total volume. At some intermediate pressure, the two corrections have opposing influences and the gas appears to follow the relationship given by $PV = nRT$ over a small range of pressures. This behavior is reflected by the "dips" in several of the compressibility curves shown in Figure 9.35. The attractive force between molecules initially makes the gas more compressible than an ideal gas, as pressure is raised (Z decreases with increasing *P*). At very high pressures, the gas becomes less compressible (Z increases with *P*), as the gas molecules begin to occupy an increasingly significant fraction of the total gas volume.

Strictly speaking, the ideal gas equation functions well when intermolecular attractions between gas molecules are negligible and the gas molecules themselves do not occupy an appreciable part of the whole volume. These criteria are satisfied under conditions of *low pressure and high temperature*. Under such conditions, the gas is said to behave ideally, and deviations from the gas laws are small enough that they may be disregarded—this is, however, very often not the case.

 **EXAMPLE 9.24**

Comparison of Ideal Gas Law and van der Waals Equation

A 4.25-L flask contains 3.46 mol CO_2 at 229 °C. Calculate the pressure of this sample of CO_2:

(a) from the ideal gas law

(b) from the van der Waals equation

(c) Explain the reason(s) for the difference.

Solution

(a) From the ideal gas law:

$$P = \frac{nRT}{V} = \frac{3.46\,\cancel{\text{mol}} \times 0.08206\,\cancel{\text{L}}\,\text{atm}\,\cancel{\text{mol}^{-1}}\,\cancel{\text{K}^{-1}} \times 502\,\cancel{\text{K}}}{4.25\,\cancel{\text{L}}} = 33.5\text{ atm}$$

(b) From the van der Waals equation:

$$\left(P + \frac{n^2a}{V^2}\right) \times (V - nb) = nRT \longrightarrow P = \frac{nRT}{(V - nb)} - \frac{n^2a}{V^2}$$

$$P = \frac{3.46\text{ mol} \times 0.08206\text{ L atm mol}^{-1}\text{ K}^{-1} \times 502\text{ K}}{\left(4.25\text{ L} - 3.46\text{ mol} \times 0.0427\text{ L mol}^{-1}\right)} - \frac{(3.46\text{ mol})^2 \times 3.59\text{ L}^2\text{ atm mol}^2}{(4.25\text{ L})^2}$$

This finally yields $P = 32.4$ atm.

(c) This is not very different from the value from the ideal gas law because the pressure is not very high and the temperature is not very low. The value is somewhat different because CO_2 molecules do have some volume and attractions between molecules, and the ideal gas law assumes they do not have volume or attractions.

Check your Learning

A 560-mL flask contains 21.3 g N_2 at 145 °C. Calculate the pressure of N_2:

(a) from the ideal gas law

(b) from the van der Waals equation

(c) Explain the reason(s) for the difference.

Answer:

(a) 46.562 atm; (b) 46.594 atm; (c) The van der Waals equation takes into account the volume of the gas molecules themselves as well as intermolecular attractions.

Key Terms

absolute zero temperature at which the volume of a gas would be zero according to Charles's law.

Amontons's law (also, Gay-Lussac's law) pressure of a given number of moles of gas is directly proportional to its kelvin temperature when the volume is held constant

atmosphere (atm) unit of pressure; 1 atm = 101,325 Pa

Avogadro's law volume of a gas at constant temperature and pressure is proportional to the number of gas molecules

bar (bar or b) unit of pressure; 1 bar = 100,000 Pa

barometer device used to measure atmospheric pressure

Boyle's law volume of a given number of moles of gas held at constant temperature is inversely proportional to the pressure under which it is measured

Charles's law volume of a given number of moles of gas is directly proportional to its kelvin temperature when the pressure is held constant

compressibility factor (Z) ratio of the experimentally measured molar volume for a gas to its molar volume as computed from the ideal gas equation

Dalton's law of partial pressures total pressure of a mixture of ideal gases is equal to the sum of the partial pressures of the component gases

diffusion movement of an atom or molecule from a region of relatively high concentration to one of relatively low concentration (discussed in this chapter with regard to gaseous species, but applicable to species in any phase)

effusion transfer of gaseous atoms or molecules from a container to a vacuum through very small openings

Graham's law of effusion rates of diffusion and effusion of gases are inversely proportional to the square roots of their molecular masses

hydrostatic pressure pressure exerted by a fluid due to gravity

ideal gas hypothetical gas whose physical properties are perfectly described by the gas laws

ideal gas constant (*R*) constant derived from the ideal gas equation $R = 0.08206$ L atm mol^{-1} K^{-1} or 8.314 L kPa mol^{-1} K^{-1}

ideal gas law relation between the pressure, volume, amount, and temperature of a gas under conditions derived by combination of the simple gas laws

kinetic molecular theory theory based on simple principles and assumptions that effectively explains ideal gas behavior

manometer device used to measure the pressure of a gas trapped in a container

mean free path average distance a molecule travels between collisions

mole fraction (*X*) concentration unit defined as the ratio of the molar amount of a mixture component to the total number of moles of all mixture components

partial pressure pressure exerted by an individual gas in a mixture

pascal (Pa) SI unit of pressure; 1 Pa = 1 N/m^2

pounds per square inch (psi) unit of pressure common in the US

pressure force exerted per unit area

rate of diffusion amount of gas diffusing through a given area over a given time

root mean square speed (u_{rms}) measure of average speed for a group of particles calculated as the square root of the average squared speed

standard conditions of temperature and pressure (STP) 273.15 K (0 °C) and 1 atm (101.325 kPa)

standard molar volume volume of 1 mole of gas at STP, approximately 22.4 L for gases behaving ideally

torr unit of pressure; 1 torr = $\frac{1}{760}$ atm

van der Waals equation modified version of the ideal gas equation containing additional terms to account for non-ideal gas behavior

vapor pressure of water pressure exerted by water vapor in equilibrium with liquid water in a closed container at a specific temperature

Key Equations

$$P = \frac{F}{A}$$

$$p = h\rho g$$

$$PV = nRT$$

$$P_{Total} = P_A + P_B + P_C + \ldots = \Sigma_i P_i$$

$$P_A = X_A P_{Total}$$

$$X_A = \frac{n_A}{n_{Total}}$$

$$\text{rate of diffusion} = \frac{\text{amount of gas passing through an area}}{\text{unit of time}}$$

$$\frac{\text{rate of effusion of gas A}}{\text{rate of effusion of gas B}} = \frac{\sqrt{m_B}}{\sqrt{m_A}} = \frac{\sqrt{\mathcal{M}_B}}{\sqrt{\mathcal{M}_A}}$$

$$u_{\text{rms}} = \sqrt{\overline{u^2}} = \sqrt{\frac{u_1^2+u_2^2+u_3^2+u_4^2+\ldots}{n}}$$

$$\text{KE}_{\text{avg}} = \frac{3}{2}RT$$

$$u_{\text{rms}} = \sqrt{\frac{3RT}{M}}$$

$$\text{Z} = \frac{\text{molar volume of gas at same } T \text{ and } P}{\text{molar volume of ideal gas at same } T \text{ and } P} = \left(\frac{P \times V_m}{R \times T}\right)_{\text{measured}}$$

$$\left(P + \frac{n^2 a}{V^2}\right) \times (V - nb) = nRT$$

Summary

9.1 Gas Pressure

Gases exert pressure, which is force per unit area. The pressure of a gas may be expressed in the SI unit of pascal or kilopascal, as well as in many other units including torr, atmosphere, and bar. Atmospheric pressure is measured using a barometer; other gas pressures can be measured using one of several types of manometers.

9.2 Relating Pressure, Volume, Amount, and Temperature: The Ideal Gas Law

The behavior of gases can be described by several laws based on experimental observations of their properties. The pressure of a given amount of gas is directly proportional to its absolute temperature, provided that the volume does not change (Amontons's law). The volume of a given gas sample is directly proportional to its absolute temperature at constant pressure (Charles's law). The volume of a given amount of gas is inversely proportional to its pressure when temperature is held constant (Boyle's law). Under the same conditions of temperature and pressure, equal volumes of all gases contain the same number of molecules (Avogadro's law).

The equations describing these laws are special cases of the ideal gas law, $PV = nRT$, where P is the pressure of the gas, V is its volume, n is the number of moles of the gas, T is its kelvin temperature, and R is the ideal (universal) gas constant.

9.3 Stoichiometry of Gaseous Substances, Mixtures, and Reactions

The ideal gas law can be used to derive a number of convenient equations relating directly measured quantities to properties of interest for gaseous substances and mixtures. Appropriate rearrangement of the ideal gas equation may be made to permit the calculation of gas densities and molar masses. Dalton's law of partial pressures may be used to relate measured gas pressures for gaseous mixtures to their compositions. Avogadro's law may be used in stoichiometric computations for chemical reactions involving gaseous reactants or products.

9.4 Effusion and Diffusion of Gases

Gaseous atoms and molecules move freely and randomly through space. Diffusion is the process whereby gaseous atoms and molecules are transferred from regions of relatively high concentration to regions of relatively low concentration. Effusion is a similar process in which gaseous species pass from a container to a vacuum through very small orifices. The rates of effusion of gases are inversely proportional to the square roots of their densities or to the square roots of their atoms/molecules' masses (Graham's law).

9.5 The Kinetic-Molecular Theory

The kinetic molecular theory is a simple but very effective model that effectively explains ideal gas behavior. The theory assumes that gases consist of widely separated molecules of negligible volume that are in constant motion, colliding elastically with one another and the walls of their container with average speeds determined by their absolute temperatures. The individual molecules of a gas exhibit a range of speeds, the distribution of these speeds being dependent on the temperature of the gas and the mass of its molecules.

9.6 Non-Ideal Gas Behavior

Gas molecules possess a finite volume and experience forces of attraction for one another. Consequently, gas behavior is not necessarily described well by the ideal gas law. Under conditions of low pressure and high temperature, these factors are negligible, the ideal gas equation is an accurate description of gas behavior, and the gas is said to exhibit ideal behavior. However, at lower temperatures and higher pressures, corrections for molecular volume and molecular attractions are required to account for finite molecular size and attractive forces. The van der Waals equation is a modified version of the ideal gas law that can be used to account for the non-ideal behavior of gases under these conditions.

Exercises

9.1 Gas Pressure

1. Why are sharp knives more effective than dull knives? (Hint: Think about the definition of pressure.)
2. Why do some small bridges have weight limits that depend on how many wheels or axles the crossing vehicle has?
3. Why should you roll or belly-crawl rather than walk across a thinly-frozen pond?
4. A typical barometric pressure in Redding, California, is about 750 mm Hg. Calculate this pressure in atm and kPa.
5. A typical barometric pressure in Denver, Colorado, is 615 mm Hg. What is this pressure in atmospheres and kilopascals?
6. A typical barometric pressure in Kansas City is 740 torr. What is this pressure in atmospheres, in millimeters of mercury, and in kilopascals?
7. Canadian tire pressure gauges are marked in units of kilopascals. What reading on such a gauge corresponds to 32 psi?
8. During the Viking landings on Mars, the atmospheric pressure was determined to be on the average about 6.50 millibars (1 bar = 0.987 atm). What is that pressure in torr and kPa?
9. The pressure of the atmosphere on the surface of the planet Venus is about 88.8 atm. Compare that pressure in psi to the normal pressure on earth at sea level in psi.
10. A medical laboratory catalog describes the pressure in a cylinder of a gas as 14.82 MPa. What is the pressure of this gas in atmospheres and torr?
11. Consider this scenario and answer the following questions: On a mid-August day in the northeastern United States, the following information appeared in the local newspaper: atmospheric pressure at sea level 29.97 in. Hg, 1013.9 mbar.
    (a) What was the pressure in kPa?
    (b) The pressure near the seacoast in the northeastern United States is usually reported near 30.0 in. Hg. During a hurricane, the pressure may fall to near 28.0 in. Hg. Calculate the drop in pressure in torr.
12. Why is it necessary to use a nonvolatile liquid in a barometer or manometer?

13. The pressure of a sample of gas is measured at sea level with a closed-end manometer. The liquid in the manometer is mercury. Determine the pressure of the gas in:

(a) torr
(b) Pa
(c) bar

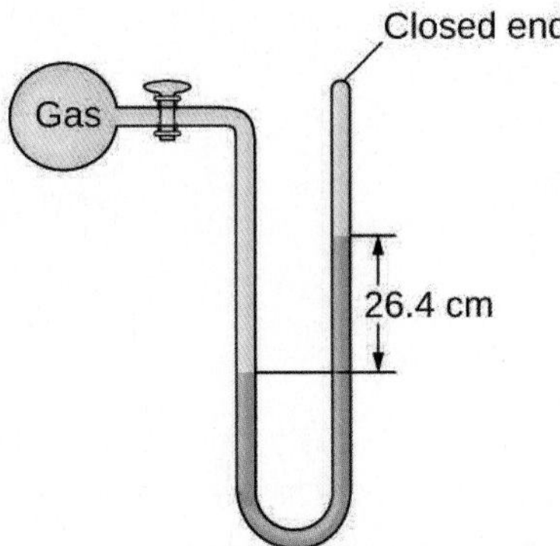

14. The pressure of a sample of gas is measured with an open-end manometer, partially shown to the right. The liquid in the manometer is mercury. Assuming atmospheric pressure is 29.92 in. Hg, determine the pressure of the gas in:

(a) torr
(b) Pa
(c) bar

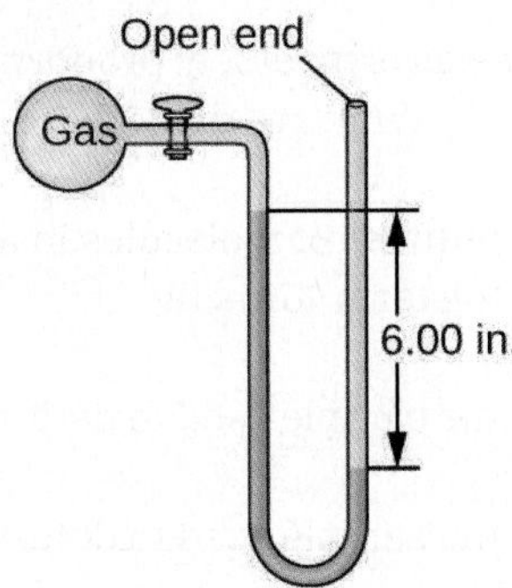

15. The pressure of a sample of gas is measured at sea level with an open-end mercury manometer. Assuming atmospheric pressure is 760.0 mm Hg, determine the pressure of the gas in:

(a) mm Hg
(b) atm
(c) kPa

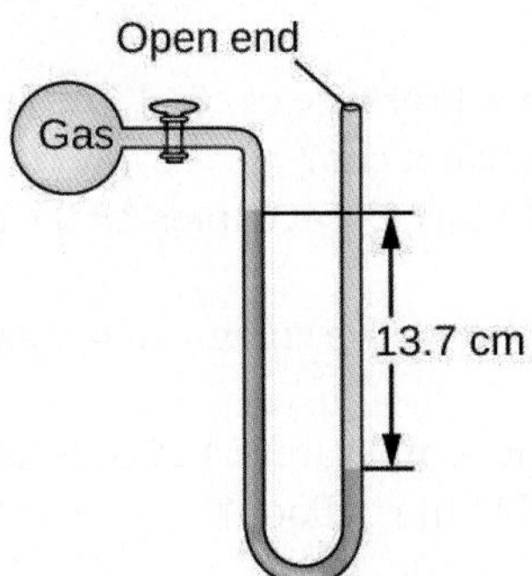

16. The pressure of a sample of gas is measured at sea level with an open-end mercury manometer. Assuming atmospheric pressure is 760 mm Hg, determine the pressure of the gas in:
(a) mm Hg
(b) atm
(c) kPa

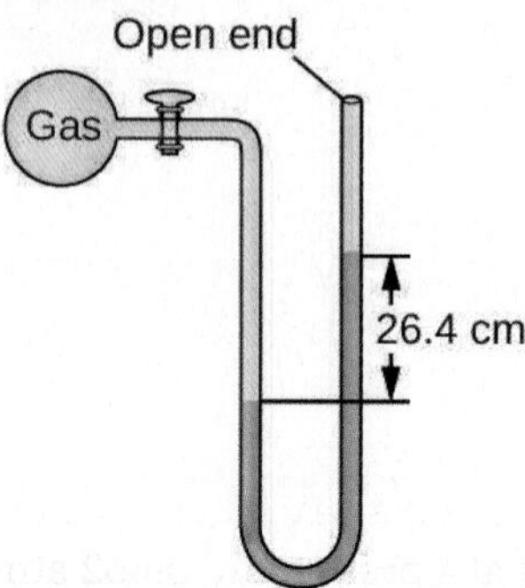

17. How would the use of a volatile liquid affect the measurement of a gas using open-ended manometers vs. closed-end manometers?

9.2 Relating Pressure, Volume, Amount, and Temperature: The Ideal Gas Law

18. Sometimes leaving a bicycle in the sun on a hot day will cause a blowout. Why?
19. Explain how the volume of the bubbles exhausted by a scuba diver (Figure 9.16) change as they rise to the surface, assuming that they remain intact.
20. One way to state Boyle's law is "All other things being equal, the pressure of a gas is inversely proportional to its volume." (a) What is the meaning of the term "inversely proportional?" (b) What are the "other things" that must be equal?
21. An alternate way to state Avogadro's law is "All other things being equal, the number of molecules in a gas is directly proportional to the volume of the gas." (a) What is the meaning of the term "directly proportional?" (b) What are the "other things" that must be equal?
22. How would the graph in Figure 9.12 change if the number of moles of gas in the sample used to determine the curve were doubled?
23. How would the graph in Figure 9.13 change if the number of moles of gas in the sample used to determine the curve were doubled?
24. In addition to the data found in Figure 9.13, what other information do we need to find the mass of the sample of air used to determine the graph?
25. Determine the volume of 1 mol of CH_4 gas at 150 K and 1 atm, using Figure 9.12.
26. Determine the pressure of the gas in the syringe shown in Figure 9.13 when its volume is 12.5 mL, using:
(a) the appropriate graph
(b) Boyle's law
27. A spray can is used until it is empty except for the propellant gas, which has a pressure of 1344 torr at 23 °C. If the can is thrown into a fire (T = 475 °C), what will be the pressure in the hot can?
28. What is the temperature of an 11.2-L sample of carbon monoxide, CO, at 744 torr if it occupies 13.3 L at 55 °C and 744 torr?
29. A 2.50-L volume of hydrogen measured at –196 °C is warmed to 100 °C. Calculate the volume of the gas at the higher temperature, assuming no change in pressure.
30. A balloon inflated with three breaths of air has a volume of 1.7 L. At the same temperature and pressure, what is the volume of the balloon if five more same-sized breaths are added to the balloon?

31. A weather balloon contains 8.80 moles of helium at a pressure of 0.992 atm and a temperature of 25 °C at ground level. What is the volume of the balloon under these conditions?

32. The volume of an automobile air bag was 66.8 L when inflated at 25 °C with 77.8 g of nitrogen gas. What was the pressure in the bag in kPa?
33. How many moles of gaseous boron trifluoride, BF_3, are contained in a 4.3410-L bulb at 788.0 K if the pressure is 1.220 atm? How many grams of BF_3?
34. Iodine, I_2, is a solid at room temperature but sublimes (converts from a solid into a gas) when warmed. What is the temperature in a 73.3-mL bulb that contains 0.292 g of I_2 vapor at a pressure of 0.462 atm?
35. How many grams of gas are present in each of the following cases?
(a) 0.100 L of CO_2 at 307 torr and 26 °C
(b) 8.75 L of C_2H_4, at 378.3 kPa and 483 K
(c) 221 mL of Ar at 0.23 torr and −54 °C
36. A high altitude balloon is filled with 1.41×10^4 L of hydrogen at a temperature of 21 °C and a pressure of 745 torr. What is the volume of the balloon at a height of 20 km, where the temperature is −48 °C and the pressure is 63.1 torr?
37. A cylinder of medical oxygen has a volume of 35.4 L, and contains O_2 at a pressure of 151 atm and a temperature of 25 °C. What volume of O_2 does this correspond to at normal body conditions, that is, 1 atm and 37 °C?
38. A large scuba tank (Figure 9.16) with a volume of 18 L is rated for a pressure of 220 bar. The tank is filled at 20 °C and contains enough air to supply 1860 L of air to a diver at a pressure of 2.37 atm (a depth of 45 feet). Was the tank filled to capacity at 20 °C?
39. A 20.0-L cylinder containing 11.34 kg of butane, C_4H_{10}, was opened to the atmosphere. Calculate the mass of the gas remaining in the cylinder if it were opened and the gas escaped until the pressure in the cylinder was equal to the atmospheric pressure, 0.983 atm, and a temperature of 27 °C.
40. While resting, the average 70-kg human male consumes 14 L of pure O_2 per hour at 25 °C and 100 kPa. How many moles of O_2 are consumed by a 70 kg man while resting for 1.0 h?
41. For a given amount of gas showing ideal behavior, draw labeled graphs of:
(a) the variation of P with V
(b) the variation of V with T
(c) the variation of P with T
(d) the variation of $\frac{1}{P}$ with V
42. A liter of methane gas, CH_4, at STP contains more atoms of hydrogen than does a liter of pure hydrogen gas, H_2, at STP. Using Avogadro's law as a starting point, explain why.
43. The effect of chlorofluorocarbons (such as CCl_2F_2) on the depletion of the ozone layer is well known. The use of substitutes, such as $CH_3CH_2F(g)$, for the chlorofluorocarbons, has largely corrected the problem. Calculate the volume occupied by 10.0 g of each of these compounds at STP:
(a) $CCl_2F_2(g)$
(b) $CH_3CH_2F(g)$
44. As 1 g of the radioactive element radium decays over 1 year, it produces 1.16×10^{18} alpha particles (helium nuclei). Each alpha particle becomes an atom of helium gas. What is the pressure in pascal of the helium gas produced if it occupies a volume of 125 mL at a temperature of 25 °C?
45. A balloon with a volume of 100.21 L at 21 °C and 0.981 atm is released and just barely clears the top of Mount Crumpit in British Columbia. If the final volume of the balloon is 144.53 L at a temperature of 5.24 °C, what is the pressure experienced by the balloon as it clears Mount Crumpet?
46. If the temperature of a fixed amount of a gas is doubled at constant volume, what happens to the pressure?
47. If the volume of a fixed amount of a gas is tripled at constant temperature, what happens to the pressure?

9.3 Stoichiometry of Gaseous Substances, Mixtures, and Reactions

48. What is the density of laughing gas, dinitrogen monoxide, N_2O, at a temperature of 325 K and a pressure of 113.0 kPa?

49. Calculate the density of Freon 12, CF_2Cl_2, at 30.0 °C and 0.954 atm.

50. Which is denser at the same temperature and pressure, dry air or air saturated with water vapor? Explain.

51. A cylinder of $O_2(g)$ used in breathing by patients with emphysema has a volume of 3.00 L at a pressure of 10.0 atm. If the temperature of the cylinder is 28.0 °C, what mass of oxygen is in the cylinder?

52. What is the molar mass of a gas if 0.0494 g of the gas occupies a volume of 0.100 L at a temperature 26 °C and a pressure of 307 torr?

53. What is the molar mass of a gas if 0.281 g of the gas occupies a volume of 125 mL at a temperature 126 °C and a pressure of 777 torr?

54. How could you show experimentally that the molecular formula of propene is C_3H_6, not CH_2?

55. The density of a certain gaseous fluoride of phosphorus is 3.93 g/L at STP. Calculate the molar mass of this fluoride and determine its molecular formula.

56. Consider this question: What is the molecular formula of a compound that contains 39% C, 45% N, and 16% H if 0.157 g of the compound occupies 125 mL with a pressure of 99.5 kPa at 22 °C?
(a) Outline the steps necessary to answer the question.
(b) Answer the question.

57. A 36.0–L cylinder of a gas used for calibration of blood gas analyzers in medical laboratories contains 350 g CO_2, 805 g O_2, and 4,880 g N_2. At 25 degrees C, what is the pressure in the cylinder in atmospheres?

58. A cylinder of a gas mixture used for calibration of blood gas analyzers in medical laboratories contains 5.0% CO_2, 12.0% O_2, and the remainder N_2 at a total pressure of 146 atm. What is the partial pressure of each component of this gas? (The percentages given indicate the percent of the total pressure that is due to each component.)

59. A sample of gas isolated from unrefined petroleum contains 90.0% CH_4, 8.9% C_2H_6, and 1.1% C_3H_8 at a total pressure of 307.2 kPa. What is the partial pressure of each component of this gas? (The percentages given indicate the percent of the total pressure that is due to each component.)

60. A mixture of 0.200 g of H_2, 1.00 g of N_2, and 0.820 g of Ar is stored in a closed container at STP. Find the volume of the container, assuming that the gases exhibit ideal behavior.

61. Most mixtures of hydrogen gas with oxygen gas are explosive. However, a mixture that contains less than 3.0 % O_2 is not. If enough O_2 is added to a cylinder of H_2 at 33.2 atm to bring the total pressure to 34.5 atm, is the mixture explosive?

62. A commercial mercury vapor analyzer can detect, in air, concentrations of gaseous Hg atoms (which are poisonous) as low as 2×10^{-6} mg/L of air. At this concentration, what is the partial pressure of gaseous mercury if the atmospheric pressure is 733 torr at 26 °C?

63. A sample of carbon monoxide was collected over water at a total pressure of 756 torr and a temperature of 18 °C. What is the pressure of the carbon monoxide? (See Table 9.2 for the vapor pressure of water.)

64. In an experiment in a general chemistry laboratory, a student collected a sample of a gas over water. The volume of the gas was 265 mL at a pressure of 753 torr and a temperature of 27 °C. The mass of the gas was 0.472 g. What was the molar mass of the gas?

65. Joseph Priestley first prepared pure oxygen by heating mercuric oxide, HgO:
$2HgO(s) \longrightarrow 2Hg(l) + O_2(g)$
(a) Outline the steps necessary to answer the following question: What volume of O_2 at 23 °C and 0.975 atm is produced by the decomposition of 5.36 g of HgO?
(b) Answer the question.

66. Cavendish prepared hydrogen in 1766 by the novel method of passing steam through a red-hot gun barrel:
$4H_2O(g) + 3Fe(s) \longrightarrow Fe_3O_4(s) + 4H_2(g)$
(a) Outline the steps necessary to answer the following question: What volume of H_2 at a pressure of 745 torr and a temperature of 20 °C can be prepared from the reaction of 15.0 g of H_2O?
(b) Answer the question.

67. The chlorofluorocarbon CCl_2F_2 can be recycled into a different compound by reaction with hydrogen to produce $CH_2F_2(g)$, a compound useful in chemical manufacturing:
$$CCl_2F_2(g) + 4H_2(g) \longrightarrow CH_2F_2(g) + 2HCl(g)$$
(a) Outline the steps necessary to answer the following question: What volume of hydrogen at 225 atm and 35.5 °C would be required to react with 1 ton (1.000×10^3 kg) of CCl_2F_2?
(b) Answer the question.

68. Automobile air bags are inflated with nitrogen gas, which is formed by the decomposition of solid sodium azide (NaN_3). The other product is sodium metal. Calculate the volume of nitrogen gas at 27 °C and 756 torr formed by the decomposition of 125 g of sodium azide.

69. Lime, CaO, is produced by heating calcium carbonate, $CaCO_3$; carbon dioxide is the other product.
(a) Outline the steps necessary to answer the following question: What volume of carbon dioxide at 875 K and 0.966 atm is produced by the decomposition of 1 ton (1.000×10^3 kg) of calcium carbonate?
(b) Answer the question.

70. Before small batteries were available, carbide lamps were used for bicycle lights. Acetylene gas, C_2H_2, and solid calcium hydroxide were formed by the reaction of calcium carbide, CaC_2, with water. The ignition of the acetylene gas provided the light. Currently, the same lamps are used by some cavers, and calcium carbide is used to produce acetylene for carbide cannons.
(a) Outline the steps necessary to answer the following question: What volume of C_2H_2 at 1.005 atm and 12.2 °C is formed by the reaction of 15.48 g of CaC_2 with water?
(b) Answer the question.

71. Calculate the volume of oxygen required to burn 12.00 L of ethane gas, C_2H_6, to produce carbon dioxide and water, if the volumes of C_2H_6 and O_2 are measured under the same conditions of temperature and pressure.

72. What volume of O_2 at STP is required to oxidize 8.0 L of NO at STP to NO_2? What volume of NO_2 is produced at STP?

73. Consider the following questions:
(a) What is the total volume of the $CO_2(g)$ and $H_2O(g)$ at 600 °C and 0.888 atm produced by the combustion of 1.00 L of $C_2H_6(g)$ measured at STP?
(b) What is the partial pressure of H_2O in the product gases?

74. Methanol, CH_3OH, is produced industrially by the following reaction:
$$CO(g) + 2H_2(g) \xrightarrow{\text{copper catalyst 300 °C, 300 atm}} CH_3OH(g)$$
Assuming that the gases behave as ideal gases, find the ratio of the total volume of the reactants to the final volume.

75. What volume of oxygen at 423.0 K and a pressure of 127.4 kPa is produced by the decomposition of 129.7 g of BaO_2 to BaO and O_2?

76. A 2.50-L sample of a colorless gas at STP decomposed to give 2.50 L of N_2 and 1.25 L of O_2 at STP. What is the colorless gas?

77. Ethanol, C_2H_5OH, is produced industrially from ethylene, C_2H_4, by the following sequence of reactions:
$$3C_2H_4 + 2H_2SO_4 \longrightarrow C_2H_5HSO_4 + (C_2H_5)_2SO_4$$
$$C_2H_5HSO_4 + (C_2H_5)_2SO_4 + 3H_2O \longrightarrow 3C_2H_5OH + 2H_2SO_4$$
What volume of ethylene at STP is required to produce 1.000 metric ton (1000 kg) of ethanol if the overall yield of ethanol is 90.1%?

78. One molecule of hemoglobin will combine with four molecules of oxygen. If 1.0 g of hemoglobin combines with 1.53 mL of oxygen at body temperature (37 °C) and a pressure of 743 torr, what is the molar mass of hemoglobin?

79. A sample of a compound of xenon and fluorine was confined in a bulb with a pressure of 18 torr. Hydrogen was added to the bulb until the pressure was 72 torr. Passage of an electric spark through the mixture produced Xe and HF. After the HF was removed by reaction with solid KOH, the final pressure of xenon and unreacted hydrogen in the bulb was 36 torr. What is the empirical formula of the xenon fluoride in the original sample? (Note: Xenon fluorides contain only one xenon atom per molecule.)

80. One method of analyzing amino acids is the van Slyke method. The characteristic amino groups ($-NH_2$) in protein material are allowed to react with nitrous acid, HNO_2, to form N_2 gas. From the volume of the gas, the amount of amino acid can be determined. A 0.0604-g sample of a biological sample containing glycine, $CH_2(NH_2)COOH$, was analyzed by the van Slyke method and yielded 3.70 mL of N_2 collected over water at a pressure of 735 torr and 29 °C. What was the percentage of glycine in the sample?
$$CH_2\,(NH_2)\,CO_2H + HNO_2 \longrightarrow CH_2\,(OH)\,CO_2H + H_2O + N_2$$

9.4 Effusion and Diffusion of Gases

81. A balloon filled with helium gas takes 6 hours to deflate to 50% of its original volume. How long will it take for an identical balloon filled with the same volume of hydrogen gas (instead of helium) to decrease its volume by 50%?
82. Explain why the numbers of molecules are not identical in the left- and right-hand bulbs shown in the center illustration of Figure 9.27.
83. Starting with the definition of rate of effusion and Graham's finding relating rate and molar mass, show how to derive the Graham's law equation, relating the relative rates of effusion for two gases to their molecular masses.
84. Heavy water, D_2O (molar mass = 20.03 g mol^{-1}), can be separated from ordinary water, H_2O (molar mass = 18.01), as a result of the difference in the relative rates of diffusion of the molecules in the gas phase. Calculate the relative rates of diffusion of H_2O and D_2O.
85. Which of the following gases diffuse more slowly than oxygen? F_2, Ne, N_2O, C_2H_2, NO, Cl_2, H_2S
86. During the discussion of gaseous diffusion for enriching uranium, it was claimed that $^{235}UF_6$ diffuses 0.4% faster than $^{238}UF_6$. Show the calculation that supports this value. The molar mass of $^{235}UF_6$ = 235.043930 + 6 × 18.998403 = 349.034348 g/mol, and the molar mass of $^{238}UF_6$ = 238.050788 + 6 × 18.998403 = 352.041206 g/mol.
87. Calculate the relative rate of diffusion of $^1H_2$ (molar mass 2.0 g/mol) compared with $^2H_2$ (molar mass 4.0 g/mol) and the relative rate of diffusion of O_2 (molar mass 32 g/mol) compared with O_3 (molar mass 48 g/mol).
88. A gas of unknown identity diffuses at a rate of 83.3 mL/s in a diffusion apparatus in which carbon dioxide diffuses at the rate of 102 mL/s. Calculate the molecular mass of the unknown gas.
89. When two cotton plugs, one moistened with ammonia and the other with hydrochloric acid, are simultaneously inserted into opposite ends of a glass tube that is 87.0 cm long, a white ring of NH_4Cl forms where gaseous NH_3 and gaseous HCl first come into contact. $NH_3\,(g) + HCl\,(g) \longrightarrow NH_4Cl\,(s)$ At approximately what distance from the ammonia moistened plug does this occur? (Hint: Calculate the rates of diffusion for both NH_3 and HCl, and find out how much faster NH_3 diffuses than HCl.)

9.5 The Kinetic-Molecular Theory

90. Using the postulates of the kinetic molecular theory, explain why a gas uniformly fills a container of any shape.
91. Can the speed of a given molecule in a gas double at constant temperature? Explain your answer.
92. Describe what happens to the average kinetic energy of ideal gas molecules when the conditions are changed as follows:
(a) The pressure of the gas is increased by reducing the volume at constant temperature.
(b) The pressure of the gas is increased by increasing the temperature at constant volume.
(c) The average speed of the molecules is increased by a factor of 2.
93. The distribution of molecular speeds in a sample of helium is shown in Figure 9.34. If the sample is cooled, will the distribution of speeds look more like that of H_2 or of H_2O? Explain your answer.
94. What is the ratio of the average kinetic energy of a SO_2 molecule to that of an O_2 molecule in a mixture of two gases? What is the ratio of the root mean square speeds, u_{rms}, of the two gases?

95. A 1-L sample of CO initially at STP is heated to 546 K, and its volume is increased to 2 L.
(a) What effect do these changes have on the number of collisions of the molecules of the gas per unit area of the container wall?
(b) What is the effect on the average kinetic energy of the molecules?
(c) What is the effect on the root mean square speed of the molecules?

96. The root mean square speed of H_2 molecules at 25 °C is about 1.6 km/s. What is the root mean square speed of a N_2 molecule at 25 °C?

97. Answer the following questions:
(a) Is the pressure of the gas in the hot-air balloon shown at the opening of this chapter greater than, less than, or equal to that of the atmosphere outside the balloon?
(b) Is the density of the gas in the hot-air balloon shown at the opening of this chapter greater than, less than, or equal to that of the atmosphere outside the balloon?
(c) At a pressure of 1 atm and a temperature of 20 °C, dry air has a density of 1.2256 g/L. What is the (average) molar mass of dry air?
(d) The average temperature of the gas in a hot-air balloon is 1.30×10^2 °F. Calculate its density, assuming the molar mass equals that of dry air.
(e) The lifting capacity of a hot-air balloon is equal to the difference in the mass of the cool air displaced by the balloon and the mass of the gas in the balloon. What is the difference in the mass of 1.00 L of the cool air in part (c) and the hot air in part (d)?
(f) An average balloon has a diameter of 60 feet and a volume of 1.1×10^5 ft^3. What is the lifting power of such a balloon? If the weight of the balloon and its rigging is 500 pounds, what is its capacity for carrying passengers and cargo?
(g) A balloon carries 40.0 gallons of liquid propane (density 0.5005 g/L). What volume of CO_2 and H_2O gas is produced by the combustion of this propane?
(h) A balloon flight can last about 90 minutes. If all of the fuel is burned during this time, what is the approximate rate of heat loss (in kJ/min) from the hot air in the bag during the flight?

98. Show that the ratio of the rate of diffusion of Gas 1 to the rate of diffusion of Gas 2, $\frac{R_1}{R_2}$, is the same at 0 °C and 100 °C.

9.6 Non-Ideal Gas Behavior

99. Graphs showing the behavior of several different gases follow. Which of these gases exhibit behavior significantly different from that expected for ideal gases?

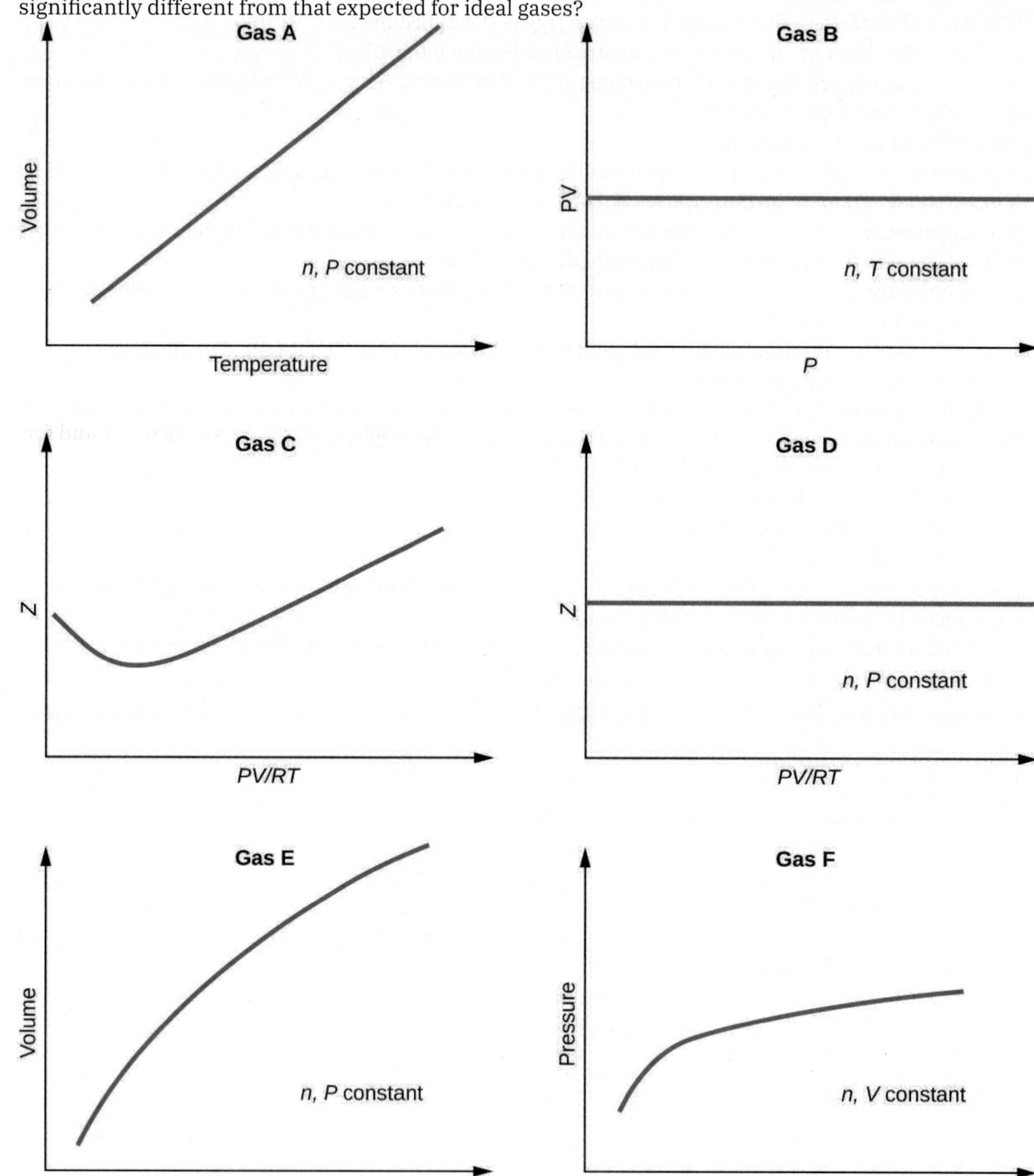

100. Explain why the plot of *PV* for CO_2 differs from that of an ideal gas.

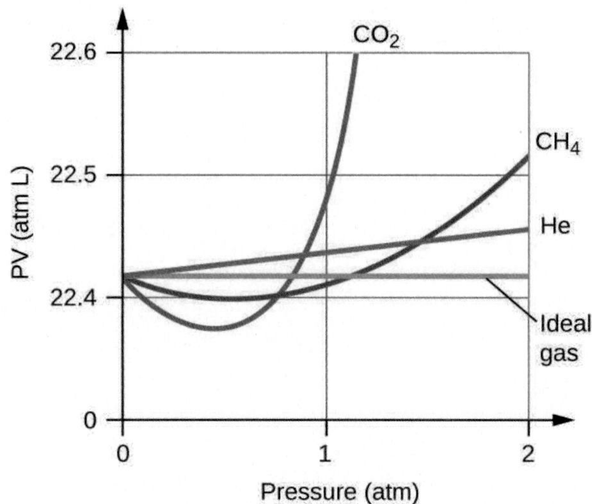

101. Under which of the following sets of conditions does a real gas behave most like an ideal gas, and for which conditions is a real gas expected to deviate from ideal behavior? Explain.
(a) high pressure, small volume
(b) high temperature, low pressure
(c) low temperature, high pressure

102. Describe the factors responsible for the deviation of the behavior of real gases from that of an ideal gas.

103. For which of the following gases should the correction for the molecular volume be largest: CO, CO_2, H_2, He, NH_3, SF_6?

104. A 0.245-L flask contains 0.467 mol CO_2 at 159 °C. Calculate the pressure:
(a) using the ideal gas law
(b) using the van der Waals equation
(c) Explain the reason for the difference.
(d) Identify which correction (that for P or V) is dominant and why.

105. Answer the following questions:
(a) If XX behaved as an ideal gas, what would its graph of Z vs. P look like?
(b) For most of this chapter, we performed calculations treating gases as ideal. Was this justified?
(c) What is the effect of the volume of gas molecules on Z? Under what conditions is this effect small? When is it large? Explain using an appropriate diagram.
(d) What is the effect of intermolecular attractions on the value of Z? Under what conditions is this effect small? When is it large? Explain using an appropriate diagram.
(e) In general, under what temperature conditions would you expect Z to have the largest deviations from the Z for an ideal gas?

CHAPTER 10
Liquids and Solids

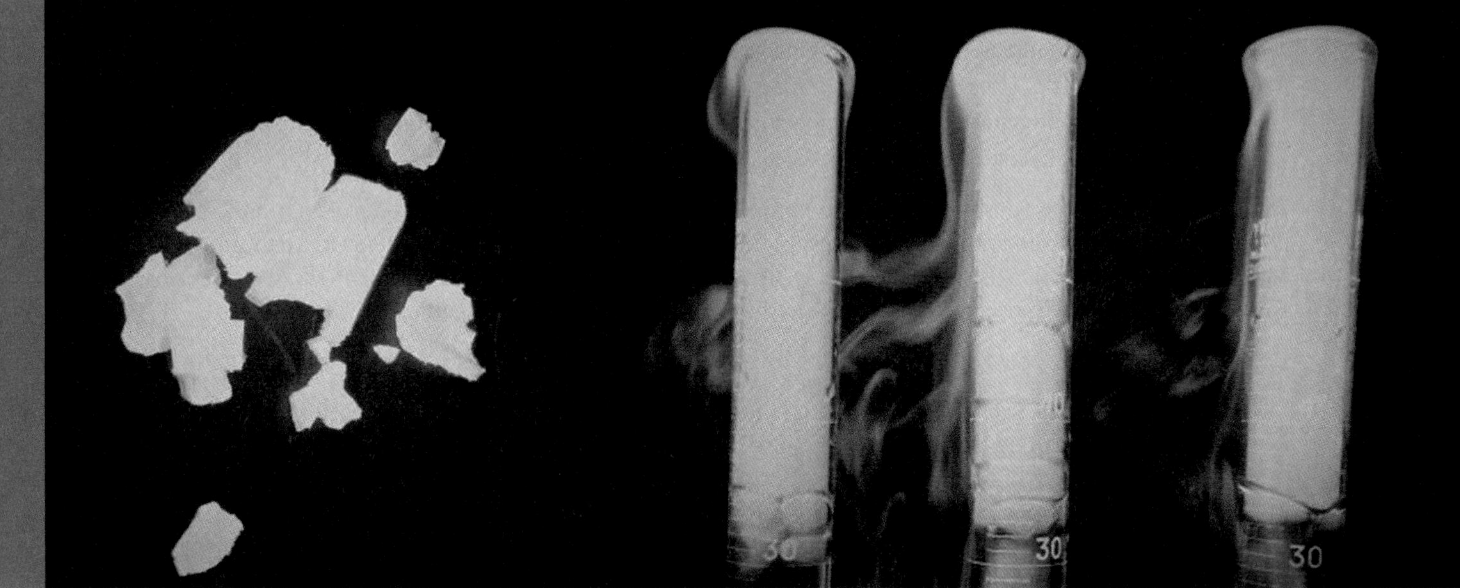

Figure 10.1 Solid carbon dioxide (“dry ice”, left) sublimes vigorously when placed in a liquid (right), cooling the liquid and generating a dense mist of water above the cylinder. (credit: modification of work by Paul Flowers)

CHAPTER OUTLINE

10.1 Intermolecular Forces
10.2 Properties of Liquids
10.3 Phase Transitions
10.4 Phase Diagrams
10.5 The Solid State of Matter
10.6 Lattice Structures in Crystalline Solids

INTRODUCTION Leprosy has been a devastating disease throughout much of human history. Aside from the symptoms and complications of the illness, its social stigma led sufferers to be cast out of communities and isolated in colonies; in some regions this practice lasted well into the twentieth century. At that time, the best potential treatment for leprosy was oil from the chaulmoogra tree, but the oil was extremely thick, causing blisters and making usage painful and ineffective. Healthcare professionals seeking a better application contacted Alice Ball, a young chemist at the University of Hawaii, who had focused her masters thesis on a similar plant. Ball initiated a sequence of procedures (repeated acidification and purification to change the characteristics of the oil and isolate the active substances (esters, discussed later in this text). The "Ball Method" as it later came to be called, became the standard treatment for leprosy for decades. In the liquid and solid states, atomic and molecular interactions are of considerable strength and play an important role in determining a number of physical properties of the substance. For example, the thickness, or viscosity, of the chaulmoogra oil was due to its intermolecular forces. In this chapter, the nature of these interactions and their effects on various physical properties of liquid and solid phases will be examined.

10.1 Intermolecular Forces

LEARNING OBJECTIVES

By the end of this section, you will be able to:

- Describe the types of intermolecular forces possible between atoms or molecules in condensed phases (dispersion forces, dipole-dipole attractions, and hydrogen bonding)
- Identify the types of intermolecular forces experienced by specific molecules based on their structures
- Explain the relation between the intermolecular forces present within a substance and the temperatures associated with changes in its physical state

As was the case for gaseous substances, the kinetic molecular theory may be used to explain the behavior of solids and liquids. In the following description, the term *particle* will be used to refer to an atom, molecule, or ion. Note that we will use the popular phrase "intermolecular attraction" to refer to attractive forces between the particles of a substance, regardless of whether these particles are molecules, atoms, or ions.

Consider these two aspects of the molecular-level environments in solid, liquid, and gaseous matter:

- Particles in a solid are tightly packed together and often arranged in a regular pattern; in a liquid, they are close together with no regular arrangement; in a gas, they are far apart with no regular arrangement.
- Particles in a solid vibrate about fixed positions and do not generally move in relation to one another; in a liquid, they move past each other but remain in essentially constant contact; in a gas, they move independently of one another except when they collide.

The differences in the properties of a solid, liquid, or gas reflect the strengths of the attractive forces between the atoms, molecules, or ions that make up each phase. The phase in which a substance exists depends on the relative extents of its **intermolecular forces** (IMFs) and the kinetic energies (KE) of its molecules. IMFs are the various forces of attraction that may exist between the atoms and molecules of a substance due to electrostatic phenomena, as will be detailed in this module. These forces serve to hold particles close together, whereas the particles' KE provides the energy required to overcome the attractive forces and thus increase the distance between particles. Figure 10.2 illustrates how changes in physical state may be induced by changing the temperature, hence, the average KE, of a given substance.

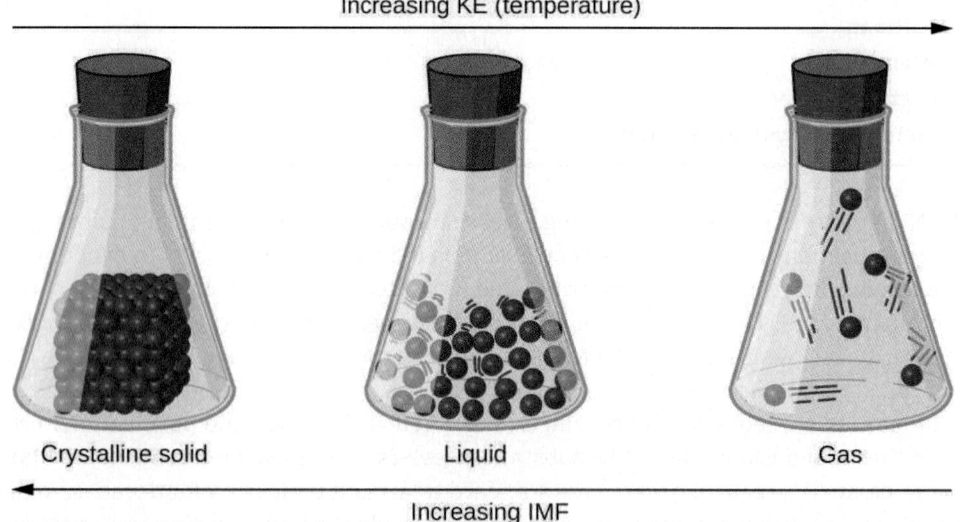

FIGURE 10.2 Transitions between solid, liquid, and gaseous states of a substance occur when conditions of temperature or pressure favor the associated changes in intermolecular forces. (Note: The space between particles in the gas phase is much greater than shown.)

As an example of the processes depicted in this figure, consider a sample of water. When gaseous water is cooled sufficiently, the attractions between H_2O molecules will be capable of holding them together when they come into contact with each other; the gas condenses, forming liquid H_2O. For example, liquid water forms on the outside of a cold glass as the water vapor in the air is cooled by the cold glass, as seen in Figure 10.3.

(a) (b)

FIGURE 10.3 Condensation forms when water vapor in the air is cooled enough to form liquid water, such as (a) on the outside of a cold beverage glass or (b) in the form of fog. (credit a: modification of work by Jenny Downing; credit b: modification of work by Cory Zanker)

We can also liquefy many gases by compressing them, if the temperature is not too high. The increased pressure brings the molecules of a gas closer together, such that the attractions between the molecules become strong relative to their KE. Consequently, they form liquids. Butane, C_4H_{10}, is the fuel used in disposable lighters and is a gas at standard temperature and pressure. Inside the lighter's fuel compartment, the butane is compressed to a pressure that results in its condensation to the liquid state, as shown in Figure 10.4.

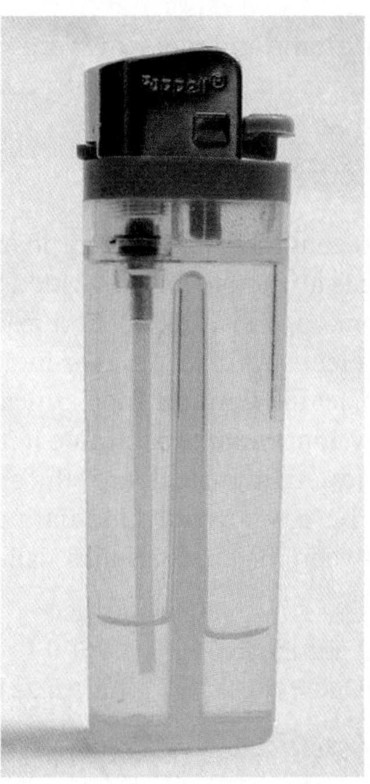

FIGURE 10.4 Gaseous butane is compressed within the storage compartment of a disposable lighter, resulting in its condensation to the liquid state. (credit: modification of work by "Sam-Cat"/Flickr)

Finally, if the temperature of a liquid becomes sufficiently low, or the pressure on the liquid becomes sufficiently high, the molecules of the liquid no longer have enough KE to overcome the IMF between them, and a solid forms. A more thorough discussion of these and other changes of state, or phase transitions, is provided in a later module of this chapter.

LINK TO LEARNING

Access this interactive simulation (http://openstax.org/l/16phetvisual) on states of matter, phase transitions, and intermolecular forces. This simulation is useful for visualizing concepts introduced throughout this chapter.

Forces between Molecules

Under appropriate conditions, the attractions between all gas molecules will cause them to form liquids or solids. This is due to intermolecular forces, not *intra*molecular forces. *Intra*molecular forces are those *within* the molecule that keep the molecule together, for example, the bonds between the atoms. *Inter*molecular forces are the attractions *between* molecules, which determine many of the physical properties of a substance. Figure 10.5 illustrates these different molecular forces. The strengths of these attractive forces vary widely, though usually the IMFs between small molecules are weak compared to the intramolecular forces that bond atoms together within a molecule. For example, to overcome the IMFs in one mole of liquid HCl and convert it into gaseous HCl requires only about 17 kilojoules. However, to break the covalent bonds between the hydrogen and chlorine atoms in one mole of HCl requires about 25 times more energy—430 kilojoules.

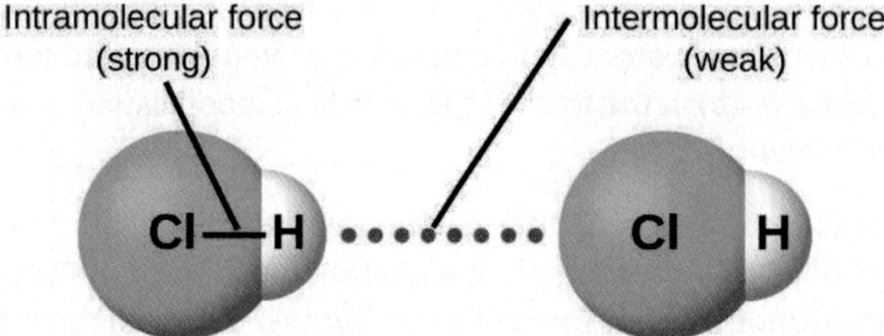

FIGURE 10.5 *Intra*molecular forces keep a molecule intact. *Inter*molecular forces hold multiple molecules together and determine many of a substance's properties.

All of the attractive forces between neutral atoms and molecules are known as **van der Waals forces**, although they are usually referred to more informally as intermolecular attraction. We will consider the various types of IMFs in the next three sections of this module.

Dispersion Forces

One of the three van der Waals forces is present in all condensed phases, regardless of the nature of the atoms or molecules composing the substance. This attractive force is called the London dispersion force in honor of German-born American physicist Fritz London who, in 1928, first explained it. This force is often referred to as simply the **dispersion force**. Because the electrons of an atom or molecule are in constant motion (or, alternatively, the electron's location is subject to quantum-mechanical variability), at any moment in time, an atom or molecule can develop a temporary, **instantaneous dipole** if its electrons are distributed asymmetrically. The presence of this dipole can, in turn, distort the electrons of a neighboring atom or molecule, producing an **induced dipole**. These two rapidly fluctuating, temporary dipoles thus result in a relatively weak electrostatic attraction between the species—a so-called dispersion force like that illustrated in Figure 10.6.

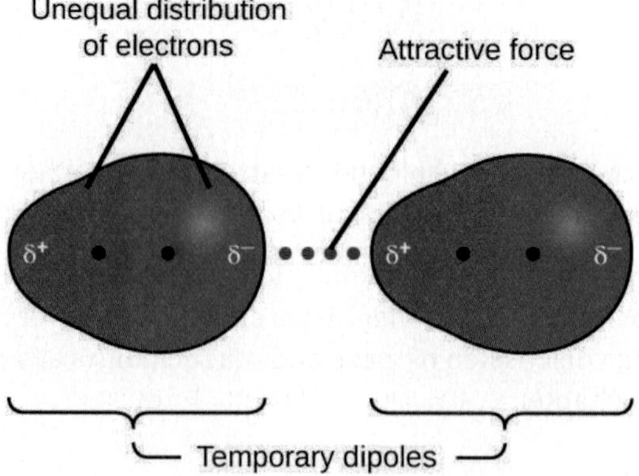

FIGURE 10.6 Dispersion forces result from the formation of temporary dipoles, as illustrated here for two nonpolar diatomic molecules.

Dispersion forces that develop between atoms in different molecules can attract the two molecules to each other. The forces are relatively weak, however, and become significant only when the molecules are very close.

Larger and heavier atoms and molecules exhibit stronger dispersion forces than do smaller and lighter atoms and molecules. F_2 and Cl_2 are gases at room temperature (reflecting weaker attractive forces); Br_2 is a liquid, and I_2 is a solid (reflecting stronger attractive forces). Trends in observed melting and boiling points for the halogens clearly demonstrate this effect, as seen in Table 10.1.

Melting and Boiling Points of the Halogens

Halogen	Molar Mass	Atomic Radius	Melting Point	Boiling Point
fluorine, F_2	38 g/mol	72 pm	53 K	85 K
chlorine, Cl_2	71 g/mol	99 pm	172 K	238 K
bromine, Br_2	160 g/mol	114 pm	266 K	332 K
iodine, I_2	254 g/mol	133 pm	387 K	457 K
astatine, At_2	420 g/mol	150 pm	575 K	610 K

TABLE 10.1

The increase in melting and boiling points with increasing atomic/molecular size may be rationalized by considering how the strength of dispersion forces is affected by the electronic structure of the atoms or molecules in the substance. In a larger atom, the valence electrons are, on average, farther from the nuclei than in a smaller atom. Thus, they are less tightly held and can more easily form the temporary dipoles that produce the attraction. The measure of how easy or difficult it is for another electrostatic charge (for example, a nearby ion or polar molecule) to distort a molecule's charge distribution (its electron cloud) is known as **polarizability**. A molecule that has a charge cloud that is easily distorted is said to be very polarizable and will have large dispersion forces; one with a charge cloud that is difficult to distort is not very polarizable and will have small dispersion forces.

London Forces and Their Effects

Order the following compounds of a group 14 element and hydrogen from lowest to highest boiling point: CH_4, SiH_4, GeH_4, and SnH_4. Explain your reasoning.

Solution

Applying the skills acquired in the chapter on chemical bonding and molecular geometry, all of these compounds are predicted to be nonpolar, so they may experience only dispersion forces: the smaller the molecule, the less polarizable and the weaker the dispersion forces; the larger the molecule, the larger the dispersion forces. The molar masses of CH_4, SiH_4, GeH_4, and SnH_4 are approximately 16 g/mol, 32 g/mol, 77 g/mol, and 123 g/mol, respectively. Therefore, CH_4 is expected to have the lowest boiling point and SnH_4 the highest boiling point. The ordering from lowest to highest boiling point is expected to be $CH_4 < SiH_4 < GeH_4 < SnH_4$.

A graph of the actual boiling points of these compounds versus the period of the group 14 element shows this prediction to be correct:

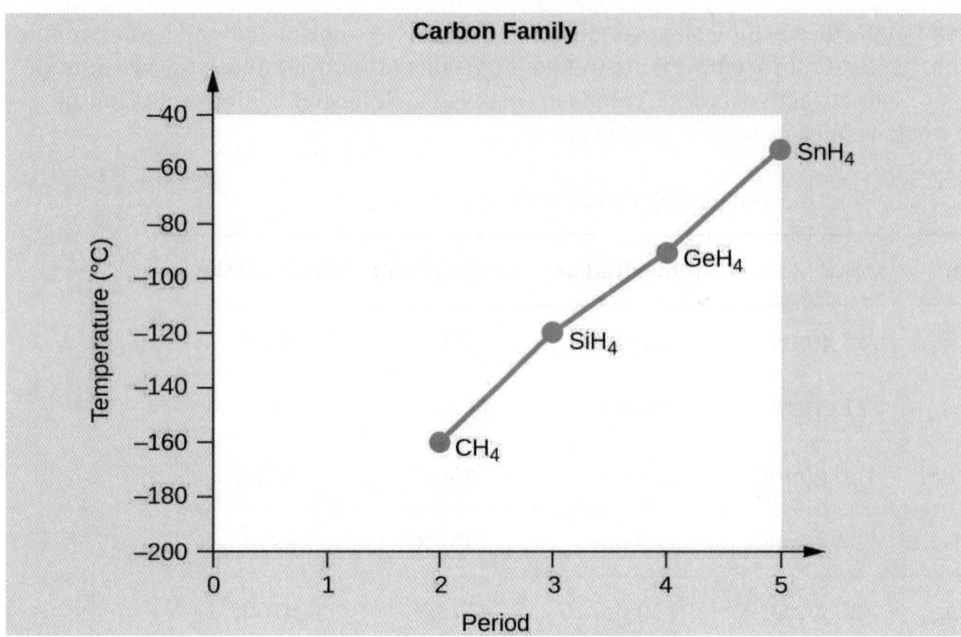

Check Your Learning

Order the following hydrocarbons from lowest to highest boiling point: C_2H_6, C_3H_8, and C_4H_{10}.

Answer:

$C_2H_6 < C_3H_8 < C_4H_{10}$. All of these compounds are nonpolar and only have London dispersion forces: the larger the molecule, the larger the dispersion forces and the higher the boiling point. The ordering from lowest to highest boiling point is therefore $C_2H_6 < C_3H_8 < C_4H_{10}$.

The shapes of molecules also affect the magnitudes of the dispersion forces between them. For example, boiling points for the isomers *n*-pentane, isopentane, and neopentane (shown in Figure 10.7) are 36 °C, 27 °C, and 9.5 °C, respectively. Even though these compounds are composed of molecules with the same chemical formula, C_5H_{12}, the difference in boiling points suggests that dispersion forces in the liquid phase are different, being greatest for *n*-pentane and least for neopentane. The elongated shape of *n*-pentane provides a greater surface area available for contact between molecules, resulting in correspondingly stronger dispersion forces. The more compact shape of isopentane offers a smaller surface area available for intermolecular contact and, therefore, weaker dispersion forces. Neopentane molecules are the most compact of the three, offering the least available surface area for intermolecular contact and, hence, the weakest dispersion forces. This behavior is analogous to the connections that may be formed between strips of VELCRO brand fasteners: the greater the area of the strip's contact, the stronger the connection.

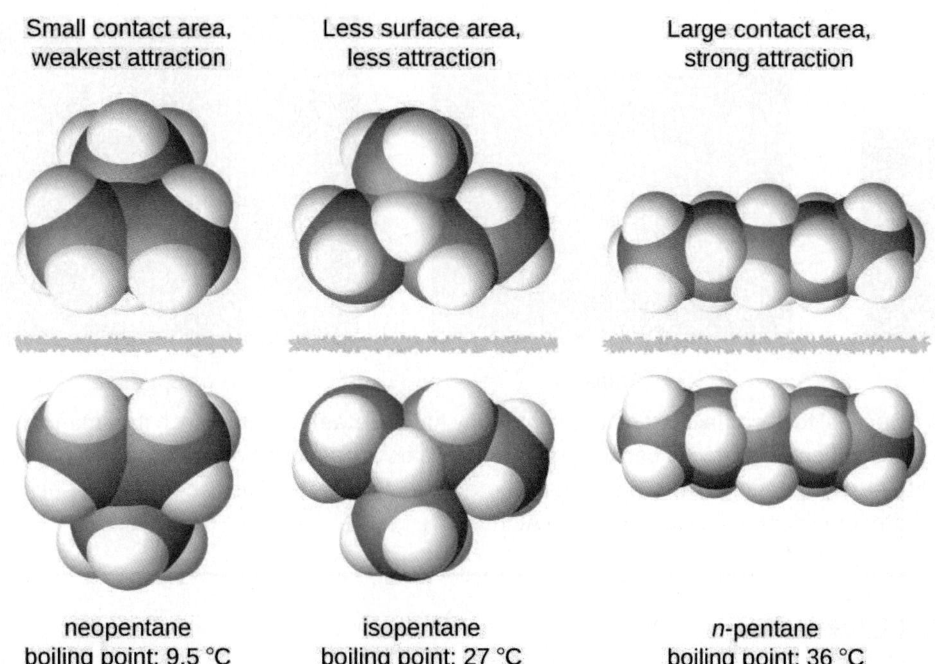

FIGURE 10.7 The strength of the dispersion forces increases with the contact area between molecules, as demonstrated by the boiling points of these pentane isomers.

Chemistry in Everyday Life

Geckos and Intermolecular Forces

Geckos have an amazing ability to adhere to most surfaces. They can quickly run up smooth walls and across ceilings that have no toe-holds, and they do this without having suction cups or a sticky substance on their toes. And while a gecko can lift its feet easily as it walks along a surface, if you attempt to pick it up, it sticks to the surface. How are geckos (as well as spiders and some other insects) able to do this? Although this phenomenon has been investigated for hundreds of years, scientists only recently uncovered the details of the process that allows geckos' feet to behave this way.

Geckos' toes are covered with hundreds of thousands of tiny hairs known as *setae*, with each seta, in turn, branching into hundreds of tiny, flat, triangular tips called *spatulae*. The huge numbers of spatulae on its setae provide a gecko, shown in Figure 10.8, with a large total surface area for sticking to a surface. In 2000, Kellar Autumn, who leads a multi-institutional gecko research team, found that geckos adhered equally well to both polar silicon dioxide and nonpolar gallium arsenide. This proved that geckos stick to surfaces because of dispersion forces—weak intermolecular attractions arising from temporary, synchronized charge distributions between adjacent molecules. Although dispersion forces are very weak, the total attraction over millions of spatulae is large enough to support many times the gecko's weight.

In 2014, two scientists developed a model to explain how geckos can rapidly transition from "sticky" to "non-sticky." Alex Greaney and Congcong Hu at Oregon State University described how geckos can achieve this by changing the angle between their spatulae and the surface. Geckos' feet, which are normally nonsticky, become sticky when a small shear force is applied. By curling and uncurling their toes, geckos can alternate between sticking and unsticking from a surface, and thus easily move across it. Later research led by Alyssa Stark at University of Akron showed that geckos can maintain their hold on hydrophobic surfaces (similar to the leaves in their habitats) equally well whether the surfaces were wet or dry. Stark's experiment used a ribbon to gently pull the geckos until they slipped, so that the researchers could determine the geckos' ability to hold various surfaces under wet and dry conditions. Further investigations may eventually lead to the development of better adhesives and other applications.

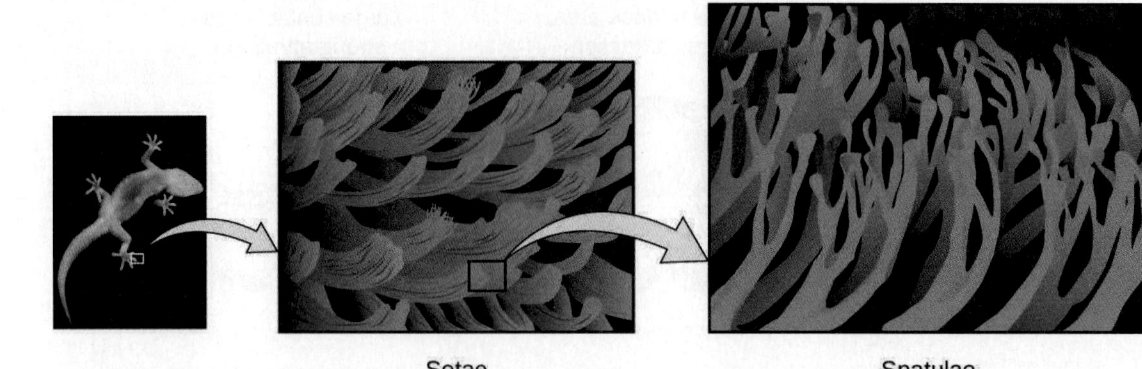

FIGURE 10.8 Geckos' toes contain large numbers of tiny hairs (setae), which branch into many triangular tips (spatulae). Geckos adhere to surfaces because of van der Waals attractions between the surface and a gecko's millions of spatulae. By changing how the spatulae contact the surface, geckos can turn their stickiness "on" and "off." (credit photo: modification of work by "JC*+A!"/Flickr)

LINK TO LEARNING

Watch this video (http://openstax.org/l/16kellaraut) to learn more about Kellar Autumn's research that determined that van der Waals forces are responsible for a gecko's ability to cling and climb.

Dipole-Dipole Attractions

Recall from the chapter on chemical bonding and molecular geometry that *polar* molecules have a partial positive charge on one side and a partial negative charge on the other side of the molecule—a separation of charge called a *dipole*. Consider a polar molecule such as hydrogen chloride, HCl. In the HCl molecule, the more electronegative Cl atom bears the partial negative charge, whereas the less electronegative H atom bears the partial positive charge. An attractive force between HCl molecules results from the attraction between the positive end of one HCl molecule and the negative end of another. This attractive force is called a **dipole-dipole attraction**—the electrostatic force between the partially positive end of one polar molecule and the partially negative end of another, as illustrated in Figure 10.9.

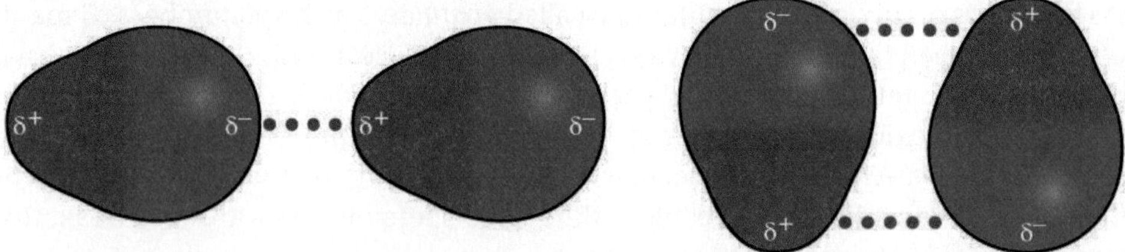

FIGURE 10.9 This image shows two arrangements of polar molecules, such as HCl, that allow an attraction between the partial negative end of one molecule and the partial positive end of another.

The effect of a dipole-dipole attraction is apparent when we compare the properties of HCl molecules to nonpolar F_2 molecules. Both HCl and F_2 consist of the same number of atoms and have approximately the same molecular mass. At a temperature of 150 K, molecules of both substances would have the same average KE. However, the dipole-dipole attractions between HCl molecules are sufficient to cause them to "stick together" to form a liquid, whereas the relatively weaker dispersion forces between nonpolar F_2 molecules are not, and so this substance is gaseous at this temperature. The higher normal boiling point of HCl (188 K) compared to F_2 (85 K) is a reflection of the greater strength of dipole-dipole attractions between HCl molecules, compared to the attractions between nonpolar F_2 molecules. We will often use values such as boiling or freezing points, or enthalpies of vaporization or fusion, as indicators of the relative strengths of IMFs of attraction present within different substances.

Dipole-Dipole Forces and Their Effects

Predict which will have the higher boiling point: N_2 or CO. Explain your reasoning.

Solution

CO and N_2 are both diatomic molecules with masses of about 28 amu, so they experience similar London dispersion forces. Because CO is a polar molecule, it experiences dipole-dipole attractions. Because N_2 is nonpolar, its molecules cannot exhibit dipole-dipole attractions. The dipole-dipole attractions between CO molecules are comparably stronger than the dispersion forces between nonpolar N_2 molecules, so CO is expected to have the higher boiling point.

Check Your Learning

Predict which will have the higher boiling point: ICl or Br_2. Explain your reasoning.

Answer:

ICl. ICl and Br_2 have similar masses (~160 amu) and therefore experience similar London dispersion forces. ICl is polar and thus also exhibits dipole-dipole attractions; Br_2 is nonpolar and does not. The relatively stronger dipole-dipole attractions require more energy to overcome, so ICl will have the higher boiling point.

Hydrogen Bonding

Nitrosyl fluoride (ONF, molecular mass 49 amu) is a gas at room temperature. Water (H_2O, molecular mass 18 amu) is a liquid, even though it has a lower molecular mass. We clearly cannot attribute this difference between the two compounds to dispersion forces. Both molecules have about the same shape and ONF is the heavier and larger molecule. It is, therefore, expected to experience more significant dispersion forces. Additionally, we cannot attribute this difference in boiling points to differences in the dipole moments of the molecules. Both molecules are polar and exhibit comparable dipole moments. The large difference between the boiling points is due to a particularly strong dipole-dipole attraction that may occur when a molecule contains a hydrogen atom bonded to a fluorine, oxygen, or nitrogen atom (the three most electronegative elements). The very large difference in electronegativity between the H atom (2.1) and the atom to which it is bonded (4.0 for an F atom, 3.5 for an O atom, or 3.0 for a N atom), combined with the very small size of a H atom and the relatively small sizes of F, O, or N atoms, leads to *highly concentrated partial charges* with these atoms. Molecules with F-H, O-H, or N-H moieties are very strongly attracted to similar moieties in nearby molecules, a particularly strong type of dipole-dipole attraction called **hydrogen bonding**. Examples of hydrogen bonds include $HF\cdots HF$, $H_2O\cdots HOH$, and $H_3N\cdots HNH_2$, in which the hydrogen bonds are denoted by dots. Figure 10.10 illustrates hydrogen bonding between water molecules.

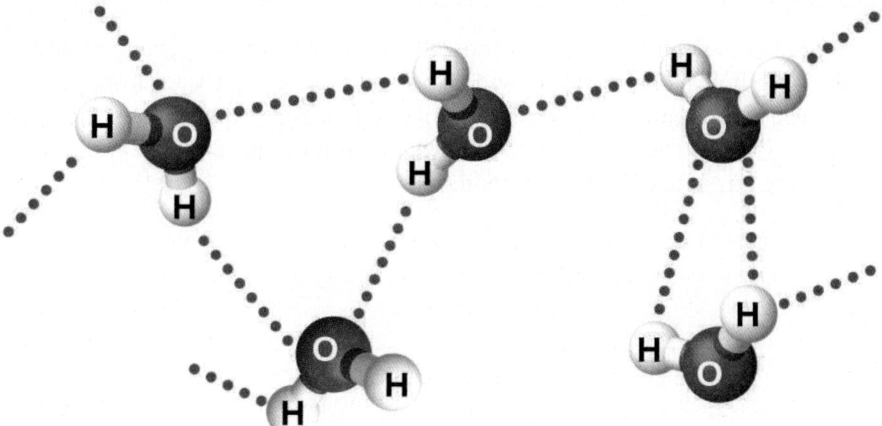

FIGURE 10.10 Water molecules participate in multiple hydrogen-bonding interactions with nearby water molecules.

Despite use of the word “bond,” keep in mind that hydrogen bonds are *intermolecular* attractive forces, not *intramolecular* attractive forces (covalent bonds). Hydrogen bonds are much weaker than covalent bonds, only about 5 to 10% as strong, but are generally much stronger than other dipole-dipole attractions and dispersion forces.

Hydrogen bonds have a pronounced effect on the properties of condensed phases (liquids and solids). For example, consider the trends in boiling points for the binary hydrides of group 15 (NH_3, PH_3, AsH_3, and SbH_3), group 16 hydrides (H_2O, H_2S, H_2Se, and H_2Te), and group 17 hydrides (HF, HCl, HBr, and HI). The boiling points of the heaviest three hydrides for each group are plotted in Figure 10.11. As we progress down any of these groups, the polarities of the molecules decrease slightly, whereas the sizes of the molecules increase substantially. The effect of increasingly stronger dispersion forces dominates that of increasingly weaker dipole-dipole attractions, and the boiling points are observed to increase steadily.

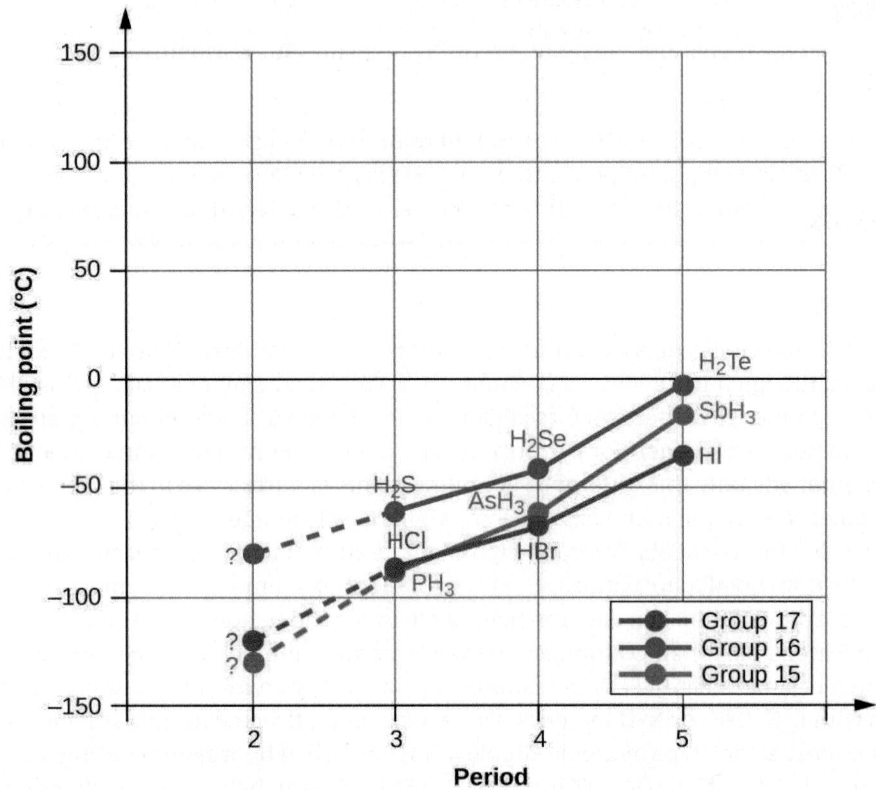

FIGURE 10.11 For the group 15, 16, and 17 hydrides, the boiling points for each class of compounds increase with increasing molecular mass for elements in periods 3, 4, and 5.

If we use this trend to predict the boiling points for the lightest hydride for each group, we would expect NH_3 to boil at about −120 °C, H_2O to boil at about −80 °C, and HF to boil at about −110 °C. However, when we measure the boiling points for these compounds, we find that they are dramatically higher than the trends would predict, as shown in Figure 10.12. The stark contrast between our naïve predictions and reality provides compelling evidence for the strength of hydrogen bonding.

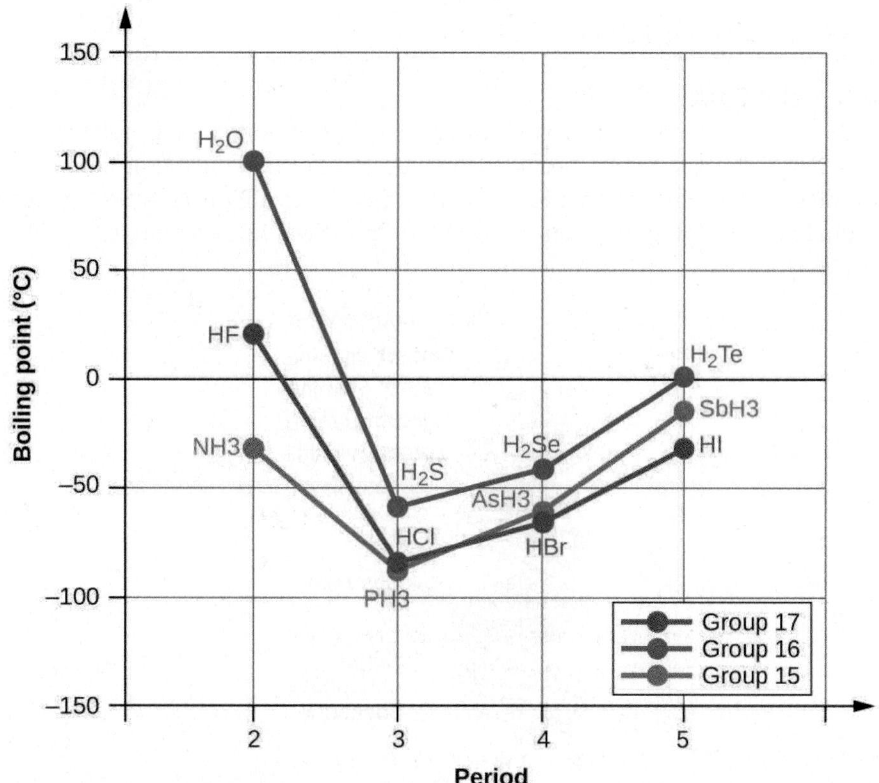

FIGURE 10.12 In comparison to periods 3–5, the binary hydrides of period 2 elements in groups 17, 16 and 15 (F, O and N, respectively) exhibit anomalously high boiling points due to hydrogen bonding.

EXAMPLE 10.3

Effect of Hydrogen Bonding on Boiling Points

Consider the compounds dimethylether (CH_3OCH_3), ethanol (CH_3CH_2OH), and propane ($CH_3CH_2CH_3$). Their boiling points, not necessarily in order, are −42.1 °C, −24.8 °C, and 78.4 °C. Match each compound with its boiling point. Explain your reasoning.

Solution

The VSEPR-predicted shapes of CH_3OCH_3, CH_3CH_2OH, and $CH_3CH_2CH_3$ are similar, as are their molar masses (46 g/mol, 46 g/mol, and 44 g/mol, respectively), so they will exhibit similar dispersion forces. Since $CH_3CH_2CH_3$ is nonpolar, it may exhibit *only* dispersion forces. Because CH_3OCH_3 is polar, it will also experience dipole-dipole attractions. Finally, CH_3CH_2OH has an −OH group, and so it will experience the uniquely strong dipole-dipole attraction known as hydrogen bonding. So the ordering in terms of strength of IMFs, and thus boiling points, is $CH_3CH_2CH_3 < CH_3OCH_3 < CH_3CH_2OH$. The boiling point of propane is −42.1 °C, the boiling point of dimethylether is −24.8 °C, and the boiling point of ethanol is 78.5 °C.

Check Your Learning

Ethane (CH_3CH_3) has a melting point of −183 °C and a boiling point of −89 °C. Predict the melting and boiling points for methylamine (CH_3NH_2). Explain your reasoning.

Answer:

The melting point and boiling point for methylamine are predicted to be significantly greater than those of ethane. CH_3CH_3 and CH_3NH_2 are similar in size and mass, but methylamine possesses an −NH group and therefore may exhibit hydrogen bonding. This greatly increases its IMFs, and therefore its melting and boiling points. It is difficult to predict values, but the known values are a melting point of −93 °C and a boiling point of −6 °C.

HOW SCIENCES INTERCONNECT

Hydrogen Bonding and DNA

Deoxyribonucleic acid (DNA) is found in every living organism and contains the genetic information that determines the organism's characteristics, provides the blueprint for making the proteins necessary for life, and serves as a template to pass this information on to the organism's offspring. A DNA molecule consists of two (anti-)parallel chains of repeating nucleotides, which form its well-known double helical structure, as shown in Figure 10.13.

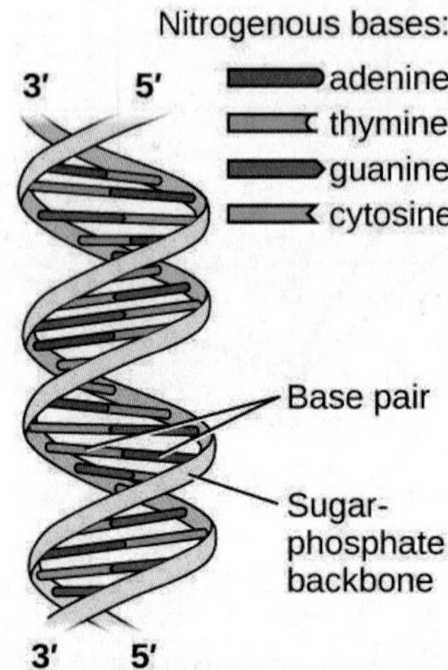

FIGURE 10.13 Two separate DNA molecules form a double-stranded helix in which the molecules are held together via hydrogen bonding. (credit: modification of work by Jerome Walker, Dennis Myts)

Each nucleotide contains a (deoxyribose) sugar bound to a phosphate group on one side, and one of four nitrogenous bases on the other. Two of the bases, cytosine (C) and thymine (T), are single-ringed structures known as pyrimidines. The other two, adenine (A) and guanine (G), are double-ringed structures called purines. These bases form complementary base pairs consisting of one purine and one pyrimidine, with adenine pairing with thymine, and cytosine with guanine. Each base pair is held together by hydrogen bonding. A and T share two hydrogen bonds, C and G share three, and both pairings have a similar shape and structure Figure 10.14.

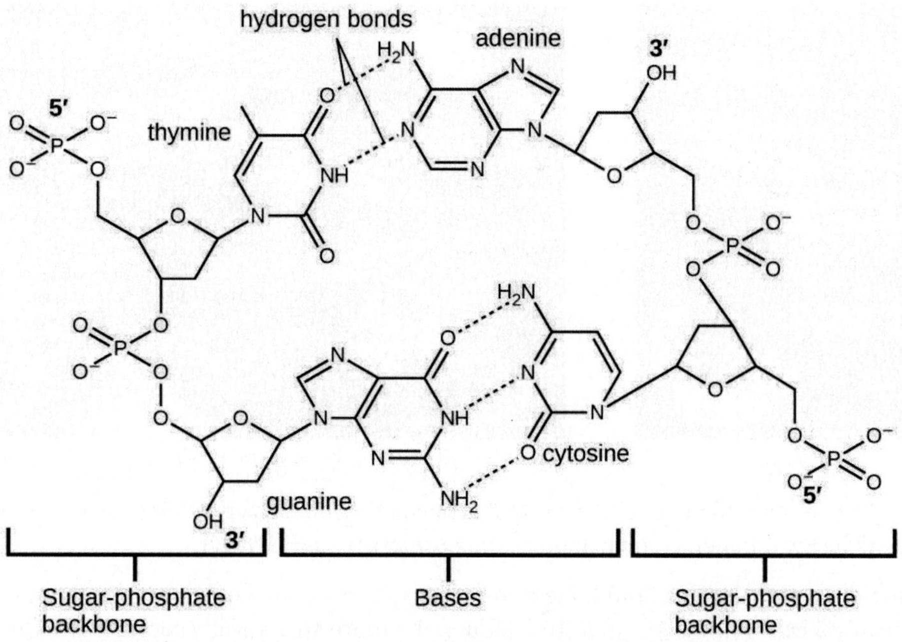

FIGURE 10.14 The geometries of the base molecules result in maximum hydrogen bonding between adenine and thymine (AT) and between guanine and cytosine (GC), so-called "complementary base pairs."

The cumulative effect of millions of hydrogen bonds effectively holds the two strands of DNA together. Importantly, the two strands of DNA can relatively easily "unzip" down the middle since hydrogen bonds are relatively weak compared to the covalent bonds that hold the atoms of the individual DNA molecules together. This allows both strands to function as a template for replication.

10.2 Properties of Liquids

LEARNING OBJECTIVES

By the end of this section, you will be able to:

- Distinguish between adhesive and cohesive forces
- Define viscosity, surface tension, and capillary rise
- Describe the roles of intermolecular attractive forces in each of these properties/phenomena

When you pour a glass of water, or fill a car with gasoline, you observe that water and gasoline flow freely. But when you pour syrup on pancakes or add oil to a car engine, you note that syrup and motor oil do not flow as readily. The **viscosity** of a liquid is a measure of its resistance to flow. Water, gasoline, and other liquids that flow freely have a low viscosity. Honey, syrup, motor oil, and other liquids that do not flow freely, like those shown in Figure 10.15, have higher viscosities. We can measure viscosity by measuring the rate at which a metal ball falls through a liquid (the ball falls more slowly through a more viscous liquid) or by measuring the rate at which a liquid flows through a narrow tube (more viscous liquids flow more slowly).

(a) (b)

FIGURE 10.15 (a) Honey and (b) motor oil are examples of liquids with high viscosities; they flow slowly. (credit a: modification of work by Scott Bauer; credit b: modification of work by David Nagy)

The IMFs between the molecules of a liquid, the size and shape of the molecules, and the temperature determine how easily a liquid flows. As Table 10.2 shows, the more structurally complex are the molecules in a liquid and the stronger the IMFs between them, the more difficult it is for them to move past each other and the greater is the viscosity of the liquid. As the temperature increases, the molecules move more rapidly and their kinetic energies are better able to overcome the forces that hold them together; thus, the viscosity of the liquid decreases.

Viscosities of Common Substances at 25 °C

Substance	Formula	Viscosity (mPa·s)
water	H_2O	0.890
mercury	Hg	1.526
ethanol	C_2H_5OH	1.074
octane	C_8H_{18}	0.508
ethylene glycol	$CH_2(OH)CH_2(OH)$	16.1
honey	variable	~2,000–10,000
motor oil	variable	~50–500

TABLE 10.2

The various IMFs between identical molecules of a substance are examples of **cohesive forces**. The molecules within a liquid are surrounded by other molecules and are attracted equally in all directions by the cohesive forces within the liquid. However, the molecules on the surface of a liquid are attracted only by about one-half as many molecules. Because of the unbalanced molecular attractions on the surface molecules, liquids contract to form a shape that minimizes the number of molecules on the surface—that is, the shape with the minimum surface area. A small drop of liquid tends to assume a spherical shape, as shown in Figure 10.16, because in a sphere, the ratio of surface area to volume is at a minimum. Larger drops are more greatly affected by gravity, air resistance, surface interactions, and so on, and as a result, are less spherical.

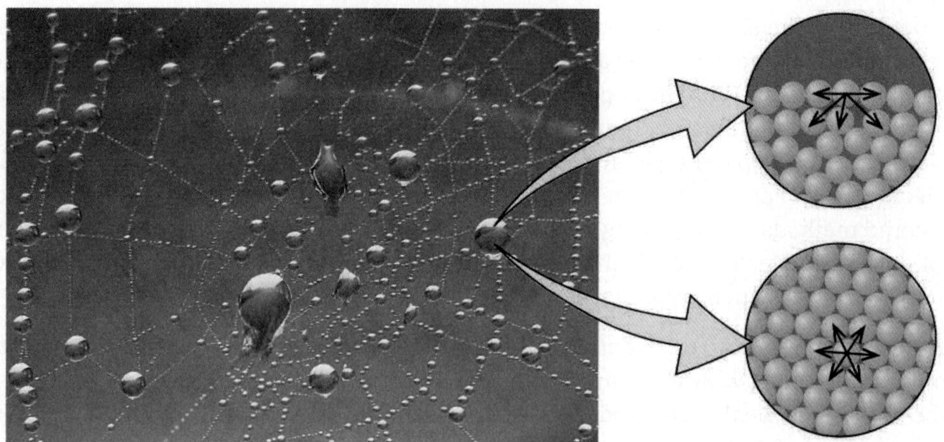

FIGURE 10.16 Attractive forces result in a spherical water drop that minimizes surface area; cohesive forces hold the sphere together; adhesive forces keep the drop attached to the web. (credit photo: modification of work by "OliBac"/Flickr)

Surface tension is defined as the energy required to increase the surface area of a liquid, or the force required to increase the length of a liquid surface by a given amount. This property results from the cohesive forces between molecules at the surface of a liquid, and it causes the surface of a liquid to behave like a stretched rubber membrane. Surface tensions of several liquids are presented in Table 10.3. Among common liquids, water exhibits a distinctly high surface tension due to strong hydrogen bonding between its molecules. As a result of this high surface tension, the surface of water represents a relatively "tough skin" that can withstand considerable force without breaking. A steel needle carefully placed on water will float. Some insects, like the one shown in Figure 10.17, even though they are denser than water, move on its surface because they are supported by the surface tension.

Surface Tensions of Common Substances at 25 °C

Substance	Formula	Surface Tension (mN/m)
water	H_2O	71.99
mercury	Hg	458.48
ethanol	C_2H_5OH	21.97
octane	C_8H_{18}	21.14
ethylene glycol	$CH_2(OH)CH_2(OH)$	47.99

TABLE 10.3

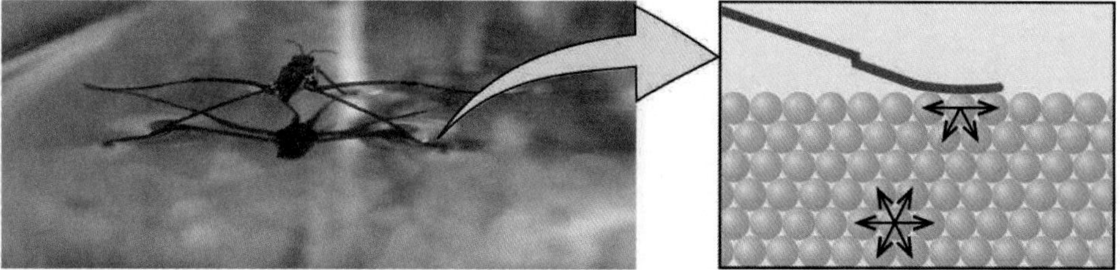

FIGURE 10.17 Surface tension (right) prevents this insect, a "water strider," from sinking into the water.

Surface tension is affected by a variety of variables, including the introduction of additional substances on the surface. In the late 1800s, Agnes Pockels, who was initially blocked from pursuing a scientific career but studied on her own, began investigating the impact and characteristics of soapy and greasy films in water. Using homemade materials, she developed an instrument known as a trough for measuring surface contaminants and their effects. With the support of renowned scientist Lord Rayleigh, her 1891 paper showed that surface contamination significantly reduces surface tension, and also that changing the characteristics of the surface (compressing or expanding it) also affects surface tension. Decades later, Irving Langmuir and Katharine Blodgett built on Pockels' work in their own trough and important advances in surface chemistry. Langmuir pioneered methods for producing single-molecule layers of film; Blodgett applied these to the development of non-reflective glass (critical for film-making and other applications), and also studied methods related to cleaning surfaces, which are important in semiconductor fabrication.

The IMFs of attraction between two *different* molecules are called **adhesive forces**. Consider what happens when water comes into contact with some surface. If the adhesive forces between water molecules and the molecules of the surface are weak compared to the cohesive forces between the water molecules, the water does not "wet" the surface. For example, water does not wet waxed surfaces or many plastics such as polyethylene. Water forms drops on these surfaces because the cohesive forces within the drops are greater than the adhesive forces between the water and the plastic. Water spreads out on glass because the adhesive force between water and glass is greater than the cohesive forces within the water. When water is confined in a glass tube, its meniscus (surface) has a concave shape because the water wets the glass and creeps up the side of the tube. On the other hand, the cohesive forces between mercury atoms are much greater than the adhesive forces between mercury and glass. Mercury therefore does not wet glass, and it forms a convex meniscus when confined in a tube because the cohesive forces within the mercury tend to draw it into a drop (Figure 10.18).

FIGURE 10.18 Differences in the relative strengths of cohesive and adhesive forces result in different meniscus shapes for mercury (left) and water (right) in glass tubes. (credit: Mark Ott)

If you place one end of a paper towel in spilled wine, as shown in Figure 10.19, the liquid wicks up the paper towel. A similar process occurs in a cloth towel when you use it to dry off after a shower. These are examples of **capillary action**—when a liquid flows within a porous material due to the attraction of the liquid molecules to the surface of the material and to other liquid molecules. The adhesive forces between the liquid and the porous material, combined with the cohesive forces within the liquid, may be strong enough to move the liquid upward against gravity.

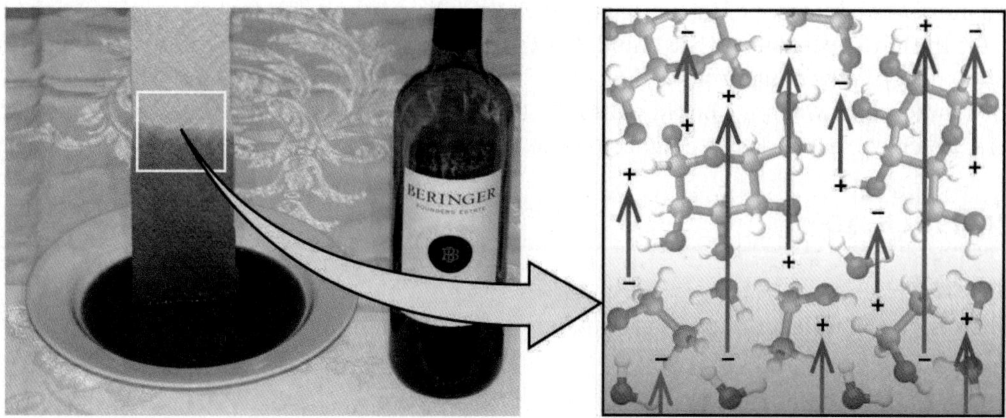

FIGURE 10.19 Wine wicks up a paper towel (left) because of the strong attractions of water (and ethanol) molecules to the –OH groups on the towel's cellulose fibers and the strong attractions of water molecules to other water (and ethanol) molecules (right). (credit photo: modification of work by Mark Blaser)

Towels soak up liquids like water because the fibers of a towel are made of molecules that are attracted to water molecules. Most cloth towels are made of cotton, and paper towels are generally made from paper pulp. Both consist of long molecules of cellulose that contain many –OH groups. Water molecules are attracted to these –OH groups and form hydrogen bonds with them, which draws the H_2O molecules up the cellulose molecules. The water molecules are also attracted to each other, so large amounts of water are drawn up the cellulose fibers.

Capillary action can also occur when one end of a small diameter tube is immersed in a liquid, as illustrated in Figure 10.20. If the liquid molecules are strongly attracted to the tube molecules, the liquid creeps up the inside of the tube until the weight of the liquid and the adhesive forces are in balance. The smaller the diameter of the tube is, the higher the liquid climbs. It is partly by capillary action occurring in plant cells called xylem that water and dissolved nutrients are brought from the soil up through the roots and into a plant. Capillary action is the basis for thin layer chromatography, a laboratory technique commonly used to separate small quantities of mixtures. You depend on a constant supply of tears to keep your eyes lubricated and on capillary action to pump tear fluid away.

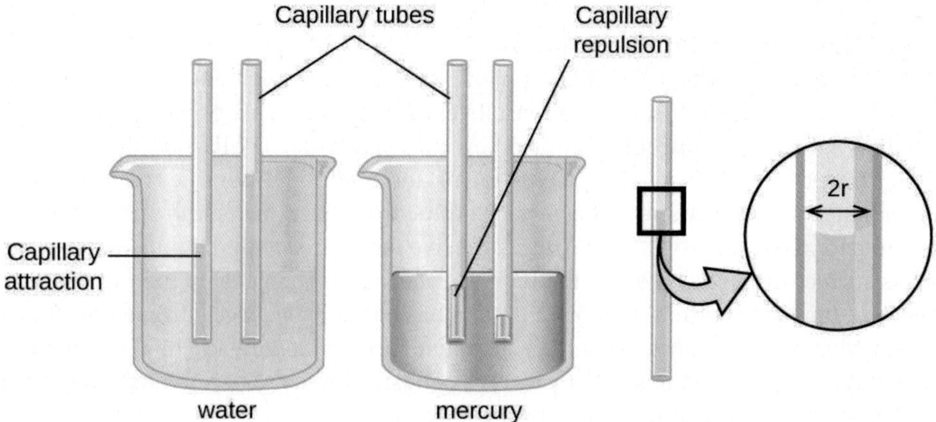

FIGURE 10.20 Depending upon the relative strengths of adhesive and cohesive forces, a liquid may rise (such as water) or fall (such as mercury) in a glass capillary tube. The extent of the rise (or fall) is directly proportional to the surface tension of the liquid and inversely proportional to the density of the liquid and the radius of the tube.

The height to which a liquid will rise in a capillary tube is determined by several factors as shown in the following equation:

$$h = \frac{2T\cos\theta}{r\rho g}$$

In this equation, h is the height of the liquid inside the capillary tube relative to the surface of the liquid outside the tube, T is the surface tension of the liquid, θ is the contact angle between the liquid and the tube, r is the radius of the tube, ρ is the density of the liquid, and g is the acceleration due to gravity, 9.8 m/s^2. When the tube is made of a material to which the liquid molecules are strongly attracted, they will spread out completely on the surface, which corresponds to a contact angle of 0°. This is the situation for water rising in a glass tube.

EXAMPLE 10.4

Capillary Rise

At 25 °C, how high will water rise in a glass capillary tube with an inner diameter of 0.25 mm?

For water, T = 71.99 mN/m and ρ = 1.0 g/cm^3.

Solution

The liquid will rise to a height h given by: $h = \frac{2T\cos\theta}{r\rho g}$

The Newton is defined as a kg m/s^2, and so the provided surface tension is equivalent to 0.07199 kg/s^2. The provided density must be converted into units that will cancel appropriately: ρ = 1000 kg/m^3. The diameter of the tube in meters is 0.00025 m, so the radius is 0.000125 m. For a glass tube immersed in water, the contact angle is θ = 0°, so cos θ = 1. Finally, acceleration due to gravity on the earth is g = 9.8 m/s^2. Substituting these values into the equation, and cancelling units, we have:

$$h = \frac{2\left(0.07199\ \text{kg/s}^2\right)}{(0.000125\ \text{m})\left(1000\ \text{kg/m}^3\right)\left(9.8\ \text{m/s}^2\right)} = 0.12\ \text{m} = 12\ \text{cm}$$

Check Your Learning

Water rises in a glass capillary tube to a height of 8.4 cm. What is the diameter of the capillary tube?

Answer:

diameter = 0.36 mm

Chemistry in Everyday Life

Biomedical Applications of Capillary Action

Many medical tests require drawing a small amount of blood, for example to determine the amount of glucose in someone with diabetes or the hematocrit level in an athlete. This procedure can be easily done because of capillary action, the ability of a liquid to flow up a small tube against gravity, as shown in Figure 10.21. When your finger is pricked, a drop of blood forms and holds together due to surface tension—the unbalanced intermolecular attractions at the surface of the drop. Then, when the open end of a narrow-diameter glass tube touches the drop of blood, the adhesive forces between the molecules in the blood and those at the glass surface draw the blood up the tube. How far the blood goes up the tube depends on the diameter of the tube (and the type of fluid). A small tube has a relatively large surface area for a given volume of blood, which results in larger (relative) attractive forces, allowing the blood to be drawn farther up the tube. The liquid itself is held together by its own cohesive forces. When the weight of the liquid in the tube generates a downward force equal to the upward force associated with capillary action, the liquid stops rising.

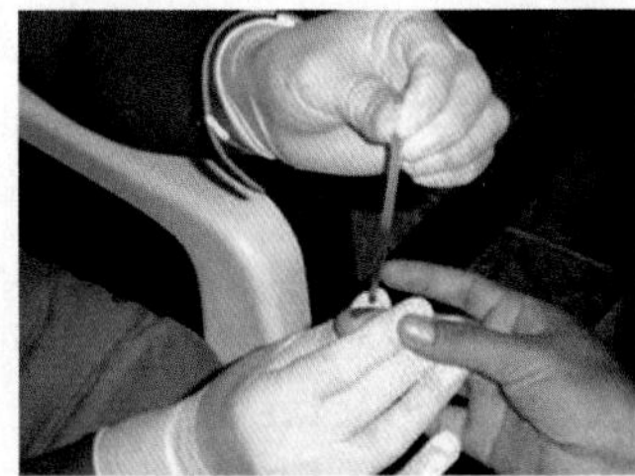

FIGURE 10.21 Blood is collected for medical analysis by capillary action, which draws blood into a small diameter glass tube. (credit: modification of work by Centers for Disease Control and Prevention)

10.3 Phase Transitions

LEARNING OBJECTIVES

By the end of this section, you will be able to:

- Define phase transitions and phase transition temperatures
- Explain the relation between phase transition temperatures and intermolecular attractive forces
- Describe the processes represented by typical heating and cooling curves, and compute heat flows and enthalpy changes accompanying these processes

We witness and utilize changes of physical state, or phase transitions, in a great number of ways. As one example of global significance, consider the evaporation, condensation, freezing, and melting of water. These changes of state are essential aspects of our earth's water cycle as well as many other natural phenomena and technological processes of central importance to our lives. In this module, the essential aspects of phase transitions are explored.

Vaporization and Condensation

When a liquid vaporizes in a closed container, gas molecules cannot escape. As these gas phase molecules move randomly about, they will occasionally collide with the surface of the condensed phase, and in some cases, these collisions will result in the molecules re-entering the condensed phase. The change from the gas phase to the liquid is called **condensation**. When the rate of condensation becomes equal to the rate of **vaporization**, neither the amount of the liquid nor the amount of the vapor in the container changes. The vapor in the container is then said to be *in equilibrium* with the liquid. Keep in mind that this is not a static situation, as molecules are continually exchanged between the condensed and gaseous phases. Such is an example of a **dynamic equilibrium**, the status of a system in which reciprocal processes (for example, vaporization and condensation) occur at equal rates. The pressure exerted by the vapor in equilibrium with a liquid in a closed container at a given temperature is called the liquid's **vapor pressure** (or equilibrium vapor pressure). The area of the surface of the liquid in contact with a vapor and the size of the vessel have no effect on the vapor pressure, although they do affect the time required for the equilibrium to be reached. We can measure the vapor pressure of a liquid by placing a sample in a closed container, like that illustrated in Figure 10.22, and using a manometer to measure the increase in pressure that is due to the vapor in equilibrium with the condensed phase.

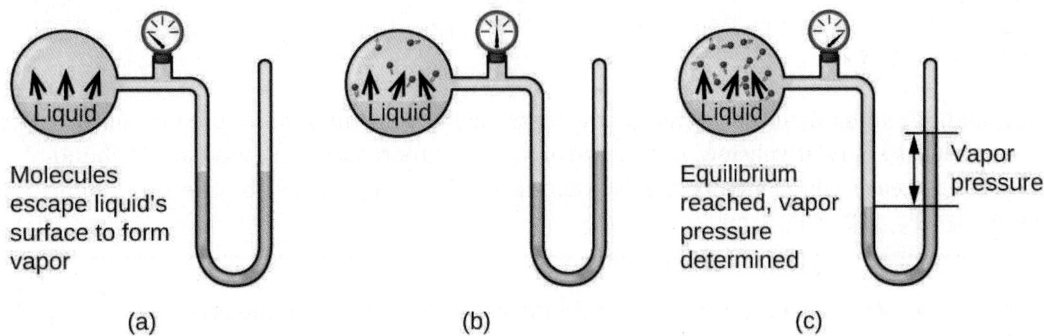

FIGURE 10.22 In a closed container, dynamic equilibrium is reached when (a) the rate of molecules escaping from

the liquid to become the gas (b) increases and eventually (c) equals the rate of gas molecules entering the liquid. When this equilibrium is reached, the vapor pressure of the gas is constant, although the vaporization and condensation processes continue.

The chemical identities of the molecules in a liquid determine the types (and strengths) of intermolecular attractions possible; consequently, different substances will exhibit different equilibrium vapor pressures. Relatively strong intermolecular attractive forces will serve to impede vaporization as well as favoring "recapture" of gas-phase molecules when they collide with the liquid surface, resulting in a relatively low vapor pressure. Weak intermolecular attractions present less of a barrier to vaporization, and a reduced likelihood of gas recapture, yielding relatively high vapor pressures. The following example illustrates this dependence of vapor pressure on intermolecular attractive forces.

EXAMPLE 10.5

Explaining Vapor Pressure in Terms of IMFs

Given the shown structural formulas for these four compounds, explain their relative vapor pressures in terms of types and extents of IMFs:

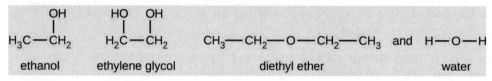

OH HO OH

$H_3C—CH_2$ $H_2C—CH_2$ $CH_3—CH_2—O—CH_2—CH_3$ and $H—O—H$

ethanol ethylene glycol diethyl ether water

Solution

Diethyl ether has a very small dipole and most of its intermolecular attractions are London forces. Although this molecule is the largest of the four under consideration, its IMFs are the weakest and, as a result, its molecules most readily escape from the liquid. It also has the highest vapor pressure. Due to its smaller size, ethanol exhibits weaker dispersion forces than diethyl ether. However, ethanol is capable of hydrogen bonding and, therefore, exhibits stronger overall IMFs, which means that fewer molecules escape from the liquid at any given temperature, and so ethanol has a lower vapor pressure than diethyl ether. Water is much smaller than either of the previous substances and exhibits weaker dispersion forces, but its extensive hydrogen bonding provides stronger intermolecular attractions, fewer molecules escaping the liquid, and a lower vapor pressure than for either diethyl ether or ethanol. Ethylene glycol has two –OH groups, so, like water, it exhibits extensive hydrogen bonding. It is much larger than water and thus experiences larger London forces. Its overall IMFs are the largest of these four substances, which means its vaporization rate will be the slowest and, consequently, its vapor pressure the lowest.

Check Your Learning

At 20 °C, the vapor pressures of several alcohols are given in this table. Explain these vapor pressures in terms of types and extents of IMFs for these alcohols:

Compound	methanol CH_3OH	ethanol C_2H_5OH	propanol C_3H_7OH	butanol C_4H_9OH
Vapor Pressure at 20 °C	11.9 kPa	5.95 kPa	2.67 kPa	0.56 kPa

Answer:

All these compounds exhibit hydrogen bonding; these strong IMFs are difficult for the molecules to overcome, so the vapor pressures are relatively low. As the size of molecule increases from methanol to butanol, dispersion forces increase, which means that the vapor pressures decrease as observed: $P_{methanol} > P_{ethanol} > P_{propanol} > P_{butanol}$.

As temperature increases, the vapor pressure of a liquid also increases due to the increased average KE of its molecules. Recall that at any given temperature, the molecules of a substance experience a range of kinetic energies, with a certain fraction of molecules having a sufficient energy to overcome IMF and escape the liquid (vaporize). At a higher temperature, a greater fraction of molecules have enough energy to escape from the liquid, as shown in Figure 10.23. The escape of more molecules per unit of time and the greater average speed of the molecules that escape both contribute to the higher vapor pressure.

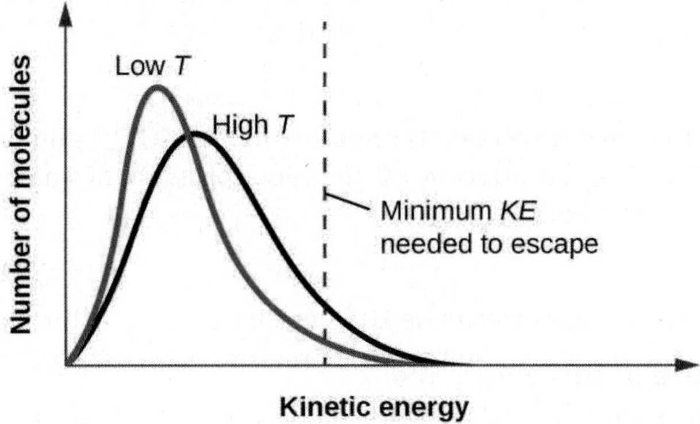

FIGURE 10.23 Temperature affects the distribution of kinetic energies for the molecules in a liquid. At the higher temperature, more molecules have the necessary kinetic energy, KE, to escape from the liquid into the gas phase.

Boiling Points

When the vapor pressure increases enough to equal the external atmospheric pressure, the liquid reaches its boiling point. The **boiling point** of a liquid is the temperature at which its equilibrium vapor pressure is equal to the pressure exerted on the liquid by its gaseous surroundings. For liquids in open containers, this pressure is that due to the earth's atmosphere. The **normal boiling point** of a liquid is defined as its boiling point when surrounding pressure is equal to 1 atm (101.3 kPa). Figure 10.24 shows the variation in vapor pressure with temperature for several different substances. Considering the definition of boiling point, these curves may be seen as depicting the dependence of a liquid's boiling point on surrounding pressure.

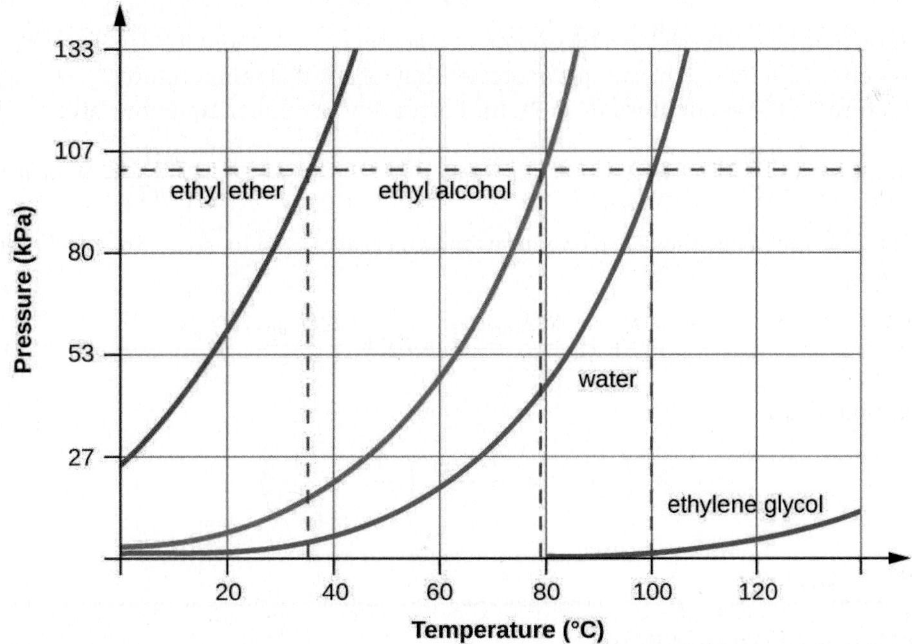

FIGURE 10.24 The boiling points of liquids are the temperatures at which their equilibrium vapor pressures equal the pressure of the surrounding atmosphere. Normal boiling points are those corresponding to a pressure of 1 atm (101.3 kPa.)

EXAMPLE 10.6

A Boiling Point at Reduced Pressure

A typical atmospheric pressure in Leadville, Colorado (elevation 10,200 feet) is 68 kPa. Use the graph in Figure 10.24 to determine the boiling point of water at this elevation.

Solution

The graph of the vapor pressure of water versus temperature in Figure 10.24 indicates that the vapor pressure of water is 68 kPa at about 90 °C. Thus, at about 90 °C, the vapor pressure of water will equal the atmospheric pressure in Leadville, and water will boil.

Check Your Learning

The boiling point of ethyl ether was measured to be 10 °C at a base camp on the slopes of Mount Everest. Use Figure 10.24 to determine the approximate atmospheric pressure at the camp.

Answer:

Approximately 40 kPa (0.4 atm)

The quantitative relation between a substance's vapor pressure and its temperature is described by the **Clausius-Clapeyron equation**:

$$P = Ae^{-\Delta H_{\text{vap}}/RT}$$

where ΔH_{vap} is the enthalpy of vaporization for the liquid, R is the gas constant, and A is a constant whose value depends on the chemical identity of the substance. Temperature T must be in Kelvin in this equation. This equation is often rearranged into logarithmic form to yield the linear equation:

$$\ln P = -\frac{\Delta H_{\text{vap}}}{RT} + \ln A$$

This linear equation may be expressed in a two-point format that is convenient for use in various computations, as demonstrated in the example exercises that follow. If at temperature T_1, the vapor pressure is P_1, and at temperature T_2, the vapor pressure is P_2, the corresponding linear equations are:

$$\ln P_1 = -\frac{\Delta H_{\text{vap}}}{RT_1} + \ln A \qquad \text{and} \qquad \ln P_2 = -\frac{\Delta H_{\text{vap}}}{RT_2} + \ln A$$

Since the constant, A, is the same, these two equations may be rearranged to isolate $\ln A$ and then set them equal to one another:

$$\ln P_1 + \frac{\Delta H_{\text{vap}}}{RT_1} = \ln P_2 + \frac{\Delta H_{\text{vap}}}{RT_2}$$

which can be combined into:

$$\ln\left(\frac{P_2}{P_1}\right) = \frac{\Delta H_{\text{vap}}}{R}\left(\frac{1}{T_1} - \frac{1}{T_2}\right)$$

EXAMPLE 10.7

Estimating Enthalpy of Vaporization

Isooctane (2,2,4-trimethylpentane) has an octane rating of 100. It is used as one of the standards for the octane-rating system for gasoline. At 34.0 °C, the vapor pressure of isooctane is 10.0 kPa, and at 98.8 °C, its vapor pressure is 100.0 kPa. Use this information to estimate the enthalpy of vaporization for isooctane.

Solution

The enthalpy of vaporization, ΔH_{vap}, can be determined by using the Clausius-Clapeyron equation:

$$\ln\left(\frac{P_2}{P_1}\right) = \frac{\Delta H_{vap}}{R}\left(\frac{1}{T_1} - \frac{1}{T_2}\right)$$

Since we have two vapor pressure-temperature values (T_1 = 34.0 °C = 307.2 K, P_1 = 10.0 kPa and T_2 = 98.8 °C = 372.0 K, P_2 = 100 kPa), we can substitute them into this equation and solve for ΔH_{vap}. Rearranging the Clausius-Clapeyron equation and solving for ΔH_{vap} yields:

$$\Delta H_{vap} = \frac{R \cdot \ln\left(\frac{P_2}{P_1}\right)}{\left(\frac{1}{T_1} - \frac{1}{T_2}\right)} = \frac{(8.3145\ \text{J/mol·K}) \cdot \ln\left(\frac{100\ \text{kPa}}{10.0\ \text{kPa}}\right)}{\left(\frac{1}{307.2\ \text{K}} - \frac{1}{372.0\ \text{K}}\right)} = 33{,}800\ \text{J/mol} = 33.8\ \text{kJ/mol}$$

Note that the pressure can be in any units, so long as they agree for both P values, but the temperature must be in kelvin for the Clausius-Clapeyron equation to be valid.

Check Your Learning

At 20.0 °C, the vapor pressure of ethanol is 5.95 kPa, and at 63.5 °C, its vapor pressure is 53.3 kPa. Use this information to estimate the enthalpy of vaporization for ethanol.

Answer:

41,360 J/mol or 41.4 kJ/mol

EXAMPLE 10.8

Estimating Temperature (or Vapor Pressure)

For benzene (C_6H_6), the normal boiling point is 80.1 °C and the enthalpy of vaporization is 30.8 kJ/mol. What is the boiling point of benzene in Denver, where atmospheric pressure = 83.4 kPa?

Solution

If the temperature and vapor pressure are known at one point, along with the enthalpy of vaporization, ΔH_{vap}, then the temperature that corresponds to a different vapor pressure (or the vapor pressure that corresponds to a different temperature) can be determined by using the Clausius-Clapeyron equation:

$$\ln\left(\frac{P_2}{P_1}\right) = \frac{\Delta H_{vap}}{R}\left(\frac{1}{T_1} - \frac{1}{T_2}\right)$$

Since the normal boiling point is the temperature at which the vapor pressure equals atmospheric pressure at sea level, we know one vapor pressure-temperature value (T_1 = 80.1 °C = 353.3 K, P_1 = 101.3 kPa, ΔH_{vap} = 30.8 kJ/mol) and want to find the temperature (T_2) that corresponds to vapor pressure P_2 = 83.4 kPa. We can substitute these values into the Clausius-Clapeyron equation and then solve for T_2. Rearranging the Clausius-Clapeyron equation and solving for T_2 yields:

$$T_2 = \left(\frac{-R \cdot \ln\left(\frac{P_2}{P_1}\right)}{\Delta H_{vap}} + \frac{1}{T_1}\right)^{-1} = \left(\frac{-(8.3145\ \text{J/mol·K}) \cdot \ln\left(\frac{83.4\ \text{kPa}}{101.3\ \text{kPa}}\right)}{30{,}800\ \text{J/mol}} + \frac{1}{353.3\ \text{K}}\right)^{-1} = 346.9\ \text{K or } 73.8^\circ\text{C}$$

Check Your Learning

For acetone $(CH_3)_2CO$, the normal boiling point is 56.5 °C and the enthalpy of vaporization is 31.3 kJ/mol. What is the vapor pressure of acetone at 25.0 °C?

Answer:

30.1 kPa

Enthalpy of Vaporization

Vaporization is an endothermic process. The cooling effect can be evident when you leave a swimming pool or a shower. When the water on your skin evaporates, it removes heat from your skin and causes you to feel cold. The energy change associated with the vaporization process is the enthalpy of vaporization, ΔH_{vap}. For example, the vaporization of water at standard temperature is represented by:

$$H_2O(l) \longrightarrow H_2O(g) \qquad \Delta H_{vap} = 44.01 \text{ kJ/mol}$$

As described in the chapter on thermochemistry, the reverse of an endothermic process is exothermic. And so, the condensation of a gas releases heat:

$$H_2O(g) \longrightarrow H_2O(l) \qquad \Delta H_{con} = -\Delta H_{vap} = -44.01 \text{ kJ/mol}$$

EXAMPLE 10.9

Using Enthalpy of Vaporization

One way our body is cooled is by evaporation of the water in sweat (Figure 10.25). In very hot climates, we can lose as much as 1.5 L of sweat per day. Although sweat is not pure water, we can get an approximate value of the amount of heat removed by evaporation by assuming that it is. How much heat is required to evaporate 1.5 L of water (1.5 kg) at T = 37 °C (normal body temperature); ΔH_{vap} = 43.46 kJ/mol at 37 °C.

FIGURE 10.25 Evaporation of sweat helps cool the body. (credit: "Kullez"/Flickr)

Solution

We start with the known volume of sweat (approximated as just water) and use the given information to convert to the amount of heat needed:

$$1.5 \cancel{\text{L}} \times \frac{1000 \cancel{\text{g}}}{1 \cancel{\text{L}}} \times \frac{1 \cancel{\text{mol}}}{18 \cancel{\text{g}}} \times \frac{43.46 \text{ kJ}}{1 \cancel{\text{mol}}} = 3.6 \times 10^3 \text{ kJ}$$

Thus, 3600 kJ of heat are removed by the evaporation of 1.5 L of water.

Check Your Learning

How much heat is required to evaporate 100.0 g of liquid ammonia, NH_3, at its boiling point if its enthalpy of vaporization is 4.8 kJ/mol?

Answer:

28 kJ

Melting and Freezing

When we heat a crystalline solid, we increase the average energy of its atoms, molecules, or ions and the solid gets hotter. At some point, the added energy becomes large enough to partially overcome the forces holding the molecules or ions of the solid in their fixed positions, and the solid begins the process of transitioning to the liquid state, or **melting**. At this point, the temperature of the solid stops rising, despite the continual input of heat, and it remains constant until all of the solid is melted. Only after all of the solid has melted will continued heating increase the temperature of the liquid (Figure 10.26).

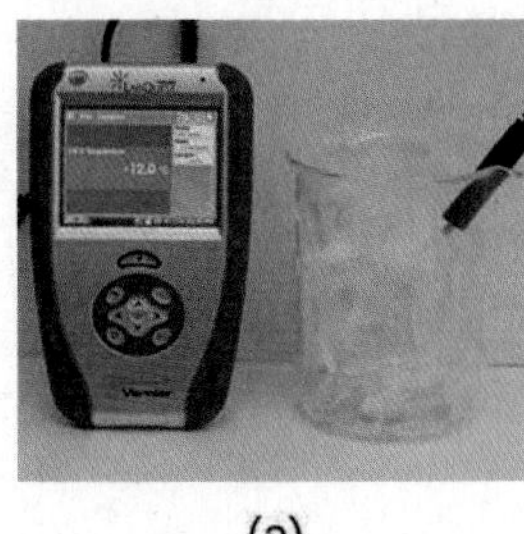
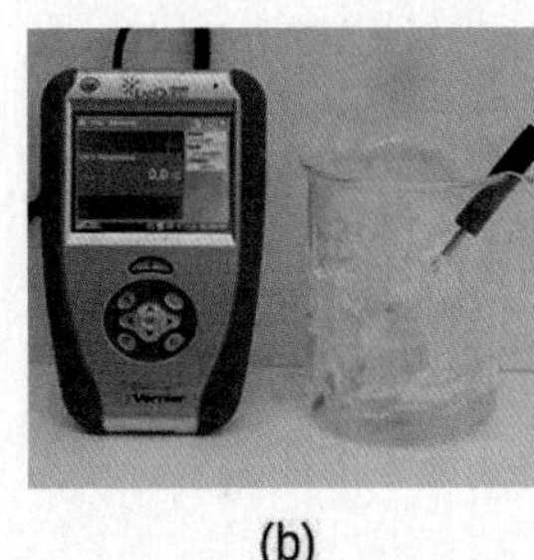
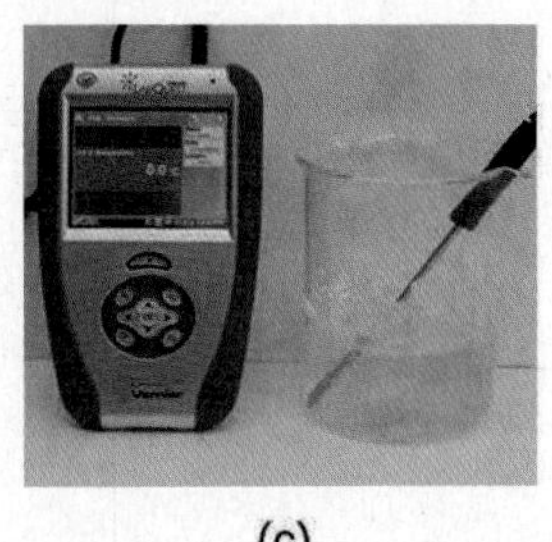
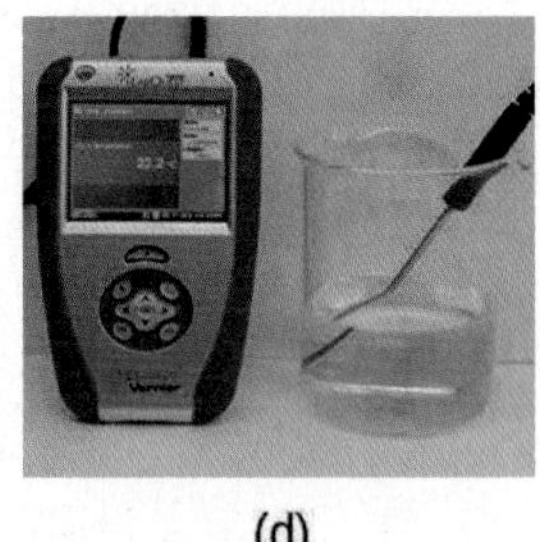

(a) (b) (c) (d)

FIGURE 10.26 (a) This beaker of ice has a temperature of −12.0 °C. (b) After 10 minutes the ice has absorbed enough heat from the air to warm to 0 °C. A small amount has melted. (c) Thirty minutes later, the ice has absorbed more heat, but its temperature is still 0 °C. The ice melts without changing its temperature. (d) Only after all the ice has melted does the heat absorbed cause the temperature to increase to 22.2 °C. (credit: modification of work by Mark Ott)

If we stop heating during melting and place the mixture of solid and liquid in a perfectly insulated container so no heat can enter or escape, the solid and liquid phases remain in equilibrium. This is almost the situation with a mixture of ice and water in a very good thermos bottle; almost no heat gets in or out, and the mixture of solid ice and liquid water remains for hours. In a mixture of solid and liquid at equilibrium, the reciprocal processes of melting and **freezing** occur at equal rates, and the quantities of solid and liquid therefore remain constant. The temperature at which the solid and liquid phases of a given substance are in equilibrium is called the **melting point** of the solid or the **freezing point** of the liquid. Use of one term or the other is normally dictated by the direction of the phase transition being considered, for example, solid to liquid (melting) or liquid to solid (freezing).

The enthalpy of fusion and the melting point of a crystalline solid depend on the strength of the attractive forces between the units present in the crystal. Molecules with weak attractive forces form crystals with low melting points. Crystals consisting of particles with stronger attractive forces melt at higher temperatures.

The amount of heat required to change one mole of a substance from the solid state to the liquid state is the enthalpy of fusion, ΔH_{fus} of the substance. The enthalpy of fusion of ice is 6.0 kJ/mol at 0 °C. Fusion (melting) is an endothermic process:

$$H_2O(s) \longrightarrow H_2O(l) \qquad \Delta H_{fus} = 6.01 \text{ kJ/mol}$$

The reciprocal process, freezing, is an exothermic process whose enthalpy change is −6.0 kJ/mol at 0 °C:

$$H_2O(l) \longrightarrow H_2O(s) \qquad \Delta H_{frz} = -\Delta H_{fus} = -6.01 \text{ kJ/mol}$$

Sublimation and Deposition

Some solids can transition directly into the gaseous state, bypassing the liquid state, via a process known as **sublimation**. At room temperature and standard pressure, a piece of dry ice (solid CO_2) sublimes, appearing to gradually disappear without ever forming any liquid. Snow and ice sublime at temperatures below the melting point of water, a slow process that may be accelerated by winds and the reduced atmospheric pressures at high altitudes. When solid iodine is warmed, the solid sublimes and a vivid purple vapor forms (Figure 10.27). The reverse of sublimation is called **deposition**, a process in which gaseous substances condense directly into the solid state, bypassing the liquid state. The formation of frost is an example of deposition.

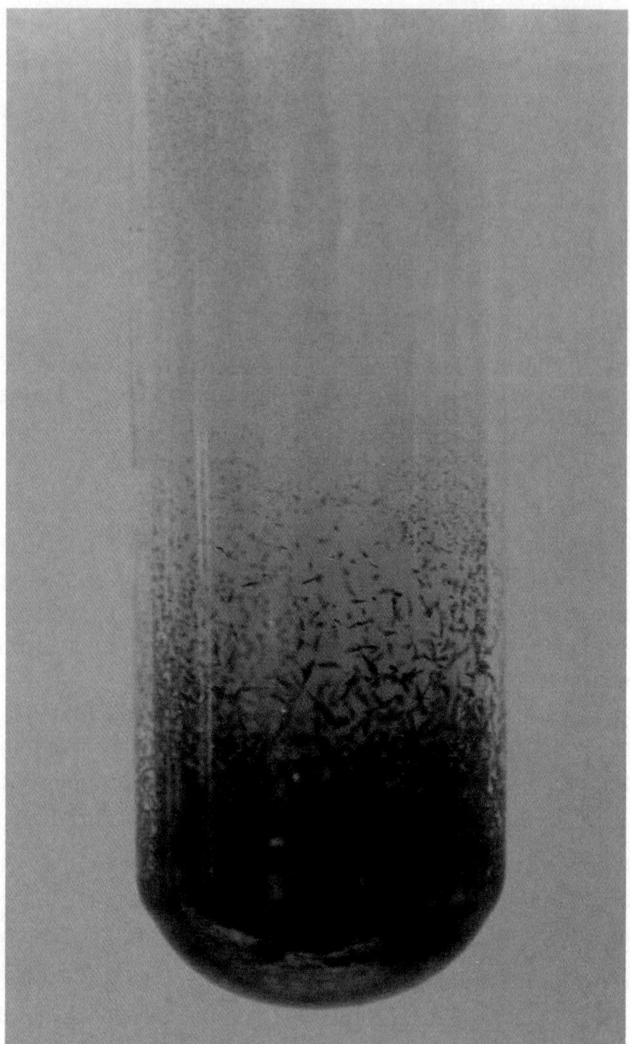

FIGURE 10.27 Sublimation of solid iodine in the bottom of the tube produces a purple gas that subsequently deposits as solid iodine on the colder part of the tube above. (credit: modification of work by Mark Ott)

Like vaporization, the process of sublimation requires an input of energy to overcome intermolecular attractions. The enthalpy of sublimation, ΔH_{sub}, is the energy required to convert one mole of a substance from the solid to the gaseous state. For example, the sublimation of carbon dioxide is represented by:

$$CO_2(s) \longrightarrow CO_2(g) \qquad \Delta H_{sub} = 26.1 \text{ kJ/mol}$$

Likewise, the enthalpy change for the reverse process of deposition is equal in magnitude but opposite in sign to that for sublimation:

$$CO_2(g) \longrightarrow CO_2(s) \qquad \Delta H_{dep} = -\Delta H_{sub} = -26.1 \text{ kJ/mol}$$

Consider the extent to which intermolecular attractions must be overcome to achieve a given phase transition. Converting a solid into a liquid requires that these attractions be only partially overcome; transition to the gaseous state requires that they be completely overcome. As a result, the enthalpy of fusion for a substance is less than its enthalpy of vaporization. This same logic can be used to derive an approximate relation between the enthalpies of all phase changes for a given substance. Though not an entirely accurate description, sublimation may be conveniently modeled as a sequential two-step process of melting followed by vaporization in order to apply Hess's Law. Viewed in this manner, the enthalpy of sublimation for a substance may be estimated as the sum of its enthalpies of fusion and vaporization, as illustrated in Figure 10.28. For example:

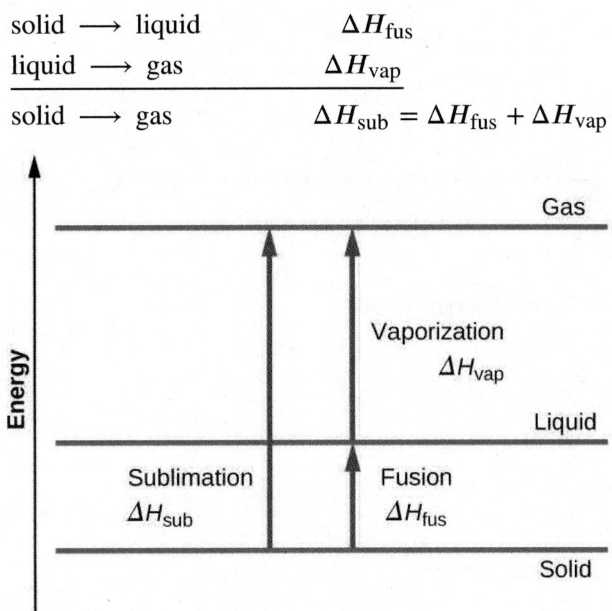

FIGURE 10.28 For a given substance, the sum of its enthalpy of fusion and enthalpy of vaporization is approximately equal to its enthalpy of sublimation.

Heating and Cooling Curves

In the chapter on thermochemistry, the relation between the amount of heat absorbed or released by a substance, q, and its accompanying temperature change, ΔT, was introduced:

$$q = mc\Delta T$$

where m is the mass of the substance and c is its specific heat. The relation applies to matter being heated or cooled, but not undergoing a change in state. When a substance being heated or cooled reaches a temperature corresponding to one of its phase transitions, further gain or loss of heat is a result of diminishing or enhancing intermolecular attractions, instead of increasing or decreasing molecular kinetic energies. While a substance is undergoing a change in state, its temperature remains constant. Figure 10.29 shows a typical heating curve.

Consider the example of heating a pot of water to boiling. A stove burner will supply heat at a roughly constant rate; initially, this heat serves to increase the water's temperature. When the water reaches its boiling point, the temperature remains constant despite the continued input of heat from the stove burner. This same temperature is maintained by the water as long as it is boiling. If the burner setting is increased to provide heat at a greater rate, the water temperature does not rise, but instead the boiling becomes more vigorous (rapid). This behavior is observed for other phase transitions as well: For example, temperature remains constant while the change of state is in progress.

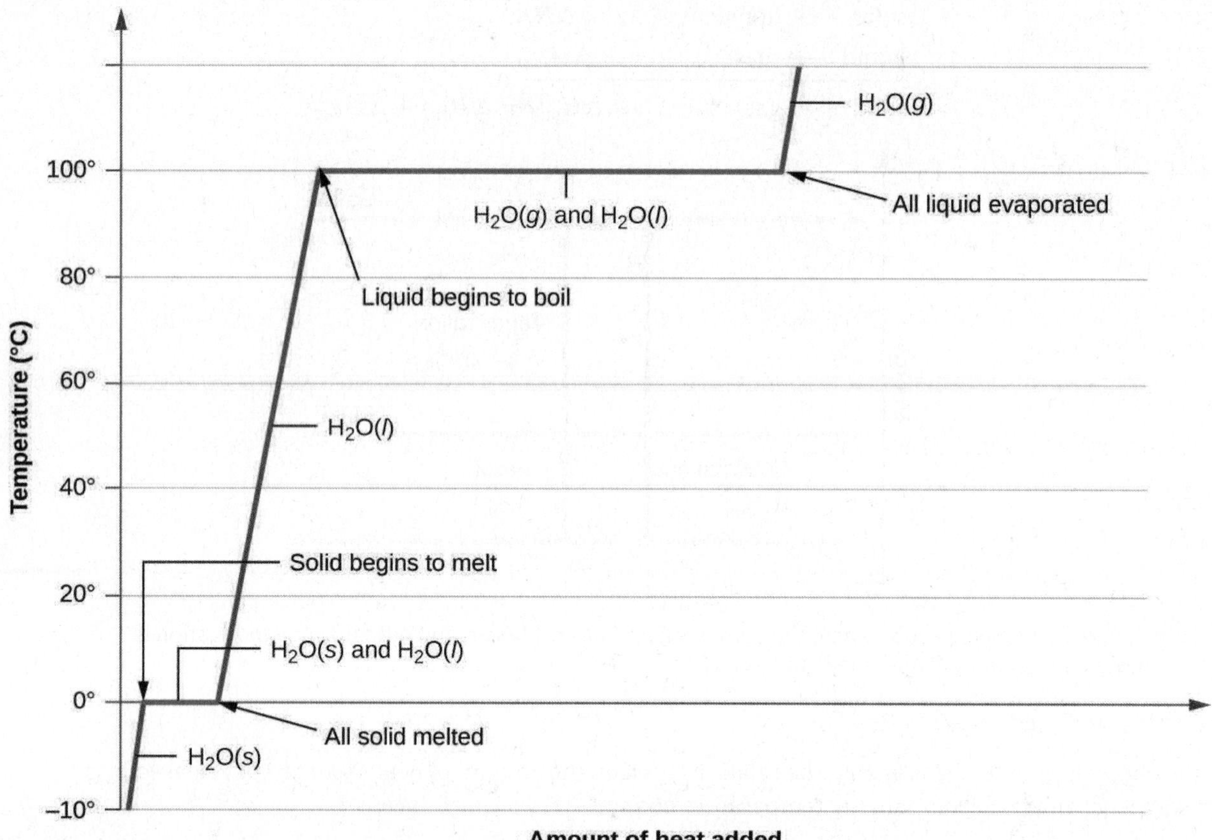

FIGURE 10.29 A typical heating curve for a substance depicts changes in temperature that result as the substance absorbs increasing amounts of heat. Plateaus in the curve (regions of constant temperature) are exhibited when the substance undergoes phase transitions.

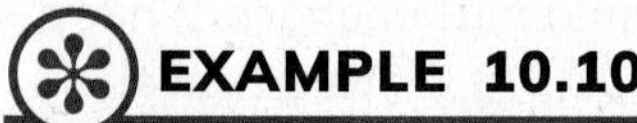

EXAMPLE 10.10

Total Heat Needed to Change Temperature and Phase for a Substance

How much heat is required to convert 135 g of ice at −15 °C into water vapor at 120 °C?

Solution

The transition described involves the following steps:

1. Heat ice from −15 °C to 0 °C
2. Melt ice
3. Heat water from 0 °C to 100 °C
4. Boil water
5. Heat steam from 100 °C to 120 °C

The heat needed to change the temperature of a given substance (with no change in phase) is: $q = m \times c \times \Delta T$ (see previous chapter on thermochemistry). The heat needed to induce a given change in phase is given by $q = n \times \Delta H$.

Using these equations with the appropriate values for specific heat of ice, water, and steam, and enthalpies of fusion and vaporization, we have:

$$q_{\text{total}} = (m{\cdot}c{\cdot}\Delta T)_{\text{ice}} + n{\cdot}\Delta H_{\text{fus}} + (m{\cdot}c{\cdot}\Delta T)_{\text{water}} + n{\cdot}\Delta H_{\text{vap}} + (m{\cdot}c{\cdot}\Delta T)_{\text{steam}}$$

$$= (135\ \text{g}\cdot 2.09\ \text{J/g}\cdot{}^\circ\text{C}\cdot 15^\circ\text{C}) + \left(135\cdot\frac{1\ \text{mol}}{18.02\ \text{g}}\cdot 6.01\ \text{kJ/mol}\right)$$
$$+ (135\ \text{g}\cdot 4.18\ \text{J/g}\cdot{}^\circ\text{C}\cdot 100^\circ\text{C}) + \left(135\ \text{g}\cdot\frac{1\ \text{mol}}{18.02\ \text{g}}\cdot 40.67\ \text{kJ/mol}\right)$$
$$+ (135\ \text{g}\cdot 1.84\ \text{J/g}\cdot{}^\circ\text{C}\cdot 20^\circ\text{C})$$
$$= 4230\ \text{J} + 45.0\ \text{kJ} + 56{,}500\ \text{J} + 305\ \text{kJ} + 4970\ \text{J}$$

Converting the quantities in J to kJ permits them to be summed, yielding the total heat required:

$$= 4.23\ \text{kJ} + 45.0\ \text{kJ} + 56.5\ \text{kJ} + 305\ \text{kJ} + 4.97\ \text{kJ} = 416\ \text{kJ}$$

Check Your Learning

What is the total amount of heat released when 94.0 g water at 80.0 °C cools to form ice at −30.0 °C?

Answer:

68.7 kJ

10.4 Phase Diagrams

LEARNING OBJECTIVES

By the end of this section, you will be able to:

- Explain the construction and use of a typical phase diagram
- Use phase diagrams to identify stable phases at given temperatures and pressures, and to describe phase transitions resulting from changes in these properties
- Describe the supercritical fluid phase of matter

In the previous module, the variation of a liquid's equilibrium vapor pressure with temperature was described. Considering the definition of boiling point, plots of vapor pressure versus temperature represent how the boiling point of the liquid varies with pressure. Also described was the use of heating and cooling curves to determine a substance's melting (or freezing) point. Making such measurements over a wide range of pressures yields data that may be presented graphically as a phase diagram. A **phase diagram** combines plots of pressure versus temperature for the liquid-gas, solid-liquid, and solid-gas phase-transition equilibria of a substance. These diagrams indicate the physical states that exist under specific conditions of pressure and temperature, and also provide the pressure dependence of the phase-transition temperatures (melting points, sublimation points, boiling points). A typical phase diagram for a pure substance is shown in Figure 10.30.

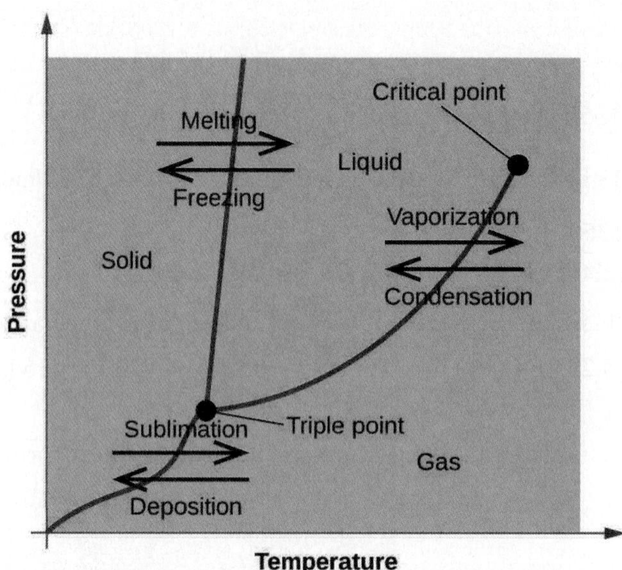

FIGURE 10.30 The physical state of a substance and its phase-transition temperatures are represented graphically in a phase diagram.

To illustrate the utility of these plots, consider the phase diagram for water shown in Figure 10.31.

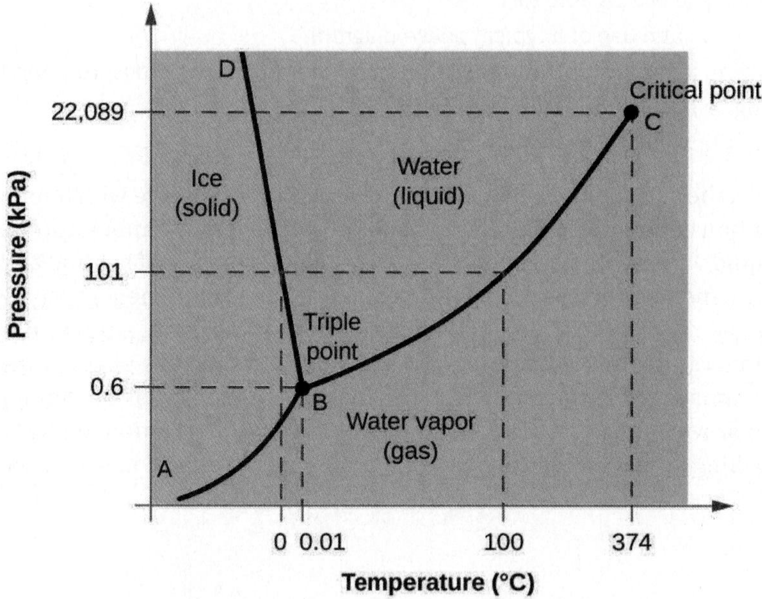

FIGURE 10.31 The pressure and temperature axes on this phase diagram of water are not drawn to constant scale in order to illustrate several important properties.

We can use the phase diagram to identify the physical state of a sample of water under specified conditions of pressure and temperature. For example, a pressure of 50 kPa and a temperature of −10 °C correspond to the region of the diagram labeled “ice.” Under these conditions, water exists only as a solid (ice). A pressure of 50 kPa and a temperature of 50 °C correspond to the “water” region—here, water exists only as a liquid. At 25 kPa and 200 °C, water exists only in the gaseous state. Note that on the H_2O phase diagram, the pressure and temperature axes are not drawn to a constant scale in order to permit the illustration of several important features as described here.

The curve BC in Figure 10.31 is the plot of vapor pressure versus temperature as described in the previous module of this chapter. This “liquid-vapor” curve separates the liquid and gaseous regions of the phase diagram and provides the boiling point for water at any pressure. For example, at 1 atm, the boiling point is 100 °C. Notice that the liquid-vapor curve terminates at a temperature of 374 °C and a pressure of 218 atm, indicating that water cannot exist as a liquid above this temperature, regardless of the pressure. The physical properties of water under these conditions are intermediate between those of its liquid and gaseous phases. This unique state of matter is called a supercritical fluid, a topic that will be described in the next section of this module.

The solid-vapor curve, labeled AB in Figure 10.31, indicates the temperatures and pressures at which ice and water vapor are in equilibrium. These temperature-pressure data pairs correspond to the sublimation, or deposition, points for water. If we could zoom in on the solid-gas line in Figure 10.31, we would see that ice has a vapor pressure of about 0.20 kPa at −10 °C. Thus, if we place a frozen sample in a vacuum with a pressure less than 0.20 kPa, ice will sublime. This is the basis for the "freeze-drying" process often used to preserve foods, such as the ice cream shown in Figure 10.32.

FIGURE 10.32 Freeze-dried foods, like this ice cream, are dehydrated by sublimation at pressures below the triple point for water. (credit: "lwao"/Flickr)

The solid-liquid curve labeled BD shows the temperatures and pressures at which ice and liquid water are in equilibrium, representing the melting/freezing points for water. Note that this curve exhibits a slight negative slope (greatly exaggerated for clarity), indicating that the melting point for water decreases slightly as pressure increases. Water is an unusual substance in this regard, as most substances exhibit an increase in melting point with increasing pressure. This behavior is partly responsible for the movement of glaciers, like the one shown in Figure 10.33. The bottom of a glacier experiences an immense pressure due to its weight that can melt some of the ice, forming a layer of liquid water on which the glacier may more easily slide.

FIGURE 10.33 The immense pressures beneath glaciers result in partial melting to produce a layer of water that provides lubrication to assist glacial movement. This satellite photograph shows the advancing edge of the Perito Moreno glacier in Argentina. (credit: NASA)

The point of intersection of all three curves is labeled B in Figure 10.31. At the pressure and temperature represented by this point, all three phases of water coexist in equilibrium. This temperature-pressure data pair is called the **triple point**. At pressures lower than the triple point, water cannot exist as a liquid, regardless of the temperature.

EXAMPLE 10.11

Determining the State of Water

Using the phase diagram for water given in Figure 10.31, determine the state of water at the following temperatures and pressures:

(a) −10 °C and 50 kPa

(b) 25 °C and 90 kPa

(c) 50 °C and 40 kPa

(d) 80 °C and 5 kPa

(e) −10 °C and 0.3 kPa

(f) 50 °C and 0.3 kPa

Solution

Using the phase diagram for water, we can determine that the state of water at each temperature and pressure given are as follows: (a) solid; (b) liquid; (c) liquid; (d) gas; (e) solid; (f) gas.

Check Your Learning

What phase changes can water undergo as the temperature changes if the pressure is held at 0.3 kPa? If the pressure is held at 50 kPa?

Answer:

At 0.3 kPa: $\text{s} \longrightarrow \text{g}$ at −58 °C. At 50 kPa: $\text{s} \longrightarrow \text{l}$ at 0 °C, $\text{l} \longrightarrow \text{g}$ at 78 °C

Consider the phase diagram for carbon dioxide shown in Figure 10.34 as another example. The solid-liquid curve exhibits a positive slope, indicating that the melting point for CO_2 increases with pressure as it does for most substances (water being a notable exception as described previously). Notice that the triple point is well above 1 atm, indicating that carbon dioxide cannot exist as a liquid under ambient pressure conditions. Instead, cooling gaseous carbon dioxide at 1 atm results in its deposition into the solid state. Likewise, solid carbon dioxide does not melt at 1 atm pressure but instead sublimes to yield gaseous CO_2. Finally, notice that the critical point for carbon dioxide is observed at a relatively modest temperature and pressure in comparison to water.

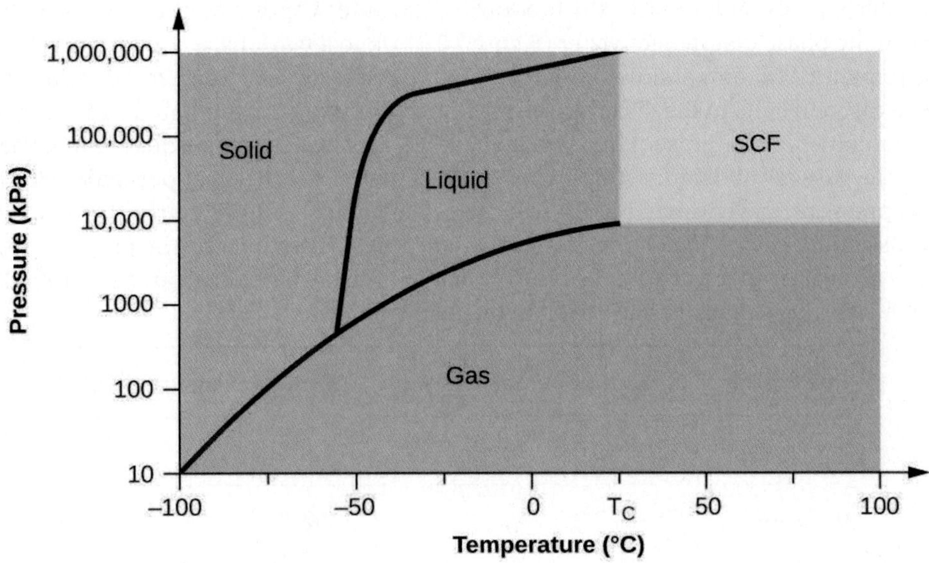

FIGURE 10.34 A phase diagram for carbon dioxide is shown. The pressure axis is plotted on a logarithmic scale to accommodate the large range of values.

EXAMPLE 10.12

Determining the State of Carbon Dioxide

Using the phase diagram for carbon dioxide shown in Figure 10.34, determine the state of CO_2 at the following temperatures and pressures:

(a) −30 °C and 2000 kPa

(b) −90 °C and 1000 kPa

(c) −60 °C and 100 kPa

(d) −40 °C and 1500 kPa

(e) 0 °C and 100 kPa

(f) 20 °C and 100 kPa

Solution

Using the phase diagram for carbon dioxide provided, we can determine that the state of CO_2 at each temperature and pressure given are as follows: (a) liquid; (b) solid; (c) gas; (d) liquid; (e) gas; (f) gas.

Check Your Learning

Identify the phase changes that carbon dioxide will undergo as its temperature is increased from −100 °C while holding its pressure constant at 1500 kPa. At 50 kPa. At what approximate temperatures do these phase changes occur?

Answer:

at 1500 kPa: s $\longrightarrow$ l at −55 °C, l $\longrightarrow$ g at −10 °C;

at 50 kPa: s $\longrightarrow$ g at −60 °C

Supercritical Fluids

If we place a sample of water in a sealed container at 25 °C, remove the air, and let the vaporization-condensation equilibrium establish itself, we are left with a mixture of liquid water and water vapor at a pressure of 0.03 atm. A distinct boundary between the more dense liquid and the less dense gas is clearly observed. As we increase the temperature, the pressure of the water vapor increases, as described by the liquid-gas curve in the phase diagram for water (Figure 10.31), and a two-phase equilibrium of liquid and gaseous phases remains. At a temperature of 374 °C, the vapor pressure has risen to 218 atm, and any further increase in temperature results in the disappearance of the boundary between liquid and vapor phases. All of the water in the container is now present in a single phase whose physical properties are intermediate between those of the gaseous and liquid states. This phase of matter is called a **supercritical fluid**, and the temperature and pressure above which this phase exists is the **critical point** (Figure 10.35). Above its critical temperature, a gas cannot be liquefied no matter how much pressure is applied. The pressure required to liquefy a gas at its critical temperature is called the critical pressure. The critical temperatures and critical pressures of some common substances are given in the following table.

Substance	Critical Temperature (°C)	Critical Pressure (kPa)
hydrogen	−240.0	1300
nitrogen	−147.2	3400
oxygen	−118.9	5000
carbon dioxide	31.1	7400
ammonia	132.4	11,300
sulfur dioxide	157.2	7800
water	374.0	22,000

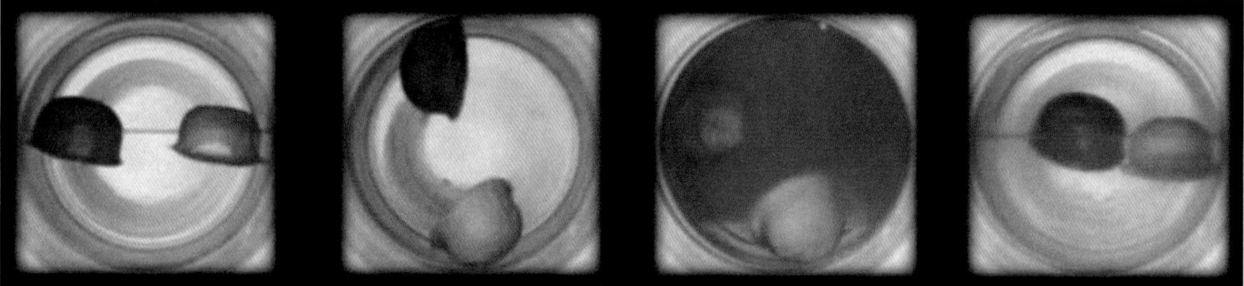

FIGURE 10.35 (a) A sealed container of liquid carbon dioxide slightly below its critical point is heated, resulting in (b) the formation of the supercritical fluid phase. Cooling the supercritical fluid lowers its temperature and pressure below the critical point, resulting in the reestablishment of separate liquid and gaseous phases (c and d). Colored floats illustrate differences in density between the liquid, gaseous, and supercritical fluid states. (credit: modification of work by "mrmrobin"/YouTube)

LINK TO LEARNING

Observe the liquid-to-supercritical fluid transition (http://openstax.org/l/16supercrit) for carbon dioxide.

Like a gas, a supercritical fluid will expand and fill a container, but its density is much greater than typical gas densities, typically being close to those for liquids. Similar to liquids, these fluids are capable of dissolving nonvolatile solutes. They exhibit essentially no surface tension and very low viscosities, however, so they can more effectively penetrate very small openings in a solid mixture and remove soluble components. These properties make supercritical fluids extremely useful solvents for a wide range of applications. For example, supercritical carbon dioxide has become a very popular solvent in the food industry, being used to decaffeinate coffee, remove fats from potato chips, and extract flavor and fragrance compounds from citrus oils. It is nontoxic, relatively inexpensive, and not considered to be a pollutant. After use, the CO_2 can be easily recovered by reducing the pressure and collecting the resulting gas.

EXAMPLE 10.13

The Critical Temperature of Carbon Dioxide

If we shake a carbon dioxide fire extinguisher on a cool day (18 °C), we can hear liquid CO_2 sloshing around inside the cylinder. However, the same cylinder appears to contain no liquid on a hot summer day (35 °C). Explain these observations.

Solution

On the cool day, the temperature of the CO_2 is below the critical temperature of CO_2, 304 K or 31 °C, so liquid CO_2 is present in the cylinder. On the hot day, the temperature of the CO_2 is greater than its critical temperature of 31 °C. Above this temperature no amount of pressure can liquefy CO_2 so no liquid CO_2 exists in the fire extinguisher.

Check Your Learning

Ammonia can be liquefied by compression at room temperature; oxygen cannot be liquefied under these conditions. Why do the two gases exhibit different behavior?

Answer:

The critical temperature of ammonia is 405.5 K, which is higher than room temperature. The critical temperature of oxygen is below room temperature; thus oxygen cannot be liquefied at room temperature.

Chemistry in Everyday Life

Decaffeinating Coffee Using Supercritical CO_2

Coffee is the world's second most widely traded commodity, following only petroleum. Across the globe, people love coffee's aroma and taste. Many of us also depend on one component of coffee—caffeine—to help us get going in the morning or stay alert in the afternoon. But late in the day, coffee's stimulant effect can keep you from sleeping, so you may choose to drink decaffeinated coffee in the evening.

Since the early 1900s, many methods have been used to decaffeinate coffee. All have advantages and disadvantages, and all depend on the physical and chemical properties of caffeine. Because caffeine is a somewhat polar molecule, it dissolves well in water, a polar liquid. However, since many of the other 400-plus compounds that contribute to coffee's taste and aroma also dissolve in H_2O, hot water decaffeination processes can also remove some of these compounds, adversely affecting the smell and taste of the decaffeinated coffee. Dichloromethane (CH_2Cl_2) and ethyl acetate ($CH_3CO_2C_2H_5$) have similar polarity to caffeine, and are therefore very effective solvents for caffeine extraction, but both also remove some flavor and aroma components, and their use requires long extraction and cleanup times. Because both of these solvents are toxic, health concerns have been raised regarding the effect of residual solvent remaining in the decaffeinated coffee.

Supercritical fluid extraction using carbon dioxide is now being widely used as a more effective and environmentally friendly decaffeination method (Figure 10.36). At temperatures above 304.2 K and pressures above 7376 kPa, CO_2 is a supercritical fluid, with properties of both gas and liquid. Like a gas, it penetrates deep into the coffee beans; like a liquid, it effectively dissolves certain substances. Supercritical carbon dioxide extraction of steamed coffee beans removes 97–99% of the caffeine, leaving coffee's flavor and aroma compounds intact. Because CO_2 is a gas under standard conditions, its removal from the extracted coffee beans is easily accomplished, as is the recovery of the caffeine from the extract. The caffeine recovered from coffee beans via this process is a valuable product that can be used subsequently as an additive to other foods or drugs.

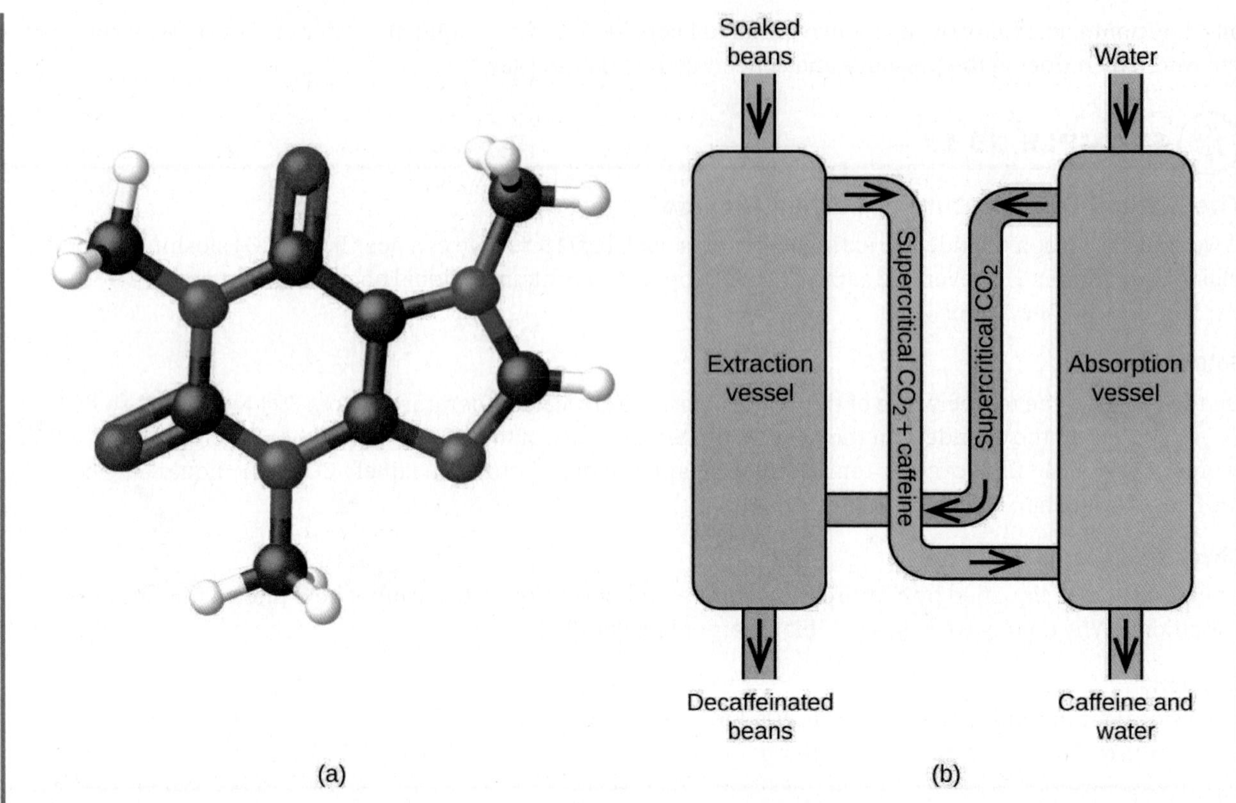

FIGURE 10.36 (a) Caffeine molecules have both polar and nonpolar regions, making it soluble in solvents of varying polarities. (b) The schematic shows a typical decaffeination process involving supercritical carbon dioxide.

10.5 The Solid State of Matter

LEARNING OBJECTIVES

By the end of this section, you will be able to:

- Define and describe the bonding and properties of ionic, molecular, metallic, and covalent network crystalline solids
- Describe the main types of crystalline solids: ionic solids, metallic solids, covalent network solids, and molecular solids
- Explain the ways in which crystal defects can occur in a solid

When most liquids are cooled, they eventually freeze and form **crystalline solids**, solids in which the atoms, ions, or molecules are arranged in a definite repeating pattern. It is also possible for a liquid to freeze before its molecules become arranged in an orderly pattern. The resulting materials are called **amorphous solids** or noncrystalline solids (or, sometimes, glasses). The particles of such solids lack an ordered internal structure and are randomly arranged (Figure 10.37).

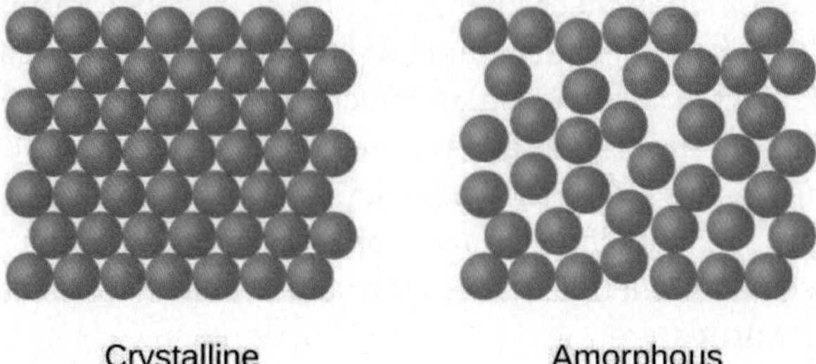

FIGURE 10.37 The entities of a solid phase may be arranged in a regular, repeating pattern (crystalline solids) or

randomly (amorphous).

Metals and ionic compounds typically form ordered, crystalline solids. Substances that consist of large molecules, or a mixture of molecules whose movements are more restricted, often form amorphous solids. For examples, candle waxes are amorphous solids composed of large hydrocarbon molecules. Some substances, such as silicon dioxide (shown in Figure 10.38), can form either crystalline or amorphous solids, depending on the conditions under which it is produced. Also, amorphous solids may undergo a transition to the crystalline state under appropriate conditions.

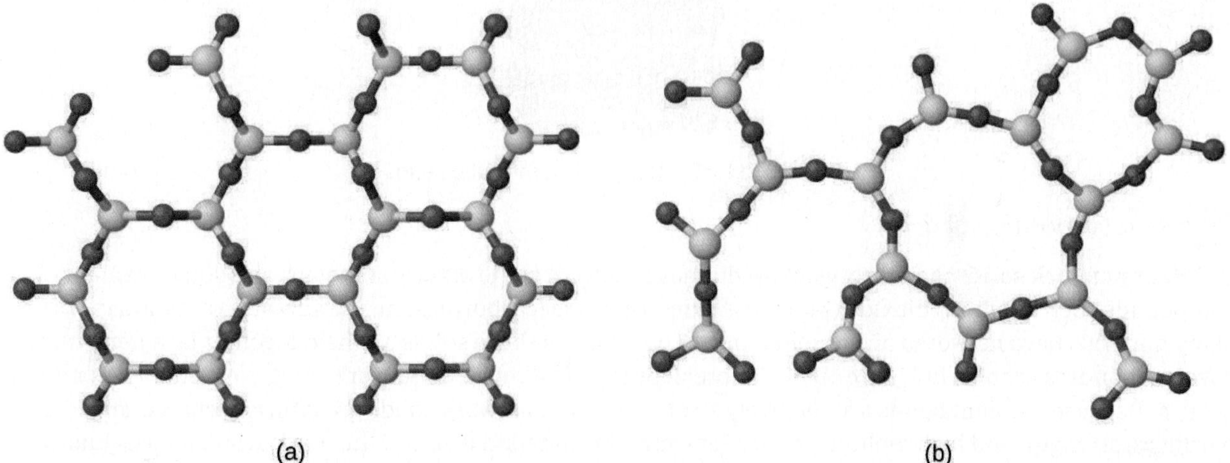

FIGURE 10.38 (a) Silicon dioxide, SiO_2, is abundant in nature as one of several crystalline forms of the mineral quartz. (b) Rapid cooling of molten SiO_2 yields an amorphous solid known as "fused silica".

Crystalline solids are generally classified according to the nature of the forces that hold its particles together. These forces are primarily responsible for the physical properties exhibited by the bulk solids. The following sections provide descriptions of the major types of crystalline solids: ionic, metallic, covalent network, and molecular.

Ionic Solids

Ionic solids, such as sodium chloride and nickel oxide, are composed of positive and negative ions that are held together by electrostatic attractions, which can be quite strong (Figure 10.39). Many ionic crystals also have high melting points. This is due to the very strong attractions between the ions—in ionic compounds, the attractions between full charges are (much) larger than those between the partial charges in polar molecular compounds. This will be looked at in more detail in a later discussion of lattice energies. Although they are hard, they also tend to be brittle, and they shatter rather than bend. Ionic solids do not conduct electricity; however, they do conduct when molten or dissolved because their ions are free to move. Many simple compounds formed by the reaction of a metallic element with a nonmetallic element are ionic.

FIGURE 10.39 Sodium chloride is an ionic solid.

Metallic Solids

Metallic solids such as crystals of copper, aluminum, and iron are formed by metal atoms Figure 10.40. The structure of metallic crystals is often described as a uniform distribution of atomic nuclei within a "sea" of delocalized electrons. The atoms within such a metallic solid are held together by a unique force known as *metallic bonding* that gives rise to many useful and varied bulk properties. All exhibit high thermal and electrical conductivity, metallic luster, and malleability. Many are very hard and quite strong. Because of their malleability (the ability to deform under pressure or hammering), they do not shatter and, therefore, make useful construction materials. The melting points of the metals vary widely. Mercury is a liquid at room temperature, and the alkali metals melt below 200 °C. Several post-transition metals also have low melting points, whereas the transition metals melt at temperatures above 1000 °C. These differences reflect differences in strengths of metallic bonding among the metals.

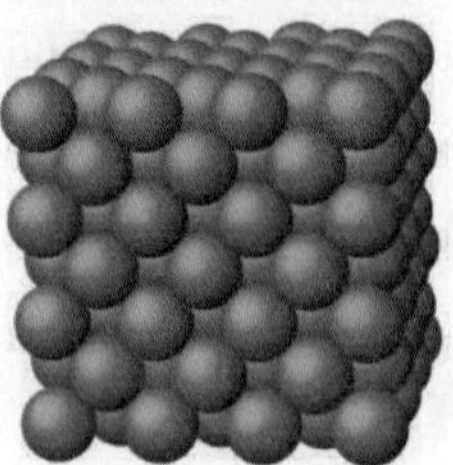

FIGURE 10.40 Copper is a metallic solid.

Covalent Network Solid

Covalent network solids include crystals of diamond, silicon, some other nonmetals, and some covalent compounds such as silicon dioxide (sand) and silicon carbide (carborundum, the abrasive on sandpaper). Many minerals have networks of covalent bonds. The atoms in these solids are held together by a network of covalent bonds, as shown in Figure 10.41. To break or to melt a covalent network solid, covalent bonds must be broken. Because covalent bonds are relatively strong, covalent network solids are typically characterized by hardness, strength, and high melting points. For example, diamond is one of the hardest substances known and melts above 3500 °C.

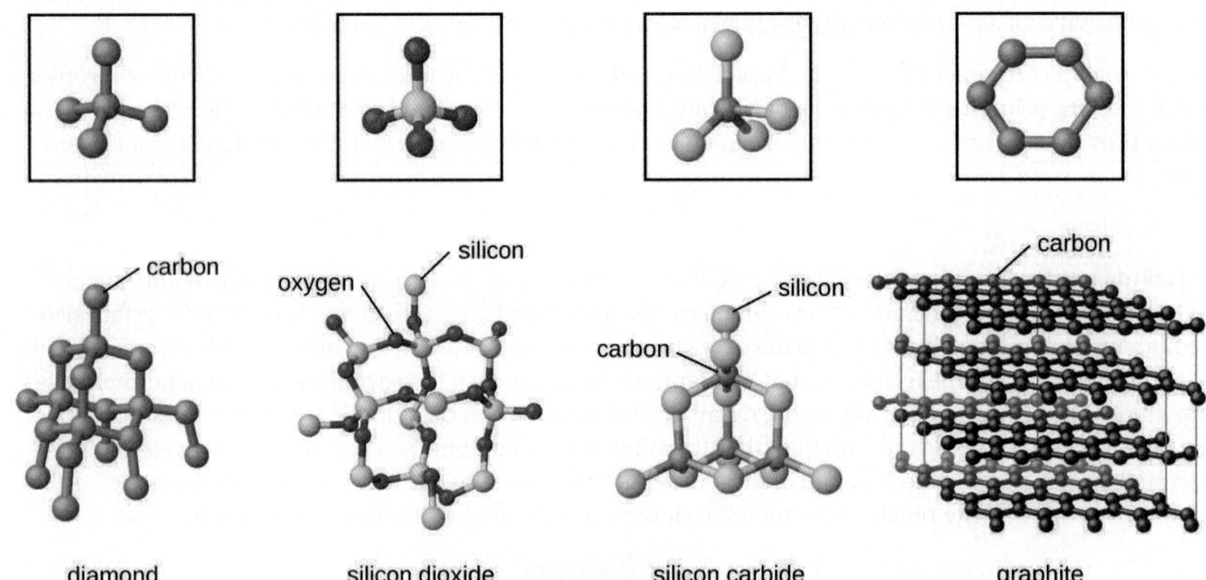

FIGURE 10.41 A covalent crystal contains a three-dimensional network of covalent bonds, as illustrated by the structures of diamond, silicon dioxide, silicon carbide, and graphite. Graphite is an exceptional example, composed of planar sheets of covalent crystals that are held together in layers by noncovalent forces. Unlike typical covalent solids, graphite is very soft and electrically conductive.

Molecular Solid

Molecular solids, such as ice, sucrose (table sugar), and iodine, as shown in Figure 10.42, are composed of neutral molecules. The strengths of the attractive forces between the units present in different crystals vary widely, as indicated by the melting points of the crystals. Small symmetrical molecules (nonpolar molecules), such as H_2, N_2, O_2, and F_2, have weak attractive forces and form molecular solids with very low melting points (below −200 °C). Substances consisting of larger, nonpolar molecules have larger attractive forces and melt at higher temperatures. Molecular solids composed of molecules with permanent dipole moments (polar molecules) melt at still higher temperatures. Examples include ice (melting point, 0 °C) and table sugar (melting point, 185 °C).

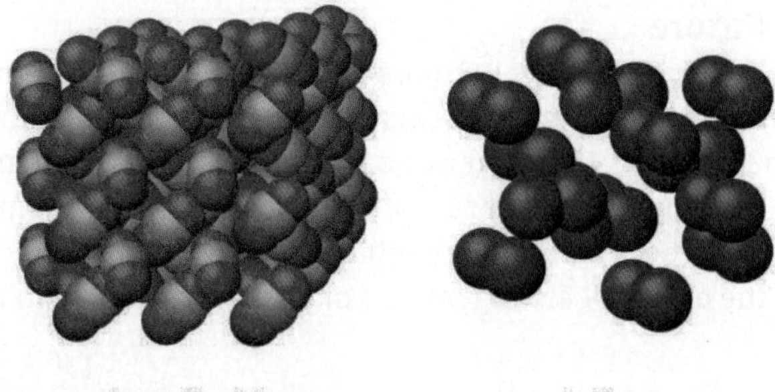

FIGURE 10.42 Carbon dioxide (CO_2) consists of small, nonpolar molecules and forms a molecular solid with a melting point of −78 °C. Iodine (I_2) consists of larger, nonpolar molecules and forms a molecular solid that melts at 114 °C.

Properties of Solids

A crystalline solid, like those listed in Table 10.4, has a precise melting temperature because each atom or molecule of the same type is held in place with the same forces or energy. Thus, the attractions between the units that make up the crystal all have the same strength and all require the same amount of energy to be broken. The gradual softening of an amorphous material differs dramatically from the distinct melting of a crystalline solid. This results from the structural nonequivalence of the molecules in the amorphous solid. Some forces are weaker than others, and when an amorphous material is heated, the weakest intermolecular attractions break first. As the temperature is increased further, the stronger attractions are broken. Thus amorphous materials soften over a range of temperatures.

Types of Crystalline Solids and Their Properties

Type of Solid	Type of Particles	Type of Attractions	Properties	Examples
ionic	ions	ionic bonds	hard, brittle, conducts electricity as a liquid but not as a solid, high to very high melting points	NaCl, Al_2O_3
metallic	atoms of electropositive elements	metallic bonds	shiny, malleable, ductile, conducts heat and electricity well, variable hardness and melting temperature	Cu, Fe, Ti, Pb, U
covalent network	atoms of electronegative elements	covalent bonds	very hard, not conductive, very high melting points	C (diamond), SiO_2, SiC
molecular	molecules (or atoms)	IMFs	variable hardness, variable brittleness, not conductive, low melting points	H_2O, CO_2, I_2, $C_{12}H_{22}O_{11}$

TABLE 10.4

HOW SCIENCES INTERCONNECT

Graphene: Material of the Future

Carbon is an essential element in our world. The unique properties of carbon atoms allow the existence of carbon-based life forms such as ourselves. Carbon forms a huge variety of substances that we use on a daily basis, including those shown in Figure 10.43. You may be familiar with diamond and graphite, the two most common *allotropes* of carbon. (Allotropes are different structural forms of the same element.) Diamond is one of the hardest-known substances, whereas graphite is soft enough to be used as pencil lead. These very different properties stem from the different arrangements of the carbon atoms in the different allotropes.

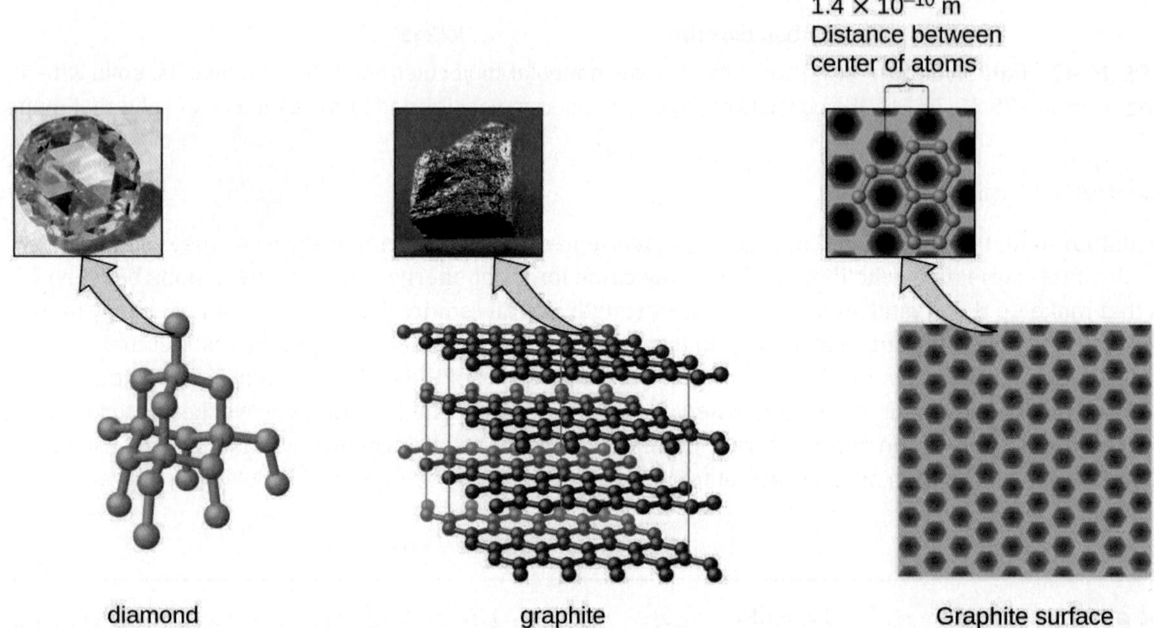

FIGURE 10.43 Diamond is extremely hard because of the strong bonding between carbon atoms in all directions. Graphite (in pencil lead) rubs off onto paper due to the weak attractions between the carbon layers. An image of a graphite surface shows the distance between the centers of adjacent carbon atoms. (credit left photo: modification of work by Steve Jurvetson; credit middle photo: modification of work by United States Geological Survey)

You may be less familiar with a recently discovered form of carbon: graphene. Graphene was first isolated in 2004 by using tape to peel off thinner and thinner layers from graphite. It is essentially a single sheet (one atom thick) of graphite. Graphene, illustrated in Figure 10.44, is not only strong and lightweight, but it is also an excellent conductor of electricity and heat. These properties may prove very useful in a wide range of applications, such as vastly improved computer chips and circuits, better batteries and solar cells, and stronger and lighter structural materials. The 2010 Nobel Prize in Physics was awarded to Andre Geim and Konstantin Novoselov for their pioneering work with graphene.

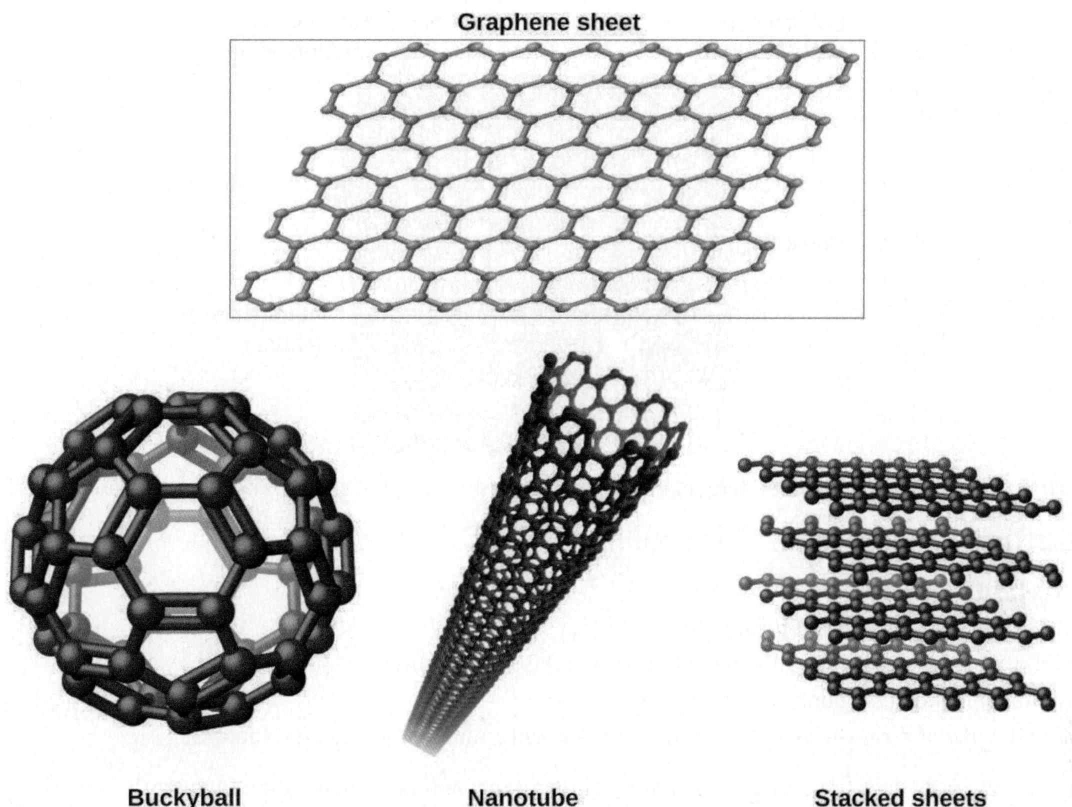

FIGURE 10.44 Graphene sheets can be formed into buckyballs, nanotubes, and stacked layers.

Crystal Defects

In a crystalline solid, the atoms, ions, or molecules are arranged in a definite repeating pattern, but occasional defects may occur in the pattern. Several types of defects are known, as illustrated in Figure 10.45. **Vacancies** are defects that occur when positions that should contain atoms or ions are vacant. Less commonly, some atoms or ions in a crystal may occupy positions, called **interstitial sites**, located between the regular positions for atoms. Other distortions are found in impure crystals, as, for example, when the cations, anions, or molecules of the impurity are too large to fit into the regular positions without distorting the structure. Trace amounts of impurities are sometimes added to a crystal (a process known as *doping)* in order to create defects in the structure that yield desirable changes in its properties. For example, silicon crystals are doped with varying amounts of different elements to yield suitable electrical properties for their use in the manufacture of semiconductors and computer chips.

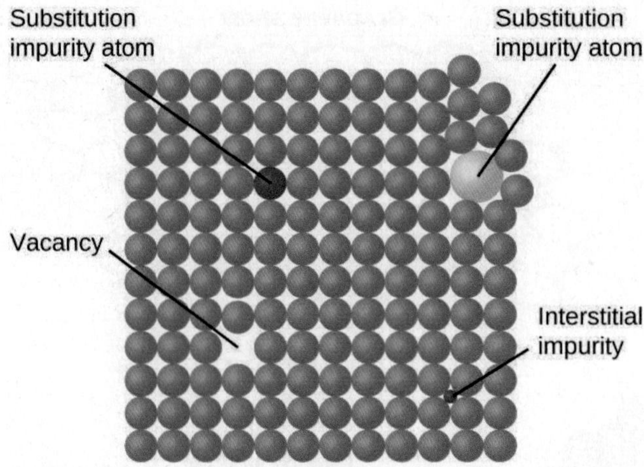

FIGURE 10.45 Types of crystal defects include vacancies, interstitial atoms, and substitutions impurities.

10.6 Lattice Structures in Crystalline Solids

LEARNING OBJECTIVES

By the end of this section, you will be able to:

- Describe the arrangement of atoms and ions in crystalline structures
- Compute ionic radii using unit cell dimensions
- Explain the use of X-ray diffraction measurements in determining crystalline structures

Over 90% of naturally occurring and man-made solids are crystalline. Most solids form with a regular arrangement of their particles because the overall attractive interactions between particles are maximized, and the total intermolecular energy is minimized, when the particles pack in the most efficient manner. The regular arrangement at an atomic level is often reflected at a macroscopic level. In this module, we will explore some of the details about the structures of metallic and ionic crystalline solids, and learn how these structures are determined experimentally.

The Structures of Metals

We will begin our discussion of crystalline solids by considering elemental metals, which are relatively simple because each contains only one type of atom. A pure metal is a crystalline solid with metal atoms packed closely together in a repeating pattern. Some of the properties of metals in general, such as their malleability and ductility, are largely due to having identical atoms arranged in a regular pattern. The different properties of one metal compared to another partially depend on the sizes of their atoms and the specifics of their spatial arrangements. We will explore the similarities and differences of four of the most common metal crystal geometries in the sections that follow.

Unit Cells of Metals

The structure of a crystalline solid, whether a metal or not, is best described by considering its simplest repeating unit, which is referred to as its **unit cell**. The unit cell consists of lattice points that represent the locations of atoms or ions. The entire structure then consists of this unit cell repeating in three dimensions, as illustrated in Figure 10.46.

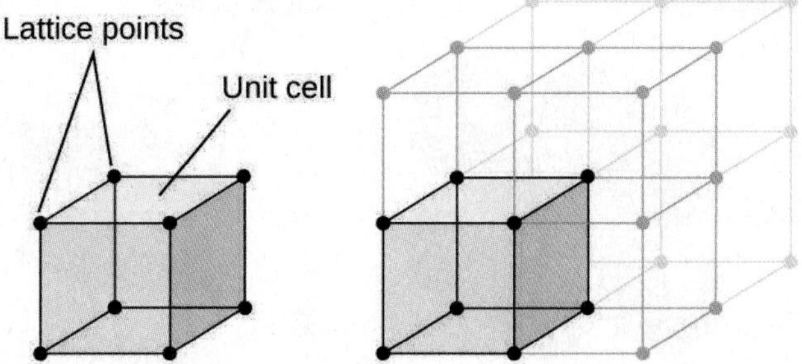

FIGURE 10.46 A unit cell shows the locations of lattice points repeating in all directions.

Let us begin our investigation of crystal lattice structure and unit cells with the most straightforward structure and the most basic unit cell. To visualize this, imagine taking a large number of identical spheres, such as tennis balls, and arranging them uniformly in a container. The simplest way to do this would be to make layers in which the spheres in one layer are directly above those in the layer below, as illustrated in Figure 10.47. This arrangement is called **simple cubic structure**, and the unit cell is called the **simple cubic unit cell** or primitive cubic unit cell.

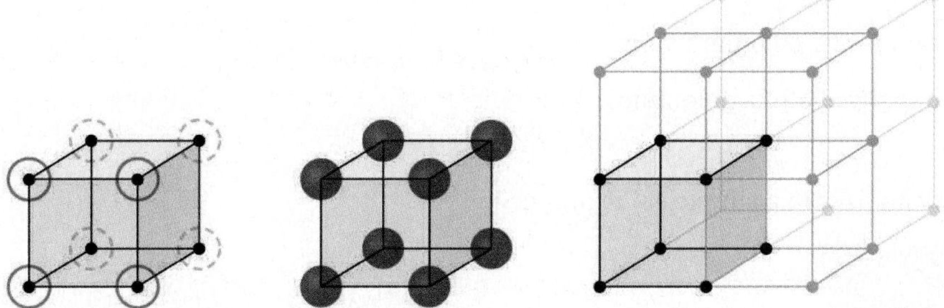

FIGURE 10.47 When metal atoms are arranged with spheres in one layer directly above or below spheres in another layer, the lattice structure is called simple cubic. Note that the spheres are in contact.

In a simple cubic structure, the spheres are not packed as closely as they could be, and they only "fill" about 52% of the volume of the container. This is a relatively inefficient arrangement, and only one metal (polonium, Po) crystallizes in a simple cubic structure. As shown in Figure 10.48, a solid with this type of arrangement consists of planes (or layers) in which each atom contacts only the four nearest neighbors in its layer; one atom directly above it in the layer above; and one atom directly below it in the layer below. The number of other particles that each particle in a crystalline solid contacts is known as its **coordination number**. For a polonium atom in a simple cubic array, the coordination number is, therefore, six.

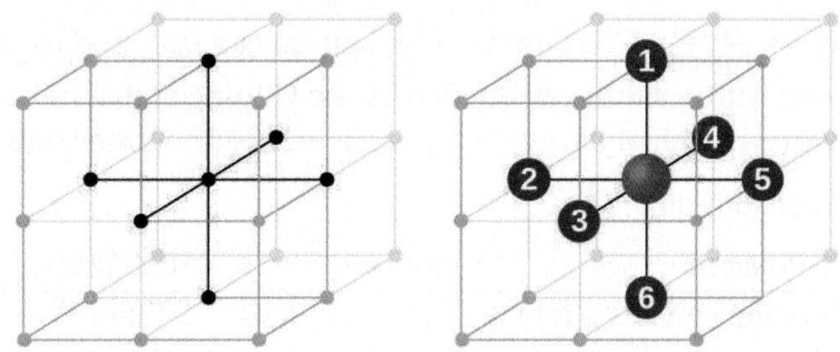

FIGURE 10.48 An atom in a simple cubic lattice structure contacts six other atoms, so it has a coordination number of six.

In a simple cubic lattice, the unit cell that repeats in all directions is a cube defined by the centers of eight atoms, as shown in Figure 10.49. Atoms at adjacent corners of this unit cell contact each other, so the edge length of this cell is equal to two atomic radii, or one atomic diameter. A cubic unit cell contains only the parts of these atoms that are within it. Since an atom at a corner of a simple cubic unit cell is contained by a total of eight unit cells, only one-eighth of that atom is within a specific unit cell. And since each simple cubic unit cell has one atom at each of its eight "corners," there is $8 \times \frac{1}{8} = 1$ atom within one simple cubic unit cell.

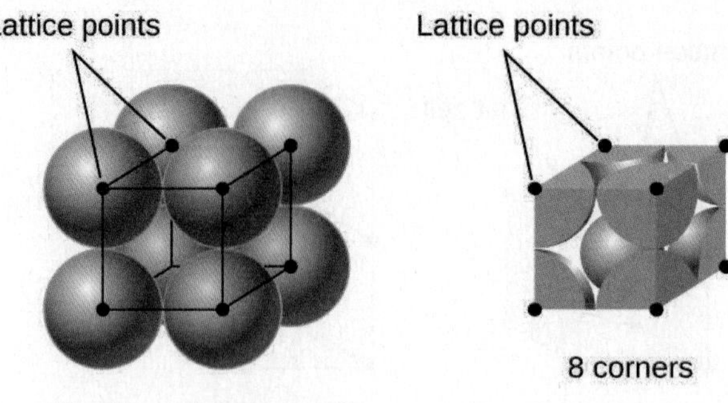

FIGURE 10.49 A simple cubic lattice unit cell contains one-eighth of an atom at each of its eight corners, so it contains one atom total.

EXAMPLE 10.14

Calculation of Atomic Radius and Density for Metals, Part 1

The edge length of the unit cell of alpha polonium is 336 pm.

(a) Determine the radius of a polonium atom.

(b) Determine the density of alpha polonium.

Solution

Alpha polonium crystallizes in a simple cubic unit cell:

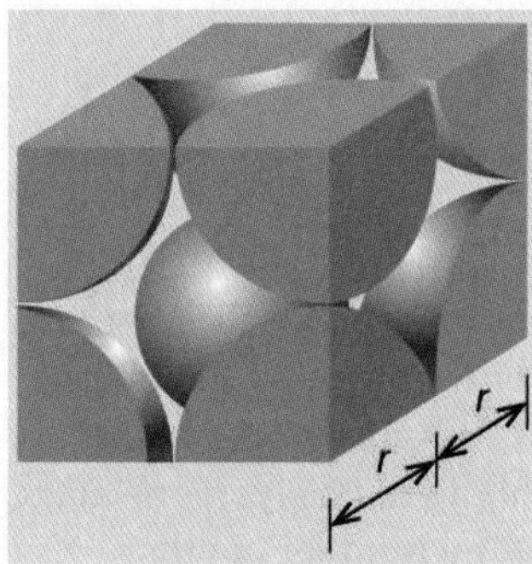

(a) Two adjacent Po atoms contact each other, so the edge length of this cell is equal to two Po atomic radii: l = 2r. Therefore, the radius of Po is $r = \frac{1}{2} = \frac{336\text{ pm}}{2} = 168\text{ pm}$.

(b) Density is given by density $= \frac{\text{mass}}{\text{volume}}$. The density of polonium can be found by determining the density of its unit cell (the mass contained within a unit cell divided by the volume of the unit cell). Since a Po unit cell contains one-eighth of a Po atom at each of its eight corners, a unit cell contains one Po atom.

The mass of a Po unit cell can be found by:

$$1\text{ Po unit cell} \times \frac{1\text{ Po atom}}{1\text{ Po unit cell}} \times \frac{1\text{ mol Po}}{6.022 \times 10^{23}\text{ Po atoms}} \times \frac{208.998\text{ g}}{1\text{ mol Po}} = 3.47 \times 10^{-22}\text{ g}$$

The volume of a Po unit cell can be found by:

$$V = l^3 = \left(336 \times 10^{-10}\text{ cm}\right)^3 = 3.79 \times 10^{-23}\text{ cm}^3$$

(Note that the edge length was converted from pm to cm to get the usual volume units for density.)

Therefore, the density of Po $= \frac{3.471 \times 10^{-22}\text{ g}}{3.79 \times 10^{-23}\text{ cm}^3} = 9.16\text{ g/cm}^3$

Check Your Learning

The edge length of the unit cell for nickel is 0.3524 nm. The density of Ni is 8.90 g/cm^3. Does nickel crystallize in a simple cubic structure? Explain.

Answer:

No. If Ni was simple cubic, its density would be given by:

$$1 \text{ Ni atom} \times \frac{1 \text{ mol Ni}}{6.022 \times 10^{23} \text{ Ni atoms}} \times \frac{58.693 \text{ g}}{1 \text{ mol Ni}} = 9.746 \times 10^{-23} \text{ g}$$

$$V = l^3 = \left(3.524 \times 10^{-8} \text{ cm}\right)^3 = 4.376 \times 10^{-23} \text{ cm}^3$$

Then the density of Ni would be $= \frac{9.746 \times 10^{-23} \text{ g}}{4.376 \times 10^{-23} \text{ cm}^3} = 2.23 \text{ g/cm}^3$

Since the actual density of Ni is not close to this, Ni does not form a simple cubic structure.

Most metal crystals are one of the four major types of unit cells. For now, we will focus on the three cubic unit cells: simple cubic (which we have already seen), **body-centered cubic unit cell**, and **face-centered cubic unit cell**—all of which are illustrated in Figure 10.50. (Note that there are actually seven different lattice systems, some of which have more than one type of lattice, for a total of 14 different types of unit cells. We leave the more complicated geometries for later in this module.)

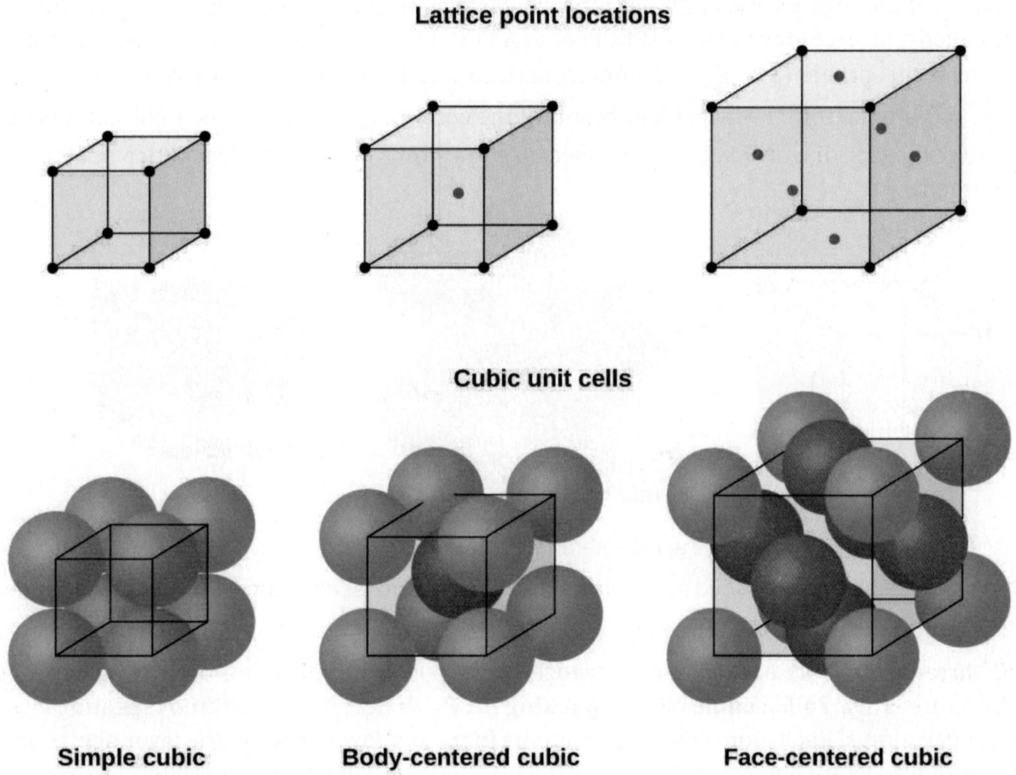

FIGURE 10.50 Cubic unit cells of metals show (in the upper figures) the locations of lattice points and (in the lower figures) metal atoms located in the unit cell.

Some metals crystallize in an arrangement that has a cubic unit cell with atoms at all of the corners and an atom in the center, as shown in Figure 10.51. This is called a **body-centered cubic (BCC) solid**. Atoms in the corners of a BCC unit cell do not contact each other but contact the atom in the center. A BCC unit cell contains two atoms: one-eighth of an atom at each of the eight corners ($8 \times \frac{1}{8} = 1$ atom from the corners) plus one atom from the center. Any atom in this structure touches four atoms in the layer above it and four atoms in the layer below it. Thus, an atom in a BCC structure has a coordination number of eight.

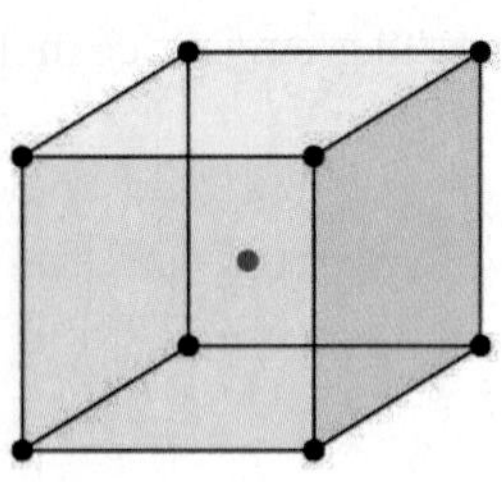
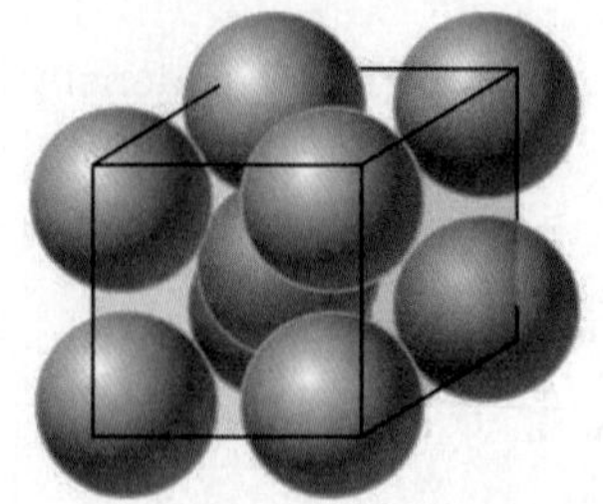

Body-centered cubic structure

FIGURE 10.51 In a body-centered cubic structure, atoms in a specific layer do not touch each other. Each atom touches four atoms in the layer above it and four atoms in the layer below it.

Atoms in BCC arrangements are much more efficiently packed than in a simple cubic structure, occupying about 68% of the total volume. Isomorphous metals with a BCC structure include K, Ba, Cr, Mo, W, and Fe at room temperature. (Elements or compounds that crystallize with the same structure are said to be **isomorphous**.)

Many other metals, such as aluminum, copper, and lead, crystallize in an arrangement that has a cubic unit cell with atoms at all of the corners and at the centers of each face, as illustrated in Figure 10.52. This arrangement is called a **face-centered cubic (FCC) solid**. A FCC unit cell contains four atoms: one-eighth of an atom at each of the eight corners ($8 \times \frac{1}{8} = 1$ atom from the corners) and one-half of an atom on each of the six faces ($6 \times \frac{1}{2} = 3$ atoms from the faces). The atoms at the corners touch the atoms in the centers of the adjacent faces along the face diagonals of the cube. Because the atoms are on identical lattice points, they have identical environments.

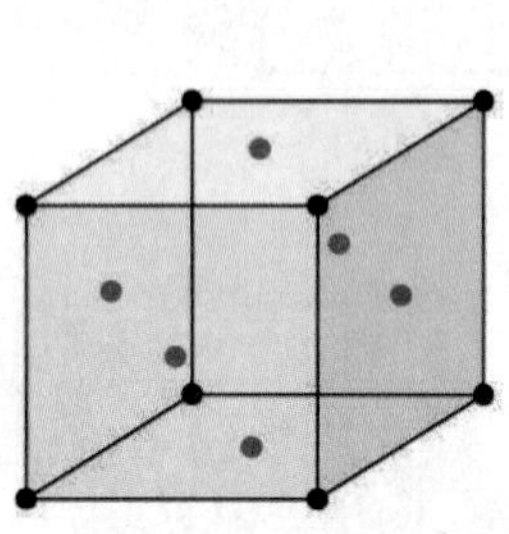
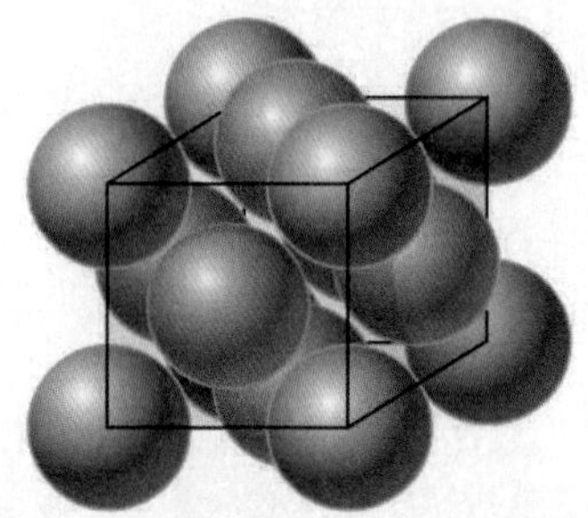

Face-centered cubic structure

FIGURE 10.52 A face-centered cubic solid has atoms at the corners and, as the name implies, at the centers of the faces of its unit cells.

Atoms in an FCC arrangement are packed as closely together as possible, with atoms occupying 74% of the volume. This structure is also called **cubic closest packing (CCP)**. In CCP, there are three repeating layers of hexagonally arranged atoms. Each atom contacts six atoms in its own layer, three in the layer above, and three in the layer below. In this arrangement, each atom touches 12 near neighbors, and therefore has a coordination number of 12. The fact that FCC and CCP arrangements are equivalent may not be immediately obvious, but why they are actually the same structure is illustrated in Figure 10.53.

FIGURE 10.53 A CCP arrangement consists of three repeating layers (ABCABC...) of hexagonally arranged atoms. Atoms in a CCP structure have a coordination number of 12 because they contact six atoms in their layer, plus three atoms in the layer above and three atoms in the layer below. By rotating our perspective, we can see that a CCP structure has a unit cell with a face containing an atom from layer A at one corner, atoms from layer B across a diagonal (at two corners and in the middle of the face), and an atom from layer C at the remaining corner. This is the same as a face-centered cubic arrangement.

Because closer packing maximizes the overall attractions between atoms and minimizes the total intermolecular energy, the atoms in most metals pack in this manner. We find two types of closest packing in simple metallic crystalline structures: CCP, which we have already encountered, and **hexagonal closest packing (HCP)** shown in Figure 10.54. Both consist of repeating layers of hexagonally arranged atoms. In both types, a second layer (B) is placed on the first layer (A) so that each atom in the second layer is in contact with three atoms in the first layer. The third layer is positioned in one of two ways. In HCP, atoms in the third layer are directly above atoms in the first layer (i.e., the third layer is also type A), and the stacking consists of alternating type A and type B close-packed layers (i.e., ABABAB⋯). In CCP, atoms in the third layer are not above atoms in either of the first two layers (i.e., the third layer is type C), and the stacking consists of alternating type A, type B, and type C close-packed layers (i.e., ABCABCABC⋯). About two–thirds of all metals crystallize in closest-packed arrays with coordination numbers of 12. Metals that crystallize in an HCP structure include Cd, Co, Li, Mg, Na, and Zn, and metals that crystallize in a CCP structure include Ag, Al, Ca, Cu, Ni, Pb, and Pt.

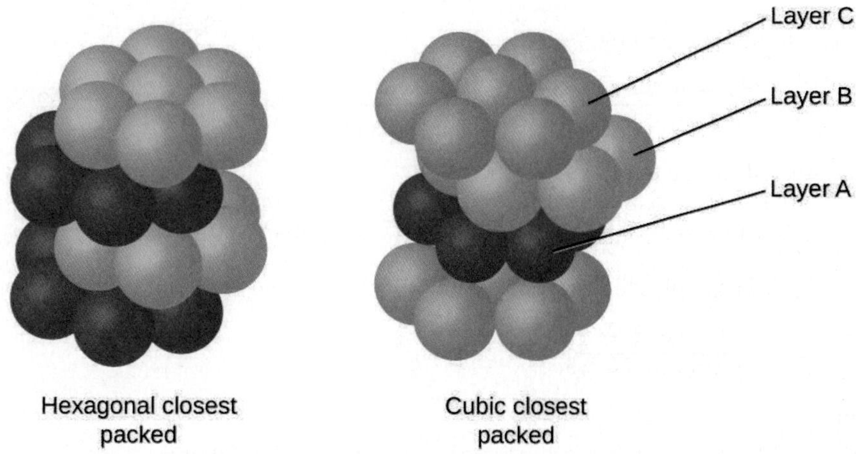

FIGURE 10.54 In both types of closest packing, atoms are packed as compactly as possible. Hexagonal closest packing consists of two alternating layers (ABABAB...). Cubic closest packing consists of three alternating layers (ABCABCABC...).

EXAMPLE 10.15

Calculation of Atomic Radius and Density for Metals, Part 2

Calcium crystallizes in a face-centered cubic structure. The edge length of its unit cell is 558.8 pm.

(a) What is the atomic radius of Ca in this structure?

(b) Calculate the density of Ca.

Solution

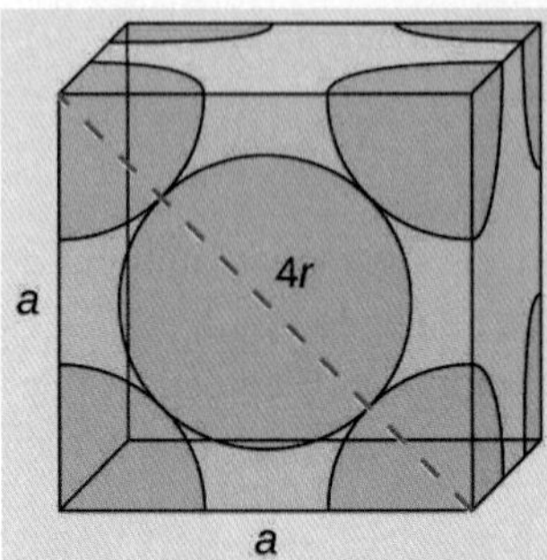

(a) In an FCC structure, Ca atoms contact each other across the diagonal of the face, so the length of the diagonal is equal to four Ca atomic radii (d = 4*r*). Two adjacent edges and the diagonal of the face form a right triangle, with the length of each side equal to 558.8 pm and the length of the hypotenuse equal to four Ca atomic radii:

$$a^2 + a^2 = d^2 \longrightarrow (558.8\text{ pm})^2 + (558.5\text{ pm})^2 = (4r)^2$$

Solving this gives $r = \sqrt{\frac{(558.8\text{ pm})^2+(558.5\text{ pm})^2}{16}} = 197.6\text{ pm}$ for a Ca radius.

(b) Density is given by density $= \frac{\text{mass}}{\text{volume}}$. The density of calcium can be found by determining the density of its unit cell: for example, the mass contained within a unit cell divided by the volume of the unit cell. A face-centered Ca unit cell has one-eighth of an atom at each of the eight corners ($8 \times \frac{1}{8} = 1$ atom) and one-half of an atom on each of the six faces $6 \times \frac{1}{2} = 3$ atoms), for a total of four atoms in the unit cell.

The mass of the unit cell can be found by:

$$1\text{ Ca unit cell} \times \frac{4\text{ Ca atoms}}{1\text{ Ca unit cell}} \times \frac{1\text{ mol Ca}}{6.022 \times 10^{23}\text{ Ca atoms}} \times \frac{40.078\text{ g}}{1\text{ mol Ca}} = 2.662 \times 10^{-22}\text{ g}$$

The volume of a Ca unit cell can be found by:

$$V = a^3 = \left(558.8 \times 10^{-10}\text{ cm}\right)^3 = 1.745 \times 10^{-22}\text{ cm}^3$$

(Note that the edge length was converted from pm to cm to get the usual volume units for density.)

Then, the density of Ca $= \frac{2.662 \times 10^{-22}\text{ g}}{1.745 \times 10^{-22}\text{ cm}^3} = 1.53\text{ g/cm}^3$

Check Your Learning

Silver crystallizes in an FCC structure. The edge length of its unit cell is 409 pm.

(a) What is the atomic radius of Ag in this structure?

(b) Calculate the density of Ag.

Answer:

(a) 144 pm; (b) 10.5 g/cm^3

In general, a unit cell is defined by the lengths of three axes (*a*, *b*, and *c*) and the angles (α, β, and γ) between them, as illustrated in Figure 10.55. The axes are defined as being the lengths between points in the space lattice. Consequently, unit cell axes join points with identical environments.

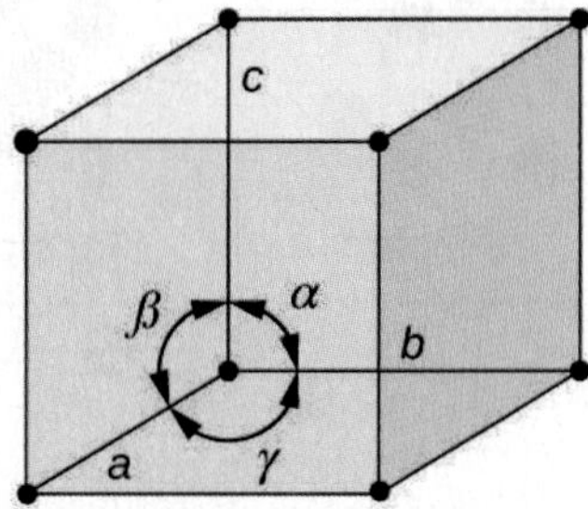

FIGURE 10.55 A unit cell is defined by the lengths of its three axes (*a*, *b*, and *c*) and the angles (α, β, and γ) between the axes.

There are seven different lattice systems, some of which have more than one type of lattice, for a total of fourteen different unit cells, which have the shapes shown in Figure 10.56.

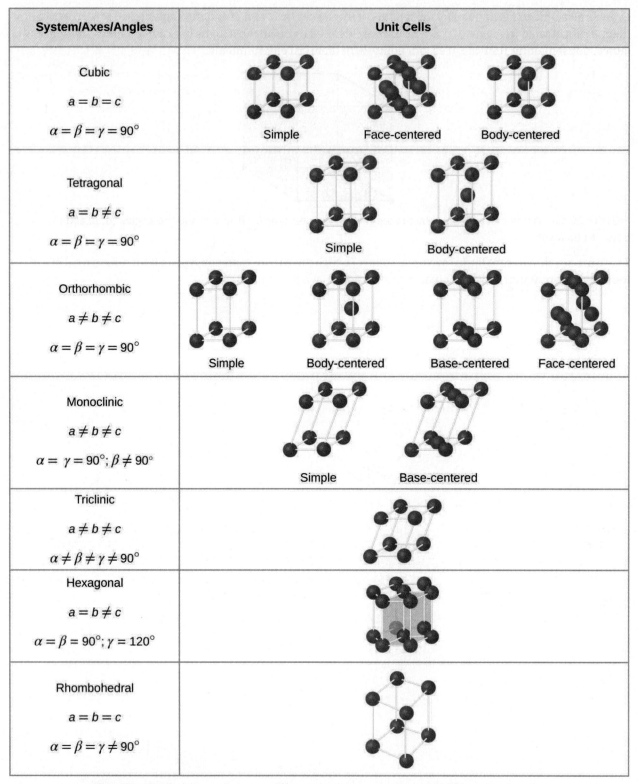

System/Axes/Angles	Unit Cells
Cubic $a = b = c$ $\alpha = \beta = \gamma = 90°$	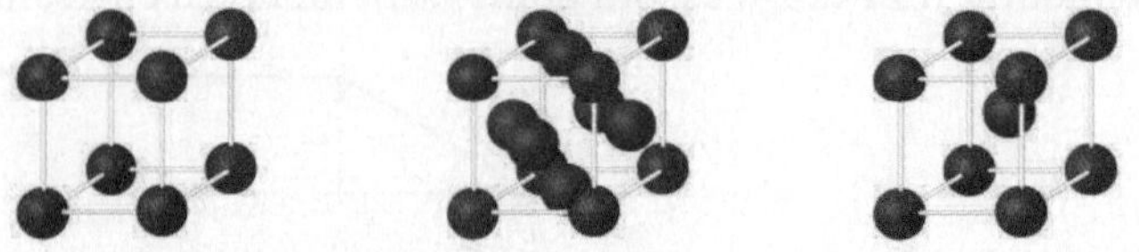 Simple, Face-centered, Body-centered
Tetragonal $a = b \neq c$ $\alpha = \beta = \gamma = 90°$	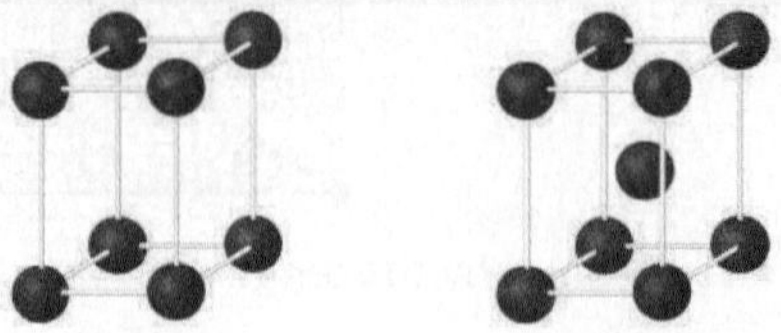 Simple, Body-centered
Orthorhombic $a \neq b \neq c$ $\alpha = \beta = \gamma = 90°$	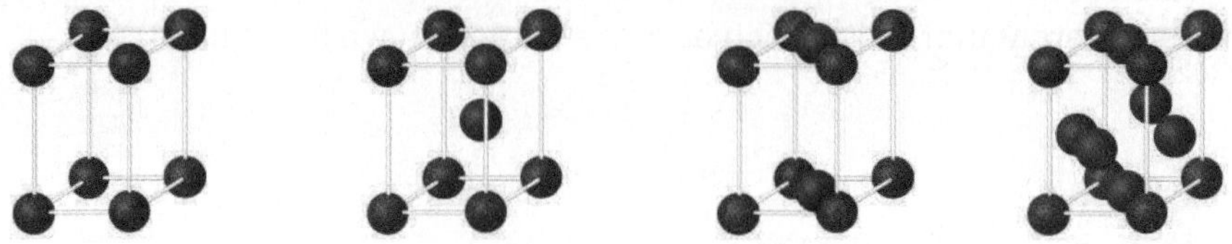 Simple, Body-centered, Base-centered, Face-centered
Monoclinic $a \neq b \neq c$ $\alpha = \gamma = 90°; \beta \neq 90°$	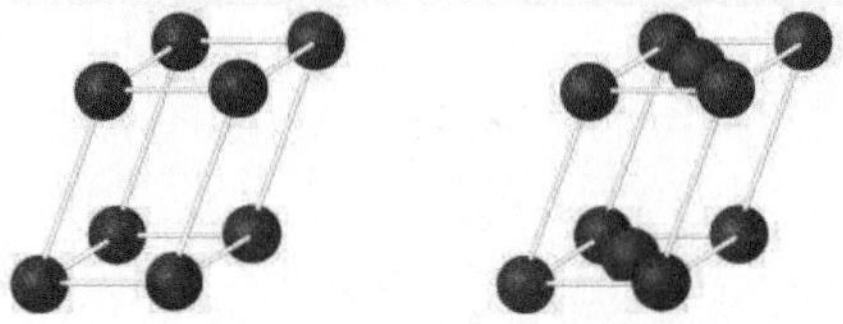 Simple, Base-centered
Triclinic $a \neq b \neq c$ $\alpha \neq \beta \neq \gamma \neq 90°$	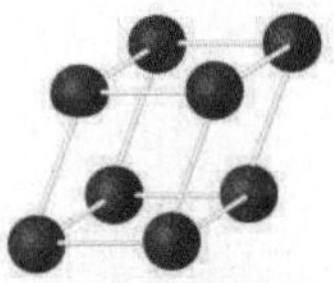
Hexagonal $a = b \neq c$ $\alpha = \beta = 90°; \gamma = 120°$	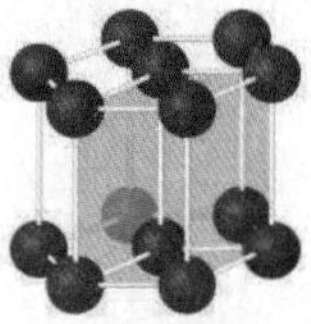
Rhombohedral $a = b = c$ $\alpha = \beta = \gamma \neq 90°$	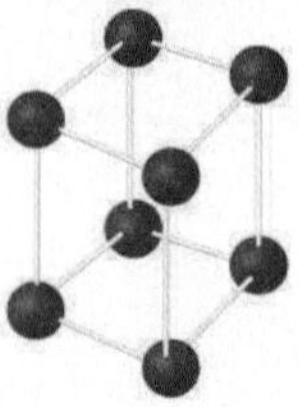

FIGURE 10.56 There are seven different lattice systems and 14 different unit cells.

The Structures of Ionic Crystals

Ionic crystals consist of two or more different kinds of ions that usually have different sizes. The packing of these ions into a crystal structure is more complex than the packing of metal atoms that are the same size.

Most monatomic ions behave as charged spheres, and their attraction for ions of opposite charge is the same in every direction. Consequently, stable structures for ionic compounds result (1) when ions of one charge are surrounded by as many ions as possible of the opposite charge and (2) when the cations and anions are in contact with each other. Structures are determined by two principal factors: the relative sizes of the ions and the ratio of the numbers of positive and negative ions in the compound.

In simple ionic structures, we usually find the anions, which are normally larger than the cations, arranged in a closest-packed array. (As seen previously, additional electrons attracted to the same nucleus make anions larger and fewer electrons attracted to the same nucleus make cations smaller when compared to the atoms from which they are formed.) The smaller cations commonly occupy one of two types of **holes** (or interstices) remaining between the anions. The smaller of the holes is found between three anions in one plane and one anion in an adjacent plane. The four anions surrounding this hole are arranged at the corners of a tetrahedron, so the hole is called a **tetrahedral hole**. The larger type of hole is found at the center of six anions (three in one layer and three in an adjacent layer) located at the corners of an octahedron; this is called an **octahedral hole**. Figure 10.57 illustrates both of these types of holes.

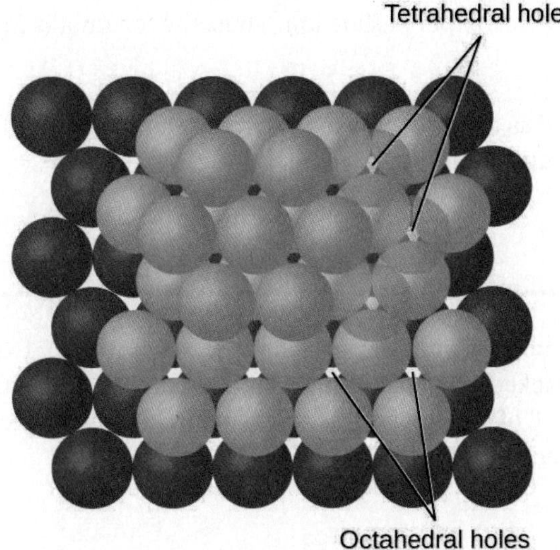

FIGURE 10.57 Cations may occupy two types of holes between anions: octahedral holes or tetrahedral holes.

Depending on the relative sizes of the cations and anions, the cations of an ionic compound may occupy tetrahedral or octahedral holes, as illustrated in Figure 10.58. Relatively small cations occupy tetrahedral holes, and larger cations occupy octahedral holes. If the cations are too large to fit into the octahedral holes, the anions may adopt a more open structure, such as a simple cubic array. The larger cations can then occupy the larger cubic holes made possible by the more open spacing.

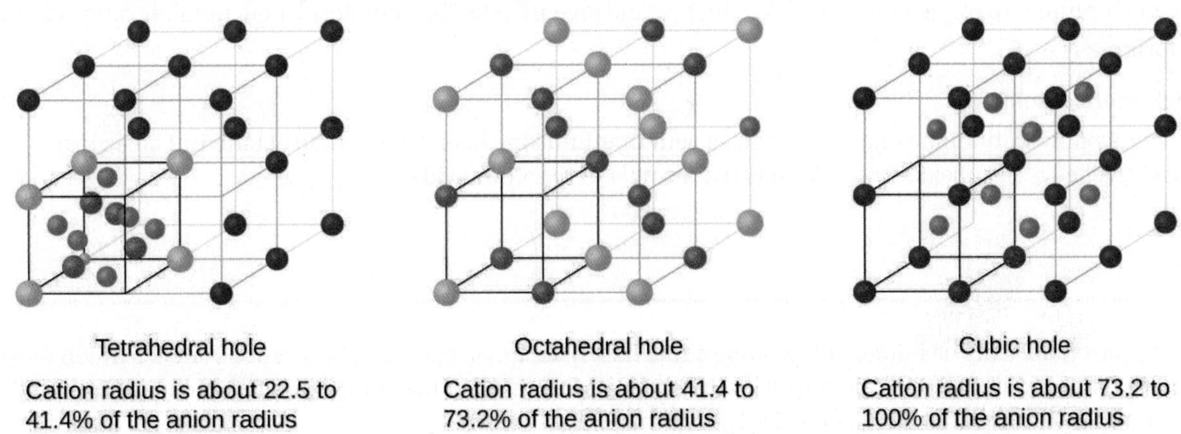

FIGURE 10.58 A cation's size and the shape of the hole occupied by the compound are directly related.

There are two tetrahedral holes for each anion in either an HCP or CCP array of anions. A compound that crystallizes in a closest-packed array of anions with cations in the tetrahedral holes can have a maximum cation:anion ratio of 2:1; all of the tetrahedral holes are filled at this ratio. Examples include Li_2O, Na_2O, Li_2S, and Na_2S. Compounds with a ratio of less than 2:1 may also crystallize in a closest-packed array of anions with cations in the tetrahedral holes, if the ionic sizes fit. In these compounds, however, some of the tetrahedral holes remain vacant.

 **EXAMPLE 10.16**

Occupancy of Tetrahedral Holes

Zinc sulfide is an important industrial source of zinc and is also used as a white pigment in paint. Zinc sulfide crystallizes with zinc ions occupying one-half of the tetrahedral holes in a closest-packed array of sulfide ions. What is the formula of zinc sulfide?

Solution

Because there are two tetrahedral holes per anion (sulfide ion) and one-half of these holes are occupied by zinc ions, there must be $\frac{1}{2} \times 2$, or 1, zinc ion per sulfide ion. Thus, the formula is ZnS.

Check Your Learning

Lithium selenide can be described as a closest-packed array of selenide ions with lithium ions in all of the tetrahedral holes. What it the formula of lithium selenide?

Answer:

Li_2Se

The ratio of octahedral holes to anions in either an HCP or CCP structure is 1:1. Thus, compounds with cations in octahedral holes in a closest-packed array of anions can have a maximum cation:anion ratio of 1:1. In NiO, MnS, NaCl, and KH, for example, all of the octahedral holes are filled. Ratios of less than 1:1 are observed when some of the octahedral holes remain empty.

 **EXAMPLE 10.17**

Stoichiometry of Ionic Compounds

Sapphire is aluminum oxide. Aluminum oxide crystallizes with aluminum ions in two-thirds of the octahedral holes in a closest-packed array of oxide ions. What is the formula of aluminum oxide?

Solution

Because there is one octahedral hole per anion (oxide ion) and only two-thirds of these holes are occupied, the ratio of aluminum to oxygen must be $\frac{2}{3}$:1, which would give $Al_{2/3}O$. The simplest whole number ratio is 2:3, so the formula is Al_2O_3.

Check Your Learning

The white pigment titanium oxide crystallizes with titanium ions in one-half of the octahedral holes in a closest-packed array of oxide ions. What is the formula of titanium oxide?

Answer:

TiO_2

In a simple cubic array of anions, there is one cubic hole that can be occupied by a cation for each anion in the array. In CsCl, and in other compounds with the same structure, all of the cubic holes are occupied. Half of the cubic holes are occupied in SrH_2, UO_2, $SrCl_2$, and CaF_2.

Different types of ionic compounds often crystallize in the same structure when the relative sizes of their ions and their stoichiometries (the two principal features that determine structure) are similar.

Unit Cells of Ionic Compounds

Many ionic compounds crystallize with cubic unit cells, and we will use these compounds to describe the general features of ionic structures.

When an ionic compound is composed of cations and anions of similar size in a 1:1 ratio, it typically forms a simple cubic structure. Cesium chloride, CsCl, (illustrated in Figure 10.59) is an example of this, with Cs^+ and Cl^- having radii of 174 pm and 181 pm, respectively. We can think of this as chloride ions forming a simple cubic unit cell, with a cesium ion in the center; or as cesium ions forming a unit cell with a chloride ion in the center; or as simple cubic unit cells formed by Cs^+ ions overlapping unit cells formed by Cl^- ions. Cesium ions and chloride ions touch along the body diagonals of the unit cells. One cesium ion and one chloride ion are present per unit cell, giving the l:l stoichiometry required by the formula for cesium chloride. Note that there is no lattice point in the center of the cell, and CsCl is not a BCC structure because a cesium ion is not identical to a chloride ion.

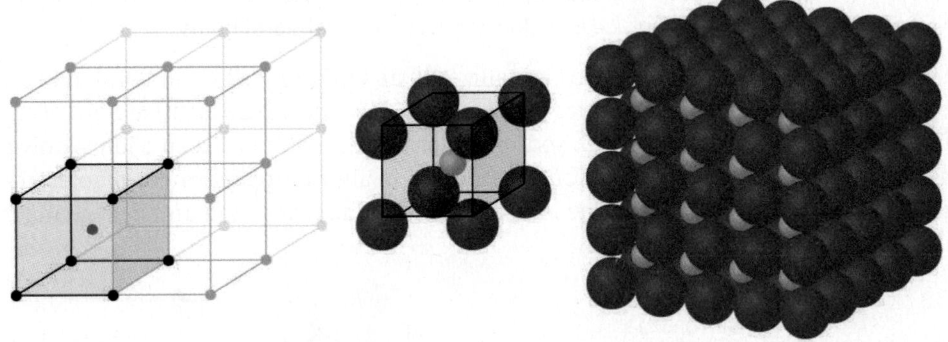

FIGURE 10.59 Ionic compounds with similar-sized cations and anions, such as CsCl, usually form a simple cubic structure. They can be described by unit cells with either cations at the corners or anions at the corners.

We have said that the location of lattice points is arbitrary. This is illustrated by an alternate description of the CsCl structure in which the lattice points are located in the centers of the cesium ions. In this description, the cesium ions are located on the lattice points at the corners of the cell, and the chloride ion is located at the center of the cell. The two unit cells are different, but they describe identical structures.

When an ionic compound is composed of a 1:1 ratio of cations and anions that differ significantly in size, it typically crystallizes with an FCC unit cell, like that shown in Figure 10.60. Sodium chloride, NaCl, is an example of this, with Na^+ and Cl^- having radii of 102 pm and 181 pm, respectively. We can think of this as chloride ions forming an FCC cell, with sodium ions located in the octahedral holes in the middle of the cell edges and in the center of the cell. The sodium and chloride ions touch each other along the cell edges. The unit cell contains four sodium ions and four chloride ions, giving the 1:1 stoichiometry required by the formula, NaCl.

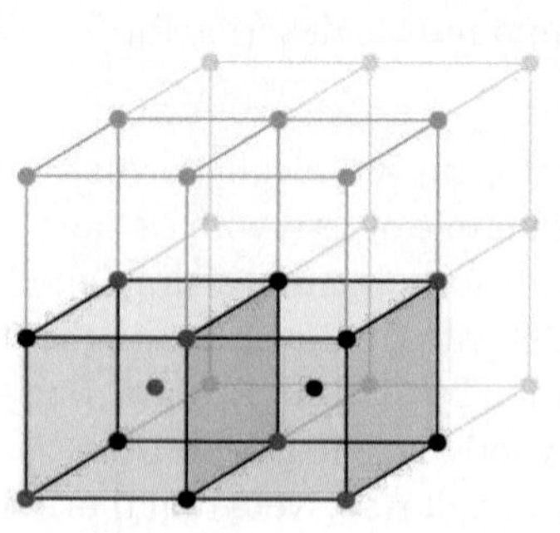
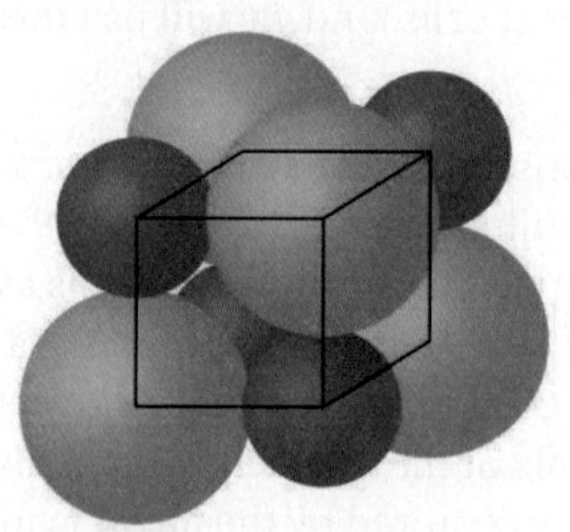

Face-centered simple cubic structure

FIGURE 10.60 Ionic compounds with anions that are much larger than cations, such as NaCl, usually form an FCC structure. They can be described by FCC unit cells with cations in the octahedral holes.

The cubic form of zinc sulfide, zinc blende, also crystallizes in an FCC unit cell, as illustrated in Figure 10.61. This structure contains sulfide ions on the lattice points of an FCC lattice. (The arrangement of sulfide ions is identical to the arrangement of chloride ions in sodium chloride.) The radius of a zinc ion is only about 40% of the radius of a sulfide ion, so these small Zn^{2+} ions are located in alternating tetrahedral holes, that is, in one half of the tetrahedral holes. There are four zinc ions and four sulfide ions in the unit cell, giving the empirical formula ZnS.

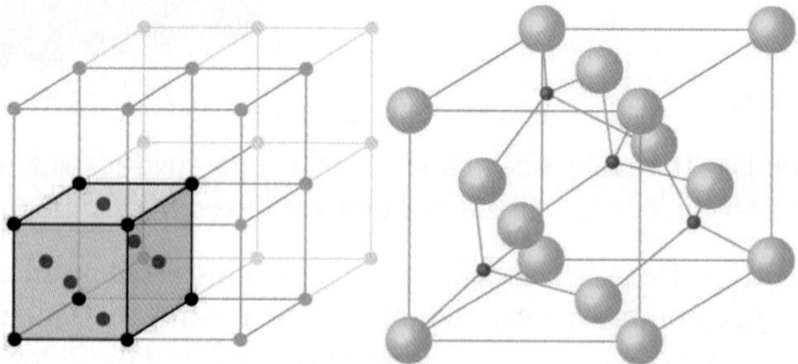

ZnS face-centered unit cell

FIGURE 10.61 ZnS, zinc sulfide (or zinc blende) forms an FCC unit cell with sulfide ions at the lattice points and much smaller zinc ions occupying half of the tetrahedral holes in the structure.

A calcium fluoride unit cell, like that shown in Figure 10.62, is also an FCC unit cell, but in this case, the cations are located on the lattice points; equivalent calcium ions are located on the lattice points of an FCC lattice. All of the tetrahedral sites in the FCC array of calcium ions are occupied by fluoride ions. There are four calcium ions and eight fluoride ions in a unit cell, giving a calcium:fluorine ratio of 1:2, as required by the chemical formula, CaF_2. Close examination of Figure 10.62 will reveal a simple cubic array of fluoride ions with calcium ions in one half of the cubic holes. The structure cannot be described in terms of a **space lattice** of points on the fluoride ions because the fluoride ions do not all have identical environments. The orientation of the four calcium ions about the fluoride ions differs.

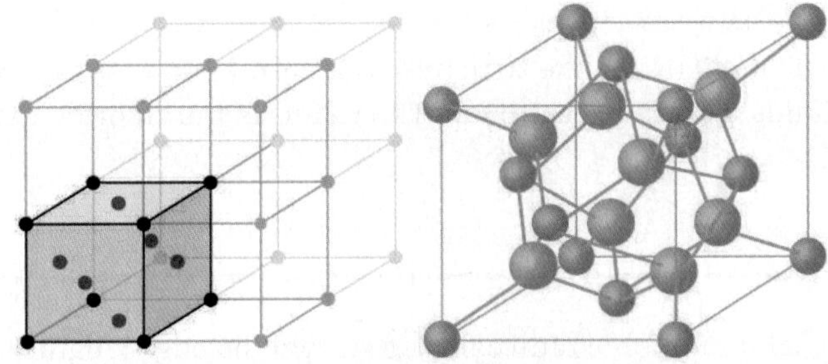

FIGURE 10.62 Calcium fluoride, CaF_2, forms an FCC unit cell with calcium ions (green) at the lattice points and fluoride ions (red) occupying all of the tetrahedral sites between them.

Calculation of Ionic Radii

If we know the edge length of a unit cell of an ionic compound and the position of the ions in the cell, we can calculate ionic radii for the ions in the compound if we make assumptions about individual ionic shapes and contacts.

EXAMPLE 10.18

Calculation of Ionic Radii

The edge length of the unit cell of LiCl (NaCl-like structure, FCC) is 0.514 nm or 5.14 Å. Assuming that the lithium ion is small enough so that the chloride ions are in contact, as in Figure 10.60, calculate the ionic radius for the chloride ion.

Note: The length unit angstrom, Å, is often used to represent atomic-scale dimensions and is equivalent to 10^{-10} m.

Solution

On the face of a LiCl unit cell, chloride ions contact each other across the diagonal of the face:

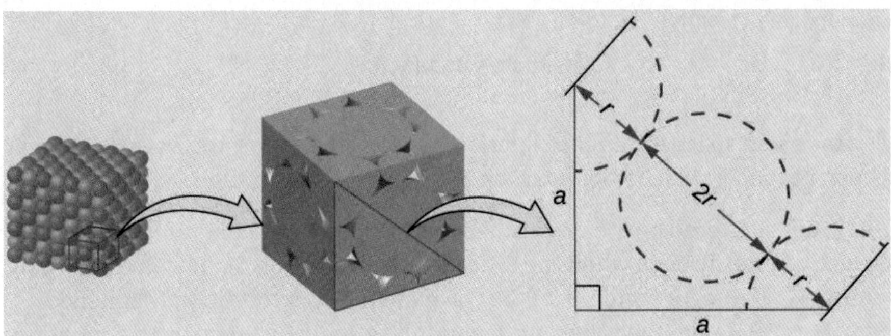

Drawing a right triangle on the face of the unit cell, we see that the length of the diagonal is equal to four chloride radii (one radius from each corner chloride and one diameter—which equals two radii—from the chloride ion in the center of the face), so $d = 4r$. From the Pythagorean theorem, we have:

$$a^2 + a^2 = d^2$$

which yields:

$$(0.514\text{ nm})^2 + (0.514\text{ nm})^2 = (4r)^2 = 16r^2$$

Solving this gives:

$$r = \sqrt{\frac{(0.514\text{ nm})^2 + (0.514\text{ nm})^2}{16}} = 0.182\text{ nm (1.82 Å) for a Cl}^- \text{ radius.}$$

Check Your Learning

The edge length of the unit cell of KCl (NaCl-like structure, FCC) is 6.28 Å. Assuming anion-cation contact along the cell edge, calculate the radius of the potassium ion. The radius of the chloride ion is 1.82 Å.

Answer:

The radius of the potassium ion is 1.33 Å.

It is important to realize that values for ionic radii calculated from the edge lengths of unit cells depend on numerous assumptions, such as a perfect spherical shape for ions, which are approximations at best. Hence, such calculated values are themselves approximate and comparisons cannot be pushed too far. Nevertheless, this method has proved useful for calculating ionic radii from experimental measurements such as X-ray crystallographic determinations.

X-Ray Crystallography

The size of the unit cell and the arrangement of atoms in a crystal may be determined from measurements of the *diffraction* of X-rays by the crystal, termed **X-ray crystallography**. **Diffraction** is the change in the direction of travel experienced by an electromagnetic wave when it encounters a physical barrier whose dimensions are comparable to those of the wavelength of the light. X-rays are electromagnetic radiation with wavelengths about as long as the distance between neighboring atoms in crystals (on the order of a few Å).

When a beam of monochromatic X-rays strikes a crystal, its rays are scattered in all directions by the atoms within the crystal. When scattered waves traveling in the same direction encounter one another, they undergo *interference*, a process by which the waves combine to yield either an increase or a decrease in amplitude (intensity) depending upon the extent to which the combining waves' maxima are separated (see Figure 10.63).

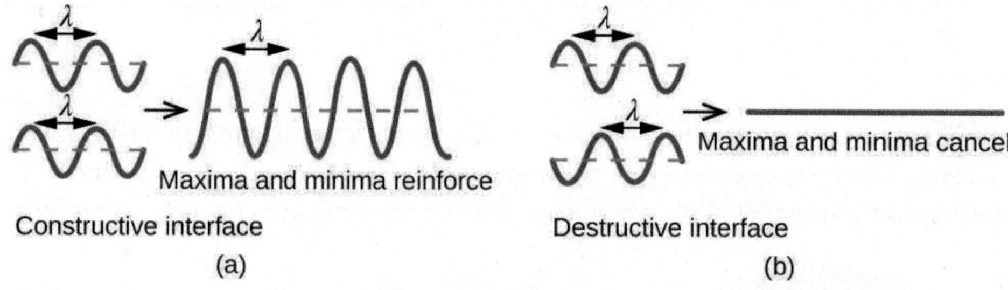

FIGURE 10.63 Light waves occupying the same space experience interference, combining to yield waves of greater (a) or lesser (b) intensity, depending upon the separation of their maxima and minima.

When X-rays of a certain wavelength, λ, are scattered by atoms in adjacent crystal planes separated by a distance, *d*, they may undergo constructive interference when the difference between the distances traveled by the two waves prior to their combination is an integer factor, *n*, of the wavelength. This condition is satisfied when the angle of the diffracted beam, θ, is related to the wavelength and interatomic distance by the equation:

$$n\lambda = 2d \sin \theta$$

This relation is known as the **Bragg equation** in honor of W. H. Bragg, the English physicist who first explained this phenomenon. Figure 10.64 illustrates two examples of diffracted waves from the same two crystal planes. The figure on the left depicts waves diffracted at the Bragg angle, resulting in constructive interference, while that on the right shows diffraction and a different angle that does not satisfy the Bragg condition, resulting in destructive interference.

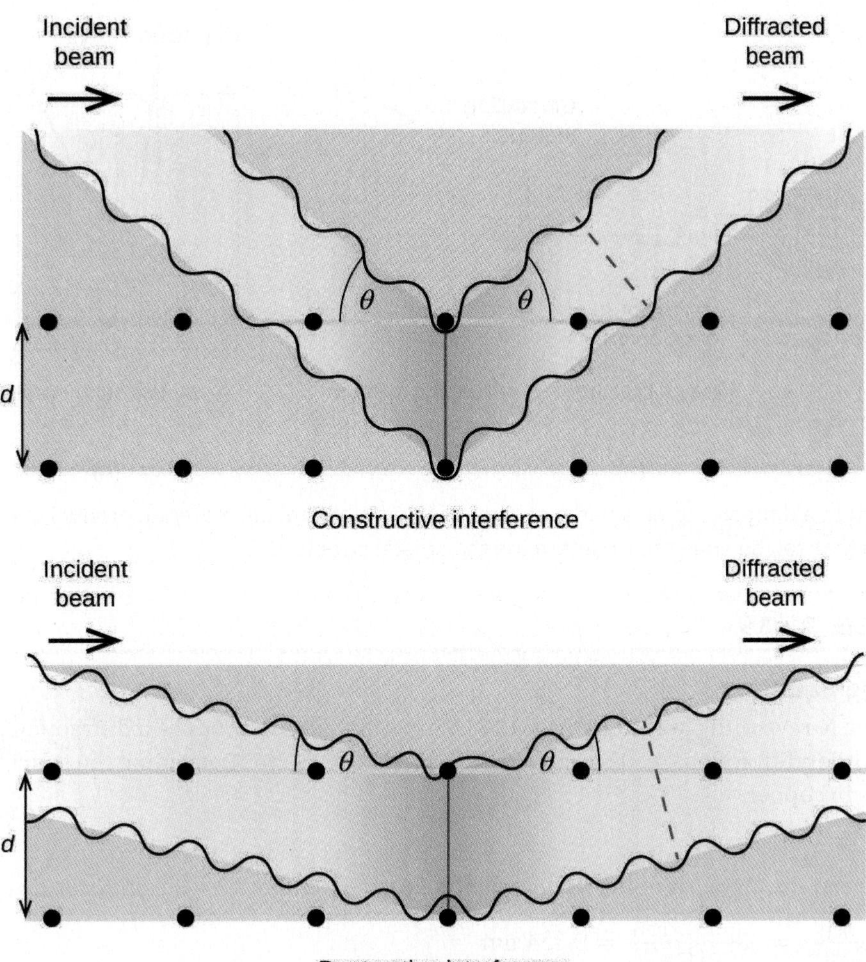

FIGURE 10.64 The diffraction of X-rays scattered by the atoms within a crystal permits the determination of the distance between the atoms. The top image depicts constructive interference between two scattered waves and a resultant diffracted wave of high intensity. The bottom image depicts destructive interference and a low intensity diffracted wave.

LINK TO LEARNING

Visit this site (http://openstax.org/l/16bragg) for more details on the Bragg equation and a simulator that allows you to explore the effect of each variable on the intensity of the diffracted wave.

An X-ray diffractometer, such as the one illustrated in Figure 10.65, may be used to measure the angles at which X-rays are diffracted when interacting with a crystal as described earlier. From such measurements, the Bragg equation may be used to compute distances between atoms as demonstrated in the following example exercise.

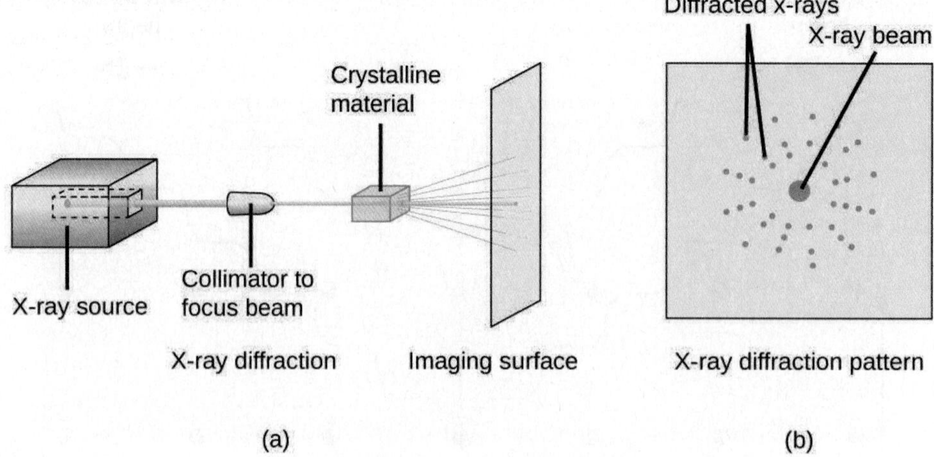

FIGURE 10.65 (a) In a diffractometer, a beam of X-rays strikes a crystalline material, producing (b) an X-ray diffraction pattern that can be analyzed to determine the crystal structure.

EXAMPLE 10.19

Using the Bragg Equation

In a diffractometer, X-rays with a wavelength of 0.1315 nm were used to produce a diffraction pattern for copper. The first order diffraction ($n = 1$) occurred at an angle $\theta = 25.25°$. Determine the spacing between the diffracting planes in copper.

Solution

The distance between the planes is found by solving the Bragg equation, $n\lambda = 2d \sin \theta$, for d.

This gives: $d = \frac{n\lambda}{2 \sin \theta} = \frac{1(0.1315 \text{ nm})}{2 \sin (25.25°)} = 0.154 \text{ nm}$

Check Your Learning

A crystal with spacing between planes equal to 0.394 nm diffracts X-rays with a wavelength of 0.147 nm. What is the angle for the first order diffraction?

Answer:

10.8°

Portrait of a Chemist

X-ray Crystallographer Rosalind Franklin

The discovery of the structure of DNA in 1953 by Francis Crick and James Watson is one of the great achievements in the history of science. They were awarded the 1962 Nobel Prize in Physiology or Medicine, along with Maurice Wilkins, who provided experimental proof of DNA's structure. British chemist Rosalind Franklin made invaluable contributions to this monumental achievement through her work in measuring X-ray diffraction images of DNA. Early in her career, Franklin's research on the structure of coals proved helpful to the British war effort. After shifting her focus to biological systems in the early 1950s, Franklin and doctoral student Raymond Gosling discovered that DNA consists of two forms: a long, thin fiber formed when wet (type "B") and a short, wide fiber formed when dried (type "A"). Her X-ray diffraction images of DNA (Figure 10.66) provided the crucial information that allowed Watson and Crick to confirm that DNA forms a double helix, and to determine details of its size and structure. Franklin also conducted pioneering research on viruses and the RNA that contains their genetic information, uncovering new information that radically changed the body of knowledge in the field. After developing ovarian cancer, Franklin continued to work until her death in 1958 at age 37. Among many posthumous recognitions of her work, the Chicago Medical School of Finch University of Health Sciences changed its name to the Rosalind Franklin University of Medicine and Science in 2004, and adopted an image of her famous X-ray diffraction image of DNA as its official university logo.

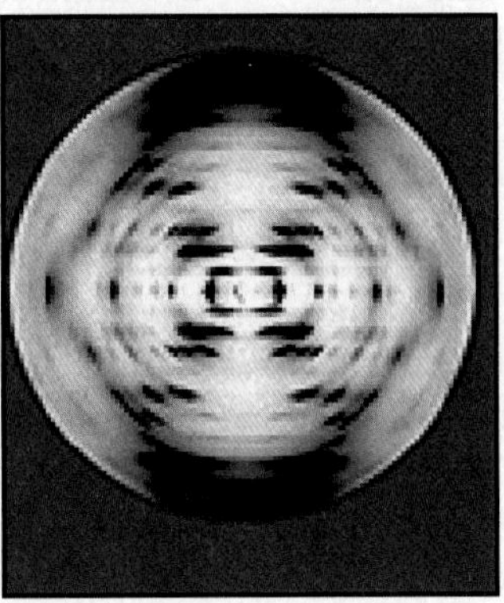

FIGURE 10.66 This illustration shows an X-ray diffraction image similar to the one Franklin found in her research. (credit: National Institutes of Health)

Key Terms

adhesive force force of attraction between molecules of different chemical identities

amorphous solid (also, noncrystalline solid) solid in which the particles lack an ordered internal structure

body-centered cubic (BCC) solid crystalline structure that has a cubic unit cell with lattice points at the corners and in the center of the cell

body-centered cubic unit cell simplest repeating unit of a body-centered cubic crystal; it is a cube containing lattice points at each corner and in the center of the cube

boiling point temperature at which the vapor pressure of a liquid equals the pressure of the gas above it

Bragg equation equation that relates the angles at which X-rays are diffracted by the atoms within a crystal

capillary action flow of liquid within a porous material due to the attraction of the liquid molecules to the surface of the material and to other liquid molecules

Clausius-Clapeyron equation mathematical relationship between the temperature, vapor pressure, and enthalpy of vaporization for a substance

cohesive force force of attraction between identical molecules

condensation change from a gaseous to a liquid state

coordination number number of atoms closest to any given atom in a crystal or to the central metal atom in a complex

covalent network solid solid whose particles are held together by covalent bonds

critical point temperature and pressure above which a gas cannot be condensed into a liquid

crystalline solid solid in which the particles are arranged in a definite repeating pattern

cubic closest packing (CCP) crystalline structure in which planes of closely packed atoms or ions are stacked as a series of three alternating layers of different relative orientations (ABC)

deposition change from a gaseous state directly to a solid state

diffraction redirection of electromagnetic radiation that occurs when it encounters a physical barrier of appropriate dimensions

dipole-dipole attraction intermolecular attraction between two permanent dipoles

dispersion force (also, London dispersion force) attraction between two rapidly fluctuating, temporary dipoles; significant only when particles are very close together

dynamic equilibrium state of a system in which reciprocal processes are occurring at equal rates

face-centered cubic (FCC) solid crystalline structure consisting of a cubic unit cell with lattice points on the corners and in the center of each face

face-centered cubic unit cell simplest repeating unit of a face-centered cubic crystal; it is a cube containing lattice points at each corner and in the center of each face

freezing change from a liquid state to a solid state

freezing point temperature at which the solid and liquid phases of a substance are in equilibrium; see also *melting point*

hexagonal closest packing (HCP) crystalline structure in which close packed layers of atoms or ions are stacked as a series of two alternating layers of different relative orientations (AB)

hole (also, interstice) space between atoms within a crystal

hydrogen bonding occurs when exceptionally strong dipoles attract; bonding that exists when hydrogen is bonded to one of the three most electronegative elements: F, O, or N

induced dipole temporary dipole formed when the electrons of an atom or molecule are distorted by the instantaneous dipole of a neighboring atom or molecule

instantaneous dipole temporary dipole that occurs for a brief moment in time when the electrons of an atom or molecule are distributed asymmetrically

intermolecular force noncovalent attractive force between atoms, molecules, and/or ions

interstitial sites spaces between the regular particle positions in any array of atoms or ions

ionic solid solid composed of positive and negative ions held together by strong electrostatic attractions

isomorphous possessing the same crystalline structure

melting change from a solid state to a liquid state

melting point temperature at which the solid and liquid phases of a substance are in equilibrium; see also *freezing point*

metallic solid solid composed of metal atoms

molecular solid solid composed of neutral molecules held together by intermolecular forces of attraction

normal boiling point temperature at which a

liquid's vapor pressure equals 1 atm (760 torr)

octahedral hole open space in a crystal at the center of six particles located at the corners of an octahedron

phase diagram pressure-temperature graph summarizing conditions under which the phases of a substance can exist

polarizability measure of the ability of a charge to distort a molecule's charge distribution (electron cloud)

simple cubic structure crystalline structure with a cubic unit cell with lattice points only at the corners

simple cubic unit cell (also, primitive cubic unit cell) unit cell in the simple cubic structure

space lattice all points within a crystal that have identical environments

sublimation change from solid state directly to gaseous state

supercritical fluid substance at a temperature and pressure higher than its critical point; exhibits properties intermediate between those of gaseous and liquid states

surface tension energy required to increase the area, or length, of a liquid surface by a given amount

tetrahedral hole tetrahedral space formed by four atoms or ions in a crystal

triple point temperature and pressure at which the vapor, liquid, and solid phases of a substance are in equilibrium

unit cell smallest portion of a space lattice that is repeated in three dimensions to form the entire lattice

vacancy defect that occurs when a position that should contain an atom or ion is vacant

van der Waals force attractive or repulsive force between molecules, including dipole-dipole, dipole-induced dipole, and London dispersion forces; does not include forces due to covalent or ionic bonding, or the attraction between ions and molecules

vapor pressure (also, equilibrium vapor pressure) pressure exerted by a vapor in equilibrium with a solid or a liquid at a given temperature

vaporization change from liquid state to gaseous state

viscosity measure of a liquid's resistance to flow

X-ray crystallography experimental technique for determining distances between atoms in a crystal by measuring the angles at which X-rays are diffracted when passing through the crystal

Key Equations

$$h = \frac{2T\cos\theta}{r\rho g}$$

$$P = Ae^{-\Delta H_{\text{vap}}/RT}$$

$$\ln P = -\frac{\Delta H_{\text{vap}}}{RT} + \ln A$$

$$\ln\left(\frac{P_2}{P_1}\right) = \frac{\Delta H_{\text{vap}}}{R}\left(\frac{1}{T_1} - \frac{1}{T_2}\right)$$

$$n\lambda = 2d\sin\theta$$

Summary

10.1 Intermolecular Forces

The physical properties of condensed matter (liquids and solids) can be explained in terms of the kinetic molecular theory. In a liquid, intermolecular attractive forces hold the molecules in contact, although they still have sufficient KE to move past each other.

Intermolecular attractive forces, collectively referred to as van der Waals forces, are responsible for the behavior of liquids and solids and are electrostatic in nature. Dipole-dipole attractions result from the electrostatic attraction of the partial negative end of one polar molecule for the partial positive end of another. The temporary dipole that results from the motion of the electrons in an atom can induce a dipole in an adjacent atom and give rise to the London dispersion force. London forces increase with increasing molecular size. Hydrogen bonds are a special type of dipole-dipole attraction that results when hydrogen is bonded to one of the three most electronegative elements: F, O, or N.

10.2 Properties of Liquids

The intermolecular forces between molecules in the liquid state vary depending upon their chemical identities and result in corresponding variations in various physical properties. Cohesive forces between like molecules are responsible for a liquid's viscosity (resistance to flow) and surface tension (elasticity of a liquid surface). Adhesive forces between the molecules of a liquid and different molecules composing a surface in contact with the liquid are responsible for phenomena such as surface wetting and capillary rise.

10.3 Phase Transitions

Phase transitions are processes that convert matter from one physical state into another. There are six phase transitions between the three phases of matter. Melting, vaporization, and sublimation are all endothermic processes, requiring an input of heat to overcome intermolecular attractions. The reciprocal transitions of freezing, condensation, and deposition are all exothermic processes, involving heat as intermolecular attractive forces are established or strengthened. The temperatures at which phase transitions occur are determined by the relative strengths of intermolecular attractions and are, therefore, dependent on the chemical identity of the substance.

10.4 Phase Diagrams

The temperature and pressure conditions at which a substance exists in solid, liquid, and gaseous states are summarized in a phase diagram for that substance. Phase diagrams are combined plots of three pressure-temperature equilibrium curves: solid-liquid, liquid-gas, and solid-gas. These curves represent the relationships between phase-transition temperatures and pressures. The point of intersection of all three curves represents the substance's triple point—the temperature and pressure at which all three phases are in equilibrium. At pressures below the triple point, a substance cannot exist in the liquid state, regardless of its temperature. The terminus of the liquid-gas curve represents the substance's critical point, the pressure and temperature above which a liquid phase cannot exist.

10.5 The Solid State of Matter

Some substances form crystalline solids consisting of particles in a very organized structure; others form amorphous (noncrystalline) solids with an internal structure that is not ordered. The main types of crystalline solids are ionic solids, metallic solids, covalent network solids, and molecular solids. The properties of the different kinds of crystalline solids are due to the types of particles of which they consist, the arrangements of the particles, and the strengths of the attractions between them. Because their particles experience identical attractions, crystalline solids have distinct melting temperatures; the particles in amorphous solids experience a range of interactions, so they soften gradually and melt over a range of temperatures. Some crystalline solids have defects in the definite repeating pattern of their particles. These defects (which include vacancies, atoms or ions not in the regular positions, and impurities) change physical properties such as electrical conductivity, which is exploited in the silicon crystals used to manufacture computer chips.

10.6 Lattice Structures in Crystalline Solids

The structures of crystalline metals and simple ionic compounds can be described in terms of packing of spheres. Metal atoms can pack in hexagonal closest-packed structures, cubic closest-packed structures, body-centered structures, and simple cubic structures. The anions in simple ionic structures commonly adopt one of these structures, and the cations occupy the spaces remaining between the anions. Small cations usually occupy tetrahedral holes in a closest-packed array of anions. Larger cations usually occupy octahedral holes. Still larger cations can occupy cubic holes in a simple cubic array of anions. The structure of a solid can be described by indicating the size and shape of a unit cell and the contents of the cell. The type of structure and dimensions of the unit cell can be determined by X-ray diffraction measurements.

Exercises

10.1 Intermolecular Forces

1. In terms of their bulk properties, how do liquids and solids differ? How are they similar?
2. In terms of the kinetic molecular theory, in what ways are liquids similar to solids? In what ways are liquids different from solids?
3. In terms of the kinetic molecular theory, in what ways are liquids similar to gases? In what ways are liquids different from gases?

4. Explain why liquids assume the shape of any container into which they are poured, whereas solids are rigid and retain their shape.
5. What is the evidence that all neutral atoms and molecules exert attractive forces on each other?
6. Open the PhET States of Matter Simulation (http://openstax.org/l/16phetvisual) to answer the following questions:
   (a) Select the Solid, Liquid, Gas tab. Explore by selecting different substances, heating and cooling the systems, and changing the state. What similarities do you notice between the four substances for each phase (solid, liquid, gas)? What differences do you notice?
   (b) For each substance, select each of the states and record the given temperatures. How do the given temperatures for each state correlate with the strengths of their intermolecular attractions? Explain.
   (c) Select the Interaction Potential tab, and use the default neon atoms. Move the Ne atom on the right and observe how the potential energy changes. Select the Total Force button, and move the Ne atom as before. When is the total force on each atom attractive and large enough to matter? Then select the Component Forces button, and move the Ne atom. When do the attractive (van der Waals) and repulsive (electron overlap) forces balance? How does this relate to the potential energy versus the distance between atoms graph? Explain.
7. Define the following and give an example of each:
   (a) dispersion force
   (b) dipole-dipole attraction
   (c) hydrogen bond
8. The types of intermolecular forces in a substance are identical whether it is a solid, a liquid, or a gas. Why then does a substance change phase from a gas to a liquid or to a solid?
9. Why do the boiling points of the noble gases increase in the order He < Ne < Ar < Kr < Xe?
10. Neon and HF have approximately the same molecular masses.
    (a) Explain why the boiling points of Neon and HF differ.
    (b) Compare the change in the boiling points of Ne, Ar, Kr, and Xe with the change of the boiling points of HF, HCl, HBr, and HI, and explain the difference between the changes with increasing atomic or molecular mass.
11. Arrange each of the following sets of compounds in order of increasing boiling point temperature:
    (a) HCl, H_2O, SiH_4
    (b) F_2, Cl_2, Br_2
    (c) CH_4, C_2H_6, C_3H_8
    (d) O_2, NO, N_2
12. The molecular mass of butanol, C_4H_9OH, is 74.14; that of ethylene glycol, $CH_2(OH)CH_2OH$, is 62.08, yet their boiling points are 117.2 °C and 174 °C, respectively. Explain the reason for the difference.
13. On the basis of intermolecular attractions, explain the differences in the boiling points of *n*–butane (−1 °C) and chloroethane (12 °C), which have similar molar masses.
14. On the basis of dipole moments and/or hydrogen bonding, explain in a qualitative way the differences in the boiling points of acetone (56.2 °C) and 1-propanol (97.4 °C), which have similar molar masses.
15. The melting point of $H_2O(s)$ is 0 °C. Would you expect the melting point of $H_2S(s)$ to be −85 °C, 0 °C, or 185 °C? Explain your answer.
16. Silane (SiH_4), phosphine (PH_3), and hydrogen sulfide (H_2S) melt at −185 °C, −133 °C, and −85 °C, respectively. What does this suggest about the polar character and intermolecular attractions of the three compounds?
17. Explain why a hydrogen bond between two water molecules is weaker than a hydrogen bond between two hydrogen fluoride molecules.

18. Under certain conditions, molecules of acetic acid, CH_3COOH, form "dimers," pairs of acetic acid molecules held together by strong intermolecular attractions:

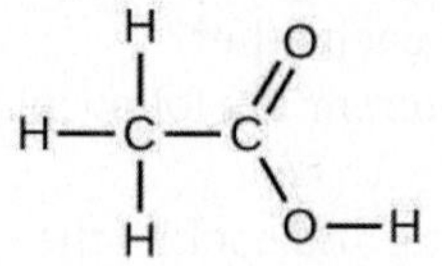

Draw a dimer of acetic acid, showing how two CH_3COOH molecules are held together, and stating the type of IMF that is responsible.

19. Proteins are chains of amino acids that can form in a variety of arrangements, one of which is a helix. What kind of IMF is responsible for holding the protein strand in this shape? On the protein image, show the locations of the IMFs that hold the protein together:

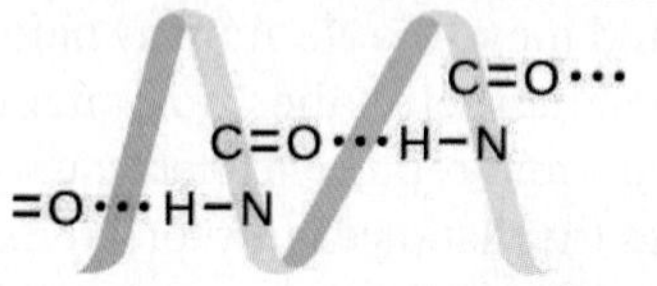

20. The density of liquid NH_3 is 0.64 g/mL; the density of gaseous NH_3 at STP is 0.0007 g/mL. Explain the difference between the densities of these two phases.

21. Identify the intermolecular forces present in the following solids:
(a) CH_3CH_2OH
(b) $CH_3CH_2CH_3$
(c) CH_3CH_2Cl

10.2 Properties of Liquids

22. The test tubes shown here contain equal amounts of the specified motor oils. Identical metal spheres were dropped at the same time into each of the tubes, and a brief moment later, the spheres had fallen to the heights indicated in the illustration.
Rank the motor oils in order of increasing viscosity, and explain your reasoning:

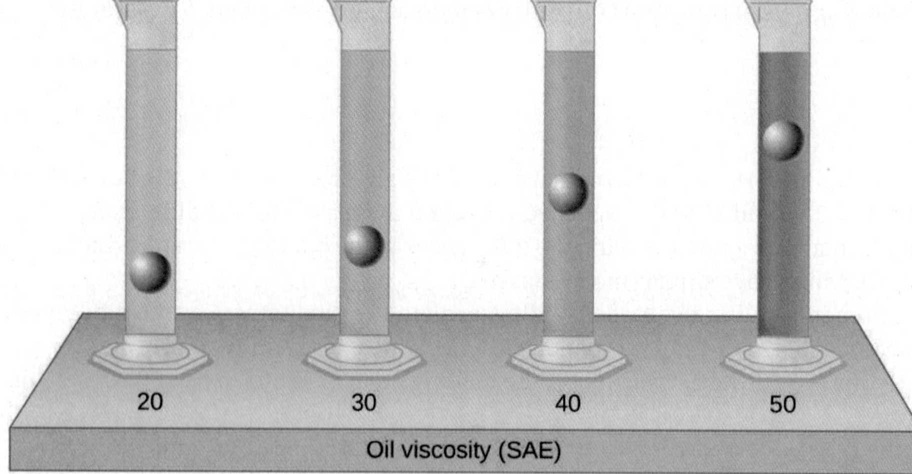

23. Although steel is denser than water, a steel needle or paper clip placed carefully lengthwise on the surface of still water can be made to float. Explain at a molecular level how this is possible.

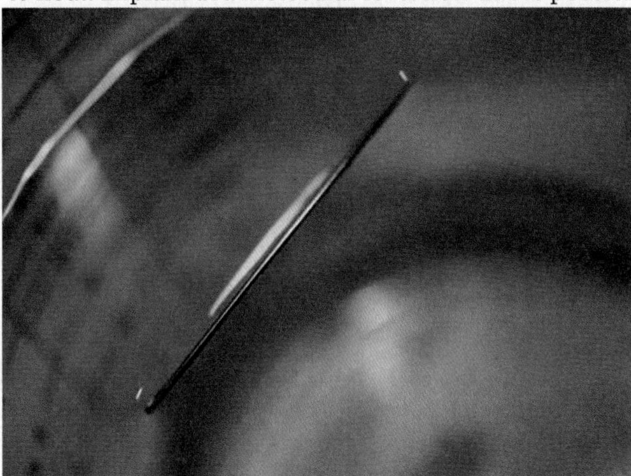

FIGURE 10.67 (credit: Cory Zanker)

24. The surface tension and viscosity values for diethyl ether, acetone, ethanol, and ethylene glycol are shown here.

Compound	Molecule	Surface Tension (mN/m)	Viscosity (mPa s)
diethyl ether $C_2H_5OC_2H_5$		17	0.22
acetone CH_3COCH_3		23	0.31
ethanol C_2H_5OH		22	1.07
ethylene glycol $CH_2(OH)CH_2(OH)$		48	16.1

(a) Explain their differences in viscosity in terms of the size and shape of their molecules and their IMFs.
(b) Explain their differences in surface tension in terms of the size and shape of their molecules and their IMFs:

25. You may have heard someone use the figure of speech "slower than molasses in winter" to describe a process that occurs slowly. Explain why this is an apt idiom, using concepts of molecular size and shape, molecular interactions, and the effect of changing temperature.

26. It is often recommended that you let your car engine run idle to warm up before driving, especially on cold winter days. While the benefit of prolonged idling is dubious, it is certainly true that a warm engine is more fuel efficient than a cold one. Explain the reason for this.

27. The surface tension and viscosity of water at several different temperatures are given in this table.

Water	Surface Tension (mN/m)	Viscosity (mPa s)
0 °C	75.6	1.79
20 °C	72.8	1.00
60 °C	66.2	0.47
100 °C	58.9	0.28

(a) As temperature increases, what happens to the surface tension of water? Explain why this occurs, in terms of molecular interactions and the effect of changing temperature.
(b) As temperature increases, what happens to the viscosity of water? Explain why this occurs, in terms of molecular interactions and the effect of changing temperature.
28. At 25 °C, how high will water rise in a glass capillary tube with an inner diameter of 0.63 mm? Refer to Example 10.4 for the required information.
29. Water rises in a glass capillary tube to a height of 17 cm. What is the diameter of the capillary tube?

10.3 Phase Transitions

30. Heat is added to boiling water. Explain why the temperature of the boiling water does not change. What does change?
31. Heat is added to ice at 0 °C. Explain why the temperature of the ice does not change. What does change?
32. What feature characterizes the dynamic equilibrium between a liquid and its vapor in a closed container?
33. Identify two common observations indicating some liquids have sufficient vapor pressures to noticeably evaporate?
34. Identify two common observations indicating some solids, such as dry ice and mothballs, have vapor pressures sufficient to sublime?
35. What is the relationship between the intermolecular forces in a liquid and its vapor pressure?
36. What is the relationship between the intermolecular forces in a solid and its melting temperature?
37. Why does spilled gasoline evaporate more rapidly on a hot day than on a cold day?
38. Carbon tetrachloride, CCl_4, was once used as a dry cleaning solvent, but is no longer used because it is carcinogenic. At 57.8 °C, the vapor pressure of CCl_4 is 54.0 kPa, and its enthalpy of vaporization is 33.05 kJ/mol. Use this information to estimate the normal boiling point for CCl_4.
39. When is the boiling point of a liquid equal to its normal boiling point?
40. How does the boiling of a liquid differ from its evaporation?
41. Use the information in Figure 10.24 to estimate the boiling point of water in Denver when the atmospheric pressure is 83.3 kPa.
42. A syringe at a temperature of 20 °C is filled with liquid ether in such a way that there is no space for any vapor. If the temperature is kept constant and the plunger is withdrawn to create a volume that can be occupied by vapor, what would be the approximate pressure of the vapor produced?
43. Explain the following observations:
(a) It takes longer to cook an egg in Ft. Davis, Texas (altitude, 5000 feet above sea level) than it does in Boston (at sea level).
(b) Perspiring is a mechanism for cooling the body.
44. The enthalpy of vaporization of water is larger than its enthalpy of fusion. Explain why.
45. Explain why the molar enthalpies of vaporization of the following substances increase in the order $CH_4 < C_2H_6 < C_3H_8$, even though the type of IMF (dispersion) is the same.
46. Explain why the enthalpies of vaporization of the following substances increase in the order $CH_4 < NH_3 < H_2O$, even though all three substances have approximately the same molar mass.

47. The enthalpy of vaporization of $CO_2(l)$ is 9.8 kJ/mol. Would you expect the enthalpy of vaporization of $CS_2(l)$ to be 28 kJ/mol, 9.8 kJ/mol, or –8.4 kJ/mol? Discuss the plausibility of each of these answers.
48. The hydrogen fluoride molecule, HF, is more polar than a water molecule, H_2O (for example, has a greater dipole moment), yet the molar enthalpy of vaporization for liquid hydrogen fluoride is lesser than that for water. Explain.
49. Ethyl chloride (boiling point, 13 °C) is used as a local anesthetic. When the liquid is sprayed on the skin, it cools the skin enough to freeze and numb it. Explain the cooling effect of liquid ethyl chloride.
50. Which contains the compounds listed correctly in order of increasing boiling points?
    (a) $N_2 < CS_2 < H_2O < KCl$
    (b) $H_2O < N_2 < CS_2 < KCl$
    (c) $N_2 < KCl < CS_2 < H_2O$
    (d) $CS_2 < N_2 < KCl < H_2O$
    (e) $KCl < H_2O < CS_2 < N_2$
51. How much heat is required to convert 422 g of liquid H_2O at 23.5 °C into steam at 150 °C?
52. Evaporation of sweat requires energy and thus take excess heat away from the body. Some of the water that you drink may eventually be converted into sweat and evaporate. If you drink a 20-ounce bottle of water that had been in the refrigerator at 3.8 °C, how much heat is needed to convert all of that water into sweat and then to vapor? (Note: Your body temperature is 36.6 °C. For the purpose of solving this problem, assume that the thermal properties of sweat are the same as for water.)
53. Titanium tetrachloride, $TiCl_4$, has a melting point of –23.2 °C and has a ΔH_{fusion} = 9.37 kJ/mol.
    (a) How much energy is required to melt 263.1 g $TiCl_4$?
    (b) For $TiCl_4$, which will likely have the larger magnitude: ΔH_{fusion} or $\Delta H_{vaporization}$? Explain your reasoning.

10.4 Phase Diagrams

54. From the phase diagram for water (Figure 10.31), determine the state of water at:
    (a) 35 °C and 85 kPa
    (b) –15 °C and 40 kPa
    (c) –15 °C and 0.1 kPa
    (d) 75 °C and 3 kPa
    (e) 40 °C and 0.1 kPa
    (f) 60 °C and 50 kPa
55. What phase changes will take place when water is subjected to varying pressure at a constant temperature of 0.005 °C? At 40 °C? At –40 °C?
56. Pressure cookers allow food to cook faster because the higher pressure inside the pressure cooker increases the boiling temperature of water. A particular pressure cooker has a safety valve that is set to vent steam if the pressure exceeds 3.4 atm. What is the approximate maximum temperature that can be reached inside this pressure cooker? Explain your reasoning.
57. From the phase diagram for carbon dioxide in Figure 10.34, determine the state of CO_2 at:
    (a) 20 °C and 1000 kPa
    (b) 10 °C and 2000 kPa
    (c) 10 °C and 100 kPa
    (d) –40 °C and 500 kPa
    (e) –80 °C and 1500 kPa
    (f) –80 °C and 10 kPa
58. Determine the phase changes that carbon dioxide undergoes as pressure is increased at a constant temperature of (a) –50 °C and (b) 50 °C. If the temperature is held at –40 °C? At 20 °C? (See the phase diagram in Figure 10.34.)
59. Consider a cylinder containing a mixture of liquid carbon dioxide in equilibrium with gaseous carbon dioxide at an initial pressure of 65 atm and a temperature of 20 °C. Sketch a plot depicting the change in the cylinder pressure with time as gaseous carbon dioxide is released at constant temperature.

60. Dry ice, $CO_2(s)$, does not melt at atmospheric pressure. It sublimes at a temperature of −78 °C. What is the lowest pressure at which $CO_2(s)$ will melt to give $CO_2(l)$? At approximately what temperature will this occur? (See Figure 10.34 for the phase diagram.)

61. If a severe storm results in the loss of electricity, it may be necessary to use a clothesline to dry laundry. In many parts of the country in the dead of winter, the clothes will quickly freeze when they are hung on the line. If it does not snow, will they dry anyway? Explain your answer.

62. Is it possible to liquefy nitrogen at room temperature (about 25 °C)? Is it possible to liquefy sulfur dioxide at room temperature? Explain your answers.

63. Elemental carbon has one gas phase, one liquid phase, and two different solid phases, as shown in the phase diagram:

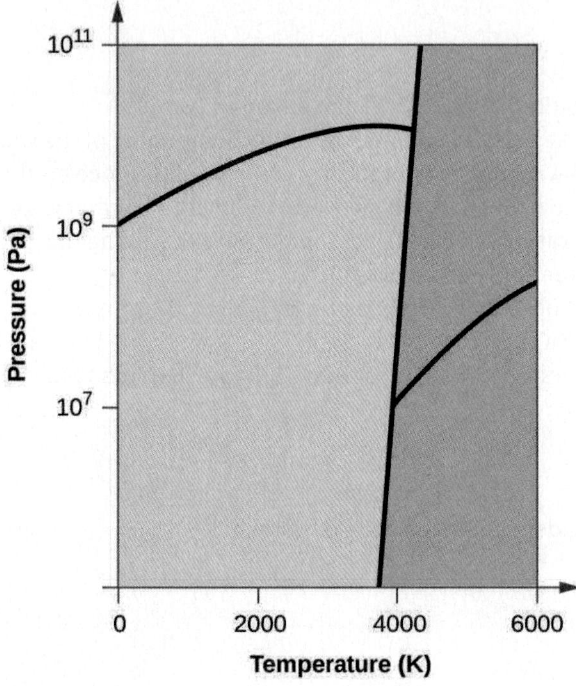

(a) On the phase diagram, label the gas and liquid regions.
(b) Graphite is the most stable phase of carbon at normal conditions. On the phase diagram, label the graphite phase.
(c) If graphite at normal conditions is heated to 2500 K while the pressure is increased to 10^{10} Pa, it is converted into diamond. Label the diamond phase.
(d) Circle each triple point on the phase diagram.
(e) In what phase does carbon exist at 5000 K and 10^8 Pa?
(f) If the temperature of a sample of carbon increases from 3000 K to 5000 K at a constant pressure of 10^6 Pa, which phase transition occurs, if any?

10.5 The Solid State of Matter

64. What types of liquids typically form amorphous solids?

65. At very low temperatures oxygen, O_2, freezes and forms a crystalline solid. Which best describes these crystals?
(a) ionic
(b) covalent network
(c) metallic
(d) amorphous
(e) molecular crystals

66. As it cools, olive oil slowly solidifies and forms a solid over a range of temperatures. Which best describes the solid?
(a) ionic
(b) covalent network
(c) metallic
(d) amorphous
(e) molecular crystals

67. Explain why ice, which is a crystalline solid, has a melting temperature of 0 °C, whereas butter, which is an amorphous solid, softens over a range of temperatures.

68. Identify the type of crystalline solid (metallic, network covalent, ionic, or molecular) formed by each of the following substances:
(a) SiO_2
(b) KCl
(c) Cu
(d) CO_2
(e) C (diamond)
(f) $BaSO_4$
(g) NH_3
(h) NH_4F
(i) C_2H_5OH

69. Identify the type of crystalline solid (metallic, network covalent, ionic, or molecular) formed by each of the following substances:
(a) $CaCl_2$
(b) SiC
(c) N_2
(d) Fe
(e) C (graphite)
(f) $CH_3CH_2CH_2CH_3$
(g) HCl
(h) NH_4NO_3
(i) K_3PO_4

70. Classify each substance in the table as either a metallic, ionic, molecular, or covalent network solid:

Substance	Appearance	Melting Point	Electrical Conductivity	Solubility in Water
X	lustrous, malleable	1500 °C	high	insoluble
Y	soft, yellow	113 °C	none	insoluble
Z	hard, white	800 °C	only if melted/ dissolved	soluble

71. Classify each substance in the table as either a metallic, ionic, molecular, or covalent network solid:

Substance	Appearance	Melting Point	Electrical Conductivity	Solubility in Water
X	brittle, white	800 °C	only if melted/ dissolved	soluble
Y	shiny, malleable	1100 °C	high	insoluble
Z	hard, colorless	3550 °C	none	insoluble

72. Identify the following substances as ionic, metallic, covalent network, or molecular solids: Substance A is malleable, ductile, conducts electricity well, and has a melting point of 1135 °C. Substance B is brittle, does not conduct electricity as a solid but does when molten, and has a melting point of 2072 °C. Substance C is very hard, does not conduct electricity, and has a melting point of 3440 °C. Substance D is soft, does not conduct electricity, and has a melting point of 185 °C.
73. Substance A is shiny, conducts electricity well, and melts at 975 °C. Substance A is likely a(n):
(a) ionic solid
(b) metallic solid
(c) molecular solid
(d) covalent network solid
74. Substance B is hard, does not conduct electricity, and melts at 1200 °C. Substance B is likely a(n):
(a) ionic solid
(b) metallic solid
(c) molecular solid
(d) covalent network solid

10.6 Lattice Structures in Crystalline Solids

75. Describe the crystal structure of iron, which crystallizes with two equivalent metal atoms in a cubic unit cell.
76. Describe the crystal structure of Pt, which crystallizes with four equivalent metal atoms in a cubic unit cell.
77. What is the coordination number of a chromium atom in the body-centered cubic structure of chromium?
78. What is the coordination number of an aluminum atom in the face-centered cubic structure of aluminum?
79. Cobalt metal crystallizes in a hexagonal closest packed structure. What is the coordination number of a cobalt atom?
80. Nickel metal crystallizes in a cubic closest packed structure. What is the coordination number of a nickel atom?
81. Tungsten crystallizes in a body-centered cubic unit cell with an edge length of 3.165 Å.
(a) What is the atomic radius of tungsten in this structure?
(b) Calculate the density of tungsten.
82. Platinum (atomic radius = 1.38 Å) crystallizes in a cubic closely packed structure. Calculate the edge length of the face-centered cubic unit cell and the density of platinum.
83. Barium crystallizes in a body-centered cubic unit cell with an edge length of 5.025 Å
(a) What is the atomic radius of barium in this structure?
(b) Calculate the density of barium.
84. Aluminum (atomic radius = 1.43 Å) crystallizes in a cubic closely packed structure. Calculate the edge length of the face-centered cubic unit cell and the density of aluminum.
85. The density of aluminum is 2.7 g/cm^3; that of silicon is 2.3 g/cm^3. Explain why Si has the lower density even though it has heavier atoms.

86. The free space in a metal may be found by subtracting the volume of the atoms in a unit cell from the volume of the cell. Calculate the percentage of free space in each of the three cubic lattices if all atoms in each are of equal size and touch their nearest neighbors. Which of these structures represents the most efficient packing? That is, which packs with the least amount of unused space?

87. Cadmium sulfide, sometimes used as a yellow pigment by artists, crystallizes with cadmium, occupying one-half of the tetrahedral holes in a closest packed array of sulfide ions. What is the formula of cadmium sulfide? Explain your answer.

88. A compound of cadmium, tin, and phosphorus is used in the fabrication of some semiconductors. It crystallizes with cadmium occupying one-fourth of the tetrahedral holes and tin occupying one-fourth of the tetrahedral holes in a closest packed array of phosphide ions. What is the formula of the compound? Explain your answer.

89. What is the formula of the magnetic oxide of cobalt, used in recording tapes, that crystallizes with cobalt atoms occupying one-eighth of the tetrahedral holes and one-half of the octahedral holes in a closely packed array of oxide ions?

90. A compound containing zinc, aluminum, and sulfur crystallizes with a closest-packed array of sulfide ions. Zinc ions are found in one-eighth of the tetrahedral holes and aluminum ions in one-half of the octahedral holes. What is the empirical formula of the compound?

91. A compound of thallium and iodine crystallizes in a simple cubic array of iodide ions with thallium ions in all of the cubic holes. What is the formula of this iodide? Explain your answer.

92. Which of the following elements reacts with sulfur to form a solid in which the sulfur atoms form a closest-packed array with all of the octahedral holes occupied: Li, Na, Be, Ca, or Al?

93. What is the percent by mass of titanium in rutile, a mineral that contains titanium and oxygen, if structure can be described as a closest packed array of oxide ions with titanium ions in one-half of the octahedral holes? What is the oxidation number of titanium?

94. Explain why the chemically similar alkali metal chlorides NaCl and CsCl have different structures, whereas the chemically different NaCl and MnS have the same structure.

95. As minerals were formed from the molten magma, different ions occupied the same cites in the crystals. Lithium often occurs along with magnesium in minerals despite the difference in the charge on their ions. Suggest an explanation.

96. Rubidium iodide crystallizes with a cubic unit cell that contains iodide ions at the corners and a rubidium ion in the center. What is the formula of the compound?

97. One of the various manganese oxides crystallizes with a cubic unit cell that contains manganese ions at the corners and in the center. Oxide ions are located at the center of each edge of the unit cell. What is the formula of the compound?

98. NaH crystallizes with the same crystal structure as NaCl. The edge length of the cubic unit cell of NaH is 4.880 Å.

(a) Calculate the ionic radius of H^-. (The ionic radius of Li^+ is 0.0.95 Å.)

(b) Calculate the density of NaH.

99. Thallium(I) iodide crystallizes with the same structure as CsCl. The edge length of the unit cell of TlI is 4.20 Å. Calculate the ionic radius of TI^+. (The ionic radius of I^- is 2.16 Å.)

100. A cubic unit cell contains manganese ions at the corners and fluoride ions at the center of each edge.

(a) What is the empirical formula of this compound? Explain your answer.

(b) What is the coordination number of the Mn^{3+} ion?

(c) Calculate the edge length of the unit cell if the radius of a Mn^{3+} ion is 0.65 A.

(d) Calculate the density of the compound.

101. What is the spacing between crystal planes that diffract X-rays with a wavelength of 1.541 nm at an angle θ of 15.55° (first order reflection)?

102. A diffractometer using X-rays with a wavelength of 0.2287 nm produced first order diffraction peak for a crystal angle $\theta = 16.21°$. Determine the spacing between the diffracting planes in this crystal.

103. A metal with spacing between planes equal to 0.4164 nm diffracts X-rays with a wavelength of 0.2879 nm. What is the diffraction angle for the first order diffraction peak?

104. Gold crystallizes in a face-centered cubic unit cell. The second-order reflection (n = 2) of X-rays for the planes that make up the tops and bottoms of the unit cells is at $\theta = 22.20°$. The wavelength of the X-rays is 1.54 Å. What is the density of metallic gold?

105. When an electron in an excited molybdenum atom falls from the L to the K shell, an X-ray is emitted. These X-rays are diffracted at an angle of 7.75° by planes with a separation of 2.64 Å. What is the difference in energy between the K shell and the L shell in molybdenum assuming a first order diffraction?

CHAPTER 11
Solutions and Colloids

Figure 11.1 Coral reefs, such as this one at the Palmyra Atoll National Wildlife Refuge, are vital to the ecosystem of earth's oceans. The health of coral reefs and all marine life depends on the specific chemical composition of the complex mixture known as seawater. (credit: modification of work by "USFWS – Pacific Region"/Wikimedia Commons)

CHAPTER OUTLINE

11.1 The Dissolution Process
11.2 Electrolytes
11.3 Solubility
11.4 Colligative Properties
11.5 Colloids

INTRODUCTION Coral reefs are home to about 25% of all marine species. They are being threatened by climate change, oceanic acidification, and water pollution, all of which change the composition of the solution known as seawater. Dissolved oxygen in seawater is critical for sea creatures, but as the oceans warm, oxygen becomes less soluble. As the concentration of carbon dioxide in the atmosphere increases, the concentration of carbon dioxide in the oceans increases, contributing to oceanic acidification. Coral reefs are particularly sensitive to the acidification of the ocean, since the exoskeletons of the coral polyps are soluble in acidic solutions. Humans contribute to the changing of seawater composition by allowing agricultural runoff and other forms of pollution to affect our oceans.

Solutions are crucial to the processes that sustain life and to many other processes involving chemical reactions. This chapter considers the nature of solutions and examines factors that determine whether a solution will form and what properties it may have. The properties of colloids—mixtures containing dispersed particles larger than the molecules and ions of typical solutions—are also discussed.

11.1 The Dissolution Process

LEARNING OBJECTIVES

By the end of this section, you will be able to:

- Describe the basic properties of solutions and how they form
- Predict whether a given mixture will yield a solution based on molecular properties of its components
- Explain why some solutions either produce or absorb heat when they form

An earlier chapter of this text introduced *solutions*, defined as homogeneous mixtures of two or more substances. Often, one component of a solution is present at a significantly greater concentration, in which case it is called the *solvent*. The other components of the solution present in relatively lesser concentrations are called *solutes*. Sugar is a covalent solid composed of sucrose molecules, $C_{12}H_{22}O_{11}$. When this compound dissolves in water, its molecules become uniformly distributed among the molecules of water:

$$C_{12}H_{22}O_{11}(s) \longrightarrow C_{12}H_{22}O_{11}(aq)$$

The subscript "*aq*" in the equation signifies that the sucrose molecules are solutes and are therefore *individually dispersed* throughout the *aqueous solution* (water is the solvent). Although sucrose molecules are heavier than water molecules, they remain dispersed throughout the solution; gravity does not cause them to "settle out" over time.

Potassium dichromate, $K_2Cr_2O_7$, is an ionic compound composed of colorless potassium ions, K^+, and orange dichromate ions, $Cr_2O_7{}^{2-}$. When a small amount of solid potassium dichromate is added to water, the compound dissolves and dissociates to yield potassium ions and dichromate ions uniformly distributed throughout the mixture (Figure 11.2), as indicated in this equation:

$$K_2Cr_2O_7(s) \longrightarrow 2K^+(aq) + Cr_2O_7{}^{2-}(aq)$$

As with the mixture of sugar and water, this mixture is also an aqueous solution. Its solutes, potassium and dichromate ions, remain individually dispersed among the solvent (water) molecules.

FIGURE 11.2 When potassium dichromate ($K_2Cr_2O_7$) is mixed with water, it forms a homogeneous orange solution. (credit: modification of work by Mark Ott)

LINK TO LEARNING

Visit this virtual lab (http://openstax.org/l/16Phetsugar) to view simulations of the dissolution of common covalent and ionic substances (sugar and salt) in water.

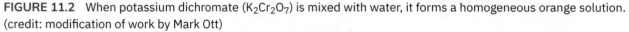

Water is used so often as a solvent that the word solution has come to imply an aqueous solution to many people. However, almost any gas, liquid, or solid can act as a solvent. Many **alloys** are solid solutions of one metal dissolved in another; for example, US five-cent coins contain nickel dissolved in copper. Air is a gaseous solution, a homogeneous mixture of nitrogen, oxygen, and several other gases. Oxygen (a gas), alcohol (a liquid), and sugar (a solid) all dissolve in water (a liquid) to form liquid solutions. Table 11.1 gives examples of several different solutions and the phases of the solutes and solvents.

Different Types of Solutions

Solution	Solute	Solvent
air	$O_2(g)$	$N_2(g)$
soft drinks[1]	$CO_2(g)$	$H_2O(l)$
hydrogen in palladium	$H_2(g)$	$Pd(s)$
rubbing alcohol	$H_2O(l)$	$C_3H_8O(l)$ (2-propanol)
saltwater	$NaCl(s)$	$H_2O(l)$
brass	$Zn(s)$	$Cu(s)$

TABLE 11.1

Solutions exhibit these defining traits:

- They are homogeneous; after a solution is mixed, it has the same composition at all points throughout (its composition is uniform).
- The physical state of a solution—solid, liquid, or gas—is typically the same as that of the solvent, as demonstrated by the examples in Table 11.1.
- The components of a solution are dispersed on a molecular scale; they consist of a mixture of separated solute particles (molecules, atoms, and/or ions) each closely surrounded by solvent species.
- The dissolved solute in a solution will not settle out or separate from the solvent.
- The composition of a solution, or the concentrations of its components, can be varied continuously (within limits determined by the *solubility* of the components, discussed in detail later in this chapter).

The Formation of Solutions

The formation of a solution is an example of a **spontaneous process**, a process that occurs under specified conditions without the requirement of energy from some external source. Sometimes a mixture is stirred to speed up the dissolution process, but this is not necessary; a homogeneous solution will form eventually. The topic of spontaneity is critically important to the study of chemical thermodynamics and is treated more thoroughly in a later chapter of this text. For purposes of this chapter's discussion, it will suffice to consider two criteria that *favor*, but do not guarantee, the spontaneous formation of a solution:

1. a decrease in the internal energy of the system (an exothermic change, as discussed in the previous chapter on thermochemistry)
2. an increased dispersal of matter in the system (which indicates an increase in the *entropy* of the system, as you will learn about in the later chapter on thermodynamics)

In the process of dissolution, an internal energy change often, but not always, occurs as heat is absorbed or evolved. An increase in matter dispersal always results when a solution forms from the uniform distribution of solute molecules throughout a solvent.

When the strengths of the intermolecular forces of attraction between solute and solvent species in a solution are no different than those present in the separated components, the solution is formed with no accompanying energy change. Such a solution is called an **ideal solution**. A mixture of ideal gases (or gases such as helium and argon, which closely approach ideal behavior) is an example of an ideal solution, since the entities comprising these gases experience no significant intermolecular attractions.

1 If bubbles of gas are observed within the liquid, the mixture is not homogeneous and, thus, not a solution.

When containers of helium and argon are connected, the gases spontaneously mix due to diffusion and form a solution (Figure 11.3). The formation of this solution clearly involves an increase in matter dispersal, since the helium and argon atoms occupy a volume twice as large as that which each occupied before mixing.

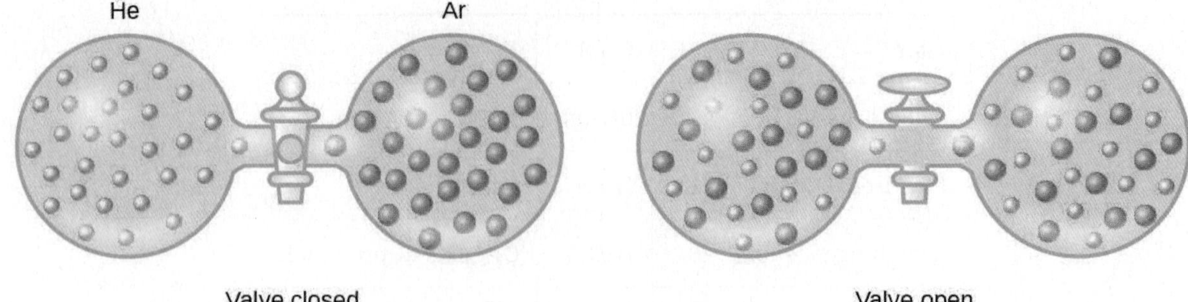

FIGURE 11.3 Samples of helium and argon spontaneously mix to give a solution.

Ideal solutions may also form when structurally similar liquids are mixed. For example, mixtures of the alcohols methanol (CH_3OH) and ethanol (C_2H_5OH) form ideal solutions, as do mixtures of the hydrocarbons pentane, C_5H_{12}, and hexane, C_6H_{14}. Placing methanol and ethanol, or pentane and hexane, in the bulbs shown in Figure 11.3 will result in the same diffusion and subsequent mixing of these liquids as is observed for the He and Ar gases (although at a much slower rate), yielding solutions with no significant change in energy. Unlike a mixture of gases, however, the components of these liquid-liquid solutions do, indeed, experience intermolecular attractive forces. But since the molecules of the two substances being mixed are structurally very similar, the intermolecular attractive forces between like and unlike molecules are essentially the same, and the dissolution process, therefore, does not entail any appreciable increase or decrease in energy. These examples illustrate how increased matter dispersal alone can provide the driving force required to cause the spontaneous formation of a solution. In some cases, however, the relative magnitudes of intermolecular forces of attraction between solute and solvent species may prevent dissolution.

Three types of intermolecular attractive forces are relevant to the dissolution process: solute-solute, solvent-solvent, and solute-solvent. As illustrated in Figure 11.4, the formation of a solution may be viewed as a stepwise process in which energy is consumed to overcome solute-solute and solvent-solvent attractions (endothermic processes) and released when solute-solvent attractions are established (an exothermic process referred to as **solvation**). The relative magnitudes of the energy changes associated with these stepwise processes determine whether the dissolution process overall will release or absorb energy. In some cases, solutions do not form because the energy required to separate solute and solvent species is so much greater than the energy released by solvation.

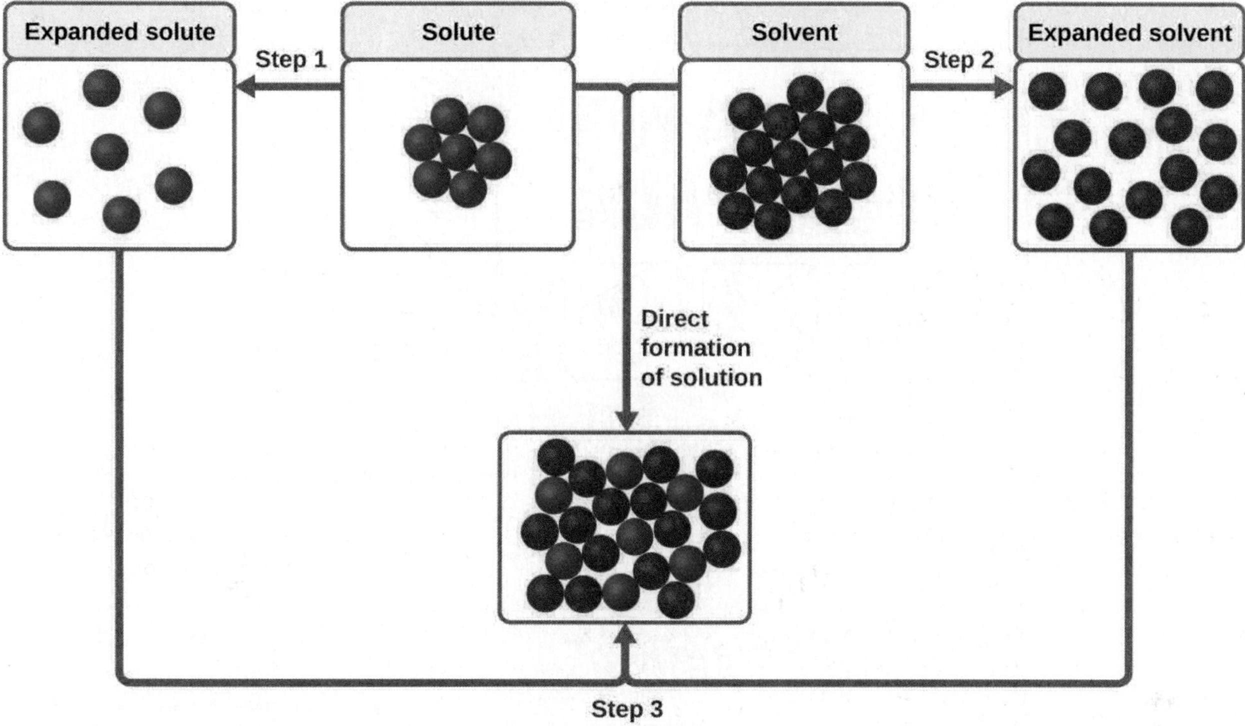

FIGURE 11.4 This schematic representation of dissolution shows a stepwise process involving the endothermic separation of solute and solvent species (Steps 1 and 2) and exothermic solvation (Step 3).

Consider the example of an ionic compound dissolving in water. Formation of the solution requires the electrostatic forces between the cations and anions of the compound (solute–solute) be overcome completely as attractive forces are established between these ions and water molecules (solute–solvent). Hydrogen bonding between a relatively small fraction of the water molecules must also be overcome to accommodate any dissolved solute. If the solute's electrostatic forces are significantly greater than the solvation forces, the dissolution process is significantly endothermic and the compound may not dissolve to an appreciable extent. Calcium carbonate, the major component of coral reefs, is one example of such an "insoluble" ionic compound (see Figure 11.1). On the other hand, if the solvation forces are much stronger than the compound's electrostatic forces, the dissolution is significantly exothermic and the compound may be highly soluble. A common example of this type of ionic compound is sodium chloride, commonly known as table salt.

As noted at the beginning of this module, spontaneous solution formation is favored, but not guaranteed, by exothermic dissolution processes. While many soluble compounds do, indeed, dissolve with the release of heat, some dissolve endothermically. Ammonium nitrate (NH_4NO_3) is one such example and is used to make instant cold packs, like the one pictured in Figure 11.5, which are used for treating injuries. A thin-walled plastic bag of water is sealed inside a larger bag with solid NH_4NO_3. When the smaller bag is broken, a solution of NH_4NO_3 forms, absorbing heat from the surroundings (the injured area to which the pack is applied) and providing a cold compress that decreases swelling. Endothermic dissolutions such as this one require a greater energy input to separate the solute species than is recovered when the solutes are solvated, but they are spontaneous nonetheless due to the increase in disorder that accompanies formation of the solution.

FIGURE 11.5 An instant cold pack gets cold when certain salts, such as ammonium nitrate, dissolve in water—an endothermic process.

LINK TO LEARNING

Watch this brief video (http://openstax.org/l/16endoexo) illustrating endothermic and exothermic dissolution processes.

11.2 Electrolytes

LEARNING OBJECTIVES

By the end of this section, you will be able to:

- Define and give examples of electrolytes
- Distinguish between the physical and chemical changes that accompany dissolution of ionic and covalent electrolytes
- Relate electrolyte strength to solute-solvent attractive forces

When some substances are dissolved in water, they undergo either a physical or a chemical change that yields ions in solution. These substances constitute an important class of compounds called **electrolytes**. Substances that do not yield ions when dissolved are called **nonelectrolytes**. If the physical or chemical process that generates the ions is essentially 100% efficient (all of the dissolved compound yields ions), then the substance is known as a **strong electrolyte**. If only a relatively small fraction of the dissolved substance undergoes the ion-producing process, it is called a **weak electrolyte**.

Substances may be identified as strong, weak, or nonelectrolytes by measuring the electrical conductance of an aqueous solution containing the substance. To conduct electricity, a substance must contain freely mobile, charged species. Most familiar is the conduction of electricity through metallic wires, in which case the mobile, charged entities are electrons. Solutions may also conduct electricity if they contain dissolved ions, with conductivity increasing as ion concentration increases. Applying a voltage to electrodes immersed in a solution permits assessment of the relative concentration of dissolved ions, either quantitatively, by measuring the electrical current flow, or qualitatively, by observing the brightness of a light bulb included in the circuit

(Figure 11.6).

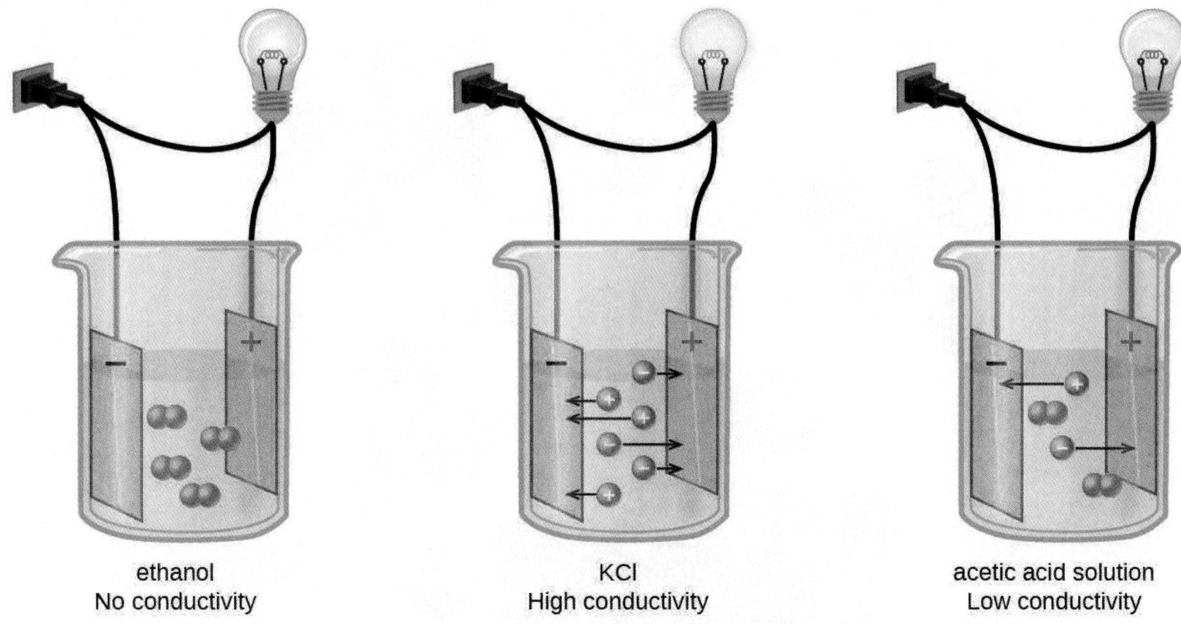

FIGURE 11.6 Solutions of nonelectrolytes such as ethanol do not contain dissolved ions and cannot conduct electricity. Solutions of electrolytes contain ions that permit the passage of electricity. The conductivity of an electrolyte solution is related to the strength of the electrolyte.

Ionic Electrolytes

Water and other polar molecules are attracted to ions, as shown in Figure 11.7. The electrostatic attraction between an ion and a molecule with a dipole is called an **ion-dipole attraction**. These attractions play an important role in the dissolution of ionic compounds in water.

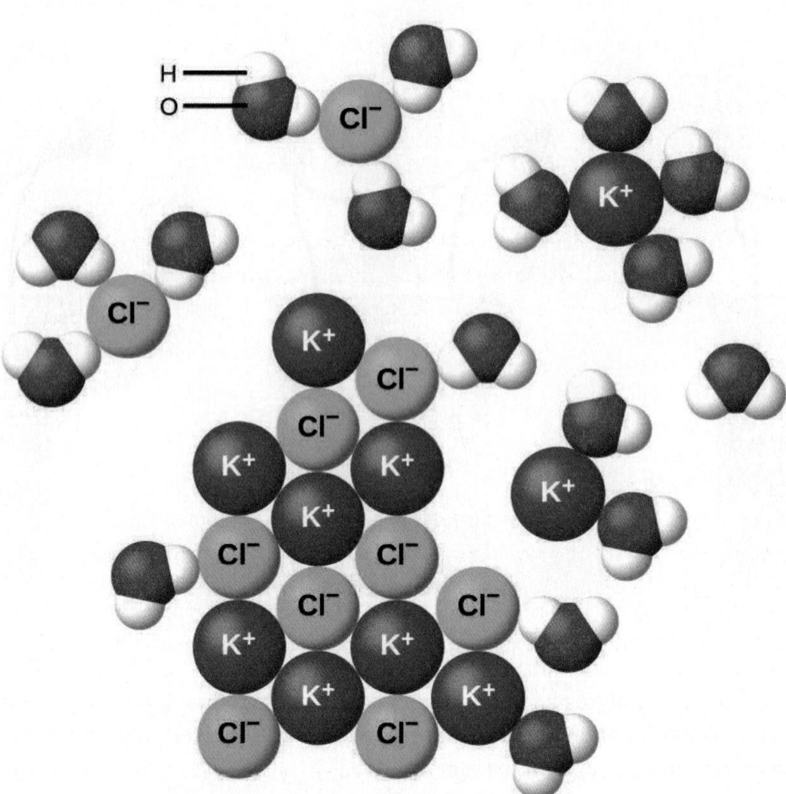

FIGURE 11.7 As potassium chloride (KCl) dissolves in water, the ions are hydrated. The polar water molecules are attracted by the charges on the K^+ and Cl^- ions. Water molecules in front of and behind the ions are not shown.

When ionic compounds dissolve in water, the ions in the solid separate and disperse uniformly throughout the solution because water molecules surround and solvate the ions, reducing the strong electrostatic forces between them. This process represents a physical change known as **dissociation**. Under most conditions, ionic compounds will dissociate nearly completely when dissolved, and so they are classified as strong electrolytes. Even sparingly, soluble ionic compounds are strong electrolytes, since the small amount that does dissolve will dissociate completely.

Consider what happens at the microscopic level when solid KCl is added to water. Ion-dipole forces attract the positive (hydrogen) end of the polar water molecules to the negative chloride ions at the surface of the solid, and they attract the negative (oxygen) ends to the positive potassium ions. The water molecules surround individual K^+ and Cl^- ions, reducing the strong interionic forces that bind the ions together and letting them move off into solution as solvated ions, as Figure 11.7 shows. Overcoming the electrostatic attraction permits the independent motion of each hydrated ion in a dilute solution as the ions transition from fixed positions in the undissolved compound to widely dispersed, solvated ions in solution.

Covalent Electrolytes

Pure water is an extremely poor conductor of electricity because it is only very slightly ionized—only about two out of every 1 billion molecules ionize at 25 °C. Water ionizes when one molecule of water gives up a proton (H^+ ion) to another molecule of water, yielding hydronium and hydroxide ions.

$$H_2O(l) + H_2O(l) \rightleftharpoons H_3O^+(aq) + OH^-(aq)$$

In some cases, solutions prepared from covalent compounds conduct electricity because the solute molecules react chemically with the solvent to produce ions. For example, pure hydrogen chloride is a gas consisting of covalent HCl molecules. This gas contains no ions. However, an aqueous solution of HCl is a very good conductor, indicating that an appreciable concentration of ions exists within the solution.

Because HCl is an *acid*, its molecules react with water, transferring H^+ ions to form hydronium ions (H_3O^+) and chloride ions (Cl^-):

$$H{-}\ddot{\underset{\underset{H}{|}}{O}}{:} + H{-}\ddot{\underset{\cdot\cdot}{Cl}}{:} \longrightarrow \left[H{-}\ddot{\underset{\underset{H}{|}}{O}}{-}H\right]^+ + {:}\ddot{\underset{\cdot\cdot}{Cl}}{:}^-$$

This reaction is essentially 100% complete for HCl (i.e., it is a *strong acid* and, consequently, a strong electrolyte). Likewise, weak acids and bases that only react partially generate relatively low concentrations of ions when dissolved in water and are classified as weak electrolytes. The reader may wish to review the discussion of strong and weak acids provided in the earlier chapter of this text on reaction classes and stoichiometry.

11.3 Solubility

LEARNING OBJECTIVES

By the end of this section, you will be able to:

- Describe the effects of temperature and pressure on solubility
- State Henry's law and use it in calculations involving the solubility of a gas in a liquid
- Explain the degrees of solubility possible for liquid-liquid solutions

Imagine adding a small amount of sugar to a glass of water, stirring until all the sugar has dissolved, and then adding a bit more. You can repeat this process until the sugar concentration of the solution reaches its natural limit, a limit determined primarily by the relative strengths of the solute-solute, solute-solvent, and solvent-solvent attractive forces discussed in the previous two modules of this chapter. You can be certain that you have reached this limit because, no matter how long you stir the solution, undissolved sugar remains. The concentration of sugar in the solution at this point is known as its solubility.

The **solubility** of a solute in a particular solvent is the maximum concentration that may be achieved under given conditions when the dissolution process is *at equilibrium*.

When a solute's concentration is equal to its solubility, the solution is said to be **saturated** with that solute. If the solute's concentration is less than its solubility, the solution is said to be **unsaturated**. A solution that contains a relatively low concentration of solute is called dilute, and one with a relatively high concentration is called concentrated.

LINK TO LEARNING

Use this interactive simulation (http://openstax.org/l/16Phetsoluble) to prepare various saturated solutions.

Solutions may be prepared in which a solute concentration *exceeds* its solubility. Such solutions are said to be **supersaturated**, and they are interesting examples of *nonequilibrium* states (a detailed treatment of this important concept is provided in the text chapters on equilibrium). For example, the carbonated beverage in an open container that has not yet "gone flat" is supersaturated with carbon dioxide gas; given time, the CO_2 concentration will decrease until it reaches its solubility.

LINK TO LEARNING

Watch this impressive video (http://openstax.org/l/16NaAcetate) showing the precipitation of sodium acetate from a supersaturated solution.

Solutions of Gases in Liquids

As for any solution, the solubility of a gas in a liquid is affected by the intermolecular attractive forces between solute and solvent species. Unlike solid and liquid solutes, however, there is no solute-solute intermolecular attraction to overcome when a gaseous solute dissolves in a liquid solvent (see Figure 11.4) since the atoms or molecules comprising a gas are far separated and experience negligible interactions. Consequently, solute-solvent interactions are the sole energetic factor affecting solubility. For example, the water solubility of oxygen is approximately three times greater than that of helium (there are greater dispersion forces between water and the larger oxygen molecules) but 100 times less than the solubility of chloromethane, $CHCl_3$ (polar chloromethane molecules experience dipole–dipole attraction to polar water molecules). Likewise note the solubility of oxygen in hexane, C_6H_{14}, is approximately 20 times greater than it is in water because greater dispersion forces exist between oxygen and the larger hexane molecules.

Temperature is another factor affecting solubility, with gas solubility typically decreasing as temperature increases (Figure 11.8). This inverse relation between temperature and dissolved gas concentration is responsible for one of the major impacts of thermal pollution in natural waters.

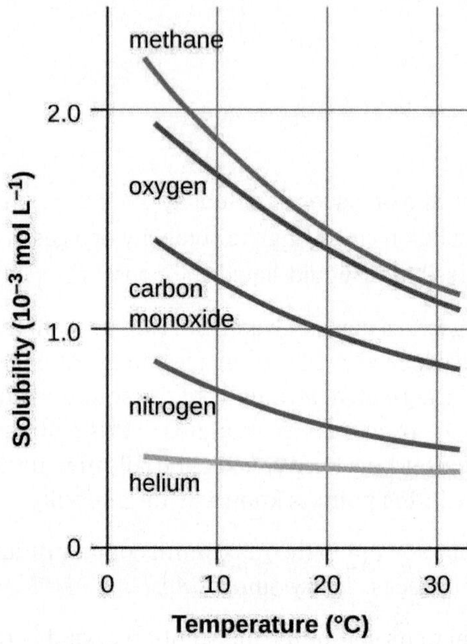

FIGURE 11.8 The solubilities of these gases in water decrease as the temperature increases. All solubilities were measured with a constant pressure of 101.3 kPa (1 atm) of gas above the solutions.

When the temperature of a river, lake, or stream is raised, the solubility of oxygen in the water is decreased. Decreased levels of dissolved oxygen may have serious consequences for the health of the water's ecosystems and, in severe cases, can result in large-scale fish kills (Figure 11.9).

FIGURE 11.9 (a) The small bubbles of air in this glass of chilled water formed when the water warmed to room temperature and the solubility of its dissolved air decreased. (b) The decreased solubility of oxygen in natural waters subjected to thermal pollution can result in large-scale fish kills. (credit a: modification of work by Liz West; credit b: modification of work by U.S. Fish and Wildlife Service)

The solubility of a gaseous solute is also affected by the partial pressure of solute in the gas to which the solution is exposed. Gas solubility increases as the pressure of the gas increases. Carbonated beverages provide a nice illustration of this relationship. The carbonation process involves exposing the beverage to a relatively high pressure of carbon dioxide gas and then sealing the beverage container, thus saturating the beverage with CO_2 at this pressure. When the beverage container is opened, a familiar hiss is heard as the carbon dioxide gas pressure is released, and some of the dissolved carbon dioxide is typically seen leaving solution in the form of small bubbles (Figure 11.10). At this point, the beverage is *supersaturated* with carbon dioxide and, with time, the dissolved carbon dioxide concentration will decrease to its equilibrium value and the beverage will become "flat."

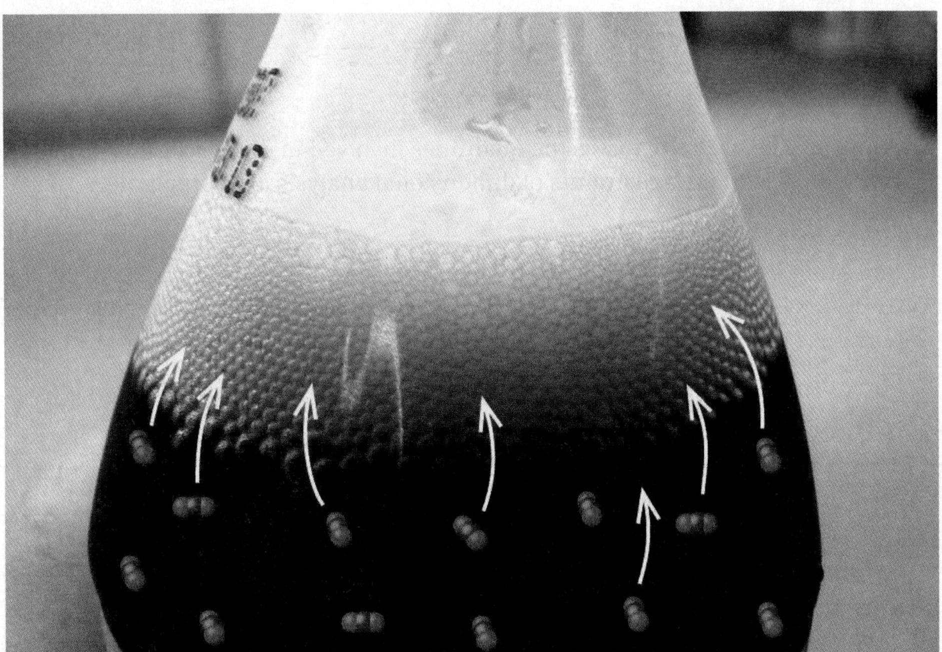

FIGURE 11.10 Opening the bottle of carbonated beverage reduces the pressure of the *gaseous* carbon dioxide above the beverage. The solubility of CO_2 is thus lowered, and some *dissolved* carbon dioxide may be seen leaving the solution as small gas bubbles. (credit: modification of work by Derrick Coetzee)

For many gaseous solutes, the relation between solubility, C_g, and partial pressure, P_g, is a proportional one:

$$C_g = kP_g$$

where k is a proportionality constant that depends on the identities of the gaseous solute and solvent, and on the solution temperature. This is a mathematical statement of **Henry's law**: *The quantity of an ideal gas that dissolves in a definite volume of liquid is directly proportional to the pressure of the gas.*

EXAMPLE 11.1

Application of Henry's Law

At 20 °C, the concentration of dissolved oxygen in water exposed to gaseous oxygen at a partial pressure of 101.3 kPa is 1.38×10^{-3} mol L^{-1}. Use Henry's law to determine the solubility of oxygen when its partial pressure is 20.7 kPa, the approximate pressure of oxygen in earth's atmosphere.

Solution

According to Henry's law, for an ideal solution the solubility, C_g, of a gas (1.38×10^{-3} mol L^{-1}, in this case) is directly proportional to the pressure, P_g, of the undissolved gas above the solution (101.3 kPa in this case). Because both C_g and P_g are known, this relation can be rearragned and used to solve for k.

$$\begin{aligned} C_g &= kP_g \\ k &= \frac{C_g}{P_g} \\ &= \frac{1.38 \times 10^{-3} \text{ mol L}^{-1}}{101.3 \text{ kPa}} \\ &= 1.36 \times 10^{-5} \text{ mol L}^{-1} \text{ kPa}^{-1} \end{aligned}$$

Now, use k to find the solubility at the lower pressure.

$$\begin{aligned} C_g &= kP_g \\ & 1.36 \times 10^{-5} \text{ mol L}^{-1} \text{ kPa}^{-1} \times 20.7 \text{ kPa} \\ &= 2.82 \times 10^{-4} \text{ mol L}^{-1} \end{aligned}$$

Note that various units may be used to express the quantities involved in these sorts of computations. Any combination of units that yield to the constraints of dimensional analysis are acceptable.

Check Your Learning

Exposing a 100.0 mL sample of water at 0 °C to an atmosphere containing a gaseous solute at 152 torr resulted in the dissolution of 1.45×10^{-3} g of the solute. Use Henry's law to determine the solubility of this gaseous solute when its pressure is 760 torr.

Answer:

7.25×10^{-3} in 100.0 mL or 0.0725 g/L

EXAMPLE 11.2

Thermal Pollution and Oxygen Solubility

A certain species of freshwater trout requires a dissolved oxygen concentration of 7.5 mg/L. Could these fish thrive in a thermally polluted mountain stream (water temperature is 30.0 °C, partial pressure of atmospheric oxygen is 0.17 atm)? Use the data in Figure 11.8 to estimate a value for the Henry's law constant at this temperature.

Solution

First, estimate the Henry's law constant for oxygen in water at the specified temperature of 30.0 °C (Figure 11.8 indicates the solubility at this temperature is approximately ~1.2 mol/L).

$$k = \frac{C_g}{P_g} = 1.2 \times 10^{-3} \text{ mol/L/1.00 atm} = 1.2 \times 10^{-3} \text{ mol/L atm}$$

Then, use this k value to compute the oxygen solubility at the specified oxygen partial pressure, 0.17 atm.

$$C_g = kP_g = (1.2 \times 10^{-3} \text{ mol/L atm})\,(0.17 \text{ atm}) = 2.0 \times 10^{-4} \text{ mol/L}$$

Finally, convert this dissolved oxygen concentration from mol/L to mg/L.

$$(2.0 \times 10^{-4} \text{ mol/L})\,(32.0 \text{ g/1 mol})\,(1000 \text{ mg/g}) = 6.4 \text{ mg/L}.$$

This concentration is lesser than the required minimum value of 7.5 mg/L, and so these trout would likely not thrive in the polluted stream.

Check Your Learning

What dissolved oxygen concentration is expected for the stream above when it returns to a normal summer time temperature of 15 °C?

Answer:

8.2 mg/L

Chemistry in Everyday Life

Decompression Sickness or “The Bends”

Decompression sickness (DCS), or “the bends,” is an effect of the increased pressure of the air inhaled by scuba divers when swimming underwater at considerable depths. In addition to the pressure exerted by the atmosphere, divers are subjected to additional pressure due to the water above them, experiencing an increase of approximately 1 atm for each 10 m of depth. Therefore, the air inhaled by a diver while submerged contains gases at the corresponding higher ambient pressure, and the concentrations of the gases dissolved in the diver’s blood are proportionally higher per Henry’s law.

As the diver ascends to the surface of the water, the ambient pressure decreases and the dissolved gases becomes less soluble. If the ascent is too rapid, the gases escaping from the diver’s blood may form bubbles that can cause a variety of symptoms ranging from rashes and joint pain to paralysis and death. To avoid DCS, divers must ascend from depths at relatively slow speeds (10 or 20 m/min) or otherwise make several decompression stops, pausing for several minutes at given depths during the ascent. When these preventive measures are unsuccessful, divers with DCS are often provided hyperbaric oxygen therapy in pressurized vessels called decompression (or recompression) chambers (Figure 11.11). Researchers are also investigating related body reactions and defenses in order to develop better testing and treatment for decompression sicknetss. For example, Ingrid Eftedal, a barophysiologist specializing in bodily reactions to diving, has shown that white blood cells undergo chemical and genetic changes as a result of the condition; these can potentially be used to create biomarker tests and other methods to manage decompression sickness.

(a)

(b)

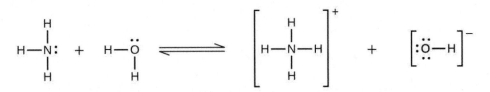

FIGURE 11.11 (a) US Navy divers undergo training in a recompression chamber. (b) Divers receive hyperbaric oxygen therapy.

Deviations from Henry’s law are observed when a chemical reaction takes place between the gaseous solute and the solvent. Thus, for example, the solubility of ammonia in water increases more rapidly with increasing pressure than predicted by the law because ammonia, being a base, reacts to some extent with water to form ammonium ions and hydroxide ions.

$$\mathrm{H{-}\ddot{N}(H)_2} + \mathrm{H{-}\ddot{O}{-}H} \rightleftharpoons \left[\mathrm{H{-}N(H)_2{-}H}\right]^+ + \left[:\ddot{O}{-}H\right]^-$$

Gases can form supersaturated solutions. If a solution of a gas in a liquid is prepared either at low temperature or under pressure (or both), then as the solution warms or as the gas pressure is reduced, the solution may become supersaturated. In 1986, more than 1700 people in Cameroon were killed when a cloud of gas, almost certainly carbon dioxide, bubbled from Lake Nyos (Figure 11.12), a deep lake in a volcanic crater. The water at the bottom of Lake Nyos is saturated with carbon dioxide by volcanic activity beneath the lake. It is believed that the lake underwent a turnover due to gradual heating from below the lake, and the warmer, less-dense water saturated with carbon dioxide reached the surface. Consequently, tremendous quantities of dissolved CO_2 were released, and the colorless gas, which is denser than air, flowed down the valley below the lake and suffocated humans and animals living in the valley.

(a)

(b)

FIGURE 11.12 (a) It is believed that the 1986 disaster that killed more than 1700 people near Lake Nyos in Cameroon resulted when a large volume of carbon dioxide gas was released from the lake. (b) A CO_2 vent has since been installed to help outgas the lake in a slow, controlled fashion and prevent a similar catastrophe from happening in the future. (credit a: modification of work by Jack Lockwood; credit b: modification of work by Bill Evans)

Solutions of Liquids in Liquids

Some liquids may be mixed in any proportions to yield solutions; in other words, they have infinite mutual solubility and are said to be **miscible**. Ethanol, sulfuric acid, and ethylene glycol (popular for use as antifreeze, pictured in Figure 11.13) are examples of liquids that are completely miscible with water. Two-cycle motor oil is miscible with gasoline, mixtures of which are used as lubricating fuels for various types of outdoor power equipment (chainsaws, leaf blowers, and so on).

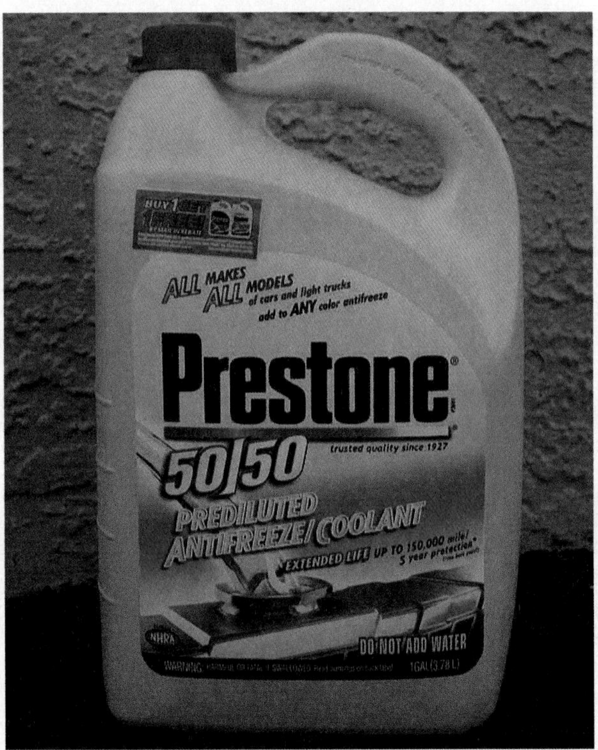

FIGURE 11.13 Water and antifreeze are miscible; mixtures of the two are homogeneous in all proportions. (credit: “dno1967”/Wikimedia commons)

Miscible liquids are typically those with very similar polarities. Consider, for example, liquids that are polar or capable of hydrogen bonding. For such liquids, the dipole-dipole attractions (or hydrogen bonding) of the solute molecules with the solvent molecules are at least as strong as those between molecules in the pure solute or in the pure solvent. Hence, the two kinds of molecules mix easily. Likewise, nonpolar liquids are miscible with each other because there is no appreciable difference in the strengths of solute-solute, solvent-solvent, and solute-solvent intermolecular attractions. The solubility of polar molecules in polar solvents and of nonpolar molecules in nonpolar solvents is, again, an illustration of the chemical axiom “like dissolves like.”

Two liquids that do not mix to an appreciable extent are called **immiscible**. Separate layers are formed when immiscible liquids are poured into the same container. Gasoline, oil (Figure 11.14), benzene, carbon tetrachloride, some paints, and many other nonpolar liquids are immiscible with water. Relatively weak attractive forces between the polar water molecules and the nonpolar liquid molecules are not adequate to overcome much stronger hydrogen bonding between water molecules. The distinction between immiscibility and miscibility is really one of extent, so that miscible liquids are of infinite mutual solubility, while liquids said to be immiscible are of very low (though not zero) mutual solubility.

FIGURE 11.14 Water and oil are immiscible. Mixtures of these two substances will form two separate layers with the less dense oil floating on top of the water. (credit: "Yortw"/Flickr)

Two liquids, such as bromine and water, that are of *moderate* mutual solubility are said to be **partially miscible**. Two partially miscible liquids usually form two layers when mixed. In the case of the bromine and water mixture, the upper layer is water, saturated with bromine, and the lower layer is bromine saturated with water. Since bromine is nonpolar, and, thus, not very soluble in water, the water layer is only slightly discolored by the bright orange bromine dissolved in it. Since the solubility of water in bromine is very low, there is no noticeable effect on the dark color of the bromine layer (Figure 11.15).

FIGURE 11.15 Bromine (the deep orange liquid on the left) and water (the clear liquid in the middle) are partially miscible. The top layer in the mixture on the right is a saturated solution of bromine in water; the bottom layer is a saturated solution of water in bromine. (credit: Paul Flowers)

Solutions of Solids in Liquids

The dependence of solubility on temperature for a number of solids in water is shown by the solubility curves in Figure 11.16. Reviewing these data indicates a general trend of increasing solubility with temperature, although there are exceptions, as illustrated by the ionic compound cerium sulfate.

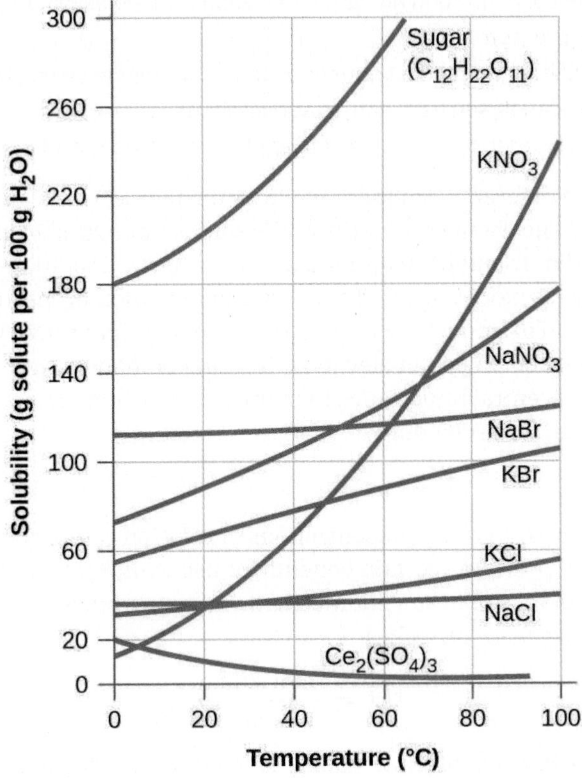

FIGURE 11.16 This graph shows how the solubility of several solids changes with temperature.

The temperature dependence of solubility can be exploited to prepare supersaturated solutions of certain compounds. A solution may be saturated with the compound at an elevated temperature (where the solute is more soluble) and subsequently cooled to a lower temperature without precipitating the solute. The resultant solution contains solute at a concentration greater than its equilibrium solubility at the lower temperature (i.e., it is supersaturated) and is relatively stable. Precipitation of the excess solute can be initiated by adding a seed crystal (see the video in the Link to Learning earlier in this module) or by mechanically agitating the solution. Some hand warmers, such as the one pictured in Figure 11.17, take advantage of this behavior.

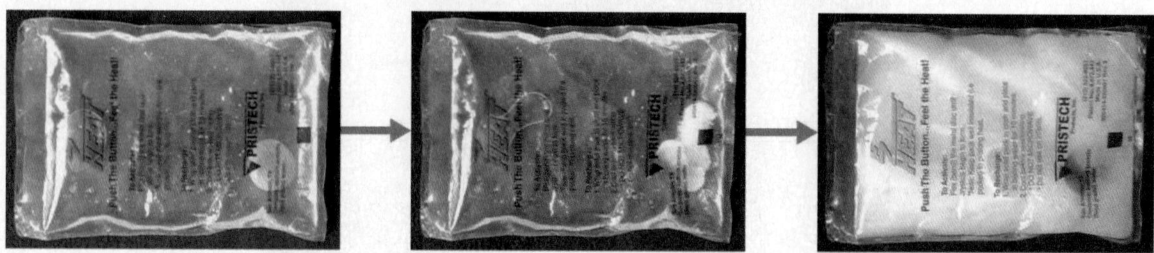

FIGURE 11.17 This hand warmer produces heat when the sodium acetate in a supersaturated solution precipitates. Precipitation of the solute is initiated by a mechanical shockwave generated when the flexible metal disk within the solution is "clicked." (credit: modification of work by "Velela"/Wikimedia Commons)

LINK TO LEARNING

This video (http://openstax.org/l/16handwarmer) shows the crystallization process occurring in a hand warmer.

11.4 Colligative Properties

LEARNING OBJECTIVES

By the end of this section, you will be able to:

- Express concentrations of solution components using mole fraction and molality
- Describe the effect of solute concentration on various solution properties (vapor pressure, boiling point, freezing point, and osmotic pressure)
- Perform calculations using the mathematical equations that describe these various colligative effects
- Describe the process of distillation and its practical applications
- Explain the process of osmosis and describe how it is applied industrially and in nature

The properties of a solution are different from those of either the pure solute(s) or solvent. Many solution properties are dependent upon the chemical identity of the solute. Compared to pure water, a solution of hydrogen chloride is more acidic, a solution of ammonia is more basic, a solution of sodium chloride is more dense, and a solution of sucrose is more viscous. There are a few solution properties, however, that depend *only* upon the total concentration of solute species, regardless of their identities. These **colligative properties** include vapor pressure lowering, boiling point elevation, freezing point depression, and osmotic pressure. This small set of properties is of central importance to many natural phenomena and technological applications, as will be described in this module.

Mole Fraction and Molality

Several units commonly used to express the concentrations of solution components were introduced in an earlier chapter of this text, each providing certain benefits for use in different applications. For example, molarity (M) is a convenient unit for use in stoichiometric calculations, since it is defined in terms of the molar amounts of solute species:

$$M = \frac{\text{mol solute}}{\text{L solution}}$$

Because solution volumes vary with temperature, molar concentrations will likewise vary. When expressed as molarity, the concentration of a solution with identical numbers of solute and solvent species will be different at different temperatures, due to the contraction/expansion of the solution. More appropriate for calculations involving many colligative properties are mole-based concentration units whose values are not dependent on temperature. Two such units are *mole fraction* (introduced in the previous chapter on gases) and *molality*.

The mole fraction, X, of a component is the ratio of its molar amount to the total number of moles of all solution components:

$$X_A = \frac{\text{mol A}}{\text{total mol of all components}}$$

By this definition, the sum of mole fractions for all solution components (the solvent and all solutes) is equal to one.

Molality is a concentration unit defined as the ratio of the numbers of moles of solute to the mass of the solvent in kilograms:

$$m = \frac{\text{mol solute}}{\text{kg solvent}}$$

Since these units are computed using only masses and molar amounts, they do not vary with temperature and, thus, are better suited for applications requiring temperature-independent concentrations, including several colligative properties, as will be described in this chapter module.

Calculating Mole Fraction and Molality

The antifreeze in most automobile radiators is a mixture of equal volumes of ethylene glycol and water, with minor amounts of other additives that prevent corrosion. What are the (a) mole fraction and (b) molality of ethylene glycol, $C_2H_4(OH)_2$, in a solution prepared from 2.22×10^3 g of ethylene glycol and 2.00×10^3 g of water (approximately 2 L of glycol and 2 L of water)?

Solution

(a) The mole fraction of ethylene glycol may be computed by first deriving molar amounts of both solution components and then substituting these amounts into the definition of mole fraction.

$$\text{mol } C_2H_4(OH)_2 = 2.22 \times 10^3 \text{ g} \times \frac{1 \text{ mol } C_2H_4(OH)_2}{62.07 \text{ g } C_2H_4(OH)_2} = 35.8 \text{ mol } C_2H_4(OH)_2$$

$$\text{mol } H_2O = 2.00 \times 10^3 \text{ g} \times \frac{1 \text{ mol } H_2O}{18.02 \text{ g } H_2O} = 111 \text{ mol } H_2O$$

$$X_{\text{ethylene glycol}} = \frac{35.8 \text{ mol } C_2H_4(OH)_2}{(35.8+111) \text{ mol total}} = 0.244$$

Notice that mole fraction is a dimensionless property, being the ratio of properties with identical units (moles).

(b) Derive moles of solute and mass of solvent (in kg).

First, use the given mass of ethylene glycol and its molar mass to find the moles of solute:

$$2.22 \times 10^3 \text{ g } C_2H_4(OH)_2 \left(\frac{\text{mol } C_2H_2(OH)_2}{62.07 \text{ g}}\right) = 35.8 \text{ mol } C_2H_4(OH)_2$$

Then, convert the mass of the water from grams to kilograms:

$$2.00 \times 10^3 \text{ g } H_2O \left(\frac{1 \text{ kg}}{1000 \text{ g}}\right) = 2.00 \text{ kg } H_2O$$

Finally, calculate molality per its definition:

$$\text{molality} = \frac{\text{mol solute}}{\text{kg solvent}}$$

$$\text{molality} = \frac{35.8 \text{ mol } C_2H_4(OH)_2}{2 \text{ kg } H_2O}$$

$$\text{molality} = 17.9\ m$$

Check Your Learning

What are the mole fraction and molality of a solution that contains 0.850 g of ammonia, NH_3, dissolved in 125 g of water?

Answer:

7.14×10^{-3}; 0.399 *m*

EXAMPLE 11.4

Converting Mole Fraction and Molal Concentrations

Calculate the mole fraction of solute and solvent in a 3.0 *m* solution of sodium chloride.

Solution

Converting from one concentration unit to another is accomplished by first comparing the two unit definitions. In this case, both units have the same numerator (moles of solute) but different denominators. The provided molal concentration may be written as:

$$\frac{3.0 \text{ mol NaCl}}{1 \text{ kg } H_2O}$$

The numerator for this solution's mole fraction is, therefore, 3.0 mol NaCl. The denominator may be computed by deriving the molar amount of water corresponding to 1.0 kg

$$1.0 \text{ kg } H_2O \left(\frac{1000 \text{ g}}{1 \text{ kg}}\right)\left(\frac{\text{mol } H_2O}{18.02 \text{ g}}\right) = 55 \text{ mol } H_2O$$

and then substituting these molar amounts into the definition for mole fraction.

$$X_{H_2O} = \frac{\text{mol } H_2O}{\text{mol NaCl} + \text{mol } H_2O}$$

$$X_{H_2O} = \frac{55 \text{ mol } H_2O}{3.0 \text{ mol NaCl} + 55 \text{ mol } H_2O}$$

$$X_{H_2O} = 0.95$$

$$X_{NaCl} = \frac{\text{mol NaCl}}{\text{mol NaCl} + \text{mol } H_2O}$$

$$X_{NaCl} = \frac{3.0 \text{ mol NaCl}}{3.0 \text{ mol NaCl} + 55 \text{ mol } H_2O}$$

$$X_{NaCl} = 0.052$$

Check Your Learning

The mole fraction of iodine, I_2, dissolved in dichloromethane, CH_2Cl_2, is 0.115. What is the molal concentration, *m*, of iodine in this solution?

Answer:

1.50 *m*

Molality and Molarity Conversions

Intravenous infusion of a 0.556 *M* aqueous solution of glucose (density of 1.04 g/mL) is part of some post-operative recovery therapies. What is the molal concentration of glucose in this solution?

Solution

The provided molal concentration may be explicitly written as:

$$M = 0.556 \text{ mol glucose/1 L solution}$$

Consider the definition of molality:

$$m = \text{mol solute/kg solvent}$$

The amount of glucose in 1-L of this solution is 0.556 mol, so the mass of water in this volume of solution is needed.

First, compute the mass of 1.00 L of the solution:

$$(1.0 \text{ L soln})\,(1.04 \text{ g/mL})\,(1000 \text{ mL/1L})\,(1 \text{ kg/1000 g}) = 1.04 \text{ kg soln}$$

This is the mass of both the water and its solute, glucose, and so the mass of glucose must be subtracted. Compute the mass of glucose from its molar amount:

$$(0.556 \text{ mol glucose})\,(180.2 \text{ g/1 mol}) = 100.2 \text{ g or } 0.1002 \text{ kg}$$

Subtracting the mass of glucose yields the mass of water in the solution:

$$1.04 \text{ kg solution} - 0.1002 \text{ kg glucose} = 0.94 \text{ kg water}$$

Finally, the molality of glucose in this solution is computed as:

$$m = 0.556 \text{ mol glucose/0.94 kg water} = 0.59 \text{ m}$$

Check Your Learning

Nitric acid, HNO_3*(aq)*, is commercially available as a 33.7 m aqueous solution (density = 1.35 g/mL). What is the molarity of this solution?

Answer:

14.6 M

Vapor Pressure Lowering

As described in the chapter on liquids and solids, the equilibrium vapor pressure of a liquid is the pressure exerted by its gaseous phase when vaporization and condensation are occurring at equal rates:

$$\text{liquid} \rightleftharpoons \text{gas}$$

Dissolving a nonvolatile substance in a volatile liquid results in a lowering of the liquid's vapor pressure. This phenomenon can be rationalized by considering the effect of added solute molecules on the liquid's vaporization and condensation processes. To vaporize, solvent molecules must be present at the surface of the solution. The presence of solute decreases the surface area available to solvent molecules and thereby reduces the rate of solvent vaporization. Since the rate of condensation is unaffected by the presence of solute, the net result is that the vaporization-condensation equilibrium is achieved with fewer solvent molecules in the vapor phase (i.e., at a lower vapor pressure) (Figure 11.18). While this interpretation is useful, it does not account for several important aspects of the colligative nature of vapor pressure lowering. A more rigorous explanation involves the property of *entropy*, a topic of discussion in a later text chapter on thermodynamics. For purposes of understanding the lowering of a liquid's vapor pressure, it is adequate to note that the more dispersed nature of matter in a solution, compared to separate solvent and solute phases, serves to effectively stabilize the solvent molecules and hinder their vaporization. A lower vapor pressure results, and a correspondingly higher boiling point as described in the next section of this module.

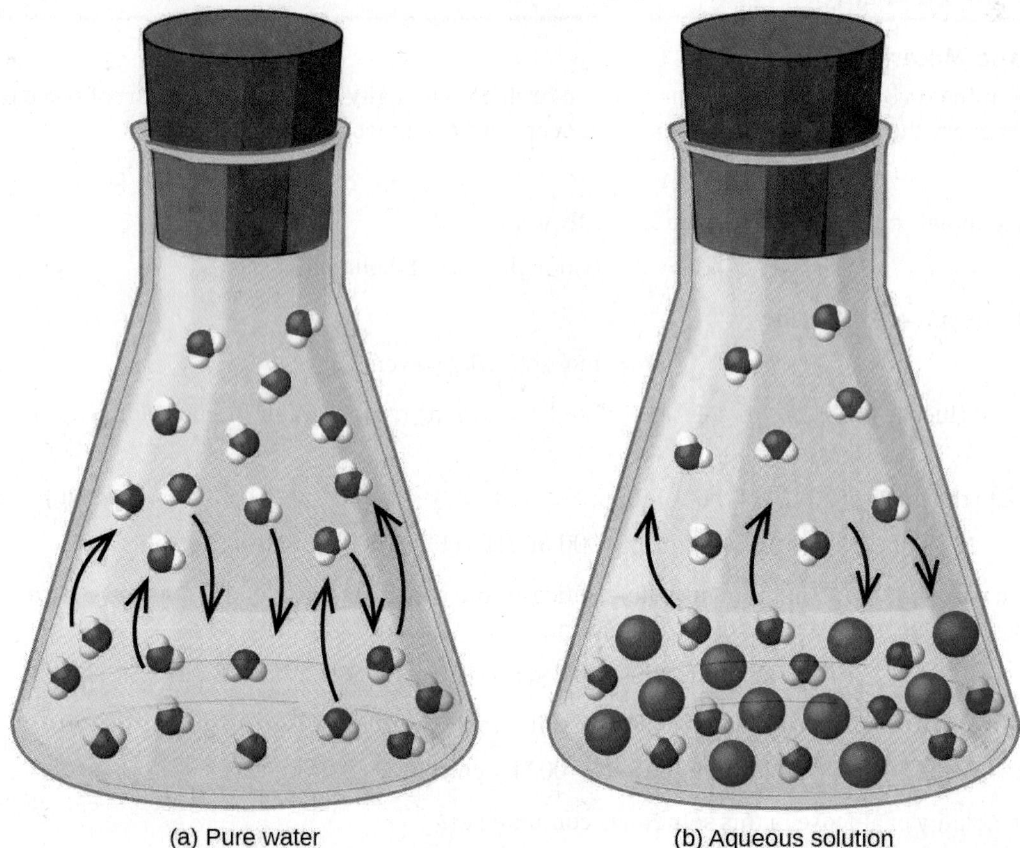

FIGURE 11.18 The presence of nonvolatile solutes lowers the vapor pressure of a solution by impeding the evaporation of solvent molecules.

The relationship between the vapor pressures of solution components and the concentrations of those components is described by **Raoult's law**: *The partial pressure exerted by any component of an ideal solution is equal to the vapor pressure of the pure component multiplied by its mole fraction in the solution.*

$$P_A = X_A P_A^*$$

where P_A is the partial pressure exerted by component A in the solution, P_A^* is the vapor pressure of pure A, and X_A is the mole fraction of A in the solution.

Recalling that the total pressure of a gaseous mixture is equal to the sum of partial pressures for all its components (Dalton's law of partial pressures), the total vapor pressure exerted by a solution containing i components is

$$P_{\text{solution}} = \sum_i P_i = \sum_i X_i P_i^*$$

A nonvolatile substance is one whose vapor pressure is negligible ($P^* \approx 0$), and so the vapor pressure above a solution containing only nonvolatile solutes is due only to the solvent:

$$P_{\text{solution}} = X_{\text{solvent}} P_{\text{solvent}}^*$$

EXAMPLE 11.6

Calculation of a Vapor Pressure

Compute the vapor pressure of an ideal solution containing 92.1 g of glycerin, $C_3H_5(OH)_3$, and 184.4 g of ethanol, C_2H_5OH, at 40 °C. The vapor pressure of pure ethanol is 0.178 atm at 40 °C. Glycerin is essentially nonvolatile at this temperature.

Solution

Since the solvent is the only volatile component of this solution, its vapor pressure may be computed per Raoult's law as:

$$P_{\text{solution}} = X_{\text{solvent}} P^*_{\text{solvent}}$$

First, calculate the molar amounts of each solution component using the provided mass data.

$$92.1 \, \cancel{\text{g } C_3H_5(OH)_3} \times \frac{1 \text{ mol } C_3H_5(OH)_3}{92.094 \, \cancel{\text{g } C_3H_5(OH)_3}} = 1.00 \text{ mol } C_3H_5(OH)_3$$

$$184.4 \, \cancel{\text{g } C_2H_5OH} \times \frac{1 \text{ mol } C_2H_5OH}{46.069 \, \cancel{\text{g } C_2H_5OH}} = 4.000 \text{ mol } C_2H_5OH$$

Next, calculate the mole fraction of the solvent (ethanol) and use Raoult's law to compute the solution's vapor pressure.

$$X_{C_2H_5OH} = \frac{4.000 \text{ mol}}{(1.00 \text{ mol} + 4.000 \text{ mol})} = 0.800$$

$$P_{\text{solv}} = X_{\text{solv}} P^*_{\text{solv}} = 0.800 \times 0.178 \text{ atm} = 0.142 \text{ atm}$$

Check Your Learning

A solution contains 5.00 g of urea, $CO(NH_2)_2$ (a nonvolatile solute) and 0.100 kg of water. If the vapor pressure of pure water at 25 °C is 23.7 torr, what is the vapor pressure of the solution assuming ideal behavior?

Answer:

23.4 torr

Distillation of Solutions

Solutions whose components have significantly different vapor pressures may be separated by a selective vaporization process known as distillation. Consider the simple case of a mixture of two volatile liquids, A and B, with A being the more volatile liquid. Raoult's law can be used to show that the vapor above the solution is enriched in component A, that is, the mole fraction of A in the vapor is greater than the mole fraction of A in the liquid (see end-of-chapter Exercise 65). By appropriately heating the mixture, component A may be vaporized, condensed, and collected—effectively separating it from component B.

Distillation is widely applied in both laboratory and industrial settings, being used to refine petroleum, to isolate fermentation products, and to purify water. A typical apparatus for laboratory-scale distillations is shown in Figure 11.19.

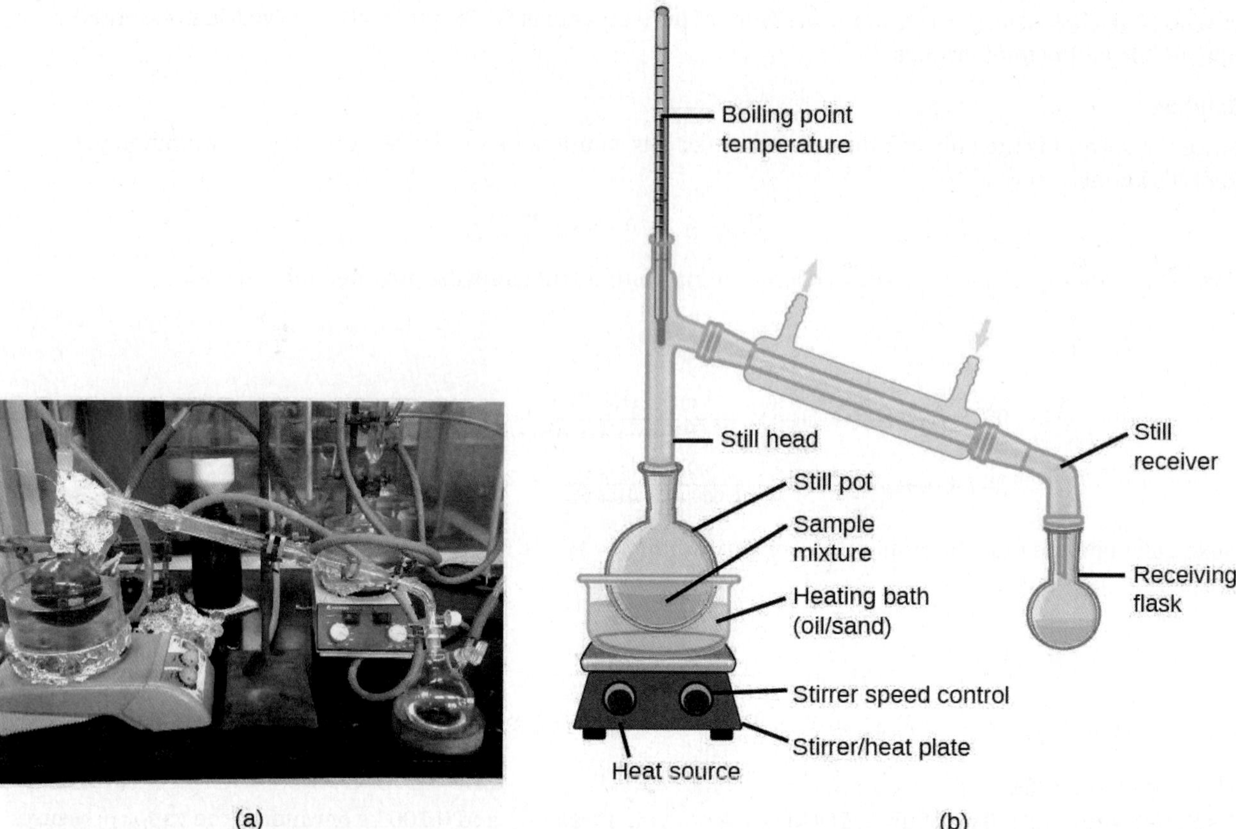

FIGURE 11.19 A typical laboratory distillation unit is shown in (a) a photograph and (b) a schematic diagram of the components. (credit a: modification of work by "Rifleman82"/Wikimedia commons; credit b: modification of work by "Slashme"/Wikimedia Commons)

Oil refineries use large-scale *fractional distillation* to separate the components of crude oil. The crude oil is heated to high temperatures at the base of a tall *fractionating column*, vaporizing many of the components that rise within the column. As vaporized components reach adequately cool zones during their ascent, they condense and are collected. The collected liquids are simpler mixtures of hydrocarbons and other petroleum compounds that are of appropriate composition for various applications (e.g., diesel fuel, kerosene, gasoline), as depicted in Figure 11.20.

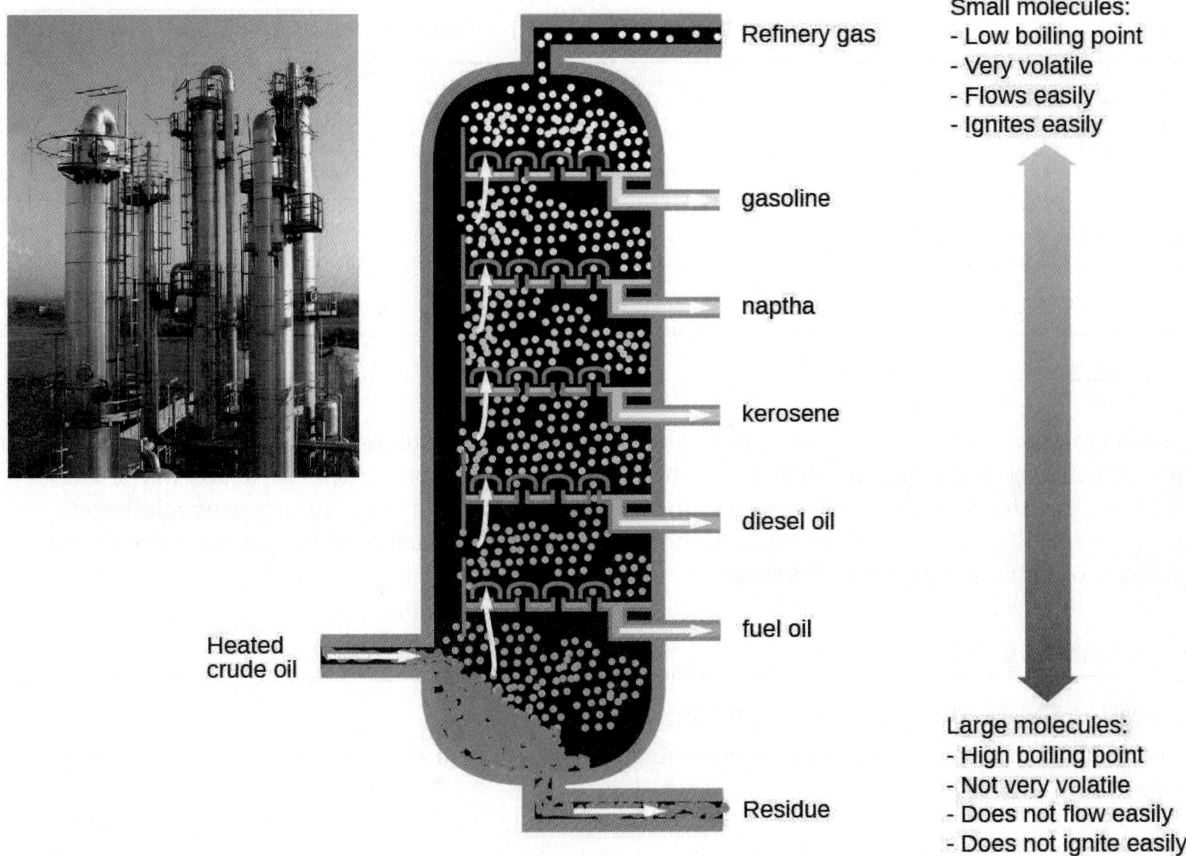

FIGURE 11.20 Crude oil is a complex mixture that is separated by large-scale fractional distillation to isolate various simpler mixtures.

Boiling Point Elevation

As described in the chapter on liquids and solids, the *boiling point* of a liquid is the temperature at which its vapor pressure is equal to ambient atmospheric pressure. Since the vapor pressure of a solution is lowered due to the presence of nonvolatile solutes, it stands to reason that the solution's boiling point will subsequently be increased. Vapor pressure increases with temperature, and so a solution will require a higher temperature than will pure solvent to achieve any given vapor pressure, including one equivalent to that of the surrounding atmosphere. The increase in boiling point observed when nonvolatile solute is dissolved in a solvent, ΔT_b, is called **boiling point elevation** and is directly proportional to the molal concentration of solute species:

$$\Delta T_b = K_b m$$

where K_b is the **boiling point elevation constant**, or the *ebullioscopic constant* and m is the molal concentration (molality) of all solute species.

Boiling point elevation constants are characteristic properties that depend on the identity of the solvent. Values of K_b for several solvents are listed in Table 11.2.

Boiling Point Elevation and Freezing Point Depression Constants for Several Solvents

Solvent	Boiling Point (°C at 1 atm)	K_b (°Cm^{-1})	Freezing Point (°C at 1 atm)	K_f (°Cm^{-1})
water	100.0	0.512	0.0	1.86

TABLE 11.2

Solvent	Boiling Point (°C at 1 atm)	K_b (°Cm^{-1})	Freezing Point (°C at 1 atm)	K_f (°Cm^{-1})
hydrogen acetate	118.1	3.07	16.6	3.9
benzene	80.1	2.53	5.5	5.12
chloroform	61.26	3.63	−63.5	4.68
nitrobenzene	210.9	5.24	5.67	8.1

TABLE 11.2

The extent to which the vapor pressure of a solvent is lowered and the boiling point is elevated depends on the total number of solute particles present in a given amount of solvent, not on the mass or size or chemical identities of the particles. A 1 *m* aqueous solution of sucrose (342 g/mol) and a 1 *m* aqueous solution of ethylene glycol (62 g/mol) will exhibit the same boiling point because each solution has one mole of solute particles (molecules) per kilogram of solvent.

EXAMPLE 11.7

Calculating the Boiling Point of a Solution

Assuming ideal solution behavior, what is the boiling point of a 0.33 *m* solution of a nonvolatile solute in benzene?

Solution

Use the equation relating boiling point elevation to solute molality to solve this problem in two steps.

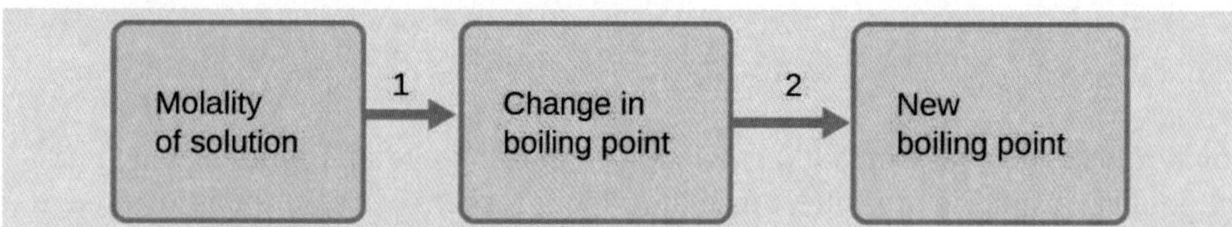

Step 1. *Calculate the change in boiling point.*

$$\Delta T_b = K_b m = 2.53\ °C\ m^{-1} \times 0.33\ m = 0.83\ °C$$

Step 2. *Add the boiling point elevation to the pure solvent's boiling point.*

$$\text{Boiling temperature} = 80.1\ °C + 0.83\ °C = 80.9\ °C$$

Check Your Learning

Assuming ideal solution behavior, what is the boiling point of the antifreeze described in Example 11.3?

Answer:

109.2 °C

EXAMPLE 11.8

The Boiling Point of an Iodine Solution

Find the boiling point of a solution of 92.1 g of iodine, I_2, in 800.0 g of chloroform, $CHCl_3$, assuming that the iodine is nonvolatile and that the solution is ideal.

Solution

A four-step approach to solving this problem is outlined below.

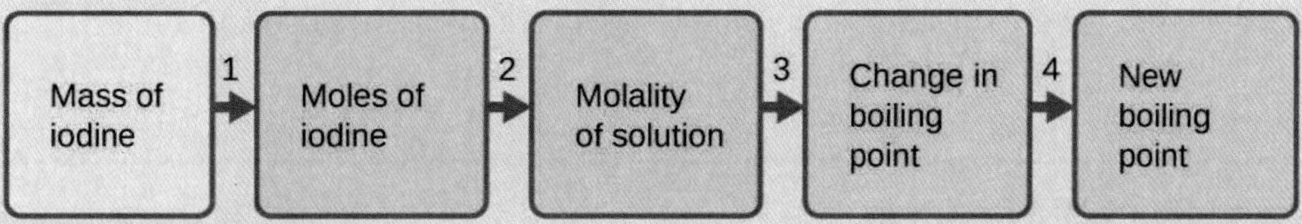

Step 1. *Convert from grams to moles of* I_2 *using the molar mass of* I_2 *in the unit conversion factor.*
Result: 0.363 mol
Step 2. *Determine the molality of the solution from the number of moles of solute and the mass of solvent, in kilograms.*
Result: 0.454 *m*
Step 3. *Use the direct proportionality between the change in boiling point and molal concentration to determine how much the boiling point changes.*
Result: 1.65 °C
Step 4. *Determine the new boiling point from the boiling point of the pure solvent and the change.*
Result: 62.91 °C
Check each result as a self-assessment.

Check Your Learning

What is the boiling point of a solution of 1.0 g of glycerin, $C_3H_5(OH)_3$, in 47.8 g of water? Assume an ideal solution.

Answer:

100.12 °C

Freezing Point Depression

Solutions freeze at lower temperatures than pure liquids. This phenomenon is exploited in "de-icing" schemes that use salt (Figure 11.21), calcium chloride, or urea to melt ice on roads and sidewalks, and in the use of ethylene glycol as an "antifreeze" in automobile radiators. Seawater freezes at a lower temperature than fresh water, and so the Arctic and Antarctic oceans remain unfrozen even at temperatures below 0 °C (as do the body fluids of fish and other cold-blooded sea animals that live in these oceans).

FIGURE 11.21 Rock salt (NaCl), calcium chloride ($CaCl_2$), or a mixture of the two are used to melt ice. (credit: modification of work by Eddie Welker)

The decrease in freezing point of a dilute solution compared to that of the pure solvent, ΔT_f, is called the **freezing point depression** and is directly proportional to the molal concentration of the solute

$$\Delta T_f = K_f m$$

where m is the molal concentration of the solute and K_f is called the **freezing point depression constant** (or *cryoscopic constant*). Just as for boiling point elevation constants, these are characteristic properties whose values depend on the chemical identity of the solvent. Values of K_f for several solvents are listed in Table 11.2.

EXAMPLE 11.9

Calculation of the Freezing Point of a Solution

Assuming ideal solution behavior, what is the freezing point of the 0.33 *m* solution of a nonvolatile nonelectrolyte solute in benzene described in Example 11.4?

Solution

Use the equation relating freezing point depression to solute molality to solve this problem in two steps.

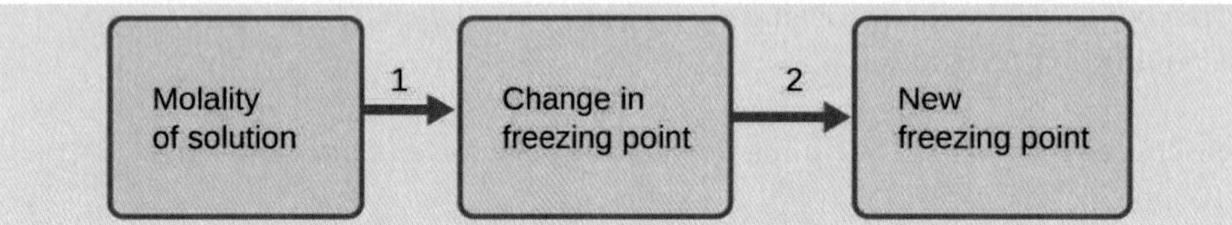

Step 1. *Calculate the change in freezing point.*

$$\Delta T_f = K_f m = 5.12 \text{ °C } m^{-1} \times 0.33 \; m = 1.7 \text{ °C}$$

Step 2. *Subtract the freezing point change observed from the pure solvent's freezing point.*

$$\text{Freezing Temperature} = 5.5 \text{ °C} - 1.7 \text{ °C} = 3.8 \text{ °C}$$

Check Your Learning

Assuming ideal solution behavior, what is the freezing point of a 1.85 *m* solution of a nonvolatile nonelectrolyte solute in nitrobenzene?

Answer:

–9.3 °C

Chemistry in Everyday Life

Colligative Properties and De-Icing

Sodium chloride and its group 2 analogs calcium and magnesium chloride are often used to de-ice roadways and sidewalks, due to the fact that a solution of any one of these salts will have a freezing point lower than 0 °C, the freezing point of pure water. The group 2 metal salts are frequently mixed with the cheaper and more readily available sodium chloride ("rock salt") for use on roads, since they tend to be somewhat less corrosive than the NaCl, and they provide a larger depression of the freezing point, since they dissociate to yield three particles per formula unit, rather than two particles like the sodium chloride.

Because these ionic compounds tend to hasten the corrosion of metal, they would not be a wise choice to use in antifreeze for the radiator in your car or to de-ice a plane prior to takeoff. For these applications, covalent compounds, such as ethylene or propylene glycol, are often used. The glycols used in radiator fluid not only lower the freezing point of the liquid, but they elevate the boiling point, making the fluid useful in both winter and summer. Heated glycols are often sprayed onto the surface of airplanes prior to takeoff in inclement weather in the winter to remove ice that has already formed and prevent the formation of more ice, which would be particularly dangerous if formed on the control surfaces of the aircraft (Figure 11.22).

(a)

(b)

FIGURE 11.22 Freezing point depression is exploited to remove ice from (a) roadways and (b) the control surfaces of aircraft.

Phase Diagram for a Solution

The colligative effects on vapor pressure, boiling point, and freezing point described in the previous section are conveniently summarized by comparing the phase diagrams for a pure liquid and a solution derived from that liquid (Figure 11.23).

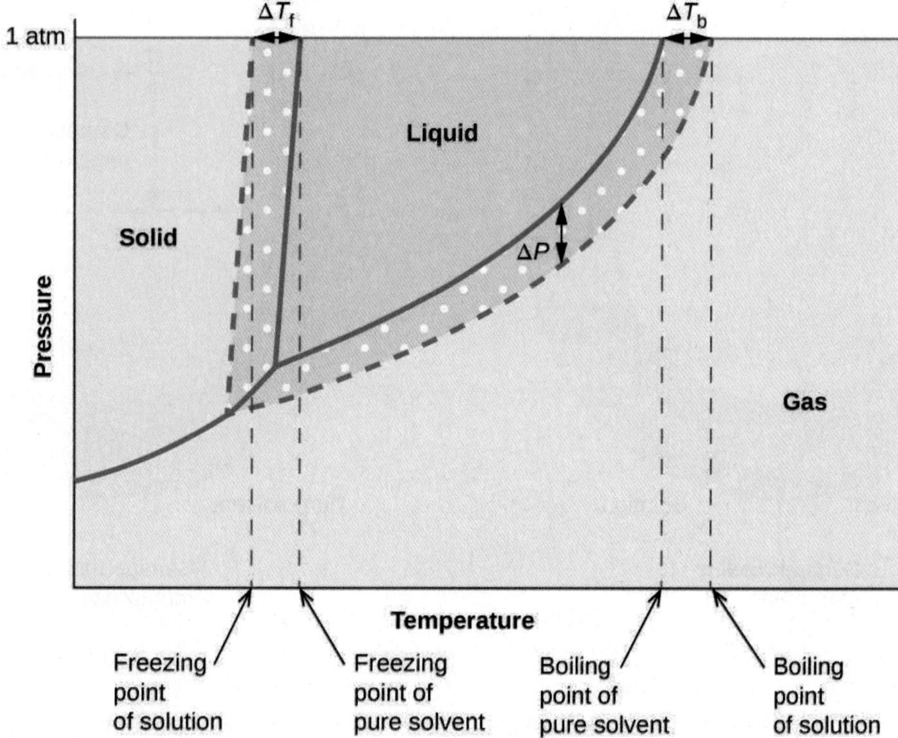

FIGURE 11.23 Phase diagrams for a pure solvent (solid curves) and a solution formed by dissolving nonvolatile solute in the solvent (dashed curves).

The liquid-vapor curve for the solution is located *beneath* the corresponding curve for the solvent, depicting the vapor pressure *lowering*, ΔP, that results from the dissolution of nonvolatile solute. Consequently, at any given pressure, the solution's boiling point is observed at a higher temperature than that for the pure solvent, reflecting the boiling point elevation, ΔT_b, associated with the presence of nonvolatile solute. The solid-liquid curve for the solution is displaced left of that for the pure solvent, representing the freezing point depression, ΔT_f, that accompanies solution formation. Finally, notice that the solid-gas curves for the solvent and its solution are identical. This is the case for many solutions comprising liquid solvents and nonvolatile solutes. Just as for vaporization, when a solution of this sort is frozen, it is actually just the *solvent* molecules that undergo the liquid-to-solid transition, forming pure solid solvent that excludes solute species. The solid and gaseous phases, therefore, are composed of solvent only, and so transitions between these phases are not subject to colligative effects.

Osmosis and Osmotic Pressure of Solutions

A number of natural and synthetic materials exhibit *selective permeation*, meaning that only molecules or ions of a certain size, shape, polarity, charge, and so forth, are capable of passing through (permeating) the material. Biological cell membranes provide elegant examples of selective permeation in nature, while dialysis tubing used to remove metabolic wastes from blood is a more simplistic technological example. Regardless of how they may be fabricated, these materials are generally referred to as **semipermeable membranes**.

Consider the apparatus illustrated in Figure 11.24, in which samples of pure solvent and a solution are separated by a membrane that only solvent molecules may permeate. Solvent molecules will diffuse across the membrane in both directions. Since the concentration of *solvent* is greater in the pure solvent than the solution, these molecules will diffuse from the solvent side of the membrane to the solution side at a faster rate than they will in the reverse direction. The result is a net transfer of solvent molecules from the pure solvent to the solution. Diffusion-driven transfer of solvent molecules through a semipermeable membrane is a process known as **osmosis**.

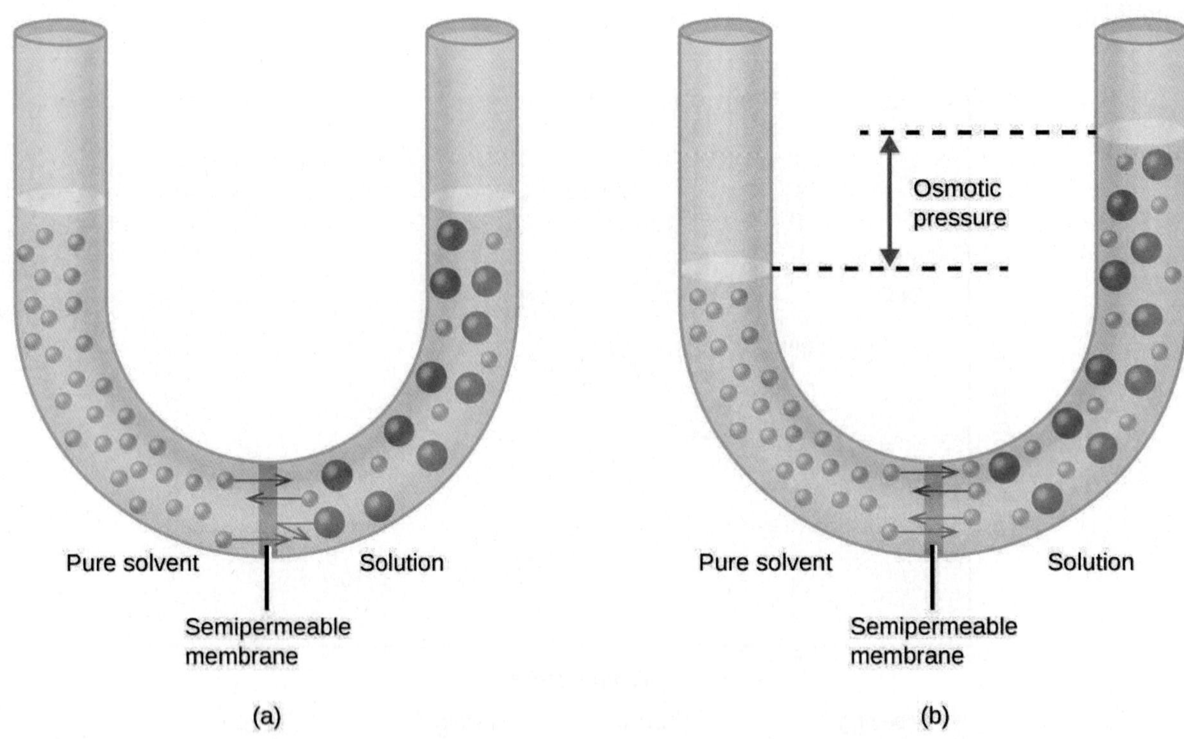

FIGURE 11.24 (a) A solution and pure solvent are initially separated by an osmotic membrane. (b) Net transfer of solvent molecules to the solution occurs until its osmotic pressure yields equal rates of transfer in both directions.

When osmosis is carried out in an apparatus like that shown in Figure 11.24, the volume of the solution increases as it becomes diluted by accumulation of solvent. This causes the level of the solution to rise, increasing its hydrostatic pressure (due to the weight of the column of solution in the tube) and resulting in a faster transfer of solvent molecules back to the pure solvent side. When the pressure reaches a value that yields a reverse solvent transfer rate equal to the osmosis rate, bulk transfer of solvent ceases. This pressure is called the **osmotic pressure (Π)** of the solution. The osmotic pressure of a dilute solution is related to its solute molarity, *M*, and absolute temperature, *T*, according to the equation

$$\Pi = MRT$$

where *R* is the universal gas constant.

EXAMPLE 11.10

Calculation of Osmotic Pressure

Assuming ideal solution behavior, what is the osmotic pressure (atm) of a 0.30 *M* solution of glucose in water that is used for intravenous infusion at body temperature, 37 °C?

Solution

Find the osmotic pressure, Π, using the formula $\Pi = MRT$, where T is on the Kelvin scale (310 K) and the value of R is expressed in appropriate units (0.08206 L atm/mol K).

$$\begin{aligned} \Pi &= MRT \\ &= 0.30 \text{ mol/L} \times 0.08206 \text{ L atm/mol K} \times 310 \text{ K} \\ &= 7.6 \text{ atm} \end{aligned}$$

Check Your Learning

Assuming ideal solution behavior, what is the osmotic pressure (atm) a solution with a volume of 0.750 L that contains 5.0 g of methanol, CH_3OH, in water at 37 °C?

Answer:

5.3 atm

If a solution is placed in an apparatus like the one shown in Figure 11.25, applying pressure greater than the osmotic pressure of the solution reverses the osmosis and pushes solvent molecules from the solution into the pure solvent. This technique of reverse osmosis is used for large-scale desalination of seawater and on smaller scales to produce high-purity tap water for drinking.

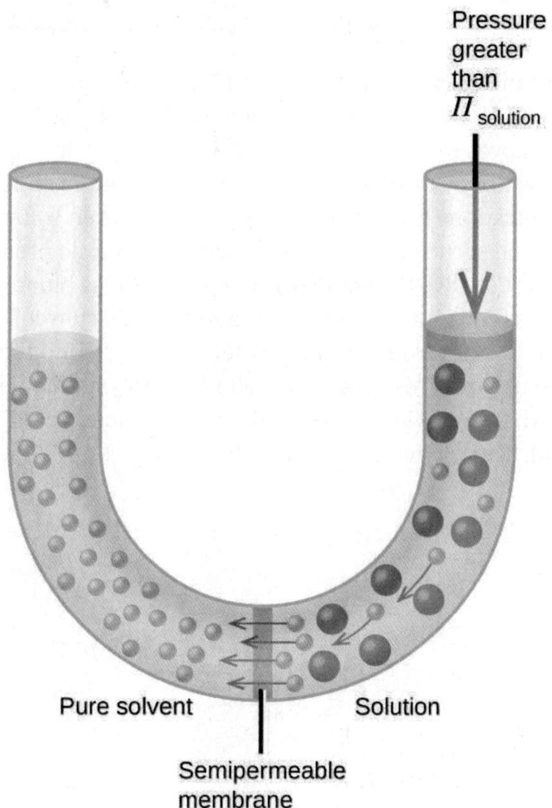

FIGURE 11.25 Applying a pressure greater than the osmotic pressure of a solution will reverse osmosis. Solvent molecules from the solution are pushed into the pure solvent.

Chemistry in Everyday Life

Reverse Osmosis Water Purification

In the process of osmosis, diffusion serves to move water through a semipermeable membrane from a less concentrated solution to a more concentrated solution. Osmotic pressure is the amount of pressure that must be applied to the more concentrated solution to cause osmosis to stop. If greater pressure is applied, the water will go from the more concentrated solution to a less concentrated (more pure) solution. This is called reverse osmosis. Reverse osmosis (RO) is used to purify water in many applications, from desalination plants in coastal cities, to water-purifying machines in grocery stores (Figure 11.26), and smaller reverse-osmosis household units. With a hand-operated pump, small RO units can be used in third-world countries, disaster areas, and in lifeboats. Our military forces have a variety of generator-operated RO units that can be transported in vehicles to remote locations.

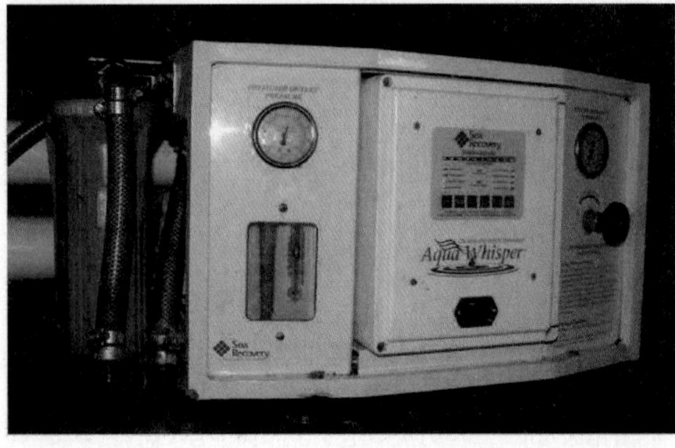

(a)

(b)

FIGURE 11.26 Reverse osmosis systems for purifying drinking water are shown here on (a) small and (b) large scales. (credit a: modification of work by Jerry Kirkhart; credit b: modification of work by Willard J. Lathrop)

Examples of osmosis are evident in many biological systems because cells are surrounded by semipermeable membranes. Carrots and celery that have become limp because they have lost water can be made crisp again by placing them in water. Water moves into the carrot or celery cells by osmosis. A cucumber placed in a concentrated salt solution loses water by osmosis and absorbs some salt to become a pickle. Osmosis can also affect animal cells. Solute concentrations are particularly important when solutions are injected into the body. Solutes in body cell fluids and blood serum give these solutions an osmotic pressure of approximately 7.7 atm. Solutions injected into the body must have the same osmotic pressure as blood serum; that is, they should be **isotonic** with blood serum. If a less concentrated solution, a **hypotonic** solution, is injected in sufficient quantity to dilute the blood serum, water from the diluted serum passes into the blood cells by osmosis, causing the cells to expand and rupture. This process is called **hemolysis**. When a more concentrated solution, a **hypertonic** solution, is injected, the cells lose water to the more concentrated solution, shrivel, and possibly die in a process called **crenation**. These effects are illustrated in Figure 11.27.

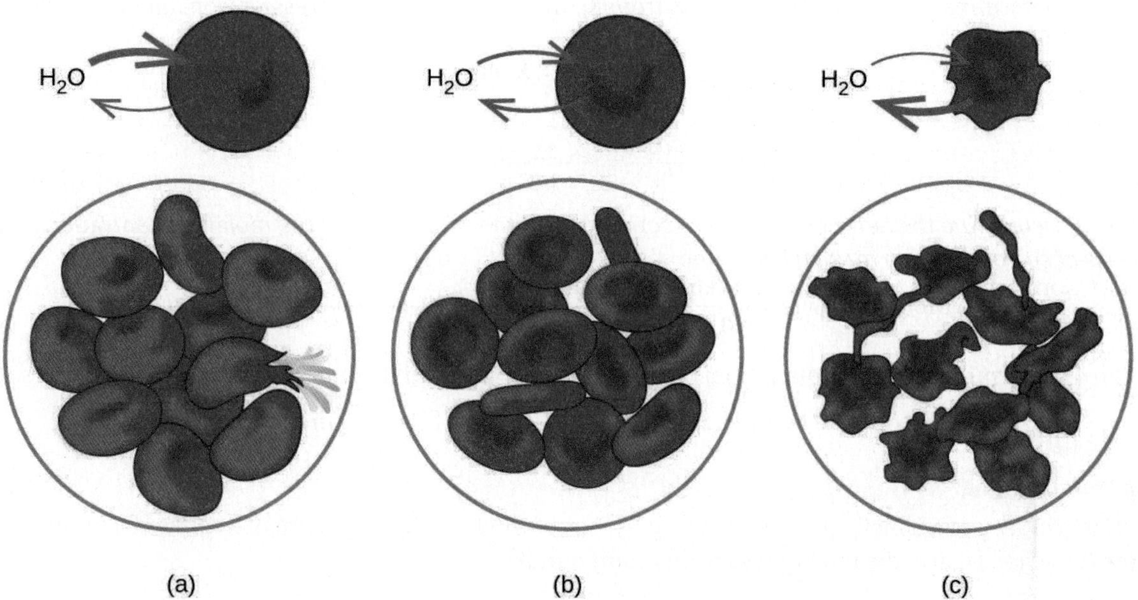

FIGURE 11.27 Red blood cell membranes are water permeable and will (a) swell and possibly rupture in a hypotonic solution; (b) maintain normal volume and shape in an isotonic solution; and (c) shrivel and possibly die in a hypertonic solution. (credit a/b/c: modifications of work by "LadyofHats"/Wikimedia commons)

Determination of Molar Masses

Osmotic pressure and changes in freezing point, boiling point, and vapor pressure are directly proportional to the number of solute species present in a given amount of solution. Consequently, measuring one of these properties for a solution prepared using a known mass of solute permits determination of the solute's molar mass.

EXAMPLE 11.11

Determination of a Molar Mass from a Freezing Point Depression

A solution of 4.00 g of a nonelectrolyte dissolved in 55.0 g of benzene is found to freeze at 2.32 °C. Assuming ideal solution behavior, what is the molar mass of this compound?

Solution

Solve this problem using the following steps.

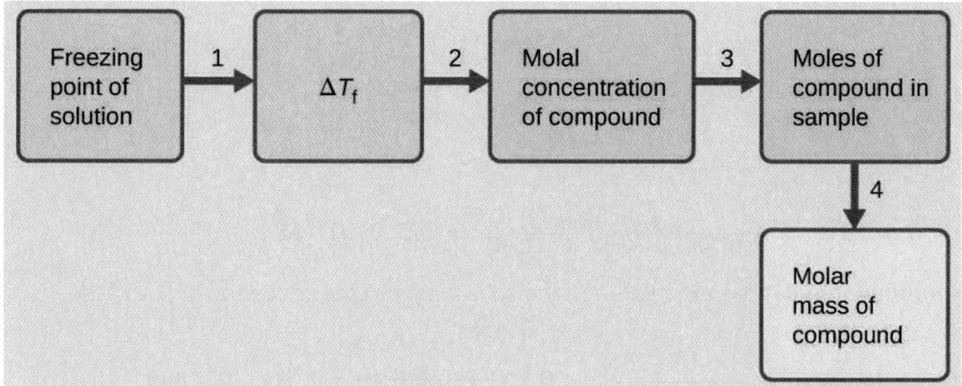

Step 1. *Determine the change in freezing point from the observed freezing point and the freezing point of pure benzene* (Table 11.2).

$$\Delta T_f = 5.5\ °C - 2.32\ °C = 3.2\ °C$$

Step 2. *Determine the molal concentration from* K_f*, the freezing point depression constant for benzene* (Table 11.2)*, and* ΔT_f.

$$\Delta T_f = K_f m$$

$$m = \frac{\Delta T_f}{K_f} = \frac{3.2\ °C}{5.12\ °C\ m^{-1}} = 0.63\ m$$

Step 3. *Determine the number of moles of compound in the solution from the molal concentration and the mass of solvent used to make the solution.*

$$\text{Moles of solute} = \frac{0.63\ \text{mol solute}}{1.00\ \cancel{\text{kg solvent}}} \times 0.0550\ \cancel{\text{kg solvent}} = 0.035\ \text{mol}$$

Step 4. *Determine the molar mass from the mass of the solute and the number of moles in that mass.*

$$\text{Molar mass} = \frac{4.00\ \text{g}}{0.035\ \text{mol}} = 1.1 \times 10^2\ \text{g/mol}$$

Check Your Learning

A solution of 35.7 g of a nonelectrolyte in 220.0 g of chloroform has a boiling point of 64.5 °C. Assuming ideal solution behavior, what is the molar mass of this compound?

Answer:

1.8×10^2 g/mol

EXAMPLE 11.12

Determination of a Molar Mass from Osmotic Pressure

A 0.500 L sample of an aqueous solution containing 10.0 g of hemoglobin has an osmotic pressure of 5.9 torr at 22 °C. Assuming ideal solution behavior, what is the molar mass of hemoglobin?

Solution

Here is one set of steps that can be used to solve the problem:

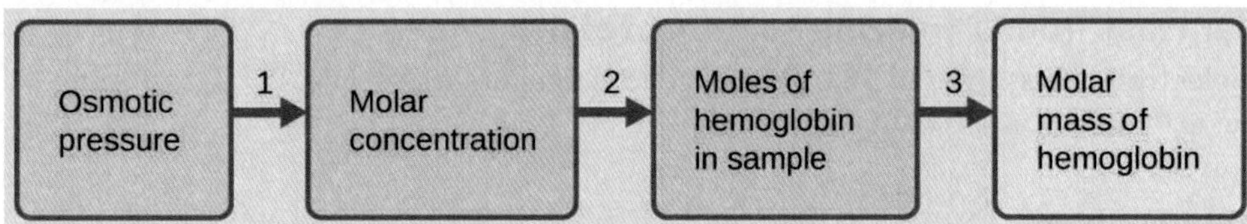

Step 1. *Convert the osmotic pressure to atmospheres, then determine the molar concentration from the osmotic pressure.*

$$\Pi = \frac{5.9\ \text{torr} \times 1\ \text{atm}}{760\ \text{torr}} = 7.8 \times 10^{-3}\ \text{atm}$$

$$\Pi = MRT$$

$$M = \frac{\Pi}{RT} = \frac{7.8 \times 10^{-3}\ \text{atm}}{(0.08206\ \text{L atm/mol K})(295\ \text{K})} = 3.2 \times 10^{-4}\ \text{M}$$

Step 2. *Determine the number of moles of hemoglobin in the solution from the concentration and the volume of the solution.*

$$\text{moles of hemoglobin} = \frac{3.2 \times 10^{-4}\ \text{mol}}{1\ \cancel{\text{L solution}}} \times 0.500\ \cancel{\text{L solution}} = 1.6 \times 10^{-4}\ \text{mol}$$

Step 3. *Determine the molar mass from the mass of hemoglobin and the number of moles in that mass.*

$$\text{molar mass} = \frac{10.0\ \text{g}}{1.6 \times 10^{-4}\ \text{mol}} = 6.2 \times 10^4\ \text{g/mol}$$

Check Your Learning

Assuming ideal solution behavior, what is the molar mass of a protein if a solution of 0.02 g of the protein in 25.0 mL of solution has an osmotic pressure of 0.56 torr at 25 °C?

Answer:

3×10^4 g/mol

Colligative Properties of Electrolytes

As noted previously in this module, the colligative properties of a solution depend only on the number, not on the identity, of solute species dissolved. The concentration terms in the equations for various colligative properties (freezing point depression, boiling point elevation, osmotic pressure) pertain to *all solute species present in the solution*. For the solutions considered thus far in this chapter, the solutes have been nonelectrolytes that dissolve physically without dissociation or any other accompanying process. Each molecule that dissolves yields one dissolved solute molecule. The dissolution of an electroyte, however, is not this simple, as illustrated by the two common examples below:

$$\begin{array}{ll} \textit{dissociation} & \text{NaCl}(s) \longrightarrow \text{Na}^+(aq) + \text{Cl}^-(aq) \\ \textit{ionization} & \text{HCl}(aq) + \text{H}_2\text{O}(l) \longrightarrow \text{Cl}^-(aq) + \text{H}_3\text{O}^+(aq) \end{array}$$

Considering the first of these examples, and assuming complete dissociation, a 1.0 *m* aqueous solution of NaCl contains 2.0 mole of ions (1.0 mol Na^+ and 1.0 mol Cl^-) per each kilogram of water, and its freezing point depression is expected to be

$$\Delta T_f = 2.0 \text{ mol ions/kg water} \times 1.86\ °\text{C kg water/mol ion} = 3.7\ °\text{C}.$$

When this solution is actually prepared and its freezing point depression measured, however, a value of 3.4 °C is obtained. Similar discrepancies are observed for other ionic compounds, and the differences between the measured and expected colligative property values typically become more significant as solute concentrations increase. These observations suggest that the ions of sodium chloride (and other strong electrolytes) are not completely dissociated in solution.

To account for this and avoid the errors accompanying the assumption of total dissociation, an experimentally measured parameter named in honor of Nobel Prize-winning German chemist Jacobus Henricus van't Hoff is used. The **van't Hoff factor (*i*)** is defined as the ratio of solute particles in solution to the number of formula units dissolved:

$$i = \frac{\text{moles of particles in solution}}{\text{moles of formula units dissolved}}$$

Values for measured van't Hoff factors for several solutes, along with predicted values assuming complete dissociation, are shown in Table 11.3.

Predicated and Measured van't Hoff Factors for Several 0.050 *m* Aqueous Solutions

Formula unit	Classification	Dissolution products	*i* (predicted)	*i* (measured)
$C_{12}H_{22}O_{11}$ (glucose)	Nonelectrolyte	$C_{12}H_{22}O_{11}$	1	1.0
NaCl	Strong electrolyte	Na^+, Cl^-	2	1.9
HCl	Strong electrolyte (acid)	H_3O^+, Cl^-	2	1.9
$MgSO_4$	Strong electrolyte	Mg^{2+}, $SO_4{}^{2-}$,	2	1.3

TABLE 11.3

Formula unit	Classification	Dissolution products	*i* (predicted)	*i* (measured)
$MgCl_2$	Strong electrolyte	Mg^{2+}, $2Cl^-$	3	2.7
$FeCl_3$	Strong electrolyte	Fe^{3+}, $3Cl^-$	4	3.4

TABLE 11.3

In 1923, the chemists Peter Debye and Erich Hückel proposed a theory to explain the apparent incomplete ionization of strong electrolytes. They suggested that although interionic attraction in an aqueous solution is very greatly reduced by solvation of the ions and the insulating action of the polar solvent, it is not completely nullified. The residual attractions prevent the ions from behaving as totally independent particles (Figure 11.28). In some cases, a positive and negative ion may actually touch, giving a solvated unit called an ion pair. Thus, the **activity**, or the effective concentration, of any particular kind of ion is less than that indicated by the actual concentration. Ions become more and more widely separated the more dilute the solution, and the residual interionic attractions become less and less. Thus, in extremely dilute solutions, the effective concentrations of the ions (their activities) are essentially equal to the actual concentrations. Note that the van't Hoff factors for the electrolytes in Table 11.3 are for 0.05 *m* solutions, at which concentration the value of *i* for NaCl is 1.9, as opposed to an ideal value of 2.

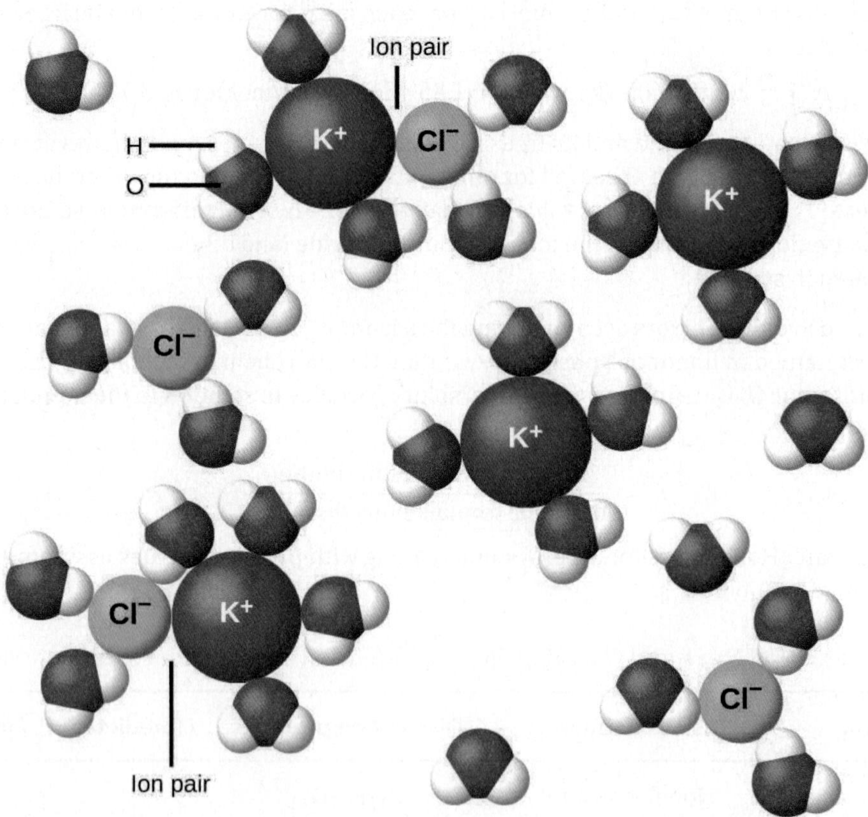

FIGURE 11.28 Dissociation of ionic compounds in water is not always complete due to the formation of ion pairs.

EXAMPLE 11.13

The Freezing Point of a Solution of an Electrolyte

The concentration of ions in seawater is approximately the same as that in a solution containing 4.2 g of NaCl dissolved in 125 g of water. Use this information and a predicted value for the van't Hoff factor (Table 11.3) to determine the freezing temperature the solution (assume ideal solution behavior).

Solution

Solve this problem using the following series of steps.

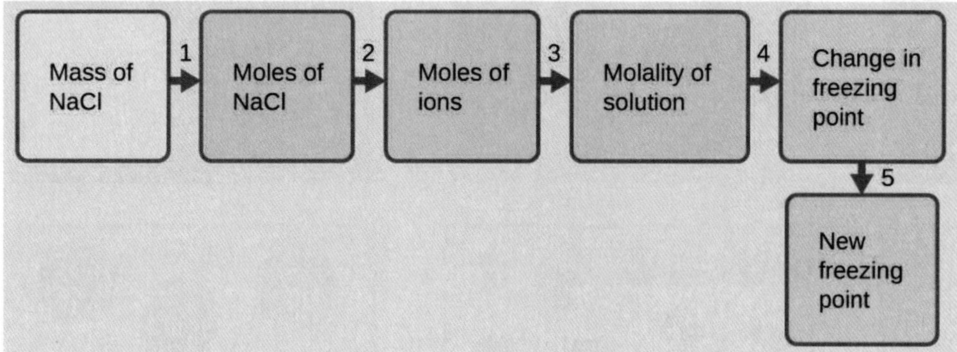

Step 1. *Convert from grams to moles of NaCl using the molar mass of NaCl in the unit conversion factor.*
Result: 0.072 mol NaCl
Step 2. *Determine the number of moles of ions present in the solution using the number of moles of ions in 1 mole of NaCl as the conversion factor (2 mol ions/1 mol NaCl).*
Result: 0.14 mol ions
Step 3. *Determine the molality of the ions in the solution from the number of moles of ions and the mass of solvent, in kilograms.*
Result: 1.2 *m*
Step 4. *Use the direct proportionality between the change in freezing point and molal concentration to determine how much the freezing point changes.*
Result: 2.1 °C
Step 5. *Determine the new freezing point from the freezing point of the pure solvent and the change.*
Result: −2.1 °C
Check each result as a self-assessment, taking care to avoid rounding errors by retaining guard digits in each step's result for computing the next step's result.

Check Your Learning

Assuming complete dissociation and ideal solution behavior, calculate the freezing point of a solution of 0.724 g of $CaCl_2$ in 175 g of water.

Answer:

−0.208 °C

11.5 Colloids

LEARNING OBJECTIVES

By the end of this section, you will be able to:

- Describe the composition and properties of colloidal dispersions
- List and explain several technological applications of colloids

As a child, you may have made suspensions such as mixtures of mud and water, flour and water, or a suspension of solid pigments in water, known as tempera paint. These **suspensions** are heterogeneous mixtures composed of relatively large particles that are visible (or that can be seen with a magnifying glass). They are cloudy, and the suspended particles settle out after mixing. On the other hand, a solution is a homogeneous mixture in which no settling occurs and in which the dissolved species are molecules or ions. Solutions exhibit completely different behavior from suspensions. A solution may be colored, but it is transparent, the molecules or ions are invisible, and they do not settle out on standing. Another class of mixtures called **colloids** (or **colloidal dispersions**) exhibit properties intermediate between those of suspensions and solutions (Figure 11.29). The particles in a colloid are larger than most simple molecules;

however, colloidal particles are small enough that they do not settle out upon standing.

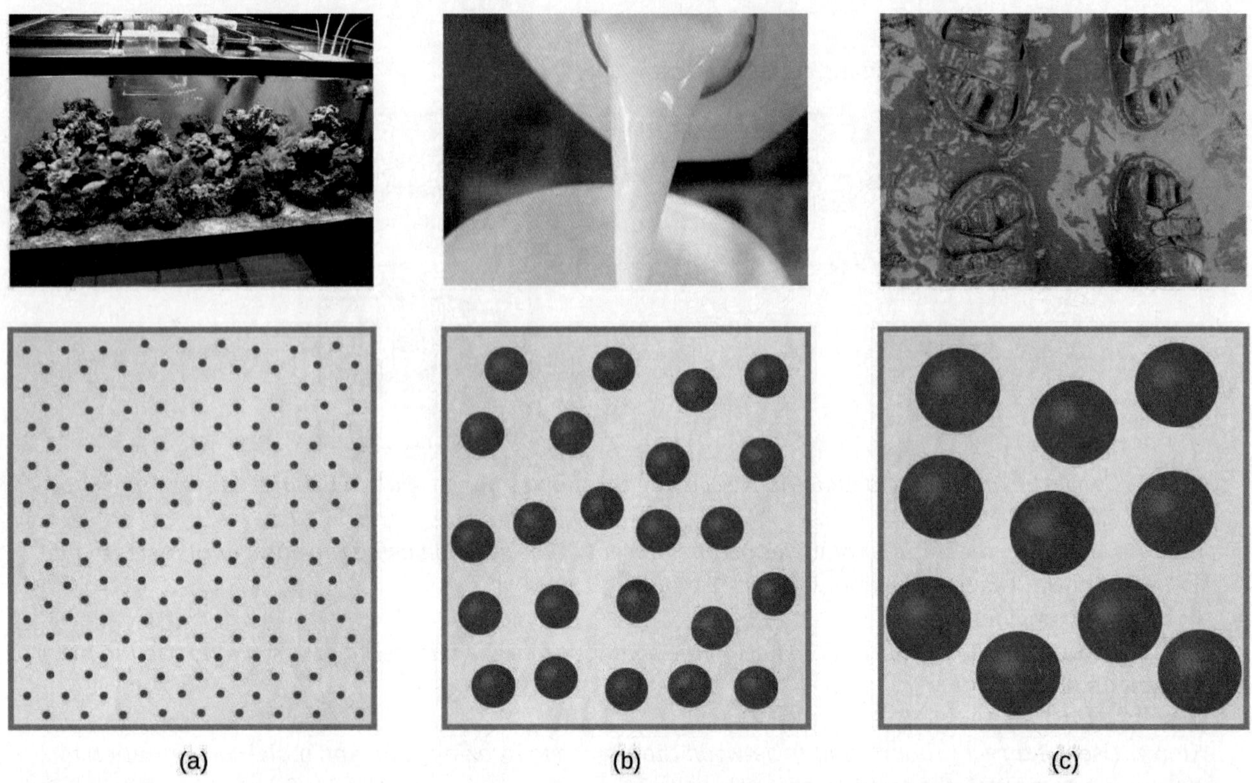

FIGURE 11.29 (a) A solution is a homogeneous mixture that appears clear, such as the saltwater in this aquarium. (b) In a colloid, such as milk, the particles are much larger but remain dispersed and do not settle. (c) A suspension, such as mud, is a heterogeneous mixture of suspended particles that appears cloudy and in which the particles can settle. (credit a photo: modification of work by Adam Wimsatt; credit b photo: modification of work by Melissa Wiese; credit c photo: modification of work by Peter Burgess)

The particles in a colloid are large enough to scatter light, a phenomenon called the **Tyndall effect**. This can make colloidal mixtures appear cloudy or opaque, such as the searchlight beams shown in Figure 11.30. Clouds are colloidal mixtures. They are composed of water droplets that are much larger than molecules, but that are small enough that they do not settle out.

FIGURE 11.30 The paths of searchlight beams are made visible when light is scattered by colloidal-size particles in the air (fog, smoke, etc.). (credit: "Bahman"/Wikimedia Commons)

The term "colloid"—from the Greek words *kolla*, meaning "glue," and *eidos*, meaning "like"—was first used in 1861 by Thomas Graham to classify mixtures such as starch in water and gelatin. Many colloidal particles are aggregates of hundreds or thousands of molecules, but others (such as proteins and polymer molecules) consist of a single extremely large molecule. The protein and synthetic polymer molecules that form colloids may have molecular masses ranging from a few thousand to many million atomic mass units.

Analogous to the identification of solution components as "solute" and "solvent," the components of a colloid are likewise classified according to their relative amounts. The particulate component typically present in a relatively minor amount is called the **dispersed phase** and the substance or solution throughout which the particulate is dispersed is called the **dispersion medium**. Colloids may involve virtually any combination of physical states (gas in liquid, liquid in solid, solid in gas, etc.), as illustrated by the examples of colloidal systems given in Table 11.4.

Examples of Colloidal Systems

Dispersed Phase	Dispersion Medium	Common Examples	Name
solid	gas	smoke, dust	—
solid	liquid	starch in water, some inks, paints, milk of magnesia	sol
solid	solid	some colored gems, some alloys	—
liquid	gas	clouds, fogs, mists, sprays	aerosol
liquid	liquid	milk, mayonnaise, butter	emulsion
liquid	solid	jellies, gels, pearl, opal (H_2O in SiO_2)	gel

TABLE 11.4

Dispersed Phase	Dispersion Medium	Common Examples	Name
gas	liquid	foams, whipped cream, beaten egg whites	foam
gas	solid	pumice, floating soaps	—

TABLE 11.4

Preparation of Colloidal Systems

Colloids are prepared by producing particles of colloidal dimensions and distributing these particles throughout a dispersion medium. Particles of colloidal size are formed by two methods:

1. Dispersion methods: breaking down larger particles. For example, paint pigments are produced by dispersing large particles by grinding in special mills.
2. Condensation methods: growth from smaller units, such as molecules or ions. For example, clouds form when water molecules condense and form very small droplets.

A few solid substances, when brought into contact with water, disperse spontaneously and form colloidal systems. Gelatin, glue, starch, and dehydrated milk powder behave in this manner. The particles are already of colloidal size; the water simply disperses them. Powdered milk particles of colloidal size are produced by dehydrating milk spray. Some atomizers produce colloidal dispersions of a liquid in air.

An **emulsion** may be prepared by shaking together or blending two immiscible liquids. This breaks one liquid into droplets of colloidal size, which then disperse throughout the other liquid. Oil spills in the ocean may be difficult to clean up, partly because wave action can cause the oil and water to form an emulsion. In many emulsions, however, the dispersed phase tends to coalesce, form large drops, and separate. Therefore, emulsions are usually stabilized by an **emulsifying agent**, a substance that inhibits the coalescence of the dispersed liquid. For example, a little soap will stabilize an emulsion of kerosene in water. Milk is an emulsion of butterfat in water, with the protein casein serving as the emulsifying agent. Mayonnaise is an emulsion of oil in vinegar, with egg yolk components as the emulsifying agents.

Condensation methods form colloidal particles by aggregation of molecules or ions. If the particles grow beyond the colloidal size range, drops or precipitates form, and no colloidal system results. Clouds form when water molecules aggregate and form colloid-sized particles. If these water particles coalesce to form adequately large water drops of liquid water or crystals of solid water, they settle from the sky as rain, sleet, or snow. Many condensation methods involve chemical reactions. A red colloidal suspension of iron(III) hydroxide may be prepared by mixing a concentrated solution of iron(III) chloride with hot water:

$$Fe^{3+}(aq) + 3Cl^-(aq) + 6H_2O(l) \longrightarrow Fe(OH)_3(s) + H_3O^+(aq) + 3Cl^-(aq).$$

A colloidal gold sol results from the reduction of a very dilute solution of gold(III) chloride by a reducing agent such as formaldehyde, tin(II) chloride, or iron(II) sulfate:

$$Au^{3+} + 3e^- \longrightarrow Au$$

Some gold sols prepared in 1857 are still intact (the particles have not coalesced and settled), illustrating the long-term stability of many colloids.

Soaps and Detergents

Pioneers made soap by boiling fats with a strongly basic solution made by leaching potassium carbonate, K_2CO_3, from wood ashes with hot water. Animal fats contain polyesters of fatty acids (long-chain carboxylic acids). When animal fats are treated with a base like potassium carbonate or sodium hydroxide, glycerol and salts of fatty acids such as palmitic, oleic, and stearic acid are formed. The salts of fatty acids are called *soaps*. The sodium salt of stearic acid, sodium stearate, has the formula $C_{17}H_{35}CO_2Na$ and contains an uncharged nonpolar hydrocarbon chain, the $C_{17}H_{35}$— unit, and an ionic carboxylate group, the $—CO_2{}^-$ unit (Figure 11.31).

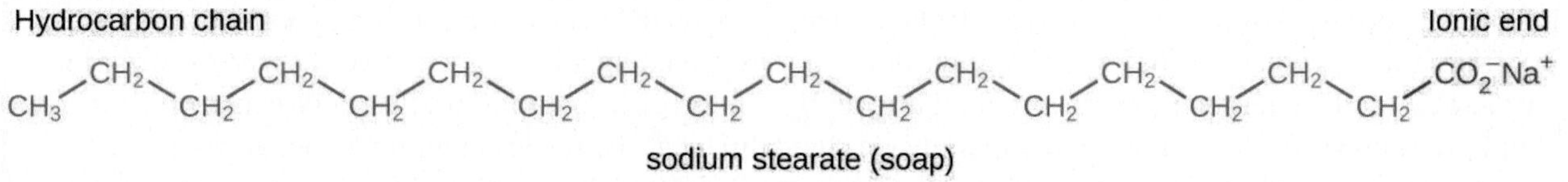

Hydrocarbon chain — Ionic end

$$CH_3-CH_2-CH_2-CH_2-CH_2-CH_2-CH_2-CH_2-CH_2-CH_2-CH_2-CH_2-CH_2-CH_2-CH_2-CH_2-CH_2-CO_2^-Na^+$$

sodium stearate (soap)

FIGURE 11.31 Soaps contain a nonpolar hydrocarbon end (blue) and an ionic end (red). The ionic end is a carboxylate group. The length of the hydrocarbon end can vary from soap to soap.

Detergents (soap substitutes) also contain nonpolar hydrocarbon chains, such as $C_{12}H_{25}-$, and an ionic group, such as a sulfate—$OSO_3{}^-$, or a sulfonate—$SO_3{}^-$ (Figure 11.32). Soaps form insoluble calcium and magnesium compounds in hard water; detergents form water-soluble products—a definite advantage for detergents.

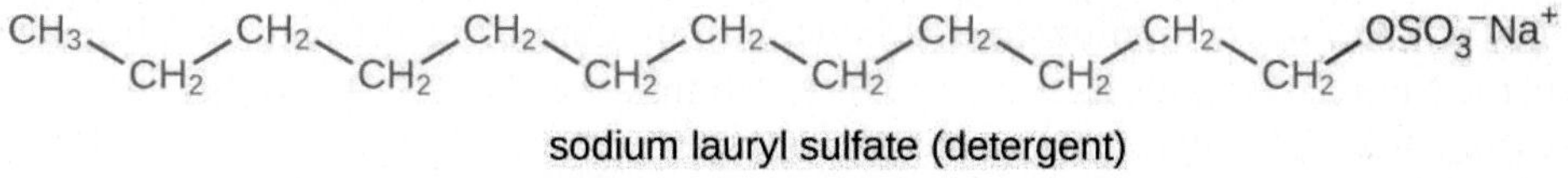

$$CH_3-CH_2-CH_2-CH_2-CH_2-CH_2-CH_2-CH_2-CH_2-CH_2-CH_2-CH_2-OSO_3^-Na^+$$

sodium lauryl sulfate (detergent)

FIGURE 11.32 Detergents contain a nonpolar hydrocarbon end (blue) and an ionic end (red). The ionic end can be either a sulfate or a sulfonate. The length of the hydrocarbon end can vary from detergent to detergent.

The cleaning action of soaps and detergents can be explained in terms of the structures of the molecules involved. The hydrocarbon (nonpolar) end of a soap or detergent molecule dissolves in, or is attracted to, nonpolar substances such as oil, grease, or dirt particles. The ionic end is attracted by water (polar), illustrated in Figure 11.33. As a result, the soap or detergent molecules become oriented at the interface between the dirt particles and the water so they act as a kind of bridge between two different kinds of matter, nonpolar and polar. Molecules such as this are termed **amphiphilic** since they have both a hydrophobic ("water-fearing") part and a hydrophilic ("water-loving") part. As a consequence, dirt particles become suspended as colloidal particles and are readily washed away.

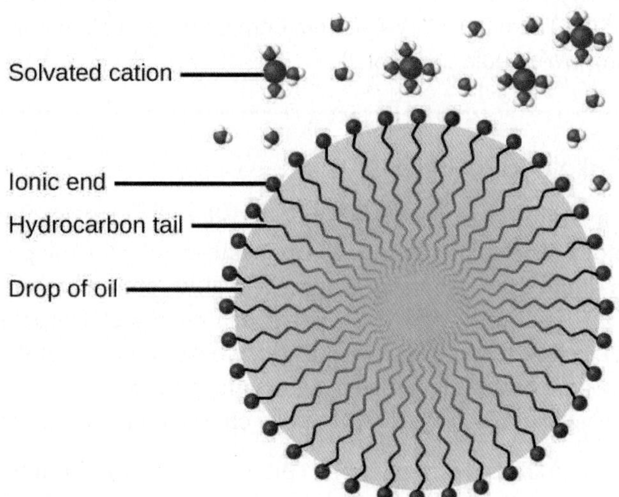

FIGURE 11.33 This diagrammatic cross section of an emulsified drop of oil in water shows how soap or detergent acts as an emulsifier.

Chemistry in Everyday Life

Deepwater Horizon Oil Spill

The blowout of the Deepwater Horizon oil drilling rig on April 20, 2010, in the Gulf of Mexico near Mississippi began the largest marine oil spill in the history of the petroleum industry. In the 87 days following the blowout, an estimated 4.9 million barrels (210 million gallons) of oil flowed from the ruptured well 5000 feet below the water's surface. The well was finally declared sealed on September 19, 2010.

Crude oil is immiscible with and less dense than water, so the spilled oil rose to the surface of the water.

Floating booms, skimmer ships, and controlled burns were used to remove oil from the water's surface in an attempt to protect beaches and wetlands along the Gulf coast. In addition to removal of the oil, attempts were also made to lessen its environmental impact by rendering it "soluble" (in the loose sense of the term) and thus allowing it to be diluted to hopefully less harmful levels by the vast volume of ocean water. This approach used 1.84 million gallons of the oil dispersant Corexit 9527, most of which was injected underwater at the site of the leak, with small amounts being sprayed on top of the spill. Corexit 9527 contains 2-butoxyethanol ($C_6H_{14}O_2$), an amphiphilic molecule whose polar and nonpolar ends are useful for emulsifying oil into small droplets, increasing the surface area of the oil and making it more available to marine bacteria for digestion (Figure 11.34). While this approach avoids many of the immediate hazards that bulk oil poses to marine and coastal ecosystems, it introduces the possibility of long-term effects resulting from the introduction of the complex and potential toxic components of petroleum into the ocean's food chain. A number of organizations are involved in monitoring the extended impact of this oil spill, including the National Oceanic and Atmospheric Administration (visit this website (http://openstax.org/l/16gulfspill) for additional details).

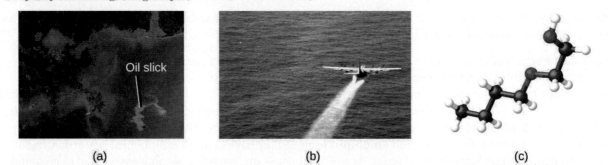

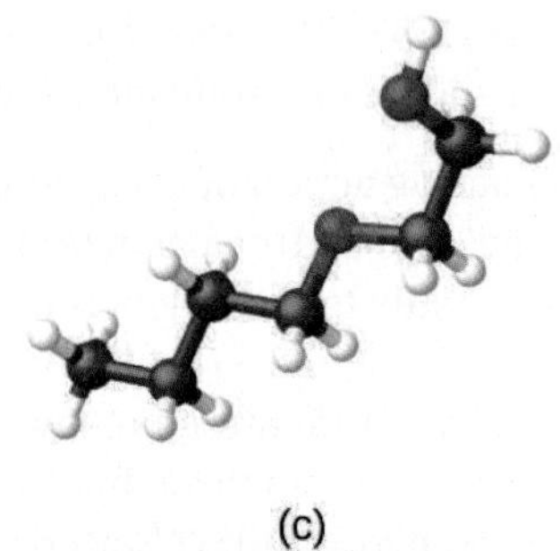

FIGURE 11.34 (a) This NASA satellite image shows the oil slick from the Deepwater Horizon spill. (b) A US Air Force plane sprays Corexit, a dispersant. (c) The molecular structure of 2-butoxyethanol is shown. (credit a: modification of work by "NASA, FT2, demis.nl"/Wikimedia Commons; credit b: modification of work by "NASA/MODIS Rapid Response Team"/Wikimedia Commons)

Electrical Properties of Colloidal Particles

Dispersed colloidal particles are often electrically charged. A colloidal particle of iron(III) hydroxide, for example, does not contain enough hydroxide ions to compensate exactly for the positive charges on the iron(III) ions. Thus, each individual colloidal particle bears a positive charge, and the colloidal dispersion consists of charged colloidal particles and some free hydroxide ions, which keep the dispersion electrically neutral. Most metal hydroxide colloids have positive charges, whereas most metals and metal sulfides form negatively charged dispersions. All colloidal particles in any one system have charges of the same sign. This helps keep them dispersed because particles containing like charges repel each other.

The charged nature of some colloidal particles may be exploited to remove them from a variety of mixtures. For example, the particles comprising smoke are often colloidally dispersed and electrically charged. Frederick Cottrell, an American chemist, developed a process to remove these particles. The charged particles are attracted to highly charged electrodes, where they are neutralized and deposited as dust (Figure 11.36). This is one of the important methods used to clean up the smoke from a variety of industrial processes. The process is also important in the recovery of valuable products from the smoke and flue dust of smelters, furnaces, and kilns. There are also similar electrostatic air filters designed for home use to improve indoor air quality.

Portrait of a Chemist

Frederick Gardner Cottrell

(a) (b)

FIGURE 11.35 (a) Frederick Cottrell developed (b) the electrostatic precipitator, a device designed to curb air pollution by removing colloidal particles from air. (credit b: modification of work by "SpLot"/Wikimedia Commons)

Born in Oakland, CA, in 1877, Frederick Cottrell devoured textbooks as if they were novels and graduated from high school at the age of 16. He then entered the University of California (UC), Berkeley, completing a Bachelor's degree in three years. He saved money from his $1200 annual salary as a chemistry teacher at Oakland High School to fund his studies in chemistry in Berlin with Nobel prize winner Jacobus Henricus van't Hoff, and in Leipzig with Wilhelm Ostwald, another Nobel awardee. After earning his PhD in physical chemistry, he returned to the United States to teach at UC Berkeley. He also consulted for the DuPont Company, where he developed the electrostatic precipitator, a device designed to curb air pollution by removing colloidal particles from air. Cottrell used the proceeds from his invention to fund a nonprofit research corporation to finance scientific research.

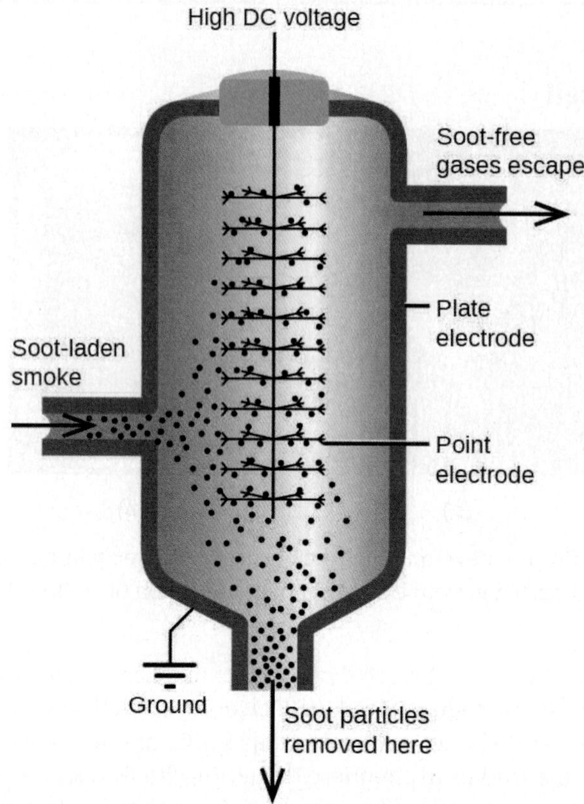

FIGURE 11.36 In a Cottrell precipitator, positively and negatively charged particles are attracted to highly charged electrodes, where they are neutralized and deposited as dust.

Gels

Gelatin desserts, such as Jell-O, are a type of colloid (Figure 11.37). Gelatin sets on cooling because the hot aqueous mixture of gelatin coagulates as it cools, yielding an extremely viscous body known as a **gel**. A gel is a colloidal dispersion of a liquid phase throughout a solid phase. It appears that the fibers of the dispersing medium form a complex three-dimensional network, the interstices being filled with the liquid medium or a dilute solution of the dispersing medium.

FIGURE 11.37 Gelatin desserts are colloids in which an aqueous solution of sweeteners and flavors is dispersed throughout a medium of solid proteins. (credit photo: modification of work by Steven Depolo)

Pectin, a carbohydrate from fruit juices, is a gel-forming substance important in jelly making. Silica gel, a colloidal dispersion of hydrated silicon dioxide, is formed when dilute hydrochloric acid is added to a dilute solution of sodium silicate. Canned Heat is a flammable gel made by mixing alcohol and a saturated aqueous solution of calcium acetate.

Key Terms

alloy solid mixture of a metallic element and one or more additional elements

amphiphilic molecules possessing both hydrophobic (nonpolar) and a hydrophilic (polar) parts

boiling point elevation elevation of the boiling point of a liquid by addition of a solute

boiling point elevation constant the proportionality constant in the equation relating boiling point elevation to solute molality; also known as the ebullioscopic constant

colligative property property of a solution that depends only on the concentration of a solute species

colloid (also, colloidal dispersion) mixture in which relatively large solid or liquid particles are dispersed uniformly throughout a gas, liquid, or solid

crenation process whereby biological cells become shriveled due to loss of water by osmosis

dispersed phase substance present as relatively large solid or liquid particles in a colloid

dispersion medium solid, liquid, or gas in which colloidal particles are dispersed

dissociation physical process accompanying the dissolution of an ionic compound in which the compound's constituent ions are solvated and dispersed throughout the solution

electrolyte substance that produces ions when dissolved in water

emulsifying agent amphiphilic substance used to stabilize the particles of some emulsions

emulsion colloid formed from immiscible liquids

freezing point depression lowering of the freezing point of a liquid by addition of a solute

freezing point depression constant (also, cryoscopic constant) proportionality constant in the equation relating freezing point depression to solute molality

gel colloidal dispersion of a liquid in a solid

hemolysis rupture of red blood cells due to the accumulation of excess water by osmosis

Henry's law the proportional relationship between the concentration of dissolved gas in a solution and the partial pressure of the gas in contact with the solution

hypertonic of greater osmotic pressure

hypotonic of less osmotic pressure

ideal solution solution that forms with no accompanying energy change

immiscible of negligible mutual solubility; typically refers to liquid substances

ion pair solvated anion/cation pair held together by moderate electrostatic attraction

ion-dipole attraction electrostatic attraction between an ion and a polar molecule

isotonic of equal osmotic pressure

miscible mutually soluble in all proportions; typically refers to liquid substances

molality (m) a concentration unit defined as the ratio of the numbers of moles of solute to the mass of the solvent in kilograms

nonelectrolyte substance that does not produce ions when dissolved in water

osmosis diffusion of solvent molecules through a semipermeable membrane

osmotic pressure (Π) opposing pressure required to prevent bulk transfer of solvent molecules through a semipermeable membrane

partially miscible of moderate mutual solubility; typically refers to liquid substances

Raoult's law the relationship between a solution's vapor pressure and the vapor pressures and concentrations of its components

saturated of concentration equal to solubility; containing the maximum concentration of solute possible for a given temperature and pressure

semipermeable membrane a membrane that selectively permits passage of certain ions or molecules

solubility extent to which a solute may be dissolved in water, or any solvent

solvation exothermic process in which intermolecular attractive forces between the solute and solvent in a solution are established

spontaneous process physical or chemical change that occurs without the addition of energy from an external source

strong electrolyte substance that dissociates or ionizes completely when dissolved in water

supersaturated of concentration that exceeds solubility; a nonequilibrium state

suspension heterogeneous mixture in which relatively large component particles are temporarily dispersed but settle out over time

Tyndall effect scattering of visible light by a colloidal dispersion

unsaturated of concentration less than solubility

van't Hoff factor (i) the ratio of the number of moles of particles in a solution to the number of moles of formula units dissolved in the solution

weak electrolyte substance that ionizes only partially when dissolved in water

Key Equations

$$C_g = kP_g$$

$$(P_A = X_A P_A^*)$$

$$P_{solution} = \sum_i P_i = \sum_i X_i P_i^*$$

$$P_{solution} = X_{solvent} P_{solvent}^*$$

$$\Delta T_b = K_b m$$

$$\Delta T_f = K_f m$$

$$\Pi = MRT$$

Summary

11.1 The Dissolution Process

A solution forms when two or more substances combine physically to yield a mixture that is homogeneous at the molecular level. The solvent is the most concentrated component and determines the physical state of the solution. The solutes are the other components typically present at concentrations less than that of the solvent. Solutions may form endothermically or exothermically, depending upon the relative magnitudes of solute and solvent intermolecular attractive forces. Ideal solutions form with no appreciable change in energy.

11.2 Electrolytes

Substances that dissolve in water to yield ions are called electrolytes. Electrolytes may be covalent compounds that chemically react with water to produce ions (for example, acids and bases), or they may be ionic compounds that dissociate to yield their constituent cations and anions, when dissolved. Dissolution of an ionic compound is facilitated by ion-dipole attractions between the ions of the compound and the polar water molecules. Soluble ionic substances and strong acids ionize completely and are strong electrolytes, while weak acids and bases ionize to only a small extent and are weak electrolytes. Nonelectrolytes are substances that do not produce ions when dissolved in water.

11.3 Solubility

The extent to which one substance will dissolve in another is determined by several factors, including the types and relative strengths of intermolecular attractive forces that may exist between the substances' atoms, ions, or molecules. This tendency to dissolve is quantified as a substance's solubility, its maximum concentration in a solution at equilibrium under specified conditions. A saturated solution contains solute at a concentration equal to its solubility. A supersaturated solution is one in which a solute's concentration exceeds its solubility—a nonequilibrium (unstable) condition that will result in solute precipitation when the solution is appropriately perturbed. Miscible liquids are soluble in all proportions, and immiscible liquids exhibit very low mutual solubility. Solubilities for gaseous solutes decrease with increasing temperature, while those for most, but not all, solid solutes increase with temperature. The concentration of a gaseous solute in a solution is proportional to the partial pressure of the gas to which the solution is exposed, a relation known as Henry's law.

11.4 Colligative Properties

Properties of a solution that depend only on the concentration of solute particles are called colligative properties. They include changes in the vapor pressure, boiling point, and freezing point of the solvent in the solution. The magnitudes of these properties depend only on the total concentration of solute particles in solution, not on the type of particles. The total concentration of solute particles in a solution also determines its osmotic pressure. This is the pressure that must be applied to the solution to prevent diffusion of molecules of pure solvent through a semipermeable membrane into the solution. Ionic compounds may not completely dissociate in solution due to activity effects, in which case observed colligative effects may be less than predicted.

11.5 Colloids

Colloids are mixtures in which one or more substances are dispersed as relatively large solid particles or liquid droplets throughout a solid, liquid, or gaseous medium. The particles of a colloid remain dispersed and do not settle due to gravity, and they are often electrically charged. Colloids are widespread in nature and are involved in many technological applications.

Exercises

11.1 The Dissolution Process

1. How do solutions differ from compounds? From other mixtures?
2. Which of the principal characteristics of solutions are evident in the solutions of $K_2Cr_2O_7$ shown in Figure 11.2?
3. When KNO_3 is dissolved in water, the resulting solution is significantly colder than the water was originally.
   (a) Is the dissolution of KNO_3 an endothermic or an exothermic process?
   (b) What conclusions can you draw about the intermolecular attractions involved in the process?
   (c) Is the resulting solution an ideal solution?
4. Give an example of each of the following types of solutions:
   (a) a gas in a liquid
   (b) a gas in a gas
   (c) a solid in a solid
5. Indicate the most important types of intermolecular attractions in each of the following solutions:
   (a) The solution in Figure 11.2.
   (b) $NO(l)$ in $CO(l)$
   (c) $Cl_2(g)$ in $Br_2(l)$
   (d) $HCl(g)$ in benzene $C_6H_6(l)$
   (e) Methanol $CH_3OH(l)$ in $H_2O(l)$
6. Predict whether each of the following substances would be more soluble in water (polar solvent) or in a hydrocarbon such as heptane (C_7H_{16}, nonpolar solvent):
   (a) vegetable oil (nonpolar)
   (b) isopropyl alcohol (polar)
   (c) potassium bromide (ionic)
7. Heat is released when some solutions form; heat is absorbed when other solutions form. Provide a molecular explanation for the difference between these two types of spontaneous processes.
8. Solutions of hydrogen in palladium may be formed by exposing Pd metal to H_2 gas. The concentration of hydrogen in the palladium depends on the pressure of H_2 gas applied, but in a more complex fashion than can be described by Henry's law. Under certain conditions, 0.94 g of hydrogen gas is dissolved in 215 g of palladium metal (solution density = 10.8 g cm^3).
   (a) Determine the molarity of this solution.
   (b) Determine the molality of this solution.
   (c) Determine the percent by mass of hydrogen atoms in this solution.

11.2 Electrolytes

9. Explain why the ions Na^+ and Cl^- are strongly solvated in water but not in hexane, a solvent composed of nonpolar molecules.
10. Explain why solutions of HBr in benzene (a nonpolar solvent) are nonconductive, while solutions in water (a polar solvent) are conductive.

11. Consider the solutions presented:
(a) Which of the following sketches best represents the ions in a solution of $Fe(NO_3)_3(aq)$?

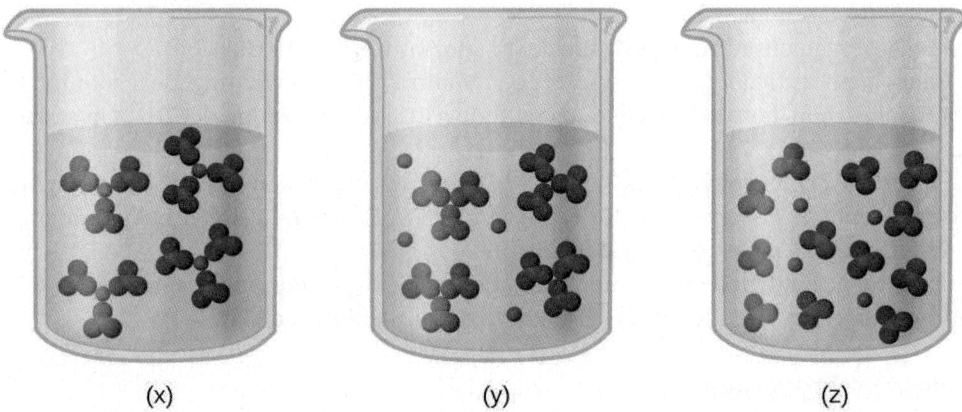

(b) Write a balanced chemical equation showing the products of the dissolution of $Fe(NO_3)_3$.

12. Compare the processes that occur when methanol (CH_3OH), hydrogen chloride (HCl), and sodium hydroxide (NaOH) dissolve in water. Write equations and prepare sketches showing the form in which each of these compounds is present in its respective solution.

13. What is the expected electrical conductivity of the following solutions?
(a) NaOH(*aq*)
(b) HCl(*aq*)
(c) $C_6H_{12}O_6(aq)$ (glucose)
(d) $NH_3(aq)$

14. Why are most *solid* ionic compounds electrically nonconductive, whereas aqueous solutions of ionic compounds are good conductors? Would you expect a *liquid* (molten) ionic compound to be electrically conductive or nonconductive? Explain.

15. Indicate the most important type of intermolecular attraction responsible for solvation in each of the following solutions:
(a) the solutions in Figure 11.7
(b) methanol, CH_3OH, dissolved in ethanol, C_2H_5OH
(c) methane, CH_4, dissolved in benzene, C_6H_6
(d) the polar halocarbon CF_2Cl_2 dissolved in the polar halocarbon $CF_2ClCFCl_2$
(e) $O_2(l)$ in $N_2(l)$

11.3 Solubility

16. Suppose you are presented with a clear solution of sodium thiosulfate, $Na_2S_2O_3$. How could you determine whether the solution is unsaturated, saturated, or supersaturated?

17. Supersaturated solutions of most solids in water are prepared by cooling saturated solutions. Supersaturated solutions of most gases in water are prepared by heating saturated solutions. Explain the reasons for the difference in the two procedures.

18. Suggest an explanation for the observations that ethanol, C_2H_5OH, is completely miscible with water and that ethanethiol, C_2H_5SH, is soluble only to the extent of 1.5 g per 100 mL of water.

19. Calculate the percent by mass of KBr in a saturated solution of KBr in water at 10 °C. See Figure 11.16 for useful data, and report the computed percentage to one significant digit.

20. Which of the following gases is expected to be most soluble in water? Explain your reasoning.
(a) CH_4
(b) CCl_4
(c) $CHCl_3$

21. At 0 °C and 1.00 atm, as much as 0.70 g of O_2 can dissolve in 1 L of water. At 0 °C and 4.00 atm, how many grams of O_2 dissolve in 1 L of water?

22. Refer to Figure 11.10.
(a) How did the concentration of dissolved CO_2 in the beverage change when the bottle was opened?
(b) What caused this change?
(c) Is the beverage unsaturated, saturated, or supersaturated with CO_2?
23. The Henry's law constant for CO_2 is 3.4×10^{-2} *M*/atm at 25 °C. Assuming ideal solution behavior, what pressure of carbon dioxide is needed to maintain a CO_2 concentration of 0.10 *M* in a can of lemon-lime soda?
24. The Henry's law constant for O_2 is 1.3×10^{-3} *M*/atm at 25 °C. Assuming ideal solution behavior, what mass of oxygen would be dissolved in a 40-L aquarium at 25 °C, assuming an atmospheric pressure of 1.00 atm, and that the partial pressure of O_2 is 0.21 atm?
25. Assuming ideal solution behavior, how many liters of HCl gas, measured at 30.0 °C and 745 torr, are required to prepare 1.25 L of a 3.20-*M* solution of hydrochloric acid?

11.4 Colligative Properties

26. Which is/are part of the macroscopic domain of solutions and which is/are part of the microscopic domain: boiling point elevation, Henry's law, hydrogen bond, ion-dipole attraction, molarity, nonelectrolyte, nonstoichiometric compound, osmosis, solvated ion?
27. What is the microscopic explanation for the macroscopic behavior illustrated in Figure 11.14?
28. Sketch a qualitative graph of the pressure versus time for water vapor above a sample of pure water and a sugar solution, as the liquids evaporate to half their original volume.
29. A solution of potassium nitrate, an electrolyte, and a solution of glycerin ($C_3H_5(OH)_3$), a nonelectrolyte, both boil at 100.3 °C. What other physical properties of the two solutions are identical?
30. What are the mole fractions of H_3PO_4 and water in a solution of 14.5 g of H_3PO_4 in 125 g of water?
(a) Outline the steps necessary to answer the question.
(b) Answer the question.
31. What are the mole fractions of HNO_3 and water in a concentrated solution of nitric acid (68.0% HNO_3 by mass)?
(a) Outline the steps necessary to answer the question.
(b) Answer the question.
32. Calculate the mole fraction of each solute and solvent:
(a) 583 g of H_2SO_4 in 1.50 kg of water—the acid solution used in an automobile battery
(b) 0.86 g of NaCl in 1.00×10^2 g of water—a solution of sodium chloride for intravenous injection
(c) 46.85 g of codeine, $C_{18}H_{21}NO_3$, in 125.5 g of ethanol, C_2H_5OH
(d) 25 g of I_2 in 125 g of ethanol, C_2H_5OH
33. Calculate the mole fraction of each solute and solvent:
(a) 0.710 kg of sodium carbonate (washing soda), Na_2CO_3, in 10.0 kg of water—a saturated solution at 0 °C
(b) 125 g of NH_4NO_3 in 275 g of water—a mixture used to make an instant ice pack
(c) 25 g of Cl_2 in 125 g of dichloromethane, CH_2Cl_2
(d) 0.372 g of tetrahydropyridine, C_5H_9N, in 125 g of chloroform, $CHCl_3$
34. Calculate the mole fractions of methanol, CH_3OH; ethanol, C_2H_5OH; and water in a solution that is 40% methanol, 40% ethanol, and 20% water by mass. (Assume the data are good to two significant figures.)
35. What is the difference between a 1 *M* solution and a 1 *m* solution?
36. What is the molality of phosphoric acid, H_3PO_4, in a solution of 14.5 g of H_3PO_4 in 125 g of water?
(a) Outline the steps necessary to answer the question.
(b) Answer the question.
37. What is the molality of nitric acid in a concentrated solution of nitric acid (68.0% HNO_3 by mass)?
(a) Outline the steps necessary to answer the question.
(b) Answer the question.
38. Calculate the molality of each of the following solutions:
(a) 583 g of H_2SO_4 in 1.50 kg of water—the acid solution used in an automobile battery
(b) 0.86 g of NaCl in 1.00×10^2 g of water—a solution of sodium chloride for intravenous injection
(c) 46.85 g of codeine, $C_{18}H_{21}NO_3$, in 125.5 g of ethanol, C_2H_5OH
(d) 25 g of I_2 in 125 g of ethanol, C_2H_5OH

39. Calculate the molality of each of the following solutions:
(a) 0.710 kg of sodium carbonate (washing soda), Na_2CO_3, in 10.0 kg of water—a saturated solution at 0°C
(b) 125 g of NH_4NO_3 in 275 g of water—a mixture used to make an instant ice pack
(c) 25 g of Cl_2 in 125 g of dichloromethane, CH_2Cl_2
(d) 0.372 g of tetrahydropyridine, C_5H_9N, in 125 g of chloroform, $CHCl_3$
40. The concentration of glucose, $C_6H_{12}O_6$, in normal spinal fluid is $\frac{75 \text{ mg}}{100 \text{ g}}$. What is the molality of the solution?
41. A 13.0% solution of K_2CO_3 by mass has a density of 1.09 g/cm^3. Calculate the molality of the solution.
42. Why does 1 mol of sodium chloride depress the freezing point of 1 kg of water almost twice as much as 1 mol of glycerin?
43. Assuming ideal solution behavior, what is the boiling point of a solution of 115.0 g of nonvolatile sucrose, $C_{12}H_{22}O_{11}$, in 350.0 g of water?
(a) Outline the steps necessary to answer the question
(b) Answer the question
44. Assuming ideal solution behavior, what is the boiling point of a solution of 9.04 g of I_2 in 75.5 g of benzene, assuming the I_2 is nonvolatile?
(a) Outline the steps necessary to answer the question.
(b) Answer the question.
45. Assuming ideal solution behavior, what is the freezing temperature of a solution of 115.0 g of sucrose, $C_{12}H_{22}O_{11}$, in 350.0 g of water?
(a) Outline the steps necessary to answer the question.
(b) Answer the question.
46. Assuming ideal solution behavior, what is the freezing point of a solution of 9.04 g of I_2 in 75.5 g of benzene?
(a) Outline the steps necessary to answer the following question.
(b) Answer the question.
47. Assuming ideal solution behavior, what is the osmotic pressure of an aqueous solution of 1.64 g of $Ca(NO_3)_2$ in water at 25 °C? The volume of the solution is 275 mL.
(a) Outline the steps necessary to answer the question.
(b) Answer the question.
48. Assuming ideal solution behavior, what is osmotic pressure of a solution of bovine insulin (molar mass, 5700 g mol^{-1}) at 18 °C if 100.0 mL of the solution contains 0.103 g of the insulin?
(a) Outline the steps necessary to answer the question.
(b) Answer the question.
49. Assuming ideal solution behavior, what is the molar mass of a solution of 5.00 g of a compound in 25.00 g of carbon tetrachloride (bp 76.8 °C; K_b = 5.02 °C/m) that boils at 81.5 °C at 1 atm?
(a) Outline the steps necessary to answer the question.
(b) Solve the problem.
50. A sample of an organic compound (a nonelectrolyte) weighing 1.35 g lowered the freezing point of 10.0 g of benzene by 3.66 °C. Assuming ideal solution behavior, calculate the molar mass of the compound.
51. A 1.0 m solution of HCl in benzene has a freezing point of 0.4 °C. Is HCl an electrolyte in benzene? Explain.
52. A solution contains 5.00 g of urea, $CO(NH_2)_2$, a nonvolatile compound, dissolved in 0.100 kg of water. If the vapor pressure of pure water at 25 °C is 23.7 torr, what is the vapor pressure of the solution (assuming ideal solution behavior)?
53. A 12.0-g sample of a nonelectrolyte is dissolved in 80.0 g of water. The solution freezes at −1.94 °C. Assuming ideal solution behavior, calculate the molar mass of the substance.
54. Arrange the following solutions in order by their decreasing freezing points: 0.1 m Na_3PO_4, 0.1 m C_2H_5OH, 0.01 m CO_2, 0.15 m NaCl, and 0.2 m $CaCl_2$.
55. Calculate the boiling point elevation of 0.100 kg of water containing 0.010 mol of NaCl, 0.020 mol of Na_2SO_4, and 0.030 mol of $MgCl_2$, assuming complete dissociation of these electrolytes and ideal solution behavior.
56. How could you prepare a 3.08 m aqueous solution of glycerin, $C_3H_8O_3$? Assuming ideal solution behavior, what is the freezing point of this solution?

57. A sample of sulfur weighing 0.210 g was dissolved in 17.8 g of carbon disulfide, CS_2 (K_b = 2.34 °C/m). If the boiling point elevation was 0.107 °C, what is the formula of a sulfur molecule in carbon disulfide (assuming ideal solution behavior)?
58. In a significant experiment performed many years ago, 5.6977 g of cadmium iodide in 44.69 g of water raised the boiling point 0.181 °C. What does this suggest about the nature of a solution of CdI_2?
59. Lysozyme is an enzyme that cleaves cell walls. A 0.100-L sample of a solution of lysozyme that contains 0.0750 g of the enzyme exhibits an osmotic pressure of 1.32×10^{-3} atm at 25 °C. Assuming ideal solution behavior, what is the molar mass of lysozyme?
60. The osmotic pressure of a solution containing 7.0 g of insulin per liter is 23 torr at 25 °C. Assuming ideal solution behavior, what is the molar mass of insulin?
61. The osmotic pressure of human blood is 7.6 atm at 37 °C. What mass of glucose, $C_6H_{12}O_6$, is required to make 1.00 L of aqueous solution for intravenous feeding if the solution must have the same osmotic pressure as blood at body temperature, 37 °C (assuming ideal solution behavior)?
62. Assuming ideal solution behavior, what is the freezing point of a solution of dibromobenzene, $C_6H_4Br_2$, in 0.250 kg of benzene, if the solution boils at 83.5 °C?
63. Assuming ideal solution behavior, what is the boiling point of a solution of NaCl in water if the solution freezes at –0.93 °C?
64. The sugar fructose contains 40.0% C, 6.7% H, and 53.3% O by mass. A solution of 11.7 g of fructose in 325 g of ethanol has a boiling point of 78.59 °C. The boiling point of ethanol is 78.35 °C, and K_b for ethanol is 1.20 °C/m. Assuming ideal solution behavior, what is the molecular formula of fructose?
65. The vapor pressure of methanol, CH_3OH, is 94 torr at 20 °C. The vapor pressure of ethanol, C_2H_5OH, is 44 torr at the same temperature.
    (a) Calculate the mole fraction of methanol and of ethanol in a solution of 50.0 g of methanol and 50.0 g of ethanol.
    (b) Ethanol and methanol form a solution that behaves like an ideal solution. Calculate the vapor pressure of methanol and of ethanol above the solution at 20 °C.
    (c) Calculate the mole fraction of methanol and of ethanol in the vapor above the solution.
66. The triple point of air-free water is defined as 273.16 K. Why is it important that the water be free of air?
67. Meat can be classified as fresh (not frozen) even though it is stored at –1 °C. Why wouldn't meat freeze at this temperature?
68. An organic compound has a composition of 93.46% C and 6.54% H by mass. A solution of 0.090 g of this compound in 1.10 g of camphor melts at 158.4 °C. The melting point of pure camphor is 178.4 °C. K_f for camphor is 37.7 °C/m. Assuming ideal solution behavior, what is the molecular formula of the solute? Show your calculations.
69. A sample of $HgCl_2$ weighing 9.41 g is dissolved in 32.75 g of ethanol, C_2H_5OH (K_b = 1.20 °C/m). The boiling point elevation of the solution is 1.27 °C. Is $HgCl_2$ an electrolyte in ethanol? Show your calculations.
70. A salt is known to be an alkali metal fluoride. A quick approximate determination of freezing point indicates that 4 g of the salt dissolved in 100 g of water produces a solution that freezes at about –1.4 °C. Assuming ideal solution behavior, what is the formula of the salt? Show your calculations.

11.5 Colloids

71. Identify the dispersed phase and the dispersion medium in each of the following colloidal systems: starch dispersion, smoke, fog, pearl, whipped cream, floating soap, jelly, milk, and ruby.
72. Distinguish between dispersion methods and condensation methods for preparing colloidal systems.
73. How do colloids differ from solutions with regard to dispersed particle size and homogeneity?
74. Explain the cleansing action of soap.
75. How can it be demonstrated that colloidal particles are electrically charged?

CHAPTER 12
Kinetics

Figure 12.1 An agama lizard basks in the sun. As its body warms, the chemical reactions of its metabolism speed up.

CHAPTER OUTLINE

12.1 Chemical Reaction Rates
12.2 Factors Affecting Reaction Rates
12.3 Rate Laws
12.4 Integrated Rate Laws
12.5 Collision Theory
12.6 Reaction Mechanisms
12.7 Catalysis

INTRODUCTION The lizard in the photograph is not simply enjoying the sunshine or working on its tan. The heat from the sun's rays is critical to the lizard's survival. A warm lizard can move faster than a cold one because the chemical reactions that allow its muscles to move occur more rapidly at higher temperatures. A cold lizard is a slower lizard and an easier meal for predators.

From baking a cake to determining the useful lifespan of a bridge, rates of chemical reactions play important roles in our understanding of processes that involve chemical changes. Two questions are typically posed when planning to carry out a chemical reaction. The first is: "Will the reaction produce the desired products in useful quantities?" The second question is: "How rapidly will the reaction occur?" A third question is often asked when investigating reactions in greater detail: "What specific molecular-level processes take place as the reaction occurs?" Knowing the answer to this question is of practical importance when the yield or rate of a reaction needs to be controlled.

The study of chemical kinetics concerns the second and third questions—that is, the rate at which a reaction yields products and the molecular-scale means by which a reaction occurs. This chapter examines the factors that influence the rates of chemical reactions, the mechanisms by which reactions proceed, and the quantitative techniques used to describe the rates at which reactions occur.

12.1 Chemical Reaction Rates

LEARNING OBJECTIVES

By the end of this section, you will be able to:

- Define chemical reaction rate
- Derive rate expressions from the balanced equation for a given chemical reaction
- Calculate reaction rates from experimental data

A *rate* is a measure of how some property varies with time. Speed is a familiar rate that expresses the distance traveled by an object in a given amount of time. Wage is a rate that represents the amount of money earned by a person working for a given amount of time. Likewise, the rate of a chemical reaction is a measure of how much reactant is consumed, or how much product is produced, by the reaction in a given amount of time.

The **rate of reaction** is the change in the amount of a reactant or product per unit time. Reaction rates are therefore determined by measuring the time dependence of some property that can be related to reactant or product amounts. Rates of reactions that consume or produce gaseous substances, for example, are conveniently determined by measuring changes in volume or pressure. For reactions involving one or more colored substances, rates may be monitored via measurements of light absorption. For reactions involving aqueous electrolytes, rates may be measured via changes in a solution's conductivity.

For reactants and products in solution, their relative amounts (concentrations) are conveniently used for purposes of expressing reaction rates. For example, the concentration of hydrogen peroxide, H_2O_2, in an aqueous solution changes slowly over time as it decomposes according to the equation:

$$2H_2O_2(aq) \longrightarrow 2H_2O(l) + O_2(g)$$

The rate at which the hydrogen peroxide decomposes can be expressed in terms of the rate of change of its concentration, as shown here:

$$\begin{aligned}\text{rate of decomposition of } H_2O_2 &= -\frac{\text{change in concentration of reactant}}{\text{time interval}} \\ &= -\frac{[H_2O_2]_{t_2} - [H_2O_2]_{t_1}}{t_2 - t_1} \\ &= -\frac{\Delta[H_2O_2]}{\Delta t}\end{aligned}$$

This mathematical representation of the change in species concentration over time is the **rate expression** for the reaction. The brackets indicate molar concentrations, and the symbol delta (Δ) indicates "change in." Thus, $[H_2O_2]_{t_1}$ represents the molar concentration of hydrogen peroxide at some time t_1; likewise,$[H_2O_2]_{t_2}$ represents the molar concentration of hydrogen peroxide at a later time t_2; and $\Delta[H_2O_2]$ represents the change in molar concentration of hydrogen peroxide during the time interval Δt (that is, $t_2 - t_1$). Since the reactant concentration decreases as the reaction proceeds, $\Delta[H_2O_2]$ is a negative quantity. Reaction rates are, by convention, positive quantities, and so this negative change in concentration is multiplied by −1. Figure 12.2 provides an example of data collected during the decomposition of H_2O_2.

Time (h)	$[H_2O_2]$ (mol L^{-1})	$\Delta[H_2O_2]$ (mol L^{-1})	Δt (h)	Rate of Decomposition, (mol L^{-1} h^{-1})
0.00	1.000			
		−0.500	6.00	0.0833
6.00	0.500			
		−0.250	6.00	0.0417
12.00	0.250			
		−0.125	6.00	0.0208
18.00	0.125			
		−0.062	6.00	0.010
24.00	0.0625			

FIGURE 12.2 The rate of decomposition of H_2O_2 in an aqueous solution decreases as the concentration of H_2O_2 decreases.

To obtain the tabulated results for this decomposition, the concentration of hydrogen peroxide was measured every 6 hours over the course of a day at a constant temperature of 40 °C. Reaction rates were computed for each time interval by dividing the change in concentration by the corresponding time increment, as shown here for the first 6-hour period:

$$\frac{-\Delta[H_2O_2]}{\Delta t} = \frac{-(0.500 \text{ mol/L} - 1.000 \text{ mol/L})}{(6.00 \text{ h} - 0.00 \text{ h})} = 0.0833 \text{ mol L}^{-1} \text{ h}^{-1}$$

Notice that the reaction rates vary with time, decreasing as the reaction proceeds. Results for the last 6-hour period yield a reaction rate of:

$$\frac{-\Delta[H_2O_2]}{\Delta t} = \frac{-(0.0625 \text{ mol/L} - 0.125 \text{ mol/L})}{(24.00 \text{ h} - 18.00 \text{ h})} = 0.010 \text{ mol L}^{-1} \text{ h}^{-1}$$

This behavior indicates the reaction continually slows with time. Using the concentrations at the beginning and end of a time period over which the reaction rate is changing results in the calculation of an **average rate** for the reaction over this time interval. At any specific time, the rate at which a reaction is proceeding is known as its **instantaneous rate**. The instantaneous rate of a reaction at "time zero," when the reaction commences, is its **initial rate**. Consider the analogy of a car slowing down as it approaches a stop sign. The vehicle's initial rate—analogous to the beginning of a chemical reaction—would be the speedometer reading at the moment the driver begins pressing the brakes (t_0). A few moments later, the instantaneous rate at a specific moment—call it t_1—would be somewhat slower, as indicated by the speedometer reading at that point in time. As time passes, the instantaneous rate will continue to fall until it reaches zero, when the car (or reaction) stops. Unlike instantaneous speed, the car's average speed is not indicated by the speedometer; but it can be calculated as the ratio of the distance traveled to the time required to bring the vehicle to a complete stop (Δt). Like the decelerating car, the average rate of a chemical reaction will fall somewhere between its initial and final rates.

The instantaneous rate of a reaction may be determined one of two ways. If experimental conditions permit the measurement of concentration changes over very short time intervals, then average rates computed as described earlier provide reasonably good approximations of instantaneous rates. Alternatively, a graphical procedure may be used that, in effect, yields the results that would be obtained if short time interval measurements were possible. In a plot of the concentration of hydrogen peroxide against time, the instantaneous rate of decomposition of H_2O_2 at any time t is given by the slope of a straight line that is tangent to the curve at that time (Figure 12.3). These tangent line slopes may be evaluated using calculus, but the procedure for doing so is beyond the scope of this chapter.

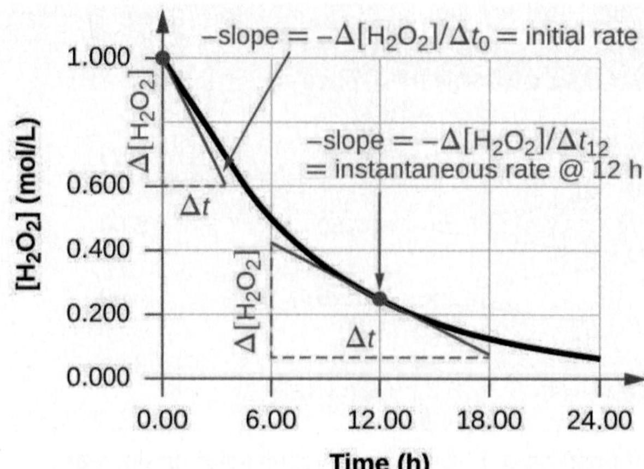

FIGURE 12.3 This graph shows a plot of concentration versus time for a 1.000 *M* solution of H_2O_2. The rate at any time is equal to the negative of the slope of a line tangent to the curve at that time. Tangents are shown at t = 0 h ("initial rate") and at t = 12 h ("instantaneous rate" at 12 h).

Chemistry in Everyday Life

Reaction Rates in Analysis: Test Strips for Urinalysis

Physicians often use disposable test strips to measure the amounts of various substances in a patient's urine (Figure 12.4). These test strips contain various chemical reagents, embedded in small pads at various locations along the strip, which undergo changes in color upon exposure to sufficient concentrations of specific substances. The usage instructions for test strips often stress that proper read time is critical for optimal results. This emphasis on read time suggests that kinetic aspects of the chemical reactions occurring on the test strip are important considerations.

The test for urinary glucose relies on a two-step process represented by the chemical equations shown here:

$$C_6H_{12}O_6 + O_2 \xrightarrow{\text{catalyst}} C_6H_{10}O_6 + H_2O_2$$

$$2H_2O_2 + 2I^- \xrightarrow{\text{catalyst}} I_2 + 2H_2O + O_2$$

The first equation depicts the oxidation of glucose in the urine to yield glucolactone and hydrogen peroxide. The hydrogen peroxide produced subsequently oxidizes colorless iodide ion to yield brown iodine, which may be visually detected. Some strips include an additional substance that reacts with iodine to produce a more distinct color change.

The two test reactions shown above are inherently very slow, but their rates are increased by special enzymes embedded in the test strip pad. This is an example of *catalysis*, a topic discussed later in this chapter. A typical glucose test strip for use with urine requires approximately 30 seconds for completion of the color-forming reactions. Reading the result too soon might lead one to conclude that the glucose concentration of the urine sample is lower than it actually is (a *false-negative* result). Waiting too long to assess the color change can lead to a *false positive* due to the slower (not catalyzed) oxidation of iodide ion by other substances found in urine.

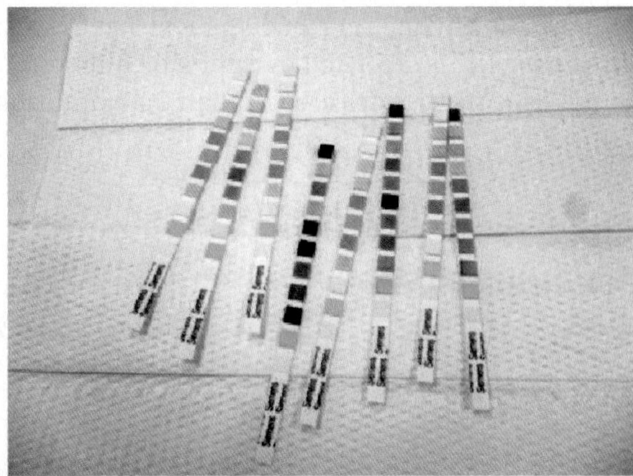

FIGURE 12.4 Test strips are commonly used to detect the presence of specific substances in a person's urine. Many test strips have several pads containing various reagents to permit the detection of multiple substances on a single strip. (credit: Iqbal Osman)

Relative Rates of Reaction

The rate of a reaction may be expressed as the change in concentration of any reactant or product. For any given reaction, these rate expressions are all related simply to one another according to the reaction stoichiometry. The rate of the general reaction

$$\mathrm{aA} \longrightarrow \mathrm{bB}$$

can be expressed in terms of the decrease in the concentration of A or the increase in the concentration of B. These two rate expressions are related by the stoichiometry of the reaction:

$$\text{rate} = -\left(\frac{1}{\mathrm{a}}\right)\left(\frac{\Delta \mathrm{A}}{\Delta t}\right) = \left(\frac{1}{\mathrm{b}}\right)\left(\frac{\Delta \mathrm{B}}{\Delta t}\right)$$

Consider the reaction represented by the following equation:

$$2\mathrm{NH_3}(g) \longrightarrow \mathrm{N_2}(g) + 3\mathrm{H_2}(g)$$

The relation between the reaction rates expressed in terms of nitrogen production and ammonia consumption, for example, is:

$$-\frac{\Delta \text{mol NH}_3}{\Delta t} \times \frac{1 \text{ mol N}_2}{2 \text{ mol NH}_3} = \frac{\Delta \text{mol N}_2}{\Delta t}$$

This may be represented in an abbreviated format by omitting the units of the stoichiometric factor:

$$-\frac{1}{2}\frac{\Delta \text{mol NH}_3}{\Delta t} = \frac{\Delta \text{mol N}_2}{\Delta t}$$

Note that a negative sign has been included as a factor to account for the opposite signs of the two amount changes (the reactant amount is decreasing while the product amount is increasing). For homogeneous reactions, both the reactants and products are present in the same solution and thus occupy the same volume, so the molar amounts may be replaced with molar concentrations:

$$-\frac{1}{2}\frac{\Delta[\mathrm{NH_3}]}{\Delta t} = \frac{\Delta[\mathrm{N_2}]}{\Delta t}$$

Similarly, the rate of formation of H_2 is three times the rate of formation of N_2 because three moles of H_2 are produced for each mole of N_2 produced.

$$\frac{1}{3}\frac{\Delta[\mathrm{H_2}]}{\Delta t} = \frac{\Delta[\mathrm{N_2}]}{\Delta t}$$

Figure 12.5 illustrates the change in concentrations over time for the decomposition of ammonia into nitrogen and hydrogen at 1100 °C. Slopes of the tangent lines at t = 500 s show that the instantaneous rates derived from all three species involved in the reaction are related by their stoichiometric factors. The rate of hydrogen production, for example, is observed to be three times greater than that for nitrogen production:

$$\frac{2.91 \times 10^{-6}\ M/\text{s}}{9.70 \times 10^{-7}\ M/\text{s}} \approx 3$$

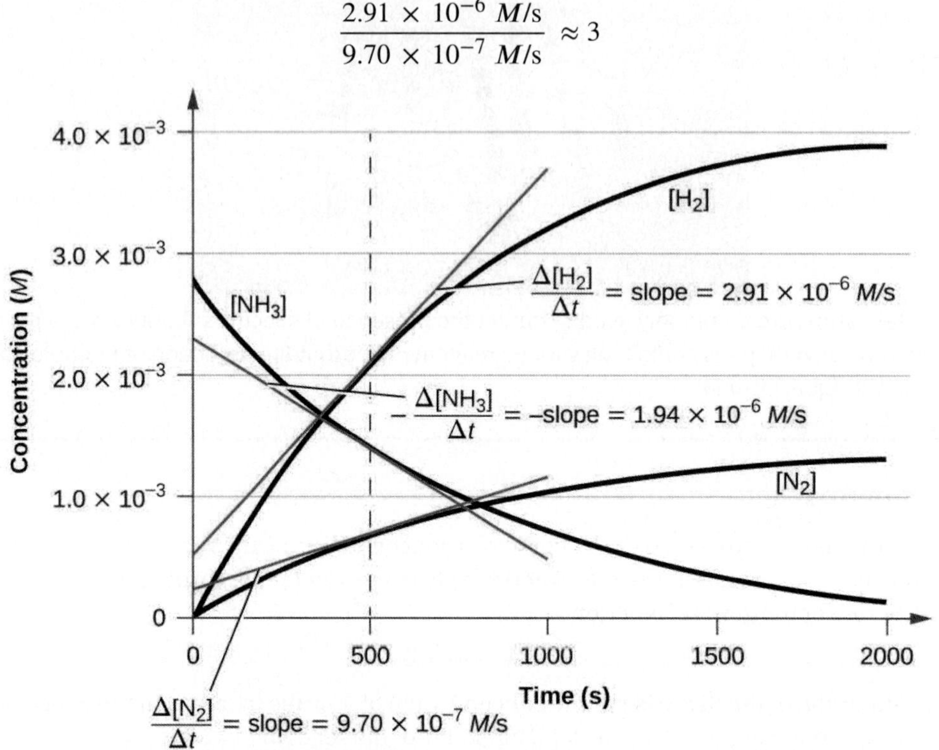

FIGURE 12.5 Changes in concentrations of the reactant and products for the reaction $2NH_3 \longrightarrow N_2 + 3H_2$. The rates of change of the three concentrations are related by the reaction stoichiometry, as shown by the different slopes of the tangents at t = 500 s.

EXAMPLE 12.1

Expressions for Relative Reaction Rates

The first step in the production of nitric acid is the combustion of ammonia:

$$4NH_3(g) + 5O_2(g) \longrightarrow 4NO(g) + 6H_2O(g)$$

Write the equations that relate the rates of consumption of the reactants and the rates of formation of the products.

Solution

Considering the stoichiometry of this homogeneous reaction, the rates for the consumption of reactants and formation of products are:

$$-\frac{1}{4}\frac{\Delta[NH_3]}{\Delta t} = -\frac{1}{5}\frac{\Delta[O_2]}{\Delta t} = \frac{1}{4}\frac{\Delta[NO]}{\Delta t} = \frac{1}{6}\frac{\Delta[H_2O]}{\Delta t}$$

Check Your Learning

The rate of formation of Br_2 is 6.0×10^{-6} mol/L/s in a reaction described by the following net ionic equation:

$$5Br^- + BrO_3^- + 6H^+ \longrightarrow 3Br_2 + 3H_2O$$

Write the equations that relate the rates of consumption of the reactants and the rates of formation of the products.

Answer:

$-\frac{1}{5}\frac{\Delta[Br^-]}{\Delta t} = -\frac{\Delta[BrO_3{}^-]}{\Delta t} = -\frac{1}{6}\frac{\Delta[H^+]}{\Delta t} = \frac{1}{3}\frac{\Delta[Br_2]}{\Delta t} = \frac{1}{3}\frac{\Delta[H_2O]}{\Delta t}$

EXAMPLE 12.2

Reaction Rate Expressions for Decomposition of H_2O_2

The graph in Figure 12.3 shows the rate of the decomposition of H_2O_2 over time:

$$2H_2O_2 \longrightarrow 2H_2O + O_2$$

Based on these data, the instantaneous rate of decomposition of H_2O_2 at t = 11.1 h is determined to be 3.20×10^{-2} mol/L/h, that is:

$$-\frac{\Delta[H_2O_2]}{\Delta t} = 3.20 \times 10^{-2} \text{mol L}^{-1}\text{h}^{-1}$$

What is the instantaneous rate of production of H_2O and O_2?

Solution

The reaction stoichiometry shows that

$$-\frac{1}{2}\frac{\Delta[H_2O_2]}{\Delta t} = \frac{1}{2}\frac{\Delta[H_2O]}{\Delta t} = \frac{\Delta[O_2]}{\Delta t}$$

Therefore:

$$\frac{1}{2} \times 3.20 \times 10^{-2} \text{ mol L}^{-1}\text{h}^{-1} = \frac{\Delta[O_2]}{\Delta t}$$

and

$$\frac{\Delta[O_2]}{\Delta t} = 1.60 \times 10^{-2} \text{ mol L}^{-1}\text{h}^{-1}$$

Check Your Learning

If the rate of decomposition of ammonia, NH_3, at 1150 K is 2.10×10^{-6} mol/L/s, what is the rate of production of nitrogen and hydrogen?

Answer:

1.05×10^{-6} mol/L/s, N_2 and 3.15×10^{-6} mol/L/s, H_2.

12.2 Factors Affecting Reaction Rates

LEARNING OBJECTIVES

By the end of this section, you will be able to:

- Describe the effects of chemical nature, physical state, temperature, concentration, and catalysis on reaction rates

The rates at which reactants are consumed and products are formed during chemical reactions vary greatly. Five factors typically affecting the rates of chemical reactions will be explored in this section: the chemical nature of the reacting substances, the state of subdivision (one large lump versus many small particles) of the reactants, the temperature of the reactants, the concentration of the reactants, and the presence of a catalyst.

The Chemical Nature of the Reacting Substances

The rate of a reaction depends on the nature of the participating substances. Reactions that appear similar may have different rates under the same conditions, depending on the identity of the reactants. For example, when small pieces of the metals iron and sodium are exposed to air, the sodium reacts completely with air overnight, whereas the iron is barely affected. The active metals calcium and sodium both react with water to form hydrogen gas and a base. Yet calcium reacts at a moderate rate, whereas sodium reacts so rapidly that the reaction is almost explosive.

The Physical States of the Reactants

A chemical reaction between two or more substances requires intimate contact between the reactants. When reactants are in different physical states, or phases (solid, liquid, gaseous, dissolved), the reaction takes place only at the interface between the phases. Consider the heterogeneous reaction between a solid phase and either a liquid or gaseous phase. Compared with the reaction rate for large solid particles, the rate for smaller particles will be greater because the surface area in contact with the other reactant phase is greater. For example, large pieces of iron react more slowly with acids than they do with finely divided iron powder (Figure 12.6). Large pieces of wood smolder, smaller pieces burn rapidly, and saw dust burns explosively.

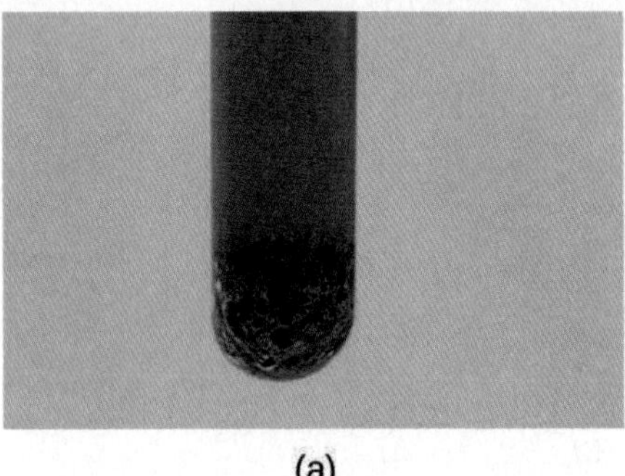

(a)

(b)

FIGURE 12.6 (a) Iron powder reacts rapidly with dilute hydrochloric acid and produces bubbles of hydrogen gas: $2Fe(s) + 6HCl(aq) \longrightarrow 2FeCl_3(aq) + 3H_2(g)$. (b) An iron nail reacts more slowly because the surface area exposed to the acid is much less.

LINK TO LEARNING

Watch this video (http://openstax.org/l/16cesium) to see the reaction of cesium with water in slow motion and a discussion of how the state of reactants and particle size affect reaction rates.

Temperature of the Reactants

Chemical reactions typically occur faster at higher temperatures. Food can spoil quickly when left on the kitchen counter. However, the lower temperature inside of a refrigerator slows that process so that the same food remains fresh for days. Gas burners, hot plates, and ovens are often used in the laboratory to increase the speed of reactions that proceed slowly at ordinary temperatures. For many chemical processes, reaction rates are approximately doubled when the temperature is raised by 10 °C.

Concentrations of the Reactants

The rates of many reactions depend on the concentrations of the reactants. Rates usually increase when the concentration of one or more of the reactants increases. For example, calcium carbonate ($CaCO_3$) deteriorates as a result of its reaction with the pollutant sulfur dioxide. The rate of this reaction depends on the amount of sulfur dioxide in the air (Figure 12.7). An acidic oxide, sulfur dioxide combines with water vapor in the air to produce sulfurous acid in the following reaction:

$$SO_2(g) + H_2O(g) \longrightarrow H_2SO_3(aq)$$

Calcium carbonate reacts with sulfurous acid as follows:

$$CaCO_3(s) + H_2SO_3(aq) \longrightarrow CaSO_3(aq) + CO_2(g) + H_2O(l)$$

In a polluted atmosphere where the concentration of sulfur dioxide is high, calcium carbonate deteriorates more rapidly than in less polluted air. Similarly, phosphorus burns much more rapidly in an atmosphere of pure oxygen than in air, which is only about 20% oxygen.

FIGURE 12.7 Statues made from carbonate compounds such as limestone and marble typically weather slowly over time due to the actions of water, and thermal expansion and contraction. However, pollutants like sulfur dioxide can accelerate weathering. As the concentration of air pollutants increases, deterioration of limestone occurs more rapidly. (credit: James P Fisher III)

LINK TO LEARNING

Phosphorus burns rapidly in air, but it will burn even more rapidly if the concentration of oxygen is higher. Watch this video (http://openstax.org/l/16phosphor) to see an example.

The Presence of a Catalyst

Relatively dilute aqueous solutions of hydrogen peroxide, H_2O_2, are commonly used as topical antiseptics. Hydrogen peroxide decomposes to yield water and oxygen gas according to the equation:

$$2H_2O_2(aq) \longrightarrow 2H_2O(l) + O_2(g)$$

Under typical conditions, this decomposition occurs very slowly. When dilute $H_2O_2(aq)$ is poured onto an open wound, however, the reaction occurs rapidly and the solution foams because of the vigorous production of oxygen gas. This dramatic difference is caused by the presence of substances within the wound's exposed tissues that accelerate the decomposition process. Substances that function to increase the rate of a reaction are called **catalysts**, a topic treated in greater detail later in this chapter.

LINK TO LEARNING

Chemical reactions occur when molecules collide with each other and undergo a chemical transformation. Before physically performing a reaction in a laboratory, scientists can use molecular modeling simulations to predict how the parameters discussed earlier will influence the rate of a reaction. Use the PhET Reactions & Rates interactive (http://openstax.org/l/16PHETreaction) to explore how temperature, concentration, and the nature of the reactants affect reaction rates.

12.3 Rate Laws

LEARNING OBJECTIVES

By the end of this section, you will be able to:

- Explain the form and function of a rate law
- Use rate laws to calculate reaction rates
- Use rate and concentration data to identify reaction orders and derive rate laws

As described in the previous module, the rate of a reaction is often affected by the concentrations of reactants. **Rate laws** (sometimes called *differential rate laws*) or **rate equations** are mathematical expressions that describe the relationship between the rate of a chemical reaction and the concentration of its reactants. As an example, consider the reaction described by the chemical equation

$$aA + bB \longrightarrow products$$

where *a* and *b* are stoichiometric coefficients. The rate law for this reaction is written as:

$$\text{rate} = k[A]^m[B]^n$$

in which [*A*] and [*B*] represent the molar concentrations of reactants, and *k* is the **rate constant**, which is specific for a particular reaction at a particular temperature. The exponents *m* and *n* are the **reaction orders** and are typically positive integers, though they can be fractions, negative, or zero. The rate constant *k* and the reaction orders *m* and *n* must be determined experimentally by observing how the rate of a reaction changes as the concentrations of the reactants are changed. The rate constant *k* is independent of the reactant concentrations, but it does vary with temperature.

The reaction orders in a rate law describe the mathematical dependence of the rate on reactant concentrations. Referring to the generic rate law above, the reaction is *m* order with respect to *A* and *n* order with respect to *B*. For example, if $m = 1$ and $n = 2$, the reaction is first order in *A* and second order in *B*. The **overall reaction order** is simply the sum of orders for each reactant. For the example rate law here, the reaction is third order overall (1 + 2 = 3). A few specific examples are shown below to further illustrate this concept.

The rate law:

$$\text{rate} = k[H_2O_2]$$

describes a reaction that is first order in hydrogen peroxide and first order overall. The rate law:

$$\text{rate} = k[C_4H_6]^2$$

describes a reaction that is second order in C_4H_6 and second order overall. The rate law:

$$\text{rate} = k[H^+][OH^-]$$

describes a reaction that is first order in H^+, first order in OH^-, and second order overall.

EXAMPLE 12.3

Writing Rate Laws from Reaction Orders

An experiment shows that the reaction of nitrogen dioxide with carbon monoxide:

$$NO_2(g) + CO(g) \longrightarrow NO(g) + CO_2(g)$$

is second order in NO_2 and zero order in CO at 100 °C. What is the rate law for the reaction?

Solution

The reaction will have the form:

$$\text{rate} = k[NO_2]^m[CO]^n$$

The reaction is second order in NO_2; thus $m = 2$. The reaction is zero order in CO; thus $n = 0$. The rate law is:

$$\text{rate} = k[NO_2]^2[CO]^0 = k[NO_2]^2$$

Remember that a number raised to the zero power is equal to 1, thus $[CO]^0 = 1$, which is why the CO concentration term may be omitted from the rate law: the rate of reaction is solely dependent on the concentration of NO_2. A later chapter section on reaction mechanisms will explain how a reactant's concentration can have no effect on a reaction rate despite being involved in the reaction.

Check Your Learning

The rate law for the reaction:

$$H_2(g) + 2NO(g) \longrightarrow N_2O(g) + H_2O(g)$$

has been determined to be rate = $k[NO]^2[H_2]$. What are the orders with respect to each reactant, and what is the overall order of the reaction?

Answer:
order in NO = 2; order in H_2 = 1; overall order = 3

Check Your Learning

In a transesterification reaction, a triglyceride reacts with an alcohol to form an ester and glycerol. Many students learn about the reaction between methanol (CH_3OH) and ethyl acetate ($CH_3CH_2OCOCH_3$) as a sample reaction before studying the chemical reactions that produce biodiesel:

$$CH_3OH + CH_3CH_2OCOCH_3 \longrightarrow CH_3OCOCH_3 + CH_3CH_2OH$$

The rate law for the reaction between methanol and ethyl acetate is, under certain conditions, determined to be:

$$\text{rate} = k\,[CH_3OH]$$

What is the order of reaction with respect to methanol and ethyl acetate, and what is the overall order of reaction?

Answer:
order in CH_3OH = 1; order in $CH_3CH_2OCOCH_3$ = 0; overall order = 1

A common experimental approach to the determination of rate laws is the **method of initial rates**. This method involves measuring reaction rates for multiple experimental trials carried out using different initial reactant concentrations. Comparing the measured rates for these trials permits determination of the reaction orders and, subsequently, the rate constant, which together are used to formulate a rate law. This approach is illustrated in the next two example exercises.

EXAMPLE 12.4

Determining a Rate Law from Initial Rates

Ozone in the upper atmosphere is depleted when it reacts with nitrogen oxides. The rates of the reactions of nitrogen oxides with ozone are important factors in deciding how significant these reactions are in the formation of the ozone hole over Antarctica (Figure 12.8). One such reaction is the combination of nitric oxide, NO, with ozone, O_3:

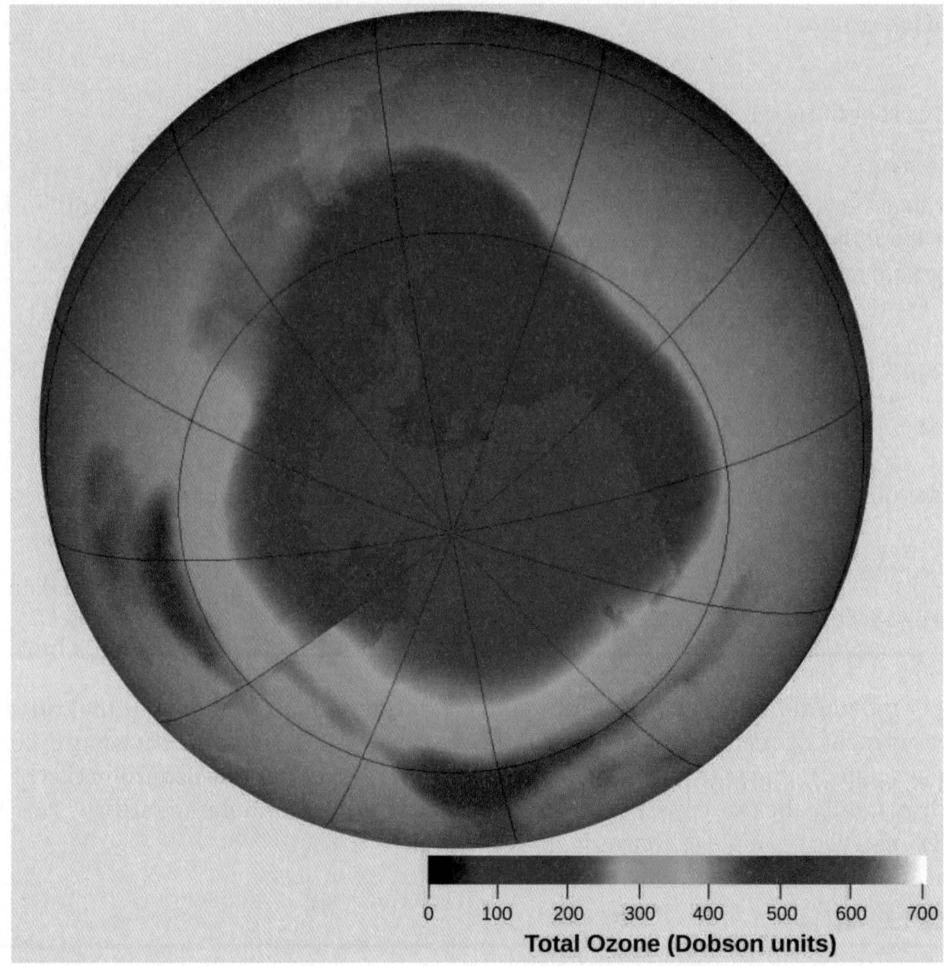

FIGURE 12.8 A contour map showing stratospheric ozone concentration and the “ozone hole” that occurs over Antarctica during its spring months. (credit: modification of work by NASA)

$$NO(g) + O_3(g) \longrightarrow NO_2(g) + O_2(g)$$

This reaction has been studied in the laboratory, and the following rate data were determined at 25 °C.

Trial	[NO] (mol/L)	[O$_3$] (mol/L)	$\frac{\Delta[NO_2]}{\Delta t}$ (mol L^{-1} s^{-1})
1	1.00×10^{-6}	3.00×10^{-6}	6.60×10^{-5}
2	1.00×10^{-6}	6.00×10^{-6}	1.32×10^{-4}
3	1.00×10^{-6}	9.00×10^{-6}	1.98×10^{-4}
4	2.00×10^{-6}	9.00×10^{-6}	3.96×10^{-4}
5	3.00×10^{-6}	9.00×10^{-6}	5.94×10^{-4}

Determine the rate law and the rate constant for the reaction at 25 °C.

Solution

The rate law will have the form:

$$\text{rate} = k[\text{NO}]^m[\text{O}_3]^n$$

Determine the values of *m*, *n*, and *k* from the experimental data using the following three-part process:

Step 1.
Determine the value of m *from the data in which [NO] varies and [O_3] is constant.* In the last three experiments, [NO] varies while [O_3] remains constant. When [NO] doubles from trial 3 to 4, the rate doubles, and when [NO] triples from trial 3 to 5, the rate also triples. Thus, the rate is also directly proportional to [NO], and *m* in the rate law is equal to 1.

Step 2.
Determine the value of n *from data in which [O_3] varies and [NO] is constant.* In the first three experiments, [NO] is constant and [O_3] varies. The reaction rate changes in direct proportion to the change in [O_3]. When [O_3] doubles from trial 1 to 2, the rate doubles; when [O_3] triples from trial 1 to 3, the rate increases also triples. Thus, the rate is directly proportional to [O_3], and *n* is equal to 1.The rate law is thus:

$$\text{rate} = k[\text{NO}]^1[\text{O}_3]^1 = k[\text{NO}][\text{O}_3]$$

Step 3.
Determine the value of k *from one set of concentrations and the corresponding rate.* The data from trial 1 are used below:

$$\begin{aligned} k &= \frac{\text{rate}}{[\text{NO}][\text{O}_3]} \\ &= \frac{6.60 \times 10^{-5}\ \cancel{\text{mol L}^{-1}}\ \text{s}^{-1}}{(1.00 \times 10^{-6}\ \cancel{\text{mol L}^{-1}})(3.00 \times 10^{-6}\ \text{mol L}^{-1})} \\ &= 2.20 \times 10^7\ \text{L mol}^{-1}\text{s}^{-1} \end{aligned}$$

Check Your Learning

Acetaldehyde decomposes when heated to yield methane and carbon monoxide according to the equation:

$$\text{CH}_3\text{CHO}(g) \longrightarrow \text{CH}_4(g) + \text{CO}(g)$$

Determine the rate law and the rate constant for the reaction from the following experimental data:

Trial	[CH_3CHO] (mol/L)	$-\frac{\Delta[\text{CH}_3\text{CHO}]}{\Delta t}$ (mol L^{-1} s^{-1})
1	1.75×10^{-3}	2.06×10^{-11}
2	3.50×10^{-3}	8.24×10^{-11}
3	7.00×10^{-3}	3.30×10^{-10}

Answer:

rate = $k[\text{CH}_3\text{CHO}]^2$ with $k = 6.73 \times 10^{-6}$ L/mol/s

 **EXAMPLE 12.5**

Determining Rate Laws from Initial Rates

Using the initial rates method and the experimental data, determine the rate law and the value of the rate constant for this reaction:

$$2\text{NO}(g) + \text{Cl}_2(g) \longrightarrow 2\text{NOCl}(g)$$

Trial	[NO] (mol/L)	$[Cl_2]$ (mol/L)	$-\frac{\Delta[NO]}{\Delta t}$ (mol L^{-1} s^{-1})
1	0.10	0.10	0.00300
2	0.10	0.15	0.00450
3	0.15	0.10	0.00675

Solution

The rate law for this reaction will have the form:

$$\text{rate} = k[NO]^m[Cl_2]^n$$

As in Example 12.4, approach this problem in a stepwise fashion, determining the values of m and n from the experimental data and then using these values to determine the value of k. In this example, however, an explicit algebraic approach (vs. the implicit approach of the previous example) will be used to determine the values of m and n:

Step 1.

Determine the value of m *from the data in which [NO] varies and [Cl_2] is constant.* Write the ratios with the subscripts x and y to indicate data from two different trials:

$$\frac{\text{rate}_x}{\text{rate}_y} = \frac{k[NO]_x^m[Cl_2]_x^n}{k[NO]_y^m[Cl_2]_y^n}$$

Using the third trial and the first trial, in which [Cl_2] does not vary, gives:

$$\frac{\text{rate 3}}{\text{rate 1}} = \frac{0.00675}{0.00300} = \frac{k(0.15)^m(0.10)^n}{k(0.10)^m(0.10)^n}$$

Canceling equivalent terms in the numerator and denominator leaves:

$$\frac{0.00675}{0.00300} = \frac{(0.15)^m}{(0.10)^m}$$

which simplifies to:

$$2.25 = (1.5)^m$$

Use logarithms to determine the value of the exponent m:

$$\begin{aligned} \ln(2.25) &= m\ln(1.5) \\ \frac{\ln(2.25)}{\ln(1.5)} &= m \\ 2 &= m \end{aligned}$$

Confirm the result

$$1.5^2 = 2.25$$

Step 2.

Determine the value of n *from data in which [Cl_2] varies and [NO] is constant.*

$$\frac{\text{rate 2}}{\text{rate 1}} = \frac{0.00450}{0.00300} = \frac{k(0.10)^m(0.15)^n}{k(0.10)^m(0.10)^n}$$

Cancelation gives:

$$\frac{0.0045}{0.0030} = \frac{(0.15)^n}{(0.10)^n}$$

which simplifies to:

$$1.5 = (1.5)^n$$

Thus n must be 1, and the form of the rate law is:

$$\text{rate} = k[\text{NO}]^m[\text{Cl}_2]^n = k[\text{NO}]^2[\text{Cl}_2]$$

Step 3.

Determine the numerical value of the rate constant k *with appropriate units.* The units for the rate of a reaction are mol/L/s. The units for k are whatever is needed so that substituting into the rate law expression affords the appropriate units for the rate. In this example, the concentration units are mol^3/L^3. The units for k should be $\text{mol}^{-2}\ \text{L}^2/\text{s}$ so that the rate is in terms of mol/L/s.

To determine the value of k once the rate law expression has been solved, simply plug in values from the first experimental trial and solve for k:

$$\begin{aligned} 0.00300\ \text{mol L}^{-1}\ \text{s}^{-1} &= k\left(0.10\ \text{mol L}^{-1}\right)^2\left(0.10\ \text{mol L}^{-1}\right)^1 \\ k &= 3.0\ \text{mol}^{-2}\ \text{L}^2\ \text{s}^{-1} \end{aligned}$$

Check Your Learning

Use the provided initial rate data to derive the rate law for the reaction whose equation is:

$$\text{OCl}^-(aq) + \text{I}^-(aq) \longrightarrow \text{OI}^-(aq) + \text{Cl}^-(aq)$$

Trial	[OCl⁻] (mol/L)	[I⁻] (mol/L)	Initial Rate (mol/L/s)
1	0.0040	0.0020	0.00184
2	0.0020	0.0040	0.00092
3	0.0020	0.0020	0.00046

Determine the rate law expression and the value of the rate constant k with appropriate units for this reaction.

Answer:

$$\frac{\text{rate 2}}{\text{rate 3}} = \frac{0.00092}{0.00046} = \frac{k(0.0020)^x(0.0040)^y}{k(0.0020)^x(0.0020)^y}$$

$2.00 = 2.00^y$

$y = 1$

$$\frac{\text{rate 1}}{\text{rate 2}} = \frac{0.00184}{0.00092} = \frac{k(0.0040)^x(0.0020)^y}{k(0.0020)^x(0.0040)^y}$$

$$\begin{aligned} 2.00 &= \frac{2^x}{2^y} \\ 2.00 &= \frac{2^x}{2^1} \\ 4.00 &= 2^x \\ x &= 2 \end{aligned}$$

Substituting the concentration data from trial 1 and solving for k yields:

$$\begin{aligned} \text{rate} &= k\left[\text{OCl}^-\right]^2\left[\text{I}^-\right]^1 \\ 0.00184 &= k(0.0040)^2(0.0020)^1 \\ k &= 5.75 \times 10^4\ \text{mol}^{-2}\ \text{L}^2\text{s}^{-1} \end{aligned}$$

Reaction Order and Rate Constant Units

In some of our examples, the reaction orders in the rate law happen to be the same as the coefficients in the chemical equation for the reaction. This is merely a coincidence and very often not the case.

Rate laws may exhibit fractional orders for some reactants, and negative reaction orders are sometimes observed when an increase in the concentration of one reactant causes a decrease in reaction rate. A few examples illustrating these points are provided:

$$\begin{array}{ll} NO_2 + CO \longrightarrow NO + CO_2 & \text{rate} = k[NO_2]^2 \\ CH_3CHO \longrightarrow CH_4 + CO & \text{rate} = k[CH_3CHO]^2 \\ 2N_2O_5 \longrightarrow NO_2 + O_2 & \text{rate} = k\left[N_2O_5\right] \\ 2NO_2 + F_2 \longrightarrow 2NO_2F & \text{rate} = k\,[NO_2]\,[F_2] \\ 2NO_2Cl \longrightarrow 2NO_2 + Cl_2 & \text{rate} = k\,[NO_2Cl] \end{array}$$

It is important to note that *rate laws are determined by experiment only and are not reliably predicted by reaction stoichiometry.*

The units for a rate constant will vary as appropriate to accommodate the overall order of the reaction. The unit of the rate constant for the second-order reaction described in Example 12.4 was determined to be $L\ mol^{-1}\ s^{-1}$. For the third-order reaction described in Example 12.5, the unit for k was derived to be $L^2\ mol^{-2}\ s^{-1}$. Dimensional analysis requires the rate constant unit for a reaction whose overall order is x to be $L^{x-1}\ mol^{1-x}\ s^{-1}$. Table 12.1 summarizes the rate constant units for common reaction orders.

Rate Constant Units for Common Reaction Orders

Overall Reaction Order (x)	Rate Constant Unit ($L^{x-1}\ mol^{1-x}\ s^{-1}$)
0 (zero)	$mol\ L^{-1}\ s^{-1}$
1 (first)	s^{-1}
2 (second)	$L\ mol^{-1}\ s^{-1}$
3 (third)	$L^2\ mol^{-2}\ s^{-1}$

TABLE 12.1

Note that the units in this table were derived using specific units for concentration (mol/L) and time (s), though any valid units for these two properties may be used.

12.4 Integrated Rate Laws

LEARNING OBJECTIVES

By the end of this section, you will be able to:

- Explain the form and function of an integrated rate law
- Perform integrated rate law calculations for zero-, first-, and second-order reactions
- Define half-life and carry out related calculations
- Identify the order of a reaction from concentration/time data

The rate laws discussed thus far relate the rate and the concentrations of reactants. We can also determine a second form of each rate law that relates the concentrations of reactants and time. These are called **integrated rate laws**. We can use an integrated rate law to determine the amount of reactant or product present after a period of time or to estimate the time required for a reaction to proceed to a certain extent. For example, an integrated rate law is used to determine the length of time a radioactive material must be stored for its radioactivity to decay to a safe level.

Using calculus, the differential rate law for a chemical reaction can be integrated with respect to time to give an equation that relates the amount of reactant or product present in a reaction mixture to the elapsed time of the reaction. This process can either be very straightforward or very complex, depending on the complexity of the differential rate law. For purposes of discussion, we will focus on the resulting integrated rate laws for first-, second-, and zero-order reactions.

First-Order Reactions

Integration of the rate law for a simple first-order reaction (rate = $k[A]$) results in an equation describing how the reactant concentration varies with time:

$$[A]_t = [A]_0 \, e^{-kt}$$

where $[A]t$ is the concentration of A at any time t, $[A]_0$ is the initial concentration of A, and k is the first-order rate constant.

For mathematical convenience, this equation may be rearranged to other formats, including direct and indirect proportionalities:

$$\ln\left(\frac{[A]_t}{[A]_0}\right) = -kt \qquad \text{or} \qquad \ln\left(\frac{[A]_0}{[A]_t}\right) = kt$$

and a format showing a linear dependence of concentration in time:

$$\ln[A]_t = \ln[A]_0 - kt$$

 **EXAMPLE 12.6**

The Integrated Rate Law for a First-Order Reaction

The rate constant for the first-order decomposition of cyclobutane, C_4H_8 at 500 °C is $9.2 \times 10^{-3}\ s^{-1}$:

$$C_4H_8 \longrightarrow 2C_2H_4$$

How long will it take for 80.0% of a sample of C_4H_8 to decompose?

Solution

Since the relative change in reactant concentration is provided, a convenient format for the integrated rate law is:

$$\ln\left(\frac{[A]_0}{[A]_t}\right) = kt$$

The initial concentration of C_4H_8, $[A]_0$, is not provided, but the provision that 80.0% of the sample has decomposed is enough information to solve this problem. Let x be the initial concentration, in which case the concentration after 80.0% decomposition is 20.0% of x or $0.200x$. Rearranging the rate law to isolate t and substituting the provided quantities yields:

$$\begin{aligned} t &= \ln \frac{[x]}{[0.200x]} \times \frac{1}{k} \\ &= \ln 5 \times \frac{1}{9.2 \times 10^{-3}\ s^{-1}} \\ &= 1.609 \times \frac{1}{9.2 \times 10^{-3}\ s^{-1}} \\ &= 1.7 \times 10^2\ s \end{aligned}$$

Check Your Learning

Iodine-131 is a radioactive isotope that is used to diagnose and treat some forms of thyroid cancer. Iodine-131 decays to xenon-131 according to the equation:

$$\text{I-131} \longrightarrow \text{Xe-131} + \text{electron}$$

The decay is first-order with a rate constant of 0.138 d^{-1}. How many days will it take for 90% of the iodine−131 in a 0.500 *M* solution of this substance to decay to Xe-131?

Answer:

16.7 days

In the next example exercise, a linear format for the integrated rate law will be convenient:

$$\begin{aligned} \ln[A]_t &= (-k)(t) + \ln[A]_0 \\ y &= mx + b \end{aligned}$$

A plot of $\ln[A]_t$ versus t for a first-order reaction is a straight line with a slope of $-k$ and a y-intercept of $\ln[A]_0$. If a set of rate data are plotted in this fashion but do *not* result in a straight line, the reaction is not first order in *A*.

EXAMPLE 12.7

Graphical Determination of Reaction Order and Rate Constant

Show that the data in Figure 12.2 can be represented by a first-order rate law by graphing $\ln[H_2O_2]$ versus time. Determine the rate constant for the decomposition of H_2O_2 from these data.

Solution

The data from Figure 12.2 are tabulated below, and a plot of $\ln[H_2O_2]$ is shown in Figure 12.9.

Trial	Time (h)	$[H_2O_2]$ (*M*)	$\ln[H_2O_2]$
1	0.00	1.000	0.000
2	6.00	0.500	−0.693
3	12.00	0.250	−1.386
4	18.00	0.125	−2.079
5	24.00	0.0625	−2.772

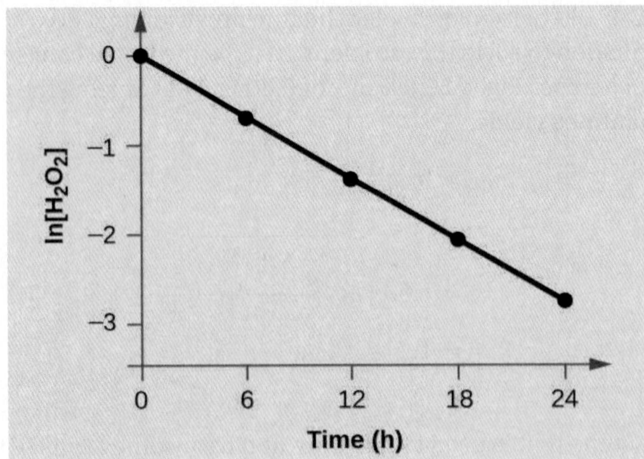

FIGURE 12.9 A linear relationship between $\ln[H_2O_2]$ and time suggests the decomposition of hydrogen peroxide is a first-order reaction.

The plot of ln[H_2O_2] versus time is linear, indicating that the reaction may be described by a first-order rate law.

According to the linear format of the first-order integrated rate law, the rate constant is given by the negative of this plot's slope.

$$\text{slope} = \frac{\text{change in } y}{\text{change in } x} = \frac{\Delta y}{\Delta x} = \frac{\Delta \ln[\text{H}_2\text{O}_2]}{\Delta t}$$

The slope of this line may be derived from two values of ln[H_2O_2] at different values of t (one near each end of the line is preferable). For example, the value of ln[H_2O_2] when t is 0.00 h is 0.000; the value when t = 24.00 h is −2.772

$$\begin{aligned} \text{slope} &= \frac{-2.772-0.000}{24.00-0.00 \text{ h}} \\ &= \frac{-2.772}{24.00 \text{ h}} \\ &= -0.116 \text{ h}^{-1} \\ k &= -\text{slope} = -\left(-0.116 \text{ h}^{-1}\right) = 0.116 \text{ h}^{-1} \end{aligned}$$

Check Your Learning

Graph the following data to determine whether the reaction $A \longrightarrow B + C$ is first order.

Trial	Time (s)	[A]
1	4.0	0.220
2	8.0	0.144
3	12.0	0.110
4	16.0	0.088
5	20.0	0.074

Answer:

The plot of ln$[A]_t$ vs. t is not linear, indicating the reaction is not first order:

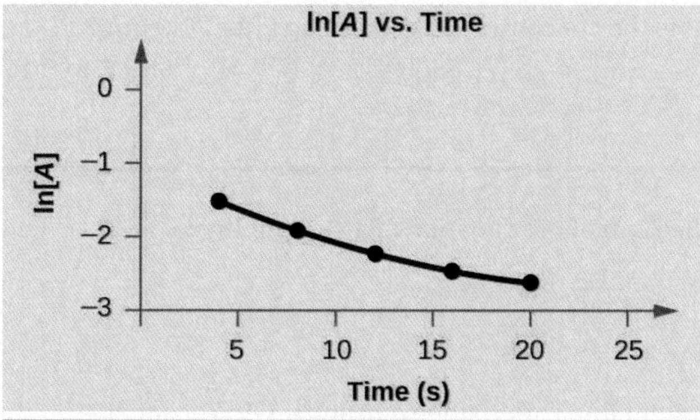

Second-Order Reactions

The equations that relate the concentrations of reactants and the rate constant of second-order reactions can be fairly complicated. To illustrate the point with minimal complexity, only the simplest second-order reactions will be described here, namely, those whose rates depend on the concentration of just one reactant. For these types of reactions, the differential rate law is written as:

$$\text{rate} = k[A]^2$$

For these second-order reactions, the integrated rate law is:

$$\frac{1}{[A]_t} = kt + \frac{1}{[A]_0}$$

where the terms in the equation have their usual meanings as defined earlier.

EXAMPLE 12.8

The Integrated Rate Law for a Second-Order Reaction

The reaction of butadiene gas (C_4H_6) to yield C_8H_{12} gas is described by the equation:

$$2C_4H_6(g) \longrightarrow C_8H_{12}(g)$$

This "dimerization" reaction is second order with a rate constant equal to 5.76×10^{-2} L mol^{-1} min^{-1} under certain conditions. If the initial concentration of butadiene is 0.200 *M*, what is the concentration after 10.0 min?

Solution

For a second-order reaction, the integrated rate law is written

$$\frac{1}{[A]_t} = kt + \frac{1}{[A]_0}$$

We know three variables in this equation: $[A]_0 = 0.200$ mol/L, $k = 5.76 \times 10^{-2}$ L/mol/min, and $t = 10.0$ min. Therefore, we can solve for $[A]$, the fourth variable:

$$\begin{aligned}
\frac{1}{[A]_t} &= \left(5.76 \times 10^{-2}\ \text{L mol}^{-1}\ \text{min}^{-1}\right)(10\ \text{min}) + \frac{1}{0.200\ \text{mol}^{-1}} \\
\frac{1}{[A]_t} &= \left(5.76 \times 10^{-1}\ \text{L mol}^{-1}\right) + 5.00\ \text{L mol}^{-1} \\
\frac{1}{[A]_t} &= 5.58\ \text{L mol}^{-1} \\
[A]_t &= 1.79 \times 10^{-1}\ \text{mol L}^{-1}
\end{aligned}$$

Therefore 0.179 mol/L of butadiene remain at the end of 10.0 min, compared to the 0.200 mol/L that was originally present.

Check Your Learning

If the initial concentration of butadiene is 0.0200 *M*, what is the concentration remaining after 20.0 min?

Answer:

0.0195 mol/L

The integrated rate law for second-order reactions has the form of the equation of a straight line:

$$\begin{aligned}
\frac{1}{[A]_t} &= kt + \frac{1}{[A]_0} \\
y &= mx + b
\end{aligned}$$

A plot of $\frac{1}{[A]_t}$ versus t for a second-order reaction is a straight line with a slope of k and a y-intercept of $\frac{1}{[A]_0}$. If the plot is not a straight line, then the reaction is not second order.

EXAMPLE 12.9

Graphical Determination of Reaction Order and Rate Constant

The data below are for the same reaction described in Example 12.8. Prepare and compare two appropriate data plots to identify the reaction as being either first or second order. After identifying the reaction order, estimate a value for the rate constant.

Solution

Trial	Time (s)	$[C_4H_6]$ (M)
1	0	1.00×10^{-2}
2	1600	5.04×10^{-3}
3	3200	3.37×10^{-3}
4	4800	2.53×10^{-3}
5	6200	2.08×10^{-3}

In order to distinguish a first-order reaction from a second-order reaction, prepare a plot of $\ln[C_4H_6]_t$ versus t and compare it to a plot of $\frac{1}{[C_4H_6]_t}$ versus t. The values needed for these plots follow.

Time (s)	$\frac{1}{[C_4H_6]}$ (M^{-1})	$\ln[C_4H_6]$
0	100	-4.605
1600	198	-5.289
3200	296	-5.692
4800	395	-5.978
6200	481	-6.175

The plots are shown in Figure 12.10, which clearly shows the plot of $\ln[C_4H_6]_t$ versus t is not linear, therefore the reaction is not first order. The plot of $\frac{1}{[C_4H_6]_t}$ versus t is linear, indicating that the reaction is second order.

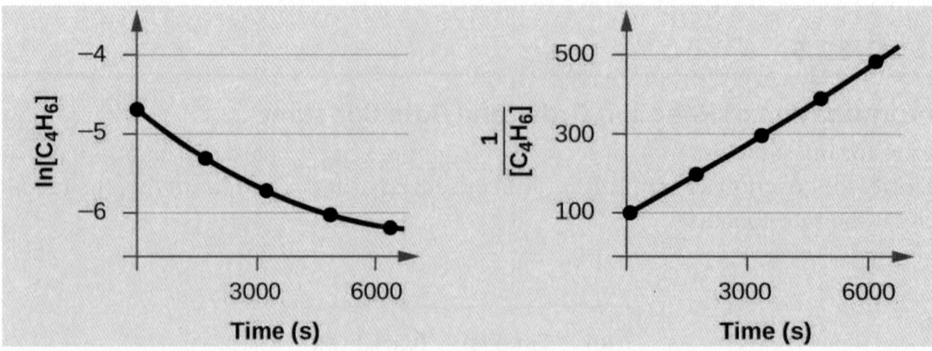

FIGURE 12.10 These two graphs show first- and second-order plots for the dimerization of C_4H_6. The linear trend in the second-order plot (right) indicates that the reaction follows second-order kinetics.

According to the second-order integrated rate law, the rate constant is equal to the slope of the $\frac{1}{[A]_t}$ versus t plot. Using the data for $t = 0$ s and $t = 6200$ s, the rate constant is estimated as follows:

$$k = \text{slope} = \frac{(481\ M^{-1} - 100\ M^{-1})}{(6200\ \text{s} - 0\ \text{s})} = 0.0614\ \text{M}^{-1}\ \text{s}^{-1}$$

Check Your Learning

Do the following data fit a second-order rate law?

Trial	Time (s)	[A] (M)
1	5	0.952
2	10	0.625
3	15	0.465
4	20	0.370
5	25	0.308
6	35	0.230

Answer:

Yes. The plot of $\frac{1}{[A]_t}$ vs. t is linear:

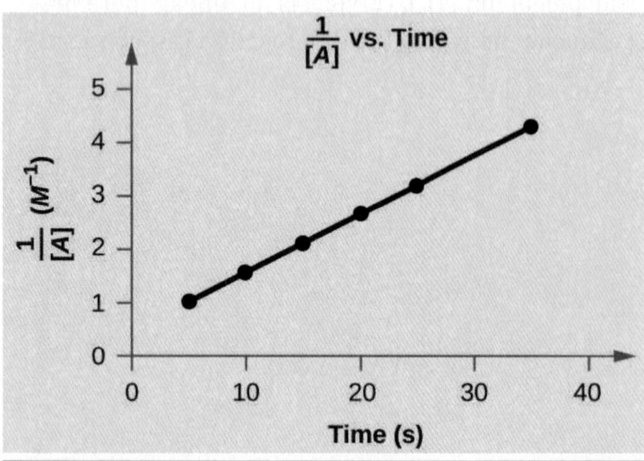

Zero-Order Reactions

For zero-order reactions, the differential rate law is:

$$\text{rate} = k$$

A zero-order reaction thus exhibits a constant reaction rate, regardless of the concentration of its reactant(s). This may seem counterintuitive, since the reaction rate certainly can't be finite when the reactant concentration is zero. For purposes of this introductory text, it will suffice to note that zero-order kinetics are observed for some reactions only under certain specific conditions. These same reactions exhibit different kinetic behaviors when the specific conditions aren't met, and for this reason the more prudent term *pseudo-zero-order* is sometimes used.

The integrated rate law for a zero-order reaction is a linear function:

$$\begin{aligned} [A]_t &= -kt + [A]_0 \\ y &= mx + b \end{aligned}$$

A plot of $[A]$ versus t for a zero-order reaction is a straight line with a slope of $-k$ and a y-intercept of $[A]_0$. Figure 12.11 shows a plot of $[NH_3]$ versus t for the thermal decomposition of ammonia at the surface of two different heated solids. The decomposition reaction exhibits first-order behavior at a quartz (SiO_2) surface, as suggested by the exponentially decaying plot of concentration versus time. On a tungsten surface, however, the plot is linear, indicating zero-order kinetics.

 **EXAMPLE 12.10**

Graphical Determination of Zero-Order Rate Constant

Use the data plot in Figure 12.11 to graphically estimate the zero-order rate constant for ammonia decomposition at a tungsten surface.

Solution

The integrated rate law for zero-order kinetics describes a linear plot of reactant concentration, $[A]_t$, versus time, t, with a slope equal to the negative of the rate constant, $-k$. Following the mathematical approach of previous examples, the slope of the linear data plot (for decomposition on W) is estimated from the graph. Using the ammonia concentrations at $t = 0$ and $t = 1000$ s:

$$k = -\text{slope} = -\frac{(0.0015\ \text{mol L}^{-1} - 0.0028\ \text{mol L}^{-1})}{(1000\ \text{s} - 0\ \text{s})} = 1.3 \times 10^{-6}\ \text{mol L}^{-1}\ \text{s}^{-1}$$

Check Your Learning

The zero-order plot in Figure 12.11 shows an initial ammonia concentration of 0.0028 mol L^{-1} decreasing linearly with time for 1000 s. Assuming no change in this zero-order behavior, at what time (min) will the concentration reach 0.0001 mol L^{-1}?

Answer:

35 min

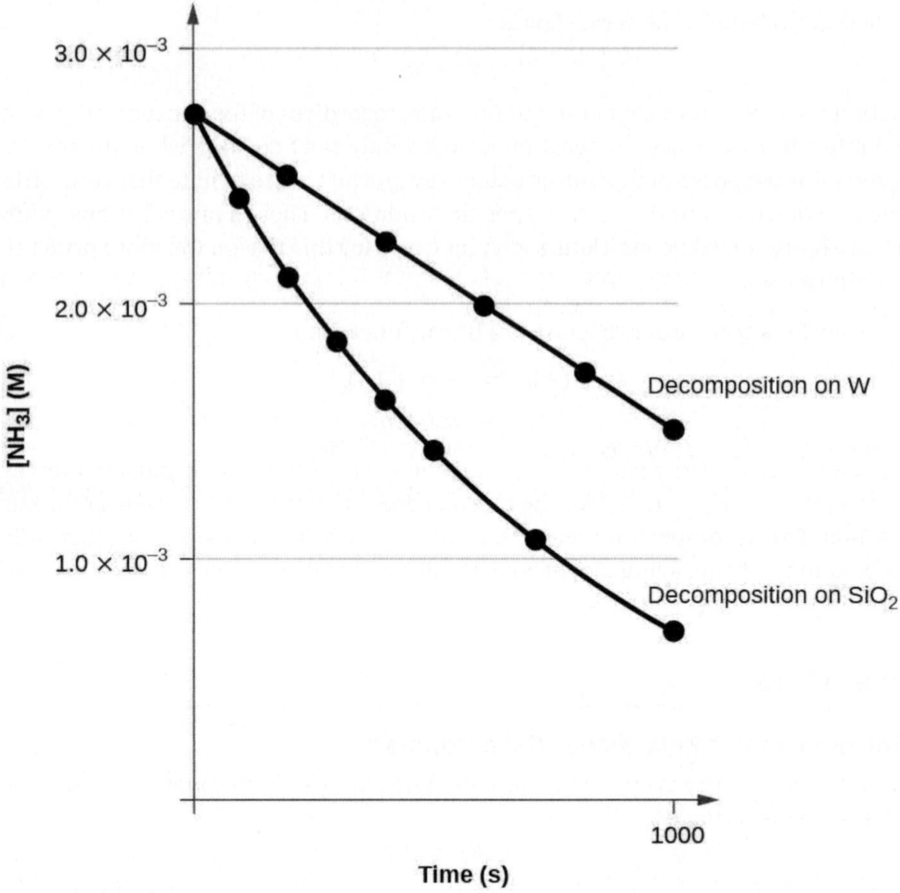

FIGURE 12.11 The decomposition of NH_3 on a tungsten (W) surface is a zero-order reaction, whereas on a quartz (SiO_2) surface, the reaction is first order.

The Half-Life of a Reaction

The **half-life of a reaction ($t_{1/2}$)** is the time required for one-half of a given amount of reactant to be consumed. In each succeeding half-life, half of the remaining concentration of the reactant is consumed. Using the decomposition of hydrogen peroxide (Figure 12.2) as an example, we find that during the first half-life (from 0.00 hours to 6.00 hours), the concentration of H_2O_2 decreases from 1.000 *M* to 0.500 *M*. During the second half-life (from 6.00 hours to 12.00 hours), it decreases from 0.500 *M* to 0.250 *M*; during the third half-life, it decreases from 0.250 *M* to 0.125 *M*. The concentration of H_2O_2 decreases by half during each successive period of 6.00 hours. The decomposition of hydrogen peroxide is a first-order reaction, and, as can be shown, the half-life of a first-order reaction is independent of the concentration of the reactant. However, half-lives of reactions with other orders depend on the concentrations of the reactants.

First-Order Reactions

An equation relating the half-life of a first-order reaction to its rate constant may be derived from the integrated rate law as follows:

$$\ln \frac{[A]_0}{[A]_t} = kt$$

$$t = \ln \frac{[A]_0}{[A]_t} \times \frac{1}{k}$$

Invoking the definition of half-life, symbolized $t_{1/2}$, requires that the concentration of A at this point is one-half its initial concentration: $t = t_{1/2}$, $[A]_t = \frac{1}{2}[A]_0$.

Substituting these terms into the rearranged integrated rate law and simplifying yields the equation for half-life:

$$\begin{aligned} t_{1/2} &= \ln \frac{[A]_0}{\frac{1}{2}[A]_0} \times \frac{1}{k} \\ &= \ln 2 \times \frac{1}{k} = 0.693 \times \frac{1}{k} \\ t_{1/2} &= \frac{0.693}{k} \end{aligned}$$

This equation describes an expected inverse relation between the half-life of the reaction and its rate constant, *k*. Faster reactions exhibit larger rate constants and correspondingly shorter half-lives. Slower reactions exhibit smaller rate constants and longer half-lives.

 **EXAMPLE 12.11**

Calculation of a First-order Rate Constant using Half-Life

Calculate the rate constant for the first-order decomposition of hydrogen peroxide in water at 40 °C, using the data given in Figure 12.12.

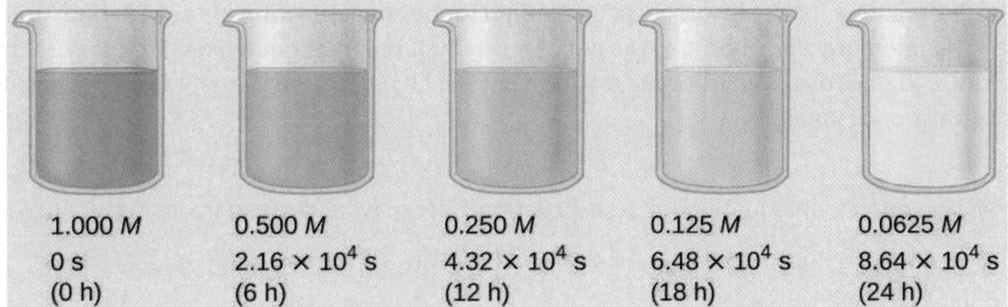

FIGURE 12.12 The decomposition of H_2O_2 ($2H_2O_2 \longrightarrow 2H_2O + O_2$) at 40 °C is illustrated. The intensity of the color symbolizes the concentration of H_2O_2 at the indicated times; H_2O_2 is actually colorless.

Solution

Inspecting the concentration/time data in Figure 12.12 shows the half-life for the decomposition of H_2O_2 is 2.16×10^4 s:

$$\begin{aligned} t_{1/2} &= \frac{0.693}{k} \\ k &= \frac{0.693}{t_{1/2}} = \frac{0.693}{2.16 \times 10^4 \text{ s}} = 3.21 \times 10^{-5} \text{ s}^{-1} \end{aligned}$$

Check Your Learning

The first-order radioactive decay of iodine-131 exhibits a rate constant of 0.138 d^{-1}. What is the half-life for this decay?

Answer:

5.02 d.

Second-Order Reactions

Following the same approach as used for first-order reactions, an equation relating the half-life of a second-order reaction to its rate constant and initial concentration may be derived from its integrated rate law:

$$\frac{1}{[A]_t} = kt + \frac{1}{[A]_0}$$

or

$$\frac{1}{[A]} - \frac{1}{[A]_0} = kt$$

Restrict t to $t_{1/2}$

$$t = t_{1/2}$$

define $[A]_t$ as one-half $[A]_0$

$$[A]_t = \frac{1}{2}[A]_0$$

and then substitute into the integrated rate law and simplify:

$$\begin{aligned} \frac{1}{\frac{1}{2}[A]_0} - \frac{1}{[A]_0} &= kt_{1/2} \\ \frac{2}{[A]_0} - \frac{1}{[A]_0} &= kt_{1/2} \\ \frac{1}{[A]_0} &= kt_{1/2} \\ t_{1/2} &= \frac{1}{k[A]_0} \end{aligned}$$

For a second-order reaction, $t_{1/2}$ is inversely proportional to the concentration of the reactant, and the half-life increases as the reaction proceeds because the concentration of reactant decreases. Unlike with first-order reactions, the rate constant of a second-order reaction cannot be calculated directly from the half-life unless the initial concentration is known.

Zero-Order Reactions

As for other reaction orders, an equation for zero-order half-life may be derived from the integrated rate law:

$$[A] = -kt + [A]_0$$

Restricting the time and concentrations to those defined by half-life: $t = t_{1/2}$ and $[A] = \frac{[A]_0}{2}$. Substituting these terms into the zero-order integrated rate law yields:

$$\begin{aligned} \frac{[\mathrm{A}]_0}{2} &= -kt_{1/2} + [\mathrm{A}]_0 \\ kt_{1/2} &= \frac{[\mathrm{A}]_0}{2} \\ t_{1/2} &= \frac{[\mathrm{A}]_0}{2k} \end{aligned}$$

As for all reaction orders, the half-life for a zero-order reaction is inversely proportional to its rate constant. However, the half-life of a zero-order reaction increases as the initial concentration increases.

Equations for both differential and integrated rate laws and the corresponding half-lives for zero-, first-, and second-order reactions are summarized in Table 12.2.

Summary of Rate Laws for Zero-, First-, and Second-Order Reactions

	Zero-Order	**First-Order**	**Second-Order**
rate law	rate = k	rate = $k[A]$	rate = $k[A]^2$
units of rate constant	$M\,\mathrm{s}^{-1}$	s^{-1}	$M^{-1}\,\mathrm{s}^{-1}$
integrated rate law	$[A] = -kt + [A]_0$	$\ln[A] = -kt + \ln[A]_0$	$\frac{1}{[A]} = kt + \left(\frac{1}{[A]_0}\right)$

TABLE 12.2

	Zero-Order	First-Order	Second-Order
plot needed for linear fit of rate data	$[A]$ vs. t	$\ln[A]$ vs. t	$\frac{1}{[A]}$ vs. t
relationship between slope of linear plot and rate constant	$k = -\text{slope}$	$k = -\text{slope}$	$k = \text{slope}$
half-life	$t_{1/2} = \frac{[A]_0}{2k}$	$t_{1/2} = \frac{0.693}{k}$	$t_{1/2} = \frac{1}{[A]_0 k}$

TABLE 12.2

 **EXAMPLE 12.12**

Half-Life for Zero-Order and Second-Order Reactions

What is the half-life for the butadiene dimerization reaction described in Example 12.8?

Solution

The reaction in question is second order, is initiated with a 0.200 mol L^{-1} reactant solution, and exhibits a rate constant of 0.0576 L mol^{-1} min^{-1}. Substituting these quantities into the second-order half-life equation:

$$t_{1/2} = \frac{1}{[(0.0576 \text{ L mol}^{-1} \text{ min}^{-1})(0.200 \text{ mol L}^{-1})]} = 18 \text{ min}$$

Check Your Learning

What is the half-life (min) for the thermal decomposition of ammonia on tungsten (see Figure 12.11)?

Answer:

87 min

12.5 Collision Theory

LEARNING OBJECTIVES

By the end of this section, you will be able to:

- Use the postulates of collision theory to explain the effects of physical state, temperature, and concentration on reaction rates
- Define the concepts of activation energy and transition state
- Use the Arrhenius equation in calculations relating rate constants to temperature

We should not be surprised that atoms, molecules, or ions must collide before they can react with each other. Atoms must be close together to form chemical bonds. This simple premise is the basis for a very powerful theory that explains many observations regarding chemical kinetics, including factors affecting reaction rates.

Collision theory is based on the following postulates:

1. The rate of a reaction is proportional to the rate of reactant collisions:

$$\text{reaction rate} \propto \frac{\#\,\text{collisions}}{\text{time}}$$

2. The reacting species must collide in an orientation that allows contact between the atoms that will become bonded together in the product.
3. The collision must occur with adequate energy to permit mutual penetration of the reacting species' valence shells so that the electrons can rearrange and form new bonds (and new chemical species).

We can see the importance of the two physical factors noted in postulates 2 and 3, the orientation and energy of collisions, when we consider the reaction of carbon monoxide with oxygen:

$$2\,CO(g) + O_2(g) \longrightarrow 2\,CO_2(g)$$

Carbon monoxide is a pollutant produced by the combustion of hydrocarbon fuels. To reduce this pollutant, automobiles have catalytic converters that use a catalyst to carry out this reaction. It is also a side reaction of the combustion of gunpowder that results in muzzle flash for many firearms. If carbon monoxide and oxygen are present in sufficient amounts, the reaction will occur at high temperature and pressure.

The first step in the gas-phase reaction between carbon monoxide and oxygen is a collision between the two molecules:

$$CO(g) + O_2(g) \longrightarrow CO_2(g) + O(g)$$

Although there are many different possible orientations the two molecules can have relative to each other, consider the two presented in Figure 12.13. In the first case, the oxygen side of the carbon monoxide molecule collides with the oxygen molecule. In the second case, the carbon side of the carbon monoxide molecule collides with the oxygen molecule. The second case is clearly more likely to result in the formation of carbon dioxide, which has a central carbon atom bonded to two oxygen atoms ($O = C = O$). This is a rather simple example of how important the orientation of the collision is in terms of creating the desired product of the reaction.

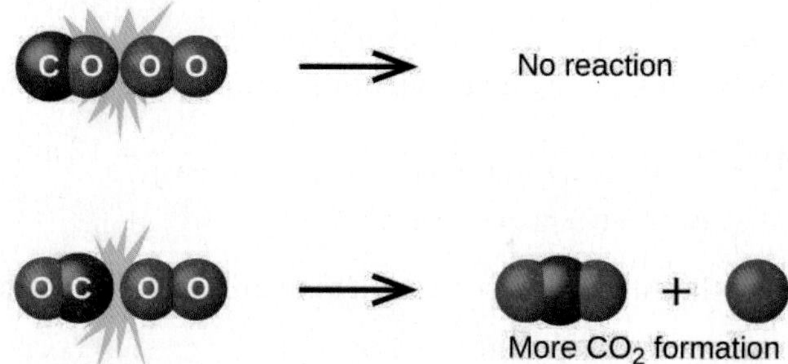

FIGURE 12.13 Illustrated are two collisions that might take place between carbon monoxide and oxygen molecules. The orientation of the colliding molecules partially determines whether a reaction between the two molecules will occur.

If the collision does take place with the correct orientation, there is still no guarantee that the reaction will proceed to form carbon dioxide. In addition to a proper orientation, the collision must also occur with sufficient energy to result in product formation. When reactant species collide with both proper orientation and adequate energy, they combine to form an unstable species called an **activated complex** or a **transition state**. These species are very short lived and usually undetectable by most analytical instruments. In some cases, sophisticated spectral measurements have been used to observe transition states.

Collision theory explains why most reaction rates increase as concentrations increase. With an increase in the concentration of any reacting substance, the chances for collisions between molecules are increased because there are more molecules per unit of volume. More collisions mean a faster reaction rate, assuming the energy of the collisions is adequate.

Activation Energy and the Arrhenius Equation

The minimum energy necessary to form a product during a collision between reactants is called the **activation energy (E_a)**. How this energy compares to the kinetic energy provided by colliding reactant molecules is a primary factor affecting the rate of a chemical reaction. If the activation energy is much larger than the average kinetic energy of the molecules, the reaction will occur slowly since only a few fast-moving molecules will have enough energy to react. If the activation energy is much smaller than the average kinetic energy of the molecules, a large fraction of molecules will be adequately energetic and the reaction will proceed rapidly.

Figure 12.14 shows how the energy of a chemical system changes as it undergoes a reaction converting reactants to products according to the equation

$$A + B \longrightarrow C + D$$

These **reaction diagrams** are widely used in chemical kinetics to illustrate various properties of the reaction of interest. Viewing the diagram from left to right, the system initially comprises reactants only, $A + B$. Reactant molecules with sufficient energy can collide to form a high-energy activated complex or transition state. The unstable transition state can then subsequently decay to yield stable products, $C + D$. The diagram depicts the reaction's activation energy, E_a, as the energy difference between the reactants and the transition state. Using a specific energy, the *enthalpy* (see chapter on thermochemistry), the enthalpy change of the reaction, ΔH, is estimated as the energy difference between the reactants and products. In this case, the reaction is exothermic ($\Delta H < 0$) since it yields a decrease in system enthalpy.

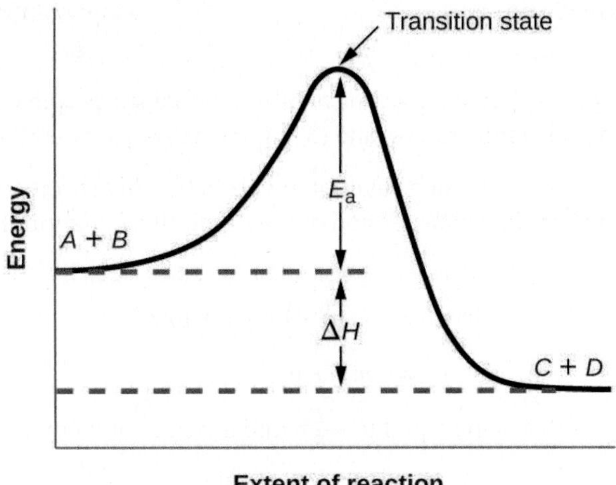

FIGURE 12.14 Reaction diagram for the exothermic reaction $A + B \longrightarrow C + D$.

The **Arrhenius equation** relates the activation energy and the rate constant, k, for many chemical reactions:

$$k = Ae^{-E_a/RT}$$

In this equation, R is the ideal gas constant, which has a value 8.314 J/mol/K, T is temperature on the Kelvin scale, E_a is the activation energy in joules per mole, e is the constant 2.7183, and A is a constant called the **frequency factor**, which is related to the frequency of collisions and the orientation of the reacting molecules.

Postulates of collision theory are nicely accommodated by the Arrhenius equation. The frequency factor, A, reflects how well the reaction conditions favor properly oriented collisions between reactant molecules. An increased probability of effectively oriented collisions results in larger values for A and faster reaction rates.

The exponential term, $e^{-Ea/RT}$, describes the effect of activation energy on reaction rate. According to kinetic molecular theory (see chapter on gases), the temperature of matter is a measure of the average kinetic energy of its constituent atoms or molecules. The distribution of energies among the molecules composing a sample of matter at any given temperature is described by the plot shown in Figure 12.15(**a**). Two shaded areas under the curve represent the numbers of molecules possessing adequate energy (RT) to overcome the activation barriers (E_a). A lower activation energy results in a greater fraction of adequately energized molecules and a faster reaction.

The exponential term also describes the effect of temperature on reaction rate. A higher temperature represents a correspondingly greater fraction of molecules possessing sufficient energy (RT) to overcome the activation barrier (E_a), as shown in Figure 12.15(**b**). This yields a greater value for the rate constant and a correspondingly faster reaction rate.

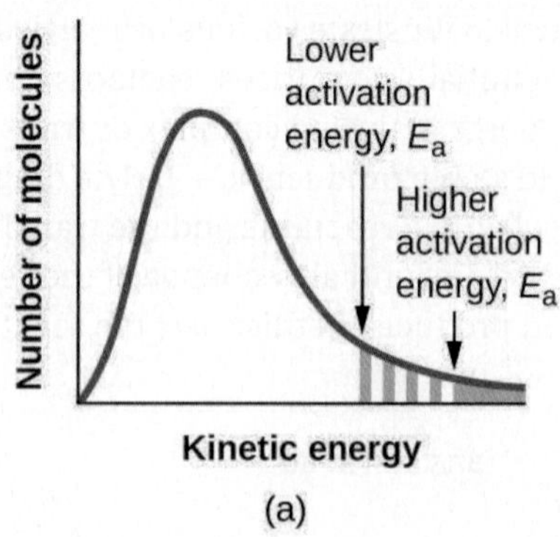

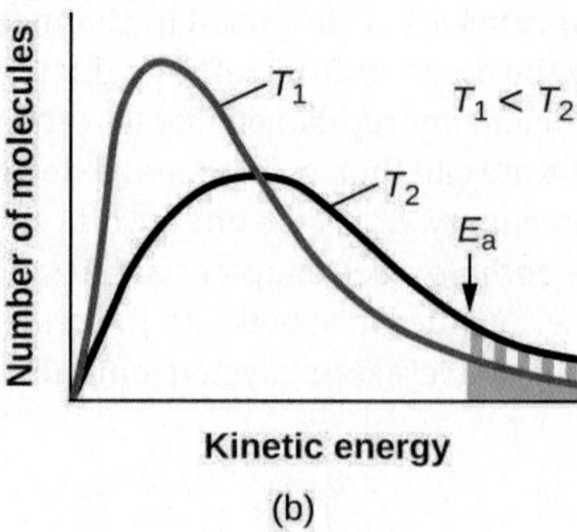

FIGURE 12.15 Molecular energy distributions showing numbers of molecules with energies exceeding (a) two different activation energies at a given temperature, and (b) a given activation energy at two different temperatures.

A convenient approach for determining E_a for a reaction involves the measurement of k at two or more different temperatures and using an alternate version of the Arrhenius equation that takes the form of a linear equation

$$\ln k = \left(\frac{-E_a}{R}\right)\left(\frac{1}{T}\right) + \ln A$$

$$y = mx + b$$

A plot of ln k versus $\frac{1}{T}$ is linear with a slope equal to $\frac{-E_a}{R}$ and a y-intercept equal to ln A.

 **EXAMPLE 12.13**

Determination of E_a

The variation of the rate constant with temperature for the decomposition of HI(g) to H_2(g) and I_2(g) is given here. What is the activation energy for the reaction?

$$2HI(g) \longrightarrow H_2(g) + I_2(g)$$

T (K)	k (L/mol/s)
555	3.52×10^{-7}
575	1.22×10^{-6}
645	8.59×10^{-5}
700	1.16×10^{-3}
781	3.95×10^{-2}

Solution

Use the provided data to derive values of $\frac{1}{T}$ and ln k:

$\frac{1}{T}$ (K^{-1})	ln k
1.80×10^{-3}	-14.860
1.74×10^{-3}	-13.617

$\frac{1}{T}$ (K^{-1})	ln *k*
1.55×10^{-3}	-9.362
1.43×10^{-3}	-6.759
1.28×10^{-3}	-3.231

Figure 12.16 is a graph of ln *k* versus $\frac{1}{T}$. In practice, the equation of the line (slope and *y*-intercept) that best fits these plotted data points would be derived using a statistical process called regression. This is helpful for most experimental data because a perfect fit of each data point with the line is rarely encountered. For the data here, the fit is nearly perfect and the slope may be estimated using any two of the provided data pairs. Using the first and last data points permits estimation of the slope.

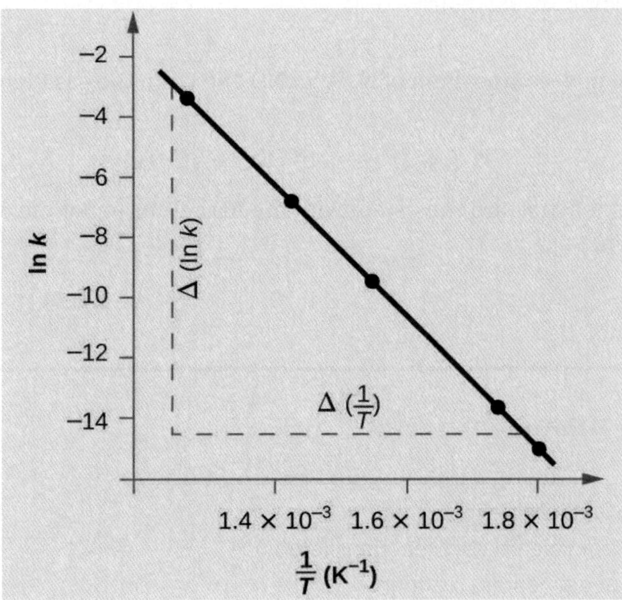

FIGURE 12.16 This graph shows the linear relationship between ln *k* and $\frac{1}{T}$ for the reaction $2HI \longrightarrow H_2 + I_2$ according to the Arrhenius equation.

$$\begin{aligned}
\text{Slope} &= \frac{\Delta(\ln k)}{\Delta\left(\frac{1}{T}\right)} \\
&= \frac{(-14.860) - (-3.231)}{\left(1.80 \times 10^{-3}\ \text{K}^{-1}\right) - \left(1.28 \times 10^{-3}\ \text{K}^{-1}\right)} \\
&= \frac{-11.629}{0.52 \times 10^{-3}\ \text{K}^{-1}} = -2.2 \times 10^{4}\ \text{K} \\
&= -\frac{E_a}{R} \\
E_a &= -\text{slope} \times R = -(-2.2 \times 10^{4}\ \text{K} \times 8.314\ \text{J mol}^{-1}\ \text{K}^{-1}) \\
& \quad 1.8 \times 10^{5}\ \text{J mol}^{-1} \text{ or } 180\ \text{kJ mol}^{-1}
\end{aligned}$$

Alternative approach: A more expedient approach involves deriving activation energy from measurements of the rate constant at just two temperatures. In this approach, the Arrhenius equation is rearranged to a convenient two-point form:

$$\ln \frac{k_1}{k_2} = \frac{E_a}{R}\left(\frac{1}{T_2} - \frac{1}{T_1}\right)$$

Rearranging this equation to isolate activation energy yields:

$$E_a = -R\left(\frac{\ln k_2 - \ln k_1}{\left(\frac{1}{T_2}\right) - \left(\frac{1}{T_1}\right)}\right)$$

Any two data pairs may be substituted into this equation—for example, the first and last entries from the above data table:

$$E_a = -8.314\,\text{J mol}^{-1}\,\text{K}^{-1}\left(\frac{-3.231 - (-14.860)}{1.28 \times 10^{-3}\,\text{K}^{-1} - 1.80 \times 10^{-3}\,\text{K}^{-1}}\right)$$

and the result is $E_a = 1.8 \times 10^5$ J mol^{-1} or 180 kJ mol^{-1}

This approach yields the same result as the more rigorous graphical approach used above, as expected. In practice, the graphical approach typically provides more reliable results when working with actual experimental data.

Check Your Learning

The rate constant for the rate of decomposition of N_2O_5 to NO and O_2 in the gas phase is 1.66 L/mol/s at 650 K and 7.39 L/mol/s at 700 K:

$$2N_2O_5(g) \longrightarrow 4NO(g) + 3O_2(g)$$

Assuming the kinetics of this reaction are consistent with the Arrhenius equation, calculate the activation energy for this decomposition.

Answer:

1.1×10^5 J mol^{-1} or 110 kJ mol^{-1}

12.6 Reaction Mechanisms

LEARNING OBJECTIVES

By the end of this section, you will be able to:

- Distinguish net reactions from elementary reactions (steps)
- Identify the molecularity of elementary reactions
- Write a balanced chemical equation for a process given its reaction mechanism
- Derive the rate law consistent with a given reaction mechanism

Chemical reactions very often occur in a step-wise fashion, involving two or more distinct reactions taking place in sequence. A balanced equation indicates what is reacting and what is produced, but it reveals no details about how the reaction actually takes place. The **reaction mechanism** (or reaction path) provides details regarding the precise, step-by-step process by which a reaction occurs.

The decomposition of ozone, for example, appears to follow a mechanism with two steps:

$$O_3(g) \longrightarrow O_2(g) + O$$
$$O + O_3(g) \longrightarrow 2O_2(g)$$

Each of the steps in a reaction mechanism is an **elementary reaction**. These elementary reactions occur precisely as represented in the step equations, and they must sum to yield the balanced chemical equation representing the overall reaction:

$$2O_3(g) \longrightarrow 3O_2(g)$$

Notice that the oxygen atom produced in the first step of this mechanism is consumed in the second step and therefore does not appear as a product in the overall reaction. Species that are produced in one step and consumed in a subsequent step are called **intermediates**.

While the overall reaction equation for the decomposition of ozone indicates that two molecules of ozone react to give three molecules of oxygen, the mechanism of the reaction *does not involve the direct collision and reaction of two ozone molecules.* Instead, one O_3 decomposes to yield O_2 and an oxygen atom, and a second O_3 molecule subsequently reacts with the oxygen atom to yield two additional O_2 molecules.

Unlike balanced equations representing an overall reaction, the equations for elementary reactions are explicit representations of the chemical change taking place. The reactant(s) in an elementary reaction's equation undergo only the bond-breaking and/or making events depicted to yield the product(s). For this reason, *the rate law for an elementary reaction may be derived directly from the balanced chemical equation describing the reaction.* This is not the case for typical chemical reactions, for which rate laws may be reliably determined only via experimentation.

Unimolecular Elementary Reactions

The **molecularity** of an elementary reaction is the number of reactant species (atoms, molecules, or ions). For example, a **unimolecular reaction** involves the reaction of a *single* reactant species to produce one or more molecules of product:

$$A \longrightarrow \text{products}$$

The rate law for a unimolecular reaction is first order:

$$\text{rate} = k[A]$$

A unimolecular reaction may be one of several elementary reactions in a complex mechanism. For example, the reaction:

$$O_3 \longrightarrow O_2 + O$$

illustrates a unimolecular elementary reaction that occurs as one part of a two-step reaction mechanism as described above. However, some unimolecular reactions may be the only step of a single-step reaction mechanism. (In other words, an "overall" reaction may also be an elementary reaction in some cases.) For example, the gas-phase decomposition of cyclobutane, C_4H_8, to ethylene, C_2H_4, is represented by the following chemical equation:

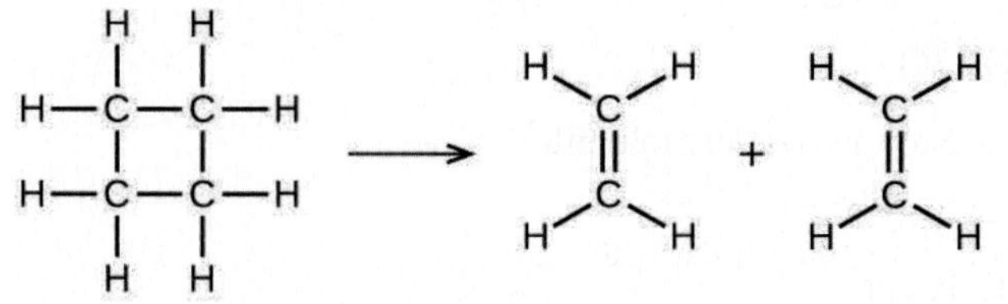

This equation represents the overall reaction observed, and it might also represent a legitimate unimolecular elementary reaction. The rate law predicted from this equation, assuming it is an elementary reaction, turns out to be the same as the rate law derived experimentally for the overall reaction, namely, one showing first-order behavior:

$$\text{rate} = -\frac{\Delta[C_4H_8]}{\Delta t} = k[C_4H_8]$$

This agreement between observed and predicted rate laws is interpreted to mean that the proposed unimolecular, single-step process is a reasonable mechanism for the butadiene reaction.

Bimolecular Elementary Reactions

A **bimolecular reaction** involves two reactant species, for example:

$$A + B \longrightarrow \text{products}$$

and

$$2A \longrightarrow \text{products}$$

For the first type, in which the two reactant molecules are different, the rate law is first-order in A and first order in B (second-order overall):

$$\text{rate} = k[A][B]$$

For the second type, in which two identical molecules collide and react, the rate law is second order in *A*:

$$\text{rate} = k[A][A] = k[A]^2$$

Some chemical reactions occur by mechanisms that consist of a single bimolecular elementary reaction. One example is the reaction of nitrogen dioxide with carbon monoxide:

$$NO_2(g) + CO(g) \longrightarrow NO(g) + CO_2(g)$$

(see Figure 12.17)

FIGURE 12.17 The probable mechanism for the reaction between NO_2 and CO to yield NO and CO_2.

Bimolecular elementary reactions may also be involved as steps in a multistep reaction mechanism. The reaction of atomic oxygen with ozone is the second step of the two-step ozone decomposition mechanism discussed earlier in this section:

$$O(g) + O_3(g) \longrightarrow 2O_2(g)$$

Termolecular Elementary Reactions

An elementary **termolecular reaction** involves the simultaneous collision of three atoms, molecules, or ions. Termolecular elementary reactions are uncommon because the probability of three particles colliding simultaneously is less than one one-thousandth of the probability of two particles colliding. There are, however, a few established termolecular elementary reactions. The reaction of nitric oxide with oxygen appears to involve termolecular steps:

$$2NO + O_2 \longrightarrow 2NO_2$$
$$\text{rate} = k[NO]^2[O_2]$$

Likewise, the reaction of nitric oxide with chlorine appears to involve termolecular steps:

$$2NO + Cl_2 \longrightarrow 2NOCl$$
$$\text{rate} = k[NO]^2[Cl_2]$$

Relating Reaction Mechanisms to Rate Laws

It's often the case that one step in a multistep reaction mechanism is significantly slower than the others. Because a reaction cannot proceed faster than its slowest step, this step will limit the rate at which the overall reaction occurs. The slowest step is therefore called the **rate-limiting step** (or rate-determining step) of the reaction Figure 12.18.

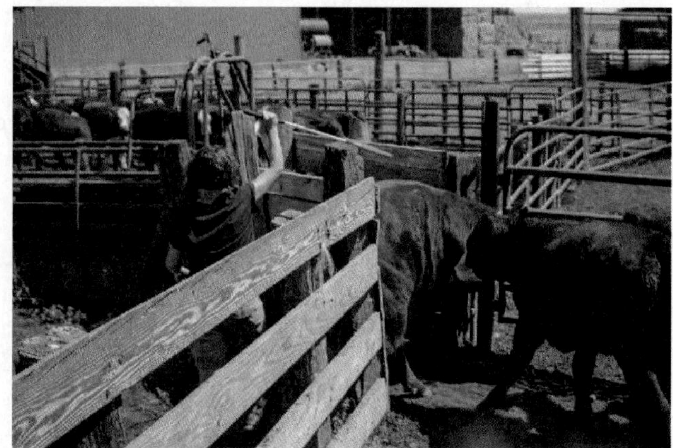

FIGURE 12.18 A cattle chute is a nonchemical example of a rate-determining step. Cattle can only be moved from one holding pen to another as quickly as one animal can make its way through the chute. (credit: Loren Kerns)

As described earlier, rate laws may be derived directly from the chemical equations for elementary reactions. This is not the case, however, for ordinary chemical reactions. The balanced equations most often encountered represent the overall change for some chemical system, and very often this is the result of some multistep reaction mechanisms. In every case, the rate law must be determined from experimental data and the reaction mechanism subsequently deduced from the rate law (and sometimes from other data). The reaction of NO_2 and CO provides an illustrative example:

$$NO_2(g) + CO(g) \longrightarrow CO_2(g) + NO(g)$$

For temperatures above 225 °C, the rate law has been found to be:

$$\text{rate} = k\,[NO_2]\,[CO]$$

The reaction is first order with respect to NO_2 and first-order with respect to CO. This is consistent with a single-step bimolecular mechanism and it is *possible* that this is the mechanism for this reaction at high temperatures.

At temperatures below 225 °C, the reaction is described by a rate law that is second order with respect to NO_2:

$$\text{rate} = k[NO_2]^2$$

This rate law is not consistent with the single-step mechanism, but is consistent with the following two-step mechanism:

$$NO_2(g) + NO_2(g) \longrightarrow NO_3(g) + NO(g)\ \text{(slow)}$$
$$NO_3(g) + CO(g) \longrightarrow NO_2(g) + CO_2(g)\ \text{(fast)}$$

The rate-determining (slower) step gives a rate law showing second-order dependence on the NO_2 concentration, and the sum of the two equations gives the net overall reaction.

In general, when the rate-determining (slower) step is the first step in a mechanism, the rate law for the overall reaction is the same as the rate law for this step. However, when the rate-determining step is preceded by a step involving a rapidly reversible reaction the rate law for the overall reaction may be more difficult to derive.

As discussed in several chapters of this text, a reversible reaction is at *equilibrium* when the rates of the forward and reverse processes are equal. Consider the reversible elementary reaction in which NO dimerizes to yield an intermediate species N_2O_2. When this reaction is at equilibrium:

$$NO + NO \rightleftharpoons N_2O_2$$
$$\text{rate}_{\text{forward}} = \text{rate}_{\text{reverse}}$$
$$k_1\,[NO]^2 = k_{-1}\,[N_2O_2]$$

This expression may be rearranged to express the concentration of the intermediate in terms of the reactant NO:

$$\left(\frac{k_1[NO]^2}{k_{-1}}\right) = [N_2O_2]$$

Since intermediate species concentrations are not used in formulating rate laws for overall reactions, this approach is sometimes necessary, as illustrated in the following example exercise.

EXAMPLE 12.14

Deriving a Rate Law from a Reaction Mechanism

The two-step mechanism below has been proposed for a reaction between nitrogen monoxide and molecular chlorine:

$$\begin{array}{ll}\text{Step 1: } NO(g) + Cl_2(g) \rightleftharpoons NOCl_2(g) & \text{fast} \\ \text{Step 2: } NOCl_2(g) + NO(g) \longrightarrow 2NOCl(g) & \text{slow}\end{array}$$

Use this mechanism to derive the equation and predicted rate law for the overall reaction.

Solution

The equation for the overall reaction is obtained by adding the two elementary reactions:

$$2NO(g) + Cl_2(g) \longrightarrow 2NOCl(g)$$

To derive a rate law from this mechanism, first write rates laws for each of the two steps.

$$\begin{aligned}\text{rate}_1 &= k_1[NO][Cl_2] \text{ for the forward reaction of step 1} \\ \text{rate}_{-1} &= k_{-1}[NOCl_2] \text{ for the reverse reaction of step 1} \\ \text{rate}_2 &= k_2[NOCl_2][NO] \text{ for step 2}\end{aligned}$$

Step 2 is the rate-determining step, and so the rate law for the overall reaction should be the same as for this step. However, the step 2 rate law, as written, contains an intermediate species concentration, $[NOCl_2]$. To remedy this, use the first step's rate laws to derive an expression for the intermediate concentration in terms of the reactant concentrations.

Assuming step 1 is at equilibrium:

$$\begin{aligned}\text{rate}_1 &= \text{rate}_{-1} \\ k_1\,[NO]\,[Cl_2] &= k_{-1}\,[NOCl_2] \\ [NOCl_2] &= \left(\frac{k_1}{k_{-1}}\right)\,[NO]\,[Cl_2]\end{aligned}$$

Substituting this expression into the rate law for step 2 yields:

$$\text{rate}_2 = \text{rate}_{\text{overall}} = \left(\frac{k_2 k_1}{k_{-1}}\right)[NO]^2[Cl_2]$$

Check Your Learning

The first step of a proposed multistep mechanism is:

$$F_2(g) \rightleftharpoons 2F(g) \text{ fast}$$

Derive the equation relating atomic fluorine concentration to molecular fluorine concentration.

Answer:

$$[F] = \left(\frac{k_1[F_2]}{k_{-1}}\right)^{1/2}$$

12.7 Catalysis

LEARNING OBJECTIVES

By the end of this section, you will be able to:

- Explain the function of a catalyst in terms of reaction mechanisms and potential energy diagrams
- List examples of catalysis in natural and industrial processes

Among the factors affecting chemical reaction rates discussed earlier in this chapter was the presence of a *catalyst*, a substance that can increase the reaction rate without being consumed in the reaction. The concepts introduced in the previous section on reaction mechanisms provide the basis for understanding how catalysts are able to accomplish this very important function.

Figure 12.19 shows reaction diagrams for a chemical process in the absence and presence of a catalyst. Inspection of the diagrams reveals several traits of these reactions. Consistent with the fact that the two diagrams represent the same overall reaction, both curves begin and end at the same energies (in this case, because products are more energetic than reactants, the reaction is endothermic). The reaction mechanisms, however, are clearly different. The uncatalyzed reaction proceeds via a one-step mechanism (one transition state observed), whereas the catalyzed reaction follows a two-step mechanism (two transition states observed) with *a notably lesser activation energy*. This difference illustrates the means by which a catalyst functions to accelerate reactions, namely, by providing an alternative reaction mechanism with a lower activation energy. Although the catalyzed reaction mechanism for a reaction needn't necessarily involve a different number of steps than the uncatalyzed mechanism, it must provide a reaction path whose rate determining step is faster (lower E_a).

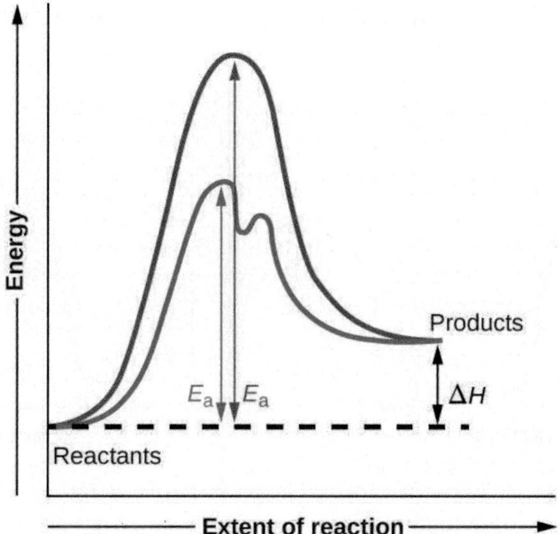

FIGURE 12.19 Reaction diagrams for an endothermic process in the absence (red curve) and presence (blue curve) of a catalyst. The catalyzed pathway involves a two-step mechanism (note the presence of two transition states) and an intermediate species (represented by the valley between the two transitions states).

EXAMPLE 12.15

Reaction Diagrams for Catalyzed Reactions

The two reaction diagrams here represent the same reaction: one without a catalyst and one with a catalyst. Estimate the activation energy for each process, and identify which one involves a catalyst.

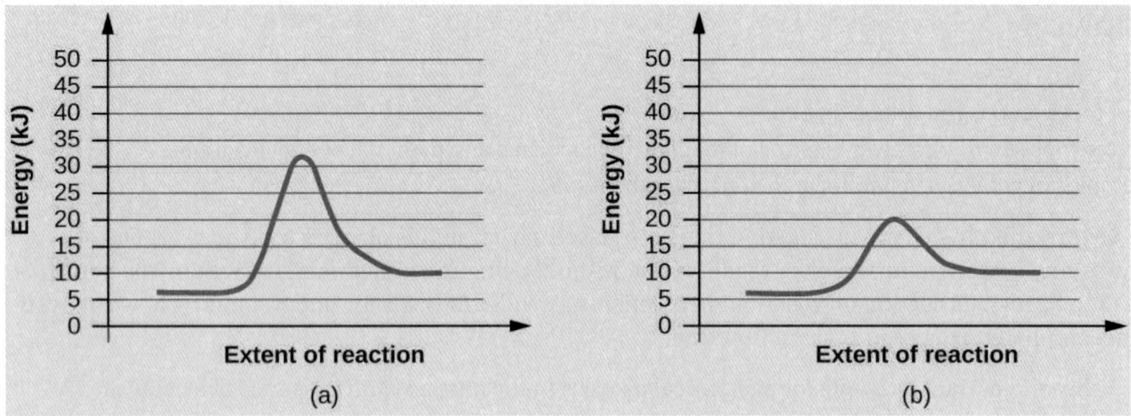

Solution

Activation energies are calculated by subtracting the reactant energy from the transition state energy.

$$\text{diagram (a): } E_a = 32\text{ kJ} - 6\text{ kJ} = 26\text{ kJ}$$
$$\text{diagram (b): } E_a = 20\text{ kJ} - 6\text{ kJ} = 14\text{ kJ}$$

The catalyzed reaction is the one with lesser activation energy, in this case represented by diagram b.

Check Your Learning

Reaction diagrams for a chemical process with and without a catalyst are shown below. Both reactions involve a two-step mechanism with a rate-determining first step. Compute activation energies for the first step of each mechanism, and identify which corresponds to the catalyzed reaction. How do the second steps of these two mechanisms compare?

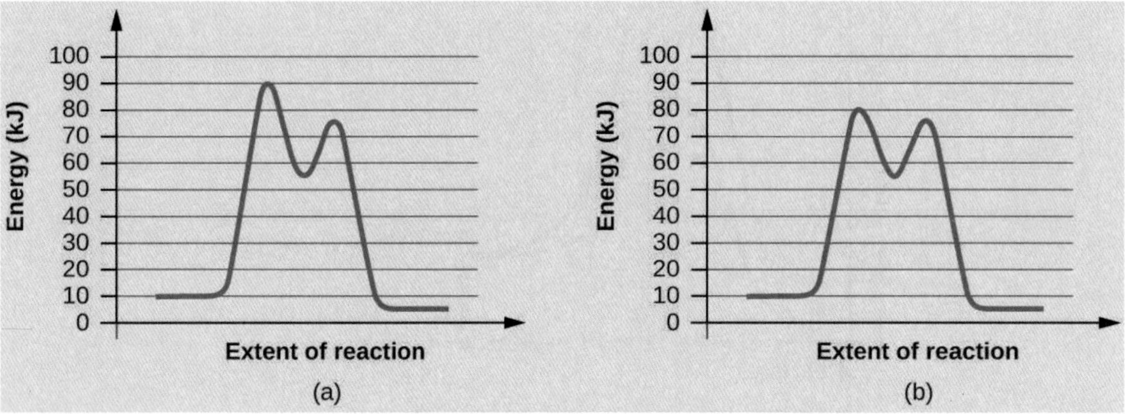

Answer:

For the first step, E_a = 80 kJ for (a) and 70 kJ for (b), so diagram (b) depicts the catalyzed reaction. Activation energies for the second steps of both mechanisms are the same, 20 kJ.

Homogeneous Catalysts

A **homogeneous catalyst** is present in the same phase as the reactants. It interacts with a reactant to form an intermediate substance, which then decomposes or reacts with another reactant in one or more steps to regenerate the original catalyst and form product.

As an important illustration of homogeneous catalysis, consider the earth's ozone layer. Ozone in the upper atmosphere, which protects the earth from ultraviolet radiation, is formed when oxygen molecules absorb ultraviolet light and undergo the reaction:

$$3O_2(g) \xrightarrow{h\upsilon} 2O_3(g)$$

Ozone is a relatively unstable molecule that decomposes to yield diatomic oxygen by the reverse of this equation. This decomposition reaction is consistent with the following two-step mechanism:

$$O_3 \longrightarrow O_2 + O$$
$$O + O_3 \longrightarrow 2O_2$$

A number of substances can catalyze the decomposition of ozone. For example, the nitric oxide–catalyzed decomposition of ozone is believed to occur via the following three-step mechanism:

$$NO(g) + O_3(g) \longrightarrow NO_2(g) + O_2(g)$$
$$O_3(g) \longrightarrow O_2(g) + O(g)$$
$$NO_2(g) + O(g) \longrightarrow NO(g) + O_2(g)$$

As required, the overall reaction is the same for both the two-step uncatalyzed mechanism and the three-step NO-catalyzed mechanism:

$$2O_3(g) \longrightarrow 3O_2(g)$$

Notice that NO is a reactant in the first step of the mechanism and a product in the last step. This is another characteristic trait of a catalyst: Though it participates in the chemical reaction, it is not consumed by the reaction.

Portrait of a Chemist

Mario J. Molina

The 1995 Nobel Prize in Chemistry was shared by Paul J. Crutzen, Mario J. Molina (Figure 12.20), and F. Sherwood Rowland "for their work in atmospheric chemistry, particularly concerning the formation and decomposition of ozone."[1] Molina, a Mexican citizen, carried out the majority of his work at the Massachusetts Institute of Technology (MIT).

(a)

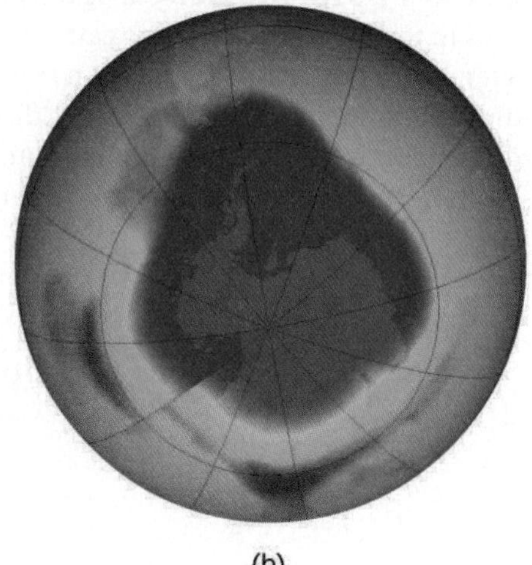

(b)

FIGURE 12.20 (a) Mexican chemist Mario Molina (1943 –) shared the Nobel Prize in Chemistry in 1995 for his research on (b) the Antarctic ozone hole. (credit a: courtesy of Mario Molina; credit b: modification of work by NASA)

In 1974, Molina and Rowland published a paper in the journal *Nature* detailing the threat of chlorofluorocarbon gases to the stability of the ozone layer in earth's upper atmosphere. The ozone layer protects earth from solar radiation by absorbing ultraviolet light. As chemical reactions deplete the amount of ozone in the upper atmosphere, a measurable "hole" forms above Antarctica, and an increase in the

1 "The Nobel Prize in Chemistry 1995," Nobel Prize.org, accessed February 18, 2015, http://www.nobelprize.org/nobel_prizes/chemistry/laureates/1995/.

amount of solar ultraviolet radiation– strongly linked to the prevalence of skin cancers—reaches earth's surface. The work of Molina and Rowland was instrumental in the adoption of the Montreal Protocol, an international treaty signed in 1987 that successfully began phasing out production of chemicals linked to ozone destruction.

Molina and Rowland demonstrated that chlorine atoms from human-made chemicals can catalyze ozone destruction in a process similar to that by which NO accelerates the depletion of ozone. Chlorine atoms are generated when chlorocarbons or chlorofluorocarbons—once widely used as refrigerants and propellants—are photochemically decomposed by ultraviolet light or react with hydroxyl radicals. A sample mechanism is shown here using methyl chloride:

$$CH_3Cl + OH \longrightarrow Cl + \text{other products}$$

Chlorine radicals break down ozone and are regenerated by the following catalytic cycle:

$$Cl + O_3 \longrightarrow ClO + O_2$$
$$ClO + O \longrightarrow Cl + O_2$$
$$\text{overall Reaction: } O_3 + O \longrightarrow 2O_2$$

A single monatomic chlorine can break down thousands of ozone molecules. Luckily, the majority of atmospheric chlorine exists as the catalytically inactive forms Cl_2 and $ClONO_2$.

Since receiving his portion of the Nobel Prize, Molina has continued his work in atmospheric chemistry at MIT.

HOW SCIENCES INTERCONNECT

Glucose-6-Phosphate Dehydrogenase Deficiency

Enzymes in the human body act as catalysts for important chemical reactions in cellular metabolism. As such, a deficiency of a particular enzyme can translate to a life-threatening disease. G6PD (glucose-6-phosphate dehydrogenase) deficiency, a genetic condition that results in a shortage of the enzyme glucose-6-phosphate dehydrogenase, is the most common enzyme deficiency in humans. This enzyme, shown in Figure 12.21, is the rate-limiting enzyme for the metabolic pathway that supplies NADPH to cells (Figure 12.22).

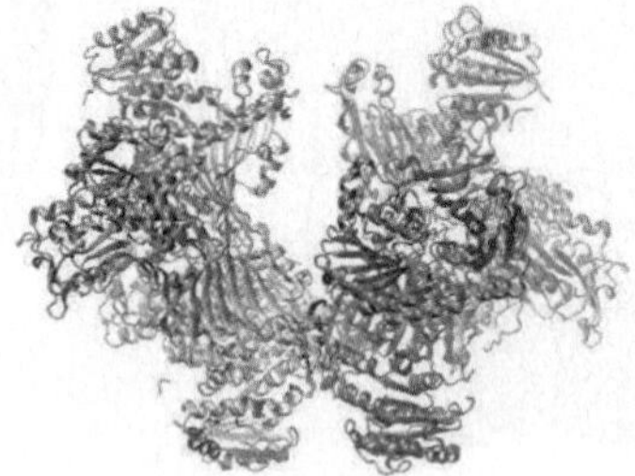

FIGURE 12.21 Glucose-6-phosphate dehydrogenase is a rate-limiting enzyme for the metabolic pathway that supplies NADPH to cells.

A disruption in this pathway can lead to reduced glutathione in red blood cells; once all glutathione is consumed, enzymes and other proteins such as hemoglobin are susceptible to damage. For example, hemoglobin can be metabolized to bilirubin, which leads to jaundice, a condition that can become severe. People who suffer from G6PD deficiency must avoid certain foods and medicines containing chemicals that can trigger damage their glutathione-deficient red blood cells.

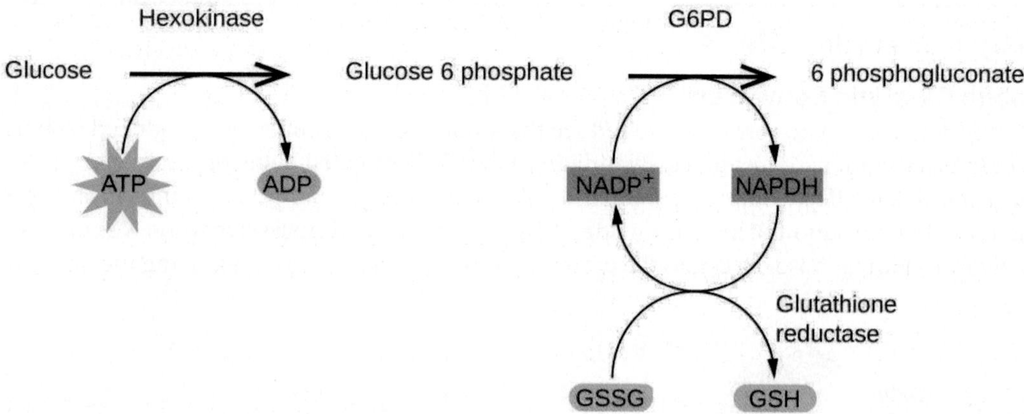

FIGURE 12.22 In the mechanism for the pentose phosphate pathway, G6PD catalyzes the reaction that regulates NADPH, a co-enzyme that regulates glutathione, an antioxidant that protects red blood cells and other cells from oxidative damage.

Heterogeneous Catalysts

A **heterogeneous catalyst** is a catalyst that is present in a different phase (usually a solid) than the reactants. Such catalysts generally function by furnishing an active surface upon which a reaction can occur. Gas and liquid phase reactions catalyzed by heterogeneous catalysts occur on the surface of the catalyst rather than within the gas or liquid phase.

Heterogeneous catalysis typically involves the following processes:

1. Adsorption of the reactant(s) onto the surface of the catalyst
2. Activation of the adsorbed reactant(s)
3. Reaction of the adsorbed reactant(s)
4. Desorption of product(s) from the surface of the catalyst

Figure 12.23 illustrates the steps of a mechanism for the reaction of compounds containing a carbon–carbon double bond with hydrogen on a nickel catalyst. Nickel is the catalyst used in the hydrogenation of polyunsaturated fats and oils (which contain several carbon–carbon double bonds) to produce saturated fats and oils (which contain only carbon–carbon single bonds).

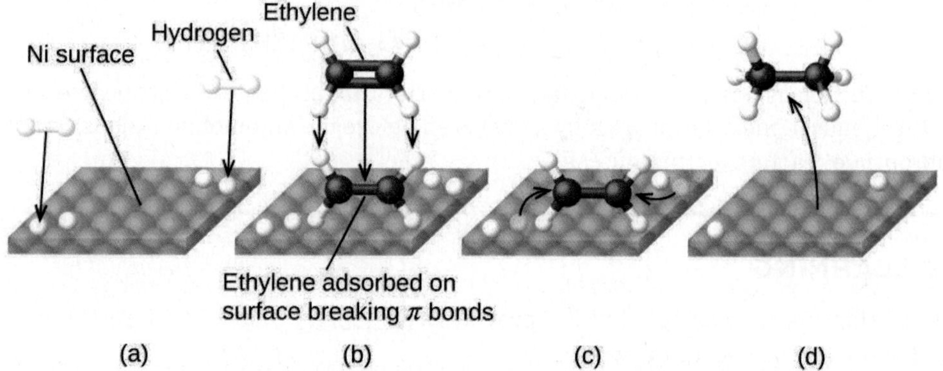

FIGURE 12.23 Mechanism for the Ni-catalyzed reaction $C_2H_4 + H_2 \longrightarrow C_2H_6$. (a) Hydrogen is adsorbed on the surface, breaking the H–H bonds and forming Ni–H bonds. (b) Ethylene is adsorbed on the surface, breaking the C–C π-bond and forming Ni–C bonds. (c) Atoms diffuse across the surface and form new C–H bonds when they collide. (d) C_2H_6 molecules desorb from the Ni surface.

Many important chemical products are prepared via industrial processes that use heterogeneous catalysts, including ammonia, nitric acid, sulfuric acid, and methanol. Heterogeneous catalysts are also used in the catalytic converters found on most gasoline-powered automobiles (Figure 12.24).

Chemistry in Everyday Life

Automobile Catalytic Converters

Scientists developed catalytic converters to reduce the amount of toxic emissions produced by burning gasoline in internal combustion engines. By utilizing a carefully selected blend of catalytically active metals, it is possible to effect complete combustion of all carbon-containing compounds to carbon dioxide while also reducing the output of nitrogen oxides. This is particularly impressive when we consider that one step involves adding more oxygen to the molecule and the other involves removing the oxygen (Figure 12.24).

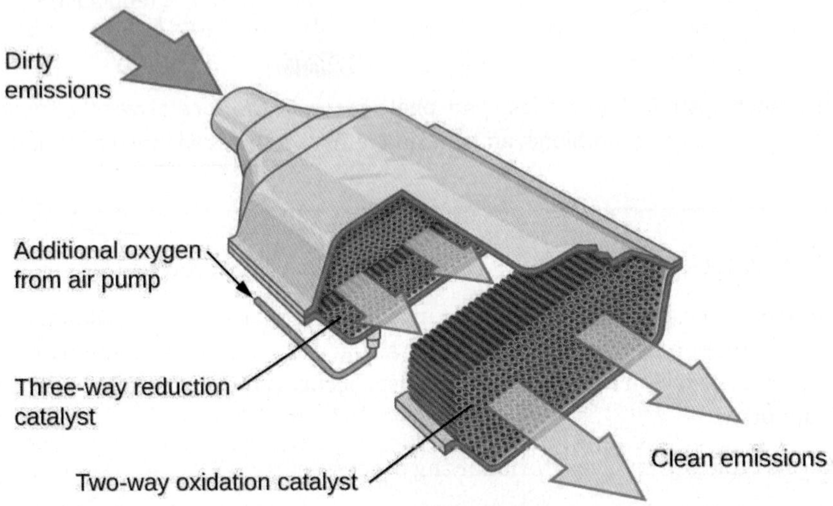

FIGURE 12.24 A catalytic converter allows for the combustion of all carbon-containing compounds to carbon dioxide, while at the same time reducing the output of nitrogen oxide and other pollutants in emissions from gasoline-burning engines.

Most modern, three-way catalytic converters possess a surface impregnated with a platinum-rhodium catalyst, which catalyzes the conversion of nitric oxide into dinitrogen and oxygen as well as the conversion of carbon monoxide and hydrocarbons such as octane into carbon dioxide and water vapor:

$$2NO_2(g) \longrightarrow N_2(g) + 2O_2(g)$$
$$2CO(g) + O_2(g) \longrightarrow 2CO_2(g)$$
$$2C_8H_{18}(g) + 25O_2(g) \longrightarrow 16CO_2(g) + 18H_2O(g)$$

In order to be as efficient as possible, most catalytic converters are preheated by an electric heater. This ensures that the metals in the catalyst are fully active even before the automobile exhaust is hot enough to maintain appropriate reaction temperatures.

LINK TO LEARNING

The University of California at Davis' "ChemWiki" provides a thorough explanation (http://openstax.org/l/16catconvert) of how catalytic converters work.

HOW SCIENCES INTERCONNECT

Enzyme Structure and Function

The study of enzymes is an important interconnection between biology and chemistry. Enzymes are usually proteins (polypeptides) that help to control the rate of chemical reactions between biologically important compounds, particularly those that are involved in cellular metabolism. Different classes of enzymes perform a variety of functions, as shown in Table 12.3.

Classes of Enzymes and Their Functions

Class	Function
oxidoreductases	redox reactions
transferases	transfer of functional groups
hydrolases	hydrolysis reactions
lyases	group elimination to form double bonds
isomerases	isomerization
ligases	bond formation with ATP hydrolysis

TABLE 12.3

Enzyme molecules possess an active site, a part of the molecule with a shape that allows it to bond to a specific substrate (a reactant molecule), forming an enzyme-substrate complex as a reaction intermediate. There are two models that attempt to explain how this active site works. The most simplistic model is referred to as the lock-and-key hypothesis, which suggests that the molecular shapes of the active site and substrate are complementary, fitting together like a key in a lock. The induced fit hypothesis, on the other hand, suggests that the enzyme molecule is flexible and changes shape to accommodate a bond with the substrate. This is not to suggest that an enzyme's active site is completely malleable, however. Both the lock-and-key model and the induced fit model account for the fact that enzymes can only bind with specific substrates, since in general a particular enzyme only catalyzes a particular reaction (Figure 12.25).

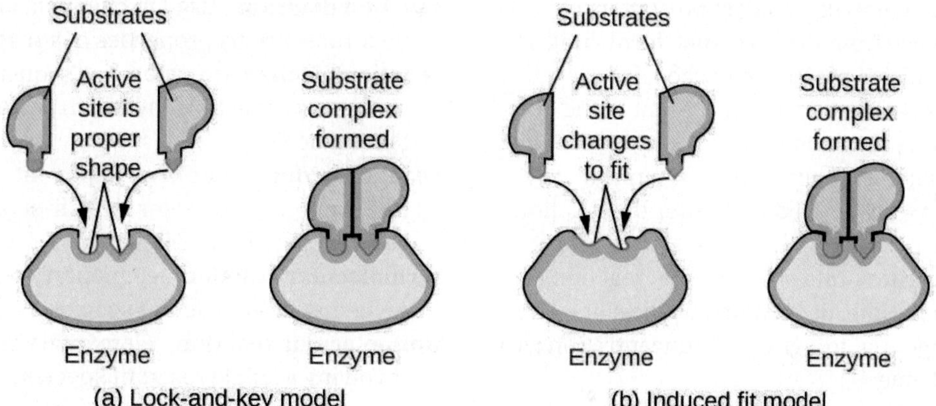

FIGURE 12.25 (a) According to the lock-and-key model, the shape of an enzyme's active site is a perfect fit for the substrate. (b) According to the induced fit model, the active site is somewhat flexible, and can change shape in order to bond with the substrate.

LINK TO LEARNING

The Royal Society of Chemistry (http://openstax.org/l/16enzymes) provides an excellent introduction to enzymes for students and teachers.

Key Terms

activated complex (also, transition state) unstable combination of reactant species formed during a chemical reaction

activation energy (E_a) minimum energy necessary in order for a reaction to take place

Arrhenius equation mathematical relationship between a reaction's rate constant, activation energy, and temperature

average rate rate of a chemical reaction computed as the ratio of a measured change in amount or concentration of substance to the time interval over which the change occurred

bimolecular reaction elementary reaction involving two reactant species

catalyst substance that increases the rate of a reaction without itself being consumed by the reaction

collision theory model that emphasizes the energy and orientation of molecular collisions to explain and predict reaction kinetics

elementary reaction reaction that takes place in a single step, precisely as depicted in its chemical equation

frequency factor (A) proportionality constant in the Arrhenius equation, related to the relative number of collisions having an orientation capable of leading to product formation

half-life of a reaction ($t_{1/2}$) time required for half of a given amount of reactant to be consumed

heterogeneous catalyst catalyst present in a different phase from the reactants, furnishing a surface at which a reaction can occur

homogeneous catalyst catalyst present in the same phase as the reactants

initial rate instantaneous rate of a chemical reaction at $t = 0$ s (immediately after the reaction has begun)

instantaneous rate rate of a chemical reaction at any instant in time, determined by the slope of the line tangential to a graph of concentration as a function of time

integrated rate law equation that relates the concentration of a reactant to elapsed time of reaction

intermediate species produced in one step of a reaction mechanism and consumed in a subsequent step

method of initial rates common experimental approach to determining rate laws that involves measuring reaction rates at varying initial reactant concentrations

molecularity number of reactant species involved in an elementary reaction

overall reaction order sum of the reaction orders for each substance represented in the rate law

rate constant (k) proportionality constant in a rate law

rate expression mathematical representation defining reaction rate as change in amount, concentration, or pressure of reactant or product species per unit time

rate law (also, rate equation) (also, differential rate laws) mathematical equation showing the dependence of reaction rate on the rate constant and the concentration of one or more reactants

rate of reaction measure of the speed at which a chemical reaction takes place

rate-determining step (also, rate-limiting step) slowest elementary reaction in a reaction mechanism; determines the rate of the overall reaction

reaction diagram used in chemical kinetics to illustrate various properties of a reaction

reaction mechanism stepwise sequence of elementary reactions by which a chemical change takes place

reaction order value of an exponent in a rate law (for example, zero order for 0, first order for 1, second order for 2, and so on)

termolecular reaction elementary reaction involving three reactant species

unimolecular reaction elementary reaction involving a single reactant species

Key Equations

relative reaction rates for $a\text{A} \longrightarrow b\text{B} = -\frac{1}{a}\frac{\Delta[\text{A}]}{\Delta t} = \frac{1}{b}\frac{\Delta[\text{B}]}{\Delta t}$

integrated rate law for zero-order reactions: $[A]_t = -kt + [A]_0$,

half-life for a zero-order reaction $t_{1/2} = \frac{[A]_0}{2k}$

integrated rate law for first-order reactions: $\ln[A]_t = -kt + \ln[A]_0$,

half-life for a first-order reaction $t_{1/2} = \frac{0.693}{k}$

integrated rate law for second-order reactions: $\frac{1}{[A]_t} = kt + \frac{1}{[A]_0}$,

half-life for a second-order reaction $t_{1/2} = \frac{1}{[A]_0 k}$

$k = Ae^{-E_a/RT}$

$\ln k = \left(\frac{-E_a}{R}\right)\left(\frac{1}{T}\right) + \ln A$

$\ln \frac{k_1}{k_2} = \frac{E_a}{R}\left(\frac{1}{T_2} - \frac{1}{T_1}\right)$

Summary

12.1 Chemical Reaction Rates

The rate of a reaction can be expressed either in terms of the decrease in the amount of a reactant or the increase in the amount of a product per unit time. Relations between different rate expressions for a given reaction are derived directly from the stoichiometric coefficients of the equation representing the reaction.

12.2 Factors Affecting Reaction Rates

The rate of a chemical reaction is affected by several parameters. Reactions involving two phases proceed more rapidly when there is greater surface area contact. If temperature or reactant concentration is increased, the rate of a given reaction generally increases as well. A catalyst can increase the rate of a reaction by providing an alternative pathway with a lower activation energy.

12.3 Rate Laws

Rate laws (*differential rate laws*) provide a mathematical description of how changes in the concentration of a substance affect the rate of a chemical reaction. Rate laws are determined experimentally and cannot be predicted by reaction stoichiometry. The order of reaction describes how much a change in the concentration of each substance affects the overall rate, and the overall order of a reaction is the sum of the orders for each substance present in the reaction. Reaction orders are typically first order, second order, or zero order, but fractional and even negative orders are possible.

12.4 Integrated Rate Laws

Integrated rate laws are mathematically derived from differential rate laws, and they describe the time dependence of reactant and product concentrations.

The half-life of a reaction is the time required to decrease the amount of a given reactant by one-half. A reaction's half-life varies with rate constant and, for some reaction orders, reactant concentration. The half-life of a zero-order reaction decreases as the initial concentration of the reactant in the reaction decreases. The half-life of a first-order reaction is independent of concentration, and the half-life of a second-order reaction decreases as the concentration increases.

12.5 Collision Theory

Chemical reactions typically require collisions between reactant species. These reactant collisions must be of proper orientation and sufficient energy in order to result in product formation. Collision theory provides a simple but effective explanation for the effect of many experimental parameters on reaction rates. The Arrhenius equation describes the relation between a reaction's rate constant, activation energy, temperature, and dependence on collision orientation.

12.6 Reaction Mechanisms

The sequence of individual steps, or elementary reactions, by which reactants are converted into products during the course of a reaction is called the reaction mechanism. The molecularity of an elementary reaction is the number of reactant species involved, typically one (unimolecular), two (bimolecular), or, less commonly, three (termolecular). The overall rate of a reaction is determined by the rate of the slowest in its mechanism, called the rate-determining step. Unimolecular elementary reactions have first-order rate laws, while bimolecular elementary reactions have second-order rate laws. By comparing the rate laws derived from a reaction mechanism to that determined experimentally, the mechanism may be deemed either incorrect or plausible.

12.7 Catalysis

Catalysts affect the rate of a chemical reaction by altering its mechanism to provide a lower activation energy. Catalysts can be homogenous (in the same phase as the reactants) or heterogeneous (a different phase than the reactants).

Exercises

12.1 Chemical Reaction Rates

1. What is the difference between average rate, initial rate, and instantaneous rate?
2. Ozone decomposes to oxygen according to the equation $2O_3(g) \longrightarrow 3O_2(g)$. Write the equation that relates the rate expressions for this reaction in terms of the disappearance of O_3 and the formation of oxygen.
3. In the nuclear industry, chlorine trifluoride is used to prepare uranium hexafluoride, a volatile compound of uranium used in the separation of uranium isotopes. Chlorine trifluoride is prepared by the reaction $Cl_2(g) + 3F_2(g) \longrightarrow 2ClF_3(g)$. Write the equation that relates the rate expressions for this reaction in terms of the disappearance of Cl_2 and F_2 and the formation of ClF_3.
4. A study of the rate of dimerization of C_4H_6 gave the data shown in the table:
$2C_4H_6 \longrightarrow C_8H_{12}$

Time (s)	0	1600	3200	4800	6200
$[C_4H_6]$ (M)	1.00×10^{-2}	5.04×10^{-3}	3.37×10^{-3}	2.53×10^{-3}	2.08×10^{-3}

(a) Determine the average rate of dimerization between 0 s and 1600 s, and between 1600 s and 3200 s.
(b) Estimate the instantaneous rate of dimerization at 3200 s from a graph of time versus $[C_4H_6]$. What are the units of this rate?
(c) Determine the average rate of formation of C_8H_{12} at 1600 s and the instantaneous rate of formation at 3200 s from the rates found in parts (a) and (b).
5. A study of the rate of the reaction represented as $2A \longrightarrow B$ gave the following data:

Time (s)	0.0	5.0	10.0	15.0	20.0	25.0	35.0
$[A]$ (M)	1.00	0.775	0.625	0.465	0.360	0.285	0.230

(a) Determine the average rate of disappearance of A between 0.0 s and 10.0 s, and between 10.0 s and 20.0 s.
(b) Estimate the instantaneous rate of disappearance of A at 15.0 s from a graph of time versus $[A]$. What are the units of this rate?
(c) Use the rates found in parts (a) and (b) to determine the average rate of formation of B between 0.00 s and 10.0 s, and the instantaneous rate of formation of B at 15.0 s.
6. Consider the following reaction in aqueous solution:
$5Br^-(aq) + BrO_3{}^-(aq) + 6H^+(aq) \longrightarrow 3Br_2(aq) + 3H_2O(l)$
If the rate of disappearance of $Br^-(aq)$ at a particular moment during the reaction is 3.5×10^{-4} mol L^{-1} s^{-1}, what is the rate of appearance of $Br_2(aq)$ at that moment?

12.2 Factors Affecting Reaction Rates

7. Describe the effect of each of the following on the rate of the reaction of magnesium metal with a solution of hydrochloric acid: the molarity of the hydrochloric acid, the temperature of the solution, and the size of the pieces of magnesium.
8. Explain why an egg cooks more slowly in boiling water in Denver than in New York City. (Hint: Consider the effect of temperature on reaction rate and the effect of pressure on boiling point.)

9. Go to the PhET Reactions & Rates (http://openstax.org/l/16PHETreaction) interactive. Use the Single Collision tab to represent how the collision between monatomic oxygen (O) and carbon monoxide (CO) results in the breaking of one bond and the formation of another. Pull back on the red plunger to release the atom and observe the results. Then, click on "Reload Launcher" and change to "Angled shot" to see the difference.
(a) What happens when the angle of the collision is changed?
(b) Explain how this is relevant to rate of reaction.
10. In the PhET Reactions & Rates (http://openstax.org/l/16PHETreaction) interactive, use the "Many Collisions" tab to observe how multiple atoms and molecules interact under varying conditions. Select a molecule to pump into the chamber. Set the initial temperature and select the current amounts of each reactant. Select "Show bonds" under Options. How is the rate of the reaction affected by concentration and temperature?
11. In the PhET Reactions & Rates (http://openstax.org/l/16PHETreaction) interactive, on the Many Collisions tab, set up a simulation with 15 molecules of A and 10 molecules of BC. Select "Show Bonds" under Options.
(a) Leave the Initial Temperature at the default setting. Observe the reaction. Is the rate of reaction fast or slow?
(b) Click "Pause" and then "Reset All," and then enter 15 molecules of A and 10 molecules of BC once again. Select "Show Bonds" under Options. This time, increase the initial temperature until, on the graph, the total average energy line is completely above the potential energy curve. Describe what happens to the reaction.

12.3 Rate Laws

12. How do the rate of a reaction and its rate constant differ?
13. Doubling the concentration of a reactant increases the rate of a reaction four times. With this knowledge, answer the following questions:
(a) What is the order of the reaction with respect to that reactant?
(b) Tripling the concentration of a different reactant increases the rate of a reaction three times. What is the order of the reaction with respect to that reactant?
14. Tripling the concentration of a reactant increases the rate of a reaction nine-fold. With this knowledge, answer the following questions:
(a) What is the order of the reaction with respect to that reactant?
(b) Increasing the concentration of a reactant by a factor of four increases the rate of a reaction four-fold. What is the order of the reaction with respect to that reactant?
15. How will the rate of reaction change for the process: $CO(g) + NO_2(g) \longrightarrow CO_2(g) + NO(g)$ if the rate law for the reaction is rate $= k[NO_2]^2$?
(a) Decreasing the pressure of NO_2 from 0.50 atm to 0.250 atm.
(b) Increasing the concentration of CO from 0.01 *M* to 0.03 *M*.
16. How will each of the following affect the rate of the reaction: $CO(g) + NO_2(g) \longrightarrow CO_2(g) + NO(g)$ if the rate law for the reaction is rate $= k\,[NO_2]\,[CO]$?
(a) Increasing the pressure of NO_2 from 0.1 atm to 0.3 atm
(b) Increasing the concentration of CO from 0.02 *M* to 0.06 *M*.
17. Regular flights of supersonic aircraft in the stratosphere are of concern because such aircraft produce nitric oxide, NO, as a byproduct in the exhaust of their engines. Nitric oxide reacts with ozone, and it has been suggested that this could contribute to depletion of the ozone layer. The reaction $NO + O_3 \longrightarrow NO_2 + O_2$ is first order with respect to both NO and O_3 with a rate constant of 2.20×10^7 L/mol/s. What is the instantaneous rate of disappearance of NO when $[NO] = 3.3 \times 10^{-6}$ *M* and $[O_3] = 5.9 \times 10^{-7}$ *M*?

18. Radioactive phosphorus is used in the study of biochemical reaction mechanisms because phosphorus atoms are components of many biochemical molecules. The location of the phosphorus (and the location of the molecule it is bound in) can be detected from the electrons (beta particles) it produces:

$${}^{32}_{15}\text{P} \longrightarrow {}^{32}_{16}\text{S} + \text{e}^-$$

$$\text{rate} = 4.85 \times 10^{-2}\ \text{day}^{-1}\ [{}^{32}\text{P}]$$

What is the instantaneous rate of production of electrons in a sample with a phosphorus concentration of 0.0033 *M*?

19. The rate constant for the radioactive decay of ^{14}C is 1.21×10^{-4} year^{-1}. The products of the decay are nitrogen atoms and electrons (beta particles):

$${}^{14}_{6}\text{C} \longrightarrow {}^{14}_{7}\text{N} + \text{e}^-$$

$$\text{rate} = k[{}^{14}_{6}\text{C}]$$

What is the instantaneous rate of production of N atoms in a sample with a carbon-14 content of 6.5×10^{-9} *M*?

20. The decomposition of acetaldehyde is a second order reaction with a rate constant of 4.71×10^{-8} L mol^{-1} s^{-1}. What is the instantaneous rate of decomposition of acetaldehyde in a solution with a concentration of 5.55×10^{-4} *M*?

21. Alcohol is removed from the bloodstream by a series of metabolic reactions. The first reaction produces acetaldehyde; then other products are formed. The following data have been determined for the rate at which alcohol is removed from the blood of an average male, although individual rates can vary by 25–30%. Women metabolize alcohol a little more slowly than men:

$[C_2H_5OH]$ (*M*)	4.4×10^{-2}	3.3×10^{-2}	2.2×10^{-2}
Rate (mol L^{-1} h^{-1})	2.0×10^{-2}	2.0×10^{-2}	2.0×10^{-2}

Determine the rate law, the rate constant, and the overall order for this reaction.

22. Under certain conditions the decomposition of ammonia on a metal surface gives the following data:

$[NH_3]$ (*M*)	1.0×10^{-3}	2.0×10^{-3}	3.0×10^{-3}
Rate (mol L^{-1} h^{-1})	1.5×10^{-6}	1.5×10^{-6}	1.5×10^{-6}

Determine the rate law, the rate constant, and the overall order for this reaction.

23. Nitrosyl chloride, NOCl, decomposes to NO and Cl_2.

$$2\text{NOCl}(g) \longrightarrow 2\text{NO}(g) + \text{Cl}_2(g)$$

Determine the rate law, the rate constant, and the overall order for this reaction from the following data:

[NOCl] (*M*)	0.10	0.20	0.30
Rate (mol L^{-1} h^{-1})	8.0×10^{-10}	3.2×10^{-9}	7.2×10^{-9}

24. From the following data, determine the rate law, the rate constant, and the order with respect to *A* for the reaction $A \longrightarrow 2C$.

[*A*] (*M*)	1.33×10^{-2}	2.66×10^{-2}	3.99×10^{-2}
Rate (mol L^{-1} h^{-1})	3.80×10^{-7}	1.52×10^{-6}	3.42×10^{-6}

25. Nitrogen monoxide reacts with chlorine according to the equation:
$2NO(g) + Cl_2(g) \longrightarrow 2NOCl(g)$
The following initial rates of reaction have been observed for certain reactant concentrations:

[NO] (mol/L)	[Cl_2] (mol/L)	Rate (mol L^{-1} h^{-1})
0.50	0.50	1.14
1.00	0.50	4.56
1.00	1.00	9.12

What is the rate law that describes the rate's dependence on the concentrations of NO and Cl_2? What is the rate constant? What are the orders with respect to each reactant?

26. Hydrogen reacts with nitrogen monoxide to form dinitrogen monoxide (laughing gas) according to the equation: $H_2(g) + 2NO(g) \longrightarrow N_2O(g) + H_2O(g)$
Determine the rate law, the rate constant, and the orders with respect to each reactant from the following data:

[NO] (*M*)	0.30	0.60	0.60
[H_2] (*M*)	0.35	0.35	0.70
Rate (mol L^{-1} s^{-1})	2.835×10^{-3}	1.134×10^{-2}	2.268×10^{-2}

27. For the reaction $A \longrightarrow B + C$, the following data were obtained at 30 °C:

[*A*] (*M*)	0.230	0.356	0.557
Rate (mol L^{-1} s^{-1})	4.17×10^{-4}	9.99×10^{-4}	2.44×10^{-3}

(a) What is the order of the reaction with respect to [*A*], and what is the rate law?
(b) What is the rate constant?

28. For the reaction $Q \longrightarrow W + X$, the following data were obtained at 30 °C:

$[Q]_{initial}$ (*M*)	0.170	0.212	0.357
Rate (mol L^{-1} s^{-1})	6.68×10^{-3}	1.04×10^{-2}	2.94×10^{-2}

(a) What is the order of the reaction with respect to [*Q*], and what is the rate law?
(b) What is the rate constant?

29. The rate constant for the first-order decomposition at 45 °C of dinitrogen pentoxide, N_2O_5, dissolved in chloroform, $CHCl_3$, is 6.2×10^{-4} min^{-1}.
$2N_2O_5 \longrightarrow 4NO_2 + O_2$
What is the rate of the reaction when $[N_2O_5] = 0.40$ *M*?

30. The annual production of HNO_3 in 2013 was 60 million metric tons Most of that was prepared by the following sequence of reactions, each run in a separate reaction vessel.
(a) $4NH_3(g) + 5O_2(g) \longrightarrow 4NO(g) + 6H_2O(g)$
(b) $2NO(g) + O_2(g) \longrightarrow 2NO_2(g)$
(c) $3NO_2(g) + H_2O(l) \longrightarrow 2HNO_3(aq) + NO(g)$
The first reaction is run by burning ammonia in air over a platinum catalyst. This reaction is fast. The reaction in equation (c) is also fast. The second reaction limits the rate at which nitric acid can be prepared from ammonia. If equation (b) is second order in NO and first order in O_2, what is the rate of formation of NO_2 when the oxygen concentration is 0.50 *M* and the nitric oxide concentration is 0.75 *M*? The rate constant for the reaction is 5.8×10^{-6} L^2 mol^{-2} s^{-1}.

31. The following data have been determined for the reaction:
$I^- + OCl^- \longrightarrow IO^- + Cl^-$

	1	2	3
$[I^-]_{initial}$ (*M*)	0.10	0.20	0.30
$[OCl^-]_{initial}$ (*M*)	0.050	0.050	0.010
Rate (mol L^{-1} s^{-1})	3.05×10^{-4}	6.20×10^{-4}	1.83×10^{-4}

Determine the rate law and the rate constant for this reaction.

12.4 Integrated Rate Laws

32. Describe how graphical methods can be used to determine the order of a reaction and its rate constant from a series of data that includes the concentration of *A* at varying times.

33. Use the data provided to graphically determine the order and rate constant of the following reaction:
$SO_2Cl_2 \longrightarrow SO_2 + Cl_2$

Time (s)	0	5.00×10^3	1.00×10^4	1.50×10^4
$[SO_2Cl_2]$ (*M*)	0.100	0.0896	0.0802	0.0719
Time (s)	2.50×10^4	3.00×10^4	4.00×10^4	
$[SO_2Cl_2]$ (*M*)	0.0577	0.0517	0.0415	

34. Pure ozone decomposes slowly to oxygen, $2O_3(g) \longrightarrow 3O_2(g)$. Use the data provided in a graphical method and determine the order and rate constant of the reaction.

Time (h)	0	2.0×10^3	7.6×10^3	1.00×10^4
$[O_3]$ (*M*)	1.00×10^{-5}	4.98×10^{-6}	2.07×10^{-6}	1.66×10^{-6}
Time (h)	1.23×10^4	1.43×10^4	1.70×10^4	
$[O_3]$ (*M*)	1.39×10^{-6}	1.22×10^{-6}	1.05×10^{-6}	

35. From the given data, use a graphical method to determine the order and rate constant of the following reaction:
$2X \longrightarrow Y + Z$

Time (s)	5.0	10.0	15.0	20.0	25.0	30.0	35.0	40.0
[X] (M)	0.0990	0.0497	0.0332	0.0249	0.0200	0.0166	0.0143	0.0125

36. What is the half-life for the first-order decay of phosphorus-32? $\left({}^{32}_{15}\text{P} \longrightarrow {}^{32}_{16}\text{S} + \text{e}^-\right)$ The rate constant for the decay is 4.85×10^{-2} day^{-1}.
37. What is the half-life for the first-order decay of carbon-14? $\left({}^{14}_{6}\text{C} \longrightarrow {}^{14}_{7}\text{N} + \text{e}^-\right)$ The rate constant for the decay is 1.21×10^{-4} year^{-1}.
38. What is the half-life for the decomposition of NOCl when the concentration of NOCl is 0.15 *M*? The rate constant for this second-order reaction is 8.0×10^{-8} L mol^{-1} s^{-1}.
39. What is the half-life for the decomposition of O_3 when the concentration of O_3 is 2.35×10^{-6} *M*? The rate constant for this second-order reaction is 50.4 L mol^{-1} h^{-1}.
40. The reaction of compound *A* to give compounds *C* and *D* was found to be second-order in *A*. The rate constant for the reaction was determined to be 2.42 L mol^{-1} s^{-1}. If the initial concentration is 0.500 mol/L, what is the value of $t_{1/2}$?
41. The half-life of a reaction of compound *A* to give compounds *D* and *E* is 8.50 min when the initial concentration of *A* is 0.150 M. How long will it take for the concentration to drop to 0.0300 M if the reaction is (a) first order with respect to *A* or (b) second order with respect to *A*?
42. Some bacteria are resistant to the antibiotic penicillin because they produce penicillinase, an enzyme with a molecular weight of 3×10^4 g/mol that converts penicillin into inactive molecules. Although the kinetics of enzyme-catalyzed reactions can be complex, at low concentrations this reaction can be described by a rate law that is first order in the catalyst (penicillinase) and that also involves the concentration of penicillin. From the following data: 1.0 L of a solution containing 0.15 µg (0.15×10^{-6} g) of penicillinase, determine the order of the reaction with respect to penicillin and the value of the rate constant.

[Penicillin] (*M*)	Rate (mol L^{-1} min^{-1})
2.0×10^{-6}	1.0×10^{-10}
3.0×10^{-6}	1.5×10^{-10}
4.0×10^{-6}	2.0×10^{-10}

43. Both technetium-99 and thallium-201 are used to image heart muscle in patients with suspected heart problems. The half-lives are 6 h and 73 h, respectively. What percent of the radioactivity would remain for each of the isotopes after 2 days (48 h)?
44. There are two molecules with the formula C_3H_6. Propene, $CH_3CH = CH_2$, is the monomer of the polymer polypropylene, which is used for indoor-outdoor carpets. Cyclopropane is used as an anesthetic:

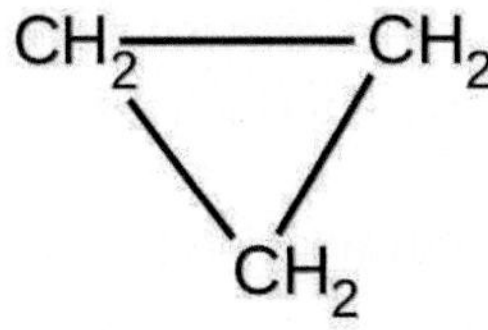

When heated to 499 °C, cyclopropane rearranges (isomerizes) and forms propene with a rate constant of 5.95×10^{-4} s^{-1}. What is the half-life of this reaction? What fraction of the cyclopropane remains after 0.75 h at 499 °C?

45. Fluorine-18 is a radioactive isotope that decays by positron emission to form oxygen-18 with a half-life of 109.7 min. (A positron is a particle with the mass of an electron and a single unit of positive charge; the equation is ${}^{18}_{9}F \longrightarrow {}^{18}_{8}O + {}^{0}_{+1}e$) Physicians use ^{18}F to study the brain by injecting a quantity of fluoro-substituted glucose into the blood of a patient. The glucose accumulates in the regions where the brain is active and needs nourishment.
(a) What is the rate constant for the decomposition of fluorine-18?
(b) If a sample of glucose containing radioactive fluorine-18 is injected into the blood, what percent of the radioactivity will remain after 5.59 h?
(c) How long does it take for 99.99% of the ^{18}F to decay?

46. Suppose that the half-life of steroids taken by an athlete is 42 days. Assuming that the steroids biodegrade by a first-order process, how long would it take for $\frac{1}{64}$ of the initial dose to remain in the athlete's body?

47. Recently, the skeleton of King Richard III was found under a parking lot in England. If tissue samples from the skeleton contain about 93.79% of the carbon-14 expected in living tissue, what year did King Richard III die? The half-life for carbon-14 is 5730 years.

48. Nitroglycerine is an extremely sensitive explosive. In a series of carefully controlled experiments, samples of the explosive were heated to 160 °C and their first-order decomposition studied. Determine the average rate constants for each experiment using the following data:

Initial $[C_3H_5N_3O_9]$ (*M*)	4.88	3.52	2.29	1.81	5.33	4.05	2.95	1.72
t (s)	300	300	300	300	180	180	180	180
% Decomposed	52.0	52.9	53.2	53.9	34.6	35.9	36.0	35.4

49. For the past 10 years, the unsaturated hydrocarbon 1,3-butadiene ($CH_2 = CH–CH = CH_2$) has ranked 38th among the top 50 industrial chemicals. It is used primarily for the manufacture of synthetic rubber. An isomer exists also as cyclobutene:

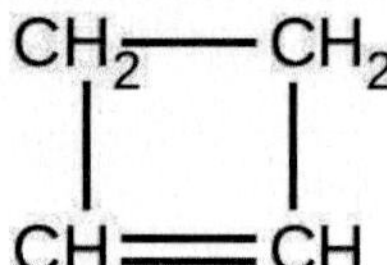

The isomerization of cyclobutene to butadiene is first-order and the rate constant has been measured as $2.0 \times 10^{-4}\ s^{-1}$ at 150 °C in a 0.53-L flask. Determine the partial pressure of cyclobutene and its concentration after 30.0 minutes if an isomerization reaction is carried out at 150 °C with an initial pressure of 55 torr.

12.5 Collision Theory

50. Chemical reactions occur when reactants collide. What are two factors that may prevent a collision from producing a chemical reaction?

51. When every collision between reactants leads to a reaction, what determines the rate at which the reaction occurs?

52. What is the activation energy of a reaction, and how is this energy related to the activated complex of the reaction?

53. Account for the relationship between the rate of a reaction and its activation energy.

54. Describe how graphical methods can be used to determine the activation energy of a reaction from a series of data that includes the rate of reaction at varying temperatures.

55. How does an increase in temperature affect rate of reaction? Explain this effect in terms of the collision theory of the reaction rate.

56. The rate of a certain reaction doubles for every 10 °C rise in temperature.
(a) How much faster does the reaction proceed at 45 °C than at 25 °C?
(b) How much faster does the reaction proceed at 95 °C than at 25 °C?

57. In an experiment, a sample of $NaClO_3$ was 90% decomposed in 48 min. Approximately how long would this decomposition have taken if the sample had been heated 20 °C higher? (Hint: Assume the rate doubles for each 10 °C rise in temperature.)

58. The rate constant at 325 °C for the decomposition reaction $C_4H_8 \longrightarrow 2C_2H_4$ is 6.1×10^{-8} s^{-1}, and the activation energy is 261 kJ per mole of C_4H_8. Determine the frequency factor for the reaction.

59. The rate constant for the decomposition of acetaldehyde, CH_3CHO, to methane, CH_4, and carbon monoxide, CO, in the gas phase is 1.1×10^{-2} L mol^{-1} s^{-1} at 703 K and 4.95 L mol^{-1} s^{-1} at 865 K. Determine the activation energy for this decomposition.

60. An elevated level of the enzyme alkaline phosphatase (ALP) in human serum is an indication of possible liver or bone disorder. The level of serum ALP is so low that it is very difficult to measure directly. However, ALP catalyzes a number of reactions, and its relative concentration can be determined by measuring the rate of one of these reactions under controlled conditions. One such reaction is the conversion of p-nitrophenyl phosphate (PNPP) to p-nitrophenoxide ion (PNP) and phosphate ion. Control of temperature during the test is very important; the rate of the reaction increases 1.47 times if the temperature changes from 30 °C to 37 °C. What is the activation energy for the ALP–catalyzed conversion of PNPP to PNP and phosphate?

61. In terms of collision theory, to which of the following is the rate of a chemical reaction proportional?
(a) the change in free energy per second
(b) the change in temperature per second
(c) the number of collisions per second
(d) the number of product molecules

62. Hydrogen iodide, HI, decomposes in the gas phase to produce hydrogen, H_2, and iodine, I_2. The value of the rate constant, k, for the reaction was measured at several different temperatures and the data are shown here:

Temperature (K)	k (L mol^{-1} s^{-1})
555	6.23×10^{-7}
575	2.42×10^{-6}
645	1.44×10^{-4}
700	2.01×10^{-3}

What is the value of the activation energy (in kJ/mol) for this reaction?

63. The element Co exists in two oxidation states, Co(II) and Co(III), and the ions form many complexes. The rate at which one of the complexes of Co(III) was reduced by Fe(II) in water was measured. Determine the activation energy of the reaction from the following data:

T (K)	k (s^{-1})
293	0.054
298	0.100

64. The hydrolysis of the sugar sucrose to the sugars glucose and fructose,
$$C_{12}H_{22}O_{11} + H_2O \longrightarrow C_6H_{12}O_6 + C_6H_{12}O_6$$
follows a first-order rate law for the disappearance of sucrose: rate = $k[C_{12}H_{22}O_{11}]$ (The products of the reaction, glucose and fructose, have the same molecular formulas but differ in the arrangement of the atoms in their molecules.)
(a) In neutral solution, $k = 2.1 \times 10^{-11}$ s^{-1} at 27 °C and 8.5×10^{-11} s^{-1} at 37 °C. Determine the activation energy, the frequency factor, and the rate constant for this equation at 47 °C (assuming the kinetics remain consistent with the Arrhenius equation at this temperature).
(b) When a solution of sucrose with an initial concentration of 0.150 *M* reaches equilibrium, the concentration of sucrose is 1.65×10^{-7} *M*. How long will it take the solution to reach equilibrium at 27 °C in the absence of a catalyst? Because the concentration of sucrose at equilibrium is so low, assume that the reaction is irreversible.
(c) Why does assuming that the reaction is irreversible simplify the calculation in part (b)?

65. Use the PhET Reactions & Rates interactive simulation (http://openstax.org/l/16PHETreaction) to simulate a system. On the "Single collision" tab of the simulation applet, enable the "Energy view" by clicking the "+" icon. Select the first $A + BC \longrightarrow AB + C$ reaction (A is yellow, B is purple, and C is navy blue). Using the "straight shot" default option, try launching the *A* atom with varying amounts of energy. What changes when the Total Energy line at launch is below the transition state of the Potential Energy line? Why? What happens when it is above the transition state? Why?

66. Use the PhET Reactions & Rates interactive simulation (http://openstax.org/l/16PHETreaction) to simulate a system. On the "Single collision" tab of the simulation applet, enable the "Energy view" by clicking the "+" icon. Select the first $A + BC \longrightarrow AB + C$ reaction (A is yellow, B is purple, and C is navy blue). Using the "angled shot" option, try launching the *A* atom with varying angles, but with more Total energy than the transition state. What happens when the *A* atom hits the *BC* molecule from different directions? Why?

12.6 Reaction Mechanisms

67. Why are elementary reactions involving three or more reactants very uncommon?

68. In general, can we predict the effect of doubling the concentration of *A* on the rate of the overall reaction $A + B \longrightarrow C$? Can we predict the effect if the reaction is known to be an elementary reaction?

69. Define these terms:
(a) unimolecular reaction
(b) bimolecular reaction
(c) elementary reaction
(d) overall reaction

70. What is the rate law for the elementary termolecular reaction $A + 2B \longrightarrow$ products? For $3A \longrightarrow$ products?

71. Given the following reactions and the corresponding rate laws, in which of the reactions might the elementary reaction and the overall reaction be the same?
(a) $Cl_2 + CO \longrightarrow Cl_2CO$
rate = $k[Cl_2]^{3/2}[CO]$
(b) $PCl_3 + Cl_2 \longrightarrow PCl_5$
rate = $k[PCl_3][Cl_2]$
(c) $2NO + H_2 \longrightarrow N_2 + H_2O_2$
rate = $k[NO][H_2]$
(d) $2NO + O_2 \longrightarrow 2NO_2$
rate = $k[NO]^2[O_2]$
(e) $NO + O_3 \longrightarrow NO_2 + O_2$
rate = $k[NO][O_3]$

72. Write the rate law for each of the following elementary reactions:
(a) $O_3 \xrightarrow{\text{sunlight}} O_2 + O$
(b) $O_3 + Cl \longrightarrow O_2 + ClO$
(c) $ClO + O \longrightarrow Cl + O_2$
(d) $O_3 + NO \longrightarrow NO_2 + O_2$
(e) $NO_2 + O \longrightarrow NO + O_2$

73. Nitrogen monoxide, NO, reacts with hydrogen, H_2, according to the following equation:
$2NO + 2H_2 \longrightarrow N_2 + 2H_2O$
What would the rate law be if the mechanism for this reaction were:
$2NO + H_2 \longrightarrow N_2 + H_2O_2$ (slow)
$H_2O_2 + H_2 \longrightarrow 2H_2O$ (fast)

74. Experiments were conducted to study the rate of the reaction represented by this equation.[2]
$2NO(g) + 2H_2(g) \longrightarrow N_2(g) + 2H_2O(g)$
Initial concentrations and rates of reaction are given here.

Experiment	Initial Concentration [NO] (mol L^{-1})	Initial Concentration, [H_2] (mol L^{-1} min^{-1})	Initial Rate of Formation of N_2 (mol L^{-1} min^{-1})
1	0.0060	0.0010	1.8×10^{-4}
2	0.0060	0.0020	3.6×10^{-4}
3	0.0010	0.0060	0.30×10^{-4}
4	0.0020	0.0060	1.2×10^{-4}

Consider the following questions:
(a) Determine the order for each of the reactants, NO and H_2, from the data given and show your reasoning.
(b) Write the overall rate law for the reaction.
(c) Calculate the value of the rate constant, *k*, for the reaction. Include units.
(d) For experiment 2, calculate the concentration of NO remaining when exactly one-half of the original amount of H_2 had been consumed.
(e) The following sequence of elementary steps is a proposed mechanism for the reaction.
Step 1: $NO + NO \rightleftharpoons N_2O_2$
Step 2: $N_2O_2 + H_2 \rightleftharpoons H_2O + N_2O$
Step 3: $N_2O + H_2 \rightleftharpoons N_2 + H_2O$
Based on the data presented, which of these is the rate determining step? Show that the mechanism is consistent with the observed rate law for the reaction and the overall stoichiometry of the reaction.

75. The reaction of CO with Cl_2 gives phosgene ($COCl_2$), a nerve gas that was used in World War I. Use the mechanism shown here to complete the following exercises:
$Cl_2(g) \rightleftharpoons 2Cl(g)$ (fast, k_1 represents the forward rate constant, k_{-1} the reverse rate constant)
$CO(g) + Cl(g) \longrightarrow COCl(g)$ (slow, k_2 the rate constant)
$COCl(g) + Cl(g) \longrightarrow COCl_2(g)$ (fast, k_3 the rate constant)
(a) Write the overall reaction.
(b) Identify all intermediates.
(c) Write the rate law for each elementary reaction.
(d) Write the overall rate law expression.

12.7 Catalysis

76. Account for the increase in reaction rate brought about by a catalyst.

2 This question is taken from the Chemistry Advanced Placement Examination and is used with the permission of the Educational Testing Service.

77. Compare the functions of homogeneous and heterogeneous catalysts.

78. Consider this scenario and answer the following questions: Chlorine atoms resulting from decomposition of chlorofluoromethanes, such as CCl_2F_2, catalyze the decomposition of ozone in the atmosphere. One simplified mechanism for the decomposition is:

$$O_3 \xrightarrow{\text{sunlight}} O_2 + O$$
$$O_3 + Cl \longrightarrow O_2 + ClO$$
$$ClO + O \longrightarrow Cl + O_2$$

(a) Explain why chlorine atoms are catalysts in the gas-phase transformation:

$$2O_3 \longrightarrow 3O_2$$

(b) Nitric oxide is also involved in the decomposition of ozone by the mechanism:

$$O_3 \xrightarrow{\text{sunlight}} O_2 + O$$
$$O_3 + NO \longrightarrow NO_2 + O_2$$
$$NO_2 + O \longrightarrow NO + O_2$$

Is NO a catalyst for the decomposition? Explain your answer.

79. For each of the following pairs of reaction diagrams, identify which of the pair is catalyzed:

(a)

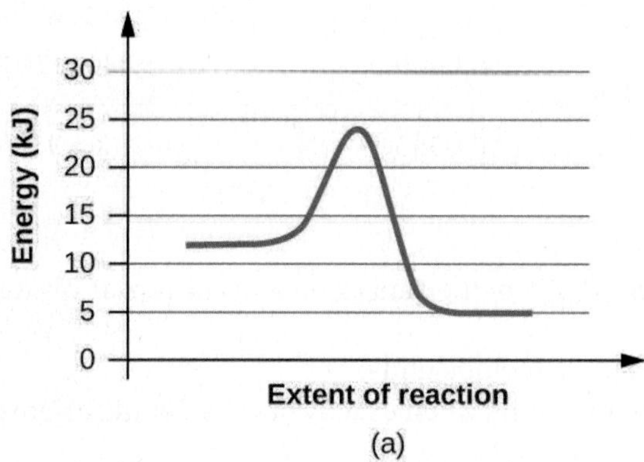

(a)

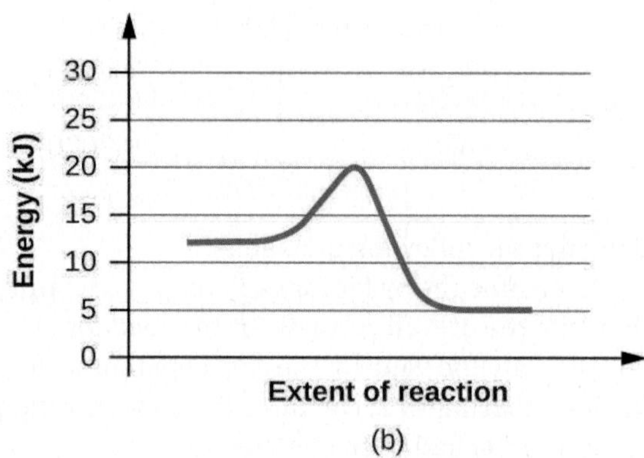

(b)

(b)

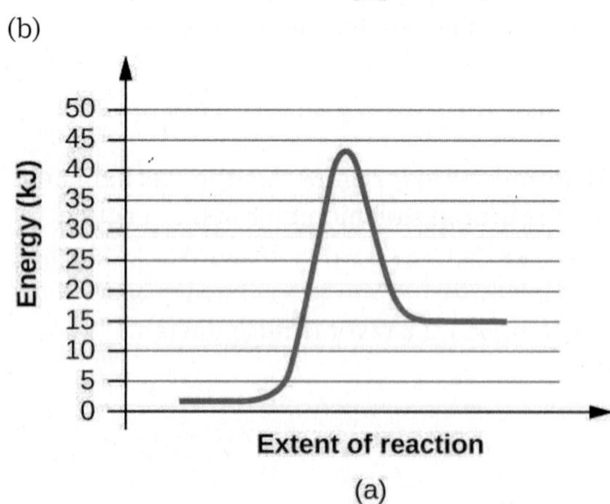

(a)

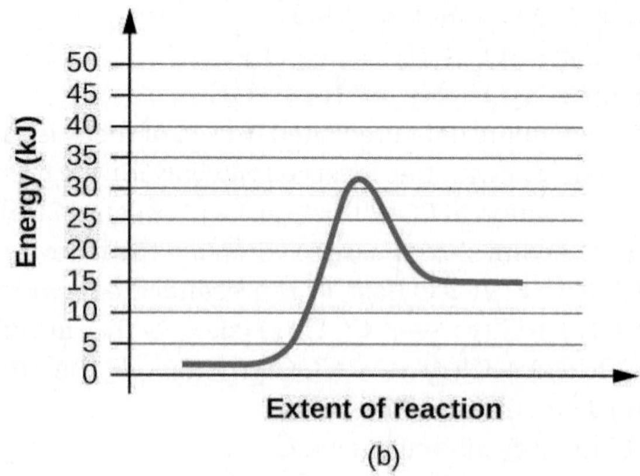

(b)

80. For each of the following pairs of reaction diagrams, identify which of the pairs is catalyzed:

(a)

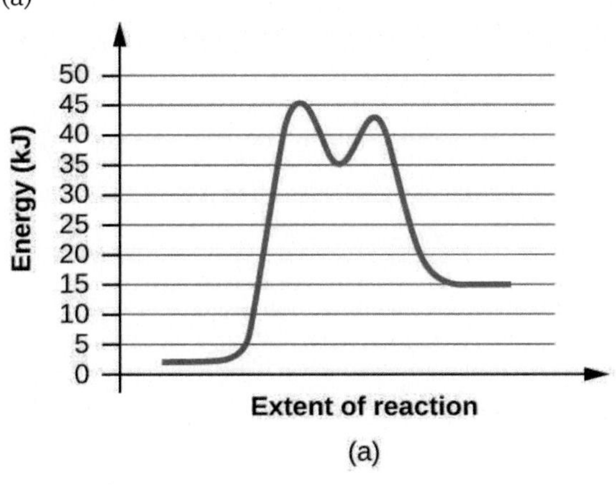

(a)

Energy (kJ)

Extent of reaction

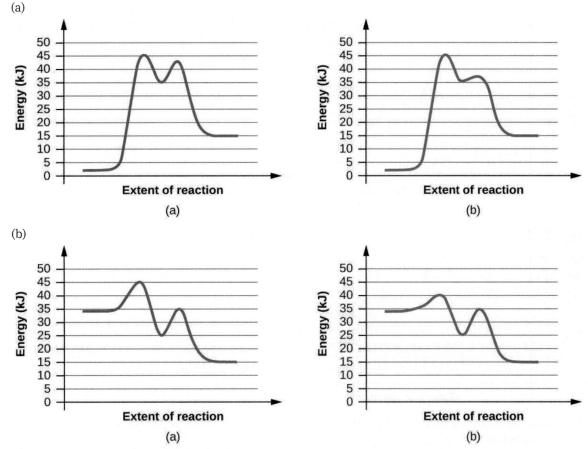

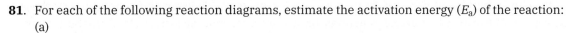

(b)

(b)

Energy (kJ)

Extent of reaction

(a)

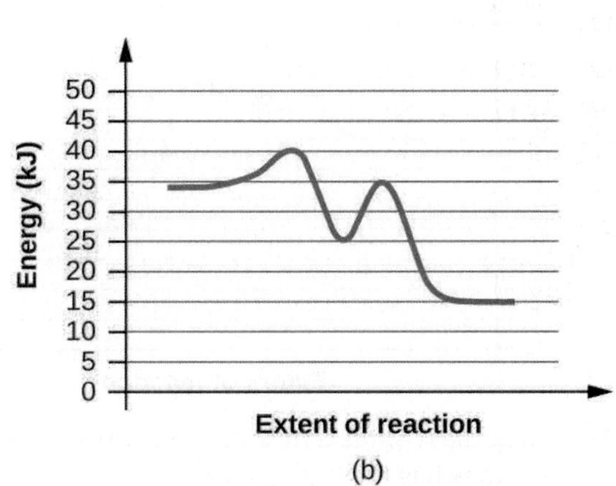

(b)

81. For each of the following reaction diagrams, estimate the activation energy (E_a) of the reaction:

(a)

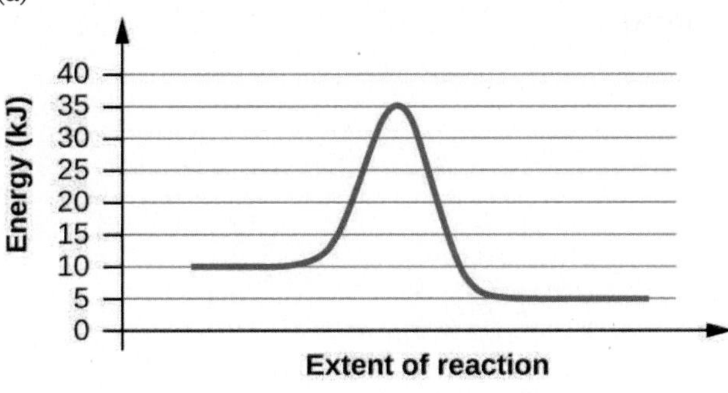

(b)

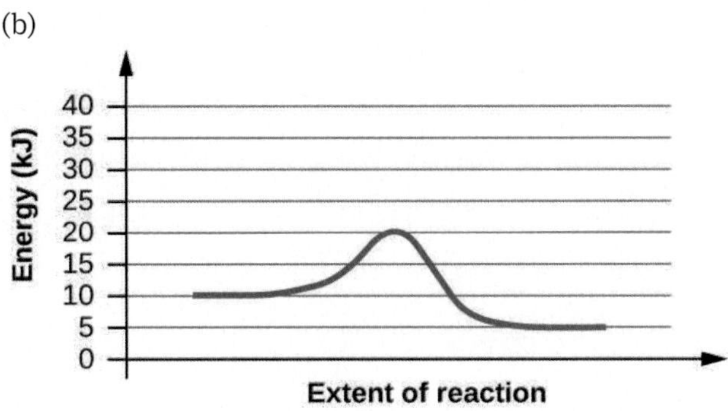

82. For each of the following reaction diagrams, estimate the activation energy (E_a) of the reaction:

(a)

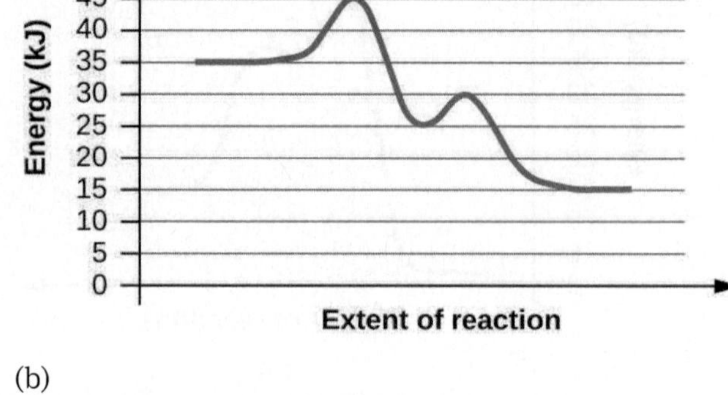

(b)

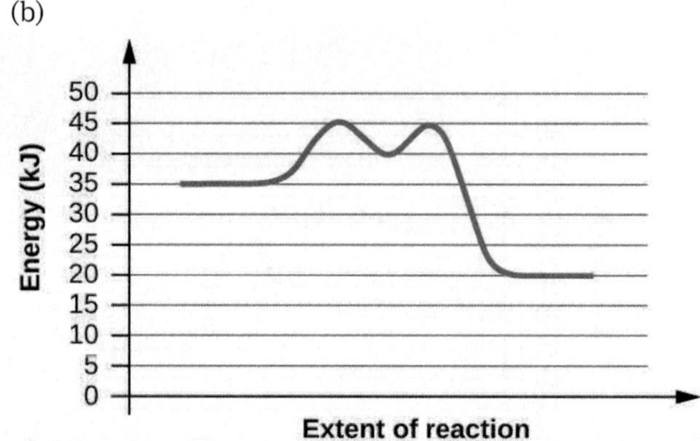

83. Assuming the diagrams in Exercise 12.81 represent different mechanisms for the same reaction, which of the reactions has the faster rate?

84. Consider the similarities and differences in the two reaction diagrams shown in Exercise 12.82. Do these diagrams represent two different overall reactions, or do they represent the same overall reaction taking place by two different mechanisms? Explain your answer.

CHAPTER 13
Fundamental Equilibrium Concepts

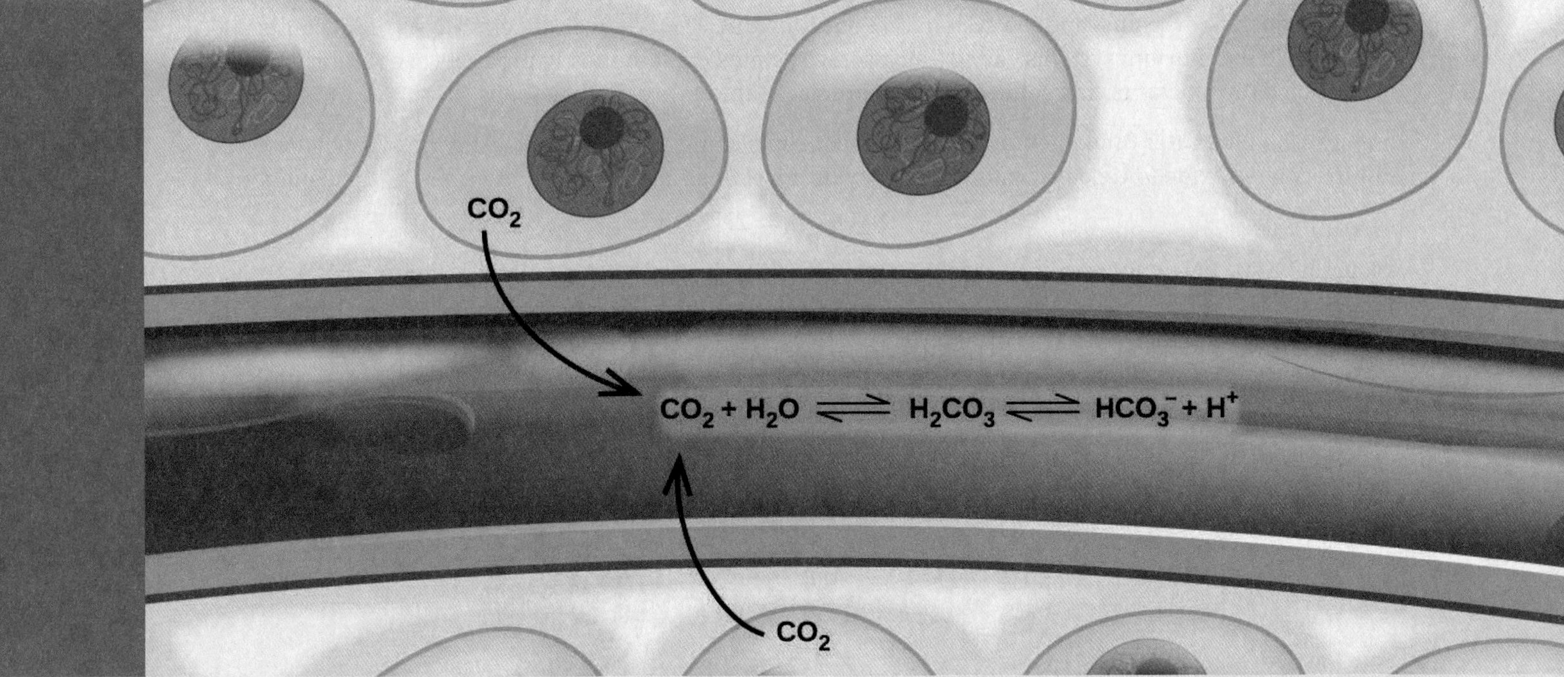

Figure 13.1 Transport of carbon dioxide in the body involves several reversible chemical reactions, including hydrolysis and acid ionization (among others).

CHAPTER OUTLINE

13.1 Chemical Equilibria
13.2 Equilibrium Constants
13.3 Shifting Equilibria: Le Châtelier's Principle
13.4 Equilibrium Calculations

INTRODUCTION Imagine a beach populated with sunbathers and swimmers. As those basking in the sun get too hot, they enter the surf to swim and cool off. As the swimmers tire, they return to the beach to rest. If the rate at which sunbathers enter the surf were to equal the rate at which swimmers return to the sand, then the numbers (though not the identities) of sunbathers and swimmers would remain constant. This scenario illustrates a dynamic phenomenon known as *equilibrium*, in which opposing processes occur at equal rates. Chemical and physical processes are subject to this phenomenon; these processes are at equilibrium when the forward and reverse reaction rates are equal. Equilibrium systems are pervasive in nature; the various reactions involving carbon dioxide dissolved in blood are examples (see Figure 13.1). This chapter provides a thorough introduction to the essential aspects of chemical equilibria.

13.1 Chemical Equilibria

LEARNING OBJECTIVES

By the end of this section, you will be able to:

- Describe the nature of equilibrium systems
- Explain the dynamic nature of a chemical equilibrium

The convention for writing chemical equations involves placing reactant formulas on the left side of a reaction arrow and product formulas on the right side. By this convention, and the definitions of "reactant" and "product," a chemical equation represents the reaction in question as proceeding from left to right. **Reversible reactions**, however, may proceed in both forward (left to right) and reverse (right to left) directions. When the rates of the forward and reverse reactions are equal, the concentrations of the reactant and product species remain constant over time and the system is at **equilibrium**. The relative concentrations of reactants and products in equilibrium systems vary greatly; some systems contain mostly products at equilibrium, some contain mostly reactants, and some contain appreciable amounts of both.

Figure 13.2 illustrates fundamental equilibrium concepts using the reversible decomposition of colorless dinitrogen tetroxide to yield brown nitrogen dioxide, an elementary reaction described by the equation:

$$N_2O_4(g) \rightleftharpoons 2NO_2(g)$$

Note that a special double arrow is used to emphasize the reversible nature of the reaction.

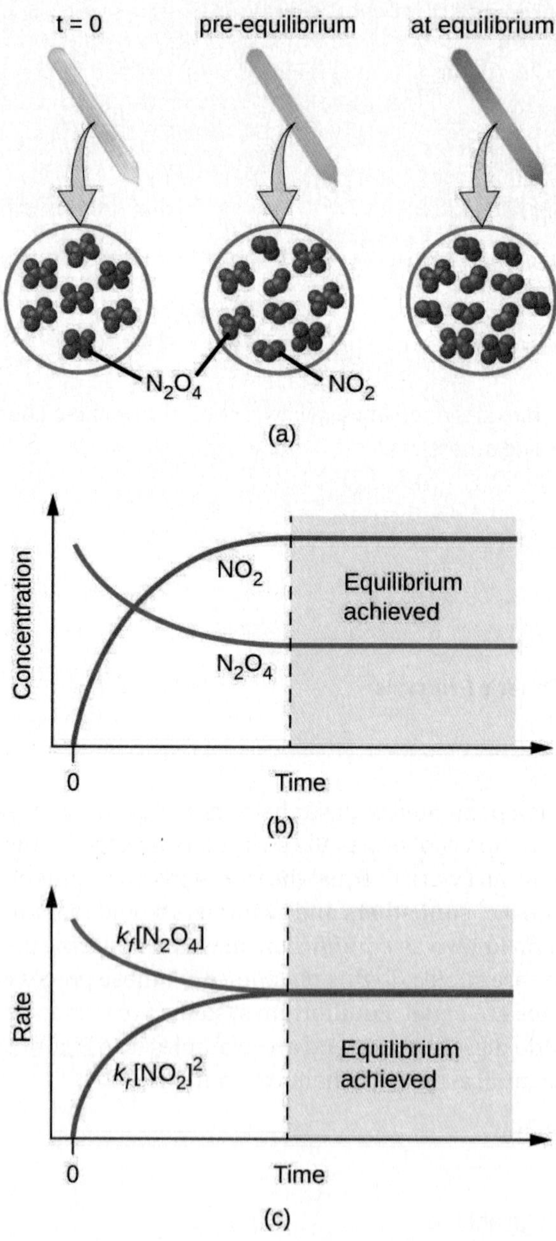

FIGURE 13.2 (a) A sealed tube containing colorless N_2O_4 darkens as it decomposes to yield brown NO_2. (b) Changes in concentration over time as the decomposition reaction achieves equilibrium. (c) At equilibrium, the forward and reverse reaction rates are equal.

For this elementary process, rate laws for the forward and reverse reactions may be derived directly from the reaction stoichiometry:

$$\text{rate}_f = k_f[N_2O_4]$$

$$\text{rate}_r = k_r[NO_2]^2$$

As the reaction begins ($t = 0$), the concentration of the N_2O_4 reactant is finite and that of the NO_2 product is zero, so the forward reaction proceeds at a finite rate while the reverse reaction rate is zero. As time passes, N_2O_4 is consumed and its concentration falls, while NO_2 is produced and its concentration increases (Figure 13.2**b**). The decreasing concentration of the reactant slows the forward reaction rate, and the increasing product concentration speeds the reverse reaction rate (Figure 13.2**c**). This process continues until *the forward and reverse reaction rates become equal*, at which time the reaction has reached equilibrium, as characterized by constant concentrations of its reactants and products (shaded areas of Figure 13.2**b** and Figure 13.2**c**). It's important to emphasize that chemical equilibria are dynamic; a reaction at equilibrium has not "stopped," but is proceeding in the forward and reverse directions at the same rate. This dynamic nature is essential to understanding equilibrium behavior as discussed in this and subsequent chapters of the text.

FIGURE 13.3 A two-person juggling act illustrates the dynamic aspect of chemical equilibria. Each person is throwing and catching clubs at the same rate, and each holds a (approximately) constant number of clubs.

Physical changes, such as phase transitions, are also reversible and may establish equilibria. This concept was introduced in another chapter of this text through discussion of the vapor pressure of a condensed phase (liquid or solid). As one example, consider the vaporization of bromine:

$$Br_2(l) \rightleftharpoons Br_2(g)$$

When liquid bromine is added to an otherwise empty container and the container is sealed, the forward process depicted above (vaporization) will commence and continue at a roughly constant rate as long as the exposed surface area of the liquid and its temperature remain constant. As increasing amounts of gaseous bromine are produced, the rate of the reverse process (condensation) will increase until it equals the rate of vaporization and equilibrium is established. A photograph showing this phase transition equilibrium is provided in Figure 13.4.

FIGURE 13.4 A sealed tube containing an equilibrium mixture of liquid and gaseous bromine. (credit: http://images-of-elements.com/bromine.php)

13.2 Equilibrium Constants

LEARNING OBJECTIVES

By the end of this section, you will be able to:

- Derive reaction quotients from chemical equations representing homogeneous and heterogeneous reactions
- Calculate values of reaction quotients and equilibrium constants, using concentrations and pressures
- Relate the magnitude of an equilibrium constant to properties of the chemical system

The status of a reversible reaction is conveniently assessed by evaluating its **reaction quotient (*Q*)**. For a reversible reaction described by

$$m\text{A} + n\text{B} + \rightleftharpoons x\text{C} + y\text{D}$$

the reaction quotient is derived directly from the stoichiometry of the balanced equation as

$$Q_c = \frac{[\text{C}]^x [\text{D}]^y}{[\text{A}]^m [\text{B}]^n}$$

where the subscript c denotes the use of molar concentrations in the expression. If the reactants and products are gaseous, a reaction quotient may be similarly derived using partial pressures:

$$Q_p = \frac{P_\text{C}{}^x \; P_\text{D}{}^y}{P_\text{A}{}^m \; P_\text{B}{}^n}$$

Note that the reaction quotient equations above are a simplification of more rigorous expressions that use *relative* values for concentrations and pressures rather than *absolute* values. These relative concentration and pressure values are dimensionless (they have no units); consequently, so are the reaction quotients. For purposes of this introductory text, it will suffice to use the simplified equations and to disregard units when computing *Q*. In most cases, this will introduce only modest errors in calculations involving reaction quotients.

EXAMPLE 13.1

Writing Reaction Quotient Expressions

Write the concentration-based reaction quotient expression for each of the following reactions:

(a) $3O_2(g) \rightleftharpoons 2O_3(g)$

(b) $N_2(g) + 3H_2(g) \rightleftharpoons 2NH_3(g)$

(c) $4NH_3(g) + 7O_2(g) \rightleftharpoons 4NO_2(g) + 6H_2O(g)$

Solution

(a) $Q_c = \frac{[O_3]^2}{[O_2]^3}$

(b) $Q_c = \frac{[NH_3]^2}{[N_2][H_2]^3}$

(c) $Q_c = \frac{[NO_2]^4[H_2O]^6}{[NH_3]^4[O_2]^7}$

Check Your Learning

Write the concentration-based reaction quotient expression for each of the following reactions:

(a) $2SO_2(g) + O_2(g) \rightleftharpoons 2SO_3(g)$

(b) $C_4H_8(g) \rightleftharpoons 2C_2H_4(g)$

(c) $2C_4H_{10}(g) + 13O_2(g) \rightleftharpoons 8CO_2(g) + 10H_2O(g)$

Answer:

(a) $Q_c = \frac{[SO_3]^2}{[SO_2]^2[O_2]}$; (b) $Q_c = \frac{[C_2H_4]^2}{[C_4H_8]}$; (c) $Q_c = \frac{[CO_2]^8[H_2O]^{10}}{[C_4H_{10}]^2[O_2]^{13}}$

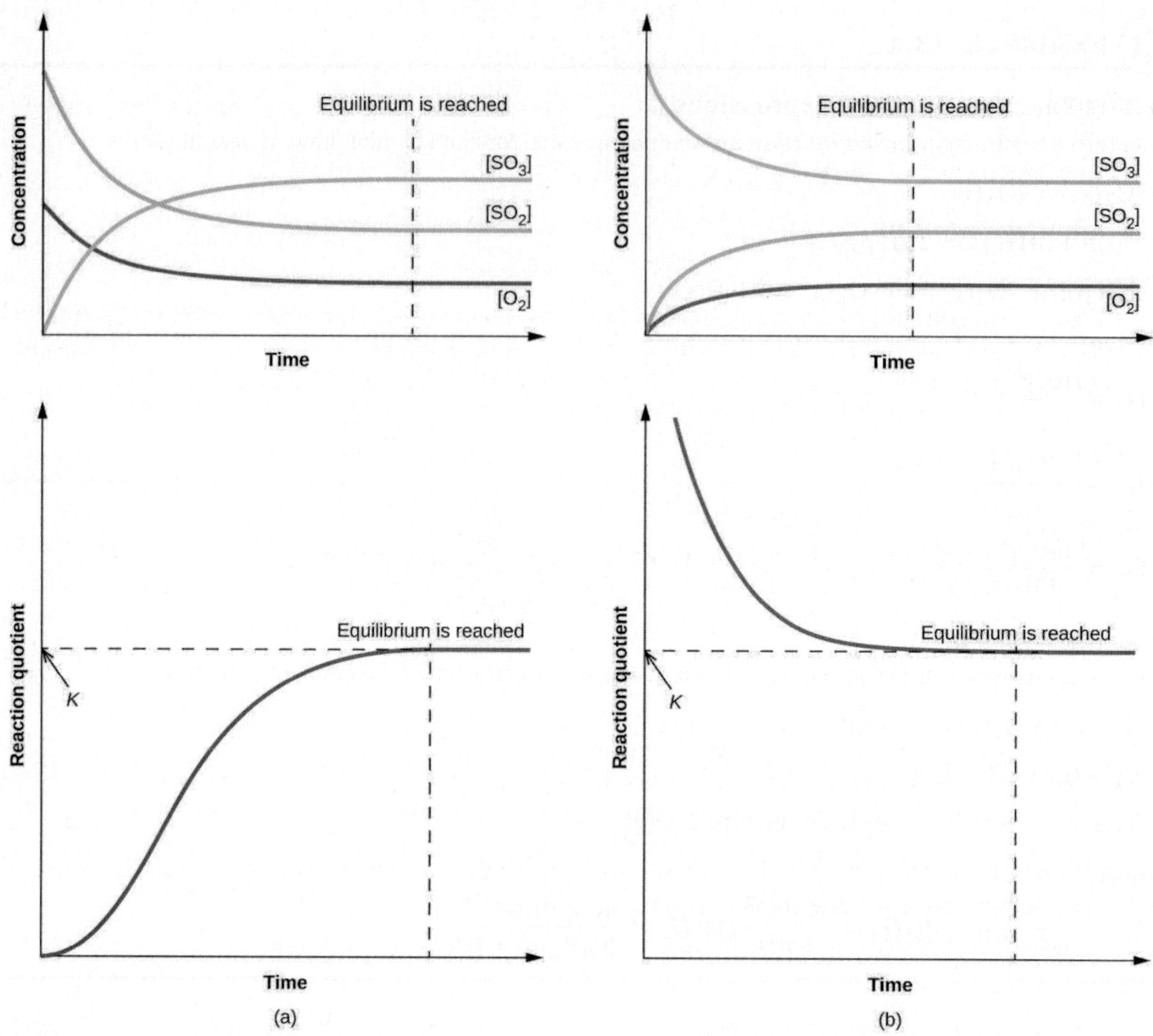

FIGURE 13.5 Changes in concentrations and Q_c for a chemical equilibrium achieved beginning with (a) a mixture of reactants only and (b) products only.

The numerical value of Q varies as a reaction proceeds towards equilibrium; therefore, it can serve as a useful indicator of the reaction's status. To illustrate this point, consider the oxidation of sulfur dioxide:

$$2SO_2(g) + O_2(g) \rightleftharpoons 2SO_3(g)$$

Two different experimental scenarios are depicted in Figure 13.5, one in which this reaction is initiated with a mixture of reactants only, SO_2 and O_2, and another that begins with only product, SO_3. For the reaction that begins with a mixture of reactants only, Q is initially equal to zero:

$$Q_c = \frac{[SO_3]^2}{[SO_2]^2\,[O_2]} = \frac{0^2}{[SO_2]^2\,[O_2]} = 0$$

As the reaction proceeds toward equilibrium in the forward direction, reactant concentrations decrease (as does the denominator of Q_c), product concentration increases (as does the numerator of Q_c), and the reaction quotient consequently increases. When equilibrium is achieved, the concentrations of reactants and product remain constant, as does the value of Q_c.

If the reaction begins with only product present, the value of Q_c is initially undefined (immeasurably large, or infinite):

$$Q_c = \frac{[SO_3]^2}{[SO_2]^2\,[O_2]} = \frac{[SO_3]^2}{0} \to \infty$$

In this case, the reaction proceeds toward equilibrium in the reverse direction. The product concentration and the numerator of Q_c decrease with time, the reactant concentrations and the denominator of Q_c increase, and the reaction quotient consequently decreases until it becomes constant at equilibrium.

The constant value of Q exhibited by a system at equilibrium is called the **equilibrium constant, *K***:

$$K \equiv Q \text{ at equilibrium}$$

Comparison of the data plots in Figure 13.5 shows that both experimental scenarios resulted in the same value for the equilibrium constant. This is a general observation for all equilibrium systems, known as the **law of mass action**: At a given temperature, the reaction quotient for a system at equilibrium is constant.

EXAMPLE 13.2

Evaluating a Reaction Quotient

Gaseous nitrogen dioxide forms dinitrogen tetroxide according to this equation:

$$2NO_2(g) \rightleftharpoons N_2O_4(g)$$

When 0.10 mol NO_2 is added to a 1.0-L flask at 25 °C, the concentration changes so that at equilibrium, $[NO_2]$ = 0.016 *M* and $[N_2O_4]$ = 0.042 *M*.

(a) What is the value of the reaction quotient before any reaction occurs?

(b) What is the value of the equilibrium constant for the reaction?

Solution

As for all equilibrium calculations in this text, use the simplified equations for Q and K and disregard any concentration or pressure units, as noted previously in this section.

(a) Before any product is formed, $[NO_2] = \frac{0.10\text{ mol}}{1.0\text{ L}} = 0.10\ M$, and $[N_2O_4]$ = 0 *M*. Thus,

$$Q_c = \frac{[N_2O_4]}{[NO_2]^2} = \frac{0}{0.10^2} = 0$$

(b) At equilibrium, $K_c = Q_c = \frac{[N_2O_4]}{[NO_2]^2} = \frac{0.042}{0.016^2} = 1.6 \times 10^2$. The equilibrium constant is 1.6×10^2.

Check Your Learning

For the reaction $2SO_2(g) + O_2(g) \rightleftharpoons 2SO_3(g)$, the concentrations at equilibrium are $[SO_2]$ = 0.90 *M*, $[O_2]$ = 0.35 *M*, and $[SO_3]$ = 1.1 *M*. What is the value of the equilibrium constant, K_c?

Answer:

$K_c = 4.3$

By its definition, the magnitude of an equilibrium constant explicitly reflects the composition of a reaction mixture at equilibrium, and it may be interpreted with regard to the extent of the forward reaction. A reaction exhibiting a large K will reach equilibrium when most of the reactant has been converted to product, whereas a small K indicates the reaction achieves equilibrium after very little reactant has been converted. It's important to keep in mind that the magnitude of K does *not* indicate how rapidly or slowly equilibrium will be reached. Some equilibria are established so quickly as to be nearly instantaneous, and others so slowly that no perceptible change is observed over the course of days, years, or longer.

The equilibrium constant for a reaction can be used to predict the behavior of mixtures containing its reactants and/or products. As demonstrated by the sulfur dioxide oxidation process described above, a chemical reaction will proceed in whatever direction is necessary to achieve equilibrium. Comparing Q to K for an equilibrium system of interest allows prediction of what reaction (forward or reverse), if any, will occur.

To further illustrate this important point, consider the reversible reaction shown below:

$$CO(g) + H_2O(g) \rightleftharpoons CO_2(g) + H_2(g) \qquad K_c = 0.640 \qquad T = 800\ °C$$

The bar charts in Figure 13.6 represent changes in reactant and product concentrations for three different reaction mixtures. The reaction quotients for mixtures 1 and 3 are initially lesser than the reaction's equilibrium constant, so each of these mixtures will experience a net forward reaction to achieve equilibrium. The reaction quotient for mixture 2 is initially greater than the equilibrium constant, so this mixture will proceed in the reverse direction until equilibrium is established.

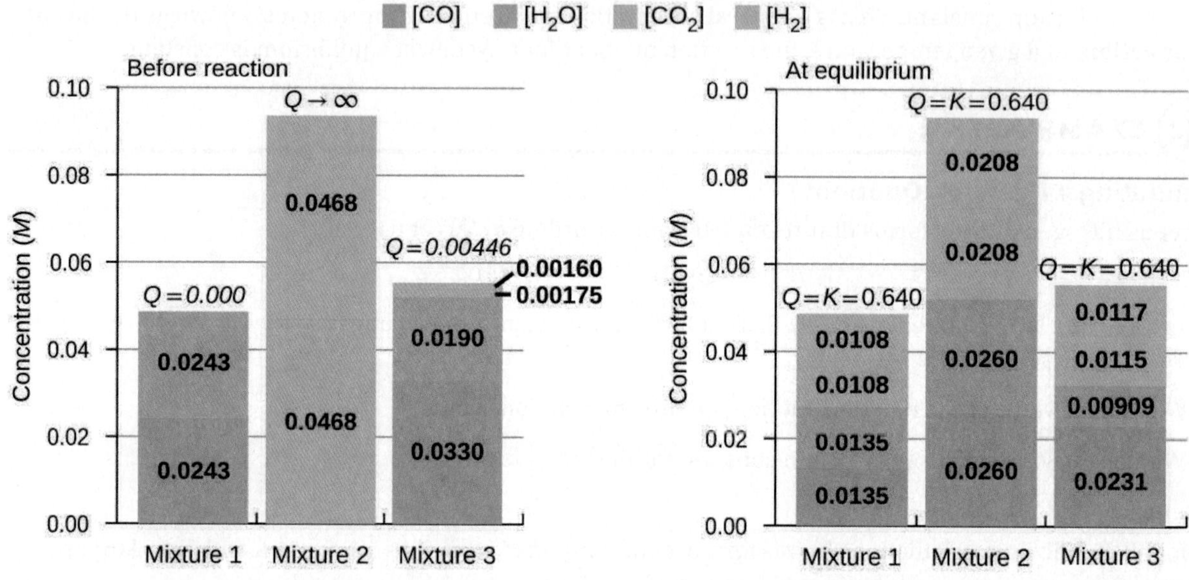

FIGURE 13.6 Compositions of three mixtures before ($Q_c \neq K_c$) and after ($Q_c = K_c$) equilibrium is established for the reaction $CO(g) + H_2O(g) \rightleftharpoons CO_2(g) + H_2(g)$.

EXAMPLE 13.3

Predicting the Direction of Reaction

Given here are the starting concentrations of reactants and products for three experiments involving this reaction:

$$CO(g) + H_2O(g) \rightleftharpoons CO_2(g) + H_2(g)$$
$$K_c = 0.64$$

Determine in which direction the reaction proceeds as it goes to equilibrium in each of the three experiments shown.

Reactants/Products	Experiment 1	Experiment 2	Experiment 3
$[CO]_i$	0.020 *M*	0.011 *M*	0.0094 *M*
$[H_2O]_i$	0.020 *M*	0.0011 *M*	0.0025 *M*
$[CO_2]_i$	0.0040 *M*	0.037 *M*	0.0015 *M*
$[H_2]_i$	0.0040 *M*	0.046 *M*	0.0076 *M*

Solution

Experiment 1:

$$Q_c = \frac{[CO_2]\,[H_2]}{[CO]\,[H_2O]} = \frac{(0.0040)\,(0.0040)}{(0.020)\,(0.020)} = 0.040.$$

$Q_c < K_c$ (0.040 < 0.64)

The reaction will proceed in the forward direction.

Experiment 2:

$$Q_c = \frac{[CO_2]\,[H_2]}{[CO]\,[H_2O]} = \frac{(0.037)\,(0.046)}{(0.011)\,(0.0011)} = 1.4 \times 10^2$$

$Q_c > K_c$ (140 > 0.64)

The reaction will proceed in the reverse direction.

Experiment 3:

$$Q_c = \frac{[CO_2]\,[H_2]}{[CO]\,[H_2O]} = \frac{(0.0015)\,(0.0076)}{(0.0094)\,(0.0025)} = 0.48$$

$Q_c < K_c$ (0.48 < 0.64)

The reaction will proceed in the forward direction.

Check Your Learning

Calculate the reaction quotient and determine the direction in which each of the following reactions will proceed to reach equilibrium.

(a) A 1.00-L flask containing 0.0500 mol of NO(g), 0.0155 mol of $Cl_2(g)$, and 0.500 mol of NOCl:

$$2NO(g) + Cl_2(g) \rightleftharpoons 2NOCl(g) \qquad K_c = 4.6 \times 10^4$$

(b) A 5.0-L flask containing 17 g of NH_3, 14 g of N_2, and 12 g of H_2:

$$N_2(g) + 3H_2(g) \rightleftharpoons 2NH_3(g) \qquad K_c = 0.060$$

(c) A 2.00-L flask containing 230 g of $SO_3(g)$:

$$2SO_3(g) \rightleftharpoons 2SO_2(g) + O_2(g) \qquad K_c = 0.230$$

Answer:

(a) $Q_c = 6.45 \times 10^3$, forward. (b) $Q_c = 0.23$, reverse. (c) $Q_c = 0$, forward.

Homogeneous Equilibria

A **homogeneous equilibrium** is one in which all reactants and products (and any catalysts, if applicable) are present in the same phase. By this definition, homogeneous equilibria take place in *solutions*. These solutions are most commonly either liquid or gaseous phases, as shown by the examples below:

$$C_2H_2(aq) + 2Br_2(aq) \rightleftharpoons C_2H_2Br_4(aq) \qquad K_c = \frac{[C_2H_2Br_4]}{[C_2H_2]\,[Br_2]^2}$$

$$I_2(aq) + I^-(aq) \rightleftharpoons I_3{}^-(aq) \qquad K_c = \frac{[I_3{}^-]}{[I_2][I^-]}$$

$$HF(aq) + H_2O(l) \rightleftharpoons H_3O^+(aq) + F^-(aq) \qquad K_c = \frac{[H_3O^+][F^-]}{[HF]}$$

$$NH_3(aq) + H_2O(l) \rightleftharpoons NH_4{}^+(aq) + OH^-(aq) \qquad K_c = \frac{[NH_4{}^+][OH^-]}{[NH_3]}$$

These examples all involve aqueous solutions, those in which water functions as the solvent. In the last two examples, water also functions as a reactant, but its concentration is *not* included in the reaction quotient. The reason for this omission is related to the more rigorous form of the Q (or K) expression mentioned previously in this chapter, in which *relative concentrations for liquids and solids are equal to 1 and needn't be included.* Consequently, reaction quotients include concentration or pressure terms only for gaseous and solute species.

The equilibria below all involve gas-phase solutions:

$$C_2H_6(g) \rightleftharpoons C_2H_4(g) + H_2(g) \qquad K_c = \frac{[C_2H_4][H_2]}{[C_2H_6]}$$

$$3O_2(g) \rightleftharpoons 2O_3(g) \qquad K_c = \frac{[O_3]^2}{[O_2]^3}$$

$$N_2(g) + 3H_2(g) \rightleftharpoons 2NH_3(g) \qquad K_c = \frac{[NH_3]^2}{[N_2][H_2]^3}$$

$$C_3H_8(g) + 5O_2(g) \rightleftharpoons 3CO_2(g) + 4H_2O(g) \qquad K_c = \frac{[CO_2]^3[H_2O]^4}{[C_3H_8][O_2]^5}$$

For gas-phase solutions, the equilibrium constant may be expressed in terms of either the molar concentrations (K_c) or partial pressures (K_p) of the reactants and products. A relation between these two K values may be simply derived from the ideal gas equation and the definition of molarity:

$$PV = nRT$$

$$P = \left(\frac{n}{V}\right) RT$$

$$= MRT$$

where P is partial pressure, V is volume, n is molar amount, R is the gas constant, T is temperature, and M is molar concentration.

For the gas-phase reaction $m\text{A} + n\text{B} \rightleftharpoons x\text{C} + y\text{D}$:

$$K_P = \frac{(P_C)^x(P_D)^y}{(P_A)^m(P_B)^n}$$

$$= \frac{([C] \times RT)^x([D] \times RT)^y}{([A] \times RT)^m([B] \times RT)^n}$$

$$= \frac{[C]^x[D]^y}{[A]^m[B]^n} \times \frac{(RT)^{x+y}}{(RT)^{m+n}}$$

$$= K_c(RT)^{(x+y)-(m+n)}$$

$$= K_c(RT)^{\Delta n}$$

And so, the relationship between K_c and K_P is

$$K_P = K_c(RT)^{\Delta n}$$

where Δn is the difference in the molar amounts of product and reactant gases, in this case:

$$\Delta n = (x+y) - (m+n)$$

EXAMPLE 13.4

Calculation of K_P

Write the equations relating K_c to K_P for each of the following reactions:

(a) $C_2H_6(g) \rightleftharpoons C_2H_4(g) + H_2(g)$

(b) $CO(g) + H_2O(g) \rightleftharpoons CO_2(g) + H_2(g)$

(c) $N_2(g) + 3H_2(g) \rightleftharpoons 2NH_3(g)$

(d) K_c is equal to 0.28 for the following reaction at 900 °C:

$$CS_2(g) + 4H_2(g) \rightleftharpoons CH_4(g) + 2H_2S(g)$$

What is K_P at this temperature?

Solution

(a) $\Delta n = (2) - (1) = 1$
$K_P = K_c (RT)^{\Delta n} = K_c (RT)^1 = K_c (RT)$

(b) $\Delta n = (2) - (2) = 0$
$K_P = K_c (RT)^{\Delta n} = K_c (RT)^0 = K_c$

(c) $\Delta n = (2) - (1 + 3) = -2$
$K_P = K_c (RT)^{\Delta n} = K_c (RT)^{-2} = \frac{K_c}{(RT)^2}$

(d) $K_P = K_c (RT)^{\Delta n} = (0.28)[(0.0821)(1173)]^{-2} = 3.0 \times 10^{-5}$

Check Your Learning

Write the equations relating K_c to K_P for each of the following reactions:

(a) $2SO_2(g) + O_2(g) \rightleftharpoons 2SO_3(g)$

(b) $N_2O_4(g) \rightleftharpoons 2NO_2(g)$

(c) $C_3H_8(g) + 5O_2(g) \rightleftharpoons 3CO_2(g) + 4H_2O(g)$

(d) At 227 °C, the following reaction has $K_c = 0.0952$:

$$CH_3OH(g) \rightleftharpoons CO(g) + 2H_2(g)$$

What would be the value of K_P at this temperature?

Answer:
(a) $K_P = K_c (RT)^{-1}$; (b) $K_P = K_c (RT)$; (c) $K_P = K_c (RT)$; (d) 160 or 1.6×10^2

Heterogeneous Equilibria

A **heterogeneous equilibrium** involves reactants and products in two or more different phases, as illustrated by the following examples:

$$\begin{aligned}
PbCl_2(s) &\rightleftharpoons Pb^{2+}(aq) + 2Cl^-(aq) & K_c &= [Pb^{2+}]\left[Cl^-\right]^2 \\
CaO(s) + CO_2(g) &\rightleftharpoons CaCO_3(s) & K_c &= \frac{1}{[CO_2]} \\
C(s) + 2S(g) &\rightleftharpoons CS_2(g) & K_c &= \frac{[CS_2]}{[S]^2} \\
Br_2(l) &\rightleftharpoons Br_2(g) & K_c &= [Br_2(g)]
\end{aligned}$$

Again, note that concentration terms are only included for gaseous and solute species, as discussed previously.

Two of the above examples include terms for gaseous species only in their equilibrium constants, and so K_p expressions may also be written:

$$\begin{aligned}
CaO(s) + CO_2(g) &\rightleftharpoons CaCO_3(s) & K_P &= \frac{1}{P_{CO_2}} \\
C(s) + 2S(g) &\rightleftharpoons CS_2(g) & K_P &= \frac{P_{CS_2}}{(P_S)^2}
\end{aligned}$$

Coupled Equilibria

The equilibrium systems discussed so far have all been relatively simple, involving just single reversible reactions. Many systems, however, involve two or more *coupled* equilibrium reactions, those which have in common one or more reactant or product species. Since the law of mass action allows for a straightforward derivation of equilibrium constant expressions from balanced chemical equations, the K value for a system involving coupled equilibria can be related to the K values of the individual reactions. Three basic manipulations are involved in this approach, as described below.

1. Changing the direction of a chemical equation essentially swaps the identities of "reactants" and "products," and so the equilibrium constant for the reversed equation is simply the reciprocal of that for the forward equation.

$$\begin{aligned} &A \rightleftharpoons B && K_c = \frac{[B]}{[A]} \\ &B \rightleftharpoons A && K_{c'} = \frac{[A]}{[B]} \end{aligned}$$

$$K_{c'} = \frac{1}{K_c}$$

2. Changing the stoichiometric coefficients in an equation by some factor *x* results in an exponential change in the equilibrium constant by that same factor:

$$\begin{aligned} &A \rightleftharpoons B && K_c = \frac{[B]}{[A]} \\ &xA \rightleftharpoons xB && K_{c'} = \frac{[B]^x}{[A]^x} \end{aligned}$$

$$K_{c'} = K_c{}^x$$

3. Adding two or more equilibrium equations together yields an overall equation whose equilibrium constant is the mathematical product of the individual reaction's K values:

$$\begin{aligned} &A \rightleftharpoons B && K_{c1} = \frac{[B]}{[A]} \\ &B \rightleftharpoons C && K_{c2} = \frac{[C]}{[B]} \end{aligned}$$

The net reaction for these coupled equilibria is obtained by summing the two equilibrium equations and canceling any redundancies:

$$\begin{aligned} &A + B \rightleftharpoons B + C \\ &A + \cancel{B} \rightleftharpoons \cancel{B} + C \\ &A \rightleftharpoons C && K_{c'} = \frac{[C]}{[A]} \end{aligned}$$

Comparing the equilibrium constant for the net reaction to those for the two coupled equilibrium reactions reveals the following relationship:

$$K_{c1}\,K_{c2} = \frac{[B]}{[A]} \times \frac{[C]}{[B]} = \frac{\cancel{[B]}[C]}{[A]\cancel{[B]}} = \frac{[C]}{[A]} = K_{c'}$$

$$K_{c'} = K_{c1}\,K_{c2}$$

Example 13.5 demonstrates the use of this strategy in describing coupled equilibrium processes.

EXAMPLE 13.5

Equilibrium Constants for Coupled Reactions

A mixture containing nitrogen, hydrogen, and iodine established the following equilibrium at 400 °C:

$$2NH_3(g) + 3I_2(g) \rightleftharpoons N_2(g) + 6HI(g)$$

Use the information below to calculate Kc for this reaction.

$$N_2(g) + 3H_2(g) \rightleftharpoons 2NH_3(g) \qquad K_{c1} = 0.50 \text{ at } 400\ °C$$
$$H_2(g) + I_2(g) \rightleftharpoons 2HI(g) \qquad K_{c2} = 50 \text{ at } 400\ °C$$

Solution

The equilibrium equation of interest and its K value may be derived from the equations for the two coupled reactions as follows.

Reverse the first coupled reaction equation:

$$2NH_3(g) \rightleftharpoons N_2(g) + 3H_2(g) \qquad K_{c1'} = \frac{1}{K_{c1}} = \frac{1}{0.50} = 2.0$$

Multiply the second coupled reaction by 3:

$$3H_2(g) + 3I_2(g) \rightleftharpoons 6HI(g) \qquad K_{c2'} = K_{c2}^3 = 50^3 = 1.2 \times 10^5$$

Finally, add the two revised equations:

$$2NH_3(g) + \cancel{3H_2(g)} + 3I_2(g) \rightleftharpoons N_2(g) + \cancel{3H_2(g)} + 6HI(g)$$
$$2NH_3(g) + 3I_2(g) \rightleftharpoons N_2(g) + 6HI(g)$$
$$K_c = K_{c1'}\ K_{c2'} = (2.0)(1.2 \times 10^5) = 2.5 \times 10^5$$

Check Your Learning

Use the provided information to calculate Kc for the following reaction at 550 °C:

$$H_2(g) + CO_2(g) \rightleftharpoons CO(g) + H_2O(g) \qquad K_c = ?$$
$$CoO(s) + CO(g) \rightleftharpoons Co(s) + CO_2(g) \qquad K_{c1} = 490$$
$$CoO(s) + H_2(g) \rightleftharpoons Co(s) + H_2O(g) \qquad K_{c1} = 67$$

Answer:

$K_c = 0.14$

13.3 Shifting Equilibria: Le Châtelier's Principle

LEARNING OBJECTIVES

By the end of this section, you will be able to:

- Describe the ways in which an equilibrium system can be stressed
- Predict the response of a stressed equilibrium using Le Châtelier's principle

A system at equilibrium is in a state of dynamic balance, with forward and reverse reactions taking place at equal rates. If an equilibrium system is subjected to a change in conditions that affects these reaction rates differently (a *stress*), then the rates are no longer equal and the system is not at equilibrium. The system will subsequently experience a net reaction in the direction of greater rate (a *shift*) that will re-establish the equilibrium. This phenomenon is summarized by **Le Châtelier's principle**: *if an equilibrium system is stressed, the system will experience a shift in response to the stress that re-establishes equilibrium.*

Reaction rates are affected primarily by concentrations, as described by the reaction's rate law, and temperature, as described by the Arrhenius equation. Consequently, changes in concentration and temperature are the two stresses that can shift an equilibrium.

Effect of a Change in Concentration

If an equilibrium system is subjected to a change in the concentration of a reactant or product species, the rate of either the forward or the reverse reaction will change. As an example, consider the equilibrium reaction

$$H_2(g) + I_2(g) \rightleftharpoons 2HI(g) \qquad K_c = 50.0 \text{ at } 400\ °C$$

The rate laws for the forward and reverse reactions are

$$\begin{array}{ll} \textit{forward}\ H_2(g) + I_2(g) \rightarrow 2HI(g) & \text{rate}_f = k_f[H_2]^m[I_2]^n \\ \textit{reverse}\ 2HI(g) \rightarrow H_2(g) + I_2(g) & \text{rate}_r = k_r[HI]^x \end{array}$$

When this system is at equilibrium, the forward and reverse reaction rates are equal.

$$\text{rate}_f = \text{rate}_r$$

If the system is stressed by adding reactant, either H_2 or I_2, the resulting increase in concentration causes the rate of the forward reaction to increase, exceeding that of the reverse reaction:

$$\text{rate}_f > \text{rate}_r$$

The system will experience a temporary net reaction in the forward direction to re-establish equilibrium (*the equilibrium will shift right*). This same shift will result if some product HI is removed from the system, which decreases the rate of the reverse reaction, again resulting in the same imbalance in rates.

The same logic can be used to explain the left shift that results from either removing reactant or adding product to an equilibrium system. These stresses both result in an increased rate for the reverse reaction

$$\text{rate}_f < \text{rate}_r$$

and a temporary net reaction in the reverse direction to re-establish equilibrium.

As an alternative to this kinetic interpretation, the effect of changes in concentration on equilibria can be rationalized in terms of reaction quotients. When the system is at equilibrium,

$$Q_c = \frac{[HI]^2}{[H_2][I_2]} = K_c$$

If reactant is added (increasing the denominator of the reaction quotient) or product is removed (decreasing the numerator), then $Q_c < K_c$ and the equilibrium will shift right. Note that the three different ways of inducing this stress result in three different changes in the composition of the equilibrium mixture. If H_2 is added, the right shift will consume I_2 and produce HI as equilibrium is re-established, yielding a mixture with a greater concentrations of H_2 and HI and a lesser concentration of I_2 than was present before. If I_2 is added, the new equilibrium mixture will have greater concentrations of I_2 and HI and a lesser concentration of H_2. Finally, if HI is removed, the concentrations of all three species will be lower when equilibrium is reestablished. Despite these differences in composition, *the value of the equilibrium constant will be the same after the stress as it was before* (per the law of mass action). The same logic may be applied for stresses involving removing reactants or adding product, in which case $Q_c > K_c$ and the equilibrium will shift left.

For gas-phase equilibria such as this one, some additional perspectives on changing the concentrations of reactants and products are worthy of mention. The partial pressure *P* of an ideal gas is proportional to its molar concentration *M*,

$$M = \frac{n}{V} = \frac{P}{RT}$$

and so changes in the partial pressures of any reactant or product are essentially changes in concentrations and thus yield the same effects on equilibria. Aside from adding or removing reactant or product, the pressures (concentrations) of species in a gas-phase equilibrium can also be changed by *changing the volume occupied by the system*. Since all species of a gas-phase equilibrium occupy the same volume, a given change in volume will cause the same change in concentration for both reactants and products. In order to discern what shift, if any, this type of stress will induce the stoichiometry of the reaction must be considered.

At equilibrium, the reaction $H_2(g) + I_2(g) \rightleftharpoons 2HI(g)$ is described by the reaction quotient

$$Q_P = \frac{{P_{HI}}^2}{P_{H_2}\ P_{I_2}} = K_P$$

If the volume occupied by an equilibrium mixture of these species is decreased by a factor of 3, the partial pressures of all three species will be increased by a factor of 3:

$$Q_P{}' = \frac{(3P_{HI})^2}{3P_{H_2}\ 3P_{I_2}} = \frac{9P_{HI}{}^2}{9P_{H_2}\ P_{I_2}} = \frac{P_{HI}{}^2}{P_{H_2}\ P_{I_2}} = Q_P = K_P$$

$$Q_P{}' = Q_P = K_P$$

And so, changing the volume of this gas-phase equilibrium mixture does not result in a shift of the equilibrium.

A similar treatment of a different system, $2NO_2(g) \rightleftharpoons 2\ NO(g) + O_2(g)$, however, yields a different result:

$$Q_P = \frac{{P_{NO}}^2\ P_{O_2}}{{P_{NO_2}}^2}$$

$$Q_P{}' = \frac{(3P_{NO})^2\ 3P_{O_2}}{(3P_{NO_2})^2} = \frac{9{P_{NO}}^2\ 3P_{O_2}}{9P_{NO_2}{}^2} = \frac{27{P_{NO}}^2\ P_{O_2}}{9{P_{NO_2}}^2} = 3Q_P > K_P$$

$$Q_P{}' = 3Q_P > K_P$$

In this case, the change in volume results in a reaction quotient greater than the equilibrium constant, and so the equilibrium will shift left.

These results illustrate the relationship between the stoichiometry of a gas-phase equilibrium and the effect of a volume-induced pressure (concentration) change. If the total molar amounts of reactants and products are equal, as in the first example, a change in volume does not shift the equilibrium. If the molar amounts of reactants and products are different, a change in volume will shift the equilibrium in a direction that better "accommodates" the volume change. In the second example, two moles of reactant (NO_2) yield three moles of product ($2NO + O_2$), and so decreasing the system volume causes the equilibrium to shift left since the reverse reaction produces less gas (2 mol) than the forward reaction (3 mol). Conversely, increasing the volume of this equilibrium system would result in a shift towards products.

LINK TO LEARNING

Check out this link (http://openstax.org/l/16equichange) to see a dramatic visual demonstration of how equilibrium changes with pressure changes.

Chemistry in Everyday Life

Equilibrium and Soft Drinks

The connection between chemistry and carbonated soft drinks goes back to 1767, when Joseph Priestley (1733–1804) developed a method of infusing water with carbon dioxide to make carbonated water. Priestley's approach involved production of carbon dioxide by reacting oil of vitriol (sulfuric acid) with chalk (calcium carbonate).

The carbon dioxide was then dissolved in water, reacting to produce hydrogen carbonate, a weak acid that subsequently ionized to yield bicarbonate and hydrogen ions:

$$\text{dissolution } CO_2(g) \rightleftharpoons CO_2(aq)$$
$$\text{hydrolysis } CO_2(aq) + H_2O(l) \rightleftharpoons H_2CO_3(aq)$$
$$\text{ionization } H_2CO_3(aq) \rightleftharpoons HCO_3^-(aq) + H^+(aq)$$

These same equilibrium reactions are the basis of today's soft-drink carbonation process. Beverages are exposed to a high pressure of gaseous carbon dioxide during the process to shift the first equilibrium above to the right, resulting in desirably high concentrations of dissolved carbon dioxide and, per similar shifts in the other two equilibria, its hydrolysis and ionization products. A bottle or can is then nearly filled with the carbonated beverage, leaving a relatively small volume of air in the container above the beverage surface (the *headspace*) before it is sealed. The pressure of carbon dioxide in the container headspace is very low immediately after sealing, but it rises as the dissolution equilibrium is re-established by shifting to the left. Since the volume of the beverage is significantly greater than the volume of the headspace, only a relatively small amount of dissolved carbon dioxide is lost to the headspace.

When a carbonated beverage container is opened, a hissing sound is heard as pressurized CO_2 escapes from the headspace. This causes the dissolution equilibrium to shift left, resulting in a decrease in the concentration of dissolved CO_2 and subsequent left-shifts of the hydrolysis and ionization equilibria. Fortunately for the consumer, the dissolution equilibrium is usually re-established slowly, and so the beverage may be enjoyed while its dissolved carbon dioxide concentration remains palatably high. Once the equilibria are re-established, the $CO_2(aq)$ concentration will be significantly lowered, and the beverage acquires a characteristic taste referred to as "flat."

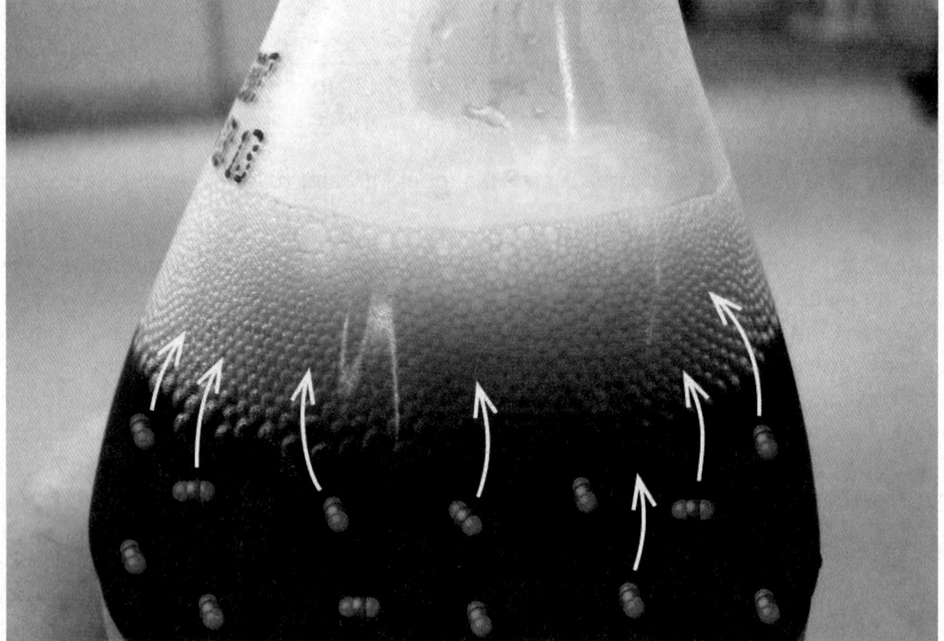

FIGURE 13.7 Opening a soft-drink bottle lowers the CO_2 pressure above the beverage, shifting the dissolution equilibrium and releasing dissolved CO_2 from the beverage. (credit: modification of work by "D Coetzee"/Flickr)

Effect of a Change in Temperature

Consistent with the law of mass action, an equilibrium stressed by a change in concentration will shift to re-establish equilibrium without any change in the value of the equilibrium constant, *K*. When an equilibrium shifts in response to a temperature change, however, it is re-established with a different relative composition that exhibits a different value for the equilibrium constant.

To understand this phenomenon, consider the elementary reaction

$$\text{A} \rightleftharpoons \text{B}$$

Since this is an elementary reaction, the rates laws for the forward and reverse may be derived directly from the balanced equation's stoichiometry:

$$\text{rate}_f = k_f[\text{A}]$$
$$\text{rate}_r = k_r[\text{B}]$$

When the system is at equilibrium,

$$\text{rate}_r = \text{rate}_f$$

Substituting the rate laws into this equality and rearranging gives

$$k_f[\text{A}] = k_r[\text{B}]$$
$$\frac{[B]}{[A]} = \frac{k_f}{k_r} = K_c$$

The equilibrium constant is seen to be a mathematical function of the rate constants for the forward and reverse reactions. Since the rate constants vary with temperature as described by the Arrhenius equation, is stands to reason that the equilibrium constant will likewise vary with temperature (assuming the rate constants are affected to different extents by the temperature change). For more complex reactions involving multistep reaction mechanisms, a similar but more complex mathematical relation exists between the equilibrium constant and the rate constants of the steps in the mechanism. Regardless of how complex the reaction may be, the temperature-dependence of its equilibrium constant persists.

Predicting the shift an equilibrium will experience in response to a change in temperature is most conveniently accomplished by considering the enthalpy change of the reaction. For example, the decomposition of dinitrogen tetroxide is an endothermic (heat-consuming) process:

$$\text{N}_2\text{O}_4(g) \rightleftharpoons 2\text{NO}_2(g) \qquad \Delta H = +57.20 \text{ kJ}$$

For purposes of applying Le Chatelier's principle, heat (q) may be viewed as a reactant:

$$\text{heat} + \text{N}_2\text{O}_4(g) \rightleftharpoons 2\text{NO}_2(g)$$

Raising the temperature of the system is akin to increasing the amount of a reactant, and so the equilibrium will shift to the right. Lowering the system temperature will likewise cause the equilibrium to shift left. For exothermic processes, heat is viewed as a product of the reaction and so the opposite temperature dependence is observed.

Effect of a Catalyst

The kinetics chapter of this text identifies a *catalyst* as a substance that enables a reaction to proceed via a different mechanism with an accelerated rate. The catalyzed reaction mechanism involves a lower energy transition state than the uncatalyzed reaction, resulting in a lower activation energy, E_a, and a correspondingly greater rate constant.

To discern the effect of catalysis on an equilibrium system, consider the reaction diagram for a simple one-step (elementary) reaction shown in Figure 13.8. The lowered transition state energy of the catalyzed reaction results in lowered activation energies for both the forward and the reverse reactions. Consequently, both forward and reverse reactions are accelerated, and equilibrium is achieved more quickly *but without a change in the equilibrium constant.*

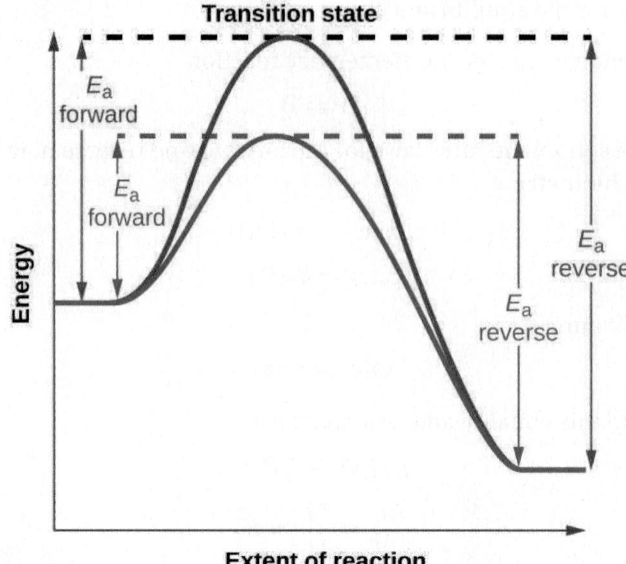

FIGURE 13.8 Reaction diagrams for an elementary process in the absence (red) and presence (blue) of a catalyst. The presence of catalyst lowers the activation energies of both the forward and reverse reactions but does not affect the value of the equilibrium constant.

An interesting case study highlighting these equilibrium concepts is the industrial production of ammonia, NH_3. This substance is among the "top 10" industrial chemicals with regard to production, with roughly two billion pounds produced annually in the US. Ammonia is used as a chemical feedstock to synthesize a wide range of commercially useful compounds, including fertilizers, plastics, dyes, and explosives.

Most industrial production of ammonia uses the *Haber-Bosch process* based on the following equilibrium reaction:

$$N_2(g) + 3H_2(g) \rightleftharpoons 2NH_3(g) \qquad \Delta H = -92.2 \text{ kJ}$$

The traits of this reaction present challenges to its use in an efficient industrial process. The equilibrium constant is relatively small (K_p on the order of 10^{-5} at 25 °C), meaning very little ammonia is present in an equilibrium mixture. Also, the rate of this reaction is relatively slow at low temperatures. To raise the yield of ammonia, the industrial process is designed to operate under conditions favoring product formation:

- High pressures (concentrations) of reactants are used, ~150–250 atm, to shift the equilibrium right, favoring product formation.
- Ammonia is continually removed (collected) from the equilibrium mixture during the process, lowering its concentration and also shifting the equilibrium right.
- Although low temperatures favor product formation for this exothermic process, the reaction rate at low temperatures is inefficiently slow. A catalyst is used to accelerate the reaction to reasonable rates at relatively moderate temperatures (400–500 °C).

A diagram illustrating a typical industrial setup for production of ammonia via the Haber-Bosch process is shown in Figure 13.9.

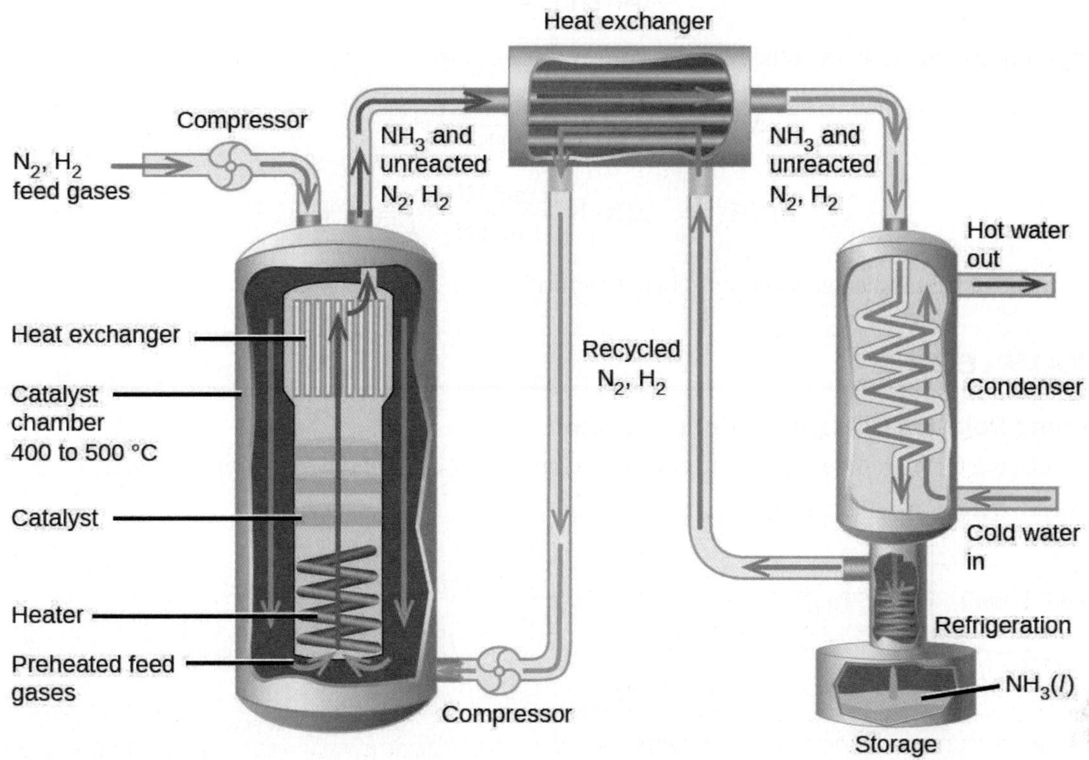

FIGURE 13.9 The figure shows a typical industrial setup for the commercial production of ammonia by the Haber-Bosch process. The process operates under conditions that stress the chemical equilibrium to favor product formation.

13.4 Equilibrium Calculations

LEARNING OBJECTIVES

By the end of this section, you will be able to:

- Identify the changes in concentration or pressure that occur for chemical species in equilibrium systems
- Calculate equilibrium concentrations or pressures and equilibrium constants, using various algebraic approaches

Having covered the essential concepts of chemical equilibria in the preceding sections of this chapter, this final section will demonstrate the more practical aspect of using these concepts and appropriate mathematical strategies to perform various equilibrium calculations. These types of computations are essential to many areas of science and technology—for example, in the formulation and dosing of pharmaceutical products. After a drug is ingested or injected, it is typically involved in several chemical equilibria that affect its ultimate concentration in the body system of interest. Knowledge of the quantitative aspects of these equilibria is required to compute a dosage amount that will solicit the desired therapeutic effect.

Many of the useful equilibrium calculations that will be demonstrated here require terms representing changes in reactant and product concentrations. These terms are derived from the stoichiometry of the reaction, as illustrated by decomposition of ammonia:

$$2NH_3(g) \rightleftharpoons N_2(g) + 3H_2(g)$$

As shown earlier in this chapter, this equilibrium may be established within a sealed container that initially contains either NH_3 only, or a mixture of any two of the three chemical species involved in the equilibrium. Regardless of its initial composition, a reaction mixture will show the same relationships between changes in the concentrations of the three species involved, as dictated by the reaction stoichiometry (see also the related content on expressing reaction rates in the chapter on kinetics). For example, if the nitrogen concentration increases by an amount *x*:

$$\Delta[N_2] = + \quad x$$

the corresponding changes in the other species concentrations are

$$\Delta[H_2] = \Delta[N_2]\left(\frac{3 \text{ mol } H_2}{1 \text{ mol } N_2}\right) = +3x$$

$$\Delta[NH_3] = -\Delta[N_2]\left(\frac{2 \text{ mol } NH_3}{1 \text{ mol } N_2}\right) = -2x$$

where the negative sign indicates a decrease in concentration.

EXAMPLE 13.6

Determining Relative Changes in Concentration

Derive the missing terms representing concentration changes for each of the following reactions.

(a)
$$\begin{array}{lllll} C_2H_2(g) + & 2Br_2(g) & \rightleftharpoons & C_2H_2Br_4(g) \\ x & \_\_\_\_ & & \_\_\_\_ \end{array}$$

(b)
$$\begin{array}{lllll} I_2(aq) + & I^-(aq) & \rightleftharpoons & I_3{}^-(aq) \\ \_\_\_\_ & \_\_\_\_ & & x \end{array}$$

(c)
$$\begin{array}{llllll} C_3H_8(g) + & 5O_2(g) & \rightleftharpoons & 3CO_2(g) + & 4H_2O(g) \\ x & \_\_\_\_ & & \_\_\_\_ & \_\_\_\_ \end{array}$$

Solution

(a)
$$\begin{array}{lllll} C_2H_2(g) + & 2Br_2(g) & \rightleftharpoons & C_2H_2Br_4(g) \\ x & 2x & & -x \end{array}$$

(b)
$$\begin{array}{lllll} I_2(aq) + & I^-(aq) & \rightleftharpoons & I_3{}^-(aq) \\ -x & -x & & x \end{array}$$

(c)
$$\begin{array}{llllll} C_3H_8(g) + & 5O_2(g) & \rightleftharpoons & 3CO_2(g) + & 4H_2O(g) \\ x & 5x & & -3x & -4x \end{array}$$

Check Your Learning

Complete the changes in concentrations for each of the following reactions:

(a)
$$\begin{array}{lllll} 2SO_2(g) + & O_2(g) & \rightleftharpoons & 2SO_3(g) \\ \_\_\_\_ & x & & \_\_\_\_ \end{array}$$

(b)
$$\begin{array}{lll} C_4H_8(g) & \rightleftharpoons & 2C_2H_4(g) \\ \_\_\_\_ & & -2x \end{array}$$

(c)
$$\begin{array}{llllll} 4NH_3(g) + & 7O_2(g) & \rightleftharpoons & 4NO_2(g) + & 6H_2O(g) \\ \_\_\_\_ & \_\_\_\_ & & \_\_\_\_ & \_\_\_\_ \end{array}$$

Answer:

(a) $2x$, x, $-2x$; (b) x, $-2x$; (c) $4x$, $7x$, $-4x$, $-6x$ or $-4x$, $-7x$, $4x$, $6x$

Calculation of an Equilibrium Constant

The equilibrium constant for a reaction is calculated from the equilibrium concentrations (or pressures) of its reactants and products. If these concentrations are known, the calculation simply involves their substitution into the K expression, as was illustrated by Example 13.2. A slightly more challenging example is provided next, in which the reaction stoichiometry is used to derive equilibrium concentrations from the information provided. The basic strategy of this computation is helpful for many types of equilibrium computations and relies on the use of terms for the reactant and product concentrations *initially* present, for how they *change* as the reaction proceeds, and for what they are when the system reaches *equilibrium*. The acronym ICE is commonly used to refer to this mathematical approach, and the concentrations terms are usually gathered in a tabular format called an ICE table.

EXAMPLE 13.7

Calculation of an Equilibrium Constant

Iodine molecules react reversibly with iodide ions to produce triiodide ions.

$$I_2(aq) + I^-(aq) \rightleftharpoons I_3{}^-(aq)$$

If a solution with the concentrations of I_2 and I^- both equal to 1.000×10^{-3} *M* before reaction gives an equilibrium concentration of I_2 of 6.61×10^{-4} *M*, what is the equilibrium constant for the reaction?

Solution

To calculate the equilibrium constants, equilibrium concentrations are needed for all the reactants and products:

$$K_C = \frac{[I_3{}^-]}{[I_2][I^-]}$$

Provided are the initial concentrations of the reactants and the equilibrium concentration of the product. Use this information to derive terms for the equilibrium concentrations of the reactants, presenting all the information in an ICE table.

	I_2 +	I^- ⇌	I_3^-
Initial concentration (*M*)	1.000×10^{-3}	1.000×10^{-3}	0
Change (*M*)	$-x$	$-x$	$+x$
Equilibrium concentration (*M*)	$1.000 \times 10^{-3} - x$	$1.000 \times 10^{-3} - x$	x

At equilibrium the concentration of I_2 is 6.61×10^{-4} *M* so that

$$1.000 \times 10^{-3} - x = 6.61 \times 10^{-4}$$
$$x = 1.000 \times 10^{-3} - 6.61 \times 10^{-4}$$
$$= 3.39 \times 10^{-4}\ M$$

The ICE table may now be updated with numerical values for all its concentrations:

	I_2 +	I^- ⇌	I_3^-
Initial concentration (*M*)	1.000×10^{-3}	1.000×10^{-3}	0
Change (*M*)	-3.39×10^{-4}	-3.39×10^{-4}	$+3.39 \times 10^{-4}$
Equilibrium concentration (*M*)	6.61×10^{-4}	6.61×10^{-4}	3.39×10^{-4}

Finally, substitute the equilibrium concentrations into the *K* expression and solve:

$$K_c = \frac{[I_3{}^-]}{[I_2][I^-]}$$

$$= \frac{3.39 \times 10^{-4}\ M}{(6.61 \times 10^{-4}\ M)(6.61 \times 10^{-4}\ M)} = 776$$

Check Your Learning

Ethanol and acetic acid react and form water and ethyl acetate, the solvent responsible for the odor of some nail polish removers.

$$C_2H_5OH + CH_3CO_2H \rightleftharpoons CH_3CO_2C_2H_5 + H_2O$$

When 1 mol each of C_2H_5OH and CH_3CO_2H are allowed to react in 1 L of the solvent dioxane, equilibrium is established when $\frac{1}{3}$ mol of each of the reactants remains. Calculate the equilibrium constant for the reaction. (Note: Water is a solute in this reaction.)

Answer:

$K_c = 4$

Calculation of a Missing Equilibrium Concentration

When the equilibrium constant and all but one equilibrium concentration are provided, the other equilibrium concentration(s) may be calculated. A computation of this sort is illustrated in the next example exercise.

EXAMPLE 13.8

Calculation of a Missing Equilibrium Concentration

Nitrogen oxides are air pollutants produced by the reaction of nitrogen and oxygen at high temperatures. At 2000 °C, the value of the K_c for the reaction, $N_2(g) + O_2(g) \rightleftharpoons 2NO(g)$, is 4.1×10^{-4}. Calculate the equilibrium concentration of NO(*g*) in air at 1 atm pressure and 2000 °C. The equilibrium concentrations of N_2 and O_2 at this pressure and temperature are 0.036 M and 0.0089 M, respectively.

Solution

Substitute the provided quantities into the equilibrium constant expression and solve for [NO]:

$$K_c = \frac{[NO]^2}{[N_2]\ [O_2]}$$

$$[NO]^2 = K_c\ [N_2]\ [O_2]$$

$$[NO] = \sqrt{K_c\ [N_2]\ [O_2]}$$

$$= \sqrt{(4.1 \times 10^{-4})(0.036)(0.0089)}$$

$$= \sqrt{1.31 \times 10^{-7}}$$

$$= 3.6 \times 10^{-4}$$

Thus [NO] is 3.6×10^{-4} mol/L at equilibrium under these conditions.

To confirm this result, it may be used along with the provided equilibrium concentrations to calculate a value for *K*:

$$K_c = \frac{[NO]^2}{[N_2]\ [O_2]}$$

$$= \frac{(3.6 \times 10^{-4})^2}{(0.036)(0.0089)}$$

$$= 4.0 \times 10^{-4}$$

This result is consistent with the provided value for K within nominal uncertainty, differing by just 1 in the least significant digit's place.

Check Your Learning

The equilibrium constant K_c for the reaction of nitrogen and hydrogen to produce ammonia at a certain temperature is 6.00×10^{-2}. Calculate the equilibrium concentration of ammonia if the equilibrium concentrations of nitrogen and hydrogen are 4.26 M and 2.09 M, respectively.

Answer:

1.53 mol/L

Calculation of Equilibrium Concentrations from Initial Concentrations

Perhaps the most challenging type of equilibrium calculation can be one in which equilibrium concentrations are derived from initial concentrations and an equilibrium constant. For these calculations, a four-step approach is typically useful:

1. Identify the direction in which the reaction will proceed to reach equilibrium.
2. Develop an ICE table.
3. Calculate the concentration changes and, subsequently, the equilibrium concentrations.
4. Confirm the calculated equilibrium concentrations.

The last two example exercises of this chapter demonstrate the application of this strategy.

EXAMPLE 13.9

Calculation of Equilibrium Concentrations

Under certain conditions, the equilibrium constant K_c for the decomposition of $PCl_5(g)$ into $PCl_3(g)$ and $Cl_2(g)$ is 0.0211. What are the equilibrium concentrations of PCl_5, PCl_3, and Cl_2 in a mixture that initially contained only PCl_5 at a concentration of 1.00 M?

Solution

Use the stepwise process described earlier.

Step 1.

Determine the direction the reaction proceeds.

The balanced equation for the decomposition of PCl_5 is

$$PCl_5(g) \rightleftharpoons PCl_3(g) + Cl_2(g)$$

Because only the reactant is present initially $Q_c = 0$ and the reaction will proceed to the right.

Step 2.

Develop an ICE table.

	PCl_5 $\rightleftharpoons$	PCl_3 +	Cl_2
Initial concentration (M)	1.00	0	0
Change (M)	$-x$	$+x$	$+x$
Equilibrium concentration (M)	$1.00 - x$	x	x

Step 3.

Solve for the change and the equilibrium concentrations.

Substituting the equilibrium concentrations into the equilibrium constant equation gives

$$K_c = \frac{[PCl_3][Cl_2]}{[PCl_5]} = 0.0211$$
$$= \frac{(x)(x)}{(1.00 - x)}$$
$$0.0211 = \frac{(x)(x)}{(1.00 - x)}$$
$$0.0211(1.00 - x) = x^2$$
$$x^2 + 0.0211x - 0.0211 = 0$$

Appendix B shows an equation of the form $ax^2 + bx + c = 0$ can be rearranged to solve for x:

$$x = \frac{-b \pm \sqrt{b^2 - 4ac}}{2a}$$

In this case, $a = 1$, $b = 0.0211$, and $c = -0.0211$. Substituting the appropriate values for a, b, and c yields:

$$x = \frac{-0.0211 \pm \sqrt{(0.0211)^2 - 4(1)(-0.0211)}}{2(1)}$$
$$= \frac{-0.0211 \pm \sqrt{(4.45 \times 10^{-4}) + (8.44 \times 10^{-2})}}{2}$$
$$= \frac{-0.0211 \pm 0.291}{2}$$

The two roots of the quadratic are, therefore,

$$x = \frac{-0.0211 + 0.291}{2} = 0.135$$

and

$$x = \frac{-0.0211 - 0.291}{2} = -0.156$$

For this scenario, only the positive root is physically meaningful (concentrations are either zero or positive), and so $x = 0.135\ M$.

The equilibrium concentrations are

$$[PCl_5] = 1.00 - 0.135 = 0.87\ M$$
$$[PCl_3] = x = 0.135\ M$$
$$[Cl_2] = x = 0.135\ M$$

Step 4.
Confirm the calculated equilibrium concentrations.

Substitution into the expression for K_c (to check the calculation) gives

$$K_c = \frac{[PCl_3][Cl_2]}{[PCl_5]} = \frac{(0.135)(0.135)}{0.87} = 0.021$$

The equilibrium constant calculated from the equilibrium concentrations is equal to the value of K_c given in the problem (when rounded to the proper number of significant figures).

Check Your Learning

Acetic acid, CH_3CO_2H, reacts with ethanol, C_2H_5OH, to form water and ethyl acetate, $CH_3CO_2C_2H_5$.

$$CH_3CO_2H + C_2H_5OH \rightleftharpoons CH_3CO_2C_2H_5 + H_2O$$

The equilibrium constant for this reaction with dioxane as a solvent is 4.0. What are the equilibrium concentrations for a mixture that is initially 0.15 *M* in CH_3CO_2H, 0.15 *M* in C_2H_5OH, 0.40 *M* in $CH_3CO_2C_2H_5$, and 0.40 *M* in H_2O?

Answer:

$[CH_3CO_2H] = 0.18$ *M*, $[C_2H_5OH] = 0.18$ *M*, $[CH_3CO_2C_2H_5] = 0.37$ *M*, $[H_2O] = 0.37$ *M*

Check Your Learning

A 1.00-L flask is filled with 1.00 mole of H_2 and 2.00 moles of I_2. The value of the equilibrium constant for the reaction of hydrogen and iodine reacting to form hydrogen iodide is 50.5 under the given conditions. What are the equilibrium concentrations of H_2, I_2, and HI in moles/L?

$$H_2(g) + I_2(g) \rightleftharpoons 2HI(g)$$

Answer:

$[H_2] = 0.06$ *M*, $[I_2] = 1.06$ *M*, $[HI] = 1.88$ *M*

EXAMPLE 13.10

Calculation of Equilibrium Concentrations Using an Algebra-Simplifying Assumption

What are the concentrations at equilibrium of a 0.15 *M* solution of HCN?

$$HCN(aq) \rightleftharpoons H^+(aq) + CN^-(aq) \qquad K_c = 4.9 \times 10^{-10}$$

Solution

Using "*x*" to represent the concentration of each product at equilibrium gives this ICE table.

	HCN(*aq*) ⇌	H⁺(*aq*) +	CN⁻(*aq*)
Initial concentration (*M*)	0.15	0	0
Change (*M*)	−*x*	+*x*	+*x*
Equilibrium concentration (*M*)	0.15 − *x*	*x*	*x*

Substitute the equilibrium concentration terms into the K_c expression

$$K_c = \frac{(x)(x)}{0.15 - x}$$

rearrange to the quadratic form and solve for *x*

$$x^2 + 4.9 \times 10^{-10} - 7.35 \times 10^{-11} = 0$$

$$x = 8.56 \times 10^{-6}\ M \text{ (3 sig. figs.)} = 8.6 \times 10^{-6}\ M \text{ (2 sig. figs.)}$$

Thus $[H^+] = [CN^-] = x = 8.6 \times 10^{-6}$ *M* and $[HCN] = 0.15 - x = 0.15$ *M*.

Note in this case that the change in concentration is significantly less than the initial concentration (a consequence of the small *K*), and so the initial concentration experiences a negligible change:

$$\text{if } x \ll 0.15 \text{ M, then } (0.15 - x) \approx 0.15$$

This approximation allows for a more expedient mathematical approach to the calculation that avoids the need to solve for the roots of a quadratic equation:

$$K_c = \frac{(x)(x)}{0.15 - x} \approx \frac{x^2}{0.15}$$

$$4.9 \times 10^{-10} = \frac{x^2}{0.15}$$

$$x^2 = (0.15)(4.9 \times 10^{-10}) = 7.4 \times 10^{-11}$$

$$x = \sqrt{7.4 \times 10^{-11}} = 8.6 \times 10^{-6}\ M$$

The value of x calculated is, indeed, much less than the initial concentration

$$8.6 \times 10^{-6} \ll 0.15$$

and so the approximation was justified. If this simplified approach were to yield a value for x that did *not* justify the approximation, the calculation would need to be repeated without making the approximation.

Check Your Learning

What are the equilibrium concentrations in a 0.25 M NH_3 solution?

$$NH_3(aq) + H_2O(l) \rightleftharpoons NH_4^{+}(aq) + OH^{-}(aq) \qquad K_c = 1.8 \times 10^{-5}$$

Answer:

$[OH^-] = [NH_4^+] = 0.0021\ M$; $[NH_3] = 0.25\ M$

Key Terms

equilibrium state of a reversible reaction in which the forward and reverse processes occur at equal rates

equilibrium constant (*K*) value of the reaction quotient for a system at equilibrium; may be expressed using concentrations (K_c) or partial pressures (K_p)

heterogeneous equilibria equilibria in which reactants and products occupy two or more different phases

homogeneous equilibria equilibria in which all reactants and products occupy the same phase

law of mass action when a reversible reaction has attained equilibrium at a given temperature, the reaction quotient remains constant

Le Châtelier's principle an equilibrium subjected to stress will shift in a way to counter the stress and re-establish equilibrium

reaction quotient (*Q*) mathematical function describing the relative amounts of reactants and products in a reaction mixture; may be expressed in terms of concentrations (Q_c) or pressures (Q_p)

reversible reaction chemical reaction that can proceed in both the forward and reverse directions under given conditions

Key Equations

$Q_c = \frac{[C]^x[D]^y}{[A]^m[B]^n}$ for the reaction $mA + nB \rightleftharpoons xC + yD$

$Q_P = \frac{(P_C)^x(P_D)^y}{(P_A)^m(P_B)^n}$ for the reaction $mA + nB \rightleftharpoons xC + yD$

$P = MRT$

$K_c = Q_c$ at equilibrium

$K_p = Q_p$ at equilibrium

$K_P = K_c (RT)^{\Delta n}$

Summary

13.1 Chemical Equilibria

A reversible reaction is at equilibrium when the forward and reverse processes occur at equal rates. Chemical equilibria are dynamic processes characterized by constant amounts of reactant and product species.

13.2 Equilibrium Constants

The composition of a reaction mixture may be represented by a mathematical function known as the reaction quotient, *Q*. For a reaction at equilibrium, the composition is constant, and *Q* is called the equilibrium constant, *K*.

A homogeneous equilibrium is an equilibrium in which all components are in the same phase. A heterogeneous equilibrium is an equilibrium in which components are in two or more phases.

13.3 Shifting Equilibria: Le Châtelier's Principle

Systems at equilibrium can be disturbed by changes to temperature, concentration, and, in some cases, volume and pressure. The system's response to these disturbances is described by Le Châtelier's principle: An equilibrium system subjected to a disturbance will shift in a way that counters the disturbance and re-establishes equilibrium. A catalyst will increase the rate of both the forward and reverse reactions of a reversible process, increasing the rate at which equilibrium is reached but not altering the equilibrium mixture's composition (*K* does not change).

13.4 Equilibrium Calculations

Calculating values for equilibrium constants and/or equilibrium concentrations is of practical benefit to many applications. A mathematical strategy that uses initial concentrations, changes in concentrations, and equilibrium concentrations (and goes by the acronym ICE) is useful for several types of equilibrium calculations.

Exercises

13.1 Chemical Equilibria

1. What does it mean to describe a reaction as "reversible"?
2. When writing an equation, how is a reversible reaction distinguished from a nonreversible reaction?
3. If a reaction is reversible, when can it be said to have reached equilibrium?
4. Is a system at equilibrium if the rate constants of the forward and reverse reactions are equal?
5. If the concentrations of products and reactants are equal, is the system at equilibrium?

13.2 Equilibrium Constants

6. Explain why there may be an infinite number of values for the reaction quotient of a reaction at a given temperature but there can be only one value for the equilibrium constant at that temperature.
7. Explain why an equilibrium between $Br_2(l)$ and $Br_2(g)$ would not be established if the container were not a closed vessel shown in Figure 13.4.
8. If you observe the following reaction at equilibrium, is it possible to tell whether the reaction started with pure NO_2 or with pure N_2O_4?
   $2NO_2(g) \rightleftharpoons N_2O_4(g)$
9. Among the solubility rules previously discussed is the statement: All chlorides are soluble except Hg_2Cl_2, AgCl, $PbCl_2$, and CuCl.
   (a) Write the expression for the equilibrium constant for the reaction represented by the equation $AgCl(s) \rightleftharpoons Ag^+(aq) + Cl^-(aq)$. Is $K_c > 1$, < 1, or ≈ 1? Explain your answer.
   (b) Write the expression for the equilibrium constant for the reaction represented by the equation $Pb^{2+}(aq) + 2Cl^-(aq) \rightleftharpoons PbCl_2(s)$. Is $K_c > 1$, < 1, or ≈ 1? Explain your answer.
10. Among the solubility rules previously discussed is the statement: Carbonates, phosphates, borates, and arsenates—except those of the ammonium ion and the alkali metals—are insoluble.
    (a) Write the expression for the equilibrium constant for the reaction represented by the equation $CaCO_3(s) \rightleftharpoons Ca^{2+}(aq) + CO_3{}^{2-}(aq)$. Is $K_c > 1$, < 1, or ≈ 1? Explain your answer.
    (b) Write the expression for the equilibrium constant for the reaction represented by the equation $3Ba^{2+}(aq) + 2PO_4{}^{3-}(aq) \rightleftharpoons Ba_3(PO_4)_2(s)$. Is $K_c > 1$, < 1, or ≈ 1? Explain your answer.
11. Benzene is one of the compounds used as octane enhancers in unleaded gasoline. It is manufactured by the catalytic conversion of acetylene to benzene: $3C_2H_2(g) \rightleftharpoons C_6H_6(g)$. Which value of K_c would make this reaction most useful commercially? $K_c \approx 0.01$, $K_c \approx 1$, or $K_c \approx 10$. Explain your answer.
12. Show that the complete chemical equation, the total ionic equation, and the net ionic equation for the reaction represented by the equation $KI(aq) + I_2(aq) \rightleftharpoons KI_3(aq)$ give the same expression for the reaction quotient. KI_3 is composed of the ions K^+ and $I_3{}^-$.
13. For a titration to be effective, the reaction must be rapid and the yield of the reaction must essentially be 100%. Is $K_c > 1$, < 1, or ≈ 1 for a titration reaction?
14. For a precipitation reaction to be useful in a gravimetric analysis, the product of the reaction must be insoluble. Is $K_c > 1$, < 1, or ≈ 1 for a useful precipitation reaction?
15. Write the mathematical expression for the reaction quotient, Q_c, for each of the following reactions:
    (a) $CH_4(g) + Cl_2(g) \rightleftharpoons CH_3Cl(g) + HCl(g)$
    (b) $N_2(g) + O_2(g) \rightleftharpoons 2NO(g)$
    (c) $2SO_2(g) + O_2(g) \rightleftharpoons 2SO_3(g)$
    (d) $BaSO_3(s) \rightleftharpoons BaO(s) + SO_2(g)$
    (e) $P_4(g) + 5O_2(g) \rightleftharpoons P_4O_{10}(s)$
    (f) $Br_2(g) \rightleftharpoons 2Br(g)$
    (g) $CH_4(g) + 2O_2(g) \rightleftharpoons CO_2(g) + 2H_2O(l)$
    (h) $CuSO_4{\cdot}5H_2O(s) \rightleftharpoons CuSO_4(s) + 5H_2O(g)$

16. Write the mathematical expression for the reaction quotient, Q_c, for each of the following reactions:
(a) $N_2(g) + 3H_2(g) \rightleftharpoons 2NH_3(g)$
(b) $4NH_3(g) + 5O_2(g) \rightleftharpoons 4NO(g) + 6H_2O(g)$
(c) $N_2O_4(g) \rightleftharpoons 2NO_2(g)$
(d) $CO_2(g) + H_2(g) \rightleftharpoons CO(g) + H_2O(g)$
(e) $NH_4Cl(s) \rightleftharpoons NH_3(g) + HCl(g)$
(f) $2Pb(NO_3)_2(s) \rightleftharpoons 2PbO(s) + 4NO_2(g) + O_2(g)$
(g) $2H_2(g) + O_2(g) \rightleftharpoons 2H_2O(l)$
(h) $S_8(g) \rightleftharpoons 8S(g)$

17. The initial concentrations or pressures of reactants and products are given for each of the following systems. Calculate the reaction quotient and determine the direction in which each system will proceed to reach equilibrium.
(a) $2NH_3(g) \rightleftharpoons N_2(g) + 3H_2(g)$ $K_c = 17$; $[NH_3] = 0.20\ M$, $[N_2] = 1.00\ M$, $[H_2] = 1.00\ M$
(b) $2NH_3(g) \rightleftharpoons N_2(g) + 3H_2(g)$ $K_P = 6.8 \times 10^4$; $NH_3 = 3.0$ atm, $N_2 = 2.0$ atm, $H_2 = 1.0$ atm
(c) $2SO_3(g) \rightleftharpoons 2SO_2(g) + O_2(g)$ $K_c = 0.230$; $[SO_3] = 0.00\ M$, $[SO_2] = 1.00\ M$, $[O_2] = 1.00\ M$
(d) $2SO_3(g) \rightleftharpoons 2SO_2(g) + O_2(g)$ $K_P = 16.5$; $SO_3 = 1.00$ atm, $SO_2 = 1.00$ atm, $O_2 = 1.00$ atm
(e) $2NO(g) + Cl_2(g) \rightleftharpoons 2NOCl(g)$ $K_c = 4.6 \times 10^4$; $[NO] = 1.00\ M$, $[Cl_2] = 1.00\ M$, $[NOCl] = 0\ M$
(f) $N_2(g) + O_2(g) \rightleftharpoons 2NO(g)$ $K_P = 0.050$; $NO = 10.0$ atm, $N_2 = O_2 = 5$ atm

18. The initial concentrations or pressures of reactants and products are given for each of the following systems. Calculate the reaction quotient and determine the direction in which each system will proceed to reach equilibrium.
(a) $2NH_3(g) \rightleftharpoons N_2(g) + 3H_2(g)$ $K_c = 17$; $[NH_3] = 0.50\ M$, $[N_2] = 0.15\ M$, $[H_2] = 0.12\ M$
(b) $2NH_3(g) \rightleftharpoons N_2(g) + 3H_2(g)$ $K_P = 6.8 \times 10^4$; $NH_3 = 2.00$ atm, $N_2 = 10.00$ atm, $H_2 = 10.00$ atm
(c) $2SO_3(g) \rightleftharpoons 2SO_2(g) + O_2(g)$ $K_c = 0.230$; $[SO_3] = 2.00\ M$, $[SO_2] = 2.00\ M$, $[O_2] = 2.00\ M$
(d) $2SO_3(g) \rightleftharpoons 2SO_2(g) + O_2(g)$ $K_P = 6.5$ atm; $SO_2 = 1.00$ atm, $O_2 = 1.130$ atm, $SO_3 = 0$ atm
(e) $2NO(g) + Cl_2(g) \rightleftharpoons 2NOCl(g)$ $K_P = 2.5 \times 10^3$; $NO = 1.00$ atm, $Cl_2 = 1.00$ atm, $NOCl = 0$ atm
(f) $N_2(g) + O_2(g) \rightleftharpoons 2NO(g)$ $K_c = 0.050$; $[N_2] = 0.100\ M$, $[O_2] = 0.200\ M$, $[NO] = 1.00\ M$

19. The following reaction has $K_P = 4.50 \times 10^{-5}$ at 720 K.
$N_2(g) + 3H_2(g) \rightleftharpoons 2NH_3(g)$
If a reaction vessel is filled with each gas to the partial pressures listed, in which direction will it shift to reach equilibrium? $P(NH_3) = 93$ atm, $P(N_2) = 48$ atm, and $P(H_2) = 52$ atm

20. Determine if the following system is at equilibrium. If not, in which direction will the system need to shift to reach equilibrium?
$SO_2Cl_2(g) \rightleftharpoons SO_2(g) + Cl_2(g)$
$[SO_2Cl_2] = 0.12\ M$, $[Cl_2] = 0.16\ M$ and $[SO_2] = 0.050\ M$. K_c for the reaction is 0.078.

21. Which of the systems described in Exercise 13.15 are homogeneous equilibria? Which are heterogeneous equilibria?

22. Which of the systems described in Exercise 13.16 are homogeneous equilibria? Which are heterogeneous equilibria?

23. For which of the reactions in Exercise 13.15 does K_c (calculated using concentrations) equal K_P (calculated using pressures)?

24. For which of the reactions in Exercise 13.16 does K_c (calculated using concentrations) equal K_P (calculated using pressures)?

25. Convert the values of K_c to values of K_P or the values of K_P to values of K_c.
(a) $N_2(g) + 3H_2(g) \rightleftharpoons 2NH_3(g)$ $K_c = 0.50$ at 400 °C
(b) $H_2(g) + I_2(g) \rightleftharpoons 2HI(g)$ $K_c = 50.2$ at 448 °C
(c) $Na_2SO_4 \cdot 10H_2O(s) \rightleftharpoons Na_2SO_4(s) + 10H_2O(g)$ $K_P = 4.08 \times 10^{-25}$ at 25 °C
(d) $H_2O(l) \rightleftharpoons H_2O(g)$ $K_P = 0.122$ at 50 °C

26. Convert the values of K_c to values of K_P or the values of K_P to values of K_c.
(a) $Cl_2(g) + Br_2(g) \rightleftharpoons 2BrCl(g)$ $K_c = 4.7 \times 10^{-2}$ at 25 °C
(b) $2SO_2(g) + O_2(g) \rightleftharpoons 2SO_3(g)$ $K_P = 48.2$ at 500 °C
(c) $CaCl_2{\cdot}6H_2O(s) \rightleftharpoons CaCl_2(s) + 6H_2O(g)$ $K_P = 5.09 \times 10^{-44}$ at 25 °C
(d) $H_2O(l) \rightleftharpoons H_2O(g)$ $K_P = 0.196$ at 60 °C

27. What is the value of the equilibrium constant expression for the change $H_2O(l) \rightleftharpoons H_2O(g)$ at 30 °C? (See Appendix E.)

28. Write the expression of the reaction quotient for the ionization of HOCN in water.

29. Write the reaction quotient expression for the ionization of NH_3 in water.

30. What is the approximate value of the equilibrium constant K_P for the change $C_2H_5OC_2H_5(l) \rightleftharpoons C_2H_5OC_2H_5(g)$ at 25 °C. (The equilibrium vapor pressure for this substance is 570 torr at 25 °C.)

13.3 Shifting Equilibria: Le Châtelier's Principle

31. The following equation represents a reversible decomposition:
$CaCO_3(s) \rightleftharpoons CaO(s) + CO_2(g)$
Under what conditions will decomposition in a closed container proceed to completion so that no $CaCO_3$ remains?

32. Explain how to recognize the conditions under which changes in volume will affect gas-phase systems at equilibrium.

33. What property of a reaction can we use to predict the effect of a change in temperature on the value of an equilibrium constant?

34. The following reaction occurs when a burner on a gas stove is lit:
$CH_4(g) + 2O_2(g) \rightleftharpoons CO_2(g) + 2H_2O(g)$
Is an equilibrium among CH_4, O_2, CO_2, and H_2O established under these conditions? Explain your answer.

35. A necessary step in the manufacture of sulfuric acid is the formation of sulfur trioxide, SO_3, from sulfur dioxide, SO_2, and oxygen, O_2, shown here. At high temperatures, the rate of formation of SO_3 is higher, but the equilibrium amount (concentration or partial pressure) of SO_3 is lower than it would be at lower temperatures.
$2SO_2(g) + O_2(g) \rightleftharpoons 2SO_3(g)$
(a) Does the equilibrium constant for the reaction increase, decrease, or remain about the same as the temperature increases?
(b) Is the reaction endothermic or exothermic?

36. Suggest four ways in which the concentration of hydrazine, N_2H_4, could be increased in an equilibrium described by the following equation:
$N_2(g) + 2H_2(g) \rightleftharpoons N_2H_4(g)$ $\Delta H = 95$ kJ

37. Suggest four ways in which the concentration of PH_3 could be increased in an equilibrium described by the following equation:
$P_4(g) + 6H_2(g) \rightleftharpoons 4PH_3(g)$ $\Delta H = 110.5$ kJ

38. How will an increase in temperature affect each of the following equilibria? How will a decrease in the volume of the reaction vessel affect each?
(a) $2NH_3(g) \rightleftharpoons N_2(g) + 3H_2(g)$ $\Delta H = 92$ kJ
(b) $N_2(g) + O_2(g) \rightleftharpoons 2NO(g)$ $\Delta H = 181$ kJ
(c) $2O_3(g) \rightleftharpoons 3O_2(g)$ $\Delta H = -285$ kJ
(d) $CaO(s) + CO_2(g) \rightleftharpoons CaCO_3(s)$ $\Delta H = -176$ kJ

39. How will an increase in temperature affect each of the following equilibria? How will a decrease in the volume of the reaction vessel affect each?
(a) $2H_2O(g) \rightleftharpoons 2H_2(g) + O_2(g)$ $\Delta H = 484$ kJ
(b) $N_2(g) + 3H_2(g) \rightleftharpoons 2NH_3(g)$ $\Delta H = -92.2$ kJ
(c) $2Br(g) \rightleftharpoons Br_2(g)$ $\Delta H = -224$ kJ
(d) $H_2(g) + I_2(s) \rightleftharpoons 2HI(g)$ $\Delta H = 53$ kJ

40. Methanol can be prepared from carbon monoxide and hydrogen at high temperature and pressure in the presence of a suitable catalyst.
(a) Write the expression for the equilibrium constant (K_c) for the reversible reaction
$2H_2(g) + CO(g) \rightleftharpoons CH_3OH(g) \qquad \Delta H = -90.2 \text{ kJ}$
(b) What will happen to the concentrations of H_2, CO, and CH_3OH at equilibrium if more H_2 is added?
(c) What will happen to the concentrations of H_2, CO, and CH_3OH at equilibrium if CO is removed?
(d) What will happen to the concentrations of H_2, CO, and CH_3OH at equilibrium if CH_3OH is added?
(e) What will happen to the concentrations of H_2, CO, and CH_3OH at equilibrium if the temperature of the system is increased?
(f) What will happen to the concentrations of H_2, CO, and CH_3OH at equilibrium if more catalyst is added?

41. Nitrogen and oxygen react at high temperatures.
(a) Write the expression for the equilibrium constant (K_c) for the reversible reaction
$N_2(g) + O_2(g) \rightleftharpoons 2NO(g) \qquad \Delta H = 181 \text{ kJ}$
(b) What will happen to the concentrations of N_2, O_2, and NO at equilibrium if more O_2 is added?
(c) What will happen to the concentrations of N_2, O_2, and NO at equilibrium if N_2 is removed?
(d) What will happen to the concentrations of N_2, O_2, and NO at equilibrium if NO is added?
(e) What will happen to the concentrations of N_2, O_2, and NO at equilibrium if the volume of the reaction vessel is decreased?
(f) What will happen to the concentrations of N_2, O_2, and NO at equilibrium if the temperature of the system is increased?
(g) What will happen to the concentrations of N_2, O_2, and NO at equilibrium if a catalyst is added?

42. Water gas, a mixture of H_2 and CO, is an important industrial fuel produced by the reaction of steam with red hot coke, essentially pure carbon.
(a) Write the expression for the equilibrium constant for the reversible reaction
$C(s) + H_2O(g) \rightleftharpoons CO(g) + H_2(g) \qquad \Delta H = 131.30 \text{ kJ}$
(b) What will happen to the concentration of each reactant and product at equilibrium if more C is added?
(c) What will happen to the concentration of each reactant and product at equilibrium if H_2O is removed?
(d) What will happen to the concentration of each reactant and product at equilibrium if CO is added?
(e) What will happen to the concentration of each reactant and product at equilibrium if the temperature of the system is increased?

43. Pure iron metal can be produced by the reduction of iron(III) oxide with hydrogen gas.
(a) Write the expression for the equilibrium constant (K_c) for the reversible reaction
$Fe_2O_3(s) + 3H_2(g) \rightleftharpoons 2Fe(s) + 3H_2O(g) \qquad \Delta H = 98.7 \text{ kJ}$
(b) What will happen to the concentration of each reactant and product at equilibrium if more Fe is added?
(c) What will happen to the concentration of each reactant and product at equilibrium if H_2O is removed?
(d) What will happen to the concentration of each reactant and product at equilibrium if H_2 is added?
(e) What will happen to the concentration of each reactant and product at equilibrium if the volume of the reaction vessel is decreased?
(f) What will happen to the concentration of each reactant and product at equilibrium if the temperature of the system is increased?

44. Ammonia is a weak base that reacts with water according to this equation:
$NH_3(aq) + H_2O(l) \rightleftharpoons NH_4^+(aq) + OH^-(aq)$
Will any of the following increase the percent of ammonia that is converted to the ammonium ion in water?
(a) Addition of NaOH
(b) Addition of HCl
(c) Addition of NH_4Cl

45. Acetic acid is a weak acid that reacts with water according to this equation:
$CH_3CO_2H(aq) + H_2O(aq) \rightleftharpoons H_3O^+(aq) + CH_3CO_2^-(aq)$
Will any of the following increase the percent of acetic acid that reacts and produces $CH_3CO_2^-$ ion?
(a) Addition of HCl
(b) Addition of NaOH
(c) Addition of $NaCH_3CO_2$

46. Suggest two ways in which the equilibrium concentration of Ag^+ can be reduced in a solution of Na^+, Cl^-, Ag^+, and NO_3^-, in contact with solid AgCl.
$Na^+(aq) + Cl^-(aq) + Ag^+(aq) + NO_3^-(aq) \rightleftharpoons AgCl(s) + Na^+(aq) + NO_3^-(aq)$
$\Delta H = -65.9\ \text{kJ}$

47. How can the pressure of water vapor be increased in the following equilibrium?
$H_2O(l) \rightleftharpoons H_2O(g) \qquad \Delta H = 41\ \text{kJ}$

48. A solution is saturated with silver sulfate and contains excess solid silver sulfate:
$Ag_2SO_4(s) \rightleftharpoons 2Ag^+(aq) + SO_4^{2-}(aq)$
A small amount of solid silver sulfate containing a radioactive isotope of silver is added to this solution. Within a few minutes, a portion of the solution phase is sampled and tests positive for radioactive Ag^+ ions. Explain this observation.

49. When equal molar amounts of HCl and HOCl are dissolved separately in equal amounts of water, the solution of HCl freezes at a lower temperature. Which compound has the larger equilibrium constant for acid ionization?
(a) HCl
(b) $H^+ + Cl^-$
(c) HOCl
(d) $H^+ + OCl^-$

13.4 Equilibrium Calculations

50. A reaction is represented by this equation: $A(aq) + 2B(aq) \rightleftharpoons 2C(aq) \qquad K_c = 1 \times 10^3$
(a) Write the mathematical expression for the equilibrium constant.
(b) Using concentrations ≤1 *M*, identify two sets of concentrations that describe a mixture of A, B, and C at equilibrium.

51. A reaction is represented by this equation: $2W(aq) \rightleftharpoons X(aq) + 2Y(aq) \qquad K_c = 5 \times 10^{-4}$
(a) Write the mathematical expression for the equilibrium constant.
(b) Using concentrations of ≤1 *M*, identify two sets of concentrations that describe a mixture of W, X, and Y at equilibrium.

52. What is the value of the equilibrium constant at 500 °C for the formation of NH_3 according to the following equation?
$N_2(g) + 3H_2(g) \rightleftharpoons 2NH_3(g)$
An equilibrium mixture of $NH_3(g)$, $H_2(g)$, and $N_2(g)$ at 500 °C was found to contain 1.35 *M* H_2, 1.15 *M* N_2, and 4.12×10^{-1} *M* NH_3.

53. Hydrogen is prepared commercially by the reaction of methane and water vapor at elevated temperatures.
$CH_4(g) + H_2O(g) \rightleftharpoons 3H_2(g) + CO(g)$
What is the equilibrium constant for the reaction if a mixture at equilibrium contains gases with the following concentrations: CH_4, 0.126 *M*; H_2O, 0.242 *M*; CO, 0.126 *M*; H_2 1.15 *M*, at a temperature of 760 °C?

54. A 0.72-mol sample of PCl_5 is put into a 1.00-L vessel and heated. At equilibrium, the vessel contains 0.40 mol of $PCl_3(g)$ and 0.40 mol of $Cl_2(g)$. Calculate the value of the equilibrium constant for the decomposition of PCl_5 to PCl_3 and Cl_2 at this temperature.

55. At 1 atm and 25 °C, NO_2 with an initial concentration of 1.00 *M* is 0.0033% decomposed into NO and O_2. Calculate the value of the equilibrium constant for the reaction.
$2NO_2(g) \rightleftharpoons 2NO(g) + O_2(g)$

56. Calculate the value of the equilibrium constant K_P for the reaction $2NO(g) + Cl_2(g) \rightleftharpoons 2NOCl(g)$ from these equilibrium pressures: NO, 0.050 atm; Cl_2, 0.30 atm; NOCl, 1.2 atm.

57. When heated, iodine vapor dissociates according to this equation:
$I_2(g) \rightleftharpoons 2I(g)$
At 1274 K, a sample exhibits a partial pressure of I_2 of 0.1122 atm and a partial pressure due to I atoms of 0.1378 atm. Determine the value of the equilibrium constant, K_P, for the decomposition at 1274 K.

58. A sample of ammonium chloride was heated in a closed container.

$NH_4Cl(s) \rightleftharpoons NH_3(g) + HCl(g)$

At equilibrium, the pressure of $NH_3(g)$ was found to be 1.75 atm. What is the value of the equilibrium constant K_P for the decomposition at this temperature?

59. At a temperature of 60 °C, the vapor pressure of water is 0.196 atm. What is the value of the equilibrium constant K_P for the vaporization equilibrium at 60 °C?

$H_2O(l) \rightleftharpoons H_2O(g)$

60. Complete the following partial ICE tables.

(a)

	$2SO_3(g)$	$\rightleftharpoons$	$2SO_2(g)+$	$O_2(g)$
change	___		___	$+x$

(b)

	$4NH_3(g)$	$+3O_2(g)$	$\rightleftharpoons$	$2N_2(g)+$	$6H_2O(g)$
change	___	$+x$		___	___

(c)

	$2CH_4(g)$	$\rightleftharpoons$	$C_2H_2(g)+$	$3H_2(g)$
change	___		$+x$	___

(d)

	$CH_4(g)+$	$H_2O(g)$	$\rightleftharpoons$	$CO(g)+$	$3H_2(g)$
change	___	$+x$		___	___

(e)

$NH_4Cl(s)$	$\rightleftharpoons$	$NH_3(g)+$	$HCl(g)$
change		$+x$	___

(f)

$Ni(s)+$	$4CO(g)$	$\rightleftharpoons$	$Ni(CO)_4(g)$
change	$+x$		___

61. Complete the following partial ICE tables.

(a)

	$2H_2(g)+$	$O_2(g)$	$\rightleftharpoons$	$2H_2O(g)$
change	___	___		$+x$

(b)

	$CS_2(g)+$	$4H_2(g)$	$\rightleftharpoons$	$CH_4(g)+$	$2H_2S(g)$
change	$+x$	___		___	___

(c)

	$H_2(g)+$	$Cl_2(g)$	$\rightleftharpoons$	$2HCl(g)$
change	$+x$	___		___

(d)

	$2NH_3(g)$	$+2O_2(g)$	$\rightleftharpoons$	$N_2O(g)+$	$3H_2O(g)$
change	___	___		___	$+x$

(e)

$NH_4HS(s)$	$\rightleftharpoons$	$NH_3(g)+$	$H_2S(g)$
change		$+x$	___

(f)

$Fe(s)+$	$5CO(g)$	$\rightleftharpoons$	$Fe(CO)_5(g)$
change	___		$+x$

62. Why are there no changes specified for Ni in Exercise 13.60, part (f)? What property of Ni does change?

63. Why are there no changes specified for NH_4HS in Exercise 13.61, part (e)? What property of NH_4HS does change?

64. Analysis of the gases in a sealed reaction vessel containing NH_3, N_2, and H_2 at equilibrium at 400 °C established the concentration of N_2 to be 1.2 *M* and the concentration of H_2 to be 0.24 *M*.
$N_2(g) + 3H_2(g) \rightleftharpoons 2NH_3(g) \qquad K_c = 0.50 \text{ at } 400\ °\text{C}$
Calculate the equilibrium molar concentration of NH_3.

65. Calculate the number of moles of HI that are at equilibrium with 1.25 mol of H_2 and 1.25 mol of I_2 in a 5.00–L flask at 448 °C.
$H_2 + I_2 \rightleftharpoons 2HI \qquad K_c = 50.2 \text{ at } 448\ °\text{C}$

66. What is the pressure of BrCl in an equilibrium mixture of Cl_2, Br_2, and BrCl if the pressure of Cl_2 in the mixture is 0.115 atm and the pressure of Br_2 in the mixture is 0.450 atm?
$Cl_2(g) + Br_2(g) \rightleftharpoons 2BrCl(g) \qquad K_P = 4.7 \times 10^{-2}$

67. What is the pressure of CO_2 in a mixture at equilibrium that contains 0.50 atm H_2, 2.0 atm of H_2O, and 1.0 atm of CO at 990 °C?
$H_2(g) + CO_2(g) \rightleftharpoons H_2O(g) + CO(g) \qquad K_P = 1.6 \text{ at } 990\ °\text{C}$

68. Cobalt metal can be prepared by reducing cobalt(II) oxide with carbon monoxide.
$CoO(s) + CO(g) \rightleftharpoons Co(s) + CO_2(g) \qquad K_c = 4.90 \times 10^2 \text{ at } 550\ °\text{C}$
What concentration of CO remains in an equilibrium mixture with $[CO_2]$ = 0.100 *M*?

69. Carbon reacts with water vapor at elevated temperatures.
$C(s) + H_2O(g) \rightleftharpoons CO(g) + H_2(g) \qquad K_c = 0.2 \text{ at } 1000\ °\text{C}$
Assuming a reaction mixture initially contains only reactants, what is the concentration of CO in an equilibrium mixture with $[H_2O]$ = 0.500 *M* at 1000 °C?

70. Sodium sulfate 10–hydrate, $Na_2SO_4{\cdot}10H_2O$, dehydrates according to the equation
$Na_2SO_4{\cdot}10H_2O(s) \rightleftharpoons Na_2SO_4(s) + 10H_2O(g) \qquad K_P = 4.08 \times 10^{-25} \text{ at } 25\ °\text{C}$
What is the pressure of water vapor at equilibrium with a mixture of $Na_2SO_4{\cdot}10H_2O$ and $NaSO_4$?

71. Calcium chloride 6–hydrate, $CaCl_2{\cdot}6H_2O$, dehydrates according to the equation
$CaCl_2{\cdot}6H_2O(s) \rightleftharpoons CaCl_2(s) + 6H_2O(g) \qquad K_P = 5.09 \times 10^{-44} \text{ at } 25\ °\text{C}$
What is the pressure of water vapor at equilibrium with a mixture of $CaCl_2{\cdot}6H_2O$ and $CaCl_2$ at 25 °C?

72. A student solved the following problem and found the equilibrium concentrations to be $[SO_2]$ = 0.590 *M*, $[O_2]$ = 0.0450 *M*, and $[SO_3]$ = 0.260 *M*. How could this student check the work without reworking the problem? The problem was: For the following reaction at 600 °C:
$2SO_2(g) + O_2(g) \rightleftharpoons 2SO_3(g) \qquad K_c = 4.32$

73. A student solved the following problem and found $[N_2O_4]$ = 0.16 *M* at equilibrium. How could this student recognize that the answer was wrong without reworking the problem? The problem was: What is the equilibrium concentration of N_2O_4 in a mixture formed from a sample of NO_2 with a concentration of 0.10 *M*?
$2NO_2(g) \rightleftharpoons N_2O_4(g) \qquad K_c = 160$

74. Assume that the change in concentration of N_2O_4 is small enough to be neglected in the following problem.
(a) Calculate the equilibrium concentration of both species in 1.00 L of a solution prepared from 0.129 mol of N_2O_4 with chloroform as the solvent.
$N_2O_4(g) \rightleftharpoons 2NO_2(g) \qquad K_c = 1.07 \times 10^{-5} \text{ in chloroform}$
(b) Confirm that the change is small enough to be neglected.

75. Assume that the change in concentration of $COCl_2$ is small enough to be neglected in the following problem.
(a) Calculate the equilibrium concentration of all species in an equilibrium mixture that results from the decomposition of $COCl_2$ with an initial concentration of 0.3166 *M*.
$COCl_2(g) \rightleftharpoons CO(g) + Cl_2(g) \qquad K_c = 2.2 \times 10^{-10}$
(b) Confirm that the change is small enough to be neglected.

76. Assume that the change in pressure of H_2S is small enough to be neglected in the following problem.
(a) Calculate the equilibrium pressures of all species in an equilibrium mixture that results from the decomposition of H_2S with an initial pressure of 0.824 atm.
$2H_2S(g) \rightleftharpoons 2H_2(g) + S_2(g) \qquad K_P = 2.2 \times 10^{-6}$
(b) Confirm that the change is small enough to be neglected.

77. What are all concentrations after a mixture that contains $[H_2O] = 1.00\ M$ and $[Cl_2O] = 1.00\ M$ comes to equilibrium at 25 °C?
$H_2O(g) + Cl_2O(g) \rightleftharpoons 2HOCl(g) \qquad K_c = 0.0900$

78. What are the concentrations of PCl_5, PCl_3, and Cl_2 in an equilibrium mixture produced by the decomposition of a sample of pure PCl_5 with $[PCl_5] = 2.00\ M$?
$PCl_5(g) \rightleftharpoons PCl_3(g) + Cl_2(g) \qquad K_c = 0.0211$

79. Calculate the number of grams of HI that are at equilibrium with 1.25 mol of H_2 and 63.5 g of iodine at 448 °C.
$H_2 + I_2 \rightleftharpoons 2HI \qquad K_c = 50.2$ at 448 °C

80. Butane exists as two isomers, *n*–butane and isobutane.

$CH_3CH_2CH_2CH_3 \rightleftharpoons CH_3CH(CH_3)CH_3$

n-butane isobutane

$K_P = 2.5$ at 25 °C
What is the pressure of isobutane in a container of the two isomers at equilibrium with a total pressure of 1.22 atm?

81. What is the minimum mass of $CaCO_3$ required to establish equilibrium at a certain temperature in a 6.50-L container if the equilibrium constant (K_c) is 0.50 for the decomposition reaction of $CaCO_3$ at that temperature?
$CaCO_3(s) \rightleftharpoons CaO(s) + CO_2(g)$

82. The equilibrium constant (K_c) for this reaction is 1.60 at 990 °C:
$H_2(g) + CO_2(g) \rightleftharpoons H_2O(g) + CO(g)$
Calculate the number of moles of each component in the final equilibrium mixture obtained from adding 1.00 mol of H_2, 2.00 mol of CO_2, 0.750 mol of H_2O, and 1.00 mol of CO to a 5.00-L container at 990 °C.

83. In a 3.0-L vessel, the following equilibrium partial pressures are measured: N_2, 190 torr; H_2, 317 torr; NH_3, 1.00×10^3 torr.
$N_2(g) + 3H_2(g) \rightleftharpoons 2NH_3(g)$
(a) How will the partial pressures of H_2, N_2, and NH_3 change if H_2 is removed from the system? Will they increase, decrease, or remain the same?
(b) Hydrogen is removed from the vessel until the partial pressure of nitrogen, at equilibrium, is 250 torr. Calculate the partial pressures of the other substances under the new conditions.

84. The equilibrium constant (K_c) for this reaction is 5.0 at a given temperature.
$CO(g) + H_2O(g) \rightleftharpoons CO_2(g) + H_2(g)$
(a) On analysis, an equilibrium mixture of the substances present at the given temperature was found to contain 0.20 mol of CO, 0.30 mol of water vapor, and 0.90 mol of H_2 in a liter. How many moles of CO_2 were there in the equilibrium mixture?
(b) Maintaining the same temperature, additional H_2 was added to the system, and some water vapor was removed by drying. A new equilibrium mixture was thereby established containing 0.40 mol of CO, 0.30 mol of water vapor, and 1.2 mol of H_2 in a liter. How many moles of CO_2 were in the new equilibrium mixture? Compare this with the quantity in part (a), and discuss whether the second value is reasonable. Explain how it is possible for the water vapor concentration to be the same in the two equilibrium solutions even though some vapor was removed before the second equilibrium was established.

85. Antimony pentachloride decomposes according to this equation:
$SbCl_5(g) \rightleftharpoons SbCl_3(g) + Cl_2(g)$
An equilibrium mixture in a 5.00-L flask at 448 °C contains 3.85 g of $SbCl_5$, 9.14 g of $SbCl_3$, and 2.84 g of Cl_2. How many grams of each will be found if the mixture is transferred into a 2.00-L flask at the same temperature?

86. Consider the equilibrium

$$4NO_2(g) + 6H_2O(g) \rightleftharpoons 4NH_3(g) + 7O_2(g)$$

(a) What is the expression for the equilibrium constant (K_c) of the reaction?

(b) How must the concentration of NH_3 change to reach equilibrium if the reaction quotient is less than the equilibrium constant?

(c) If the reaction were at equilibrium, how would an increase in the volume of the reaction vessel affect the pressure of NO_2?

(d) If the change in the pressure of NO_2 is 28 torr as a mixture of the four gases reaches equilibrium, how much will the pressure of O_2 change?

87. The binding of oxygen by hemoglobin (Hb), giving oxyhemoglobin (HbO_2), is partially regulated by the concentration of H_3O^+ and dissolved CO_2 in the blood. Although the equilibrium is complicated, it can be summarized as

$$HbO_2(aq) + H_3O^+(aq) + CO_2(g) \rightleftharpoons CO_2{-}Hb{-}H^+ + O_2(g) + H_2O(l)$$

(a) Write the equilibrium constant expression for this reaction.

(b) Explain why the production of lactic acid and CO_2 in a muscle during exertion stimulates release of O_2 from the oxyhemoglobin in the blood passing through the muscle.

88. Liquid N_2O_3 is dark blue at low temperatures, but the color fades and becomes greenish at higher temperatures as the compound decomposes to NO and NO_2. At 25 °C, a value of $K_P = 1.91$ has been established for this decomposition. If 0.236 moles of N_2O_3 are placed in a 1.52-L vessel at 25 °C, calculate the equilibrium partial pressures of $N_2O_3(g)$, $NO_2(g)$, and $NO(g)$.

89. A 1.00-L vessel at 400 °C contains the following equilibrium concentrations: N_2, 1.00 *M*; H_2, 0.50 *M*; and NH_3, 0.25 *M*. How many moles of hydrogen must be removed from the vessel to increase the concentration of nitrogen to 1.1 *M*? The equilibrium reaction is

$$N_2(g) + 3H_2(g) \rightleftharpoons 2NH_3(g)$$

CHAPTER 14
Acid-Base Equilibria

Figure 14.1 Sinkholes such as this are the result of reactions between acidic groundwaters and basic rock formations, like limestone. (credit: modification of work by Emil Kehnel)

CHAPTER OUTLINE

14.1 Brønsted-Lowry Acids and Bases
14.2 pH and pOH
14.3 Relative Strengths of Acids and Bases
14.4 Hydrolysis of Salts
14.5 Polyprotic Acids
14.6 Buffers
14.7 Acid-Base Titrations

INTRODUCTION Liquid water is essential to life on our planet, and chemistry involving the characteristic ions of water, H^+ and OH^-, is widely encountered in nature and society. As introduced in another chapter of this text, acid-base chemistry involves the transfer of hydrogen ions from donors (acids) to acceptors (bases). These H+ transfer reactions are reversible, and the equilibria established by acid-base systems are essential aspects of phenomena ranging from sinkhole formation (Figure 14.1) to oxygen transport in the human body. This chapter will further explore acid-base chemistry with an emphasis on the equilibrium aspects of this important reaction class.

14.1 Brønsted-Lowry Acids and Bases

LEARNING OBJECTIVES

By the end of this section, you will be able to:

- Identify acids, bases, and conjugate acid-base pairs according to the Brønsted-Lowry definition
- Write equations for acid and base ionization reactions
- Use the ion-product constant for water to calculate hydronium and hydroxide ion concentrations
- Describe the acid-base behavior of amphiprotic substances

The acid-base reaction class has been studied for quite some time. In 1680, Robert Boyle reported traits of acid solutions that included their ability to dissolve many substances, to change the colors of certain natural dyes, and to lose these traits after coming in contact with alkali (base) solutions. In the eighteenth century, it was recognized that acids have a sour taste, react with limestone to liberate a gaseous substance (now known to be CO_2), and interact with alkalis to form neutral substances. In 1815, Humphry Davy contributed greatly to the development of the modern acid-base concept by demonstrating that hydrogen is the essential constituent of acids. Around that same time, Joseph Louis Gay-Lussac concluded that acids are substances that can neutralize bases and that these two classes of substances can be defined only in terms of each other. The significance of hydrogen was reemphasized in 1884 when Svante Arrhenius defined an acid as a compound that dissolves in water to yield hydrogen cations (now recognized to be hydronium ions) and a base as a compound that dissolves in water to yield hydroxide anions.

Johannes Brønsted and Thomas Lowry proposed a more general description in 1923 in which acids and bases were defined in terms of the transfer of hydrogen ions, H^+. (Note that these hydrogen ions are often referred to simply as *protons*, since that subatomic particle is the only component of cations derived from the most abundant hydrogen isotope, ^1H.) A compound that donates a proton to another compound is called a **Brønsted-Lowry acid**, and a compound that accepts a proton is called a **Brønsted-Lowry base**. An acid-base reaction is, thus, the transfer of a proton from a donor (acid) to an acceptor (base).

The concept of *conjugate pairs* is useful in describing Brønsted-Lowry acid-base reactions (and other reversible reactions, as well). When an acid donates H^+, the species that remains is called the **conjugate base** of the acid because it reacts as a proton acceptor in the reverse reaction. Likewise, when a base accepts H^+, it is converted to its **conjugate acid**. The reaction between water and ammonia illustrates this idea. In the forward direction, water acts as an acid by donating a proton to ammonia and subsequently becoming a hydroxide ion, OH^-, the conjugate base of water. The ammonia acts as a base in accepting this proton, becoming an ammonium ion, $NH_4{}^+$, the conjugate acid of ammonia. In the reverse direction, a hydroxide ion acts as a base in accepting a proton from ammonium ion, which acts as an acid.

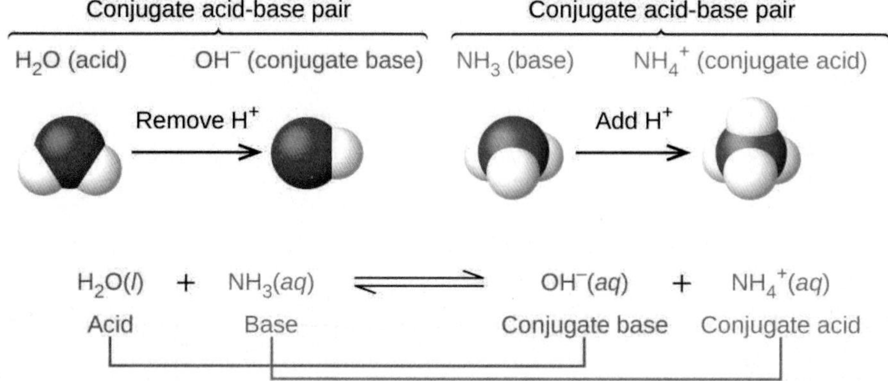

The reaction between a Brønsted-Lowry acid and water is called **acid ionization**. For example, when hydrogen fluoride dissolves in water and ionizes, protons are transferred from hydrogen fluoride molecules to water molecules, yielding hydronium ions and fluoride ions:

$$\underset{\text{Acid}}{HF} + \underset{\text{Base}}{H_2O} \rightleftharpoons \underset{\text{Acid}}{H_3O^+} + \underset{\text{Base}}{F^-}$$

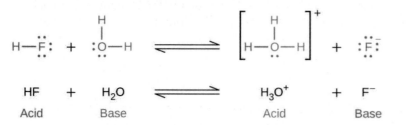

Base ionization of a species occurs when it accepts protons from water molecules. In the example below, pyridine molecules, C_5NH_5, undergo base ionization when dissolved in water, yielding hydroxide and pyridinium ions:

$$H_2O + C_5NH_5 \rightleftharpoons C_5NH_6^+ + OH^-$$

Acid Base Acid Base

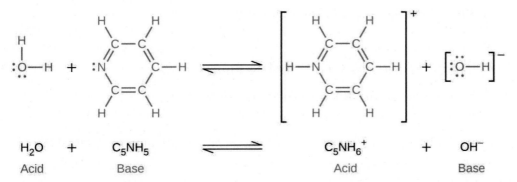

The preceding ionization reactions suggest that water may function as both a base (as in its reaction with hydrogen fluoride) and an acid (as in its reaction with ammonia). Species capable of either donating or accepting protons are called **amphiprotic**, or more generally, **amphoteric**, a term that may be used for acids and bases per definitions other than the Brønsted-Lowry one. The equations below show the two possible acid-base reactions for two amphiprotic species, bicarbonate ion and water:

$$HCO_3^-(aq) + H_2O(l) \qquad CO_3^{2-}(aq) + H_3O^+(aq)$$

$$HCO_3^-(aq) + H_2O(l) \qquad H_2CO_3(aq) + OH^-(aq)$$

The first equation represents the reaction of bicarbonate as an acid with water as a base, whereas the second represents reaction of bicarbonate as a base with water as an acid. When bicarbonate is added to water, both these equilibria are established simultaneously and the composition of the resulting solution may be determined through appropriate equilibrium calculations, as described later in this chapter.

In the liquid state, molecules of an amphiprotic substance can react with one another as illustrated for water in the equations below:

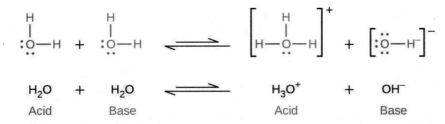

$$H_2O + H_2O \rightleftharpoons H_3O^+ + OH^-$$

Acid Base Acid Base

The process in which like molecules react to yield ions is called **autoionization**. Liquid water undergoes autoionization to a very slight extent; at 25 °C, approximately two out of every billion water molecules are ionized. The extent of the water autoionization process is reflected in the value of its equilibrium constant, the **ion-product constant for water, K_w**:

$$H_2O(l) + H_2O(l) \rightleftharpoons H_3O^+(aq) + OH^-(aq) \qquad K_w = [H_3O^+][OH^-]$$

The slight ionization of pure water is reflected in the small value of the equilibrium constant; at 25 °C, K_w has a value of 1.0×10^{-14}. The process is endothermic, and so the extent of ionization and the resulting concentrations of hydronium ion and hydroxide ion increase with temperature. For example, at 100 °C, the value for K_w is about 5.6×10^{-13}, roughly 50 times larger than the value at 25 °C.

EXAMPLE 14.1

Ion Concentrations in Pure Water

What are the hydronium ion concentration and the hydroxide ion concentration in pure water at 25 °C?

Solution

The autoionization of water yields the same number of hydronium and hydroxide ions. Therefore, in pure water, $[H_3O^+] = [OH^-] = x$. At 25 °C:

$$K_w = [H_3O^+][OH^-] = (x)(x) = x^2 = 1.0 \times 10^{-14}$$

So:

$$x = [H_3O^+] = [OH^-] = \sqrt{1.0 \times 10^{-14}} = 1.0 \times 10^{-7}\ M$$

The hydronium ion concentration and the hydroxide ion concentration are the same, 1.0×10^{-7} *M*.

Check Your Learning

The ion product of water at 80 °C is 2.4×10^{-13}. What are the concentrations of hydronium and hydroxide ions in pure water at 80 °C?

Answer:

$[H_3O^+] = [OH^-] = 4.9 \times 10^{-7}\ M$

EXAMPLE 14.2

The Inverse Relation between $[H_3O^+]$ and $[OH^-]$

A solution of an acid in water has a hydronium ion concentration of 2.0×10^{-6} *M*. What is the concentration of hydroxide ion at 25 °C?

Solution

Use the value of the ion-product constant for water at 25 °C

$$2H_2O(l) \rightleftharpoons H_3O^+(aq) + OH^-(aq) \qquad K_w = [H_3O^+][OH^-] = 1.0 \times 10^{-14}$$

to calculate the missing equilibrium concentration.

Rearrangement of the K_w expression shows that $[OH^-]$ is inversely proportional to $[H_3O^+]$:

$$[OH^-] = \frac{K_w}{[H_3O^+]} = \frac{1.0 \times 10^{-14}}{2.0 \times 10^{-6}} = 5.0 \times 10^{-9}$$

Compared with pure water, a solution of acid exhibits a higher concentration of hydronium ions (due to ionization of the acid) and a proportionally lower concentration of hydroxide ions. This may be explained via Le Châtelier's principle as a left shift in the water autoionization equilibrium resulting from the stress of increased hydronium ion concentration.

Substituting the ion concentrations into the K_w expression confirms this calculation, resulting in the expected value:

$$K_w = [H_3O^+][OH^-] = (2.0 \times 10^{-6})(5.0 \times 10^{-9}) = 1.0 \times 10^{-14}$$

Check Your Learning

What is the hydronium ion concentration in an aqueous solution with a hydroxide ion concentration of 0.001 *M* at 25 °C?

Answer:

$[H_3O^+] = 1 \times 10^{-11}\ M$

EXAMPLE 14.3

Representing the Acid-Base Behavior of an Amphoteric Substance

Write separate equations representing the reaction of $HSO_3{}^-$

(a) as an acid with OH^-

(b) as a base with HI

Solution

(a) $HSO_3^-(aq) + OH^-(aq) \rightleftharpoons SO_3^{2-}(aq) + H_2O(l)$

(b) $HSO_3^-(aq) + HI(aq) \rightleftharpoons H_2SO_3(aq) + I^-(aq)$

Check Your Learning

Write separate equations representing the reaction of $H_2PO_4^-$

(a) as a base with HBr

(b) as an acid with OH^-

Answer:

(a) $H_2PO_4^-(aq) + HBr(aq) \rightleftharpoons H_3PO_4(aq) + Br^-(aq)$; (b) $H_2PO_4^-(aq) + OH^-(aq) \rightleftharpoons HPO_4^{2-}(aq) + H_2O(l)$

14.2 pH and pOH

LEARNING OBJECTIVES

By the end of this section, you will be able to:

- Explain the characterization of aqueous solutions as acidic, basic, or neutral
- Express hydronium and hydroxide ion concentrations on the pH and pOH scales
- Perform calculations relating pH and pOH

As discussed earlier, hydronium and hydroxide ions are present both in pure water and in all aqueous solutions, and their concentrations are inversely proportional as determined by the ion product of water (K_w). The concentrations of these ions in a solution are often critical determinants of the solution's properties and the chemical behaviors of its other solutes, and specific vocabulary has been developed to describe these concentrations in relative terms. A solution is **neutral** if it contains equal concentrations of hydronium and hydroxide ions; **acidic** if it contains a greater concentration of hydronium ions than hydroxide ions; and **basic** if it contains a lesser concentration of hydronium ions than hydroxide ions.

A common means of expressing quantities that may span many orders of magnitude is to use a logarithmic scale. One such scale that is very popular for chemical concentrations and equilibrium constants is based on the p-function, defined as shown where "X" is the quantity of interest and "log" is the base-10 logarithm:

$$pX = -\log X$$

The **pH** of a solution is therefore defined as shown here, where $[H_3O^+]$ is the molar concentration of hydronium ion in the solution:

$$pH = -\log [H_3O^+]$$

Rearranging this equation to isolate the hydronium ion molarity yields the equivalent expression:

$$[H_3O^+] = 10^{-pH}$$

Likewise, the hydroxide ion molarity may be expressed as a p-function, or **pOH**:

$$pOH = -\log [OH^-]$$

or

$$\left[OH^-\right] = 10^{-pOH}$$

Finally, the relation between these two ion concentration expressed as p-functions is easily derived from the K_w expression:

$$K_w = [H_3O^+]\left[OH^-\right]$$

$$-\log K_w = -\log\left([H_3O^+][OH^-]\right) = -\log[H_3O^+] + -\log[OH^-]$$

$$pK_w = pH + pOH$$

At 25 °C, the value of K_w is 1.0×10^{-14}, and so:

$$14.00 = pH + pOH$$

As was shown in Example 14.1, the hydronium ion molarity in pure water (or any neutral solution) is 1.0×10^{-7} *M* at 25 °C. The pH and pOH of a neutral solution at this temperature are therefore:

$$pH = -\log[H_3O^+] = -\log(1.0 \times 10^{-7}) = 7.00$$

$$pOH = -\log[OH^-] = -\log(1.0 \times 10^{-7}) = 7.00$$

And so, *at this temperature*, acidic solutions are those with hydronium ion molarities greater than 1.0×10^{-7} *M* and hydroxide ion molarities less than 1.0×10^{-7} *M* (corresponding to pH values less than 7.00 and pOH values greater than 7.00). Basic solutions are those with hydronium ion molarities less than 1.0×10^{-7} *M* and hydroxide ion molarities greater than 1.0×10^{-7} *M* (corresponding to pH values greater than 7.00 and pOH values less than 7.00).

Since the autoionization constant K_w is temperature dependent, these correlations between pH values and the acidic/neutral/basic adjectives will be different at temperatures other than 25 °C. For example, the "Check Your Learning" exercise accompanying Example 14.1 showed the hydronium molarity of pure water at 80 °C is 4.9×10^{-7} *M*, which corresponds to pH and pOH values of:

$$pH = -\log[H_3O^+] = -\log(4.9 \times 10^{-7}) = 6.31$$

$$pOH = -\log[OH^-] = -\log(4.9 \times 10^{-7}) = 6.31$$

At this temperature, then, neutral solutions exhibit pH = pOH = 6.31, acidic solutions exhibit pH less than 6.31 and pOH greater than 6.31, whereas basic solutions exhibit pH greater than 6.31 and pOH less than 6.31. This distinction can be important when studying certain processes that occur at other temperatures, such as enzyme reactions in warm-blooded organisms at a temperature around 36–40 °C. Unless otherwise noted, references to pH values are presumed to be those at 25 °C (Table 14.1).

Summary of Relations for Acidic, Basic and Neutral Solutions

Classification	Relative Ion Concentrations	pH at 25 °C
acidic	$[H_3O^+] > [OH^-]$	$pH < 7$
neutral	$[H_3O^+] = [OH^-]$	$pH = 7$
basic	$[H_3O^+] < [OH^-]$	$pH > 7$

TABLE 14.1

Figure 14.2 shows the relationships between $[H_3O^+]$, $[OH^-]$, pH, and pOH for solutions classified as acidic, basic, and neutral.

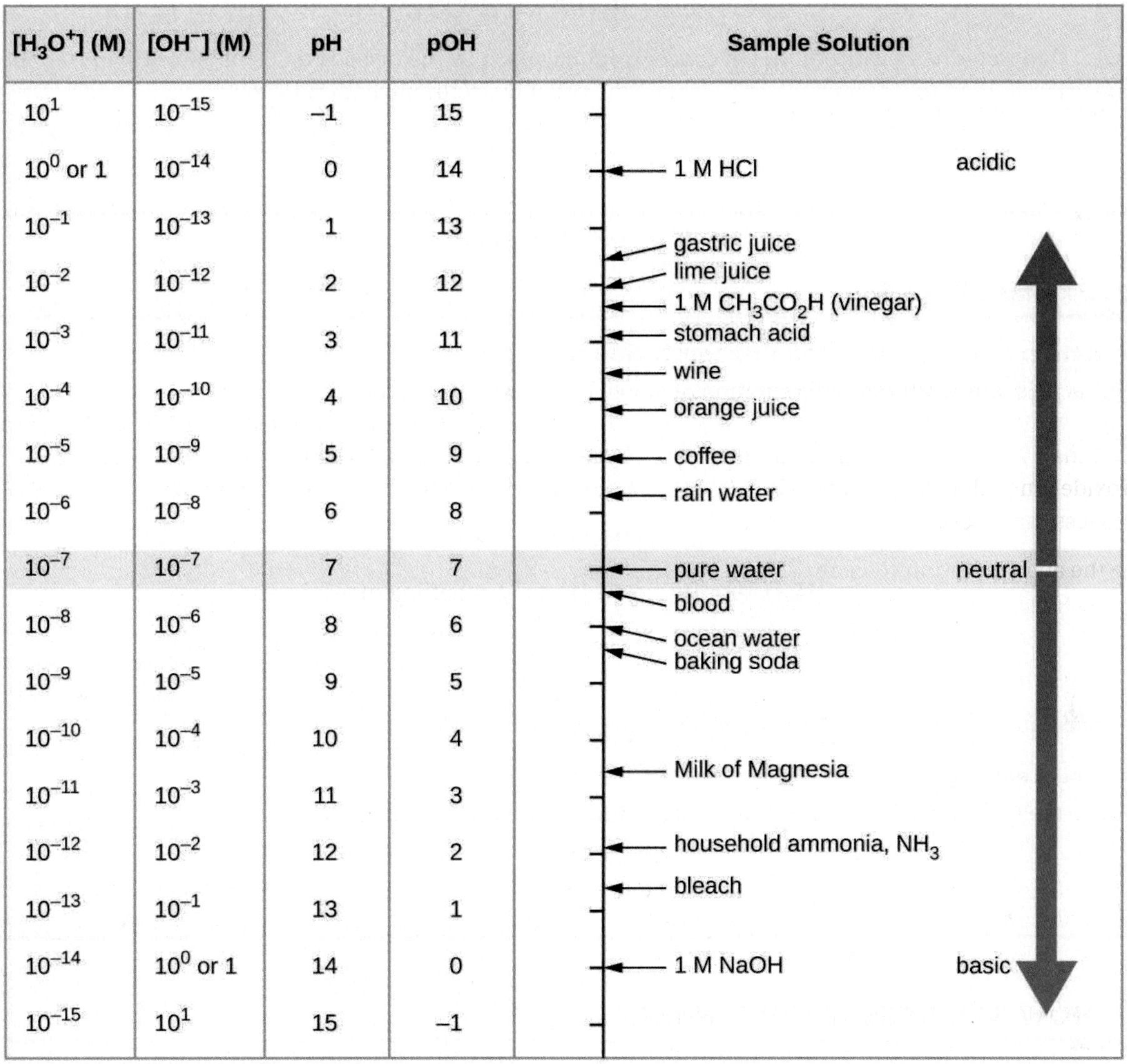

$[H_3O^+]$ (M)	$[OH^-]$ (M)	pH	pOH	Sample Solution
10^1	10^{-15}	–1	15	
10^0 or 1	10^{-14}	0	14	1 M HCl — acidic
10^{-1}	10^{-13}	1	13	gastric juice
10^{-2}	10^{-12}	2	12	lime juice; 1 M CH_3CO_2H (vinegar)
10^{-3}	10^{-11}	3	11	stomach acid
10^{-4}	10^{-10}	4	10	wine; orange juice
10^{-5}	10^{-9}	5	9	coffee
10^{-6}	10^{-8}	6	8	rain water
10^{-7}	10^{-7}	7	7	pure water — neutral
10^{-8}	10^{-6}	8	6	blood; ocean water; baking soda
10^{-9}	10^{-5}	9	5	
10^{-10}	10^{-4}	10	4	Milk of Magnesia
10^{-11}	10^{-3}	11	3	
10^{-12}	10^{-2}	12	2	household ammonia, NH_3
10^{-13}	10^{-1}	13	1	bleach
10^{-14}	10^0 or 1	14	0	1 M NaOH — basic
10^{-15}	10^1	15	–1	

FIGURE 14.2 The pH and pOH scales represent concentrations of H_3O^+ and OH^-, respectively. The pH and pOH values of some common substances at 25 °C are shown in this chart.

 **EXAMPLE 14.4**

Calculation of pH from $[H_3O^+]$

What is the pH of stomach acid, a solution of HCl with a hydronium ion concentration of 1.2×10^{-3} *M*?

Solution

$$\begin{aligned} \text{pH} &= -\log [H_3O^+] \\ &= -\log(1.2 \times 10^{-3}) \\ &= -(-2.92) = 2.92 \end{aligned}$$

(The use of logarithms is explained in Appendix B. When taking the log of a value, keep as many decimal places in the result as there are significant figures in the value.)

Check Your Learning

Water exposed to air contains carbonic acid, H_2CO_3, due to the reaction between carbon dioxide and water:

$$CO_2(aq) + H_2O(l) \rightleftharpoons H_2CO_3(aq)$$

Air-saturated water has a hydronium ion concentration caused by the dissolved CO_2 of 2.0×10^{-6} *M*, about 20-times larger than that of pure water. Calculate the pH of the solution at 25 °C.

Answer:

5.70

EXAMPLE 14.5

Calculation of Hydronium Ion Concentration from pH

Calculate the hydronium ion concentration of blood, the pH of which is 7.3.

Solution

$$\text{pH} = -\log[H_3O^+] = 7.3$$

$$\log[H_3O^+] = -7.3$$

$$[H_3O^+] = 10^{-7.3} \text{ or } [H_3O^+] = \text{antilog of } -7.3$$

$$[H_3O^+] = 5 \times 10^{-8}\ M$$

(On a calculator take the antilog, or the "inverse" log, of −7.3, or calculate $10^{-7.3}$.)

Check Your Learning

Calculate the hydronium ion concentration of a solution with a pH of −1.07.

Answer:

12 *M*

HOW SCIENCES INTERCONNECT

Environmental Science

Normal rainwater has a pH between 5 and 6 due to the presence of dissolved CO_2 which forms carbonic acid:

$$H_2O(l) + CO_2(g) \longrightarrow H_2CO_3(aq)$$

$$H_2CO_3(aq) \rightleftharpoons H^+(aq) + HCO_3^-(aq)$$

Acid rain is rainwater that has a pH of less than 5, due to a variety of nonmetal oxides, including CO_2, SO_2, SO_3, NO, and NO_2 being dissolved in the water and reacting with it to form not only carbonic acid, but sulfuric acid and nitric acid. The formation and subsequent ionization of sulfuric acid are shown here:

$$H_2O(l) + SO_3(g) \longrightarrow H_2SO_4(aq)$$

$$H_2SO_4(aq) \longrightarrow H^+(aq) + HSO_4^-(aq)$$

Carbon dioxide is naturally present in the atmosphere because most organisms produce it as a waste product of metabolism. Carbon dioxide is also formed when fires release carbon stored in vegetation or fossil fuels. Sulfur trioxide in the atmosphere is naturally produced by volcanic activity, but it also originates from burning fossil fuels, which have traces of sulfur, and from the process of "roasting" ores of metal sulfides in metal-refining processes. Oxides of nitrogen are formed in internal combustion engines where the high temperatures make it possible for the nitrogen and oxygen in air to chemically combine.

Acid rain is a particular problem in industrial areas where the products of combustion and smelting are released into the air without being stripped of sulfur and nitrogen oxides. In North America and Europe until the 1980s, it was responsible for the destruction of forests and freshwater lakes, when the acidity of the rain actually killed trees, damaged soil, and made lakes uninhabitable for all but the most acid-tolerant species. Acid rain also corrodes statuary and building facades that are made of marble and limestone (Figure 14.3). Regulations limiting the amount of sulfur and nitrogen oxides that can be released into the atmosphere by industry and automobiles have reduced the severity of acid damage to both natural and manmade environments in North America and Europe. It is now a growing problem in industrial areas of China and India.

For further information on acid rain, visit this website (http://openstax.org/l/16EPA) hosted by the US Environmental Protection Agency.

(a)

(b)

FIGURE 14.3 (a) Acid rain makes trees more susceptible to drought and insect infestation, and depletes nutrients in the soil. (b) It also is corrodes statues that are carved from marble or limestone. (credit a: modification of work by Chris M Morris; credit b: modification of work by "Eden, Janine and Jim"/Flickr)

EXAMPLE 14.6

Calculation of pOH

What are the pOH and the pH of a 0.0125-*M* solution of potassium hydroxide, KOH?

Solution

Potassium hydroxide is a highly soluble ionic compound and completely dissociates when dissolved in dilute solution, yielding $[OH^-] = 0.0125\ M$:

$$\begin{aligned} \text{pOH} &= -\log [\text{OH}^-] = -\log 0.0125 \\ &= -(-1.903) = 1.903 \end{aligned}$$

The pH can be found from the pOH:

$$\text{pH} + \text{pOH} = 14.00$$

$$\text{pH} = 14.00 - \text{pOH} = 14.00 - 1.903 = 12.10$$

Check Your Learning

The hydronium ion concentration of vinegar is approximately $4 \times 10^{-3}\ M$. What are the corresponding values of pOH and pH?

Answer:

pOH = 11.6, pH = 2.4

The acidity of a solution is typically assessed experimentally by measurement of its pH. The pOH of a solution is not usually measured, as it is easily calculated from an experimentally determined pH value. The pH of a solution can be directly measured using a pH meter (Figure 14.4).

(a) (b)

FIGURE 14.4 (a) A research-grade pH meter used in a laboratory can have a resolution of 0.001 pH units, an accuracy of ± 0.002 pH units, and may cost in excess of $1000. (b) A portable pH meter has lower resolution (0.01 pH units), lower accuracy (± 0.2 pH units), and a far lower price tag. (credit b: modification of work by Jacopo Werther)

The pH of a solution may also be visually estimated using colored indicators (Figure 14.5). The acid-base equilibria that enable use of these indicator dyes for pH measurements are described in a later section of this chapter.

(a) (b)

FIGURE 14.5 (a) A solution containing a dye mixture, called universal indicator, takes on different colors depending upon its pH. (b) Convenient test strips, called pH paper, contain embedded indicator dyes that yield pH-dependent color changes on contact with aqueous solutions.(credit: modification of work by Sahar Atwa)

14.3 Relative Strengths of Acids and Bases

LEARNING OBJECTIVES

By the end of this section, you will be able to:

- Assess the relative strengths of acids and bases according to their ionization constants
- Rationalize trends in acid–base strength in relation to molecular structure
- Carry out equilibrium calculations for weak acid–base systems

Acid and Base Ionization Constants

The relative strength of an acid or base is the extent to which it ionizes when dissolved in water. If the ionization reaction is essentially complete, the acid or base is termed *strong*; if relatively little ionization occurs, the acid or base is weak. As will be evident throughout the remainder of this chapter, there are many more weak acids and bases than strong ones. The most common strong acids and bases are listed in Figure 14.6.

6 Strong Acids		6 Strong Bases	
$HClO_4$	perchloric acid	LiOH	lithium hydroxide
HCl	hydrochloric acid	NaOH	sodium hydroxide
HBr	hydrobromic acid	KOH	potassium hydroxide
HI	hydroiodic acid	$Ca(OH)_2$	calcium hydroxide
HNO_3	nitric acid	$Sr(OH)_2$	strontium hydroxide
H_2SO_4	sulfuric acid	$Ba(OH)_2$	barium hydroxide

FIGURE 14.6 Some of the common strong acids and bases are listed here.

The relative strengths of acids may be quantified by measuring their equilibrium constants in aqueous solutions. In solutions of the same concentration, stronger acids ionize to a greater extent, and so yield higher concentrations of hydronium ions than do weaker acids. The equilibrium constant for an acid is called the **acid-ionization constant, K_a**. For the reaction of an acid HA:

$$HA(aq) + H_2O(l) \rightleftharpoons H_3O^+(aq) + A^-(aq),$$

the acid ionization constant is written

$$K_a = \frac{[H_3O^+][A^-]}{[HA]}$$

where the concentrations are those at equilibrium. Although water is a reactant in the reaction, it is the solvent as well, so we do not include $[H_2O]$ in the equation. The larger the K_a of an acid, the larger the concentration of H_3O^+ and A^- relative to the concentration of the nonionized acid, HA, in an equilibrium mixture, and the stronger the acid. An acid is classified as "strong" when it undergoes complete ionization, in which case the concentration of HA is zero and the acid ionization constant is immeasurably large ($K_a \approx \infty$). Acids that are partially ionized are called "weak," and their acid ionization constants may be experimentally measured. A table of ionization constants for weak acids is provided in Appendix H.

To illustrate this idea, three acid ionization equations and K_a values are shown below. The ionization constants increase from first to last of the listed equations, indicating the relative acid strength increases in the order $CH_3CO_2H < HNO_2 < HSO_4^-$:

$$CH_3CO_2H(aq) + H_2O(l) \rightleftharpoons H_3O^+(aq) + CH_3CO_2^-(aq) \qquad K_a = 1.8 \times 10^{-5}$$

$$HNO_2(aq) + H_2O(l) \rightleftharpoons H_3O^+(aq) + NO_2^-(aq) \qquad K_a = 4.6 \times 10^{-4}$$

$$HSO_4^-(aq) + H_2O(aq) \rightleftharpoons H_3O^+(aq) + SO_4^{2-}(aq) \qquad K_a = 1.2 \times 10^{-2}$$

Another measure of the strength of an acid is its percent ionization. The **percent ionization** of a weak acid is defined in terms of the composition of an equilibrium mixture:

$$\%\ \text{ionization} = \frac{[H_3O^+]_{eq}}{[HA]_0} \times 100$$

where the numerator is equivalent to the concentration of the acid's conjugate base (per stoichiometry, $[A^-] =$

$[H_3O^+]$). Unlike the K_a value, the percent ionization of a weak acid varies with the initial concentration of acid, typically decreasing as concentration increases. Equilibrium calculations of the sort described later in this chapter can be used to confirm this behavior.

EXAMPLE 14.7

Calculation of Percent Ionization from pH

Calculate the percent ionization of a 0.125-*M* solution of nitrous acid (a weak acid), with a pH of 2.09.

Solution

The percent ionization for an acid is:

$$\frac{[H_3O^+]_{eq}}{[HNO_2]_0} \times 100$$

Converting the provided pH to hydronium ion molarity yields

$$[H_3O^+] = 10^{-2.09} = 0.0081\ M$$

Substituting this value and the provided initial acid concentration into the percent ionization equation gives

$$\frac{8.1 \times 10^{-3}}{0.125} \times 100 = 6.5\%$$

(Recall the provided pH value of 2.09 is logarithmic, and so it contains just two significant digits, limiting the certainty of the computed percent ionization.)

Check Your Learning

Calculate the percent ionization of a 0.10-*M* solution of acetic acid with a pH of 2.89.

Answer:

1.3% ionized

LINK TO LEARNING

View the simulation (http://openstax.org/l/16AcidBase) of strong and weak acids and bases at the molecular level.

Just as for acids, the relative strength of a base is reflected in the magnitude of its **base-ionization constant** **(K_b)** in aqueous solutions. In solutions of the same concentration, stronger bases ionize to a greater extent, and so yield higher hydroxide ion concentrations than do weaker bases. A stronger base has a larger ionization constant than does a weaker base. For the reaction of a base, B:

$$B(aq) + H_2O(l) \rightleftharpoons HB^+(aq) + OH^-(aq),$$

the ionization constant is written as

$$K_b = \frac{[HB^+]\left[OH^-\right]}{[B]}$$

Inspection of the data for three weak bases presented below shows the base strength increases in the order $NO_2{}^- < CH_2CO_2{}^- < NH_3$.

$$NO_2{}^-(aq) + H_2O(l) \rightleftharpoons HNO_2(aq) + OH^-(aq) \qquad K_b = 2.17 \times 10^{-11}$$

$$CH_3CO_2{}^-(aq) + H_2O(l) \rightleftharpoons CH_3CO_2H(aq) + OH^-(aq) \qquad K_b = 5.6 \times 10^{-10}$$

$$NH_3(aq) + H_2O(l) \rightleftharpoons NH_4{}^+(aq) + OH^-(aq) \qquad K_b = 1.8 \times 10^{-5}$$

A table of ionization constants for weak bases appears in Appendix I. As for acids, the relative strength of a base is also reflected in its percent ionization, computed as

$$\%\ \text{ionization} = [OH^-]_{eq}/[B]_0 \times 100\%$$

but will vary depending on the base ionization constant and the initial concentration of the solution.

Relative Strengths of Conjugate Acid-Base Pairs

Brønsted-Lowry acid-base chemistry is the transfer of protons; thus, logic suggests a relation between the relative strengths of conjugate acid-base pairs. The strength of an acid or base is quantified in its ionization constant, K_a or K_b, which represents the extent of the acid or base ionization reaction. For the conjugate acid-base pair HA / A^-, ionization equilibrium equations and ionization constant expressions are

$$HA(aq) + H_2O(l) \rightleftharpoons H_3O^+(aq) + A^-(aq) \qquad K_a = \frac{[H_3O^+][A^-]}{[HA]}$$

$$A^-(aq) + H_2O(l) \rightleftharpoons OH^-(aq) + HA(aq) \qquad K_b = \frac{[HA][OH^-]}{[A^-]}$$

Adding these two chemical equations yields the equation for the autoionization for water:

$$\cancel{HA(aq)} + H_2O(l) + \cancel{A^-(aq)} + H_2O(l) \rightleftharpoons H_3O^+(aq) + \cancel{A^-(aq)} + OH^-(aq) + \cancel{HA(aq)}$$

$$2H_2O(l) \rightleftharpoons H_3O^+(aq) + OH^-(aq)$$

As discussed in another chapter on equilibrium, the equilibrium constant for a summed reaction is equal to the mathematical product of the equilibrium constants for the added reactions, and so

$$K_a \times K_b = \frac{[H_3O^+][A^-]}{[HA]} \times \frac{[HA][OH^-]}{[A^-]} = [H_3O^+][OH^-] = K_w$$

This equation states the relation between ionization constants for any conjugate acid-base pair, namely, their mathematical product is equal to the ion product of water, K_w. By rearranging this equation, a reciprocal relation between the strengths of a conjugate acid-base pair becomes evident:

$$K_a = K_w/K_b \text{ or } K_b = K_w/K_a$$

The inverse proportional relation between K_a and K_b means *the stronger the acid or base, the weaker its conjugate partner.* Figure 14.7 illustrates this relation for several conjugate acid-base pairs.

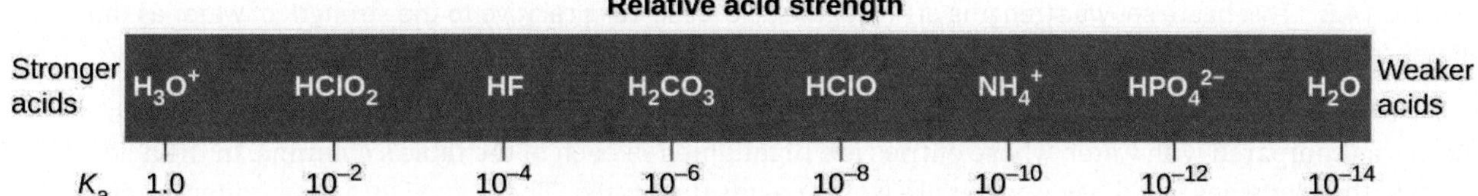

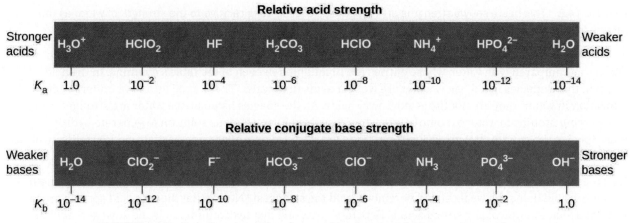

FIGURE 14.7 Relative strengths of several conjugate acid-base pairs are shown.

Acid			Base		
perchloric acid	$HClO_4$	Undergo complete acid ionization in water	Do not undergo base ionization in water	ClO_4^-	perchlorate ion
sulfuric acid	H_2SO_4			HSO_4^-	hydrogen sulfate ion
hydrogen iodide	HI			I^-	iodide ion
hydrogen bromide	HBr			Br^-	bromide ion
hydrogen chloride	HCl			Cl^-	chloride ion
nitric acid	HNO_3			NO_3^-	nitrate ion
hydronium ion	H_3O^+			H_2O	water
hydrogen sulfate ion	HSO_4^-			SO_4^{2-}	sulfate ion
phosphoric acid	H_3PO_4			$H_2PO_4^-$	dihydrogen phosphate ion
hydrogen fluoride	HF			F^-	fluoride ion
nitrous acid	HNO_2			NO_2^-	nitrite ion
acetic acid	CH_3CO_2H			$CH_3CO_2^-$	acetate ion
carbonic acid	H_2CO_3			HCO_3^-	hydrogen carbonate ion
hydrogen sulfide	H_2S			HS^-	hydrogen sulfide ion
ammonium ion	NH_4^+			NH_3	ammonia
hydrogen cyanide	HCN			CN^-	cyanide ion
hydrogen carbonate ion	HCO_3^-			CO_3^{2-}	carbonate ion
water	H_2O			OH^-	hydroxide ion
hydrogen sulfide ion	HS^-	Do not undergo acid ionization in water	Undergo complete base ionization in water	S^{2-}	sulfide ion
ethanol	C_2H_5OH			$C_2H_5O^-$	ethoxide ion
ammonia	NH_3			NH_2^-	amide ion
hydrogen	H_2			H^-	hydride ion
methane	CH_4			CH_3^-	methide ion

Increasing acid strength (upward); Increasing base strength (downward)

FIGURE 14.8 This figure shows strengths of conjugate acid-base pairs relative to the strength of water as the reference substance.

The listing of conjugate acid–base pairs shown in Figure 14.8 is arranged to show the relative strength of each species as compared with water, whose entries are highlighted in each of the table's columns. In the acid column, those species listed below water are weaker acids than water. These species do not undergo acid ionization in water; they are not Bronsted-Lowry acids. All the species listed above water are stronger acids, transferring protons to water to some extent when dissolved in an aqueous solution to generate hydronium ions. Species above water but below hydronium ion are *weak acids*, undergoing partial acid ionization, wheres those above hydronium ion are *strong acids* that are completely ionized in aqueous solution.

If all these strong acids are completely ionized in water, why does the column indicate they vary in strength, with nitric acid being the weakest and perchloric acid the strongest? Notice that the sole acid species present in an aqueous solution of any strong acid is $H_3O^+(aq)$, meaning that hydronium ion is the strongest acid that may exist in water; any stronger acid will react completely with water to generate hydronium ions. This limit on the acid strength of solutes in a solution is called a **leveling effect**. To measure the differences in acid strength for "strong" acids, the acids must be dissolved in a solvent that is *less basic* than water. In such solvents, the acids will be "weak," and so any differences in the extent of their ionization can be determined. For example, the binary hydrogen halides HCl, HBr, and HI are strong acids in water but weak acids in ethanol (strength increasing HCl < HBr < HI).

The right column of Figure 14.8 lists a number of substances in order of increasing base strength from top to bottom. Following the same logic as for the left column, species listed above water are weaker bases and so they don't undergo base ionization when dissolved in water. Species listed between water and its conjugate base, hydroxide ion, are weak bases that partially ionize. Species listed below hydroxide ion are strong bases that completely ionize in water to yield hydroxide ions (i.e., they are *leveled* to hydroxide). A comparison of the acid and base columns in this table supports the reciprocal relation between the strengths of conjugate acid-base pairs. For example, the conjugate bases of the strong acids (top of table) are all of negligible strength. A strong acid exhibits an immeasurably large K_a, and so its conjugate base will exhibit a K_b that is essentially zero:

$$\text{strong acid}: \qquad K_a \approx \infty$$

$$\text{conjugate base}: \; K_b = K_w/K_a = K_w/\infty \approx 0$$

A similar approach can be used to support the observation that conjugate acids of strong bases ($K_b \approx \infty$) are of negligible strength ($K_a \approx 0$).

 **EXAMPLE 14.8**

Calculating Ionization Constants for Conjugate Acid-Base Pairs

Use the K_b for the nitrite ion, NO_2^-, to calculate the K_a for its conjugate acid.

Solution

K_b for NO_2^- is given in this section as 2.17×10^{-11}. The conjugate acid of NO_2^- is HNO_2; K_a for HNO_2 can be calculated using the relationship:

$$K_a \times K_b = 1.0 \times 10^{-14} = K_w$$

Solving for K_a yields

$$K_a = \frac{K_w}{K_b} = \frac{1.0 \times 10^{-14}}{2.17 \times 10^{-11}} = 4.6 \times 10^{-4}$$

This answer can be verified by finding the K_a for HNO_2 in Appendix H.

Check Your Learning

Determine the relative acid strengths of NH_4^+ and HCN by comparing their ionization constants. The ionization constant of HCN is given in Appendix H as 4.9×10^{-10}. The ionization constant of NH_4^+ is not listed, but the ionization constant of its conjugate base, NH_3, is listed as 1.8×10^{-5}.

Answer:

NH_4^+ is the slightly stronger acid (K_a for $NH_4^+ = 5.6 \times 10^{-10}$).

Acid-Base Equilibrium Calculations

The chapter on chemical equilibria introduced several types of equilibrium calculations and the various mathematical strategies that are helpful in performing them. These strategies are generally useful for equilibrium systems regardless of chemical reaction class, and so they may be effectively applied to acid-base equilibrium problems. This section presents several example exercises involving equilibrium calculations for acid-base systems.

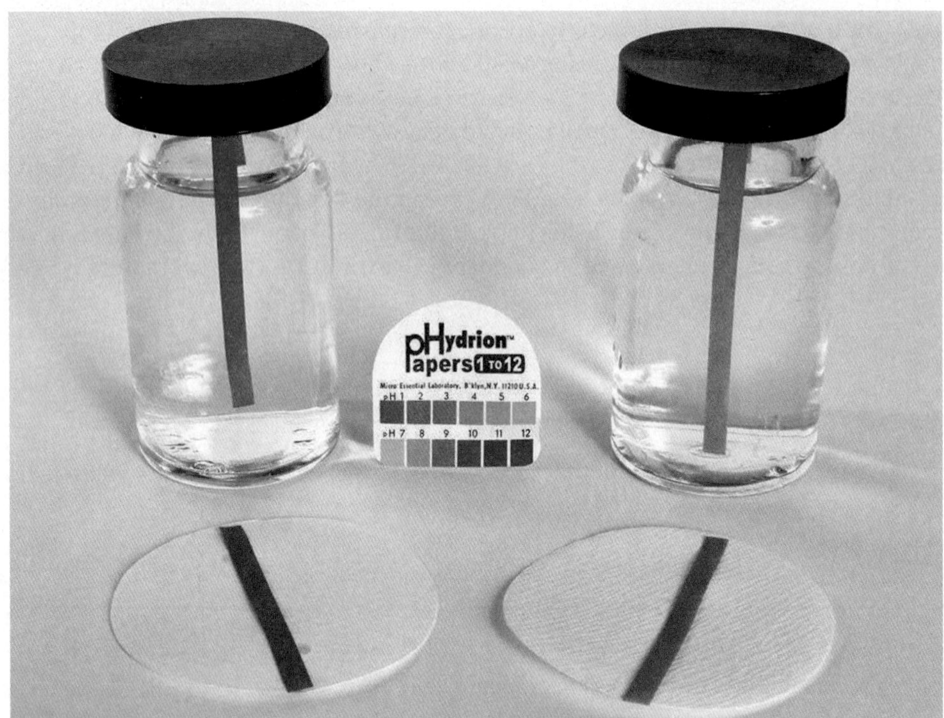

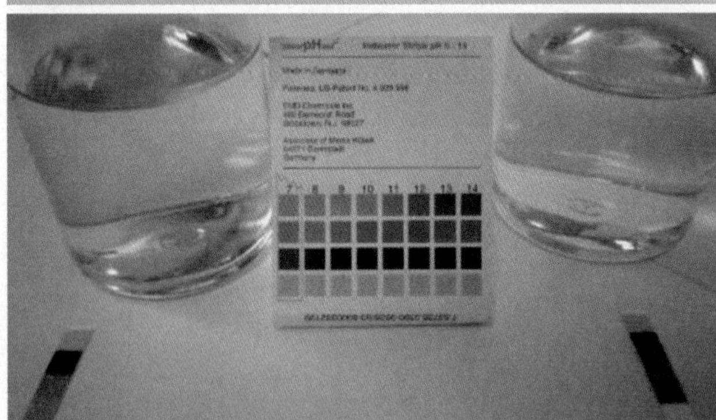

 **EXAMPLE 14.9**

Determination of K_a from Equilibrium Concentrations

Acetic acid is the principal ingredient in vinegar (Figure 14.9) that provides its sour taste. At equilibrium, a solution contains $[CH_3CO_2H] = 0.0787\ M$ and $[H_3O^+] = [CH_3CO_2^-] = 0.00118\ M$. What is the value of K_a for acetic acid?

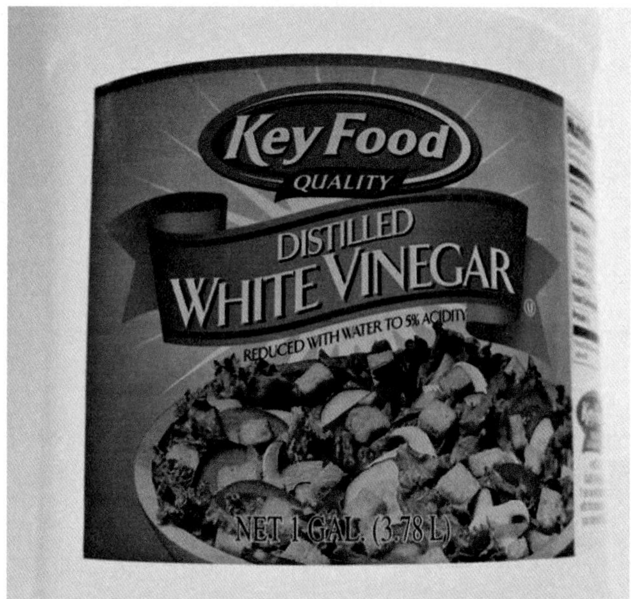

FIGURE 14.9 Vinegar contains acetic acid, a weak acid. (credit: modification of work by "HomeSpot HQ"/Flickr)

Solution

The relevant equilibrium equation and its equilibrium constant expression are shown below. Substitution of the provided equilibrium concentrations permits a straightforward calculation of the K_a for acetic acid.

$$CH_3CO_2H(aq) + H_2O(l) \rightleftharpoons H_3O^+(aq) + CH_3CO_2^-(aq)$$

$$K_a = \frac{[H_3O^+][CH_3CO_2^-]}{[CH_3CO_2H]} = \frac{(0.00118)(0.00118)}{0.0787} = 1.77 \times 10^{-5}$$

Check Your Learning

The HSO_4^- ion, weak acid used in some household cleansers:

$$HSO_4^-(aq) + H_2O(l) \rightleftharpoons H_3O^+(aq) + SO_4^{2-}(aq)$$

What is the acid ionization constant for this weak acid if an equilibrium mixture has the following composition: $[H_3O^+] = 0.027\ M$; $[HSO_4^-] = 0.29\ M$; and $[SO_4^{2-}] = 0.13\ M$?

Answer:

K_a for $HSO_4^- = 1.2 \times 10^{-2}$

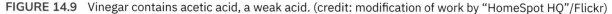

EXAMPLE 14.10

Determination of K_b from Equilibrium Concentrations

Caffeine, $C_8H_{10}N_4O_2$ is a weak base. What is the value of K_b for caffeine if a solution at equilibrium has $[C_8H_{10}N_4O_2] = 0.050\ M$, $[C_8H_{10}N_4O_2H^+] = 5.0 \times 10^{-3}\ M$, and $[OH^-] = 2.5 \times 10^{-3}\ M$?

Solution

The relevant equilibrium equation and its equilibrium constant expression are shown below. Substitution of the provided equilibrium concentrations permits a straightforward calculation of the K_b for caffeine.

$$C_8H_{10}N_4O_2(aq) + H_2O(l) \rightleftharpoons C_8H_{10}N_4O_2H^+(aq) + OH^-(aq)$$

$$K_b = \frac{[C_8H_{10}N_4O_2H^+][OH^-]}{[C_8H_{10}N_4O_2]} = \frac{(5.0 \times 10^{-3})(2.5 \times 10^{-3})}{0.050} = 2.5 \times 10^{-4}$$

Check Your Learning

What is the equilibrium constant for the ionization of the $HPO_4{}^{2-}$ ion, a weak base

$$HPO_4{}^{2-}(aq) + H_2O(l) \rightleftharpoons H_2PO_4{}^-(aq) + OH^-(aq)$$

if the composition of an equilibrium mixture is as follows: $[OH^-] = 1.3 \times 10^{-6}$ M; $[H_2PO_4{}^-] = 0.042$ M; and $[HPO_4{}^{2-}] = 0.341$ M?

Answer:

K_b for $HPO_4{}^{2-} = 1.6 \times 10^{-7}$

 **EXAMPLE 14.11**

Determination of K_a or K_b from pH

The pH of a 0.0516-M solution of nitrous acid, HNO_2, is 2.34. What is its K_a?

$$HNO_2(aq) + H_2O(l) \rightleftharpoons H_3O^+(aq) + NO_2{}^-(aq)$$

Solution

The nitrous acid concentration provided is a *formal* concentration, one that does not account for any chemical equilibria that may be established in solution. Such concentrations are treated as "initial" values for equilibrium calculations using the ICE table approach. Notice the initial value of hydronium ion is listed as *approximately* zero because a small concentration of H_3O^+ is present (1×10^{-7} M) due to the autoprotolysis of water. In many cases, such as all the ones presented in this chapter, this concentration is much less than that generated by ionization of the acid (or base) in question and may be neglected.

The pH provided is a logarithmic measure of the hydronium ion concentration resulting from the acid ionization of the nitrous acid, and so it represents an "equilibrium" value for the ICE table:

$$[H_3O^+] = 10^{-2.34} = 0.0046\ M$$

The ICE table for this system is then

	HNO_2 +	H_2O ⇌	H_3O^+ +	$NO_2{}^-$
Initial concentration (M)	0.0516		~0	0
Change (M)	−0.0046		+0.0046	+0.0046
Equilibrium concentration (M)	0.0470		0.0046	0.0046

Finally, calculate the value of the equilibrium constant using the data in the table:

$$K_a = \frac{[H_3O^+][NO_2{}^-]}{[HNO_2]} = \frac{(0.0046)(0.0046)}{(0.0470)} = 4.6 \times 10^{-4}$$

Check Your Learning.

The pH of a solution of household ammonia, a 0.950-M solution of NH_3, is 11.612. What is K_b for NH_3.

Answer:

$K_b = 1.8 \times 10^{-5}$

EXAMPLE 14.12

Calculating Equilibrium Concentrations in a Weak Acid Solution

Formic acid, HCO_2H, is one irritant that causes the body's reaction to some ant bites and stings (Figure 14.10).

FIGURE 14.10 The pain of some ant bites and stings is caused by formic acid. (credit: John Tann)

What is the concentration of hydronium ion and the pH of a 0.534-*M* solution of formic acid?

$$HCO_2H(aq) + H_2O(l) \rightleftharpoons H_3O^+(aq) + HCO_2{}^-(aq) \qquad K_a = 1.8 \times 10^{-4}$$

Solution

The ICE table for this system is

	HCO_2H	+ H_2O	$\rightleftharpoons$ H_3O^+	+ HCO_2^-
Initial concentration (M)	0.534		~0	0
Change (M)			$+x$	$+x$
Equilibrium concentration (M)	$0.534 - x$		x	x

Substituting the equilibrium concentration terms into the K_a expression gives

$$K_a = 1.8 \times 10^{-4} = \frac{[H_3O^+][HCO_2{}^-]}{[HCO_2H]}$$

$$= \frac{(x)(x)}{0.534 - x} = 1.8 \times 10^{-4}$$

The relatively large initial concentration and small equilibrium constant permits the simplifying assumption that x will be much lesser than 0.534, and so the equation becomes

$$K_a = 1.8 \times 10^{-4} = \frac{x^2}{0.534}$$

Solving the equation for x yields

$$x^2 = 0.534 \times \left(1.8 \times 10^{-4}\right) = 9.6 \times 10^{-5}$$

$$x = \sqrt{9.6 \times 10^{-5}}$$

$$= 9.8 \times 10^{-3}\ M$$

To check the assumption that x is small compared to 0.534, its relative magnitude can be estimated:

$$\frac{x}{0.534} = \frac{9.8 \times 10^{-3}}{0.534} = 1.8 \times 10^{-2}\ (1.8\%\ \text{of}\ 0.534)$$

Because x is less than 5% of the initial concentration, the assumption is valid.

As defined in the ICE table, x is equal to the equilibrium concentration of hydronium ion:

$$x = [H_3O^+] = 0.0098\ M$$

Finally, the pH is calculated to be

$$\text{pH} = -\log[H_3O^+] = -\log(0.0098) = 2.01$$

Check Your Learning

Only a small fraction of a weak acid ionizes in aqueous solution. What is the percent ionization of a 0.100-M solution of acetic acid, CH_3CO_2H?

$$CH_3CO_2H(aq) + H_2O(l) \rightleftharpoons H_3O^+(aq) + CH_3CO_2^-(aq) \qquad K_a = 1.8 \times 10^{-5}$$

Answer:

percent ionization = 1.3%

EXAMPLE 14.13

Calculating Equilibrium Concentrations in a Weak Base Solution

Find the concentration of hydroxide ion, the pOH, and the pH of a 0.25-M solution of trimethylamine, a weak base:

$$(CH_3)_3N(aq) + H_2O(l) \rightleftharpoons (CH_3)_3NH^+(aq) + OH^-(aq) \qquad K_b = 6.3 \times 10^{-5}$$

Solution

The ICE table for this system is

	$(CH_3)_3N$	+ H_2O ⇌	$(CH_3)_3NH^+$	+ OH^-
Initial concentration (M)	0.25		0	~0
Change (M)	$-x$		x	x
Equilibrium concentration (M)	$0.25 + (-x)$		$0 + x$	$\sim 0 + x$

Substituting the equilibrium concentration terms into the K_b expression gives

$$K_b = \frac{[(CH_3)_3NH^+][OH^-]}{[(CH_3)_3N]} = \frac{(x)(x)}{0.25 - x} = 6.3 \times 10^{-5}$$

Assuming $x << 0.25$ and solving for x yields

$$x = 4.0 \times 10^{-3}\ M$$

This value is less than 5% of the initial concentration (0.25), so the assumption is justified.

As defined in the ICE table, x is equal to the equilibrium concentration of hydroxide ion:

$$[OH^-] = \sim 0 + x = x = 4.0 \times 10^{-3}\ M$$

$$= 4.0 \times 10^{-3}\ M$$

The pOH is calculated to be

$$pOH = -\log\left(4.0 \times 10^{-3}\right) = 2.40$$

Using the relation introduced in the previous section of this chapter:

$$pH + pOH = pK_w = 14.00$$

permits the computation of pH:

$$pH = 14.00 - pOH = 14.00 - 2.40 = 11.60$$

Check Your Learning

Calculate the hydroxide ion concentration and the percent ionization of a 0.0325-M solution of ammonia, a weak base with a K_b of 1.76×10^{-5}.

Answer:

7.56×10^{-4} M, 2.33%

In some cases, the strength of the weak acid or base and its formal (initial) concentration result in an appreciable ionization. Though the ICE strategy remains effective for these systems, the algebra is a bit more involved because the simplifying assumption that x is negligible cannot be made. Calculations of this sort are demonstrated in Example 14.14 below.

 **EXAMPLE 14.14**

Calculating Equilibrium Concentrations without Simplifying Assumptions

Sodium bisulfate, $NaHSO_4$, is used in some household cleansers as a source of the HSO_4^- ion, a weak acid. What is the pH of a 0.50-M solution of HSO_4^-?

$$HSO_4^-(aq) + H_2O(l) \rightleftharpoons H_3O^+(aq) + SO_4^{2-}(aq) \qquad K_a = 1.2 \times 10^{-2}$$

Solution

The ICE table for this system is

	HSO_4^-	+ H_2O ⇌	H_3O^+	+ SO_4^{2-}
Initial concentration (M)	0.50		~0	0
Change (M)	$-x$		$+x$	$+x$
Equilibrium concentration (M)	$0.50 - x$		x	x

Substituting the equilibrium concentration terms into the K_a expression gives

$$K_a = 1.2 \times 10^{-2} = \frac{[H_3O^+][SO_4{}^{2-}]}{[HSO_4{}^-]} = \frac{(x)(x)}{0.50 - x}$$

If the assumption that $x << 0.5$ is made, simplifying and solving the above equation yields

$$x = 0.077\ M$$

This value of x is clearly not significantly less than 0.50 M; rather, it is approximately 15% of the initial concentration:
When we check the assumption, we calculate:

$$\frac{x}{[HSO_4{}^-]_i}$$

$$\frac{x}{0.50} = \frac{7.7 \times 10^{-2}}{0.50} = 0.15\ (15\%)$$

Because the simplifying assumption is not valid for this system, the equilibrium constant expression is solved as follows:

$$K_a = 1.2 \times 10^{-2} = \frac{(x)(x)}{0.50 - x}$$

Rearranging this equation yields

$$6.0 \times 10^{-3} - 1.2 \times 10^{-2}x = x^2$$

Writing the equation in quadratic form gives

$$x^2 + 1.2 \times 10^{-2}x - 6.0 \times 10^{-3} = 0$$

Solving for the two roots of this quadratic equation results in a negative value that may be discarded as physically irrelevant and a positive value equal to x. As defined in the ICE table, x is equal to the hydronium concentration.

$$x = [H_3O^+] = 0.072\ M$$
$$\text{pH} = -\log[H_3O^+] = -\log(0.072) = 1.14$$

Check Your Learning

Calculate the pH in a 0.010-M solution of caffeine, a weak base:

$$C_8H_{10}N_4O_2(aq) + H_2O(l) \rightleftharpoons C_8H_{10}N_4O_2H^+(aq) + OH^-(aq) \qquad K_b = 2.5 \times 10^{-4}$$

Answer:

pH 11.16

Effect of Molecular Structure on Acid-Base Strength

Binary Acids and Bases

In the absence of any leveling effect, the acid strength of binary compounds of hydrogen with nonmetals (A) increases as the H-A bond strength decreases down a group in the periodic table. For group 17, the order of increasing acidity is HF < HCl < HBr < HI. Likewise, for group 16, the order of increasing acid strength is H_2O < H_2S < H_2Se < H_2Te.

Across a row in the periodic table, the acid strength of binary hydrogen compounds increases with increasing electronegativity of the nonmetal atom because the polarity of the H-A bond increases. Thus, the order of increasing acidity (for removal of one proton) across the second row is $CH_4 < NH_3 < H_2O < HF$; across the third row, it is $SiH_4 < PH_3 < H_2S < HCl$ (see Figure 14.11).

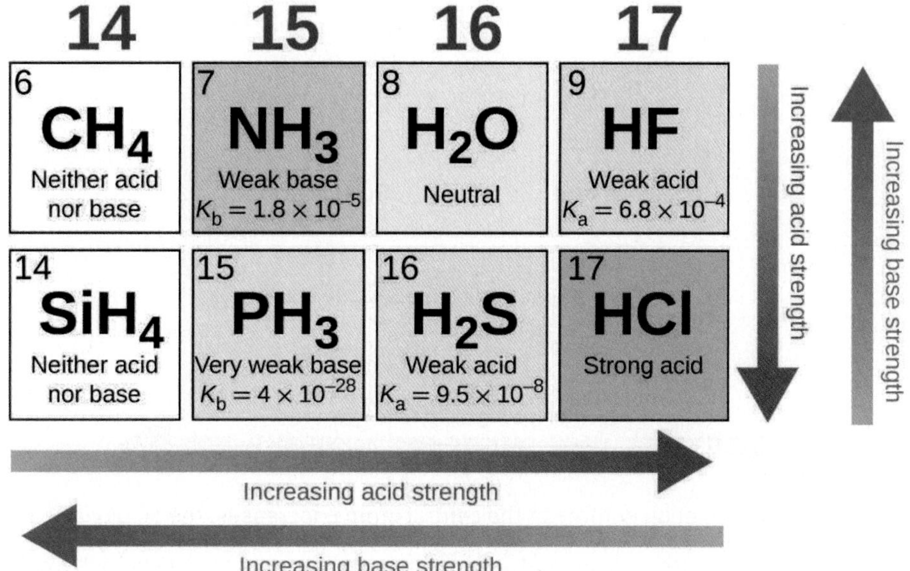

FIGURE 14.11 The figure shows trends in the strengths of binary acids and bases.

Ternary Acids and Bases

Ternary compounds composed of hydrogen, oxygen, and some third element ("E") may be structured as depicted in the image below. In these compounds, the central E atom is bonded to one or more O atoms, and at least one of the O atoms is also bonded to an H atom, corresponding to the general molecular formula $O_mE(OH)_n$. These compounds may be acidic, basic, or amphoteric depending on the properties of the central E atom. Examples of such compounds include sulfuric acid, $O_2S(OH)_2$, sulfurous acid, $OS(OH)_2$, nitric acid, O_2NOH, perchloric acid, O_3ClOH, aluminum hydroxide, $Al(OH)_3$, calcium hydroxide, $Ca(OH)_2$, and potassium hydroxide, KOH:

—E—O—H
Bond *b*
Bond *a*

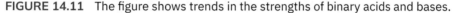

If the central atom, E, has a low electronegativity, its attraction for electrons is low. Little tendency exists for the central atom to form a strong covalent bond with the oxygen atom, and bond *a* between the element and oxygen is more readily broken than bond *b* between oxygen and hydrogen. Hence bond *a* is ionic, hydroxide ions are released to the solution, and the material behaves as a base—this is the case with $Ca(OH)_2$ and KOH. Lower electronegativity is characteristic of the more metallic elements; hence, the metallic elements form ionic hydroxides that are by definition basic compounds.

If, on the other hand, the atom E has a relatively high electronegativity, it strongly attracts the electrons it shares with the oxygen atom, making bond *a* relatively strongly covalent. The oxygen-hydrogen bond, bond *b*, is thereby weakened because electrons are displaced toward E. Bond *b* is polar and readily releases hydrogen ions to the solution, so the material behaves as an acid. High electronegativities are characteristic of the more nonmetallic elements. Thus, nonmetallic elements form covalent compounds containing acidic −OH groups that are called **oxyacids**.

Increasing the oxidation number of the central atom E also increases the acidity of an oxyacid because this increases the attraction of E for the electrons it shares with oxygen and thereby weakens the O-H bond. Sulfuric acid, H_2SO_4, or $O_2S(OH)_2$ (with a sulfur oxidation number of +6), is more acidic than sulfurous acid,

H_2SO_3, or $OS(OH)_2$ (with a sulfur oxidation number of +4). Likewise nitric acid, HNO_3, or O_2NOH (N oxidation number = +5), is more acidic than nitrous acid, HNO_2, or ONOH (N oxidation number = +3). In each of these pairs, the oxidation number of the central atom is larger for the stronger acid (Figure 14.12).

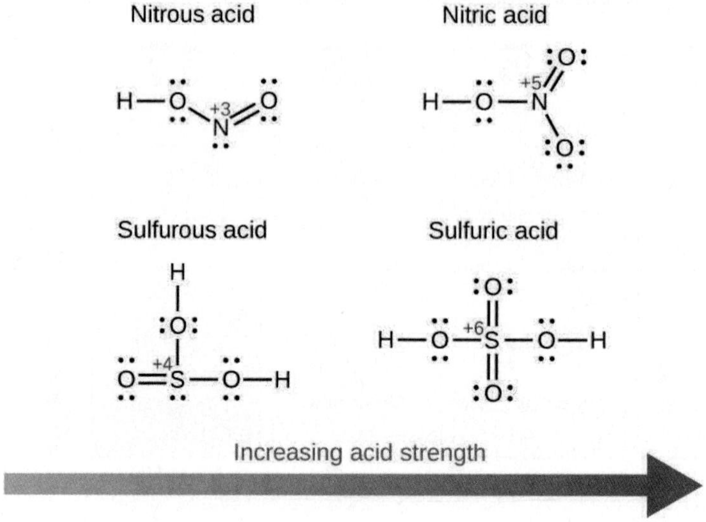

FIGURE 14.12 As the oxidation number of the central atom E increases, the acidity also increases.

Hydroxy compounds of elements with intermediate electronegativities and relatively high oxidation numbers (for example, elements near the diagonal line separating the metals from the nonmetals in the periodic table) are usually amphoteric. This means that the hydroxy compounds act as acids when they react with strong bases and as bases when they react with strong acids. The amphoterism of aluminum hydroxide, which commonly exists as the hydrate $Al(H_2O)_3(OH)_3$, is reflected in its solubility in both strong acids and strong bases. In strong bases, the relatively insoluble hydrated aluminum hydroxide, $Al(H_2O)_3(OH)_3$, is converted into the soluble ion, $[Al(H_2O)_2(OH)_4]^-$, by reaction with hydroxide ion:

$$Al(H_2O)_3(OH)_3(aq) + OH^-(aq) \rightleftharpoons H_2O(l) + [Al(H_2O)_2(OH)_4]^-(aq)$$

In this reaction, a proton is transferred from one of the aluminum-bound H_2O molecules to a hydroxide ion in solution. The $Al(H_2O)_3(OH)_3$ compound thus acts as an acid under these conditions. On the other hand, when dissolved in strong acids, it is converted to the soluble ion $[Al(H_2O)_6]^{3+}$ by reaction with hydronium ion:

$$3H_3O^+(aq) + Al(H_2O)_3(OH)_3(aq) \rightleftharpoons Al(H_2O)_6{}^{3+}(aq) + 3H_2O(l)$$

In this case, protons are transferred from hydronium ions in solution to $Al(H_2O)_3(OH)_3$, and the compound functions as a base.

14.4 Hydrolysis of Salts

LEARNING OBJECTIVES

By the end of this section, you will be able to:

- Predict whether a salt solution will be acidic, basic, or neutral
- Calculate the concentrations of the various species in a salt solution
- Describe the acid ionization of hydrated metal ions

Salts with Acidic Ions

Salts are ionic compounds composed of cations and anions, either of which may be capable of undergoing an acid or base ionization reaction with water. Aqueous salt solutions, therefore, may be acidic, basic, or neutral, depending on the relative acid-base strengths of the salt's constituent ions. For example, dissolving ammonium chloride in water results in its dissociation, as described by the equation

$$NH_4Cl(s) \rightleftharpoons NH_4{}^+(aq) + Cl^-(aq)$$

The ammonium ion is the conjugate acid of the base ammonia, NH_3; its acid ionization (or acid hydrolysis) reaction is represented by

$$NH_4{}^+(aq) + H_2O(l) \rightleftharpoons H_3O^+(aq) + NH_3(aq) \qquad K_a = K_w/K_b$$

Since ammonia is a weak base, K_b is measurable and $K_a > 0$ (ammonium ion is a weak acid).

The chloride ion is the conjugate base of hydrochloric acid, and so its base ionization (or *base hydrolysis*) reaction is represented by

$$Cl^-(aq) + H_2O(l) \rightleftharpoons HCl(aq) + OH^-(aq) \qquad K_b = K_w/K_a$$

Since HCl is a strong acid, K_a is immeasurably large and $K_b \approx 0$ (chloride ions don't undergo appreciable hydrolysis).

Thus, dissolving ammonium chloride in water yields a solution of weak acid cations ($NH_4{}^+$) and inert anions (Cl^-), resulting in an acidic solution.

 **EXAMPLE 14.15**

Calculating the pH of an Acidic Salt Solution

Aniline is an amine that is used to manufacture dyes. It is isolated as anilinium chloride, $[C_6H_5NH_3]Cl$, a salt prepared by the reaction of the weak base aniline and hydrochloric acid. What is the pH of a 0.233 *M* solution of anilinium chloride

$$C_6H_5NH_3{}^+(aq) + H_2O(l) \rightleftharpoons H_3O^+(aq) + C_6H_5NH_2(aq)$$

Solution

The K_a for anilinium ion is derived from the K_b for its conjugate base, aniline (see Appendix H):

$$K_a = \frac{K_w}{K_b} = \frac{1.0 \times 10^{-14}}{4.3 \times 10^{-10}} = 2.3 \times 10^{-5}$$

Using the provided information, an ICE table for this system is prepared:

	$C_6H_5NH_3{}^+$ +	H_2O $\rightleftharpoons$	$C_6H_5NH_2$ +	H_3O^+
Initial concentration (*M*)	0.233		0	~0
Change (*M*)	$-x$		$+x$	$+x$
Equilibrium concentration (*M*)	$0.233 - x$		x	x

Substituting these equilibrium concentration terms into the K_a expression gives

$$K_a = [C_6H_5NH_2][H_3O^+]/[C_6H_5NH_3{}^+]$$
$$2.3 \times 10^{-5} = (x)(x)/0.233 - x)$$

Assuming $x << 0.233$, the equation is simplified and solved for x:

$$2.3 \times 10^{-5} = x^2/0.233$$
$$x = 0.0023\ M$$

The ICE table defines x as the hydronium ion molarity, and so the pH is computed as

$$\text{pH} = -\log[H_3O^+] = -\log(0.0023) = 2.64$$

Check Your Learning

What is the hydronium ion concentration in a 0.100-*M* solution of ammonium nitrate, NH_4NO_3, a salt composed of the ions NH_4^+ and NO_3^-. Which is the stronger acid $C_6H_5NH_3^+$ or NH_4^+?

Answer:
$[H_3O^+] = 7.5 \times 10^{-6}$ M; $C_6H_5NH_3^+$ is the stronger acid.

Salts with Basic Ions

As another example, consider dissolving sodium acetate in water:

$$NaCH_3CO_2(s) \rightleftharpoons Na^+(aq) + CH_3CO_2^-(aq)$$

The sodium ion does not undergo appreciable acid or base ionization and has no effect on the solution pH. This may seem obvious from the ion's formula, which indicates no hydrogen or oxygen atoms, but some dissolved metal ions function as weak acids, as addressed later in this section.

The acetate ion, $CH_3CO_2^-$, is the conjugate base of acetic acid, CH_3CO_2H, and so its base ionization (or *base hydrolysis*) reaction is represented by

$$CH_3CO_2^-(aq) + H_2O(l) \rightleftharpoons CH_3CO_2H(aq) + OH-(aq) \qquad K_b = K_w/K_a$$

Because acetic acid is a weak acid, its K_a is measurable and $K_b > 0$ (acetate ion is a weak base).

Dissolving sodium acetate in water yields a solution of inert cations (Na^+) and weak base anions ($CH_3CO_2^-$), resulting in a basic solution.

 **EXAMPLE 14.16**

Equilibrium in a Solution of a Salt of a Weak Acid and a Strong Base

Determine the acetic acid concentration in a solution with $[CH_3CO_2^-] = 0.050$ M and $[OH^-] = 2.5 \times 10^{-6}$ M at equilibrium. The reaction is:

$$CH_3CO_2^-(aq) + H_2O(l) \rightleftharpoons CH_3CO_2H(aq) + OH^-(aq)$$

Solution

The provided equilibrium concentrations and a value for the equilibrium constant will permit calculation of the missing equilibrium concentration. The process in question is the base ionization of acetate ion, for which

$$K_b \text{ (for } CH_3CO_2^-) = \frac{K_w}{K_a \text{ (for } CH_3CO_2H)} = \frac{1.0 \times 10^{-14}}{1.8 \times 10^{-5}} = 5.6 \times 10^{-10}$$

Substituting the available values into the K_b expression gives

$$K_b = \frac{[CH_3CO_2H][OH^-]}{[CH_3CO_2^-]} = 5.6 \times 10^{-10}$$

$$= \frac{[CH_3CO_2H](2.5 \times 10^{-6})}{(0.050)} = 5.6 \times 10^{-10}$$

Solving the above equation for the acetic acid molarity yields $[CH_3CO_2H] = 1.1 \times 10^{-5}$ M.

Check Your Learning

What is the pH of a 0.083-M solution of NaCN?

Answer:
11.11

Salts with Acidic and Basic Ions

Some salts are composed of both acidic and basic ions, and so the pH of their solutions will depend on the relative strengths of these two species. Likewise, some salts contain a single ion that is amphiprotic, and so the relative strengths of this ion's acid and base character will determine its effect on solution pH. For both types of salts, a comparison of the K_a and K_b values allows prediction of the solution's acid-base status, as illustrated in the following example exercise.

EXAMPLE 14.17

Determining the Acidic or Basic Nature of Salts

Determine whether aqueous solutions of the following salts are acidic, basic, or neutral:

(a) KBr

(b) $NaHCO_3$

(c) Na_2HPO_4

(d) NH_4F

Solution

Consider each of the ions separately in terms of its effect on the pH of the solution, as shown here:

(a) The K^+ cation is inert and will not affect pH. The bromide ion is the conjugate base of a strong acid, and so it is of negligible base strength (no appreciable base ionization). The solution is neutral.

(b) The Na^+ cation is inert and will not affect the pH of the solution; while the $HCO_3{}^-$ anion is amphiprotic. The K_a of $HCO_3{}^-$ is 4.7×10^{-11},and its K_b is $\frac{1.0 \times 10^{-14}}{4.3 \times 10^{-7}} = 2.3 \times 10^{-8}$.

Since $K_b >> K_a$, the solution is basic.

(c) The Na^+ cation is inert and will not affect the pH of the solution, while the $HPO_4{}^{2-}$ anion is amphiprotic. The K_a of $HPO_4{}^{2-}$ is 4.2×10^{-13},

and its K_b is $\frac{1.0 \times 10^{-14}}{6.2 \times 10^{-8}} = 1.6 \times 10^{-7}$. Because $K_b >> K_a$, the solution is basic.

(d) The $NH_4{}^+$ ion is acidic (see above discussion) and the F^- ion is basic (conjugate base of the weak acid HF). Comparing the two ionization constants: K_a of $NH_4{}^+$ is 5.6×10^{-10} and the K_b of F^- is 1.6×10^{-11}, so the solution is acidic, since $K_a > K_b$.

Check Your Learning

Determine whether aqueous solutions of the following salts are acidic, basic, or neutral:

(a) K_2CO_3

(b) $CaCl_2$

(c) KH_2PO_4

(d) $(NH_4)_2CO_3$

Answer:

(a) basic; (b) neutral; (c) acidic; (d) basic

The Ionization of Hydrated Metal Ions

Unlike the group 1 and 2 metal ions of the preceding examples (Na^+, Ca^{2+}, etc.), some metal ions function as acids in aqueous solutions. These ions are not just loosely solvated by water molecules when dissolved, instead they are covalently bonded to a fixed number of water molecules to yield a complex ion (see chapter on coordination chemistry). As an example, the dissolution of aluminum nitrate in water is typically represented as

$$Al(NO_3)(s) \rightleftharpoons Al^3 + (aq) + 3NO_3{}^-(aq)$$

However, the aluminum(III) ion actually reacts with six water molecules to form a stable complex ion, and so the more explicit representation of the dissolution process is

$$Al(NO_3)_3(s) + 6H_2O(l) \rightleftharpoons Al(H_2O)_6{}^{3+}(aq) + 3NO_3{}^-(aq)$$

As shown in Figure 14.13, the $Al(H_2O)_6{}^{3+}$ ions involve bonds between a central Al atom and the O atoms of the six water molecules. Consequently, the bonded water molecules' O–H bonds are more polar than in nonbonded water molecules, making the bonded molecules more prone to donation of a hydrogen ion:

$$Al(H_2O)_6{}^{3+}(aq) + H_2O(l) \rightleftharpoons H_3O^+(aq) + Al(H_2O)_5(OH)^{2+}(aq) \qquad K_a = 1.4 \times 10^{-5}$$

The conjugate base produced by this process contains five other bonded water molecules capable of acting as acids, and so the sequential or step-wise transfer of protons is possible as depicted in few equations below:

$$Al(H_2O)_6{}^{3+}(aq) + H_2O(l) \rightleftharpoons H_3O^+(aq) + Al(H_2O)_5(OH)^{2+}(aq)$$

$$Al(H_2O)_5(OH)^{2+}(aq) + H_2O(l) \rightleftharpoons H_3O^+(aq) + Al(H_2O)_4(OH)_2{}^+(aq)$$

$$Al(H_2O)_4(OH)_2{}^+(aq) + H_2O(l) \rightleftharpoons H_3O^+(aq) + Al(H_2O)_3(OH)_3(aq)$$

This is an example of a polyprotic acid, the topic of discussion in a later section of this chapter.

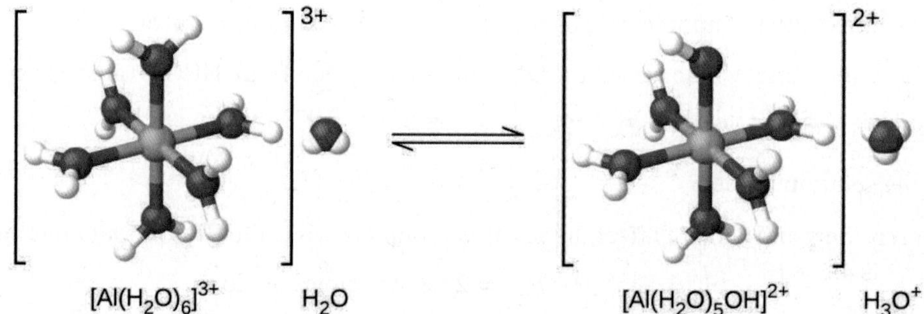

FIGURE 14.13 When an aluminum ion reacts with water, the hydrated aluminum ion becomes a weak acid.

Aside from the alkali metals (group 1) and some alkaline earth metals (group 2), most other metal ions will undergo acid ionization to some extent when dissolved in water. The acid strength of these complex ions typically increases with increasing charge and decreasing size of the metal ions. The first-step acid ionization equations for a few other acidic metal ions are shown below:

$$Fe(H_2O)_6{}^{3+}(aq) + H_2O(l) \rightleftharpoons H_3O^+(aq) + Fe(H_2O)_5(OH)^{2+}(aq) \qquad pK_a = 2.74$$

$$Cu(H_2O)_6{}^{2+}(aq) + H_2O(l) \rightleftharpoons H_3O^+(aq) + Cu(H_2O)_5(OH)^+(aq) \qquad pK_a = \text{~}6.3$$

$$Zn(H_2O)_4{}^{2+}(aq) + H_2O(l) \rightleftharpoons H_3O^+(aq) + Zn(H_2O)_3(OH)^+(aq) \qquad pK_a = 9.6$$

EXAMPLE 14.18

Hydrolysis of $[Al(H_2O)_6]^{3+}$

Calculate the pH of a 0.10-*M* solution of aluminum chloride, which dissolves completely to give the hydrated aluminum ion $[Al(H_2O)_6]^{3+}$ in solution.

Solution

The equation for the reaction and K_a are:

$$Al(H_2O)_6{}^{3+}(aq) + H_2O(l) \rightleftharpoons H_3O^+(aq) + Al(H_2O)_5(OH)^{2+}(aq) \qquad K_a = 1.4 \times 10^{-5}$$

An ICE table with the provided information is

	$Al(H_2O)_6^{3+} + H_2O \rightleftharpoons$	H_3O^+ +	$Al(H_2O)_5(OH)^{2+}$
Initial concentration (*M*)	0.10	~0	0
Change (*M*)	$-x$	$+x$	$+x$
Equilibrium concentration (*M*)	$0.10 - x$	x	x

Substituting the expressions for the equilibrium concentrations into the equation for the ionization constant yields:

$$K_a = \frac{[H_3O^+][Al(H_2O)_5(OH)^{2+}]}{[Al(H_2O)_6{}^{3+}]}$$

$$= \frac{(x)(x)}{0.10 - x} = 1.4 \times 10^{-5}$$

Assuming $x << 0.10$ and solving the simplified equation gives:

$$x = 1.2 \times 10^{-3}\ M$$

The ICE table defined x as equal to the hydronium ion concentration, and so the pH is calculated to be

$$[H_3O^+] = 0 + x = 1.2 \times 10^{-3}\ M$$

$$pH = -\log[H_3O^+] = 2.92 \text{ (an acidic solution)}$$

Check Your Learning

What is $[Al(H_2O)_5(OH)^{2+}]$ in a 0.15-*M* solution of $Al(NO_3)_3$ that contains enough of the strong acid HNO_3 to bring $[H_3O^+]$ to 0.10 *M*?

Answer:

$2.1 \times 10^{-5}\ M$

14.5 Polyprotic Acids

LEARNING OBJECTIVES

By the end of this section, you will be able to:

- Extend previously introduced equilibrium concepts to acids and bases that may donate or accept more than one proton

Acids are classified by the number of protons per molecule that they can give up in a reaction. Acids such as HCl, HNO_3, and HCN that contain one ionizable hydrogen atom in each molecule are called **monoprotic acids**. Their reactions with water are:

$$HCl(aq) + H_2O(l) \longrightarrow H_3O^+(aq) + Cl^-(aq)$$
$$HNO_3(aq) + H_2O(l) \longrightarrow H_3O^+(aq) + NO_3{}^-(aq)$$
$$HCN(aq) + H_2O(l) \rightleftharpoons H_3O^+(aq) + CN^-(aq)$$

Even though it contains four hydrogen atoms, acetic acid, CH_3CO_2H, is also monoprotic because only the hydrogen atom from the carboxyl group (COOH) reacts with bases:

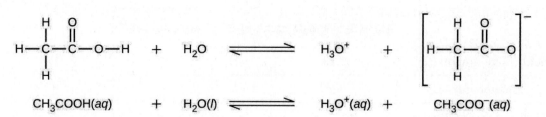

$$CH_3COOH(aq) + H_2O(l) \rightleftharpoons H_3O^+(aq) + CH_3COO^-(aq)$$

Similarly, monoprotic bases are bases that will accept a single proton.

Diprotic acids contain two ionizable hydrogen atoms per molecule; ionization of such acids occurs in two steps. The first ionization always takes place to a greater extent than the second ionization. For example, sulfuric acid, a strong acid, ionizes as follows:

$$\text{First ionization: } H_2SO_4(aq) + H_2O(l) \rightleftharpoons H_3O^+(aq) + HSO_4^-(aq) \qquad K_{a1} = \text{more than } 10^2\text{; complete dissociation}$$

$$\text{Second ionization: } HSO_4^-(aq) + H_2O(l) \rightleftharpoons H_3O^+(aq) + SO_4^{2-}(aq) \qquad K_{a2} = 1.2 \times 10^{-2}$$

This **stepwise ionization** process occurs for all polyprotic acids. Carbonic acid, H_2CO_3, is an example of a weak diprotic acid. The first ionization of carbonic acid yields hydronium ions and bicarbonate ions in small amounts.

First ionization:

$$H_2CO_3(aq) + H_2O(l) \rightleftharpoons H_3O^+(aq) + HCO_3^-(aq) \qquad K_{H_2CO_3} = \frac{[H_3O^+][HCO_3^-]}{[H_2CO_3]} = 4.3 \times 10^{-7}$$

The bicarbonate ion can also act as an acid. It ionizes and forms hydronium ions and carbonate ions in even smaller quantities.

Second ionization:

$$HCO_3^-(aq) + H_2O(l) \rightleftharpoons H_3O^+(aq) + CO_3^{2-}(aq) \qquad K_{HCO_3^-} = \frac{[H_3O^+][CO_3^{2-}]}{[HCO_3^-]} = 4.7 \times 10^{-11}$$

$K_{H_2CO_3}$ is larger than $K_{HCO_3^-}$ by a factor of 10^4, so H_2CO_3 is the dominant producer of hydronium ion in the solution. This means that little of the HCO_3^- formed by the ionization of H_2CO_3 ionizes to give hydronium ions (and carbonate ions), and the concentrations of H_3O^+ and HCO_3^- are practically equal in a pure aqueous solution of H_2CO_3.

If the first ionization constant of a weak diprotic acid is larger than the second by a factor of at least 20, it is appropriate to treat the first ionization separately and calculate concentrations resulting from it before calculating concentrations of species resulting from subsequent ionization. This approach is demonstrated in the following example exercise.

EXAMPLE 14.19

Ionization of a Diprotic Acid

"Carbonated water" contains a palatable amount of dissolved carbon dioxide. The solution is acidic because CO_2 reacts with water to form carbonic acid, H_2CO_3. What are $[H_3O^+]$, $[HCO_3^-]$, and $[CO_3^{2-}]$ in a saturated solution of CO_2 with an initial $[H_2CO_3] = 0.033$ *M*?

$$H_2CO_3(aq) + H_2O(l) \rightleftharpoons H_3O^+(aq) + HCO_3^-(aq) \qquad K_{a1} = 4.3 \times 10^{-7}$$

$$HCO_3^-(aq) + H_2O(l) \rightleftharpoons H_3O^+(aq) + CO_3^{2-}(aq) \qquad K_{a2} = 4.7 \times 10^{-11}$$

Solution

As indicated by the ionization constants, H_2CO_3 is a much stronger acid than HCO_3^-, so the stepwise ionization reactions may be treated separately.

The first ionization reaction is

$$H_2CO_3(aq) + H_2O(l) \rightleftharpoons H_3O^+(aq) + HCO_3^-(aq) \qquad K_{a1} = 4.3 \times 10^{-7}$$

Using provided information, an ICE table for this first step is prepared:

	H_2CO_3	+ H_2O	$\rightleftharpoons$ H_3O^+	+ HCO_3^-
Initial concentration (*M*)	0.033		~0	0
Change (*M*)	−*x*		+*x*	+*x*
Equilibrium concentration (*M*)	0.033 − *x*		*x*	*x*

Substituting the equilibrium concentrations into the equilibrium equation gives

$$K_{H_2CO_3} = \frac{[H_3O^+][HCO_3^-]}{[H_2CO_3]} = \frac{(x)(x)}{0.033 - x} = 4.3 \times 10^{-7}$$

Assuming $x << 0.033$ and solving the simplified equation yields

$$x = 1.2 \times 10^{-4}$$

The ICE table defined *x* as equal to the bicarbonate ion molarity and the hydronium ion molarity:

$$[H_2CO_3] = 0.033\ M$$

$$[H_3O^+] = [HCO_3^-] = 1.2 \times 10^{-4}\ M$$

Using the bicarbonate ion concentration computed above, the second ionization is subjected to a similar equilibrium calculation:

$$HCO_3^-(aq) + H_2O(l) \rightleftharpoons H_3O^+(aq) + CO_3^{2-}(aq)$$

$$K_{HCO_3^-} = \frac{[H_3O^+][CO_3^{2-}]}{[HCO_3^-]} = \frac{(1.2 \times 10^{-4})[CO_3^{2-}]}{1.2 \times 10^{-4}}$$

$$[CO_3^{2-}] = \frac{(4.7 \times 10^{-11})(1.2 \times 10^{-4})}{1.2 \times 10^{-4}} = 4.7 \times 10^{-11}\ M$$

To summarize: at equilibrium $[H_2CO_3] = 0.033\ M$; $[H_3O^+] = 1.2 \times 10^{-4}$; $[HCO_3^-] = 1.2 \times 10^{-4}\ M$; $[CO_3^{2-}] = 4.7 \times 10^{-11}\ M$.

Check Your Learning

The concentration of H_2S in a saturated aqueous solution at room temperature is approximately 0.1 *M*. Calculate $[H_3O^+]$, $[HS^-]$, and $[S^{2-}]$ in the solution:

$$H_2S(aq) + H_2O(l) \rightleftharpoons H_3O^+(aq) + HS^-(aq) \qquad K_{a1} = 8.9 \times 10^{-8}$$

$$HS^-(aq) + H_2O(l) \rightleftharpoons H_3O^+(aq) + S^{2-}(aq) \qquad K_{a2} = 1.0 \times 10^{-19}$$

Answer:

$[H_2S] = 0.1\ M$; $[H_3O^+] = [HS^-] = 0.000094\ M$; $[S^{2-}] = 1 \times 10^{-19}\ M$

A **triprotic acid** is an acid that has three ionizable H atoms. Phosphoric acid is one example:

$$\text{First ionization: } H_3PO_4(aq) + H_2O(l) \rightleftharpoons H_3O^+(aq) + H_2PO_4^-(aq) \qquad K_{a1} = 7.5 \times 10^{-3}$$
$$\text{Second ionization: } H_2PO_4^-(aq) + H_2O(l) \rightleftharpoons H_3O^+(aq) + HPO_4^{2-}(aq) \qquad K_{a2} = 6.2 \times 10^{-8}$$
$$\text{Third ionization: } HPO_4^{2-}(aq) + H_2O(l) \rightleftharpoons H_3O^+(aq) + PO_4^{3-}(aq) \qquad K_{a3} = 4.2 \times 10^{-13}$$

As for the diprotic acid examples, each successive ionization reaction is less extensive than the former, reflected in decreasing values for the stepwise acid ionization constants. This is a general characteristic of polyprotic acids and successive ionization constants often differ by a factor of about 10^5 to 10^6.

This set of three dissociation reactions may appear to make calculations of equilibrium concentrations in a solution of H_3PO_4 complicated. However, because the successive ionization constants differ by a factor of 10^5 to 10^6, large differences exist in the small changes in concentration accompanying the ionization reactions. This allows the use of math-simplifying assumptions and processes, as demonstrated in the examples above.

Polyprotic bases are capable of accepting more than one hydrogen ion. The carbonate ion is an example of a **diprotic base**, because it can accept two protons, as shown below. Similar to the case for polyprotic acids, note the ionization constants decrease with ionization step. Likewise, equilibrium calculations involving polyprotic bases follow the same approaches as those for polyprotic acids.

$$H_2O(l) + CO_3^{2-}(aq) \rightleftharpoons HCO_3^-(aq) + OH^-(aq) \qquad K_{b1} = 2.1 \times 10^{-4}$$
$$H_2O(l) + HCO_3^-(aq) \rightleftharpoons H_2CO_3(aq) + OH^-(aq) \qquad K_{b2} = 2.3 \times 10^{-8}$$

14.6 Buffers

LEARNING OBJECTIVES

By the end of this section, you will be able to:

- Describe the composition and function of acid–base buffers
- Calculate the pH of a buffer before and after the addition of added acid or base

A solution containing appreciable amounts of a weak conjugate acid-base pair is called a buffer solution, or a **buffer**. Buffer solutions resist a change in pH when small amounts of a strong acid or a strong base are added (Figure 14.14). A solution of acetic acid and sodium acetate (CH_3COOH + CH_3COONa) is an example of a buffer that consists of a weak acid and its salt. An example of a buffer that consists of a weak base and its salt is a solution of ammonia and ammonium chloride ($NH_3(aq)$ + $NH_4Cl(aq)$).

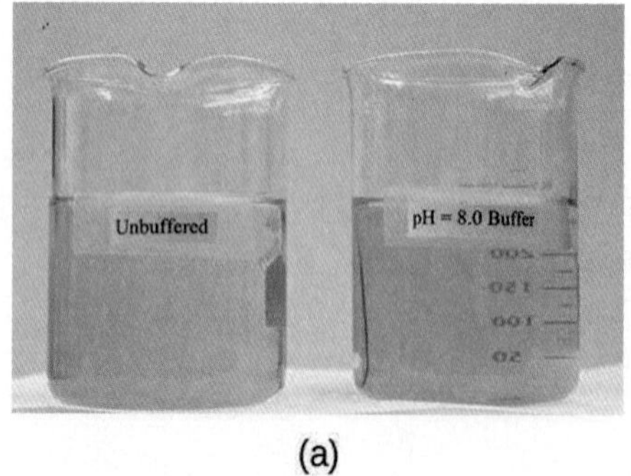

(a)

(b)

FIGURE 14.14 (a) The unbuffered solution on the left and the buffered solution on the right have the same pH (pH 8); they are basic, showing the yellow color of the indicator methyl orange at this pH. (b) After the addition of 1 mL of a 0.01-*M* HCl solution, the buffered solution has not detectably changed its pH but the unbuffered solution has become acidic, as indicated by the change in color of the methyl orange, which turns red at a pH of about 4. (credit: modification of work by Mark Ott)

How Buffers Work

To illustrate the function of a buffer solution, consider a mixture of roughly equal amounts of acetic acid and sodium acetate. The presence of a weak conjugate acid-base pair in the solution imparts the ability to neutralize modest amounts of added strong acid or base. For example, strong base added to this solution will neutralize hydronium ion, causing the acetic acid ionization equilibrium to shift to the right and generate additional amounts of the weak conjugate base (acetate ion):

$$CH_3CO_2H(aq) + H_2O(l) \rightleftharpoons H_3O^+(aq) + CH_3CO_2{}^-(aq)$$

Likewise, strong acid added to this buffer solution will shift the above ionization equilibrium left, producing additional amounts of the weak conjugate acid (acetic acid). Figure 14.15 provides a graphical illustration of the changes in conjugate-partner concentration that occur in this buffer solution when strong acid and base are added. The buffering action of the solution is essentially a result of the added strong acid and base being converted to the weak acid and base that make up the buffer's conjugate pair. The weaker acid and base undergo only slight ionization, as compared with the complete ionization of the strong acid and base, and the solution pH, therefore, changes much less drastically than it would in an unbuffered solution.

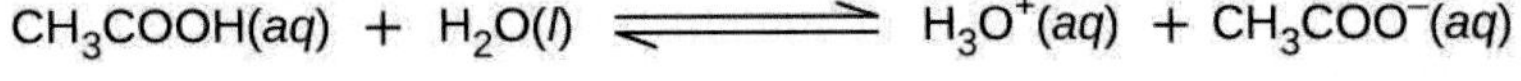

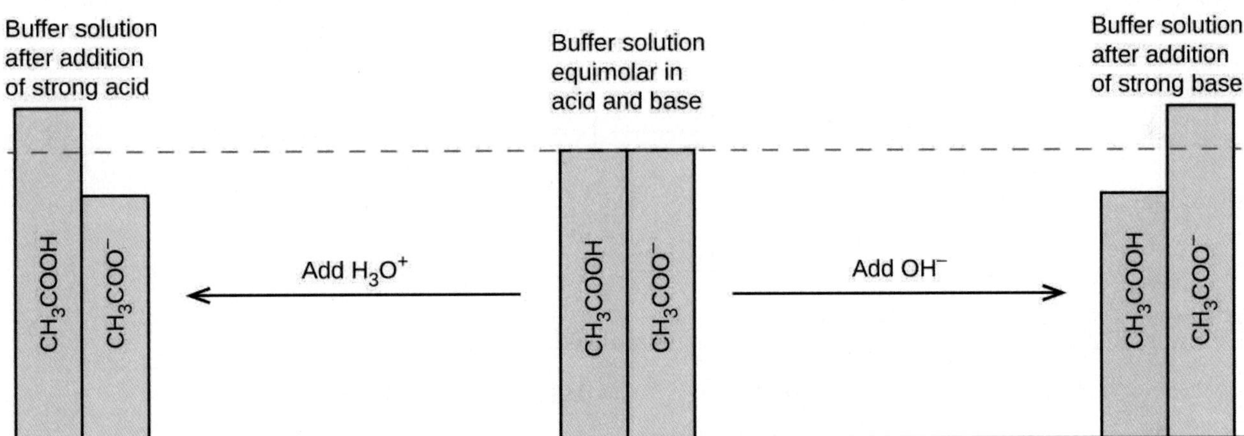

FIGURE 14.15 Buffering action in a mixture of acetic acid and acetate salt.

EXAMPLE 14.20

pH Changes in Buffered and Unbuffered Solutions

Acetate buffers are used in biochemical studies of enzymes and other chemical components of cells to prevent pH changes that might affect the biochemical activity of these compounds.

(a) Calculate the pH of an acetate buffer that is a mixture with 0.10 *M* acetic acid and 0.10 *M* sodium acetate.

(b) Calculate the pH after 1.0 mL of 0.10 NaOH is added to 100 mL of this buffer.

(c) For comparison, calculate the pH after 1.0 mL of 0.10 *M* NaOH is added to 100 mL of a solution of an unbuffered solution with a pH of 4.74.

Solution

(a) Following the ICE approach to this equilibrium calculation yields the following:

	CH_3CO_2H	+ H_2O	$\rightleftharpoons$ H_3O^+	+ $CH_3CO_2^-$
Initial concentration (*M*)	0.10		~0	0.10
Change (*M*)	$-x$		$+x$	$+x$
Equilibrium concentration (*M*)	$0.10 - x$		x	$0.10 + x$

Substituting the equilibrium concentration terms into the K_a expression, assuming $x << 0.10$, and solving the simplified equation for x yields

$$x = 1.8 \times 10^{-5}\, M$$

$$[H_3O^+] = 0 + x = 1.8 \times 10^{-5}\, M$$

$$pH = -\log [H_3O^+] = -\log (1.8 \times 10^{-5})$$

$$= 4.74$$

(b) Calculate the pH after 1.0 mL of 0.10 *M* NaOH is added to 100 mL of this buffer.

Adding strong base will neutralize some of the acetic acid, yielding the conjugate base acetate ion. Compute the new concentrations of these two buffer components, then repeat the equilibrium calculation of part (a) using these new concentrations.

$$0.0010\,\cancel{L} \times \left(\frac{0.10 \text{ mol NaOH}}{1\,\cancel{L}}\right) = 1.0 \times 10^{-4} \text{ mol NaOH}$$

The initial molar amount of acetic acid is

$$0.100\,\cancel{L} \times \left(\frac{0.100 \text{ mol } CH_3CO_2H}{1\,\cancel{L}}\right) = 1.00 \times 10^{-2} \text{ mol } CH_3CO_2H$$

The amount of acetic acid remaining after some is neutralized by the added base is

$$(1.0 \times 10^{-2}) - (0.01 \times 10^{-2}) = 0.99 \times 10^{-2} \text{ mol } CH_3CO_2H$$

The newly formed acetate ion, along with the initially present acetate, gives a final acetate concentration of

$$(1.0 \times 10^{-2}) + (0.01 \times 10^{-2}) = 1.01 \times 10^{-2} \text{mol } NaCH_3CO_2$$

Compute molar concentrations for the two buffer components:

$$[CH_3CO_2H] = \frac{9.9 \times 10^{-3} \text{ mol}}{0.101 \text{ L}} = 0.098\, M$$

$$[NaCH_3CO_2] = \frac{1.01 \times 10^{-2} \text{ mol}}{0.101 \text{ L}} = 0.100\, M$$

Using these concentrations, the pH of the solution may be computed as in part (a) above, yielding pH = 4.75 (only slightly different from that prior to adding the strong base).

(c) For comparison, calculate the pH after 1.0 mL of 0.10 *M* NaOH is added to 100 mL of a solution of an unbuffered solution with a pH of 4.74.

The amount of hydronium ion initially present in the solution is

$$[H_3O^+] = 10^{-4.74} = 1.8 \times 10^{-5}\, M$$

$$\text{mol } H_3O^+ = (0.100\, L)(1.8 \times 10^{-5}\, M) = 1.8 \times 10^{-6} \text{ mol } H_3O^+$$

The amount of hydroxide ion added to the solution is

$$\text{mol } OH^- = (0.0010\, L)(0.10\, M) = 1.0 \times 10^{-4} \text{ mol } OH^-$$

The added hydroxide will neutralize hydronium ion via the reaction

$$H_3O^+(aq) + OH^-(aq) \leftrightharpoons 2H_2O(l)$$

The 1:1 stoichiometry of this reaction shows that an excess of hydroxide has been added (greater molar amount than the initially present hydronium ion).

The amount of hydroxide ion remaining is

$$1.0 \times 10^{-4} \text{ mol} - 1.8 \times 10^{-6} \text{mol} = 9.8 \times 10^{-5} \text{ mol OH}^-$$

corresponding to a hydroxide molarity of

$$9.8 \times 10^{-5} \text{ mol OH}^-/0.101\ L = 9.7 \times 10^{-4}\ M$$

The pH of the solution is then calculated to be

$$\text{pH} = 14.00 - \text{pOH} = 14.00 - -\log(9.7 \times 10^{-4}) = 10.99$$

In this unbuffered solution, addition of the base results in a significant rise in pH (from 4.74 to 10.99) compared with the very slight increase observed for the buffer solution in part (b) (from 4.74 to 4.75).

Check Your Learning

Show that adding 1.0 mL of 0.10 M HCl changes the pH of 100 mL of a 1.8×10^{-5} M HCl solution from 4.74 to 3.00.

Answer:

Initial pH of 1.8×10^{-5} M HCl; pH = $-\log[H_3O^+] = -\log[1.8 \times 10^{-5}] = 4.74$
Moles of H_3O^+ in 100 mL 1.8×10^{-5} M HCl; 1.8×10^{-5} moles/L $\times$ 0.100 L = 1.8×10^{-6}
Moles of H_3O^+ added by addition of 1.0 mL of 0.10 M HCl: 0.10 moles/L $\times$ 0.0010 L = 1.0×10^{-4} moles; final pH after addition of 1.0 mL of 0.10 M HCl:

$$\text{pH} = -\log[H_3O^+] = -\log\left(\frac{\text{total moles } H_3O^+}{\text{total volume}}\right) = -\log\left(\frac{1.0 \times 10^{-4} \text{ mol} + 1.8 \times 10^{-6} \text{ mol}}{101 \text{ mL}\left(\frac{1 \text{ L}}{1000 \text{ mL}}\right)}\right) = 3.00$$

Buffer Capacity

Buffer solutions do not have an unlimited capacity to keep the pH relatively constant (Figure 14.16). Instead, the ability of a buffer solution to resist changes in pH relies on the presence of appreciable amounts of its conjugate weak acid-base pair. When enough strong acid or base is added to substantially lower the concentration of either member of the buffer pair, the buffering action within the solution is compromised.

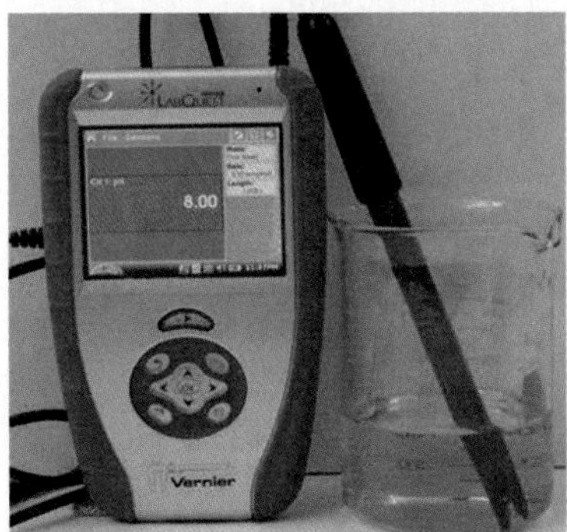

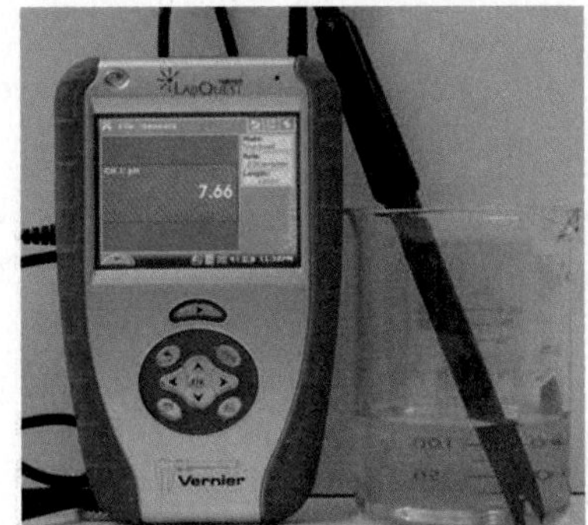

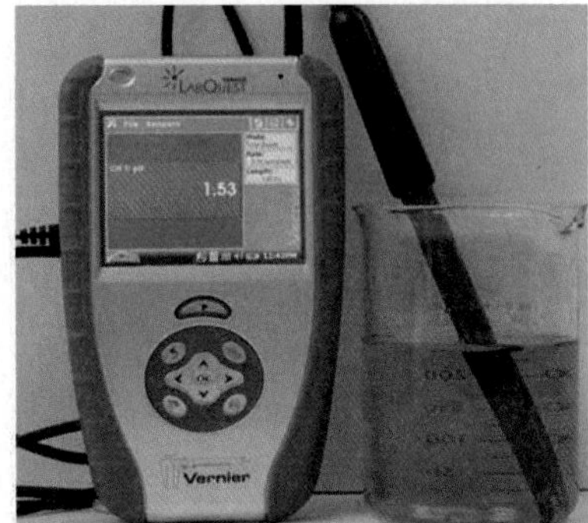

FIGURE 14.16 The indicator color (methyl orange) shows that a small amount of acid added to a buffered solution of pH 8 (beaker on the left) has little affect on the buffered system (middle beaker). However, a large amount of acid exhausts the buffering capacity of the solution and the pH changes dramatically (beaker on the right). (credit: modification of work by Mark Ott)

The **buffer capacity** is the amount of acid or base that can be added to a given volume of a buffer solution before the pH changes significantly, usually by one unit. Buffer capacity depends on the amounts of the weak acid and its conjugate base that are in a buffer mixture. For example, 1 L of a solution that is 1.0 *M* in acetic acid and 1.0 *M* in sodium acetate has a greater buffer capacity than 1 L of a solution that is 0.10 *M* in acetic acid and 0.10 *M* in sodium acetate even though both solutions have the same pH. The first solution has more buffer capacity because it contains more acetic acid and acetate ion.

Selection of Suitable Buffer Mixtures

There are two useful rules of thumb for selecting buffer mixtures:

1. A good buffer mixture should have about equal concentrations of both of its components. A buffer solution has generally lost its usefulness when one component of the buffer pair is less than about 10% of the other. Figure 14.17 shows how pH changes for an acetic acid-acetate ion buffer as base is added. The initial pH is 4.74. A change of 1 pH unit occurs when the acetic acid concentration is reduced to 11% of the acetate ion concentration.

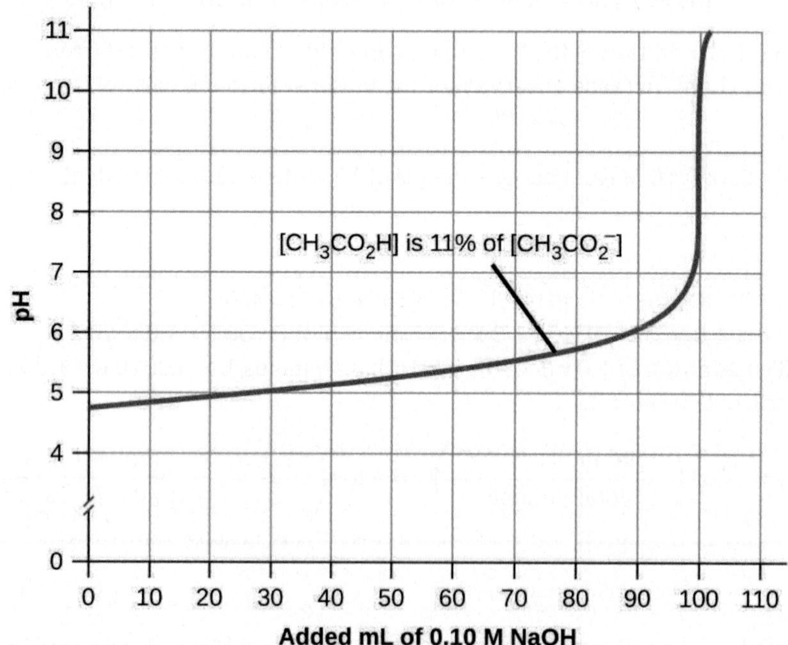

FIGURE 14.17 Change in pH as an increasing amount of a 0.10-*M* NaOH solution is added to 100 mL of a buffer solution in which, initially, $[CH_3CO_2H]$ = 0.10 *M* and $[CH_3CO_2{}^-] = 0.10\ M$. Note the greatly diminished buffering action occurring after the buffer capacity has been reached, resulting in drastic rises in pH on adding more strong base.

2. Weak acids and their salts are better as buffers for pHs less than 7; weak bases and their salts are better as buffers for pHs greater than 7.

Blood is an important example of a buffered solution, with the principal acid and ion responsible for the buffering action being carbonic acid, H_2CO_3, and the bicarbonate ion, $HCO_3{}^-$. When a hydronium ion is introduced to the blood stream, it is removed primarily by the reaction:

$$H_3O^+(aq) + HCO_3{}^-(aq) \longrightarrow H_2CO_3(aq) + H_2O(l)$$

An added hydroxide ion is removed by the reaction:

$$OH^-(aq) + H_2CO_3(aq) \longrightarrow HCO_3{}^-(aq) + H_2O(l)$$

The added strong acid or base is thus effectively converted to the much weaker acid or base of the buffer pair (H_3O^+ is converted to H_2CO_3 and OH^- is converted to $HCO_3{}^-$). The pH of human blood thus remains very near the value determined by the buffer pairs pKa, in this case, 7.35. Normal variations in blood pH are usually less than 0.1, and pH changes of 0.4 or greater are likely to be fatal.

The Henderson-Hasselbalch Equation

The ionization-constant expression for a solution of a weak acid can be written as:

$$K_a = \frac{[H_3O^+][A^-]}{[HA]}$$

Rearranging to solve for $[H_3O^+]$ yields:

$$[H_3O^+] = K_a \times \frac{[HA]}{[A^-]}$$

Taking the negative logarithm of both sides of this equation gives

$$-\log[H_3O^+] = -\log K_a - \log\frac{[HA]}{[A^-]},$$

which can be written as

$$pH = pK_a + \log\frac{[A^-]}{[HA]}$$

where pK_a is the negative of the logarithm of the ionization constant of the weak acid ($pK_a = -\log K_a$). This equation relates the pH, the ionization constant of a weak acid, and the concentrations of the weak conjugate acid-base pair in a buffered solution. Scientists often use this expression, called the **Henderson-Hasselbalch equation**, to calculate the pH of buffer solutions. It is important to note that the "*x* is small" assumption must be valid to use this equation.

Portrait of a Chemist

Lawrence Joseph Henderson and Karl Albert Hasselbalch

Lawrence Joseph Henderson (1878–1942) was an American physician, biochemist and physiologist, to name only a few of his many pursuits. He obtained a medical degree from Harvard and then spent 2 years studying in Strasbourg, then a part of Germany, before returning to take a lecturer position at Harvard. He eventually became a professor at Harvard and worked there his entire life. He discovered that the acid-base balance in human blood is regulated by a buffer system formed by the dissolved carbon dioxide in blood. He wrote an equation in 1908 to describe the carbonic acid-carbonate buffer system in blood. Henderson was broadly knowledgeable; in addition to his important research on the physiology of blood, he also wrote on the adaptations of organisms and their fit with their environments, on sociology and on university education. He also founded the Fatigue Laboratory, at the Harvard Business School, which examined human physiology with specific focus on work in industry, exercise, and nutrition.

In 1916, Karl Albert Hasselbalch (1874–1962), a Danish physician and chemist, shared authorship in a paper with Christian Bohr in 1904 that described the Bohr effect, which showed that the ability of hemoglobin in the blood to bind with oxygen was inversely related to the acidity of the blood and the concentration of carbon dioxide. The pH scale was introduced in 1909 by another Dane, Sørensen, and in 1912, Hasselbalch published measurements of the pH of blood. In 1916, Hasselbalch expressed Henderson's equation in logarithmic terms, consistent with the logarithmic scale of pH, and thus the Henderson-Hasselbalch equation was born.

HOW SCIENCES INTERCONNECT

Medicine: The Buffer System in Blood

The normal pH of human blood is about 7.4. The carbonate buffer system in the blood uses the following equilibrium reaction:

$$CO_2(g) + 2H_2O(l) \rightleftharpoons H_2CO_3(aq) \rightleftharpoons HCO_3{}^-(aq) + H_3O^+(aq)$$

The concentration of carbonic acid, H_2CO_3 is approximately 0.0012 *M*, and the concentration of the hydrogen carbonate ion, $HCO_3{}^-$, is around 0.024 *M*. Using the Henderson-Hasselbalch equation and the pK_a of carbonic acid at body temperature, we can calculate the pH of blood:

$$\text{pH} = \text{p}K_a + \log \frac{[\text{base}]}{[\text{acid}]} = 6.4 + \log \frac{0.024}{0.0012} = 7.7$$

The fact that the H_2CO_3 concentration is significantly lower than that of the $HCO_3{}^-$ ion may seem unusual, but this imbalance is due to the fact that most of the by-products of our metabolism that enter our bloodstream are acidic. Therefore, there must be a larger proportion of base than acid, so that the capacity of the buffer will not be exceeded.

Lactic acid is produced in our muscles when we exercise. As the lactic acid enters the bloodstream, it is neutralized by the $HCO_3{}^-$ ion, producing H_2CO_3. An enzyme then accelerates the breakdown of the excess carbonic acid to carbon dioxide and water, which can be eliminated by breathing. In fact, in addition to the regulating effects of the carbonate buffering system on the pH of blood, the body uses breathing to regulate blood pH. If the pH of the blood decreases too far, an increase in breathing removes CO_2 from the blood through the lungs driving the equilibrium reaction such that $[H_3O^+]$ is lowered. If the blood is too alkaline, a lower breath rate increases CO_2 concentration in the blood, driving the equilibrium reaction the other way, increasing $[H^+]$ and restoring an appropriate pH.

LINK TO LEARNING

View information (http://openstax.org/l/16BufferSystem) on the buffer system encountered in natural waters.

14.7 Acid-Base Titrations

LEARNING OBJECTIVES

By the end of this section, you will be able to:

- Interpret titration curves for strong and weak acid-base systems
- Compute sample pH at important stages of a titration
- Explain the function of acid-base indicators

As seen in the chapter on the stoichiometry of chemical reactions, titrations can be used to quantitatively analyze solutions for their acid or base concentrations. In this section, we will explore the underlying chemical equilibria that make acid-base titrimetry a useful analytical technique.

Titration Curves

A **titration curve** is a plot of some solution property versus the amount of added titrant. For acid-base titrations, solution pH is a useful property to monitor because it varies predictably with the solution composition and, therefore, may be used to monitor the titration's progress and detect its end point. The following example exercise demonstrates the computation of pH for a titration solution after additions of several specified titrant volumes. The first example involves a strong acid titration that requires only stoichiometric calculations to derive the solution pH. The second example addresses a weak acid titration requiring equilibrium calculations.

EXAMPLE 14.21

Calculating pH for Titration Solutions: Strong Acid/Strong Base

A titration is carried out for 25.00 mL of 0.100 *M* HCl (strong acid) with 0.100 *M* of a strong base NaOH (the titration curve is shown in Figure 14.18). Calculate the pH at these volumes of added base solution:

(a) 0.00 mL

(b) 12.50 mL

(c) 25.00 mL

(d) 37.50 mL

Solution

(a) Titrant volume = 0 mL. The solution pH is due to the acid ionization of HCl. Because this is a strong acid, the ionization is complete and the hydronium ion molarity is 0.100 *M*. The pH of the solution is then

$$\text{pH} = -\log\ (0.100) = 1.000$$

(b) Titrant volume = 12.50 mL. Since the acid sample and the base titrant are both monoprotic and equally concentrated, this titrant addition involves less than a stoichiometric amount of base, and so it is completely consumed by reaction with the excess acid in the sample. The concentration of acid remaining is computed by subtracting the consumed amount from the intial amount and then dividing by the solution volume:

$$[\text{H}_3\text{O}^+] = \frac{\text{n(H}^+)}{V} = \frac{0.002500\ \text{mol} \times \left(\frac{1000\ \text{mL}}{1\ \text{L}}\right) - 0.100\ M \times 12.50\ \text{mL}}{25.00\ \text{mL} + 12.50\ \text{mL}} = 0.0333\ M$$

(c) Titrant volume = 25.00 mL. This titrant addition involves a stoichiometric amount of base (the *equivalence point*), and so only products of the neutralization reaction are in solution (water and NaCl). Neither the cation nor the anion of this salt undergo acid-base ionization; the only process generating hydronium ions is the autoprotolysis of water. The solution is neutral, having a pH = 7.00.

(d) Titrant volume = 37.50 mL. This involves the addition of titrant in excess of the equivalence point. The solution pH is then calculated using the concentration of hydroxide ion:

$$\text{n(OH}^-)_0 > \text{n(H}^+)_0$$

$$[\text{OH}^-] = \frac{\text{n(OH}^-)}{V} = \frac{0.100\ M \times 37.50\ \text{mL} - 0.002500\ \text{mol} \times \left(\frac{1000\ \text{mL}}{1\ \text{L}}\right)}{25.00\ \text{mL} + 37.50\ \text{mL}} = 0.0200\ M$$

pH = 14 − pOH = 14 + log([OH^-]) = 14 + log(0.0200) = 12.30

Check Your Learning

Calculate the pH for the strong acid/strong base titration between 50.0 mL of 0.100 *M* $HNO_3(aq)$ and 0.200 *M* NaOH (titrant) at the listed volumes of added base: 0.00 mL, 15.0 mL, 25.0 mL, and 40.0 mL.

Answer:

0.00: 1.000; 15.0: 1.5111; 25.0: 7; 40.0: 12.523

 **EXAMPLE 14.22**

Titration of a Weak Acid with a Strong Base

Consider the titration of 25.00 mL of 0.100 *M* CH_3CO_2H with 0.100 *M* NaOH. The reaction can be represented as:

$$\text{CH}_3\text{CO}_2\text{H} + \text{OH}^- \longrightarrow \text{CH}_3\text{CO}_2{}^- + \text{H}_2\text{O}$$

Calculate the pH of the titration solution after the addition of the following volumes of NaOH titrant:

(a) 0.00 mL

(b) 25.00 mL

(c) 12.50 mL

(d) 37.50 mL

Solution

(a) The initial pH is computed for the acetic acid solution in the usual ICE approach:

$$K_a = \frac{[H_3O^+][CH_3CO_2^-]}{[CH_3CO_2H]} \approx \frac{[H_3O^+]^2}{[CH_3CO_2H]_0}, \text{ and}$$

$$[H_3O^+] = \sqrt{K_a \times [CH_3CO_2H]} = \sqrt{1.8 \times 10^{-5} \times 0.100} = 1.3 \times 10^{-3}$$

$$pH = -\log(1.3 \times 10^{-3}) = 2.87$$

(b) The acid and titrant are both monoprotic and the sample and titrant solutions are equally concentrated; thus, this volume of titrant represents the equivalence point. Unlike the strong-acid example above, however, the reaction mixture in this case contains a weak conjugate base (acetate ion). The solution pH is computed considering the base ionization of acetate, which is present at a concentration of

$$\frac{0.00250 \text{ mol}}{0.0500 \text{ L}} = 0.0500 \, M CH_3CO_2^-$$

Base ionization of acetate is represented by the equation

$$CH_3CO_2^-(aq) + H_2O(l) \rightleftharpoons CH_3CO_2H(aq) + OH^-(aq)$$

$$K_b = \frac{[H^+][OH^-]}{K_a} = \frac{K_w}{K_a} = \frac{1.0 \times 10^{-14}}{1.8 \times 10^{-5}} = 5.6 \times 10^{-10}$$

Assuming $x << 0.0500$, the pH may be calculated via the usual ICE approach: $K_b = \frac{x^2}{0.0500 \, M}$

$$x = [OH^-] = 5.3 \times 10^{-6}$$

$$pOH = -\log(5.3 \times 10^{-6}) = 5.28$$

$$pH = 14.00 - 5.28 = 8.72$$

Note that the pH at the equivalence point of this titration is significantly greater than 7, as expected when titrating a weak acid with a strong base.

(c) Titrant volume = 12.50 mL. This volume represents one-half of the stoichiometric amount of titrant, and so one-half of the acetic acid has been neutralized to yield an equivalent amount of acetate ion. The concentrations of these conjugate acid-base partners, therefore, are equal. A convenient approach to computing the pH is use of the Henderson-Hasselbalch equation:

$$pH = pK_a + \log\frac{[Base]}{[Acid]} = -\log(K_a) + \log\frac{[CH_3CO_2^-]}{[CH_3CO_2H]} = -\log(1.8 \times 10^{-5}) + \log(1)$$

$$pH = -\log(1.8 \times 10^{-5}) = 4.74$$

($pH = pK_a$ at the half-equivalence point in a titration of a weak acid)

(d) Titrant volume = 37.50 mL. This volume represents a stoichiometric excess of titrant, and a reaction solution containing both the titration product, acetate ion, and the excess strong titrant. In such solutions, the solution pH is determined primarily by the amount of excess strong base:

$$[OH^-] = \frac{(0.003750 \text{ mol} - 0.00250 \text{ mol})}{0.06250 \text{ L}} = 2.00 \times 10^{-2} \, M$$

$$pOH = -\log\left(2.00 \times 10^{-2}\right) = 1.70, \text{ and } pH = 14.00 - 1.70 = 12.30$$

Check Your Learning

Calculate the pH for the weak acid/strong base titration between 50.0 mL of 0.100 M HCOOH(aq) (formic acid) and 0.200 M NaOH (titrant) at the listed volumes of added base: 0.00 mL, 15.0 mL, 25.0 mL, and 30.0 mL.

Answer:

0.00 mL: 2.37; 15.0 mL: 3.92; 25.00 mL: 8.29; 30.0 mL: 12.097

Performing additional calculations similar to those in the preceding example permits a more full assessment of titration curves. A summary of pH/volume data pairs for the strong and weak acid titrations is provided in Table 14.2 and plotted as titration curves in Figure 14.18. A comparison of these two curves illustrates several important concepts that are best addressed by identifying the four stages of a titration:

initial state (added titrant volume = 0 mL): pH is determined by the acid being titrated; because the two acid samples are equally concentrated, the weak acid will exhibit a greater initial pH

pre-equivalence point (0 mL < V < 25 mL): solution pH increases gradually and the acid is consumed by reaction with added titrant; composition includes unreacted acid and the reaction product, its conjugate base

equivalence point (V = 25 mL): a drastic rise in pH is observed as the solution composition transitions from acidic to either neutral (for the strong acid sample) or basic (for the weak acid sample), with pH determined by ionization of the conjugate base of the acid

postequivalence point (V > 25 mL): pH is determined by the amount of excess strong base titrant added; since both samples are titrated with the same titrant, both titration curves appear similar at this stage.

pH Values in the Titrations of a Strong Acid and of a Weak Acid

Volume of 0.100 *M* NaOH Added (mL)	Moles of NaOH Added	pH Values 0.100 *M* HCl[1]	pH Values 0.100 *M* CH_3CO_2H[2]
0.0	0.0	1.00	2.87
5.0	0.00050	1.18	4.14
10.0	0.00100	1.37	4.57
15.0	0.00150	1.60	4.92
20.0	0.00200	1.95	5.35
22.0	0.00220	2.20	5.61
24.0	0.00240	2.69	6.13
24.5	0.00245	3.00	6.44
24.9	0.00249	3.70	7.14
25.0	0.00250	7.00	8.72
25.1	0.00251	10.30	10.30
25.5	0.00255	11.00	11.00
26.0	0.00260	11.29	11.29

1 Titration of 25.00 mL of 0.100 *M* HCl (0.00250 mol of HCl) with 0.100 *M* NaOH.
2 Titration of 25.00 mL of 0.100 *M* CH_3CO_2H (0.00250 mol of CH_3CO_2H) with 0.100 *M* NaOH.

Volume of 0.100 *M* NaOH Added (mL)	Moles of NaOH Added	pH Values 0.100 *M* HCl[1]	pH Values 0.100 *M* CH_3CO_2H[2]
28.0	0.00280	11.75	11.75
30.0	0.00300	11.96	11.96
35.0	0.00350	12.22	12.22
40.0	0.00400	12.36	12.36
45.0	0.00450	12.46	12.46
50.0	0.00500	12.52	12.52

TABLE 14.2

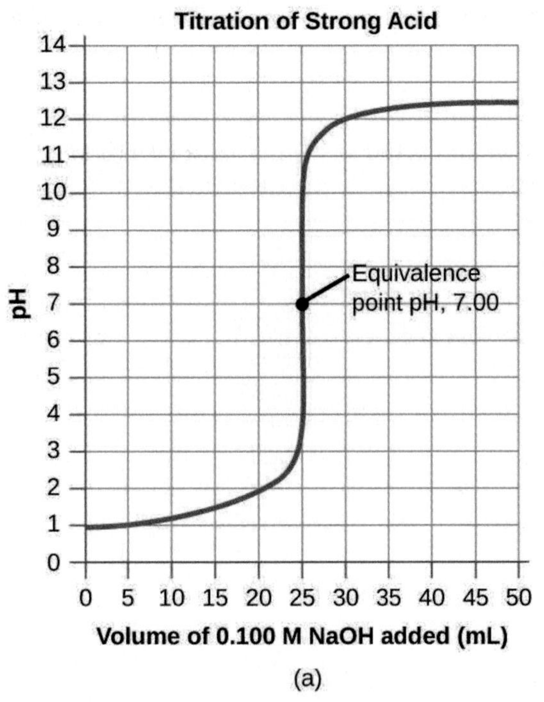

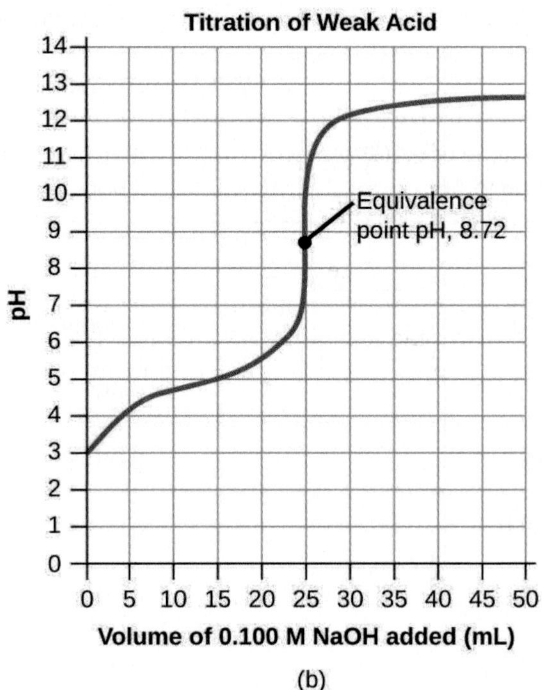

FIGURE 14.18 (a) The titration curve for the titration of 25.00 mL of 0.100 *M* HCl (strong acid) with 0.100 *M* NaOH (strong base) has an equivalence point of 7.00 pH. (b) The titration curve for the titration of 25.00 mL of 0.100 *M* acetic acid (weak acid) with 0.100 *M* NaOH (strong base) has an equivalence point of 8.72 pH.

Acid-Base Indicators

Certain organic substances change color in dilute solution when the hydronium ion concentration reaches a particular value. For example, phenolphthalein is a colorless substance in any aqueous solution with a hydronium ion concentration greater than 5.0×10^{-9} *M* (pH < 8.3). In more basic solutions where the hydronium ion concentration is less than 5.0×10^{-9} *M* (pH > 8.3), it is red or pink. Substances such as phenolphthalein, which can be used to determine the pH of a solution, are called **acid-base indicators**. Acid-base indicators are either weak organic acids or weak organic bases.

The equilibrium in a solution of the acid-base indicator methyl orange, a weak acid, can be represented by an equation in which we use HIn as a simple representation for the complex methyl orange molecule:

$$\underset{\text{red}}{HIn(aq)} + H_2O(l) \rightleftharpoons H_3O^+(aq) + \underset{\text{yellow}}{In^-(aq)}$$

$$K_a = \frac{[H_3O^+][In^-]}{[HIn]} = 4.0 \times 10^{-4}$$

The anion of methyl orange, In^-, is yellow, and the nonionized form, HIn, is red. When we add acid to a solution of methyl orange, the increased hydronium ion concentration shifts the equilibrium toward the nonionized red form, in accordance with Le Châtelier's principle. If we add base, we shift the equilibrium towards the yellow form. This behavior is completely analogous to the action of buffers.

The perceived color of an indicator solution is determined by the ratio of the concentrations of the two species In^- and HIn. If most of the indicator (typically about 60−90% or more) is present as In^-, the perceived color of the solution is yellow. If most is present as HIn, then the solution color appears red. The Henderson-Hasselbalch equation is useful for understanding the relationship between the pH of an indicator solution and its composition (thus, perceived color):

$$\text{pH} = \text{p}Ka + \log\left(\frac{[In^-]}{[HIn]}\right)$$

In solutions where pH > pK_a, the logarithmic term must be positive, indicating an excess of the conjugate base form of the indicator (yellow solution). When pH < pK_a, the log term must be negative, indicating an excess of the conjugate acid (red solution). When the solution pH is close to the indicator pKa, appreciable amounts of both conjugate partners are present, and the solution color is that of an additive combination of each (yellow and red, yielding orange). The **color change interval** (or *pH interval*) for an acid-base indicator is defined as the range of pH values over which a change in color is observed, and for most indicators this range is approximately $pK_a \pm 1$.

There are many different acid-base indicators that cover a wide range of pH values and can be used to determine the approximate pH of an unknown solution by a process of elimination. Universal indicators and pH paper contain a mixture of indicators and exhibit different colors at different pHs. Figure 14.19 presents several indicators, their colors, and their color-change intervals.

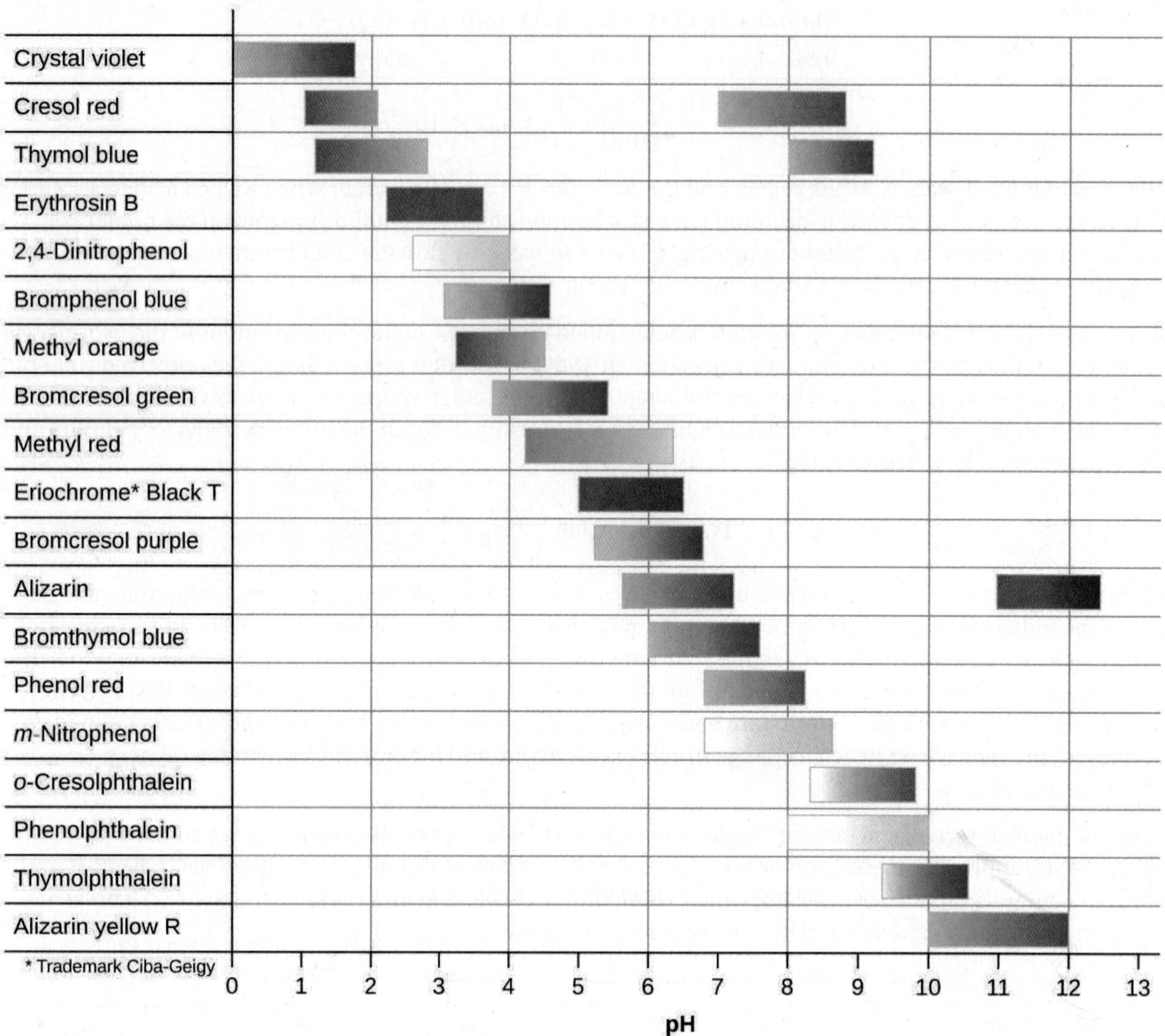

FIGURE 14.19 This chart illustrates the color change intervals for several acid-base indicators.

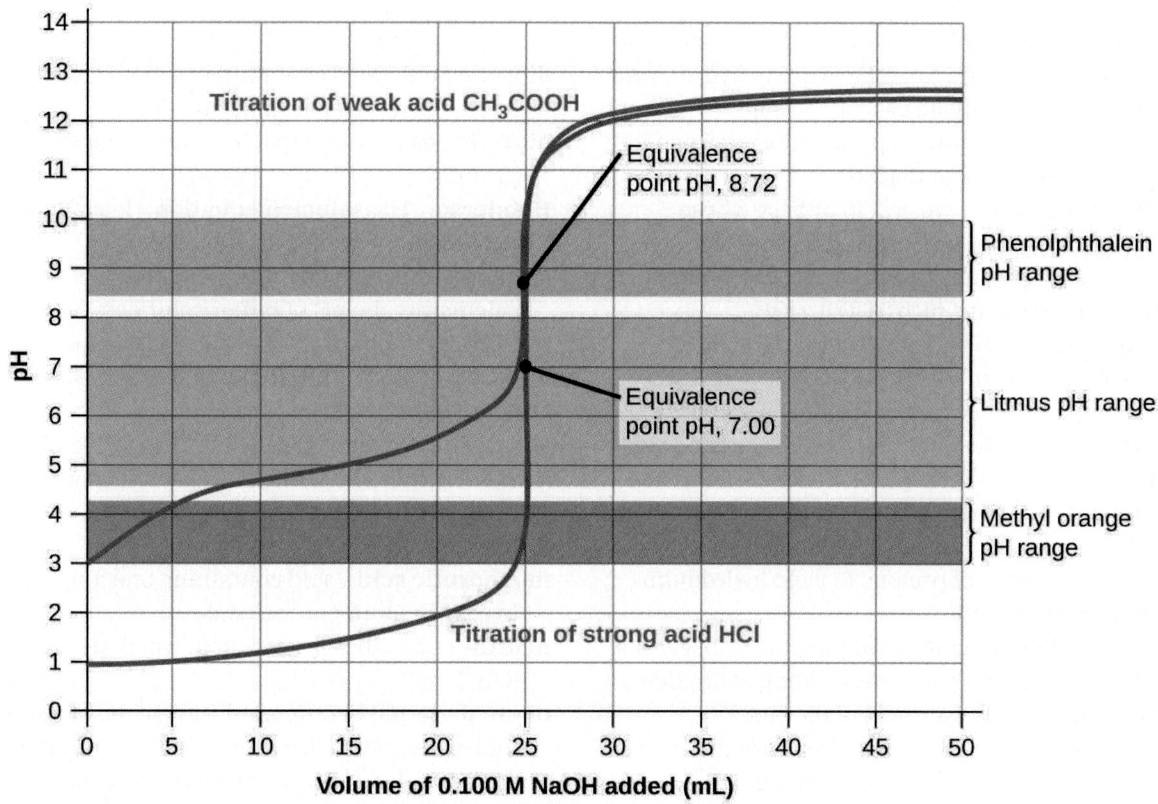

FIGURE 14.20 Titration curves for strong and weak acids illustrating the proper choice of acid-base indicator. Any of the three indicators will exhibit a reasonably sharp color change at the equivalence point of the strong acid titration, but only phenolphthalein is suitable for use in the weak acid titration.

The titration curves shown in Figure 14.20 illustrate the choice of a suitable indicator for specific titrations. In the strong acid titration, use of any of the three indicators should yield reasonably sharp color changes and accurate end point determinations. For this titration, the solution pH reaches the lower limit of the methyl orange color change interval after addition of ~24 mL of titrant, at which point the initially red solution would begin to appear orange. When 25 mL of titrant has been added (the equivalence point), the pH is well above the upper limit and the solution will appear yellow. The titration's end point may then be estimated as the volume of titrant that yields a distinct orange-to-yellow color change. This color change would be challenging for most human eyes to precisely discern. More-accurate estimates of the titration end point are possible using either litmus or phenolphthalein, both of which exhibit color change intervals that are encompassed by the steep rise in pH that occurs around the 25.00 mL equivalence point.

The weak acid titration curve in Figure 14.20 shows that only one of the three indicators is suitable for end point detection. If methyl orange is used in this titration, the solution will undergo a gradual red-to-orange-to-yellow color change over a relatively large volume interval (0–6 mL), completing the color change well before the equivalence point (25 mL) has been reached. Use of litmus would show a color change that begins after adding 7–8 mL of titrant and ends just before the equivalence point. Phenolphthalein, on the other hand, exhibits a color change interval that nicely brackets the abrupt change in pH occurring at the titration's equivalence point. A sharp color change from colorless to pink will be observed within a very small volume interval around the equivalence point.

Key Terms

acid ionization reaction involving the transfer of a proton from an acid to water, yielding hydronium ions and the conjugate base of the acid

acid ionization constant (K_a) equilibrium constant for an acid ionization reaction

acid-base indicator weak acid or base whose conjugate partner imparts a different solution color; used in visual assessments of solution pH

acidic a solution in which $[H_3O^+] > [OH^-]$

amphiprotic species that may either donate or accept a proton in a Bronsted-Lowry acid-base reaction

amphoteric species that can act as either an acid or a base

autoionization reaction between identical species yielding ionic products; for water, this reaction involves transfer of protons to yield hydronium and hydroxide ions

base ionization reaction involving the transfer of a proton from water to a base, yielding hydroxide ions and the conjugate acid of the base

base ionization constant (K_b) equilibrium constant for a base ionization reaction

basic a solution in which $[H_3O^+] < [OH^-]$

Brønsted-Lowry acid proton donor

Brønsted-Lowry base proton acceptor

buffer mixture of appreciable amounts of a weak acid-base pair the pH of a buffer resists change when small amounts of acid or base are added

buffer capacity amount of an acid or base that can be added to a volume of a buffer solution before its pH changes significantly (usually by one pH unit)

color-change interval range in pH over which the color change of an indicator is observed

conjugate acid substance formed when a base gains a proton

conjugate base substance formed when an acid loses a proton

diprotic acid acid containing two ionizable hydrogen atoms per molecule

diprotic base base capable of accepting two protons

Henderson-Hasselbalch equation logarithmic version of the acid ionization constant expression, conveniently formatted for calculating the pH of buffer solutions

ion-product constant for water (K_w) equilibrium constant for the autoionization of water

leveling effect observation that acid-base strength of solutes in a given solvent is limited to that of the solvent's characteristic acid and base species (in water, hydronium and hydroxide ions, respectively)

monoprotic acid acid containing one ionizable hydrogen atom per molecule

neutral describes a solution in which $[H_3O^+] = [OH^-]$

oxyacid ternary compound with acidic properties, molecules of which contain a central nonmetallic atom bonded to one or more O atoms, at least one of which is bonded to an ionizable H atom

percent ionization ratio of the concentration of ionized acid to initial acid concentration expressed as a percentage

pH logarithmic measure of the concentration of hydronium ions in a solution

pOH logarithmic measure of the concentration of hydroxide ions in a solution

stepwise ionization process in which a polyprotic acid is ionized by losing protons sequentially

titration curve plot of some sample property (such as pH) versus volume of added titrant

triprotic acid acid that contains three ionizable hydrogen atoms per molecule

Key Equations

$K_w = [H_3O^+][OH^-] = 1.0 \times 10^{-14}$ (at 25 °C)

$pH = -\log[H_3O^+]$

$pOH = -\log[OH^-]$

$[H_3O^+] = 10^{-pH}$

$[OH^-] = 10^{-pOH}$

$pH + pOH = pK_w = 14.00$ at 25 °C

$$K_a = \frac{[H_3O^+][A^-]}{[HA]}$$

$$K_b = \frac{[HB^+][OH^-]}{[B]}$$

$K_a \times K_b = 1.0 \times 10^{-14} = K_w$

$$\text{Percent ionization} = \frac{[H_3O^+]_{eq}}{[HA]_0} \times 100$$

$pK_a = -\log K_a$

$pK_b = -\log K_b$

$$pH = pK_a + \log \frac{[A^-]}{[HA]}$$

Summary

14.1 Brønsted-Lowry Acids and Bases

A compound that can donate a proton (a hydrogen ion) to another compound is called a Brønsted-Lowry acid. The compound that accepts the proton is called a Brønsted-Lowry base. The species remaining after a Brønsted-Lowry acid has lost a proton is the conjugate base of the acid. The species formed when a Brønsted-Lowry base gains a proton is the conjugate acid of the base. Thus, an acid-base reaction occurs when a proton is transferred from an acid to a base, with formation of the conjugate base of the reactant acid and formation of the conjugate acid of the reactant base. Amphiprotic species can act as both proton donors and proton acceptors. Water is the most important amphiprotic species. It can form both the hydronium ion, H_3O^+, and the hydroxide ion, OH^- when it undergoes autoionization:

$$2H_2O(l) \rightleftharpoons H_3O^+(aq) + OH^-(aq)$$

The ion product of water, K_w is the equilibrium constant for the autoionization reaction:

$$K_w = [H_3O^+][OH^-] = 1.0 \times 10^{-14} \text{ at } 25\ °C$$

14.2 pH and pOH

Concentrations of hydronium and hydroxide ions in aqueous media are often represented as logarithmic pH and pOH values, respectively. At 25 °C, the autoprotolysis equilibrium for water requires the sum of pH and pOH to equal 14 for any aqueous solution. The relative concentrations of hydronium and hydroxide ion in a solution define its status as acidic ($[H_3O^+] > [OH^-]$), basic ($[H_3O^+] < [OH^-]$), or neutral ($[H_3O^+] = [OH^-]$). At 25 °C, a pH < 7 indicates an acidic solution, a pH > 7 a basic solution, and a pH = 7 a neutral solution.

14.3 Relative Strengths of Acids and Bases

The relative strengths of acids and bases are reflected in the magnitudes of their ionization constants; the stronger the acid or base, the larger its ionization constant. A reciprocal relation exists between the strengths of a conjugate acid-base pair: the stronger the acid, the weaker its conjugate base. Water exerts a leveling effect on dissolved acids or bases, reacting completely to generate its characteristic hydronium and hydroxide ions (the strongest acid and base that may exist in water). The strengths of the binary acids increase from left to right across a period of the periodic table ($CH_4 < NH_3 < H_2O < HF$), and they increase down a group ($HF < HCl < HBr < HI$). The strengths of oxyacids that contain the same central element increase as the oxidation number of the element increases ($H_2SO_3 < H_2SO_4$). The strengths of oxyacids also increase as the electronegativity of the central element increases [$H_2SeO_4 < H_2SO_4$].

14.4 Hydrolysis of Salts

The ions composing salts may possess acidic or basic character, ionizing when dissolved in water to yield acidic or basic solutions. Acidic cations are typically the conjugate partners of weak bases, and basic anions are the conjugate partners of weak acids. Many metal ions bond to water molecules when dissolved to yield complex ions that may function as acids.

14.5 Polyprotic Acids

An acid that contains more than one ionizable proton is a polyprotic acid. These acids undergo stepwise ionization reactions involving the transfer of single protons. The ionization constants for polyprotic acids decrease with each subsequent step; these decreases typically are large enough to permit simple equilibrium calculations that treat each step separately.

14.6 Buffers

Solutions that contain appreciable amounts of a weak conjugate acid-base pair are called buffers. A buffered solution will experience only slight changes in pH when small amounts of acid or base are added. Addition of large amounts of acid or base can exceed the buffer capacity, consuming most of one conjugate partner and preventing further buffering action.

14.7 Acid-Base Titrations

The titration curve for an acid-base titration is typically a plot of pH versus volume of added titrant. These curves are useful in selecting appropriate acid-base indicators that will permit accurate determinations of titration end points.

Exercises

14.1 Brønsted-Lowry Acids and Bases

1. Write equations that show NH_3 as both a conjugate acid and a conjugate base.
2. Write equations that show $H_2PO_4{}^-$ acting both as an acid and as a base.
3. Show by suitable net ionic equations that each of the following species can act as a Brønsted-Lowry acid:
   (a) H_3O^+
   (b) HCl
   (c) NH_3
   (d) CH_3CO_2H
   (e) $NH_4{}^+$
   (f) $HSO_4{}^-$
4. Show by suitable net ionic equations that each of the following species can act as a Brønsted-Lowry acid:
   (a) HNO_3
   (b) $PH_4{}^+$
   (c) H_2S
   (d) CH_3CH_2COOH
   (e) $H_2PO_4{}^-$
   (f) HS^-
5. Show by suitable net ionic equations that each of the following species can act as a Brønsted-Lowry base:
   (a) H_2O
   (b) OH^-
   (c) NH_3
   (d) CN^-
   (e) S^{2-}
   (f) $H_2PO_4{}^-$
6. Show by suitable net ionic equations that each of the following species can act as a Brønsted-Lowry base:
   (a) HS^-
   (b) $PO_4{}^{3-}$
   (c) $NH_2{}^-$
   (d) C_2H_5OH
   (e) O^{2-}
   (f) $H_2PO_4{}^-$
7. What is the conjugate acid of each of the following? What is the conjugate base of each?
   (a) OH^-
   (b) H_2O
   (c) $HCO_3{}^-$
   (d) NH_3
   (e) $HSO_4{}^-$
   (f) H_2O_2
   (g) HS^-
   (h) $H_5N_2{}^+$

8. What is the conjugate acid of each of the following? What is the conjugate base of each?
   (a) H_2S
   (b) $H_2PO_4^-$
   (c) PH_3
   (d) HS^-
   (e) HSO_3^-
   (f) $H_3O_2^+$
   (g) H_4N_2
   (h) CH_3OH
9. Identify and label the Brønsted-Lowry acid, its conjugate base, the Brønsted-Lowry base, and its conjugate acid in each of the following equations:
   (a) $HNO_3 + H_2O \longrightarrow H_3O^+ + NO_3^-$
   (b) $CN^- + H_2O \longrightarrow HCN + OH^-$
   (c) $H_2SO_4 + Cl^- \longrightarrow HCl + HSO_4^-$
   (d) $HSO_4^- + OH^- \longrightarrow SO_4^{2-} + H_2O$
   (e) $O^{2-} + H_2O \longrightarrow 2OH^-$
   (f) $[Cu(H_2O)_3(OH)]^+ + [Al(H_2O)_6]^{3+} \longrightarrow [Cu(H_2O)_4]^{2+} + [Al(H_2O)_5(OH)]^{2+}$
   (g) $H_2S + NH_2^- \longrightarrow HS^- + NH_3$
10. Identify and label the Brønsted-Lowry acid, its conjugate base, the Brønsted-Lowry base, and its conjugate acid in each of the following equations:
   (a) $NO_2^- + H_2O \longrightarrow HNO_2 + OH^-$
   (b) $HBr + H_2O \longrightarrow H_3O^+ + Br^-$
   (c) $HS^- + H_2O \longrightarrow H_2S + OH^-$
   (d) $H_2PO_4^- + OH^- \longrightarrow HPO_4^{2-} + H_2O$
   (e) $H_2PO_4^- + HCl \longrightarrow H_3PO_4 + Cl^-$
   (f) $[Fe(H_2O)_5(OH)]^{2+} + [Al(H_2O)_6]^{3+} \longrightarrow [Fe(H_2O)_6]^{3+} + [Al(H_2O)_5(OH)]^{2+}$
   (g) $CH_3OH + H^- \longrightarrow CH_3O^- + H_2$
11. What are amphiprotic species? Illustrate with suitable equations.
12. State which of the following species are amphiprotic and write chemical equations illustrating the amphiprotic character of these species:
   (a) H_2O
   (b) $H_2PO_4^-$
   (c) S^{2-}
   (d) CO_3^{2-}
   (e) HSO_4^-
13. State which of the following species are amphiprotic and write chemical equations illustrating the amphiprotic character of these species.
   (a) NH_3
   (b) HPO_4^-
   (c) Br^-
   (d) NH_4^+
   (e) ASO_4^{3-}
14. Is the self-ionization of water endothermic or exothermic? The ionization constant for water (K_w) is 2.9×10^{-14} at 40 °C and 9.3×10^{-14} at 60 °C.

14.2 pH and pOH

15. Explain why a sample of pure water at 40 °C is neutral even though $[H_3O^+] = 1.7 \times 10^{-7}$ *M*. K_w is 2.9×10^{-14} at 40 °C.
16. The ionization constant for water (K_w) is 2.9×10^{-14} at 40 °C. Calculate $[H_3O^+]$, $[OH^-]$, pH, and pOH for pure water at 40 °C.
17. The ionization constant for water (K_w) is 9.311×10^{-14} at 60 °C. Calculate $[H_3O^+]$, $[OH^-]$, pH, and pOH for pure water at 60 °C.

18. Calculate the pH and the pOH of each of the following solutions at 25 °C for which the substances ionize completely:
(a) 0.200 M HCl
(b) 0.0143 M NaOH
(c) 3.0 M HNO_3
(d) 0.0031 M $Ca(OH)_2$
19. Calculate the pH and the pOH of each of the following solutions at 25 °C for which the substances ionize completely:
(a) 0.000259 M $HClO_4$
(b) 0.21 M NaOH
(c) 0.000071 M $Ba(OH)_2$
(d) 2.5 M KOH
20. What are the pH and pOH of a solution of 2.0 M HCl, which ionizes completely?
21. What are the hydronium and hydroxide ion concentrations in a solution whose pH is 6.52?
22. Calculate the hydrogen ion concentration and the hydroxide ion concentration in wine from its pH. See Figure 14.2 for useful information.
23. Calculate the hydronium ion concentration and the hydroxide ion concentration in lime juice from its pH. See Figure 14.2 for useful information.
24. The hydronium ion concentration in a sample of rainwater is found to be 1.7×10^{-6} M at 25 °C. What is the concentration of hydroxide ions in the rainwater?
25. The hydroxide ion concentration in household ammonia is 3.2×10^{-3} M at 25 °C. What is the concentration of hydronium ions in the solution?

14.3 Relative Strengths of Acids and Bases

26. Explain why the neutralization reaction of a strong acid and a weak base gives a weakly acidic solution.
27. Explain why the neutralization reaction of a weak acid and a strong base gives a weakly basic solution.
28. Use this list of important industrial compounds (and Figure 14.8) to answer the following questions regarding: $Ca(OH)_2$, CH_3CO_2H, HCl, H_2CO_3, HF, HNO_2, HNO_3, H_3PO_4, H_2SO_4, NH_3, NaOH, Na_2CO_3.
(a) Identify the strong Brønsted-Lowry acids and strong Brønsted-Lowry bases.
(b) Identify the compounds that can behave as Brønsted-Lowry acids with strengths lying between those of H_3O^+ and H_2O.
(c) Identify the compounds that can behave as Brønsted-Lowry bases with strengths lying between those of H_2O and OH^-.
29. The odor of vinegar is due to the presence of acetic acid, CH_3CO_2H, a weak acid. List, in order of descending concentration, all of the ionic and molecular species present in a 1-M aqueous solution of this acid.
30. Household ammonia is a solution of the weak base NH_3 in water. List, in order of descending concentration, all of the ionic and molecular species present in a 1-M aqueous solution of this base.
31. Explain why the ionization constant, K_a, for H_2SO_4 is larger than the ionization constant for H_2SO_3.
32. Explain why the ionization constant, K_a, for HI is larger than the ionization constant for HF.
33. Gastric juice, the digestive fluid produced in the stomach, contains hydrochloric acid, HCl. Milk of Magnesia, a suspension of solid $Mg(OH)_2$ in an aqueous medium, is sometimes used to neutralize excess stomach acid. Write a complete balanced equation for the neutralization reaction, and identify the conjugate acid-base pairs.
34. Nitric acid reacts with insoluble copper(II) oxide to form soluble copper(II) nitrate, $Cu(NO_3)_2$, a compound that has been used to prevent the growth of algae in swimming pools. Write the balanced chemical equation for the reaction of an aqueous solution of HNO_3 with CuO.
35. What is the ionization constant at 25 °C for the weak acid $CH_3NH_3^+$, the conjugate acid of the weak base CH_3NH_2, $K_b = 4.4 \times 10^{-4}$.
36. What is the ionization constant at 25 °C for the weak acid $(CH_3)_2NH_2^+$, the conjugate acid of the weak base $(CH_3)_2NH$, $K_b = 5.9 \times 10^{-4}$?
37. Which base, CH_3NH_2 or $(CH_3)_2NH$, is the stronger base? Which conjugate acid, $(CH_3)_2NH_2^+$ or $CH_3NH_3^+$, is the stronger acid?

38. Which is the stronger acid, NH_4^+ or HBrO?
39. Which is the stronger base, $(CH_3)_3N$ or $H_2BO_3^-$?
40. Predict which acid in each of the following pairs is the stronger and explain your reasoning for each.
    (a) H_2O or HF
    (b) $B(OH)_3$ or $Al(OH)_3$
    (c) HSO_3^- or HSO_4^-
    (d) NH_3 or H_2S
    (e) H_2O or H_2Te
41. Predict which compound in each of the following pairs of compounds is more acidic and explain your reasoning for each.
    (a) HSO_4^- or $HSeO_4^-$
    (b) NH_3 or H_2O
    (c) PH_3 or HI
    (d) NH_3 or PH_3
    (e) H_2S or HBr
42. Rank the compounds in each of the following groups in order of increasing acidity or basicity, as indicated, and explain the order you assign.
    (a) acidity: HCl, HBr, HI
    (b) basicity: H_2O, OH^-, H^-, Cl^-
    (c) basicity: $Mg(OH)_2$, $Si(OH)_4$, $ClO_3(OH)$ (Hint: Formula could also be written as $HClO_4$.)
    (d) acidity: HF, H_2O, NH_3, CH_4
43. Rank the compounds in each of the following groups in order of increasing acidity or basicity, as indicated, and explain the order you assign.
    (a) acidity: $NaHSO_3$, $NaHSeO_3$, $NaHSO_4$
    (b) basicity: BrO_2^-, ClO_2^-, IO_2^-
    (c) acidity: HOCl, HOBr, HOI
    (d) acidity: HOCl, HOClO, $HOClO_2$, $HOClO_3$
    (e) basicity: NH_2^-, HS^-, HTe^-, PH_2^-
    (f) basicity: BrO^-, BrO_2^-, BrO_3^-, BrO_4^-
44. Both HF and HCN ionize in water to a limited extent. Which of the conjugate bases, F^- or CN^-, is the stronger base?
45. The active ingredient formed by aspirin in the body is salicylic acid, $C_6H_4OH(CO_2H)$. The carboxyl group ($-CO_2H$) acts as a weak acid. The phenol group (an OH group bonded to an aromatic ring) also acts as an acid but a much weaker acid. List, in order of descending concentration, all of the ionic and molecular species present in a 0.001-*M* aqueous solution of $C_6H_4OH(CO_2H)$.
46. Are the concentrations of hydronium ion and hydroxide ion in a solution of an acid or a base in water directly proportional or inversely proportional? Explain your answer.
47. What two common assumptions can simplify calculation of equilibrium concentrations in a solution of a weak acid or base?
48. Which of the following will increase the percent of NH_3 that is converted to the ammonium ion in water?
    (a) addition of NaOH
    (b) addition of HCl
    (c) addition of NH_4Cl
49. Which of the following will increase the percentage of HF that is converted to the fluoride ion in water?
    (a) addition of NaOH
    (b) addition of HCl
    (c) addition of NaF

50. What is the effect on the concentrations of NO_2^-, HNO_2, and OH^- when the following are added to a solution of KNO_2 in water:
(a) HCl
(b) HNO_2
(c) NaOH
(d) NaCl
(e) KNO

51. What is the effect on the concentration of hydrofluoric acid, hydronium ion, and fluoride ion when the following are added to separate solutions of hydrofluoric acid?
(a) HCl
(b) KF
(c) NaCl
(d) KOH
(e) HF

52. Why is the hydronium ion concentration in a solution that is 0.10 *M* in HCl and 0.10 *M* in HCOOH determined by the concentration of HCl?

53. From the equilibrium concentrations given, calculate K_a for each of the weak acids and K_b for each of the weak bases.
(a) CH_3CO_2H: $[H_3O^+] = 1.34 \times 10^{-3}$ *M*;
$[CH_3CO_2^-] = 1.34 \times 10^{-3}$ *M*;
$[CH_3CO_2H] = 9.866 \times 10^{-2}$ *M*;
(b) ClO^-: $[OH^-] = 4.0 \times 10^{-4}$ *M*;
$[HClO] = 2.38 \times 10^{-4}$ *M*;
$[ClO^-] = 0.273$ *M*;
(c) HCO_2H: $[HCO_2H] = 0.524$ *M*;
$[H_3O^+] = 9.8 \times 10^{-3}$ *M*;
$[HCO_2^-] = 9.8 \times 10^{-3}$ *M*;
(d) $C_6H_5NH_3^+$: $[C_6H_5NH_3^+] = 0.233$ *M*;
$[C_6H_5NH_2] = 2.3 \times 10^{-3}$ *M*;
$[H_3O^+] = 2.3 \times 10^{-3}$ *M*

54. From the equilibrium concentrations given, calculate K_a for each of the weak acids and K_b for each of the weak bases.
(a) NH_3: $[OH^-] = 3.1 \times 10^{-3}$ *M*;
$[NH_4^+] = 3.1 \times 10^{-3}$ *M*;
$[NH_3] = 0.533$ *M*;
(b) HNO_2: $[H_3O^+] = 0.011$ *M*;
$[NO_2^-] = 0.0438$ *M*;
$[HNO_2] = 1.07$ *M*;
(c) $(CH_3)_3N$: $[(CH_3)_3N] = 0.25$ *M*;
$[(CH_3)_3NH^+] = 4.3 \times 10^{-3}$ *M*;
$[OH^-] = 3.7 \times 10^{-3}$ *M*;
(d) NH_4^+ : $[NH_4^+] = 0.100$ *M*;
$[NH_3] = 7.5 \times 10^{-6}$ *M*;
$[H_3O^+] = 7.5 \times 10^{-6}$ *M*

55. Determine K_b for the nitrite ion, NO_2^-. In a 0.10-*M* solution this base is 0.0015% ionized.

56. Determine K_a for hydrogen sulfate ion, HSO_4^-. In a 0.10-*M* solution the acid is 29% ionized.

57. Calculate the ionization constant for each of the following acids or bases from the ionization constant of its conjugate base or conjugate acid:
(a) F^-
(b) $NH_4{}^+$
(c) $AsO_4{}^{3-}$
(d) $(CH_3)_2NH_2{}^+$
(e) $NO_2{}^-$
(f) $HC_2O_4{}^-$ (as a base)

58. Calculate the ionization constant for each of the following acids or bases from the ionization constant of its conjugate base or conjugate acid:
(a) HTe^- (as a base)
(b) $(CH_3)_3NH^+$
(c) $HAsO_4{}^{2-}$ (as a base)
(d) $HO_2{}^-$ (as a base)
(e) $C_6H_5NH_3{}^+$
(f) $HSO_3{}^-$ (as a base)

59. Using the K_a value of 1.4×10^{-5}, place $Al(H_2O)_6{}^{3+}$ in the correct location in Figure 14.7.

60. Calculate the concentration of all solute species in each of the following solutions of acids or bases. Assume that the ionization of water can be neglected, and show that the change in the initial concentrations can be neglected.
(a) 0.0092 M HClO, a weak acid
(b) 0.0784 M $C_6H_5NH_2$, a weak base
(c) 0.0810 M HCN, a weak acid
(d) 0.11 M $(CH_3)_3N$, a weak base
(e) 0.120 M $Fe(H_2O)_6{}^{2+}$ a weak acid, $K_a = 1.6 \times 10^{-7}$

61. Propionic acid, $C_2H_5CO_2H$ ($K_a = 1.34 \times 10^{-5}$), is used in the manufacture of calcium propionate, a food preservative. What is the pH of a 0.698-M solution of $C_2H_5CO_2H$?

62. White vinegar is a 5.0% by mass solution of acetic acid in water. If the density of white vinegar is 1.007 g/cm^3, what is the pH?

63. The ionization constant of lactic acid, $CH_3CH(OH)CO_2H$, an acid found in the blood after strenuous exercise, is 1.36×10^{-4}. If 20.0 g of lactic acid is used to make a solution with a volume of 1.00 L, what is the concentration of hydronium ion in the solution?

64. Nicotine, $C_{10}H_{14}N_2$, is a base that will accept two protons ($K_{b1} = 7 \times 10^{-7}$, $K_{b2} = 1.4 \times 10^{-11}$). What is the concentration of each species present in a 0.050-M solution of nicotine?

65. The pH of a 0.23-M solution of HF is 1.92. Determine K_a for HF from these data.

66. The pH of a 0.15-M solution of $HSO_4{}^-$ is 1.43. Determine K_a for $HSO_4{}^-$ from these data.

67. The pH of a 0.10-M solution of caffeine is 11.70. Determine K_b for caffeine from these data:
$C_8H_{10}N_4O_2(aq) + H_2O(l) \rightleftharpoons C_8H_{10}N_4O_2H^+(aq) + OH^-(aq)$

68. The pH of a solution of household ammonia, a 0.950 M solution of NH_3, is 11.612. Determine K_b for NH_3 from these data.

14.4 Hydrolysis of Salts

69. Determine whether aqueous solutions of the following salts are acidic, basic, or neutral:
(a) $Al(NO_3)_3$
(b) RbI
(c) $KHCO_2$
(d) CH_3NH_3Br

70. Determine whether aqueous solutions of the following salts are acidic, basic, or neutral:
(a) $FeCl_3$
(b) K_2CO_3
(c) NH_4Br
(d) $KClO_4$

71. Novocaine, $C_{13}H_{21}O_2N_2Cl$, is the salt of the base procaine and hydrochloric acid. The ionization constant for procaine is 7×10^{-6}. Is a solution of novocaine acidic or basic? What are $[H_3O^+]$, $[OH^-]$, and pH of a 2.0% solution by mass of novocaine, assuming that the density of the solution is 1.0 g/mL.

14.5 Polyprotic Acids

72. Which of the following concentrations would be practically equal in a calculation of the equilibrium concentrations in a 0.134-*M* solution of H_2CO_3, a diprotic acid: $[H_3O^+]$, $[OH^-]$, $[H_2CO_3]$, $[HCO_3^-]$, $[CO_3^{2-}]$? No calculations are needed to answer this question.
73. Calculate the concentration of each species present in a 0.050-*M* solution of H_2S.
74. Calculate the concentration of each species present in a 0.010-*M* solution of phthalic acid, $C_6H_4(CO_2H)_2$.

$$C_6H_4(CO_2H)_2(aq) + H_2O(l) \rightleftharpoons H_3O^+(aq) + C_6H_4\left(CO_2H\right)\left(CO_2\right)^-(aq) \qquad K_a = 1.1 \times 10^{-3}$$

$$C_6H_4\left(CO_2H\right)\left(CO_2\right)(aq) + H_2O(l) \rightleftharpoons H_3O^+(aq) + C_6H_4(CO_2)_2{}^{2-}(aq) \qquad K_a = 3.9 \times 10^{-6}$$

75. Salicylic acid, $HOC_6H_4CO_2H$, and its derivatives have been used as pain relievers for a long time. Salicylic acid occurs in small amounts in the leaves, bark, and roots of some vegetation (most notably historically in the bark of the willow tree). Extracts of these plants have been used as medications for centuries. The acid was first isolated in the laboratory in 1838.
(a) Both functional groups of salicylic acid ionize in water, with $K_a = 1.0 \times 10^{-3}$ for the—CO_2H group and 4.2×10^{-13} for the –OH group. What is the pH of a saturated solution of the acid (solubility = 1.8 g/L).
(b) Aspirin was discovered as a result of efforts to produce a derivative of salicylic acid that would not be irritating to the stomach lining. Aspirin is acetylsalicylic acid, $CH_3CO_2C_6H_4CO_2H$. The $-CO_2H$ functional group is still present, but its acidity is reduced, $K_a = 3.0 \times 10^{-4}$. What is the pH of a solution of aspirin with the same concentration as a saturated solution of salicylic acid (See Part a).
76. The ion HTe^- is an amphiprotic species; it can act as either an acid or a base.
(a) What is K_a for the acid reaction of HTe^- with H_2O?
(b) What is K_b for the reaction in which HTe^- functions as a base in water?
(c) Demonstrate whether or not the second ionization of H_2Te can be neglected in the calculation of $[HTe^-]$ in a 0.10 M solution of H_2Te.

14.6 Buffers

77. Explain why a buffer can be prepared from a mixture of NH_4Cl and NaOH but not from NH_3 and NaOH.
78. Explain why the pH does not change significantly when a small amount of an acid or a base is added to a solution that contains equal amounts of the acid H_3PO_4 and a salt of its conjugate base NaH_2PO_4.
79. Explain why the pH does not change significantly when a small amount of an acid or a base is added to a solution that contains equal amounts of the base NH_3 and a salt of its conjugate acid NH_4Cl.
80. What is $[H_3O^+]$ in a solution of 0.25 *M* CH_3CO_2H and 0.030 *M* $NaCH_3CO_2$?
$CH_3CO_2H(aq) + H_2O(l) \rightleftharpoons H_3O^+(aq) + CH_3CO_2{}^-(aq) \qquad K_a = 1.8 \times 10^{-5}$
81. What is $[H_3O^+]$ in a solution of 0.075 *M* HNO_2 and 0.030 *M* $NaNO_2$?
$HNO_2(aq) + H_2O(l) \rightleftharpoons H_3O^+(aq) + NO_2{}^-(aq) \qquad K_a = 4.5 \times 10^{-5}$
82. What is $[OH^-]$ in a solution of 0.125 *M* CH_3NH_2 and 0.130 *M* CH_3NH_3Cl?
$CH_3NH_2(aq) + H_2O(l) \rightleftharpoons CH_3NH_3{}^+(aq) + OH^-(aq) \qquad K_b = 4.4 \times 10^{-4}$
83. What is $[OH^-]$ in a solution of 1.25 *M* NH_3 and 0.78 *M* NH_4NO_3?
$NH_3(aq) + H_2O(l) \rightleftharpoons NH_4{}^+(aq) + OH^-(aq) \qquad K_b = 1.8 \times 10^{-5}$
84. What is the effect on the concentration of acetic acid, hydronium ion, and acetate ion when the following are added to an acidic buffer solution of equal concentrations of acetic acid and sodium acetate:
(a) HCl
(b) KCH_3CO_2
(c) NaCl
(d) KOH
(e) CH_3CO_2H

85. What is the effect on the concentration of ammonia, hydroxide ion, and ammonium ion when the following are added to a basic buffer solution of equal concentrations of ammonia and ammonium nitrate:
(a) KI
(b) NH_3
(c) HI
(d) NaOH
(e) NH_4Cl

86. What will be the pH of a buffer solution prepared from 0.20 mol NH_3, 0.40 mol NH_4NO_3, and just enough water to give 1.00 L of solution?

87. Calculate the pH of a buffer solution prepared from 0.155 mol of phosphoric acid, 0.250 mole of KH_2PO_4, and enough water to make 0.500 L of solution.

88. How much solid $NaCH_3CO_2•3H_2O$ must be added to 0.300 L of a 0.50-*M* acetic acid solution to give a buffer with a pH of 5.00? (Hint: Assume a negligible change in volume as the solid is added.)

89. What mass of NH_4Cl must be added to 0.750 L of a 0.100-*M* solution of NH_3 to give a buffer solution with a pH of 9.26? (Hint: Assume a negligible change in volume as the solid is added.)

90. A buffer solution is prepared from equal volumes of 0.200 *M* acetic acid and 0.600 *M* sodium acetate. Use 1.80×10^{-5} as K_a for acetic acid.
(a) What is the pH of the solution?
(b) Is the solution acidic or basic?
(c) What is the pH of a solution that results when 3.00 mL of 0.034 *M* HCl is added to 0.200 L of the original buffer?

91. A 5.36–g sample of NH_4Cl was added to 25.0 mL of 1.00 *M* NaOH and the resulting solution diluted to 0.100 L.
(a) What is the pH of this buffer solution?
(b) Is the solution acidic or basic?
(c) What is the pH of a solution that results when 3.00 mL of 0.034 *M* HCl is added to the solution?

14.7 Acid-Base Titrations

92. Explain how to choose the appropriate acid-base indicator for the titration of a weak base with a strong acid.

93. Explain why an acid-base indicator changes color over a range of pH values rather than at a specific pH.

94. Calculate the pH at the following points in a titration of 40 mL (0.040 L) of 0.100 *M* barbituric acid ($K_a = 9.8 \times 10^{-5}$) with 0.100 *M* KOH.
(a) no KOH added
(b) 20 mL of KOH solution added
(c) 39 mL of KOH solution added
(d) 40 mL of KOH solution added
(e) 41 mL of KOH solution added

95. The indicator dinitrophenol is an acid with a K_a of 1.1×10^{-4}. In a 1.0×10^{-4}-*M* solution, it is colorless in acid and yellow in base. Calculate the pH range over which it goes from 10% ionized (colorless) to 90% ionized (yellow).

CHAPTER 15
Equilibria of Other Reaction Classes

Figure 15.1 The mineral fluorite (CaF_2) is formed when dissolved calcium and fluoride ions precipitate from groundwater within the Earth's crust. Note that pure fluorite is colorless, and that the color in this sample is due to the presence of other metal ions in the crystal.

CHAPTER OUTLINE

15.1 Precipitation and Dissolution
15.2 Lewis Acids and Bases
15.3 Coupled Equilibria

INTRODUCTION The mineral fluorite, CaF_2 Figure 15.1, is commonly used as a semiprecious stone in many types of jewelry because of its striking appearance. Deposits of fluorite are formed through a process called hydrothermal precipitation in which calcium and fluoride ions dissolved in groundwater combine to produce insoluble CaF_2 in response to some change in solution conditions. For example, a decrease in temperature may trigger fluorite precipitation if its solubility is exceeded at the lower temperature. Because fluoride ion is a weak base, its solubility is also affected by solution pH, and so geologic or other processes that change groundwater pH will also affect the precipitation of fluorite. This chapter extends the equilibrium discussion of other chapters by addressing some additional reaction classes (including precipitation) and systems involving coupled equilibrium reactions.

15.1 Precipitation and Dissolution

LEARNING OBJECTIVES

By the end of this section, you will be able to:

- Write chemical equations and equilibrium expressions representing solubility equilibria
- Carry out equilibrium computations involving solubility, equilibrium expressions, and solute concentrations

Solubility equilibria are established when the dissolution and precipitation of a solute species occur at equal rates. These equilibria underlie many natural and technological processes, ranging from tooth decay to water purification. An understanding of the factors affecting compound solubility is, therefore, essential to the effective management of these processes. This section applies previously introduced equilibrium concepts and tools to systems involving dissolution and precipitation.

The Solubility Product

Recall from the chapter on solutions that the solubility of a substance can vary from essentially zero (*insoluble* or *sparingly soluble*) to infinity (*miscible*). A solute with finite solubility can yield a *saturated* solution when it is added to a solvent in an amount exceeding its solubility, resulting in a heterogeneous mixture of the saturated solution and the excess, undissolved solute. For example, a saturated solution of silver chloride is one in which the equilibrium shown below has been established.

$$\text{AgCl}(s) \underset{\text{precipitation}}{\overset{\text{dissolution}}{\rightleftharpoons}} \text{Ag}^+(aq) + \text{Cl}^-(aq)$$

In this solution, an excess of solid AgCl dissolves and dissociates to produce aqueous Ag^+ and Cl^- ions at the same rate that these aqueous ions combine and precipitate to form solid AgCl (Figure 15.2). Because silver chloride is a sparingly soluble salt, the equilibrium concentration of its dissolved ions in the solution is relatively low.

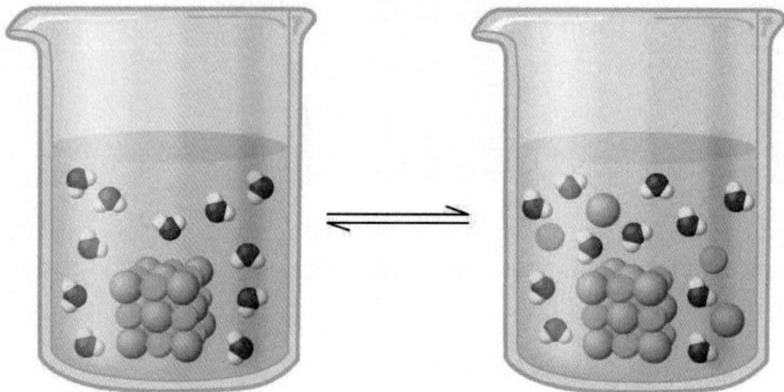

FIGURE 15.2 Silver chloride is a sparingly soluble ionic solid. When it is added to water, it dissolves slightly and produces a mixture consisting of a very dilute solution of Ag^+ and Cl^- ions in equilibrium with undissolved silver chloride.

The equilibrium constant for solubility equilibria such as this one is called the **solubility product constant,** $\boldsymbol{K_{sp}}$, in this case

$$\text{AgCl}(s) \rightleftharpoons \text{Ag}^+(aq) + \text{Cl}^-(aq) \qquad K_{sp} = [\text{Ag}^+(aq)][\text{Cl}^-(aq)]$$

Recall that only gases and solutes are represented in equilibrium constant expressions, so the K_{sp} does not include a term for the undissolved AgCl. A listing of solubility product constants for several sparingly soluble compounds is provided in Appendix J.

 **EXAMPLE 15.1**

Writing Equations and Solubility Products

Write the dissolution equation and the solubility product expression for each of the following slightly soluble ionic compounds:

(a) AgI, silver iodide, a solid with antiseptic properties

(b) $CaCO_3$, calcium carbonate, the active ingredient in many over-the-counter chewable antacids

(c) $Mg(OH)_2$, magnesium hydroxide, the active ingredient in Milk of Magnesia

(d) $Mg(NH_4)PO_4$, magnesium ammonium phosphate, an essentially insoluble substance used in tests for magnesium

(e) $Ca_5(PO_4)_3OH$, the mineral apatite, a source of phosphate for fertilizers

Solution

(a) $AgI(s) \rightleftharpoons Ag^+(aq) + I^-(aq) \qquad K_{sp} = [Ag^+][I^-]$

(b) $CaCO_3(s) \rightleftharpoons Ca^{2+}(aq) + CO_3{}^{2-}(aq) \qquad K_{sp} = [Ca^{2+}][CO_3{}^{2-}]$

(c) $Mg(OH)_2(s) \rightleftharpoons Mg^{2+}(aq) + 2OH^-(aq) \qquad K_{sp} = [Mg^{2+}][OH^-]^2$

(d) $Mg(NH_4)PO_4(s) \rightleftharpoons Mg^{2+}(aq) + NH_4{}^+(aq) + PO_4{}^{3-}(aq) \qquad K_{sp} = [Mg^{2+}][NH_4{}^+][PO_4{}^{3-}]$

(e) $Ca_5(PO_4)3OH(s) \rightleftharpoons 5Ca^{2+}(aq) + 3PO_4{}^{3-}(aq) + OH^-(aq) \qquad K_{sp} = [Ca^{2+}]^5[PO_4{}^{3-}]^3[OH^-]$

Check Your Learning

Write the dissolution equation and the solubility product for each of the following slightly soluble compounds:

(a) $BaSO_4$

(b) Ag_2SO_4

(c) $Al(OH)_3$

(d) Pb(OH)Cl

Answer:

(a) $BaSO_4(s) \rightleftharpoons Ba^{2+}(aq) + SO_4{}^{2-}(aq) \qquad K_{sp} = [Ba^{2+}][SO_4{}^{2-}];$

(b) $Ag_2SO_4(s) \rightleftharpoons 2Ag^+(aq) + SO_4{}^{2-}(aq) \qquad K_{sp} = [Ag^+]^2[SO_4{}^{2-}];$

(c) $Al(OH)_3(s) \rightleftharpoons Al^{3+}(aq) + 3OH^-(aq) \qquad K_{sp} = [Al^{3+}][OH^-]^3;$

(d) $Pb(OH)Cl(s) \rightleftharpoons Pb^{2+}(aq) + OH^-(aq) + Cl^-(aq) \qquad K_{sp} = [Pb^{2+}][OH^-][Cl^-]$

K_{sp} and Solubility

The K_{sp} of a slightly soluble ionic compound may be simply related to its measured solubility provided the dissolution process involves only dissociation and solvation, for example:

$$M_pX_q(s) \rightleftharpoons pM^{m+}(aq) + qX^{n-}(aq)$$

For cases such as these, one may derive K_{sp} values from provided solubilities, or vice-versa. Calculations of this sort are most conveniently performed using a compound's molar solubility, measured as moles of dissolved solute per liter of saturated solution.

EXAMPLE 15.2

Calculation of K_{sp} from Equilibrium Concentrations

Fluorite, CaF_2, is a slightly soluble solid that dissolves according to the equation:

$$CaF_2(s) \rightleftharpoons Ca^{2+}(aq) + 2F^-(aq)$$

The concentration of Ca^{2+} in a saturated solution of CaF_2 is 2.15×10^{-4} *M*. What is the solubility product of fluorite?

Solution

According to the stoichiometry of the dissolution equation, the fluoride ion molarity of a CaF_2 solution is equal to twice its calcium ion molarity:

$$[F^-] = (2 \text{ mol } F^- / 1 \text{ mol } Ca^{2+}) = (2)(2.15 \times 10^{-4}\ M) = 4.30 \times 10^{-4}\ M$$

Substituting the ion concentrations into the K_{sp} expression gives

$$K_{sp} = [Ca^{2+}][F^-]^2 = (2.15 \times 10^{-4})(4.30 \times 10^{-4})^2 = 3.98 \times 10^{-11}$$

Check Your Learning

In a saturated solution of $Mg(OH)_2$, the concentration of Mg^{2+} is 1.31×10^{-4} *M*. What is the solubility product for $Mg(OH)_2$?

$$Mg(OH)_2(s) \rightleftharpoons Mg^{2+}(aq) + 2OH^-(aq)$$

Answer:

8.99×10^{-12}

EXAMPLE 15.3

Determination of Molar Solubility from K_{sp}

The K_{sp} of copper(I) bromide, CuBr, is 6.3×10^{-9}. Calculate the molar solubility of copper bromide.

Solution

The dissolution equation and solubility product expression are

$$CuBr(s) \rightleftharpoons Cu^+(aq) + Br^-(aq)$$

$$K_{sp} = [Cu^+][Br^-]$$

Following the ICE approach to this calculation yields the table

	$CuBr(s) \rightleftharpoons$	$Cu^+(aq)$ +	$Br^-(aq)$
Initial concentration (*M*)		0	0
Change (*M*)		$+x$	$+x$
Equilibrium concentration (*M*)		x	x

Substituting the equilibrium concentration terms into the solubility product expression and solving for x yields

$$K_{sp} = [Cu^+][Br^-]$$

$$6.3 \times 10^{-9} = (x)(x) = x^2$$

$$x = \sqrt{(6.3 \times 10^{-9})} = 7.9 \times 10^{-5}\ M$$

Since the dissolution stoichiometry shows one mole of copper(I) ion and one mole of bromide ion are produced for each moles of Br dissolved, the molar solubility of CuBr is 7.9×10^{-5} *M*.

Check Your Learning

The K_{sp} of AgI is 1.5×10^{-16}. Calculate the molar solubility of silver iodide.

Answer:

1.2×10^{-8} *M*

EXAMPLE 15.4

Determination of Molar Solubility from K_{sp}

The K_{sp} of calcium hydroxide, $Ca(OH)_2$, is 1.3×10^{-6}. Calculate the molar solubility of calcium hydroxide.

Solution

The dissolution equation and solubility product expression are

$$Ca(OH)_2(s) \rightleftharpoons Ca^{2+}(aq) + 2OH^-(aq)$$

$$K_{sp} = [Ca^{2+}][OH^-]^2$$

The ICE table for this system is

	$Ca(OH)_2(s) \rightleftharpoons$	$Ca^{2+}(aq)$ +	$2OH^-(aq)$
Initial concentration (*M*)		0	0
Change (*M*)		$+x$	$+2x$
Equilibrium concentration (*M*)		x	$2x$

Substituting terms for the equilibrium concentrations into the solubility product expression and solving for x gives

$$K_{sp} = [Ca^{2+}][OH^-]^2$$

$$1.3 \times 10^{-6} = (x)(2x)^2 = (x)(4x^2) = 4x^3$$

$$x = \sqrt[3]{\frac{1.3 \times 10^{-6}}{4}} = 6.9 \times 10^{-3}\ M$$

As defined in the ICE table, x is the molarity of calcium ion in the saturated solution. The dissolution stoichiometry shows a 1:1 relation between moles of calcium ion in solution and moles of compound dissolved, and so, the molar solubility of $Ca(OH)_2$ is $6.9 \times 10^{-3}\ M$.

Check Your Learning

The K_{sp} of PbI_2 is 1.4×10^{-8}. Calculate the molar solubility of lead(II) iodide.

Answer:

$1.5 \times 10^{-3}\ M$

EXAMPLE 15.5

Determination of K_{sp} from Gram Solubility

Many of the pigments used by artists in oil-based paints (Figure 15.3) are sparingly soluble in water. For example, the solubility of the artist's pigment chrome yellow, $PbCrO_4$, is 4.6×10^{-6} g/L. Determine the solubility product for $PbCrO_4$.

FIGURE 15.3 Oil paints contain pigments that are very slightly soluble in water. In addition to chrome yellow ($PbCrO_4$), examples include Prussian blue ($Fe_7(CN)_{18}$), the reddish-orange color vermilion (HgS), and green color veridian (Cr_2O_3). (credit: Sonny Abesamis)

Solution

Before calculating the solubility product, the provided solubility must be converted to molarity:

$$\begin{aligned}[\text{PbCrO}_4] &= \frac{4.6 \times 10^{-6}\text{g PbCrO}_4}{1\text{ L}} \times \frac{1\text{ mol PbCrO}_4}{323.2\text{ g PbCrO}_4} \\ &= \frac{1.4 \times 10^{-8}\text{ mol PbCrO}_4}{1\text{ L}} \\ &= 1.4 \times 10^{-8}\,M\end{aligned}$$

The dissolution equation for this compound is

$$\text{PbCrO}_4(s) \rightleftharpoons \text{Pb}^{2+}(aq) + \text{CrO}_4{}^{2-}(aq)$$

The dissolution stoichiometry shows a 1:1 relation between the molar amounts of compound and its two ions, and so both $[Pb^{2+}]$ and $[CrO_4{}^{2-}]$ are equal to the molar solubility of $PbCrO_4$:

$$[\text{Pb}^{2+}] = [\text{CrO}_4{}^{2-}] = 1.4 \times 10^{-8}\,M$$

$K_{sp} = [Pb^{2+}][CrO_4{}^{2-}] = (1.4 \times 10^{-8})(1.4 \times 10^{-8}) = 2.0 \times 10^{-16}$

Check Your Learning

The solubility of TlCl [thallium(I) chloride], an intermediate formed when thallium is being isolated from ores, is 3.12 grams per liter at 20 °C. What is its solubility product?

Answer:

1.69×10^{-4}

EXAMPLE 15.6

Calculating the Solubility of Hg_2Cl_2

Calomel, Hg_2Cl_2, is a compound composed of the diatomic ion of mercury(I), $Hg_2{}^{2+}$, and chloride ions, Cl^-. Although most mercury compounds are now known to be poisonous, eighteenth-century physicians used calomel as a medication. Their patients rarely suffered any mercury poisoning from the treatments because calomel has a very low solubility, as suggested by its very small K_{sp}:

$$\text{Hg}_2\text{Cl}_2(s) \rightleftharpoons \text{Hg}_2{}^{2+}(aq) + 2\text{Cl}^-(aq) \qquad K_{sp} = 1.1 \times 10^{-18}$$

Calculate the molar solubility of Hg_2Cl_2.

Solution

The dissolution stoichiometry shows a 1:1 relation between the amount of compound dissolved and the amount of mercury(I) ions, and so the molar solubility of Hg_2Cl_2 is equal to the concentration of $Hg_2{}^{2+}$ ions

Following the ICE approach results in

	$Hg_2Cl_2(s) \rightleftharpoons$	$Hg_2{}^{2+}(aq)$ +	$2Cl^-(aq)$
Initial concentration (*M*)		0	0
Change (*M*)		$+x$	$+2x$
Equilibrium concentration (*M*)		x	$2x$

Substituting the equilibrium concentration terms into the solubility product expression and solving for *x* gives

$$K_{sp} = [Hg_2{}^{2+}][Cl^-]^2$$
$$1.1 \times 10^{-18} = (x)(2x)^2$$
$$4x^3 = 1.1 \times 10^{-18}$$
$$x = \sqrt[3]{\left(\frac{1.1 \times 10^{-18}}{4}\right)} = 6.5 \times 10^{-7}\ M$$
$$[Hg_2{}^{2+}] = 6.5 \times 10^{-7}\ M = 6.5 \times 10^{-7}\ M$$
$$[Cl^-] = 2x = 2(6.5 \times 10^{-7}) = 1.3 \times 10^{-6}\ M$$

The dissolution stoichiometry shows the molar solubility of Hg_2Cl_2 is equal to $[Hg_2{}^{2+}]$, or 6.5×10^{-7} *M*.

Check Your Learning

Determine the molar solubility of MgF_2 from its solubility product: $K_{sp} = 6.4 \times 10^{-9}$.

Answer:

1.2×10^{-3} *M*

HOW SCIENCES INTERCONNECT

Using Barium Sulfate for Medical Imaging

Various types of medical imaging techniques are used to aid diagnoses of illnesses in a noninvasive manner. One such technique utilizes the ingestion of a barium compound before taking an X-ray image. A suspension of barium sulfate, a chalky powder, is ingested by the patient. Since the K_{sp} of barium sulfate is 2.3×10^{-8}, very little of it dissolves as it coats the lining of the patient's intestinal tract. Barium-coated areas of the digestive tract then appear on an X-ray as white, allowing for greater visual detail than a traditional X-ray (Figure 15.4).

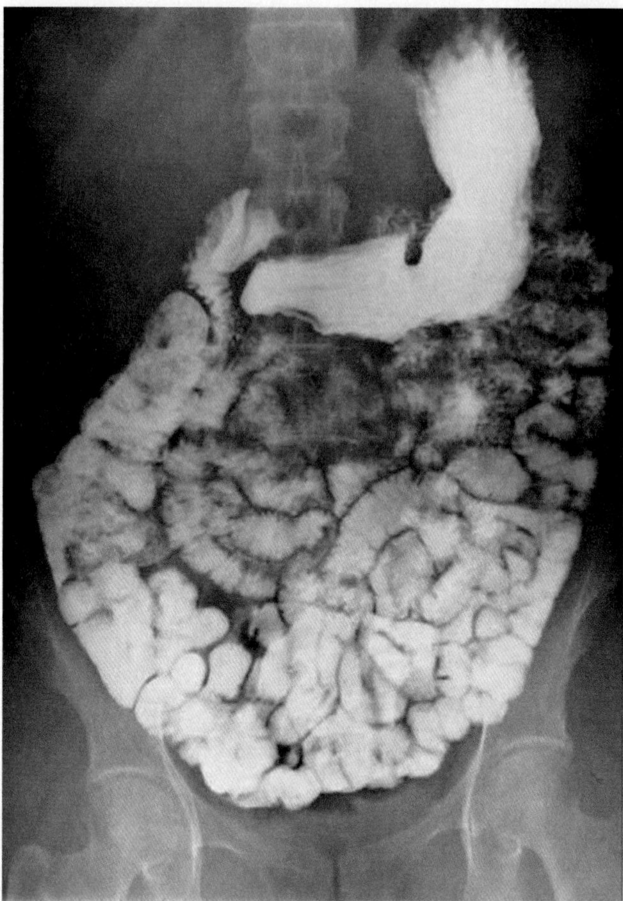

FIGURE 15.4 A suspension of barium sulfate coats the intestinal tract, permitting greater visual detail than a traditional X-ray. (credit modification of work by "glitzy queen00"/Wikimedia Commons)

Medical imaging using barium sulfate can be used to diagnose acid reflux disease, Crohn's disease, and ulcers in addition to other conditions.

Visit this website (http://openstax.org/l/16barium) for more information on how barium is used in medical diagnoses and which conditions it is used to diagnose.

Predicting Precipitation

The equation that describes the equilibrium between solid calcium carbonate and its solvated ions is:

$$\text{CaCO}_3(s) \rightleftharpoons \text{Ca}^{2+}(aq) + \text{CO}_3{}^{2-}(aq) \qquad K_{sp} = [\text{Ca}^{2+}][\text{CO}_3{}^{2-}] = 8.7 \times 10^{-9}$$

It is important to realize that this equilibrium is established in any aqueous solution containing Ca^{2+} and $CO_3{}^{2-}$ ions, not just in a solution formed by saturating water with calcium carbonate. Consider, for example, mixing aqueous solutions of the soluble compounds sodium carbonate and calcium nitrate. If the concentrations of calcium and carbonate ions in the mixture do not yield a reaction quotient, Q_{sp}, that exceeds the solubility product, K_{sp}, then no precipitation will occur. If the ion concentrations yield a reaction quotient greater than the solubility product, then precipitation will occur, lowering those concentrations until equilibrium is established ($Q_{sp} = K_{\text{sp}}$). The comparison of Q_{sp} to K_{sp} to predict precipitation is an example of the general approach to predicting the direction of a reaction first introduced in the chapter on equilibrium. For the specific case of solubility equilibria:

$Q_{sp} < K_{\text{sp}}$: the reaction proceeds in the forward direction (solution is not saturated; no precipitation observed)

$Q_{sp} > K_{\text{sp}}$: the reaction proceeds in the reverse direction (solution is supersaturated; precipitation will occur)

This predictive strategy and related calculations are demonstrated in the next few example exercises.

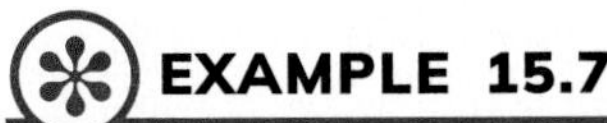

EXAMPLE 15.7

Precipitation of $Mg(OH)_2$

The first step in the preparation of magnesium metal is the precipitation of $Mg(OH)_2$ from sea water by the addition of lime, $Ca(OH)_2$, a readily available inexpensive source of OH^- ion:

$$Mg(OH)_2(s) \rightleftharpoons Mg^{2+}(aq) + 2OH^-(aq) \qquad K_{sp} = 8.9 \times 10^{-12}$$

The concentration of $Mg^{2+}(aq)$ in sea water is 0.0537 *M*. Will $Mg(OH)_2$ precipitate when enough $Ca(OH)_2$ is added to give a $[OH^-]$ of 0.0010 *M*?

Solution

Calculation of the reaction quotient under these conditions is shown here:

$$Q = [Mg^{2+}][OH^-]^2 = (0.0537)(0.0010)^2 = 5.4 \times 10^{-8}$$

Because Q is greater than K_{sp} ($Q = 5.4 \times 10^{-8}$ is larger than $K_{sp} = 8.9 \times 10^{-12}$), the reverse reaction will proceed, precipitating magnesium hydroxide until the dissolved ion concentrations have been sufficiently lowered, so that $Q_{sp} = K_{sp}$.

Check Your Learning

Predict whether $CaHPO_4$ will precipitate from a solution with $[Ca^{2+}] = 0.0001$ *M* and $[HPO_4{}^{2-}] = 0.001$ *M*.

Answer:

No precipitation of $CaHPO_4$; $Q = 1 \times 10^{-7}$, which is less than K_{sp} (7×10^{-7})

EXAMPLE 15.8

Precipitation of AgCl

Does silver chloride precipitate when equal volumes of a 2.0×10^{-4}-*M* solution of $AgNO_3$ and a 2.0×10^{-4}-*M* solution of NaCl are mixed?

Solution

The equation for the equilibrium between solid silver chloride, silver ion, and chloride ion is:

$$AgCl(s) \rightleftharpoons Ag^+(aq) + Cl^-(aq)$$

The solubility product is 1.6×10^{-10} (see Appendix J).

AgCl will precipitate if the reaction quotient calculated from the concentrations in the mixture of $AgNO_3$ and NaCl is greater than K_{sp}. Because the volume doubles when equal volumes of $AgNO_3$ and NaCl solutions are mixed, each concentration is reduced to half its initial value

$$\frac{1}{2}(2.0 \times 10^{-4})\ M = 1.0 \times 10^{-4}\ M$$

The reaction quotient, Q, is greater than K_{sp} for AgCl, so a supersaturated solution is formed:

$$Q = [Ag^+][Cl^-] = (1.0 \times 10^{-4})(1.0 \times 10^{-4}) = 1.0 \times 10^{-8} > K_{sp}$$

AgCl will precipitate from the mixture until the dissolution equilibrium is established, with Q equal to K_{sp}.

Check Your Learning

Will $KClO_4$ precipitate when 20 mL of a 0.050-*M* solution of K^+ is added to 80 mL of a 0.50-*M* solution of $ClO_4{}^-$? (Hint: Use the dilution equation to calculate the concentrations of potassium and perchlorate ions in the mixture.)

Answer:

No, $Q = 4.0 \times 10^{-3}$, which is less than $K_{sp} = 1.05 \times 10^{-2}$

EXAMPLE 15.9

Precipitation of Calcium Oxalate

Blood will not clot if calcium ions are removed from its plasma. Some blood collection tubes contain salts of the oxalate ion, $C_2O_4{}^{2-}$, for this purpose (Figure 15.5). At sufficiently high concentrations, the calcium and oxalate ions form solid, $CaC_2O_4 \cdot H_2O$ (calcium oxalate monohydrate). The concentration of Ca^{2+} in a sample of blood serum is 2.2×10^{-3} *M*. What concentration of $C_2O_4{}^{2-}$ ion must be established before $CaC_2O_4 \cdot H_2O$ begins to precipitate?

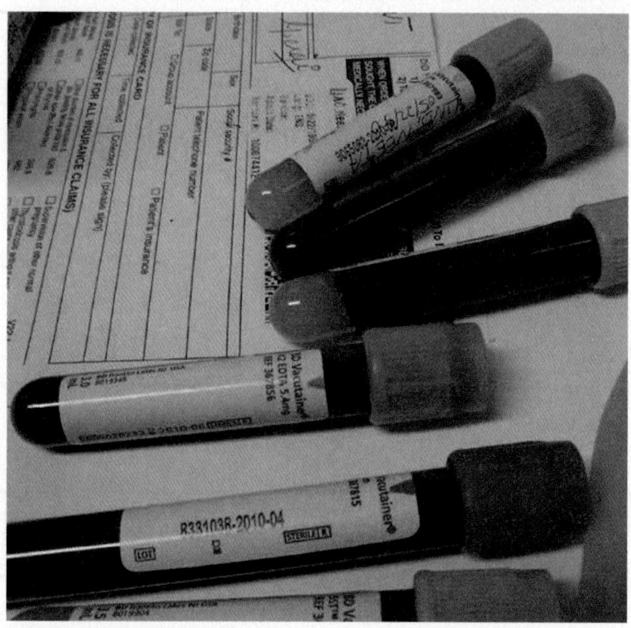

FIGURE 15.5 Anticoagulants can be added to blood that will combine with the Ca^{2+} ions in blood serum and prevent the blood from clotting. (credit: modification of work by Neeta Lind)

Solution

The equilibrium expression is:

$$CaC_2O_4(s) \rightleftharpoons Ca^{2+}(aq) + C_2O_4{}^{2-}(aq)$$

For this reaction:

$$K_{sp} = [Ca^{2+}][C_2O_4{}^{2-}] = 1.96 \times 10^{-8}$$

(see Appendix J)

Substitute the provided calcium ion concentration into the solubility product expression and solve for oxalate concentration:

$$Q = K_{sp} = [Ca^{2+}][C_2O_4{}^{2-}] = 1.96 \times 10^{-8}$$

$$(2.2 \times 10^{-3})[C_2O_4{}^{2-}] = 1.96 \times 10^{-8}$$

$$[C_2O_4{}^{2-}] = \frac{1.96 \times 10^{-8}}{2.2 \times 10^{-3}} = 8.9 \times 10^{-6}\ M$$

A concentration of $[C_2O_4{}^{2-}] = 8.9 \times 10^{-6}$ *M* is necessary to initiate the precipitation of CaC_2O_4 under these conditions.

Check Your Learning

If a solution contains 0.0020 mol of $CrO_4{}^{2-}$ per liter, what concentration of Ag^+ ion must be reached by adding solid $AgNO_3$ before Ag_2CrO_4 begins to precipitate? Neglect any increase in volume upon adding the solid silver nitrate.

Answer:

6.7×10^{-5} *M*

 **EXAMPLE 15.10**

Concentrations Following Precipitation

Clothing washed in water that has a manganese [Mn^{2+}(*aq*)] concentration exceeding 0.1 mg/L (1.8×10^{-6} *M*) may be stained by the manganese upon oxidation, but the amount of Mn^{2+} in the water can be decreased by adding a base to precipitate $Mn(OH)_2$. What pH is required to keep [Mn^{2+}] equal to 1.8×10^{-6} *M*?

Solution

The dissolution of $Mn(OH)_2$ is described by the equation:

$$Mn(OH)_2(s) \rightleftharpoons Mn^{2+}(aq) + 2OH^-(aq) \qquad K_{sp} = 2 \times 10^{-13}$$

At equilibrium:

$$K_{sp} = [Mn^{2+}][OH^-]^2$$

or

$$(1.8 \times 10^{-6})[OH^-]^2 = 2 \times 10^{-13}$$

so

$$[OH^-] = 3.3 \times 10^{-4}\ M$$

Calculate the pH from the pOH:

$$\text{pOH} = -\log[OH^-] = -\log(3.3 \times 10 - 4) = 3.48$$
$$\text{pH} = 14.00 - \text{pOH} = 14.00 - 3.48 = 10.52$$

(final result rounded to one significant digit, limited by the certainty of the K_{sp})

Check Your Learning

The first step in the preparation of magnesium metal is the precipitation of $Mg(OH)_2$ from sea water by the addition of $Ca(OH)_2$. The concentration of Mg^{2+}(*aq*) in sea water is 5.37×10^{-2} *M*. Calculate the pH at which [Mg^{2+}] is decreased to 1.0×10^{-5} *M*

Answer:

10.97

In solutions containing two or more ions that may form insoluble compounds with the same counter ion, an experimental strategy called **selective precipitation** may be used to remove individual ions from solution. By increasing the counter ion concentration in a controlled manner, ions in solution may be precipitated individually, assuming their compound solubilities are adequately different. In solutions with equal concentrations of target ions, the ion forming the least soluble compound will precipitate first (at the lowest concentration of counter ion), with the other ions subsequently precipitating as their compound's solubilities are reached. As an illustration of this technique, the next example exercise describes separation of a two halide ions via precipitation of one as a silver salt.

Chemistry in Everyday Life

The Role of Precipitation in Wastewater Treatment

Solubility equilibria are useful tools in the treatment of wastewater carried out in facilities that may treat the municipal water in your city or town (Figure 15.6). Specifically, selective precipitation is used to remove contaminants from wastewater before it is released back into natural bodies of water. For example, phosphate ions ($PO_4{}^{3-}$) are often present in the water discharged from manufacturing facilities. An abundance of phosphate causes excess algae to grow, which impacts the amount of oxygen available for marine life as well as making water unsuitable for human consumption.

FIGURE 15.6 Wastewater treatment facilities, such as this one, remove contaminants from wastewater before the water is released back into the natural environment. (credit: "eutrophication&hypoxia"/Wikimedia Commons)

One common way to remove phosphates from water is by the addition of calcium hydroxide, or lime, $Ca(OH)_2$. As the water is made more basic, the calcium ions react with phosphate ions to produce hydroxylapatite, $Ca_5(PO4)_3OH$, which then precipitates out of the solution:

$$5Ca^{2+} + 3PO_4{}^{3-} + OH^- \rightleftharpoons Ca_5(PO_4)_3{\cdot}OH(s)$$

Because the amount of calcium ion added does not result in exceeding the solubility products for other calcium salts, the anions of those salts remain behind in the wastewater. The precipitate is then removed by filtration and the water is brought back to a neutral pH by the addition of CO_2 in a recarbonation process. Other chemicals can also be used for the removal of phosphates by precipitation, including iron(III) chloride and aluminum sulfate.

View this site (http://openstax.org/l/16Wastewater) for more information on how phosphorus is removed from wastewater.

 **EXAMPLE 15.11**

Precipitation of Silver Halides

A solution contains 0.00010 mol of KBr and 0.10 mol of KCl per liter. $AgNO_3$ is gradually added to this solution. Which forms first, solid AgBr or solid AgCl?

Solution

The two equilibria involved are:

$$\begin{aligned} AgCl(s) &\rightleftharpoons Ag^+(aq) + Cl^-(aq) & K_{sp} &= 1.6 \times 10^{-10} \\ AgBr(s) &\rightleftharpoons Ag^+(aq) + Br^-(aq) & K_{sp} &= 5.0 \times 10^{-13} \end{aligned}$$

If the solution contained about *equal* concentrations of Cl^- and Br^-, then the silver salt with the smaller K_{sp} (AgBr) would precipitate first. The concentrations are not equal, however, so the $[Ag^+]$ at which AgCl begins to precipitate and the $[Ag^+]$ at which AgBr begins to precipitate must be calculated. The salt that forms at the lower $[Ag^+]$ precipitates first.

AgBr precipitates when Q equals K_{sp} for AgBr

$$Q_{sp} = K_{sp} = [Ag^+][Br^-] = [Ag^+](0.00010) = 5.0 \times 10^{-13}$$

$$[Ag^+] = \frac{5.0 \times 10^{-13}}{0.00010} = 5.0 \times 10^{-9}\ M$$

AgBr begins to precipitate when $[Ag^+]$ is 5.0×10^{-9} *M*.

For AgCl: AgCl precipitates when Q equals K_{sp} for AgCl (1.6×10^{-10}). When $[Cl^-] = 0.10$ *M*:

$$Q_{sp} = K_{sp} = [Ag^+][Cl^-] = [Ag^+](0.10) = 1.6 \times 10^{-10}$$

$$[Ag^+] = \frac{1.6 \times 10^{-10}}{0.10} = 1.6 \times 10^{-9}\ M$$

AgCl begins to precipitate when $[Ag^+]$ is 1.6×10^{-9} *M*.

AgCl begins to precipitate at a lower $[Ag^+]$ than AgBr, so AgCl begins to precipitate first. Note the chloride ion concentration of the initial mixture was significantly greater than the bromide ion concentration, and so silver chloride precipitated first despite having a K_{sp} greater than that of silver bromide.

Check Your Learning

If silver nitrate solution is added to a solution which is 0.050 *M* in both Cl^- and Br^- ions, at what $[Ag^+]$ would precipitation begin, and what would be the formula of the precipitate?

Answer:

$[Ag^+] = 1.0 \times 10^{-11}$ *M*; AgBr precipitates first

Common Ion Effect

Compared with pure water, the solubility of an ionic compound is less in aqueous solutions containing a *common ion* (one also produced by dissolution of the ionic compound). This is an example of a phenomenon known as the **common ion effect**, which is a consequence of the law of mass action that may be explained using Le ChÂtelier's principle. Consider the dissolution of silver iodide:

$$AgI(s) \rightleftharpoons Ag^+(aq) + I^-(aq)$$

This solubility equilibrium may be shifted left by the addition of either silver(I) or iodide ions, resulting in the precipitation of AgI and lowered concentrations of dissolved Ag^+ and I^-. In solutions that already contain either of these ions, less AgI may be dissolved than in solutions without these ions.

This effect may also be explained in terms of mass action as represented in the solubility product expression:

$$K_{sp} = [Ag^+][I^-]$$

The mathematical product of silver(I) and iodide ion molarities is constant in an equilibrium mixture *regardless of the source of the ions*, and so an increase in one ion's concentration must be balanced by a proportional decrease in the other.

LINK TO LEARNING

View this simulation (http://openstax.org/l/16solublesalts) to explore various aspects of the common ion effect.

EXAMPLE 15.12

Common Ion Effect on Solubility

What is the effect on the amount of solid $Mg(OH)_2$ and the concentrations of Mg^{2+} and OH^- when each of the following are added to a saturated solution of $Mg(OH)_2$?

(a) $MgCl_2$

(b) KOH

(c) $NaNO_3$

(d) $Mg(OH)_2$

Solution

The solubility equilibrium is

$$Mg(OH)_2(s) \rightleftharpoons Mg^{2+}(aq) + 2OH^-(aq)$$

(a) Adding a common ion, Mg^{2+}, will increase the concentration of this ion and shift the solubility equilibrium to the left, decreasing the concentration of hydroxide ion and increasing the amount of undissolved magnesium hydroxide.

(b) Adding a common ion, OH^-, will increase the concentration of this ion and shift the solubility equilibrium to the left, decreasing the concentration of magnesium ion and increasing the amount of undissolved magnesium hydroxide.

(c) The added compound does not contain a common ion, and no effect on the magnesium hydroxide solubility equilibrium is expected.

(d) Adding more solid magnesium hydroxide will increase the amount of undissolved compound in the mixture. The solution is already saturated, though, so the concentrations of dissolved magnesium and hydroxide ions will remain the same.

$$Q = [Mg^{2+}][OH^-]^2$$

Thus, changing the amount of solid magnesium hydroxide in the mixture has no effect on the value of Q, and no shift is required to restore Q to the value of the equilibrium constant.

Check Your Learning

What is the effect on the amount of solid $NiCO_3$ and the concentrations of Ni^{2+} and ${CO_3}^{2-}$ when each of the following are added to a saturated solution of $NiCO_3$

(a) $Ni(NO_3)_2$

(b) $KClO_4$

(c) $NiCO_3$

(d) K_2CO_3

Answer:

(a) mass of $NiCO_3(s)$ increases, $[Ni^{2+}]$ increases, $[{CO_3}^{2-}]$ decreases; (b) no appreciable effect; (c) no effect except to increase the amount of solid $NiCO_3$; (d) mass of $NiCO_3(s)$ increases, $[Ni^{2+}]$ decreases, $[{CO_3}^{2-}]$ increases;

EXAMPLE 15.13

Common Ion Effect

Calculate the molar solubility of cadmium sulfide (CdS) in a 0.010-*M* solution of cadmium bromide ($CdBr_2$). The K_{sp} of CdS is 1.0×10^{-28}.

Solution

This calculation can be performed using the ICE approach:

$$CdS(s) \rightleftharpoons Cd^{2+}(aq) + S^{2-}(aq)$$

	CdS (s) ⇌	Cd^{2+} (aq) +	S^{2-} (aq)
Initial concentration (*M*)		0.010	0
Change (*M*)		$+x$	$+x$
Equilibrium concentration (*M*)		$0.010 + x$	x

$$K_{sp} = [Cd^{2+}][S^{2-}] = 1.0 \times 10^{-28}$$

$$(0.010 + x)(x) = 1.0 \times 10^{-28}$$

Because K_{sp} is very small, assume $x << 0.010$ and solve the simplified equation for x:

$$(0.010)(x) = 1.0 \times 10^{-28}$$

$$x = 1.0 \times 10^{-26}\ M$$

The molar solubility of CdS in this solution is 1.0×10^{-26} *M*.

Check Your Learning

Calculate the molar solubility of aluminum hydroxide, $Al(OH)_3$, in a 0.015-*M* solution of aluminum nitrate, $Al(NO_3)_3$. The K_{sp} of $Al(OH)_3$ is 2×10^{-32}.

Answer:

4×10^{-11} *M*

15.2 Lewis Acids and Bases

LEARNING OBJECTIVES

By the end of this section, you will be able to:

- Explain the Lewis model of acid-base chemistry
- Write equations for the formation of adducts and complex ions
- Perform equilibrium calculations involving formation constants

In 1923, G. N. Lewis proposed a generalized definition of acid-base behavior in which acids and bases are identified by their ability to accept or to donate a pair of electrons and form a coordinate covalent bond.

A **coordinate covalent bond** (or dative bond) occurs when one of the atoms in the bond provides both bonding electrons. For example, a coordinate covalent bond occurs when a water molecule combines with a hydrogen ion to form a hydronium ion. A coordinate covalent bond also results when an ammonia molecule combines with a hydrogen ion to form an ammonium ion. Both of these equations are shown here.

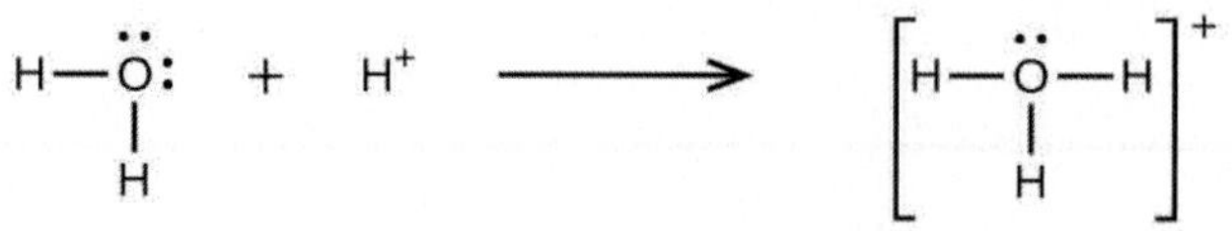

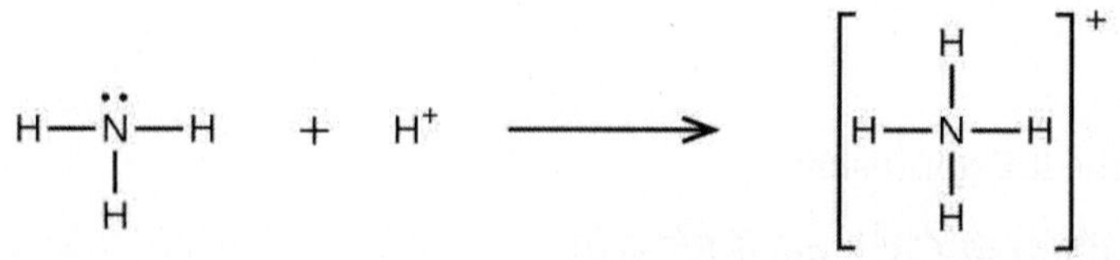

Reactions involving the formation of coordinate covalent bonds are classified as **Lewis acid-base chemistry**. The species donating the electron pair that compose the bond is a **Lewis base**, the species accepting the electron pair is a **Lewis acid**, and the product of the reaction is a **Lewis acid-base adduct**. As the two examples above illustrate, Brønsted-Lowry acid-base reactions represent a subcategory of Lewis acid reactions, specifically, those in which the acid species is H^+. A few examples involving other Lewis acids and bases are described below.

The boron atom in boron trifluoride, BF_3, has only six electrons in its valence shell. Being short of the preferred octet, BF_3 is a very good Lewis acid and reacts with many Lewis bases; a fluoride ion is the Lewis base in this reaction, donating one of its lone pairs:

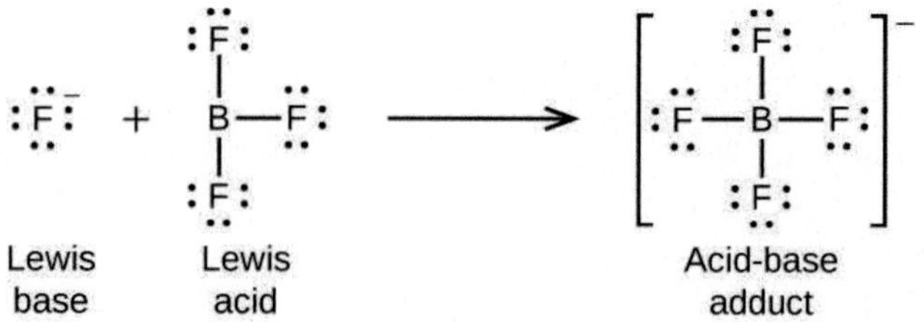

Lewis base | Lewis acid | Acid-base adduct

In the following reaction, each of two ammonia molecules, Lewis bases, donates a pair of electrons to a silver ion, the Lewis acid:

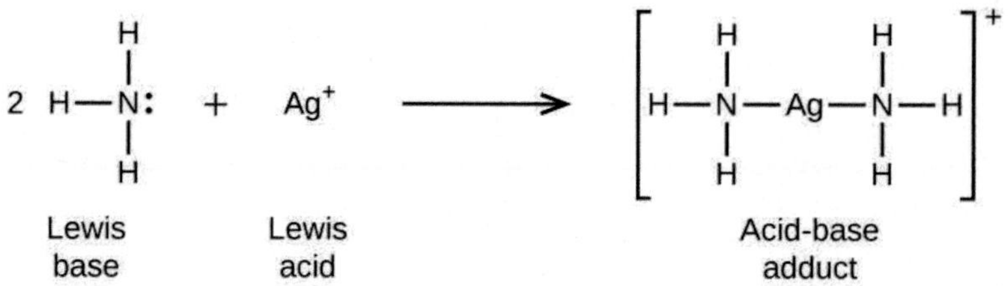

Lewis base | Lewis acid | Acid-base adduct

Nonmetal oxides act as Lewis acids and react with oxide ions, Lewis bases, to form oxyanions:

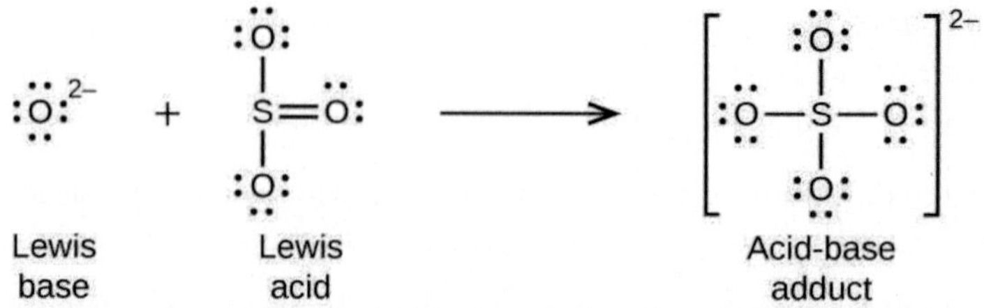

Lewis base | Lewis acid | Acid-base adduct

Many Lewis acid-base reactions are displacement reactions in which one Lewis base displaces another Lewis base from an acid-base adduct, or in which one Lewis acid displaces another Lewis acid:

$$\left[\mathrm{H_3N{-}Ag{-}NH_3}\right]^+ + 2\left[:\mathrm{C{\equiv}N}:\right]^- \longrightarrow \left[:\mathrm{N{\equiv}C{-}Ag{-}C{\equiv}N}:\right]^- + 2\ :\mathrm{NH_3}$$

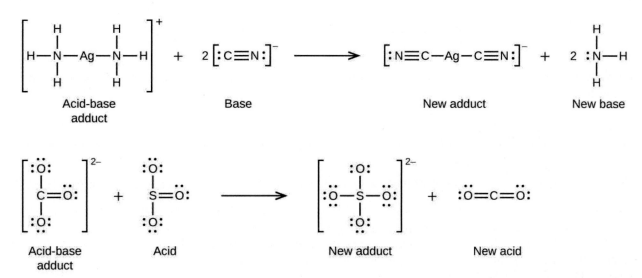

Acid-base adduct | Base | New adduct | New base

$$\left[\mathrm{CO_3}\right]^{2-} + \mathrm{SO_2} \longrightarrow \left[\mathrm{SO_3}\right]^{2-} + \mathrm{O{=}C{=}O}$$

Acid-base adduct | Acid | New adduct | New acid

Another type of Lewis acid-base chemistry involves the formation of a complex ion (or a coordination complex) comprising a central atom, typically a transition metal cation, surrounded by ions or molecules called **ligands**. These ligands can be neutral molecules like H_2O or NH_3, or ions such as CN^- or OH^-. Often, the ligands act as Lewis bases, donating a pair of electrons to the central atom. These types of Lewis acid-base reactions are examples of a broad subdiscipline called *coordination chemistry*—the topic of another chapter in this text.

The equilibrium constant for the reaction of a metal ion with one or more ligands to form a coordination complex is called a **formation constant (K_f)** (sometimes called a stability constant). For example, the complex ion $Cu(CN)_2{}^-$

$$\left[:\mathrm{N{\equiv}C{-}Cu{-}C{\equiv}N}:\right]^-$$

is produced by the reaction

$$\mathrm{Cu^+}(aq) + \mathrm{2CN^-}(aq) \rightleftharpoons \mathrm{Cu(CN)_2{}^-}(aq)$$

The formation constant for this reaction is

$$K_\mathrm{f} = \frac{[\mathrm{Cu(CN)_2{}^-}]}{[\mathrm{Cu^+}]\,[\mathrm{CN^-}]^2}$$

Alternatively, the reverse reaction (decomposition of the complex ion) can be considered, in which case the equilibrium constant is a **dissociation constant (K_d)**. Per the relation between equilibrium constants for reciprocal reactions described, the dissociation constant is the mathematical inverse of the formation constant, $K_d = K_f{}^{-1}$. A tabulation of formation constants is provided in Appendix K.

As an example of dissolution by complex ion formation, let us consider what happens when we add aqueous ammonia to a mixture of silver chloride and water. Silver chloride dissolves slightly in water, giving a small concentration of Ag^+ ($[Ag^+] = 1.3 \times 10^{-5}$ *M*):

$$\mathrm{AgCl}(s) \rightleftharpoons \mathrm{Ag^+}(aq) + \mathrm{Cl^-}(aq)$$

However, if NH_3 is present in the water, the complex ion, $Ag(NH_3)_2{}^+$, can form according to the equation:

$$\mathrm{Ag^+}(aq) + \mathrm{2NH_3}(aq) \rightleftharpoons \mathrm{Ag(NH_3)_2{}^+}(aq)$$

with

$$K_\mathrm{f} = \frac{[\mathrm{Ag(NH_3)_2{}^+}]}{[\mathrm{Ag^+}][\mathrm{NH_3}]^2} = 1.7 \times 10^7$$

The large size of this formation constant indicates that most of the free silver ions produced by the dissolution of AgCl combine with NH_3 to form $Ag(NH_3)_2{}^+$. As a consequence, the concentration of silver ions, $[Ag^+]$, is reduced, and the reaction quotient for the dissolution of silver chloride, $[Ag^+][Cl^-]$, falls below the solubility product of AgCl:

$$Q = [\text{Ag}^+]\left[\text{Cl}^-\right] < K_{sp}$$

More silver chloride then dissolves. If the concentration of ammonia is great enough, all of the silver chloride dissolves.

EXAMPLE 15.14

Dissociation of a Complex Ion

Calculate the concentration of the silver ion in a solution that initially is 0.10 *M* with respect to $\text{Ag(NH}_3)_2{}^+$.

Solution

Applying the standard ICE approach to this reaction yields the following:

	Ag^+ +	2NH_3 ⇌	$\text{Ag(NH}_3)_2{}^+$
Initial concentration (*M*)	0	0	0.10
Change (*M*)	$+x$	$+2x$	$-x$
Equilibrium concentration (*M*)	x	$2x$	$0.10 - x$

Substituting these equilibrium concentration terms into the K_f expression gives

$$K_f = \frac{[\text{Ag(NH}_3)_2{}^+]}{[\text{Ag}^+][\text{NH}_3]^2}$$

$$1.7 \times 10^7 = \frac{0.10 - x}{(x)(2x)^2}$$

The very large equilibrium constant means the amount of the complex ion that will dissociate, *x*, will be very small. Assuming $x << 0.1$ permits simplifying the above equation:

$$1.7 \times 10^7 = \frac{0.10}{(x)(2x)^2}$$

$$x^3 = \frac{0.10}{4(1.7 \times 10^7)} = 1.5 \times 10^{-9}$$

$$x = \sqrt[3]{1.5 \times 10^{-9}} = 1.1 \times 10^{-3}$$

Because only 1.1% of the $\text{Ag(NH}_3)_2{}^+$ dissociates into Ag^+ and NH_3, the assumption that *x* is small is justified.

Using this value of *x* and the relations in the above ICE table allows calculation of all species' equilibrium concentrations:

$$[\text{Ag}^+] = 0 + x = 1.1 \times 10^{-3}\ M$$

$$[\text{NH}_3] = 0 + 2x = 2.2 \times 10^{-3}\ M$$

$$[\text{Ag(NH}_3)_2{}^+] = 0.10 - x = 0.10 - 0.0011 = 0.099$$

The concentration of free silver ion in the solution is 0.0011 *M*.

Check Your Learning

Calculate the silver ion concentration, $[\text{Ag}^+]$, of a solution prepared by dissolving 1.00 g of AgNO_3 and 10.0 g of KCN in sufficient water to make 1.00 L of solution. (Hint: Because K_f is very large, assume the reaction goes to completion then calculate the $[\text{Ag}^+]$ produced by dissociation of the complex.)

Answer:

2.9×10^{-22} *M*

15.3 Coupled Equilibria

LEARNING OBJECTIVES

By the end of this section, you will be able to:

- Describe examples of systems involving two (or more) coupled chemical equilibria
- Calculate reactant and product concentrations for coupled equilibrium systems

As discussed in preceding chapters on equilibrium, *coupled equilibria* involve two or more separate chemical reactions that share one or more reactants or products. This section of this chapter will address solubility equilibria coupled with acid-base and complex-formation reactions.

An environmentally relevant example illustrating the coupling of solubility and acid-base equilibria is the impact of ocean acidification on the health of the ocean's coral reefs. These reefs are built upon skeletons of sparingly soluble calcium carbonate excreted by colonies of corals (small marine invertebrates). The relevant dissolution equilibrium is

$$CaCO_3(s) \rightleftharpoons Ca^{2+}(aq) + CO_3{}^{-2}(aq) \qquad K_{sp} = 8.7 \times 10^{-9}$$

Rising concentrations of atmospheric carbon dioxide contribute to an increased acidity of ocean waters due to the dissolution, hydrolysis, and acid ionization of carbon dioxide:

$$CO_2(g) \rightleftharpoons CO_2(aq)$$

$$CO_2(aq) + H_2O(l) \rightleftharpoons H_2CO_3(aq)$$

$$H_2CO_3(aq) + H_2O(l) \rightleftharpoons HCO_3{}^-(aq) + H_3O^+(aq) \qquad K_{a1} = 4.3 \times 10^{-7}$$

$$HCO_3{}^-(aq) + H_2O(l) \rightleftharpoons CO_3{}^{2-}(aq) + H_3O^+(aq) \qquad K_{a2} = 4.7 \times 10^{-11}$$

Inspection of these equilibria shows the carbonate ion is involved in the calcium carbonate dissolution and the acid hydrolysis of bicarbonate ion. Combining the dissolution equation with the reverse of the acid hydrolysis equation yields

$$CaCO_3(s) + H_3O^+(aq) \rightleftharpoons Ca^{2+}(aq) + HCO_3{}^-(aq) + H_2O(l) \qquad K = K_{sp} / K_{a2} = 180$$

The equilibrium constant for this net reaction is much greater than the K_{sp} for calcium carbonate, indicating its solubility is markedly increased in acidic solutions. As rising carbon dioxide levels in the atmosphere increase the acidity of ocean waters, the calcium carbonate skeletons of coral reefs become more prone to dissolution and subsequently less healthy (Figure 15.7).

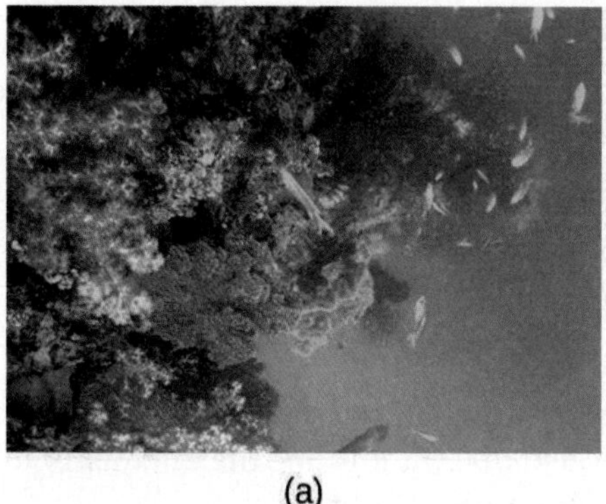

(a)

(b)

FIGURE 15.7 Healthy coral reefs (a) support a dense and diverse array of sea life across the ocean food chain. But when coral are unable to adequately build and maintain their calcium carbonate skeletons because of excess ocean acidification, the unhealthy reef (b) is only capable of hosting a small fraction of the species as before, and the local food chain starts to collapse. (credit a: modification of work by NOAA Photo Library; credit b: modification of work by "prilfish"/Flickr)

LINK TO LEARNING

Learn more about ocean acidification (http://openstax.org/l/16acidicocean) and how it affects other marine creatures.

This site (http://openstax.org/l/16coralreef) has detailed information about how ocean acidification specifically affects coral reefs.

The dramatic increase in solubility with increasing acidity described above for calcium carbonate is typical of salts containing basic anions (e.g., carbonate, fluoride, hydroxide, sulfide). Another familiar example is the formation of dental cavities in tooth enamel. The major mineral component of enamel is calcium hydroxyapatite (Figure 15.8), a sparingly soluble ionic compound whose dissolution equilibrium is

$$Ca_5(PO_4)_3OH(s) \rightleftharpoons 5Ca^{2+}(aq) + 3PO_4{}^{3-}(aq) + OH^-(aq)$$

FIGURE 15.8 Crystal of the mineral hydroxyapatite, $Ca_5(PO_4)_3OH$, is shown here. The pure compound is white, but like many other minerals, this sample is colored because of the presence of impurities.

This compound dissolved to yield two different basic ions: triprotic phosphate ions

$$PO_4{}^{3-}(aq) + H_3O^+(aq) \longrightarrow H_2PO_4{}^{2-}(aq) + H_2O(l)$$

$$H_2PO_4{}^{2-}(aq) + H_3O^+(aq) \longrightarrow H_2PO_4{}^{-}(aq) + H_2O(l)$$

$$H_2PO_4{}^{-}(aq) + H_3O^+(aq) \longrightarrow H_3PO_4(aq) + H_2O(l)$$

and monoprotic hydroxide ions:

$$OH^-(aq) + H_3O^+ \longrightarrow 2H_2O$$

Of the two basic productions, the hydroxide is, of course, by far the stronger base (it's the strongest base that can exist in aqueous solution), and so it is the dominant factor providing the compound an acid-dependent solubility. Dental cavities form when the acid waste of bacteria growing on the surface of teeth hastens the dissolution of tooth enamel by reacting completely with the strong base hydroxide, shifting the hydroxyapatite solubility equilibrium to the right. Some toothpastes and mouth rinses contain added NaF or SnF_2 that make enamel more acid resistant by replacing the strong base hydroxide with the weak base fluoride:

$$NaF + Ca_5(PO_4)_3OH \rightleftharpoons Ca_5(PO_4)_3F + Na^+ + OH^-$$

The weak base fluoride ion reacts only partially with the bacterial acid waste, resulting in a less extensive shift in the solubility equilibrium and an increased resistance to acid dissolution. See the Chemistry in Everyday Life feature on the role of fluoride in preventing tooth decay for more information.

Chemistry in Everyday Life

Role of Fluoride in Preventing Tooth Decay

As we saw previously, fluoride ions help protect our teeth by reacting with hydroxylapatite to form fluorapatite, $Ca_5(PO_4)_3F$. Since it lacks a hydroxide ion, fluorapatite is more resistant to attacks by acids in our mouths and is thus less soluble, protecting our teeth. Scientists discovered that naturally fluorinated water could be beneficial to your teeth, and so it became common practice to add fluoride to drinking water. Toothpastes and mouthwashes also contain amounts of fluoride (Figure 15.9).

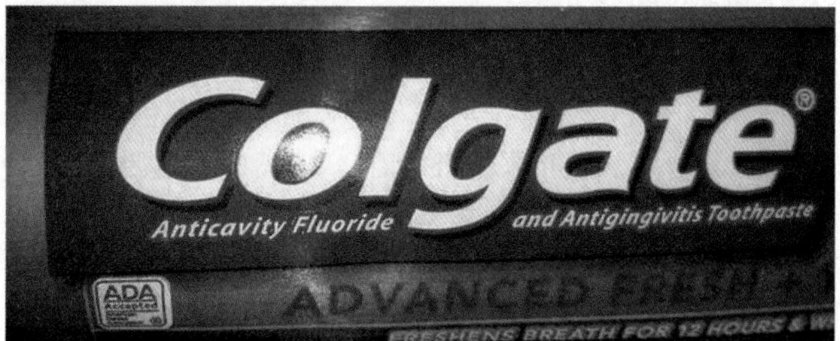

FIGURE 15.9 Fluoride, found in many toothpastes, helps prevent tooth decay (credit: Kerry Ceszyk).

Unfortunately, excess fluoride can negate its advantages. Natural sources of drinking water in various parts of the world have varying concentrations of fluoride, and places where that concentration is high are prone to certain health risks when there is no other source of drinking water. The most serious side effect of excess fluoride is the bone disease, skeletal fluorosis. When excess fluoride is in the body, it can cause the joints to stiffen and the bones to thicken. It can severely impact mobility and can negatively affect the thyroid gland. Skeletal fluorosis is a condition that over 2.7 million people suffer from across the world. So while fluoride can protect our teeth from decay, the US Environmental Protection Agency sets a maximum level of 4 ppm (4 mg/L) of fluoride in drinking water in the US. Fluoride levels in water are not regulated in all countries, so fluorosis is a problem in areas with high levels of fluoride in the groundwater.

The solubility of ionic compounds may also be increased when dissolution is coupled to the formation of a complex ion. For example, aluminum hydroxide dissolves in a solution of sodium hydroxide or another strong base because of the formation of the complex ion $Al(OH)_4{}^-$.

The equations for the dissolution of aluminum hydroxide, the formation of the complex ion, and the combined (net) equation are shown below. As indicated by the relatively large value of K for the net reaction, coupling complex formation with dissolution drastically increases the solubility of $Al(OH)_3$.

$$Al(OH)_3(s) \rightleftharpoons Al^{3+}(aq) + 3OH^-(aq) \qquad K_{sp} = 2 \times 10^{-32}$$

$$Al^{3+}(aq) + 4OH^-(aq) \rightleftharpoons Al(OH)_4{}^-(aq) \qquad K_f = 1.1 \times 10^{33}$$

$$\text{Net: } Al(OH)_3(s) + OH^-(aq) \rightleftharpoons Al(OH)_4{}^-(aq) \qquad K = K_{sp}K_f = 22$$

 **EXAMPLE 15.15**

Increased Solubility in Acidic Solutions

Compute and compare the molar solublities for aluminum hydroxide, $Al(OH)_3$, dissolved in (a) pure water and (b) a buffer containing 0.100 *M* acetic acid and 0.100 *M* sodium acetate.

Solution

(a) The molar solubility of aluminum hydroxide in water is computed considering the dissolution equilibrium only as demonstrated in several previous examples:

$$Al(OH)_3(s) \rightleftharpoons Al^{3+}(aq) + 3OH^-(aq) \qquad K_{sp} = 2 \times 10^{-32}$$

$$\text{molar solubility in water} = [Al^{3+}] = (2 \times 10^{-32} / 27)^{1/4} = 5 \times 10^{-9}\ M$$

(b) The concentration of hydroxide ion of the buffered solution is conveniently calculated by the Henderson-Hasselbalch equation:

$$pH = pK_a + \log [CH_3COO^-] / [CH_3COOH]$$

$$pH = 4.74 + \log (0.100 / 0.100) = 4.74$$

At this pH, the concentration of hydroxide ion is

$$pOH = 14.00 - 4.74 = 9.26$$

$$[OH^-] = 10^{-9.26} = 5.5 \times 10^{-10}$$

The solubility of $Al(OH)_3$ in this buffer is then calculated from its solubility product expressions:

$$K_{sp} = [Al^{3+}][OH^-]^3$$

$$\text{molar solubility in buffer} = [Al^{3+}] = K_{sp} / [OH^-]^3 = (2 \times 10^{-32}) / (5.5 \times 10^{-10})^3 = 1.2 \times 10^{-4}\ M$$

Compared to pure water, the solubility of aluminum hydroxide in this mildly acidic buffer is approximately ten million times greater (though still relatively low).

Check Your Learning

What is the solubility of aluminum hydroxide in a buffer comprised of 0.100 *M* formic acid and 0.100 *M* sodium formate?

Answer:

0.1 *M*

EXAMPLE 15.16

Multiple Equilibria

Unexposed silver halides are removed from photographic film when they react with sodium thiosulfate ($Na_2S_2O_3$, called hypo) to form the complex ion $Ag(S_2O_3)_2{}^{3-}$ ($K_f = 4.7 \times 10^{13}$).

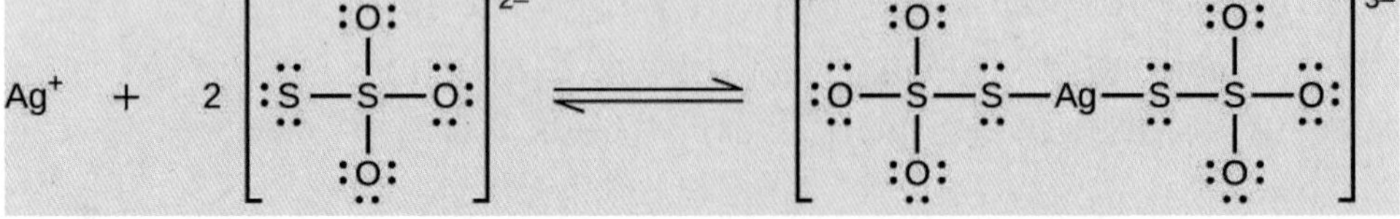

What mass of $Na_2S_2O_3$ is required to prepare 1.00 L of a solution that will dissolve 1.00 g of AgBr by the formation of $Ag(S_2O_3)_2{}^{3-}$?

Solution

Two equilibria are involved when silver bromide dissolves in an aqueous thiosulfate solution containing the $S_2O_3{}^{2-}$ ion:

dissolution: $AgBr(s) \rightleftharpoons Ag^+(aq) + Br^-(aq)$ $K_{sp} = 5.0 \times 10^{-13}$

complexation: $Ag^+(aq) + 2S_2O_3{}^{2-}(aq) \rightleftharpoons Ag(S_2O_3)_2{}^{3-}(aq)$ $K_f = 4.7 \times 10^{13}$

Combining these two equilibrium equations yields

$$AgBr(s) + 2S_2O_3{}^{2-}(aq) \qquad Ag(S_2O_3)_2{}^{3-}(aq)$$

$$\frac{K=[Ag(S_2O_3)_2{}^{3-}][Br^-]}{[S_2O_3{}^{2-}]^2} = K_{sp}K_f = 24$$

The concentration of bromide resulting from dissolution of 1.00 g of AgBr in 1.00 L of solution is

$$[Br^-] = \frac{1.00 \text{ g AgBr} \times \frac{1 \text{ mol AgBr}}{187.77 \text{ g/mol}} \times \frac{1 \text{ mol Br}^-}{1 \text{ mol AgBr}}}{1.00 \text{ L}} = 0.00532 \ M$$

The stoichiometry of the dissolution equilibrium indicates the same concentration of aqueous silver ion will result, 0.00532 M, and the very large value of K_f ensures that essentially all the dissolved silver ion will be complexed by thiosulfate ion:

$$[Ag(S_2O_3)_2{}^{3-}] = 0.00532 \ M$$

Rearranging the K expression for the combined equilibrium equations and solving for the concentration of thiosulfate ion yields

$$[S_2O_3{}^{2-}] = \frac{[Ag(S_2O_3)_2{}^{3-}][Br^-]}{K} = \frac{(0.00532 \ M)(0.00532 \ M)}{24} = 0.0011 \ M$$

Finally, the total mass of $Na_2S_2O_3$ required to provide enough thiosulfate to yield the concentrations cited above can be calculated.

Mass of $Na_2S_2O_3$ required to yield 0.00532 M $Ag\left(S_2O_2\right)_2{}^{3-}$

$$0.00532 \frac{\text{mol } Ag(S_2O_3)_2{}^{3-}}{1.00 \text{ L}} \times \frac{2 \text{ mol } S_2O_3{}^{2-}}{1 \text{ mol } Ag(S_2O_3)_2{}^{3-}} \times \frac{1 \text{ mol } NaS_2O_3}{1 \text{ mol } S_2O_3{}^{2-}} \times \frac{158.1 \text{ g } NaS_2O_3}{1 \text{ mol } NaS_2O_3} = 1.68 \text{ g}$$

Mass of $Na_2S_2O_3$ required to yield 0.00110 M $S_2O_3{}^{2-}$

$$0.0011 \frac{\text{mol } S_2O_3{}^{2-}}{1.00 \text{ L}} \times \frac{1 \text{ mol } S_2O_3{}^{2-}}{1 \text{ mol } Na_2S_2O_3} \times \frac{158.1 \text{ g } Na_2S_2O_3}{1 \text{ mol } Na_2S_2O_3} = 0.17 \text{ g}$$

The mass of $Na_2S_2O_3$ required to dissolve 1.00 g of AgBr in 1.00 L of water is thus 1.68 g + 0.17 g = 1.85 g

Check Your Learning

AgCl(*s*), silver chloride, has a very low solubility: $AgCl(s) \rightleftharpoons Ag^+(aq) + Cl^-(aq)$, $K_{sp} = 1.6 \times 10^{-10}$. Adding ammonia significantly increases the solubility of AgCl because a complex ion is formed:
$Ag^+(aq) + 2NH_3(aq) \rightleftharpoons Ag(NH_3)_2{}^+(aq)$, $K_f = 1.7 \times 10^7$. What mass of NH_3 is required to prepare 1.00 L of solution that will dissolve 2.00 g of AgCl by formation of $Ag(NH_3)_2{}^+$?

Answer:

1.00 L of a solution prepared with 4.81 g NH_3 dissolves 2.0 g of AgCl.

Key Terms

common ion effect effect on equilibrium when a substance with an ion in common with the dissolved species is added to the solution; causes a decrease in the solubility of an ionic species, or a decrease in the ionization of a weak acid or base

complex ion ion consisting of a central atom surrounding molecules or ions called ligands via coordinate covalent bonds

coordinate covalent bond (also, dative bond) covalent bond in which both electrons originated from the same atom

coupled equilibria system characterized the simultaneous establishment of two or more equilibrium reactions sharing one or more reactant or product

dissociation constant (**K_d**) equilibrium constant for the decomposition of a complex ion into its components

formation constant (**K_f**) (also, stability constant) equilibrium constant for the formation of a complex ion from its components

Lewis acid any species that can accept a pair of electrons and form a coordinate covalent bond

Lewis acid-base adduct compound or ion that contains a coordinate covalent bond between a Lewis acid and a Lewis base

Lewis acid-base chemistry reactions involving the formation of coordinate covalent bonds

Lewis base any species that can donate a pair of electrons and form a coordinate covalent bond

ligand molecule or ion acting as a Lewis base in complex ion formation; bonds to the central atom of the complex

molar solubility solubility of a compound expressed in units of moles per liter (mol/L)

selective precipitation process in which ions are separated using differences in their solubility with a given precipitating reagent

solubility product constant (K_{sp}**)** equilibrium constant for the dissolution of an ionic compound

Key Equations

$M_pX_q(s) \rightleftharpoons pM^{m+}(aq) + qX^{n-}(aq)$

$K_{sp} = [M^{m+}]^p[X^{n-}]^q$

Summary

15.1 Precipitation and Dissolution

The equilibrium constant for an equilibrium involving the precipitation or dissolution of a slightly soluble ionic solid is called the solubility product, K_{sp}, of the solid. For a heterogeneous equilibrium involving the slightly soluble solid M_pX_q and its ions M^{m+} and X^{n-}:

$$M_pX_q(s) \rightleftharpoons pM^{m+}(aq) + qX^{n-}(aq)$$

the solubility product expression is:

$$K_{sp} = [M^{m+}]^p[X^{n-}]^q$$

The solubility product of a slightly soluble electrolyte can be calculated from its solubility; conversely, its solubility can be calculated from its K_{sp}, provided the only significant reaction that occurs when the solid dissolves is the formation of its ions.

A slightly soluble electrolyte begins to precipitate when the magnitude of the reaction quotient for the dissolution reaction exceeds the magnitude of the solubility product. Precipitation continues until the reaction quotient equals the solubility product.

15.2 Lewis Acids and Bases

A Lewis acid is a species that can accept an electron pair, whereas a Lewis base has an electron pair available for donation to a Lewis acid. Complex ions are examples of Lewis acid-base adducts and comprise central metal atoms or ions acting as Lewis acids bonded to molecules or ions called ligands that act as Lewis bases. The equilibrium constant for the reaction between a metal ion and ligands produces a complex ion called a formation constant; for the reverse reaction, it is called a dissociation constant.

15.3 Coupled Equilibria

Systems involving two or more chemical equilibria that share one or more reactant or product are called coupled equilibria. Common examples of coupled equilibria include the increased solubility of some compounds in acidic solutions (coupled dissolution and neutralization equilibria) and in solutions containing ligands (coupled dissolution and complex formation). The equilibrium tools from other chapters may be applied to describe and perform calculations on these systems.

Exercises

15.1 Precipitation and Dissolution

1. Complete the changes in concentrations for each of the following reactions:
(a)
$AgI(s) \longrightarrow Ag^{+}(aq) + I^{-}(aq)$

x ______

(b)
$CaCO_3(s) \longrightarrow Ca^{2+}(aq) + CO_3^{2-}(aq)$

______ x

(c)
$Mg(OH)_2(s) \longrightarrow Mg^{2+}(aq) + 2OH^{-}(aq)$

x ______

(d)
$Mg_3(PO_4)_2(s) \longrightarrow 3Mg^{2+}(aq) + 2PO_4^{3-}(aq)$

x ______

(e)
$Ca_5(PO_4)_3OH(s) \longrightarrow 5Ca^{2+}(aq) + 3PO_4^{3-}(aq) + OH^{-}(aq)$

______ ______ x

2. Complete the changes in concentrations for each of the following reactions:
(a)
$BaSO_4(s) \longrightarrow Ba^{2+}(aq) + SO_4^{2-}(aq)$

x ______

(b
$Ag_2SO_4(s) \longrightarrow 2Ag^{+}(aq) + SO_4^{2-}(aq)$

______ x

(c)
$Al(OH)_3(s) \longrightarrow Al^{3+}(aq) + 3OH^{-}(aq)$

x ______

(d)
$Pb(OH)Cl(s) \longrightarrow Pb^{2+}(aq) + OH^{-}(aq) + Cl^{-}(aq)$

______ x ______

(e)
$Ca_3(AsO_4)_2(s) \longrightarrow 3Ca^{2+}(aq) + 2AsO_4^{3-}(aq)$

$3x$ ______

3. How do the concentrations of Ag^{+} and CrO_4^{2-} in a saturated solution above 1.0 g of solid Ag_2CrO_4 change when 100 g of solid Ag_2CrO_4 is added to the system? Explain.
4. How do the concentrations of Pb^{2+} and S^{2-} change when K_2S is added to a saturated solution of PbS?

5. What additional information do we need to answer the following question: How is the equilibrium of solid silver bromide with a saturated solution of its ions affected when the temperature is raised?
6. Which of the following slightly soluble compounds has a solubility greater than that calculated from its solubility product because of hydrolysis of the anion present: $CoSO_3$, CuI, $PbCO_3$, $PbCl_2$, Tl_2S, $KClO_4$?
7. Which of the following slightly soluble compounds has a solubility greater than that calculated from its solubility product because of hydrolysis of the anion present: AgCl, $BaSO_4$, CaF_2, Hg_2I_2, $MnCO_3$, and ZnS?
8. Write the ionic equation for dissolution and the solubility product (K_{sp}) expression for each of the following slightly soluble ionic compounds:
(a) $PbCl_2$
(b) Ag_2S
(c) $Sr_3(PO_4)_2$
(d) $SrSO_4$
9. Write the ionic equation for the dissolution and the K_{sp} expression for each of the following slightly soluble ionic compounds:
(a) LaF_3
(b) $CaCO_3$
(c) Ag_2SO_4
(d) $Pb(OH)_2$
10. The Handbook of Chemistry and Physics (http://openstax.org/l/16Handbook) gives solubilities of the following compounds in grams per 100 mL of water. Because these compounds are only slightly soluble, assume that the volume does not change on dissolution and calculate the solubility product for each.
(a) $BaSiF_6$, 0.026 g/100 mL (contains $SiF_6{}^{2-}$ ions)
(b) $Ce(IO_3)_4$, 1.5×10^{-2} g/100 mL
(c) $Gd_2(SO_4)_3$, 3.98 g/100 mL
(d) $(NH_4)_2PtBr_6$, 0.59 g/100 mL (contains $PtBr_6{}^{2-}$ ions)
11. The Handbook of Chemistry and Physics (http://openstax.org/l/16Handbook) gives solubilities of the following compounds in grams per 100 mL of water. Because these compounds are only slightly soluble, assume that the volume does not change on dissolution and calculate the solubility product for each.
(a) $BaSeO_4$, 0.0118 g/100 mL
(b) $Ba(BrO_3)_2{\cdot}H_2O$, 0.30 g/100 mL
(c) $NH_4MgAsO_4{\cdot}6H_2O$, 0.038 g/100 mL
(d) $La_2(MoO_4)_3$, 0.00179 g/100 mL
12. Use solubility products and predict which of the following salts is the most soluble, in terms of moles per liter, in pure water: CaF_2, Hg_2Cl_2, PbI_2, or $Sn(OH)_2$.
13. Assuming that no equilibria other than dissolution are involved, calculate the molar solubility of each of the following from its solubility product:
(a) $KHC_4H_4O_6$
(b) PbI_2
(c) $Ag_4[Fe(CN)_6]$, a salt containing the $Fe(CN)_6{}^{4-}$ ion
(d) Hg_2I_2
14. Assuming that no equilibria other than dissolution are involved, calculate the molar solubility of each of the following from its solubility product:
(a) Ag_2SO_4
(b) $PbBr_2$
(c) AgI
(d) $CaC_2O_4{\cdot}H_2O$
15. Assuming that no equilibria other than dissolution are involved, calculate the concentration of all solute species in each of the following solutions of salts in contact with a solution containing a common ion. Show that changes in the initial concentrations of the common ions can be neglected.
(a) $AgCl(s)$ in 0.025 M NaCl
(b) $CaF_2(s)$ in 0.00133 M KF
(c) $Ag_2SO_4(s)$ in 0.500 L of a solution containing 19.50 g of K_2SO_4
(d) $Zn(OH)_2(s)$ in a solution buffered at a pH of 11.45

16. Assuming that no equilibria other than dissolution are involved, calculate the concentration of all solute species in each of the following solutions of salts in contact with a solution containing a common ion. Show that changes in the initial concentrations of the common ions can be neglected.
(a) $TlCl(s)$ in 1.250 *M* HCl
(b) $PbI_2(s)$ in 0.0355 *M* CaI_2
(c) $Ag_2CrO_4(s)$ in 0.225 L of a solution containing 0.856 g of K_2CrO_4
(d) $Cd(OH)_2(s)$ in a solution buffered at a pH of 10.995

17. Assuming that no equilibria other than dissolution are involved, calculate the concentration of all solute species in each of the following solutions of salts in contact with a solution containing a common ion. Show that it is not appropriate to neglect the changes in the initial concentrations of the common ions.
(a) $TlCl(s)$ in 0.025 *M* $TlNO_3$
(b) $BaF_2(s)$ in 0.0313 *M* KF
(c) MgC_2O_4 in 2.250 L of a solution containing 8.156 g of $Mg(NO_3)_2$
(d) $Ca(OH)_2(s)$ in an unbuffered solution initially with a pH of 12.700

18. Explain why the changes in concentrations of the common ions in Exercise 15.17 can be neglected.

19. Explain why the changes in concentrations of the common ions in Exercise 15.18 cannot be neglected.

20. Calculate the solubility of aluminum hydroxide, $Al(OH)_3$, in a solution buffered at pH 11.00.

21. Refer to Appendix J for solubility products for calcium salts. Determine which of the calcium salts listed is most soluble in moles per liter and which is most soluble in grams per liter.

22. Most barium compounds are very poisonous; however, barium sulfate is often administered internally as an aid in the X-ray examination of the lower intestinal tract (Figure 15.4). This use of $BaSO_4$ is possible because of its low solubility. Calculate the molar solubility of $BaSO_4$ and the mass of barium present in 1.00 L of water saturated with $BaSO_4$.

23. Public Health Service standards for drinking water set a maximum of 250 mg/L (2.60×10^{-3} *M*) of $SO_4{}^{2-}$ because of its cathartic action (it is a laxative). Does natural water that is saturated with $CaSO_4$ ("gyp" water) as a result or passing through soil containing gypsum, $CaSO_4{\cdot}2H_2O$, meet these standards? What is the concentration of $SO_4{}^{2-}$ in such water?

24. Perform the following calculations:
(a) Calculate $[Ag^+]$ in a saturated aqueous solution of AgBr.
(b) What will $[Ag^+]$ be when enough KBr has been added to make $[Br^-] = 0.050$ *M*?
(c) What will $[Br^-]$ be when enough $AgNO_3$ has been added to make $[Ag^+] = 0.020$ *M*?

25. The solubility product of $CaSO_4{\cdot}2H_2O$ is 2.4×10^{-5}. What mass of this salt will dissolve in 1.0 L of 0.010 *M* $SO4^{2-}$?

26. Assuming that no equilibria other than dissolution are involved, calculate the concentrations of ions in a saturated solution of each of the following (see Appendix J for solubility products).
(a) TlCl
(b) BaF_2
(c) Ag_2CrO_4
(d) $CaC_2O_4{\cdot}H_2O$
(e) the mineral anglesite, $PbSO_4$

27. Assuming that no equilibria other than dissolution are involved, calculate the concentrations of ions in a saturated solution of each of the following (see Appendix J for solubility products):
(a) AgI
(b) Ag_2SO_4
(c) $Mn(OH)_2$
(d) $Sr(OH)_2{\cdot}8H_2O$
(e) the mineral brucite, $Mg(OH)_2$

28. The following concentrations are found in mixtures of ions in equilibrium with slightly soluble solids. From the concentrations given, calculate K_{sp} for each of the slightly soluble solids indicated:
(a) AgBr: $[Ag^+] = 5.7 \times 10^{-7}$ *M*, $[Br^-] = 5.7 \times 10^{-7}$ *M*
(b) $CaCO_3$: $[Ca^{2+}] = 5.3 \times 10^{-3}$ *M*, $[CO_3{}^{2-}] = 9.0 \times 10^{-7}$ *M*
(c) PbF_2: $[Pb^{2+}] = 2.1 \times 10^{-3}$ *M*, $[F^-] = 4.2 \times 10^{-3}$ *M*
(d) Ag_2CrO_4: $[Ag^+] = 5.3 \times 10^{-5}$ *M*, 3.2×10^{-3} *M*
(e) InF_3: $[In^{3+}] = 2.3 \times 10^{-3}$ *M*, $[F^-] = 7.0 \times 10^{-3}$ *M*

29. The following concentrations are found in mixtures of ions in equilibrium with slightly soluble solids. From the concentrations given, calculate K_{sp} for each of the slightly soluble solids indicated:
(a) TlCl: $[Tl^+] = 1.21 \times 10^{-2}$ *M*, $[Cl^-] = 1.2 \times 10^{-2}$ *M*
(b) $Ce(IO_3)_4$: $[Ce^{4+}] = 1.8 \times 10^{-4}$ *M*, $[IO_3{}^-] = 2.6 \times 10^{-13}$ *M*
(c) $Gd_2(SO_4)_3$: $[Gd^{3+}] = 0.132$ *M*, $[SO_4{}^{2-}] = 0.198$ *M*
(d) Ag_2SO_4: $[Ag^+] = 2.40 \times 10^{-2}$ *M*, $[SO_4{}^{2-}] = 2.05 \times 10^{-2}$ *M*
(e) $BaSO_4$: $[Ba^{2+}] = 0.500$ *M*, $[SO_4{}^{2-}] = 4.6 \times 10^{-8}$ *M*

30. Which of the following compounds precipitates from a solution that has the concentrations indicated? (See Appendix J for K_{sp} values.)
(a) $KClO_4$: $[K^+] = 0.01$ *M*, $[ClO_4{}^-] = 0.01$ *M*
(b) K_2PtCl_6: $[K^+] = 0.01$ *M*, $[PtCl_6{}^{2-}] = 0.01$ *M*
(c) PbI_2: $[Pb^{2+}] = 0.003$ *M*, $[I^-] = 1.3 \times 10^{-3}$ *M*
(d) Ag_2S: $[Ag^+] = 1 \times 10^{-10}$ *M*, $[S^{2-}] = 1 \times 10^{-13}$ *M*

31. Which of the following compounds precipitates from a solution that has the concentrations indicated? (See Appendix J for K_{sp} values.)
(a) $CaCO_3$: $[Ca^{2+}] = 0.003$ *M*, $[CO_3{}^{2-}] = 0.003$ *M*
(b) $Co(OH)_2$: $[Co^{2+}] = 0.01$ *M*, $[OH^-] = 1 \times 10^{-7}$ *M*
(c) $CaHPO_4$: $[Ca^{2+}] = 0.01$ *M*, $[HPO_4{}^{2-}] = 2 \times 10^{-6}$ *M*
(d) $Pb_3(PO_4)_2$: $[Pb^{2+}] = 0.01$ *M*, $[PO_4{}^{3-}] = 1 \times 10^{-13}$ *M*

32. Calculate the concentration of Tl^+ when TlCl just begins to precipitate from a solution that is 0.0250 *M* in Cl^-.

33. Calculate the concentration of sulfate ion when $BaSO_4$ just begins to precipitate from a solution that is 0.0758 *M* in Ba^{2+}.

34. Calculate the concentration of Sr^{2+} when $SrCrO_4$ starts to precipitate from a solution that is 0.0025 *M* in $CrO_4{}^{2-}$.

35. Calculate the concentration of $PO_4{}^{3-}$ when Ag_3PO_4 starts to precipitate from a solution that is 0.0125 *M* in Ag^+.

36. Calculate the concentration of F^- required to begin precipitation of CaF_2 in a solution that is 0.010 *M* in Ca^{2+}.

37. Calculate the concentration of Ag^+ required to begin precipitation of Ag_2CO_3 in a solution that is 2.50×10^{-6} *M* in $CO_3{}^{2-}$.

38. What $[Ag^+]$ is required to reduce $[CO_3{}^{2-}]$ to 8.2×10^{-4} *M* by precipitation of Ag_2CO_3?

39. What $[F^-]$ is required to reduce $[Ca^{2+}]$ to 1.0×10^{-4} *M* by precipitation of CaF_2?

40. A volume of 0.800 L of a 2×10^{-4}-*M* $Ba(NO_3)_2$ solution is added to 0.200 L of 5×10^{-4} *M* Li_2SO_4. Does $BaSO_4$ precipitate? Explain your answer.

41. Perform these calculations for nickel(II) carbonate. (a) With what volume of water must a precipitate containing $NiCO_3$ be washed to dissolve 0.100 g of this compound? Assume that the wash water becomes saturated with $NiCO_3$ ($K_{sp} = 1.36 \times 10^{-7}$).
(b) If the $NiCO_3$ were a contaminant in a sample of $CoCO_3$ ($K_{sp} = 1.0 \times 10^{-12}$), what mass of $CoCO_3$ would have been lost? Keep in mind that both $NiCO_3$ and $CoCO_3$ dissolve in the same solution.

42. Iron concentrations greater than 5.4×10^{-6} *M* in water used for laundry purposes can cause staining. What $[OH^-]$ is required to reduce $[Fe^{2+}]$ to this level by precipitation of $Fe(OH)_2$?

43. A solution is 0.010 *M* in both Cu^{2+} and Cd^{2+}. What percentage of Cd^{2+} remains in the solution when 99.9% of the Cu^{2+} has been precipitated as CuS by adding sulfide?

44. A solution is 0.15 *M* in both Pb^{2+} and Ag^+. If Cl^- is added to this solution, what is $[Ag^+]$ when $PbCl_2$ begins to precipitate?

45. What reagent might be used to separate the ions in each of the following mixtures, which are 0.1 *M* with respect to each ion? In some cases it may be necessary to control the pH. (Hint: Consider the K_{sp} values given in Appendix J.)
(a) $Hg_2{}^{2+}$ and Cu^{2+}
(b) $SO_4{}^{2-}$ and Cl^-
(c) Hg^{2+} and Co^{2+}
(d) Zn^{2+} and Sr^{2+}
(e) Ba^{2+} and Mg^{2+}
(f) $CO_3{}^{2-}$ and OH^-
46. A solution contains 1.0×10^{-5} mol of KBr and 0.10 mol of KCl per liter. $AgNO_3$ is gradually added to this solution. Which forms first, solid AgBr or solid AgCl?
47. A solution contains 1.0×10^{-2} mol of KI and 0.10 mol of KCl per liter. $AgNO_3$ is gradually added to this solution. Which forms first, solid AgI or solid AgCl?
48. The calcium ions in human blood serum are necessary for coagulation (Figure 15.5). Potassium oxalate, $K_2C_2O_4$, is used as an anticoagulant when a blood sample is drawn for laboratory tests because it removes the calcium as a precipitate of $CaC_2O_4{\cdot}H_2O$. It is necessary to remove all but 1.0% of the Ca^{2+} in serum in order to prevent coagulation. If normal blood serum with a buffered pH of 7.40 contains 9.5 mg of Ca^{2+} per 100 mL of serum, what mass of $K_2C_2O_4$ is required to prevent the coagulation of a 10 mL blood sample that is 55% serum by volume? (All volumes are accurate to two significant figures. Note that the volume of serum in a 10-mL blood sample is 5.5 mL. Assume that the K_{sp} value for CaC_2O_4 in serum is the same as in water.)
49. About 50% of urinary calculi (kidney stones) consist of calcium phosphate, $Ca_3(PO_4)_2$. The normal mid range calcium content excreted in the urine is 0.10 g of Ca^{2+} per day. The normal mid range amount of urine passed may be taken as 1.4 L per day. What is the maximum concentration of phosphate ion that urine can contain before a calculus begins to form?
50. The pH of normal urine is 6.30, and the total phosphate concentration ($[PO_4{}^{3-}] + [HPO_4{}^{2-}] + [H_2PO_4{}^{-}] + [H_3PO_4]$) is 0.020 *M*. What is the minimum concentration of Ca^{2+} necessary to induce kidney stone formation? (See Exercise 15.49 for additional information.)
51. Magnesium metal (a component of alloys used in aircraft and a reducing agent used in the production of uranium, titanium, and other active metals) is isolated from sea water by the following sequence of reactions:
$$Mg^{2+}(aq) + Ca(OH)_2(aq) \longrightarrow Mg(OH)_2(s) + Ca^{2+}(aq)$$
$$Mg(OH)_2(s) + 2HCl(aq) \longrightarrow MgCl_2(s) + 2H_2O(l)$$
$$MgCl_2(l) \xrightarrow{\text{electrolysis}} Mg(s) + Cl_2(g)$$
Sea water has a density of 1.026 g/cm^3 and contains 1272 parts per million of magnesium as $Mg^{2+}(aq)$ by mass. What mass, in kilograms, of $Ca(OH)_2$ is required to precipitate 99.9% of the magnesium in 1.00×10^3 L of sea water?
52. Hydrogen sulfide is bubbled into a solution that is 0.10 *M* in both Pb^{2+} and Fe^{2+} and 0.30 *M* in HCl. After the solution has come to equilibrium it is saturated with H_2S ($[H_2S] = 0.10$ *M*). What concentrations of Pb^{2+} and Fe^{2+} remain in the solution? For a saturated solution of H_2S we can use the equilibrium:
$$H_2S(aq) + 2H_2O(l) \rightleftharpoons 2H_3O^+(aq) + S^{2-}(aq) \qquad K = 1.0 \times 10^{-26}$$
(Hint: The $[H_3O^+]$ changes as metal sulfides precipitate.)
53. Perform the following calculations involving concentrations of iodate ions:
(a) The iodate ion concentration of a saturated solution of $La(IO_3)_3$ was found to be 3.1×10^{-3} mol/L. Find the K_{sp}.
(b) Find the concentration of iodate ions in a saturated solution of $Cu(IO_3)_2$ ($K_{sp} = 7.4 \times 10^{-8}$).
54. Calculate the molar solubility of AgBr in 0.035 *M* NaBr ($K_{sp} = 5 \times 10^{-13}$).
55. How many grams of $Pb(OH)_2$ will dissolve in 500 mL of a 0.050-*M* $PbCl_2$ solution ($K_{sp} = 1.2 \times 10^{-15}$)?
56. Use the simulation (http://openstax.org/l/16solublesalts) from the earlier Link to Learning to complete the following exercise. Using 0.01 g CaF_2, give the K_{sp} values found in a 0.2-*M* solution of each of the salts. Discuss why the values change as you change soluble salts.

57. How many grams of Milk of Magnesia, $Mg(OH)_2$ (*s*) (58.3 g/mol), would be soluble in 200 mL of water. K_{sp} = 7.1×10^{-12}. Include the ionic reaction and the expression for K_{sp} in your answer. ($K_w = 1 \times 10^{-14} = [H_3O^+][OH^-]$)
58. Two hypothetical salts, LM_2 and LQ, have the same molar solubility in H_2O. If K_{sp} for LM_2 is 3.20×10^{-5}, what is the K_{sp} value for LQ?
59. The carbonate ion concentration is gradually increased in a solution containing equal concentrations of the divalent cations of magnesium, calcium, strontium, barium, and manganese. Which of the following carbonates will precipitate first? Which will precipitate last? Explain.
    (a) $MgCO_3 \bullet 3H_2O$ $K_{sp} = 1 \times 10^{-5}$
    (b) $CaCO_3$ $K_{sp} = 8.7 \times 10^{-9}$
    (c) $SrCO_3$ $K_{sp} = 7 \times 10^{-10}$
    (d) $BaCO_3$ $K_{sp} = 1.6 \times 10^{-9}$
    (e) $MnCO_3$ $K_{sp} = 8.8 \times 10^{-11}$
60. How many grams of $Zn(CN)_2(s)$ (117.44 g/mol) would be soluble in 100 mL of H_2O? Include the balanced reaction and the expression for K_{sp} in your answer. The K_{sp} value for $Zn(CN)_2(s)$ is 3.0×10^{-16}.

15.2 Lewis Acids and Bases

61. Even though $Ca(OH)_2$ is an inexpensive base, its limited solubility restricts its use. What is the pH of a saturated solution of $Ca(OH)_2$?
62. Under what circumstances, if any, does a sample of solid AgCl completely dissolve in pure water?
63. Explain why the addition of NH_3 or HNO_3 to a saturated solution of Ag_2CO_3 in contact with solid Ag_2CO_3 increases the solubility of the solid.
64. Calculate the cadmium ion concentration, $[Cd^{2+}]$, in a solution prepared by mixing 0.100 L of 0.0100 *M* $Cd(NO_3)_2$ with 0.150 L of 0.100 $NH_3(aq)$.
65. Explain why addition of NH_3 or HNO_3 to a saturated solution of $Cu(OH)_2$ in contact with solid $Cu(OH)_2$ increases the solubility of the solid.
66. Sometimes equilibria for complex ions are described in terms of dissociation constants, K_d. For the complex ion $AlF_6{}^{3-}$ the dissociation reaction is:
    $$AlF_6{}^{3-} \rightleftharpoons Al^{3+} + 6F^- \text{ and } K_d = \frac{[Al^{3+}][F^-]^6}{[AlF_6{}^{3-}]} = 2 \times 10^{-24}$$
    Calculate the value of the formation constant, K_f, for $AlF_6{}^{3-}$.
67. Using the value of the formation constant for the complex ion $Co(NH_3)_6{}^{2+}$, calculate the dissociation constant.
68. Using the dissociation constant, $K_d = 7.8 \times 10^{-18}$, calculate the equilibrium concentrations of Cd^{2+} and CN^- in a 0.250-*M* solution of $Cd(CN)_4{}^{2-}$.
69. Using the dissociation constant, $K_d = 3.4 \times 10^{-15}$, calculate the equilibrium concentrations of Zn^{2+} and OH^- in a 0.0465-*M* solution of $Zn(OH)_4{}^{2-}$.
70. Using the dissociation constant, $K_d = 2.2 \times 10^{-34}$, calculate the equilibrium concentrations of Co^{3+} and NH_3 in a 0.500-*M* solution of $Co(NH_3)_6{}^{3+}$.
71. Using the dissociation constant, $K_d = 1 \times 10^{-44}$, calculate the equilibrium concentrations of Fe^{3+} and CN^- in a 0.333 M solution of $Fe(CN)_6{}^{3-}$.
72. Calculate the mass of potassium cyanide ion that must be added to 100 mL of solution to dissolve 2.0×10^{-2} mol of silver cyanide, AgCN.
73. Calculate the minimum concentration of ammonia needed in 1.0 L of solution to dissolve 3.0×10^{-3} mol of silver bromide.
74. A roll of 35-mm black and white photographic film contains about 0.27 g of unexposed AgBr before developing. What mass of $Na_2S_2O_3{\cdot}5H_2O$ (sodium thiosulfate pentahydrate or hypo) in 1.0 L of developer is required to dissolve the AgBr as $Ag(S_2O_3)_2{}^{3-}$ ($K_f = 4.7 \times 10^{13}$)?

75. We have seen an introductory definition of an acid: An acid is a compound that reacts with water and increases the amount of hydronium ion present. In the chapter on acids and bases, we saw two more definitions of acids: a compound that donates a proton (a hydrogen ion, H^+) to another compound is called a Brønsted-Lowry acid, and a Lewis acid is any species that can accept a pair of electrons. Explain why the introductory definition is a macroscopic definition, while the Brønsted-Lowry definition and the Lewis definition are microscopic definitions.
76. Write the Lewis structures of the reactants and product of each of the following equations, and identify the Lewis acid and the Lewis base in each:
(a) $CO_2 + OH^- \longrightarrow HCO_3^-$
(b) $B(OH)_3 + OH^- \longrightarrow B(OH)_4^-$
(c) $I^- + I_2 \longrightarrow I_3^-$
(d) $AlCl_3 + Cl^- \longrightarrow AlCl_4^-$ (use Al-Cl single bonds)
(e) $O^{2-} + SO_3 \longrightarrow SO_4^{2-}$
77. Write the Lewis structures of the reactants and product of each of the following equations, and identify the Lewis acid and the Lewis base in each:
(a) $CS_2 + SH^- \longrightarrow HCS_3^-$
(b) $BF_3 + F^- \longrightarrow BF_4^-$
(c) $I^- + SnI_2 \longrightarrow SnI_3^-$
(d) $Al(OH)_3 + OH^- \longrightarrow Al(OH)_4^-$
(e) $F^- + SO_3 \longrightarrow SFO_3^-$
78. Using Lewis structures, write balanced equations for the following reactions:
(a) $HCl(g) + PH_3(g) \longrightarrow$
(b) $H_3O^+ + CH_3^- \longrightarrow$
(c) $CaO + SO_3 \longrightarrow$
(d) $NH_4^+ + C_2H_5O^- \longrightarrow$
79. Calculate $[HgCl_4^{2-}]$ in a solution prepared by adding 0.0200 mol of NaCl to 0.250 L of a 0.100-M $HgCl_2$ solution.
80. In a titration of cyanide ion, 28.72 mL of 0.0100 M $AgNO_3$ is added before precipitation begins. [The reaction of Ag^+ with CN^- goes to completion, producing the $Ag(CN)_2^-$ complex.] Precipitation of solid AgCN takes place when excess Ag^+ is added to the solution, above the amount needed to complete the formation of $Ag(CN)_2^-$. How many grams of NaCN were in the original sample?
81. What are the concentrations of Ag^+, CN^-, and $Ag(CN)_2^-$ in a saturated solution of AgCN?
82. In dilute aqueous solution HF acts as a weak acid. However, pure liquid HF (boiling point = 19.5 °C) is a strong acid. In liquid HF, HNO_3 acts like a base and accepts protons. The acidity of liquid HF can be increased by adding one of several inorganic fluorides that are Lewis acids and accept F^- ion (for example, BF_3 or SbF_5). Write balanced chemical equations for the reaction of pure HNO_3 with pure HF and of pure HF with BF_3.
83. The simplest amino acid is glycine, $H_2NCH_2CO_2H$. The common feature of amino acids is that they contain the functional groups: an amine group, $-NH_2$, and a carboxylic acid group, $-CO_2H$. An amino acid can function as either an acid or a base. For glycine, the acid strength of the carboxyl group is about the same as that of acetic acid, CH_3CO_2H, and the base strength of the amino group is slightly greater than that of ammonia, NH_3.
(a) Write the Lewis structures of the ions that form when glycine is dissolved in 1 M HCl and in 1 M KOH.
(b) Write the Lewis structure of glycine when this amino acid is dissolved in water. (Hint: Consider the relative base strengths of the $-NH_2$ and $-CO_2^-$ groups.)
84. Boric acid, H_3BO_3, is not a Brønsted-Lowry acid but a Lewis acid.
(a) Write an equation for its reaction with water.
(b) Predict the shape of the anion thus formed.
(c) What is the hybridization on the boron consistent with the shape you have predicted?

15.3 Coupled Equilibria

85. A saturated solution of a slightly soluble electrolyte in contact with some of the solid electrolyte is said to be a system in equilibrium. Explain. Why is such a system called a heterogeneous equilibrium?

86. Calculate the equilibrium concentration of Ni^{2+} in a 1.0-M solution $[Ni(NH_3)_6](NO_3)_2$.

87. Calculate the equilibrium concentration of Zn^{2+} in a 0.30-M solution of $Zn(CN)_4{}^{2-}$.

88. Calculate the equilibrium concentration of Cu^{2+} in a solution initially with 0.050 M Cu^{2+} and 1.00 M NH_3.

89. Calculate the equilibrium concentration of Zn^{2+} in a solution initially with 0.150 M Zn^{2+} and 2.50 M CN^-.

90. Calculate the Fe^{3+} equilibrium concentration when 0.0888 mole of $K_3[Fe(CN)_6]$ is added to a solution with 0.0.00010 M CN^-.

91. Calculate the Co^{2+} equilibrium concentration when 0.010 mole of $[Co(NH_3)_6](NO_3)_2$ is added to a solution with 0.25 M NH_3. Assume the volume is 1.00 L.

92. Calculate the molar solubility of $Sn(OH)_2$ in a buffer solution containing equal concentrations of NH_3 and $NH_4{}^+$.

93. Calculate the molar solubility of $Al(OH)_3$ in a buffer solution with 0.100 M NH_3 and 0.400 M $NH_4{}^+$.

94. What is the molar solubility of CaF_2 in a 0.100-M solution of HF? K_a for HF = 6.4×10^{-4}.

95. What is the molar solubility of $BaSO_4$ in a 0.250-M solution of $NaHSO_4$? K_a for $HSO_4{}^-$ = 1.2×10^{-2}.

96. What is the molar solubility of $Tl(OH)_3$ in a 0.10-M solution of NH_3?

97. What is the molar solubility of $Pb(OH)_2$ in a 0.138-M solution of CH_3NH_2?

98. A solution of 0.075 M $CoBr_2$ is saturated with H_2S ($[H_2S]$ = 0.10 M). What is the minimum pH at which CoS begins to precipitate?

$$CoS(s) \rightleftharpoons Co^{2+}(aq) + S^{2-}(aq) \qquad K_{sp} = 2.3 \times 10^{27}$$

$$H_2S(aq) + 2H_2O(l) \rightleftharpoons 2H_3O^+(aq) + S^{2-}(aq) \qquad K = 8.9 \times 10^{-27}$$

99. A 0.125-M solution of $Mn(NO_3)_2$ is saturated with H_2S ($[H_2S]$ = 0.10 M). At what pH does MnS begin to precipitate?

$$MnS(s) \rightleftharpoons Mn^{2+}(aq) + S^{2-}(aq) \qquad K_{sp} = 2.3 \times 10^{-13}$$

$$H_2S(aq) + 2H_2O(l) \rightleftharpoons 2H_3O^+(aq) + S^{2-}(aq) \qquad K = 1.0 \times 10^{-26}$$

100. Both AgCl and AgI dissolve in NH_3.

(a) What mass of AgI dissolves in 1.0 L of 1.0 M NH_3?

(b) What mass of AgCl dissolves in 1.0 L of 1.0 M NH_3?

101. The following question is taken from a Chemistry Advanced Placement Examination and is used with the permission of the Educational Testing Service.

Solve the following problem:

$$MgF_2(s) \rightleftharpoons Mg^{2+}(aq) + 2F^-(aq)$$

In a saturated solution of MgF_2 at 18 °C, the concentration of Mg^{2+} is 1.21×10^{-3} M. The equilibrium is represented by the preceding equation.

(a) Write the expression for the solubility-product constant, K_{sp}, and calculate its value at 18 °C.

(b) Calculate the equilibrium concentration of Mg^{2+} in 1.000 L of saturated MgF_2 solution at 18 °C to which 0.100 mol of solid KF has been added. The KF dissolves completely. Assume the volume change is negligible.

(c) Predict whether a precipitate of MgF_2 will form when 100.0 mL of a 3.00×10^{-3}-M solution of $Mg(NO_3)_2$ is mixed with 200.0 mL of a 2.00×10^{-3}-M solution of NaF at 18 °C. Show the calculations to support your prediction.

(d) At 27 °C the concentration of Mg^{2+} in a saturated solution of MgF_2 is 1.17×10^{-3} M. Is the dissolving of MgF_2 in water an endothermic or an exothermic process? Give an explanation to support your conclusion.

102. Which of the following compounds, when dissolved in a 0.01-M solution of $HClO_4$, has a solubility greater than in pure water: CuCl, $CaCO_3$, MnS, $PbBr_2$, CaF_2? Explain your answer.

103. Which of the following compounds, when dissolved in a 0.01-M solution of $HClO_4$, has a solubility greater than in pure water: AgBr, BaF_2, $Ca_3(PO_4)_2$, ZnS, PbI_2? Explain your answer.

104. What is the effect on the amount of solid $Mg(OH)_2$ that dissolves and the concentrations of Mg^{2+} and OH^- when each of the following are added to a mixture of solid $Mg(OH)_2$ and water at equilibrium?
(a) $MgCl_2$
(b) KOH
(c) $HClO_4$
(d) $NaNO_3$
(e) $Mg(OH)_2$

105. What is the effect on the amount of $CaHPO_4$ that dissolves and the concentrations of Ca^{2+} and $HPO_4{}^{2-}$ when each of the following are added to a mixture of solid $CaHPO_4$ and water at equilibrium?
(a) $CaCl_2$
(b) HCl
(c) $KClO_4$
(d) NaOH
(e) $CaHPO_4$

106. Identify all chemical species present in an aqueous solution of $Ca_3(PO_4)_2$ and list these species in decreasing order of their concentrations. (Hint: Remember that the $PO_4{}^{3-}$ ion is a weak base.)

CHAPTER 16
Thermodynamics

Figure 16.1 Geysers are a dramatic display of thermodynamic principles in nature. Water deep within the underground channels of the geyser is under high pressure and heated to high temperature by magma. When a pocket of water near the surface reaches boiling point and is expelled, the resulting drop in pressure causes larger volumes of water to flash boil, forcefully ejecting steam and water in an impressive eruption. (credit: modification of work by Yellowstone National Park)

CHAPTER OUTLINE

16.1 Spontaneity
16.2 Entropy
16.3 The Second and Third Laws of Thermodynamics
16.4 Free Energy

INTRODUCTION Among the many capabilities of chemistry is its ability to predict if a process will occur under specified conditions. Thermodynamics, the study of relationships between the energy and work associated with chemical and physical processes, provides this predictive ability. Previous chapters in this text have described various applications of thermochemistry, an important aspect of thermodynamics concerned with the heat flow accompanying chemical reactions and phase transitions. This chapter will introduce additional thermodynamic concepts, including those that enable the prediction of any chemical or physical changes under a given set of conditions.

16.1 Spontaneity

LEARNING OBJECTIVES

By the end of this section, you will be able to:

- Distinguish between spontaneous and nonspontaneous processes
- Describe the dispersal of matter and energy that accompanies certain spontaneous processes

Processes have a natural tendency to occur in one direction under a given set of conditions. Water will naturally flow downhill, but uphill flow requires outside intervention such as the use of a pump. Iron exposed to the earth's atmosphere will corrode, but rust is not converted to iron without intentional chemical treatment. A **spontaneous process** is one that occurs naturally under certain conditions. A **nonspontaneous process**, on the other hand, will not take place unless it is "driven" by the continual input of energy from an external source. A process that is spontaneous in one direction under a particular set of conditions is nonspontaneous in the reverse direction. At room temperature and typical atmospheric pressure, for example, ice will spontaneously melt, but water will not spontaneously freeze.

The spontaneity of a process is *not* correlated to the speed of the process. A spontaneous change may be so rapid that it is essentially instantaneous or so slow that it cannot be observed over any practical period of time. To illustrate this concept, consider the decay of radioactive isotopes, a topic more thoroughly treated in the chapter on nuclear chemistry. Radioactive decay is by definition a spontaneous process in which the nuclei of unstable isotopes emit radiation as they are converted to more stable nuclei. All the decay processes occur spontaneously, but the rates at which different isotopes decay vary widely. Technetium-99m is a popular radioisotope for medical imaging studies that undergoes relatively rapid decay and exhibits a half-life of about six hours. Uranium-238 is the most abundant isotope of uranium, and its decay occurs much more slowly, exhibiting a half-life of more than four billion years (Figure 16.2).

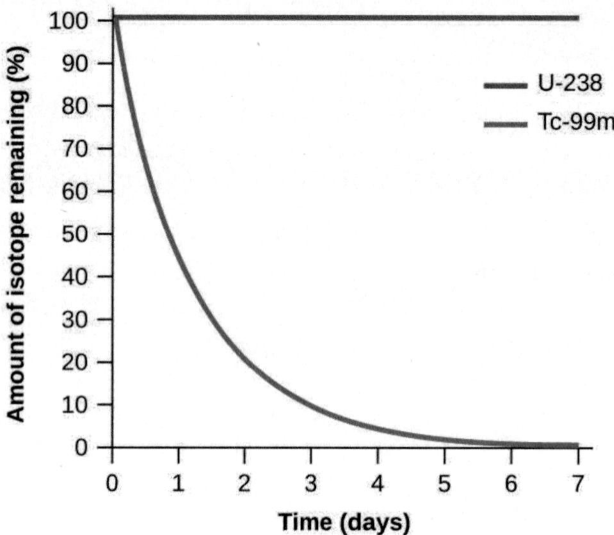

FIGURE 16.2 Both U-238 and Tc-99m undergo spontaneous radioactive decay, but at drastically different rates. Over the course of one week, essentially all of a Tc-99m sample and none of a U-238 sample will have decayed.

As another example, consider the conversion of diamond into graphite (Figure 16.3).

$$\text{C}(s,\ \text{diamond}) \longrightarrow \text{C}(s,\ \text{graphite})$$

The phase diagram for carbon indicates that graphite is the stable form of this element under ambient atmospheric pressure, while diamond is the stable allotrope at very high pressures, such as those present during its geologic formation. Thermodynamic calculations of the sort described in the last section of this chapter indicate that the conversion of diamond to graphite at ambient pressure occurs spontaneously, yet diamonds are observed to exist, and persist, under these conditions. Though the process is spontaneous under typical ambient conditions, its rate is extremely slow; so, for all practical purposes diamonds are indeed "forever." Situations such as these emphasize the important distinction between the thermodynamic and the kinetic aspects of a process. In this particular case, diamonds are said to be *thermodynamically unstable* but *kinetically stable* under ambient conditions.

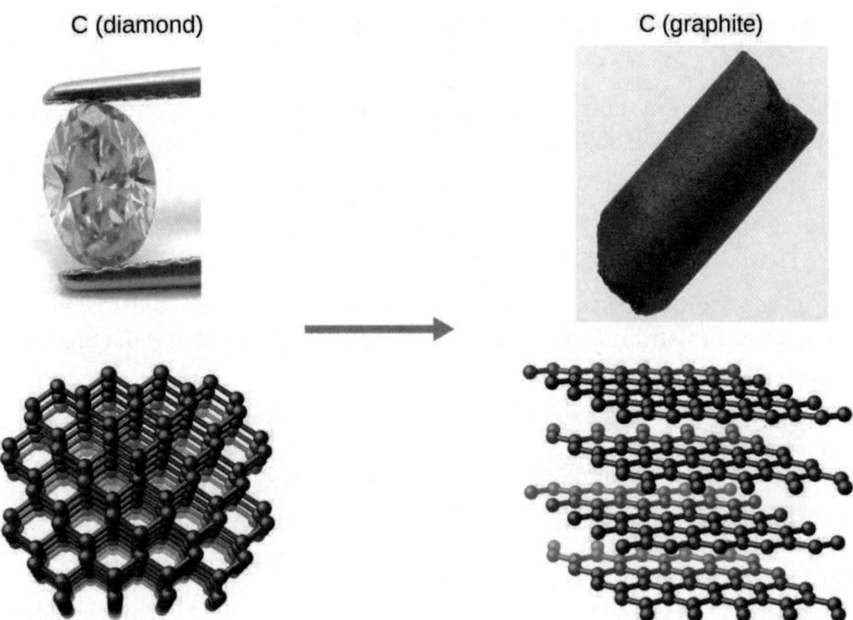

FIGURE 16.3 The conversion of carbon from the diamond allotrope to the graphite allotrope is spontaneous at ambient pressure, but its rate is immeasurably slow at low to moderate temperatures. This process is known as *graphitization*, and its rate can be increased to easily measurable values at temperatures in the 1000–2000 K range. (credit "diamond" photo: modification of work by "Fancy Diamonds"/Flickr; credit "graphite" photo: modification of work by images-of-elements.com/carbon.php)

Dispersal of Matter and Energy

Extending the discussion of thermodynamic concepts toward the objective of predicting spontaneity, consider now an isolated system consisting of two flasks connected with a closed valve. Initially there is an ideal gas in one flask and the other flask is empty ($P = 0$). (Figure 16.4). When the valve is opened, the gas spontaneously expands to fill both flasks equally. Recalling the definition of pressure-volume work from the chapter on thermochemistry, note that no work has been done because the pressure in a vacuum is zero.

$$w = -P\Delta V = 0 \qquad\qquad (P = 0 \text{ in a vacuum})$$

Note as well that since the system is isolated, no heat has been exchanged with the surroundings ($q = 0$). The *first law of thermodynamics* confirms that there has been no change in the system's internal energy as a result of this process.

$$\Delta U = q + w = 0 + 0 = 0$$

The spontaneity of this process is therefore not a consequence of any change in energy that accompanies the process. Instead, the driving force appears to be related to the *greater, more uniform dispersal of matter* that results when the gas is allowed to expand. Initially, the system was comprised of one flask containing matter and another flask containing nothing. After the spontaneous expansion took place, the matter was distributed both more widely (occupying twice its original volume) and more uniformly (present in equal amounts in each flask).

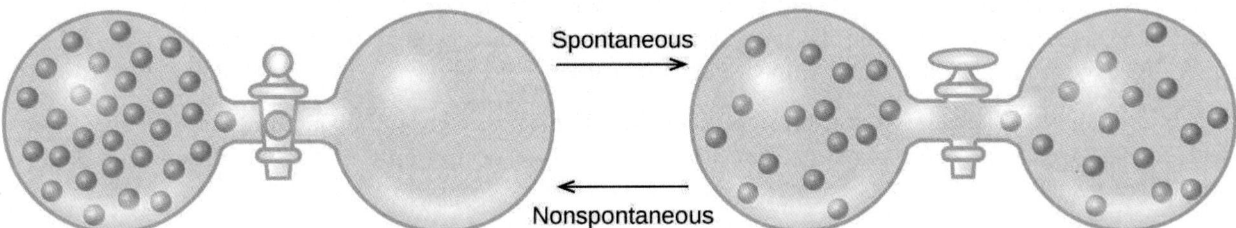

FIGURE 16.4 An isolated system consists of an ideal gas in one flask that is connected by a closed valve to a

second flask containing a vacuum. Once the valve is opened, the gas spontaneously becomes evenly distributed between the flasks.

Now consider two objects at different temperatures: object X at temperature T_X and object Y at temperature T_Y, with $T_X > T_Y$ (Figure 16.5). When these objects come into contact, heat spontaneously flows from the hotter object (X) to the colder one (Y). This corresponds to a loss of thermal energy by X and a gain of thermal energy by Y.

$$q_X < 0 \qquad \text{and} \qquad q_Y = -q_X > 0$$

From the perspective of this two-object system, there was no net gain or loss of thermal energy, rather the available thermal energy was redistributed among the two objects. This spontaneous process resulted in a *more uniform dispersal of energy.*

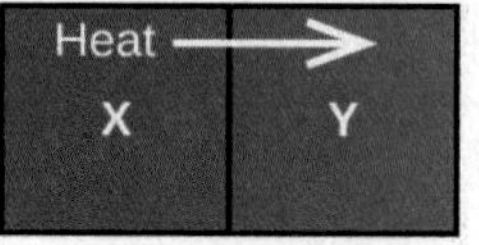

X and Y in contact

FIGURE 16.5 When two objects at different temperatures come in contact, heat spontaneously flows from the hotter to the colder object.

As illustrated by the two processes described, an important factor in determining the spontaneity of a process is the extent to which it changes the dispersal or distribution of matter and/or energy. In each case, a spontaneous process took place that resulted in a more uniform distribution of matter or energy.

EXAMPLE 16.1

Redistribution of Matter during a Spontaneous Process

Describe how matter is redistributed when the following spontaneous processes take place:

(a) A solid sublimes.

(b) A gas condenses.

(c) A drop of food coloring added to a glass of water forms a solution with uniform color.

Solution

FIGURE 16.6 (credit a: modification of work by Jenny Downing; credit b: modification of work by "Fuzzy Gerdes"/Flickr; credit c: modification of work by Paul A. Flowers)

(a) Sublimation is the conversion of a solid (relatively high density) to a gas (much lesser density). This process yields a much greater dispersal of matter, since the molecules will occupy a much greater volume after the solid-to-gas transition.

(b) Condensation is the conversion of a gas (relatively low density) to a liquid (much greater density). This process yields a much lesser dispersal of matter, since the molecules will occupy a much lesser volume after the gas-to-liquid transition.

(c) The process in question is diffusion. This process yields a more uniform dispersal of matter, since the initial state of the system involves two regions of different dye concentrations (high in the drop of dye, zero in the water), and the final state of the system contains a single dye concentration throughout.

Check Your Learning

Describe how energy is redistributed when a spoon at room temperature is placed in a cup of hot coffee.

Answer:

Heat will spontaneously flow from the hotter object (coffee) to the colder object (spoon), resulting in a more uniform distribution of thermal energy as the spoon warms and the coffee cools.

16.2 Entropy

LEARNING OBJECTIVES

By the end of this section, you will be able to:

- Define entropy
- Explain the relationship between entropy and the number of microstates
- Predict the sign of the entropy change for chemical and physical processes

In 1824, at the age of 28, Nicolas Léonard Sadi Carnot (Figure 16.7) published the results of an extensive study regarding the efficiency of steam heat engines. A later review of Carnot's findings by Rudolf Clausius introduced a new thermodynamic property that relates the spontaneous heat flow accompanying a process to the temperature at which the process takes place. This new property was expressed as the ratio of the *reversible* heat (q_{rev}) and the kelvin temperature (T). In thermodynamics, a **reversible process** is one that takes place at such a slow rate that it is always at equilibrium and its direction can be changed (it can be "reversed") by an infinitesimally small change in some condition. Note that the idea of a reversible process is a formalism required to support the development of various thermodynamic concepts; no real processes are truly reversible, rather they are classified as *irreversible*.

(a)

(b)

FIGURE 16.7 (a) Nicholas Léonard Sadi Carnot's research into steam-powered machinery and (b) Rudolf Clausius's later study of those findings led to groundbreaking discoveries about spontaneous heat flow processes.

Similar to other thermodynamic properties, this new quantity is a state function, so its change depends only upon the initial and final states of a system. In 1865, Clausius named this property **entropy (S)** and defined its change for any process as the following:

$$\Delta S = \frac{q_{\text{rev}}}{T}$$

The entropy change for a real, irreversible process is then equal to that for the theoretical reversible process that involves the same initial and final states.

Entropy and Microstates

Following the work of Carnot and Clausius, Ludwig Boltzmann developed a molecular-scale statistical model that related the entropy of a system to the *number of microstates* (W) possible for the system. A **microstate** is a specific configuration of all the locations and energies of the atoms or molecules that make up a system. The relation between a system's entropy and the number of possible microstates is

$$S = k \ln W$$

where k is the Boltzmann constant, 1.38×10^{-23} J/K.

As for other state functions, the change in entropy for a process is the difference between its final (S_f) and initial (S_i) values:

$$\Delta S = S_f - S_i = k \ln W_f - k \ln W_i = k \ln \frac{W_f}{W_i}$$

For processes involving an increase in the number of microstates, $W_f > W_i$, the entropy of the system increases and $\Delta S > 0$. Conversely, processes that reduce the number of microstates, $W_f < W_i$, yield a decrease in system entropy, $\Delta S < 0$. This molecular-scale interpretation of entropy provides a link to the probability that a process will occur as illustrated in the next paragraphs.

Consider the general case of a system comprised of N particles distributed among n boxes. The number of microstates possible for such a system is n^N. For example, distributing four particles among two boxes will result in $2^4 = 16$ different microstates as illustrated in Figure 16.8. Microstates with equivalent particle arrangements (not considering individual particle identities) are grouped together and are called *distributions*. The probability that a system will exist with its components in a given distribution is proportional to the number of microstates within the distribution. Since entropy increases logarithmically with the number of microstates, *the most probable distribution is therefore the one of greatest entropy.*

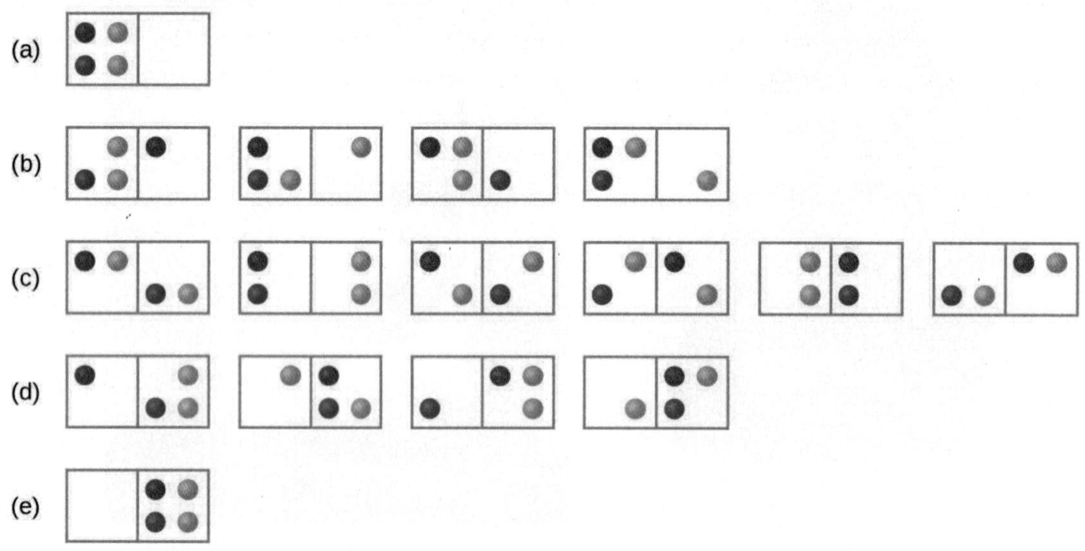

FIGURE 16.8 The sixteen microstates associated with placing four particles in two boxes are shown. The microstates are collected into five distributions—(a), (b), (c), (d), and (e)—based on the numbers of particles in each box.

For this system, the most probable configuration is one of the six microstates associated with distribution (c) where the particles are evenly distributed between the boxes, that is, a configuration of two particles in each box. The probability of finding the system in this configuration is $\frac{6}{16}$ or $\frac{3}{8}$. The least probable configuration of the system is one in which all four particles are in one box, corresponding to distributions (a) and (e), each with a probability of $\frac{1}{16}$. The probability of finding all particles in only one box (either the left box or right box) is then $\left(\frac{1}{16} + \frac{1}{16}\right) = \frac{2}{16}$ or $\frac{1}{8}$.

As you add more particles to the system, the number of possible microstates increases exponentially (2^N). A macroscopic (laboratory-sized) system would typically consist of moles of particles ($N \sim 10^{23}$), and the corresponding number of microstates would be staggeringly huge. Regardless of the number of particles in the system, however, the distributions in which roughly equal numbers of particles are found in each box are always the most probable configurations.

This matter dispersal model of entropy is often described qualitatively in terms of the *disorder* of the system. By this description, microstates in which all the particles are in a single box are the most ordered, thus possessing the least entropy. Microstates in which the particles are more evenly distributed among the boxes are more disordered, possessing greater entropy.

The previous description of an ideal gas expanding into a vacuum (Figure 16.4) is a macroscopic example of this particle-in-a-box model. For this system, the most probable distribution is confirmed to be the one in which the matter is most uniformly dispersed or distributed between the two flasks. Initially, the gas molecules are confined to just one of the two flasks. Opening the valve between the flasks increases the volume available to the gas molecules and, correspondingly, the number of microstates possible for the system. Since $W_f > W_i$, the expansion process involves an increase in entropy ($\Delta S > 0$) and is spontaneous.

A similar approach may be used to describe the spontaneous flow of heat. Consider a system consisting of two objects, each containing two particles, and two units of thermal energy (represented as "*") in Figure 16.9. The hot object is comprised of particles **A** and **B** and initially contains both energy units. The cold object is comprised of particles **C** and **D**, which initially has no energy units. Distribution (a) shows the three microstates possible for the initial state of the system, with both units of energy contained within the hot object. If one of the two energy units is transferred, the result is distribution (b) consisting of four microstates. If both energy units are transferred, the result is distribution (c) consisting of three microstates. Thus, we may describe this system by a total of ten microstates. The probability that the heat does not flow when the two objects are brought into contact, that is, that the system remains in distribution (a), is $\frac{3}{10}$. More likely is the flow of heat to yield one of the other two distribution, the combined probability being $\frac{7}{10}$. The most likely result is the flow of heat to yield the uniform dispersal of energy represented by distribution (b), the probability of this configuration being $\frac{4}{10}$. This supports the common observation that placing hot and cold objects in contact results in spontaneous heat flow that ultimately equalizes the objects' temperatures. And, again, this spontaneous process is also characterized by an increase in system entropy.

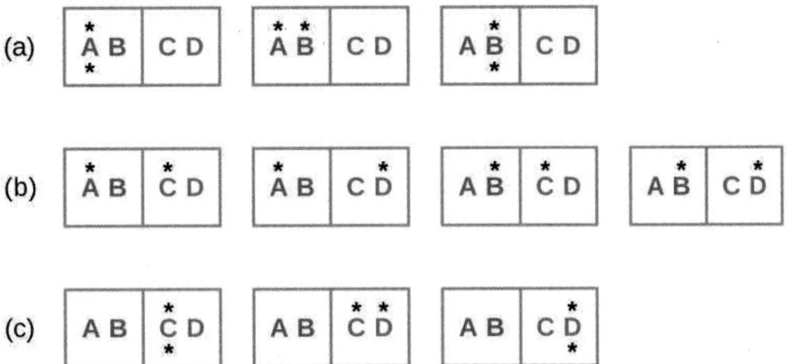

FIGURE 16.9 This shows a microstate model describing the flow of heat from a hot object to a cold object. (a) Before the heat flow occurs, the object comprised of particles **A** and **B** contains both units of energy and as represented by a distribution of three microstates. (b) If the heat flow results in an even dispersal of energy (one energy unit transferred), a distribution of four microstates results. (c) If both energy units are transferred, the resulting distribution has three microstates.

EXAMPLE 16.2

Determination of ΔS

Calculate the change in entropy for the process depicted below.

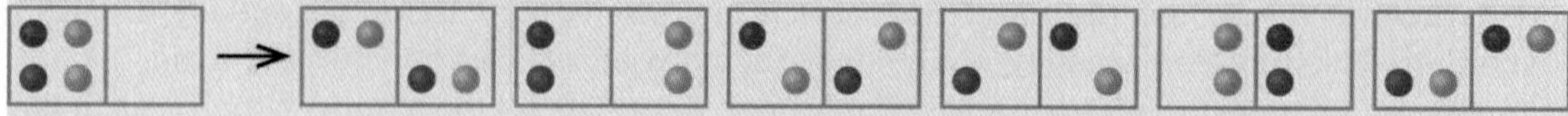

Solution

The initial number of microstates is one, the final six:

$$\Delta S = k \ln \frac{W_c}{W_a} = 1.38 \times 10^{-23}\ \text{J/K} \times \ln \frac{6}{1} = 2.47 \times 10^{-23}\ \text{J/K}$$

The sign of this result is consistent with expectation; since there are more microstates possible for the final state than for the initial state, the change in entropy should be positive.

Check Your Learning

Consider the system shown in Figure 16.9. What is the change in entropy for the process where *all* the energy is transferred from the hot object (**AB**) to the cold object (**CD**)?

Answer:

0 J/K

Predicting the Sign of ΔS

The relationships between entropy, microstates, and matter/energy dispersal described previously allow us to make generalizations regarding the relative entropies of substances and to predict the sign of entropy changes for chemical and physical processes. Consider the phase changes illustrated in Figure 16.10. In the solid phase, the atoms or molecules are restricted to nearly fixed positions with respect to each other and are capable of only modest oscillations about these positions. With essentially fixed locations for the system's component particles, the number of microstates is relatively small. In the liquid phase, the atoms or molecules are free to move over and around each other, though they remain in relatively close proximity to one another. This increased freedom of motion results in a greater variation in possible particle locations, so the number of microstates is correspondingly greater than for the solid. As a result, $S_{\text{liquid}} > S_{\text{solid}}$ and the process of converting a substance from solid to liquid (melting) is characterized by an increase in entropy, $\Delta S > 0$. By the same logic, the reciprocal process (freezing) exhibits a decrease in entropy, $\Delta S < 0$.

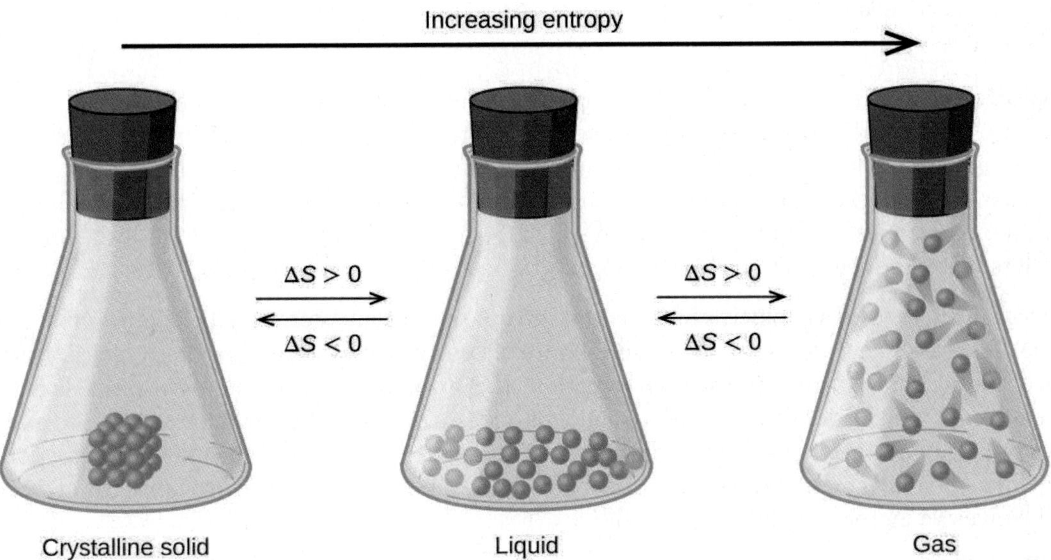

FIGURE 16.10 The entropy of a substance increases ($\Delta S > 0$) as it transforms from a relatively ordered solid, to a less-ordered liquid, and then to a still less-ordered gas. The entropy decreases ($\Delta S < 0$) as the substance transforms from a gas to a liquid and then to a solid.

Now consider the gaseous phase, in which a given number of atoms or molecules occupy a *much* greater volume than in the liquid phase. Each atom or molecule can be found in many more locations, corresponding to a much greater number of microstates. Consequently, for any substance, $S_{gas} > S_{liquid} > S_{solid}$, and the processes of vaporization and sublimation likewise involve increases in entropy, $\Delta S > 0$. Likewise, the reciprocal phase transitions, condensation and deposition, involve decreases in entropy, $\Delta S < 0$.

According to kinetic-molecular theory, the temperature of a substance is proportional to the average kinetic energy of its particles. Raising the temperature of a substance will result in more extensive vibrations of the particles in solids and more rapid translations of the particles in liquids and gases. At higher temperatures, the distribution of kinetic energies among the atoms or molecules of the substance is also broader (more dispersed) than at lower temperatures. Thus, the entropy for any substance increases with temperature (Figure 16.11).

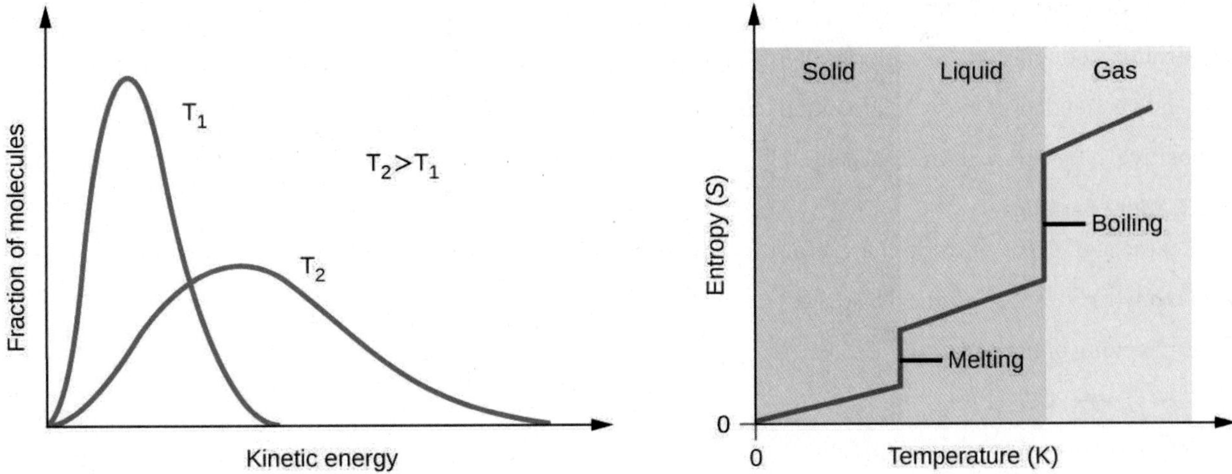

FIGURE 16.11 Entropy increases as the temperature of a substance is raised, which corresponds to the greater spread of kinetic energies. When a substance undergoes a phase transition, its entropy changes significantly.

LINK TO LEARNING

Try this simulator (http://openstax.org/l/16freemotion) with interactive visualization of the dependence of particle location and freedom of motion on physical state and temperature.

The entropy of a substance is influenced by the structure of the particles (atoms or molecules) that comprise the substance. With regard to atomic substances, heavier atoms possess greater entropy at a given temperature than lighter atoms, which is a consequence of the relation between a particle's mass and the spacing of quantized translational energy levels (a topic beyond the scope of this text). For molecules, greater numbers of atoms increase the number of ways in which the molecules can vibrate and thus the number of possible microstates and the entropy of the system.

Finally, variations in the types of particles affects the entropy of a system. Compared to a pure substance, in which all particles are identical, the entropy of a mixture of two or more different particle types is greater. This is because of the additional orientations and interactions that are possible in a system comprised of nonidentical components. For example, when a solid dissolves in a liquid, the particles of the solid experience both a greater freedom of motion and additional interactions with the solvent particles. This corresponds to a more uniform dispersal of matter and energy and a greater number of microstates. The process of dissolution therefore involves an increase in entropy, $\Delta S > 0$.

Considering the various factors that affect entropy allows us to make informed predictions of the sign of ΔS for various chemical and physical processes as illustrated in Example 16.3.

EXAMPLE 16.3

Predicting the Sign of ΔS

Predict the sign of the entropy change for the following processes. Indicate the reason for each of your predictions.

(a) One mole liquid water at room temperature $\longrightarrow$ one mole liquid water at 50 °C

(b) $Ag^+(aq) + Cl^-(aq) \longrightarrow AgCl(s)$

(c) $C_6H_6(l) + \frac{15}{2}O_2(g) \longrightarrow 6CO_2(g) + 3H_2O(l)$

(d) $NH_3(s) \longrightarrow NH_3(l)$

Solution

(a) positive, temperature increases

(b) negative, reduction in the number of ions (particles) in solution, decreased dispersal of matter

(c) negative, net decrease in the amount of gaseous species

(d) positive, phase transition from solid to liquid, net increase in dispersal of matter

Check Your Learning

Predict the sign of the entropy change for the following processes. Give a reason for your prediction.

(a) $NaNO_3(s) \longrightarrow Na^+(aq) + NO_3{}^-(aq)$

(b) the freezing of liquid water

(c) $CO_2(s) \longrightarrow CO_2(g)$

(d) $CaCO_3(s) \longrightarrow CaO(s) + CO_2(g)$

Answer:

(a) Positive; The solid dissolves to give an increase of mobile ions in solution. (b) Negative; The liquid becomes a more ordered solid. (c) Positive; The relatively ordered solid becomes a gas. (d) Positive; There is a net increase in the amount of gaseous species.

16.3 The Second and Third Laws of Thermodynamics

LEARNING OBJECTIVES

By the end of this section, you will be able to:

- State and explain the second and third laws of thermodynamics
- Calculate entropy changes for phase transitions and chemical reactions under standard conditions

The Second Law of Thermodynamics

In the quest to identify a property that may reliably predict the spontaneity of a process, a promising candidate has been identified: entropy. Processes that involve an increase in entropy *of the system* ($\Delta S > 0$) are very often spontaneous; however, examples to the contrary are plentiful. By expanding consideration of entropy changes to include *the surroundings*, we may reach a significant conclusion regarding the relation between this property and spontaneity. In thermodynamic models, the system and surroundings comprise everything, that is, the universe, and so the following is true:

$$\Delta S_{\text{univ}} = \Delta S_{\text{sys}} + \Delta S_{\text{surr}}$$

To illustrate this relation, consider again the process of heat flow between two objects, one identified as the system and the other as the surroundings. There are three possibilities for such a process:

1. The objects are at different temperatures, and heat flows from the hotter to the cooler object. *This is always observed to occur spontaneously.* Designating the hotter object as the system and invoking the definition of entropy yields the following:

   $$\Delta S_{\text{sys}} = \frac{-q_{\text{rev}}}{T_{\text{sys}}} \quad \text{and} \quad \Delta S_{\text{surr}} = \frac{q_{\text{rev}}}{T_{\text{surr}}}$$

   The magnitudes of $-q_{\text{rev}}$ and q_{rev} are equal, their opposite arithmetic signs denoting loss of heat by the system and gain of heat by the surroundings. Since $T_{\text{sys}} > T_{\text{surr}}$ in this scenario, the entropy *decrease* of the system will be less than the entropy *increase* of the surroundings, and so *the entropy of the universe will increase*:

   $$|\Delta S_{\text{sys}}| < |\Delta S_{\text{surr}}|$$
   $$\Delta S_{\text{univ}} = \Delta S_{\text{sys}} + \Delta S_{\text{surr}} > 0$$

2. The objects are at different temperatures, and heat flows from the cooler to the hotter object. *This is never observed to occur spontaneously.* Again designating the hotter object as the system and invoking the definition of entropy yields the following:

   $$\Delta S_{\text{sys}} = \frac{q_{\text{rev}}}{T_{\text{sys}}} \quad \text{and} \quad \Delta S_{\text{surr}} = \frac{-q_{\text{rev}}}{T_{\text{surr}}}$$

   The arithmetic signs of q_{rev} denote the gain of heat by the system and the loss of heat by the surroundings. The magnitude of the entropy change for the surroundings will again be greater than that for the system, but in this case, the signs of the heat changes (that is, *the direction of the heat flow*) will yield a negative value for ΔS_{univ}. *This process involves a decrease in the entropy of the universe.*
3. The objects are at essentially the same temperature, $T_{\text{sys}} \approx T_{\text{surr}}$, and so the magnitudes of the entropy changes are essentially the same for both the system and the surroundings. In this case, the entropy change of the universe is zero, and the system is *at equilibrium*.

   $$|\Delta S_{\text{sys}}| \approx |\Delta S_{\text{surr}}|$$
   $$\Delta S_{\text{univ}} = \Delta S_{\text{sys}} + \Delta S_{\text{surr}} = 0$$

These results lead to a profound statement regarding the relation between entropy and spontaneity known as the **second law of thermodynamics**: *all spontaneous changes cause an increase in the entropy of the universe.* A summary of these three relations is provided in Table 16.1.

The Second Law of Thermodynamics

$\Delta S_{univ} > 0$	spontaneous
$\Delta S_{univ} < 0$	nonspontaneous (spontaneous in opposite direction)
$\Delta S_{univ} = 0$	at equilibrium

TABLE 16.1

For many realistic applications, the surroundings are vast in comparison to the system. In such cases, the heat gained or lost by the surroundings as a result of some process represents a very small, nearly infinitesimal, fraction of its total thermal energy. For example, combustion of a fuel in air involves transfer of heat from a system (the fuel and oxygen molecules undergoing reaction) to surroundings that are infinitely more massive (the earth's atmosphere). As a result, q_{surr} is a good approximation of q_{rev}, and the second law may be stated as the following:

$$\Delta S_{univ} = \Delta S_{sys} + \Delta S_{surr} = \Delta S_{sys} + \frac{q_{surr}}{T}$$

We may use this equation to predict the spontaneity of a process as illustrated in Example 16.4.

EXAMPLE 16.4

Will Ice Spontaneously Melt?

The entropy change for the process

$$H_2O(s) \longrightarrow H_2O(l)$$

is 22.1 J/K and requires that the surroundings transfer 6.00 kJ of heat to the system. Is the process spontaneous at −10.00 °C? Is it spontaneous at +10.00 °C?

Solution

We can assess the spontaneity of the process by calculating the entropy change of the universe. If ΔS_{univ} is positive, then the process is spontaneous. At both temperatures, ΔS_{sys} = 22.1 J/K and q_{surr} = −6.00 kJ.

At −10.00 °C (263.15 K), the following is true:

$$\begin{aligned}\Delta S_{univ} &= \Delta S_{sys} + \Delta S_{surr} = \Delta S_{sys} + \frac{q_{surr}}{T} \\ &= 22.1\text{ J/K} + \frac{-6.00 \times 10^3\text{ J}}{263.15\text{ K}} = -0.7\text{ J/K}\end{aligned}$$

$S_{univ} < 0$, so melting is nonspontaneous (*not* spontaneous) at −10.0 °C.

At 10.00 °C (283.15 K), the following is true:

$$\Delta S_{univ} = \Delta S_{sys} + \frac{q_{surr}}{T}$$

$$= 22.1\text{ J/K} + \frac{-6.00 \times 10^3\text{ J}}{283.15\text{ K}} = +0.9\text{ J/K}$$

$S_{univ} > 0$, so melting *is* spontaneous at 10.00 °C.

Check Your Learning

Using this information, determine if liquid water will spontaneously freeze at the same temperatures. What can you say about the values of S_{univ}?

Answer:

Entropy is a state function, so $\Delta S_{freezing} = -\Delta S_{melting} = -22.1$ J/K and $q_{surr} = +6.00$ kJ. At −10.00 °C spontaneous, +0.7 J/K; at +10.00 °C nonspontaneous, −0.9 J/K.

The Third Law of Thermodynamics

The previous section described the various contributions of matter and energy dispersal that contribute to the entropy of a system. With these contributions in mind, consider the entropy of a pure, perfectly crystalline solid possessing no kinetic energy (that is, at a temperature of absolute zero, 0 K). This system may be described by a single microstate, as its purity, perfect crystallinity and complete lack of motion means there is but one possible location for each identical atom or molecule comprising the crystal ($W = 1$). According to the Boltzmann equation, the entropy of this system is zero.

$$S = k \ln W = k \ln(1) = 0$$

This limiting condition for a system's entropy represents the **third law of thermodynamics**: *the entropy of a pure, perfect crystalline substance at 0 K is zero.*

Careful calorimetric measurements can be made to determine the temperature dependence of a substance's entropy and to derive absolute entropy values under specific conditions. **Standard entropies ($S°$)** are for one mole of substance under standard conditions (a pressure of 1 bar and a temperature of 298.15 K; see details regarding standard conditions in the thermochemistry chapter of this text). The **standard entropy change ($\Delta S°$)** for a reaction may be computed using standard entropies as shown below:

$$\Delta S° = \sum \nu S°(\text{products}) - \sum \nu S°(\text{reactants})$$

where ν represents stoichiometric coefficients in the balanced equation representing the process. For example, $\Delta S°$ for the following reaction at room temperature

$$m\text{A} + n\text{B} \longrightarrow x\text{C} + y\text{D},$$

is computed as:

$$= [xS°(\text{C}) + yS°(\text{D})] - [mS°(\text{A}) + nS°(\text{B})]$$

A partial listing of standard entropies is provided in Table 16.2, and additional values are provided in Appendix G. The example exercises that follow demonstrate the use of $S°$ values in calculating standard entropy changes for physical and chemical processes.

Substance	$S°$ (J mol^{-1} K^{-1})
carbon	
C(*s*, graphite)	5.740
C(*s*, diamond)	2.38
CO(*g*)	197.7
CO_2(*g*)	213.8
CH_4(*g*)	186.3
C_2H_4(*g*)	219.5
C_2H_6(*g*)	229.5

$CH_3OH(l)$	126.8
$C_2H_5OH(l)$	160.7
hydrogen	
$H_2(g)$	130.57
$H(g)$	114.6
$H_2O(g)$	188.71
$H_2O(l)$	69.91
$HCl(g)$	186.8
$H_2S(g)$	205.7
oxygen	
$O_2(g)$	205.03

TABLE 16.2 Standard entropies for selected substances measured at 1 atm and 298.15 K. (Values are approximately equal to those measured at 1 bar, the currently accepted standard state pressure.)

EXAMPLE 16.5

Determination of $\Delta S°$

Calculate the standard entropy change for the following process:

$$H_2O(g) \longrightarrow H_2O(l)$$

Solution

Calculate the entropy change using standard entropies as shown above:

$$\Delta S° = (1\ \text{mol})(70.0\ \text{J mol}^{-1}\ \text{K}^{-1}) - (1\ \text{mol})(188.8\ \text{J mol}^{-1}\ \text{K}^{-1}) = -118.8\ \text{J/K}$$

The value for $\Delta S°$ is negative, as expected for this phase transition (condensation), which the previous section discussed.

Check Your Learning

Calculate the standard entropy change for the following process:

$$H_2(g) + C_2H_4(g) \longrightarrow C_2H_6(g)$$

Answer:

$-120.6\ \text{J K}^{-1}\ \text{mol}^{-1}$

EXAMPLE 16.6

Determination of $\Delta S°$

Calculate the standard entropy change for the combustion of methanol, CH_3OH:

$$2CH_3OH(l) + 3O_2(g) \longrightarrow 2CO_2(g) + 4H_2O(l)$$

Solution

Calculate the entropy change using standard entropies as shown above:

$$\Delta S° = \sum \nu S°(\text{products}) - \sum \nu S°(\text{reactants})$$

$$[2\text{ mol} \times S°(CO_2(g)) + 4\text{ mol} \times S°(H_2O(l))] - [2\text{ mol} \times S°(CH_3OH(l)) + 3\text{ mol} \times S°(O_2(g))]$$

$$= \{[2(213.8) + 4 \times 70.0] - [2(126.8) + 3(205.03)]\} = -161.1\text{ J/K}$$

Check Your Learning

Calculate the standard entropy change for the following reaction:

$$Ca(OH)_2(s) \longrightarrow CaO(s) + H_2O(l)$$

Answer:

24.7 J/K

16.4 Free Energy

LEARNING OBJECTIVES

By the end of this section, you will be able to:

- Define Gibbs free energy, and describe its relation to spontaneity
- Calculate free energy change for a process using free energies of formation for its reactants and products
- Calculate free energy change for a process using enthalpies of formation and the entropies for its reactants and products
- Explain how temperature affects the spontaneity of some processes
- Relate standard free energy changes to equilibrium constants

One of the challenges of using the second law of thermodynamics to determine if a process is spontaneous is that it requires measurements of the entropy change for the system *and* the entropy change for the surroundings. An alternative approach involving a new thermodynamic property defined in terms of system properties only was introduced in the late nineteenth century by American mathematician Josiah Willard Gibbs. This new property is called the **Gibbs free energy (*G*)** (or simply the *free energy*), and it is defined in terms of a system's enthalpy and entropy as the following:

$$G = H - TS$$

Free energy is a state function, and at constant temperature and pressure, the **free energy change (Δ*G*)** may be expressed as the following:

$$\Delta G = \Delta H - T\Delta S$$

(For simplicity's sake, the subscript "sys" will be omitted henceforth.)

The relationship between this system property and the spontaneity of a process may be understood by recalling the previously derived second law expression:

$$\Delta S_{\text{univ}} = \Delta S + \frac{q_{\text{surr}}}{T}$$

The first law requires that $q_{\text{surr}} = -q_{\text{sys}}$, and at constant pressure $q_{\text{sys}} = \Delta H$, so this expression may be rewritten as:

$$\Delta S_{\text{univ}} = \Delta S - \frac{\Delta H}{T}$$

Multiplying both sides of this equation by $-T$, and rearranging yields the following:

$$-T\Delta S_{\text{univ}} = \Delta H - T\Delta S$$

Comparing this equation to the previous one for free energy change shows the following relation:

$$\Delta G = -T\Delta S_{univ}$$

The free energy change is therefore a reliable indicator of the spontaneity of a process, being directly related to the previously identified spontaneity indicator, ΔS_{univ}. Table 16.3 summarizes the relation between the spontaneity of a process and the arithmetic signs of these indicators.

Relation between Process Spontaneity and Signs of Thermodynamic Properties

$\Delta S_{univ} > 0$	$\Delta G < 0$	spontaneous
$\Delta S_{univ} < 0$	$\Delta G > 0$	nonspontaneous
$\Delta S_{univ} = 0$	$\Delta G = 0$	at equilibrium

TABLE 16.3

What's "Free" about ΔG?

In addition to indicating spontaneity, the free energy change also provides information regarding the amount of useful work (w) that may be accomplished by a spontaneous process. Although a rigorous treatment of this subject is beyond the scope of an introductory chemistry text, a brief discussion is helpful for gaining a better perspective on this important thermodynamic property.

For this purpose, consider a spontaneous, exothermic process that involves a decrease in entropy. The free energy, as defined by

$$\Delta G = \Delta H - T\Delta S$$

may be interpreted as representing the difference between the energy produced by the process, ΔH, and the energy lost to the surroundings, $T\Delta S$. The difference between the energy produced and the energy lost is the energy available (or "free") to do useful work by the process, ΔG. If the process somehow could be made to take place under conditions of thermodynamic reversibility, the amount of work that could be done would be maximal:

$$\Delta G = w_{max}$$

However, as noted previously in this chapter, such conditions are not realistic. In addition, the technologies used to extract work from a spontaneous process (e.g., automobile engine, steam turbine) are never 100% efficient, and so the work done by these processes is always less than the theoretical maximum. Similar reasoning may be applied to a nonspontaneous process, for which the free energy change represents the *minimum* amount of work that must be done *on* the system to carry out the process.

Calculating Free Energy Change

Free energy is a state function, so its value depends only on the conditions of the initial and final states of the system. A convenient and common approach to the calculation of free energy changes for physical and chemical reactions is by use of widely available compilations of standard state thermodynamic data. One method involves the use of standard enthalpies and entropies to compute **standard free energy changes, $\Delta G°$**, according to the following relation.

$$\Delta G° = \Delta H° - T\Delta S°$$

EXAMPLE 16.7

Using Standard Enthalpy and Entropy Changes to Calculate $\Delta G°$

Use standard enthalpy and entropy data from Appendix G to calculate the standard free energy change for the vaporization of water at room temperature (298 K). What does the computed value for $\Delta G°$ say about the spontaneity of this process?

Solution

The process of interest is the following:

$$H_2O(l) \longrightarrow H_2O(g)$$

The standard change in free energy may be calculated using the following equation:

$$\Delta G° = \Delta H° - T\Delta S°$$

From Appendix G:

Substance	ΔH_f°(kJ/mol)	$S°$(J/K·mol)
$H_2O(l)$	-285.83	70.0
$H_2O(g)$	-241.82	188.8

Using the appendix data to calculate the standard enthalpy and entropy changes yields:

$$\Delta H° = \Delta H_f^\circ\,(H_2O(g)) - \Delta H_f^\circ\,(H_2O(l))$$
$$= [-241.82\text{ kJ/mol} - (-285.83)]\text{ kJ/mol} = 44.01\text{ kJ}$$
$$\Delta S° = 1\text{ mol} \times S°\,(H_2O(g)) - 1\text{ mol} \times S°\,(H_2O(l))$$
$$= (1\text{ mol})188.8\text{ J/mol·K} - (1\text{ mol})70.0\text{ J/mol K} = 118.8\text{ J/K}$$
$$\Delta G° = \Delta H° - T\Delta S°$$

Substitution into the standard free energy equation yields:

$$\Delta G° = \Delta H° - T\Delta S°$$

$$= 44.01\text{ kJ} - (298\text{ K} \times 118.8\text{ J/K}) \times \frac{1\text{ kJ}}{1000\text{ J}}$$

$$44.01\text{ kJ} - 35.4\text{ kJ} = 8.6\text{ kJ}$$

At 298 K (25 °C) $\Delta G° > 0$, so boiling is nonspontaneous (*not* spontaneous).

Check Your Learning

Use standard enthalpy and entropy data from Appendix G to calculate the standard free energy change for the reaction shown here (298 K). What does the computed value for $\Delta G°$ say about the spontaneity of this process?

$$C_2H_6(g) \longrightarrow H_2(g) + C_2H_4(g)$$

Answer:

$\Delta G° = 102.0$ kJ/mol; the reaction is nonspontaneous (*not* spontaneous) at 25 °C.

The standard free energy change for a reaction may also be calculated from **standard free energy of formation $\Delta G_f°$** values of the reactants and products involved in the reaction. The standard free energy of formation is the free energy change that accompanies the formation of one mole of a substance from its elements in their standard states. Similar to the standard enthalpy of formation, ΔG_f° is by definition zero for elemental substances in their standard states. The approach used to calculate ΔG° for a reaction from ΔG_f° values is the same as that demonstrated previously for enthalpy and entropy changes. For the reaction

$$m\text{A} + n\text{B} \longrightarrow x\text{C} + y\text{D},$$

the standard free energy change at room temperature may be calculated as

$$\Delta G° = \sum \nu \Delta G°(\text{products}) - \sum \nu \Delta G°(\text{reactants})$$

$$= \left[x\Delta G_f^\circ (\text{C}) + y\Delta G_f^\circ (\text{D})\right] - \left[m\Delta G_f^\circ (\text{A}) + n\Delta G_f^\circ (\text{B})\right].$$

EXAMPLE 16.8

Using Standard Free Energies of Formation to Calculate ΔG°

Consider the decomposition of yellow mercury(II) oxide.

$$\text{HgO}(s,\ \text{yellow}) \longrightarrow \text{Hg}(l) + \frac{1}{2}\text{O}_2(g)$$

Calculate the standard free energy change at room temperature, $\Delta G°$, using (a) standard free energies of formation and (b) standard enthalpies of formation and standard entropies. Do the results indicate the reaction to be spontaneous or nonspontaneous under standard conditions?

Solution

The required data are available in Appendix G and are shown here.

Compound	ΔG_f° (kJ/mol)	ΔH_f° (kJ/mol)	$S°$ (J/K·mol)
HgO (*s*, yellow)	−58.43	−90.46	71.13
Hg(*l*)	0	0	75.9
$O_2(g)$	0	0	205.2

(a) Using free energies of formation:

$$\Delta G° = \sum \nu G_f^\circ(\text{products}) - \sum \nu \Delta G_f^\circ(\text{reactants})$$

$$= \left[1\Delta G_f^\circ \text{Hg}(l) + \frac{1}{2}\Delta G_f^\circ \text{O}_2(g)\right] - 1\Delta G_f^\circ \text{HgO}(s,\ \text{yellow})$$

$$= \left[1\ \text{mol}(0\ \text{kJ/mol}) + \frac{1}{2}\ \text{mol}(0\ \text{kJ/mol})\right] - 1\ \text{mol}(-58.43\ \text{kJ/mol}) = 58.43\ \text{kJ/mol}$$

(b) Using enthalpies and entropies of formation:

$$\Delta H° = \sum \nu \Delta H_f^\circ(\text{products}) - \sum \nu \Delta H_f^\circ(\text{reactants})$$

$$= \left[1\Delta H_f^\circ \text{Hg}(l) + \frac{1}{2}\Delta H_f^\circ \text{O}_2(g)\right] - 1\Delta H_f^\circ \text{HgO}(s,\ \text{yellow})$$

$$= \left[1\ \text{mol}(0\ \text{kJ/mol}) + \frac{1}{2}\ \text{mol}(0\ \text{kJ/mol})\right] - 1\ \text{mol}(-90.46\ \text{kJ/mol}) = 90.46\ \text{kJ/mol}$$

$$\Delta S° = \sum \nu \Delta S°(\text{products}) - \sum \nu \Delta S°(\text{reactants})$$

$$= \left[1\Delta S° \text{Hg}(l) + \frac{1}{2}\Delta S° \text{O}_2(g)\right] - 1\Delta S° \text{HgO}(s,\ \text{yellow})$$

$$= \left[1 \text{ mol } (75.9 \text{ J/mol K}) + \frac{1}{2} \text{ mol}(205.2 \text{ J/mol K})\right] - 1 \text{ mol}(71.13 \text{ J/mol K}) = 107.4 \text{ J/mol K}$$

$$\Delta G° = \Delta H° - T\Delta S° = 90.46 \text{ kJ} - 298.15 \text{ K} \times 107.4 \text{ J/K·mol} \times \frac{1 \text{ kJ}}{1000 \text{ J}}$$

$$\Delta G° = (90.46 - 32.01) \text{ kJ/mol} = 58.45 \text{ kJ/mol}$$

Both ways to calculate the standard free energy change at 25 °C give the same numerical value (to three significant figures), and both predict that the process is nonspontaneous (*not* spontaneous) at room temperature.

Check Your Learning

Calculate $\Delta G°$ using (a) free energies of formation and (b) enthalpies of formation and entropies (Appendix G). Do the results indicate the reaction to be spontaneous or nonspontaneous at 25 °C?

$$C_2H_4(g) \longrightarrow H_2(g) + C_2H_2(g)$$

Answer:

(a) 140.8 kJ/mol, nonspontaneous

(b) 141.5 kJ/mol, nonspontaneous

Free Energy Changes for Coupled Reactions

The use of free energies of formation to compute free energy changes for reactions as described above is possible because ΔG is a state function, and the approach is analogous to the use of Hess' Law in computing enthalpy changes (see the chapter on thermochemistry). Consider the vaporization of water as an example:

$$H_2O(l) \rightarrow H_2O(g)$$

An equation representing this process may be derived by adding the formation reactions for the two phases of water (necessarily reversing the reaction for the liquid phase). The free energy change for the sum reaction is the sum of free energy changes for the two added reactions:

$$H_2(g) + \tfrac{1}{2}O_2(g) \rightarrow H_2O(g) \qquad \Delta G_f^\circ \text{gas}$$

$$H_2O(l) \rightarrow H_2(g) + \tfrac{1}{2}O_2(g) \qquad -\Delta G_f^\circ \text{liquid}$$

$$H_2O(l) \rightarrow H_2O(g) \qquad \Delta G° = \Delta G_f^\circ \text{gas} - \Delta G_f^\circ \text{liquid}$$

This approach may also be used in cases where a nonspontaneous reaction is enabled by coupling it to a spontaneous reaction. For example, the production of elemental zinc from zinc sulfide is thermodynamically unfavorable, as indicated by a positive value for $\Delta G°$:

$$ZnS(s) \rightarrow Zn(s) + S(s) \qquad \Delta G_1^\circ = 201.3 \text{ kJ}$$

The industrial process for production of zinc from sulfidic ores involves coupling this decomposition reaction to the thermodynamically favorable oxidation of sulfur:

$$S(s) + O_2(g) \rightarrow SO_2(g) \qquad \Delta G_2^\circ = -300.1 \text{ kJ}$$

The coupled reaction exhibits a negative free energy change and is spontaneous:

$$ZnS(s) + O_2(g) \rightarrow Zn(s) + SO_2(g) \qquad \Delta G° = 201.3 \text{ kJ} + -300.1 \text{ kJ} = -98.8 \text{ kJ}$$

This process is typically carried out at elevated temperatures, so this result obtained using standard free energy values is just an estimate. The gist of the calculation, however, holds true.

Calculating Free Energy Change for a Coupled Reaction

Is a reaction coupling the decomposition of ZnS to the formation of H2S expected to be spontaneous under standard conditions?

Solution

Following the approach outlined above and using free energy values from Appendix G:

Decomposition of zinc sulfide:	$\text{ZnS}(s) \rightarrow \text{Zn}(s) + \text{S}(s)$	$\Delta G_1^\circ = 201.3\text{ kJ}$
Formation of hydrogen sulfide:	$\text{S}(s) + \text{H}_2(g) \rightarrow \text{H}_2\text{S}(g)$	$\Delta G_2^\circ = -33.4\text{ kJ}$
Coupled reaction:	$\text{ZnS}(s) + \text{H}_2(g) \rightarrow \text{Zn}(s) + \text{H}_2\text{S}(g)$	$\Delta G^\circ = 201.3\text{ kJ} + -33.4\text{ kJ} = 167.9\text{ kJ}$

The coupled reaction exhibits a positive free energy change and is thus nonspontaneous.

Check Your Learning

What is the standard free energy change for the reaction below? Is the reaction expected to be spontaneous under standard conditions?

$$\text{FeS}(s) + \text{O}_2(g) \rightarrow \text{Fe}(s) + \text{SO}_2(g)$$

Answer:

−199.7 kJ; spontaneous

Temperature Dependence of Spontaneity

As was previously demonstrated in this chapter's section on entropy, the spontaneity of a process may depend upon the temperature of the system. Phase transitions, for example, will proceed spontaneously in one direction or the other depending upon the temperature of the substance in question. Likewise, some chemical reactions can also exhibit temperature dependent spontaneities. To illustrate this concept, the equation relating free energy change to the enthalpy and entropy changes for the process is considered:

$$\Delta G = \Delta H - T\Delta S$$

The spontaneity of a process, as reflected in the arithmetic sign of its free energy change, is then determined by the signs of the enthalpy and entropy changes and, in some cases, the absolute temperature. Since T is the absolute (kelvin) temperature, it can only have positive values. Four possibilities therefore exist with regard to the signs of the enthalpy and entropy changes:

1. **Both ΔH and ΔS are positive.** This condition describes an endothermic process that involves an increase in system entropy. In this case, ΔG will be negative if the magnitude of the $T\Delta S$ term is greater than ΔH. If the $T\Delta S$ term is less than ΔH, the free energy change will be positive. Such a process is *spontaneous at high temperatures and nonspontaneous at low temperatures.*
2. **Both ΔH and ΔS are negative.** This condition describes an exothermic process that involves a decrease in system entropy. In this case, ΔG will be negative if the magnitude of the $T\Delta S$ term is less than ΔH. If the $T\Delta S$ term's magnitude is greater than ΔH, the free energy change will be positive. Such a process is *spontaneous at low temperatures and nonspontaneous at high temperatures.*
3. **ΔH is positive and ΔS is negative.** This condition describes an endothermic process that involves a decrease in system entropy. In this case, ΔG will be positive regardless of the temperature. Such a process is *nonspontaneous at all temperatures.*
4. **ΔH is negative and ΔS is positive.** This condition describes an exothermic process that involves an increase in system entropy. In this case, ΔG will be negative regardless of the temperature. Such a process is *spontaneous at all temperatures.*

These four scenarios are summarized in Figure 16.12.

Summary of the Four Scenarios for Enthalpy and Entropy Changes

	ΔH > 0 (endothermic)	ΔH < 0 (exothermic)
ΔS > 0 (increase in entropy)	ΔG < 0 at high temperature ΔG > 0 at low temperature Process is spontaneous at high temperature	ΔG < 0 at any temperature Process is spontaneous at any temperature
ΔS < 0 (decrease in entropy)	ΔG > 0 at any temperature Process is nonspontaneous at any temperature	ΔG < 0 at low temperature ΔG > 0 at high temperature Process is spontaneous at low temperature

FIGURE 16.12 There are four possibilities regarding the signs of enthalpy and entropy changes.

EXAMPLE 16.10

Predicting the Temperature Dependence of Spontaneity

The incomplete combustion of carbon is described by the following equation:

$$2C(s) + O_2(g) \longrightarrow 2CO(g)$$

How does the spontaneity of this process depend upon temperature?

Solution

Combustion processes are exothermic ($\Delta H < 0$). This particular reaction involves an increase in entropy due to the accompanying increase in the amount of gaseous species (net gain of one mole of gas, $\Delta S > 0$). The reaction is therefore spontaneous ($\Delta G < 0$) at all temperatures.

Check Your Learning

Popular chemical hand warmers generate heat by the air-oxidation of iron:

$$4Fe(s) + 3O_2(g) \longrightarrow 2Fe_2O_3(s)$$

How does the spontaneity of this process depend upon temperature?

Answer:

ΔH and ΔS are negative; the reaction is spontaneous at low temperatures.

When considering the conclusions drawn regarding the temperature dependence of spontaneity, it is important to keep in mind what the terms "high" and "low" mean. Since these terms are adjectives, the temperatures in question are deemed high or low relative to some reference temperature. A process that is nonspontaneous at one temperature but spontaneous at another will necessarily undergo a change in "spontaneity" (as reflected by its ΔG) as temperature varies. This is clearly illustrated by a graphical presentation of the free energy change equation, in which ΔG is plotted on the y axis versus T on the x axis:

$$\Delta G = \Delta H - T\Delta S$$

$$y = b + mx$$

Such a plot is shown in Figure 16.13. A process whose enthalpy and entropy changes are of the same arithmetic sign will exhibit a temperature-dependent spontaneity as depicted by the two yellow lines in the plot. Each line crosses from one spontaneity domain (positive or negative ΔG) to the other at a temperature that is characteristic of the process in question. This temperature is represented by the x-intercept of the line, that is, the value of T for which ΔG is zero:

$$\Delta G = 0 = \Delta H - T\Delta S$$

$$T = \frac{\Delta H}{\Delta S}$$

So, saying a process is spontaneous at “high” or “low” temperatures means the temperature is above or below, respectively, that temperature at which ΔG for the process is zero. As noted earlier, the condition of $\Delta G = 0$ describes a system at equilibrium.

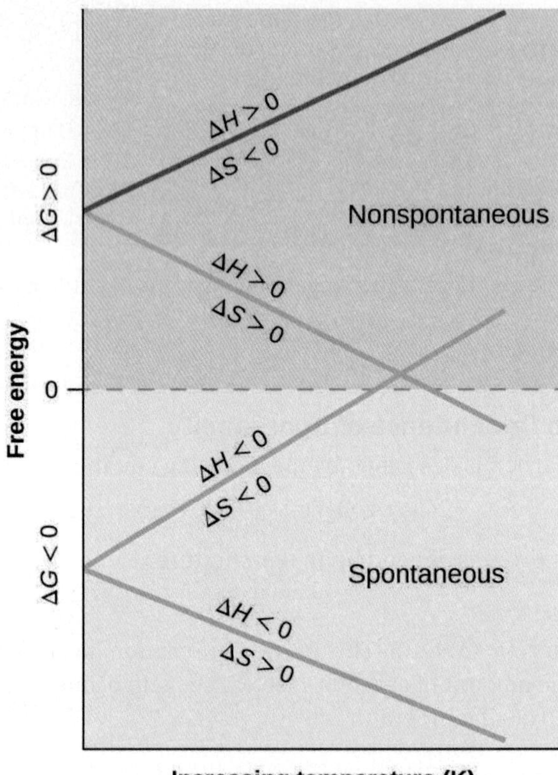

FIGURE 16.13 These plots show the variation in ΔG with temperature for the four possible combinations of arithmetic sign for ΔH and ΔS.

EXAMPLE 16.11

Equilibrium Temperature for a Phase Transition

As defined in the chapter on liquids and solids, the boiling point of a liquid is the temperature at which its liquid and gaseous phases are in equilibrium (that is, when vaporization and condensation occur at equal rates). Use the information in Appendix G to estimate the boiling point of water.

Solution

The process of interest is the following phase change:

$$H_2O(l) \longrightarrow H_2O(g)$$

When this process is at equilibrium, $\Delta G = 0$, so the following is true:

$$0 = \Delta H^\circ - T\Delta S^\circ \quad \text{or} \quad T = \frac{\Delta H^\circ}{\Delta S^\circ}$$

Using the standard thermodynamic data from Appendix G,

$$\begin{aligned} \Delta H^\circ &= 1 \text{ mol} \times \Delta H_f^\circ (H_2O(g)) - 1 \text{ mol} \times \Delta H_f^\circ (H_2O(l)) \\ &= (1 \text{ mol}) - 241.82 \text{ kJ/mol} - (1 \text{ mol})(-286.83 \text{ kJ/mol}) = 44.01 \text{ kJ} \end{aligned}$$

$$\Delta S^\circ = 1 \text{ mol} \times \Delta S^\circ(H_2O(g)) - 1 \text{ mol} \times \Delta S^\circ(H_2O(l))$$
$$= (1 \text{ mol})\ 188.8 \text{ J/K·mol} - (1 \text{ mol})\ 70.0 \text{ J/K·mol} = 118.8 \text{ J/K}$$

$$T = \frac{\Delta H^\circ}{\Delta S^\circ} = \frac{44.01 \times 10^3 \text{ J}}{118.8 \text{ J/K}} = 370.5 \text{ K} = 97.3\ ^\circ\text{C}$$

The accepted value for water's normal boiling point is 373.2 K (100.0 °C), and so this calculation is in reasonable agreement. Note that the values for enthalpy and entropy changes data used were derived from standard data at 298 K (Appendix G). If desired, you could obtain more accurate results by using enthalpy and entropy changes determined at (or at least closer to) the actual boiling point.

Check Your Learning

Use the information in Appendix G to estimate the boiling point of CS_2.

Answer:

313 K (accepted value 319 K)

Free Energy and Equilibrium

The free energy change for a process may be viewed as a measure of its driving force. A negative value for ΔG represents a driving force for the process in the forward direction, while a positive value represents a driving force for the process in the reverse direction. When ΔG is zero, the forward and reverse driving forces are equal, and the process occurs in both directions at the same rate (the system is at equilibrium).

In the chapter on equilibrium the *reaction quotient*, Q, was introduced as a convenient measure of the status of an equilibrium system. Recall that Q is the numerical value of the mass action expression for the system, and that you may use its value to identify the direction in which a reaction will proceed in order to achieve equilibrium. When Q is lesser than the equilibrium constant, K, the reaction will proceed in the forward direction until equilibrium is reached and $Q = K$. Conversely, if $Q > K$, the process will proceed in the reverse direction until equilibrium is achieved.

The free energy change for a process taking place with reactants and products present under *nonstandard conditions* (pressures other than 1 bar; concentrations other than 1 M) is related to the standard free energy change according to this equation:

$$\Delta G = \Delta G^\circ + RT \ln Q$$

R is the gas constant (8.314 J/K mol), T is the kelvin or absolute temperature, and Q is the reaction quotient. For gas phase equilibria, the pressure-based reaction quotient, Q_P, is used. The concentration-based reaction quotient, Q_C, is used for condensed phase equilibria. This equation may be used to predict the spontaneity for a process under any given set of conditions as illustrated in Example 16.12.

 **EXAMPLE 16.12**

Calculating ΔG under Nonstandard Conditions

What is the free energy change for the process shown here under the specified conditions?

$T = 25\ ^\circ\text{C}$, $P_{N_2} = 0.870 \text{ atm}$, $P_{H_2} = 0.250 \text{ atm}$, and $P_{NH_3} = 12.9 \text{ atm}$

$$2NH_3(g) \longrightarrow 3H_2(g) + N_2(g) \qquad \Delta G^\circ = 33.0 \text{ kJ/mol}$$

Solution

The equation relating free energy change to standard free energy change and reaction quotient may be used directly:

$$\Delta G = \Delta G° + RT \ln Q = 33.0 \frac{\text{kJ}}{\text{mol}} + \left(8.314 \frac{\text{J}}{\text{mol K}} \times 298 \text{ K} \times \ln \frac{(0.250^3) \times 0.870}{12.9^2}\right)$$

$$= 9680 \frac{\text{J}}{\text{mol}} \text{ or } 9.68 \text{ kJ/mol}$$

Since the computed value for ΔG is positive, the reaction is nonspontaneous under these conditions.

Check Your Learning

Calculate the free energy change for this same reaction at 875 °C in a 5.00 L mixture containing 0.100 mol of each gas. Is the reaction spontaneous under these conditions?

Answer:

$\Delta G = -123.5$ kJ/mol; yes

For a system at equilibrium, $Q = K$ and $\Delta G = 0$, and the previous equation may be written as

$$0 = \Delta G° + RT \ln K \qquad \text{(at equilibrium)}$$

$$\Delta G° = -RT \ln K \qquad \text{or} \qquad K = e^{-\frac{\Delta G°}{RT}}$$

This form of the equation provides a useful link between these two essential thermodynamic properties, and it can be used to derive equilibrium constants from standard free energy changes and vice versa. The relations between standard free energy changes and equilibrium constants are summarized in Table 16.4.

Relations between Standard Free Energy Changes and Equilibrium Constants

K	$\Delta G°$	Composition of an Equilibrium Mixture
> 1	< 0	Products are more abundant
< 1	> 0	Reactants are more abundant
= 1	= 0	Reactants and products are comparably abundant

TABLE 16.4

 **EXAMPLE 16.13**

Calculating an Equilibrium Constant using Standard Free Energy Change

Given that the standard free energies of formation of $Ag^+(aq)$, $Cl^-(aq)$, and $AgCl(s)$ are 77.1 kJ/mol, −131.2 kJ/mol, and −109.8 kJ/mol, respectively, calculate the solubility product, K_{sp}, for AgCl.

Solution

The reaction of interest is the following:

$$\text{AgCl}(s) \rightleftharpoons \text{Ag}^+(aq) + \text{Cl}^-(aq) \qquad K_{sp} = [\text{Ag}^+][\text{Cl}^-]$$

The standard free energy change for this reaction is first computed using standard free energies of formation for its reactants and products:

$$\Delta G° = \left[\Delta G_f^\circ\left(\text{Ag}^+(aq)\right) + \Delta G_f^\circ\left(\text{Cl}^-(aq)\right)\right] - \left[\Delta G_f^\circ\left(\text{AgCl}(s)\right)\right]$$
$$= [77.1 \text{ kJ/mol} - 131.2 \text{ kJ/mol}] - [-109.8 \text{ kJ/mol}] = 55.7 \text{ kJ/mol}$$

The equilibrium constant for the reaction may then be derived from its standard free energy change:

$$K_{sp} = e^{-\frac{\Delta G°}{RT}} = \exp\left(-\frac{\Delta G°}{RT}\right) = \exp\left(-\frac{55.7 \times 10^3\ \text{J/mol}}{8.314\ \text{J/mol·K} \times 298.15\ \text{K}}\right) = \exp(-22.470) = e^{-22.470} = 1.74 \times 10^{-10}$$

This result is in reasonable agreement with the value provided in Appendix J.

Check Your Learning

Use the thermodynamic data provided in Appendix G to calculate the equilibrium constant for the dissociation of dinitrogen tetroxide at 25 °C.

$$2NO_2(g) \rightleftharpoons N_2O_4(g)$$

Answer:

$K = 6.9$

To further illustrate the relation between these two essential thermodynamic concepts, consider the observation that reactions spontaneously proceed in a direction that ultimately establishes equilibrium. As may be shown by plotting the free energy versus the extent of the reaction (for example, as reflected in the value of Q), equilibrium is established when the system's free energy is minimized (Figure 16.14). If a system consists of reactants and products in nonequilibrium amounts ($Q \neq K$), the reaction will proceed spontaneously in the direction necessary to establish equilibrium.

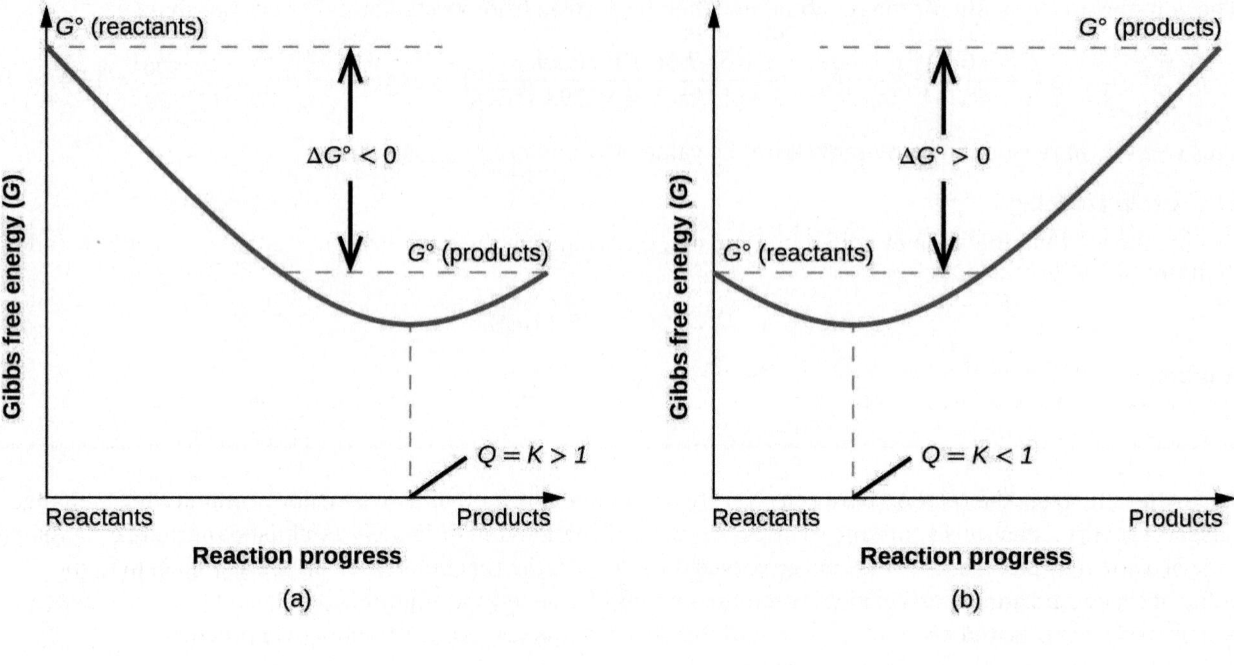

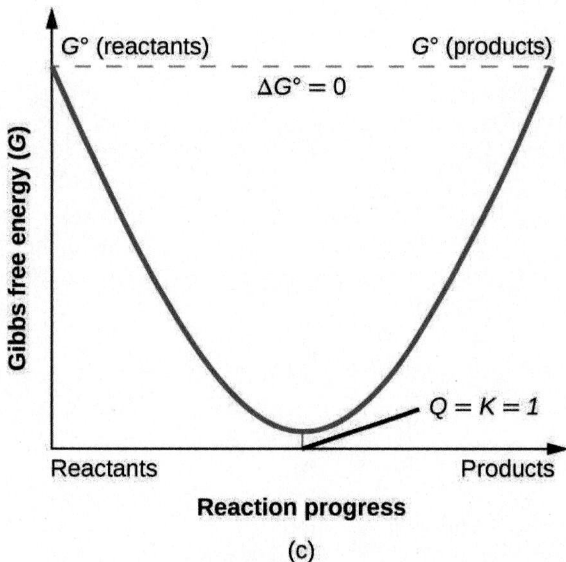

FIGURE 16.14 These plots show the free energy versus reaction progress for systems whose standard free energy changes are (a) negative, (b) positive, and (c) zero. Nonequilibrium systems will proceed spontaneously in whatever direction is necessary to minimize free energy and establish equilibrium.

Key Terms

entropy (S) state function that is a measure of the matter and/or energy dispersal within a system, determined by the number of system microstates; often described as a measure of the disorder of the system

Gibbs free energy change (G) thermodynamic property defined in terms of system enthalpy and entropy; all spontaneous processes involve a decrease in G

microstate possible configuration or arrangement of matter and energy within a system

nonspontaneous process process that requires continual input of energy from an external source

reversible process process that takes place so slowly as to be capable of reversing direction in response to an infinitesimally small change in conditions; hypothetical construct that can only be approximated by real processes

second law of thermodynamics all spontaneous processes involve an increase in the entropy of the universe

spontaneous change process that takes place without a continuous input of energy from an external source

standard entropy ($S°$) entropy for one mole of a substance at 1 bar pressure; tabulated values are usually determined at 298.15 K

standard entropy change ($\Delta S°$) change in entropy for a reaction calculated using the standard entropies

standard free energy change ($\Delta G°$) change in free energy for a process occurring under standard conditions (1 bar pressure for gases, 1 M concentration for solutions)

standard free energy of formation (ΔG_f°) change in free energy accompanying the formation of one mole of substance from its elements in their standard states

third law of thermodynamics entropy of a perfect crystal at absolute zero (0 K) is zero

Key Equations

$$\Delta S = \frac{q_{rev}}{T}$$

$$S = k \ln W$$

$$\Delta S = k \ln \frac{W_f}{W_i}$$

$$\Delta S° = \sum \nu S°(\text{products}) - \sum \nu S°(\text{reactants})$$

$$\Delta S = \frac{q_{rev}}{T}$$

$$\Delta S_{univ} = \Delta S_{sys} + \Delta S_{surr}$$

$$\Delta S_{univ} = \Delta S_{sys} + \Delta S_{surr} = \Delta S_{sys} + \frac{q_{surr}}{T}$$

$$\Delta G = \Delta H - T\Delta S$$

Summary

16.1 Spontaneity

Chemical and physical processes have a natural tendency to occur in one direction under certain conditions. A spontaneous process occurs without the need for a continual input of energy from some external source, while a nonspontaneous process requires such. Systems undergoing a spontaneous process may or may not experience a gain or loss of energy, but they will experience a change in the way matter and/or energy is distributed within the system.

16.2 Entropy

Entropy (S) is a state function that can be related to the number of microstates for a system (the number of ways the system can be arranged) and to the ratio of reversible heat to kelvin temperature. It may be interpreted as a measure of the dispersal or distribution of matter and/or energy in a system, and it is often described as representing the "disorder" of the system.

For a given substance, entropy depends on phase with $S_{solid} < S_{liquid} < S_{gas}$. For different substances in the same physical state at a given temperature, entropy is typically greater for heavier atoms or more complex molecules. Entropy increases when a system is heated and when solutions form. Using these guidelines, the sign of entropy changes for some chemical reactions and physical changes may be reliably predicted.

16.3 The Second and Third Laws of Thermodynamics

The second law of thermodynamics states that a spontaneous process increases the entropy of the universe, $S_{univ} > 0$. If $\Delta S_{univ} < 0$, the process is nonspontaneous, and if $\Delta S_{univ} = 0$, the system is at equilibrium. The third law of thermodynamics establishes the zero for entropy as that of a perfect, pure crystalline solid at 0 K. With only one possible microstate, the entropy is zero. We may compute the standard entropy change for a process by using standard entropy values for the reactants and products involved in the process.

16.4 Free Energy

Gibbs free energy (*G*) is a state function defined with regard to system quantities only and may be used to predict the spontaneity of a process. A negative value for ΔG indicates a spontaneous process; a positive ΔG indicates a nonspontaneous process; and a ΔG of zero indicates that the system is at equilibrium. A number of approaches to the computation of free energy changes are possible.

Exercises

16.1 Spontaneity

1. What is a spontaneous reaction?
2. What is a nonspontaneous reaction?
3. Indicate whether the following processes are spontaneous or nonspontaneous.
   (a) Liquid water freezing at a temperature below its freezing point
   (b) Liquid water freezing at a temperature above its freezing point
   (c) The combustion of gasoline
   (d) A ball thrown into the air
   (e) A raindrop falling to the ground
   (f) Iron rusting in a moist atmosphere
4. A helium-filled balloon spontaneously deflates overnight as He atoms diffuse through the wall of the balloon. Describe the redistribution of matter and/or energy that accompanies this process.
5. Many plastic materials are organic polymers that contain carbon and hydrogen. The oxidation of these plastics in air to form carbon dioxide and water is a spontaneous process; however, plastic materials tend to persist in the environment. Explain.

16.2 Entropy

6. In Figure 16.8 all possible distributions and microstates are shown for four different particles shared between two boxes. Determine the entropy change, ΔS, if the particles are initially evenly distributed between the two boxes, but upon redistribution all end up in Box (b).
7. In Figure 16.8 all of the possible distributions and microstates are shown for four different particles shared between two boxes. Determine the entropy change, ΔS, for the system when it is converted from distribution (b) to distribution (d).
8. How does the process described in the previous item relate to the system shown in Figure 16.4?
9. Consider a system similar to the one in Figure 16.8, except that it contains six particles instead of four. What is the probability of having all the particles in only one of the two boxes in the case? Compare this with the similar probability for the system of four particles that we have derived to be equal to $\frac{1}{8}$. What does this comparison tell us about even larger systems?
10. Consider the system shown in Figure 16.9. What is the change in entropy for the process where the energy is initially associated only with particle A, but in the final state the energy is distributed between two different particles?
11. Consider the system shown in Figure 16.9. What is the change in entropy for the process where the energy is initially associated with particles A and B, and the energy is distributed between two particles in different boxes (one in A-B, the other in C-D)?

12. Arrange the following sets of systems in order of increasing entropy. Assume one mole of each substance and the same temperature for each member of a set.
(a) $H_2(g)$, $HBrO_4(g)$, $HBr(g)$
(b) $H_2O(l)$, $H_2O(g)$, $H_2O(s)$
(c) $He(g)$, $Cl_2(g)$, $P_4(g)$
13. At room temperature, the entropy of the halogens increases from I_2 to Br_2 to Cl_2. Explain.
14. Consider two processes: sublimation of $I_2(s)$ and melting of $I_2(s)$ (Note: the latter process can occur at the same temperature but somewhat higher pressure).
$I_2(s) \longrightarrow I_2(g)$
$I_2(s) \longrightarrow I_2(l)$
Is ΔS positive or negative in these processes? In which of the processes will the magnitude of the entropy change be greater?
15. Indicate which substance in the given pairs has the higher entropy value. Explain your choices.
(a) $C_2H_5OH(l)$ or $C_3H_7OH(l)$
(b) $C_2H_5OH(l)$ or $C_2H_5OH(g)$
(c) $2H(g)$ or $H(g)$
16. Predict the sign of the entropy change for the following processes.
(a) An ice cube is warmed to near its melting point.
(b) Exhaled breath forms fog on a cold morning.
(c) Snow melts.
17. Predict the sign of the entropy change for the following processes. Give a reason for your prediction.
(a) $Na^+(aq) + Cl^-(aq) \longrightarrow NaCl(s)$
(b) $2Fe(s) + \frac{3}{2}O_2(g) \longrightarrow Fe_2O_2(s)$
(c) $2C_6H_{14}(l) + 19O_2(g) \longrightarrow 14H_2O(g) + 12CO_2(g)$
18. Write the balanced chemical equation for the combustion of methane, $CH_4(g)$, to give carbon dioxide and water vapor. Explain why it is difficult to predict whether ΔS is positive or negative for this chemical reaction.
19. Write the balanced chemical equation for the combustion of benzene, $C_6H_6(l)$, to give carbon dioxide and water vapor. Would you expect ΔS to be positive or negative in this process?

16.3 The Second and Third Laws of Thermodynamics

20. What is the difference between ΔS and $\Delta S°$ for a chemical change?
21. Calculate $\Delta S°$ for the following changes.
(a) $SnCl_4(l) \longrightarrow SnCl_4(g)$
(b) $CS_2(g) \longrightarrow CS_2(l)$
(c) $Cu(s) \longrightarrow Cu(g)$
(d) $H_2O(l) \longrightarrow H_2O(g)$
(e) $2H_2(g) + O_2(g) \longrightarrow 2H_2O(l)$
(f) $2HCl(g) + Pb(s) \longrightarrow PbCl_2(s) + H_2(g)$
(g) $Zn(s) + CuSO_4(s) \longrightarrow Cu(s) + ZnSO_4(s)$
22. Determine the entropy change for the combustion of liquid ethanol, C_2H_5OH, under the standard conditions to give gaseous carbon dioxide and liquid water.
23. Determine the entropy change for the combustion of gaseous propane, C_3H_8, under the standard conditions to give gaseous carbon dioxide and water.
24. "Thermite" reactions have been used for welding metal parts such as railway rails and in metal refining. One such thermite reaction is $Fe_2O_3(s) + 2Al(s) \longrightarrow Al_2O_3(s) + 2Fe(s)$. Is the reaction spontaneous at room temperature under standard conditions? During the reaction, the surroundings absorb 851.8 kJ/mol of heat.
25. Using the relevant $S°$ values listed in Appendix G, calculate $\Delta S°_{298}$ for the following changes:
(a) $N_2(g) + 3H_2(g) \longrightarrow 2NH_3(g)$
(b) $N_2(g) + \frac{5}{2}O_2(g) \longrightarrow N_2O_5(g)$

26. From the following information, determine ΔS° for the following:

$N(g) + O(g) \longrightarrow NO(g) \quad \Delta S^\circ = ?$

$N_2(g) + O_2(g) \longrightarrow 2NO(g) \quad \Delta S^\circ = 24.8 \text{ J/K}$

$N_2(g) \longrightarrow 2N(g) \quad \Delta S^\circ = 115.0 \text{ J/K}$

$O_2(g) \longrightarrow 2O(g) \quad \Delta S^\circ = 117.0 \text{ J/K}$

27. By calculating ΔS_{univ} at each temperature, determine if the melting of 1 mole of NaCl(*s*) is spontaneous at 500 °C and at 700 °C.

$S^\circ_{\text{NaCl}(s)} = 72.11 \frac{\text{J}}{\text{mol·K}} \quad S^\circ_{\text{NaCl}(l)} = 95.06 \frac{\text{J}}{\text{mol·K}} \quad \Delta H^\circ_{\text{fusion}} = 27.95 \text{ kJ/mol}$

What assumptions are made about the thermodynamic information (entropy and enthalpy values) used to solve this problem?

28. Use the standard entropy data in Appendix G to determine the change in entropy for each of the following reactions. All the processes occur at the standard conditions and 25 °C.

(a) $MnO_2(s) \longrightarrow Mn(s) + O_2(g)$

(b) $H_2(g) + Br_2(l) \longrightarrow 2HBr(g)$

(c) $Cu(s) + S(g) \longrightarrow CuS(s)$

(d) $2LiOH(s) + CO_2(g) \longrightarrow Li_2CO_3(s) + H_2O(g)$

(e) $CH_4(g) + O_2(g) \longrightarrow C(s, \text{ graphite}) + 2H_2O(g)$

(f) $CS_2(g) + 3Cl_2(g) \longrightarrow CCl_4(g) + S_2Cl_2(g)$

29. Use the standard entropy data in Appendix G to determine the change in entropy for each of the following reactions. All the processes occur at the standard conditions and 25 °C.

(a) $C(s, \text{ graphite}) + O_2(g) \longrightarrow CO_2(g)$

(b) $O_2(g) + N_2(g) \longrightarrow 2NO(g)$

(c) $2Cu(s) + S(g) \longrightarrow Cu_2S(s)$

(d) $CaO(s) + H_2O(l) \longrightarrow Ca(OH)_2(s)$

(e) $Fe_2O_3(s) + 3CO(g) \longrightarrow 2Fe(s) + 3CO_2(g)$

(f) $CaSO_4{\cdot}2H_2O(s) \longrightarrow CaSO_4(s) + 2H_2O(g)$

16.4 Free Energy

30. What is the difference between ΔG and ΔG° for a chemical change?

31. A reaction has $\Delta H^\circ = 100$ kJ/mol and $\Delta S^\circ = 250$ J/mol·K. Is the reaction spontaneous at room temperature? If not, under what temperature conditions will it become spontaneous?

32. Explain what happens as a reaction starts with $\Delta G < 0$ (negative) and reaches the point where $\Delta G = 0$.

33. Use the standard free energy of formation data in Appendix G to determine the free energy change for each of the following reactions, which are run under standard state conditions and 25 °C. Identify each as either spontaneous or nonspontaneous at these conditions.

(a) $MnO_2(s) \longrightarrow Mn(s) + O_2(g)$

(b) $H_2(g) + Br_2(l) \longrightarrow 2HBr(g)$

(c) $Cu(s) + S(g) \longrightarrow CuS(s)$

(d) $2LiOH(s) + CO_2(g) \longrightarrow Li_2CO_3(s) + H_2O(g)$

(e) $CH_4(g) + O_2(g) \longrightarrow C(s, \text{ graphite}) + 2H_2O(g)$

(f) $CS_2(g) + 3Cl_2(g) \longrightarrow CCl_4(g) + S_2Cl_2(g)$

34. Use the standard free energy data in Appendix G to determine the free energy change for each of the following reactions, which are run under standard state conditions and 25 °C. Identify each as either spontaneous or nonspontaneous at these conditions.

(a) $C(s, \text{ graphite}) + O_2(g) \longrightarrow CO_2(g)$

(b) $O_2(g) + N_2(g) \longrightarrow 2NO(g)$

(c) $2Cu(s) + S(g) \longrightarrow Cu_2S(s)$

(d) $CaO(s) + H_2O(l) \longrightarrow Ca(OH)_2(s)$

(e) $Fe_2O_3(s) + 3CO(g) \longrightarrow 2Fe(s) + 3CO_2(g)$

(f) $CaSO_4{\cdot}2H_2O(s) \longrightarrow CaSO_4(s) + 2H_2O(g)$

35. Given:

$P_4(s) + 5O_2(g) \longrightarrow P_4O_{10}(s)$ $\Delta G° = -2697.0$ kJ/mol

$2H_2(g) + O_2(g) \longrightarrow 2H_2O(g)$ $\Delta G° = -457.18$ kJ/mol

$6H_2O(g) + P_4O_{10}(s) \longrightarrow 4H_3PO_4(l)$ $\Delta G° = -428.66$ kJ/mol

(a) Determine the standard free energy of formation, $\Delta G_f°$, for phosphoric acid.

(b) How does your calculated result compare to the value in Appendix G? Explain.

36. Is the formation of ozone ($O_3(g)$) from oxygen ($O_2(g)$) spontaneous at room temperature under standard state conditions?

37. Consider the decomposition of red mercury(II) oxide under standard state conditions.

$2HgO(s, \text{red}) \longrightarrow 2Hg(l) + O_2(g)$

(a) Is the decomposition spontaneous under standard state conditions?

(b) Above what temperature does the reaction become spontaneous?

38. Among other things, an ideal fuel for the control thrusters of a space vehicle should decompose in a spontaneous exothermic reaction when exposed to the appropriate catalyst. Evaluate the following substances under standard state conditions as suitable candidates for fuels.

(a) Ammonia: $2NH_3(g) \longrightarrow N_2(g) + 3H_2(g)$

(b) Diborane: $B_2H_6(g) \longrightarrow 2B(g) + 3H_2(g)$

(c) Hydrazine: $N_2H_4(g) \longrightarrow N_2(g) + 2H_2(g)$

(d) Hydrogen peroxide: $H_2O_2(l) \longrightarrow H_2O(g) + \frac{1}{2}O_2(g)$

39. Calculate $\Delta G°$ for each of the following reactions from the equilibrium constant at the temperature given.

(a) $N_2(g) + O_2(g) \longrightarrow 2NO(g)$ T = 2000 °C $K_p = 4.1 \times 10^{-4}$

(b) $H_2(g) + I_2(g) \longrightarrow 2HI(g)$ T = 400 °C $K_p = 50.0$

(c) $CO_2(g) + H_2(g) \longrightarrow CO(g) + H_2O(g)$ T = 980 °C $K_p = 1.67$

(d) $CaCO_3(s) \longrightarrow CaO(s) + CO_2(g)$ T = 900 °C $K_p = 1.04$

(e) $HF(aq) + H_2O(l) \longrightarrow H_3O^+(aq) + F^-(aq)$ T = 25 °C $K_p = 7.2 \times 10^{-4}$

(f) $AgBr(s) \longrightarrow Ag^+(aq) + Br^-(aq)$ T = 25 °C $K_p = 3.3 \times 10^{-13}$

40. Calculate $\Delta G°$ for each of the following reactions from the equilibrium constant at the temperature given.

(a) $Cl_2(g) + Br_2(g) \longrightarrow 2BrCl(g)$ T = 25 °C $K_p = 4.7 \times 10^{-2}$

(b) $2SO_2(g) + O_2(g) \rightleftharpoons 2SO_3(g)$ T = 500 °C $K_p = 48.2$

(c) $H_2O(l) \rightleftharpoons H_2O(g)$ T = 60 °C $K_p = 0.196$

(d) $CoO(s) + CO(g) \rightleftharpoons Co(s) + CO_2(g)$ T = 550 °C $K_p = 4.90 \times 10^2$

(e)

$CH_3NH_2(aq) + H_2O(l) \longrightarrow CH_3NH_3{}^+(aq) + OH^-(aq)$ T = 25 °C $K_p = 4.4 \times 10^{-4}$

(f) $PbI_2(s) \longrightarrow Pb^{2+}(aq) + 2I^-(aq)$ T = 25 °C $K_p = 8.7 \times 10^{-9}$

41. Calculate the equilibrium constant at 25 °C for each of the following reactions from the value of $\Delta G°$ given.

(a) $O_2(g) + 2F_2(g) \longrightarrow 2OF_2(g)$ $\Delta G° = -9.2$ kJ

(b) $I_2(s) + Br_2(l) \longrightarrow 2IBr(g)$ $\Delta G° = 7.3$ kJ

(c) $2LiOH(s) + CO_2(g) \longrightarrow Li_2CO_3(s) + H_2O(g)$ $\Delta G° = -79$ kJ

(d) $N_2O_3(g) \longrightarrow NO(g) + NO_2(g)$ $\Delta G° = -1.6$ kJ

(e) $SnCl_4(l) \longrightarrow SnCl_4(l)$ $\Delta G° = 8.0$ kJ

42. Calculate the equilibrium constant at 25 °C for each of the following reactions from the value of $\Delta G°$ given.

(a) $I_2(s) + Cl_2(g) \longrightarrow 2ICl(g)$ $\Delta G° = -10.88$ kJ

(b) $H_2(g) + I_2(s) \longrightarrow 2HI(g)$ $\Delta G° = 3.4$ kJ

(c) $CS_2(g) + 3Cl_2(g) \longrightarrow CCl_4(g) + S_2Cl_2(g)$ $\Delta G° = -39$ kJ

(d) $2SO_2(g) + O_2(g) \longrightarrow 2SO_3(g)$ $\Delta G° = -141.82$ kJ

(e) $CS_2(g) \longrightarrow CS_2(l)$ $\Delta G° = -1.88$ kJ

43. Calculate the equilibrium constant at the temperature given.

(a) $O_2(g) + 2F_2(g) \longrightarrow 2F_2O(g)$ (T = 100 °C)

(b) $I_2(s) + Br_2(l) \longrightarrow 2IBr(g)$ (T = 0.0 °C)

(c) $2LiOH(s) + CO_2(g) \longrightarrow Li_2CO_3(s) + H_2O(g)$ (T = 575 °C)

(d) $N_2O_3(g) \longrightarrow NO(g) + NO_2(g)$ (T = −10.0 °C)

(e) $SnCl_4(l) \longrightarrow SnCl_4(g)$ (T = 200 °C)

44. Calculate the equilibrium constant at the temperature given.
(a) $I_2(s) + Cl_2(g) \longrightarrow 2ICl(g)$ (T = 100 °C)
(b) $H_2(g) + I_2(s) \longrightarrow 2HI(g)$ (T = 0.0 °C)
(c) $CS_2(g) + 3Cl_2(g) \longrightarrow CCl_4(g) + S_2Cl_2(g)$ (T = 125 °C)
(d) $2SO_2(g) + O_2(g) \longrightarrow 2SO_3(g)$ (T = 675 °C)
(e) $CS_2(g) \longrightarrow CS_2(l)$ (T = 90 °C)

45. Consider the following reaction at 298 K:
$N_2O_4(g) \rightleftharpoons 2NO_2(g) \qquad K_P = 0.142$
What is the standard free energy change at this temperature? Describe what happens to the initial system, where the reactants and products are in standard states, as it approaches equilibrium.

46. Determine the normal boiling point (in kelvin) of dichloromethane, CH_2Cl_2. Find the actual boiling point using the Internet or some other source, and calculate the percent error in the temperature. Explain the differences, if any, between the two values.

47. Under what conditions is $N_2O_3(g) \longrightarrow NO(g) + NO_2(g)$ spontaneous?

48. At room temperature, the equilibrium constant (K_w) for the self-ionization of water is 1.00×10^{-14}. Using this information, calculate the standard free energy change for the aqueous reaction of hydrogen ion with hydroxide ion to produce water. (Hint: The reaction is the reverse of the self-ionization reaction.)

49. Hydrogen sulfide is a pollutant found in natural gas. Following its removal, it is converted to sulfur by the reaction $2H_2S(g) + SO_2(g) \rightleftharpoons \frac{3}{8}S_8(s, \text{rhombic}) + 2H_2O(l)$. What is the equilibrium constant for this reaction? Is the reaction endothermic or exothermic?

50. Consider the decomposition of $CaCO_3(s)$ into $CaO(s)$ and $CO_2(g)$. What is the equilibrium partial pressure of CO_2 at room temperature?

51. In the laboratory, hydrogen chloride ($HCl(g)$) and ammonia ($NH_3(g)$) often escape from bottles of their solutions and react to form the ammonium chloride ($NH_4Cl(s)$), the white glaze often seen on glassware. Assuming that the number of moles of each gas that escapes into the room is the same, what is the maximum partial pressure of HCl and NH_3 in the laboratory at room temperature? (Hint: The partial pressures will be equal and are at their maximum value when at equilibrium.)

52. Benzene can be prepared from acetylene. $3C_2H_2(g) \rightleftharpoons C_6H_6(g)$. Determine the equilibrium constant at 25 °C and at 850 °C. Is the reaction spontaneous at either of these temperatures? Why is all acetylene not found as benzene?

53. Carbon dioxide decomposes into CO and O_2 at elevated temperatures. What is the equilibrium partial pressure of oxygen in a sample at 1000 °C for which the initial pressure of CO_2 was 1.15 atm?

54. Carbon tetrachloride, an important industrial solvent, is prepared by the chlorination of methane at 850 K.
$CH_4(g) + 4Cl_2(g) \longrightarrow CCl_4(g) + 4HCl(g)$
What is the equilibrium constant for the reaction at 850 K? Would the reaction vessel need to be heated or cooled to keep the temperature of the reaction constant?

55. Acetic acid, CH_3CO_2H, can form a dimer, $(CH_3CO_2H)_2$, in the gas phase.
$2CH_3CO_2H(g) \longrightarrow (CH_3CO_2H)_2(g)$
The dimer is held together by two hydrogen bonds with a total strength of 66.5 kJ per mole of dimer.

O•••H—O
CH3—C C—CH3
O—H•••O

At 25 °C, the equilibrium constant for the dimerization is 1.3×10^3 (pressure in atm). What is $\Delta S°$ for the reaction?

56. Determine $\Delta G°$ for the following reactions.
(a) Antimony pentachloride decomposes at 448 °C. The reaction is:
$SbCl_5(g) \longrightarrow SbCl_3(g) + Cl_2(g)$
An equilibrium mixture in a 5.00 L flask at 448 °C contains 3.85 g of $SbCl_5$, 9.14 g of $SbCl_3$, and 2.84 g of Cl_2.
(b) Chlorine molecules dissociate according to this reaction:
$Cl_2(g) \longrightarrow 2Cl(g)$
1.00% of Cl_2 molecules dissociate at 975 K and a pressure of 1.00 atm.

57. Given that the ΔG_f° for $Pb^{2+}(aq)$ and $Cl^-(aq)$ is −24.3 kJ/mole and −131.2 kJ/mole respectively, determine the solubility product, K_{sp}, for $PbCl_2(s)$.

58. Determine the standard free energy change, ΔG_f°, for the formation of $S^{2-}(aq)$ given that the ΔG_f° for $Ag^+(aq)$ and $Ag_2S(s)$ are 77.1 kJ/mole and −39.5 kJ/mole respectively, and the solubility product for $Ag_2S(s)$ is 8×10^{-51}.

59. Determine the standard enthalpy change, entropy change, and free energy change for the conversion of diamond to graphite. Discuss the spontaneity of the conversion with respect to the enthalpy and entropy changes. Explain why diamond spontaneously changing into graphite is not observed.

60. The evaporation of one mole of water at 298 K has a standard free energy change of 8.58 kJ.
$H_2O(l) \rightleftharpoons H_2O(g) \qquad \Delta G^\circ = 8.58 \text{ kJ}$
(a) Is the evaporation of water under standard thermodynamic conditions spontaneous?
(b) Determine the equilibrium constant, K_P, for this physical process.
(c) By calculating ΔG, determine if the evaporation of water at 298 K is spontaneous when the partial pressure of water, P_{H_2O}, is 0.011 atm.
(d) If the evaporation of water were always nonspontaneous at room temperature, wet laundry would never dry when placed outside. In order for laundry to dry, what must be the value of P_{H_2O} in the air?

61. In glycolysis, the reaction of glucose (Glu) to form glucose-6-phosphate (G6P) requires ATP to be present as described by the following equation:
$\text{Glu} + \text{ATP} \longrightarrow \text{G6P} + \text{ADP} \qquad \Delta G^\circ = -17 \text{ kJ}$
In this process, ATP becomes ADP summarized by the following equation:
$\text{ATP} \longrightarrow \text{ADP} \qquad \Delta G^\circ = -30 \text{ kJ}$
Determine the standard free energy change for the following reaction, and explain why ATP is necessary to drive this process:
$\text{Glu} \longrightarrow \text{G6P} \qquad \Delta G^\circ = ?$

62. One of the important reactions in the biochemical pathway glycolysis is the reaction of glucose-6-phosphate (G6P) to form fructose-6-phosphate (F6P):
$\text{G6P} \rightleftharpoons \text{F6P} \qquad \Delta G^\circ = 1.7 \text{ kJ}$
(a) Is the reaction spontaneous or nonspontaneous under standard thermodynamic conditions?
(b) Standard thermodynamic conditions imply the concentrations of G6P and F6P to be 1 *M*, however, in a typical cell, they are not even close to these values. Calculate ΔG when the concentrations of G6P and F6P are 120 μ*M* and 28 μ*M* respectively, and discuss the spontaneity of the forward reaction under these conditions. Assume the temperature is 37 °C.

63. Without doing a numerical calculation, determine which of the following will reduce the free energy change for the reaction, that is, make it less positive or more negative, when the temperature is increased. Explain.
(a) $N_2(g) + 3H_2(g) \longrightarrow 2NH_3(g)$
(b) $HCl(g) + NH_3(g) \longrightarrow NH_4Cl(s)$
(c) $(NH_4)_2Cr_2O_7(s) \longrightarrow Cr_2O_3(s) + 4H_2O(g) + N_2(g)$
(d) $2Fe(s) + 3O_2(g) \longrightarrow Fe_2O_3(s)$

64. When ammonium chloride is added to water and stirred, it dissolves spontaneously and the resulting solution feels cold. Without doing any calculations, deduce the signs of ΔG, ΔH, and ΔS for this process, and justify your choices.

65. An important source of copper is from the copper ore, chalcocite, a form of copper(I) sulfide. When heated, the Cu_2S decomposes to form copper and sulfur described by the following equation:
$Cu_2S(s) \longrightarrow Cu(s) + S(s)$
(a) Determine $\Delta G°$ for the decomposition of $Cu_2S(s)$.
(b) The reaction of sulfur with oxygen yields sulfur dioxide as the only product. Write an equation that describes this reaction, and determine $\Delta G°$ for the process.
(c) The production of copper from chalcocite is performed by roasting the Cu_2S in air to produce the Cu. By combining the equations from Parts (a) and (b), write the equation that describes the roasting of the chalcocite, and explain why coupling these reactions together makes for a more efficient process for the production of the copper.

66. What happens to ΔG (becomes more negative or more positive) for the following chemical reactions when the partial pressure of oxygen is increased?
(a) $S(s) + O_2(g) \longrightarrow SO_2(g)$
(b) $2SO_2(g) + O_2(g) \longrightarrow 2SO_3(g)$
(c) $HgO(s) \longrightarrow Hg(l) + O_2(g)$

CHAPTER 17
Electrochemistry

Figure 17.1 Electric vehicles are powered by batteries, devices that harness the energy of spontaneous redox reactions. (credit: modification of work by Robert Couse-Baker)

CHAPTER OUTLINE

17.1 Review of Redox Chemistry
17.2 Galvanic Cells
17.3 Electrode and Cell Potentials
17.4 Potential, Free Energy, and Equilibrium
17.5 Batteries and Fuel Cells
17.6 Corrosion
17.7 Electrolysis

INTRODUCTION Another chapter in this text introduced the chemistry of reduction-oxidation (redox) reactions. This important reaction class is defined by changes in oxidation states for one or more reactant elements, and it includes a subset of reactions involving the transfer of electrons between reactant species. Around the turn of the nineteenth century, chemists began exploring ways these electrons could be transferred *indirectly* via an external circuit rather than directly via intimate contact of redox reactants. In the two centuries since, the field of *electrochemistry* has evolved to yield significant insights on the fundamental aspects of redox chemistry as well as a wealth of technologies ranging from industrial-scale metallurgical processes to robust, rechargeable batteries for electric vehicles (Figure 17.1). In this chapter, the essential concepts of electrochemistry will be addressed.

17.1 Review of Redox Chemistry

LEARNING OBJECTIVES

By the end of this section, you will be able to:

- Describe defining traits of redox chemistry
- Identify the oxidant and reductant of a redox reaction
- Balance chemical equations for redox reactions using the half-reaction method

Since reactions involving electron transfer are essential to the topic of electrochemistry, a brief review of redox chemistry is provided here that summarizes and extends the content of an earlier text chapter (see chapter on reaction stoichiometry). Readers wishing additional review are referred to the text chapter on reaction stoichiometry.

Oxidation Numbers

By definition, a redox reaction is one that entails changes in *oxidation number* (or *oxidation state*) for one or more of the elements involved. The oxidation number of an element in a compound is essentially an assessment of how the electronic environment of its atoms is different in comparison to atoms of the pure element. By this description, *the oxidation number of an atom in an element is equal to zero.* For an atom in a compound, *the oxidation number is equal to the charge the atom would have in the compound if the compound were ionic.* Consequential to these rules, *the sum of oxidation numbers for all atoms in a molecule is equal to the charge on the molecule.* To illustrate this formalism, examples from the two compound classes, ionic and covalent, will be considered.

Simple ionic compounds present the simplest examples to illustrate this formalism, since by definition the elements' oxidation numbers are numerically equivalent to ionic charges. Sodium chloride, NaCl, is comprised of Na^+ cations and Cl^- anions, and so oxidation numbers for sodium and chlorine are, +1 and −1, respectively. Calcium fluoride, CaF_2, is comprised of Ca^{2+} cations and F^- anions, and so oxidation numbers for calcium and fluorine are, +2 and −1, respectively.

Covalent compounds require a more challenging use of the formalism. Water is a covalent compound whose molecules consist of two H atoms bonded separately to a central O atom via polar covalent O–H bonds. The shared electrons comprising an O–H bond are more strongly attracted to the more electronegative O atom, and so it acquires a partial negative charge in the water molecule (relative to an O atom in elemental oxygen). Consequently, H atoms in a water molecule exhibit partial positive charges compared to H atoms in elemental hydrogen. The sum of the partial negative and partial positive charges for each water molecule is zero, and the water molecule is neutral.

Imagine that the polarization of shared electrons within the O–H bonds of water were 100% complete—the result would be *transfer* of electrons from H to O, and water would be an ionic compound comprised of O^{2-} anions and H^+ cations. And so, the oxidations numbers for oxygen and hydrogen in water are −2 and +1, respectively. Applying this same logic to carbon tetrachloride, CCl_4, yields oxidation numbers of +4 for carbon and −1 for chlorine. In the nitrate ion, NO_3^-, the oxidation number for nitrogen is +5 and that for oxygen is −2, summing to equal the 1− charge on the molecule:

$$(1\text{N atom})\left(\frac{+5}{\text{N atom}}\right) + (3\text{ O atoms})\left(\frac{-2}{\text{O atom}}\right) = +5 + -6 = -1$$

Balancing Redox Equations

The unbalanced equation below describes the decomposition of molten sodium chloride:

$$\text{NaCl}(l) \longrightarrow \text{Na}(l) + \text{Cl}_2(g) \qquad \textit{unbalanced}$$

This reaction satisfies the criterion for redox classification, since the oxidation number for Na is decreased from +1 to 0 (it undergoes *reduction*) and that for Cl is increased from −1 to 0 (it undergoes *oxidation*). The equation in this case is easily balanced by inspection, requiring stoichiometric coefficients of 2 for the NaCl and Na:

$$2NaCl(l) \longrightarrow 2Na(l) + Cl_2(g) \qquad \textit{balanced}$$

Redox reactions that take place in aqueous solutions are commonly encountered in electrochemistry, and many involve water or its characteristic ions, $H^+(aq)$ and $OH^-(aq)$, as reactants or products. In these cases, equations representing the redox reaction can be very challenging to balance by inspection, and the use of a systematic approach called the *half-reaction method* is helpful. This approach involves the following steps:

1. Write skeletal equations for the oxidation and reduction half-reactions.
2. Balance each half-reaction for all elements except H and O.
3. Balance each half-reaction for O by adding H_2O.
4. Balance each half-reaction for H by adding H^+.
5. Balance each half-reaction for charge by adding electrons.
6. If necessary, multiply one or both half-reactions so that the number of electrons consumed in one is equal to the number produced in the other.
7. Add the two half-reactions and simplify.
8. If the reaction takes place in a basic medium, add OH^- ions the equation obtained in step 7 to neutralize the H^+ ions (add in equal numbers to both sides of the equation) and simplify.

The examples below demonstrate the application of this method to balancing equations for aqueous redox reactions.

 **EXAMPLE 17.1**

Balancing Equations for Redox Reactions in Acidic Solutions

Write the balanced equation representing reaction between solid copper and nitric acid to yield aqueous copper(II) ions and nitrogen monoxide gas.

Solution

Following the steps of the half-reaction method:

1. *Write skeletal equations for the oxidation and reduction half-reactions.*
   oxidation: $Cu(s) \longrightarrow Cu^{2+}(aq)$
   reduction: $HNO_3(aq) \longrightarrow NO(g)$
2. *Balance each half-reaction for all elements except H and O.*
   oxidation: $Cu(s) \longrightarrow Cu^{2+}(aq)$
   reduction: $HNO_3(aq) \longrightarrow NO(g)$
3. *Balance each half-reaction for O by adding H_2O.*
   oxidation: $Cu(s) \longrightarrow Cu^{2+}(aq)$
   reduction: $HNO_3(aq) \longrightarrow NO(g) + \mathbf{2H_2O}(l)$
4. *Balance each half-reaction for H by adding H^+.*
   oxidation: $Cu(s) \longrightarrow Cu^{2+}(aq)$
   reduction: $\mathbf{3H^+}(aq) + HNO_3(aq) \longrightarrow NO(g) + 2H_2O(l)$
5. *Balance each half-reaction for charge by adding electrons.*
   oxidation: $Cu(s) \longrightarrow Cu^{2+}(aq) + \mathbf{2e^-}$
   reduction: $\mathbf{3e^-} + 3H^+(aq) + HNO_3(aq) \longrightarrow NO(g) + 2H_2O(l)$
6. *If necessary, multiply one or both half-reactions so that the number of electrons consumed in one is equal to the number produced in the other.*
   oxidation (×3): $\mathbf{3}Cu(s) \longrightarrow \mathbf{3}Cu^{2+}(aq) + \mathbf{6}\cancel{2}e^-$
   reduction (×2): $\mathbf{6}\cancel{3}e^- + \mathbf{6}\cancel{3}H^+(aq) + \mathbf{2}HNO_3(aq) \longrightarrow \mathbf{2}NO(g) + \mathbf{4}\cancel{2}H_2O(l)$
7. *Add the two half-reactions and simplify.*
   $3Cu(s) + \cancel{\mathbf{6e^-}} + 6H^+(aq) + 2HNO_3(aq) \longrightarrow 3Cu^{2+}(aq) + \cancel{\mathbf{6e^-}} + 2NO(g) + 4H_2O(l)$
   $3Cu(s) + 6H^+(aq) + 2HNO_3(aq) \longrightarrow 3Cu^{2+}(aq) + 2NO(g) + 4H_2O(l)$
8. *If the reaction takes place in a basic medium, add OH^- ions the equation obtained in step 7 to neutralize the H^+ ions (add in equal numbers to both sides of the equation) and simplify.*

This step not necessary since the solution is stipulated to be acidic.

The balanced equation for the reaction in an acidic solution is then

$$3Cu(s) + 6H^+(aq) + 2HNO_3(aq) \longrightarrow 3Cu^{2+}(aq) + 2NO(g) + 4H_2O(l)$$

Check Your Learning

The reaction above results when using relatively diluted nitric acid. If concentrated nitric acid is used, nitrogen dioxide is produced instead of nitrogen monoxide. Write a balanced equation for this reaction.

Answer:

$Cu(s) + 2H^+(aq) + 2HNO_3(aq) \longrightarrow Cu^{2+}(aq) + 2NO_2(g) + 2H_2O(l)$

EXAMPLE 17.2

Balancing Equations for Redox Reactions in Basic Solutions

Write the balanced equation representing reaction between aqueous permanganate ion, MnO_4^-, and solid chromium(III) hydroxide, $Cr(OH)_3$, to yield solid manganese(IV) oxide, MnO_2, and aqueous chromate ion, CrO_4^{2-} The reaction takes place in a basic solution.

Solution

Following the steps of the half-reaction method:

1. *Write skeletal equations for the oxidation and reduction half-reactions.*
   oxidation: $Cr(OH)_3(s) \longrightarrow CrO_4^{2-}(aq)$
   reduction: $MnO_4^-(aq) \longrightarrow MnO_2(s)$
2. *Balance each half-reaction for all elements except H and O.*
   oxidation: $Cr(OH)_3(s) \longrightarrow CrO_4^{2-}(aq)$
   reduction: $MnO_4^-(aq) \longrightarrow MnO_2(s)$
3. *Balance each half-reaction for O by adding H_2O.*
   oxidation: $\mathbf{H_2O(\mathit{l})}+Cr(OH)_3(s) \longrightarrow CrO_4^{2-}(aq)$
   reduction: $MnO_4^-(aq) \longrightarrow MnO_2(s) + \mathbf{2H_2O(\mathit{l})}$
4. *Balance each half-reaction for H by adding H^+.*
   oxidation: $H_2O(l) + Cr(OH)_3(s) \longrightarrow CrO_4^{2-}(aq) + \mathbf{5H^+(\mathit{aq})}$
   reduction: $\mathbf{4H^+(\mathit{aq})}+MnO_4^-(aq) \longrightarrow MnO_2(s) + 2H_2O(l)$
5. *Balance each half-reaction for charge by adding electrons.*
   oxidation: $H_2O(l) + Cr(OH)_3(s) \longrightarrow CrO_4^{2-}(aq) + 5H^+(aq) + \mathbf{3e^-}$
   reduction: $\mathbf{3e^-} + 4H^+(aq) + MnO_4^-(aq) \longrightarrow MnO_2(s) + 2H_2O(l)$
6. *If necessary, multiply one or both half-reactions so that the number of electrons consumed in one is equal to the number produced in the other.*
   This step is not necessary since the number of electrons is already in balance.
7. *Add the two half-reactions and simplify.*
   $\cancel{H_2O(l)} + Cr(OH)_3(s) + \cancel{3e^-} + \cancel{4H^+(aq)} + MnO_4^-(aq) \longrightarrow CrO_4^{2-}(aq) + \cancel{5}H^+(aq) + \cancel{3e^-} + MnO_2(s) + 2H_2O(l)$
   $Cr(OH)_3(s) + MnO_4^-(aq) \longrightarrow CrO_4^{2-}(aq) + H^+(aq) + MnO_2(s) + H_2O(l)$
8. *If the reaction takes place in a basic medium, add OH^- ions the equation obtained in step 7 to neutralize the H^+ ions (add in equal numbers to both sides of the equation) and simplify.*
   $\mathbf{OH^-(\mathit{aq})}+Cr(OH)_3(s) + MnO_4^-(aq) \longrightarrow CrO_4^{2-}(aq) + H^+(aq) + \mathbf{OH^-(\mathit{aq})} + MnO_2(s) + H_2O(l)$
   $OH^-(aq) + Cr(OH)_3(s) + MnO_4^-(aq) \longrightarrow CrO_4^{2-}(aq) + MnO_2(s) + 2H_2O(l)$

Check Your Learning

Aqueous permanganate ion may also be reduced using aqueous bromide ion, Br^-, the products of this reaction being solid manganese(IV) oxide and aqueous bromate ion, BrO_3^-. Write the balanced equation for this reaction occurring in a basic medium.

Answer:

$H_2O(l) + 2MnO_4^-(aq) + Br^-(aq) \longrightarrow 2MnO_2(s) + BrO_3^-(aq) + 2OH^-(aq)$

17.2 Galvanic Cells

LEARNING OBJECTIVES

By the end of this section, you will be able to:

- Describe the function of a galvanic cell and its components
- Use cell notation to symbolize the composition and construction of galvanic cells

As demonstration of spontaneous chemical change, Figure 17.2 shows the result of immersing a coiled wire of copper into an aqueous solution of silver nitrate. A gradual but visually impressive change spontaneously occurs as the initially colorless solution becomes increasingly blue, and the initially smooth copper wire becomes covered with a porous gray solid.

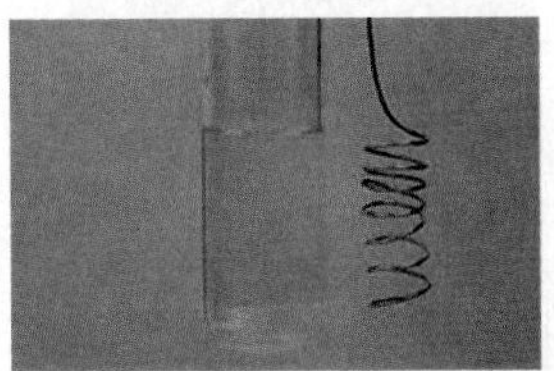
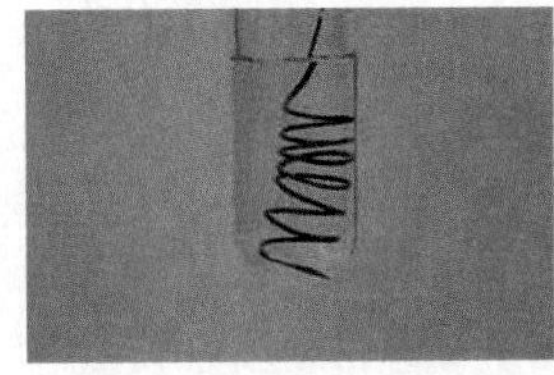

FIGURE 17.2 A copper wire and an aqueous solution of silver nitrate (left) are brought into contact (center) and a spontaneous transfer of electrons occurs, creating blue $Cu^{2+}(aq)$ and gray Ag(*s*) (right).

These observations are consistent with (i) the oxidation of elemental copper to yield copper(II) ions, $Cu^{2+}(aq)$, which impart a blue color to the solution, and (ii) the reduction of silver(I) ions to yield elemental silver, which deposits as a fluffy solid on the copper wire surface. And so, *the direct transfer of electrons from the copper wire to the aqueous silver ions is spontaneous* under the employed conditions. A summary of this redox system is provided by these equations:

$$\begin{array}{ll} \text{overall reaction:} & Cu(s) + 2Ag^+(aq) \longrightarrow Cu^{2+}(aq) + 2Ag(s) \\ \text{oxidation half-reaction:} & Cu(s) \longrightarrow Cu^{2+}(aq) + 2e^- \\ \text{reduction half-reaction:} & 2Ag^+(aq) + 2e^- \longrightarrow 2Ag(s) \end{array}$$

Consider the construction of a device that contains all the reactants and products of a redox system like the one here, but prevents physical contact between the reactants. Direct transfer of electrons is, therefore, prevented; transfer, instead, takes place indirectly through an external circuit that contacts the separated reactants. Devices of this sort are generally referred to as *electrochemical cells*, and those in which a spontaneous redox reaction takes place are called **galvanic cells** (or **voltaic cells**).

A galvanic cell based on the spontaneous reaction between copper and silver(I) is depicted in Figure 17.3. The cell is comprised of two **half-cells**, each containing the redox conjugate pair ("couple") of a single reactant. The half-cell shown at the left contains the Cu(0)/Cu(II) couple in the form of a solid copper foil and an aqueous solution of copper nitrate. The right half-cell contains the Ag(I)/Ag(0) couple as solid silver foil and an aqueous silver nitrate solution. An external circuit is connected to each half-cell at its solid foil, meaning the Cu and Ag foil each function as an *electrode*. By definition, the **anode** of an electrochemical cell is the electrode at which oxidation occurs (in this case, the Cu foil) and the **cathode** is the electrode where reduction occurs (the Ag foil). The redox reactions in a galvanic cell occur only at the interface between each half-cell's reaction mixture and its electrode. To keep the reactants separate while maintaining charge-balance, the two half-cell solutions are connected by a tube filled with inert electrolyte solution called a **salt bridge**. The spontaneous reaction in this cell produces Cu^{2+} cations in the anode half-cell and consumes Ag^+ ions in the cathode half-cell, resulting in a compensatory flow of inert ions from the salt bridge that maintains charge balance. Increasing concentrations of Cu^{2+} in the anode half-cell are balanced by an influx of NO_3^- from the salt bridge, while a flow of Na^+ into the cathode half-cell compensates for the decreasing Ag^+ concentration.

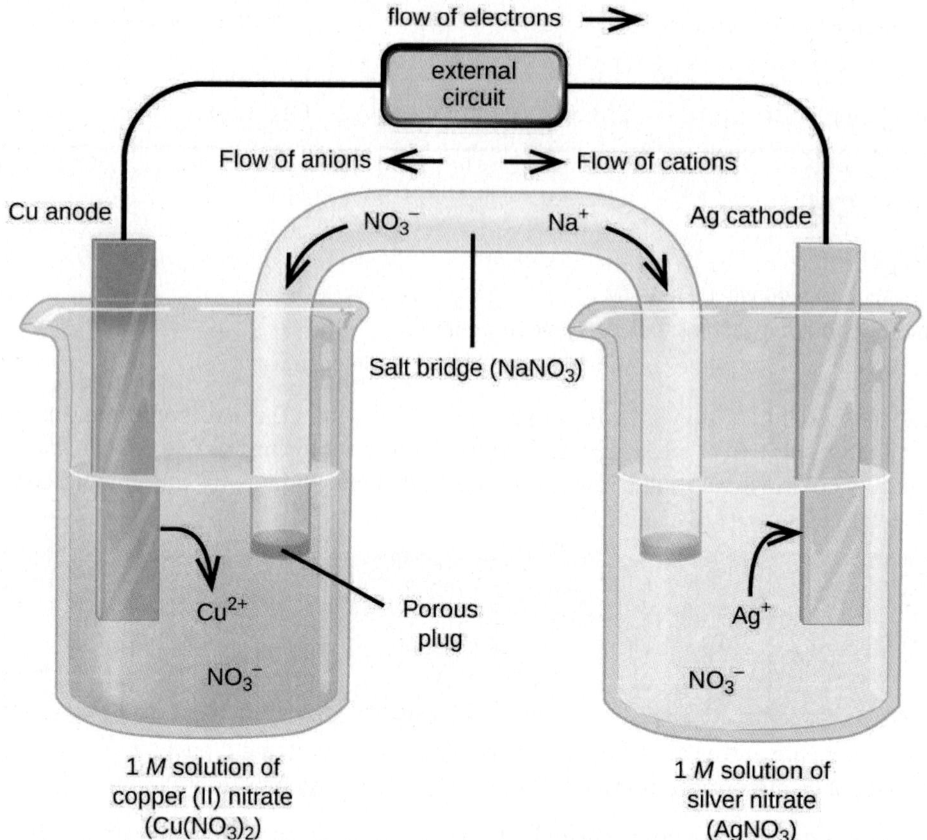

FIGURE 17.3 A galvanic cell based on the spontaneous reaction between copper and silver(I) ions.

Cell Notation

Abbreviated symbolism is commonly used to represent a galvanic cell by providing essential information on its composition and structure. These symbolic representations are called **cell notations** or **cell schematics**, and they are written following a few guidelines:

- The relevant components of each half-cell are represented by their chemical formulas or element symbols
- All interfaces between component phases are represented by vertical parallel lines; if two or more components are present in the same phase, their formulas are separated by commas
- By convention, the schematic begins with the anode and proceeds left-to-right identifying phases and interfaces encountered within the cell, ending with the cathode

A verbal description of the cell as viewed from anode-to-cathode is often a useful first-step in writing its schematic. For example, the galvanic cell shown in Figure 17.3 consists of a solid copper anode immersed in an aqueous solution of copper(II) nitrate that is connected via a salt bridge to an aqueous silver(I) nitrate solution, immersed in which is a solid silver cathode. Converting this statement to symbolism following the above guidelines results in the cell schematic:

$$\text{Cu}(s)\,\middle|\,1\ M\ \text{Cu(NO}_3)_2(aq)\,\middle\|\,1\ M\ \text{AgNO}_3(aq)\,\middle|\,\text{Ag}(s)$$

Consider a different galvanic cell (see Figure 17.4) based on the spontaneous reaction between solid magnesium and aqueous iron(III) ions:

net cell reaction:	$\text{Mg}(s) + 2\text{Fe}^{3+}(aq) \longrightarrow \text{Mg}^{2+}(aq) + 2\text{Fe}^{2+}(aq)$
oxidation half-reaction:	$\text{Mg}(s) \longrightarrow \text{Mg}^{2+}(aq) + 2\text{e}^-$
reduction half-reaction:	$2\text{Fe}^{3+}(aq) + 2\text{e}^- \longrightarrow 2\text{Fe}^{2+}(aq)$

In this cell, a solid magnesium anode is immersed in an aqueous solution of magnesium chloride that is connected via a salt bridge to an aqueous solution containing a mixture of iron(III) chloride and iron(II) chloride, immersed in which is a platinum cathode. The cell schematic is then written as

$$\text{Mg}(s)\,|\,0.1\,M\,\text{MgCl}_2(aq)\,\|\,0.2\,M\,\text{FeCl}_3(aq),0.3\,M\,\text{FeCl}_2(aq)\,|\,\text{Pt}(s)$$

Notice the cathode half-cell is different from the others considered thus far in that its electrode is comprised of a substance (Pt) that is neither a reactant nor a product of the cell reaction. This is required when neither member of the half-cell's redox couple can reasonably function as an electrode, which must be electrically conductive and in a phase separate from the half-cell solution. In this case, both members of the redox couple are solute species, and so Pt is used as an **inert electrode** that can simply provide or accept electrons to redox species in solution. Electrodes constructed from a member of the redox couple, such as the Mg anode in this cell, are called **active electrodes**.

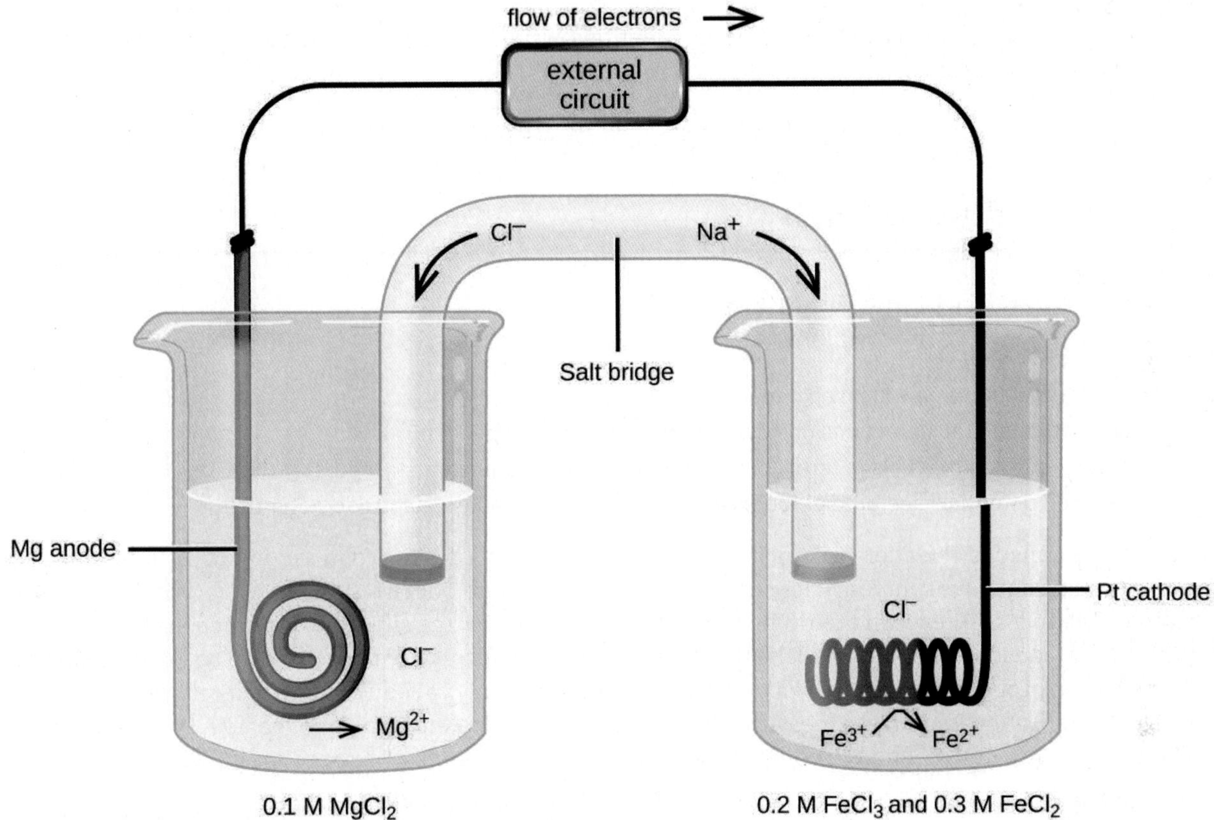

FIGURE 17.4 A galvanic cell based on the spontaneous reaction between magnesium and iron(III) ions.

EXAMPLE 17.3

Writing Galvanic Cell Schematics

A galvanic cell is fabricated by connecting two half-cells with a salt bridge, one in which a chromium wire is immersed in a 1 M $CrCl_3$ solution and another in which a copper wire is immersed in 1 M $CuCl_2$. Assuming the chromium wire functions as an anode, write the schematic for this cell along with equations for the anode half-reaction, the cathode half-reaction, and the overall cell reaction.

Solution

Since the chromium wire is stipulated to be the anode, the schematic begins with it and proceeds left-to-right, symbolizing the other cell components until ending with the copper wire cathode:

$$\text{Cr}(s)\,|\,1\,M\,\text{CrCl}_3(aq)\,\|\,1\,M\,\text{CuCl}_2(aq)\,|\,\text{Cu}(s)$$

The half-reactions for this cell are

$$\text{anode (oxidation):} \quad Cr(s) \longrightarrow Cr^{3+}(aq) + 3e^-$$
$$\text{cathode (reduction):} \quad Cu^{2+}(aq) + 2e^- \longrightarrow Cu(s)$$

Multiplying to make the number of electrons lost by Cr and gained by Cu^{2+} equal yields

$$\text{anode (oxidation):} \quad 2Cr(s) \longrightarrow 2Cr^{3+}(aq) + 6e^-$$
$$\text{cathode (reduction):} \quad 3Cu^{2+}(aq) + 6e^- \longrightarrow 3Cu(s)$$

Adding the half-reaction equations and simplifying yields an equation for the cell reaction:

$$2Cr(s) + 3Cu^{2+}(aq) \longrightarrow 2Cr^{3+}(aq) + 3Cu(s)$$

Check Your Learning

Omitting solute concentrations and spectator ion identities, write the schematic for a galvanic cell whose net cell reaction is shown below.

$$Sn^{4+}(aq) + Zn(s) \longrightarrow Sn^{2+}(aq) + Zn^{2+}(aq)$$

Answer:

$Zn(s) \mid Zn^{2+}(aq) \parallel Sn^{4+}(aq), Sn^{2+}(aq) \mid Pt(s)$

17.3 Electrode and Cell Potentials

LEARNING OBJECTIVES

By the end of this section, you will be able to:

- Describe and relate the definitions of electrode and cell potentials
- Interpret electrode potentials in terms of relative oxidant and reductant strengths
- Calculate cell potentials and predict redox spontaneity using standard electrode potentials

Unlike the spontaneous oxidation of copper by aqueous silver(I) ions described in section 17.2, immersing a copper wire in an aqueous solution of lead(II) ions yields no reaction. The two species, Ag^+*(aq)* and Pb^{2+}*(aq)*, thus show a distinct difference in their redox activity towards copper: the silver ion spontaneously oxidized copper, but the lead ion did not. Electrochemical cells permit this relative redox activity to be quantified by an easily measured property, *potential*. This property is more commonly called *voltage* when referenced in regard to electrical applications, and it is a measure of energy accompanying the transfer of charge. Potentials are measured in the volt unit, defined as one joule of energy per one coulomb of charge, V = J/C.

When measured for purposes of electrochemistry, a potential reflects the driving force for a specific type of charge transfer process, namely, the transfer of electrons between redox reactants. Considering the nature of potential in this context, it is clear that the potential of a single half-cell or a single electrode can't be measured; "transfer" of electrons requires both a donor and recipient, in this case a reductant and an oxidant, respectively. Instead, a half-cell potential may only be assessed relative to that of another half-cell. It is only the *difference in potential* between two half-cells that may be measured, and these measured potentials are called **cell potentials, E_{cell}**, defined as

$$E_{cell} = E_{cathode} - E_{anode}$$

where $E_{cathode}$ and E_{anode} are the potentials of two different half-cells functioning as specified in the subscripts. As for other thermodynamic quantities, the **standard cell potential, $E°_{cell}$**, is a cell potential measured when both half-cells are under standard-state conditions (1 *M* concentrations, 1 bar pressures, 298 K):

$$E^\circ_{cell} = E^\circ_{cathode} - E^\circ_{anode}$$

To simplify the collection and sharing of potential data for half-reactions, the scientific community has designated one particular half-cell to serve as a universal reference for cell potential measurements, assigning it a potential of exactly 0 V. This half-cell is the **standard hydrogen electrode (SHE)** and it is based on half-reaction below:

$$2H^+(aq) + 2e^- \longrightarrow H_2(g)$$

A typical SHE contains an inert platinum electrode immersed in precisely 1 *M* aqueous H^+ and a stream of bubbling H_2 gas at 1 bar pressure, all maintained at a temperature of 298 K (see Figure 17.5).

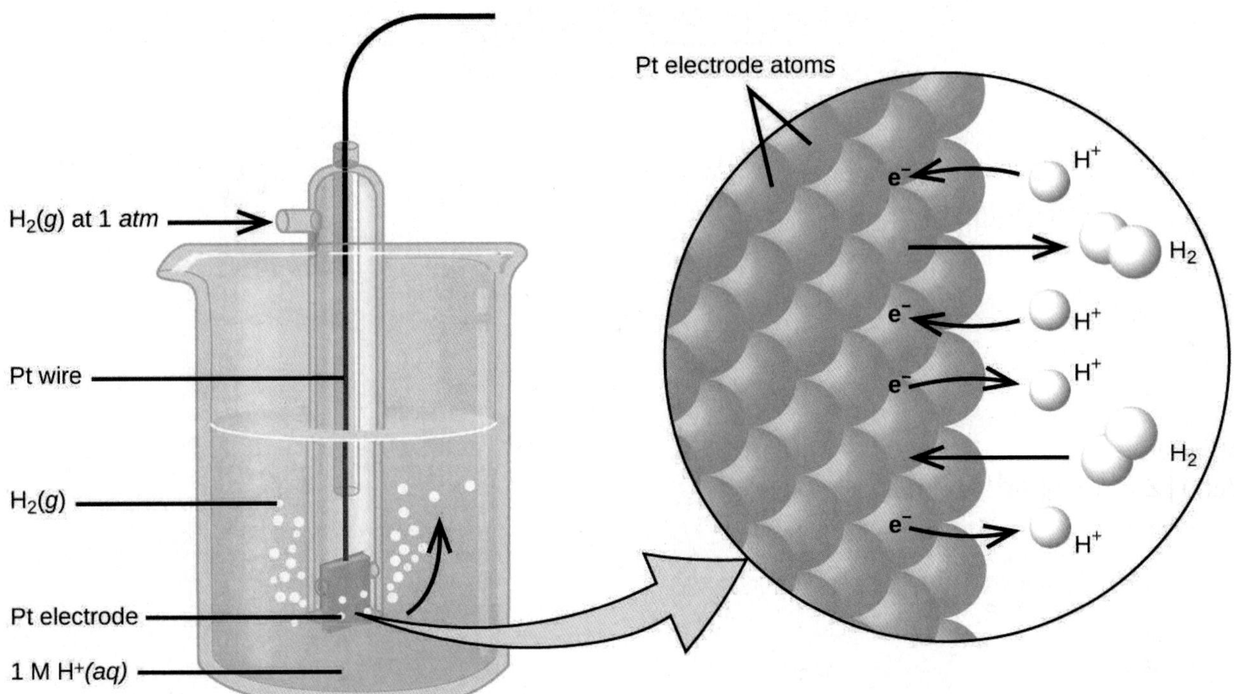

FIGURE 17.5 A standard hydrogen electrode (SHE).

The assigned potential of the SHE permits the definition of a conveniently measured potential for a single half-cell. The **electrode potential (E_X)** for a half-cell *X* is defined as *the potential measured for a cell comprised of X acting as cathode and the SHE acting as anode*:

$$\begin{aligned} E_{cell} &= E_X - E_{SHE} \\ E_{SHE} &= 0 \text{ V (defined)} \\ E_{cell} &= E_X \end{aligned}$$

When the half-cell *X* is under standard-state conditions, its potential is the **standard electrode potential, $E°_X$**. Since the definition of cell potential requires the half-cells function as cathodes, these potentials are sometimes called *standard reduction potentials*.

This approach to measuring electrode potentials is illustrated in Figure 17.6, which depicts a cell comprised of an SHE connected to a copper(II)/copper(0) half-cell under standard-state conditions. A voltmeter in the external circuit allows measurement of the potential difference between the two half-cells. Since the Cu half-cell is designated as the cathode in the definition of cell potential, it is connected to the red (positive) input of the voltmeter, while the designated SHE anode is connected to the black (negative) input. These connections insure that the sign of the measured potential will be consistent with the sign conventions of electrochemistry per the various definitions discussed above. A cell potential of +0.337 V is measured, and so

$$E^\circ_{cell} = E^\circ_{Cu} = +0.337 \text{ V}$$

Tabulations of E° values for other half-cells measured in a similar fashion are available as reference literature to permit calculations of cell potentials and the prediction of the spontaneity of redox processes.

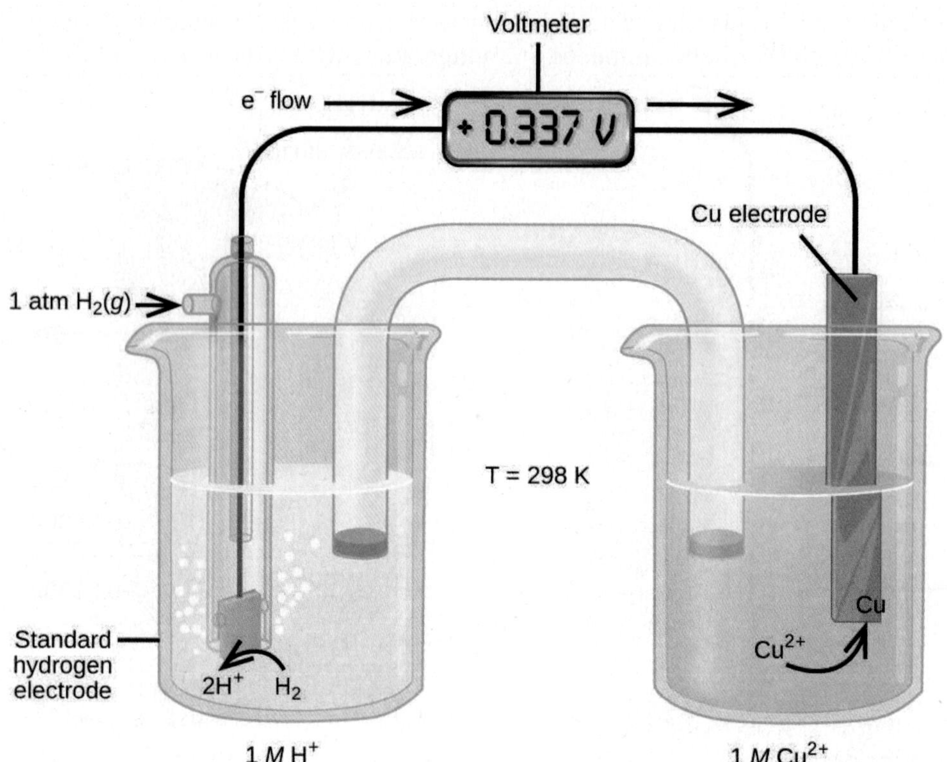

FIGURE 17.6 A cell permitting experimental measurement of the standard electrode potential for the half-reaction $Cu^{2+}(aq) + 2e^- \longrightarrow Cu(s)$

Table 17.1 provides a listing of standard electrode potentials for a selection of half-reactions in numerical order, and a more extensive alphabetical listing is given in Appendix L.

Selected Standard Reduction Potentials at 25 °C

Half-Reaction	$E°$ (V)
$F_2(g) + 2e^- \longrightarrow 2F^-(aq)$	+2.866
$PbO_2(s) + SO_4{}^{2-}(aq) + 4H^+(aq) + 2e^- \longrightarrow PbSO_4(s) + 2H_2O(l)$	+1.69
$MnO_4{}^-(aq) + 8H^+(aq) + 5e^- \longrightarrow Mn^{2+}(aq) + 4H_2O(l)$	+1.507
$Au^{3+}(aq) + 3e^- \longrightarrow Au(s)$	+1.498
$Cl_2(g) + 2e^- \longrightarrow 2Cl^-(aq)$	+1.35827
$O_2(g) + 4H^+(aq) + 4e^- \longrightarrow 2H_2O(l)$	+1.229
$Pt^{2+}(aq) + 2e^- \longrightarrow Pt(s)$	+1.20
$Br_2(aq) + 2e^- \longrightarrow 2Br^-(aq)$	+1.0873
$Ag^+(aq) + e^- \longrightarrow Ag(s)$	+0.7996

TABLE 17.1

Half-Reaction	$E°$ (V)
$Hg_2{}^{2+}(aq) + 2e^- \longrightarrow 2Hg(l)$	+0.7973
$Fe^{3+}(aq) + e^- \longrightarrow Fe^{2+}(aq)$	+0.771
$MnO_4{}^-(aq) + 2H_2O(l) + 3e^- \longrightarrow MnO_2(s) + 4OH^-(aq)$	+0.558
$I_2(s) + 2e^- \longrightarrow 2I^-(aq)$	+0.5355
$NiO_2(s) + 2H_2O(l) + 2e^- \longrightarrow Ni(OH)_2(s) + 2OH^-(aq)$	+0.49
$Cu^{2+}(aq) + 2e^- \longrightarrow Cu(s)$	+0.34
$Hg_2Cl_2(s) + 2e^- \longrightarrow 2Hg(l) + 2Cl^-(aq)$	+0.26808
$AgCl(s) + e^- \longrightarrow Ag(s) + Cl^-(aq)$	+0.22233
$Sn^{4+}(aq) + 2e^- \longrightarrow Sn^{2+}(aq)$	+0.151
$2H^+(aq) + 2e^- \longrightarrow H_2(g)$	0.00
$Pb^{2+}(aq) + 2e^- \longrightarrow Pb(s)$	−0.1262
$Sn^{2+}(aq) + 2e^- \longrightarrow Sn(s)$	−0.1375
$Ni^{2+}(aq) + 2e^- \longrightarrow Ni(s)$	−0.257
$Co^{2+}(aq) + 2e^- \longrightarrow Co(s)$	−0.28
$PbSO_4(s) + 2e^- \longrightarrow Pb(s) + SO_4{}^{2-}(aq)$	−0.3505
$Cd^{2+}(aq) + 2e^- \longrightarrow Cd(s)$	−0.4030
$Fe^{2+}(aq) + 2e^- \longrightarrow Fe(s)$	−0.447
$Cr^{3+}(aq) + 3e^- \longrightarrow Cr(s)$	−0.744
$Mn^{2+}(aq) + 2e^- \longrightarrow Mn(s)$	−1.185
$Zn(OH)_2(s) + 2e^- \longrightarrow Zn(s) + 2OH^-(aq)$	−1.245
$Zn^{2+}(aq) + 2e^- \longrightarrow Zn(s)$	−0.7618
$Al^{3+}(aq) + 3e^- \longrightarrow Al(s)$	−1.662
$Mg^{2}(aq) + 2e^- \longrightarrow Mg(s)$	−2.372
$Na^+(aq) + e^- \longrightarrow Na(s)$	−2.71

TABLE 17.1

Half-Reaction	$E°$ (V)
$Ca^{2+}(aq) + 2e^- \longrightarrow Ca(s)$	−2.868
$Ba^{2+}(aq) + 2e^- \longrightarrow Ba(s)$	−2.912
$K^+(aq) + e^- \longrightarrow K(s)$	−2.931
$Li^+(aq) + e^- \longrightarrow Li(s)$	−3.04

TABLE 17.1

EXAMPLE 17.4

Calculating Standard Cell Potentials

What is the standard potential of the galvanic cell shown in Figure 17.3?

Solution

The cell in Figure 17.3 is galvanic, the spontaneous cell reaction involving oxidation of its copper anode and reduction of silver(I) ions at its silver cathode:

$$\begin{array}{ll} \text{cell reaction:} & Cu(s) + 2Ag^+(aq) \longrightarrow Cu^{2+}(aq) + 2Ag(s) \\ \text{anode half-reaction:} & Cu(s) \longrightarrow Cu^{2+}(aq) + 2e^- \\ \text{cathode half-reaction:} & 2Ag^+(aq) + 2e^- \longrightarrow 2Ag(s) \end{array}$$

The standard cell potential computed as

$$\begin{aligned} E^\circ_{cell} &= E^\circ_{cathode} - E^\circ_{anode} \\ &= E^\circ_{Ag} - E^\circ_{Cu} \\ &= 0.7996\text{ V} - 0.34\text{ V} \\ &= +0.46\text{ V} \end{aligned}$$

Check Your Learning

What is the standard cell potential expected if the silver cathode half-cell in Figure 17.3 is replaced with a lead half-cell: $Pb^{2+}(aq) + 2e^- \longrightarrow Pb(s)$?

Answer:

−0. 47 V

Intrepreting Electrode and Cell Potentials

Thinking carefully about the definitions of cell and electrode potentials and the observations of spontaneous redox change presented thus far, a significant relation is noted. The previous section described the spontaneous oxidation of copper by aqueous silver(I) ions, but no observed reaction with aqueous lead(II) ions. Results of the calculations in Example 17.4 have just shown *the spontaneous process is described by a positive cell potential* while *the nonspontaneous process exhibits a negative cell potential.* And so, with regard to the relative effectiveness ("strength") with which aqueous Ag^+ and Pb^{2+} ions oxidize Cu under standard conditions, *the stronger oxidant is the one exhibiting the greater standard electrode potential, $E°$.* Since by convention electrode potentials are for reduction processes, an increased value of $E°$ corresponds to an increased driving force behind the reduction of the species (hence increased effectiveness of its action as an *oxidizing agent* on some other species). Negative values for electrode potentials are simply a consequence of assigning a value of 0 V to the SHE, indicating the reactant of the half-reaction is a weaker oxidant than aqueous hydrogen ions.

Applying this logic to the numerically ordered listing of standard electrode potentials in Table 17.1 shows this listing to be likewise in order of the oxidizing strength of the half-reaction's reactant species, decreasing from strongest oxidant (most positive $E°$) to weakest oxidant (most negative $E°$). Predictions regarding the spontaneity of redox reactions under standard state conditions can then be easily made by simply comparing the relative positions of their table entries. By definition, $E°_{cell}$ is positive when $E°_{cathode} > E°_{anode}$, and so any redox reaction in which the oxidant's entry is above the reductant's entry is predicted to be spontaneous.

Reconsideration of the two redox reactions in Example 17.4 provides support for this fact. The entry for the silver(I)/silver(0) half-reaction is above that for the copper(II)/copper(0) half-reaction, and so the oxidation of Cu by Ag^+ is predicted to be spontaneous ($E°_{cathode} > E°_{anode}$ and so $E°_{cell} > 0$). Conversely, the entry for the lead(II)/lead(0) half-cell is beneath that for copper(II)/copper(0), and the oxidation of Cu by Pb^{2+} is nonspontaneous ($E°_{cathode} < E°_{anode}$ and so $E°_{cell} < 0$).

Recalling the chapter on thermodynamics, the spontaneities of the forward and reverse reactions of a reversible process show a reciprocal relationship: if a process is spontaneous in one direction, it is non-spontaneous in the opposite direction. As an indicator of spontaneity for redox reactions, the potential of a cell reaction shows a consequential relationship in its arithmetic sign. The spontaneous oxidation of copper by lead(II) ions is *not* observed,

$$\text{Cu}(s) + \text{Pb}^{2+}(aq) \longrightarrow \text{Cu}^{2+}(aq) + \text{Pb}(s) E^{\circ}{}_{forward} = -0.47 \text{ V (negative, non-spontaneous)}$$

and so the reverse reaction, the oxidation of lead by copper(II) ions, is predicted to occur spontaneously:

$$\text{Pb}(s) + \text{Cu}^{2+}(aq) \longrightarrow \text{Pb}^{2+}(aq) + \text{Cu}(s) E^{\circ}{}_{forward} = +0.47 \text{ V (positive, spontaneous)}$$

Note that reversing the direction of a redox reaction effectively interchanges the identities of the cathode and anode half-reactions, and so the cell potential is calculated from electrode potentials in the reverse subtraction order than that for the forward reaction. In practice, a voltmeter would report a potential of −0.47 V with its red and black inputs connected to the Pb and Cu electrodes, respectively. If the inputs were swapped, the reported voltage would be +0.47 V.

 **EXAMPLE 17.5**

Predicting Redox Spontaneity

Are aqueous iron(II) ions predicted to spontaneously oxidize elemental chromium under standard state conditions? Assume the half-reactions to be those available in Table 17.1.

Solution

Referring to the tabulated half-reactions, the redox reaction in question can be represented by the equations below:

$$\text{Cr}(s) + \text{Fe}^{2+}(aq) \longrightarrow \text{Cr}^{3+}(aq) + \text{Fe}(s)$$

The entry for the putative oxidant, Fe^{2+}, appears *above* the entry for the reductant, Cr, and so a spontaneous reaction is predicted per the quick approach described above. Supporting this predication by calculating the standard cell potential for this reaction gives

$$\begin{aligned} \text{E}^{\circ}{}_{\text{cell}} &= \text{E}^{\circ}{}_{\text{cathode}} - \text{E}^{\circ}{}_{\text{anode}} \\ &= \text{E}^{\circ}{}_{\text{Fe(II)}} - \text{E}^{\circ}{}_{\text{Cr}} \\ &= -0.447 \text{ V} - -0.744 \text{ V} = +0.297 \text{ V} \end{aligned}$$

The positive value for the standard cell potential indicates the process is spontaneous under standard state conditions.

Check Your Learning

Use the data in Table 17.1 to predict the spontaneity of the oxidation of bromide ion by molecular iodine under standard state conditions, supporting the prediction by calculating the standard cell potential for the reaction.

Repeat for the oxidation of iodide ion by molecular bromine.

Answer:

$$I_2(s) + 2Br^-(aq) \longrightarrow 2I^-(aq) + Br_2(l) \qquad E^\circ_{cell} = -0.5518\text{ V (nonspontaneous)}$$

$$Br_2(s) + 2I^-(aq) \longrightarrow 2Br^-(aq) + I_2(l) \qquad E^\circ_{cell} = +0.5518\text{ V (spontaneous)}$$

17.4 Potential, Free Energy, and Equilibrium

LEARNING OBJECTIVES

By the end of this section, you will be able to:

- Explain the relations between potential, free energy change, and equilibrium constants
- Perform calculations involving the relations between cell potentials, free energy changes, and equilibrium
- Use the Nernst equation to determine cell potentials under nonstandard conditions

So far in this chapter, the relationship between the cell potential and reaction *spontaneity* has been described, suggesting a link to the free energy change for the reaction (see chapter on thermodynamics). The interpretation of potentials as measures of oxidant *strength* was presented, bringing to mind similar measures of acid-base strength as reflected in equilibrium constants (see the chapter on acid-base equilibria). This section provides a summary of the relationships between potential and the related thermodynamic properties ΔG and K.

E° and ΔG°

The standard free energy change of a process, $\Delta G°$, was defined in a previous chapter as the maximum work that could be performed by a system, w_{max}. In the case of a redox reaction taking place within a galvanic cell under standard state conditions, essentially all the work is associated with transferring the electrons from reductant-to-oxidant, w_{elec}:

$$\Delta G° = w_{max} = w_{elec}$$

The work associated with transferring electrons is determined by the total amount of charge (coulombs) transferred and the cell potential:

$$\begin{aligned} \Delta G° = w_{elec} &= -nFE^\circ_{cell} \\ \Delta G° &= -nFE^\circ_{cell} \end{aligned}$$

where n is the number of moles of electrons transferred, F is **Faraday's constant**, and $E°_{cell}$ is the standard cell potential. The relation between free energy change and standard cell potential confirms the sign conventions and spontaneity criteria previously discussed for both of these properties: spontaneous redox reactions exhibit positive potentials and negative free energy changes.

E° and K

Combining a previously derived relation between ΔG° and K (see the chapter on thermodynamics) and the equation above relating ΔG° and $E°_{cell}$ yields the following:

$$\begin{aligned} \Delta G° &= -RT\ln K = -nFE^\circ_{cell} \\ E^\circ_{cell} &= \left(\frac{RT}{nF}\right)\ln K \end{aligned}$$

This equation indicates redox reactions with large (positive) standard cell potentials will proceed far towards completion, reaching equilibrium when the majority of reactant has been converted to product. A summary of the relations between $E°$, $\Delta G°$ and K is depicted in Figure 17.7, and a table correlating reaction spontaneity to values of these properties is provided in Table 17.2.

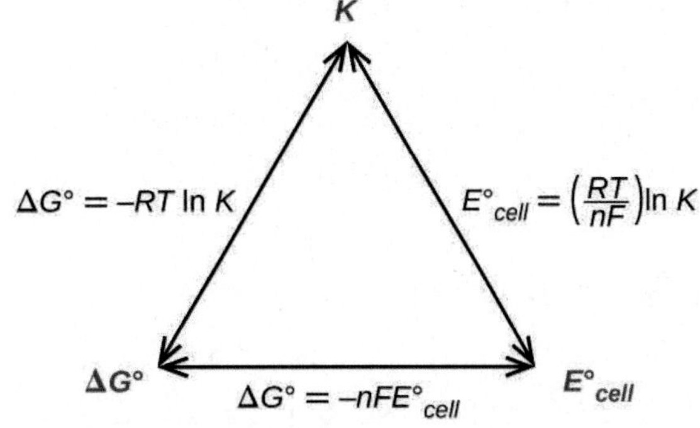

FIGURE 17.7 Graphic depicting the relation between three important thermodynamic properties.

K	$\Delta G°$	$E°_{cell}$	
> 1	< 0	> 0	Reaction is spontaneous under standard conditions Products more abundant at equilibrium
< 1	> 0	< 0	Reaction is non-spontaneous under standard conditions Reactants more abundant at equilibrium
= 1	= 0	= 0	Reaction is at equilibrium under standard conditions Reactants and products equally abundant

TABLE 17.2

 **EXAMPLE 17.6**

Equilibrium Constants, Standard Cell Potentials, and Standard Free Energy Changes

Use data from Appendix L to calculate the standard cell potential, standard free energy change, and equilibrium constant for the following reaction at 25 °C. Comment on the spontaneity of the forward reaction and the composition of an equilibrium mixture of reactants and products.

$$2Ag^{+}(aq) + Fe(s) \rightleftharpoons 2Ag(s) + Fe^{2+}(aq)$$

Solution

The reaction involves an oxidation-reduction reaction, so the standard cell potential can be calculated using the data in Appendix L.

$$\begin{array}{llrl} \text{anode (oxidation):} & Fe(s) \longrightarrow Fe^{2+}(aq) + 2e^{-} & E°_{Fe^{2+}/Fe} & = -0.447 \text{ V} \\ \text{cathode (reduction):} & 2 \times (Ag^{+}(aq) + e^{-} \longrightarrow Ag(s)) & E°_{Ag^{+}/Ag} & = 0.7996 \text{ V} \\ & E°_{cell} = E°_{cathode} - E°_{anode} = & E°_{Ag^{+}/Ag} - E°_{Fe^{2+}/Fe} & = +1.247 \text{ V} \end{array}$$

With $n = 2$, the equilibrium constant is then

$$\begin{aligned} E^\circ_{\text{cell}} &= \frac{0.0592\text{ V}}{n} \log K \\ K &= 10^{n \times E^\circ_{\text{cell}}/0.0592\text{ V}} \\ K &= 10^{2 \times 1.247\text{ V}/0.0592\text{ V}} \\ K &= 10^{42.128} \\ K &= 1.3 \times 10^{42} \end{aligned}$$

The standard free energy is then

$$\begin{aligned} \Delta G^\circ &= -nFE^\circ_{\text{cell}} \\ \Delta G^\circ &= -2 \times 96{,}485 \ \frac{\text{C}}{\text{mol}} \times 1.247 \ \frac{\text{J}}{\text{C}} = -240.6 \ \frac{\text{kJ}}{\text{mol}} \end{aligned}$$

The reaction is spontaneous, as indicated by a negative free energy change and a positive cell potential. The K value is very large, indicating the reaction proceeds to near completion to yield an equilibrium mixture containing mostly products.

Check Your Learning

What is the standard free energy change and the equilibrium constant for the following reaction at room temperature? Is the reaction spontaneous?

$$\text{Sn}(s) + 2\text{Cu}^{2+}(aq) \rightleftharpoons \text{Sn}^{2+}(aq) + 2\text{Cu}^{+}(aq)$$

Answer:

Spontaneous; $n = 2$; $E^\circ_{\text{cell}} = +0.291\text{ V}$; $\Delta G^\circ = -56.2 \ \frac{\text{kJ}}{\text{mol}}$; $K = 6.8 \times 10^9$.

Potentials at Nonstandard Conditions: The Nernst Equation

Most of the redox processes that interest science and society do not occur under standard state conditions, and so the potentials of these systems under nonstandard conditions are a property worthy of attention. Having established the relationship between potential and free energy change in this section, the previously discussed relation between free energy change and reaction mixture composition can be used for this purpose.

$$\Delta G = \Delta G^\circ + RT \ln Q$$

Notice the reaction quotient, Q, appears in this equation, making the free energy change dependent upon the composition of the reaction mixture. Substituting the equation relating free energy change to cell potential yields the **Nernst equation**:

$$-nFE_{\text{cell}} = -nFE^\circ_{\text{cell}} + RT \ln Q$$

$$E_{\text{cell}} = E^\circ_{\text{cell}} - \frac{RT}{nF} \ln Q$$

This equation describes how the potential of a redox system (such as a galvanic cell) varies from its standard state value, specifically, showing it to be a function of the number of electrons transferred, n, the temperature, T, and the reaction mixture composition as reflected in Q. A convenient form of the Nernst equation for most work is one in which values for the fundamental constants (R and F) and standard temperature (298) K), along with a factor converting from natural to base-10 logarithms, have been included:

$$E_{\text{cell}} = E^\circ_{\text{cell}} - \frac{0.0592\text{V}}{n} \log Q$$

 **EXAMPLE 17.7**

Predicting Redox Spontaneity Under Nonstandard Conditions

Use the Nernst equation to predict the spontaneity of the redox reaction shown below.

$$\text{Co}(s) + \text{Fe}^{2+}(aq,\ 1.94\ M) \longrightarrow \text{Co}^{2+}(aq, 0.15\ M) + \text{Fe}(s)$$

Solution

Collecting information from Appendix L and the problem,

$$\begin{array}{lllll} \text{Anode (oxidation):} & \text{Co}(s) \longrightarrow \text{Co}^{2+}(aq) + 2\text{e}^- & E^\circ_{\text{Co}^{2+}/\text{Co}} & = & -0.28\ \text{V} \\ \text{Cathode (reduction):} & \text{Fe}^{2+}(aq) + 2\text{e}^- \longrightarrow \text{Fe}(s) & E^\circ_{\text{Fe}^{2+}/\text{Fe}} & = & -0.447\ \text{V} \end{array}$$

$$E^\circ_{\text{cell}} = E^\circ_{\text{cathode}} - E^\circ_{\text{anode}} = -0.447\ \text{V} - (-0.28\ \text{V}) = -0.17\ \text{V}$$

Notice the negative value of the standard cell potential indicates the process is not spontaneous under standard conditions. Substitution of the Nernst equation terms for the nonstandard conditions yields:

$$\begin{aligned} Q &= \frac{[\text{Co}^{2+}]}{[\text{Fe}^{2+}]} = \frac{0.15\ M}{1.94\ M} = 0.077 \\ E_{\text{cell}} &= E^\circ_{\text{cell}} - \frac{0.0592\ \text{V}}{n} \log Q \\ E_{\text{cell}} &= -0.17\ \text{V} - \frac{0.0592\ \text{V}}{2} \log 0.077 \\ E_{\text{cell}} &= -0.17\ \text{V} + 0.033\ \text{V} = -0.14\ \text{V} \end{aligned}$$

The cell potential remains negative (slightly) under the specified conditions, and so the reaction remains nonspontaneous.

Check Your Learning

For the cell schematic below, identify values for n and Q, and calculate the cell potential, E_{cell}.

$$\text{Al}(s) \,|\, \text{Al}^{3+}(aq,\ 0.15\ M) \,\|\, \text{Cu}^{2+}(aq,\ 0.025\ M) \,|\, \text{Cu}(s)$$

Answer:

$n = 6$; $Q = 1440$; $E_{\text{cell}} = +1.97$ V, spontaneous.

A **concentration cell** is constructed by connecting two nearly identical half-cells, each based on the same half-reaction and using the same electrode, varying only in the concentration of one redox species. The potential of a concentration cell, therefore, is determined only by the difference in concentration of the chosen redox species. The example problem below illustrates the use of the Nernst equation in calculations involving concentration cells.

EXAMPLE 17.8

Concentration Cells

What is the cell potential of the concentration cell described by

$$\text{Zn}(s) \,|\, \text{Zn}^{2+}(aq, 0.10\ M) \,\|\, \text{Zn}^{2+}(aq, 0.50\ M) \,|\, \text{Zn}(s)$$

Solution

From the information given:

$$\begin{array}{lllll} \text{Anode:} & \text{Zn}(s) \longrightarrow \text{Zn}^{2+}(aq, 0.10\ M) + 2\text{e}^- & E^\circ_{\text{anode}} & = & -0.7618\ \text{V} \\ \text{Cathode:} & \text{Zn}^{2+}(aq, 0.50\ M) + 2\text{e}^- \longrightarrow \text{Zn}(s) & E^\circ_{\text{cathode}} & = & -0.7618\ \text{V} \\ \hline \text{Overall:} & \text{Zn}^{2+}(aq, 0.50\ M) \longrightarrow \text{Zn}^{2+}(aq, 0.10\ M) & E^\circ_{\text{cell}} & = & 0.000\ \text{V} \end{array}$$

Substituting into the Nernst equation,

$$E_{cell} = 0.000\text{ V} - \frac{0.0592\text{ V}}{2}\log\frac{0.10}{0.50} = +0.021\text{ V}$$

The positive value for cell potential indicates the overall cell reaction (see above) is spontaneous. This spontaneous reaction is one in which the zinc ion concentration in the cathode falls (it is reduced to elemental zinc) while that in the anode rises (it is produced by oxidation of the zinc anode). A greater driving force for zinc reduction is present in the cathode, where the zinc(II) ion concentration is greater ($E_{cathode} > E_{anode}$).

Check Your Learning

The concentration cell above was allowed to operate until the cell reaction reached equilibrium. What are the cell potential and the concentrations of zinc(II) in each half-cell for the cell now?

Answer:

$E_{cell} = 0.000$ V; $[Zn^{2+}]_{cathode} = [Zn^{2+}]_{anode} = 0.30\ M$

17.5 Batteries and Fuel Cells

LEARNING OBJECTIVES

By the end of this section, you will be able to:

- Describe the electrochemistry associated with several common batteries
- Distinguish the operation of a fuel cell from that of a battery

There are many technological products associated with the past two centuries of electrochemistry research, none more immediately obvious than the battery. A **battery** is a galvanic cell that has been specially designed and constructed in a way that best suits its intended use a source of electrical power for specific applications. Among the first successful batteries was the *Daniell cell*, which relied on the spontaneous oxidation of zinc by copper(II) ions (Figure 17.8):

$$Zn(s) + Cu^{2+}(aq) \longrightarrow Zn^{2+}(aq) + Cu(s)$$

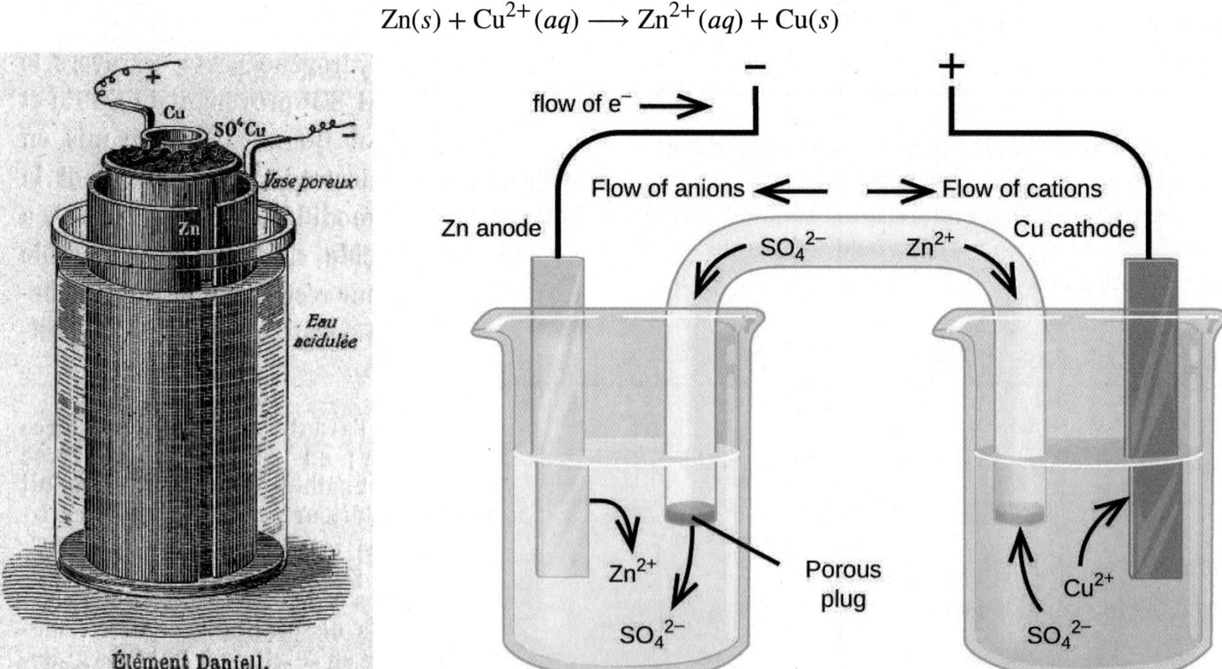

FIGURE 17.8 Illustration of a Daniell cell taken from a 1904 journal publication (left) along with a simplified illustration depicting the electrochemistry of the cell (right). The 1904 design used a porous clay pot to both contain one of the half-cell's content and to serve as a salt bridge to the other half-cell.

Modern batteries exist in a multitude of forms to accommodate various applications, from tiny button batteries that provide the modest power needs of a wristwatch to the very large batteries used to supply backup energy to municipal power grids. Some batteries are designed for single-use applications and cannot be recharged (**primary cells**), while others are based on conveniently reversible cell reactions that allow recharging by an external power source (**secondary cells**). This section will provide a summary of the basic electrochemical aspects of several batteries familiar to most consumers, and will introduce a related electrochemical device called a *fuel cell* that can offer improved performance in certain applications.

LINK TO LEARNING

Visit this site (http://openstax.org/l/16batteries) to learn more about batteries.

Single-Use Batteries

A common primary battery is the **dry cell**, which uses a zinc can as both container and anode ("–" terminal) and a graphite rod as the cathode ("+" terminal). The Zn can is filled with an electrolyte paste containing manganese(IV) oxide, zinc(II) chloride, ammonium chloride, and water. A graphite rod is immersed in the electrolyte paste to complete the cell. The spontaneous cell reaction involves the oxidation of zinc:

$$\text{anode reaction: Zn}(s) \longrightarrow \text{Zn}^{2+}(aq) + 2\text{e}^-$$

and the reduction of manganese(IV)

$$\text{reduction reaction: } 2\text{MnO}_2(s) + 2\text{NH}_4\text{Cl}(aq) + 2\,\text{e}^- \longrightarrow \text{Mn}_2\text{O}_3(s) + 2\text{NH}_3(aq) + \text{H}_2\text{O}(l) + 2\text{Cl}^-$$

which together yield the cell reaction:

$$\text{cell reaction: } 2\text{MnO}_2(s) + 2\text{NH}_4\text{Cl}(aq) + \text{Zn}(s) \longrightarrow \text{Zn}^{2+}(aq) + \text{Mn}_2\text{O}_3(s) + 2\text{NH}_3(aq) + \text{H}_2\text{O}(l) + 2\text{Cl}^- \;\; E_\text{cell}\text{\textasciitilde}1.5\text{ V}$$

The voltage (*cell potential*) of a dry cell is approximately 1.5 V. Dry cells are available in various sizes (e.g., D, C, AA, AAA). All sizes of dry cells comprise the same components, and so they exhibit the same voltage, but larger cells contain greater amounts of the redox reactants and therefore are capable of transferring correspondingly greater amounts of charge. Like other galvanic cells, dry cells may be connected in series to yield batteries with greater voltage outputs, if needed.

FIGURE 17.9 A schematic diagram shows a typical dry cell.

LINK TO LEARNING

Visit this site (http://openstax.org/l/16zinccarbon) to learn more about zinc-carbon batteries.

Alkaline batteries (Figure 17.10) were developed in the 1950s to improve on the performance of the dry cell, and they were designed around the same redox couples. As their name suggests, these types of batteries use alkaline electrolytes, often potassium hydroxide. The reactions are

$$\begin{array}{llll} \text{anode:} & Zn(s) + 2OH^-(aq) \longrightarrow ZnO(s) + H_2O(l) + 2e^- & & \\ \text{cathode:} & 2MnO_2(s) + H_2O(l) + 2e^- \longrightarrow Mn_2O_3(s) + 2OH^-(aq) & & \\ \hline \text{cell:} & Zn(s) + 2MnO_2(s) \longrightarrow ZnO(s) + Mn_2O_3(s) & & E_{cell} = +1.43\ V \end{array}$$

An alkaline battery can deliver about three to five times the energy of a zinc-carbon dry cell of similar size. Alkaline batteries are prone to leaking potassium hydroxide, so they should be removed from devices for long-term storage. While some alkaline batteries are rechargeable, most are not. Attempts to recharge an alkaline battery that is not rechargeable often leads to rupture of the battery and leakage of the potassium hydroxide electrolyte.

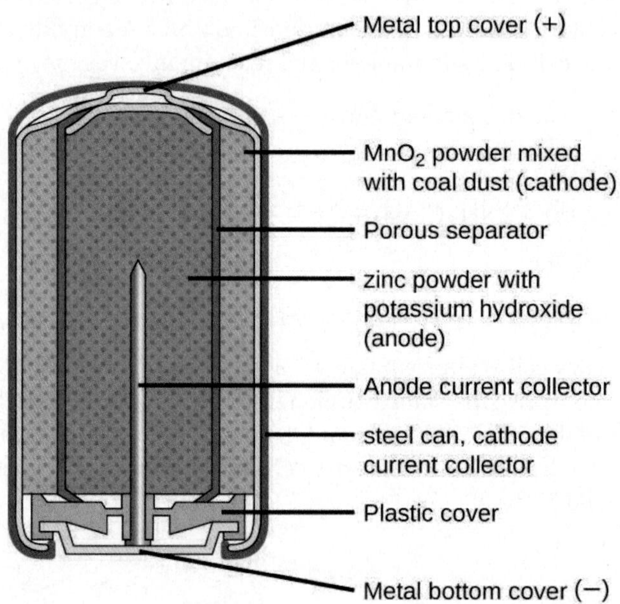

FIGURE 17.10 Alkaline batteries were designed as improved replacements for zinc-carbon (dry cell) batteries.

LINK TO LEARNING

Visit this site (http://openstax.org/l/16alkaline) to learn more about alkaline batteries.

Rechargeable (Secondary) Batteries

Nickel-cadmium, or NiCd, batteries (Figure 17.11) consist of a nickel-plated cathode, cadmium-plated anode, and a potassium hydroxide electrode. The positive and negative plates, which are prevented from shorting by the separator, are rolled together and put into the case. This is a "jelly-roll" design and allows the NiCd cell to deliver much more current than a similar-sized alkaline battery. The reactions are

$$\begin{array}{llll} \text{anode:} & Cd(s) + 2OH^-(aq) \longrightarrow Cd(OH)_2(s) + 2e^- & & \\ \text{cathode:} & NiO_2(s) + 2H_2O(l) + 2e^- \longrightarrow Ni(OH)_2(s) + 2OH^-(aq) & & \\ \hline \text{cell:} & Cd(s) + NiO_2(s) + 2H_2O(l) \longrightarrow Cd(OH)_2(s) + Ni(OH)_2(s) & & E_{cell} \sim 1.2\ V \end{array}$$

When properly treated, a NiCd battery can be recharged about 1000 times. Cadmium is a toxic heavy metal so NiCd batteries should never be ruptured or incinerated, and they should be disposed of in accordance with relevant toxic waste guidelines.

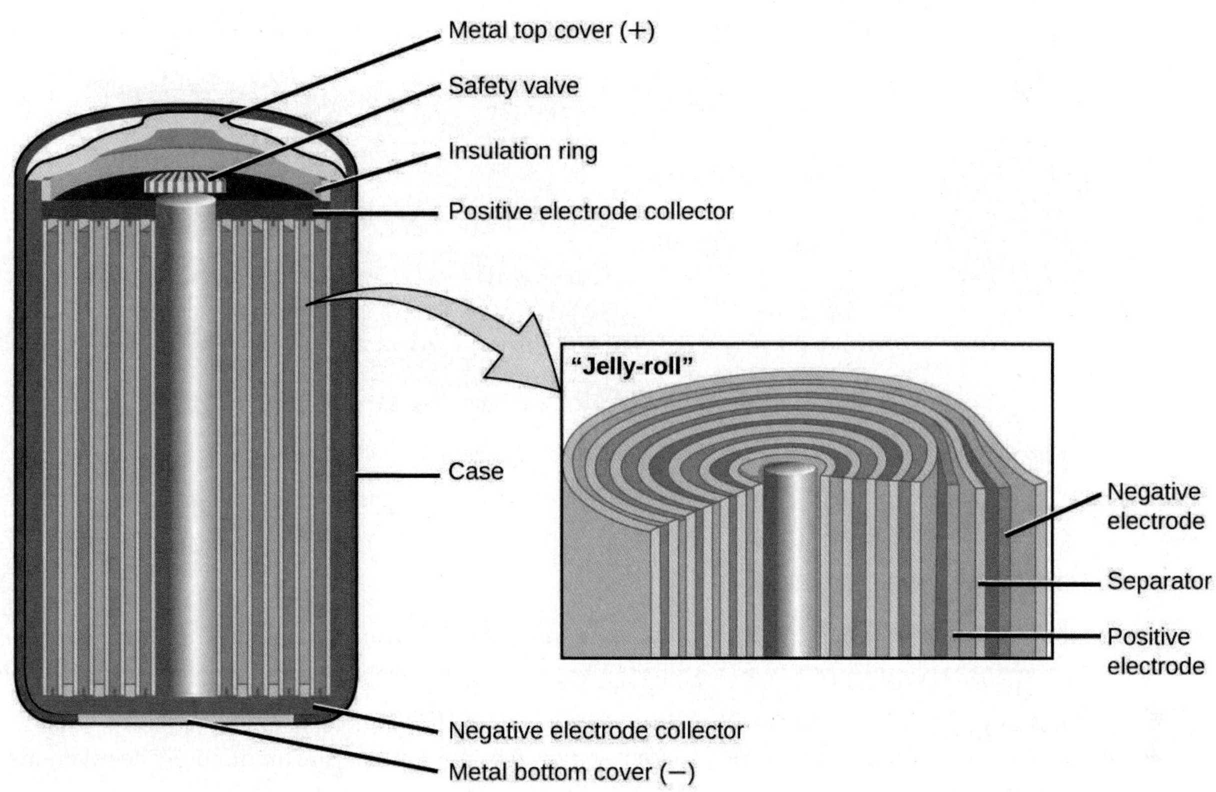

FIGURE 17.11 NiCd batteries use a "jelly-roll" design that significantly increases the amount of current the battery can deliver as compared to a similar-sized alkaline battery.

LINK TO LEARNING

Visit this site (http://openstax.org/l/16NiCdrecharge) for more information about nickel cadmium rechargeable batteries.

Lithium ion batteries (Figure 17.12) are among the most popular rechargeable batteries and are used in many portable electronic devices. The reactions are

$$\begin{array}{ll} \text{anode:} & \text{LiCoO}_2 \rightleftharpoons \text{Li}_{1-x}\text{CoO}_2 + x\,\text{Li}^+ + x\,\text{e}^- \\ \text{cathode:} & x\,\text{Li}^+ + x\,\text{e}^- + x\,\text{C}_6 \rightleftharpoons x\,\text{LiC}_6 \\ \hline \text{cell:} & \text{LiCoO}_2 + x\,\text{C}_6 \rightleftharpoons \text{Li}_{1-x}\text{CoO}_2 + x\,\text{LiC}_6 \qquad E_\text{cell}\text{\textasciitilde}3.7\text{ V} \end{array}$$

The variable stoichiometry of the cell reaction leads to variation in cell voltages, but for typical conditions, x is usually no more than 0.5 and the cell voltage is approximately 3.7 V. Lithium batteries are popular because they can provide a large amount current, are lighter than comparable batteries of other types, produce a nearly constant voltage as they discharge, and only slowly lose their charge when stored.

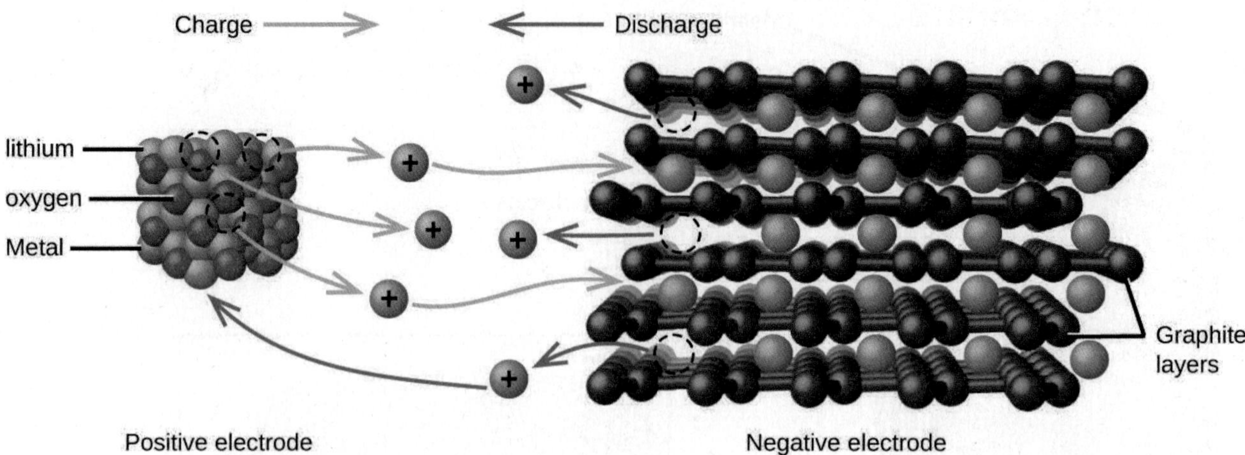

FIGURE 17.12 In a lithium ion battery, charge flows as the lithium ions are transferred between the anode and cathode.

LINK TO LEARNING

Visit this site (http://openstax.org/l/16lithiumion) for more information about lithium ion batteries.

The **lead acid battery** (Figure 17.13) is the type of secondary battery commonly used in automobiles. It is inexpensive and capable of producing the high current required by automobile starter motors. The reactions for a lead acid battery are

$$\begin{array}{llll} \text{anode:} & Pb(s) + HSO_4^-(aq) & \longrightarrow PbSO_4(s) + H^+(aq) + 2e^- & \\ \text{cathode:} & PbO_2(s) + HSO_4^-(aq) + 3H^+(aq) + 2e^- & \longrightarrow PbSO_4(s) + 2H_2O(l) & \\ \hline \text{cell:} & Pb(s) + PbO_2(s) + 2H_2SO_4(aq) & \longrightarrow 2PbSO_4(s) + 2H_2O(l) & E_{cell} \sim 2\text{ V} \end{array}$$

Each cell produces 2 V, so six cells are connected in series to produce a 12-V car battery. Lead acid batteries are heavy and contain a caustic liquid electrolyte, $H_2SO_4(aq)$, but are often still the battery of choice because of their high current density. Since these batteries contain a significant amount of lead, they must always be disposed of properly.

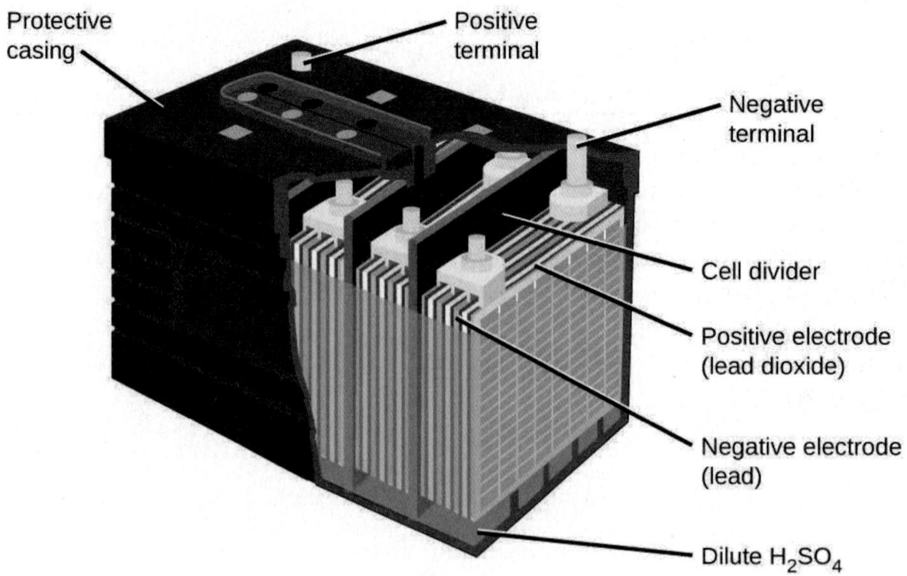

FIGURE 17.13 The lead acid battery in your automobile consists of six cells connected in series to give 12 V.

LINK TO LEARNING

Visit this site (http://openstax.org/l/16leadacid) for more information about lead acid batteries.

Fuel Cells

A **fuel cell** is a galvanic cell that uses traditional combustive fuels, most often hydrogen or methane, that are continuously fed into the cell along with an oxidant. (An alternative, but not very popular, name for a fuel cell is a *flow battery*.) Within the cell, fuel and oxidant undergo the same redox chemistry as when they are combusted, but via a catalyzed electrochemical that is significantly more efficient. For example, a typical hydrogen fuel cell uses graphite electrodes embedded with platinum-based catalysts to accelerate the two half-cell reactions:

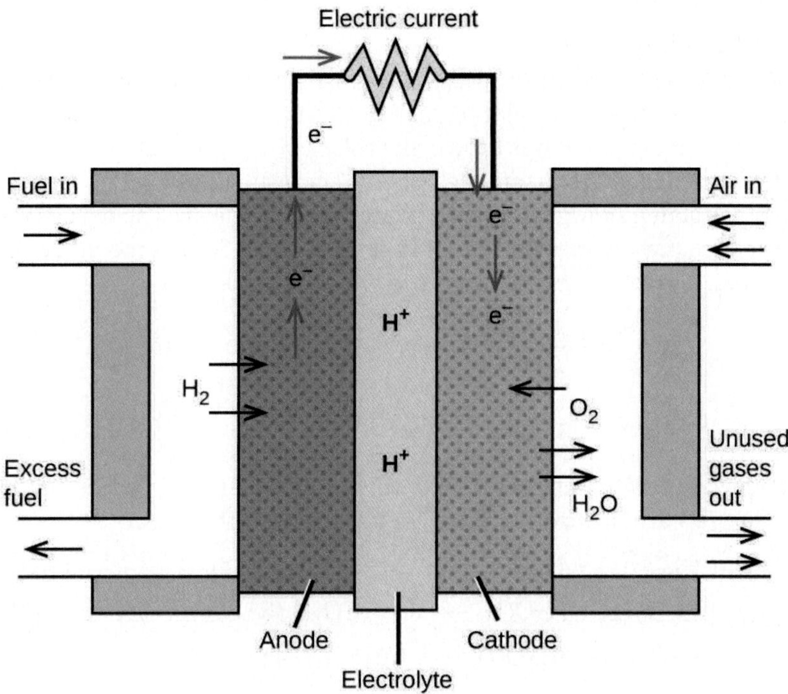

FIGURE 17.14 In this hydrogen fuel cell, oxygen from the air reacts with hydrogen, producing water and electricity.

$$\begin{array}{ll} \text{Anode:} & 2\text{H}_2(g) \longrightarrow 4\text{H}^+(aq) + 4\text{e}^- \\ \text{Cathode:} & \text{O}_2(g) + 4\text{H}^+(aq) + 4\text{e}^- \longrightarrow 2\text{H}_2\text{O}(g) \\ \hline \text{Cell:} & 2\text{H}_2(g) + \text{O}_2(g) \longrightarrow 2\text{H}_2\text{O}(g) \qquad E_{\text{cell}} \sim 1.2 \text{ V} \end{array}$$

These types of fuel cells generally produce voltages of approximately 1.2 V. Compared to an internal combustion engine, the energy efficiency of a fuel cell using the same redox reaction is typically more than double (~20%–25% for an engine versus ~50%–75% for a fuel cell). Hydrogen fuel cells are commonly used on extended space missions, and prototypes for personal vehicles have been developed, though the technology remains relatively immature.

LINK TO LEARNING

Check out this link (http://openstax.org/l/16fuelcells) to learn more about fuel cells.

17.6 Corrosion

LEARNING OBJECTIVES

By the end of this section, you will be able to:

- Define corrosion
- List some of the methods used to prevent or slow corrosion

Corrosion is usually defined as the degradation of metals by a naturally occurring electrochemical process. The formation of rust on iron, tarnish on silver, and the blue-green patina that develops on copper are all examples of corrosion. The total cost of corrosion remediation in the United States is significant, with estimates in excess of half a trillion dollars a year.

Chemistry in Everyday Life

Statue of Liberty: Changing Colors

The Statue of Liberty is a landmark every American recognizes. The Statue of Liberty is easily identified by its height, stance, and unique blue-green color (Figure 17.15). When this statue was first delivered from France, its appearance was not green. It was brown, the color of its copper "skin." So how did the Statue of Liberty change colors? The change in appearance was a direct result of corrosion. The copper that is the primary component of the statue slowly underwent oxidation from the air. The oxidation-reduction reactions of copper metal in the environment occur in several steps. Copper metal is oxidized to copper(I) oxide (Cu_2O), which is red, and then to copper(II) oxide, which is black

$$2Cu(s) + \frac{1}{2}O_2(g) \longrightarrow Cu_2O(s) \qquad \text{(red)}$$

$$Cu_2O(s) + \frac{1}{2}O_2(g) \longrightarrow 2CuO(s) \qquad \text{(black)}$$

Coal, which was often high in sulfur, was burned extensively in the early part of the last century. As a result, atmospheric sulfur trioxide, carbon dioxide, and water all reacted with the CuO

$$2CuO(s) + CO_2(g) + H_2O(l) \longrightarrow Cu_2CO_3(OH)_2(s) \qquad \text{(green)}$$

$$3CuO(s) + 2CO_2(g) + H_2O(l) \longrightarrow Cu_2(CO_3)_2(OH)_2(s) \qquad \text{(blue)}$$

$$4CuO(s) + SO_3(g) + 3H_2O(l) \longrightarrow Cu_4SO_4(OH)_6(s) \qquad \text{(green)}$$

These three compounds are responsible for the characteristic blue-green patina seen on the Statue of Liberty (and other outdoor copper structures). Fortunately, formation of patina creates a protective layer on the copper surface, preventing further corrosion of the underlying copper. The formation of the protective layer is called *passivation*, a phenomenon discussed further in another chapter of this text.

(a)

(b)

FIGURE 17.15 (a) The Statue of Liberty is covered with a copper skin, and was originally brown, as shown in this painting. (b) Exposure to the elements has resulted in the formation of the blue-green patina seen today.

Perhaps the most familiar example of corrosion is the formation of rust on iron. Iron will rust when it is exposed to oxygen and water. Rust formation involves the creation of a galvanic cell at an iron surface, as illustrated in Figure 17.15. The relevant redox reactions are described by the following equations:

$$\begin{array}{llrl} \text{anode:} & \text{Fe}(s) \longrightarrow \text{Fe}^{2+}(aq) + 2\text{e}^- & E^\circ_{\text{Fe}^{2+}/\text{Fe}} & = -0.44 \text{ V} \\ \text{cathode:} & \text{O}_2(g) + 4\text{H}^+(aq) + 4\text{e}^- \longrightarrow 2\text{H}_2\text{O}(l) & E^\circ_{\text{O}_2/\text{O}2} & = +1.23 \text{ V} \\ \text{overall:} & 2\text{Fe}(s) + \text{O}_2(g) + 4\text{H}^+(aq) \longrightarrow 2\text{Fe}^{2+}(aq) + 2\text{H}_2\text{O}(l) & E^\circ_{\text{cell}} & = +1.67 \text{ V} \end{array}$$

Further reaction of the iron(II) product in humid air results in the production of an iron(III) oxide hydrate known as rust:

$$4\text{Fe}^{2+}(aq) + \text{O}_2(g) + (4 + 2x)\,\text{H}_2\text{O}(l) \longrightarrow 2\text{Fe}_2\text{O}_3{\cdot}x\text{H}_2\text{O}(s) + 8\text{H}^+(aq)$$

The stoichiometry of the hydrate varies, as indicated by the use of x in the compound formula. Unlike the patina on copper, the formation of rust does not create a protective layer and so corrosion of the iron continues as the rust flakes off and exposes fresh iron to the atmosphere.

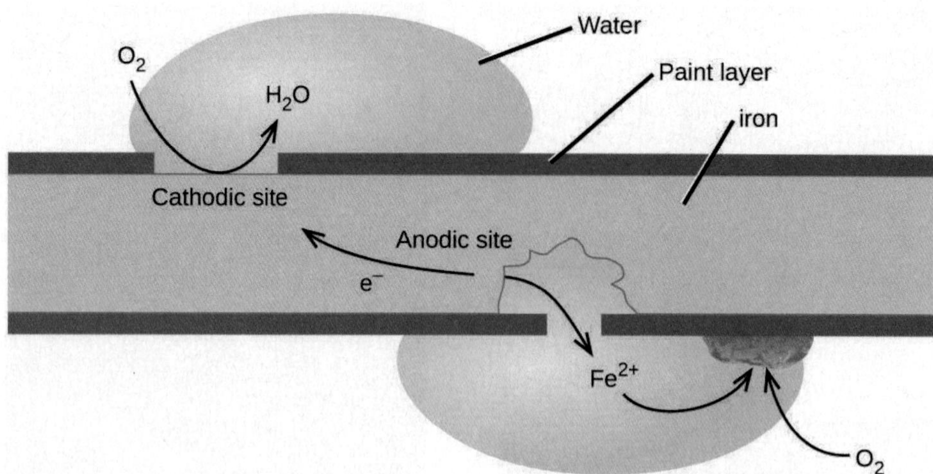

FIGURE 17.16 Corrosion can occur when a painted iron or steel surface is exposed to the environment by a scratch through the paint. A galvanic cell results that may be approximated by the simplified cell schematic Fe(*s*) | Fe^{2+}(*aq*) ||O_2(*aq*), H_2O(*l*) | Fe(*s*).

One way to keep iron from corroding is to keep it painted. The layer of paint prevents the water and oxygen necessary for rust formation from coming into contact with the iron. As long as the paint remains intact, the iron is protected from corrosion.

Other strategies include alloying the iron with other metals. For example, stainless steel is an alloy of iron containing a small amount of chromium. The chromium tends to collect near the surface, where it corrodes and forms a passivating an oxide layer that protects the iron.

Iron and other metals may also be protected from corrosion by **galvanization**, a process in which the metal to be protected is coated with a layer of a more readily oxidized metal, usually zinc. When the zinc layer is intact, it prevents air from contacting the underlying iron and thus prevents corrosion. If the zinc layer is breached by either corrosion or mechanical abrasion, the iron may still be protected from corrosion by a *cathodic protection* process, which is described in the next paragraph.

Another important way to protect metal is to make it the cathode in a galvanic cell. This is **cathodic protection** and can be used for metals other than just iron. For example, the rusting of underground iron storage tanks and pipes can be prevented or greatly reduced by connecting them to a more active metal such as zinc or magnesium (Figure 17.17). This is also used to protect the metal parts in water heaters. The more active metals (lower reduction potential) are called **sacrificial anodes** because as they get used up as they corrode (oxidize) at the anode. The metal being protected serves as the cathode for the reduction of oxygen in air, and so it simply serves to conduct (not react with) the electrons being transferred. When the anodes are properly monitored and periodically replaced, the useful lifetime of the iron storage tank can be greatly extended.

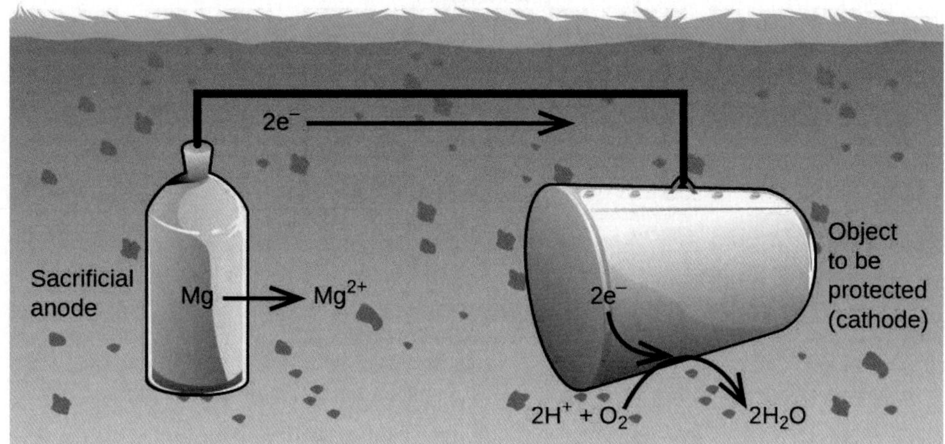

FIGURE 17.17 Cathodic protection is a useful approach to electrochemically preventing corrosion of underground storage tanks.

17.7 Electrolysis

LEARNING OBJECTIVES

By the end of this section, you will be able to:

- Describe the process of electrolysis
- Compare the operation of electrolytic cells with that of galvanic cells
- Perform stoichiometric calculations for electrolytic processes

Electrochemical cells in which spontaneous redox reactions take place (*galvanic cells*) have been the topic of discussion so far in this chapter. In these cells, *electrical work is done by a redox system on its surroundings* as electrons produced by the redox reaction are transferred through an external circuit. This final section of the chapter will address an alternative scenario in which *an external circuit does work on a redox system* by imposing a voltage sufficient to drive an otherwise nonspontaneous reaction, a process known as **electrolysis**. A familiar example of electrolysis is recharging a battery, which involves use of an external power source to drive the spontaneous (discharge) cell reaction in the reverse direction, restoring to some extent the composition of the half-cells and the voltage of the battery. Perhaps less familiar is the use of electrolysis in the refinement of metallic ores, the manufacture of commodity chemicals, and the *electroplating* of metallic coatings on various products (e.g., jewelry, utensils, auto parts). To illustrate the essential concepts of electrolysis, a few specific processes will be considered.

The Electrolysis of Molten Sodium Chloride

Metallic sodium, Na, and chlorine gas, Cl_2, are used in numerous applications, and their industrial production relies on the large-scale electrolysis of molten sodium chloride, NaCl(*l*). The industrial process typically uses a *Downs cell* similar to the simplified illustration shown in Figure 17.18. The reactions associated with this process are:

$$\begin{array}{llcl} \text{anode:} & 2\text{Cl}^-(l) & \longrightarrow & \text{Cl}_2(g) + 2\text{e}^- \\ \text{cathode:} & \text{Na}^+(l) + \text{e}^- & \longrightarrow & \text{Na}(l) \\ \hline \text{cell:} & 2\text{Na}^+(l) + 2\text{Cl}^-(l) & \longrightarrow & 2\text{Na}(l) + \text{Cl}_2(g) \end{array}$$

The cell potential for the above process is negative, indicating the reaction as written (decomposition of liquid NaCl) is not spontaneous. To force this reaction, *a positive potential of magnitude greater than the negative cell potential* must be applied to the cell.

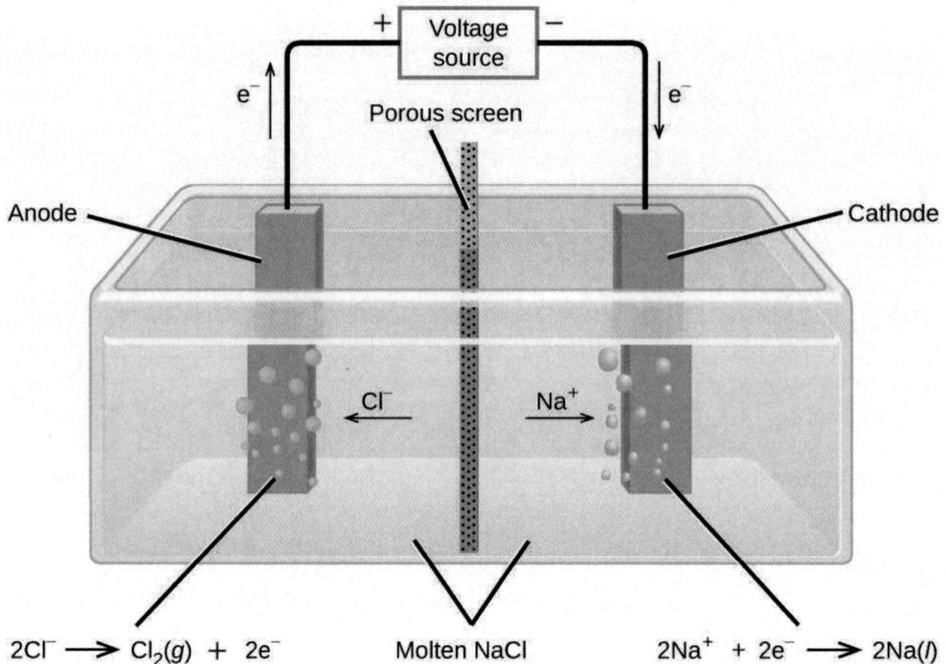

FIGURE 17.18 Cells of this sort (a cell for the electrolysis of molten sodium chloride) are used in the *Downs process* for production of sodium and chlorine, and they typically use iron cathodes and carbon anodes.

The Electrolysis of Water

Water may be electrolytically decomposed in a cell similar to the one illustrated in Figure 17.19. To improve electrical conductivity without introducing a different redox species, the hydrogen ion concentration of the water is typically increased by addition of a strong acid. The redox processes associated with this cell are

$$\begin{array}{llll} \text{anode:} & 2H_2O(l) \longrightarrow O_2(g) + 4H^+(aq) + 4e^- & E^\circ_{\text{anode}} & = +1.229\text{ V} \\ \text{cathode:} & 2H^+(aq) + 2e^- \longrightarrow H_2(g) & E^\circ_{\text{cathode}} & = 0\text{ V} \\ \hline \text{cell:} & 2H_2O(l) \longrightarrow 2H_2(g) + O_2(g) & E^\circ_{\text{cell}} & = -1.229\text{ V} \end{array}$$

Again, the cell potential as written is negative, indicating a nonspontaneous cell reaction that must be driven by imposing a cell voltage greater than +1.229 V. Keep in mind that *standard* electrode potentials are used to inform thermodynamic predictions here, though the cell is *not* operating under standard state conditions. Therefore, at best, calculated cell potentials should be considered ballpark estimates.

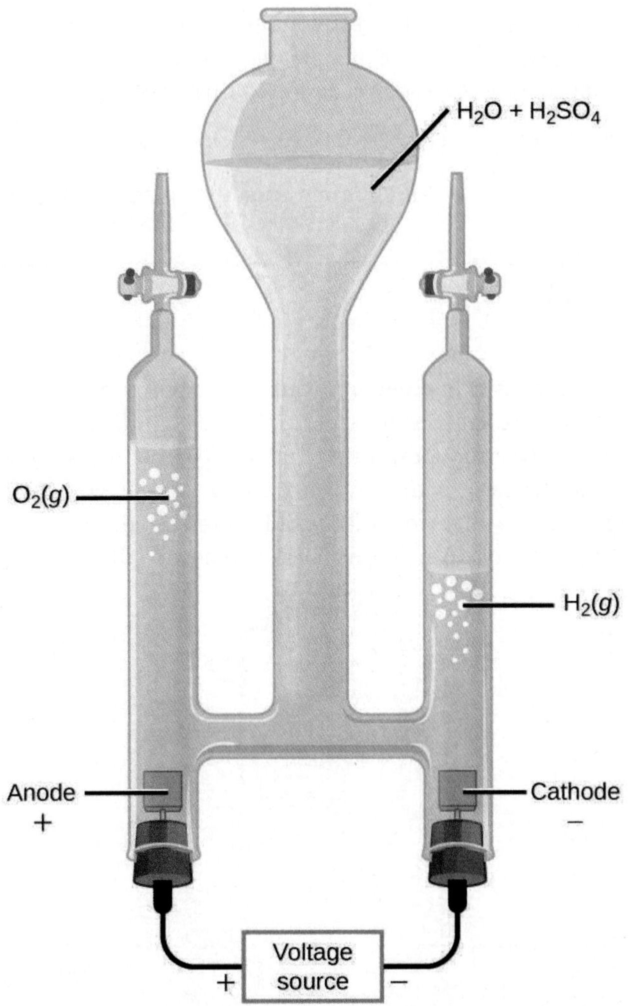

FIGURE 17.19 The electrolysis of water produces stoichiometric amounts of oxygen gas at the anode and hydrogen at the anode.

The Electrolysis of Aqueous Sodium Chloride

When aqueous solutions of ionic compounds are electrolyzed, the anode and cathode half-reactions may involve the electrolysis of either water species (H_2O, H^+, OH^-) or solute species (the cations and anions of the compound). As an example, the electrolysis of aqueous sodium chloride could involve either of these two anode reactions:

$$\begin{array}{llll} \text{(i) } 2\text{Cl}^-(aq) & \longrightarrow \text{Cl}_2(g) + 2\,\text{e}^- & E^\circ_{\text{anode}} & = +1.35827 \text{ V} \\ \text{(ii) } 2\text{H}_2\text{O}(l) & \longrightarrow \text{O}_2(g) + 4\text{H}^+(aq) + 4\text{e}^- & E^\circ_{\text{anode}} & = +1.229 \text{ V} \end{array}$$

The standard electrode (*reduction*) potentials of these two half-reactions indicate water may be *oxidized* at a less negative/more positive potential (–1.229 V) than chloride ion (–1.358 V). Thermodynamics thus predicts that water would be more readily oxidized, though in practice it is observed that both water and chloride ion are oxidized under typical conditions, producing a mixture of oxygen and chlorine gas.

Turning attention to the cathode, the possibilities for reduction are:

$$\begin{array}{llll} \text{(iii) } 2\text{H}^+(aq) + 2\text{e}^- & \longrightarrow \text{H}_2(g) & E^\circ_{\text{cathode}} & = 0 \text{ V} \\ \text{(iv) } 2\text{H}_2\text{O}(l) + 2\text{e}^- & \longrightarrow \text{H}_2(g) + 2\text{OH}^-(aq) & E^\circ_{\text{cathode}} & = -0.8277 \text{ V} \\ \text{(v) } \text{Na}^+(aq) + \text{e}^- & \longrightarrow \text{Na}(s) & E^\circ_{\text{cathode}} & = -2.71 \text{ V} \end{array}$$

Comparison of these *standard* half-reaction potentials suggests the reduction of hydrogen ion is thermodynamically favored. However, in a neutral aqueous sodium chloride solution, the concentration of hydrogen ion is far below the standard state value of 1 *M* (approximately 10^{-7} *M*), and so the observed cathode reaction is actually reduction of water. The net cell reaction in this case is then

$$\text{cell: } 2H_2O(l) + 2Cl^-(aq) \longrightarrow H_2(g) + Cl_2(g) + 2OH^-(aq) \qquad E^\circ_{\text{cell}} = -2.186 \text{ V}$$

This electrolysis reaction is part of the *chlor-alkali process* used by industry to produce chlorine and sodium hydroxide (lye).

Chemistry in Everyday Life

Electroplating

An important use for electrolytic cells is in **electroplating**. Electroplating results in a thin coating of one metal on top of a conducting surface. Reasons for electroplating include making the object more corrosion resistant, strengthening the surface, producing a more attractive finish, or for purifying metal. The metals commonly used in electroplating include cadmium, chromium, copper, gold, nickel, silver, and tin. Common consumer products include silver-plated or gold-plated tableware, chrome-plated automobile parts, and jewelry. The silver plating of eating utensils is used here to illustrate the process. (Figure 17.20).

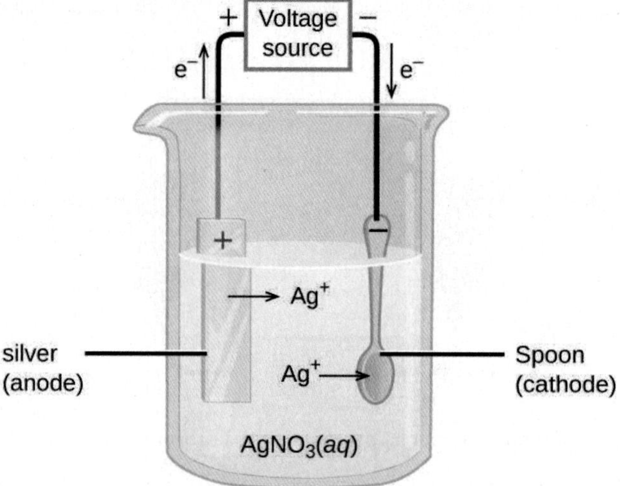

FIGURE 17.20 This schematic shows an electrolytic cell for silver plating eating utensils.

In the figure, the anode consists of a silver electrode, shown on the left. The cathode is located on the right and is the spoon, which is made from inexpensive metal. Both electrodes are immersed in a solution of silver nitrate. Applying a sufficient potential results in the oxidation of the silver anode

$$\text{anode: } Ag(s) \longrightarrow Ag^+(aq) + e^-$$

and reduction of silver ion at the (spoon) cathode:

$$\text{cathode: } Ag^+(aq) + e^- \longrightarrow Ag(s)$$

The net result is the transfer of silver metal from the anode to the cathode. Several experimental factors must be carefully controlled to obtain high-quality silver coatings, including the exact composition of the electrolyte solution, the cell voltage applied, and the rate of the electrolysis reaction (electrical current).

Quantitative Aspects of Electrolysis

Electrical current is defined as the rate of flow for any charged species. Most relevant to this discussion is the flow of electrons. Current is measured in a composite unit called an ampere, defined as one coulomb per second (A = 1 C/s). The charge transferred, *Q*, by passage of a constant current, *I*, over a specified time interval, *t*, is then given by the simple mathematical product

$$Q = It$$

When electrons are transferred during a redox process, the stoichiometry of the reaction may be used to derive the total amount of (electronic) charge involved. For example, the generic reduction process

$$M^{n+}(aq) + ne^- \longrightarrow M(s)$$

involves the transfer of n mole of electrons. The charge transferred is, therefore,

$$Q = nF$$

where F is Faraday's constant, the charge in coulombs for one mole of electrons. If the reaction takes place in an electrochemical cell, the current flow is conveniently measured, and it may be used to assist in stoichiometric calculations related to the cell reaction.

 **EXAMPLE 17.9**

Converting Current to Moles of Electrons

In one process used for electroplating silver, a current of 10.23 A was passed through an electrolytic cell for exactly 1 hour. How many moles of electrons passed through the cell? What mass of silver was deposited at the cathode from the silver nitrate solution?

Solution

Faraday's constant can be used to convert the charge (Q) into moles of electrons (n). The charge is the current (I) multiplied by the time

$$n = \frac{Q}{F} = \frac{\frac{10.23\text{ C}}{\text{s}} \times 1\text{ hr} \times \frac{60\text{ min}}{\text{hr}} \times \frac{60\text{ s}}{\text{min}}}{96{,}485\text{ C/mol e}^-} = \frac{36{,}830\text{ C}}{96{,}485\text{ C/mol e}^-} = 0.3817\text{ mol e}^-$$

From the problem, the solution contains $AgNO_3$, so the reaction at the cathode involves 1 mole of electrons for each mole of silver

$$\text{cathode: } Ag^+(aq) + e^- \longrightarrow Ag(s)$$

The atomic mass of silver is 107.9 g/mol, so

$$\text{mass Ag} = 0.3817\text{ mol e}^- \times \frac{1\text{ mol Ag}}{1\text{ mol e}^-} \times \frac{107.9\text{ g Ag}}{1\text{ mol Ag}} = 41.19\text{ g Ag}$$

Check Your Learning

Aluminum metal can be made from aluminum(III) ions by electrolysis. What is the half-reaction at the cathode? What mass of aluminum metal would be recovered if a current of 25.0 A passed through the solution for 15.0 minutes?

Answer:

$Al^{3+}(aq) + 3\,e^- \longrightarrow Al(s)$; 0.0777 mol Al = 2.10 g Al.

 **EXAMPLE 17.10**

Time Required for Deposition

In one application, a 0.010-mm layer of chromium must be deposited on a part with a total surface area of 3.3 m^2 from a solution of containing chromium(III) ions. How long would it take to deposit the layer of chromium if the current was 33.46 A? The density of chromium (metal) is 7.19 g/cm^3.

Solution

First, compute the volume of chromium that must be produced (equal to the product of surface area and thickness):

$$\text{volume} = \left(0.010\ \text{mm} \times \frac{1\ \text{cm}}{10\ \text{mm}}\right) \times \left(3.3\ \text{m}^2 \times \left(\frac{10{,}000\ \text{cm}^2}{1\ \text{m}^2}\right)\right) = 33\ \text{cm}^3$$

Use the computed volume and the provided density to calculate the molar amount of chromium required:

$$\text{mass} = \text{volume} \times \text{density} = 33\ \cancel{\text{cm}^3} \times \frac{7.19\ \text{g}}{\cancel{\text{cm}^3}} = 237\ \text{g Cr}$$

$$\text{mol Cr} = 237\ \text{g Cr} \times \frac{1\ \text{mol Cr}}{52.00\ \text{g Cr}} = 4.56\ \text{mol Cr}$$

The stoichiometry of the chromium(III) reduction process requires three moles of electrons for each mole of chromium(0) produced, and so the total charge required is:

$$Q = 4.56\ \text{mol Cr} \times \frac{3\ \text{mol e}^-}{1\ \text{mol Cr}} \times \frac{96485\ \text{C}}{\text{mol e}^-} = 1.32 \times 10^6\ \text{C}$$

Finally, if this charge is passed at a rate of 33.46 C/s, the required time is:

$$t = \frac{Q}{I} = \frac{1.32 \times 10^6\ \text{C}}{33.46\ \text{C/s}} = 3.95 \times 10^4\ \text{s} = 11.0\ \text{hr}$$

Check Your Learning

What mass of zinc is required to galvanize the top of a 3.00 m × 5.50 m sheet of iron to a thickness of 0.100 mm of zinc? If the zinc comes from a solution of $Zn(NO_3)_2$ and the current is 25.5 A, how long will it take to galvanize the top of the iron? The density of zinc is 7.140 g/cm^3.

Answer:

11.8 kg Zn requires 382 hours.

Key Terms

active electrode electrode that participates as a reactant or product in the oxidation-reduction reaction of an electrochemical cell; the mass of an active electrode changes during the oxidation-reduction reaction

alkaline battery primary battery similar to a *dry cell* that uses an alkaline (often potassium hydroxide) electrolyte; designed to be an improved replacement for the dry cell, but with more energy storage and less electrolyte leakage than typical dry cell

anode electrode in an electrochemical cell at which oxidation occurs

battery single or series of galvanic cells designed for use as a source of electrical power

cathode electrode in an electrochemical cell at which reduction occurs

cathodic protection approach to preventing corrosion of a metal object by connecting it to a *sacrificial anode* composed of a more readily oxidized metal

cell notation (schematic) symbolic representation of the components and reactions in an electrochemical cell

cell potential (E_{cell}) difference in potential of the cathode and anode half-cells

concentration cell galvanic cell comprising half-cells of identical composition but for the concentration of one redox reactant or product

corrosion degradation of metal via a natural electrochemical process

dry cell primary battery, also called a zinc-carbon battery, based on the spontaneous oxidation of zinc by manganese(IV)

electrode potential (E_X) the potential of a cell in which the half-cell of interest acts as a cathode when connected to the standard hydrogen electrode

electrolysis process using electrical energy to cause a nonspontaneous process to occur

electrolytic cell electrochemical cell in which an external source of electrical power is used to drive an otherwise nonspontaneous process

Faraday's constant (F) charge on 1 mol of electrons; $F = 96{,}485$ C/mol e^-

fuel cell devices similar to galvanic cells that require a continuous feed of redox reactants; also called a *flow battery*

galvanic (voltaic) cell electrochemical cell in which a spontaneous redox reaction takes place; also called a *voltaic cell*

galvanization method of protecting iron or similar metals from corrosion by coating with a thin layer of more easily oxidized zinc.

half cell component of a cell that contains the redox conjugate pair ("couple") of a single reactant

inert electrode electrode that conducts electrons to and from the reactants in a half-cell but that is not itself oxidized or reduced

lead acid battery rechargeable battery commonly used in automobiles; it typically comprises six galvanic cells based on Pb half-reactions in acidic solution

lithium ion battery widely used rechargeable battery commonly used in portable electronic devices, based on lithium ion transfer between the anode and cathode

Nernst equation relating the potential of a redox system to its composition

nickel-cadmium battery rechargeable battery based on Ni/Cd half-cells with applications similar to those of lithium ion batteries

primary cell nonrechargeable battery, suitable for single use only

sacrificial anode electrode constructed from an easily oxidized metal, often magnesium or zinc, used to prevent corrosion of metal objects via cathodic protection

salt bridge tube filled with inert electrolyte solution

secondary cell battery designed to allow recharging

standard cell potential (E°_{cell}) the cell potential when all reactants and products are in their standard states (1 bar or 1 atm or gases; 1 *M* for solutes), usually at 298.15 K

standard electrode potential ((E°_X)) electrode potential measured under standard conditions (1 bar or 1 atm for gases; 1 *M* for solutes) usually at 298.15 K

standard hydrogen electrode (SHE) half-cell based on hydrogen ion production, assigned a potential of exactly 0 V under standard state conditions, used as the universal reference for measuring electrode potential

Key Equations

$$E^\circ_{cell} = E^\circ_{cathode} - E^\circ_{anode}$$

$$E^\circ_{cell} = \frac{RT}{nF} \ln K$$

$$E^\circ_{cell} = \frac{0.0257 \text{ V}}{n} \ln K = \frac{0.0592 \text{ V}}{n} \log K \qquad \text{(at 298.15 } K\text{)}$$

$$E_{cell} = E^\circ_{cell} - \frac{RT}{nF} \ln Q \qquad \text{(Nernst equation)}$$

$$E_{cell} = E^\circ_{cell} - \frac{0.0592 \text{ V}}{n} \log Q \qquad \text{(at 298.15 } K\text{)}$$

$$\Delta G = -nFE_{cell}$$

$$\Delta G^\circ = -nFE^\circ_{cell}$$

$$w_{ele} = w_{max} = -nFE_{cell}$$

$$Q = I \times t = n \times F$$

Summary

17.1 Review of Redox Chemistry

Redox reactions are defined by changes in reactant oxidation numbers, and those most relevant to electrochemistry involve actual transfer of electrons. Aqueous phase redox processes often involve water or its characteristic ions, H^+ and OH^-, as reactants in addition to the oxidant and reductant, and equations representing these reactions can be challenging to balance. The half-reaction method is a systematic approach to balancing such equations that involves separate treatment of the oxidation and reduction half-reactions.

17.2 Galvanic Cells

Galvanic cells are devices in which a spontaneous redox reaction occurs indirectly, with the oxidant and reductant redox couples contained in separate half-cells. Electrons are transferred from the reductant (in the anode half-cell) to the oxidant (in the cathode half-cell) through an external circuit, and inert solution phase ions are transferred between half-cells, through a salt bridge, to maintain charge neutrality. The construction and composition of a galvanic cell may be succinctly represented using chemical formulas and others symbols in the form of a cell schematic (cell notation).

17.3 Electrode and Cell Potentials

The property of potential, *E*, is the energy associated with the separation/transfer of charge. In electrochemistry, the potentials of cells and half-cells are thermodynamic quantities that reflect the driving force or the spontaneity of their redox processes. The cell potential of an electrochemical cell is the difference in between its cathode and anode. To permit easy sharing of half-cell potential data, the standard hydrogen electrode (SHE) is assigned a potential of exactly 0 V and used to define a single electrode potential for any given half-cell. The electrode potential of a half-cell, E_X, is the cell potential of said half-cell acting as a cathode when connected to a SHE acting as an anode. When the half-cell is operating under standard state conditions, its potential is the standard electrode potential, E°_X. Standard electrode potentials reflect the relative oxidizing strength of the half-reaction's reactant, with stronger oxidants exhibiting larger (more positive) E°_X values. Tabulations of standard electrode potentials may be used to compute standard cell potentials, E°_{cell}, for many redox reactions. The arithmetic sign of a cell potential indicates the spontaneity of the cell reaction, with positive values for spontaneous reactions and negative values for nonspontaneous reactions (spontaneous in the reverse direction).

17.4 Potential, Free Energy, and Equilibrium

Potential is a thermodynamic quantity reflecting the intrinsic driving force of a redox process, and it is directly related to the free energy change and equilibrium constant for the process. For redox processes taking place in electrochemical cells, the maximum (electrical) work done by the system is easily computed from the cell potential and the reaction stoichiometry and is equal to the free energy change for the process. The equilibrium constant for a redox reaction is logarithmically related to the reaction's cell potential, with larger (more positive) potentials indicating reactions with greater driving force that equilibrate when the reaction has proceeded far towards completion (large value of *K*). Finally, the potential of a redox process varies with the composition of the reaction mixture, being related to the reactions standard potential and the value of its reaction quotient, *Q*, as described by the Nernst equation.

17.5 Batteries and Fuel Cells

Galvanic cells designed specifically to function as electrical power supplies are called batteries. A variety of both single-use batteries (primary cells) and rechargeable batteries (secondary cells) are commercially available to serve a variety of applications, with important specifications including voltage, size, and lifetime. Fuel cells, sometimes called flow batteries, are devices that harness the energy of spontaneous redox reactions normally associated with combustion processes. Like batteries, fuel cells enable the reaction's electron transfer via an external circuit, but they require continuous input of the redox reactants (fuel and oxidant) from an external reservoir. Fuel cells are typically much more efficient in converting the energy released by the reaction to useful work in comparison to internal combustion engines.

17.6 Corrosion

Spontaneous oxidation of metals by natural electrochemical processes is called corrosion, familiar examples including the rusting of iron and the tarnishing of silver. Corrosion process involve the creation of a galvanic cell in which different sites on the metal object function as anode and cathode, with the corrosion taking place at the anodic site. Approaches to preventing corrosion of metals include use of a protective coating of zinc (galvanization) and the use of sacrificial anodes connected to the metal object (cathodic protection).

17.7 Electrolysis

Nonspontaneous redox processes may be forced to occur in electrochemical cells by the application of an appropriate potential using an external power source—a process known as electrolysis. Electrolysis is the basis for certain ore refining processes, the industrial production of many chemical commodities, and the electroplating of metal coatings on various products. Measurement of the current flow during electrolysis permits stoichiometric calculations.

Exercises

17.1 Review of Redox Chemistry

1. Identify each half-reaction below as either oxidation or reduction.
   (a) $Fe^{3+} + 3e^- \longrightarrow Fe$
   (b) $Cr \longrightarrow Cr^{3+} + 3e^-$
   (c) $MnO_4{}^{2-} \longrightarrow MnO_4{}^- + e^-$
   (d) $Li^+ + e^- \longrightarrow Li$
2. Identify each half-reaction below as either oxidation or reduction.
   (a) $Cl^- \longrightarrow Cl_2$
   (b) $Mn^{2+} \longrightarrow MnO_2$
   (c) $H_2 \longrightarrow H^+$
   (d) $NO_3{}^- \longrightarrow NO$
3. Assuming each pair of half-reactions below takes place in an acidic solution, write a balanced equation for the overall reaction.
   (a) $Ca \longrightarrow Ca^{2+} + 2e^-, F_2 + 2e^- \longrightarrow 2F^-$
   (b) $Li \longrightarrow Li^+ + e^-, Cl_2 + 2e^- \longrightarrow 2Cl^-$
   (c) $Fe \longrightarrow Fe^{3+} + 3e^-, Br_2 + 2e^- \longrightarrow 2Br^-$
   (d) $Ag \longrightarrow Ag^+ + e^-, MnO_4{}^- + 4H^+ + 3e^- \longrightarrow MnO_2 + 2H_2O$
4. Balance the equations below assuming they occur in an acidic solution.
   (a) $H_2O_2 + Sn^{2+} \longrightarrow H_2O + Sn^{4+}$
   (b) $PbO_2 + Hg \longrightarrow Hg_2{}^{2+} + Pb^{2+}$
   (c) $Al + Cr_2O_7{}^{2-} \longrightarrow Al^{3+} + Cr^{3+}$
5. Identify the oxidant and reductant of each reaction of the previous exercise.

6. Balance the equations below assuming they occur in a basic solution.
   (a) $SO_3{}^{2-}(aq) + Cu(OH)_2(s) \longrightarrow SO_4{}^{2-}(aq) + Cu(OH)(s)$
   (b) $O_2(g) + Mn(OH)_2(s) \longrightarrow MnO_2(s)$
   (c) $NO_3{}^-(aq) + H_2(g) \longrightarrow NO(g)$
   (d) $Al(s) + CrO_4{}^{2-}(aq) \longrightarrow Al(OH)_3(s) + Cr(OH)_4{}^-(aq)$
7. Identify the oxidant and reductant of each reaction of the previous exercise.
8. Why don't hydroxide ions appear in equations for half-reactions occurring in acidic solution?
9. Why don't hydrogen ions appear in equations for half-reactions occurring in basic solution?
10. Why must the charge balance in oxidation-reduction reactions?

17.2 Galvanic Cells

11. Write cell schematics for the following cell reactions, using platinum as an inert electrode as needed.
    (a) $Mg(s) + Ni^{2+}(aq) \longrightarrow Mg^{2+}(aq) + Ni(s)$
    (b) $2Ag^+(aq) + Cu(s) \longrightarrow Cu^{2+}(aq) + 2Ag(s)$
    (c) $Mn(s) + Sn(NO_3)_2(aq) \longrightarrow Mn(NO_3)_2(aq) + Sn(s)$
    (d) $3CuNO_3(aq) + Au(NO_3)_3(aq) \longrightarrow 3Cu(NO_3)_2(aq) + Au(s)$
12. Assuming the schematics below represent galvanic cells as written, identify the half-cell reactions occurring in each.
    (a) $Mg(s) \mid Mg^{2+}(aq) \parallel Cu^{2+}(aq) \mid Cu(s)$
    (b) $Ni(s) \mid Ni^{2+}(aq) \parallel Ag^+(aq) \mid Ag(s)$
13. Write a balanced equation for the cell reaction of each cell in the previous exercise.
14. Balance each reaction below, and write a cell schematic representing the reaction as it would occur in a galvanic cell.
    (a) $Al(s) + Zr^{4+}(aq) \longrightarrow Al^{3+}(aq) + Zr(s)$
    (b) $Ag^+(aq) + NO(g) \longrightarrow Ag(s) + NO_3{}^-(aq)$ (acidic solution)
    (c) $SiO_3{}^{2-}(aq) + Mg(s) \longrightarrow Si(s) + Mg(OH)_2(s)$ (basic solution)
    (d) $ClO_3{}^-(aq) + MnO_2(s) \longrightarrow Cl^-(aq) + MnO_4{}^-(aq)$ (basic solution)
15. Identify the oxidant and reductant in each reaction of the previous exercise.
16. From the information provided, use cell notation to describe the following systems:
    (a) In one half-cell, a solution of $Pt(NO_3)_2$ forms Pt metal, while in the other half-cell, Cu metal goes into a $Cu(NO_3)_2$ solution with all solute concentrations 1 *M*.
    (b) The cathode consists of a gold electrode in a 0.55 *M* $Au(NO_3)_3$ solution and the anode is a magnesium electrode in 0.75 *M* $Mg(NO_3)_2$ solution.
    (c) One half-cell consists of a silver electrode in a 1 *M* $AgNO_3$ solution, and in the other half-cell, a copper electrode in 1 *M* $Cu(NO_3)_2$ is oxidized.
17. Why is a salt bridge necessary in galvanic cells like the one in Figure 17.3?
18. An active (metal) electrode was found to gain mass as the oxidation-reduction reaction was allowed to proceed. Was the electrode an anode or a cathode? Explain.
19. An active (metal) electrode was found to lose mass as the oxidation-reduction reaction was allowed to proceed. Was the electrode an anode or a cathode? Explain.
20. The masses of three electrodes (A, B, and C), each from three different galvanic cells, were measured before and after the cells were allowed to pass current for a while. The mass of electrode A increased, that of electrode B was unchanged, and that of electrode C decreased. Identify each electrode as active or inert, and note (if possible) whether it functioned as anode or cathode.

17.3 Electrode and Cell Potentials

21. Calculate the standard cell potential for each reaction below, and note whether the reaction is spontaneous under standard state conditions.
(a) $Mg(s) + Ni^{2+}(aq) \longrightarrow Mg^{2+}(aq) + Ni(s)$
(b) $2Ag^{+}(aq) + Cu(s) \longrightarrow Cu^{2+}(aq) + 2Ag(s)$
(c) $Mn(s) + Sn(NO_3)_2(aq) \longrightarrow Mn(NO_3)_2(aq) + Sn(s)$
(d) $3Fe(NO_3)_2(aq) + Au(NO_3)_3(aq) \longrightarrow 3Fe(NO_3)_3(aq) + Au(s)$

22. Calculate the standard cell potential for each reaction below, and note whether the reaction is spontaneous under standard state conditions.
(a) $Mn(s) + Ni^{2+}(aq) \longrightarrow Mn^{2+}(aq) + Ni(s)$
(b) $3Cu^{2+}(aq) + 2Al(s) \longrightarrow 2Al^{3+}(aq) + 3Cu(s)$
(c) $Na(s) + LiNO_3(aq) \longrightarrow NaNO_3(aq) + Li(s)$
(d) $Ca(NO_3)_2(aq) + Ba(s) \longrightarrow Ba(NO_3)_2(aq) + Ca(s)$

23. Write the balanced cell reaction for the cell schematic below, calculate the standard cell potential, and note whether the reaction is spontaneous under standard state conditions.
$Cu(s) \mid Cu^{2+}(aq) \parallel Au^{3+}(aq) \mid Au(s)$

24. Determine the cell reaction and standard cell potential at 25 °C for a cell made from a cathode half-cell consisting of a silver electrode in 1 *M* silver nitrate solution and an anode half-cell consisting of a zinc electrode in 1 *M* zinc nitrate. Is the reaction spontaneous at standard conditions?

25. Determine the cell reaction and standard cell potential at 25 °C for a cell made from an anode half-cell containing a cadmium electrode in 1 *M* cadmium nitrate and a cathode half-cell consisting of an aluminum electrode in 1 *M* aluminum nitrate solution. Is the reaction spontaneous at standard conditions?

26. Write the balanced cell reaction for the cell schematic below, calculate the standard cell potential, and note whether the reaction is spontaneous under standard state conditions.
$Pt(s) \mid H_2(g) \mid H^{+}(aq) \parallel Br_2(aq),\ Br^{-}(aq) \mid Pt(s)$

17.4 Potential, Free Energy, and Equilibrium

27. For each pair of standard cell potential and electron stoichiometry values below, calculate a corresponding standard free energy change (kJ).
(a) 0.000 V, n = 2
(b) +0.434 V, n = 2
(c) −2.439 V, n = 1

28. For each pair of standard free energy change and electron stoichiometry values below, calculate a corresponding standard cell potential.
(a) 12 kJ/mol, n = 3
(b) −45 kJ/mol, n = 1

29. Determine the standard cell potential and the cell potential under the stated conditions for the electrochemical reactions described here. State whether each is spontaneous or nonspontaneous under each set of conditions at 298.15 K.
(a) $Hg(l) + S^{2-}(aq, 0.10\ M) + 2Ag^{+}(aq, 0.25\ M) \longrightarrow 2Ag(s) + HgS(s)$
(b) The cell made from an anode half-cell consisting of an aluminum electrode in 0.015 *M* aluminum nitrate solution and a cathode half-cell consisting of a nickel electrode in 0.25 *M* nickel(II) nitrate solution.
(c) The cell comprised of a half-cell in which aqueous bromine (1.0 *M*) is being oxidized to bromide ion (0.11 *M*) and a half-cell in which Al^{3+} (0.023 *M*) is being reduced to aluminum metal.

30. Determine ΔG and $\Delta G°$ for each of the reactions in the previous problem.

31. Use the data in Appendix L to calculate equilibrium constants for the following reactions. Assume 298.15 K if no temperature is given.
(a) $AgCl(s) \rightleftharpoons Ag^{+}(aq) + Cl^{-}(aq)$
(b) $CdS(s) \rightleftharpoons Cd^{2+}(aq) + S^{2-}(aq)$ at 377 K
(c) $Hg^{2+}(aq) + 4Br^{-}(aq) \rightleftharpoons [HgBr_4]^{2-}(aq)$
(d) $H_2O(l) \rightleftharpoons H^{+}(aq) + OH^{-}(aq)$ at 25 °C

17.5 Batteries and Fuel Cells

32. Consider a battery made from one half-cell that consists of a copper electrode in 1 *M* $CuSO_4$ solution and another half-cell that consists of a lead electrode in 1 *M* $Pb(NO_3)_2$ solution.
(a) What is the standard cell potential for the battery?
(b) What are the reactions at the anode, cathode, and the overall reaction?
(c) Most devices designed to use dry-cell batteries can operate between 1.0 and 1.5 V. Could this cell be used to make a battery that could replace a dry-cell battery? Why or why not.
(d) Suppose sulfuric acid is added to the half-cell with the lead electrode and some $PbSO_4(s)$ forms. Would the cell potential increase, decrease, or remain the same?

33. Consider a battery with the overall reaction: $Cu(s) + 2Ag^+(aq) \longrightarrow 2Ag(s) + Cu^{2+}(aq)$.
(a) What is the reaction at the anode and cathode?
(b) A battery is "dead" when its cell potential is zero. What is the value of *Q* when this battery is dead?
(c) If a particular dead battery was found to have $[Cu^{2+}] = 0.11$ *M*, what was the concentration of silver ion?

34. Why do batteries go dead, but fuel cells do not?

35. Use the Nernst equation to explain the drop in voltage observed for some batteries as they discharge.

36. Using the information thus far in this chapter, explain why battery-powered electronics perform poorly in low temperatures.

17.6 Corrosion

37. Which member of each pair of metals is more likely to corrode (oxidize)?
(a) Mg or Ca
(b) Au or Hg
(c) Fe or Zn
(d) Ag or Pt

38. Consider the following metals: Ag, Au, Mg, Ni, and Zn. Which of these metals could be used as a sacrificial anode in the cathodic protection of an underground steel storage tank? Steel is an alloy composed mostly of iron, so use −0.447 V as the standard reduction potential for steel.

39. Aluminum ($E^\circ_{Al^{3+}/Al} = -2.07$ V) is more easily oxidized than iron ($E^\circ_{Fe^{3+}/Fe} = -0.477$ V), and yet when both are exposed to the environment, untreated aluminum has very good corrosion resistance while the corrosion resistance of untreated iron is poor. What might explain this observation?

40. If a sample of iron and a sample of zinc come into contact, the zinc corrodes but the iron does not. If a sample of iron comes into contact with a sample of copper, the iron corrodes but the copper does not. Explain this phenomenon.

41. Suppose you have three different metals, A, B, and C. When metals A and B come into contact, B corrodes and A does not corrode. When metals A and C come into contact, A corrodes and C does not corrode. Based on this information, which metal corrodes and which metal does not corrode when B and C come into contact?

42. Why would a sacrificial anode made of lithium metal be a bad choice

17.7 Electrolysis

43. If a 2.5 A current flows through a circuit for 35 minutes, how many coulombs of charge moved through the circuit?

44. For the scenario in the previous question, how many electrons moved through the circuit?

45. Write the half-reactions and cell reaction occurring during electrolysis of each molten salt below.
(a) $CaCl_2$
(b) LiH
(c) $AlCl_3$
(d) $CrBr_3$

46. What mass of each product is produced in each of the electrolytic cells of the previous problem if a total charge of 3.33×10^5 C passes through each cell?

47. How long would it take to reduce 1 mole of each of the following ions using the current indicated?
(a) Al^{3+}, 1.234 A
(b) Ca^{2+}, 22.2 A
(c) Cr^{5+}, 37.45 A
(d) Au^{3+}, 3.57 A

48. A current of 2.345 A passes through the cell shown in Figure 17.19 for 45 minutes. What is the volume of the hydrogen collected at room temperature if the pressure is exactly 1 atm? (Hint: Is hydrogen the only gas present above the water?)

49. An irregularly shaped metal part made from a particular alloy was galvanized with zinc using a $Zn(NO_3)_2$ solution. When a current of 2.599 A was used, it took exactly 1 hour to deposit a 0.01123-mm layer of zinc on the part. What was the total surface area of the part? The density of zinc is 7.140 g/cm^3.

CHAPTER 18
Representative Metals, Metalloids, and Nonmetals

Figure 18.1 Purity is extremely important when preparing silicon wafers. Technicians in a cleanroom prepare silicon without impurities (left). The CEO of VLSI Research, Don Hutcheson, shows off a pure silicon wafer (center). A silicon wafer covered in Pentium chips is an enlarged version of the silicon wafers found in many electronics used today (right). (credit middle: modification of work by "Intel Free Press"/Flickr; credit right: modification of work by Naotake Murayama)

CHAPTER OUTLINE

18.1 Periodicity
18.2 Occurrence and Preparation of the Representative Metals
18.3 Structure and General Properties of the Metalloids
18.4 Structure and General Properties of the Nonmetals
18.5 Occurrence, Preparation, and Compounds of Hydrogen
18.6 Occurrence, Preparation, and Properties of Carbonates
18.7 Occurrence, Preparation, and Properties of Nitrogen
18.8 Occurrence, Preparation, and Properties of Phosphorus
18.9 Occurrence, Preparation, and Compounds of Oxygen
18.10 Occurrence, Preparation, and Properties of Sulfur
18.11 Occurrence, Preparation, and Properties of Halogens
18.12 Occurrence, Preparation, and Properties of the Noble Gases

INTRODUCTION The development of the periodic table in the mid-1800s came from observations that there was a periodic relationship between the properties of the elements. Chemists, who have an understanding of the variations of these properties, have been able to use this knowledge to solve a wide variety of technical challenges. For example, silicon and other semiconductors form the backbone of modern electronics because of our ability to fine-tune the electrical properties of these materials. This chapter explores important properties of representative metals, metalloids, and nonmetals in the periodic table.

18.1 Periodicity

LEARNING OBJECTIVES

By the end of this section, you will be able to:

- Classify elements
- Make predictions about the periodicity properties of the representative elements

We begin this section by examining the behaviors of representative metals in relation to their positions in the periodic table. The primary focus of this section will be the application of periodicity to the representative metals.

It is possible to divide elements into groups according to their electron configurations. The **representative elements** are elements where the *s* and *p* orbitals are filling. The transition elements are elements where the *d* orbitals (groups 3–11 on the periodic table) are filling, and the inner transition metals are the elements where the *f* orbitals are filling. The *d* orbitals fill with the elements in group 11; therefore, the elements in group 12 qualify as representative elements because the last electron enters an *s* orbital. Metals among the representative elements are the **representative metals**. Metallic character results from an element's ability to lose its outer valence electrons and results in high thermal and electrical conductivity, among other physical and chemical properties. There are 20 nonradioactive representative metals in groups 1, 2, 3, 12, 13, 14, and 15 of the periodic table (the elements shaded in yellow in Figure 18.2). The radioactive elements copernicium, flerovium, polonium, and livermorium are also metals but are beyond the scope of this chapter.

In addition to the representative metals, some of the representative elements are metalloids. A **metalloid** is an element that has properties that are between those of metals and nonmetals; these elements are typically semiconductors.

The remaining representative elements are nonmetals. Unlike **metals**, which typically form cations and ionic compounds (containing ionic bonds), nonmetals tend to form anions or molecular compounds. In general, the combination of a metal and a nonmetal produces a salt. A salt is an ionic compound consisting of cations and anions.

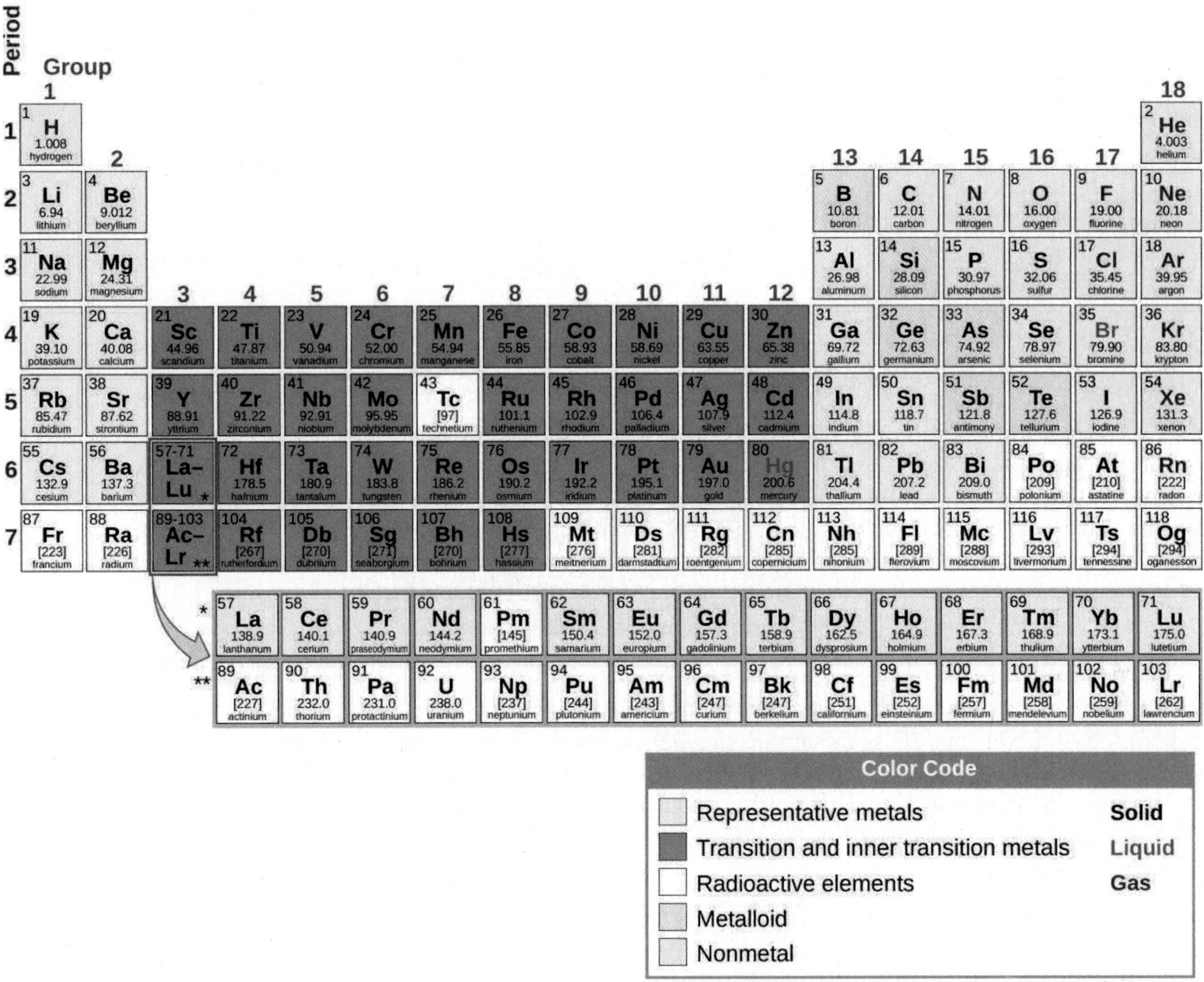

Period
Group
1
2
3
4
5
6
7
8
9
10
11
12
13
14
15
16
17
18
1 H 1.008 hydrogen
2 He 4.003 helium
3 Li 6.94 lithium
4 Be 9.012 beryllium
5 B 10.81 boron
6 C 12.01 carbon
7 N 14.01 nitrogen
8 O 16.00 oxygen
9 F 19.00 fluorine
10 Ne 20.18 neon
11 Na 22.99 sodium
12 Mg 24.31 magnesium
13 Al 26.98 aluminum
14 Si 28.09 silicon
15 P 30.97 phosphorus
16 S 32.06 sulfur
17 Cl 35.45 chlorine
18 Ar 39.95 argon
19 K 39.10 potassium
20 Ca 40.08 calcium
21 Sc 44.96 scandium
22 Ti 47.87 titanium
23 V 50.94 vanadium
24 Cr 52.00 chromium
25 Mn 54.94 manganese
26 Fe 55.85 iron
27 Co 58.93 cobalt
28 Ni 58.69 nickel
29 Cu 63.55 copper
30 Zn 65.38 zinc
31 Ga 69.72 gallium
32 Ge 72.63 germanium
33 As 74.92 arsenic
34 Se 78.97 selenium
35 Br 79.90 bromine
36 Kr 83.80 krypton
37 Rb 85.47 rubidium
38 Sr 87.62 strontium
39 Y 88.91 yttrium
40 Zr 91.22 zirconium
41 Nb 92.91 niobium
42 Mo 95.95 molybdenum
43 Tc [97] technetium
44 Ru 101.1 ruthenium
45 Rh 102.9 rhodium
46 Pd 106.4 palladium
47 Ag 107.9 silver
48 Cd 112.4 cadmium
49 In 114.8 indium
50 Sn 118.7 tin
51 Sb 121.8 antimony
52 Te 127.6 tellurium
53 I 126.9 iodine
54 Xe 131.3 xenon
55 Cs 132.9 cesium
56 Ba 137.3 barium
57-71 La–Lu *
72 Hf 178.5 hafnium
73 Ta 180.9 tantalum
74 W 183.8 tungsten
75 Re 186.2 rhenium
76 Os 190.2 osmium
77 Ir 192.2 iridium
78 Pt 195.1 platinum
79 Au 197.0 gold
80 Hg 200.6 mercury
81 Tl 204.4 thallium
82 Pb 207.2 lead
83 Bi 209.0 bismuth
84 Po [209] polonium
85 At [210] astatine
86 Rn [222] radon
87 Fr [223] francium
88 Ra [226] radium
89-103 Ac–Lr **
104 Rf [267] rutherfordium
105 Db [270] dubnium
106 Sg [271] seaborgium
107 Bh [270] bohrium
108 Hs [277] hassium
109 Mt [276] meitnerium
110 Ds [281] darmstadtium
111 Rg [282] roentgenium
112 Cn [285] copernicium
113 Nh [285] nihonium
114 Fl [289] flerovium
115 Mc [288] moscovium
116 Lv [293] livermorium
117 Ts [294] tennessine
118 Og [294] oganesson
*
57 La 138.9 lanthanum
58 Ce 140.1 cerium
59 Pr 140.9 praseodymium
60 Nd 144.2 neodymium
61 Pm [145] promethium
62 Sm 150.4 samarium
63 Eu 152.0 europium
64 Gd 157.3 gadolinium
65 Tb 158.9 terbium
66 Dy 162.5 dysprosium
67 Ho 164.9 holmium
68 Er 167.3 erbium
69 Tm 168.9 thulium
70 Yb 173.1 ytterbium
71 Lu 175.0 lutetium
**
89 Ac [227] actinium
90 Th 232.0 thorium
91 Pa 231.0 protactinium
92 U 238.0 uranium
93 Np [237] neptunium
94 Pu [244] plutonium
95 Am [243] americium
96 Cm [247] curium
97 Bk [247] berkelium
98 Cf [251] californium
99 Es [252] einsteinium
100 Fm [257] fermium
101 Md [258] mendelevium
102 No [259] nobelium
103 Lr [262] lawrencium
Color Code
Representative metals
Transition and inner transition metals
Radioactive elements
Metalloid
Nonmetal
Solid
Liquid
Gas

FIGURE 18.2 The location of the representative metals is shown in the periodic table. Nonmetals are shown in green, metalloids in purple, and the transition metals and inner transition metals in blue.

Most of the representative metals do not occur naturally in an uncombined state because they readily react with water and oxygen in the air. However, it is possible to isolate elemental beryllium, magnesium, zinc, cadmium, mercury, aluminum, tin, and lead from their naturally occurring minerals and use them because they react very slowly with air. Part of the reason why these elements react slowly is that these elements react with air to form a protective coating. The formation of this protective coating is **passivation**. The coating is a nonreactive film of oxide or some other compound. Elemental magnesium, aluminum, zinc, and tin are important in the fabrication of many familiar items, including wire, cookware, foil, and many household and personal objects. Although beryllium, cadmium, mercury, and lead are readily available, there are limitations in their use because of their toxicity.

Group 1: The Alkali Metals

The alkali metals lithium, sodium, potassium, rubidium, cesium, and francium constitute group 1 of the periodic table. Although hydrogen is in group 1 (and also in group 17), it is a nonmetal and deserves separate consideration later in this chapter. The name alkali metal is in reference to the fact that these metals and their oxides react with water to form very basic (alkaline) solutions.

The properties of the alkali metals are similar to each other as expected for elements in the same family. The alkali metals have the largest atomic radii and the lowest first ionization energy in their periods. This combination makes it very easy to remove the single electron in the outermost (valence) shell of each. The easy loss of this valence electron means that these metals readily form stable cations with a charge of 1+. Their reactivity increases with increasing atomic number due to the ease of losing the lone valence electron (decreasing ionization energy). Since oxidation is so easy, the reverse, reduction, is difficult, which explains why it is hard to isolate the elements. The solid alkali metals are very soft; lithium, shown in Figure 18.3, has the lowest density of any metal (0.5 g/cm^3).

The alkali metals all react vigorously with water to form hydrogen gas and a basic solution of the metal hydroxide. This means they are easier to oxidize than is hydrogen. As an example, the reaction of lithium with water is:

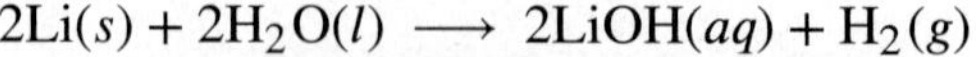

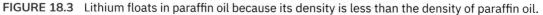

$$2\text{Li}(s) + 2\text{H}_2\text{O}(l) \longrightarrow 2\text{LiOH}(aq) + \text{H}_2(g)$$

FIGURE 18.3 Lithium floats in paraffin oil because its density is less than the density of paraffin oil.

Alkali metals react directly with all the nonmetals (except the noble gases) to yield binary ionic compounds containing 1+ metal ions. These metals are so reactive that it is necessary to avoid contact with both moisture and oxygen in the air. Therefore, they are stored in sealed containers under mineral oil, as shown in Figure 18.4, to prevent contact with air and moisture. The pure metals never exist free (uncombined) in nature due to their high reactivity. In addition, this high reactivity makes it necessary to prepare the metals by electrolysis of alkali metal compounds.

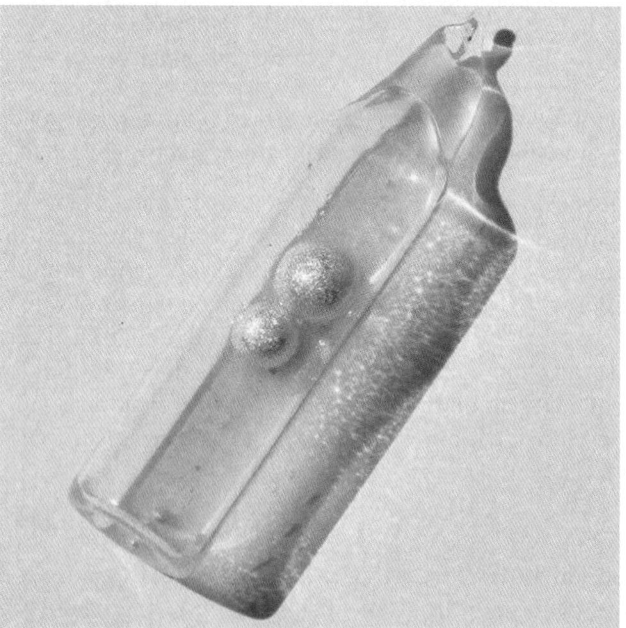

FIGURE 18.4 To prevent contact with air and water, potassium for laboratory use comes as sticks or beads stored under kerosene or mineral oil, or in sealed containers. (credit: http://images-of-elements.com/potassium.php)

Unlike many other metals, the reactivity and softness of the alkali metals make these metals unsuitable for structural applications. However, there are applications where the reactivity of the alkali metals is an advantage. For example, the production of metals such as titanium and zirconium relies, in part, on the ability of sodium to reduce compounds of these metals. The manufacture of many organic compounds, including certain dyes, drugs, and perfumes, utilizes reduction by lithium or sodium.

Sodium and its compounds impart a bright yellow color to a flame, as seen in Figure 18.5. Passing an electrical discharge through sodium vapor also produces this color. In both cases, this is an example of an emission spectrum as discussed in the chapter on electronic structure. Streetlights sometime employ sodium vapor lights because the sodium vapor penetrates fog better than most other light. This is because the fog does not scatter yellow light as much as it scatters white light. The other alkali metals and their salts also impart color to a flame. Lithium creates a bright, crimson color, whereas the others create a pale, violet color.

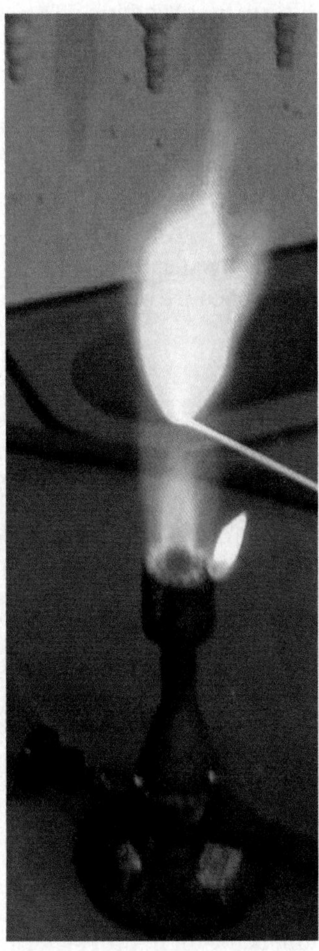

FIGURE 18.5 Dipping a wire into a solution of a sodium salt and then heating the wire causes emission of a bright yellow light, characteristic of sodium.

LINK TO LEARNING

This video (http://openstax.org/l/16alkalih2o) demonstrates the reactions of the alkali metals with water.

Group 2: The Alkaline Earth Metals

The **alkaline earth metals** (beryllium, magnesium, calcium, strontium, barium, and radium) constitute group 2 of the periodic table. The name alkaline metal comes from the fact that the oxides of the heavier members of the group react with water to form alkaline solutions. The nuclear charge increases when going from group 1 to group 2. Because of this charge increase, the atoms of the alkaline earth metals are smaller and have higher first ionization energies than the alkali metals within the same period. The higher ionization energy makes the alkaline earth metals less reactive than the alkali metals; however, they are still very reactive elements. Their reactivity increases, as expected, with increasing size and decreasing ionization energy. In chemical reactions, these metals readily lose both valence electrons to form compounds in which they exhibit an oxidation state of 2+. Due to their high reactivity, it is common to produce the alkaline earth metals, like the alkali metals, by electrolysis. Even though the ionization energies are low, the two metals with the highest ionization energies (beryllium and magnesium) do form compounds that exhibit some covalent characters. Like the alkali metals, the heavier alkaline earth metals impart color to a flame. As in the case of the alkali metals, this is part of the emission spectrum of these elements. Calcium and strontium produce shades of red, whereas barium produces a green color.

Magnesium is a silver-white metal that is malleable and ductile at high temperatures. Passivation decreases the reactivity of magnesium metal. Upon exposure to air, a tightly adhering layer of magnesium oxycarbonate forms on the surface of the metal and inhibits further reaction. (The carbonate comes from the reaction of carbon dioxide in the atmosphere.) Magnesium is the lightest of the widely used structural metals, which is why most magnesium production is for lightweight alloys.

Magnesium (shown in Figure 18.6), calcium, strontium, and barium react with water and air. At room temperature, barium shows the most vigorous reaction. The products of the reaction with water are hydrogen and the metal hydroxide. The formation of hydrogen gas indicates that the heavier alkaline earth metals are better reducing agents (more easily oxidized) than is hydrogen. As expected, these metals react with both acids and nonmetals to form ionic compounds. Unlike most salts of the alkali metals, many of the common salts of the alkaline earth metals are insoluble in water because of the high lattice energies of these compounds, containing a divalent metal ion.

FIGURE 18.6 From left to right: Mg(*s*), warm water at pH 7, and the resulting solution with a pH greater than 7, as indicated by the pink color of the phenolphthalein indicator. (credit: modification of work by Sahar Atwa)

The potent reducing power of hot magnesium is useful in preparing some metals from their oxides. Indeed, magnesium's affinity for oxygen is so great that burning magnesium reacts with carbon dioxide, producing elemental carbon:

$$2Mg(s) + CO_2(g) \longrightarrow 2MgO(s) + C(s)$$

For this reason, a CO_2 fire extinguisher will not extinguish a magnesium fire. Additionally, the brilliant white light emitted by burning magnesium makes it useful in flares and fireworks.

Group 12

The elements in group 12 are transition elements; however, the last electron added is not a *d* electron, but an *s* electron. Since the last electron added is an *s* electron, these elements qualify as representative metals, or post-transition metals. The group 12 elements behave more like the alkaline earth metals than transition metals. Group 12 contains the four elements zinc, cadmium, mercury, and copernicium. Each of these elements has two electrons in its outer shell (ns^2). When atoms of these metals form cations with a charge of 2+, where the two outer electrons are lost, they have pseudo-noble gas electron configurations. Mercury is sometimes an exception because it also exhibits an oxidation state of 1+ in compounds that contain a diatomic ${Hg_2}^{2+}$ ion. In their elemental forms and in compounds, cadmium and mercury are both toxic.

Zinc is the most reactive in group 12, and mercury is the least reactive. (This is the reverse of the reactivity trend of the metals of groups 1 and 2, in which reactivity increases down a group. The increase in reactivity with increasing atomic number only occurs for the metals in groups 1 and 2.) The decreasing reactivity is due to the formation of ions with a pseudo-noble gas configuration and to other factors that are beyond the scope of this discussion. The chemical behaviors of zinc and cadmium are quite similar to each other but differ from that of mercury.

Zinc and cadmium have lower reduction potentials than hydrogen, and, like the alkali metals and alkaline earth metals, they will produce hydrogen gas when they react with acids. The reaction of zinc with hydrochloric acid, shown in Figure 18.7, is:

$$\mathrm{Zn}(s) + 2\mathrm{H_3O^+}(aq) + 2\mathrm{Cl^-}(aq) \longrightarrow \mathrm{H_2}(g) + \mathrm{Zn^{2+}}(aq) + 2\mathrm{Cl^-}(aq) + 2\mathrm{H_2O}(l)$$

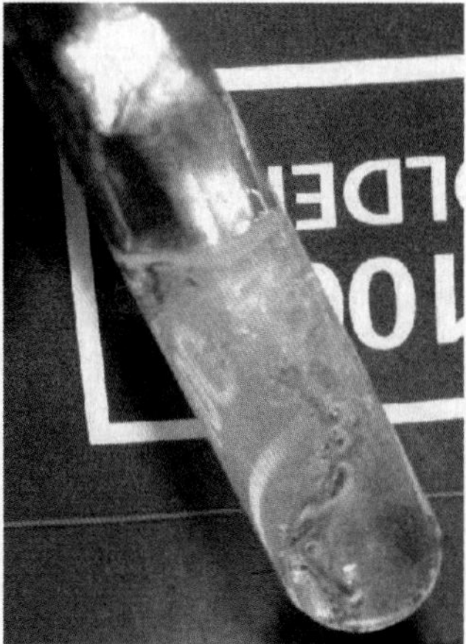

FIGURE 18.7 Zinc is an active metal. It dissolves in hydrochloric acid, forming a solution of colorless Zn^{2+} ions, Cl^- ions, and hydrogen gas.

Zinc is a silvery metal that quickly tarnishes to a blue-gray appearance. This change in color is due to an adherent coating of a basic carbonate, $Zn_2(OH)_2CO_3$, which passivates the metal to inhibit further corrosion. Dry cell and alkaline batteries contain a zinc anode. Brass (Cu and Zn) and some bronze (Cu, Sn, and sometimes Zn) are important zinc alloys. About half of zinc production serves to protect iron and other metals from corrosion. This protection may take the form of a sacrificial anode (also known as a galvanic anode, which is a means of providing cathodic protection for various metals) or as a thin coating on the protected metal. Galvanized steel is steel with a protective coating of zinc.

Chemistry in Everyday Life

Sacrificial Anodes

A sacrificial anode, or galvanic anode, is a means of providing cathodic protection of various metals. Cathodic protection refers to the prevention of corrosion by converting the corroding metal into a cathode. As a cathode, the metal resists corrosion, which is an oxidation process. Corrosion occurs at the sacrificial anode instead of at the cathode.

The construction of such a system begins with the attachment of a more active metal (more negative reduction potential) to the metal needing protection. Attachment may be direct or via a wire. To complete the circuit, a *salt bridge* is necessary. This salt bridge is often seawater or ground water. Once the circuit is complete, oxidation (corrosion) occurs at the anode and not the cathode.

The commonly used sacrificial anodes are magnesium, aluminum, and zinc. Magnesium has the most negative reduction potential of the three and serves best when the salt bridge is less efficient due to a low electrolyte concentration such as in freshwater. Zinc and aluminum work better in saltwater than does magnesium. Aluminum is lighter than zinc and has a higher capacity; however, an oxide coating may passivate the aluminum. In special cases, other materials are useful. For example, iron will protect copper.

Mercury is very different from zinc and cadmium. Mercury is the only metal that is liquid at 25 °C. Many metals dissolve in mercury, forming solutions called amalgams (see the feature on Amalgams), which are alloys of mercury with one or more other metals. Mercury, shown in Figure 18.8, is a nonreactive element that is more difficult to oxidize than hydrogen. Thus, it does not displace hydrogen from acids; however, it will react with strong oxidizing acids, such as nitric acid:

$$\text{Hg}(l) + \text{HCl}(aq) \longrightarrow \text{no reaction}$$

$$3\text{Hg}(l) + 8\text{HNO}_3(aq) \longrightarrow 3\text{Hg}(\text{NO}_3)_2(aq) + 4\text{H}_2\text{O}(l) + 2\text{NO}(g)$$

The clear NO initially formed quickly undergoes further oxidation to the reddish brown NO_2.

FIGURE 18.8 From left to right: Hg(*l*), Hg + concentrated HCl, Hg + concentrated HNO_3. (credit: Sahar Atwa)

Most mercury compounds decompose when heated. Most mercury compounds contain mercury with a 2+-oxidation state. When there is a large excess of mercury, it is possible to form compounds containing the ${Hg_2}^{2+}$ ion. All mercury compounds are toxic, and it is necessary to exercise great care in their synthesis.

Chemistry in Everyday Life

Amalgams

An amalgam is an alloy of mercury with one or more other metals. This is similar to considering steel to be an alloy of iron with other metals. Most metals will form an amalgam with mercury, with the main exceptions being iron, platinum, tungsten, and tantalum.

Due to toxicity issues with mercury, there has been a significant decrease in the use of amalgams. Historically, amalgams were important in electrolytic cells and in the extraction of gold. Amalgams of the alkali metals still find use because they are strong reducing agents and easier to handle than the pure alkali metals.

Prospectors had a problem when they found finely divided gold. They learned that adding mercury to their pans collected the gold into the mercury to form an amalgam for easier collection. Unfortunately, losses of small amounts of mercury over the years left many streams in California polluted with mercury.

Dentists use amalgams containing silver and other metals to fill cavities. There are several reasons to use an amalgam including low cost, ease of manipulation, and longevity compared to alternate materials. Dental amalgams are approximately 50% mercury by weight, which, in recent years, has become a concern due to the toxicity of mercury.

After reviewing the best available data, the Food and Drug Administration (FDA) considers amalgam-based fillings to be safe for adults and children over six years of age. Even with multiple fillings, the mercury levels in the patients remain far below the lowest levels associated with harm. Clinical studies have found no link between dental amalgams and health problems. Health issues may not be the same in cases of children under six or pregnant women. The FDA conclusions are in line with the opinions of the Environmental Protection Agency (EPA) and Centers for Disease Control (CDC). The only health consideration noted is that some people are allergic to the amalgam or one of its components.

Group 13

Group 13 contains the metalloid boron and the metals aluminum, gallium, indium, and thallium. The lightest element, boron, is semiconducting, and its binary compounds tend to be covalent and not ionic. The remaining elements of the group are metals, but their oxides and hydroxides change characters. The oxides and hydroxides of aluminum and gallium exhibit both acidic and basic behaviors. A substance, such as these two, that will react with both acids and bases is amphoteric. This characteristic illustrates the combination of nonmetallic and metallic behaviors of these two elements. Indium and thallium oxides and hydroxides exhibit only basic behavior, in accordance with the clearly metallic character of these two elements. The melting point of gallium is unusually low (about 30 °C) and will melt in your hand.

Aluminum is amphoteric because it will react with both acids and bases. A typical reaction with an acid is:

$$2\text{Al}(s) + 6\text{HCl}(aq) \longrightarrow 2\text{AlCl}_3(aq) + 3\text{H}_2(g)$$

The products of the reaction of aluminum with a base depend upon the reaction conditions, with the following being one possibility:

$$2\text{Al}(s) + 2\text{NaOH}(aq) + 6\text{H}_2\text{O}(l) \longrightarrow 2\text{Na}\left[\text{Al(OH)}_4\right](aq) + 3\text{H}_2(g)$$

With both acids and bases, the reaction with aluminum generates hydrogen gas.

The group 13 elements have a valence shell electron configuration of ns^2np^1. Aluminum normally uses all of its valence electrons when it reacts, giving compounds in which it has an oxidation state of 3+. Although many of these compounds are covalent, others, such as AlF_3 and $Al_2(SO_4)_3$, are ionic. Aqueous solutions of aluminum salts contain the cation $\left[\text{Al(H}_2\text{O)}_6\right]^{3+}$, abbreviated as $Al^{3+}(aq)$. Gallium, indium, and thallium also form ionic compounds containing M^{3+} ions. These three elements exhibit not only the expected oxidation state of 3+ from the three valence electrons but also an oxidation state (in this case, 1+) that is two below the expected value. This phenomenon, the inert pair effect, refers to the formation of a stable ion with an oxidation state two lower than expected for the group. The pair of electrons is the valence *s* orbital for those elements. In general, the inert pair effect is important for the lower *p*-block elements. In an aqueous solution, the $Tl^+(aq)$ ion is more stable than is $Tl^{3+}(aq)$. In general, these metals will react with air and water to form 3+ ions; however, thallium reacts to give thallium(I) derivatives. The metals of group 13 all react directly with nonmetals such as sulfur, phosphorus, and the halogens, forming binary compounds.

The metals of group 13 (Al, Ga, In, and Tl) are all reactive. However, passivation occurs as a tough, hard, thin film of the metal oxide forms upon exposure to air. Disruption of this film may counter the passivation, allowing the metal to react. One way to disrupt the film is to expose the passivated metal to mercury. Some of the metal dissolves in the mercury to form an amalgam, which sheds the protective oxide layer to expose the metal to further reaction. The formation of an amalgam allows the metal to react with air and water.

LINK TO LEARNING

Although easily oxidized, the passivation of aluminum makes it very useful as a strong, lightweight building material. Because of the formation of an amalgam, mercury is corrosive to structural materials made of aluminum. This video (http://openstax.org/l/16aluminumhg) demonstrates how the integrity of an aluminum beam can be destroyed by the addition of a small amount of elemental mercury.

The most important uses of aluminum are in the construction and transportation industries, and in the manufacture of aluminum cans and aluminum foil. These uses depend on the lightness, toughness, and strength of the metal, as well as its resistance to corrosion. Because aluminum is an excellent conductor of heat and resists corrosion, it is useful in the manufacture of cooking utensils.

Aluminum is a very good reducing agent and may replace other reducing agents in the isolation of certain metals from their oxides. Although more expensive than reduction by carbon, aluminum is important in the isolation of Mo, W, and Cr from their oxides.

Group 14

The metallic members of group 14 are tin, lead, and flerovium. Carbon is a typical nonmetal. The remaining elements of the group, silicon and germanium, are examples of semimetals or metalloids. Tin and lead form the stable divalent cations, Sn^{2+} and Pb^{2+}, with oxidation states two below the group oxidation state of 4+. The stability of this oxidation state is a consequence of the inert pair effect. Tin and lead also form covalent compounds with a formal 4+-oxidation state. For example, $SnCl_4$ and $PbCl_4$ are low-boiling covalent liquids.

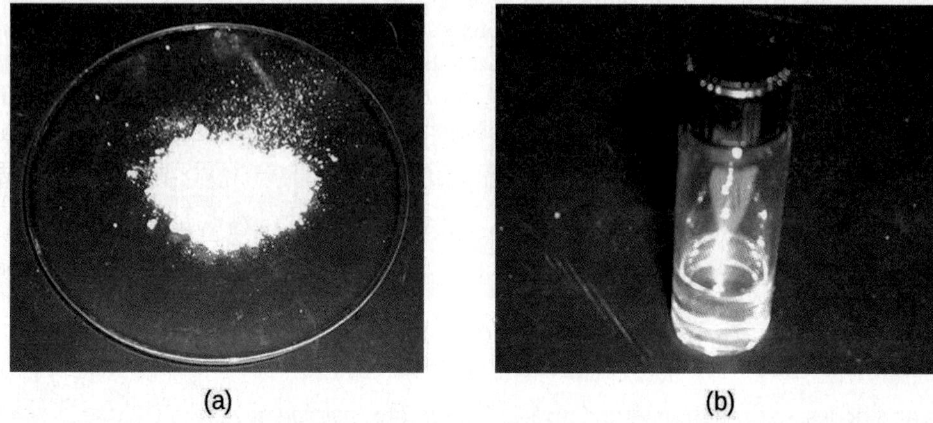

FIGURE 18.9 (a) Tin(II) chloride is an ionic solid; (b) tin(IV) chloride is a covalent liquid.

Tin reacts readily with nonmetals and acids to form tin(II) compounds (indicating that it is more easily oxidized than hydrogen) and with nonmetals to form either tin(II) or tin(IV) compounds (shown in Figure 18.9), depending on the stoichiometry and reaction conditions. Lead is less reactive. It is only slightly easier to oxidize than hydrogen, and oxidation normally requires a hot concentrated acid.

Many of these elements exist as allotropes. **Allotropes** are two or more forms of the same element in the same physical state with different chemical and physical properties. There are two common allotropes of tin. These allotropes are grey (brittle) tin and white tin. As with other allotropes, the difference between these forms of tin is in the arrangement of the atoms. White tin is stable above 13.2 °C and is malleable like other metals. At low temperatures, gray tin is the more stable form. Gray tin is brittle and tends to break down to a powder. Consequently, articles made of tin will disintegrate in cold weather, particularly if the cold spell is lengthy. The change progresses slowly from the spot of origin, and the gray tin that is first formed catalyzes further change. In a way, this effect is similar to the spread of an infection in a plant or animal body, leading people to call this process tin disease or tin pest.

The principal use of tin is in the coating of steel to form tin plate-sheet iron, which constitutes the tin in tin cans. Important tin alloys are bronze (Cu and Sn) and solder (Sn and Pb). Lead is important in the lead storage batteries in automobiles.

Group 15

Bismuth, the heaviest member of group 15, is a less reactive metal than the other representative metals. It readily gives up three of its five valence electrons to active nonmetals to form the tri-positive ion, Bi^{3+}. It forms compounds with the group oxidation state of 5+ only when treated with strong oxidizing agents. The stability of the 3+-oxidation state is another example of the inert pair effect.

18.2 Occurrence and Preparation of the Representative Metals

LEARNING OBJECTIVES

By the end of this section, you will be able to:

- Identify natural sources of representative metals
- Describe electrolytic and chemical reduction processes used to prepare these elements from natural sources

Because of their reactivity, we do not find most representative metals as free elements in nature. However, compounds that contain ions of most representative metals are abundant. In this section, we will consider the two common techniques used to isolate the metals from these compounds—electrolysis and chemical reduction.

These metals primarily occur in minerals, with lithium found in silicate or phosphate minerals, and sodium and potassium found in salt deposits from evaporation of ancient seas and in silicates. The alkaline earth metals occur as silicates and, with the exception of beryllium, as carbonates and sulfates. Beryllium occurs as the mineral beryl, $Be_3Al_2Si_6O_{18}$, which, with certain impurities, may be either the gemstone emerald or aquamarine. Magnesium is in seawater and, along with the heavier alkaline earth metals, occurs as silicates, carbonates, and sulfates. Aluminum occurs abundantly in many types of clay and in bauxite, an impure aluminum oxide hydroxide. The principle tin ore is the oxide cassiterite, SnO_2, and the principle lead and thallium ores are the sulfides or the products of weathering of the sulfides. The remaining representative metals occur as impurities in zinc or aluminum ores.

Electrolysis

Ions of metals in of groups 1 and 2, along with aluminum, are very difficult to reduce; therefore, it is necessary to prepare these elements by electrolysis, an important process discussed in the chapter on electrochemistry. Briefly, electrolysis involves using electrical energy to drive unfavorable chemical reactions to completion; it is useful in the isolation of reactive metals in their pure forms. Sodium, aluminum, and magnesium are typical examples.

The Preparation of Sodium

The most important method for the production of sodium is the electrolysis of molten sodium chloride; the set-up is a **Downs cell**, shown in Figure 18.10. The reaction involved in this process is:

$$2\text{NaCl}(l) \xrightarrow[600\ °\text{C}]{\text{electrolysis}} 2\text{Na}(l) + \text{Cl}_2(g)$$

The electrolysis cell contains molten sodium chloride (melting point 801 °C), to which calcium chloride has been added to lower the melting point to 600 °C (a colligative effect). The passage of a direct current through the cell causes the sodium ions to migrate to the negatively charged cathode and pick up electrons, reducing the ions to sodium metal. Chloride ions migrate to the positively charged anode, lose electrons, and undergo oxidation to chlorine gas. The overall cell reaction comes from adding the following reactions:

$$\begin{aligned}
&\text{at the cathode: } 2\text{Na}^+ + 2\text{e}^- \longrightarrow 2\text{Na}(l) \\
&\text{at the anode: } 2\text{Cl}^- \longrightarrow \text{Cl}_2(g) + 2\text{e}^- \\
&\text{overall change: } 2\text{Na}^+ + 2\text{Cl}^- \longrightarrow 2\text{Na}(l) + \text{Cl}_2(g)
\end{aligned}$$

Separation of the molten sodium and chlorine prevents recombination. The liquid sodium, which is less dense than molten sodium chloride, floats to the surface and flows into a collector. The gaseous chlorine goes to storage tanks. Chlorine is also a valuable product.

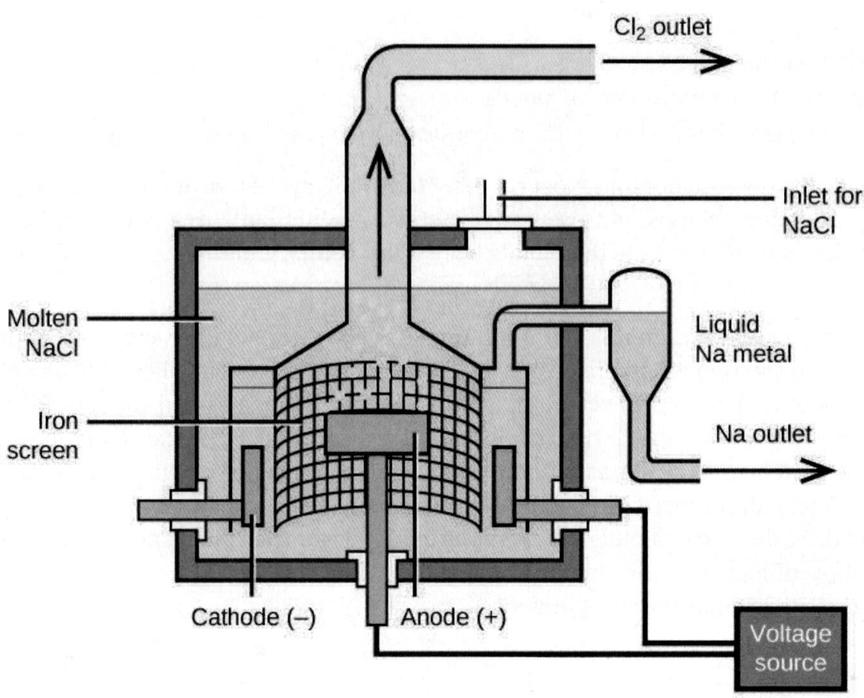

FIGURE 18.10 Pure sodium metal is isolated by electrolysis of molten sodium chloride using a Downs cell. It is not possible to isolate sodium by electrolysis of aqueous solutions of sodium salts because hydrogen ions are more easily reduced than are sodium ions; as a result, hydrogen gas forms at the cathode instead of the desired sodium metal. The high temperature required to melt NaCl means that liquid sodium metal forms.

The Preparation of Aluminum

The preparation of aluminum utilizes a process invented in 1886 by Charles M. Hall, who began to work on the problem while a student at Oberlin College in Ohio. Paul L. T. Héroult discovered the process independently a month or two later in France. In honor to the two inventors, this electrolysis cell is known as the **Hall–Héroult cell**. The Hall–Héroult cell is an electrolysis cell for the production of aluminum. Figure 18.11 illustrates the Hall–Héroult cell.

The production of aluminum begins with the purification of bauxite, the most common source of aluminum. The reaction of bauxite, AlO(OH), with hot sodium hydroxide forms soluble sodium aluminate, while clay and other impurities remain undissolved:

$$\text{AlO(OH)}(s) + \text{NaOH}(aq) + \text{H}_2\text{O}(l) \longrightarrow \text{Na[Al(OH)}_4](aq)$$

After the removal of the impurities by filtration, the addition of acid to the aluminate leads to the reprecipitation of aluminum hydroxide:

$$\text{Na[Al(OH)}_4](aq) + \text{H}_3\text{O}^+(aq) \longrightarrow \text{Al(OH)}_3(s) + \text{Na}^+(aq) + 2\text{H}_2\text{O}(l)$$

The next step is to remove the precipitated aluminum hydroxide by filtration. Heating the hydroxide produces aluminum oxide, Al_2O_3, which dissolves in a molten mixture of cryolite, Na_3AlF_6, and calcium fluoride, CaF_2. Electrolysis of this solution takes place in a cell like that shown in Figure 18.11. Reduction of aluminum ions to the metal occurs at the cathode, while oxygen, carbon monoxide, and carbon dioxide form at the anode.

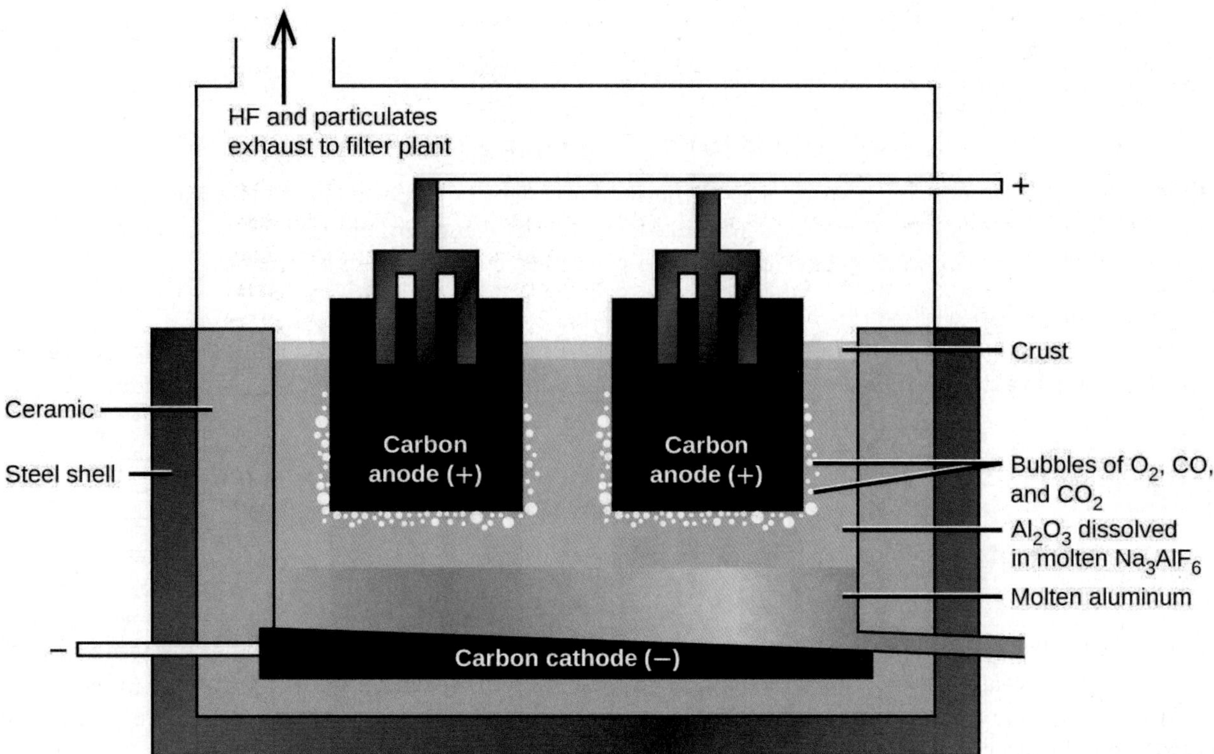

FIGURE 18.11 An electrolytic cell is used for the production of aluminum. The electrolysis of a solution of cryolite and calcium fluoride results in aluminum metal at the cathode, and oxygen, carbon monoxide, and carbon dioxide at the anode.

The Preparation of Magnesium

Magnesium is the other metal that is isolated in large quantities by electrolysis. Seawater, which contains approximately 0.5% magnesium chloride, serves as the major source of magnesium. Addition of calcium hydroxide to seawater precipitates magnesium hydroxide. The addition of hydrochloric acid to magnesium hydroxide, followed by evaporation of the resultant aqueous solution, leaves pure magnesium chloride. The electrolysis of molten magnesium chloride forms liquid magnesium and chlorine gas:

$$MgCl_2(aq) + Ca(OH)_2(aq) \longrightarrow Mg(OH)_2(s) + CaCl_2(aq)$$

$$Mg(OH)_2(s) + 2HCl(aq) \longrightarrow MgCl_2(aq) + 2H_2O(l)$$

$$MgCl_2(l) \longrightarrow Mg(l) + Cl_2(g)$$

Some production facilities have moved away from electrolysis completely. In the next section, we will see how the Pidgeon process leads to the chemical reduction of magnesium.

Chemical Reduction

It is possible to isolate many of the representative metals by **chemical reduction** using other elements as reducing agents. In general, chemical reduction is much less expensive than electrolysis, and for this reason, chemical reduction is the method of choice for the isolation of these elements. For example, it is possible to produce potassium, rubidium, and cesium by chemical reduction, as it is possible to reduce the molten chlorides of these metals with sodium metal. This may be surprising given that these metals are more reactive than sodium; however, the metals formed are more volatile than sodium and can be distilled for collection. The removal of the metal vapor leads to a shift in the equilibrium to produce more metal (see how reactions can be driven in the discussions of Le Châtelier's principle in the chapter on fundamental equilibrium concepts).

The production of magnesium, zinc, and tin provide additional examples of chemical reduction.

The Preparation of Magnesium

The **Pidgeon process** involves the reaction of magnesium oxide with elemental silicon at high temperatures to form pure magnesium:

$$\text{Si}(s) + 2\text{MgO}(s) \xrightarrow{\Delta} \text{SiO}_2(s) + 2\text{Mg}(g)$$

Although this reaction is unfavorable in terms of thermodynamics, the removal of the magnesium vapor produced takes advantage of Le Châtelier's principle to continue the forward progress of the reaction. Over 75% of the world's production of magnesium, primarily in China, comes from this process.

The Preparation of Zinc

Zinc ores usually contain zinc sulfide, zinc oxide, or zinc carbonate. After separation of these compounds from the ores, heating in air converts the ore to zinc oxide by one of the following reactions:

$$2\text{ZnS}(s) + 3\text{O}_2(g) \xrightarrow{\Delta} 2\text{ZnO}(s) + 2\text{SO}_2(g)$$

$$\text{ZnCO}_3(s) \xrightarrow{\Delta} \text{ZnO}(s) + \text{CO}_2(g)$$

Carbon, in the form of coal, reduces the zinc oxide to form zinc vapor:

$$\text{ZnO}(s) + \text{C}(s) \longrightarrow \text{Zn}(g) + \text{CO}(g)$$

The zinc can be distilled (boiling point 907 °C) and condensed. This zinc contains impurities of cadmium (767 °C), iron (2862 °C), lead (1750 °C), and arsenic (613 °C). Careful redistillation produces pure zinc. Arsenic and cadmium are distilled from the zinc because they have lower boiling points. At higher temperatures, the zinc is distilled from the other impurities, mainly lead and iron.

The Preparation of Tin

The ready reduction of tin(IV) oxide by the hot coals of a campfire accounts for the knowledge of tin in the ancient world. In the modern process, the roasting of tin ores containing SnO_2 removes contaminants such as arsenic and sulfur as volatile oxides. Treatment of the remaining material with hydrochloric acid removes the oxides of other metals. Heating the purified ore with carbon at temperature above 1000 °C produces tin:

$$\text{SnO}_2(s) + 2\text{C}(s) \xrightarrow{\Delta} \text{Sn}(s) + 2\text{CO}(g)$$

The molten tin collects at the bottom of the furnace and is drawn off and cast into blocks.

18.3 Structure and General Properties of the Metalloids

LEARNING OBJECTIVES

By the end of this section, you will be able to:

- Describe the general preparation, properties, and uses of the metalloids
- Describe the preparation, properties, and compounds of boron and silicon

A series of six elements called the metalloids separate the metals from the nonmetals in the periodic table. The metalloids are boron, silicon, germanium, arsenic, antimony, and tellurium. These elements look metallic; however, they do not conduct electricity as well as metals so they are semiconductors. They are semiconductors because their electrons are more tightly bound to their nuclei than are those of metallic conductors. Their chemical behavior falls between that of metals and nonmetals. For example, the pure metalloids form covalent crystals like the nonmetals, but like the metals, they generally do not form monatomic anions. This intermediate behavior is in part due to their intermediate electronegativity values. In this section, we will briefly discuss the chemical behavior of metalloids and deal with two of these elements—boron and silicon—in more detail.

The metalloid boron exhibits many similarities to its neighbor carbon and its diagonal neighbor silicon. All three elements form covalent compounds. However, boron has one distinct difference in that its $2s^2 2p^1$ outer electron structure gives it one less valence electron than it has valence orbitals. Although boron exhibits an oxidation state of 3+ in most of its stable compounds, this electron deficiency provides boron with the ability to form other, sometimes fractional, oxidation states, which occur, for example, in the boron hydrides.

Silicon has the valence shell electron configuration $3s^2 3p^2$, and it commonly forms tetrahedral structures in which it is sp^3 hybridized with a formal oxidation state of 4+. The major differences between the chemistry of carbon and silicon result from the relative strength of the carbon-carbon bond, carbon's ability to form stable bonds to itself, and the presence of the empty 3*d* valence-shell orbitals in silicon. Silicon's empty *d* orbitals and boron's empty *p* orbital enable tetrahedral silicon compounds and trigonal planar boron compounds to act as Lewis acids. Carbon, on the other hand, has no available valence shell orbitals; tetrahedral carbon compounds cannot act as Lewis acids. Germanium is very similar to silicon in its chemical behavior.

Arsenic and antimony generally form compounds in which an oxidation state of 3+ or 5+ is exhibited; however, arsenic can form arsenides with an oxidation state of 3−. These elements tarnish only slightly in dry air but readily oxidize when warmed.

Tellurium combines directly with most elements. The most stable tellurium compounds are the tellurides—salts of Te^{2-} formed with active metals and lanthanides—and compounds with oxygen, fluorine, and chlorine, in which tellurium normally exhibits an oxidation state 2+ or 4+. Although tellurium(VI) compounds are known (for example, TeF_6), there is a marked resistance to oxidation to this maximum group oxidation state.

Structures of the Metalloids

Covalent bonding is the key to the crystal structures of the metalloids. In this regard, these elements resemble nonmetals in their behavior.

Elemental silicon, germanium, arsenic, antimony, and tellurium are lustrous, metallic-looking solids. Silicon and germanium crystallize with a diamond structure. Each atom within the crystal has covalent bonds to four neighboring atoms at the corners of a regular tetrahedron. Single crystals of silicon and germanium are giant, three-dimensional molecules. There are several allotropes of arsenic with the most stable being layer like and containing puckered sheets of arsenic atoms. Each arsenic atom forms covalent bonds to three other atoms within the sheet. The crystal structure of antimony is similar to that of arsenic, both shown in Figure 18.12. The structures of arsenic and antimony are similar to the structure of graphite, covered later in this chapter. Tellurium forms crystals that contain infinite spiral chains of tellurium atoms. Each atom in the chain bonds to two other atoms.

LINK TO LEARNING

Explore a cubic diamond (http://openstax.org/l/16crystal) crystal structure.

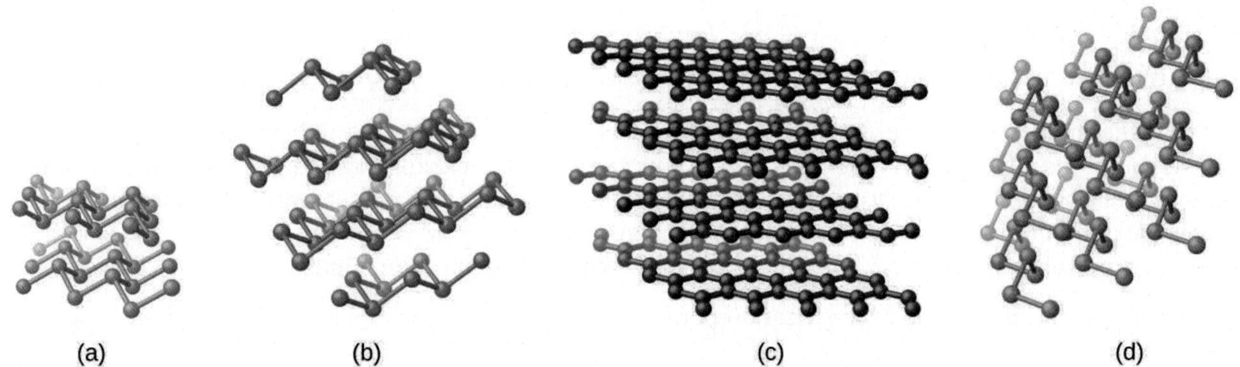

FIGURE 18.12 (a) Arsenic and (b) antimony have a layered structure similar to that of (c) graphite, except that the layers are puckered rather than planar. (d) Elemental tellurium forms spiral chains.

Pure crystalline boron is transparent. The crystals consist of icosahedra, as shown in Figure 18.13, with a boron atom at each corner. In the most common form of boron, the icosahedra pack together in a manner similar to the cubic closest packing of spheres. All boron-boron bonds within each icosahedron are identical and are approximately 176 pm in length. In the different forms of boron, there are different arrangements and connections between the icosahedra.

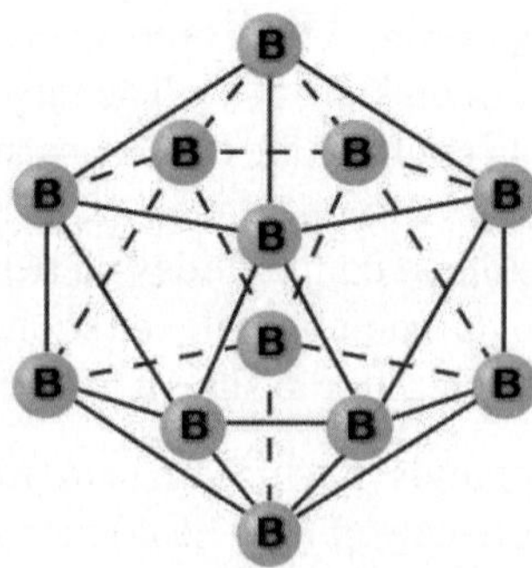

FIGURE 18.13 An icosahedron is a symmetrical, solid shape with 20 faces, each of which is an equilateral triangle. The faces meet at 12 corners.

The name silicon is derived from the Latin word for flint, *silex*. The metalloid silicon readily forms compounds containing Si-O-Si bonds, which are of prime importance in the mineral world. This bonding capability is in contrast to the nonmetal carbon, whose ability to form carbon-carbon bonds gives it prime importance in the plant and animal worlds.

Occurrence, Preparation, and Compounds of Boron and Silicon

Boron constitutes less than 0.001% by weight of the earth's crust. In nature, it only occurs in compounds with oxygen. Boron is widely distributed in volcanic regions as boric acid, $B(OH)_3$, and in dry lake regions, including the desert areas of California, as borates and salts of boron oxyacids, such as borax, $Na_2B_4O_7{\cdot}10H_2O$.

Elemental boron is chemically inert at room temperature, reacting with only fluorine and oxygen to form boron trifluoride, BF_3, and boric oxide, B_2O_3, respectively. At higher temperatures, boron reacts with all nonmetals, except tellurium and the noble gases, and with nearly all metals; it oxidizes to B_2O_3 when heated with concentrated nitric or sulfuric acid. Boron does not react with nonoxidizing acids. Many boron compounds react readily with water to give boric acid, $B(OH)_3$ (sometimes written as H_3BO_3).

Reduction of boric oxide with magnesium powder forms boron (95–98.5% pure) as a brown, amorphous powder:

$$\mathrm{B_2O_3}(s) + 3\mathrm{Mg}(s) \longrightarrow 2\mathrm{B}(s) + 3\mathrm{MgO}(s)$$

An **amorphous** substance is a material that appears to be a solid, but does not have a long-range order like a true solid. Treatment with hydrochloric acid removes the magnesium oxide. Further purification of the boron begins with conversion of the impure boron into boron trichloride. The next step is to heat a mixture of boron trichloride and hydrogen:

$$2\mathrm{BCl_3}(g) + 3\mathrm{H_2}(g) \xrightarrow{1500\ ^\circ\mathrm{C}} 2\mathrm{B}(s) + 6\mathrm{HCl}(g) \qquad \Delta H^\circ = 253.7\ \mathrm{kJ}$$

Silicon makes up nearly one-fourth of the mass of the earth's crust—second in abundance only to oxygen. The crust is composed almost entirely of minerals in which the silicon atoms are at the center of the silicon-oxygen tetrahedron, which connect in a variety of ways to produce, among other things, chains, layers, and three-dimensional frameworks. These minerals constitute the bulk of most common rocks, soil, and clays. In addition, materials such as bricks, ceramics, and glasses contain silicon compounds.

It is possible to produce silicon by the high-temperature reduction of silicon dioxide with strong reducing agents, such as carbon and magnesium:

$$\mathrm{SiO_2}(s) + 2\mathrm{C}(s) \xrightarrow{\Delta} \mathrm{Si}(s) + 2\mathrm{CO}(g)$$

$$\mathrm{SiO_2}(s) + 2\mathrm{Mg}(s) \xrightarrow{\Delta} \mathrm{Si}(s) + 2\mathrm{MgO}(s)$$

Extremely pure silicon is necessary for the manufacture of semiconductor electronic devices. This process begins with the conversion of impure silicon into silicon tetrahalides, or silane (SiH_4), followed by decomposition at high temperatures. Zone refining, illustrated in Figure 18.14, completes the purification. In this method, a rod of silicon is heated at one end by a heat source that produces a thin cross-section of molten silicon. Slowly lowering the rod through the heat source moves the molten zone from one end of the rod to other. As this thin, molten region moves, impurities in the silicon dissolve in the liquid silicon and move with the molten region. Ultimately, the impurities move to one end of the rod, which is then cut off.

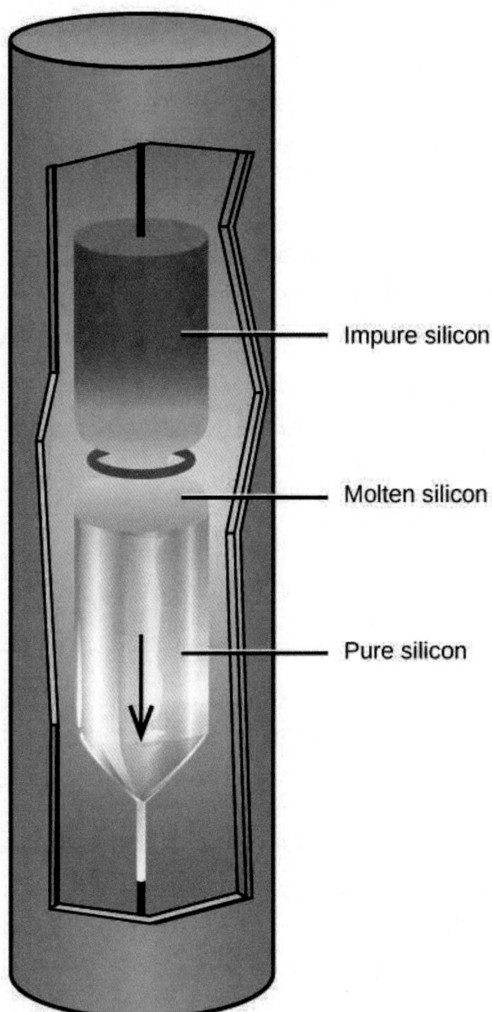

FIGURE 18.14 A zone-refining apparatus used to purify silicon.

This highly purified silicon, containing no more than one part impurity per million parts of silicon, is the most important element in the computer industry. Pure silicon is necessary in semiconductor electronic devices such as transistors, computer chips, and solar cells.

Like some metals, passivation of silicon occurs due the formation of a very thin film of oxide (primarily silicon dioxide, SiO_2). Silicon dioxide is soluble in hot aqueous base; thus, strong bases destroy the passivation. Removal of the passivation layer allows the base to dissolve the silicon, forming hydrogen gas and silicate anions. For example:

$$\mathrm{Si}(s) + 4\mathrm{OH}^-(aq) \longrightarrow \mathrm{SiO_4}^{4-}(aq) + 2\mathrm{H_2}(g)$$

Silicon reacts with halogens at high temperatures, forming volatile tetrahalides, such as SiF_4.

Unlike carbon, silicon does not readily form double or triple bonds. Silicon compounds of the general formula SiX_4, where X is a highly electronegative group, can act as Lewis acids to form six-coordinate silicon. For example, silicon tetrafluoride, SiF_4, reacts with sodium fluoride to yield $Na_2[SiF_6]$, which contains the octahedral $\left[\mathrm{SiF_6}\right]^{2-}$ ion in which silicon is sp^3d^2 hybridized:

$$2\text{NaF}(s) + \text{SiF}_4(g) \longrightarrow \text{Na}_2\text{SiF}_6(s)$$

Antimony reacts readily with stoichiometric amounts of fluorine, chlorine, bromine, or iodine, yielding trihalides or, with excess fluorine or chlorine, forming the pentahalides SbF_5 and $SbCl_5$. Depending on the stoichiometry, it forms antimony(III) sulfide, Sb_2S_3, or antimony(V) sulfide when heated with sulfur. As expected, the metallic nature of the element is greater than that of arsenic, which lies immediately above it in group 15.

Boron and Silicon Halides

Boron trihalides—BF_3, BCl_3, BBr_3, and BI_3—can be prepared by the direct reaction of the elements. These nonpolar molecules contain boron with sp^2 hybridization and a trigonal planar molecular geometry. The fluoride and chloride compounds are colorless gasses, the bromide is a liquid, and the iodide is a white crystalline solid.

Except for boron trifluoride, the boron trihalides readily hydrolyze in water to form boric acid and the corresponding hydrohalic acid. Boron trichloride reacts according to the equation:

$$\text{BCl}_3(g) + 3\text{H}_2\text{O}(l) \longrightarrow \text{B(OH)}_3(aq) + 3\text{HCl}(aq)$$

Boron trifluoride reacts with hydrofluoric acid, to yield a solution of fluoroboric acid, HBF_4:

$$\text{BF}_3(aq) + \text{HF}(aq) + \text{H}_2\text{O}(l) \longrightarrow \text{H}_3\text{O}^+(aq) + \text{BF}_4{}^-(aq)$$

In this reaction, the BF_3 molecule acts as the Lewis acid (electron pair acceptor) and accepts a pair of electrons from a fluoride ion:

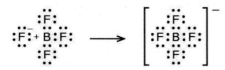

All the tetrahalides of silicon, SiX_4, have been prepared. Silicon tetrachloride can be prepared by direct chlorination at elevated temperatures or by heating silicon dioxide with chlorine and carbon:

$$\text{SiO}_2(s) + 2\text{C}(s) + 2\text{Cl}_2(g) \xrightarrow{\Delta} \text{SiCl}_4(g) + 2\text{CO}(g)$$

Silicon tetrachloride is a covalent tetrahedral molecule, which is a nonpolar, low-boiling (57 °C), colorless liquid.

It is possible to prepare silicon tetrafluoride by the reaction of silicon dioxide with hydrofluoric acid:

$$\text{SiO}_2(s) + 4\text{HF}(g) \longrightarrow \text{SiF}_4(g) + 2\text{H}_2\text{O}(l) \qquad \Delta H^\circ = -191.2 \text{ kJ}$$

Hydrofluoric acid is the only common acid that will react with silicon dioxide or silicates. This reaction occurs because the silicon-fluorine bond is the only bond that silicon forms that is stronger than the silicon-oxygen bond. For this reason, it is possible to store all common acids, other than hydrofluoric acid, in glass containers.

Except for silicon tetrafluoride, silicon halides are extremely sensitive to water. Upon exposure to water, $SiCl_4$ reacts rapidly with hydroxide groups, replacing all four chlorine atoms to produce unstable orthosilicic acid, $Si(OH)_4$ or H_4SiO_4, which slowly decomposes into SiO_2.

Boron and Silicon Oxides and Derivatives

Boron burns at 700 °C in oxygen, forming boric oxide, B_2O_3. Boric oxide is necessary for the production of heat-resistant borosilicate glass, like that shown in Figure 18.15 and certain optical glasses. Boric oxide dissolves in hot water to form boric acid, $B(OH)_3$:

$$\text{B}_2\text{O}_3(s) + 3\text{H}_2\text{O}(l) \longrightarrow 2\text{B(OH)}_3(aq)$$

FIGURE 18.15 Laboratory glassware, such as Pyrex and Kimax, is made of borosilicate glass because it does not break when heated. The inclusion of borates in the glass helps to mediate the effects of thermal expansion and contraction. This reduces the likelihood of thermal shock, which causes silicate glass to crack upon rapid heating or cooling. (credit: "Tweenk"/Wikimedia Commons)

The boron atom in $B(OH)_3$ is sp^2 hybridized and is located at the center of an equilateral triangle with oxygen atoms at the corners. In solid $B(OH)_3$, hydrogen bonding holds these triangular units together. Boric acid, shown in Figure 18.16, is a very weak acid that does not act as a proton donor but rather as a Lewis acid, accepting an unshared pair of electrons from the Lewis base OH^-:

$$B(OH)_3(aq) + 2H_2O(l) \rightleftharpoons B(OH)_4{}^-(aq) + H_3O^+(aq) \qquad K_a = 5.8 \times 10^{-10}$$

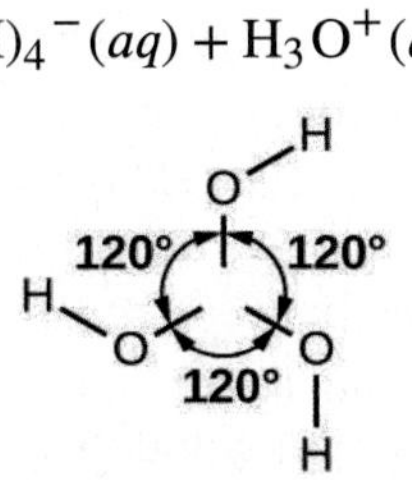

FIGURE 18.16 Boric acid has a planar structure with three –OH groups spread out equally at 120° angles from each other.

Heating boric acid to 100 °C causes molecules of water to split out between pairs of adjacent –OH groups to form metaboric acid, HBO_2. At about 150 °C, additional B-O-B linkages form, connecting the BO_3 groups together with shared oxygen atoms to form tetraboric acid, $H_2B_4O_7$. Complete water loss, at still higher temperatures, results in boric oxide.

Borates are salts of the oxyacids of boron. Borates result from the reactions of a base with an oxyacid or from the fusion of boric acid or boric oxide with a metal oxide or hydroxide. Borate anions range from the simple trigonal planar $BO_3{}^{3-}$ ion to complex species containing chains and rings of three- and four-coordinated boron atoms. The structures of the anions found in CaB_2O_4, $K[B_5O_6(OH)_4]·2H_2O$ (commonly written $KB_5O_8·4H_2O$) and $Na_2[B_4O_5(OH)_4]·8H_2O$ (commonly written $Na_2B_4O_7·10H_2O$) are shown in Figure 18.17. Commercially, the most important borate is borax, $Na_2[B_4O_5(OH)_4]·8H_2O$, which is an important component of some laundry detergents. Most of the supply of borax comes directly from dry lakes, such as Searles Lake in California, or is prepared from kernite, $Na_2B_4O_7·4H_2O$.

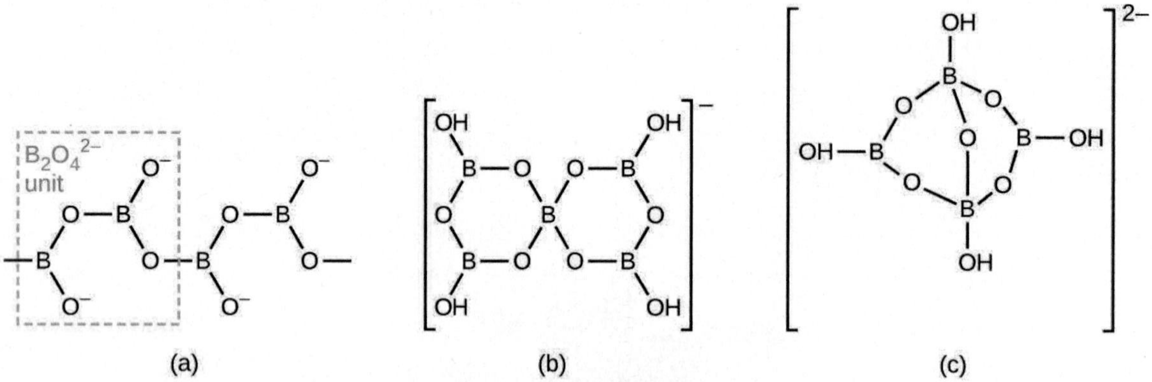

FIGURE 18.17 The borate anions are (a) CaB_2O_4, (b) $KB_5O_8{\cdot}4H_2O$, and (c) $Na_2B_4O_7{\cdot}10H_2O$. The anion in CaB_2O_4 is an "infinite" chain.

Silicon dioxide, silica, occurs in both crystalline and amorphous forms. The usual crystalline form of silicon dioxide is quartz, a hard, brittle, clear, colorless solid. It is useful in many ways—for architectural decorations, semiprecious jewels, and frequency control in radio transmitters. Silica takes many crystalline forms, or **polymorphs**, in nature. Trace amounts of Fe^{3+} in quartz give amethyst its characteristic purple color. The term *quartz* is also used for articles such as tubing and lenses that are manufactured from amorphous silica. Opal is a naturally occurring form of amorphous silica.

The contrast in structure and physical properties between silicon dioxide and carbon dioxide is interesting, as illustrated in Figure 18.18. Solid carbon dioxide (dry ice) contains single CO_2 molecules with each of the two oxygen atoms attached to the carbon atom by double bonds. Very weak intermolecular forces hold the molecules together in the crystal. The volatility of dry ice reflect these weak forces between molecules. In contrast, silicon dioxide is a covalent network solid. In silicon dioxide, each silicon atom links to four oxygen atoms by single bonds directed toward the corners of a regular tetrahedron, and SiO_4 tetrahedra share oxygen atoms. This arrangement gives a three dimensional, continuous, silicon-oxygen network. A quartz crystal is a macromolecule of silicon dioxide. The difference between these two compounds is the ability of the group 14 elements to form strong π bonds. Second-period elements, such as carbon, form very strong π bonds, which is why carbon dioxide forms small molecules with strong double bonds. Elements below the second period, such as silicon, do not form π bonds as readily as second-period elements, and when they do form, the π bonds are weaker than those formed by second-period elements. For this reason, silicon dioxide does not contain π bonds but only σ bonds.

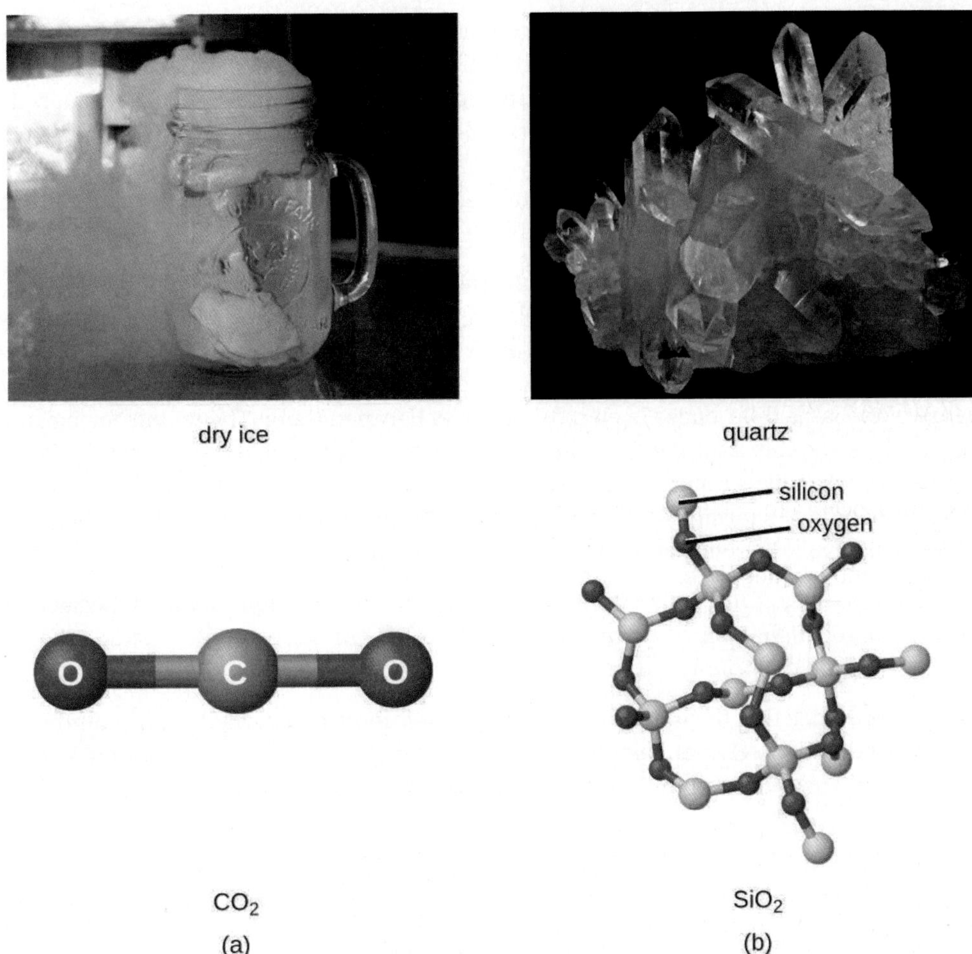

FIGURE 18.18 Because carbon tends to form double and triple bonds and silicon does not, (a) carbon dioxide is a discrete molecule with two C=O double bonds and (b) silicon dioxide is an infinite network of oxygen atoms bridging between silicon atoms with each silicon atom possessing four Si-O single bonds. (credit a photo: modification of work by Erica Gerdes; credit b photo: modification of work by Didier Descouens)

At 1600 °C, quartz melts to yield a viscous liquid. When the liquid cools, it does not crystallize readily but usually supercools and forms a glass, also called silica. The SiO_4 tetrahedra in glassy silica have a random arrangement characteristic of supercooled liquids, and the glass has some very useful properties. Silica is highly transparent to both visible and ultraviolet light. For this reason, it is important in the manufacture of lamps that give radiation rich in ultraviolet light and in certain optical instruments that operate with ultraviolet light. The coefficient of expansion of silica glass is very low; therefore, rapid temperature changes do not cause it to fracture. CorningWare and other ceramic cookware contain amorphous silica.

Silicates are salts containing anions composed of silicon and oxygen. In nearly all silicates, sp^3-hybridized silicon atoms occur at the centers of tetrahedra with oxygen at the corners. There is a variation in the silicon-to-oxygen ratio that occurs because silicon-oxygen tetrahedra may exist as discrete, independent units or may share oxygen atoms at corners in a variety of ways. In addition, the presence of a variety of cations gives rise to the large number of silicate minerals.

Many ceramics are composed of silicates. By including small amounts of other compounds, it is possible to modify the physical properties of the silicate materials to produce ceramics with useful characteristics.

18.4 Structure and General Properties of the Nonmetals

LEARNING OBJECTIVES

By the end of this section, you will be able to:

- Describe structure and properties of nonmetals

The nonmetals are elements located in the upper right portion of the periodic table. Their properties and behavior are quite different from those of metals on the left side. Under normal conditions, more than half of the nonmetals are gases, one is a liquid, and the rest include some of the softest and hardest of solids. The nonmetals exhibit a rich variety of chemical behaviors. They include the most reactive and least reactive of elements, and they form many different ionic and covalent compounds. This section presents an overview of the properties and chemical behaviors of the nonmetals, as well as the chemistry of specific elements. Many of these nonmetals are important in biological systems.

In many cases, trends in electronegativity enable us to predict the type of bonding and the physical states in compounds involving the nonmetals. We know that electronegativity decreases as we move down a given group and increases as we move from left to right across a period. The nonmetals have higher electronegativities than do metals, and compounds formed between metals and nonmetals are generally ionic in nature because of the large differences in electronegativity between them. The metals form cations, the nonmetals form anions, and the resulting compounds are solids under normal conditions. On the other hand, compounds formed between two or more nonmetals have small differences in electronegativity between the atoms, and covalent bonding—sharing of electrons—results. These substances tend to be molecular in nature and are gases, liquids, or volatile solids at room temperature and pressure.

In normal chemical processes, nonmetals do not form monatomic positive ions (cations) because their ionization energies are too high. All monatomic nonmetal ions are anions; examples include the chloride ion, Cl^-, the nitride ion, N^{3-}, and the selenide ion, Se^{2-}.

The common oxidation states that the nonmetals exhibit in their ionic and covalent compounds are shown in Figure 18.19. Remember that an element exhibits a positive oxidation state when combined with a more electronegative element and that it exhibits a negative oxidation state when combined with a less electronegative element.

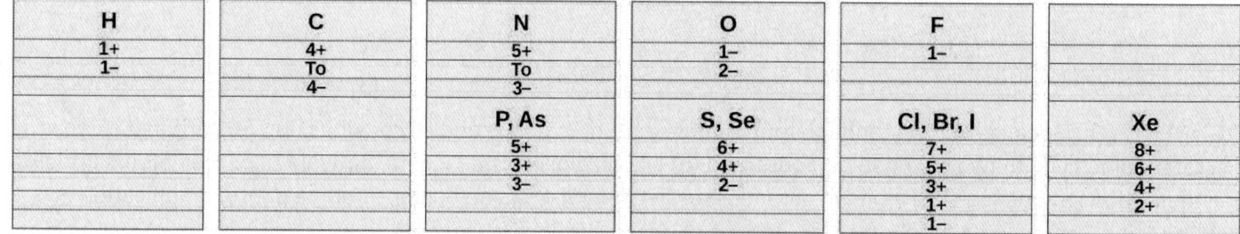

H
1+
1–
C
4+
To
4–
N
5+
To
3–
P, As
5+
3+
3–
O
1–
2–
S, Se
6+
4+
2–
F
1–
Cl, Br, I
7+
5+
3+
1+
1–
Xe
8+
6+
4+
2+

FIGURE 18.19 Nonmetals exhibit these common oxidation states in ionic and covalent compounds.

The first member of each nonmetal group exhibits different behaviors, in many respects, from the other group members. The reasons for this include smaller size, greater ionization energy, and (most important) the fact that the first member of each group has only four valence orbitals (one 2*s* and three 2*p*) available for bonding, whereas other group members have empty *d* orbitals in their valence shells, making possible five, six, or even more bonds around the central atom. For example, nitrogen forms only NF_3, whereas phosphorus forms both PF_3 and PF_5.

Another difference between the first group member and subsequent members is the greater ability of the first member to form *π* bonds. This is primarily a function of the smaller size of the first member of each group, which allows better overlap of atomic orbitals. Nonmetals, other than the first member of each group, rarely form π bonds to nonmetals that are the first member of a group. For example, sulfur-oxygen π bonds are well known, whereas sulfur does not normally form stable π bonds to itself.

The variety of oxidation states displayed by most of the nonmetals means that many of their chemical reactions involve changes in oxidation state through oxidation-reduction reactions. There are five general aspects of the oxidation-reduction chemistry:

1. Nonmetals oxidize most metals. The oxidation state of the metal becomes positive as it undergoes oxidation and that of the nonmetal becomes negative as it undergoes reduction. For example:

$$\underset{0}{4Fe(s)} + \underset{0}{3O_2(g)} \longrightarrow \underset{+3\ -2}{2Fe_2O_3(s)}$$

2. With the exception of nitrogen and carbon, which are poor oxidizing agents, a more electronegative nonmetal oxidizes a less electronegative nonmetal or the anion of the nonmetal:

$$\underset{0}{S(s)} + \underset{0}{O_2(g)} \longrightarrow \underset{+4\ -2}{2SO_2(s)}$$

$$\underset{0}{Cl_2(g)} + 2I^-(aq) \longrightarrow \underset{0}{I_2(s)} + 2Cl^-(aq)$$

3. Fluorine and oxygen are the strongest oxidizing agents within their respective groups; each oxidizes all the elements that lie below it in the group. Within any period, the strongest oxidizing agent is in group 17. A nonmetal often oxidizes an element that lies to its left in the same period. For example:

$$\underset{0}{2As(s)} + \underset{0}{3Br_2(l)} \longrightarrow \underset{+3\ -1}{2AsBr_3(s)}$$

4. The stronger a nonmetal is as an oxidizing agent, the more difficult it is to oxidize the anion formed by the nonmetal. This means that the most stable negative ions are formed by elements at the top of the group or in group 17 of the period.
5. Fluorine and oxygen are the strongest oxidizing elements known. Fluorine does not form compounds in which it exhibits positive oxidation states; oxygen exhibits a positive oxidation state only when combined with fluorine. For example:

$$\underset{0}{2F_2(g)} + 2OH^-(aq) \longrightarrow \underset{+2\quad -1}{OF_2(g)} + 2F^-(aq) + H_2O(l)$$

With the exception of most of the noble gases, all nonmetals form compounds with oxygen, yielding covalent oxides. Most of these oxides are acidic, that is, they react with water to form oxyacids. Recall from the acid-base chapter that an oxyacid is an acid consisting of hydrogen, oxygen, and some other element. Notable exceptions are carbon monoxide, CO, nitrous oxide, N_2O, and nitric oxide, NO. There are three characteristics of these acidic oxides:

1. Oxides such as SO_2 and N_2O_5, in which the nonmetal exhibits one of its common oxidation states, are **acid anhydrides** and react with water to form acids with no change in oxidation state. The product is an oxyacid. For example:

$$SO_2(g) + H_2O(l) \longrightarrow H_2SO_3(aq)$$

$$N_2O_5(s) + H_2O(l) \longrightarrow 2HNO_3(aq)$$

2. Those oxides such as NO_2 and ClO_2, in which the nonmetal does not exhibit one of its common oxidation states, also react with water. In these reactions, the nonmetal is both oxidized and reduced. For example:

$$\underset{+4}{3NO_2(g)} + H_2O(l) \longrightarrow \underset{+5}{2HNO_3(aq)} + \underset{+2}{NO(g)}$$

   Reactions in which the same element is both oxidized and reduced are called **disproportionation reactions**.
3. The acid strength increases as the electronegativity of the central atom increases. To learn more, see the discussion in the chapter on acid-base chemistry.

The binary hydrogen compounds of the nonmetals also exhibit an acidic behavior in water, although only HCl, HBr, and HI are strong acids. The acid strength of the nonmetal hydrogen compounds increases from left to right across a period and down a group. For example, ammonia, NH_3, is a weaker acid than is water, H_2O, which is weaker than is hydrogen fluoride, HF. Water, H_2O, is also a weaker acid than is hydrogen sulfide, H_2S, which is weaker than is hydrogen selenide, H_2Se. Weaker acidic character implies greater basic character.

Structures of the Nonmetals

The structures of the nonmetals differ dramatically from those of metals. Metals crystallize in closely packed arrays that do not contain molecules or covalent bonds. Nonmetal structures contain covalent bonds, and many nonmetals consist of individual molecules. The electrons in nonmetals are localized in covalent bonds, whereas in a metal, there is delocalization of the electrons throughout the solid.

The noble gases are all monatomic, whereas the other nonmetal gases—hydrogen, nitrogen, oxygen, fluorine, and chlorine—normally exist as the diatomic molecules H_2, N_2, O_2, F_2, and Cl_2. The other halogens are also diatomic; Br_2 is a liquid and I_2 exists as a solid under normal conditions. The changes in state as one moves down the halogen family offer excellent examples of the increasing strength of intermolecular London forces with increasing molecular mass and increasing polarizability.

Oxygen has two allotropes: O_2, dioxygen, and O_3, ozone. Phosphorus has three common allotropes, commonly referred to by their colors: white, red, and black. Sulfur has several allotropes. There are also many carbon allotropes. Most people know of diamond, graphite, and charcoal, but fewer people know of the recent discovery of fullerenes, carbon nanotubes, and graphene.

Descriptions of the physical properties of three nonmetals that are characteristic of molecular solids follow.

Carbon

Carbon occurs in the uncombined (elemental) state in many forms, such as diamond, graphite, charcoal, coke, carbon black, graphene, and fullerene.

Diamond, shown in Figure 18.20, is a very hard crystalline material that is colorless and transparent when pure. Each atom forms four single bonds to four other atoms at the corners of a tetrahedron (sp^3 hybridization); this makes the diamond a giant molecule. Carbon-carbon single bonds are very strong, and, because they extend throughout the crystal to form a three-dimensional network, the crystals are very hard and have high melting points (~4400 °C).

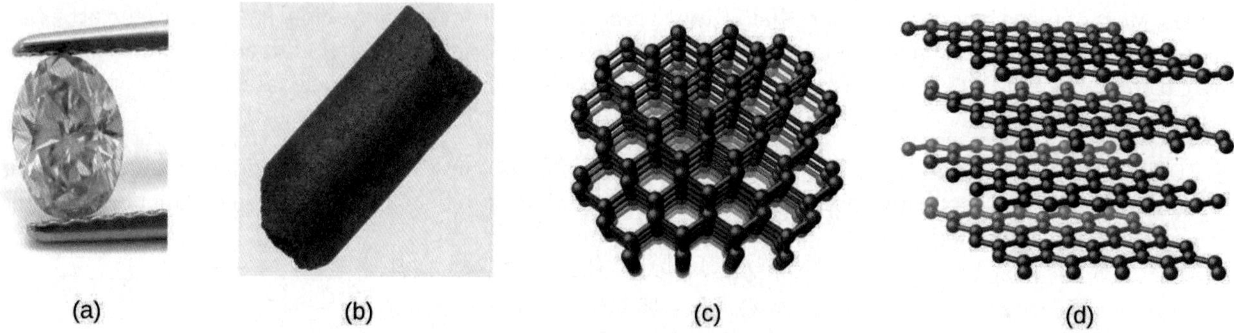

FIGURE 18.20 (a) Diamond and (b) graphite are two forms of carbon. (c) In the crystal structure of diamond, the covalent bonds form three-dimensional tetrahedrons. (d) In the crystal structure of graphite, each planar layer is composed of six-membered rings. (credit a: modification of work by “Fancy Diamonds”/Flickr; credit b: modification of work from http://images-of-elements.com/carbon.php)

Graphite, also shown in Figure 18.20, is a soft, slippery, grayish-black solid that conducts electricity. These properties relate to its structure, which consists of layers of carbon atoms, with each atom surrounded by three other carbon atoms in a trigonal planar arrangement. Each carbon atom in graphite forms three σ bonds, one to each of its nearest neighbors, by means of sp^2-hybrid orbitals. The unhybridized *p* orbital on each carbon atom will overlap unhybridized orbitals on adjacent carbon atoms in the same layer to form π bonds. Many resonance forms are necessary to describe the electronic structure of a graphite layer; Figure 18.21 illustrates two of these forms.

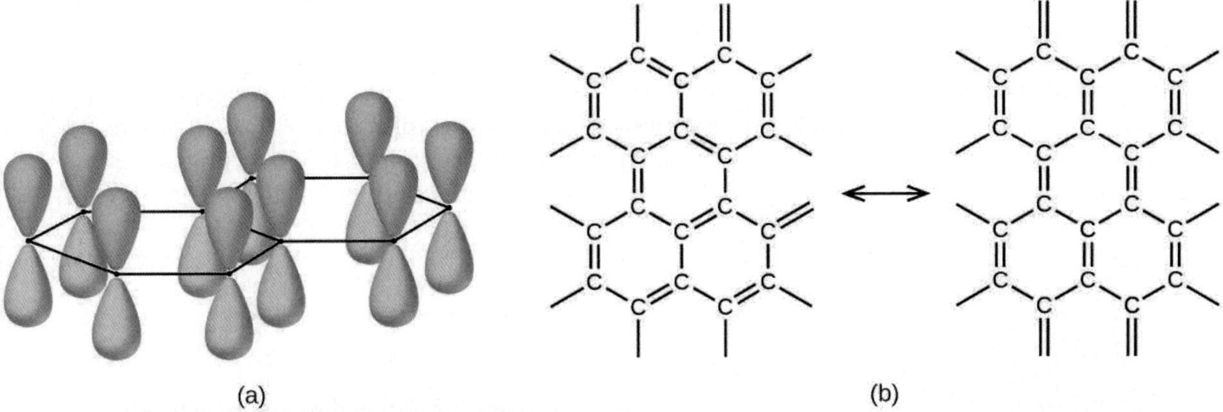

FIGURE 18.21 (a) Carbon atoms in graphite have unhybridized *p* orbitals. Each *p* orbital is perpendicular to the plane of carbon atoms. (b) These are two of the many resonance forms of graphite necessary to describe its electronic structure as a resonance hybrid.

Atoms within a graphite layer are bonded together tightly by the σ and π bonds; however, the forces between layers are weak. London dispersion forces hold the layers together. To learn more, see the discussion of these weak forces in the chapter on liquids and solids. The weak forces between layers give graphite the soft, flaky character that makes it useful as the so-called "lead" in pencils and the slippery character that makes it useful as a lubricant. The loosely held electrons in the resonating π bonds can move throughout the solid and are responsible for the electrical conductivity of graphite.

Other forms of elemental carbon include carbon black, charcoal, and coke. Carbon black is an amorphous form of carbon prepared by the incomplete combustion of natural gas, CH_4. It is possible to produce charcoal and coke by heating wood and coal, respectively, at high temperatures in the absence of air.

Recently, new forms of elemental carbon molecules have been identified in the soot generated by a smoky flame and in the vapor produced when graphite is heated to very high temperatures in a vacuum or in helium. One of these new forms, first isolated by Professor Richard Smalley and coworkers at Rice University, consists of icosahedral (soccer-ball-shaped) molecules that contain 60 carbon atoms, C_{60}. This is buckminsterfullerene (often called bucky balls) after the architect Buckminster Fuller, who designed domed structures, which have a similar appearance (Figure 18.22).

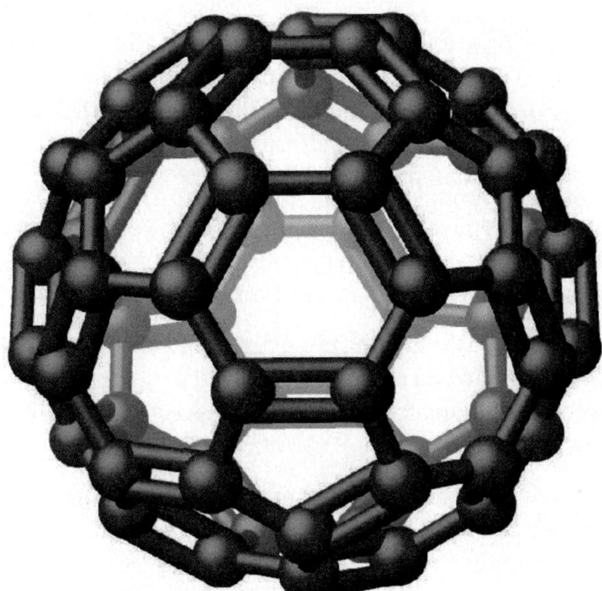

FIGURE 18.22 The molecular structure of C_{60}, buckminsterfullerene, is icosahedral.

Chemistry in Everyday Life

Nanotubes and Graphene

Graphene and carbon nanotubes are two recently discovered allotropes of carbon. Both of the forms bear some relationship to graphite. Graphene is a single layer of graphite (one atom thick), as illustrated in Figure 18.23, whereas carbon nanotubes roll the layer into a small tube, as illustrated in Figure 18.23.

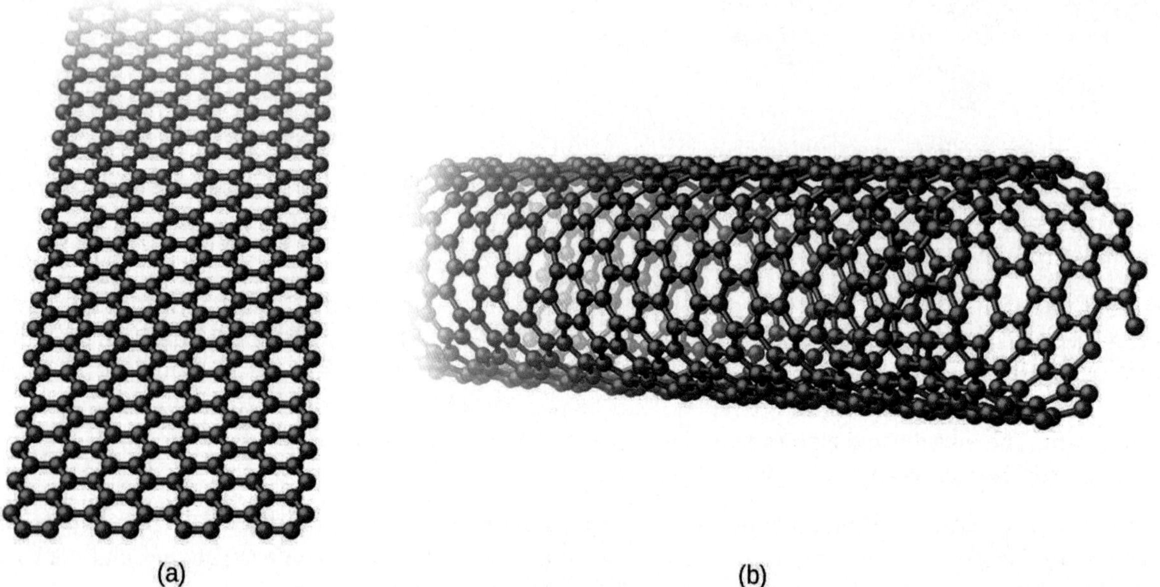

FIGURE 18.23 (a) Graphene and (b) carbon nanotubes are both allotropes of carbon.

Graphene is a very strong, lightweight, and efficient conductor of heat and electricity discovered in 2003. As in graphite, the carbon atoms form a layer of six-membered rings with sp^2-hybridized carbon atoms at the corners. Resonance stabilizes the system and leads to its conductivity. Unlike graphite, there is no stacking of the layers to give a three-dimensional structure. Andre Geim and Kostya Novoselov at the University of Manchester won the 2010 Nobel Prize in Physics for their pioneering work characterizing graphene.

The simplest procedure for preparing graphene is to use a piece of adhesive tape to remove a single layer of graphene from the surface of a piece of graphite. This method works because there are only weak London dispersion forces between the layers in graphite. Alternative methods are to deposit a single layer of carbon atoms on the surface of some other material (ruthenium, iridium, or copper) or to synthesize it at the surface of silicon carbide via the sublimation of silicon.

There currently are no commercial applications of graphene. However, its unusual properties, such as high electron mobility and thermal conductivity, should make it suitable for the manufacture of many advanced electronic devices and for thermal management applications.

Carbon nanotubes are carbon allotropes, which have a cylindrical structure. Like graphite and graphene, nanotubes consist of rings of sp^2-hybridized carbon atoms. Unlike graphite and graphene, which occur in layers, the layers wrap into a tube and bond together to produce a stable structure. The walls of the tube may be one atom or multiple atoms thick.

Carbon nanotubes are extremely strong materials that are harder than diamond. Depending upon the shape of the nanotube, it may be a conductor or semiconductor. For some applications, the conducting form is preferable, whereas other applications utilize the semiconducting form.

The basis for the synthesis of carbon nanotubes is the generation of carbon atoms in a vacuum. It is possible to produce carbon atoms by an electrical discharge through graphite, vaporization of graphite with a laser, and the decomposition of a carbon compound.

The strength of carbon nanotubes will eventually lead to some of their most exciting applications, as a thread produced from several nanotubes will support enormous weight. However, the current applications only employ bulk nanotubes. The addition of nanotubes to polymers improves the mechanical, thermal, and electrical properties of the bulk material. There are currently nanotubes in some bicycle parts, skis, baseball bats, fishing rods, and surfboards.

Phosphorus

The name *phosphorus* comes from the Greek words meaning *light bringing.* When phosphorus was first isolated, scientists noted that it glowed in the dark and burned when exposed to air. Phosphorus is the only member of its group that does not occur in the uncombined state in nature; it exists in many allotropic forms. We will consider two of those forms: white phosphorus and red phosphorus.

White phosphorus is a white, waxy solid that melts at 44.2 °C and boils at 280 °C. It is insoluble in water (in which it is stored—see Figure 18.24), is very soluble in carbon disulfide, and bursts into flame in air. As a solid, as a liquid, as a gas, and in solution, white phosphorus exists as P_4 molecules with four phosphorus atoms at the corners of a regular tetrahedron, as illustrated in Figure 18.24. Each phosphorus atom covalently bonds to the other three atoms in the molecule by single covalent bonds. White phosphorus is the most reactive allotrope and is very toxic.

FIGURE 18.24 (a) Because white phosphorus bursts into flame in air, it is stored in water. (b) The structure of white phosphorus consists of P_4 molecules arranged in a tetrahedron. (c) Red phosphorus is much less reactive than is white phosphorus. (d) The structure of red phosphorus consists of networks of P_4 tetrahedra joined by P-P single bonds. (credit a: modification of work from http://images-of-elements.com/phosphorus.php)

Heating white phosphorus to 270–300 °C in the absence of air yields red phosphorus. Red phosphorus (shown in Figure 18.24) is denser, has a higher melting point (~600 °C), is much less reactive, is essentially nontoxic, and is easier and safer to handle than is white phosphorus. Its structure is highly polymeric and appears to contain three-dimensional networks of P_4 tetrahedra joined by P-P single bonds. Red phosphorus is insoluble in solvents that dissolve white phosphorus. When red phosphorus is heated, P_4 molecules sublime from the solid.

Sulfur

The allotropy of sulfur is far greater and more complex than that of any other element. Sulfur is the brimstone referred to in the Bible and other places, and references to sulfur occur throughout recorded history—right up to the relatively recent discovery that it is a component of the atmospheres of Venus and of Io, a moon of Jupiter. The most common and most stable allotrope of sulfur is yellow, rhombic sulfur, so named because of the shape of its crystals. Rhombic sulfur is the form to which all other allotropes revert at room temperature. Crystals of rhombic sulfur melt at 113 °C. Cooling this liquid gives long needles of monoclinic sulfur. This form is stable from 96 °C to the melting point, 119 °C. At room temperature, it gradually reverts to the rhombic form.

Both rhombic sulfur and monoclinic sulfur contain S_8 molecules in which atoms form eight-membered, puckered rings that resemble crowns, as illustrated in Figure 18.25. Each sulfur atom is bonded to each of its two neighbors in the ring by covalent S-S single bonds.

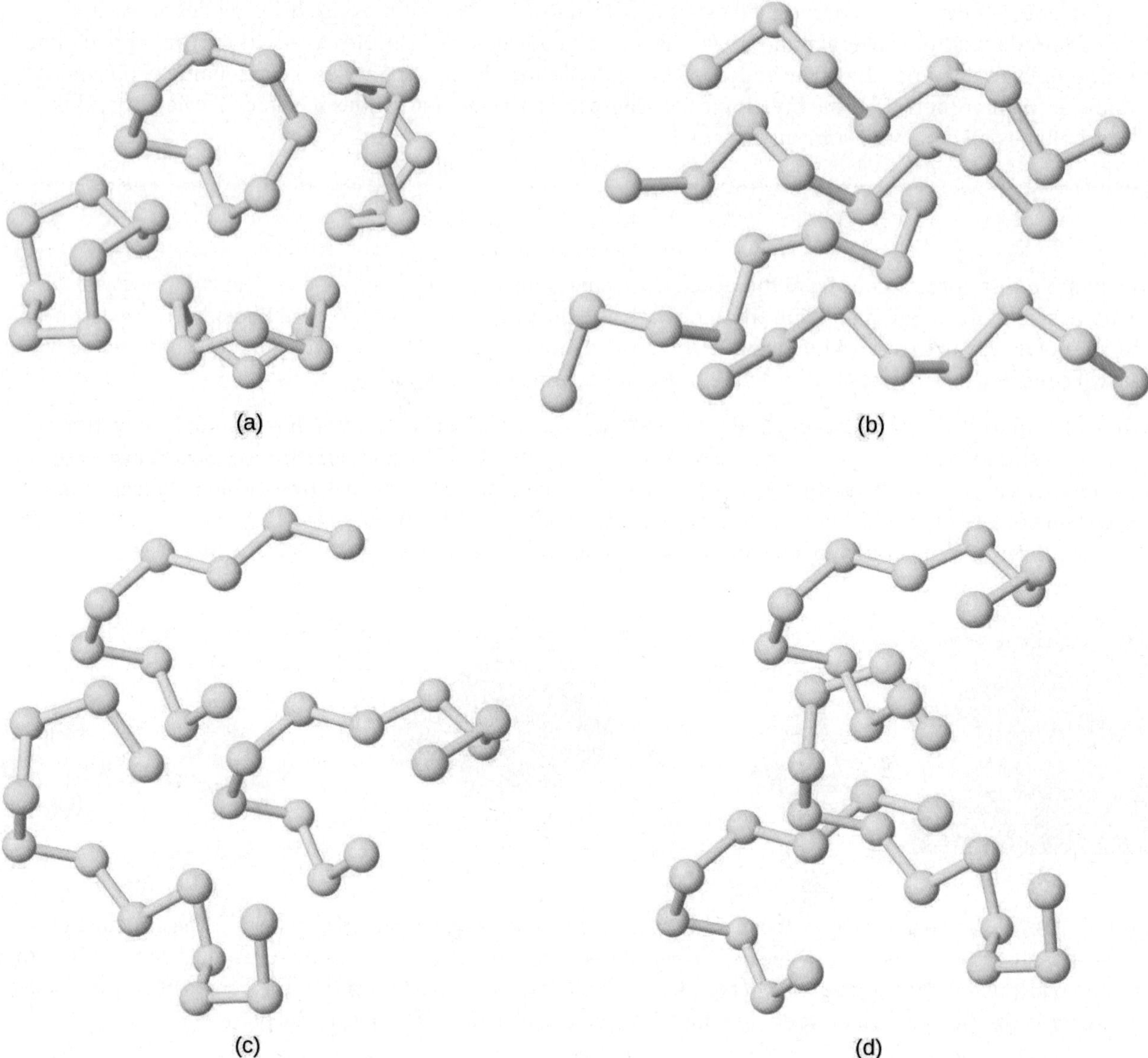

FIGURE 18.25 These four sulfur allotropes show eight-membered, puckered rings. Each sulfur atom bonds to each of its two neighbors in the ring by covalent S-S single bonds. Here are (a) individual S_8 rings, (b) S_8 chains formed when the rings open, (c) longer chains formed by adding sulfur atoms to S_8 chains, and (d) part of the very long sulfur chains formed at higher temperatures.

When rhombic sulfur melts, the straw-colored liquid is quite mobile; its viscosity is low because S_8 molecules are essentially spherical and offer relatively little resistance as they move past each other. As the temperature rises, S-S bonds in the rings break, and polymeric chains of sulfur atoms result. These chains combine end to end, forming still longer chains that tangle with one another. The liquid gradually darkens in color and becomes so viscous that finally (at about 230 °C) it does not pour easily. The dangling atoms at the ends of the chains of sulfur atoms are responsible for the dark red color because their electronic structure differs from those of sulfur atoms that have bonds to two adjacent sulfur atoms. This causes them to absorb light differently and results in a different visible color. Cooling the liquid rapidly produces a rubberlike amorphous mass, called plastic sulfur.

Sulfur boils at 445 °C and forms a vapor consisting of S_2, S_6, and S_8 molecules; at about 1000 °C, the vapor density corresponds to the formula S_2, which is a paramagnetic molecule like O_2 with a similar electronic structure and a weak sulfur-sulfur double bond.

As seen in this discussion, an important feature of the structural behavior of the nonmetals is that the elements usually occur with eight electrons in their valence shells. If necessary, the elements form enough covalent bonds to supplement the electrons already present to possess an octet. For example, members of group 15 have five valence electrons and require only three additional electrons to fill their valence shells. These elements form three covalent bonds in their free state: triple bonds in the N_2 molecule or single bonds to three different atoms in arsenic and phosphorus. The elements of group 16 require only two additional electrons. Oxygen forms a double bond in the O_2 molecule, and sulfur, selenium, and tellurium form two single bonds in various rings and chains. The halogens form diatomic molecules in which each atom is involved in only one bond. This provides the electron required necessary to complete the octet on the halogen atom. The noble gases do not form covalent bonds to other noble gas atoms because they already have a filled outer shell.

18.5 Occurrence, Preparation, and Compounds of Hydrogen

LEARNING OBJECTIVES

By the end of this section, you will be able to:

- Describe the properties, preparation, and compounds of hydrogen

Hydrogen is the most abundant element in the universe. The sun and other stars are composed largely of hydrogen. Astronomers estimate that 90% of the atoms in the universe are hydrogen atoms. Hydrogen is a component of more compounds than any other element. Water is the most abundant compound of hydrogen found on earth. Hydrogen is an important part of petroleum, many minerals, cellulose and starch, sugar, fats, oils, alcohols, acids, and thousands of other substances.

At ordinary temperatures, hydrogen is a colorless, odorless, tasteless, and nonpoisonous gas consisting of the diatomic molecule H_2. Hydrogen is composed of three isotopes, and unlike other elements, these isotopes have different names and chemical symbols: protium, ^1H, deuterium, ^2H (or "D"), and tritium ^3H (or "T"). In a naturally occurring sample of hydrogen, there is one atom of deuterium for every 7000 H atoms and one atom of radioactive tritium for every 10^{18} H atoms. The chemical properties of the different isotopes are very similar because they have identical electron structures, but they differ in some physical properties because of their differing atomic masses. Elemental deuterium and tritium have lower vapor pressure than ordinary hydrogen. Consequently, when liquid hydrogen evaporates, the heavier isotopes are concentrated in the last portions to evaporate. Electrolysis of heavy water, D_2O, yields deuterium. Most tritium originates from nuclear reactions.

Preparation of Hydrogen

Elemental hydrogen must be prepared from compounds by breaking chemical bonds. The most common methods of preparing hydrogen follow.

From Steam and Carbon or Hydrocarbons

Water is the cheapest and most abundant source of hydrogen. Passing steam over coke (an impure form of elemental carbon) at 1000 °C produces a mixture of carbon monoxide and hydrogen known as water gas:

$$\mathrm{C}(s) + \mathrm{H_2O}(g) \xrightarrow{1000\ °\mathrm{C}} \underset{\text{water gas}}{\mathrm{CO}(g) + \mathrm{H_2}(g)}$$

Water gas is as an industrial fuel. It is possible to produce additional hydrogen by mixing the water gas with steam in the presence of a catalyst to convert the CO to CO_2. This reaction is the water gas shift reaction.

It is also possible to prepare a mixture of hydrogen and carbon monoxide by passing hydrocarbons from natural gas or petroleum and steam over a nickel-based catalyst. Propane is an example of a hydrocarbon reactant:

$$\mathrm{C_3H_8}(g) + 3\mathrm{H_2O}(g) \xrightarrow[\text{catalyst}]{900\ °\mathrm{C}} 3\mathrm{CO}(g) + 7\mathrm{H_2}(g)$$

Electrolysis

Hydrogen forms when direct current electricity passes through water containing an electrolyte such as H_2SO_4, as illustrated in Figure 18.26. Bubbles of hydrogen form at the cathode, and oxygen evolves at the anode. The net reaction is:

$$2H_2O(l) + \text{electrical energy} \longrightarrow 2H_2(g) + O_2(g)$$

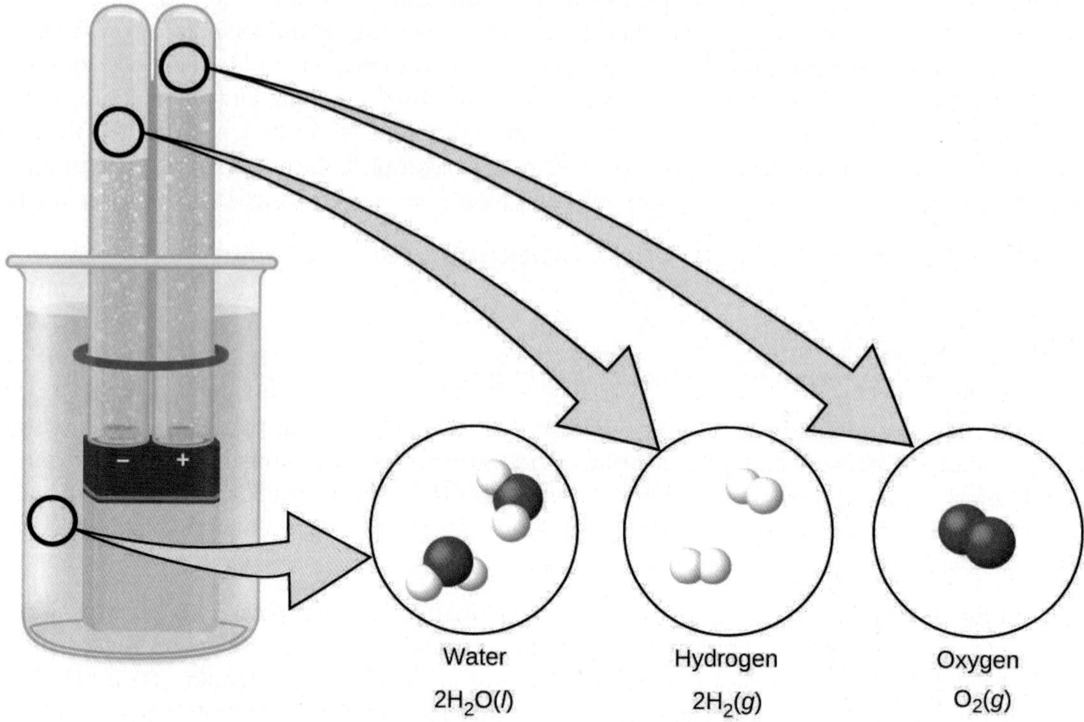

FIGURE 18.26 The electrolysis of water produces hydrogen and oxygen. Because there are twice as many hydrogen atoms as oxygen atoms and both elements are diatomic, there is twice the volume of hydrogen produced at the cathode as there is oxygen produced at the anode.

Reaction of Metals with Acids

This is the most convenient laboratory method of producing hydrogen. Metals with lower reduction potentials reduce the hydrogen ion in dilute acids to produce hydrogen gas and metal salts. For example, as shown in Figure 18.27, iron in dilute hydrochloric acid produces hydrogen gas and iron(II) chloride:

$$Fe(s) + 2H_3O^+(aq) + 2Cl^-(aq) \longrightarrow Fe^{2+}(aq) + 2Cl^-(aq) + H_2(g) + 2H_2O(l)$$

FIGURE 18.27 The reaction of iron with an acid produces hydrogen. Here, iron reacts with hydrochloric acid. (credit: Mark Ott)

Reaction of Ionic Metal Hydrides with Water

It is possible to produce hydrogen from the reaction of hydrides of the active metals, which contain the very strongly basic H^- anion, with water:

$$CaH_2(s) + 2H_2O(l) \longrightarrow Ca^{2+}(aq) + 2OH^-(aq) + 2H_2(g)$$

Metal hydrides are expensive but convenient sources of hydrogen, especially where space and weight are important factors. They are important in the inflation of life jackets, life rafts, and military balloons.

Reactions

Under normal conditions, hydrogen is relatively inactive chemically, but when heated, it enters into many chemical reactions.

Two thirds of the world's hydrogen production is devoted to the manufacture of ammonia, which is a fertilizer and used in the manufacture of nitric acid. Large quantities of hydrogen are also important in the process of **hydrogenation**, discussed in the chapter on organic chemistry.

It is possible to use hydrogen as a nonpolluting fuel. The reaction of hydrogen with oxygen is a very exothermic reaction, releasing 286 kJ of energy per mole of water formed. Hydrogen burns without explosion under controlled conditions. The oxygen-hydrogen torch, because of the high heat of combustion of hydrogen, can achieve temperatures up to 2800 °C. The hot flame of this torch is useful in cutting thick sheets of many metals. Liquid hydrogen is also an important rocket fuel (Figure 18.28).

FIGURE 18.28 Before the fleet's retirement in 2011, liquid hydrogen and liquid oxygen were used in the three main engines of a space shuttle. Two compartments in the large tank held these liquids until the shuttle was launched. (credit: "reynermedia"/Flickr)

An uncombined hydrogen atom consists of a nucleus and one valence electron in the 1*s* orbital. The $n = 1$ valence shell has a capacity for two electrons, and hydrogen can rightfully occupy two locations in the periodic table. It is possible to consider hydrogen a group 1 element because hydrogen can lose an electron to form the cation, H^+. It is also possible to consider hydrogen to be a group 17 element because it needs only one electron to fill its valence orbital to form a hydride ion, H^-, or it can share an electron to form a single, covalent bond. In reality, hydrogen is a unique element that almost deserves its own location in the periodic table.

Reactions with Elements

When heated, hydrogen reacts with the metals of group 1 and with Ca, Sr, and Ba (the more active metals in group 2). The compounds formed are crystalline, ionic hydrides that contain the hydride anion, H^-, a strong reducing agent and a strong base, which reacts vigorously with water and other acids to form hydrogen gas.

The reactions of hydrogen with nonmetals generally produce *acidic* hydrogen compounds with hydrogen in the 1+ oxidation state. The reactions become more exothermic and vigorous as the electronegativity of the nonmetal increases. Hydrogen reacts with nitrogen and sulfur only when heated, but it reacts explosively with fluorine (forming HF) and, under some conditions, with chlorine (forming HCl). A mixture of hydrogen and oxygen explodes if ignited. Because of the explosive nature of the reaction, it is necessary to exercise caution when handling hydrogen (or any other combustible gas) to avoid the formation of an explosive mixture in a confined space. Although most hydrides of the nonmetals are acidic, ammonia and phosphine (PH_3) are very, very weak acids and generally function as bases. There is a summary of these reactions of hydrogen with the elements in Table 18.1.

Chemical Reactions of Hydrogen with Other Elements

General Equation	Comments
MH or $MH_2 \longrightarrow MOH$ or $M(OH)_2 + H_2$	ionic hydrides with group 1 and Ca, Sr, and Ba
$H_2 + C \longrightarrow$ (no reaction)	
$3H_2 + N_2 \longrightarrow 2NH_3$	requires high pressure and temperature; low yield
$2H_2 + O_2 \longrightarrow 2H_2O$	exothermic and potentially explosive
$H_2 + S \longrightarrow H_2S$	requires heating; low yield
$H_2 + X_2 \longrightarrow 2HX$	X = F, Cl, Br, and I; explosive with F_2; low yield with I_2

TABLE 18.1

Reaction with Compounds

Hydrogen reduces the heated oxides of many metals, with the formation of the metal and water vapor. For example, passing hydrogen over heated CuO forms copper and water.

Hydrogen may also reduce the metal ions in some metal oxides to lower oxidation states:

$$H_2(g) + MnO_2(s) \xrightarrow{\Delta} MnO(s) + H_2O(g)$$

Hydrogen Compounds

Other than the noble gases, each of the nonmetals forms compounds with hydrogen. For brevity, we will discuss only a few hydrogen compounds of the nonmetals here.

Nitrogen Hydrogen Compounds

Ammonia, NH_3, forms naturally when any nitrogen-containing organic material decomposes in the absence of air. The laboratory preparation of ammonia is by the reaction of an ammonium salt with a strong base such as sodium hydroxide. The acid-base reaction with the weakly acidic ammonium ion gives ammonia, illustrated in Figure 18.29. Ammonia also forms when ionic nitrides react with water. The nitride ion is a much stronger base than the hydroxide ion:

$$Mg_3N_2(s) + 6H_2O(l) \longrightarrow 3Mg(OH)_2(s) + 2NH_3(g)$$

The commercial production of ammonia is by the direct combination of the elements in the **Haber process**:

$$N_2(g) + 3H_2(g) \overset{\text{catalyst}}{\rightleftharpoons} 2NH_3(g) \qquad \Delta H^\circ = -92\ \text{kJ}$$

FIGURE 18.29 The structure of ammonia is shown with a central nitrogen atom and three hydrogen atoms.

Ammonia is a colorless gas with a sharp, pungent odor. Smelling salts utilize this powerful odor. Gaseous ammonia readily liquefies to give a colorless liquid that boils at −33 °C. Due to intermolecular hydrogen bonding, the enthalpy of vaporization of liquid ammonia is higher than that of any other liquid except water, so ammonia is useful as a refrigerant. Ammonia is quite soluble in water (658 L at STP dissolves in 1 L H_2O).

The chemical properties of ammonia are as follows:

1. Ammonia acts as a Brønsted base, as discussed in the chapter on acid-base chemistry. The ammonium ion is similar in size to the potassium ion; compounds of the two ions exhibit many similarities in their structures and solubilities.
2. Ammonia can display acidic behavior, although it is a much weaker acid than water. Like other acids, ammonia reacts with metals, although it is so weak that high temperatures are necessary. Hydrogen and (depending on the stoichiometry) amides (salts of NH_2^-), imides (salts of NH^{2-}), or nitrides (salts of N^{3-}) form.
3. The nitrogen atom in ammonia has its lowest possible oxidation state (3−) and thus is not susceptible to reduction. However, it can be oxidized. Ammonia burns in air, giving NO and water. Hot ammonia and the ammonium ion are active reducing agents. Of particular interest are the oxidations of ammonium ion by nitrite ion, NO_2^-, to yield pure nitrogen and by nitrate ion to yield nitrous oxide, N_2O.
4. There are a number of compounds that we can consider derivatives of ammonia through the replacement of one or more hydrogen atoms with some other atom or group of atoms. Inorganic derivations include chloramine, NH_2Cl, and hydrazine, N_2H_4:

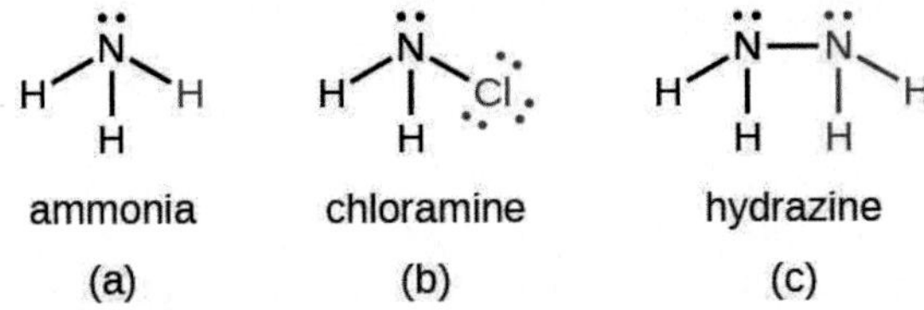

Chloramine, NH_2Cl, results from the reaction of sodium hypochlorite, NaOCl, with ammonia in basic solution. In the presence of a large excess of ammonia at low temperature, the chloramine reacts further to produce hydrazine, N_2H_4:

$$NH_3(aq) + OCl^-(aq) \longrightarrow NH_2Cl(aq) + OH^-(aq)$$

$$NH_2Cl(aq) + NH_3(aq) + OH^-(aq) \longrightarrow N_2H_4(aq) + Cl^-(aq) + H_2O(l)$$

Anhydrous hydrazine is relatively stable in spite of its positive free energy of formation:

$$N_2(g) + 2H_2(g) \longrightarrow N_2H_4(l) \qquad \Delta G_f^\circ = 149.2 \text{ kJ mol}^{-1}$$

Hydrazine is a fuming, colorless liquid that has some physical properties remarkably similar to those of H_2O (it melts at 2 °C, boils at 113.5 °C, and has a density at 25 °C of 1.00 g/mL). It burns rapidly and completely in air with substantial evolution of heat:

$$N_2H_4(l) + O_2(g) \longrightarrow N_2(g) + 2H_2O(l) \qquad \Delta H^\circ = -621.5 \text{ kJ mol}^{-1}$$

Like ammonia, hydrazine is both a Brønsted base and a Lewis base, although it is weaker than ammonia. It reacts with strong acids and forms two series of salts that contain the $N_2H_5^+$ and $N_2H_6^{2+}$ ions, respectively. Some rockets use hydrazine as a fuel.

Phosphorus Hydrogen Compounds

The most important hydride of phosphorus is phosphine, PH_3, a gaseous analog of ammonia in terms of both formula and structure. Unlike ammonia, it is not possible to form phosphine by direct union of the elements. There are two methods for the preparation of phosphine. One method is by the action of an acid on an ionic phosphide. The other method is the disproportionation of white phosphorus with hot concentrated base to produce phosphine and the hydrogen phosphite ion:

$$AlP(s) + 3H_3O^+(aq) \longrightarrow PH_3(g) + Al^{3+}(aq) + 3H_2O(l)$$

$$P_4(s) + 4OH^-(aq) + 2H_2O(l) \longrightarrow 2HPO_3^{2-}(aq) + 2PH_3(g)$$

Phosphine is a colorless, very poisonous gas, which has an odor like that of decaying fish. Heat easily decomposes phosphine ($4PH_3 \longrightarrow P_4 + 6H_2$), and the compound burns in air. The major uses of phosphine are as a fumigant for grains and in semiconductor processing. Like ammonia, gaseous phosphine unites with gaseous hydrogen halides, forming phosphonium compounds like PH_4Cl and PH_4I. Phosphine is a much weaker base than ammonia; therefore, these compounds decompose in water, and the insoluble PH_3 escapes from solution.

Sulfur Hydrogen Compounds

Hydrogen sulfide, H_2S, is a colorless gas that is responsible for the offensive odor of rotten eggs and of many hot springs. Hydrogen sulfide is as toxic as hydrogen cyanide; therefore, it is necessary to exercise great care in handling it. Hydrogen sulfide is particularly deceptive because it paralyzes the olfactory nerves; after a short exposure, one does not smell it.

The production of hydrogen sulfide by the direct reaction of the elements (H_2 + S) is unsatisfactory because the yield is low. A more effective preparation method is the reaction of a metal sulfide with a dilute acid. For example:

$$FeS(s) + 2H_3O^+(aq) \longrightarrow Fe^{2+}(aq) + H_2S(g) + 2H_2O(l)$$

It is easy to oxidize the sulfur in metal sulfides and in hydrogen sulfide, making metal sulfides and H_2S good reducing agents. In acidic solutions, hydrogen sulfide reduces Fe^{3+} to Fe^{2+}, $MnO_4{}^-$ to Mn^{2+}, $Cr_2O_7{}^{2-}$ to Cr^{3+}, and HNO_3 to NO_2. The sulfur in H_2S usually oxidizes to elemental sulfur, unless a large excess of the oxidizing agent is present. In which case, the sulfide may oxidize to $SO_3{}^{2-}$ or $SO_4{}^{2-}$ (or to SO_2 or SO_3 in the absence of water):

$$2H_2S(g) + O_2(g) \longrightarrow 2S(s) + 2H_2O(l)$$

This oxidation process leads to the removal of the hydrogen sulfide found in many sources of natural gas. The deposits of sulfur in volcanic regions may be the result of the oxidation of H_2S present in volcanic gases.

Hydrogen sulfide is a weak diprotic acid that dissolves in water to form hydrosulfuric acid. The acid ionizes in two stages, yielding hydrogen sulfide ions, HS^-, in the first stage and sulfide ions, S^{2-}, in the second. Since hydrogen sulfide is a weak acid, aqueous solutions of soluble sulfides and hydrogen sulfides are basic:

$$S^{2-}(aq) + H_2O(l) \rightleftharpoons HS^-(aq) + OH^-(aq)$$

$$HS^-(aq) + H_2O(l) \rightleftharpoons H_2S(g) + OH^-(aq)$$

Halogen Hydrogen Compounds

Binary compounds containing only hydrogen and a halogen are **hydrogen halides**. At room temperature, the pure hydrogen halides HF, HCl, HBr, and HI are gases.

In general, it is possible to prepare the halides by the general techniques used to prepare other acids. Fluorine, chlorine, and bromine react directly with hydrogen to form the respective hydrogen halide. This is a commercially important reaction for preparing hydrogen chloride and hydrogen bromide.

The acid-base reaction between a nonvolatile strong acid and a metal halide will yield a hydrogen halide. The escape of the gaseous hydrogen halide drives the reaction to completion. For example, the usual method of preparing hydrogen fluoride is by heating a mixture of calcium fluoride, CaF_2, and concentrated sulfuric acid:

$$CaF_2(s) + H_2SO_4(aq) \longrightarrow CaSO_4(s) + 2HF(g)$$

Gaseous hydrogen fluoride is also a by-product in the preparation of phosphate fertilizers by the reaction of fluoroapatite, $Ca_5(PO_4)_3F$, with sulfuric acid. The reaction of concentrated sulfuric acid with a chloride salt produces hydrogen chloride both commercially and in the laboratory.

In most cases, sodium chloride is the chloride of choice because it is the least expensive chloride. Hydrogen bromide and hydrogen iodide cannot be prepared using sulfuric acid because this acid is an oxidizing agent capable of oxidizing both bromide and iodide. However, it is possible to prepare both hydrogen bromide and hydrogen iodide using an acid such as phosphoric acid because it is a weaker oxidizing agent. For example:

$$H_3PO_4(l) + Br^-(aq) \longrightarrow HBr(g) + H_2PO_4{}^-(aq)$$

All of the hydrogen halides are very soluble in water, forming hydrohalic acids. With the exception of hydrogen fluoride, which has a strong hydrogen-fluoride bond, they are strong acids. Reactions of hydrohalic acids with metals, metal hydroxides, oxides, or carbonates produce salts of the halides. Most chloride salts are soluble in water. AgCl, $PbCl_2$, and Hg_2Cl_2 are the commonly encountered exceptions.

The halide ions give the substances the properties associated with $X^-(aq)$. The heavier halide ions (Cl^-, Br^-, and I^-) can act as reducing agents, and the lighter halogens or other oxidizing agents will oxidize them:

$$Cl_2(aq) + 2e^- \longrightarrow 2Cl^-(aq) \qquad E° = 1.36\text{ V}$$

$$Br_2(aq) + 2e^- \longrightarrow 2Br^-(aq) \qquad E° = 1.09\text{ V}$$

$$I_2(aq) + 2e^- \longrightarrow 2I^-(aq) \qquad E° = 0.54\text{ V}$$

For example, bromine oxidizes iodine:

$$Br_2(aq) + 2HI(aq) \longrightarrow 2HBr(aq) + I_2(aq) \qquad E° = 0.55\text{ V}$$

Hydrofluoric acid is unique in its reactions with sand (silicon dioxide) and with glass, which is a mixture of silicates:

$$SiO_2(s) + 4HF(aq) \longrightarrow SiF_4(g) + 2H_2O(l)$$

$$CaSiO_3(s) + 6HF(aq) \longrightarrow CaF_2(s) + SiF_4(g) + 3H_2O(l)$$

The volatile silicon tetrafluoride escapes from these reactions. Because hydrogen fluoride attacks glass, it can frost or etch glass and is used to etch markings on thermometers, burets, and other glassware.

The largest use for hydrogen fluoride is in production of hydrochlorofluorocarbons for refrigerants, in plastics, and in propellants. The second largest use is in the manufacture of cryolite, Na_3AlF_6, which is important in the production of aluminum. The acid is also important in the production of other inorganic fluorides (such as BF_3), which serve as catalysts in the industrial synthesis of certain organic compounds.

Hydrochloric acid is relatively inexpensive. It is an important and versatile acid in industry and is important for the manufacture of metal chlorides, dyes, glue, glucose, and various other chemicals. A considerable amount is also important for the activation of oil wells and as pickle liquor—an acid used to remove oxide coating from iron or steel that is to be galvanized, tinned, or enameled. The amounts of hydrobromic acid and hydroiodic acid used commercially are insignificant by comparison.

18.6 Occurrence, Preparation, and Properties of Carbonates

LEARNING OBJECTIVES

By the end of this section, you will be able to:

- Describe the preparation, properties, and uses of some representative metal carbonates

The chemistry of carbon is extensive; however, most of this chemistry is not relevant to this chapter. The other aspects of the chemistry of carbon will appear in the chapter covering organic chemistry. In this chapter, we will focus on the carbonate ion and related substances. The metals of groups 1 and 2, as well as zinc, cadmium, mercury, and lead(II), form ionic **carbonates**—compounds that contain the carbonate anions, ${CO_3}^{2-}$. The metals of group 1, magnesium, calcium, strontium, and barium also form **hydrogen carbonates**—compounds that contain the hydrogen carbonate anion, ${HCO_3}^-$, also known as the **bicarbonate anion**.

With the exception of magnesium carbonate, it is possible to prepare carbonates of the metals of groups 1 and 2 by the reaction of carbon dioxide with the respective oxide or hydroxide. Examples of such reactions include:

$$Na_2O(s) + CO_2(g) \longrightarrow Na_2CO_3(s)$$

$$Ca(OH)_2(s) + CO_2(g) \longrightarrow CaCO_3(s) + H_2O(l)$$

The carbonates of the alkaline earth metals of group 12 and lead(II) are not soluble. These carbonates precipitate upon mixing a solution of soluble alkali metal carbonate with a solution of soluble salts of these metals. Examples of net ionic equations for the reactions are:

$$Ca^{2+}(aq) + {CO_3}^{2-}(aq) \longrightarrow CaCO_3(s)$$

$$Pb^{2+}(aq) + {CO_3}^{2-}(aq) \longrightarrow PbCO_3(s)$$

Pearls and the shells of most mollusks are calcium carbonate. Tin(II) or one of the trivalent or tetravalent ions such as Al^{3+} or Sn^{4+} behave differently in this reaction as carbon dioxide and the corresponding oxide form instead of the carbonate.

Alkali metal hydrogen carbonates such as $NaHCO_3$ and $CsHCO_3$ form by saturating a solution of the hydroxides with carbon dioxide. The net ionic reaction involves hydroxide ion and carbon dioxide:

$$OH^-(aq) + CO_2(aq) \longrightarrow HCO_3{}^-(aq)$$

It is possible to isolate the solids by evaporation of the water from the solution.

Although they are insoluble in pure water, alkaline earth carbonates dissolve readily in water containing carbon dioxide because hydrogen carbonate salts form. For example, caves and sinkholes form in limestone when $CaCO_3$ dissolves in water containing dissolved carbon dioxide:

$$CaCO_3(s) + CO_2(aq) + H_2O(l) \longrightarrow Ca^{2+}(aq) + 2HCO_3{}^-(aq)$$

Hydrogen carbonates of the alkaline earth metals remain stable only in solution; evaporation of the solution produces the carbonate. Stalactites and stalagmites, like those shown in Figure 18.30, form in caves when drops of water containing dissolved calcium hydrogen carbonate evaporate to leave a deposit of calcium carbonate.

(a)

(b)

FIGURE 18.30 (a) Stalactites and (b) stalagmites are cave formations of calcium carbonate. (credit a: modification of work by Arvind Govindaraj; credit b: modification of work by the National Park Service.)

The two carbonates used commercially in the largest quantities are sodium carbonate and calcium carbonate. In the United States, sodium carbonate is extracted from the mineral trona, $Na_3(CO_3)(HCO_3)(H_2O)_2$. Following recrystallization to remove clay and other impurities, heating the recrystallized trona produces Na_2CO_3:

$$2Na_3(CO_3)(HCO_3)(H_2O)_2(s) \longrightarrow 3Na_2CO_3(s) + 5H_2O(l) + CO_2(g)$$

Carbonates are moderately strong bases. Aqueous solutions are basic because the carbonate ion accepts hydrogen ion from water in this reversible reaction:

$$CO_3{}^{2-}(aq) + H_2O(l) \rightleftharpoons HCO_3{}^-(aq) + OH^-(aq)$$

Carbonates react with acids to form salts of the metal, gaseous carbon dioxide, and water. The reaction of calcium carbonate, the active ingredient of the antacid Tums, with hydrochloric acid (stomach acid), as shown in Figure 18.31, illustrates the reaction:

$$CaCO_3(s) + 2HCl(aq) \longrightarrow CaCl_2(aq) + CO_2(g) + H_2O(l)$$

FIGURE 18.31 The reaction of calcium carbonate with hydrochloric acid is shown. (credit: Mark Ott)

Other applications of carbonates include glass making—where carbonate ions serve as a source of oxide ions—and synthesis of oxides.

Hydrogen carbonates are amphoteric because they act as both weak acids and weak bases. Hydrogen carbonate ions act as acids and react with solutions of soluble hydroxides to form a carbonate and water:

$$\mathrm{KHCO_3}(aq) + \mathrm{KOH}(aq) \longrightarrow \mathrm{K_2CO_3}(aq) + \mathrm{H_2O}(l)$$

With acids, hydrogen carbonates form a salt, carbon dioxide, and water. Baking soda (bicarbonate of soda or sodium bicarbonate) is sodium hydrogen carbonate. Baking powder contains baking soda and a solid acid such as potassium hydrogen tartrate (cream of tartar), $KHC_4H_4O_6$. As long as the powder is dry, no reaction occurs; immediately after the addition of water, the acid reacts with the hydrogen carbonate ions to form carbon dioxide:

$$\mathrm{HC_4H_4O_6}^-(aq) + \mathrm{HCO_3}^-(aq) \longrightarrow \mathrm{C_4H_4O_6}^{2-}(aq) + \mathrm{CO_2}(g) + \mathrm{H_2O}(l)$$

Dough will trap the carbon dioxide, causing it to expand during baking, producing the characteristic texture of baked goods.

18.7 Occurrence, Preparation, and Properties of Nitrogen

LEARNING OBJECTIVES

By the end of this section, you will be able to:

- Describe the properties, preparation, and uses of nitrogen

Most pure nitrogen comes from the fractional distillation of liquid air. The atmosphere consists of 78% nitrogen by volume. This means there are more than 20 million tons of nitrogen over every square mile of the earth's surface. Nitrogen is a component of proteins and of the genetic material (DNA/RNA) of all plants and animals.

Under ordinary conditions, nitrogen is a colorless, odorless, and tasteless gas. It boils at 77 K and freezes at 63 K. Liquid nitrogen is a useful coolant because it is inexpensive and has a low boiling point. Nitrogen is very unreactive because of the very strong triple bond between the nitrogen atoms. The only common reactions at room temperature occur with lithium to form Li_3N, with certain transition metal complexes, and with hydrogen or oxygen in nitrogen-fixing bacteria. The general lack of reactivity of nitrogen makes the remarkable ability of some bacteria to synthesize nitrogen compounds using atmospheric nitrogen gas as the source one of the most exciting chemical events on our planet. This process is one type of **nitrogen fixation**. In this case, nitrogen fixation is the process where organisms convert atmospheric nitrogen into biologically useful chemicals. Nitrogen fixation also occurs when lightning passes through air, causing molecular nitrogen to react with oxygen to form nitrogen oxides, which are then carried down to the soil.

Chemistry in Everyday Life

Nitrogen Fixation

All living organisms require nitrogen compounds for survival. Unfortunately, most of these organisms cannot absorb nitrogen from its most abundant source—the atmosphere. Atmospheric nitrogen consists of N_2 molecules, which are very unreactive due to the strong nitrogen-nitrogen triple bond. However, a few organisms can overcome this problem through a process known as nitrogen fixation, illustrated in Figure 18.32.

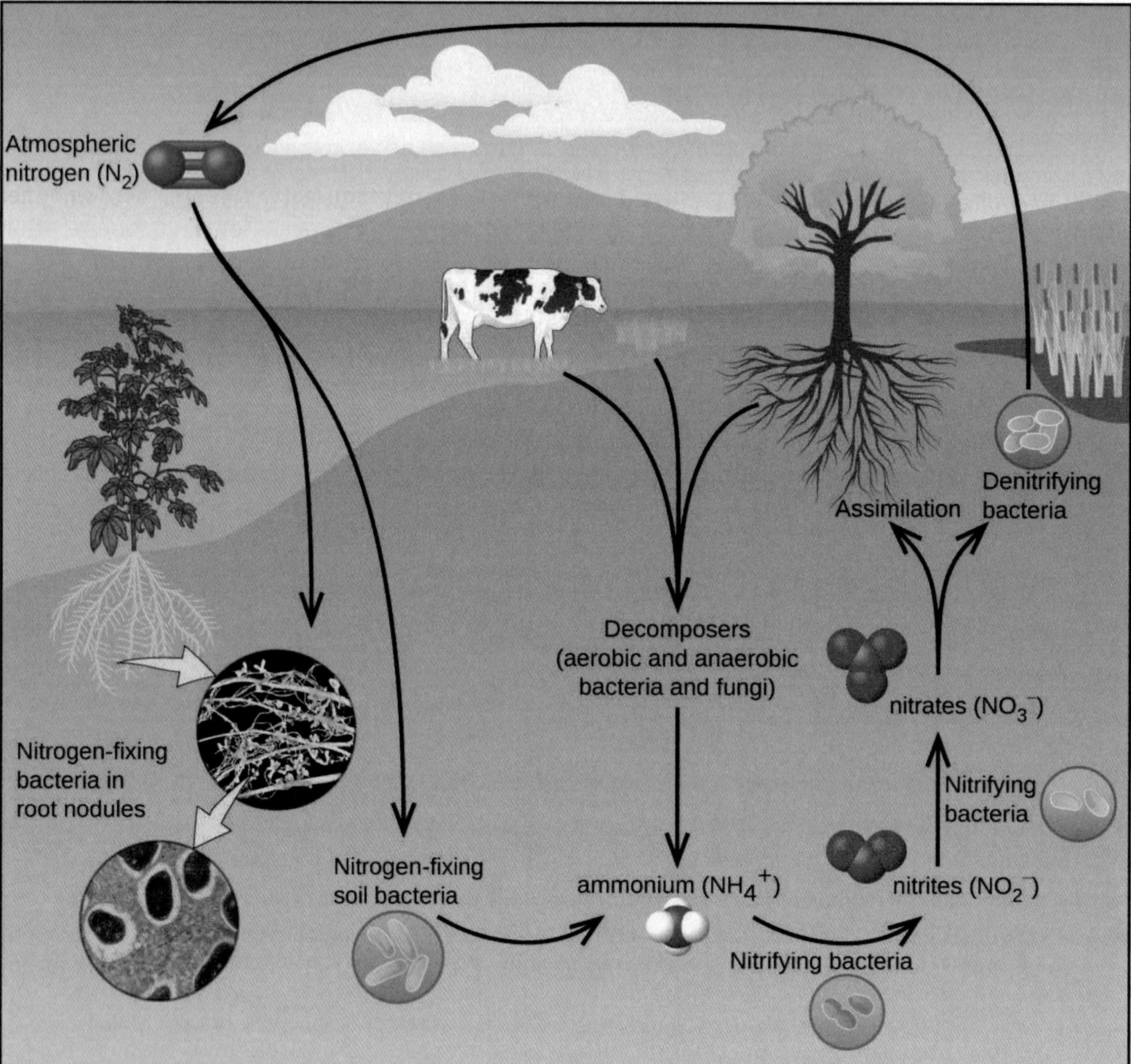

FIGURE 18.32 All living organisms require nitrogen. A few microorganisms are able to process atmospheric nitrogen using nitrogen fixation. (credit "roots": modification of work by the United States Department of Agriculture; credit "root nodules": modification of work by Louisa Howard)

Nitrogen fixation is the process where organisms convert atmospheric nitrogen into biologically useful chemicals. To date, the only known kind of biological organisms capable of nitrogen fixation are microorganisms. These organisms employ enzymes called nitrogenases, which contain iron and molybdenum. Many of these microorganisms live in a symbiotic relationship with plants, with the best-known example being the presence of rhizobia in the root nodules of legumes.

Large volumes of atmospheric nitrogen are necessary for making ammonia—the principal starting material used for preparation of large quantities of other nitrogen-containing compounds. Most other uses for elemental nitrogen depend on its inactivity. It is helpful when a chemical process requires an inert atmosphere. Canned foods and luncheon meats cannot oxidize in a pure nitrogen atmosphere, so they retain a better flavor and color, and spoil less rapidly, when sealed in nitrogen instead of air. This technology allows fresh produce to be available year-round, regardless of growing season.

There are compounds with nitrogen in all of its oxidation states from 3− to 5+. Much of the chemistry of nitrogen involves oxidation-reduction reactions. Some active metals (such as alkali metals and alkaline earth metals) can reduce nitrogen to form metal nitrides. In the remainder of this section, we will examine nitrogen-oxygen chemistry.

There are well-characterized nitrogen oxides in which nitrogen exhibits each of its positive oxidation numbers from 1+ to 5+. When ammonium nitrate is carefully heated, nitrous oxide (dinitrogen oxide) and water vapor form. Stronger heating generates nitrogen gas, oxygen gas, and water vapor. No one should ever attempt this reaction—it can be very explosive. In 1947, there was a major ammonium nitrate explosion in Texas City, Texas, and, in 2013, there was another major explosion in West, Texas. In the last 100 years, there were nearly 30 similar disasters worldwide, resulting in the loss of numerous lives. In this oxidation-reduction reaction, the nitrogen in the nitrate ion oxidizes the nitrogen in the ammonium ion. Nitrous oxide, shown in Figure 18.33, is a colorless gas possessing a mild, pleasing odor and a sweet taste. It finds application as an anesthetic for minor operations, especially in dentistry, under the name "laughing gas."

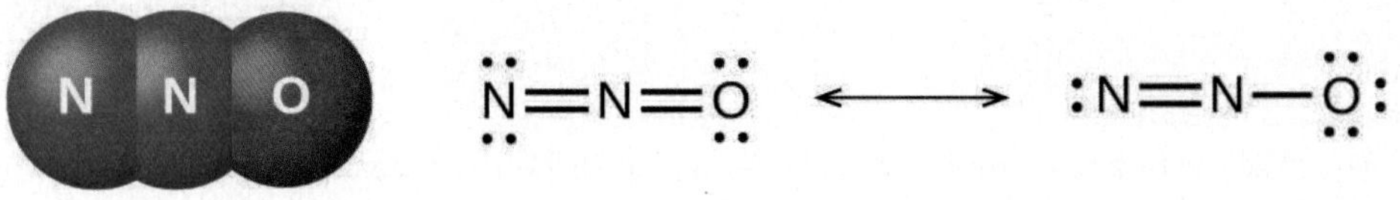

FIGURE 18.33 Nitrous oxide, N_2O, is an anesthetic that has these molecular (left) and resonance (right) structures.

Low yields of nitric oxide, NO, form when heating nitrogen and oxygen together. NO also forms when lightning passes through air during thunderstorms. Burning ammonia is the commercial method of preparing nitric oxide. In the laboratory, the reduction of nitric acid is the best method for preparing nitric oxide. When copper reacts with dilute nitric acid, nitric oxide is the principal reduction product:

$$3Cu(s) + 8HNO_3(aq) \longrightarrow 2NO(g) + 3Cu(NO_3)_2(aq) + 4H_2O(l)$$

Gaseous nitric oxide is the most thermally stable of the nitrogen oxides and is the simplest known thermally stable molecule with an unpaired electron. It is one of the air pollutants generated by internal combustion engines, resulting from the reaction of atmospheric nitrogen and oxygen during the combustion process.

At room temperature, nitric oxide is a colorless gas consisting of diatomic molecules. As is often the case with molecules that contain an unpaired electron, two molecules combine to form a dimer by pairing their unpaired electrons to form a bond. Liquid and solid NO both contain N_2O_2 dimers, like that shown in Figure 18.34. Most substances with unpaired electrons exhibit color by absorbing visible light; however, NO is colorless because the absorption of light is not in the visible region of the spectrum.

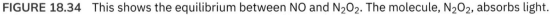

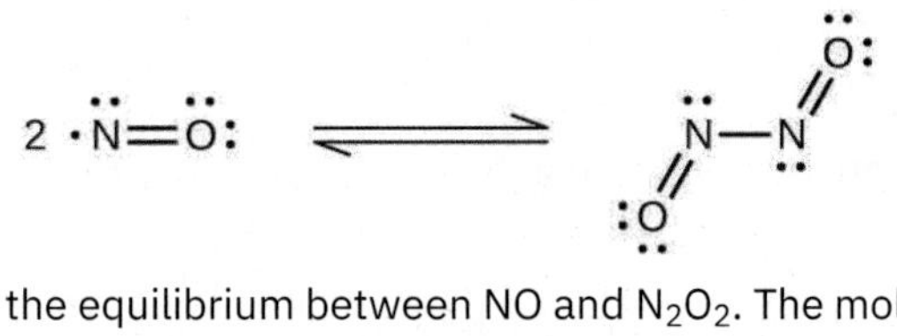

FIGURE 18.34 This shows the equilibrium between NO and N_2O_2. The molecule, N_2O_2, absorbs light.

Cooling a mixture of equal parts nitric oxide and nitrogen dioxide to −21 °C produces dinitrogen trioxide, a blue liquid consisting of N_2O_3 molecules (shown in Figure 18.35). Dinitrogen trioxide exists only in the liquid and solid states. When heated, it reverts to a mixture of NO and NO_2.

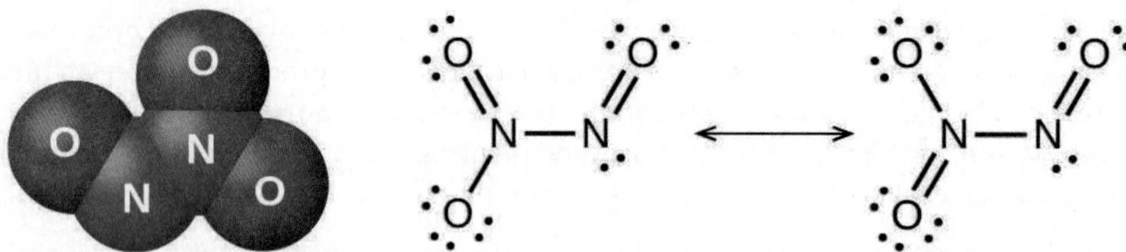

FIGURE 18.35 Dinitrogen trioxide, N_2O_3, only exists in liquid or solid states and has these molecular (left) and resonance (right) structures.

It is possible to prepare nitrogen dioxide in the laboratory by heating the nitrate of a heavy metal, or by the reduction of concentrated nitric acid with copper metal, as shown in Figure 18.36. Commercially, it is possible to prepare nitrogen dioxide by oxidizing nitric oxide with air.

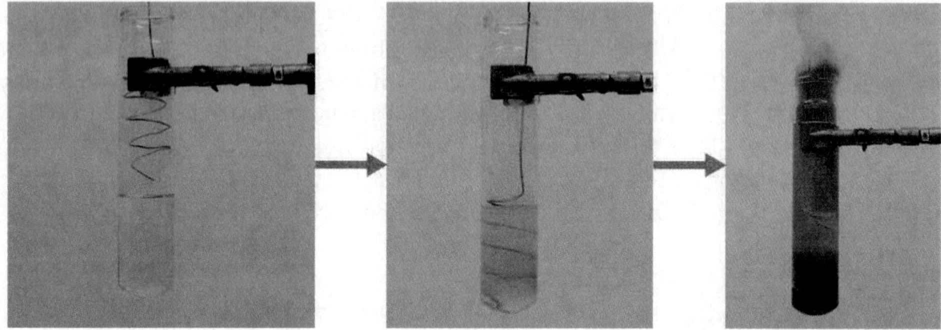

FIGURE 18.36 The reaction of copper metal with concentrated HNO_3 produces a solution of $Cu(NO_3)_2$ and brown fumes of NO_2. (credit: modification of work by Mark Ott)

The nitrogen dioxide molecule (illustrated in Figure 18.37) contains an unpaired electron, which is responsible for its color and paramagnetism. It is also responsible for the dimerization of NO_2. At low pressures or at high temperatures, nitrogen dioxide has a deep brown color that is due to the presence of the NO_2 molecule. At low temperatures, the color almost entirely disappears as dinitrogen tetraoxide, N_2O_4, forms. At room temperature, an equilibrium exists:

$$2NO_2(g) \rightleftharpoons N_2O_4(g) \qquad K_P = 6.86$$

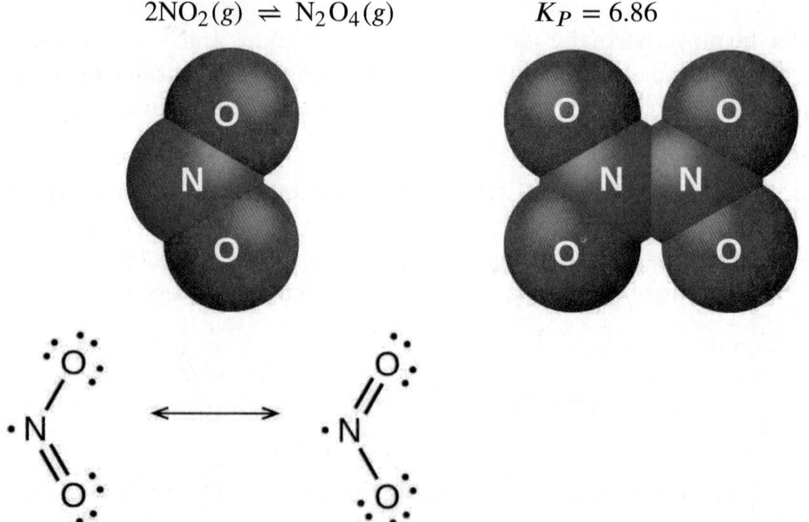

FIGURE 18.37 The molecular and resonance structures for nitrogen dioxide (NO_2, left) and dinitrogen tetraoxide (N_2O_4, right) are shown.

Dinitrogen pentaoxide, N_2O_5 (illustrated in Figure 18.38), is a white solid that is formed by the dehydration of nitric acid by phosphorus(V) oxide (tetraphosphorus decoxide):

$$P_4O_{10}(s) + 4HNO_3(l) \longrightarrow 4HPO_3(s) + 2N_2O_5(s)$$

It is unstable above room temperature, decomposing to N_2O_4 and O_2.

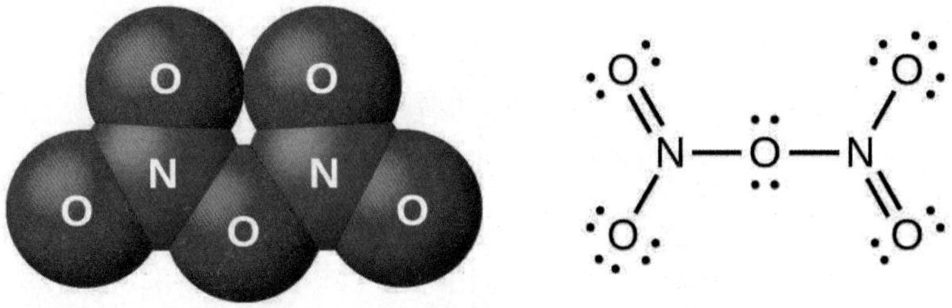

FIGURE 18.38 This image shows the molecular structure and one resonance structure of a molecule of dinitrogen pentaoxide, N_2O_5.

The oxides of nitrogen(III), nitrogen(IV), and nitrogen(V) react with water and form nitrogen-containing oxyacids. Nitrogen(III) oxide, N_2O_3, is the anhydride of nitrous acid; HNO_2 forms when N_2O_3 reacts with water. There are no stable oxyacids containing nitrogen with an oxidation state of 4+; therefore, nitrogen(IV) oxide, NO_2, disproportionates in one of two ways when it reacts with water. In cold water, a mixture of HNO_2 and HNO_3 forms. At higher temperatures, HNO_3 and NO will form. Nitrogen(V) oxide, N_2O_5, is the anhydride of nitric acid; HNO_3 is produced when N_2O_5 reacts with water:

$$N_2O_5(s) + H_2O(l) \longrightarrow 2HNO_3(aq)$$

The nitrogen oxides exhibit extensive oxidation-reduction behavior. Nitrous oxide resembles oxygen in its behavior when heated with combustible substances. N_2O is a strong oxidizing agent that decomposes when heated to form nitrogen and oxygen. Because one-third of the gas liberated is oxygen, nitrous oxide supports combustion better than air (one-fifth oxygen). A glowing splinter bursts into flame when thrust into a bottle of this gas. Nitric oxide acts both as an oxidizing agent and as a reducing agent. For example:

$$\text{oxidizing agent: } P_4(s) + 6NO(g) \longrightarrow P_4O_6(s) + 3N_2(g)$$

$$\text{reducing agent: } Cl_2(g) + 2NO(g) \longrightarrow 2ClNO(g)$$

Nitrogen dioxide (or dinitrogen tetraoxide) is a good oxidizing agent. For example:

$$NO_2(g) + CO(g) \longrightarrow NO(g) + CO_2(g)$$

$$NO_2(g) + 2HCl(aq) \longrightarrow NO(g) + Cl_2(g) + H_2O(l)$$

18.8 Occurrence, Preparation, and Properties of Phosphorus

LEARNING OBJECTIVES

By the end of this section, you will be able to:

- Describe the properties, preparation, and uses of phosphorus

The industrial preparation of phosphorus is by heating calcium phosphate, obtained from phosphate rock, with sand and coke:

$$2Ca_3(PO_4)_2(s) + 6SiO_2(s) + 10C(s) \xrightarrow{\Delta} 6CaSiO_3(l) + 10CO(g) + P_4(g)$$

The phosphorus distills out of the furnace and is condensed into a solid or burned to form P_4O_{10}. The preparation of many other phosphorus compounds begins with P_4O_{10}. The acids and phosphates are useful as fertilizers and in the chemical industry. Other uses are in the manufacture of special alloys such as ferrophosphorus and phosphor bronze. Phosphorus is important in making pesticides, matches, and some plastics. Phosphorus is an active nonmetal. In compounds, phosphorus usually occurs in oxidation states of 3−, 3+, and 5+. Phosphorus exhibits oxidation numbers that are unusual for a group 15 element in compounds that contain phosphorus-phosphorus bonds; examples include diphosphorus tetrahydride, $H_2P\text{-}PH_2$, and tetraphosphorus trisulfide, P_4S_3, illustrated in Figure 18.39.

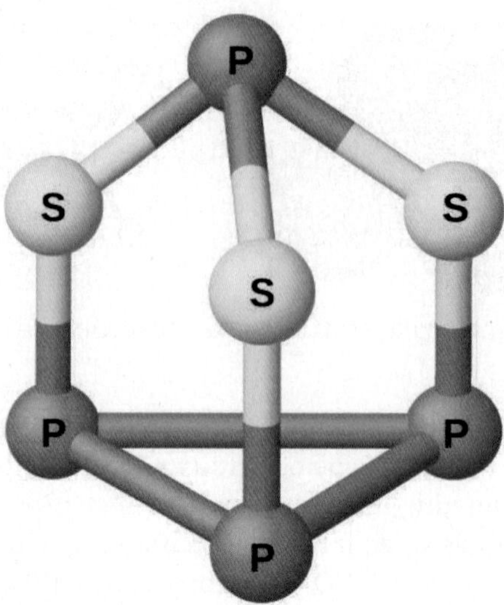

FIGURE 18.39 P_4S_3 is a component of the heads of strike-anywhere matches.

Phosphorus Oxygen Compounds

Phosphorus forms two common oxides, phosphorus(III) oxide (or tetraphosphorus hexaoxide), P_4O_6, and phosphorus(V) oxide (or tetraphosphorus decaoxide), P_4O_{10}, both shown in Figure 18.40. Phosphorus(III) oxide is a white crystalline solid with a garlic-like odor. Its vapor is very poisonous. It oxidizes slowly in air and inflames when heated to 70 °C, forming P_4O_{10}. Phosphorus(III) oxide dissolves slowly in cold water to form phosphorous acid, H_3PO_3.

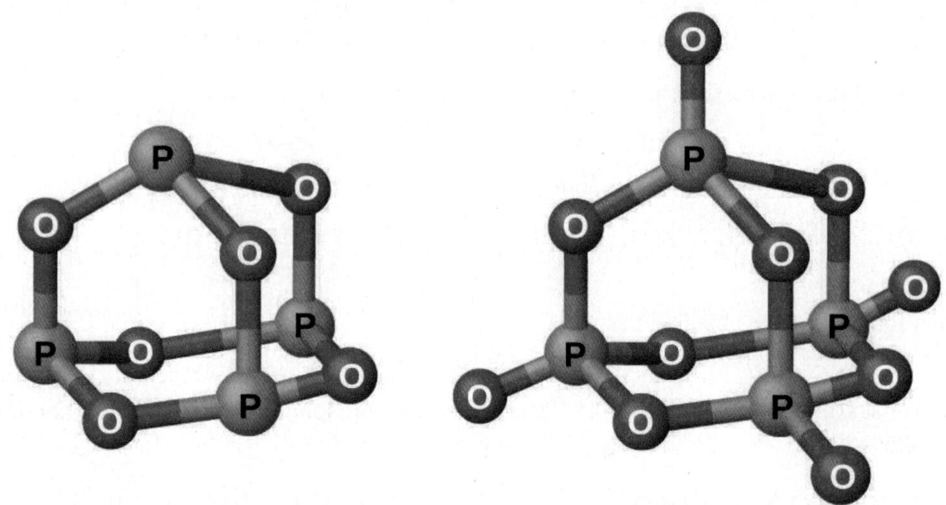

FIGURE 18.40 This image shows the molecular structures of P_4O_6 (left) and P_4O_{10} (right).

Phosphorus(V) oxide, P_4O_{10}, is a white powder that is prepared by burning phosphorus in excess oxygen. Its enthalpy of formation is very high (−2984 kJ), and it is quite stable and a very poor oxidizing agent. Dropping P_4O_{10} into water produces a hissing sound, heat, and orthophosphoric acid:

$$P_4O_{10}(s) + 6H_2O(l) \longrightarrow 4H_3PO_4(aq)$$

Because of its great affinity for water, phosphorus(V) oxide is an excellent drying agent for gases and solvents, and for removing water from many compounds.

Phosphorus Halogen Compounds

Phosphorus will react directly with the halogens, forming trihalides, PX_3, and pentahalides, PX_5. The trihalides are much more stable than the corresponding nitrogen trihalides; nitrogen pentahalides do not form because of nitrogen's inability to form more than four bonds.

The chlorides PCl_3 and PCl_5, both shown in Figure 18.41, are the most important halides of phosphorus. Phosphorus trichloride is a colorless liquid that is prepared by passing chlorine over molten phosphorus. Phosphorus pentachloride is an off-white solid that is prepared by oxidizing the trichloride with excess chlorine. The pentachloride sublimes when warmed and forms an equilibrium with the trichloride and chlorine when heated.

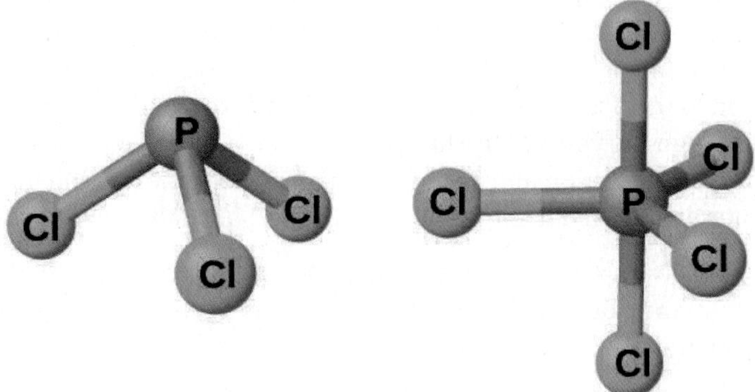

FIGURE 18.41 This image shows the molecular structure of PCl_3 (left) and PCl_5 (right) in the gas phase.

Like most other nonmetal halides, both phosphorus chlorides react with an excess of water and yield hydrogen chloride and an oxyacid: PCl_3 yields phosphorous acid H_3PO_3 and PCl_5 yields phosphoric acid, H_3PO_4.

The pentahalides of phosphorus are Lewis acids because of the empty valence *d* orbitals of phosphorus. These compounds readily react with halide ions (Lewis bases) to give the anion $PX_6{}^-$. Whereas phosphorus pentafluoride is a molecular compound in all states, X-ray studies show that solid phosphorus pentachloride is an ionic compound, $[PCl_4{}^+][PCl_6{}^-]$, as are phosphorus pentabromide, $[PBr_4{}^+][Br^-]$, and phosphorus pentaiodide, $[PI_4{}^+][I^-]$.

18.9 Occurrence, Preparation, and Compounds of Oxygen

LEARNING OBJECTIVES

By the end of this section, you will be able to:

- Describe the properties, preparation, and compounds of oxygen
- Describe the preparation, properties, and uses of some representative metal oxides, peroxides, and hydroxides

Oxygen is the most abundant element on the earth's crust. The earth's surface is composed of the crust, atmosphere, and hydrosphere. About 50% of the mass of the earth's crust consists of oxygen (combined with other elements, principally silicon). Oxygen occurs as O_2 molecules and, to a limited extent, as O_3 (ozone) molecules in air. It forms about 20% of the mass of the air. About 89% of water by mass consists of combined oxygen. In combination with carbon, hydrogen, and nitrogen, oxygen is a large part of plants and animals.

Oxygen is a colorless, odorless, and tasteless gas at ordinary temperatures. It is slightly denser than air. Although it is only slightly soluble in water (49 mL of gas dissolves in 1 L at STP), oxygen's solubility is very important to aquatic life.

Most of the oxygen isolated commercially comes from air and the remainder from the electrolysis of water. The separation of oxygen from air begins with cooling and compressing the air until it liquefies. As liquid air warms, oxygen with its higher boiling point (90 K) separates from nitrogen, which has a lower boiling point (77 K). It is possible to separate the other components of air at the same time based on differences in their boiling points.

Oxygen is essential in combustion processes such as the burning of fuels. Plants and animals use the oxygen from the air in respiration. The administration of oxygen-enriched air is an important medical practice when a patient is receiving an inadequate supply of oxygen because of shock, pneumonia, or some other illness.

The chemical industry employs oxygen for oxidizing many substances. A significant amount of oxygen produced commercially is important in the removal of carbon from iron during steel production. Large quantities of pure oxygen are also necessary in metal fabrication and in the cutting and welding of metals with oxyhydrogen and oxyacetylene torches.

Liquid oxygen is important to the space industry. It is an oxidizing agent in rocket engines. It is also the source of gaseous oxygen for life support in space.

As we know, oxygen is very important to life. The energy required for the maintenance of normal body functions in human beings and in other organisms comes from the slow oxidation of chemical compounds. Oxygen is the final oxidizing agent in these reactions. In humans, oxygen passes from the lungs into the blood, where it combines with hemoglobin, producing oxyhemoglobin. In this form, blood transports the oxygen to tissues, where it is transferred to the tissues. The ultimate products are carbon dioxide and water. The blood carries the carbon dioxide through the veins to the lungs, where the blood releases the carbon dioxide and collects another supply of oxygen. Digestion and assimilation of food regenerate the materials consumed by oxidation in the body; the energy liberated is the same as if the food burned outside the body.

Green plants continually replenish the oxygen in the atmosphere by a process called **photosynthesis**. The products of photosynthesis may vary, but, in general, the process converts carbon dioxide and water into glucose (a sugar) and oxygen using the energy of light:

$$\underset{\text{carbon dioxide}}{6CO_2(g)} + \underset{\text{water}}{6H_2O(l)} \xrightarrow[\text{light}]{\text{chlorophyll}} \underset{\text{glucose}}{C_6H_{12}O_6(aq)} + \underset{\text{oxygen}}{6O_2(g)}$$

Thus, the oxygen that became carbon dioxide and water by the metabolic processes in plants and animals returns to the atmosphere by photosynthesis.

When dry oxygen is passed between two electrically charged plates, **ozone** (O_3, illustrated in Figure 18.42), an allotrope of oxygen possessing a distinctive odor, forms. The formation of ozone from oxygen is an endothermic reaction, in which the energy comes from an electrical discharge, heat, or ultraviolet light:

$$3O_2(g) \xrightarrow{\text{electric discharge}} 2O_3(g) \qquad \Delta H^\circ = 287 \text{ kJ}$$

The sharp odor associated with sparking electrical equipment is due, in part, to ozone.

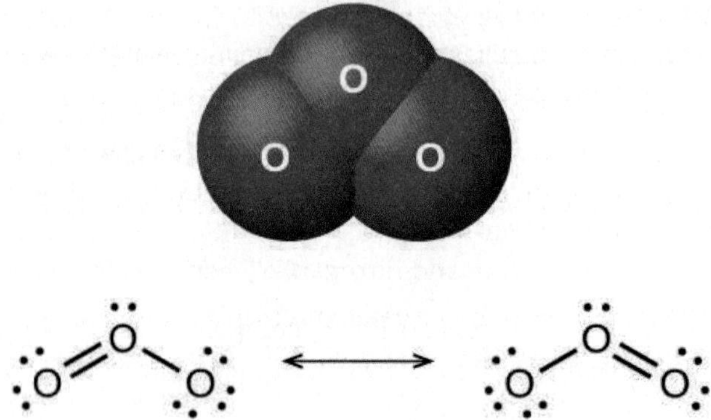

FIGURE 18.42 The image shows the bent ozone (O_3) molecule and the resonance structures necessary to describe its bonding.

Ozone forms naturally in the upper atmosphere by the action of ultraviolet light from the sun on the oxygen there. Most atmospheric ozone occurs in the stratosphere, a layer of the atmosphere extending from about 10 to 50 kilometers above the earth's surface. This ozone acts as a barrier to harmful ultraviolet light from the sun by absorbing it via a chemical decomposition reaction:

$$O_3(g) \xrightarrow{\text{ultraviolet light}} O(g) + O_2(g)$$

The reactive oxygen atoms recombine with molecular oxygen to complete the ozone cycle. The presence of stratospheric ozone decreases the frequency of skin cancer and other damaging effects of ultraviolet radiation. It has been clearly demonstrated that chlorofluorocarbons, CFCs (known commercially as Freons), which were present as aerosol propellants in spray cans and as refrigerants, caused depletion of ozone in the stratosphere. This occurred because ultraviolet light also causes CFCs to decompose, producing atomic chlorine. The chlorine atoms react with ozone molecules, resulting in a net removal of O_3 molecules from stratosphere. This process is explored in detail in our coverage of chemical kinetics. There is a worldwide effort to reduce the amount of CFCs used commercially, and the ozone hole is already beginning to decrease in size as atmospheric concentrations of atomic chlorine decrease. While ozone in the stratosphere helps protect us, ozone in the troposphere is a problem. This ozone is a toxic component of photochemical smog.

The uses of ozone depend on its reactivity with other substances. It can be used as a bleaching agent for oils, waxes, fabrics, and starch: It oxidizes the colored compounds in these substances to colorless compounds. It is an alternative to chlorine as a disinfectant for water.

Reactions

Elemental oxygen is a strong oxidizing agent. It reacts with most other elements and many compounds.

Reaction with Elements

Oxygen reacts directly at room temperature or at elevated temperatures with all other elements except the noble gases, the halogens, and few second- and third-row transition metals of low reactivity (those with higher reduction potentials than copper). Rust is an example of the reaction of oxygen with iron. The more active metals form peroxides or superoxides. Less active metals and the nonmetals give oxides. Two examples of these reactions are:

$$2Mg(s) + O_2(g) \longrightarrow 2MgO(s)$$

$$P_4(s) + 5O_2(g) \longrightarrow P_4O_{10}(s)$$

The oxides of halogens, at least one of the noble gases, and metals with higher reduction potentials than copper do not form by the direct action of the elements with oxygen.

Reaction with Compounds

Elemental oxygen also reacts with some compounds. If it is possible to oxidize any of the elements in a given compound, further oxidation by oxygen can occur. For example, hydrogen sulfide, H_2S, contains sulfur with an oxidation state of 2−. Because the sulfur does not exhibit its maximum oxidation state, we would expect H_2S to react with oxygen. It does, yielding water and sulfur dioxide. The reaction is:

$$2H_2S(g) + 3O_2(g) \longrightarrow 2H_2O(l) + 2SO_2(g)$$

It is also possible to oxidize oxides such as CO and P_4O_6 that contain an element with a lower oxidation state. The ease with which elemental oxygen picks up electrons is mirrored by the difficulty of removing electrons from oxygen in most oxides. Of the elements, only the very reactive fluorine can oxidize oxides to form oxygen gas.

Oxides, Peroxides, and Hydroxides

Compounds of the representative metals with oxygen fall into three categories: (1) **oxides**, containing oxide ions, O^{2-}; (2) **peroxides**, containing peroxides ions, ${O_2}^{2-}$, with oxygen-oxygen covalent single bonds and a very limited number of **superoxides**, containing superoxide ions, ${O_2}^{-}$, with oxygen-oxygen covalent bonds that have a bond order of $1\frac{1}{2}$, In addition, there are (3) **hydroxides**, containing hydroxide ions, OH^-. All representative metals form oxides. Some of the metals of group 2 also form peroxides, MO_2, and the metals of group 1 also form peroxides, M_2O_2, and superoxides, MO_2.

Oxides

It is possible to produce the oxides of most representative metals by heating the corresponding hydroxides (forming the oxide and gaseous water) or carbonates (forming the oxide and gaseous CO_2). Equations for example reactions are:

$$2Al(OH)_3(s) \xrightarrow{\Delta} Al_2O_3(s) + 3H_2O(g)$$

$$CaCO_3(s) \xrightarrow{\Delta} CaO(s) + CO_2(g)$$

However, alkali metal salts generally are very stable and do not decompose easily when heated. Alkali metal oxides result from the oxidation-reduction reactions created by heating nitrates or hydroxides with the metals. Equations for sample reactions are:

$$2KNO_3(s) + 10K(s) \xrightarrow{\Delta} 6K_2O(s) + N_2(g)$$

$$2LiOH(s) + 2Li(s) \xrightarrow{\Delta} 2Li_2O(s) + H_2(g)$$

With the exception of mercury(II) oxide, it is possible to produce the oxides of the metals of groups 2–15 by burning the corresponding metal in air. The heaviest member of each group, the member for which the inert pair effect is most pronounced, forms an oxide in which the oxidation state of the metal ion is two less than the group oxidation state (inert pair effect). Thus, Tl_2O, PbO, and Bi_2O_3 form when burning thallium, lead, and bismuth, respectively. The oxides of the lighter members of each group exhibit the group oxidation state. For example, SnO_2 forms from burning tin. Mercury(II) oxide, HgO, forms slowly when mercury is warmed below 500 °C; it decomposes at higher temperatures.

Burning the members of groups 1 and 2 in air is not a suitable way to form the oxides of these elements. These metals are reactive enough to combine with nitrogen in the air, so they form mixtures of oxides and ionic nitrides. Several also form peroxides or superoxides when heated in air.

Ionic oxides all contain the oxide ion, a very powerful hydrogen ion acceptor. With the exception of the very insoluble aluminum oxide, Al_2O_3, tin(IV), SnO_2, and lead(IV), PbO_2, the oxides of the representative metals react with acids to form salts. Some equations for these reactions are:

$$Na_2O + 2HNO_3(aq) \longrightarrow 2NaNO_3(aq) + H_2O(l)$$

$$CaO(s) + 2HCL(aq) \longrightarrow CaCl_2(aq) + H_2O(l)$$

$$SnO(s) + 2HClO_4(aq) \longrightarrow Sn(ClO_4)_2(aq) + H_2O(l)$$

The oxides of the metals of groups 1 and 2 and of thallium(I) oxide react with water and form hydroxides. Examples of such reactions are:

$$Na_2O(s) + H_2O(l) \longrightarrow NaOH(aq)$$

$$CaO(s) + H_2O(l) \longrightarrow Ca(OH)_2(aq)$$

$$Tl_2O(s) + H_2O(aq) \longrightarrow 2TlOH(aq)$$

The oxides of the alkali metals have little industrial utility, unlike magnesium oxide, calcium oxide, and aluminum oxide. Magnesium oxide is important in making firebrick, crucibles, furnace linings, and thermal insulation—applications that require chemical and thermal stability. Calcium oxide, sometimes called *quicklime* or lime in the industrial market, is very reactive, and its principal uses reflect its reactivity. Pure calcium oxide emits an intense white light when heated to a high temperature (as illustrated in Figure 18.43). Blocks of calcium oxide heated by gas flames were the stage lights in theaters before electricity was available. This is the source of the phrase “in the limelight.”

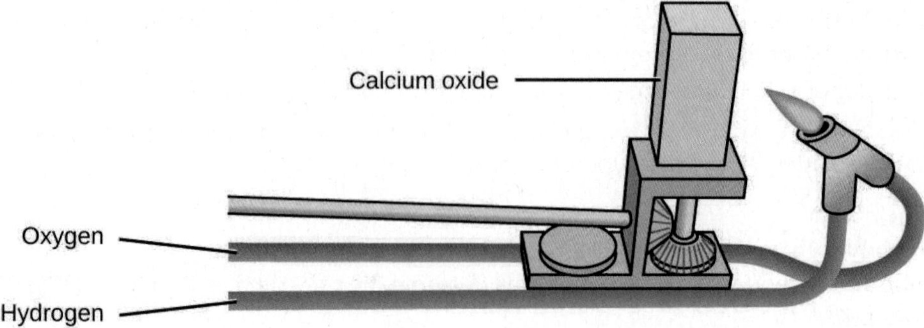

FIGURE 18.43 Calcium oxide has many industrial uses. When it is heated at high temperatures, it emits an intense white light.

Calcium oxide and calcium hydroxide are inexpensive bases used extensively in chemical processing, although most of the useful products prepared from them do not contain calcium. Calcium oxide, CaO, is made by heating calcium carbonate, $CaCO_3$, which is widely and inexpensively available as limestone or oyster shells:

$$CaCO_3(s) \longrightarrow CaO(s) + CO_2(g)$$

Although this decomposition reaction is reversible, it is possible to obtain a 100% yield of CaO by allowing the CO_2 to escape. It is possible to prepare calcium hydroxide by the familiar acid-base reaction of a soluble metal oxide with water:

$$CaO(s) + H_2O(l) \longrightarrow Ca(OH)_2(s)$$

Both CaO and $Ca(OH)_2$ are useful as bases; they accept protons and neutralize acids.

Alumina (Al_2O_3) occurs in nature as the mineral corundum, a very hard substance used as an abrasive for grinding and polishing. Corundum is important to the jewelry trade as ruby and sapphire. The color of ruby is due to the presence of a small amount of chromium; other impurities produce the wide variety of colors possible for sapphires. Artificial rubies and sapphires are now manufactured by melting aluminum oxide (melting point = 2050 °C) with small amounts of oxides to produce the desired colors and cooling the melt in such a way as to produce large crystals. Ruby lasers use synthetic ruby crystals.

Zinc oxide, ZnO, was a useful white paint pigment; however, pollutants tend to discolor the compound. The compound is also important in the manufacture of automobile tires and other rubber goods, and in the preparation of medicinal ointments. For example, zinc-oxide-based sunscreens, as shown in Figure 18.44, help prevent sunburn. The zinc oxide in these sunscreens is present in the form of very small grains known as nanoparticles. Lead dioxide is a constituent of charged lead storage batteries. Lead(IV) tends to revert to the more stable lead(II) ion by gaining two electrons, so lead dioxide is a powerful oxidizing agent.

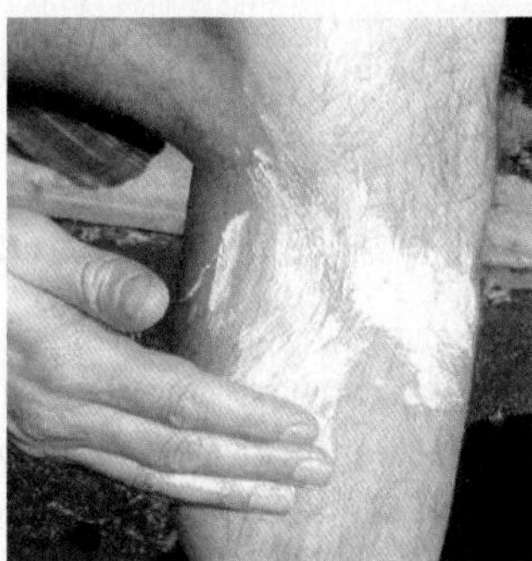

FIGURE 18.44 Zinc oxide protects exposed skin from sunburn. (credit: modification of work by "osseous"/Flickr)

Peroxides and Superoxides

Peroxides and superoxides are strong oxidizers and are important in chemical processes. Hydrogen peroxide, H_2O_2, prepared from metal peroxides, is an important bleach and disinfectant. Peroxides and superoxides form when the metal or metal oxides of groups 1 and 2 react with pure oxygen at elevated temperatures. Sodium peroxide and the peroxides of calcium, strontium, and barium form by heating the corresponding metal or metal oxide in pure oxygen:

$$2Na(s) + O_2(g) \xrightarrow{\Delta} Na_2O_2(s)$$

$$2Na_2O(s) + O_2(g) \xrightarrow{\Delta} 2Na_2O_2(s)$$

$$2SrO(s) + O_2(g) \xrightarrow{\Delta} 2SrO_2(s)$$

The peroxides of potassium, rubidium, and cesium can be prepared by heating the metal or its oxide in a carefully controlled amount of oxygen:

$$2K(s) + O_2(g) \longrightarrow K_2O_2(s) \qquad (2\text{ mol K per mol } O_2)$$

With an excess of oxygen, the superoxides KO_2, RbO_2, and CsO_2 form. For example:

$$K(s) + O_2(g) \longrightarrow KO_2(s) \qquad (1\text{ mol K per mol } O_2)$$

The stability of the peroxides and superoxides of the alkali metals increases as the size of the cation increases.

Hydroxides

Hydroxides are compounds that contain the OH^- ion. It is possible to prepare these compounds by two general types of reactions. Soluble metal hydroxides can be produced by the reaction of the metal or metal oxide with water. Insoluble metal hydroxides form when a solution of a soluble salt of the metal combines with a solution containing hydroxide ions.

With the exception of beryllium and magnesium, the metals of groups 1 and 2 react with water to form hydroxides and hydrogen gas. Examples of such reactions include:

$$2\text{Li}(s) + 2\text{H}_2\text{O}(l) \longrightarrow 2\text{LiOH}(aq) + \text{H}_2(g)$$

$$\text{Ca}(s) + 2\text{H}_2\text{O}(l) \longrightarrow \text{Ca(OH)}_2(aq) + \text{H}_2(g)$$

However, these reactions can be violent and dangerous; therefore, it is preferable to produce soluble metal hydroxides by the reaction of the respective oxide with water:

$$\text{Li}_2\text{O}(s) + \text{H}_2\text{O}(l) \longrightarrow 2\text{LiOH}(aq)$$

$$\text{CaO}(s) + \text{H}_2\text{O}(l) \longrightarrow \text{Ca(OH)}_2(aq)$$

Most metal oxides are **base anhydrides**. This is obvious for the soluble oxides because they form metal hydroxides. Most other metal oxides are insoluble and do not form hydroxides in water; however, they are still base anhydrides because they will react with acids.

It is possible to prepare the insoluble hydroxides of beryllium, magnesium, and other representative metals by the addition of sodium hydroxide to a solution of a salt of the respective metal. The net ionic equations for the reactions involving a magnesium salt, an aluminum salt, and a zinc salt are:

$$\text{Mg}^{2+}(aq) + 2\text{OH}^-(aq) \longrightarrow \text{Mg(OH)}_2(s)$$

$$\text{Al}^{3+}(aq) + 3\text{OH}^-(aq) \longrightarrow \text{Al(OH)}_3(s)$$

$$\text{Zn}^{2+}(aq) + 2\text{OH}^-(aq) \longrightarrow \text{Zn(OH)}_2(s)$$

An excess of hydroxide must be avoided when preparing aluminum, gallium, zinc, and tin(II) hydroxides, or the hydroxides will dissolve with the formation of the corresponding complex ions: $\text{Al(OH)}_4{}^-$, $\text{Ga(OH)}_4{}^-$, $\text{Zn(OH)}_4{}^{2-}$, and $\text{Sn(OH)}_3{}^-$ (see Figure 18.45). The important aspect of complex ions for this chapter is that they form by a Lewis acid-base reaction with the metal being the Lewis acid.

(a)

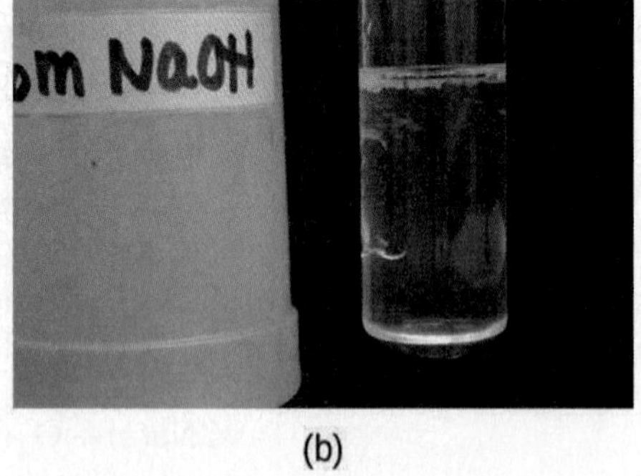

(b)

FIGURE 18.45 (a) Mixing solutions of NaOH and $\text{Zn(NO}_3)_2$ produces a white precipitate of Zn(OH)_2. (b) Addition of an excess of NaOH results in dissolution of the precipitate. (credit: modification of work by Mark Ott)

Industry uses large quantities of sodium hydroxide as a cheap, strong base. Sodium chloride is the starting material for the production of NaOH because NaCl is a less expensive starting material than the oxide. Sodium hydroxide is among the top 10 chemicals in production in the United States, and this production was almost entirely by electrolysis of solutions of sodium chloride. This process is the **chlor-alkali process**, and it is the primary method for producing chlorine.

Sodium hydroxide is an ionic compound and melts without decomposition. It is very soluble in water, giving off a great deal of heat and forming very basic solutions: 40 grams of sodium hydroxide dissolves in only 60 grams of water at 25 °C. Sodium hydroxide is employed in the production of other sodium compounds and is used to neutralize acidic solutions during the production of other chemicals such as petrochemicals and polymers.

Many of the applications of hydroxides are for the neutralization of acids (such as the antacid shown in Figure 18.46) and for the preparation of oxides by thermal decomposition. An aqueous suspension of magnesium hydroxide constitutes the antacid milk of magnesia. Because of its ready availability (from the reaction of water with calcium oxide prepared by the decomposition of limestone, $CaCO_3$), low cost, and activity, calcium hydroxide is used extensively in commercial applications needing a cheap, strong base. The reaction of hydroxides with appropriate acids is also used to prepare salts.

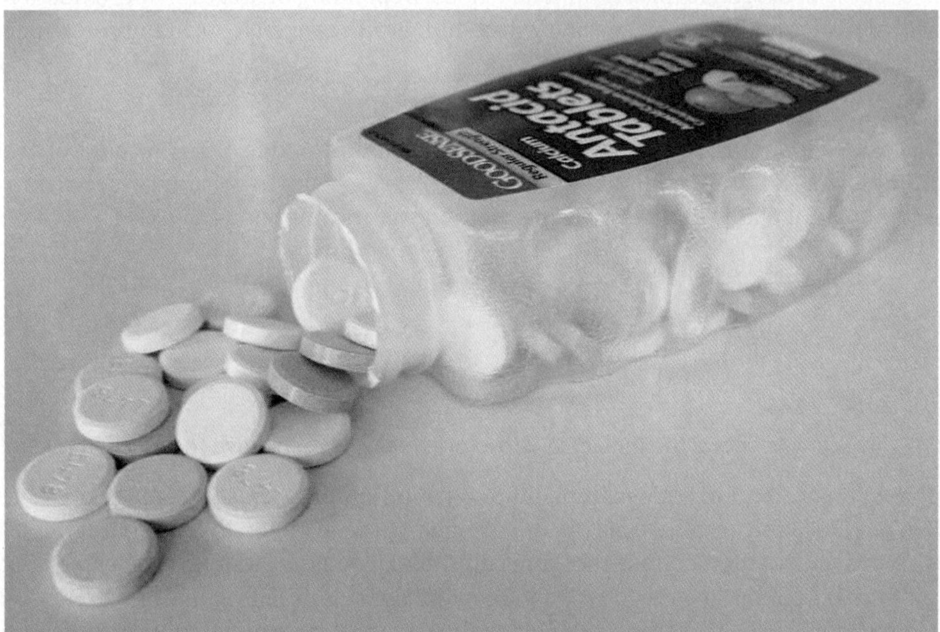

FIGURE 18.46 Calcium carbonate, $CaCO_3$, can be consumed in the form of an antacid to neutralize the effects of acid in your stomach. (credit: “Midnightcomm”/Wikimedia Commons)

Chemistry in Everyday Life

The Chlor-Alkali Process

Although they are very different chemically, there is a link between chlorine and sodium hydroxide because there is an important electrochemical process that produces the two chemicals simultaneously. The process known as the chlor-alkali process, utilizes sodium chloride, which occurs in large deposits in many parts of the world. This is an electrochemical process to oxidize chloride ion to chlorine and generate sodium hydroxide.

Passing a direct current of electricity through a solution of NaCl causes the chloride ions to migrate to the positive electrode where oxidation to gaseous chlorine occurs when the ion gives up an electron to the electrode:

$$2Cl^-(aq) \longrightarrow Cl_2(g) + 2e^- \qquad \text{(at the positive electrode)}$$

The electrons produced travel through the outside electrical circuit to the negative electrode. Although the positive sodium ions migrate toward this negative electrode, metallic sodium does not form because sodium ions are too difficult to reduce under the conditions used. (Recall that metallic sodium is active enough to react with water and hence, even if produced, would immediately react with water to produce sodium ions again.) Instead, water molecules pick up electrons from the electrode and undergo reduction to form hydrogen gas and hydroxide ions:

$$2H_2O(l) + 2e^- \text{ (from the negative electrode)} \longrightarrow H_2(g) + 2OH^-(aq)$$

The overall result is the conversion of the aqueous solution of NaCl to an aqueous solution of NaOH, gaseous Cl_2, and gaseous H_2:

$$2Na^+(aq) + 2Cl^-(aq) + 2H_2O(l) \xrightarrow{\text{electrolysis}} 2Na^+(aq) + 2OH^-(aq) + Cl_2(g) + H_2(g)$$

Nonmetal Oxygen Compounds

Most nonmetals react with oxygen to form nonmetal oxides. Depending on the available oxidation states for the element, a variety of oxides might form. Fluorine will combine with oxygen to form fluorides such as OF_2, where the oxygen has a 2+-oxidation state.

Sulfur Oxygen Compounds

The two common oxides of sulfur are sulfur dioxide, SO_2, and sulfur trioxide, SO_3. The odor of burning sulfur comes from sulfur dioxide. Sulfur dioxide, shown in Figure 18.47, occurs in volcanic gases and in the atmosphere near industrial plants that burn fuel containing sulfur compounds.

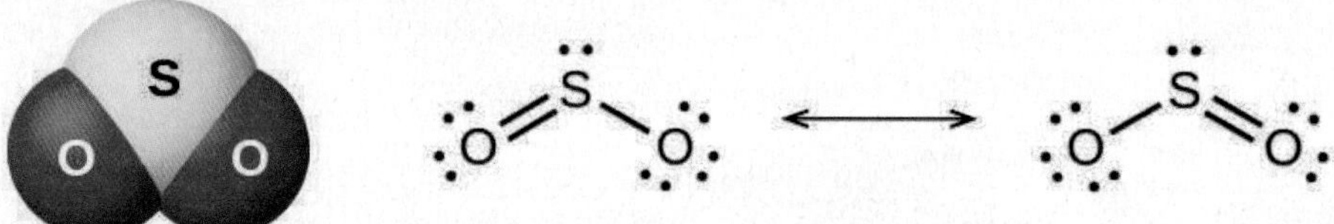

FIGURE 18.47 This image shows the molecular structure (left) and resonance forms (right) of sulfur dioxide.

Commercial production of sulfur dioxide is from either burning sulfur or roasting sulfide ores such as ZnS, FeS_2, and Cu_2S in air. (Roasting, which forms the metal oxide, is the first step in the separation of many metals from their ores.) A convenient method for preparing sulfur dioxide in the laboratory is by the action of a strong acid on either sulfite salts containing the $SO_3{}^{2-}$ ion or hydrogen sulfite salts containing $HSO_3{}^-$. Sulfurous acid, H_2SO_3, forms first, but quickly decomposes into sulfur dioxide and water. Sulfur dioxide also forms when many reducing agents react with hot, concentrated sulfuric acid. Sulfur trioxide forms slowly when heating sulfur dioxide and oxygen together, and the reaction is exothermic:

$$2SO_2(g) + O_2(g) \longrightarrow 2SO_3(g) \qquad \Delta H° = -197.8 \text{ kJ}$$

Sulfur dioxide is a gas at room temperature, and the SO_2 molecule is bent. Sulfur trioxide melts at 17 °C and boils at 43 °C. In the vapor state, its molecules are single SO_3 units (shown in Figure 18.48), but in the solid state, SO_3 exists in several polymeric forms.

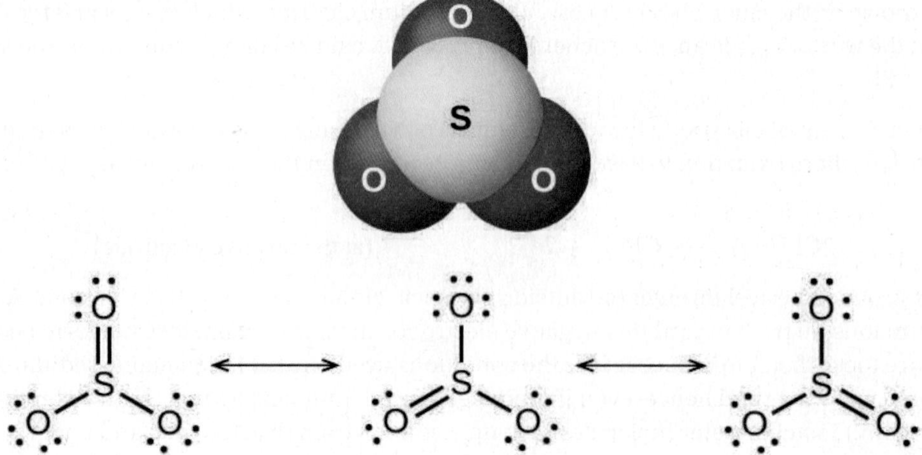

FIGURE 18.48 This image shows the structure (top) of sulfur trioxide in the gas phase and its resonance forms (bottom).

The sulfur oxides react as Lewis acids with many oxides and hydroxides in Lewis acid-base reactions, with the formation of **sulfites** or **hydrogen sulfites**, and **sulfates** or **hydrogen sulfates**, respectively.

Halogen Oxygen Compounds

The halogens do not react directly with oxygen, but it is possible to prepare binary oxygen-halogen compounds by the reactions of the halogens with oxygen-containing compounds. Oxygen compounds with chlorine, bromine, and iodine are oxides because oxygen is the more electronegative element in these compounds. On the other hand, fluorine compounds with oxygen are fluorides because fluorine is the more electronegative element.

As a class, the oxides are extremely reactive and unstable, and their chemistry has little practical importance. Dichlorine oxide, formally called dichlorine monoxide, and chlorine dioxide, both shown in Figure 18.49, are the only commercially important compounds. They are important as bleaching agents (for use with pulp and flour) and for water treatment.

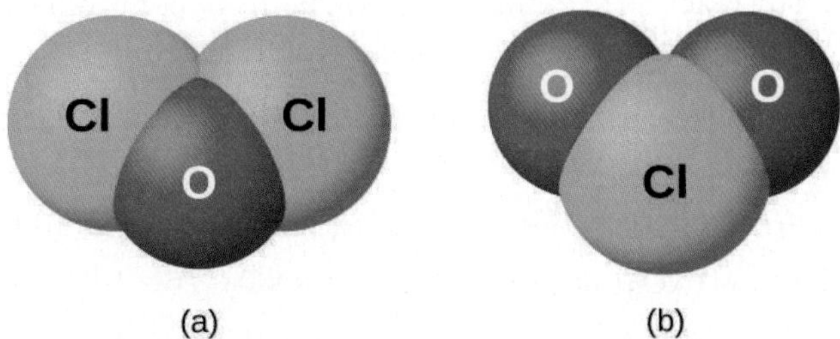

FIGURE 18.49 This image shows the structures of the (a) Cl_2O and (b) ClO_2 molecules.

Nonmetal Oxyacids and Their Salts

Nonmetal oxides form acids when allowed to react with water; these are acid anhydrides. The resulting oxyanions can form salts with various metal ions.

Nitrogen Oxyacids and Salts

Nitrogen pentaoxide, N_2O_5, and NO_2 react with water to form nitric acid, HNO_3. Alchemists, as early as the eighth century, knew nitric acid (shown in Figure 18.50) as *aqua fortis* (meaning "strong water"). The acid was useful in the separation of gold from silver because it dissolves silver but not gold. Traces of nitric acid occur in the atmosphere after thunderstorms, and its salts are widely distributed in nature. There are tremendous deposits of Chile saltpeter, $NaNO_3$, in the desert region near the boundary of Chile and Peru. Bengal saltpeter, KNO_3, occurs in India and in other countries of the Far East.

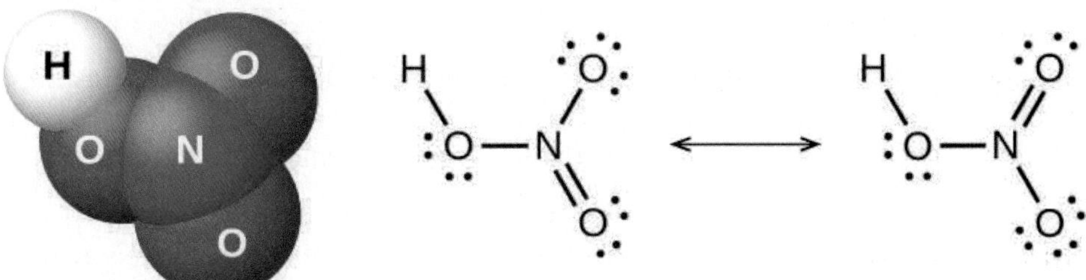

FIGURE 18.50 This image shows the molecular structure (left) of nitric acid, HNO_3 and its resonance forms (right).

In the laboratory, it is possible to produce nitric acid by heating a nitrate salt (such as sodium or potassium nitrate) with concentrated sulfuric acid:

$$NaNO_3(s) + H_2SO_4(l) \xrightarrow{\Delta} NaHSO_4(s) + HNO_3(g)$$

The **Ostwald process** is the commercial method for producing nitric acid. This process involves the oxidation of ammonia to nitric oxide, NO; oxidation of nitric oxide to nitrogen dioxide, NO_2; and further oxidation and hydration of nitrogen dioxide to form nitric acid:

$$4NH_3(g) + 5O_2(g) \longrightarrow 4NO(g) + 6H_2O(g)$$

$$2NO(g) + O_2(g) \longrightarrow 2NO_2(g)$$
$$3NO_2(g) + H_2O(l) \longrightarrow 2HNO_3(aq) + NO(g)$$

Or

$$4NO_2(g) + O_2(g) + 2H_2O(g) \longrightarrow 4HNO_3(l)$$

Pure nitric acid is a colorless liquid. However, it is often yellow or brown in color because NO_2 forms as the acid decomposes. Nitric acid is stable in aqueous solution; solutions containing 68% of the acid are commercially available concentrated nitric acid. It is both a strong oxidizing agent and a strong acid.

The action of nitric acid on a metal rarely produces H_2 (by reduction of H^+) in more than small amounts. Instead, the reduction of nitrogen occurs. The products formed depend on the concentration of the acid, the activity of the metal, and the temperature. Normally, a mixture of nitrates, nitrogen oxides, and various reduction products form. Less active metals such as copper, silver, and lead reduce concentrated nitric acid primarily to nitrogen dioxide. The reaction of dilute nitric acid with copper produces NO. In each case, the nitrate salts of the metals crystallize upon evaporation of the resultant solutions.

Nonmetallic elements, such as sulfur, carbon, iodine, and phosphorus, undergo oxidation by concentrated nitric acid to their oxides or oxyacids, with the formation of NO_2:

$$S(s) + 6HNO_3(aq) \longrightarrow H_2SO_4(aq) + 6NO_2(g) + 2H_2O(l)$$
$$C(s) + 4HNO_3(aq) \longrightarrow CO_2(g) + 4NO_2(g) + 2H_2O(l)$$

Nitric acid oxidizes many compounds; for example, concentrated nitric acid readily oxidizes hydrochloric acid to chlorine and chlorine dioxide. A mixture of one part concentrated nitric acid and three parts concentrated hydrochloric acid (called *aqua regia*, which means royal water) reacts vigorously with metals. This mixture is particularly useful in dissolving gold, platinum, and other metals that are more difficult to oxidize than hydrogen. A simplified equation to represent the action of *aqua regia* on gold is:

$$Au(s) + 4HCl(aq) + 3HNO_3(aq) \longrightarrow HAuCl_4(aq) + 3NO_2(g) + 3H_2O(l)$$

LINK TO LEARNING

Although gold is generally unreactive, you can watch a video (http://openstax.org/l/16gold) of the complex mixture of compounds present in *aqua regia* dissolving it into solution.

Nitrates, salts of nitric acid, form when metals, oxides, hydroxides, or carbonates react with nitric acid. Most nitrates are soluble in water; indeed, one of the significant uses of nitric acid is to prepare soluble metal nitrates.

Nitric acid finds extensive use in the laboratory and in chemical industries as a strong acid and strong oxidizing agent. It is important in the manufacture of explosives, dyes, plastics, and drugs. Salts of nitric acid (nitrates) are valuable as fertilizers. Gunpowder is a mixture of potassium nitrate, sulfur, and charcoal.

The reaction of N_2O_3 with water gives a pale blue solution of nitrous acid, HNO_2. However, HNO_2 (shown in Figure 18.51) is easier to prepare by the addition of an acid to a solution of nitrite; nitrous acid is a weak acid, so the nitrite ion is basic in aqueous solution:

$$NO_2{}^-(aq) + H_3O^+(aq) \longrightarrow HNO_2(aq) + H_2O(l)$$

Nitrous acid is very unstable and exists only in solution. It disproportionates slowly at room temperature (rapidly when heated) into nitric acid and nitric oxide. Nitrous acid is an active oxidizing agent with strong reducing agents, and strong oxidizing agents oxidize it to nitric acid.

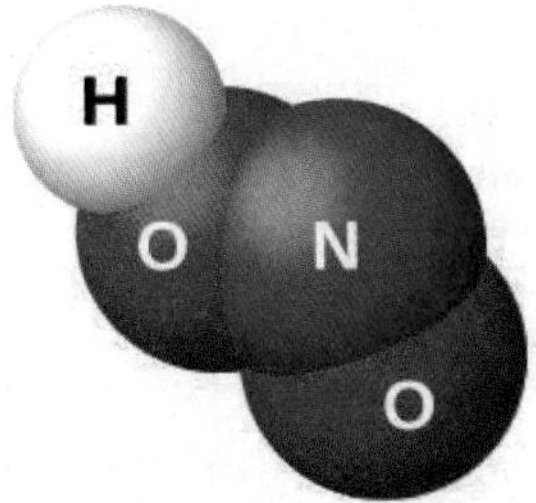

FIGURE 18.51 This image shows the molecular structure of a molecule of nitrous acid, HNO_2.

Sodium nitrite, $NaNO_2$, is an additive to meats such as hot dogs and cold cuts. The nitrite ion has two functions. It limits the growth of bacteria that can cause food poisoning, and it prolongs the meat's retention of its red color. The addition of sodium nitrite to meat products is controversial because nitrous acid reacts with certain organic compounds to form a class of compounds known as nitrosamines. Nitrosamines produce cancer in laboratory animals. This has prompted the FDA to limit the amount of $NaNO_2$ in foods.

The nitrites are much more stable than the acid, but nitrites, like nitrates, can explode. Nitrites, like nitrates, are also soluble in water ($AgNO_2$ is only slightly soluble).

Phosphorus Oxyacids and Salts

Pure orthophosphoric acid, H_3PO_4 (shown in Figure 18.52), forms colorless, deliquescent crystals that melt at 42 °C. The common name of this compound is phosphoric acid, and is commercially available as a viscous 82% solution known as syrupy phosphoric acid. One use of phosphoric acid is as an additive to many soft drinks.

One commercial method of preparing orthophosphoric acid is to treat calcium phosphate rock with concentrated sulfuric acid:

$$Ca_3(PO_4)_2(s) + 3H_2SO_4(aq) \longrightarrow 2H_3PO_4(aq) + 3CaSO_4(s)$$

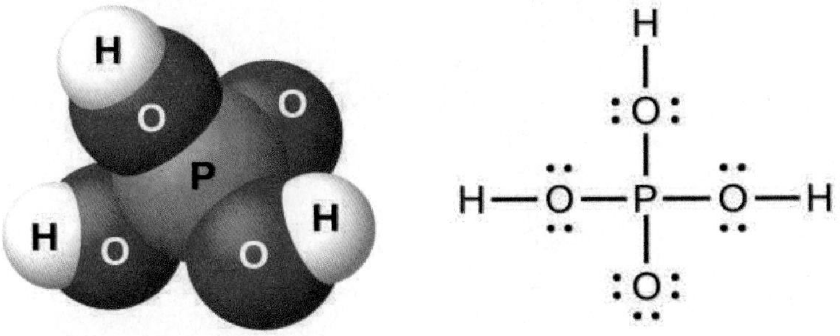

FIGURE 18.52 Orthophosphoric acid, H_3PO_4, is colorless when pure and has this molecular (left) and Lewis structure (right).

Dilution of the products with water, followed by filtration to remove calcium sulfate, gives a dilute acid solution contaminated with calcium dihydrogen phosphate, $Ca(H_2PO_4)_2$, and other compounds associated with calcium phosphate rock. It is possible to prepare pure orthophosphoric acid by dissolving P_4O_{10} in water.

The action of water on P_4O_6, PCl_3, PBr_3, or PI_3 forms phosphorous acid, H_3PO_3 (shown in Figure 18.53). The best method for preparing pure phosphorous acid is by hydrolyzing phosphorus trichloride:

$$PCl_3(l) + 3H_2O(l) \longrightarrow H_3PO_3(aq) + 3HCl(g)$$

Heating the resulting solution expels the hydrogen chloride and leads to the evaporation of water. When sufficient water evaporates, white crystals of phosphorous acid will appear upon cooling. The crystals are deliquescent, very soluble in water, and have an odor like that of garlic. The solid melts at 70.1 °C and decomposes at about 200 °C by disproportionation into phosphine and orthophosphoric acid:

$$4H_3PO_3(l) \longrightarrow PH_3(g) + 3H_3PO_4(l)$$

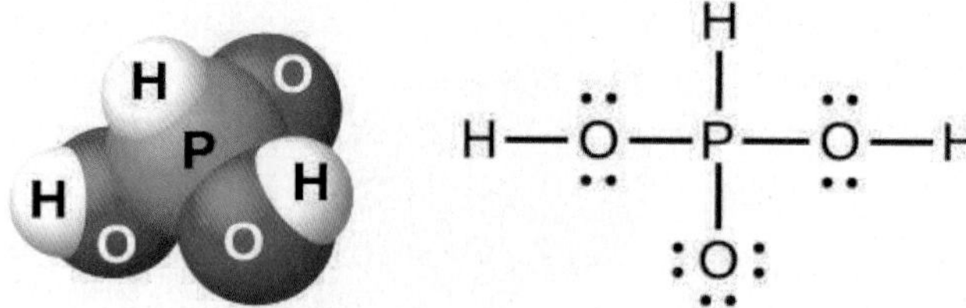

FIGURE 18.53 In a molecule of phosphorous acid, H_3PO_3, only the two hydrogen atoms bonded to an oxygen atom are acidic.

Phosphorous acid forms only two series of salts, which contain the dihydrogen phosphite ion, $H_2PO_3{}^-$, or the hydrogen phosphate ion, $HPO_3{}^{2-}$, respectively. It is not possible to replace the third atom of hydrogen because it is not very acidic, as it is not easy to ionize the P-H bond.

Sulfur Oxyacids and Salts

The preparation of sulfuric acid, H_2SO_4 (shown in Figure 18.54), begins with the oxidation of sulfur to sulfur trioxide and then converting the trioxide to sulfuric acid. Pure sulfuric acid is a colorless, oily liquid that freezes at 10.5 °C. It fumes when heated because the acid decomposes to water and sulfur trioxide. The heating process causes the loss of more sulfur trioxide than water, until reaching a concentration of 98.33% acid. Acid of this concentration boils at 338 °C without further change in concentration (a constant boiling solution) and is commercially concentrated H_2SO_4. The amount of sulfuric acid used in industry exceeds that of any other manufactured compound.

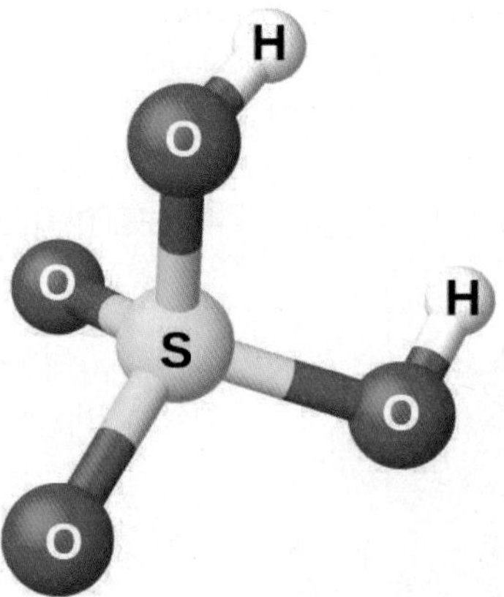

FIGURE 18.54 Sulfuric acid has a tetrahedral molecular structure.

The strong affinity of concentrated sulfuric acid for water makes it a good dehydrating agent. It is possible to dry gases and immiscible liquids that do not react with the acid by passing them through the acid.

Sulfuric acid is a strong diprotic acid that ionizes in two stages. In aqueous solution, the first stage is essentially complete. The secondary ionization is not nearly so complete, and $HSO_4{}^-$ is a moderately strong acid (about 25% ionized in solution of a $HSO_4{}^-$ salt: $K_a = 1.2 \times 10^{-2}$).

Being a diprotic acid, sulfuric acid forms both sulfates, such as Na_2SO_4, and hydrogen sulfates, such as $NaHSO_4$. Most sulfates are soluble in water; however, the sulfates of barium, strontium, calcium, and lead are only slightly soluble in water.

Among the important sulfates are $Na_2SO_4{\cdot}10H_2O$ and Epsom salts, $MgSO_4{\cdot}7H_2O$. Because the $HSO_4{}^-$ ion is an acid, hydrogen sulfates, such as $NaHSO_4$, exhibit acidic behavior, and this compound is the primary ingredient in some household cleansers.

Hot, concentrated sulfuric acid is an oxidizing agent. Depending on its concentration, the temperature, and the strength of the reducing agent, sulfuric acid oxidizes many compounds and, in the process, undergoes reduction to SO_2, $HSO_3{}^-$, $SO_3{}^{2-}$, S, H_2S, or S^{2-}.

Sulfur dioxide dissolves in water to form a solution of sulfurous acid, as expected for the oxide of a nonmetal.

Sulfurous acid is unstable, and it is not possible to isolate anhydrous H_2SO_3. Heating a solution of sulfurous acid expels the sulfur dioxide. Like other diprotic acids, sulfurous acid ionizes in two steps: The hydrogen sulfite ion, HSO_3^{-}, and the sulfite ion, SO_3^{2-}, form. Sulfurous acid is a moderately strong acid. Ionization is about 25% in the first stage, but it is much less in the second ($K_{a1} = 1.2 \times 10^{-2}$ and $K_{a2} = 6.2 \times 10^{-8}$).

In order to prepare solid sulfite and hydrogen sulfite salts, it is necessary to add a stoichiometric amount of a base to a sulfurous acid solution and then evaporate the water. These salts also form from the reaction of SO_2 with oxides and hydroxides. Heating solid sodium hydrogen sulfite forms sodium sulfite, sulfur dioxide, and water:

$$2NaHSO_3(s) \xrightarrow{\Delta} Na_2SO_3(s) + SO_2(g) + H_2O(l)$$

Strong oxidizing agents can oxidize sulfurous acid. Oxygen in the air oxidizes it slowly to the more stable sulfuric acid:

$$2H_2SO_3(aq) + O_2(g) + 2H_2O(l) \xrightarrow{\Delta} 2H_3O^+(aq) + 2HSO_4^-(aq)$$

Solutions of sulfites are also very susceptible to air oxidation to produce sulfates. Thus, solutions of sulfites always contain sulfates after exposure to air.

Halogen Oxyacids and Their Salts

The compounds HXO, HXO_2, HXO_3, and HXO_4, where X represents Cl, Br, or I, are the hypohalous, halous, halic, and perhalic acids, respectively. The strengths of these acids increase from the hypohalous acids, which are very weak acids, to the perhalic acids, which are very strong. Table 18.2 lists the known acids, and, where known, their pK_a values are given in parentheses.

Oxyacids of the Halogens

Name	Fluorine	Chlorine	Bromine	Iodine
hypohalous	HOF	HOCl (7.5)	HOBr (8.7)	HOI (11)
halous		$HClO_2$ (2.0)		
halic		$HClO_3$	$HBrO_3$	HIO_3 (0.8)
perhalic		$HClO_4$	$HBrO_4$	HIO_4 (1.6)
paraperhalic				H_5IO_6 (1.6)

TABLE 18.2

The only known oxyacid of fluorine is the very unstable hypofluorous acid, HOF, which is prepared by the reaction of gaseous fluorine with ice:

$$F_2(g) + H_2O(s) \longrightarrow HOF(g) + HF(g)$$

The compound is very unstable and decomposes above −40 °C. This compound does not ionize in water, and there are no known salts. It is uncertain whether the name hypofluorous acid is even appropriate for HOF; a more appropriate name might be hydrogen hypofluorite.

The reactions of chlorine and bromine with water are analogous to that of fluorine with ice, but these reactions do not go to completion, and mixtures of the halogen and the respective hypohalous and hydrohalic acids result. Other than HOF, the hypohalous acids only exist in solution. The hypohalous acids are all very weak acids; however, HOCl is a stronger acid than HOBr, which, in turn, is stronger than HOI.

The addition of base to solutions of the hypohalous acids produces solutions of salts containing the basic hypohalite ions, OX^-. It is possible to isolate these salts as solids. All of the hypohalites are unstable with respect to disproportionation in solution, but the reaction is slow for hypochlorite. Hypobromite and hypoiodite disproportionate rapidly, even in the cold:

$$3XO^-(aq) \longrightarrow 2X^-(aq) + XO_3{}^-(aq)$$

Sodium hypochlorite is an inexpensive bleach (Clorox) and germicide. The commercial preparation involves the electrolysis of cold, dilute, aqueous sodium chloride solutions under conditions where the resulting chlorine and hydroxide ion can react. The net reaction is:

$$Cl^-(aq) + H_2O(l) \xrightarrow{\text{electrical energy}} ClO^-(aq) + H_2(g)$$

The only definitely known halous acid is chlorous acid, $HClO_2$, obtained by the reaction of barium chlorite with dilute sulfuric acid:

$$Ba(ClO_2)_2(aq) + H_2SO_4(aq) \longrightarrow BaSO_4(s) + 2HClO_2(aq)$$

Filtering the insoluble barium sulfate leaves a solution of $HClO_2$. Chlorous acid is not stable; it slowly decomposes in solution to yield chlorine dioxide, hydrochloric acid, and water. Chlorous acid reacts with bases to give salts containing the chlorite ion (shown in Figure 18.55). Sodium chlorite finds an extensive application in the bleaching of paper because it is a strong oxidizing agent and does not damage the paper.

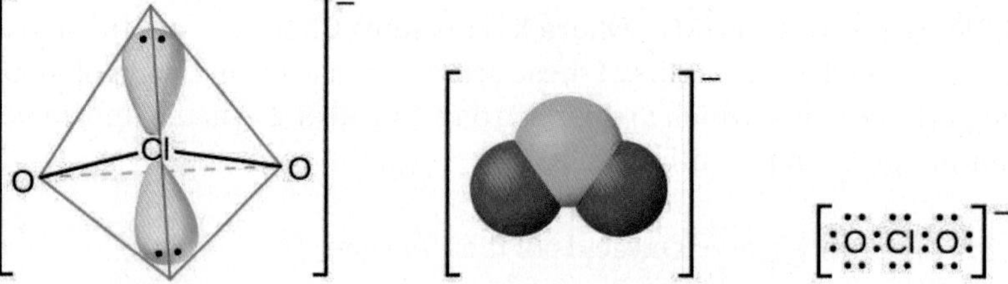

FIGURE 18.55 Chlorite ions, $ClO_2{}^-$, are produced when chlorous acid reacts with bases.

Chloric acid, $HClO_3$, and bromic acid, $HBrO_3$, are stable only in solution. The reaction of iodine with concentrated nitric acid produces stable white iodic acid, HIO_3:

$$I_2(s) + 10HNO_3(aq) \longrightarrow 2HIO_3(s) + 10NO_2(g) + 4H_2O(l)$$

It is possible to obtain the lighter halic acids from their barium salts by reaction with dilute sulfuric acid. The reaction is analogous to that used to prepare chlorous acid. All of the halic acids are strong acids and very active oxidizing agents. The acids react with bases to form salts containing chlorate ions (shown in Figure 18.56). Another preparative method is the electrochemical oxidation of a hot solution of a metal halide to form the appropriate metal chlorates. Sodium chlorate is a weed killer; potassium chlorate is used as an oxidizing agent.

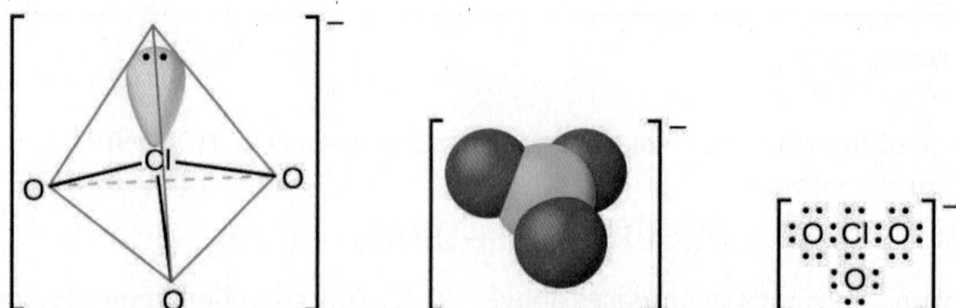

FIGURE 18.56 Chlorate ions, $ClO_3{}^-$, are produced when halic acids react with bases.

Perchloric acid, $HClO_4$, forms when treating a perchlorate, such as potassium perchlorate, with sulfuric acid under reduced pressure. The $HClO_4$ can be distilled from the mixture:

$$KClO_4(s) + H_2SO_4(aq) \longrightarrow HClO_4(g) + KHSO_4(s)$$

Dilute aqueous solutions of perchloric acid are quite stable thermally, but concentrations above 60% are unstable and dangerous. Perchloric acid and its salts are powerful oxidizing agents, as the very electronegative chlorine is more stable in a lower oxidation state than 7+. Serious explosions have occurred when heating concentrated solutions with easily oxidized substances. However, its reactions as an oxidizing agent are slow when perchloric acid is cold and dilute. The acid is among the strongest of all acids. Most salts containing the perchlorate ion (shown in Figure 18.57) are soluble. It is possible to prepare them from reactions of bases with perchloric acid and, commercially, by the electrolysis of hot solutions of their chlorides.

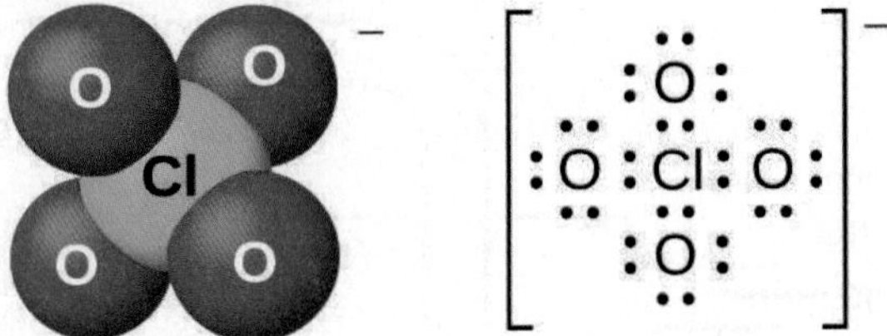

FIGURE 18.57 Perchlorate ions, ClO_4^-, can be produced when perchloric acid reacts with a base or by electrolysis of hot solutions of their chlorides.

Perbromate salts are difficult to prepare, and the best syntheses currently involve the oxidation of bromates in basic solution with fluorine gas followed by acidification. There are few, if any, commercial uses of this acid or its salts.

There are several different acids containing iodine in the 7+-oxidation state; they include metaperiodic acid, HIO_4, and paraperiodic acid, H_5IO_6. These acids are strong oxidizing agents and react with bases to form the appropriate salts.

18.10 Occurrence, Preparation, and Properties of Sulfur

LEARNING OBJECTIVES

By the end of this section, you will be able to:

- Describe the properties, preparation, and uses of sulfur

Sulfur exists in nature as elemental deposits as well as sulfides of iron, zinc, lead, and copper, and sulfates of sodium, calcium, barium, and magnesium. Hydrogen sulfide is often a component of natural gas and occurs in many volcanic gases, like those shown in Figure 18.58. Sulfur is a constituent of many proteins and is essential for life.

FIGURE 18.58 Volcanic gases contain hydrogen sulfide. (credit: Daniel Julie/Wikimedia Commons)

The **Frasch process**, illustrated in Figure 18.59, is important in the mining of free sulfur from enormous underground deposits in Texas and Louisiana. Superheated water (170 °C and 10 atm pressure) is forced down the outermost of three concentric pipes to the underground deposit. The hot water melts the sulfur. The innermost pipe conducts compressed air into the liquid sulfur. The air forces the liquid sulfur, mixed with air, to flow up through the outlet pipe. Transferring the mixture to large settling vats allows the solid sulfur to separate upon cooling. This sulfur is 99.5% to 99.9% pure and requires no purification for most uses.

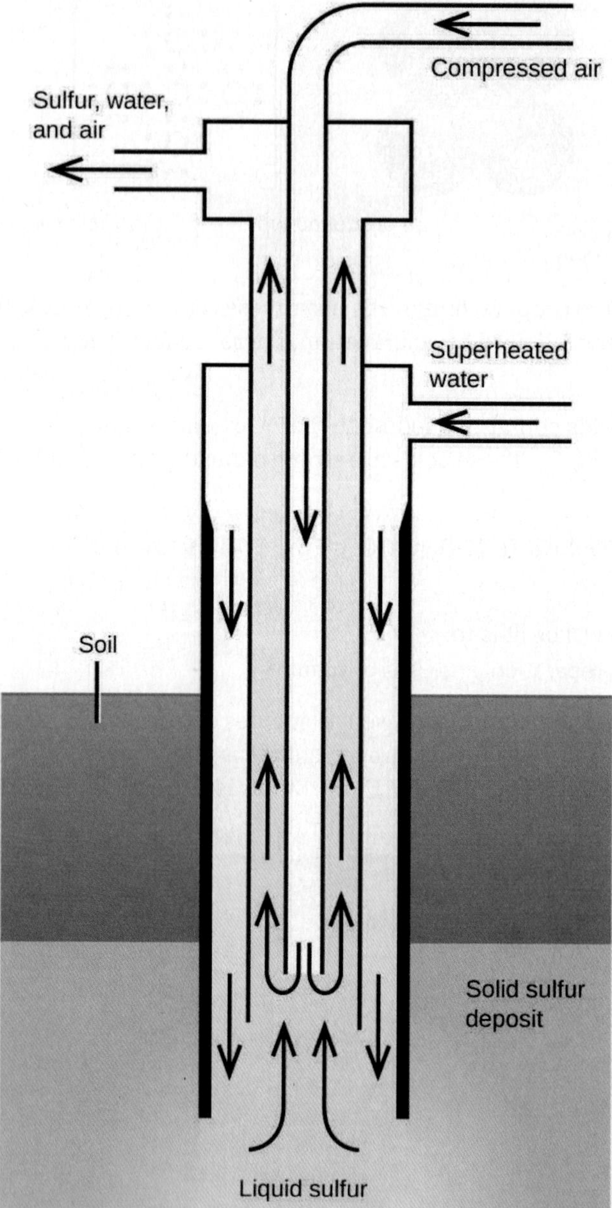

FIGURE 18.59 The Frasch process is used to mine sulfur from underground deposits.

Larger amounts of sulfur also come from hydrogen sulfide recovered during the purification of natural gas.

Sulfur exists in several allotropic forms. The stable form at room temperature contains eight-membered rings, and so the true formula is S_8. However, chemists commonly use S to simplify the coefficients in chemical equations; we will follow this practice in this book.

Like oxygen, which is also a member of group 16, sulfur exhibits a distinctly nonmetallic behavior. It oxidizes metals, giving a variety of binary sulfides in which sulfur exhibits a negative oxidation state (2−). Elemental sulfur oxidizes less electronegative nonmetals, and more electronegative nonmetals, such as oxygen and the halogens, will oxidize it. Other strong oxidizing agents also oxidize sulfur. For example, concentrated nitric acid oxidizes sulfur to the sulfate ion, with the concurrent formation of nitrogen(IV) oxide:

$$S(s) + 6HNO_3(aq) \longrightarrow 2H_3O^+(aq) + SO_4{}^{2-}(aq) + 6NO_2(g)$$

The chemistry of sulfur with an oxidation state of 2− is similar to that of oxygen. Unlike oxygen, however, sulfur forms many compounds in which it exhibits positive oxidation states.

18.11 Occurrence, Preparation, and Properties of Halogens

LEARNING OBJECTIVES

By the end of this section, you will be able to:

- Describe the preparation, properties, and uses of halogens
- Describe the properties, preparation, and uses of halogen compounds

The elements in group 17 are the halogens. These are the elements fluorine, chlorine, bromine, iodine, and astatine. These elements are too reactive to occur freely in nature, but their compounds are widely distributed. Chlorides are the most abundant; although fluorides, bromides, and iodides are less common, they are reasonably available. In this section, we will examine the occurrence, preparation, and properties of halogens. Next, we will examine halogen compounds with the representative metals followed by an examination of the interhalogens. This section will conclude with some applications of halogens.

Occurrence and Preparation

All of the halogens occur in seawater as halide ions. The concentration of the chloride ion is 0.54 *M*; that of the other halides is less than 10^{-4} *M*. Fluoride also occurs in minerals such as CaF_2, $Ca(PO_4)_3F$, and Na_3AlF_6. Chloride also occurs in the Great Salt Lake and the Dead Sea, and in extensive salt beds that contain NaCl, KCl, or $MgCl_2$. Part of the chlorine in your body is present as hydrochloric acid, which is a component of stomach acid. Bromine compounds occur in the Dead Sea and underground brines. Iodine compounds are found in small quantities in Chile saltpeter, underground brines, and sea kelp. Iodine is essential to the function of the thyroid gland.

The best sources of halogens (except iodine) are halide salts. It is possible to oxidize the halide ions to free diatomic halogen molecules by various methods, depending on the ease of oxidation of the halide ion. Fluoride is the most difficult to oxidize, whereas iodide is the easiest.

The major method for preparing fluorine is electrolytic oxidation. The most common electrolysis procedure is to use a molten mixture of potassium hydrogen fluoride, KHF_2, and anhydrous hydrogen fluoride. Electrolysis causes HF to decompose, forming fluorine gas at the anode and hydrogen at the cathode. It is necessary to keep the two gases separated to prevent their explosive recombination to reform hydrogen fluoride.

Most commercial chlorine comes from the electrolysis of the chloride ion in aqueous solutions of sodium chloride; this is the chlor-alkali process discussed previously. Chlorine is also a product of the electrolytic production of metals such as sodium, calcium, and magnesium from their fused chlorides. It is also possible to prepare chlorine by the chemical oxidation of the chloride ion in acid solution with strong oxidizing agents such as manganese dioxide (MnO_2) or sodium dichromate ($Na_2Cr_2O_7$). The reaction with manganese dioxide is:

$$MnO_2(s) + 2Cl^-(aq) + 4H_3O^+(aq) \longrightarrow Mn^{2+}(aq) + Cl_2(g) + 6H_2O(l)$$

The commercial preparation of bromine involves the oxidation of bromide ion by chlorine:

$$2Br^-(aq) + Cl_2(g) \longrightarrow Br_2(l) + 2Cl^-(aq)$$

Chlorine is a stronger oxidizing agent than bromine. This method is important for the production of essentially all domestic bromine.

Some iodine comes from the oxidation of iodine chloride, ICl, or iodic acid, HlO_3. The commercial preparation of iodine utilizes the reduction of sodium iodate, $NaIO_3$, an impurity in deposits of Chile saltpeter, with sodium hydrogen sulfite:

$$2IO_3{}^-(aq) + 5HSO_3{}^-(aq) \longrightarrow 3HSO_4{}^-(aq) + 2SO_4{}^{2-}(aq) + H_2O(l) + I_2(s)$$

Properties of the Halogens

Fluorine is a pale yellow gas, chlorine is a greenish-yellow gas, bromine is a deep reddish-brown liquid, and iodine is a grayish-black crystalline solid. Liquid bromine has a high vapor pressure, and the reddish vapor is readily visible in Figure 18.60. Iodine crystals have a noticeable vapor pressure. When gently heated, these crystals sublime and form a beautiful deep violet vapor.

FIGURE 18.60 Chlorine is a pale yellow-green gas (left), gaseous bromine is deep orange (center), and gaseous iodine is purple (right). (Fluorine is so reactive that it is too dangerous to handle.) (credit: Sahar Atwa)

Bromine is only slightly soluble in water, but it is miscible in all proportions in less polar (or nonpolar) solvents such as chloroform, carbon tetrachloride, and carbon disulfide, forming solutions that vary from yellow to reddish-brown, depending on the concentration.

Iodine is soluble in chloroform, carbon tetrachloride, carbon disulfide, and many hydrocarbons, giving violet solutions of I_2 molecules. Iodine dissolves only slightly in water, giving brown solutions. It is quite soluble in aqueous solutions of iodides, with which it forms brown solutions. These brown solutions result because iodine molecules have empty valence *d* orbitals and can act as weak Lewis acids towards the iodide ion. The equation for the reversible reaction of iodine (Lewis acid) with the iodide ion (Lewis base) to form triiodide ion, $I_3{}^-$, is:

$$I_2(s) + I^-(aq) \longrightarrow I_3{}^-(aq)$$

The easier it is to oxidize the halide ion, the more difficult it is for the halogen to act as an oxidizing agent. Fluorine generally oxidizes an element to its highest oxidation state, whereas the heavier halogens may not. For example, when excess fluorine reacts with sulfur, SF_6 forms. Chlorine gives SCl_2 and bromine, S_2Br_2. Iodine does not react with sulfur.

Fluorine is the most powerful oxidizing agent of the known elements. It spontaneously oxidizes most other elements; therefore, the reverse reaction, the oxidation of fluorides, is very difficult to accomplish. Fluorine reacts directly and forms binary fluorides with all of the elements except the lighter noble gases (He, Ne, and Ar). Fluorine is such a strong oxidizing agent that many substances ignite on contact with it. Drops of water inflame in fluorine and form O_2, OF_2, H_2O_2, O_3, and HF. Wood and asbestos ignite and burn in fluorine gas. Most hot metals burn vigorously in fluorine. However, it is possible to handle fluorine in copper, iron, or nickel containers because an adherent film of the fluoride salt passivates their surfaces. Fluorine is the only element that reacts directly with the noble gas xenon.

Although it is a strong oxidizing agent, chlorine is less active than fluorine. Mixing chlorine and hydrogen in the dark makes the reaction between them to be imperceptibly slow. Exposure of the mixture to light causes the two to react explosively. Chlorine is also less active towards metals than fluorine, and oxidation reactions usually require higher temperatures. Molten sodium ignites in chlorine. Chlorine attacks most nonmetals (C, N_2, and O_2 are notable exceptions), forming covalent molecular compounds. Chlorine generally reacts with compounds that contain only carbon and hydrogen (hydrocarbons) by adding to multiple bonds or by substitution.

In cold water, chlorine undergoes a disproportionation reaction:

$$Cl_2(aq) + 2H_2O(l) \longrightarrow HOCl(aq) + H_3O^+(aq) + Cl^-(aq)$$

Half the chlorine atoms oxidize to the 1+ oxidation state (hypochlorous acid), and the other half reduce to the 1− oxidation state (chloride ion). This disproportionation is incomplete, so chlorine water is an equilibrium mixture of chlorine molecules, hypochlorous acid molecules, hydronium ions, and chloride ions. When exposed to light, this solution undergoes a photochemical decomposition:

$$2HOCl(aq) + 2H_2O(l) \xrightarrow{\text{sunlight}} 2H_3O^+(aq) + 2Cl^-(aq) + O_2(g)$$

The nonmetal chlorine is more electronegative than any other element except fluorine, oxygen, and nitrogen. In general, very electronegative elements are good oxidizing agents; therefore, we would expect elemental chlorine to oxidize all of the other elements except for these three (and the nonreactive noble gases). Its oxidizing property, in fact, is responsible for its principal use. For example, phosphorus(V) chloride, an important intermediate in the preparation of insecticides and chemical weapons, is manufactured by oxidizing the phosphorus with chlorine:

$$P_4(s) + 10Cl_2(g) \longrightarrow 4PCl_5(l)$$

A great deal of chlorine is also used to oxidize, and thus to destroy, organic or biological materials in water purification and in bleaching.

The chemical properties of bromine are similar to those of chlorine, although bromine is the weaker oxidizing agent and its reactivity is less than that of chlorine.

Iodine is the least reactive of the halogens. It is the weakest oxidizing agent, and the iodide ion is the most easily oxidized halide ion. Iodine reacts with metals, but heating is often required. It does not oxidize other halide ions.

Compared with the other halogens, iodine reacts only slightly with water. Traces of iodine in water react with a mixture of starch and iodide ion, forming a deep blue color. This reaction is a very sensitive test for the presence of iodine in water.

Halides of the Representative Metals

Thousands of salts of the representative metals have been prepared. The binary halides are an important subclass of salts. A salt is an ionic compound composed of cations and anions, other than hydroxide or oxide ions. In general, it is possible to prepare these salts from the metals or from oxides, hydroxides, or carbonates. We will illustrate the general types of reactions for preparing salts through reactions used to prepare binary halides.

The binary compounds of a metal with the halogens are the **halides**. Most binary halides are ionic. However, mercury, the elements of group 13 with oxidation states of 3+, tin(IV), and lead(IV) form covalent binary halides.

The direct reaction of a metal and a halogen produce the halide of the metal. Examples of these oxidation-reduction reactions include:

$$Cd(s) + Cl_2(g) \longrightarrow CdCl_2(s)$$
$$2Ga(l) + 3Br_2(l) \longrightarrow 2GaBr_3(s)$$

LINK TO LEARNING

Reactions of the alkali metals with elemental halogens are very exothermic and often quite violent. Under controlled conditions, they provide exciting demonstrations for budding students of chemistry. You can view the initial heating (http://openstax.org/l/16sodium) of the sodium that removes the coating of sodium hydroxide, sodium peroxide, and residual mineral oil to expose the reactive surface. The reaction with chlorine gas then proceeds very nicely.

If a metal can exhibit two oxidation states, it may be necessary to control the stoichiometry in order to obtain the halide with the lower oxidation state. For example, preparation of tin(II) chloride requires a 1:1 ratio of Sn to Cl_2, whereas preparation of tin(IV) chloride requires a 1:2 ratio:

$$Sn(s) + Cl_2(g) \longrightarrow SnCl_2(s)$$
$$Sn(s) + 2Cl_2(g) \longrightarrow SnCl_4(l)$$

The active representative metals—those that are easier to oxidize than hydrogen—react with gaseous hydrogen halides to produce metal halides and hydrogen. The reaction of zinc with hydrogen fluoride is:

$$Zn(s) + 2HF(g) \longrightarrow ZnF_2(s) + H_2(g)$$

The active representative metals also react with solutions of hydrogen halides to form hydrogen and solutions of the corresponding halides. Examples of such reactions include:

$$Cd(s) + 2HBr(aq) \longrightarrow CdBr_2(aq) + H_2(g)$$
$$Sn(s) + 2HI(aq) \longrightarrow SnI_2(aq) + H_2(g)$$

Hydroxides, carbonates, and some oxides react with solutions of the hydrogen halides to form solutions of halide salts. It is possible to prepare additional salts by the reaction of these hydroxides, carbonates, and oxides with aqueous solution of other acids:

$$CaCo_3(s) + 2HCl(aq) \longrightarrow CaCl_2(aq) + CO_2(g) + H_2O(l)$$
$$TlOH(aq) + HF(aq) \longrightarrow TlF(aq) + H_2O(l)$$

A few halides and many of the other salts of the representative metals are insoluble. It is possible to prepare these soluble salts by metathesis reactions that occur when solutions of soluble salts are mixed (see Figure 18.61). Metathesis reactions are examined in the chapter on the stoichiometry of chemical reactions.

FIGURE 18.61 Solid HgI_2 forms when solutions of KI and $Hg(NO_3)_2$ are mixed. (credit: Sahar Atwa)

Several halides occur in large quantities in nature. The ocean and underground brines contain many halides. For example, magnesium chloride in the ocean is the source of magnesium ions used in the production of magnesium. Large underground deposits of sodium chloride, like the salt mine shown in Figure 18.62, occur in many parts of the world. These deposits serve as the source of sodium and chlorine in almost all other compounds containing these elements. The chlor-alkali process is one example.

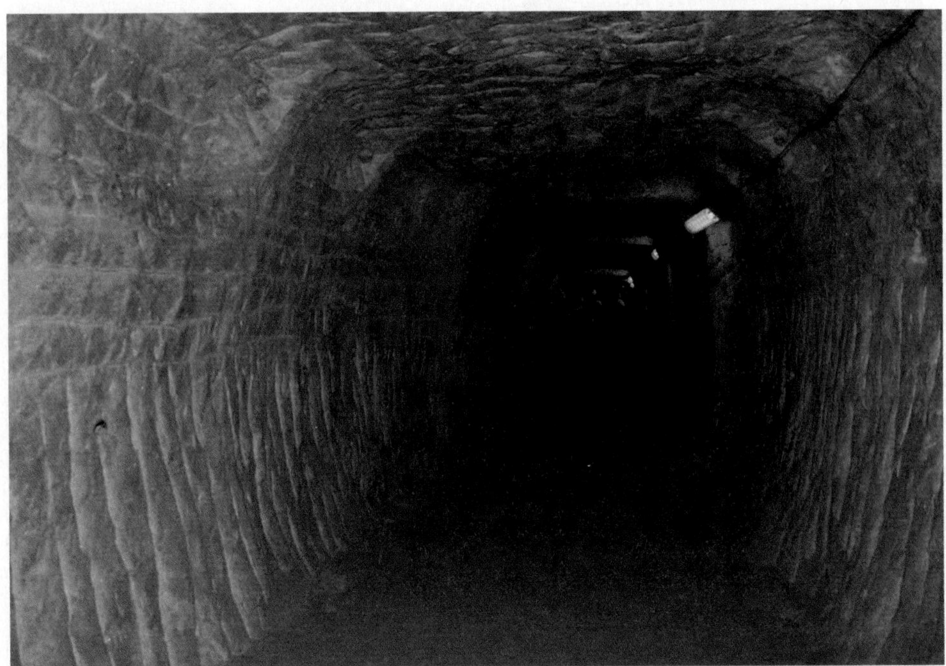

FIGURE 18.62 Underground deposits of sodium chloride are found throughout the world and are often mined. This is a tunnel in the Kłodawa salt mine in Poland. (credit: Jarek Zok)

Interhalogens

Compounds formed from two or more different halogens are **interhalogens**. Interhalogen molecules consist of one atom of the heavier halogen bonded by single bonds to an odd number of atoms of the lighter halogen. The structures of IF_3, IF_5, and IF_7 are illustrated in Figure 18.63. Formulas for other interhalogens, each of which comes from the reaction of the respective halogens, are in Table 18.3.

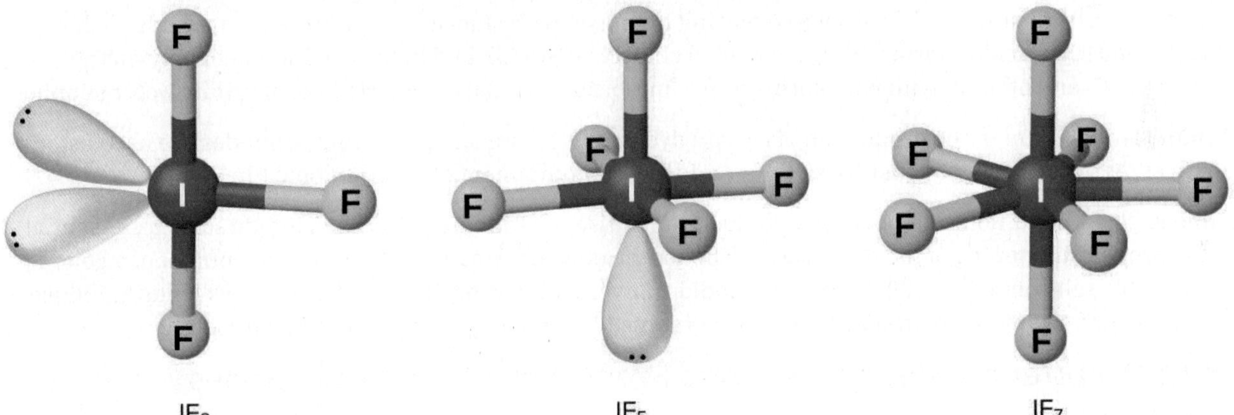

FIGURE 18.63 The structure of IF_3 is T-shaped (left), IF_5 is square pyramidal (center), and IF_7 is pentagonal bipyramidal (right).

Note from Table 18.3 that fluorine is able to oxidize iodine to its maximum oxidation state, 7+, whereas bromine and chlorine, which are more difficult to oxidize, achieve only the 5+-oxidation state. A 7+-oxidation state is the limit for the halogens. Because smaller halogens are grouped about a larger one, the maximum number of smaller atoms possible increases as the radius of the larger atom increases. Many of these compounds are unstable, and most are extremely reactive. The interhalogens react like their component halides; halogen fluorides, for example, are stronger oxidizing agents than are halogen chlorides.

The ionic polyhalides of the alkali metals, such as KI_3, $KICl_2$, $KICl_4$, $CsIBr_2$, and $CsBrCl_2$, which contain an anion composed of at least three halogen atoms, are closely related to the interhalogens. As seen previously, the formation of the polyhalide anion $I_3{}^-$ is responsible for the solubility of iodine in aqueous solutions containing an iodide ion.

Interhalogens

YX	YX_3	YX_5	YX_7
$ClF(g)$	$ClF_3(g)$	$ClF_5(g)$	
$BrF(g)$	$BrF_3(l)$	$BrF_5(l)$	
$BrCl(g)$			
$IF(s)$	$IF_3(s)$	$IF_5(l)$	$IF_7(g)$
$ICl(l)$	$ICl_3(s)$		
$IBr(s)$			

TABLE 18.3

Applications

The fluoride ion and fluorine compounds have many important uses. Compounds of carbon, hydrogen, and fluorine are replacing Freons (compounds of carbon, chlorine, and fluorine) as refrigerants. Teflon is a polymer composed of $–CF_2CF_2–$ units. Fluoride ion is added to water supplies and to some toothpastes as SnF_2 or NaF to fight tooth decay. Fluoride partially converts teeth from $Ca_5(PO_4)_3(OH)$ into $Ca_5(PO_4)_3F$.

Chlorine is important to bleach wood pulp and cotton cloth. The chlorine reacts with water to form hypochlorous acid, which oxidizes colored substances to colorless ones. Large quantities of chlorine are important in chlorinating hydrocarbons (replacing hydrogen with chlorine) to produce compounds such as tetrachloride (CCl_4), chloroform ($CHCl_3$), and ethyl chloride (C_2H_5Cl), and in the production of polyvinyl chloride (PVC) and other polymers. Chlorine is also important to kill the bacteria in community water supplies.

Bromine is important in the production of certain dyes, and sodium and potassium bromides are used as sedatives. At one time, light-sensitive silver bromide was a component of photographic film.

Iodine in alcohol solution with potassium iodide is an antiseptic (tincture of iodine). Iodide salts are essential for the proper functioning of the thyroid gland; an iodine deficiency may lead to the development of a goiter. Iodized table salt contains 0.023% potassium iodide. Silver iodide is useful in the seeding of clouds to induce rain; it was important in the production of photographic film and iodoform, CHI_3, is an antiseptic.

18.12 Occurrence, Preparation, and Properties of the Noble Gases

LEARNING OBJECTIVES

By the end of this section, you will be able to:

- Describe the properties, preparation, and uses of the noble gases

The elements in group 18 are the noble gases (helium, neon, argon, krypton, xenon, and radon). They earned the name “noble” because they were assumed to be nonreactive since they have filled valence shells. In 1962, Dr. Neil Bartlett at the University of British Columbia proved this assumption to be false.

These elements are present in the atmosphere in small amounts. Some natural gas contains 1–2% helium by mass. Helium is isolated from natural gas by liquefying the condensable components, leaving only helium as a gas. The United States possesses most of the world’s commercial supply of this element in its helium-bearing gas fields. Argon, neon, krypton, and xenon come from the fractional distillation of liquid air. Radon comes from other radioactive elements. More recently, it was observed that this radioactive gas is present in very small amounts in soils and minerals. Its accumulation in well-insulated, tightly sealed buildings, however, constitutes a health hazard, primarily lung cancer.

The boiling points and melting points of the noble gases are extremely low relative to those of other substances of comparable atomic or molecular masses. This is because only weak London dispersion forces are present, and these forces can hold the atoms together only when molecular motion is very slight, as it is at very low temperatures. Helium is the only substance known that does not solidify on cooling at normal pressure. It remains liquid close to absolute zero (0.001 K) at ordinary pressures, but it solidifies under elevated pressure.

Helium is used for filling balloons and lighter-than-air craft because it does not burn, making it safer to use than hydrogen. Helium at high pressures is not a narcotic like nitrogen. Thus, mixtures of oxygen and helium are important for divers working under high pressures. Using a helium-oxygen mixture avoids the disoriented mental state known as nitrogen narcosis, the so-called rapture of the deep. Helium is important as an inert atmosphere for the melting and welding of easily oxidizable metals and for many chemical processes that are sensitive to air.

Liquid helium (boiling point, 4.2 K) is an important coolant to reach the low temperatures necessary for cryogenic research, and it is essential for achieving the low temperatures necessary to produce superconduction in traditional superconducting materials used in powerful magnets and other devices. This cooling ability is necessary for the magnets used for magnetic resonance imaging, a common medical diagnostic procedure. The other common coolant is liquid nitrogen (boiling point, 77 K), which is significantly cheaper.

Neon is a component of neon lamps and signs. Passing an electric spark through a tube containing neon at low pressure generates the familiar red glow of neon. It is possible to change the color of the light by mixing argon or mercury vapor with the neon or by utilizing glass tubes of a special color.

Argon was useful in the manufacture of gas-filled electric light bulbs, where its lower heat conductivity and chemical inertness made it preferable to nitrogen for inhibiting the vaporization of the tungsten filament and prolonging the life of the bulb. Fluorescent tubes commonly contain a mixture of argon and mercury vapor. Argon is the third most abundant gas in dry air.

Krypton-xenon flash tubes are used to take high-speed photographs. An electric discharge through such a tube gives a very intense light that lasts only $\frac{1}{50{,}000}$ of a second. Krypton forms a difluoride, KrF_2, which is thermally unstable at room temperature.

Stable compounds of xenon form when xenon reacts with fluorine. Xenon difluoride, XeF_2, forms after heating an excess of xenon gas with fluorine gas and then cooling. The material forms colorless crystals, which are stable at room temperature in a dry atmosphere. Xenon tetrafluoride, XeF_4, and xenon hexafluoride, XeF_6, are prepared in an analogous manner, with a stoichiometric amount of fluorine and an excess of fluorine, respectively. Compounds with oxygen are prepared by replacing fluorine atoms in the xenon fluorides with oxygen.

When XeF_6 reacts with water, a solution of XeO_3 results and the xenon remains in the 6+-oxidation state:

$$XeF_6(s) + 3H_2O(l) \longrightarrow XeO_3(aq) + 6HF(aq)$$

Dry, solid xenon trioxide, XeO_3, is extremely explosive—it will spontaneously detonate. Both XeF_6 and XeO_3 disproportionate in basic solution, producing xenon, oxygen, and salts of the perxenate ion, $XeO_6{}^{4-}$, in which xenon reaches its maximum oxidation sate of 8+.

Radon apparently forms RnF_2—evidence of this compound comes from radiochemical tracer techniques.

Unstable compounds of argon form at low temperatures, but stable compounds of helium and neon are not known.

Key Terms

acid anhydride compound that reacts with water to form an acid or acidic solution

alkaline earth metal any of the metals (beryllium, magnesium, calcium, strontium, barium, and radium) occupying group 2 of the periodic table; they are reactive, divalent metals that form basic oxides

allotropes two or more forms of the same element, in the same physical state, with different chemical structures

amorphous solid material such as a glass that does not have a regular repeating component to its three-dimensional structure; a solid but not a crystal

base anhydride metal oxide that behaves as a base towards acids

bicarbonate anion salt of the hydrogen carbonate ion, $HCO_3{}^-$

bismuth heaviest member of group 15; a less reactive metal than other representative metals

borate compound containing boron-oxygen bonds, typically with clusters or chains as a part of the chemical structure

carbonate salt of the anion $CO_3{}^{2-}$; often formed by the reaction of carbon dioxide with bases

chemical reduction method of preparing a representative metal using a reducing agent

chlor-alkali process electrolysis process for the synthesis of chlorine and sodium hydroxide

disproportionation reaction chemical reaction where a single reactant is simultaneously reduced and oxidized; it is both the reducing agent and the oxidizing agent

Downs cell electrochemical cell used for the commercial preparation of metallic sodium (and chlorine) from molten sodium chloride

Frasch process important in the mining of free sulfur from enormous underground deposits

Haber process main industrial process used to produce ammonia from nitrogen and hydrogen; involves the use of an iron catalyst and elevated temperatures and pressures

halide compound containing an anion of a group 17 element in the 1− oxidation state (fluoride, F^-; chloride, Cl^-; bromide, Br^-; and iodide, I^-)

Hall–Héroult cell electrolysis apparatus used to isolate pure aluminum metal from a solution of alumina in molten cryolite

hydrogen carbonate salt of carbonic acid, H_2CO_3 (containing the anion $HCO_3{}^-$) in which one hydrogen atom has been replaced; an acid carbonate; also known as *bicarbonate ion*

hydrogen halide binary compound formed between hydrogen and the halogens: HF, HCl, HBr, and HI

hydrogen sulfate $HSO_4{}^-$ ion

hydrogen sulfite $HSO_3{}^-$ ion

hydrogenation addition of hydrogen (H_2) to reduce a compound

hydroxide compound of a metal with the hydroxide ion OH^- or the group −OH

interhalogen compound formed from two or more different halogens

metal (representative) atoms of the metallic elements of groups 1, 2, 12, 13, 14, 15, and 16, which form ionic compounds by losing electrons from their outer *s* or *p* orbitals

metalloid element that has properties that are between those of metals and nonmetals; these elements are typically semiconductors

nitrate $NO_3{}^-$ ion; salt of nitric acid

nitrogen fixation formation of nitrogen compounds from molecular nitrogen

Ostwald process industrial process used to convert ammonia into nitric acid

oxide binary compound of oxygen with another element or group, typically containing O^{2-} ions or the group −O− or =O

ozone allotrope of oxygen; O_3

passivation metals with a protective nonreactive film of oxide or other compound that creates a barrier for chemical reactions; physical or chemical removal of the passivating film allows the metals to demonstrate their expected chemical reactivity

peroxide molecule containing two oxygen atoms bonded together or as the anion, $O_2{}^{2-}$

photosynthesis process whereby light energy promotes the reaction of water and carbon dioxide to form carbohydrates and oxygen; this allows photosynthetic organisms to store energy

Pidgeon process chemical reduction process used to produce magnesium through the thermal reaction of magnesium oxide with silicon

polymorph variation in crystalline structure that results in different physical properties for the resulting compound

representative element element where the *s* and *p* orbitals are filling

representative metal metal among the representative elements

silicate compound containing silicon-oxygen bonds, with silicate tetrahedra connected in rings, sheets, or three-dimensional networks, depending on the other elements involved in the formation of the compounds

sulfate SO_4^{2-} ion

sulfite SO_3^{2-} ion

superoxide oxide containing the anion O_2^{-}

Summary

18.1 Periodicity

This section focuses on the periodicity of the representative elements. These are the elements where the electrons are entering the *s* and *p* orbitals. The representative elements occur in groups 1, 2, and 12–18. These elements are representative metals, metalloids, and nonmetals. The alkali metals (group 1) are very reactive, readily form ions with a charge of 1+ to form ionic compounds that are usually soluble in water, and react vigorously with water to form hydrogen gas and a basic solution of the metal hydroxide. The outermost electrons of the alkaline earth metals (group 2) are more difficult to remove than the outer electron of the alkali metals, leading to the group 2 metals being less reactive than those in group 1. These elements easily form compounds in which the metals exhibit an oxidation state of 2+. Zinc, cadmium, and mercury (group 12) commonly exhibit the group oxidation state of 2+ (although mercury also exhibits an oxidation state of 1+ in compounds that contain Hg_2^{2+}). Aluminum, gallium, indium, and thallium (group 13) are easier to oxidize than is hydrogen. Aluminum, gallium, and indium occur with an oxidation state 3+ (however, thallium also commonly occurs as the Tl^+ ion). Tin and lead form stable divalent cations and covalent compounds in which the metals exhibit the 4+-oxidation state.

18.2 Occurrence and Preparation of the Representative Metals

Because of their chemical reactivity, it is necessary to produce the representative metals in their pure forms by reduction from naturally occurring compounds. Electrolysis is important in the production of sodium, potassium, and aluminum. Chemical reduction is the primary method for the isolation of magnesium, zinc, and tin. Similar procedures are important for the other representative metals.

18.3 Structure and General Properties of the Metalloids

The elements boron, silicon, germanium, arsenic, antimony, and tellurium separate the metals from the nonmetals in the periodic table. These elements, called metalloids or sometimes semimetals, exhibit properties characteristic of both metals and nonmetals. The structures of these elements are similar in many ways to those of nonmetals, but the elements are electrical semiconductors.

18.4 Structure and General Properties of the Nonmetals

Nonmetals have structures that are very different from those of the metals, primarily because they have greater electronegativity and electrons that are more tightly bound to individual atoms. Most nonmetal oxides are acid anhydrides, meaning that they react with water to form acidic solutions. Molecular structures are common for most of the nonmetals, and several have multiple allotropes with varying physical properties.

18.5 Occurrence, Preparation, and Compounds of Hydrogen

Hydrogen is the most abundant element in the universe and its chemistry is truly unique. Although it has some chemical reactivity that is similar to that of the alkali metals, hydrogen has many of the same chemical properties of a nonmetal with a relatively low electronegativity. It forms ionic hydrides with active metals, covalent compounds in which it has an oxidation state of 1− with less electronegative elements, and covalent compounds in which it has an oxidation state of 1+ with more electronegative nonmetals. It reacts explosively with oxygen, fluorine, and chlorine, less readily with bromine, and much less readily with iodine, sulfur, and nitrogen. Hydrogen reduces the oxides of metals with lower reduction potentials than chromium to form the metal and water. The hydrogen halides are all acidic when dissolved in water.

18.6 Occurrence, Preparation, and Properties of Carbonates

The usual method for the preparation of the carbonates of the alkali and alkaline earth metals is by reaction of an oxide or hydroxide with carbon dioxide. Other carbonates form by precipitation. Metal carbonates or hydrogen carbonates such as limestone ($CaCO_3$), the antacid Tums ($CaCO_3$), and baking soda ($NaHCO_3$) are common examples. Carbonates and hydrogen carbonates decompose in the presence of acids and most decompose on heating.

18.7 Occurrence, Preparation, and Properties of Nitrogen

Nitrogen exhibits oxidation states ranging from 3− to 5+. Because of the stability of the N≡N triple bond, it requires a great deal of energy to make compounds from molecular nitrogen. Active metals such as the alkali metals and alkaline earth metals can reduce nitrogen to form metal nitrides. Nitrogen oxides and nitrogen hydrides are also important substances.

18.8 Occurrence, Preparation, and Properties of Phosphorus

Phosphorus (group 15) commonly exhibits oxidation states of 3− with active metals and of 3+ and 5+ with more electronegative nonmetals. The halogens and oxygen will oxidize phosphorus. The oxides are phosphorus(V) oxide, P_4O_{10}, and phosphorus(III) oxide, P_4O_6. The two common methods for preparing orthophosphoric acid, H_3PO_4, are either the reaction of a phosphate with sulfuric acid or the reaction of water with phosphorus(V) oxide. Orthophosphoric acid is a triprotic acid that forms three types of salts.

18.9 Occurrence, Preparation, and Compounds of Oxygen

Oxygen is one of the most reactive elements. This reactivity, coupled with its abundance, makes the chemistry of oxygen very rich and well understood.

Compounds of the representative metals with oxygen exist in three categories (1) oxides, (2) peroxides and superoxides, and (3) hydroxides. Heating the corresponding hydroxides, nitrates, or carbonates is the most common method for producing oxides. Heating the metal or metal oxide in oxygen may lead to the formation of peroxides and superoxides. The soluble oxides dissolve in water to form solutions of hydroxides. Most metals oxides are base anhydrides and react with acids. The hydroxides of the representative metals react with acids in acid-base reactions to form salts and water. The hydroxides have many commercial uses.

All nonmetals except fluorine form multiple oxides. Nearly all of the nonmetal oxides are acid anhydrides. The acidity of oxyacids requires that the hydrogen atoms bond to the oxygen atoms in the molecule rather than to the other nonmetal atom. Generally, the strength of the oxyacid increases with the number of oxygen atoms bonded to the nonmetal atom and not to a hydrogen.

18.10 Occurrence, Preparation, and Properties of Sulfur

Sulfur (group 16) reacts with almost all metals and readily forms the sulfide ion, S^{2-}, in which it has as oxidation state of 2−. Sulfur reacts with most nonmetals.

18.11 Occurrence, Preparation, and Properties of Halogens

The halogens form halides with less electronegative elements. Halides of the metals vary from ionic to covalent; halides of nonmetals are covalent. Interhalogens form by the combination of two or more different halogens.

All of the representative metals react directly with elemental halogens or with solutions of the hydrohalic acids (HF, HCl, HBr, and HI) to produce representative metal halides. Other laboratory preparations involve the addition of aqueous hydrohalic acids to compounds that contain such basic anions, such as hydroxides, oxides, or carbonates.

18.12 Occurrence, Preparation, and Properties of the Noble Gases

The most significant property of the noble gases (group 18) is their inactivity. They occur in low concentrations in the atmosphere. They find uses as inert atmospheres, neon signs, and as coolants. The three heaviest noble gases react with fluorine to form fluorides. The xenon fluorides are the best characterized as the starting materials for a few other noble gas compounds.

Exercises

18.1 Periodicity

1. How do alkali metals differ from alkaline earth metals in atomic structure and general properties?
2. Why does the reactivity of the alkali metals decrease from cesium to lithium?

3. Predict the formulas for the nine compounds that may form when each species in column 1 of the table reacts with each species in column 2.

1	2
Na	I
Sr	Se
Al	O

4. Predict the best choice in each of the following. You may wish to review the chapter on electronic structure for relevant examples.
   (a) the most metallic of the elements Al, Be, and Ba
   (b) the most covalent of the compounds NaCl, $CaCl_2$, and $BeCl_2$
   (c) the lowest first ionization energy among the elements Rb, K, and Li
   (d) the smallest among Al, Al^{+}, and Al^{3+}
   (e) the largest among Cs^{+}, Ba^{2+}, and Xe
5. Sodium chloride and strontium chloride are both white solids. How could you distinguish one from the other?
6. The reaction of quicklime, CaO, with water produces slaked lime, $Ca(OH)_2$, which is widely used in the construction industry to make mortar and plaster. The reaction of quicklime and water is highly exothermic:
   $CaO(s) + H_2O(l) \longrightarrow Ca(OH)_2(s) \qquad \Delta H = -350 \text{ kJ mol}^{-1}$
   (a) What is the enthalpy of reaction per gram of quicklime that reacts?
   (b) How much heat, in kilojoules, is associated with the production of 1 ton of slaked lime?
7. Write a balanced equation for the reaction of elemental strontium with each of the following:
   (a) oxygen
   (b) hydrogen bromide
   (c) hydrogen
   (d) phosphorus
   (e) water
8. How many moles of ionic species are present in 1.0 L of a solution marked 1.0 *M* mercury(I) nitrate?
9. What is the mass of fish, in kilograms, that one would have to consume to obtain a fatal dose of mercury, if the fish contains 30 parts per million of mercury by weight? (Assume that all the mercury from the fish ends up as mercury(II) chloride in the body and that a fatal dose is 0.20 g of $HgCl_2$.) How many pounds of fish is this?
10. The elements sodium, aluminum, and chlorine are in the same period.
    (a) Which has the greatest electronegativity?
    (b) Which of the atoms is smallest?
    (c) Write the Lewis structure for the simplest covalent compound that can form between aluminum and chlorine.
    (d) Will the oxide of each element be acidic, basic, or amphoteric?
11. Does metallic tin react with HCl?
12. What is tin pest, also known as tin disease?
13. Compare the nature of the bonds in $PbCl_2$ to that of the bonds in $PbCl_4$.
14. Is the reaction of rubidium with water more or less vigorous than that of sodium? How does the rate of reaction of magnesium compare?

18.2 Occurrence and Preparation of the Representative Metals

15. Write an equation for the reduction of cesium chloride by elemental calcium at high temperature.

16. Why is it necessary to keep the chlorine and sodium, resulting from the electrolysis of sodium chloride, separate during the production of sodium metal?
17. Give balanced equations for the overall reaction in the electrolysis of molten lithium chloride and for the reactions occurring at the electrodes. You may wish to review the chapter on electrochemistry for relevant examples.
18. The electrolysis of molten sodium chloride or of aqueous sodium chloride produces chlorine. Calculate the mass of chlorine produced from 3.00 kg sodium chloride in each case. You may wish to review the chapter on electrochemistry for relevant examples.
19. What mass, in grams, of hydrogen gas forms during the complete reaction of 10.01 g of calcium with water?
20. How many grams of oxygen gas are necessary to react completely with 3.01×10^{21} atoms of magnesium to yield magnesium oxide?
21. Magnesium is an active metal; it burns in the form of powder, ribbons, and filaments to provide flashes of brilliant light. Why is it possible to use magnesium in construction?
22. Why is it possible for an active metal like aluminum to be useful as a structural metal?
23. Describe the production of metallic aluminum by electrolytic reduction.
24. What is the common ore of tin and how is tin separated from it?
25. A chemist dissolves a 1.497-g sample of a type of metal (an alloy of Sn, Pb, Sb, and Cu) in nitric acid, and metastannic acid, H_2SnO_3, is precipitated. She heats the precipitate to drive off the water, which leaves 0.4909 g of tin(IV) oxide. What was the percentage of tin in the original sample?
26. Consider the production of 100 kg of sodium metal using a current of 50,000 A, assuming a 100% yield.
(a) How long will it take to produce the 100 kg of sodium metal?
(b) What volume of chlorine at 25 °C and 1.00 atm forms?
27. What mass of magnesium forms when 100,000 A is passed through a $MgCl_2$ melt for 1.00 h if the yield of magnesium is 85% of the theoretical yield?

18.3 Structure and General Properties of the Metalloids

28. Give the hybridization of the metalloid and the molecular geometry for each of the following compounds or ions. You may wish to review the chapters on chemical bonding and advanced covalent bonding for relevant examples.
(a) GeH_4
(b) SbF_3
(c) $Te(OH)_6$
(d) H_2Te
(e) GeF_2
(f) $TeCl_4$
(g) $SiF_6{}^{2-}$
(h) $SbCl_5$
(i) TeF_6
29. Write a Lewis structure for each of the following molecules or ions. You may wish to review the chapter on chemical bonding.
(a) H_3BPH_3
(b) $BF_4{}^-$
(c) BBr_3
(d) $B(CH_3)_3$
(e) $B(OH)_3$
30. Describe the hybridization of boron and the molecular structure about the boron in each of the following:
(a) H_3BPH_3
(b) $BF_4{}^-$
(c) BBr_3
(d) $B(CH_3)_3$
(e) $B(OH)_3$

31. Using only the periodic table, write the complete electron configuration for silicon, including any empty orbitals in the valence shell. You may wish to review the chapter on electronic structure.
32. Write a Lewis structure for each of the following molecules and ions:
(a) $(CH_3)_3SiH$
(b) $SiO_4{}^{4-}$
(c) Si_2H_6
(d) $Si(OH)_4$
(e) $SiF_6{}^{2-}$
33. Describe the hybridization of silicon and the molecular structure of the following molecules and ions:
(a) $(CH_3)_3SiH$
(b) $SiO_4{}^{4-}$
(c) Si_2H_6
(d) $Si(OH)_4$
(e) $SiF_6{}^{2-}$
34. Describe the hybridization and the bonding of a silicon atom in elemental silicon.
35. Classify each of the following molecules as polar or nonpolar. You may wish to review the chapter on chemical bonding.
(a) SiH_4
(b) Si_2H_6
(c) $SiCl_3H$
(d) SiF_4
(e) $SiCl_2F_2$
36. Silicon reacts with sulfur at elevated temperatures. If 0.0923 g of silicon reacts with sulfur to give 0.3030 g of silicon sulfide, determine the empirical formula of silicon sulfide.
37. Name each of the following compounds:
(a) TeO_2
(b) Sb_2S_3
(c) GeF_4
(d) SiH_4
(e) GeH_4
38. Write a balanced equation for the reaction of elemental boron with each of the following (most of these reactions require high temperature):
(a) F_2
(b) O_2
(c) S
(d) Se
(e) Br_2
39. Why is boron limited to a maximum coordination number of four in its compounds?
40. Write a formula for each of the following compounds:
(a) silicon dioxide
(b) silicon tetraiodide
(c) silane
(d) silicon carbide
(e) magnesium silicide
41. From the data given in Appendix G, determine the standard enthalpy change and the standard free energy change for each of the following reactions:
(a) $BF_3(g) + 3H_2O(l) \longrightarrow B(OH)_3(s) + 3HF(g)$
(b) $BCl_3(g) + 3H_2O(l) \longrightarrow B(OH)_3(s) + 3HCl(g)$
(c) $B_2H_6(g) + 6H_2O(l) \longrightarrow 2B(OH)_3(s) + 6H_2(g)$
42. A hydride of silicon prepared by the reaction of Mg_2Si with acid exerted a pressure of 306 torr at 26 °C in a bulb with a volume of 57.0 mL. If the mass of the hydride was 0.0861 g, what is its molecular mass? What is the molecular formula for the hydride?

43. Suppose you discovered a diamond completely encased in a silicate rock. How would you chemically free the diamond without harming it?

18.4 Structure and General Properties of the Nonmetals

44. Carbon forms a number of allotropes, two of which are graphite and diamond. Silicon has a diamond structure. Why is there no allotrope of silicon with a graphite structure?
45. Nitrogen in the atmosphere exists as very stable diatomic molecules. Why does phosphorus form less stable P_4 molecules instead of P_2 molecules?
46. Write balanced chemical equations for the reaction of the following acid anhydrides with water:
    (a) SO_3
    (b) N_2O_3
    (c) Cl_2O_7
    (d) P_4O_{10}
    (e) NO_2
47. Determine the oxidation number of each element in each of the following compounds:
    (a) HCN
    (b) OF_2
    (c) $AsCl_3$
48. Determine the oxidation state of sulfur in each of the following:
    (a) SO_3
    (b) SO_2
    (c) ${SO_3}^{2-}$
49. Arrange the following in order of increasing electronegativity: F; Cl; O; and S.
50. Why does white phosphorus consist of tetrahedral P_4 molecules while nitrogen consists of diatomic N_2 molecules?

18.5 Occurrence, Preparation, and Compounds of Hydrogen

51. Why does hydrogen not exhibit an oxidation state of 1− when bonded to nonmetals?
52. The reaction of calcium hydride, CaH_2, with water can be characterized as a Lewis acid-base reaction:
    $CaH_2(s) + 2H_2O(l) \longrightarrow Ca(OH)_2(aq) + 2H_2(g)$
    Identify the Lewis acid and the Lewis base among the reactants. The reaction is also an oxidation-reduction reaction. Identify the oxidizing agent, the reducing agent, and the changes in oxidation number that occur in the reaction.
53. In drawing Lewis structures, we learn that a hydrogen atom forms only one bond in a covalent compound. Why?
54. What mass of CaH_2 is necessary to react with water to provide enough hydrogen gas to fill a balloon at 20 °C and 0.8 atm pressure with a volume of 4.5 L? The balanced equation is:
    $CaH_2(s) + 2H_2O(l) \longrightarrow Ca(OH)_2(aq) + 2H_2(g)$
55. What mass of hydrogen gas results from the reaction of 8.5 g of KH with water?
    $KH + H_2O \longrightarrow KOH + H_2$

18.6 Occurrence, Preparation, and Properties of Carbonates

56. Carbon forms the ${CO_3}^{2-}$ ion, yet silicon does not form an analogous ${SiO_3}^{2-}$ ion. Why?
57. Complete and balance the following chemical equations:
    (a) hardening of plaster containing slaked lime
    $Ca(OH)_2 + CO_2 \longrightarrow$
    (b) removal of sulfur dioxide from the flue gas of power plants
    $CaO + SO_2 \longrightarrow$
    (c) the reaction of baking powder that produces carbon dioxide gas and causes bread to rise
    $NaHCO_3 + NaH_2PO_4 \longrightarrow$

58. Heating a sample of $Na_2CO_3{\cdot}xH_2O$ weighing 4.640 g until the removal of the water of hydration leaves 1.720 g of anhydrous Na_2CO_3. What is the formula of the hydrated compound?

18.7 Occurrence, Preparation, and Properties of Nitrogen

59. Write the Lewis structures for each of the following:
(a) NH^{2-}
(b) N_2F_4
(c) NH_2^-
(d) NF_3
(e) N_3^-

60. For each of the following, indicate the hybridization of the nitrogen atom (for N_3^-, the central nitrogen).
(a) N_2F_4
(b) NH_2^-
(c) NF_3
(d) N_3^-

61. Explain how ammonia can function both as a Brønsted base and as a Lewis base.

62. Determine the oxidation state of nitrogen in each of the following. You may wish to review the chapter on chemical bonding for relevant examples.
(a) NCl_3
(b) ClNO
(c) N_2O_5
(d) N_2O_3
(e) NO_2^-
(f) N_2O_4
(g) N_2O
(h) NO_3^-
(i) HNO_2
(j) HNO_3

63. For each of the following, draw the Lewis structure, predict the ONO bond angle, and give the hybridization of the nitrogen. You may wish to review the chapters on chemical bonding and advanced theories of covalent bonding for relevant examples.
(a) NO_2
(b) NO_2^-
(c) NO_2^+

64. How many grams of gaseous ammonia will the reaction of 3.0 g hydrogen gas and 3.0 g of nitrogen gas produce?

65. Although PF_5 and AsF_5 are stable, nitrogen does not form NF_5 molecules. Explain this difference among members of the same group.

66. The equivalence point for the titration of a 25.00-mL sample of CsOH solution with 0.1062 M HNO_3 is at 35.27 mL. What is the concentration of the CsOH solution?

18.8 Occurrence, Preparation, and Properties of Phosphorus

67. Write the Lewis structure for each of the following. You may wish to review the chapter on chemical bonding and molecular geometry.
(a) PH_3
(b) PH_4^+
(c) P_2H_4
(d) PO_4^{3-}
(e) PF_5

68. Describe the molecular structure of each of the following molecules or ions listed. You may wish to review the chapter on chemical bonding and molecular geometry.
(a) PH_3
(b) $PH_4{}^+$
(c) P_2H_4
(d) $PO_4{}^{3-}$
69. Complete and balance each of the following chemical equations. (In some cases, there may be more than one correct answer.)
(a) $P_4 + Al \longrightarrow$
(b) $P_4 + Na \longrightarrow$
(c) $P_4 + F_2 \longrightarrow$
(d) $P_4 + Cl_2 \longrightarrow$
(e) $P_4 + O_2 \longrightarrow$
(f) $P_4O_6 + O_2 \longrightarrow$
70. Describe the hybridization of phosphorus in each of the following compounds: P_4O_{10}, P_4O_6, PH_4I (an ionic compound), PBr_3, H_3PO_4, H_3PO_3, PH_3, and P_2H_4. You may wish to review the chapter on advanced theories of covalent bonding.
71. What volume of 0.200 *M* NaOH is necessary to neutralize the solution produced by dissolving 2.00 g of PCl_3 is an excess of water? Note that when H_3PO_3 is titrated under these conditions, only one proton of the acid molecule reacts.
72. How much $POCl_3$ can form from 25.0 g of PCl_5 and the appropriate amount of H_2O?
73. How many tons of $Ca_3(PO_4)_2$ are necessary to prepare 5.0 tons of phosphorus if the yield is 90%?
74. Write equations showing the stepwise ionization of phosphorous acid.
75. Draw the Lewis structures and describe the geometry for the following:
(a) $PF_4{}^+$
(b) PF_5
(c) $PF_6{}^-$
(d) POF_3
76. Why does phosphorous acid form only two series of salts, even though the molecule contains three hydrogen atoms?
77. Assign an oxidation state to phosphorus in each of the following:
(a) NaH_2PO_3
(b) PF_5
(c) P_4O_6
(d) K_3PO_4
(e) Na_3P
(f) $Na_4P_2O_7$
78. Phosphoric acid, one of the acids used in some cola drinks, is produced by the reaction of phosphorus(V) oxide, an acidic oxide, with water. Phosphorus(V) oxide is prepared by the combustion of phosphorus.
(a) Write the empirical formula of phosphorus(V) oxide.
(b) What is the molecular formula of phosphorus(V) oxide if the molar mass is about 280.
(c) Write balanced equations for the production of phosphorus(V) oxide and phosphoric acid.
(d) Determine the mass of phosphorus required to make 1.00×10^4 kg of phosphoric acid, assuming a yield of 98.85%.

18.9 Occurrence, Preparation, and Compounds of Oxygen

79. Predict the product of burning francium in air.
80. Using equations, describe the reaction of water with potassium and with potassium oxide.
81. Write balanced chemical equations for the following reactions:
(a) zinc metal heated in a stream of oxygen gas
(b) zinc carbonate heated until loss of mass stops
(c) zinc carbonate added to a solution of acetic acid, CH_3CO_2H
(d) zinc added to a solution of hydrobromic acid

82. Write balanced chemical equations for the following reactions:
(a) cadmium burned in air
(b) elemental cadmium added to a solution of hydrochloric acid
(c) cadmium hydroxide added to a solution of acetic acid, CH_3CO_2H
83. Illustrate the amphoteric nature of aluminum hydroxide by citing suitable equations.
84. Write balanced chemical equations for the following reactions:
(a) metallic aluminum burned in air
(b) elemental aluminum heated in an atmosphere of chlorine
(c) aluminum heated in hydrogen bromide gas
(d) aluminum hydroxide added to a solution of nitric acid
85. Write balanced chemical equations for the following reactions:
(a) sodium oxide added to water
(b) cesium carbonate added to an excess of an aqueous solution of HF
(c) aluminum oxide added to an aqueous solution of $HClO_4$
(d) a solution of sodium carbonate added to solution of barium nitrate
(e) titanium metal produced from the reaction of titanium tetrachloride with elemental sodium
86. What volume of 0.250 M H_2SO_4 solution is required to neutralize a solution that contains 5.00 g of $CaCO_3$?
87. Which is the stronger acid, $HClO_4$ or $HBrO_4$? Why?
88. Write a balanced chemical equation for the reaction of an excess of oxygen with each of the following. Remember that oxygen is a strong oxidizing agent and tends to oxidize an element to its maximum oxidation state.
(a) Mg
(b) Rb
(c) Ga
(d) C_2H_2
(e) CO
89. Which is the stronger acid, H_2SO_4 or H_2SeO_4? Why? You may wish to review the chapter on acid-base equilibria.

18.10 Occurrence, Preparation, and Properties of Sulfur

90. Explain why hydrogen sulfide is a gas at room temperature, whereas water, which has a lower molecular mass, is a liquid.
91. Give the hybridization and oxidation state for sulfur in SO_2, in SO_3, and in H_2SO_4.
92. Which is the stronger acid, $NaHSO_3$ or $NaHSO_4$?
93. Determine the oxidation state of sulfur in SF_6, SO_2F_2, and KHS.
94. Which is a stronger acid, sulfurous acid or sulfuric acid? Why?
95. Oxygen forms double bonds in O_2, but sulfur forms single bonds in S_8. Why?
96. Give the Lewis structure of each of the following:
(a) SF_4
(b) K_2SO_4
(c) SO_2Cl_2
(d) H_2SO_3
(e) SO_3
97. Write two balanced chemical equations in which sulfuric acid acts as an oxidizing agent.
98. Explain why sulfuric acid, H_2SO_4, which is a covalent molecule, dissolves in water and produces a solution that contains ions.
99. How many grams of Epsom salts ($MgSO_4 \cdot 7H_2O$) will form from 5.0 kg of magnesium?

18.11 Occurrence, Preparation, and Properties of Halogens

100. What does it mean to say that mercury(II) halides are weak electrolytes?
101. Why is $SnCl_4$ not classified as a salt?

102. The following reactions are all similar to those of the industrial chemicals. Complete and balance the equations for these reactions:
(a) reaction of a weak base and a strong acid
$NH_3 + HClO_4 \longrightarrow$
(b) preparation of a soluble silver salt for silver plating
$Ag_2CO_3 + HNO_3 \longrightarrow$
(c) preparation of strontium hydroxide by electrolysis of a solution of strontium chloride
$SrCl_2(aq) + H_2O(l) \xrightarrow{\text{electrolysis}}$

103. Which is the stronger acid, $HClO_3$ or $HBrO_3$? Why?

104. What is the hybridization of iodine in IF_3 and IF_5?

105. Predict the molecular geometries and draw Lewis structures for each of the following. You may wish to review the chapter on chemical bonding and molecular geometry.
(a) IF_5
(b) $I_3{}^-$
(c) PCl_5
(d) SeF_4
(e) ClF_3

106. Which halogen has the highest ionization energy? Is this what you would predict based on what you have learned about periodic properties?

107. Name each of the following compounds:
(a) BrF_3
(b) $NaBrO_3$
(c) PBr_5
(d) $NaClO_4$
(e) $KClO$

108. Explain why, at room temperature, fluorine and chlorine are gases, bromine is a liquid, and iodine is a solid.

109. What is the oxidation state of the halogen in each of the following?
(a) H_5IO_6
(b) $IO_4{}^-$
(c) ClO_2
(d) ICl_3
(e) F_2

110. Physiological saline concentration—that is, the sodium chloride concentration in our bodies—is approximately 0.16 *M*. A saline solution for contact lenses is prepared to match the physiological concentration. If you purchase 25 mL of contact lens saline solution, how many grams of sodium chloride have you bought?

18.12 Occurrence, Preparation, and Properties of the Noble Gases

111. Give the hybridization of xenon in each of the following. You may wish to review the chapter on the advanced theories of covalent bonding.
(a) XeF_2
(b) XeF_4
(c) XeO_3
(d) XeO_4
(e) $XeOF_4$

112. What is the molecular structure of each of the following molecules? You may wish to review the chapter on chemical bonding and molecular geometry.
(a) XeF_2
(b) XeF_4
(c) XeO_3
(d) XeO_4
(e) $XeOF_4$

113. Indicate whether each of the following molecules is polar or nonpolar. You may wish to review the chapter on chemical bonding and molecular geometry.
(a) XeF_2
(b) XeF_4
(c) XeO_3
(d) XeO_4
(e) $XeOF_4$

114. What is the oxidation state of the noble gas in each of the following? You may wish to review the chapter on chemical bonding and molecular geometry.
(a) XeO_2F_2
(b) KrF_2
(c) $XeF_3{}^+$
(d) $XeO_6{}^{4-}$
(e) XeO_3

115. A mixture of xenon and fluorine was heated. A sample of the white solid that formed reacted with hydrogen to yield 81 mL of xenon (at STP) and hydrogen fluoride, which was collected in water, giving a solution of hydrofluoric acid. The hydrofluoric acid solution was titrated, and 68.43 mL of 0.3172 *M* sodium hydroxide was required to reach the equivalence point. Determine the empirical formula for the white solid and write balanced chemical equations for the reactions involving xenon.

116. Basic solutions of Na_4XeO_6 are powerful oxidants. What mass of $Mn(NO_3)_2{\cdot}6H_2O$ reacts with 125.0 mL of a 0.1717 *M* basic solution of Na_4XeO_6 that contains an excess of sodium hydroxide if the products include Xe and solution of sodium permanganate?

CHAPTER 19
Transition Metals and Coordination Chemistry

Figure 19.1 Transition metals often form vibrantly colored complexes. The minerals malachite (green), azurite (blue), and proustite (red) are some examples. (credit left: modification of work by James St. John; credit middle: modification of work by Stephanie Clifford; credit right: modification of work by Terry Wallace)

CHAPTER OUTLINE

19.1 Occurrence, Preparation, and Properties of Transition Metals and Their Compounds
19.2 Coordination Chemistry of Transition Metals
19.3 Spectroscopic and Magnetic Properties of Coordination Compounds

INTRODUCTION We have daily contact with many transition metals. Iron occurs everywhere—from the rings in your spiral notebook and the cutlery in your kitchen to automobiles, ships, buildings, and in the hemoglobin in your blood. Titanium is useful in the manufacture of lightweight, durable products such as bicycle frames, artificial hips, and jewelry. Chromium is useful as a protective plating on plumbing fixtures and automotive detailing.

In addition to being used in their pure elemental forms, many compounds containing transition metals have numerous other applications. Silver nitrate is used to create mirrors, zirconium silicate provides friction in automotive brakes, and many important cancer-fighting agents, like the drug cisplatin and related species, are platinum compounds.

The variety of properties exhibited by transition metals is due to their complex valence shells. Unlike most main group metals where one oxidation state is normally observed, the valence shell structure of transition metals means that they usually occur in several different stable oxidation states. In addition, electron transitions in these elements can correspond with absorption of photons in the visible electromagnetic spectrum, leading to colored compounds. Because of these behaviors, transition metals exhibit a rich and fascinating chemistry.

19.1 Occurrence, Preparation, and Properties of Transition Metals and Their Compounds

Transition metals are defined as those elements that have (or readily form) partially filled *d* orbitals. As shown in Figure 19.2, the ***d*-block elements** in groups 3–11 are transition elements. The ***f*-block elements**, also called *inner transition metals* (the lanthanides and actinides), also meet this criterion because the *d* orbital is partially occupied before the *f* orbitals. The *d* orbitals fill with the copper family (group 11); for this reason, the next family (group 12) are technically not transition elements. However, the group 12 elements do display some of the same chemical properties and are commonly included in discussions of transition metals. Some chemists do treat the group 12 elements as transition metals.

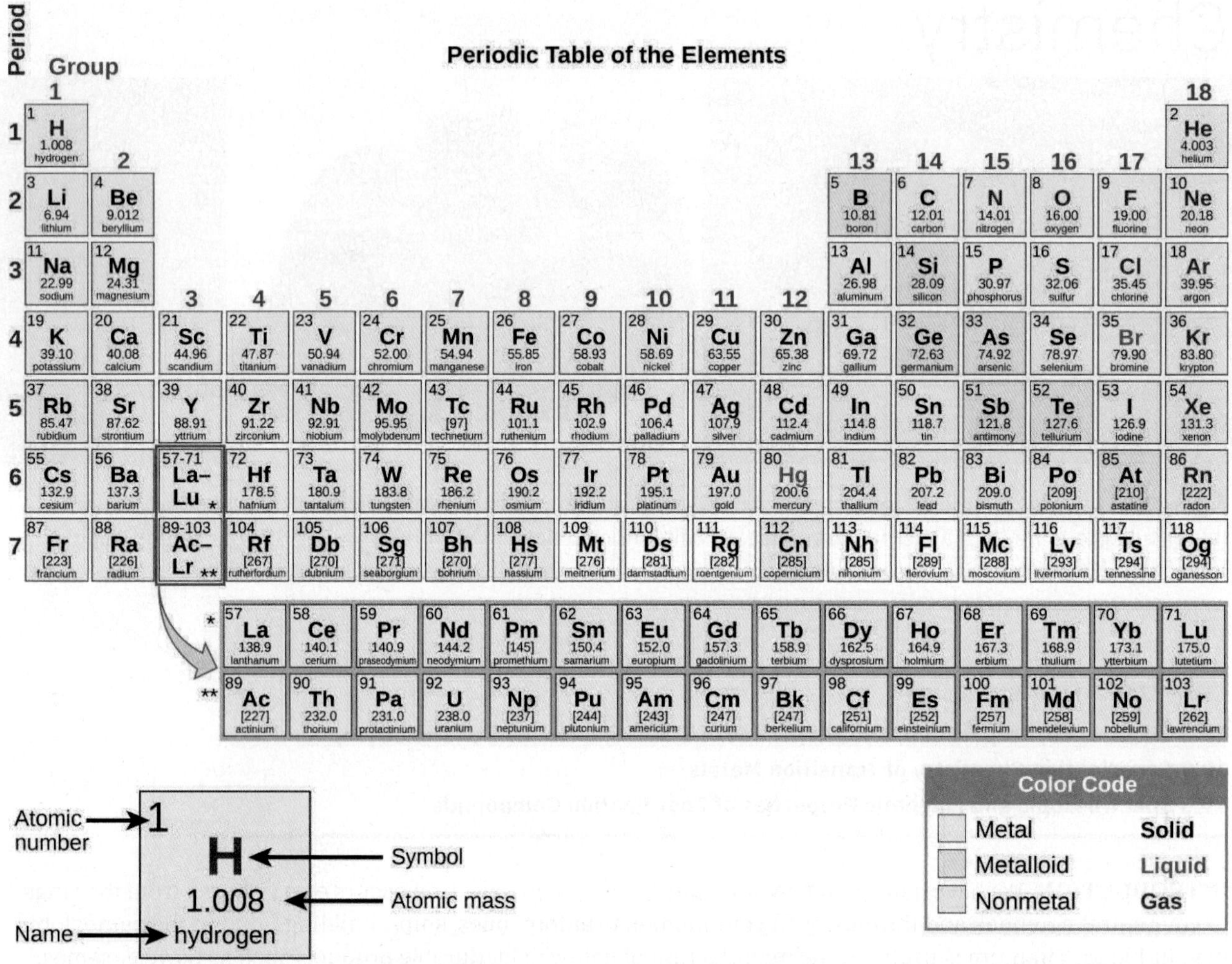

Periodic Table of the Elements

Period / Group	1	2	3	4	5	6	7	8	9	10	11	12	13	14	15	16	17	18
1	1 H 1.008 hydrogen																	2 He 4.003 helium
2	3 Li 6.94 lithium	4 Be 9.012 beryllium											5 B 10.81 boron	6 C 12.01 carbon	7 N 14.01 nitrogen	8 O 16.00 oxygen	9 F 19.00 fluorine	10 Ne 20.18 neon
3	11 Na 22.99 sodium	12 Mg 24.31 magnesium											13 Al 26.98 aluminum	14 Si 28.09 silicon	15 P 30.97 phosphorus	16 S 32.06 sulfur	17 Cl 35.45 chlorine	18 Ar 39.95 argon
4	19 K 39.10 potassium	20 Ca 40.08 calcium	21 Sc 44.96 scandium	22 Ti 47.87 titanium	23 V 50.94 vanadium	24 Cr 52.00 chromium	25 Mn 54.94 manganese	26 Fe 55.85 iron	27 Co 58.93 cobalt	28 Ni 58.69 nickel	29 Cu 63.55 copper	30 Zn 65.38 zinc	31 Ga 69.72 gallium	32 Ge 72.63 germanium	33 As 74.92 arsenic	34 Se 78.97 selenium	35 Br 79.90 bromine	36 Kr 83.80 krypton
5	37 Rb 85.47 rubidium	38 Sr 87.62 strontium	39 Y 88.91 yttrium	40 Zr 91.22 zirconium	41 Nb 92.91 niobium	42 Mo 95.95 molybdenum	43 Tc [97] technetium	44 Ru 101.1 ruthenium	45 Rh 102.9 rhodium	46 Pd 106.4 palladium	47 Ag 107.9 silver	48 Cd 112.4 cadmium	49 In 114.8 indium	50 Sn 118.7 tin	51 Sb 121.8 antimony	52 Te 127.6 tellurium	53 I 126.9 iodine	54 Xe 131.3 xenon
6	55 Cs 132.9 cesium	56 Ba 137.3 barium	57-71 La–Lu *	72 Hf 178.5 hafnium	73 Ta 180.9 tantalum	74 W 183.8 tungsten	75 Re 186.2 rhenium	76 Os 190.2 osmium	77 Ir 192.2 iridium	78 Pt 195.1 platinum	79 Au 197.0 gold	80 Hg 200.6 mercury	81 Tl 204.4 thallium	82 Pb 207.2 lead	83 Bi 209.0 bismuth	84 Po [209] polonium	85 At [210] astatine	86 Rn [222] radon
7	87 Fr [223] francium	88 Ra [226] radium	89-103 Ac–Lr **	104 Rf [267] rutherfordium	105 Db [270] dubnium	106 Sg [271] seaborgium	107 Bh [270] bohrium	108 Hs [277] hassium	109 Mt [276] meitnerium	110 Ds [281] darmstadtium	111 Rg [282] roentgenium	112 Cn [285] copernicium	113 Nh [285] nihonium	114 Fl [289] flerovium	115 Mc [288] moscovium	116 Lv [293] livermorium	117 Ts [294] tennessine	118 Og [294] oganesson

*	57 La 138.9 lanthanum	58 Ce 140.1 cerium	59 Pr 140.9 praseodymium	60 Nd 144.2 neodymium	61 Pm [145] promethium	62 Sm 150.4 samarium	63 Eu 152.0 europium	64 Gd 157.3 gadolinium	65 Tb 158.9 terbium	66 Dy 162.5 dysprosium	67 Ho 164.9 holmium	68 Er 167.3 erbium	69 Tm 168.9 thulium	70 Yb 173.1 ytterbium	71 Lu 175.0 lutetium
**	89 Ac [227] actinium	90 Th 232.0 thorium	91 Pa 231.0 protactinium	92 U 238.0 uranium	93 Np [237] neptunium	94 Pu [244] plutonium	95 Am [243] americium	96 Cm [247] curium	97 Bk [247] berkelium	98 Cf [251] californium	99 Es [252] einsteinium	100 Fm [257] fermium	101 Md [258] mendelevium	102 No [259] nobelium	103 Lr [262] lawrencium

Atomic number → 1
H ← Symbol
1.008 ← Atomic mass
Name → hydrogen

Color Code	
Metal	Solid
Metalloid	Liquid
Nonmetal	Gas

FIGURE 19.2 The transition metals are located in groups 3–11 of the periodic table. The inner transition metals are in the two rows below the body of the table.

The *d*-block elements are divided into the **first transition series** (the elements Sc through Cu), the **second transition series** (the elements Y through Ag), and the **third transition series** (the element La and the elements Hf through Au). Actinium, Ac, is the first member of the **fourth transition series**, which also includes Rf through Rg.

The *f*-block elements are the elements Ce through Lu, which constitute the **lanthanide series** (or **lanthanoid series**), and the elements Th through Lr, which constitute the **actinide series** (or **actinoid series**). Because lanthanum behaves very much like the lanthanide elements, it is considered a lanthanide element, even though its electron configuration makes it the first member of the third transition series. Similarly, the behavior of actinium means it is part of the actinide series, although its electron configuration makes it the first member of the fourth transition series.

EXAMPLE 19.1

Valence Electrons in Transition Metals

Review how to write electron configurations, covered in the chapter on electronic structure and periodic properties of elements. Recall that for the transition and inner transition metals, it is necessary to remove the *s* electrons before the *d* or *f* electrons. Then, for each ion, give the electron configuration:

(a) cerium(III)

(b) lead(II)

(c) Ti^{2+}

(d) Am^{3+}

(e) Pd^{2+}

For the examples that are transition metals, determine to which series they belong.

Solution

For ions, the *s*-valence electrons are lost prior to the *d* or *f* electrons.

(a) $Ce^{3+}[Xe]4f^1$; Ce^{3+} is an inner transition element in the lanthanide series.

(b) $Pb^{2+}[Xe]6s^25d^{10}4f^{14}$; the electrons are lost from the *p* orbital. This is a main group element.
(c) titanium(II) $[Ar]3d^2$; first transition series

(d) americium(III) $[Rn]5f^6$; actinide

(e) palladium(II) $[Kr]4d^8$; second transition series

Check Your Learning

Give an example of an ion from the first transition series with no *d* electrons.

Answer:

V^{5+} is one possibility. Other examples include Sc^{3+}, Ti^{4+}, Cr^{6+}, and Mn^{7+}.

Chemistry in Everyday Life

Uses of Lanthanides in Devices

Lanthanides (elements 57–71) are fairly abundant in the earth's crust, despite their historic characterization as **rare earth elements**. Thulium, the rarest naturally occurring lanthanoid, is more common in the earth's crust than silver ($4.5 \times 10^{-5}\%$ versus $0.79 \times 10^{-5}\%$ by mass). There are 17 rare earth elements, consisting of the 15 lanthanoids plus scandium and yttrium. They are called rare because they were once difficult to extract economically, so it was rare to have a pure sample; due to similar chemical properties, it is difficult to separate any one lanthanide from the others. However, newer separation methods, such as ion exchange resins similar to those found in home water softeners, make the separation of these elements easier and more economical. Most ores that contain these elements have low concentrations of all the rare earth elements mixed together.

The commercial applications of lanthanides are growing rapidly. For example, europium is important in flat screen displays found in computer monitors, cell phones, and televisions. Neodymium is useful in laptop hard drives and in the processes that convert crude oil into gasoline (Figure 19.3). Holmium is found in dental and medical equipment. In addition, many alternative energy technologies rely heavily on lanthanoids. Neodymium and dysprosium are key components of hybrid vehicle engines and the magnets used in wind turbines.

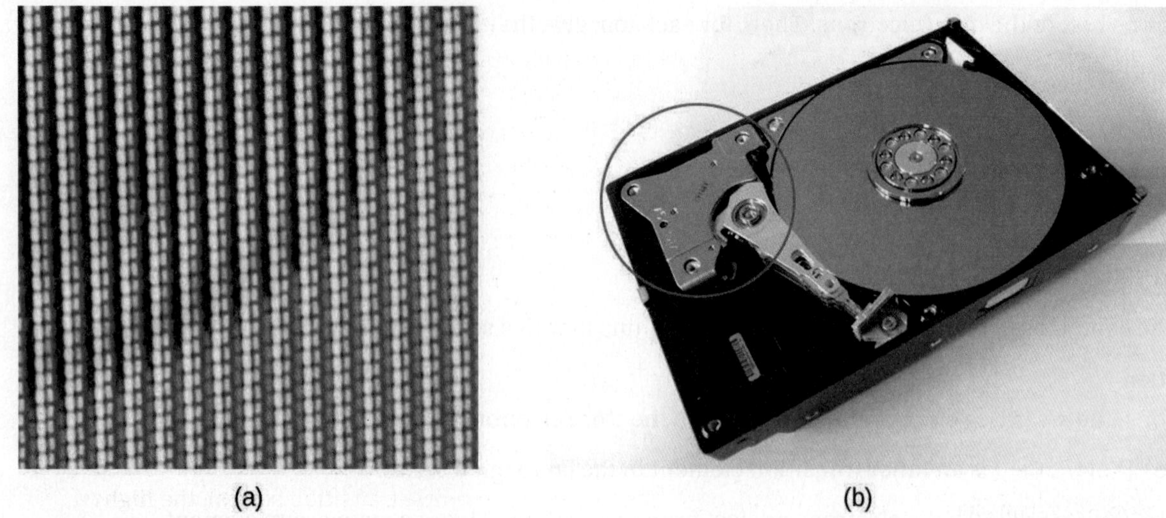

FIGURE 19.3 (a) Europium is used in display screens for televisions, computer monitors, and cell phones. (b) Neodymium magnets are commonly found in computer hard drives. (credit b: modification of work by "KUERT Datenrettung"/Flickr)

As the demand for lanthanide materials has increased faster than supply, prices have also increased. In 2008, dysprosium cost \$110/kg; by 2014, the price had increased to \$470/kg. Increasing the supply of lanthanoid elements is one of the most significant challenges facing the industries that rely on the optical and magnetic properties of these materials.

The transition elements have many properties in common with other metals. They are almost all hard, high-melting solids that conduct heat and electricity well. They readily form alloys and lose electrons to form stable cations. In addition, transition metals form a wide variety of stable **coordination compounds**, in which the central metal atom or ion acts as a Lewis acid and accepts one or more pairs of electrons. Many different molecules and ions can donate lone pairs to the metal center, serving as Lewis bases. In this chapter, we shall focus primarily on the chemical behavior of the elements of the first transition series.

Properties of the Transition Elements

Transition metals demonstrate a wide range of chemical behaviors. As can be seen from their reduction potentials (see Appendix H), some transition metals are strong reducing agents, whereas others have very low reactivity. For example, the lanthanides all form stable 3+ aqueous cations. The driving force for such oxidations is similar to that of alkaline earth metals such as Be or Mg, forming Be^{2+} and Mg^{2+}. On the other hand, materials like platinum and gold have much higher reduction potentials. Their ability to resist oxidation makes them useful materials for constructing circuits and jewelry.

Ions of the lighter *d*-block elements, such as Cr^{3+}, Fe^{3+}, and Co^{2+}, form colorful hydrated ions that are stable in water. However, ions in the period just below these (Mo^{3+}, Ru^{3+}, and Ir^{2+}) are unstable and react readily with oxygen from the air. The majority of simple, water-stable ions formed by the heavier *d*-block elements are oxyanions such as $MoO_4{}^{2-}$ and $ReO_4{}^{-}$.

Ruthenium, osmium, rhodium, iridium, palladium, and platinum are the **platinum metals**. With difficulty, they form simple cations that are stable in water, and, unlike the earlier elements in the second and third transition series, they do not form stable oxyanions.

Both the *d*- and *f*-block elements react with nonmetals to form binary compounds; heating is often required. These elements react with halogens to form a variety of halides ranging in oxidation state from 1+ to 6+. On heating, oxygen reacts with all of the transition elements except palladium, platinum, silver, and gold. The oxides of these latter metals can be formed using other reactants, but they decompose upon heating. The *f*-block elements, the elements of group 3, and the elements of the first transition series except copper react with aqueous solutions of acids, forming hydrogen gas and solutions of the corresponding salts.

Transition metals can form compounds with a wide range of oxidation states. Some of the observed oxidation states of the elements of the first transition series are shown in Figure 19.4. As we move from left to right across the first transition series, we see that the number of common oxidation states increases at first to a maximum towards the middle of the table, then decreases. The values in the table are typical values; there are other known values, and it is possible to synthesize new additions. For example, in 2014, researchers were successful in synthesizing a new oxidation state of iridium (9+).

21 Sc	22 Ti	23 V	24 Cr	25 Mn	26 Fe	27 Co	28 Ni	29 Cu	30 Zn
								1+	
		2+	2+	2+	2+	2+	2+	2+	2+
3+	3+	3+	3+	3+	3+	3+	3+	3+	
	4+	4+	4+	4+					
		5+							
			6+	6+	6+				
				7+					

FIGURE 19.4 Transition metals of the first transition series can form compounds with varying oxidation states.

For the elements scandium through manganese (the first half of the first transition series), the highest oxidation state corresponds to the loss of all of the electrons in both the *s* and *d* orbitals of their valence shells. The titanium(IV) ion, for example, is formed when the titanium atom loses its two 3*d* and two 4*s* electrons. These highest oxidation states are the most stable forms of scandium, titanium, and vanadium. However, it is not possible to continue to remove all of the valence electrons from metals as we continue through the series. Iron is known to form oxidation states from 2+ to 6+, with iron(II) and iron(III) being the most common. Most of the elements of the first transition series form ions with a charge of 2+ or 3+ that are stable in water, although those of the early members of the series can be readily oxidized by air.

The elements of the second and third transition series generally are more stable in higher oxidation states than are the elements of the first series. In general, the atomic radius increases down a group, which leads to the ions of the second and third series being larger than are those in the first series. Removing electrons from orbitals that are located farther from the nucleus is easier than removing electrons close to the nucleus. For example, molybdenum and tungsten, members of group 6, are limited mostly to an oxidation state of 6+ in aqueous solution. Chromium, the lightest member of the group, forms stable Cr^{3+} ions in water and, in the absence of air, less stable Cr^{2+} ions. The sulfide with the highest oxidation state for chromium is Cr_2S_3, which contains the Cr^{3+} ion. Molybdenum and tungsten form sulfides in which the metals exhibit oxidation states of 4+ and 6+.

EXAMPLE 19.2

Activity of the Transition Metals

Which is the strongest oxidizing agent in acidic solution: dichromate ion, which contains chromium(VI), permanganate ion, which contains manganese(VII), or titanium dioxide, which contains titanium(IV)?

Solution

First, we need to look up the reduction half reactions (in Appendix L) for each oxide in the specified oxidation state:

$$Cr_2O_7{}^{2-} + 14H^+ + 6e^- \longrightarrow 2Cr^{3+} + 7H_2O \qquad +1.33\text{ V}$$

$$MnO_4{}^- + 8H^+ + 5e^- \longrightarrow Mn^{2+} + H_2O \qquad +1.51\text{ V}$$

$$TiO_2 + 4H^+ + 2e^- \longrightarrow Ti^{2+} + 2H_2O \qquad -0.50\text{ V}$$

A larger reduction potential means that it is easier to reduce the reactant. Permanganate, with the largest reduction potential, is the strongest oxidizer under these conditions. Dichromate is next, followed by titanium dioxide as the weakest oxidizing agent (the hardest to reduce) of this set.

Check Your Learning

Predict what reaction (if any) will occur between HCl and Co(*s*), and between HBr and Pt(*s*). You will need to use the standard reduction potentials from Appendix L.

Answer:
$Co(s) + 2HCl \longrightarrow H_2 + CoCl_2(aq)$; no reaction because Pt(*s*) will not be oxidized by H^+

Preparation of the Transition Elements

Ancient civilizations knew about iron, copper, silver, and gold. The time periods in human history known as the Bronze Age and Iron Age mark the advancements in which societies learned to isolate certain metals and use them to make tools and goods. Naturally occurring ores of copper, silver, and gold can contain high concentrations of these metals in elemental form (Figure 19.5). Iron, on the other hand, occurs on earth almost exclusively in oxidized forms, such as rust (Fe_2O_3). The earliest known iron implements were made from iron meteorites. Surviving iron artifacts dating from approximately 4000 to 2500 BC are rare, but all known examples contain specific alloys of iron and nickel that occur only in extraterrestrial objects, not on earth. It took thousands of years of technological advances before civilizations developed iron **smelting**, the ability to extract a pure element from its naturally occurring ores and for iron tools to become common.

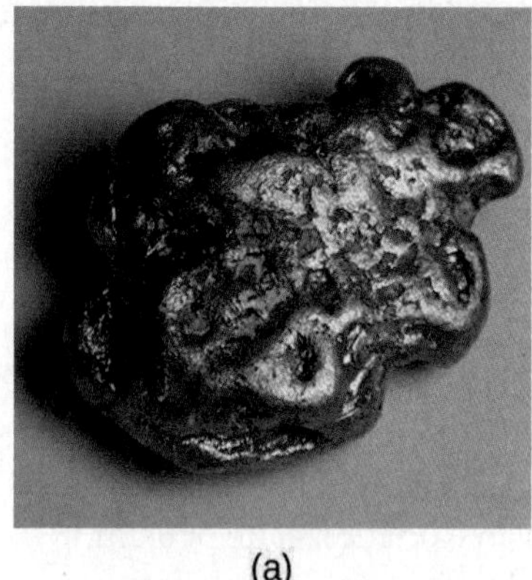
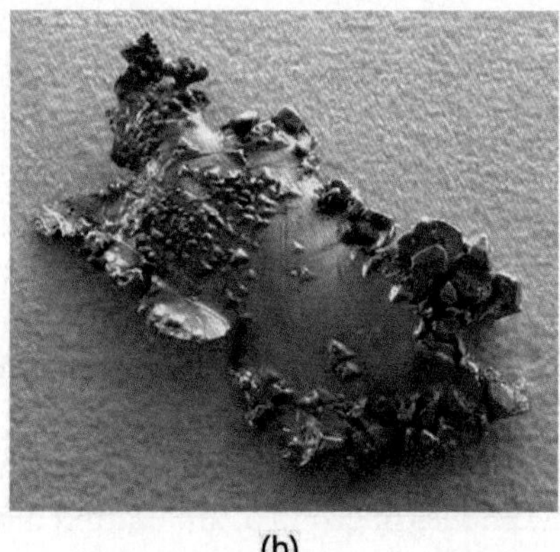

(a) (b) (c)

FIGURE 19.5 Transition metals occur in nature in various forms. Examples include (a) a nugget of copper, (b) a deposit of gold, and (c) an ore containing oxidized iron. (credit a: modification of work by http://images-of-elements.com/copper-2.jpg; credit c: modification of work by http://images-of-elements.com/iron-ore.jpg)

Generally, the transition elements are extracted from minerals found in a variety of ores. However, the ease of their recovery varies widely, depending on the concentration of the element in the ore, the identity of the other elements present, and the difficulty of reducing the element to the free metal.

In general, it is not difficult to reduce ions of the *d*-block elements to the free element. Carbon is a sufficiently strong reducing agent in most cases. However, like the ions of the more active main group metals, ions of the *f*-block elements must be isolated by electrolysis or by reduction with an active metal such as calcium.

We shall discuss the processes used for the isolation of iron, copper, and silver because these three processes illustrate the principal means of isolating most of the *d*-block metals. In general, each of these processes involves three principal steps: preliminary treatment, smelting, and refining.

1. Preliminary treatment. In general, there is an initial treatment of the ores to make them suitable for the extraction of the metals. This usually involves crushing or grinding the ore, concentrating the metal-bearing components, and sometimes treating these substances chemically to convert them into compounds that are easier to reduce to the metal.
2. Smelting. The next step is the extraction of the metal in the molten state, a process called smelting, which includes reduction of the metallic compound to the metal. Impurities may be removed by the addition of a compound that forms a slag—a substance with a low melting point that can be readily separated from the molten metal.
3. Refining. The final step in the recovery of a metal is refining the metal. Low boiling metals such as zinc and mercury can be refined by distillation. When fused on an inclined table, low melting metals like tin flow away from higher-melting impurities. Electrolysis is another common method for refining metals.

Isolation of Iron

The early application of iron to the manufacture of tools and weapons was possible because of the wide distribution of iron ores and the ease with which iron compounds in the ores could be reduced by carbon. For a long time, charcoal was the form of carbon used in the reduction process. The production and use of iron became much more widespread about 1620, when coke was introduced as the reducing agent. Coke is a form of carbon formed by heating coal in the absence of air to remove impurities.

The first step in the metallurgy of iron is usually roasting the ore (heating the ore in air) to remove water, decomposing carbonates into oxides, and converting sulfides into oxides. The oxides are then reduced in a blast furnace that is 80–100 feet high and about 25 feet in diameter (Figure 19.6) in which the roasted ore, coke, and limestone (impure $CaCO_3$) are introduced continuously into the top. Molten iron and slag are withdrawn at the bottom. The entire stock in a furnace may weigh several hundred tons.

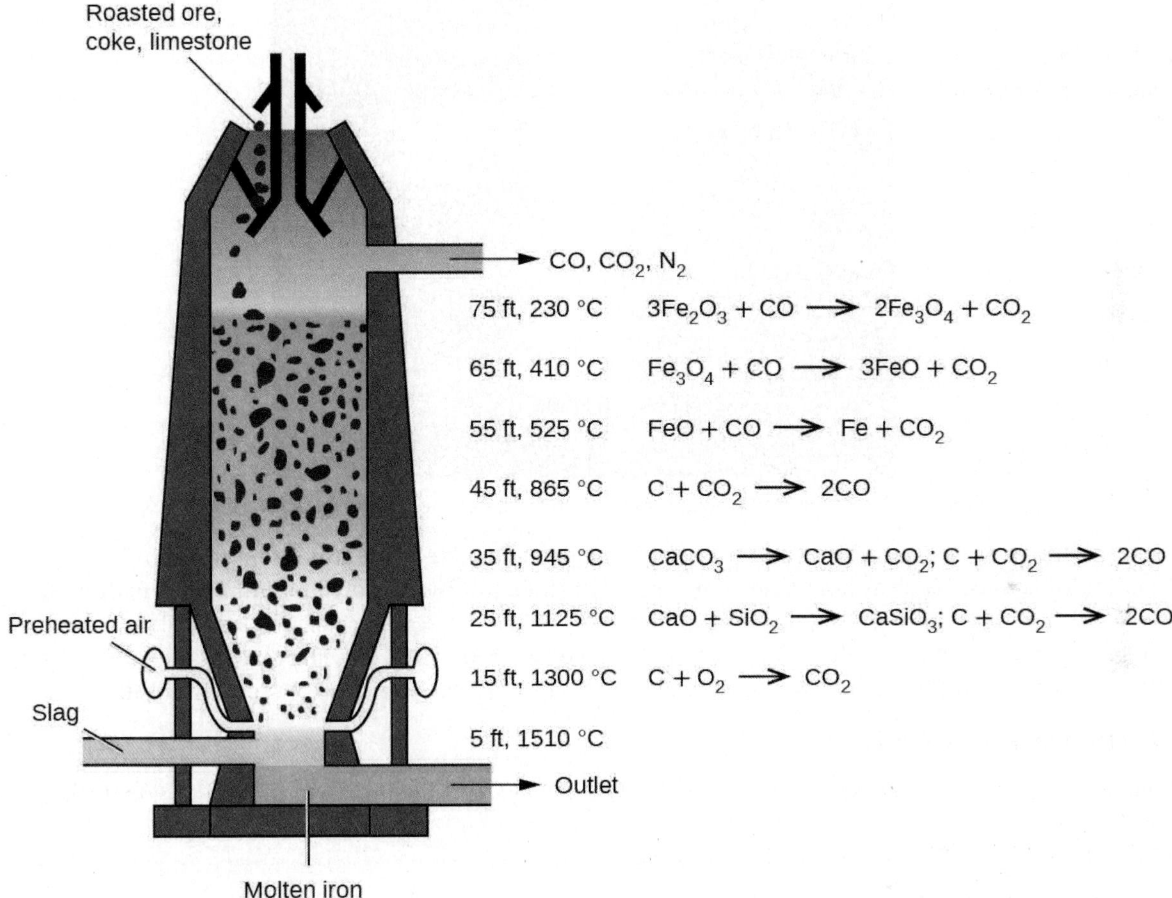

FIGURE 19.6 Within a blast furnace, different reactions occur in different temperature zones. Carbon monoxide is generated in the hotter bottom regions and rises upward to reduce the iron oxides to pure iron through a series of reactions that take place in the upper regions.

Near the bottom of a furnace are nozzles through which preheated air is blown into the furnace. As soon as the air enters, the coke in the region of the nozzles is oxidized to carbon dioxide with the liberation of a great deal of heat. The hot carbon dioxide passes upward through the overlying layer of white-hot coke, where it is reduced to carbon monoxide:

$$CO_2(g) + C(s) \longrightarrow 2CO(g)$$

The carbon monoxide serves as the reducing agent in the upper regions of the furnace. The individual reactions are indicated in Figure 19.6.

The iron oxides are reduced in the upper region of the furnace. In the middle region, limestone (calcium carbonate) decomposes, and the resulting calcium oxide combines with silica and silicates in the ore to form slag. The slag is mostly calcium silicate and contains most of the commercially unimportant components of the ore:

$$CaO(s) + SiO_2(s) \longrightarrow CaSiO_3(l)$$

Just below the middle of the furnace, the temperature is high enough to melt both the iron and the slag. They collect in layers at the bottom of the furnace; the less dense slag floats on the iron and protects it from oxidation. Several times a day, the slag and molten iron are withdrawn from the furnace. The iron is transferred to casting machines or to a steelmaking plant (Figure 19.7).

FIGURE 19.7 Molten iron is shown being cast as steel. (credit: Clint Budd)

Much of the iron produced is refined and converted into steel. **Steel** is made from iron by removing impurities and adding substances such as manganese, chromium, nickel, tungsten, molybdenum, and vanadium to produce alloys with properties that make the material suitable for specific uses. Most steels also contain small but definite percentages of carbon (0.04%–2.5%). However, a large part of the carbon contained in iron must be removed in the manufacture of steel; otherwise, the excess carbon would make the iron brittle.

LINK TO LEARNING

You can watch an animation of steelmaking (http://openstax.org/l/16steelmaking) that walks you through the process.

Isolation of Copper

The most important ores of copper contain copper sulfides (such as covellite, CuS), although copper oxides (such as tenorite, CuO) and copper hydroxycarbonates [such as malachite, $Cu_2(OH)_2CO_3$] are sometimes found. In the production of copper metal, the concentrated sulfide ore is roasted to remove part of the sulfur as sulfur dioxide. The remaining mixture, which consists of Cu_2S, FeS, FeO, and SiO_2, is mixed with limestone, which serves as a flux (a material that aids in the removal of impurities), and heated. Molten slag forms as the iron and silica are removed by Lewis acid-base reactions:

$$CaCO_3(s) + SiO_2(s) \longrightarrow CaSiO_3(l) + CO_2(g)$$
$$FeO(s) + SiO_2(s) \longrightarrow FeSiO_3(l)$$

In these reactions, the silicon dioxide behaves as a Lewis acid, which accepts a pair of electrons from the Lewis base (the oxide ion).

Reduction of the Cu_2S that remains after smelting is accomplished by blowing air through the molten material.

The air converts part of the Cu_2S into Cu_2O. As soon as copper(I) oxide is formed, it is reduced by the remaining copper(I) sulfide to metallic copper:

$$2Cu_2S(l) + 3O_2(g) \longrightarrow 2Cu_2O(l) + 2SO_2(g)$$
$$2Cu_2O(l) + Cu_2S(l) \longrightarrow 6Cu(l) + SO_2(g)$$

The copper obtained in this way is called blister copper because of its characteristic appearance, which is due to the air blisters it contains (Figure 19.8). This impure copper is cast into large plates, which are used as anodes in the electrolytic refining of the metal (which is described in the chapter on electrochemistry).

FIGURE 19.8 Blister copper is obtained during the conversion of copper-containing ore into pure copper. (credit: "Tortie tude"/Wikimedia Commons)

Isolation of Silver

Silver sometimes occurs in large nuggets (Figure 19.9) but more frequently in veins and related deposits. At one time, panning was an effective method of isolating both silver and gold nuggets. Due to their low reactivity, these metals, and a few others, occur in deposits as nuggets. The discovery of platinum was due to Spanish explorers in Central America mistaking platinum nuggets for silver. When the metal is not in the form of nuggets, it often useful to employ a process called **hydrometallurgy** to separate silver from its ores. Hydrology involves the separation of a metal from a mixture by first converting it into soluble ions and then extracting and reducing them to precipitate the pure metal. In the presence of air, alkali metal cyanides readily form the soluble dicyanoargentate(I) ion, $\left[Ag(CN)_2\right]^-$, from silver metal or silver-containing compounds such as Ag_2S and AgCl. Representative equations are:

$$4Ag(s) + 8CN^-(aq) + O_2(g) + 2H_2O(l) \longrightarrow 4\left[Ag(CN)_2\right]^-(aq) + 4OH^-(aq)$$
$$2Ag_2S(s) + 8CN^-(aq) + O_2(g) + 2H_2O(l) \longrightarrow 4\left[Ag(CN)_2\right]^-(aq) + 2S(s) + 4OH^-(aq)$$
$$AgCl(s) + 2CN^-(aq) \longrightarrow \left[Ag(CN)_2\right]^-(aq) + Cl^-(aq)$$

(a)

(b)

FIGURE 19.9 Naturally occurring free silver may be found as nuggets (a) or in veins (b). (credit a: modification of work by “Teravolt”/Wikimedia Commons; credit b: modification of work by James St. John)

The silver is precipitated from the cyanide solution by the addition of either zinc or iron(II) ions, which serves as the reducing agent:

$$2\left[Ag(CN)_2\right]^-(aq) + Zn(s) \longrightarrow 2Ag(s) + \left[Zn(CN)_4\right]^{2-}(aq)$$

EXAMPLE 19.3

Refining Redox

One of the steps for refining silver involves converting silver into dicyanoargenate(I) ions:

$$4Ag(s) + 8CN^-(aq) + O_2(g) + 2H_2O(l) \longrightarrow 4\left[Ag(CN)_2\right]^-(aq) + 4OH^-(aq)$$

Explain why oxygen must be present to carry out the reaction. Why does the reaction not occur as:

$$4Ag(s) + 8CN^-(aq) \longrightarrow 4\left[Ag(CN)_2\right]^-(aq)?$$

Solution

The charges, as well as the atoms, must balance in reactions. The silver atom is being oxidized from the 0 oxidation state to the 1+ state. Whenever something loses electrons, something must also gain electrons (be reduced) to balance the equation. Oxygen is a good oxidizing agent for these reactions because it can gain electrons to go from the 0 oxidation state to the 2− state.

Check Your Learning

During the refining of iron, carbon must be present in the blast furnace. Why is carbon necessary to convert iron oxide into iron?

Answer:

The carbon is converted into CO, which is the reducing agent that accepts electrons so that iron(III) can be reduced to iron(0).

Transition Metal Compounds

The bonding in the simple compounds of the transition elements ranges from ionic to covalent. In their lower oxidation states, the transition elements form ionic compounds; in their higher oxidation states, they form covalent compounds or polyatomic ions. The variation in oxidation states exhibited by the transition elements gives these compounds a metal-based, oxidation-reduction chemistry. The chemistry of several classes of compounds containing elements of the transition series follows.

Halides

Anhydrous halides of each of the transition elements can be prepared by the direct reaction of the metal with halogens. For example:

$$2Fe(s) + 3Cl_2(g) \longrightarrow 2FeCl_3(s)$$

Heating a metal halide with additional metal can be used to form a halide of the metal with a lower oxidation state:

$$Fe(s) + 2FeCl_3(s) \longrightarrow 3FeCl_2(s)$$

The stoichiometry of the metal halide that results from the reaction of the metal with a halogen is determined by the relative amounts of metal and halogen and by the strength of the halogen as an oxidizing agent. Generally, fluorine forms fluoride-containing metals in their highest oxidation states. The other halogens may not form analogous compounds.

In general, the preparation of stable water solutions of the halides of the metals of the first transition series is by the addition of a hydrohalic acid to carbonates, hydroxides, oxides, or other compounds that contain basic anions. Sample reactions are:

$$NiCO_3(s) + 2HF(aq) \longrightarrow NiF_2(aq) + H_2O(l) + CO_2(g)$$
$$Co(OH)_2(s) + 2HBr(aq) \longrightarrow CoBr_2(aq) + 2H_2O(l)$$

Most of the first transition series metals also dissolve in acids, forming a solution of the salt and hydrogen gas. For example:

$$Cr(s) + 2HCl(aq) \longrightarrow CrCl_2(aq) + H_2(g)$$

The polarity of bonds with transition metals varies based not only upon the electronegativities of the atoms involved but also upon the oxidation state of the transition metal. Remember that bond polarity is a continuous spectrum with electrons being shared evenly (covalent bonds) at one extreme and electrons being transferred completely (ionic bonds) at the other. No bond is ever 100% ionic, and the degree to which the electrons are evenly distributed determines many properties of the compound. Transition metal halides with low oxidation numbers form more ionic bonds. For example, titanium(II) chloride and titanium(III) chloride ($TiCl_2$ and $TiCl_3$) have high melting points that are characteristic of ionic compounds, but titanium(IV) chloride ($TiCl_4$) is a volatile liquid, consistent with having covalent titanium-chlorine bonds. All halides of the heavier *d*-block elements have significant covalent characteristics.

The covalent behavior of the transition metals with higher oxidation states is exemplified by the reaction of the metal tetrahalides with water. Like covalent silicon tetrachloride, both the titanium and vanadium tetrahalides react with water to give solutions containing the corresponding hydrohalic acids and the metal oxides:

$$SiCl_4(l) + 2H_2O(l) \longrightarrow SiO_2(s) + 4HCl(aq)$$
$$TiCl_4(l) + 2H_2O(l) \longrightarrow TiO_2(s) + 4HCl(aq)$$

Oxides

As with the halides, the nature of bonding in oxides of the transition elements is determined by the oxidation state of the metal. Oxides with low oxidation states tend to be more ionic, whereas those with higher oxidation states are more covalent. These variations in bonding are because the electronegativities of the elements are not fixed values. The electronegativity of an element increases with increasing oxidation state. Transition metals in low oxidation states have lower electronegativity values than oxygen; therefore, these metal oxides are ionic. Transition metals in very high oxidation states have electronegativity values close to that of oxygen, which leads to these oxides being covalent.

The oxides of the first transition series can be prepared by heating the metals in air. These oxides are Sc_2O_3, TiO_2, V_2O_5, Cr_2O_3, Mn_3O_4, Fe_3O_4, Co_3O_4, NiO, and CuO.

Alternatively, these oxides and other oxides (with the metals in different oxidation states) can be produced by heating the corresponding hydroxides, carbonates, or oxalates in an inert atmosphere. Iron(II) oxide can be prepared by heating iron(II) oxalate, and cobalt(II) oxide is produced by heating cobalt(II) hydroxide:

$$FeC_2O_4(s) \longrightarrow FeO(s) + CO(g) + CO_2(g)$$
$$Co(OH)_2(s) \longrightarrow CoO(s) + H_2O(g)$$

With the exception of CrO_3 and Mn_2O_7, transition metal oxides are not soluble in water. They can react with acids and, in a few cases, with bases. Overall, oxides of transition metals with the lowest oxidation states are basic (and react with acids), the intermediate ones are amphoteric, and the highest oxidation states are primarily acidic. Basic metal oxides at a low oxidation state react with aqueous acids to form solutions of salts and water. Examples include the reaction of cobalt(II) oxide accepting protons from nitric acid, and scandium(III) oxide accepting protons from hydrochloric acid:

$$CoO(s) + 2HNO_3(aq) \longrightarrow Co(NO_3)_2(aq) + H_2O(l)$$
$$Sc_2O_3(s) + 6HCl(aq) \longrightarrow 2ScCl_3(aq) + 3H_2O(l)$$

The oxides of metals with oxidation states of 4+ are amphoteric, and most are not soluble in either acids or bases. Vanadium(V) oxide, chromium(VI) oxide, and manganese(VII) oxide are acidic. They react with solutions of hydroxides to form salts of the oxyanions $VO_4{}^{3-}$, $CrO_4{}^{2-}$, and $MnO_4{}^{-}$. For example, the complete ionic equation for the reaction of chromium(VI) oxide with a strong base is given by:

$$CrO_3(s) + 2Na^+(aq) + 2OH^-(aq) \longrightarrow 2Na^+(aq) + CrO_4{}^{2-}(aq) + H_2O(l)$$

Chromium(VI) oxide and manganese(VII) oxide react with water to form the acids H_2CrO_4 and $HMnO_4$, respectively.

Hydroxides

When a soluble hydroxide is added to an aqueous solution of a salt of a transition metal of the first transition series, a gelatinous precipitate forms. For example, adding a solution of sodium hydroxide to a solution of cobalt sulfate produces a gelatinous pink or blue precipitate of cobalt(II) hydroxide. The net ionic equation is:

$$Co^{2+}(aq) + 2OH^-(aq) \longrightarrow Co(OH)_2(s)$$

In this and many other cases, these precipitates are hydroxides containing the transition metal ion, hydroxide ions, and water coordinated to the transition metal. In other cases, the precipitates are hydrated oxides composed of the metal ion, oxide ions, and water of hydration:

$$4Fe^{3+}(aq) + 6OH^-(aq) + n\,H_2O(l) \longrightarrow 2Fe_2O_3{\cdot}(n + 3)H_2O(s)$$

These substances do not contain hydroxide ions. However, both the hydroxides and the hydrated oxides react with acids to form salts and water. When precipitating a metal from solution, it is necessary to avoid an excess of hydroxide ion, as this may lead to complex ion formation as discussed later in this chapter. The precipitated metal hydroxides can be separated for further processing or for waste disposal.

Carbonates

Many of the elements of the first transition series form insoluble carbonates. It is possible to prepare these carbonates by the addition of a soluble carbonate salt to a solution of a transition metal salt. For example, nickel carbonate can be prepared from solutions of nickel nitrate and sodium carbonate according to the following net ionic equation:

$$Ni^{2+}(aq) + CO_3{}^{2-} \longrightarrow NiCO_3(s)$$

The reactions of the transition metal carbonates are similar to those of the active metal carbonates. They react with acids to form metals salts, carbon dioxide, and water. Upon heating, they decompose, forming the transition metal oxides.

Other Salts

In many respects, the chemical behavior of the elements of the first transition series is very similar to that of the main group metals. In particular, the same types of reactions that are used to prepare salts of the main group metals can be used to prepare simple ionic salts of these elements.

A variety of salts can be prepared from metals that are more active than hydrogen by reaction with the corresponding acids: Scandium metal reacts with hydrobromic acid to form a solution of scandium bromide:

$$2Sc(s) + 6HBr(aq) \longrightarrow 2ScBr_3(aq) + 3H_2(g)$$

The common compounds that we have just discussed can also be used to prepare salts. The reactions involved include the reactions of oxides, hydroxides, or carbonates with acids. For example:

$$Ni(OH)_2(s) + 2H_3O^+(aq) + 2ClO_4{}^-(aq) \longrightarrow Ni^{2+}(aq) + 2ClO_4{}^-(aq) + 4H_2O(l)$$

Substitution reactions involving soluble salts may be used to prepare insoluble salts. For example:

$$Ba^{2+}(aq) + 2Cl^-(aq) + 2K^+(aq) + CrO_4{}^{2-}(aq) \longrightarrow BaCrO_4(s) + 2K^+(aq) + 2Cl^-(aq)$$

In our discussion of oxides in this section, we have seen that reactions of the covalent oxides of the transition elements with hydroxides form salts that contain oxyanions of the transition elements.

 HOW SCIENCES INTERCONNECT

High Temperature Superconductors

A **superconductor** is a substance that conducts electricity with no resistance. This lack of resistance means that there is no energy loss during the transmission of electricity. This would lead to a significant reduction in the cost of electricity.

Most currently used, commercial superconducting materials, such as NbTi and Nb_3Sn, do not become superconducting until they are cooled below 23 K (−250 °C). This requires the use of liquid helium, which has a boiling temperature of 4 K and is expensive and difficult to handle. The cost of liquid helium has deterred the widespread application of superconductors.

One of the most exciting scientific discoveries of the 1980s was the characterization of compounds that exhibit superconductivity at temperatures above 90 K. (Compared to liquid helium, 90 K is a high temperature.) Typical among the high-temperature superconducting materials are oxides containing yttrium (or one of several rare earth elements), barium, and copper in a 1:2:3 ratio. The formula of the ionic yttrium compound is $YBa_2Cu_3O_7$.

The new materials become superconducting at temperatures close to 90 K (Figure 19.10), temperatures that can be reached by cooling with liquid nitrogen (boiling temperature of 77 K). Not only are liquid nitrogen-cooled materials easier to handle, but the cooling costs are also about 1000 times lower than for liquid helium.

Further advances during the same period included materials that became superconducting at even higher temperatures and with a wider array of materials. The DuPont team led by Uma Chowdry and Arthur Sleight identified Bismouth-Strontium-Copper-Oxides that became superconducting at temperatures as high as 110 K and, importantly, did not contain rare earth elements. Advances continued through the subsequent decades until, in 2020, a team led by Ranga Dias at University of Rochester announced the development of a room-temperature superconductor, opening doors to widespread applications. More research and development is needed to realize the potential of these materials, but the possibilities are very promising.

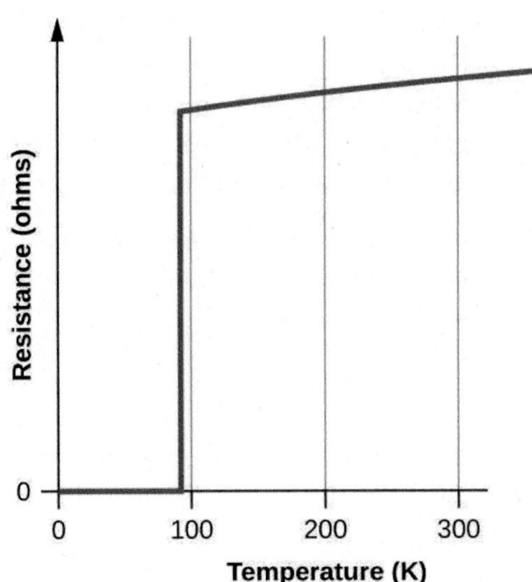

FIGURE 19.10 The resistance of the high-temperature superconductor $YBa_2Cu_3O_7$ varies with temperature. Note how the resistance falls to zero below 92 K, when the substance becomes superconducting.

Although the brittle, fragile nature of these materials presently hampers their commercial applications, they have tremendous potential that researchers are hard at work improving their processes to help realize. Superconducting transmission lines would carry current for hundreds of miles with no loss of power due to resistance in the wires. This could allow generating stations to be located in areas remote from population centers and near the natural resources necessary for power production. The first project demonstrating the viability of high-temperature superconductor power transmission was established in New York in 2008.

Researchers are also working on using this technology to develop other applications, such as smaller and more powerful microchips. In addition, high-temperature superconductors can be used to generate magnetic fields for applications such as medical devices, magnetic levitation trains, and containment fields for nuclear fusion reactors (Figure 19.11).

(a)

(b)

FIGURE 19.11 (a) This magnetic levitation train (or maglev) uses superconductor technology to move along its tracks. (b) A magnet can be levitated using a dish like this as a superconductor. (credit a: modification of work by Alex Needham; credit b: modification of work by Kevin Jarrett)

LINK TO LEARNING

Watch how a high-temperature superconductor (http://openstax.org/l/16supercond) levitates around a magnetic racetrack in the video.

19.2 Coordination Chemistry of Transition Metals

LEARNING OBJECTIVES

By the end of this section, you will be able to:

- List the defining traits of coordination compounds
- Describe the structures of complexes containing monodentate and polydentate ligands
- Use standard nomenclature rules to name coordination compounds
- Explain and provide examples of geometric and optical isomerism
- Identify several natural and technological occurrences of coordination compounds

The hemoglobin in your blood, the chlorophyll in green plants, vitamin B-12, and the catalyst used in the manufacture of polyethylene all contain coordination compounds. Ions of the metals, especially the transition metals, are likely to form complexes. Many of these compounds are highly colored (Figure 19.12). In the remainder of this chapter, we will consider the structure and bonding of these remarkable compounds.

FIGURE 19.12 Metal ions that contain partially filled *d* subshell usually form colored complex ions; ions with empty *d* subshell (d^0) or with filled *d* subshells (d^{10}) usually form colorless complexes. This figure shows, from left to right, solutions containing $[M(H_2O)_6]^{n+}$ ions with $M = Sc^{3+}(d^0)$, $Cr^{3+}(d^3)$, $Co^{2+}(d^7)$, $Ni^{2+}(d^8)$, $Cu^{2+}(d^9)$, and $Zn^{2+}(d^{10})$. (credit: Sahar Atwa)

Remember that in most main group element compounds, the valence electrons of the isolated atoms combine to form chemical bonds that satisfy the octet rule. For instance, the four valence electrons of carbon overlap with electrons from four hydrogen atoms to form CH_4. The one valence electron leaves sodium and adds to the seven valence electrons of chlorine to form the ionic formula unit NaCl (Figure 19.13). Transition metals do not normally bond in this fashion. They primarily form coordinate covalent bonds, a form of the Lewis acid-base interaction in which both of the electrons in the bond are contributed by a donor (Lewis base) to an electron acceptor (Lewis acid). The Lewis acid in coordination complexes, often called a **central metal** ion (or atom), is often a transition metal or inner transition metal, although main group elements can also form **coordination compounds**. The Lewis base donors, called **ligands**, can be a wide variety of chemicals—atoms, molecules, or ions. The only requirement is that they have one or more electron pairs, which can be donated to the central metal. Most often, this involves a **donor atom** with a lone pair of electrons that can form a coordinate bond to the metal.

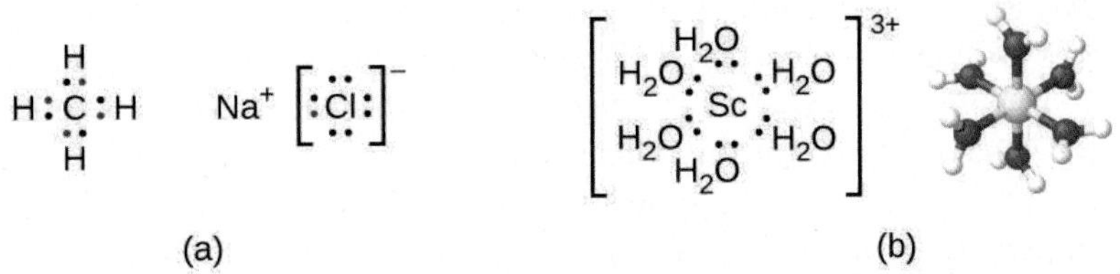

FIGURE 19.13 (a) Covalent bonds involve the sharing of electrons, and ionic bonds involve the transferring of electrons associated with each bonding atom, as indicated by the colored electrons. (b) However, coordinate covalent bonds involve electrons from a Lewis base being donated to a metal center. The lone pairs from six water molecules form bonds to the scandium ion to form an octahedral complex. (Only the donated pairs are shown.)

The **coordination sphere** consists of the central metal ion or atom plus its attached ligands. Brackets in a formula enclose the coordination sphere; species outside the brackets are not part of the coordination sphere. The **coordination number** of the central metal ion or atom is the number of donor atoms bonded to it. The coordination number for the silver ion in $[Ag(NH_3)_2]^+$ is two (Figure 19.14). For the copper(II) ion in $[CuCl_4]^{2-}$, the coordination number is four, whereas for the cobalt(II) ion in $[Co(H_2O)_6]^{2+}$ the coordination number is six. Each of these ligands is **monodentate**, from the Greek for "one toothed," meaning that they connect with the central metal through only one atom. In this case, the number of ligands and the coordination number are equal.

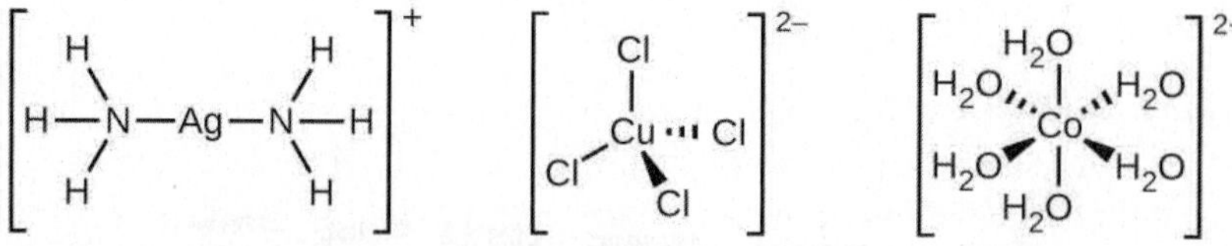

FIGURE 19.14 The complexes (a) $[Ag(NH_3)_2]^+$, (b) $[Cu(Cl)_4]^{2-}$, and (c) $[Co(H_2O)_6]^{2+}$ have coordination numbers of two, four, and six, respectively. The geometries of these complexes are the same as we have seen with VSEPR theory for main group elements: linear, tetrahedral, and octahedral.

Many other ligands coordinate to the metal in more complex fashions. **Bidentate ligands** are those in which two atoms coordinate to the metal center. For example, ethylenediamine (en, $H_2NCH_2CH_2NH_2$) contains two nitrogen atoms, each of which has a lone pair and can serve as a Lewis base (Figure 19.15). Both of the atoms can coordinate to a single metal center. In the complex $[Co(en)_3]^{3+}$, there are three bidentate en ligands, and the coordination number of the cobalt(III) ion is six. The most common coordination numbers are two, four, and six, but examples of all coordination numbers from 1 to 15 are known.

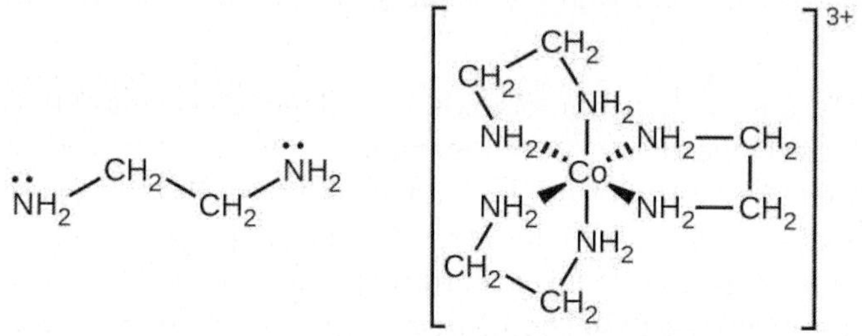

(a) (b)

FIGURE 19.15 (a) The ethylenediamine (en) ligand contains two atoms with lone pairs that can coordinate to the metal center. (b) The cobalt(III) complex $\left[Co(en)_3\right]^{3+}$ contains three of these ligands, each forming two bonds to the cobalt ion.

Any ligand that bonds to a central metal ion by more than one donor atom is a **polydentate ligand** (or "many teeth") because it can bite into the metal center with more than one bond. The term **chelate** (pronounced "KEY-late") from the Greek for "claw" is also used to describe this type of interaction. Many polydentate ligands are **chelating ligands**, and a complex consisting of one or more of these ligands and a central metal is a chelate. A chelating ligand is also known as a chelating agent. A chelating ligand holds the metal ion rather like a crab's claw would hold a marble. Figure 19.15 showed one example of a chelate. The heme complex in hemoglobin is another important example (Figure 19.16). It contains a polydentate ligand with four donor atoms that coordinate to iron.

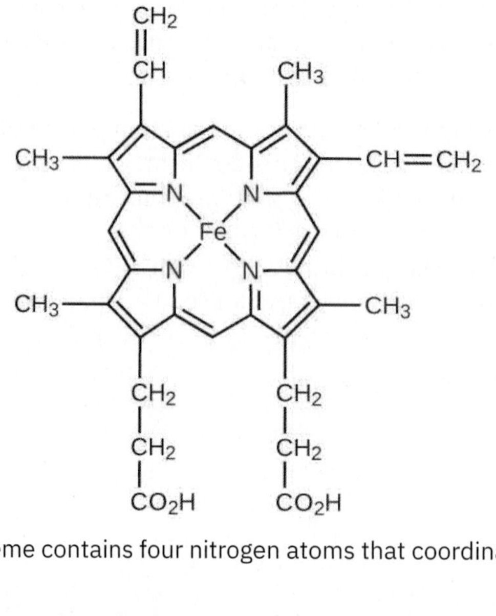

FIGURE 19.16 The single ligand heme contains four nitrogen atoms that coordinate to iron in hemoglobin to form a

chelate.

Polydentate ligands are sometimes identified with prefixes that indicate the number of donor atoms in the ligand. As we have seen, ligands with one donor atom, such as NH_3, Cl^-, and H_2O, are monodentate ligands. Ligands with two donor groups are bidentate ligands. Ethylenediamine, $H_2NCH_2CH_2NH_2$, and the anion of the acid glycine, $NH_2CH_2CO_2^-$ (Figure 19.17) are examples of bidentate ligands. Tridentate ligands, tetradentate ligands, pentadentate ligands, and hexadentate ligands contain three, four, five, and six donor atoms, respectively. The ligand in heme (Figure 19.16) is a tetradentate ligand.

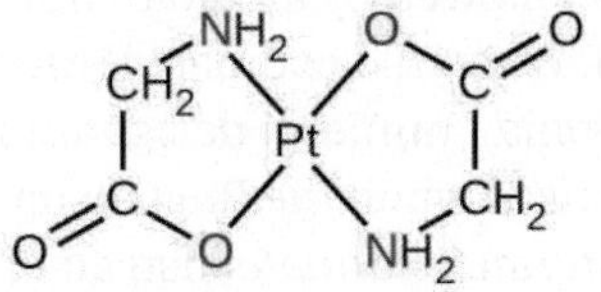

FIGURE 19.17 Each of the anionic ligands shown attaches in a bidentate fashion to platinum(II), with both a nitrogen and oxygen atom coordinating to the metal.

The Naming of Complexes

The nomenclature of the complexes is patterned after a system suggested by Alfred Werner, a Swiss chemist and Nobel laureate, whose outstanding work more than 100 years ago laid the foundation for a clearer understanding of these compounds. The following five rules are used for naming complexes:

1. If a coordination compound is ionic, name the cation first and the anion second, in accordance with the usual nomenclature.
2. Name the ligands first, followed by the central metal. Name the ligands alphabetically. Negative ligands (anions) have names formed by adding *-o* to the stem name of the group. For examples, see Table 19.1. For most neutral ligands, the name of the molecule is used. The four common exceptions are *aqua* (H_2O), *ammine* (NH_3), *carbonyl* (CO), and *nitrosyl* (NO). For example, name $[Pt(NH_3)_2Cl_4]$ as diamminetetrachloroplatinum(IV).

Examples of Anionic Ligands

Anionic Ligand	Name
F^-	fluoro
Cl^-	chloro
Br^-	bromo
I^-	iodo
CN^-	cyano
NO_3^-	nitrato
OH^-	hydroxo
O^{2-}	oxo
$C_2O_4^{2-}$	oxalato
CO_3^{2-}	carbonato

TABLE 19.1

3. If more than one ligand of a given type is present, the number is indicated by the prefixes *di-* (for two), *tri-* (for three), *tetra-* (for four), *penta-* (for five), and *hexa-* (for six). Sometimes, the prefixes *bis-* (for two), *tris-* (for three), and *tetrakis-* (for four) are used when the name of the ligand already includes *di-*, *tri-*, or *tetra-*, or when the ligand name begins with a vowel. For example, the ion bis(bipyridyl)osmium(II) uses bis- to signify that there are two ligands attached to Os, and each bipyridyl ligand contains two pyridine groups (C_5H_4N).

When the complex is either a cation or a neutral molecule, the name of the central metal atom is spelled exactly like the name of the element and is followed by a Roman numeral in parentheses to indicate its oxidation state (Table 19.2 and Table 19.3). When the complex is an anion, the suffix -ate is added to the stem of the name of the metal, followed by the Roman numeral designation of its oxidation state (Table 19.4). Sometimes, the Latin name of the metal is used when the English name is clumsy. For example, *ferrate* is used instead of *ironate*, *plumbate* instead *leadate*, and *stannate* instead of *tinate*. The oxidation state of the metal is determined based on the charges of each ligand and the overall charge of the coordination compound. For example, in $[Cr(H_2O)_4Cl_2]Br$, the coordination sphere (in brackets) has a charge of 1+ to balance the bromide ion. The water ligands are neutral, and the chloride ligands are anionic with a charge of 1− each. To determine the oxidation state of the metal, we set the overall charge equal to the sum of the ligands and the metal: $+1 = -2 + x$, so the oxidation state (x) is equal to 3+.

Examples in Which the Complex Is a Cation

$[Co(NH_3)_6]Cl_3$	hexaamminecobalt(III) chloride
$[Pt(NH_3)_4Cl_2]^{2+}$	tetraamminedichloroplatinum(IV) ion
$[Ag(NH_3)_2]^+$	diamminesilver(I) ion
$[Cr(H_2O)_4Cl_2]Cl$	tetraaquadichlorochromium(III) chloride
$[Co(H_2NCH_2CH_2NH_2)_3]_2(SO_4)_3$	tris(ethylenediamine)cobalt(III) sulfate

TABLE 19.2

Examples in Which the Complex Is Neutral

$[Pt(NH_3)_2Cl_4]$	diamminetetrachloroplatinum(IV)
$[Ni(H_2NCH_2CH_2NH_2)_2Cl_2]$	dichlorobis(ethylenediamine)nickel(II)

TABLE 19.3

Examples in Which the Complex Is an Anion

$[PtCl_6]^{2-}$	hexachloroplatinate(IV) ion
$Na_2[SnCl_6]$	sodium hexachlorostannate(IV)

TABLE 19.4

LINK TO LEARNING

Do you think you understand naming coordination complexes? You can look over more examples and test yourself with online quizzes (http://openstax.org/l/16namingcomps) at the University of Sydney's site.

EXAMPLE 19.4

Coordination Numbers and Oxidation States

Determine the name of the following complexes and give the coordination number of the central metal atom.

(a) $Na_2[PtCl_6]$

(b) $K_3[Fe(C_2O_4)_3]$

(c) $[Co(NH_3)_5Cl]Cl_2$

Solution

(a) There are two Na^+ ions, so the coordination sphere has a negative two charge: $[PtCl_6]^{2-}$. There are six anionic chloride ligands, so $-2 = -6 + x$, and the oxidation state of the platinum is 4+. The name of the complex is sodium hexachloroplatinate(IV), and the coordination number is six. (b) The coordination sphere has a charge of 3− (based on the potassium) and the oxalate ligands each have a charge of 2−, so the metal oxidation state is given by $-3 = -6 + x$, and this is an iron(III) complex. The name is potassium trisoxalatoferrate(III) (note that tris is used instead of tri because the ligand name starts with a vowel). Because oxalate is a bidentate ligand, this complex has a coordination number of six. (c) In this example, the coordination sphere has a cationic charge of 2+. The NH_3 ligand is neutral, but the chloro ligand has a charge of 1−. The oxidation state is found by $+2 = -1 + x$ and is 3+, so the complex is pentaamminechlorocobalt(III) chloride and the coordination number is six.

Check Your Learning

The complex potassium dicyanoargenate(I) is used to make antiseptic compounds. Give the formula and coordination number.

Answer:

$K[Ag(CN)_2]$; coordination number two

The Structures of Complexes

The most common structures of the complexes in coordination compounds are octahedral, tetrahedral, and square planar (see Figure 19.18). For transition metal complexes, the coordination number determines the geometry around the central metal ion. Table 19.5 compares coordination numbers to the molecular geometry:

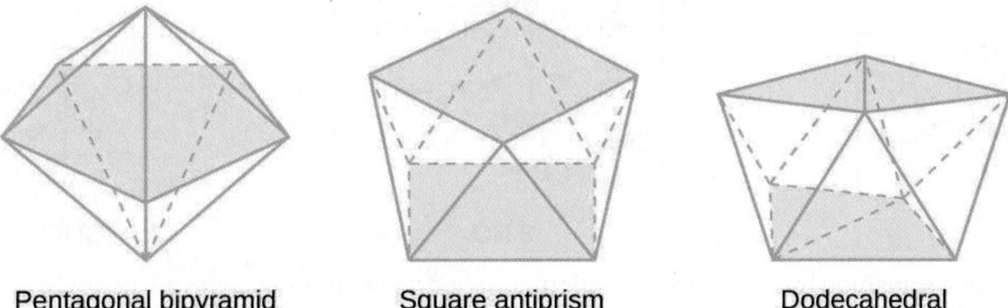

FIGURE 19.18 These are geometries of some complexes with coordination numbers of seven and eight.

Coordination Numbers and Molecular Geometry

Coordination Number	Molecular Geometry	Example
2	linear	$[Ag(NH_3)_2]^+$
3	trigonal planar	$[Cu(CN)_3]^{2-}$
4	tetrahedral(d^0 or d^{10}), low oxidation states for M	$[Ni(CO)_4]$
4	square planar (d^8)	$[Ni(CN)_4]^{2-}$
5	trigonal bipyramidal	$[CoCl_5]^{2-}$
5	square pyramidal	$[VO(CN)_4]^{2-}$
6	octahedral	$[CoCl_6]^{3-}$
7	pentagonal bipyramid	$[ZrF_7]^{3-}$
8	square antiprism	$[ReF_8]^{2-}$
8	dodecahedron	$[Mo(CN)_8]^{4-}$
9 and above	more complicated structures	$[ReH_9]^{2-}$

TABLE 19.5

Unlike main group atoms in which both the bonding and nonbonding electrons determine the molecular shape, the nonbonding *d*-electrons do not change the arrangement of the ligands. Octahedral complexes have a coordination number of six, and the six donor atoms are arranged at the corners of an octahedron around the central metal ion. Examples are shown in Figure 19.19. The chloride and nitrate anions in $[Co(H_2O)_6]Cl_2$ and $[Cr(en)_3](NO_3)_3$, and the potassium cations in $K_2[PtCl_6]$, are outside the brackets and are not bonded to the metal ion.

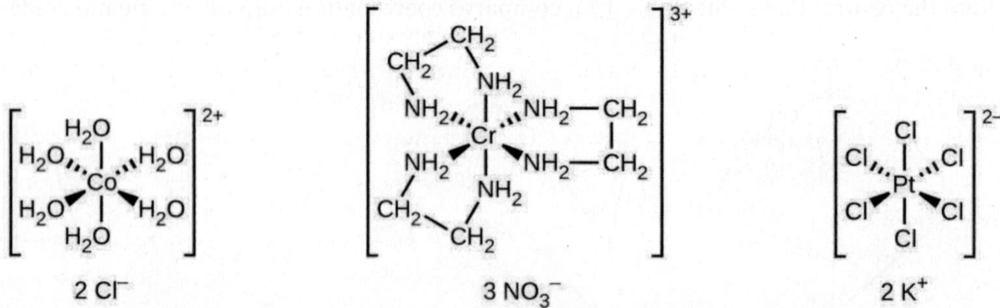

FIGURE 19.19 Many transition metal complexes adopt octahedral geometries, with six donor atoms forming bond angles of 90° about the central atom with adjacent ligands. Note that only ligands within the coordination sphere affect the geometry around the metal center.

For transition metals with a coordination number of four, two different geometries are possible: tetrahedral or square planar. Unlike main group elements, where these geometries can be predicted from VSEPR theory, a more detailed discussion of transition metal orbitals (discussed in the section on Crystal Field Theory) is required to predict which complexes will be tetrahedral and which will be square planar. In tetrahedral complexes such as $[Zn(CN)_4]^{2-}$ (Figure 19.20), each of the ligand pairs forms an angle of 109.5°. In square planar complexes, such as $[Pt(NH_3)_2Cl_2]$, each ligand has two other ligands at 90° angles (called the *cis* positions) and one additional ligand at an 180° angle, in the *trans* position.

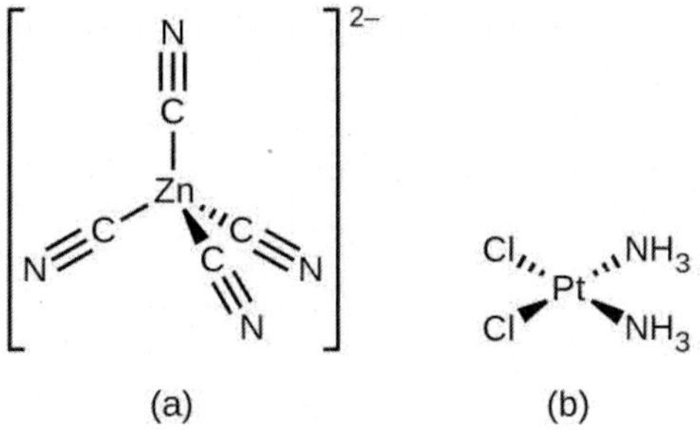

(a) (b)

FIGURE 19.20 Transition metals with a coordination number of four can adopt a tetrahedral geometry (a) as in $K_2[Zn(CN)_4]$ or a square planar geometry (b) as shown in $[Pt(NH_3)_2Cl_2]$.

Isomerism in Complexes

Isomers are different chemical species that have the same chemical formula. Transition metal complexes often exist as **geometric isomers**, in which the same atoms are connected through the same types of bonds but with differences in their orientation in space. Coordination complexes with two different ligands in the *cis* and *trans* positions from a ligand of interest form isomers. For example, the octahedral $[Co(NH_3)_4Cl_2]^+$ ion has two isomers. In the ***cis* configuration**, the two chloride ligands are adjacent to each other (Figure 19.21). The other isomer, the ***trans* configuration**, has the two chloride ligands directly across from one another.

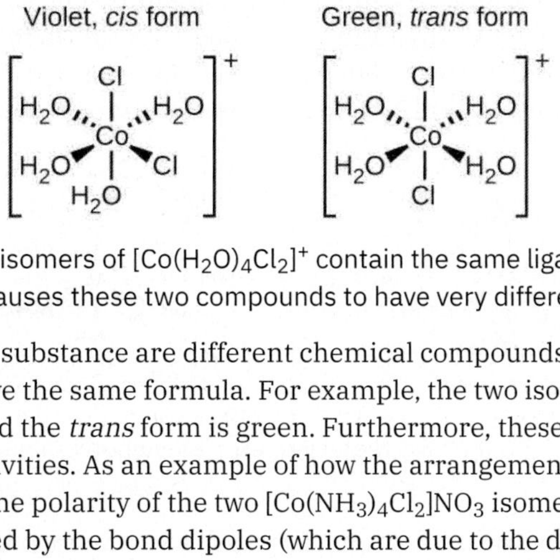

Violet, *cis* form Green, *trans* form

FIGURE 19.21 The *cis* and *trans* isomers of $[Co(H_2O)_4Cl_2]^+$ contain the same ligands attached to the same metal ion, but the spatial arrangement causes these two compounds to have very different properties.

Different geometric isomers of a substance are different chemical compounds. They exhibit different properties, even though they have the same formula. For example, the two isomers of $[Co(NH_3)_4Cl_2]NO_3$ differ in color; the *cis* form is violet, and the *trans* form is green. Furthermore, these isomers have different dipole moments, solubilities, and reactivities. As an example of how the arrangement in space can influence the molecular properties, consider the polarity of the two $[Co(NH_3)_4Cl_2]NO_3$ isomers. Remember that the polarity of a molecule or ion is determined by the bond dipoles (which are due to the difference in electronegativity of the bonding atoms) and their arrangement in space. In one isomer, *cis* chloride ligands cause more electron density on one side of the molecule than on the other, making it polar. For the *trans* isomer, each ligand is directly across from an identical ligand, so the bond dipoles cancel out, and the molecule is nonpolar.

EXAMPLE 19.5

Geometric Isomers

Identify which geometric isomer of $[Pt(NH_3)_2Cl_2]$ is shown in Figure 19.20. Draw the other geometric isomer and give its full name.

Solution

In the Figure 19.20, the two chlorine ligands occupy *cis* positions. The other form is shown in Figure 19.22. When naming specific isomers, the descriptor is listed in front of the name. Therefore, this complex is *trans*-diamminedichloroplatinum(II).

FIGURE 19.22 The *trans* isomer of $[Pt(NH_3)_2Cl_2]$ has each ligand directly across from an adjacent ligand.

Check Your Learning

Draw the ion *trans*-diaqua-*trans*-dibromo-*trans*-dichlorocobalt(II).

Answer:

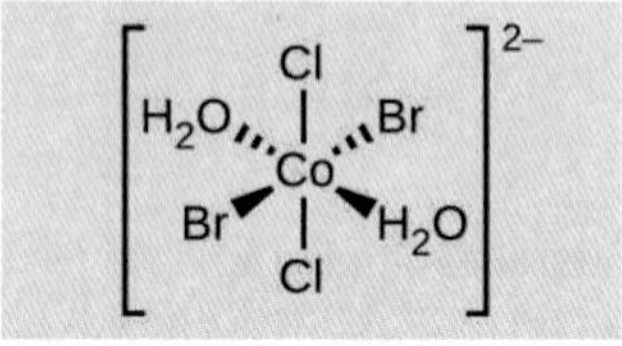

Another important type of isomers are **optical isomers**, or **enantiomers**, in which two objects are exact mirror images of each other but cannot be lined up so that all parts match. This means that optical isomers are nonsuperimposable mirror images. A classic example of this is a pair of hands, in which the right and left hand are mirror images of one another but cannot be superimposed. Optical isomers are very important in organic and biochemistry because living systems often incorporate one specific optical isomer and not the other. Unlike geometric isomers, pairs of optical isomers have identical properties (boiling point, polarity, solubility, etc.). Optical isomers differ only in the way they affect polarized light and how they react with other optical isomers. For coordination complexes, many coordination compounds such as $[M(en)_3]^{n+}$ [in which M^{n+} is a central metal ion such as iron(III) or cobalt(II)] form enantiomers, as shown in Figure 19.23. These two isomers will react differently with other optical isomers. For example, DNA helices are optical isomers, and the form that occurs in nature (right-handed DNA) will bind to only one isomer of $[M(en)_3]^{n+}$ and not the other.

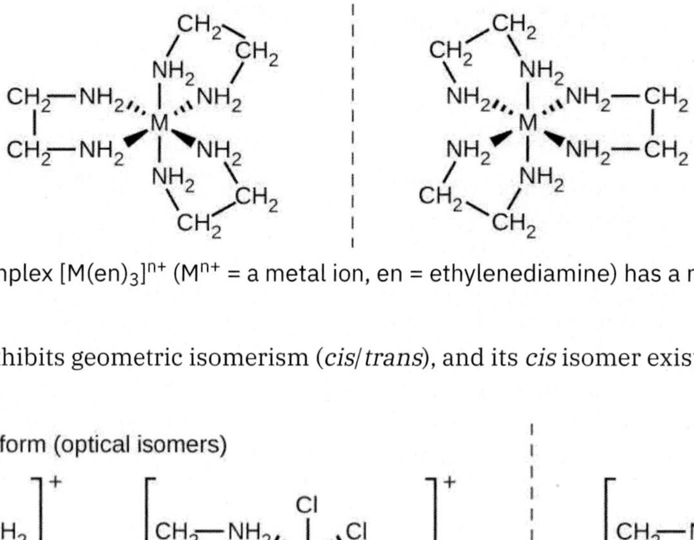

FIGURE 19.23 The complex $[M(en)_3]^{n+}$ (M^{n+} = a metal ion, en = ethylenediamine) has a nonsuperimposable mirror image.

The $[Co(en)_2Cl_2]^+$ ion exhibits geometric isomerism (*cis*/*trans*), and its *cis* isomer exists as a pair of optical isomers (Figure 19.24).

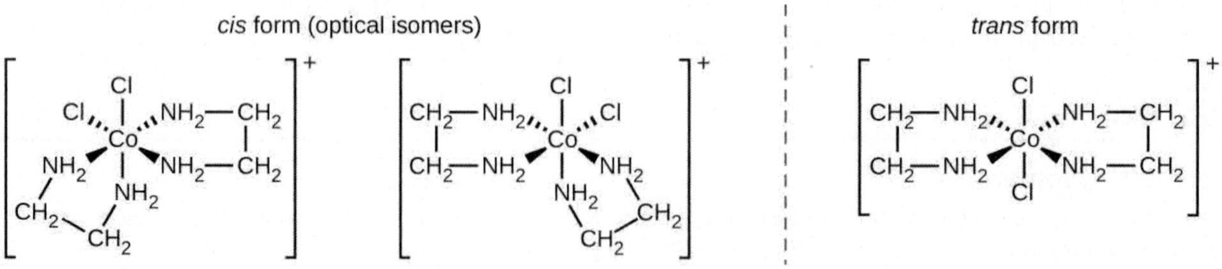

FIGURE 19.24 Three isomeric forms of $[Co(en)_2Cl_2]^+$ exist. The *trans* isomer, formed when the chlorines are positioned at a 180° angle, has very different properties from the *cis* isomers. The mirror images of the *cis* isomer form a pair of optical isomers, which have identical behavior except when reacting with other enantiomers.

Linkage isomers occur when the coordination compound contains a ligand that can bind to the transition metal center through two different atoms. For example, the CN ligand can bind through the carbon atom (cyano) or through the nitrogen atom (isocyano). Similarly, SCN– can be bound through the sulfur or nitrogen atom, affording two distinct compounds ($[Co(NH_3)_5SCN]^{2+}$ or $[Co(NH_3)_5NCS]^{2+}$).

Ionization isomers (or **coordination isomers**) occur when one anionic ligand in the inner coordination sphere is replaced with the counter ion from the outer coordination sphere. A simple example of two ionization isomers are $[CoCl_6][Br]$ and $[CoCl_5Br][Cl]$.

Coordination Complexes in Nature and Technology

Chlorophyll, the green pigment in plants, is a complex that contains magnesium (Figure 19.25). This is an example of a main group element in a coordination complex. Plants appear green because chlorophyll absorbs red and purple light; the reflected light consequently appears green. The energy resulting from the absorption of light is used in photosynthesis.

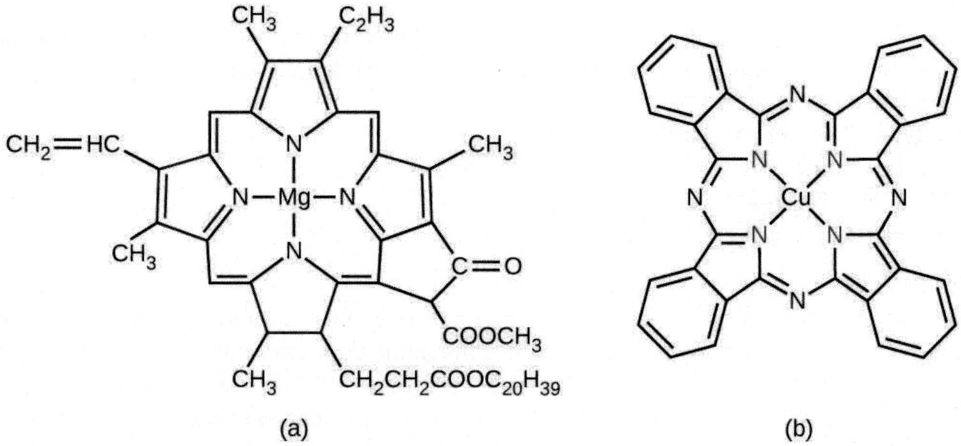

FIGURE 19.25 (a) Chlorophyll comes in several different forms, which all have the same basic structure around the magnesium center. (b) Copper phthalocyanine blue, a square planar copper complex, is present in some blue dyes.

Chemistry in Everyday Life

Transition Metal Catalysts

One of the most important applications of transition metals is as industrial catalysts. As you recall from the chapter on kinetics, a catalyst increases the rate of reaction by lowering the activation energy and is regenerated in the catalytic cycle. Over 90% of all manufactured products are made with the aid of one or more catalysts. The ability to bind ligands and change oxidation states makes transition metal catalysts well suited for catalytic applications. Vanadium oxide is used to produce 230,000,000 tons of sulfuric acid worldwide each year, which in turn is used to make everything from fertilizers to cans for food. Plastics are made with the aid of transition metal catalysts, along with detergents, fertilizers, paints, and more (see Figure 19.26). Very complicated pharmaceuticals are manufactured with catalysts that are selective, reacting with one specific bond out of a large number of possibilities. Catalysts allow processes to be more economical and more environmentally friendly. Developing new catalysts and better understanding of existing systems are important areas of current research.

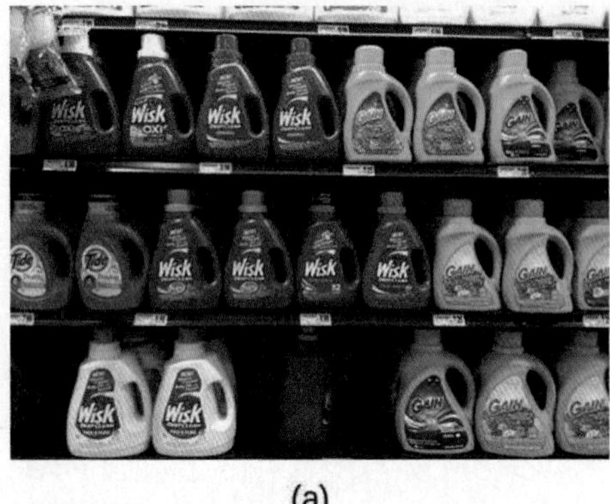

(a)

(b)

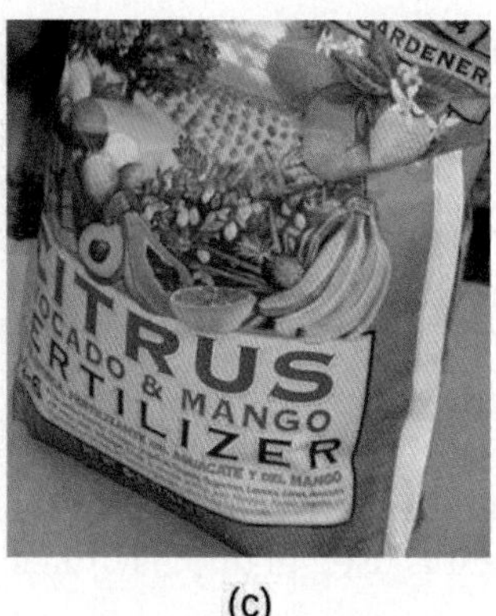

(c)

FIGURE 19.26 (a) Detergents, (b) paints, and (c) fertilizers are all made using transition metal catalysts. (credit a: modification of work by "Mr. Brian"/Flickr; credit b: modification of work by Ewen Roberts; credit c: modification of work by "osseous"/Flickr)

Portrait of a Chemist

Deanna D'Alessandro

Dr. Deanna D'Alessandro develops new metal-containing materials that demonstrate unique electronic, optical, and magnetic properties. Her research combines the fields of fundamental inorganic and physical chemistry with materials engineering. She is working on many different projects that rely on transition metals. For example, one type of compound she is developing captures carbon dioxide waste from power plants and catalytically converts it into useful products (see Figure 19.27).

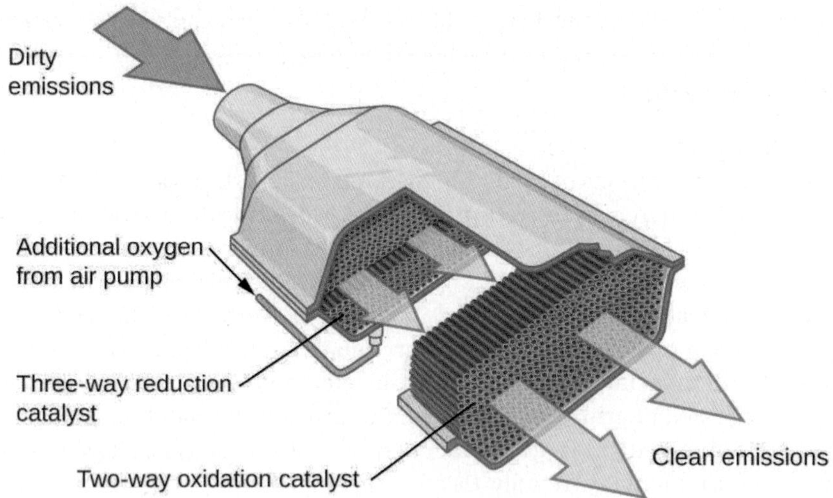

FIGURE 19.27 Catalytic converters change carbon dioxide emissions from power plants into useful products, and, like the one shown here, are also found in cars.

Another project involves the development of porous, sponge-like materials that are "photoactive." The absorption of light causes the pores of the sponge to change size, allowing gas diffusion to be controlled. This has many potential useful applications, from powering cars with hydrogen fuel cells to making better electronics components. Although not a complex, self-darkening sunglasses are an example of a photoactive substance.

Watch this video (http://openstax.org/l/16DeannaD) to learn more about this research and listen to Dr. D'Alessandro (shown in Figure 19.28) describe what it is like being a research chemist.

FIGURE 19.28 Dr. Deanna D'Alessandro is a functional materials researcher. Her work combines the inorganic and physical chemistry fields with engineering, working with transition metals to create new systems to power cars and convert energy (credit: image courtesy of Deanna D'Alessandro).

Many other coordination complexes are also brightly colored. The square planar copper(II) complex phthalocyanine blue (from Figure 19.25) is one of many complexes used as pigments or dyes. This complex is used in blue ink, blue jeans, and certain blue paints.

The structure of heme (Figure 19.29), the iron-containing complex in hemoglobin, is very similar to that in chlorophyll. In hemoglobin, the red heme complex is bonded to a large protein molecule (globin) by the attachment of the protein to the heme ligand. Oxygen molecules are transported by hemoglobin in the blood by being bound to the iron center. When the hemoglobin loses its oxygen, the color changes to a bluish red. Hemoglobin will only transport oxygen if the iron is Fe^{2+}; oxidation of the iron to Fe^{3+} prevents oxygen transport.

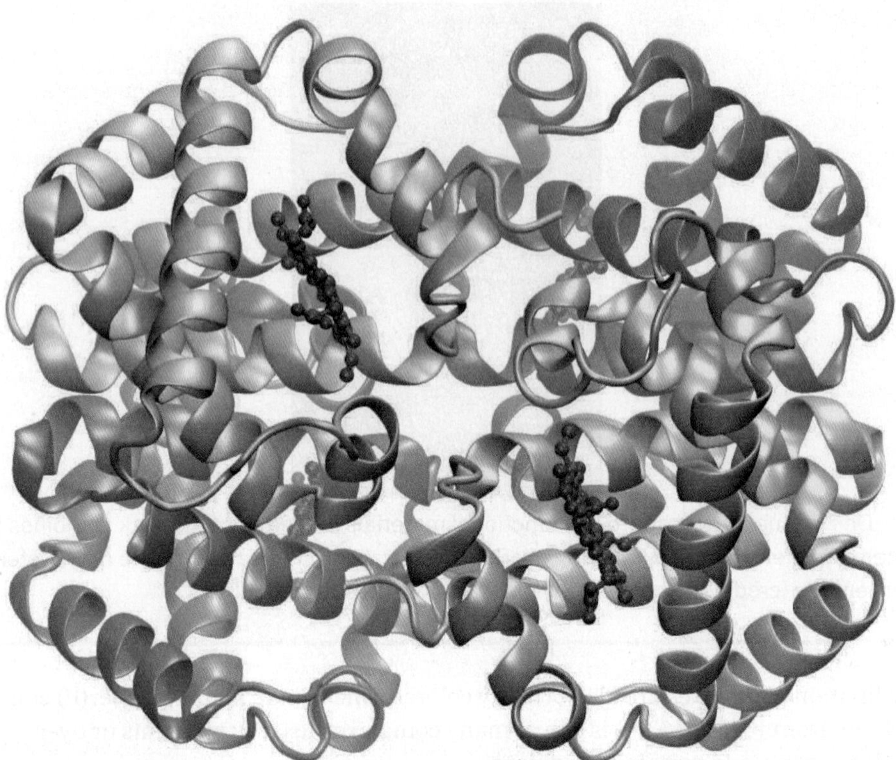

FIGURE 19.29 Hemoglobin contains four protein subunits, each of which has an iron center attached to a heme ligand (shown in red), which is coordinated to a globin protein. Each subunit is shown in a different color.

Complexing agents often are used for water softening because they tie up such ions as Ca^{2+}, Mg^{2+}, and Fe^{2+}, which make water hard. Many metal ions are also undesirable in food products because these ions can catalyze reactions that change the color of food. Coordination complexes are useful as preservatives. For example, the ligand EDTA, $(HO_2CCH_2)_2NCH_2CH_2N(CH_2CO_2H)_2$, coordinates to metal ions through six donor atoms and prevents the metals from reacting (Figure 19.30). This ligand also is used to sequester metal ions in paper production, textiles, and detergents, and has pharmaceutical uses.

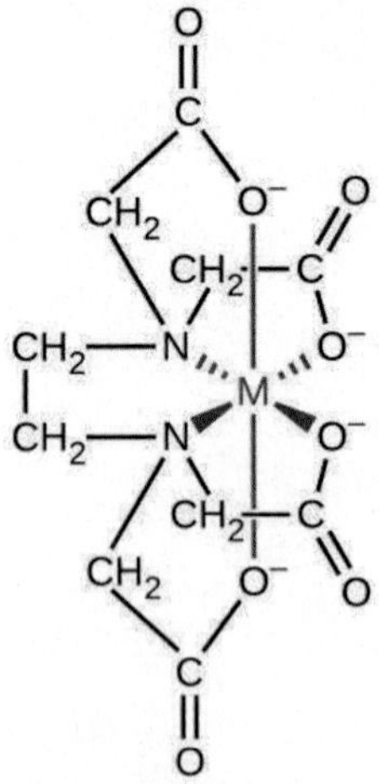

FIGURE 19.30 The ligand EDTA binds tightly to a variety of metal ions by forming hexadentate complexes.

Complexing agents that tie up metal ions are also used as drugs. British Anti-Lewisite (BAL), $HSCH_2CH(SH)CH_2OH$, is a drug developed during World War I as an antidote for the arsenic-based war gas Lewisite. BAL is now used to treat poisoning by heavy metals, such as arsenic, mercury, thallium, and chromium. The drug is a ligand and functions by making a water-soluble chelate of the metal; the kidneys eliminate this metal chelate (Figure 19.31). Another polydentate ligand, enterobactin, which is isolated from certain bacteria, is used to form complexes of iron and thereby to control the severe iron buildup found in patients suffering from blood diseases such as Cooley's anemia, who require frequent transfusions. As the transfused blood breaks down, the usual metabolic processes that remove iron are overloaded, and excess iron can build up to fatal levels. Enterobactin forms a water-soluble complex with excess iron, and the body can safely eliminate this complex.

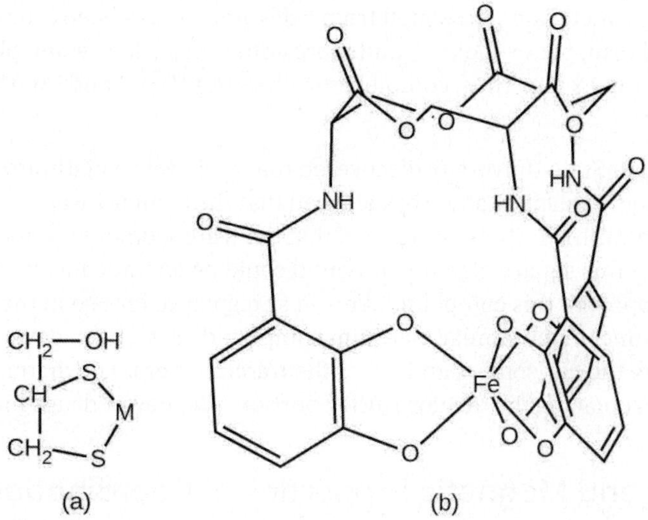

FIGURE 19.31 Coordination complexes are used as drugs. (a) British Anti-Lewisite is used to treat heavy metal poisoning by coordinating metals (M), and enterobactin (b) allows excess iron in the blood to be removed.

EXAMPLE 19.6

Chelation Therapy

Ligands like BAL and enterobactin are important in medical treatments for heavy metal poisoning. However, chelation therapies can disrupt the normal concentration of ions in the body, leading to serious side effects, so researchers are searching for new chelation drugs. One drug that has been developed is dimercaptosuccinic acid (DMSA), shown in Figure 19.32. Identify which atoms in this molecule could act as donor atoms.

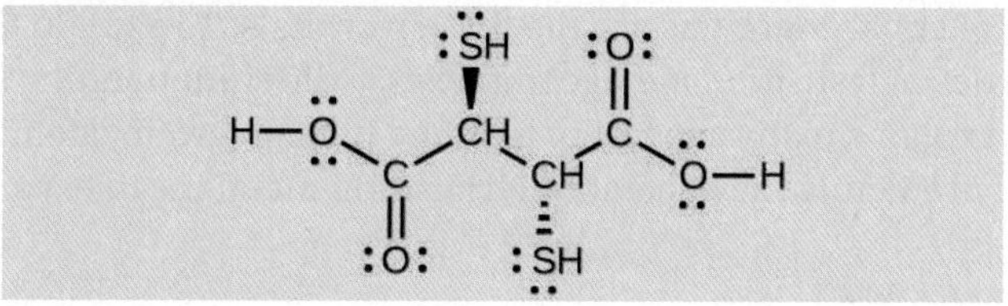

FIGURE 19.32 Dimercaptosuccinic acid is used to treat heavy metal poisoning.

Solution

All of the oxygen and sulfur atoms have lone pairs of electrons that can be used to coordinate to a metal center, so there are six possible donor atoms. Geometrically, only two of these atoms can be coordinated to a metal at once. The most common binding mode involves the coordination of one sulfur atom and one oxygen atom, forming a five-member ring with the metal.

Check Your Learning

Some alternative medicine practitioners recommend chelation treatments for ailments that are not clearly related to heavy metals, such as cancer and autism, although the practice is discouraged by many scientific organizations.[1] Identify at least two biologically important metals that could be disrupted by chelation therapy.

1 National Council against Health Fraud, *NCAHF Policy Statement on Chelation Therapy*, (Peabody, MA, 2002).

Answer:

Ca, Fe, Zn, and Cu

Ligands are also used in the electroplating industry. When metal ions are reduced to produce thin metal coatings, metals can clump together to form clusters and nanoparticles. When metal coordination complexes are used, the ligands keep the metal atoms isolated from each other. It has been found that many metals plate out as a smoother, more uniform, better-looking, and more adherent surface when plated from a bath containing the metal as a complex ion. Thus, complexes such as $[Ag(CN)_2]^-$ and $[Au(CN)_2]^-$ are used extensively in the electroplating industry.

In 1965, scientists at Michigan State University discovered that there was a platinum complex that inhibited cell division in certain microorganisms. Later work showed that the complex was *cis*-diamminedichloroplatinum(II), $[Pt(NH_3)_2(Cl)_2]$, and that the *trans* isomer was not effective. The inhibition of cell division indicated that this square planar compound could be an anticancer agent. In 1978, the US Food and Drug Administration approved this compound, known as cisplatin, for use in the treatment of certain forms of cancer. Since that time, many similar platinum compounds have been developed for the treatment of cancer. In all cases, these are the *cis* isomers and never the *trans* isomers. The diammine $(NH_3)_2$ portion is retained with other groups, replacing the dichloro $[(Cl)_2]$ portion. The newer drugs include carboplatin, oxaliplatin, and satraplatin.

19.3 Spectroscopic and Magnetic Properties of Coordination Compounds

LEARNING OBJECTIVES

By the end of this section, you will be able to:

- Outline the basic premise of crystal field theory (CFT)
- Identify molecular geometries associated with various d-orbital splitting patterns
- Predict electron configurations of split d orbitals for selected transition metal atoms or ions
- Explain spectral and magnetic properties in terms of CFT concepts

The behavior of coordination compounds cannot be adequately explained by the same theories used for main group element chemistry. The observed geometries of coordination complexes are not consistent with hybridized orbitals on the central metal overlapping with ligand orbitals, as would be predicted by valence bond theory. The observed colors indicate that the *d* orbitals often occur at different energy levels rather than all being degenerate, that is, of equal energy, as are the three *p* orbitals. To explain the stabilities, structures, colors, and magnetic properties of transition metal complexes, a different bonding model has been developed. Just as valence bond theory explains many aspects of bonding in main group chemistry, crystal field theory is useful in understanding and predicting the behavior of transition metal complexes.

Crystal Field Theory

To explain the observed behavior of transition metal complexes (such as how colors arise), a model involving electrostatic interactions between the electrons from the ligands and the electrons in the unhybridized *d* orbitals of the central metal atom has been developed. This electrostatic model is **crystal field theory** (CFT). It allows us to understand, interpret, and predict the colors, magnetic behavior, and some structures of coordination compounds of transition metals.

CFT focuses on the nonbonding electrons on the central metal ion in coordination complexes not on the metal-ligand bonds. Like valence bond theory, CFT tells only part of the story of the behavior of complexes. However, it tells the part that valence bond theory does not. In its pure form, CFT ignores any covalent bonding between ligands and metal ions. Both the ligand and the metal are treated as infinitesimally small point charges.

All electrons are negative, so the electrons donated from the ligands will repel the electrons of the central metal. Let us consider the behavior of the electrons in the unhybridized *d* orbitals in an octahedral complex. The five *d* orbitals consist of lobe-shaped regions and are arranged in space, as shown in Figure 19.33. In an octahedral complex, the six ligands coordinate along the axes.

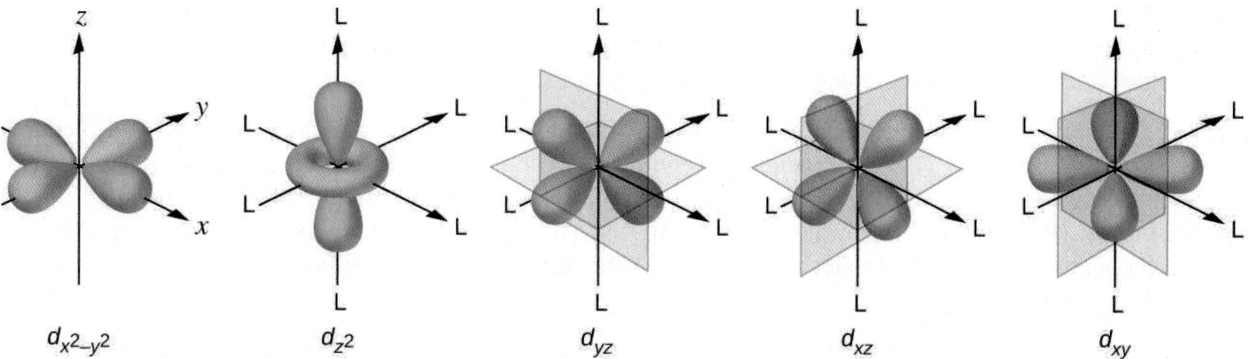

FIGURE 19.33 The directional characteristics of the five *d* orbitals are shown here. The shaded portions indicate the phase of the orbitals. The ligands (L) coordinate along the axes. For clarity, the ligands have been omitted from the $d_{x^2-y^2}$ orbital so that the axis labels could be shown.

In an uncomplexed metal ion in the gas phase, the electrons are distributed among the five *d* orbitals in accord with Hund's rule because the orbitals all have the same energy. However, when ligands coordinate to a metal ion, the energies of the *d* orbitals are no longer the same.

In octahedral complexes, the lobes in two of the five *d* orbitals, the d_{z^2} and $d_{x^2-y^2}$ orbitals, point toward the ligands (Figure 19.33). These two orbitals are called the **e_g orbitals** (the symbol actually refers to the symmetry of the orbitals, but we will use it as a convenient name for these two orbitals in an octahedral complex). The other three orbitals, the d_{xy}, d_{xz}, and d_{yz} orbitals, have lobes that point between the ligands and are called the **t_{2g} orbitals** (again, the symbol really refers to the symmetry of the orbitals). As six ligands approach the metal ion along the axes of the octahedron, their point charges repel the electrons in the *d* orbitals of the metal ion. However, the repulsions between the electrons in the e_g orbitals (the d_{z^2} and $d_{x^2-y^2}$ orbitals) and the ligands are greater than the repulsions between the electrons in the t_{2g} orbitals (the d_{zy}, d_{xz}, and d_{yz} orbitals) and the ligands. This is because the lobes of the e_g orbitals point directly at the ligands, whereas the lobes of the t_{2g} orbitals point between them. Thus, electrons in the e_g orbitals of the metal ion in an octahedral complex have higher potential energies than those of electrons in the t_{2g} orbitals. The difference in energy may be represented as shown in Figure 19.34.

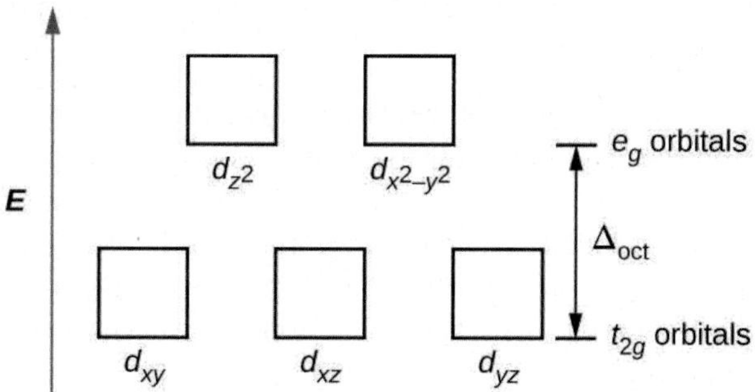

FIGURE 19.34 In octahedral complexes, the e_g orbitals are destabilized (higher in energy) compared to the t_{2g} orbitals because the ligands interact more strongly with the *d* orbitals at which they are pointed directly.

The difference in energy between the e_g and the t_{2g} orbitals is called the **crystal field splitting** and is symbolized by $\mathbf{\Delta_{oct}}$, where oct stands for octahedral.

The magnitude of Δ_{oct} depends on many factors, including the nature of the six ligands located around the central metal ion, the charge on the metal, and whether the metal is using 3*d*, 4*d*, or 5*d* orbitals. Different ligands produce different crystal field splittings. The increasing crystal field splitting produced by ligands is expressed in the **spectrochemical series**, a short version of which is given here:

$$\xrightarrow[\text{a few ligands of the spectrochemical series, in order of increasing field strength of the ligand}]{I^- < Br^- < Cl^- < F^- < H_2O < C_2O_4{}^{2-} < NH_3 < en < NO_2{}^- < CN^-}$$

In this series, ligands on the left cause small crystal field splittings and are **weak-field ligands**, whereas those on the right cause larger splittings and are **strong-field ligands**. Thus, the Δ_{oct} value for an octahedral complex with iodide ligands (I^-) is much smaller than the Δ_{oct} value for the same metal with cyanide ligands (CN^-).

Electrons in the *d* orbitals follow the aufbau ("filling up") principle, which says that the orbitals will be filled to give the lowest total energy, just as in main group chemistry. When two electrons occupy the same orbital, the like charges repel each other. The energy needed to pair up two electrons in a single orbital is called the **pairing energy (P)**. Electrons will always singly occupy each orbital in a degenerate set before pairing. P is similar in magnitude to Δ_{oct}. When electrons fill the *d* orbitals, the relative magnitudes of Δ_{oct} and P determine which orbitals will be occupied.

In $[Fe(CN)_6]^{4-}$, the strong field of six cyanide ligands produces a large Δ_{oct}. Under these conditions, the electrons require less energy to pair than they require to be excited to the e_g orbitals ($\Delta_{oct} > P$). The six 3*d* electrons of the Fe^{2+} ion pair in the three t_{2g} orbitals (Figure 19.35). Complexes in which the electrons are paired because of the large crystal field splitting are called **low-spin complexes** because the number of unpaired electrons (spins) is minimized.

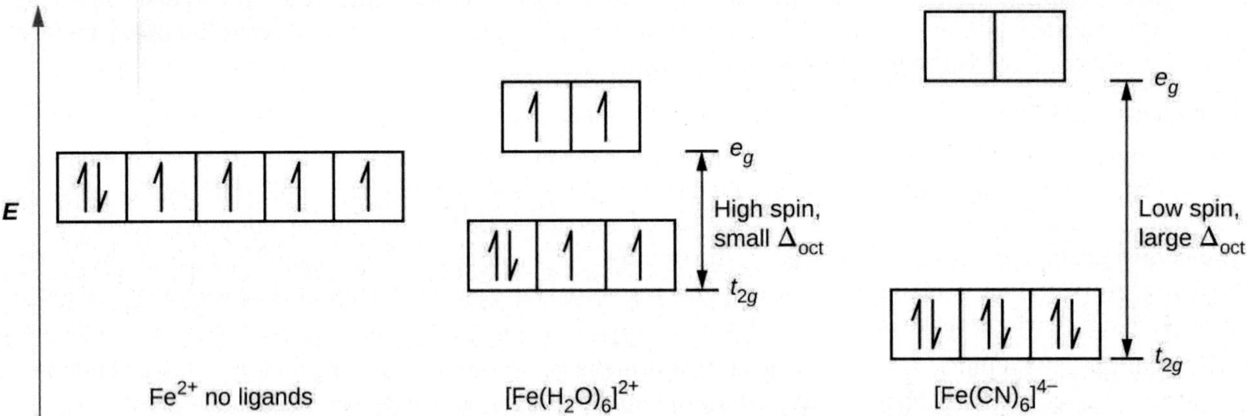

FIGURE 19.35 Iron(II) complexes have six electrons in the 5*d* orbitals. In the absence of a crystal field, the orbitals are degenerate. For coordination complexes with strong-field ligands such as $[Fe(CN)_6]^{4-}$, Δ_{oct} is greater than P, and the electrons pair in the lower energy t_{2g} orbitals before occupying the *eg* orbitals. With weak-field ligands such as H_2O, the ligand field splitting is less than the pairing energy, Δ_{oct} less than P, so the electrons occupy all *d* orbitals singly before any pairing occurs.

In $[Fe(H_2O)_6]^{2+}$, on the other hand, the weak field of the water molecules produces only a small crystal field splitting ($\Delta_{oct} < P$). Because it requires less energy for the electrons to occupy the e_g orbitals than to pair together, there will be an electron in each of the five 3*d* orbitals before pairing occurs. For the six *d* electrons on the iron(II) center in $[Fe(H_2O)_6]^{2+}$, there will be one pair of electrons and four unpaired electrons (Figure 19.35). Complexes such as the $[Fe(H_2O)_6]^{2+}$ ion, in which the electrons are unpaired because the crystal field splitting is not large enough to cause them to pair, are called **high-spin complexes** because the number of unpaired electrons (spins) is maximized.

A similar line of reasoning shows why the $[Fe(CN)_6]^{3-}$ ion is a low-spin complex with only one unpaired electron, whereas both the $[Fe(H_2O)_6]^{3+}$ and $[FeF_6]^{3-}$ ions are high-spin complexes with five unpaired electrons.

EXAMPLE 19.7

High- and Low-Spin Complexes

Predict the number of unpaired electrons.

(a) $K_3[CrI_6]$

(b) $[Cu(en)_2(H_2O)_2]Cl_2$

(c) $Na_3[Co(NO_2)_6]$

Solution

The complexes are octahedral.

(a) Cr^{3+} has a d^3 configuration. These electrons will all be unpaired.

(b) Cu^{2+} is d^9, so there will be one unpaired electron.

(c) Co^{3+} has d^6 valence electrons, so the crystal field splitting will determine how many are paired. Nitrite is a strong-field ligand, so the complex will be low spin. Six electrons will go in the t_{2g} orbitals, leaving 0 unpaired.

Check Your Learning

The size of the crystal field splitting only influences the arrangement of electrons when there is a choice between pairing electrons and filling the higher-energy orbitals. For which *d*-electron configurations will there be a difference between high- and low-spin configurations in octahedral complexes?

Answer:

d^4, d^5, d^6, and d^7

 **EXAMPLE 19.8**

CFT for Other Geometries

CFT is applicable to molecules in geometries other than octahedral. In octahedral complexes, remember that the lobes of the e_g set point directly at the ligands. For tetrahedral complexes, the *d* orbitals remain in place, but now we have only four ligands located between the axes (Figure 19.36). None of the orbitals points directly at the tetrahedral ligands. However, the e_g set (along the Cartesian axes) overlaps with the ligands less than does the t_{2g} set. By analogy with the octahedral case, predict the energy diagram for the *d* orbitals in a tetrahedral crystal field. To avoid confusion, the octahedral e_g set becomes a tetrahedral *e* set, and the octahedral t_{2g} set becomes a t_2 set.

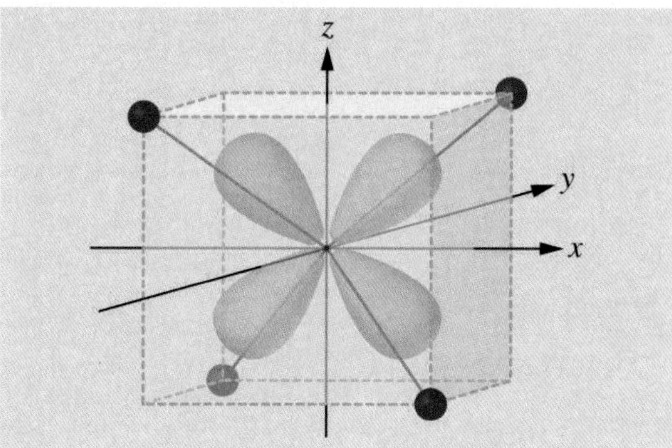

FIGURE 19.36 This diagram shows the orientation of the tetrahedral ligands with respect to the axis system for the orbitals.

Solution

Since CFT is based on electrostatic repulsion, the orbitals closer to the ligands will be destabilized and raised in energy relative to the other set of orbitals. The splitting is less than for octahedral complexes because the overlap is less, so Δ_{tet} is usually small $\left(\Delta_{tet} = \frac{4}{9}\Delta_{oct}\right)$:

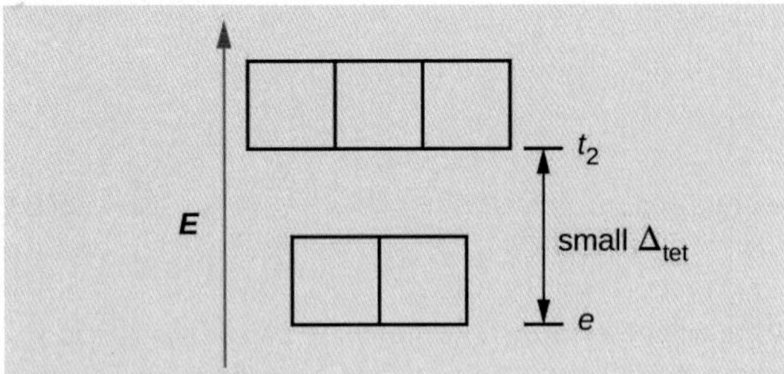

Check Your Learning

Explain how many unpaired electrons a tetrahedral d^4 ion will have.

Answer:

4; because Δ_{tet} is small, all tetrahedral complexes are high spin and the electrons go into the t_2 orbitals before pairing

---

The other common geometry is square planar. It is possible to consider a square planar geometry as an octahedral structure with a pair of *trans* ligands removed. The removed ligands are assumed to be on the *z*-axis. This changes the distribution of the *d* orbitals, as orbitals on or near the *z*-axis become more stable, and those on or near the *x*- or *y*-axes become less stable. This results in the octahedral t_{2g} and the e_g sets splitting and gives a more complicated pattern, as depicted below:

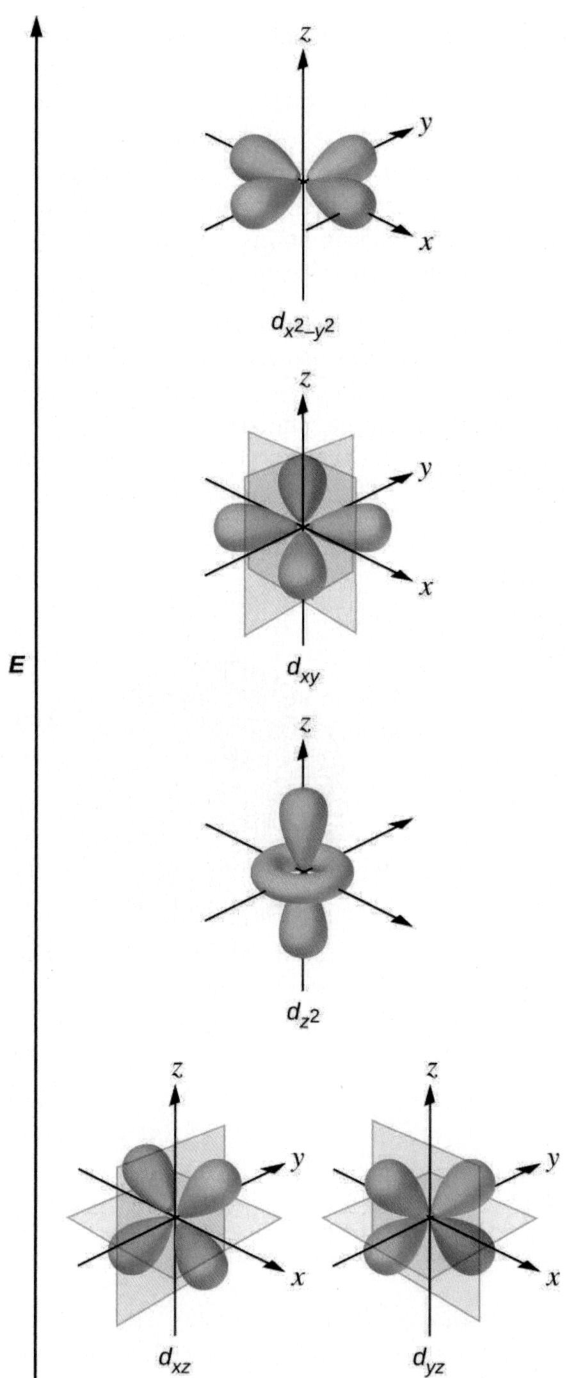

Magnetic Moments of Molecules and Ions

Experimental evidence of magnetic measurements supports the theory of high- and low-spin complexes. Remember that molecules such as O_2 that contain unpaired electrons are paramagnetic. Paramagnetic substances are attracted to magnetic fields. Many transition metal complexes have unpaired electrons and hence are paramagnetic. Molecules such as N_2 and ions such as Na^+ and $[Fe(CN)_6]^{4-}$ that contain no unpaired electrons are diamagnetic. Diamagnetic substances have a slight tendency to be repelled by magnetic fields.

When an electron in an atom or ion is unpaired, the magnetic moment due to its spin makes the entire atom or ion paramagnetic. The size of the magnetic moment of a system containing unpaired electrons is related directly to the number of such electrons: the greater the number of unpaired electrons, the larger the magnetic moment. Therefore, the observed magnetic moment is used to determine the number of unpaired electrons present. The measured magnetic moment of low-spin d^6 $[Fe(CN)_6]^{4-}$ confirms that iron is diamagnetic, whereas high-spin d^6 $[Fe(H_2O)_6]^{2+}$ has four unpaired electrons with a magnetic moment that confirms this arrangement.

Colors of Transition Metal Complexes

When atoms or molecules absorb light at the proper frequency, their electrons are excited to higher-energy orbitals. For many main group atoms and molecules, the absorbed photons are in the ultraviolet range of the electromagnetic spectrum, which cannot be detected by the human eye. For coordination compounds, the energy difference between the *d* orbitals often allows photons in the visible range to be absorbed.

The human eye perceives a mixture of all the colors, in the proportions present in sunlight, as white light. Complementary colors, those located across from each other on a color wheel, are also used in color vision. The eye perceives a mixture of two complementary colors, in the proper proportions, as white light. Likewise, when a color is missing from white light, the eye sees its complement. For example, when red photons are absorbed from white light, the eyes see the color green. When violet photons are removed from white light, the eyes see lemon yellow. The blue color of the $[Cu(NH_3)_4]^{2+}$ ion results because this ion absorbs orange and red light, leaving the complementary colors of blue and green (Figure 19.37).

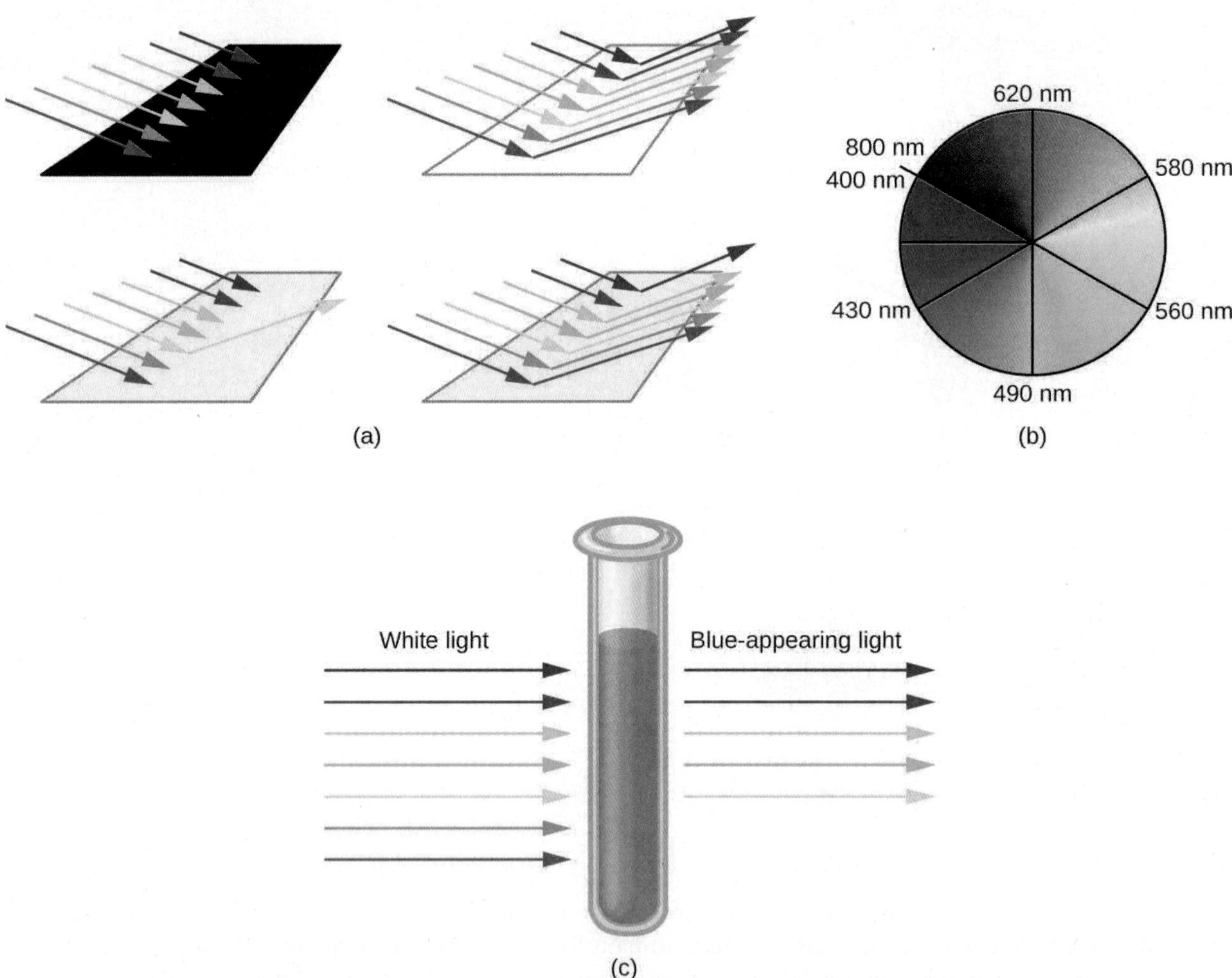

FIGURE 19.37 (a) An object is black if it absorbs all colors of light. If it reflects all colors of light, it is white. An object has a color if it absorbs all colors except one, such as this yellow strip. The strip also appears yellow if it absorbs the complementary color from white light (in this case, indigo). (b) Complementary colors are located directly across from one another on the color wheel. (c) A solution of $[Cu(NH_3)_4]^{2+}$ ions absorbs red and orange light, so the transmitted light appears as the complementary color, blue.

EXAMPLE 19.9

Colors of Complexes

The octahedral complex $[Ti(H_2O)_6]^{3+}$ has a single *d* electron. To excite this electron from the ground state t_{2g} orbital to the e_g orbital, this complex absorbs light from 450 to 600 nm. The maximum absorbance corresponds to Δ_{oct} and occurs at 499 nm. Calculate the value of Δ_{oct} in Joules and predict what color the solution will appear.

Solution

Using Planck's equation (refer to the section on electromagnetic energy), we calculate:

$$\upsilon = \frac{c}{\lambda} \text{ so } \frac{3.00 \times 10^8 \text{m/s}}{\frac{499 \text{ nm} \times 1 \text{ m}}{10^9 \text{nm}}} = 6.01 \times 10^{14} \text{ Hz}$$

$$E = h\nu \text{ so } 6.63 \times 10^{-34} \text{ J·s} \times 6.01 \times 10^{14} \text{ Hz} = 3.99 \times 10^{-19} \text{ Joules/ion}$$

Because the complex absorbs 600 nm (orange) through 450 (blue), the indigo, violet, and red wavelengths will be transmitted, and the complex will appear purple.

Check Your Learning

A complex that appears green, absorbs photons of what wavelengths?

Answer:

red, 620–800 nm

Small changes in the relative energies of the orbitals that electrons are transitioning between can lead to drastic shifts in the color of light absorbed. Therefore, the colors of coordination compounds depend on many factors. As shown in Figure 19.38, different aqueous metal ions can have different colors. In addition, different oxidation states of one metal can produce different colors, as shown for the vanadium complexes in the link below.

FIGURE 19.38 The partially filled *d* orbitals of the stable ions $Cr^{3+}(aq)$, $Fe^{3+}(aq)$, and $Co^{2+}(aq)$ (left, center and right, respectively) give rise to various colors. (credit: Sahar Atwa)

The specific ligands coordinated to the metal center also influence the color of coordination complexes. For example, the iron(II) complex $[Fe(H_2O)_6]SO_4$ appears blue-green because the high-spin complex absorbs photons in the red wavelengths (Figure 19.39). In contrast, the low-spin iron(II) complex $K_4[Fe(CN)_6]$ appears pale yellow because it absorbs higher-energy violet photons.

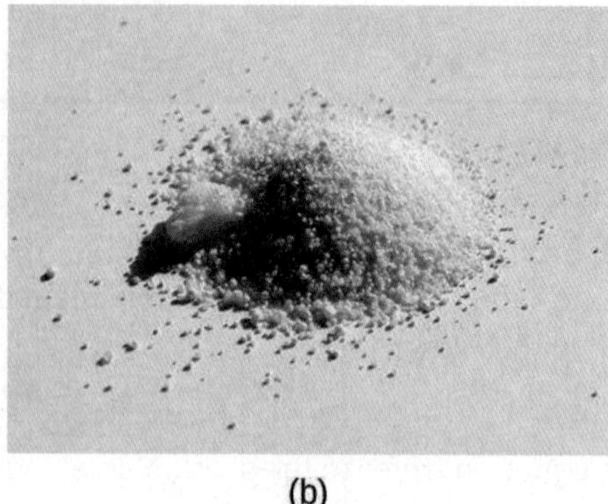

(a) (b)

FIGURE 19.39 Both (a) hexaaquairon(II) sulfate and (b) potassium hexacyanoferrate(II) contain d^6 iron(II) octahedral metal centers, but they absorb photons in different ranges of the visible spectrum.

LINK TO LEARNING

Watch this video (http://openstax.org/l/16vanadium) of the reduction of vanadium complexes to observe the colorful effect of changing oxidation states.

In general, strong-field ligands cause a large split in the energies of *d* orbitals of the central metal atom (large Δ_{oct}). Transition metal coordination compounds with these ligands are yellow, orange, or red because they absorb higher-energy violet or blue light. On the other hand, coordination compounds of transition metals with weak-field ligands are often blue-green, blue, or indigo because they absorb lower-energy yellow, orange, or red light.

A coordination compound of the Cu^+ ion has a d^{10} configuration, and all the e_g orbitals are filled. To excite an electron to a higher level, such as the 4*p* orbital, photons of very high energy are necessary. This energy corresponds to very short wavelengths in the ultraviolet region of the spectrum. No visible light is absorbed, so the eye sees no change, and the compound appears white or colorless. A solution containing $[Cu(CN)_2]^-$, for example, is colorless. On the other hand, octahedral Cu^{2+} complexes have a vacancy in the e_g orbitals, and electrons can be excited to this level. The wavelength (energy) of the light absorbed corresponds to the visible part of the spectrum, and Cu^{2+} complexes are almost always colored—blue, blue-green violet, or yellow (Figure 19.40). Although CFT successfully describes many properties of coordination complexes, molecular orbital explanations (beyond the introductory scope provided here) are required to understand fully the behavior of coordination complexes.

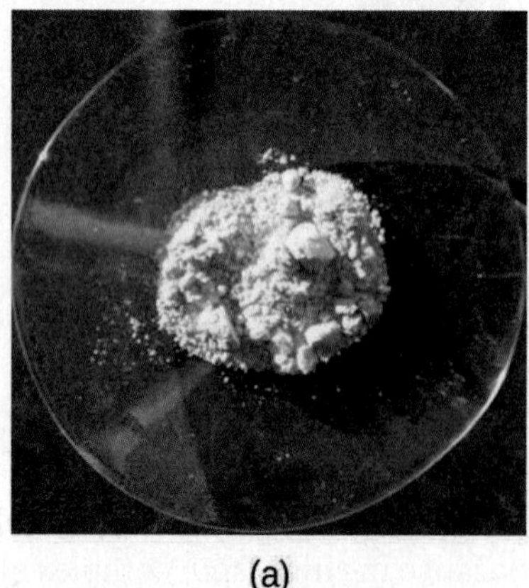

(a) (b)

FIGURE 19.40 (a) Copper(I) complexes with d^{10} configurations such as CuI tend to be colorless, whereas (b) d^9 copper(II) complexes such as $Cu(NO_3)_2{\cdot}5H_2O$ are brightly colored.

Key Terms

actinide series (also, actinoid series) actinium and the elements in the second row or the *f*-block, atomic numbers 89–103

bidentate ligand ligand that coordinates to one central metal through coordinate bonds from two different atoms

central metal ion or atom to which one or more ligands is attached through coordinate covalent bonds

chelate complex formed from a polydentate ligand attached to a central metal

chelating ligand ligand that attaches to a central metal ion by bonds from two or more donor atoms

***cis* configuration** configuration of a geometrical isomer in which two similar groups are on the same side of an imaginary reference line on the molecule

coordination compound stable compound in which the central metal atom or ion acts as a Lewis acid and accepts one or more pairs of electrons

coordination compound substance consisting of atoms, molecules, or ions attached to a central atom through Lewis acid-base interactions

coordination number number of coordinate covalent bonds to the central metal atom in a complex or the number of closest contacts to an atom in a crystalline form

coordination sphere central metal atom or ion plus the attached ligands of a complex

crystal field splitting (Δ_{oct}) difference in energy between the t_{2g} and e_g sets or *t* and *e* sets of orbitals

crystal field theory model that explains the energies of the orbitals in transition metals in terms of electrostatic interactions with the ligands but does not include metal ligand bonding

***d*-block element** one of the elements in groups 3–11 with valence electrons in *d* orbitals

donor atom atom in a ligand with a lone pair of electrons that forms a coordinate covalent bond to a central metal

e_g orbitals set of two *d* orbitals that are oriented on the Cartesian axes for coordination complexes; in octahedral complexes, they are higher in energy than the t_{2g} orbitals

***f*-block element** (also, inner transition element) one of the elements with atomic numbers 58–71 or 90–103 that have valence electrons in *f* orbitals; they are frequently shown offset below the periodic table

first transition series transition elements in the fourth period of the periodic table (first row of the *d*-block), atomic numbers 21–29

fourth transition series transition elements in the seventh period of the periodic table (fourth row of the *d*-block), atomic numbers 89 and 104–111

geometric isomers isomers that differ in the way in which atoms are oriented in space relative to each other, leading to different physical and chemical properties

high-spin complex complex in which the electrons maximize the total electron spin by singly populating all of the orbitals before pairing two electrons into the lower-energy orbitals

hydrometallurgy process in which a metal is separated from a mixture by first converting it into soluble ions, extracting the ions, and then reducing the ions to precipitate the pure metal

ionization isomer (or coordination isomer) isomer in which an anionic ligand is replaced by the counter ion in the inner coordination sphere

lanthanide series (also, lanthanoid series) lanthanum and the elements in the first row or the *f*-block, atomic numbers 57–71

ligand ion or neutral molecule attached to the central metal ion in a coordination compound

linkage isomer coordination compound that possesses a ligand that can bind to the transition metal in two different ways (CN^- vs. NC^-)

low-spin complex complex in which the electrons minimize the total electron spin by pairing in the lower-energy orbitals before populating the higher-energy orbitals

monodentate ligand that attaches to a central metal through just one coordinate covalent bond

optical isomer (also, enantiomer) molecule that is a nonsuperimposable mirror image with identical chemical and physical properties, except when it reacts with other optical isomers

pairing energy (P) energy required to place two electrons with opposite spins into a single orbital

platinum metals group of six transition metals consisting of ruthenium, osmium, rhodium, iridium, palladium, and platinum that tend to occur in the same minerals and demonstrate similar chemical properties

polydentate ligand ligand that is attached to a central metal ion by bonds from two or more donor atoms, named with prefixes specifying how many donors are present (e.g., hexadentate = six coordinate bonds formed)

rare earth element collection of 17 elements including the lanthanides, scandium, and yttrium that often occur together and have similar chemical properties, making separation difficult

second transition series transition elements in the fifth period of the periodic table (second row of the *d*-block), atomic numbers 39–47

smelting process of extracting a pure metal from a molten ore

spectrochemical series ranking of ligands according to the magnitude of the crystal field splitting they induce

steel material made from iron by removing impurities in the iron and adding substances that produce alloys with properties suitable for specific uses

strong-field ligand ligand that causes larger crystal field splittings

superconductor material that conducts electricity with no resistance

t_{2g} orbitals set of three *d* orbitals aligned between the Cartesian axes for coordination complexes; in octahedral complexes, they are lowered in energy compared to the e_g orbitals according to CFT

third transition series transition elements in the sixth period of the periodic table (third row of the *d*-block), atomic numbers 57 and 72–79

***trans* configuration** configuration of a geometrical isomer in which two similar groups are on opposite sides of an imaginary reference line on the molecule

weak-field ligand ligand that causes small crystal field splittings

Summary

19.1 Occurrence, Preparation, and Properties of Transition Metals and Their Compounds

The transition metals are elements with partially filled *d* orbitals, located in the *d*-block of the periodic table. The reactivity of the transition elements varies widely from very active metals such as scandium and iron to almost inert elements, such as the platinum metals. The type of chemistry used in the isolation of the elements from their ores depends upon the concentration of the element in its ore and the difficulty of reducing ions of the elements to the metals. Metals that are more active are more difficult to reduce.

Transition metals exhibit chemical behavior typical of metals. For example, they oxidize in air upon heating and react with elemental halogens to form halides. Those elements that lie above hydrogen in the activity series react with acids, producing salts and hydrogen gas. Oxides, hydroxides, and carbonates of transition metal compounds in low oxidation states are basic. Halides and other salts are generally stable in water, although oxygen must be excluded in some cases. Most transition metals form a variety of stable oxidation states, allowing them to demonstrate a wide range of chemical reactivity.

19.2 Coordination Chemistry of Transition Metals

The transition elements and main group elements can form coordination compounds, or complexes, in which a central metal atom or ion is bonded to one or more ligands by coordinate covalent bonds. Ligands with more than one donor atom are called polydentate ligands and form chelates. The common geometries found in complexes are tetrahedral and square planar (both with a coordination number of four) and octahedral (with a coordination number of six). *Cis* and *trans* configurations are possible in some octahedral and square planar complexes. In addition to these geometrical isomers, optical isomers (molecules or ions that are mirror images but not superimposable) are possible in certain octahedral complexes. Coordination complexes have a wide variety of uses including oxygen transport in blood, water purification, and pharmaceutical use.

19.3 Spectroscopic and Magnetic Properties of Coordination Compounds

Crystal field theory treats interactions between the electrons on the metal and the ligands as a simple electrostatic effect. The presence of the ligands near the metal ion changes the energies of the metal *d* orbitals relative to their energies in the free ion. Both the color and the magnetic properties of a complex can be attributed to this crystal field splitting. The magnitude of the splitting (Δ_{oct}) depends on the nature of the ligands bonded to the metal. Strong-field ligands produce large splitting and favor low-spin complexes, in which the t_{2g} orbitals are completely filled before any electrons occupy the e_g orbitals. Weak-field ligands favor formation of high-spin complexes. The t_{2g} and the e_g orbitals are singly occupied before any are doubly occupied.

Exercises

19.1 Occurrence, Preparation, and Properties of Transition Metals and Their Compounds

1. Write the electron configurations for each of the following elements:
   (a) Sc
   (b) Ti
   (c) Cr
   (d) Fe
   (e) Ru
2. Write the electron configurations for each of the following elements and its ions:
   (a) Ti
   (b) Ti^{2+}
   (c) Ti^{3+}
   (d) Ti^{4+}
3. Write the electron configurations for each of the following elements and its 3+ ions:
   (a) La
   (b) Sm
   (c) Lu
4. Why are the lanthanoid elements not found in nature in their elemental forms?
5. Which of the following elements is most likely to be used to prepare La by the reduction of La_2O_3: Al, C, or Fe? Why?
6. Which of the following is the strongest oxidizing agent: $VO_4{}^{3}$, $CrO_4{}^{2-}$, or $MnO_4{}^{-}$?
7. Which of the following elements is most likely to form an oxide with the formula MO_3: Zr, Nb, or Mo?
8. The following reactions all occur in a blast furnace. Which of these are redox reactions?
   (a) $3Fe_2O_3(s) + CO(g) \longrightarrow 2Fe_3O_4(s) + CO_2(g)$
   (b) $Fe_3O_4(s) + CO(g) \longrightarrow 3FeO(s) + CO_2(g)$
   (c) $FeO(s) + CO(g) \longrightarrow Fe(l) + CO_2(g)$
   (d) $C(s) + O_2(g) \longrightarrow CO_2(g)$
   (e) $C(s) + CO_2(g) \longrightarrow 2CO(g)$
   (f) $CaCO_3(s) \longrightarrow CaO(s) + CO_2(g)$
   (g) $CaO(s) + SiO_2(s) \longrightarrow CaSiO_3(l)$
9. Why is the formation of slag useful during the smelting of iron?
10. Would you expect an aqueous manganese(VII) oxide solution to have a pH greater or less than 7.0? Justify your answer.
11. Iron(II) can be oxidized to iron(III) by dichromate ion, which is reduced to chromium(III) in acid solution. A 2.5000-g sample of iron ore is dissolved and the iron converted into iron(II). Exactly 19.17 mL of 0.0100 M $Na_2Cr_2O_7$ is required in the titration. What percentage of the ore sample was iron?
12. How many cubic feet of air at a pressure of 760 torr and 0 °C is required per ton of Fe_2O_3 to convert that Fe_2O_3 into iron in a blast furnace? For this exercise, assume air is 19% oxygen by volume.
13. Find the potentials of the following electrochemical cell:
    $Cd \mid Cd^{2+}, M = 0.10 \parallel Ni^{2+}, M = 0.50 \mid Ni$
14. A 2.5624-g sample of a pure solid alkali metal chloride is dissolved in water and treated with excess silver nitrate. The resulting precipitate, filtered and dried, weighs 3.03707 g. What was the percent by mass of chloride ion in the original compound? What is the identity of the salt?
15. The standard reduction potential for the reaction $[Co(H_2O)_6]^{3+}(aq) + e^- \longrightarrow [Co(H_2O)_6]^{2+}(aq)$ is about 1.8 V. The reduction potential for the reaction $[Co(NH_3)_6]^{3+}(aq) + e^- \longrightarrow [Co(NH_3)_6]^{2+}(aq)$ is +0.1 V. Calculate the cell potentials to show whether the complex ions, $[Co(H_2O)_6]^{2+}$ and/or $[Co(NH_3)_6]^{2+}$, can be oxidized to the corresponding cobalt(III) complex by oxygen.

16. Predict the products of each of the following reactions. (Note: In addition to using the information in this chapter, also use the knowledge you have accumulated at this stage of your study, including information on the prediction of reaction products.)
(a) $MnCO_3(s) + HI(aq) \longrightarrow$
(b) $CoO(s) + O_2(g) \longrightarrow$
(c) $La(s) + O_2(g) \longrightarrow$
(d) $V(s) + VCl_4(s) \longrightarrow$
(e) $Co(s) + xsF_2(g) \longrightarrow$
(f) $CrO_3(s) + CsOH(aq) \longrightarrow$
17. Predict the products of each of the following reactions. (Note: In addition to using the information in this chapter, also use the knowledge you have accumulated at this stage of your study, including information on the prediction of reaction products.)
(a) $Fe(s) + H_2SO_4(aq) \longrightarrow$
(b) $FeCl_3(aq) + NaOH(aq) \longrightarrow$
(c) $Mn(OH)_2(s) + HBr(aq) \longrightarrow$
(d) $Cr(s) + O_2(g) \longrightarrow$
(e) $Mn_2O_3(s) + HCl(aq) \longrightarrow$
(f) $Ti(s) + xsF_2(g) \longrightarrow$
18. Describe the electrolytic process for refining copper.
19. Predict the products of the following reactions and balance the equations.
(a) Zn is added to a solution of $Cr_2(SO_4)_3$ in acid.
(b) $FeCl_2$ is added to a solution containing an excess of $Cr_2O_7{}^{2-}$ in hydrochloric acid.
(c) Cr^{2+} is added to $Cr_2O_7{}^{2-}$ in acid solution.
(d) Mn is heated with CrO_3.
(e) CrO is added to $2HNO_3$ in water.
(f) $FeCl_3$ is added to an aqueous solution of NaOH.
20. What is the gas produced when iron(II) sulfide is treated with a nonoxidizing acid?
21. Predict the products of each of the following reactions and then balance the chemical equations.
(a) Fe is heated in an atmosphere of steam.
(b) NaOH is added to a solution of $Fe(NO_3)_3$.
(c) $FeSO_4$ is added to an acidic solution of $KMnO_4$.
(d) Fe is added to a dilute solution of H_2SO_4.
(e) A solution of $Fe(NO_3)_2$ and HNO_3 is allowed to stand in air.
(f) $FeCO_3$ is added to a solution of $HClO_4$.
(g) Fe is heated in air.
22. Balance the following equations by oxidation-reduction methods; note that *three* elements change oxidation state.
$Co(NO_3)_2(s) \longrightarrow Co_2O_3(s) + NO_2(g) + O_2(g)$
23. Dilute sodium cyanide solution is slowly dripped into a slowly stirred silver nitrate solution. A white precipitate forms temporarily but dissolves as the addition of sodium cyanide continues. Use chemical equations to explain this observation. Silver cyanide is similar to silver chloride in its solubility.
24. Predict which will be more stable, $[CrO_4]^{2-}$ or $[WO_4]^{2-}$, and explain.

25. Give the oxidation state of the metal for each of the following oxides of the first transition series. (Hint: Oxides of formula M_3O_4 are examples of *mixed valence compounds* in which the metal ion is present in more than one oxidation state. It is possible to write these compound formulas in the equivalent format $MO{\cdot}M_2O_3$, to permit estimation of the metal's two oxidation states.)
(a) Sc_2O_3
(b) TiO_2
(c) V_2O_5
(d) CrO_3
(e) MnO_2
(f) Fe_3O_4
(g) Co_3O_4
(h) NiO
(i) Cu_2O

19.2 Coordination Chemistry of Transition Metals

26. Indicate the coordination number for the central metal atom in each of the following coordination compounds:
(a) $[Pt(H_2O)_2Br_2]$
(b) $[Pt(NH_3)(py)(Cl)(Br)]$ (py = pyridine, C_5H_5N)
(c) $[Zn(NH_3)_2Cl_2]$
(d) $[Zn(NH_3)(py)(Cl)(Br)]$
(e) $[Ni(H_2O)_4Cl_2]$
(f) $[Fe(en)_2(CN)_2]^+$ (en = ethylenediamine, $C_2H_8N_2$)

27. Give the coordination numbers and write the formulas for each of the following, including all isomers where appropriate:
(a) tetrahydroxozincate(II) ion (tetrahedral)
(b) hexacyanopalladate(IV) ion
(c) dichloroaurate(I) ion (note that *aurum* is Latin for "gold")
(d) diamminedichloroplatinum(II)
(e) potassium diamminetetrachlorochromate(III)
(f) hexaamminecobalt(III) hexacyanochromate(III)
(g) dibromobis(ethylenediamine) cobalt(III) nitrate

28. Give the coordination number for each metal ion in the following compounds:
(a) $[Co(CO_3)_3]^{3-}$ (note that $CO_3{}^{2-}$ is bidentate in this complex)
(b) $[Cu(NH_3)_4]^{2+}$
(c) $[Co(NH_3)_4Br_2]_2(SO_4)_3$
(d) $[Pt(NH_3)_4][PtCl_4]$
(e) $[Cr(en)_3](NO_3)_3$
(f) $[Pd(NH_3)_2Br_2]$ (square planar)
(g) $K_3[Cu(Cl)_5]$
(h) $[Zn(NH_3)_2Cl_2]$

29. Sketch the structures of the following complexes. Indicate any *cis, trans,* and optical isomers.
(a) $[Pt(H_2O)_2Br_2]$ (square planar)
(b) $[Pt(NH_3)(py)(Cl)(Br)]$ (square planar, py = pyridine, C_5H_5N)
(c) $[Zn(NH_3)_3Cl]^+$ (tetrahedral)
(d) $[Pt(NH_3)_3Cl]^+$ (square planar)
(e) $[Ni(H_2O)_4Cl_2]$
(f) $[Co(C_2O_4)_2Cl_2]^{3-}$ (note that $C_2O_4{}^{2-}$ is the bidentate oxalate ion, $-O_2CCO_2{}^-$)

30. Draw diagrams for any *cis*, *trans*, and optical isomers that could exist for the following (en is ethylenediamine):
(a) $[Co(en)_2(NO_2)Cl]^+$
(b) $[Co(en)_2Cl_2]^+$
(c) $[Pt(NH_3)_2Cl_4]$
(d) $[Cr(en)_3]^{3+}$
(e) $[Pt(NH_3)_2Cl_2]$
31. Name each of the compounds or ions given in Exercise 19.28, including the oxidation state of the metal.
32. Name each of the compounds or ions given in Exercise 19.30.
33. Specify whether the following complexes have isomers.
(a) tetrahedral $[Ni(CO)_2(Cl)_2]$
(b) trigonal bipyramidal $[Mn(CO)_4NO]$
(c) $[Pt(en)_2Cl_2]Cl_2$
34. Predict whether the carbonate ligand ${CO_3}^{2-}$ will coordinate to a metal center as a monodentate, bidentate, or tridentate ligand.
35. Draw the geometric, linkage, and ionization isomers for $[CoCl_5CN][CN]$.

19.3 Spectroscopic and Magnetic Properties of Coordination Compounds

36. Determine the number of unpaired electrons expected for $[Fe(NO_2)_6]^{3-}$ and for $[FeF_6]^{3-}$ in terms of crystal field theory.
37. Draw the crystal field diagrams for $[Fe(NO_2)_6]^{4-}$ and $[FeF_6]^{3-}$. State whether each complex is high spin or low spin, paramagnetic or diamagnetic, and compare Δ_{oct} to P for each complex.
38. Give the oxidation state of the metal, number of *d* electrons, and the number of unpaired electrons predicted for $[Co(NH_3)_6]Cl_3$.
39. The solid anhydrous solid $CoCl_2$ is blue in color. Because it readily absorbs water from the air, it is used as a humidity indicator to monitor if equipment (such as a cell phone) has been exposed to excessive levels of moisture. Predict what product is formed by this reaction, and how many unpaired electrons this complex will have.
40. Is it possible for a complex of a metal in the transition series to have six unpaired electrons? Explain.
41. How many unpaired electrons are present in each of the following?
(a) $[CoF_6]^{3-}$ (high spin)
(b) $[Mn(CN)_6]^{3-}$ (low spin)
(c) $[Mn(CN)_6]^{4-}$ (low spin)
(d) $[MnCl_6]^{4-}$ (high spin)
(e) $[RhCl_6]^{3-}$ (low spin)
42. Explain how the diphosphate ion, $[O_3P{-}O{-}PO_3]^{4-}$, can function as a water softener that prevents the precipitation of Fe^{2+} as an insoluble iron salt.
43. For complexes of the same metal ion with no change in oxidation number, the stability increases as the number of electrons in the t_{2g} orbitals increases. Which complex in each of the following pairs of complexes is more stable?
(a) $[Fe(H_2O)_6]^{2+}$ or $[Fe(CN)_6]^{4-}$
(b) $[Co(NH_3)_6]^{3+}$ or $[CoF_6]^{3-}$
(c) $[Mn(CN)_6]^{4-}$ or $[MnCl_6]^{4-}$
44. Trimethylphosphine, $P(CH_3)_3$, can act as a ligand by donating the lone pair of electrons on the phosphorus atom. If trimethylphosphine is added to a solution of nickel(II) chloride in acetone, a blue compound that has a molecular mass of approximately 270 g and contains 21.5% Ni, 26.0% Cl, and 52.5% $P(CH_3)_3$ can be isolated. This blue compound does not have any isomeric forms. What are the geometry and molecular formula of the blue compound?
45. Would you expect the complex $[Co(en)_3]Cl_3$ to have any unpaired electrons? Any isomers?
46. Would you expect the $Mg_3[Cr(CN)_6]_2$ to be diamagnetic or paramagnetic? Explain your reasoning.
47. Would you expect salts of the gold(I) ion, Au^+, to be colored? Explain.
48. $[CuCl_4]^{2-}$ is green. $[Cu(H_2O)_6]^{2+}$ is blue. Which absorbs higher-energy photons? Which is predicted to have a larger crystal field splitting?

CHAPTER 20
Organic Chemistry

Figure 20.1 All organic compounds contain carbon and most are formed by living things, although they are also formed by geological and artificial processes. (credit left: modification of work by Jon Sullivan; credit left middle: modification of work by Deb Tremper; credit right middle: modification of work by “annszyp”/Wikimedia Commons; credit right: modification of work by George Shuklin)

CHAPTER OUTLINE

20.1 Hydrocarbons
20.2 Alcohols and Ethers
20.3 Aldehydes, Ketones, Carboxylic Acids, and Esters
20.4 Amines and Amides

INTRODUCTION All living things on earth are formed mostly of carbon compounds. The prevalence of carbon compounds in living things has led to the epithet “carbon-based” life. The truth is we know of no other kind of life. Early chemists regarded substances isolated from *organisms* (plants and animals) as a different type of matter that could not be synthesized artificially, and these substances were thus known as *organic compounds*. The widespread belief called vitalism held that organic compounds were formed by a vital force present only in living organisms. The German chemist Friedrich Wohler was one of the early chemists to refute this aspect of vitalism, when, in 1828, he reported the synthesis of urea, a component of many body fluids, from nonliving materials. Since then, it has been recognized that organic molecules obey the same natural laws as inorganic substances, and the category of organic compounds has evolved to include both natural and synthetic compounds that contain carbon. Some carbon-containing compounds are *not* classified as organic, for example, carbonates and cyanides, and simple oxides, such as CO and CO_2. Although a single, precise definition has yet to be identified by the chemistry community, most agree that a defining trait of organic molecules is the presence of carbon as the principal element, bonded to hydrogen and other carbon atoms.

Today, organic compounds are key components of plastics, soaps, perfumes, sweeteners, fabrics, pharmaceuticals, and many other substances that we use every day. The value to us of organic compounds ensures that organic chemistry is an important discipline within the general field of chemistry. In this chapter, we discuss why the element carbon gives rise to a vast number and variety of compounds, how those compounds are classified, and the role of organic compounds in representative biological and industrial settings.

20.1 Hydrocarbons

LEARNING OBJECTIVES

By the end of this section, you will be able to:

- Explain the importance of hydrocarbons and the reason for their diversity
- Name saturated and unsaturated hydrocarbons, and molecules derived from them
- Describe the reactions characteristic of saturated and unsaturated hydrocarbons
- Identify structural and geometric isomers of hydrocarbons

The largest database[1] of organic compounds lists about 10 million substances, which include compounds originating from living organisms and those synthesized by chemists. The number of potential organic compounds has been estimated[2] at 10^{60}—an astronomically high number. The existence of so many organic molecules is a consequence of the ability of carbon atoms to form up to four strong bonds to other carbon atoms, resulting in chains and rings of many different sizes, shapes, and complexities.

The simplest **organic compounds** contain only the elements carbon and hydrogen, and are called hydrocarbons. Even though they are composed of only two types of atoms, there is a wide variety of hydrocarbons because they may consist of varying lengths of chains, branched chains, and rings of carbon atoms, or combinations of these structures. In addition, hydrocarbons may differ in the types of carbon-carbon bonds present in their molecules. Many hydrocarbons are found in plants, animals, and their fossils; other hydrocarbons have been prepared in the laboratory. We use hydrocarbons every day, mainly as fuels, such as natural gas, acetylene, propane, butane, and the principal components of gasoline, diesel fuel, and heating oil. The familiar plastics polyethylene, polypropylene, and polystyrene are also hydrocarbons. We can distinguish several types of hydrocarbons by differences in the bonding between carbon atoms. This leads to differences in geometries and in the hybridization of the carbon orbitals.

Alkanes

Alkanes, or **saturated hydrocarbons**, contain only single covalent bonds between carbon atoms. Each of the carbon atoms in an alkane has sp^3 hybrid orbitals and is bonded to four other atoms, each of which is either carbon or hydrogen. The Lewis structures and models of methane, ethane, and pentane are illustrated in Figure 20.2. Carbon chains are usually drawn as straight lines in Lewis structures, but one has to remember that Lewis structures are not intended to indicate the geometry of molecules. Notice that the carbon atoms in the structural models (the ball-and-stick and space-filling models) of the pentane molecule do not lie in a straight line. Because of the sp^3 hybridization, the bond angles in carbon chains are close to 109.5°, giving such chains in an alkane a zigzag shape.

The structures of alkanes and other organic molecules may also be represented in a less detailed manner by condensed structural formulas (or simply, *condensed formulas*). Instead of the usual format for chemical formulas in which each element symbol appears just once, a condensed formula is written to suggest the bonding in the molecule. These formulas have the appearance of a Lewis structure from which most or all of the bond symbols have been removed. Condensed structural formulas for ethane and pentane are shown at the bottom of Figure 20.2, and several additional examples are provided in the exercises at the end of this chapter.

1 This is the Beilstein database, now available through the Reaxys site (www.elsevier.com/online-tools/reaxys).
2 Peplow, Mark. "Organic Synthesis: The Robo-Chemist," *Nature* 512 (2014): 20–2.

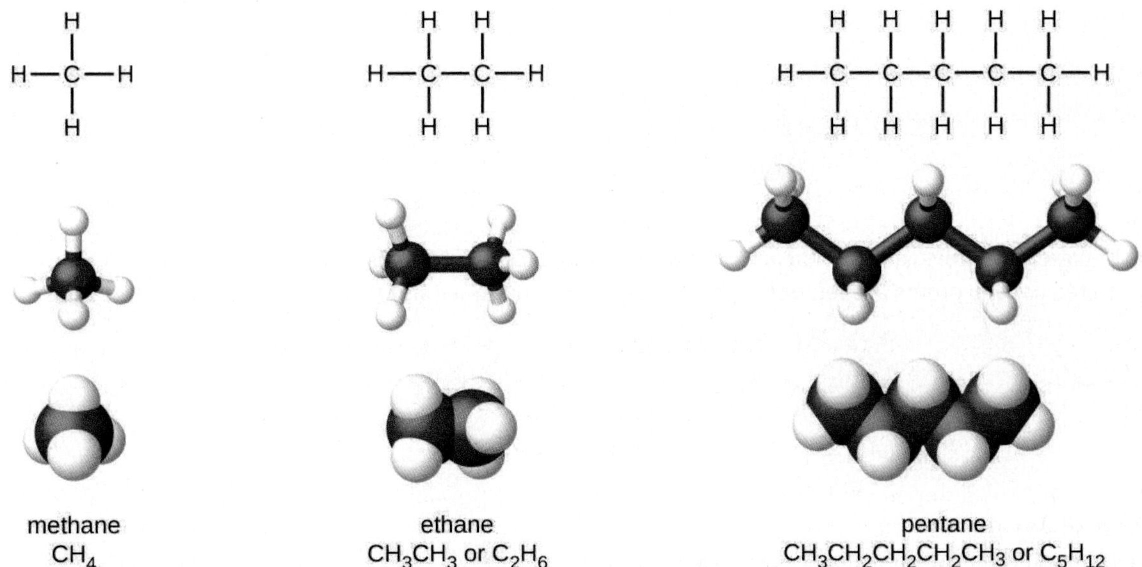

FIGURE 20.2 Pictured are the Lewis structures, ball-and-stick models, and space-filling models for molecules of methane, ethane, and pentane.

A common method used by organic chemists to simplify the drawings of larger molecules is to use a **skeletal structure** (also called a line-angle structure). In this type of structure, carbon atoms are not symbolized with a C, but represented by each end of a line or bend in a line. Hydrogen atoms are not drawn if they are attached to a carbon. Other atoms besides carbon and hydrogen are represented by their elemental symbols. Figure 20.3 shows three different ways to draw the same structure.

Expanded formula

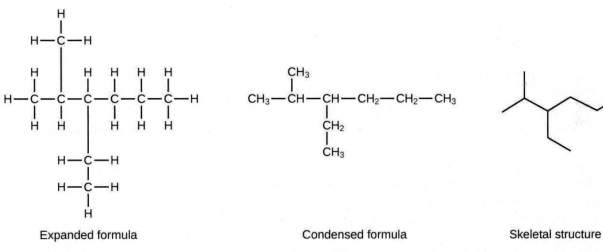

Condensed formula

Skeletal structure

FIGURE 20.3 The same structure can be represented three different ways: an expanded formula, a condensed formula, and a skeletal structure.

EXAMPLE 20.1

Drawing Skeletal Structures

Draw the skeletal structures for these two molecules:

CH3 CH3
CH3—C—CH
CH3 CH3
$CH_3CH_2CH_2CH_2CH_2CH_2CH_3$
(a) (b)

Solution

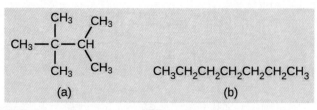

Each carbon atom is converted into the end of a line or the place where lines intersect. All hydrogen atoms attached to the carbon atoms are left out of the structure (although we still need to recognize they are there):

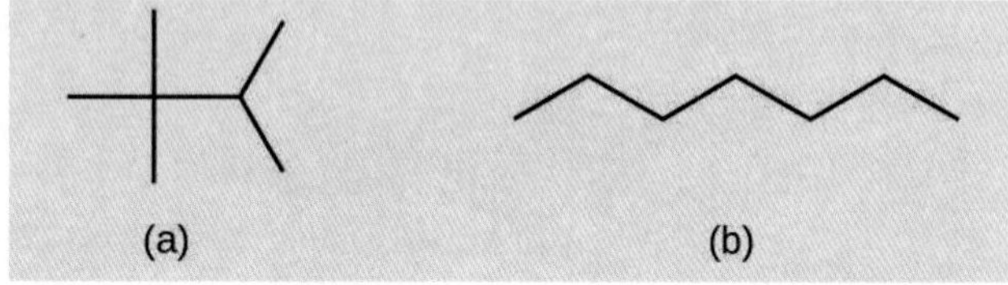

Check Your Learning

Draw the skeletal structures for these two molecules:

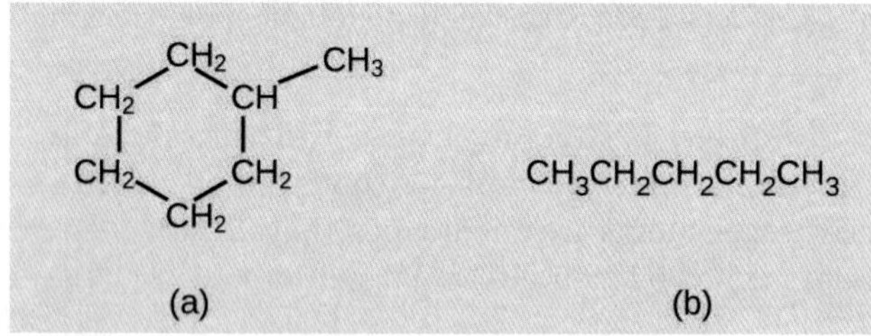

Answer:

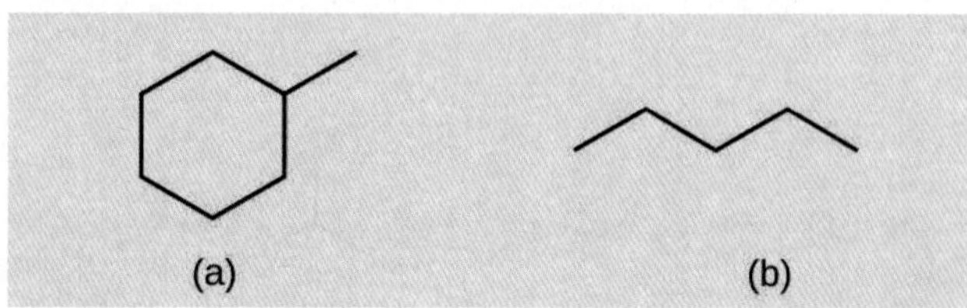

EXAMPLE 20.2

Interpreting Skeletal Structures

Identify the chemical formula of the molecule represented here:

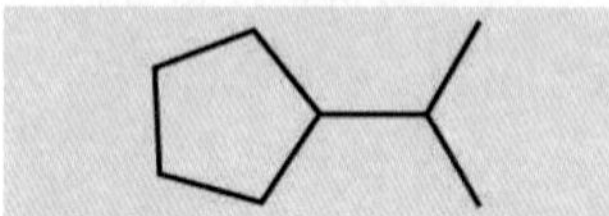

Solution

There are eight places where lines intersect or end, meaning that there are eight carbon atoms in the molecule. Since we know that carbon atoms tend to make four bonds, each carbon atom will have the number of hydrogen atoms that are required for four bonds. This compound contains 16 hydrogen atoms for a molecular formula of C_8H_{16}.

Location of the hydrogen atoms:

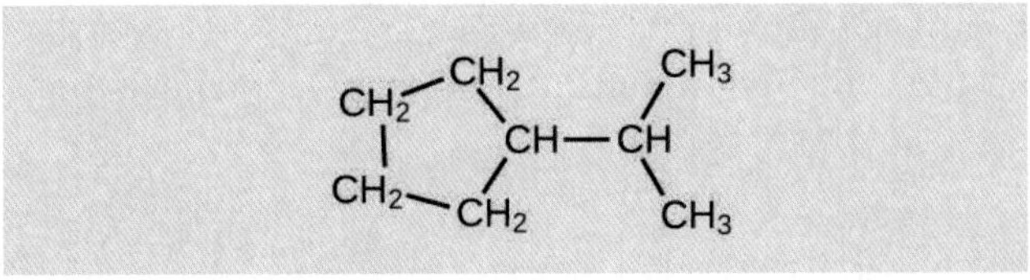

Check Your Learning

Identify the chemical formula of the molecule represented here:

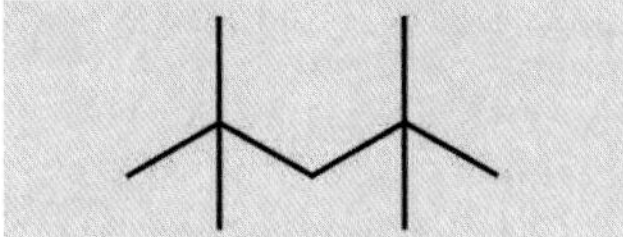

Answer:

C_9H_{20}

All alkanes are composed of carbon and hydrogen atoms, and have similar bonds, structures, and formulas; noncyclic alkanes all have a formula of C_nH_{2n+2}. The number of carbon atoms present in an alkane has no limit. Greater numbers of atoms in the molecules will lead to stronger intermolecular attractions (dispersion forces) and correspondingly different physical properties of the molecules. Properties such as melting point and boiling point (Table 20.1) usually change smoothly and predictably as the number of carbon and hydrogen atoms in the molecules change.

Properties of Some Alkanes[3]

Alkane	Molecular Formula	Melting Point (°C)	Boiling Point (°C)	Phase at STP[4]	Number of Structural Isomers
methane	CH_4	–182.5	–161.5	gas	1
ethane	C_2H_6	–183.3	–88.6	gas	1
propane	C_3H_8	–187.7	–42.1	gas	1
butane	C_4H_{10}	–138.3	–0.5	gas	2
pentane	C_5H_{12}	–129.7	36.1	liquid	3
hexane	C_6H_{14}	–95.3	68.7	liquid	5
heptane	C_7H_{16}	–90.6	98.4	liquid	9
octane	C_8H_{18}	–56.8	125.7	liquid	18
nonane	C_9H_{20}	–53.6	150.8	liquid	35
decane	$C_{10}H_{22}$	–29.7	174.0	liquid	75
tetradecane	$C_{14}H_{30}$	5.9	253.5	solid	1858
octadecane	$C_{18}H_{38}$	28.2	316.1	solid	60,523

3 Physical properties for C_4H_{10} and heavier molecules are those of the *normal isomer*, *n*-butane, *n*-pentane, etc.

Hydrocarbons with the same formula, including alkanes, can have different structures. For example, two alkanes have the formula C_4H_{10}: They are called *n*-butane and 2-methylpropane (or isobutane), and have the following Lewis structures:

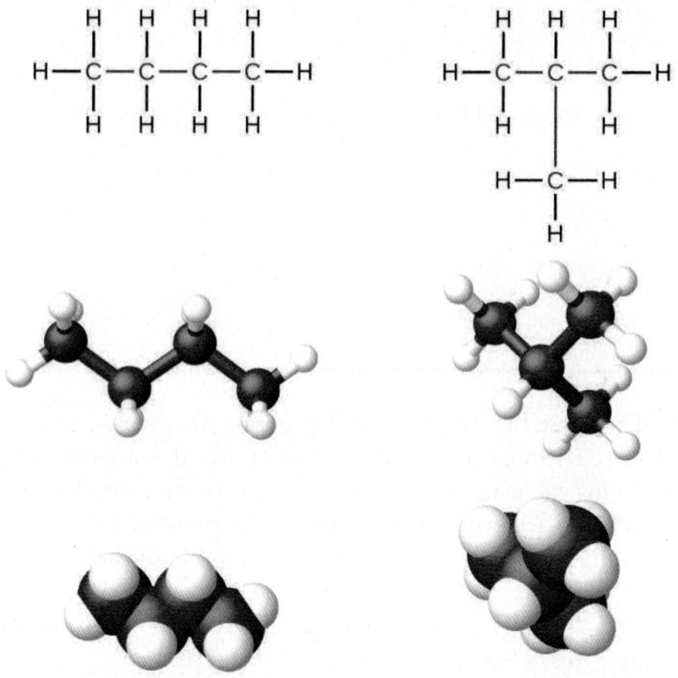

n-butane 2-methylpropane

The compounds *n*-butane and 2-methylpropane are structural isomers (the term constitutional isomers is also commonly used). Constitutional isomers have the same molecular formula but different spatial arrangements of the atoms in their molecules. The *n*-butane molecule contains an *unbranched chain*, meaning that no carbon atom is bonded to more than two other carbon atoms. We use the term *normal*, or the prefix *n*, to refer to a chain of carbon atoms without branching. The compound 2–methylpropane has a branched chain (the carbon atom in the center of the Lewis structure is bonded to three other carbon atoms)

Identifying isomers from Lewis structures is not as easy as it looks. Lewis structures that look different may actually represent the same isomers. For example, the three structures in Figure 20.4 all represent the same molecule, *n*-butane, and hence are not different isomers. They are identical because each contains an unbranched chain of four carbon atoms.

4 STP indicates a temperature of 0 °C and a pressure of 1 atm.

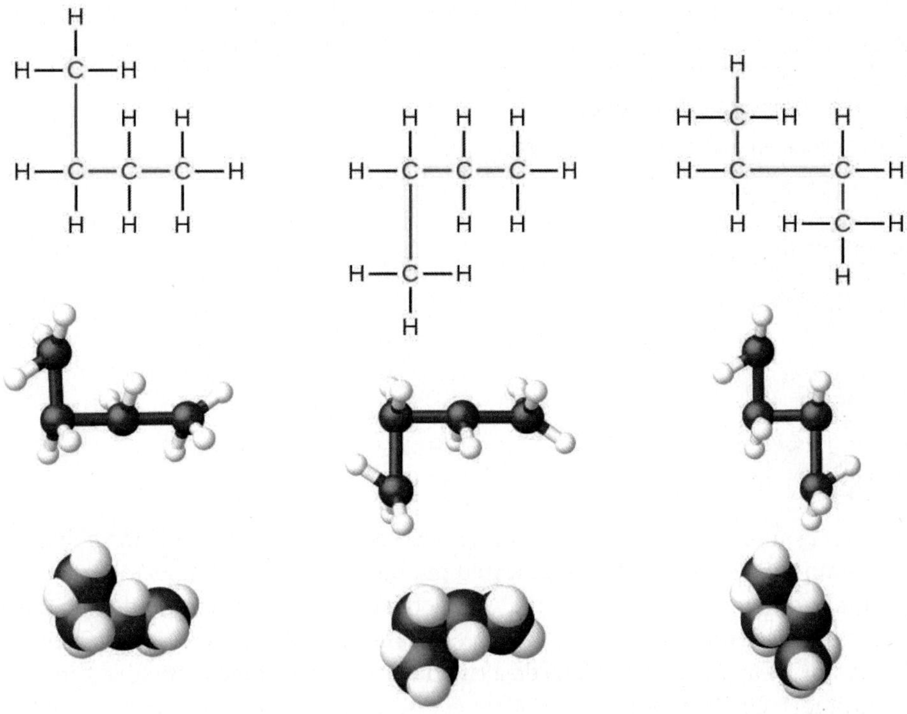

FIGURE 20.4 These three representations of the structure of n-butane are not isomers because they all contain the same arrangement of atoms and bonds.

The Basics of Organic Nomenclature: Naming Alkanes

The International Union of Pure and Applied Chemistry (IUPAC) has devised a system of nomenclature that begins with the names of the alkanes and can be adjusted from there to account for more complicated structures. The nomenclature for alkanes is based on two rules:

1. To name an alkane, first identify the longest chain of carbon atoms in its structure. A two-carbon chain is called ethane; a three-carbon chain, propane; and a four-carbon chain, butane. Longer chains are named as follows: pentane (five-carbon chain), hexane (6), heptane (7), octane (8), nonane (9), and decane (10). These prefixes can be seen in the names of the alkanes described in Table 20.1.
2. Add prefixes to the name of the longest chain to indicate the positions and names of **substituents**. Substituents are branches or functional groups that replace hydrogen atoms on a chain. The position of a substituent or branch is identified by the number of the carbon atom it is bonded to in the chain. We number the carbon atoms in the chain by counting from the end of the chain nearest the substituents. Multiple substituents are named individually and placed in alphabetical order at the front of the name.

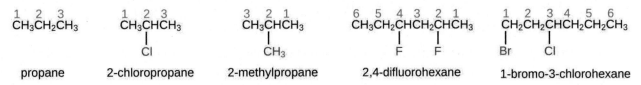

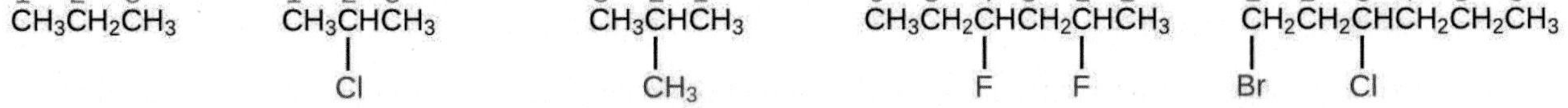

Numbering	Formula	Substituents	Name
1 2 3	$CH_3CH_2CH_3$		propane
1 2 3	CH_3CHCH_3	Cl (on C2)	2-chloropropane
3 2 1	CH_3CHCH_3	CH_3 (on C2)	2-methylpropane
6 5 4 3 2 1	$CH_3CH_2CHCH_2CHCH_3$	F, F (on C4, C2)	2,4-difluorohexane
1 2 3 4 5 6	$CH_2CH_2CHCH_2CH_2CH_3$	Br (on C1), Cl (on C3)	1-bromo-3-chlorohexane

When more than one substituent is present, either on the same carbon atom or on different carbon atoms, the substituents are listed alphabetically. Because the carbon atom numbering begins at the end closest to a substituent, the longest chain of carbon atoms is numbered in such a way as to produce the lowest number for the substituents. The ending *-o* replaces *-ide* at the end of the name of an electronegative substituent (in ionic compounds, the negatively charged ion ends with *-ide* like chloride; in organic compounds, such atoms are treated as substituents and the *-o* ending is used). The number of substituents of the same type is indicated by the prefixes *di-* (two), *tri-* (three), *tetra-* (four), and so on (for example, *difluoro-* indicates two fluoride substituents).

EXAMPLE 20.3

Naming Halogen-substituted Alkanes

Name the molecule whose structure is shown here:

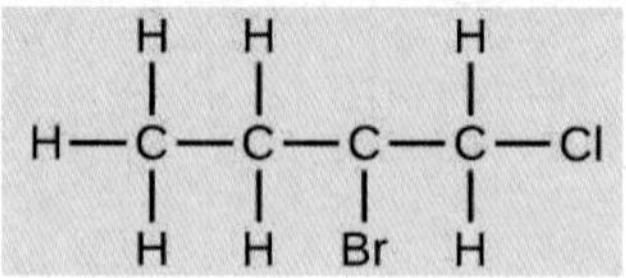

Solution

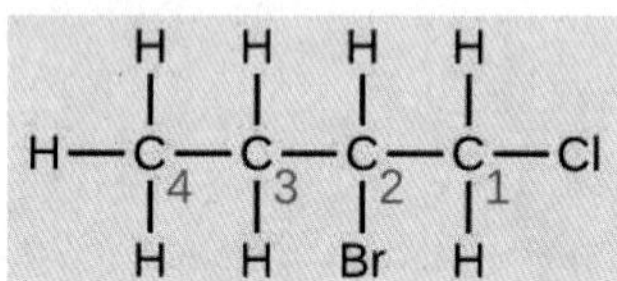

The four-carbon chain is numbered from the end with the chlorine atom. This puts the substituents on positions 1 and 2 (numbering from the other end would put the substituents on positions 3 and 4). Four carbon atoms means that the base name of this compound will be butane. The bromine at position 2 will be described by adding 2-bromo-; this will come at the beginning of the name, since bromo- comes before chloro- alphabetically. The chlorine at position 1 will be described by adding 1-chloro-, resulting in the name of the molecule being 2-bromo-1-chlorobutane.

Check Your Learning

Name the following molecule:

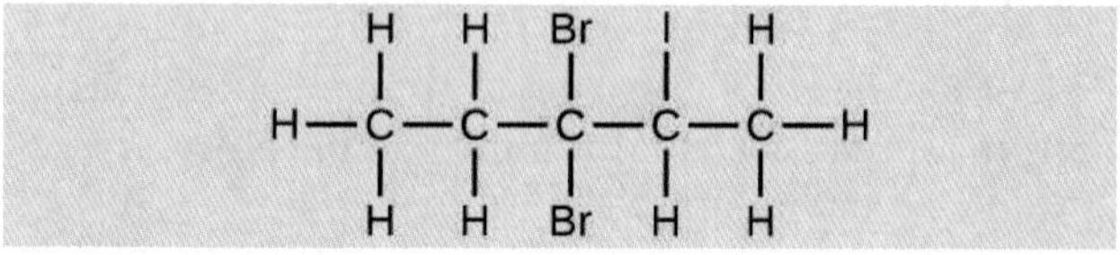

Answer:

3,3-dibromo-2-iodopentane

We call a substituent that contains one less hydrogen than the corresponding alkane an alkyl group. The name of an **alkyl group** is obtained by dropping the suffix *-ane* of the alkane name and adding *-yl*:

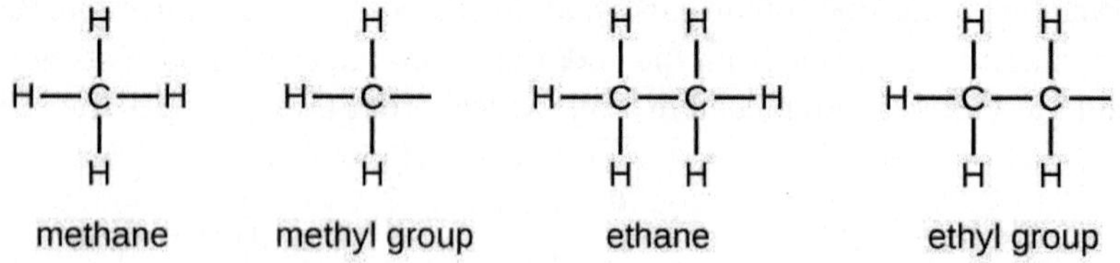

methane methyl group ethane ethyl group

The open bonds in the methyl and ethyl groups indicate that these alkyl groups are bonded to another atom.

EXAMPLE 20.4

Naming Substituted Alkanes

Name the molecule whose structure is shown here:

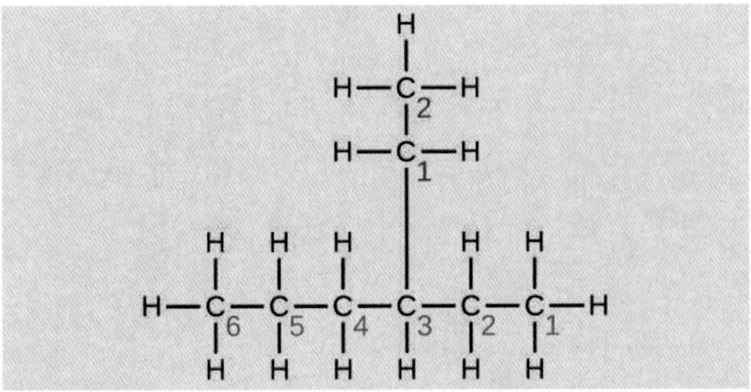

Solution

The longest carbon chain runs horizontally across the page and contains six carbon atoms (this makes the base of the name hexane, but we will also need to incorporate the name of the branch). In this case, we want to number from right to left (as shown by the blue numbers) so the branch is connected to carbon 3 (imagine the numbers from left to right—this would put the branch on carbon 4, violating our rules). The branch attached to position 3 of our chain contains two carbon atoms (numbered in red)—so we take our name for two carbons *eth-* and attach *-yl* at the end to signify we are describing a branch. Putting all the pieces together, this molecule is 3-ethylhexane.

Check Your Learning

Name the following molecule:

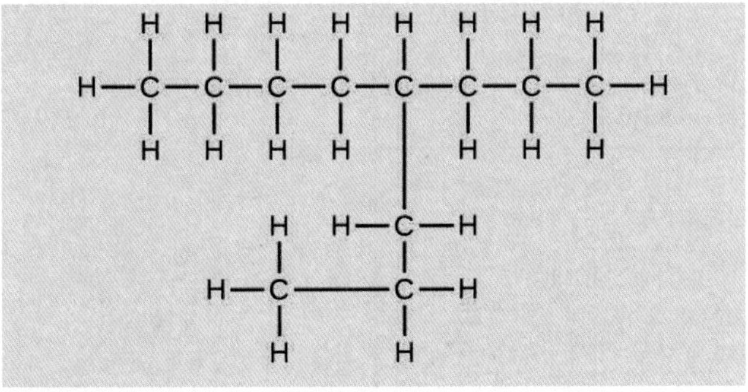

Answer:

4-propyloctane

Some hydrocarbons can form more than one type of alkyl group when the hydrogen atoms that would be removed have different "environments" in the molecule. This diversity of possible alkyl groups can be identified in the following way: The four hydrogen atoms in a methane molecule are equivalent; they all have the same environment. They are equivalent because each is bonded to a carbon atom (the same carbon atom) that is bonded to three hydrogen atoms. (It may be easier to see the equivalency in the ball and stick models in Figure 20.2. Removal of any one of the four hydrogen atoms from methane forms a methyl group. Likewise, the six hydrogen atoms in ethane are equivalent (Figure 20.2) and removing any one of these hydrogen atoms produces an ethyl group. Each of the six hydrogen atoms is bonded to a carbon atom that is bonded to two other hydrogen atoms and a carbon atom. However, in both propane and 2–methylpropane, there are hydrogen atoms in two different environments, distinguished by the adjacent atoms or groups of atoms:

```
    H   H   H              H   H   H
    |   |   |              |   |   |
H — C — C — C — H      H — C — C — C — H
    |   |   |              |   |   |
    H   H   H              H   |   H
                               |
                           H — C — H
                               |
                               H
```

propane 2-methylpropane

Each of the six equivalent hydrogen atoms of the first type in propane and each of the nine equivalent hydrogen atoms of that type in 2-methylpropane (all shown in black) are bonded to a carbon atom that is bonded to only one other carbon atom. The two purple hydrogen atoms in propane are of a second type. They differ from the six hydrogen atoms of the first type in that they are bonded to a carbon atom bonded to two other carbon atoms. The green hydrogen atom in 2-methylpropane differs from the other nine hydrogen atoms in that molecule and from the purple hydrogen atoms in propane. The green hydrogen atom in 2-methylpropane is bonded to a carbon atom bonded to three other carbon atoms. Two different alkyl groups can be formed from each of these molecules, depending on which hydrogen atom is removed. The names and structures of these and several other alkyl groups are listed in Figure 20.5.

Alkyl Group	Structure
methyl	$CH_3—$
ethyl	$CH_3CH_2—$
n-propyl	$CH_3CH_2CH_2—$
isopropyl	$CH_3\overset{|}{C}HCH_3$
n-butyl	$CH_3CH_2CH_2CH_2—$
sec-butyl	$CH_3CH_2\overset{|}{C}HCH_3$
isobutyl	$CH_3\underset{\underset{CH_3}{|}}{C}HCH_2—$
tert-butyl	$CH_3\overset{|}{\underset{\underset{CH_3}{|}}{C}}CH_3$

FIGURE 20.5 This listing gives the names and formulas for various alkyl groups formed by the removal of hydrogen atoms from different locations.

Note that alkyl groups do not exist as stable independent entities. They are always a part of some larger molecule. The location of an alkyl group on a hydrocarbon chain is indicated in the same way as any other substituent:

7 6 5 4 3 2 1
$CH_3CH_2CH_2CH_2CHCH_2CH_3$
with CH_2CH_3 on carbon 3

3-ethylheptane

CH_3 on carbons 2 and 4
1 2 3 4 5
$CH_3C{-}CH_2CHCH_3$
CH_3 on carbon 2

2,2,4-trimethylpentane

CH_3 CH_3
CH
$CH_3CH_2CH_2CHCH_2CH_2CH_3$

4-isopropylheptane

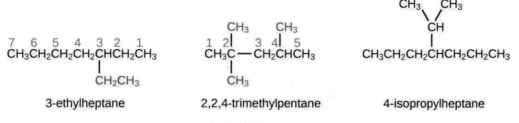

Alkanes are relatively stable molecules, but heat or light will activate reactions that involve the breaking of C–H or C–C single bonds. Combustion is one such reaction:

$$CH_4(g) + 2O_2(g) \longrightarrow CO_2(g) + 2H_2O(g)$$

Alkanes burn in the presence of oxygen, a highly exothermic oxidation-reduction reaction that produces carbon dioxide and water. As a consequence, alkanes are excellent fuels. For example, methane, CH_4, is the principal component of natural gas. Butane, C_4H_{10}, used in camping stoves and lighters is an alkane. Gasoline is a liquid mixture of continuous- and branched-chain alkanes, each containing from five to nine carbon atoms, plus various additives to improve its performance as a fuel. Kerosene, diesel oil, and fuel oil are primarily mixtures of alkanes with higher molecular masses. The main source of these liquid alkane fuels is crude oil, a complex mixture that is separated by fractional distillation. Fractional distillation takes advantage of differences in the boiling points of the components of the mixture (see Figure 20.6). You may recall that boiling point is a function of intermolecular interactions, which was discussed in the chapter on solutions and colloids.

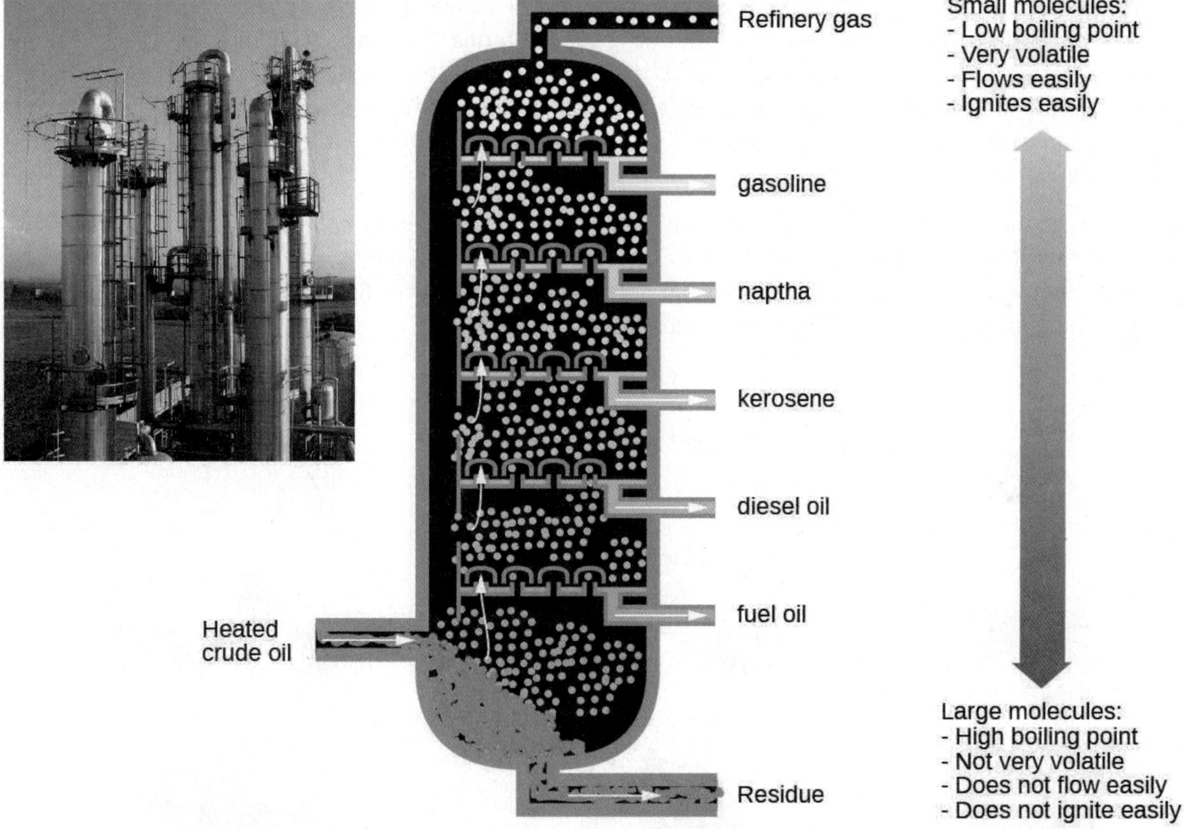

FIGURE 20.6 In a column for the fractional distillation of crude oil, oil heated to about 425 °C in the furnace vaporizes when it enters the base of the tower. The vapors rise through bubble caps in a series of trays in the tower. As the vapors gradually cool, fractions of higher, then of lower, boiling points condense to liquids and are drawn off. (credit left: modification of work by Luigi Chiesa)

In a **substitution reaction**, another typical reaction of alkanes, one or more of the alkane's hydrogen atoms is replaced with a different atom or group of atoms. No carbon-carbon bonds are broken in these reactions, and the hybridization of the carbon atoms does not change. For example, the reaction between ethane and molecular chlorine depicted here is a substitution reaction:

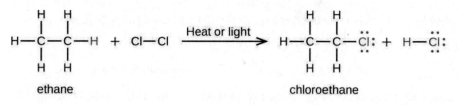

$$\underset{\text{ethane}}{H_3C{-}CH_3} + Cl{-}Cl \xrightarrow{\text{Heat or light}} \underset{\text{chloroethane}}{H_3C{-}CH_2{-}Cl} + H{-}Cl$$

The C–Cl portion of the chloroethane molecule is an example of a **functional group**, the part or moiety of a molecule that imparts a specific chemical reactivity. The types of functional groups present in an organic molecule are major determinants of its chemical properties and are used as a means of classifying organic compounds as detailed in the remaining sections of this chapter.

LINK TO LEARNING

Want more practice naming alkanes? Watch this brief video tutorial (http://openstax.org/l/16alkanes) to review the nomenclature process.

Alkenes

Organic compounds that contain one or more double or triple bonds between carbon atoms are described as unsaturated. You have likely heard of unsaturated fats. These are complex organic molecules with long chains of carbon atoms, which contain at least one double bond between carbon atoms. Unsaturated hydrocarbon molecules that contain one or more double bonds are called **alkenes**. Carbon atoms linked by a double bond are bound together by two bonds, one σ bond and one π bond. Double and triple bonds give rise to a different geometry around the carbon atom that participates in them, leading to important differences in molecular shape and properties. The differing geometries are responsible for the different properties of unsaturated versus saturated fats.

Ethene, C_2H_4, is the simplest alkene. Each carbon atom in ethene, commonly called ethylene, has a trigonal planar structure. The second member of the series is propene (propylene) (Figure 20.7); the butene isomers follow in the series. Four carbon atoms in the chain of butene allows for the formation of isomers based on the position of the double bond, as well as a new form of isomerism.

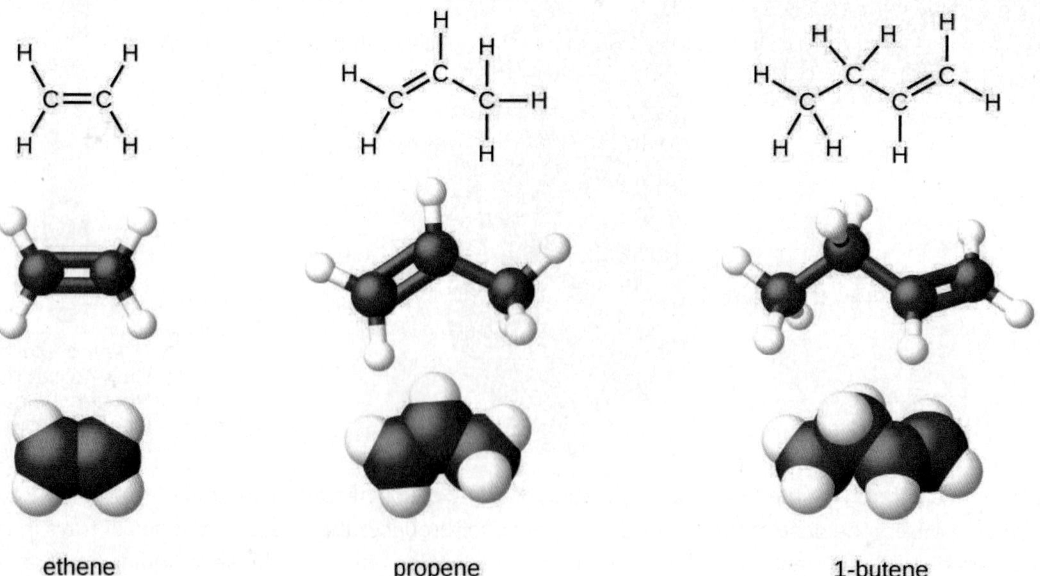

FIGURE 20.7 Expanded structures, ball-and-stick structures, and space-filling models for the alkenes ethene, propene, and 1-butene are shown.

Ethylene (the common industrial name for ethene) is a basic raw material in the production of polyethylene and other important compounds. Over 135 million tons of ethylene were produced worldwide in 2010 for use in the polymer, petrochemical, and plastic industries. Ethylene is produced industrially in a process called cracking, in which the long hydrocarbon chains in a petroleum mixture are broken into smaller molecules.

Chemistry in Everyday Life

Recycling Plastics

Polymers (from Greek words *poly* meaning "many" and *mer* meaning "parts") are large molecules made up of repeating units, referred to as monomers. Polymers can be natural (starch is a polymer of sugar residues and proteins are polymers of amino acids) or synthetic [like polyethylene, polyvinyl chloride (PVC), and polystyrene]. The variety of structures of polymers translates into a broad range of properties and uses that make them integral parts of our everyday lives. Adding functional groups to the structure of a polymer can result in significantly different properties (see the discussion about Kevlar later in this chapter).

An example of a polymerization reaction is shown in Figure 20.8. The monomer ethylene (C_2H_4) is a gas at room temperature, but when polymerized, using a transition metal catalyst, it is transformed into a solid material made up of long chains of $-CH_2-$ units called polyethylene. Polyethylene is a commodity plastic used primarily for packaging (bags and films).

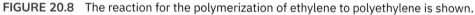

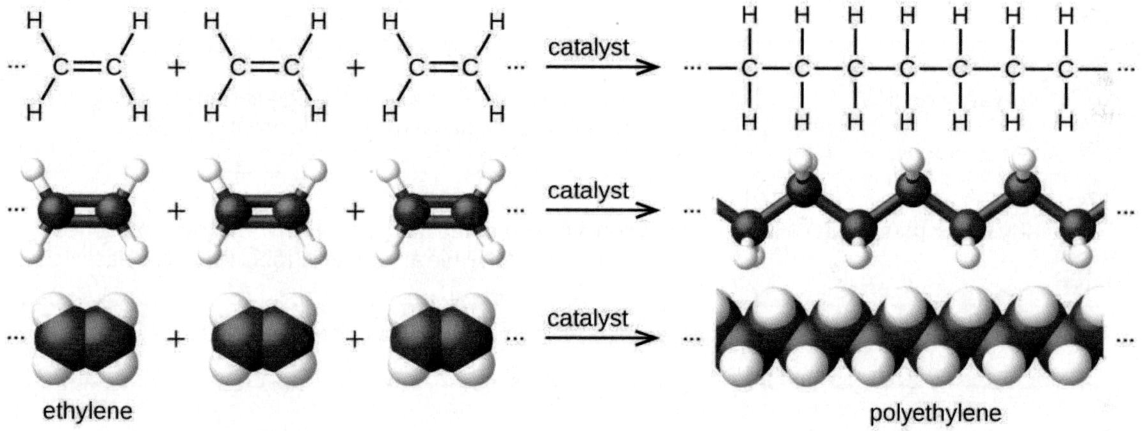

FIGURE 20.8 The reaction for the polymerization of ethylene to polyethylene is shown.

Polyethylene is a member of one subset of synthetic polymers classified as plastics. Plastics are synthetic organic solids that can be molded; they are typically organic polymers with high molecular masses. Most of the monomers that go into common plastics (ethylene, propylene, vinyl chloride, styrene, and ethylene terephthalate) are derived from petrochemicals and are not very biodegradable, making them candidate materials for recycling. Recycling plastics helps minimize the need for using more of the petrochemical supplies and also minimizes the environmental damage caused by throwing away these nonbiodegradable materials.

Plastic recycling is the process of recovering waste, scrap, or used plastics, and reprocessing the material into useful products. For example, polyethylene terephthalate (soft drink bottles) can be melted down and used for plastic furniture, in carpets, or for other applications. Other plastics, like polyethylene (bags) and polypropylene (cups, plastic food containers), can be recycled or reprocessed to be used again. Many areas of the country have recycling programs that focus on one or more of the commodity plastics that have been assigned a recycling code (see Figure 20.9). These operations have been in effect since the 1970s and have made the production of some plastics among the most efficient industrial operations today.

Code	Plastic	Uses
1 PETE	polyethylene terephthalate (PETE)	Soda bottles and oven-ready food trays
2 HDPE	high-density polyethylene (HDPE)	Bottles for milk and dishwashing liquids
3 V	polyvinyl chloride (PVC)	Food trays, plastic wrap, bottles for mineral water and shampoo
4 LDPE	low density polyethylene (LDPE)	Shopping bags and garbage bags
5 PP	polypropylene (PP)	Margarine tubs, microwaveable food trays
6 PS	polystyrene (PS)	Yogurt tubs, foam meat trays, egg cartons, vending cups, plastic cutlery, packaging for electronics and toys
7 OTHER	any other plastics (OTHER)	Plastics that do not fall into any of the above categories One example is melamine resin (plastic plates, plastic cups)

FIGURE 20.9 Each type of recyclable plastic is imprinted with a code for easy identification.

The name of an alkene is derived from the name of the alkane with the same number of carbon atoms. The presence of the double bond is signified by replacing the suffix *-ane* with the suffix *-ene*. The location of the double bond is identified by naming the smaller of the numbers of the carbon atoms participating in the double bond:

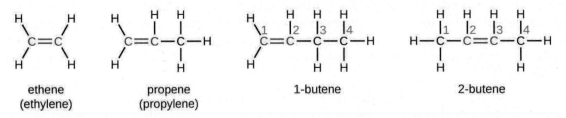

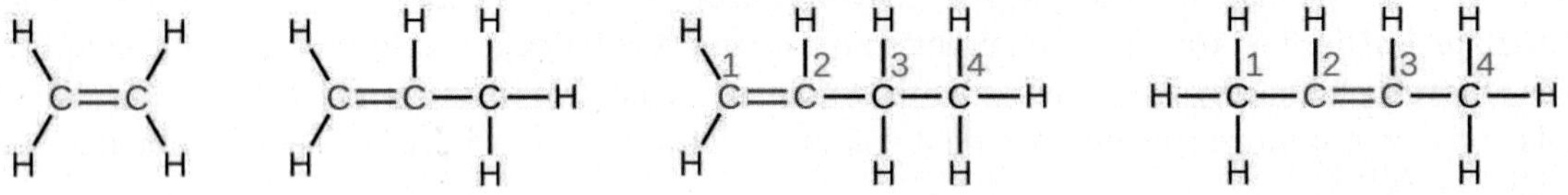

ethene (ethylene) propene (propylene) 1-butene 2-butene

Isomers of Alkenes

Molecules of 1-butene and 2-butene are structural isomers; the arrangement of the atoms in these two molecules differs. As an example of arrangement differences, the first carbon atom in 1-butene is bonded to two hydrogen atoms; the first carbon atom in 2-butene is bonded to three hydrogen atoms.

The compound 2-butene and some other alkenes also form a second type of isomer called a geometric isomer. In a set of geometric isomers, the same types of atoms are attached to each other in the same order, but the geometries of the two molecules differ. Geometric isomers of alkenes differ in the orientation of the groups on either side of a $C = C$ bond.

Carbon atoms are free to rotate around a single bond but not around a double bond; a double bond is rigid.

This makes it possible to have two isomers of 2-butene, one with both methyl groups on the same side of the double bond and one with the methyl groups on opposite sides. When structures of butene are drawn with 120° bond angles around the sp^2-hybridized carbon atoms participating in the double bond, the isomers are apparent. The 2-butene isomer in which the two methyl groups are on the same side is called a *cis*-isomer; the one in which the two methyl groups are on opposite sides is called a *trans*-isomer (Figure 20.10). The different geometries produce different physical properties, such as boiling point, that may make separation of the isomers possible:

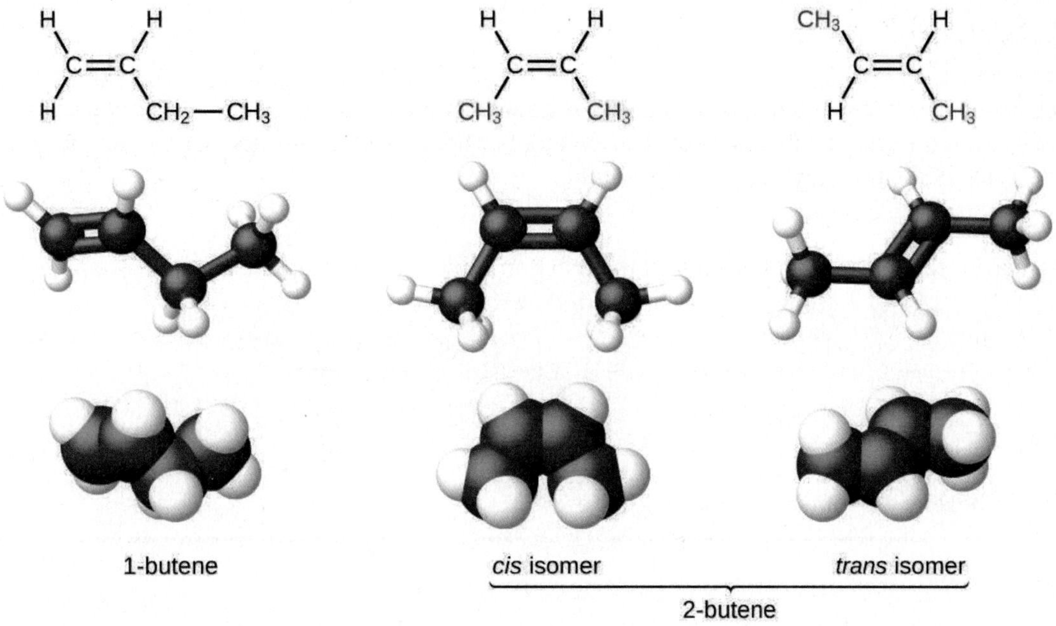

1-butene *cis* isomer *trans* isomer

2-butene

FIGURE 20.10 These molecular models show the structural and geometric isomers of butene.

Alkenes are much more reactive than alkanes because the C = C moiety is a reactive functional group. A π bond, being a weaker bond, is disrupted much more easily than a σ bond. Thus, alkenes undergo a characteristic reaction in which the π bond is broken and replaced by two σ bonds. This reaction is called an **addition reaction**. The hybridization of the carbon atoms in the double bond in an alkene changes from sp^2 to sp^3 during an addition reaction. For example, halogens add to the double bond in an alkene instead of replacing hydrogen, as occurs in an alkane:

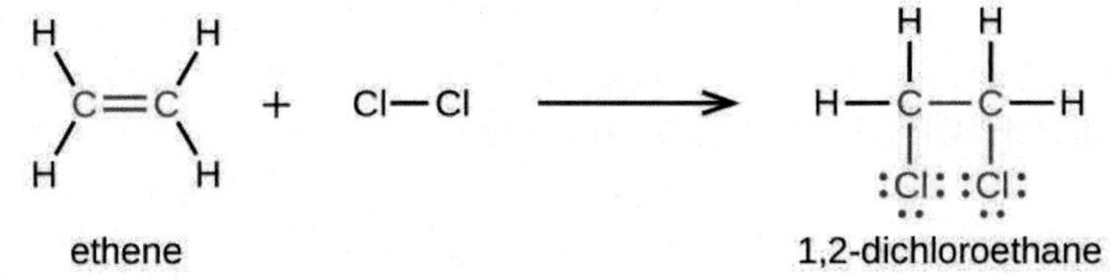

ethene 1,2-dichloroethane

EXAMPLE 20.5

Alkene Reactivity and Naming

Provide the IUPAC names for the reactant and product of the halogenation reaction shown here:

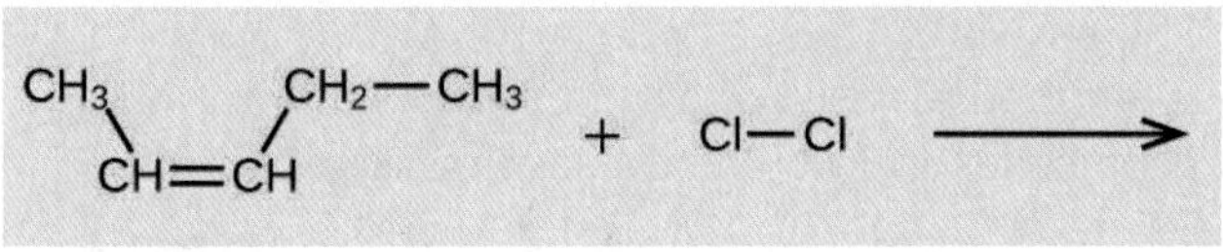

CH_3–CH=CH–CH_2–CH_3 + Cl–Cl ⟶

Solution

The reactant is a five-carbon chain that contains a carbon-carbon double bond, so the base name will be pentene. We begin counting at the end of the chain closest to the double bond—in this case, from the left—the double bond spans carbons 2 and 3, so the name becomes 2-pentene. Since there are two carbon-containing groups attached to the two carbon atoms in the double bond—and they are on the same side of the double bond—this molecule is the *cis*-isomer, making the name of the starting alkene *cis*-2-pentene. The product of the halogenation reaction will have two chlorine atoms attached to the carbon atoms that were a part of the carbon-carbon double bond:

```
CH3       CH2—CH3
   \     /
   CH—CH
   /     \
 Cl       Cl
```

This molecule is now a substituted alkane and will be named as such. The base of the name will be pentane. We will count from the end that numbers the carbon atoms where the chlorine atoms are attached as 2 and 3, making the name of the product 2,3-dichloropentane.

Check Your Learning

Provide names for the reactant and product of the reaction shown:

```
    H  H  H  H  H  H
    |  |  |  |  |  |
H—C—C—C=C—C—C—H  +  Cl—Cl  ⟶
    |  |        |  |
    H  H        H  H
```

Answer:

reactant: cis-3-hexene product: 3,4-dichlorohexane

Alkynes

Hydrocarbon molecules with one or more triple bonds are called **alkynes**; they make up another series of unsaturated hydrocarbons. Two carbon atoms joined by a triple bond are bound together by one σ bond and two π bonds. The *sp*-hybridized carbons involved in the triple bond have bond angles of 180°, giving these types of bonds a linear, rod-like shape.

The simplest member of the alkyne series is ethyne, C_2H_2, commonly called acetylene. The Lewis structure for ethyne, a linear molecule, is:

H—C≡C—H

ethyne (acetylene)

The IUPAC nomenclature for alkynes is similar to that for alkenes except that the suffix *-yne* is used to indicate a triple bond in the chain. For example, $CH_3CH_2C \equiv CH$ is called 1-butyne.

EXAMPLE 20.6

Structure of Alkynes

Describe the geometry and hybridization of the carbon atoms in the following molecule:

```
 1     2   3   4
CH3—C≡C—CH3
```

Solution

Carbon atoms 1 and 4 have four single bonds and are thus tetrahedral with sp^3 hybridization. Carbon atoms 2 and 3 are involved in the triple bond, so they have linear geometries and would be classified as *sp* hybrids.

Check Your Learning

Identify the hybridization and bond angles at the carbon atoms in the molecule shown:

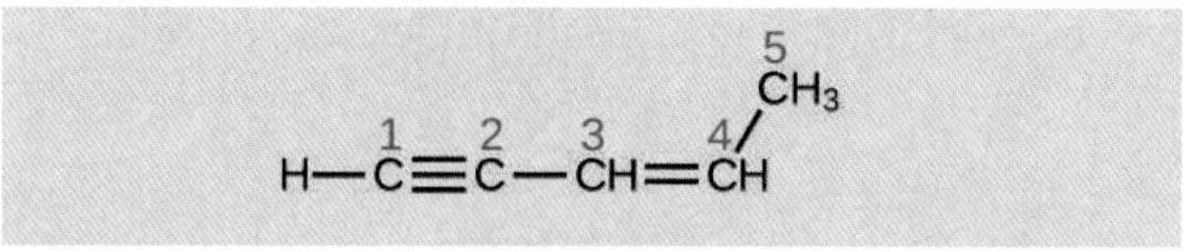

$$\overset{1}{H-C}\equiv\overset{2}{C}-\overset{3}{CH}=\overset{4}{CH}-\overset{5}{CH_3}$$

Answer:

carbon 1: *sp*, 180°; carbon 2: *sp*, 180°; carbon 3: sp^2, 120°; carbon 4: sp^2, 120°; carbon 5: sp^3, 109.5°

Chemically, the alkynes are similar to the alkenes. Since the $C \equiv C$ functional group has two π bonds, alkynes typically react even more readily, and react with twice as much reagent in addition reactions. The reaction of acetylene with bromine is a typical example:

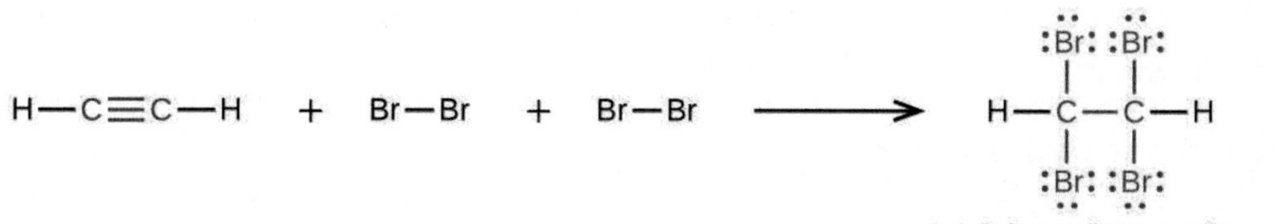

$$H-C\equiv C-H \;+\; Br-Br \;+\; Br-Br \longrightarrow CHBr_2-CHBr_2$$

1,1,2,2-tetrabromoethane

Acetylene and the other alkynes also burn readily. An acetylene torch takes advantage of the high heat of combustion for acetylene.

Aromatic Hydrocarbons

Benzene, C_6H_6, is the simplest member of a large family of hydrocarbons, called **aromatic hydrocarbons**. These compounds contain ring structures and exhibit bonding that must be described using the resonance hybrid concept of valence bond theory or the delocalization concept of molecular orbital theory. (To review these concepts, refer to the earlier chapters on chemical bonding). The resonance structures for benzene, C_6H_6, are:

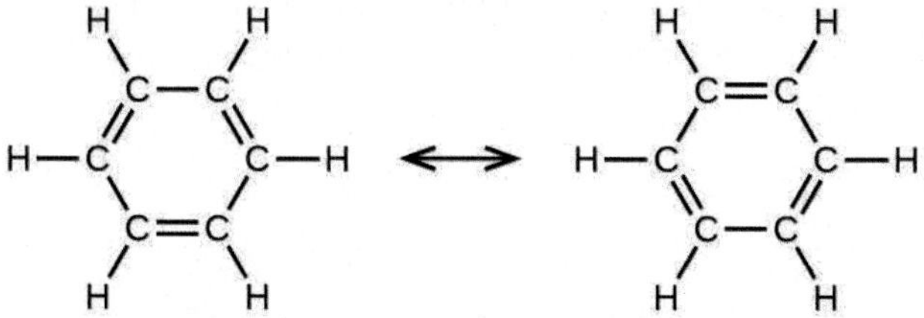

Valence bond theory describes the benzene molecule and other planar aromatic hydrocarbon molecules as hexagonal rings of sp^2-hybridized carbon atoms with the unhybridized *p* orbital of each carbon atom perpendicular to the plane of the ring. Three valence electrons in the sp^2 hybrid orbitals of each carbon atom and the valence electron of each hydrogen atom form the framework of σ bonds in the benzene molecule. The fourth valence electron of each carbon atom is shared with an adjacent carbon atom in their unhybridized *p* orbitals to yield the π bonds. Benzene does not, however, exhibit the characteristics typical of an alkene. Each of the six bonds between its carbon atoms is equivalent and exhibits properties that are intermediate between those of a C–C single bond and a $C = C$ double bond. To represent this unique bonding, structural formulas for benzene and its derivatives are typically drawn with single bonds between the carbon atoms and a circle within the ring as shown in Figure 20.11.

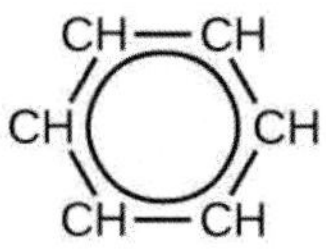

FIGURE 20.11 This condensed formula shows the unique bonding structure of benzene.

There are many derivatives of benzene. The hydrogen atoms can be replaced by many different substituents. Aromatic compounds more readily undergo substitution reactions than addition reactions; replacement of one of the hydrogen atoms with another substituent will leave the delocalized double bonds intact. The following are typical examples of substituted benzene derivatives:

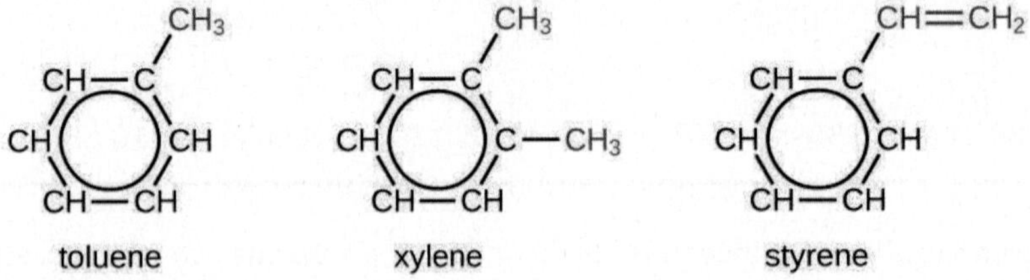

toluene xylene styrene

Toluene and xylene are important solvents and raw materials in the chemical industry. Styrene is used to produce the polymer polystyrene.

EXAMPLE 20.7

Structure of Aromatic Hydrocarbons

One possible isomer created by a substitution reaction that replaces a hydrogen atom attached to the aromatic ring of toluene with a chlorine atom is shown here. Draw two other possible isomers in which the chlorine atom replaces a different hydrogen atom attached to the aromatic ring:

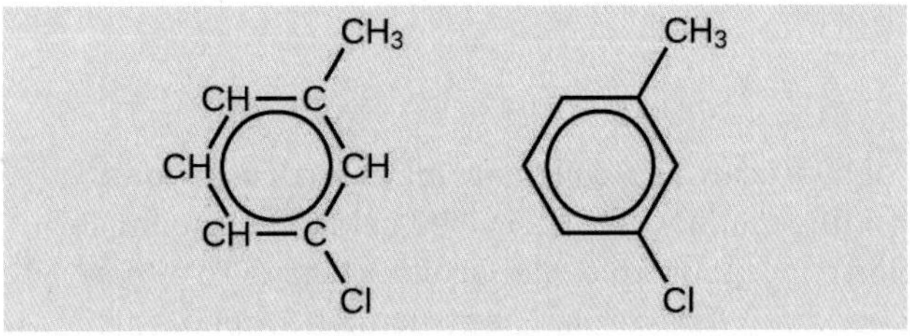

Solution

Since the six-carbon ring with alternating double bonds is necessary for the molecule to be classified as aromatic, appropriate isomers can be produced only by changing the positions of the chloro-substituent relative to the methyl-substituent:

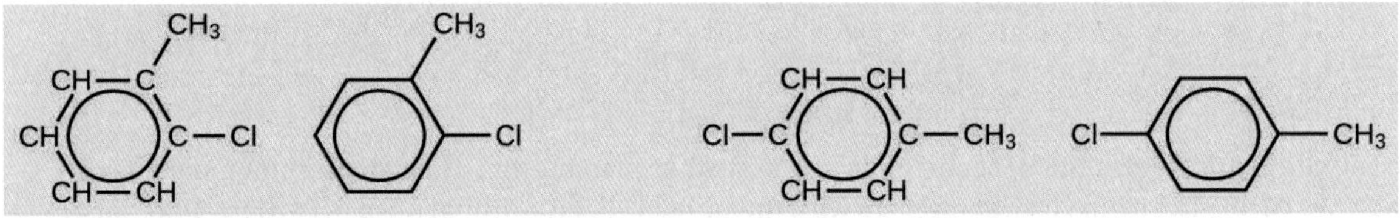

Check Your Learning

Draw three isomers of a six-membered aromatic ring compound substituted with two bromines.

Answer:

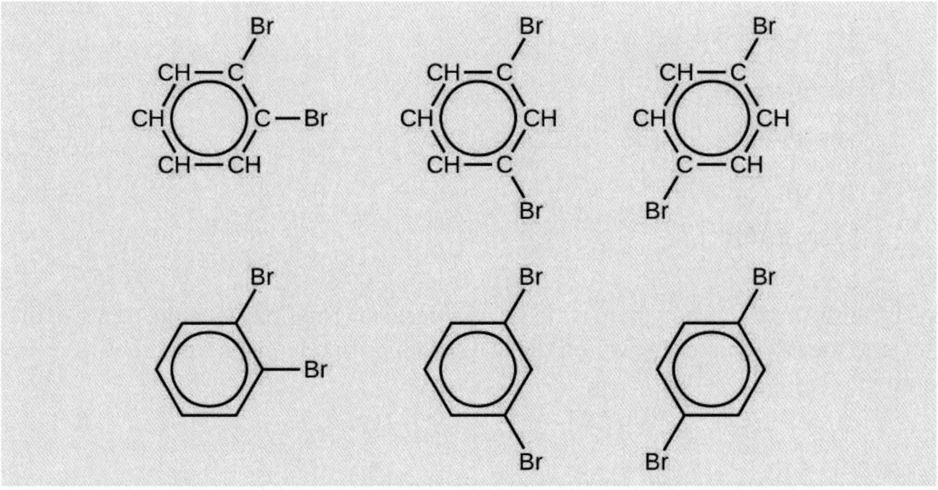

Br, CH—C, CH, C—Br, CH—CH; Br, CH—C, CH, CH, CH—C, Br; Br, CH—C, CH, CH, C—CH, Br

20.2 Alcohols and Ethers

LEARNING OBJECTIVES

By the end of this section, you will be able to:

- Describe the structure and properties of alcohols
- Describe the structure and properties of ethers
- Name and draw structures for alcohols and ethers

In this section, we will learn about alcohols and ethers.

Alcohols

Incorporation of an oxygen atom into carbon- and hydrogen-containing molecules leads to new functional groups and new families of compounds. When the oxygen atom is attached by single bonds, the molecule is either an alcohol or ether.

Alcohols are derivatives of hydrocarbons in which an –OH group has replaced a hydrogen atom. Although all alcohols have one or more hydroxyl (–OH) functional groups, they do not behave like bases such as NaOH and KOH. NaOH and KOH are ionic compounds that contain OH^- ions. Alcohols are covalent molecules; the –OH group in an alcohol molecule is attached to a carbon atom by a covalent bond.

Ethanol, CH_3CH_2OH, also called ethyl alcohol, is a particularly important alcohol for human use. Ethanol is the alcohol produced by some species of yeast that is found in wine, beer, and distilled drinks. It has long been prepared by humans harnessing the metabolic efforts of yeasts in fermenting various sugars:

$$\underset{\text{glucose}}{C_6H_{12}O_6(aq)} \xrightarrow{\text{Yeast}} \underset{\text{ethanol}}{2C_2H_5OH(aq)} + 2CO_2(g)$$

Large quantities of ethanol are synthesized from the addition reaction of water with ethylene using an acid as a catalyst:

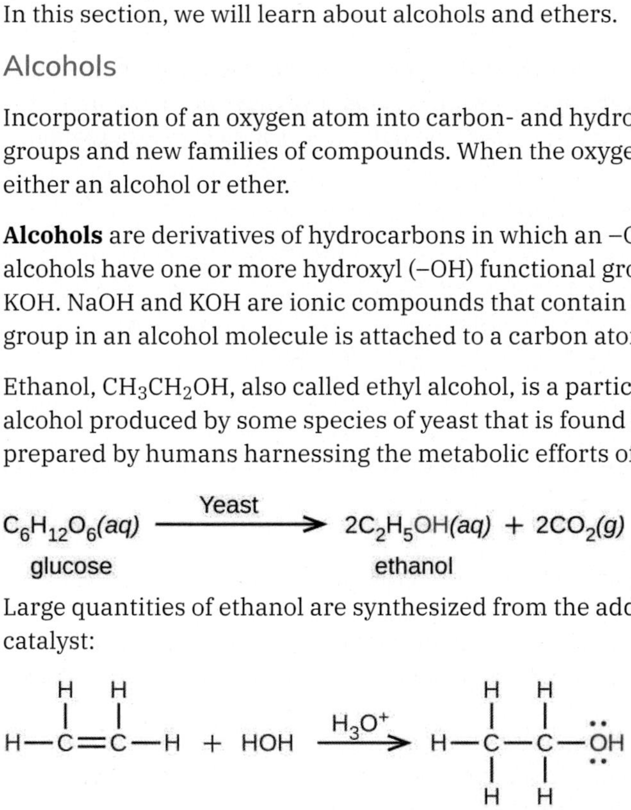

$$H_2C{=}CH_2 + HOH \xrightarrow{H_3O^+} H_3C{-}CH_2{-}\ddot{\underset{\cdot\cdot}{O}}H$$

Alcohols containing two or more hydroxyl groups can be made. Examples include 1,2-ethanediol (ethylene glycol, used in antifreeze) and 1,2,3-propanetriol (glycerine, used as a solvent for cosmetics and medicines):

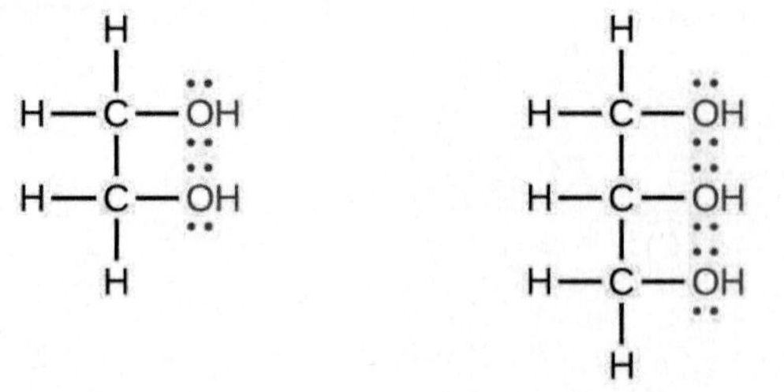

1,2-ethanediol   1,2,3-propanetriol

Naming Alcohols

The name of an alcohol comes from the hydrocarbon from which it was derived. The final *-e* in the name of the hydrocarbon is replaced by *-ol,* and the carbon atom to which the –OH group is bonded is indicated by a number placed before the name.[5]

EXAMPLE 20.8

Naming Alcohols

Consider the following example. How should it be named?

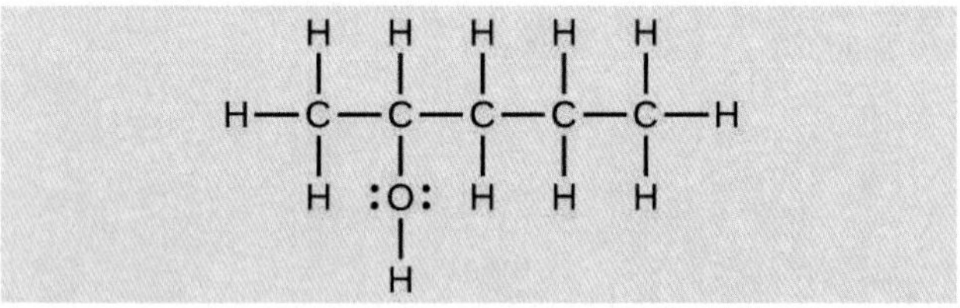

Solution

The carbon chain contains five carbon atoms. If the hydroxyl group was not present, we would have named this molecule pentane. To address the fact that the hydroxyl group is present, we change the ending of the name to *-ol.* In this case, since the –OH is attached to carbon 2 in the chain, we would name this molecule 2-pentanol.

Check Your Learning

Name the following molecule:

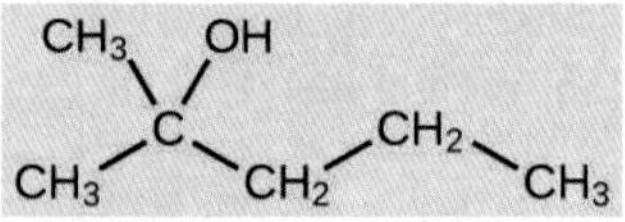

Answer:

2-methyl-2-pentanol

Ethers

Ethers are compounds that contain the functional group –O–. Ethers do not have a designated suffix like the other types of molecules we have named so far. In the IUPAC system, the oxygen atom and the smaller carbon branch are named as an alkoxy substituent and the remainder of the molecule as the base chain, as in alkanes. As shown in the following compound, the red symbols represent the smaller alkyl group and the oxygen atom, which would be named "methoxy." The larger carbon branch would be ethane, making the molecule methoxyethane. Many ethers are referred to with common names instead of the IUPAC system names. For common names, the two branches connected to the oxygen atom are named separately and followed by "ether." The common name for the compound shown in Example 20.9 is ethylmethyl ether:

5 The IUPAC adopted new nomenclature guidelines in 2013 that require this number to be placed as an "infix" rather than a prefix. For example, the new name for 2-propanol would be propan-2-ol. Widespread adoption of this new nomenclature will take some time, and students are encouraged to be familiar with both the old and new naming protocols.

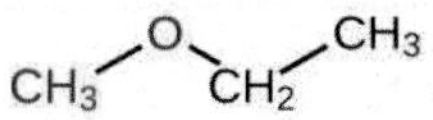

EXAMPLE 20.9

Naming Ethers

Provide the IUPAC and common name for the ether shown here:

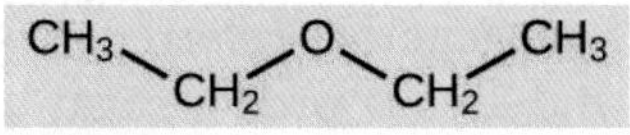

$CH_3CH_2OCH_2CH_3$

Solution

IUPAC: The molecule is made up of an ethoxy group attached to an ethane chain, so the IUPAC name would be ethoxyethane.

Common: The groups attached to the oxygen atom are both ethyl groups, so the common name would be diethyl ether.

Check Your Learning

Provide the IUPAC and common name for the ether shown:

$CH_3OCH(CH_3)CH_3$

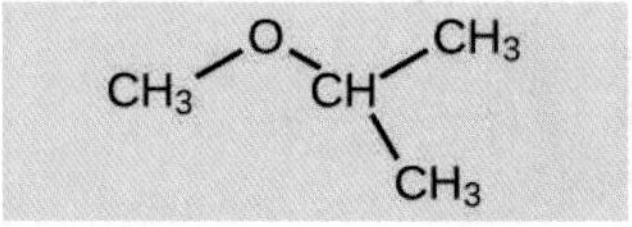

Answer:

IUPAC: 2-methoxypropane; common: isopropylmethyl ether

Ethers can be obtained from alcohols by the elimination of a molecule of water from two molecules of the alcohol. For example, when ethanol is treated with a limited amount of sulfuric acid and heated to 140 °C, diethyl ether and water are formed:

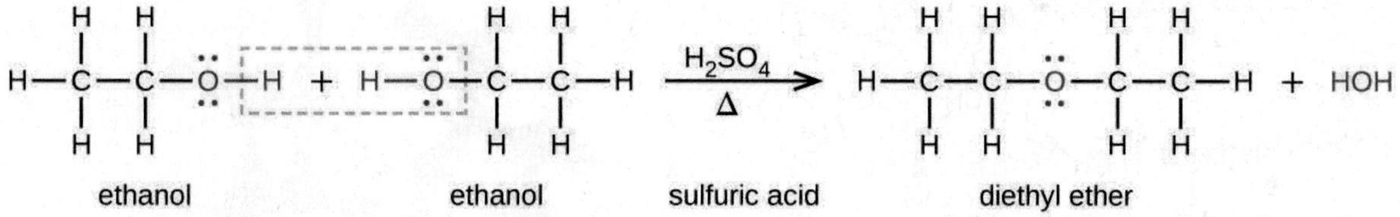

$$\underset{\text{ethanol}}{H{-}CH_2{-}CH_2{-}O{-}H} + \underset{\text{ethanol}}{H{-}O{-}CH_2{-}CH_2{-}H} \xrightarrow[\Delta]{H_2SO_4} \underset{\text{diethyl ether}}{H{-}CH_2{-}CH_2{-}O{-}CH_2{-}CH_2{-}H} + HOH$$

ethanol ethanol sulfuric acid diethyl ether

In the general formula for ethers, R–**O**–R, the hydrocarbon groups (R) may be the same or different. Diethyl ether, the most widely used compound of this class, is a colorless, volatile liquid that is highly flammable. It was first used in 1846 as an anesthetic, but better anesthetics have now largely taken its place. Diethyl ether and other ethers are presently used primarily as solvents for gums, fats, waxes, and resins. *Tertiary*-butyl methyl ether, $C_4H_9OCH_3$ (abbreviated MTBE—italicized portions of names are not counted when ranking the groups alphabetically—so butyl comes before methyl in the common name), is used as an additive for gasoline. MTBE belongs to a group of chemicals known as oxygenates due to their capacity to increase the oxygen content of gasoline.

LINK TO LEARNING

Want more practice naming ethers? This brief video review (http://openstax.org/l/16ethers) summarizes the nomenclature for ethers.

Chemistry in Everyday Life

Carbohydrates and Diabetes

Carbohydrates are large biomolecules made up of carbon, hydrogen, and oxygen. The dietary forms of carbohydrates are foods rich in these types of molecules, like pastas, bread, and candy. The name "carbohydrate" comes from the formula of the molecules, which can be described by the general formula $C_m(H_2O)_n$, which shows that they are in a sense "carbon and water" or "hydrates of carbon." In many cases, *m* and *n* have the same value, but they can be different. The smaller carbohydrates are generally referred to as "sugars," the biochemical term for this group of molecules is "saccharide" from the Greek word for sugar (Figure 20.12). Depending on the number of sugar units joined together, they may be classified as monosaccharides (one sugar unit), disaccharides (two sugar units), oligosaccharides (a few sugars), or polysaccharides (the polymeric version of sugars—polymers were described in the feature box earlier in this chapter on recycling plastics). The scientific names of sugars can be recognized by the suffix *-ose* at the end of the name (for instance, fruit sugar is a monosaccharide called "fructose" and milk sugar is a disaccharide called lactose composed of two monosaccharides, glucose and galactose, connected together). Sugars contain some of the functional groups we have discussed: Note the alcohol groups present in the structures and how monosaccharide units are linked to form a disaccharide by formation of an ether.

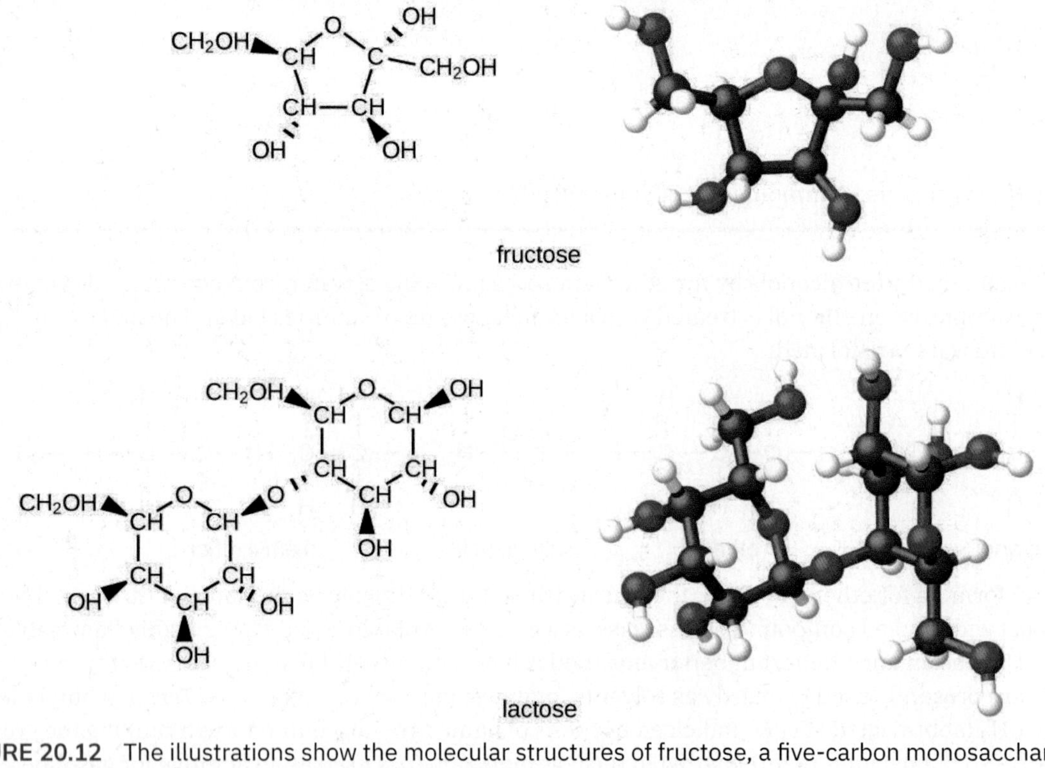

fructose

lactose

FIGURE 20.12 The illustrations show the molecular structures of fructose, a five-carbon monosaccharide, and of lactose, a disaccharide composed of two isomeric, six-carbon sugars.

Organisms use carbohydrates for a variety of functions. Carbohydrates can store energy, such as the polysaccharides glycogen in animals or starch in plants. They also provide structural support, such as the polysaccharide cellulose in plants and the modified polysaccharide chitin in fungi and animals. The sugars ribose and deoxyribose are components of the backbones of RNA and DNA, respectively. Other sugars play key roles in the function of the immune system, in cell-cell recognition, and in many other biological roles.

Diabetes is a group of metabolic diseases in which a person has a high sugar concentration in their blood (Figure 20.13). Diabetes may be caused by insufficient insulin production by the pancreas or by the body's cells not responding properly to the insulin that is produced. In a healthy person, insulin is produced when it is needed and functions to transport glucose from the blood into the cells where it can be used for energy.

The long-term complications of diabetes can include loss of eyesight, heart disease, and kidney failure.

In 2013, it was estimated that approximately 3.3% of the world's population (~380 million people) suffered from diabetes, resulting in over a million deaths annually. Prevention involves eating a healthy diet, getting plenty of exercise, and maintaining a normal body weight. Treatment involves all of these lifestyle practices and may require injections of insulin.

Even after treatment protocols were introduced, the need to continually monitor their glucose levels posed a challenge for people with diabetes. The first tests required a doctor or lab, and therefore limited access and frequency. Eventually, researchers developed small tablets that would react to the presence of glucose in urine, but these still required a relatively complex process. Chemist Helen Free, who was working on improvements to the tablets, conceived a simpler device: a small test strip. With her husband and research partner, Alfred Free, she produced the first such product for measuring glucose; soon after, she expanded the technology to provide test strips for other compounds and conditions. While very recent advances (such as breath tests, discussed earlier in the text) have shown promise in replacing test strips, they have been widely used for decades and remain a primary method today.

FIGURE 20.13 Diabetes is a disease characterized by high concentrations of glucose in the blood. Treating diabetes involves making lifestyle changes, monitoring blood-sugar levels, and sometimes insulin injections. (credit: "Blausen Medical Communications"/Wikimedia Commons)

20.3 Aldehydes, Ketones, Carboxylic Acids, and Esters

LEARNING OBJECTIVES

By the end of this section, you will be able to:

- Describe the structure and properties of aldehydes, ketones, carboxylic acids and esters

Another class of organic molecules contains a carbon atom connected to an oxygen atom by a double bond, commonly called a carbonyl group. The trigonal planar carbon in the carbonyl group can attach to two other substituents leading to several subfamilies (aldehydes, ketones, carboxylic acids and esters) described in this section.

Aldehydes and Ketones

Both **aldehydes** and **ketones** contain a **carbonyl group**, a functional group with a carbon-oxygen double bond. The names for aldehyde and ketone compounds are derived using similar nomenclature rules as for alkanes and alcohols, and include the class-identifying suffixes *-al* and *-one*, respectively:

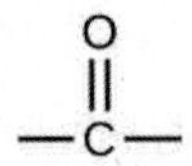

$$-\overset{\overset{\large O}{||}}{C}-$$

In an aldehyde, the carbonyl group is bonded to at least one hydrogen atom. In a ketone, the carbonyl group is bonded to two carbon atoms:

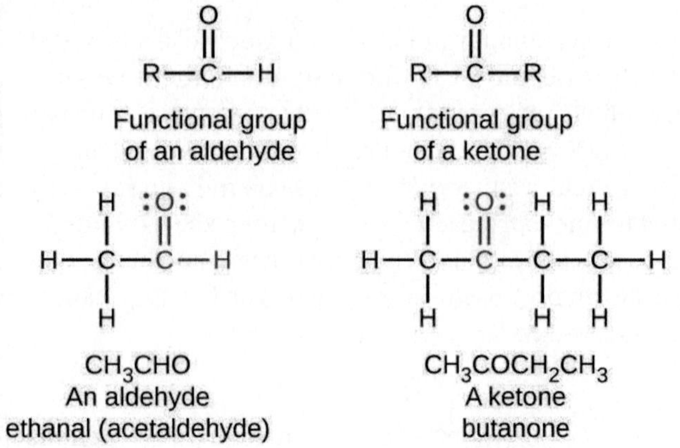

$$R-\overset{\overset{\large O}{||}}{C}-H \qquad R-\overset{\overset{\large O}{||}}{C}-R$$

Functional group of an aldehyde   Functional group of a ketone

CH_3CHO
An aldehyde
ethanal (acetaldehyde)

$CH_3COCH_2CH_3$
A ketone
butanone

As text, an aldehyde group is represented as –CHO; a ketone is represented as –C(O)– or –CO–.

In both aldehydes and ketones, the geometry around the carbon atom in the carbonyl group is trigonal planar; the carbon atom exhibits sp^2 hybridization. Two of the sp^2 orbitals on the carbon atom in the carbonyl group are used to form σ bonds to the other carbon or hydrogen atoms in a molecule. The remaining sp^2 hybrid orbital forms a σ bond to the oxygen atom. The unhybridized *p* orbital on the carbon atom in the carbonyl group overlaps a *p* orbital on the oxygen atom to form the π bond in the double bond.

Like the $C = O$ bond in carbon dioxide, the $C = O$ bond of a carbonyl group is polar (recall that oxygen is significantly more electronegative than carbon, and the shared electrons are pulled toward the oxygen atom and away from the carbon atom). Many of the reactions of aldehydes and ketones start with the reaction between a Lewis base and the carbon atom at the positive end of the polar $C = O$ bond to yield an unstable intermediate that subsequently undergoes one or more structural rearrangements to form the final product (Figure 20.14).

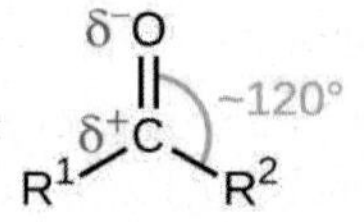

FIGURE 20.14 The carbonyl group is polar, and the geometry of the bonds around the central carbon is trigonal planar.

The importance of molecular structure in the reactivity of organic compounds is illustrated by the reactions that produce aldehydes and ketones. We can prepare a carbonyl group by oxidation of an alcohol—for organic molecules, oxidation of a carbon atom is said to occur when a carbon-hydrogen bond is replaced by a carbon-oxygen bond. The reverse reaction—replacing a carbon-oxygen bond by a carbon-hydrogen bond—is a reduction of that carbon atom. Recall that oxygen is generally assigned a –2 oxidation number unless it is elemental or attached to a fluorine. Hydrogen is generally assigned an oxidation number of +1 unless it is attached to a metal. Since carbon does not have a specific rule, its oxidation number is determined algebraically by factoring the atoms it is attached to and the overall charge of the molecule or ion. In general, a carbon atom attached to an oxygen atom will have a more positive oxidation number and a carbon atom attached to a hydrogen atom will have a more negative oxidation number. This should fit nicely with your understanding of the polarity of C–O and C–H bonds. The other reagents and possible products of these reactions are beyond the scope of this chapter, so we will focus only on the changes to the carbon atoms:

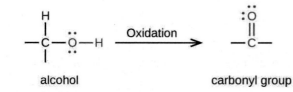

Oxidation

:O:
C

alcohol carbonyl group

EXAMPLE 20.10

Oxidation and Reduction in Organic Chemistry

Methane represents the completely reduced form of an organic molecule that contains one carbon atom. Sequentially replacing each of the carbon-hydrogen bonds with a carbon-oxygen bond would lead to an alcohol, then an aldehyde, then a carboxylic acid (discussed later), and, finally, carbon dioxide:

$$CH_4 \longrightarrow CH_3OH \longrightarrow CH_2O \longrightarrow HCO_2H \longrightarrow CO_2$$

What are the oxidation numbers for the carbon atoms in the molecules shown here?

Solution

In this example, we can calculate the oxidation number (review the chapter on oxidation-reduction reactions if necessary) for the carbon atom in each case (note how this would become difficult for larger molecules with additional carbon atoms and hydrogen atoms, which is why organic chemists use the definition dealing with replacing C–H bonds with C–O bonds described). For CH_4, the carbon atom carries a –4 oxidation number (the hydrogen atoms are assigned oxidation numbers of +1 and the carbon atom balances that by having an oxidation number of –4). For the alcohol (in this case, methanol), the carbon atom has an oxidation number of –2 (the oxygen atom is assigned –2, the four hydrogen atoms each are assigned +1, and the carbon atom balances the sum by having an oxidation number of –2; note that compared to the carbon atom in CH_4, this carbon atom has lost two electrons so it was oxidized); for the aldehyde, the carbon atom's oxidation number is 0 (–2 for the oxygen atom and +1 for each hydrogen atom already balances to 0, so the oxidation number for the carbon atom is 0); for the carboxylic acid, the carbon atom's oxidation number is +2 (two oxygen atoms each at –2 and two hydrogen atoms at +1); and for carbon dioxide, the carbon atom's oxidation number is +4 (here, the carbon atom needs to balance the –4 sum from the two oxygen atoms).

Check Your Learning

Indicate whether the marked carbon atoms in the three molecules here are oxidized or reduced relative to the marked carbon atom in ethanol:

CH_3 CH_2 OH

There is no need to calculate oxidation states in this case; instead, just compare the types of atoms bonded to the marked carbon atoms:

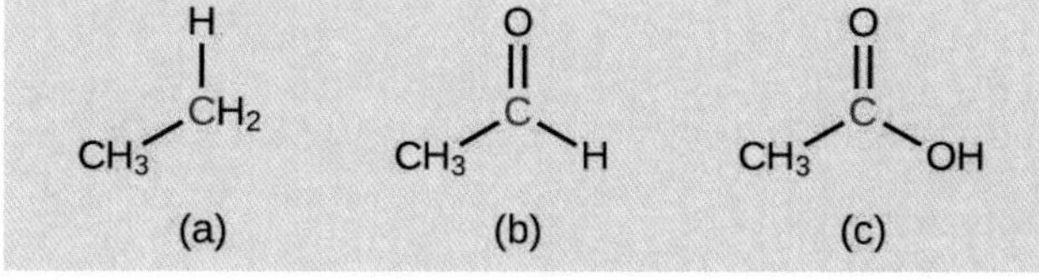

Answer:

(a) reduced (bond to oxygen atom replaced by bond to hydrogen atom); (b) oxidized (one bond to hydrogen atom replaced by one bond to oxygen atom); (c) oxidized (2 bonds to hydrogen atoms have been replaced by bonds to an oxygen atom)

Aldehydes are commonly prepared by the oxidation of alcohols whose –OH functional group is located on the carbon atom at the end of the chain of carbon atoms in the alcohol:

$$\underset{\text{alcohol}}{CH_3CH_2CH_2OH} \longrightarrow \underset{\text{aldehyde}}{CH_3CH_2CHO}$$

Alcohols that have their –OH groups in the middle of the chain are necessary to synthesize a ketone, which requires the carbonyl group to be bonded to two other carbon atoms:

$$\underset{\text{alcohol}}{CH_3CH(OH)CH_3} \longrightarrow \underset{\text{ketone}}{CH_3COCH_3}$$

An alcohol with its –OH group bonded to a carbon atom that is bonded to no or one other carbon atom will form an aldehyde. An alcohol with its –OH group attached to two other carbon atoms will form a ketone. If three carbons are attached to the carbon bonded to the –OH, the molecule will not have a C–H bond to be replaced, so it will not be susceptible to oxidation.

Formaldehyde, an aldehyde with the formula HCHO, is a colorless gas with a pungent and irritating odor. It is sold in an aqueous solution called formalin, which contains about 37% formaldehyde by weight. Formaldehyde causes coagulation of proteins, so it kills bacteria (and any other living organism) and stops many of the biological processes that cause tissue to decay. Thus, formaldehyde is used for preserving tissue specimens and embalming bodies. It is also used to sterilize soil or other materials. Formaldehyde is used in the manufacture of Bakelite, a hard plastic having high chemical and electrical resistance.

Dimethyl ketone, CH_3COCH_3, commonly called acetone, is the simplest ketone. It is made commercially by fermenting corn or molasses, or by oxidation of 2-propanol. Acetone is a colorless liquid. Among its many uses are as a solvent for lacquer (including fingernail polish), cellulose acetate, cellulose nitrate, acetylene, plastics, and varnishes; as a paint and varnish remover; and as a solvent in the manufacture of pharmaceuticals and chemicals.

Carboxylic Acids and Esters

The odor of vinegar is caused by the presence of acetic acid, a carboxylic acid, in the vinegar. The odor of ripe bananas and many other fruits is due to the presence of esters, compounds that can be prepared by the reaction of a carboxylic acid with an alcohol. Because esters do not have hydrogen bonds between molecules, they have lower vapor pressures than the alcohols and carboxylic acids from which they are derived (see Figure 20.15).

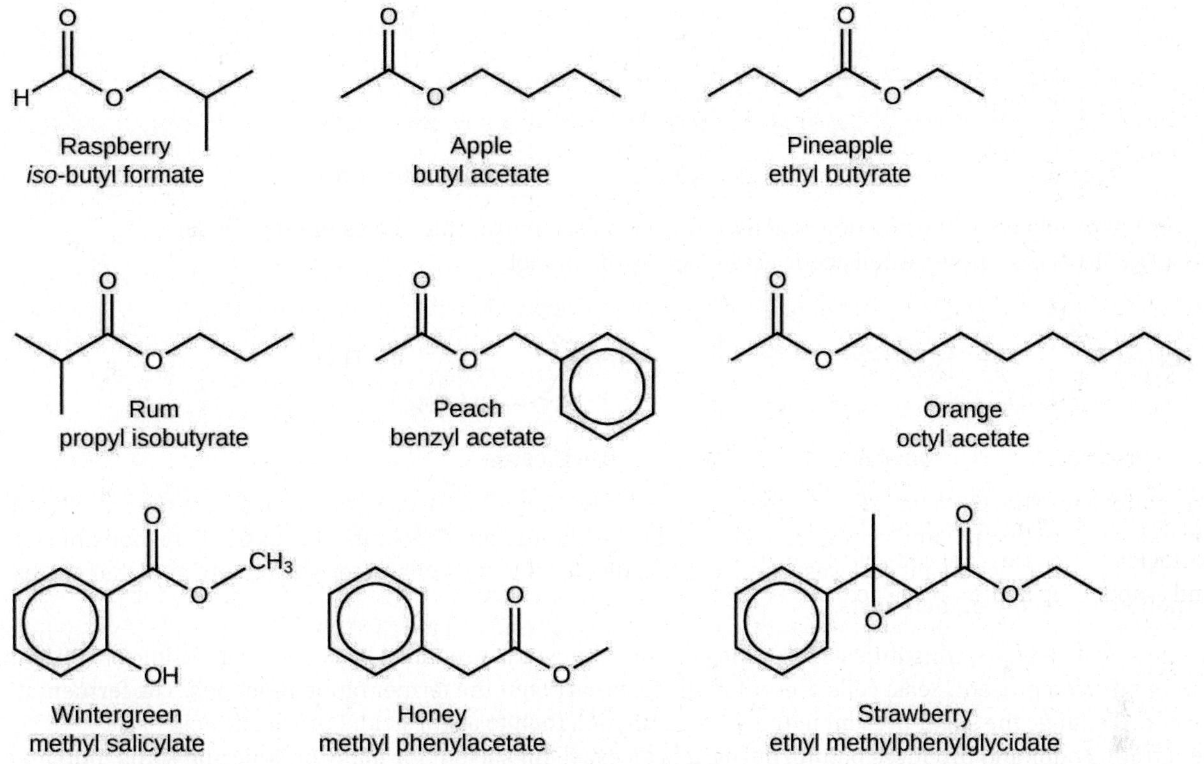

Raspberry
iso-butyl formate

Apple
butyl acetate

Pineapple
ethyl butyrate

Rum
propyl isobutyrate

Peach
benzyl acetate

Orange
octyl acetate

Wintergreen
methyl salicylate

Honey
methyl phenylacetate

Strawberry
ethyl methylphenylglycidate

FIGURE 20.15 Esters are responsible for the odors associated with various plants and their fruits.

Both **carboxylic acids** and **esters** contain a carbonyl group with a second oxygen atom bonded to the carbon atom in the carbonyl group by a single bond. In a carboxylic acid, the second oxygen atom also bonds to a hydrogen atom. In an ester, the second oxygen atom bonds to another carbon atom. The names for carboxylic acids and esters include prefixes that denote the lengths of the carbon chains in the molecules and are derived following nomenclature rules similar to those for inorganic acids and salts (see these examples):

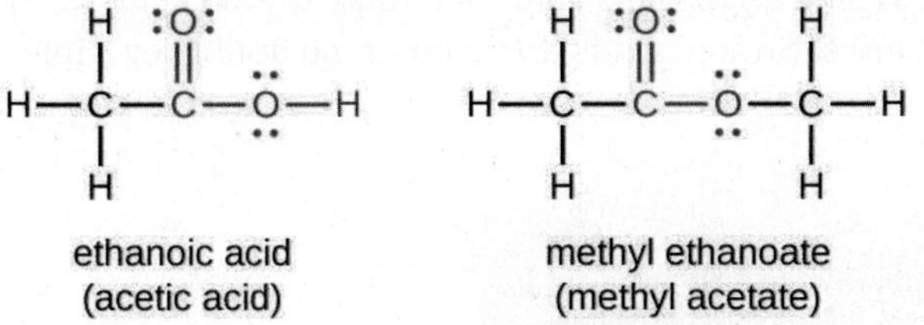

ethanoic acid
(acetic acid)

methyl ethanoate
(methyl acetate)

The functional groups for an acid and for an ester are shown in red in these formulas.

The hydrogen atom in the functional group of a carboxylic acid will react with a base to form an ionic salt:

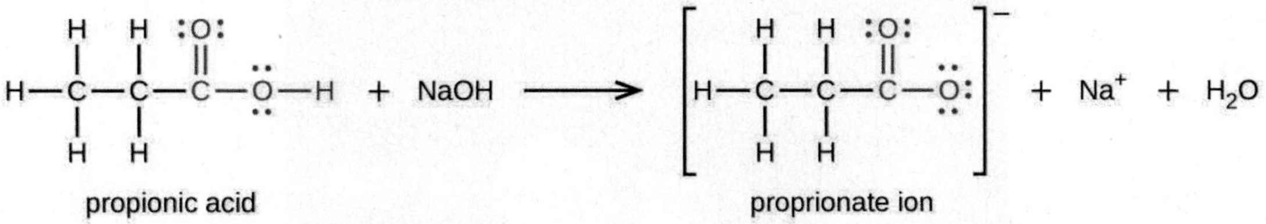

$$\text{propionic acid} + NaOH \longrightarrow \text{proprionate ion} + Na^+ + H_2O$$

propionic acid

proprionate ion

Carboxylic acids are weak acids (see the chapter on acids and bases), meaning they are not 100% ionized in water. Generally only about 1% of the molecules of a carboxylic acid dissolved in water are ionized at any given time. The remaining molecules are undissociated in solution.

We prepare carboxylic acids by the oxidation of aldehydes or alcohols whose –OH functional group is located on the carbon atom at the end of the chain of carbon atoms in the alcohol:

$$R{-}CH_2{-}\ddot{O}{-}H \longrightarrow R{-}C(H){=}\ddot{O} \longrightarrow R{-}C(\ddot{O}{-}H){=}\ddot{O}$$

alcohol aldehyde carboxylic acid

Esters are produced by the reaction of acids with alcohols. For example, the ester ethyl acetate, $CH_3CO_2CH_2CH_3$, is formed when acetic acid reacts with ethanol:

$$CH_3{-}C(=\ddot{O}){-}\ddot{O}{-}H + HOCH_2CH_3 \longrightarrow CH_3{-}C(=\ddot{O}){-}\ddot{O}{-}CH_2CH_3 + H_2O$$

acetic acid ethanol ethyl acetate

The simplest carboxylic acid is formic acid, HCO_2H, known since 1670. Its name comes from the Latin word *formicus*, which means “ant”; it was first isolated by the distillation of red ants. It is partially responsible for the pain and irritation of ant and wasp stings, and is responsible for a characteristic odor of ants that can be sometimes detected in their nests.

Acetic acid, CH_3CO_2H, constitutes 3–6% vinegar. Cider vinegar is produced by allowing apple juice to ferment without oxygen present. Yeast cells present in the juice carry out the fermentation reactions. The fermentation reactions change the sugar present in the juice to ethanol, then to acetic acid. Pure acetic acid has a penetrating odor and produces painful burns. It is an excellent solvent for many organic and some inorganic compounds, and it is essential in the production of cellulose acetate, a component of many synthetic fibers such as rayon.

The distinctive and attractive odors and flavors of many flowers, perfumes, and ripe fruits are due to the presence of one or more esters (Figure 20.16). Among the most important of the natural esters are fats (such as lard, tallow, and butter) and oils (such as linseed, cottonseed, and olive oils), which are esters of the trihydroxyl alcohol glycerine, $C_3H_5(OH)_3$, with large carboxylic acids, such as palmitic acid, $CH_3(CH_2)_{14}CO_2H$, stearic acid, $CH_3(CH_2)_{16}CO_2H$, and oleic acid, $CH_3(CH_2)_7CH = CH(CH_2)_7CO_2H$. Oleic acid is an unsaturated acid; it contains a $C = C$ double bond. Palmitic and stearic acids are saturated acids that contain no double or triple bonds.

FIGURE 20.16 Over 350 different volatile molecules (many members of the ester family) have been identified in strawberries. (credit: Rebecca Siegel)

20.4 Amines and Amides

LEARNING OBJECTIVES

By the end of this section, you will be able to:

- Describe the structure and properties of an amine
- Describe the structure and properties of an amide

Amines are molecules that contain carbon-nitrogen bonds. The nitrogen atom in an amine has a lone pair of electrons and three bonds to other atoms, either carbon or hydrogen. Various nomenclatures are used to derive names for amines, but all involve the class-identifying suffix *–ine* as illustrated here for a few simple examples:

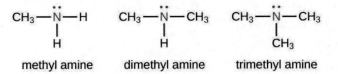

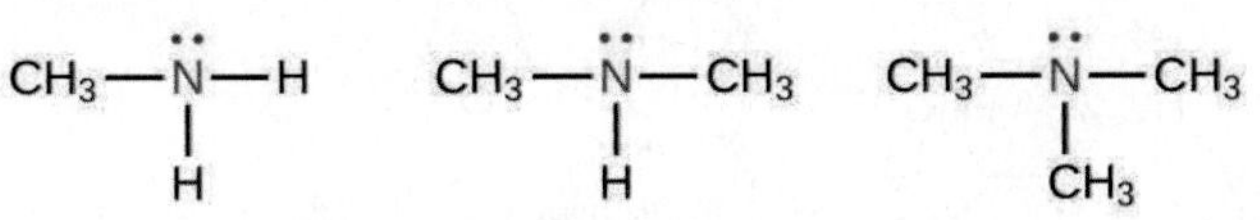

$CH_3—\ddot{N}(H)—H$	$CH_3—\ddot{N}(H)—CH_3$	$CH_3—\ddot{N}(CH_3)—CH_3$
methyl amine	dimethyl amine	trimethyl amine

In some amines, the nitrogen atom replaces a carbon atom in an aromatic hydrocarbon. Pyridine (Figure 20.17) is one such heterocyclic amine. A heterocyclic compound contains atoms of two or more different elements in its ring structure.

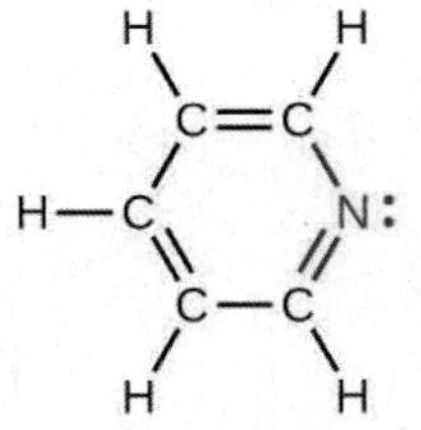

FIGURE 20.17 The illustration shows one of the resonance structures of pyridine.

HOW SCIENCES INTERCONNECT

DNA in Forensics and Paternity

The genetic material for all living things is a polymer of four different molecules, which are themselves a combination of three subunits. The genetic information, the code for developing an organism, is contained in the specific sequence of the four molecules, similar to the way the letters of the alphabet can be sequenced to form words that convey information. The information in a DNA sequence is used to form two other types of polymers, one of which are proteins. The proteins interact to form a specific type of organism with individual characteristics.

A genetic molecule is called DNA, which stands for deoxyribonucleic acid. The four molecules that make up DNA are called nucleotides. Each nucleotide consists of a single- or double-ringed molecule containing nitrogen, carbon, oxygen, and hydrogen called a nitrogenous base. Each base is bonded to a five-carbon sugar called deoxyribose. The sugar is in turn bonded to a phosphate group ($–PO_4{}^{3-}$) When new DNA is made, a polymerization reaction occurs that binds the phosphate group of one nucleotide to the sugar group of a second nucleotide. The nitrogenous bases of each nucleotide stick out from this sugar-phosphate backbone. DNA is actually formed from two such polymers coiled around each other and held together by hydrogen bonds between the nitrogenous bases. Thus, the two backbones are on the outside of the coiled pair of strands, and the bases are on the inside. The shape of the two strands wound around each other is called a double helix (see Figure 20.18).

It probably makes sense that the sequence of nucleotides in the DNA of a cat differs from those of a dog. But it is also true that the sequences of the DNA in the cells of two individual pugs differ. Likewise, the sequences of DNA in you and a sibling differ (unless your sibling is an identical twin), as do those between you and an unrelated individual. However, the DNA sequences of two related individuals are more similar than the sequences of two unrelated individuals, and these similarities in sequence can be observed in various ways.

This is the principle behind DNA fingerprinting, which is a method used to determine whether two DNA samples came from related (or the same) individuals or unrelated individuals.

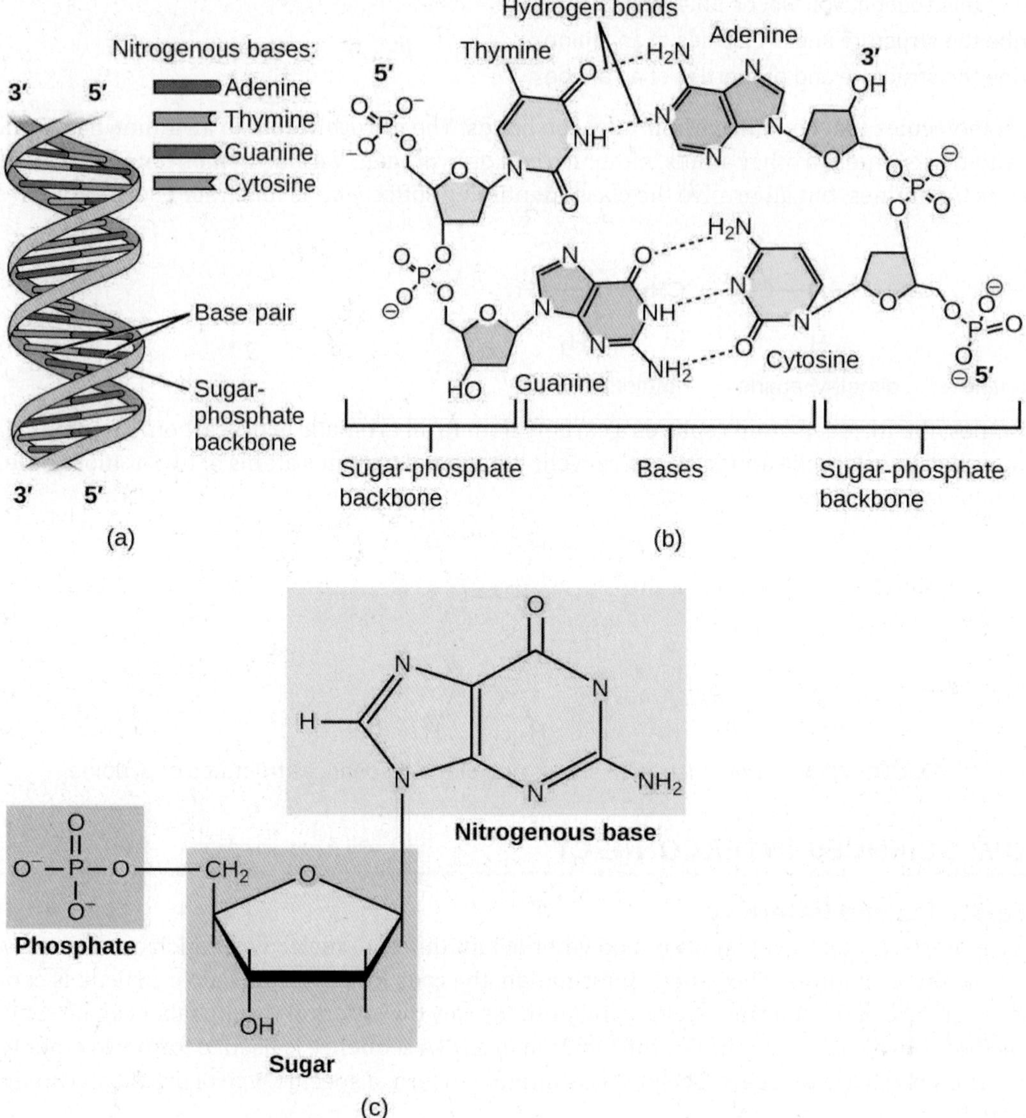

FIGURE 20.18 DNA is an organic molecule and the genetic material for all living organisms. (a) DNA is a double helix consisting of two single DNA strands hydrogen bonded together at each nitrogenous base. (b) This detail shows the hydrogen bonding (dotted lines) between nitrogenous bases on each DNA strand and the way in which each nucleotide is joined to the next, forming a backbone of sugars and phosphate groups along each strand. (c) This detail shows the structure of one of the four nucleotides that makes up the DNA polymer. Each nucleotide consists of a nitrogenous base (a double-ring molecule, in this case), a five-carbon sugar (deoxyribose), and a phosphate group.

Using similarities in sequences, technicians can determine whether a man is the father of a child (the identity of the mother is rarely in doubt, except in the case of an adopted child and a potential birth mother). Likewise, forensic geneticists can determine whether a crime scene sample of human tissue, such as blood or skin cells, contains DNA that matches exactly the DNA of a suspect.

LINK TO LEARNING

Watch this video animation (http://openstax.org/l/16dnapackaging) of how DNA is packaged for a visual lesson

in its structure.

Like ammonia, amines are weak bases due to the lone pair of electrons on their nitrogen atoms:

$$NH_3 + H^+ \longrightarrow [NH_4]^+$$

ammonia → ammonium ion

$$CH_3NH_2 + H^+ \longrightarrow [CH_3NH_3]^+$$

methyl amine → methyl ammonium ion

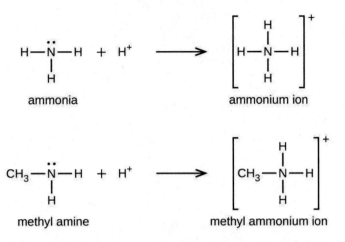

The basicity of an amine's nitrogen atom plays an important role in much of the compound's chemistry. Amine functional groups are found in a wide variety of compounds, including natural and synthetic dyes, polymers, vitamins, and medications such as penicillin and codeine. They are also found in many molecules essential to life, such as amino acids, hormones, neurotransmitters, and DNA.

HOW SCIENCES INTERCONNECT

Addictive Alkaloids

Since ancient times, plants have been used for medicinal purposes. One class of substances, called *alkaloids*, found in many of these plants has been isolated and found to contain cyclic molecules with an amine functional group. These amines are bases. They can react with H_3O^+ in a dilute acid to form an ammonium salt, and this property is used to extract them from the plant:

$$R_3N + H_3O^+ + Cl^- \longrightarrow \left[R_3NH^+\right] Cl^- + H_2O$$

The name alkaloid means "like an alkali." Thus, an alkaloid reacts with acid. The free compound can be recovered after extraction by reaction with a base:

$$\left[R_3NH^+\right] Cl^- + OH^- \longrightarrow R_3N + H_2O + Cl^-$$

The structures of many naturally occurring alkaloids have profound physiological and psychotropic effects in humans. Examples of these drugs include nicotine, morphine, codeine, and heroin. The plant produces these substances, collectively called secondary plant compounds, as chemical defenses against the numerous pests that attempt to feed on the plant:

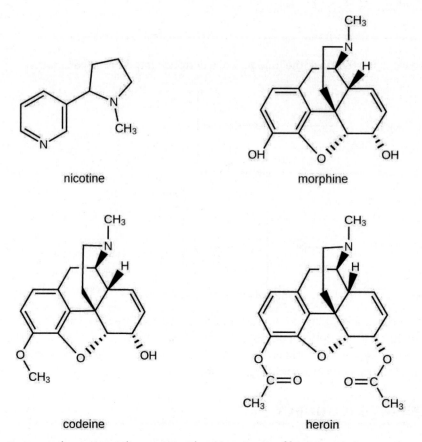

nicotine morphine

codeine heroin

In these diagrams, as is common in representing structures of large organic compounds, carbon atoms in the rings and the hydrogen atoms bonded to them have been omitted for clarity. The solid wedges indicate bonds that extend out of the page. The dashed wedges indicate bonds that extend into the page. Notice that small changes to a part of the molecule change the properties of morphine, codeine, and heroin. Morphine, a strong narcotic used to relieve pain, contains two hydroxyl functional groups, located at the bottom of the molecule in this structural formula. Changing one of these hydroxyl groups to a methyl ether group forms codeine, a less potent drug used as a local anesthetic. If both hydroxyl groups are converted to esters of acetic acid, the powerfully addictive drug heroin results (Figure 20.19).

FIGURE 20.19 Poppies can be used in the production of opium, a plant latex that contains morphine from which other opiates, such as heroin, can be synthesized. (credit: Karen Roe)

Amides are molecules that contain nitrogen atoms connected to the carbon atom of a carbonyl group. Like amines, various nomenclature rules may be used to name amides, but all include use of the class-specific suffix *-amide*:

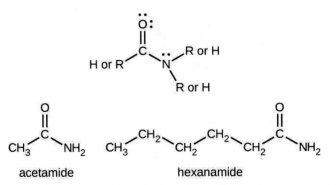

H or R—C(=O)—N(R or H)(R or H)

$CH_3C(=O)NH_2$ $CH_3CH_2CH_2CH_2CH_2C(=O)NH_2$

acetamide hexanamide

Amides can be produced when carboxylic acids react with amines or ammonia in a process called amidation. A water molecule is eliminated from the reaction, and the amide is formed from the remaining pieces of the carboxylic acid and the amine (note the similarity to formation of an ester from a carboxylic acid and an alcohol discussed in the previous section):

$$\underset{\text{carboxylic acid}}{CH_3-CH_2-\overset{O}{\overset{\|}{C}}-OH} + \underset{\text{amine}}{H-\underset{H}{\underset{|}{N}}-CH_3} \longrightarrow \underset{\text{amide}}{CH_3-CH_2-\overset{O}{\overset{\|}{C}}-\underset{H}{\underset{|}{N}}-CH_3} + \underset{\text{water}}{H_2O}$$

The reaction between amines and carboxylic acids to form amides is biologically important. It is through this reaction that amino acids (molecules containing both amine and carboxylic acid substituents) link together in a polymer to form proteins.

HOW SCIENCES INTERCONNECT

Proteins and Enzymes

Proteins are large biological molecules made up of long chains of smaller molecules called amino acids. Organisms rely on proteins for a variety of functions—proteins transport molecules across cell membranes, replicate DNA, and catalyze metabolic reactions, to name only a few of their functions. The properties of proteins are functions of the combination of amino acids that compose them and can vary greatly. Interactions between amino acid sequences in the chains of proteins result in the folding of the chain into specific, three-dimensional structures that determine the protein's activity.

Amino acids are organic molecules that contain an amine functional group ($-NH_2$), a carboxylic acid functional group ($-COOH$), and a side chain (that is specific to each individual amino acid). Most living things build proteins from the same 20 different amino acids. Amino acids connect by the formation of a peptide bond, which is a covalent bond formed between two amino acids when the carboxylic acid group of one amino acid reacts with the amine group of the other amino acid. The formation of the bond results in the production of a molecule of water (in general, reactions that result in the production of water when two other molecules combine are referred to as condensation reactions). The resulting bond—between the carbonyl group carbon atom and the amine nitrogen atom is called a peptide link or peptide bond. Since each of the original amino acids has an unreacted group (one has an unreacted amine and the other an unreacted carboxylic acid), more peptide bonds can form to other amino acids, extending the structure. (Figure 20.20) A chain of connected amino acids is called a polypeptide. Proteins contain at least one long polypeptide chain.

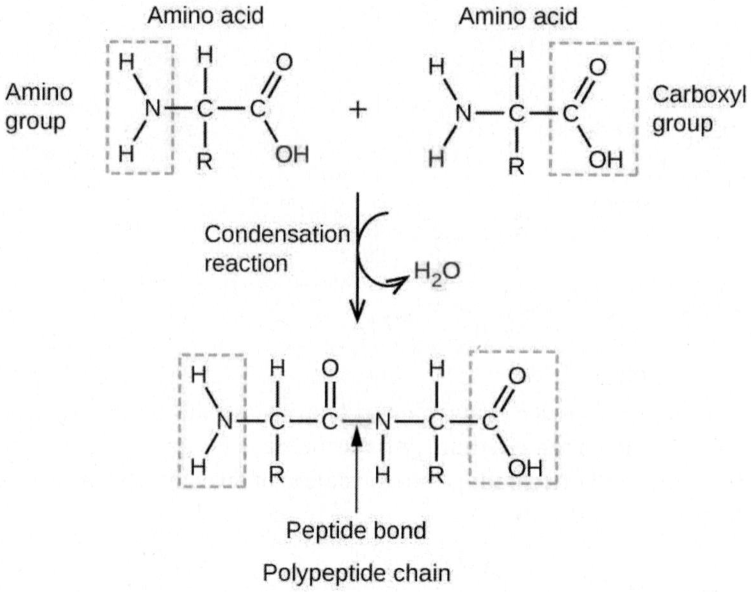

FIGURE 20.20 This condensation reaction forms a dipeptide from two amino acids and leads to the formation of water.

Enzymes are large biological molecules, mostly composed of proteins, which are responsible for the thousands of metabolic processes that occur in living organisms. Enzymes are highly specific catalysts; they speed up the rates of certain reactions. Enzymes function by lowering the activation energy of the reaction they are catalyzing, which can dramatically increase the rate of the reaction. Most reactions catalyzed by enzymes have rates that are millions of times faster than the noncatalyzed version. Like all catalysts, enzymes are not consumed during the reactions that they catalyze. Enzymes do differ from other catalysts in how specific they are for their substrates (the molecules that an enzyme will convert into a different product). Each enzyme is only capable of speeding up one or a few very specific reactions or types of reactions. Since the function of enzymes is so specific, the lack or malfunctioning of an enzyme can lead to serious health consequences. One disease that is the result of an enzyme malfunction is phenylketonuria. In this disease, the enzyme that catalyzes the first step in the degradation of the amino acid phenylalanine is not functional (Figure 20.21). Untreated, this can lead to an accumulation of phenylalanine, which can lead to intellectual disabilities.

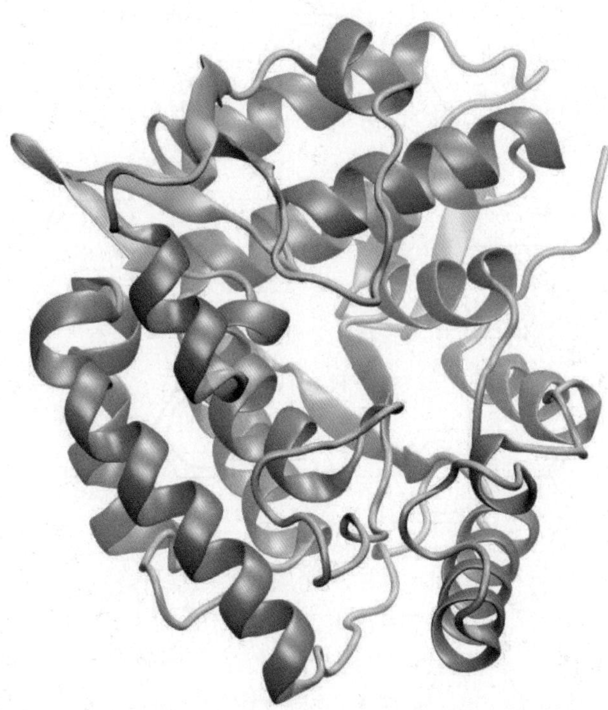

FIGURE 20.21 A computer rendering shows the three-dimensional structure of the enzyme phenylalanine hydroxylase. In the disease phenylketonuria, a defect in the shape of phenylalanine hydroxylase causes it to lose its function in breaking down phenylalanine.

Chemistry in Everyday Life

Kevlar

Kevlar (Figure 20.22) is a synthetic polymer made from two monomers 1,4-phenylene-diamine and terephthaloyl chloride (Kevlar is a registered trademark of DuPont). The material was developed by Susan Kwolek while she worked to find a replacement for steel in tires. Kwolek's work involved synthesizing polyamides and dissolving them in solvents, then spinning the resulting solution into fibers. One of her solutions proved to be quite different in initial appearance and structure. And once spun, the resulting fibers were particularly strong. From this initial discovery, Kevlar was created. The material has a high tensile strength-to-weight ratio (it is about 5 times stronger than an equal weight of steel), making it useful for many applications from bicycle tires to sails to body armor.

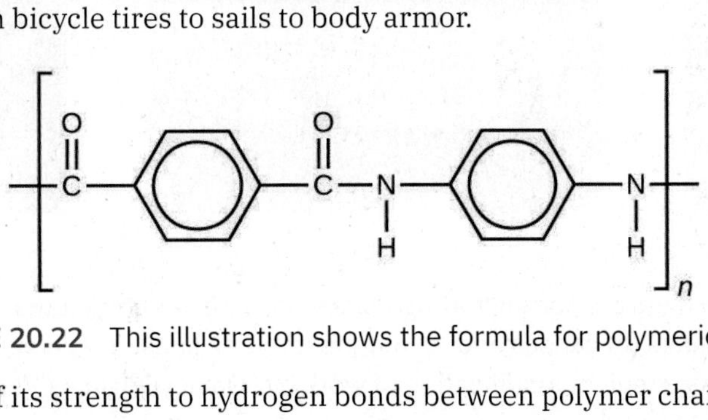

FIGURE 20.22 This illustration shows the formula for polymeric Kevlar.

The material owes much of its strength to hydrogen bonds between polymer chains (refer back to the chapter on intermolecular interactions). These bonds form between the carbonyl group oxygen atom (which has a partial negative charge due to oxygen's electronegativity) on one monomer and the partially positively charged hydrogen atom in the N–H bond of an adjacent monomer in the polymer structure (see dashed line in Figure 20.23). There is additional strength derived from the interaction between the unhybridized *p* orbitals in the six-membered rings, called aromatic stacking.

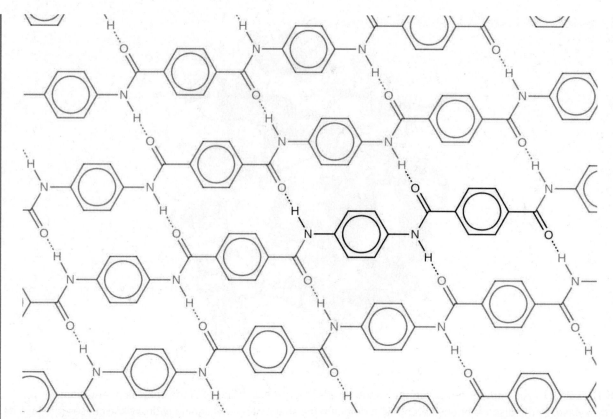

FIGURE 20.23 The diagram shows the polymer structure of Kevlar, with hydrogen bonds between polymer chains represented by dotted lines.

Kevlar may be best known as a component of body armor, combat helmets, and face masks. Since the 1980s, the US military has used Kevlar as a component of the PASGT (personal armor system for ground troops) helmet and vest. Kevlar is also used to protect armored fighting vehicles and aircraft carriers. Civilian applications include protective gear for emergency service personnel such as body armor for police officers and heat-resistant clothing for fire fighters. Kevlar based clothing is considerably lighter and thinner than equivalent gear made from other materials (Figure 20.24). Beyond Kevlar, Susan Kwolek was instrumental in the development of Nomex, a fireproof material, and was also involved in the creation of Lycra. She became just the fourth woman inducted into the National Inventors Hall of Fame, and received a number of other awards for her significant contributions to science and society.

(a)

(b)

(c)

FIGURE 20.24 (a) These soldiers are sorting through pieces of a Kevlar helmet that helped absorb a grenade blast. Kevlar is also used to make (b) canoes and (c) marine mooring lines. (credit a: modification of work by "Cla68"/Wikimedia Commons; credit b: modification of work by "OakleyOriginals"/Flickr; credit c: modification of work by Casey H. Kyhl)

In addition to its better-known uses, Kevlar is also often used in cryogenics for its very low thermal conductivity (along with its high strength). Kevlar maintains its high strength when cooled to the temperature of liquid nitrogen (–196 °C).

The table here summarizes the structures discussed in this chapter:

Compound Name	Structure of Compound and Functional Group (red)	Example	
		Formula	Name
alkene	C=C	C_2H_4	ethene
alkyne	C≡C	C_2H_2	ethyne
alcohol	R—O—H	CH_3CH_2OH	ethanol
ether	R—O—R′	$(C_2H_5)_2O$	diethyl ether
aldehyde	R—C(=O)—H	CH_3CHO	ethanal
ketone	R—C(=O)—R′	$CH_3COCH_2CH_3$	methyl ethyl ketone
carboxylic acid	R—C(=O)—O—H	CH_3COOH	acetic acid
ester	R—C(=O)—O—R′	$CH_3CO_2CH_2CH_3$	ethyl acetate
amine	R—N(H)—H R—N(R′)—H R—N(R′)—R″	$C_2H_5NH_2$	ethylamine
amide	R—C(=O)—N(H)—R′	CH_3CONH_2	acetamide

Key Terms

addition reaction reaction in which a double carbon-carbon bond forms a single carbon-carbon bond by the addition of a reactant. Typical reaction for an alkene.

alcohol organic compound with a hydroxyl group (–OH) bonded to a carbon atom

aldehyde organic compound containing a carbonyl group bonded to two hydrogen atoms or a hydrogen atom and a carbon substituent

alkane molecule consisting of only carbon and hydrogen atoms connected by single (σ) bonds

alkene molecule consisting of carbon and hydrogen containing at least one carbon-carbon double bond

alkyl group substituent, consisting of an alkane missing one hydrogen atom, attached to a larger structure

alkyne molecule consisting of carbon and hydrogen containing at least one carbon-carbon triple bond

amide organic molecule that features a nitrogen atom connected to the carbon atom in a carbonyl group

amine organic molecule in which a nitrogen atom is bonded to one or more alkyl group

aromatic hydrocarbon cyclic molecule consisting of carbon and hydrogen with delocalized alternating carbon-carbon single and double bonds, resulting in enhanced stability

carbonyl group carbon atom double bonded to an oxygen atom

carboxylic acid organic compound containing a carbonyl group with an attached hydroxyl group

ester organic compound containing a carbonyl group with an attached oxygen atom that is bonded to a carbon substituent

ether organic compound with an oxygen atom that is bonded to two carbon atoms

functional group part of an organic molecule that imparts a specific chemical reactivity to the molecule

ketone organic compound containing a carbonyl group with two carbon substituents attached to it

organic compound natural or synthetic compound that contains carbon

saturated hydrocarbon molecule containing carbon and hydrogen that has only single bonds between carbon atoms

skeletal structure shorthand method of drawing organic molecules in which carbon atoms are represented by the ends of lines and bends in between lines, and hydrogen atoms attached to the carbon atoms are not shown (but are understood to be present by the context of the structure)

substituent branch or functional group that replaces hydrogen atoms in a larger hydrocarbon chain

substitution reaction reaction in which one atom replaces another in a molecule

Summary

20.1 Hydrocarbons

Strong, stable bonds between carbon atoms produce complex molecules containing chains, branches, and rings. The chemistry of these compounds is called organic chemistry. Hydrocarbons are organic compounds composed of only carbon and hydrogen. The alkanes are saturated hydrocarbons—that is, hydrocarbons that contain only single bonds. Alkenes contain one or more carbon-carbon double bonds. Alkynes contain one or more carbon-carbon triple bonds. Aromatic hydrocarbons contain ring structures with delocalized π electron systems.

20.2 Alcohols and Ethers

Many organic compounds that are not hydrocarbons can be thought of as derivatives of hydrocarbons. A hydrocarbon derivative can be formed by replacing one or more hydrogen atoms of a hydrocarbon by a functional group, which contains at least one atom of an element other than carbon or hydrogen. The properties of hydrocarbon derivatives are determined largely by the functional group. The –OH group is the functional group of an alcohol. The –R–O–R– group is the functional group of an ether.

20.3 Aldehydes, Ketones, Carboxylic Acids, and Esters

Functional groups related to the carbonyl group include the –CHO group of an aldehyde, the –CO– group of a ketone, the $-CO_2H$ group of a carboxylic acid, and the $-CO_2R$ group of an ester. The carbonyl group, a carbon-oxygen double bond, is the key structure in these classes of organic molecules: Aldehydes contain at least one hydrogen atom attached to the carbonyl carbon atom, ketones contain two carbon groups attached to the carbonyl carbon atom, carboxylic acids contain a hydroxyl group attached to the carbonyl carbon atom, and esters contain an oxygen atom attached to another carbon group connected to the carbonyl carbon atom. All of these compounds contain oxidized carbon atoms relative to the carbon atom of an alcohol group.

20.4 Amines and Amides

The addition of nitrogen into an organic framework leads to two families of molecules. Compounds containing a nitrogen atom bonded in a hydrocarbon framework are classified as amines. Compounds that have a nitrogen atom bonded to one side of a carbonyl group are classified as amides. Amines are a basic functional group. Amines and carboxylic acids can combine in a condensation reaction to form amides.

Exercises

20.1 Hydrocarbons

1. Write the chemical formula and Lewis structure of the following, each of which contains five carbon atoms:
(a) an alkane
(b) an alkene
(c) an alkyne

2. What is the difference between the hybridization of carbon atoms' valence orbitals in saturated and unsaturated hydrocarbons?

3. On a microscopic level, how does the reaction of bromine with a saturated hydrocarbon differ from its reaction with an unsaturated hydrocarbon? How are they similar?

4. On a microscopic level, how does the reaction of bromine with an alkene differ from its reaction with an alkyne? How are they similar?

5. Explain why unbranched alkenes can form geometric isomers while unbranched alkanes cannot. Does this explanation involve the macroscopic domain or the microscopic domain?

6. Explain why these two molecules are not isomers:

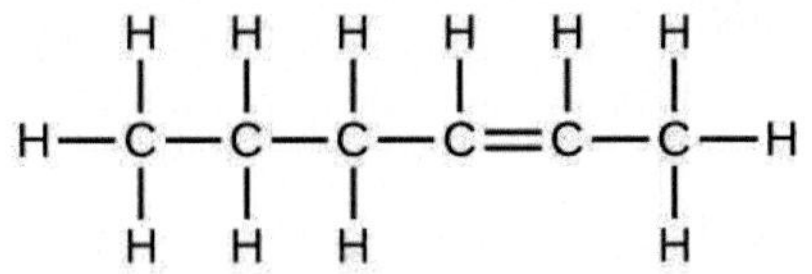

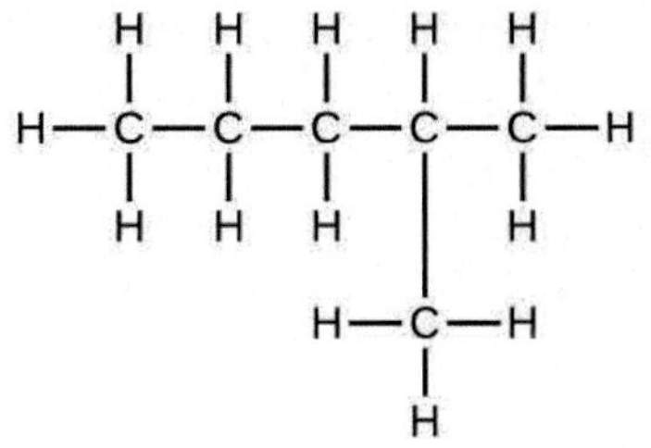

7. Explain why these two molecules are not isomers:

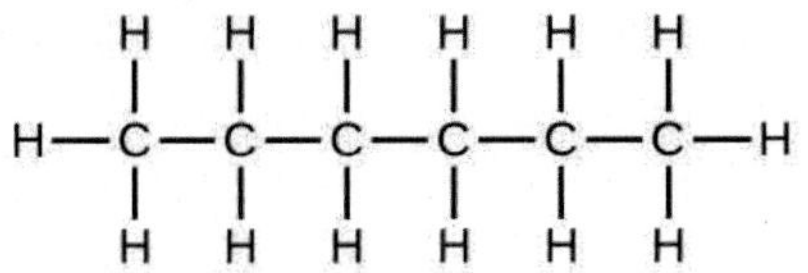

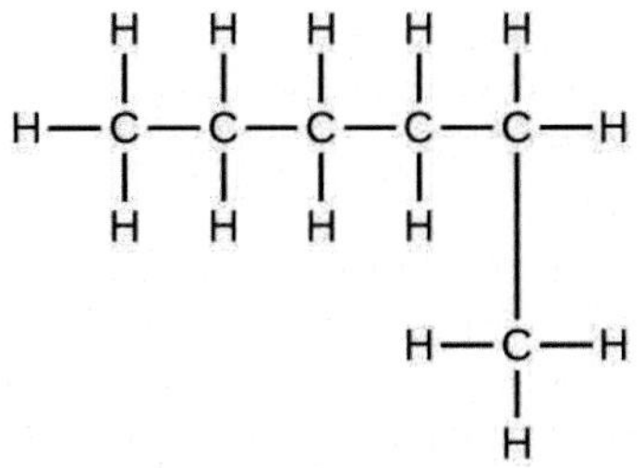

8. How does the carbon-atom hybridization change when polyethylene is prepared from ethylene?

9. Write the Lewis structure and molecular formula for each of the following hydrocarbons:
(a) hexane
(b) 3-methylpentane
(c) *cis*-3-hexene
(d) 4-methyl-1-pentene
(e) 3-hexyne
(f) 4-methyl-2-pentyne

10. Write the chemical formula, condensed formula, and Lewis structure for each of the following hydrocarbons:
(a) heptane
(b) 3-methylhexane
(c) *trans*-3-heptene
(d) 4-methyl-1-hexene
(e) 2-heptyne
(f) 3,4-dimethyl-1-pentyne

11. Give the complete IUPAC name for each of the following compounds:
(a) $CH_3CH_2CBr_2CH_3$
(b) $(CH_3)_3CCl$
(c)

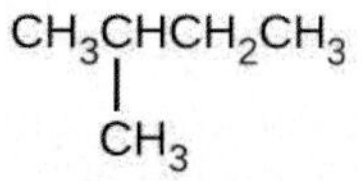

$CH_3CHCH_2CH_3$ (with CH_3 attached to the second carbon)

(d) $CH_3CH_2C \equiv CH\ CH_3CH_2C \equiv CH$
(e)

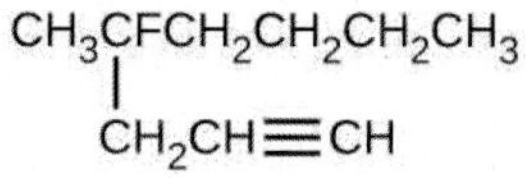

$CH_3CFCH_2CH_2CH_2CH_3$ (with $CH_2CH{\equiv}CH$ attached to the second carbon)

(f)

Cl(H)C=C(H)(CH_3): Cl and H on the first carbon, H and CH_3 on the second carbon (Cl and CH_3 on opposite sides)

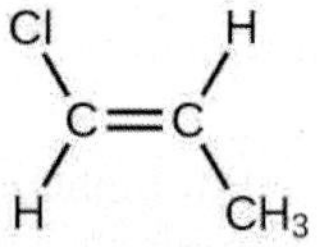

(g) $(CH_3)_2CHCH_2CH = CH_2$

12. Give the complete IUPAC name for each of the following compounds:
(a) $(CH_3)_2CHF$
(b) $CH_3CHClCHClCH_3$
(c)

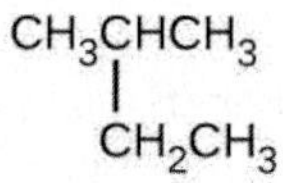

CH_3CHCH_3 (with CH_2CH_3 attached to the second carbon)

(d) $CH_3CH_2CH = CHCH_3$
(e)

$CH_3CH_2CH_2CHBrCH_2CH_3$ (with $CH_2CH{=}CH_2$ attached to the second carbon)

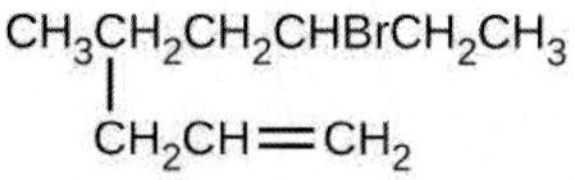

(f) $(CH_3)_3CCH_2C \equiv CH$

13. Butane is used as a fuel in disposable lighters. Write the Lewis structure for each isomer of butane.

14. Write Lewis structures and name the five structural isomers of hexane.

15. Write Lewis structures for the *cis–trans* isomers of $CH_3CH = CHCl$.

16. Write structures for the three isomers of the aromatic hydrocarbon xylene, $C_6H_4(CH_3)_2$.

17. Isooctane is the common name of the isomer of C_8H_{18} used as the standard of 100 for the gasoline octane rating:

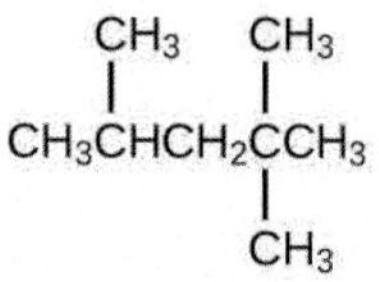

$CH_3CHCH_2CCH_3$ (with one CH_3 on the second carbon, and two CH_3 groups on the fourth carbon)

(a) What is the IUPAC name for the compound?
(b) Name the other isomers that contain a five-carbon chain with three methyl substituents.

18. Write Lewis structures and IUPAC names for the alkyne isomers of C_4H_6.

19. Write Lewis structures and IUPAC names for all isomers of C_4H_9Cl.
20. Name and write the structures of all isomers of the propyl and butyl alkyl groups.
21. Write the structures for all the isomers of the $–C_5H_{11}$ alkyl group.
22. Write Lewis structures and describe the molecular geometry at each carbon atom in the following compounds:
(a) *cis*-3-hexene
(b) *cis*-1-chloro-2-bromoethene
(c) 2-pentyne
(d) *trans*-6-ethyl-7-methyl-2-octene
23. Benzene is one of the compounds used as an octane enhancer in unleaded gasoline. It is manufactured by the catalytic conversion of acetylene to benzene:
$3C_2H_2 \longrightarrow C_6H_6$
Draw Lewis structures for these compounds, with resonance structures as appropriate, and determine the hybridization of the carbon atoms in each.
24. Teflon is prepared by the polymerization of tetrafluoroethylene. Write the equation that describes the polymerization using Lewis symbols.
25. Write two complete, balanced equations for each of the following reactions, one using condensed formulas and one using Lewis structures.
(a) 1 mol of 1-butyne reacts with 2 mol of iodine.
(b) Pentane is burned in air.
26. Write two complete, balanced equations for each of the following reactions, one using condensed formulas and one using Lewis structures.
(a) 2-butene reacts with chlorine.
(b) benzene burns in air.
27. What mass of 2-bromopropane could be prepared from 25.5 g of propene? Assume a 100% yield of product.
28. Acetylene is a very weak acid; however, it will react with moist silver(I) oxide and form water and a compound composed of silver and carbon. Addition of a solution of HCl to a 0.2352-g sample of the compound of silver and carbon produced acetylene and 0.2822 g of AgCl.
(a) What is the empirical formula of the compound of silver and carbon?
(b) The production of acetylene on addition of HCl to the compound of silver and carbon suggests that the carbon is present as the acetylide ion, $C_2{}^{2-}$. Write the formula of the compound showing the acetylide ion.
29. Ethylene can be produced by the pyrolysis of ethane:
$C_2H_6 \longrightarrow C_2H_4 + H_2$
How many kilograms of ethylene is produced by the pyrolysis of 1.000×10^3 kg of ethane, assuming a 100.0% yield?

20.2 Alcohols and Ethers

30. Why do the compounds hexane, hexanol, and hexene have such similar names?
31. Write condensed formulas and provide IUPAC names for the following compounds:
(a) ethyl alcohol (in beverages)
(b) methyl alcohol (used as a solvent, for example, in shellac)
(c) ethylene glycol (antifreeze)
(d) isopropyl alcohol (used in rubbing alcohol)
(e) glycerine

32. Give the complete IUPAC name for each of the following compounds:

(a)

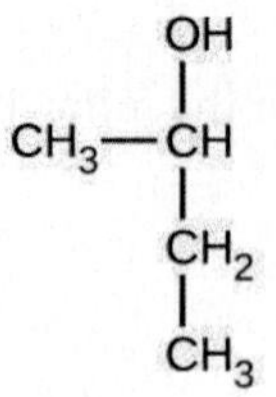

$$\begin{array}{c} \quad OH \\ \quad | \\ CH_3-CH \\ \quad | \\ \quad CH_2 \\ \quad | \\ \quad CH_3 \end{array}$$

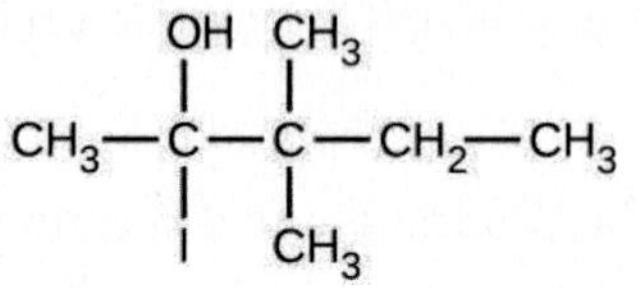

(b)

$$\begin{array}{l} \qquad\;\; OH \;\; CH_3 \\ \qquad\;\;\; | \qquad | \\ CH_3-C-C-CH_2-CH_3 \\ \qquad\;\;\; | \qquad | \\ \qquad\;\;\; I \quad\; CH_3 \end{array}$$

(c)

$$\begin{array}{l} \qquad\qquad\qquad\;\; CH_3 \\ \qquad\qquad\qquad\;\;\; | \\ \qquad\qquad OH \quad CH_2 \\ \qquad\qquad\;\; | \qquad\; | \\ CH_3-CH-C-CH_2-CH_2 \\ \qquad\qquad\qquad\;\; | \qquad\qquad | \\ \qquad\qquad\qquad\;\; Cl \qquad\quad CH_3 \end{array}$$

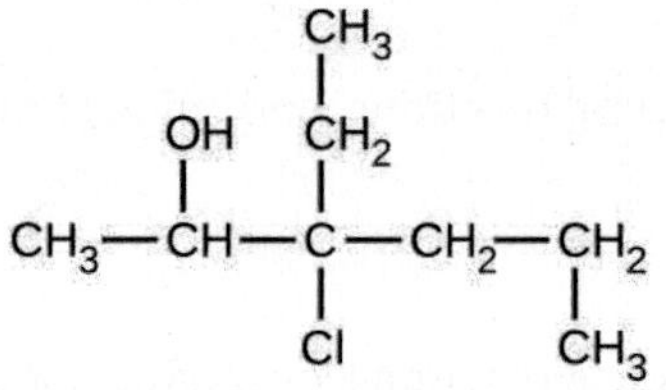

33. Give the complete IUPAC name and the common name for each of the following compounds:

(a)

$$CH_3-CH_2-O-CH_2-CH_2-CH_2-CH_3$$

(b)

$$CH_3-CH_2-O-CH_2-CH_2-CH_3$$

(c)

$$CH_3-O-CH_2-CH_2-CH_3$$

34. Write the condensed structures of both isomers with the formula C_2H_6O. Label the functional group of each isomer.

35. Write the condensed structures of all isomers with the formula $C_2H_6O_2$. Label the functional group (or groups) of each isomer.

36. Draw the condensed formulas for each of the following compounds:
(a) dipropyl ether
(b) 2,2-dimethyl-3-hexanol
(c) 2-ethoxybutane

37. MTBE, Methyl *tert*-butyl ether, $CH_3OC(CH_3)_3$, is used as an oxygen source in oxygenated gasolines. MTBE is manufactured by reacting 2-methylpropene with methanol.
(a) Using Lewis structures, write the chemical equation representing the reaction.
(b) What volume of methanol, density 0.7915 g/mL, is required to produce exactly 1000 kg of MTBE, assuming a 100% yield?

38. Write two complete balanced equations for each of the following reactions, one using condensed formulas and one using Lewis structures.
(a) propanol is converted to dipropyl ether
(b) propene is treated with water in dilute acid.

39. Write two complete balanced equations for each of the following reactions, one using condensed formulas and one using Lewis structures.
(a) 2-butene is treated with water in dilute acid
(b) ethanol is dehydrated to yield ethene

20.3 Aldehydes, Ketones, Carboxylic Acids, and Esters

40. Order the following molecules from least to most oxidized, based on the marked carbon atom:

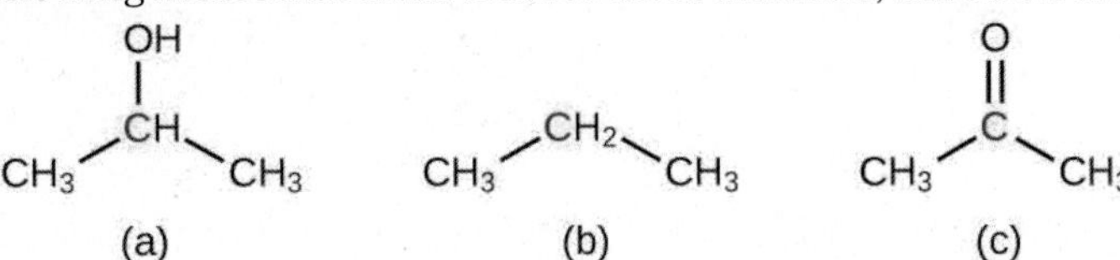

41. Predict the products of oxidizing the molecules shown in this problem. In each case, identify the product that will result from the minimal increase in oxidation state for the highlighted carbon atom:

(a)

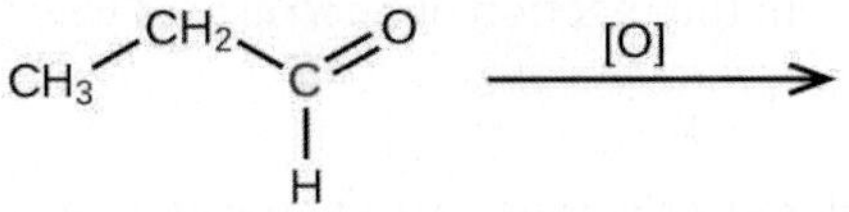

$CH_3CH_2C(=O)H \xrightarrow{[O]}$

(b)

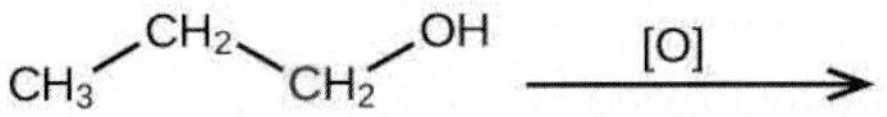

$CH_3CH_2CH_2OH \xrightarrow{[O]}$

(c)

$CH_3CH_2CH(CH_3)OH \xrightarrow{[O]}$

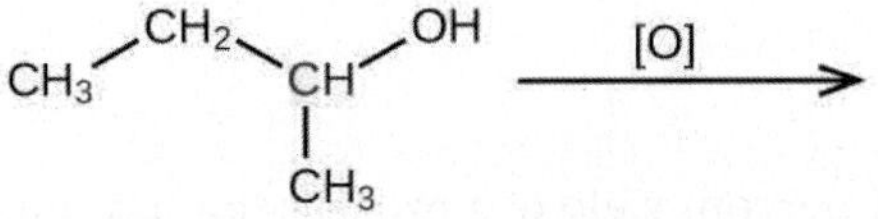

42. Predict the products of reducing the following molecules. In each case, identify the product that will result from the minimal decrease in oxidation state for the highlighted carbon atom:

(a)

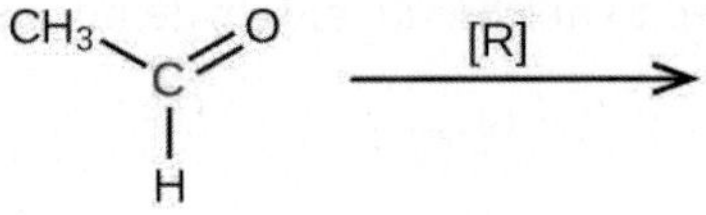

$CH_3C(=O)H \xrightarrow{[R]}$

(b)

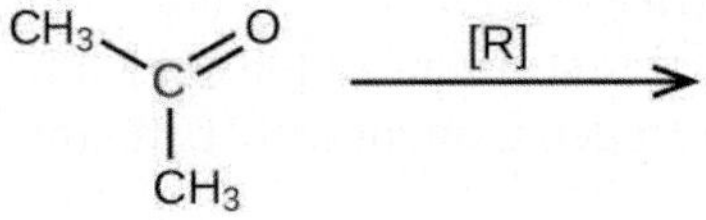

$CH_3C(=O)CH_3 \xrightarrow{[R]}$

(c)

$CH_3C(=O)OH \xrightarrow{[R]}$

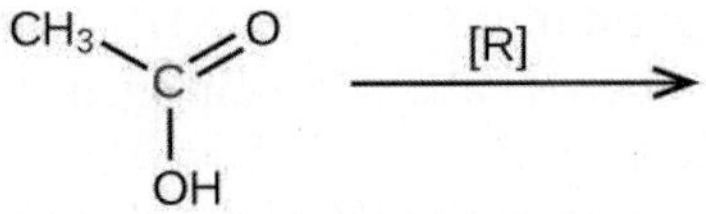

43. Explain why it is not possible to prepare a ketone that contains only two carbon atoms.

44. How does hybridization of the substituted carbon atom change when an alcohol is converted into an aldehyde? An aldehyde to a carboxylic acid?

45. Fatty acids are carboxylic acids that have long hydrocarbon chains attached to a carboxylate group. How does a saturated fatty acid differ from an unsaturated fatty acid? How are they similar?

46. Write a condensed structural formula, such as CH_3CH_3, and describe the molecular geometry at each carbon atom.

(a) propene
(b) 1-butanol
(c) ethyl propyl ether
(d) *cis*-4-bromo-2-heptene
(e) 2,2,3-trimethylhexane
(f) formaldehyde

47. Write a condensed structural formula, such as CH_3CH_3, and describe the molecular geometry at each carbon atom.
(a) 2-propanol
(b) acetone
(c) dimethyl ether
(d) acetic acid
(e) 3-methyl-1-hexene
48. The foul odor of rancid butter is caused by butyric acid, $CH_3CH_2CH_2CO_2H$.
(a) Draw the Lewis structure and determine the oxidation number and hybridization for each carbon atom in the molecule.
(b) The esters formed from butyric acid are pleasant-smelling compounds found in fruits and used in perfumes. Draw the Lewis structure for the ester formed from the reaction of butyric acid with 2-propanol.
49. Write the two-resonance structures for the acetate ion.
50. Write two complete, balanced equations for each of the following reactions, one using condensed formulas and one using Lewis structures:
(a) ethanol reacts with propionic acid
(b) benzoic acid, $C_6H_5CO_2H$, is added to a solution of sodium hydroxide
51. Write two complete balanced equations for each of the following reactions, one using condensed formulas and one using Lewis structures.
(a) 1-butanol reacts with acetic acid
(b) propionic acid is poured onto solid calcium carbonate
52. Yields in organic reactions are sometimes low. What is the percent yield of a process that produces 13.0 g of ethyl acetate from 10.0 g of CH_3CO_2H?
53. Alcohols A, B, and C all have the composition $C_4H_{10}O$. Molecules of alcohol A contain a branched carbon chain and can be oxidized to an aldehyde; molecules of alcohol B contain a linear carbon chain and can be oxidized to a ketone; and molecules of alcohol C can be oxidized to neither an aldehyde nor a ketone. Write the Lewis structures of these molecules.

20.4 Amines and Amides

54. Write the Lewis structures of both isomers with the formula C_2H_7N.
55. What is the molecular structure about the nitrogen atom in trimethyl amine and in the trimethyl ammonium ion, $(CH_3)_3NH^+$? What is the hybridization of the nitrogen atom in trimethyl amine and in the trimethyl ammonium ion?
56. Write the two resonance structures for the pyridinium ion, $C_5H_5NH^+$.
57. Draw Lewis structures for pyridine and its conjugate acid, the pyridinium ion, $C_5H_5NH^+$. What are the hybridizations, electron domain geometries, and molecular geometries about the nitrogen atoms in pyridine and in the pyridinium ion?
58. Write the Lewis structures of all isomers with the formula C_3H_7ON that contain an amide linkage.
59. Write two complete balanced equations for the following reaction, one using condensed formulas and one using Lewis structures.
Methyl amine is added to a solution of HCl.
60. Write two complete, balanced equations for each of the following reactions, one using condensed formulas and one using Lewis structures.
Ethylammonium chloride is added to a solution of sodium hydroxide.
61. Identify any carbon atoms that change hybridization and the change in hybridization during the reactions in Exercise 20.26.
62. Identify any carbon atoms that change hybridization and the change in hybridization during the reactions in Exercise 20.39.
63. Identify any carbon atoms that change hybridization and the change in hybridization during the reactions in Exercise 20.51.

CHAPTER 21
Nuclear Chemistry

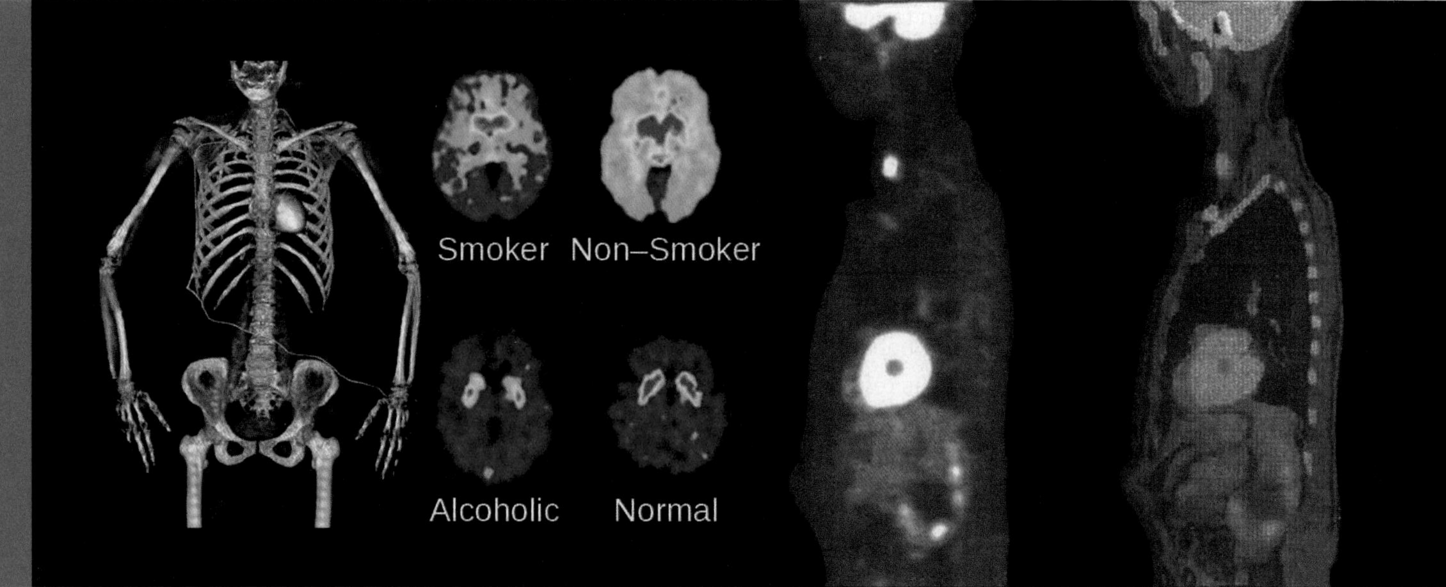

Figure 21.1 Nuclear chemistry provides the basis for many useful diagnostic and therapeutic methods in medicine, such as these positron emission tomography (PET) scans. The PET/computed tomography scan on the left shows muscle activity. The brain scans in the center show chemical differences in dopamine signaling in the brains of addicts and nonaddicts. The images on the right show an oncological application of PET scans to identify lymph node metastasis.

CHAPTER OUTLINE

21.1 Nuclear Structure and Stability
21.2 Nuclear Equations
21.3 Radioactive Decay
21.4 Transmutation and Nuclear Energy
21.5 Uses of Radioisotopes
21.6 Biological Effects of Radiation

INTRODUCTION The chemical reactions that we have considered in previous chapters involve changes in the *electronic* structure of the species involved, that is, the arrangement of the electrons around atoms, ions, or molecules. *Nuclear* structure, the numbers of protons and neutrons within the nuclei of the atoms involved, remains unchanged during chemical reactions.

This chapter will introduce the topic of nuclear chemistry, which began with the discovery of radioactivity in 1896 by French physicist Antoine Becquerel and has become increasingly important during the twentieth and twenty-first centuries, providing the basis for various technologies related to energy, medicine, geology, and many other areas.

21.1 Nuclear Structure and Stability

LEARNING OBJECTIVES

By the end of this section, you will be able to:

- Describe nuclear structure in terms of protons, neutrons, and electrons
- Calculate mass defect and binding energy for nuclei
- Explain trends in the relative stability of nuclei

Nuclear chemistry is the study of reactions that involve changes in nuclear structure. The chapter on atoms, molecules, and ions introduced the basic idea of nuclear structure, that the nucleus of an atom is composed of protons and, with the exception of ${}^{1}_{1}H$, neutrons. Recall that the number of protons in the nucleus is called the atomic number (Z) of the element, and the sum of the number of protons and the number of neutrons is the mass number (A). Atoms with the same atomic number but different mass numbers are isotopes of the same element. When referring to a single type of nucleus, we often use the term **nuclide** and identify it by the notation ${}^{A}_{Z}X$, where X is the symbol for the element, A is the mass number, and Z is the atomic number (for example, ${}^{14}_{6}C$). Often a nuclide is referenced by the name of the element followed by a hyphen and the mass number. For example, ${}^{14}_{6}C$ is called "carbon-14."

Protons and neutrons, collectively called **nucleons**, are packed together tightly in a nucleus. With a radius of about 10^{-15} meters, a nucleus is quite small compared to the radius of the entire atom, which is about 10^{-10} meters. Nuclei are extremely dense compared to bulk matter, averaging 1.8×10^{14} grams per cubic centimeter. For example, water has a density of 1 gram per cubic centimeter, and iridium, one of the densest elements known, has a density of 22.6 g/cm^3. If the earth's density were equal to the average nuclear density, the earth's radius would be only about 200 meters (earth's actual radius is approximately 6.4×10^6 meters, 30,000 times larger). Example 21.1 demonstrates just how great nuclear densities can be in the natural world.

EXAMPLE 21.1

Density of a Neutron Star

Neutron stars form when the core of a very massive star undergoes gravitational collapse, causing the star's outer layers to explode in a supernova. Composed almost completely of neutrons, they are the densest-known stars in the universe, with densities comparable to the average density of an atomic nucleus. A neutron star in a faraway galaxy has a mass equal to 2.4 solar masses (1 solar mass = $M_{\odot}$ = mass of the sun = 1.99×10^{30} kg) and a diameter of 26 km.

(a) What is the density of this neutron star?

(b) How does this neutron star's density compare to the density of a uranium nucleus, which has a diameter of about 15 fm (1 fm = 10^{-15} m)?

Solution

We can treat both the neutron star and the U-235 nucleus as spheres. Then the density for both is given by:

$$d = \frac{m}{V} \qquad \text{with} \qquad V = \frac{4}{3}\pi r^3$$

(a) The radius of the neutron star is $\frac{1}{2} \times 26\text{ km} = \frac{1}{2} \times 2.6 \times 10^4\text{ m} = 1.3 \times 10^4\text{ m}$, so the density of the neutron star is:

$$d = \frac{m}{V} = \frac{m}{\frac{4}{3}\pi r^3} = \frac{2.4\left(1.99 \times 10^{30}\text{ kg}\right)}{\frac{4}{3}\pi\left(1.3 \times 10^4\text{ m}\right)^3} = 5.2 \times 10^{17}\text{ kg/m}^3$$

(b) The radius of the U-235 nucleus is $\frac{1}{2} \times 15 \times 10^{-15}\text{ m} = 7.5 \times 10^{-15}\text{ m}$, so the density of the U-235 nucleus is:

$$d = \frac{m}{V} = \frac{m}{\frac{4}{3}\pi r^3} = \frac{235 \text{ amu} \left(\frac{1.66 \times 10^{-27} \text{kg}}{1 \text{ amu}}\right)}{\frac{4}{3}\pi(7.5 \times 10^{-15} \text{m})^3} = 2.2 \times 10^{17} \text{ kg/m}^3$$

These values are fairly similar (same order of magnitude), but the neutron star is more than twice as dense as the U-235 nucleus.

Check Your Learning

Find the density of a neutron star with a mass of 1.97 solar masses and a diameter of 13 km, and compare it to the density of a hydrogen nucleus, which has a diameter of 1.75 fm (1 fm = 1×10^{-15} m).

Answer:

The density of the neutron star is 3.4×10^{18} kg/m^3. The density of a hydrogen nucleus is 6.0×10^{17} kg/m^3. The neutron star is 5.7 times denser than the hydrogen nucleus.

To hold positively charged protons together in the very small volume of a nucleus requires very strong attractive forces because the positively charged protons repel one another strongly at such short distances. The force of attraction that holds the nucleus together is the **strong nuclear force**. (The strong force is one of the four fundamental forces that are known to exist. The others are the electromagnetic force, the gravitational force, and the nuclear weak force.) This force acts between protons, between neutrons, and between protons and neutrons. It is very different from the electrostatic force that holds negatively charged electrons around a positively charged nucleus (the attraction between opposite charges). Over distances less than 10^{-15} meters and within the nucleus, the strong nuclear force is much stronger than electrostatic repulsions between protons; over larger distances and outside the nucleus, it is essentially nonexistent.

LINK TO LEARNING

Visit this website (http://openstax.org/l/16fourfund) for more information about the four fundamental forces.

Nuclear Binding Energy

As a simple example of the energy associated with the strong nuclear force, consider the helium atom composed of two protons, two neutrons, and two electrons. The total mass of these six subatomic particles may be calculated as:

$$\underset{\text{protons}}{(2 \times 1.0073 \text{ amu})} + \underset{\text{neutrons}}{(2 \times 1.0087 \text{ amu})} + \underset{\text{electrons}}{(2 \times 0.00055 \text{ amu})} = 4.0331 \text{ amu}$$

However, mass spectrometric measurements reveal that the mass of an ${}^{4}_{2}\text{He}$ atom is 4.0026 amu, less than the combined masses of its six constituent subatomic particles. This difference between the calculated and experimentally measured masses is known as the **mass defect** of the atom. In the case of helium, the mass defect indicates a "loss" in mass of 4.0331 amu – 4.0026 amu = 0.0305 amu. The loss in mass accompanying the formation of an atom from protons, neutrons, and electrons is due to the conversion of that mass into energy that is evolved as the atom forms. The **nuclear binding energy** is the energy produced when the atoms' nucleons are bound together; this is also the energy needed to break a nucleus into its constituent protons and neutrons. In comparison to chemical bond energies, nuclear binding energies are *vastly* greater, as we will learn in this section. Consequently, the energy changes associated with nuclear reactions are vastly greater than are those for chemical reactions.

The conversion between mass and energy is most identifiably represented by the **mass-energy equivalence equation** as stated by Albert Einstein:

$$E = mc^2$$

where E is energy, m is mass of the matter being converted, and c is the speed of light in a vacuum. This equation can be used to find the amount of energy that results when matter is converted into energy. Using this mass-energy equivalence equation, the nuclear binding energy of a nucleus may be calculated from its mass defect, as demonstrated in Example 21.2. A variety of units are commonly used for nuclear binding energies, including **electron volts (eV)**, with 1 eV equaling the amount of energy necessary to the move the charge of an electron across an electric potential difference of 1 volt, making 1 eV = 1.602×10^{-19} J.

EXAMPLE 21.2

Calculation of Nuclear Binding Energy

Determine the binding energy for the nuclide $^{4}_{2}\text{He}$ in:

(a) joules per mole of nuclei

(b) joules per nucleus

(c) MeV per nucleus

Solution

The mass defect for a $^{4}_{2}\text{He}$ nucleus is 0.0305 amu, as shown previously. Determine the binding energy in joules per nuclide using the mass-energy equivalence equation. To accommodate the requested energy units, the mass defect must be expressed in kilograms (recall that 1 J = 1 kg m^2/s^2).

(a) First, express the mass defect in g/mol. This is easily done considering the *numerical equivalence* of atomic mass (amu) and molar mass (g/mol) that results from the definitions of the amu and mole units (refer to the previous discussion in the chapter on atoms, molecules, and ions if needed). The mass defect is therefore 0.0305 g/mol. To accommodate the units of the other terms in the mass-energy equation, the mass must be expressed in kg, since 1 J = 1 kg m^2/s^2. Converting grams into kilograms yields a mass defect of 3.05×10^{-5} kg/mol. Substituting this quantity into the mass-energy equivalence equation yields:

$$E = mc^2 = \frac{3.05 \times 10^{-5}\text{ kg}}{\text{mol}} \times \left(\frac{2.998 \times 10^{8}\text{ m}}{\text{s}}\right)^2 = 2.74 \times 10^{12}\text{ kg m}^2\text{s}^{-2}\text{mol}^{-1}$$
$$= 2.74 \times 10^{12}\text{ J mol}^{-1} = 2.74\text{ TJ mol}^{-1}$$

Note that this tremendous amount of energy is associated with the conversion of a very small amount of matter (about 30 mg, roughly the mass of typical drop of water).

(b) The binding energy for a single nucleus is computed from the molar binding energy using Avogadro's number:

$$E = 2.74 \times 10^{12}\text{ J mol}^{-1} \times \frac{1\text{ mol}}{6.022 \times 10^{23}\text{ nuclei}} = 4.55 \times 10^{-12}\text{ J} = 4.55\text{ pJ}$$

(c) Recall that 1 eV = 1.602×10^{-19} J. Using the binding energy computed in part (b):

$$E = 4.55 \times 10^{-12}\text{ J} \times \frac{1\text{ eV}}{1.602 \times 10^{-19}\text{ J}} = 2.84 \times 10^{7}\text{ eV} = 28.4\text{ MeV}$$

Check Your Learning

What is the binding energy for the nuclide $^{19}_{9}\text{F}$ (atomic mass: 18.9984 amu) in MeV per nucleus?

Answer:

148.4 MeV

Because the energy changes for breaking and forming bonds are so small compared to the energy changes for breaking or forming nuclei, the changes in mass during all ordinary chemical reactions are virtually undetectable. As described in the chapter on thermochemistry, the most energetic chemical reactions exhibit enthalpies on the order of *thousands* of kJ/mol, which is equivalent to mass differences in the nanogram range (10^{-9} g). On the other hand, nuclear binding energies are typically on the order of *billions* of kJ/mol, corresponding to mass differences in the milligram range (10^{-3} g).

Nuclear Stability

A nucleus is stable if it cannot be transformed into another configuration without adding energy from the outside. Of the thousands of nuclides that exist, about 250 are stable. A plot of the number of neutrons versus the number of protons for stable nuclei reveals that the stable isotopes fall into a narrow band. This region is known as the **band of stability** (also called the belt, zone, or valley of stability). The straight line in Figure 21.2 represents nuclei that have a 1:1 ratio of protons to neutrons (n:p ratio). Note that the lighter stable nuclei, in general, have equal numbers of protons and neutrons. For example, nitrogen-14 has seven protons and seven neutrons. Heavier stable nuclei, however, have increasingly more neutrons than protons. For example: iron-56 has 30 neutrons and 26 protons, an n:p ratio of 1.15, whereas the stable nuclide lead-207 has 125 neutrons and 82 protons, an n:p ratio equal to 1.52. This is because larger nuclei have more proton-proton repulsions, and require larger numbers of neutrons to provide compensating strong forces to overcome these electrostatic repulsions and hold the nucleus together.

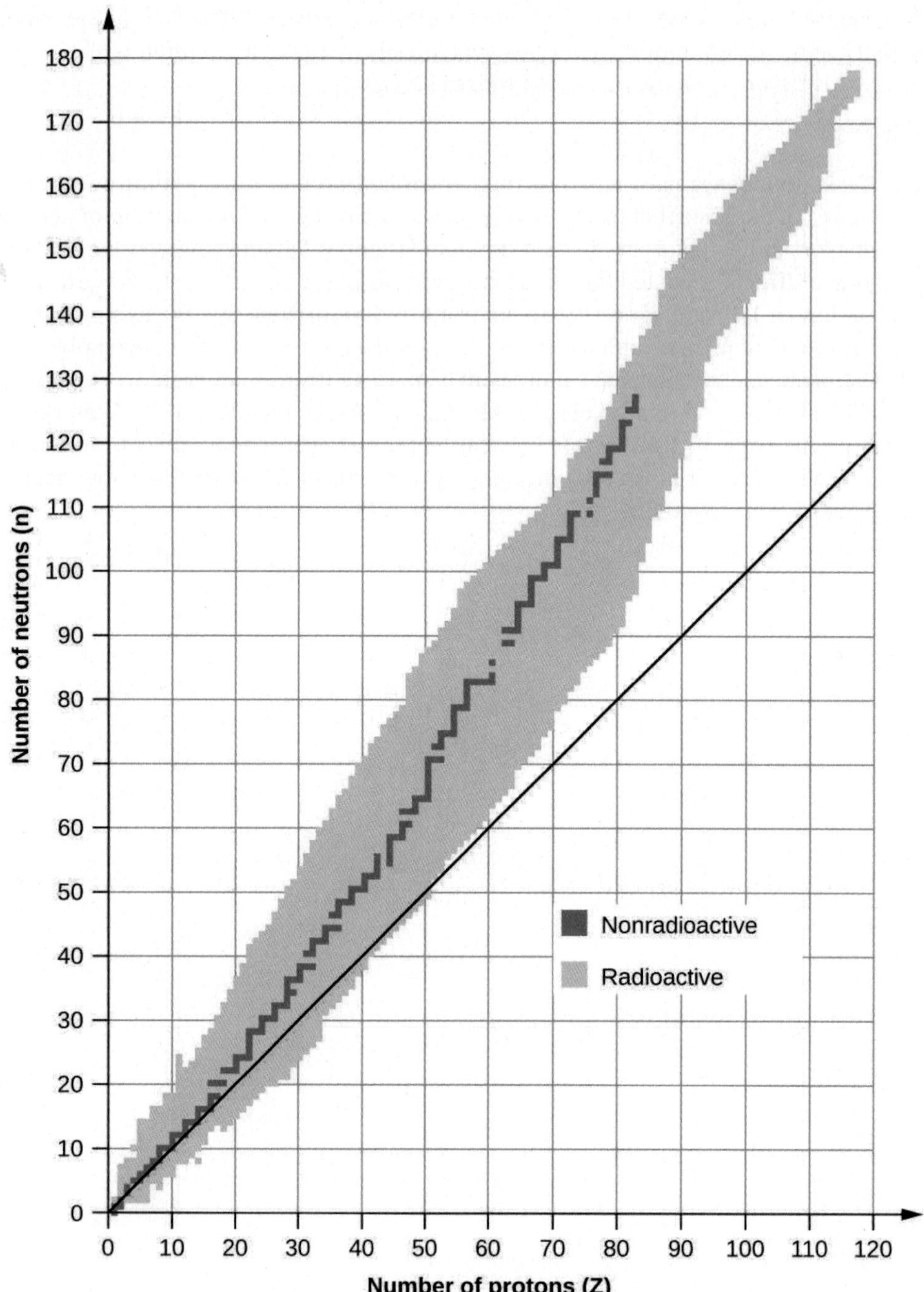

FIGURE 21.2 This plot shows the nuclides that are known to exist and those that are stable. The stable nuclides are indicated in blue, and the unstable nuclides are indicated in green. Note that all isotopes of elements with atomic numbers greater than 83 are unstable. The solid line is the line where n = Z.

The nuclei that are to the left or to the right of the band of stability are unstable and exhibit **radioactivity**. They change spontaneously (decay) into other nuclei that are either in, or closer to, the band of stability. These nuclear decay reactions convert one unstable isotope (or **radioisotope**) into another, more stable, isotope. We will discuss the nature and products of this radioactive decay in subsequent sections of this chapter.

Several observations may be made regarding the relationship between the stability of a nucleus and its structure. Nuclei with even numbers of protons, neutrons, or both are more likely to be stable (see Table 21.1). Nuclei with certain numbers of nucleons, known as **magic numbers**, are stable against nuclear decay. These numbers of protons or neutrons (2, 8, 20, 28, 50, 82, and 126) make complete shells in the nucleus. These are similar in concept to the stable electron shells observed for the noble gases. Nuclei that have magic numbers of both protons and neutrons, such as $^{4}_{2}He$, $^{16}_{8}O$, $^{40}_{20}Ca$, and $^{208}_{82}Pb$, are called "double magic" and are particularly stable. These trends in nuclear stability may be rationalized by considering a quantum mechanical model of nuclear energy states analogous to that used to describe electronic states earlier in this textbook. The details of this model are beyond the scope of this chapter.

Stable Nuclear Isotopes

Number of Stable Isotopes	Proton Number	Neutron Number
157	even	even
53	even	odd
50	odd	even
5	odd	odd

TABLE 21.1

The relative stability of a nucleus is correlated with its **binding energy per nucleon**, the total binding energy for the nucleus divided by the number or nucleons in the nucleus. For instance, we saw in Example 21.2 that the binding energy for a $^{4}_{2}He$ nucleus is 28.4 MeV. The binding energy *per nucleon* for a $^{4}_{2}He$ nucleus is therefore:

$$\frac{28.4 \text{ MeV}}{4 \text{ nucleons}} = 7.10 \text{ MeV/nucleon}$$

In Example 21.3, we learn how to calculate the binding energy per nucleon of a nuclide on the curve shown in Figure 21.3.

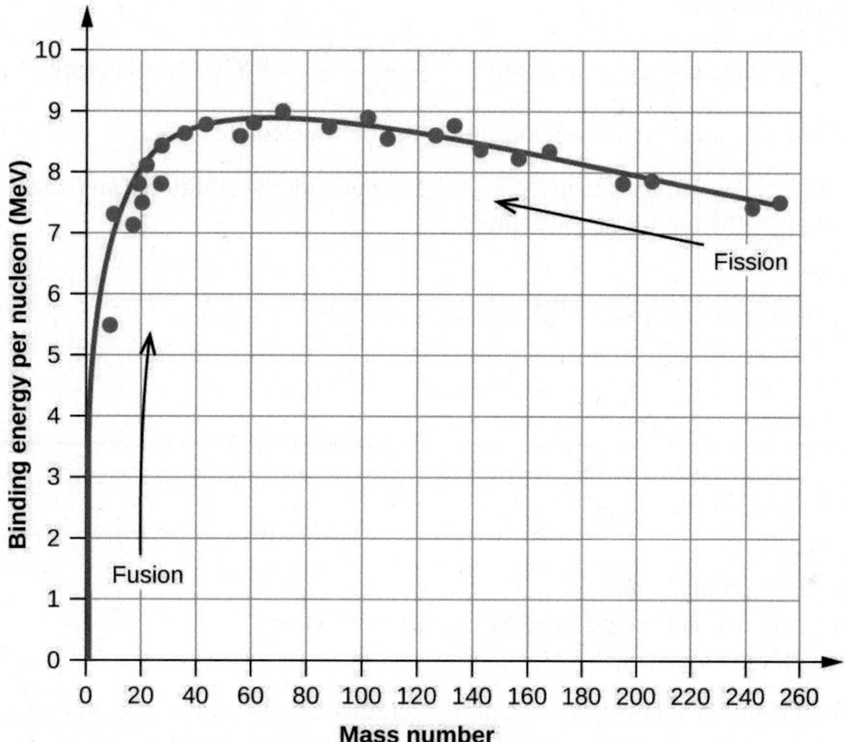

FIGURE 21.3 The binding energy per nucleon is largest for nuclides with mass number of approximately 56.

EXAMPLE 21.3

Calculation of Binding Energy per Nucleon

The iron nuclide ${}^{56}_{26}\text{Fe}$ lies near the top of the binding energy curve (Figure 21.3) and is one of the most stable nuclides. What is the binding energy per nucleon (in MeV) for the nuclide ${}^{56}_{26}\text{Fe}$ (atomic mass of 55.9349 amu)?

Solution

As in Example 21.2, we first determine the mass defect of the nuclide, which is the difference between the mass of 26 protons, 30 neutrons, and 26 electrons, and the observed mass of an ${}^{56}_{26}\text{Fe}$ atom:

$$\begin{aligned}\text{Mass defect} &= [(26 \times 1.0073\text{ amu}) + (30 \times 1.0087\text{ amu}) + (26 \times 0.00055\text{ amu})] - 55.9349\text{ amu}\\ &= 56.4651\text{ amu} - 55.9349\text{ amu}\\ &= 0.5302\text{ amu}\end{aligned}$$

We next calculate the binding energy for one nucleus from the mass defect using the mass-energy equivalence equation:

$$\begin{aligned}E = mc^2 &= 0.5302\text{ amu} \times \frac{1.6605 \times 10^{-27}\text{ kg}}{1\text{ amu}} \times \left(2.998 \times 10^8\text{ m/s}\right)^2\\ &= 7.913 \times 10^{-11}\text{ kg·m/s}^2\\ &= 7.913 \times 10^{-11}\text{J}\end{aligned}$$

We then convert the binding energy in joules per nucleus into units of MeV per nuclide:

$$7.913 \times 10^{-11}\text{ J} \times \frac{1\text{ MeV}}{1.602 \times 10^{-13}\text{ J}} = 493.9\text{ MeV}$$

Finally, we determine the binding energy per nucleon by dividing the total nuclear binding energy by the number of nucleons in the atom:

$$\text{Binding energy per nucleon} = \frac{493.9\text{ MeV}}{56} = 8.820\text{ MeV/nucleon}$$

Note that this is almost 25% larger than the binding energy per nucleon for ${}^{4}_{2}\text{He}$.

(Note also that this is the same process as in Example 21.1, but with the additional step of dividing the total nuclear binding energy by the number of nucleons.)

Check Your Learning

What is the binding energy per nucleon in ${}^{19}_{9}\text{F}$ (atomic mass, 18.9984 amu)?

Answer:

7.810 MeV/nucleon

21.2 Nuclear Equations

LEARNING OBJECTIVES

By the end of this section, you will be able to:

- Identify common particles and energies involved in nuclear reactions
- Write and balance nuclear equations

Changes of nuclei that result in changes in their atomic numbers, mass numbers, or energy states are **nuclear reactions**. To describe a nuclear reaction, we use an equation that identifies the nuclides involved in the reaction, their mass numbers and atomic numbers, and the other particles involved in the reaction.

Types of Particles in Nuclear Reactions

Many entities can be involved in nuclear reactions. The most common are protons, neutrons, alpha particles, beta particles, positrons, and gamma rays, as shown in Figure 21.4. Protons $\left({}_{1}^{1}\text{p}\right.$, also represented by the symbol $\left.{}_{1}^{1}\text{H}\right)$ and neutrons $\left({}_{0}^{1}\text{n}\right)$ are the constituents of atomic nuclei, and have been described previously. **Alpha particles** $\left({}_{2}^{4}\text{He}\right.$, also represented by the symbol $\left.{}_{2}^{4}\alpha\right)$ are high-energy helium nuclei. **Beta particles** $\left({}_{-1}^{0}\beta\right.$, also represented by the symbol $\left.{}_{-1}^{0}\text{e}\right)$ are high-energy electrons, and gamma rays are photons of very high-energy electromagnetic radiation. **Positrons** $\left({}_{+1}^{0}\text{e}\right.$, also represented by the symbol $\left.{}_{+1}^{0}\beta\right)$ are positively charged electrons ("anti-electrons"). The subscripts and superscripts are necessary for balancing nuclear equations, but are usually optional in other circumstances. For example, an alpha particle is a helium nucleus (He) with a charge of +2 and a mass number of 4, so it is symbolized ${}_{2}^{4}\text{He}$. This works because, in general, the ion charge is not important in the balancing of nuclear equations.

Name	Symbol(s)	Representation	Description
Alpha particle	${}_{2}^{4}\text{He}$ or ${}_{2}^{4}\alpha$		(High-energy) helium nuclei consisting of two protons and two neutrons
Beta particle	${}_{-1}^{0}\text{e}$ or ${}_{-1}^{0}\beta$		(High-energy) electrons
Positron	${}_{+1}^{0}\text{e}$ or ${}_{+1}^{0}\beta$		Particles with the same mass as an electron but with 1 unit of positive charge
Proton	${}_{1}^{1}\text{H}$ or ${}_{1}^{1}\text{p}$		Nuclei of hydrogen atoms
Neutron	${}_{0}^{1}\text{n}$		Particles with a mass approximately equal to that of a proton but with no charge
Gamma ray	γ	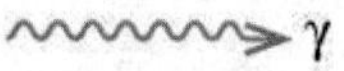 γ	Very high-energy electromagnetic radiation

FIGURE 21.4 Although many species are encountered in nuclear reactions, this table summarizes the names, symbols, representations, and descriptions of the most common of these.

Note that positrons are exactly like electrons, except they have the opposite charge. They are the most common example of **antimatter**, particles with the same mass but the opposite state of another property (for example, charge) than ordinary matter. When antimatter encounters ordinary matter, both are annihilated and their mass is converted into energy in the form of **gamma rays (γ)**—and other much smaller subnuclear particles, which are beyond the scope of this chapter—according to the mass-energy equivalence equation $E = mc^2$, seen in the preceding section. For example, when a positron and an electron collide, both are annihilated and two gamma ray photons are created:

$${}_{-1}^{0}\text{e} + {}_{+1}^{0}\text{e} \longrightarrow \gamma + \gamma$$

As seen in the chapter discussing light and electromagnetic radiation, gamma rays compose short wavelength, high-energy electromagnetic radiation and are (much) more energetic than better-known X-rays that can behave as particles in the wave-particle duality sense. Gamma rays are a type of high energy electromagnetic radiation produced when a nucleus undergoes a transition from a higher to a lower energy state, similar to how a photon is produced by an electronic transition from a higher to a lower energy level. Due to the much larger energy differences between nuclear energy shells, gamma rays emanating from a nucleus have energies that are typically millions of times larger than electromagnetic radiation emanating from electronic transitions.

Balancing Nuclear Reactions

A balanced chemical reaction equation reflects the fact that during a chemical reaction, bonds break and form, and atoms are rearranged, but the total numbers of atoms of each element are conserved and do not change. A balanced nuclear reaction equation indicates that there is a rearrangement during a nuclear reaction, but of nucleons (subatomic particles within the atoms' nuclei) rather than atoms. Nuclear reactions also follow conservation laws, and they are balanced in two ways:

1. The sum of the mass numbers of the reactants equals the sum of the mass numbers of the products.
2. The sum of the charges of the reactants equals the sum of the charges of the products.

If the atomic number and the mass number of all but one of the particles in a nuclear reaction are known, we can identify the particle by balancing the reaction. For instance, we could determine that ${}^{17}_{8}O$ is a product of the nuclear reaction of ${}^{14}_{7}N$ and ${}^{4}_{2}He$ if we knew that a proton, ${}^{1}_{1}H$, was one of the two products. Example 21.4 shows how we can identify a nuclide by balancing the nuclear reaction.

 **EXAMPLE 21.4**

Balancing Equations for Nuclear Reactions

The reaction of an α particle with magnesium-25 (${}^{25}_{12}Mg$) produces a proton and a nuclide of another element. Identify the new nuclide produced.

Solution

The nuclear reaction can be written as:

$${}^{25}_{12}Mg + {}^{4}_{2}He \longrightarrow {}^{1}_{1}H + {}^{A}_{Z}X$$

where A is the mass number and Z is the atomic number of the new nuclide, X. Because the sum of the mass numbers of the reactants must equal the sum of the mass numbers of the products:

$$25 + 4 = A + 1, \text{ or } A = 28$$

Similarly, the charges must balance, so:

$$12 + 2 = Z + 1, \text{ and } Z = 13$$

Check the periodic table: The element with nuclear charge = +13 is aluminum. Thus, the product is ${}^{28}_{13}Al$.

Check Your Learning

The nuclide ${}^{125}_{53}I$ combines with an electron and produces a new nucleus and no other massive particles. What is the equation for this reaction?

Answer:

${}^{125}_{53}I + {}^{0}_{-1}e \longrightarrow {}^{125}_{52}Te$

Following are the equations of several nuclear reactions that have important roles in the history of nuclear chemistry:

- The first naturally occurring unstable element that was isolated, polonium, was discovered by the Polish scientist Marie Curie and her husband Pierre in 1898. It decays, emitting α particles:
$${}^{212}_{84}Po \longrightarrow {}^{208}_{82}Pb + {}^{4}_{2}He$$
- The first nuclide to be prepared by artificial means was an isotope of oxygen, ${}^{17}O$. It was made by Ernest Rutherford in 1919 by bombarding nitrogen atoms with α particles:
$${}^{14}_{7}N + {}^{4}_{2}He \longrightarrow {}^{17}_{8}O + {}^{1}_{1}H$$
- James Chadwick discovered the neutron in 1932, as a previously unknown neutral particle produced along with ${}^{12}C$ by the nuclear reaction between ${}^{9}Be$ and ${}^{4}He$:

$$^{9}_{4}\text{Be} + ^{4}_{2}\text{He} \longrightarrow ^{12}_{6}\text{C} + ^{1}_{0}\text{n}$$

- The first element to be prepared that does not occur naturally on the earth, technetium, was created by bombardment of molybdenum by deuterons (heavy hydrogen, $^{2}_{1}\text{H}$), by Emilio Segre and Carlo Perrier in 1937:
$$^{2}_{1}\text{H} + ^{97}_{42}\text{Mo} \longrightarrow 2^{1}_{0}\text{n} + ^{97}_{43}\text{Tc}$$
- The first controlled nuclear chain reaction was carried out in a reactor at the University of Chicago in 1942. One of the many reactions involved was:
$$^{235}_{92}\text{U} + ^{1}_{0}\text{n} \longrightarrow ^{87}_{35}\text{Br} + ^{146}_{57}\text{La} + 3^{1}_{0}\text{n}$$

21.3 Radioactive Decay

LEARNING OBJECTIVES

By the end of this section, you will be able to:

- Recognize common modes of radioactive decay
- Identify common particles and energies involved in nuclear decay reactions
- Write and balance nuclear decay equations
- Calculate kinetic parameters for decay processes, including half-life
- Describe common radiometric dating techniques

Following the somewhat serendipitous discovery of radioactivity by Becquerel, many prominent scientists began to investigate this new, intriguing phenomenon. Among them were Marie Curie (the first woman to win a Nobel Prize, and the only person to win two Nobel Prizes in different sciences—chemistry and physics), who was the first to coin the term "radioactivity," and Ernest Rutherford (of gold foil experiment fame), who investigated and named three of the most common types of radiation. During the beginning of the twentieth century, many radioactive substances were discovered, the properties of radiation were investigated and quantified, and a solid understanding of radiation and nuclear decay was developed.

The spontaneous change of an unstable nuclide into another is **radioactive decay**. The unstable nuclide is called the **parent nuclide**; the nuclide that results from the decay is known as the **daughter nuclide**. The daughter nuclide may be stable, or it may decay itself. The radiation produced during radioactive decay is such that the daughter nuclide lies closer to the band of stability than the parent nuclide, so the location of a nuclide relative to the band of stability can serve as a guide to the kind of decay it will undergo (Figure 21.5).

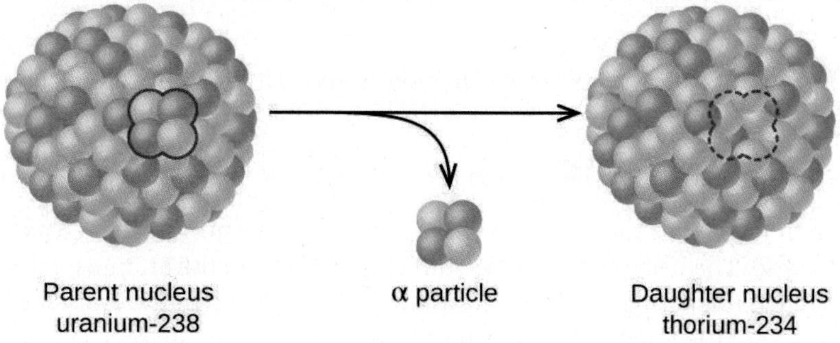

FIGURE 21.5 A nucleus of uranium-238 (the parent nuclide) undergoes α decay to form thorium-234 (the daughter nuclide). The alpha particle removes two protons (green) and two neutrons (gray) from the uranium-238 nucleus.

LINK TO LEARNING

Although the radioactive decay of a nucleus is too small to see with the naked eye, we can indirectly view radioactive decay in an environment called a cloud chamber. Click here (http://openstax.org/l/16cloudchamb) to learn about cloud chambers and to view an interesting Cloud Chamber Demonstration from the Jefferson Lab.

Types of Radioactive Decay

Ernest Rutherford's experiments involving the interaction of radiation with a magnetic or electric field (Figure 21.6) helped him determine that one type of radiation consisted of positively charged and relatively massive α particles; a second type was made up of negatively charged and much less massive β particles; and a third was uncharged electromagnetic waves, γ rays. We now know that α particles are high-energy helium nuclei, β particles are high-energy electrons, and γ radiation compose high-energy electromagnetic radiation. We classify different types of radioactive decay by the radiation produced.

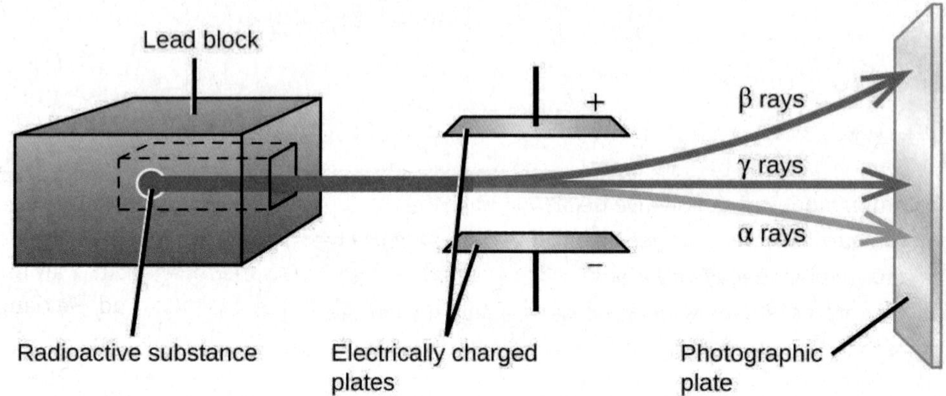

FIGURE 21.6 Alpha particles, which are attracted to the negative plate and deflected by a relatively small amount, must be positively charged and relatively massive. Beta particles, which are attracted to the positive plate and deflected a relatively large amount, must be negatively charged and relatively light. Gamma rays, which are unaffected by the electric field, must be uncharged.

Alpha (α) decay is the emission of an α particle from the nucleus. For example, polonium-210 undergoes α decay:

$$^{210}_{84}\text{Po} \longrightarrow {}^{4}_{2}\text{He} + {}^{206}_{82}\text{Pb} \qquad \text{or} \qquad ^{210}_{84}\text{Po} \longrightarrow {}^{4}_{2}\alpha + {}^{206}_{82}\text{Pb}$$

Alpha decay occurs primarily in heavy nuclei (A > 200, Z > 83). Because the loss of an α particle gives a daughter nuclide with a mass number four units smaller and an atomic number two units smaller than those of the parent nuclide, the daughter nuclide has a larger n:p ratio than the parent nuclide. If the parent nuclide undergoing α decay lies below the band of stability (refer to Figure 21.2), the daughter nuclide will lie closer to the band.

Beta (β) decay is the emission of an electron from a nucleus. Iodine-131 is an example of a nuclide that undergoes β decay:

$$^{131}_{53}\text{I} \longrightarrow {}^{0}_{-1}\text{e} + {}^{131}_{54}\text{Xe} \qquad \text{or} \qquad ^{131}_{53}\text{I} \longrightarrow {}^{0}_{-1}\beta + {}^{131}_{54}\text{Xe}$$

Beta decay, which can be thought of as the conversion of a neutron into a proton and a β particle, is observed in nuclides with a large n:p ratio. The beta particle (electron) emitted is from the atomic nucleus and is not one of the electrons surrounding the nucleus. Such nuclei lie above the band of stability. Emission of an electron does not change the mass number of the nuclide but does increase the number of its protons and decrease the number of its neutrons. Consequently, the n:p ratio is decreased, and the daughter nuclide lies closer to the band of stability than did the parent nuclide.

Gamma emission (γ emission) is observed when a nuclide is formed in an excited state and then decays to its ground state with the emission of a γ ray, a quantum of high-energy electromagnetic radiation. The presence of a nucleus in an excited state is often indicated by an asterisk (*). Cobalt-60 emits γ radiation and is used in many applications including cancer treatment:

$$^{60}_{27}\text{Co}^{*} \longrightarrow {}^{0}_{0}\gamma + {}^{60}_{27}\text{Co}$$

There is no change in mass number or atomic number during the emission of a γ ray unless the γ emission accompanies one of the other modes of decay.

Positron emission (β^+ decay) is the emission of a positron from the nucleus. Oxygen-15 is an example of a nuclide that undergoes positron emission:

$$^{15}_{8}\text{O} \longrightarrow {}^{0}_{+1}\text{e} + {}^{15}_{7}\text{N} \qquad \text{or} \qquad {}^{15}_{8}\text{O} \longrightarrow {}^{0}_{+1}\beta + {}^{15}_{7}\text{N}$$

Positron emission is observed for nuclides in which the n:p ratio is low. These nuclides lie below the band of stability. Positron decay is the conversion of a proton into a neutron with the emission of a positron. The n:p ratio increases, and the daughter nuclide lies closer to the band of stability than did the parent nuclide.

Electron capture occurs when one of the inner electrons in an atom is captured by the atom's nucleus. For example, potassium-40 undergoes electron capture:

$$^{40}_{19}\text{K} + {}^{0}_{-1}\text{e} \longrightarrow {}^{40}_{18}\text{Ar}$$

Electron capture occurs when an inner shell electron combines with a proton and is converted into a neutron. The loss of an inner shell electron leaves a vacancy that will be filled by one of the outer electrons. As the outer electron drops into the vacancy, it will emit energy. In most cases, the energy emitted will be in the form of an X-ray. Like positron emission, electron capture occurs for "proton-rich" nuclei that lie below the band of stability. Electron capture has the same effect on the nucleus as does positron emission: The atomic number is decreased by one and the mass number does not change. This increases the n:p ratio, and the daughter nuclide lies closer to the band of stability than did the parent nuclide. Whether electron capture or positron emission occurs is difficult to predict. The choice is primarily due to kinetic factors, with the one requiring the smaller activation energy being the one more likely to occur.

Figure 21.7 summarizes these types of decay, along with their equations and changes in atomic and mass numbers.

Type	Nuclear equation	Representation	Change in mass/atomic numbers
Alpha decay	$^{A}_{Z}\text{X} \rightarrow {}^{4}_{2}\text{He} + {}^{A-4}_{Z-2}\text{Y}$	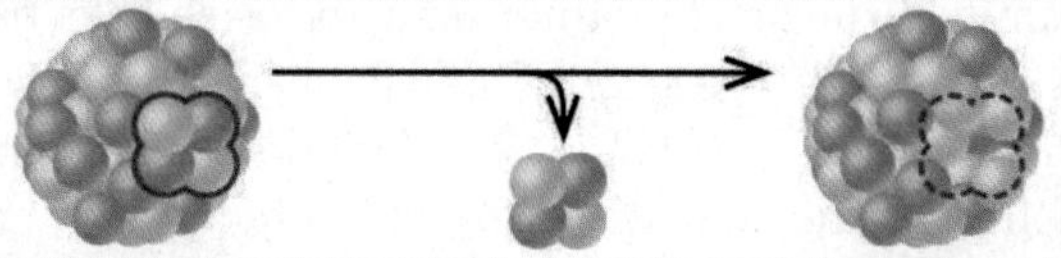	A: decrease by 4 Z: decrease by 2
Beta decay	$^{A}_{Z}\text{X} \rightarrow {}^{0}_{-1}\text{e} + {}^{A}_{Z+1}\text{Y}$	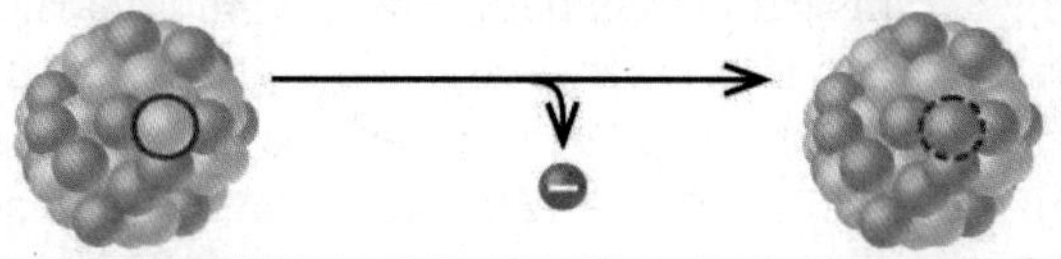	A: unchanged Z: increase by 1
Gamma decay	$^{A}_{Z}\text{X} \rightarrow {}^{0}_{0}\gamma + {}^{A}_{Z}\text{Y}$	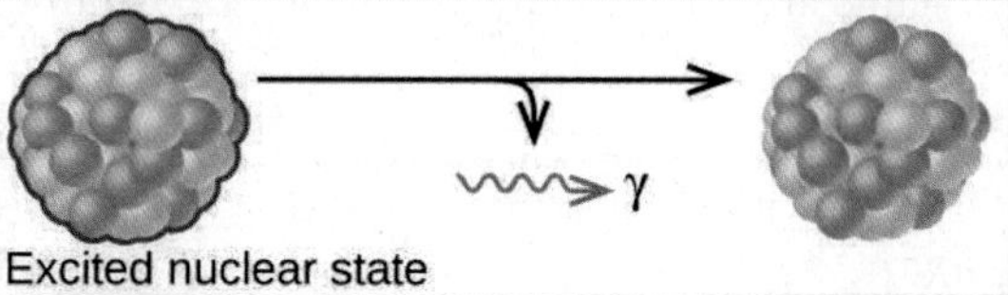 γ Excited nuclear state	A: unchanged Z: unchanged
Positron emission	$^{A}_{Z}\text{X} \rightarrow {}^{0}_{+1}\text{e} + {}^{A}_{Y-1}\text{Y}$	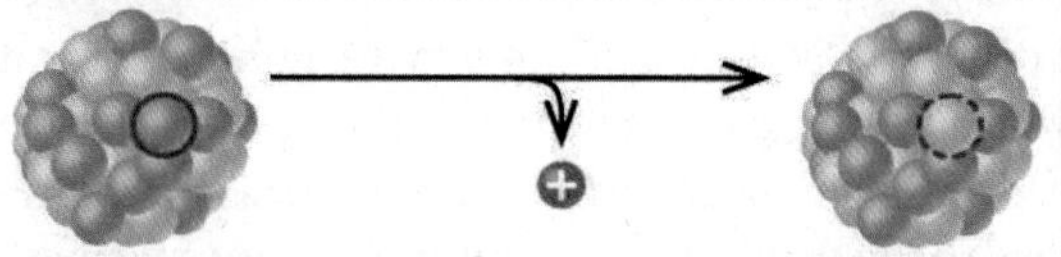	A: unchanged Z: decrease by 1
Electron capture	$^{A}_{Z}\text{X} \rightarrow {}^{0}_{-1}\text{e} + {}^{A}_{Y-1}\text{Y}$	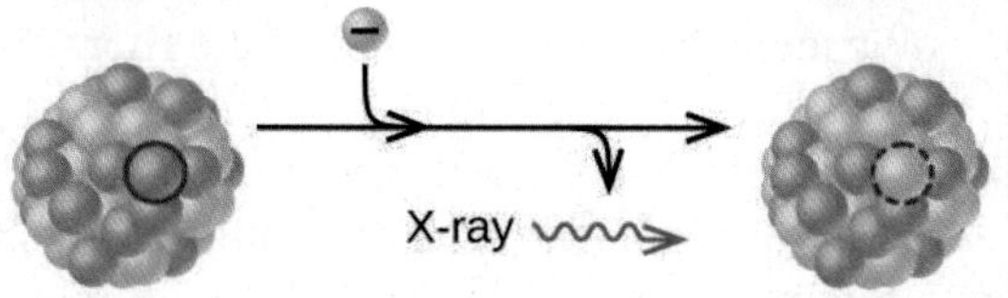 X-ray	A: unchanged Z: decrease by 1

FIGURE 21.7 This table summarizes the type, nuclear equation, representation, and any changes in the mass or atomic numbers for various types of decay.

Chemistry in Everyday Life

PET Scan

Positron emission tomography (PET) scans use radiation to diagnose and track health conditions and monitor medical treatments by revealing how parts of a patient's body function (Figure 21.8). To perform a PET scan, a positron-emitting radioisotope is produced in a cyclotron and then attached to a substance that is used by the part of the body being investigated. This "tagged" compound, or radiotracer, is then put into the patient (injected via IV or breathed in as a gas), and how it is used by the tissue reveals how that organ or other area of the body functions.

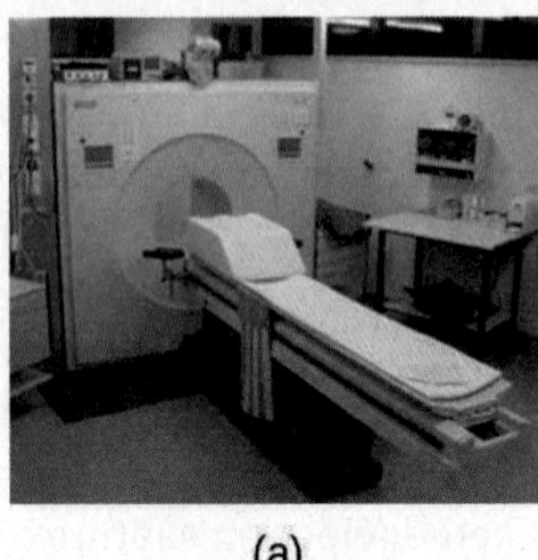
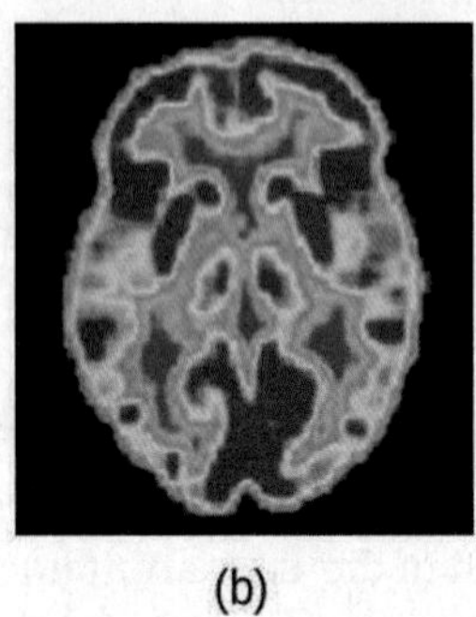
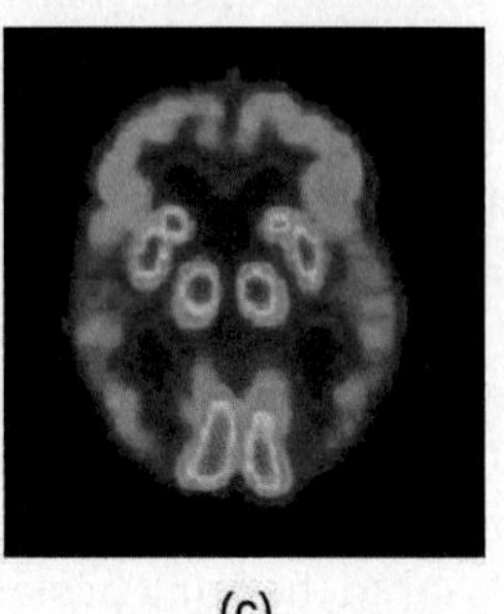

(a) (b) (c)

FIGURE 21.8 A PET scanner (a) uses radiation to provide an image of how part of a patient's body functions. The scans it produces can be used to image a healthy brain (b) or can be used for diagnosing medical conditions such as Alzheimer's disease (c). (credit a: modification of work by Jens Maus)

For example, F-18 is produced by proton bombardment of ^{18}O ($^{18}_{8}O + ^{1}_{1}p \longrightarrow ^{18}_{9}F + ^{1}_{0}n$) and incorporated into a glucose analog called fludeoxyglucose (FDG). How FDG is used by the body provides critical diagnostic information; for example, since cancers use glucose differently than normal tissues, FDG can reveal cancers. The ^{18}F emits positrons that interact with nearby electrons, producing a burst of gamma radiation. This energy is detected by the scanner and converted into a detailed, three-dimensional, color image that shows how that part of the patient's body functions. Different levels of gamma radiation produce different amounts of brightness and colors in the image, which can then be interpreted by a radiologist to reveal what is going on. PET scans can detect heart damage and heart disease, help diagnose Alzheimer's disease, indicate the part of a brain that is affected by epilepsy, reveal cancer, show what stage it is, and how much it has spread, and whether treatments are effective. Unlike magnetic resonance imaging and X-rays, which only show how something looks, the big advantage of PET scans is that they show how something functions. PET scans are now usually performed in conjunction with a computed tomography scan.

Radioactive Decay Series

The naturally occurring radioactive isotopes of the heaviest elements fall into chains of successive disintegrations, or decays, and all the species in one chain constitute a radioactive family, or **radioactive decay series**. Three of these series include most of the naturally radioactive elements of the periodic table. They are the uranium series, the actinide series, and the thorium series. The neptunium series is a fourth series, which is no longer significant on the earth because of the short half-lives of the species involved. Each series is characterized by a parent (first member) that has a long half-life and a series of daughter nuclides that ultimately lead to a stable end-product—that is, a nuclide on the band of stability (Figure 21.9). In all three series, the end-product is a stable isotope of lead. The neptunium series, previously thought to terminate with bismuth-209, terminates with thallium-205.

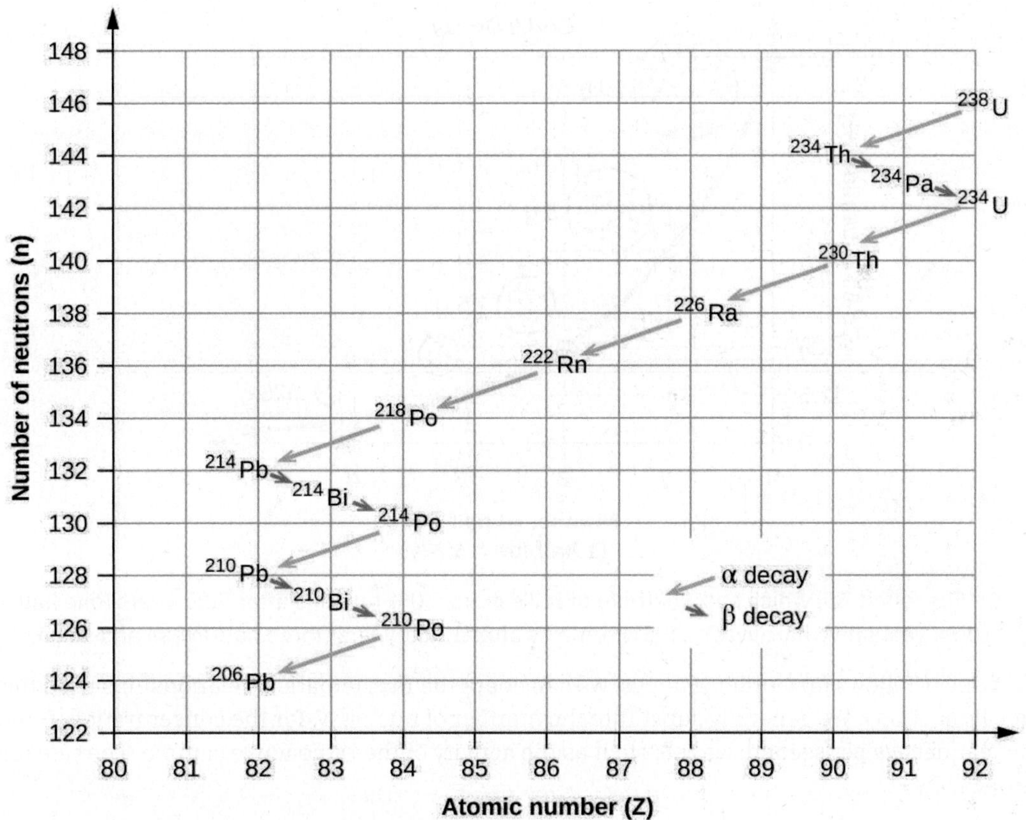

FIGURE 21.9 Uranium-238 undergoes a radioactive decay series consisting of 14 separate steps before producing stable lead-206. This series consists of eight α decays and six β decays.

Radioactive Half-Lives

Radioactive decay follows first-order kinetics. Since first-order reactions have already been covered in detail in the kinetics chapter, we will now apply those concepts to nuclear decay reactions. Each radioactive nuclide has a characteristic, constant **half-life** ($t_{1/2}$), the time required for half of the atoms in a sample to decay. An isotope's half-life allows us to determine how long a sample of a useful isotope will be available, and how long a sample of an undesirable or dangerous isotope must be stored before it decays to a low-enough radiation level that is no longer a problem.

For example, cobalt-60, an isotope that emits gamma rays used to treat cancer, has a half-life of 5.27 years (Figure 21.10). In a given cobalt-60 source, since half of the ${}^{60}_{27}Co$ nuclei decay every 5.27 years, both the amount of material and the intensity of the radiation emitted is cut in half every 5.27 years. (Note that for a given substance, the intensity of radiation that it produces is directly proportional to the rate of decay of the substance and the amount of the substance.) This is as expected for a process following first-order kinetics. Thus, a cobalt-60 source that is used for cancer treatment must be replaced regularly to continue to be effective.

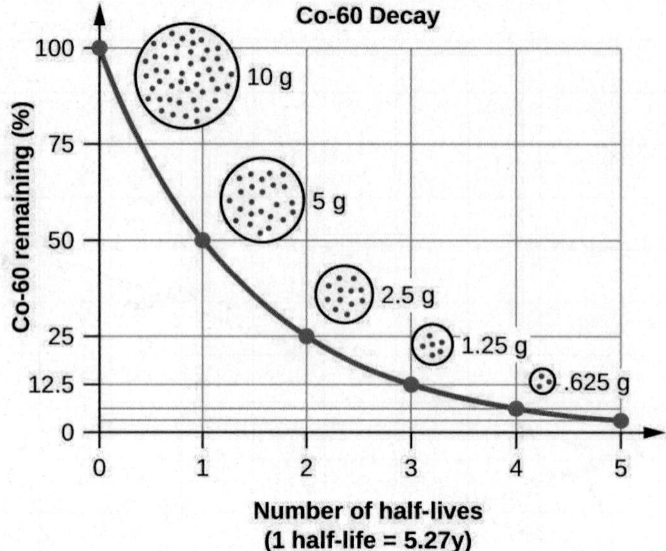

FIGURE 21.10 For cobalt-60, which has a half-life of 5.27 years, 50% remains after 5.27 years (one half-life), 25% remains after 10.54 years (two half-lives), 12.5% remains after 15.81 years (three half-lives), and so on.

Since nuclear decay follows first-order kinetics, we can adapt the mathematical relationships used for first-order chemical reactions. We generally substitute the number of nuclei, *N*, for the concentration. If the rate is stated in nuclear decays per second, we refer to it as the activity of the radioactive sample. The rate for radioactive decay is:

decay rate = λN with λ = the decay constant for the particular radioisotope

The decay constant, λ, which is the same as a rate constant discussed in the kinetics chapter. It is possible to express the decay constant in terms of the half-life, $t_{1/2}$:

$$\lambda = \frac{\ln 2}{t_{1/2}} = \frac{0.693}{t_{1/2}} \qquad \text{or} \qquad t_{1/2} = \frac{\ln 2}{\lambda} = \frac{0.693}{\lambda}$$

The first-order equations relating amount, *N*, and time are:

$$N_t = N_0 e^{-\lambda t} \qquad \text{or} \qquad t = -\frac{1}{\lambda} \ln \left(\frac{N_t}{N_0} \right)$$

where N_0 is the initial number of nuclei or moles of the isotope, and N_t is the number of nuclei/moles remaining at time *t*. Example 21.5 applies these calculations to find the rates of radioactive decay for specific nuclides.

EXAMPLE 21.5

Rates of Radioactive Decay

${}^{60}_{27}\text{Co}$ decays with a half-life of 5.27 years to produce ${}^{60}_{28}\text{Ni}$.

(a) What is the decay constant for the radioactive disintegration of cobalt-60?

(b) Calculate the fraction of a sample of the ${}^{60}_{27}\text{Co}$ isotope that will remain after 15 years.

(c) How long does it take for a sample of ${}^{60}_{27}\text{Co}$ to disintegrate to the extent that only 2.0% of the original amount remains?

Solution

(a) The value of the rate constant is given by:

$$\lambda = \frac{\ln 2}{t_{1/2}} = \frac{0.693}{5.27\ \text{y}} = 0.132\ \text{y}^{-1}$$

(b) The fraction of ${}^{60}_{27}\text{Co}$ that is left after time t is given by $\frac{N_t}{N_0}$. Rearranging the first-order relationship $N_t = N_0 e^{-\lambda t}$ to solve for this ratio yields:

$$\frac{N_t}{N_0} = e^{-\lambda t} = e^{-(0.132/\text{y})(15\times\text{y})} = 0.138$$

The fraction of ${}^{60}_{27}\text{Co}$ that will remain after 15.0 years is 0.138. Or put another way, 13.8% of the ${}^{60}_{27}\text{Co}$ originally present will remain after 15 years.

(c) 2.00% of the original amount of ${}^{60}_{27}\text{Co}$ is equal to $0.0200 \times N_0$. Substituting this into the equation for time for first-order kinetics, we have:

$$t = -\frac{1}{\lambda}\ln\left(\frac{N_t}{N_0}\right) = -\frac{1}{0.132\ \text{y}^{-1}}\ln\left(\frac{0.0200 \times N_0}{N_0}\right) = 29.6\ \text{y}$$

Check Your Learning

Radon-222, ${}^{222}_{86}\text{Rn}$, has a half-life of 3.823 days. How long will it take a sample of radon-222 with a mass of 0.750 g to decay into other elements, leaving only 0.100 g of radon-222?

Answer:

11.1 days

Because each nuclide has a specific number of nucleons, a particular balance of repulsion and attraction, and its own degree of stability, the half-lives of radioactive nuclides vary widely. For example: the half-life of ${}^{209}_{83}\text{Bi}$ is 1.9×10^{19} years; ${}^{239}_{94}\text{Ra}$ is 24,000 years; ${}^{222}_{86}\text{Rn}$ is 3.82 days; and element-111 (Rg for roentgenium) is 1.5×10^{-3} seconds. The half-lives of a number of radioactive isotopes important to medicine are shown in Table 21.2, and others are listed in Appendix M.

Half-lives of Radioactive Isotopes Important to Medicine

Type[1]	Decay Mode	Half-Life	Uses
F-18	β^+ decay	110. minutes	PET scans
Co-60	β decay, γ decay	5.27 years	cancer treatment
Tc-99m	γ decay	8.01 hours	scans of brain, lung, heart, bone
I-131	β decay	8.02 days	thyroid scans and treatment
Tl-201	electron capture	73 hours	heart and arteries scans; cardiac stress tests

TABLE 21.2

Radiometric Dating

Several radioisotopes have half-lives and other properties that make them useful for purposes of "dating" the origin of objects such as archaeological artifacts, formerly living organisms, or geological formations. This process is **radiometric dating** and has been responsible for many breakthrough scientific discoveries about

1 The "m" in Tc-99m stands for "metastable," indicating that this is an unstable, high-energy state of Tc-99. Metastable isotopes emit γ radiation to rid themselves of excess energy and become (more) stable.

the geological history of the earth, the evolution of life, and the history of human civilization. We will explore some of the most common types of radioactive dating and how the particular isotopes work for each type.

Radioactive Dating Using Carbon-14

The radioactivity of carbon-14 provides a method for dating objects that were a part of a living organism. This method of radiometric dating, which is also called **radiocarbon dating** or carbon-14 dating, is accurate for dating carbon-containing substances that are up to about 30,000 years old, and can provide reasonably accurate dates up to a maximum of about 50,000 years old.

Naturally occurring carbon consists of three isotopes: ${}^{12}_{6}C$, which constitutes about 99% of the carbon on earth; ${}^{13}_{6}C$, about 1% of the total; and trace amounts of ${}^{14}_{6}C$. Carbon-14 forms in the upper atmosphere by the reaction of nitrogen atoms with neutrons from cosmic rays in space:

$$ {}^{14}_{7}N + {}^{1}_{0}n \longrightarrow {}^{14}_{6}C + {}^{1}_{1}H $$

All isotopes of carbon react with oxygen to produce CO_2 molecules. The ratio of ${}^{14}_{6}CO_2$ to ${}^{12}_{6}CO_2$ depends on the ratio of ${}^{14}_{6}CO$ to ${}^{12}_{6}CO$ in the atmosphere. The natural abundance of ${}^{14}_{6}CO$ in the atmosphere is approximately 1 part per trillion; until recently, this has generally been constant over time, as seen is gas samples found trapped in ice. The incorporation of ${}^{14}_{6}C{}^{14}_{6}CO_2$ and ${}^{12}_{6}CO_2$ into plants is a regular part of the photosynthesis process, which means that the ${}^{14}_{6}C$: ${}^{12}_{6}C$ ratio found in a living plant is the same as the ${}^{14}_{6}C$: ${}^{12}_{6}C$ ratio in the atmosphere. But when the plant dies, it no longer traps carbon through photosynthesis. Because ${}^{12}_{6}C$ is a stable isotope and does not undergo radioactive decay, its concentration in the plant does not change. However, carbon-14 decays by β emission with a half-life of 5730 years:

$$ {}^{14}_{6}C \longrightarrow {}^{14}_{7}N + {}^{0}_{-1}e $$

Thus, the ${}^{14}_{6}C$: ${}^{12}_{6}C$ ratio gradually decreases after the plant dies. The decrease in the ratio with time provides a measure of the time that has elapsed since the death of the plant (or other organism that ate the plant). Figure 21.11 visually depicts this process.

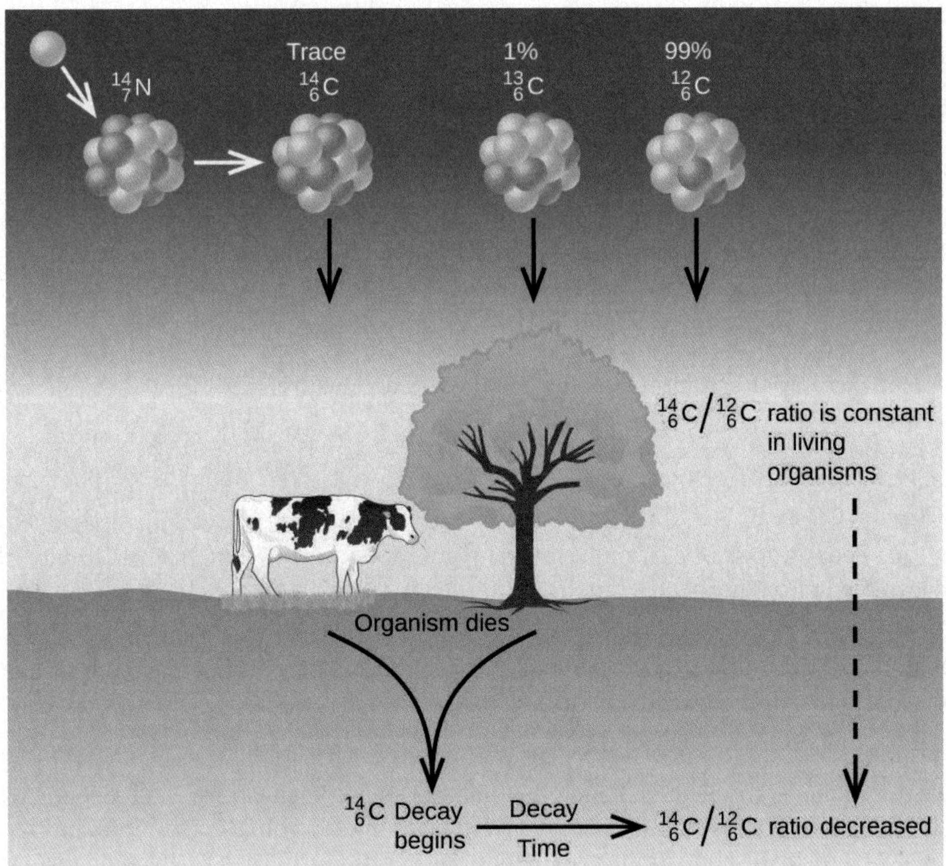

FIGURE 21.11 Along with stable carbon-12, radioactive carbon-14 is taken in by plants and animals, and remains at a constant level within them while they are alive. After death, the C-14 decays and the C-14:C-12 ratio in the remains decreases. Comparing this ratio to the C-14:C-12 ratio in living organisms allows us to determine how long ago the organism lived (and died).

For example, with the half-life of $^{14}_{6}C$ being 5730 years, if the $^{14}_{6}C$: $^{12}_{6}C$ ratio in a wooden object found in an archaeological dig is half what it is in a living tree, this indicates that the wooden object is 5730 years old. Highly accurate determinations of $^{14}_{6}C$: $^{12}_{6}C$ ratios can be obtained from very small samples (as little as a milligram) by the use of a mass spectrometer.

LINK TO LEARNING

Visit this website (http://openstax.org/l/16phetradiom) to perform simulations of radiometric dating.

EXAMPLE 21.6

Radiocarbon Dating

A tiny piece of paper (produced from formerly living plant matter) taken from the Dead Sea Scrolls has an activity of 10.8 disintegrations per minute per gram of carbon. If the initial C-14 activity was 13.6 disintegrations/min/g of C, estimate the age of the Dead Sea Scrolls.

Solution

The rate of decay (number of disintegrations/minute/gram of carbon) is proportional to the amount of radioactive C-14 left in the paper, so we can substitute the rates for the amounts, *N*, in the relationship:

$$t = -\frac{1}{\lambda}\ln\left(\frac{N_t}{N_0}\right) \longrightarrow t = -\frac{1}{\lambda}\ln\left(\frac{\text{Rate}_t}{\text{Rate}_0}\right)$$

where the subscript 0 represents the time when the plants were cut to make the paper, and the subscript t represents the current time.

The decay constant can be determined from the half-life of C-14, 5730 years:

$$\lambda = \frac{\ln 2}{t_{1/2}} = \frac{0.693}{5730\text{ y}} = 1.21 \times 10^{-4}\text{ y}^{-1}$$

Substituting and solving, we have:

$$t = -\frac{1}{\lambda}\ln\left(\frac{\text{Rate}_t}{\text{Rate}_0}\right) = -\frac{1}{1.21 \times 10^{-4}\text{ y}^{-1}}\ln\left(\frac{10.8\text{ dis/min/g C}}{13.6\text{ dis/min/g C}}\right) = 1910\text{ y}$$

Therefore, the Dead Sea Scrolls are approximately 1900 years old (Figure 21.12).

FIGURE 21.12 Carbon-14 dating has shown that these pages from the Dead Sea Scrolls were written or copied on paper made from plants that died between 100 BC and AD 50.

Check Your Learning

More accurate dates of the reigns of ancient Egyptian pharaohs have been determined recently using plants that were preserved in their tombs. Samples of seeds and plant matter from King Tutankhamun's tomb have a C-14 decay rate of 9.07 disintegrations/min/g of C. How long ago did King Tut's reign come to an end?

Answer:

about 3350 years ago, or approximately 1340 BC

There have been some significant, well-documented changes to the ${}^{14}_{6}\text{C}$: ${}^{12}_{6}\text{C}$ ratio. The accuracy of a straightforward application of this technique depends on the ${}^{14}_{6}\text{C}$: ${}^{12}_{6}\text{C}$ ratio in a living plant being the same now as it was in an earlier era, but this is not always valid. Due to the increasing accumulation of CO_2 molecules (largely ${}^{12}_{6}\text{CO}_2$) in the atmosphere caused by combustion of fossil fuels (in which essentially all of the ${}^{14}_{6}\text{C}$ has decayed), the ratio of ${}^{14}_{6}\text{C}$: ${}^{12}_{6}\text{C}$ in the atmosphere may be changing. This manmade increase in ${}^{12}_{6}\text{CO}_2$ in the atmosphere causes the ${}^{14}_{6}\text{C}$: ${}^{12}_{6}\text{C}$ ratio to decrease, and this in turn affects the ratio in currently living organisms on the earth. Fortunately, however, we can use other data, such as tree dating via examination of annual growth rings, to calculate correction factors. With these correction factors, accurate dates can be determined. In general, radioactive dating only works for about 10 half-lives; therefore, the limit for carbon-14 dating is about 57,000 years.

Radioactive Dating Using Nuclides Other than Carbon-14

Radioactive dating can also use other radioactive nuclides with longer half-lives to date older events. For example, uranium-238 (which decays in a series of steps into lead-206) can be used for establishing the age of rocks (and the approximate age of the oldest rocks on earth). Since U-238 has a half-life of 4.5 billion years, it takes that amount of time for half of the original U-238 to decay into Pb-206. In a sample of rock that does not contain appreciable amounts of Pb-208, the most abundant isotope of lead, we can assume that lead was not present when the rock was formed. Therefore, by measuring and analyzing the ratio of U-238:Pb-206, we can determine the age of the rock. This assumes that all of the lead-206 present came from the decay of uranium-238. If there is additional lead-206 present, which is indicated by the presence of other lead isotopes in the sample, it is necessary to make an adjustment. Potassium-argon dating uses a similar method. K-40 decays by positron emission and electron capture to form Ar-40 with a half-life of 1.25 billion years. If a rock sample is crushed and the amount of Ar-40 gas that escapes is measured, determination of the Ar-40:K-40 ratio yields the age of the rock. Other methods, such as rubidium-strontium dating (Rb-87 decays into Sr-87 with a half-life of 48.8 billion years), operate on the same principle. To estimate the lower limit for the earth's age, scientists determine the age of various rocks and minerals, making the assumption that the earth is older than the oldest rocks and minerals in its crust. As of 2014, the oldest known rocks on earth are the Jack Hills zircons from Australia, found by uranium-lead dating to be almost 4.4 billion years old.

EXAMPLE 21.7

Radioactive Dating of Rocks

An igneous rock contains 9.58×10^{-5} g of U-238 and 2.51×10^{-5} g of Pb-206, and much, much smaller amounts of Pb-208. Determine the approximate time at which the rock formed.

Solution

The sample of rock contains very little Pb-208, the most common isotope of lead, so we can safely assume that all the Pb-206 in the rock was produced by the radioactive decay of U-238. When the rock formed, it contained all of the U-238 currently in it, plus some U-238 that has since undergone radioactive decay.

The amount of U-238 currently in the rock is:

$$9.58 \times 10^{-5} \cancel{\text{g U}} \times \left(\frac{1 \text{ mol U}}{238 \cancel{\text{g U}}}\right) = 4.03 \times 10^{-7} \text{ mol U}$$

Because when one mole of U-238 decays, it produces one mole of Pb-206, the amount of U-238 that has undergone radioactive decay since the rock was formed is:

$$2.51 \times 10^{-5} \cancel{\text{g Pb}} \times \left(\frac{1 \cancel{\text{mol Pb}}}{206 \cancel{\text{g Pb}}}\right) \times \left(\frac{1 \text{ mol U}}{1 \cancel{\text{mol Pb}}}\right) = 1.22 \times 10^{-7} \text{ mol U}$$

The total amount of U-238 originally present in the rock is therefore:

$$4.03 \times 10^{-7} \text{ mol} + 1.22 \times 10^{-7} \text{ mol} = 5.25 \times 10^{-7} \text{ mol U}$$

The amount of time that has passed since the formation of the rock is given by:

$$t = -\frac{1}{\lambda} \ln\left(\frac{N_t}{N_0}\right)$$

with N_0 representing the original amount of U-238 and N_t representing the present amount of U-238.

U-238 decays into Pb-206 with a half-life of 4.5×10^9 y, so the decay constant λ is:

$$\lambda = \frac{\ln 2}{t_{1/2}} = \frac{0.693}{4.5 \times 10^9 \text{ y}} = 1.54 \times 10^{-10} \text{ y}^{-1}$$

Substituting and solving, we have:

$$t = -\frac{1}{1.54 \times 10^{-10} \text{ y}^{-1}} \ln\left(\frac{4.03 \times 10^{-7} \cancel{\text{mol U}}}{5.25 \times 10^{-7} \cancel{\text{mol U}}}\right) = 1.7 \times 10^9 \text{ y}$$

Therefore, the rock is approximately 1.7 billion years old.

Check Your Learning

A sample of rock contains 6.14×10^{-4} g of Rb-87 and 3.51×10^{-5} g of Sr-87. Calculate the age of the rock. (The half-life of the β decay of Rb-87 is 4.7×10^{10} y.)

Answer:

3.7×10^9 y

21.4 Transmutation and Nuclear Energy

LEARNING OBJECTIVES

By the end of this section, you will be able to:

- Describe the synthesis of transuranium nuclides
- Explain nuclear fission and fusion processes
- Relate the concepts of critical mass and nuclear chain reactions
- Summarize basic requirements for nuclear fission and fusion reactors

After the discovery of radioactivity, the field of nuclear chemistry was created and developed rapidly during the early twentieth century. A slew of new discoveries in the 1930s and 1940s, along with World War II, combined to usher in the Nuclear Age in the mid-twentieth century. Scientists learned how to create new substances, and certain isotopes of certain elements were found to possess the capacity to produce unprecedented amounts of energy, with the potential to cause tremendous damage during war, as well as produce enormous amounts of power for society's needs during peace.

Synthesis of Nuclides

Nuclear transmutation is the conversion of one nuclide into another. It can occur by the radioactive decay of a nucleus, or the reaction of a nucleus with another particle. The first manmade nucleus was produced in Ernest Rutherford's laboratory in 1919 by a **transmutation** reaction, the bombardment of one type of nuclei with other nuclei or with neutrons. Rutherford bombarded nitrogen atoms with high-speed α particles from a natural radioactive isotope of radium and observed protons resulting from the reaction:

$$^{14}_{7}N + ^{4}_{2}He \longrightarrow ^{17}_{8}O + ^{1}_{1}H$$

The $^{17}_{8}O$ and $^{1}_{1}H$ nuclei that are produced are stable, so no further (nuclear) changes occur.

To reach the kinetic energies necessary to produce transmutation reactions, devices called **particle accelerators** are used. These devices use magnetic and electric fields to increase the speeds of nuclear particles. In all accelerators, the particles move in a vacuum to avoid collisions with gas molecules. When neutrons are required for transmutation reactions, they are usually obtained from radioactive decay reactions or from various nuclear reactions occurring in nuclear reactors. The Chemistry in Everyday Life feature that follows discusses a famous particle accelerator that made worldwide news.

Chemistry in Everyday Life

CERN Particle Accelerator

Located near Geneva, the CERN ("Conseil Européen pour la Recherche Nucléaire," or European Council for Nuclear Research) Laboratory is the world's premier center for the investigations of the fundamental particles that make up matter. It contains the 27-kilometer (17 mile) long, circular Large Hadron Collider (LHC), the largest particle accelerator in the world (Figure 21.13). In the LHC, particles are boosted to high energies and are then made to collide with each other or with stationary targets at nearly the speed of light. Superconducting electromagnets are used to produce a strong magnetic field that guides the particles around the ring. Specialized, purpose-built detectors observe and record the results of these collisions, which are then analyzed by CERN scientists using powerful computers.

FIGURE 21.13 A small section of the LHC is shown with workers traveling along it. (credit: Christophe Delaere)

In 2012, CERN announced that experiments at the LHC showed the first observations of the Higgs boson, an elementary particle that helps explain the origin of mass in fundamental particles. This long-anticipated discovery made worldwide news and resulted in the awarding of the 2013 Nobel Prize in Physics to François Englert and Peter Higgs, who had predicted the existence of this particle almost 50 years previously.

LINK TO LEARNING

Famous physicist Brian Cox talks about his work on the Large Hadron Collider at CERN, providing an entertaining and engaging tour (http://openstax.org/l/16tedCERN) of this massive project and the physics behind it.

View a short video (http://openstax.org/l/16CERNvideo) from CERN, describing the basics of how its particle accelerators work.

Prior to 1940, the heaviest-known element was uranium, whose atomic number is 92. Now, many artificial elements have been synthesized and isolated, including several on such a large scale that they have had a profound effect on society. One of these—element 93, neptunium (Np)—was first made in 1940 by McMillan and Abelson by bombarding uranium-238 with neutrons. The reaction creates unstable uranium-239, with a half-life of 23.5 minutes, which then decays into neptunium-239. Neptunium-239 is also radioactive, with a half-life of 2.36 days, and it decays into plutonium-239. The nuclear reactions are:

$$ {}^{238}_{92}\text{U} + {}^{1}_{0}\text{n} \longrightarrow {}^{239}_{92}\text{U} $$

$$ {}^{239}_{92}\text{U} \longrightarrow {}^{239}_{93}\text{Np} + {}^{0}_{-1}\text{e} \qquad \text{half-life} = 23.5\text{ min} $$

$$ {}^{239}_{93}\text{Np} \longrightarrow {}^{239}_{94}\text{Pu} + {}^{0}_{-1}\text{e} \qquad \text{half-life} = 2.36\text{ days} $$

Plutonium is now mostly formed in nuclear reactors as a byproduct during the fission of U-235. Additional neutrons are released during this fission process (see the next section), some of which combine with U-238 nuclei to form uranium-239; this undergoes β decay to form neptunium-239, which in turn undergoes β decay to form plutonium-239 as illustrated in the preceding three equations. These processes are summarized in the equation:

$$ {}^{238}_{92}\text{U} + {}^{1}_{0}\text{n} \longrightarrow {}^{239}_{92}\text{U} \xrightarrow{\beta^-} {}^{239}_{93}\text{Np} \xrightarrow{\beta^-} {}^{239}_{94}\text{Pu} $$

Heavier isotopes of plutonium—Pu-240, Pu-241, and Pu-242—are also produced when lighter plutonium nuclei capture neutrons. Some of this highly radioactive plutonium is used to produce military weapons, and the rest presents a serious storage problem because they have half-lives from thousands to hundreds of thousands of years.

Although they have not been prepared in the same quantity as plutonium, many other synthetic nuclei have been produced. Nuclear medicine has developed from the ability to convert atoms of one type into other types of atoms. Radioactive isotopes of several dozen elements are currently used for medical applications. The radiation produced by their decay is used to image or treat various organs or portions of the body, among other uses.

The elements beyond element 92 (uranium) are called **transuranium elements**. As of this writing, 22 transuranium elements have been produced and officially recognized by IUPAC; several other elements have formation claims that are waiting for approval. Some of these elements are shown in Table 21.3.

Preparation of Some of the Transuranium Elements

Name	Symbol	Atomic Number	Reaction
americium	Am	95	${}^{239}_{94}Pu + {}^{1}_{0}n \longrightarrow {}^{240}_{95}Am + {}^{0}_{-1}e$
curium	Cm	96	${}^{239}_{94}Pu + {}^{4}_{2}He \longrightarrow {}^{242}_{96}Cm + {}^{1}_{0}n$
californium	Cf	98	${}^{242}_{96}Cm + {}^{4}_{2}He \longrightarrow {}^{245}_{98}Cf + {}^{1}_{0}n$
einsteinium	Es	99	${}^{238}_{92}U + 15{}^{1}_{0}n \longrightarrow {}^{253}_{99}Es + 7{}^{0}_{-1}e$
mendelevium	Md	101	${}^{253}_{99}Es + {}^{4}_{2}He \longrightarrow {}^{256}_{101}Md + {}^{1}_{0}n$
nobelium	No	102	${}^{246}_{96}Cm + {}^{12}_{6}C \longrightarrow {}^{254}_{102}No + 4{}^{1}_{0}n$
rutherfordium	Rf	104	${}^{249}_{98}Cf + {}^{12}_{6}C \longrightarrow {}^{257}_{104}Rf + 4{}^{1}_{0}n$
seaborgium	Sg	106	${}^{206}_{82}Pb + {}^{54}_{24}Cr \longrightarrow {}^{257}_{106}Sg + 3{}^{1}_{0}n$ ${}^{249}_{98}Cf + {}^{18}_{8}O \longrightarrow {}^{263}_{106}Sg + 4{}^{1}_{0}n$
meitnerium	Mt	107	${}^{209}_{83}Bi + {}^{58}_{26}Fe \longrightarrow {}^{266}_{109}Mt + {}^{1}_{0}n$

TABLE 21.3

Nuclear Fission

Many heavier elements with smaller binding energies per nucleon can decompose into more stable elements that have intermediate mass numbers and larger binding energies per nucleon—that is, mass numbers and binding energies per nucleon that are closer to the "peak" of the binding energy graph near 56 (see Figure 21.3). Sometimes neutrons are also produced. This decomposition is called **fission**, the breaking of a large nucleus into smaller pieces. The breaking is rather random with the formation of a large number of different products. Fission usually does not occur naturally, but is induced by bombardment with neutrons. The first reported nuclear fission occurred in 1939 when three German scientists, Lise Meitner, Otto Hahn, and Fritz Strassman, bombarded uranium-235 atoms with slow-moving neutrons that split the U-238 nuclei into smaller fragments that consisted of several neutrons and elements near the middle of the periodic table. Since then, fission has been observed in many other isotopes, including most actinide isotopes that have an odd number of neutrons. A typical nuclear fission reaction is shown in Figure 21.14.

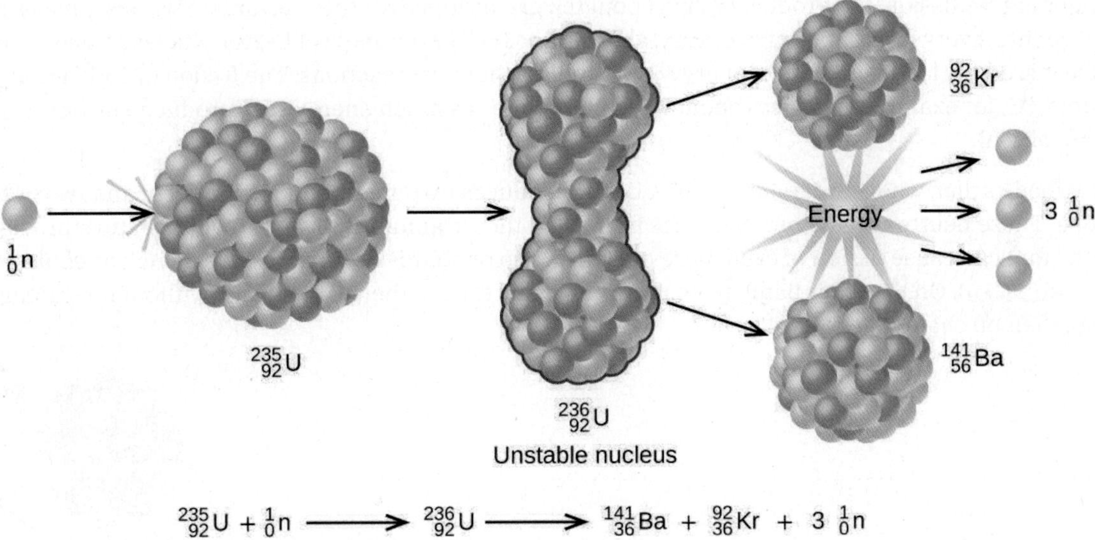

FIGURE 21.14 When a slow neutron hits a fissionable U-235 nucleus, it is absorbed and forms an unstable U-236 nucleus. The U-236 nucleus then rapidly breaks apart into two smaller nuclei (in this case, Ba-141 and Kr-92) along with several neutrons (usually two or three), and releases a very large amount of energy.

Among the products of Meitner, Hahn, and Strassman's fission reaction were barium, krypton, lanthanum, and cerium, all of which have nuclei that are more stable than uranium-235. Since then, hundreds of different isotopes have been observed among the products of fissionable substances. A few of the many reactions that occur for U-235, and a graph showing the distribution of its fission products and their yields, are shown in Figure 21.15. Similar fission reactions have been observed with other uranium isotopes, as well as with a variety of other isotopes such as those of plutonium.

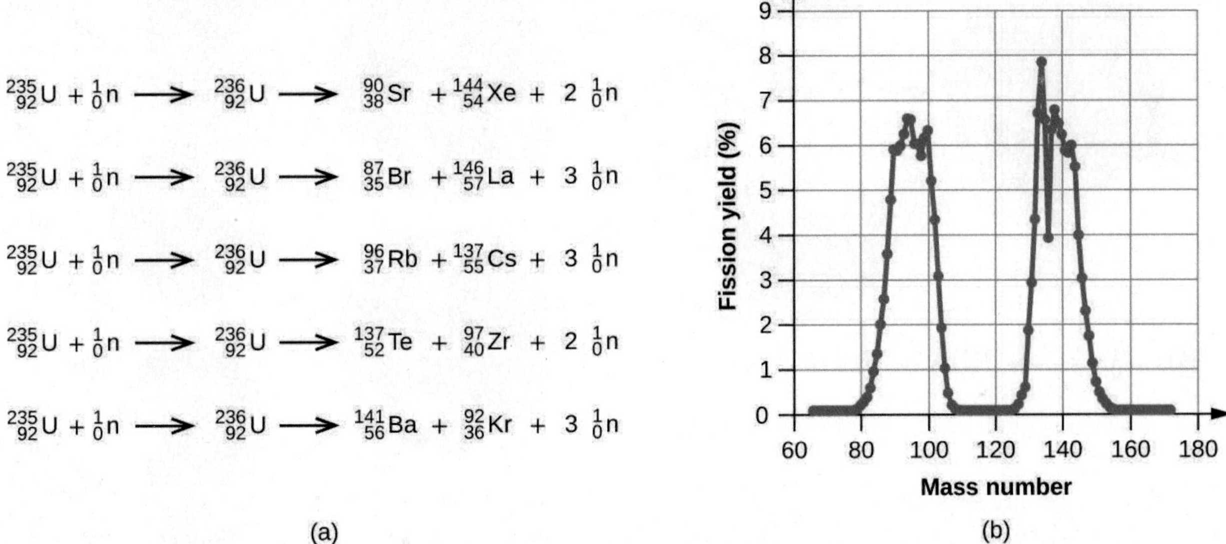

FIGURE 21.15 (a) Nuclear fission of U-235 produces a range of fission products. (b) The larger fission products of U-235 are typically one isotope with a mass number around 85–105, and another isotope with a mass number that is about 50% larger, that is, about 130–150.

LINK TO LEARNING

View this link (http://openstax.org/l/16fission) to see a simulation of nuclear fission.

A tremendous amount of energy is produced by the fission of heavy elements. For instance, when one mole of

U-235 undergoes fission, the products weigh about 0.2 grams less than the reactants; this "lost" mass is converted into a very large amount of energy, about 1.8×10^{10} kJ per mole of U-235. Nuclear fission reactions produce incredibly large amounts of energy compared to chemical reactions. The fission of 1 kilogram of uranium-235, for example, produces about 2.5 million times as much energy as is produced by burning 1 kilogram of coal.

As described earlier, when undergoing fission U-235 produces two "medium-sized" nuclei, and two or three neutrons. These neutrons may then cause the fission of other uranium-235 atoms, which in turn provide more neutrons that can cause fission of even more nuclei, and so on. If this occurs, we have a nuclear **chain reaction** (see Figure 21.16). On the other hand, if too many neutrons escape the bulk material without interacting with a nucleus, then no chain reaction will occur.

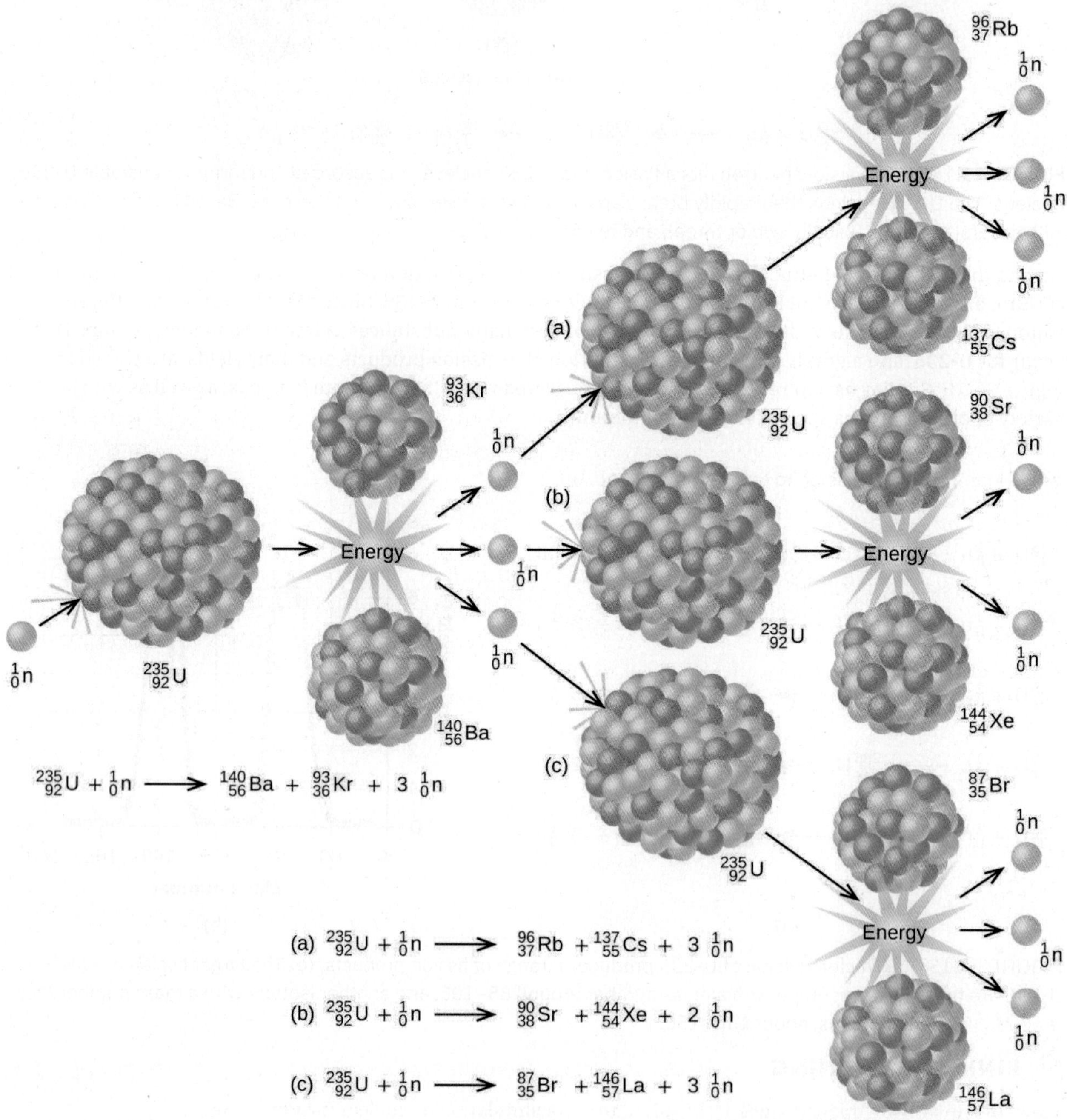

FIGURE 21.16 The fission of a large nucleus, such as U-235, produces two or three neutrons, each of which is capable of causing fission of another nucleus by the reactions shown. If this process continues, a nuclear chain reaction occurs.

Material that can sustain a nuclear fission chain reaction is said to be **fissile** or **fissionable**. (Technically, fissile material can undergo fission with neutrons of any energy, whereas fissionable material requires high-energy neutrons.) Nuclear fission becomes self-sustaining when the number of neutrons produced by fission equals or exceeds the number of neutrons absorbed by splitting nuclei plus the number that escape into the surroundings. The amount of a fissionable material that will support a self-sustaining chain reaction is a **critical mass**. An amount of fissionable material that cannot sustain a chain reaction is a **subcritical mass**. An amount of material in which there is an increasing rate of fission is known as a **supercritical mass**. The critical mass depends on the type of material: its purity, the temperature, the shape of the sample, and how the neutron reactions are controlled (Figure 21.17).

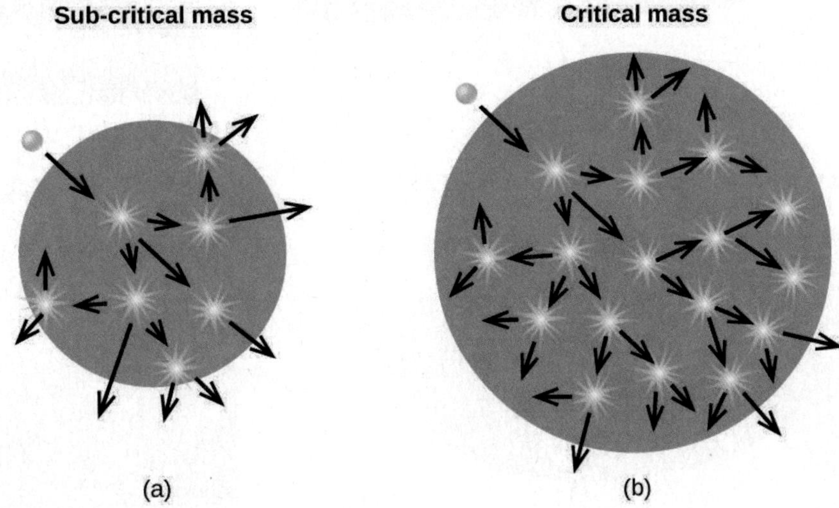

FIGURE 21.17 (a) In a subcritical mass, the fissile material is too small and allows too many neutrons to escape the material, so a chain reaction does not occur. (b) In a critical mass, a large enough number of neutrons in the fissile material induce fission to create a chain reaction.

An atomic bomb (Figure 21.18) contains several pounds of fissionable material, ${}^{235}_{92}\text{U}$ or ${}^{239}_{94}\text{Pu}$, a source of neutrons, and an explosive device for compressing it quickly into a small volume. When fissionable material is in small pieces, the proportion of neutrons that escape through the relatively large surface area is great, and a chain reaction does not take place. When the small pieces of fissionable material are brought together quickly to form a body with a mass larger than the critical mass, the relative number of escaping neutrons decreases, and a chain reaction and explosion result.

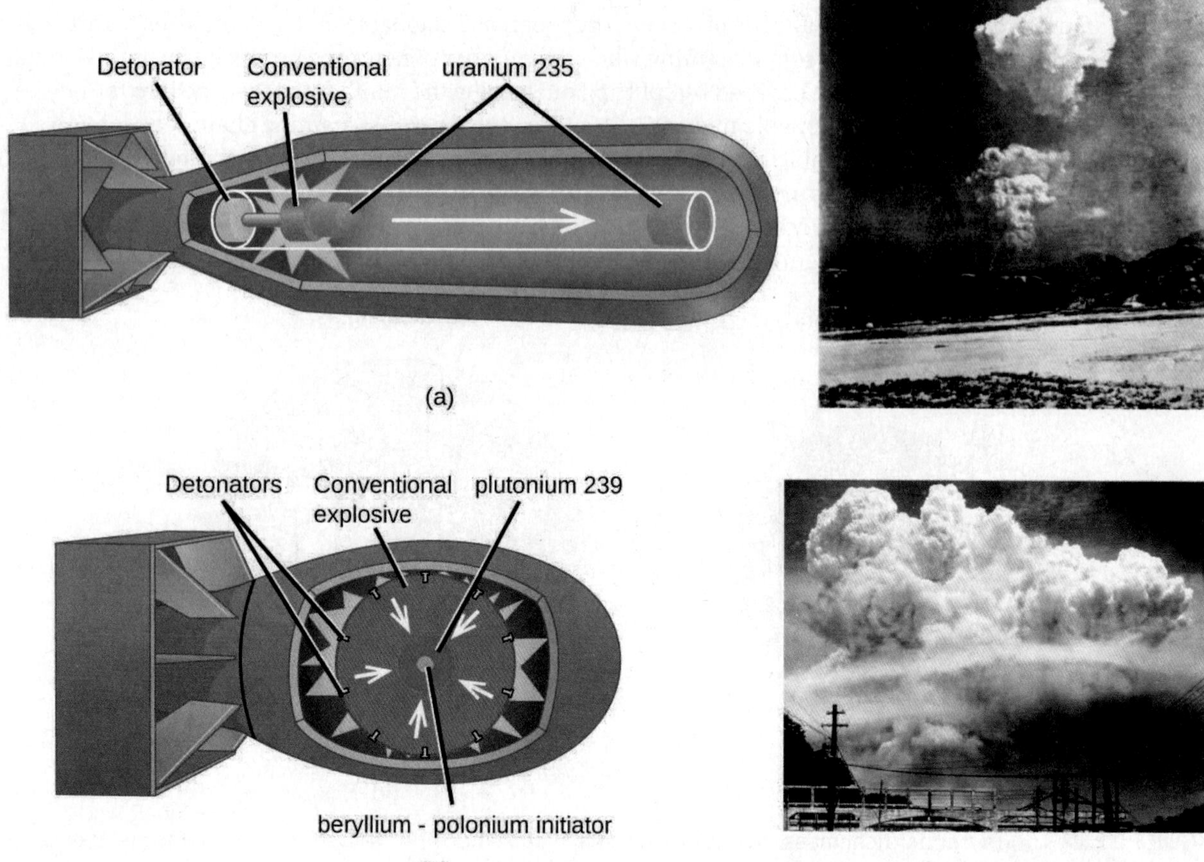

FIGURE 21.18 (a) The nuclear fission bomb that destroyed Hiroshima on August 6, 1945, consisted of two subcritical masses of U-235, where conventional explosives were used to fire one of the subcritical masses into the other, creating the critical mass for the nuclear explosion. (b) The plutonium bomb that destroyed Nagasaki on August 9, 1945, consisted of a hollow sphere of plutonium that was rapidly compressed by conventional explosives. This led to a concentration of plutonium in the center that was greater than the critical mass necessary for the nuclear explosion.

Fission Reactors

Chain reactions of fissionable materials can be controlled and sustained without an explosion in a **nuclear reactor** (Figure 21.19). Any nuclear reactor that produces power via the fission of uranium or plutonium by bombardment with neutrons must have at least five components: nuclear fuel consisting of fissionable material, a nuclear moderator, reactor coolant, control rods, and a shield and containment system. We will discuss these components in greater detail later in the section. The reactor works by separating the fissionable nuclear material such that a critical mass cannot be formed, controlling both the flux and absorption of neutrons to allow shutting down the fission reactions. In a nuclear reactor used for the production of electricity, the energy released by fission reactions is trapped as thermal energy and used to boil water and produce steam. The steam is used to turn a turbine, which powers a generator for the production of electricity.

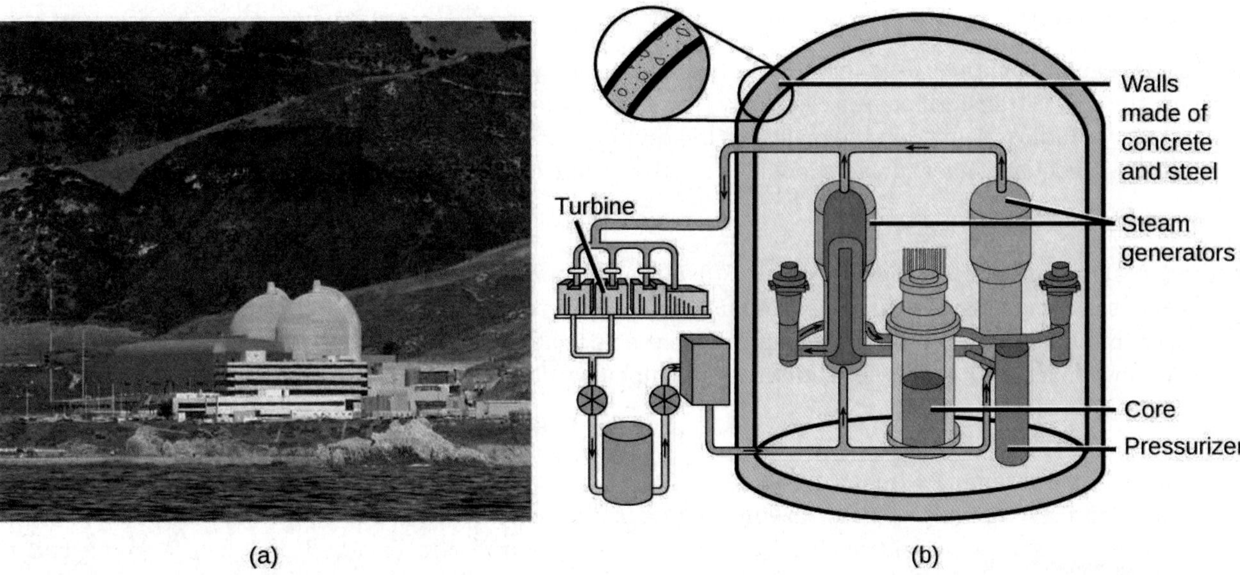

FIGURE 21.19 (a) The Diablo Canyon Nuclear Power Plant near San Luis Obispo is the only nuclear power plant currently in operation in California. The domes are the containment structures for the nuclear reactors, and the brown building houses the turbine where electricity is generated. Ocean water is used for cooling. (b) The Diablo Canyon uses a pressurized water reactor, one of a few different fission reactor designs in use around the world, to produce electricity. Energy from the nuclear fission reactions in the core heats water in a closed, pressurized system. Heat from this system produces steam that drives a turbine, which in turn produces electricity. (credit a: modification of work by "Mike" Michael L. Baird; credit b: modification of work by the Nuclear Regulatory Commission)

Nuclear Fuels

Nuclear fuel consists of a fissionable isotope, such as uranium-235, which must be present in sufficient quantity to provide a self-sustaining chain reaction. In the United States, uranium ores contain from 0.05–0.3% of the uranium oxide U_3O_8; the uranium in the ore is about 99.3% nonfissionable U-238 with only 0.7% fissionable U-235. Nuclear reactors require a fuel with a higher concentration of U-235 than is found in nature; it is normally enriched to have about 5% of uranium mass as U-235. At this concentration, it is not possible to achieve the supercritical mass necessary for a nuclear explosion. Uranium can be enriched by gaseous diffusion (the only method currently used in the US), using a gas centrifuge, or by laser separation.

In the gaseous diffusion enrichment plant where U-235 fuel is prepared, UF_6 (uranium hexafluoride) gas at low pressure moves through barriers that have holes just barely large enough for UF_6 to pass through. The slightly lighter $^{235}UF_6$ molecules diffuse through the barrier slightly faster than the heavier $^{238}UF_6$ molecules. This process is repeated through hundreds of barriers, gradually increasing the concentration of $^{235}UF_6$ to the level needed by the nuclear reactor. The basis for this process, Graham's law, is described in the chapter on gases. The enriched UF_6 gas is collected, cooled until it solidifies, and then taken to a fabrication facility where it is made into fuel assemblies. Each fuel assembly consists of fuel rods that contain many thimble-sized, ceramic-encased, enriched uranium (usually UO_2) fuel pellets. Modern nuclear reactors may contain as many as 10 million fuel pellets. The amount of energy in each of these pellets is equal to that in almost a ton of coal or 150 gallons of oil.

Nuclear Moderators

Neutrons produced by nuclear reactions move too fast to cause fission (refer back to Figure 21.17). They must first be slowed to be absorbed by the fuel and produce additional nuclear reactions. A **nuclear moderator** is a substance that slows the neutrons to a speed that is low enough to cause fission. Early reactors used high-purity graphite as a moderator. Modern reactors in the US exclusively use heavy water ($^2_1H_2O$) or light water (ordinary H_2O), whereas some reactors in other countries use other materials, such as carbon dioxide, beryllium, or graphite.

Reactor Coolants

A nuclear **reactor coolant** is used to carry the heat produced by the fission reaction to an external boiler and turbine, where it is transformed into electricity. Two overlapping coolant loops are often used; this counteracts the transfer of radioactivity from the reactor to the primary coolant loop. All nuclear power plants in the US use water as a coolant. Other coolants include molten sodium, lead, a lead-bismuth mixture, or molten salts.

Control Rods

Nuclear reactors use **control rods** (Figure 21.20) to control the fission rate of the nuclear fuel by adjusting the number of slow neutrons present to keep the rate of the chain reaction at a safe level. Control rods are made of boron, cadmium, hafnium, or other elements that are able to absorb neutrons. Boron-10, for example, absorbs neutrons by a reaction that produces lithium-7 and alpha particles:

$$^{10}_{5}\text{B} + ^{1}_{0}\text{n} \longrightarrow ^{7}_{3}\text{Li} + ^{4}_{2}\text{He}$$

When control rod assemblies are inserted into the fuel element in the reactor core, they absorb a larger fraction of the slow neutrons, thereby slowing the rate of the fission reaction and decreasing the power produced. Conversely, if the control rods are removed, fewer neutrons are absorbed, and the fission rate and energy production increase. In an emergency, the chain reaction can be shut down by fully inserting all of the control rods into the nuclear core between the fuel rods.

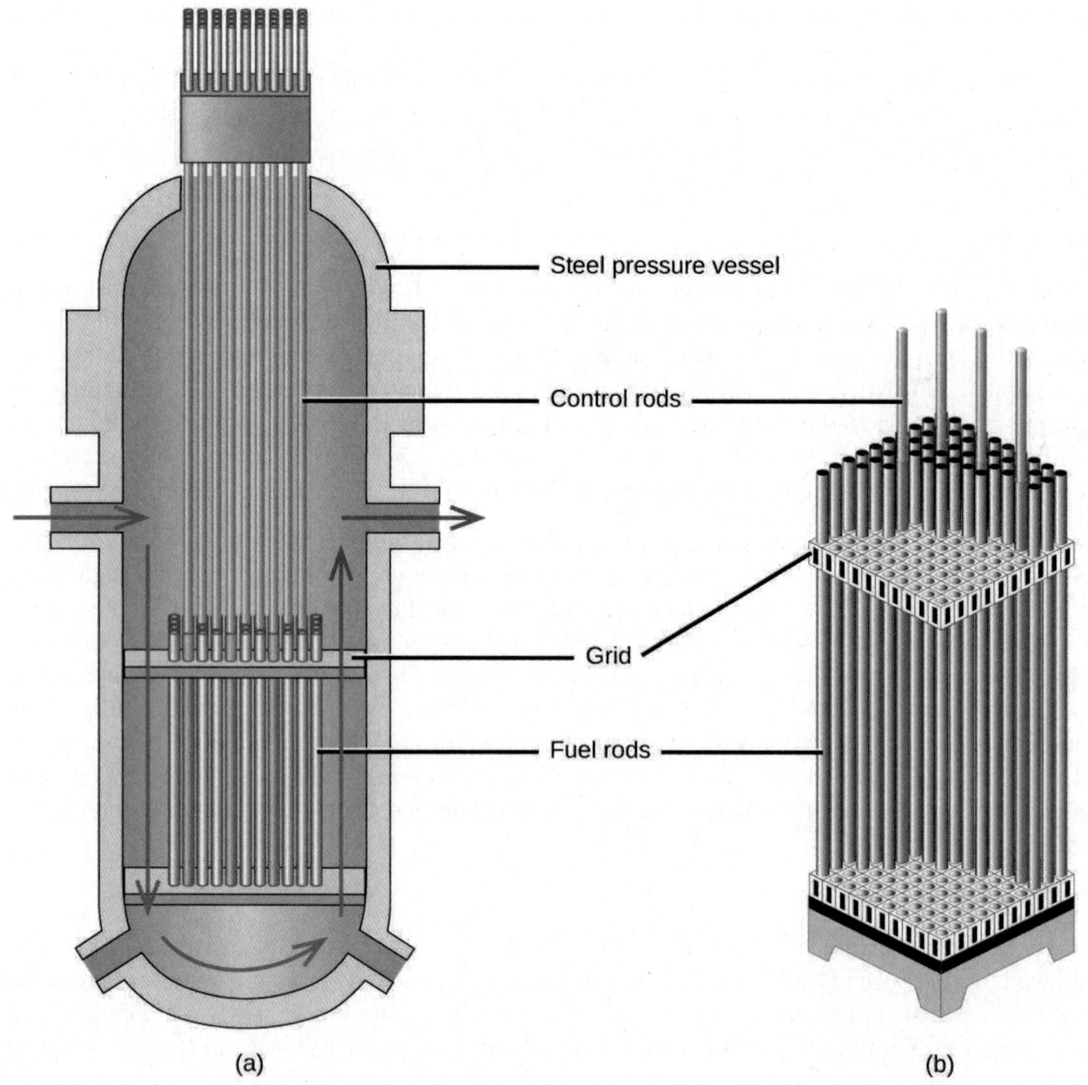

FIGURE 21.20 The nuclear reactor core shown in (a) contains the fuel and control rod assembly shown in (b).

(credit: modification of work by E. Generalic, http://glossary.periodni.com/glossary.php?en=control+rod)

Shield and Containment System

During its operation, a nuclear reactor produces neutrons and other radiation. Even when shut down, the decay products are radioactive. In addition, an operating reactor is thermally very hot, and high pressures result from the circulation of water or another coolant through it. Thus, a reactor must withstand high temperatures and pressures, and must protect operating personnel from the radiation. Reactors are equipped with a **containment system** (or shield) that consists of three parts:

1. The reactor vessel, a steel shell that is 3–20-centimeters thick and, with the moderator, absorbs much of the radiation produced by the reactor
2. A main shield of 1–3 meters of high-density concrete
3. A personnel shield of lighter materials that protects operators from γ rays and X-rays

In addition, reactors are often covered with a steel or concrete dome that is designed to contain any radioactive materials might be released by a reactor accident.

LINK TO LEARNING

Click here to watch a 3-minute video (http://openstax.org/l/16nucreactors) from the Nuclear Energy Institute on how nuclear reactors work.

Nuclear power plants are designed in such a way that they cannot form a supercritical mass of fissionable material and therefore cannot create a nuclear explosion. But as history has shown, failures of systems and safeguards can cause catastrophic accidents, including chemical explosions and nuclear meltdowns (damage to the reactor core from overheating). The following Chemistry in Everyday Life feature explores three infamous meltdown incidents.

Chemistry in Everyday Life

Nuclear Accidents

The importance of cooling and containment are amply illustrated by three major accidents that occurred with the nuclear reactors at nuclear power generating stations in the United States (Three Mile Island), the former Soviet Union (Chernobyl), and Japan (Fukushima).

In March 1979, the cooling system of the Unit 2 reactor at Three Mile Island Nuclear Generating Station in Pennsylvania failed, and the cooling water spilled from the reactor onto the floor of the containment building. After the pumps stopped, the reactors overheated due to the high radioactive decay heat produced in the first few days after the nuclear reactor shut down. The temperature of the core climbed to at least 2200 °C, and the upper portion of the core began to melt. In addition, the zirconium alloy cladding of the fuel rods began to react with steam and produced hydrogen:

$$\mathrm{Zr}(s) + 2\mathrm{H_2O}(g) \longrightarrow \mathrm{ZrO_2}(s) + 2\mathrm{H_2}(g)$$

The hydrogen accumulated in the confinement building, and it was feared that there was danger of an explosion of the mixture of hydrogen and air in the building. Consequently, hydrogen gas and radioactive gases (primarily krypton and xenon) were vented from the building. Within a week, cooling water circulation was restored and the core began to cool. The plant was closed for nearly 10 years during the cleanup process.

Although zero discharge of radioactive material is desirable, the discharge of radioactive krypton and xenon, such as occurred at the Three Mile Island plant, is among the most tolerable. These gases readily disperse in the atmosphere and thus do not produce highly radioactive areas. Moreover, they are noble gases and are not incorporated into plant and animal matter in the food chain. Effectively none of the heavy elements of the core of the reactor were released into the environment, and no cleanup of the area outside of the containment building was necessary (Figure 21.21).

(a) (b)

FIGURE 21.21 (a) In this 2010 photo of Three Mile Island, the remaining structures from the damaged Unit 2 reactor are seen on the left, whereas the separate Unit 1 reactor, unaffected by the accident, continues generating power to this day (right). (b) President Jimmy Carter visited the Unit 2 control room a few days after the accident in 1979.

Another major nuclear accident involving a reactor occurred in April 1986, at the Chernobyl Nuclear Power Plant in Ukraine, which was still a part of the former Soviet Union. While operating at low power during an unauthorized experiment with some of its safety devices shut off, one of the reactors at the plant became unstable. Its chain reaction became uncontrollable and increased to a level far beyond what the reactor was designed for. The steam pressure in the reactor rose to between 100 and 500 times the full power pressure and ruptured the reactor. Because the reactor was not enclosed in a containment building, a large amount of radioactive material spewed out, and additional fission products were released, as the graphite (carbon) moderator of the core ignited and burned. The fire was controlled, but over 200 plant workers and firefighters developed acute radiation sickness and at least 32 soon died from the effects of the radiation. It is predicted that about 4000 more deaths will occur among emergency workers and former Chernobyl residents from radiation-induced cancer and leukemia. The reactor has since been encapsulated in steel and concrete, a now-decaying structure known as the sarcophagus. Almost 30 years later, significant radiation problems still persist in the area, and Chernobyl largely remains a wasteland.

In 2011, the Fukushima Daiichi Nuclear Power Plant in Japan was badly damaged by a 9.0-magnitude earthquake and resulting tsunami. Three reactors up and running at the time were shut down automatically, and emergency generators came online to power electronics and coolant systems. However, the tsunami quickly flooded the emergency generators and cut power to the pumps that circulated coolant water through the reactors. High-temperature steam in the reactors reacted with zirconium alloy to produce hydrogen gas. The gas escaped into the containment building, and the mixture of hydrogen and air exploded. Radioactive material was released from the containment vessels as the result of deliberate venting to reduce the hydrogen pressure, deliberate discharge of coolant water into the sea, and accidental or uncontrolled events.

An evacuation zone around the damaged plant extended over 12.4 miles away, and an estimated 200,000 people were evacuated from the area. All 48 of Japan's nuclear power plants were subsequently shut down, remaining shuttered as of December 2014. Since the disaster, public opinion has shifted from largely favoring to largely opposing increasing the use of nuclear power plants, and a restart of Japan's atomic energy program is still stalled (Figure 21.22).

(a)

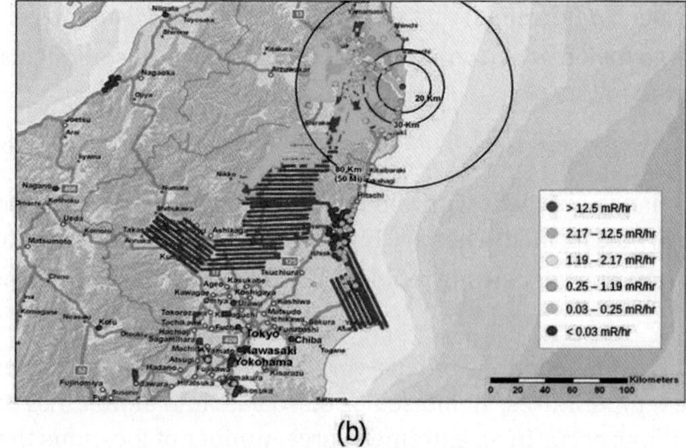

(b)

FIGURE 21.22 (a) After the accident, contaminated waste had to be removed, and (b) an evacuation zone was set up around the plant in areas that received heavy doses of radioactive fallout. (credit a: modification of work by "Live Action Hero"/Flickr)

The energy produced by a reactor fueled with enriched uranium results from the fission of uranium as well as from the fission of plutonium produced as the reactor operates. As discussed previously, the plutonium forms from the combination of neutrons and the uranium in the fuel. In any nuclear reactor, only about 0.1% of the mass of the fuel is converted into energy. The other 99.9% remains in the fuel rods as fission products and unused fuel. All of the fission products absorb neutrons, and after a period of several months to a few years, depending on the reactor, the fission products must be removed by changing the fuel rods. Otherwise, the concentration of these fission products would increase and absorb more neutrons until the reactor could no longer operate.

Spent fuel rods contain a variety of products, consisting of unstable nuclei ranging in atomic number from 25 to 60, some transuranium elements, including plutonium and americium, and unreacted uranium isotopes. The unstable nuclei and the transuranium isotopes give the spent fuel a dangerously high level of radioactivity. The long-lived isotopes require thousands of years to decay to a safe level. The ultimate fate of the nuclear reactor as a significant source of energy in the United States probably rests on whether or not a politically and scientifically satisfactory technique for processing and storing the components of spent fuel rods can be developed.

LINK TO LEARNING

Explore the information in this link (http://openstax.org/l/16wastemgmt) to learn about the approaches to nuclear waste management.

Nuclear Fusion and Fusion Reactors

The process of converting very light nuclei into heavier nuclei is also accompanied by the conversion of mass into large amounts of energy, a process called **fusion**. The principal source of energy in the sun is a net fusion reaction in which four hydrogen nuclei fuse and produce one helium nucleus and two positrons. This is a net reaction of a more complicated series of events:

$$4\,{}^{1}_{1}\text{H} \longrightarrow {}^{4}_{2}\text{He} + 2\,{}^{0}_{+1}\text{e}^{+}$$

A helium nucleus has a mass that is 0.7% less than that of four hydrogen nuclei; this lost mass is converted into energy during the fusion. This reaction produces about 3.6×10^{11} kJ of energy per mole of ${}^{4}_{2}\text{He}$ produced. This is somewhat larger than the energy produced by the nuclear fission of one mole of U-235 (1.8×10^{10} kJ), and over 3 million times larger than the energy produced by the (chemical) combustion of one mole of octane (5471 kJ).

It has been determined that the nuclei of the heavy isotopes of hydrogen, a deuteron, ${}^{2}_{1}\text{H}$ and a triton, ${}^{3}_{1}\text{H}$, undergo fusion at extremely high temperatures (thermonuclear fusion). They form a helium nucleus and a neutron:

$$ {}^{2}_{1}\text{H} + {}^{3}_{1}\text{H} \longrightarrow {}^{4}_{2}\text{He} + {}^{1}_{0}\text{n} $$

This change proceeds with a mass loss of 0.0188 amu, corresponding to the release of 1.69×10^{9} kilojoules per mole of ${}^{4}_{2}\text{He}$ formed. The very high temperature is necessary to give the nuclei enough kinetic energy to overcome the very strong repulsive forces resulting from the positive charges on their nuclei so they can collide.

Useful fusion reactions require very high temperatures for their initiation—about 15,000,000 K or more. At these temperatures, all molecules dissociate into atoms, and the atoms ionize, forming plasma. These conditions occur in an extremely large number of locations throughout the universe—stars are powered by fusion. Humans have already figured out how to create temperatures high enough to achieve fusion on a large scale in thermonuclear weapons. A thermonuclear weapon such as a hydrogen bomb contains a nuclear fission bomb that, when exploded, gives off enough energy to produce the extremely high temperatures necessary for fusion to occur.

Another much more beneficial way to create fusion reactions is in a **fusion reactor**, a nuclear reactor in which fusion reactions of light nuclei are controlled. Because no solid materials are stable at such high temperatures, mechanical devices cannot contain the plasma in which fusion reactions occur. Two techniques to contain plasma at the density and temperature necessary for a fusion reaction are currently the focus of intensive research efforts: containment by a magnetic field and by the use of focused laser beams (Figure 21.23). A number of large projects are working to attain one of the biggest goals in science: getting hydrogen fuel to ignite and produce more energy than the amount supplied to achieve the extremely high temperatures and pressures that are required for fusion. At the time of this writing, there are no self-sustaining fusion reactors operating in the world, although small-scale controlled fusion reactions have been run for very brief periods.

(a)

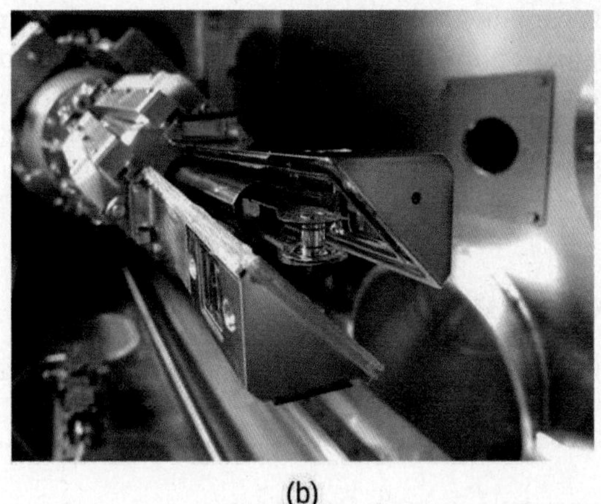

(b)

FIGURE 21.23 (a) This model is of the International Thermonuclear Experimental Reactor (ITER) reactor. Currently under construction in the south of France with an expected completion date of 2027, the ITER will be the world's largest experimental Tokamak nuclear fusion reactor with a goal of achieving large-scale sustained energy production. (b) In 2012, the National Ignition Facility at Lawrence Livermore National Laboratory briefly produced over 500,000,000,000 watts (500 terawatts, or 500 TW) of peak power and delivered 1,850,000 joules (1.85 MJ) of energy, the largest laser energy ever produced and 1000 times the power usage of the entire United States in any given moment. Although lasting only a few billionths of a second, the 192 lasers attained the conditions needed for nuclear fusion ignition. This image shows the target prior to the laser shot. (credit a: modification of work by Stephan Mosel)

21.5 Uses of Radioisotopes

LEARNING OBJECTIVES

By the end of this section, you will be able to:

- List common applications of radioactive isotopes

Radioactive isotopes have the same chemical properties as stable isotopes of the same element, but they emit radiation, which can be detected. If we replace one (or more) atom(s) with radioisotope(s) in a compound, we can track them by monitoring their radioactive emissions. This type of compound is called a **radioactive tracer** (or **radioactive label**). Radioisotopes are used to follow the paths of biochemical reactions or to determine how a substance is distributed within an organism. Radioactive tracers are also used in many medical applications, including both diagnosis and treatment. They are used to measure engine wear, analyze the geological formation around oil wells, and much more.

Radioimmunossays (RIA), for example, rely on radioisotopes to detect the presence and/or concentration of certain antigens. Developed by Rosalyn Sussman Yalow and Solomon Berson in the 1950s, the technique is known for extreme sensitivity, meaning that it can detect and measure very small quantities of a substance. Prior to its discovery, most similar detection relied on large enough quantities to produce visible outcomes. RIA revolutionized and expanded entire fields of study, most notably endocrinology, and is commonly used in narcotics detection, blood bank screening, early cancer screening, hormone measurement, and allergy diagnosis. Based on her significant contribution to medicine, Yalow received a Nobel Prize, making her the second woman to be awarded the prize for medicine.

Radioisotopes have revolutionized medical practice (see Appendix M), where they are used extensively. Over 10 million nuclear medicine procedures and more than 100 million nuclear medicine tests are performed annually in the United States. Four typical examples of radioactive tracers used in medicine are technetium-99 ($^{99}_{43}Tc$), thallium-201 ($^{201}_{81}Tl$), iodine-131 ($^{131}_{53}I$), and sodium-24 ($^{24}_{11}Na$) . Damaged tissues in the heart, liver, and lungs absorb certain compounds of technetium-99 preferentially. After it is injected, the location of the technetium compound, and hence the damaged tissue, can be determined by detecting the γ rays emitted by the Tc-99 isotope. Thallium-201 (Figure 21.24) becomes concentrated in healthy heart tissue, so the two isotopes, Tc-99 and Tl-201, are used together to study heart tissue. Iodine-131 concentrates in the thyroid gland, the liver, and some parts of the brain. It can therefore be used to monitor goiter and treat thyroid conditions, such as Grave's disease, as well as liver and brain tumors. Salt solutions containing compounds of sodium-24 are injected into the bloodstream to help locate obstructions to the flow of blood.

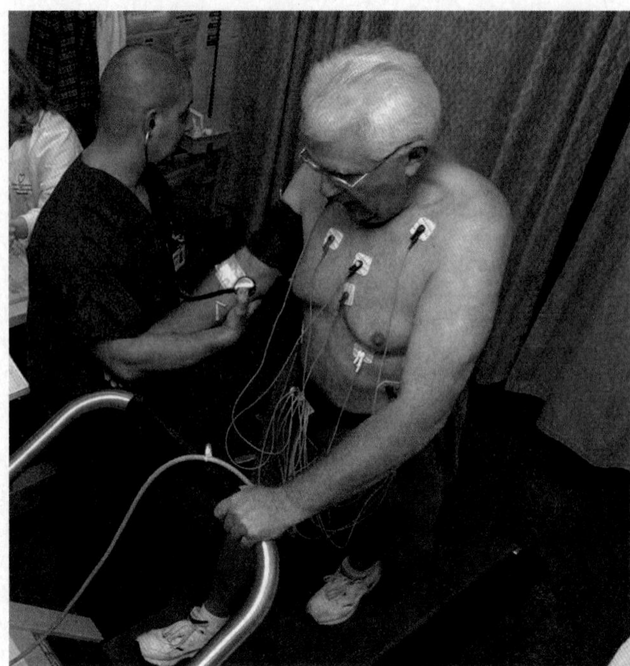

FIGURE 21.24 Administering thallium-201 to a patient and subsequently performing a stress test offer medical professionals an opportunity to visually analyze heart function and blood flow. (credit: modification of work by "BlueOctane"/Wikimedia Commons)

Radioisotopes used in medicine typically have short half-lives—for example, the ubiquitous Tc-99m has a half-life of 6.01 hours. This makes Tc-99m essentially impossible to store and prohibitively expensive to transport, so it is made on-site instead. Hospitals and other medical facilities use Mo-99 (which is primarily extracted from U-235 fission products) to generate Tc-99. Mo-99 undergoes β decay with a half-life of 66 hours, and the Tc-99 is then chemically extracted (Figure 21.25). The parent nuclide Mo-99 is part of a molybdate ion, $MoO_4{}^{2-}$; when it decays, it forms the pertechnetate ion, $TcO_4{}^{-}$. These two water-soluble ions are separated by column chromatography, with the higher charge molybdate ion adsorbing onto the alumina in the column, and the lower charge pertechnetate ion passing through the column in the solution. A few micrograms of Mo-99 can produce enough Tc-99 to perform as many as 10,000 tests.

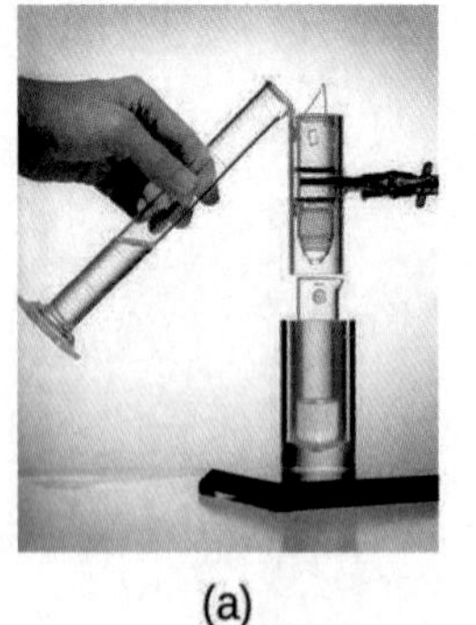

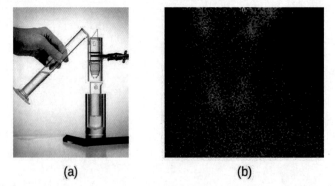

(a)

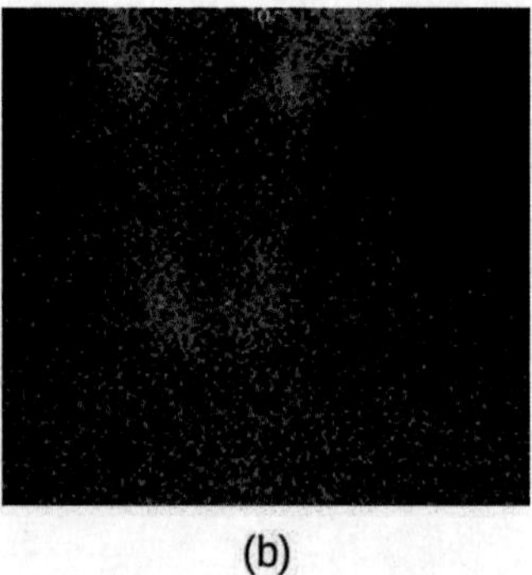

(b)

FIGURE 21.25 (a) The first Tc-99m generator (circa 1958) is used to separate Tc-99 from Mo-99. The $MoO_4{}^{2-}$ is retained by the matrix in the column, whereas the $TcO_4{}^{-}$ passes through and is collected. (b) Tc-99 was used in this scan of the neck of a patient with Grave's disease. The scan shows the location of high concentrations of Tc-99. (credit a: modification of work by the Department of Energy; credit b: modification of work by "MBq"/Wikimedia Commons)

Radioisotopes can also be used, typically in higher doses than as a tracer, as treatment. **Radiation therapy** is the use of high-energy radiation to damage the DNA of cancer cells, which kills them or keeps them from dividing (Figure 21.26). A cancer patient may receive **external beam radiation therapy** delivered by a machine outside the body, or **internal radiation therapy (brachytherapy)** from a radioactive substance that has been introduced into the body. Note that **chemotherapy** is similar to internal radiation therapy in that the cancer treatment is injected into the body, but differs in that chemotherapy uses chemical rather than radioactive substances to kill the cancer cells.

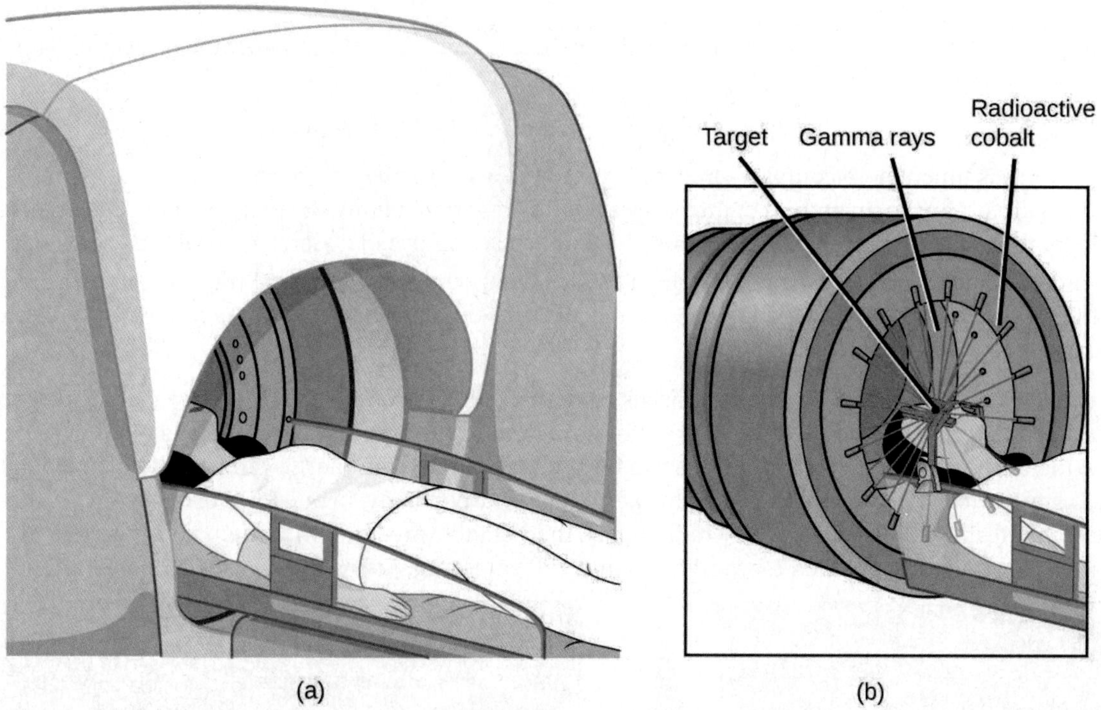

FIGURE 21.26 The cartoon in (a) shows a cobalt-60 machine used in the treatment of cancer. The diagram in (b) shows how the gantry of the Co-60 machine swings through an arc, focusing radiation on the targeted region (tumor) and minimizing the amount of radiation that passes through nearby regions.

Cobalt-60 is a synthetic radioisotope produced by the neutron activation of Co-59, which then undergoes β decay to form Ni-60, along with the emission of γ radiation. The overall process is:

$$^{59}_{27}\text{Co} + ^{1}_{0}\text{n} \longrightarrow ^{60}_{27}\text{Co} \longrightarrow ^{60}_{28}\text{Ni} + ^{0}_{-1}\beta + 2^{0}_{0}\gamma$$

The overall decay scheme for this is shown graphically in Figure 21.27.

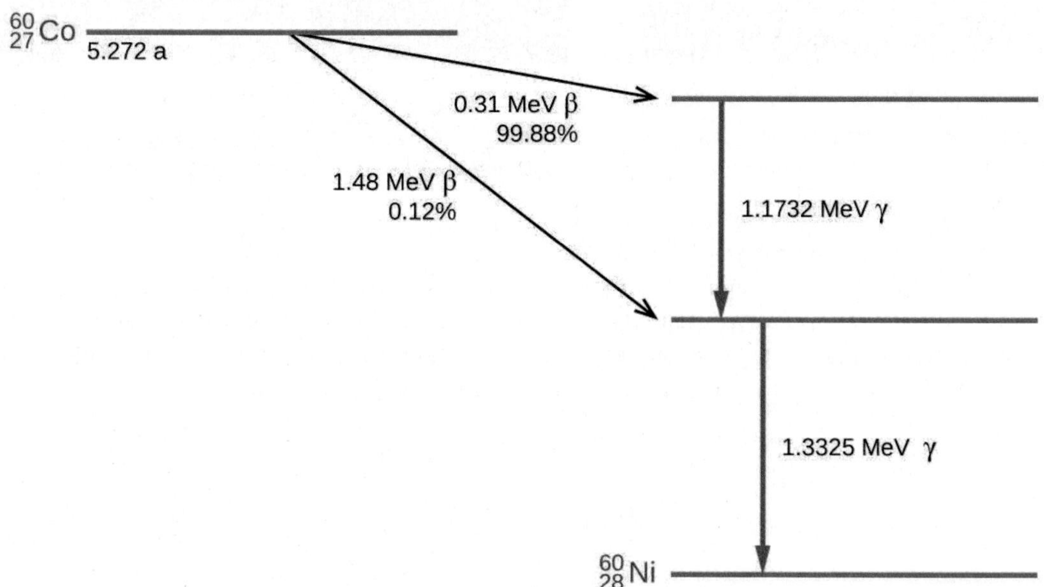

FIGURE 21.27 Co-60 undergoes a series of radioactive decays. The γ emissions are used for radiation therapy.

Radioisotopes are used in diverse ways to study the mechanisms of chemical reactions in plants and animals. These include labeling fertilizers in studies of nutrient uptake by plants and crop growth, investigations of digestive and milk-producing processes in cows, and studies on the growth and metabolism of animals and plants.

For example, the radioisotope C-14 was used to elucidate the details of how photosynthesis occurs. The overall reaction is:

$$6CO_2(g) + 6H_2O(l) \longrightarrow C_6H_{12}O_6(s) + 6O_2(g),$$

but the process is much more complex, proceeding through a series of steps in which various organic compounds are produced. In studies of the pathway of this reaction, plants were exposed to CO_2 containing a high concentration of ${}^{14}_{6}C$. At regular intervals, the plants were analyzed to determine which organic compounds contained carbon-14 and how much of each compound was present. From the time sequence in which the compounds appeared and the amount of each present at given time intervals, scientists learned more about the pathway of the reaction.

Commercial applications of radioactive materials are equally diverse (Figure 21.28). They include determining the thickness of films and thin metal sheets by exploiting the penetration power of various types of radiation. Flaws in metals used for structural purposes can be detected using high-energy gamma rays from cobalt-60 in a fashion similar to the way X-rays are used to examine the human body. In one form of pest control, flies are controlled by sterilizing male flies with γ radiation so that females breeding with them do not produce offspring. Many foods are preserved by radiation that kills microorganisms that cause the foods to spoil.

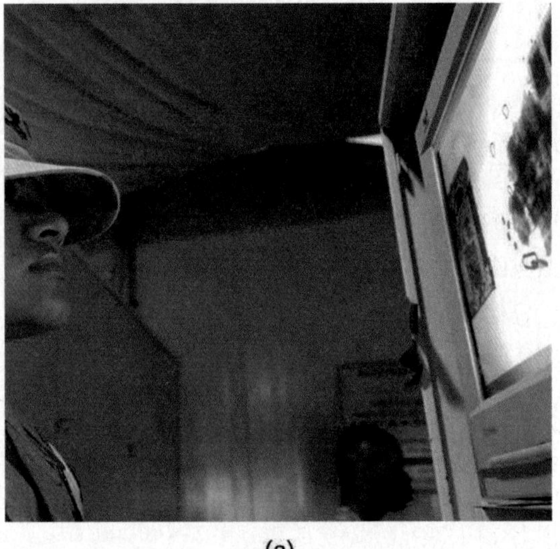

(a)

(b)

FIGURE 21.28 Common commercial uses of radiation include (a) X-ray examination of luggage at an airport and (b) preservation of food. (credit a: modification of work by the Department of the Navy; credit b: modification of work by the US Department of Agriculture)

Americium-241, an α emitter with a half-life of 458 years, is used in tiny amounts in ionization-type smoke detectors (Figure 21.29). The α emissions from Am-241 ionize the air between two electrode plates in the ionizing chamber. A battery supplies a potential that causes movement of the ions, thus creating a small electric current. When smoke enters the chamber, the movement of the ions is impeded, reducing the conductivity of the air. This causes a marked drop in the current, triggering an alarm.

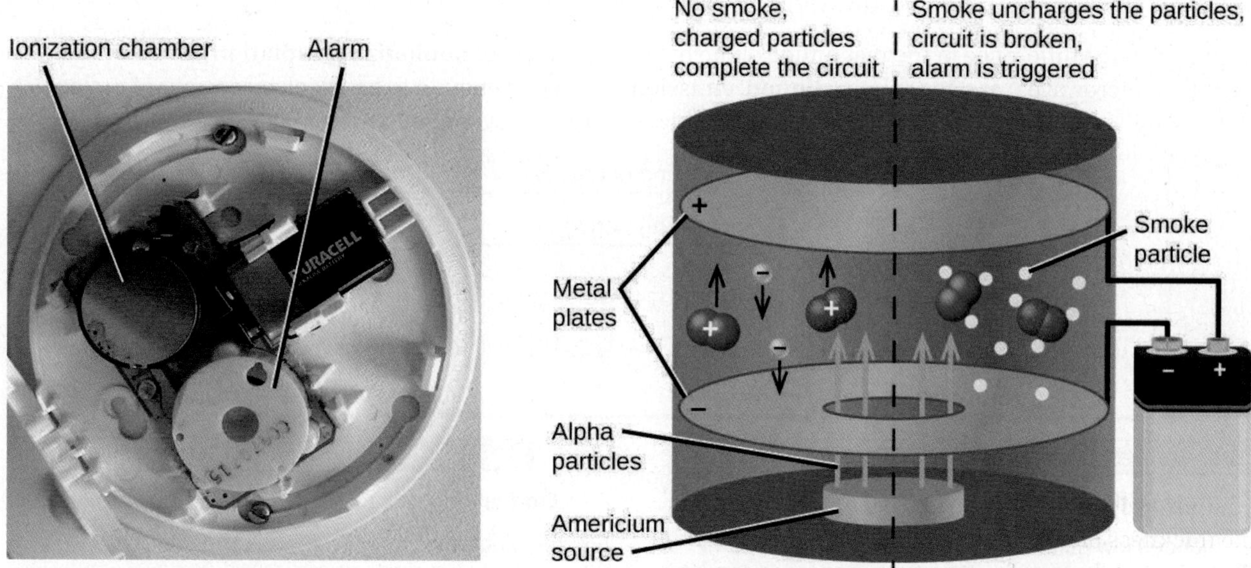

FIGURE 21.29 Inside a smoke detector, Am-241 emits α particles that ionize the air, creating a small electric current. During a fire, smoke particles impede the flow of ions, reducing the current and triggering an alarm. (credit a: modification of work by "Muffet"/Wikimedia Commons)

21.6 Biological Effects of Radiation

LEARNING OBJECTIVES

By the end of this section, you will be able to:

- Describe the biological impact of ionizing radiation
- Define units for measuring radiation exposure
- Explain the operation of common tools for detecting radioactivity
- List common sources of radiation exposure in the US

The increased use of radioisotopes has led to increased concerns over the effects of these materials on biological systems (such as humans). All radioactive nuclides emit high-energy particles or electromagnetic waves. When this radiation encounters living cells, it can cause heating, break chemical bonds, or ionize molecules. The most serious biological damage results when these radioactive emissions fragment or ionize molecules. For example, alpha and beta particles emitted from nuclear decay reactions possess much higher energies than ordinary chemical bond energies. When these particles strike and penetrate matter, they produce ions and molecular fragments that are extremely reactive. The damage this does to biomolecules in living organisms can cause serious malfunctions in normal cell processes, taxing the organism's repair mechanisms and possibly causing illness or even death (Figure 21.30).

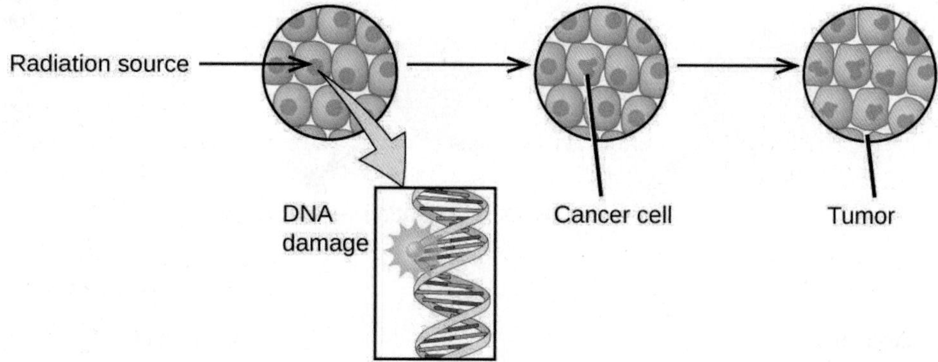

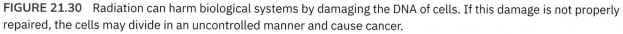

FIGURE 21.30 Radiation can harm biological systems by damaging the DNA of cells. If this damage is not properly repaired, the cells may divide in an uncontrolled manner and cause cancer.

Ionizing and Nonionizing Radiation

There is a large difference in the magnitude of the biological effects of **nonionizing radiation** (for example, light and microwaves) and **ionizing radiation**, emissions energetic enough to knock electrons out of molecules (for example, α and β particles, γ rays, X-rays, and high-energy ultraviolet radiation) (Figure 21.31).

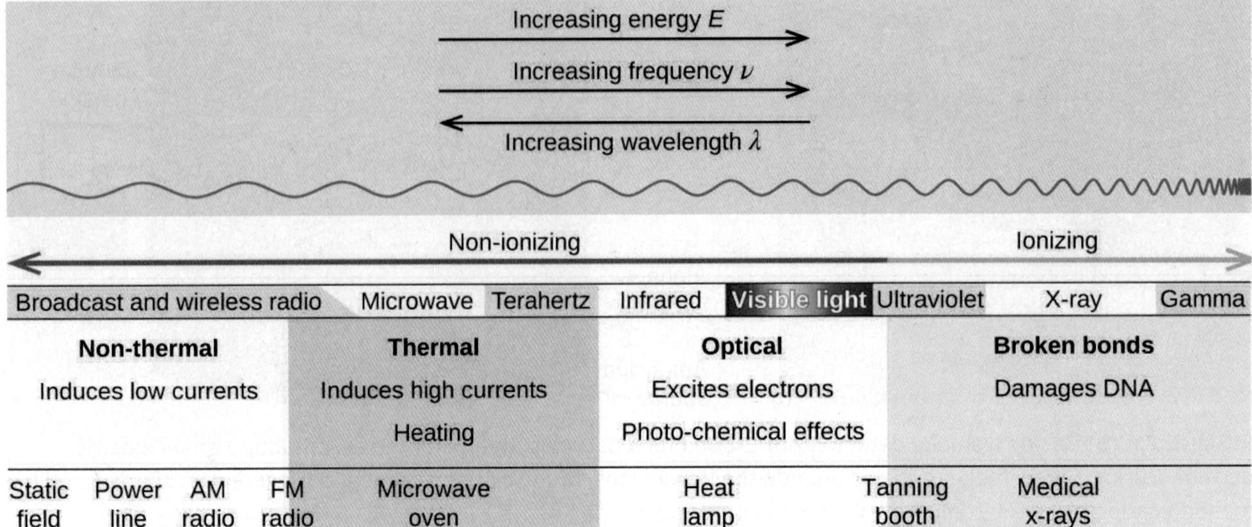

FIGURE 21.31 Lower frequency, lower-energy electromagnetic radiation is nonionizing, and higher frequency, higher-energy electromagnetic radiation is ionizing.

Energy absorbed from nonionizing radiation speeds up the movement of atoms and molecules, which is equivalent to heating the sample. Although biological systems are sensitive to heat (as we might know from touching a hot stove or spending a day at the beach in the sun), a large amount of nonionizing radiation is necessary before dangerous levels are reached. Ionizing radiation, however, may cause much more severe damage by breaking bonds or removing electrons in biological molecules, disrupting their structure and function. The damage can also be done indirectly, by first ionizing H_2O (the most abundant molecule in living organisms), which forms a H_2O^+ ion that reacts with water, forming a hydronium ion and a hydroxyl radical:

$$H_2O + \text{radiation} \xrightarrow{e^-} H_2O^+ + H_2O \longrightarrow H_3O^+ + OH^{\bullet}$$

Because the hydroxyl radical has an unpaired electron, it is highly reactive. (This is true of any substance with unpaired electrons, known as a free radical.) This hydroxyl radical can react with all kinds of biological molecules (DNA, proteins, enzymes, and so on), causing damage to the molecules and disrupting physiological processes. Examples of direct and indirect damage are shown in Figure 21.32.

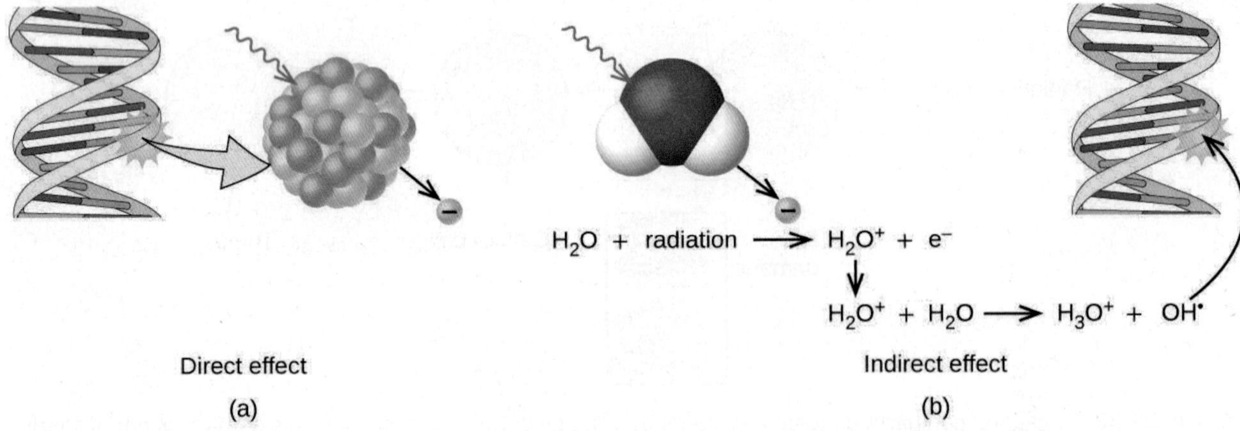

FIGURE 21.32 Ionizing radiation can (a) directly damage a biomolecule by ionizing it or breaking its bonds, or (b) create an H_2O^+ ion, which reacts with H_2O to form a hydroxyl radical, which in turn reacts with the biomolecule,

causing damage indirectly.

Biological Effects of Exposure to Radiation

Radiation can harm either the whole body (somatic damage) or eggs and sperm (genetic damage). Its effects are more pronounced in cells that reproduce rapidly, such as the stomach lining, hair follicles, bone marrow, and embryos. This is why patients undergoing radiation therapy often feel nauseous or sick to their stomach, lose hair, have bone aches, and so on, and why particular care must be taken when undergoing radiation therapy during pregnancy.

Different types of radiation have differing abilities to pass through material (Figure 21.33). A very thin barrier, such as a sheet or two of paper, or the top layer of skin cells, usually stops alpha particles. Because of this, alpha particle sources are usually not dangerous if outside the body, but are quite hazardous if ingested or inhaled (see the Chemistry in Everyday Life feature on Radon Exposure). Beta particles will pass through a hand, or a thin layer of material like paper or wood, but are stopped by a thin layer of metal. Gamma radiation is very penetrating and can pass through a thick layer of most materials. Some high-energy gamma radiation is able to pass through a few feet of concrete. Certain dense, high atomic number elements (such as lead) can effectively attenuate gamma radiation with thinner material and are used for shielding. The ability of various kinds of emissions to cause ionization varies greatly, and some particles have almost no tendency to produce ionization. Alpha particles have about twice the ionizing power of fast-moving neutrons, about 10 times that of β particles, and about 20 times that of γ rays and X-rays.

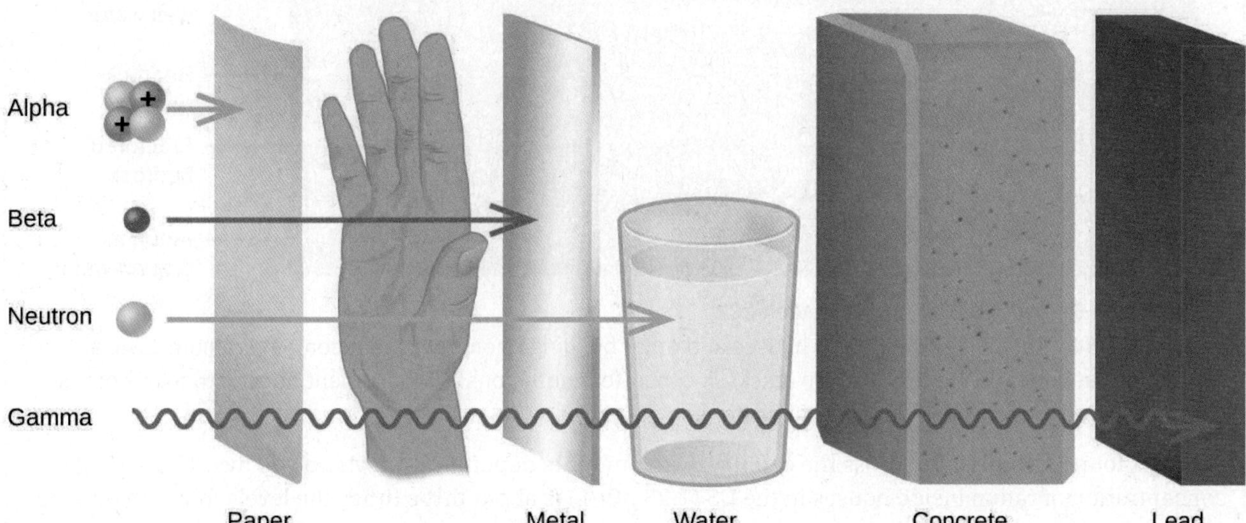

FIGURE 21.33 The ability of different types of radiation to pass through material is shown. From least to most penetrating, they are alpha < beta < neutron < gamma.

Chemistry in Everyday Life

Radon Exposure

For many people, one of the largest sources of exposure to radiation is from radon gas (Rn-222). Radon-222 is an α emitter with a half–life of 3.82 days. It is one of the products of the radioactive decay series of U-238 (Figure 21.9), which is found in trace amounts in soil and rocks. The radon gas that is produced slowly escapes from the ground and gradually seeps into homes and other structures above. Since it is about eight times more dense than air, radon gas accumulates in basements and lower floors, and slowly diffuses throughout buildings (Figure 21.34).

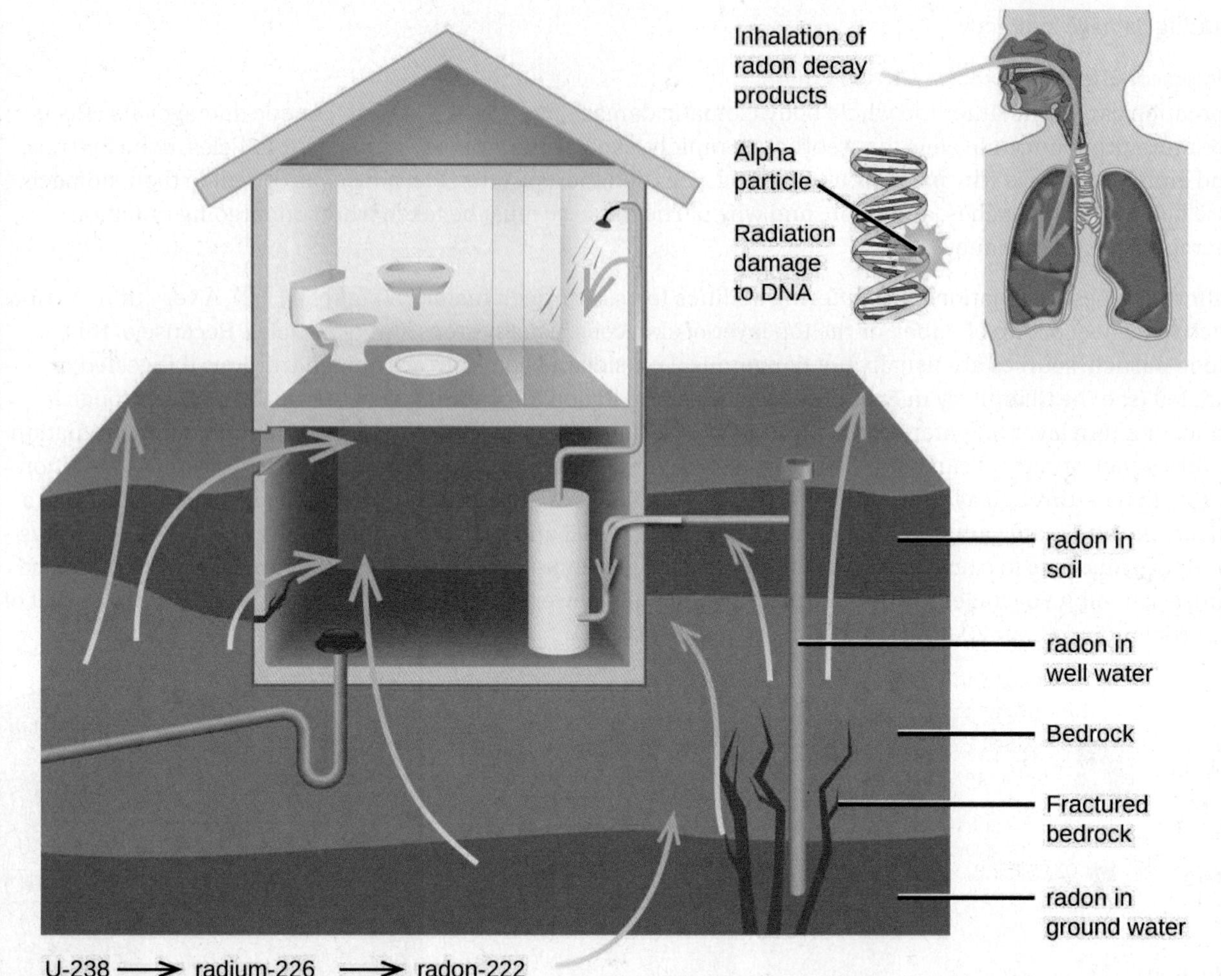

FIGURE 21.34 Radon-222 seeps into houses and other buildings from rocks that contain uranium-238, a radon emitter. The radon enters through cracks in concrete foundations and basement floors, stone or porous cinderblock foundations, and openings for water and gas pipes.

Radon is found in buildings across the country, with amounts depending on where you live. The average concentration of radon inside houses in the US (1.25 pCi/L) is about three times the levels found in outside air, and about one in six houses have radon levels high enough that remediation efforts to reduce the radon concentration are recommended. Exposure to radon increases one's risk of getting cancer (especially lung cancer), and high radon levels can be as bad for health as smoking a carton of cigarettes a day. Radon is the number one cause of lung cancer in nonsmokers and the second leading cause of lung cancer overall. Radon exposure is believed to cause over 20,000 deaths in the US per year.

Measuring Radiation Exposure

Several different devices are used to detect and measure radiation, including Geiger counters, scintillation counters (scintillators), and radiation dosimeters (Figure 21.35). Probably the best-known radiation instrument, the **Geiger counter** (also called the Geiger-Müller counter) detects and measures radiation. Radiation causes the ionization of the gas in a Geiger-Müller tube. The rate of ionization is proportional to the amount of radiation. A **scintillation counter** contains a scintillator—a material that emits light (luminesces) when excited by ionizing radiation—and a sensor that converts the light into an electric signal. **Radiation dosimeters** also measure ionizing radiation and are often used to determine personal radiation exposure. Commonly used types are electronic, film badge, thermoluminescent, and quartz fiber dosimeters.

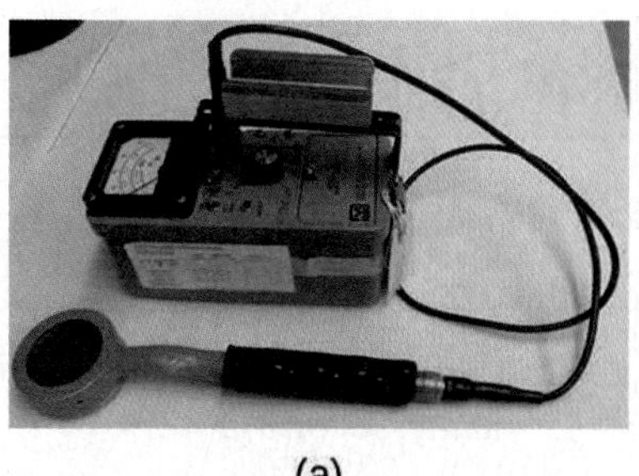
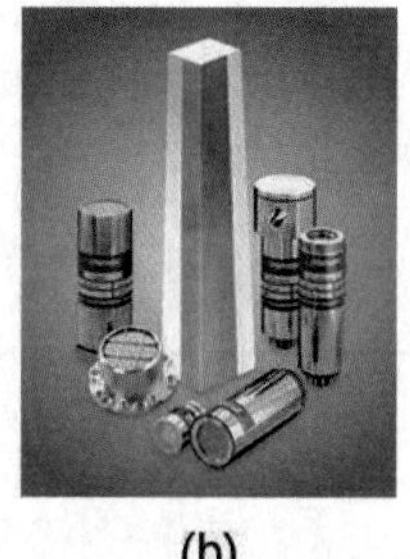
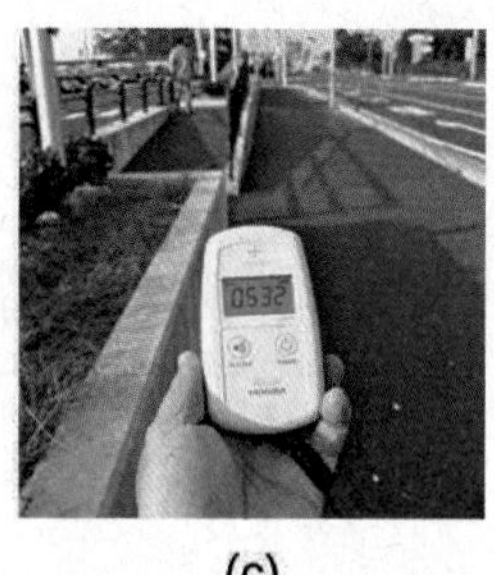

(a) (b) (c)

FIGURE 21.35 Devices such as (a) Geiger counters, (b) scintillators, and (c) dosimeters can be used to measure radiation. (credit c: modification of work by "osaMu"/Wikimedia commons)

A variety of units are used to measure various aspects of radiation (Figure 21.36). The SI unit for rate of radioactive decay is the **becquerel (Bq)**, with 1 Bq = 1 disintegration per second. The **curie (Ci)** and **millicurie (mCi)** are much larger units and are frequently used in medicine (1 curie = 1 Ci = 3.7×10^{10} disintegrations per second). The SI unit for measuring radiation dose is the **gray (Gy)**, with 1 Gy = 1 J of energy absorbed per kilogram of tissue. In medical applications, the **radiation absorbed dose (rad)** is more often used (1 rad = 0.01 Gy; 1 rad results in the absorption of 0.01 J/kg of tissue). The SI unit measuring tissue damage caused by radiation is the **sievert (Sv)**. This takes into account both the energy and the biological effects of the type of radiation involved in the radiation dose. The **roentgen equivalent for man (rem)** is the unit for radiation damage that is used most frequently in medicine (100 rem = 1 Sv). Note that the tissue damage units (rem or Sv) includes the energy of the radiation dose (rad or Gy) along with a biological factor referred to as the **RBE** (for **relative biological effectiveness**) that is an approximate measure of the relative damage done by the radiation. These are related by:

$$\text{number of rems} = \text{RBE} \times \text{number of rads}$$

with RBE approximately 10 for α radiation, 2(+) for protons and neutrons, and 1 for β and γ radiation.

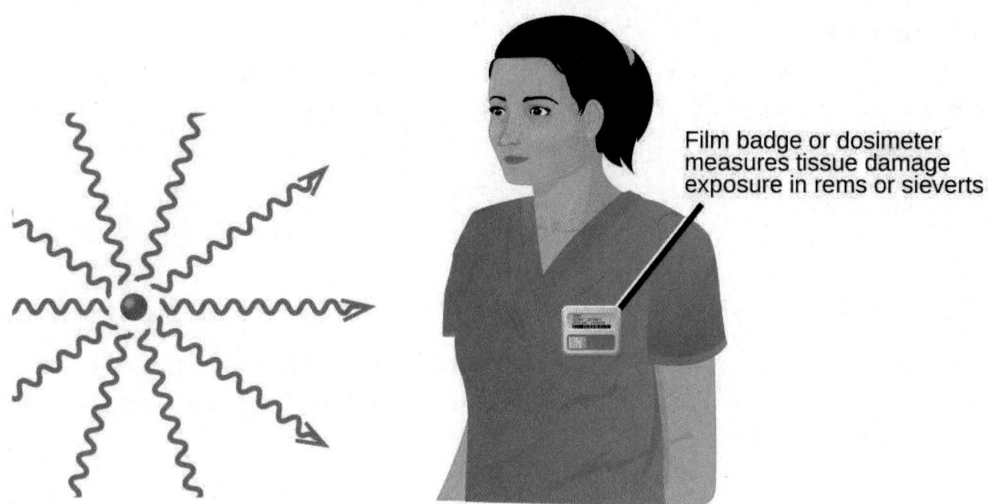

FIGURE 21.36 Different units are used to measure the rate of emission from a radioactive source, the energy that is absorbed from the source, and the amount of damage the absorbed radiation does.

Units of Radiation Measurement

Table 21.4 summarizes the units used for measuring radiation.

Units Used for Measuring Radiation

Measurement Purpose	Unit	Quantity Measured	Description
activity of source	becquerel (Bq)	radioactive decays or emissions	amount of sample that undergoes 1 decay/second
	curie (Ci)		amount of sample that undergoes 3.7 $\times 10^{10}$ decays/second
absorbed dose	gray (Gy)	energy absorbed per kg of tissue	1 Gy = 1 J/kg tissue
	radiation absorbed dose (rad)		1 rad = 0.01 J/kg tissue
biologically effective dose	sievert (Sv)	tissue damage	Sv = RBE × Gy
	roentgen equivalent for man (rem)		Rem = RBE × rad

TABLE 21.4

EXAMPLE 21.8

Amount of Radiation

Cobalt-60 ($t_{1/2}$ = 5.26 y) is used in cancer therapy since the γ rays it emits can be focused in small areas where the cancer is located. A 5.00-g sample of Co-60 is available for cancer treatment.

(a) What is its activity in Bq?

(b) What is its activity in Ci?

Solution

The activity is given by:

$$\text{Activity} = \lambda N = \left(\frac{\ln 2}{t_{1/2}}\right) N = \left(\frac{\ln 2}{5.26\text{ y}}\right) \times 5.00\text{ g} = 0.659\ \frac{\text{g}}{\text{y}}\ \text{of Co}-60\text{ that decay}$$

And to convert this to decays per second:

$$0.659\ \frac{\text{g}}{\text{y}} \times \frac{1\text{ y}}{365\text{ d}} \times \frac{1\text{ d}}{24\text{ h}} \times \frac{1\text{ h}}{3600\text{ s}} \times \frac{1\text{ mol}}{59.9\text{ g}} \times \frac{6.02 \times 10^{23}\text{ atoms}}{1\text{ mol}} \times \frac{1\text{ decay}}{1\text{ atom}}$$

$$= 2.10 \times 10^{14}\ \frac{\text{decay}}{\text{s}}$$

(a) Since 1 Bq = $\frac{1\text{ decay}}{\text{s}}$, the activity in Becquerel (Bq) is:

$$2.10 \times 10^{14}\ \frac{\text{decay}}{\text{s}} \times \left(\frac{1\text{ Bq}}{1\ \frac{\text{decay}}{\text{s}}}\right) = 2.10 \times 10^{14}\text{ Bq}$$

(b) Since 1 Ci = $\frac{3.7 \times 10^{11}\text{ decay}}{\text{s}}$, the activity in curie (Ci) is:

$$2.10 \times 10^{14}\ \frac{\text{decay}}{\text{s}} \times \left(\frac{1\text{ Ci}}{\frac{3.7 \times 10^{11}\text{ decay}}{\text{s}}}\right) = 5.7 \times 10^{2}\text{ Ci}$$

Check Your Learning

Tritium is a radioactive isotope of hydrogen ($t_{1/2}$ = 12.32 y) that has several uses, including self-powered lighting, in which electrons emitted in tritium radioactive decay cause phosphorus to glow. Its nucleus contains one proton and two neutrons, and the atomic mass of tritium is 3.016 amu. What is the activity of a sample containing 1.00mg of tritium (a) in Bq and (b) in Ci?

Answer:

(a) 3.56×10^{11} Bq; (b) 0.962 Ci

Effects of Long-term Radiation Exposure on the Human Body

The effects of radiation depend on the type, energy, and location of the radiation source, and the length of exposure. As shown in Figure 21.37, the average person is exposed to background radiation, including cosmic rays from the sun and radon from uranium in the ground (see the Chemistry in Everyday Life feature on Radon Exposure); radiation from medical exposure, including CAT scans, radioisotope tests, X-rays, and so on; and small amounts of radiation from other human activities, such as airplane flights (which are bombarded by increased numbers of cosmic rays in the upper atmosphere), radioactivity from consumer products, and a variety of radionuclides that enter our bodies when we breathe (for example, carbon-14) or through the food chain (for example, potassium-40, strontium-90, and iodine-131).

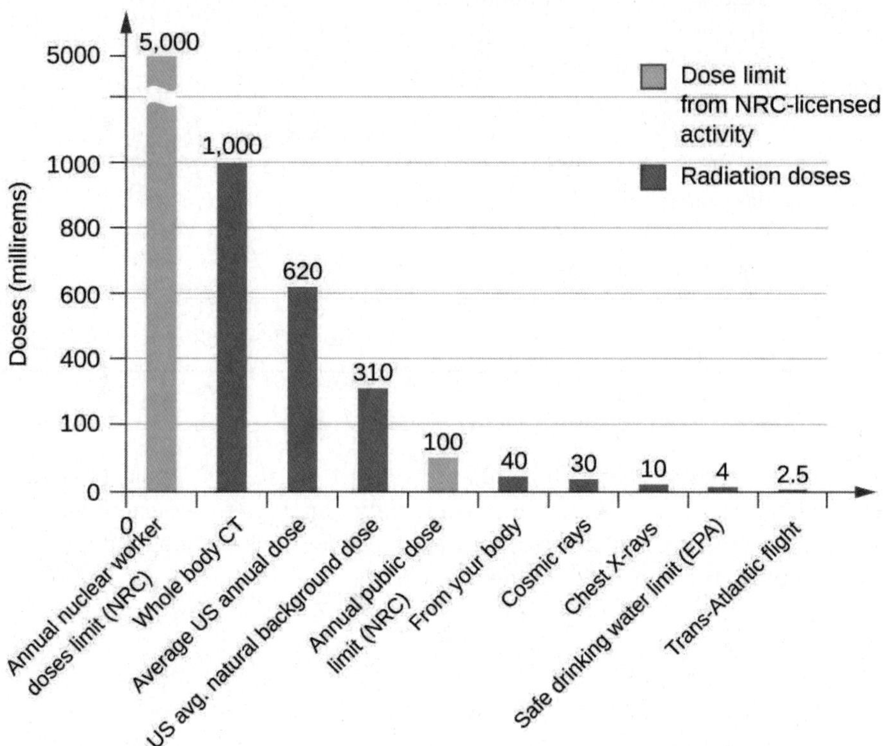

FIGURE 21.37 The total annual radiation exposure for a person in the US is about 620 mrem. The various sources and their relative amounts are shown in this bar graph. (source: U.S. Nuclear Regulatory Commission)

A short-term, sudden dose of a large amount of radiation can cause a wide range of health effects, from changes in blood chemistry to death. Short-term exposure to tens of rems of radiation will likely cause very noticeable symptoms or illness; a dose of about 500 rems is estimated to have a 50% probability of causing the death of the victim within 30 days of exposure. Exposure to radioactive emissions has a cumulative effect on the body during a person's lifetime, which is another reason why it is important to avoid any unnecessary exposure to radiation. Health effects of short-term exposure to radiation are shown in Table 21.5.

Health Effects of Radiation[2]

Exposure (rem)	Health Effect	Time to Onset (without treatment)
5–10	changes in blood chemistry	—
50	nausea	hours
55	fatigue	—
70	vomiting	—
75	hair loss	2–3 weeks
90	diarrhea	—
100	hemorrhage	—
400	possible death	within 2 months
	destruction of intestinal lining	—
1000	internal bleeding	—
	death	1–2 weeks
	damage to central nervous system	—
2000	loss of consciousness;	minutes
	death	hours to days

TABLE 21.5

It is impossible to avoid some exposure to ionizing radiation. We are constantly exposed to background radiation from a variety of natural sources, including cosmic radiation, rocks, medical procedures, consumer products, and even our own atoms. We can minimize our exposure by blocking or shielding the radiation, moving farther from the source, and limiting the time of exposure.

2 Source: US Environmental Protection Agency

Key Terms

alpha (α) decay loss of an alpha particle during radioactive decay

alpha particle (α or ${}^{4}_{2}\text{He}$ or ${}^{4}_{2}\alpha$) high-energy helium nucleus; a helium atom that has lost two electrons and contains two protons and two neutrons

antimatter particles with the same mass but opposite properties (such as charge) of ordinary particles

band of stability (also, belt of stability, zone of stability, or valley of stability) region of graph of number of protons versus number of neutrons containing stable (nonradioactive) nuclides

becquerel (Bq) SI unit for rate of radioactive decay; 1 Bq = 1 disintegration/s

beta (β) decay breakdown of a neutron into a proton, which remains in the nucleus, and an electron, which is emitted as a beta particle

beta particle (β or ${}^{0}_{-1}\text{e}$ or ${}^{0}_{-1}\beta$) high-energy electron

binding energy per nucleon total binding energy for the nucleus divided by the number of nucleons in the nucleus

chain reaction repeated fission caused when the neutrons released in fission bombard other atoms

chemotherapy similar to internal radiation therapy, but chemical rather than radioactive substances are introduced into the body to kill cancer cells

containment system (also, shield) a three-part structure of materials that protects the exterior of a nuclear fission reactor and operating personnel from the high temperatures, pressures, and radiation levels inside the reactor

control rod material inserted into the fuel assembly that absorbs neutrons and can be raised or lowered to adjust the rate of a fission reaction

critical mass amount of fissionable material that will support a self-sustaining (nuclear fission) chain reaction

curie (Ci) larger unit for rate of radioactive decay frequently used in medicine; 1 Ci = 3.7×10^{10} disintegrations/s

daughter nuclide nuclide produced by the radioactive decay of another nuclide; may be stable or may decay further

electron capture combination of a core electron with a proton to yield a neutron within the nucleus

electron volt (eV) measurement unit of nuclear binding energies, with 1 eV equaling the amount energy due to the moving an electron across an electric potential difference of 1 volt

external beam radiation therapy radiation delivered by a machine outside the body

fissile (or fissionable) when a material is capable of sustaining a nuclear fission reaction

fission splitting of a heavier nucleus into two or more lighter nuclei, usually accompanied by the conversion of mass into large amounts of energy

fusion combination of very light nuclei into heavier nuclei, accompanied by the conversion of mass into large amounts of energy

fusion reactor nuclear reactor in which fusion reactions of light nuclei are controlled

gamma (γ) emission decay of an excited-state nuclide accompanied by emission of a gamma ray

gamma ray (γ or ${}^{0}_{0}\gamma$) short wavelength, high-energy electromagnetic radiation that exhibits wave-particle duality

Geiger counter instrument that detects and measures radiation via the ionization produced in a Geiger-Müller tube

gray (Gy) SI unit for measuring radiation dose; 1 Gy = 1 J absorbed/kg tissue

half-life ($t_{1/2}$) time required for half of the atoms in a radioactive sample to decay

internal radiation therapy (also, brachytherapy) radiation from a radioactive substance introduced into the body to kill cancer cells

ionizing radiation radiation that can cause a molecule to lose an electron and form an ion

magic number nuclei with specific numbers of nucleons that are within the band of stability

mass defect difference between the mass of an atom and the summed mass of its constituent subatomic particles (or the mass "lost" when nucleons are brought together to form a nucleus)

mass-energy equivalence equation Albert Einstein's relationship showing that mass and energy are equivalent

millicurie (mCi) larger unit for rate of radioactive decay frequently used in medicine; 1 Ci = 3.7×10^{10} disintegrations/s

nonionizing radiation radiation that speeds up the movement of atoms and molecules; it is equivalent to heating a sample, but is not energetic enough to cause the ionization of molecules

nuclear binding energy energy lost when an

atom's nucleons are bound together (or the energy needed to break a nucleus into its constituent protons and neutrons)

nuclear chemistry study of the structure of atomic nuclei and processes that change nuclear structure

nuclear fuel fissionable isotope present in sufficient quantities to provide a self-sustaining chain reaction in a nuclear reactor

nuclear moderator substance that slows neutrons to a speed low enough to cause fission

nuclear reaction change to a nucleus resulting in changes in the atomic number, mass number, or energy state

nuclear reactor environment that produces energy via nuclear fission in which the chain reaction is controlled and sustained without explosion

nuclear transmutation conversion of one nuclide into another nuclide

nucleon collective term for protons and neutrons in a nucleus

nuclide nucleus of a particular isotope

parent nuclide unstable nuclide that changes spontaneously into another (daughter) nuclide

particle accelerator device that uses electric and magnetic fields to increase the kinetic energy of nuclei used in transmutation reactions

positron (${}^{0}_{+1}\beta$ **or** ${}^{0}_{+1}e$) antiparticle to the electron; it has identical properties to an electron, except for having the opposite (positive) charge

positron emission (also, β^+ decay) conversion of a proton into a neutron, which remains in the nucleus, and a positron, which is emitted

radiation absorbed dose (rad) SI unit for measuring radiation dose, frequently used in medical applications; 1 rad = 0.01 Gy

radiation dosimeter device that measures ionizing radiation and is used to determine personal radiation exposure

radiation therapy use of high-energy radiation to damage the DNA of cancer cells, which kills them or keeps them from dividing

radioactive decay spontaneous decay of an unstable nuclide into another nuclide

radioactive decay series chains of successive disintegrations (radioactive decays) that ultimately lead to a stable end-product

radioactive tracer (also, radioactive label) radioisotope used to track or follow a substance by monitoring its radioactive emissions

radioactivity phenomenon exhibited by an unstable nucleon that spontaneously undergoes change into a nucleon that is more stable; an unstable nucleon is said to be radioactive

radiocarbon dating highly accurate means of dating objects 30,000–50,000 years old that were derived from once-living matter; achieved by calculating the ratio of ${}^{14}_{6}C : {}^{12}_{6}C$ in the object vs. the ratio of ${}^{14}_{6}C : {}^{12}_{6}C$ in the present-day atmosphere

radioisotope isotope that is unstable and undergoes conversion into a different, more stable isotope

radiometric dating use of radioisotopes and their properties to date the formation of objects such as archeological artifacts, formerly living organisms, or geological formations

reactor coolant assembly used to carry the heat produced by fission in a reactor to an external boiler and turbine where it is transformed into electricity

relative biological effectiveness (RBE) measure of the relative damage done by radiation

roentgen equivalent man (rem) unit for radiation damage, frequently used in medicine; 100 rem = 1 Sv

scintillation counter instrument that uses a scintillator—a material that emits light when excited by ionizing radiation—to detect and measure radiation

sievert (Sv) SI unit measuring tissue damage caused by radiation; takes into account energy and biological effects of radiation

strong nuclear force force of attraction between nucleons that holds a nucleus together

subcritical mass amount of fissionable material that cannot sustain a chain reaction; less than a critical mass

supercritical mass amount of material in which there is an increasing rate of fission

transmutation reaction bombardment of one type of nuclei with other nuclei or neutrons

transuranium element element with an atomic number greater than 92; these elements do not occur in nature

Key Equations

$E = mc^2$

decay rate = λN

$t_{1/2} = \frac{\ln 2}{\lambda} = \frac{0.693}{\lambda}$

rem = RBE × rad

Sv = RBE × Gy

Summary

21.1 Nuclear Structure and Stability

An atomic nucleus consists of protons and neutrons, collectively called nucleons. Although protons repel each other, the nucleus is held tightly together by a short-range, but very strong, force called the strong nuclear force. A nucleus has less mass than the total mass of its constituent nucleons. This "missing" mass is the mass defect, which has been converted into the binding energy that holds the nucleus together according to Einstein's mass-energy equivalence equation, $E = mc^2$. Of the many nuclides that exist, only a small number are stable. Nuclides with even numbers of protons or neutrons, or those with magic numbers of nucleons, are especially likely to be stable. These stable nuclides occupy a narrow band of stability on a graph of number of protons versus number of neutrons. The binding energy per nucleon is largest for the elements with mass numbers near 56; these are the most stable nuclei.

21.2 Nuclear Equations

Nuclei can undergo reactions that change their number of protons, number of neutrons, or energy state. Many different particles can be involved in nuclear reactions. The most common are protons, neutrons, positrons (which are positively charged electrons), alpha (α) particles (which are high-energy helium nuclei), beta (β) particles (which are high-energy electrons), and gamma (γ) rays (which compose high-energy electromagnetic radiation). As with chemical reactions, nuclear reactions are always balanced. When a nuclear reaction occurs, the total mass (number) and the total charge remain unchanged.

21.3 Radioactive Decay

Nuclei that have unstable n:p ratios undergo spontaneous radioactive decay. The most common types of radioactivity are α decay, β decay, γ emission, positron emission, and electron capture. Nuclear reactions also often involve γ rays, and some nuclei decay by electron capture. Each of these modes of decay leads to the formation of a new nucleus with a more stable n:p ratio. Some substances undergo radioactive decay series, proceeding through multiple decays before ending in a stable isotope. All nuclear decay processes follow first-order kinetics, and each radioisotope has its own characteristic half-life, the time that is required for half of its atoms to decay. Because of the large differences in stability among nuclides, there is a very wide range of half-lives of radioactive substances. Many of these substances have found useful applications in medical diagnosis and treatment, determining the age of archaeological and geological objects, and more.

21.4 Transmutation and Nuclear Energy

It is possible to produce new atoms by bombarding other atoms with nuclei or high-speed particles. The products of these transmutation reactions can be stable or radioactive. A number of artificial elements, including technetium, astatine, and the transuranium elements, have been produced in this way.

Nuclear power as well as nuclear weapon detonations can be generated through fission (reactions in which a heavy nucleus is split into two or more lighter nuclei and several neutrons). Because the neutrons may induce additional fission reactions when they combine with other heavy nuclei, a chain reaction can result. Useful power is obtained if the fission process is carried out in a nuclear reactor. The conversion of light nuclei into heavier nuclei (fusion) also produces energy. At present, this energy has not been contained adequately and is too expensive to be feasible for commercial energy production.

21.5 Uses of Radioisotopes

Compounds known as radioactive tracers can be used to follow reactions, track the distribution of a substance, diagnose and treat medical conditions, and much more. Other radioactive substances are helpful for controlling pests, visualizing structures, providing fire warnings, and for many other applications. Hundreds of millions of nuclear medicine tests and procedures, using a wide variety of radioisotopes with relatively short half-lives, are performed every year in the US. Most of these radioisotopes have relatively short half-lives; some are short enough that the radioisotope must be made on-site at medical facilities. Radiation therapy uses high-energy radiation to kill cancer cells by damaging their DNA. The radiation used for this treatment may be delivered externally or internally.

21.6 Biological Effects of Radiation

We are constantly exposed to radiation from a variety of naturally occurring and human-produced sources. This radiation can affect living organisms. Ionizing radiation is the most harmful because it can ionize molecules or break chemical bonds, which damages the molecule and causes malfunctions in cell processes. It can also create reactive hydroxyl radicals that damage biological molecules and disrupt physiological processes. Radiation can cause somatic or genetic damage, and is most harmful to rapidly reproducing cells. Types of radiation differ in their ability to penetrate material and damage tissue, with alpha particles the least penetrating but potentially most damaging and gamma rays the most penetrating.

Various devices, including Geiger counters, scintillators, and dosimeters, are used to detect and measure radiation, and monitor radiation exposure. We use several units to measure radiation: becquerels or curies for rates of radioactive decay; gray or rads for energy absorbed; and rems or sieverts for biological effects of radiation. Exposure to radiation can cause a wide range of health effects, from minor to severe, and including death. We can minimize the effects of radiation by shielding with dense materials such as lead, moving away from the source, and limiting time of exposure.

Exercises

21.1 Nuclear Structure and Stability

1. Write the following isotopes in hyphenated form (e.g., “carbon-14”)
   (a) ${}^{24}_{11}\text{Na}$
   (b) ${}^{29}_{13}\text{Al}$
   (c) ${}^{73}_{36}\text{Kr}$
   (d) ${}^{194}_{77}\text{Ir}$
2. Write the following isotopes in nuclide notation (e.g., "${}^{14}_{6}\text{C}$")
   (a) oxygen-14
   (b) copper-70
   (c) tantalum-175
   (d) francium-217
3. For the following isotopes that have missing information, fill in the missing information to complete the notation
   (a) ${}^{34}_{14}\text{X}$
   (b) ${}^{36}_{\text{X}}\text{P}$
   (c) ${}^{57}_{\text{X}}\text{Mn}$
   (d) ${}^{121}_{56}\text{X}$
4. For each of the isotopes in Exercise 21.1, determine the numbers of protons, neutrons, and electrons in a neutral atom of the isotope.
5. Write the nuclide notation, including charge if applicable, for atoms with the following characteristics:
   (a) 25 protons, 20 neutrons, 24 electrons
   (b) 45 protons, 24 neutrons, 43 electrons
   (c) 53 protons, 89 neutrons, 54 electrons
   (d) 97 protons, 146 neutrons, 97 electrons
6. Calculate the density of the ${}^{24}_{12}\text{Mg}$ nucleus in g/mL, assuming that it has the typical nuclear diameter of 1×10^{-13} cm and is spherical in shape.
7. What are the two principal differences between nuclear reactions and ordinary chemical changes?

8. The mass of the atom $^{23}_{11}Na$ is 22.9898 amu.
   (a) Calculate its binding energy per atom in millions of electron volts.
   (b) Calculate its binding energy per nucleon.
9. Which of the following nuclei lie within the band of stability shown in Figure 21.2?
   (a) chlorine-37
   (b) calcium-40
   (c) ^{204}Bi
   (d) ^{56}Fe
   (e) ^{206}Pb
   (f) ^{211}Pb
   (g) ^{222}Rn
   (h) carbon-14
10. Which of the following nuclei lie within the band of stability shown in Figure 21.2?
   (a) argon-40
   (b) oxygen-16
   (c) ^{122}Ba
   (d) ^{58}Ni
   (e) ^{205}Tl
   (f) ^{210}Tl
   (g) ^{226}Ra
   (h) magnesium-24

21.2 Nuclear Equations

11. Write a brief description or definition of each of the following:
   (a) nucleon
   (b) α particle
   (c) β particle
   (d) positron
   (e) γ ray
   (f) nuclide
   (g) mass number
   (h) atomic number
12. Which of the various particles (α particles, β particles, and so on) that may be produced in a nuclear reaction are actually nuclei?
13. Complete each of the following equations by adding the missing species:
   (a) $^{27}_{13}Al + ^{4}_{2}He \longrightarrow ? + ^{1}_{0}n$
   (b) $^{239}_{94}Pu + ? \longrightarrow ^{242}_{96}Cm + ^{1}_{0}n$
   (c) $^{14}_{7}N + ^{4}_{2}He \longrightarrow ? + ^{1}_{1}H$
   (d) $^{235}_{92}U \longrightarrow ? + ^{135}_{55}Cs + 4^{1}_{0}n$
14. Complete each of the following equations:
   (a) $^{7}_{3}Li + ? \longrightarrow 2^{4}_{2}He$
   (b) $^{14}_{6}C \longrightarrow ^{14}_{7}N + ?$
   (c) $^{27}_{13}Al + ^{4}_{2}He \longrightarrow ? + ^{1}_{0}n$
   (d) $^{250}_{96}Cm \longrightarrow ? + ^{98}_{38}Sr + 4^{1}_{0}n$
15. Write a balanced equation for each of the following nuclear reactions:
   (a) the production of ^{17}O from ^{14}N by α particle bombardment
   (b) the production of ^{14}C from ^{14}N by neutron bombardment
   (c) the production of ^{233}Th from ^{232}Th by neutron bombardment
   (d) the production of ^{239}U from ^{238}U by $^{2}_{1}H$ bombardment

16. Technetium-99 is prepared from ^{98}Mo. Molybdenum-98 combines with a neutron to give molybdenum-99, an unstable isotope that emits a β particle to yield an excited form of technetium-99, represented as $^{99}Tc^*$. This excited nucleus relaxes to the ground state, represented as ^{99}Tc, by emitting a γ ray. The ground state of ^{99}Tc then emits a β particle. Write the equations for each of these nuclear reactions.
17. The mass of the atom $^{19}_{9}F$ is 18.99840 amu.
    (a) Calculate its binding energy per atom in millions of electron volts.
    (b) Calculate its binding energy per nucleon.
18. For the reaction $^{14}_{6}C \longrightarrow {}^{14}_{7}N + ?$, if 100.0 g of carbon reacts, what volume of nitrogen gas (N_2) is produced at 273K and 1 atm?

21.3 Radioactive Decay

19. What are the types of radiation emitted by the nuclei of radioactive elements?
20. What changes occur to the atomic number and mass of a nucleus during each of the following decay scenarios?
    (a) an α particle is emitted
    (b) a β particle is emitted
    (c) γ radiation is emitted
    (d) a positron is emitted
    (e) an electron is captured
21. What is the change in the nucleus that results from the following decay scenarios?
    (a) emission of a β particle
    (b) emission of a β^+ particle
    (c) capture of an electron
22. Many nuclides with atomic numbers greater than 83 decay by processes such as electron emission. Explain the observation that the emissions from these unstable nuclides also normally include α particles.
23. Why is electron capture accompanied by the emission of an X-ray?
24. Explain, in terms of Figure 21.2, how unstable heavy nuclides (atomic number > 83) may decompose to form nuclides of greater stability (a) if they are below the band of stability and (b) if they are above the band of stability.
25. Which of the following nuclei is most likely to decay by positron emission? Explain your choice.
    (a) chromium-53
    (b) manganese-51
    (c) iron-59
26. The following nuclei do not lie in the band of stability. How would they be expected to decay? Explain your answer.
    (a) $^{34}_{15}P$
    (b) $^{239}_{92}U$
    (c) $^{38}_{20}Ca$
    (d) $^{3}_{1}H$
    (e) $^{245}_{94}Pu$
27. The following nuclei do not lie in the band of stability. How would they be expected to decay?
    (a) $^{28}_{15}P$
    (b) $^{235}_{92}U$
    (c) $^{37}_{20}Ca$
    (d) $^{9}_{3}Li$
    (e) $^{245}_{96}Cm$

28. Predict by what mode(s) of spontaneous radioactive decay each of the following unstable isotopes might proceed:
(a) ${}^{6}_{2}He$
(b) ${}^{60}_{30}Zn$
(c) ${}^{235}_{91}Pa$
(d) ${}^{241}_{94}Np$
(e) ${}^{18}F$
(f) ${}^{129}Ba$
(g) ${}^{237}Pu$

29. Write a nuclear reaction for each step in the formation of ${}^{218}_{84}Po$ from ${}^{238}_{98}U$, which proceeds by a series of decay reactions involving the step-wise emission of α, β, β, α, α, α particles, in that order.

30. Write a nuclear reaction for each step in the formation of ${}^{208}_{82}Pb$ from ${}^{228}_{90}Th$, which proceeds by a series of decay reactions involving the step-wise emission of α, α, α, α, β, β, α particles, in that order.

31. Define the term half-life and illustrate it with an example.

32. A 1.00×10^{-6}-g sample of nobelium, ${}^{254}_{102}No$, has a half-life of 55 seconds after it is formed. What is the percentage of ${}^{254}_{102}No$ remaining at the following times?
(a) 5.0 min after it forms
(b) 1.0 h after it forms

33. ${}^{239}Pu$ is a nuclear waste byproduct with a half-life of 24,000 y. What fraction of the ${}^{239}Pu$ present today will be present in 1000 y?

34. The isotope ${}^{208}Tl$ undergoes β decay with a half-life of 3.1 min.
(a) What isotope is produced by the decay?
(b) How long will it take for 99.0% of a sample of pure ${}^{208}Tl$ to decay?
(c) What percentage of a sample of pure ${}^{208}Tl$ remains un-decayed after 1.0 h?

35. If 1.000 g of ${}^{226}_{88}Ra$ produces 0.0001 mL of the gas ${}^{222}_{86}Rn$ at STP (standard temperature and pressure) in 24 h, what is the half-life of ${}^{226}Ra$ in years?

36. The isotope ${}^{90}_{38}Sr$ is one of the extremely hazardous species in the residues from nuclear power generation. The strontium in a 0.500-g sample diminishes to 0.393 g in 10.0 y. Calculate the half-life.

37. Technetium-99 is often used for assessing heart, liver, and lung damage because certain technetium compounds are absorbed by damaged tissues. It has a half-life of 6.0 h. Calculate the rate constant for the decay of ${}^{99}_{43}Tc$.

38. What is the age of mummified primate skin that contains 8.25% of the original quantity of ${}^{14}C$?

39. A sample of rock was found to contain 8.23 mg of rubidium-87 and 0.47 mg of strontium-87.
(a) Calculate the age of the rock if the half-life of the decay of rubidium by β emission is 4.7×10^{10} y.
(b) If some ${}^{87}_{38}Sr$ was initially present in the rock, would the rock be younger, older, or the same age as the age calculated in (a)? Explain your answer.

40. A laboratory investigation shows that a sample of uranium ore contains 5.37 mg of ${}^{238}_{92}U$ and 2.52 mg of ${}^{206}_{82}Pb$. Calculate the age of the ore. The half-life of ${}^{238}_{92}U$ is 4.5×10^{9} yr.

41. Plutonium was detected in trace amounts in natural uranium deposits by Glenn Seaborg and his associates in 1941. They proposed that the source of this ${}^{239}Pu$ was the capture of neutrons by ${}^{238}U$ nuclei. Why is this plutonium not likely to have been trapped at the time the solar system formed 4.7×10^{9} years ago?

42. A ${}^{7}_{4}Be$ atom (mass = 7.0169 amu) decays into a ${}^{7}_{3}Li$ atom (mass = 7.0160 amu) by electron capture. How much energy (in millions of electron volts, MeV) is produced by this reaction?

43. A ${}^{8}_{5}B$ atom (mass = 8.0246 amu) decays into a ${}^{8}_{4}B$ atom (mass = 8.0053 amu) by loss of a β^{+} particle (mass = 0.00055 amu) or by electron capture. How much energy (in millions of electron volts) is produced by this reaction?

44. Isotopes such as ^{26}Al (half-life: 7.2×10^5 years) are believed to have been present in our solar system as it formed, but have since decayed and are now called extinct nuclides.
(a) ^{26}Al decays by β^+ emission or electron capture. Write the equations for these two nuclear transformations.
(b) The earth was formed about 4.7×10^9 (4.7 billion) years ago. How old was the earth when 99.999999% of the ^{26}Al originally present had decayed?
45. Write a balanced equation for each of the following nuclear reactions:
(a) bismuth-212 decays into polonium-212
(b) beryllium-8 and a positron are produced by the decay of an unstable nucleus
(c) neptunium-239 forms from the reaction of uranium-238 with a neutron and then spontaneously converts into plutonium-239
(d) strontium-90 decays into yttrium-90
46. Write a balanced equation for each of the following nuclear reactions:
(a) mercury-180 decays into platinum-176
(b) zirconium-90 and an electron are produced by the decay of an unstable nucleus
(c) thorium-232 decays and produces an alpha particle and a radium-228 nucleus, which decays into actinium-228 by beta decay
(d) neon-19 decays into fluorine-19

21.4 Transmutation and Nuclear Energy

47. Write the balanced nuclear equation for the production of the following transuranium elements:
(a) berkelium-244, made by the reaction of Am-241 and He-4
(b) fermium-254, made by the reaction of Pu-239 with a large number of neutrons
(c) lawrencium-257, made by the reaction of Cf-250 and B-11
(d) dubnium-260, made by the reaction of Cf-249 and N-15
48. How does nuclear fission differ from nuclear fusion? Why are both of these processes exothermic?
49. Both fusion and fission are nuclear reactions. Why is a very high temperature required for fusion, but not for fission?
50. Cite the conditions necessary for a nuclear chain reaction to take place. Explain how it can be controlled to produce energy, but not produce an explosion.
51. Describe the components of a nuclear reactor.
52. In usual practice, both a moderator and control rods are necessary to operate a nuclear chain reaction safely for the purpose of energy production. Cite the function of each and explain why both are necessary.
53. Describe how the potential energy of uranium is converted into electrical energy in a nuclear power plant.
54. The mass of a hydrogen atom (^1_1H) is 1.007825 amu; that of a tritium atom (^3_1H) is 3.01605 amu; and that of an α particle is 4.00150 amu. How much energy in kilojoules per mole of ^4_2He produced is released by the following fusion reaction: $^1_1H + ^3_1H \longrightarrow ^4_2He$.

21.5 Uses of Radioisotopes

55. How can a radioactive nuclide be used to show that the equilibrium:
$AgCl(s) \rightleftharpoons Ag^+(aq) + Cl^-(aq)$
is a dynamic equilibrium?
56. Technetium-99m has a half-life of 6.01 hours. If a patient injected with technetium-99m is safe to leave the hospital once 75% of the dose has decayed, when is the patient allowed to leave?
57. Iodine that enters the body is stored in the thyroid gland from which it is released to control growth and metabolism. The thyroid can be imaged if iodine-131 is injected into the body. In larger doses, I-133 is also used as a means of treating cancer of the thyroid. I-131 has a half-life of 8.70 days and decays by β^- emission.
(a) Write an equation for the decay.
(b) How long will it take for 95.0% of a dose of I-131 to decay?

21.6 Biological Effects of Radiation

58. If a hospital were storing radioisotopes, what is the minimum containment needed to protect against:
(a) cobalt-60 (a strong γ emitter used for irradiation)
(b) molybdenum-99 (a beta emitter used to produce technetium-99 for imaging)

59. Based on what is known about Radon-222's primary decay method, why is inhalation so dangerous?

60. Given specimens uranium-232 ($t_{1/2}$ = 68.9 y) and uranium-233 ($t_{1/2}$ = 159,200 y) of equal mass, which one would have greater activity and why?

61. A scientist is studying a 2.234 g sample of thorium-229 ($t_{1/2}$ = 7340 y) in a laboratory.
(a) What is its activity in Bq?
(b) What is its activity in Ci?

62. Given specimens neon-24 ($t_{1/2}$ = 3.38 min) and bismuth-211 ($t_{1/2}$ = 2.14 min) of equal mass, which one would have greater activity and why?

APPENDIX A

The Periodic Table

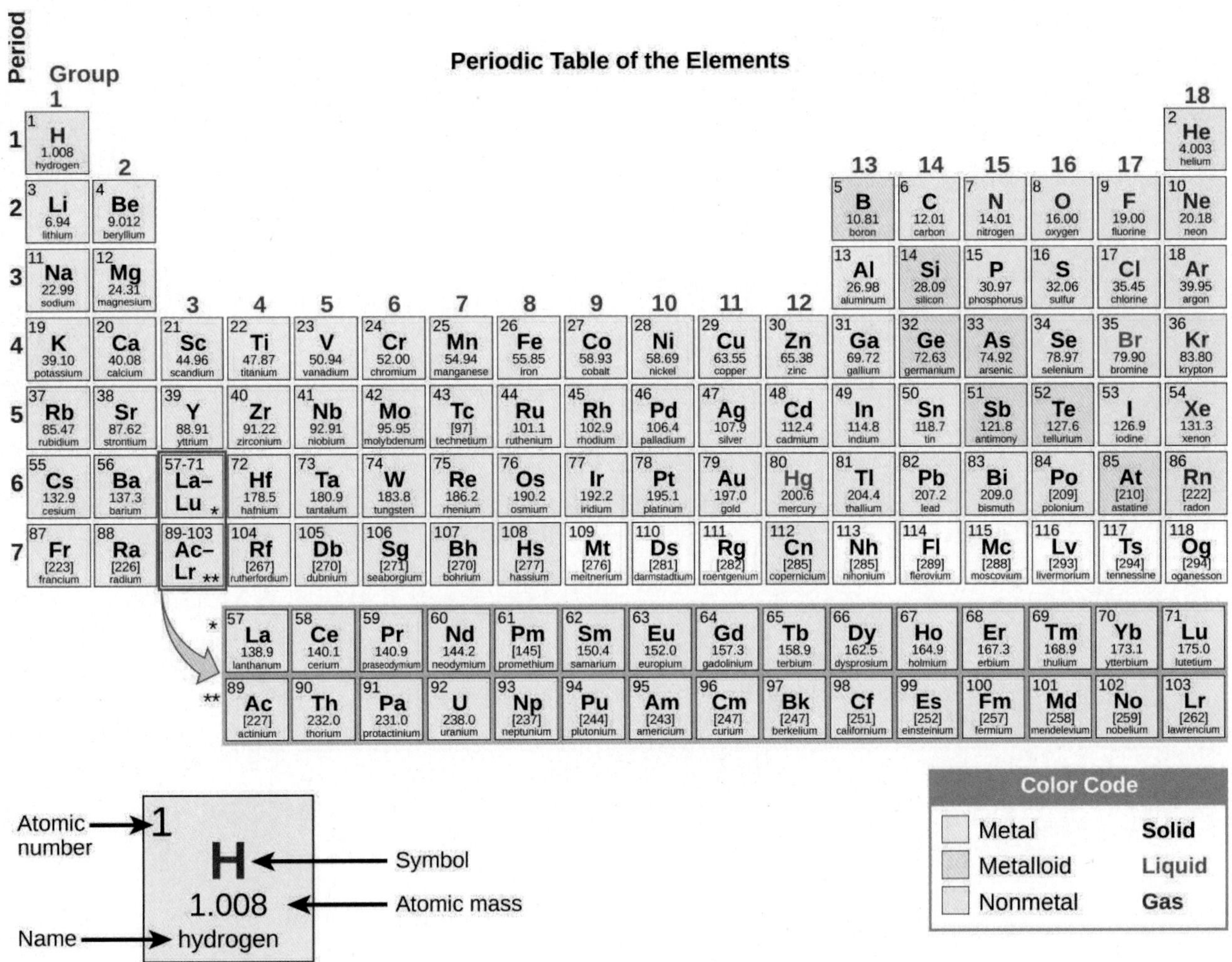

FIGURE A1

APPENDIX B

Essential Mathematics

Exponential Arithmetic

Exponential notation is used to express very large and very small numbers as a product of two numbers. The first number of the product, the *digit term*, is usually a number not less than 1 and not equal to or greater than 10. The second number of the product, the *exponential term*, is written as 10 with an exponent. Some examples of exponential notation are:

$$\begin{aligned} 1000 &= 1 \times 10^3 \\ 100 &= 1 \times 10^2 \\ 10 &= 1 \times 10^1 \\ 1 &= 1 \times 10^0 \\ 0.1 &= 1 \times 10^{-1} \\ 0.001 &= 1 \times 10^{-3} \\ 2386 &= 2.386 \times 1000 = 2.386 \times 10^3 \\ 0.123 &= 1.23 \times 0.1 = 1.23 \times 10^{-1} \end{aligned}$$

The power (exponent) of 10 is equal to the number of places the decimal is shifted to give the digit number. The exponential method is particularly useful notation for very large and very small numbers. For example, $1,230,000,000 = 1.23 \times 10^9$, and $0.00000000036 = 3.6 \times 10^{-10}$.

Addition of Exponentials

Convert all numbers to the same power of 10, add the digit terms of the numbers, and if appropriate, convert the digit term back to a number between 1 and 10 by adjusting the exponential term.

EXAMPLE B1

Adding Exponentials

Add 5.00×10^{-5} and 3.00×10^{-3}.

Solution

$$\begin{aligned} 3.00 \times 10^{-3} &= 300 \times 10^{-5} \\ (5.00 \times 10^{-5}) + (300 \times 10^{-5}) &= 305 \times 10^{-5} = 3.05 \times 10^{-3} \end{aligned}$$

Subtraction of Exponentials

Convert all numbers to the same power of 10, take the difference of the digit terms, and if appropriate, convert the digit term back to a number between 1 and 10 by adjusting the exponential term.

EXAMPLE B2

Subtracting Exponentials

Subtract 4.0×10^{-7} from 5.0×10^{-6}.

Solution

$$4.0 \times 10^{-7} = 0.40 \times 10^{-6}$$
$$(5.0 \times 10^{-6}) - (0.40 \times 10^{-6}) = 4.6 \times 10^{-6}$$

Multiplication of Exponentials

Multiply the digit terms in the usual way and add the exponents of the exponential terms.

EXAMPLE B3

Multiplying Exponentials

Multiply 4.2×10^{-8} by 2.0×10^{3}.

Solution

$$(4.2 \times 10^{-8}) \times (2.0 \times 10^{3}) = (4.2 \times 2.0) \times 10^{(-8)+(+3)} = 8.4 \times 10^{-5}$$

Division of Exponentials

Divide the digit term of the numerator by the digit term of the denominator and subtract the exponents of the exponential terms.

EXAMPLE B4

Dividing Exponentials

Divide 3.6×10^{-5} by 6.0×10^{-4}.

Solution

$$\frac{3.6 \times 10^{-5}}{6.0 \times 10^{-4}} = \left(\frac{3.6}{6.0}\right) \times 10^{(-5)-(-4)} = 0.60 \times 10^{-1} = 6.0 \times 10^{-2}$$

Squaring of Exponentials

Square the digit term in the usual way and multiply the exponent of the exponential term by 2.

EXAMPLE B5

Squaring Exponentials

Square the number 4.0×10^{-6}.

Solution

$$(4.0 \times 10^{-6})^2 = 4 \times 4 \times 10^{2 \times (-6)} = 16 \times 10^{-12} = 1.6 \times 10^{-11}$$

Cubing of Exponentials

Cube the digit term in the usual way and multiply the exponent of the exponential term by 3.

EXAMPLE B6

Cubing Exponentials

Cube the number 2×10^4.

Solution

$$(2 \times 10^4)^3 = 2 \times 2 \times 2 \times 10^{3 \times 4} = 8 \times 10^{12}$$

Taking Square Roots of Exponentials

If necessary, decrease or increase the exponential term so that the power of 10 is evenly divisible by 2. Extract the square root of the digit term and divide the exponential term by 2.

 **EXAMPLE B7**

Finding the Square Root of Exponentials

Find the square root of 1.6×10^{-7}.

Solution

$$1.6 \times 10^{-7} = 16 \times 10^{-8}$$

$$\sqrt{16 \times 10^{-8}} = \sqrt{16} \times \sqrt{10^{-8}} = \sqrt{16} \times 10^{-\frac{8}{2}} = 4.0 \times 10^{-4}$$

Significant Figures

A beekeeper reports that he has 525,341 bees. The last three figures of the number are obviously inaccurate, for during the time the keeper was counting the bees, some of them died and others hatched; this makes it quite difficult to determine the exact number of bees. It would have been more reasonable if the beekeeper had reported the number 525,000. In other words, the last three figures are not significant, except to set the position of the decimal point. Their exact values have no useful meaning in this situation. When reporting quantities, use only as many significant figures as the accuracy of the measurement warrants.

The importance of significant figures lies in their application to fundamental computation. In addition and subtraction, the sum or difference should contain as many digits to the right of the decimal as that in the least certain of the numbers used in the computation (indicated by underscoring in the following example).

 **EXAMPLE B8**

Addition and Subtraction with Significant Figures

Add 4.383 g and 0.0023 g.

Solution

$$\begin{array}{r} 4.38\underline{3}\text{ g} \\ 0.002\underline{3}\text{ g} \\ \hline 4.38\underline{5}\text{ g} \end{array}$$

In multiplication and division, the product or quotient should contain no more digits than that in the factor containing the least number of significant figures.

 **EXAMPLE B9**

Multiplication and Division with Significant Figures

Multiply 0.6238 by 6.6.

Solution

$$0.623\underline{8} \times 6.\underline{6} = 4.\underline{1}$$

When rounding numbers, increase the retained digit by 1 if it is followed by a number larger than 5 ("round up"). Do not change the retained digit if the digits that follow are less than 5 ("round down"). If the retained digit is followed by 5, round up if the retained digit is odd, or round down if it is even (after rounding, the retained digit will thus always be even).

The Use of Logarithms and Exponential Numbers

The common logarithm of a number (log) is the power to which 10 must be raised to equal that number. For example, the common logarithm of 100 is 2, because 10 must be raised to the second power to equal 100. Additional examples follow.

Logarithms and Exponential Numbers

Number	Number Expressed Exponentially	Common Logarithm
1000	10^3	3
10	10^1	1
1	10^0	0
0.1	10^{-1}	−1
0.001	10^{-3}	−3

TABLE B1

What is the common logarithm of 60? Because 60 lies between 10 and 100, which have logarithms of 1 and 2, respectively, the logarithm of 60 is 1.7782; that is,

$$60 = 10^{1.7782}$$

The common logarithm of a number less than 1 has a negative value. The logarithm of 0.03918 is −1.4069, or

$$0.03918 = 10^{-1.4069} = \frac{1}{10^{1.4069}}$$

To obtain the common logarithm of a number, use the *log* button on your calculator. To calculate a number from its logarithm, take the inverse log of the logarithm, or calculate 10^x (where x is the logarithm of the number).

The natural logarithm of a number (ln) is the power to which e must be raised to equal the number; e is the constant 2.7182818. For example, the natural logarithm of 10 is 2.303; that is,

$$10 = e^{2.303} = 2.7182818^{2.303}$$

To obtain the natural logarithm of a number, use the *ln* button on your calculator. To calculate a number from its natural logarithm, enter the natural logarithm and take the inverse ln of the natural logarithm, or calculate e^x (where x is the natural logarithm of the number).

Logarithms are exponents; thus, operations involving logarithms follow the same rules as operations involving exponents.

1. The logarithm of a product of two numbers is the sum of the logarithms of the two numbers.
$$\log xy = \log x + \log y, \text{ and } \ln xy = \ln x + \ln y$$

2. The logarithm of the number resulting from the division of two numbers is the difference between the logarithms of the two numbers.
$$\log \frac{x}{y} = \log x - \log y, \text{ and } \ln \frac{x}{y} = \ln x - \ln y$$

3. The logarithm of a number raised to an exponent is the product of the exponent and the logarithm of the number.
$$\log x^n = n\log x \text{ and } \ln x^n = n\ln x$$

The Solution of Quadratic Equations

Mathematical functions of this form are known as second-order polynomials or, more commonly, quadratic functions.

$$ax^2 + bx + c = 0$$

The solution or roots for any quadratic equation can be calculated using the following formula:

$$x = \frac{-b \pm \sqrt{b^2 - 4ac}}{2a}$$

 **EXAMPLE B10**

Solving Quadratic Equations

Solve the quadratic equation $3x^2 + 13x - 10 = 0$.

Solution

Substituting the values $a = 3$, $b = 13$, $c = -10$ in the formula, we obtain

$$x = \frac{-13 \pm \sqrt{(13)^2 - 4 \times 3 \times (-10)}}{2 \times 3}$$

$$x = \frac{-13 \pm \sqrt{169 + 120}}{6} = \frac{-13 \pm \sqrt{289}}{6} = \frac{-13 \pm 17}{6}$$

The two roots are therefore

$$x = \frac{-13 + 17}{6} = \frac{2}{3} \text{ and } x = \frac{-13 - 17}{6} = -5$$

Quadratic equations constructed on physical data always have real roots, and of these real roots, often only those having positive values are of any significance.

Two-Dimensional (*x*-*y*) Graphing

The relationship between any two properties of a system can be represented graphically by a two-dimensional data plot. Such a graph has two axes: a horizontal one corresponding to the independent variable, or the variable whose value is being controlled (x), and a vertical axis corresponding to the dependent variable, or the variable whose value is being observed or measured (y).

When the value of y is changing as a function of x (that is, different values of x correspond to different values of y), a graph of this change can be plotted or sketched. The graph can be produced by using specific values for (x,y) data pairs.

 **EXAMPLE B11**

Graphing the Dependence of *y* on *x*

x	*y*
1	5
2	10
3	7
4	14

This table contains the following points: (1,5), (2,10), (3,7), and (4,14). Each of these points can be plotted on a graph and connected to produce a graphical representation of the dependence of *y* on *x*.

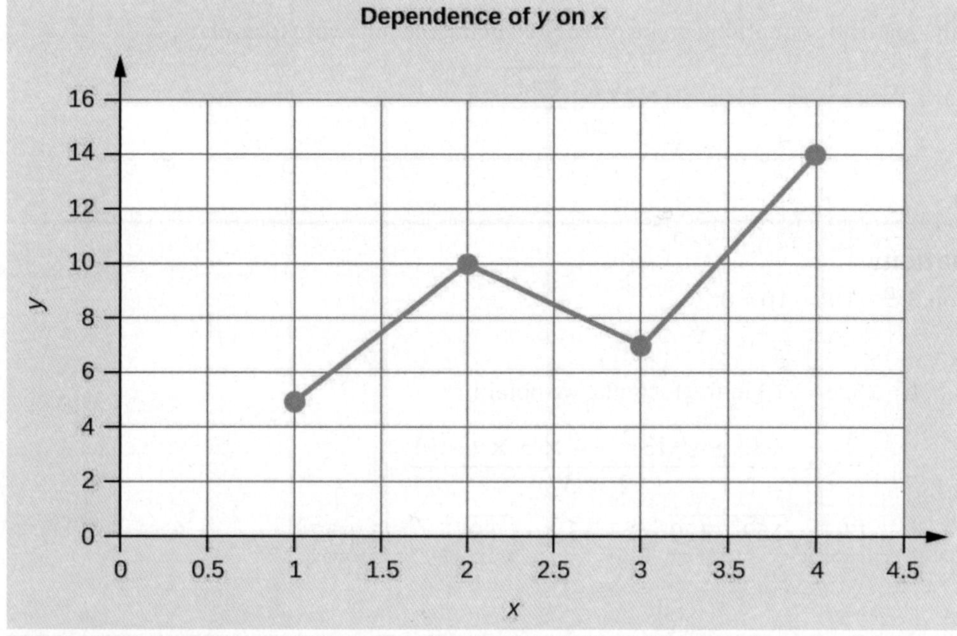

If the function that describes the dependence of *y* on *x* is known, it may be used to compute x,y data pairs that may subsequently be plotted.

EXAMPLE B12

Plotting Data Pairs

If we know that $y = x^2 + 2$, we can produce a table of a few (*x*,*y*) values and then plot the line based on the data shown here.

x	$y = x^2 + 2$
1	3
2	6

x	$y = x^2 + 2$
3	11
4	18

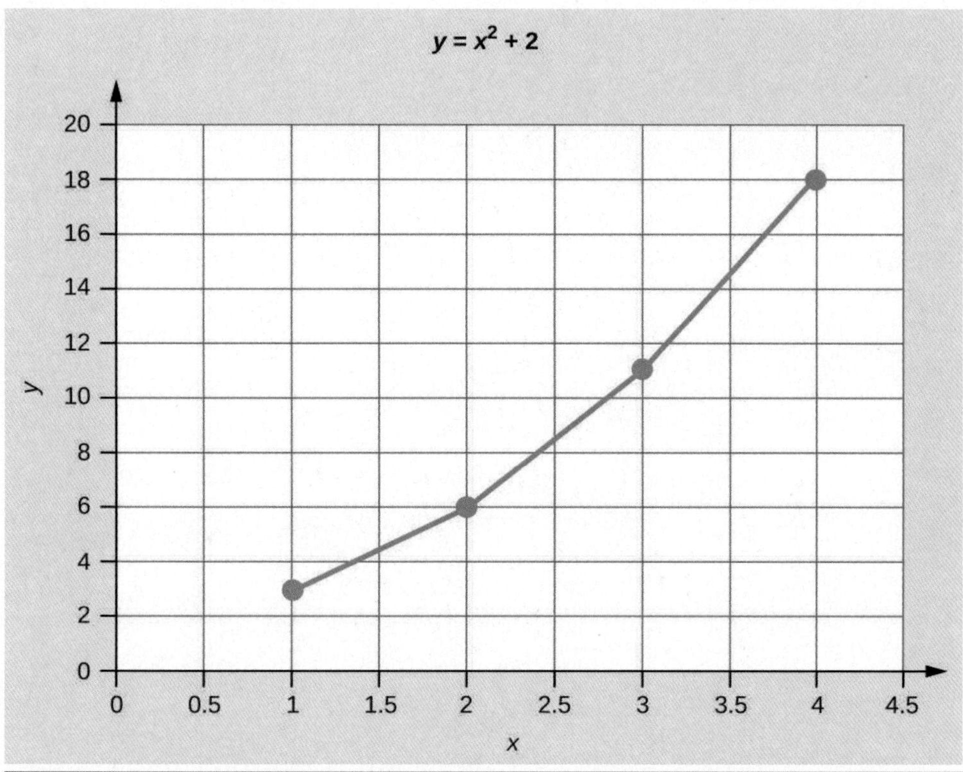

APPENDIX C

Units and Conversion Factors

Units of Length

meter (m)	= 39.37 inches (in.) = 1.094 yards (yd)
centimeter (cm)	= 0.01 m (exact, definition)
millimeter (mm)	= 0.001 m (exact, definition)
kilometer (km)	= 1000 m (exact, definition)
angstrom (Å)	= 10^{-8} cm (exact, definition) = 10^{-10} m (exact, definition)
yard (yd)	= 0.9144 m
inch (in.)	= 2.54 cm (exact, definition)
mile (US)	= 1.60934 km

TABLE C1

Units of Volume

liter (L)	= 0.001 m^3 (exact, definition) = 1000 cm^3 (exact, definition) = 1.057 (US) quarts
milliliter (mL)	= 0.001 L (exact, definition) = 1 cm^3 (exact, definition)
microliter (μL)	= 10^{-6} L (exact, definition) = 10^{-3} cm^3 (exact, definition)
liquid quart (US)	= 32 (US) liquid ounces (exact, definition) = 0.25 (US) gallon (exact, definition) = 0.9463 L
dry quart	= 1.1012 L
cubic foot (US)	= 28.316 L

TABLE C2

Units of Mass

gram (g)	= 0.001 kg (exact, definition)
milligram (mg)	= 0.001 g (exact, definition)
kilogram (kg)	= 1000 g (exact, definition) = 2.205 lb
ton (metric)	=1000 kg (exact, definition) = 2204.62 lb
ounce (oz)	= 28.35 g
pound (lb)	= 0.4535924 kg
ton (short)	=2000 lb (exact, definition) = 907.185 kg
ton (long)	= 2240 lb (exact, definition) = 1.016 metric ton

TABLE C3

Units of Energy

4.184 joule (J)	= 1 thermochemical calorie (cal)
1 thermochemical calorie (cal)	= 4.184×10^7 erg
erg	= 10^{-7} J (exact, definition)
electron-volt (eV)	= 1.60218×10^{-19} J = 23.061 kcal mol^{-1}
liter•atmosphere	= 24.217 cal = 101.325 J (exact, definition)
nutritional calorie (Cal)	= 1000 cal (exact, definition) = 4184 J
British thermal unit (BTU)	= 1054.804 J[1]

TABLE C4

Units of Pressure

torr	= 1 mm Hg (exact, definition)

TABLE C5

1 BTU is the amount of energy needed to heat one pound of water by one degree Fahrenheit. Therefore, the exact relationship of BTU to joules and other energy units depends on the temperature at which BTU is measured. 59 °F (15 °C) is the most widely used reference temperature for BTU definition in the United States. At this temperature, the conversion factor is the one provided in this table.

pascal (Pa)	= N m^{-2} (exact, definition) = kg m^{-1} s^{-2} (exact, definition)
atmosphere (atm)	= 760 mm Hg (exact, definition) = 760 torr (exact, definition) = 101,325 N m^{-2} (exact, definition) = 101,325 Pa (exact, definition)
bar	= 10^5 Pa (exact, definition) = 10^5 kg m^{-1} s^{-2} (exact, definition)

TABLE C5

APPENDIX D

Fundamental Physical Constants

Fundamental Physical Constants

Name and Symbol	Value
atomic mass unit (amu)	$1.6605402 \times 10^{-27}$ kg
Avogadro's number	$6.02214076 \times 10^{23}$ mol^{-1}
Boltzmann's constant (k)	1.380649×10^{-23} J K^{-1}
charge-to-mass ratio for electron (e/m_e)	$1.75881962 \times 10^{11}$ C kg^{-1}
fundamental unit of charge (e)	$1.602176634 \times 10^{-19}$ C
electron rest mass (m_e)	$9.1093897 \times 10^{-31}$ kg
Faraday's constant (F)	9.6485309×10^{4} C mol^{-1}
gas constant (R)	8.205784×10^{-2} L atm mol^{-1} K^{-1} = 8.314510 J mol^{-1} K^{-1}
molar volume of an ideal gas, 1 atm, 0 °C	22.41409 L mol^{-1}
molar volume of an ideal gas, 1 bar, 0 °C	22.71108 L mol^{-1}
neutron rest mass (m_n)	$1.6749274 \times 10^{-27}$ kg
Planck's constant (h)	$6.62607015 \times 10^{-34}$ J s
proton rest mass (m_p)	$1.6726231 \times 10^{-27}$ kg
Rydberg constant (R)	$1.0973731534 \times 10^{7}$ m^{-1} = $2.1798736 \times 10^{-18}$ J
speed of light (in vacuum) (c)	2.99792458×10^{8} m s^{-1}

TABLE D1

APPENDIX E

Water Properties

Water Density (g/mL) at Different Temperatures (°C)

Temperature	Density (g/mL)
0	0.9998395
4	0.9999720 (density maximum)
10	0.9997026
15	0.9991026
20	0.9982071
22	0.9977735
25	0.9970479
30	0.9956502
40	0.9922
60	0.9832
80	0.9718
100	0.9584

TABLE E1

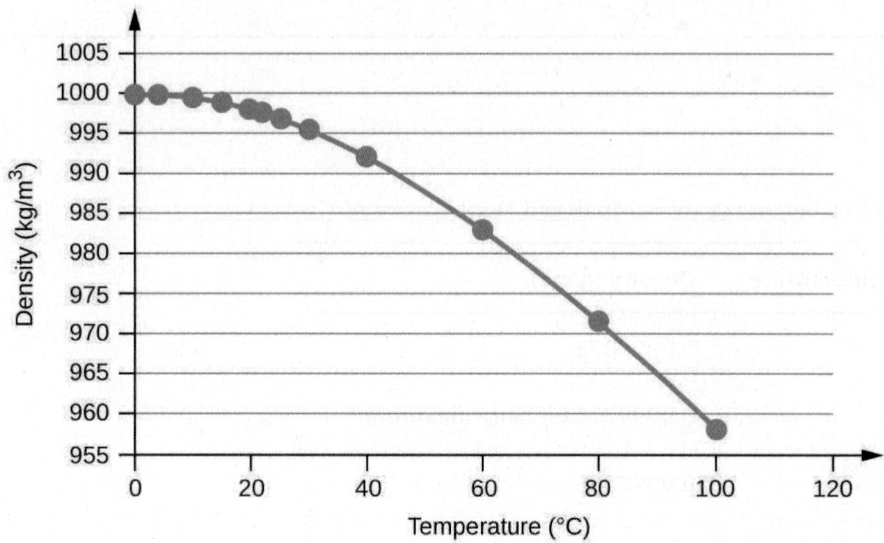

Water Vapor Pressure at Different Temperatures (°C)

Temperature	Vapor Pressure (torr)	Vapor Pressure (Pa)
0	4.6	613.2812
4	6.1	813.2642
10	9.2	1226.562
15	12.8	1706.522
20	17.5	2333.135
22	19.8	2639.776
25	23.8	3173.064
30	31.8	4239.64
35	42.2	5626.188
40	55.3	7372.707
45	71.9	9585.852
50	92.5	12332.29
55	118.0	15732

TABLE E2

Temperature	Vapor Pressure (torr)	Vapor Pressure (Pa)
60	149.4	19918.31
65	187.5	24997.88
70	233.7	31157.35
75	289.1	38543.39
80	355.1	47342.64
85	433.6	57808.42
90	525.8	70100.71
95	633.9	84512.82
100	760.0	101324.7

TABLE E2

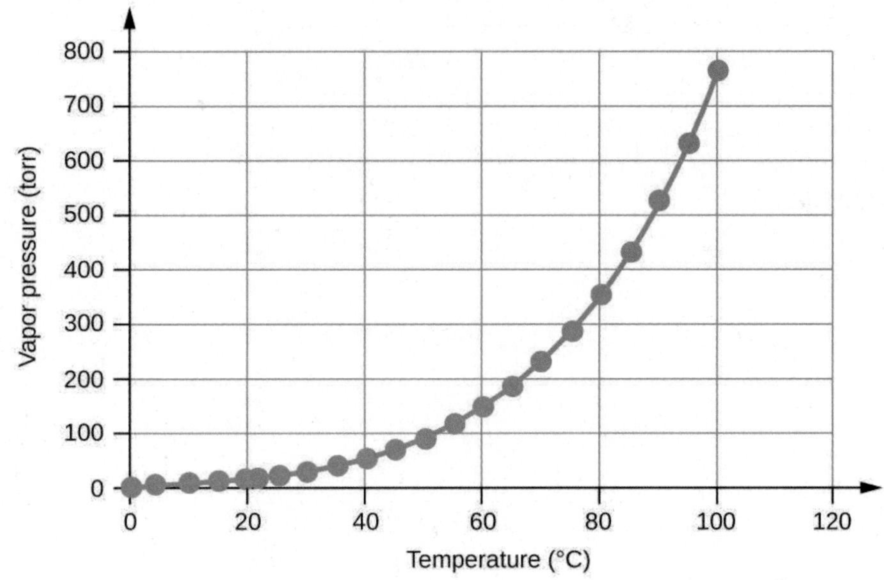

Water K_w and pK_w at Different Temperatures (°C)

Temperature	K_w 10^{-14}	pK_w[1]
0	0.112	14.95

TABLE E3

1 $pK_w = -\log_{10}(K_w)$

Temperature	K_w 10^{-14}	pK_w[1]
5	0.182	14.74
10	0.288	14.54
15	0.465	14.33
20	0.671	14.17
25	0.991	14.00
30	1.432	13.84
35	2.042	13.69
40	2.851	13.55
45	3.917	13.41
50	5.297	13.28
55	7.080	13.15
60	9.311	13.03
75	19.95	12.70
100	56.23	12.25

TABLE E3

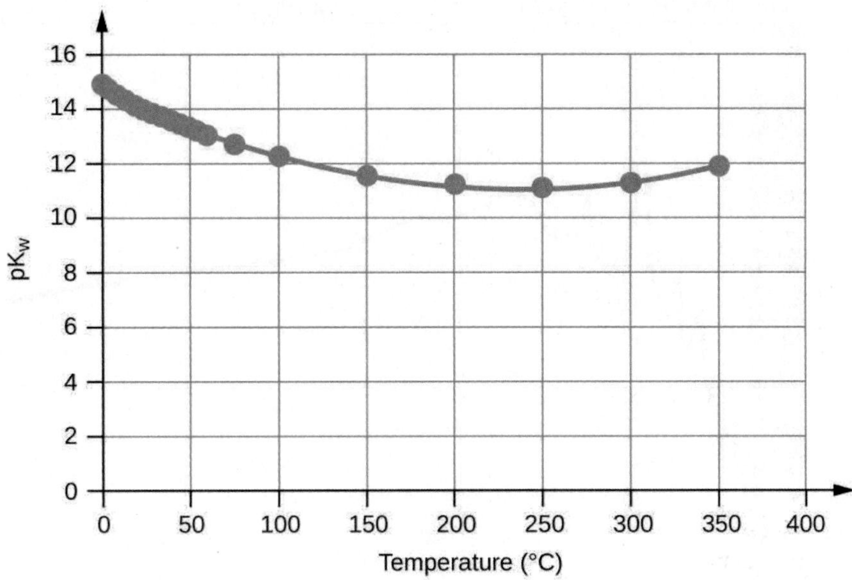

Specific Heat Capacity for Water
$C°(H_2O(l)) = 4.184\ J•g^{-1}•°C^{-1}$
$C°(H_2O(s)) = 1.864\ J•K^{-1}•g^{-1}$
$C°(H_2O(g)) = 2.093\ J•K^{-1}•g^{-1}$

TABLE E4

Standard Water Melting and Boiling Temperatures and Enthalpies of the Transitions

	Temperature (K)	ΔH (kJ/mol)
melting	273.15	6.088
boiling	373.15	40.656 (44.016 at 298 K)

TABLE E5

Water Cryoscopic (Freezing Point Depression) and Ebullioscopic (Boiling Point Elevation) Constants
$K_f = 1.86°C•kg•mol^{-1}$ (cryoscopic constant)
$K_b = 0.51°C•kg•mol^{-1}$ (ebullioscopic constant)

TABLE E6

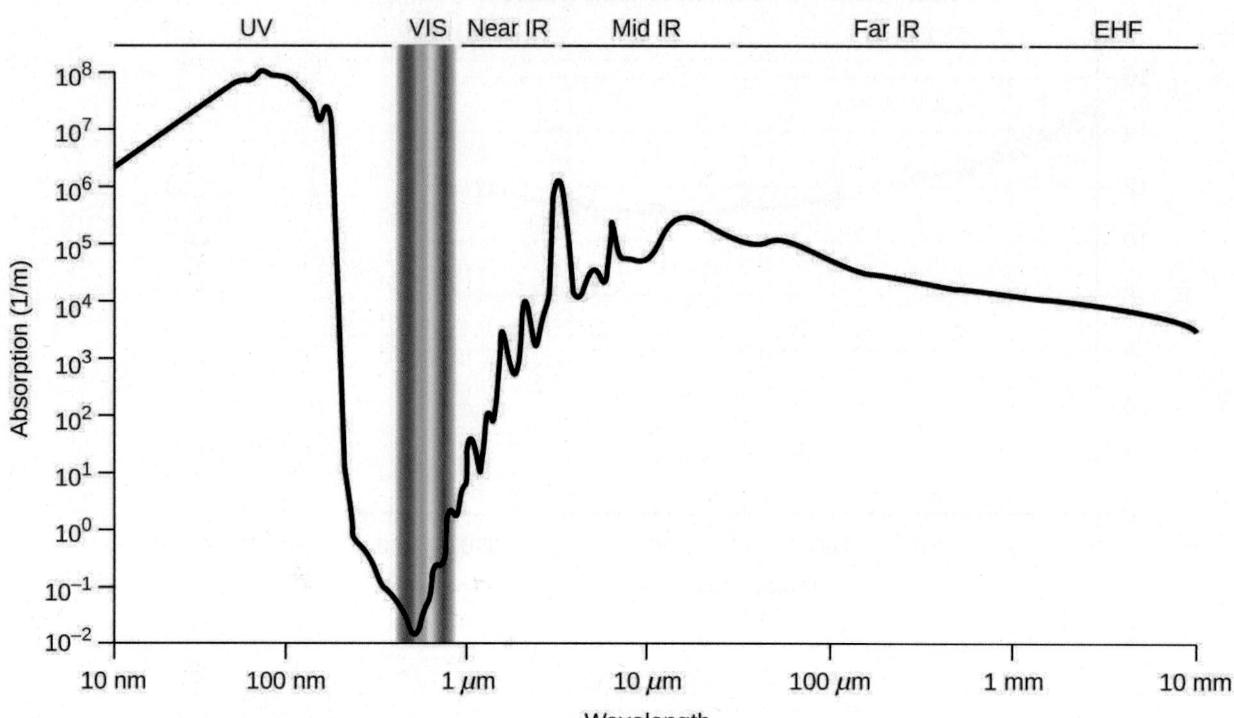

FIGURE E1 The plot shows the extent of light absorption versus wavelength for water. Absorption is reported in reciprocal meters and corresponds to the inverse of the distance light may travel through water before its intensity is diminished by $1/e$ (~37%).

APPENDIX F

Composition of Commercial Acids and Bases

Composition of Commercial Acids and Bases

Acid or Base[1]	Density (g/mL)[2]	Percentage by Mass	Molarity
acetic acid, glacial	1.05	99.5%	17.4
aqueous ammonia[3]	0.90	28%	14.8
hydrochloric acid	1.18	36%	11.6
nitric acid	1.42	71%	16.0
perchloric acid	1.67	70%	11.65
phosphoric acid	1.70	85%	14.7
sodium hydroxide	1.53	50%	19.1
sulfuric acid	1.84	96%	18.0

TABLE F1

1 Acids and bases are commercially available as aqueous solutions. This table lists properties (densities and concentrations) of common acid and base solutions. Nominal values are provided in cases where the manufacturer cites a range of concentrations and densities.

2 This column contains specific gravity data. In the case of this table, specific gravity is the ratio of density of a substance to the density of pure water at the same conditions. Specific gravity is often cited on commercial labels.

3 This solution is sometimes called "ammonium hydroxide," although this term is not chemically accurate.

APPENDIX G

Standard Thermodynamic Properties for Selected Substances

Standard Thermodynamic Properties for Selected Substances

Substance	ΔH_f° (kJ mol^{-1})	ΔG_f° (kJ mol^{-1})	S° (J K^{-1} mol^{-1})
aluminum			
$Al(s)$	0	0	28.3
$Al(g)$	324.4	285.7	164.54
$Al^{3+}(aq)$	−531	−485	−321.7
$Al_2O_3(s)$	−1676	−1582	50.92
$AlF_3(s)$	−1510.4	−1425	66.5
$AlCl_3(s)$	−704.2	−628.8	110.67
$AlCl_3 \cdot 6H_2O(s)$	−2691.57	−2269.40	376.56
$Al_2S_3(s)$	−724.0	−492.4	116.9
$Al_2(SO_4)_3(s)$	−3445.06	−3506.61	239.32
antimony			
$Sb(s)$	0	0	45.69
$Sb(g)$	262.34	222.17	180.16
$Sb_4O_6(s)$	−1440.55	−1268.17	220.92
$SbCl_3(g)$	−313.8	−301.2	337.80
$SbCl_5(g)$	−394.34	−334.29	401.94
$Sb_2S_3(s)$	−174.89	−173.64	182.00
$SbCl_3(s)$	−382.17	−323.72	184.10
$SbOCl(s)$	−374.0	—	—

TABLE G1

Substance	ΔH_f° (kJ mol^{-1})	ΔG_f° (kJ mol^{-1})	S° (J K^{-1} mol^{-1})
arsenic			
As(*s*)	0	0	35.1
As(*g*)	302.5	261.0	174.21
As_4(*g*)	143.9	92.4	314
As_4O_6(*s*)	–1313.94	–1152.52	214.22
As_2O_5(*s*)	–924.87	–782.41	105.44
$AsCl_3$(*g*)	–261.50	–248.95	327.06
As_2S_3(*s*)	–169.03	–168.62	163.59
AsH_3(*g*)	66.44	68.93	222.78
H_3AsO_4(*s*)	–906.3	—	—
barium			
Ba(*s*)	0	0	62.5
Ba(*g*)	180	146	170.24
Ba^{2+}(*aq*)	–537.6	–560.8	9.6
BaO(*s*)	–548.0	–520.3	72.1
$BaCl_2$(*s*)	–855.0	–806.7	123.7
$BaSO_4$(*s*)	–1473.2	–1362.3	132.2
beryllium			
Be(*s*)	0	0	9.50
Be(*g*)	324.3	286.6	136.27
BeO(*s*)	–609.4	–580.1	13.8
bismuth			
Bi(*s*)	0	0	56.74
Bi(*g*)	207.1	168.2	187.00

TABLE G1

Substance	ΔH_f° (kJ mol^{-1})	ΔG_f° (kJ mol^{-1})	S° (J K^{-1} mol^{-1})
$Bi_2O_3(s)$	−573.88	−493.7	151.5
$BiCl_3(s)$	−379.07	−315.06	176.98
$Bi_2S_3(s)$	−143.1	−140.6	200.4
boron			
$B(s)$	0	0	5.86
$B(g)$	565.0	521.0	153.4
$B_2O_3(s)$	−1273.5	−1194.3	53.97
$B_2H_6(g)$	36.4	87.6	232.1
$H_3BO_3(s)$	−1094.33	−968.92	88.83
$BF_3(g)$	−1136.0	−1119.4	254.4
$BCl_3(g)$	−403.8	−388.7	290.1
$B_3N_3H_6(l)$	−540.99	−392.79	199.58
$HBO_2(s)$	−794.25	−723.41	37.66
bromine			
$Br_2(l)$	0	0	152.23
$Br_2(g)$	30.91	3.142	245.5
$Br(g)$	111.88	82.429	175.0
$Br^-(aq)$	−120.9	−102.82	80.71
$BrF_3(g)$	−255.60	−229.45	292.42
$HBr(g)$	−36.3	−53.43	198.7
cadmium			
$Cd(s)$	0	0	51.76
$Cd(g)$	112.01	77.41	167.75
$Cd^{2+}(aq)$	−75.90	−77.61	−73.2

TABLE G1

Substance	ΔH_f° (kJ mol^{-1})	ΔG_f° (kJ mol^{-1})	S° (J K^{-1} mol^{-1})
$CdO(s)$	−258.2	−228.4	54.8
$CdCl_2(s)$	−391.5	−343.9	115.3
$CdSO_4(s)$	−933.3	−822.7	123.0
$CdS(s)$	−161.9	−156.5	64.9
calcium			
$Ca(s)$	0	0	41.6
$Ca(g)$	178.2	144.3	154.88
$Ca^{2+}(aq)$	−542.96	−553.04	−55.2
$CaO(s)$	−634.9	−603.3	38.1
$Ca(OH)_2(s)$	−985.2	−897.5	83.4
$CaSO_4(s)$	−1434.5	−1322.0	106.5
$CaSO_4{\cdot}2H_2O(s)$	−2022.63	−1797.45	194.14
$CaCO_3(s)$ (calcite)	−1220.0	−1081.4	110.0
$CaSO_3{\cdot}H_2O(s)$	−1752.68	−1555.19	184.10
carbon			
$C(s)$ (graphite)	0	0	5.740
$C(s)$ (diamond)	1.89	2.90	2.38
$C(g)$	716.681	671.2	158.1
$CO(g)$	−110.52	−137.15	197.7
$CO_2(g)$	−393.51	−394.36	213.8
$CO_3{}^{2-}(aq)$	−677.1	−527.8	−56.9
$CH_4(g)$	−74.6	−50.5	186.3
$CH_3OH(l)$	−239.2	−166.6	126.8
$CH_3OH(g)$	−201.0	−162.3	239.9

TABLE G1

Substance	ΔH_f° (kJ mol^{-1})	ΔG_f° (kJ mol^{-1})	S° (J K^{-1} mol^{-1})
$CCl_4(l)$	−128.2	−62.5	214.4
$CCl_4(g)$	−95.7	−58.2	309.7
$CHCl_3(l)$	−134.1	−73.7	201.7
$CHCl_3(g)$	−103.14	−70.34	295.71
$CS_2(l)$	89.70	65.27	151.34
$CS_2(g)$	116.9	66.8	238.0
$C_2H_2(g)$	227.4	209.2	200.9
$C_2H_4(g)$	52.4	68.4	219.3
$C_2H_6(g)$	−84.0	−32.0	229.2
$CH_3CO_2H(l)$	−484.3	−389.9	159.8
$CH_3CO_2H(g)$	−434.84	−376.69	282.50
$C_2H_5OH(l)$	−277.6	−174.8	160.7
$C_2H_5OH(g)$	−234.8	−167.9	281.6
$HCO_3{}^-(aq)$	−691.11	−587.06	95
$C_3H_8(g)$	−103.8	−23.4	270.3
$C_6H_6(g)$	82.927	129.66	269.2
$C_6H_6(l)$	49.1	124.50	173.4
$CH_2Cl_2(l)$	−124.2	−63.2	177.8
$CH_2Cl_2(g)$	−95.4	−65.90	270.2
$CH_3Cl(g)$	−81.9	−60.2	234.6
$C_2H_5Cl(l)$	−136.52	−59.31	190.79
$C_2H_5Cl(g)$	−112.17	−60.39	276.00
$C_2N_2(g)$	308.98	297.36	241.90
$HCN(l)$	108.9	125.0	112.8

TABLE G1

Substance	ΔH_f° (kJ mol^{-1})	ΔG_f° (kJ mol^{-1})	S° (J K^{-1} mol^{-1})
$HCN(g)$	135.5	124.7	201.8
cesium			
$Cs^+(aq)$	−248	−282.0	133
chlorine			
$Cl_2(g)$	0	0	223.1
$Cl(g)$	121.3	105.70	165.2
$Cl^-(aq)$	−167.2	−131.2	56.5
$ClF(g)$	−54.48	−55.94	217.78
$ClF_3(g)$	−158.99	−118.83	281.50
$Cl_2O(g)$	80.3	97.9	266.2
$Cl_2O_7(l)$	238.1	—	—
$Cl_2O_7(g)$	272.0	—	—
$HCl(g)$	−92.307	−95.299	186.9
$HClO_4(l)$	−40.58	—	—
chromium			
$Cr(s)$	0	0	23.77
$Cr(g)$	396.6	351.8	174.50
$CrO_4{}^{2-}(aq)$	−881.2	−727.8	50.21
$Cr_2O_7{}^{2-}(aq)$	−1490.3	−1301.1	261.9
$Cr_2O_3(s)$	−1139.7	−1058.1	81.2
$CrO_3(s)$	−589.5	—	—
$(NH_4)_2Cr_2O_7(s)$	−1806.7	—	—
cobalt			
$Co(s)$	0	0	30.0

TABLE G1

Substance	ΔH_f° (kJ mol^{-1})	ΔG_f° (kJ mol^{-1})	S° (J K^{-1} mol^{-1})
$Co^{2+}(aq)$	−67.4	−51.5	−155
$Co^{3+}(aq)$	92	134	−305.0
$CoO(s)$	−237.9	−214.2	52.97
$Co_3O_4(s)$	−910.02	−794.98	114.22
$Co(NO_3)_2(s)$	−420.5	—	—
copper			
$Cu(s)$	0	0	33.15
$Cu(g)$	338.32	298.58	166.38
$Cu^{+}(aq)$	51.9	50.2	−26
$Cu^{2+}(aq)$	64.77	65.49	−99.6
$CuO(s)$	−157.3	−129.7	42.63
$Cu_2O(s)$	−168.6	−146.0	93.14
$CuS(s)$	−53.1	−53.6	66.5
$Cu_2S(s)$	−79.5	−86.2	120.9
$CuSO_4(s)$	−771.36	−662.2	109.2
$Cu(NO_3)_2(s)$	−302.9	—	—
fluorine			
$F_2(g)$	0	0	202.8
$F(g)$	79.4	62.3	158.8
$F^{-}(aq)$	−332.6	−278.8	−13.8
$F_2O(g)$	24.7	41.9	247.43
$HF(g)$	−273.3	−275.4	173.8
hydrogen			
$H_2(g)$	0	0	130.7

TABLE G1

Substance	ΔH_f° (kJ mol^{-1})	ΔG_f° (kJ mol^{-1})	S° (J K^{-1} mol^{-1})
$H(g)$	217.97	203.26	114.7
$H^+(aq)$	0	0	0
$OH^-(aq)$	−230.0	−157.2	−10.75
$H_3O^+(aq)$	−285.8		69.91
$H_2O(l)$	−285.83	−237.1	70.0
$H_2O(g)$	−241.82	−228.59	188.8
$H_2O_2(l)$	−187.78	−120.35	109.6
$H_2O_2(g)$	−136.3	−105.6	232.7
$HF(g)$	−273.3	−275.4	173.8
$HCl(g)$	−92.307	−95.299	186.9
$HBr(g)$	−36.3	−53.43	198.7
$HI(g)$	26.48	1.70	206.59
$H_2S(g)$	−20.6	−33.4	205.8
$H_2Se(g)$	29.7	15.9	219.0
HNO_3	−206.64	–	–
iodine			
$I_2(s)$	0	0	116.14
$I_2(g)$	62.438	19.3	260.7
$I(g)$	106.84	70.2	180.8
$I^-(aq)$	−55.19	−51.57	11.13
$IF(g)$	95.65	−118.49	236.06
$ICl(g)$	17.78	−5.44	247.44
$IBr(g)$	40.84	3.72	258.66
$IF_7(g)$	−943.91	−818.39	346.44

TABLE G1

Substance	ΔH_f° (kJ mol^{-1})	ΔG_f° (kJ mol^{-1})	S° (J K^{-1} mol^{-1})
HI(*g*)	26.48	1.70	206.59
iron			
Fe(*s*)	0	0	27.3
Fe(*g*)	416.3	370.7	180.5
Fe^{2+}(*aq*)	−89.1	−78.90	−137.7
Fe^{3+}(*aq*)	−48.5	−4.7	−315.9
Fe_2O_3(*s*)	−824.2	−742.2	87.40
Fe_3O_4(*s*)	−1118.4	−1015.4	146.4
$Fe(CO)_5$(*l*)	−774.04	−705.42	338.07
$Fe(CO)_5$(*g*)	−733.87	−697.26	445.18
$FeCl_2$(*s*)	−341.79	−302.30	117.95
$FeCl_3$(*s*)	−399.49	−334.00	142.3
FeO(*s*)	−272.0	−255.2	60.75
$Fe(OH)_2$(*s*)	−569.0	−486.5	88.
$Fe(OH)_3$(*s*)	−823.0	−696.5	106.7
FeS(*s*)	−100.0	−100.4	60.29
Fe_3C(*s*)	25.10	20.08	104.60
lead			
Pb(*s*)	0	0	64.81
Pb(*g*)	195.2	162.	175.4
Pb^{2+}(*aq*)	−1.7	−24.43	10.5
PbO(*s*) (yellow)	−217.32	−187.89	68.70
PbO(*s*) (red)	−218.99	−188.93	66.5
$Pb(OH)_2$(*s*)	−515.9	—	—

TABLE G1

Substance	ΔH_f° (kJ mol^{-1})	ΔG_f° (kJ mol^{-1})	S° (J K^{-1} mol^{-1})
$PbS(s)$	−100.4	−98.7	91.2
$Pb(NO_3)_2(s)$	−451.9	—	—
$PbO_2(s)$	−277.4	−217.3	68.6
$PbCl_2(s)$	−359.4	−314.1	136.0
lithium			
$Li(s)$	0	0	29.1
$Li(g)$	159.3	126.6	138.8
$Li^+(aq)$	−278.5	−293.3	13.4
$LiH(s)$	−90.5	−68.3	20.0
$Li(OH)(s)$	−487.5	−441.5	42.8
$LiF(s)$	−616.0	−587.5	35.7
$Li_2CO_3(s)$	−1216.04	−1132.19	90.17
magnesium			
$Mg^{2+}(aq)$	−466.9	−454.8	−138.1
manganese			
$Mn(s)$	0	0	32.0
$Mn(g)$	280.7	238.5	173.7
$Mn^{2+}(aq)$	−220.8	−228.1	−73.6
$MnO(s)$	−385.2	−362.9	59.71
$MnO_2(s)$	−520.03	−465.1	53.05
$Mn_2O_3(s)$	−958.97	−881.15	110.46
$Mn_3O_4(s)$	−1378.83	−1283.23	155.64
$MnO_4^-(aq)$	−541.4	−447.2	191.2
$MnO_4^{2-}(aq)$	−653.0	−500.7	59

TABLE G1

Substance	ΔH_f° (kJ mol^{-1})	ΔG_f° (kJ mol^{-1})	S° (J K^{-1} mol^{-1})
mercury			
$Hg(l)$	0	0	75.9
$Hg(g)$	61.4	31.8	175.0
$Hg^{2+}(aq)$		164.8	
$Hg^{2+}(aq)$	172.4	153.9	84.5
$HgO(s)$ (red)	−90.83	−58.5	70.29
$HgO(s)$ (yellow)	−90.46	−58.43	71.13
$HgCl_2(s)$	−224.3	−178.6	146.0
$Hg_2Cl_2(s)$	−265.4	−210.7	191.6
$HgS(s)$ (red)	−58.16	−50.6	82.4
$HgS(s)$ (black)	−53.56	−47.70	88.28
$HgSO_4(s)$	−707.51	−594.13	0.00
nickel			
$Ni^{2+}(aq)$	−64.0	−46.4	−159
nitrogen			
$N_2(g)$	0	0	191.6
$N(g)$	472.704	455.5	153.3
$NO(g)$	90.25	87.6	210.8
$NO_2(g)$	33.2	51.30	240.1
$N_2O(g)$	81.6	103.7	220.0
$N_2O_3(g)$	83.72	139.41	312.17
$NO_3^{\ -}(aq)$	−205.0	−108.7	146.4
$N_2O_4(g)$	11.1	99.8	304.4
$N_2O_5(g)$	11.3	115.1	355.7

TABLE G1

Substance	ΔH_f° (kJ mol^{-1})	ΔG_f° (kJ mol^{-1})	S° (J K^{-1} mol^{-1})
$NH_3(g)$	−45.9	−16.5	192.8
$NH_4{}^+(aq)$	−132.5	−79.31	113.4
$N_2H_4(l)$	50.63	149.43	121.21
$N_2H_4(g)$	95.4	159.4	238.5
$NH_4NO_3(s)$	−365.56	−183.87	151.08
$NH_4Cl(s)$	−314.43	−202.87	94.6
$NH_4Br(s)$	−270.8	−175.2	113.0
$NH_4I(s)$	−201.4	−112.5	117.0
$NH_4NO_2(s)$	−256.5	—	—
$HNO_3(l)$	−174.1	−80.7	155.6
$HNO_3(g)$	−133.9	−73.5	266.9
$HNO_3(aq)$	−207.4	−110.5	146
oxygen			
$O_2(g)$	0	0	205.2
$O(g)$	249.17	231.7	161.1
$O_3(g)$	142.7	163.2	238.9
phosphorus			
$P_4(s)$	0	0	164.4
$P_4(g)$	58.91	24.4	280.0
$P(g)$	314.64	278.25	163.19
$PH_3(g)$	5.4	13.5	210.2
$PCl_3(g)$	−287.0	−267.8	311.78
$PCl_5(g)$	−374.9	−305.0	364.4
$P_4O_6(s)$	−1640.1	—	—

TABLE G1

Substance	ΔH_f° (kJ mol^{-1})	ΔG_f° (kJ mol^{-1})	S° (J K^{-1} mol^{-1})
$P_4O_{10}(s)$	−2984.0	−2697.0	228.86
$PO_4{}^{3-}(aq)$	−1277	−1019	−222
$HPO_3(s)$	−948.5	—	—
$HPO_4{}^{2-}(aq)$	−1292.1	−1089.3	−33
$H_2PO_4{}^{2-}(aq)$	−1296.3	−1130.4	90.4
$H_3PO_2(s)$	−604.6	—	—
$H_3PO_3(s)$	−964.4	—	—
$H_3PO_4(s)$	−1279.0	−1119.1	110.50
$H_3PO_4(l)$	−1266.9	−1124.3	110.5
$H_4P_2O_7(s)$	−2241.0	—	—
$POCl_3(l)$	−597.1	−520.8	222.5
$POCl_3(g)$	−558.5	−512.9	325.5
potassium			
$K(s)$	0	0	64.7
$K(g)$	89.0	60.5	160.3
$K^+(aq)$	−252.4	−283.3	102.5
$KF(s)$	−576.27	−537.75	66.57
$KCl(s)$	−436.5	−408.5	82.6
rubidium			
$Rb^+(aq)$	−246	−282.2	124
silicon			
$Si(s)$	0	0	18.8
$Si(g)$	450.0	405.5	168.0
$SiO_2(s)$	−910.7	−856.3	41.5

TABLE G1

Substance	ΔH_f° (kJ mol^{-1})	ΔG_f° (kJ mol^{-1})	S° (J K^{-1} mol^{-1})
$SiH_4(g)$	34.3	56.9	204.6
$H_2SiO_3(s)$	−1188.67	−1092.44	133.89
$H_4SiO_4(s)$	−1481.14	−1333.02	192.46
$SiF_4(g)$	−1615.0	−1572.8	282.8
$SiCl_4(l)$	−687.0	−619.8	239.7
$SiCl_4(g)$	−662.75	−622.58	330.62
SiC(*s, beta cubic*)	−73.22	−70.71	16.61
SiC(*s, alpha hexagonal*)	−71.55	−69.04	16.48
silver			
Ag(*s*)	0	0	42.55
Ag(*g*)	284.9	246.0	172.89
$Ag^+(aq)$	105.6	77.11	72.68
$Ag_2O(s)$	−31.05	−11.20	121.3
AgCl(*s*)	−127.0	−109.8	96.3
$Ag_2S(s)$	−32.6	−40.7	144.0
sodium			
Na(*s*)	0	0	51.3
Na(*g*)	107.5	77.0	153.7
$Na^+(aq)$	−240.1	−261.9	59
$Na_2O(s)$	−414.2	−375.5	75.1
NaCl(*s*)	−411.2	−384.1	72.1
strontium			
$Sr^{2+}(aq)$	−545.8	−557.3	−32.6
sulfur			

TABLE G1

Substance	ΔH_f° (kJ mol^{-1})	ΔG_f° (kJ mol^{-1})	S° (J K^{-1} mol^{-1})
$S_8(s)$ (rhombic)	0	0	256.8
$S(g)$	278.81	238.25	167.82
$S^{2-}(aq)$	41.8	83.7	22
$SO_2(g)$	−296.83	−300.1	248.2
$SO_3(g)$	−395.72	−371.06	256.76
$SO_4{}^{2-}(aq)$	−909.3	−744.5	20.1
$S_2O_3{}^{2-}(aq)$	−648.5	−522.5	67
$H_2S(g)$	−20.6	−33.4	205.8
$HS^-(aq)$	−17.7	12.6	61.1
$H_2SO_4(l)$	−813.989	−690.00	156.90
$HSO_4{}^{2-}(aq)$	−885.75	−752.87	126.9
$H_2S_2O_7(s)$	−1273.6	—	—
$SF_4(g)$	−728.43	−684.84	291.12
$SF_6(g)$	−1220.5	−1116.5	291.5
$SCl_2(l)$	−50	—	—
$SCl_2(g)$	−19.7	—	—
$S_2Cl_2(l)$	−59.4	—	—
$S_2Cl_2(g)$	−19.50	−29.25	319.45
$SOCl_2(g)$	−212.55	−198.32	309.66
$SOCl_2(l)$	−245.6	—	—
$SO_2Cl_2(l)$	−394.1	—	—
$SO_2Cl_2(g)$	−354.80	−310.45	311.83
tin			
$Sn(s)$	0	0	51.2

TABLE G1

Substance	ΔH_f° (kJ mol^{-1})	ΔG_f° (kJ mol^{-1})	S° (J K^{-1} mol^{-1})
Sn(*g*)	301.2	266.2	168.5
SnO(*s*)	–285.8	–256.9	56.5
SnO_2(*s*)	–577.6	–515.8	49.0
$SnCl_4$(*l*)	–511.3	–440.1	258.6
$SnCl_4$(*g*)	–471.5	–432.2	365.8
titanium			
Ti(*s*)	0	0	30.7
Ti(*g*)	473.0	428.4	180.3
TiO_2(*s*)	–944.0	–888.8	50.6
$TiCl_4$(*l*)	–804.2	–737.2	252.4
$TiCl_4$(*g*)	–763.2	–726.3	353.2
tungsten			
W(*s*)	0	0	32.6
W(*g*)	849.4	807.1	174.0
WO_3(*s*)	–842.9	–764.0	75.9
zinc			
Zn(*s*)	0	0	41.6
Zn(*g*)	130.73	95.14	160.98
Zn^{2+}(*aq*)	–153.9	–147.1	–112.1
ZnO(*s*)	–350.5	–320.5	43.7
$ZnCl_2$(*s*)	–415.1	–369.43	111.5
ZnS(*s*)	–206.0	–201.3	57.7
$ZnSO_4$(*s*)	–982.8	–871.5	110.5
$ZnCO_3$(*s*)	–812.78	–731.57	82.42

TABLE G1

Substance	ΔH°_f (kJ mol^{-1})	ΔG°_f (kJ mol^{-1})	S° (J K^{-1} mol^{-1})
complexes			
$[Co(NH_3)_4(NO_2)_2]NO_3$, *cis*	−898.7	—	—
$[Co(NH_3)_4(NO_2)_2]NO_3$, *trans*	−896.2	—	—
$NH_4[Co(NH_3)_2(NO_2)_4]$	−837.6	—	—
$[Co(NH_3)_6][Co(NH_3)_2(NO_2)_4]_3$	−2733.0	—	—
$[Co(NH_3)_4Cl_2]Cl$, *cis*	−874.9	—	—
$[Co(NH_3)_4Cl_2]Cl$, *trans*	−877.4	—	—
$[Co(en)_2(NO_2)_2]NO_3$, *cis*	−689.5	—	—
$[Co(en)_2Cl_2]Cl$, *cis*	−681.2	—	—
$[Co(en)_2Cl_2]Cl$, *trans*	−677.4	—	—
$[Co(en)_3](ClO_4)_3$	−762.7	—	—
$[Co(en)_3]Br_2$	−595.8	—	—
$[Co(en)_3]I_2$	−475.3	—	—
$[Co(en)_3]I_3$	−519.2	—	—
$[Co(NH_3)_6](ClO_4)_3$	−1034.7	−221.1	615
$[Co(NH_3)_5NO_2](NO_3)_2$	−1088.7	−412.9	331
$[Co(NH_3)_6](NO_3)_3$	−1282.0	−524.5	448
$[Co(NH_3)_5Cl]Cl_2$	−1017.1	−582.5	366.1
$[Pt(NH_3)_4]Cl_2$	−725.5	—	—
$[Ni(NH_3)_6]Cl_2$	−994.1	—	—
$[Ni(NH_3)_6]Br_2$	−923.8	—	—
$[Ni(NH_3)_6]I_2$	−808.3	—	—

TABLE G1

APPENDIX H

Ionization Constants of Weak Acids

Ionization Constants of Weak Acids

Acid	Formula	K_a at 25 °C	Lewis Structure
acetic	$CH_3CO_2\mathbf{H}$	1.8×10^{-5}	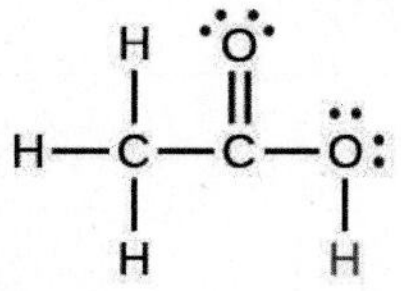
arsenic	$\mathbf{H_3}AsO_4$	5.5×10^{-3}	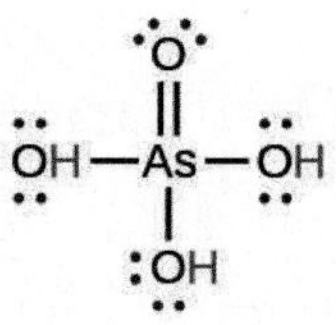
	$H_2AsO_4^-$	1.7×10^{-7}	
	$HAsO_4^{2-}$	3.0×10^{-12}	
arsenous	$\mathbf{H_3}AsO_3$	5.1×10^{-10}	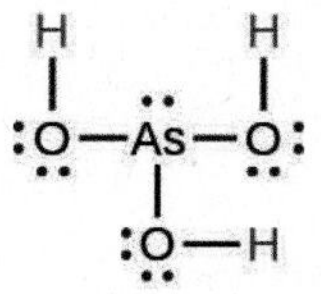
boric	$\mathbf{H_3}BO_3$	5.4×10^{-10}	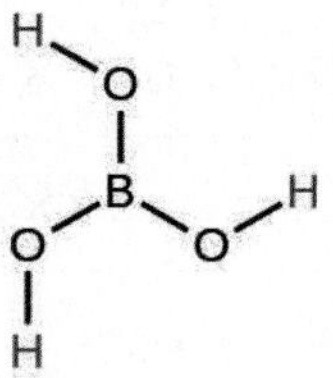
carbonic	$\mathbf{H_2}CO_3$	4.3×10^{-7}	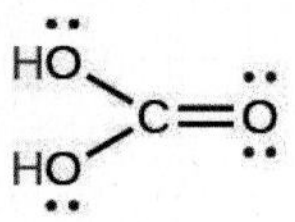
	HCO_3^-	4.7×10^{-11}	
cyanic	$\mathbf{H}CNO$	2×10^{-4}	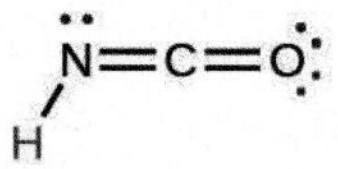
formic	$HCO_2\mathbf{H}$	1.8×10^{-4}	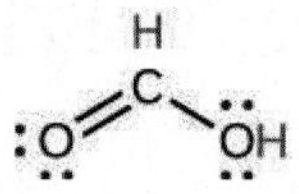

TABLE H1

Acid	Formula	K_a at 25 °C	Lewis Structure
hydrazoic	HN_3	2.5×10^{-5}	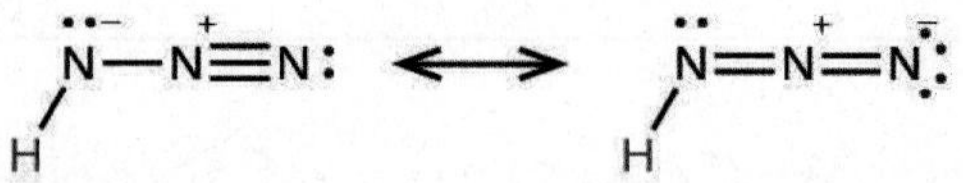
hydrocyanic	HCN	4.9×10^{-10}	
hydrofluoric	HF	6.4×10^{-4}	
hydrogen peroxide	H_2O_2	2.4×10^{-12}	
hydrogen selenide	H_2Se	1.29×10^{-4}	
	HSe^-	1×10^{-12}	
hydrogen sulfate ion	HSO_4^-	1.2×10^{-2}	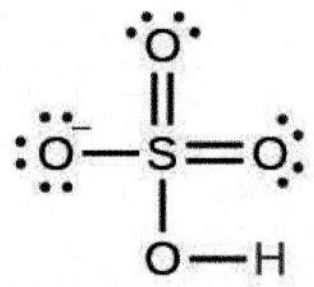
hydrogen sulfide	H_2S	8.9×10^{-8}	
	HS^-	1.0×10^{-19}	
hydrogen telluride	H_2Te	2.3×10^{-3}	
	HTe^-	1.6×10^{-11}	
hypobromous	HBrO	2.8×10^{-9}	
hypochlorous	HClO	2.9×10^{-8}	
nitrous	HNO_2	4.6×10^{-4}	
oxalic	$H_2C_2O_4$	6.0×10^{-2}	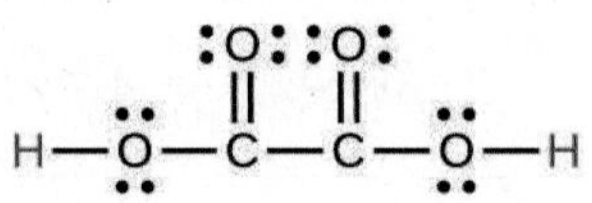
	$HC_2O_4^-$	6.1×10^{-5}	
phosphoric	H_3PO_4	7.5×10^{-3}	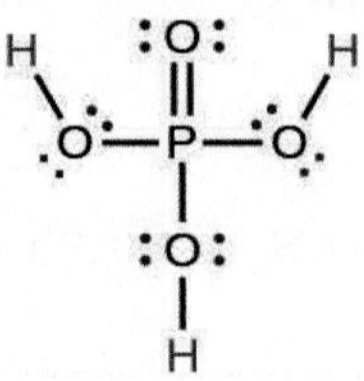
	$H_2PO_4^-$	6.2×10^{-8}	
	HPO_4^{2-}	4.2×10^{-13}	

TABLE H1

Acid	Formula	K_a at 25 °C	Lewis Structure
phosphorous	H_3PO_3	5×10^{-2}	:OH HO—P—OH
	$H_2PO_3^-$	2.0×10^{-7}	
sulfurous	H_2SO_3	1.6×10^{-2}	:OH :O—S—OH
	HSO_3^-	6.4×10^{-8}	

TABLE H1

APPENDIX I

Ionization Constants of Weak Bases

Ionization Constants of Weak Bases

Base	Lewis Structure	K_b at 25 °C
ammonia	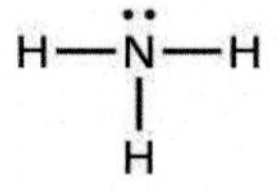	1.8×10^{-5}
dimethylamine	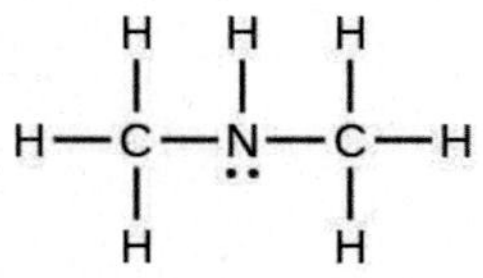	5.9×10^{-4}
methylamine	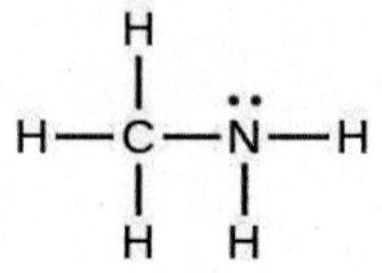	4.4×10^{-4}
phenylamine (aniline)	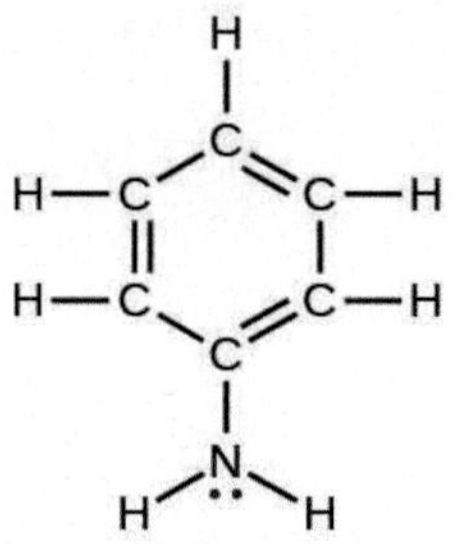	4.3×10^{-10}
trimethylamine	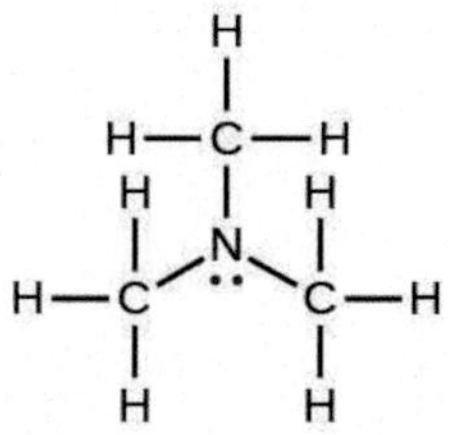	6.3×10^{-5}

TABLE I1

APPENDIX J

Solubility Products

Solubility Products

Substance	K_{sp} at 25 °C
aluminum	
$Al(OH)_3$	2×10^{-32}
barium	
$BaCO_3$	1.6×10^{-9}
$BaC_2O_4 \cdot 2H_2O$	1.1×10^{-7}
$BaSO_4$	2.3×10^{-8}
$BaCrO_4$	8.5×10^{-11}
BaF_2	2.4×10^{-5}
$Ba(OH)_2 \cdot 8H_2O$	5.0×10^{-3}
$Ba_3(PO_4)_2$	6×10^{-39}
$Ba_3(AsO_4)_2$	1.1×10^{-13}
bismuth	
$BiO(OH)$	4×10^{-10}
$BiOCl$	1.8×10^{-31}
Bi_2S_3	1×10^{-97}
cadmium	
$Cd(OH)_2$	5.9×10^{-15}
CdS	1.0×10^{-28}
$CdCO_3$	5.2×10^{-12}

TABLE J1

Substance	K_{sp} at 25 °C
calcium	
$Ca(OH)_2$	1.3×10^{-6}
$CaCO_3$	8.7×10^{-9}
$CaSO4{\cdot}2H_2O$	6.1×10^{-5}
$CaC_2O_4{\cdot}H_2O$	1.96×10^{-8}
$Ca_3(PO_4)_2$	1.3×10^{-32}
$CaHPO_4$	7×10^{-7}
CaF_2	4.0×10^{-11}
chromium	
$Cr(OH)_3$	6.7×10^{-31}
cobalt	
$Co(OH)_2$	2.5×10^{-16}
$CoS(\alpha)$	5×10^{-22}
$CoS(\beta)$	3×10^{-26}
$CoCO_3$	1.4×10^{-13}
$Co(OH)_3$	2.5×10^{-43}
copper	
CuCl	1.2×10^{-6}
CuBr	6.27×10^{-9}
CuI	1.27×10^{-12}
CuSCN	1.6×10^{-11}
Cu_2S	2.5×10^{-48}
$Cu(OH)_2$	2.2×10^{-20}
CuS	8.5×10^{-45}

TABLE J1

Substance	K_{sp} at 25 °C
$CuCO_3$	2.5×10^{-10}
iron	
$Fe(OH)_2$	1.8×10^{-15}
$FeCO_3$	2.1×10^{-11}
FeS	3.7×10^{-19}
$Fe(OH)_3$	4×10^{-38}
lead	
$Pb(OH)_2$	1.2×10^{-15}
PbF_2	4×10^{-8}
$PbCl_2$	1.6×10^{-5}
$PbBr_2$	4.6×10^{-6}
PbI_2	1.4×10^{-8}
$PbCO_3$	1.5×10^{-15}
PbS	7×10^{-29}
$PbCrO_4$	2×10^{-16}
$PbSO_4$	1.3×10^{-8}
$Pb_3(PO_4)_2$	1×10^{-54}
magnesium	
$Mg(OH)_2$	8.9×10^{-12}
$MgCO_3{\cdot}3H_2O$	*ca* 1×10^{-5}
$MgNH_4PO_4$	3×10^{-13}
MgF_2	6.4×10^{-9}
MgC_2O_4	7×10^{-7}
manganese	

TABLE J1

Substance	K_{sp} at 25 °C
$Mn(OH)_2$	2×10^{-13}
$MnCO_3$	8.8×10^{-11}
MnS	2.3×10^{-13}
mercury	
$Hg_2O{\cdot}H_2O$	3.6×10^{-26}
Hg_2Cl_2	1.1×10^{-18}
Hg_2Br_2	1.3×10^{-22}
Hg_2I_2	4.5×10^{-29}
Hg_2CO_3	9×10^{-15}
Hg_2SO_4	7.4×10^{-7}
Hg_2S	1.0×10^{-47}
Hg_2CrO_4	2×10^{-9}
HgS	1.6×10^{-54}
nickel	
$Ni(OH)_2$	1.6×10^{-16}
$NiCO_3$	1.4×10^{-7}
$NiS(\alpha)$	4×10^{-20}
$NiS(\beta)$	1.3×10^{-25}
potassium	
$KClO_4$	1.05×10^{-2}
K_2PtCl_6	7.48×10^{-6}
$KHC_4H_4O_6$	3×10^{-4}
silver	
$\frac{1}{2}Ag_2O(Ag^+ + OH^-)$	2×10^{-8}

TABLE J1

Substance	K_{sp} at 25 °C
$AgCl$	1.6×10^{-10}
$AgBr$	5.0×10^{-13}
AgI	1.5×10^{-16}
$AgCN$	1.2×10^{-16}
$AgSCN$	1.0×10^{-12}
Ag_2S	1.6×10^{-49}
Ag_2CO_3	8.1×10^{-12}
Ag_2CrO_4	9.0×10^{-12}
$Ag_4Fe(CN)_6$	1.55×10^{-41}
Ag_2SO_4	1.2×10^{-5}
Ag_3PO_4	1.8×10^{-18}
strontium	
$Sr(OH)_2{\cdot}8H_2O$	3.2×10^{-4}
$SrCO_3$	7×10^{-10}
$SrCrO_4$	3.6×10^{-5}
$SrSO_4$	3.2×10^{-7}
$SrC_2O_4{\cdot}H_2O$	4×10^{-7}
thallium	
$TlCl$	1.7×10^{-4}
$TlSCN$	1.6×10^{-4}
Tl_2S	6×10^{-22}
$Tl(OH)_3$	6.3×10^{-46}
tin	
$Sn(OH)_2$	3×10^{-27}

TABLE J1

Substance	K_{sp} at 25 °C
SnS	1×10^{-26}
$Sn(OH)_4$	1.0×10^{-57}
zinc	
$ZnCO_3$	2×10^{-10}

TABLE J1

APPENDIX K

Formation Constants for Complex Ions

Formation Constants for Complex Ions

Equilibrium	K_f
$Al^{3+} + 6F^- \rightleftharpoons [AlF_6]^{3-}$	7×10^{19}
$Cd^{2+} + 4NH_3 \rightleftharpoons [Cd(NH_3)_4]^{2+}$	1.3×10^7
$Cd^{2+} + 4CN^- \rightleftharpoons [Cd(CN)_4]^{2-}$	3×10^{18}
$Co^{2+} + 6NH_3 \rightleftharpoons [Co(NH_3)_6]^{2+}$	1.3×10^5
$Co^{3+} + 6NH_3 \rightleftharpoons [Co(NH_3)_6]^{3+}$	2.3×10^{33}
$Cu^+ + 2CN \rightleftharpoons [Cu(CN)_2]^-$	1.0×10^{16}
$Cu^{2+} + 4NH_3 \rightleftharpoons [Cu(NH_3)_4]^{2+}$	1.7×10^{13}
$Fe^{2+} + 6CN^- \rightleftharpoons [Fe(CN)_6]^{4-}$	1.5×10^{35}
$Fe^{3+} + 6CN^- \rightleftharpoons [Fe(CN)_6]^{3-}$	2×10^{43}
$Fe^{3+} + 6SCN^- \rightleftharpoons [Fe(SCN)_6]^{3-}$	3.2×10^3
$Hg^{2+} + 4Cl^- \rightleftharpoons [HgCl_4]^{2-}$	1.1×10^{16}
$Ni^{2+} + 6NH_3 \rightleftharpoons [Ni(NH_3)_6]^{2+}$	2.0×10^8
$Ag^+ + 2Cl^- \rightleftharpoons [AgCl_2]^-$	1.8×10^5
$Ag^+ + 2CN^- \rightleftharpoons [Ag(CN)_2]^-$	1×10^{21}
$Ag^+ + 2NH_3 \rightleftharpoons [Ag(NH_3)_2]^+$	1.7×10^7
$Zn^{2+} + 4CN^- \rightleftharpoons [Zn(CN)_4]^{2-}$	2.1×10^{19}
$Zn^{2+} + 4OH^- \rightleftharpoons [Zn(OH)_4]^{2-}$	2×10^{15}

TABLE K1

Equilibrium	K_f
$Fe^{3+} + SCN^- \rightleftharpoons [Fe(SCN)]^{2+}$	8.9×10^2
$Ag^+ + 4SCN^- \rightleftharpoons [Ag(SCN)_4]^{3-}$	1.2×10^{10}
$Pb^{2+} + 4I^- \rightleftharpoons [PbI_4]^{2-}$	3.0×10^4
$Pt^{2+} + 4Cl^- \rightleftharpoons [PtCl_4]^{2-}$	1×10^{16}
$Cu^{2+} + 4CN \rightleftharpoons [Cu(CN)_4]^{2-}$	1.0×10^{25}
$Co^{2+} + 4SCN^- \rightleftharpoons [Co(SCN)_4]^{2-}$	1×10^3

TABLE K1

APPENDIX L

Standard Electrode (Half-Cell) Potentials

Standard Electrode (Half-Cell) Potentials

Half-Reaction	$E°$ (V)
$Ag^+ + e^- \longrightarrow Ag$	+0.7996
$AgCl + e^- \longrightarrow Ag + Cl^-$	+0.22233
$[Ag(CN)_2]^- + e^- \longrightarrow Ag + 2CN^-$	−0.31
$Ag_2CrO_4 + 2e^- \longrightarrow 2Ag + CrO_4{}^{2-}$	+0.45
$[Ag(NH_3)_2]^+ + e^- \longrightarrow Ag + 2NH_3$	+0.373
$[Ag(S_2O_3)_2]^{3+} + e^- \longrightarrow Ag + 2S_2O_3{}^{2-}$	+0.017
$[AlF_6]^{3-} + 3e^- \longrightarrow Al + 6F^-$	−2.07
$Al^{3+} + 3e^- \longrightarrow Al$	−1.662
$Am^{3+} + 3e^- \longrightarrow Am$	−2.048
$Au^{3+} + 3e^- \longrightarrow Au$	+1.498
$Au^+ + e^- \longrightarrow Au$	+1.692
$Ba^{2+} + 2e^- \longrightarrow Ba$	−2.912
$Be^{2+} + 2e^- \longrightarrow Be$	−1.847
$Br_2(aq) + 2e^- \longrightarrow 2Br^-$	+1.0873
$Ca^{2+} + 2e^- \longrightarrow Ca$	−2.868
$Ce^3 + 3e^- \longrightarrow Ce$	−2.483
$Ce^{4+} + e^- \longrightarrow Ce^{3+}$	+1.61
$Cd^{2+} + 2e^- \longrightarrow Cd$	−0.4030
$[Cd(CN)_4]^{2-} + 2e^- \longrightarrow Cd + 4CN^-$	−1.09

TABLE L1

Half-Reaction	$E°$ (V)
$[Cd(NH_3)_4]^{2+} + 2e^- \longrightarrow Cd + 4NH_3$	−0.61
$CdS + 2e^- \longrightarrow Cd + S^{2-}$	−1.17
$Cl_2 + 2e^- \longrightarrow 2Cl^-$	+1.35827
$ClO_4{}^- + H_2O + 2e^- \longrightarrow ClO_3{}^- + 2OH^-$	+0.36
$ClO_3{}^- + H_2O + 2e^- \longrightarrow ClO_2{}^- + 2OH^-$	+0.33
$ClO_2{}^- + H_2O + 2e^- \longrightarrow ClO^- + 2OH^-$	+0.66
$ClO^- + H_2O + 2e^- \longrightarrow Cl^- + 2OH^-$	+0.89
$ClO_4{}^- + 2H_3O^+ + 2e^- \longrightarrow ClO_3{}^- + 3H_2O$	+1.189
$ClO_3{}^- + 3H_3O^+ + 2e^- \longrightarrow HClO_2 + 4H_2O$	+1.21
$HClO + H_3O^+ + 2e^- \longrightarrow Cl^- + 2H_2O$	+1.482
$HClO + H_3O^+ + e^- \longrightarrow \frac{1}{2}Cl_2 + 2H_2O$	+1.611
$HClO_2 + 2H_3O^+ + 2e^- \longrightarrow HClO + 3H_2O$	+1.628
$Co^{3+} + e^- \longrightarrow Co^{2+}$ (2 mol // H_2SO_4)	+1.83
$Co^{2+} + 2e^- \longrightarrow Co$	−0.28
$[Co(NH_3)_6]^{3+} + e^- \longrightarrow [Co(NH_3)_6]^{2+}$	+0.1
$Co(OH)_3 + e^- \longrightarrow Co(OH)_2 + OH^-$	+0.17
$Cr^3 + 3e^- \longrightarrow Cr$	−0.744
$Cr^{3+} + e^- \longrightarrow Cr^{2+}$	−0.407
$Cr^{2+} + 2e^- \longrightarrow Cr$	−0.913
$[Cu(CN)_2]^- + e^- \longrightarrow Cu + 2CN^-$	−0.43
$CrO_4{}^{2-} + 4H_2O + 3e^- \longrightarrow Cr(OH)_3 + 5OH^-$	−0.13
$Cr_2O_7{}^{2-} + 14H_3O^+ + 6e^- \longrightarrow 2Cr^{3+} + 21H_2O$	+1.232
$[Cr(OH)_4]^- + 3e^- \longrightarrow Cr + 4OH^-$	−1.2

TABLE L1

Half-Reaction	$E°$ (V)
$Cr(OH)_3 + 3e^- \longrightarrow Cr + 3OH^-$	−1.48
$Cu^{2+} + e^- \longrightarrow Cu^+$	+0.153
$Cu^{2+} + 2e^- \longrightarrow Cu$	+0.34
$Cu^+ + e^- \longrightarrow Cu$	+0.521
$F_2 + 2e^- \longrightarrow 2F^-$	+2.866
$Fe^{2+} + 2e^- \longrightarrow Fe$	−0.447
$Fe^{3+} + e^- \longrightarrow Fe^{2+}$	+0.771
$[Fe(CN)_6]^{3-} + e^- \longrightarrow [Fe(CN)_6]^{4-}$	+0.36
$Fe(OH)_2 + 2e^- \longrightarrow Fe + 2OH^-$	−0.88
$FeS + 2e^- \longrightarrow Fe + S^{2-}$	−1.01
$Ga^{3+} + 3e^- \longrightarrow Ga$	−0.549
$Gd^{3+} + 3e^- \longrightarrow Gd$	−2.279
$\frac{1}{2}H_2 + e^- \longrightarrow H^-$	−2.23
$2H_2O + 2e^- \longrightarrow H_2 + 2OH^-$	−0.8277
$H_2O_2 + 2H_3O^+ + 2e^- \longrightarrow 4H_2O$	+1.776
$2H_3O^+ + 2e^- \longrightarrow H_2 + 2H_2O$	0.00
$HO_2{}^- + H_2O + 2e^- \longrightarrow 3OH^-$	+0.878
$Hf^{4+} + 4e^- \longrightarrow Hf$	−1.55
$Hg^{2+} + 2e^- \longrightarrow Hg$	+0.851
$2Hg^{2+} + 2e^- \longrightarrow Hg_2{}^{2+}$	+0.92
$Hg_2{}^{2+} + 2e^- \longrightarrow 2Hg$	+0.7973
$[HgBr_4]^{2-} + 2e^- \longrightarrow Hg + 4Br^-$	+0.21
$Hg_2Cl_2 + 2e^- \longrightarrow 2Hg + 2Cl^-$	+0.26808

TABLE L1

Half-Reaction	$E°$ (V)
$[Hg(CN)_4]^{2-} + 2e^- \longrightarrow Hg + 4CN^-$	−0.37
$[HgI_4]^{2-} + 2e^- \longrightarrow Hg + 4I^-$	−0.04
$HgS + 2e^- \longrightarrow Hg + S^{2-}$	−0.70
$I_2 + 2e^- \longrightarrow 2I^-$	+0.5355
$In^{3+} + 3e^- \longrightarrow In$	−0.3382
$K^+ + e^- \longrightarrow K$	−2.931
$La^{3+} + 3e^- \longrightarrow La$	−2.52
$Li^+ + e^- \longrightarrow Li$	−3.04
$Lu^{3+} + 3e^- \longrightarrow Lu$	−2.28
$Mg^{2+} + 2e^- \longrightarrow Mg$	−2.372
$Mn^{2+} + 2e^- \longrightarrow Mn$	−1.185
$MnO_2 + 2H_2O + 2e^- \longrightarrow Mn(OH)_2 + 2OH^-$	−0.05
$MnO_4{}^- + 2H_2O + 3e^- \longrightarrow MnO_2 + 4OH^-$	+0.558
$MnO_2 + 4H^+ + 2e^- \longrightarrow Mn^{2+} + 2H_2O$	+1.23
$MnO_4{}^- + 8H^+ + 5e^- \longrightarrow Mn^{2+} + 4H_2O$	+1.507
$Na^+ + e^- \longrightarrow Na$	−2.71
$Nd^{3+} + 3e^- \longrightarrow Nd$	−2.323
$Ni^{2+} + 2e^- \longrightarrow Ni$	−0.257
$[Ni(NH_3)_6]^{2+} + 2e^- \longrightarrow Ni + 6NH_3$	−0.49
$NiO_2 + 4H^+ + 2e^- \longrightarrow Ni^{2+} + 2H_2O$	+1.593
$NiO_2 + 2H_2O + 2e^- \longrightarrow Ni(OH)_2 + 2OH^-$	+0.49
$NiS + 2e^- \longrightarrow Ni + S^{2-}$	+0.76
$NO_3{}^- + 4H^+ + 3e^- \longrightarrow NO + 2H_2O$	+0.957
$NO_3{}^- + 3H^+ + 2e^- \longrightarrow HNO_2 + H_2O$	+0.92

TABLE L1

Half-Reaction	E° (V)
$NO_3^- + H_2O + 2e^- \longrightarrow NO_2^- + 2OH^-$	+0.10
$Np^{3+} + 3e^- \longrightarrow Np$	−1.856
$O_2 + 2H_2O + 4e^- \longrightarrow 4OH^-$	+0.401
$O_2 + 2H^+ + 2e^- \longrightarrow H_2O_2$	+0.695
$O_2 + 4H^+ + 4e^- \longrightarrow 2H_2O$	+1.229
$Pb^{2+} + 2e^- \longrightarrow Pb$	−0.1262
$PbO_2 + SO_4^{2-} + 4H^+ + 2e^- \longrightarrow PbSO_4 + 2H_2O$	+1.69
$PbS + 2e^- \longrightarrow Pb + S^{2-}$	−0.95
$PbSO_4 + 2e^- \longrightarrow Pb + SO_4^{2-}$	−0.3505
$Pd^{2+} + 2e^- \longrightarrow Pd$	+0.987
$[PdCl_4]^{2-} + 2e^- \longrightarrow Pd + 4Cl^-$	+0.591
$Pt^{2+} + 2e^- \longrightarrow Pt$	+1.20
$[PtBr_4]^{2-} + 2e^- \longrightarrow Pt + 4Br^-$	+0.58
$[PtCl_4]^{2-} + 2e^- \longrightarrow Pt + 4Cl^-$	+0.755
$\left[PtCl_6\right]^{2-} + 2e^- \longrightarrow [PtCl_4]^{2-} + 2Cl^-$	+0.68
$Pu^{3} + 3e^- \longrightarrow Pu$	−2.03
$Ra^{2+} + 2e^- \longrightarrow Ra$	−2.92
$Rb^+ + e^- \longrightarrow Rb$	−2.98
$\left[RhCl_6\right]^{3-} + 3e^- \longrightarrow Rh + 6Cl^-$	+0.44
$S + 2e^- \longrightarrow S^{2-}$	−0.47627
$S + 2H^+ + 2e^- \longrightarrow H_2S$	+0.142
$Sc^{3+} + 3e^- \longrightarrow Sc$	−2.09
$Se + 2H^+ + 2e^- \longrightarrow H_2Se$	−0.399
$\left[SiF_6\right]^{2-} + 4e^- \longrightarrow Si + 6F^-$	−1.2

TABLE L1

Half-Reaction	E° (V)
$SiO_3{}^{2-} + 3H_2O + 4e^- \longrightarrow Si + 6OH^-$	−1.697
$SiO_2 + 4H^+ + 4e^- \longrightarrow Si + 2H_2O$	−0.86
$Sm^{3+} + 3e^- \longrightarrow Sm$	−2.304
$Sn^{4+} + 2e^- \longrightarrow Sn^{2+}$	+0.151
$Sn^{2+} + 2e^- \longrightarrow Sn$	−0.1375
$\left[SnF_6\right]^{2-} + 4e^- \longrightarrow Sn + 6F^-$	−0.25
$SnS + 2e^- \longrightarrow Sn + S^{2-}$	−0.94
$Sr^{2+} + 2e^- \longrightarrow Sr$	−2.89
$TeO_2 + 4H^+ + 4e^- \longrightarrow Te + 2H_2O$	+0.593
$Th^{4+} + 4e^- \longrightarrow Th$	−1.90
$Ti^{2+} + 2e^- \longrightarrow Ti$	−1.630
$U^{3+} + 3e^- \longrightarrow U$	−1.79
$V^{2+} + 2e^- \longrightarrow V$	−1.19
$Y^{3+} + 3e^- \longrightarrow Y$	−2.37
$Zn^{2+} + 2e^- \longrightarrow Zn$	−0.7618
$\left[Zn(CN)_4\right]^{2-} + 2e^- \longrightarrow Zn + 4CN^-$	−1.26
$\left[Zn(NH_3)_4\right]^{2+} + 2e^- \longrightarrow Zn + 4NH_3$	−1.04
$Zn(OH)_2 + 2e^- \longrightarrow Zn + 2OH^-$	−1.245
$\left[Zn(OH)_4\right]^{2} + 2e^- \longrightarrow Zn + 4OH^-$	−1.199
$ZnS + 2e^- \longrightarrow Zn + S^{2-}$	−1.40
$Zr^{4} + 4e^- \longrightarrow Zr$	−1.539

TABLE L1

APPENDIX M

Half-Lives for Several Radioactive Isotopes

Half-Lives for Several Radioactive Isotopes

Isotope	Half-Life[1]	Type of Emission[2]	Isotope	Half-Life[3]	Type of Emission[4]
${}^{14}_{6}C$	5730 y	(β^-)	${}^{210}_{83}Bi$	5.01 d	(β^-)
${}^{13}_{7}N$	9.97 m	(β^+)	${}^{212}_{83}Bi$	60.55 m	$(\alpha$ or $\beta^-)$
${}^{15}_{9}F$	4.1×10^{-22} s	(p)	${}^{210}_{84}Po$	138.4 d	(α)
${}^{24}_{11}Na$	15.00 h	(β^-)	${}^{212}_{84}Po$	3×10^{-7} s	(α)
${}^{32}_{15}P$	14.29 d	(β^-)	${}^{216}_{84}Po$	0.15 s	(α)
${}^{40}_{19}K$	1.27×10^{9} y	$(\beta$ or $E.C.)$	${}^{218}_{84}Po$	3.05 m	(α)
${}^{49}_{26}Fe$	0.08 s	(β^+)	${}^{215}_{85}At$	1.0×10^{-4} s	(α)
${}^{60}_{26}Fe$	2.6×10^{6} y	(β^-)	${}^{218}_{85}At$	1.6 s	(α)
${}^{60}_{27}Co$	5.27 y	(β^-)	${}^{220}_{86}Rn$	55.6 s	(α)
${}^{87}_{37}Rb$	4.7×10^{10} y	(β^-)	${}^{222}_{86}Rn$	3.82 d	(α)
${}^{90}_{38}Sr$	29 y	(β^-)	${}^{224}_{88}Ra$	3.66 d	(α)
${}^{115}_{49}In$	5.1×10^{15} y	(β^-)	${}^{226}_{88}Ra$	1600 y	(α)
${}^{131}_{53}I$	8.040 d	(β^-)	${}^{228}_{88}Ra$	5.75 y	(β^-)
${}^{142}_{58}Ce$	5×10^{15} y	(α)	${}^{228}_{89}Ac$	6.13 h	(β^-)
${}^{208}_{81}Tl$	3.07 m	(β^-)	${}^{228}_{90}Th$	1.913 y	(α)
${}^{210}_{82}Pb$	22.3 y	(β^-)	${}^{232}_{90}Th$	1.4×10^{10} y	(α)
${}^{212}_{82}Pb$	10.6 h	(β^-)	${}^{233}_{90}Th$	22 m	(β^-)
${}^{214}_{82}Pb$	26.8 m	(β^-)	${}^{234}_{90}Th$	24.10 d	(β^-)

1 y = years, d = days, h = hours, m = minutes, s = seconds
2 *E.C.* = electron capture, *S.F.* = Spontaneous fission
3 y = years, d = days, h = hours, m = minutes, s = seconds

Isotope	Half-Life[1]	Type of Emission[2]	Isotope	Half-Life[3]	Type of Emission[4]
$^{206}_{83}\text{Bi}$	6.243 d	$(E.C.)$	$^{233}_{91}\text{Pa}$	27 d	(β^-)
$^{233}_{92}\text{U}$	1.59×10^5 y	(α)	$^{242}_{96}\text{Cm}$	162.8 d	(α)
$^{234}_{92}\text{U}$	2.45×10^5 y	(α)	$^{243}_{97}\text{Bk}$	4.5 h	$(\alpha \text{ or } E.C.)$
$^{235}_{92}\text{U}$	7.03×10^8 y	(α)	$^{253}_{99}\text{Es}$	20.47 d	(α)
$^{238}_{92}\text{U}$	4.47×10^9 y	(α)	$^{254}_{100}\text{Fm}$	3.24 h	$(\alpha \text{ or } S.F.)$
$^{239}_{92}\text{U}$	23.54 m	(β^-)	$^{255}_{100}\text{Fm}$	20.1 h	(α)
$^{239}_{93}\text{Np}$	2.3 d	(β^-)	$^{256}_{101}\text{Md}$	76 m	$(\alpha \text{ or } E.C.)$
$^{239}_{94}\text{Pu}$	2.407×10^4 y	(α)	$^{254}_{102}\text{No}$	55 s	(α)
$^{239}_{94}\text{Pu}$	6.54×10^3 y	(α)	$^{257}_{103}\text{Lr}$	0.65 s	(α)
$^{241}_{94}\text{Pu}$	14.4 y	$(\alpha \text{ or } \beta^-)$	$^{260}_{105}\text{Ha}$	1.5 s	$(\alpha \text{ or } S.F.)$
$^{241}_{95}\text{Am}$	432.2 y	(α)	$^{263}_{106}\text{Sg}$	0.8 s	$(\alpha \text{ or } S.F.)$

TABLE M1

4 *E.C.* = electron capture, *S.F.* = Spontaneous fission

ANSWER KEY

Chapter 1

1. Place a glass of water outside. It will freeze if the temperature is below 0 °C.
3. (a) law (states a consistently observed phenomenon, can be used for prediction); (b) theory (a widely accepted explanation of the behavior of matter); (c) hypothesis (a tentative explanation, can be investigated by experimentation)
5. (a) symbolic, microscopic; (b) macroscopic; (c) symbolic, macroscopic; (d) microscopic
7. Macroscopic. The heat required is determined from macroscopic properties.
9. Liquids can change their shape (flow); solids can't. Gases can undergo large volume changes as pressure changes; liquids do not. Gases flow and change volume; solids do not.
11. The mixture can have a variety of compositions; a pure substance has a definite composition. Both have the same composition from point to point.
13. Molecules of elements contain only one type of atom; molecules of compounds contain two or more types of atoms. They are similar in that both are comprised of two or more atoms chemically bonded together.
15. Answers will vary. Sample answer: Gatorade contains water, sugar, dextrose, citric acid, salt, sodium chloride, monopotassium phosphate, and sucrose acetate isobutyrate.
17. (a) element; (b) element; (c) compound; (d) mixture; (e) compound; (f) compound; (g) compound; (h) mixture
19. In each case, a molecule consists of two or more combined atoms. They differ in that the types of atoms change from one substance to the next.
21. Gasoline (a mixture of compounds), oxygen, and to a lesser extent, nitrogen are consumed. Carbon dioxide and water are the principal products. Carbon monoxide and nitrogen oxides are produced in lesser amounts.
23. (a) Increased as it would have combined with oxygen in the air thus increasing the amount of matter and therefore the mass. (b) 0.9 g
25. (a) 200.0 g; (b) The mass of the container and contents would decrease as carbon dioxide is a gaseous product and would leave the container. (c) 102.3 g
27. (a) physical; (b) chemical; (c) chemical; (d) physical; (e) physical
29. physical
31. The value of an extensive property depends upon the amount of matter being considered, whereas the value of an intensive property is the same regardless of the amount of matter being considered.
33. Being extensive properties, both mass and volume are directly proportional to the amount of substance under study. Dividing one extensive property by another will in effect "cancel" this dependence on amount, yielding a ratio that is independent of amount (an intensive property).
35. about a yard
37. (a) kilograms; (b) meters; (c) meters/second; (d) kilograms/cubic meter; (e) kelvin; (f) square meters; (g) cubic meters
39. (a) centi-, $\times 10^{-2}$; (b) deci-, $\times 10^{-1}$; (c) Giga-, $\times 10^{9}$; (d) kilo-, $\times 10^{3}$; (e) milli-, $\times 10^{-3}$; (f) nano-, $\times 10^{-9}$; (g) pico-, $\times 10^{-12}$; (h) tera-, $\times 10^{12}$
41. (a) m = 18.58 g, V = 5.7 mL. (b) d = 3.3 g/mL (c) dioptase (copper cyclosilicate, d = 3.28—3.31 g/mL); malachite (basic copper carbonate, d = 3.25—4.10 g/mL); Paraiba tourmaline (sodium lithium boron silicate with copper, d = 2.82—3.32 g/mL)
43. (a) displaced water volume = 2.8 mL; (b) displaced water mass = 2.8 g; (c) The block mass is 2.76 g, essentially equal to the mass of displaced water (2.8 g) and consistent with Archimedes' principle of buoyancy.
45. (a) 7.04×10^{2}; (b) 3.344×10^{-2}; (c) 5.479×10^{2}; (d) 2.2086×10^{4}; (e) 1.00000×10^{3}; (f) 6.51×10^{-8}; (g) 7.157×10^{-3}
47. (a) exact; (b) exact; (c) uncertain; (d) exact; (e) uncertain; (f) uncertain

49. (a) two; (b) three; (c) five; (d) four; (e) six; (f) two; (g) five
51. (a) 0.44; (b) 9.0; (c) 27; (d) 140; (e) 1.5×10^{-3}; (f) 0.44
53. (a) 2.15×10^5; (b) 4.2×10^6; (c) 2.08; (d) 0.19; (e) 27,440; (f) 43.0
55. (a) Archer X; (b) Archer W; (c) Archer Y
57. (a) $\frac{1.0936 \text{ yd}}{1 \text{ m}}$; (b) $\frac{0.94635 \text{ L}}{1 \text{ qt}}$; (c) $\frac{2.2046 \text{ lb}}{1 \text{ kg}}$
59. $\frac{2.0 \text{ L}}{67.6 \text{ fl oz}} = \frac{0.030 \text{ L}}{1 \text{ fl oz}}$
Only two significant figures are justified.
61. 68–71 cm; 400–450 g
63. 355 mL
65. 8×10^{-4} cm
67. yes; weight = 89.4 kg
69. 5.0×10^{-3} mL
71. (a) 1.3×10^{-4} kg; (b) 2.32×10^8 kg; (c) 5.23×10^{-12} m; (d) 8.63×10^{-5} kg; (e) 3.76×10^{-1} m; (f) 5.4×10^{-5} m; (g) 1×10^{12} s; (h) 2.7×10^{-11} s; (i) 1.5×10^{-4} K
73. 45.4 L
75. 1.0160×10^3 kg
77. (a) 394 ft; (b) 5.9634 km; (c) 6.0×10^2; (d) 2.64 L; (e) 5.1×10^{18} kg; (f) 14.5 kg; (g) 324 mg
79. 0.46 m; 1.5 ft/cubit
81. Yes, the acid's volume is 123 mL.
83. 62.6 in (about 5 ft 3 in.) and 101 lb
85. (a) 3.81 cm × 8.89 cm × 2.44 m; (b) 40.6 cm
87. 2.70 g/cm^3
89. (a) 81.6 g; (b) 17.6 g
91. (a) 5.1 mL; (b) 37 L
93. 5371 °F, 3239 K
95. −23 °C, 250 K
97. −33.4 °C, 239.8 K
99. 113 °F

Chapter 2

1. The starting materials consist of one green sphere and two purple spheres. The products consist of two green spheres and two purple spheres. This violates Dalton's postulate that that atoms are not created during a chemical change, but are merely redistributed.
3. This statement violates Dalton's fourth postulate: In a given compound, the numbers of atoms of each type (and thus also the percentage) always have the same ratio.
5. Dalton originally thought that all atoms of a particular element had identical properties, including mass. Thus, the concept of isotopes, in which an element has different masses, was a violation of the original idea. To account for the existence of isotopes, the second postulate of his atomic theory was modified to state that atoms of the same element must have identical chemical properties.
7. Both are subatomic particles that reside in an atom's nucleus. Both have approximately the same mass. Protons are positively charged, whereas neutrons are uncharged.
9. (a) The Rutherford atom has a small, positively charged nucleus, so most α particles will pass through empty space far from the nucleus and be undeflected. Those α particles that pass near the nucleus will be deflected from their paths due to positive-positive repulsion. The more directly toward the nucleus the α particles are headed, the larger the deflection angle will be. (b) Higher-energy α particles that pass near the nucleus will still undergo deflection, but the faster they travel, the less the expected angle of deflection. (c) If the nucleus is smaller, the positive charge is smaller and the expected deflections are smaller—both in terms of how closely the α particles pass by the nucleus undeflected and the angle of deflection. If the nucleus is larger, the positive charge is larger and the expected deflections are larger—more α particles will be deflected, and the deflection angles will be larger. (d) The paths followed by the α particles match the predictions from (a), (b), and (c).
11. (a) $^{133}Cs^+$; (b) $^{127}I^-$; (c) $^{31}P^{3-}$; (d) $^{57}Co^{3+}$

13. (a) Carbon-12, ^{12}C; (b) This atom contains six protons and six neutrons. There are six electrons in a neutral ^{12}C atom. The net charge of such a neutral atom is zero, and the mass number is 12. (c) The preceding answers are correct. (d) The atom will be stable since C-12 is a stable isotope of carbon. (e) The preceding answer is correct. Other answers for this exercise are possible if a different element of isotope is chosen.
15. (a) Lithium-6 contains three protons, three neutrons, and three electrons. The isotope symbol is ^{6}Li or $^{6}_{3}Li$. (b) $^{6}Li^{+}$ or $^{6}_{3}Li^{+}$
17. (a) Iron, 26 protons, 24 electrons, and 32 neutrons; (b) iodine, 53 protons, 54 electrons, and 74 neutrons
19. (a) 3 protons, 3 electrons, 4 neutrons; (b) 52 protons, 52 electrons, 73 neutrons; (c) 47 protons, 47 electrons, 62 neutrons; (d) 7 protons, 7 electrons, 8 neutrons; (e) 15 protons, 15 electrons, 16 neutrons
21. Let us use neon as an example. Since there are three isotopes, there is no way to be sure to accurately predict the abundances to make the total of 20.18 amu average atomic mass. Let us guess that the abundances are 9% Ne-22, 91% Ne-20, and only a trace of Ne-21. The average mass would be 20.18 amu. Checking the nature's mix of isotopes shows that the abundances are 90.48% Ne-20, 9.25% Ne-22, and 0.27% Ne-21, so our guessed amounts have to be slightly adjusted.
23. 79.90 amu
25. Turkey source: 20.3% (of 10.0129 amu isotope); US source: 19.1% (of 10.0129 amu isotope)
27. The symbol for the element oxygen, O, represents both the element and one atom of oxygen. A molecule of oxygen, O_2, contains two oxygen atoms; the subscript 2 in the formula must be used to distinguish the diatomic molecule from two single oxygen atoms.
29. (a) molecular CO_2, empirical CO_2; (b) molecular C_2H_2, empirical CH; (c) molecular C_2H_4, empirical CH_2; (d) molecular H_2SO_4, empirical H_2SO_4
31. (a) $C_4H_5N_2O$; (b) $C_{12}H_{22}O_{11}$; (c) HO; (d) CH_2O; (e) $C_3H_4O_3$
33. (a) CH_2O; (b) C_2H_4O
35. (a) ethanol

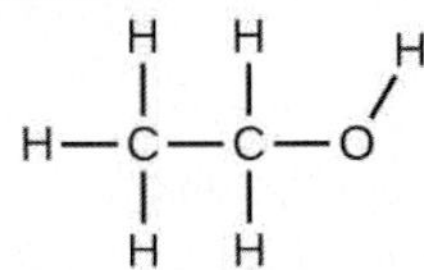

(b) methoxymethane, more commonly known as dimethyl ether

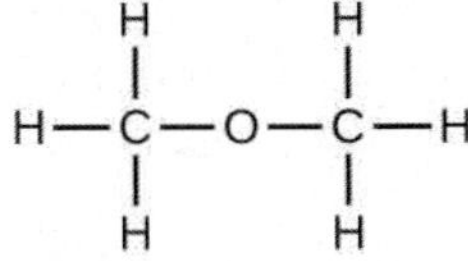

(c) These molecules have the same chemical composition (types and number of atoms) but different chemical structures. They are structural isomers.
37. (a) metal, inner transition metal; (b) nonmetal, representative element; (c) metal, representative element; (d) nonmetal, representative element; (e) metal, transition metal; (f) metal, inner transition metal; (g) metal, transition metal; (h) nonmetal, representative element; (i) nonmetal, representative element; (j) metal, representative element
39. (a) He; (b) Be; (c) Li; (d) O
41. (a) krypton, Kr; (b) calcium, Ca; (c) fluorine, F; (d) tellurium, Te
43. (a) $^{23}_{11}Na$; (b) $^{129}_{54}Xe$; (c) $^{73}_{33}As$; (d) $^{226}_{88}Ra$
45. Ionic: KCl, $MgCl_2$; Covalent: NCl_3, ICl, PCl_5, CCl_4
47. (a) covalent; (b) ionic, Ba^{2+}, O^{2-}; (c) ionic, NH_4^{+}, CO_3^{2-}; (d) ionic, Sr^{2+}, $H_2PO_4^{-}$; (e) covalent; (f) ionic, Na^{+}, O^{2-}
49. (a) CaS; (b) $(NH_4)_2SO_4$; (c) $AlBr_3$; (d) Na_2HPO_4; (e) $Mg_3(PO_4)_2$
51. (a) cesium chloride; (b) barium oxide; (c) potassium sulfide; (d) beryllium chloride; (e) hydrogen bromide; (f) aluminum fluoride
53. (a) RbBr; (b) MgSe; (c) Na_2O; (d) $CaCl_2$; (e) HF; (f) GaP; (g) $AlBr_3$; (h) $(NH_4)_2SO_4$
55. (a) ClO_2; (b) N_2O_4; (c) K_3P; (d) Ag_2S; (e) $AlF_3 \cdot 3H_2O$; (f) SiO_2
57. (a) chromium(III) oxide; (b) iron(II) chloride; (c) chromium(VI) oxide; (d) titanium(IV) chloride; (e)

cobalt(II) chloride hexahydrate; (f) molybdenum(IV) sulfide

59. (a) K_3PO_4; (b) $CuSO_4$; (c) $CaCl_2$; (d) TiO_2; (e) NH_4NO_3; (f) $NaHSO_4$

61. (a) manganese(IV) oxide; (b) mercury(I) chloride; (c) iron(III) nitrate; (d) titanium(IV) chloride; (e) copper(II) bromide

Chapter 3

1. (a) 12.01 amu; (b) 12.01 amu; (c) 144.12 amu; (d) 60.05 amu
3. (a) 123.896 amu; (b) 18.015 amu; (c) 164.086 amu; (d) 60.052 amu; (e) 342.297 amu
5. (a) 56.107 amu; (b) 54.091 amu; (c) 199.9976 amu; (d) 97.9950 amu
7. Use the molecular formula to find the molar mass; to obtain the number of moles, divide the mass of compound by the molar mass of the compound expressed in grams.
9. Formic acid. Its formula has twice as many oxygen atoms as the other two compounds (one each). Therefore, 0.60 mol of formic acid would be equivalent to 1.20 mol of a compound containing a single oxygen atom.
11. The two masses have the same numerical value, but the units are different: The molecular mass is the mass of 1 molecule while the molar mass is the mass of 6.022×10^{23} molecules.
13. (a) 256.48 g/mol; (b) 72.150 g mol^{-1}; (c) 378.103 g mol^{-1}; (d) 58.080 g mol^{-1}; (e) 180.158 g mol^{-1}
15. (a) 197.382 g mol^{-1}; (b) 257.163 g mol^{-1}; (c) 194.193 g mol^{-1}; (d) 60.056 g mol^{-1}; (e) 306.464 g mol^{-1}
17. (a) 0.819 g; (b) 307 g; (c) 0.23 g; (d) 1.235×10^6 g (1235 kg); (e) 765 g
19. (a) 99.41 g; (b) 2.27 g; (c) 3.5 g; (d) 222 kg; (e) 160.1 g
21. (a) 9.60 g; (b) 19.2 g; (c) 28.8 g
23. zirconium: 2.038×10^{23} atoms; 30.87 g; silicon: 2.038×10^{23} atoms; 9.504 g; oxygen: 8.151×10^{23} atoms; 21.66 g
25. $AlPO_4$: 1.000 mol, or 26.98 g Al; Al_2Cl_6: 1.994 mol, or 53.74 g Al; Al_2S_3: 3.00 mol, or 80.94 g Al; The Al_2S_3 sample thus contains the greatest mass of Al.
27. 3.113×10^{25} C atoms
29. 0.865 servings, or about 1 serving.
31. 20.0 g H_2O represents the least number of molecules since it has the least number of moles.
33. (a) % N = 82.24%, % H = 17.76%; (b) % Na = 29.08%, % S = 40.56%, % O = 30.36%; (c) % Ca^{2+} = 38.76%
35. % NH_3 = 38.2%
37. (a) CS_2; (b) CH_2O
39. C_6H_6
41. $Mg_3Si_2H_3O_8$ (empirical formula), $Mg_6Si_4H_6O_{16}$ (molecular formula)
43. $C_{15}H_{15}N_3$
45. We need to know the number of moles of sulfuric acid dissolved in the solution and the volume of the solution.
47. (a) 0.679 *M*; (b) 1.00 *M*; (c) 0.06998 *M*; (d) 1.75 *M*; (e) 0.070 *M*; (f) 6.6 *M*
49. (a) determine the number of moles of glucose in 0.500 L of solution; determine the molar mass of glucose; determine the mass of glucose from the number of moles and its molar mass; (b) 27 g
51. (a) 37.0 mol H_2SO_4, 3.63×10^3 g H_2SO_4; (b) 3.8×10^{-7} mol NaCN, 1.9×10^{-5} g NaCN; (c) 73.2 mol H_2CO, 2.20 kg H_2CO; (d) 5.9×10^{-7} mol $FeSO_4$, 8.9×10^{-5} g $FeSO_4$
53. (a) Determine the molar mass of $KMnO_4$; determine the number of moles of $KMnO_4$ in the solution; from the number of moles and the volume of solution, determine the molarity; (b) 1.15×10^{-3} *M*
55. (a) 5.04×10^{-3} *M*; (b) 0.499 *M*; (c) 9.92 *M*; (d) 1.1×10^{-3} *M*
57. 0.025 *M*
59. 0.5000 L
61. 1.9 mL
63. (a) 0.125 *M*; (b) 0.04888 *M*; (c) 0.206 *M*; (d) 0.0056 *M*
65. 11.9 *M*
67. 1.6 L
69. (a) The dilution equation can be used, appropriately modified to accommodate mass-based concentration units: $\%mass_1 \times mass_1 = \%mass_2 \times mass_2$. This equation can be rearranged to isolate $mass_1$ and the given quantities substituted into this equation. (b) 58.8 g

71. 114 g
73. 1.75×10^{-3} *M*
75. 95 mg/dL
77. 2.38×10^{-4} mol
79. 0.29 mol

Chapter 4

1. An equation is balanced when the same number of each element is represented on the reactant and product sides. Equations must be balanced to accurately reflect the law of conservation of matter.
3. (a) $PCl_5(s) + H_2O(l) \longrightarrow POCl_3(l) + 2HCl(aq)$; (b) $3Cu(s) + 8HNO_3(aq) \longrightarrow 3Cu(NO_3)_2(aq) + 4H_2O(l) + 2NO(g)$; (c) $H_2(g) + I_2(s) \longrightarrow 2HI(s)$; (d) $4Fe(s) + 3O_2(g) \longrightarrow 2Fe_2O_3(s)$; (e) $2Na(s) + 2H_2O(l) \longrightarrow 2NaOH(aq) + H_2(g)$; (f) $(NH_4)_2Cr_2O_7(s) \longrightarrow Cr_2O_3(s) + N_2(g) + 4H_2O(g)$; (g) $P_4(s) + 6Cl_2(g) \longrightarrow 4PCl_3(l)$; (h) $PtCl_4(s) \longrightarrow Pt(s) + 2Cl_2(g)$
5. (a) $CaCO_3(s) \longrightarrow CaO(s) + CO_2(g)$; (b) $2C_4H_{10}(g) + 13O_2(g) \longrightarrow 8CO_2(g) + 10H_2O(g)$; (c) $MgCl_2(aq) + 2NaOH(aq) \longrightarrow Mg(OH)_2(s) + 2NaCl(aq)$; (d) $2H_2O(g) + 2Na(s) \longrightarrow 2NaOH(s) + H_2(g)$
7. (a) $Ba(NO_3)_2$, $KClO_3$; (b) $2KClO_3(s) \longrightarrow 2KCl(s) + 3O_2(g)$; (c) $2Ba(NO_3)_2(s) \longrightarrow 2BaO(s) + 2N_2(g) + 5O_2(g)$; (d) $2Mg(s) + O_2(g) \longrightarrow 2MgO(s)$; $4Al(s) + 3O_2(g) \longrightarrow 2Al_2O_3(s)$; $4Fe(s) + 3O_2(g) \longrightarrow 2Fe_2O_3(s)$
9. (a) $4HF(aq) + SiO_2(s) \longrightarrow SiF_4(g) + 2H_2O(l)$; (b) complete ionic equation: $2Na^+(aq) + 2F^-(aq) + Ca^{2+}(aq) + 2Cl^-(aq) \longrightarrow CaF_2(s) + 2Na^+(aq) + 2Cl^-(aq)$, net ionic equation: $2F^-(aq) + Ca^{2+}(aq) \longrightarrow CaF_2(s)$
11. (a)

$$2K^+(aq) + C_2O_4{}^{2-}(aq) + Ba^{2+}(aq) + 2OH^-(aq) \longrightarrow 2K^+(aq) + 2OH^-(aq) + BaC_2O_4(s) \quad \text{(complete)}$$
$$Ba^{2+}(aq) + C_2O_4{}^{2-}(aq) \longrightarrow BaC_2O_4(s) \quad \text{(net)}$$

(b)

$$Pb^{2+}(aq) + 2NO_3{}^-(aq) + 2H^+(aq) + SO_4{}^{2-}(aq) \longrightarrow PbSO_4(s) + 2H^+(aq) + 2NO_3{}^-(aq) \quad \text{(complete)}$$
$$Pb^{2+}(aq) + SO_4{}^{2-}(aq) \longrightarrow PbSO_4(s) \quad \text{(net)}$$

(c) $$CaCO_3(s) + 2H^+(aq) + SO_4{}^{2-}(aq) \longrightarrow CaSO_4(s) + CO_2(g) + H_2O(l) \quad \text{(complete)}$$
$$CaCO_3(s) + 2H^+(aq) + SO_4{}^{2-}(aq) \longrightarrow CaSO_4(s) + CO_2(g) + H_2O(l) \quad \text{(net)}$$

13. (a) oxidation-reduction (addition); (b) acid-base (neutralization); (c) oxidation-reduction (combustion)
15. It is an oxidation-reduction reaction because the oxidation state of the silver changes during the reaction.
17. (a) H +1, P +5, O −2; (b) Al +3, H +1, O −2; (c) Se +4, O −2; (d) K +1, N +3, O −2; (e) In +3, S −2; (f) P +3, O −2
19. (a) acid-base; (b) oxidation-reduction: Na is oxidized, H^+ is reduced; (c) oxidation-reduction: Mg is oxidized, Cl_2 is reduced; (d) acid-base; (e) oxidation-reduction: P^{3-} is oxidized, O_2 is reduced; (f) acid-base
21. (a) $2HCl(g) + Ca(OH)_2(s) \longrightarrow CaCl_2(s) + 2H_2O(l)$; (b) $Sr(OH)_2(aq) + 2HNO_3(aq) \longrightarrow Sr(NO_3)_2(aq) + 2H_2O(l)$
23. (a) $2Al(s) + 3F_2(g) \longrightarrow 2AlF_3(s)$; (b) $2Al(s) + 3CuBr_2(aq) \longrightarrow 3Cu(s) + 2AlBr_3(aq)$; (c) $P_4(s) + 5O_2(g) \longrightarrow P_4O_{10}(s)$; (d) $Ca(s) + 2H_2O(l) \longrightarrow Ca(OH)_2(aq) + H_2(g)$
25. (a) $Mg(OH)_2(s) + 2HClO_4(aq) \longrightarrow Mg^{2+}(aq) + 2ClO_4{}^-(aq) + 2H_2O(l)$; (b) $SO_3(g) + 2H_2O(l) \longrightarrow H_3O^+(aq) + HSO_4{}^-(aq)$, (a solution of H_2SO_4); (c) $SrO(s) + H_2SO_4(l) \longrightarrow SrSO_4(s) + H_2O$
27. $H_2(g) + F_2(g) \longrightarrow 2HF(g)$
29. $2NaBr(aq) + Cl_2(g) \longrightarrow 2NaCl(aq) + Br_2(l)$
31. $2LiOH(aq) + CO_2(g) \longrightarrow Li_2CO_3(aq) + H_2O(l)$
33. (a) $Ca(OH)_2(s) + H_2S(g) \longrightarrow CaS(s) + 2H_2O(l)$; (b) $Na_2CO_3(aq) + H_2S(g) \longrightarrow Na_2S(aq) + CO_2(g) + H_2O(l)$
35. (a) step 1: $N_2(g) + 3H_2(g) \longrightarrow 2NH_3(g)$, step 2: $NH_3(g) + HNO_3(aq) \longrightarrow NH_4NO_3(aq) \longrightarrow NH_4NO_3(s)$ (after drying); (b)

$H_2(g) + Br_2(l) \longrightarrow 2HBr(g)$; (c) $Zn(s) + S(s) \longrightarrow ZnS(s)$ and $ZnS(s) + 2HCl(aq) \longrightarrow ZnCl_2(aq) + H_2S(g)$

37. (a) $Sn^{4+}(aq) + 2e^- \longrightarrow Sn^{2+}(aq)$, (b) $[Ag(NH_3)_2]^+(aq) + e^- \longrightarrow Ag(s) + 2NH_3(aq)$; (c) $Hg_2Cl_2(s) + 2e^- \longrightarrow 2Hg(l) + 2Cl^-(aq)$; (d) $2H_2O(l) \longrightarrow O_2(g) + 4H^+(aq) + 4e^-$; (e) $6H_2O(l) + 2IO_3^-(aq) + 10e^- \longrightarrow I_2(s) + 12OH^-(aq)$; (f) $H_2O(l) + SO_3^{2-}(aq) \longrightarrow SO_4^{2-}(aq) + 2H^+(aq) + 2e^-$; (g) $8H^+(aq) + MnO_4^-(aq) + 5e^- \longrightarrow Mn^{2+}(aq) + 4H_2O(l)$; (h) $Cl^-(aq) + 6OH^-(aq) \longrightarrow ClO_3^-(aq) + 3H_2O(l) + 6e^-$

39. (a) $Sn^{2+}(aq) + 2Cu^{2+}(aq) \longrightarrow Sn^{4+}(aq) + 2Cu^+(aq)$; (b) $H_2S(g) + Hg_2^{2+}(aq) + 2H_2O(l) \longrightarrow 2Hg(l) + S(s) + 2H_3O^+(aq)$; (c) $5CN^-(aq) + 2ClO_2(aq) + 3H_2O(l) \longrightarrow 5CNO^-(aq) + 2Cl^-(aq) + 2H_3O^+(aq)$; (d) $Fe^{2+}(aq) + Ce^{4+}(aq) \longrightarrow Fe^{3+}(aq) + Ce^{3+}(aq)$; (e) $2HBrO(aq) + 2H_2O(l) \longrightarrow 2H_3O^+(aq) + 2Br^-(aq) + O_2(g)$

41. (a) $2MnO_4^-(aq) + 3NO_2^-(aq) + H_2O(l) \longrightarrow 2MnO_2(s) + 3NO_3^-(aq) + 2OH^-(aq)$; (b) $3MnO_4^{2-}(aq) + 2H_2O(l) \longrightarrow 2MnO_4^-(aq) + 4OH^-(aq) + MnO_2(s)$ (in base); (c) $Br_2(l) + SO_2(g) + 2H_2O(l) \longrightarrow 4H^+(aq) + 2Br^-(aq) + SO_4^{2-}(aq)$

43. (a) 0.435 mol Na, 0.217 mol Cl_2, 15.4 g Cl_2; (b) 0.005780 mol HgO, 2.890×10^{-3} mol O_2, 9.248×10^{-2} g O_2; (c) 8.00 mol $NaNO_3$, 6.8×10^2 g $NaNO_3$; (d) 1665 mol CO_2, 73.3 kg CO_2; (e) 18.86 mol CuO, 2.330 kg $CuCO_3$; (f) 0.4580 mol $C_2H_4Br_2$, 86.05 g $C_2H_4Br_2$

45. (a) 0.0686 mol Mg, 1.67 g Mg; (b) 2.701×10^{-3} mol O_2, 0.08644 g O_2; (c) 6.43 mol $MgCO_3$, 542 g $MgCO_3$ (d) 768 mol H_2O, 13.8 kg H_2O; (e) 16.31 mol BaO_2, 2762 g BaO_2; (f) 0.207 mol C_2H_4, 5.81 g C_2H_4

47. (a) volume HCl solution $\longrightarrow$ mol HCl $\longrightarrow$ mol $GaCl_3$; (b) 1.25 mol $GaCl_3$, 2.2×10^2 g $GaCl_3$

49. (a) 5.337×10^{22} molecules; (b) 10.41 g $Zn(CN)_2$

51. $SiO_2 + 3C \longrightarrow SiC + 2CO$, 4.50 kg SiO_2

53. 5.00×10^3 kg

55. 1.28×10^5 g CO_2

57. 161.4 mL KI solution

59. 176 g TiO_2

61. The limiting reactant is Cl2.

63. Percent yield = 31%

65. g $CCl_4 \longrightarrow$ mol $CCl_4 \longrightarrow$ mol $CCl_2F_2 \longrightarrow$ g CCl_2F_2, percent yield = 48.3%

67. percent yield = 91.3%

69. Convert mass of ethanol to moles of ethanol; relate the moles of ethanol to the moles of ether produced using the stoichiometry of the balanced equation. Convert moles of ether to grams; divide the actual grams of ether (determined through the density) by the theoretical mass to determine the percent yield; 87.6%

71. The conversion needed is mol Cr $\longrightarrow$ mol H_3PO_4. Then compare the amount of Cr to the amount of acid present. Cr is the limiting reactant.

73. $Na_2C_2O_4$ is the limiting reactant. percent yield = 86.56%

75. Only four molecules can be made.

77. This amount cannot be weighted by ordinary balances and is worthless.

79. 3.4×10^{-3} M H_2SO_4

81. 9.6×10^{-3} M Cl^-

83. 22.4%

85. The empirical formula is BH_3. The molecular formula is B_2H_6.

87. 49.6 mL

89. 13.64 mL

91. 0.0122 M

93. 34.99 mL KOH

95. The empirical formula is WCl_4.

Chapter 5

1. The temperature of 1 gram of burning wood is approximately the same for both a match and a bonfire. This is an intensive property and depends on the material (wood). However, the overall amount of produced heat depends on the amount of material; this is an extensive property. The amount of wood in a bonfire is much greater than that in a match; the total amount of produced heat is also much greater, which is why we can sit around a bonfire to stay warm, but a match would not provide enough heat to keep us from getting cold.
3. Heat capacity refers to the heat required to raise the temperature of the mass of the substance 1 degree; specific heat refers to the heat required to raise the temperature of 1 gram of the substance 1 degree. Thus, heat capacity is an extensive property, and specific heat is an intensive one.
5. (a) 47.6 J/°C; 11.38 cal $°C^{-1}$; (b) 407 J/°C; 97.3 cal $°C^{-1}$
7. 1310 J; 313 cal
9. 7.15 °C
11. (a) 0.390 J/g °C; (b) Copper is a likely candidate.
13. We assume that the density of water is 1.0 g/cm^3(1 g/mL) and that it takes as much energy to keep the water at 85 °F as to heat it from 72 °F to 85 °F. We also assume that only the water is going to be heated. Energy required = 7.47 kWh
15. lesser; more heat would be lost to the coffee cup and the environment and so ΔT for the water would be lesser and the calculated q would be lesser
17. greater, since taking the calorimeter's heat capacity into account will compensate for the thermal energy transferred to the solution from the calorimeter; this approach includes the calorimeter itself, along with the solution, as "surroundings": $q_{rxn} = -(q_{solution} + q_{calorimeter})$; since both $q_{solution}$ and $q_{calorimeter}$ are negative, including the latter term (q_{rxn}) will yield a greater value for the heat of the dissolution
19. The temperature of the coffee will drop 1 degree.
21. 5.7×10^2 kJ
23. 38.5 °C
25. −2.2 kJ; The heat produced shows that the reaction is exothermic.
27. 1.4 kJ
29. 22.6. Since the mass and the heat capacity of the solution is approximately equal to that of the water, the two-fold increase in the amount of water leads to a two-fold decrease of the temperature change.
31. 11.7 kJ
33. 30%
35. 0.24 g
37. 1.4×10^2 Calories
39. The enthalpy change of the indicated reaction is for exactly 1 mol HCL and 1 mol NaOH; the heat in the example is produced by 0.0500 mol HCl and 0.0500 mol NaOH.
41. 25 kJ mol^{-1}
43. 81 kJ mol^{-1}
45. 5204.4 kJ
47. 1.83×10^{-2} mol
49. −802 kJ mol^{-1}
51. 15.5 kJ/°C
53. 7.43 g
55. Yes.
57. 459.6 kJ
59. −494 kJ/mol
61. 44.01 kJ/mol
63. −394 kJ
65. 265 kJ
67. 90.3 kJ/mol
69. (a) −1615.0 kJ mol^{-1}; (b) −484.3 kJ mol^{-1}; (c) 164.2 kJ; (d) −232.1 kJ
71. −54.04 kJ mol^{-1}
73. −2660 kJ mol^{-1}

75. −66.4 kJ
77. −122.8 kJ
79. 3.7 kg
81. On the assumption that the best rocket fuel is the one that gives off the most heat, B_2H_6 is the prime candidate.
83. −88.2 kJ
85. (a) $C_3H_8(g) + 5O_2(g) \longrightarrow 3CO_2(g) + 4H_2O(l)$; (b) 1570 L air; (c) −104.5 kJ mol^{-1}; (d) 75.4 °C

Chapter 6

1. The spectrum consists of colored lines, at least one of which (probably the brightest) is red.
3. 3.15 m
5. 3.233×10^{-19} J; 2.018 eV
7. $\nu = 4.568 \times 10^{14}$ s; $\lambda = 656.3$ nm; Energy mol^{-1} = 1.823×10^5 J mol^{-1}; red
9. (a) $\lambda = 8.69 \times 10^{-7}$ m; $E = 2.29 \times 10^{-19}$ J; (b) $\lambda = 4.59 \times 10^{-7}$ m; $E = 4.33 \times 10^{-19}$ J; The color of (a) is red; (b) is blue.
11. $E = 9.502 \times 10^{-15}$ J; $\nu = 1.434 \times 10^{19}$ s^{-1}
13. Red: 660 nm; 4.54×10^{14} Hz; 3.01×10^{-19} J. Green: 520 nm; 5.77×10^{14} Hz; 3.82×10^{-19} J. Blue: 440 nm; 6.81×10^{14} Hz; 4.51×10^{-19} J. Somewhat different numbers are also possible.
15. 5.49×10^{14} s^{-1}; no
17. Quantized energy means that the electrons can possess only certain discrete energy values; values between those quantized values are not permitted.
19. 2.856 eV
21. -8.716×10^{-18} J
23. -3.405×10^{-20} J
25. 33.9 Å
27. 1.471×10^{-17} J
29. Both involve a relatively heavy nucleus with electrons moving around it, although strictly speaking, the Bohr model works only for one-electron atoms or ions. According to classical mechanics, the Rutherford model predicts a miniature “solar system” with electrons moving about the nucleus in circular or elliptical orbits that are confined to planes. If the requirements of classical electromagnetic theory that electrons in such orbits would emit electromagnetic radiation are ignored, such atoms would be stable, having constant energy and angular momentum, but would not emit any visible light (contrary to observation). If classical electromagnetic theory is applied, then the Rutherford atom would emit electromagnetic radiation of continually increasing frequency (contrary to the observed discrete spectra), thereby losing energy until the atom collapsed in an absurdly short time (contrary to the observed long-term stability of atoms). The Bohr model retains the classical mechanics view of circular orbits confined to planes having constant energy and angular momentum, but restricts these to quantized values dependent on a single quantum number, n. The orbiting electron in Bohr’s model is assumed not to emit any electromagnetic radiation while moving about the nucleus in its stationary orbits, but the atom can emit or absorb electromagnetic radiation when the electron changes from one orbit to another. Because of the quantized orbits, such “quantum jumps” will produce discrete spectra, in agreement with observations.
31. Both models have a central positively charged nucleus with electrons moving about the nucleus in accordance with the Coulomb electrostatic potential. The Bohr model *assumes* that the electrons move in circular orbits that have quantized energies, angular momentum, and radii that are specified by a single quantum number, n = 1, 2, 3, …, but this quantization is an ad hoc assumption made by Bohr to incorporate quantization into an essentially classical mechanics description of the atom. Bohr also assumed that electrons orbiting the nucleus normally do not emit or absorb electromagnetic radiation, but do so when the electron switches to a different orbit. In the quantum mechanical model, the electrons do not move in precise orbits (such orbits violate the Heisenberg uncertainty principle) and, instead, a probabilistic interpretation of the electron’s position at any given instant is used, with a mathematical function ψ called a wavefunction that can be used to determine the electron’s spatial probability distribution. These wavefunctions, or orbitals, are three-dimensional stationary waves that can be specified by three quantum numbers that arise naturally from their underlying mathematics (no ad hoc

assumptions required): the principal quantum number, n (the same one used by Bohr), which specifies shells such that orbitals having the same n all have the same energy and approximately the same spatial extent; the angular momentum quantum number l, which is a measure of the orbital's angular momentum and corresponds to the orbitals' general shapes, as well as specifying subshells such that orbitals having the same l (and n) all have the same energy; and the orientation quantum number m, which is a measure of the z component of the angular momentum and corresponds to the orientations of the orbitals. The Bohr model gives the same expression for the energy as the quantum mechanical expression and, hence, both properly account for hydrogen's discrete spectrum (an example of getting the right answers for the wrong reasons, something that many chemistry students can sympathize with), but gives the wrong expression for the angular momentum (Bohr orbits necessarily all have non-zero angular momentum, but some quantum orbitals [s orbitals] can have zero angular momentum).

33. n determines the general range for the value of energy and the probable distances that the electron can be from the nucleus. l determines the shape of the orbital. m_1 determines the orientation of the orbitals of the same l value with respect to one another. m_s determines the spin of an electron.

35. (a) 2p; (b) 4d; (c) 6s

37. (a) 3d; (b) 1s; (c) 4f

39.

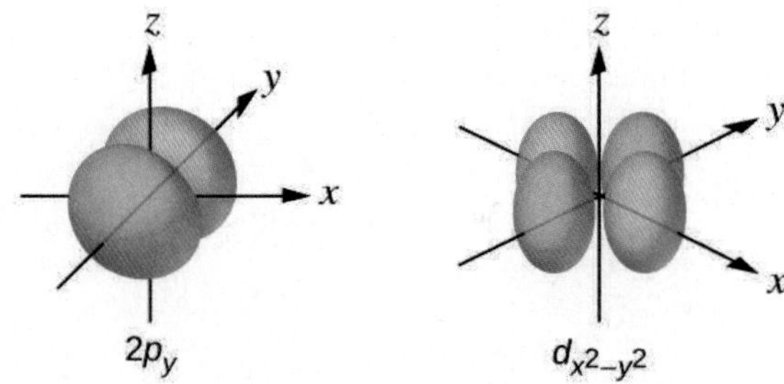

41. (a) x. 2, y. 2, z. 2; (b) x. 1, y. 3, z. 0; (c) x. 4 0 0 $\frac{1}{2}$, y. 2 1 0 $\frac{1}{2}$, z. 3 2 0 $\frac{1}{2}$; (d) x. 1, y. 2, z. 3; (e) x. $l = 0$, $m_l = 0$, y. $l = 1$, $m_l = -1$, 0, or +1, z. $l = 2$, $m_l = -2, -1, 0, +1, +2$

43. 12

45.

n	l	m_l	s
4	0	0	$+\frac{1}{2}$
4	0	0	$-\frac{1}{2}$
4	1	−1	$+\frac{1}{2}$
4	1	0	$+\frac{1}{2}$
4	1	+1	$+\frac{1}{2}$
4	1	−1	$-\frac{1}{2}$

47. For example, Na^+: $1s^22s^22p^6$; Ca^{2+}: $1s^22s^22p^63s^23p^6$; Sn^{2+}: $1s^22s^22p^63s^23p^63d^{10}4s^24p^64d^{10}5s^2$; F^-: $1s^22s^22p^6$; O^{2-}: $1s^22s^22p^6$; Cl^-: $1s^22s^22p^63s^23p^6$.

49. (a) $1s^22s^22p^3$; (b) $1s^22s^22p^63s^23p^2$; (c) $1s^22s^22p^63s^23p^64s^23d^6$; (d) $1s^22s^22p^63s^23p^64s^23d^{10}4p^65s^24d^{10}5p^4$; (e) $1s^22s^22p^63s^23p^64s^23d^{10}4p^65s^24d^{10}5p^66s^24f^9$

51. The charge on the ion.

53. (a)

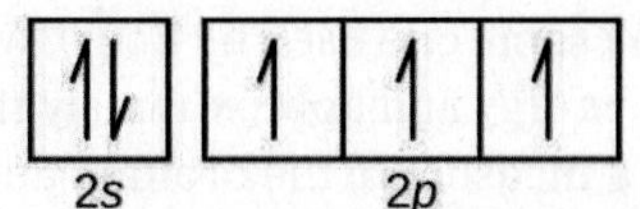

(b)

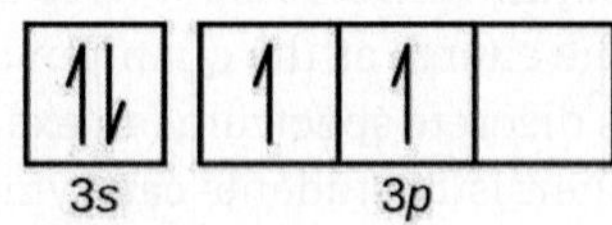

(c)

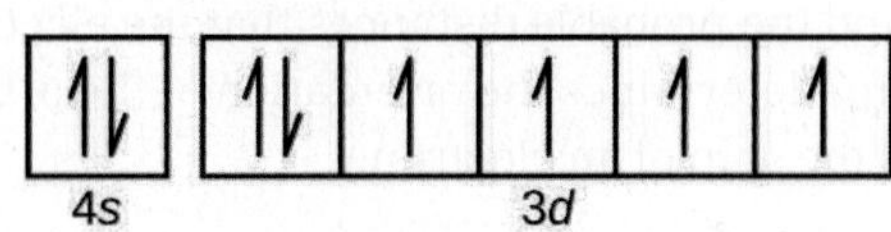

(d)

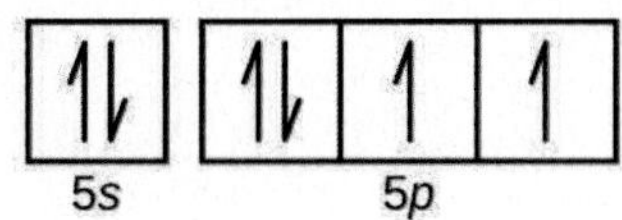

(e)

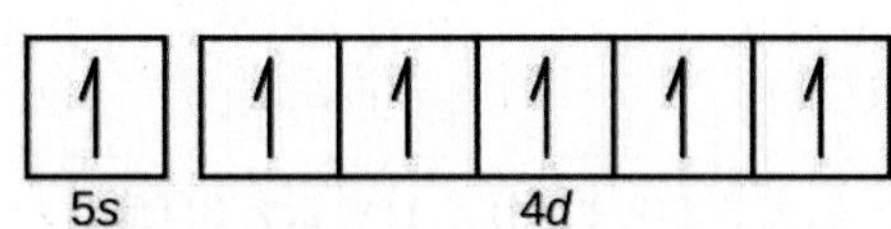

55. Zr
57. Rb^+, Se^{2-}
59. Although both (b) and (c) are correct, (e) encompasses both and is the best answer.
61. K
63. $1s^2 2s^2 2p^6 3s^2 3p^6 4s^2 3d^{10} 4p^6 5s^2 4d^{10} 5p^6 6s^2 4f^{14} 5d^{10}$
65. Co has 27 protons, 27 electrons, and 33 neutrons: $1s^2 2s^2 2p^6 3s^2 3p^6 4s^2 3d^7$. I has 53 protons, 53 electrons, and 78 neutrons: $1s^2 2s^2 2p^6 3s^2 3p^6 3d^{10} 4s^2 4p^6 4d^{10} 5s^2 5p^5$.
67. Cl
69. O
71. Rb < Li < N < F
73. 15 (5A)
75. Mg < Ca < Rb < Cs
77. $Si^{4+} < Al^{3+} < Ca^{2+} < K^+$
79. Se, As^-
81. $Mg^{2+} < K^+ < Br^- < As^{3-}$
83. O, IE_1
85. Ra

Chapter 7

1. The protons in the nucleus do not change during normal chemical reactions. Only the outer electrons move. Positive charges form when electrons are lost.
3. P, I, Cl, and O would form anions because they are nonmetals. Mg, In, Cs, Pb, and Co would form cations because they are metals.
5. (a) P^{3-}; (b) Mg^{2+}; (c) Al^{3+}; (d) O^{2-}; (e) Cl^-; (f) Cs^+
7. (a) $[Ar]4s^2 3d^{10} 4p^6$; (b) $[Kr]4d^{10} 5s^2 5p^6$ (c) $1s^2$ (d) $[Kr]4d^{10}$; (e) $[He]2s^2 2p^6$; (f) $[Ar]3d^{10}$; (g) $1s^2$ (h) $[He]2s^2 2p^6$ (i) $[Kr]4d^{10} 5s^2$ (j) $[Ar]3d^7$ (k) $[Ar]3d^6$, (l) $[Ar]3d^{10} 4s^2$

9. (a) $1s^2 2s^2 2p^6 3s^2 3p^1$; Al^{3+}: $1s^2 2s^2 2p^6$; (b) $1s^2 2s^2 2p^6 3s^2 3p^6 3d^{10} 4s^2 4p^5$; $1s^2 2s^2 2p^6 3s^2 3p^6 3d^{10} 4s^2 4p^6$; (c) $1s^2 2s^2 2p^6 3s^2 3p^6 3d^{10} 4s^2 4p^6 5s^2$; Sr^{2+}: $1s^2 2s^2 2p^6 3s^2 3p^6 3d^{10} 4s^2 4p^6$; (d) $1s^2 2s^1$; Li^+: $1s^2$; (e) $1s^2 2s^2 2p^6 3s^2 3p^6 3d^{10} 4s^2 4p^3$; $1s^2 2s^2 2p^6 3s^2 3p^6 3d^{10} 4s^2 4p^6$; (f) $1s^2 2s^2 2p^6 3s^2 3p^4$; $1s^2 2s^2 2p^6 3s^2 3p^6$

11. NaCl consists of discrete ions arranged in a crystal lattice, not covalently bonded molecules.

13. ionic: (b), (d), (e), (g), and (i); covalent: (a), (c), (f), (h), (j), and (k)

15. (a) Cl; (b) O; (c) O; (d) S; (e) N; (f) P; (g) N

17. (a) H, C, N, O, F; (b) H, I, Br, Cl, F; (c) H, P, S, O, F; (d) Na, Al, H, P, O; (e) Ba, H, As, N, O

19. N, O, F, and Cl

21. (a) HF; (b) CO; (c) OH; (d) PCl; (e) NH; (f) PO; (g) CN

23. (a) eight electrons:

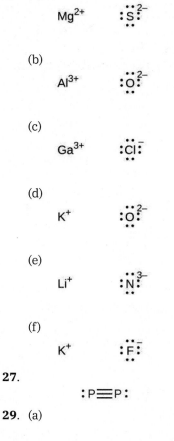

:As:$^{3-}$

(b) eight electrons:

:I:$^{-}$

(c) no electrons Be^{2+}

(d) eight electrons:

:O:$^{2-}$

(e) no electrons Ga^{3+}

(f) no electrons Li^+

(g) eight electrons:

:N:$^{3-}$

25. (a)

Mg^{2+} :S:$^{2-}$

(b)

Al^{3+} :O:$^{2-}$

(c)

Ga^{3+} :Cl:$^{-}$

(d)

K^+ :O:$^{2-}$

(e)

Li^+ :N:$^{3-}$

(f)

K^+ :F:$^{-}$

27.

:P≡P:

29. (a)

:O=O:

In this case, the Lewis structure is inadequate to depict the fact that experimental studies have shown two unpaired electrons in each oxygen molecule.

(b)

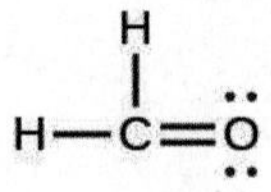

H—C(H)=O:

(c)

:F—As(—F:)—F:

(d)

:Cl—N=O:

(e)

:Cl—Si(Cl)(Cl)—Cl:

(f)

[H—O(H)—H]$^+$

(g)

[H—N(H)(H)—H]$^+$

(h)

[:F—B(F)(F)—F:]$^-$

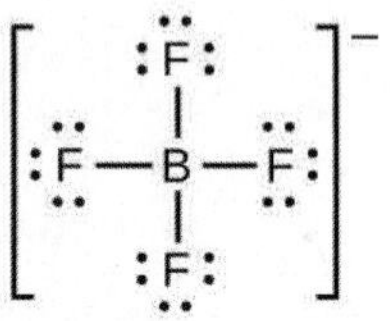

(i)

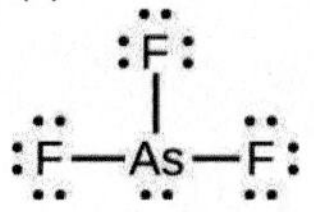

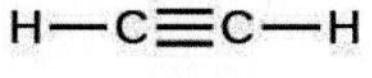

H—C≡C—H

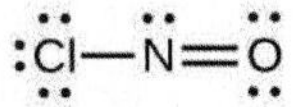

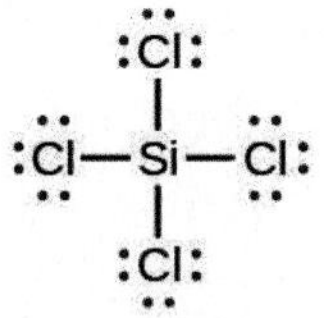

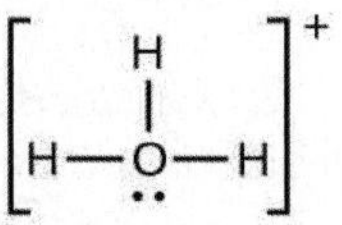

(j)

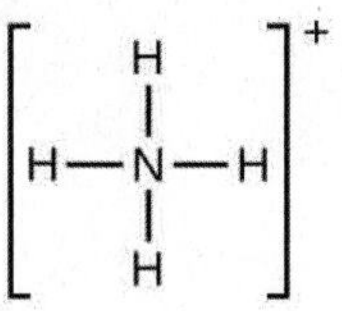

:Cl—C≡N:

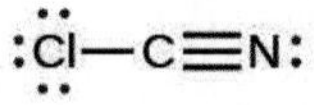

(k)

C≡C^{2+}

31. (a) SeF_6:

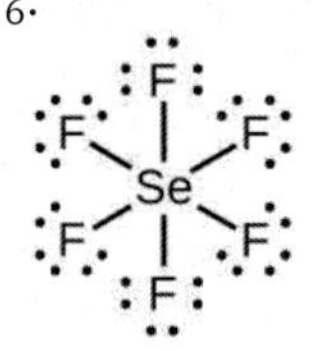

(b) XeF_4:

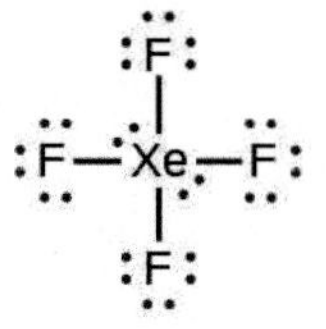

(c) $SeCl_3{}^+$:

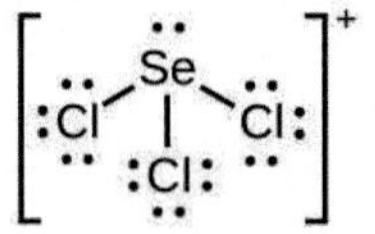

(d) Cl_2BBCl_2:

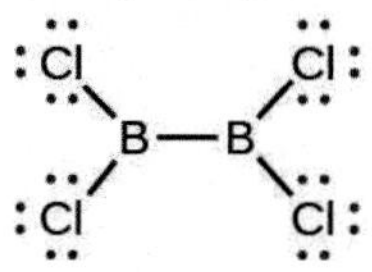

33. Two valence electrons per Pb atom are transferred to Cl atoms; the resulting Pb^{2+} ion has a $6s^2$ valence shell configuration. Two of the valence electrons in the HCl molecule are shared, and the other six are located on the Cl atom as lone pairs of electrons.

35.

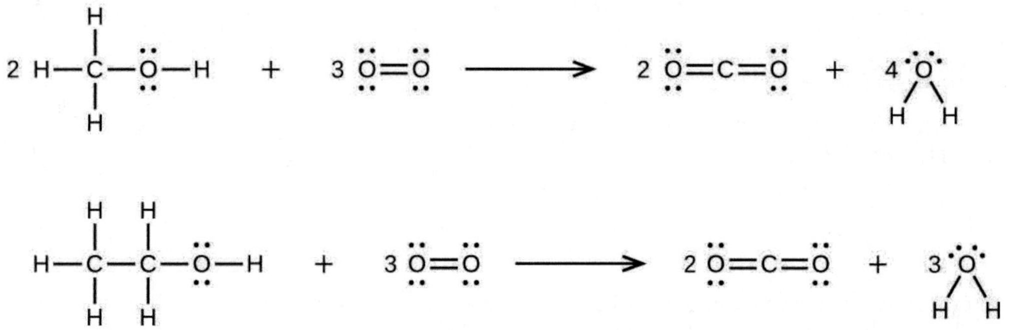

37.

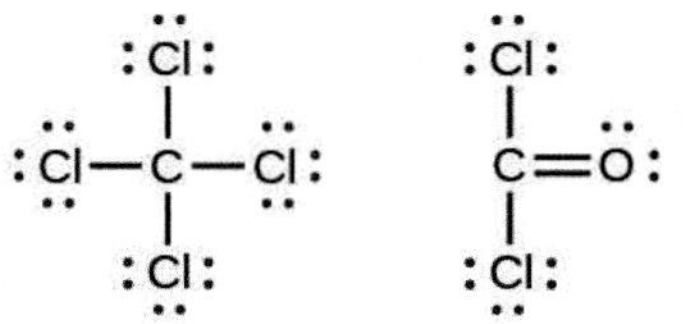

39. (a)

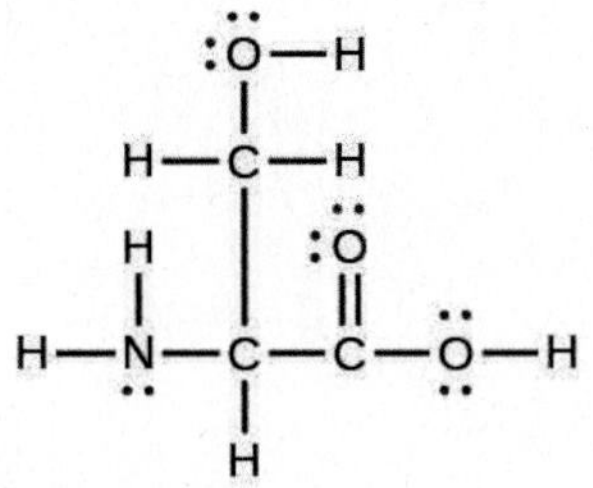

(b)

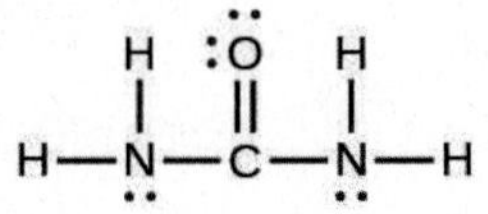

(c)

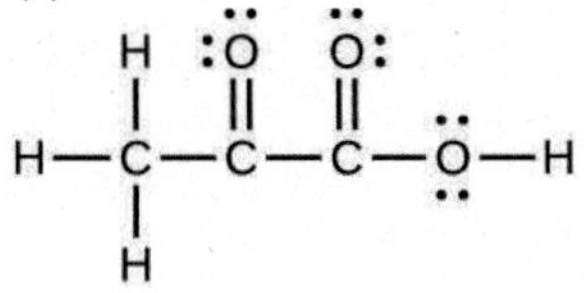

(d)

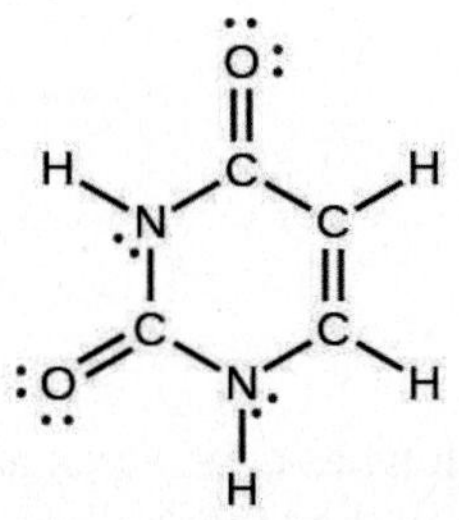

(e)

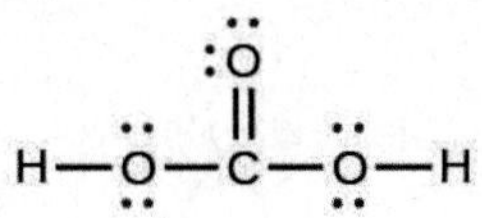

41.

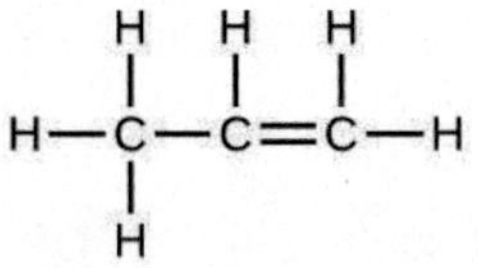

43. Each bond includes a sharing of electrons between atoms. Two electrons are shared in a single bond; four electrons are shared in a double bond; and six electrons are shared in a triple bond.

45. (a)

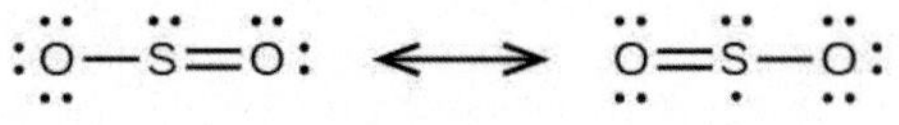

(b)

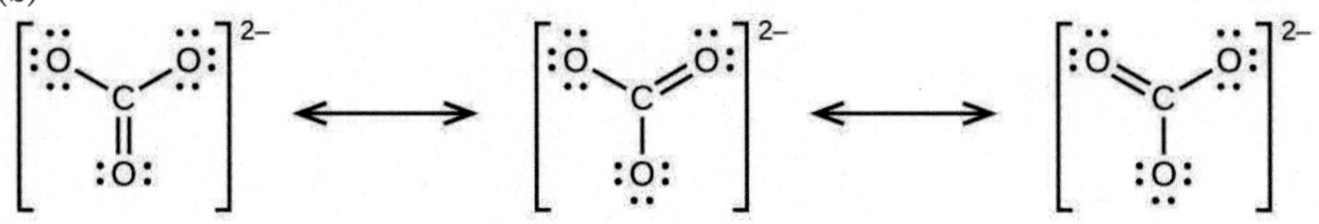

(c)

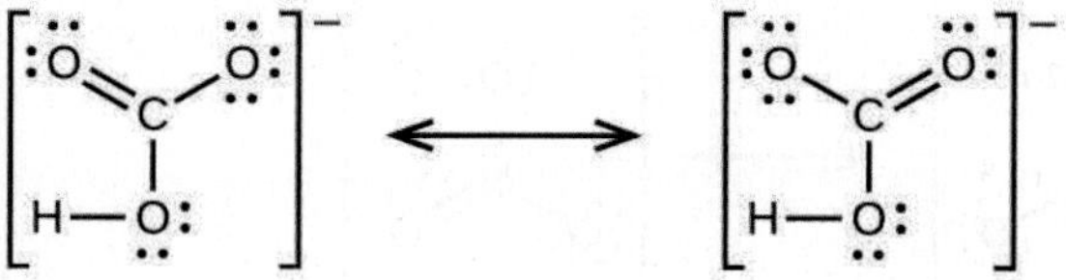

(d)

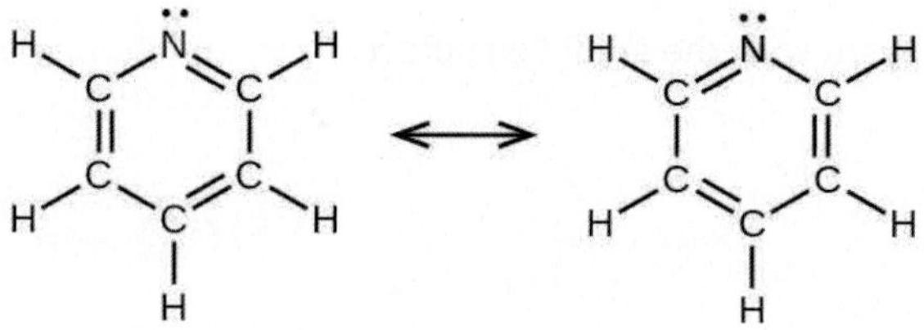

(e)

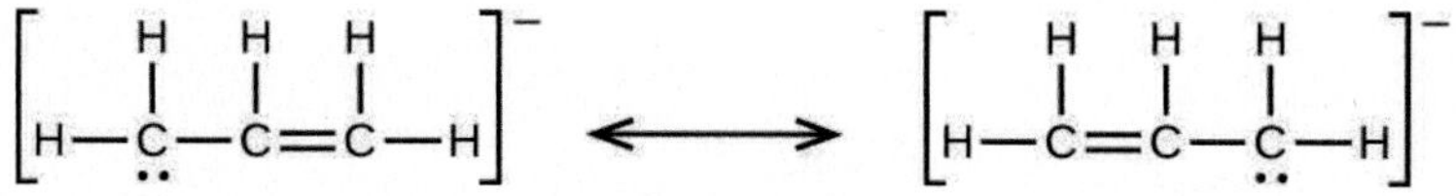

47.

For NO_2^-

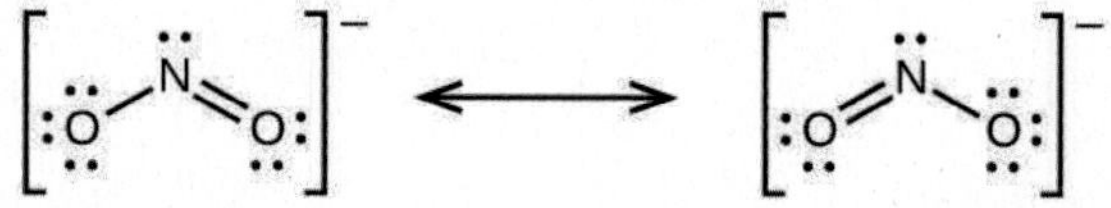

49. (a)

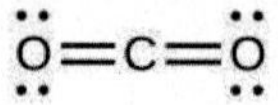

(b)

CO has the strongest carbon-oxygen bond because there is a triple bond joining C and O. CO_2 has double bonds.

51. (a) H: 0, Cl: 0; (b) C: 0, F: 0; (c) P: 0, Cl 0; (d) P: 0, F: 0

53. Cl in Cl_2: 0; Cl in $BeCl_2$: 0; Cl in ClF_5: 0

55. (a)

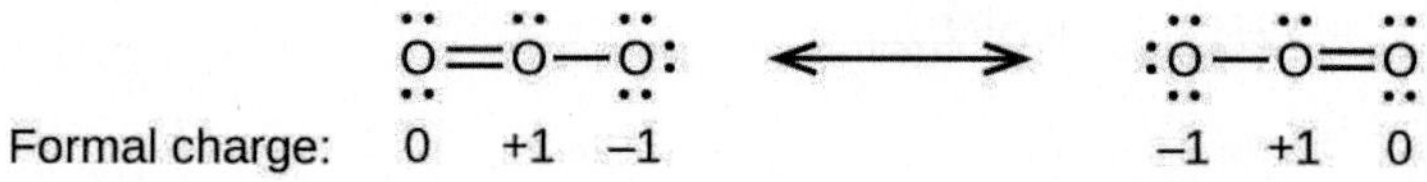

Formal charge: 0 +1 −1 | −1 +1 0

(b)

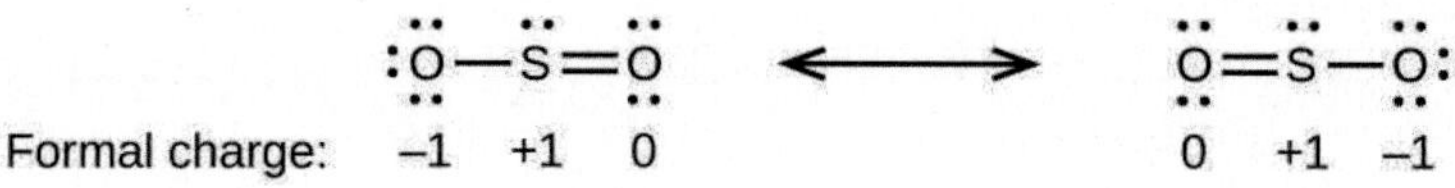

Formal charge: −1 +1 0 | 0 +1 −1

(c)

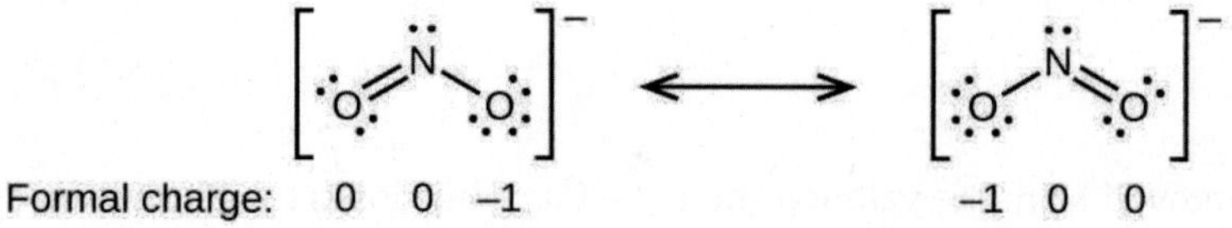

Formal charge: 0 0 −1 | −1 0 0

(d)

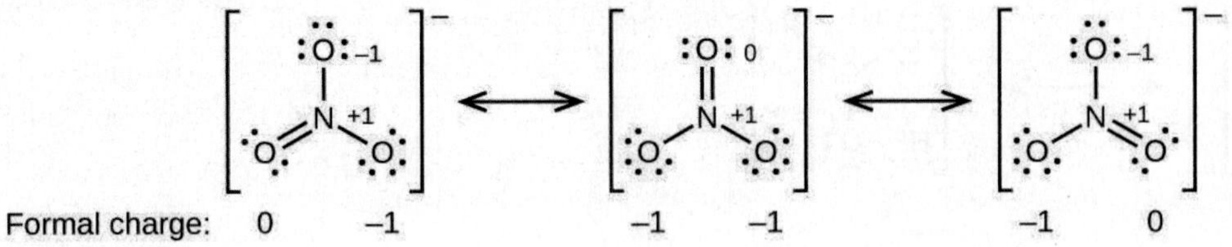

Formal charge: 0 −1 −1 −1 −1 0

57. HOCl

59. The structure that gives zero formal charges is consistent with the actual structure:

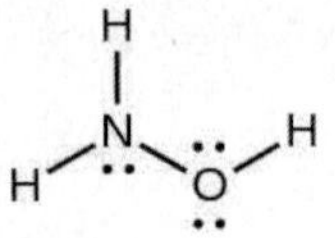

61. NF_3;

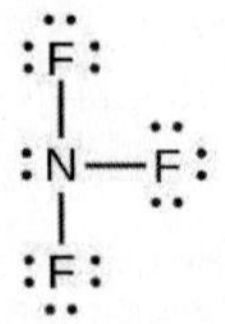

63.

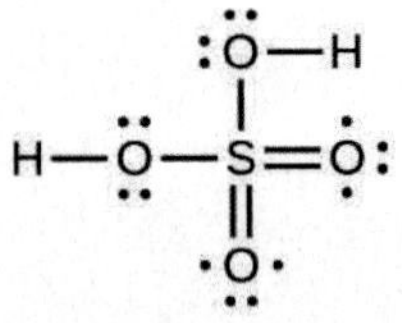

65. (a) −114 kJ; (b) 30 kJ; (c) −1055 kJ

67. The greater bond energy is in the figure on the left. It is the more stable form.

69.

$$\begin{array}{lll} \text{HCl}(g) \longrightarrow \frac{1}{2}\text{H}_2(g) + \frac{1}{2}\text{Cl}_2(g) & & \Delta H_1^\circ = -\Delta H^\circ_{\text{f[HCl}(g)]} \\ \frac{1}{2}\text{H}_2(g) \longrightarrow \text{H}(g) & & \Delta H_2^\circ = \Delta H^\circ_{\text{f[H}(g)]} \\ \frac{1}{2}\text{Cl}_2(g) \longrightarrow \text{Cl}(g) & & \Delta H_3^\circ = \Delta H^\circ_{\text{f[Cl}(g)]} \\ \hline \text{HCl}(g) \longrightarrow \text{H}(g) + \text{Cl}(g) & & \Delta H^\circ = \Delta H_1^\circ + \Delta H_2^\circ + \Delta H_3^\circ \end{array}$$

$$D_{\text{HCl}} = \Delta H^\circ = \Delta H^\circ_{\text{f[HCl}(g)]} + \Delta H^\circ_{\text{f[H}(g)]} + \Delta H^\circ_{\text{f[Cl}(g)]}$$

$$= -(-92.307\text{ kJ}) + 217.97\text{ kJ} + 121.3\text{ kJ}$$

$$= 431.6\text{ kJ}$$

71. The S–F bond in SF_4 is stronger.

73.

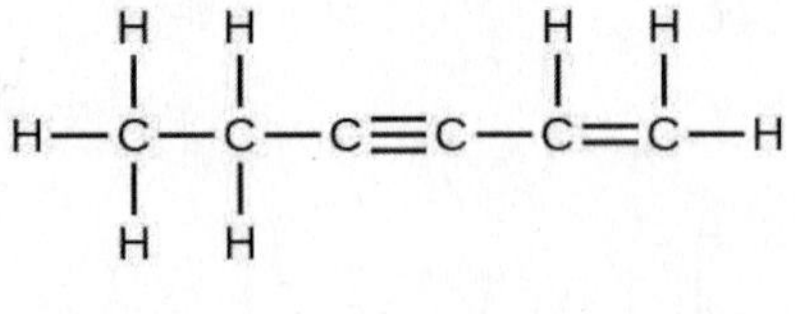

The C–C single bonds are longest.

75. (a) When two electrons are removed from the valence shell, the Ca radius loses the outermost energy level and reverts to the lower $n = 3$ level, which is much smaller in radius. (b) The +2 charge on calcium pulls the oxygen much closer compared with K, thereby increasing the lattice energy relative to a less charged ion. (c) Removal of the 4*s* electron in Ca requires more energy than removal of the 4*s* electron in K because of the stronger attraction of the nucleus and the extra energy required to break the pairing of the

electrons. The second ionization energy for K requires that an electron be removed from a lower energy level, where the attraction is much stronger from the nucleus for the electron. In addition, energy is required to unpair two electrons in a full orbital. For Ca, the second ionization potential requires removing only a lone electron in the exposed outer energy level. (d) In Al, the removed electron is relatively unprotected and unpaired in a *p* orbital. The higher energy for Mg mainly reflects the unpairing of the 2*s* electron.

77. (d)

79. 4008 kJ/mol; both ions in MgO have twice the charge of the ions in LiF; the bond length is very similar and both have the same structure; a quadrupling of the energy is expected based on the equation for lattice energy

81. (a) Na_2O; Na^+ has a smaller radius than K^+; (b) BaS; Ba has a larger charge than K; (c) BaS; Ba and S have larger charges; (d) BaS; S has a larger charge

83. (e)

85. The placement of the two sets of unpaired electrons in water forces the bonds to assume a tetrahedral arrangement, and the resulting HOH molecule is bent. The HBeH molecule (in which Be has only two electrons to bond with the two electrons from the hydrogens) must have the electron pairs as far from one another as possible and is therefore linear.

87. Space must be provided for each pair of electrons whether they are in a bond or are present as lone pairs. Electron-pair geometry considers the placement of all electrons. Molecular structure considers only the bonding-pair geometry.

89. As long as the polar bonds are compensated (for example. two identical atoms are found directly across the central atom from one another), the molecule can be nonpolar.

91. (a) Both the electron geometry and the molecular structure are octahedral. (b) Both the electron geometry and the molecular structure are trigonal bipyramid. (c) Both the electron geometry and the molecular structure are linear. (d) Both the electron geometry and the molecular structure are trigonal planar.

93. (a) electron-pair geometry: octahedral, molecular structure: square pyramidal; (b) electron-pair geometry: tetrahedral, molecular structure: bent; (c) electron-pair geometry: octahedral, molecular structure: square planar; (d) electron-pair geometry: tetrahedral, molecular structure: trigonal pyramidal; (e) electron-pair geometry: trigonal bypyramidal, molecular structure: seesaw; (f) electron-pair geometry: tetrahedral, molecular structure: bent (109°)

95. (a) electron-pair geometry: trigonal planar, molecular structure: bent (120°); (b) electron-pair geometry: linear, molecular structure: linear; (c) electron-pair geometry: trigonal planar, molecular structure: trigonal planar; (d) electron-pair geometry: tetrahedral, molecular structure: trigonal pyramidal; (e) electron-pair geometry: tetrahedral, molecular structure: tetrahedral; (f) electron-pair geometry: trigonal bipyramidal, molecular structure: seesaw; (g) electron-pair geometry: tetrahedral, molecular structure: trigonal pyramidal

97. All of these molecules and ions contain polar bonds. Only ClF_5, $ClO_2{}^-$, PCl_3, SeF_4, and $PH_2{}^-$ have dipole moments.

99. SeS_2, CCl_2F_2, PCl_3, and ClNO all have dipole moments.

101. P

103. nonpolar

105. (a) tetrahedral; (b) trigonal pyramidal; (c) bent (109°); (d) trigonal planar; (e) bent (109°); (f) bent (109°); (g) $\underline{C}H_3CCH$ tetrahedral, $CH_3\underline{CC}H$ linear; (h) tetrahedral; (i) $H_2C\underline{C}CH_2$ linear; $H_2\underline{C}C\underline{C}H_2$ trigonal planar

107.

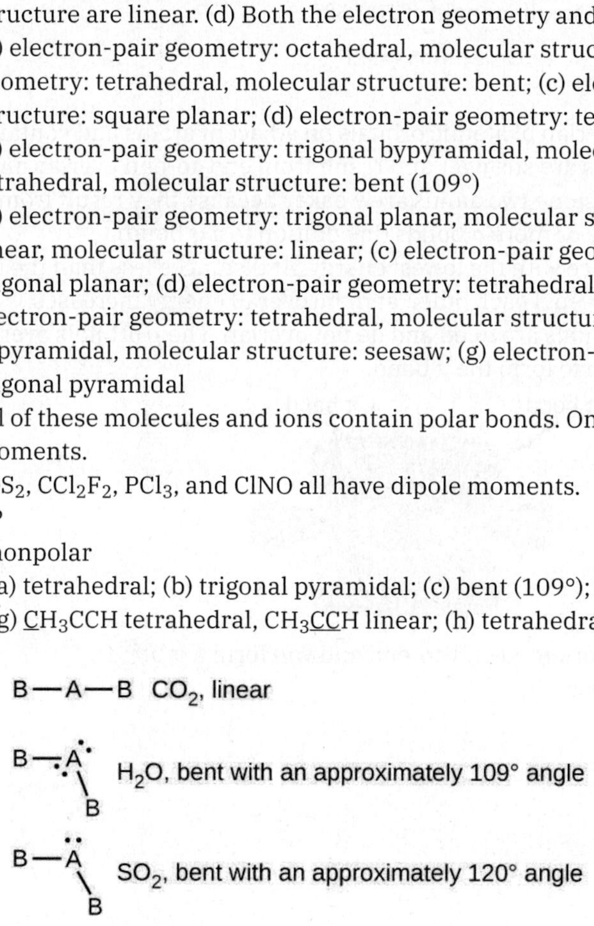

B—A—B CO_2, linear

B—A(—B) H_2O, bent with an approximately 109° angle

B—A(—B) SO_2, bent with an approximately 120° angle

109. (a)

$[:S-C(=S)-S:]^{2-} \longleftrightarrow [:S=C(-S)-S:]^{2-} \longleftrightarrow [:S-C(-S)=S:]^{2-}$

(b)

S=C=S

(c)

:C≡S:

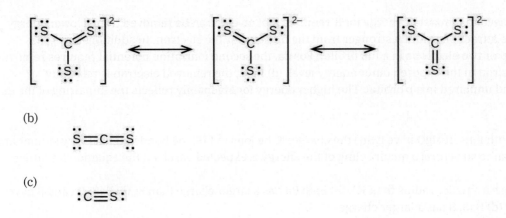

(d) $CS_3{}^{2-}$ includes three regions of electron density (all are bonds with no lone pairs); the shape is trigonal planar; CS_2 has only two regions of electron density (all bonds with no lone pairs); the shape is linear

111. The Lewis structure is made from three units, but the atoms must be rearranged:

```
    H   H   H
    |   |   |
H—C—C=C—H
    |
    H
```

113. The molecular dipole points away from the hydrogen atoms.

115. The structures are very similar. In the model mode, each electron group occupies the same amount of space, so the bond angle is shown as 109.5°. In the "real" mode, the lone pairs are larger, causing the hydrogens to be compressed. This leads to the smaller angle of 104.5°.

Chapter 8

1. Similarities: Both types of bonds result from overlap of atomic orbitals on adjacent atoms and contain a maximum of two electrons. Differences: σ bonds are stronger and result from end-to-end overlap and all single bonds are σ bonds; π bonds between the same two atoms are weaker because they result from side-by-side overlap, and multiple bonds contain one or more π bonds (in addition to a σ bond).

3. The specific average bond distance is the distance with the lowest energy. At distances less than the bond distance, the positive charges on the two nuclei repel each other, and the overall energy increases.

5. Bonding: One σ bond and one π bond. The *s* orbitals are filled and do not overlap. The *p* orbitals overlap along the axis to form a σ bond and side-by-side to form the π bond.

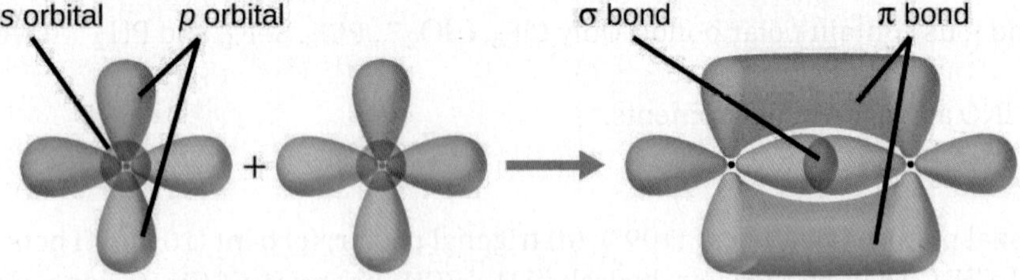

7. No, two of the *p* orbitals (one on each N) will be oriented end-to-end and will form a σ bond.

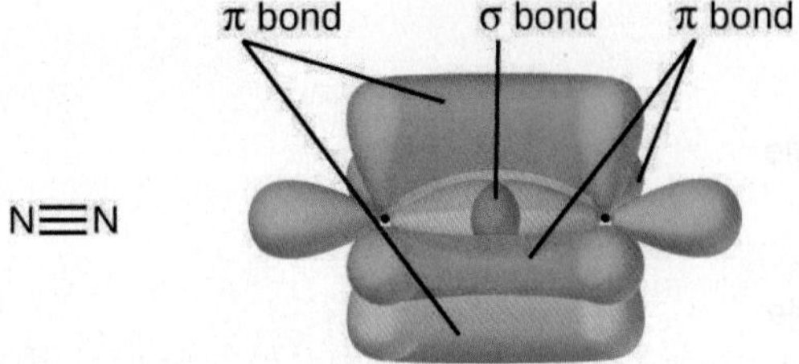

9. Hybridization is introduced to explain the geometry of bonding orbitals in valance bond theory.

11. There are no *d* orbitals in the valence shell of carbon.

13. trigonal planar, sp^2; trigonal pyramidal (one lone pair on A) sp^3; T-shaped (two lone pairs on A sp^3d, or (three lone pairs on A) sp^3d^2

15. (a) Each S has a bent (109°) geometry, sp^3

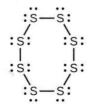

(b) Bent (120°), sp^2

(c) Trigonal planar, sp^2

(d) Tetrahedral, sp^3

17. (a) XeF_2

(b)

(c) linear (d) sp^3d

19. (a)

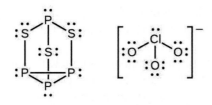

(b) P atoms, trigonal pyramidal; S atoms, bent, with two lone pairs; Cl atoms, trigonal pyramidal; (c) Hybridization about P, S, and Cl is, in all cases, sp^3; (d) Oxidation states P +1, S$-1\frac{1}{3}$, Cl +5, O −2. Formal charges: P 0; S 0; Cl +2: O −1

21.

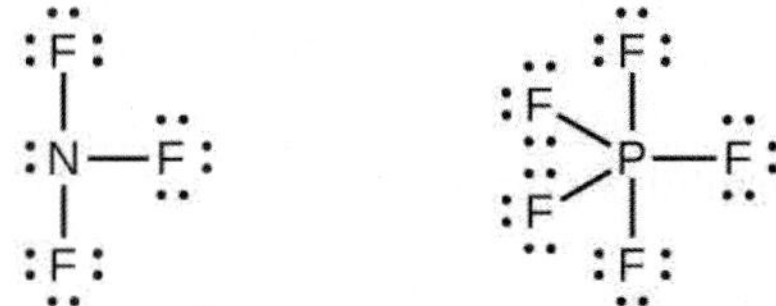

Phosphorus and nitrogen can form sp^3 hybrids to form three bonds and hold one lone pair in PF_3 and NF_3, respectively. However, nitrogen has no valence *d* orbitals, so it cannot form a set of sp^3d hybrid orbitals to bind five fluorine atoms in NF_5. Phosphorus has *d* orbitals and can bind five fluorine atoms with sp^3d

hybrid orbitals in PF_5.

23. A triple bond consists of one σ bond and two π bonds. A σ bond is stronger than a π bond due to greater overlap.

25. (a)

H
|
H—C—C≡N:
|
H ⟶ dipole moment

(b) The terminal carbon atom uses sp^3 hybrid orbitals, while the central carbon atom is *sp* hybridized. (c) Each of the two π bonds is formed by overlap of a 2*p* orbital on carbon and a nitrogen 2*p* orbital.

27. (a) sp^2; (b) *sp*; (c) sp^2; (d) sp^3; (e) sp^3; (f) sp^3d; (g) sp^3

29. (a) sp^2, delocalized; (b) *sp*, localized; (c) sp^2, delocalized; (d) sp^3, delocalized

31.

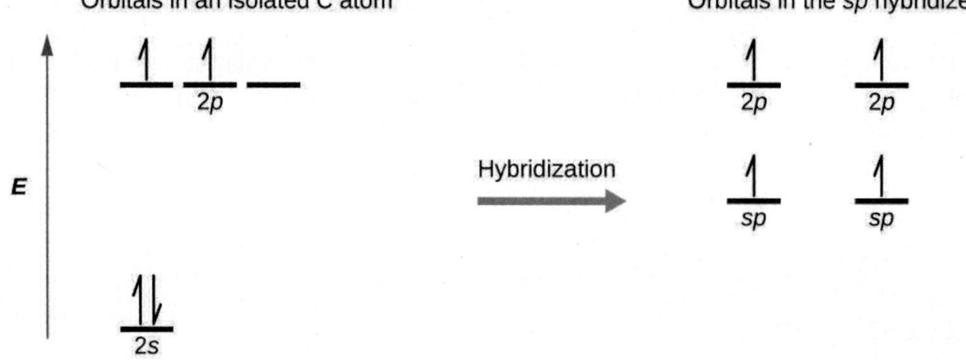

Each of the four electrons is in a separate orbital and overlaps with an electron on an oxygen atom.

33. (a) Similarities: Both are bonding orbitals that can contain a maximum of two electrons. Differences: σ orbitals are end-to-end combinations of atomic orbitals, whereas π orbitals are formed by side-by-side overlap of orbitals. (b) Similarities: Both are quantum-mechanical constructs that represent the probability of finding the electron about the atom or the molecule. Differences: ψ for an atomic orbital describes the behavior of only one electron at a time based on the atom. For a molecule, ψ represents a mathematical combination of atomic orbitals. (c) Similarities: Both are orbitals that can contain two electrons. Differences: Bonding orbitals result in holding two or more atoms together. Antibonding orbitals have the effect of destabilizing any bonding that has occurred.

35. An odd number of electrons can never be paired, regardless of the arrangement of the molecular orbitals. It will always be paramagnetic.

37. Bonding orbitals have electron density in close proximity to more than one nucleus. The interaction between the bonding positively charged nuclei and negatively charged electrons stabilizes the system.

39. The pairing of the two bonding electrons lowers the energy of the system relative to the energy of the nonbonded electrons.

41. (a) H_2 bond order = 1, $H_2{}^+$ bond order = 0.5, $H_2{}^-$ bond order = 0.5, strongest bond is H_2; (b) O_2 bond order = 2, $O_2{}^{2+}$ bond order = 3; $O_2{}^{2-}$ bond order = 1, strongest bond is $O_2{}^{2+}$; (c) Li_2 bond order = 1, $Be_2{}^+$ bond order = 0.5, Be_2 bond order = 0, strongest bond is Li_2;(d) F_2 bond order = 1, $F_2{}^+$ bond order = 1.5, $F_2{}^-$ bond order = 0.5, strongest bond is $F_2{}^+$; (e) N_2 bond order = 3, $N_2{}^+$ bond order = 2.5, $N_2{}^-$ bond order = 2.5, strongest bond is N_2

43. (a) H_2; (b) N_2; (c) O; (d) C_2; (e) B_2

45. Yes, fluorine is a smaller atom than Li, so atoms in the 2*s* orbital are closer to the nucleus and more stable.

47. 2+

49. N_2 has s-p mixing, so the π orbitals are the last filled in $N_2{}^{2+}$. O_2 does not have s-p mixing, so the σ_p orbital fills before the π orbitals.

Chapter 9

1. The cutting edge of a knife that has been sharpened has a smaller surface area than a dull knife. Since pressure is force per unit area, a sharp knife will exert a higher pressure with the same amount of force and cut through material more effectively.
3. Lying down distributes your weight over a larger surface area, exerting less pressure on the ice compared to standing up. If you exert less pressure, you are less likely to break through thin ice.
5. 0.809 atm; 82.0 kPa
7. 2.2×10^2 kPa
9. Earth: 14.7 lb in^{-2}; Venus: 1.30×10^3 lb in^{-2}
11. (a) 101.5 kPa; (b) 51 torr drop
13. (a) 264 torr; (b) 35,200 Pa; (c) 0.352 bar
15. (a) 623 mm Hg; (b) 0.820 atm; (c) 83.1 kPa
17. With a closed-end manometer, no change would be observed, since the vaporized liquid would contribute equal, opposing pressures in both arms of the manometer tube. However, with an open-ended manometer, a higher pressure reading of the gas would be obtained than expected, since $P_{gas} = P_{atm} + P_{vol\ liquid}$.
19. As the bubbles rise, the pressure decreases, so their volume increases as suggested by Boyle's law.
21. (a) The number of particles in the gas increases as the volume increases. (b) temperature, pressure
23. The curve would be farther to the right and higher up, but the same basic shape.
25. About 12.5 L
27. 3.40×10^3 torr
29. 12.1 L
31. 217 L
33. 8.190×10^{-2} mol; 5.553 g
35. (a) 7.24×10^{-2} g; (b) 23.1 g; (c) 1.5×10^{-4} g
37. 5561 L
39. 46.4 g
41. For a gas exhibiting ideal behavior:

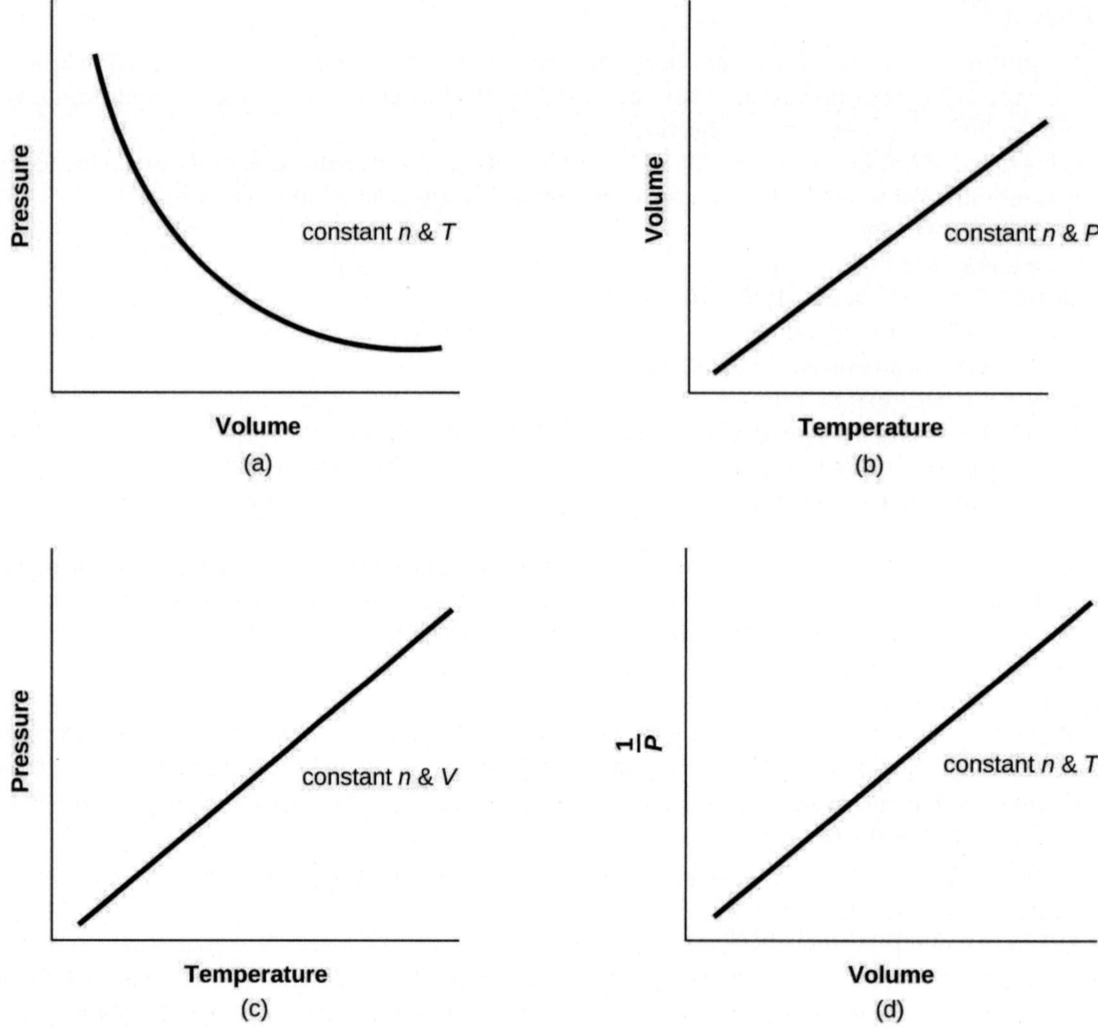

43. (a) 1.85 L CCl_2F_2; (b) 4.66 L CH_3CH_2F

45. 0.644 atm

47. The pressure decreases by a factor of 3.

49. 4.64 g L^{-1}

51. 38.8 g

53. 72.0 g mol^{-1}

55. 88.1 g mol^{-1}; PF_3

57. 141 atm, 107,000 torr, 14,300 kPa

59. CH_4: 276 kPa; C_2H_6: 27 kPa; C_3H_8: 3.4 kPa

61. Yes

63. 740 torr

65. (a) Determine the moles of HgO that decompose; using the chemical equation, determine the moles of O_2 produced by decomposition of this amount of HgO; and determine the volume of O_2 from the moles of O_2, temperature, and pressure. (b) 0.308 L

67. (a) Determine the molar mass of CCl_2F_2. From the balanced equation, calculate the moles of H_2 needed for the complete reaction. From the ideal gas law, convert moles of H_2 into volume. (b) 3.72×10^3 L

69. (a) Balance the equation. Determine the grams of CO_2 produced and the number of moles. From the ideal gas law, determine the volume of gas. (b) 7.43×10^5 L

71. 42.00 L

73. (a) 18.0 L; (b) 0.533 atm

75. 10.57 L O_2

77. 5.40×10^5 L
79. XeF_4
81. 4.2 hours
83. Effusion can be defined as the process by which a gas escapes through a pinhole into a vacuum. Graham's law states that with a mixture of two gases A and B: $\left(\frac{\text{rate A}}{\text{rate B}}\right) = \left(\frac{\text{molar mass of B}}{\text{molar mass of A}}\right)^{1/2}$. Both A and B are in the same container at the same temperature, and therefore will have the same kinetic energy:
$KE_A = KE_B \; KE = \frac{1}{2}mv^2$
Therefore, $\frac{1}{2}m_A v_A^2 = \frac{1}{2}m_B v_B^2$
$$\frac{v_A^2}{v_B^2} = \frac{m_B}{m_A}$$
$$\left(\frac{v_A^2}{v_B^2}\right)^{1/2} = \left(\frac{m_B}{m_A}\right)^{1/2}$$
$$\frac{v_A}{v_B} = \left(\frac{m_B}{m_A}\right)^{1/2}$$
85. F_2, N_2O, Cl_2, H_2S
87. 1.4; 1.2
89. 51.7 cm
91. Yes. At any given instant, there are a range of values of molecular speeds in a sample of gas. Any single molecule can speed up or slow down as it collides with other molecules. The average speed of all the molecules is constant at constant temperature.
93. H_2O. Cooling slows the speeds of the He atoms, causing them to behave as though they were heavier.
95. (a) The number of collisions per unit area of the container wall is constant. (b) The average kinetic energy doubles. (c) The root mean square speed increases to $\sqrt{2}$ times its initial value; u_{rms} is proportional to $\sqrt{KE_{avg}}$.
97. (a) equal; (b) less than; (c) 29.48 g mol^{-1}; (d) 1.0966 g L^{-1}; (e) 0.129 g/L; (f) 4.01×10^5 g; net lifting capacity = 384 lb; (g) 270 L; (h) 39.1 kJ min^{-1}
99. Gases C, E, and F
101. The gas behavior most like an ideal gas will occur under the conditions in (b). Molecules have high speeds and move through greater distances between collision; they also have shorter contact times and interactions are less likely. Deviations occur with the conditions described in (a) and (c). Under conditions of (a), some gases may liquefy. Under conditions of (c), most gases will liquefy.
103. SF_6
105. (a) A straight horizontal line at 1.0; (b) When real gases are at low pressures and high temperatures, they behave close enough to ideal gases that they are approximated as such; however, in some cases, we see that at a high pressure and temperature, the ideal gas approximation breaks down and is significantly different from the pressure calculated by the ideal gas equation. (c) The greater the compressibility, the more the volume matters. At low pressures, the correction factor for intermolecular attractions is more significant, and the effect of the volume of the gas molecules on Z would be a small lowering compressibility. At higher pressures, the effect of the volume of the gas molecules themselves on Z would increase compressibility (see Figure 9.35). (d) Once again, at low pressures, the effect of intermolecular attractions on Z would be more important than the correction factor for the volume of the gas molecules themselves, though perhaps still small. At higher pressures and low temperatures, the effect of intermolecular attractions would be larger. See Figure 9.35. (e) Low temperatures

Chapter 10

1. Liquids and solids are similar in that they are matter composed of atoms, ions, or molecules. They are incompressible and have similar densities that are both much larger than those of gases. They are different in that liquids have no fixed shape, and solids are rigid.
3. They are similar in that the atoms or molecules are free to move from one position to another. They differ in that the particles of a liquid are confined to the shape of the vessel in which they are placed. In contrast, a gas will expand without limit to fill the space into which it is placed.

5. All atoms and molecules will condense into a liquid or solid in which the attractive forces exceed the kinetic energy of the molecules, at sufficiently low temperature.
7. (a) Dispersion forces occur as an atom develops a temporary dipole moment when its electrons are distributed asymmetrically about the nucleus. This structure is more prevalent in large atoms such as argon or radon. A second atom can then be distorted by the appearance of the dipole in the first atom. The electrons of the second atom are attracted toward the positive end of the first atom, which sets up a dipole in the second atom. The net result is rapidly fluctuating, temporary dipoles that attract one another (e.g., Ar). (b) A dipole-dipole attraction is a force that results from an electrostatic attraction of the positive end of one polar molecule for the negative end of another polar molecule (e.g., ICI molecules attract one another by dipole-dipole interaction). (c) Hydrogen bonds form whenever a hydrogen atom is bonded to one of the more electronegative atoms, such as a fluorine, oxygen, or nitrogen atom. The electrostatic attraction between the partially positive hydrogen atom in one molecule and the partially negative atom in another molecule gives rise to a strong dipole-dipole interaction called a hydrogen bond (e.g., $HF\cdots HF$).
9. The London forces typically increase as the number of electrons increase.
11. (a) $SiH_4 < HCl < H_2O$; (b) $F_2 < Cl_2 < Br_2$; (c) $CH_4 < C_2H_6 < C_3H_8$; (d) $N_2 < O_2 < NO$
13. Only rather small dipole-dipole interactions from C-H bonds are available to hold *n*-butane in the liquid state. Chloroethane, however, has rather large dipole interactions because of the Cl-C bond; the interaction, therefore, is stronger, leading to a higher boiling point.
15. −85 °C. Water has stronger hydrogen bonds, so it melts at a higher temperature.
17. The hydrogen bond between two hydrogen fluoride molecules is stronger than that between two water molecules because the electronegativity of F is greater than that of O. Consequently, the partial negative charge on F is greater than that on O. The hydrogen bond between the partially positive H and the larger partially negative F will be stronger than that formed between H and O.
19. H-bonding is the principle IMF holding the protein strands together. The H-bonding is between the $N-H$ and $C=O$.
21. (a) hydrogen bonding, dipole-dipole attraction, and dispersion forces; (b) dispersion forces; (c) dipole-dipole attraction and dispersion forces
23. The water molecules have strong intermolecular forces of hydrogen bonding. The water molecules are thus attracted strongly to one another and exhibit a relatively large surface tension, forming a type of "skin" at its surface. This skin can support a bug or paper clip if gently placed on the water.
25. Temperature has an effect on intermolecular forces: The higher the temperature, the greater the kinetic energies of the molecules and the greater the extent to which their intermolecular forces are overcome, and so the more fluid (less viscous) the liquid. The lower the temperature, the less the intermolecular forces are overcome, and so the less viscous the liquid.
27. (a) As the water reaches higher temperatures, the increased kinetic energies of its molecules are more effective in overcoming hydrogen bonding, and so its surface tension decreases. Surface tension and intermolecular forces are directly related. (b) The same trend in viscosity is seen as in surface tension, and for the same reason.
29. 1.7×10^{-4} m
31. The heat is absorbed by the ice, providing the energy required to partially overcome intermolecular attractive forces in the solid and causing a phase transition to liquid water. The solution remains at 0 °C until all the ice is melted. Only the amount of water existing as ice changes until the ice disappears. Then the temperature of the water can rise.
33. We can see the amount of liquid in an open container decrease and we can smell the vapor of some liquids.
35. The vapor pressure of a liquid decreases as the strength of its intermolecular forces increases.
37. As the temperature increases, the average kinetic energy of the molecules of gasoline increases and so a greater fraction of molecules have sufficient energy to escape from the liquid than at lower temperatures.
39. They are equal when the pressure of gas above the liquid is exactly 1 atm.
41. approximately 95 °C
43. (a) At 5000 feet, the atmospheric pressure is lower than at sea level, and water will therefore boil at a lower temperature. This lower temperature will cause the physical and chemical changes involved in cooking the egg to proceed more slowly, and a longer time is required to fully cook the egg. (b) As long as the air

surrounding the body contains less water vapor than the maximum that air can hold at that temperature, perspiration will evaporate, thereby cooling the body by removing the heat of vaporization required to vaporize the water.

45. Dispersion forces increase with molecular mass or size. As the number of atoms composing the molecules in this homologous series increases, so does the extent of intermolecular attraction via dispersion forces and, consequently, the energy required to overcome these forces and vaporize the liquids.

47. The boiling point of CS_2 is higher than that of CO_2 partially because of the higher molecular weight of CS_2; consequently, the attractive forces are stronger in CS_2. It would be expected, therefore, that the heat of vaporization would be greater than that of 9.8 kJ/mol for CO_2. A value of 28 kJ/mol would seem reasonable. A value of −8.4 kJ/mol would indicate a release of energy upon vaporization, which is clearly implausible.

49. The thermal energy (heat) needed to evaporate the liquid is removed from the skin.

51. 1130 kJ

53. (a) 13.0 kJ; (b) It is likely that the heat of vaporization will have a larger magnitude since in the case of vaporization the intermolecular interactions have to be completely overcome, while melting weakens or destroys only some of them.

55. At low pressures and 0.005 °C, the water is a gas. As the pressure increases to 4.6 torr, the water becomes a solid; as the pressure increases still more, it becomes a liquid. At 40 °C, water at low pressure is a vapor; at pressures higher than about 75 torr, it converts into a liquid. At −40 °C, water goes from a gas to a solid as the pressure increases above very low values.

57. (a) gas; (b) gas; (c) gas; (d) gas; (e) solid; (f) gas

59.

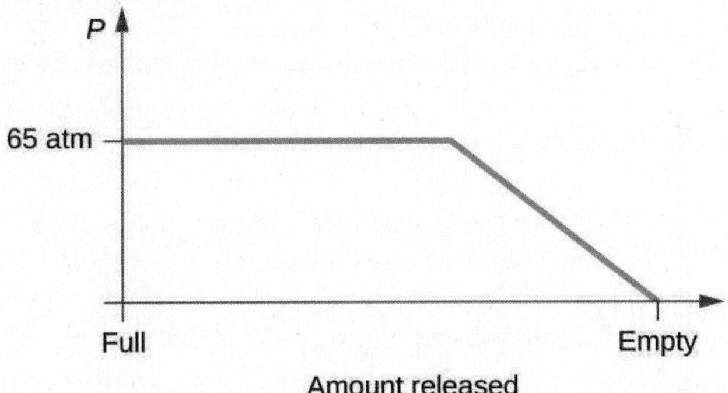

61. Yes, ice will sublime, although it may take it several days. Ice has a small vapor pressure, and some ice molecules form gas and escape from the ice crystals. As time passes, more and more solid converts to gas until eventually the clothes are dry.

63. (a)

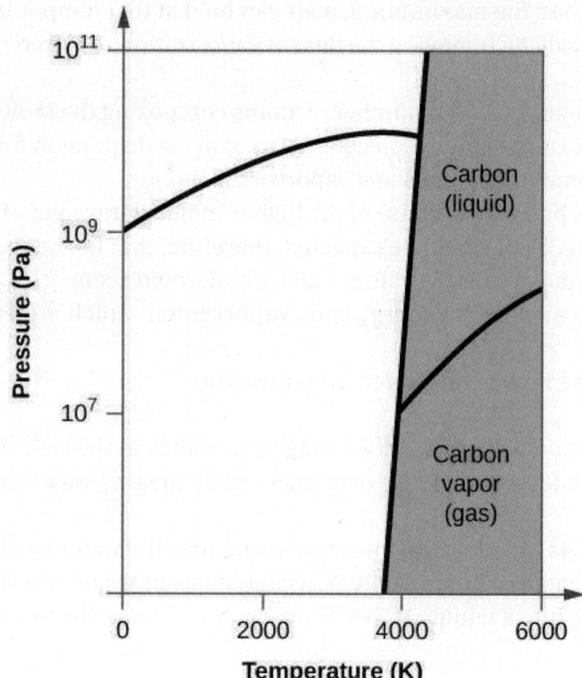
10^11
10^9
10^7
Pressure (Pa)
Carbon (liquid)
Carbon vapor (gas)
0
2000
4000
6000
Temperature (K)

(b)

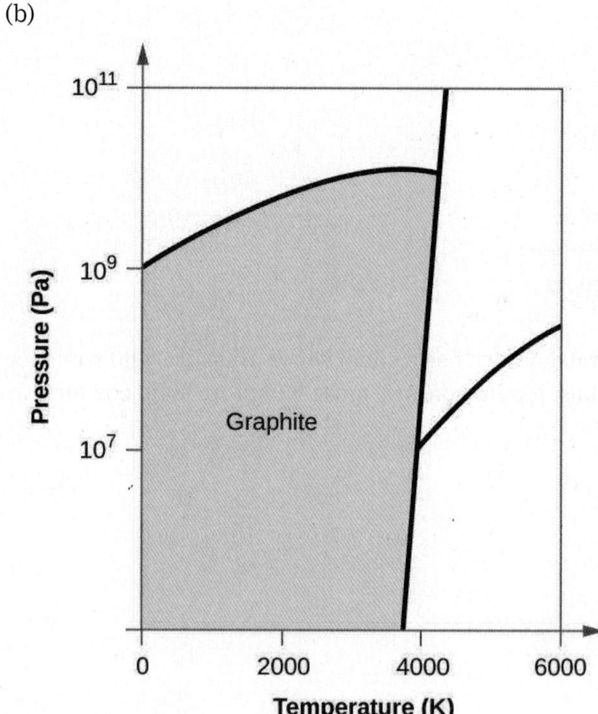
10^11
10^9
10^7
Pressure (Pa)
Graphite
0
2000
4000
6000
Temperature (K)

(c)

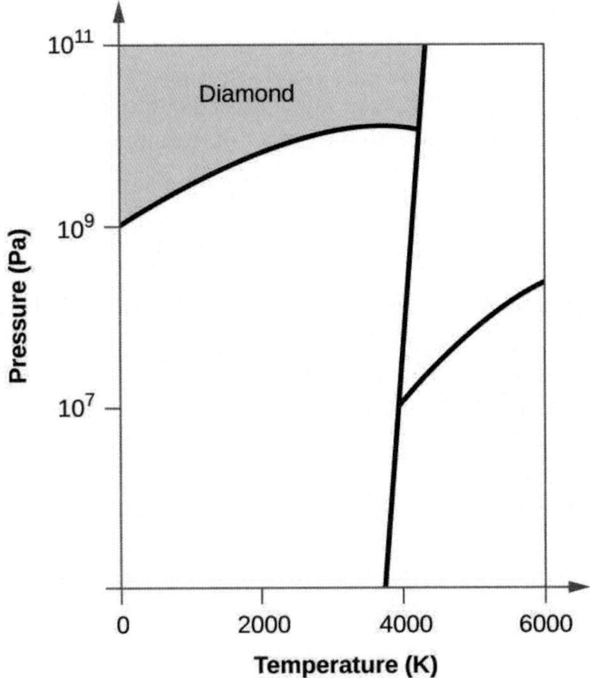

(d)

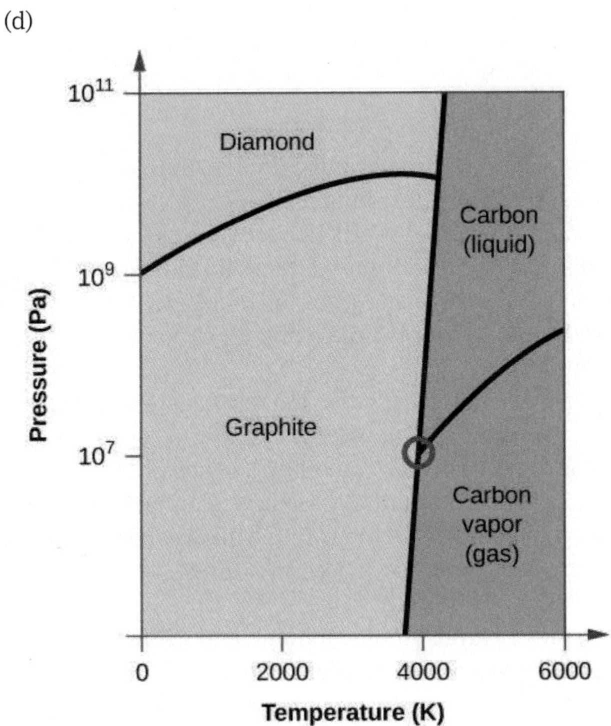

(e) liquid phase (f) sublimation

65. (e) molecular crystals

67. Ice has a crystalline structure stabilized by hydrogen bonding. These intermolecular forces are of comparable strength and thus require the same amount of energy to overcome. As a result, ice melts at a single temperature and not over a range of temperatures. The various, very large molecules that compose butter experience varied van der Waals attractions of various strengths that are overcome at various temperatures, and so the melting process occurs over a wide temperature range.

69. (a) ionic; (b) covalent network; (c) molecular; (d) metallic; (e) covalent network; (f) molecular; (g) molecular; (h) ionic; (i) ionic

71. X = ionic; Y = metallic; Z = covalent network

73. (b) metallic solid

75. The structure of this low-temperature form of iron (below 910 °C) is body-centered cubic. There is one-eighth atom at each of the eight corners of the cube and one atom in the center of the cube.

77. eight

79. 12

81. (a) 1.370 Å; (b) 19.26 g/cm

83. (a) 2.176 Å; (b) 3.595 g/cm^3

85. The crystal structure of Si shows that it is less tightly packed (coordination number 4) in the solid than Al (coordination number 12).

87. In a closest-packed array, two tetrahedral holes exist for each anion. If only half the tetrahedral holes are occupied, the numbers of anions and cations are equal. The formula for cadmium sulfide is CdS.

89. Co_3O_4

91. In a simple cubic array, only one cubic hole can be occupied be a cation for each anion in the array. The ratio of thallium to iodide must be 1:1; therefore, the formula for thallium is TlI.

93. 59.95%; The oxidation number of titanium is +4.

95. Both ions are close in size: Mg, 0.65; Li, 0.60. This similarity allows the two to interchange rather easily. The difference in charge is generally compensated by the switch of Si^{4+} for Al^{3+}.

97. Mn_2O_3

99. 1.48 Å

101. 2.874 Å

103. 20.2°

105. 1.74×10^4 eV

Chapter 11

1. A solution can vary in composition, while a compound cannot vary in composition. Solutions are homogeneous at the molecular level, while other mixtures are heterogeneous.

3. (a) The process is endothermic as the solution is consuming heat. (b) Attraction between the K^+ and NO_3^- ions is stronger than between the ions and water molecules (the ion-ion interactions have a lower, more negative energy). Therefore, the dissolution process increases the energy of the molecular interactions, and it consumes the thermal energy of the solution to make up for the difference. (c) No, an ideal solution is formed with no appreciable heat release or consumption.

5. (a) ion-dipole forces; (b) dipole-dipole forces; (c) dispersion forces; (d) dispersion forces; (e) hydrogen bonding

7. Heat is released when the total intermolecular forces (IMFs) between the solute and solvent molecules are stronger than the total IMFs in the pure solute and in the pure solvent: Breaking weaker IMFs and forming stronger IMFs releases heat. Heat is absorbed when the total IMFs in the solution are weaker than the total of those in the pure solute and in the pure solvent: Breaking stronger IMFs and forming weaker IMFs absorbs heat.

9. Crystals of NaCl dissolve in water, a polar liquid with a very large dipole moment, and the individual ions become strongly solvated. Hexane is a nonpolar liquid with a dipole moment of zero and, therefore, does not significantly interact with the ions of the NaCl crystals.

11. (a) $Fe(NO_3)_3$ is a strong electrolyte, thus it should completely dissociate into Fe^{3+} and NO_3^- ions. Therefore, (z) best represents the solution. (b) $Fe(NO_3)_3(s) \longrightarrow Fe^{3+}(aq) + 3NO_3^-(aq)$

13. (a) high conductivity (solute is an ionic compound that will dissociate when dissolved); (b) high conductivity (solute is a strong acid and will ionize completely when dissolved); (c) nonconductive (solute is a covalent compound, neither acid nor base, unreactive towards water); (d) low conductivity (solute is a weak base and will partially ionize when dissolved)

15. (a) ion-dipole; (b) hydrogen bonds; (c) dispersion forces; (d) dipole-dipole attractions; (e) dispersion forces

17. The solubility of solids usually decreases upon cooling a solution, while the solubility of gases usually decreases upon heating.

19. 40%

21. 2.8 g
23. 2.9 atm
25. 102 L HCl
27. The strength of the bonds between like molecules is stronger than the strength between unlike molecules. Therefore, some regions will exist in which the water molecules will exclude oil molecules and other regions will exist in which oil molecules will exclude water molecules, forming a heterogeneous region.
29. Both form homogeneous solutions; their boiling point elevations are the same, as are their lowering of vapor pressures. Osmotic pressure and the lowering of the freezing point are also the same for both solutions.
31. (a) Find number of moles of HNO_3 and H_2O in 100 g of the solution. Find the mole fractions for the components. (b) The mole fraction of HNO_3 is 0.378. The mole fraction of H_2O is 0.622.
33. (a) $X_{Na_2CO_3} = 0.0119$; $X_{H_2O} = 0.988$; (b) $X_{NH_4NO_3} = 0.0928$; $X_{H_2O} = 0.907$; (c) $X_{Cl_2} = 0.192$; $X_{CH_2Cl_2} = 0.808$; (d) $X_{C_5H_9N} = 0.00426$; $X_{CHCl_3} = 0.997$
35. In a 1 *M* solution, the mole is contained in exactly 1 L of solution. In a 1 *m* solution, the mole is contained in exactly 1 kg of solvent.
37. (a) Determine the molar mass of HNO_3. Determine the number of moles of acid in the solution. From the number of moles and the mass of solvent, determine the molality. (b) 33.7 *m*
39. (a) 6.70×10^{-1} *m*; (b) 5.67 *m*; (c) 2.8 *m*; (d) 0.0358 *m*
41. 1.08 *m*
43. (a) Determine the molar mass of sucrose; determine the number of moles of sucrose in the solution; convert the mass of solvent to units of kilograms; from the number of moles and the mass of solvent, determine the molality; determine the difference between the boiling point of water and the boiling point of the solution; determine the new boiling point. (b) 100.5 °C
45. (a) Determine the molar mass of sucrose; determine the number of moles of sucrose in the solution; convert the mass of solvent to units of kilograms; from the number of moles and the mass of solvent, determine the molality; determine the difference between the freezing temperature of water and the freezing temperature of the solution; determine the new freezing temperature. (b) −1.8 °C
47. (a) Determine the molar mass of $Ca(NO_3)_2$; determine the number of moles of $Ca(NO_3)_2$ in the solution; determine the number of moles of ions in the solution; determine the molarity of ions, then the osmotic pressure. (b) 2.67 atm
49. (a) Determine the molal concentration from the change in boiling point and K_b; determine the moles of solute in the solution from the molal concentration and mass of solvent; determine the molar mass from the number of moles and the mass of solute. (b) 2.1×10^2 g mol^{-1}
51. No. Pure benzene freezes at 5.5 °C, and so the observed freezing point of this solution is depressed by ΔT_f = 5.5 − 0.4 = 5.1 °C. The value computed, assuming no ionization of HCl, is ΔT_f = (1.0 m)(5.14 °C/*m*) = 5.1 °C. Agreement of these values supports the assumption that HCl is not ionized.
53. 144 g mol^{-1}
55. 0.870 °C
57. S_8
59. 1.39×10^4 g mol^{-1}
61. 54 g
63. 100.26 °C
65. (a) $X_{CH_3OH} = 0.590$; $X_{C_2H_5OH} = 0.410$; (b) Vapor pressures are: CH_3OH: 55 torr; C_2H_5OH: 18 torr; (c) CH_3OH: 0.75; C_2H_5OH: 0.25
67. The ions and compounds present in the water in the beef lower the freezing point of the beef below −1 °C.
69. $\Delta bp = K_b m = (1.20\ °C/m)\left(\dfrac{9.41\text{g} \times \dfrac{1\text{mol Hg Cl}_2}{271.496\text{ g}}}{0.03275\text{ kg}}\right) = 1.27\ °C$

The observed change equals the theoretical change; therefore, no dissociation occurs.

71.

Colloidal System	Dispersed Phase	Dispersion Medium
starch dispersion	starch	water
smoke	solid particles	air
fog	water	air
pearl	water	calcium carbonate ($CaCO_3$)
whipped cream	air	cream
floating soap	air	soap
jelly	fruit juice	pectin gel
milk	butterfat	water
ruby	chromium(III) oxide (Cr_2O_3)	aluminum oxide (Al_2O_3)

73. Colloidal dispersions consist of particles that are much bigger than the solutes of typical solutions. Colloidal particles are either very large molecules or aggregates of smaller species that usually are big enough to scatter light. Colloids are homogeneous on a macroscopic (visual) scale, while solutions are homogeneous on a microscopic (molecular) scale.

75. If they are placed in an electrolytic cell, dispersed particles will move toward the electrode that carries a charge opposite to their own charge. At this electrode, the charged particles will be neutralized and will coagulate as a precipitate.

Chapter 12

1. The instantaneous rate is the rate of a reaction at any particular point in time, a period of time that is so short that the concentrations of reactants and products change by a negligible amount. The initial rate is the instantaneous rate of reaction as it starts (as product just begins to form). Average rate is the average of the instantaneous rates over a time period.
3. rate $= +\frac{1}{2}\frac{\Delta[\text{ClF}_3]}{\Delta t} = -\frac{\Delta[\text{Cl}_2]}{\Delta t} = -\frac{1}{3}\frac{\Delta[\text{F}_2]}{\Delta t}$
5. (a) average rate, 0 – 10 s = 0.0375 mol L^{-1} s^{-1}; average rate, 10 – 20 s = 0.0265 mol L^{-1} s^{-1}; (b) instantaneous rate, 15 s = 0.023 mol L^{-1} s^{-1}; (c) average rate for B formation = 0.0188 mol L^{-1} s^{-1}; instantaneous rate for B formation = 0.012 mol L^{-1} s^{-1}
7. Higher molarity increases the rate of the reaction. Higher temperature increases the rate of the reaction. Smaller pieces of magnesium metal will react more rapidly than larger pieces because more reactive surface exists.
9. (a) Depending on the angle selected, the atom may take a long time to collide with the molecule and, when a collision does occur, it may not result in the breaking of the bond and the forming of the other. (b) Particles of reactant must come into contact with each other before they can react.
11. (a) very slow; (b) As the temperature is increased, the reaction proceeds at a faster rate. The amount of reactants decreases, and the amount of products increases. After a while, there is a roughly equal amount of *BC*, *AB*, and *C* in the mixture and a slight excess of *A*.
13. (a) 2; (b) 1
15. (a) The process reduces the rate by a factor of 4. (b) Since CO does not appear in the rate law, the rate is not affected.
17. 4.3×10^{-5} mol/L/s

19. 7.9×10^{-13} mol/L/year
21. rate = k; $k = 2.0 \times 10^{-2}$ mol L^{-1} h^{-1} (about 0.9 g L^{-1} h^{-1} for the average male); The reaction is zero order.
23. rate = $k[NOCl]^2$; $k = 8.0 \times 10^{-8}$ L/mol/h; second order
25. rate = $k[NO]^2[Cl_2]$; $k = 9.1$ L^2 mol^{-2} h^{-1}; second order in NO; first order in Cl_2
27. (a) The rate law is second order in A and is written as rate = $k[A]^2$. (b) $k = 7.88 \times 10^{-3}$ L mol^{-1} s^{-1}
29. (a) 2.5×10^{-4} mol/L/min
31. rate = $k[I^-][OCl^-]$; $k = 6.1 \times 10^{-2}$ L mol $^{-1}$ s^{-1}
33. Plotting a graph of ln[SO_2Cl_2] versus t reveals a linear trend; therefore we know this is a first-order reaction:

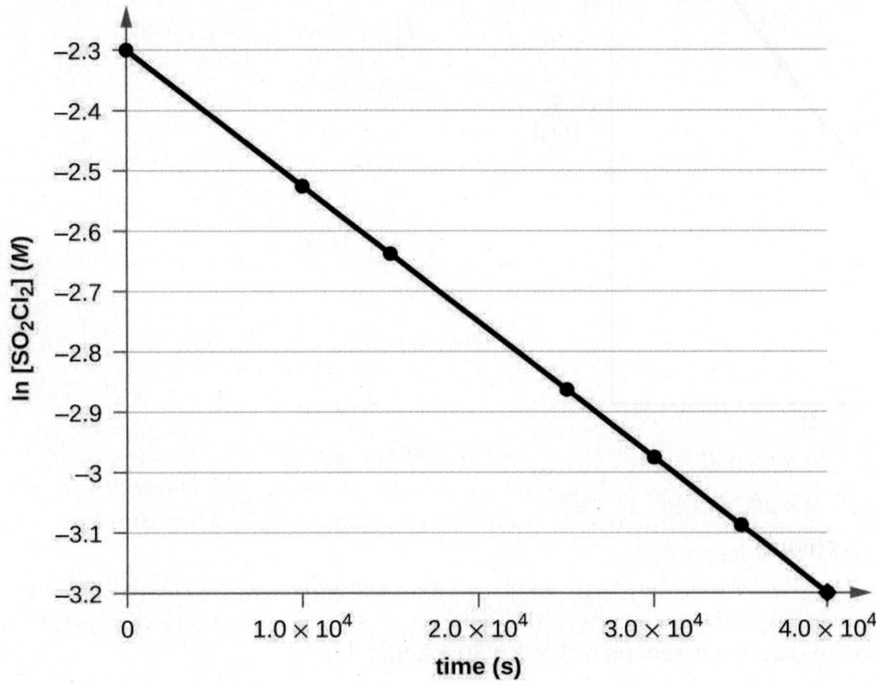

$k = 2.20 \times 10^{-5}$ s^{-1}

34.

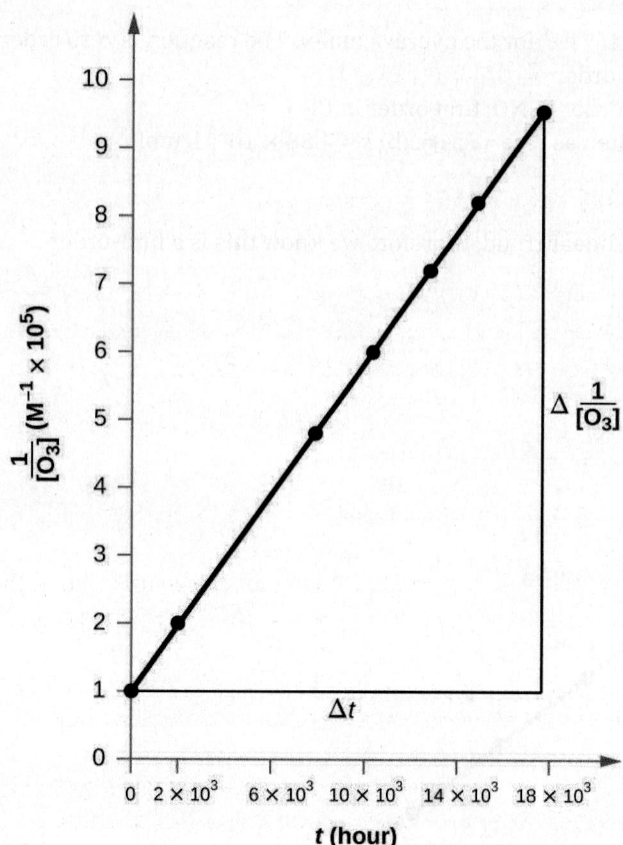

The plot is nicely linear, so the reaction is second order. $k = 50.1 \text{ L mol}^{-1} \text{ h}^{-1}$

36. 14.3 d

38. 8.3×10^7 s

40. 0.826 s

42. The reaction is first order. $k = 1.0 \times 10^7 \text{ L mol}^{-1} \text{ min}^{-1}$

44. 1.16×10^3 s ; 20% remains

46. 252 days

48.

$[A]_0$ (*M*)	$k \times 10^3$ (s^{-1})
4.88	2.45
3.52	2.51
2.29	2.53
1.81	2.58
5.33	2.36
4.05	2.47
2.95	2.48

$[A]_0$ (*M*)	$k \times 10^3$ (s^{-1})
1.72	2.43

50. The reactants either may be moving too slowly to have enough kinetic energy to exceed the activation energy for the reaction, or the orientation of the molecules when they collide may prevent the reaction from occurring.

52. The activation energy is the minimum amount of energy necessary to form the activated complex in a reaction. It is usually expressed as the energy necessary to form one mole of activated complex.

54. After finding *k* at several different temperatures, a plot of ln *k* versus $\frac{1}{T}$, gives a straight line with the slope $\frac{-E_a}{R}$ from which E_a may be determined.

56. (a) 4-times faster (b) 128-times faster

58. $3.9 \times 10^{15}\ s^{-1}$

60. 43.0 kJ/mol

62. 177 kJ/mol

64. E_a = 108 kJ; $A = 2.0 \times 10^8\ s^{-1}$; $k = 3.2 \times 10^{-10}\ s^{-1}$; (b) 1.81×10^8 h or 7.6×10^6 day; (c) Assuming that the reaction is irreversible simplifies the calculation because we do not have to account for any reactant that, having been converted to product, returns to the original state.

66. The *A* atom has enough energy to react with *BC*; however, the different angles at which it bounces off of *BC* without reacting indicate that the orientation of the molecule is an important part of the reaction kinetics and determines whether a reaction will occur.

68. No. In general, for the overall reaction, we cannot predict the effect of changing the concentration without knowing the rate law. Yes. If the reaction is an elementary reaction, then doubling the concentration of *A* doubles the rate.

70. Rate = $k[A][B]^2$; Rate = $k[A]^3$

72. (a) $\text{Rate}_1 = k[O_3]$; (b) $\text{Rate}_2 = k[O_3][Cl]$; (c) $\text{Rate}_3 = k[ClO][O]$; (d) $\text{Rate}_2 = k[O_3][NO]$; (e) $\text{Rate}_3 = k[NO_2][O]$

74. (a) Doubling $[H_2]$ doubles the rate. $[H_2]$ must enter the rate law to the first power. Doubling [NO] increases the rate by a factor of 4. [NO] must enter the rate law to the second power. (b) Rate = $k\,[NO]^2[H_2]$; (c) $k = 5.0 \times 10^3\ mol^{-2}\ L^{-2}\ min^{-1}$; (d) 0.0050 mol/L; (e) Step II is the rate-determining step. If step I gives N_2O_2 in adequate amount, steps 1 and 2 combine to give $2NO + H_2 \longrightarrow H_2O + N_2O$. This reaction corresponds to the observed rate law. Combine steps 1 and 2 with step 3, which occurs by supposition in a rapid fashion, to give the appropriate stoichiometry.

76. The general mode of action for a catalyst is to provide a mechanism by which the reactants can unite more readily by taking a path with a lower reaction energy. The rates of both the forward and the reverse reactions are increased, leading to a faster achievement of equilibrium.

78. (a) Chlorine atoms are a catalyst because they react in the second step but are regenerated in the third step. Thus, they are not used up, which is a characteristic of catalysts. (b) NO is a catalyst for the same reason as in part (a).

80. The lowering of the transition state energy indicates the effect of a catalyst. (a) B; (b) B

82. The energy needed to go from the initial state to the transition state is (a) 10 kJ; (b) 10 kJ.

84. Both diagrams describe two-step, exothermic reactions, but with different changes in enthalpy, suggesting the diagrams depict two different overall reactions.

Chapter 13

1. The reaction can proceed in both the forward and reverse directions.

3. When a system has reached equilibrium, no further changes in the reactant and product concentrations occur; the forward and reverse reactions continue to proceed, but at equal rates.

5. Not necessarily. A system at equilibrium is characterized by *constant* reactant and product concentrations, but the values of the reactant and product concentrations themselves need not be equal.

7. Equilibrium cannot be established between the liquid and the gas phase if the top is removed from the bottle because the system is not closed; one of the components of the equilibrium, the Br_2 vapor, would escape from the bottle until all liquid disappeared. Thus, more liquid would evaporate than can condense back from the gas phase to the liquid phase.

9. (a) $K_c = [Ag^+][Cl^-] < 1$. AgCl is insoluble; thus, the concentrations of ions are much less than 1 *M*; (b) $K_c = \frac{1}{[Pb^{2+}][Cl^-]^2} > 1$ because $PbCl_2$ is insoluble and formation of the solid will reduce the concentration of ions to a low level (<1 *M*).

11. Since $K_c = \frac{[C_6H_6]}{[C_2H_2]^3}$, a value of $K_c \approx 10$ means that C_6H_6 predominates over C_2H_2. In such a case, the reaction would be commercially feasible if the rate to equilibrium is suitable.

13. $K_c > 1$

15. (a) $Q_c = \frac{[CH_3Cl][HCl]}{[CH_4][Cl_2]}$; (b) $Q_c = \frac{[NO]^2}{[N_2][O_2]}$; (c) $Q_c = \frac{[SO_3]^2}{[SO_2]^2[O_2]}$; (d) $Q_c = [SO_2]$; (e) $Q_c = \frac{1}{[P_4][O_2]^5}$; (f) $Q_c = \frac{[Br]^2}{[Br_2]}$; (g) $Q_c = \frac{[CO_2]}{[CH_4][O_2]^2}$; (h) $Q_c = [H_2O]^5$

17. (a) Q_c 25 proceeds left; (b) Q_P 0.22 proceeds right; (c) Q_c undefined proceeds left; (d) Q_P 1.00 proceeds right; (e) Q_P 0 proceeds right; (f) Q_c 4 proceeds left

19. The system will shift toward the reactants to reach equilibrium.

21. (a) homogenous; (b) homogenous; (c) homogenous; (d) heterogeneous; (e) heterogeneous; (f) homogenous; (g) heterogeneous; (h) heterogeneous

23. This situation occurs in (a) and (b).

25. (a) $K_P = 1.6 \times 10^{-4}$; (b) $K_P = 50.2$; (c) $K_c = 5.34 \times 10^{-39}$; (d) $K_c = 4.60 \times 10^{-3}$

27. $K_P = P_{H_2O} = 0.042$.

29. $Q_c = \frac{[NH_4^+][OH^-]}{[NH_3]}$

31. The amount of $CaCO_3$ must be so small that P_{CO_2} is less than K_P when the $CaCO_3$ has completely decomposed. In other words, the starting amount of $CaCO_3$ cannot completely generate the full P_{CO_2} required for equilibrium.

33. The change in enthalpy may be used. If the reaction is exothermic, the heat produced can be thought of as a product. If the reaction is endothermic the heat added can be thought of as a reactant. Additional heat would shift an exothermic reaction back to the reactants but would shift an endothermic reaction to the products. Cooling an exothermic reaction causes the reaction to shift toward the product side; cooling an endothermic reaction would cause it to shift to the reactants' side.

34. No, it is not at equilibrium. Because the system is not confined, products continuously escape from the region of the flame; reactants are also added continuously from the burner and surrounding atmosphere.

36. Add N_2; add H_2; decrease the container volume; heat the mixture.

38. (a) *T* increase = shift right, *V* decrease = shift left; (b) *T* increase = shift right, *V* = no effect; (c) *T* increase = shift left, *V* decrease = shift left; (d) *T* increase = shift left, *V* decrease = shift right.

40. (a) $K_c = \frac{[CH_3OH]}{[H_2]^2[CO]}$; (b) $[H_2]$ increases, [CO] decreases, $[CH_3OH]$ increases; (c), $[H_2]$ increases, [CO] decreases, $[CH_3OH]$ decreases; (d), $[H_2]$ increases, [CO] increases, $[CH_3OH]$ increases; (e), $[H_2]$ increases, [CO] increases, $[CH_3OH]$ decreases; (f), no changes.

42. (a) $K_c = \frac{[CO][H_2]}{[H_2O]}$; (b) $[H_2O]$ no change, [CO] no change, $[H_2]$ no change; (c) $[H_2O]$ decreases, [CO] decreases, $[H_2]$ decreases; (d) $[H_2O]$ increases, [CO] increases, $[H_2]$ decreases; (e) $[H_2O]$ decreases, [CO] increases, $[H_2]$ increases. In (b), (c), (d), and (e), the mass of carbon will change, but its concentration (activity) will not change.

44. Only (b)

46. Add NaCl or some other salt that produces Cl^- to the solution. Cooling the solution forces the equilibrium to the right, precipitating more AgCl(*s*).

48. Though the solution is saturated, the dynamic nature of the solubility equilibrium means the opposing processes of solid dissolution and precipitation continue to occur (just at equal rates, meaning the dissolved ion concentrations and the amount of undissolved solid remain constant). The radioactive Ag^+

ions detected in the solution phase come from dissolution of the added solid, and their presence is countered by precipitation of nonradioactive Ag^+.

50. $K_c = \frac{[C]^2}{[A][B]^2}$. [A] = 0.1 *M*, [B] = 0.1 *M*, [C] = 1 *M*; and [A] = 0.01, [B] = 0.250, [C] = 0.791.

52. $K_c = 6.00 \times 10^{-2}$

54. $K_c = 0.50$

56. $K_P = 1.9 \times 10^3$

58. $K_P = 3.06$

60. (a) $-2x$, $+2x$; (b) $\frac{4}{3}x, -\frac{2}{3}x, -2x$; (c) $-2x$, $3x$; (d) $x, -x, -3x$; (e) $+x$; (f) $-\frac{1}{4}x$

62. Activities of pure crystalline solids equal 1 and are constant; however, the mass of Ni does change.

64. $[NH_3] = 9.1 \times 10^{-2}$ *M*

66. $P_{BrCl} = 4.9 \times 10^{-2}$ atm

68. $[CO] = 2.04 \times 10^{-4}$ *M*

70. $P_{H_2O} = 3.64 \times 10^{-3}$ atm

72. Calculate *Q* based on the calculated concentrations and see if it is equal to K_c. Because *Q* does equal 4.32, the system must be at equilibrium.

74. (a) $[NO_2] = 1.17 \times 10^{-3}$ *M*; $[N_2O_4] = 0.128$ *M*; (b) The assumption that *x* is negligibly small compared to 0.129 is confirmed by comparing the initial concentration of the N_2O_4 to its concentration at equilibrium (they differ by just 1 in the least significant digit's place).

76. (a) $[H_2S] = 0.810$ atm, $[H_2] = 0.014$ atm, $[S_2] = 0.0072$ atm; (b) The assumption that $2x$ is negligibly small compared to 0.824 is confirmed by comparing the initial concentration of the H_2S to its concentration at equilibrium (0.824 atm versus 0.810 atm, a difference of less than 2%).

78. $[PCl_5] = 1.80$ *M*; $[Cl_2] = 0.195$ *M*; $[PCl_3] = 0.195$ *M*.

79. 507 g

81. 330 g

84. (a) 0.33 mol. (b) $[CO_2] = 0.50$ *M*. Added H_2 forms some water as a result of a shift to the left after H_2 is added.

86. (a) $K_c = \frac{[NH_3]^4[O_2]^7}{[NO_2]^4[H_2O]^6}$. (b) $[NH_3]$ must increase for Q_c to reach K_c. (c) The increase in system volume would lower the partial pressures of all reactants (including NO_2). (d) $P_{O_2} = 49$ torr

88. $P_{N_2O_3} = 1.90$ atm and $P_{NO} = P_{NO_2} = 1.90$ atm

Chapter 14

1. One example for NH_3 as a conjugate acid: $NH_2^- + H^+ \longrightarrow NH_3$; as a conjugate base: $NH_4^+(aq) + OH^-(aq) \longrightarrow NH_3(aq) + H_2O(l)$

3. (a) $H_3O^+(aq) \longrightarrow H^+(aq) + H_2O(l)$; (b) $HCl(aq) \longrightarrow H^+(aq) + Cl^-(aq)$; (c) $NH_3(aq) \longrightarrow H^+(aq) + NH_2^-(aq)$; (d) $CH_3CO_2H(aq) \longrightarrow H^+(aq) + CH_3CO_2^-(aq)$; (e) $NH_4^+(aq) \longrightarrow H^+(aq) + NH_3(aq)$; (f) $HSO_4^-(aq) \longrightarrow H^+(aq) + SO_4^{2-}(aq)$

5. (a) $H_2O(l) + H^+(aq) \longrightarrow H_3O^+(aq)$; (b) $OH^-(aq) + H^+(aq) \longrightarrow H_2O(l)$; (c) $NH_3(aq) + H^+(aq) \longrightarrow NH_4^+(aq)$; (d) $CN^-(aq) + H^+(aq) \longrightarrow HCN(aq)$; (e) $S^{2-}(aq) + H^+(aq) \longrightarrow HS^-(aq)$; (f) $H_2PO_4^-(aq) + H^+(aq) \longrightarrow H_3PO_4(aq)$

7. (a) H_2O, O^{2-}; (b) H_3O^+, OH^-; (c) H_2CO_3, CO_3^{2-}; (d) NH_4^+, NH_2^-; (e) H_2SO_4, SO_4^{2-}; (f) $H_3O_2^+$, HO_2^-; (g) H_2S; S^{2-}; (h) $H_6N_2^{2+}$, H_4N_2

9. The labels are Brønsted-Lowry acid = BA; its conjugate base = CB; Brønsted-Lowry base = BB; its conjugate acid = CA. (a) HNO_3(BA), H_2O(BB), H_3O^+(CA), NO_3^-(CB); (b) CN^-(BB), H_2O(BA), HCN(CA), OH^-(CB); (c) H_2SO_4(BA), Cl^-(BB), HCl(CA), HSO_4^-(CB); (d) HSO_4^-(BA), OH^-(BB), SO_4^{2-}(CB), H_2O(CA); (e) O^{2-}(BB), H_2O(BA) OH^-(CB and CA); (f) $[Cu(H_2O)_3(OH)]^+$(BB), $[Al(H_2O)_6]^{3+}$(BA), $[Cu(H_2O)_4]^{2+}$(CA), $[Al(H_2O)_5(OH)]^{2+}$(CB); (g) H_2S(BA), NH_2^-(BB), HS^-(CB), NH_3(CA)

11. Amphiprotic species may either gain or lose a proton in a chemical reaction, thus acting as a base or an acid. An example is H_2O. As an acid: $H_2O(aq) + NH_3(aq) \rightleftharpoons NH_4^+(aq) + OH^-(aq)$. As a base: $H_2O(aq) + HCl(aq) \rightleftharpoons H_3O^+(aq) + Cl^-(aq)$

13. amphiprotic: (a) $NH_3 + H_3O^+ \longrightarrow NH_4OH + H_2O$, $NH_3 + OCH_3^- \longrightarrow NH_2^- + CH_3OH$; (b)

$HPO_4{}^{2-} + OH^- \longrightarrow PO_4{}^{3-} + H_2O, HPO_4{}^{2-} + HClO_4 \longrightarrow H_2PO_4{}^- + ClO_4{}^-$; not amphiprotic: (c) Br^-; (d) $NH_4{}^+$; (e) $AsO_4{}^{3-}$

15. In a neutral solution $[H_3O^+] = [OH^-]$. At 40 °C, $[H_3O^+] = [OH^-] = (2.910 \times 10^{-14})^{1/2} = 1.7 \times 10^{-7}$.
17. $x = 3.051 \times 10^{-7}\ M = [H_3O^+] = [OH^-]$; pH = −log3.051 × 10
19. (a) pH = 3.587; pOH = 10.413; (b) pOH = 0.68; pH = 13.32; (c) pOH = 3.85; pH = 10.15; (d) pOH = −0.40; pH = 14.4
21. $[H_3O^+] = 3.0 \times 10^{-7}\ M$; $[OH^-] = 3.3 \times 10^{-8}\ M$
23. $[H_3O^+] = 1 \times 10^{-2}\ M$; $[OH^-] = 1 \times 10^{-12}\ M$
25. $[OH^-] = 3.1 \times 10^{-12}\ M$
27. The salt ionizes in solution, but the anion slightly reacts with water to form the weak acid. This reaction also forms OH^-, which causes the solution to be basic.
29. $[H_2O] > [CH_3CO_2H] > [H_3O^+] \approx [CH_3CO_2{}^-] > [OH^-]$
31. The oxidation state of the sulfur in H_2SO_4 is greater than the oxidation state of the sulfur in H_2SO_3.
33. $\underset{BB}{Mg(OH)_2(s)} + \underset{BA}{2HCl(aq)} \longrightarrow Mg^{2+}(aq) + \underset{CB}{2Cl^-(aq)} + \underset{CA}{2H_2O(l)}$
35. $K_a = 2.3 \times 10^{-11}$
37. The stronger base or stronger acid is the one with the larger K_b or K_a, respectively. In these two examples, they are $(CH_3)_2NH$ and $CH_3NH_3{}^+$.
39. triethylamine
41. (a) $HSO_4{}^-$; higher electronegativity of the central ion. (b) H_2O; NH_3 is a base and water is neutral, or decide on the basis of K_a values. (c) HI; PH_3 is weaker than HCl; HCl is weaker than HI. Thus, PH_3 is weaker than HI. (d) PH_3; in binary compounds of hydrogen with nonmetals, the acidity increases for the element lower in a group. (e) HBr; in a period, the acidity increases from left to right; in a group, it increases from top to bottom. Br is to the left and below S, so HBr is the stronger acid.
43. (a) $NaHSeO_3 < NaHSO_3 < NaHSO_4$; in polyoxy acids, the more electronegative central element—S, in this case—forms the stronger acid. The larger number of oxygen atoms on the central atom (giving it a higher oxidation state) also creates a greater release of hydrogen atoms, resulting in a stronger acid. As a salt, the acidity increases in the same manner. (b) $ClO_2{}^- < BrO_2{}^- < IO_2{}^-$; the basicity of the anions in a series of acids will be the opposite of the acidity in their oxyacids. The acidity increases as the electronegativity of the central atom increases. Cl is more electronegative than Br, and I is the least electronegative of the three. (c) HOI < HOBr < HOCl; in a series of the same form of oxyacids, the acidity increases as the electronegativity of the central atom increases. Cl is more electronegative than Br, and I is the least electronegative of the three. (d) $HOCl < HOClO < HOClO_2 < HOClO_3$; in a series of oxyacids of the same central element, the acidity increases as the number of oxygen atoms increases (or as the oxidation state of the central atom increases). (e) $HTe^- < HS^- << PH_2{}^- < NH_2{}^-$; $PH_2{}^-$ and $NH_2{}^-$ are anions of weak bases, so they act as strong bases toward H^+. HTe^- and HS^- are anions of weak acids, so they have less basic character. In a periodic group, the more electronegative element has the more basic anion. (f) $BrO_4{}^- < BrO_3{}^- < BrO_2{}^- < BrO^-$; with a larger number of oxygen atoms (that is, as the oxidation state of the central ion increases), the corresponding acid becomes more acidic and the anion consequently less basic.
45. $[H_2O] > [C_6H_4OH(CO_2H)] > [H^+]0 > [C_6H_4OH(CO_2)^-] \gg [C_6H_4O(CO_2H)^-] > [OH^-]$
47. 1. Assume that the change in initial concentration of the acid as the equilibrium is established can be neglected, so this concentration can be assumed constant and equal to the initial value of the total acid concentration. 2. Assume we can neglect the contribution of water to the equilibrium concentration of H_3O^+.
48. (b) The addition of HCl
50. (a) Adding HCl will add H_3O^+ ions, which will then react with the OH^- ions, lowering their concentration. The equilibrium will shift to the right, increasing the concentration of HNO_2, and decreasing the concentration of $NO_2{}^-$ ions. (b) Adding HNO_2 increases the concentration of HNO_2 and shifts the equilibrium to the left, increasing the concentration of $NO_2{}^-$ ions and decreasing the concentration of OH^- ions. (c) Adding NaOH adds OH^- ions, which shifts the equilibrium to the left, increasing the concentration of $NO_2{}^-$ ions and decreasing the concentrations of HNO_2. (d) Adding NaCl has no effect on the concentrations of the ions. (e) Adding KNO_2 adds $NO_2{}^-$ ions and shifts the equilibrium to the right,

increasing the HNO_2 and OH^- ion concentrations.

52. This is a case in which the solution contains a mixture of acids of different ionization strengths. In solution, the HCO_2H exists primarily as HCO_2H molecules because the ionization of the weak acid is suppressed by the strong acid. Therefore, the HCO_2H contributes a negligible amount of hydronium ions to the solution. The stronger acid, HCl, is the dominant producer of hydronium ions because it is completely ionized. In such a solution, the stronger acid determines the concentration of hydronium ions, and the ionization of the weaker acid is fixed by the $[H_3O^+]$ produced by the stronger acid.

54. (a) $K_b = 1.8 \times 10^{-5}$; (b) $K_a = 4.5 \times 10^{-4}$; (c) $K_b = 6.4 \times 10^{-5}$; (d) $K_a = 5.6 \times 10^{-10}$

56. $K_a = 1.2 \times 10^{-2}$

58. (a) $K_b = 4.3 \times 10^{-12}$ (b) $K_a = 1.6 \times 10^{10}$ (c) $K_b = 5.9 \times 10^{8}$ (d) $K_b = 4.2 \times 10^{-3}$ (e) $K_b = 2.3 \times 10^{5}$ (f) $K_b = 6.3 \times 10^{-13}$

60. (a) $\frac{[H_3O^+][ClO^-]}{[HClO]} = \frac{(x)(x)}{(0.0092-x)} \approx \frac{(x)(x)}{0.0092} = 2.9 \times 10^{-8}$

Solving for x gives 1.63×10^{-5} M. This value is less than 5% of 0.0092, so the assumption that it can be neglected is valid. Thus, the concentrations of solute species at equilibrium are:

$[H_3O^+] = [ClO^-] = 1.6 \times 10^{-5}$ M

$[HClO^-] = 0.0092$ M

$[OH^-] = 6.1 \times 10^{-10}$ M;

(b) $\frac{[C_6H_5NH_3{}^+][OH^-]}{[C_6H_5NH_2]} = \frac{(x)(x)}{(0.0784-x)} \approx \frac{(x)(x)}{0.0784} = 4.3 \times 10^{-10}$

Solving for x gives 5.81×10^{-6} M. This value is less than 5% of 0.0784, so the assumption that it can be neglected is valid. Thus, the concentrations of solute species at equilibrium are:

$[C_6H5NH_3{}^+] = [OH^-] = 5.8 \times 10^{-6}$ M

$[C_6H_5NH_2] = 0.0784$ M

$[H_3O^+] = 1.7 \times 10^{-9}$ M;

(c) $\frac{[H_3O^+][CN^-]}{[HCN]} = \frac{(x)(x)}{(0.0810-x)} \approx \frac{(x)(x)}{0.0810} = 4.9 \times 10^{-10}$

Solving for x gives 6.30×10^{-6} M. This value is less than 5% of 0.0810, so the assumption that it can be neglected is valid. Thus, the concentrations of solute species at equilibrium are:

$[H_3O^+] = [CN^-] = 6.3 \times 10^{-6}$ M

$[HCN] = 0.0810$ M

$[OH^-] = 1.6 \times 10^{-9}$ M;

(d) $\frac{[(CH_3)_3NH^+][OH^-]}{[(CH_3)_3N]} = \frac{(x)(x)}{(0.11-x)} \approx \frac{(x)(x)}{0.11} = 6.3 \times 10^{-5}$

Solving for x gives 2.63×10^{-3} M. This value is less than 5% of 0.11, so the assumption that it can be neglected is valid. Thus, the concentrations of solute species at equilibrium are:

$[(CH_3)_3NH^+] = [OH^-] = 2.6 \times 10^{-3}$ M

$[(CH_3)_3N] = 0.11$ M

$[H_3O^+] = 3.8 \times 10^{-12}$ M;

(e) $\frac{[Fe(H_2O)_5(OH)^+][H_3O^+]}{[Fe(H_2O)_6{}^{2+}]} = \frac{(x)(x)}{(0.120-x)} \approx \frac{(x)(x)}{0.120} = 1.6 \times 10^{-7}$

Solving for x gives 1.39×10^{-4} M. This value is less than 5% of 0.120, so the assumption that it can be neglected is valid. Thus, the concentrations of solute species at equilibrium are:

$[Fe(H_2O)_5(OH)^+] = [H_3O^+] = 1.4 \times 10^{-4}$ M

$[Fe(H_2O)_6{}^{2+}] = 0.120$ M

$[OH^-] = 7.2 \times 10^{-11}$ M

62. pH = 2.41

64. $[C_{10}H_{14}N_2] = 0.049$ M; $[C_{10}H_{14}N_2H^+] = 1.9 \times 10^{-4}$ M; $[C_{10}H_{14}N_2H_2{}^{2+}] = 1.4 \times 10^{-11}$ M; $[OH^-] = 1.9 \times 10^{-4}$ M; $[H_3O^+] = 5.3 \times 10^{-11}$ M

66. $K_a = 1.2 \times 10^{-2}$

68. $K_b = 1.77 \times 10^{-5}$

70. (a) acidic; (b) basic; (c) acidic; (d) neutral

72. $[H_3O^+]$ and $[HCO_3{}^-]$ are practically equal

74. $[C_6H_4(CO_2H)_2]$ 7.2×10^{-3} M, $[C_6H_4(CO_2H)(CO_2)^-] = [H_3O^+]$ 2.8×10^{-3} M, $[C_6H_4(CO_2)_2{}^{2-}]$$3.9 \times 10^{-6}$ M,

$[OH^-]$ 3.6 × 10^{-12} M

76. (a) $K_{a2} = 1.5 \times 10^{-11}$;
(b) $K_b = 4.3 \times 10^{-12}$;
(c) $\frac{[Te^{2-}][H_3O^+]}{[HTe^-]} = \frac{(x)(0.0141+x)}{(0.0141-x)} \approx \frac{(x)(0.0141)}{0.0141} = 1.5 \times 10^{-11}$
Solving for x gives 1.5 × 10^{-11} M. Therefore, compared with 0.014 M, this value is negligible (1.1 × 10^{-7}%).

78. Excess H_3O^+ is removed primarily by the reaction: $H_3O^+(aq) + H_2PO_4^-(aq) \longrightarrow H_3PO_4(aq) + H_2O(l)$
Excess base is removed by the reaction: $OH^-(aq) + H_3PO_4(aq) \longrightarrow H_2PO_4^-(aq) + H_2O(l)$

80. $[H_3O^+] = 1.5 \times 10^{-4}$ M

82. $[OH^-] = 4.2 \times 10^{-4}$ M

84. (a) The added HCl will increase the concentration of H_3O^+ slightly, which will react with $CH_3CO_2^-$ and produce CH_3CO_2H in the process. Thus, $[CH_3CO_2^-]$ decreases and $[CH_3CO_2H]$ increases. (b) The added KCH_3CO_2 will increase the concentration of $[CH_3CO_2^-]$ which will react with H_3O^+ and produce CH_3CO_2 H in the process. Thus, $[H_3O^+]$ decreases slightly and $[CH_3CO_2H]$ increases. (c) The added NaCl will have no effect on the concentration of the ions. (d) The added KOH will produce OH^- ions, which will react with the H_3O^+, thus reducing $[H_3O^+]$. Some additional CH_3CO_2H will dissociate, producing $[CH_3CO_2^-]$ ions in the process. Thus, $[CH_3CO_2H]$ decreases slightly and $[CH_3CO_2^-]$ increases. (e) The added CH_3CO_2H will increase its concentration, causing more of it to dissociate and producing more $[CH_3CO_2^-]$ and H_3O^+ in the process. Thus, $[H_3O^+]$ increases slightly and $[CH_3CO_2^-]$ increases.

86. pH = 8.95

88. 37 g (0.27 mol)

90. (a) pH = 5.222; (b) The solution is acidic. (c) pH = 5.220

92. At the equivalence point in the titration of a weak base with a strong acid, the resulting solution is slightly acidic due to the presence of the conjugate acid. Thus, pick an indicator that changes color in the acidic range and brackets the pH at the equivalence point. Methyl orange is a good example.

94. (a) pH = 2.50; (b) pH = 4.01; (c) pH = 5.60; (d) pH = 8.35; (e) pH = 11.08

Chapter 15

1. (a)

$$AgI(s) \rightleftharpoons Ag^+(aq) + I^-(aq)$$
$$x \qquad \underline{x}$$

(b)

$$CaCO_3(s) \rightleftharpoons Ca^{2+}(aq) + CO_3^{2-}(aq)$$
$$\underline{x} \qquad x$$

(c)

$$Mg(OH)_2(s) \rightleftharpoons Mg^{2+}(aq) + 2OH^-(aq)$$
$$x \qquad \underline{2x}$$

(d)

$$Mg_3(PO_4)_2(s) \rightleftharpoons 3Mg^{2+}(aq) + 2PO_4^{3-}(aq)$$
$$\underline{x} \qquad \tfrac{2}{3}x$$

(e)

$$Ca_5(PO_4)_3OH(s) \rightleftharpoons 5Ca^{2+}(aq) + 3PO_4^{3-}(aq) + OH^-(aq)$$
$$\underline{5x} \qquad \underline{3x} \qquad x$$

3. There is no change. A solid has an activity of 1 whether there is a little or a lot.

5. The solubility of silver bromide at the new temperature must be known. Normally the solubility increases and some of the solid silver bromide will dissolve.

7. CaF_2, $MnCO_3$, and ZnS
9. (a) $LaF_3(s) \rightleftharpoons La^{3+}(aq) + 3F^-(aq)$ $K_{sp} = [La^{3+}][F^-]^3$;
(b) $CaCO_3(s) \rightleftharpoons Ca^{2+}(aq) + CO_3{}^{2-}(aq)$ $K_{sp} = [Ca^{2+}][CO_3{}^{2-}]$;
(c) $Ag_2SO_4(s) \rightleftharpoons 2Ag^+(aq) + SO_4{}^{2-}(aq)$ $K_{sp} = [Ag^+]^2[SO_4{}^{2-}]$;
(d) $Pb(OH)_2(s) \rightleftharpoons Pb^{2+}(aq) + 2OH^-(aq)$ $K_{sp} = [Pb^{2+}][OH^-]^2$
11. (a)1.77×10^{-7}; (b) 1.6×10^{-6}; (c) 2.2×10^{-9}; (d) 7.91×10^{-22}
13. (a) 2×10^{-2} *M*; (b) 1.5×10^{-3} *M*; (c) 2.27×10^{-9} *M*; (d) 2.2×10^{-10} *M*
15. (a) 6.4×10^{-9} *M* = [Ag⁺], [Cl⁻] = 0.025 *M*. Check: $\frac{6.4 \times 10^{-9}\ M}{0.025\ M} \times 100\% = 2.6 \times 10^{-5}\%$,an insignificant change;
(b) 2.2×10^{-5} *M* = $[Ca^{2+}]$, $[F^-]$ = 0.0013 *M*. Check: $\frac{2.26 \times 10^{-5}\ M}{0.00133\ M} \times 100\% = 1.70\%$. This value is less than 5% and can be ignored.
(c) 0.2238 *M* = $[SO_4{}^{2-}]$; $[Ag^+] = 7.4 \times 10^{-3}$ *M*. Check: $\frac{3.7 \times 10^{-3}}{0.2238} \times 100\% = 1.64 \times 10^{-2}$; the condition is satisfied.
(d) $[OH^-] = 2.8 \times 10^{-3}$ *M*; 5.7×10^{-12} *M* = $[Zn^{2+}]$. Check: $\frac{5.7 \times 10^{-12}}{2.8 \times 10^{-3}} \times 100\% = 2.0 \times 10^{-7}\%$; *x* is less than 5% of $[OH^-]$ and is, therefore, negligible.
17. (a) $[Cl^-] = 7.6 \times 10^{-3}$ *M*
Check: $\frac{7.6 \times 10^{-3}}{0.025} \times 100\% = 30\%$
This value is too large to drop *x*. Therefore solve by using the quadratic equation:
$[Ti^+] = 3.1 \times 10^{-2}$ *M*
$[Cl^-] = 6.1 \times 10^{-3}$
(b) $[Ba^{2+}] = 7.7 \times 10^{-4}$ *M*
Check: $\frac{7.7 \times 10^{-4}}{0.0313} \times 100\% = 2.4\%$
Therefore, the condition is satisfied.
$[Ba^{2+}] = 7.7 \times 10^{-4}$ *M*
$[F^-]$ = 0.0321 *M*;
(c) $Mg(NO_3)_2$ = 0.02444 *M*
$[C_2O_4{}^{2-}] = 2.9 \times 10^{-5}$
Check: $\frac{2.9 \times 10^{-5}}{0.02444} \times 100\% = 0.12\%$
The condition is satisfied; the above value is less than 5%.
$[C_2O_4{}^{2-}] = 2.9 \times 10^{-5}$ *M*
$[Mg^{2+}]$ = 0.0244 *M*
(d) $[OH^-]$ = 0.0501 *M*
$[Ca^{2+}] = 3.15 \times 10^{-3}$
Check: $\frac{3.15 \times 10^{-3}}{0.050} \times 100\% = 6.28\%$
This value is greater than 5%, so a more exact method, such as successive approximations, must be used.
$[Ca^{2+}] = 2.8 \times 10^{-3}$ *M*
$[OH^-] = 0.053 \times 10^{-2}$ *M*
19. The changes in concentration are greater than 5% and thus exceed the maximum value for disregarding the change.
21. $CaSO_4{\bullet}2H_2O$ is the most soluble Ca salt in mol/L, and it is also the most soluble Ca salt in g/L.
23. 4.8×10^{-3} *M* = $[SO_4{}^{2-}] = [Ca^{2+}]$; Since this concentration is higher than 2.60×10^{-3} *M*, "gyp" water does not meet the standards.
25. Mass ($CaSO_4{\cdot}2H_2O$) = 0.72 g/L
27. (a) $[Ag^+] = [I^-] = 1.3 \times 10^{-5}$ *M*; (b) $[Ag^+] = 2.88 \times 10^{-2}$ *M*, $[SO_4{}^{2-}] = 1.44 \times 10^{-2}$ *M*; (c) $[Mn^{2+}] = 3.7 \times 10^{-5}$ *M*, $[OH^-] = 7.4 \times 10^{-5}$ *M*; (d) $[Sr^{2+}] = 4.3 \times 10^{-2}$ *M*, $[OH^-] = 8.6 \times 10^{-2}$ *M*; (e) $[Mg^{2+}] = 1.3 \times 10^{-4}$ *M*, $[OH^-] = 2.6 \times 10^{-4}$ *M*.
29. (a) 1.45×10^{-4}; (b) 8.2×10^{-55}; (c) 1.35×10^{-4}; (d) 1.18×10^{-5}; (e) 1.08×10^{-10}
31. (a) $CaCO_3$ does precipitate. (b) The compound does not precipitate. (c) The compound does not precipitate.

(d) The compound precipitates.

33. 3.03×10^{-7} *M*

35. 9.2×10^{-13} *M*

37. $[Ag^+] = 1.8 \times 10^{-3}$ *M*

39. 6.3×10^{-4}

41. (a) 2.25 L; (b) 7.2×10^{-7} g

43. 100% of it is dissolved

45. (a) Hg_2^{2+} and Cu^{2+}: Add SO_4^{2-}. (b) SO_4^{2-} and Cl^-: Add Ba^{2+}. (c) Hg^{2+} and Co^{2+}: Add S^{2-}. (d) Zn^{2+} and Sr^{2+}: Add OH^- until $[OH^-] = 0.050$ *M*. (e) Ba^{2+} and Mg^{2+}: Add SO_4^{2-}. (f) CO_3^{2-} and OH^-: Add Ba^{2+}.

47. AgI will precipitate first.

49. 1.5×10^{-12} *M*

51. 3.99 kg

53. (a) 3.1×10^{-11}; (b) $[Cu^{2+}] = 2.6 \times 10^{-3}$; $[IO_3^-] = 5.3 \times 10^{-3}$

55. 1.8×10^{-5} g $Pb(OH)_2$

57. $Mg(OH)_2(s) \rightleftharpoons Mg^{2+} + 2OH^- \qquad K_{sp} = [Mg^{2+}][OH^-]^2$

1.23×10^{-3} g $Mg(OH)_2$

59. $MnCO_3$ will form first since it has the smallest K_{sp} value among these homologous compounds and is therefore the least soluble. $MgCO_3{\cdot}3H_2O$ will be the last to precipitate since it has the largest K_sp value and is the most soluble. K_{sp} value.

62. when the amount of solid is so small that a saturated solution is not produced

64. 1.8×10^{-5} *M*

66. 5×10^{23}

68.

	$[Cd(CN)_4^{2-}]$	$[CN^-]$	$[Cd^{2+}]$
Initial concentration (*M*)	0.250	0	0
Equilibrium (*M*)	$0.250 - x$	$4x$	x

$[Cd^{2+}] = 9.5 \times 10^{-5}$ *M*; $[CN^-] = 3.8 \times 10^{-4}$ *M*

70. $[Co^{3+}] = 3.0 \times 10^{-6}$ *M*; $[NH_3] = 1.8 \times 10^{-5}$ *M*

72. 1.3 g

74. 0.79 g

76. (a)

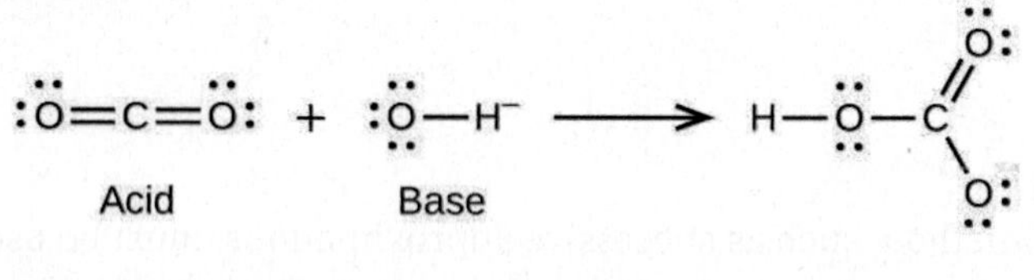

:Ö=C=Ö: + :Ö—H⁻ ⟶ H—Ö—C(=O:)—O:

Acid Base

(b)

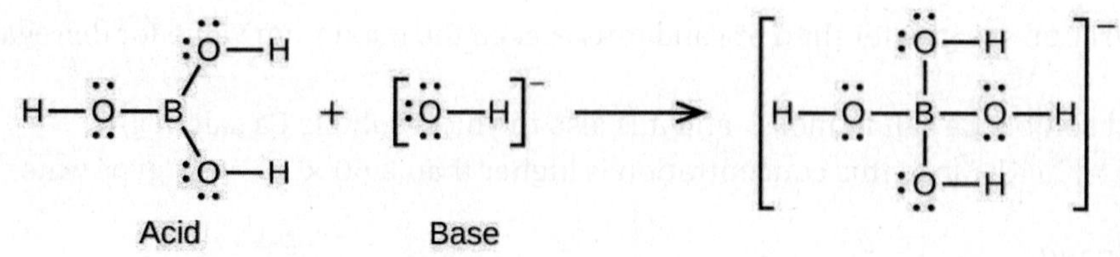

$H-O-B(OH)_2$ + $[:O-H]^-$ ⟶ $[H-O-B(OH)_2-O-H]^-$

Acid Base

(c)

:I—I: + $[:I:]^-$ ⟶ $[:I—I—I:]^-$

Acid Base

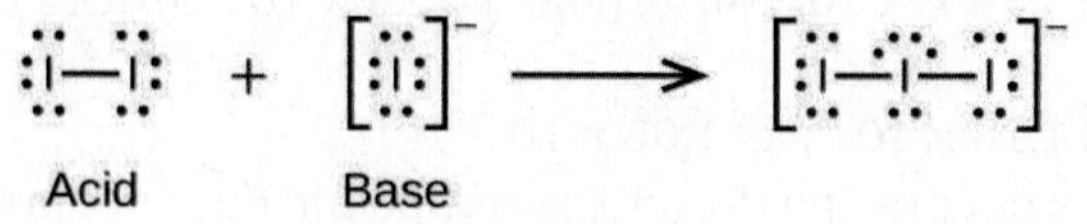

(d)

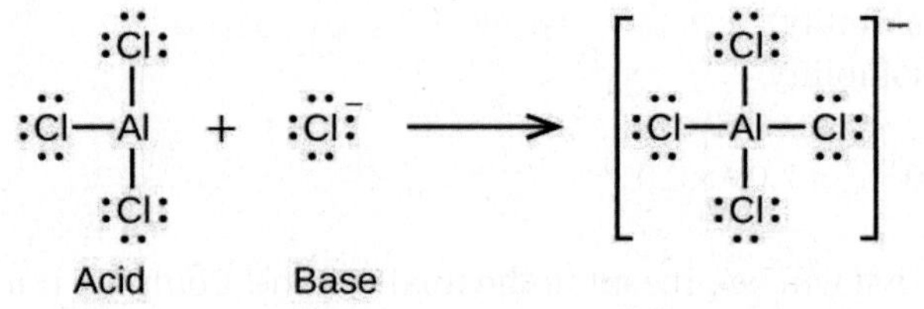

Acid Base

(e)

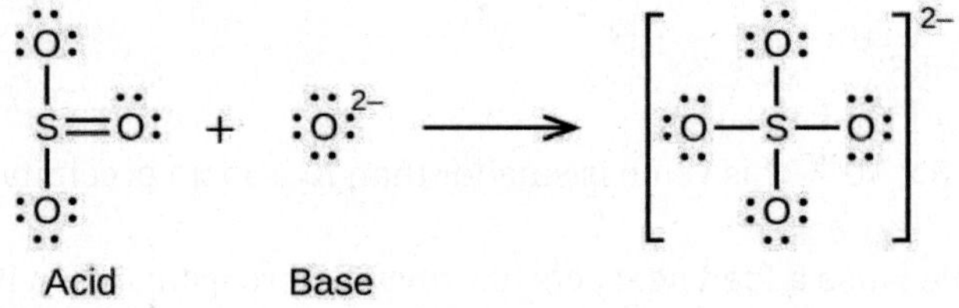

Acid Base

78. (a)

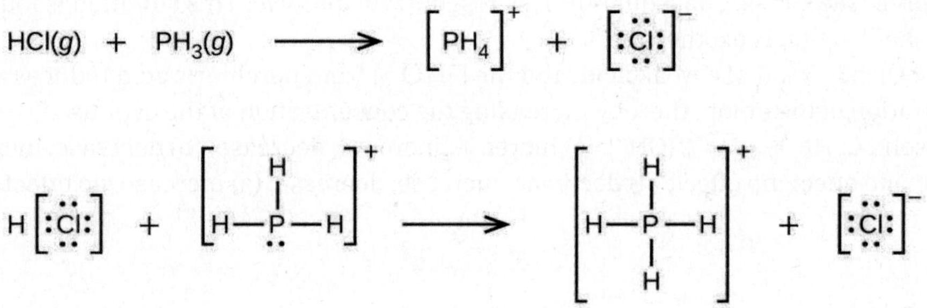

$HCl(g) + PH_3(g) \longrightarrow [PH_4]^+ + [:Cl:]^-$

(b) $H_3O^+ + CH_3^- \longrightarrow CH_4 + H_2O$

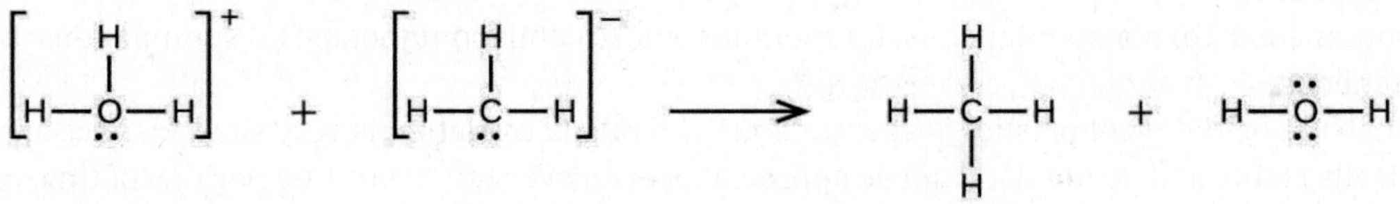

(c) $CaO + SO_3 \longrightarrow CaSO4$

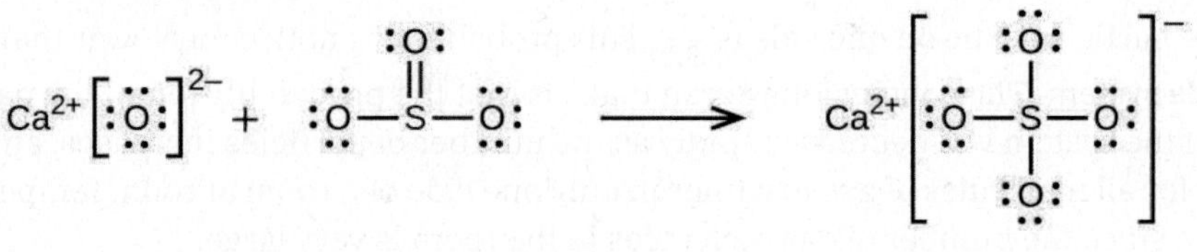

(d) $NH_4{}^+ + C_2H_5O^- \longrightarrow C_2H_5OH + NH_3$

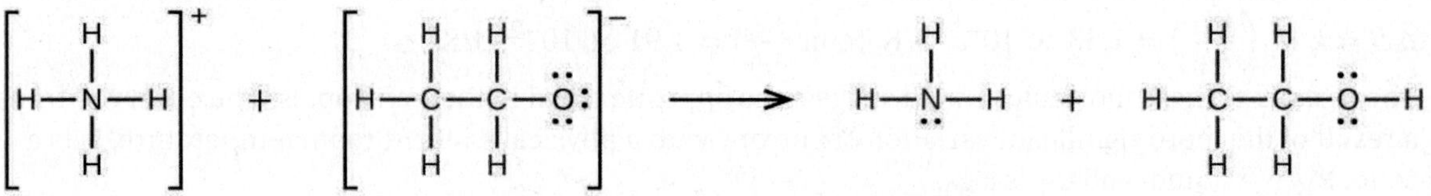

80. 0.0281 g

82. $HNO_3(l) + HF(l) \longrightarrow H_2NO_3{}^+ + F^-$; $HF(l) + BF_3(g) \longrightarrow H^+ + BF_4$

84. (a) $H_3BO_3 + H_2O \longrightarrow H_4BO_4{}^- + H^+$; (b) The electronic and molecular shapes are the same—both tetrahedral. (c) The tetrahedral structure is consistent with sp^3 hybridization.

86. 0.014 M

88. 7.2×10^{-15} M

90. 4.4×10^{-22} *M*
93. $[OH^-] = 4.5 \times 10^{-6}$; $[Al^{3+}] = 2 \times 10^{-16}$ (molar solubility)
95. $[SO_4^{2-}] = 0.049$ *M*; $[Ba^{2+}] = 4.7 \times 10^{-7}$ (molar solubility)
97. $[OH^-] = 7.6 \times 10^{-3}$ *M*; $[Pb^{2+}] = 2.1 \times 10^{-11}$ (molar solubility)
99. 7.66
101. (a) $K_{sp} = [Mg^{2+}][F^-]^2 = (1.21 \times 10^{-3})(2 \times 1.21 \times 10^{-3})^2 = 7.09 \times 10^{-9}$
(b) 7.09×10^{-7} *M*
(c) Determine the concentration of Mg^{2+} and F^- that will be present in the final volume. Compare the value of the ion product $[Mg^{2+}][F^-]^2$ with K_{sp}. If this value is larger than K_{sp}, precipitation will occur.
$0.1000\ L \times 3.00 \times 10^{-3}\ M\ Mg(NO_3)_2 = 0.3000\ L \times M\ Mg(NO_3)_2$
$M\ Mg(NO_3)_2 = 1.00 \times 10^{-3}\ M$
$0.2000\ L \times 2.00 \times 10^{-3}\ M\ NaF = 0.3000\ L \times M\ NaF$
$M\ NaF = 1.33 \times 10^{-3}\ M$
ion product $= (1.00 \times 10^{-3})(1.33 \times 10^{-3})^2 = 1.77 \times 10^{-9}$ This value is smaller than K_{sp}, so no precipitation will occur.
(d) MgF_2 is less soluble at 27 °C than at 18 °C. Because added heat acts like an added reagent, when it appears on the product side, the Le Châtelier's principle states that the equilibrium will shift to the reactants' side to counter the stress. Consequently, less reagent will dissolve. This situation is found in our case. Therefore, the reaction is exothermic.
103. BaF_2, $Ca_3(PO_4)_2$, ZnS; each is a salt of a weak acid, and the $[H_3O^+]$ from perchloric acid reduces the equilibrium concentration of the anion, thereby increasing the concentration of the cations
105. Effect on amount of solid $CaHPO_4$, $[Ca^{2+}]$, $[OH^-]$: (a) increase, increase, decrease; (b) decrease, increase, decrease; (c) no effect, no effect, no effect; (d) decrease, increase, decrease; (e) increase, no effect, no effect

Chapter 16

1. A reaction has a natural tendency to occur and takes place without the continual input of energy from an external source.
3. (a) spontaneous; (b) nonspontaneous; (c) spontaneous; (d) nonspontaneous; (e) spontaneous; (f) spontaneous
5. Although the oxidation of plastics is spontaneous, the rate of oxidation is very slow. Plastics are therefore kinetically stable and do not decompose appreciably even over relatively long periods of time.
7. There are four initial microstates and four final microstates.
$\Delta S = k \ln \frac{W_f}{W_i} = 1.38 \times 10^{-23}\ J/K \times \ln \frac{4}{4} = 0$
9. The probability for all the particles to be on one side is $\frac{1}{32}$. This probability is noticeably lower than the $\frac{1}{8}$ result for the four-particle system. The conclusion we can make is that the probability for all the particles to stay in only one part of the system will decrease rapidly as the number of particles increases, and, for instance, the probability for all molecules of gas to gather in only one side of a room at room temperature and pressure is negligible since the number of gas molecules in the room is very large.
11. There is only one initial state. For the final state, the energy can be contained in pairs A-C, A-D, B-C, or B-D. Thus, there are four final possible states.
$\Delta S = k \ln \left(\frac{W_f}{W_i}\right) = 1.38 \times 10^{-23}\ J/K \times \ln \left(\frac{4}{1}\right) = 1.91 \times 10^{-23}\ J/K$
13. The masses of these molecules would suggest the opposite trend in their entropies. The observed trend is a result of the more significant variation of entropy with a physical state. At room temperature, I_2 is a solid, Br_2 is a liquid, and Cl_2 is a gas.
15. (a) $C_3H_7OH(l)$ as it is a larger molecule (more complex and more massive), and so more microstates describing its motions are available at any given temperature. (b) $C_2H_5OH(g)$ as it is in the gaseous state. (c) 2H(*g*), since entropy is an extensive property, and so two H atoms (or two moles of H atoms) possess twice as much entropy as one atom (or one mole of atoms).
17. (a) Negative. The relatively ordered solid precipitating decreases the number of mobile ions in solution. (b) Negative. There is a net loss of three moles of gas from reactants to products. (c) Positive. There is a net

increase of seven moles of gas from reactants to products.

19. $C_6H_6(l) + 7.5O_2(g) \longrightarrow 3H_2O(g) + 6CO_2(g)$
There are 7.5 moles of gas initially, and 3 + 6 = 9 moles of gas in the end. Therefore, it is likely that the entropy increases as a result of this reaction, and ΔS is positive.

21. (a) 107 J/K; (b) −86.4 J/K; (c) 133.2 J/K; (d) 118.8 J/K; (e) −326.6 J/K; (f) −171.9 J/K; (g) −7.2 J/K

23. 100.6 J/K

25. (a) −198.1 J/K; (b) −348.9 J/K

27. As $\Delta S_{univ} < 0$ at each of these temperatures, melting is not spontaneous at either of them. The given values for entropy and enthalpy are for NaCl at 298 K. It is assumed that these do not change significantly at the higher temperatures used in the problem.

29. (a) 2.86 J/K; (b) 24.8 J/K; (c) −113.2 J/K; (d) −24.7 J/K; (e) 15.5 J/K; (f) 290.0 J/K

31. The reaction is nonspontaneous at room temperature.
Above 400 K, ΔG will become negative, and the reaction will become spontaneous.

33. (a) 465.1 kJ nonspontaneous; (b) −106.86 kJ spontaneous; (c) −291.9 kJ spontaneous; (d) −83.4 kJ spontaneous; (e) −406.7 kJ spontaneous; (f) −154.3 kJ spontaneous

35. (a) The standard free energy of formation is −1124.3 kJ/mol. (b) The calculation agrees with the value in Appendix G because free energy is a state function (just like the enthalpy and entropy), so its change depends only on the initial and final states, not the path between them.

37. (a) The reaction is nonspontaneous; (b) Above 566 °C the process is spontaneous.

39. (a) 1.5×10^2 kJ; (b) −21.9 kJ; (c) −5.34 kJ; (d) −0.383 kJ; (e) 18 kJ; (f) 71 kJ

41. (a) $K = 41$; (b) $K = 0.053$; (c) $K = 6.9 \times 10^{13}$; (d) $K = 1.9$; (e) $K = 0.04$

43. In each of the following, the value of ΔG is not given at the temperature of the reaction. Therefore, we must calculate ΔG from the values $\Delta H°$ and ΔS and then calculate ΔG from the relation $\Delta G = \Delta H° - T\Delta S°$. (a) $K = 1.07 \times 10^{-13}$; (b) $K = 2.51 \times 10^{-3}$; (c) $K = 4.83 \times 10^3$; (d) $K = 0.219$; (e) $K = 16.1$

45. The standard free energy change is $\Delta G° = -RT \ln K = 4.84$ kJ/mol. When reactants and products are in their standard states (1 bar or 1 atm), $Q = 1$. As the reaction proceeds toward equilibrium, the reaction shifts left (the amount of products drops while the amount of reactants increases): $Q < 1$, and ΔG becomes less positive as it approaches zero. At equilibrium, $Q = K$, and $\Delta G = 0$.

47. The reaction will be spontaneous at temperatures greater than 287 K.

49. $K = 5.35 \times 10^{15}$; The process is exothermic.

51. 1.0×10^{-8} atm. This is the maximum pressure of the gases under the stated conditions.

53. $x = 1.29 \times 10^{-5}\text{ atm} = P_{O_2}$

55. −0.16 kJ

56. (a) 22.1 kJ; (b) 98.9 kJ/mol

58. 90 kJ/mol

60. (a) Under standard thermodynamic conditions, the evaporation is nonspontaneous; (b) $K_p = 0.031$; (c) The evaporation of water is spontaneous; (d) P_{H_2O} must always be less than K_p or less than 0.031 atm. 0.031 atm represents air saturated with water vapor at 25 °C, or 100% humidity.

62. (a) Nonspontaneous as $\Delta G° > 0$; (b) $\Delta G = \Delta G° + RT \ln Q$,
$\Delta G = 1.7 \times 10^3 + \left(8.314 \times 310 \times \ln \frac{28}{120}\right) = -2.1$ kJ. The forward reaction to produce F6P is spontaneous under these conditions.

64. ΔG is negative as the process is spontaneous. ΔH is positive as with the solution becoming cold, the dissolving must be endothermic. ΔS must be positive as this drives the process, and it is expected for the dissolution of any soluble ionic compound.

66. (a) Increasing the oxygen partial pressure will yield a decrease in Q and ΔG thus becomes more negative. (b) Increasing the oxygen partial pressure will yield a decrease in Q and ΔG thus becomes more negative. (c) Increasing the oxygen partial pressure will yield an increase in Q and ΔG thus becomes more positive.

Chapter 17

1. (a) reduction; (b) oxidation; (c) oxidation; (d) reduction

3. (a) $F_2 + Ca \longrightarrow 2F^- + Ca^{2+}$; (b) $Cl_2 + 2Li \longrightarrow 2Li^+ + 2Cl^-$; (c) $3Br_2 + 2Fe \longrightarrow 2Fe^{3+} + 6Br^-$; (d) $MnO_4{}^- + 4H^+ + 3Ag \longrightarrow 3Ag^+ + MnO_2 + 2H_2O$

5. Oxidized: (a) Sn^{2+}; (b) Hg; (c) Al; reduced: (a) H_2O_2; (b) PbO_2; (c) $Cr_2O_7{}^{2-}$; oxidizing agent: (a) H_2O_2; (b) PbO_2; (c) $Cr_2O_7{}^{2-}$; reducing agent: (a) Sn^{2+}; (b) Hg; (c) Al
7. Oxidized = reducing agent: (a) $SO_3{}^{2-}$; (b) $Mn(OH)_2$; (c) H_2; (d) Al; reduced = oxidizing agent: (a) $Cu(OH)_2$; (b) O_2; (c) $NO_3{}^-$; (d) $CrO_4{}^{2-}$
9. In basic solution, $[OH^-] > 1 \times 10^{-7}\ M > [H^+]$. Hydrogen ion cannot appear as a reactant because its concentration is essentially zero. If it were produced, it would instantly react with the excess hydroxide ion to produce water. Thus, hydrogen ion should *not* appear as a reactant or product in basic solution.
11. (a) $Mg(s) \mid Mg^{2+}(aq) \| Ni^{2+}(aq) \mid Ni(s)$; (b) $Cu(s) \mid Cu^{2+}(aq) \| Ag^{+}(aq) \mid Ag(s)$; (c) $Mn(s) \mid Mn^{2+}(aq) \| Sn^{2+}(aq) \mid Sn(s)$; (d) $Pt(s) \mid Cu^{+}(aq), Cu^{2+}(aq) \| Au^{3+}(aq) \mid Au(s)$
13. (a) $Mg(s) + Cu^{2+}(aq) \longrightarrow Mg^{2+}(aq) + Cu(s)$; (b) $2Ag^{+}(aq) + Ni(s) \longrightarrow Ni^{2+}(aq) + 2Ag(s)$
15. Species oxidized = reducing agent: (a) Al(*s*); (b) NO(*g*); (c) Mg(*s*); and (d) $MnO_2(s)$; Species reduced = oxidizing agent: (a) $Zr^{4+}(aq)$; (b) $Ag^{+}(aq)$; (c) $SiO_3{}^{2-}(aq)$; and (d) $ClO_3{}^-(aq)$
17. Without the salt bridge, the circuit would be open (or broken) and no current could flow. With a salt bridge, each half-cell remains electrically neutral and current can flow through the circuit.
19. Active electrodes participate in the oxidation-reduction reaction. Since metals form cations, the electrode would lose mass if metal atoms in the electrode were to oxidize and go into solution. Oxidation occurs at the anode.
21. (a) +2.115 V (spontaneous); (b) +0.4626 V (spontaneous); (c) +1.0589 V (spontaneous); (d) +0.727 V (spontaneous)
23. $3Cu(s) + 2Au^{3+}(aq) \longrightarrow 3Cu^{2+}(aq) + 2Au(s)$; +1.16 V; spontaneous
25. $3Cd(s) + 2Al^{3+}(aq) \longrightarrow 3Cd^{2+}(aq) + 2Al(s)$; −1.259 V; nonspontaneous
27. (a) 0 kJ/mol; (b) −83.7 kJ/mol; (c) +235.3 kJ/mol
29. (a) standard cell potential: 1.50 V, spontaneous; cell potential under stated conditions: 1.43 V, spontaneous; (b) standard cell potential: 1.405 V, spontaneous; cell potential under stated conditions: 1.423 V, spontaneous; (c) standard cell potential: −2.749 V, nonspontaneous; cell potential under stated conditions: −2.757 V, nonspontaneous
31. (a) 1.7×10^{-10}; (b) 2.6×10^{-21}; (c) 4.693×10^{21}; (d) 1.0×10^{-14}

33. (a) $\begin{array}{ll} \text{anode: } Cu(s) \longrightarrow Cu^{2+}(aq) + 2e^- & E^\circ_{\text{anode}} = 0.34\ \text{V} \\ \text{cathode: } 2 \times \left(Ag^{+}(aq) + e^- \longrightarrow Ag(s)\right) & E^\circ_{\text{cathode}} = 0.7996\ \text{V} \end{array}$; (b) 3.5×10^{15}; (c) 5.6×10^{-9} *M*
34. Batteries are self-contained and have a limited supply of reagents to expend before going dead. Alternatively, battery reaction byproducts accumulate and interfere with the reaction. Because a fuel cell is constantly resupplied with reactants and products are expelled, it can continue to function as long as reagents are supplied.
36. E_{cell}, as described in the Nernst equation, has a term that is directly proportional to temperature. At low temperatures, this term is decreased, resulting in a lower cell voltage provided by the battery to the device—the same effect as a battery running dead.
38. Mg and Zn
40. Both examples involve cathodic protection. The (sacrificial) anode is the metal that corrodes (oxidizes or reacts). In the case of iron (−0.447 V) and zinc (−0.7618 V), zinc has a more negative standard reduction potential and so serves as the anode. In the case of iron and copper (0.34 V), iron has the smaller standard reduction potential and so corrodes (serves as the anode).
42. While the reduction potential of lithium would make it capable of protecting the other metals, this high potential is also indicative of how reactive lithium is; it would have a spontaneous reaction with most substances. This means that the lithium would react quickly with other substances, even those that would not oxidize the metal it is attempting to protect. Reactivity like this means the sacrificial anode would be depleted rapidly and need to be replaced frequently. (Optional additional reason: fire hazard in the presence of water.)

46. (a) $\begin{array}{l} \text{mass Ca} = 69.1\ \text{g} \\ \text{mass } Cl_2 = 122\ \text{g} \end{array}$; (b) $\begin{array}{l} \text{mass Li} = 23.9\ \text{g} \\ \text{mass } H_2 = 3.48\ \text{g} \end{array}$; (c) $\begin{array}{l} \text{mass Al} = 31.0\ \text{g} \\ \text{mass } Cl_2 = 122\ \text{g} \end{array}$; (d) $\begin{array}{l} \text{mass Cr} = 59.8\ \text{g} \\ \text{mass } Br_2 = 276\ \text{g} \end{array}$

48. 0.79 L

Chapter 18

1. The alkali metals all have a single *s* electron in their outermost shell. In contrast, the alkaline earth metals have a completed *s* subshell in their outermost shell. In general, the alkali metals react faster and are more reactive than the corresponding alkaline earth metals in the same period.

3.

$$\text{Na} + \text{I}_2 \longrightarrow 2\text{NaI}$$
$$2\text{Na} + \text{Se} \longrightarrow \text{Na}_2\text{Se}$$
$$2\text{Na} + \text{O}_2 \longrightarrow \text{Na}_2\text{O}_2$$
$$\text{Sr} + \text{I}_2 \longrightarrow \text{SrI}_2$$
$$\text{Sr} + \text{Se} \longrightarrow \text{SrSe}$$
$$2\text{Sr} + \text{O}_2 \longrightarrow 2\text{SrO}$$
$$2\text{Al} + 3\text{I}_2 \longrightarrow 2\text{AlI}_3$$
$$2\text{Al} + 3\text{Se} \longrightarrow \text{Al}_2\text{Se}_3$$
$$4\text{Al} + 3\text{O}_2 \longrightarrow 2\text{Al}_2\text{O}_3$$

5. The possible ways of distinguishing between the two include infrared spectroscopy by comparison of known compounds, a flame test that gives the characteristic yellow color for sodium (strontium has a red flame), or comparison of their solubilities in water. At 20 °C, NaCl dissolves to the extent of $\frac{35.7\text{ g}}{100\text{ mL}}$ compared with $\frac{53.8\text{ g}}{100\text{ mL}}$ for $SrCl_2$. Heating to 100 °C provides an easy test, since the solubility of NaCl is $\frac{39.12\text{ g}}{100\text{ mL}}$, but that of $SrCl_2$ is $\frac{100.8\text{ g}}{100\text{ mL}}$. Density determination on a solid is sometimes difficult, but there is enough difference (2.165 g/mL NaCl and 3.052 g/mL $SrCl_2$) that this method would be viable and perhaps the easiest and least expensive test to perform.

7. (a) $2\text{Sr}(s) + \text{O}_2(g) \longrightarrow 2\text{SrO}(s)$; (b) $\text{Sr}(s) + 2\text{HBr}(g) \longrightarrow \text{SrBr}_2(s) + \text{H}_2(g)$; (c) $\text{Sr}(s) + \text{H}_2(g) \longrightarrow \text{SrH}_2(s)$; (d) $6\text{Sr}(s) + \text{P}_4(s) \longrightarrow 2\text{Sr}_3\text{P}_2(s)$; (e) $\text{Sr}(s) + 2\text{H}_2\text{O}(l) \longrightarrow \text{Sr(OH)}_2(aq) + \text{H}_2(g)$

9. 11 lb

11. Yes, tin reacts with hydrochloric acid to produce hydrogen gas.

13. In $PbCl_2$, the bonding is ionic, as indicated by its melting point of 501 °C. In $PbCl_4$, the bonding is covalent, as evidenced by it being an unstable liquid at room temperature.

15. $2\text{CsCl}(l) + \text{Ca}(g) \xrightarrow[\text{tower}]{\text{countercurrent fractionating}} 2\text{Cs}(g) + \text{CaCl}_2(l)$

17. Cathode (reduction): $2\text{Li}^+ + 2\text{e}^- \longrightarrow 2\text{Li}(l)$; Anode (oxidation): $2\text{Cl}^- \longrightarrow \text{Cl}_2(g) + 2\text{e}^-$; Overall reaction: $2\text{Li}^+ + 2\text{Cl}^- \longrightarrow 2\text{Li}(l) + \text{Cl}_2(g)$

19. 0.5035 g H_2

21. Despite its reactivity, magnesium can be used in construction even when the magnesium is going to come in contact with a flame because a protective oxide coating is formed, preventing gross oxidation. Only if the metal is finely subdivided or present in a thin sheet will a high-intensity flame cause its rapid burning.

23. Extract from ore: $\text{AlO(OH)}(s) + \text{NaOH}(aq) + \text{H}_2\text{O}(l) \longrightarrow \text{Na}\left[\text{Al(OH)}_4\right](aq)$
Recover: $2\text{Na}\left[\text{Al(OH)}_4\right](s) + \text{H}_2\text{SO}_4(aq) \longrightarrow 2\text{Al(OH)}_3(s) + \text{Na}_2\text{SO}_4(aq) + 2\text{H}_2\text{O}(l)$
Sinter: $2\text{Al(OH)}_3(s) \longrightarrow \text{Al}_2\text{O}_3(s) + 3\text{H}_2\text{O}(g)$
Dissolve in $\text{Na}_3\text{AlF}_6(l)$ and electrolyze: $\text{Al}^{3+} + 3\text{e}^- \longrightarrow \text{Al}(s)$

25. 25.83%

27. 39 kg

29. (a) H_3BPH_3:

```
    H   H
    |   |
H — B — P — H
    |   |
    H   H
```

(b) $BF_4{}^-$:

```
[      ..      ]-
[     :F:      ]
[      |       ]
[ ..       ..  ]
[:F — B — F:   ]
[ ..       ..  ]
[      |       ]
[     :F:      ]
[      ..      ]
```

(c) BBr_3:

```
 ..
:Br:
 |     ..
 B — Br:
 |     ..
:Br:
 ..
```

(d) $B(CH_3)_3$:

```
     H
     |
H —  C — H    H
     |        |
     B ——— C — H
     |        |
H —  C — H    H
     |
     H
```

(e) $B(OH)_3$:

```
 H
 |
:O:
 |   ..
 B — O — H
 |   ..
:O:
 |
 H
```

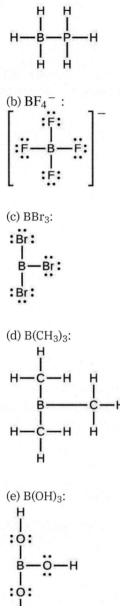

31. $1s^2 2s^2 2p^6 3s^2 3p^2 3d^0$.
33. (a) $(CH_3)_3SiH$: sp^3 bonding about Si; the structure is tetrahedral; (b) $SiO_4{}^{4-}$: sp^3 bonding about Si; the structure is tetrahedral; (c) Si_2H_6: sp^3 bonding about each Si; the structure is linear along the Si-Si bond; (d) $Si(OH)_4$: sp^3 bonding about Si; the structure is tetrahedral; (e) $SiF_6{}^{2-}$: sp^3d^2 bonding about Si; the structure is octahedral
35. (a) nonpolar; (b) nonpolar; (c) polar; (d) nonpolar; (e) polar
37. (a) tellurium dioxide or tellurium(IV) oxide; (b) antimony(III) sulfide; (c) germanium(IV) fluoride; (d) silane or silicon(IV) hydride; (e) germanium(IV) hydride
39. Boron has only *s* and *p* orbitals available, which can accommodate a maximum of four electron pairs. Unlike silicon, no *d* orbitals are available in boron.
41. (a) $\Delta H° = 87$ kJ; $\Delta G° = 44$ kJ; (b) $\Delta H° = -109.9$ kJ; $\Delta G° = -154.7$ kJ; (c) $\Delta H° = -510$ kJ; $\Delta G° = -601.5$ kJ
43. A mild solution of hydrofluoric acid would dissolve the silicate and would not harm the diamond.
45. In the N_2 molecule, the nitrogen atoms have an σ bond and two π bonds holding the two atoms together. The presence of three strong bonds makes N_2 a very stable molecule. Phosphorus is a third-period element, and as such, does not form π bonds efficiently; therefore, it must fulfill its bonding requirement by forming three σ bonds.

47. (a) H = 1+, C = 2+, and N = 3−; (b) O = 2+ and F = 1−; (c) As = 3+ and Cl = 1−

49. S < Cl < O < F

51. The electronegativity of the nonmetals is greater than that of hydrogen. Thus, the negative charge is better represented on the nonmetal, which has the greater tendency to attract electrons in the bond to itself.

53. Hydrogen has only one orbital with which to bond to other atoms. Consequently, only one two-electron bond can form.

55. 0.43 g H_2

57. (a) $Ca(OH)_2(aq) + CO_2(g) \longrightarrow CaCO_3(s) + H_2O(l)$; (b) $CaO(s) + SO_2(g) \longrightarrow CaSO_3(s)$; (c) $2NaHCO_3(s) + NaH_2PO_4(aq) \longrightarrow Na_3PO_4(aq) + 2CO_2(g) + 2H_2O(l)$

59. (a) NH^{2-}:

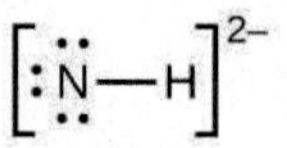

(b) N_2F_4:

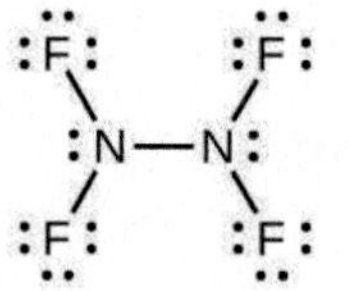

(c) $NH_2{}^-$:

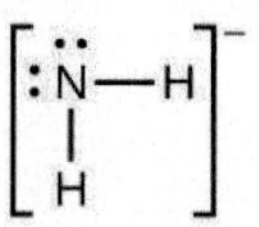

(d) NF_3:

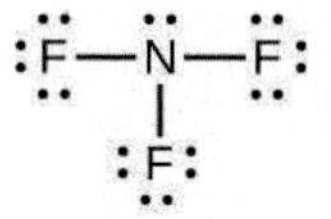

(e) $N_3{}^-$:

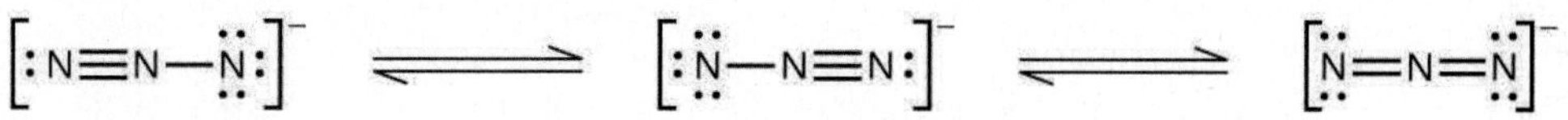

61. Ammonia acts as a Brønsted base because it readily accepts protons and as a Lewis base in that it has an electron pair to donate.

Brønsted base: $NH_3 + H_3O^+ \longrightarrow NH_4{}^+ + H_2O$

Lewis base: $2NH_3 + Ag^+ \longrightarrow [H_3N - Ag - NH_3]^+$

63. (a) NO_2:

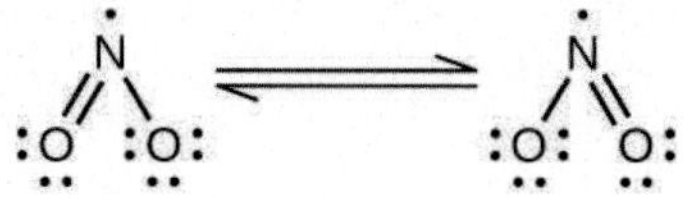

Nitrogen is sp^2 hybridized. The molecule has a bent geometry with an ONO bond angle of approximately 120°.

(b) $NO_2{}^-$:

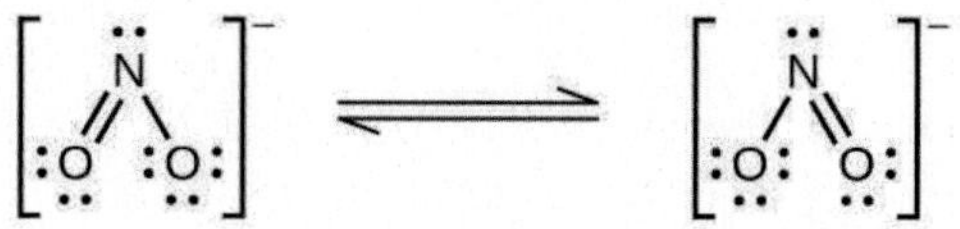

Nitrogen is sp^2 hybridized. The molecule has a bent geometry with an ONO bond angle slightly less than

120°.

(c) $NO_2{}^+$:

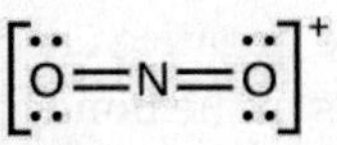

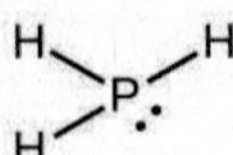

Nitrogen is *sp* hybridized. The molecule has a linear geometry with an ONO bond angle of 180°.

65. Nitrogen cannot form a NF_5 molecule because it does not have *d* orbitals to bond with the additional two fluorine atoms.

67. (a)

(b)

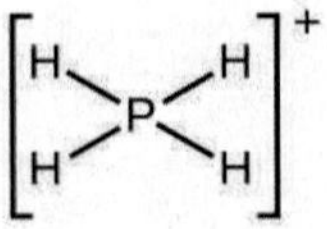

(c)

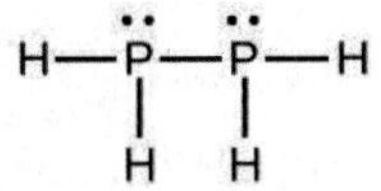

(d)

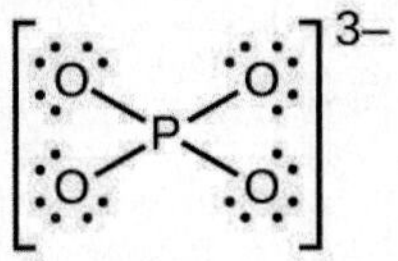

(e)

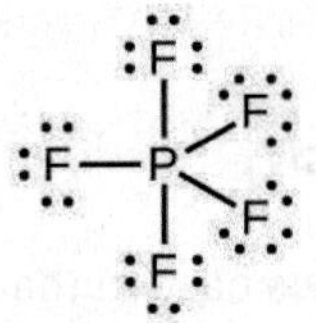

69. (a) $P_4(s) + 4Al(s) \longrightarrow 4AlP(s)$; (b) $P_4(s) + 12Na(s) \longrightarrow 4Na_3P(s)$; (c) $P_4(s) + 10F_2(g) \longrightarrow 4PF_5(l)$; (d) $P_4(s) + 6Cl_2(g) \longrightarrow 4PCl_3(l)$ or $P_4(s) + 10Cl_2(g) \longrightarrow 4PCl_5(l)$; (e) $P_4(s) + 3O_2(g) \longrightarrow P_4O_6(s)$ or $P_4(s) + 5O_2(g) \longrightarrow P_4O_{10}(s)$; (f) $P_4O_6(s) + 2O_2(g) \longrightarrow P_4O_{10}(s)$

71. 291 mL

73. 28 tons

75. (a)

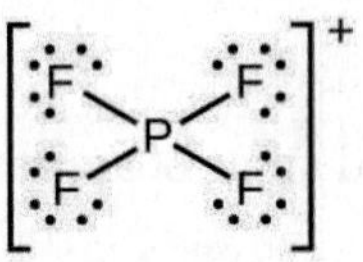

Tetrahedral

(b)

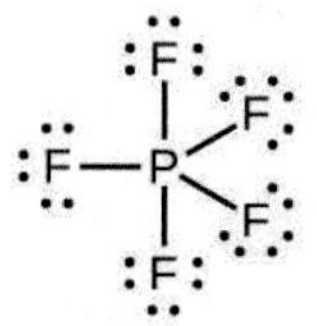

Trigonal bipyramid

(c)

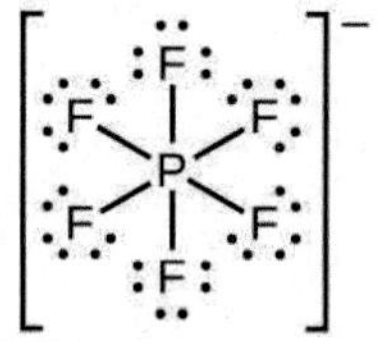

Octahedral

(d)

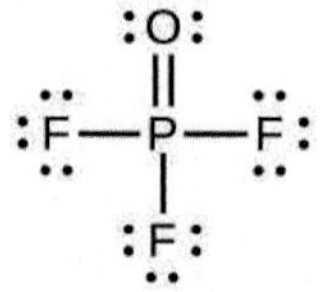

Tetrahedral

77. (a) P = 3+; (b) P = 5+; (c) P = 3+; (d) P = 5+; (e) P = 3−; (f) P = 5+

79. FrO_2

81. (a) $2Zn(s) + O_2(g) \longrightarrow 2ZnO(s)$; (b) $ZnCO_3(s) \longrightarrow ZnO(s) + CO_2(g)$; (c) $ZnCO_3(s) + 2CH_3COOH(aq) \longrightarrow Zn(CH_3COO)_2(aq) + CO_2(g) + H_2O(l)$; (d) $Zn(s) + 2HBr(aq) \longrightarrow ZnBr_2(aq) + H_2(g)$

83. $Al(OH)_3(s) + 3H^+(aq) \longrightarrow Al^{3+} + 3H_2O(l)$; $Al(OH)_3(s) + OH^- \longrightarrow \left[Al(OH)_4\right]^-(aq)$

85. (a) $Na_2O(s) + H_2O(l) \longrightarrow 2NaOH(aq)$; (b) $Cs_2CO_3(s) + 2HF(aq) \longrightarrow 2CsF(aq) + CO_2(g) + H_2O(l)$; (c) $Al_2O_3(s) + 6HClO_4(aq) \longrightarrow 2Al(ClO_4)_3(aq) + 3H_2O(l)$; (d) $Na_2CO_3(aq) + Ba(NO_3)_2(aq) \longrightarrow 2NaNO_3(aq) + BaCO_3(s)$; (e) $TiCl_4(l) + 4Na(s) \longrightarrow Ti(s) + 4NaCl(s)$

87. $HClO_4$ is the stronger acid because, in a series of oxyacids with similar formulas, the higher the electronegativity of the central atom, the stronger is the attraction of the central atom for the electrons of the oxygen(s). The stronger attraction of the oxygen electron results in a stronger attraction of oxygen for the electrons in the O-H bond, making the hydrogen more easily released. The weaker this bond, the stronger the acid.

89. As H_2SO_4 and H_2SeO_4 are both oxyacids and their central atoms both have the same oxidation number, the acid strength depends on the relative electronegativity of the central atom. As sulfur is more electronegative than selenium, H_2SO_4 is the stronger acid.

91. SO_2, sp^2 4+; SO_3, sp^2, 6+; H_2SO_4, sp^3, 6+

93. SF_6: S = 6+; SO_2F_2: S = 6+; KHS: S = 2−

95. Sulfur is able to form double bonds only at high temperatures (substantially endothermic conditions), which is not the case for oxygen.

97. There are many possible answers including: $Cu(s) + 2H_2SO_4(l) \longrightarrow CuSO_4(aq) + SO_2(g) + 2H_2O(l)$ and $C(s) + 2H_2SO_4(l) \longrightarrow CO_2(g) + 2SO_2(g) + 2H_2O(l)$

99. 5.1×10^4 g

101. $SnCl_4$ is not a salt because it is covalently bonded. A salt must have ionic bonds.

103. In oxyacids with similar formulas, the acid strength increases as the electronegativity of the central atom increases. $HClO_3$ is stronger than $HBrO_3$; Cl is more electronegative than Br.

105. (a)

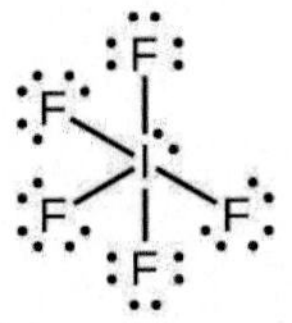

Square pyramidal

(b)

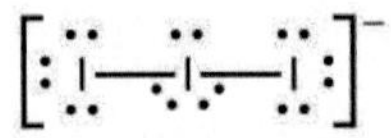

Linear

(c)

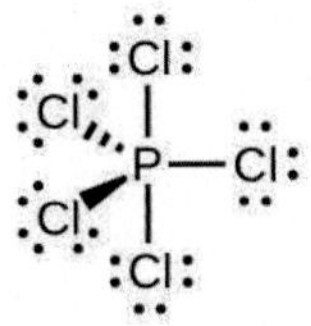

Trigonal bipyramidal

(d)

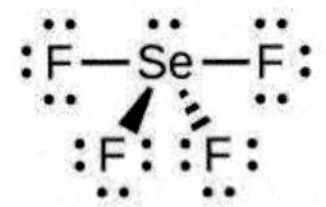

Seesaw

(e)

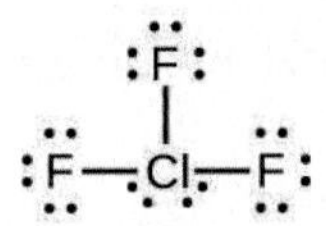

T-shaped

107. (a) bromine trifluoride; (b) sodium bromate; (c) phosphorus pentabromide; (d) sodium perchlorate; (e) potassium hypochlorite

109. (a) I: 7+; (b) I: 7+; (c) Cl: 4+; (d) I: 3+; Cl: 1−; (e) F: 0

111. (a) sp^3d hybridized; (b) sp^3d^2 hybridized; (c) sp^3 hybridized; (d) sp^3 hybridized; (e) sp^3d^2 hybridized;

113. (a) nonpolar; (b) nonpolar; (c) polar; (d) nonpolar; (e) polar

115. The empirical formula is XeF_6, and the balanced reactions are:

$$\text{Xe}(g) + 3\text{F}_2(g) \xrightarrow{\Delta} \text{XeF}_6(s)$$
$$\text{XeF}_6(s) + 3\text{H}_2(g) \longrightarrow 6\text{HF}(g) + \text{Xe}(g)$$

Chapter 19

1. (a) Sc: $[Ar]4s^23d^1$; (b) Ti: $[Ar]4s^23d^2$; (c) Cr: $[Ar]4s^13d^5$; (d) Fe: $[Ar]4s^23d^6$; (e) Ru: $[Kr]5s^24d^6$

3. (a) La: $[Xe]6s^25d^1$, La^{3+}: [Xe]; (b) Sm: $[Xe]6s^24f^6$, Sm^{3+}: $[Xe]4f^5$; (c) Lu: $[Xe]6s^24f^{14}5d^1$, Lu^{3+}: $[Xe]4f^{14}$

5. Al is used because it is the strongest reducing agent and the only option listed that can provide sufficient driving force to convert La(III) into La.

7. Mo

9. The $CaSiO_3$ slag is less dense than the molten iron, so it can easily be separated. Also, the floating slag layer creates a barrier that prevents the molten iron from exposure to O_2, which would oxidize the Fe back to Fe_2O_3.

11. 2.57%

13. 0.167 V

15. $E° = -0.6$ V, $E°$ is negative so this reduction is not spontaneous. $E° = +1.1$ V

17. (a) $Fe(s) + 2H_3O^+(aq) + SO_4{}^{2-}(aq) \longrightarrow Fe^{2+}(aq) + SO_4{}^{2-}(aq) + H_2(g) + 2H_2O(l)$; (b) $FeCl_3(aq) + 3Na^+(aq) + 3OH^-(aq) \longrightarrow Fe(OH)_3(s) + 3Na^+(aq) + 3Cl^+(aq)$; (c) $Mn(OH)_2(s) + 2H_3O^+(aq) + 2Br^-(aq) \longrightarrow Mn^{2+}(aq) + 2Br^-(aq) + 4H_2O(l)$; (d) $4Cr(s) + 3O_2(g) \longrightarrow 2Cr_2O_3(s)$; (e) $Mn_2O_3(s) + 6H_3O^+(aq) + 6Cl^-(aq) \longrightarrow 2MnCl_3(s) + 9H_2O(l)$; (f) $Ti(s) + xsF_2(g) \longrightarrow TiF_4(g)$

19. (a) $Cr_2(SO_4)_3(aq) + 2Zn(s) + 2H_3O^+(aq) \longrightarrow 2Zn^{2+}(aq) + H_2(g) + 2H_2O(l) + 2Cr^{2+}(aq) + 3SO_4{}^{2-}(aq)$; (b) $4TiCl_3(s) + CrO_4{}^{2-}(aq) + 8H^+(aq) \longrightarrow 4Ti^{4+}(aq) + Cr(s) + 4H_2O(l) + 12Cl^-(aq)$; (c) In acid solution between pH 2 and pH 6, $CrO_4{}^{2-}$ forms $HCrO_4{}^-$, which is in equilibrium with dichromate ion. The reaction is $2HCrO_4{}^-(aq) \longrightarrow Cr_2O_7{}^{2-}(aq) + H_2O(l)$. At other acidic pHs, the reaction is $3Cr^{2+}(aq) + CrO_4{}^{2-}(aq) + 8H_3O^+(aq) \longrightarrow 4Cr^{3+}(aq) + 12H_2O(l)$; (d) $8CrO_3(s) + 9Mn(s) \xrightarrow{\Delta} 4Cr_2O_3(s) + 3Mn_3O_4(s)$; (e) $CrO(s) + 2H_3O^+(aq) + 2NO_3{}^-(aq) \longrightarrow Cr^{2+}(aq) + 2NO_3{}^-(aq) + 3H_2O(l)$; (f) $CrCl_3(s) + 3NaOH(aq) \longrightarrow Cr(OH)_3(s) + 3Na^+(aq) + 3Cl^-(aq)$

21. (a) $3Fe(s) + 4H_2O(g) \longrightarrow Fe_3O_4(s) + 4H_2(g)$; (b) $3NaOH(aq) + Fe(NO_3)_3(aq) \xrightarrow{H_2O} Fe(OH)_3(s) + 3Na^+(aq) + 3NO_3{}^-(aq)$; (c) $MnO^{4-} + 5Fe^{2+} + 8H^+ \longrightarrow Mn^{2+} + 5Fe_3 + 4H_2O$; (d) $Fe(s) + 2H_3O^+(aq) + SO_4{}^{2-}(aq) \longrightarrow Fe^{2+}(aq) + SO_4{}^{2-}(aq) + H_2(g) + 2H_2O(l)$; (e) $4Fe^{2+}(aq) + O_2(g) + 4HNO_3(aq) \longrightarrow 4Fe^{3+}(aq) + 2H_2O(l) + 4NO_3{}^-(aq)$; (f) $FeCO_3(s) + 2HClO_4(aq) \longrightarrow Fe(ClO_4)_2(aq) + H_2O(l) + CO_2(g)$; (g) $3Fe(s) + 2O_2(g) \xrightarrow{\Delta} Fe_3O_4(s)$

23. As CN^- is added,

$Ag^+(aq) + CN^-(aq) \longrightarrow AgCN(s)$

As more CN^- is added,

$Ag^+(aq) + 2CN^-(aq) \longrightarrow [Ag(CN)_2]^-(aq)$

$AgCN(s) + CN^-(aq) \longrightarrow [Ag(CN)_2]^-(aq)$

25. (a) Sc^{3+}; (b) Ti^{4+}; (c) V^{5+}; (d) Cr^{6+}; (e) Mn^{4+}; (f) Fe^{2+} and Fe^{3+}; (g) Co^{2+} and Co^{3+}; (h) Ni^{2+}; (i) Cu^+

27. (a) 4, $[Zn(OH)_4]^{2-}$; (b) 6, $[Pd(CN)_6]^{2-}$; (c) 2, $[AuCl_2]^-$; (d) 4, $[Pt(NH_3)_2Cl_2]$; (e) 6, $K[Cr(NH_3)_2Cl_4]$; (f) 6, $[Co(NH_3)_6][Cr(CN)_6]$; (g) 6, $[Co(en)_2Br_2]NO_3$

29. (a) $[Pt(H_2O)_2Br_2]$:

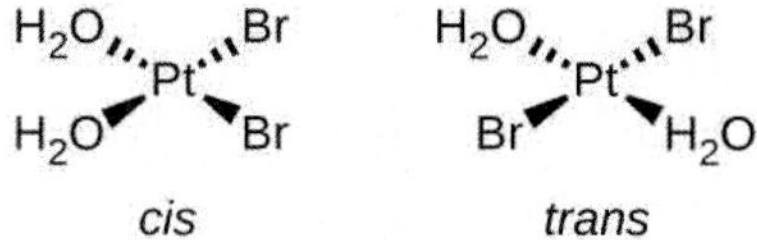

(b) $[Pt(NH_3)(py)(Cl)(Br)]$:

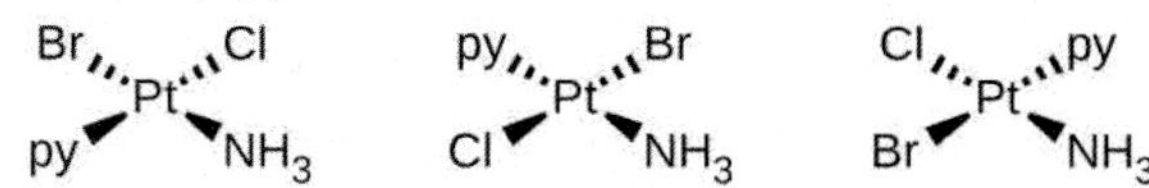

(c) $[Zn(NH_3)_3Cl]^+$:

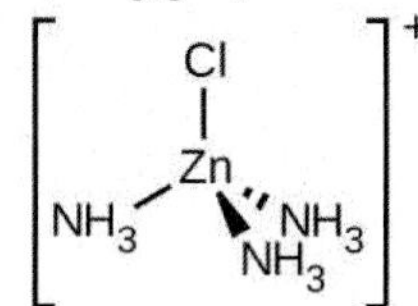

(d) $[Pt(NH_3)_3Cl]^+$:

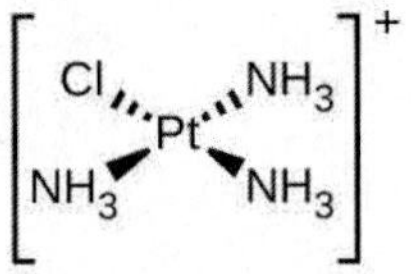

(e) $[Ni(H_2O)_4Cl_2]$:

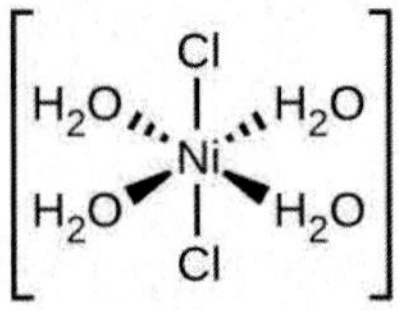

trans

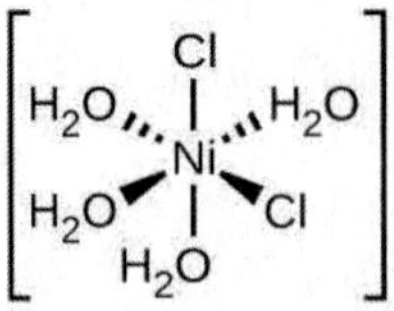

cis

(f) $[Co(C_2O_4)_2Cl_2]^{3-}$:

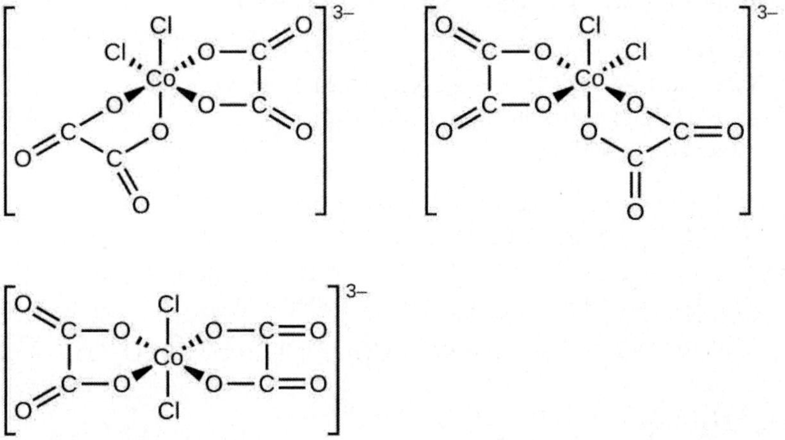

31. (a) tricarbonatocobaltate(III) ion; (b) tetraaminecopper(II) ion; (c) tetraaminedibromocobalt(III) sulfate; (d) tetraamineplatinum(II) tetrachloroplatinate(II); (e) *tris*-(ethylenediamine)chromium(III) nitrate; (f) diaminedibromopalladium(II); (g) potassium pentachlorocuprate(II); (h) diaminedichlorozinc(II)

33. (a) none; (b) none; (c) The two Cl ligands can be *cis* or *trans*. When they are *cis*, there will also be an optical isomer.

35.

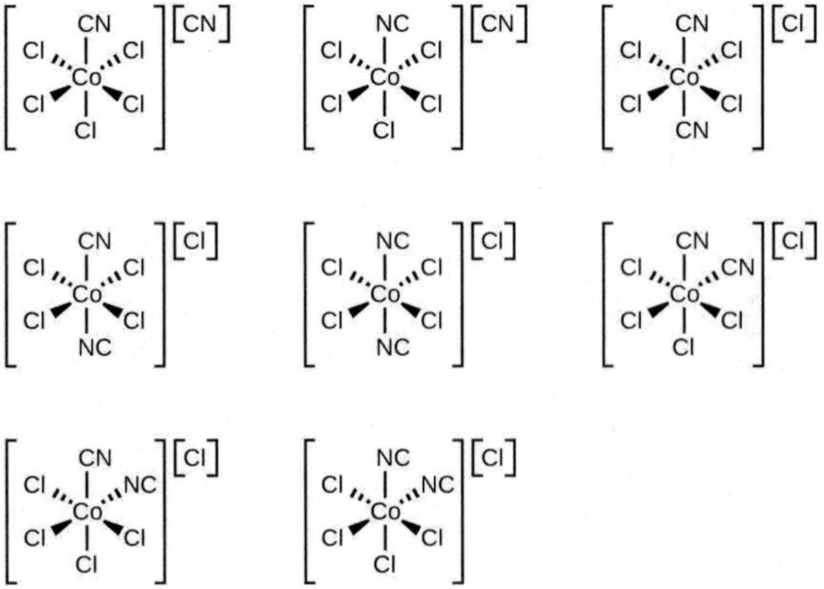

37.

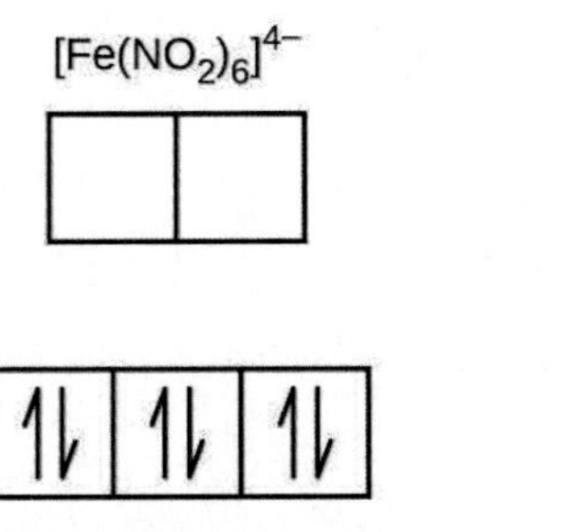

$[FeF_6]^{3-}$

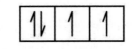

Low spin, diamagnetic, $P<\Delta_{oct}$ High spin, paramagnetic, $P>\Delta_{oct}$

39. $[Co(H_2O)_6]Cl_2$ with three unpaired electrons.

41. (a) 4; (b) 2; (c) 1; (d) 5; (e) 0

43. (a) $[Fe(CN)_6]^{4-}$; (b) $[Co(NH_3)_6]^{3+}$; (c) $[Mn(CN)_6]^{4-}$

45. The complex does not have any unpaired electrons. The complex does not have any geometric isomers, but the mirror image is nonsuperimposable, so it has an optical isomer.

47. No. Au^+ has a complete 5*d* sublevel.

Chapter 20

1. There are several sets of answers; one is:

(a) C_5H_{12}

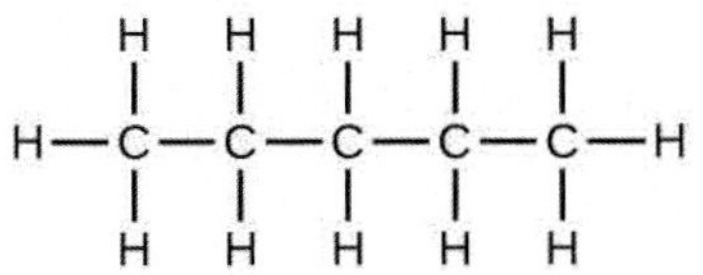

```
    H   H   H   H   H
    |   |   |   |   |
H — C — C — C — C — C — H
    |   |   |   |   |
    H   H   H   H   H
```

(b) C_5H_{10}

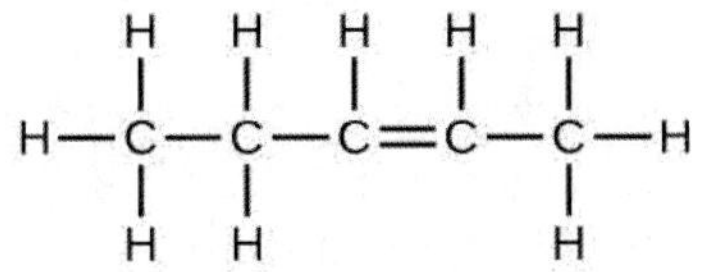

```
    H   H   H   H   H
    |   |   |   |   |
H — C — C — C = C — C — H
    |   |           |
    H   H           H
```

(c) C_5H_8

```
    H   H           H
    |   |           |
H — C — C — C ≡ C — C — H
    |   |           |
    H   H           H
```

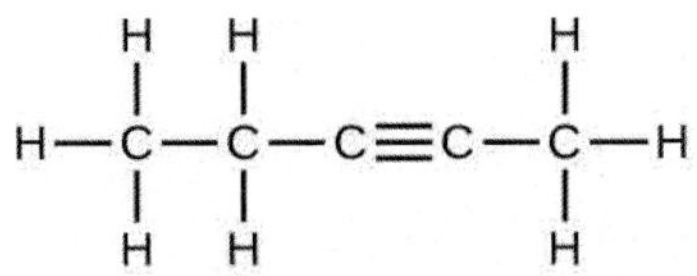

3. Both reactions result in bromine being incorporated into the structure of the product. The difference is the way in which that incorporation takes place. In the saturated hydrocarbon, an existing C–H bond is broken, and a bond between the C and the Br can then be formed. In the unsaturated hydrocarbon, the only bond broken in the hydrocarbon is the π bond whose electrons can be used to form a bond to one of the bromine atoms in Br_2 (the electrons from the Br–Br bond form the other C–Br bond on the other carbon that was part of the π bond in the starting unsaturated hydrocarbon).

5. Unbranched alkanes have free rotation about the C–C bonds, yielding all orientations of the substituents about these bonds equivalent, interchangeable by rotation. In the unbranched alkenes, the inability to rotate about the $C = C$ bond results in fixed (unchanging) substituent orientations, thus permitting different isomers. Since these concepts pertain to phenomena at the molecular level, this explanation involves the microscopic domain.

7. They are the same compound because each is a saturated hydrocarbon containing an unbranched chain of six carbon atoms.

9. (a) C_6H_{14}

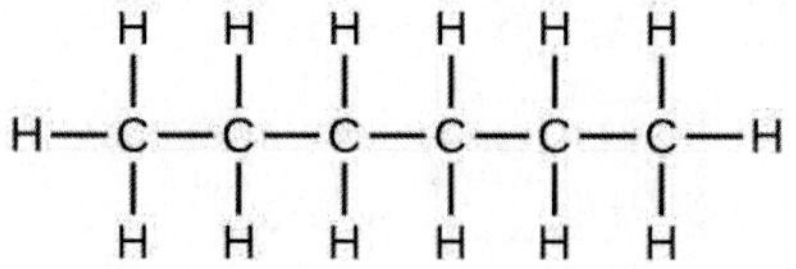

(b) C_6H_{14}

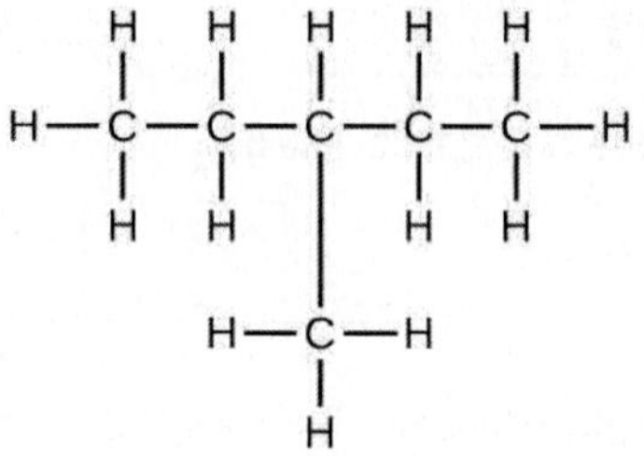

(c) C_6H_{12}

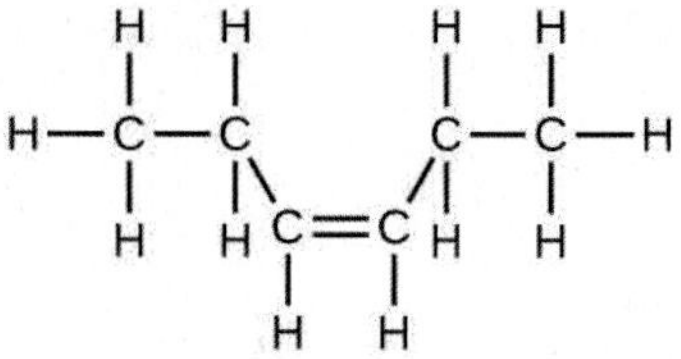

(d) C_6H_{12}

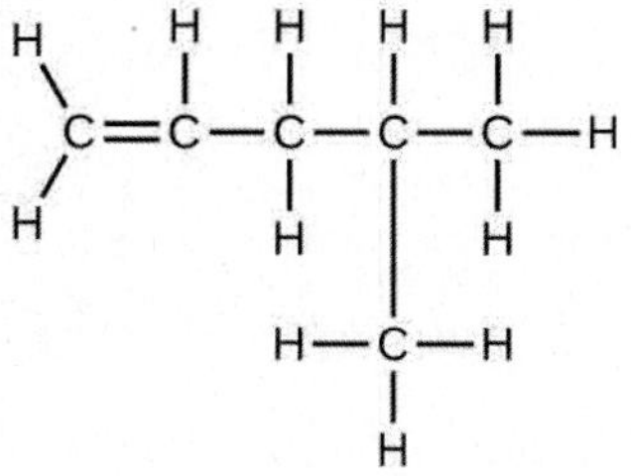

(e) C_6H_{10}

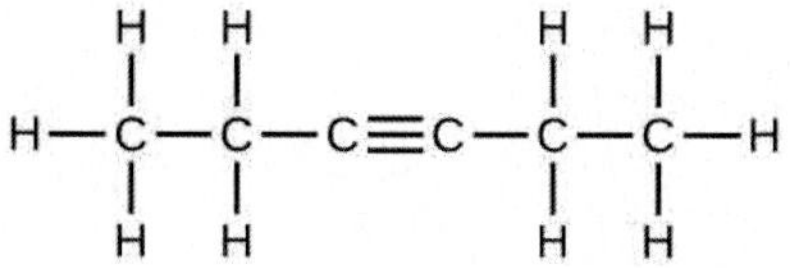

(f) C_6H_{10}

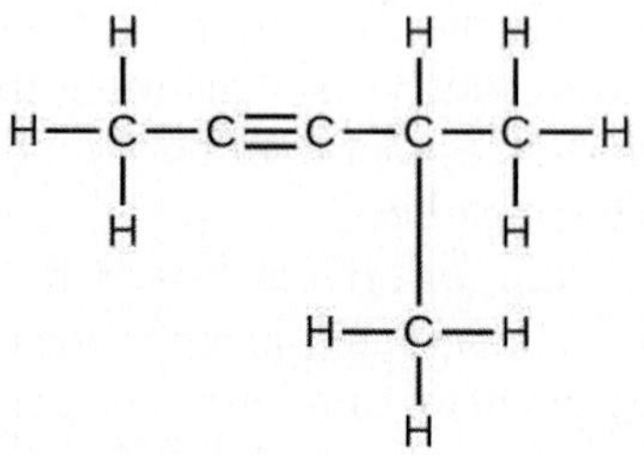

11. (a) 2,2-dibromobutane; (b) 2-chloro-2-methylpropane; (c) 2-methylbutane; (d) 1-butyne; (e) 4-fluoro-4-methyl-1-octyne; (f) *trans*-1-chloropropene; (g) 4-methyl-1-pentene

13.

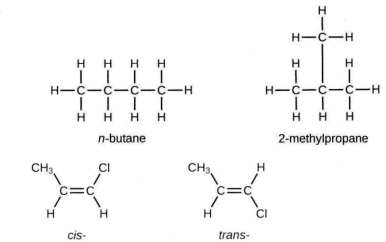

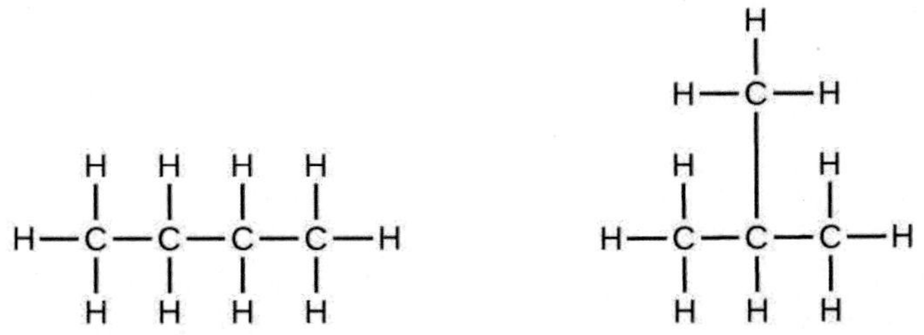

n-butane

2-methylpropane

15.

cis-

trans-

17. (a) 2,2,4-trimethylpentane; (b) 2,2,3-trimethylpentane, 2,3,4-trimethylpentane, and 2,3,3-trimethylpentane:

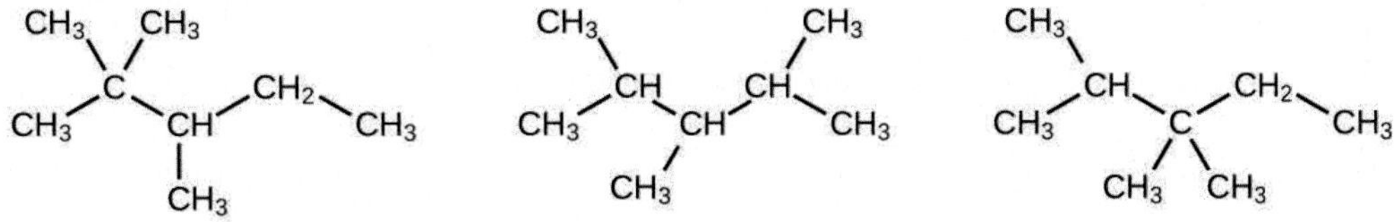

19.

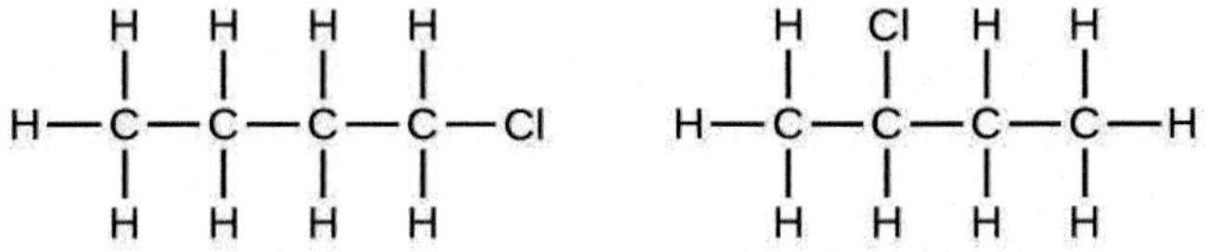

1-chlorobutane

2-chlorobutane

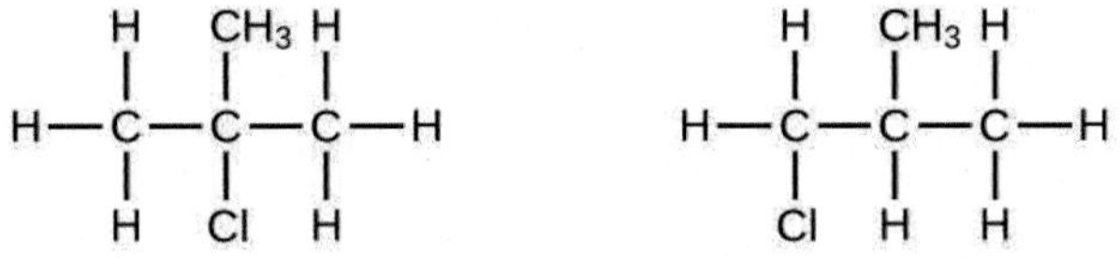

2-chloro-2-methylpropane

1-chloro-2-methylpropane
(1-chloro-2-methylpropane)

21. In the following, the carbon backbone and the appropriate number of hydrogen atoms are shown in condensed form:

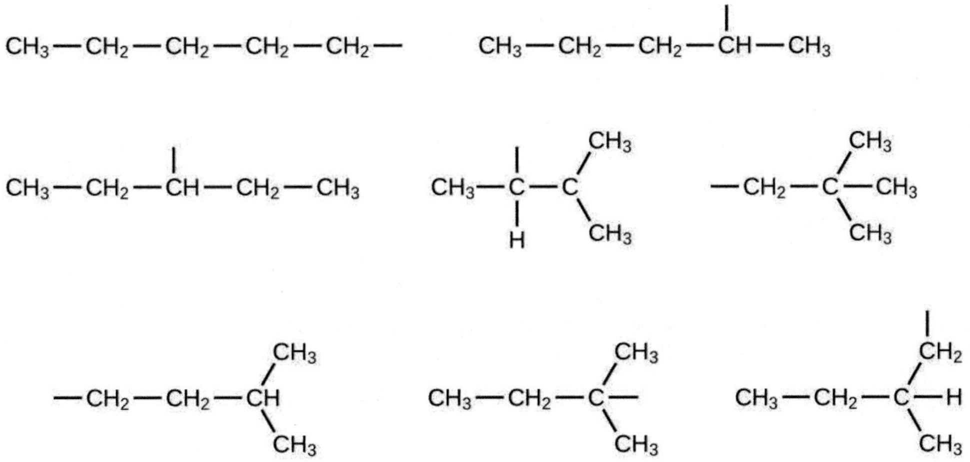

23.

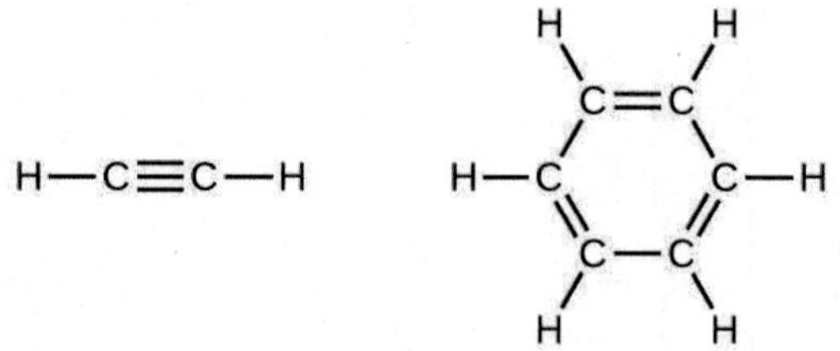

In acetylene, the bonding uses *sp* hybrids on carbon atoms and *s* orbitals on hydrogen atoms. In benzene, the carbon atoms are sp^2 hybridized.

25. (a) $CH \equiv CCH_2CH_3 + 2I_2 \longrightarrow CHI_2CI_2CH_2CH_3$

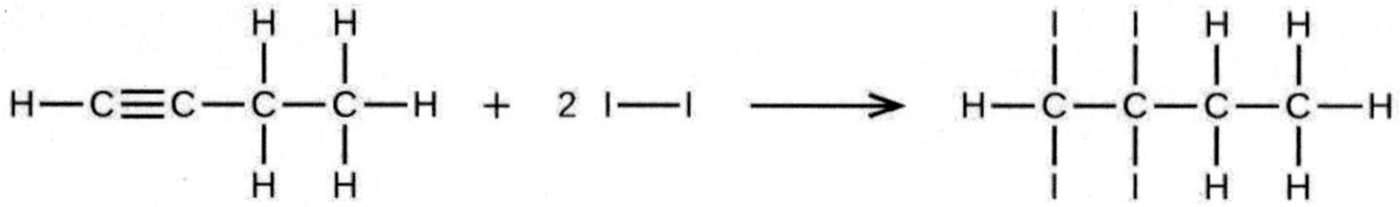

(b) $CH_3CH_2CH_2CH_2CH_3 + 8O_2 \longrightarrow 5CO_2 + 6H_2O$

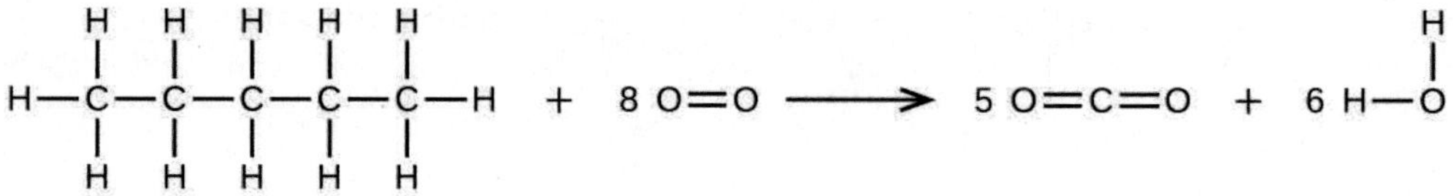

27. 65.2 g

29. 9.328×10^2 kg

31. (a) ethyl alcohol, ethanol: CH_3CH_2OH; (b) methyl alcohol, methanol: CH_3OH; (c) ethylene glycol, ethanediol: $HOCH_2CH_2OH$; (d) isopropyl alcohol, 2-propanol: $CH_3CH(OH)CH_3$; (e) glycerine, 1,2,3-trihydroxypropane: $HOCH_2CH(OH)CH_2OH$

33. (a) 1-ethoxybutane, butyl ethyl ether; (b) 1-ethoxypropane, ethyl propyl ether; (c) 1-methoxypropane, methyl propyl ether

35. $HOCH_2CH_2OH$, two alcohol groups; CH_3OCH_2OH, ether and alcohol groups

37. (a)

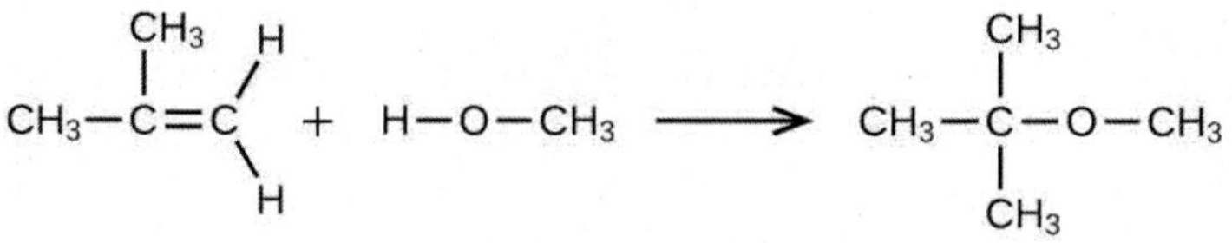

(b) 4.593×10^2 L

39. (a) $CH_3CH = CHCH_3 + H_2O \longrightarrow CH_3CH_2CH(OH)CH_3$

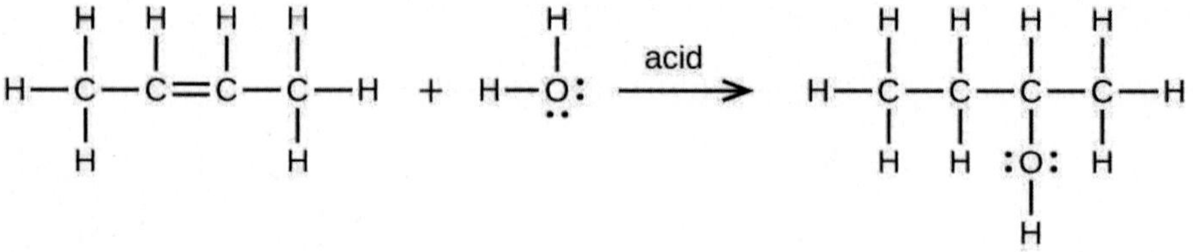

(b) $CH_3CH_2OH \longrightarrow CH_2 = CH_2 + H_2O$

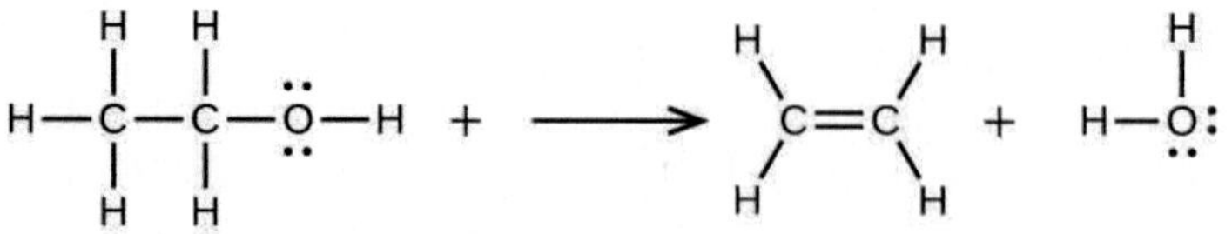

41. (a)

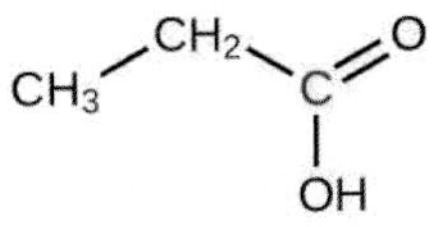

(b)

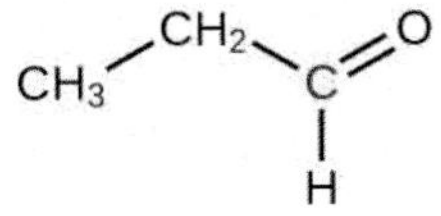

(c)

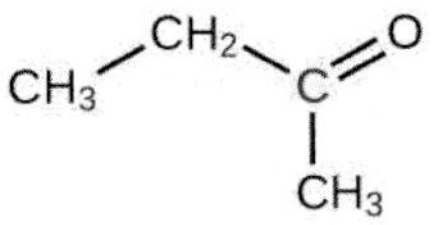

43. A ketone contains a group bonded to two additional carbon atoms; thus, a minimum of three carbon atoms are needed.

45. Since they are both carboxylic acids, they each contain the –COOH functional group and its characteristics. The difference is the hydrocarbon chain in a saturated fatty acid contains no double or triple bonds, whereas the hydrocarbon chain in an unsaturated fatty acid contains one or more multiple bonds.

47. (a) $CH_3CH(OH)CH_3$: all carbons are tetrahedral; (b) CH_3COCH_3: the end carbons are tetrahedral and the central carbon is trigonal planar; (c) CH_3OCH_3: all are tetrahedral; (d) CH_3COOH: the methyl carbon is tetrahedral and the acid carbon is trigonal planar; (e) $CH_3CH_2CH_2CH(CH_3)CHCH_2$: all are tetrahedral except the right-most two carbons, which are trigonal planar

49.

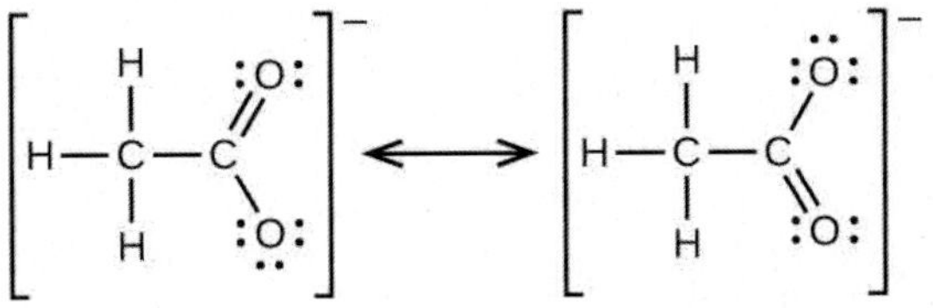

51. (a) $CH_3CH_2CH_2CH_2OH + CH_3C(O)OH \longrightarrow CH_3C(O)OCH_2CH_2CH_2CH_3 + H_2O$:

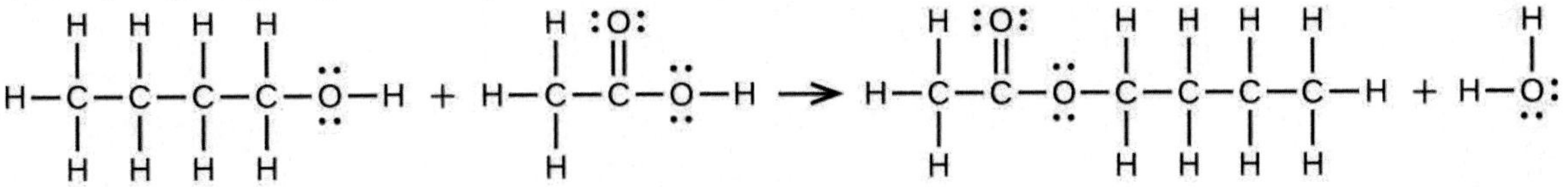

(b) $2CH_3CH_2COOH + CaCO_3 \longrightarrow (CH_3CH_2COO)_2Ca + CO_2 + H_2O$:

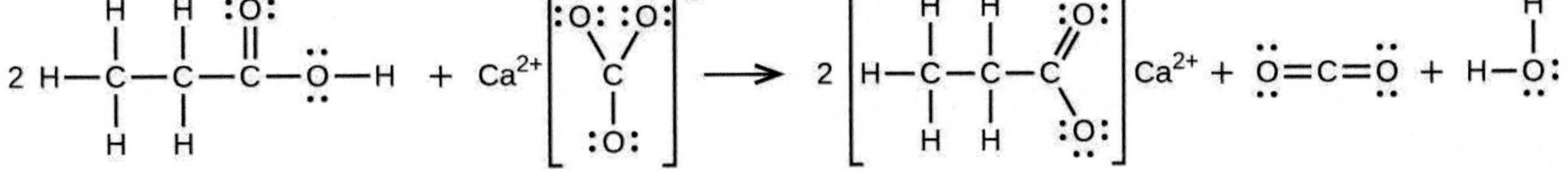

53.

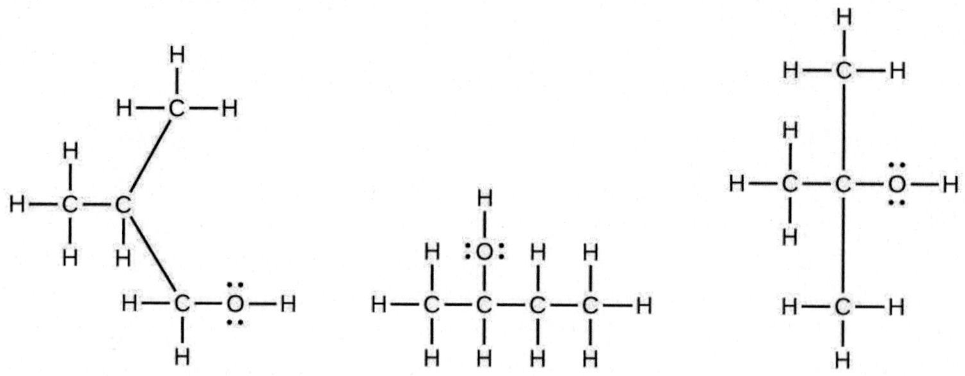

Compound A

Compound B

Compound C

55. Trimethyl amine: trigonal pyramidal, sp^3; trimethyl ammonium ion: tetrahedral, sp^3

57.

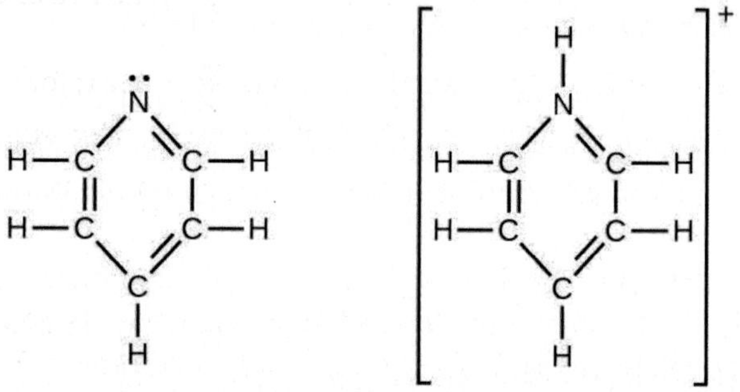

pyridine,
trigonal planar, sp^2
(nonlinear or bent)

pyridinium ion,
trigonal planar, sp^2
(trigonal planar)

59. $CH_3NH_2 + H_3O^+ \longrightarrow CH_3NH_3{}^+ + H_2O$

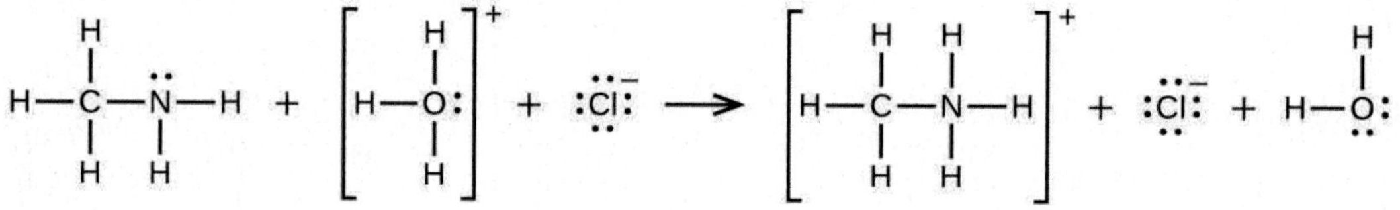

61. $CH_3\underline{C}H = \underline{C}HCH_3(sp^2) + Cl \longrightarrow CH_3\underline{C}H(Cl)H(Cl)CH_3(sp^3)$; $2\underline{C}_6H_6(sp^2) + 15O_2 \longrightarrow 12\underline{C}O_2(sp) + 6H_2O$

63. The carbon in $CO_3{}^{2-}$, initially at sp^2, changes hybridization to sp in CO_2.

Chapter 21

1. (a) sodium-24; (b) aluminum-29; (c) krypton-73; (d) iridium-194

3. (a) ${}^{34}_{14}Si$; (b) ${}^{36}_{15}P$; (c) ${}^{57}_{25}Mn$; (d) ${}^{121}_{56}Ba$

5. (a) ${}^{45}_{25}Mn^{+1}$; (b) ${}^{69}_{45}Rh^{+2}$; (c) ${}^{142}_{53}I^{-1}$; (d) ${}^{243}_{97}Bk$

7. Nuclear reactions usually change one type of nucleus into another; chemical changes rearrange atoms. Nuclear reactions involve much larger energies than chemical reactions and have measureable mass changes.

9. (a), (b), (c), (d), and (e)

11. (a) A nucleon is any particle contained in the nucleus of the atom, so it can refer to protons and neutrons. (b) An α particle is one product of natural radioactivity and is the nucleus of a helium atom. (c) A β particle is a product of natural radioactivity and is a high-speed electron. (d) A positron is a particle with the same mass as an electron but with a positive charge. (e) Gamma rays compose electromagnetic radiation of high energy and short wavelength. (f) Nuclide is a term used when referring to a single type of nucleus. (g) The mass number is the sum of the number of protons and the number of neutrons in an element. (h) The atomic number is the number of protons in the nucleus of an element.

13. (a) $^{27}_{13}Al + ^{4}_{2}He \longrightarrow ^{30}_{15}P + ^{1}_{0}n$; (b) $^{239}_{94}Pu + ^{4}_{2}He \longrightarrow ^{242}_{96}Cm + ^{1}_{0}n$; (c) $^{14}_{7}N + ^{4}_{2}He \longrightarrow ^{17}_{8}O + ^{1}_{1}H$; (d) $^{235}_{92}U \longrightarrow ^{96}_{37}Rb + ^{135}_{55}Cs + 4^{1}_{0}n$

15. (a) $^{14}_{7}N + ^{4}_{2}He \longrightarrow ^{17}_{8}O + ^{1}_{1}H$; (b) $^{14}_{7}C + ^{1}_{0}n \longrightarrow ^{14}_{6}C + ^{1}_{1}H$; (c) $^{232}_{90}Th + ^{1}_{0}n \longrightarrow ^{233}_{90}Th$; (d) $^{238}_{92}U + ^{2}_{1}H \longrightarrow ^{239}_{92}U + ^{1}_{1}H$

17. (a) 148.8 MeV per atom; (b) 7.808 MeV/nucleon

19. α (helium nuclei), β (electrons), β⁺ (positrons), and η (neutrons) may be emitted from a radioactive element, all of which are particles; γ rays also may be emitted.

21. (a) conversion of a neutron to a proton: $^{1}_{0}n \longrightarrow ^{1}_{1}p + ^{0}_{+1}e$; (b) conversion of a proton to a neutron; the positron has the same mass as an electron and the same magnitude of positive charge as the electron has negative charge; when the n:p ratio of a nucleus is too low, a proton is converted into a neutron with the emission of a positron: $^{1}_{1}p \longrightarrow ^{1}_{0}n + ^{0}_{+1}e$; (c) In a proton-rich nucleus, an inner atomic electron can be absorbed. In simplest form, this changes a proton into a neutron: $^{1}_{1}p + ^{0}_{-1}e \longrightarrow ^{1}_{0}p$

23. The electron pulled into the nucleus was most likely found in the 1*s* orbital. As an electron falls from a higher energy level to replace it, the difference in the energy of the replacement electron in its two energy levels is given off as an X-ray.

25. Manganese-51 is most likely to decay by positron emission. The n:p ratio for Cr-53 is $\frac{29}{24}$ = 1.21; for Mn-51, it is $\frac{26}{25}$ = 1.04; for Fe-59, it is $\frac{33}{26}$ = 1.27. Positron decay occurs when the n:p ratio is low. Mn-51 has the lowest n:p ratio and therefore is most likely to decay by positron emission. Besides, $^{53}_{24}Cr$ is a stable isotope, and $^{59}_{26}Fe$ decays by beta emission.

27. (a) β decay; (b) α decay; (c) positron emission; (d) β decay; (e) α decay

29. $^{238}_{92}U \longrightarrow ^{234}_{90}Th + ^{4}_{2}He$; $^{234}_{90}Th \longrightarrow ^{234}_{91}Pa + ^{0}_{-1}e$; $^{234}_{91}Pa \longrightarrow ^{234}_{92}U + ^{0}_{-1}e$; $^{234}_{92}U \longrightarrow ^{230}_{90}Th + ^{4}_{2}He$ $^{230}_{90}Th \longrightarrow ^{226}_{88}Ra + ^{4}_{2}He$ $^{226}_{88}Ra \longrightarrow ^{222}_{86}Rn + ^{4}_{2}He$; $^{222}_{86}Rn \longrightarrow ^{218}_{84}Po + ^{4}_{2}He$

31. Half-life is the time required for half the atoms in a sample to decay. Example (answers may vary): For C-14, the half-life is 5770 years. A 10-g sample of C-14 would contain 5 g of C-14 after 5770 years; a 0.20-g sample of C-14 would contain 0.10 g after 5770 years.

33. $\left(\frac{1}{2}\right)^{0.04} = 0.973$ or 97.3%

35. 2×10^3 y

37. 0.12 h^{-1}

39. (a) 3.8 billion years; (b) The rock would be younger than the age calculated in part (a). If Sr was originally in the rock, the amount produced by radioactive decay would equal the present amount minus the initial amount. As this amount would be smaller than the amount used to calculate the age of the rock and the age is proportional to the amount of Sr, the rock would be younger.

41. $c = 0$; This shows that no Pu-239 could remain since the formation of the earth. Consequently, the plutonium now present could not have been formed with the uranium.

43. 17.5 MeV

45. (a) $^{212}_{83}Bi \longrightarrow ^{212}_{84}Po + ^{0}_{-1}e$; (b) $^{8}_{5}B \longrightarrow ^{8}_{4}Be + ^{0}_{-1}e$; (c) $^{238}_{92}U + ^{1}_{0}n \longrightarrow ^{239}_{93}Np + ^{0}_{-1}Np$, $^{239}_{93}Np \longrightarrow ^{239}_{94}Pu + ^{0}_{-1}e$; (d) $^{90}_{38}Sr \longrightarrow ^{90}_{39}Y + ^{0}_{-1}e$

47. (a) $^{241}_{95}Am + ^{4}_{2}He \longrightarrow ^{244}_{97}Bk + ^{1}_{0}n$; (b) $^{239}_{94}Pu + 15^{1}_{0}n \longrightarrow ^{254}_{100}Fm + 6^{0}_{-1}e$; (c) $^{250}_{98}Cf + ^{11}_{5}B \longrightarrow ^{257}_{103}Lr + 4^{1}_{0}n$; (d) $^{249}_{98}Cf + ^{15}_{7}N \longrightarrow ^{260}_{105}Db + 4^{1}_{0}n$

49. Two nuclei must collide for fusion to occur. High temperatures are required to give the nuclei enough kinetic energy to overcome the very strong repulsion resulting from their positive charges.

51. A nuclear reactor consists of the following:
    1. A nuclear fuel. A fissionable isotope must be present in large enough quantities to sustain a controlled chain reaction. The radioactive isotope is contained in tubes called fuel rods.
    2. A moderator. A moderator slows neutrons produced by nuclear reactions so that they can be absorbed by the fuel and cause additional nuclear reactions.
    3. A coolant. The coolant carries heat from the fission reaction to an external boiler and turbine where it is transformed into electricity.
    4. A control system. The control system consists of control rods placed between fuel rods to absorb neutrons and is used to adjust the number of neutrons and keep the rate of the chain reaction at a safe

level.

5. A shield and containment system. The function of this component is to protect workers from radiation produced by the nuclear reactions and to withstand the high pressures resulting from high-temperature reactions.

53. The fission of uranium generates heat, which is carried to an external steam generator (boiler). The resulting steam turns a turbine that powers an electrical generator.

55. Introduction of either radioactive Ag^+ or radioactive Cl^- into the solution containing the stated reaction, with subsequent time given for equilibration, will produce a radioactive precipitate that was originally devoid of radiation.

57. (a) ${}^{133}_{53}\text{I} \longrightarrow {}^{133}_{54}\text{Xe} + {}^{0}_{-1}\text{e}$; (b) 37.6 days

59. Alpha particles can be stopped by very thin shielding but have much stronger ionizing potential than beta particles, X-rays, and γ-rays. When inhaled, there is no protective skin covering the cells of the lungs, making it possible to damage the DNA in those cells and cause cancer.

61. (a) 7.64×10^9 Bq; (b) 2.06×10^{-2} Ci

INDEX

Symbols

Δoct 963
π* antibonding molecular orbital 396
σs molecular orbital 395
σs*σs* molecular orbital 396

A

absolute zero 427
accuracy 40
acid 169
acid anhydrides 879
acid ionization 694
acid-base indicators 734
acid-base reaction 169
acid-ionization constant, Ka 703
acidic 697
acids 96
actinide series 936
actinides 87
actinoid series 936
activated complex 626
activation energy (Ea) 626
active electrodes 823
activity 582
actual yield 188
addition reaction 991
adhesive forces 490
Alcohols 995
aldehydes 999
alkali metals 87
Alkaline batteries 835
alkaline earth metals 87, 861
Alkanes 978
alkenes 988
alkyl group 984
alkynes 992
Allotropes 866
alloys 548
Alpha (α) decay 1032
Alpha particles 1029
alpha particles (α particles) 69
Amides 1008
Amines 1005
Amontons's law 426
amorphous 872
amorphous solids 510
amphiphilic 587
amphiprotic 695
amphoteric 695
amplitude 259
analyte 191
anion 72
anode 821
antibonding orbitals 396
antimatter 1029
aqueous solution 137
aromatic hydrocarbons 993
Arrhenius 694
Arrhenius equation 627
atmosphere (atm) 417
atom 19, 62
atomic mass 77
atomic mass unit (amu) 71
atomic number (Z) 72
atomic orbital 280
Atwater system 232
Aufbau principle 288
autoionization 695
Autumn 481
average rate 601
Avogadro's law 433
Avogadro's number (NA) 121
axial position 347

B

balanced 161
Balmer 269
band of stability 1025
bar 417
barometer 418
Bartlett 920
base 171
base anhydrides 904
Base ionization 694
base-ionization constant (Kb) 704
basic 697
battery 834
becquerel (Bq) 1063
Beta (β) decay 1032
Beta particles 1029
bicarbonate anion 891
Bidentate ligands 950
bimolecular reaction 631
binary acid 102
binary compounds 96
binding energy per nucleon 1027
biofuel 239
Bismuth 866
blackbody 264
body-centered cubic (BCC) solid 519
body-centered cubic unit cell 519
Bohr 269, 270
Bohr's model 270
boiling point 495
boiling point elevation 571
boiling point elevation constant 571
Boltzmann 788
bomb calorimeter 230
bond angle 343
bond dipole moment 354
bond distance 343
bond length 317
bond order 400
bonding orbitals 396
Borates 875
Born 279
Born-Haber cycle 341
Boyle 694
Boyle's law 431
Bragg 530
Bragg equation 530
Brønsted-Lowry acid 694
Brønsted-Lowry base 694
buffer 724
buffer capacity 727

buret 191

C

calories (cal) 216
calorimeter 221
calorimetry 221
capillary action 490
carbonates 891
carbonyl group 999
carboxylic acids 1003
Carnot 787
catalysts 607
cathode 821
cathode ray 66
cathodic protection 842
cations 72
cell notations 822
cell potentials, Ecell 824
cell schematics 822
Celsius (°C) 30
central metal 949
Chadwick 70, 1030
chain reaction 1046
chalcogens 87
Charles's law 428
chelate 950
chelating ligands 950
chemical change 24
chemical equation 160
chemical property 24
chemical reduction 869
chemical symbol 73
chemical thermodynamics 233
chemistry 11
chemotherapy 1057
chlor-alkali process 904
cis configuration 955
Clausius 787
Clausius-Clapeyron equation 496
coefficients 160
cohesive forces 488
colligative properties 564
Collision theory 625
colloidal dispersions 583
colloids 583
color change interval 735
combustion analysis 195
combustion reactions 177
complete ionic equation 165
compounds 16
compressibility factor (Z) 459
concentrated 137
concentration 137
concentration cell 833
condensation 493
containment system 1051
continuous spectrum 264
control rods 1050
coordinate covalent bond 763
coordination compounds 938, 949
coordination isomers 957
coordination number 517, 949
coordination sphere 949
core electrons 291
Corrosion 840
Cottrell 589
covalent bonds 92, 317
Covalent network solids 512
covalent radius 296
crenation 578
Crick 532
critical mass 1047
critical point 508
Cronin 82
crystal field splitting 963
crystal field theory 962
crystalline solids 510
cubic centimeter (cm3) 31
cubic closest packing (CCP) 520
cubic meter (m3) 31
Curie 1030
curie (Ci) 1063

D

d orbitals 280
d-block elements 935
Dalton 19
Dalton (Da) 71
Dalton's atomic theory 62
Dalton's law of partial pressures 440
daughter nuclide 1031
Davisson 276
Davy 694
de Broglie 275
Debye 582
degenerate orbitals 283, 397
density 31
deposition 499
diamagnetic 394
Diffraction 530
diffusion 449
dilute 137
Dilution 141
dimensional analysis 42
dipole moment 354
dipole-dipole attraction 482
Diprotic acids 722
diprotic base 724
dispersed phase 585
dispersion force 478
dispersion medium 585
disproportionation reactions 879
dissociation 554
dissociation constant (Kd) 765
dissolved 137
donor atom 949
double bond 325
Downs cell 867
dry cell 835
dynamic equilibrium 493

E

effective nuclear charge, Zeff 298
effusion 450
eg orbitals 963
electrode potential (EX) 825
electrolysis 843
electrolytes 552
electromagnetic radiation 258
electromagnetic spectrum 259
electron 67
electron affinity 302
Electron capture 1033
electron configuration 288
electron volts (eV) 1024
electron-pair geometry 345
electronegativity 319
electroplating 846
element 62
elementary reaction 630
elements 16
empirical formula 81

empirical formula mass 135
emulsifying agent 586
emulsion 586
enantiomers 956
end point 191
endothermic process 215
Energy 213
enthalpy (H) 234
enthalpy change (ΔH) 234
entropy (S) 787
equatorial position 347
equilibrium 658
equilibrium constant, K 663
equivalence point 191
esters 1003
Ethers 996
exact number 34
excess reactant 186
excited electronic state 271
exothermic process 215
expansion work 233
extensive property 25
external beam radiation therapy 1056

F

f orbitals 281
f-block elements 935
face-centered cubic (FCC) solid 520
face-centered cubic unit cell 519
factor-label method 42
Fahrenheit 45
first law of thermodynamics 233
first transition series 936
fissile 1047
fission 1044
fissionable 1047
formal charge 332
formation constant (Kf) 765
formula mass 118
fourth transition series 936
Franklin 532
Frasch process 913
free energy change (ΔG) 797
free radicals 330
freezing 499
freezing point 499
freezing point depression 573
freezing point depression constant 573
frequency 259
frequency factor 627
fuel cell 839
Fuller 313, 881
functional group 988
fundamental unit of charge (e) 71
fusion 1053
fusion reactor 1054

G

galvanic cells 821
galvanization 842
Gamma emission (γ emission) 1032
gamma rays (γ) 1029
gas 14
Gay-Lussac's law 426
Geiger 69
Geiger counter 1062
Geim 514, 882
gel 590
Germer 276
Gibbs 797
Gibbs free energy (G) 797
Gouy 394
Graham's law of effusion 450
gravimetric analysis 193
gray (Gy) 1063
Greaney 481
ground electronic state 271
groups 85

H

Haber process 888
half-cells 821
half-life 1035
half-life of a reaction (t1/2) 622
half-reaction 174
halides 917
Hall 868
Hall–Héroult cell 868
halogens 87
Hasselbalch 729
Heat (q) 215
heat capacity (C) 216
Heisenberg uncertainty principle 278
hemolysis 578
Henderson 729
Henderson-Hasselbalch equation 729
Henry's law 557, 559
Héroult 868
hertz (Hz) 259
Hess's law 242, 245
heterogeneous catalyst 639
heterogeneous equilibrium 667
heterogeneous mixture 17
hexagonal closest packing (HCP) 521
high-spin complexes 964
holes 525
homogeneous catalyst 636
homogeneous equilibrium 665
homogeneous mixture 17
homonuclear diatomic molecules 395
Hu 481
Hückel 582
Hund's rule 290
Huygens 258
hybrid orbitals 380
hybridization 380
hydrates 98
hydrocarbons 237
hydrogen bonding 483
hydrogen carbonates 891
hydrogen halides 890
hydrogen sulfates 907
hydrogen sulfites 907
hydrogenation 887
hydrometallurgy 943
hydrostatic pressure 419
hydroxides 901
hypertonic 578
hypervalent molecules 331
hypothesis 11
hypotonic 578

I

ideal gas 433
ideal gas constant 433
ideal gas law 433

ideal solution 549
immiscible 561
indicators 191
induced dipole 478
inert electrode 823
inert gases 87
inert pair effect 315
initial rate 601
inner transition metals 87
insoluble 166
instantaneous dipole 478
instantaneous rate 601
integrated rate laws 614
intensive property 25
interference patterns 262
interhalogens 919
intermediates 630
intermolecular forces 476
internal energy (U) 233
internal radiation therapy (brachytherapy) 1056
International System of Units 28
interstitial sites 515
ion 72
ion-dipole attraction 553
ion-product constant for water, Kw 695
ionic bonds 92, 314
ionic compound 93
Ionic solids 511
ionization energy 299
Ionization isomers 957
ionizing radiation 1060
isoelectronic 299
isomers 83
isomorphous 520
isotonic 578
isotopes 70
IUPAC 30, 983

J

joule (J) 216

K

kelvin (K) 30
ketones 999
kilogram (kg) 30
kinetic energy 213
kinetic molecular theory 454
Kohn 398

L

lanthanide series 936
lanthanides 87
lanthanoid series 936
lattice energy (ΔHlattice) 340
Lavoisier 437
law of conservation of matter 15
law of constant composition 64
law of definite proportions 64
law of mass action 663
law of multiple proportions 64
laws 12
Le Châtelier's principle 669
lead acid battery 838
length 29
Lewis 763
Lewis acid 764
Lewis acid-base adduct 764
Lewis acid-base chemistry 764
Lewis base 764
Lewis structures 324
Lewis symbol 323
ligands 765, 949
limiting reactant 186
line spectra 268
linear 344
linear combination of atomic orbitals (LCAO) 395
Linkage isomers 956
liquid 14
liter (L) 31
Lithium ion batteries 837
London 478
London dispersion force 478
lone pairs 324
low-spin complexes 964

M

macroscopic domain 12
magic numbers 1026
magnetic quantum number 282
main-group elements 87
manometer 420
Marsden 69
mass 15
mass defect 1023
mass number (A) 72
mass percentage 144
mass-energy equivalence equation 1023
mass-volume percent 146
Matter 14
Maxwell 258
mean free path 449
melting 498
melting point 499
Mendeleev 85
Metallic solids 511
metalloid 858
metalloids 86
metals 86, 858
meter (m) 29
method of initial rates 609
Meyer 85
microscopic domain 12
microstate 788
millicurie (mCi) 1063
Millikan 67
milliliter (mL) 31
miscible 560
mixture 17
Molality 565
molar mass 121
Molarity (M) 137
mole 121
mole fraction (X) 441
molecular compounds 95
molecular equation 164
molecular formula 79
molecular mass 81
molecular orbital (Ψ2) 395
molecular orbital diagram 399
Molecular orbital theory 395
Molecular solids 512
molecular structure 334, 345
molecularity 631
molecule 20
Molina 637
monatomic ions 91
monodentate 949
monoprotic acids 721
ms 283

N

Nagaoka 68
Nernst equation 832
net ionic equation 165
neutral 697
neutralization reaction 172
neutrons 70
Newton 258
Nickel-cadmium 836
Nitrates 908
nitrogen fixation 893
noble gases 87
node 378
nodes 263
Nomenclature 96
nonelectrolytes 552
nonionizing radiation 1060
nonmetals 86
nonspontaneous process 784
normal boiling point 495
Novoselov 514, 882
nuclear binding energy 1023
Nuclear chemistry 1022
Nuclear fuel 1049
nuclear moderator 1049
nuclear reactions 1028
nuclear reactor 1048
Nuclear transmutation 1042
nucleons 1022
nucleus 69
nuclide 1022
nutritional calorie (Calorie) 231

O

octahedral 344
octahedral hole 525
octet rule 324
optical isomers 956
Orbital diagrams 289
organic compounds 978
osmosis 576
osmotic pressure (Π) 576
Ostwald process 907
overall reaction order 608
overlap 376
oxidation 174
oxidation number 175
oxidation state 175
Oxidation-reduction (redox)
reactions 176
oxides 901
oxidizing agent (oxidant) 175
oxyacids 103, 715
Oxyanions 91
ozone 900

P

p orbitals 280
pairing energy (P) 964
paramagnetism 393
parent nuclide 1031
partial pressure 440
partially miscible 562
particle accelerators 1042
parts per billion (ppb) 147
parts per million (ppm) 147
pascal (Pa) 417
passivation 859
Pauli exclusion principle 284
Pauling 320
percent composition 129
percent ionization 703
percent yield 188
periodic law 85
periodic table 85
periods 85
peroxides 901
Perrier 1031
pH 697
phase diagram 503
photons 266
photosynthesis 900
physical change 23
physical property 23
pi (π) bonding molecular orbital 396
pi bond (π bond) 378
Pidgeon process 870
plasma 14
platinum metals 938
pnictogens 87
pOH 697
polar covalent bond 319
polar molecule 354
polarizability 479
polyatomic ions 91
polydentate ligand 950
polymorphs 876
Positron emission (β+ decay 1033
Positrons 1029
potential energy 213
pounds per square inch (psi) 417
precipitate 166
precipitation reaction 166
precision 40
pressure 416
primary cells 835
principal quantum number 279
products 160
proton 70
pure covalent bond 318
pure substance 16

Q

quantitative analysis 190
quantization 263
quantum mechanics 279
quantum numbers 274

R

radiation absorbed dose (rad) 1063
Radiation dosimeters 1062
Radiation therapy 1056
radioactive decay 1031
radioactive decay series 1034
radioactive label 1055
radioactive tracer 1055
radioactivity 1026
radiocarbon dating 1038
radioisotope 1026
radiometric dating 1037
Raoult's law 568
rare earth elements 937
rate constant 608
rate equations 607
rate expression 600
Rate laws 607
rate of diffusion 450
rate of reaction 600
rate-limiting step 632
RBE 1063
reactants 160
reaction diagrams 627

reaction mechanism 630
reaction orders 608
reaction quotient (Q) 660
reactor coolant 1050
reducing agent (reductant) 175
reduction 175
relative biological effectiveness 1063
representative elements 87, 858
representative metals 858
resonance 335
resonance forms 335
resonance hybrid 335
reversible process 787
Reversible reactions 658
roentgen equivalent for man (rem) 1063
root mean square speed 456
rounding 36
Rutherford 69, 270, 1030
Rydberg 269

S

s orbitals 280
s-p mixing 402
sacrificial anodes 842
salt 172
salt bridge 821
saturated 555
saturated hydrocarbons 978
scientific method 12
scintillation counter 1062
second (s) 31
second law of thermodynamics 793
second transition series 936
secondary (angular momentum) quantum number 280
secondary cells 835
Segre 1031
selective precipitation 759
semipermeable membranes 576
series 85
shells 279
SI Units 28
sievert (Sv) 1063
sigma bonds (σ bonds) 378
significant digits 35
significant figures 35
Silicates 877
simple cubic structure 517
simple cubic unit cell 517
single bond 324
Single-displacement (replacement) reactions 177
skeletal structure 979
Smalley 329, 881
smelting 940
Soddy 70
solid 14
Solomon 448
solubility 166, 555
soluble 166
solute 137
solution 17
solvation 550
solvent 137
sp hybrid orbitals 381
sp2 hybrid orbitals 382
sp3 hybrid orbitals 384
sp3d hybrid orbitals 386
sp3d2 hybrid orbitals 386
space lattice 528
spatial isomers 84
specific heat capacity (c) 216
spectator ions 165
spectrochemical series 963
spin quantum number 283
spontaneous process 549, 784
standard cell potential, E°cell 824
standard electrode potential, E°X 825
Standard enthalpy of combustion 237
standard enthalpy of formation ΔHf°ΔHf° 240
Standard entropies (S°) 795
standard entropy change (ΔS°) 795
standard free energy changes, ΔG° 798
standard free energy of formation ΔGf° 799
standard hydrogen electrode (SHE) 824
standard molar volume 436
standard state 237
standard temperature and pressure (STP) 436
Standing waves 263
state function 234
stationary waves 263
Steel 942
stepwise ionization 722
stoichiometric factors 181
stoichiometry 181
strong acids 170
strong bases 171
strong electrolyte 552
strong nuclear force 1023
strong-field ligands 964
structural formula 80
structural isomers 83
subcritical mass 1047
sublimation 499
subshell 280
substituents 983
substitution reaction 987
sulfates 907
sulfites 907
superconductor 947
supercritical fluid 508
supercritical mass 1047
superoxides 901
supersaturated 555
Surface tension 489
surroundings 221
suspensions 583
symbolic domain 13
system 221

T

t2g orbitals 963
temperature 45, 213
termolecular reaction 632
tetrahedral 344
tetrahedral hole 525
theoretical yield 188
theories 12
Thermal energy 213
thermochemistry 212
third law of thermodynamics 795
third transition series 936

Thomson 66
titrant 191
titration analysis 191
titration curve 730
torr 419
trans configuration 955
transition metals 87
transition state 626
transmutation 1042
transuranium elements 1044
trigonal bipyramidal 344
trigonal planar 344
triple bond 325
triple point 506
triprotic acid 724
Tyndall effect 584

U

uncertainty 34
unified atomic mass unit (u) 71
unimolecular reaction 631
unit cell 516
unit conversion factor 42
Units 28
unsaturated 555
urms 456

V

Vacancies 515
Valence bond theory 376
valence electrons 291
valence shell 294
Valence shell electron-pair repulsion theory (VSEPR theory) 343
van der Waals equation 460
van der Waals forces 478
van't Hoff factor (i) 581
vapor pressure 493
vapor pressure of water 443
vaporization 493
vector 354
viscosity 487
voltaic cells 821
Volume 31
volume percentage 146

W

Watson 532
wave 258
wave-particle duality 266
wavefunctions 279
wavelength 259
weak acids 170
weak bases 171
weak electrolyte 552
weak-field ligands 964
Weight 15
Wilkins 532
Wohler 977
work (w) 213

X

X-ray crystallography 530

Y

Young 258